Parkinson's Disease

Parkinson's Disease

Edited by

Manuchair Ebadi, Ph.D., FACCP
ster Fritz Distinguished Professor of Pharmacology
Associate Vice President for Medical Research
University of North Dakota
School of Medicine and Health Sciences

Ronald F. Pfeiffer, M.D.
Professor and Vice Chairman
University of Tennessee
Health Science Center
College of Medicine
Department of Neurology

CRC PRESS

Boca Raton   London   New York   Washington, D.C.

Library of Congress Cataloging-in-Publication Data

Parkinson's disease / [edited by] Manuchair Ebadi, Ronald F. Pfeiffer.
   p. cm.
   Includes bibliographical references and index.
   ISBN 0-8493-1590-5 (alk. paper)
   1. Parkinson's disease. I. Ebadi, Manuchair S. II. Pfeiffer, Ronald.
[DNLM: 1. Parkinson's Disease. WL 359 P24666 2004]
RC382.P242 2004
616.8'33—dc22   2004054497

Visit the CRC Press Web site at www.crcpress.com

© 2005 by CRC Press

No claim to original U.S. Government works
International Standard Book Number 0-8493-1590-5
Library of Congress Card Number 2004054497
Printed in the United States of America 1 2 3 4 5 6 7 8 9 0
Printed on acid-free paper

Dedication

We dedicate this book humbly and reverently to the individuals who have suffered, are suffering, or will suffer from Parkinson's disease. We have high hope that the etiology of this mysterious disease will be discovered soon and a cure and prevention will be found in the very near future.

In books lie the souls of the whole past time,
The articulate audible voice of the past,
When the body and material substances
Of it have altogether vanished like a dream.

—Thomas Carlyle

In Memoriam

In loving memory of Dr. Benjamin Pfeiffer, who faced the trials of Parkinson's disease with tremendous courage and served as an inspiration to so many, and Irene Pfeiffer, who filled the often lonely role of caregiver with unfailing dedication and love.

Preface

What we know today as Parkinson's disease (PD) was first described by James Parkinson in his 1817 *Essay on the Shaking Palsy.* However, it is probable that PD was present long before this landmark description. A disease known as *kampavata,* consisting of shaking (kampa) and lack of muscular movement (vata), existed in ancient India as long as 4500 years ago.

It was not until more than 100 years after Parkinson's original description that the loss of dopamine-containing cells in the substantia nigra (SN), characteristic of PD, was recognized. Although neuropathological examination documented the distinctive presence of *Lewy bodies (LB)* and degeneration of the SN as hallmarks of PD, no definitive clinical test or procedure to diagnose PD exists, and the diagnosis must be made on the basis of clinical features alone.

The progressive loss of dopaminergic and other neurons that characterizes PD neuropathologically leads to a sometimes bewildering array of clinical features whose identification and management can challenge even the most astute clinician. Four cardinal signs constitute the core clinical complex of parkinsonism: tremor, akinesia or bradykinesia, rigidity, and loss of postural reflexes.

In addition to these cardinal signs, a variety of additional motor features may develop in PD, produced at least in part by combinations of the four cardinal features. Speech becomes both soft and poorly articulated. Dysphagia is often present and may lead to aspiration. Handwriting becomes micrographic, sometimes displaying a fatigue-like quality in which it starts out at normal size but becomes progressively smaller in prolonged writing. Posture becomes flexed and gait is characterized by small, shuffling steps on a narrow base, sometimes with a propulsive, or festinating, quality to it. With advancing parkinsonism, patients may experience transient *freezing*, typically upon initiation of gait, but sometimes also in narrow confines, such as doorways.

Although they have received less attention, a number of nonmotor features also characterize PD. Autonomic abnormalities may include bowel dysfunction, urinary difficulties, sexual disturbances, cardiovascular changes, and thermoregulatory alterations. Behavioral changes, such as depression and anxiety, are frequently present in PD, while, with advancing PD, cognitive impairment may also become evident. A variety of sleep disturbances also may appear in PD, as can fatigue and progressive weight loss. To further complicate matters, most of these nonmotor features can also be triggered or accentuated by the medications used to treat PD.

While no preventive or curative treatment for PD has been discovered to date, the evolution of treatment for PD has been characterized by a fascinating, and in many respects dramatic, progression to more effective symptomatic therapies. In tandem with these advances, therapeutic attention has also begun to focus on treatment that might actually alter or slow progression of the disease process itself.

A common cause of parkinsonism in 1920 was encephalitis, which prompted an attempt to develop a vaccine that would prevent the development of postencephalitic parkinsonism. This earlier effort to develop *protective immunologic therapy,* has now been replaced by trials investigating agents that may represent *neuroprotective therapy.* Potential examples include *coenzyme Q_{10}* (antioxidant and mitochondrial stabilizer) and *minocycline* (anti-inflammatory/antiapoptotic).

Advances in our understanding of the neuropathological processes and genetic factors that produce PD have also unfolded at a dizzying pace. Over the past few years, several genes for monogenically inherited forms of PD have been mapped or cloned. In a small number of families with autosomal dominant inheritance and typical LB pathology, mutations have been identified in the gene for α-*synuclein.* Aggregation of this protein in LB may be a crucial step in the molecular pathogenesis of both familial and sporadic PD. On the other hand, mutations in the *parkin gene* result in an autosomal recessive form of parkinsonism, which appears to be a relatively frequent cause of PD in patients with very early onset. In this form of PD, nigral degeneration is not accompanied by LB formation. Parkin has been implicated in cellular protein degradation pathways, as it has been shown that it functions as a *ubiquitin ligase.* The potential importance of this pathway is also highlighted by the finding of a mutation in the gene for *ubiquitin C-terminal hydrolase L1* in another small family with PD. Other loci have been mapped on chromosome 2p and 4p, respectively, in a small number of families with dominantly inherited PD, but those genes have not yet been identified. The identification of specific mutations that cause a parkinsonian phenotype has caused confusion and raised some valid questions as to whether the term PD, which implies a single unitary disease process, is still valid.

The contribution of genetic factors and a defect of *complex I* of the mitochondrial respiratory chain have been confirmed at the biochemical level. Disease specificity of this defect has been demonstrated for the parkinsonian SN. These findings and the observation that the neurotoxin, 1-methyl-4-phenyl-1, 2, 3, 6-tetrahydropyridine *(MPTP)*, which causes a parkinson-like syndrome in humans, acts via inhibition of complex I, have triggered research interest in the *mitochondrial genetics of PD.*

Oxidative phosphorylation consists of five protein-lipid enzyme complexes located in the mitochondrial inner membrane that contain flavins (FMN, FAD), quinoid compounds (coenzyme Q_{10}, CoQ_{10}), and transition metal compounds (iron-sulfur clusters, hemes, protein-bound copper). These enzymes are designated *complex I* (NADH:ubiquinone oxidoreductase, EC 1.6.5.3), *complex II* (succinate:ubiquinone oxidoreductase, EC 1.3.5.1), *complex III* (ubiquinol:ferrocytochrome c oxidoreductase, EC 1.10.2.2), *complex IV* (ferrocytochrome c:oxygen oxidoreductase or cytochrome c oxidase, EC 1.9.3.1), and *complex V* (ATP synthase, EC 3.6.1.34). A defect in mitochondrial oxidative phosphorylation, in terms of a reduction in the activity of NADH CoQ reductase (complex I) has been reported in the striatum of patients with PD. The reduction in the activity of complex I is found in the SN, but not in other areas of the brain, such as globus pallidus or cerebral cortex. Therefore, the specificity of mitochondrial impairment may play a role in the degeneration of nigrostriatal dopaminergic neurons. This view is supported by the fact that MPTP, generating 1-methyl-4-phenylpyridine (MPP^+), destroys dopaminergic neurons in the SN. Although the serum levels of CoQ_{10} are normal in patients with PD, CoQ_{10} is able to attenuate the MPTP-induced loss of striatal dopaminergic neurons.

Oxidative stress is believed to play a key role in the degeneration of dopaminergic neurons in the SN of PD patients. An important biochemical feature of PD is a significant early depletion in levels of the thiol antioxidant compound *glutathione* (GSH), which may lead to the generation of reactive oxygen species, mitochondrial dysfunction, and ultimately to subsequent neuronal cell death. GSH has been reported to be markedly reduced in PD, particularly in patients with advanced disease. Furthermore, the GSH decrease seems to appear before neurodegeneration in presymptomatic PD and is not a consequence thereof. This suggests that a link may exist between these two events, although it remains to be established whether or not the loss of GSH can induce neurodegeneration. *R-lipoic acid* acts to prevent depletion of GSH content and preserve the mitochondrial complex I activity, which normally is impaired as a consequence of GSH loss.

Nitric oxide (NO) has been also implicated in neurodegenerative disease. Several studies have reported markers that suggest a NO overproduction in PD brains. *NO radicals* have been detected in PD SN, as have increased *nitrosilated proteins,* such as α-synuclein. Increased nitrite concentration has also been described in cerebrospinal fluid. Finally, the core of the LB is immunoreactive for *nitrotyrosine.*

Decreased GSH may predispose cells to the toxicity of other insults that selectively target dopaminergic neurons. GSH depletion synergistically increases the selective toxicity of MPP^+ in dopamine (DA) cell cultures and the toxicity of 6-hyroxydopamine (6OHDA) and MPTP *in vivo*. GSH peroxidase *(GPx)-knockout mice* show increased vulnerability to MPTP. There is evidence that NO may play an important role in DA cell death and functionality. A redox-based mechanism for the neuroprotective effects of NO and related nitroso-compounds has been postulated. In this regard, GSH is an endogenous thiol that reacts with NO to form *S-nitrosoglutathione* and which protects DA neurons from oxidative stress.

Metallothioneins, low-molecular-weight zinc-binding proteins, are able to scavenge free radicals, including *hydroxyl radicals* implicated in PD. In addition, metallothionein averts α-synuclein nitration, enhances the elaboration of coenzyme Q10, increases the activity of complex I and the synthesis of ATP, and as an antioxidant is 50 times more potent than GSH.

This growing recognition of the mechanisms and pathways potentially involved in the death of dopaminergic neurons in PD is leading to an ever-expanding array of investigative approaches whose aim is to achieve effective neuroprotective or even neurorestorative treatment. Avenues being investigated encompass not only traditional (and nontraditional) pharmacological approaches, but also innovative and frontier-crossing surgical and other modalities such as gene therapy, stem cell therapy, and neurotransplantation.

In this volume, an impressive armada of authorities has been assembled to address the challenging topic of PD. Historical background, basic neuropathological and neurophysiological characteristics, epidemiological considerations, clinical features, current treatment approaches, and potential future therapeutic modalities of PD are addressed in a fashion that is, we hope, both comprehensive and comprehensible.

The editors also fervently hope that in 2017, when the 200th anniversary of James Parkinson's original description of the disease that now bears his name is marked, it will be firmly established fact that the etiology of the disease has been unraveled, unfailingly effective treatment established, and its cause(s) averted.

M. Ebadi, Ph.D.
Grand Forks, North Dakota

R. Pfeiffer, M.D.
Memphis, Tennessee

Acknowledgments

The authors express their appreciation to Barbara Norwitz, Life Sciences publisher for CRC Press, for her gracious invitation to prepare a book on Parkinson's disease.

The authors acknowledge the effort of Patricia Roberson, the project coordinator assigned to bring this manuscript to completion.

The authors salute the magnificent contributions of Gail Renard, production editor, and her capable staff for polishing and refining this book.

The authors remain grateful to Jeffrey K. Eckert for his diligence in preparing and typesetting all chapters and producing the book in a timely fashion.

The authors express their delight and joy in the artistic talent of Victoria Swift in completing various art works for the book.

The authors appreciate the secretarial assistance of Lori Wagner and JoAnn Johnson (UND) and Sharon Williams (UT) in helping in the various stages of preparing this book.

The authors remain indebted to Dani Stramer for her relentless dedication to her job, unmatched work ethics, and incredible secretarial skills in typing various chapters and supervising the composition of this project.

Ronald Pfeiffer, M.D., expresses his sincere appreciation to William Pulsinelli, M.D., Ph.D., Chairman of the Department of Neurology, University of Tennessee, College of Medicine, for his visionary leadership and support of the Division of Neurodegenerative Diseases.

Manuchair Ebadi, Ph.D., pays an affectionate tribute and extends his heartfelt gratitude to H. David Wilson, M.D., Dean and Vice President for Health Affairs, University of North Dakota School of Medicine and Health Sciences, for his unyielding support, solomonic wisdom, and genuine friendship in facilitating the completion of this book.

About the Editors

Manuchair Ebadi earned a B.S. degree in chemistry from Park University (Parkville, Missouri, 1960), an M.S. degree in pharmacology from the University of Missouri College of Pharmacy (Kansas City, 1962), and a Ph.D. degree in pharmacology from the University of Missouri College of Medicine (Columbia, 1967). He completed his postdoctoral training in the Laboratory of Preclinical Pharmacology at the National Institute of Mental Health (Washington, D.C., 1970), under the able direction of Erminio Costa, M.D., an eminent member of the National Academy of Sciences.

Dr. Ebadi served as Chairman of the Department of Pharmacology at the University of Nebraska College of Medicine from 1970 until 1988, and subsequently as Professor of Pharmacology, Neurology, and Psychiatry from 1988 through 1999. In July 1999, he was appointed Professor and Chairman of the Department of Pharmacology and Toxicology at the University of North Dakota School of Medicine and Health Sciences. In September 1999, Dr. Ebadi became Professor and Chairman of the newly created Department of Pharmacology, Physiology, and Therapeutics; in November 1999, he became Professor of Neuroscience; and in December 1999, he was appointed Associate Dean for Research and Program Development. In September 2000, Dr. Ebadi was appointed Director of the Center of Excellence in Neurosciences at the University of North Dakota School of Medicine and Health Sciences, and in March 2002, Associate Vice President for Medical Research at the University of North Dakota.

During his academic career, Professor Ebadi has received 36 awards, including the Burlington Northern Faculty Achievement Award (1987), the University of Nebraska's system-wide Outstanding Teaching and Creative Activity Award (1995); and was inducted into the *Golden Apple Hall of Fame* (1995) for having received 11 Golden Apple awards. He is a member of 18 research and scholarly societies including Alpha Omega Alpha Honor Medical Society.

In 1976, Dr. Ebadi became the Mid-America State Universities Association (MASUA) honor lecturer; in 1987, he received an award for Meritorious Contributions to Pharmaceutical Sciences from the University of Missouri Alumni Association; in 1995, he was honored by a Resolution and Commendation of the Board of Regents of the University of Nebraska for having developed a sustained record of excellence in teaching, including creative instructional methodology; and in 1996, he received the *Distinguished Alumni Award* from Park University, his alma mater. In November 2002, Dr. Ebadi received a Recognition Award in appreciation of his outstanding contribution to the UND School of Medicine. In May 2003, Dr. Ebadi received the Outstanding Block Instructor Award for outstanding performance "in the encouragement, enrichment, and education of tomorrow's physicians." In 2003, Dr. Ebadi was elected to the Prestigious *Cosmos Club* (Washington, D.C.) for individuals who have distinguished themselves in art, literature, or science.

Professor Ebadi discovered and characterized brain metallothioneins isoforms in 1983 and subsequently showed that they are able to scavenge free radicals implicated in Parkinson's disease. In addition, he showed that metallothionein averts α-synuclein nitration, enhances the elaboration of coenzyme Q10, increases the activity of complex I, enhances the synthesis of ATP, and as an antioxidant is fifty times more potent than glutathione. His research programs have been supported in the past and currently by the National Institute on Aging (AG 17059-06); the National Institute of Environmental Health Sciences (NIEHS 03949); the National Institute of Child Health and Human Development (NICHD 00370); the National Institute of Neurological Disorders and Stroke (NINDS 08932, NINDS 34566, and NINDS 40160); and the Office of National Drug Control Policy, Counter Drug Technology Assessments Center (DATM 05-02-C-1252).

Professor Ebadi has written seven textbooks. The *Pharmacology* text was translated into Japanese in 1987 (Medical Science International LTD, Tokyo); the *Core Concepts in Pharmacology* was translated into Chinese in 2002 (Ho-Chi Book Publishing of Taiwan); and the *Pharmacodynamic Basis of Herbal Medicine* (CRC Press 2002) became a best seller.

On February 26, 2004, Dr. Ebadi received the University of North Dakota Foundation's *Thomas J. Clifford Faculty Achievement Award for Excellence in Research* and, on September 7, 2004, he received from President Charles E. Kupchella, the designation of *Chester Fritz Distinguished Professor of Pharmacology,* the highest honor bestowed by the University of North Dakota.

Ronald Frederick Pfeiffer earned a B.S. degree from the University of Nebraska (Lincoln, 1969), graduating with honors and becoming a member of Phi Beta Kappa. He completed his M.D. degree at the University of Nebraska College of Medicine (Omaha, 1973). Dr. Pfeiffer completed his internship in internal medicine (1974) and residency in neurology (1977) at Walter Reed Army Medical Center in Washington, D.C.

Dr. Pfeiffer completed a student research fellowship in the Laboratory of Preclinical Pharmacology (1972) under the able direction of Erminio Costa, an eminent member of the National Academy of Sciences. From 1975 to 1977, during his neurology residency, he was a guest fellow at the Experimental Therapeutics Branch, NINDS, participating in research programs under the tutelage of D. B. Calne, D.M. In 2001, Professor Pfeiffer received an honorary Doctor of Laws degree from Concordia University in Nebraska.

Dr. Pfeiffer served as Professor and Chief of the Section of Neurology (1987–1993) at the University of Nebraska College of Medicine. Thereafter, he was appointed Professor of Neurology (1994–present) and then Vice Chairman of the Department of Neurology (1996–present) at the University of Tennessee College of Medicine.

Professor Pfeiffer is board certified in psychiatry and neurology (1979–present) and is a member of various medical and scientific societies, including the American Neurological Association. He has participated in numerous clinical trials of experimental agents for the treatment of PD and has written and lectured extensively about nonmotor aspects of PD, especially GI dysfunction.

Contributors

John R. Adams, M.D., M.Sc.
Neurodegenerative Disorders Centre
Vancouver, BC, Canada

J. Eric Ahlskog, M.D., Ph.D.
Department of Neurology
Mayo Clinic
Rochester, MN

Amornpan Ajjimaporn, Ph.D.
School of Medicine and Health Sciences
University of North Dakota
Grand Forks, ND

Andrea Antal, Ph.D.
Department of Clinical Neurophysiology
Georg-August University of Göttingen
Göttingen, Germany

Yasuhiko Baba, M.D.
Department of Neurology
Mayo Clinic
Jacksonville, FL

Yacov Balash, M.D., Ph.D.
Movement Disorders Unit
Department of Neurology
Tel Aviv Sourasky Medical Center
Tel Aviv, Israel

Anne L. Barba, Ph.D.
Parkinson's Disease and Movement Disorders Center
Albany Medical Center
Albany, NY

John Bertoni, M.D., Ph.D.
Department of Neurology
Creighton University
Omaha, NE

Pierre J. Blanchet, M.D., Ph.D.
Department of Stomatology
University of Montreal
Montreal, Canada

Ivan Bodis-Wollner, M.D., DSC
Department of Neurology
SUNY Health Science Center at Brooklyn
Brooklyn, NY

Daryl Bohac, Ph.D.
Department of Psychiatry
University of Nebraska Medical Center
Omaha, NE

William Burke, M.D.
Department of Psychiatry
University of Nebraska Medical Center
Omaha, NE

L. Cartwright, B.Sc.
Department of Medicine
Edgbaston, Birmingham, UK

Donald B. Calne, D.M., F.R.S.C.
Pacific Parkinson's Research Centre
Vancouver Hospital and Health Sciences Centre
Vancouver, BC, Canada

Susan Calne, C.M., R.N.
Pacific Parkinson's Research Centre
University Hospital UBC
Vancouver, BC, Canada

M. Angela Cenci, M.D., Ph.D.
Wallenberg Neurosciences Center
Neurobiology Division
University of Lund
Lund, Sweden

Jaturaporn Chagkutip, Ph.D.
University of North Dakota
School of Medicine and Health Sciences
Grand Forks, ND

Kelvin L. Chou, M.D.
Department of Clinical Neurosciences
Brown University School of Medicine, and
Division of Neurology
Memorial Hospital of Rhode Island
Pawtucket, RI

Thomas L. Davis, M.D.
Division of Movement Disorders
Vanderbilt University School of Medicine
Nashville, TN

Ruth Djaldetti, M.D.
Department of Neurology
Sackler Medical School
Tel Aviv University
Tel Aviv, Israel

John E. Duda, M.D.
Department of Neurology
University of Pennsylvania School of Medicine and
    Parkinson's Disease Research, Education and Clinical
    Center
Philadelphia Veterans Affairs Medical Center
Philadelphia, PA

Manuchair Ebadi, Ph.D.
Department of Pharmacology, Physiology and
    Therapeutics
University of North Dakota
School of Medicine & Health Sciences
Grand Forks, ND

Larry Elmer, M.D., Ph.D.
Department of Neurology
Medical College of Ohio
Toledo, OH

Stewart A. Factor, D.O.
Department of Neurology
Albany Medical College
Albany, NY

Ciaran J. Faherty, Ph.D.
Department of Developmental Neurobiology
St. Jude Children's Research Hospital
Memphis, TN

Stanley Fahn, M.D.
Columbia University College of Physicians and Surgeons
New York, NY

Robert G. Feldman, M.D.
Department of Neurology and Pharmacology and
    Environmental Health
Boston University Schools of Medicine and Public Health
Boston, MA

Blair Ford, M.D.
Center for Parkinson's Disease
Columbia-Presbyterian Medical Center
New York, NY

Tatiana Foroud, Ph.D.
Department of Medical and Molecular Genetics
Indiana University School of Medicine
Indianapolis, IN

Joseph H. Friedman, M.D.
Department of Clinical Neurosciences
Brown University School of Medicine
Providence, RI

Raul de la Fuente-Fernandez, M.D.
Neurodegenerative Disorders Centre
University of British Columbia
Vancouver, BC, Canada

Carol Ewing Garber, Ph.D.
Department of Cardiopulmonary and Exercise Sciences
Bouve College of Health Sciences
Northeastern University
Boston, MA

Gaëtan Garraux, M.D.
Human Motor Control Section
NINDS, NIH
Bethesda, MD

Thomas Gasser, M.D.
Department of Neurodegenerative Disorders
Hertie-Institute for Clinical Brain Research and Program
    Development
University of Tubingen
Tubingen, Germany

Nir Giladi, M.D.
Movement Disorders Unit
Department of Neurology
Tel Aviv Sourasky Medical Center
Tel Aviv, Israel

Christopher G. Goetz, M.D.
Rush University Medical Center
Chicago, IL

Lawrence I. Golbe, M.D.
Department of Neurology
UMDNJ—Robert Wood Johnson Medical School
New Brunswick, NJ

Jennifer G. Goldman, M.D.
Rush University Medical Center
Chicago, IL

John L. Goudreau, D.O., Ph.D.
Department of Neurology
Department of Pharmacology and Toxicology
Michigan State University
East Lansing, MI

J. Timothy Greenamyre, M.D., Ph.D.
Department of Neurology
Emory University
Atlanta, GA

James G. Greene, M.D., Ph.D.
Department of Neurology
Emory University
Atlanta, GA

Ruth A. Hagestuen, R.N., M.A.
National Parkinson Foundation
Miami, FL

Mark Hallett, M.D.
Human Motor Control Section
NINDS, NIH
Bethesda, MD

Svenja Happe, M.D.
Department of Clinical Neurophysiology
University of Göttingen
Göttingen, Germany

Jeffrey M. Hausdorff, Ph.D.
Sackler School of Medicine
Tel Aviv University
Tel Aviv, Israel

Robert A. Hauser, M.D., M.B.A.
Department of Neurology
Movement Disorder Center
University of South Florida
Tampa, FL

Donald S. Higgins, M.D.
Parkinson's Disease and Movement Disorders Center
Albany Medical Center
Albany, NY

S. L. Ho, M.D.
Division of Neurology
Department of Medicine
University of Hong Kong, China

Robert G. Holloway, M.D., M.P.H.
Department of Neurology
University of Rochester School of Medicine and Dentistry
Rochester, NY

Sandra L. Holten, M.T., B.C.
Struthers Parkinson's Center
Golden Valley, MN

Zhigao Huang, M.D., Ph.D.
Pacific Parkinson's Research Centre
Vancouver Hospital and Health Sciences Centre
Vancouver, BC, Canada

Serena W. Hung, M.D.
Toronto West Hospital
Movement Disorders Clinic
Division of Neurology
Toronto, ON, Canada

Howard I. Hurtig, M.D.
Department of Neurology
University of Pennsylvania Health Systems
Parkinson's Disease and Movement Disorders Center
Pennsylvania Hospital
Philadelphia, PA

Bahman Jabbari, M.D.
Yale University School of Medicine
New Haven, CT

Michael W. Jakowec, Ph.D.
Department of Neurology
Keck School of Medicine
University of Southern California
Los Angeles, CA

Joseph Jankovic, M.D.
Department of Neurology
Parkinson Disease Center
Baylor College of Medicine
Houston, TX

Danna Jennings, M.D.
Department of Neurology
The Institute for Neurodegenerative Disorders
New Haven, CT

Monica Korell, M.Ph.
The Parkinson's Institute
Sunnyvale, CA

Ajit Kumar, D.M.
Pacific Parkinson's Research Centre
Vancouver Hospital and Health Sciences Centre
Vancouver, BC, Canada

Sandra Kuniyoshi, M.D., Ph.D.
Department of Neurology
Parkinson Disease Center and Movement Disorders Clinic
Baylor College of Medicine
Houston, TX

Roger Kurlan, M.D.
Department of Neurology
University of Rochester
Rochester, NY

Anthony E. Lang, M.D., F.R.C.P.
Toronto West Hospital
Movement Disorders Clinic
Division of Neurology
Toronto, ON, Canada

Yuen-Sum Lau, Ph.D.
Division of Pharmacology
School of Pharmacy
University of Missouri, Kansas City
Kansas City, MO

Mark S. LeDoux, M.D., Ph.D.
Department of Neurology
University of Tennessee Health Science Center
Memphis, TN

Stuart E. Leff, Ph.D.
Emory University
Atlanta, GA

Sarah C. Lidstone, B.Sc.
Pacific Parkinson's Research Centre
Vancouver, BC, Canada

Kelly E. Lyons, Ph.D.
Department of Neurology
University of Kansas Medical Center
Kansas City, KS

Scott Maanum
Department of Pharmacology
School of Medicine and Health Services
University of North Dakota
Grand Forks, ND

Kirsten Maier, M.A.
Pacific Parkinson's Research Centre
Vancouver Hospital and Health Sciences Centre
Vancouver, BC, Canada

Ronald J. Mandel, Ph.D.
Department of Neuroscience
Powell Gene Therapy Center
McKnight Brain Institute
University of Florida College of Medicine
Gainesville, FL

Fredric P. Manfredsson, B.S.
Department of Neuroscience
Powell Gene Therapy Center
McKnight Brain Institute
University of Florida College of Medicine
Gainesville, FL

Ken Marek, M.D.
Department of Neurology
The Institute for Neurodegenerative Disorders
New Haven, CT

Katerina Markopoulou, M.D., Ph.D.
Department of Neurology
University of Nebraska Medical Center
Omaha, NE

Christopher J. Mathias, D.Phil., D.S.C., F.R.C.P.
Imperial College School of Medicine
St. Mary's Hospital
London, UK

Eldad Melamed, M.D.
Department of Neurology
Rabin Medical Center, Beilinson Campus
Sackler Medical School
Tel Aviv University
Tel Aviv, Israel

Yoshikuni Mizuno, M.D.
Department of Neurology
Juntendo University School of Medicine
Tokyo, Japan

Eric S. Molho, M.D.
Department of Neurology
Albany Medical College
Albany, NY

Erwin B. Montgomery, Jr., M.D.
Department of Neurology
National Regional Primate Center
University of Wisconsin—Madison
Madison, WI

John C. Morgan, M.D., Ph.D.
Department of Neurology
Medical College of Georgia
Augusta, GA

L. Charles Murrin, Ph.D.
Department of Pharmacology
Nebraska Medical Center
Omaha, NE

Lisa A. Newman, Sc.D.
Army Audiology and Speech Center
Walter Reed Army Medical Center
Washington, DC

Ann Nolen, Psy., O.T.R.
University of Tennessee Health Science Center, CAHS
Memphis, TN

Katia Noyes, Ph.D., M.P.H.
Department of Neurology
University of Rochester School of Medicine and Dentistry
Rochester, NY

Padraig O'Suilleabhain, M.B., B.Ch.
Department of Neurology
University of Texas, Southwestern Medical School
Dallas, TX

Rajesh Pahwa, M.D.
Department of Neurology
University of Kansas Medical Center
Kansas City, KS

Pramod Kumar Pal, D.M.
National Institute of Mental Health and Neurosciences
  (NIMHANS)
Bangalore, India

Nathan Pankratz, Ph.D.
Department of Medical and Molecular Genetics
Indiana University School of Medicine
Indianapolis, IN

R.B. Parsons, Ph.D.
School of Biosciences
Edgbaston, Birmingham, UK

Walter Paulus, M.D.
Department of Clinical Neurophysiology
Georg-August University of Göttingen
Göttingen, Germany

Rene Pazdan, M.D.
Department of Neurology
Uniformed Services University
Bethesda, MD

Carmen S. Peden, Ph.D.
Department of Neuroscience
Powell Gene Therapy Center
McKnight Brain Institute
University of Florida College of Medicine
Gainesville, FL

Ronald F. Pfeiffer, M.D.
Department of Neurology
University of Tennessee Health Science Center
Memphis, TN

Giselle M. Petzinger, M.D.
Department of Neurology
Keck School of Medicine
University of Southern California
Los Angeles, CA

Brad A. Racette, M.D.
Department of Neurology
Washington University School of Medicine
St. Louis, MO

Ali H. Rajput, M.D.
Royal University Hospital
University of Saskatchewan
Saskatoon, SK, Canada

D.B. Ramsden, Ph.D.
Queen Elizabeth Hospital
Edgbaston, Birmingham, UK

Marcia H. Ratner, Ph.D.
Departments of Neurology and Pharmacology
Boston University School of Medicine
Boston, MA

Christopher A. Robinson, B.Sc., M.Sc., M.D., F.R.C.P.C.
Department of Pathology
Royal University Hospital
University of Saskatchewan and Saskatoon Health Region
Saskatoon, SK, Canada

Robert L. Rodnitzky, M.D.
Department of Neurology
University of Iowa Hospital
Iowa City, IA

Edgardo Rodriguez, Ph.D.
Department of Neuroscience
Powell Gene Therapy Center
McKnight Brain Institute
University of Florida College of Medicine
Gainesville, FL

Anthony J. Santiago, M.D.
Parkinson's Disease and Movement Disorders Center
Albany Medical Center
Albany, NY

John Seibyl, M.D.
Department of Neurology
The Institute for Neurodegenerative Disorders
New Haven, CT

Kapil D. Sethi, M.D.
Department of Neurology
Medical College of Georgia
Augusta, GA

Surya Shah, Ph.D.
College of Allied Health Sciences
University of Tennessee Health Science Center
Memphis, TN

Kathleen M. Shannon, M.D.
Department Neurological Sciences
Rush Presbyterian St. Luke's Medical Center
Chicago, IL

Sushil K. Sharma, Ph.D.
University of North Dakota
School of Medicine and Health Sciences
Grand Forks, ND

Shaik Shavali, Ph.D.
University of North Dakota
School of Medicine and Health Sciences
Grand Forks, ND

Holly Shill, M.D.
Muhammad Ali Parkinson Research Center
Phoenix, AZ

Andrew Siderowf, M.D.
Department of Neurology
University of Pennsylvania
Philadelphia, PA

Carlos Singer, M.D.
Department of Neurology
University of Miami
Miami, FL

Richard J. Smeyne, Ph.D.
Department of Developmental Neurobiology
St. Jude Children's Research Hospital
Memphis, TN

S.W. Smith, M.B.Ch.B.
Department of Medicine
Edgbaston, Birmingham, UK

Janice Smolowitz, R.N., Ed.D., ANP
Neurological Institute
Columbia University
New York, NY

Dennis A. Steindler, Ph.D.
Department of Neuroscience
McKnight Brain Institute
University of Florida College of Medicine
Gainesville, FL

Matthew B. Stern, M.D.
Parkinson's Disease Research, Education and Clinical
  Center
Philadelphia Veteran's Affairs Medical Center
University of Pennsylvania School of Medicine
Parkinson's Disease and Movement Disorders Center
Pennsylvania Hospital
Philadelphia, PA

A. Jon Stoessl, M.D., F.R.C.P.C.
Neurodegenerative Disorders Centre
University of British Columbia
Vancouver, BC, Canada

Daniel Strickland, MSPH, Ph.D.
Kaiser Permanente, Southern California
Pasadena, CA

Oksana Suchowersky, M.D., F.R.C.P.
Department of Clinical Neurosciences
University of Calgary
Calgary, AB, Canada

Caroline M. Tanner, M.D., Ph.D.
The Parkinson's Institute
Sunnyvale, CA

Daniel Tarsy, M.D.
Department of Neurology
Harvard Medical School
Movement Disorders Center
Boston, MA

James W. Tetrud, M.D.
Movement Disorders Treatment Center
The Parkinson's Institute
Sunnyvale, CA

Claudia M. Trenkwalder, M.D.
Center of Parkinsonism and Movement Disorders
University of Göttingen
Göttingen, Germany

Michael Trew, M.D., F.R.C.P.C.
Department of Psychiatry
University of Calgary
Calgary Health Region
Calgary, AB, Canada

Joel M. Trugman, M.D.
Department of Neurology
University of Virginia Health System
Charlottesville, VA

Ryan J. Uitti, M.D.
Department of Neurology
Mayo Clinic
Jacksonville, FL
Mayo Clinic College of Medicine
Rochester, MN

Leo Verhagen Metman, M.D., Ph.D.
Department of Neurological Sciences
Rush University Medical Center
Chicago, IL

Mervat Wahba, M.D.
Movement Disorder Center
Department of Neurology
University of South Florida
Tampa, FL

Sawitri Wanpen, Ph.D.
University of North Dakota
School of Medicine and Health Sciences
Grand Forks, ND

R. H. Waring, D.Sc.
School of Biosciences
Edgbaston, Birmingham, UK

Cheryl Waters, M.D., F.R.C.P.
Neurological Institute
Columbia University
New York, NY

Mickie D. Welsh, R.N., D.N.Sc.
Keck School of Medicine
University of Southern California
Los Angeles, CA

Steven P. Wengel, M.D.
Department of Psychiatry
University of Nebraska Medical Center
Omaha, NE

Robert E. Wharen, Jr. M.D.
Department of Neurology
Mayo Clinic
Jacksonville, FL
Mayo Clinic College of Medicine
Rochester, MN

Rose Wichmann, P.T.
Struthers Parkinson's Center
Golden Valley, MN

A. C. Williams, M.D., F.R.C.P.
Queen Elizabeth Hospital
Edgbaston, Birmingham, UK

James B. Wood, M.D.
Department of Radiology
Veterans Affairs Medical Center
Memphis, TN

Zbigniew K. Wszolek, M.D.
Department of Neurology
Mayo Clinic
Jacksonville, FL
Mayo Clinic College of Medicine
Rochester, MN

Theresa A. Zesiewicz, M.D.
Movement Disorder Center
Department of Neurology
University of South Florida
Tampa, FL

Table of Contents

1 James Parkinson

Jennifer G. Goldman and Christopher G. Goetz
Rush University Medical Center

CONTENTS

Although James Parkinson may be best remembered in the medical profession for his *Essay on the Shaking Palsy* (1817) describing cases of paralysis agitans, he was also a prolific and respected writer in other fields such as politics, social reform, mental health, chemistry, and geology. His writings demonstrate his keen sense of observation, breadth of knowledge, and devotion to humanity. This introductory chapter explores the background of James Parkinson, his medical training, his writings on diverse subjects, and lastly, the contributions of his *Essay on the Shaking Palsy*.

FAMILY MEDICAL TRADITION

James Parkinson was born on April 11, 1755, at No. 1 Hoxton Square in the parish of St. Leonard's, Shoreditch, England, to John and Mary Parkinson. Hoxton, now a London neighborhood but then a separate village, grew from a medieval town to a place of gardens and large residential homes in the seventeenth and eighteenth centuries. Subsequently, Hoxton ceded to industrial development, overcrowding, and poverty in the eighteenth and nineteenth centuries.[1-4] The parish church of St. Leonard's remained a focal point, and here Parkinson was baptized, married, and buried. Parkinson practiced medicine nearby in a two-storied house behind the main house at No. 1 Hoxton Square.

He was born into the medical tradition, as his father John Parkinson was an apothecary and surgeon in Hoxton for many years. While James was a child, his father received the Grand Diploma of the Corporation of Surgeons of London in 1765 and served as Anatomical Warden of the Company at Surgeon's Hall from 1775 to 1776.[1,5] It is possible that James Parkinson's apprenticeship to his father included anatomical studies.[6,7] While an apprentice, James often accompanied his father on resuscitation and recovery operations for the Royal Humane Society. The father and son team published several cases illustrating their resuscitative measures.[1]

Regarding family life, it is known that James Parkinson had a younger brother, William, and a sister, Mary Sedgwick, who married Parkinson's close friend, John Keys. During his apprenticeship, James Parkinson married Mary Dale in 1781 in St. Leonard's church. James and Mary Parkinson had six children, of which one, John William Keys (b. 1785), became a physician.[1,2] James Parkinson served as an apprentice to his father from 1802 to 1808 and then received his diploma from the Royal College of Surgeons. He joined his father in medical practice at No. 1 Hoxton Square, where he practiced until 12 years after his father's death. In turn, his son, James Keys, became a Licentiate of the Society of Apothecaries in 1834 and practiced with his father in Hoxton until his father's death in 1838.

James Parkinson died from a stroke at the age of 69 on December 21, 1824, at No. 3 Pleasant Row, Kingsland

Road, Hoxton. Addressing the Board of Trustees of the Poor of the Parish of St. Leonard's, John William Keys Parkinson reported his father's sudden onset of severe right-sided paralysis and inability to speak. Despite attempts made for his recovery, Parkinson died three days later.

Medical Education

Parkinson studied at the London Hospital Medical College for six months in 1776 as one of the school's earliest medical students.[1] While an apprentice, he performed rescues for drownings in London Waterways. He received the Honorary Silver Medal of the Royal Humane Society in 1777 for the rescue of a Hoxton man who had hanged himself, and this case was reported by his father. Parkinson obtained his diploma of the Company of Surgeons in April 1784 shortly after his father's death. His election to Fellow of the Medical Society of London in 1787 followed the delivery of his first paper to the society. This paper, "Some Account of the Effects of Lightning," (1789) describes injuries sustained by two men whose house was struck by lightning.[1,8] The description focused on the dermatological and neurological sequelae of lightning injuries.

Parkinson attended the surgical lectures of the English surgeon and experimentalist, John Hunter (1728–1793) in 1785.[1,5,9] His shorthand notes of the lectures were later transcribed and published by his son, John William Keys, in a volume entitled, *Hunterian Reminiscences* (1833). Whether Parkinson attended Hunter's lectures on tremor and paralysis remains speculative. In his notes, Parkinson quotes Hunter's illustrations of tremor, but these examples date from 1776 and 1786, and his attendance at these particular lectures is not established. Parkinson's notes cite Hunter's case on the "wrong actions of parts or tremor."

> A lady, at the age of seventy-one, had universal palsy: every part of the body shook which was not fully supported. The muscles of respiration were so affected, that respiration was with difficulty effected; but in sleep the vibratory motions of the muscles ceased, and the respiration was performed more equably: any endeavor of the will to alter these morbid actions increased them.[10]

In his Croonian lecture on muscular motion in 1776, Hunter portrays the case of Lord L.

> For instance, Lord L's hands are almost perpetually in motion, and he never feels the sensation in them of being tired. When he is asleep his hands, &c., are perfectly at rest; but when he wakes in a little time they begin to move.[10]

Since sources do not indicate that Parkinson was among those pupils identified by Hunter, their acquaintance is not established.[10–12]

Overall, little is known about Parkinson's early medical training, but much information is inferred from his outline of a "sound liberal education" in *The Hospital Pupil* (1800).[13] He was likely versed in Latin and Greek, shorthand, and drawing. In *The Hospital Pupil* (1800), Parkinson offers opinions on medical education and prerequisites for medical or surgical careers. Written as a letter advising an anonymous friend whose son contemplates a medical career, this book provides a glimpse of medical education in eighteenth century England. In this system, on completion of a common school education, students apprenticed for seven years and then went to a metropolitan hospital to attend lectures and witness the practice for a year or less. Parkinson argues that this model does not adequately train physicians in the community and neglects studies of observation, anatomy, and physiology. Traditionally, the first four or five years were spent "almost entirely appropriated to the compounding of medicines; the art of which, with every habit of necessary exactness, might be just as well obtained in as many months."[13] The remainder of the apprenticeship focused on the "art of bleeding, of dressing a blister, and, for the completion of the climax—of exhibiting an enema."[13] Instead, Parkinson proposes the following curriculum: anatomy, natural philosophy, physiology, chemistry, physics, French, and German during the first two years; clinical lectures during the third year; morbid anatomy and clinical work during the fourth year; and clinical work as a dressing pupil and lectures during the fifth year. Personal characteristics of "sympathetic concern, and a tender interest for the sufferings of others… the object of which should be to mitigate or remove, one great portion of the calamities to which humanity is subject" were deemed important.[13] Parkinson even outlines strategies for good study habits, taking notes, maintaining concentration, and optimizing one's education. He concludes with advice on patient relations and business and legal aspects in medicine. Many of these principles remain true in the current study and practice of medicine.

Political Writings and Plots

Parkinson's publications in the decade after his medical career began focused on politics. Dramatic political changes, reforms, and revolutions were occurring in England and France at the time. He belonged to the political societies, Society for Constitutional Information (est. 1780) and the London Corresponding Society (est. 1792), which espoused parliamentary reform, representation of the people, and universal suffrage.[1,9]

Parkinson's political beliefs emerged in multiple pamphlets written between 1793 and 1796 under the pseudonym of "Old Hubert." Several notable works by "Old Hubert" included *Pearls cast before Swine* (1793) and *An Address to the Honorable Edmund Burke from the*

Swinish Multitude (1793), which refuted antireform sentiments presented in Edmund Burke's "Reflections on the Revolution in France" (1790).[14–16]

His political endeavors involved acting as witness before the Privy Council in a trial for high treason regarding the Pop-Gun Plot in 1795. This plot implicated several members of the London Corresponding Society in an attempted assassination of King George III with an air gun. As a witness, Parkinson tried to prevent incrimination of himself in his radical political roles but eventually confessed to being "Old Hubert," the mysterious political pamphlet author. Despite this revelation, he provided pertinent trial information and helped a prisoner-friend obtain necessary medical attention. After the Pop-Gun Plot trial, Parkinson published several political works but shifted his focus to other medical and scientific matters.

MEDICAL WRITINGS

Parkinson's publications embrace a multitude of scientific and medical topics. An early work, *Observations on Dr. Hugh Smith's Philosophy of Physic* (1780), although published anonymously, already demonstrates his inquisitiveness and writing skills as he questions current scientific theories.[1] Merely a young student in medicine, Parkinson challenges Dr. Smith's ideas on the definition of glands, the role of "Vital Air" in circulation, and respiratory function.[1,5] In a politely apologetic but determined, scientific style that recurs in many of his works, Parkinson states in the preface,

> These observations, Sir, are dedicated to you, with that earnestness, which the subject demands, that deference, which is due you, and that diffidence, which ought to accompany an opposition to opinions, which are said to be founded on experiments and confirmed by physiological researching and the closest method of reasoning.[5]

Several common themes of environmental injuries and accidents, approaches to common ailments for the lay person, social issues, and medical cases arise in Parkinson's medical works.

Accidents and Dangers

Writings on accidents and resuscitation stem from his service as rescuer for the Royal Humane Society. A later work, *Dangerous Sports, a Tale Addressed to Children* (1808), addresses potential dangers and injuries associated with childhood play and pranks.[17] In contrast to other works on injuries, this book assumes a literary quality as a tale for children presented by an old cripple named Millson. The reader meets old lame Millson after he saves the life of a young boy found in the snow with a head wound and hypothermia. The story unfolds as Millson, a guest at the rescued boy's birthday party, lectures the mischievous children on jumping from high places, throwing stones, swimming, throwing snowballs with pebbles hidden in them, walking on frozen ponds, tasting unknown medications, and playing with gun-powder and pistols. One child's prank of altering Millson's crutches causes him to fall. To this child, Millson states,

> Before you determine on a frolic, consider first the probable consequences; if then you discover it is innocent, and cannot injure any person, or even hurt their feelings, go through it with spirit; but if you see that any one may really suffer by it, give it up at once, for where, for instance, would have been the joke of breaking the legs of such a poor old cripple as I am.[17d]

Millson offers constructive suggestions for safe and productive play. Rather than climbing great heights, he recommends that the leader

> …perform some act of real ingenuity, calling the powers of the mind into action; or some feat of useful dexterity, and let him then be imitated… may it be tried who can get most lines by rote in a certain space of time; who can spell the most difficult words, or who can most readily find the corresponding words, in French, or Latin.[17]

Question and answer games might incorporate subjects with

> …some curious circumstance in the natural history of animals and vegetables. In the summer time the exercise might be, to discover the names of the various plants in the fields, and of the trees in the woods. In the evening… the talk might now be to mark the constellations, the planets, and the larger stars.[17]

Through his protagonist, Parkinson's curiosity in nature and science shines. Millson amazes the children with a microscope magnifying common objects and electrical experiments that provoke shock-like sensations to the servant, who thinks him a conjuror. Despite concerns that he is a conjuror, Millson instills in the children, as well as the reader, not only a sense of safety and awareness but also an interest in science and nature.

Medical Advice for the Public

Parkinson wrote several medical handbooks for the lay public. *Medical Admonitions* (first published in 1799) instructs families to recognize symptoms of both minor and major illnesses.[18] The first section includes a table of common symptoms listed alphabetically from *Anxiety*, "When fever is accompanied by extreme anxiety, the patient sustaining, at the same time, a considerable loss of spirits and strength, the fever may be judged to be of a malignant kind, and to require the most powerful aid," to *Yawning*, "Generally occurs at the commencement of the ague fit."[18] In addition to general medical topics such as breathing, palpitations, and swelling, he defines certain neurologic conditions:

> Convulsions: Of the whole body, with frothing at the mouth, and total loss of sensibility, characterize Epilepsy, or the Falling Sickness; so termed from the subjects of this disease falling suddenly on the coming on of the fit" and "With a sensation as if a ball was rising in the throat, flutterings and rumbling in the bowel, show the disease to be *Hysterics.*

> Stupor: After wounds, or blows on the head, requires particular attention.

> Tremor: In fever, a sign of great disability.[18]

The majority of the text includes medical information regarding symptoms and treatments with the aim,

> To prevent you, on the one hand, from unnecessarily incurring the expense of medical attendance in the various trifling ails to which you and your family may be subjected; and, on the other, from sacrificing a friend, or perhaps a beloved child, by delay or improper interference, in some insidious disease.[18]

Parkinson seeks to expose evils arising from quackery and from the wealthy acting as "dispensers of physic to all their poor neighbors;" he would rather that the rich contribute to the community by supporting public hospitals.[18] He describes symptoms, treatments, and prevention of a spectrum of diseases ranging from inflammation of different organs; infections with croup, mumps, measles, and smallpox; dental problems of toothaches and teething; to pulmonary consumption; hydrocephalus; cancer; and fractures. Parkinson reiterates his beliefs on studying anatomy, physiology, pathology, and chemistry to understand human disease. He recommends several texts including Dr. Gregory's *Oeconomy of Nature and of the Medical Extracts,* the lectures of Dr. A. F. M. Willich, information for nurses by Dr. Hamilton, Physician to the General Dispensary, and instructional material for parents in Dr. Darwin's *Essay on the Education of Females.*

He includes a section entitled "Observations on the Excessive Indulgence of Children," which discloses harmful effects of indulgence on health and difficulties occasioned in treatment of illness. He writes of those children,

> ...continually undergoing either disappointment or punishment; or engaged in extorting gratifications, which he often triumphs at having gained by an artful display of passion; his time passes on, until at last the poor child manifests ill nature sufficient to render him odious to all around him, and acquires pride and meanness sufficient to render him the little hated tyrant of his playfellows and

inferiors. Can the duties of a parent have been fulfilled in this case? Can the child owe any duty, in return for such conduct? Certainly not.[18]

Parkinson asserts that indulgence of a child leads to certain diseases and affects a child's health care "when the expressions of impatience magnify one particular symptom, and conceal the rest; the nicest investigation may prove insufficient to obtain the necessary information."[18] Moreover, indulgent parents often restrict treatments for their children: "The medicines he shall prescribe, he will, very likely, be told, must not only not be ill-flavored, but, if he expects they shall be gotten down by his patient, they must be absolutely without any taste."[18]

The *Villager's Friend and Physician* (1800) offers another medical resource for the lay public on disease and health preservation.[19] He proposes the following thoughts on exercise, labor, and drunkenness:

> ...moderate and regular labour [which] coils up the main spring of life, but wild and irregular sallies may break it. He that is steady is ever ready. Regular exercise will demand regular rest.[19]
> ...would it be if we all knew, as well, the mischiefs arising, from taking a little too frequently, what is called a little drop, so that we might be sufficiently on our guard against that insidious enemy, the love of drink. This is an enemy against whom you should always be on your guard, for he uses every trick of war: sometimes he comes on by flow and unheeded approaches; sometimes his attacks are open and violent; and oftentimes will he fight under false colours, and whilst he is received as a friend, cruelly deprive those he has deluded of every comfort, and at last of life itself.[19]

Parkinson teaches how to assess the severity of fevers and pain. For example,

> If pain in the head, light-headedness, fever, redness of the eyes, and impatience at viewing much light, or hearing loud noises, succeed to shiverings, INFLAMMATION OF THE BRAIN, OR ITS MEMBRANES may be feared to exist. This must be followed with death in a very few days, if not opposed by the exertions of some skilful person. Bleeding profusely, blisters, the strictest regimen and proper medicines must be here employed, with that degree of firmness and decision, as cannot be hoped for, but where they are directed by a person of real skill, and where the attendants are impressed with the danger of the smallest deviation from orders.[19]

In contrast, he describes the common cold,

> ...pointed out by the ticking, which occasions a frequent troublesome cough. This may in general be removed by obtaining a copious perspiration at the commencement of the complaint. By drinking freely of treacle posset, vine-

gar or orange whey, barley water or gruel; but without having recourse to any considerable increase of bed clothes, or of the temperature of the room. Bleeding in general, is not here necessary.[19]

Many passages resemble those presented in his *Medical Admonitions*. Nevertheless, the two books complement each other by elaborating on different aspects of diseases and therapeutic regimens.

Social Issues of Child Abuse and Education

Although no longer known as "Old Hubert," Parkinson still expressed his social viewpoints, particularly when related to medical care and reform. Comments on child abuse and parenting appear in *The Villager's Friend and Physician* (1800) under the section on "Dropsy of the brain, or watery head," since vigorous correction of children was another cause of dropsy:

> Parents too often forget the weight of their hands and the delicate structure of a child. You must excuse the digression— it was but yesterday I passed the cottage of one you all know to have always neglected his children; I heard the plaintive and suppliant cries of a child, and rushed into the cottage; where I saw the father, whose countenance was dreadful, from the strong marks of passion and cruelty which it bore, beating most unmercifully his son, about ten years old.[19,20]

In *Remarks on Mr. Whitbread's Plan for the Education of the Poor; with Observations of Sunday School, and on the State of the Apprenticed Poor* (1807), Parkinson opposes a bill in the House of Commons to establish parochial schools for the education of poor children in the countryside.[21] Parkinson, secretary to St. Leonard's Sunday school, worries that attendance, and therefore religious and moral education, at the Sunday Schools would suffer. The proposal would "crush those institutions to which religion, morality, and the good order of society are already most highly indebted, and that the establishments which are to succeed these will fail in two most material points—the promoting of industry, and the inculcation of religious sentiments."[21] He anticipates arguments between justices and parish officers in governing the locations, curriculum, and management of the schools. He also discusses the dilemma of eliminating a day of the child's labor, and thus, family's income, for the purpose of education. He argues that since Sunday School

> …prevents no labour, nor the obtaining of the necessary earning, so it does not break in upon nor destroy those habits, on the powerful influence of which chiefly depends the prosperity of the possessor. The varied employments of the Sunday School, the change of scene, and the visit to the church, all excite sufficient interest to make the return of the day of instruction wished for through the succeeding week; whilst, as it opposes not useful habits, it throws no obstacles in the way of industrious exertions.[21]

Reform measures by Parkinson included the creation of a Register Book for St. Leonard's Church of children educated in Sunday School who were willing and fit for service. This register provided names, ages, qualifications, and behaviors of the children and whether they might be employed. Often, the unprotected, poor children apprenticed in the parish were "left to almost the unrestrained caprice of their masters, no law existing by which the duties of the master are defined, or any inspectors of his conduct appointed."[21] As a result, Parkinson supported measures to monitor working conditions:

> …children be apprenticed in the parish whenever proper masters can be obtained… with the direction to apply to him (Vestry clerk) in case of actual ill treatment—that children already apprenticed out of the parish, and within the bills of mortality, be visited by a committee of the trustees and overseers of the poor twice every year.[21]

Mental Health

As a visiting doctor for 30 years to a madhouse in Hoxton, Parkinson gained experience in the area of mental health.[22,23] In the late seventeenth and early eighteenth centuries, three private madhouses in Hoxton housed many of the lunatics in London and its surroundings, and conditions were often squalid. The case of Mary Daintree brought Parkinson back to the court of law in a trial in 1810. He had signed a certificate of insanity for Mary Daintree who was confined in Holly House in Hoxton, but she later charged that her relatives conspired to confine her illegally. Parkinson was accused of declaring her insane based on her relatives' testimony rather than the patient's arguments of sanity. He was much criticized by the newspapers, but after the trial, the newspapers published his account of the case and views on the regulation of madhouses. In *Observations on the Act for Regulating Mad-Houses* (1811), Parkinson argues that only physicians and surgeons rather than apothecaries should sign orders for confinement to madhouses. One should provide "not merely that such person is a *lunatic*, but that such person is proper to be received *into such house or place as a lunatic.*"[22] Not all lunatics require confinement, and other maladies of the mind such as delirium or dementia, apart from madness, necessitate confinement.

Obtaining direct evidence of insanity by history and declaration from the patient was often difficult. Testimony from relations of the patient was sometimes needed to determine a patient's insanity and safety risks to themselves and others. Parkinson suggests that

> …perhaps the evidence of the relatives, where the medical examiner cannot himself obtain proof, ought to be

required upon oath; and as it is a case in which the safety of society is concerned, the justice of peace administering the oath, might, if he thinks that evidence sufficiently strong, either give his order for the confinement of the party, or add his signature to the certificate.[22]

Scientific Writings on Chemistry and Geology

Parkinson published *The Chemical Pocket Book* (1800) as a short book for beginners to complement the well-known chemistry texts of Lavoisier, Fourcroy, Chaptal, and Nicholson.[1,24] His book provided useful information for novices and inspired questions for more advanced readers. *The Chemical Pocket Book* presents information on elements and compounds that was up to date with early nineteenth century scientific literature. Chemical properties of earths, calorics, light, gases, alkalis, acids, metallic substances, stones, and vegetable and animal substances are outlined. Twenty-one different metals ranging from platina, gold, and silver to arsenic, molybdenite, and tungstenite are described. For example, the entry for gold begins,

> Its colour is orange red, or reddish yellow… Melts at 32°… It may be volatilized and calcined, in high and long continued heats. It is the most perfect, ductile, tenacious, and unchangeable of all the known metals. Not being combinable with Oxygen, Sulphur, &c. in low heats, it can never be found, strictly speaking, mineralized.[24]

He describes bile, blood, urine and other components such as fat, tooth enamel, synovia, and feathers of animals. It is possible that knowledge of these animal substances corresponded with his understanding of human physiology and disease.

Parkinson was a renowned geologist or oryctologist (known today as a *paleontologist*). His museum at No. 1 Hoxton Square housed a notable collection of fossils, shells, metals, coins, and medals. Parkinson's interests in oryctology paralleled scientific advancements in the eighteenth and nineteenth centuries, spanning the period from 1790 to 1820, known as "The Heroic Age of Geology," during which fossils and rocks were studied individually and then in the context of their respective strata.[1] Parkinson's interests in geology and chemistry were intertwined, and his knowledge of chemical properties likely influenced his discovery of using muriatic acid to demonstrate presence of animal membranes in marble fossils.

Parkinson published his first book on geology, *Organic Remains of a Former World. An Examination of the mineralized remains of the vegetables and animals of the antediluvian world generally termed extraneous fossils,* in 1804. It contained about 42 plates with 700 figures, many of which came from his specimen collection. This book became a standard text on paleontology for half a century. The second and third quarto volumes were devoted to "The Fossil Zoophytes" and "Fossil starfish, Echini, Shells, Insects, Amphibia, Mammalia, &c.," respectively. Parkinson also was one of 13 founding members of the Geological Society in 1807. He published "Observations on some of the Strata in the neighbourhood of London, and on the Fossil Remains contained in them" (1811) in the first volume of the *Transactions of the Geological Society*. Two years before his death, Parkinson wrote *Outlines of Oryctology. An Introduction to the Study of Fossil Organic Remains* (1822), which he humbly considered an adjunct to the valuable work, *Outlines of the Geology of England and Wales,* by Rev. W. D. Conybeare and W. Phillips, published several months before his book.[1,25]

Essay on the Shaking Palsy

Parkinson's *Essay on the Shaking Palsy* (1817) is often thought to represent his greatest contribution to medicine.[26,27] The study is based on six cases, some never actually examined by Parkinson but observed on the streets. The five chapters of this 66-page octavo volume include

> I. Definition—History—Illustrative Cases, II. Pathognomic symptoms examined—*Tremor Coactus—Scelotyrbe Festinians,* III. Shaking Palsy distinguished from other disease with which it may be confounded, IV. Proximate cause—Remote causes—Illustrative cases, V. Considerations respecting the means of cure.

Apologizing for mere conjecture regarding the etiology of the shaking palsy, Parkinson states his "duty to submit his opinions to the examination of others, even in their present state of immaturity and imperfection" and mission to inspire research on this disease.[26]

Parkinson recognized the long duration and slowly progressive nature of the disease. His first chapter commences with an often-quoted definition of shaking palsy:

> …involuntary tremulous motion, with lessened muscular power, in parts not in action and even when supported; with a propensity to bend the trunk forward, and to pass from a walking to a running pace: the senses and intellect being uninjured.[26]

Similar tremors were noted years before by Galen, Sylvius de la Boe, Juncker, and Cullen, and these different types of tremor are further discussed in his second chapter. Parkinson illustrates the insidious onset and progression of this condition:

> So slight and nearly imperceptible are the first inroads of this malady, and so extremely slow is its progress, that it rarely happens, that the patient can form any recollection of the precise period of its commencement. The first symptoms perceived are, a slight sense of weakness, with

a proneness to trembling in some particular part; sometimes in the head, but most commonly in one of the hands and arms.[26]

As the disease progresses, other features appear:

After a few more months the patient is found to be less strict than usual in preserving an upright posture: this being the most observable whilst walking, but sometimes whilst sitting or standing. Sometime after the appearance of this symptom, and during its slow increase, one of the legs is discovered slightly to tremble, and is also found to suffer fatigue sooner than the leg of the other side.[26]

Inevitably, a state of immobility and dependence occurs with disturbances of sleep and bodily functions of bowels, speech, and swallowing. Parkinson's descriptions quite accurately elaborate on the cardinal symptoms associated with the disease. He also comments on the asymmetric onset of disease, patients' perceptions of weakness, and problems with sleep, constipation, hypophonia, and sialorrhea. Observations in years following his publication would lead to recognition of other features such as hypomimia, rigidity, and dementia.

The six cases reported differ in severity as well as depth of Parkinson's observation. The first case features a man older than 50 years, with left upper extremity tremor, who we are told succinctly had almost all symptoms reported in Parkinson's first chapter. Parkinson actually examined only three of the six cases directly. Of those observed casually in the street, he includes the following cases:

…a sixty-two year old man with an eight to ten year history of symptoms of tremor, interrupted speech, flexed posture, and gait impairment; a sixty-five year old man with agitation of his whole body, flexed posture, and festinating gait; and a man with "inability for motion except in a running pace."[26]

Case four, a 55-year-old man with trembling of his arms for 5 years and costal inflammation necessitating drainage, was examined but lost to follow-up. The sixth case provides a more comprehensive account of the patient's afflictions—gradually progressive tremor, interrupted speech, constipation, intelligible handwriting, gait disturbance, and inability to feed himself. More striking was the occurrence of a stroke in this patient, which suppressed his tremor while his affected arm was paralyzed.

Parkinson systematically dissects each symptom described in his first chapter. Past distinctions between tremors occurring during terror, anger, advanced age, or palsy were explored. Parkinson comments on the useful classification of tremor at rest (tremor coactus) and action by Sylvius de la Boe (1614–1672) and Sauvages (1706–1767). It is possible that he learned about tremors during his study of Latin and Greek and attendance at John Hunter's lectures.[10] He examines the origins of flexed posture and running gait. Sauvages described this gait as *Scelotyrbe festinans,* "a peculiar species of scelotyrbe, in which the patients, whilst wishing to walk in the ordinary mode, are forced to run" and differentiated it from *Chorea Viti:*

…the patients make shorter steps, and strive with a more than common exertion or impetus to overcome the resistance; walking with a quick and hastened step, as if hurried along against their will. Chorea Viti… attacks the youth of both sexes, but this disease only those advanced in years.[26]

Parkinson differentiated tremor observed in the shaking palsy from that seen in apoplexy, epilepsy, worms, alcohol and caffeine use, and advanced age. He used the term "palsy" as a synonym for weakness and did not appreciate the unique quality of bradykinesia. He did not discuss rigidity. These distinctions were added later by Trousseau in his "Fifteenth Lecture on Clinical Medicine."[28]

The *medulla spinalis* and *medulla oblongata* were the proposed neuroanatomical localization of this disease. Cases with features similar to the shaking palsy were suspected to have involvement of the medulla with "some slow morbid change in the structure of the medulla, or its investing membranes, or theca, occasioned by simple inflammation, or rheumatic or scrophulous affection."[26] Despite contributions of spine fractures, venereal disease, and rheumatism in these cases, Parkinson suggests how pathology in the medulla might underlie the weakness, gait disturbance, and bulbar symptoms seen in these examples. Parkinson's arguments for involvement of the medulla spinalis and oblongata reflect understanding of the nervous system in the eighteenth and nineteenth centuries. It was not until the latter part of the nineteenth century, with observations and anatomo-clinical correlates on amyotrophic lateral sclerosis, tabes dorsalis, and multiple sclerosis from physicians such as Charcot, that functions of the brain and spinal cord were better understood.[29] The nigral degeneration implicit to Parkinson's disease was not suggested until the late nineteenth century and not systematically studied until Tretiakoff in his 1919 thesis.[1,30] Parkinson's belief, however, in clinical and pathological correlations to understand disease mechanisms is clearly stated:

Before concluding these pages, it may be proper to observe once more, that an important object proposed to be obtained by them is, the leading of the attention of those who humanely employ anatomical examination in detecting the causes and nature of diseases, particularly to this malady. By their benevolent labours its real nature may be ascertained, and appropriate modes, of relief, or even of cure, pointed out.[26]

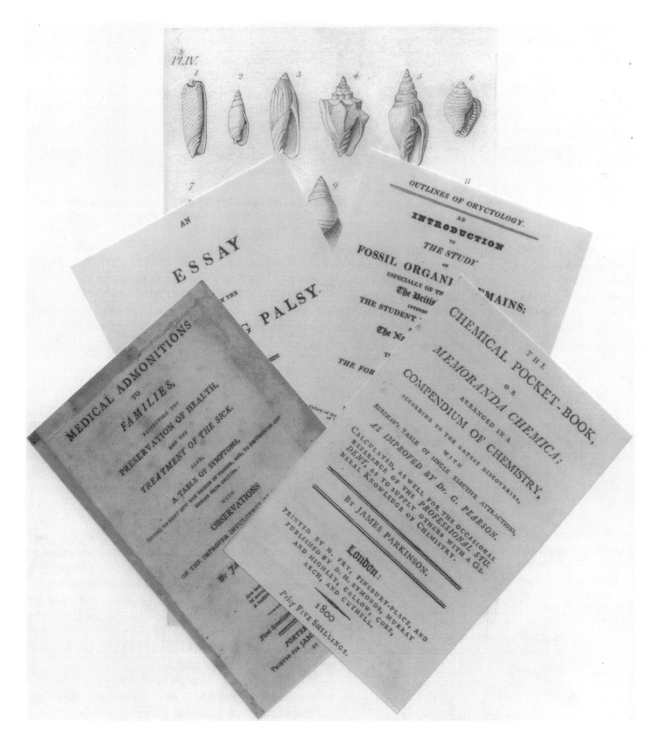

FIGURE 1.1   Title pages from Parkinson's *Chemical Pocket Book, Medical Admonitions, Essay on the Shaking Palsy,* and *Outlines of Oryctology,* and Plate IV from *Outlines of Oryctology,* from the collection of Dr. Christopher Goetz (gift of Dr. Robert Currier).

In his final chapter, Parkinson enumerates treatments for the shaking palsy. He expresses the potential for neuroprotection as

…it seldom happens that the agitation extends beyond the arms within the first two years… in this period, it is very probable, that remedial means might be employed with success: and even, if unfortunately deferred to a later period, they might then arrest the farther progress of the disease, although the removing of the effects already produced, might be hardly to be expected.[26]

Recommended treatments were bleeding from the upper part of the neck, application of vesicatories, and resultant drainage of purulent discharge; the use of internal medicines was not justified until more knowledge of the disease was available.

Overall, his essay was well received in the English medical community. The first review, appearing in *The London Medical and Physical Journal* in 1818, did criticize his speculation on localization and causes but largely recommended it for reading.[31] Other reviews in *The London Medical Repository* and *The Medico-chirurgical Journal* praised his observations and excused his speculations on the basis of his respectable reputation.[31] It was not until the 1860s that Charcot and Vulpian coined Parkinson's disease as an eponym for paralysis agitans.

Parkinson's Legacy

Parkinson's legacy encompasses not only medical works such as the well known *Essay on the Shaking Palsy,* but also a diverse assortment of writings on politics, medical care, chemistry, and geology. Although these writings provide insight into Parkinson's character, no portrait of him exists. A plaque designates his house in Hoxton Square, now a factory site, and an inscribed marble tablet, a gift by St. Leonard's Hospital to commemorate his bicentennial in 1955, can be seen in St. Leonard's Church.[1,32] Several pieces from his fossil collection are in possession of the British Museum of Natural History.[32] His *Essay on the Shaking Palsy* is nearly impossible to find in original, although several reprints exist. Other writings can be found in libraries and among antiquarians. These writings share the same sense of combined clarity, humility, and competence revealed in the celebrated *Essay.* Charcot's proclivity for eponyms introduced the term Parkinson's disease, a name that has retained its place of continued honor in modern neurology.

REFERENCES

1. Morris A.D. *James Parkinson: His life and times.* Birkhauser, Boston, 1989.
2. Morris A.D. James Parkinson. *The Lancet* 1955; 761-763.
3. A. Hoxton and Shoreditch Walk: Route and what to see. From www.london-footprints.co.uk
4. Shoreditch and Hoxton – Medieaval Village. From http://learningcurve.pro.gov.uk
5. Rowntree L.G. James Parkinson. *Bulletin of the Johns Hopkins Hospital* 1912; 23: 33-45.
6. Critchley, M. (Ed.). James Parkinson (1755-1824). Macmillan and Co., Ltd., New York, 1955.
7. Jefferson M. James Parkinson 1755-1824. *British Medical Journal* 1973; 2: 601-603.
8. Parkinson J. *"Some account of the effects of lightning."* Memoirs of the Medical Society of London, 1789; Vol. 2 pp.193; 493-503. In Morris A.D. *James Parkinson: His life and times.* Birkhauser, Boston, 1989.
9. Bett, W. R., James Parkinson: Practitioner, pamphleteer, politician and pioneer in Neurology. *Proceedings of the Royal Society of Medicine* 1955; 48: 865.
10. Currier, R. D., Did John Hunter give James Parkinson an idea? Archives of Neurology 1996; 53: 377-378.
11. Allen, E., Turk J. L., Murley R. (Eds.). The case books of John Hunter FRS. Royal Society of Medicine Services Limited, London, 1993.
12. Gloyne, S. R., John Hunter. E & S Livingstone, Ltd., Edinburgh, 1950.
13. Parkinson, J., *The Hospital pupil; Or an essay intended to facilitate the study of Medicine and Surgery. In Four Letters.* H. D. Symonds, London, 1800.
14. Parkinson J., *Pearls cast before swine,* by Edmund Burke, scraped together by Old Hubert. D. I. Eaton, London, 1793.
15. Parkinson J., *An address to the Hon. Edmund Burke from the swinish multitude by Old Hubert.* J. Ridgeway, London, 1793.
16. Tyler, K. L., Tyler, H. R. The secret life of James Parkinson (1755-1824): *The writings of Old Hubert.* Neurology 1986; 36: 222-224.
17. Parkinson, J., *Dangerous sports: A tale addressed to children.* H. D. Symonds, London, 1808.
18. Parkinson, J., *Medical admonitions to families, respecting the preservation of health, and the treatment of the sick.* 4th ed., H. D. Symonds, London, 1801.
19. Parkinson, J., *The Villager's friend and physician; or a familiar address on the preservation of health.* H.D. Symonds, London, 1800.
20. Currier, R. D., Currier, M. M., James Parkinson: on child abuse and other things. *Archives of Neurology* 1991; 48: 95-97.
21. Parkinson, J., *Remarks of Mr. Whitbread's plan for the education of the poor.* H. D. Symonds, London, 1807.
22. Parkinson, J., *Mad-Houses: Observations of the act for regulating mad-houses.* Whittingham and Rowland, London, 1811.
23. McMenemey, W. H., A note on James Parkinson as a reformer of the Lunacy Acts. *Proceedings of the Royal Society of Medicine* 1955; 48: 593-594.
24. Parkinson, J., *The Chemical pocket book; Or memoranda chemica; Arranged in a compendia of Chemistry.* D.H., Symonds, London, 1800.
25. Parkinson, J., *Outlines of Oryctology: An introduction to the study of fossil organic remains.* London, 1822.
26. Parkinson, J., *An Essay on the Shaking Palsy* (1817). In Medical Classics 1938; 2:10.
27. Williamson, R. T., James Parkinson and his essay on paralysis agitans. *Janus* 1925; 29: 193-197.
28. Pearce, J. M. S., Aspects of the history of Parkinson's Disease. *Journal of Neurology, Neurosurgery, and Psychiatry* 1989; 6-10.
29. Goetz, C. G., Bonduelle, M., Gelfand, T., Charcot: *Constructing Neurology.* Oxford University Press, New York, 1995.
30. Duvoisin, R. C., A brief history of parkinsonism. *Neurologic Clinics* 1992; 10: 301-316.

31. Herzberg, L., An essay on the shaking palsy: Reviews and notes on the journals in which they appeared. *Movement Disorders* 1990; 5: 162-166.

32. Eyles, J. M., James Parkinson (1755-1824). *Nature* 1955; 176: 580-581.

2 Paralysis Agitans—Refining the Diagnosis and Treatment

Lawrence Elmer

Department of Neurology, Medical College of Ohio

CONTENTS

During the decades after the publication of James Parkinson's seminal article,[1] the definition of paralysis agitans was debated, discussed, and further refined through observations and publications in the medical community. This process, common to all newly described disease entities, clarified the essential clinical features of paralysis agitans, the treatment of this disorder, and the pathology underlying the syndrome. Historically, progress in understanding Dr. Parkinson's namesake was not without misunderstandings and confusion before a more distinct and accurate picture of this disease developed.

Physicians worldwide tried to diagnose and describe cases of paralysis agitans in light of the description given by Dr. Parkinson. Initially, most of these manuscripts were simply restatements of the original clinical description and appeared in textbooks, sometimes quoting directly from Parkinson's treatise. Other articles described cases that were clearly not parkinsonian in nature, reflecting other neurological diseases. However, some of the clinical descriptions met the diagnostic criteria of Parkinson's disease and thus give insight into the clinical acumen of this age.[2] In this chapter, examination will first be made of the early reports and descriptions of cases corresponding to paralysis agitans prior to Charcot. Subsequently, the contributions of Charcot in clarifying and expanding on the diagnosis of paralysis agitans will be reviewed. Finally, the contributions of neurologists and other physicians to the diagnosis and management of Parkinson's disease after Charcot and before the advent of levodopa therapy will be discussed.

THE EARLY YEARS (1817–1861)

One of the most prolific writers during this time period was Dr. John Elliotson, a practitioner from St. Thomas' Hospital in London. While Dr. Elliotson described some cases that were clearly not parkinsonian, his writings and lectures reveal much about the diagnosis and treatment of neurological disorders in the first two decades after Dr. Parkinson's publication.

In his earliest article, Elliotson first comments on a common practice in the management of neurological diseases, the oral administration of subcarbonate of iron. A multitude of neurological disorders were reportedly cured by the administration of 15 to 30 gr of the iron subcarbonate per dose repeated until the desired benefit was obtained. For example, for the treatment of neuralgia, the usual quantity was described as 90 gr over a period of 24 h. (Elliotson mentioned that Dr. Sydenham's favorite form of the iron subcarbonate for administration to patients was iron filings but others suggested that simple rust was just as effective as any other preparation and could actually be easier on the stomach!)[3]

Elliotson described multiple cases of nonparkinsonian movement disorders, commonly chorea in children and young adults. These patients were given the typical prescription of subcarbonate of iron, leeches, blistering agents, and purgatives. In many of these cases, the patients were described as exhibiting significant improvement in their choreiform disorders. However, Dr. Elliotson went on to lament,

> I have failed with the largest quantities of iron in epilepsy, cancer, and lupus, but found it very beneficial in chronic neuralgia, and various chronic ulcerations and chronic pustular diseases, as well as those diseases of debility in which it is so justly celebrated.[3]

Elliotson's lecture then elaborates on his first case of paralysis agitans, which he treated with the iron subcarbonate regimen. The patient was described as a male, 28 years of age, who had symptoms approximately 1 year before his encounter with Elliotson. The clinical symptoms consisted of constant shaking of the legs and arms of variable intensity. According to Elliotson,

> Till within the last week, the agitation would sometimes cease for a few hours or even the whole day, but for the last week (it) has been constant. At first, sometimes only one leg was affected, sometimes both. He has pain in the head, loins, and legs, and vertigo, and cannot fix his attention.[3]

The first treatment described for this unfortunate young man involved the application of tartarized antimony ointment to produce pustules on both legs. He then received orally administered oil of turpentine followed by zinc sulfate, but this was stopped due to nausea. Finally, he received the subcarbonate of iron prescription accompanied by the administration of leeches to his temples daily. Within four days of the subcarbonate of iron and leech administration, the patient was described as feeling better—the dose of the subcarbonate of iron was increased and the leeches were continued. Within 28 days following the initiation of this prescription, the patient was described as dramatically improved, at least insofar as the shaking was concerned. Unfortunately, the pustules, leeches, and treatments other than the subcarbonate of iron had failed to relieve the patient's headache, vertigo, other pain, and smarting of his eyes. Therefore, all treatments other than the subcarbonate of iron were discontinued. The patient significantly improved. He continued to receive the subcarbonate of iron prescription for a total of six weeks. At the end of this time, the patient was discharged from the hospital and was seen several months later in perfect health.[3] In this case, the presumptive diagnosis was "paralysis agitans," while the true diagnosis almost certainly was not!

Dr. Elliotson went on to describe multiple other cases referred to as paralysis agitans. Most of these cases probably did not represent paralysis agitans at all, which confused the diagnosis for students of neurology during that day. Several of these cases are reviewed here and in other historical reviews.[4] Unfortunately, the apparent effectiveness of the treatments administered may have resulted from the fact that the cases did not represent true paralysis agitans. Elliotson himself comments on the lack of efficacy of these treatments in some individuals, most likely those with true paralysis agitans.[3]

In a separate publication,[5] Dr. Elliotson attends to another presumed positive case of paralysis agitans. This monograph describes an individual 54 years of age admitted under the care of Dr. Elliotson. His symptoms began after a frightening episode in which he fell into water. Following that all of his extremities were described as a "continual state of tremor: head and jaw also affected." He was given subcarbonate of iron four times daily with alternating cold and warm baths. Reportedly, he recovered but then relapsed after a fortnight, "when he was suddenly seized with pain in the head, and giddiness, soon followed by his old complaint; since then, articulation has been indistinct, and his superior and inferior extremities have been in a constant state of tremor, and he has constantly complained of pain in the head." He was readmitted to the hospital, treated extensively with various preparations and was felt to have significant improvement. He was subsequently discharged from the hospital however, due to his frequent habit of drinking alcohol on short jaunts away from a hospital![5]

In another case presentation from March 21, 1831, Dr. Elliotson relates a case of paralysis agitans that resulted from fright.

> I spoke of this disease before. The patient was a man 50 years of age; and usually, I believe, it arises at such an age from an organic cause, for I have never been able to cure a person of it at or after middle life. I cured one between 30 and 40 years of age, but he was the oldest. In this case it came on from fright, and therefore there may be nothing organic, and perhaps I shall cure him. He is taking the carbonate of iron, and is much better. The man regularly receives me with a smile, and fancies he is getting well. It would not be right at present to give a decided prognosis.[6]

It is likely that these cases represented essential tremor or tremulousness associated with toxins or psychiatric symptoms, simply misdiagnosed as paralysis agitans. However, Dr. Elliotson did attempt to distinguish paralysis agitans from the tremulousness seen in alcoholics.

> This disease is to be carefully distinguished from the tremulous motions with which drunkards are affected. It is entirely distinct from the effect produced by habitual intoxication . . . . The shaking is continuous, and it is only

by discontinuing their use that the tremors cease. This trembling, too, will be produced temporarily by occasional strong doses only. It generally, also, affects both hands, and is seen chiefly when any effort is made by the individual; if, for instance, a pen is taken in the hand, a shaking comes on the moments an attempt is made to write; or if a cup or glass be lifted, the contents are spilled over. The greater, too, the effort which is made, the more excessive is the tremor that follows. But in paralysis agitans exactly the reverse of this is observable, for a strong effort will, for the time, overcome the disease. By this, and by the affection occurring pretty equally in both hands, you may distinguish nervous trembling from paralysis agitans. You are aware that strong passions, as fear and rage, will also, like strong tea, coffee, or tobacco, produce a trembling.[7]

The misdiagnoses described above should not discount the value of this physician's contributions. It is likely that many other physicians attempted to accurately diagnose this newly described syndrome of "paralysis agitans." It is also likely that while many incorrect diagnoses followed for those physicians as well, Elliotson had the courage to record his observations and opinions permanently for future evaluation. Indeed, Elliotson did observe and correctly diagnose some cases of "paralysis agitans." In one of his clinical lectures on paralysis agitans, delivered on October 11, 1830, Elliotson described a patient of age 38 who developed symptoms that were interpreted as representing the shaking palsy. In this clinical discussion, Dr. Elliotson expounds on the symptoms of shaking palsy, describing characteristic features of the tremors, gait, festination, dysphagia, hypophonia, and ultimately death.[7]

Now this disease usually commences in some one part of the frame, as, for instance in the head; but it more frequently begins in one hand, or in the arm; there it will sometimes remain for many months, and even for years, before it spreads, and perhaps it never spreads at all. Sometimes, however, it increases in degree and extent, and other parts become affected, until, at last, the whole body is in a constant shake. Though the tremulous motions in this disease are involuntary, yet they may be checked by an effort of the will. The effort exerted, however, must be of a powerful nature, and then it will for a few moments stop the shaking. As the disease extends, first one extremity and then another becomes affected, at length the head and trunk bend forwards, the individual walks in some measure upon his toes, the motion of walking becomes gradually quickened, at last it is altogether lost, and the man unconsciously gets into a trot, and has all the appearance of a person in a most violent hurry. This change is owing to the disease being slightly under the will. The individual who was afflicted, finds that a powerful exertion of the muscles will stop the tremors, and as running requires more effort than walking, running answers better to control them; or we may say, that when he is walking,

the same effort which he makes to check them, forces him to run, which state he continues, because he finds that he thus partly conquers the tremulous motions,—that they do not so much get the better of him and impede him. In the usual progress of paralysis agitans, the voice is not affected until the muscles of the upper extremity and head have been so for a long time. At last, however, speech becomes involved, and the muscles employed in the acts of deglutition and mastication are affected, and speaking, chewing, and swallowing, are extremely difficult to be performed. By and by the urine and feces pass away involuntarily, general emaciation ensues, entire decay of the powers, and ultimately death.[7]

In some cases, the physicians of this era did record their observations. As mentioned above, many cases probably did not reflect "paralysis agitans" but rather involved another neurological disorder. For example, Dr. Thomas Gowry describes a case of paralysis agitans "intermittens" in which the symptoms were intermittent in nature and cured by treatment. A woman of age 26 presented as a patient to Dr. Gowry on June 18, 1831. She was experiencing involuntary tremors of the upper and lower extremities which would continue for about five or six minutes at a time and occurred up to three times per hour. During that time she had complete loss of function of her limbs. Her lips and mouth were also involved, the tongue protruded intermittently, and the paroxysm was terminated with a heavy sigh. The patient also complained of vertigo, heaviness of her head, and great difficulty supporting her head on her shoulders. Her treatment consisted of cathartics, following which she developed syncopal episodes. A prescription change occurred and she underwent other therapy including leeches resulting in complete resolution of her symptoms![8]

In a case presentation dated September 6, 1839, Dr. Matthew Gibson described a case of "paralysis agitans." A girl, age 14, developed "constant and violent involuntary motion or shaking of the right forearm, and slightly of the arm; the motion is so violent that it cannot be stopped, though held down." The patient was treated with blistering agents over her spinal column (described as tender) and was given a combination of calomel, opium, and cinnamon. Four days later this case of "paralysis agitans" was reportedly almost completely cured.[9]

Another popular treatment for diseases of the nervous system in the mid 1800s was the application of direct galvanic current. Again, it is unclear whether the symptoms undergoing treatment truly represent "paralysis agitans" at all. One such case is described in a publication of the Lancet dated December 3, 1859, where Dr. Russell Reynolds treated a case of "paralysis agitans." The patient was a male 57 years old, who for the prior 2 years before his examination had experienced intermittent tremors of the right arm and leg. These tremors were present with multiple inciting variables including exasperation, excessive physical exertion, cold temperatures, attempting to drink liquid substances, or fully extending the arm and forearm. He came to the attention of Dr. Reynolds when he had an episode of severe vertigo, aching in his knee joints, and a feeling of general disturbance. At this point, the shaking was reported as violent in the right upper extremity.

The tremors occurred with activity and were aggravated by emotional upset or any attempt at voluntary movement of the extremity. The treatment for this tremor was the application of direct galvanic current to the right arm and forearm. After five minutes of electrical stimulation, the tremor was noted to be absent and emotional excitement failed to reproduce the exacerbation of tremor seen previously. The tremors returned after three hours following the first treatment with electrical stimulation. Subsequent application of electrical stimulation was applied to the arm on a daily basis followed by an every other day basis for a total of almost two months. At this point, the patient's tremor was felt "cured" except for a very slight tremulousness when he raised something to his mouth.

This dramatic improvement in the subject's symptoms was regarded as evidence that electricity could cure a case of "paralysis agitans." Dr. Reynolds proceeded to write,

> ...the above case requires, I think, no comment. It is more important that a fact of this character should be placed on record than that any speculation should be advanced in regard to the pathology of paralysis agitans, or the modus operandi of the continuous galvanic current. The term which I have employed to denote the case involves no theory; it is but the name of a prominent symptom—a symptom which, in this instance, constituted almost the whole of the affection, and which, after a fortnight's duration without the slightest tendency to improvement, was quickly, but progressively and effectually, removed by a special form of treatment.

Clearly, it was the opinion of Dr. Reynolds that the application of electrical current to the limbs of patients with "paralysis agitans" was the definitive treatment requiring no further explanation. Unfortunately, the clinical description appears far more typical for a patient with essential or familial tremor.[10]

Some physicians concentrated solely on the gait abnormalities in paralysis agitans, using that symptom as the only criterion for a diagnosis of "paralysis agitans." For example, in a clinical description from 1855, Dr. Paget outlines the case of a 41-year-old male who developed giddiness 6 weeks before his first examination. He subsequently experienced loss of vision and was found by his wife at home reclining in a chair, unresponsive, with one side of his mouth drawn and both arms and legs rigid and immovable. His coma lasted about 24 h, after which

time he recovered use of his arms and legs. When he presented to the hospital, he was weak and could not walk without assistance. He couldn't feed himself and had bowel and bladder incontinence. He also experienced emotional incontinence and apparently had some difficulty with language. Due to his involuntary tendency to fall precipitately forwards, this patient was considered to have a tentative diagnosis of "paralysis agitans."[11] Again, unfortunately, the diagnosis is certainly in question and reflects a measurable amount of uncertainty during this time period about what truly constituted a diagnosis of "paralysis agitans."

THE CONTRIBUTIONS OF CHARCOT

While multiple authors attempted to refine the clinical description of "paralysis agitans," the diagnosis was still confounded until J.-M. Charcot and colleagues evaluated, diagnosed, and accurately revised, refined, and reiterated the clinical description of James Parkinson. These physicians contributed a wealth of insight and clarification to the original treatise, mostly as a result of exposure to large numbers of invalid patients at the La Salpetriere in Paris. Charcot provided clarification of the differential diagnosis between paralysis agitans and other tremor disorders. Charcot was also the first to refer to this disorder as "Parkinson's disease."

Charcot's lectures and publications were all in French but fortunately were translated in 1877 by G. Sigerson.[12] One of Charcot's greatest contributions was the differentiation of Parkinson's rest tremor from the kinetic tremor associated with essential and familial tremor as well as tremors associated with multiple sclerosis.

CHARCOT ON TREMOR

Initially, Charcot distinguishes between the rest tremor and the kinetic tremor.[12] As mentioned in the paragraphs above, accurate diagnosis of Parkinson's disease was confounded by the presence of multiple other tremorogenic disorders, including essential tremor, tremor associated with multiple sclerosis, chorea, as well as abnormal movements from intoxicating substances including alcohol. Charcot eloquently differentiates between the tremors of Parkinson's disease and the tremors of multiple sclerosis,

> Let it suffice that we have put prominently forward those characters which can be recognized by the simplest observation, irrespective of any theoretical prepossession. It is because these have not been considered, that the two affections which are to form the object of our first clinical studies— paralysis agitans and disseminated sclerosis— have remained until today, confounded under the same rubric, although they are, in every respect, perfectly independent of each other. Both, indeed, reckon tremor

amongst their most important symptoms; but, in the first, the rhythmical oscillations of the limbs are nearly quite permanent, whilst in the second they only supervene on the attempt to execute intended movements.[12, pp. 132,133]

CHARCOT ON THE FUNDAMENTAL CHARACTERISTICS OF PARKINSON'S DISEASE

In the next section of his lectures, Charcot gives an overview of Parkinson's disease with respect to its etiology and diagnostic features. He begins by proposing that the disease itself is a neurosis, absent of a true physiological and pathological basis. This conclusion Charcot based on the absence of clear neuropathological abnormalities seen in postmortem examination of patients who died with Parkinson's. He states, "Paralysis agitans, separated from foreign elements, is, gentlemen, at present a neurosis, in this sense that it possesses no proper lesion."[12, p. 134]

Charcot next turned his attention to the age of onset and the potential causes of Parkinson's disease.

It assails persons already advanced in age, those especially who have passed their fortieth or fiftieth year. Frequently the causes remain unknown. However, of the etiological data two deserve to be cited: 1, damp cold, such as that arising from a prolonged sojourn in a badly ventilated apartment, or in a low dark dwelling on the ground floor; 2, acute moral emotions. The latter cause appears to be tolerably common.[12, pp. 134,135]

The discussion of the pertinence and applicability of symptoms in the diagnosis of Parkinson's disease serves as the next topic of discussion. In this discourse, Dr. Charcot refers to the tremors, postural instability, and bradykinesia.

The symptoms of paralysis agitans are not all of the same value. The most striking symptom consists of the tremor, existing even when the individual reposes, limited at first to one member, then little by little becoming generalized, whilst respecting, however, the head. To this phenomenon is superadded sooner or later an apparent diminution of muscular strength.... A singular symptom is that which, frequently at an early, but usually at a late period, comes to complicate the situation—the patient loses the faculty of preserving equilibrium whilst walking. In some patients also we notice a tendency to propulsion or to retropulsion.... A peculiar attitude of the body and its members, a fixed look, and immobile features should also be enumerated among the more important symptoms of this disease.[12, p. 135]

Charcot then turned his attention to the progressive nature of Parkinson's disease as well as the circumstances surrounding the demise of individuals afflicted with this particular disorder.

The march of paralysis agitans is slow, and progressive. Its duration is long—sometimes it has gone on for 30 years. The fatal term supervenes either by the advance of age, or because of intercurrent diseases. In the first case, an acute disease, such as pneumonia, occurs. In the second, death takes place from a sort of nervous exhaustion, nutrition degenerates, the patient cannot sleep, eschars are formed and conclude the morbid scene.[12, pp. 135,136]

CHARCOT ON THE PROGRESSIVE NATURE OF PARKINSON'S DISEASE

What follows next in Charcot's discourse is a more elaborate and detailed discussion on the "manner of its invasion," when referring to the shaking palsy.

In the immense majority of cases, the invasion is insidious, the disease first showing itself as slight and benignant. The tremor is circumscribed to the foot, the hand, or the thumb. At this stage of the disease the tremor may be merely passing and transitory. It breaks out when least expected, the patient enjoying complete repose of mind and body, and it frequently occurs without his being conscious of it. The act of walking (even where the upper extremities are affected), the act of grasping, lifting, taking a pen, writing, any effort at all of the will, may at this epoch often suffice to suspend the tremor. Later on, it will be no longer so. Moreover, as it augments in intensity and persistence, the tremor invades little by little, and not without observing certain rules in its progress, the parts which have hitherto remained sound. If, for instance, it first affected the right hand, at the end of some months or of some years, the turn of the right foot will come; next the left hand, and after that the left foot will be, successively, assailed.[12, p. 136]

Charcot deliberately pointed out that the symptoms of Parkinson's usually began unilaterally, only later to cross over to the unaffected side. However, he had seen cases where the symptoms of Parkinson's began on opposite sides of the body. He also elaborated on the nonmotor aspects of paralysis agitans, such as fatigue and pain.

Decussated invasion is more rare. I have, however, at least twice seen the affection first seize the right upper extremity, and at next to the left lower extremity. It is much more common to see the tremor confined for a long time to the members of one side of the body (hemiplegic type), or to the two lower extremities (paraplegic type). The tremor is not absolutely the first symptom recorded. It may possibly be preceded sometimes by a very remarkable feeling of fatigue, sometimes by rheumatoid or neuralgic pains, which are occasionally most severe, occupying the member or the regions of the members which shall soon be seized, but secondarily, by the convulsive agitation.[12, pp. 136,137]

Charcot distinguishes the slow invasion from the abrupt invasion. The abrupt invasion usually followed an emotional trauma, and frequently only involved one limb. The symptoms usually would persist only for days and then depart just as quickly. However, most of these patients would later manifest the symptoms of Parkinson's in a more persistent manner.

> After persisting for a few days it may possibly improve or even vanish. But, later on, after a series of alternate improvements and exacerbations, it takes up its abode in a permanent manner.[12, p. 137]

Finally, Charcot defines the period of stationary intensity. This is best described as a period of time when the tremor has been completely established in the patient's life but fluctuations in the intensity of that tremor still occur. Emotional upset and other crises were noted as situational protagonists of the tremor.[12, pp. 137,138]

CHARCOT ON BRADYKINESIA AND RIGIDITY

Charcot clearly describes masked facies as an essential component of Parkinson's disease. While James Parkinson mentioned face and voice abnormalities in his clinical cases and definitively described bradykinesia in his patients, he did not elaborate on this symptom nor did he include masked facies in his definition of shaking palsy. Charcot elaborates extensively on this topic.

> Far from trembling, the muscles of the face are motionless, there is even a remarkable fixity of look, and the features present a permanent expression of mournfulness, sometimes of stolidness or stupidity. There is no real difficulty of speech, but the utterance is slow, jerky, and short of phrase: the pronunciation of each word appears to cost a considerable effort of the will. Finally, the patients seem to speak between their teeth. Deglutition is accomplished with ease, though perhaps slowly; frequently in cases of somewhat old standing the saliva, accumulated in the mouth, is involuntarily allowed to escape."[12, pp. 139,140]

Charcot is recognized as the first prominent physician-scientist to describe and document the rigidity seen in Parkinson's disease. This symptom was either missed or not documented by James Parkinson. Regardless, Charcot contributed significantly to the accurate diagnosis of Parkinson's disease by his observations and clear description of the change in muscle tone. "

> We shall now point out the characteristic which, we believe, was overlooked by Parkinson as well as by most of his successors: we allude to the rigidity to be found, at a certain stage of the disease, in the muscles of the extrem-

ities, of the body, and, for the most part, in those of the neck also. When this symptom declares itself, the patients complain of cramps, followed by stiffness, which, at first transient, is afterwards more or less lasting, and is subject to exacerbations. Thus on account of the rigidity of the anterior muscles of the neck, the head, as Parkinson remarked, is greatly bent forward, and, as one might say, fixed in that position; for the patient cannot, without much effort, raise it up, or turn it to the right or left. The body also was almost always slightly inclined forward, when the patient is standing.[12, p. 140]

The posture of the hands and the upper extremities was documented in his lectures and recorded in drawings.

> The elbows are habitually held a little apart from the chest, the forearms being slightly flexed upon the arms; the hands, flexed upon the forearms, rest upon the stomach …Commonly, the thumb and index are extended and apposed, as if to hold a pen; the fingers, slightly inclined towards the palm, are all deviated outwards to the ulnar side.[12, p. 141]

In reference to the bradykinesia, Dr. Charcot states,

> You will readily discover, in some of the patients whom I have shown you, that laboriousness in the execution of movements which is dependent neither on the existence of tremors, nor on that of muscular rigidity; and is somewhat attentive examination will enable you to recognize the significant fact that, in such cases, there is a rather retardation in the execution of movements than real enfeeblement of the motor powers. The patient is still able to accomplish most of the motor acts, in spite of the trembling, but goes about performing them with extreme slowness. We noticed this fact a few moments ago, in its relation to the faculty of speech; there is a comparatively considerable lapse of time between the thought and the act.[12, p. 144]

CHARCOT ON GAIT DISTURBANCES IN PARKINSON'S DISEASE

Charcot also described freezing of gait, festination, propulsion, and retropulsion.

> Yet a word upon the gait peculiar to patients affected by paralysis agitans. You have seen some of our patients get up slowly and laboriously from their seats, hesitate for some seconds to step out, then, once started, go off in spite of themselves at a rapid rate. Several times they threatened to fall heavily forward. Does this irresistible tendency to adopt a running pace depend exclusively on the center of gravity being displaced forward by the inclination of the head and body? There are, in fact, certain patients who, in contradistinction to those described, tend to run backwards when in motion, and to fall backwards,

although their bodies are manifestly inclined forward. Besides, propulsion, like retropulsion, is not absolutely connected with the bent attitude of the body, for it is sometimes seen at an early period of the disease, even before there is any inclination of the body at all.[12, p. 145,146]

CHARCOT ON NONMOTOR SYMPTOMS OF PARKINSON'S DISEASE

There is also elaboration on the more subtle physical, emotional and psychiatric nuances of paralysis agitans.

Paralysis agitans is not merely one of the saddest of diseases, in as much as it deprives the patients of the use of their limbs, and sooner or later reduces them to almost absolute inaction; it is also a cruel affection, because of the unpleasant sensations which the sufferers experience. Usually, indeed, (the neuralgic cases which we have already described being excepted), they are not affected by acute pains, but by disagreeable sensations of a special order. They complain of cramps, or rather of a nearly permanent sensation of tension and traction in most of the muscles. There is also a feeling of utter prostration, of fatigue, which comes on especially after the fits of trembling; in short, an indefinable uneasiness, which shows itself in a perpetual desire for change of posture. Seated, the patients every moment feel obliged to get up; standing, after a few steps they require to sit down. This need for change of position is principally exhibited at night in bed by the more infirm, who are incapable of attending on themselves. The nurses charged with their care will tell you: "They must be turned now on the right side, now on the left, now on the back." Half an hour, a quarter of an hour, has scarcely elapsed until they require to be turned again, and if their wish be not immediately gratified they give vent to moans, which sufficiently testify to the intense uneasiness they experience.[12, p. 147]

Charcot described symptoms in his patients most consistent with autonomic dysfunction and temperature deregulation, which are nonetheless described as extremely uncomfortable in nature from the standpoint of the patient.

But there is one very troublesome sensation which the patients experience, in which I have not found mentioned in any description; this is an habitual sensation of excessive heat, so that you shall see them in the heart of winter throw off the bed clothes, and in the daytime only retain the lightest garments. All the cases under our charge give evidence in favor of this assertion. It appears to obtain its maximum after the paroxysms of trembling, and is then frequently accompanied by profuse perspiration, which is sometimes so great as to necessitate a change of linen; but it may also be found in patients who do not thus perspire and who are but little troubled with tremor.[12, p. 147]

CHARCOT ON THE TERMINAL EVENTS OF PARKINSON'S DISEASE

Charcot details the events leading to death in patients with Parkinson's disease.

The affection pursuing its course, the difficulty of movement increases, and the patients are obliged to remain, the whole day long, seated on a chair, or are altogether confined to the bed. Then, nutrition suffers, especially the nutrition of the muscular system. There may supervene, as I have twice observed, a genuine fatty wasting of the muscles. At a given moment, the mind becomes clouded and the memory is lost. General prostration sets in, the urine and feces are passed unconsciously, and eschars appear upon the sacrum. In such cases, the patients succumb to the mere progress of their disease, by a sort of exhaustion of the nervous system; and it is perfectly true, as several authors have remarked, that at this terminal period the tremor, however intense it was before, is frequently seem to diminish and even to disappear.[12, p. 149]

CHARCOT ON TREATMENT ALTERNATIVES FOR PARKINSON'S DISEASE

Lastly, Charcot comments on some of the treatment alternatives, although he seems to have taken a skeptical view of its effectiveness.

Everything, or almost everything, has been tried against this disease. Among the medicinal substances that have been extolled, in which I have administered without any beneficial effect, I need only enumerate a few. Strychnine, praised by Trousseau, appears to me rather to exasperate the trembling than to calm it. Ergot of rye and belladonna, recommended on account of their anticonvulsive qualities, have not yielded any very profitable results. The same verdict must be given in reference to opium, which, on the contrary, augments reflex excitability, and which was supposed capable of moderating the tremor because of diminishing the pain. Latterly I have made use of hyoscyamine, from which some patients have obtained relief; its action, however, is simply palliative.[12, p. 155]

This is one of the first descriptions of a centrally acting anticholinergic medication used for the treatment of Parkinson's disease. Subsequent practitioners drew from Charcot's experience and treated patients for years with various preparations of other derivatives of hyoscyamine.

While Charcot speculated extensively on the pathology and etiology of paralysis agitans, no significant contribution was made at this time.[12, pp. 150–152] However, his work allowed future neuropathologists to distinguish Parkinson's disease from the multitude of other neurological disorders with which this disorder was confused. In a very real sense, then, the careful scientific documentation of

Charcot set the stage for the determination of the precise pathological abnormality causing Parkinson's.

PARKINSON'S DISEASE AFTER CHARCOT

It is not clear that Dr. Parkinson distinguished the rigidity of paralysis agitans from the spasticity associated with spinal cord damage, mentioning repeatedly his belief that the pathology of paralysis agitans resided in the medulla. In 1871, Dr. Meynert described damage to the corpus striatum and the lenticular nucleus.[13] This may have been the first suggestion that the tremor of paralysis agitans as well as chorea might involve the basal ganglia. Drs. Murchison and Cayley described a case in 1871 of paralysis agitans.[14] There was shrinkage of the cerebral hemispheres, thickening of the spinal cord and infiltration of the spinal cord with connective tissue and inflammatory cells probably related to typhus. In 1878, Dr. Dowse described a case of a patient who died at age 43.[15] No gross lesions were found the central nervous system; however, there was pigmented granular degeneration of nerve cells along the spinal cord and diffuse sclerosis in white matter tracts. Miliary degeneration was seen throughout the dentate nucleus, cerebellum, corpora striata and the thalamus.

Further progress in elucidating the pathological hallmarks of paralysis agitans was limited by the pervasive opinion that sclerosis of the spinal cord was the only definite pathological change. Dr. Gowers remarked that Hughlings Jackson favored the cerebellum as the source of pathological change.[16] In 1899, Dr. Gowers stated, "whatever anatomical changes may underlie the symptoms they are too minute to be at present within the reach even of the microscope."

Dr. Gowers felt that the motor cortex was the cause of paralysis agitans and postulated "chronic senile change in the nutrition of the branching processes of the motor nerve cells."[17] As late as 1910, Dr. Gowers still felt that the cortex was involved but noted,

> The most careful search has failed to reveal any microscopic changes peculiar to paralysis agitans. Those which have been found are such as are common at the time of life at which its subjects die.

Due to the position of the hands in paralysis agitans, and the similarity of this position to patients experiencing tetany, some authors speculated about the possible role of the thyroid and parathyroid glands in the etiology of paralysis agitans. Other authors, including Charcot and others considered paralysis agitans a neurosis, the disease without pathology. However, in light of its progressive nature and the inability to cure it by psychotherapy, most neurologists rejected this theory. In 1926, Byrnes suggested that the rigidity seen in paralysis agitans was actually due to

degeneration of the muscle spindles.[18] This theory gradually lost favor at the lack of evidence according a corticospinal or peripheral etiology that failed to materialize. There was a short report published in 1893 and 1894 of a patient with unilateral symptoms of parkinsonism. The doctors were Blocq and Marinesco.[19] Autopsy on a 38-year-old man with unilateral parkinsonism revealed a tuberculoma in the midbrain destroying the substantia nigra. This report encouraged Dr. Brissaud in 1895 to suggest the substantia nigra as the source of Parkinson's disease.[20] His writings however were largely ignored until 1919, when Tretiakoff published his thesis regarding the involvement of the substantia nigra in the etiology of Parkinson's disease and encephalitis lethargica.[21]

From this point forward, the role of the substantia nigra grew in importance until the time of Cotzias,[22] when the dopamine deficiency hypothesis, confirmed by biochemical and neuropathological studies, came full circle resulting in the first remarkably effective treatment of Parkinson's disease. This development is discussed further in subsequent chapters.

The author gratefully acknowledges the editorial and technical assistance of David Velliquette, friend and colleague.

REFERENCES

1. Parkinson, J., *An Essay on the Shaking Palsy,* Whittingham and Rowland, London, 1817.
2. Louis, E. D., Paralysis Agitans in the nineteenth century, in *Parkinson's Disease Diagnosis and Clinical Management,* Factor, S. A. and Weiner, W. J., Eds., Demos, New York, 2002, Chapter 2.
3. Elliotson, J., On the medical properties of the subcarbonate of iron, *Medico. Chirurgic. Transact.* 13, 232, 1827.
4. Louis, E. D., Paralysis agitans in the nineteenth century, in *Parkinson's Disease: Diagnosis and Clinical Management,* Factor S. A. and Weiner, W. J., Eds., Demos, New York, 2002, Chapter 2.
5. Hospital reports St. Thomas's Hospital, Paralysis agitans, *London Med. Surg. J.,* 11, 605, 1832.
6. Elliotson J., Clinical lecture, *The Lancet,* I:289-297, 1831.
7. Elliotson, J., Clinical lecture on paralysis agitans, *The Lancet,* 119, 1880.
8. Gowry, T. C., Case of paralysis agitans intermittens, *The Lancet,* II,1831, 651.
9. Gibson, M., On spinal irritation, *The Lancet,* 567, 1839.
10. Reynolds, J. R., Report of a case of paralysis agitans removed by continuous galvanic current, *The Lancet,* II, 558, 1859.
11. Paget, G. E., Case of involuntary tendency to fall precipitately forwards, with remarks. *Med. Times Gaz.,* 10, 178, 1855.

12. Charcot, J. M., Lectures on diseases of the nervous system Lecture V, The New Sydenham Society, London 1877, 129.

13. Meynert, Beitage zur differentiel Dianose des paralytischen Irrsinns, *Wien. Med. Pr.,* Vol. XII, p. 645, 1871.

14. Murchison, C. and Cayley, W., Case of paralysis agitans, *Trans. Path Soc.,* Vol. XXII, p. 24, 1871.

15. Dowse, T. S. The pathology of the case of paralysis agitans or Parkinson's disease, *Trans. Path. Soc.,* Vol. XXIXX, p. 17, 1878.

16. Gowers, W. R., *A manual of the disease of the nervous system,* Vol. II, p. 589, Churchill, London, 1888.

17. Gowers, W. R., *A system of medicine,* Vol. VII, 2nd ed., p. 473, Macmillan & Co., London, 1910.

18. Byrnes, C. M., A contribution to the pathology of paralysis agitans, *Arch. Neurol. Psychiat.,* Vol. XV, p. 407, 1926.

19. Blocq, P. and Marinesco, G., Sur un cas de tremblent parkinsonien hemiplegique symtomatique dune tumeur du peduncule cerebrale, *Bull. Et Mem. Soc.*

20. Brissaud, *Lecons sur les maladies nereuses,* Vols. XXII and XXIII, Masson, Paris 1895.

21. Tretiakoff, C., Contribution a l'etude de l'anatomie pathologique du locus niger, *These de Paris,* 1919.

22. Cotzias, G. C., Van Woert, M. H., and Schiffer, L. M., Aromatic amino acids and modifications of parkinsons, *N. Eng. J. Med.,* pp. 276, 374, 1967.

3 The Role of Dopamine in Parkinson's Disease: A Historical Review

L. Charles Murrin
Department of Pharmacology, Nebraska Medical Center

CONTENTS

Dopamine is universally known as the neurotransmitter most intimately involved with Parkinson's disease and the severe loss of this neurotransmitter has been shown to be associated with most of the primary symptoms of the disease. However, even though James Parkinson described the neurological disease named after him almost two centuries ago, it required the development of relatively modern techniques to identify dopamine as the critical neurotransmitter in Parkinson's disease and to realize the consequent therapeutic advances that were based on this discovery. In this review, I have focused on the discoveries, both in animals and man, that presented us with the concept of a nigrostriatal pathway that uses dopamine as neurotransmitter, that plays a key role in control of motor function, and that is central to Parkinson's disease. More detailed and personal accounts of some aspects of this subject are available.[1,2] In addition, after dopamine became a focus of research related to Parkinson's disease, many advances have been made concerning the dopamine receptors and the other neurotransmitter pathways involved in Parkinson's disease. These subjects are covered in other chapters in this volume.

DISCOVERY OF THE NIGROSTRIATAL PATHWAY

One of the most dramatic aspects of the pathology of Parkinson's disease, the loss of neuronal cell bodies containing neuromelanin in the substantia nigra zona compacta, was reported as early as 1895 and has been confirmed by numerous investigators since then (see Ref. 3). At this early time, however, the terminal fields and even the function of the substantia nigra were unknown. For decades, most anatomists thought that the nigrostriatal neurons projected primarily to the globus pallidus. Numerous attempts were made to determine the anatomy of the nigral projection systems using retrograde axonal degeneration.[4-6] Ferraro postulated that the major terminal field for the substantia nigra was the corpus striatum, with the greatest projection to the putamen.[5] Mettler, in a more extensive study using similar techniques, came to the same conclusion.[6] However, these studies were characterized by the inability to demonstrate the entire axonal pathway and so unequivocally establish a direct connection between the substantia nigra and the corpus striatum. This inability to demonstrate the complete pathway, probably due to the very fine unmyelinated nature of these neurons, left room for doubt.

The first clear demonstration of a pathway from the substantia nigra to the corpus striatum was published by Rosegay in 1944 using a combination of Nissl and Marchi techniques to allow cross-checking and so identification of very fine fibers with more confidence.[7] Since then, improvements in anatomic techniques have allowed numerous investigators to confirm this work (e.g., Refs. 8, 9). The many studies delineating the nigrostriatal pathway laid critical groundwork for subsequently connecting the known pathology of Parkinson's disease with a loss of dopaminergic neurons.

DISCOVERY OF DOPAMINE AS A NEUROTRANSMITTER

The concept of neurohumoral transmission arose around the turn of the twentieth century with the work of Lewandowsky, Langley, and Elliott (see Ref. 10), and epineph-rine was the first candidate neurotransmitter. A little over a decade later, norepinephrine began to emerge as the leading candidate for the neurohumoral substance released by the sympathetic nervous system. In subsequent years, the role of norepinephrine was clearly delineated,[11] and research in this area led to the awarding of the Nobel Prize in Medicine in 1970 to U. S. von Euler and Julius Axelrod, along with Sir Bernard Katz. In the course of these studies, it became clear that the immediate precursor to norepi-nephrine is 3,4-dihydroxyphenylethylamine, or dopamine.

The discovery that dopamine itself is a neurotransmit-ter, similar to norepinephrine and acetylcholine, proved to be a major scientific breakthrough for the understanding of several neurological and psychiatric diseases, includ-ing Parkinson's disease and schizophrenia, but it required another 20 years work to establish this. Dopamine was known to be an intermediate in the synthesis of norepi-nephrine and epinephrine, the product of aromatic amino acid decarboxylase acting on L-dihydroxyphenylalanine. Initially, dopamine was thought to be simply a precursor to norepinephrine and epinephrine, since it was found in extremely low levels and norepinephrine had been shown to be the primary transmitter of the sympathetic nervous system.[12] The presence of dopamine in greater than trace amounts was demonstrated by Holtz and von Euler in urine, and in the adrenal gland and the heart of sheep by Goodall.[13–15] This began to raise the question of whether dopamine was only a metabolic intermediate. However, the techniques available for these studies were relatively insensitive for dopamine and made study in other tissues difficult. A further step in uncovering dopamine as worthy of study in its own right was the demonstration by Montagu[16] of a compound in whole brain from several species, including man, that was tentatively identified as dopamine (hydroxytyramine). Carlsson and Waldeck, in a similar time frame, discovered that a change in pH pro-vided a marked improvement to the fluorescence assay of Weil-Malherbe and Bone, and this allowed a more defini-tive detection of dopamine in tissue samples.[17] Based on this, they demonstrated that dopamine (3-hydroxy-tyramine) was indeed present in whole brain of rabbits at levels equivalent to those of norepinephrine.[18] The fact that dopamine concentrations equalled those of norepi-nephrine suggested to them that dopamine might have an independent function beyond being the precursor for norepinephrine. In this paper, Carlsson and co-workers carried out an experiment that foreshadowed the future treatment of Parkinson's disease when they administered high doses of dopa, albeit the racemic form, to rabbits and found that this led to a marked increase in the levels of dopamine in brain while having far less impact on norepi-nephrine levels.

There quickly followed a series of papers from a strongly collaborating group of Swedish scientists that provided further evidence for dopamine as a neurotrans-mitter, independent of its role in norepinephrine synthe-sis. Using the technique developed by Carlsson and Waldeck, Bertler and Rosengren examined the brains of numerous mammalian species and found dopamine present in all and at concentrations similar to those of norepinephrine.[19,20] They also demonstrated that most of the dopamine present in brain (about 80%) was in the cor-pus striatum. In general, they found that, in areas with high concentrations of dopamine, norepinephrine was usually at low levels. Conversely, in regions where nore-pinephrine concentration was highest, dopamine was quite low. These data supported the idea that dopamine had a function of its own beyond being a precursor for norepinephrine. These authors also suggested that, since dopamine was found concentrated in the corpus striatum, it probably was important for the function of that region, i.e., control of motor function. This idea was supported by the fact that reserpine, which depleted the brain of dopamine (among other neurotransmitters) produced motor hypoactivity, while administration of high doses of levodopa produced motor hyperactivity,[21] presumably due at least in part to production of excess dopamine. Interest-ingly, the authors make a comparison between the motor effects of reserpine and the symptoms of Parkinson's dis-ease. Based on these and other studies, Carlsson[22] mar-shaled three arguments supporting dopamine's role in controlling motor functions:

1. Large amounts of dopamine are present in the corpus striatum, known to be an important com-ponent of the extrapyramidal system.
2. Hypokinesis is produced by administration of reserpine, which depletes dopamine from the corpus striatum.
3. Levodopa, the immediate precursor to dopam-ine, is able to counteract the hypokinetic effects produced by reserpine.

All of these ultimately pointed to a role for dopamine in Parkinson's disease.

At about the same time that the fluorescence bio-chemical assay to quantitatively measure dopamine (and other monoamines) was being established, fluorescence histochemical procedures were developed that allowed semiquantitative analysis of these same monoamines in tissue sections.[23,24] This approach, known as the Falck-Hillarp technique, provided high-resolution anatomical data to complement the biochemical and pharmacological data.

Based on the data demonstrating that the substantia nigra sends fibers to the neostriatum, that the neostriatum contains very high levels of dopamine, and that the sub-stantia nigra has dopamine-containing nerve cells, Andén and colleagues postulated that the substantia nigra neurons were the source of dopamine in the neostriatum.[3] Using a series of lesions of either the substantia nigra or of the neostriatum, they found support for their ideas. Lesion of the substantia nigra led to a loss of fluorescence in the neostriatum that correlated well with the extent of loss of catecholamine nerve cells in substantia nigra, particularly the cells in the pars compacta. Conversely, removal of the neostriatum allowed demonstration of fluorescent nerve fibers from the substantia nigra up to the neostriatum due to the apparent backup of dopamine and its synthetic enzymes in the lesioned axons. Given these data and the findings that the pars compacta has a much higher dopamine content than the pars reticulata in man[25] and that the characteristic lesion in Parkinson's disease is loss of neuromelanin containing cells in the pars compacta [see Ref. 3], the authors suggested that parkinsonism is due to selective degeneration of nigro-neostriatal dopamine neurons.

Later animal studies, including those using the glyoxylic acid fluorescence technique,[26] provided more detailed analyses of catecholamine pathways, terminal regions, and cell bodies in both brain and spinal cord.[27–33] They supplied evidence for dopamine-rich areas outside the neostriatum, such as the olfactory tubercles, nucleus accumbens, hypothalamus, and frontal cortex. Other studies substantiated the idea that the catecholamine cell bodies give rise to catecholamine-containing terminal regions,[34] analogous to those previously described in the peripheral nervous system, and that terminal regions contained dopamine in varicosities at very high concentrations.[35]

In the years following the initial suggestion that dopamine functions as a neurotransmitter, this catecholamine has been the subject of intense research. Numerous laboratories have provided evidence that dopamine fulfills the classic criteria for a neurotransmitter: localization and synthesis in nerve terminals, release from neurons on stimulation, production of the same effects as the endogenous compound released by nerve stimulation, metabolic machinery in the appropriate locations, and appropriate pharmacology.[36, pp. 225–270] These basic science studies provided the foundation for and, at the same time, were chronologically intertwined with similar studies on human tissue that substantiated the key role dopamine plays Parkinson's disease. Research on dopamine played a major role in the Nobel Prize in Medicine and Physiology awarded in 2000 to Carlsson, Greengard, and Kandel.

DOPAMINE AND HUMAN STUDIES

As mentioned above, it has been known for 100 years that a cardinal feature of Parkinson's disease is loss of neuromelanin-containing neurons in the substantia nigra, particularly in the zona compacta. A nigrostriatal pathway was postulated in humans by Ferraro, but this proved difficult to substantiate because of the necessity of using postmortem tissue. Today, however, the existence of this and other related pathways[37] is widely accepted and is the basis for our understanding of the anatomy and pathology of Parkinson's disease.[38]

The early work demonstrating that dopamine was a probable neurotransmitter in mammals and that a dopaminergic nigrostriatal pathway appeared to be a major component of the extrapyramidal pathway led to investigations of human tissue. It was reasoned that a similar pathway existed in humans, analogous to that found in many other species. In addition, since it was known that Parkinson's disease was characterized by loss of nigral neurons and a primary feature was loss of motor control, it seemed likely that loss of dopamine neurons would be an important pathological feature of this disease. The presence of high concentrations of dopamine in the neostriatum and in the substantia nigra of human brain[39–41] provided evidence for the existence of this pathway. The dramatic loss of dopamine in neostriatum and substantia nigra of Parkinson's patients demonstrated by Hornykiewicz and colleagues[25,41] helped confirm the involvement of these neurons in the disease process. These findings were further supported by the strong correlation between the loss of dopamine and the loss of nigral neurons in humans, by a similar correlation between severity of the disease and the degree of loss of dopamine neurons, and by the finding that in hemiparkinsonism the dopamine deficiency was much more severe in the neostriatum contralateral to the side of the symptoms.[42] Since these early studies, loss of the neuromelanin-containing dopamine neurons in the substantia nigra zona compacta has become the diagnostic hallmark of Parkinson's disease.

One of the drawbacks of the studies examining dopamine levels in human brain was the fact that postmortem tissue samples were the source of all the data. As a result, the possibility of artifacts due to tissue storage and handling or due to postmortem changes could not be ruled out completely. More recently, though, the development of positron emission tomography (PET) and single positron emission computed tomography (SPECT) have provided methods that allow examination of markers for dopamine neurons in living brain.

The development of [¹⁸F]fluorodopa as a marker for dopamine neurons that is applicable to PET studies in humans[43] allowed further *in vivo* examination in Parkinson's patients. These studies confirmed a severe loss of dopamine terminals in living patients,[44,45] in agreement with previous studies. As would be expected, other markers specific to dopamine neurons, such as the dopamine transporter, were also dramatically reduced in Parkinson's patients when compared to control subjects.[46–48] Thus, these studies with technically far more sophisticated techniques confirmed in living patients the early findings in this field.

DOPAMINE AND TREATMENT OF PARKINSON'S DISEASE

Given this pathology, a logical approach to treating Parkinson's disease would be to try to restore the levels of dopamine in the CNS. It was well known that dopamine does not cross the blood-brain barrier because of its positive charge. However, zwitterionic amino acids had the advantage of broadly specific transport systems to carry them across this barrier into the brain. Based on this, a top candidate for increasing dopamine levels in the CNS would be its immediate precursor, L-dihydroxyphenylalanine or L-dopa. Not only could L-dopa be transported across the blood-brain barrier, it also had the advantage of by-passing the rate limiting step in the synthesis of catecholamines, tyrosine hydroxylase. This, in turn, provided two advantages. First, it avoided the slowest step in the synthetic process. In addition, tyrosine hydroxylase had been shown to be present only in catecholamine-containing neurons in the CNS, the very neurons that were disappearing in Parkinson's disease. Second, aromatic amino acid decarboxylase (dopa decarboxylase), the enzyme converting L-dopa to dopamine, although shown to be primarily in dopamine neurons in the corpus striatum, was also clearly found in non-dopamine cells (i.e., serotonin and norepinephrine terminals).[19,49] As a result, the loss of this enzyme in Parkinson's disease was not as dramatic as the loss of tyrosine hydroxylase,[50] and so use of dopa decarboxylase as a critical enzymatic step in treatment would be expected to be more successful than going through tyrosine hydroxylase.

Several groups tried administering large doses of L-dopa to Parkinson's patients and found promising results, particularly in combating akinesia.[51–53] There were, however, significant side effects associated with this therapy, and they threatened to severely limit its usefulness. Several significant improvements were made in an extensive study by Cotzias and colleagues.[54] In their own initial studies, they had used racemic dopa and had encountered serious side effects.[53] In the later study,[54] they used only L-dopa, as had others previously, and found that the use of the stereoisomer instead of the racemic mixture reduced the incidence of side effects greatly, including avoiding some of the most problematic, such as reversible granulocytopenia. They found that slowly increasing the dose also reduced the incidence of side effects. Perhaps most important, they introduced the use of a peripheral dopa decarboxylase inhibitor. They reasoned that many of the side effects that had been encountered could be explained by the conversion of L-dopa to dopamine and norepinephrine in the periphery. If this peripheral conversion could be reduced or stopped, the side effects should be reduced, and more L-dopa would be available for transport into the CNS. Animal studies had provided support for this notion.[55,56] Using this approach, Cotzias

and colleagues were very successful in reducing side effects and in allowing a reduction in the dosage of L-dopa necessary to produce beneficial effects in Parkinson's patients.[54] These early studies in humans introducing L-dopa therapy as a means of partially restoring central dopamine levels, and hence countering the symptoms of Parkinson's disease, initiated what is currently the most common therapeutic approach to treating Parkinson's disease.

CONCLUSION

Since this early work, our understanding of the pathology of Parkinson's disease and of potential approaches to treatment have become much more detailed and sophisticated. While this review has focused on the nigrostriatal dopamine pathway as being central to Parkinson's disease, it is becoming increasingly clear that the basal ganglia circuitry is quite complex, with multiple parallel circuits subserving specific functions.[37,57] Indeed, not only are there regional differences in the loss of neurons within the substantia nigra,[58,59] but dopamine neuronal systems besides the nigrostriatal pathway are also affected,[60] although usually to a lesser extent. The loss of these other dopamine neurons contributes to a number of the characteristic symptoms of Parkinson's disease. It also is clear that the dopamine neuronal system is not the only neurotransmitter system affected in Parkinson's disease[61–63] and that loss of other neurotransmitter neurons probably plays a significant role in Parkinson's disease. These must now be taken into consideration in our understanding of the symptomatology and, hopefully, the treatment of Parkinson's disease. It has been suggested that dopamine itself may be, in a certain sense, a contributor to the development of Parkinson's disease,[64,65] an idea that is disputed.[66] Our understanding of these areas is covered in detail in other chapters of this volume. Nevertheless, it is clear that our current knowledge of the pathology, symptomatology and therapy of Parkinson's disease has, as its base, the early studies in animals and man that provided evidence there is a nigrostriatal pathway critical for motor function, that dopamine is a neurotransmitter in the central nervous system and specifically in the nigrostriatal pathway, and that degeneration of this pathway is the hallmark of Parkinson's disease.

REFERENCES

1. Dahlstrom, A. and Carlsson, A., Making visible the invisible. Recollections of the first experiences with the histochemical fluorescence method for visualization of tissue monoamines, in *Discoveries in Pharmacology. Vol. 3. Chemical Pharmacology and Chemotherapy,* M.

J. Parnham, J. Bruinvels, Eds., Elsevier: Amsterdam, 1986j, pp. 97–125.

2. Carlsson, A., *Annual Review of Neuroscience,* 10, 19, 1987.

3. Anden, N. E., Carlsson, A., Dahlstrom, A., Fuxe, K., Hillarp, N. A., and Larsson, K., *Life Sciences.,* 3, 8, 523, 1964.

4. von Monakow, C., *Arch. f. Psychiat. u. Nervenkr.,* 27, 1, 1895.

5. Ferraro, A., *Arch.Neurol.Psychiat.,* 19, 177, 1928.

6. Mettler, F. A., *Journal of Comparative Neurology,* 79, 185, 1943.

7. Rosegay, H., *Journal of Comparative Neurology,* 80, 293, 1944.

8. Carpenter, M. B., McMaster, R. E., *American Journal of Anatomy,* 114, 293, 1964.

9. Moore, R. Y., Bhatnager, R. K., Heller, A., *Brain Research,* 30, 119, 1971.

10. Hoffman, B. B. and Taylor, P., Neurotransmission. The autonomic and somatic motor nervous systems, in *Goodman and Gilman's, The Pharmacological Basis of Therapeutics,* J. G.Hardman, and L. E. Limbird, Eds., McGraw-Hill, New York, 2001, Chapter 6, pp. 115-153.

11. von Euler, U. S., *Noradrenaline: Chemistry, Physiology, Pharmacology and Clinical Aspects,* Thomas: Springfield, IL, 1–382, 1956.

12. von Euler, U. S., *Acta Physiologica Scandinavica.* 12, 73, 1946.

13. Holtz, P., Credner, K., and Kroneberg, G., *Naunyn-Schmiedeberg's Arch.Exp.Pathol.Pharmakol.,* 204, 228, 1947.

14. von Euler, U. S., Hamberg, U., and Hellner, S., *Biochemical Journal,* 49, 655, 1951.

15. Goodall, M., *Acta Physiologica Scandinavica,* 24, Suppl. 85, 1, 1951.

16. Montagu, K. A., *Nature,* 180, 244, 1957.

17. Carlsson, A., Waldeck, B., *Acta Physiologica Scandinavica,* 44, 293, 1958.

18. Carlsson, A., Lindqvist, M., Magnusson, T., Waldeck, B., *Science,* 127, 471, 1958.

19. Bertler, A., Rosengren, E., *Acta Physiologica Scandinavica.* 47, 350, 1959.

20. Bertler, A., Rosengren, E., *Experientia,* 15, 10, 1959.

21. Carlsson, A., Lindqvist, M., Magnusson, T., *Nature,* 180, 1200, 1957.

22. Carlsson, A., *Pharmacological Reviews,* 11, 490, 1959.

23. Falck, B., *Acta Physiologica Scandinavica,* 56, Suppl. 197, 1, 1962.

24. Falck, B., Hillarp, N. A., Thieme, G., Torp, A., *Journal of Histochemistry and Cytochemistry.* 10, 348, 1962.

25. Hornykiewicz, O., *Wien. klin. Wsch.,* 75, 309, 1963.

26. Lindvall, O., Bjorklund, A., *Histochemistry,* 39, 97, 1974.

27. Dahlström, A., Fuxe, K., *Acta Physiologica Scandinavica,* 62, 1, 1964.

28. Fuxe, K., *Acta Physiologica Scandinavica,* S247, 37, 1965.

29. Anden, N. E., Dahlstrom, A., Fuxe, K., Larsson, K., Olson, L., Ungerstedt, U., *Acta Physiologica Scandinavica,* 67, 313, 1966.

30. Ungerstedt, U., *Acta Physiologica Scandinavica,* Supp., 367, 1, 1971.

31. Lindvall, O. and Björklund, A., *Acta Physiologica Scandinavica,* Supp. 412, 1, 1974.

32. Lindvall, O., Bjorklund, A., Divac, I., *Brain Research,* 142, 1, 1978.

33. Björklund, A., Lindvall, O., Dopamine-containing systems in the CNS, in *Handbook of Chemical Neuroanatomy,* A. Björklund, Ed., Elsevier: Amsterdam, pp. 55–122, 1984.

34. Fuxe, K., *Zeitschrift fur Zellforschung und Mikroskopische Anatomie,* 65, 573, 1965.

35. Anden, N. E., Fuxe, K., Hamberger, B., Hokfelt, T., *Acta Physiologica Scandinavica,* 67, 306, 1966.

36. Cooper, J. R., Bloom, F. E., Roth, R. H., *The Biochemical Basis of Neuropharmacology,* Oxford Univ. Press: Oxford, pp. 1–405, 2003.

37. Alexander, G. E., DeLong, M. R., Strick, P. L., *Annual Review of Neuroscience,* 9, 357, 1986.

38. Fletcher, N., Movement disorders, in *Brain's Diseases of the Nervous System,* M. Donaghy, Ed., Oxford University Press: Oxford, Chapter 32, p. 1015–1095, 2001.

39. Bertler, A., *Acta Physiologica Scandinavica,* 51, 97, 1961.

40. Sano, I., Gamo, T., Kakimoto, Y., Taniguchi, K., Takesada, M., Nishinuma, K., *Biochimica et Biophysica Acta,* 32, 586, 1959.

41. Ehringer, H., Hornykiewicz, O., *Klin.Wschr,* 38, 1236, 1960.

42. Hornykiewicz, O., *Research Publication of the Association for Research in Nervous & Mental Disease,* 50, 390, 1972.

43. Garnett, E. S., Firnau, G., Nahmias, C., *Nature.* 305, 137, 1983.

44. Leenders, K. L., Palmer, A. J., Quinn, N., Clark, J. C., Firnau, G., Garnett, E. S., Nahmias, C., Jones, T. and Marsden, C. D., *Journal of Neurology, Neurosurgery and Psychiatry,* 49 (8), 853, 1986.

45. Leenders, K. L., Salmon, E. P., Tyrrell, P., Perani, D., Brooks, D. J., Sager, H., Jones, T., Marsden, C. D., and Frackowiak, R. S. J., *Archives of Neurology,* 47, 1290, 1990.

46. Frost, J. J., Rosier, A. J., Reich, S. G., Smith, J. S., Ehlers, M. D., Snyder, S. H., Ravert, H. T. and Dannals, R. F., *Annals of Neurology,* 34, 423, 1993.

47. Seibyl, J. P., Marek, K. L., Quinlan, D., Sheff, K., Zoghbi, S., Zea-Ponce, Y., Baldwin, R. M., Fussell, B., Smith, E. O., Charney, D. S., Hoffer, P. B. and Innis, R. B., *Annals of Neurology,* 38, 589, 1995.

48. Innis, R. B., Seibyl, J. P., Scanley, B. E., Laruelle, M., Abi-Dargham, A., Wallace, E., Baldwin, R. M., Zea-Ponce, Y., Zoghbi, S. and Wang, S., *Proc. Natl. Acad. Sci. USA,* 90 (24), 11965, 1993.

49. Lloyd, K., Hornykiewicz, O., *Brain Research,* 22, 426, 1970.

50. Lloyd, K.; Hornykiewicz, O., *Science,* 170, 1212, 1970.

51. Birkmayer, W., Hornykiewicz, O., *Wien. klin. Wsch,* 73, 787, 1961.

52. Birkmayer, W., Hornykiewicz, O., *Arch.Psychiat.Nervenkr,* 203, 560, 1962.

53. Cotzias, G. C., Van Woert, M. H., Schiffer, L. M., *New England Journal of Medicine,* 276 (7), 374, 1967.

54. Cotzias, G. C., Papavasiliou, P. S., Gellene, R., *New England Journal of Medicine,* 280 (7), 337, 1969.

55. Sjoerdsma, A., Vendsalu, A., Engelman, K., *Circulation,* 28, 492, 1963.

56. Udenfriend, S., Zaltzman-Nirenberg, P., Gordon, R. and Spector, S., *Molecular Pharmacology,* 2, 95, 1966.

57. Parent, A., Sato, F., Wu, Y., Gauthier, J., Levesque, M. and Parent, M., *Trends in Neuroscience,* 23, Suppl., S20, 2003.

58. Fearnley, J. M., Lees, A. J., *Brain,* 114 (5), 2283, 1991.

59. Gibb, W. R. G., Lees, A. J., *Journal of Neurology, Neurosurgery and Psychiatry,* 54, 388, 1991.

60. Rakshi, J. S., Uema, T., Ito, K., Bailey, D. L., Morrish, P. K., Ashburner, J., Dagher, A., Jenkins, I. H., Friston, K. J. and Brooks, D. J., *Brain,* 122 (9), 1637, 1999.

61. Przuntek, H., Müller, T., *Journal of Neurology,* 247, Suppl. 2, II/2, 2000.

62. Braak, H., Braak, E., *Journal of Neurology,* 247, Suppl. 2, II/3, 2000.

63. Stokes, A. H., Hastings, T. G., Vrana, K. E., *Journal of Neuroscience Research,* 55, 659, 1999.

64. Fahn, S., *Advances In Neurology,* 69, 477, 1996.

65. Olanow, C. W., Stocchi, F., *European Journal of Neurology,* 1, Suppl. 7, 3, 2000.

66. Mytilineou, C., Walker, R. H., JnoBaptiste, R., Olanow, C. W., *Journal of Pharmacology and Experimental Therapeutics,* 304 (2), 792, 2003.

4 Parkinson's Disease: Where Are We?

Ajit Kumar, Zhigao Huang, and Donald B. Calne
Pacific Parkinson's Research Centre, Vancouver Hospital and Health Sciences Centre, University of British Columbia

CONTENTS

Since the original description of the "Shaking Palsy" by James Parkinson in 1817, knowledge about Parkinson's disease (PD) made slow progress for over a century. Beginning in the late 1950s, our understanding of PD has progressed by leaps and bounds, largely through the advent of better biomedical technology and extraordinary progress in allied medical disciplines such as neuropharmacology, neurochemistry, neuropathology, molecular biology, and genetics. The advances made thus far can be discussed from the broad perspectives of etiopathogenesis and management.

ETIOPATHOGENESIS

The definition of PD is rather difficult. A practical definition with the recognition that the brunt of the pathology falls on the dopaminergic nigrostriatal pathway suffices for most clinicians. There is evidence to suggest that PD may actually be a syndrome with many causes.[1] Phenotypes identical to sporadic PD have been described in patients with the parkin mutation, spinocerebellar ataxia type 2, and mutations in the alpha-synuclein gene.[2–5] Similarly, parkinsonism indistinguishable from PD has been described after viral encephalitis. These observations suggest that PD may not be a single homogenous entity. This view assumes particular importance with regard to causation, as there may be several causes for the PD phenotype.

The cause(s) of PD are still largely unknown. There has been considerable debate about the relative importance of genetic versus environmental factors. These factors are not mutually exclusive, as genetic susceptibility may confer selective vulnerability to specific environmental factors. The role of heredity in the etiology of PD has been fortified by the discovery of some kindreds with rare genetic forms of PD. However, the cause remains unresolved in the vast majority of patients.[6] Environmental risk factors, including exposure to pesticides and metals, viruses, well-water drinking, rural living, and farming, have been investigated in many recent case-control and epidemiological studies.[7–10] In addition, there is evidence that aging is also a likely contributory factor.[11]

GENETIC FACTORS

The discovery of families with inherited forms of parkinsonism has generated considerable interest over the past few years. These inherited forms account only for a miniscule proportion of PD patients. However, their discovery has provided insight into the possible pathogenesis of PD, particularly with regard to the role of abnormal protein processing at the subcellular level.[6] Parkinsonism has been associated with mutations in four genes thus far: α-synuclein, on locus 4q21-23,[4,5] parkin on locus 6q25-27,[3] ubiquitin C-terminal hydrolase L1 (UCH-L1) on locus 4p14[12] and DJ-1 on 1p36.[13] Other gene loci with linkage to inherited parkinsonism identified include 2p13,[14] 4p14–16[15] and 12p11-q13[16] (See Table 4.1). There is evidence for additional loci being involved as well.[17]

Two disease-causing mutations, A53T and A30P, have been identified in the α-synuclein gene.[4–5] Multiple mutations have been described in the parkin gene.[18,19] Mutations in the α-synuclein and UCH-L1 genes are associated with a phenotype resembling sporadic PD (young onset form is common with the former), whereas mutations in the parkin gene are usually associated with a juvenile form of PD with no Lewy bodies in the brain.[6,20] Monogenic forms of parkinsonism due to genetic mutations result in abnormalities of protein processing, particularly the ubiquitin mediated pathway of protein degradation. These abnormalities presumably set off a cascade of adverse events at the cellular level that eventually culminate in cell death. Abnormal protein processing is further discussed later under 'pathogenetic mechanisms' (see discussion below).

ENVIRONMENTAL FACTORS

Many epidemiological studies have shown a tenuous link between environmental factors and PD. This, together with the observation that the vast majority of PD patients lack the genetic abnormalities described earlier, makes a case for examining the environmental theory of PD causation. Familial occurrence of PD does not necessarily imply genetic causation. A large survey comparing monozygotic and dizygotic twins failed to reveal a higher concordance rate in the monozygotic group in the age range when PD usually starts.[21] Family members share their environment as well as their genes. Another study demonstrated that the risk of developing PD for a child in a parent-child cluster depended on the child's age when the parent started to show symptoms rather than the age of the parent; the younger the child, the greater the risk.[22] This suggests a significant role for shared environment. An older study showed that PD is commoner in the north compared to the south in the United State, irrespective of race, again suggesting a role for environmental factors.[23]

The discovery that 1-methyl-4-phenyl-1,2,3,6-tetrahydropyridine (MPTP) leads to selective destruction of the nigrostriatal pathway opened up a new line of thinking among neuroscientists on the potential role of toxins in PD causation.[24] Transient exposure to MPTP can also lead to delayed death of nigral neurons after a long latent period.[25] There is no evidence to date that MPTP is involved in the causation of sporadic PD. There have been several reports on the relation of exposure to herbicides, fungicides and pesticides, and the development of PD.[8–10,26,27] A recent meta-analysis seems to lend some credibility to these reports.[28] It is interesting that the pesticide rotenone, which has a somewhat similar structure to MPTP, can cause alpha synuclein accumulation and neuronal death in animals.[29] Rotenone-induced dopaminergic neuronal degeneration is thought to result from selective dysfunction of mitochondrial Complex I, as is the case for the other selective dopaminergic neurotoxin, MPTP. Other environmental factors proposed to be associated with a higher risk of developing PD include rural living, well-water drinking, and farming.[9] Young-onset parkinsonism in particular has been associated with exposure to well water.[30] No toxic constituents have been identified, and well-water drinking may simply be a marker for rural

TABLE 4.1
Genes Associated with Parkinsonism

Gene	Locus	Inheritance	Phenotype	Reference
α-synuclein	4q21-q23	Autosomal dominant	Early onset PD	Kruger et al., 1998 [4]
Parkin	6q25.2-q27	Autosomal recessive	Juvenile onset PD	Kitada et al., 1998 [3]
UCH-L1	4p14	Autosomal dominant	Typical PD	Leroy et al., 1998 [12]
DJ-1	1p36	Autosomal recessive	Early onset	Bonifati et al., 2003 [13]
PARK3	2p13	Autosomal dominant	Typical PD	Gasser et al., 1998 [14]
PARK4	4p14-p16	Autosomal dominant	PD/Essential tremor	Farrer et al., 1999 [15]
PARK6	1p35-p36	Autosomal recessive	Early onset	Valente et al., 2001 [15a]
PARK8	12p11.2-q13.1	Autosomal dominant	Typical PD	Funayama et al., 2002 [16]

environment, which may in turn point to pesticide exposure.[6] Associations between PD and exposure to plastic or epoxy resins, and metals such as manganese have also been reported.[7,31] An inverse risk between smoking and the risk for developing PD has been reported by some studies, including one on monozygotic twins.[32] There is controversy as to whether this relationship is real or merely reflects a "rigid" premorbid personality trait described in PD patients[33] that prevents them from smoking. Recent studies suggest that nicotine may play a protective role by acting on toxin-neutralizing enzymatic pathways or by inducing neurotrophic factors in the striatum.[34,35] It has also been proposed that caffeine intake has a protective effect and prevents PD independent of the protective effect of smoking.[36] This effect could possibly be mediated through the neuroprotective effect of adenosine receptor blockade as demonstrated in animal models of parkinsonism.[37,38]

A possible role for infection, particularly viruses, has been speculated about since the epidemic of Von Economo's encephalitis in the early years of the last century. Parkinsonism was frequently a sequel, sometimes after a delay of several years.[39] Symptom progression has been documented in these patients despite the absence of markers of persisting viral infection.[40] There has been renewed interest in the possible role of viruses in PD causation. Positron emission tomography (PET) scanning has demonstrated selective lesions of the nigrostriatal pathway in human subjects with parkinsonism following viral encephalitis.[41] Japanese workers have shown that certain strains of Influenza A virus are selectively tropic to the nigral neurons and can gain access to the brain via the nasal passages in mice.[42-44] Another study showed that antibodies to the Epstein-Barr virus cross react with alpha synuclein in the brains of patients with PD.[45] Though no markers of viral infection have been shown in autopsy studies in PD, a persistent atypical inflammatory reaction has been demonstrated in the substantia nigra.[46-48] A recent epidemiological report suggests that the prevalence of PD is increased in teachers, medical workers, loggers and miners.[9] One explanation for this would be an infectious etiology as teachers and medical workers come in contact with large numbers of the public. In the case of miners and loggers, increased prevalence could be related to the cramped poorly ventilated living quarters they share. Another recent study examined occupational risk factors in pairs of monozygotic twins discordant for PD and found a highly significantly increased risk in the teaching and health care professions.[49]

PATHOGENETIC MECHANISMS

Abnormal protein processing, oxidative stress, mitochondrial dysfunction, apoptosis, excitotoxicity, and inflammation have all been thought to contribute to cell death in PD.[6] These mechanisms are not mutually exclusive and may indeed be intimately related. These pathogenetic mechanisms are ostensibly set off by a trigger that, as discussed earlier, may be genetic, environmental, or the result of a complex gene-environment interaction.

Abnormal Protein Processing

Abnormalities in the ubiquitin-mediated pathway of protein degradation are associated with some monogenic forms of parkinsonism. Proteins destined to be degraded by the ubiquitin-mediated pathway are labeled with polyubiquitin chains through a series of enzymatic reactions and then degraded by the proteasome, a multicatalytic complex, or by the lysosomal system.[50,51] α-synuclein and ubiquitin are major components of the filaments associated with Lewy bodies.[52] Parkinsonism associated with mutations of the α-synuclein gene is characterized by the presence of Lewy bodies in surviving neurons of the substantia nigra. Mutant A53T or A30P α-synuclein are associated with the formation of small ubiquitinated aggregates and autophagic cellular degeneration.[4,53] Defects occur in the lysosomal and proteasomal degradation systems and may be a consequence of the above effects.[51,54] Accumulation of mutant α-synuclein in the neuron possibly contributes to cell death. There is experimental evidence supporting this notion.[55,56] Degeneration of dopaminergic terminals is seen in mice and in Drosophila with transgenic expression of human mutant α-synuclein.[57] The degeneration is particularly marked in the Drosophila model when mutant human α-synuclein (either A30P or A53T) is expressed and is associated with the formation of abnormal inclusion bodies resembling Lewy bodies.[57] Dopamine-dependent apoptosis is enhanced by α-synuclein accumulation, and this may be one explanation for the selective vulnerability seen in PD.[56]

Mutations of the parkin gene result in a juvenile form of PD characterized by selective loss of nigral dopamine neurons without Lewy bodies.[3,20] The gene product, parkin, functions as an E3 ubiquitin–protein ligase and is responsible for the attachment of ubiquitin to substrates such as synaptic vesicle-associated protein, PNUTL1 (drosophila peanut-like gene 1 protein)/CDCrel-1,[58] parkin-associated endothelin receptor-like receptor (Pael-R),[59] and a glycosylated form of α-synuclein.[60] Mutations in the parkin gene ostensibly result in abnormal accumulation of substrate proteins leading to insoluble Pael-R mediated cell death.[59]

UCH-L1 is an enzyme that hydrolyzes small C-terminal adducts of ubiquitin to generate ubiquitin monomers, which can then be recycled and used to label other proteins. Mutation in the UCH-L1 gene results in an abnormal form of the enzyme with reduced activity and this in turn leads to impaired protein processing by the ubiquitin-proteasome pathway.[50]

The recently discovered DJ-1 gene encodes a ubiquitous highly conserved protein whose function remains largely unknown though it has been suggested that it may be involved in the oxidative stress response.[13] Mutations in this gene result in a young onset form of PD.

Oxidative Stress

Indicators for a role for oxidative stress in PD include changes in the substantia nigra in the form of increased lipid peroxidation, reduced glutathione (GSH) levels, and high concentrations of iron and reactive oxygen free radicals (ROS).[50,61,62] While there is evidence for increased lipid peroxidase and abnormally oxidized DNA in PD, these findings are not restricted to the substantia nigra.[63,64] Deficiency of reduced GSH has been shown in the substantia nigra of parkinsonian subjects and this may contribute to reduced clearance of hydrogen peroxide.[65] GSH deficiency appears to result at least in part from increased activity of the degradative enzyme γ-glutamyltranspeptidase. Experimental evidence that shows reduction of the neurotoxic effect of 6-hydroxydopamine (6-OHDA) in the laboratory with exogenous administration of antioxidants such as cysteine, N-acetyl cysteine or glutathione lends some support to the oxidative hypothesis.[66]

Dopamine itself can undergo both enzymatic and nonenzymatic reactions resulting in the formation of toxic radicals.[67] While high concentrations of levodopa in artificial conditions can result in oxidative cell death,[68] there is no *in vivo* evidence that levodopa can be toxic to substantia nigra neurons.[69] On the contrary, studies have shown that levodopa could actually be neuroprotective not only in rodent models but also in humans.[70,71]

Mitochondrial Dysfunction

The elucidation of the mechanism by which MPTP produces parkinsonism in experimental animals has contributed to the understanding of the possible role of mitochondrial dysfunction in the pathogenesis of PD. MPTP is first deaminated by MAO-B in glial cells resulting in the formation of the active moiety, the 1-methyl-4-phenylpyridinium ion (MPP+). MPP+ is then selectively accumulated in dopamine nerve terminals by the plasma membrane dopamine transporter. Once inside the dopamine nerve terminals, MPP+ generates hydrogen peroxide and other free radicals that interfere with mitochondrial respiration. MPP+ is concentrated in mitochondria, where it impairs mitochondrial respiration by inhibiting complex I of the electron transfer complex and consequently causing cell death.[72] Complex I deficiency specific to the substantia nigra has been reported in human PD brains.[73,74] Also, selective nigral death following chronic exposure to rotenone, a well-known inhibitor of complex I, has been reported.[29] No such defect of oxidative phosphorylation has been found in multiple system atrophy.[75] Similarly, no complex I abnormality has been shown in the Lewy body rich cingulate cortex of diffuse Lewy body dementia brains.[73]

Apoptosis

Altered expression of pro-apoptotic genes has been reported to be associated with PD. Activated caspase-3, which is the major downstream caspase involved in the execution phase of neuronal cell death, has been detected in the substantia nigra of PD patients.[76] Other studies have shown that activated forms of caspase-8 and caspase-9, upstream caspases that are known to cleave and activate caspase-3, are present in dopaminergic neurons of the substantia nigra in MPTP-treated mice.[77] Caspase-mediated parkin cleavage that compromises parkin function has also been demonstrated in cell lines.[78] The significance of these observations is at yet indeterminate, especially as there are studies that do not support the notion of active apoptosis in PD.[79,80]

Other Mechanisms

Excitatory neurotransmission may result in neurotoxicity through impaired mitochondrial function.[81] Prolonged survival in PD has been claimed with the use of amantadine, which is a weak N-methyl-D-aspartate (NMDA) antagonist.[82]

Activated microglia have been shown in the substantia nigra in PD.[46,83] Inflammatory and glial responses have also been observed in the substantia nigra of patients exposed to MPTP and in MPTP-treated primates.[84] The pathogenetic role of atypical inflammation is unclear at present, especially whether it represents a primary or secondary phenomenon.

MANAGEMENT

To date, no intervention has been convincingly shown to slow down, arrest, or reverse the progression of PD. However, there are indicators from recent research that effective neuroprotective agents for PD may be available in the not too distant future as our understanding of the pathogenetic mechanisms steadily grows. There have also been considerable advances in the symptomatic treatment of PD over the past few years. These are discussed below under the broad headings of pharmacological, surgical, and nonpharmacological interventions.

PHARMACOLOGICAL

The Placebo Effect

There is a prominent placebo effect associated with PD. Placebo treatment significantly increases the release of endogenous DA in the striatum.[85] The degree of DA release induced by a placebo in PD is comparable to that induced by antiparkinson medication.[85] DA is involved in mediating the expectation of, as well as the response to reward, and increased expectation of therapeutic benefit results in greater DA release in the striatum. This study reinforces the necessity of having a placebo arm while evaluating therapies for PD.

Dopamine Agonists

Since loss of nigrostriatal dopaminergic function is the basic underlying pathophysiology in PD, drugs that enhance dopminergic function in the striatum remain the cornerstone treatment. Levodopa is still the most widely used and the most effective drug for PD, though there has been a rapidly increasing trend for the use of DA agonists in the past few years. One of the main reasons for this trend has been the concern that early use of levodopa may predispose patients to developing long-term motor complications such as "wearing off," "on-off," and dyskinesia. The reported prevalence of these complications ranges from 50 to 75% or higher after five years of treatment.[86–90] Motor fluctuations and dyskinesias have a negative impact on the quality of life of patients and are often difficult to manage. Ideally, the goal of treatment would be reduction of parkinsonian symptoms without risk of long-term side effects. The theoretical benefits of DA agonists over levodopa are a longer half-life, resulting in less pulsatile stimulation of dopamine receptors, and no dependence on degenerating presynaptic DA nerve terminals, thereby reducing the risk of development of motor fluctuations and dyskinesias.

Once functional disability in PD requires treatment with a dopaminergic agent, the choice of levodopa versus a dopamine agonist has largely been arbitrary. Results from more recent clinical trials claim to shed some light on this contentious issue. The study of cabergoline versus levodopa by Rinne et al. found that the Unified Parkinson's Disease Rating Scale (UPDRS) motor scores decreased by 40 to 50% with both drugs during the first year of therapy.[91] Levodopa appeared to be better than cabergoline for improvement in UPDRS motor scores as well as activities of daily living (ADL). After four years in the clinical trial, levodopa treated subjects still showed an average of 30% improvement in motor disability, while patients treated with cabergoline showed a 22% improvement. There was a risk reduction of 12% for the development of "motor complications" in patients on cabergoline compared to levodopa. The study of ropinirole versus levodopa by Rascol et al. found that levodopa treatment resulted in a significantly greater increase in motor improvement than ropinirole in patients who completed the study (five years).[92] They also reported that there was no significant difference in ADL scores between the two groups at five years. The risk reduction for motor compli-

cations was 14% in the ropinirole group. The pramipexole versus levodopa study by the Parkinson Study Group (PSG) found that levodopa resulted in a significantly greater improvement than pramipexole in both the UPDRS motor scores as well as in ADL scores after 23.5 months of treatment.[93] Motor complications, defined as dyskinesias, wearing off, and on-off motor fluctuations, were significantly less common in the pramipexole group (28%) versus levodopa-treated patients (51%) at the end of 23.5 months. However, the incidence of hallucinations, peripheral edema, and somnolence were significantly higher in the pramipexole group than in the levodopa group.

The above studies suggest that the incidence of motor complications is lower with DA agonists compared to levodopa. However, both motor scores and activities of daily living are better with levodopa. It is unclear whether the higher incidence of complications in the levodopa treated groups is in part simply a reflection of greater therapeutic efficacy. Also, DA agonists have a longer half-life than levodopa and the lower incidence of complications may reflect a smoother pattern of receptor stimulation.

COMT Inhibitors

In early PD, the motor response to levodopa administration lasts longer than would be inferred from the plasma half-life of levodopa. Presumably, this phenomenon is related to surviving nigrostriatal neurons being able to store dopamine (DA) synthesized from exogenous levodopa, thus serving a buffer function. In more advanced PD, especially with the appearance of "wearing-off," the motor response tends to correlate more with plasma levodopa levels.[94] Degeneration of nigrostriatal DA nerve terminals leading to loss of buffering capacity has been suggested as one mechanism for this effect.[95] With the onset of fluctuations, there is steepening of the dose-response curve resulting in narrowing of the dose range of levodopa that produces a clinically significant antiparkinsonian response, resembling an "all-or-none" response.[96] In such a situation, even slight variations of plasma levodopa levels, resulting from pharmacokinetic factors such as absorption and changes in transport across the blood-brain barrier, can result in a highly variable antiparkinsonian response. This mechanism may contribute to the "on-off" effect.[96] Increased DA turnover has also been implicated in the pathogenesis of motor fluctuations. Uptake and decarboxylation of levodopa, release in to the synaptic cleft, and reuptake by the presynaptic neuron are all accelerated resulting in high synaptic levels of DA for a very brief period in the synaptic cleft. This has been demonstrated in a PET study where synaptic DA levels were estimated using the ^{11}C-raclopride binding paradigm. One hour after administration of a dose of levodopa, fluctuators not only had a synaptic DA level three times higher than stable responders, but also faster clearance of DA from the synaptic cleft.[97] This suggests that increased DA turnover leading to marked swings in synaptic DA levels contributes to motor fluctuations. Whatever the mechanism, maintenance of steady plasma levels of levodopa in the therapeutic range over a sustained period is helpful in PD, especially when fluctuations set in. Strategies for achieving this goal include more frequent dosing, sustained-release preparations of levodopa, long-acting DA agonists, monoamine oxidase (MAO-B) inhibitors to inhibit DA metabolism, and catechol-O-methyltransferase (COMT) inhibitors to increase bioavailability of levodopa. Current evidence indicates that use of COMT inhibitors is sometimes an effective option to deal with fluctuations, particularly "wearing-off."[94]

COMT catalyzes transfer of a methyl group from S-adenosyl-methionine to endogenous catechols.[98] Levodopa is O-methylated by COMT to form 3-O-methyldopa (3-OMD).[99] Peripheral decarboxylation and methylation result in only 1% of an oral dose of levodopa reaching the brain.[100] Furthermore, 3-OMD may itself contribute to decreased brain levels of levodopa by competing with levodopa for absorption from the GI tract and transport into the brain. When COMT inhibition is added to levodopa/AADC (aromatic L-amino acid decarboxylase) inhibitor therapy (carbidopa and benserazide), peripheral levodopa metabolism is further reduced. Consequently, a greater level of levodopa becomes available for entry into the brain and conversion to DA. Another potential benefit of adding a COMT inhibitor to antiparkinson drugs is the sparing of S-adenosylmethionine, the methyl donor for O-methylation reactions. Low concentrations of S-adenosyl-methionine in the cerebrospinal fluid have been associated with depression and dementia; both common comorbidities associated with PD.[101]

Entacapone, a selective and reversible COMT inhibitor, increases both the peripheral (and thus central) concentration of levodopa by preventing its biotransformation to 3-OMD. When entacapone is added to levodopa/AADC-inhibitor therapy, its ability to inhibit COMT results in greater and more sustained plasma and central nervous system levels of DA than with levodopa/AADC-inhibitor alone.[102] This leads to a prolonged duration of antiparkinsonian action, which translates into improvement in motor function and better ability to perform activities of daily living. Also, [^{18}F] 6-fluorodopa (^{18}FD) PET studies have shown a significant increase in striatal uptake of ^{18}FD in PD patients when coadministered with COMT inhibitors, as compared to without.[103] This indicates the effectiveness of COMT inhibitors in indirectly enhancing striatal levodopa uptake by increasing the peripheral concentration of levodopa.

Results from a number of trials in PD patients who experience motor fluctuations while on levodopa have shown an increase in levodopa bioavailability resulting from COMT inhibition.[102,104,105] These patients show prolongation of the clinical benefit of levodopa therapy with specific improvement in "on" time. Correspondingly, decreases in the daily "off" time, and to a lesser extent in UPDRS motor scores, have been observed. Another COMT inhibitor, tolcapone, has been studied in stable PD patients.[106] The required levodopa doses were much less in stable patients than in fluctuators when combining treatment with tolcapone. There was significant improvement in activities of daily living and motor function in stable PD patients treated with a combination of levodopa and tolcapone. Adverse effects of COMT inhibitors are usually related to increased plasma levels of levodopa and include increased risk of dyskinesias and neuropsychiatric disturbances[102]. Diarrhea is a specific side effect of entacapone. Tolcapone is seldom used now because of hepatotoxicity.

Neuroprotective Agents

Over the past few years, there has been renewed interest in designing interventions that potentially slow, stop, or reverse the neurodegenerative process in PD. This has in part been due to the proliferation of theories of the pathogenetic processes involved in cell death in PD such as oxidative stress, mitochondrial dysfunction, excitotoxicity, and apoptosis.[107]

The DATATOP study and other similar later clinical trials suggested that selegiline may have a neuroprotective effect in PD.[108–112] However, these studies have been confounded by the symptomatic effect of selegiline as changes in the UPDRS motor score were used as primary outcome measures.[107] Even a study with a two-week washout of selegiline[111] that suggested a neuroprotective effect has been criticized as a recent report suggests that even this time frame may be insufficient to eliminate the symptomatic effect of antiparkinson medication.[113] A study with rasagiline, which is a selective and irreversible inhibitor of MAO B, showed significant clinical benefit, but again this study may have been confounded by a symptomatic effect.[114] It has been suggested, based on both *in vitro* and *in vivo* studies, that propargylamines such as selegiline and rasagiline may have a protective and antiapoptotic effect independent of their monoamino oxidase B (MAO B) inhibiting capability.[114–117] This effect may be mediated through altered expression of genes involved in apoptosis such as superoxide dismutase (SOD) 1, SOD 2, BCL 2, BAX, c-JUN, glutathione peroxidase, and glyceraldehydes-3-phosphate dehydrogenase (GAPDH).[116,118] Rasagiline is currently undergoing testing for neuroprotection in PD.

Glutamate mediated excitotoxic cell damage in dopaminergic neurons has been reported to be prevented by N-methyl-D-aspartate (NMDA) antagonists *in vitro* and in animal models of PD.[119–121] One report suggests that amantadine, which has weak NMDA antagonist activity, may be neuroprotective.[82] Other NMDA antagonists such as remacemide and riluzole have not shown any neuroprotective effect.[122,123]

Exposure to rotenone, a pesticide that inhibits complex 1 of the mitochondria, has been associated with PD.[29] A trial last year with coenzyme Q_{10}, a bioenergetic agent, showed a trend for clinical improvement with doses of 1200 mg or more.[124] The results have to be interpreted with caution, as the study was a small one.

DA agonists have been shown to have a neuroprotective effect both *in vitro* and *in vivo* in experimental setting. The CALM-PD SPECT study evaluated the neuroprotective effect of pramipexole using striatal β-CIT (a marker for the DA transporter) uptake as measured by single photon emission tomography (SPECT) as an outcome measure and concluded that pramipexole had a neuroprotective effect.[93] There have been technical concerns with the interpretation of surrogate markers of nigrostriatal function. More importantly, there was no clinical benefit observed in the patients treated with the DA agonists though they had better PET scans. On the contrary, the levodopa treated patients had better motor scores. Another study evaluated the putative neuroprotective effect of ropinirole using striatal [18]fluorodopa uptake measured by PET as an outcome measure and concluded that ropinirole did not have a neuroprotective effect in PD.[125]

— Glial-derived neurotrophic factor (GDNF) has been shown to protect dopaminergic neurons from toxins *in vitro* as well as restore dopaminergic function in MPTP lesioned primates.[126,127] A trial with intraventricular GDNF in human PD subjects showed no benefit.[128] One of the reasons for this is believed to be inability of GDNF to penetrate into brain tissue. Significant improvement has been noted in MPTP lesioned primates that received GDNF using a lenti virus vector to deliver the drug into the striatum.[129]

In summary, none of the interventions thus far have convincingly shown a neuroprotective effect sufficient to make clinical claims.

SURGICAL THERAPIES

The past decade has seen renewed interest in surgery for the treatment of PD.[130] The limitations of drug treatment, a better understanding of disordered basal ganglia physiology, and the significant clinical benefits of surgery have made this option more attractive. Advances in neuroimaging, stereotactic surgery, and better physiological localization with techniques such as microelectrode recording and macrostimulation have also made surgery more accurate and therefore safer.[130] The two techniques employed are ablation and deep brain stimulation. The latter has become popular because of reversibility and the ability to adjust the stimulus and thus potentially "follow" the disease. Disadvantages include technical difficulties, high cost, and potential surgical complications such as infection and migration of electrodes.[131] Neural structures targeted include the ventral intermediate nucleus of the thalamus (for tremor), the posteroventral pallidum (for bradykinesia and drug-induced dyskinesia), and the subthalamic nucleus (for bradykinesia, and dyskinesia indirectly by virtue of reduction in the dose of antiparkinson medication following surgery). There remain unresolved issues such as guidelines for optimal patient selection and timing of the surgery.[132]

Restorative surgery with transplantation of fetal nigral cells into the striatum has been claimed to show clinical benefit and evidence of increased striatal[18] fluorodopa uptake in PD patients. No improvement has been noted in quality of life measures.[133,134] Also, these patients have have severe persistent "off" period dyskinesia, the pathophysiology of which is not well understood.[135,136] A xenotransplant study using porcine fetal nigral cells showed no benefit.[137] Embryonic neural stem cell transplants have shown promise in rodent models of PD though there are no human studies to date.[138,139] Another promising approach has been intrastriatal implantation of cultured human retinal pigment epithelial cells capable of producing levodopa and DA. A preliminary report with six patients indicates significant improvement in UPDRS motor scores at one and two years after surgery without any "off-state" dyskinesia.[140]

CONCLUSION

Advances over the last few years have resulted in a better understanding of PD and consequently better patient care. Specific genetic mutations result in the PD phenotype in a minority of instances. In the vast majority, the etiology remains unknown. Environmental and possibly complex genetic-environmental interactions may be involved. Recent research indicates that abnormal protein processing leading to aberrant protein accumulation is a major pathogenetic mechanism. Other processes possibly implicated in cell death in PD include oxidative stress, impaired mitochondrial function, apoptosis and excitotoxicity. The availability of a larger repertoire of drugs together with a better understanding of their action has greatly improved the symptomatic treatment of PD. Refinement in stereotactic surgery has provided an effective option for symptomatic relief in PD. Neuroprotective agents and restorative techniques such as transplant surgery have yet to translate into meaningful clinical benefit for human PD. Growing understanding of the pathogenesis of PD makes it likely that more effective treatment is likely in the future, for symptoms and perhaps even for neuroprotection.

ACKNOWLEDGMENTS

The authors wish to thank the Canadian Institutes of Health Research, the Parkinson Foundation of Canada, the National Parkinson Foundation, and the Pacific Parkinson Research Institute for supporting this work.

REFERENCES

1. Calne, D. B., Parkinson's disease is not one disease, *Parkinsonism Relat. Disord.* 2001; 7:3–7.

2. Furtado, S., Farrer, M., Tsuboi, Y., Klimek, M. L., Fuente-Fernandez, R, Hussey, J., et al., SCA-2 presenting as parkinsonism in an Alberta family: clinical, genetic, and PET findings, *Neurology,* 59(10):1625–1627, 2002.

3. Kitada, T., Asakawa, S., Hattori, N., Matsumine, H., Yamamura, Y., Minoshima, S., et al., Mutations in the parkin gene cause autosomal recessive juvenile parkinsonism, *Nature,* 392(6676):605–608, 1998.

4. Kruger, R., Kuhn, W., Muller, T., Woitalla, D., Graeber, M., Kosel, S., et al., Ala30Pro mutation in the gene encoding alpha-synuclein in Parkinson's disease, *Nat Genet*; 18(2):106–108, 1998.

5. Polymeropoulos, M. H., Lavedan, C., Leroy, E., Ide, S. E., Dehejia, A., Dutra, A., et al., Mutation in the alpha-synuclein gene identified in families with Parkinson's disease, *Science,* 276(5321):2045–2047, 1997.

6. Huang, Z., Fuente-Fernandez, R., Stoessl, A. J., Etiology of Parkinson's disease, *Can. J. Neurol. Sci.,* 30 Suppl. 1:S10–S18, 2003.

7. Gorell, J. M., Johnson, C. C., Rybicki, B. A., Peterson, E. L., Kortsha, G. X., Brown, G. G., et al., Occupational exposures to metals as risk factors for Parkinson's disease. *Neurology,* 48(3):650–658, 1997.

8. Gorell, J. M., Johnson, C. C., Rybicki, B. A., Peterson, E. L., Richardson, R. J., The risk of Parkinson's disease with exposure to pesticides, farming, well water, and rural living, *Neurology,* 50(5):1346–1350, 1998.

9. Lai, B. C. L., Marion, S. A., Teschke, K., Tsui, J. K. C., Occupational and environmental risk factors for Parkinson's disease, *Parkinsonism Relat. Disord.,* 8:297–309, 2002.

10. Zorzon, M., Capus, L., Pellegrino, A., Cazzato, G., Zivadinov, R., Familial and environmental risk factors in Parkinson's disease: a case-control study in north-east Italy, *Acta Neurol. Scand.,* 105(2):77–82, 2002.

11. Calne, D. B., Langston, J. W., Aetiology of Parkinson's disease, *Lancet,* 2(8365–66):1457–1459, 1983.

12. Leroy, E., Boyer, R., Auburger, G., Leube, B., Ulm, G., Mezey, E., et al., The ubiquitin pathway in Parkinson's disease, *Nature,* 395(6701):451–452, 1998.

13. Bonifati, V., Rizzu, P., van Baren, M. J., Schaap, O., Breedveld, G. J., Krieger, E., et al., Mutations in the DJ-1 gene associated with autosomal recessive early-onset parkinsonism, *Science,* 299(5604):256–259, 2003.

14. Gasser, T., Muller-Myhsok, B., Wszolek, Z. K., Oehlmann, R, Calne, D. B., Bonifati, V., et al., A suscepti-bility locus for Parkinson's disease maps to chromosome 2p13, *Nat Genet,* 18(3):262–265, 1998.

15. Farrer, M., Gwinn-Hardy, K., Muenter, M., DeVrieze, F. W., Crook, R., Perez-Tur, J., et al., A chromosome 4p haplotype segregating with Parkinson's disease and postural tremor, *Hum. Mol. Genet.,* 8(1):81–85, 1999.

15a. Valente, E. M., Bentivoglio, A. R., Dixon, P. H., Ferraris, A., Ialongo, T., Frontali, M., Albanese, A., and Wood, N. W., Localization of a novel locus for autosomal recessive early-onset parkinsonism, PARK6, on human chromosome 1p35–p36, *Am. J. Hum. Genet.,* 68(4):895–900, 2001.

16. Funayama, M., Hasegawa, K., Kowa, H., Saito, M., Tsuji, S., Obata, F., A new locus for Parkinson's disease (PARK8) maps to chromosome 12p11.2–q13.1. *Ann. Neurol.,* 51(3):296-301, 2001.

17. Scott, W. K., Nance, M. A., Watts, R. L., Hubble, J. P., Koller, W. C., Lyons, K., et al., Complete genomic screen in Parkinson disease: evidence for multiple genes. *JAMA,* 286(18):2239–2244, 2001.

18. Mouradian, M. M., Recent advances in the genetics and pathogenesis of Parkinson disease, *Neurology,* 58(2):179–185, 2002.

19. Shastry, B. S., Parkinson disease: etiology, pathogenesis and future of gene therapy, *Neurosci. Res.,* 41(1):5–12, 2001.

20. Takahashi, H., Ohama, E., Suzuki, S., Horikawa, Y., Ishikawa, A., Morita, T., et al., Familial juvenile parkinsonism: clinical and pathologic study in a family, *Neurology,* 44(3 Pt. 1):437–441, 1994.

21. Tanner, C. M., Ottman, R., Goldman, S. M., Ellenberg, J., Chan, P., Mayeux, R., et al., Parkinson disease in twins: an etiologic study, *JAMA,* 281(4):341–346, 1999.

22. de la Fuente-Fernandez, R., Calne, D. B., Evidence for environmental causation of Parkinson's disease, *Parkinsonism Relat. Disord.,* 8(4):235–241, 2001.

23. Kurtzke, J. F., Goldberg, I. D., Parkinsonism death rates by race, sex, and geography, *Neurology,* 38(10):1558–1561, 1998.

24. Langston, J. W., Forno, L. S., Rebert, C. S., Irwin, I., Selective nigral toxicity after systemic administration of 1-methyl-4- phenyl-1,2,5,6-tetrahydropyrine (MPTP) in the squirrel monkey, *Brain Res.,* 292(2):390–394, 1984.

25. Vingerhoets, F. J., Snow, B.J., Tetrud, J. W., Langston, J. W., Schulzer, M., Calne, D. B., Positron emission tomographic evidence for progression of human MPTP-induced dopaminergic lesions, *Ann. Neurol.,* 36(5):765–770, 1994.

26. Menegon, A, Board, P. G., Blackburn, A. C., Mellick, G. D., Le Couteur, D. G., Parkinson's disease, pesticides, and glutathione transferase polymorphisms, *Lancet,* 352(9137):1344–1346, 1998.

27. Semchuk, K. M., Love, E. J, Effects of agricultural work and other proxy-derived case-control data on Parkinson's disease risk estimates, *Am. J. Epidemiol.,* 141(8):747–754, 1995.

28. Priyadarshi, A., Khuder, S. A., Schaub, E. A., Shrivastava, S., A meta-analysis of Parkinson's disease and exposure to pesticides, *Neurotoxicology,* 21(4):435–440, 2000.

29. Betarbet, R., Sherer, T. B., MacKenzie, G., Garcia-Osuna, M., Panov, A. V., Greenamyre, J. T., Chronic systemic pesticide exposure reproduces features of Parkinson's disease, *Nat. Neurosci.*, 3(12):1301–1306, 2000.

30. Rajput, A. H., Uitti, R. J., Stern, W., Laverty, W., Early onset Parkinson's disease in Saskatchewan—environmental considerations for etiolog., *Can. J. Neurol. Sci.*, 13(4):312–316, 1986.

31. Gorell, J. M., Johnson, C. C., Rybicki, B. A., Peterson, E. L., Kortsha, G. X., Brown, G. G., et al., Occupational exposure to manganese, copper, lead, iron, mercury and zinc and the risk of Parkinson's disease, *Neurotoxicology*, 20(2–3):239–247, 1999.

32. Tanner, C. M., Goldman, S. M., Aston, D. A., Ottman, R., Ellenberg, J., Mayeux, R., et al., Smoking and Parkinson's disease in twins, *Neurology*, 58(4):581–588, 2002.

33. Golbe, L. I., Langston, J. W., The etiology of Parkinson's disease: new directions for research. Jankovic, J., Tolosa, E., Eds., *Parkinson's disease and movement disorders*, Williams and Wilkins, Baltimore, MD, 93–101, 1993.

34. Baron, J. A., Cigarette smoking and Parkinson's disease, *Neurology*, 36(11):1490–1496, 1986.

35. Maggio, R., Riva, M., Vaglini, F., Fornai, F., Molteni, R., Armogida, M., et al., Nicotine prevents experimental parkinsonism in rodents and induces striatal increase of neurotrophic factors, *J. Neurochem.*, 71(6):2439–2446, 1998.

36. Ross, G. W., Abbott, R. D., Petrovitch, H., Morens, D. M., Grandinetti, A., Tung, K. H., et al., Association of coffee and caffeine intake with the risk of Parkinson disease, *JAMA*, 283(20):2674–2679, 2000.

37. Chen, J. F., Xu, K., Petzer, J. P., Staal, R., Xu, YH., Beilstein, M., et al., Neuroprotection by caffeine and A(2A) adenosine receptor inactivation in a model of Parkinson's disease, *J. Neurosci.*, 21(10):RC143, 2001.

38. Ross, G. W., Petrovitch, H., Current evidence for neuroprotective effects of nicotine and caffeine against Parkinson's disease, *Drugs Aging*, 18(11):797–806, 2001.

39. Duvoisin, R. C., Yahr, M. D., Encephalitis and parkinsonism, *Arch. Neurol.*, 112:227–239, 1985.

40. Calne, D. B., Lees, A. J., Late progression of post-encephalitic Parkinson's syndrome, *Can. J. Neurol. Sci.*, 15(2):135–138, 1988.

41. Lin, S. K., Lu, C. S., Vingerhoets, F. J., Snow, B. J., Schulzer, M., Wai, Y. Y., et al., Isolated involvement of substantia nigra in acute transient parkinsonism: MRI and PET observations, *Parkinsonism Relat. Disord.*, 1(2):67–73, 1995.

42. Takahashi, M., Yamada, T., Viral etiology for Parkinson's disease—a possible role of influenza A virus infection, *Jpn. J. Infect. Dis.*, 52(3):89–98, 1999.

43. Yamada, T., Yamanaka, I., Takahashi, M., Namajima, S., Invasion of brain by neurovirulent influenza A virus after intranasal inoculation, *Parkinsonism Relat. Disord.*, 2(4):187–193, 1996.

44. Yamada, T., Yamanaka, I., Takahashi, M., Namajima, S., Invasion of brain by neurovirulent influenza A virus after intranasal inoculation, *Parkinsonism Relat. Disord.*, 2(4):187–193, 1996.

45. Woulfe, J., Hoogendoorn, H., Tarnopolsky, M., Munoz, D. G., Monoclonal antibodies against Epstein-Barr virus cross-react with alpha- synuclein in human brain, *Neurology*, 55(9):1398–1401, 2000.

46. McGeer, P. L., Yasojima, K., McGeer, E. G., Inflammation in Parkinson's disease, *Adv. Neurol.*, 86:83–89, 2001.

47. McGeer, P. L., Yasojima, K., McGeer, E. G., Association of interleukin 1a polymorphisms with idiopathic Parkinson's disease, *Neurosc. Lett.*, 326(1):67–69, 2002.

48. McGeer, P. L. and McGeer, E. G., Inflammation and neurodegeneration in Parkinson's disease, *Parkinsonism Relat. Disord.*, 10(Suppl. 1):S3–7, 2004.

49. Tanner, C. M., Goldman, S. M., Quinlan, P., Field, R., Aston, D. A., Comyns, K., et al., Occupation and risk of Parkinson's disease: A preliminary investigation of standard occupational codes in twins discordant for disease, *Neurology*, 60(Suppl. 1):A 415 (Abstract), 2003.

50. Ciechanover, A., Schwartz, A. L., The ubiquitin-proteasome pathway: the complexity and myriad functions of proteins death, *Proc. Natl. Acad. Sci., USA*, 95(6):2727–2730, 1998.

51. Jenner, P., Olanow, C. W., Understanding cell death in Parkinson's disease, *Ann. Neurol.*, 44(3 Suppl. 1):S72–S84, 1998.

52. Spillantini, M. G., Crowther, R. A., Jakes, R., Hasegawa, M., Goedert, M., alpha-synuclein in filamentous inclusions of Lewy bodies from Parkinson's disease and dementia with lewy bodies, *Proc. Natl. Acad. Sci., USA*, 95(11):6469–6473, 1998.

53. Stefanis, L., Larsen, K. E., Rideout, H. J., Sulzer, D., Greene, L. A., Expression of A53T mutant but not wild-type alpha-synuclein in PC12 cells induces alterations of the ubiquitin-dependent degradation system, loss of dopamine release, and autophagic cell death, *J. Neurosci.*, 21(24):9549–9560, 2001.

54. Tofaris, G. K., Layfield, R., Spillantini, M.G., alpha-synuclein metabolism and aggregation is linked to ubiquitin- independent degradation by the proteasome, *FEBS Lett.*, 509(1):22–26, 2001.

55. Kanda, S., Bishop, J. F., Eglitis, M. A., Yang, Y., Mouradian, M. M., Enhanced vulnerability to oxidative stress by alpha-synuclein mutations and C-terminal truncation, *Neuroscience*, 97(2):279–284, 2000.

56. Xu, J., Kao, S. Y., Lee, F. J., Song, W., Jin, L. W., Yankner, B. A., Dopamine-dependent neurotoxicity of alpha-synuclein: a mechanism for selective neurodegeneration in Parkinson disease, *Nat. Med.*, 8(6):600–606, 2002.

57. Feany, M. B., Bender, W. W., A Drosophila model of Parkinson's disease, *Nature*, 404(6776):394–398, 2000.

58. Zhang, Y., Gao, J., Chung, K. K., Huang, H., Dawson, V. L., Dawson, T. M., Parkin functions as an E2-dependent ubiquitin- protein ligase and promotes the degradation of the synaptic vesicle-associated protein, CDCrel-1, *Proc. Nat.l Acad. Sci., USA*, 97(24):13354–13359, 2000.

59. Imai, Y., Soda, M., Inoue, H., Hattori, N., Mizuno, Y., Takahashi, R., An unfolded putative transmembrane polypeptide, which can lead to endoplasmic reticulum stress, is a substrate of Parkin, *Cell*, 105(7):891–902, 2001.

60. Shimura, H., Schlossmacher, M. G., Hattori, N., Frosch, M. P., Trockenbacher, A., Schneider, R., et al., Ubiquitination of a new form of alpha-synuclein by parkin from human brain: implications for Parkinson's disease, *Science*, 293(5528):263–269, 2001.

61. Jha, N., Jurma, O., Lalli, G., Liu, Y., Pettus, E. H., Greenamyre, J. T., et al., Glutathione depletion in PC12 results in selective inhibition of mitochondrial complex I activity. Implications for Parkinson's disease, *J. Biol. Chem.*, 275(34):26096–26101, 2000.

62. Merad-Boudia, M., Nicole, A., Santiard-Baron, D., Saille, C., Ceballos-Picot, I., Mitochondrial impairment as an early event in the process of apoptosis induced by glutathione depletion in neuronal cells: relevance to Parkinson's disease, *Biochem. Pharmacol.*, 56(5):645–655, 1998.

63. Alam, Z. I., Jenner, A., Daniel, S. E., Lees, A. J., Cairns, N., Marsden, C. D., et al., Oxidative DNA damage in the parkinsonian brain: an apparent selective increase in 8-hydroxyguanine levels in substantia nigra, *J. Neurochem.*, 69(3):1196–1203, 1997.

64. Sanchez-Ramos, J., Overvik, E., Overvik, A. B. N., A marker of oxyradical-mediated DNA damage (8-hydroxy-2'deoxyguanosine)is increased in the nigrostriatum of Parkinson's disease brain, *Neurodegeneration*, 3(197):204, 1994.

65. Sian, J., Dexter, D. T., Lees, A. J., Daniel, S., Jenner, P., Marsden, C. D., Glutathione-related enzymes in brain in Parkinson's disease. *Ann. Neurol.*, 36(3):356–361, 1994.

66. Soto-Otero, R., Mendez-Alvarez, E., Hermida-Ameijeiras, A., Munoz-Patino, A. M., Labandeira-Garcia, J. L., Autoxidation and neurotoxicity of 6-hydroxydopamine in the presence of some antioxidants: potential implication in relation to the pathogenesis of Parkinson's disease, *J. Neurochem.* 74(4):1605–1612, 2000.

67. Jenner, P., Olanow, C.W., Oxidative stress and the pathogenesis of Parkinson's disease, *Neurology*, 47(6 Suppl. 3):S161–S170, 1996.

68. Agid, Y., Chase, T., Marsden, D., Adverse reactions to levodopa: drug toxicity or progression of disease? *Lancet*, 351(9106):851–852, 1998.

69. Calne, D. B., The free radical hypothesis in idiopathic parkinsonism: evidence against it, *Ann. Neurol.*, 32(6):799–803, 1992.

70. Murer, M. G., Dziewczapolski, G., Menalled, L. B., Garcia, M. C., Agid, Y., Gershanik, O., et al., Chronic levodopa is not toxic for remaining dopamine neurons, but instead promotes their recovery, in rats with moderate nigrostriatal lesions, *Ann. Neurol.*, 43(5):561–575, 1998.

71. Rajput, A. H., The protective role of levodopa in the human substantia nigra, *Adv. Neurol.*, 86:327–336, 2001.

72. Ebadi, M., Muralikrishnan, D., Pellett, L. J., Murphy, T., Drees, K., Ubiquinone (coenzyme Q10) and complex I in mitochondrial oxidative disorder of Parkinson's disease, *Proc. West Pharmacol. Soc.*, 43:55–63, 2000.

73. Gu, M., Cooper, J. M., Taanman, J. W., Schapira, A. H., Mitochondrial DNA transmission of the mitochondrial defect in Parkinson's disease, *Ann. Neurol.*, 44(2):177–186, 1998.

74. Janetzky, B., Hauck, S., Youdim, M. B., Riederer, P., Jellinger, K., Pantucek, F., et al., Unaltered aconitase activity, but decreased complex I activity in substantia nigra pars compacta of patients with Parkinson's disease, *Neurosci. Lett.*, 169(1–2):126–128, 1994.

75. Gu, M., Gash, M. T., Cooper, J. M., Wenning, G. K., Daniel, S. E., Quinn, N. P., et al., Mitochondrial respiratory chain function in multiple system atrophy, *Mov. Disord.*, 12(3):418–422, 1997.

76. Hartmann, A., Hunot, S., Michel, P. P., Muriel, M. P., Vyas, S., Faucheux, B. A., et al., Caspase-3: A vulnerability factor and final effector in apoptotic death of dopaminergic neurons in Parkinson's disease, *Proc. Natl. Acad. Sci.*, USA, 97(6):2875–2880, 2000.

77. Viswanath, V., Wu, Y., Boonplueang, R., Chen, S., Stevenson, F. F., Yantiri, F., et al., Caspase-9 activation results in downstream caspase-8 activation and bid cleavage in 1-methyl-4-phenyl-1,2,3,6-tetrahydropyridine-induced Parkinson's disease, *J. Neurosci.*, 21(24):9519–9528, 2000.

78. Kahns, S., Lykkebo, S., Jakobsen, L. D., Nielsen, M. S., Jensen, P. H., Caspase-mediated parkin cleavage in apoptotic cell death, *J. Biol. Chem.*, 277(18):15303–15308, 2002.

79. Banati, R. B., Daniel, S. E., Blunt, S. B., Glial pathology but absence of apoptotic nigral neurons in long- standing Parkinson's disease, *Mov. Disord.*, 13(2):221-227, 1998.

80. Jellinger, K. A., Stadelmann, C. H., The enigma of cell death in neurodegenerative disorders, *J. Neural. Transm., Suppl.*, (60):21–36, 2000.

81. Beal, M. F., Energetics in the pathogenesis of neurodegenerative disease, *Trends Neurosci.* 23(7):298–304, 2000.

82. Uitti, R. J., Rajput, A. H., Ahlskog, J. E., Offord, K. P., Schroeder, D. R., Ho, M. M., et al., Amantadine treatment is an independent predictor of improved survival in Parkinson's disease, *Neurology*, 46(6):1551–1556, 1996.

83. McGeer, P. L., Itagaki, S., Boyes, B. E., McGeer, E.G., Reactive microglia are positive for HLA-DR in the substantia nigra of Parkinson's and Alzheimer's disease brains, *Neurology*, 38(8):1285–1291, 1988.

84. Langston, J. W., Forno, L. S., Tetrud, J., Reeves, A. G., Kaplan, J. A., Karluk, D., Evidence of active nerve cell degeneration in the substantia nigra of humans years after 1-methyl-4-phenyl-1,2,3,6-tetrahydropyridine exposure, *Ann. Neurol.*, 46(4):598–605, 1999.

85. de la Fuente-Fernandez, R., Ruth, T. J., Sossi, V., Schulzer, M., Calne, D. B., Stoessl, A. J., Expectation and dopamine release: mechanism of the placebo effect in Parkinson's disease, *Science*, 293(5532):1164–1166, 2001.

86. Marsden, C. D., Parkes, J. D., "On-off" effects in patients with Parkinson's disease on chronic levodopa therapy, *Lancet,* 1(7954):292–296, 1976.

87. Quinn, N., Critchley, P., Marsden, C. D., Young onset Parkinson's disease, *Mov. Disord.,* 2(2):73–91, 1987.

88. Rinne, U. K., Problems associated with long-term levodopa treatment of Parkinson's disease, *Acta Neurol. Scand. Suppl.,* 95:19–26, 1983.

89. Stocchi, F., Nordera, G., Marsden, C. D., Strategies for treating patients with advanced Parkinson's disease with disastrous fluctuations and dyskinesias, *Clin. Neuropharmacol.,* 20(2):95–115, 1997.

90. Sweet, R. D., McDowell, F. H., The "on-off" response to chronic L-DOPA treatment of Parkinsonism, *Adv. Neurol.,* 5:331–338, 1974.

91. Rinne, U. K., Bracco, F., Chouza, C., Dupont, E., Gershanik, O., Marti, Masso, J. F., et al., Early treatment of Parkinson's disease with cabergoline delays the onset of motor complications. Results of a double-blind levodopa controlled trial, The PKDS009 Study Group, *Drugs,* 55 Suppl. 1:23–30, 1998.

92. Rascol, O., Brooks, D. J., Korczyn, A. D., De Deyn, P. P., Clarke, C. E., Lang, A. E., A five-year study of the incidence of dyskinesia in patients with early Parkinson's disease who were treated with ropinirole or levodopa, 056 Study Group, *New Engl. J. Med.,* 342(20):1484–1491, 2000.

93. A randomized controlled trial comparing pramipexole with levodopa in early Parkinson's disease: design and methods of the CALM-PD Study, Parkinson Study Group, *Clin. Neuropharmacol.,* 23(1):34–44, 2000.

94. Huang, Z., Kumar, A., Tsui, J. K. C., COMT inhibition in Parkinson's disease, *Geriatrics and Aging,* 5(4):32–35, 2002.

95. Saint-Hilaire, M. H., Feldman, R. G., The "on-off" phenomenon in Parkinson's disease, In Joseph, A. B., Young, R. R., EEEd., *Movement disorders in neurology and neuropsychiatry.* Boston: Blackwell Scientific Publications, 180–184, 1999.

96. Chase, T. N., Oh, J. D., Striatal mechanisms and pathogenesis of parkinsonian signs and motor complications, *Ann. Neurol.,* 47(4 Suppl. 1):S122–S129, 2001.

97. de la Fuente-Fernandez, R., Lu, J. Q., Sossi, V., Jivan, S., Schulzer, M., Holden, J. E., et al., Biochemical variations in the synaptic level of dopamine precede motor fluctuations in Parkinson's disease: PET evidence of increased dopamine turnover, *Ann. Neurol.,* 49(3):298–303, 2001.

98. Guldberg, H. C., Marsden, C. A., Catechol-O-methyl transferase: pharmacological aspects and physiological role, *Pharmacol. Rev.,* 27(2):135–206, 1975.

99. Nutt, J. G., Woodward, W. R., Beckner, R. M., Stone, C. K., Berggren, K., Carter, J. H., et al., Effect of peripheral catechol-O-methyltransferase inhibition on the pharmacokinetics and pharmacodynamics of levodopa in parkinsonian patients, *Neurology,* 44(5):913–919, 1994.

100. Mannisto, P. T., Tuomainen, P., Toivonen, M., Tornwall, M., Kaakkola, S., Effect of acute levodopa on brain catecholamines after selective MAO and COMT inhibition in male rats, *J. Neural. Transm.,* Park Dis Dement Sect, 2(1):31–43, 1990.

101. Nutt, J. G., Catechol-O-methyltransferase inhibitors for treatment of Parkinson's disease, *Lancet,* 351(9111):1221–1222, 1998.

102. Entacapone improves motor fluctuations in levodopa-treated Parkinson's disease patients, Parkinson Study Group, *Ann. Neurol.,* 42(5):747–755, 1997.

103. Ruottinen, H. M., Niinivirta, M., Bergman, J., Oikonen, V., Solin, O., Eskola, O., et al., Detection of response to COMT inhibition in FDOPA PET in advanced Parkinson's disease requires prolonged imaging, *Synapse,* 40(1):19–26, 2001.

104. Merello, M., Lees, A. J., Webster, R., Bovingdon, M., Gordin, A., Effect of entacapone, a peripherally acting catechol-O- methyltransferase inhibitor, on the motor response to acute treatment with levodopa in patients with Parkinson's disease, *J. Neurol. Neurosurg. Psychiatry,* 57(2):186–189, 1994.

105. Rinne, U. K., Larsen, J. P., Siden, A., Worm-Petersen, J., Entacapone enhances the response to levodopa in parkinsonian patients with motor fluctuations. Nomecomt Study Group, *Neurology,* 51(5):1309—1314, 1998.

106. Waters, C. H., Kurth, M., Bailey, P., Shulman, L. M., LeWitt, P., Dorflinger, E., et al., Tolcapone in stable Parkinson's disease: efficacy and safety of long- term treatment, The Tolcapone Stable Study Group, *Neurology,* 49(3):665–671, 1997.

107. Stocchi, F., Olanow, C. W., Neuroprotection in Parkinson's disease: clinical trials, *Ann. Neurol.,* 53 Suppl. 3:S87–S97, 2003.

108. DATATOP: a multicenter controlled clinical trial in early Parkinson's disease, Parkinson Study Group, *Arch. Neurol.,* 46(10):1052–1060, 1989.

109. Effect of deprenyl on the progression of disability in early Parkinson's disease, The Parkinson Study Group, *N. Engl. J. Med.,* 321(20):1364–1371, 1989.

110. Effects of tocopherol and deprenyl on the progression of disability in early Parkinson's disease, The Parkinson Study Group, *N. Engl. J. Med.,* 328(3):176–183, 1993.

111. Olanow, C. W., Hauser, R. A., Gauger, L., Malapira, T., Koller, W., Hubble, J., et al., The effect of deprenyl and levodopa on the progression of Parkinson's disease, *Ann. Neurol.,* 38(5):771–777, 1995.

112. Przuntek, H., Conrad, B., Dichgans, J., Kraus, P. H,. Krauseneck, P., Pergande, G., et al., SELEDO: a 5-year long-term trial on the effect of selegiline in early Parkinsonian patients treated with levodopa, *Eur. J. Neurol.,* 6(2):141–150, 1999.

113. Hauser, R. A., Koller, W. C., Hubble, J. P., Malapira, T., Busenbark, K., Olanow, C. W., Time course of loss of clinical benefit following withdrawal of levodopa/carbidopa and bromocriptine in early Parkinson' s disease, *Mov. Disord.,* 15(3):485–489, 2000.

114. Youdim, M. B., Weinstock, M., Molecular basis of neuroprotective activities of rasagiline and the anti-Alzheimer drug TV3326 [(N-propargyl-(3R)aminoindan-5-YL)-ethyl methyl carbamate], *Cell Mol. Neurobiol.,* 21(6):555–573, 2001.

115. Mytilineou, C., Radcliffe, P., Leonardi, E. K., Werner, P., Olanow, C. W., L-deprenyl protects mesencephalic dopamine neurons from glutamate receptor-mediated toxicity in vitro, *J. Neurochem.*, 68(1):33–39, 1997.

116. Tatton, W. G., Chalmers-Redman, R. M., Ju,W. J., Mammen, M., Carlile, G. W., Pong, A. W., et al., Propargylamines induce antiapoptotic new protein synthesis in serum- and nerve growth factor (NGF)-withdrawn, NGF-differentiated PC-12 cells, *J. Pharmacol. Exp. Ther.*, 301(2):753–764, 2002.

117. Carlile, G. W., Chalmers-Redman, R. M., Tatton, N. A., Pong, A., Borden, K. E., Tatton, W. G., Reduced apoptosis after nerve growth factor and serum withdrawal: conversion of tetrameric glyceraldehyde-3-phosphate dehydrogenase to a dimer, *Mol. Pharmacol.*, 57(1):2–12, 2000.

118. Tatton, W. G., Ju, W. Y., Holland, D. P., Tai, C., Kwan, M., (-)-Deprenyl reduces PC12 cell apoptosis by inducing new protein synthesis, *J. Neurochem.*,63(4):1572–1575, 1994.

119. Doble, A., The role of excitotoxicity in neurodegenerative disease: implications for therapy, *Pharmacol. Ther,* 81(3):163–221, 1999.

120. Turski, L., Bressler, K., Rettig, K. J., Loschmann, P. A., Wachtel, H., Protection of substantia nigra from MPP+ neurotoxicity by N-methyl-D- aspartate antagonists, *Nature*, 349(6308):414–418, 1991.

121. Greenamyre, J. T., Eller, R. V., Zhang, Z., Ovadia, A., Kurlan, R., Gash, D. M., Antiparkinsonian effects of remacemide hydrochloride, a glutamate antagonist, in rodent and primate models of Parkinson's disease, *Ann. Neurol.*, 35(6):655–661, 1994.

122. A multicenter randomized controlled trial of remacemide hydrochloride as monotherapy for PD, Parkinson Study Group, *Neurology,* 54(8):1583–1588, 2000.

123. Rascol, O., Olanow, C. W., Brooks, D. J., et al., A 2-year multicenter placebo-controlled, double-blind parallel group study of the effect of riluzole on Parkinson's disease progression, *Mov. Disord.*, 17:39, 2003.

124. Shults, C. W., Oakes, D., Kieburtz, K., Beal, M. F., Haas, R., Plumb, S., et al., Effects of coenzyme Q10 in early Parkinson disease: evidence of slowing of the functional decline, *Arch. Neurol.*, 59(10):1541–1550, 2002.

125. Rakshi, J. S., Pavese, N., Uema, T., Ito, K., Morrish, P. K., Bailey, D. L., et al., A comparison of the progression of early Parkinson's disease in patients started on ropinirole or L-dopa: an 18F-dopa PET stud, *J. Neural. Transm.*, 109(12):1433–1443, 2002.

126. Lin, L. F., Doherty, D. H., Lile, J. D., Bektesh, S., Collins, F., GDNF: a glial cell line-derived neurotrophic factor for midbrain dopaminergic neurons, *Science,* 260(5111):1130–1132, 1993.

127. Herbert, M. A., Hoffer, B. J., Zhang, Z., et al., Functional effects of GDNF in normal and parkinsonian rats and monkeys, In: Tuszynski, M., Kordower, J. H., Eds., CNS regeneration: basic science and clinical advances, Academic Press, NY, 419–436, 1999.

128. Kordower, J. H., Palfi, S., Chen, E. Y., Ma, S. Y., Sendera, T., Cochran, E. J., et al., Clinicopathological findings following intraventricular glial-derived neurotrophic factor treatment in a patient with Parkinson's disease, *Ann. Neurol.*, 46(3):419–424, 1999.

129. Kordower, J. H., Emborg, M. E., Bloch, J., Ma, S. Y., Chu, Y., Leventhal, L., et al., Neurodegeneration prevented by lentiviral vector delivery of GDNF in primate models of Parkinson's disease, *Science,* 290(5492):767–773, 2000.

130. Koller, W. C., Wilkinson, S., Pahwa, R., Miyawaki, E. K., Surgical treatment options in Parkinson's disease, *Neurosurg. Clin. N. Am.*, 9(2):295–306, 1998.

131. Hariz, M. I., Complications of deep brain stimulation surgery, *Mov. Disord.*, 17 Suppl. 3:S162–S166, 2002.

132. Lang, A. E., Widner, H., Deep brain stimulation for Parkinson's disease: patient selection and evaluation, *Mov. Disord.*, 17 Suppl. 3:S94–101, 2002.

133. Olanow, C. W., Kordower, J. H., Freeman, T. B., Fetal nigral transplantation as a therapy for Parkinson's disease, *Trends Neurosci.*, 19(3):102–109, 1996.

134. Freed, C. R., Greene, P. E., Breeze, R. E., Tsai, W. Y., DuMouchel, W., Kao, R., et al., Transplantation of embryonic dopamine neurons for severe Parkinson's diseas,. *N. Engl. J. Med.*, 344(10):710–719, 2001.

135. Hagell, P., Piccini, P., Bjorklund, A., Brundin, P., Rehncrona, S., Widner, H., et al., Dyskinesias following neural transplantation in Parkinson's disease, *Nat. Neurosci.*, 5(7):627–628, 2002.

136. Huang, Z., de la Fuente-Fernandez, R., Hauser, R. A., Freeman, T. B., Sossi, V., Olanow, C. W., et al., Dopaminergic alteration in Parkinson's patients with off-period dyskinesia following striatal embryonic mesencephalic transplant, *Neurology,* 60 (S1):A127, 2003.

137. Fink, J. S., Schumacher, J. M., Ellias, S. L., Palmer, E. P., Saint-Hilaire, M., Shannon, K., et al., Porcine xenografts in Parkinson's disease and Huntington's disease patients: preliminary results, *Cell Transplant.*, 9(2):273–278, 2000.

138. Bjorklund, L. M., Sanchez-Pernaute, R., Chung, S., Andersson, T., Chen, I. Y., McNaught, K. S., et al., Embryonic stem cells develop into functional dopaminergic neurons after transplantation in a Parkinson rat model, *Proc. Natl. Acad. Sci.,* U S A, 99(4):2344–2349, 2002.

139. Kim, J. H., Auerbach, J. M., Rodriguez-Gomez, J. A., Velasco, I., Gavin, D., Lumelsky, N., et al., Dopamine neurons derived from embryonic stem cells function in an animal model of Parkinson's disease, *Nature,* 418(6893):50–56, 2002.

140. Watts, R. L., Raiser, C. D., Stover, N. P., Cornfeldt, M. L., Schweikert, A. W., Allen, R. C., et al., Stereotaxic intrastriatal implantation of retinal pigment epithelial cells attached to microcarriers in six advanced parkinson disease patients; two year follow-up, *Neurology,* 60(S1):A164–A165, 2003.

5 Epidemiology of Parkinson's Disease: An Overview

Monica Korell and Caroline M. Tanner
The Parkinson's Institute

CONTENTS

This chapter provides a brief overview of the epidemiology of Parkinson's disease. The primary focus is a review of descriptive epidemiological findings and a discussion of how these patterns might help us understand the causes of disease. The last section summarizes risk factors that will be covered in greater detail in later chapters.

INTRODUCTION

The first step to a better understanding of Parkinson's disease is to identify and describe the people who have the disease. The ability to identify shared characteristics such as age, gender, occupation, residence, or family membership among persons with Parkinson's disease will bring us closer to solving the mystery of what causes the disease. These observations are most informative when they apply to all cases of disease within a population, referred to as *complete ascertainment*. When ascertainment is not complete, conclusions may be misleading, because those persons missed may be different from those identified. However, there are many challenges faced when trying to determine the frequency and distribution of Parkinson's disease within populations. The first challenge lies in quantifying the number of Parkinson's disease cases. Because there is not a diagnostic test for Parkinson's disease, epidemiological studies must rely on the clinical examination to determine the number of cases. Accuracy of the clinical diagnosis of Parkinson's disease depends on both the experience of the examiner and the duration of disease in each individual examined. Inexperienced investigators may confuse Parkinson's disease with other disorders such as essential tremor or atypical parkinsonism, or even with senescent changes in movement and balance. Mutch et al. found that 57 out of 393 of their original Parkinson's disease referrals had a different diagnosis when examined, most commonly essential tremor.[1] Similarly, a study conducted in Finland found that 26% of the subjects identified as having Parkinson's disease based on medical record review instead were found to

0-8493-1590-5/05/$0.00+$1.50
© 2005 by CRC Press

have essential tremor upon physical examination.[2] Even experienced examiners can misdiagnose Parkinson's disease early in the course of disease, when distinguishing features of other disorders may not yet be present. Therefore, estimates of incident disease may have more misclassification than prevalence estimates.

Comparisons across studies are hampered not only by differences in the experience of the diagnostician, but by differences in the study diagnostic criteria. These may vary greatly over time but can also be quite different among contemporaneous studies. For example, some early studies grouped all forms of parkinsonism together, combining atypical parkinsonism, secondary parkinsonism (such as post-encephalitic and drug-induced parkinsonism), and Parkinson's disease. Anderson and colleagues reviewed several prevalence surveys of Parkinson's disease and concluded that the prevalence comparisons between surveys can have diminished value if the surveys used different diagnostic criteria for Parkinson's disease.[3] They found that the differences in prevalence estimates between surveys could be explained by the differences in diagnostic criteria in some instances.

Three general approaches have been used to identify Parkinson's disease cases for epidemiologic studies: (1) evaluating clinic patients, (2) searching health utilization records (medical charts, prescription registries, disease registries, billing databases), and (3) directly screening a population to identify persons with Parkinson's disease living within a defined area. Each approach has strengths and weaknesses. The first method, relying on information derived from clinics, while relatively inexpensive and easy to perform, may be influenced by social and economic factors determining attendance at the facility studied. This could result in mistaken conclusions about disease frequency—for example, patients at a referral center and those at a neighborhood clinic may differ in many ways (socioeconomic status, race/ethnicity, gender), but none of these differences may be specific to Parkinson's disease. The second approach, relying on the review of health utilization records, while subject to some of the same biases, can provide good estimates where health care is universally available. Both of these methods will miss those cases of Parkinson's disease who have never been diagnosed.[4–6] The proportion of undiagnosed cases will also vary across populations, reflecting variations in such factors as health resources and disease awareness. The third method, identifying persons with Parkinson's disease within a geographic area, typically employs a staged, community-based ascertainment method such as a door-to-door survey. This study design attempts to minimize undercounting of previously undiagnosed cases by surveying all households within a targeted area. When all households are surveyed in a community using a screening interview followed up by examination of individuals suspected of having disease,

this method is more likely to identify all prevalent cases of Parkinson's disease in a community. Expert application of stringent diagnostic criteria is important, however, to avoid overestimation of cases. In addition, the large amount of time and great expense involved in the latter method limit its application.

DESCRIPTIVE EPIDEMIOLOGY

INCIDENCE

Incidence, the number of new cases of a disease occurring in a specific population during a given period of time, is the best measure of disease frequency, because it is not affected by survival or migration. Parkinson's disease is a relatively rare disorder. Therefore, large numbers of people must be studied to obtain reliable estimates of incidence. Because Parkinson's disease is rare before age 50 and increases with increasing age thereafter, the age distribution of the sample population can influence the number of cases observed. For this reason, direct comparison of crude incidence estimated from different populations may be misleading. For example, estimated crude incidence ranges from 5 to 20/100,000/year in different reports.[7–9] Because the age distribution of these populations differed, as did case ascertainment methods and diagnostic criteria among these studies, it is possible that this fourfold difference reflects these factors, rather than a true difference in disease frequency. Estimated incidence is more similar in studies including only those with Parkinson's disease. Some examples of recent studies using similar diagnostic criteria, adjusted for age to the 1990 U.S. census to allow direct comparison, are shown in Table 5.1.

PREVALENCE

Prevalence measures the total number of individuals in a population who have disease at a specific point in time. Even greater differences are observed in estimates of crude prevalence. When Zhang and Roman reviewed studies published through 1991,[10] they found that crude prevalence ranged from 10/100,000 in Igbo-Ora, Nigeria, to 405/100,000 in Uruguay, Montevideo. While this 40-fold difference across populations is reduced by age adjustment, the range of estimated PD prevalence remained broad. Some of these differences likely reflect variations in ascertainment and diagnostic criteria. However, real differences in prevalence may be due to shortened survival in some populations. Some examples of more recent studies of PD prevalence are shown in Table 5.2.

MORTALITY

Parkinson's disease is not a direct cause of death *per se*, although death may occur as a secondary result of severe

TABLE 5.1
Age-Adjusted Total[a] and Age-Specific Incidence of Parkinson's Disease from Selected Studies

Reference	Population Studied	Total Age-Adjusted[a] Incidence/100,000 Person-Years	Age Strata										
			<40	40-44	45-49	50-54	55-59	60-64	65-69	70-74	75-79	80-84	>85
Bower 1999[18]	Olmsted County, MN, USA	13.8		0.44		17.4		52.5		93.1		79.1	
Mayeux 1995[15]	New York, NY, USA	13.5		0			10.7		54.2		136.6		180.9
Van Den Eden 2003[22]	Northern, CA, USA	13.9	0.15		2.5	9.8		38.8		107.2		119.0	
Fall 1996[17]	SE Sweden	9.7	1.6		3.3	9.0		22.4		59.4		79.5	
Chen 2001[12]	Ilan County, Taiwan	11.3	NA		0		18.5	47.4		100.2		0	
Kusumi 1996[26]	Yanago City, Japan	11.7	0	0	4.2	2.3	16.7	27.2	51.5	81.1	76.8	113.7	26.0

a. Age-adjusted to the 1990 U.S. Census. NA = not available.

TABLE 5.2
Age-Adjusted Total[a]

Reference	Population Studied	Total Age-Adjusted* Prevalence/100,000	Age Strata										
			<40	40-44	45-49	50-54	55-59	60-64	65-69	70-74	75-79	80-84	>85
Mayeux 1995[15]	New York, NY, USA	114.6	1.3			99.3		254.0	509.5	839.6	1192.9		823.8
Svenson 1991[123]	Alberta, Canada	NA	NA	46.6		77.9						1925	
Morgante 1992[b]	Sicily, Italy	258.8		0		115.6		621.4		1978.3		3055	
De Rijk 1995[6,b]	Netherlands	NA	NA		NA		300		1000		3100		4300
Chen 2001[12,b]	Ilan County, Taiwan	168.8[a]	NA		37.8		122.5		546.7		819.7		2197.8
Wang 1996[34,b]	Kinmen, Taiwan	NA	NA			273		535		565		1839	
Kusumi 1996[26]	Yanago City, Japan	104.7		0	8.4	41.8	23.3	71.5	210.0	457.9	669.1	850.5	750.0
Zhang 2003[35,b]	Beijing, China	NA	NA					289.7		1157.2		3534.0	3472.2

a. Age-adjusted to the 1990 U.S. Census.

b. Assumes no cases <40.

c. Door-to-door survey.

motor dysfunction, causing aspiration or falls in advanced Parkinson's disease. Parkinson's disease patients have an increased risk of death as compared to other persons of the same age and gender.[11] For example, Chen found a threefold risk of death for Parkinson's disease subjects in Taiwan.[12] Similarly, investigators in Hawaii found a two- to threefold increased risk of death for persons with Parkinson's disease between the ages of 70 and 89 that corresponded to an 8-year decrease in life expectancy.[13] Mortality from Parkinson's disease has been suggested to be reduced in more recent years, following the use of more effective therapies, such as levodopa, but little information is available from the prelevodopa era, making this assertion controversial.[14]

AGE

Age is the only unequivocal risk factor for Parkinson's disease. Overall risk of Parkinson's disease increases with advancing age.[6,12,13,15] A few studies have reported an apparent decrease in the incidence of Parkinson's disease in the oldest age groups: over 80[16,17] and men over 80.[13,18] However, rather than representing a true drop in incidence, this reported decrease likely reflects incomplete ascertainment of Parkinson's disease in the very old. Especially in those over 80, comorbid conditions may complicate diagnosis or lower participation rates, causing under ascertainment. Furthermore, the small number of living persons in the oldest age groups could lead to unstable estimates of Parkinson's disease incidence and prevalence within this group. Other studies, such as the Northern Manhattan Study, the Olmsted County Study (among women only), the Italian Longitudinal Study on Aging, and the Netherlands Study, have reported continued increases in incidence rates with advancing age.[15,18–20] The association between increasing age and Parkinson's disease occurrence may reflect an age-related neuronal vulnerability or a causal mechanism dependent on the passage of time.[21] If this theory is correct, Parkinson's disease risk would be expected to continuously increase with increasing age even among the oldest of the old.

GENDER

Gender-specific differences in Parkinson's disease have been observed in incidence and prevalence studies. Because the distribution of men and women will differ among populations, crude estimates cannot be compared directly. Instead, rates should be adjusted to a standard reference for comparison. A preponderance of Parkinson's disease among males has been reported in many prevalence[8,15–17] and incidence[15,16,18,19,22] surveys. This finding is less robust than the association with increasing age, and there is variability across studies. A few studies

have reported similar incidence or prevalence for males and females.[4,6,9,23,24,25] Two studies conducted in Japan found higher prevalence of Parkinson's disease in women.[26,27] The higher prevalence of Parkinson's disease among females in Japan could be an artifact and may reflect longer survival among females in Japan, differential access to medical services, or an ascertainment bias that leads to an under ascertainment of male cases. In sharp contrast, the incidence of Parkinson's disease among men was approximately twofold higher than the incidence among women in the Northern California population and in the Italian Longitudinal Study on Aging population.[19,22] If there is a true preponderance of Parkinson's disease among males in some populations, this could be due to differential exposure to risk factors (such as occupational exposure to toxicants) in men and women. Alternatively, a male preponderance could also suggest an X-linked genetic predisposition. Or sex hormones could confer different risks in men and women—for example, female sex hormones have been proposed to reduce disease risk.[28]

RACE

Parkinson's disease prevalence is reported to vary internationally. This apparent difference may reflect differences in study methods, diagnostic patterns, or survival. Alternatively, these variations may reflect actual differences in Parkinson's disease risk due to exposure to environmental factors, cultural factors, or a genetic predisposition of populations surveyed. Lower Parkinson's disease prevalence among blacks as compared to whites has been suggested, but remains controversial.[15,24] Mayeux and colleagues[15] found a difference in the prevalence and incidence rates for African-Americans living in Manhattan. The prevalence rate for Parkinson's disease was lower for African-Americans as compared with whites and Hispanics, but the incidence rate was highest for African-American men. The investigators speculated that the lower prevalence among African-Americans may be related to diminished survival after diagnosis, because significantly more deaths occurred in the incident cohort among African-Americans than in the other ethnic groups. Interestingly, among African-American men under age 65, the incidence was more than four times that of white men. However, even more remarkable is the fact that no cases of Parkinson's disease were identified among the white men in this age group. Whether these unexpected results represent the actual frequency in other communities cannot be known. However, the absence of cases in white men is in contrast to estimates in many other populations. While intriguing, some of these findings may be due to design features, rather than "true" incidence. In contrast, a second study investigating incidence of Parkinson's disease in a multiethnic population found that incidence rates were highest among Hispanics, followed by non-Hispanic whites, Asians, and blacks.[22] This study identified 588 incident cases of Parkinson's disease in two years, by several-fold the largest number of cases to date on which an estimate of incidence has been based. Nonetheless, precision was still low for estimates by race/ethnicity, so it remains important to further investigate multiethnic populations to determine Parkinson's disease incidence. An alternative approach is to compare populations that are genetically similar but separated geographically. In a study of Japanese and Okinawan men living in Hawaii, the incidence of Parkinson's disease was similar to published rates for Caucasian men in Europe and the U.S., and higher than incidence rates published for Asian men living in Asian nations.[13] Such an observation may be due to environmental differences causing an increased frequency of Parkinson's disease in Hawaii, although differences in study methods are always a possible contributing factor.

Others have investigated the prevalence of Parkinson's disease among different racial groups. A prevalence estimate derived using a door-to-door survey in Copiah County, Mississippi, found similar age-adjusted prevalence ratios of Parkinson's disease for African-Americans and whites when the least stringent diagnostic criterion was used to define Parkinson's disease. However, when the analysis was limited to only definite cases of Parkinson's disease, a definition more likely to include only those with Parkinson's disease, investigators found age-adjusted prevalence to be lower for African-Americans than for whites.[4] In another study, Schoenberg et al.[29] utilized the same door-to-door methods to compare the prevalence of Parkinson's disease among African-Americans living in rural Copiah County, Mississippi, and Africans living in rural Igbo-Ora, Nigeria. The age-adjusted prevalence ratio was 67/100,000 for Africans living in Nigeria and 341/100,000 for African-Americans living in Mississippi. The findings were consistent with the hypothesis that an environmental agent (or agents) may be responsible for the observed differences. In these prevalence estimates, as for the incidence estimates, precision is low, because the nonwhite populations observed were small.

TIME

If the frequency of Parkinson's disease changes over time, this would suggest a change in risk factors for the disease, particularly if the change is in incidence. If changes are in prevalence, changes in survival in those with the disease may also contribute to changes in frequency estimates. Investigations of temporal changes and parallel changes in the source populations can provide useful clues to the cause of Parkinson's disease. However, such investigation can be explored in only a few locations, as accurate inci-

dence and prevalence estimates are not available for most populations, and typically those available are for a relatively recent time period. The population of Olmsted County, Minnesota, has been studied for the longest period of time, beginning in the middle of the last century.[9] To minimize differences in diagnostic criteria over time, an analysis investigating rates of change in Parkinson's disease incidence over a 15-year period was conducted in the Olmsted County population.[30] For the 15-year period of 1976 to 1990, the investigators found no important trends, suggesting that major environmental risk factors for Parkinson's disease had neither been introduced nor removed from the population. When Parkinson's disease incidence was compared for the same population from 1935 to 1988, incidence gradually increased from 9.2/100,000 for the interval 1935 to 1944 to 16.3/100,000 for the interval 1975 to 1984.[31] In all reports, the actual number of incident cases was few, and precision was poor. Whether these patterns represent differences in diagnosis or access to care over time, or actual differences in Parkinson's disease frequency remains uncertain.

A survey of Parkinson's disease in Finland found that prevalence and incidence increased among men between 1971 and 1992. Among women, the prevalence remained stable and the incidence decreased during the same time period.[16] The increased prevalence among men over time could reflect decreased mortality due to levodopa therapy. The increased incidence in men may be explained by the presence of an environmental factor not present in 1971 to which men are more exposed than are women. It could also reflect a greater genetic susceptibility among men to the environmental factor relative to women.

Two estimates of Parkinson's disease prevalence in mainland China based on door-to-door ascertainment methods found rates much lower than those reported in western countries.[32,33] More recently, two door-to-door studies in Taiwan found Parkinson's disease prevalence to be much closer to that of Western countries (Table 5.2).[12,34] Similarly, a recent survey in Beijing reported age-specific prevalence ratios that were similar to those found in Western countries[35] (Table 5.2). This change in Parkinson's disease frequency over time in the racially homogeneous Chinese population could reflect differences in environmental factors. Moreover, study methods were similar, arguing against methodological explanations for the differences in estimates. Because many environmental factors have been changing rapidly in China over the last several decades, a similar change in Parkinson's disease frequency might be predicted if environmental causes are important. Further studies of these populations could provide important insights into the causes of Parkinson's disease. As with all such estimates, however, it is still possible that undetectable differences in study conduct could instead have resulted in this apparent difference in prevalence over time.

GEOGRAPHY

Differences in the geographic distribution of Parkinson's disease have been observed within as well as among countries. Such differences, if accurate, may reflect geographic differences in environmental or genetic risk factors. Alternatively, they may be due to differences in study methods. Several groups have addressed the pattern of difference internationally by adjusting reported incidence or prevalence from many countries to a standard population. For recent incidence studies, Twelves et al. found adjusted rates in some countries to be more than twice those reported for others (from 8/100,000 in Italy to 18/100,000 in the U.K.).[7] Zhang and Roman found even greater differences in adjusted prevalence rates, reporting more than tenfold differences internationally (ranging from 18/100,000 in China to 234/100,000 in Montevideo, Uruguay).[10] While studies are too few to provide definitive patterns, some have been suggested. These include the suggestion that Mediterranean populations are at lower risk for Parkinson's disease, given the low incidence or prevalence found in Sardinia, Italy, and Benghazi, Libya.[8,24] Similarly, in China and Nigeria, Parkinson's disease prevalence determined by door-to-door surveys conducted in the 1980s was much lower than rates obtained contemporaneously using similar methods in western countries.[32,33]

ANALYTIC EPIDEMIOLOGY

RISK FACTORS

Many factors have been associated with an increased risk for Parkinson's disease (Table 5.3). The causes of Parkinson's disease are likely to be multifactorial, a combination of age related, genetic, and environmental risk factors. As discussed earlier, demographic factors that may influence the risk of developing Parkinson's disease include age, gender, and race.

GENETIC AND FAMILIAL FACTORS

Parkinsonism due to purely genetic causes remains unusual. Most genetic parkinsonism is found in those with younger age at disease onset, typically before age 50. To date, parkinsonism-causing mutations in three genes have been identified: α-synuclein,[36–38] parkin,[39,40] and DJ-1.[41] An additional seven autosomal loci[42–48] and a family with a pattern suggesting a defect in the mitochondrial genome have been observed.[49,50] Taken together, the known mutations account for a small number of persons with parkinsonism.[51–54] Of these, the most common is the parkin mutation. These genetic forms of parkinsonism are not addressed here, as they will be discussed in detail elsewhere in this volume.

TABLE 5.3
Factors Associated with Increased Risk for Parkinson's Disease

Demographic Factors
  Increasing age
  Male gender
  White race
  Family history of Parkinson's disease
  Lifestyle
  Head trauma
  Emotional stress
  Personality traits (shyness, risk averse)
Environmental Exposures
  MPTP and MPTP like compounds
  Pesticides
  Industrial agents
  Carbon monoxide
  Metals (manganese, mercury, iron)
  Drinking well water
  Pulp mills
  Farming
  Rural residence
  Occupation (health care, teaching, construction work)
Infections
  Encephalitis
  Nocardia asteroides

In case-control studies, persons with Parkinson's disease consistently report more affected family members than do controls.[55,56] However, biased recall has been found to contribute to this in at least one study in which reported history was verified by an examination.[57] Twin studies may also provide clues to the relative contribution of genetic factors to the cause of disease. If the cause of a disease were primarily genetic, then the rate of disease in monozygotic (MZ) twins would be greater than that in dizygotic (DZ) twins. Twin studies of Parkinson's disease have failed to support a major genetic effect. A recent large, population based twin study of 163 pairs showed similar rates of concordance in MZ and DZ twin pairs.[58] Two small follow-up studies had contradictory results, one finding increased MZ concordance and the other failing to find a difference in MZ and DZ concordance.[59,60] A prospective follow-up of the population-based cohort is now under way.

While genetic causes do not appear to be primary in typical, sporadic Parkinson's disease, investigating the mechanisms underlying the genetic causes of parkinsonism can provide important clues to the common pathogenesis of all parkinsonism. The relative paucity of evidence supporting a genetic cause for typical Parkinson's disease has sparked interest in investigating environmental risk factors for Parkinson's disease.

ENVIRONMENTAL FACTORS

In the early 1980s, interest in environmental causes of Parkinson's disease was ignited by the description of a cluster of parkinsonism produced by the neurotoxic pyridine,1-methyl-4-phenyl-1,2,3,6-tetrahydropyridine (MPTP), a relatively simple pyridine compound that induces most if not all of the features of Parkinson's disease in humans[61] and experimental animals.[62,63] While MPTP is rare outside of the laboratory, related chemicals more commonly present in the environment were proposed as possible causes of Parkinson's disease. Subsequently, case-control studies have investigated and suggested numerous associations between environmental exposures and Parkinson's disease (Table 5.3). These specific environmental toxins and associations are mentioned here briefly and will be discussed in detail in later chapters.

Because MPTP resembles the herbicide paraquat, and several ecologic studies suggested a rural preponderance of Parkinson's disease, factors associated with the rural environment have been studied. Multiple case-control studies have detected positive associations between Parkinson's disease and exposure to pesticides,[64–66] well water,[67–69] and rural living.[67,69] Information about exposures to specific agents is limited but suggests paraquat,[70] dieldrin,[71] organochlorines,[64] alkylated phosphates,[64] and carbamate derivatives[65] may have a causal role in Parkinson's disease. In China, several decades ago, a case-control study found that exposure to industrial chemicals, printing plants, or quarries was associated with an increased risk of developing Parkinson's disease but found no relationship with agricultural work and Parkinson's disease.[72] In contrast, an investigation in Hong Kong during that time period did find such an association,[73] perhaps due to differences in farming practices or other environmental factors between the less developed mainland and the more developed island of Hong Kong at that time. Individual studies have had conflicting results, possibly due to methodological differences, small samples, and regional differences in farming practices or differences in population characteristics.

Occupational exposures and risk of Parkinson's disease has been investigated in a small number of studies. Several of these studies have shown associations of Parkinson's disease with industries using metals,[74] although results of studies of specific metals have been inconsistent.[75] Occupational exposures to metals suggested a positive association between exposure to manganese and mercury.[75] In addition to occupational exposures to pesticides and metals, a higher frequency of Parkinson's disease has been reported among teachers,[76] health workers,[76] construction workers,[77] carpenters, and cleaners.[78] The clues to the etiology of Parkinson's disease gathered by these associations will be explored in detail in later chapters.

OTHER ASSOCIATIONS

The association between head trauma and Parkinson's disease has been investigated in many case-control studies. The overall epidemiological evidence in favor of or against the role of head trauma in Parkinson's disease remains controversial. Several studies reported a positive association between head trauma and Parkinson's disease risk,[64,79–81] while others have found no relation between head trauma and Parkinson's disease risk.[68,82–84] Because most studies have relied on the report of the person with Parkinson's disease or a control, biased recall has been a concern. In a recent population-based study in Olmstead County, Minnesota, head trauma documented in the medical record was associated with an increased risk of Parkinson's disease, although this association was restricted to more severe head trauma. Head trauma could be a direct causal factor triggering or predisposing to factors causing loss of nigral neurons. Alternatively, head trauma could be a reflection of early motor problems in preclinical Parkinson's disease,[85] although in some studies the trauma preceded the onset of Parkinson's disease by many decades, making this less plausible.

Parkinsonism was a sequela in the survivors of the 1917 to 1935 epidemic encephalitis, resulting in the proposal that all Parkinson's disease was the result of this infection. This belief persisted, despite the fact that postencephalitic disease and Parkinson's disease have clear differences both clinically and pathologically,[86] until it was laid to rest when disease rates did not decrease despite the lack of exposed persons in the population. To date, an infectious agent has never been shown to cause typical Parkinson's disease. The soil pathogen *Nocardia asteroides* causes a levodopa responsive movement disorder and nigral degeneration in mice.[87] However, a serologic case-control study in humans did not support a role for *Nocardia asteroides* in Parkinson's disease.[88] Several case-control studies found an association between occupations thought to be associated with an increased risk of infection, such as teaching and health care.[89,90] Because inflammatory processes appear to contribute to nerve cell death in Parkinson's disease, investigating the role of infectious agents remains interesting, but challenging.

PROTECTIVE FACTORS

Factors proposed to protect against the development of Parkinson's disease include cigarette smoking, coffee consumption, and the use of nonsteroidal anti-inflammatory drugs (NSAIDs). Of these, the most compelling is the inverse association of cigarette smoking with Parkinson's disease risk seen in both case-control and prospective studies,[64,70,83,91–98] with multiple studies confirming an inverse dose-response pattern with regard to cumulative lifetime cigarette smoking.[96,99,100] Although not all studies confirmed the inverse association,[101,102] and a clearly defined biological basis for this finding has yet to be defined. One hypothesis is that smoking protects against the development of Parkinson's disease because of its effect on the enzyme monoamine oxidase (MAO) B. Cigarette smoke reduces MAO B activity in the animal and human brain.[103,104] MAO B activates the neurotoxin MPTP, and a MAO B inhibitor, cigarette smoke, may offer neuroprotection in Parkinson's disease.[61,105] Another hypothesis is that nicotine itself is neuroprotective,[92] given that it has antioxidant properties.[106] The nicotine in cigarette smoke may inhibit free radical formation and offer associated neuronal protection. An alternative hypothesis is that some inherent, perhaps life-long characteristic of those destined to develop Parkinson's disease also determines a constitutional lack of interest in smoking cigarettes. The relationship between dopaminergic systems and addiction lends some credence to this proposal. Evidence against an inborn "low dopamine" state is provided by the observation that an inverse effect of smoking and Parkinson's disease risk is seen in monozygotic twin pairs, one of whom has Parkinson's disease.[92] There are overwhelming health risks associated with cigarette smoking, and this behavior should be avoided. Nonetheless, the evidence that smoking is somehow protective is intriguing, and delineation of the underlying biochemical mechanism could lead not only to insights into the cause of the disease but perhaps also to useful and safe preventive approaches.

Coffee drinking or caffeine intake is a second behavior inversely associated with the risk of Parkinson's disease.[93,97,107–109] In the most methodologically compelling study, a dose-dependent reduction in Parkinson's disease risk was observed in a prospective cohort of men.[108] But further investigation of this effect will be needed, as caffeine may not affect all persons similarly. For example, Ascherio et al. reported an interesting difference among women—that caffeine reduces the risk of Parkinson's disease among women who do not use postmenopausal hormones, but it increases risk among hormone users.[110] As a result, there was not a clear overall relationship between caffeine or coffee intake and Parkinson's disease in this cohort. The latter findings suggest a potential interaction between caffeine consumption and estrogen exposure in mediating Parkinson's disease risk. An investigation of the actions of caffeine in the brain provides some plausibility for the possibility that caffeine my alter disease risk. The biologic effects of caffeine are mediated in part through its antagonist action on the adenosine A2 receptor, which in turn modulates dopaminergic neurotransmission[111,112] and protects against striatal dopamine loss caused by MPTP in laboratory studies.[113–115]

Glial cell-mediated inflammation may contribute to the nigralstriatal degeneration found in Parkinson's disease. Studies of Alzheimer's have shown that the regular use of NSAIDs may reduce the risk of Alzheimer's in humans.[116,117] Since Alzheimer's and Parkinson's disease share common pathogenic mechanisms of neuronal cell death and degeneration, investigations have looked at the use of NSAIDs and the risk of Parkinson's disease. The regular use of nonaspirin NSAIDs was associated with a 45% lower risk of Parkinson's disease in a prospective study of men and women suggesting neuroprotective effects of NSAIDs.[118] This was also the second prospective study to find that aspirin use was associated with a lower incidence of Parkinson's disease.[118,119] Although current evidence is compelling, it will be important for future studies to determine the specific compounds and mechanisms that may mediate the protective effects of smoking, coffee drinking, and NSAIDs use.

Alcohol has been found by some to be inversely associated with Parkinson's disease in a few prospective studies,[99,109,120] although the associations were of borderline significance. There is a great deal of variability across studies and the inverse association between alcohol and Parkinson's disease is weak in studies to date. Other protective factors associated with Parkinson's disease have been identified in single studies but not replicated to date. These include early childhood measles infection[121] and early or mid-life exercise.[122]

CONCLUSION

Parkinson's disease is a complex disorder that likely has several etiologies. Epidemiological studies can contribute a great deal to our understanding of the genetic, molecular, and environmental factors involved in the disease process. Investigations need to continue to document the frequency and distribution of Parkinson's disease to better understand the etiologic factors. Ideally, future studies will use similar methods of case ascertainment and case definitions to enable investigators to better compare results across studies. The collaboration of clinicians, basic scientists, epidemiologists, toxicologists, industrial hygienists, and statisticians is essential to determine which associations identified in epidemiological studies are biologically plausible and determine the relevance in human populations. As our understanding of Parkinson's disease increases, it will be important to identify populations at risk as preventative or protective treatment strategies become available.

REFERENCES

1. Mutch, W. J. et al., Parkinson's disease in a Scottish city, *Br. Med. J. (Clin. Res. Ed.)*, 292, pp. 534–536, 1986.
2. Marttila, R. J. and U. K., Rinne, Epidemiology of Parkinson's disease in Finland, *Acta Neurol. Scand.*, 53, pp. 81–102, 1976.

3. Anderson, D. W. et al., Case ascertainment uncertainties in prevalence surveys of Parkinson's disease, *Mov. Disord.*, 13, pp. 626–32, 1998.

4. Schoenberg, B. S., D.W. Anderson, and A. F. Haerer, Prevalence of Parkinson's disease in the biracial population of Copiah County, Mississippi, *Neurology*, 35, pp. 841–845, 1985.

5. Morgante, L. et al., Prevalence of Parkinson's disease and other types of parkinsonism: a door-to-door survey in three Sicilian municipalities, The Sicilian Neuro-Epidemiologic Study (SNES) Group, *Neurology*, 42, pp. 1901–1907, 1992.

6. de Rijk, M. C. et al., Prevalence of Parkinson's disease in the elderly: the Rotterdam Study, *Neurology*, 45, pp. 2143–2146, 1995.

7. Twelves, D., K. S. Perkins, and C. Counsell, Systematic review of incidence studies of Parkinson's disease, *Mov. Disord.*, 18, pp. 19–31, 2003.

8. Rosati, G. et al., The risk of Parkinson disease in Mediterranean people, *Neurology*, 30, pp. 250–255, 1980.

9. Rajput, A. H. et al., Epidemiology of parkinsonism: incidence, classification, and mortality, *Ann. Neurol.*, 16, pp. 278–282, 1984.

10. Zhang, Z. X. and G. C. Roman, Worldwide occurrence of Parkinson's disease: an updated review, *Neuroepidemiology*, 12, pp. 195–208, 1993.

11. Uitti, R. J. et al., Levodopa therapy and survival in idiopathic Parkinson's disease: Olmsted County project, *Neurology*, 43, pp. 1918–1926, 1993.

12. Chen, R. C. et al., Prevalence, incidence, and mortality of PD: a door-to-door survey in Ilan county, Taiwan, *Neurology*, 57, pp. 1679–86, 2001.

13. Morens, D. M. et al., Epidemiologic observations on Parkinson's disease: incidence and mortality in a prospective study of middle-aged men, *Neurology*, 46, pp. 1044–1050, 1996.

14. Poewe, W., The Sydney multicentre study of Parkinson's disease, *J. Neurol. Neurosurg. Psychiatry*, 67, pp. 280–281, 1999.

15. Mayeux, R. et al., The frequency of idiopathic Parkinson's disease by age, ethnic group, and sex in northern Manhattan, 1988–1993, *Am. J. Epidemiol.*, 142, pp. 820–827, 1995.

16. Kuopio, A. M. et al., Changing epidemiology of Parkinson's disease in southwestern Finland, *Neurology*, 52, pp. 302–308, 1999.

17. Fall, P. A. et al., Age-standardized incidence and prevalence of Parkinson's disease in a Swedish community, *J. Clin. Epidemiol.*, 49, pp. 637–641, 1996.

18. Bower, J. H. et al., Incidence and distribution of parkinsonism in Olmsted County, Minnesota, 1976–1990, *Neurology*, 52, pp. 1214–1220, 1999.

19. Baldereschi, M. et al., Parkinson's disease and parkinsonism in a longitudinal study: two-fold higher incidence in men. ILSA Working Group, Italian Longitudinal Study on Aging, *Neurology*, 55, pp. 1358–1363, 2000.

20. Leentjens, A. F. et al., The incidence of Parkinson's disease in the Netherlands: results from a longitudinal general practice-based registration, *Neuroepidemiology*, 22, pp. 311–312, 2003.

21. Tanner, C. M. and J. W. Langston, Do environmental toxins cause Parkinson's disease? A critical review, *Neurology*, 40, pp. suppl. 17-30; discussion 30-1, 1990.

22. Van Den Eeden, S. K., et al., Incidence of Parkinson's disease: variation by age, gender, and race/ethnicity, *Am. J. Epidemiol.*, 157, pp. 1015–1022, 2003.

23. Taba, P. and T. Asser, Incidence of Parkinson's disease in estonia, *Neuroepidemiology*, 22, pp. 41–45, 2003.

24. Ashok, P. P. et al., Epidemiology of Parkinson's disease in Benghazi, North-East Libya, *Clin. Neurol. Neurosurg.*, 88, pp. 109–113, 1986.

25. Taba, P. and T. Asser, Prevalence of Parkinson's disease in Estonia, *Acta Neurol. Scand.*, 106, pp. 276–281, 2002.

26. Kusumi, M. et al., Epidemiology of Parkinson's disease in Yonago City, Japan: comparison with a study carried out 12 years ago, *Neuroepidemiology*, 15, pp. 201–207, 1996.

27. Kimura, H. et al., Female preponderance of Parkinson's disease in Japan, *Neuroepidemiology*, 21, pp. 292, 2002.

28. Tanner, C M. and D. A. Aston, Epidemiology of Parkinson's disease and akinetic syndromes, *Curr. Opin. Neurol.*, 13, pp. 427–430, 2000.

29. Schoenberg, B. S. et al., Comparison of the prevalence of Parkinson's disease in black populations in the rural United States and in rural Nigeria: door-to-door community studies, *Neurology*, 38, pp. 645–646, 1988.

30. Rocca, W. A. et al., Time trends in the incidence of parkinsonism in Olmsted County, Minnesota, *Neurology*, 57, pp. 462–467, 2001.

31. Tanner, C. M. et al., Parkinson's disease incidence in Olmsted County, MN: 1935–1988, *Neurology*, 42, pp. 194–194, 1992.

32. Li, S. C. et al., A prevalence survey of Parkinson's disease and other movement disorders in the People's Republic of China, *Arch. Neurol.*, 42, pp. 655–657, 1985.

33. Wang, Y. S. et al., Parkinson's disease in China. Coordinational Group of Neuroepidemiology, PLA, *Chin. Med. J. (Engl.)*, 104, pp. 960–964, 1991.

34. Wang, S. J. et al., A door-to-door survey of Parkinson's disease in a Chinese population in Kinmen, *Arch. Neurol.*, 53, pp. 66–71, 1996.

35. Zhang, Z. X. et al., Prevalence of Parkinson's disease and related disorders in the elderly population of greater Beijing, China, *Mov. Disord.*, 18, pp. 764–772, 2003.

36. Polymeropoulos, M. et al., Mutation in the α-synuclein gene identified in families with Parkinson's disease, *Science*, pp. 2045–2047, 1997.

37. Golbe, L. I. et al., Clinical genetic analysis of Parkinson's disease in the Contursi kindred, *Ann. Neurol.*, 40, pp. 767–775, 1996.

38. Kruger, R., et al., Ala30Pro mutation in the gene encoding alpha-synuclein in Parkinson's disease, *Nat. Genet.*, 18, pp. 106–108, 1998.

39. Hattori, N. et al., Molecular genetic analysis of a novel Parkin gene in Japanese families with autosomal recessive juvenile parkinsonism: evidence for variable

homozygous deletions in the Parkin gene in affected individuals, *Ann. Neurol.*, 44, pp. 935–941, 1998.

40. Kitada, T. et al., Mutations in the parkin gene cause autosomal recessive juvenile parkinsonism, *Nature*, 392, pp. 605–608, 1998.

41. Bonifati, V. et al., Mutations in the DJ-1 gene associated with autosomal recessive early-onset parkinsonism, *Science*, 299, pp. 256–259, 2003.

42. Farrer, M., et al. A chromosome 4p haplotype segregating with Parkinson's disease and postural tremor, *Hum. Mol. Genet.*, 8, pp. 81–85, 1999.

43. Funayama, M. et al., A new locus for Parkinson's disease (PARK8) maps to chromosome 12p11.2-q13.1, *Ann. Neurol.*, 51, pp. 296–301, 2002.

44. Gasser, T. et al., A susceptibility locus for Parkinson's disease maps to chromosome 2p13, *Nat. Genet.*, 18, pp. 262–265, 1998.

45. Hicks, A. A. et al., A susceptibility gene for late-onset idiopathic Parkinson's disease, *Ann. Neurol.*, 52, pp. 549–555, 2002.

46. Leroy, E. et al., The ubiquitin pathway in Parkinson's disease, *Nature*, 395, pp. 451–452, 1998.

47. Valente, E. M. et al., Localization of a novel locus for autosomal recessive early-onset parkinsonism, PARK6, on human chromosome 1p35-p36, *Am. J. Hum. Genet.*, 68, pp. 895–900, 2001.

48. van Duijn, C. M. et al., Park7, a novel locus for autosomal recessive early-onset parkinsonism, on chromosome 1p36, *Am. J. Hum. Genet.*, 69, pp. 629–634, 2001.

49. Wooten, G. F. et al., Maternal inheritance in Parkinson's disease, *Ann. Neurol.*, 41, pp. 265–268, 1997.

50. Swerdlow, R. H. et al., Matrilineal inheritance of complex I dysfunction in a multigenerational Parkinson's disease family, *Ann. Neurol.*, 44, pp. 873–881, 1998.

51. Farrer, M. et al., Low frequency of alpha-synuclein mutations in familial Parkinson's disease, *Ann. Neurol.*, 43, pp. 394–397, 1998.

52. Papadimitriou, A. et al., Mutated alpha-synuclein gene in two Greek kindreds with familial PD: incomplete penetrance?, *Neurology*, 52, pp. 651–654, 1999.

53. Abbas, N. et al., A wide variety of mutations in the parkin gene are responsible for autosomal recessive parkinsonism in Europe, *Hum. Mol. Genet.*, 8, pp. 567–574, 1999.

54. Wang, M. et al., Polymorphism in the parkin gene in sporadic Parkinson's disease, *Ann. Neurol.*, 45, pp. 655–658, 1999.

55. Autere, J. M. et al., Familial aggregation of Parkinson's disease in a Finnish population, *J. Neurol. Neurosurg. Psychiatry*, 69, pp. 107–109, 2000.

56. Payami, H. et al., Increased risk of Parkinson's disease in parents and siblings of patients, *Annals of Neurology*, 36, pp. 659–661, 1994.

57. Elbaz, A. et al., Validity of family history data on PD: Evidence for a family information bias, *Neurology*, 61, pp. 11–17, 2003.

58. Tanner, C. M. et al., Parkinson disease in twins: an etiologic study, *Jama*, 281, pp. 341–346, 1999.

59. Piccini, P. et al., The role of inheritance in sporadic Parkinson's disease: evidence from a longitudinal study of dopaminergic function in twins, *Ann. Neurol.*, 45, pp. 577–582, 1999.

60. Vieregge, P. et al., Parkinson's disease in twins: a follow-up study, *Neurology*, 53, pp. 566–572, 1999.

61. Langston, J. W. et al., Chronic Parkinsonism in humans due to a product of meperidine-analog synthesis, *Science*, 219, pp. 979–980, 1983.

62. Burns, R. S. et al., A primate model of parkinsonism: selective destruction of dopaminergic neurons in the pars compacta of the substantia nigra by N-methyl-4-phenyl-1,2,3,6-tetrahydropyridine, *Proceedings of the National Academy of Sciences*, 80, pp. 4546–4550, 1983.

63. Langston, J. W. et al., Selective nigral toxicity after systemic administration of 1-methyl-4-phenyl-1,2,3,6-tetrahydropyridine (MPTP) in the squirrel monkey, *Brain Research*, 292, pp. 390–394, 1984.

64. Seidler, A. et al., Possible environmental, occupational, and other etiologic factors for Parkinson's disease: a case-control study in Germany, *Neurology*, 46, pp. 1275–1284, 1996.

65. Semchuk, K. M., Love, E. J. and Lee, R. G., Parkinson's disease and exposure to agricultural work and pesticide chemicals, *Neurology*, 42, pp. 1328–1335, 1992.

66. Gorell, J. M. et al., The risk of Parkinson's disease with exposure to pesticides, farming, well water, and rural living, *Neurology*, 50, pp. 1346–1350, 1998.

67. Koller, W. et al., Environmental risk factors in Parkinson's disease, *Neurology*, 40, pp. 1218–1221, 1990.

68. De Michele, G. et al., Environmental and genetic risk factors in Parkinson's disease: a case-control study in southern Italy, *Mov. Disord.*, 11, pp. 17–23, 1996.

69. Wong, G. F. et al., Environmental risk factors in siblings with Parkinson's disease, *Arch. Neurol.*, 48, pp. 287–289, 1991.

70. Liou, H. H. et al., Environmental risk factors and Parkinson's disease: a case-control study in Taiwan, *Neurology*, 48, pp. 1583–1588, 1997.

71. Fleming, L. et al., Parkinson's disease and brain levels of organochlorine pesticides, *Ann. Neurol.*, 36, pp. 100–103, 1994.

72. Tanner, C. M. et al., Environmental factors and Parkinson's disease: a case-control study in China, *Neurology*, 39, pp. 660–664, 1989.

73. Ho, S. C., Woo, J., and Lee, C. M., Epidemiologic study of Parkinson's disease in Hong Kong, *Neurology*, 39, pp. 1314–1318, 1989.

74. Tanner, C. M. and Goldman, S. M., Epidemiology of Parkinson's disease, *Neurol. Clin.*, 14, pp. 317–335, 1996.

75. Gorell, J. M. et al., Occupational metal exposures and the risk of Parkinson's disease, *Neuroepidemiology*, 18, pp. 303–308, 1999.

76. Schulte, P. A. et al., Neurodegenerative diseases: occupational occurrence and potential risk factors, 1982 through 1991, *Am. J. Public Health*, 86, pp. 1281–1288, 1996.

77. Herishanu, Y. O. et al., A case-control study of Parkinson's disease in urban population of southern Israel, *Can. J. Neurol. Sci.*, 28, pp. 144–147, 2001.

78. Fall, P. A. et al., Nutritional and occupational factors influencing the risk of Parkinson's disease: a case-control study in southeastern Sweden, *Mov. Disord.*, 14, pp. 28–37, 1999.

79. Semchuk, K M., Love, E. J., and Lee, R. G., Parkinson's disease: a test of the multifactorial etiologic hypothesis, *Neurology*, 43, pp. 1173–1180, 1993.

80. Stern, M. et al., The epidemiology of Parkinson's disease. A case-control study of young-onset and old-onset patients, *Arch. Neurol.*, 48, pp. 903–907, 1991.

81. Tsai, C. H. et al., Environmental risk factors of young onset Parkinson's disease: a case-control study, *Clin. Neurol. Neurosurg*, 104, pp. 328–333, 2002.

82. Tanner, C. M. et al., Environmental factors in the etiology of Parkinson's disease, *Can. J. Neurol. Sci.*, 14, pp. 419–423, 1987.

83. Hofman, A., Collette, H. J., and Bartelds, A. I., Incidence and risk factors of Parkinson's disease in The Netherlands, *Neuroepidemiology*, 8, pp. 296–269, 1989.

84. Morano, A. et al., Risk-factors for Parkinson's disease: case-control study in the province of Caceres, Spain, *Acta Neurol. Scand.*, 89, pp. 164–170, 1994.

85. Bower, J. H. et al., Head trauma preceding PD: A case-control study, *Neurology*, 60, pp. 1610–1615, 2003.

86. Poskanzer, D. C., Schwab, R. S., and Fraser, D. W., *Further observations on the cohort phenomenon in Parkinson's syndrome*, in *Progress in Neurogenetics*, A. Barbeau and J. R. Brunette, Eds., 1969, Excerpta Medica Foundation: Amsterdam, pp. 497–505.

87. Kobbata, S. and B.L. Beaman, L-dopa-responsive movement disorder caused by *Norcardia asteroides* localized in the brains of mice, *Infect Immun*, 59, pp. 181–191, 1991.

88. Hubble, J. P. et al., Nocardia species as an etiologic agent in Parkinson's disease: serological testing in a case-control study, *J. Clin. Microbiol.*, 33, pp. 2768–2769, 1995.

89. Tanner, C. et al., Occupation and risk of Parkinson's disease (PD): A preliminary investigation of Standard Occupational Codes (SOC) in twins discordant for disease, *Neurology*, 60, p. A415, 2003.

90. Tsui, J. K. et al., Occupational risk factors in Parkinson's disease, *Can J Public Health*, 90, pp. 334–337, 1999.

91. Zayed, J. et al., Environmental factors in the etiology of Parkinson's disease, *Can. J. Neurol. Sci.*, 17, pp. 286–291, 1990.

92. Tanner, C. M. et al., Smoking and Parkinson's disease in twins, *Neurology*, 58, pp. 581–588, 2002.

93. Ross, G. W. and Petrovitch, H., Current evidence for neuroprotective effects of nicotine and caffeine against Parkinson's disease, *Drugs Aging*, 18, pp. 797–806, 2001.

94. Smargiassi, A. et al., A case-control study of occupational and environmental risk factors for Parkinson's disease in the Emilia-Romagna region of Italy, *Neurotoxicology*, 19, pp. 709–712, 1998.

95. Taylor, C. A. et al., Environmental, medical, and family history risk factors for Parkinson's disease: a New England-based case control study, *Am. J. Med. Genet.*, 88, pp. 742–749, 1999.

96. Checkoway, H. et al., Parkinson's disease risks associated with cigarette smoking, alcohol consumption, and caffeine intake, *Am. J. Epidemiol.*, 155, pp. 732–738, 2002.

97. Hernan, M. A. et al., A meta-analysis of coffee drinking, cigarette smoking, and the risk of Parkinson's disease, *Ann. Neurol.*, 52, pp. 276–284, 2002.

98. Tanner, C. et al., Cigarette smoking, alcohol drinking and Parkinson's disease: cross-cultural risk assessment., *Mov. Disord.*, 5, p. 11, 1990.

99. Grandinetti, A. et al., Prospective study of cigarette smoking and the risk of developing idiopathic Parkinson's disease, *Am. J. Epidemiol.*, 139, pp. 1129–1138, 1994.

100. Gorell, J. M. et al., Smoking and Parkinson's disease: a dose-response relationship, *Neurology*, 52, pp. 115–119, 1999.

101. Mayeux, R. et al., Smoking and Parkinson's disease, *Mov Disord*, 9, pp. 207–212, 1994.

102. Rajput, A. H. et al., A case-control study of smoking habits, dementia, and other illnesses in idiopathic Parkinson's disease, *Neurology*, 37, pp. 226–232, 1987.

103. Fowler, J. et al., Inhibition of monoamine oxidase B in the brains of smokers., *Nature*, 379, pp. 733–736, 1996.

104. Mendez-Alvarez, E. et al., Inhibition of brain monoamine oxidase by adducts of 1,2,3,4-tetrahydroisoquinoline with components of cigarette smoke, *Life Sciences*, 60, pp. 1719–1727, 1997.

105. Salach, J. I. et al., Oxidation of the neurotoxic amine 1-methyl-4-phenyl-1,2,3,6-tetrahydropyridine (MPTP) by monoamine oxidases A and B and suicide inactivation of the enzymes by MPTP, *Biochem. Biophys. Res. Commun.*, 125, pp. 831–835, 1984.

106. Ferger, B. et al., Effects of nicotine on hydroxyl free radical formation *in vitro* and on MPTP-induced neurotoxicity *in vivo*, *Naunyn Schmiedebergs Arch. Pharmacol.*, 358, pp. 351–359, 1998.

107. Ascherio, A. et al., Prospective study of caffeine consumption and risk of Parkinson's disease in men and women, *Ann. Neurol.*, 50, pp. 56–63, 2001.

108. Ross, G. W. et al., Association of coffee and caffeine intake with the risk of Parkinson disease, *Jama*, 283, pp. 2674–2679, 2000.

109. Paganini-Hill, A., Risk factors for Parkinson's disease: the leisure world cohort study, *Neuroepidemiology*, 20, pp. 118–124, 2001.

110. Ascherio, A. et al., Caffeine, postmenopausal estrogen, and risk of Parkinson's disease, *Neurology*, 60, pp. 790–795, 2003.

111. Popoli, P., Caporali, M. G., and Scotti de Carolis, A., Akinesia due to catecholamine depletion in mice is prevented by caffeine. Further evidence for an involvement of adenosinergic system in the control of motility, *J. Pharm. Pharmacol.*, 43, pp. 280–281, 1991.

112. Nehlig, A., Daval, J. L. and Debry, G., Caffeine and the central nervous system: mechanisms of action, biochemical, metabolic and psychostimulant effects, *Brain Res. Brain Res. Rev.*, 17, pp. 139–170, 1992.

113. Kanda, T. et al., Adenosine A2A receptors modify motor function in MPTP-treated common marmosets, *Neuroreport*, 9, pp. 2857–2860, 1998.

114. Richardson, P., Kase, H., and Jenner, P., Adenosine A2A receptor antagonists as new agents for the treatment of Parkinson's disease., *Trends Pharmacol. Sci.*, 18, pp. 338–344, 1997.

115. Chen, J. F. et al., Neuroprotection by caffeine and A(2A) adenosine receptor inactivation in a model of Parkinson's disease, *J. Neurosci.*, 21, p. RC143, 2001.

116. in t' Veld, B. A. et al., Nonsteroidal antiinflammatory drugs and the risk of Alzheimer's disease, *N. Engl. J. Med.*, 345, pp. 1515–1521, 2001.

117. McGeer, P. L. and McGeer, E. G., The inflammatory response system of brain: implications for therapy of Alzheimer and other neurodegenerative diseases, *Brain Res. Brain Res. Rev.*, 21, pp. 195–218, 1995.

118. Chen, H. et al., Nonsteroidal anti-inflammatory drugs and the risk of Parkinson disease, *Arch. Neurol.*, 60, pp. 1059–1064, 2003.

119. Ross, G. et al., NSAID use and risk of Parkinson's disease, *Neurology*, 60, p. A416, 2003.

120. Willems-Giesbergen, P. et al., Smoking, Alcohol, and Coffee Consumption and the Risk of PD: Results from the Rotterdam Study, *Neurol.*, 54, p. A347, 2000.

121. Sasco, A. J. and Paffenbarger, Jr., R. S., Measles infection and Parkinson's disease, *Am. J. Epidemiol.*, 122, pp. 1017–1031, 1985.

122. Sasco, A. J. et al., The role of physical exercise in the occurrence of Parkinson's disease, *Arch. Neurol.*, 49, pp. 360–365, 1992.

123. Svenson, L. W., Regional disparities in the annual prevalence rates of Parkinson's disease in Canada, *Neuroepidemiology*, 10, pp. 205–210, 1991.

6 Environmental Toxins and Parkinson's Disease

Marcia H. Ratner
Departments of Neurology and Pharmacology, Boston University School of Medicine

Robert G. Feldman
Departments of Neurology, Pharmacology, and Environmental Health, Boston University Schools of Medicine and Public Health

CONTENTS

INTRODUCTION

Studies suggest that genetic and environmental factors may interact to influence the progression of idiopathic Parkinson's disease (PD). Family history of PD is a risk factor for the disease. Genetic factors have been associated with early and late onset forms of PD. Exposure to certain chemicals has been associated with parkinsonism, but no consistent association has been made between exposure to any particular chemical and the prevalence or incidence of idiopathic PD. In this chapter, we review the research to date on the role of the environment in the normal aging of the nigrostrial pathway, parkinsonism, and idiopathic Parkinson's disease.

NORMAL AGING AND THE NIGROSTRIATAL PATHWAY AND PD

A review of the literature pertaining to the normal aging of the nigrostriatal pathway suggests that, if we all live long enough, we may all develop symptoms of parkinsonism, although we probably won't develop PD. Studies indicate that striatal dopamine concentrations decrease markedly after age 60 years. Nigrostriatal neuronal loss occurs at a rate of approximately 1.4% per decade from ages 15 to 65, but the rate is accelerated to about 10% per decade after age 65 years. These findings indicate that normal aging is associated with an age-related increase in nigrostriatal degeneration.[1]

Magnetic resonance imaging (MRI) studies of normal subjects reveals a negative correlation of age with estimated midbrain volume, anteroposterior diameter through the substantia nigra, and interpeduncular distance. The linear measurements for the right and left side were found to be almost identical demonstrating symmetry of normal age-related changes between the right and left side of the brain, a laterality finding that is in contrast to idiopathic PD.[2]

Normal aging has also been associated with an age-related increase in tissue damage due to free radicals.

The role of free radicals in apoptosis, and the aging process has received considerable attention.[3] Factors that decrease free radical damage, including limiting caloric intake and the use of antioxidants, have been associated with slowing of the normal aging process and with increasing longevity.[3,4,5–8]

Iron has been implicated in the formation of free radicals via the Fenton reaction and therefore may have a role in the normal age related effects of free radicals as well as in neurodegenerative disease. MRI studies were used to determine the relationship between age and basal ganglia iron content in 20 normal individuals ranging from 24 to 79 years of age. These authors analyzed paramagnetic centers sequestered inside cellular membranes to predict local brain iron content. A strong direct relationship between age and regional iron content was found in the putamen and caudate but not in the globus pallidus or thalamus. These findings indicate that striatal iron content increases with normal aging and may play a role in age related loss of function in this brain region.[115]

Ultrastructural analysis of neurons of the substantia nigra in four normal aged subjects revealed changes characteristic of apoptosis, including cell shrinkage and chromatin condensation in 2% of melanized neurons.[116] Although the endoplasmic reticulum appears normal, mitochrondria are markedly shrunken. Fragments of melanized neurons are found in glial cells. Evidence of autophagic degeneration or necrosis are not detected in melanized neurons. Signs of oxidative stress, such as vacuolation of mitochondria, are observed in melanized neurons devoid of apoptotic features. These findings suggest that apoptosis is involved in cell death of nigral dopaminergic neurons during normal aging. However, the morphological abnormalities found in this study, such as marked mitochondrial shrinkage in apoptotic neurons, are not characteristic of those observed in patients with Parkinson's disease, suggesting that the mechanisms underlying the apoptosis associated with PD differs from that associated with normal aging.

Collectively, these finding suggest that normal aging is associated with a loss of neurons and with a loss of function in the nigrostriatal pathway, which may be due to the cumulative effects of oxidative stress, and that at least one metal (iron) may play a role in this process. These findings also suggest that, although free radicals appear to have a role in normal aging and PD, the neuropathology of PD is also distinctly different from that of normal aging. These findings also provide evidence for a putative point of interaction (i.e., free radical generation and scavenging) between normal aging, the environment, and neurodegenerative disease, which may hasten the progression of idiopathic PD, possibly leading to a younger age of onset.

IDIOPATHIC PARKINSON'S DISEASE

Idiopathic Parkinson's disease is a progressive neurodegenerative movement disorder, the etiology of which remains unknown. The clinical manifestations of the disease result from the loss of pigmented dopaminergic neurons in the pars compacta of the substantia nigra. Symptoms of Parkinson's disease include tremor, bradykinesia, gait disturbances, cogwheel rigidity, postural instability, hypomimia, hypophonia, and micrographia. The symptoms of PD are alleviated by Levo-dopa, dopamine agonists, and anticholinergics.

Parkinson's disease incidence rates of up to 190 per 100,000 persons have been reported.[9–14] A long preclinical or asymptomatic period may occur in PD. The presence of early-life risk factors is consistent with a long prodromal period. Marked degeneration of the substantia nigra and loss of striatal dopamine are necessary before clinical symptoms develop. Lewy bodies, the histological hallmark of PD, occur in 10% of normal individuals over age 50. Clinical symptoms develop slowly and are often unilateral and intermittent early in the clinical course of PD. Reduction of striatal dopamine can be detected with positron emission tomography (PET) scans in "at-risk" asymptomatic individuals. Biologic markers may eventually be able to detect subclinical PD and permit prophylactic therapeutic measures to prevent or at least forestall the onset of the disease.[15]

Although PD has been reported in relatively young persons, the likelihood of an individual developing PD increases as one ages. The mean age at death in PD patients increased from about 60 years in 1950 to 77 years in 1992 for both sexes living in Japan from 1950 to 1992.[16] In the United States, approximately 1% of the population over the age of 60 years is afflicted with PD.[17] The number of affected individuals and the cost associated with caring for affected individuals is likely to increase dramatically over the next several decades as the world's population becomes older and those individuals afflicted with PD live longer with the disease.[16,18] Projections indicate that therapies that delay disease onset will markedly reduce overall disease prevalence, whereas therapies to treat existing disease will alter the proportion of cases that are mild as opposed to moderate/severe. The public health impact of such changes would likely involve both the amount and type of health services needed.[19]

Most studies indicate that PD is a multifactorial disorder, which involves genetic and environmental factors acting together.[20,21] Evidence suggests that oxidative stress mediated by free radicals plays a central role in the pathogenesis of PD. This opinion is based in part on a decrease in the levels of glutathione found in the substantia nigra of patients with PD.

The evidence for a genetic risk factor includes reports of families presenting with a highly penetrant, ostensibly dominantly transmitted, form of PD.[22] Studies demonstrating an increased risk for PD among first degree relatives (i.e., siblings, parents, and/or children) of a patient with PD.[23] The incidence of PD in Blacks in is about one fourth of that found in Caucasians.[117,118] The greater similarity for age at onset than for year at onset among siblings with PD, together with increased risk among the subject's biological relatives compared with the subject's spouse, further supports a genetic component. However, no increase in risk was found among twins with an age of onset greater than 50 years old, suggesting that there may an interaction between environment and genetics in those PD cases with an age of onset greater than 50 years old.[24]

Genes that may be involved in the age of onset as well as the etiology of PD include α-synuclein on chromosome 4q21-23/PARK1. α-synuclein aggregation may be involved in Lewy body formation and in the pathogenesis of autosomal dominant forms of familial PD. The ubiquitin C-terminal hydrolase gene located on chromosome 4p14/PARK5 has been associated with autosomal dominant PD. The parkin gene on chromosome 6q25-27/PARK2 appears to be involved in autosomal recessive juvenile onset PD. Parkin mutations account for at least 15% (38 out of 246) of early-onset cases (≤45 years old) without family history, and this proportion decreases significantly with increasing age at onset. Loci on chromosome 1p35-36/PARK6 and 1p36/PARK7 have also been associated with autosomal recessive early-onset PD. Although the role of parkin, an E2-dependent ubiquitin protein ligase, in juvenile onset PD is well established, its role in the late-onset form of Parkinson's disease (PD) is not as clear. At least one study suggests that heterozygous mutations, especially those lying in exon 7, may act as susceptibility alleles for a later-onset form of PD.[25] Excluding exon 7 mutations, the mean age at onset among patients with parkin mutations is 31.5 years, but mutations in exon 7, are observed primarily in heterozygous PD patients with a mean age at onset of 49.2 years. Allele 174 of marker D2S1394 on chromosome 2p13/PARK3 has been associated with an older age at onset of PD (mean age: 69.8 years).[26]

PARKINSONISM

Parkinsonism is a movement disorder that clinically resembles idiopathic PD. Although parkinsonism shares many of the features of PD, the underlying pathologies as well as the etiologies are different. Parkinsonism has been associated with vascular disease, head trauma, encephalitis, and exposure to pharmaceuticals (metachlopramide), illicit drugs (MPTP), pesticides, and industrial toxins.[27,28]

NEURODEGENERATIVE PARKINSONISM VERSUS PARKINSON'S DISEASE: THE ROLE OF OXIDATIVE STRESS

Several neurodegenerative diseases resemble idiopathic PD, including progressive supranuclear palsy (PSP), multiple system atrophy (MSA), Machado-Joseph disease, and Wilson's disease. These disorders are manifested clinically by symptoms, which may include tremors, dystonic posture, gait disturbances, and cognitive disturbances. The specific clinical manifestations seen in each of these parkinsonisms reflects the specific differences in the underlying pathologies (e.g., involvement of cerebellar pathways as well as neurons in the basal ganglia). These disorders are differentiated clinically from idiopathic PD by the findings on neuroimaging studies, rates of progression, and by the therapeutic responses to levodopa and/or dopamine agonist therapy.

The consistent findings of decreased levels of the major antioxidant glutathione in the substantia nigra of patients with idiopathic PD has provided the basis for the oxidative stress hypothesis of the etiology of this neurodegenerative disease. Recent studies have explored whether the nigral glutathione deficiency seen in idiopathic PD is present in patients with parkinsonism associated with nigral damage (PSP and MSA). These studies reveal decreased nigral levels of reduced glutathione in postmortem brain of patients with PD and PSP. A similar decrease was seen in the MSA patient group, but this did not reach statistical significance. Levels of reduced glutathione were within normal limits in all unaffected brain regions and in degenerating extranigral brain areas in PSP and MSA. A trend for decreased levels of uric acid (antioxidant and product of purine catabolism) also was observed in nigra of all patient groups (−19 to −30%). These data suggest that glutathione depletion, possibly consequent to over utilization in oxidative stress reactions, could play a causal role in nigral degeneration in all nigrostriatal dopamine deficiency disorders and that minimizing oxidative stress may be relevant to slowing the progression of these diseases as well.[119]

CHEMICAL EXPOSURE-INDUCED PARKINSONISM: THE ROLE OF OXIDATIVE STRESS

There is evidence to suggest that oxidative stress in involved in the in the neuronal loss seen in the substantia nigra of patients with PD and other forms of parkinsonism. Free radicals and other metabolites, which are conjugated with glutathione are formed during the metabolism of many industrial chemicals and thus exposure to these compounds may contribute to the progression of nigral degeneration.

There are many published reports suggesting that exposure to industrial chemicals including manganese, paraquat, carbon monoxide, carbon disulfide, n-haxane, and ethylene oxide, and pharmaceuticals such as metoclo-pramide and the neuroleptics (e.g., chlorpromazine) can induce extrapyramidal syndromes resembling PD (i.e., parkinsonism).[29–32] Exposures to organophosphate insec-ticides have also been reported to induce parkinsonism. While the specific mechanisms by which these com-pounds induce neuronal loss leading to parkinsonism may differ, the net result is a loss of viable neurons in the extrapyramidal system.

In some patients, parkinsonian symptoms arise for the first time immediately following a severe acute chemical exposure,[29] while in others, the onset of symptoms is insidious and associated with chronic chemical expo-sures.[28,30] The differential diagnosis of PD versus parkin-sonism is complex and requires a review of the history of occupational and environmental exposure, on whether the onset of symptoms was unilateral or bilateral, the constel-lations of symptoms observed, and the response of the patient to levodopa and/or dopamine agonist therapy. Patients with idiopathic PD typically have a unilateral onset of symptoms and respond well to levodopa and/or dopamine agonist therapy while the symptoms seen among patients with chemically induced parkinsonism often develop bilaterally and show limited if any improve-ment with levodopa and/or dopamine agonist therapy. Patients with idiopathic PD and chemical induced parkin-sonism can also be differentiated by the presence of asso-ciated symptoms. For example, patients exposed to carbon disulfide will exhibit symptoms consistent with idiopathic PD but also typically present with peripheral neuropathy as well.[33,34]

MPTP

The most widely studied chemical that can induce parkin-sonism in humans is 1-methyl-4-phenyl-1,2,5,6-tetrahy-dropyridine (MPTP). MPTP is formed as a by-product of the synthesis of 1-methyl-4-phenyl-4-propionoxypiperi-dine (MPPP), a potent meperidine-analog. The occurrence of parkinsonism following exposure to MPTP was first reported in users of illicit "synthetic heroin" or MPPP that was contaminated with MPTP.[35] The clinical picture and neuropathology of MPTP poisoning is very similar to PD. Exposure to the MPTP, has been associated with damage to cells in the pars compacta of the substantia nigra.[35,36] Although cell loss is also seen in the pars compacta of the substantia nigra in patients with PD, other neuropatholog-ical features of PD, most notably the presence of Lewy inclusion bodies in the substantia nigra and locus ceruleus, are not seen in patients exposed to MPTP.[37,38]

Administration of either pargyline, a nonselective monoamine oxidase (MAO) inhibitor, or deprenil (sele-giline), which specifically inhibits MAO-B, has been shown to the prevent both the clinical and pathological effects of MPTP exposure.[36,39] These findings led to the

conclusion that the formation of a charged metabolite, specifically 1-methyl-4-phenyl pyridine (MPP$^+$), was dependent on the actions of MAO-B and that the produc-tion of free radicals from a redox reaction might be involved in the pathogenesis of the MPTP-induced parkin-sonism.[39] This hypothesis is further supported by studies showing that the administration of diethyldithiocarbam-ate, a potent inhibitor of superoxide dismutase (i.e., an enzyme which scavenges free radicals) can potentiate the effects of exposure to MPTP.[40,41] Nontoxic doses of MPTP produced neuronal loss when given simultaneously with diethyldithiocarbamate. These findings indicate that chemicals that deplete the activity of neuroprotective enzymes such as superoxide dismutase may put certain susceptible individuals at an increased risk for developing PD especially if they are simultaneously exposed to more than one neurotoxicant.[41]

The discovery of a chemical that could induce parkin-sonism led to the hypothesis that an environmental toxin or protoxin (a compound that is metabolized to a toxin) might cause Parkinson's disease. That MAO-B inhibitors do not have as profound an effect on preventing the pro-gression of idiopathic PD attests to the subtle differences in the underlying pathogenesis of these two and suggests that Parkinson's disease is not the result of a simple toxic effect. Nevertheless, due to the theoretical practicality of interfering with the production of free radicals, the poten-tial effectiveness of selegiline and/or antioxidants such as vitamin E and coenzyme Q_{10} for slowing the progression of PD continue to be studied and debated.[42]

Research has also demonstrated that MPP$^+$ inhibits activity of Complex I of the mitochondrial electron trans-port chain.[43,44] MPP+ is transported into dopaminergic neurons by the dopamine transporter.[45] Once inside the cell, MPP+ accumulates within mitochondria and inhibits the activity of Complex-I of the electron-transport chain.[46,47,41] Although iron-induced oxidative stress has been associated with reduced Complex-I activity in the substantia nigra, the relationship of this finding to the onset or progression of idiopathic PD has yet to be eluci-dated.[48–52] These findings have led to recent research into the therapeutic benefit of coenzyme Q_{10}, which is an endogenous electron acceptor for Complex-I and a pow-erful antioxidant. The level of coenzyme Q_{10} is reduced in the platelet mitochondria of patients with PD. Oral administration of coenzyme Q_{10} has been shown to increase Complex-I activity, but the therapeutic benefit of oral coenzyme Q_{10} has not yet been established.[42]

Paraquat

The pesticide paraquat is a bipyridyl herbicide that is metabolized by NADPH-dependent reduction to yield a free radical that reacts with molecular oxygen to form a superoxide anion, which is then converted to hydrogen peroxide by superoxide dismutase. Both the superoxide anion and hydrogen peroxide are capable of reacting with lipids to induce lipid peroxidation, thereby altering membrane permeability and disrupting cellular functioning.[53] The toxic effects of paraquat are attenuated by the conjugation of the free radicals metabolites with glutathione-S-transferases.[54]

Paraquat has been shown cross the blood-brain barrier and to induce a loss of dopaminergic neurons in the substantia nigra of rats.[55] Exposure of humans to paraquat, which is structurally similar to 1-methyl-4-phenyl-1,2-5,6-tetrahydropyridine (MPTP), has been associated with an increased risk for PD.[9,31,56] Possible synergistic effects of paraquat and iron have been proposed. Recent studies in animals suggest that paraquat potentiates the toxicity of Maneb (manganese ethylene bisdithiocarbamate).[120]

Iron

Studies on postmortem brains from patients with PD reveal elevated iron in the substantia nigra. Studies have demonstrated that iron chelation via either transgenic expression of the iron binding protein ferritin or oral administration of the bioavailable metal chelator clioquinol reduces susceptibility to the effects of MPTP in animals suggesting that iron is involved in parkinsonism induced by MPTP and that this metal may have a role in the progression of parkinsonism associated with exposures to other chemicals that are metabolized to free radicals and/or contribute to the adverse effects of oxidative stress by depleting stores of glutathione.[121,122]

Accumulation of iron within the brain is associated with specific disorders of iron metabolism and transport. The mutations responsible for hemochromatosis, a hereditary iron overload disorder, led to intracellular sequestration of iron. Studies comparing subjects with PD and parkinsonism suggest that subjects with PD are significantly more likely to be homozygous for the highly penetrant C282Y mutation associated with hemochromatosis than are healthy controls. Furthermore, subjects with parkinsonism are more often carriers of the C282Y mutation than are controls suggesting that the C282Y mutation increases the risk of PD and parkinsonism.[123]

Increased levels of iron within the brain can lead to an increase in oxidative stress mediated via the Fenton reaction. The vulnerability of the dopaminergic neurons of the substantia nigra has been related to the presence of neuromelanin in these neurons. It is hypothesized that neuromelanin may act as an endogenous storage molecule for iron, an interaction suggested to influence free-radical production.[124]

Recent studies have looked at the redox activity of neuromelanin-aggregates in parkinsonian patients who presented with a statistically significant reduction (−70%)

in the number of melanized-neurones and an increased non-heme (Fe^{3+}) iron content as compared with a group of matched-control subjects.[125] The level of redox activity detected in neuromelanin-aggregates was significantly increased (+69%) in parkinsonian patients and was highest in patients with the most severe neuronal loss. This change was not observed in tissue in the immediate vicinity of melanized-neurones. A possible consequence of an overloading of neuromelanin with redox-active elements is an increased contribution to oxidative stress and intra-neuronal damage in patients with Parkinson's disease and parkinsonism.

Manganese

Manganese (Mn) is an essential trace element necessary for normal development and normal biological functioning.[57,58] Occupational exposures to manganese typically occur among miners, welders and during the manufacture and application of Maneb.[58–62] Food is the main source of nonoccupational intake of manganese.[63,64] Dietary intake of manganese alone has not associated with toxic effects except in those individuals with decreased excretion due to liver failure.[65,64] Recent studies suggest that high dietary intakes of manganese plus iron may contribute to the risk for PD.[64]

The valence state (i.e., the number of electrons in the outermost or valence orbital) of manganese is a factor in its effect on living tissues.[66] Divalent manganese (Mn^{2+}) acts as a powerful antioxidant, while trivalent manganese (Mn^{3+}) appears to have a high affinity for those brain regions with high concentrations of neuromelanin (e.g., substantiate nigra).[27,67–79] The capacity of manganese to induce selective lesions in the substantia nigra and basal ganglia appears to be related to the neuromelanin content in these brain regions, where divalent manganese is readily oxidized to the cytotoxic trivalent species.[37,77,80] Trivalent manganese potentiates the autooxidation of catecholamines (e.g., dopamine), thereby generating toxic free radicals.[71,76] The death of the dopaminergic cells results in a concurrent release of neuromelanin and accompanying depigmentation of the affected region.[71,81–83] Loss of neuromelanin, which is also a scavenger of free radicals, may further potentiate the toxic effects of manganese poisoning.[77,84] Studies indicate that neuromelanin has a high affinity for iron, lipids, pesticides, and MPP+ as well as manganese. The affinity of neuromelanin for a variety of inorganic and organic toxicants suggests that it normally acts to protect cells from the neurotoxic effects of these compounds. The synthesis and accumulation of neuromelanin associated with normal aging are also consistent with a putative protective role which could be exceeded during conditions of neurotoxicant overload such as may occur with occupational exposures to manganese.

The adverse effects of trivalent manganese are also possible when protective scavenger enzymes such as manganese superoxide dismutase (Mn SOD) are unable to alter the oxidation potential of critical amounts of reactive oxygen species.[73,82,83,85] There is an increase in the levels of Mn SOD in patients with idiopathic PD, suggesting that synthesis of Mn SOD may be induced in response to free radical induced injury to cells.[86,87] Ironically, increased Mn SOD activity results in increased production of hydrogen peroxide, which can react with ferrous iron to yield hydroxyl radicals.[88] Mn SOD is also considered the point of contact between mitochondrial respiratory failure and oxidative stress. Experimental data have shown that treatment of dopaminergic neurons (PC 12 cells) with manganese chloride ($MnCl_2$) inhibits mitochondrial Complex I activity, while glial cells (C6) are not similarly affected. These findings suggest that environmental factors such as exposure to manganese, can induce oxidative stress and mitochondrial respiratory dysfunction and may also contribute to the pathogenesis of idiopathic PD (see Figure 6.1).[27,52,89]

Significant differences exist in the underlying neuropathology of manganese poisoning and idiopathic PD. Manganese damages cells in the basal ganglia as well as the substantia nigra, but the cell loss in the substantia nigra primarily involves the pars reticulata and is less marked than that which occurs in PD. Lewy inclusion bodies, which are found in the substantia nigra and locus ceruleus of idiopathic PD, further differentiate the two pathologically.[37,38] This difference in neuropathology is responsible for a remarkable difference in response to levodopa and/or dopamine agonist therapy, which is considerably less favorable the in those patients with manganese poisoning than it is in idiopathic PD.[58,90–96]

Aside from the reduced clinical response to dopaminergic therapy, there are other distinct differences in the clinical manifestations of manganese-induced parkinsonism that differentiate it from idiopathic PD.[38,52,81,84,89,95–99]

Perhaps the most overt clinical difference between idiopathic PD and manganese poisoning is that the extrapyramidal signs of manganese poisoning are frequently preceded or accompanied by an acute psychosis, which is not seen among patients with idiopathic PD. As manganese poisoning progresses, the acute psychosis gradually subsides, while dystonia, action tremor, and an awkward high-stepping dystonic gait appear.[58,100–103] The high-stepping dystonic gait of manganese poisoning is in stark contrast to the shuffling gait of patients with PD. Although the extrapyramidal symptoms of manganese poisoning may progress following cessation of exposure, the progression is slower than that seen in PD.[94,58]

ROLE OF ENVIRONMENTAL AND OCCUPATIONAL EXPOSURE TO CHEMICALS IN PARKINSON'S DISEASE

The possible role of environmental and occupational exposures to neurotoxicants (e.g., solvents, metals, and pesticides) in the development of idiopathic PD has received considerable attention from the medical community and public health researchers.[20,21,30,104–107] Environmental factors that have been associated with an increased risk for PD include home pesticide exposure, rural living, well-water consumption, and diet.[106,108–110] A rural predominance has been reported by several authors suggesting that environmental factors such as exposure to pesticides (e.g., organophosphates) may play a role in the etiology of PD.[13,31,110] Several researchers[84,111] have noted a correlation between living in a rural environment and drinking well water at an early age and the development of PD later in life. These authors have suggested that water is a likely vehicle for the causal agent, but that neither the concentrations of metals in the water nor any of the herbicides and pesticides used in agriculture in the areas where these studies were conducted are related to the cause. Tanner et al.[24] studied the role of environment in the development of PD in a Chinese population using a case-control method. These authors investigated the relationship between PD and place of residence, source of drinking water, and environmental and occupational exposure to various agricultural and industrial processes. Occupational or residential exposure to industrial chemicals, printing plants, or quarries was associated with an increased risk of developing PD while living in villages, and exposure to the common accompaniments of village life, wheat growing and pig raising, were associated with a decreased risk for PD. PD cases and controls did not differ with respect to other factors investigated. These findings were interpreted as consistent with the hypothesis that environmental exposure to certain industrial chemicals may be

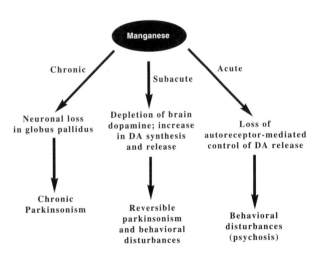

FIGURE 6.1   Effects of manganese exposure.

related to the development of PD. A case-control study of 150 PD patients and 150 age- and sex-matched controls by Koller et al.,[15] which looked at residential histories, sources of drinking water, and occupations such as farming, revealed that rural living and drinking well water (which was dependent on rural living) were significantly more common among PD patients than controls, regardless of age at onset of PD. These data were interpreted to provide further evidence that an environmental factor such as exposure to a neurotoxic agent could be involved in the etiology of PD, but a putative toxic agent to account for this has not been identified.

Despite decades of research, exposure to a specific neurotoxicant has never been shown to induce idiopathic PD. Studies suggest that there has been an increase in the incidence of PD since the beginning of the industrial revolution. However, the existence of PD prior to the industrial revolution suggests that exposure to any of the synthetic neurotoxicants recently released into the environment is unlikely to be exclusively responsible for this neurodegenerative disease.

Review of the literature reveals many studies that have looked at the incidence and/or prevalence of PD among exposed and unexposed populations, although few studies that have stratified subjects by exposure history and age at onset.[15,20,24,84,112–114] The lack of consistent findings in the research performed to date that has looked at factors that increase the incidence and/or prevalence of PD suggest that this approach may need to be reconsidered. By contrast, the few studies that have looked at factors that influence age at onset have revealed interesting results which suggest that exposure to neurotoxic chemicals may unmask latent idiopathic PD by accelerating neuronal loss in the substantia nigra due to normal aging as well as that due to the neurodegenerative process.[113,114]

A study of 15 manganese-exposed welders with a mean of 47,144 welding hours revealed that welders were younger than controls at the time of PD symptom onset (46 years old versus 63 years old; $p < 0.0001$). There was no difference in frequency of tremor, bradykinesia, rigidity, asymmetric onset, postural instability, clinical depression, dementia, or drug-induced psychosis between the welders and the two control groups. Thirteen of the welders responded favorably to levodopa therapy, while one responded to pramipexole, and one did not yet require symptomatic treatment. Motor fluctuations and dyskinesias occurred at a similar frequency in welders and controls. PET with 6-[18F]fluorodopa obtained in two of the welders showed findings typical of idiopathic PD, with greatest loss in posterior putamen. Eight of the 15 welders (53%) included in this study had a positive family history for PD. Although these authors concluded that there was no significant difference between welders and controls subjects with regard to family history of PD, this was a small study, and cases and controls were not specifically matched for family history of PD.[114]

A similar study by Pezzoli et al.[113] looking at the role of hydrocarbon exposure also revealed a younger age at onset among exposed subjects with PD. The subjects worked in various industries where exposures to petroleum products, plastics, rubber, paints, paint thinners, lacquers, degreasing solvents, glues, pesticides, dyes, and refrigerants were likely to occur. The majority of subjects reported occupational exposure to hydrocarbons. Only 14 subjects reported occupational exposure to pesticides/herbicides. The most frequently encountered substances in this study were acetone, 2-di-methyl-ethyl-ketone (MEK), n-hexane and its isomers, cyclo-hexane and its isomers, hepthane and its isomers, ethyl-acetate, isobutylacetate, butyl-acetate, dichloropropane, trichloroethylene, trichloroethane, tetrachloroethylene, freon, toluene, and 1-methoxy-2-propanol. The exposed group ($n = 188$) had a mean age at onset of 55.2 years (± 9.8 years) compared with 58.6 years (±10 years) for unexposed subjects ($n = 188$) matched for duration of disease and gender. In addition to an earlier age at onset, the exposed subjects with PD were less responsive to treatment and required a higher mean dosage of levodopa than did the unexposed subjects. Although the diminished response to levodopa treatment in this group may be construed to be consistent with a diagnosis of multiple systems atrophy (MSA) rather than idiopathic PD, none of the subjects met all the clinical criteria for the diagnosis of MSA. Furthermore, among those subjects who had PET studies, these authors found no difference in striatal glucose metabolism, which is typically diminished in MSA. The exposed group was also composed primarily of men (76.4%) and was less educated and more disabled than the unexposed subjects. The severity of symptoms was directly proportional to the duration and the intensity of the exposure. These findings were interpreted to suggest that hydrocarbons may be involved in the etiopathogenesis of PD, which does not appear to have a major genetic component. There was no difference in the number of subjects in each group with a positive family history of PD.

Each of these two studies independently concluded that exposure to a specific chemical or a type of chemicals (i.e., manganese or hydrocarbons) could influence age at onset of PD. Because the findings in these studies are complementary, these findings can be interpreted to suggest that neither manganese nor hydrocarbon exposure is responsible for the younger age at onset of PD but, rather, that a common mechanism of neurotoxic action (e.g., free radicals induced neuronal loss in the substantia nigra) may hasten the progression of idiopathic PD. Assuming this hypothesis to be true, then it could further be expected that any factor that induces neuronal loss in the substantia nigra could hasten the neuropathological process that underlies PD by contributing to the total number of neurons lost per unit time. If the loss of a critical percentage of neurons in the substantia nigra is associated with the onset of parkinsonian symptoms, then it logically follows that any factor that hastens the loss of neurons in the substantia nigra will lead to a younger age at onset of PD.

CONCLUSIONS

The material presented in this review suggests that PD can interact with environmental factors to affect age at onset of the disease and that this interaction could possibly increase the incidence of PD in an exposed population, but these data do not provide evidence for a causal relationship between the environment and prevalence of PD. However, this does provide a foundation upon which to base future research. In addition, it appears that higher levels of exposure that occur in occupational settings are more likely to significantly affect age at onset of PD, but whether subtle changes in age at onset occur at lower levels of exposure has not been determined. Large-scale epidemiological studies could provide the statistical power needed to answer this enduring question. The need remains for large retrospective studies that stratify subjects based on exposure history and genetic risk factors that can plausibly interact with exposure (e.g., glutathione-S-transferase polymorphisms). Prospective studies to ascertain the role of exposure and therapeutic measures such as antioxidants at onset and disease progression are also needed. As a our understanding of genetic modulation of enzyme activity and neurotransmitter receptor assembly and function, as well as genetic control of apoptosis and other factors that modulator neuronal viability, increases, our understanding of many of the interactions between environment and disease will likely grow as well.

Few therapeutic strategies to slow PD progression have been tested in clinical trials, but pharmaceuticals with antioxidant properties including ropinirole and pramipexole may slow PD progression. Neutraceuticals that scavenge free radicals, including antioxidant vitamins (e.g., vitamin E) and co-enzyme Q10, have also gained attention for their potential ability to slow the progression of PD.[10] Although the effects of minimizing lifetime exposure to free radicals via use of antioxidant vitamins and limiting cumulative lifetime exposure to neurotoxic chemicals have not yet been fully elucidated by prospective human studies, emerging evidence from retrospective studies suggests that these measures may slow the progression of PD and delay its onset until later in life.[10,21,27,113,114] For the time being, its seems prudent to conclude that minimizing exposure to free radicals, which appears to be the single common factor involved in normal aging of the nigrostriatal pathway, parkinsonism, and PD, is a reasonable first step in reducing the risk for developing PD.

A LOVING TRIBUTE TO ROBERT G. FELDMAN, M.D.

Robert G. Feldman, M.D., unfortunately passed away before this chapter was completed. Dr. Feldman was a dedicated physician, researcher, and mentor. He was world renowned for his work in both Parkinson's disease and neurotoxicology. His interest in these two distinct areas afforded him the unique level of understanding necessary to not only recognize but to logically investigate potential interactions between the environment and neurodegenerative disease. Although the role of the environment in the etiology of Parkinson's disease was not fully elucidated during his lifetime, his unique perspective and the contributions he made through his research and publications will undoubtedly facilitate future research in this area for years to come.

REFERENCES

1. Calne, D. B., Peppard, R. F., Aging of the nigrostriatal pathway in humans, *Can. J. Neurol. Sci.,* 14(3 Suppl):424–7, 1987.
2. Doraiswamy, P. M., Na, C., Husain, M. M., Figiel, G. S., McDonald, W. M., Ellinwood, Jr., E. H., Boyko, O. B., Krishnan, K.R., Morphometric changes of the human midbrain with normal aging: MR and stereologic findings. *AJNR Am. J. Neuroradiol.,* 13(1):383–386, 1992.
3. Pollack, M., Leeuwenburgh, C., Apoptosis and aging: role of the mitochondria. *J. Gerontol. A. Biol. Sci. Med. Sci.,* (11):B475–82, 2001.
4. Duan, W, and Mattson, M. P., Dietary restriction and 2-deoxyglucose administration improve behavioral outcome and reduce degeneration of dopaminergic neurons in models of Parkinson's disease. *J. Neurosci. Res.,* 57(2):195–206, 1999.
5. Casadesus, G., Shukitt-Hale, B., Joseph, J. A., Qualitative versus quantitative caloric intake: are they equivalent paths to successful aging? *Neurobiol. Aging,* 23(5):747–69, 2002.
6. Sreekumar, R., Unnikrishnan, J., Fu, A., Nygren, J., Short, K. R., Schimke, J., Barazzoni, R., Nair, K. S., Effects of caloric restriction on mitochondrial function and gene transcripts in rat muscle, *Am. J. Physiol. Endocrinol Metab.,* (1):E38–43, 2002.
7. Chen, S. K., Hsieh, W. A., Tsai, M. H., Chen, C. C., Hong, A. I., Wei, Y. H., Chang, W. P., Age-associated decrease of oxidative repair enzymes, human 8-oxoguanine DNA glycosylases (hOgg1), in human aging, *J. Radiat. Res.* (Tokyo), 44(1):31–35, 2003.
8. Mattson, M. P., Duan, W., Guo, Z., Meal size and frequency affect neuronal plasticity and vulnerability to disease: cellular and molecular mechanisms, *J. Neurochem,* 84(3):417–31, 2003.
9. Rajput, A. H., Uitti, R. J., Stern, W., Laverty, W., O'Donnell, K., O'Donnell, D., Yuen, W. K., Dua. Geography,

drinking water chemistry, pesticides and herbicides and the etiology of Parkinson's disease, *Can. J. Neurol. Sci.,* 14(3 Suppl):414–418, 1987.

10. Zhang, Z. X., and Roman, G. C., Worldwide occurrence of Parkinson's disease: an updated review, *Neuroepidemiol,* 12:195–208, 1993.

11. Bharucha, E. P., Bharucha, N.E., Epidemiological study of Parkinson's disease in Parsis in India, *Adv. Neurol.,* 60:352–354, 1993.

12. Mayeux, R., Marder, K., Cote, L. J., Denaro, J., Hemenegildo, N., Mejia, H., Tang, M. X., Lantigua, R., Wilder, D., Gurland, B. et al, The frequency of idiopathic Parkinson's disease by age, ethnic group, and sex in northern Manhattan, 1988–1993, *Am. J. Epidemiol.,* 142:820–827, 1995.

13. Tandberg, E., Larsen, J. P., Nessler, E. G., Riise, T., Aarli, J. A., The epidemiology of Parkinson's disease in the county of Rogaland, Norway, *Mov. Disord.,* 10:541–549, 1995.

14. Sutcliffe, R. L., Meara, J. R., Parkinson's disease epidemiology in the Northampton District, England, 1992, *Acta Neurol. Scand.,* 92:443–450, 1995.

15. Koller, W., Vetere-Overfield, B., Gray, C., Alexander, C., Chin, T., Dolezal, J., Hassanein, R., Tanner, C., Environmental risk factors in Parkinson's disease, *Neurology,* 40(8):1218–1221, 1990.

16. Imaizumi, Y., Kaneko, R., Rising mortality from Parkinson's disease in Japan, 1950–1992, *Acta Neurol. Scand.* 91:169–176, 1995.

17. Martilla, R. J., Epidemiology of Parkinson's disease, In: Koller WC (ed); *Handbook of Parkinson's Disease.* New York, Marcel Dekker, pp. 35–50, 1989.

18. Dodel, R. C., Eggert, K. M, Singer, M. S., Eichhorn, T. E., Pogarell, O., Oertel. W. H., Costs of drug treatment in Parkinson's disease, *Mov. Disord.,* 13:249–254, 1998.

19. Sloane, P. D., Zimmerman, S., Suchindran, C., Reed, P., Wang, L., Boustani, M., Sudha, S., The public health impact of Alzheimer's disease, 2000–2050: potential implication of treatment advances, *Annu. Rev. Public Health,* 23:213–231, 2002.

20. Seidler, A., Hellenbrand, W., Robra, B. P., Vieregge, P., Nischan, P., Joerg, J., Oertel, W. H., Ulm, G., Schneider, E., Possible environmental, occupational, and other etiologic factors for Parkinson's disease: a case-control study in Germany, *Neurology,* 46(5):1275–1284, 1996.

21. Feldman, R.G., and Ratner, M. H., The pathogenesis of neurodegenerative disease: neurotoxic mechanisms of action and genetics, *Current Opinion in Neurology,* 12:725–731, 1999.

22. Goble, L. I., The genetics of Parkinson's disease, *Rev. Neurosci.,* 4:1–16, 1993.

23. Lazzarini, A. M., Myers, R. H., Zimmerman, T. R., Mark, M. H., Golbe, L. I., Sage, J. I., Johnson, W. G., Duvoisin, R. C., A clinical genetic study of Parkinson's disease: evidence for dominant transmission, *Neurology,* 44:499–506, 1994.

24. Tanner, C. M., Chen, B., Wang, W., Peng, M., Liu, Z., Liang. X., Kao, L. C., Gilley, D. W., Goetz, C. G., Schoenberg, B. S., Environmental factors and Parkinson's disease: a case-control study in China, *Neurology,* 39(5):660–664, 1989.

25. Oliveira, S. A., Scott, W. K., Martin, E. R., Nance, M. A., Watts, R. L., Hubble, J. P., Koller, W. C., Pahwa, R., Stern, M. B., Hiner, B. C., Ondo, W. G., Allen, F. H., Jr., Scott, B. L., Goetz, C. G., Small, G. W., Mastaglia, F., Stajich, J. M., Zhang, F., Booze, M. W., Winn, M. P., Middleton, L. T., Haines, J. L., Pericak-Vance, M. A., Vance, J. M., Parkin mutations and susceptibility alleles in late-onset Parkinson's disease, *Ann. Neurol.,* 53(5):624–629, 2003.

26. DeStefano, A. L., Lew, M. F., Golbe, L. I., Mark, M. H., Lazzarini, A. M., Guttman, M., Montgomery, E., Waters, C. H., Singer, C., Watts, R. L., Currie, L. J., Wooten, G. F., Maher, N. E., Wilk, J. B., Sullivan, K. M., Slater, K. M., Saint-Hilaire, M. H., Feldman, R. G., Suchowersky, O., Lafontaine, A. L., Labelle, N., Growdon, J. H., Vieregge, P., Pramstaller, P. P, Klein, C., Hubble, J. P., Reider, C. R., Stacy, M., MacDonald, M. E., Gusella, J. F., Myers, R. H., PARK3 influences age at onset in Parkinson disease: a genome scan in the GenePD study, *Am. J. Hum. Genet.,* 70(5):1089–1095, 2002.

27. Feldman, R. G., and Ratner, M. H., Essentials of Metal Neurotoxicity: Mechanisms and Pathology. Clinics in Occupational and Environmental Medicine, *Neurotoxicology,* 1(3):1526–1546, 2001.

28. Ratner, M., Cabello, D., Thaler, D., and Feldman, R., Movement disorder in an adult following exposure to DEET, *J. Toxicol. Clin. Toxicol.,* 39: 477, 2001.

29. Barbosa, ER., Comerlatti, L. R., Haddad, M. S., Scaff, M., Parkinsonism secondary to ethylene oxide exposure: case report, *Arq. Neuropsiquiatr.* 50(4):531–3, 1992.

30. Pezzoli, G., Antonini, A., Barbieri, S., Canesi, M., Perbellini, L., Zecchinelli, A., Mariani, C. B., Bonetti, A., and Leenders, K. L., *n*-Hexane-induced parkinsonism: pathogenetic hypotheses, *Mov. Disord.,* 10:279–282, 1995.

31. Liou, H. H., Tsai, M. C., Chewn, C. J., Jeng, J. S., Chang, Y. C., Chen, S. Y., Chen, R. C., Environmental risk factors and Parkinson's disease: a case-control study in Taiwan, *Neurology,* 48:1583–1588, 1997.

32. Kim, Y., Kim, J. W., Ito, K., Lim, H. S., Cheong, H. K., Kim, J. Y., Shin, Y. C., Kim, K. S., Moon, Y., Idiopathic parkinsonism with superimposed manganese exposure: utility of positron emission tomography, *Neurotoxicolog,* 20(2–3):249–52, 1999.

33. Seppäläinen, A. M., and Haltia, M., 1980, Carbon disulfide, in Spencer PS, and Schaumburg HH (eds); *Experimental and Clinical Neurotoxicology,* Williams and Wilkins, Baltimore/London, pp. 356–373, 1980.

34. Laplane, D., Attal, N., Sauron, B., de Billy, A., and Dubois, B., Lesions of basal ganglia due to disulfiram neurotoxicity, *J. Neurol. Neurosurg. Psychiatr.,* 55:925–929, 1992.

35. Langston, J. W., Ballard, P., Tetrud, J. W., and Irwin, I,: Chronic parkinsonism in humans due to a product of meperidine-analog synthesis, *Science,* 219:979–980, 1983.

36. Langston, J. W., Forno, L. S., Rebert, C. S., and Irwin, I., Selective nigral toxicity after systemic administration of 1-methyl-4-phenyl-1,2,5,6-tetrahydropyridine (MPTP) in the squirrel monkey, *Brain Res.* 292:390–394, 1984.

37. Donaldson, J., Manganese and human health, In: Stokes, P. M., Campbell, P. G., Schroeder, W. H., Trick, C., France, R. L., Puckett, K. J., Lazerte, B., Speyer, M., Hanna, J. E., and Donaldson, J. (eds), *Manganese in the Canadian Environment,* NRCC No. 26193; ISSN 0316-0114. Ottawa, NRCC Associate Committee on Scientific Criteria for Environmental Quality, pp. 93–111, 1988.

38. Gibb, W. R. G., The neuropathology of parkinsonian disorders. Parkinson's disease and movement disorders, in Jankovic, J., and Tolosa, E., (Eds): *Parkinson's Disease and Movement Disorders,* Baltimore-Munich, Urban and Schwartzenberg, pp. 205–223, 1988.

39. Heikkila, R. E., Manzino, L., Cabbat, F. S., and Duvoisin, R. C., Protection against the dopaminergic neurotoxicity of 1-methyl-4-phenyl-1,2,5,6-tetrahydropyridine by monoamine oxidase inhibitors, *Nature,* 311:467–469, 1984.

40. Heikkila, R. E., Cabbat, F. S., and Cohen, G., In vivo inhibition of superoxide dismutase in mice by diethyldithiocarbamate, *J. Biol. Chem.,* 251:1451–1453, 1976.

41. Walters, T. L., Irwin, I., Delfani, K., Langston, J. W., and Janson, A. M., Diethyldithiocarbamate causes nigral cell loss and dopamine depletion with nontoxic doses of MPTP, *Exp. Neurol.,* 156:62–70, 1999.

42. Shults, C. W., Beal, M. F., Fontaine, D. et al, Absorption, tolerability, and effects on mitochondrial activity of oral coenzyme Q_{10} in parkinsonian patients, *Neurology,* 50:793–795, 1998.

43. Davey, G. P., Tipton, K. F., and Murphy, M. P., Uptake and accumulation of 1-methyl-4-phenylpyridinium by rat liver mitochondria measured using an ion selective electrode, *Biochem. Med. J.,* 288:439–443, 1992.

44. Schapira, A. H., Mann, V. M., Cooper, J. M. et al, Anatomic and disease specificity of NADH CoQ1 reductase (complex I) deficiency in Parkinson's disease, *J. Neurochem.,* 55:2142–2145, 1990.

45. Javitch, J. A., D'Amato, R. J., Strittmatter, S. M., and Snyder, S. H., Parkinsonism inducing neurotoxin, *N*-methyl-4-phenyl-1,2,3,6-tetrahydropyridine: Uptake of the metabolite *N*-methyl-4-phenylpyridine by dopamine neurons explains selective toxicity, *Proc. Natl. Acad. Sci.,* 82:2173–2177, 1985.

46. Nicklas, W. J., Vyas, I., and Heikkila, R. E., Inhibition of NADH-linked oxidation in brain mitochondria by 1-methyl-4-phenyl-1,2,3-6-tetrahydropyridine, *Life Sci.,* 36:2503–2508, 1985.

47. Ramsey, R. R., and Singer, T. P., Energy-dependent uptake of *N*-methyl-4-phenylpyridinium, the neurotoxic metabolite of 1-methyl-4-phenyl-1,2,3-6-tetrahydropyridine, by mitochondri,. 261:7585–7587, 1986.

48. Harley, A., Cooper, J. M., Schapira, A. H., Iron induced oxidative stress and mitochondrial dysfunction: relevance to Parkinson's disease, *Brain Res.* 627:349–353, 1993.

49. Janetsky, B., Hauck, S., Youdim, M. B. H. et al, Unlatered aconitase activity, but decreased complex I acvtivity in substantia pars compacta of patients with Parkinson's disease, *Neurosci. Lett.,* 169:126–128, 1994.

50. Mann, V. M., Cooper, J. M., Daniel, S. E. et al, Complex I, iron, and ferritin in Parkinson's disease substantia nigra, *Ann. Neurol.,* 36:876–881, 1994.

51. Reichmann, H., Janetsky, B., and Riederer, P., Iron-dependent enzymes in Parkinson's disease, *J. Neural. Trans.* (Suppl) 46:157–164, 1995.

52. Blake, C. I., Spitz, E., Leehy, M., Hoffer, B. J., and Boyson, S. J., Platelet mitochondrial respiratory chain function in Parkinson's disease, *Mov. Disord.,* 12:3–8, 1997.

53. Ecobichon, D., Pesticides, in Klaassen C.D. (Ed.), *Casarett and Doull's Toxicology: the Basic Science of Poisons,* 5th Ed. McGraw-Hill, N.Y., pp. 671–675, 1996.

54. Di Llio, C., Sacchetta, P., Iannarelli, V., and Aceto, A., Binding of pesticides to alpha, mu and pi class glutathione transferase, *Toxicol. Lett.,* 76:173–177, 1995.

55. Brooks, A. L., Chadwick, C. A., Gelbard, H. A., Cory-Sclechta, D. A., Federoff, H. J., Paraquat elicited neurobehavioral syndrome caused by dopaminergic neuron loss, *Brain Res.,* 823:1–10, 1999.

56. Barbeau, A., Roy, M., Bernier, G., Campanella, G., Paris, S., Ecogenetics of Parkinson's disease: prevalence and environmental aspects in rural areas, *Can. J. Neurol. Sci.,* 14:36–41, 1987.

57. Hurley, L. S., Teratogenic aspects of manganese, zinc, and copper nutrition, *Physiol. Rev.,* 61:249–295, 1981.

58. Feldman, R. G., *Occupational and Environmental Neurotoxicology.* Lippincott-Raven, Philadelphia, P.A., 1999.

59. Flinn, R. H., Neal, P. A., Reinhart, W. H., D'Allavalle, J. M., Fulton, W. B., and Dooley, A. E., Chronic manganese poisoning in an ore crushing mill. Public Health Bulletin. No. 247. U.S. Government Printing Office, Washington, D.C., pp. 1–77, 1940.

60. Huang, C. P., and Quist, G. C., The dissolution of manganese ore in dilute aqueous solution, *Environ. Int.,* 9:379–389, 1983.

61. Saric, M., Manganese, in Friberg, L., Nordberg, G. F., and Vouk, V., (Eds.): *Handbook on the Toxicology of Metals,* 2nd ed. Elsevier, Amsterdam, pp. 354–386, 1986.

62. Sjogren, B., Ingren, A., Frech, W., Hagman, M., Johansson, L., Tesarz, M., and Wennberg, A., Effects on the nervous system among welders exposed to aluminum and manganese, *Occup. Environ. Med.,* 53(1):32–40, 1996.

63. Sittig, W., Kleindienst, G., and Irmisch, R., Effects of food intake on CNS activity. psychomotor performance and mood states in healthy volunteers, *Pharmacopsychiat.,* 18(1):123–126, 1985.

64. Powers, K. M., Smith-Weller, T., Franklin, G. M., Longstreth, W. T., Jr., Swanson, P. D., Checkoway, H., Parkinson's disease risks associated with dietary iron, manganese, and other nutrient intakes, *Neurology,* 10;60(11):1761–1766, 2003.

65. Hauser, R. A., Zesiewicz, T. A., Rosemurgy, A. S. et al, Manganese intoxication and chronic liver failure, *Ann. Neurol.*, 36:871–875, 1994.

66. Cotzias, G. C., Manganese in health and disease, *Physiol. Rev.*, 38:503–532, 1958.

67. Maynard, L. S., and Cotzias, G. C., Partition of manganese among organs and intracellular organelles of the rat, *J. Biol. Chem.*, 214:489–495, 1955.

68. Kono, Y., Takahashi, M., Asada, K., Oxidation of manganous pyrophosphate by superoxide radicals and illuminated spinach chloroplasts, *Arch. Biochem. Biophys.*, 174:454–461, 1976.

69. Graham, D. G., Oxidative pathways for catecholamines in the genesis of neuromelanin and cytotoxic quinones, *Mol. Pharmacol.*, 14:633–643, 1978.

70. Graham, D. G., Tiffany, S. M., Bell, W. R., and Gutknecht, W. F., Autoxidation versus covalent binding of quinones as the mechanism of toxicity of dopamine, 6-hydroxydopamine, and related compounds towards C1300 neuroblastoma cells *in vitro*, *Mol. Pharmacol.*, 14:644–653, 1978.

71. Donaldson, J., LaBella, F. S., and Gesser, D., Enhanced autoxidation of dopamine as a possible basis of manganese neurotoxicity, *Neurotoxicol.*, 2:53–64, 1980.

72. World Health Organization (WHO), Manganese, Environmental Health Criteria, Geneva, 17:1–110, 1981.

73. Halliwell, B., Manganese ions, oxidation reactions and the superoxide radical, *Neurotoxicology*, 5(1):113–118, 1984.

74. Lydén, A., Larsson, B. S., and Lindquist, N. G., Melanin affinity of manganese, *Acta Pharmaco. et Teratol.*, 55:133–138, 1984.

75. Hintz, P., and Kalyaraman, B., Metal ion-induced activation of molecular oxygen in pigmented polymers, *Biochem. Biophys. Acta*, 883:41–45, 1986.

76. Archibald, F. S., and Tyree, C., Manganese poisoning and the attack of trivalent manganese upon catecholamines, *Arch. Biochem. Biophys.*, 256:638–650, 1987.

77. Swartz, H. M., Sarna, T., and Zecca, L., Modulation by neuromelanin of the availability and reactivity of metal ions, *Ann. Neurol.*, 32:S69–S75, 1992.

78. Ali, S. F., Duhart, H. M., Newport, G. D., Lipe, G. W., and Slikker, W., Jr., Manganese-induced reactive oxygen species: comparison between Mn+2 and Mn+3, *Neurodegeneration*, 4(3):329–334, 1995.

79. Olanow, C. W., Good, P. F., Shinotoh, H., Hewitt, K. A., Vingerhoets, F., Snow, B. J., Beal, M. F., Calne, D. B., and Perl, D. P., Manganese intoxication in the rhesus monkey: a clinical, imaging, pathologic, and biochemical study, *Neurology*, 46(2):492–498, 1996.

80. Ambani, L. M., Van Woert, M. H., and Murphy, S., Brain peroxidase and catalase in Parkinson's disease, *Arch. Neuro.*, l 32:114–118, 1975.

81. Graham, D. G., Catecholamine toxicity: A proposal for the molecular pathogenesis of manganese neurotoxicity and Parkinson's disease, *Neurotoxicology*, 5(1):83–96, 1984.

82. Donaldson, J., and Barbeau, A., Manganese neurotoxicity: possible clues to the etiology of human brain disorders, in Gabay, S. J., Harris and Ho, B. T., (Eds.): *Metal ions in Neurology and Psychiatry*, Alan R. Liss Inc., New York, pp. 259–285, 1985.

83. Segura-Aguila, J., and Lind, C., On the mechanism of the Mn^{3+}-induced neurotoxicity of dopamine: Prevention of quinone-derived oxygen toxicity by DT diaphorase and superoxide dismutase *Chem-Bio.l Interact.*, 72:309–324, 1989.

84. Barbeau, A., Manganese and extrapyramidal disorders, (A critical tribute to Dr. George C. Cotzias), *Neurotoxicol.*, 5:13–36, 1984.

85. Halliwell, B., Oxidants and the central nervous system: some fundamental questions. Is oxidant damage relevant to Parkinson's disease, Alzheimer's disease, traumatic injury and stroke? *Acta Neurol.*, scand. 126:23–33, 1989.

86. Yoritaka, A., Hattori, N., Mori et al, An immunohistochemical study on manganese superoxide dismutase in Parkinson's disease, *J. Neurol. Sci.*, 148:181–186, 1997.

87. Radunovic, A., Porto, W. G., Zeman, S., and Leigh, P. N., Increased mitochondrial superoxide dismutase activity in Parkinson's disease but not amyotrophic lateral sclerosis motor cortex, *Neurosci. Lett.*, 239:105–108, 1997.

88. Shimoda-Matsubayashi, S., Hattori, T., Matsumine, H. et al, Mn SOD activity and protein in a patient with chromosome 6-linked autosomal recessive parkinsonism in comparison with Parkinson's disease and control, *Neurology*, 49:1257–1262, 1997.

89. Feldman, R. G., Manganese as a possible ecoetiologic factor in Parkinson's disease, *Ann. N.Y. Acad. Sci.*, 648:266–267, 1992.

90. Mena, I., Court, J., Fuezalida, S. et al, Modification of chronic manganese poisoning: treatment with L-dopa or 5-OH tryptophane, *N. Eng. J. Med.*, 282(1):5–10, 1970.

91. Rosenstock, H. A., Simons, D. G., and Meyer, J. S., Chronic manganism. Neurologic and laboratory studies during treatment with levodopa, *JAMA*, 217:1354–1359, 1971.

92. Greenhouse, A. H., Manganese Intoxication in the United States. *Trans. Am. Neurol. Assoc.*, 96:248–249, 1971.

93. Cook, D. G., Fahn, S., and Brait, K. A., Chronic manganese intoxication, *Arch. Neurol.*, 30:59–64, 1974.

94. Huang, C., Lu, C., Chu, N., Hochberg, F, et al, Progression after chronic manganese exposure, *Neurology*, 43:1479–1482, 1993.

95. Calne, D. B., Chu, N. S., Huang, C. C., Lu, C. S., Olanow, W., Manganism and idiopathic parkinsonism: similarities and differences, *Neurology*, 44(9):1583–1586, 1994.

96. Lu, C. S., Huang, C. C., Chu, N. S., Calne, D. B., Levodopa failure in chronic manganism, *Neurology*, 44(9):1600–1602, 1994.

97. Wolters, E., Huang, C. C., Clark, C., Peppard, R. F., Okada, J., Chu, N. S., Adam, M. J., Ruth, T. J., Li, D., and Calne, D. B., Positron Emission Tomography in manganese intoxication, *Ann. Neurol.*, 26:647–651, 1989.

98. Tanner, C. M., and Langston, J. W., Do environmental toxins cause Parkinson's disease? A critical review, *Neurology,* 40(3):17–30, 1990.

99. Galvani, P., Fumagalli, P., and Santagostino, A., Vulnerability of mitochondrial complex I in PC12 cells exposed to manganese, *Euro. J. Pharmacol.,* 293(4):377–383, 1995.

100. Rodier, J., Manganese poisoning in Moroccan miners, *Brit. J. Ind. Med.,* 12:21–35, 1955.

101. Penalver, R., Diagnosis and treatment of manganese intoxication, *Arch. Ind. Health,* 16:64–66, 1957.

102. Mena, I., Marin, O., Fuenzalida, S., and Cotzias, G. C., Chronic manganese poisoning: Clinical picture and manganese turnover, *Neurology,* 17:128–136, 1967.

103. Chandra, S. V., Neurological consequences of manganese imbalance, in Dreosti, i.e., Smith, R. M., (eds), *Neurobiology of the Trace Elements, Vol. 2: Neurotoxicology and Neuropharmacology.* Humana Press, Clifton, pp. 167–196, 1983.

104. Menegon, A., Board, P. G., Blackburn, A. C., Mellick, G. D., Le Couteur, D. G., Parkinson's disease, pesticides, and glutathione transferase polymorphisms, *Lancet,* 352:1344–1246, 1998.

105. Savolainen, K. M., Loikkanen, J., Eerikainen, S., Naarala, J., Glutamte-stimulated ROS production in neuronal cultures: interaction with lead and cholinergic system, *Neurotoxicology,* 19:669–674, 1998.

106. Smargiassi, A., Mutti, A., De Rosa, A., De Palma, G., Negrotti, A., Calzetti, S., A case-control study of occupational and environmental risk factors for Parkinson's disease in the Emilia-Romagna region of Italy, *Neurotoxicology,* 19(4–5):709–712, 1998.

107. Priyadarshi, A., Khuder, S. A., Schaub, E. A., Shrivastava, S., A meta-analysis of Parkinson's disease and exposure to pesticides, *Neurotoxicology,* 21(4):435–440, 2000.

108. Calne, S., Schoenberg, B., Martin, W., Uitti, R. J., Spencer, P., Calne, D. B., Familial Parkinson's disease: possible role of environmental factors, *Can. J. Neurol. Sci.,* 14(3):303–305, 1987.

109. Logroscino, G., Marder, K., Cote, L., Tang, M. X., Shea, S., Mayeux, R., Dietary lipids and antioxidants in Parkinson's disease: a population-based, case-control study, *Ann. Neurol.,* 39(1):89–94, 1996.

110. Kuopio, A.M., Marttila, R. J., Helenius, H., Rinne, U. K., Changing epidemiology of Parkinson's disease in southwestern Finland, *Neurology,* 52:302–308, 1999.

111. Rajput, A. H., Offord, K. P., Beard, C. M., Kurland, L. T:, Epidemiology of Parkinsonism: Incidence, classification, and mortality, *Ann. Neurol.,* 16:278–282, 1984.

112. McCann, S. J., LeCouteur, D. G., Green, A. C., Brayne, C., Johnson, A. G., Chan, D., McManus, M. E., Pond, S. M.,The epidemiology of Parkinson's disease in an Australian population, *Neuroepidemiology,* 17(6):310–317, 1998.

113. Pezzoli, G., Canesi, M., Antonini, A., Righini, A., Perbellini, L., Barichella, M., Mariani, C. B., Tenconi, F., Tesei, S., Zecchinelli, A., Leenders, K. L., Hydrocarbon exposure and Parkinson's disease, *Neurology,* 55(5):667–673, 2000.

114. Racette, B. A., McGee-Minnich, L., Moerlein, S. M., Mink, J. W., Videen, T. O., Perlmutter, J. S., Welding-related parkinsonism: clinical features, treatment, and pathophysiology, *Neurology,* 56(1):8–13, 2001.

115. Martin, W. R., Ye, F. Q., Allen P. S., Increasing striatal iron content associated with normal aging, *Mov Disord.,* 13(2):281–286, 1998.

116. Anglade, P., Vyas, S., Hirsch, E. C., Agid, Y., Apoptosis in dopaminergic neurons of the human substantia nigra during normal aging, *Histol. Histopathol.,* 12(3): 603–610, 1997.

117. Kessler, I. I., Epidemiologic studies of Parkinson's disease, II, A hospital-based survey, *Am. J. Epidemiol.,* 95(4):308–318, 1972a.

118. Kessler, I. I., Epidemiologic studies of Parkinson's disease, 3, A community-based survey, *Am. J. Epidemiol.,* 96(4):242–254, 1972b.

119. Fitzmaurice, P. S., Ang, L., Guttman. M., Rajput, A. H., Furukawa, Y., Kish, S. J., Nigral glutathione deficiency is not specific for idiopathic Parkinson's disease, *Mov. Disord.* 18(9):969–976, 2003.

120. Thiruchelvam, M., McCormack, A., Richfield, E. K., Baggs, R. B., Tank, A.W., Di Monte, D.A., Cory-Slechta, D. A., Age-related irreversible progressive nigrostriatal dopaminergic neurotoxicity in the paraquat and maneb model of the Parkinson's disease phenotype. *Eur. J. Neurosci.,* 18(3):589–600, 2003.

121. Bharath, S., Hsu, M., Kaur, D., Rajagopalan, S., Andersen, J. K., Glutathione, iron and Parkinson's disease, *Biochem Pharmacol.,* 64(5–6):1037–1048, 2002.

122. Kaur, D., Yantiri, F., Rajagopalan, S., Kumar, J., Mo, J. Q., Boonplueang, R., Viswanath, V., Jacobs, R., Yang, L., Beal, M. F., DiMonte, D., Volitaskis, I., Ellerby, L., Cherny, R. A., Bush, A. I., Andersen, J. K., Genetic or pharmacological iron chelation prevents MPTP-induced neurotoxicity in vivo: a novel therapy for Parkinson's disease, *Neuron.,* 37(6):899–909, 2003.

123. Dekker, M. C., Giesbergen, P. C., Njajou, O. T., van Swieten, J. C., Hofman, A., Breteler, M. M., van Duijn, C. M., Mutations in the hemochromatosis gene (HFE), Parkinson's disease and parkinsonism, *Neurosci Lett.,* 348(2):117–119, 2003.

124. Gotz, M. E., Double, K., Gerlach, M., Youdim, M. B., Riederer, P., The relevance of iron in the pathogenesis of Parkinson's disease, *Ann. NY Acad. Sci.,* 1012:193–208, 2004.

125. Faucheux, B. A., Martin, M. E., Beaumont, C., Hauw J. J., Agid, Y., Hirsch, E. C., Neuromelanin associated redox-active iron is increased in the substantia nigra of patients with Parkinson's disease, *J. Neurochem.* 86(5):1142–1148, 2003.

7 Rural Environment and Parkinson's Disease

Daniel Strickland
Kaiser Permanente, Southern California

CONTENTS

INTRODUCTION

The perception and concern of a risk of Parkinson's disease associated with rural exposures was not considered until the finding of chemically generated (so to speak) acute Parkinson's disease by a garage lab-made street narcotic metabolite (MPTP) in the early 1980s.[1,2] Because of this chemical's resemblance to certain common herbicides, the essence of the investigations of "rural" exposure or "well-water" use is as a proxy measure for deleterious chemicals. Since the initial broad-stroke studies, we have moved from using that proxy exposure to see if the association exists to looking directly at the specific chemicals that MPTP story leads us to believe represents the underlying risk. Even more recently, difficulty in arriving at clear results from studies of pesticide exposure has led the PD research community to an interest in pesticide-gene interactions, but not further explorations of other agricultural exposures. Most likely, we are correct in this; that there really is no more to the story than the farming chemical or chemicals that are hazardous to dopaminergic brain cells, but since so many researchers have put so much effort into examining everything about farming and rural living, it seems wasteful not to look a little deeper and consider a little more the other possibilities. A few of those other possibilities are discussed in the course of reviewing the history of this literature, but there really is comparatively little work on rural exposures other than pesticides. Just to list them, some of the other possibilities include diet, occupational chemicals other than pesticides (such things as heavy metals, sawdust, engine fumes, paints, and so on), and zoonotic infectious agent exposures from farm animals. Farmers often have ample opportunity for exposure to all of these and, given that many farmers find it financially necessary to take supplementary jobs in the local area, they then have opportunity for more exposure to the types of industry, often quite dirty, commonly seen in rural areas.

Besides the alternative causal exposures, one might hypothesize that the conservative PD personality is the type of person who would choose to live in a rural setting, but that seems a stretch. A similar problem is that many of the case-control studies showing this association used prevalent cases, and hence one could suppose that the rural residence helps long-term PD survival somehow. A slightly more reasonable hypothesis is that, rather than rural living being harmful, there might be some protective factor operating in a nonrural environment. Perhaps city people drink significantly more coffee, for example, thus deriving protection.[3]

EARLY OBSERVATIONS

It is of note that, while PD in its later stages is readily noticeable and distinct, it was described in detail by James Parkinson in 1817,[4] as the Industrial Revolution was gathering speed. A number of other neurologic conditions had been well described before that time, so one might speculate that, until Parkinson's day, the disease was very rare.

Parkinson indeed noted conspicuous symptoms such as a gentleman who found it necessary to have a servant walk backward in front of him to keep him from falling forward. It was not long following Parkinson that PD was noted to be among the more common illnesses seen in neurologic practice.[5,6] The inference one could draw is that some environmental exposure that added considerably to the population risk of Parkinson's disease became more prevalent in the cities of the European Industrial Revolution. These may have now shifted to the rural areas as cities have become cleaner, although it may be that the rural exposures are not agriculturally related but were a result of some of the dirtier industries preferentially locating in more remote areas, whether to be near natural resources needed in that industry or to avoid the more restrictive regulatory environment of major cities. Plainly, Barbeau,[7] in his descriptive studies in Quebec described below, was concerned as much with such things as sawmills and paper plants as with agriculture. Several studies made use of anti-PD drug sales to look for associations with vegetable farming, sawmills, and metal foundries.[8,7] Rybicki and colleagues[9] used death certificates in Michigan to examine correlations with industry of varying sorts, finding increased PD mortality in counties with paper, chemical, iron, and copper related industries. There were examinations of the rural prevalence hypothesis by Granierie[10] and Svensen[11] in Italy and Alberta, Canada, with positive results in both cases.

THE MPTP STUDIES

The majority of descriptive studies imply a rural elevation of PD prevalence. Further concern about agricultural exposures arose from incidents with the designer drug breakdown product MPTP 1-methyl-4-phenyl-1,2,3,6-tetrahydropyridine), which is chemically related to the herbicide Paraquat and a number of other compounds that may be present in an agricultural or industrial environment.[1,2] MPTP is destructive to the substantia nigra, resulting in rapid onset of acute Parkinsonism, and it causes neuromuscular symptoms in animal models resulting from that destruction.[2] Those incidents have driven the search for an agricultural chemical cause for PD since.[2,12,13]

PRELIMINARY DESCRIPTIVE AND ECOLOGIC STUDIES

An early study in Finland[14] relied on the traditional two stage prevalence study design[15] to examine descriptive epidemiology issues but did not find a significant rural/urban difference, something they do not remark upon. It does not seem this was a major concern at the time, and slightly later works, in particular that of Ali Rajput,[16,17] cite the Langston work as a rationale for

believing rural living to be a concern. Later workers followed Rajput's lead in this.

Studies seeking to clarify etiologic or risk factors have therefore looked for the sources of apparent differences between urban and rural rates in descriptive studies. Following is a selection of the epidemiologic research on PD and agricultural factors that followed the MPTP incidents. The studies for the most part built upon each other and are therefore listed in chronological order. The community of PD epidemiology researchers, being small, communicates internally unusually well. The first studies were simply ecologic looks at rates of PD versus rural or industrial region (as the exposures). Earlier individual studies tended to use convenience or clinical samples of PD cases and broad definitions of exposure, typical of exploratory stages. Later studies used case-control design, more detailed exposure assessments, and broader case-finding techniques. Results are not always consistent, and design differences likely account for some of that.

Numerous case-control studies report elevation of risk in rural areas (for example, Rajput, 1986 and 1987; Ho, 1989; Koller, 1990; Tanner, 1986; and Svenson, 1993).[11,17-21] A few report elevated risk in urban areas (Tanner, 1989; Zayed, 1990; Butterfield, 1993),[22-24] and there are reports of no association (Stern, 1991, and Semchuk, 1991).[25,26] Part of these differences is likely a result of geographic differences (comparing China versus Kansas, for instance), and part is due to inadequate statistical power or other methodologic problems, but part remains unexplained. The balance of opinion favors a risk associated with rural living in developed countries and city living in poorer countries (suggested by Tanner, 1990).[13]

The most important of the early studies following the MPTP papers was by Barbeau,[12,27] who compared detailed use and sales data for pesticides in Quebec, Canada, with prevalence of PD, determined by four methods.[27] These included reports to physicians, sales data for anti-PD medications, examination of death certificates, and contact of major movement disorder clinics and neurologists practicing within 100 miles of Montreal for neurologically confirmed cases. These latter clinicians were contacted to overcome the deficiency of the first methods, namely diagnostic accuracy. A final group of clinicians, those in general practice in one of the areas of high prevalence, were also contacted to confirm the diagnosis in their PD patients. The investigators' choice of this area, Quebec Province, cited the advantages of genetic homogeneity, equal access to medical care, and an even distribution of age groups. Prevalence estimates of Parkinson's disease as defined by these three methods in the province ranged from 0.834 to 0.967 per 1,000. The areas found to have the highest relative prevalence were those with the most land in intensive high-irrigation and chemical use agriculture. Of note, the investigators reported results by hydrographic regions rather than political divisions, stating that this gives a better depiction of environmental influences. While they seem unaware of the problem of attributing group-level exposures to population risk of disease, they do discuss the possibility of other unmeasured causal agents entering into the associations observed. They found an uneven prevalence within rural areas, contrary to previous reports at the time. After examination of possible confounders such as age and gender structure or reporting and diagnostic differences, they conclude that differences in agricultural practice and chemical use explain a large proportion of the prevalence differences observed.

The next Quebec study used individual patients rather than grouped data. Zayed et al.[23] chose southern Quebec, Canada, in which to assess environmental risks in PD patients because of its stable population, mixed economy including farming, orchards, and metallurgic industries, and indications of an elevated prevalence of PD found earlier by Barbeau.[27] Cases ($n = 42$) were found by contacting local primary care practitioners and neurologic consultants by mail and by publicity in local periodicals. Controls were drawn using municipal phone books, matched two to one on age, sex, and municipality and examined for neurologic illness. Rural residence was protective [odds ratio (OR) = 0.31]. However, there was a significant tendency for risk to increase with increasing exposure (i.e., longer residence in those areas). Some elevation was seen of risk associated with pesticide use. While Barbeau's study was ecologic (group-based), Zayed's was an individual-based study. Possible interpretations of the apparent contradiction include study design differences, including the possibility that Zayed overmatched and thereby reduced the effect of residence.s

Rajput and colleagues[14] focused on early-onset (younger than age 40) PD in Saskatchewan, Canada. An important assumption that was usually explicitly stated by the investigators who studied early-onset PD was that it was etiologically the same as normal-onset disease, differing primarily in intensity of exposure or perhaps susceptibility to the exposures of interest. In any event, it was hoped that the younger subjects would be closer in time to the cause of the PD and hence better able to report such exposures. A complete assessment of drinking water, use of farm chemicals, and early onset PD was done on all early onset cases located in earlier work.[17] Agricultural chemical use records were obtained from provincial records and from experts in the field at the University of Saskatchewan. Early-onset PD did not seem to rise in incidence with the use and introduction of agricultural chemicals. The weakness of this technique is the imputation of very broad group-level exposures to individuals with disease. They did nevertheless find an association of early-onset PD with both rural living and well-water use in childhood. A strength of this study is the set-ting—Saskatchewan, Canada, which is a rural farming province with a very stable population, making these broad exposures easy to assess with accuracy. As might be expected, however, sample size was quite low, and Rajput did not describe the source of his control group other than to mention matching by age and sex and some exclusionary criteria. It might be reasonably expected that, in a place such as Saskatchewan, Parkinson's disease clinics might be so limited in number that were the controls to develop the disease, they would appear at Dr. Rajput's clinic and thereby have the opportunity to be included as case subjects in this study.

Sethi and colleagues[28] took note of the elevation of PD risk reported by Barbeau[27] in areas containing paper mills. To attempt to confirm that finding, they obtained death certificates for the U.S. state of Georgia and the total U.S.A., years 1979 through 1983. Unfortunately, they had available only primary and underlying causes of death and, as they acknowledged, the majority of people dying with PD do not have the PD noted on those first two causes. Therefore, their sample was biased, and that may have contributed to their inability to show an association of PD either with residence in counties that had paper mills or in rural areas.

A late-comer amongst the descriptive studies used administrative data in the province of Alberta, Canada[11] (Svenson). The design differed from the usual cross-sectional study in that the investigators included all enrollees in the Alberta Health Care Insurance Plan, which includes the physician billing information for all "registered residents" of Alberta. The cohort's inception was April 1, 1984, and they were followed until March 31, 1989. It stayed closed during the period, during which 6% attrition was observed. All diagnoses of PD were extracted and described demographically. The investigators estimated both prevalence and "morbidity," or incidence, as a standardized morbidity ratio. Excesses of both prevalence and morbidity were observed for census divisions outside Alberta's two large cities, while urban areas were significantly reduced. They discuss the usual precautions about misclassification of PD diagnosis and an inability to detect undiagnosed PD. As an editorial comment, it seems a little odd that this work apparently was done independently of the Semchuk[26] research, which was in preparation at about the same time, also in Alberta. These authors are located in Edmonton and are government employees, while Semchuk and her colleagues are academics located at that time in Calgary.

The balance of opinion favors a risk associated with rural living in developed countries and city living in poorer countries (suggested by Tanner, 1990.)[13] A question not fully addressed by the early studies is precisely what is meant by "rural living." Does it refer to farming, cattle ranching, or living in the deep woods? Barbeau and Zayed both differentiated rural areas into farming and areas of rural industries, but ranches have not, to this writer's knowledge, been examined for association with PD, and the size of a community before it is termed "rural" has had only cursory examination.

INTERNATIONAL STUDIES

Some of the more interesting approaches to the question entail looking at what rural living means in non-Western societies with regards to Parkinson's disease.

Tanner[21] tested Rajput's findings, questioning 95 outpatients with PD regarding well-water use. She found that those with PD onset before age 47 reported well-water use significantly more often. Following that work, she made use of recent industrialization and stable population in China,[29] specifically in areas around Beijing and Guangzhou. She hypothesized that more recent industrialization would result in lower rates of PD and that patients would have correspondingly higher exposure to those chemicals or processes that increase risk of PD. Patients were recruited from the neurology outpatient clinics of university teaching hospitals in both cities and controls were nonmovement disorder patients from the same clinics, matched on age, sex, and institution. There was a total of 23 women and 77 men among the cases, mean age of 57.2, and mean age of onset was 51.5. The population was quite stable, with a mean number of life places of residence of 2.5 for cases and 2.3 for controls. Chemical and industrial exposures were assessed by interview, and elevated risks were found of 2.39 for exposure to industrial chemical plants and 2.40 to printing plants as well as for exposure to rock quarries (4.50), although confidence limits were wide (1.35 to 15.0) in the last case. Decreased risks were associated with living in villages (OR = 0.57), exposure to pig raising (OR = 0.17), chicken raising (OR = 0.53), and wheat growing (OR = 0.40). Those PD cases with exposure to chemical, pharmaceutical, herbicide, or pesticide industries did not differ from PD cases without such exposures on age of onset or mean duration of disease. However, all such exposures were of fairly recent onset. She comments that the rural exposures found significant are ubiquitous in China and could represent village residence or some other as-yet unassessed factor, either positive or negative. Both swine and poultry are a source of many human zoonoses, and other authors have speculated on a possible protective effect of early infection with childhood diseases. However, at least one longitudinal study of a long-term cohort with careful exposure assessment[30] found childhood infections to be protective, similar to Tanner's finding for small livestock exposures.

Ho et al.[19] surveyed the residents of eight homes for the elderly in the immediate area of a university teaching hospital in Hong Kong. All patients found by this method and all prevalent patients under care at the University Geriatric Clinic were combined to form a case group and compared to controls matched three to one on age and sex. Controls were drawn from the University Geriatric Clinic and the homes for the elderly in the survey. Total study size for the case-control portion was 35 cases and 105 controls, age ranging from 65 to 87, 69% women. The authors found living in rural areas over 40 years yielded an odds ratio (OR) of 4.9 of PD, while greater than 20 years of farming was significant at 5.2. Also significantly elevated were history of pesticide and herbicide use (OR = 3.6), and the authors remark upon an odds ratio of 11 for eating raw vegetables, which had, however, a 95% confidence interval of 2.4 to 40.0. Significant trends were seen of increasing PD risk with increased years living in rural areas and years occupied in farming. The authors contrast their study with Tanner's[29] and postulate that geographic (mainland China versus Hong Kong) and age differences account for differing results. It seems likelier that case-finding differences might explain the contrast, although in fairness, Ho does devote some time to delineating farming differences that could also contribute. These include the year-round farming season, the warmer weather and thus lighter clothing including bare feet, the farming of more vegetables than grains which may have higher pesticide use, and consumption of fresh vegetables usually bought daily, typically not well washed and often eaten raw. These add up to a plausible list of alternative hypotheses, but methodologic differences could explain the majority of the variance between Ho's and Tanner's findings.

NORTH AMERICAN CASE-CONTROL STUDIES

Agricultural risk factors were investigated in a case-control study in Kansas.[20] The source for the 150 cases was the prevalent patient population of the Movement Disorder Clinic of the University of Kansas Medical Center and, for controls, the neurologic and medical clinics of the same institution. An equal number of controls were matched on age and sex. Subjects were asked to check their records of herbicide/pesticide use to verify the information. These authors used an earlier version of the Blair questionnaire that had been used in a study of soft-tissue cancers in Kansas.[31] Cases had significantly greater number of years of rural residence and well-water drinking, and it was found that these two variables interacted. The odds ratios (McNemar's) were 1.9 (rural living) and 1.7 (well-water). Most of the subjects lived in Kansas City at the time of the study. The authors report no difference for farming history and pesticide/herbicide use.

This group pursued their ideas with a different study design.[32] They chose 19 patients from their PD clinic who also had a sibling with PD, 19 more such sibling-pairs with essential tremor rather than PD, and 38 control subjects from the general neurology and internal medicine clinics. Cases and controls were age and sex matched, and the controls had diagnoses such as headache, back pain, arthritis, and heart disease. They again used the Blair questionnaire[33] and found differences between the PD cases and controls on rural residence (or 4.3, $p < 0.01$) and well-water drinking (or 2.9, $p < 0.07$). There was a significant dose-response, i.e., number of years spent in rural residence and drinking well water as well. There was no difference on farm chemical exposure, nor was there a difference between the PD patients and the essential tremor patients.

Both of the Kansas studies drew from a selected population, i.e., their PD clinic, and may suffer bias because of that—although the group appear to have tried to draw controls from the base population of the cases. The sibling study has small numbers, and the power to detect differences may be limited, but both studies found essentially the same result. Variations on the Blair questionnaire have been used by most PD investigators and indeed are still in use—for example, in the studies by Tanner and colleagues in Iowa and North Carolina. It is a well validated instrument,[31,33,34] and that the Koller group might have missed a strong pesticide association twice seems unlikely.

A comparison of young-onset versus old-onset PD patients in a case-control design was conducted by Stern.[25] Cases were 161 prevalent PD patients from the Graduate Hospital Department of Neurology in Philadelphia and the Department of Neurology of the University of Medicine and Dentistry of New Jersey in New Brunswick. Controls, matched on age, gender, race, and geographic area, were drawn using peer nomination. If nominated controls could not be matched or the case subject could not nominate any, controls were drawn from the patients and hospital volunteers at the Philadelphia hospital. The study design was stratified on young-onset (those having their first symptoms before age 40) versus old-onset (those having their first symptoms after age 59). The authors hypothesized that environmental causes would be revealed most strongly in the background of those who developed symptoms earliest. Exposures assessed included residential history, water supply, population density, and exposure to herbicides and pesticides. Rural living was associated with PD with an odds ratio of 1.7, but chance could not be excluded (95% CI—1.0, 2.9), and the same was true of well water, with an odds ratio of 1.8 (1.0, 3.4). Reporting of pesticide exposure was even across groups. The two different age groupings did not show different associations with the exposures measured. The authors hypothesize that their study subjects are too urbanized—i.e., pesticide or other rural exposures are too uncommon in this group—as an explanation for their finding of no elevation in risk for rural living, well water, or pesticides and herbicides, in contrast to other studies.

There were problems as well with possible selection bias in the case group (since these were prevalent cases assembled in clinics) and with the size of the study sample, although most of the results reported combined the two age strata. Nevertheless, the minimum sample required to detect an odds ratio of 2.0 is 130, and while the study sample was larger at 166, the odds ratios they report could not exclude chance as an explanation of their findings.

The first population-based case-control study of PD was done in Alberta, Canada, by Semchuk and colleagues.[26,35,36] Case finding techniques included hospital discharge records search for any diagnosis of PD, a search of a listing of all long-term care residents, listing of all PD patients from the Movement Disorder Clinic of the University of Calgary and the 16 neurologists practicing in Calgary, and finally a roster from the Parkinson's Society of Southern Alberta. After merging these files, chart review was done to confirm diagnosis of PD by a neurologist. After exclusions, final numbers were 130 cases and 260 controls. Two community controls matched on age and sex were drawn using random digit dialing sampling, with no matching on phone numbers other than restriction to Calgary exchanges. For those 23 cases who were in long-term care facilities, a third control was randomly selected from the total census of all 6 long-term care facilities in which the cases resided. Exposure assessment was by questionnaire, which Semchuk wisely obtained from Blair,[33] so that her results can be compared easily to most of the other PD case-control studies.

First, they examined the risk factors that previous studies had evaluated.[26] These included rural living, farm living, and well-water exposure during each five-year age period up to the 45th birthday, and during exposure periods of increasing duration (by five-year increments for the first 45 years of life). They found no differences between cases and controls on these three variables, with a large proportion of both groups responding positively to all three questions. The authors discuss the possibility that there is geographic variation in the association of PD and broad rural factors. They point out the large proportion of controls (61.1%) who had some exposure to rural living. They conclude that ruralness itself is likely not a strong risk factor, and any causative agent must be something particular to the rural environment that varies geographically (and perhaps with time as well). The difficulty with her interpretation of the weak association is that strength of association varies inversely with prevalence,[37] so that in an environment where the exposure of interest is common, the association with disease will be weak, and other rarer etiologic components will appear stronger. Thus, in the United States, the association of dietary fat consumption and cardiovascular disease is weak, and correspondingly, a family history of heart illness is a strong risk factor, whereas in environments where the usual diet is much less fatty, those relationships are reversed.[38] Semchuk is correct to think that "farm living," "rural living," and perhaps even well-water consumption represent proxy measures of a more direct hazard, and in fact the environment where farming is common is exactly the place to be able to tease out the specifics. Koutsky[39] used this to good effect when determining the association of human papilloma virus and cervical carcinoma when she used the STD clinic of a large public hospital as her sampling frame, drawing both cases and controls from among the patients of that clinic.

Our study of Parkinson's disease in Nebraska[40] used a design very similar to Semchuk's (including use of a slightly later version of the Blair questionnaire), although relying on drawing case subjects from a passive registry[41] rather than from government medical records. We found that 58% of controls and 65% of case subjects stated they had lived on a farm for at least a year, although our estimated odds ratio of 1.3 did reach borderline statistical significance. Here again, we argue that this particular question is no longer important. There is wide agreement that, in many environments, there is an increased risk of Parkinson's disease associated with rural living; the matter of greater import is teasing out what underlying exposures—and there are likely to be a number—play a direct role in contribution to that increased risk.

Semchuk next[35] looked at the more specific exposures. Univariate analysis showed increased PD risk for agricultural work, herbicide use, and pesticide use. Multivariate analysis led to the conclusion that the significant risk factor was herbicide exposure. Odds ratio estimates stayed around 3, despite adjustment for a variety of other correlates and covariates including type of farming and pesticide and fungicide use. Although only 41% of the exposed cases could recall names of herbicides, all who could recalled using compounds in the chlorophenoxy and thiocarbonate group. Semchuk cautiously does not make a statement that these compounds are causal, on the grounds that her study lacked statistical power to ascertain dose-response relation.

In the third paper examining this data set[36] Semchuk used multivariate techniques to evaluate interactions and control confounding amongst the risk factors she measured. Significantly elevated odds ratios included herbicide use (OR = 3.09). Calculation of population attributable risk showed family history of PD and head trauma together explaining 22% of cases, while herbicide use explained 10%.

Butterfield[24] developed a case group of young-onset PD patients in Oregon and Washington by contacting all American Academy of Neurology directory-listed neurologists and young-onset PD patient-support groups affiliated with Oregon Health Sciences University (located in Portland) and University of Washington (located in Seattle). The authors contacted 100 PD patients with the exposure history questionnaire, of whom 28 refused. Controls

were rheumatoid arthritis patients. Of 168 possible controls, 100 refused to participate. Ever living on a farm and well-water use both had protective odds ratios for PD in this study. However, PD patients with farm employment or residency had longer exposure than controls, and while rural residency at birth was slightly protective, rural residence at diagnosis showed an elevation of risk, at 2.72, $p = 0.0027$. Crude odds ratios were found of 3.22 for herbicide exposure and 5.75 for insecticide exposure. Multivariate analysis showed differences for insecticide exposure (OR = 4.3). Herbicide exposure was not included because of colinearity with insecticide exposure, although this is a situation where the associations should be assessed independently. Two other variables were not significant by Walds test but were by the likelihood ratio chi-square. These were rural residence at time of diagnosis (OR = 2.35) and residence in a house that was fumigated (OR = 4.57). Although findings are in reasonable accord with other studies, response bias and lack of a community control group raise concerns and may have affected the results.

The Kansas group[42] reexamined the hypotheses of rural living, well-water consumption, and pesticide exposure using two sets of cases and controls: one set urban-based and the other drawn from a rural community in western Kansas. The techniques used to recruit patients in Kansas City (PD clinic at the University Hospital and PD support groups) may be subject to referral bias, but in the rural setting, recruitment through a PD outreach clinic in a rural clinic implies that the researchers may have recruited nearly all the PD cases in that area. Unfortunately, controls were recruited in the rural setting from two media releases and a senior citizens' lunch program, ensuring nonrepresentativeness, whereas, in Kansas City, they came from the University Hospital neurology clinic and case-nominated nonfamily controls (about one third of the controls). Therefore, the city-based controls were mostly hospital controls, unlike the rural controls. They found that general pesticide exposure and family history of neurologic disease were significantly associated with PD. It appears that details about specific pesticides and herbicides were not solicited.

Likewise, the British Columbia group[43] developed a solid toolbox of techniques for examination of exposures related to orchard work, which they had found to be associated with PD in their initial study (Snow, 1989).[44] This included selecting a region where farm chemicals use is common (ensuring that respondents will know their exposures better) and where detailed farm chemical use records are kept by a provincial office with dates of marketing, and the use of cue cards and full descriptions of chemicals as memory aids. To further reduce recall bias, two control groups were selected: one from voter rolls to represent the general community and one a group with chronic cardiac disease. Exposure history to 79 agricultural chemicals was asked, and associations with PD were tested for individual chemicals and grouped in several ways. The odds ratios for occupational exposure to pesticides was significantly elevated in white males (OR versus cardiac controls: 2.03) (OR vs. community controls: 2.32), and elevated but not significantly so in women (OR versus cardiac controls: 1.11) (OR versus community controls: 1.36). Despite this, the authors were unable to demonstrate a clear risk for any farming chemicals singly or grouped. In their discussion, they cover the usual questions of possible lack of power and biases and conclude that those matters are not a major difficulty; indeed, they seem correct in that. They suggest either PD is a result of combined causes (and that is certainly where the field is going at the moment with the emphasis on gene-environment interactions) or that we have been looking at the wrong rural risk factors, giving as example a common soil pathogen, *Nocardia asteroides*, which has been demonstrated to cause an L-dopa-responsive movement disorder in a mouse model.[45] This possibility was explored further in a case-control study[46] with no difference found in *Nocardia* titer elevations between cases and controls. Very little work on this exposure has been done since.

An active group of PD researchers in Spain drew case subjects from the population of a largely rural province with residents who tended to stay in one place for very long periods.[47] The province has neurology service in only two hospitals, and newly diagnosed patients attending clinics in those hospitals served as case subjects. There were 74 such patients enrolled, and these were compared with control subjects who presented to the emergency rooms with minor non-neurologic complaints or to the neurology clinics with "functional" CNS pathology such as headache. Questionnaires were given covering a spectrum of exposures, and differences were significant for long exposure (more than 50 years) to both rural environments and well water. The authors did not estimate odds ratios but cite elevated proportions. For both exposures, odds ratios would be between 1.5 and 2.0. Other exposures that might be of concern, especially chemical toxins of various types, did not show a strong association, although pesticide use approached significance. It is of interest that these authors were able to show an association of PD with rural living despite drawing subjects from a highly rural population. However, the associations were seen only with very long residence, raising concerns about competing causes of mortality (i.e., those who lived long enough to develop PD had the long rural exposure simply by virtue of having avoided, say, cardiovascular disease) and about confounding of age and rural residence. The latter would be a result of those who choose to live for long periods in rural areas, perhaps on a specific farm, being older than those who are more mobile and hence more at risk of PD. Age adjustment should take care of some of this, but the cross-sectional

nature of the sampling may have introduced the bias discussed earlier—meaning that the case group will disproportionately represent longer-lived, older PD patients. If older age is associated with fewer changes of residence and rural residence, then it will appear that PD is associated with rural residence, even when adjustments for age are made.

Paffenbarger has the privilege of being investigator of a long-term cohort that had been particularly well-described at baseline: a group of graduates from Harvard University and a similar, smaller group from University of Pennsylvania.[30] These students, all males, were given thorough medical exams when they entered college between 1916 and 1950; there were 50,002 subjects in all. Sasco and Paffenbarger[30] chose to pursue a case-control design of PD, both for economy and to examine a number of different exposures. They were able to locate and recruit 137 PD cases, matching by age of college entry, year of college entry, and institution attended to four controls from the PD-free members of the cohort. PD was "validated" by confirmation of the self-report by a physician or mention on a death certificate. This poses obvious potential for misclassification, of which the authors are aware. They therefore take some pains to delineate the possible effects of such misclassification and conclude that the odds ratio would be pushed toward the null. The college entry exam included physiologic parameters and information on childhood illnesses, as well as health habits such as smoking, drinking, and exercise. The authors found protective effects for viral diseases, although chance could be excluded only in the case of measles—hence the title of the paper. Multivariate techniques, although not easily available at the time, were used to test independence of effects and confounding, and measles infection history proved remarkably stable with an odds ratio around 0.5. The other viral infections ranged from about 0.5 to about 0.8. The minimum sample size necessary to exclude chance (with alpha of 0.05 and power of 80%) when the true odds ratio is 2.0 or 0.5 is 130, so Sasco and Paffenbarger are on the edge of that range, especially with the probable misclassifications. Nevertheless, the evidence implies a protective effect of viral infection in childhood for PD. This could also explain the protective effect, of about the same magnitude, of chicken or hog ownership.[29,41]

Diet has been explored by several workers, with a focus usually on antioxidants, alpha-tocopherol in particular. Golbe[48,49] showed a protective effect of consumption of a diet higher in alpha-tocopherol, and this was followed by Tanner's exploration[50] of multivitamin supplement use, vitamin E use, and cod-liver oil intake also showing protection. Later prospective collection of dietary data[51] failed to show an association. A few other studies followed,[22,52,53] but the PD research community seems to have reached a consensus that determining dietary habits prior to true disease onset in an unbiased fashion is very difficult and perhaps not worth the pursuit.

CONCLUSION

Indeed, the lack of interest in pursuing the dietary hypothesis is representative of the strong focus the field currently has on the genetic and pesticide interaction hypothesis. While this seems like a very promising line of research, all the investigators in the cited hypothesis testing studies have great amounts of data on all factors of ruralness, most of which will never see publication. Perhaps this is appropriate, given the time involved in analyzing what may eventually prove to be a false trail. There may be, however, overlooked associations buried in these collections of data that could suggest other avenues for more rigorous testing. It would be most unfortunate if potentially valuable resources of information on the etiology of PD are neglected.

REFERENCES

1. Langston, J. W., Ballaro, P. and Tetrud, J. W., Chronic Parkinsonism in humans due to a product of mepridine-analog synthesis, *Science*, 219, 979, 1983.
2. Langston, J. W. and Irwin, I., MPTP: current concepts and controversies, *Clin. Neuropharmacol.*, 9, 485, 1986.
3. Ross, G. W. et al., Association of coffee and caffeine intake with the risk of Parkinson disease, *JAMA*, 283(20), 2674, 2000.
4. Parkinson, J., *An Essay On The Shaking Palsy,* London: Sherwood, Neely and Jones, 1817.
5. Charcot, J. M., *Lectures on the Diseases of The Nervous System,* Vol. 1, Sigerson, G., trans., London, The New Sydenham Society, 1878.
6. Gowers, W. R., *Diseases of the Nervous System*, American Ed., Philadelphia, P. Blakiston, Son and Co, 1888.
7. Barbeau, A. and Roy, M., Uneven prevalence of Parkinson's disease in the province of Quebec, *Can. J. Neurol. Sci.*, 12, 169, 1985b.
8. Aquilonius, S. M. and Hartvig, P., A Swedish county with unexpectedly high utilization of anti-parkinsonian drugs, *Acta Neurol. Scand.*, 74, 379, 1986.
9. Rybicki, B. A. et al., Parkinson's disease mortality and the industrial use of heavy metals in Michigan, *Mov. Disord.*, 8, 87, 1993.
10. Granieri, E. et al., Parkinson's disease in Ferrara, Italy, 1967 through 1987, *Arch. Neurol.*, 48, 854, 1991.
11. Svenson, L. W., Platt, G. H. and Woodhead, S. E., Geographic variations in the prevalence rates of Parkinson's disease in Alberta, *Can. J. Neurol. Sci.*, 20, 307, 1993.
12. Barbeau, A. et al., Environmental and genetic factors in the etiology of Parkinson's Disease, *Adv. Neurol.*, 45, 299, 1985a.
13. Tanner, C. M. and Langston, J. W., Do environmental toxins cause Parkinson's disease? A critical review, *Neurology*, 40 (suppl. 3), 17, 1990.
14. Marttila, R. J. and Rinne, U. K., Epidemiology of Parkinson's Disease in Finland, *Acta Neurol. Scandinav.*, 53, 81, 1976.
15. Bermejo, F. et al., Problems and issues with door-to-door, two-phase surveys: An illustration from central Spain, *Neuroepidemiology* 20, 225, 2001.
16. Rajput, A.H. et al., Etiology of Parkinson's disease: environmental factor(s), *Neurology*, 34 (Suppl. I, 1), 207, 1984.
17. Rajput, A. H. et al., Early onset Parkinson's disease and childhood environment, *Adv. Neurol.*, 45, 295, 1986.
18. Rajput, A.H. et al., Geography, drinking water chemistry, pesticides and herbicides and the etiology of Parkinson's Disease, *Can. J. Neurol. Sci.*, 14, 414, 1987.
19. Ho, S. C., Woo, J. and Lee, C. M., Epidemiologic study of Parkinson's disease in Hong Kong, *Neurology*, 39, 1314, 1989.
20. Koller, W. et al., Environmental risk factors in Parkinson's disease, *Neurology*, 40, 1218, 1990.
21. Tanner, C. M., Influence of environmental factors on the onset of Parkinson's disease, *Neurology*, 36 (suppl.), 215, 1986.
22. Tanner, C. M. et al., Dietary antioxidant vitamins and the risk of developing Parkinson's disease, *Neurology* 39 (suppl.), 181, 1989.
23. Zayed, J. et al., Facteurs environnementaux dans l'étiologie de la maladie de Parkinson, *Can. J. Neurol. Sci.*, 17, 286, 1990.
24. Butterfield, P. G. et al., Environmental antecedents of young-onset Parkinson's disease, *Neurology*, 43, 1150, 1993.
25. Stern, M. et al., The epidemiology of Parkinson's disease: a case-control study of young-onset and old-onset patients, *Arch. Neurol.*, 48, 903, 1991.
26. Semchuk, K. M., Love, E. J. and Lee, R. G., Parkinson's disease and exposure to rural environmental factors: A population based case-control study, *Can. J. Neurol. Sci.*, 18, 279, 1991.
27. Barbeau, A. et al., Ecogenetics of Parkinson's Disease: Prevalence and environmental aspects in rural areas, *Can. J. Neurol. Sci.*, 14, 36, 1987.
28. Sethi, K. et al., Neuroepidemiology of Parkinson's disease: Analysis of mortality data for the U.S.A. and Georgia, *Intern. J. Neuroscience*, 46, 87, 1989.
29. Tanner, C. M. et al., Environmental factors and Parkinson's disease: A case-control study in China, *Neurology*, 39, 660, 1989.
30. Sasco, A. J. and Paffenbarger, R. S., Jr., Measles infection and Parkinson's disease, *Am. J. Epidemiol.*, 122(6), 1017, 1985.
31. Hoar, S. K. et al., Agricultural herbicide use and the risk of lymphoma and soft tissue sarcoma, *JAMA*, 256, 1141, 1986.
32. Wong, G. F. et al., Environmental risk factors in siblings with Parkinson's Disease, *Arch. Neurol.*, 48, 287, 1991.
33. Blair, A. and Zahm, S. H., Methodologic issues in exposure assessment for case-control studies of cancer and herbicides, *Am. J. Ind. Med.*, 18, 285, 1990.

34. Blair, A. and Zahm, S. H., Patterns of pesticide use among farmers: implications for epidemiologic research, *Epidemiology*, 4, 55, 1993.

35. Semchuk, K. M., Love, E. J. and Lee, R. G., Parkinson's disease and exposure to agricultural work and pesticide chemicals, *Neurology*, 42, 1328, 1992.

36. Semchuk, K. M., Love, E. J. and Lee, R. G., Parkinson's disease: A test of the multifactorial etiologic hypothesis, *Neurology* 43, 1173, 1993.

37. Rothman, K. J., *Modern Epidemiology*, Little, Brown and Co., Boston/Toronto, 1986.

38. Rose, G., Sick individuals and sick populations, *Int. J. Epidemiol.*, 14 (1), 32, 1985.

39. Koutsky, L. A., Galloway, D. A. and Holmes, K. K., Epidemiology of genital human papillomavirus infection, *Epidemiol. Rev.*, 10, 122, 1988.

40. Strickland, D. and Bertoni, J. M., Epidemiology of Parkinson's Disease and Farm Risk Factors, manuscripts in preparation, 2003a.

41. Strickland, D. and Bertoni, J. M., Parkinson's Prevalence Estimated by a State Registry, *Movement Disorders*, in press 2003b.

42. Hubble, J. P. et al., Risk factors for Parkinson's disease, *Neurology*, 43, 1693, 1993.

43. Hertzman, C. et al., A case-control study of Parkinson's Disease in a horticultural region of British Columbia, *Movement Disorders*, 9, 69, 1994.

44. Snow, B. et al., A community survey of Parkinson's disease, *CMAJ*, 141, 418, 1989.

45. Kohbata, S. and Beaman, B., L-dopa-responsive movement disorder caused by *Nocardia asteroides* localized in the brains of mice, *Infect. Immun.*, 59, 181, 1991.

46. Hubble, J. P. et al., *Nocardia* species as an etiologic agent in Parkinson's disease; serologocal testing in a case-control study, *J. Clin. Microbiol.*, 33, 2768, 1995.

47. Morano, A. et al., Risk-factors for Parkinson's disease: case-control study in the province of Cáceres, Spain, *Acta Neurol. Scand.*, 89, 164, 1994.

48. Golbe, L. L., Farrell, T. M. and Davis, P. H., Case-control study of early life dietary factors in Parkinson's disease, *Arch. Neurol.*, 45, 350, 1988.

49. Golbe, L. L., Farrell, T. M. and Davis, P. H., Follow-up study of early life protective and risk factors in Parkinson's disease, *Mov. Disorder*, 566, 1990.

50. Tanner, C. M. et al., Vitamin use and Parkinson's disease, *Ann. Neurol.*, 233, 182, 1988.

51. Morens, D. M. et al., Epidemiologic observations on Parkinson's disease: incidence and mortality in a prospective study of middle-aged men, *Neurology*, 46, 1044, 1996.

52. Logroscino, G. et al., Dietary lipids and antioxidants in Parkinson's disease: a population-based, case-control study, *Ann. Neurol.* 39, 89, 1996.

53. Hellenbrand, W. et al., Diet and Parkinson's disease: A possible role for the past intake of specific nutrients: results from a self-administered food-frequency questionnaire in a case-control study, *Neurology* 47, 644, 1996.

8 Industrial and Occupational Exposures and Parkinson's Disease

Brad A. Racette
Department of Neurology, Washington University School of Medicine, and American Parkinson Disease Association Advanced Center for Parkinson Research

CONTENTS

INDUSTRIES AND OCCUPATIONS ASSOCIATED WITH PD

Epidemiologic studies have provided contradictory evidence regarding etiologic risks for PD. No area more clearly demonstrates these contradictions than in the area of occupational and environmental risk factors and PD. Numerous studies demonstrate a higher risk of PD for individuals living in a rural environment;[1,2] however, these findings are inconsistent, possibly due to lack of power in the negative studies.[3–6] On the other hand, there is some evidence that the prevalence of PD may be higher in industrialized countries. Schoenberg et al. compared the prevalence of PD in Copiah County, Mississippi, (341/100,000 over age 39) to Igbo-Ora, Nigeria (67/100,000 over age 39), using similar methodology and studying genetically similar populations. They concluded that environmental factors may be responsible for the observed higher prevalence in the industrialized US population.[7] Similarly, a door-to-door epidemiologic study of PD in China found a prevalence of 57/100,000.[8] Presumably, the degree of U.S. industrialization may account for some of this difference. However, a small study based in a health district in Canada found a lower risk of PD in industrialized areas of the district.[5] In a population based mortality study, Rybicki et al. demonstrated that counties in Michigan with a higher concentration of industries with potential for heavy-metal exposures (iron, zinc, copper, mercury, magnesium, and manganese) had a higher PD death rates.[9] Using levodopa prescription records as a surrogate for PD, two studies have shown an increased risk of PD in areas with prominent employment in wood pulp and steel alloy industries.[10] Some potential confounds to the surrogate diagnosis and study methodology include inclusion of non-PD phenocopies and inability to separate working in an environment from living in an environment. If increasing world industrialization is a risk factor for PD, the incidence should be increasing throughout the last century. Only one study has addressed the incidence of PD over time. The yearly incidence of PD has not significantly changed between 1955 and 1970 in Rochester, Minnesota.[11] However, it is unlikely that there has been a substantial change in the industrialization of this relatively rural community over that period of time. No preindustrial epidemiology studies of PD exist, and many cases of PD likely went unrecognized in the beginning of industrialization in this country. It may be possible to reconcile these contradictory data with more attention to regional differences in industrial pollution and farming practices.

SPECIFIC INDUSTRIAL OCCUPATIONS

Epidemiologic studies have also been performed to attempt to find occupations at high risk for developing PD. Fall et al. performed an occupation case control study and found an increased risk of PD in carpenters [odds ratio (OR) = 3.9], cabinet makers (OR = 11), and cleaners (OR = 6.7) compared to a population-based control group.[12] Tanner et al. performed a case control study (nonpopulation-based) of occupational exposures and PD in China and found that occupations involving industrial chemical plants (OR = 2.39), printing plants (OR = 2.40), and quarries (OR = 4.50) were associated with a higher risk of PD.[13] Unfortunately, no detailed occupational information was provided. A population-based survey of PD in British Columbia found an association between PD and working in an orchard (AOR = 2.30) or planer mill (AOR = 4.97).[14] They hypothesized that industrial chemicals, including pesticides and herbicides, could be etiologic agents. Another nonpopulation-based case-control study in the same region found that occupational categories including forestry, logging, mining, and oil/gas field work had the highest odds ratio (3.79) of any occupation studied.[15] Although referral bias of affected subjects may have influenced the results, the number of subjects studied (n = 414) was substantial. Another nonpopulation-based case control study in the Emilia-Romagna region of Italy found that occupational exposure to "industrial chemicals" was a risk factor for PD (OR = 2.13).[16] Among industrial chemicals, only organic solvents were identified as a risk factor (OR = 2.78). Limitations of this study include lack of specific information regarding occupations, small sample size, and subject selection bias. Occupational exposure to magnetic fields may be a risk factor for PD.[17] A death certificate (population-based) case control study in Colorado utilizing a tiered exposure matrix found an adjusted odds ratio of 1.76 for Parkinson's disease for subjects exposed to magnetic fields. Occupations included in this study were electronic technicians and engineers, repairers of electronic equipment, telephone and telephone line installers and repairers, electric power installers and repairers, supervisors of electricians and power transmission installers, power plant operators, motion picture projectionists, broadcast equipment operators, and electricians.[17] Another study of electrical workers in a similar group of occupations found a nonsignificant elevated odds ratio of 1.1 for PD compared to controls, but the study lacked power.[18]

Several studies have investigated the association between residential exposures to industrial toxins and PD. Using standard industrial code classifications, Rybicki and colleagues found that residential exposure to industrial chemicals, iron, and paper was significantly associated with the development of PD.[9] The counties in Michigan with the highest concentration of these industries had the greatest death rate from PD, suggesting that these individuals resided in an environmentally high-risk region. In a case-control study in China, subjects living near a rubber plant appeared to have a higher risk of PD; however, no specific data on those working in the plant was provided.[19]

Most epidemiologic studies have focused on categories of exposure and not on specific occupations. A few occupations warrant specific attention given the type of chemical exposures or the amount of data supporting these occupations in the etiology of PD.

AGRICULTURE

Studies demonstrate both an association[20,21] and lack of an association[6] between PD and farming as an occupation. Duration of plantation work demonstrated a significant dose response relationship with PD in the population-based Honolulu Heart Study, although individual deciles failed to demonstrate significant odds ratios.[22] In a nonpopulation-based case-control study, orchard workers had a higher risk of PD.[14] Exposures specific to rural environments include pesticide and herbicide exposures, livestock, and well water. Most studies of rural occupational risk factors have focused on pesticides and herbicides. If pesticide and herbicide exposures cause PD, those applying or working directly with the substances might be at higher risk. Few studies investigate the relative frequency of PD in pesticide/herbicide workers as compared to those living (but not working) in rural environments. However, this difference may be somewhat arbitrary, as family farms reside directly in the treated areas. Gorell et al. performed a population-based case-control study of 144 PD subjects with occupational exposure to pesticides and herbicides in Michigan.[23] They found that occupational exposure to pesticides (OR = 3.55) and herbicides (OR = 4.10) were significant risk factors for PD; fungicide use was not associated with PD. Ferraz et al. performed a case control study of parkinsonian features in a group of agricultural workers exposed to the manganese containing fungicide Maneb.[24] They found that the exposed workers (n = 50) were significantly more likely to have rigidity and a variety of constitutional symptoms than nonexposed workers (n = 19). There was no significant difference in other parkinsonian signs but the number of subjects was small. Isolated reports of parkinsonism developing after acute paraquat exposure[25–27] and glyphosate[28] suggest a role of the use of these pesticides as a risk factor for PD. However, the high prevalence of PD in the population makes sporadic PD a substantial confound in these reports. An outbreak of an atypical parkinsonism in grain workers implicated the fumigant, carbon disulfide, as the cause of a syndrome characterized by cerebellar signs, bradykinesia, rest tremor, sensory neuropathy.[29,30] Atypical features such as MRI abnormalities in some cases of paraquat exposure and atypical clinical features in carbon disulfide exposure (cerebellar signs and neuropathy) argue for a primary causative effect.[25]

THE STEEL INDUSTRY

There is some evidence that acute exposures to fumes in the steel industry are associated with an atypical parkinsonian syndrome.[31–33] The primary exposure in the steel industry is manganese. Wang et al. described an outbreak of parkinsonism in a Taiwanese ferromanganese smelter due to defective ventilation control system. Of those subjects with brief, high-level exposure to inhaled manganese (>28.8 mg/m^3) six of eight subjects developed parkinsonism.[33] Symptoms of affected individuals included bradykinesia, rigidity, gait abnormalities, and tremor. Only one subject developed a "cock gait," a characteristic dystonic gait disorder reportedly characteristic of manganism.[34] No details on disease asymmetry or characteristics of tremor were reported, but subjects were noted to experience 50% improvement in parkinsonism with levodopa.[33] Subsequent clinical assessment demonstrated some progression of disease at five years in four subjects in whom follow-up was available. A follow-up, double-blind, placebo-controlled study of a single dose of levodopa failed to result in improvement in the four subjects.[35] Given substantial individual patient variability in response to a single dose of levodopa in typical PD patients and the small number of subjects studied in this report, it is not entirely clear that the full spectrum of response of manganism to levodopa is known.

WELDING

Welding is the process of joining metals by electric arc or flame with a filler material.[36] The filler material, also called the *consumable*, is usually a coated electrode or wire that contributes metal to the joint. The four most common welds are the TIG (tungsten inert gas), MIG (metal inert gas), MAG (metal active gas), and MMA (manual metal arc).[37] The process of melting the parent metal and the consumable produces concentrated particulate fumes and gases; the consumable produces 80 to 95% of the fume. These fumes contain a number of elements including F, Mn, Zn, Pb, As, Ca, S, Cr, and Ni. Gases released include CO, CO_2, F, and HF.[38,39] Neurologic complications of welding exposures include encephalopathy[40] and lead poisoning from heating lead-based paint.[41]

Welding is an intriguing possible etiologic risk factor for PD, given the presence of an established neurotoxin, manganese, in the flux. Furthermore, if there is a relationship between welding and PD, the public health implications could be substantial, given that more than 500,000 people identify welding as their primary occupation.[42]

The materials safety data sheet (MSDS) for welding consumables lists parkinsonism as a potential hazard of welding, but no definitive epidemiologic evidence has been published.

There are several case reports that identified welders with PD or parkinsonism, although many patients had atypical features, including cognitive abnormalities, disturbances of sleep, peripheral nerve complaints, and mild motor slowing.[32,43–45] A pilot epidemiology study suggested that occupational welding may be more common in PD patients as compared to neurologic controls; however, this study was not population based, and the number of subjects studied was small.[46] In a study of magnetic-field exposed workers, Noonan and colleagues studied nonelectrical, magnetic-field-related workers and found that welders were overrepresented in PD deaths.[17] Blood manganese[45] and aluminum[44] levels may be elevated in welders, but no study convincingly demonstrates an association between motor signs and these metals. However, a small study suggests that welders with exposure to manganese may be slower on peg-board and finger-tapping scores as compared to welders without these exposures.[44] In a specialty clinic based, case-control study of 15 career welders compared to consecutively ascertained and age-matched PD controls, welders with PD were clinically identical to the control groups except for a significantly younger age of onset (46 years).[47] PET with 6-[^{18}F]fluorodopa imaging in two welders with PD demonstrated reduced 6-[^{18}F]fluorodopa uptake more prominent in the posterior putamen contralateral to the most affected side.[47] The authors concluded that parkinsonism in welders is distinguished clinically from idiopathic PD only by age of onset, suggesting it may accelerate the onset of the disease. These findings appeared to make manganese a less likely etiologic agent given previous reports of atypical parkinsonism in manganese exposed individuals.[33,48]

MANGANISM AS A MODEL FOR OCCUPATIONALLY INDUCED PARKINSONISM

CLINICAL FEATURES

Manganese (Mn) was first recognized as a neurotoxin in the nineteenth century with the report of four manganese ore crushers developing a syndrome of a lower extremity predominant "muscular weakness," festination, postural instability, facial masking, hypophonia, and sialorrhea.[49] The syndrome was more clearly delineated by Rodier in 1954, when he described a group of Moroccan manganese miners with a neurologic illness characterized by parkinsonism, gait disorder, dystonia, psychosis, and emotional lability.[48] All of these individuals worked underground, and the majority mined manganese ore. The latency to symptom onset from work exposure was one month to over ten years. Rodier divided the syndrome into three phases: the prodromal period, the intermediate phase, and the established phase. The first phase was characterized by akinesia and apathy followed by "manganese psychosis." During this phase the gait was described as "staggering," and patients became aggressive. Early characteristics of the intermediate phase were hypophonia with vocal "freezing," facial masking, and emotional lability. In the final phase, the patients developed rigidity, bradykinesia, tremor, flexed posture, shuffling gait, and postural instability. Some patients developed a dystonic, wide based gait described as a "cock gait." The disease progressed to total disability in most, despite discontinuing exposure.[48]

In another report, the dystonic gait was only present in 1 in 13 Chilean manganese miners,[50] although the prodromal psychiatric disorder, locally termed "locura manganica," was seen commonly. Tanaka et al. described seven cases of a neurologic illness associated with exposure to airborne manganese.[51] The clinical features of affected individuals included bradykinesia, postural instability, tremor, rigidity, and a "spastic gait" in one subject. There was no documentation of a "cock gait" in any subject. Two subjects were followed for four years. One of the subjects had a static disease course, and the other experienced improvement attributed to chelation.[51]

Many additional studies document the neurotoxic effects of manganese on the central nervous and as a result the Occupational Safety and Health Administration (OSHA) has a permissible exposure limit ceiling for manganese of 5 mg/m^3.[52] A cross-sectional epidemiologic study of workers exposed in a Mn oxide and salt producing plant found that workers exposed to low levels of manganese (approximately 1 mg/m^3) had slower simple reaction times on a standardized reaction time test and more hand tremor as measured by a standardized hand steadiness assessment.[53] A dose-response relationship existed for the quantitative tremor assessment but not the reaction time assessment. Standard neurologic examination revealed axial rigidity but no other objective neurologic findings. Manganese-exposed foundry workers in Sweden (mean Mn exposure 0.18 to 0.41 mg/m^3) demonstrated slower reaction time, reduced finger-tapping speed, reduced tapping endurance, and diadochokinesis.[54–56] A larger, population-based study of workers in a Mn alloy facility found that exposed subjects had slower computerized finger tapping scores and less hand steadiness.[57] Lucchini et al. found an exposure-dependent increase in blood and urine manganese levels and slowing of finger tapping in workers in a ferroalloy plant exposed to low-level, chronic manganese. Even nonoccupational blood elevations in manganese are also associated with a exposure-related slowing of motor tasks and difficulty with pointing tasks consistent with tremor.[58] Although many of the subjects in these studies had been exposed to manganese for many years, none of these studies provides

longitudinal follow-up to determine the natural course of these physiologic differences. The finding that low doses may be associated with parkinsonism suggests that the OSHA permissible exposure ceiling limit for manganese may need to be revised.

Although high-level acute exposures are clearly associated with parkinsonism, and lower-level exposures are associated with parkinsonian motor abnormalities, studies are contradictory in implicating manganese in the etiology of PD. In a population-based, case-control study, Gorell and colleagues found that occupational exposures to copper (OR = 2.49) and manganese (OR = 10.61) for more than 20 years were associated with the diagnosis of PD.[59] However, their study only had three cases and one control with long duration manganese exposure.[60] A variety of occupations (including pipe fitter, electrical worker, engineer, chemist, machinist, firefighter, steam fitter, process technician, and toolmaker) were included in their cohort. This study included a blinded industrial hygiene assessment of exposure but only exposure duration was quantified. Zayed et al. found an increased risk of PD in subjects exposed to a combination of manganese, iron, and aluminum for greater than 30 years.[5] This study did not analyze the effects of individual metals, nor was there a dose-response relationship (the association was only significant with the longest duration of exposure). In addition, the study sample was small, and occupational categories were broad. Using a more detailed assessment of occupational exposure, a job exposure matrix, Seidler et al. found no association between PD and occupational heavy-metal exposure in a population-based, German cohort.[61] Another population-based study using self-reported occupational exposures found no association between PD and manganese.[62] Differences in study design, populations studied, and exposures likely account for the discrepant findings in theses studies. Unfortunately, no studies have investigated the relationship between PD or parkinsonian signs in a large cohort with a specific industrial exposure.

PATHOPHYSIOLOGY AND PATHOLOGY OF MANGANISM

Functional imaging studies provide evidence of the pathophysiology of parkinsonian disorders and may be useful in clarifying pathogenesis. Previous reports have suggested that 6-[18F]fluorodopa uptake is normal in patients with manganism, indicating intact decarboxylase activity and storage in the nigrostriatal pathway.[63] This contrasts with the typical 6-[18F]fluorodopa findings in PD of asymmetric reduction of putamenal uptake in idiopathic PD.[64] These reports were consistent with direct striatal damage, preservation of nigrostriatal neurons, and parkinsonian disorder being poorly responsive to levodopa.[63] However, these previous studies used older imaging equipment that may have limited the sensitivity to detect dysfunction of the nigrostrial pathway.[47] A recent report, using the dopamine transporter ligand [[123I]]-B-CIT to measure nigrostriatal neurons in two patients with presumed manganese induced parkinsonism, suggested that manganism may be associated with reduced [[123I]]-B-CIT uptake, a pattern consistent with idiopathic PD.[65] Another recent report describes a patient with substantial occupational manganese exposure, levodopa-responsive parkinsonism, and reduced 6-[18F]fluorodopa uptake more prominent in the posterior putamen, contralateral to the most affected side.[66]

Brain MRI in patients with manganism demonstrate increased signal in the globus pallidum on T1 weighted images but normal signal on T2.[43,67–69] Although this findings appears to be relatively specific to manganese, the sensitivity is unclear. It is also unclear whether these findings represent irreversible striatal injury. Kim and colleagues performed a cross-sectional MRI study of manganese exposed workers in South Korea.[68] They performed brain MRI in randomly selected workers at six factories, including subjects with and without manganese exposure (38 welders, 39 steel smelters, and 16 welding rod manufacturing workers). They demonstrated a striking association between increased T1 signal in the globus pallidum and manganese exposure, as quantified by an industrial hygiene evaluation, in exposed compared to nonexposed clerical workers in the same plants.[68] However, they found no objective evidence of parkinsonism in their subjects. Only subjects with postural tremor were followed up at six months, and none reportedly had any clinical change.

There is very limited autopsy data on subjects with manganism, but pathology affecting primarily basal ganglia appears to be characteristic. Yamada et al. reported a 52-year-old man with "increased muscle tone," weakness, hyperreflexia, and a mood disorder attributed to working in a manganese ore crushing plant.[70] Autopsy demonstrated cell loss in the pallidum with astrocytosis but normal substantia nigra. Bernheimer et al. reported a patient with a long history of a parkinsonian illness, associated with markedly elevated blood manganese levels, that developed while exposed to manganese dioxide in a battery factory.[71] Autopsy revealed pallidal atrophy, marked degeneration of the substantia nigra pars compacta, and occasional Lewy bodies in nigral neurons. Older reports comment primarily on cell loss in the putamen and pallidum.[72,73] Clearly, additional pathologic material studied with modern techniques and stains will be necessary to more clearly define the pathologic spectrum of manganism.

CONCLUSIONS

The industrialization of societies may have increased the prevalence of PD, although longitudinal incidence data are still lacking due to the more modern classification of PD. Employment in agricultural and steel industry appears to be a risk factor for PD in multiple studies; however, there is very little data on specific occupations within these broad categories. Studies correlating detailed occupational exposures with PD in at-risk occupations may provide more definitive evidence of an association between PD and specific occupations. Although the evidence for specific occupations as risk factors for PD remains speculative, occupational manganese is a candidate neurotoxin that may play a role in the selective degeneration seen in PD. There may be a spectrum of manganese-induced neurodegeneration from mild parkinsonism in low-level exposures to more fulminant PD or atypical parkinsonism in subjects exposed to high levels of manganese. Prospective, long-term follow-up of a large cohort of manganese exposed subjects will be necessary to define more clearly the role of manganese as an occupational risk factor for PD. Finally, efforts to obtain pathologic material in exposed subjects should clarify the pathologic spectrum of toxin-induced parkinsonism.

ACKNOWLEDGMENTS

This work was supported by NIH grants K23NS43351 and the Greater St. Louis Chapter of the American Parkinson Disease Association.

REFERENCES

1. Koller, W., Vetere-Overfield, B., Gray, C., Alexander, C., Chin, T., Dolezal, J. et al., Environmental risk factors in Parkinson's disease, *Neurology*, 1990; 40(8):1218–1221.
2. Granieri, E., Carreras, M., Casetta, I., Govoni, V., Tola, M. R., Paolino, E. et al., Parkinson's disease in Ferrara, Italy, 1967 through 1987, *Arch. Neurol.*, 48(8):854–857, 1991.
3. Rocca, W. A., Anderson, D.W., Meneghini, F., Grigoletto, F., Morgante, L., Reggio, A. et al., Occupation, education, and Parkinson's disease: a case-control study in an Italian population, *Mov. Disord.*, 11(2):201–206, 1996.
4. Hubble, J. P., Cao, T., Hassanein, R. E. S., Neuberger, J. S., Koller, W. C., Risk factors in Parkinson's disease. *Neurology*, 1993; 43:1693–1697.
5. Zayed, J., Ducic, S., Campanella, G., Panisset, J. C., Andre, P., Masson, H. et al., Environmental factors in the etiology of Parkinson's disease, *Can. J. Neurol. Sci.*, 17(3):286–291, 1990.
6. Semchuk, K. M., Love, E. J., Lee, R. G., Parkinson's disease and exposure to agricultural work and pesticide chemicals, *Neurology,* 42(7):1328–1335, 1992.
7. Schoenberg, B. S., Osuntokun, B. O., Adeuja, A. O., Bademosi, O., Nottidge, V., Anderson, D. W. et al., Comparison of the prevalence of Parkinson's disease in black populations in the rural United States and in rural Nige-

ria: door-to-door community studies, *Neurology,* 38(4):645–646, 1988.

8. Li, S. C., Schoenberg, B. S., Wang, C. C., Cheng, X. M., Rui, D. Y., Bolis, C. L. et al., A prevalence survey of Parkinson's disease and other movement disorders in the People's Republic of China, *Arch. Neurol.,* 42(7):655–657, 1985.

9. Rybicki, B. A., Johnson, C. C., Uman, J., Gorell, J. M., Parkinson's disease mortality and the industrial use of heavy metals in Michigan, *Mov. Disord.,* 8(1):87–92, 1993.

10. Aquilonius, S. M., Hartvig, P. A., Swedish county with unexpectedly high utilization of anti-parkinsonian drugs, *Acta Neurol. Scand.* 74(5):379–382, 1986.

11. Rajput, A. H., Offord, K. P., Beard, C. M., Kurland, L. T., Epidemiology of parkinsonism: incidence, classification, and mortality, *Ann. Neurol.,* 16(3):278–282, 1984.

12. Fall, P. A., Fredrikson, M., Axelson, O., Granerus, A. K., Nutritional and occupational factors influencing the risk of Parkinson's disease: a case-control study in southeastern, Sweden, *Mov. Disord.,* 14(1):28–37, 1999.

13. Tanner, C. M., Chen B., Wang, W., Peng, M., Liu, Z., Liang, X. et al., Environmental factors and Parkinson's disease: a case-control study in China, *Neurology,* 39(5):660–664, 1989.

14. Hertzman, C., Wiens, M., Bowering, D., Snow, B., Calne, D., Parkinson's disease: a case-control study of occupational and environmental risk factors. *Am. J. Ind. Med.,* 17(3):349–355, 1990.

15. Tsui, J. K., Calne, D. B., Wang, Y., Schulzer, M., Marion, S. A., Occupational risk factors in Parkinson's disease, *Can. J. Public. Health.,* 90(5):334–337, 1999.

16. Smargiassi, A., Mutti, A., De Rosa, A., De Palma, G., Negrotti, A., Calzetti, S., A case-control study of occupational and environmental risk factors for Parkinson's disease in the Emilia-Romagna region of Italy, *Neurotoxicology,* 19(4-5):709–712, 1998.

17. Noonan, C. W., Reif, J. S., Yost, M., Touchstone, J., Occupational exposure to magnetic fields in case-referent studies of neurodegenerative diseases, *Scand. J. Work. Environ. Health,* 28(1):42–48, 2002.

18. Savitz, D. A., Loomis, D. P., Tse, C. K,. Electrical occupations and neurodegenerative disease: analysis of U.S. mortality data, Arch. Environ. Health, 53(1):71–74, 1998.

19. Wang, W. Z., Fang, X. H., Cheng, X. M., Jiang, D. H., Lin, Z. J., A case-control study on the environmental risk factors of Parkinson's disease in Tianjin, China, *Neuroepidemiology,* 12(4):209–218, 1993.

20. Gorell, J. M., Johnson, C. C., Rybicki, B. A., Peterson, E. L., Richardson, R. J., The risk of Parkinson's disease with exposure to pesticides, farming, well water, and rural living, *Neurology,* 50(5):1346–1350, 1998.

21. Zorzon, M., Capus, L., Pellegrino, A., Cazzato, G,. Zivadinov, R., Familial and environmental risk factors in Parkinson's disease: a case-control study in north-east Italy, *Acta Neurol. Scand.,* 105(2):77–82, 2002.

22. Petrovitch, H., Ross, G. W., Abbott, R. D., Sanderson, W. T., Sharp, D. S., Tanner, C. M. et al., Plantation work

and risk of Parkinson disease in a population-based longitudinal study, *Arch. Neurol.,* 59(11):1787–1792.

23. Gorell, J. M., Johnson, C. C., Rybicki, B. A., Peterson, E. L., Richardson, R. J., The risk of Parkinson's disease with exposure to pesticides, farming, well water, and rural living, *Neurology,* 50(5):1346–1350, 1998.

24. Ferraz, H. B., Bertolucci, P. H., Pereira, J. S., Lima, J. G., Andrade, L. A., Chronic exposure to the fungicide maneb may produce symptoms and signs of CNS manganese intoxication, *Neurology,* 38(4):550–553, 1988.

25. Sechi, G. P., Agnetti, V., Piredda, M., Canu, M,. Deserra, F., Omar, H. A. et al., Acute and persistent parkinsonism after use of diquat, *Neurology,* 42(1):261–263, 1992.

26. Bocchetta, A., Corsini, G. U., Parkinson's disease and pesticides, *Lancet,* 2(8516):1163, 1986.

27. Sanchez-Ramos, J. R., Hefti, F., Weiner, W. J., Paraquat and Parkinson's disease, *Neurology,* 37(4):728, 1987.

28. Barbosa, E. R., Leiros, D. C., Bacheschi, L.A., Scaff, M., Leite, C. C., Parkinsonism after glycine-derivate exposure, *Mov. Disord.,* 16(3):565–568, 2001.

29. Melamed, E., Lavy, S., Parkinsonism associated with chronic inhalation of carbon tetrachloride, *Lancet,* 1(8019):1015, 1977.

30. Peters, H. A., Levine, R.L., Matthews, C. G., Chapman, L. J., Extrapyramidal and other neurologic manifestations associated with carbon disulfide fumigant exposure, *Arch. Neurol.,* 45(5):537–540, 1988.

31. Cook, D. G., Fahn, S., Brait, K. A., Chronic manganese intoxication, *Arch. Neurol.,* 30(1):59–64, 1974.

32. Whitlock, C. M., Jr., Amuso, S. J., Bittenbender, J. B., Chronic neurological disease in two manganese steel workers, *Am. Ind. Hyg. Assoc. J.,* 27(5):454–459, 1966.

33. Wang, J. D., Huang, C. C., Hwang, Y. H., Chiang, J. R., Lin, J. M., Chen, J. S., Manganese induced parkinsonism: an outbreak due to an unrepaired ventilation control system in a ferromanganese smelter, *Br. J. Ind. Med.,* 46(12):856–859, 1989.

34. Huang, C. C., Lu, C. S., Chu, N. S., Hochberg, F., Lilienfeld, D., Olanow, W. et al., Progression after chronic manganese exposure, *Neurology,* 43(8):1479–1483, 1993.

35. Lu, C. S., Huang, C. C., Chu, N. S., Calne, D. B., Levodopa failure in chronic manganism, *Neurology,* 44(9):1600–1602, 1994.

36. Wallace, M. E., Fischbach, T., Kovein, R. J., In-Depth Survey Report. Control Technology Assessment for the Welding Operations Boilermaker's National Apprenticeship Training School Kansas City, Kansas, OSHA Report No., ECTB, 214-13a, 1-27-1997.

37. Hemminki, Peto, J., Stern, R. M., Summary report: International conference on health hazards and biological effects of welding fumes and gases, In: Stern, R. M., Berlin, A., Fletcher, A. C., Jarvisalo, J., Eds., *Amsterdam: Excerpta Medica,* 1–5, 1986.

38. NIOSH. Criteria for a recommended standard: occupational exposure to welding, brazing, and thermal cutting. U.S.Department of Health and Human Services, Public Health Service, Centers for Disease Control, National Institute for Occupational Safety and Health, DHHS (NIOSH), Eds., Publication No. 88-110, Cincinnati, OH, 1988.

39. AWS Safety and Health Committee. Effects of welding on health VIII, Miami, FL., American Welding Society, Safety and Health Committee, SBN: 0-87171-437-X, 1993.

40. Johnson, J. S., Kilburn, K. H., Cadmium induced metal fume fever: results of inhalation challenge, *Am. J. Ind. Med.*, 4(4):533–540, 1983.

41. NIOSH Bulletin, Preventing Lead Poisoning in Construction Workers, NIOSH ALERT Publication No. 91-116a, DHHS (NIOSH), 1992.

42. OSHA, Welding, Cutting and Brazing, OSHA archive, 2002.

43. Nelson, K., Golnick, J., Korn, T., Angle, C., Manganese encephalopathy: Utility of early magnetic resonance imaging, *Br. J. Ind. Med.*, 50(6):510–513, 1993.

44. Sjogren, B., Iregren, A., Frech, W., Hagman, M., Johansson, L., Tesarz, M. et al., Effects on the nervous system among welders exposed to aluminium and manganese, *Occup. Environ. Med.*, 53(1):32–40, 1996.

45. Chandra, S. V., Shukla, G. S., Srivastava, R. S., Singh, H., Gupta, V. P., An exploratory study of manganese exposure to welders, *Clin. Toxicol.*, 18(4):407–416, 1981.

46. Wechsler, L. S., Checkoway, H., Franklin, G. M., Costa, L. G., A pilot study of occupational and environmental risk factors for Parkinson's disease, *Neurotoxicology*, 12(3):387–392, 1991.

47. Racette, B. A., McGee-Minnich, L., Moerlein, S. M., Mink, J. W., Videen, T. O., Perlmutter, J. S., Welding-related parkinsonism: clinical features, treatment, and pathophysiology, *Neurology*, 56(1):8–13, 2001.

48. Rodier, J., Manganese poisoning in Moroccan miners, *Brit. J. Industr. Med.*, 12, 21–35, 1955.

49. Couper, J., On the effects of black oxide of manganese when inhaled into the lungs, *British Annals of Medical Pharmacology*, 1, 41–42, 1837.

50. Mena, I., Marin, O., Fuenzalida, S., Cotzias, G. C., Chronic manganese poisoning: Clinical pictures and manganese turnover, *Neurology*, 17:128–136, 1967.

51. Tanaka, S., Lieben, J., Manganese poisoning and exposure in *Pennsylvania, Arch. Environ. Health.*, 19(5):674–684, 1969.

52. http://www.osha-slc.gov/oshinfo/priorities/welding.html.

53. Roels, H., Lauwerys, R., Buchet, J. P., Genet, P., Sarhan, M. J., Hanotiau, I. et al., Epidemiological survey among workers exposed to manganese: effects on lung, central nervous system, and some biological indices, *Am. J. Ind. Med.*, 11(3):307–327, 1987.

54. Wennberg, A., Iregren, A., Struwe, G., Cizinsky, G., Hagman, M., Johansson, L., Manganese exposure in steel smelters a health hazard to the nervous system, *Sc.and J. Work. Environ. Health*, 17(4):255–262, 1991.

55. Wennberg, A., Hagman, M., Johansson, L., Preclinical neurophysiological signs of parkinsonism in occupational manganese exposure, Neurotoxicology, 13(1):271–274, 1992.

56. Iregren, A., Psychological test performance in foundry workers exposed to low levels of manganese, *Neurotoxicol. Teratol.*, 12(6):673–675, 1990.

57. Mergler, D., Huel, G., Bowler, R., Iregren, A., Belanger, S., Baldwin, M. et al., Nervous system dysfunction among workers with long-term exposure to manganese, *Environ. Res.*, 64(2):151–180, 1994.

58. Mergler, D., Baldwin, M., Belanger, S., Larribe, F., Beuter, A., Bowler, R. et al., Manganese neurotoxicity, a continuum of dysfunction: results from a community based study, *Neurotoxicology*, 20(2-3):327–342, 1999.

59. Gorell, J. M., Johnson, C. C., Rybicki, B. A., Peterson, E. L., Kortsha, G. X., Brown, G. G. et al., Occupational exposures to metals as risk factors for Parkinson's disease, *Neurology*, 48(3):650–658, 1997.

60. Gorell, J. M., Rybicki, B. A., Cole, J. C., Peterson, E. L., Occupational metal exposures and the risk of Parkinson's disease, *Neuroepidemiology*, 18(6):303–308, 1999.

61. Seidler, A., Hellenbrand, W., Robra, B. P., Vieregge, P., Nischan, P., Joerg, J. et al., Possible environmental, occupational, and other etiologic factors for Parkinson's disease: a case-control study in Germany, *Neurology*, 46(5):1275–1284, 1996.

62. Semchuk, K. M., Love, E. J., Lee, R. G., Parkinson's disease: a test of the multifactorial etiologic hypothesis, *Neurology*, 43(6):1173–1180, 1993.

63. Wolters, E. C., Huang, C. C., Clark, C., Peppard, R. F., Okada, J., Chu, N. S. et al., Positron emission tomography in manganes intoxication, *Ann, Neurol.*, 26(5):647-651, 1989.

64. Eidelberg, D., Moeller, J. R., Dhawan, V., Sidtis, J. J., Ginos, J. Z., Strother, S. C. et al., The metabolic anatomy of Parkinson's disease: Complementary [^{18}F]fluorodeoxyglucose and [^{18}F]fluorodopa positron emission tomographic studies, *Mov. Disord.*, 5(3):203–213, 1990.

65. Kim, Y., Kim, J. M., Kim, J. W., Yoo, C. I., Lee, C. R., Lee, J. H. et al., Dopamine transporter density is decreased in parkinsonian patients with a history of manganese exposure: What does it mean? *Mov. Disord.*, 17(3):568–575, 2002.

66. Racette, B., Antenor, J., Kotagal, V., Videen, T., Moerlein, S., Goldman, J. et al., [18F]FDOPA PET and clinical features in a Parkinson's disease patient with manganese exposure, *Mov. Disord.*, 17:S108 (Abstract), 2002.

67. Hauser, R. A., Zesiewicz, T. A., Rosemurgy, A. S., Martinez, C., Olanow, C. W., Manganese intoxication and chronic liver failure, *Ann. Neurol.*, 36(6):871–875, 1994.

68. Kim, Y., Kim, K. S., Yang, J. S., Park, I. J., Kim, E., Jin, Y. et al., Increase in signal intensities on T1-weighted magnetic resonance images in asymptomatic manganese-exposed workers, *Neurotoxicology*, 20(6):901–907, 1999.

69. Arjona, A., Mata, M., Bonet, M., Diagnosis of chronic manganese intoxication by magnetic resonance imaging, *N. Engl. J. Med.*, 336(13):964–965, 1997.

70. Yamada, M., Ohno, S., Okayasu, I., Okeda, R., Hatakeyama, S., Watanabe, H. et al., Chronic manganese poisoning: a neuropathological study with determination of manganese distribution in the brain, *Acta Neuropathol. (Berl)*, 70(3–4):273–278, 1986.

71. Bernheimer, H., Birkmayer, W., Hornykiewicz, O., Jellinger, K., Seitelberger, F., Brain dopamine and the syndromes of Parkinson and Huntington. Clinical, morphological and neurochemical correlations, *J. Neurol. Sci.*, 20(4):415–455, 1973.

72. Casamajor, L., An unusual form of mineral poisoning affecting the nervous system: manganese? *J. Am. Med. Assoc.,* 60:646–649, 1913.

73. Canavan, M. M., Cobb, W., Drnovsek, B,. Chronic manganese poisoning: Report of a case with autopsy, *Archives of Neurologic Psychiatry,* 32:501–512, 1934.

9 Tetrahydroisoquinolines and Parkinson's Disease

Mark S. LeDoux

Departments of Neurology and Anatomy and Neurobiology, University of Tennessee Health Science Center

CONTENTS

INTRODUCTION

The discovery that 1-methyl-4-phenyl-1,2,3,6-tetrahydro-pyridine (MPTP; Figure 9.1A) can cause parkinsonism[1,2] led to a search for MPTP analogs as possible endogenous or exogenous neurotoxins critical to the neurodegeneration seen in Parkinson's disease (PD). In the context of identifying other neurotoxins that show some degree of dopaminergic selectively, it is important to remember that (1) MPTP is converted to 1-methyl-4-phenyl-pyridinium ion (MPP+) by monoamine oxidase (MAO)-B, (2) MPP+ is actively transported into presynaptic dopaminergic nerve terminals through the plasma membrane dopamine transporter (DAT), and (3) it is MPP+, and not MPTP, that kills dopaminergic neurons. Isoquinoline (Figure 9.1B) derivatives are one group of neurotoxins that are both substrates for MAO and structurally homologous to MPTP. Several lines of evidence suggest that isoquinoline derivatives may be endogenous and/or exogenous neurotoxins that contribute to dopaminergic cell death in PD.[3–5] McNaught et al.[6] showed that isoquinoline derivates such as the tetrahydroisoquinolines (TIQs) and dihydroisoquinolines are more potent inhibitors of complex I of the respiratory chain than MPP+. TIQ (Figure 9.1C) is present in several common foods such as cheese, bananas, milk, wine, and cocoa.[7–10] TIQ has also been localized to rodent and human brain.[7,11] Nagatsu and Yoshida[12] stirred interest in TIQ by showing that systemic administration of this molecule to marmosets produced parkinsonism, despite the lack of demonstrable cell death within the substantia nigra, pars compacta (SNpc). TIQ also appears to be a much less potent neurotoxin than MPTP in mice.[13] Cell death could have been missed, however, since neither markers of apoptosis nor stereological counting methods were employed. N-methylated derivatives of TIQ (Figures 9.1C through 9.1D and Figure 9.2) and benzyl

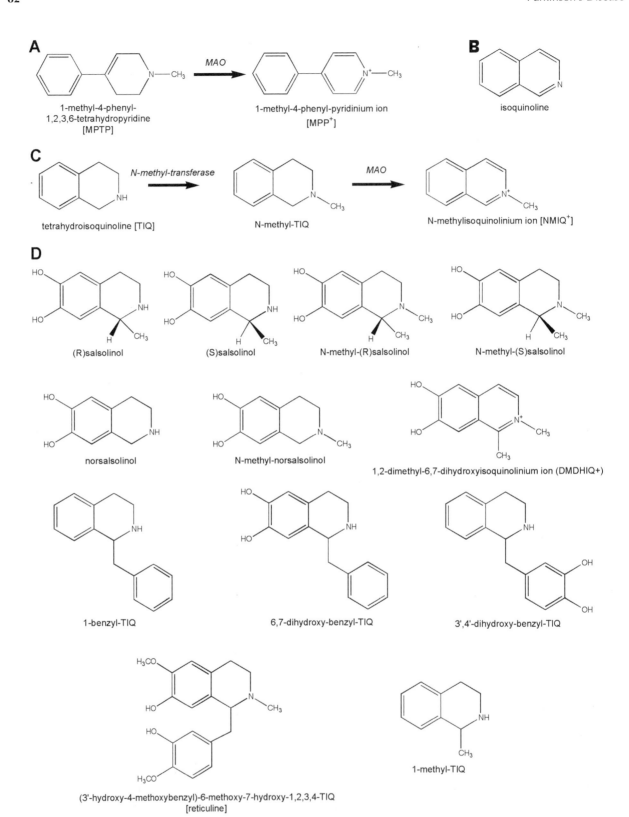

FIGURE 9.1 MPP+, isoquinoline, and tetrahydroisoquinoline derivatives. (A) MPTP crosses the blood-brain barrier and is converted to the neurotoxin MPP+ by MAO. (B) Isoquinoline [IQ]. (C) Combined exogenous-endogenous pathway for tetrahydroisoquinoline (TIQ) neurotoxicity. The TIQ present in foods and beverages is (1) absorbed from the gut, (2) crosses the blood-brain barrier, (3) converted to N-methyl-TIQ by an N-methyl-transferase and (4) converted to N-methylisoquinolinium ion [NMIQ+] by an MAO. (D) Chemical structures of selected TIQ derivatives. Reticuline is a potential exogenous neurotoxin from the leaves and bark of *Annonaceae* trees.

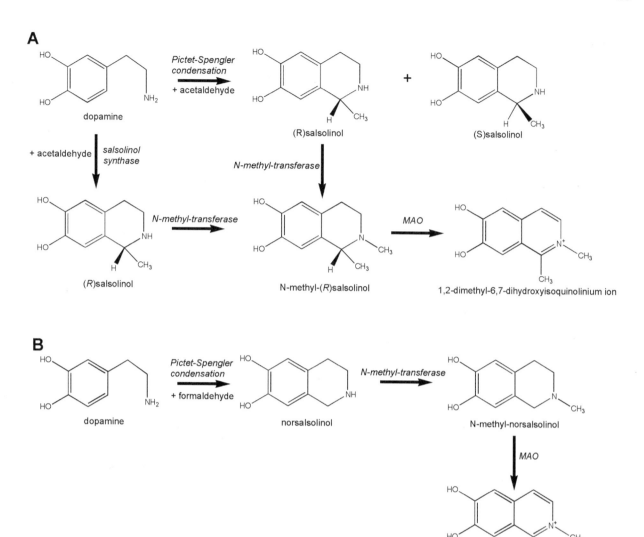

FIGURE 9.2 Putative endogenous synthetic pathways for 1,2-dimethyl-6,7-dihydroxyisoquinolinium ion (A) and N-methyl-6,7-dihydroxyisoquinolinium ion (B).

TIQs (Figures 9.1D and Figure 9.3) may be more neurotoxic than TIQ.[5,14–21] Although several TIQ derivates have been shown to destroy cultured dopaminergic neurons,[22,23] convincing and consistent anatomical proof of cell death in animals exposed to isoquinolines has been lacking.

Unlike the relatively acute effects of MPTP, dopaminergic cell loss within the SNpc in idiopathic PD is generally considered to occur gradually over the course of years or decades. Despite this reality, many neurotoxins unable to produce rapid and extensive dopaminergic cell death have been dismissed as unlikely contributors to the pathophysiology of PD. A more rational approach should instead consider all neurotoxins with relative dopaminergic selectivity as potential contributors to the pathophysiology of PD.

TIQ derivatives are one class of "weak" neurotoxins that may be important in the pathogenesis of idiopathic PD. TIQ derivatives can be formed endogenously by condensation of catecholamines with aldehydes or α-keto acids. TIQ derivatives may accumulate in neurons through numerous combinations of endogenous and exogenous pathways. Neuronal or glial enzymes may also convert exogenous chemicals present in foods such as TIQ to more potent neurotoxins (Figure 9.1C). As shown in Figures 9.2 and 9.3, "normal" metabolic pathways may "inadvertently" convert neurotransmitters such as dopamine to toxic by-products. Evolution may have not provided neurons with mechanisms to prevent accumulation of these "accidental" by-products of normal metabolism. Susceptible neuronal populations may gradually shrink in size via apoptotic cell death caused, in part, by slow neuronal accumulation of compounds that inhibit complex I of the respiratory chain.

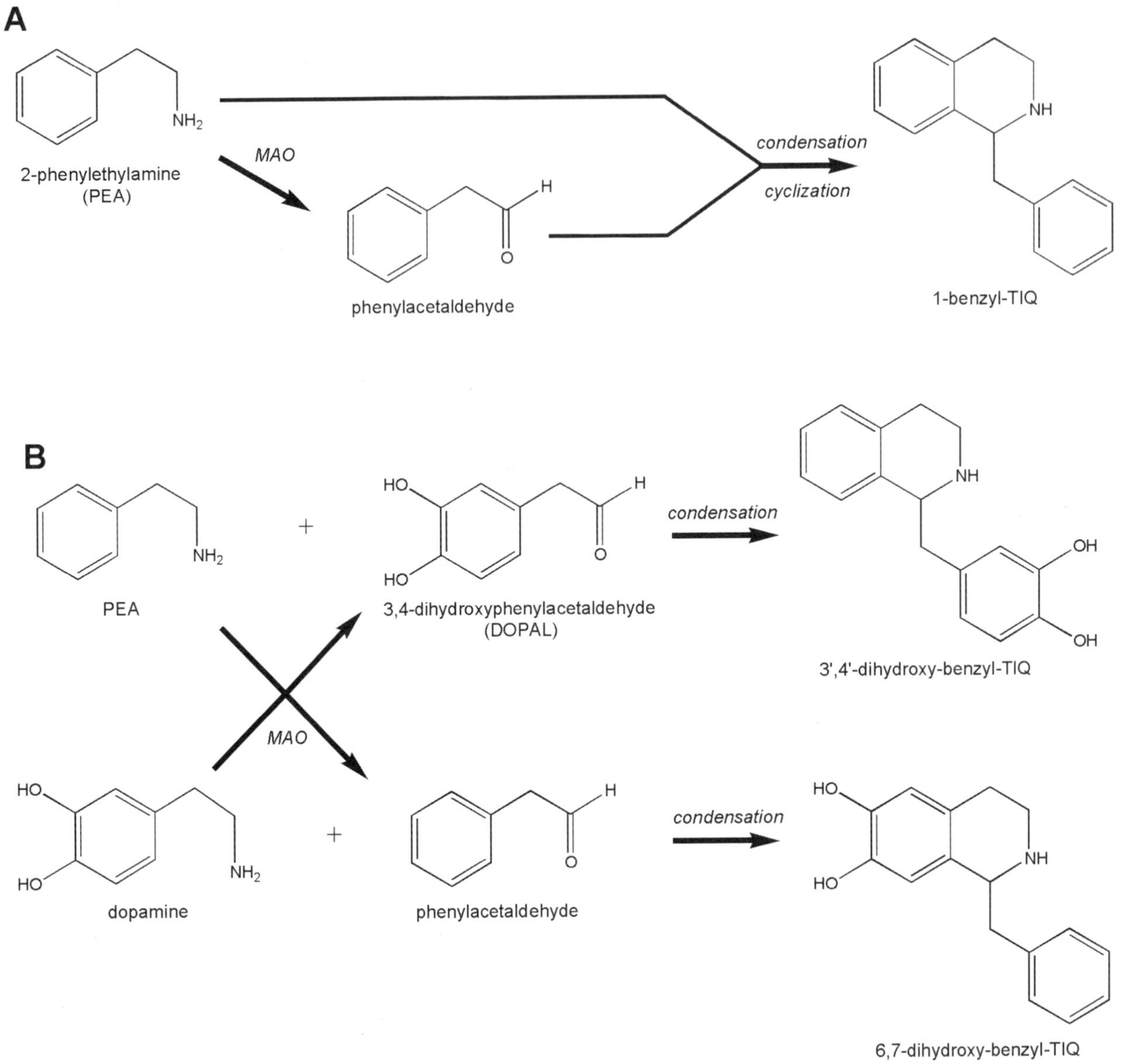

A

B

FIGURE 9.3 Putative endogenous synthetic pathways for 1-benzyl-TIQ (A), 3',4'-dihydroxy-benzyl-TIQ (B) and 6,7-dihydroxy-benzyl-TIQ (B).

ORIGIN OF ISOQUINOLINES

EXOGENOUS

Occurrence in Foodstuffs

Twins studies indicate that environmental factors probably play an important role in typical late-onset (>50 years) idiopathic PD.[24] The development of PD has been statistically associated with a number of environmental factors including exposure to pesticides, well water, herbicides, wood pulp, and rural living. The risk of PD is only slightly increased in persons exposed to these risk factors, suggesting a dose effect and/or interaction with genetic pre-

disposition. In addition, exposure to well water, herbicides, and pesticides may be surrogate markers for rural living or farming. Likewise, rural living and farming may be surrogate markers for a more potent risk factor such as consumption of fresh fruits and vegetables that concentrate one or more TIQ derivatives.

As shown in Table 9.1, TIQs have been detected in several common foods and beverages including wine, cheese, cocoa, bananas, broiled sardines, flour, eggs, beer, and milk.[7-10] Furthermore, TIQs are natural substances in many plants,[25] some of which are consumed by humans. For example, both nonmethylated and methylated TIQs are present in mucuna beans derived from the climbing

TABLE 9.1
Tetrahydroisoquinolines in Foods and Beverages

Tetrahydroisoquinoline	Food Sample	Concentration	Reference
TIQ	White wine	1.7 ng/g	Makino et al. 1988[7]
TIQ	Cheese (Emmenthal)	15.0 ng/g	Makino et al. 1988[7]
TIQ	Cocoa A	0.8 ng/g	Makino et al. 1988[7]
TIQ	Cocoa B	1.1 ng/g	Makino et al. 1988[7]
TIQ	Cheese	5.2 ng/g	Niwa et al. 1989[9]
TIQ	Banana	2.2 ng/g	Niwa et al. 1989[9]
TIQ	Broiled sardine	0.96 ng/g	Niwa et al. 1989[9]
TIQ	Broiled beef	1.3 ng/g	Niwa et al. 1989[9]
TIQ	Flour	0.52 ng/g	Niwa et al. 1989[9]
TIQ	Yolk of boiled egg	1.8 ng/g	Niwa et al. 1989[9]
TIQ	White of boiled egg	2.2 ng/g	Niwa et al. 1989[9]
TIQ	Wine	0.56 ng/ml	Niwa et al. 1989[9]
TIQ	Beer	0.36 ng/ml	Niwa et al. 1989[9]
TIQ	Whisky	0.73 ng/ml	Niwa et al. 1989[9]
TIQ	Milk	3.3 ng/ml	Niwa et al. 1989[9]
1-methyl-TIQ	White wine	354 ng/g	Makino et al. 1998[7]
1-methyl-TIQ	Cheese (Emmenthal)	0.5 ng/g	Makino et al. 1998[7]
1-methyl-TIQ	Cocoa A	6.5 ng/g	Makino et al. 1998[7]
1-methyl-TIQ	Cocoa B	12.0 ng/g	Makino et al. 1998[7]
(R)salsolinol	Dried banana	365 nmol/g	Deng et al. 1997[10]
(R)salsolinol	German wine	35 pmol/ml	Deng et al. 1997[10]
(R)salsolinol	French wine	19 pmol/ml	Deng et al. 1997[10]
(S)salsolinol	Dried banana	344 nmol/g	Deng et al. 1997[10]
(S)salsolinol	German wine	44 pmol/ml	Deng et al. 1997[10]
(S)salsolinol	French wine	28 pmol/ml	Deng et al. 1997[10]
N-methyl-(R)salsolinol	Dried banana	ND	Deng et al. 1997[10]
N-methyl-(R)ssalsolinol	German wine	ND	Deng et al. 1997[10]
N-methyl-(R)salsollinol	French wine	ND	Deng et al. 1997[10]
N-methyl-(S)salsolinol	Dried banana	ND	Deng et al. 1997[10]
N-methyl-(S)salsolinol	German wine	ND	Deng et al. 1997[10]
N-methyl-(S)salsolinol	French wine	ND	Deng et al. 1997[10]

legume commonly called nescafé (*Mucuna pruriens*).[26] In Central and South America, mucuna beans have been roasted, ground, and used as a coffee substitute for centuries. Interestingly, mucuna beans contain very high concentrations of levodopa. Finally, scientists studying neurodegenerative processes should take note that some rat chows may contain small amounts of the TIQ derivative, salsolinol.[27]

Blood-Brain Barrier Permeability

After oral consumption but before reaching the brain, TIQ and its derivatives may be metabolized in the liver. In rats, TIQ accumulation in brain is enhanced in poor debrisoquine (debrisoquine hydroxylase, CYP2D6) metabolizers.[28] In this context, it is important to mention the reported associations between the development of PD and certain cytochrome P450 polymorphisms.[29] TIQ and a number of its derivatives can cross the blood-brain barrier[7,30,31] For example, Makino et al.[7] showed that both 1-methyl-TIQ and TIQ can accumulate in rat brain after intraperitoneal

injection. Rats were injected with both 1-methyl-TIQ (5 mg/kg) and TIQ (5 mg/kg) daily for five days and killed two days after the last injection. Levels of both compounds increased over threefold in whole brain homogenate in comparison with an untreated rat. In another study, both radiolabeled TIQ and 1-methyl-TIQ were detected in brain at four hours after their oral administration.[30]

Larger, more complex TIQ derivatives can also cross the blood-brain barrier. For instance, the dihydroxylated and methylated TIQ derivative, N-methyl-norsalsolinol, was not detectable in the caudate nucleus of untreated rats but was present at high levels after intraperitoneal injection.[31] Earlier studies that questioned the blood-brain barrier transport of TIQ derivatives could have been confounded by technical factors. For example, Origitano and co-workers[32] were unable to show transport of 6,7-dihydroxy-1-methyl-TIQ (i.e., salsolinol) across the blood-brain barrier after intraperitoneal injection of racemic salsolinol. In the same publication, they also indicated that salsolinol was absent from rat brain. The results of Origitano et al.[32] should be interpreted with caution, however, since several other research groups have identified and quantified (R)-salsolinol in brain tissue.[10,33] Transport of other TIQ derivates such as 1-benzyl-TIQ across the blood-brain barrier is suggested by their biochemical and pathological effects on mouse brain after intraperitoneal injection.[34]

Tetrahydroisoquinolines and Monoamine Transporters

PD is often superficially described as a neurodegenerative disorder attributable to selective dopaminergic cell death and characterized by the presence of tremor, rigidity, and bradykinesia. PD is, however, associated with protean manifestations, many of which are due to neuronal death or dysfunction outside to confines of the SNpc. Involvement of extranigral structures, particularly monoaminergic nuclei within the brain stem, is present in the vast majority of postmortem PD brains.[35–37] Furthermore, in many cases, the pathological hallmark of PD, the Lewy body, can even be found in autonomic ganglia.[38,39] Therefore, PD is more appropriately described as a neurodegenerative syndrome associated with widespread neuronal death that preferentially affects the SNpc. Analysis of neurotoxins potentially important to the pathogenesis of PD must take the pathological spectrum of PD into consideration.

Both direct and indirect lines of evidence indicate that monoamine transporters, especially the DAT, could be fundamental to the relatively selective cell death (dopaminergic > noradrenergic > serotonergic) that characterizes PD. The DAT, for example, is mandatory for the *in vivo* dopaminergic neurotoxicity of MPTP.[40] The norepinephrine transporter (NET) is not absolutely specific for its namesake neurotransmitter and can also clear dopamine.[41] In mice, for illustration, much of the dopamine uptake in frontal cortex depends on the NET rather than the DAT.[41]

Storch and colleagues[42] used heterologous expression systems of the DAT to study the dopaminergic neurotoxicity of a large selection of isoquinoline derivatives. Human embryonic kidney (HEK-293) and mouse neuroblastoma Neuro-2A cells that were stably transfected with the mouse DAT were used for cell viability assays. The toxicity of each compound was compared between wildtype cell lines and the cell lines expressing DAT. MPP+ was used as a positive control. Only 2[N]-methylated compounds showed enhanced toxicity in DAT-expressing cell lines: MPP+ >> N-methyl-isoquinolinium ion+ > N-methyl-norsalsolinol = N-methyl-(R/S)salsolinol. In HEK-293 cell lines, MPP+ was almost 1000-fold more DAT-specific toxicity than either N-methyl-norsalsolinol or N-methyl-(R/S)salsolinol. In Neuro-2A cell lines, MPP+ showed approximately 30 times the DAT-specific toxicity of both N-methyl-norsalsolinol and N-methyl-(R/S)salsolinol. Unfortunately, the assays did not include

the active metabolites of N-methyl-norsalsolinol (i.e., N-methyl-6,7-dihydroxyisoquinilinium ion) and N-methyl-(R/S)salsolinol (i.e., 1,2-dimethyl-6,7-dihydroxyisoquinolinium ion{DMDHIQ+}). Based on the potent effects of MPP+, it is quite reasonable to expect that N-methyl-isoquinolinium ions are expected to be much more toxic than their parent compounds.

In the peripheral autonomic nervous system, 6,7-dihydroxy-TIQ is taken up by sympathetic nerve terminals.[43,44] Uptake is blocked by desmethylimipramine thereby implicating the NET in accumulation of tetrahydroisoquinlines in sympathetic synapses. In addition, 6,7-dihydroxy-TIQ can accumulate in sympathetic synaptic vesicles after uptake and potentially act as a false neurotransmitter.[44] Therefore, in mammalian systems, both central and peripheral transport mechanisms are in place for the accumulation of TIQs in dopaminergic and noradrenergic cell populations.

ENDOGENOUS

Regional Distribution

As depicted in Table 9.2, several TIQs have been detected and quantified in mammalian brain and human cerebrospinal fluid (CSF). For comparison with the values in Table 9.2, concentrations of dopamine, 3,4-dihydroxyphenylacetic acid (DOPAC), and homovanillic acid (HVA) in human striatum are in the ranges of 3, 1, and 7 μmol/g wet weight, respectively.[45] For comparisons among the TIQs listed in Table 9.2, it is worth noting that the molecular weight of TIQ is 133; therefore, 1 pmol of TIQ is equivalent to 0.133 ηg. The brain concentrations of most individual TIQs are in the range of 10 to 500 pmol/g wet weight. TIQs are not isolated to the basal ganglia. Instead, they appear to be widely distributed in brain. The study by Yamakawa and colleagues,[46] for example, detected TIQ, 1-methyl-TIQ and 1-benzyl-TIQ in both the cerebellum and thalamus in addition to components of the basal ganglia.

Synthetic Pathways

Glial cells and/or neurons may convert relatively innocuous TIQs to more toxic derivatives. For example, TIQ can be methylated to form either N- or 1-methyl-TIQ after crossing the blood-brain barrier and, as shown in Figure 9.1C, subsequent oxidation by an MAO converts N-methyl-TIQ to the more toxic N-methylisoquinolinium ion (NMIQ+). Other neurotoxins in this family may be produced endogenously from catecholamines, particularly dopamine, and either aldehydes or α-keto acids by several combinations of enzymatic and non-enzymatic pathways.

Both enzymatic and nonenzymatic synthesis of the (R)- and (S)-enantiomers of salsolinol is shown in Figure 9.2A. Endogenous formation of (R/S)salsolinol by

TABLE 9.2
Tetrahydrosisoquinolines in Brain

Compound	Species and Location	Concentration	Reference
TIQ	Human, frontal cortex	1 ng/g	Niwa et al. 1987[11]
n-methyl-TIQ	Human, frontal cortex	3 ng/g	Niwa et al. 1987[11]
TIQ	Rat, whole brain	4.2 ng/g	Makino et al. 1988[7]
1-methyl-TIQ	Rat, whole brain	1.7 ng/g	Tasaki et al. 1991[72]
TIQ	Mouse, whole brain	1.1 ng/g	Tasaki et al. 1991[72]
1-methyl-TIQ	Mouse, whole brain	9.8 ng/g	Tasaki et al. 1991[72]
TIQ	Monkey, substantia nigra	140 pmol/g	Yamakawa et al. 1999[46]
TIQ	Monkey, striatum	20 pmol/g	Yamakawa et al. 1999[46]
TIQ	Monkey, cerebrum	80 pmol/g	Yamakawa et al. 1999[46]
TIQ	Monkey, cerebellum	10 pmol/g	Yamakawa et al. 1999[46]
TIQ	Monkey, medulla	30 pmol/g	Yamakawa et al. 1999[46]
TIQ	Monkey, thalamus	25 pmol/g	Yamakawa et al. 1999[46]
1-methyl-TIQ	Monkey, substantia nigra	475 pmol/g	Yamakawa et al. 1999[46]
1-methyl-TIQ	Monkey, striatum	300 pmol/g	Yamakawa et al. 1999[46]
1-methyl-TIQ	Monkey, cerebrum	160 pmol/g	Yamakawa et al. 1999[46]
1-methyl-TIQ	Monkey, cerebellum	90 pmol/g	Yamakawa et al. 1999[46]
1-methyl-TIQ	Monkey, medulla	70 pmol/g	Yamakawa et al. 1999[46]
1-methyl-TIQ	Monkey, thalamus	65 pmol/g	Yamakawa et al. 1999[46]
1-benzyl-TIQ	Monkey, substantia nigra	110 pmol/g	Yamakawa et al. 1999[46]
1-benzyl-TIQ	Monkey, striatum	30 pmol/g	Yamakawa et al. 1999[46]
1-benzyl-TIQ	Monkey, cerebrum	55 pmol/g	Yamakawa et al. 1999[46]
1-benzyl-TIQ	Monkey, cerebellum	15 pmol/g	Yamakawa et al. 1999[46]
1-benzyl-TIQ	Monkey, medulla	10 pmol/g	Yamakawa et al. 1999[46]
1-benzyl-TIQ	Monkey, thalamus	20 pmol/g	Yamakawa et al. 1999[46]
1-benzyl-TIQ	Monkey, cerebral cortex	23.5 nmol/g[a]	Kotake et al. 1996[20]
1-benzyl-TIQ	Monkey, cerebellum	16.9 nmol/g[a]	Kotake et al. 1996[20]
1-benzyl-TIQ	Monkey, striatum	15.3 nmol/g[a]	Kotake et al. 1996[20]
1-benzyl-TIQ	Monkey, thalamus	17.2 nmol/g[a]	Kotake et al. 1996[20]
1-benzyl-TIQ	Monkey, hippocampus	26.2 nmol/g[a]	Kotake et al. 1996[20]
1-benzyl-TIQ	Monkey, spinal cord	21.2 nmol/g	Kotake et al. 1996[20]
1-benzyl-TIQ	Mouse, whole brain	Detected with GC/MS	Kotake et al. 1996[19]
3',4'-dihydroxy-benzyl TIQ	Mouse, whole brain	Detected with GC/MS	Kawai et al. 1998[14]
6,7-dihydroxy-benzyl-TIQ	Mouse, whole brain	Detected with GC/MS	Kawai et al. 1998[14]
(R)salsolinol	Human, frontal cortex	134 pmol/g	Maruyama et al. 1997[82]
(R)salsolinol	Human, caudate	73 pmol/g	Maruyama et al. 1997[82]
(R)salsolinol	Human putamen	38 pmol/g	Maruyama et al. 1997[82]
(R)salsolinol	Human, substantia nigra	95 pmol/g	Maruyama et al. 1997[82]
N-methyl-(R)salsolinol	Human, frontal cortex	ND	Maruyama et al. 1997[82]
N-methyl-(R)salsolinol	Human caudate	66 pmol/g	Maruyama et al. 1997[82]
N-methyl-(R)salsolinol	Human, putamen	110 pmol/g	Maruyama et al. 1997[82]
N-methyl-(R)salsolinol	Human, substantia nigra	77 pmol/g	Maruyama et al. 1997[82]
DMDHIQ+	Human, frontal cortex	134 pmol/g	Maruyama et al. 1997[82]
DMDHIQ+	Human, caudate	73 pmol/g	Maruyama et al. 1997[82]
DMDHIQ+	Human, putamen	38 pmol/g	Maruyama et al. 1997[82]
DMDHIQ+	Human, substantia nigra	95 pmol/g	Maruyama et al. 1997[82]
(S)salsolinol	Human; frontal cortex, caudate, putamen, substantia nigra	ND	Maruyama et al. 1997[82]
N-methyl-(S)salsolinol	Human; frontal cortex, caudate, putamen, substantia nigra	ND	Maruyama et al. 1997[82]
(R)salsolinol	Human, cerebral gray matter	71 pmol/g	Deng et al. 1997[10]
(S)salsolinol	Human, cerebral gray matter	ND	Deng et al. 1997[10]
N-methyl-(R)salsolinol	Human, cerebral gray matter	96 pmol/g	Deng et al. 1997[10]
N-methyl-(S)salsolinol	Human, cerebral gray matter	ND	Deng et al. 1997[10]
(R)salsolinol	Human, nucleus accumbens	16.5 nmol/g	Musshoff et al. 2000[33]

TABLE 9.2
Tetrahydrosisoquinolines in Brain (continued)

Compound	Species and Location	Concentration	Reference
(R)salsolinol	Human, caudate nucleus	12.5 nmol/g	Musshoff et al. 2000[33]
(R)salsolinol	Human, putamen	24.1 nmol/g	Musshoff et al. 2000[33]
(R)salsolinol	Human, substantia nigra	28.6 ng/g	Musshoff et al. 2000[33]
(R)salsolinol	Human, hypothalamus	11.9 ng/g	Musshoff et al. 2000[33]
(S)salsolinol	Human nucleus accumbens	7.8 ng/g	Musshoff et al. 2000[33]
(S)salsolinol	Human caudate nucleus	7.9 ng/g	Musshoff et al. 2000[33]
(S)salsolinol	Human, putamen	14.2 ng/g	Musshoff et al. 2000[33]
(S)salsolinol	Human, substantia nigra	18.5 ng/g	Musshoff et al. 2000[33]
(S)salsolinol	Human, hypothalamus	12.8 ng/g	Musshoff et al. 2000[33]
Norsalsolinol	Human, nucleus accumbens	55.5 ng/g	Musshoff et al. 2000[33]
Norsalsolinol	Human, caudate nucleus	42.1 ng/g	Musshoff et al. 2000[33]
Norsalsolinol	Human putamen	40.0 ng/g	Musshoff et al. 2000[33]
Norsalsolinol	Human, substantia nigra	12.3 ng/g	Musshoff et al. 2000[33]
Norsalsolinol	Human, hypothalamus	4.5 ng/g	Musshoff et al. 2000[33]
N-methyl-norsalsolinol	Human frontal cortex	1 ng/g	Niwa et al. 1991[63]
N-methyl-(R/S)salsolinol	Human, frontal cortex	1 ng/g	Niwa et al. 1991[63]

[a] 1-benzyl-TIQ was detected in a single monkey brain 5 days after a 66-day period of daily subcutaneous injections. DMDHIQ+, 1,2-dimethyl-6,7-dihydroxyisoquinolinium ion. ND, none detected.

Pictet-Spengler condensation of acetaldehyde with dopamine yields equal amounts of each enantiomer. In contrast, putative stereoselective enzymatic synthesis by way of a salsolinol synthase would generate only the (R)-enantiomer.[47] As indicated in Table 9.2, levels of (R)salsolinol tend to be higher than those of (S)salsolinol in human brain tissue, which supports the existence of both mechanisms.[33] Other TIQs may be exclusively generated by Pictet-Spengler condensation; norsalsolinol, for instance, is derived from the condensation of dopamine with formaldehyde (Figure 9.2B).

The methylated and oxidized derivatives of (R)salsolinol and norsalsolinol are much more toxic than their parent compounds. Both methylation and oxidation occur enzymatically. PD patients reportedly have increased N-methyltransferase activity that is not isolated to brain. Using lymphocytes from peripheral blood, Naoi and co-workers[48] found markedly elevated N-methyl-transferase activity in PD patients in comparison with controls. Several N-methyltransferases such as phenylethanolamine N-methyltransferase, nicotinamide N-methyltransferase, and histamine N-methyltransferase that are present in neural tissues could increase methylation of TIQs such as (R)salsolinol and norsalsolinol. In the context of PD, nicotinamide N-methyltransferase (NNMT) has received the most attention.

NNMT, a widely-expressed cytosolic enzyme that requires S-adenosyl methionine, catalyzes the catabolic N-methylation of nicotinamide to N-methylnicotinamide. Theoretically, increases NNMT activity could (1) reduce the levels of nicotinamide available for synthesis of nicotinamide adenine dinucleotide and (2) increase the forma-

tion of N-methylated TIQs.[49] Using immunoblot analysis, NNMT protein levels have been measured in the CSF from PD patients and controls.[50] The CSF levels of NNMT were higher in young PD patients than in age-matched controls. NNMT expression was also higher in PD than in control brains using quantitative immunohistochemistry on postmortem cerebella.[49]

After methylation, TIQs can be oxidized to more potent neurotoxins. As is the case for MPTP, oxidation by MAO may be critical for the neurotoxicity of TIQs. Using MAO-B isolated from liver mitochondria, Booth and co-workers[51] showed that TIQ and N-methyl-TIQ are oxidized at only 3% the rate of MPTP. Thus, TIQ, N-methyl-TIQ and, possibly, other TIQ derivatives are poor substrates for MAO and this may explain their relatively limited neurotoxicity in comparison with MPTP.

As outlined in Figure 9.3, hypothetical biosynthetic schemes for benzyl-TIQs depend on the generation of precursor compounds by MAO. For example, MAO can convert endogenous 2-phenylethylamine (PEA) to phenylacetaldehyde and dopamine to 3,4-dihydroxyphenylacetaldehyde (DOPAL). These four compounds can, via condensation, form three different benzyl-TIQs: 1-benzyl-TIQ, 3',4'-dihydroxy-benzyl-TIQ and 6,7-dihydroxy-benzyl-TIQ.[14]

TOXICITY OF ISOQUINOLINES

INHIBITION OF THE RESPIRATORY CHAIN

Neurotoxins may disrupt a variety of cellular activities such as DNA repair, protein trafficking, and neurotransmitter release. Cell death precipitated by neurotoxin inhibition of oxidative phosphorylation may, however, be the mechanism most relevant to PD. With this perspective, McKnaught and colleagues[6] examined the effects of three classes of isoquinoline derivatives (e.g., isoquinolines, dihydroisoquinolines, TIQs) or MPP[+] on respiratory chain complexes I, II-III, and IV from rat brain mitochondria. The isoquinoline derivatives or MPP[+] were added to reaction mixes at a concentration of 10 mM and complexes I, II-III, and IV activities were compared to control. Under the experimental conditions of their study, MPP[+] produced approximately 70% inhibition of complex I but had no effects on complexes II-III and IV. At 10 mM, isoquinoline and isoquinolinium ion were associated with complete inhibition of complex I, whereas the effects of TIQ (20%), N-methyl-TIQ (65%), (R/S) salsolinol (55%), and N-methyl-(R/S) salsolinol (60%) were less dramatic. The IC$_{50}$ values for N-methyl-TIQ (4.3 mM) and N-methyl-isoquinolinium ion (1.3 mM) reported by McNaught et al.[6] were similar to those provided earlier by Suzuki and co-workers.[52] For comparison, the complex I IC$_{50}$ for MPP[+] is about 4.1 mM.[6] Only isoquinoline, 6,7-dimethoxyisoquinoline and N-methyl-(R/S)salsolinol produced notable inhibition of complex II-III. None of the isoquinoline derivatives inhibited complex IV. For the series of compounds examined, there was no significant general relationship between a lipophilicity measure and the extent of complex I inhibition.

In Vitro Toxicity

The *in vitro* toxicity of TIQ derivatives is summarized in Table 9.3. Like MPTP,[53] several TIQs that have been detected in brain inhibit tyrosine hydroxylase (TH).[15,54] In a comparison of six TIQs (e.g., N-methyl-TIQ, N-methyl-IQ[+], (R/S)salsolinol, N-methyl-(R/S)salsolinol, norsalsolinol, and N-methyl-norsalsolinol), N-methyl-norsalsolinol was the most potent in reducing TH activity.[15] TH enzyme activity was virtually eliminated by 1 mM concentrations of N-methyl-norsalsolinol, N-methyl-IQ[+], and N-methyl-salsolinol. Scholz et al.[54] showed that the inhibition of TH by N-methyl-norsalsolinol was noncompetitive with an IC$_{50}$ of 10 μM. Therefore, investigators of nigral neurotoxins should bear in mind that reductions in dopamine content may be a manifestation of TH inhibition rather than an indication of dopaminergic cell death.

Early cell culture studies have provided convincing evidence of isoquinoline neurotoxicity.[22,23] More recent studies have explored mechanisms of dopaminergic cell death produced by TIQs using both molecular and morphological assays. In general, N-methylated derivatives are more toxic than their nonmethylated parent compounds, and (R)-enantiomers are more toxic than (S)-enantiomers. Among the TIQ derivates, N-methyl-(R)salsolinol may be the most potent dopaminergic neurotoxin

and produces cell death through apoptotic mechanisms. Using an assay of DNA fragmentation, the (R)-enantiomer of N-methyl-(R/S)salsolinol is at least tenfold more toxic than the (S)-enantiomer.[55] In addition to DNA fragmentation, apoptosis has been demonstrated with TUNEL labeling,[55] Hoechst 33342 staining,[56] and Western blot detection of caspase-3 activation.[56] Dopaminergic cells (SH-SY5Y) overexpressing Bcl-2 are resistant to apoptosis induced by N-methyl(R)salsolinol.[57] In addition, the MAO inhibitors deprenyl and rasagiline prevent cell death produced by N-methyl(R)salsolinol.[18,57]

In Vivo Neurotoxicity

In vivo neurotoxicological studies of TIQ and its derivatives have failed to provide convincing evidence of cell death within the SNpc; this failure may be, in part, due to methodological issues. For example, no publication to date has employed rigorous stereological counting methods and highly sensitive markers of neurodegeneration such as Fluoro-Jade® B and amino cupric silver staining after challenge with TIQ or its derivatives. Furthermore, small sample sizes, variable routes of compound administration, and wide-ranging dosing durations compromise any effort to summarize the literature on TIQ neurotoxicity. Moreover, few studies of TIQ *in vivo* neurotoxicity have combined morphological, behavioral, and biochemical endpoints into a cohesive analysis.

Two early studies in marmosets suggest that the both the behavioral and biochemical effects of TIQs could be due to TH inhibition rather than cell death within the SNpc. Nagatsu and Yoshida[12] injected marmosets with TIQ subcutaneously for 11 days. This dosing regimen was associated with parkinsonism, reduced striatal dopamine, and reduced TH activity. However, they did not describe cell death within the SNpc. In squirrel monkeys, TIQ was injected subcutaneously for up to 104 days at a dosage of 20 mg/kg-day.[3] These monkeys exhibited rigidity and bradykinesia, reduced nigral dopamine, and decreased TH activity. Again, there was no indication of anatomical changes within the substantia nigra.

Nigral cell death was also not reported in two investigations of TIQ neurotoxicity in mice;[13,58] TIQ was injected subcutaneously on a daily basis for at least four weeks in both studies. In C57BL/6J mice, reduced TH immunoreactivity in the absence of overt cell death was detected.[13] Surprisingly, there was no reduction in striatal dopamine, DOPAC or HVA in C57BL mice.[58]

The N-methylated derivative of TIQ, N-methyl-TIQ, and its oxidized product, N-methyl-IQ[+], have as well been examined for *in vivo* neurotoxicity. In a study by Perry and colleagues,[59] after N-methyl-TIQ was injected into marmosets subcutaneously, three out of four animals died; disappointingly, postmortem pathological findings were not reported. The one surviving marmoset showed

TABLE 9.3
In Vitro Toxicity of Tetrahydroisoquinolines

Compound(s)	Cell Line(s) or Tissue	Methods	Results	Reference
N-methyl-IQ⁺	Mesencephalic dopaminergic neurons	Cell counts	Death of dopaminergic neurons	Nijima et al. 1991,[22] Nishi et al. 1994[23]
N-methyl-TIQ, N-methyl-IQ⁺, (R/S)salsolinol, N-methyl-(R/S) salsolinol, norsalsolinol, N-methyl-norsalsolinol	Rat pheochromocytoma PC12h cells	Protein quantification, TH activity, ATP levels, subcellular accumulation		
(R)salsolinol, (S salsolinol, DMDHIQ⁺, N-methyl-(R)salsolinol	Dopaminergic neuroblastoma SH-SY5Y cells	Comet assay for DNA fragmentation, TUNEL staining for apoptosis	Only N-methyl-(R)salsolinol produces apoptotic cell death (5% with 100 μM concentration)	Maruyama et al. 1997[55]
N-methyl-norsalsolinol	Rat nucleus accumbens homogenate	TH activity	N-methyl-(R)salsolinol inhibited TH activity with an IC₅₀ of 10 μM	Scholz et al. 1997[54]
(R/S)salsolinol	Dopaminergic neuroblastoma SH-Sy5Y cells	MTT assay, Tyrpan blue exclusion, ATP/ADP ratio	(R/S) Salsolinol decreased cell survival and intracellular ATP content	Storch et al. 2000[83]
N-methyl-(R)salsolinol	Dopaminergic neuroblastoma SH-SY5Y cells	Markers of DNA damage and mitochondrial membrane potential	Apoptotic cell death	Naoi et al. 2000,[18] 2002[84]
(R/S) salsolinol	Nigral dopaminergic cell line, SN4741	XTT assay, DNA fragmentation, measurement of reactive oxygen species and caspase activation	Apoptotic cell death (100%) at 48 hr. produced by 1mM salsolinol	Chun et al. 2001[85]
1-benzyl-TIQ	Organotypic slice co-culture of ventral midbrain and striatum	Dopamine content, morphological changes	Dopamine content decreased in both a dose- and time-dependent manner. Morphological changes included some shrinkage and distortion of dendritic morphology.	Kotake et al. 2003[21]

DMDHIQ⁺, 1,2-dimenthyl-6,7-dihydroxyisoquinolinium ion.

no reduction in striatal dopamine, DOPAC, or HVA. Rhesus monkeys exposed to prolonged subcutaneous administration of N-methyl-TIQ showed only slight motor disturbances.[3] In mice, long-term (20-week) intraperitoneal injection of N-methyl-TIQ was associated with atrophy of the SNpc and reduced nigral TH immunoreactivity.[60] In rats, N-methyl-IQ⁺ showed toxicity similar to that of MPP⁺ when directly infused into the striatum.[51] However, in cynomolgous monkeys, N-methyl-IQ⁺ was associated with only mild bradykinesia after eight weeks of intraventricular administration.[3] The potential neurotoxicity of N-methyl-IQ⁺ should not be dismissed though, since parenchymal penetration of intraventricularly-administered drugs is frequently quite limited.

In contrast to other several other TIQs, the effects of N-methyl-(R)salsolinol have been thoroughly detailed.[17] In the study by Naoi et al.,[17] N-methyl-(R)salsolinol was delivered to rats by intrastriatal infusion with a mini-osmotic pump over a period of one week. Parkinsonian behavioral changes, including hypokinesia, stiff tail, and postural abnormalities, were associated with infusions of N-methyl-(R)salsolinol. Dopamine and norepinephrine levels were reduced in both the striatum and substantia nigra. For instance, striatal dopamine was reduced to 18% of control values. Neuronal density was determined in sections either stained with an antibody to TH or processed by the Klüver-Barrera (K-B) method; markers of apoptosis were not employed. On the side of intrastriatal N-methyl-(R)salsolinol infusion, TH-immunoreactive neurons were reduced by more than 50%. The density of K-B stained neurons was also reduced although the differences between control and injected sides did not reach statistical significance. Therefore, although intriguing, the work by Naoi and his cohort did not show that N-methyl-(R)salsolinol can cause apoptotic cell death *in vivo*.

The *in vivo* effects of several other TIQ derivatives such as 1-benzyl-TIQ and N-methyl-norsalsolinol are worth mentioning. In a study by Moser and colleagues,[61] N-methyl-norsalsolinol was injected into the left medial forebrain bundle of Wistar rats, and both the behavioral response to apomorphine and striatal catecholamine levels were determined three weeks later. There was no response to apomorphine, and the mild reductions in dopamine, DOPAC, and HVA that were noted did not reach statistical significance. In a single cynomolgous monkey, there were no behavioral or morphological effects of intraventricular N-methyl-norsalsolinol administered over an eight week period.[3]

The *in vivo* effects of 1-benzyl-TIQ and benzy-TIQ derivatives have been mixed. In a single monkey, chronic subcutaneous injection of 1-benzyl-TIQ produced parkinsonism, although morphological and biochemical effects were not reported.[20] In the original study of 1-benzyl-TIQ neurotoxicity in the mouse, motor testing was performed after short-term (i.e., five-day) intraperitoneal injections.[19] Mice showed mild parkinsonism that was corrected by carbidopa/levodopa. In a subsequent study by Abe and co-workers,[34] 1-benzyl-TIQ was injected intraperitoneally for four days prior to motor testing and quantification of catecholamines in the striatum. Even though the mice receiving 1-benzyl-TIQ were bradykinetic, they showed no reductions in striatal dopamine levels or TH+ neurons in the SNpc. Experimental findings were quite different in another study that was also published in 2001.[62] Intraperitoneal injections of 1-benzyl-TIQ were performed for 17 days in mice prior to quantification of striatal catecholamines. The 17-day protocol produced a 60% reduction in striatal dopamine levels. Chronic injection of 1-benzyl-TIQ was also associated with muscular rigidity. The benzyl-TIQ derivative 3'4'-dihydroxy-benzyl-TIQ induced parkinsonism in mice after five days of intraperitoneal administration; striatal dopamine levels and SNpc morphology were not analyzed.[14] Using the same experimental paradigm, 1-benzyl-6,7-dihydroxy-TIQ had no behavioral effects. The *in vivo* neurotoxicity of all TIQ derivatives is reviewed in Table 9.4.

TETRAHYDROISOQUINOLINES IN HUMAN PARKINSON'S DISEASE

PRESENCE IN BRAIN AND BODY FLUIDS

The body fluid and brain concentrations of several TIQ derivatives have been compared between PD patients and controls, although the data on PD brain tissue is meager (Table 9.5). For instance, single PD and control brains were used in a 1987 study by Niwa et al.[11] Therefore, no reasonable conclusions can be made about their report that TIQ was present at a tenfold greater concentration in the PD than the normal brain. In another study by the same group, only estimated wet tissue concentrations (1 ηg/g) for both N-methyl-norsalsolinol and N-methyl-(R/S)salsolinol were provided.[63] More rigorous topological analysis of TIQ derivatives in PD brain is sorely needed.

(R/S)salsolinol, N-methyl-norsalsolinol, N-methyl-(R)salsolinol, and 1-benzyl-TIQ have been detected in PD CSF. (R/S)salsolinol is also present in plasma and urine in both PD patients and controls. In a study by Müller and associates,[64] concentrations of both enantiomers of salsolinol were higher in plasma than in CSF. Furthermore, both enantiomers were, on average, at lower concentrations in PD patients than in controls, although the differences did not reach statistical significance. The presence of both enantiomers argues against enantioselective synthesis of (R)salsolinol in the body.

The CSF concentrations of two N-methylated TIQ derivates have been correlated to PD disease duration. In one publication, N-methyl-(R)salsolinol levels were reported to decrease after two years of treatment with levodopa.[65] In another study, N-methyl-norsalsolinol was detected in the CSF of PD patients but not in controls.[66] The presence of N-methyl-norsalsolinol in CSF was negatively correlated with disease duration. The results of human CSF studies suggest that (1) dying dopaminergic neurons release both *N*-methyl-(*R*)salsolinol and *N*-methyl-norsalsolinol into the CSF, (2) healthy neurons release little *N*-methyl-(*R*)salsolinol and *N*-methyl-norsalsolinol into the CSF, and (3) at later stages of PD, there are too few dying cells to release significant amounts of *N*-methyl-(*R*)salsolinol and *N*-methyl-norsalsolinol into the CSF.

RELATIONSHIP TO OTHER ENVIRONMENTAL RISK FACTORS

A substantial body of evidence suggests that the development of idiopathic PD is due to exogenous and/or endogenous toxins interacting with a genetically and/or environmentally vulnerable host. In addition, the risk of PD is influenced by nonessential host behavioral activities such as smoking cigarettes and drinking coffee.[67,68] The metabolism of TIQ highlights the complex interactions between host and environment. For instance, once accumulated in brain, some components in tobacco smoke may limit the conversion of TIQ to its more toxic derivative, N-methyl-TIQ.[69] Chronic administration of cigarette smoke solution also reduces brain tissue levels of TIQ in rats.[70] Since epidemiological studies have shown that cigarette smoking is associated with a lowered incidence of PD,[67] it is tempting to speculate that reduced brain TIQ and decreased conversion of TIQ to N-methyl-TIQ explains the epidemiological effect of cigarette smoking on PD incidence.

IS 1-METHYL-TIQ A NEUROPROTECTIVE TETRAHYDROISOQUINOLINE?

One simple TIQ derivative, 1-methyl-TIQ, which is present in foods,[8] has been detected in the brain[46] and may

TABLE 9.4
In Vivo **Neurotoxicity of Tetrahydroisoquinolines**

Compounds	Species	Route of Administration	Effects	Reference
N-methyl-TIQ	Marmosets	Subcutaneously (10–140 mg/kg) with cumulative dose of 420 mg/kg in a single surviving marmoset	Three out of four marmosets died. No reduction in striatal dopamine, DOPAC or HVA.	Perry et al., 1987[59]
TIQ	Marmosets	Subcutaneously (50 mg/kg) × 11 days	Parkinsonism, reduced nigral and striatal dopamine levels and TH activity	Nagatsu and Yoshida, 1988[12]
TIQ	C57BL mice	Subcutaneously (60–150 mg/kg) 5 ×/wk × 4 wk	No reduction in striatal dopamine, DOPAC or HVA	Perry et al., 1988[58]
TIQ	C57BL/6J mice	Subcutaneously (50 mg/kg) daily or 70 days	Reduced TH immunoreactivity but no cell death via cresyl violet staining	Ogawa et al., 1989[13]
TIQ, 1-methyl-TIQ, N-methyl-TIQ, N-methyl-IQ⁺	Sprague-Dawley rats	Intrastriatal infusion via microdialysis cannula with measurement of dopamine	N-methyl-IQ⁺ most potent toxin with similarities to MPP⁺	Booth et al., 1989[51]
TIQ	Squirrel monkeys	Subcutaneus (20 mg/kg) daily for up to 104 days	Parkinsonism: rigidity, bradykinesia. Reduced SN but not striatal dopamine and TH activity.	Yoshida et al., 1990[86]
N-methyl-(R/S)salsolinol	Wistar rats	Intrastriatal infusion	Parkinsonism, reduced striatal dopamine and norepinephrine, decreased numbers of TH(+) neurons	Naoi et al., 1996[17]
N-methyl-norsalsolinol	Wistar rats	Stereotactically into left medial forebrain bundle	Increased ipsiversive rotation with amphetamine. No SNpc cell loss.	Moser et al., 1996[61]
N-methyl-norsalsolinol	Cynomolgous monkey	Intraventricular × 8 wk	No behavioral or morphological abnormalities.	Yoshida et al., 1993[3]
N-methyl-TIQ	Rhesus and crab-eating monkeys	Subcutaneously (10 mg/k) 2 ×/day for 100 days	Slight motor disturbances	Yoshida et al., 1993[3]
N-methyl-TIQ	C57BL/6J mice	Intraperitoneally 6×/wk × 20 wk	Atrophy of SNpc with reduced TH immunoreactivity	Fukuda et al., 1994[60]
N-methyl-IQ⁺	Cynomologous monkey	Intraventricular × 8 wk	Mild bradykinesia but no morphological abnormalities	Yoshida et al., 1993[3]
1-benzyl-TIQ	C57BL/6N mice	Intraperitoneally 2×/day × 4 days	No loss of TH(+) neurons. Increased striatal dopamine.	Abe et al., 2001[34]
1-benzyl-TIQ	Cynomologous monkey	Subcutaneous (22 mg/kg) daily × 17 days	Parkinsonism: rigidity, akinesia, resting tremor	Kotake et al., 1996[20]
1-benzyl-TIQ	Wistar rats	Intraperitoneal (50 mg/kg) daily × 17 days	Decreased (60%) striatal dopamine	Antkiewicz-ichaluk et al., 2001[62]
1-methyl-TIQ	Wistar rats	Intraperitoneal (100 mg/kg) daily × 17 days	No changes in striatal dopamine. Reduced DOPAC (60%) levels	Antkiewicz-Michaluk et al., 2001[62]

be neuroprotective.[71] It is the only neuroprotective TIQ derivative that has been described to date and prevents the neurotoxic effects of MPP⁺ and other TIQ derivatives, possibly by "shielding" complex I.[71] The enantiomeric ratio (R/S) of 1-methyl-TIQ in mouse brain is 0.6.[72] Enzymatic synthesis of 1-methyl-TIQ from 2-phenylethylamine and pyruvate has been proposed as one explanation for this enantiomeric ratio.[73] The brain ratios of 1-methyl-TIQ to the potentially neurotoxic TIQs may be critical to the pathogenesis of idiopathic PD. Along these lines, Absi

et al.[74] showed decreased 1-methyl-TIQ synthesizing enzyme activity in aged rats (24 months old) with a 50% reduction in substantia nigra and striatum in comparison to young rats (3 months old).

ISOQUINOLINES IN OTHER NEURODEGENERATIVE DISEASES

In addition to epidemiological data and the striking effects of MPTP, a role for neurotoxins in the pathogenesis of

TABLE 9.5
Tetrahydroisoquinolines in Parkinson's Disease

Compound	Clinical Features	Source	Concentration PD	Concentration Controls	Reference
Salsolinol	PD on levodopa	Urine	>200 μg/12 hr	>12 μg/12 hr	Sadler et al. 1973[87]
Salsolinol	Early and late PD	CSF	Early PD-119 ng/ml Late PD-222 ng/ml	118 ng/ml	Antkiewicz-Michaluk et al. 1997[88]
(R)salsolinol	PD	Plasma	441 pg/ml	1300 pg/ml	Müller et al. 1999[64]
(S)salsolinol	PD	Plasma	418 pg/ml	851 pg/ml	Müller et al. 1999[64]
(R)salsolinol	PD	CSF	179 pg/ml	427 pg/ml	Müller et al. 1999[64]
(S)salsolinol	PD	CSF	159 pg/ml	245 pg/ml	Müller et al. 1999[64]
TIQ	PD	Frontal cortex	10 ng/g	1 ng/g	Niwa et al. 1987[11]
N-methyl-TIQ	PD	Frontal cortex	2 ng/g	3 ng/g	Niwa et al. 1987[11]
N-methyl-norsalsolinol	PD	Frontal cortex	1 ng/g	1 ng/g	Niwa et al. 1991[63]
N-methyl-(R/S)salsolinol	PD	Frontal cortex	1 ng/g	1 ng/g	Niwa et al. 1991[63]
N-methyl-norsalsolinol	PD	CSF	Presence negatively correlated with disease duration	Not detected in a control group of 15 subjects	Moser et al. 1995[66]
N-methyl-(R)salsolinol	PD on levodopa	CSF	N-methyl-(R)salsolinol: 6.15 nM, HVA: 250 nM. Levels decreased after 2 yr of treatment with levodopa	No control group	Maruyama et al. 1999[65]
1-benzyl-TIQ	PD	SF	1.2 ng/ml	0.4 ng/ml	Kotake et al. 1995[19]

idiopathic PD is also suggested by the conspicuous incidence of atypical parkinsonism on the island of Guadeloupe that may be related to the oral consumption of fruits like soursop, sweetsop, and custard-apple and teas made from the leaves and bark of trees in the *Annonaceae* family.[75] These fruits and drinks contain neurotoxins like reticuline, a benzyl-TIQ derivative.

Actually, the phenotype of Guadeloupean Parkinsonism more closely resembles progressive supranuclear palsy (PSP) than PD. Patients with Guadeloupean Parkinsonism present with relatively symmetrical rigidity and bradykinesia. They respond poorly to levodopa in the long-term despite a positive early response in some patients. Frequently, they also exhibit ophthalmoparesis, pseudo-bulbar palsy and frontal lobe signs. In a series of 96 patients with Guadeloupean parkinsonism, all were homozygous for the H1 tau haplotype.[76] When compared to PSP and Guadeloupean parkinsonism, the frequency of H1 tau homozygotes is only mildly increased in PD patients compared to controls.[77,78] Therefore, interactions between tau and neurotoxins present in *Annonaceae* and possibly other fruit trees may play a much smaller role in idiopathic PD than in Guadeloupean parkinsonism and PSP.

Several neurotoxins present in *Annonaceae* species are quite potent. Benzyl-TIQs (e.g., reticuline), acetogenins, and the tetrahydroprotoberberine corexime present in the fruit, bark, and roots of *Annona muricata* are toxic to cultured nigral neurons. After 24 h of exposure to reticuline (100 μg/ml), 50% of cultured dopaminergic neurons degenerated.[79] Cell death caused by reticuline does not require the DAT and is not specific for dopaminergic or monoaminergic neurons.

Acetogenins like annonacin are more potent inhibitors of complex I than most TIQ derivatives and produce apoptotic cell death.[80,81] Acetogenin neurotoxins do not require uptake by the dopamine transporter,[79] which is consistent with their diffuse pathological effects. When infused into rats for 28 days, annonacin induced loss of nigral dopaminergic neurons and produced neurodegenerative changes in the basal ganglia and brain stem.[80] Annonacin is a more potent dopaminergic neurotoxin than either rotenone or MPP+.

SUMMARY

The most straightforward explanation for the relatively selective dopaminergic cell death seen in PD is the accumulation of neurotoxin(s) derived from dopamine. Neurotoxins with a TIQ moiety can be formed either enzymatically or nonenzymatically from dopamine and other small molecules that are normal constituents of most cells. An aggregate of TIQ derivatives may accumulate in dopaminergic neurons and, on reaching critical levels, they could produce oxidative stress by inhibiting complex I of the respiratory chain. There has been essentially no evolutionary pressure to prevent synthesis of these unwanted by-products of normal metabolic pathways. Dopaminergic neurons in elderly persons with marginal oxidative phosphorylation capacity may not tolerate a buildup of complex I inhibitors; as a result, cells die, and Parkinsonism emerges.

ACKNOWLEDGMENTS

This work was supported, in part, by a grant from the University of Tennessee Center for Neurobiology of Brain Disease.

REFERENCES

1. Langston, J. W. et al., 1983. Chronic parkinsonism in humans due to a product of meperidine-analog synthesis, *Science*, 219, 979, 1983.
2. Ballard, P. A., Tetrud, J. W., and Langston, J. W., Permanent human parkinsonism due to 1-methyl-4-phenyl-1,2,3,6-tetrahydropyridine (MPTP): seven cases, *Neurology*, 35, 949, 1985.
3. Yoshida, M. et al., Parkinsonism produced by tetrahydroisoquinoline (TIQ) or the analogues, *Adv. Neurol.*, 60, 207, 1993.
4. Kajita, M. et al., Endogenous synthesis of *N*-methylnorsalsolinol in rat brain during *in vivo* microdialysis with epinine, *J. Chromatogr.*, B654, 263, 1993.
5. Nagatsu, T., Isoquinoline neurotoxins in the brain and Parkinson's disease, *Neurosci. Res.*, 29, 99, 1997.
6. McNaught, K. S. et al., Inhibition of complex I by isoquinoline derivatives structurally related to 1-methyl-4-phenyl-1,2,3,6-tetrahydropyridine (MPTP), *Biochem. Pharmacol.*, 50, 1903, 1995.
7. Makino, Y. et al., 1988. Presence of tetrahydroisoquinoline and 1-methyl-tetrahydroisoquinoline in foods: compounds related to Parkinson's disease, *Life Sci.*, 43, 373, 1988.
8. Makino, Y. et al., Confirmation of the enantiomers of 1-methyl-1,2,3,4-tetrahydroisoquinoline in the mouse brain and foods applying gas chromatography/mass spectrometry with negative ion chemical ionization, *Biomed. Environ. Mass Spectrom.*, 19, 415, 1990.
9. Niwa, T. et al., Presence of tetrahydroisoquinoline, a parkinsonism related compound, in foods, *J. Chromatogr.*, 493, 345, 1989.
10. Deng, Y. et al., Assay for the (R)- and (S)-enantiomers of salsolinols in biological samples and foods with ion-pair high-performance liquid chromatography using beta-cyclodextrin as a chiral mobile phase additive, *J. Chromatogr. B. Biomed. Sci. Appl.*, 689, 313, 1997.
11. Niwa, T. et al., Presence of tetrahydroisoquinoline and 2-methyl-tetrahydroisoquinoline in parkinsonian and normal human brains, *Biochem. Biophys. Res. Comm.*, 144, 1084, 1987.
12. Nagatsu, T. and Yoshida, M., An endogenous substance of the brain, tetrahydroisoquinoline, produces parkinsonism in primates with decreased dopamine, tyrosine hydroxylase and biopterine in the nigrostriatal regions, *Neurosci. Lett.*, 87, 178, 1988.
13. Ogawa, M. et al., The effect of 1,2,3,4-tetrahydroisoquinoline (TIQ) on mesencephalic dopaminergic neurons in C57BL/6J mice: immunohistochemical studies-tyrosine hydroxylase, *Biogenic Amines*, 6, 436, 1989.
14. Kawai, H. et al., Novel endogenous 1,2,3,4-tetrahydroisoquinoline derivatives: uptake by dopamine transporter and activity to induce parkinsonism, *J. Neurochem.*, 70, 745, 1998.
15. Maruyama, W. et al., Cytotoxicity of dopamine-derived 6,7-dihydroxy-1,2,3,4-tetrahydroisoquinolines, *Adv. Neurol.*, 60, 207, 1993.
16. Maruyama, W. et al., A neurotoxin N-methyl(R)salsolinol induces apoptotic cell death in differentiated human dopaminergic neuroblastoma SH-SY5Y cells, *Neurosci. Lett.*, 232, 147, 1997.
17. Naoi, M et al., Dopamine-derived endogenous 1(R),2(N)-dimethyl-6,7-dihydroxy-1,2,3,4-tetrahydroisoquinoline, N-methyl-(R)-salsolinol, induced parkinsonism in rat: biochemical, pathological, and behavioral studies, *Brain Res.*, 709, 285, 1996.
18. Naoi, M. et al., Apoptosis induced by an endogenous neurotoxin, *N*-methyl(R)salsolinol, in dopamine neurons, *Toxicology* 153, 123, 2000.
19. Kotake, Y. et al., 1-benzyl 1-2-3-4 tetrahydroisoquinoline as a parkinsonism-inducing agent: a novel endogenous amine in mouse brain and parkinsonian CSF, *J. Neurochem.* 65, 2633, 1995.
20. Kotake, Y. et al., Chronic administration of 1-benzyl-1,2,3,4-tetrahydroisoquinoline, an endogenous amine in the brain, induces parkinsonism in a primate, *Neurosci. Lett.*, 217, 69, 1996.
21. Kotake, Y. et al., Neurotoxicity of an endogenous brain amine, 1-benzyl-1,2,3,4-tetrahydroisoquinoline, in organotypic slice co-culture of mesencephalon and striatum, *Neuroscience*, 117, 63, 2003.
22. Nijima, K. et al., N-methylisoquinolinium ion (NMIQ⁺) destroys cultured mesencephalic dopamine neurons, *Biogenic Amines*, 8, 61, 1991.
23. Nishi, K. et al., Neurotoxic effects of 1-methyl-4-phenylpyridinium (MPP⁺) and tetrahydroisoquinoline derivatives on dopaminergic neurons in ventral mesencephalic-striatal co-culture, *Neurodegeneration*, 3, 33, 1994.
24. Tanner, C.M. et al., Parkinson disease in twins: an etiologic study, *JAMA*, 281, 341, 1999.
25. Rommelspacher, H. and Susilo, R., Tetrahydroisoquinolines and beta-carbolines: putative natural substances in plants and mammals, *Prog. Drug Res.*, 29, 415, 1985.
26. Siddhuraju, P., Becker, K., and Makkar, H. P., Studies on the nutritional composition and antinutritional factors of three different germplasm seed materials of an underutilized tropical legume, Mucuna pruriens var. utilis, *J. Agric. Food Chem.*, 48, 6048, 2000.

27. Collins, M. A. et al., Brain and plasma tetrahydroiso-quinolines in rats: effects of chronic ethanol intake and diet, *J. Neurochem.*, 55, 1507, 1990.

28. Ohta, S. et al., Metabolism and brain accumulation of tetrahydroisoquinoline (TIQ) a possible parkinsonism inducing substance, in an animal model of a poor debrisoquine metabolizer, *Life Sci.*, 46, **599**, 1990.

29. Bon, M. A. et al., Neurogenetic correlates of Parkinson's disease: apolipoprotein-E and cytochrome P450 2D6 genetic polymorphism, *Neurosci. Lett.*, 266, 149, 1999.

30. Kikuchi, K. et al., Metabolism and penetration through blood-brain barrier of parkinsonism-related compounds. 1,2,3,4-tetrahydroisoquinoline and 1-methyl-1,2,3,4-tet-rahydroisoquinoline, *Drug Metab. Dispos.*, 19, 257, 1991.

31. Thümen, A. et al., N-Methyl-norsalsolinol, a putative dopaminergic neurotoxin, passes through the blood-brain barrier *in vivo*, *Neuroreport*, 13, 25, 2002.

32. Origitano, T., Hannigan, J., and Collins, M. A., Rat brain salsolinol and blood-brain barrier. *Brain Res.*, 224, **446**, 1981.

33. Musshoff, F. et al., Determination of dopamine and dopamine-derived (R)-/(S)-salsolinol and norsalsolinol in various human brain areas using solid-phase extrac-tion and gas chromatography/mass spectrometry, *Foren-sic Sci. Int.*, 113, 359, 2000.

34. Abe, K. et al., Biochemical and pathological study of endogenous 1-benzyl-1,2,3,4,-tetrahydroisoquinoline-induced parkinsonism in the mouse, *Brain Res.*, 907, 134, 2001.

35. Jellinger, K.A., Post mortem studies in Parkinson's dis-ease—is it possible to detect brain areas for specific symptoms?, *J. Neural. Transm. Suppl.*, 56, 1, 1999.

36. Gesi, M. et al., The role of the locus coeruleus in the development of Parkinson's disease, *Neurosci. Biobe-hav. Rev.*, 24, 655, 2000.

37. Del Tredici, K. et al., Where does Parkinson disease pathology begin in the brain? *J. Neuropathol. Exp. Neu-rol.*, 61, 413, 2002.

38. Den Hartog Jager, W. A. and Bethlem, J., The distribu-tion of Lewy bodies in the central and autonomic ner-vous systems in idiopathic paralysis agitans, *J. Neurol. Neurosurg. Psychiatry*, 23, 283, 1960.

39. Rajput, A. H. and Rozdilsky, B., Dysautonomia in Par-kinsonism: a clinicopathological study, *J. Neurol. Neur-surg. Psychiatry*, 39, 1092, 1976.

40. Gainetdinov, R. R. et al., Dopamine transporter is required for *in vivo* MPTP neurotoxicity: evidence from mice lacking the transporter, *J. Neurochem.*, 69, 1322, 1997.

41. Moron, J.A. et al., Dopamine uptake through the nore-pinephrine transporter in brain regions with low levels of the dopamine transporter: evidence from knock-out mouse lines, *J. Neurosci.*, 22, 389, 2002.

42. Storch, A. et al., Selective dopaminergic neurotoxicity of isoquinoline derivatives related to Parkinson's dis-ease: studies using heterologous expression systems of the dopamine transporter, *Biochem. Pharmacol.*, 63, 909, 2002.

43. Cohen, G., Mytilineou, C., and Barrett, R. E., 6,7-dihy-droxytetrahydroisoquinoline: uptake and storage by peripheral sympathetic nerve of the rat, *Science*, 175, 1269, 1972.

44. Tennyson, V. M. et al., 6,7-dihydroxytetrahdroisoquin-oline: electron microscopic evidence for uptake into the amine-binding vesicles in sympathetic nerves of rat iris and pineal gland, *Brain Res.*, 51, 161, 1973.

45. Konradi, C. et al., Variations of monoamines and their metabolites in the human brain putamen, *Brain Res.*, 579, 285, 1992.

46. Yamakawa, T. et al., Regional distribution of parkin-sonism-preventing endogenous tetrahydroisoquinoline derivatives and an endogenous parkinsonism-preventing substance-synthesizing enzyme in monkey brain, *Neu-rosci. Lett.*, 276, 68, 1999.

47. Naoi, M. et al., A novel enzyme enantio-selectively syn-thesizes (R)salsolinol, a precursor of a dopaminergic neurotoxin, N-methyl-(R)salsolinol, *Neurosci. Lett.*, 212, 183, 1996.

48. Naoi, M. et al., (R)salsolinol N-methyltransferase activ-ity increases in parkinsonian lymphocytes, *Ann. Neurol.*, 43, 212, 1998.

49. Parsons, R. B. et al., High expression of nicotinamide N-methyltransferase in patients with idiopathic Parkin-son's disease, *Neurosci. Lett.*, 342, 13, 2003.

50. Aoyama, K. et al., Nicotinamide-N-methyl-transferase is higher in the lumbar cerebrospinal fluid of patients with Parkinson's disease, *Neurosci. Lett.*, 298, 78, 2001.

51. Booth, R. G., Castagnoli, N., Jr., and Rollema, H., Intracerebral microdialysis neurotoxicity studies of quinoline and isoquinoline derivatives related to MPTP/MPP+, *Neurosci. Lett.*, 100, 306, 1989.

52. Suzuki, K. et al., Selective inhibition of complex I by N-methylisoquinolinium ion and N-methyl-1,2,3,4-tet-rahydroisoquinoline in isolated mitochondria prepared from mouse brain, *J. Neurol. Sci.*, 109, 219, 1992.

53. Hirata, Y. and Nagatsu, T., Inhibition of tyrosine hydrox-ylation in tissue slices of the rat striatum by 1-methyl-4-phenyl-1,2,3,6-tetrahydropyridine, *Brain Res.*, 337, 193, 1985.

54. Scholz, J., Bamberg, H., and Moser, A., N-methyl-nor-salsolinol, an endogenous neurotoxin, inhibits tyrosine hydroxylase activity in the rat brain nucleus accumbens *in vitro*, *Neurochem. Int.*, 31, 845, 1997.

55. Maruyama, W. et al., An endogenous dopaminergic neu-rotoxin, N-methyl(R)salsolinol, induces DNA damage in human dopaminergic neuroblastoma SH-SY-SY5Y cells, *J. Neurochem.*, 69, 322, 1997.

56. Akao, Y. et al., Apoptosis induced by an endogenous neurotoxin, N-methyl(R)salsolinol, is mediated by acti-vation of caspase 3, *Neurosci. Lett.*, 267, 153, 1999.

57. Maruyama, W. et al., Transfection-enforced Bcl-2 over-expression and an antiparkinson drug, rasagiline, pre-vent nuclear accumulation of glyceraldehyde-3-phosphate dehydrogenase induced by an endogenous dopaminergic neurotoxin, N-methyl(R)salsolinol, *J. Neurochem.*, 78, 727, 2001.

58. Perry, T. L., Jones, K., and Hansen, S., Tetrahydroiso-quinoline lacks dopaminergic nigrostriatal neurotoxicity in mice, *Neurosci. Lett.*, 85, 101, 1988.

59. Perry, T. L. et al., 4-phenylpyridine and three other ana-logues of 1-methyl-4-phenyl-1,2,3,6-tetrahydropyridine lack dopaminergic nigrostriatal neurotoxicity in mice and marmosets, *Neurosci. Lett.*, 75, 65, 1987.

60. Fukuda, T., 2-methyl-1,2,3,4-tetrahydroisoquinoline does dependently reduce the number of tyrosine hydrox-ylase-immunoreactive cells in the substantia nigra and the locus ceruleus of C57BL/6J mice, *Brain Res.*, 639, 325, 1994.

61. Moser, A. et al., Rotational behaviour and neurochemi-cal changes in unilateral *N*-methyl-norsalsolinol and 6-hydroxydopamine lesioned rats, *Exp. Brain Res.*, 112, 89, 1996.

62. Antkiewicz-Michaluk, L. et al., Different action on dopamine catabolic pathways of two endogenous 1,2,3,4-tetrahydroisoquinolines with similar antidopam-inergic properties, *J. Neurochem.*, 78, 100, 2001.

63. Niwa, T. et al., Presence of 2-methyl-6,7-dihydroxy-1,2,3,4-tetrahydroisoquinoline and 1,2-dimethyl-6,7-dihydroxy-1,2,3,4-tetrahydroisoquinoline, novel endog-enous amines, in parkinsonism and normal human brains, *Biochem. Biophys. Res. Comm.*, 177, 603, 1991.

64. Müller, T. et al., No increase of synthesis of (R) salsoli-nol in Parkinson's disease, *Mov. Disord.*, 14, 514, 1999.

65. Maruyama, W. et al., An endogenous MPTP-like dopaminergic neurotoxin, N-methyl(R)salsolinol, in the cerebrospinal fluid decreases with progression of Par-kinson's disease, *Neurosci. Lett.*, 262, 13, 1999.

66. Moser, A. et al., Presence of N-methyl-norsalsolinol in the CSF: correlations with dopamine metabolites of patients with Parkinson's disease, *J. Neurol. Sci.*, 131, 183, 1995.

67. Grandinetti, A. et al., Prospective study of cigarette smoking and the risk of developing idiopathic Parkin-son's disease, *Am. J. Epidemiol.*, 139, 1129, 1994.

68. Paganini-Hill, A. et al., Risk factors for Parkinson's disease: the leisure world cohort study, *Neuroepidemi-ology*, 20, 118, 2001.

69. Soto-Otero, R. et al., Interaction of 1,2,3,4-tetrahy-droisoquinoline with some components of cigarette smoke: potential implications for Parkinson's disease. *Biochem. Biophys. Res Comm.*, 222, 607, 1996.

70. Soto-Otero, R. et al., Reduction of rat brain levels of the endogenous dopaminergic proneurotoxins 1,2,3,4-tet-rahydroisoquinoline and 1,2,3,4-tetrahydro-beta-carbo-line by cigarette smoke, *Neurosci. Lett.*, 298, 187, 2001.

71. Parrado, J. et al., The endogenous amine 1-methyl-1,2,3,4-tetrahydroisoquinoline prevents the inhibition of complex I of the respiratory chain produced by MPP+, *J. Neurochem.*, 75, 65, 2000.

72. Tasaki, Y., et al., 1-methyl-1,2,3,4-tetrahydroisoquino-line, decreasing in 1-methyl-4-phenyl-1,2,3,6-tetrahy-droisoquinoline-treated mouse, prevents parkinsonism-like behavior abnormalities, *J. Neurochem.*, 57, 1940, 1991.

73. Yamakawa, T. and Ohta, S., Isolation of 1-methyl-1,2,3,4-tetrahydroisoquinoline-synthesizing enzyme from rat brain: a possible Parkinson's disease-preventing enzyme, *Biochem. Biophys. Res. Commun.*, 236, 676, 1997.

74. Absi, E. et al., Decrease of 1-methyl-1,2,3,4-tetrahy-droisoquinoline synthesizing enzyme activity in the brain areas of aged rat, *Brain Res.*, 955, 161, 2002.

75. Caparros-Lefebvre, D., Elbaz, A., and the Caribbean Parkinsonian Study Group, Possible relation of atypical parkinsonism in the French West Indies with consump-tion of tropical plants: a case-control study, *Lancet*, 354, 281, 1999.

76. Caparros-Lefebvre, D. et al., Guadeloupean parkin-sonism: a cluster of progressive supranuclear palsy-like tauopathy, *Brain*, 125, 801, 2002.

77. de Silva, R. et al., The tau locus is not significantly associated with pathologically confirmed sporadic Par-kinson's disease, *Neurosci. Lett.*, 330, 201, 2002.

78. Farrer, M. et al., The tau H1 haplotype is associated with Parkinson's disease in the Norwegian population, *Neu-rosci. Lett.*, 322, 83, 2002.

79. Lannuzel A, et al., Toxicity of Annonaceae for dopam-inergic neurons: potential role in atypical Parkinsonism in Guadeloupe, *Mov. Disord.*, 17, 84, 2002.

80. Lannuzel, A. et al., Role of the mitochondrial complex I inhibitor annonacin in Parkinsonism, *Neurology*, 60(Suppl. 1), A331, 2003.

81. Yuan, S.-S. F. et al., Annonacin, a mono-tetrahydrofuran acetogenin, arrests cancer cells at the G1 phase and-causes cytotoxicity in a Bax- and caspase-3-related path-way, *Life Sci.*, 72, 2853, 2003.

82. Maruyama, W. et al., A dopaminergic neurotoxin, 1(R), 2(N)-dimethyl-6,7-dihydroxy-1,2,3,4-tetrahydroiso-quinoline, N-methyl(R)salsolinol, and its oxidation product, 1,2(N)-dimethyl-6,7-dihydroxyisoquinolinium ion, accumulate in the nigro-striatal system of the human brain, *Neurosci. Lett.*, 223, 61, 1997.

83. Storch, A. et al., 1-methyl-6,7-dihydroxy-1,2,3,4-tet-rahydroisoquinoline (salsolinol) is toxic to dopaminer-gic neuroblastoma SH-SY5Y cells via impairment of cellular energy metabolism, *Brain Res.*, 855, 67, 2000.

84. Naoi, M. et al., Dopamine-derived endogenous N-methyl-(R)-salsolinol. Its role in Parkinson's disease, *Neurotoxicol. Teratol.*, 24, 579, 2002.

85. Chun, H. S. et al., Dopaminergic cell death induced by MPP(+), oxidant an specific neurotoxicants shares the common molecular mechanism, *J. Neurochem.*, 76, 1010, 2001.

86. Yoshida, M., Niwa, T., and Nagatsu, T., Parkinsonism in monkeys produced by chronic administration of an endogenous substance of the brain, tetrahydroisoquino-line: the behavioral and biochemical changes, *Neurosci. Lett.*, 119, 109, 1990.

87. Sandler, M. et al., Tetrahydroisoquinoline alkaloids: *in vivo* metabolite of L-dopa in man, *Nature*, 241, 439, 1973.

88. Antkiewicz-Michaluk, L. et al., Increase in salsolinol level in the cerebrospinal fluid of parkinsonian patients is related to dementia: advantage of a new high-perfor-mance liquid chromatography methodology, *Biol. Psy-chiatry*, 42, 514, 1997.

10 Xenobiotic Metabolism and Idiopathic Parkinson's Disease

L. Cartwright, D. B. Ramsden, and S. W. Smith
Department of Medicine, University of Birmingham, U.K.

R. B. Parsons and R. H. Waring
School of Biosciences, University of Birmingham, U.K.

S. L. Ho
Division of Neurology, Department of Medicine, University of Hong Kong

A. C. Williams
Centre for Neuroscience, Queen Elizabeth Hospital, U.K.

CONTENTS

XENOBIOTIC METABOLISM AND GENETIC PREDISPOSITION

One of the most widely held hypotheses concerning the etiology of idiopathic Parkinson's disease (IPD) is the so-called "two-hit" model, which links genetic and environmental factors. Underlying this model are the facts that familial association of the disease was noted early in the previous century (the genetic influence), and there is a higher prevalence of the disease in industrialized Western societies than is seen in less developed ones (the neurotoxicant exposure). A major boost to the hypothesis came from the recognition of the effects of the protoxin, 1-methyl-4-phenyl-1,2,3,6-tetrahydropyridine (MPTP), which *in vivo* is converted to the selective dopaminergic neurotoxin, 1-methyl-4-phenylpyridinium ion (MPP+). This chemical causes a condition that is clinically indistinguishable from IPD.[1,2] Discovery of the mutation in the α-synuclein gene that gives rise to the A30P variant protein demonstrated for the first time that genetic polymorphism could be the causal defect in familial PD.[3] Unfortunately, none of the genetic loci subsequently identified with other forms of familial PD[4–13] has been shown to be a major factor in the etiology of IPD. Although mutations in the parkin gene were found in a small minority of IPD patients, the latest evidence suggests that there is no association of this gene with IPD.[14]

The concept that nongenetic factors are an important component in the etiology of IPD was strengthened by the finding of major studies of twins by Tanner et al., 1999,[15] and Vieregge et al., 1999,[16] both of whom showed that concordance in monozygotic twins was similar to that of dizygotic twins. Nevertheless, these findings should be viewed in the light of those of Piccini et al., 1999,[17] in the same year. Here, a sophisticated positron emission tomography (PET) scanning system was employed to assess clinical status rather than the more conventional methods of patient assessment used in the other two studies. PET scanning gives a picture of the number of healthy dopaminergic neurones in the living brain. The findings of this study were that the concordance between monozygotic twins was significantly higher than in dizygotic twins. The conclusions drawn were that overt clinical differences measured by conventional methods of patient assessment could be misleading and that genetics was an important, if not the major, etiological component.

A logical consequence of the two-hit hypothesis is that it has engendered searches for predisposing genes and both endogenous and exogenous neurotoxicants. The idea that an environmental agent may be involved suggests the xenobiotic metabolizing system as an area where the genetic factor might be found. Humans possess an elaborate system composed of families of genes for combating the effects of noxious chemicals in their environment. One or more of the genes encoding enzymes that make up this system would seem an obvious candidate as the genetic component. Classically, xenobiotic metabolism is thought to involve two phases. First, the toxicant is subjected to attack, usually oxidative, in which a reactive hydrophilic group is introduced into the molecule, followed by a second phase in which the molecule is conjugated to a hydrophilic residue to aid excretion. Many of the genes encoding these enzymes are highly polymorphic, giving rise to a variety of isoenzymes with, potentially, subtle differences in substrate specificity. Thus, the general population is made up of numerous subgroups, each possessing genetically determined, different characteristics in the way they metabolize xenobiotica. This diversity is sustained by a further feature—redundancy, with different enzymes capable of metabolizing the same substrate. Thus, if one particular allelic variation gives rise to a product that cannot metabolize a common

environmental agent, another will. However, redundancy may have its disadvantages, too, in that, although different enzymes may be capable of detoxifying a common chemical, they will not necessarily metabolize the same range of less common substrates, so that these less common substrates may be metabolized differently, depending on the profile of xenobiotic metabolizing enzymes a person expresses.

To support this basic system are enzymes involved in the supply of cofactors for Phase 2 metabolism, e.g., sulfate, glucuronic acid and glutathione. Additionally, the various enzymes are not simply involved in the metabolism of exogenous compounds but also endogenous ones. Indeed, the primary function of some appears to be the latter, e.g., oestradiol sulphotranferase. This leads to a fourth class of enzymes, which we class as *releasing enzymes,* the function of which is deconjugation, e.g., sulphatases, with specific ones involved in the deconjugation of hormone sulfates.

The overall system is summarized in Figure 10.1. The largest family of genes within the system is that encoding the cytochrome P450 enzymes, which contains more than 40 genes, approximately half of which are chiefly involved in xenobiotic metabolism. More than 50 allelic variants have been identified in just one of these, the *CYP2D6* gene on chromosome 22q13.1,[18] which gives rise to a spectrum of activities. The other families are smaller, however, many of the genes are also polymorphic, giving rise to enzymic isoforms. Cytosolic sulphotransferases (SULT) are encoded by ten genes. One of these, SULT1A2, encoding the enzyme phenol sulphotransferase 2, has six isoforms.[19] Given the number of genes and enzyme isoforms generated by individual genes, it can be seen that the general population is composed of a large number of subgroups, with each subgroup having its own pattern of genetic diversity.

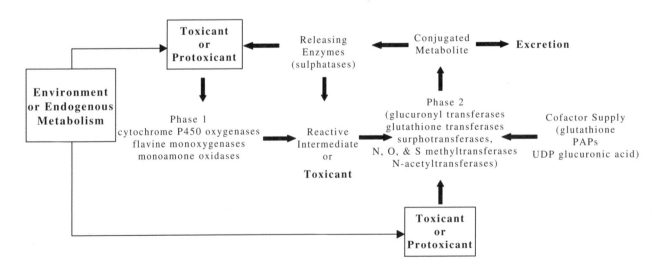

FIGURE 10.1 The xenobiotic metabolizing system.

This picture is complicated by the fact that not all toxicants are metabolized by this two-step process. Some enter directly into Phase 2 because of the presence of hydrophilic reactive groups within the toxin. A further complication is the fact that detoxification is not necessarily the result of a particular reaction. In the case of MPTP, oxidation by MAO-B leads to the creation of a toxin rather than its destruction.

Despite the appropriateness of a search for predisposing genes in the xenobiotic metabolizing system no convincing associations have been found. Genes investigated to date are listed in Table 10.1. Often, claims have been made of an association with a particular allele, then later data fail to substantiate the link. Two examples of this are *CYP 2D6* and MAO-B. Over the years, numerous studies on the association of *CYP 2D6* and MAO-B polymorphisms and IPD have been carried out. Initial results have suggested a positive association of IPD with a particular allele, only to be followed by studies that failed to confirm this. The fundamental flaw in most of these studies is the relatively small number of patients and controls in the populations investigated, which then allows the possibility of spurious statistical significance being derived. However, the fact that, to date, no data have convincingly and consistently linked with a particular allele in the xenobiotic metabolizing system does not mean that no such linkage exists. Until studies on adequately large populations are conducted, no firm conclusions can be drawn.

The search for a cause of IPD raises important questions about the nature of the disease. For the moment, assuming the "two-hit" model posed earlier, some of these questions are outlined below.

1. Is the disease a single entity with a single genetic predisposing factor and a single environmental factor?
2. Is it single disease entity due to a single genetic predisposing factor plus multiple environmental causes?
3. Is it a spectrum of diseases with multiple genetic predisposing factors and multiple possible environmental causes?

If the scenario posed in the third question approximates to the truth, it emphasizes the need for large populations to be studied and requires that the relationship between the genetic factors be modelled. Figure 10.2 illustrates possible interactions among these factors, where it may be a combination of mandatory elements

TABLE 10.1
Some Xenobiotic–Metabolizing Enzymes and Their Association with IPD

	Reference	Study Type Genotype	Phenotype
Phase 1 Enzymes			
Cytochrome P450 Monooxygenases			
CYP 1A1	20, 92	−ve	
CYP 1A2	78		−ve
CYP 2D6	79–83, 85, 86, 88, 90	−ve	
CYP 2D6[a]	84, 87, 90	+ve	
CYP 3E1	91–93	−ve	
Monoamine Oxidases (MAO)			
MAO-A	94,	+ve	
	95, 96	−ve	
MAO-B	97, 98		−ve
	99	+ve	
Phase 2 Enzymes			
Glutathione–S–Transferases (GST)			
GST M1	100, 101, 102		−ve
GST P1	100	+ve	
GST T1	100, 102	−ve	
N–Acetyltransferases			
NAT–1	103	−ve	
NAT–2	84, 103, 106		−ve
	104, 105	+ve	
Catecholamine–O–methylatransferase (COMT)			
	108	+ve	
	107		−ve

[a] Where +ve and −ve represent a positive or negative association of the disease with a particular allele.

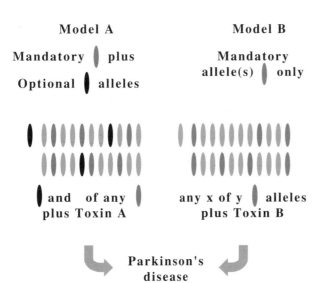

FIGURE 10.2   Possible interactions.

and permissive influences (Model A) or simply a selection of one or more mandatory elements (Model B). If more than one toxin is involved, then the genetic etiological component may well be unique to each toxin.

Two further variations on the theme of genetic influence are that of (a) allelic frequency and (b) level of expression. The frequencies of the alleles mentioned above vary greatly both intraracially and interracially. Thus, if there is more than one xenobiotic metabolizing gene making up the genetic component of IPD, the smaller the populations in any particular study, the more difficult it is to be certain that the control and patient groups are genetically the same. This was exemplified in a study of our own involving 500 control subjects drawn mainly from two different areas of the U.K. No significant differences were found in allelic frequencies between the control and IPD group. The closest approach to a statistically significant finding, although significance was not reached, was between the two sub-groups of the control population.[20]

The previous discussion has regarded genetic polymorphisms as creating structurally different isoenzymes with unique spectra of substrate specificities. However, genetic influence is not solely limited to this. Genetics may powerfully influence levels of expression without affecting any change in the structure of an enzyme. Thus, new scenarios can be envisaged where threshold levels of expression are the determining factor. Every drug user who injected MPTP did not, as a consequence, develop parkinsonism. The condition arose when the level of drug exposure exceeded the limit derived from the individual's ability to catabolize the chemical nontoxically and thus allow neurotoxic levels of MPP+ to be formed. The neurotoxic potency of MPTP is such that relatively small amounts were required for rapid development of the ill-

ness. However, IPD develops over years rather than weeks, as in the case of MPTP exposure, during which time levels of enzyme expression and toxicant exposure may well fluctuate resulting in differing times of onset, severity, and speeds of progression of the illness.

In summarizing the findings to date on genetic factors and predisposition, one may say that, in terms of the "two-hit" hypothesis, the xenobiotic-metabolizing enzymes offer a tempting target for investigation, but as yet no study has get been carried out on a sufficiently large population to show convincingly a strong association between an allele and the disease. The following sections deal with xenobiotica as potential toxins.

POTENTIAL TOXINS—COMPLEX 1 INHIBITORS

MPTP ANALOGS

The discovery of MPTP quickly led to searches for the compound in the natural world. Almost as quickly, it was realized that MPTP itself was not present; however, its analogs may be. In considering MPTP analogs as potential toxins, three salient facts about MPTP must be borne in mind. The first of these is that MPTP is not a toxin but a protoxin. The uncharged nature of MPTP allows it to cross the blood-brain barrier. Only when it has crossed this is a neurotoxic compound generated with the conversion to the charged pyridinium ion, MPP+.[21,22] The second important consideration is its mode of entry into the dopaminergic cell. This is brought about by the dopamine reuptake system.[23] Thus, the steric and charge requirements of the dopamine transporter are crucial. Using PC-12 cells (a rat, dopaminergic phaeochromocytoma cell line) and a series of MPTP analogs, we explored the steric properties of the dopamine transporter.[24] Two endpoints, inhibition of ^{3}H-dopamine uptake and cell toxicity, were used to assess interaction of the analogs with the transporter. Both endpoints indicated that substitutions in the benzene ring were tolerated reasonably well, whereas replacement of the N-methyl group with larger substituents quickly led to a decreased ability to inhibit dopamine uptake or induce cell death. A notable feature of the disturbed metabolism in these cells brought about by the analogs, even some that poorly inhibited dopamine uptake, was the formation of large lipofuscin deposits within the cytosol (Figure 10.3). The third fact is that MPP+ causes cell death by a combination of routes: (a) inhibition of mitochondrial Complex 1 and subsequent oxidative stress,[25, 26] and (b) release of dopamine, which may then go on to form oxidative free radicals,[27] both of which are features of the pathology of IPD.[28–30] As pointed out earlier, IPD is a disease that develops over years; therefore, the relative inefficiency of compounds to induce cell death compared to that of MPTP may not be important. Consequently, the chemical struc-

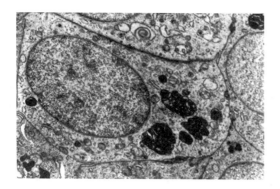

FIGURE 10.3 Rat phaeochromocytoma cell (PC-12) treated with MPTP. Electron micrograph shows large black irregular shaped particles of lipofuscin in the cytosol.

ture of potential toxins may not be instantly recognizable as closely similar to MPTP or MPP$^+$. What does appear of importance from our own studies and those of others[31–33] is the possession of a methyl group on an aromatic nitrogen.

Three potential sources of MPTP analogs have been identified: simple pyridines present in trace amounts in foodstuffs,[31] tyrosine metabolites (β-carbolines),[34] and tryptophan metabolites (tetrahydrodisoquinolines—TIQs)[35] (Figure 10.4). These last two types of compound may be present in humans because of both endogenous and exogenous metabolism. All of these may be N-methylated inside the CNS to form specific dopaminergic neurotoxins which inhibit Complex 1 and release dopamine. Enzymic activities capable of bringing about the methylation of TIQs at either of the two nitrogen atoms are present in the CNS.[36,37] Subsequently, it was shown that patients with IPD have higher levels of N-methylated β-carbolines in the CSF and higher β-carboline N-methylating enzymic activity in brain compared with control subjects.[38,39]

Our own studies in this area have concentrated on the formation of N-methylated simple pyridines. 4-Phenylpyridine is present in some foodstuffs,[31] and N-methylation of this would give rise to MPP$^+$. In considering enzymes that could bring about this reaction, our attention

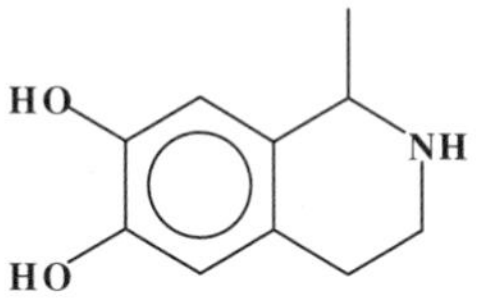

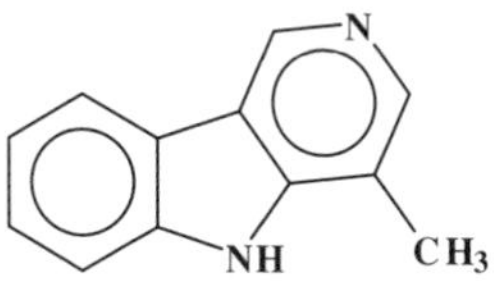

was drawn to nicotinamide-N-methyltransferase (NNMT). NNMT methylates the pyridinium nitrogen of niacin amide (vitamin B3 amide). It also has a wide substrate specificity and is capable of methylating other simple pyridines. Its activity in humans is strongly genetically determined, with 25% of the population having high hepatic levels.[40,41] This variation in activity is not the result of a polymorphism in the coding regions of the gene but is due to an as yet unidentified polymorphism causing different levels of expression of the enzyme.[41] Although, as demonstrated by MPP$^+$ synthesis, it is not essential that a N-methylpyridinium ion is formed within the dopaminergic neurone itself, we demonstrated that NNMT was expressed there.[42] However, the substantia nigra is a difficult brain region to study in humans when one wishes to demonstrate differences of protein expression between IPD patients and control subjects. Apart from the difficulties of controlling confounding factors such as comparisons between two disease groups, drug treatment prior to death, and postmortem interval between death and sample acquisition, the nature of IPD is a major difficulty. At the development of symptoms, approximately 50% of the 450,000 dopaminergic neurones in either hemisphere of the substantia nigra are dead, and the surviving neurones are underperforming. Therefore, by the time of death, on average some ten years later, this number is significantly further reduced. Consequently, any measurable reduction in the expression of a protein may be due to the consequences of the disease rather than being a cause of the disease. Any causal overexpression of a protein may be hidden by the fact that there are few surviving neurones. To overcome these difficulties, we investigated expression in a brain region largely unaffected by the disease process—the granular layer of the cerebellum. In 50 IPD patients, NNMT expression in this layer was significantly higher than in the same region in control subjects.[43] These findings were in accord with those of Akoyama et al., 2001, who showed increased levels of NNMT in the CSF of IPD patients compared to controls.[44] One can speculate that this may be causal, in that high levels of NNMT may give rise to high levels of neurotoxic N-methyl pyridinium ions. However attractive this speculation may appear, the data do not prove this. Alternative explanations may be that such increased expression may be a cause of the drugs used in the treatment of IPD, or it may occur as a consequence of particular types of illness. NNMT is elevated in certain cancers. Thus, further studies are required to clarify the physiological significance of these findings.

SULPHUR-CONTAINING TOXINS

Dryhurst and co-workers have shown that, under mild oxidizing conditions, cysteine and dopamine or other cat-

FIGURE 10.4 1-methyl-6,7-dihydroxy-1,2,3,4-tetrahydroisoquinoline and 4-methylpyridino [4,3-b]indole (a beta carboline).

echolamines react together to form a variety of compounds, including dihydrobenzothiazines (Figure 10.5), which are potent inhibitors of Complex 1, and free-radical species. They have postulated that the products of these reactions may be causal agents in IPD.[45–51] If this is the case, factors which control the intracellular concentration of cysteine in the dopaminergic neurone would be important. Within the hepatocyte, the enzyme cysteine dioxygenase, is one such factor. Therefore, we explored the possibility that this enzyme, which has been shown to be expressed in brain previously,[52] is expressed in dopaminergic neurones. To date, we have shown that, in the rat, the enzyme is present in numerous brain regions, and the highest unstimulated activity levels can be seen in the basal ganglia and olfactory bulb.[53] Initial immunohistochemistry shows that the enzyme is present in neurones,[54] and further work is underway to characterize expression in the human substantia nigra. Relatively little work has been done on the genetics of CDO regulation, but some phenotyping studies suggest that the majority of the population have a high to medium level of constitutive activity with a small percentage in a low-activity group.[55] A low activity would favor a high intracellular cysteine concentration and hence the formation of toxic reaction products.

ROTENONE

The emphasis on causal environmental agents received a boost with the recognition of the selective toxicity of levels of the pesticide rotenone (Figure 10.6), which caused symptoms similar to those of PD.[56] This is quite widely used as a safe insecticide. The compound is also a potent inhibitor of mitochondrial complex 1 and thus shares a common mode of action with 1-methyl-4-phenylpyridinium ion (MPP+) formed from *in vivo* oxidation of the protoxin, 1-methyl-4-phenyl-1,2,3,6-tetrahydropyridine (MPTP). The recognition that a new class of rotenone-like compounds might exist in the environment helps to counter one objection to the concept that environmental toxins might be involved in the etiology of IPD, simply that MPTP is not encountered in nature. In exploring the properties of rotenone, its injection into rats has been shown to cause extensive microglial activation without astrocytosis in the striatum and substantia nigra and to a much less extent in the cortex. Microglial activation was observable before marked lesioning of the substantia nigra.[57] This mimics the pathology seen in IPD. Microglia are the phagocytic cells of the CNS and on activation NADPH oxidase releases bactericidal superoxide anions into the phagosome.[58] The target of rotenone in microglia is the enzyme complex NADPH oxidase. Evidence for this comes both from the effects of NADPH oxidase inhibitors such as naloxone,[59] diphenylene iodonium and apocynin[61] and primary mesencephalic cultures from homozygous knock-out mice in which one of the components of the NADPH oxidase complex (gp91) had been deleted.[60] A possibly important insight into the etiology of IPD came from the fact that rotenone and the inflammatory stimulus, lipopolysaccharide (LPS) acted synergistically to cause dopaminergic depletion.[62] Subsequently it was shown that microglial NADPH oxidase-derived free radicals were an important component in the neurotoxicity of MPTP.[63] In addition, MPTP and LPS were also shown to act synergistically, and that inhibition of microglial activation mitigated the neurotoxic effects of this combination. This synergism may occur through formation of the peroxynitrite ion, because NO synthase inhibition attenuated neurotoxicity.[64]

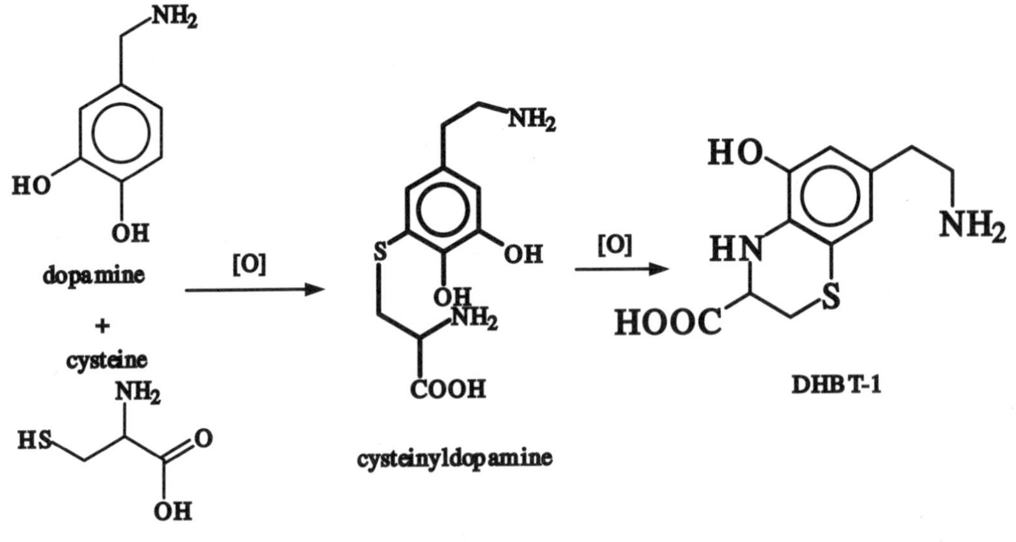

FIGURE 10.5

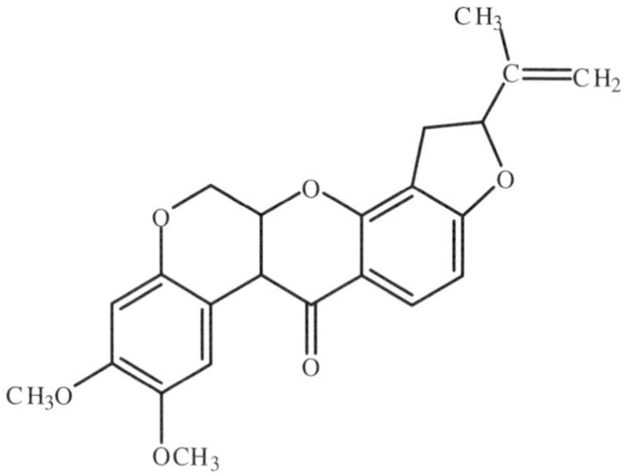

FIGURE 10.6 Rotenone (also known as Derrin; Nicouline; Rotenonum; Tubatoxin; Derril, Extrax, Mexide). Chemical name [2R-(2a, 6a alpha, 12a alpha)]–1, 2,12, 12a-tetrahydro-8,9-dimethoxy-2-(1-methylethyl)[I] benzopyranol [3,4-b] furo 12,3-h] [I]benzopyran-6(6aH)-one.

A fourth potentially neurotoxic action of rotenone, in addition to inhibiting Complex 1, activating microglia, and acting synergistically with inflammogens, is that of stimulating α-synuclein aggregation. Rotenone, in common with other pesticides, including dieldrin and paraquat, induce a conformational change in the protein, which accelerates the rate of formation of α-synuclein fibrils *in vitro*.[65] This effect is exacerbated by the presence of metal ions;[66] metal ions such as Ti^{3+}, Zn^{2+}, Al^{3+}, and Pb^{2+} were able to facilitate fibril formation of S-oxidized methionine α-synuclein, which does not undergo aggregation on its own, but which may be formed *in vivo* under conditions of oxidative stress.[67]

DITHIOCARBAMATES (E.G., MANEB) AND PARAQUAT

In the mid 1980s, the late J. Barbeau proposed that exposure to the herbicide, paraquat, which is closely related to MPP+, was an etiological factor in IPD. At the time, the subsequent evidence failed to substantiate this hypothesis. However, more recently paraquat exposure has been shown to cause selective nigrostriatal degeneration in animal models of IPD.[68] Diethyldithiocarbamate and some related compounds enhance the effects of MPTP.[69] Possibly more importantly for the understanding of risk factors, because MPTP is not encountered in the environment, the combination of paraquat with dithiocarbamate pesticides such as Maneb has been found to be more destructive. In adult animals, this combination causes selective dopaminergic neuronal loss in the substantia nigra,[70] which is accompanied by accumulation of paraquat in the neurones and delayed release of dopamine from synaptosomes.[71] A cause for even greater concern is the fact that exposure to the chemicals in early life rendered the animals more susceptible to follow-up exposure in later life.[72,73]

XENOBIOTIC NEUROPROTECTANTS AND SMOKING

The above discussion has focused on xenobiotica as neurotoxicants; however, their role is not limited simply to that. Examples of neuroprotection may be found in two unexpected areas. The first of these are TIQs, which have previously been discussed in terms of their neurotoxic potential. However, some TIQs, notably 1-methyl-1,2,3,4-tetrahydroisoquinoline, protect against the action of rotenone, possibly by affecting shifting the metabolism of dopamine toward O-methylation and away from oxidation and the release of oxidative free radicals.[74,75]

IPD is a perverse disease in numerous ways, one of which is the fact that smoking exerts a protective influence, in contrast to its carcinogenic actions. The protective agent in tobacco smoke is unidentified as yet, but much attention has been paid to nicotine as the beneficial agent. One other candidate that has received scant attention is the rather unusual hydrocarbon, azulene (Figure 10.7), which, as its name implies, is associated with the blue color of tobacco smoke. This chemical fails to protect against MPP+ toxicity in cell models.[76] However, the protection afforded by smoking is relatively weak and MPP+ *in vitro* cell models do not mimic the *in vivo* situation at all well in the speed and scale of the insult. Azulene, being a hydrocarbon, is likely to cross the blood-brain barrier, particularly in view of its pulmonary route of intake, and it is a potent reducing agent. As the relevance of microglial activation is becoming recognized, azulene's primary role may be in scavenging oxidative free radicals formed as a consequence of this process. Thus, azulene analogs, which are being investigated as drugs to treat other conditions, may also be useful in IPD.[77]

CONCLUSION

The evidence so far supports the hypothesis that the etiology of IPD is multifactorial, with both genetic and environmental components. The end result is a failure to generate adequate amounts of ATP, and this, in the long term, leads to a failure to catabolize intracellular proteins and oxidative stress in the energy-demanding dopaminergic neurones. This picture is further complicated by activation of microglia. The combination leads to destruction of the neurones and depletion of dopamine supply to the striatum.

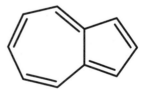

FIGURE 10.7 Azulene.

REFERENCES

1. Davis, G. C. Williams, A. C., Markey, S. P., Ebert, M., Caine, H., Reichert, E. D., Kopin, C. M., I. J. Chronic parkinsonism secondary to intravenous injection of meperidine analogues, *Psychiatry Research,* 1, 249–254, 1979.

2. Langston, J. W., Ballard, P., Tetrud, J. W. and Irwin, I., Chronic parkinsonism in humans due to a product of meperidine-analog synthesis, *Science,* 219, 979–980, 1983.

3. Polymeropoulos, M. H., Lavedan, C., Leroy, E., Ide, S. E., Dehejia, A., Dutra, A., Pike, B., Root, H., Rubenstein, J., Boyer, R., Stenroos, E. S., Chandrasekharappa, S., Athanassiadou, A., Papapetropoulos, T., Johnson, W. G., Lazzarini, A. M., Duvoisin, R. C., Di Iorio, G., Golbe, L. I. and Nussbaum, R. L., Mutation in the alpha-synuclein gene identified in families with Parkinson's disease, *Science,* 276, 2045–2047, 1997.

4. Kitada, T., Asakawa, S., Hattori, N., Matsumine, H., Yamamura, Y., Minoshima, S., Yokochi, M., Mizuno, Y. and Shimizu, N., Mutations in the parkin gene cause autosomal recessive juvenile parkinsonism, *Nature,* 392, 605–608, 1998.

5. Farrer, M., Gwinn-Hardy, K., Muenter, M., DeVrieze, F. W., Crook, R., Perez-Tur, J., Lincoln, S., Maraganore, D., Adler, C., Newman, S., MacElwee, K., McCarthy, P., Miller, C., Waters, C., Hardy, J., A chromosome 4p haplotype segregating with Parkinson's disease and postural tremor, *Human Molecular Genetics,* 8, 81–85, 1999.

6. Gasser, T., Muller-Myhsok, B., Wszolek, Z. K., Oehlmann, R., Calne, D. B., Bonifati, V., Bereznai, B., Fabrizio, E., Vieregge, P., Horstmann, R. D., A susceptibility locus for Parkinson's disease maps to chromosome 2p13, *Nature Genetics,* 18, 262–265, 1998.

7. Gwinn-Hardy K. A., Crook, R. Lincoln, S., Adler, C. H., Caviness, J. N., Hardy, J., Farrer, M., A kindred with Parkinson's disease not showing genetic linkage to established loci, *Neurology,* 54, 504–507, 2000.

8. Rana, M., de Coo, I., Diaz, F., Smeets, H., Moraes, C. T., An out-of-frame cytochrome b gene deletion from a patient with parkinsonism is associated with impaired complex III assembly and an increase in free radical production, *Annals of Neurology,* 48, 774–781, 2000.

9. Baker, M., Kwok, J. B., Kucera, S., Crook, R., Farrer, M., Houlden, H., Isaacs, A., Lincoln, S., Onstead, L., Hardy, J., Wittenberg, L., Dodd, P., Webb, S., Hayward, N., Tannenberg, T., Andreadis, A., Hallupp, M., Schofield, P., Dark, F., Hutton, M., Localization of frontotemporal dementia with parkinsonism in an Australian kindred to chromosome 17q21-22, *Annals of Neurology,* 42, 794–798, 1997.

10. Funayama, M., A new locus for Parkinson's disease (PARK8) maps to chromosome 12p11.2–q13.1, *Annals of Neurology,* 51, 296–301, 2002.

11. Valente, E. M., Bentivoglio, A. R., Dixon, P. H., Ferraris, A., Ialongo, T., Frontali, M., Albanese, A., Wood, N. W., localisation of a novel locus for autosomal recessive early-onset parkinsonism, PARK6, on human chromosome 1p35-p36, *American Journal of Human Genetics,* 68, 895–900, 2001.

12. van Duijn, C. M., Dekker, M. C., Bonifati, V., Galjaard, R. J., Houwing-Duistermaat, J. J., Snijders, P. J., Testers, L., Breedveld, G. J.,Horstink, M., Sandkuijl, L. A., van Swieten, J. C., Oostra, B. A., Heutink, P., Park7, a novel locus for autosomal recessive early-onset parkinsonism, on chromosome 1p36, *American Journal of Human Genetics,* 69, 629–634, 2001

13. Krygowska-Wajs A, Hussey JM, Hulihan, M., Farrer, M., Tsuboi, Y., Uitti, R.J., Wszolek, Z. K.,Two large Polish kindreds with levodopa-responsive Parkinsonism not linked to known Parkinsonian genes and loci. *Parkinsonism and Related Disorders,* 9, 193–200, 2003.

14. Oliveira, S. A., Scott, W. K., Nance, M. A., Watts, R. L., Hubble, J. P., Koller, W. C., Lyons, K. E., Pahwa, R., Stern, M. B., Hiner, B. C., Jankovic, J., Ondo, W. G., Allen, F. H., Jr., Scott, B. L., Goetz, C. G., Small, G. W., Mastaglia, F. L., Stajich, J. M., Zhang, F., Booze, M. W., Reaves, J. A., Middleton, L. T., Haines, J. L., Pericak-Vance, M. A., Vance, J. M., Martin, E. R., Association study of Parkin gene polymorphisms with idiopathic Parkinson disease, *Archives of Neurology,* 60, 975–980, 2003.

15. Tanner, C. M., Ottman, R., Goldman, S. M., Ellenberg, J., Chan, P., Mayeux, R., Langston, J. W., Parkinson disease in twins: an etiologic study, *JAMA,* 281, 376–378, 1999.

16. Vieregge, P., Hagenah, J., Heberlein, I., Klein, C., Ludin, H. P., Parkinson's disease in twins: a follow-up study, *Neurology,* 53, 566–572, 1999.

17. Piccini, P., Burn, D. J., Ceravolo, R., Maraganore, D., Brooks, D. J., The role of inheritance in sporadic Parkinson's disease: evidence from a longitudinal study of dopaminergic function in twins, *Annals of Neurology,* 45, 577–582, 1999.

18. Marez, D., Legrand, M., Sabbagh, N., Guidice, J. M., Spire, C., Lafitte, J. J., Meyer, U. A., Broly, F., Polymorphism of the cytochrome P450 CYP2D6 gene in a European population: characterisation of 48 mutations and 53 allele, their frequencies and evolution, *Pharmacogenetics,* 7, 193–202, 1997.

19. Glatt, H., Boeing, H., Engelke, C. E. H., Ma L., Kuhlow, A., Pabel, U., Pomplun, D., Teuber, W., Meinl, W., Human cytosolic sulphotransferases: genetics, characteristics, toxicological aspects, *Mutation Research,* 482, 27–40, 2001.

20. Nicholl, D. J., Bennett, P., Hiller, L., Bonifati, V., Vanacore, N., Fabbrini, G., Marconi, R., Colosimo, C., Stocchi, F., Boucelli, U., Vieregge, P., Ramsden, D. B., Meco, G., Williams, A. C., A study of five candidate genes in Parkinson's disease and related neurodegenerative disorders, *Neurology,* 53, 1415–1421, 1999.

21. Langston, J. W., Irwin, I., Langston, E. B., Forno, L. S., 1-Methyl-4-phenylpyridinium ion (MPP+): identification of a metabolite of MPTP, a toxin selective to the substantia nigra, *Neuroscience Letters,* 48, 87-92, 1984.

22. Mytilineou, C., Cohen, G., Heikkila, R. E., 1-Methyl-4-phenyl-pyridine (MPP+) is toxic to mesencephalic

dopamine neurones in culture, *Neuroscience Letters*, 57, 19–24, 1985.

23. Javitch, J. A., Synder, S. H., Uptake of MPP+ by dopamine neurones explains selectivity of parkinsonism-inducing neurotoxin, MPTP, *European Journal of Pharmacology*, 106, 455–456, 1985.

24. Athwall, N. S. S., Ramsden, D. B., Simpson, M., Williams, A. C., Inhibition of dopamine uptake into PC-12 cells by analogues of 1-methyl-4-phenyl-1,2,3,6-tetrahydropyridine (MPTP), *Parkinsonism and Related Disorders*, 2, 1–6, 1996.

25. Jenner, P., Schapira, A. H., Marsden, C. D., New insights into the cause of Parkinson's disease, *Neurology*, 42, 2241–2250, 1992.

26. Ramsay, R. R., Salach, J. I., Dadgar, J., Singer, T. P., Inhibition of mitochondrial NADH dehydrogenase by pyridine derivatives and its possible relation to experimental and idiopathic parkinsonism, *Biochemical and Biophysical Research Communications*, 135, 269–275, 1986.

27. Lotharius, J., O'Malley, K. L., The parkinsonian drug 1-methyl-4-phenylpyridinium triggers intracellular dopamine oxidation. A novel mechanism of toxicity, *Journal of Biological. Chemistry*, 275, 38581–38588, 2000.

28. Owen, A. D., Schapira, A. H., Jenner, P., Marsden, C. D., Indices of oxidative stress in Parkinson's disease, Alzheimer's disease and dementia with Lewy bodies, *Journal of Neural Transmission*, Suppl., 51, 167–173, 1997.

29. Schapira, A. H., Mitochondrial Complex I deficiency in Parkinson's disease, *Advances in Neurology*, 60, 288–291, 1993.

30. Schapira, A. H., Cooper, J. M., Dexter, D., Clark, J. B. and Marsden, C. D., Mitochondrial complex I deficiency in Parkinson's disease, *Journal of Neurochemistry*, 54, 823–827, 1990

31. Youngster, S. K., Nicklas, W. J, Heikkila, R. E., Structure-activity study of the mechanism of 1-methyl-4-phenyl-1,2,3,6-tetrahydropyridine (MPTP)-induced neurotoxicity. II Evaluation of biological activity of the pyridinium metabolites formed from monoamine oxidase-catalysed oxidation of MPTP analogues, *Journal of Pharmacology and Experimental Therapeutics* 249, 829–835, 1989.

32. Snyder, S. H., D'Amato, R. J., MPTP, a neurotoxin of relevance to the pathophysiology of Parkinson's disease, *Neurology*, 36, 250–258, 1986.

33. Ansher, S. S., Cadet, J. L., Jakoby, W. B., Baker, J. K., Role of N-methyltransferases in the neurotoxicity associated with the metabolites of 1-methyl-4-phenyl-1,2,3,6-tetrahydropyridine (MPTP) and other 4-substituted pyridines present in the environment, *Biochemical Pharmacology*, 35, 3359–3363, 1986.

34. Collins, M. A., Neafsey, E. J., Matsubara, K., Cobuzzi, R. J., Jr., Rollema, H. Indole-N-methylated beta-carbolinium ions as potential brain-bioactivated neurotoxins. *Brain Research*, 570, 154–160, 1992.

35. Kotake, Y., Tasaki, Y., Makino, Y., Ohta, S., Hirobe, M., 1-Benzyl-1,2,3,4-Tetrahydroisoquinoline as a Parkin-

sonism-Inducing Agent-a Novel Endogenous Amine in Mouse-Brain and Parkinsonian CSF, *Journal of Neurochemistry*, 65, 2633–2638, 1995.

36. Naoi, M., Matsuura, S., Takahashi, T., Nagatsu, T. A., N-methyltransferase in human brain catalyses N-methylation of 1,2,3,4-tetrahydroisoquinoline into N-methyl-1,2,3,4-tetrahydroisoquinoline, a precursor of a dopaminergic neurotoxin, N-methylisoquinolinium ion, *Biochemical and Biophysical Research Communications*, 161, 1213–1219, 1989.

37. Naoi, M., Dostert, P., Yoshida, M., Nagatsu, T., N-methylated tetrahydroisoquinolines as dopaminergic neurotoxins, *Advances in Neurology*, 60, 212–217, 1993.

38. Matsubara, K., Kobayashi, S., Kobayashi, Y., Yamashita, K., Koide, H., Hatta, M., Iwamoto, K., Tanaka, O., Kimura, K., Beta-Carbolinium cations, endogenous MPP+ analogs, in the lumbar cerebrospinal fluid of patients with Parkinson's disease, *Neurology*, 45, 2240–2245, 1995.

39. Gearhart, D. A., Collins, M. A., Lee, J. M., Neafsey, E. J., Increased beta-carboline N-methyltransferase activity in the frontal cortex in Parkinson's disease, *Neurobiology of Disease*, 7, 201–211, 2000.

40. Rini, J., Szumlanski, C., Guerciolini, R.,Weinshilboum, R. M., Human liver nicotinamide N-methyltransferase: ion-pairing radiochemical assay, biochemical properties and individual variation, *Clinica Chimica Acta*, 186, 359–374, 1990.

41. Smith, M. L., Burnett, D., Bennett, P., Waring, R. H., Brown, H. M., Williams, A. C. and Ramsden, D. B., A direct correlation between nicotinamide N-methyltransferase activity and protein levels in the human liver cytosol, *Biochimica et Biophysica, Acta* 1422, 238–244, 1998.

42. Parsons, R. B., Smith, M. L., Williams, A. C., Waring, R. H. and Ramsden, D. B., Expression of Nicotinamide N-methyltransferase (E.C. 2.1.1.1) in the Parkinsonian brain, *Journal of Neuropathology and Experimental Neurology*, 61, 111–124, 2002.

43. Parsons, R. B., Smith, S. W., Waring, R. H., Williams, A. C., Ramsden, D. B., High expression of nicotinamide N-methyltransferase in patients with idiopathic Parkinson's disease, *Neuroscience Letters*, 342,13–16, 2003.

44. Akoyama, K., Matsubara, K., Kondo, M., Murakawa, Y., Suno, M., Yamashita, K.,Yamaguchi, S., and Kobayashi, S., Nicotinamide N-methyltransferase is higher in the lumbar cerebrospinal fluid of patients with Parkinson's disease, *Neuroscience Letters*, 298, 78–80, 2001.

45. Zhang, F., Dryhurst, G., Effects of L-Cysteine on the Oxidation Chemistry of Dopamine: New Reaction Pathways of Potential Relevance to Idiopathic Parkinson's Disease, *Journal of Medicinal Chemistry*, 34, 1084–1098, 1994.

46. Li, H., Shen, X. M., Dryhurst, G., Brain mitochondria catalyze the oxidation of 7-(2-aminoethyl)-3,4-dihydro-5-hydroxy-2H-1,4-benzothiazine-3-carboxylic acid (DHBT-1) to intermediates that irreversibly inhibit complex I and scavenge glutathione: potential relevance to

the pathogenesis of Parkinson's disease, Journal of Neurochemistry, 71, 2049–2062, 1998.

47. Shen, X. M., Dryhurst, G., Iron- and manganese-catalyzed autoxidation of dopamine in the presence of L-cysteine: possible insights into iron- and manganese-mediated dopaminergic neurotoxicity, *Chemical Research in Toxicology,* 11, 824–837, 1998.

48. Shen, X. M., Zhang, F., Dryhurst, G., Oxidation of dopamine in the presence of cysteine: characterization of new toxic products, *Chemical Research in Toxicology,* 10, 147–155, 1997.

49. Cheng, F. C., Kuo, J. S., Chia, L. G., Dryhurst, G., Elevated 5-S-cysteinyldopamine/homovanillic acid ratio and reduced homovanillic acid in cerebrospinal fluid: possible markers or and potential insights into the pathoetiology of Parkinson's disease, *Journal of Neural Transmission (Budapest),* 103, 433–446, 1996.

50. Shen, X.M., Dryhurst, G., Further insights into the influence of L-cysteine on the oxidation chemistry of dopamine: reaction pathways of potential relevance to Parkinson's disease, Chemical Research in Toxicology, 9, 751–763, 1996.

51. Xin, W., Shen, X. M., Li, H., Dryhurst, G., Oxidative metabolites of 5-S-cysteinylnorepinephrine are irreversible inhibitors of mitochondrial complex I and the alpha-ketoglutarate dehydrogenase and pyruvate dehydrogenase complexes: possible implications for neurodegenerative brain disorders, *Chemical Research in Toxicology,* 13, 749–760, 2000.

52. Misra, C. H., *In vitro* study of cysteine oxidase in rat brain, *Neurochemistry Research,* 8, 1497-1508, 1983.

53. Parsons, R. B., Barber, P. C., Waring, R. H., Williams, A. C., and Ramsden, D, B., Human Cysteine Dioxygenase Type I (CDO-I; EC 1.13.11.20): Regional Expression of Activity in Rat Brain, *Neuroscience Letters,* 248, 101–104, 1998.

54. Williams, A. C., Smith, M. L., Waring, R. H., Ramsden, D. B., The Aetiology of Idiopathic Parkinson's Disease: A Genetic and Environmental Model, Advances in Neurology, 80, 215–218, 1999.

55. Mitchell, S. C., Waring, R. H., Haley, C. S., Idle, J. R., Smith, R. L., Genetic aspects of the polymodally distributed sulphoxidation of S-carboxymethyl-L-cysteine in man, *British Journal of Clinical Pharmacology,* 18, 507–521, 1984.

56. Betarbet, R., Sherer, T. B., MacKenzie, G., Garcia-Osuna, M., Panov, A. V., Greenamyre, J. T., Chronic systemic pesticide exposure reproduces features of Parkinson's disease, *Nature Neuroscience,* 3, 1301–1306, 2000.

57. Sherer, T. B., Betarbet, R., Kim, J. H., Greenamyre, J. T., Selective microglial activation in the rat rotenone model of Parkinson's disease, *Neuroscience Letters,* 341, 87–90, 2003.

58. Eder, C., DeCoursey, T. E.,Voltage-gated proton channels in microglia, *Progress in Neurobiology,* 64, 277-305, 2001.

59. Lui, B., Du, L., Hong, J. S., Naloxone protects rat dopaminergic neurones against inflammatory damage through inhibition of microglia activation and superox-

ide generation, *Journal of Pharmacology and Experimental Therapeutics,* 293, 607–617, 2000.

60. Gao, H. M., Hong, J. S., Zhang, W., Lui, B., Distinct role for microglia in rotenone-induced degeneration of dopaminergic neurons, *Journal of Neuroscience,* 22, 782–790, 2002.

61. Gao, H. M., Liu, B., Hong, J. S., Critical role for microglial NADPH oxidase in rotenone-induced degeneration of dopaminergic neurons, *Journal of Neuroscience,* 23, 6181–6187, 2003.

62. Gao, H. M., Hong, J. S., Zhang, W., Liu, B., Synergistic dopaminergic neurotoxicity of the pesticide rotenone and inflammogen lipopolysaccharide: relevance to the etiology of Parkinson's disease, *Journal of Neuroscience,* 23, 1228–1236, 2003.

63. Gao, H. M., Liu, B., Zhang, W., Hong, J. S., Critical role for microglial NADPH oxidase-derived free radicals in the vitro MPTP model of Parkinson's disease, *FASEB J.,* Aug (ahead of print publication), 2003.

64. Gao, H. M., Liu, B., Zhang, W., Hong, J. S., Synergistic dopaminergic neurotoxicity of MPTP and inflammogen lipopolysaccharide: relevance to the etiology of Parkinson's disease, *FASEB J.,* August 15, 2003.

65. Uversky, V. N., Li, J., Fink, A. L., Pesticides directly accelerate the rate of alpha-synuclein fibril formation: a possible factor in Parkinson's disease, *FEBS Letters,* 500, 105–108, 2001.

66. Uversky, V. N., Li, J., Bower, K., Fink, A. L., Synergisic effects of pesticides and metals on the fibrillation of alpha-synuclein: implications for Parkinson's disease, *Neurotoxicology,* 23, 527–536, 2002.

67. Yamin, G., Glaser, C. B., Uversky, V. N., Fink, A. L., Certain metals trigger fibrillation of methionine-oxidized alpha-synuclein, *Journal of Biological Chemistry,* 278, 27630–27635, 2003.

68. McCormack, A. L., Thiruchelvam, M., Manning-Bog, A. B., Thiffault, C., Langston, J. W., Cory-Slechta, D. A., Di Monte, D. A., Environmental risk factors and Parkinson's disease: selective degeneration of nigral dopaminergic neurons caused by the herbicide paraquat, *Neurobiology of Disease,* 10, 119–127, 2002.

69. McGrew, D. M., Irwin, I., Langston, J. W., Ethylenebisdithiocarbamate enhances MPTP-induced striatal dopamine depletion in mice, *Neurotoxicology,* 21, 309–312, 2000.

70. Thiruchelvam, M., Richfield, E. K., Baggs, R. B., Tank, A. W., Cory-Slechta, D. A. J., The nigrostriatal dopaminergic system as a preferential target of repeated exposures to combined paraquat and maneb: implications for Parkinson's disease, *Neuroscience,* 2000, 9207–9214, 2000.

71. Barlow, B. K., Thiruchelvam, M. J., Bennice, L., Cory-Slechta, D. A., Ballatori, N., Richfield, E. K., Increased synaptosomal dopamine content and brain concentration of paraquat produced by selective dithiocarbamates, *Journal of Neurochemistry,* 85, 1075–1086, 2003.

72. Thiruchelvam, M., Richfield, E. K., Goodman, B. M., Baggs, R. B., Cory-Slechta, D. A., Developmental exposure to the pesticides paraquat and maneb and the Par-

kinson's disease phenotype, *Neurotoxicology,* 23, 621–633, 2002.

73. Thiruchelvam, M., McCormack, A., Richfield, E. K., Baggs, R. B., Tank, A. W., Di Monte, D. A., Cory-Slechta, D. A., Age-related irreversible progressive nigrostriatal dopaminergic neurotoxicity in the paraquat and maneb model of the Parkinson's disease phenotype, *European Journal of Neuroscience,* 18, 589–600, 2003.

74. Antkiewicz-Michaluk, L., Michaluk, J., Mokrosz, M., Romanska, I., Lorenc-Koci, E., Ohta, S., Vetulani, J., Different action on dopamine catabolic pathways of two endogenous 1,2,3,4-tetrahydroisoquinolines with similar antidopaminergic properties, *Journal of Neurochemistry,* 78, 100–108, 2001.

75. Antkiewicz-Michaluk, L., Karolewicz, B., Romanska, I., Michaluk, J., Bojarski, A. J., Vetulani, J., 1-Methyl-1,2,3,4-tetrahydroisoquinoline protects against rotenone-induced mortality and biochemical changes in rat brain, *European Journal of Pharmacology,* 466, 263–269, 2003.

76. Mazzio, E., Huber, J., Darling, S., Harris, N., Soliman, K. F. A., Effect of antioxidants on L-glumate and N-methyl-4-pyridinium ion induced neurotoxicity in PC-12 cells, *Neurotoxicology,* 22, 283–288, 2001.

77. Rekka, E., Chysselis, M., Siskou, I., Kourounakis, A., Synthesis of new azulene derivatives and study of their effects on lipid peroxidation and lipoxygenase activity, *Chemical Pharmacy Bulletin,* 50, 904–907, 2002.

78. Forsyth, J. T., Grunewald, R. A., Rostami-Hodjegan, A., Lennard, M. S., Sagar, H. J., Tucker, G. T., Parkinson's disease and CYP1A2 activity, British Journal of Clinical Pharmacology, 50, 303–309, 2000.

79. Smith, C. A. D., Gough, A., C., Leigh, P. N., Summers, B. A., Harding, A. E., Maranganore, D. M., Sturman, S. G., Schapira, A. H. V., Williams, A. C., Spurr, N. K., et al. Debrisoquine hydroxylase gene polymorphism and susceptibility to Parkinson's disease, *Lancet,* 339, 1375–1377, 1992.

80. Woo, S. I., Kim, J. W., Seo, H. G., Park, C. H., Han, S. H., Kim, S. H., Kim, K. W., Jhoo, J. H., Woo, J. I., CYP2D6*4 polymorphism is not associated with Parkinson's disease and has no protective role against Alzheimer's disease in the Korean population, *Psychiatry and Clinical Neuroscience,* 55, 373–377, 2001.

81. Payami, H., Lee, N., Zareparsi, S., Gonzales McNeal, M., Camicioli, R., Bird, T. D., Sexton, G., Gancher, S., Kaye, J., Calhoun, D., Swanson, P. D., Nutt, J., Parkinson's disease, CYP2D6 polymorphism, and age, *Neurology,* 56, 1363–1370, 2001.

82. Harhangi, B. S., Oostra, B. A., Heutink, P., van Duijn, C. M., Hofman, A., Breteler, M. M., CYP2D6 polymorphism in Parkinson's disease: the Rotterdam Study, *Movement Disorders,* 16, 290–293, 2001.

83. Stefanovic, M., Topic, E., Ivanisevic, A. M., Relja, M., Korsic, M., Genotyping of CYP2D6 in Parkinson's disease, *Clinical Chemistry and Laboratory Medicine,* 38, 929–934, 2000.

84. Maraganore, D. M., Farrer, M. J., Hardy, J. A., McDonnell, S. K., Schaid, D. J., Rocca, W. A., Case-control study of debrisoquine 4-hydroxylase, N-acetyltransferase 2, and apolipoprotein E gene polymorphisms in Parkinson's disease, *Movement Disorders,* 15, 714–719, 2000.

85. Chida, M., Yokoi, T., Kosaka, Y., Chiba, K., Nakamura, H., Ishizaki, T., Yokota, J., Kinoshita, M., Sato, K., Inaba, M., Aoki, Y., Gonzalez, F. J., Kamataki, T., Genetic polymorphism of CYP2D6 in the Japanese population, *Pharmacogenetics,* 9, 601–605, 1999.

86. Joost, O., Taylor, C. A., Thomas, C. A., Cupples, L. A., Saint-Hilaire, M. H., Feldman, R. G., Baldwin, C. T., Myers, R. H., Absence of effect of seven functional mutations in the CYP2D6 gene in Parkinson's disease, *Movement Disorders,* 14, 590–595, 1999.

87. Bon, M. A., Jansen Steur, E. N., de Vos, R. A., Vermes. I., Neurogenetic correlates of Parkinson's disease: apolipoprotein-E and cytochrome P450 2D6 genetic polymorphism, *Neuroscience Letters,* 266, 149–151, 1999.

88. Ho, S. L., Kung, M. H., Li, L. S., Lauder, I. J., Ramsden, D. B., Cytochrome P4502D6 (debrisoquine 4-hydroxylase) and Parkinson's disease in Chinese and Caucasians, *European Journal of Neurology,* 6, 323–329, 1999.

89. Atkinson, A., Singleton, A. B., Steward, A., Ince, P. G., Perry, R. H., McKeith, I. G., Fairbairn, A. F., Edwardson, J. A., Daly, A. K., Morris, C. M., CYP2D6 is associated with Parkinson's disease but not with dementia with Lewy Bodies or Alzheimer's disease, *Pharmacogenetics,* 9, 31–35, 1999.

90. Sabbagh, N., Brice, A., Marez, D., Durr, A., Legrand, M., Lo Guidice, J. M., Destee, A., Agid, Y., Broly, F., CYP2D6 polymorphism and Parkinson's disease, *Movement Disorders,* 14, 230–236, 1999.

91. Wu, R. M., Cheng, C. W., Chen, K. H., Shan, D. E., Kuo, J. W., Ho, Y. F., Chern, H. D., Genetic polymorphism of the CYP2E1 gene and susceptibility to Parkinson's disease in Taiwanese, *Journal of Neural Transmission,* 109, 1403–1414, 2002.

92. Wang, J., Liu, Z., Chen, B., Association between cytochrome P-450 enzyme gene polymorphisms and Parkinson's disease, *Zhonghua Yi Xue Za Zhi,* 80, 585–587, 2000.

93. Wang, J., Liu, Z., Chan, P., Lack of association between cytochrome P450 2E1 gene polymorphisms and Parkinson's disease in a Chinese population, *Movement Disorders,* 15, 1267–1269, 2000.

94. Hotamisligil, G. S., Girmen, A. S., Fink, J. S., Tivol, E., Shalish, C., Trofatter, J., Baenziger, J., Diamond, S., Markham, C., Sullivan, J., et al., Hereditary variations in monoamine oxidase as a risk factor for Parkinson's disease, *Movement Disorders,* 9, 305–310, 1994.

95. Nanko, S., Ueki, A., Hattori, M., No association between Parkinson's disease and monoamine oxidase A and B gene polymorphisms, *Neuroscience Letters,* 204, 125–127, 1996.

96. Xie, H., Wang, X., Hao, Y., Tang, G., Xu, L., Wu, Q., Chen, L., Ren, D., The EcoR V polymorphism of human monoamine oxidase A is not associated with idiopathic Parkinson's disease in a Shanghai Han population, *Zhonghua Yi Xue Yi Chuan Xue Za Zhi,* 19, 329–331, 2002.

97. Goudreau, J. L., Maraganore, D. M., Farrer, M. J., Lesnick, T. G., Singleton, A. B., Bower, J. H., Hardy, J. A., Rocca, W. A., Case-control study of dopamine transporter-1, monoamine oxidase-B, and catechol-O-methyl transferase polymorphisms in Parkinson's disease, *Movement Disorders,* 17, 1305–1311, 2002.

98. Hernan, M. A., Checkoway, H., O'Brien, R., Costa-Mallen, P., De Vivo, I., Colditz, G. A., Hunter, D. J., Kelsey, K. T., Ascherio, A., MAOB intron 13 and COMT codon 158 polymorphisms, cigarette smoking, and the risk of PD, *Neurology,* 58, 1381–1387, 2002.

99. Mellick, G. D., Buchanan, D. D., McCann, S. J., James, K. M., Johnson, A. G., Davis, D. R., Liyou, N., Chan, D., Le Couteur, D. G., Variations in the monoamine oxidase B (MAOB) gene are associated with Parkinson's disease, *Movement Disorders,* 14, 219–224, 1999.

100. Kelada, S. N., Stapleton, P. L., Farin, F. M., Bammler, T. K., Eaton, D. L., Smith-Weller, T., Franklin, G. M., Swanson, P. D., Longstreth, W. T., Jr., Checkoway, H., Glutathione S-transferase M1, T1, and P1 polymorphisms and Parkinson's disease, *Neuroscience Letters,* 337, 5–8., 2003.

101. Harada, S., Fujii, C., Hayashi, A., Ohkoshi, N., An association between idiopathic Parkinson's disease and polymorphisms of phase II detoxification enzymes: glutathione S-transferase M1 and quinone oxidoreductase 1 and 2, *Biochemical and Biophysical Research Communications,* 288, 887–892, 2001.

102. Rahbar, A., Kempkes, M., Muller, T., Reich, S., Welter, F. L., Meves, S., Przuntek, H., Bolt, H. M., Kuhn, W., Glutathione S-transferase polymorphism in Parkinson's disease, *Journal of Neural Transmission,* 107, 331–334, 2000.

103. van der Walt, J. M., Martin, E. R., Scott, W. K., Zhang, F., Nance, M. A., Watts, R. L., Hubble, J. P., Haines, J. L., Koller, W. C., Lyons, K., Pahwa, R., Stern, M. B., Colcher, A., Hiner, B. C., Jankovic, J., Ondo, W. G., Allen, F. H., Jr., Goetz, C. G., Small, G. W., Mastaglia, F., Roses, A. D., Stajich, J. M., Booze, M. W., Fujiwara, K., Gibson, R. A., Middleton, L. T., Scott, B. L., Pericak-Vance, M. A., Vance, J. M., Genetic polymorphisms of the N-acetyltransferase genes and risk of Parkinson's disease, *Neurology,* 60, 1189–1191, 2003.

104. Bialecka, M., Gawronska-Szklarz, B., Drozdzik, M., Honczarenko, K., Stankiewicz, J., N-acetyltransferase 2 polymorphism in sporadic Parkinson's disease in a Polish population, *European Journal of Clinical Pharmacology,* 57, 857–862, 2002.

105. Agundez, J.A., Jimenez-Jimenez, F. J., Luengo, A., Molina, J. A., Orti-Pareja, M., Vazquez, A., Ramos, F., Duarte, J., Coria, F., Ladero, J. M., Alvarez-Cermeno, J. C., Benitez, J., Slow allotypic variants of the NAT2 gene and susceptibility to early-onset Parkinson's disease, *Neurology,* 51, 1587–1592, 1998.

106. Ladero, J. M., Jimenez, F. J., Benitez, J., Fernandez-Gundin, M. J., Martinez, C., Llerena, A., Cobaleda, J., Munoz, J. J., Acetylator polymorphism in Parkinson's disease. *European Journal of Clinical Pharmacology,* 37, 391–393, 1989.

107. Eerola, J., Launes, J., Hellstrom, O., Tienari, P. J., Apolipoprotein E (APOE), PARKIN and catechol-O-methyltransferase (COMT) genes and susceptibility to sporadic Parkinson's disease in Finland, *Neuroscience Letters,* 330, 296–298, 2002.

108. Kunugi, H., Nanko, S., Ueki, A., Otsuka, E., Hattori, M., Hoda, F., Vallada, H. P., Arranz, M. J., Collier, D. A., High and low activity alleles of catechol-O-methyltransferase gene: ethnic difference and possible association with Parkinson's disease, *Neuroscience Letters,* 221, 202–204, 1997.

11 Progressive Neurodegeneration in the Chronic MPTP/Probenecid Model of Parkinson's Disease

Yuen-Sum Lau

Division of Pharmacology, School of Pharmacy, University of Missouri—Kansas City

CONTENTS

INTRODUCTION

Parkinson's disease (PD) is one of the most common neurological disorders, afflicting millions of individuals worldwide. Idiopathic PD, the prevailing form of disease, has a late age onset, and results from a slow, progressive, and complex degeneration of the nigrostriatal dopaminergic neurons. Accordingly, PD patients experience a gradual deterioration of motor performance and show signs of dyskinesia, rigidity, gait imbalance, and/or uncontrollable resting tremors. When the disease reaches an advanced stage, without medical support, PD patients can virtually be debilitated and may also suffer cognitive and behavioral dysfunctions such as dementia, anxiety, and depression. These clinical manifestations are caused by the marked loss of dopamine containing neurons in the substantia nigra pars compacta and by the vast depletion of the key neurotransmitter, dopamine (DA). Although current therapeutic paradigms largely target the replenishing of the lost DA in the brain or temporarily correcting the declined motor functions by improving mobility and reducing tremors and stiffness, no protective strategies have yet been developed to clinically guard the remaining intact neurons from undergoing degeneration or to restore and remodel the degenerated neurons back to function.

In addition to our understanding of the neurological explanations for parkinsonism and our awareness of the known distinctive motor and behavioral symptoms that serve as clinical indicators for PD diagnosis,[1] prominent Lewy bodies and dystrophic neurites are also detected in the degenerated nigrostriatal dopaminergic and other cortical/subcortical neurons in the disease-afflicted postmortem brains.[2,3] The Lewy bodies found in the substantia nigra of PD are typically described as concentric, intracytoplasmic inclusions consisting of a dense granular core and surrounded by a halo of radiating filaments.[3] The Lewy bodies further exhibit abnormally phosphorylated neurofilaments with an accumulation of proteins (such as ubiquitin, α-synuclein) and lipids,[4–8] Postmortem brains from PD patients also show evidence of nigral cell apoptosis displaying fragmented nuclei and caspase activation.[9–11]

The causes of idiopathic PD and the nigrostriatal neuron degenerative processes at cellular and molecular levels are still uncertain. Although analysis of postmortem human PD brain tissues may offer some clues or confirmation showing specific neuropathological features, in many instances, such findings could occur and associate with the end stage of the disease, whereas the understanding of disease initiation processes would still remain elusive. Therefore, to uncover the course of PD neurodegeneration and to design neuroprotective and neurorestorative strategies, comparable animal models that resemble the natural history and demonstrate progressive neuronal loss as in PD are desirable for research investigation.

ANIMAL MODELS FOR PARKINSON'S DISEASE RESEARCH

In conformity with the known human PD symptomatology, a desirable experimental model for PD research should display

1. Persistent depletion of at least 80% of striatal DA and its metabolites
2. Pronounced reduction of striatal sites for DA uptake
3. Significant and progressive loss of substantia nigral cells
4. Marked impairment in the animal's motor performance
5. The formation and accumulation of pathological inclusion bodies in nigral neurons

Over decades, several general animal models have been used in PD research including the unilateral, 6-hydroxydopamine lesion rat produced by chemically induced neuroectomy of the dopaminergic tracts.[12,13] The pharmacologically induced parkinsonism was obtained either by administering the monoamine depleting agents (e.g., reserpine, α-methyl-p-tyrosine) or by blocking the postsynaptic DA receptors using neuroleptic drugs (e.g., haloperidol, phenothiazines).[14] In recent years, many specially designed models have been introduced and have shown to be particularly valuable in elucidating certain chosen biochemical or genetic targets suspected to be involved in PD neurodegeneration. Such models include the use of rotenone, a known inhibitor of the complex I enzyme of the mitochondria electron transfer chain system,[15,16] or paraquat, which generates superoxide anion radicals and causes lipid peroxidation.[17,18] Transgenic models have also been generated in the fruit fly, *Drosophila melanogaster*,[19] and in the mouse[20] overexpressing the human α-synuclein, a constituent protein deposited in the Lewy body inclusions. Moreover, chronic intranigral infusion of lipopolysaccharide, a bacterial cell wall endotoxin, produces an indirect dopaminergic neurodegeneration through the proinflammatory activation of microglia.[21] Many of the described models show either reversible or abrupt end-stage type of PD symptomatology, which would not allow for long-term animal survival studies. The chronic lipopolysaccharide-infusion model demonstrates a progressive nature of neurodegeneration; however, the experimental procedure is not a simple one.

Since 1983, 1-methyl-4-phenyl-1,2,3,6-tetrahydropyridine (MPTP) has emerged as a leading neurotoxic agent used for inducing parkinsonism in animal model.[22] The MPTP neurotoxicity in humans is irreversible, and the consequential clinical and biochemical traits closely resemble those of the idiopathic PD.[23,24] This neurotoxin has been tested in many species of animals, and the results are strikingly different. Although similar symptoms can be replicated in nonhuman primates,[25] their sensitivity to the MPTP insult is variable.[26] It has been reported that the MPTP-induced parkinsonian symptoms in primates tend to reverse spontaneously over time.[27] While the classical Lewy bodies are not detected in humans[28] or nonhuman primates[29] intoxicated with MPTP, α-synuclein aggregations are found in the substantia nigra of MPTP-treated baboons.[30] Even though the use of the nonhuman primate MPTP model may be invaluable for preclinical evaluations of new therapies for PD, it would be very expensive and unjustifiable for investigating the neurological and pathological mechanisms underlying this disease.

The mode of dopaminolytic actions of MPTP has been well studied. Following a single dose of bolus injection, MPTP is readily absorbed and crosses the blood-brain-barrier. The toxin is then converted to an active toxic metabolite, 1-methyl-4-phenylpyridinium (MPP+), by the monoamine oxidase B enzyme, located in the astroglial cells.[31] The produced MPP+ is taken up into the nigrostriatal DA nerve terminals by the DA uptake transporter, DAT.[32] The elevated cytoplasmic MPP+ level sequentially causes the release and accumulation of glutamate, an excitatory amino acid transmitter, free Ca++, and reactive oxygen species within the afflicted neurons, leading to an inhibition of complex I of the mitochondrial electron transport chain system.[33,34] Impairment of respiration and oxidative damage to the mitochondria could effect depletion of cellular ATP and eventual neuronal death.[35] Free-radical species generation and oxidative stress in the neuron could bring about damages to DNA, lipid, and protein structures[36] and contribute to pathological amassment of ubiquitin and α-synuclein.[37] Subacute MPTP treatment is shown to induce apoptosis in dopaminergic neurons[38] and elicit endogenous reactive gliogenesis in the striatum.[39,40] Nevertheless, these mechanistic pathways of MPTP are established under *in vitro* or subacute studies; it needs to be experimentally confirmed whether these same pathways are involved in the slow, progressive development of neurodegeneration in PD.

Toxic effects of many substances after an acute exposure are quite different from those produced by repeated exposures over a prolonged time course. In general, long-term exposure to small doses of toxicants tends to instigate devastating and severe clinical consequences. It depends on the dose level, time interval, and frequency of toxic exposure; chemicals can accumulate in the biological systems and continuously cause residual cellular damage. Some toxicants, when given at intervals without providing sufficient time for cells to recover between successive doses, can produce irreversible cellular damages even after they disappear and are no longer detectable within the tissues. The latter scenario is apparently demonstrated in the case of MPTP intoxicated human subjects.[28] Experimentally, when low doses of MPTP are administered following a slow and prolonged time course in chronic primate models of PD, a gradual neurodegenerative process has been observed.[41,42] It is conceivable that a chronic small animal model of PD would be most practicable for studying and testing neuroprotective and neurorestorative paradigms. In this review, I describe the chronic mouse MPTP/probenecid model of PD that has been developed in our laboratory that shows similar symptomatic characteristics to the idiopathic PD. This model would be useful for examining the cellular and molecular pathways involving the slow and progressive course of parkinsonism development.

THE CHRONIC MOUSE MPTP/PROBENECID MODEL

ANIMAL TREATMENT

In our laboratory we routinely treat male, C57BL/6 mice (8 to 10 wk old, 25 to 28 g) with 10 combined regimens of MPTP and probenecid in an extended time course. For a period of 5 wk, at a twice-per-week interval (3.5 days apart), we first inject mice intraperitoneally with 250 mg/kg probenecid (made in dimethyl sulfoxide, 30-μl injection volume using a 50-μl size Hamilton syringe with a disposable 26G3/8 tuberculin needle). Approximately 30 min later, we treat the mice subcutaneously with 25 mg/kg MPTP HCl (dissolved in saline, 0.1-ml injection volume using a standard 1-ml size tuberculin syringe with a 26G3/8 needle). The background information and pharmacological rationale behind the development of this chronic model of PD have been previously described in detail.[22,43,44]

ANIMAL SURVIVAL

To establish a dependable model of PD for research studies, it is essential to maintain a high rate of animal survival. Otherwise, the outcome of the study will be greatly compromised if results are obtained from only a few animals that survive the treatment. Rapid induction of parkinsonian symptoms with frequent and massive doses of MPTP injections often causes a high mortality rate among animals.[22] From our experience, the high animal mortality caused by short-interval, multiple-dose MPTP injections is a combined result of hypothermia, catalepsy-like immobility, deprivation of food and water intake, and severe peripheral organ toxicity. Although the multiple-dose,

acute MPTP model can effectively cause rapid and drastic depletion of striatal DA levels, because of its unpredictable and extremely high fatality rate and reversible toxicity, the use of acute and subacute MPTP model for elucidating slow and progressive DA neurodegeneration may not be suitable.

In our laboratory, the rate of fatality in mice following the chronic MPTP/probenecid treatment protocol is very low. This treatment procedure lowers the risk of acute peripheral toxicity and enhances the central neurodegenerative process. In the most recent study, we injected 46 mice with chronic MPTP and probenecid, recording only 3 deaths (<1% mortality). These deaths are attributed to accidental inaccurate injections, inherent diseases in mice, or victimization after vicious fighting with a cagemate. To circumvent these problems, we now routinely keep each mouse in a single cage, within a sanitary murine virus-free environment, and ensure accurate drug injections by an experienced technician. Immediately following each treatment, animals typically show signs of lethargy, either moving slowly or staying in one location. Checking back 12 to 15 h later, these animals are normally active and display no overt behavioral inhibitions. We also notice that the food and water intake and body weight gain in these mice are normal during and after the treatment.

PARKINSONIAN FEATURES

In this chronic MPTP/probenecid mouse model of PD, we observe an immediate and robust (over 90%) lost of striatal DA, its metabolite DOPAC, and uptake, which are the typical single-dose effects of MPTP. Strikingly, the reduction in DA, its metabolite, and uptake persists throughout a survival period of at least 6 mo. The tyrosine hydroxylase (TH) immunoreactivity in nigral cells is down-regulated.[44] The loss of DA neurons in the substantia nigra six months after treatment is estimated to be at least 60%. The first sign of motor deficit in chronic MPTP/probenecid-treated mice as demonstrated by rotarod performance test is detected at three weeks and persists for at least six months after treatment.[44] We further observed that these mice initially show slowness in movement with a "wobbling" or "shuffling" pattern of walking (Lau, unpublished observations).

Morphologically, we detect an increasing number of abnormal inclusion bodies in the cytoplasm of nigral dopaminergic and cortical neurons beginning at three weeks after chronic MPTP/probenecid treatment.[45] These inclusions are granular and filamentous in appearance and are immuno-positive to α-synuclein and ubiquitin, resembling those Lewy body structures found in the cortex of human PD. At the ultrastructural level, we further confirm that these inclusions contain the dense and granular core similar to that of the classical Lewy bodies. In addition, numerous lobulated, secondary lysosomes filled with lipofuscin are observed in the cytoplasm of α-synuclein-immunoreactive neurons, and the density of these neurons is significantly reduced.

It is recognized that neurons die at a low rate during normal aging but at an accelerated pace in cases of neuro-degenerative disorders. However, it is not clear how cell death under normal and disease state is regulated. In a preliminary study using TUNEL analysis, we observed that nigral neurons undergo an apoptotic mode of degeneration within a week after the chronic treatment of mice with MPTP/probenecid, suggesting that, in this model, neurodegeneration occurs at an early stage and progresses as cell death intensifies (Novikova and Lau, unpublished observations). Therefore, molecules that block apoptotic process might be potentially useful for preventing further neuronal loss and functional deterioration in neurodegenerative diseases. Our current understanding of apoptotic pathways has been generated from studies involving worms, flies, cell cultures, and murine knockouts of apoptosis-related genes, which have demonstrated severe abnormalities resulting in early postnatal cell death. The neuro-apoptotic concept needs to be thoroughly tested in the chronic animal disease models. Furthermore, when cells die, they should be quickly removed from the neuronal tissue. Abnormal dead cell accumulation may lead to further obstruction of neural transmission and eventually incapacitate neurological functions.

RISK AND SAFETY CONSIDERATIONS FOR LABORATORY USE OF MPTP

Since MPTP became a widely used chemical for inducing a research model of PD in a variety of laboratory animals, the sale and price of MPTP have gone up sharply. It is also noticed that the neurotoxic susceptibility of MPTP is much greater in humans than in other species. Because human subjects who are intoxicated with MPTP can develop severe cardinal signs and symptoms indistinguishable from those of the idiopathic PD,[23,24] extra caution should be exercised among the research investigators to avoid accidental injections or exposure to MPTP.

Since mice are less sensitive to MPTP than humans, larger doses are normally required to produce neuronal damage in mice, thus increasing the risk potential of exposure to investigators who handle a large quantity of the hazardous chemical. It is understandable that there are increasing concerns expressed by the animal care personnel including those in our own Laboratory Animal Center regarding how much MPTP exposure they might receive by handling the animals, changing their cages, bedding, food, and water, or even entering the room.

It should be made aware that most reported clinical cases of MPTP intoxication are due to *intravenous* injection of the chemical by drug abusers;[46] through this route,

the chemical readily crosses the blood brain barrier in massive quantities causing severe neuronal damage. Another fatal incidence occurred with a chemist who works with the compound.[47] Fortunately, no incidence has been reported among investigators who use MPTP in their research. In the laboratory, the most vulnerable chances for MPTP exposure are probably during the handling and weighing of the stock chemical and in the preparation of solutions for injection, since MPTP can be absorbed through inhalation, ingestion, direct skin contact, and self-injection. If the chemical is (1) always stored in a secure place and not left in unattended, open areas, (2) prepared by a carefully trained technician who wears a protective laboratory coat, facial mask, eyeglasses or goggles, and gloves while handling and weighing the chemical under a safety cabinet, and (3) made into solution and contained in a tightly capped vial placed in another protective container lined with an absorbant pad, the risk for self-contamination should be low to none.

During treatment, if animals are steadily held by one person and injections are done by an experienced technician, accidental self-inoculation and possible intoxication can be avoided. Since MPTP is incompatible with, and can be decomposed by, strong oxidizing agents, the unused MPTP as a powder or solution should be properly treated with oxidizing agents such as nitric acid (HNO_3), chromic acid ($H_2Cr_2O_7$), potassium chlorate ($KClO_3$), or sodium hypochlorite ($NaClO$), an active ingredient in the household bleach, before disposal.

MPTP is classified as, and should be regarded as, a hazardous toxic substance at all times. Animal care technicians should be notified about the ongoing study, and a sign, along with a copy of the material safety data sheet (MSDS) for MPTP, should be posted on the door of an isolated room designated solely for the MPTP study. On each animal cage, the date and time of the most recent injection should be indicated so the animal care personnel will know how long the chemical has been with the animals. Additional information and suggestions for the safe use of MPTP can be found in other reviews.[48,49] However, a discussion on MPTP-N-oxide, a major metabolite found in the mouse urine after MPTP injection and its significance to animal excreta handling will be presented here.

It is reasonable to first ask whether handling the animals that are treated with MPTP and their excreta will pose health hazards to researchers or animal care takers. We have demonstrated that, after subcutaneous administration of [³H]MPTP to mice,[50] there is no significant increase in the background radiation detected in the cage environment, suggesting that the tritiated compound is not released by animal exhalation. The tritiated activity is found in the body of the untreated cagemate (kept freely in the same cage with the treated mouse) but not in the untreated near-neighbor (kept in an adjacent cage separated from the treated mouse by mesh wire). The route of transmission for cross-animal contamination appears to be through urinary contact, since, after the [³H]MPTP injection, we found that urine is the major excretory source containing MPTP metabolites in comparison to the fecal waste. Approximately 42% of the total radioactivity is detected in the urine within first three hours after the [³H]MPTP injection. Chemical species analysis shows that MPTP-N-oxide, a product of the liver microsomal flavin-containing monooxygenase metabolism is the major metabolite found in the urine. Only traces of MPP$^+$ and MPTP are retrieved from urine samples.[43,51] In the chronic MPTP/probenecid model, it is interesting to report that probenecid not only potentiates MPTP neurotoxicity, it also significantly reduces urinary excretion of MPTP-N-oxide and MPTP is not noticeable.[43]

If MPP$^+$ and MPTP-N-oxide are the primary metabolites present in the urine of MPTP-treated mice and end up in the animal cage and bedding, it is logical to ask the next question: are these two metabolites neurotoxic and transmittable? It has been determined that MPP$^+$ and MPTP-N-oxide may cause DA depletion only if injected directly into the neostriatum.[52,53] Nevertheless, these two metabolites are not lipophilic; thus, when given systemically, they would not cross the blood-brain barrier and produce neurotoxic effects as the parent compound, MPTP, does.[53–55] Therefore, the major metabolites (MPTP-N-oxide and MPP$^+$) released from the mouse urine after MPTP injection may not cause neurotoxicity through peripheral contact due to their lack of CNS penetration. However, it is still highly advisable to always treat MPTP and its metabolites as hazardous materials and to follow proper safety procedures when handling them.

CONCLUSION

We have shown that the chronic MPTP/probenecid-treated mouse shares many key features of human PD. This model demonstrates improvements over the conventional, acute and subacute models and can serve as an important and a potentially useful tool for studying disease progression, neuroprotective, and neurorestorative strategies in PD. This chronic model may offer the window of opportunity for the testing of different viable hypotheses underlying the mechanisms of PD type of nigrostriatal neurodegeneration. We further reveal that this model reduces the hazardous risks of MPTP in the research environment. Not only the chemical species released from the animals are lowered after MPTP injection with probenecid; these metabolites do not have the same neurotoxic potential as the parent compound through peripheral contact.

ACKNOWLEDGMENTS

The author wishes to acknowledge and thank the following individuals who have contributed to the work described in this review: Shanon Callen, Richard Callison, James Crampton, Thomas Doyle, Eric Fung, Gloria Meredith, Elizabeth Petroske, Christopher Runice, Karen Santa Cruz, Susan Totterdell, Karen Trobough, and John Wilson. The author's research on PD is supported by the Health Future Foundation, the National Parkinson Foundation, the University of Missouri and the National Institute of Neurological Diseases and Stroke (R01 NS41799, R01 NS47920).

REFERENCES

1. Fahn, S., Description of Parkinson's disease as a clinical syndrome, *Ann. N. Y. Acad. Sci.*, 991, 1, 2003.
2. Braak, H. et al., Pattern of brain destruction in Parkinson's and Alzheimer's diseases, *J. Neural Transm.*, 103, 455, 1996.
3. Forno, L. S., Neuropathology of Parkinson's disease, *J. Neuropathol. Exp. Neurol.*, 55, 259, 1996.
4. Kuzuhara, S. et al., Lewy bodies are ubiquitinated. A light and electron microscopic immunocytochemical study, *Acta Neuropathol. (Berl.)*, 75, 345, 1988.
5. Bancher, C. et al., An antigenic profile of Lewy bodies: immunocytochemical indication for protein phosphorylation and ubiquitination, *J. Neuropathol. Exp. Neurol.*, 48, 81, 1989.
6. Spillantini, M. G. et al., Alpha-synuclein in filamentous inclusions of Lewy bodies from Parkinson's disease and dementia with Lewy bodies, *Proc. Natl. Acad. Sci. USA*, 95, 6469, 1998.
7. Gai, W. P. et al., *In situ* and *in vitro* study of colocalization and segregation of alpha-synuclein, ubiquitin, and lipids in Lewy bodies, *Exp. Neurol.*, 166, 324, 2000.
8. Shimura, H. et al., Ubiquitination of a new form of alpha-synuclein by parkin from human brain: implications for Parkinson's disease, *Science*, 293, 263, 2001.
9. Tatton, N. A. et al., A fluorescent double-labeling method to detect and confirm apoptotic nuclei in Parkinson's disease, *Ann. Neurol.*, 44, S142, 1998.
10. Hartmann, A. et al., Caspase-3: A vulnerability factor and final effector in apoptotic death of dopaminergic neurons in Parkinson's disease, *Proc. Natl. Acad. Sci. USA*, 97, 2875, 2000.
11. Tatton, N. A., Increased caspase 3 and Bax immunoreactivity accompany nuclear GAPDH translocation and neuronal apoptosis in Parkinson's disease, *Exp. Neurol.*, 166, 29, 2000.
12. Ungerstedt, U., 6-hydroxydopamine-induced degeneration of the nigrostriatal dopamine pathway: the turning syndrome, *Pharmacol. Ther.*, 2, 37, 1976.
13. Schwarting, R. K. and Huston, J. P., The unilateral 6-hydroxydopamine lesion model in behavioral brain research. Analysis of functional deficits, recovery and treatments, *Prog. Neurobiol.*, 50, 275, 1996.
14. Montastruc, J. L. et al., Drug-induced parkinsonism: a review, *Fundam. Clin. Pharmacol.* 8, 293, 1994.
15. Ferrante, R. J. et al., Systemic administration of rotenone produces selective damage in the striatum and glo-

bus pallidus, but not in the substantia nigra, *Brain Res.,* 753, 157, 1997.

16. Betarbet, R., et al., Chronic systemic pesticide exposure reproduces features of Parkinson's disease, *Nat. Neurosci.,* 3, 1301, 2000.

17. Corasaniti, M. T. et al., Paraquat: a useful tool for the *in vivo* study of mechanisms of neuronal cell death, *Pharmacol. Toxicol.,* 83, 1, 1998.

18. Thiruchelvam, M., et al., The nigrostriatal dopaminergic system as a preferential target of repeated exposures to combined paraquat and maneb: implications for Parkinson's disease, *J. Neurosci.,* 20, 9207, 2000.

19. Feany, M. B. and Bender, W. W., A Drosophila model of Parkinson's disease, *Nature,* 404, 394, 2000.

20. Masliah, E. et al., Dopaminergic loss and inclusion body formation in alpha-synuclein mice: implications for neurodegenerative disorders, *Science,* 287, 1265, 2000.

21. Gao, H. M. et al., Microglial activation-mediated delayed and progressive degeneration of rat nigral dopaminergic neurons: relevance to Parkinson's disease, *J. Neurochem.,* 81, 1285, 2002.

22. Lau, Y. S. and Meredith, G. E., From drugs of abuse to parkinsonism. The MPTP mouse model of Parkinson's disease, *Methods Mol. Med.,* 79, 103, 2003.

23. Ballard, P. A., Tetrud, J. W., and Langston, J. W., Permanent human parkinsonism due to 1-methyl-4-phenyl-1,2,3,6-tetrahydropyridine (MPTP): seven cases, *Neurology,* 35, 949, 1985.

24. Burns, R. S. et al., The clinical syndrome of striatal dopamine deficiency. Parkinsonism induced by 1-methyl-4-phenyl-1,2,3,6-tetrahydropyridine (MPTP), *N. Engl. J. Med.,* 312, 1418, 1985.

25. Burns, R. S. et al., A primate model of parkinsonism: selective destruction of dopaminergic neurons in the pars compacta of the substantia nigra by N-methyl-4-phenyl-1,2,3,6-tetrahydropyridine, *Proc. Natl. Acad. Sci. USA,* 80, 4546, 1983.

26. Gerlach, M. and Riederer, P., Animal models of Parkinson's disease: an empirical comparison with the phenomenology of the disease in man, *J. Neural Transm.,* 103, 987, 1996.

27. Eidelberg, E. et al., Variability and functional recovery in the N-methyl-4-phenyl-1,2,3,6-tetrahydropyridine model of parkinsonism in monkeys, *Neuroscience,* 18, 817, 1986.

28. Langston, J. W. et al., Evidence of active nerve cell degeneration in the substantia nigra of humans years after 1-methyl-4-phenyl-1,2,3,6-tetrahydropyridine exposure, *Ann. Neurol.,* 46, 598, 1999.

29. Forno, L. S. et al., Similarities and differences between MPTP-induced parkinsonsim and Parkinson's disease. Neuropathologic considerations, *Adv. Neurol.* 60, 600, 1993.

30. Kowall, N. W. et al., MPTP induces alpha-synuclein aggregation in the substantia nigra of baboons, *Neuroreport,* 11, 211, 2000.

31. Singer, T. P. et al., Biochemical events in the development of parkinsonism induced by 1-methyl-4-phenyl-1,2,3,6-tetrahydropyridine, *J. Neurochem.,* 49, 1, 1987.

32. Gainetdinov, R. R. et al., Dopamine transporter is required for *in vivo* MPTP neurotoxicity: evidence from mice lacking the transporter, *J. Neurochem.,* 69, 1322, 1997.

33. Leist, M. et al., 1-Methyl-4-phenylpyridinium induces autocrine excitotoxicity, protease activation, and neuronal apoptosis, *Mol. Pharmacol.,* 54, 789, 1998.

34. Cleeter, M. W., Cooper, J. M., and Schapira, A. H., Irreversible inhibition of mitochondrial complex I by 1-methyl-4-phenylpyridinium: evidence for free radical involvement, *J. Neurochem.,* 58, 786, 1992.

35. Chan, P. et al., Rapid ATP loss caused by 1-methyl-4-phenyl-1,2,3,6-tetrahydropyridine in mouse brain, *J. Neurochem.,* 57, 348, 1991.

36. Jenner, P., Oxidative mechanisms in nigral cell death in Parkinson's disease, *Mov. Disord.,* 13, 24, 1998.

37. Souza, J. M., et al., Dityrosine cross-linking promotes formation of stable alpha -synuclein polymers. Implication of nitrative and oxidative stress in the pathogenesis of neurodegenerative synucleinopathies, *J. Biol. Chem.,* 275, 18344, 2000.

38. Tatton, N. A. and Kish, S. J., In situ detection of apoptotic nuclei in the substantia nigra compacta of 1-methyl-4-phenyl-1,2,3,6-tetrahydropyridine-treated mice using terminal deoxynucleotidyl transferase labeling and acridine orange staining, *Neuroscience,* 77, 1037, 1997.

39. Kay, J. N. and Blum, M., Differential response of ventral midbrain and striatal progenitor cells to lesions of the nigrostriatal dopaminergic projection, *Dev. Neurosci.,* 22, 56, 2000.

40. Mao, L. et al., Profound astrogenesis in the striatum of adult mice following nigrostriatal dopaminergic lesion by repeated MPTP administration, *Brain Res. Dev. Brain Res.,* 131, 57, 2001.

41. Schneider, J. S. and Roeltgen, D. P., Delayed matching-to-sample, object retrieval, and discrimination reversal deficits in chronic low dose MPTP-treated monkeys, *Brain Res.,* 615, 351, 1993.

42. Bezard, E. et al., A chronic MPTP model reproducing the slow evolution of Parkinson's disease: evolution of motor symptoms in the monkey, *Brain Res.,* 766, 107, 1997.

43. Lau, Y.-S. et al., Effects of probenecid on striatal dopamine depletion in acute and long-term 1-methyl-4-phenyl-1,2,3,6-tetrahydropyridine (MPTP)-treated mice, *Gen. Pharmacol.,* 21, 181, 1990.

44. Petroske, E. et al., Mouse model of parkinsonism: A comparison between subacute MPTP and chronic MPTP/probenecid treatment, *Neuroscience,* 106, 589, 2001.

45. Meredith, G. E. et al., Lysosomal malfunction accompanies alpha-synuclein aggregation in a progressive mouse model of Parkinson's disease, *Brain Res.,* 956, 156, 2002.

46. Langston, J. W. et al., Chronic Parkinsonism in humans due to a product of meperidine-analog synthesis, *Science,* 219, 979, 1983.

47. Langston, J. W. and Ballard, P. A. Jr., Parkinson's disease in a chemist working with 1-methyl-4-phenyl-1,2,5,6-tetrahydropyridine, *N. Engl. J. Med.,* 309, 310, 1983.

48. Yang, S. C. et al., Recommended safe practices for using the neurotoxin MPTP in animal experiments, *Lab. Anim. Sci.,* 38, 563, 1988.

49. Przedborski, S. et al., The parkinsonian toxin 1-methyl-4-phenyl-1,2,3,6-tetrahydropyridine (MPTP): a technical review of its utility and safety, *J. Neurochem.,* 76, 1265, 2001.

50. Crampton, J. M. et al., MPTP in mice: treatment, distribution and possible source of contamination, *Life Sci.,* 42, 73, 1988.

51. Lau, Y. S., Crampton, J. M., and Wilson, J. A., Urinary excretion of MPTP and its primary metabolites in mice, *Life Sci.,* 43, 1459, 1988.

52. Giovanni, A., Sonsalla, P. K., and Heikkila, R. E., Studies on species sensitivity to the dopaminergic neurotoxin 1-methyl-4-phenyl-1,2,3,6-tetrahydropyridine. Part 2: Central administration of 1-methyl-4-phenylpyridinium, *J. Pharmacol. Exp. Ther.,* 270, 1008, 1994.

53. Lau, Y. S. et al., Depletion of striatal dopamine by the N-oxide of 1-methyl-4-phenyl-1,2,3,6-tetrahydropyridine (MPTP), *Neurotoxicology,* 12, 189, 1991.

54. Johannessen, J. N. et al., 1-Methyl-4-phenylpyridine (MPP+) induces oxidative stress in the rodent, *Life Sci.,* 38, 743, 1986.

55. Chiba, K. et al., Characterization of hepatic microsomal metabolism as an *in vivo* detoxication pathway of 1-methyl-4-phenyl-1,2,3,6-tetrahydropyridine in mice, *J. Pharmacol. Exp. Ther.,* 246, 1108, 1988.

12 Alpha-Synuclein and Parkinson's Disease

Lawrence I. Golbe
University of Medicine and Dentistry of New Jersey, Robert Wood Johnson Medical School

CONTENTS

INTRODUCTION

In 1997 and 1998, the protein alpha-synuclein (α-syn) was found to aggregate abnormally in sporadic and familial PD. Many observers have considered this insight into the pathogenesis of PD to be at least as important as the description of MPTP toxicity 15 years earlier. The causes of α-syn aggregation and the details of its relationship to cell loss in PD are the subject of much current work and will continue to cascade out of laboratories for years. But the result so far is a wealth of new etiological and pathogenetic insights, several animal models, and, most important, new potential targets for neuroprotective therapy. This chapter will summarize the current state of knowledge of the role of α-syn in PD.

HISTORY

α-syn was discovered in 1988 as a component of cholinergic synapses of the electric organ of a fish, *Torpedo californica*.[1] A role in synaptic plasticity and learning was found in other species, and a hint of a relationship to human disease appeared when a portion of the protein was found to be identical to the non-β-amyloid component of amyloid plaques in Alzheimer's disease[2] and to bind beta-amyloid[3] But any further role of the protein in Alzheimer's disease was elusive and a test for an association of α-syn genetic polymorphisms with AD proved negative.[4] The chromosomal location of α-syn was found in 1995 to be 4q21.[5]

Soon thereafter, in 1996, linkage analysis of a family with autosomal dominant PD, the Contursi kindred,[6,7] incriminated a genomic interval at chromosome 4q21–23.[8] Subsequent sequencing of candidate genes in that chromosomal interval revealed a point mutation substituting adenosine for guanosine at nucleotide position 209 of the gene for α-syn.[9] It translated into a substitution of threonine for alanine at amino acid position.[53]

Sequencing the α-syn gene in other families with autosomal dominant PD revealed two other point mutations, A30P in a German family[10] and E46K in a Spanish family.[11] Linkage analysis revealed a triplication of the α-syn gene as the cause of PD in the "Iowa kindred," a U.S. family of mixed northern European origin[12] and another triplication causing the disorder in a Swedish-American family.[13] α-syn mutations associated with PD are collectively dubbed "PARK1."

The A30P substitution and the two triplications have each been found in only one family, but the group that found the linkage and A53T mutation in the Contursi kindred found the same defect in three Greek families with PD clinical identical to that in the Contursi kindred.[9] Subsequently, the same mutation was discovered in a few more families of Greek origin in the U.S., Australia, and Greece.[14–18] The A53T mutation has been otherwise absent from hundreds of other families and patients with sporadic PD,[19] suggesting that it is a "private mutation" with infinitesimally small contribution to the population burden of PD.

It is tempting to speculate that a Greek traveler of centuries ago journeyed to Contursi for treatment of his A53T-related parkinsonism at that the town's sulfurous springs, which have attracted medical tourists since Roman times. Perhaps he, or just his DNA, remained.

CLINICAL PHENOTYPES CAUSED BY α-SYN MUTATIONS

The most important difference between patients with A53T PD and "idiopathic PD" is the young average age of symptom onset (in the late 40s compared with 60), the paucity of rest tremor, and the rapid progression to severe disability and death.[7] In the Contursi kindred, for example, the mean survival is only 9.6 yr, compared with approximately 25 yr for patients with "idiopathic" PD of similarly young onset age.[7] A clinical picture consistent with that of dementia with Lewy bodies, together with its pathological correlate, seems to occur more frequently among A53G PARK1 carriers than in the rest of the parkinsonian population.[20] The motor symptomatology responds to levodopa as well as "idiopathic" PD, but rest tremor is relatively rare, occurring at disease onset in only 7%, compared with 42% for controls with familial PD without α-syn mutations.[17] In the Iowa kindred, early-onset dementia is particularly common and severe.[21]

An intriguing observation is that the range of clinical variation, even among patients with the A53T mutation, is as wide as that in the population with "idiopathic" PD, with a standard deviation of approximately 12 years.[7] This suggests that other genetic or nongenetic factors influence the clinical phenotype. Identifying those factors in PARK1 carriers could provide clues to the range of variation in others with PD. In fact, preliminary data suggest that certain alleles of glutathione S-transferase, a detoxification enzyme, are disproportionately represented among younger-onset affected individuals with the α-syn A53T mutation.[22] This is only one way in which identification of rare α- mutations may shed light on all PD, in this case by pointing to possible pathogenetic collaborators that may be obscured by the mix of etiologies in the general PD population.

NORMAL STRUCTURE AND FUNCTION

α-syn is a small protein, with 140 amino acids (Figure 12.1). It is normally expressed not only in dopaminergic neurons but also in cortical and noradrenergic neurons and in endothelial cells and platelets.[23,24] Its normal function is not known with certainty, and α-syn knockout mice exhibit little behavioral or anatomic pathology.[25] This offers the hope that if inhibition of α-syn synthesis or activity in humans is found to slow or prevent PD, such treatment would produce little or no toxicity.

The physiologic function of α-syn best supported by experimental evidence is the maintenance, storage, and regulation of dopamine vesicles at synaptic terminals.[26] It may also play a role in synaptic vesicle function in non-dopaminergic neurons including those of the hippocampus.[27] This could explain the involvement of the CA2-3 area and early appearance of dementia in some carriers of the A53T mutation.

The large amino-terminal portion of α-syn, residues 1–67, consists of two narrowly separated domains. They have no secondary structure in solution but, on binding to lipid, assume α-helical conformations.[28] This structure is similar to that of apolipoproteins. Residing within this amino-terminal region are seven imperfect copies of an 11-residue motif with KTKEGV as its core, "consensus," sequence. This region is essential to the ability of α-syn to bind to lipid membranes such as synaptic vesicles. All three of the known PD-associated point mutations in α-syn occur in this region, suggesting that they disrupt the ability of the protein to assume an α-helical structure.

Residues 61–95 form a hydrophobic portion necessary to the aggregation of α-syn, a critical issue in the pathogenesis of PD. The form of aggregation is the β-sheet or β-amyloid-like fibril.[29] This region of α-syn is identical to the non-β-amyloid portion of the amyloid plaques of Alzheimer's disease.[2] This has prompted the name "non-β-amyloid component precursor" (NACP) for α-syn, a synonym that is still frequent in the literature.

The carboxy-terminal portion, residues 96–140, is acidic by virtue of an abundance of glutamic acid and aspartic acid. It is also rich in proline. The resemblance of residues 123–140 to fatty acid binding protein suggests that some aspect of the behavior of α-syn may be regulated by fatty acids.[30]

Although α-syn knockout mice display little pathology,[25] they are fully resistant to the toxicity of MPTP.[31] This suggests, broadly, that α-syn participates in the process whereby mitochondrial damage kills neurons, a notion that will be discussed below.

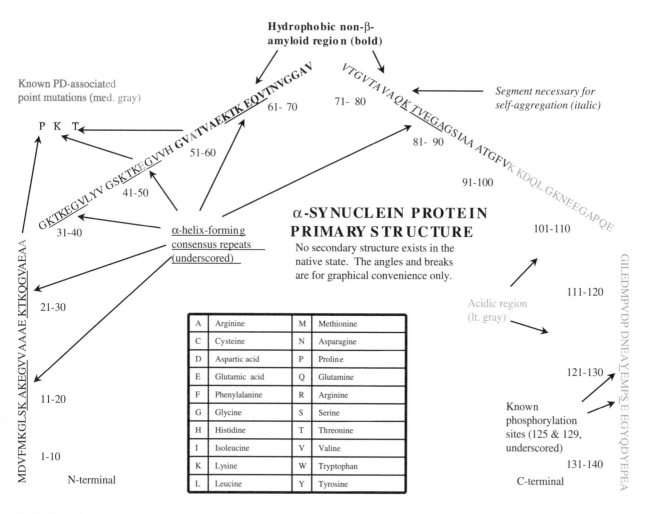

FIGURE 12.1

INTERACTIONS AND BINDING PARTNERS

The list of proteins with which α-syn interacts is long and growing, but the relevance of any of these to the pathogenesis of PD or the other α-synucleinopathies remains unclear. One interacting protein with a direct role in dopamine synthesis, if not the pathogenesis of PD, is tyrosine hydroxylase. α-syn increases the ratio of the less active phosphorylated form of TH to the more active unphosphorylated form.[32] α-syn also enhances the activity of the dopamine transporter,[33] which could increase cytoplasmic concentration of dopamine and other toxins. α-syn interacts with mitochondrial cytochrome oxidase[34] a pro-apoptotic agent that is part of the mitochondrial respiratory chain. α-syn strongly inhibits phospholipase D-2,[35] which is located on the plasma membrane and is part of the physiology of presynaptic vesicles.

Proteins associated with Alzheimer's disease that interact with α-syn include β-amyloid[3] and tau,[36,37] and apoE3 is associated with PD at the epidemiologic level.[38]

This supports the notion that the α-synucleinopathies and the tauopathies share common pathogenetic pathways, and possibly common neuroprotective strategies.

Components of the apoptotic pathway that interact with α-syn include ERK-1/2 and p38MAPK. It also interacts with ERK-2 and Elk-1, suggesting that it may regulate the mitogen-activated protein (MAP) kinase pathway and the response of cells to growth factors.[39,40]

A search for novel proteins interacting with α-syn yielded synphilin-1.[41] Its normal function may be to link α-syn to other proteins involved in vesicular transport. It does appear to promote the abnormal aggregation of α-syn, perhaps by interacting with parkin, a component of the ubiquitin-proteasome system.[42] Synphilin-1 co-localizes with α-syn to a striking degree and co-transfection of cultured cells with both the NAC fragment of α-syn and synphilin-1 produces cytoplasmic inclusions resembling Lewy bodies.[41] Unfortunately, a search for mutations or allelic associations in synphilin-1 in families with inherited PD has proved negative.[43]

AGGREGATION PROCESS

α-syn tends to aggregate with itself, especially when it is misfolded. The bulk of the evidence suggests that a central pathogenetic event in PD is excessive α-syn aggregation, failure of the cell to remove misfolded α-syn before it aggregates, failure to dispose of aggregates in an early stage of development when they are still soluble, or a combination of these. Many factors have been found to influence the rate of α-syn aggregation. The relevance of these to PD and their vulnerability to therapeutic intervention is the subject of much active research.

The first stage of α-syn aggregation is the oligomer, which in its transient β-sheet configuration is called a protofibril. These consist typically of three to 30 α-syn monomers and can take a variety of geometric forms. Some but not all of these forms form strings, termed fibrils, that join together as they mature.[44] The fibrils, in turn, form the basis of Lewy bodies and Lewy neurites,[45] well structured cytoplasmic inclusions with some two dozen chemical components, principally α-syn, ubiquitin, and tubulin.

The rate of fibril formation is accelerated by the A53T mutation but slowed by the A30P mutation.[46–48] This suggests that the fibrils themselves are unlikely to be the toxic species in PD. In this, PD would be similar to Huntington's disease and other trinucleotide repeat disorders.[49]

The α-syn fibrils in Lewy bodies are radially arranged within the usually spherical structures,[50] suggesting that they are not mere innocent bystanders caught in the process of aggregation of a different macromolecule. The fibrils are ultrastructurally similar to those formed from *in vitro* α-syn aggregation.[51] Pale bodies, which are probably a precursor of Lewy bodies, also stain strongly for α-syn.[52]

Lewy bodies display a protein content and organization similar to those of aggresomes,[53] which are sequestrations of misfolded, worn-out, or defective proteins formed by a metabolically active process involving transport of such proteins along microtubules. This suggests that Lewy bodies may be a form of aggresome, part of an adaptive strategy to keep α-syn protoaggregates away from vital cell processes.[54]

PRO-AGGREGATION FACTORS

Many chemical species promote the aggregation of α-syn. Chief among these is genetically wild-type α-syn itself. Excessive concentrations of the protein cause it to aggregate. This phenomenon has been demonstrated in multiple animal models[55,56] and in yeast.[57] It also may be the mechanism of toxicity of one form of PARK1 human PD in which α-syn of normal amino acid sequence is overexpressed by virtue of a genetic triplication.[12,13]

The process of aggregate formation by excessive concentrations of α-syn is nucleation-dependent. This means that, after a critical concentration of α-syn is attained, a particle, which may be an α-syn fibril or an aggregate of another protein, precipitates full-scale aggregation.[58] This may help explain the late-life onset and subsequent rapid progression of PD despite the early-life appearance of etiologic factors such as genetic defects and/or toxic exposures.

Tau is another important protein relevant to PD that promotes the aggregation of α-syn. In fact, α-syn promotes the aggregation of tau as well[59] and promotes tau phosphorylation, which may disrupt the normal function of tau in stabilizing microtubules.[36] The notion that tau interacts with α-syn in the pathogenesis of PD is supported by

1. The presence of tau in some Lewy bodies of sporadic PD[60]
2. The co-occurrence of α-syn-positive aggregates and tau-positive aggregates in some neurodegenerative disorders[61]
3. The co-occurrence of tau aggregates in the same brain areas as the α-syn aggregates in a Contursi kindred brain[62]
4. A slight but confirmed over-representation of the tau H1 haplotype in patients with PD[63]

An important clue to the dopaminergic predilection of cellular damage in PD arises from the observation that dopamine inhibits the maturation of toxic protoaggregates to insoluble, probably nontoxic, or less toxic, fibrils by forming adducts.[64] Other catecholamines tend to have similar activity, perhaps explaining the involvement of other catecholaminergic nuclei in PD.

In light of the reported epidemiological association of PD with exposure to industrial metals, it is intriguing that α-syn interacts with ionic forms of copper, iron, aluminum, and zinc.[65–67] Neuromelanin acts as an iron sink and may present a source of that metal to stimulate α-syn aggregation under certain conditions.[68,69] Pesticides, a heterogeneous class of chemicals epidemiologically associated with PD, also enhance α-syn aggregation.[67]

Head trauma is an even more important risk factor for PD, with an odds ratio of 3.1 to 4.2, depending on the multivariate model.[70,71] (The only stronger independent risk factor is a family history of PD, with odds ratio 5.1 to 5.8.)[70] This etiologic relationship may involve α-syn. Mice subjected to brain trauma develop an increase in α-syn concentration in axon bundles in the striatum and elsewhere.[71] The α-syn in this model is nitrated, as in human PD, suggesting that oxidative stress plays a role in this process. One may hypothesize that mechanical damage to axons causes accumulation of proteins, including α-syn, which proceeds to aggregate by virtue of its excessive concentration. Supporting this hypothesis is the gross mechanical anatomy of the human brain, where a deceleration or acceleration injury produces a torque at the point where the relatively mobile cerebrum hinges on the relative immobile brain stem. This occurs at the diencephalic-mesencephalic junction, the area that includes the nigrostriatal pathway.

GENETIC PROMOTER POLYMORPHISM

The promoter region of α-syn harbors a number of haplotypes, one of which may be associated with sporadic PD. The first report[38] of this association found that one allele of a dinucleotide repeat marker, NACP-Rep1, in the promoter was associated with a 2.2-fold increase in PD risk. A study in a U.S. population[73] and one Singaporean survey[74] confirm the association, but a British survey[75] and another Singaporean survey[74] refute it. The inconsistency of the NACP Rep-1 studies could be the result of variability in other loci with which the NACP Rep-1 itself is in linkage disequilibrium. Such loci could even lie outside the α-syn gene itself: alcohol dehydrogenase 5 and necrosis factor 6B, both of which may be involved in the pathogenesis of PD, are encoded by neighboring genes.[38]

The co-occurrence of the NACP-Rep1 marker and the ApoE,4 allele was associated with a 12.8-fold increased PD risk in the German population[38] that was the subject of the original NACP-Rep1 observation, but the British study[75] failed to confirm that genetic interaction. Such an interaction has a theoretical appeal, because both α-syn and ApoE have similar α-helical regions based on 11-residue repeats, a common structure that may allow them to cooperate in cellular processes.

How would an alteration of α-syn promoter help cause PD? The leading hypothesis is that it simply increases the transcription of the protein, leading to increased concentration and aggregation.[73] This idea has received important support by analogy with progressive supranuclear palsy, where a haplotype that spans the entire *tau* gene is overrepresented.[76] Very recently, one of the components of that H1 haplotype, located in the *tau* promoter, has been found to increase tau expression and lead to tau aggregation in a cellular system.[77] By analogy, the Rep1 polymorphism in α-syn could be doing the same. This possibility has received indirect support from reports of two families with highly penetrant, autosomal dominant PD associated with triplications in the α-syn gene.[12,13]

REACTIVE OXYGEN SPECIES

A major advance in understanding the dopaminergic neuronal predominance of α-syn aggregation and damage in PD is the insight that an oxidative environment promotes the aggregation of α-syn.[78] Such an environment predominates in dopaminergic neurons, which accumulate hydrogen peroxide and other oxidants as a result of dopamine metabolism. This causes post-translational modification of α-syn in the form of tyrosine nitration, methionine oxidation and dopamine adduct formation.[59,64,79] These modifications occur at an early stage of α-syn aggregation, suggesting that they cause that aggregation rather than appearing as epiphenomena of the oxidative environment unrelated to the aggregation process.[64,80] α-syn that has been so modified can undergo dimerization via cross-linking at the modified sites. Such dimers have been shown to be toxic in cellular models.[81]

The reverse also occurs: aggregates of α-syn aggravate the oxidative milieu of the neuron. This is probably mediated by damage to dopamine vesicles, liberating dopamine into the cytoplasm, where its degradation produces reactive oxygen species.[47] The free dopamine would also cause dopamine adduct formation, which promotes further α-syn aggregation in a vicious cycle.[47] The aggregates may also damage the mitochondria, liberating cytochrome c, a pro-apoptotic factor.[82]

The oxidative stress prevalent in neurons in PD is probably partly the result of the well established defect in mitochondrial Complex I enzyme activity in that disorder.[83] However, it is not clear whether that defect is caused by an early stage of cellular metabolic damage in PD or is further upstream, possibly the result of mutations in the mitochondrial DNA that encodes part of Complex I. An important clue has appeared in the results of cybrid experiments which, by stably transmitting mitochondrial hypofunction from patients with sporadic PD to host cells, suggest that the mitochondrial defect in fact resides in mitochondrial DNA.[84] The question then becomes whether that mitochondrial DNA defect is itself caused by metabolic insults or is inherited. The latter alternative is supported by the result of a further cybrid experiment in which mitochondria from patients with the A53T α-syn mutation, unlike those from patients with sporadic PD, fail to harm host cells.[85] In contrast to sporadic PD, A53T PD seems to require no pathogenetic contribution from mitochondrial genetic mutations.

OTHER POST-TRANSLATIONAL MODIFICATIONS

Post-translational modifications in α-syn are an important potential pathogenetic consideration in an illness such as PD, where a vast majority of patients have no α-syn genomic DNA mutation but do suffer from abnormal aggregation of that protein.

Transglutamination, where glutamines cross-link tyrosine residues, has been shown to occur in α-syn in sporadic PD,[86] as it has been in tau in Alzheimer's disease and huntingtin in Huntington's disease.[86] Intriguingly, this effect in a cellular model was inhibited by cystamine, an orally available drug with relatively little toxicity.[86]

Phosphorylation of α-syn occurs in transgenic mouse and *Drosophila* models of PD[87,88] and in human PD,[89] predominantly at the serine 129 position.[90] There is evidence that this is catalyzed by casein kinase 1 and 2, enzymes that regulate other synaptic proteins, and is subject to inhibition by a fatty acid, phosphatidylinositol 4,5-bisphosphate.[90] This insight provides a clue to the normal mode of regulation of the chemical behavior of α-syn and offers a therapeutic opportunity.

CHAPERONES

α-syn's carboxy-terminal portion is critical to its putative function as a chaperone. Chaperones are small proteins that are upregulated during conditions of metabolic stress. One of their principal functions is to return misfolded proteins to an appropriate conformation. They also help identify misfolded proteins for ubiquitination in preparation for degradation by the proteasome.[91] Two chaperones, Hsp 40 and Hsp70, are abundant in Lewy bodies.[92]

Transgenic *Drosophila* with a copy of the human wild-type α-syn gene develop important dopaminergic neuronal loss with α-syn aggregates.[56] Addition of the gene for the human chaperone Hsp70 to that model prevents the neuronal loss, presumably via its known anti-apoptotic action.[92] Hsp 70 can disassemble protein aggregates and can catalyze the unfolding of partially folded proteins, allowing them the opportunity to refold appropriately.[93] Treatment of the singly transgenic α-syn flies with geldanamycin, an anticancer drug that inhibits the activity of the chaperone Hsp 90, accomplishes the same result,[94] which is a very exciting finding.

The addition of the Hsp 70 gene or geldanamycin treatment to the α-syn fly, despite preventing neurodegeneration, does not prevent or reduce the formation of α-syn aggregates.[92] Here, the PD model is similar to models of other neurodegenerative disorders with protein aggregation.[95] This suggests that the aggregates themselves are not toxic and, more important, that pharmacologic stimulation of chaperone production could provide anti-PD prophylaxis.

α-syn itself functions as a chaperone as assessed by its ability to prevent thermally-induced aggregation of other proteins.[96] The 42-residue C-terminal amino acid sequence of α-syn is similar to that of 14-3-3, an antiapoptotic chaperone important in other neurodegenerative disorders.[97,98] It remains unclear whether the chaperone property of α-syn is physiologically relevant, and, more important to the issue of PD, whether α-syn can inhibit its own aggregation.

THE UBIQUITIN-PROTEASOME SYSTEM

Perhaps the most important interactions of α-syn are with the ubiquitin-proteasomal[99] and lysosomal degradative pathways.[80,100] Defective catabolism of misfolded or abnormally modified α-syn is central to the issue of α-syn aggregation and pathology.

Aggregates of α-syn impair the function of the proteasome,[101,102] possibly via a simple mechanism of clogging the opening of the barrel-shaped structure.[103] However, even soluble oligomers of α-syn (but not monomers) impair the proteolytic properties of the proteasome.[104] Experimental inhibition of the proteasome causes α-syn to accumulate[99] and to aggregate[105,106] and sensitizes the cell to damage by proteasome inhibitors.[101] Thus, another vicious cycle occurs in the pathogenesis of PD, where α-syn aggregation impairs proteasomal function, permitting further accumulation of abnormally folded or aggregated α-syn and rendering the neuron more vulnerable to any deleterious effects of other misfolded proteins.

Elements of the ubiquitin-proteasome system that occur in Lewy bodies are ubiquitin itself and proteasomal subunits.[107] Within Lewy bodies, α-syn appears to be ubiquitinated, but normal α-syn degradation by the proteasome does not require ubiquitination.[108–110] Our knowledge of how α-syn is degraded in the normal state, in sporadic PD, and in PD related to Mendelian mutations remains rudimentary. A clue is the observation that a glycosylated form of α-syn is ubiquitinated *in vitro* by normal but not by mutant parkin.[111] While this does provide an important clue to the pathogenesis of autosomal recessive PD related to *parkin* mutations, its relevance to non-Mendelian PD may simply be to point to the possibility of other post-translational modifications as steps in the pathogenetic process.[112]

SUMMARY AND PATHOGENETIC HYPOTHESIS

The foregoing describes many ways by which dysfunction in the cellular physiology of α-syn can damage neurons. Which of these biologically possible events actually participate in real-world PD remains conjectural and may well differ among patients depending on the individual's exogenous toxin exposures and genetic background. However, one may propose a generic pathogenetic pathway with α-syn at its center that takes account of present knowledge of that protein and the pathology and epidemiology of PD (Figure 12.2).

The evidence suggests that α-syn aggregation results from its excessive concentration caused by (a) overexpression related to genetic variants in its promoter region or to (b) impaired breakdown related to defects in the

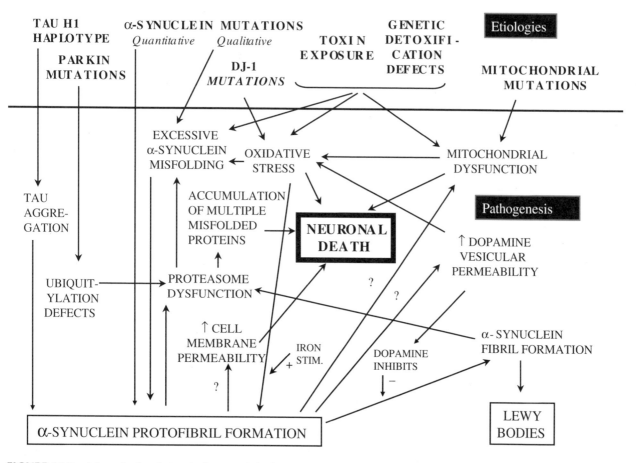

FIGURE 12.2 A hypothesis relating the known etiologic contributors to PD with the known potentially relevant pathogenetic steps. Multiple vicious cycles are evident. The question marks denote that the α-syn pore hypothesis[29,44,113] is not fully accepted. The potential of many genetic allelic associations, even those that have been found in multiple studies, are not shown. "Quantitative" α-synuclein mutations include triplications, which are very rare,[12,13] and the Rep-1 promoter mutation,[73] which is common. "Qualitative" α-synuclein mutations include the three known point mutations.[8,10,11] Note the roles of iron and dopamine, which are plentiful in dopaminergic neurons, in promoting the presence of α-syn protoaggregates.[64,67–69]

ubiquitin-proteasome system caused by genetic variants in parkin.

Aggregation is further promoted by the oxidative environment in dopaminergic neurons caused by the catabolism of dopamine, abetted by genetic defects in mitochondrial Complex I and the ready availability of iron presented by neuromelanin. Furthermore, aggregation is promoted by exposure to certain pesticides and environmental metals and by tau protein in excessive concentration, which may be abetted by a certain promoter allele. The concentration of exogenous pesticides or other toxins could be enhanced by genetically determined allelic variants in toxin catabolism.

An early, β-sheet stage of oligomeric aggregate, the protoaggregate, impairs the proteasome, permitting further accumulation of misfolded α-syn and other toxic, misfolded proteins. This may be sufficient to cause the neuronal death seen in PD, but the protoaggregate also may have pore-like activity in the manner of some bacte-

rial toxins.[29,44,113] This would impair the structure and function of mitochondria, dopaminergic vesicles, and even the cell membrane. This would create vicious cycles whereby (a) the oxidative stress so produced aggravates α-syn aggregation, and (b) the release of dopamine into the cytoplasm stabilizes the toxic protoaggregates rather than allowing them to form nontoxic fibrils

CONCLUSIONS

The literature on α-syn, particularly in relation to PD, is growing rapidly, from none in 1996 and 11 in 1997, to 166 in 2002 and 201 in 2003. Of the many potential pathogenetic properties of α-syn that have been and will be described, many will prove not to be relevant to PD. Animal models using α-syn knock-ins, discussed elsewhere in this volume, will be critical to sorting this out. The ultimate goal is to find a rate-limiting step that is vulnerable to a therapeutically convenient molecule. The animal models will be equally critical in evaluating such treatments. Another central goal is to take advantage of the newly discovered biochemistry of PD to devise a diagnostic marker to permit treatments to be prescribed before the degenerative process causes disability. The cascade of discovery initiated by the observation of a relation between α-syn and PD has only just begun.

ACKNOWLEDGMENTS

This work was supported by NS43647 (NIH/NINDS) and by a Center of Excellence grant from the American Parkinson's Disease Association.

REFERENCES

1. Maroteaux, L., Campanelli, J. T., and Scheller, R. H. Synuclein: A neuron-specific protein localized to the nucleus and presynaptic nerve terminal, *J. Neurosci.*, 8, 2804, 1988.

2. Uéda, K. et al., Molecular cloning of cDNA encoding an unrecognized component of amyloid in Alzheimer disease, *Proc. Natl. Acad. Sci., USA*, 90, 11282, 1993.

3. Jensen, P. H. et al., Binding of Abeta to α and β-synucleins: identification of segments in α-synuclein/NAC precursor that bind Aβ and NAC, *Biochem. J.*, 323, 539, 1997.

4. Campion, D. et al., The NACP/synuclein gene: chromosomal assignment and screening for alterations in Alzheimer disease, *Genomics*, 26, 254, 1995.

5. Spillantini, M. G., Divane, A. and Goedert, M., Assignment of human alpha-synuclein (SNCA) and beta-synuclein (SNCB) genes to chromosomes 4q21 and 5q35, *Genomics*, 27, 379, 1995.

6. Golbe, L. I. et al., A large kindred with autosomal dominant Parkinson's disease, *Ann. Neurol.*, 27, 276, 1990.

7. Golbe, L. I. et al., Clinical genetic analysis of Parkinson's disease in the Contursi kindred, *Ann. Neurol.*, 40, 767, 1996.

8. Polymeropoulos, M. H. et al., Mapping of a gene for Parkinson's disease to chromosome 4q21-23, *Science*, 274, 1197, 1996.

9. Polymeropoulos, M. H. et al., Mutation in the alpha-synuclein gene identified in families with Parkinson's disease, *Science*, 276, 2045, 1997.

10. Krüger, R. et al., Ala30Pro mutation in the gene encoding α-synuclein in Parkinson's disease, *Nat. Genetics*, 18, 106, 1998.

11. Zarranz, J. J. et al., Familial parkinsonism and dementia with Lewy bodies is related to a novel mutation, E46K, of the alpha-synuclein gene, in a family from Bilbao (Basque country-Spain), *Ann. Neurol.*, 55, 164, 2004.

12. Singleton, A. et al., Alpha-synuclein locus triplication causes Parkinson's disease, *Science*, 302, 841, 2003.

13. Farrer, M. et al., Comparison of kindreds with familial parkinsonism and alpha-synuclein genomic multiplications, *Ann. Neurol.*, 55, 174, 2004.

14. Athanassiadou, A. et al., Genetic analysis of families with Parkinson disease that carry the ala53thr mutation in the gene encoding α-synuclein, *Am. J. Hum. Gen.*, 65, 555, 1999.

15. Bostantjopoulou, S. et al., Clinical features of parkinsonian patients with the α-synuclein (G209A) mutation, *Mov. Disord.*, 16, 1007, 2001.

16. Papadimitriou, A. et al., Mutated alpha-synuclein gene in two Greek kindreds with familial PD: incomplete penetrance? *Neurology*, 52, 651, 1999.

17. Papapetropoulos, S. et al., Clinical characteristics of the alpha-synuclein mutation (G209A)-associated Parkinson's disease in comparison with other forms of familial Parkinson's disease in Greece, *Eur. J. Neurol.*, 10, 281, 2003.

18. Spira, P. J. et al., Clinical and pathological features of a parkinsonian syndrome in a family with an ala53thr α-synuclein mutation, *Ann. Neurol.*, 49, 313, 2001.

19. Scott, W. K. et al., The alpha-synuclein gene is not a major risk factor in familial Parkinson disease, *Neurogenetics*, 2, 191, 1999.

20. Langston, J. W. et al., Novel alpha-synuclein-immunoreactive proteins in brain samples from the Contursi kindred, Parkinson's, and Alzheimer's disease, *Exp. Neurol.*, 154, 684, 1998.

21. Muenter, M. D. et al., A familial Parkinson-dementia syndrome, *Ann. Neurol.*, 43, 768-781, 1998.

22. Golbe, L. I., unpublished.

23. Hashimoto, M. et al., NACP, a synaptic protein involved in Alzheimer's disease, is differentially regulated during megakaryocyte differentiation, *Biochem. Biophys. Res. Commun.*, 237, 611, 1997.

24. Li, J. Y. et al., Differential localization of alpha-, beta- and gamma-synucleins in the rat CNS, *Neuroscience*, 113, 463, 2002.

25. Abeliovich, A. et al., Mice lacking alpha-synuclein display functional deficits in the nigrostriatal dopamine system, *Neuron*, 25, 239, 2000.

26. Lotharius, J. and Brundin, P., Impaired dopamine storage resulting from alpha-synuclein mutations may contribute to the pathogenesis of Parkinson's disease, *Hum. Mol. Genet.*, 11, 2395, 2002.

27. Cabin, D. E. et al., Synaptic vesicle depletion correlates with attenuated synaptic responses to prolonged repetitive stimulation in mice lacking alpha-synuclein, *J. Neurosci.*, 22, 8797, 2002.

28. Perrin, R. J. et al., Interaction of human alpha-Synuclein and Parkinson's disease variants with phospholipids. Structural analysis using site-directed mutagenesis, *J. Biol. Chem.* 275, 34393, 2000.

29. Volles, M. J. et al., Vesicle permeabilization by protofibrillar alpha-synuclein is sensitive to Parkinson's disease-linked mutations and occurs by a pore-like mechanism, *Biochemistry*, 41, 4595, 2002.

30. Sharon, R. et al., The formation of highly soluble oligomers of alpha-synuclein is regulated by fatty acids and enhanced in Parkinson's disease, *Neuron*, 37, 583, 2003.

31. Dauer, W. et al., Resistance of alpha-synuclein null mice to the parkinsonian neurotoxin MPTP, *Proc. Natl. Acad. Sci. USA*, 29, 14524, 2002.

32. Perez, R. G. et al., A role for alpha-synuclein in the regulation of dopamine biosynthesis, *J. Neurosci.* 22, 3090, 2002.

33. Lee, F. J. et al., Direct binding and functional coupling of alpha-synuclein to the dopamine transporters accelerate dopamine-induced apoptosis, *FASEB J.*, 15, 916, 2001.

34. Elkon, H. et al., Mutant and wild-type alpha-synuclein interact with mitochondrial cytochrome C oxidase, *J. Mol. Neurosci.*, 18, 229, 2002.

35. Jenco, J. M. et al., Regulation of phospholipase D_2: selective inhibition of mammalian phospholipase D isoenzymes by alpha- and beta-synucleins, *Biochemistry*, 37, 4901, 1998.

36. Jensen, P. H. et al., Alpha-synuclein binds to tau and stimulates the protein kinase A-catalyzed tau phosphorylation of serine residues 262 and 356, *J. Biol. Chem.*, 274, 25481, 1999.

37. Giasson, B. I. et al., Initiation and synergistic fibrillization of tau and alpha-synuclein, *Science*, 300, 636, 2003.

38. Krüger, R. et al., Increased susceptibility to sporadic Parkinson's disease by a certain combined alpha-synuclein/apolipoprotein E genotype, *Ann. Neurol.*, 45, 611, 1999.

39. Iwata, A. et al., Alpha-Synuclein forms a complex with transcription factor Elk-1, *J. Neurochem.*, 77, 239, 2001.

40. Iwata, A. et al., Alpha-synuclein degradation by serine protease neurosin: implication for pathogenesis of synucleinopathies, *Hum. Mol. Genet.*, 12, 2625, 2003.

41. Engelender, S. et al., Synphilin-1 associates with alpha-synuclein and promotes the formation of cytosolic inclusions, *Nat. Genetics*, 22, 110, 1999.

42. Chung, K. K. et al., Parkin ubiquitinates the alpha-synuclein-interacting protein, synphilin-1: implications for Lewy-body formation in Parkinson disease, *Nature Med.*, 7, 1144, 2001.

43. Bandopadhyay, R. et al., No pathogenic mutations in the synphilin-1 gene in Parkinson's disease, *Neurosci. Lett.*, 307, 125, 2001.

44. Volles, M. J. and Lansbury, Jr., P. T., Zeroing in on the pathogenic form of alpha-synuclein and its mechanism of neurotoxicity in Parkinson's disease, *Biochemistry*, 42, 7871, 2003.

45. Spillantini, M. G. et al., Alpha-synuclein in Lewy bodies, *Nature*, 388, 839, 1997.

46. Ding, T. T. et al., Annular alpha-synuclein protofibrils are produced when spherical protofibrils are incubated in solution or bound to brain-derived membranes, *Biochemistry*, 41, 10209, 2002.

47. Conway, K. A. et al., Acceleration of oligomerization, not fibrillization, is a shared property of both alpha-synuclein mutations linked to early-onset Parkinson's disease: implications for pathogenesis and therapy, *Proc. Natl. Acad. Sci. USA*, 97, 571, 2000.

48. Li, J. Y., Uversky, V. N. and Fink, A. L., Effect of familial Parkinson's disease point mutations A30P and A53T on the structural properties, aggregation, and fibrillation of human alpha-synuclein, *Biochemistry*, 40, 11604, 2001.

49. Michalik A. and Van Broeckhoven, C., Pathogenesis of polyglutamine disorders: aggregation revisited, *Hum. Mol. Genet.*, 12, R173, 2003.

50. Spillantini, M. G. et al., Alpha-synuclein in filamentous inclusions of Lewy bodies from Parkinson's disease and dementia with Lewy bodies, *Proc. Nat. Acad. Sci. USA.*, 95, 6469, 1998.

51. El-Agnaf, O. M. et al., Effects of the mutations Ala30 to Pro and Ala53 to Thr on the physical and morphological properties of alpha-synuclein protein implicated in Parkinson's disease, *FEBS Lett.*, 440, 67, 1998.

52. Irizarry, M. C. et al., Nigral and cortical Lewy bodies and dystrophic nigral neurites in Parkinson's disease and cortical Lewy body disease contain alpha-synuclein immunoreactivity, *J. Neuropathol. Exp. Neurol.*, 57, 334, 1998.

53. McNaught, K. S. et al., Aggresome-related biogenesis of Lewy bodies, *Eur. J. Neurosci.*, 16, 2136, 2002.

54. Taylor, J. P. et al., Aggresomes protect cells by enhancing the degradation of toxic polyglutamine-containing protein, *Hum. Mol. Genet.*, 12, 749, 2003.

55. Masliah, E. et al., Dopaminergic loss and inclusion body formation in alpha-synuclein mice: implications for neurodegenerative disorders, *Science*, 287, 1265, 2000.

56. Feany, M. B. and Bender, W. W., A Drosophila model of Parkinson's disease, *Nature* 404, 394, 2000.

57. Fleming Outeiro, T. and Lindquist, S., Yeast cells provide insight into alpha-synuclein biology and pathobiology, *Science*, 302, 1772, 2003.

58. Wood, S. J. et al., Alpha-synuclein fibrillogenesis is nucleation-dependent, *J. Biol. Chem.*, 274, 19509, 1999.

59. Giasson, B. I. et al., Oxidative damage linked to neurodegeneration by a selective alpha-synuclein nitration in synucleinopathy lesions, *Science*, 290, 985, 2000.

60. Arima, K. et al., Cellular co-localization of phosphorylated tau- and NACP/alpha-synuclein-epitopes in Lewy bodies in sporadic Parkinson's disease and in dementia with Lewy bodies, *Brain Res.*, 843, 53, 1999.

61. Takeda, A. et al., Abnormal accumulation of NACP/alpha-synuclein in neurodegenerative disorders, *Am. J. Pathol.*, 152, 367, 1998.

62. Duda, J. E. et al., Concurrence of alpha-synuclein and tau brain pathology in the Contursi kindred, *Acta Neuropathol., (Berl.)*, 104, 7, 2002.

63. Golbe, L. I. et al., The tau A0 allele in Parkinson's disease, *Mov. Disord.*, 16, 442, 2001.

64. Conway, K. A. et al., Kinetic stabilization of the alpha-synuclein protofibril by a dopamine-alpha-synuclein adduct, *Science*, 294, 1346, 2001.

65. Paik, S. R. et al., Aluminum-induced structural alterations of the precursor of the non-A beta component of Alzheimer's disease amyloid, *Arch. Biochem. Biophys.* 344, 325, 1997.

66. Paik, S. R. et al., Copper(II)-induced self-oligomerization of alpha-synuclein, *Biochem. J.*, 340, 821, 1999.

67. Uversky V. N. et al., Synergistic effects of pesticides and metals on the fibrillation of alpha-synuclein: impli-

cations for Parkinson's disease, *Neurotoxicology*, 23, 527, 2002.

68. Double K. L. et al., Iron-binding characteristics of neuromelanin of the human substantia nigra, *Biochem. Pharmacol.*, 66, 489, 2003.

69. Faucheux B. A. et al., Neuromelanin associated redox-active iron is increased in the substantia nigra of patients with Parkinson's disease, *J. Neurochem.*, 86, 1142, 2003.

70. Semchuk, K. M., Love, E. J. and Lee, R. G., Parkinson's disease: a test of the multifactorial etiologic hypothesis, *Neurology*, 43, 1173, 1993.

71. Stern, M. B. et al., Head trauma as a risk factor for Parkinson's disease, *Mov. Disord.*, 6, 95, 1991.

72. Uryu, K. et al., Age-dependent synuclein pathology following traumatic brain injury in mice, *Exp. Neurol.*, 184, 214, 2003.

73. Farrer, M. et al., Alpha-Synuclein gene haplotypes are associated with Parkinson's disease, *Hum. Mol. Genet.*, 10, 1847, 2001.

74. Tan, E. K., Alpha synuclein promoter and risk of Parkinson's disease: microsatellite and allelic size variability, *Neurosci. Lett.*, 336, 70, 2003.

75. Khan, N. et al., Parkinson's disease is not associated with the combined alpha-synuclein/apolipoprotein E susceptibility genotype, *Ann. Neurol.*, 49, 665, 2001.

76. Conrad, C. et al., Genetic evidence for the involvement of tau in progressive supranuclear palsy, *Ann. Neurol.*, 41, 277, 1997.

77. Kwok, J. B. et al., Tau haplotypes regulate transcription and are associated with Parkinson's disease, *Ann. Neurol.*, 55, 329, 2004.

78. Kowall, N.W. et al., MPTP induces alpha-synuclein aggregation in the substantia nigra of baboons, *Neuroreport*, 11, 211, 2000.

79. Yamin, G. et al., Certain metals trigger fibrillation of methionine-oxidized alpha-synuclein, *J. Biol. Chem.*, 274, 33855, 1999.

80. Paxinou, E. et al., Induction of alpha-synuclein aggregation by intracellular nitrative insult, *J. Neurosci.*, 21, 8053, 2001.

81. Zhou, W. and Freed, C. R., Tyrosine-to-cysteine modification of human alpha-synuclein enhances protein aggregation and cellular toxicity, *J. Biol. Chem*, in press.

82. Junn, E. and Mouradian, M. M., Apoptotic signaling in dopamine-induced cell death: the role of oxidative stress, p38 mitogen-activated protein kinase, cytochrome c and caspases, *J. Neurochem.* 78, 374, 2001.

83. Sherer, T. B., Betarbet, R., and Greenamyre, J. T., Environment, mitochondria, and Parkinson's disease, *Neuroscientist*, 8, 192, 2002.

84. Swerdlow, R. H. et al., Origin and functional consequences of the complex I defect in Parkinson's disease, *Ann. Neurol.*, 40, 663, 1996.

85. Swerdlow, R. H. et al., Biochemical analysis of cybrids expressing mitochondrial DNA from Contursi kindred Parkinson's subjects, *Exp. Neurol.*169, 479, 2001.

86. Junn, E. et al., Tissue transglutaminase-induced aggregation of alpha-synuclein: Implications for Lewy body formation in Parkinson's disease and dementia with Lewy bodies, *Proc. Natl. Acad. Sci., USA*, 100, 2047, 2003.

87. Takahashi, M. et al., Phosphorylation of alpha-synuclein characteristic of synucleinopathy lesions is recapitulated in alpha-synuclein transgenic *Drosophila*, *Neurosci. Lett.*, 336, 155, 2003.

88. Neumann, M. et al., Misfolded proteinase K-resistant hyperphosphorylated alpha-synuclein in aged transgenic mice with locomotor deterioration and in human alpha-synucleinopathies, *J. Clin. Invest.*, 110, 1429, 2002.

89. Fujiwara, H. et al., Alpha-Synuclein is phosphorylated in synucleinopathy lesions, *Nat. Cell Biol.*, 4, 160, 2002.

90. Okochi, M. et al., Constitutive phosphorylation of the Parkinson's disease associated alpha-synuclein, *J. Biol. Chem.*, 275, 390, 2000.

91. Helfand, S. L., Chaperones take flight, *Science*, 295, 809, 2002.

92. Auluck, P. V. et al., Chaperone suppression of alpha-synuclein toxicity in a *Drosophila* model for Parkinson's disease, *Science*, 295, 865, 2002.

93. Sherman, M. Y. and Goldberg, A. L., Cellular defenses against unfolded proteins: a cell biologist thinks about neurodegenerative diseases, *Neuron*, 29, 15, 2001.

94. Auluck, P. K. and Bonini, N. M., Pharmacological prevention of Parkinson disease in *Drosophila*, *Nat. Med.*, 8, 1185, 2002.

95. Muchowsky, P. J., Protein misfolding, amyloid formation, and neurodegeneration: a critical role for molecular chaperones? *Neuron*, 35, 9, 2002.

96. Souza, J. M. et al., Chaperone-like activity of synucleins, *FEBS Lett.*, 474, 116, 2000.

97. Ostrerova, N. et al., Alpha-Synuclein shares physical and functional homology with 14-3-3 proteins, *J. Neurosci.*, 19, 5782, 1999.

98. Park, S. M. et al., Distinct roles of the N-terminal-binding domain and the C-terminal-solubilizing domain of alpha-synuclein, a molecular chaperone, *J. Biol. Chem.*, 277, 28512, 2002.

99. Bennett, M. C. et al., Degradation of alpha-synuclein by proteasome, *J. Biol. Chem.* 274, 48, 1999.

100. Iwata, A. et al., Alpha-Synuclein affects the MAPK pathway and accelerates cell death, *J. Biol.Chem.*, 30, 45320, 2001.

101. Tanaka, Y. et al., Inducible expression of mutant alpha-synuclein decreases proteasome activity and increases sensitivity to mitochondria-dependent apoptosis, *Hum. Mol. Genet.*, 10, 919, 2001.

102. Stefanis, L. et al., Expression of A53T mutant but not wild-type alpha-synuclein in PC12 cells induces alterations of the ubiquitin-dependent degradation system, loss of dopamine release, and autophagic cell death, *J. Neurosci.*, 21, 9549, 2001.

103. Bence, N. F., Sampat, R. M., and Kopito, R. R., Impairment of the ubiquitin-proteasome system by protein aggregation, *Science*, 292, 1552, 2001.

104. Lindersson, E. et al., Proteasomal inhibition by alpha-synuclein filaments and oligomers. *J. Biol. Chem.*, 279, 12924, 2004.

105. Rideout, H. J., Proateasomal inhibition leads to formation of ubiquitin/alpha-synuclein-immunireactive inclusions in PC12 cells, *J. Neurochem.*, 78, 899, 2001.

106. Sawada, H. et al., Proteasome mediates dopaminergic neuronal degeneration and inhibition causes alpha-synuclein inclusions, *J. Biol. Chem*, in press.

107. Gai, W. P., Blessing, W. W. and Blumbergs, P. C., Ubiquitin-positive degenerating neurites in the brainstem in Parkinson's disease, *Brain*, 118, 1447, 1995.

108. Tofaris, G. K. et al., Ubiquitination of alpha-synuclein in Lewy bodies is a pathological event not associated with impairment of proteasome function, *J Biol Chem.*, 278, 44405, 2003.

109. Tofaris, G. K., Layfield, R., and Spillantini, M. G., Alpha-synuclein metabolism and aggregation is linked to ubiquitin-independent degradation by the proteasome, *FEBS Lett.*, 509, 22, 2001.

110. Sampathu, D. M. et al., Ubiquitination of alpha-synuclein is not required for formation of pathological inclusions in alpha-synucleinopathies, *Am. J. Pathol.*, 163, 91, 2003.

111. Shimura, H. et al., Ubiquitination of a new form of alpha-synuclein by parkin from human brain: implications for Parkinson's disease, *Science*, 293, 263, 2001.

112. Burke, R. E., Alpha-synuclein and parkin: coming together of pieces in puzzle of Parkinson's disease, *Lancet,* 358, 1567, 2001.

113. Volles, M. J. et al., Vesicle permeabilization by protofibrillar alpha-synuclein: implications for the pathogenesis and treatment of Parkinson's disease, *Biochemistry,* 40, 7812, 2001.

13 Parkin and Its Role in Parkinson's Disease

Thomas Gasser
University of Tübingen, Hertie-Institute for Clinical Brain Research

CONTENTS

ABSTRACT

Mutations in the *parkin* gene (*PRKN)* have been recognized as a major cause of early-onset parkinsonism with recessive inheritance. Up to 50% of familial and 15 to 20% of sporadic young-onset cases may be due to *PRKN* mutations. Although the neuropathologic changes found in parkin-related parkinsonism differ from those of typical idiopathic Parkinson's disease (PD) because of the absence of the characteristic alpha-synuclein positive inclusions (Lewy-bodies and Lewy-neurites), the understanding of the underlying molecular events is thought to provide important insight into the mechanisms of selective neuronal cell death. The study of the normal function of parkin and its dysfunction in several cellular and animal models showed that defective protein degradation by the ubiquitin-proteasome complex and/or early mitochondrial dysfunction and increased oxidative cell damage may be central to the pathogenic cascade.

INTRODUCTION

In recent years, the strategies of linkage analysis and positional cloning have lead to the identification of an increasing number of loci and genes that cause monogenically inherited forms of parkinsonism. More than 10 loci have been mapped, and several of the causative genes have been cloned.[1] The analysis of the encoded gene products has greatly advanced our knowledge on the pathogenesis of the disease.

None of the monogenic forms of parkinsonism recapitulates all the clinical and pathologic features of typical sporadic Parkinson's disease (PD). However, "typical PD," in all likelihood, should probably not be thought of as a disease entity but rather as a spectrum of "parkinsonian disorders." If this is true, each of the identified monogenic variants may reflect, and help to understand, certain aspects of this common and devastating neurodegenerative disorder.

Autosomal-dominant forms of PD account for only a very small proportion of all cases, but the identification of mutations in the gene for α-synuclein in a few families with dominant PD[2] has proven extremely important, as it has focused research on this protein, which has been found to accumulate, in an insoluble and fibrillary form, not only in these rare familial cases, but also in the common sporadic form of PD, which in fact represents the pathologic hallmark of the disease.

One of the surprising developments of recent years was the recognition of the relatively high proportion of patients with a monogenically inherited form of parkinsonism following autosomal-recessive transmission. So far, three of the causative genes, *parkin PRKN* (PARK2), *DJ-1* (PARK7), and *PINK1* (PARK6) have been cloned. It is likely that several other loci exist.

This review focuses on the role of *PRKN*, the parkin gene, and its encoded gene product in Parkinson's disease. *PRKN* is undoubtedly the major gene responsible for juvenile-onset parkinsonism. Whether *PRKN* itself or the pathogenic mechanisms operative in parkin-deficiency are relevant to the more common late-onset form of PD remains to be determined.

MAPPING AND CLONING OF PRKN

Cases of parkinsonism with early disease-onset and recessive inheritance (families with affected siblings, but usually no transmission from one generation to the next) were first recognized and described as a clinical entity in Japan.[3] Clinically, these patients suffered from L-dopa-responsive parkinsonism with onset in the second to fourth decade. Some of the patients showed diurnal fluctuations with symptoms becoming worse later in the day, similar to patients with dopa-responsive dystonia (DRD), which is caused by mutations in the gene for GTP-cyclohydrolase,[4] one of the genes involved in dopamine biosythesis. In contrast to patients with DRD, however, patients with recessive parkinsonism develop early and severe levodopa-induced motor fluctuations and dyskinesias. Dystonia at onset of the disease was found to be a common feature. The disorder was called autosomal-recessive juvenile parkinsonism (AR-JP).

The genetic locus for AR-JP was mapped by linkage analysis in a set of 13 Japanese families to the long arm of chromosome 6.[5] This region was analyzed as a candidate locus bearing the gene for mitochondrial superoxide dismutase 2 (Mn-SOD2), based on hypotheses of oxidative stress and mitochondrial dysfunction being involved in the pathogenesis of PD. The locus was later confirmed and refined in different ethnic groups,[67] excluding SOD2 from the refined region.

Positional cloning eventually identified several homozygous deletions in consanguineous families with AR-JP in a large gene in this region that was named parkin.[89]

PRKN mutations turned out to be a common cause of parkinsonism with early onset, and novel mutations were reported in rapid succession.[10–12] All types of mutations, including missense mutations leading to amino acid exchanges, nonsense mutations resulting in premature termination of translation, and exonic rearrangements (deletions, duplications and triplications) were identified.[13]

To study the origin of *PRKN* mutations, Periquet et al. performed haplotype analyses using 10 microsatellite markers covering a 4.7-cM region known to contain the parkin gene in 48 families, mostly from European countries, with early-onset autosomal recessive parkinsonism. The patients carried 14 distinct mutations, and each mutation was detected in more than one family. The results indicate that exon rearrangements occurred indepen-

dently, whereas some point mutations, found in families from different geographic origins, may have been transmitted by a common founder.[14]

In the vast majority of cases, homozygous or compound heterozygous *PRKN* mutations cause AR-JP. Whether (some) heterozygous *PRKN* mutations are sufficient to cause clinically apparent parkinsonism by haplotype insufficiency or a dominant negative effect is currently a major topic of discussion (see below).

CLINICAL PICTURE OF PARKIN-ASSOCIATED PARKINSONISM

A number of studies have delineated the clinical spectrum associated with parkin mutations. By and large, the clinical picture is frequently indistinguishable from that of the sporadic disease, with the notable exception, of course, of the earlier age at onset.

For example, a large collection of European sib pairs collected by the European Consortium on Genetic Susceptibility in PD (GSPD) allowed the characterization of the clinical spectrum of *parkin*-associated parkinsonism.[12] Mean age at onset was 32 yr, progression of the disease was slow, but L-dopa associated fluctuations and dyskinesias occurred frequently and early. Brisk reflexes of the lower limbs were present in 44%.

Comparing young-onset carriers and noncarriers of *parkin* mutations, those with a mutation tended to have earlier and more symmetrical onset, slower progression of the disease and greater response to L-dopa despite lower doses. Lower limb dystonia at disease-onset occurred in about a third of patients, but this feature did not appear to be specific for *parkin*-related disease, and was more correlated with the age at onset than with genetic status.[15] Nevertheless, dystonia at disease-onset can be prominent enough to render distinction from dopa-responsive dystonia difficult or impossible.[16]

In the initial European study by Lücking and coworkers, there was no discernible difference in the clinical phenotype between patients with missense mutations, truncating point mutations, or deletions, suggesting that a complete loss of *parkin* function is associated with all of these mutations and with the full phenotype of early-onset parkinsonism. More recent phenotype-genotype studies by the same group in an extended sample, however, suggested that the type of mutation may in fact influence the clinical phenotype to a certain degree: patients with at least one missense mutation showed a faster progression of the disease with a higher UPDRS (United Parkinson's Disease Rating Scale) motor score than carriers of truncating mutations. Missense mutations in functional domains of the parkin gene resulted in earlier onset.[15] A very similar clinical picture was described in Japanese patients, but age of onset was still lower (average about 26 years).[17]

Although the majority of patients with *parkin* mutations in most series conform to this relatively well defined clinical picture, a minority will present with unusual features, including cervical dystonia; autonomic dysfunction and peripheral neuropathy; and pure exercise-induced dystonia,[18] or camptocormia.[19] Further clinical features that can be seen in *parkin* disease: focal dystonia; early postural instability; severe and early freezing; festination or retropulsion; concurrent autonomic failure; dramatic response to anticholinergics; early or atypical L-dopa-induced dyskinesias; exquisite sensitivity to small doses of L-dopa; and recurrent psychosis, even taking L-dopa alone.[18] Another unusual presenting feature may be exercise induced dystonia.[20] Psychiatric and behavioral disorders prior to the onset of parkinsonism have also been recognized in patients with *parkin* mutations[17,18] and some relatives carrying a single *parkin* mutation without extrapyramidal symptoms or signs also had psychiatric symptoms that might be related to their carrier status.[18]

Exceptionally, patients with *parkin* mutations may present with a clinical picture of levodopa unresponsive parkinsonism complicated by cerebellar and pyramidal tract dysfunction.[21] The report of a single case presenting with cerebellar ataxia several years before typical parkinsonism also extends the spectrum of parkin-related disease.[12,22]

The spectrum of parkin-related clinical signs and symptoms has been expanded further by the study of a large family from South Tyrol. Affected members show a wide range of age at onset and differences in the clinical picture, despite identical mutations and almost identical environment. Some mutation carriers in this family had onset in their 50s with tremor-dominant parkinsonism, indistinguishable from the sporadic disease.[23]

GENETICS OF PARKIN-ASSOCIATED PARKINSONISM

Parkin-related AR-JP is an autosomal-recessive disorder with high penetrance. Homozygous or compound heterozygous disease gene carriers will develop the disease with an age dependent penetrance. The vast majority of cases manifest the disease before the age of 40, very few patients with disease-onset in their 50s have been described.[25] The fact that homozygous deletions that virtually abolish the parkin protein and point mutations cause very similar (if not identical) phenotypes argue for a classical loss-of-function mechanism.

The question whether (specific?) heterozygous parkin mutations may confer a susceptibility to late-onset typical PD is still a matter of debate. It is well established that heterozygous carriers of parkin mutations have reduced [^{18}F]-Fluorodopa-uptake by PET,[26,27] which has been interpreted as a "first hit" to the dopaminergic system. Some larger multicase-pedigrees have been ascertained on the basis of typical parkin-related cases of

parkinsonism in patients with two parkin-mutations. Heterozygous carriers of parkin mutations have been described in these families either with minor signs of basal ganglia dysfunction[23] or with late-onset parkinsonism.[28,25] In at least two these cases, typical Lewy-bodies have been described, suggesting that a heterozygous parkin mutation may be a risk factor for typical LB-positive PD. However, PD is a common condition, and the causal relationship between heterozygous parkin mutations and disease status is still not clear. A recent study screening late-onset PD cases and healthy, age-matched controls for parkin mutations found a similar frequency of 3.8 and 3.1%, respectively, indicating that heterozygous parkin mutations generally do not confer a significantly increased risk for late-onset PD.[29] Apparently dominant inheritance of parkin-related disease may be found when multiple disease alleles are segregating within a single family (which may occur particularly in the context of consanguinity).[30]

To further address the possible role of parkin in sporadic PD, several association studies with coding and noncoding polymorphisms within the *parkin* gene have been performed. Overall, the results of these studies are still inconclusive. Some positive results have been reported[31–33] but not fully corroborated by other studies.[34–37]

Parkin mutations are a major cause of early-onset parkinsonism. In cases with recessive inheritance (e.g., in a group of 73 sibling pairs with PD, at least one of them having an age at onset before 45 yr), 50% were due to parkin mutations.[12] The prevalence of parkin-associated parkinsonism in a population of sporadic patients strongly depends on age at onset. Taking into account a number of different studies, *PRKN* mutations probably account for about 10 to 15% of early-onset cases (onset before 45 yr) without family history. Within this group, the proportion of mutation-carriers increases significantly with decreasing age at onset (82% before age 20 yr).[22] Rarely, the disease may manifest at a later age and disease onset in the sixth decade has been described in individual patients.[12,18] A similar distribution has been found in Japanese patients.[38]

PET STUDIES

Functional neuroimaging in parkin-linked parkinsonism consistently demonstrated a severely reduced uptake of dopamine tracer in both hemispheres in the putamen und caudate nucleus,[43] which differs at least quantitatively from the initially unilateral reduction in dopa uptake of sporadic PD patients.[44] In terms of the rostro-caudal gradient of tracer-uptake in the putamen (the decrease being more severe in the posterior portion), the pattern of uptake in YOPD patients is similar to that of patients with idiopathic Parkinson's disease and does not depend on the presence or absence of mutations of the parkin gene.[45]

Khan et al. have used [¹⁸F]dopa PET serially to study members of a family with young-onset parkinsonism who are compound heterozygous for mutations in the parkin gene, to assess disease progression. The *parkin* patients showed a significantly slower loss of putamen [¹⁸F]dopa uptake compared with a group of idiopathic Parkinson's disease (IPD) patients who had baseline putamen [¹⁸F]dopa uptake and disease severity similar to the *parkin* group. Interestingly, asymptomatic carriers of a single *parkin* mutation as well as presymptomatic carriers of compound heterozygous mutations also showed a mild but significant striatal dopaminergic dysfunction.[26,27] One study also provided evidence for an alteration of postsynaptic mechanisms, indicated by a reduced binding of the D₂-receptor ligand 11C-raclopride.[27] This finding is remarkable, as it is usually associated with an impaired response to dopaminergic medication, which is clearly not typical of *parkin*-disease.

Pathology

There is still relatively little information available on the neuropathology of molecularly confirmed cases of *parkin*-related AR-JP. Severe and selective degeneration of dopaminergic neurons and gliosis in the substantia nigra pars compacta and, to a somewhat lesser degree, of the locus coeruleus, has been described.[39–41] Lewy bodies or other alpha-synuclein-containing Lewy-pathology have usually not been described, suggesting that the disease process probably differs in some important ways from that of typical idiopathic PD. This is further supported by the finding of neurofibrillary tangles and tau-pathology in the brains of patients with *parkin* disease, a feature that is usually not associated with LB-positive PD. Lewy bodies are thought to represent a mechanism whereby cells sequester damaged and toxic proteins in an inactive form. The ubiquitine-proteasome system (UPS) is involved in the detoxification of these proteins, so complete loss of *parkin* (a component of the UPS-see below) may be sufficient to prevent LB formation. However, two cases have been described with parkinsonism, parkin-mutations and positive Lewy-bodies at autopsy[28,24] The significance of this finding is still unclear, but it may indicate that some specific parkin mutations retain sufficient activity to allow LB-formation. This has been supported by experimental studies indicating that mutations in the RING-domain retain ubiquitine ligase activity,[28] and that certain point

mutations may confer a toxic gain of function, leading to parkin protein aggregation and aggresome formation.[42]

The *Parkin* Gene and Its Function

The *parkin* gene (*PRKN*) is one of the largest genes in the genome. The gene has 12 exons, encoding 465 amino acids, with very large introns, so the entire gene spans about 1.3Mb of genomic DNA (Figure 13.1).

As mutations in *PRKN* are likely to cause parkinsonism, at least in the majority of cases, by a loss-of-function mechanism, the study of the normal function of parkin is thought to provide important insight into the molecular pathogenesis of the disorder. The distribution of expression of *PRKN* did not provide any clues, as the gene was found to be expressed in many tissues and widely in most brain regions studied. Intracellularly, the parkin protein has been found in the cytosole, the nucleus, and also associated with membranes.

The parkin protein contains several conserved domains indicating its cellular functions. A ubiquitin-like (Ubl) domain at the N' terminus, a central domain, and, in the C'-terminal half, two RING (*Really-Interesting-New-Gene*) finger domains separated by an IBR (*In-Between-Ring*) domain.[51] The motif of two RING fingers separated by an IBR domain is common to several E3 ligases, enzymes that catalyze the conjugation of activated ubiquitin to target proteins prior to their destruction via the proteasome.

In fact, several groups have now shown that parkin functions in the cellular ubiquitination/protein degradation pathway as an E3-ubiquitin ligase.[46,47] (Figure 13.2) The cell uses three different classes of proteins (E1: ubiquitin activase; E2: ubiquitin conjugase, and E3: ubiquitin ligase) to transfer a small signal peptide, ubiquitin, to target proteins. The addition of a polyubiquitin chain consisting of approximately five ubiquitin units to a protein targets this protein to the proteasome, a multi-subunit protease which efficiently degrades ubiquitinated proteins. This system is thought to play a central role in cellular protein homeostasis. In addition to its function in protein turnover and in regulatory processes, the UPS is particularly important because of its role in eliminating proteins that have become destabilized, unfolded, and consequently dysfunctional. An increased concentration of unfolded proteins, which may occur, for example, in the context of oxidative stress or other damaging conditions,

FIGURE 13.1 Structure of the parkin gene. The gene has 12 exons. The ubiquitin like domain (UBL) is thought to be instrumental in substrate recognition, while the RING-domains interact with other components of the UPS-system. IBR: in between rings.

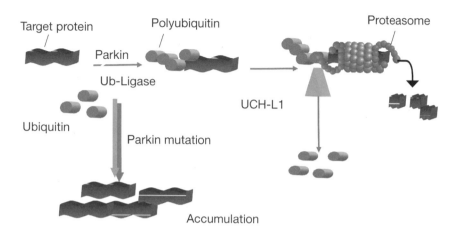

FIGURE 13.2 Parkin and its function in the ubiquitin-proteasome system (UPS). Parkin functions as a ubiquitin ligase, adding ubiquitin, a small signal polypeptide, to target proteins. The polyubiquitinated protein is directed to the proteasome, a multi-subunit protease that degrades the target protein. Ubiquitin is recycled by ubiquitin C-terminal hydrolase L-1 (UCH-L1). Parkin mutations may lead to a failure of the UPS, to accumulation of unwanted proteins, and consequently to cell damage.

may be an important trigger of apoptosis-mediated cell death. Several lines of evidence suggest that a failure of the ubiquitin-proteasome system may be one of the crucial events in the molecular pathogenesis of nigral degeneration.[48]

The specificity of the UPS is provided by a large number (several hundred) of E3 ligases, each having its own subset of target proteins. The search for parkin substrates therefore holds the promise to identify those target proteins that are involved in the specific neurodegenerative process in AR-JP. It has been hypothesized that loss of parkin may lead to the accumulation of a nonubiquitinated, potentially toxic substrate that is deleterious to the dopaminergic cell, but due to its nonubiquitinated nature, it can neither be sequestered in typical Lewy-bodies nor degraded by the proteasome. Several proteins have been shown to interact with parkin which could possibly serve as a substrate and thereby play a crucial role with regard to neurodegeneration: an O-glycosylated form of alpha-synuclein (α-Sp22),[49] a protein associated with synaptic vesicles, CDCrel-1,[47] and a transmembrane protein, called the pael-receptor,[50] SEPT5_v2,[51] synphillin-1,[52] and others. α-Sp22 and the Pael-Receptor have been shown to accumulate in the brains of patients with proven parkin mutations.[49,50]

PARKIN CELL MODELS

The function of parkin has also been studied in a number of cellular systems, either transiently or stably expressing wild-type or mutated parkin. The combined evidence from these studies suggests that parkin acts protective in a number of different paradigms thought to be pertinent in the pathogenesis of PD. In particular, these studies support the potential protective role of parkin in the setting of unfolded protein stress in the presence of damaged, misfolded, or unwanted protein.

For example, when one of the identified parkin substrates, the putative G protein-coupled transmembrane polypeptide, the Pael receptor, is overexpressed in cells, this receptor tends to become unfolded and insoluble. The insoluble Pael receptor leads to unfolded protein-induced cell death. Parkin specifically ubiquitinates this receptor in the presence of ubiquitin-conjugating enzymes resident in the endoplasmic reticulum and promotes the degradation of insoluble Pael receptor, resulting in suppression of the cell death induced by Pael receptor overexpression.[53] As the Pael receptor is expressed in dopaminergic neurons, this mechanism may contribute to the neurodegenerative process *in vivo*.

Analyzing another paradigm, Darios and co-workers have overexpressed wt and mt parkin in PC12 cells. In this cell line, neuronally differentiated by nerve growth factor, parkin overproduction protected against cell death mediated by ceramide, but not by a variety of other cell death inducers (H_2O_2, 4-hydroxynonenal, rotenone, 6-OHDA, tunicamycin, 2-mercaptoethanol and staurosporine). Protection was abrogated by the proteasome inhibitor epoxomicin and disease-causing *PRKN* variants, indicating that it was mediated by the E3 ubiquitin ligase activity of parkin.[54]

Hyun et al. demonstrated that increasing the expression of parkin by gene transfection in NT-2 and SK-N-MC cells led to increased proteasomal activity, decreased levels of protein carbonyls, 3-nitrotyrosine-containing proteins, and a trend to a reduction in ubiquitinated protein levels. Transfection of these cells with DNA encoding three mutant parkins associated with autosomal recessive juvenile parkinsonism (Del 3-5, T240R, and Q311X) was less effective.[55]

A potential protective role against α-synuclein mediated toxic effects by parkin is of particular interest as it may contribute to the understanding of parkin's role in typical sporadic PD. In a model using catecholaminergic neurons in primary midbrain cultures, overexpression of mutant alpha-synuclein was found to increase sensitivity to proteasome inhibitors by decreasing proteasome function.[56] Overexpression of parkin counteracted this effect and decreased sensitivity to proteasome inhibitors in a manner dependent on parkin's ubiquitin-protein E3 ligase activity. Antisense knockdown of parkin increases sensitivity to proteasome inhibitors. This study indicates that parkin and alpha-synuclein may be linked by common effects on a pathway associated with selective cell death in catecholaminergic neurons.

In a similar way, a protective role of parkin against the toxic effects caused by overexpression of wt alpha-synuclein in SHSY-5Y cells has been reported.[57] Although non-glycosylated alpha-synuclein is not a parkin substrate, parkin can block alpha-synuclein cytotoxicity through the activation of calpain.[58]

Parkin Animal Models

Further insight into the normal function of parkin and its role in the pathogenesis of nigral degeneration has been derived from animal models. Assuming a loss of function mechanism in the human disease, knock-out models of *PRKN* have been constructed to serve as model systems to address the molecular basis of parkin dysfunction and dopaminergic cell loss.

The classic model to study is the "knock-out" mouse. Two different lines have been reported in the literature.[59,60] Both lines were constructed by deleting exon 3 of the gene, with consequent loss of parkin protein. Neither model demonstrates alterations in gross brain morphology or dopaminergic neuron loss. However, both models do exhibit nigrostriatal deficits, with subtle behavioral phenotypes that may represent a relatively early stage of the degenerative process. For example, despite normal numbers of dopamine neurons, the level of dopamine transporter protein was reduced in these animals, suggesting a decreased density of dopamine terminals, or adaptive changes in the nigrostriatal dopamine system. GSH levels were increased in the striatum and in fetal mesencephalic neurons from parkin mutant mice, suggesting that a compensatory mechanism may protect dopamine neurons from neuronal death.[60] Extracellular dopamine concentrations were found to be increased in both lines of parkin[-/-] mice, again possibly indicating compensatory mechanisms. Interestingly, levels of CDCrel-1, synphilin-1 and alpha-synuclein, which had previously been identified as parkin substrates, were unaltered. Very recently, further analysis of one of these mouse models

has provided a possible alternative scenario linking lack of parkin protein to neuronal damage.

Palacino et al. demonstrated a down-regulation of a number of proteins involved in mitochondrial function or oxidative stress in the midbrain of parkin[-/-] mice by two-dimensional gel electrophoresis. Consistent with reductions in several subunits of complexes I and IV, functional assays showed reductions in respiratory capacity of striatal mitochondria. Parkin[-/-] mice also exhibited decreased levels of proteins involved in protection from oxidative stress and decreased serum antioxidant capacity and increased protein and lipid peroxidation.[61] These findings indicate an important role for parkin in the regulation of mitochondrial function, which is particularly striking as mitochondrial dysfunction has long been thought to be one of the key events leading to neuronal cell death in typical sporadic PD.[62]

Interestingly, mitochondrial dysfunction also appears to be central to another model of parkin disease, the PRKN-null model in drosophila.[63] Flies lacking parkin have a shorter life-span, a compromised locomotor activity, compared to its wild-type counterpart, due to apoptotic degeneration of muscle cells, and male sterility. Mitochondrial pathology is the earliest manifestation of muscle degeneration and a prominent characteristic of individualizing spermatids in parkin mutants, raising the possibility that similar mitochondrial impairment triggers the selective cell loss observed in AR-JP. [63]

However, other studies in the drosophila model have strengthened the putative role of parkin in protein degradation. Panneuronal expression of parkin substrate Pael-R was shown to cause an age-dependent selective degeneration of Drosophila dopaminergic (DA) neurons. Coexpression of parkin degrades Pael-R and suppressed its toxicity, whereas interfering with endogenous Drosophila parkin function was found to promote Pael-R accumulation and augments its toxicity.[53]

CONCLUDING REMARKS

Since the discovery in 1998 that homozygous or compound heterozygous loss of function mutations in the parkin gene cause early-onset recessive parkinsonism, enormous progress has been made by a combination of clinical, genetic, molecular, biochemical, and functional studies in patients and in cellular and animal models. None of the models examined so far reflects all aspects of PD, and the human disease caused by parkin mutations itself may well share only some features with the sporadic disease. Nevertheless, the analysis of this gene, its molecular interactions, and the consequences of its dysfunction have already greatly advanced our knowledge of the mechanisms operating in nigral degeneration and will undoubtedly further contribute to a more profound understanding of the disease mechanisms.

REFERENCES

1. Gasser T., Overview of the genetics of parkinsonism, *Adv. Neurol.*, 91:143–152, 2003.

2. Polymeropoulos, M. H., Lavedan, C., Leroy, E. et al., Mutation in the α-synuclein gene identified in families with Parkinson's disease, *Science*, 276:2045–2047, 1997.

3. Ishikawa, A., Tsuji, S., Clinical analysis of 17 patients in 12 Japanese families with autosomal-recessive type juvenile parkinsonism, *Neurology*, 47(1):160–166, 1996.

4. Ichinose, H., Ohye, T., Takahashi, E. et al., Hereditary progressive dystonia with marked diurnal fluctuation caused by mutations in the GTP cyclohydrolase I gene, *Nat. Genet.*, 8(3):236–242, 1994.

5. Matsumine, H., Saito, M., Shimoda-Matsubayashi, S. et al., Localization of a gene for an autosomal recessive form of juvenile Parkinsonism to chromosome 6q25.2–27, *Am. J. Hum. Genet.*, 60(3):588–596, 1997.

6. Tassin, J., Durr, A., de Broucker, T. et al., Chromosome 6-linked autosomal recessive early-onset parkinsonism: linkage in european and algerian families, extension of the clinical spectrum, and evidence of a small homozygous deletion in one family, *Am. J. Hum. Genet.*, 63(1):88–94, 1998.

7. Jones, A. C., Yamamura, Y., Almasy, L. et al., Autosomal recessive juvenile parkinsonism maps to 6q25.2-q27 in four ethnic groups: detailed genetic mapping of the linked region, *Am. J. Hum. Genet.*, 63(1):80–87, 1998.

8. Kitada, T., Asakawa, S., Hattori, N. et al., Mutations in the parkin gene cause autosomal recessive juvenile parkinsonism, *Nature*, 392:605–608, 1998.

9. Hattori, N., Kitada, T., Matsumine, H. et al., Molecular genetic analysis of a novel Parkin gene in Japanese families with autosomal recessive juvenile parkinsonism: evidence for variable homozygous deletions in the Parkin gene in affected individuals, *Ann. Neurol.*, 44(6):935–941, 1998.

10. Lücking, C. B., Abbas, N., Dürr, A. et al., Homozygous deletions in parkin gene in European and North African families with autosomal recessive juvenile parkinsonism. The European Consortium on Genetic Susceptibility in Parkinson's Disease and the French Parkinson's Disease Genetics Study Group, *Lancet*, 352(9137):1355–1356, 1998.

11. Abbas, N., Lücking, C. B., Ricard, S. et al., A wide variety of mutations in the parkin gene are responsible for autosomal recessive parkinsonism in Europe, *Hum. Mol. Genet.*, 8(4):567–574, 1999.

12. Lücking, C. B., Dürr, A., Bonifati, V. et al., Association between Early-Onset Parkinson's Disease and Mutations in the Parkin Gene, *N. Engl. J. Med.*, 342(21):1560–1567, 2000.

13. Lücking, C. B., Bonifati, V., Periquet, M., Vanacore, N., Brice, A., Meco, G., Pseudo-dominant inheritance and exon 2 triplication in a family with parkin gene mutations, *Neurology*, 57(5):924–7, 2001.

14. Periquet, M., Lücking, C. B., Vaughan, J. R. et al., Origin of the Mutations in the parkin Gene in Europe: Exon Rearrangements Are Independent Recurrent Events, whereas Point Mutations May Result from Founder Effects, *Am. J. Hum. Genet.*, 68(3):617–626, 2001.

15. Lohmann, E., Periquet, M., Bonifati, V. et al., How much phenotypic variation can be attributed to parkin genotype? *Ann. Neurol.*, 54(2):176–185, 2003.

16. Tassin, J., Durr, A., Bonnet, A. M. et al., Levodopa-responsive dystonia: GTP cyclohydrolase I or parkin mutations? *Brain*, 123 (Pt. 6):1112–1121, 2000.

17. Yamamura, Y., Hattori, N., Matsumine, H., Kuzuhara, S., Mizuno, Y., Autosomal recessive early-onset parkinsonism with diurnal fluctuation: clinicopathologic characteristics and molecular genetic identification, *Brain Dev.*, 22 Suppl., 1:87–91.:87–91, 2000.

18. Khan, N. L., Graham, E., Critchley, P. et al., Parkin disease: a phenotypic study of a large case series, *Brain*, 126 (Pt. 6): 1279–1292, 2003.

19. Inzelberg, R., Hattori, N., Nisipeanu, P. et al., Camptocormia, axial dystonia, and parkinsonism: Phenotypic heterogeneity of a parkin mutation, *Neurology*, 60(8):1393–1394, 2003.

20. Bruno, M. K., Ravina, B., Garraux, G. et al., Exercise-induced dystonia as a preceding symptom of familial Parkinson's disease, *Mov. Disord.*, 19(2):228–30, 2004.

21. Kuroda, Y., Mitsui, T., Akaike, M., Azuma, H., Matsumoto, T., Homozygous deletion mutation of the parkin gene in patients with atypical parkinsonism, *J. Neurol. Neurosurg. Psychiatry*, 71(2):231–234, 2001.

22. Periquet, M., Latouche, M., Lohmann, E. et al., Parkin mutations are frequent in patients with isolated early-onset parkinsonism, *Brain*, 126(Pt 6):1271–1278, 2003.

23. Pramstaller, P. P., Kis, B., Eskelson, C. et al., Phenotypic variability in a large kindred (Family LA) with deletions in the parkin gene, *Mov. Disord.*, 17(2):424–426, 2002.

24. Rawal, N., Periquet, M., Lohmann, E. et al., New parkin mutations and atypical phenotypes in families with autosomal recessive parkinsonism, *Neurology*, 60(8):1378–81, 2003.

25. Klein, C., Pramstaller, P. P., Kis, B. et al., Parkin deletions in a family with adult-onset, tremor-dominant parkinsonism: expanding the phenotype, *Ann. Neurol.*, 48(1):65–71, 2000.

26. Khan, N. L., Brooks, D. J., Pavese, N. et al., Progression of nigrostriatal dysfunction in a parkin kindred: an [18F]dopa PET and clinical study, *Brain*, 125 (Pt. 10):2248–2256, 2002.

27. Hilker, R., Klein, C., Ghaemi, M. et al., Positron emission tomographic analysis of the nigrostriatal dopaminergic system in familial parkinsonism associated with mutations in the parkin gene, *Ann. Neurol.*, 49(3):367–376, 2001.

28. Farrer, M., Chan, P., Chen, R. et al., Lewy bodies and parkinsonism in families with parkin mutations, *Ann. Neurol.*, 50(3):293–300, 2001.

29. Lincoln, S. J., Maraganore, D. M., Lesnick, T. G. et al., Parkin variants in North American Parkinson's disease: cases and controls, *Mov. Disord.*, 18(11):1306–11, 2003.

30. Maruyama, M., Ikeuchi, T., Saito, M. et al., Novel mutations, pseudo-dominant inheritance, and possible famil-

ial affects in patients with autosomal recessive juvenile parkinsonism, *Ann. Neurol.,* 48(2):245–250, 2000.

31. Satoh, J., Kuroda, Y., Association of codon 167 Ser/Asn heterozygosity in the parkin gene with sporadic Parkinson's disease, *Neuroreport.,* 10(13):2735–2739, 1999.

32. Lücking, C. B., Chesneau, V., Lohmann, E. et al., Coding polymorphisms in the parkin gene and susceptibility to Parkinson disease, *Arch. Neurol.,* 60(9):1253-1256, 2003.

33. West, A. B., Maraganore, D., Crook, J. et al., Functional association of the parkin gene promoter with idiopathic Parkinson's disease, *Hum. Mol. Genet.,* 11(22):2787–2792, 2002.

34. Klein, C., Schumacher, K., Jacobs, H. et al., Association studies of Parkinson's disease and parkin polymorphisms, *Ann. Neurol.,* 48(1):126–127, 2000.

35. Mellick, G. D., Buchanan, D. D., Hattori, N. et al., The parkin gene S/N167 polymorphism in Australian Parkinson's disease patients and controls, 7(2):89–91, 2001.

36. Oliveira, S. A., Scott, W. K., Nance, M. A. et al., Association study of Parkin gene polymorphisms with idiopathic Parkinson disease, *Arch. Neurol.,* 60(7):975–980, 2003.

37. Wang, M., Hattori, N., Matsumine, H. et al., Polymorphism in the parkin gene in sporadic Parkinson's disease, *Ann. Neurol.,* 45(5):655–658, 1999.

38. Ujike, H., Yamamoto, M., Kanzaki, A., Okumura, K., Takaki, M., Kuroda, S., Prevalence of homozygous deletions of the parkin gene in a cohort of patients with sporadic and familial Parkinson's disease, *Mov. Disord.,* 16(1):111–113, 2001.

39. Takahashi, H., Ohama, E., Suzuki, S. et al., Familial juvenile parkinsonism: clinical and pathologic study in a family, *Neurology,* 44(3 Pt. 1):437–441, 1994.

40. van De Warrenburg, B. P., Lammens, M., Lücking, C. B. et al., Clinical and pathologic abnormalities in a family with parkinsonism and parkin gene mutations, *Neurology,* 56(4):555-557, 2001.

41. Mori, H., Kondo, T., Yokochi, M. et al., Pathologic and biochemical studies of juvenile parkinsonism linked to chromosome 6q, *Neurology,* 51(3):890-892, 1998.

42. Gu, W. J., Corti, O., Araujo, F. et al., The C289G and C418R missense mutations cause rapid sequestration of human Parkin into insoluble aggregates, *Neurobiol Dis.,* 14(3):357–64, 2003.

43. Portman, A. T., Giladi, N., Leenders, K. L. et al., The nigrostriatal dopaminergic system in familial early onset parkinsonism with parkin mutations, *Neurology,* 56(12):1759–1762, 2001.

44. Leenders, K. L., Oertel, W. H., Parkinson's disease: clinical signs and symptoms, neural mechanisms, positron emission tomography, and therapeutic interventions, *Neural Plast.,* 8(1–2):99–110, 2001.

45. Thobois, S., Ribeiro, M. J., Lohmann, E. et al., Young-onset Parkinson disease with and without parkin gene mutations: a fluorodopa F 18 positron emission tomography study, *Arch. Neurol.,* 60(5):713–718, 2003.

46. Shimura, H., Hattori, N., Kubo, S. et al., Familial Parkinson disease gene product, parkin, is a ubiquitin-pro-

tein ligase [In Process Citation], *Nat. Genet.,* 25(3):302-305, 2000.

47. Zhang, Y., Gao, J., Chung, K. K., Huang, H., Dawson, V. L., Dawson, T. M., Parkin functions as an E2-dependent ubiquitin- protein ligase and promotes the degradation of the synaptic vesicle-associated protein, CDCrel-1, *Proc. Natl. Acad. Sci., U.S.A.,* 97(24):13354–13359, 2000.

48. McNaught, K. S., Olanow, C. W., Proteolytic stress: a unifying concept for the etiopathogenesis of Parkinson's disease, *Ann. Neurol.,* 53 Suppl., 3:S73–84; discussion S84–6, 2003.

49. Shimura, H., Schlossmacher, M. G., Hattori, N. et al., Ubiquitination of a new form of alpha-synuclein by parkin from human brain: implications for Parkinson's disease, *Science,* 293(5528):263–269, 2001.

50. Imai, Y., Soda, M., Inoue, H., Hattori, N., Mizuno, Y., Takahashi, R., An unfolded putative transmembrane polypeptide, which can lead to endoplasmic reticulum stress, is a substrate of Parkin, *Cell,* 105(7):891–902, 2001.

51. Choi, P., Snyder, H., Petrucelli, L. et al., SEPT5_v2 is a parkin-binding protein, *Brain Res. Mol. Brain Res.,* 117(2):179–89, 2003.

52. Chung, K. K., Zhang, Y., Lim, K. L. et al., Parkin ubiquitinates the alpha-synuclein-interacting protein, synphilin-1: implications for Lewy-body formation in Parkinson disease, *Nat. Med.,* 7(10):1144–1150, 2001.

53. Yang, Y., Nishimura, I., Imai, Y., Takahashi, R., Lu, B., Parkin suppresses dopaminergic neuron-selective neurotoxicity induced by Pael-R in Drosophila, *Neuron,* 37(6):911–924, 2003.

54. Darios, F., Corti, O., Lücking, C. B., et al. Parkin prevents mitochondrial swelling and cytochrome c release in mitochondria-dependent cell death, *Hum. Mol. Genet.,* 12(5):517–26, 2003.

55. Hyun, D. H., Lee, M., Hattori, N. et al., Effect of wild-type or mutant Parkin on oxidative damage, nitric oxide, antioxidant defenses, and the proteasome, *J. Biol. Chem.,* 277(32):28572–7, 2002.

56. Petrucelli, L., O'Farrell, C., Lockhart, P. J. et al., Parkin protects against the toxicity associated with mutant alpha-synuclein: proteasome dysfunction selectively affects catecholaminergic neurons, *Neuron,* 36(6):1007–19, 2002.

57. Oluwatosin-Chigbum Y., Robbins, A., Scott, C. W. et al., Parkin suppresses wild-type alpha-synuclein-induced toxicity in SHSY-5Y cells, *Biochem. Biophys. Res. Commun.,* 309(3):679-84, 2003.

58. Kim, S. J., Sung, J. Y., Um, J. W. et al., Parkin cleaves intracellular alpha-synuclein inclusions via the activation of calpain, *J. Biol. Chem.,* 278(43):41890-9, 2003.

59. Goldberg, M. S., Fleming, S. M., Palacino, J. J. et al., Parkin-deficient mice exhibit nigrostriatal deficits but not loss of dopaminergic neurons, *J. Biol. Chem.,* 2003.

60. Itier, J. M., Ibanez, P., Mena, M. A. et al., Parkin gene inactivation alters behaviour and dopamine neurotransmission in the mouse, *Hum. Mol. Genet.,* 12(18):2277–2291, 2003.

61. Palacino, J. J., Sagi, D., Goldberg, M. S. et al., Mitochondrial dysfunction and oxidative damage in Parkin-deficient mice, *J. Biol. Chem.,* 2004.

62. Mizuno, Y., Yoshino, H., Ikebe, S. et al., Mitochondrial dysfunction in Parkinson's disease, *Ann. Neurol.,* 44(3 Suppl., 1):S99–109, 1998.

63. Greene, J. C., Whitworth, A. J., Kuo, I., Andrews, L. A., Feany, M. B., Pallanck, L. J., Mitochondrial pathology and apoptotic muscle degeneration in Drosophila parkin mutants, *Proc. Natl. Acad. Sci., U.S.A.* 100(7): 4078–4083, 2003.

14 Heredofamilial Parkinsonism

Pramod Kumar Pal
National Institute of Mental Health and Neurosciences (NIMHANS), Bangalore, India

Zbigniew K. Wszolek
Department of Neurology, Mayo Clinic, Jacksonville, Florida, and Mayo Clinic College of Medicine, Rochester, Minnesota

CONTENTS

0-8493-1590-5/05/$0.00+$1.50
© 2005 by CRC Press

ABSTRACT

The pathogenesis of Parkinson's disease remains unknown, even 186 yr after its first description. However, major advances in molecular genetics in the past few decades have revolutionized the classification of Parkinson's disease and Parkinson-plus syndromes. A strong genetic basis has been established for early-onset Parkinson's disease, and genetic factors have been identified that play a crucial role in a subset of patients with the late-onset, sporadic form. Several genetically, clinically, and pathologically distinct forms of these disorders can be caused by mutations in *α-synuclein, parkin, UCH-L1, DJ-1, NR4A2, ND4, tau,* or as yet unknown causative genes. Molecular characterization has also provided clues regarding their pathogenesis, leading to the categorization of these disorders into polyglutamine disorders, synucleinopathies, and tauopathies. This chapter reviews the current knowledge of the genetic basis of heredofamilial parkinsonism. With growing interest in this field, additional genes and susceptibility loci will undoubtedly continue to be discovered.

INTRODUCTION

Major advances in the field of molecular genetics have revolutionized the understanding and classification of neurodegenerative conditions. Many disorders previously believed to be nonhereditary or nongenetic are now found to have genetic defects. One such group of disorders is primarily manifested by features of parkinsonism, a term loosely used to describe 1 or more of the 4 cardinal manifestations of Parkinson's disease:

1. Resting tremor
2. Bradykinesia
3. Rigidity
4. Postural instability

PHENOTYPES OF PARKINSONISM

Parkinsonism may have no identifiable cause (idiopathic), or it may be observed in conjunction with another neurodegenerative disorder or as a result of an identifiable cause.

1. Without an identifiable cause, the condition is known as Parkinson's disease or idiopathic parkinsonism.
2. As a part of other neurodegenerative disorders, it is known as a Parkinson-plus syndrome, such as progressive supranuclear palsy, corticobasal degeneration, multiple system atrophy, spinocerebellar ataxias (especially Machado-Joseph disease, also known as spinocerebellar ataxia type 3, and spinocerebellar ataxia type 2), diffuse Lewy body disease, or frontotemporal dementia and parkinsonism linked to chromosome 17 (FTDP-17).
3. With identifiable causes (e.g., viral, neuroleptic medications, vascular, toxins, tumors, metabolic, Wilson disease, Hallervorden-Spatz disease), the condition is referred to as secondary parkinsonism.

Evidence of familial clustering indicating a possible genetic disorder has been observed mostly in the first two groups, whereas metabolic disorders have been found in the third group. However, caution should be exercised when equating familial with genetic disorders, because a common environmental exposure (e.g., 1-methyl-4-phenyl-1,2,3,6-tetrahydropyridine, manganese, other toxins, or viruses) can result in the familial occurrence of parkinsonism. It is often reasonable to assume that most cases of parkinsonism involve a combination of environmental and inherited risk factors.

Lennox was probably the first to document the familial clustering of parkinsonism in 1880; the first systematic studies were conducted by Mjönes[1] in Sweden in the 1940s. With the developments in molecular genetics of the past two decades has come a flurry of reports of multigenerational kindreds who have multiple cases of parkinsonism. Many of these reports have been published in the molecular genetics literature, but good clinical, pathologic, and longitudinal follow-up data are available in only a few of them. Although earlier twin studies in Parkinson's disease were inconclusive,[2-4] a later study[5] of a large cohort of twins indicated a genetic pathogenesis in patients whose onset of disease occurred at or before 50 yr of age.

This chapter provides a review of the available evidence that implicates genetic factors in the causation of heredofamilial parkinsonism. This form of parkinsonism can be divided broadly into two groups that have some degree of overlap in their features.

Typical Parkinson's Disease Phenotype

Patients who have familial Parkinson's disease usually have phenotypic characteristics of sporadic disease that are characterized clinically by the presence of the following cardinal features: bradykinesia, rigidity, resting tremor, postural instability, or positive response to levodopa therapy.[6] However, certain features may provide clues that the proband does not have typical, idiopathic, sporadic Parkinson's disease (e.g., young age at onset of symptoms, inconspicuous resting tremor, early levodopa-induced dyskinesia, dystonia, presence of cognitive dysfunction, or familial occurrence).

Phenotypic variability or the phenomenon of anticipation may occur in the affected members of a given kindred. The pathologic hallmark is loss of pigmented neurons, gliosis in the substantia nigra and other pigmented brain stem nuclei, or the presence of Lewy bodies in the surviving neurons of these nuclei.[6–8]

PARKINSON-PLUS SYNDROME PHENOTYPE

The neurodegenerative disorders that make up Parkinson-plus syndrome have the following:

1. Some cardinal features of Parkinson's disease, as described above; resting tremor is usually absent.
2. Other neurological symptoms or signs that may antedate or follow the features of Parkinson's disease—namely, upper or lower motor neuron signs, cerebellar dysfunction, sensory impairment, gaze palsy, autonomic dysfunction not induced by medications, behavioral disturbances, or dementia.
3. Absence of response or only a brief or partial response to dopaminomimetic medications.[9]
4. Neuroimaging or neuropathologic abnormalities found in the cortex and subcortical areas: neurofibrillary tangles, senile plaques, amyloid deposits, ballooned neurons, neuronal or glial inclusions stained with anti-tau antibodies, or other histochemical markers.[10]

The Parkinson's disease phenotype of the heredofamilial parkinsonism disorders, which are more homogeneous than the Parkinson-plus syndromes of the heredofamilial parkinsonism disorders, is mostly inherited as an autosomal dominant trait,[11] whereas another group of disorders with juvenile onset is inherited as an autosomal recessive trait.[12]

DIAGNOSIS OF HEREDOFAMILIAL PARKINSONISM

For a diagnosis of heredofamilial parkinsonism, evidence may be of the following types:

1. An exact genetic defect (mutation) identified in the kindred.
2. An abnormal chromosome region or loci (but not the mutation or gene) identified in the kindred. Abnormal product resulting from a genetic defect may be identified in tissues.
3. Kindred in whom the molecular genetic status is at present unknown but who have been examined clinically.

4. A history of similar disease or a neurological illness with the features of parkinsonism in a relative of the index patient (the proband) who has not been examined clinically.
5. Apparently sporadic Parkinson's disease with a young age at onset, or an isolated case of parkinsonism with a phenotype similar to that of the genetically characterized heredofamilial parkinsonism. Although the disease seems sporadic, a characteristic phenotype in these subsets of patients suggests heredofamilial parkinsonism. Thus, testing for candidate genes of the clinical phenotype may help achieve a genetic characterization.

Positron emission tomography (PET) may be an alternative tool for establishing an antemortem diagnosis of heredofamilial parkinsonism. PET studies of clinically asymptomatic family members of a proband may show evidence of striatal dysfunction, even in the presymptomatic stage, which indicates probable heredofamilial parkinsonism. Finally, autopsy studies of brains, if available, should be helpful in identifying disorders such as tauopathies, synucleinopathies, or polyglutamine diseases.

Heredofamilial parkinsonism is transmitted in most kindred in an autosomal dominant fashion,[13] but autosomal recessive and X-linked modes of inheritance have also been described. Defects are usually single-gene defects leading either to a novel aberrant gene function or an abnormal reduction in normal gene function.[14] The available data clearly point to a genetic heterogeneity in the pathogenesis of heredofamilial parkinsonism.

FAMILIAL PARKINSONISM OF PARKINSON'S DISEASE PHENOTYPE

In the Parkinson's disease phenotype of heredofamilial parkinsonism, some families have the exact genetic defect or an identified susceptibility locus on the chromosome (Table 14.1). Others have an as yet unknown genetic status.

The pathogenesis of Parkinson's disease is largely unknown, but it is probably due to a combination of genetic and environmental factors.[15] Genetics seem to play a prominent role, especially in patients with an early onset.[5,16] Genetic linkages have been identified in families with Parkinson's disease or sporadic Parkinson's disease. Familial Parkinson's disease can be inherited as autosomal dominant or autosomal recessive trait. Patients with sporadic Parkinson's disease are most likely to carry an autosomal recessive gene, although it may be possible for Parkinson's disease that is autosomal dominant to have a sporadic presentation because of reduced penetrance or clinical expression.[17]

TABLE 14.1
Familial Parkinsonism of Parkinson's Disease Phenotype with Reported Mutations or Loci

Mode of Inheritance	Chromosome	Gene	Locus
Autosomal dominant			
Known genetic mutation	4q21–23	α-synuclein (2 mutations)	PARK1
	4p14–15	*UCH-L1*	PARK5
	2q22–23	*NR4A2 (NURR1)*	
Unknown gene	2p13	—	PARK3
	4p15	—	PARK4
	12p11.2-q13.1	—	PARK8
Autosomal recessive			
Known genetic mutation	6q25.2-27	*Parkin* (>32 mutations)	PARK2
	1p36	*DJ-1*	PARK7
Unknown gene	1p35-36	—	PARK6
	1p36	—	PARK9
Uncertain (polygenetic?)			
Unknown gene	1p32	—	PARK10

AUTOSOMAL DOMINANT PARKINSON'S DISEASE

Known Genes

α-Synuclein (PARK1)

In 1996, the genotype of a large Italian-American kindred (from the town of Contursi in Salerno, Italy) with a Parkinson's disease phenotype was mapped to the chromosomal region 4q21–23.[18,19] The following year, the analysis of candidate genes in this region led to the identification of a missense mutation in exon 4 of the α-synuclein gene in the Contursi kindred as well as in three unrelated Greek kindreds.[20] The mutation is a G-to-A transversion at position 209 (G209A), causing an alanine-to-threonine substitution at amino acid position 53 (Ala53Thr), and it is almost fully penetrant in the Contursi kindred.[20]

However, many other investigators from other parts of Europe, the United States, China, India, and Brazil have not found any α-synuclein gene mutation in familial or sporadic cases of Parkinson's disease.[21–25] Only a few families with the G209A mutation, from mainland Greece or of Greek-American descent, have been reported.[26–28]

Another mutation of the α-synuclein gene (Ala30Pro) was described in the affected members of a German kindred.[29] The latter finding was a G-to-C transversion at position 88 (G88C) in exon 3 of the α-synuclein gene.

The phenotype of the G209A mutants was that of sporadic Parkinson's disease: resting tremor, rigidity, bradykinesia, levodopa responsiveness, and levodopa-induced dyskinesia. The mean age at onset of Parkinson's disease symptoms in the Contursi kindred was 46 yr (range, 20 to 85 yr). Dementia of variable severity developed eventually in many patients in the Contursi kindred;[18] in family H (Greek-American), one elderly patient had dementia.[30] Bostantjopoulou et al.[31] recently reported the detailed clinical and neuropsychological features of eight Greek patients who had Parkinson's disease and α-synuclein (G209A) mutation. Resting tremor was rare and, despite their relatively young age, one patient was mildly demented at 44 yr of age, whereas another at 54 yr of age had abnormal scores in memory, visuoconstructive, and executive function tests. In summary, available evidence suggests that a greater cognitive impairment and relative rarity of resting tremor may be helpful in clinically distinguishing an autosomal dominant type of Parkinson's disease with the α-synuclein mutation from sporadic Parkinson's disease.

The phenomenon of anticipation was present in family H.[30] The mean age at onset was 61.2 yr in generation 3, 48.7 yr in generation 4, and 31 yr in generation 5. In contrast to what has been reported for the Contursi kindred, the allele carrying the G209A mutation in members of family H is not expressed, or its expression is significantly reduced in affected and asymptomatic at-risk persons older than the mean age at onset for their generation, whereas it is expressed in asymptomatic at-risk persons who are younger than the mean age at onset for their generation. These findings suggest that, at least in this form of parkinsonism, the timing of α-synuclein gene expression and not solely the presence of the G209A mutation may be critical for the development of parkinsonism.[30]

The functional PET study with ^{18}F-labeled dopa or ^{11}C-raclopride of members of family H was similar to that observed in sporadic Parkinson's disease.[32] In the Contursi kindred, autopsy findings confirmed the presence of Lewy bodies in the substantia nigra.

In 1988, *α-synuclein* was first described as a neuron-specific protein localized to a part of the nucleus and cholinergic presynaptic nerve terminal of *Torpedo californica*,[33] but its normal function is unknown. *α-Synuclein* is found in a wide range of vertebrates, where it is expressed in the nervous system and localized in the nucleus and presynaptic nerve terminals,[34] and has a probable role in the synaptic transport of vesicles or plasticity.[35] The presence of cellular inclusions containing *α-synuclein* (synucleinopathies) in a large number of clinically distinct neurodegenerative diseases suggests a possible pleiotropic effect of the *α-synuclein* gene. Synucleinopathies include Parkinson's disease (patients have mutations of the *α-synuclein* gene or other genes in sporadic variety), dementia with Lewy bodies, or atrophy of multiple systems.[36,37] Whether the appearance of the *α-synuclein*–positive cellular inclusions is caused by neurodegeneration or results in neurodegeneration is unknown.[38] *In vitro* studies suggest that the mutant protein is prone to fibrillogenesis, which probably leads to the formation of Lewy bodies.[39-41]

UCH-L1 (PARK5)

In the year after the identification of mutation in the *α-synuclein* gene, a mutation in the ubiquitin carboxy-terminal hydrolase-L1 (*UCH-L1*) (PARK5) on chromosome 4p14–15 was identified in a sibling pair of a small German pedigree with probable Parkinson's disease and apparent autosomal dominant inheritance.[42] The age at onset was 49 yr in one sibling and 51 yr in the other sibling, and the clinical phenotype was dopa-responsive parkinsonism resembling idiopathic Parkinson's disease. Radiologic and neuropathologic details are not available.

The *UCH-L1* gene encodes a protein that is abundant in the brain and present in Lewy bodies.[43] It belongs to a family of deubiquitinating enzymes and is thought to hydrolyze small adducts of ubiquitin and generate a free monomeric ubiquitin.[44] The deficits in the ubiquitin pathway may lead to the formation of pathogenetic intracellular inclusions.[38]

NR4A2

The *NR4A2* gene (also known as *NOT, TINUR, RNR-1, HZF-3,* or *NURR1*), which is mapped to 2q22–23,[45] is a highly conserved gene. It can activate expression of tyrosine hydroxylase[46] and enhance transcription of the dopamine transporter,[47] and it is therefore essential for differentiation of nigral dopaminergic neurons. Its mutations have been reported in persons with schizophrenia[48,49] or manic depression,[48] which are believed to be related to dopaminergic dysfunction. Recently, Le et al.[50] identified 2 mutations (–291Tdel and –245T→G) of this gene in 10 of 107 persons with familial Parkinson's disease but not in 94 patients with sporadic Parkinson's disease or in 221 unaffected controls. The mutations were mapped to the first exon (noncoding exon) of the *NR4A2* gene; they affected one allele and led to a marked reduction in levels of *NR4A2* messenger RNA. The age at onset and the clinical profile of these ten patients were similar to those of persons with typical Parkinson's disease, and all were of European descent.

Susceptibility Loci on Chromosome Sites with Unknown Gene

PARK3

The PARK3 susceptibility locus for Parkinson's disease on chromosome 2p13[51] has been assigned a parkinsonian phenotype in six kindreds: four American and Canadian (families B, C, D, and G),[52-55] one Italian (family IT-1),[56] and one German (family K).[57] The presence of a disease-associated haplotype in clinically unaffected relatives suggests that the penetrance of mutations at this locus is low (40% or less) and may be implicated in sporadic Parkinson's disease. Clinically, these kindreds have dopa-responsive parkinsonism, with a mean age at onset of 59 yr (range, 37 to 89 yr). Postural tremor in family C and dementia with or without parkinsonism in families B (Danish-North American), C (German-American), and G (Russian German-American) have been reported.[54] Some members of families B and D showed reduced striatal fluorodopa uptake (family D had asymmetric reduction with the putamen more affected than the caudate, similar to that observed on PET studies in sporadic Parkinson's disease).[58] Autopsy findings showed degeneration of dopaminergic neurons in the substantia nigra and the presence of Lewy bodies in the substantia nigra and also the presence of cortical and subcortical Lewy bodies in the patients of family G.[52]

Recently, affected members of family D have been linked to the PARK8 locus with the LOD score of 2.8.[59] Further studies of this kindred are needed to explain the possible interactions between the two implicated chromosomal locations for the genetic disorder present in family D.

PARK4

PARK4, a haplotype on chromosome 4p15.1, was reported in the kindred known as the Spellman-Muenter family (also known as the Iowa family) of Scottish-Irish-Dutch-German origin.[60,61] The clinical features in this kindred are characterized by the appearance of dopa-responsive parkinsonism in their early or mid-30s (range, third to sixth decade), early-stage weight loss, dementia in later stages, or occasionally myoclonus, frontal lobe release signs, or urinary incontinence.[38] The course of illness is rapidly progressive. Subcortical and cortical Lewy bodies are found on autopsy.[60] Postural tremor, consistent with essential tremor, has been reported in persons of this kindred who do not have a Parkinson's disease phenotype but who do have the disease haplotype.[61]

PARK8

A genomewide linkage analysis of a Japanese family from Sagamihara (Sagamihara family) with autosomal dominant parkinsonism strongly supported the mapping of the parkinsonism locus to 12p11.2–q13.1.[62] This new locus has been named PARK8. The clinical features of these patients were compatible with those of patients with typical sporadic Parkinson's disease: a mean age at onset of 51 ± 6 yr, the presence of asymmetry of parkinsonism at onset, and a good response to levodopa.[63,64] The haplotype was shared by all 15 patients and by 8 of the 12 unaffected carriers, suggesting that disease penetration in this family is incomplete. This reduced penetrance indicates that environmental or other genetic factors may modify expression of the disease. A pure nigral degeneration without Lewy bodies was the neuropathologic observation in four patients.

Affected members of another familial parkinsonism of the Parkinson-plus syndrome phenotype (family A, see below) have also been linked to PARK8.[59]

AUTOSOMAL RECESSIVE PARKINSON'S DISEASE

Known Genes

Parkin (PARK2)

Mutations in the *parkin* gene on chromosome 6q25.2–27 in kindreds with autosomal recessive Parkinson's disease were first reported in Japanese families in 1998.[65] The *parkin* gene is a highly conserved gene across species, containing 12 exons spanning about 1.5 Mb, and it encodes a novel protein of 465 amino acids with a molecular mass of about 52 kDa. This protein is a RING-type ubiquitin protein ligase that collaborates with a ubiquitin-conjugating enzyme for degradation of specific substrates through the ubiquitin-proteasome pathway.[66,67] One substrate has been reported to be a glycosylated form of α-synuclein.[68] *Parkin* is involved in the pathogenesis of several neurodegenerative disorders, and ubiquitin is a component of the paired helical filaments in Alzheimer disease[69] and of Lewy bodies in Parkinson's disease.[70] The translational products of *α-synuclein* and *UCH-L1* genes, whose mutations cause Parkinson's disease, have also been identified as major components of Lewy bodies, presumably being at least partly ubiquinated.[43,71,72] Thus, *parkin* may function similarly to ubiquitin, and its defect in autosomal recessive juvenile parkinsonism may interfere with the ubiquitin-mediated proteolytic pathway, leading to the death of nigral neurons.[65] However, although the *parkin* gene is defective in autosomal recessive juvenile parkinsonism, Lewy bodies usually do not occur.

After the initial report from Japan, several patients with early-onset autosomal recessive Parkinson's disease or sporadic Parkinson's disease have been screened worldwide, leading to reports of multiple mutations, including exon rearrangements (deletions and multiplications) and point mutations (truncating and missense).[73–78] To date, more than 32 mutations have been reported in the *parkin* gene.[79] Twenty percent to 50% of families with autosomal recessive Parkinson's disease may have *parkin* mutations.[77] Mutations were found in 50% of families with autosomal recessive Parkinson's disease whose onset occurred before 45 yr of age and in 18% of patients with Parkinson's disease who had sporadic early onset (less than 45 yr of age).[78] Moreover, in the latter subgroup, the frequency of the *parkin* mutation increased with parental consanguinity (50%) and with decreasing age at onset of symptoms (77% with an age at onset of 20 yr or younger versus 3% with an age at onset of 31 to 45 yr). In a recent report,[80] 2% of families with idiopathic late-onset Parkinson's disease had *parkin* mutations, which indicates that *parkin* mutations may contribute to the common form of Parkinson's disease. Although *parkin* disease is typically an autosomal recessive disorder with manifestations in siblings of one generation, it has been reported in multiple generations of families with[81] or without[82–84] consanguineous marriages. This finding suggests that different *parkin* mutations are frequent enough in some populations to lead to allelic heterogeneity.[84] Most *parkin*-positive persons have either homozygous deletion or compound point mutation and deletion of the *parkin* gene. However, several small kindreds have heterozygous mutations affecting only one allele. Carriers of *parkin* mutations manifesting parkinsonism may be homozygotes, heterozygotes, compound heterozygotes, or "unclear" heterozygotes, with a trend toward an increase in age at onset of disease.[80] Certain mutations in the *parkin* gene cause or constitute a risk factor for development of Parkinson's disease in the heterozygous state, and these mutations in the heterozygous form act as susceptibility alleles for Parkinson's disease.

On average, patients with *parkin* mutations begin to have symptoms in their early 30s (Table 14.2). However, there is much interfamilial and intrafamilial variability in the age of onset, ranging from 7 to 64 yr.[78,82,85] Onset usually occurs earlier in persons with isolated Parkinson's disease rather than in those with familial Parkinson's disease. Although the clinical manifestations among patients with *parkin* mutations are independent of the age at onset,[78] features such as dystonia, symmetric signs at onset, or hyperreflexia (at onset or later) seem more common in patients who have early-onset Parkinson's disease with *parkin* mutations.[73,86] Diurnal fluctuations of symptoms or sleep benefit may be present. Patients with *parkin* mutations demonstrate significantly less rigidity and bradykinesia than do those without such mutations. In contrast to patients with idiopathic Parkinson's disease, some *parkin*-positive patients may experience freezing, festination, retropulsion, or falls as early or presenting

TABLE 14.2
Characteristics of Patients with Parkin Mutation

Onset of symptoms

Usually in early 30s, with wide variability (range, 7 to 64 yr)

Usually early onset in persons with isolated rather than familial *parkin*-positive Parkinson's disease

Clinical manifestations

Usually independent of age at onset

Common features

Parkinsonism, foot dystonia, symmetric signs at onset, hyperreflexia, diurnal symptom fluctuation, sleep benefit

Atypical features

Akinetic rigid syndrome without tremor, exercise-induced dystonia, cervical dystonia and writer's cramp, leg tremor predominant on standing, autonomic dysfunction, peripheral neuropathy, pyramidal and cerebellar dysfunctions

Dementia (rare)

Excellent response to levodopa-carbidopa, to dopamine agonist with or without anticholinergics, and (in some cases) to anticholinergics alone

Dyskinesias

Early and more common occurrence, sometimes to very low doses of levodopa

Multiple parkin *mutations*

Nonpenetration or age-dependent penetration of disease

Slow progression of disease

Autopsy findings

Neuronal loss restricted to substantia nigra and locus ceruleus

Lewy bodies usually absent

features.[84] Although dementia is rare in *parkin*-positive patients, these patients may have behavioral disturbances that complicate treatment or even precede the onset of parkinsonism, and some *parkin* carriers who do not have parkinsonism may have dementia.[84]

As more patients with Parkinson's disease have had genetic testing for *parkin*, the phenotypic heterogeneity in those who are *parkin*-positive has become evident. Atypical-presenting phenotypes include Parkinson's disease with a later onset that mimics idiopathic Parkinson's disease, exercise-induced dystonia observed in dopa-responsive dystonias, focal dystonias (cervical dystonia and writer's cramp), leg tremor, akinetic rigid syndrome without tremor, autonomic or peripheral neuropathy, or cerebellar and pyramidal dysfunctions.[84] Predominant tremor of the legs on standing may be the most prominent initial manifestation in patients who are *parkin*-positive.[87] Some studies suggest that the phenotype may vary with different types of *parkin* mutations. In a large Irish kindred that was *parkin*-positive, Khan et al.[84] identified 4 siblings, 26 to 32 yr of age, who shared the same mutations and had in common an abduction-adduction oscillatory tremor in the legs; yet, in another family with 2 siblings who had the same mutation, the phenotypes were different. Among *parkin*-positive patients, those carrying truncation mutations (mutations that truncate the protein products c.202–203delAG and c.438–477del) had an earlier age at onset (mean age, 33.2 yr) than those carrying missense mutations (mutations coding for the incorrect amino acid; mean age at onset, 44.5 yr) but did not differ significantly in any other clinical characteristics. Conversely, patients who had mutations in exon 7 were older at onset (mean age, 49.2 yr) than patients with other abnormalities of the *parkin* gene (mean age, 31.5 yr).[80] A recent study[88] also confirmed phenotypic variability with type and localization of mutations in the *parkin* gene: those patients with at least one missense mutation had a higher Unified Parkinson's Disease Rating Scale than did patients carrying two truncating mutations; missense mutations in functional domains of the *parkin* gene resulted in an earlier onset of disease; and patients with a single heterozygous mutation had significantly later and more frequent levodopa-induced fluctuations and dystonia than did those with two mutations. In a kindred with 5 members who had parkinsonism or dystonia or both and marked variability in age at onset and phenotype, a heterozygous 40-base-pair deletion was found in exon 3 of the *parkin* gene.[85] Three of these five persons had mild, slowly progressive parkinsonism, whereas the other two had striking dystonia in addition to their parkinsonism.

The rate of disease progression varies among patients with *parkin* mutations, with slower evolution in those with a younger age at onset.[88] Management of *parkin*-positive patients is usually satisfactory. They have a good and sustained response to levodopa/carbidopa, and some authors have reported a similar response to dopamine-agonist monotherapy[89] or a combination of agonist with biperiden.[90] As a group, *parkin*-positive patients respond better to levodopa, but they are also more likely to have dyskinesias, especially when they have early-onset disease. Nevertheless, whether the dyskinesia occurs earlier in the course of treatment for *parkin*-positive versus *parkin*-negative patients is still controversial.[65,73,78,86]

Among patients with late-onset Parkinson's disease, those who have *parkin* mutations probably have a slower disease progression and a better response to a low dose of levodopa. Rare occurrences of dementia have been reported in patients who have *parkin* mutations.[78,91] Cognitive decline may result from a long course of illness.[89]

The frequency of mutations in the *parkin* gene in certain populations may be great enough to cause allelic heterogeneity in the same sibship. The frequency of mutations may also cause either nonpenetrance or an age-dependent presentation of disease,[89] or it may result in pseudodominant transmission.[81]

Several authors have reported similarities as well as differences in the pattern of striatal ^{18}F-labeled dopa uptake in *parkin*-positive patients with Parkinson's disease compared with that observed in patients with idiopathic Parkinson's disease.[92–97] In *parkin*-positive patients with early-onset Parkinson's disease, authors have reported a rostrocaudal gradient in the reduction of ^{18}F-labeled dopa uptake in the striatum, with severely reduced uptake in the posterior putamen and relative sparing of the caudate nucleus.[93] The reduction was asymmetrical according to the most affected hemibody for the anterior and posterior putamen, and it did not correlate with the motor disability or type of mutations. No significant difference was found in striatal ^{18}F-labeled dopa uptake in patients with early-onset Parkinson's disease, with or without the *parkin* mutations. However, other authors did not observe this asymmetry of striatal ^{18}F-labeled dopa uptake[95,96] or the relative sparing of the caudate compared with the putamen.[92,97] In a longitudinal study of *parkin*-positive patients with Parkinson's disease, the loss of ^{18}F-labeled dopa uptake in the putamen was significantly slower compared with the loss of uptake in a group of patients with idiopathic Parkinson's disease who had a similar severity of disease and baseline ^{18}F-labeled dopa uptake in the putamen.[92] In asymptomatic carriers of a single *parkin* mutation with an apparently normal allele, a mild but statistically significant decrease of mean ^{18}F-labeled dopa uptake, compared with that of control subjects, was found in all striatal regions, indicating a preclinical disease process in these patients.[95]

A fluorodeoxyglucose PET study of 2 *parkin*-positive patients with Parkinson's disease showed a tendency for relative hypermetabolism in striatal regions similar to that found in patients with idiopathic Parkinson's disease; cerebellar hypometabolism also was noted.[97] The cerebellar hypometabolism may be either a primary phenomenon of the disease or a result of striatal deafferentation,[98] supporting earlier observations of clinical cerebellar dysfunction and neuropathology in Japanese and Dutch families with *parkin*-related parkinsonism.[99,100] In *parkin*-positive patients, ^{11}C-raclopride PET studies show striatal D$_2$-receptor binding to be different from that observed in patients with sporadic Parkinson's disease. Hilker et al.[95] found a uniformly low striatal ^{11}C-raclopride binding in all five affected family members who had compound heterozygous *parkin* mutations, whereas Portman et al.[97] found a slightly increased binding only in the putamen, similar to that observed in patients with sporadic Parkinson's disease but with conspicuously low binding in the caudate.

In summary, available PET data suggest that, although the rate of progression of striatal damage is slower in *parkin*-positive patients with Parkinson's disease than in patients with sporadic Parkinson's disease, the pathology is more symmetrical and uniform.

Few neuropathologic reports exist on patients with Parkinson's disease who are positive for *parkin* mutations. Mori et al.[101] reported neuronal cell loss restricted

to the substantia nigra and locus ceruleus and the absence of Lewy bodies, whereas Farrer et al.[102] reported typical Lewy bodies at autopsy in a *parkin*-positive patient.

DJ-1 (PARK7)

The PARK7 locus on chromosome 1p36, ~25 centimorgans from the PARK6 locus, was discovered in 2001 by van Duijn et al.[103] in four patients with early-onset Parkinson's disease who belonged to a genetically isolated community in the southwestern part of the Netherlands. The following year, this locus was confirmed in an Italian family by Bonifati et al.[104] Mutations in the *DJ-1* gene were associated with parkinsonism in both kindreds.[105]

The *DJ-1* gene is an 8-exon gene spanning 24kb. The DJ-1 protein is absent in the Dutch family as a result of a homozygous 14kb deletion removing exons 1 to 5, and it is functionally inactive in the Italian family because of the homozygous missense mutation Leu[166]Pro.[105] The *DJ-1* gene encodes a ubiquitous, highly conserved protein, the function of which is unknown. Evidence suggests that its role is similar to that of an antioxidant protein[106,107] and that loss of its function leads to neurodegeneration.

Because the mutant *DJ-1* is still present in the nucleus, the loss of cytoplasmic activity is most likely pathogenic, or else nuclear activity is affected by mutation.[105] Of interest is that this locus overlaps with DYT13, a dominant locus for primary torsion dystonia, which indicates allelic function despite different modes of inheritance.[108] As in *parkin* mutations, a compound heterozygous mutation of the *DJ-1* gene has been reported, but only in 1 of 107 patients with early-onset Parkinson's disease.[109]

Phenotypically, PARK7-linked families are similar to those with *parkin*-related or PARK6-linked forms of the disease. They have an average age at onset of 32.1 yr (range, 27 to 40 yr), symptoms that are primarily asymmetrical, a good prolonged response to treatment with levodopa, and a slow progression. Resting tremor is less common in patients with PARK7 than in those with PARK6.[103] Some authors have noted behavioral or psychological disturbances (e.g., anxiety or recurrent episodes of paranoid delusions) that occurred even before the onset of parkinsonism and focal dystonias, including blepharospasm, in some patients with PARK7.[98,103,104] Interestingly, in one family, two affected persons shared haplotypes at all three regions implicated in autosomal recessive early-onset parkinsonism (PARK2, PARK6, and PARK7).[103]

Pathologic features in carriers of the *DJ-1* mutation are not yet available, but the findings of functional neuroimaging tests show a symmetrical decrease in the dopa uptake in the putamen and caudate similar to that observed in patients with *parkin* or PARK6-linked parkinsonism.[98,105] The cerebellar hypometabolism of glucose found on a fluorodeoxyglucose PET study was reported in a patient with PARK7-linked parkinsonism,[98] similar to that found in a patient with *parkin*-linked parkinsonism.[97] PET study with ^{123}I-iodobenzamide showed normal D$_2$-receptor binding in one patient with PARK7-linked parkinsonism.[98]

Susceptibility Loci on Chromosome Sites with Unknown Gene

PARK6

Linkage to chromosome 1p35-36 (PARK6) was found in a large consanguineous family from Italy (the Marsala kindred).[110] The clinical features of nine members of three unrelated Italian families affected with PARK6 were subsequently reported by the same researchers.[17] The mean age at onset, mean age of affected patients, and disease duration were 36 ± 4.6 yr, 57 ± 8.5 yr, and 21 ± 7.8 yr, respectively. All three families had inheritance suggestive of autosomal recessive Parkinson's disease. The clinical features are similar to those of idiopathic Parkinson's disease with Lewy bodies: asymmetric tremor and akinesia. However, compared with patients who have sporadic Parkinson's disease, these patients had a younger age at onset of disease and earlier onset of ON-state dyskinesias. Also, compared with autosomal dominant parkinsonians, they did not have a rapid progression of disease or cognitive abnormalities. The onset of symptoms in the legs may represent a peculiarity of PARK6. Response to treatment with medication was excellent. Patients with PARK6 disease are similar to patients with *parkin* gene mutations (PARK2), except for the symmetric involvement and occurrence of dystonia at the onset of disease commonly observed in European *parkin* patients.[78] The clinical presentation of PARK6 patients is in between idiopathic, Lewy body, classic Parkinson's disease and *parkin* disease.[17]

Recently, 8 families (16 affected persons) of the 28 *parkin*-negative families with autosomal recessive juvenile parkinsonism from European countries had positive test findings for linkage to PARK6.[111] These PARK6-positive patients reside in four countries (the Netherlands, Germany, Italy, or the United Kingdom). No common haplotype could be detected in the linked families, suggesting independent mutational events. The mean age at onset was 42.1 ± 9.4 yr, affecting 25% of the patients when they were older than 45 yr of age (latest onset, 68 yr). Clinical features were indistinguishable from those of *parkin*-positive autosomal recessive juvenile parkinsonism of non-Japanese origin (i.e., mild-to-moderate parkinsonian syndrome, good response to L-dopa, and slow progression). Half of these patients had L-dopa–induced dyskinesias, but the dystonia at onset and the sleep benefit found in Japanese patients with autosomal recessive juvenile parkinsonism were absent in these PARK6 patients.

Functional neuroimaging with ^{18}F-labeled dopa PET shows a more uniformly reduced dopa uptake in the putamen as well as in the caudate, unlike that found in idiopathic Parkinson's disease where the caudate and the anterior putamen are relatively unaffected compared with the posterior putamen.[112] There is also subclinical loss of striatal dopamine storage capacity in asymptomatic PARK6 carriers, which implies that the unidentified genes on chromosome 1p exhibit either haploinsufficiency or a dominant negative effect.[112] No neuropathologic data are available on PARK6 patients.

PARK9

Hampshire et al.[113] reported a linkage to a region of 9cM on chromosome 1p36 in Kufor-Rakeb syndrome. This linkage was subsequently assigned PARK9 status, although no official report is available yet (The Genome Database, URL: http://www.gdb.org). Kufor-Rakeb syndrome was first described in an Arab consanguineous kindred by Najim al-Din et al.[114] It is an autosomal recessive, juvenile-onset, nigrostriatal-pallidal-pyramidal neurodegeneration. Clinically, patients have levodopa-responsive parkinsonism with additional features, such as spasticity, dementia, or supranuclear gaze paralysis. Neuroimaging shows atrophy of the globus pallidus and the pyramids, which later becomes generalized. Neuropathologic data are not available.

LATE-ONSET NON-MENDELIAN PARKINSON'S DISEASE

PARK10

Unlike the previously reported genes and loci, which all exhibit a Mendelian inheritance pattern, PARK10 is a locus on chromosome 1p32 for late-onset non-Mendelian Parkinson's disease.[115] This susceptibility locus was identified in Iceland in the families of patients with Parkinson's disease. Eighty-four percent of patients had an onset of disease when they were older than 50 yr of age (mean, 65.8 yr). This phenotype has a good response to levodopa. Its pathogenesis is unknown.

FAMILIAL PARKINSONISM OF PARKINSON-PLUS SYNDROME PHENOTYPE

FRONTOTEMPORAL DEMENTIA AND PARKINSONISM LINKED TO CHROMOSOME 17 (FTDP-17)

In heredofamilial parkinsonism of the Parkinson-plus syndrome type, frontotemporal dementia and parkinsonism linked to chromosome 17 (FTDP-17) form the most important group of disorders (Table 14.3). Descriptions of a familial frontotemporal dementia date back to 1939, when Sanders et al.[116] reported a presenile-dementing disorder affecting a Dutch kindred (subsequently known as Dutch family 2) in an autosomal dominant fashion and

TABLE 14.3
Familial Parkinsonism of Parkinson-Plus Syndrome Phenotype with Reported Mutations or Loci

	Chromosome	Gene	Locus
Autosomal dominant			
Known genetic mutation	17q21–22	*Tau (>20 mutations)*	FTDP-17
	12q23–24.1	*SCA2 (Ataxin-2)*	SCA2
	14q24.3–q32	*SCA3 (Ataxin-3)*	MJD/SCA3
Unknown genetic mutation	19q13	—	DYT12
X-linked recessive			
Unknown genetic mutation	Xq13.1	—	DYT3
Mitochondrial			
Known genetic mutation	Complex 1	*ND4*	—
Unknown genetic mutation	Complex 1	—	—

characterized by prominent behavioral disturbances, including disinhibition, aggression, obsessive behavior, and hyperorality. Subsequently, other cases of familial frontotemporal dementia, as defined by the Lund-Manchester criteria,[117] were reported with varied clinical manifestations apart from dementia (e.g., personality or behavioral changes, parkinsonism, amyotrophy, dystonias). These familial disorders were given different clinical specific eponyms, such as (1) pallido-ponto-nigral degeneration,[118] (2) disinhibition-dementia-parkinsonism-amyotrophy complex,[119] (3) familial progressive subcortical gliosis,[120] (4) familial multiple system tauopathy with presenile dementia,[36] or (5) hereditary dysphasic disinhibition dementia.[121] The term "FTDP-17" was introduced during an International Consensus Conference in Ann Arbor, Michigan, in 1996,[122] when it was found that many of these families shared a common locus on chromosome 17q21–22.[123–125]

At the time, reports were presented on 13 families with relatives affected by syndromes linked to chromosome 17q21–22. With increased awareness of this disorder, reports of additional families came in from different parts of world, including the United States, the United Kingdom, Japan, the Netherlands, France, Canada, Australia, Italy, Germany, Israel, Ireland, Spain, and Sweden. About 80 families are known to have or have had relatives affected with FTDP-17. Probably fewer than 50 of these persons are still alive.

Clinical Features

The onset of symptoms is usually insidious. The average age at onset is 49 yr (range, 25 to 76 yr), and the average duration of the clinical course is 8.5 yr (range, 2 to 26 yr). There are three major categories of symptoms: (1) behavioral and personality disturbances, (2) cognitive deficits, and (3) motor dysfunction (most often signs of Parkinson-plus syndrome) (Table 14.4).[122] In the fully developed stage of the disease, affected patients have at least two of

TABLE 14.4
Symptomatology of FTDP-17 (Frontotemporal Dementia and Parkinsonism Linked to Chromosome 17) Disorders

Behavioral and personality disturbances
 Apathy, depression
 Psychosis, verbal and physical aggressiveness, family abuse
 Obsessive-compulsive stereotyped behavior
 Defective judgment
 Hyper-religiosity, alcoholism, illicit drug addiction
 Hyperorality, hyperphagia
 Loss of personal awareness, poor hygiene, disinhibition
Cognitive deficits
 Early stage
  Aphasia of nonfluent type
  Impaired executive functions
  Relatively well preserved memory, orientation, and visuospatial functions
 Late stage
  Global deterioration of cognitive functions
  Increased repetition (echolalia, palilalia)
  Verbal and vocal preservations
  Mutism (finally)
Motor dysfunctions
 Parkinsonism (early or late manifestation)
  Axial and limb rigidity
  Bradykinesia
  Postural instability poorly responding to levodopa
  Resting tremor (uncommon)
 Other motor signs
  Pyramidal signs
  Dystonia
  Supranuclear gaze palsy
  Weakness due to amyotrophy, myoclonus, dysphagia

these groups of symptoms. Depending on the clinical presentation, families with FTDP-17 are often classified into one of two broad types: (1) families with the dementia-predominant phenotype or (2) families with the parkinsonism-predominant phenotype, such as pallido-ponto-nigral degeneration, disinhibition-dementia-parkin-sonism-amyotrophy complex, or multisystem tauopathy with presenile dementia.

Although FTDP-17 is an uncommon disorder without any strict clinical diagnostic criteria, it should be considered in the differential diagnosis in the presence of one or more of the following:[126]

1. Onset of symptoms in the third to fifth decades
2. Rapid disease progression
3. Neuropsychiatric symptoms or frontotemporal dementia
4. Parkinson-plus syndrome with levodopa-nonresponsive parkinsonism, frequent early falls, supranuclear gaze palsy, or (less commonly) apraxia, dystonia, or lateralization
5. Occasionally, early progressive speech difficulties
6. Poorly controlled seizure disorder superimposed on dementia and parkinsonism
7. Positive family history suggestive of an autosomal dominant inheritance of a neurodegenerative disorder, although variability of clinical presentation may be present even in persons of the same kindred

Genetic Aspects

FTDP-17 is inherited in an autosomal dominant pattern. Mutations in the *tau* gene, which is located on chromosome 17q21–22, have been linked to some families with FTDP-17. The human *tau* gene consists of 16 exons, only 11 of which are expressed in the central nervous system. The mutations are either exonic or intronic. The exonic mutations are missense, deletion, or silent. In more than 60 separately ascertained families, more than 31 different mutations have been reported on exons 1, 9, 10, 12, and 13 and on the intron after exon 10 in the *tau* gene.[126–130] Recently, Rosso et al.[131] described a novel missense mutation, S320F, in exon 11 of the *tau* gene in a family with presenile dementia.

Phenotypic-Genotypic Correlation

Both interkindred (different mutations) and intrakindred (same mutation) variability in clinical manifestations have been observed in the FTDP-17 disorders, and a precise phenotype-genotype correlation is not yet possible. In general, a parkinsonism-predominant phenotype develops in the families with an exon 10 missense or a 5′-splice-site intronic mutation, whereas a dementia-predominant phenotype develops in those with nonexon 10 missense mutations. Neuropathologically, in the former group, there is cortical and subcortical neuronal and glial deposition of filaments containing 4R tau isoforms, whereas in the

latter, there are widespread cortical neuronal accumulations of straight filaments composed of six tau isoforms.

Pathologic Characteristics

The consistent feature in the FTDP-17 kindred is severe frontotemporal neocortical atrophy; quite frequently, there is destruction of the basal ganglia and the substantia nigra (especially in N279K kindred).[122,132] Medial temporal lobe structures are variably involved. The common denominator of all brains affected by FTDP-17 is the pathologic accumulation of the tau protein in neurons or glia (tauopathy). The tau protein is a microtubule-associated protein present in healthy brain tissue, and it promotes assembly and stabilization of microtubules responsible for axonal transport.[133] It consists of six major isoforms resulting from the splicing exon 2, 3, or 10 of the *tau* gene. In the physiologic state of the adult brain, the ratio of 3 repeat to 4 repeat is equal to 1.[134] In tauopathies, hyperphosphorylation of the tau protein leads to its reduced solubility and the formation of pathologic filaments and inclusions.[135] Tauopathies include disorders with parkinsonism, abnormal movements, and dementia in varied combinations, such as progressive supranuclear palsy, corticobasal degeneration, Alzheimer disease, Pick disease, subacute-sclerosing panencephalitis, or Niemann-Pick disease type C.[136] FTDP-17 is the most important familial tauopathy in which mutations in the *tau* gene lead to an accumulation of either 3 repeat or 4 repeat tau isoforms in the neurons and glia.

REPRESENTATIVE FTDP–17 KINDREDS

Mutations in Exon 10

N279K Missense

This mutation is the third most prevalent mutation of the *tau* gene. It was originally described in the pallido-ponto-nigral degeneration family from the United States (genealogically traced to the colonial settlement of Virginia), and this kindred is still the largest among all the FTDP-17 families. It has 43 affected members in 8 generations.[118,137,138] Three Japanese families and one French family were subsequently described with this mutation[139–142] (reviewed by Tsuboi et al.[139]).

Affected members of this kindred have a disease presentation early in the fifth decade, usually with parkinsonian features (akinetic limb and axial rigidity, most often symmetric, postural instability, absence of resting tremor), followed by behavioral changes and dementia (frontal type). On presentation, one-third of the affected persons have personality changes or dementia alone or either or both in combination with parkinsonism. Other features include dystonia, ocular abnormalities, pyramidal tract dysfunction, bladder dysfunction, and perseverative vocalizations. The parkinsonism is usually poorly responsive to levodopa, although, in the initial stages, some response may be found. A retrospective analysis of the clinical features of affected persons shows that there may be two broad phenotypes: one with corticobasal degeneration and the other with progressive supranuclear palsy.[126] PET studies show a marked reduction in striatal ^{18}F-labeled dopa uptake.[143] The course of illness is relentlessly progressive, with death occurring within eight to nine years after the onset of clinical manifestations. Autopsy shows severe destruction of the globus pallidus and the substantia nigra, variable neocortical atrophy, ballooned neurons, and widespread compact phosphorylated tau (p-tau) inclusions within neurons and striking p-tau inclusions within oligodendroglia.[138,144]

P301L Missense

This mutation has been identified in 22 separately ascertained kindreds and is the most prevalent FTDP-17 mutation, with a worldwide distribution.[125,127,145–152] The important families are the family F in Seattle (English-French-Canadian), comprising 15 affected members in 2 generations, and the Dutch 1 family (the Netherlands), comprising 49 affected members in 6 generations. The usual presenting symptoms include behavioral disturbances and personality changes, disinhibition, loss of executive function, and language abnormalities, followed later by parkinsonism. Autopsy findings include severe destruction of basal ganglia, substantia nigra, frontotemporal atrophy, neurofibrillary tangles, ballooned neurons, and neuronal and glial p-tau inclusions.

P301S Missense

Thus far, four kindreds have been reported with this mutation: Italian (family P),[153] German,[154] Japanese,[155] and Jewish-Algerian.[130] Patients present in the third or fourth decade with symptoms of either affective disorder or movement disorder, followed by a rapidly progressive dementia and parkinsonism. Distinctive features include refractory epilepsy in the German kindred, myoclonus in the German and Italian families, and early bilateral pyramidal syndrome in the Jewish-Algerian patients. Magnetic resonance imaging has documented frontal and caudate atrophy.

Mutations of Intron Following Exon 10

Seven mutations (+3, +11, +12, +13, +14, +16, +19) have been described in the intron following exon 10, clustered in the 5′-splice-site of the intron, which affects the alternative splicing of exon 10.[127,129,156–159]

The +16 intronic mutation is the second most frequent mutation after the P301L mutation of exon 10. Eight separate families with a total of 78 affected persons have been reported with this mutation. The important kindreds are Aus I (Australia) with 28 affected members in 5 generations;[127,160,161] familial, rapidly progressive subcor-

tical gliosis (U.S.A.) with 17 affected members in 5 generations;[156] and Duke family (U.S.A.) with 14 affected members in 3 generations.[162] Personality and behavioral changes and other frontal lobe dysfunctions are early and prominent manifestations, and parkinsonian features are late and not always present. Neuropathologic features are frontotemporal atrophy, ballooned neurons, and neuronal and glial p-tau inclusions. The neuronal p-tau inclusions may be diffuse or shaped like dots or Pick bodies.[156,162]

The +14 intronic mutation has been described in a family of Irish descent (U.S.A.) with disinhibition-dementia-parkinsonism-amyotrophy complex. This kindred is the first one linked to chromosome 17, and it is the only one with this mutation. The kindred has 13 affected members in 3 generations. Clinical characteristics include personality changes, dementia of frontotemporal type, parkinsonism, and amyotrophy. A long prodromal period may be present and characterized by behavioral aberrations, such as inappropriate sexual advances, overeating, shoplifting, or excessive religiosity.[126] There is frontotemporal atrophy, degeneration of the substantia nigra, ballooned neurons, and intraneuronal and glial inclusions.[163] The hippocampus is spared.

The +3 intronic mutation has been reported so far in only a single family having 41 affected members in 7 generations (U.S.A.) with multisystem tauopathy and presenile dementia. The presenting symptoms are disequilibrium and short-term memory loss, followed by parkinsonism and superior gaze palsy.[144] The reported neuropathologic findings include cortical atrophy; neuronal loss in the hippocampus, substantia nigra, cerebellum (Purkinje cell loss), and other subcortical areas; degeneration and demyelination of the spinal cord; and intraneuronal and intraglial p-tau inclusions.[129,132,164]

Mutations Not in Exon 10

Mutations outside exon 10 include those localized on exons 1, 9, 12, and 13 (reviewed in detail by Wszolek et al.[126] and Ghetti et al.[165]). The clinical features include early personality changes, behavioral and cognitive dysfunctions, psychological manifestations, and variable and late onset of parkinsonian features.[126]

OTHER TYPES OF FAMILIAL PARKINSONISM WITH PARKINSON-PLUS SYNDROME PHENOTYPE

SPINOCEREBELLAR ATAXIAS WITH LEVODOPA-RESPONSIVE PARKINSONISM

Ataxin-2 (SCA2)

The ataxin-2 (SCA2) mutation causing a CAG repeat expansion within the coding region of the cytoplasmic protein (ataxin-2) on chromosome 12q has been reported in several patients of Chinese origin who have dopa-responsive familial parkinsonism.[166–168] Members of a large ethnic Chinese family with SCA2 mutation and autosomal dominant inheritances have been described as having typical dopa-responsive asymmetric Parkinson's disease, parkinsonism-ataxia, or parkinsonism resembling progressive supranuclear palsy.[166] Shan et al.[167] also reported expanded CAG repeats in the *SCA2* gene in two patients of Chinese origin who presented with dopa-responsive familial parkinsonism at the age of 50 yr. Both of these patients presented predominantly with 4 Hz resting tremor of the legs that responded to treatment with levodopa in one patient and to alcohol, primidone, or trihexiphenidyl in the other patient. Overt signs of cerebellar dysfunction were absent, and there was mild slowing of the ocular saccades and gait hesitation suggestive of Parkinson-plus syndrome. Gait ataxia has been reported to occur as much as 25 yr after the onset of dopa-responsive parkinsonism.[168]

Neuroimaging studies have shown features typical of idiopathic Parkinson's disease, such as reduced ^{18}F-labeled dopa uptake in the striatum with a rostrocaudal gradient,[169] bilateral asymmetric reduction of striatal dopamine transporters,[168] and normal ^{11}C-raclopride binding of the striatum.[169] However, Shan et al.[167] reported a severe involvement of the caudate nucleus unlike that observed in patients with sporadic Parkinson's disease.

The dopa-responsive parkinsonism phenotype of SCA2 is observed mainly in Chinese persons. Kock et al.[170] did not find any expanded trinucleotide repeats in the *SCA2* gene in all 270 unrelated patients of mixed ethnicity who had dopa-responsive parkinsonism (young-onset familial, young-onset sporadic, or late-onset familial), and thus far there is only one published report of an *SCA2* mutation in a family of non-Chinese origin (English family in Alberta, Canada) with levodopa-responsive parkinsonism without cerebellar abnormalities.[169]

Ataxin-3 (SCA3)

Parkinsonism and ataxia have been described in patients with SCA3 or Machado-Joseph disease, both of which have identical mutation on chromosome 14q24.3–q32; currently, these two terms have become interchangeable.[171–174] Levodopa-responsive parkinsonism and levodopa-responsive motor complications have been reported in patients with the SCA3 mutation.[166,175,176]

FAMILIAL DYSTONIA PARKINSONISM

Rapid-Onset Dystonia Parkinsonism

This autosomal dominant movement disorder is characterized by the abrupt onset of dysarthria, dysphagia, dys-

tonic spasms, bradykinesia, or postural instability.[177] The onset of symptoms usually occurs in late childhood or early adulthood and may be triggered by stressful events. Treatment with levodopa/carbidopa is usually unsatisfactory. Linkage has been established to chromosome 19q13, and the locus has been named DYT12, but the mutation is unknown.[177]

X-Linked Dystonia Parkinsonism or Lubag

Lubag, reported in men from the island of Panay in the Philippines,[178] is characterized by parkinsonism, action tremor, or dystonia manifesting usually in the fourth or fifth decade but also occurring as early as adolescence.[179–182] The neuropathologic features of this form of parkinsonism, which responds poorly to levodopa, are neuronal loss restricted to the caudate and the putamen, without evidence of Lewy bodies.[180] Linkage has been established to chromosome Xq13.1, and the gene has been named *DYT3*.[183,184]

Familial Parkinsonism Related to Mitochondrial Dysfunction

Mitochondrial DNA (mtDNA) represents a well recognized non-Mendelian genetic system, abnormalities of which cause a multitude of human diseases.[185] The mtDNA codes for constituents of the mitochondrial electron transport chain, such as the nicotinamide-adenine-dinucleotide ubiquinone reductase (complex 1). A role of decreased activity of complex 1 in the pathogenesis of Parkinson's disease comes from observations that (1) 1-methyl-4-phenyl-1,2,3,6-tetrahydropyridine, which produces clinical and pathologic features of Parkinson's disease, is an inhibitor of complex 1,[186] or that (2) complex 1 patients with Parkinson's disease exhibit decreased activity.[187–191] Swerdlow et al.[191] reported a large family with multiple members in three generations who were affected with levodopa-responsive Parkinson's disease (age at onset, 35 to 79 yr; mean, 42 yr) through exclusively maternal lines. This kindred had complex 1 dysfunction in mtDNA of maternal descendants with Parkinson's disease and in asymptomatic young maternal descendants. In another large family with maternally inherited, adult-onset, multisystem degeneration and prominent parkinsonism, a G-to-A missense mutation was found at nucleotide position 11778 of the mitochondrial *ND4* gene of complex 1 that converts a highly conserved arginine to a histidine.[192] This family had variable clinical features that included levodopa-responsive parkinsonism, dementia, dystonia, dysarthria, areflexia or hyperreflexia, spasticity, ptosis, and progressive external ophthalmoplegia. Neuropathologic findings in one patient showed a marked loss of pigmented neurons in the substantia nigra, the loss of large neurons in the caudate and the putamen, and the absence of Lewy bodies, neurofibrillary tangles, or amyloid plaques.

The identification of the G11778A mutation demonstrates that adult-onset, multisystem, neurodegenerative disease with prominent parkinsonism can be associated with an inherited mtDNA mutation.[192]

Familial Parkinsonism with Central Hypoventilation (Perry Syndrome)

The clinical features of this syndrome include the onset of symptoms at 45 to 50 yr of age, autosomal dominant inheritance, parkinsonism in the form of bradykinesia and resting tremor not responding to levodopa, depression, dementia, weight loss, sleep disorders, and central hypoventilation.[193,194] After its initial description, other cases were reported in Canada, the United States, the United Kingdom, France, Turkey,[195] and Japan.[196–200] The disease progresses relentlessly, and persons die suddenly or of respiratory failure in 4 to 8 yr.[197,198] Neuropathologic studies show the presence of severe neuronal loss and gliosis mainly in the substantia nigra, with or without scarce Lewy bodies.[196–200]

Family A (German-Canadian)

The characteristic features of this family include autosomal dominant inheritance, symptom onset at a mean age of 51 yr, levodopa-responsive parkinsonism (resting tremor, bradykinesia, rigidity, and postural instability), amyotrophy, dementia, and dystonia.[201] Affected persons have a reduction in the uptake of striatal ^{18}F-labeled dopa (increased caudate–putamen ratio) and an increase in ^{11}C-raclopride binding, whereas neuropathologic findings in autopsied patients have included neuronal loss and gliosis but an absence of Lewy bodies in the substantia nigra.[202] The affected persons in the Canadian branch of this family have been found to have linkage to the PARK8 locus on chromosome 12p11.23–q13.11.[59] Members of the Sagamihara family, the first to be linked to the PARK8 locus,[62] typically have the Parkinson's disease phenotype, which indicates the marked variability of phenotypic expression in the PARK8-linked disorders.

CONCLUSIONS

The recent explosion of genetic information in familial parkinsonism, and in Parkinson's disease in particular, underscores the importance of genes in the pathogenesis of these disorders. Although family history may not always be present, Parkinson's disease is undoubtedly a complex disorder with a strong genetic basis for the early onset of disease. Several genetic studies of patients with late-onset, sporadic Parkinson's disease also indicate that genetic factors may play a crucial role in at least a subset

of such patients. There are several genetically, clinically, and pathologically distinct forms of Parkinson's disease and Parkinson-plus syndrome (e.g., synucleinopathy, tauopathy, polyglutamine disorders) that can be caused by mutations in *α-synuclein, parkin, UCH-L1, DJ-1, NR4A2, tau, ND4,* or still unknown causative genes, which initiate a cascade of events that culminate in the death of nigral neurons.

However, in spite of the identification of several mutations and susceptibility loci and of distinct forms of familial parkinsonism, the origin of Parkinson's disease in most patients is still unresolved. Moreover, phenotypic variability in persons having the same mutations suggests that Parkinson's disease is a highly heterogeneous disorder in which multiple gene-to-gene and gene-to-environment interactions may play a critical role in the onset of disease and in phenotypic variability. Apart from family history, an early age at onset of symptoms is still the most reliable clue for suspecting a genetic basis of parkinsonism. Most families with early-onset parkinsonism display clinical and pathologic features consistent with Parkinson-plus syndrome rather than with Parkinson's disease. Thus, among familial parkinsonism, the Parkinson-plus phenotype is more likely than the Parkinson's disease phenotype to have an underlying genetic disorder.

Finally, the future holds promise for the prevention of familial parkinsonism through risk prediction and genetic counseling. At the same time, neuroprotective and curative strategies are being developed to target the abnormal protein or metabolic pathways that result from genetic mutations.

REFERENCES

1. Mjönes, H., Paralysis agitans: clinical and genetic study, *Acta Psychiatr. Neurol.*, suppl. 54, 1, 1949.
2. Duvoisin, R. C. et al., Twin study of Parkinson disease, *Neurology*, 31, 77, 1981.
3. Ward, C. D. et al., Parkinson's disease in 65 pairs of twins and in a set of quadruplets, *Neurology*, 33, 815, 1983.
4. Johnson, W. G., Hodge, S. E. and Duvoisin, R., Twin studies and the genetics of Parkinson's disease—a reappraisal, *Mov. Disord.*, 5, 187, 1990.
5. Tanner, C. M. et al., Parkinson disease in twins: an etiologic study, *JAMA*, 281, 341, 1999.
6. Calne, D. B., Is idiopathic parkinsonism the consequence of an event or a process? *Neurology*, 44, 5, 1994.
7. Gelb, D. J., Oliver, E. and Gilman, S., Diagnostic criteria for Parkinson disease, *Arch. Neurol.*, 56, 33, 1999.
8. Forno, L. S., Neuropathology of Parkinson's disease, *J. Neuropathol. Exp. Neurol.*, 55, 259, 1996.
9. Jankovic, J., Parkinsonism-plus syndromes, *Mov. Disord.*, 4 suppl. 1, S95, 1989.
10. Jellinger, K. A., The neuropathologic diagnosis of secondary parkinsonian syndromes, *Adv. Neurol.*, 69, 293, 1996.

11. Maraganore, D. M., Harding, A. E., and Marsden, C. D., A clinical and genetic study of familial Parkinson's disease, *Mov. Disord.*, 6, 205, 1991.

12. Yamamura, Y. et al., Paralysis agitans of early onset with marked diurnal fluctuation of symptoms, *Neurology*, 23, 239, 1973.

13. Wszolek, Z. K. and Pfeiffer, R. F., Heredofamilial parkinsonian syndromes, in *Movement Disorders, Neurologic Principles and Practice*, Watts, R. and Koller, W. C., Eds., McGraw-Hill, New York, 1997, 351.

14. Wilkie, A. O., The molecular basis of genetic dominance, *J. Med. Genet.*, 31, 89, 1994.

15. Williams, A. C. et al., Idiopathic Parkinson's disease: a genetic and environmental model, *Adv. Neurol.*, 80, 215, 1999.

16. Zhang, Y., Dawson, V. L. and Dawson, T. M., Oxidative stress and genetics in the pathogenesis of Parkinson's disease, *Neurobiol. Dis.*, 7, 240, 2000.

17. Bentivoglio, A. R. et al., Phenotypic characterisation of autosomal recessive PARK6-linked parkinsonism in three unrelated Italian families, *Mov. Disord.*, 16, 999, 2001.

18. Golbe, L. I. et al., Clinical genetic analysis of Parkinson's disease in the Contursi kindred, *Ann. Neurol.*, 40, 767, 1996.

19. Polymeropoulos, M. H. et al., Mapping of a gene for Parkinson's disease to chromosome 4q21-q23, *Science*, 274, 1197, 1996.

20. Polymeropoulos, M. H. et al., Mutation in the α-synuclein gene identified in families with Parkinson's disease, *Science*, 276, 2045, 1997.

21. Munoz, E. et al., Identification of Spanish familial Parkinson's disease and screening for the Ala53Thr mutation of the α-synuclein gene in early onset patients, *Neurosci. Lett.*, 235, 57, 1997.

22. Vaughan, J. et al., and the European Consortium on Genetic Susceptibility in Parkinson's Disease, The α-synuclein Ala53Thr mutation is not a common cause of familial Parkinson's disease: a study of 230 European cases, *Ann. Neurol.*, 44, 270, 1998.

23. Lin, J. J. et al., Absence of G209A and G88C mutations in the alpha-synuclein gene of Parkinson's disease in a Chinese population, *Eur. Neurol.*, 42, 217, 1999.

24. Nagar, S. et al., Mutations in the α-synuclein gene in Parkinson's disease among Indians, *Acta Neurol. Scand.*, 103, 120, 2001.

25. Teive, H. A. et al., The G209A mutation in the alpha-synuclein gene in Brazilian families with Parkinson's disease, *Arq. Neuropsiquiatr.*, 59, 722, 2001.

26. Markopoulou, K. et al., Reduced expression of the G209A α-synuclein allele in familial Parkinsonism, *Ann. Neurol.*, 46, 374, 1999.

27. Athanassiadou, A. et al., Genetic analysis of families with Parkinson disease that carry the Ala53Thr mutation in the gene encoding α-synuclein, *Am. J. Hum. Genet.*, 65, 555, 1999.

28. Papadimitriou, A. et al., Mutated alpha-synuclein gene in two Greek kindreds with familial PD: incomplete penetrance? *Neurology*, 52, 651, 1999.

29. Kruger, R. et al., Ala30Pro mutation in the gene encoding alpha-synuclein in Parkinson's disease, *Nat. Genet.*, 18, 106, 1998.

30. Markopoulou, K., Wszolek, Z.K. and Pfeiffer, R.F., A Greek-American kindred with autosomal dominant, levodopa-responsive parkinsonism and anticipation, *Ann. Neurol.*, 38, 373, 1995.

31. Bostantjopoulou, S., et al., Clinical features of parkinsonian patients with the α-synuclein (G209A) mutation, *Mov. Disord.*, 16, 1007, 2001.

32. Samii, A. et al., PET studies of parkinsonism associated with mutation in the alpha-synuclein gene, *Neurology*, 53, 2097, 1999.

33. Maroteaux, L., Campanelli, J. T. and Scheller, R. H., Synuclein: a neuron-specific protein localized to the nucleus and presynaptic nerve terminal, *J. Neurosci.*, 8, 2804, 1988.

34. Maroteaux, L. and Scheller, R. H., The rat brain synucleins: family of proteins transiently associated with neuronal membrane, *Brain Res. Mol. Brain Res.*, 11, 335, 1991.

35. Clayton, D. F. and George, J. M., The synucleins: a family of proteins involved in synaptic function, plasticity, neurodegeneration and disease, *Trends Neurosci.*, 21, 249, 1998.

36. Spillantini, M. G. et al., Familial multiple system tauopathy with presenile dementia: a disease with abundant neuronal and glial tau filaments, *Proc. Natl. Acad. Sci. USA*, 94, 4113, 1997.

37. Dickson, D. W. et al., Widespread alterations of α-synuclein in multiple system atrophy, *Am. J. Pathol.*, 155, 1241, 1999.

38. Wszolek, Z. K. and Markopoulou, K., Molecular genetics of familial parkinsonism, *Parkinsonism Relat. Disord.*, 5, 145, 1999.

39. Conway, K. A., Harper, J. D. and Lansbury, P. T., Accelerated *in vitro* fibril formation by a mutant α-synuclein linked to early-onset Parkinson disease, *Nat. Med.*, 4, 1318, 1998.

40. Giasson, B. I. et al., Mutant and wild type human α-synucleins assemble into elongated filaments with distinct morphologies *in vitro*, *J. Biol. Chem.*, 274, 7619, 1999.

41. El-Agnaf, O. M. et al., Effects of the mutations Ala30 to Pro and Ala53 to Thr on the physical and morphological properties of α–synuclein protein implicated in Parkinson's disease, *FEBS Lett.*, 440, 67, 1998.

42. Leroy, E. et al., The ubiquitin pathway in Parkinson's disease, *Nature*, 395, 451, 1998.

43. Lowe, J. et al., Ubiquitin carboxyl-terminal hydrolase (PGP 9.5) is selectively present in ubiquitinated inclusion bodies characteristic of human neurodegenerative diseases, *J. Pathol.*, 161, 153, 1990.

44. Larson, C. N., Krantz, B. A. and Wilkinson, K. D., Substrate specificity of deubiquitinating enzymes: ubiquitin C-terminal hydrolases, *Biochemistry*, 37, 3358, 1998.

45. Mages, H. W. et al., NOT, a human immediate-early response gene closely related to the steroid/thyroid hor-

mone receptor NAK1/TR3, *Mol. Endocrinol.*, 8, 1583, 1994.

46. Iwawaki, T., Kohno, K. and Kobayashi, K., Identification of a potential nurr1 response element that activates the tyrosine hydroxylase gene promoter in cultured cells, *Biochem. Biophys. Res. Commun.*, 274, 590, 2000.

47. Sacchetti, P. et al., Nurr1 enhances transcription of the human dopamine transporter gene through a novel mechanism, *J. Neurochem.*, 76, 1565, 2001.

48. Buervenich, S. et al., NURR1 mutations in cases of schizophrenia and manic-depressive disorder, *Am. J. Med. Genet.*, 96, 808, 2000.

49. Chen, Y. H. et al., Mutation analysis of the human NR4A2 gene, an essential gene for midbrain dopaminergic neurogenesis, in schizophrenic patients, *Am. J. Med. Genet.*, 105, 753, 2001.

50. Le, W. D. et al., Mutations in NR4A2 associated with familial Parkinson disease, *Nat. Genet.*, 33, 85, 2003. Erratum in: *Nat. Genet.*, 33, 214, 2003.

51. Gasser, T. et al., A susceptibility locus for Parkinson's disease maps to chromosome 2p13, *Nat. Genet.*, 18, 262, 1998.

52. Denson, M. A. et al., Familial parkinsonism, dementia, and Lewy body disease: study of family G, *Ann. Neurol.*, 42, 638, 1997.

53. Wszolek, Z. K. et al., Western Nebraska family (family D) with autosomal dominant parkinsonism, *Neurology*, 45, 502, 1995.

54. Wszolek, Z. K. et al., Hereditary Parkinson disease: report of 3 families with dominant autosomal inheritance [German], *Nervenarzt.*, 64, 331, 1993.

55. Denson, M. A. and Wszolek, Z. K., Familial parkinsonism: our experience and review, *Parkinsonism Relat. Disord.*, 1, 35, 1995.

56. Bonifati, V. et al., A large Italian family with dominantly inherited levodopa-responsive parkinsonism and isolated tremors (abstract), *Mov. Disord.*, 11 suppl. 1, 86, 1996.

57. Gaser, T. et al., Genetic complexity and Parkinson's disease, *Science*, 277, 388, 1997.

58. Pal, P. K. et al., Positron emission tomography of dopamine pathways in familial Parkinsonian syndromes. *Parkinsonism Relat. Disord.*, 8, 51, 2001.

59. Wszolek, Z. K. et al., PARK8 locus is associated with late-onset autosomal dominant Parkinsonism: clinical, pathological and linkage analysis study of family A and D (abstract), *Neurology*, 60 Suppl. 1, A282, 2003.

60. Muenter, M. D. et al., Hereditary form of parkinsonism—dementia, *Ann. Neurol.*, 43, 768, 1998.

61. Farrer, M. et al., A chromosome 4p haplotype segregating with Parkinson's disease and postural tremor, *Hum. Mol. Genet.*, 8, 81, 1999.

62. Funayama, M. et al., A new locus for Parkinson's disease (PARK8) maps to chromosome 12p11.2-q13.1, *Ann. Neurol.*, 51, 296, 2002.

63. Nukada, H. et al., A big family of paralysis agitans [Japanese], *Rinsho Shinkeigaku.*, 18, 627, 1978.

64. Hasegawa, K. et al., Analysis of alpha-synuclein, parkin, tau, and UCH-L1 in a Japanese family with autosomal dominant parkinsonism, *Eur. Neurol.*, 46, 20, 2001.

65. Kitada, T. et al., Mutations in the parkin gene cause autosomal recessive juvenile parkinsonism, *Nature*, 392, 605, 1998.

66. Shimura, H. et al., Familial Parkinson disease gene product, parkin, is a ubiquitin-protein ligase, *Nat. Genet.*, 25, 302, 2000.

67. Tanaka, K. et al., Parkin is linked to the ubiquitin pathway, *J. Mol. Med.*, 79, 482, 2001.

68. Shimura, H. et al., Ubiquitination of a new form of alpha-synuclein by parkin from human brain: implications for Parkinson's disease, *Science*, 293, 263, 2001.

69. Mori, H., Kondo, J. and Ihara, Y., Ubiquitin is a component of paired helical filaments in Alzheimer's disease, *Science*, 235, 1641, 1987.

70. Iwatsubo, T. et al., Purification and characterization of Lewy bodies from the brains of patients with diffuse Lewy body disease, *Am. J. Pathol.*, 148, 1517, 1996. Erratum in: *Am. J. Pathol.*, 149, 1770, 1996; *Am. J. Pathol.*, 150, 2255, 1997.

71. Spillantini, M. G. et al., α-synuclein in Lewy bodies, *Nature*, 388, 839, 1997.

72. Baba, M. et al., Aggregation of alpha-synuclein in Lewy bodies of sporadic Parkinson's disease and dementia with Lewy bodies, *Am J. Pathol.*, 152, 879, 1998.

73. Hattori, N. et al., Molecular genetic analysis of a novel Parkin gene in Japanese families with autosomal recessive juvenile parkinsonism: evidence for variable homozygous deletions in the Parkin gene in affected individuals, *Ann. Neurol.*, 44, 935, 1998.

74. Hattori, N. et al., Point mutations (Thr240Arg and Gln311Stop) [correction of Thr240Arg and Ala311Stop] in the Parkin gene, *Biochem. Biophys. Res. Commun.*, 249, 754, 1998. Erratum in: *Biochem. Biophys. Res. Commun.*, 251, 666, 1998.

75. Leroy, E. et al., Deletions in the Parkin gene and genetic heterogeneity in a Greek family with early onset Parkinson's disease, *Hum. Genet.*, 103, 424, 1998.

76. Lucking, C. B. et al., for the European Consortium on Genetic Susceptibility in Parkinson's Disease and the French Parkinson's Disease Genetics Study Group, Homozygous deletions in parkin gene in European and North African families with autosomal recessive juvenile parkinsonism. *Lancet*, 352, 1355, 1998.

77. Abbas, N. et al., the French Parkinson's Disease Genetics Study Group and the European Consortium on Genetic Susceptibility in Parkinson's Disease, A wide variety of mutations in the parkin gene are responsible for autosomal recessive parkinsonism in Europe, *Hum. Mol. Genet.*, 8, 567, 1999.

78. Lucking, C. B. et al., for the French Parkinson's Disease Genetics Study Group, Association between early-onset Parkinson's disease and mutations in the parkin gene, *N. Engl. J. Med.*, 342, 1560, 2000.

79. Wszolek, Z. K. and Farrer, M., Genetics, in *Handbook of Parkinson's Disease*, 3rd ed., Pahwa, R., Lyons, K. E., and Koller, W. C., Eds., Marcel Dekker, New York, 2003, 325.

80. Oliveira, S. A. et al., Parkin mutations and susceptibility alleles in late-onset Parkinson's disease, *Ann. Neurol.*, 53, 624, 2003.

81. Maruyama, M. et al., Novel mutations, pseudo-dominant inheritance, and possible familial effects in patients with autosomal recessive juvenile parkinsonism, *Ann. Neurol.*, 48, 245, 2000.

82. Klein, C. et al., Parkin deletions in a family with adult-onset, tremor-dominant parkinsonism: expanding the phenotype, *Ann. Neurol.*, 48, 65, 2000.

83. Lucking, C. B. et al., Pseudo-dominant inheritance and exon 2 triplication in a family with parkin gene mutations, *Neurology*, 57, 924, 2001.

84. Khan, N. L. et al., Parkin disease: a phenotypic study of a large case series, *Brain*, 126, 1279, 2003.

85. Tan, L. C. et al., Marked variation in clinical presentation and age of onset in a family with a heterozygous parkin mutation, *Mov. Disord.*, 18, 758, 2003.

86. Ishikawa, A. and Tsuji, S., Clinical analysis of 17 patients in 12 Japanese families with autosomal-recessive type juvenile parkinsonism, *Neurology*, 47, 160, 1996.

87. Rawal, N. et al., New parkin mutations and atypical phenotypes in families with autosomal recessive parkinsonism, *Neurology*, 60, 1378, 2003.

88. Lohmann, E. et al., French Parkinson's Disease Genetics Study Group; European Consortium on Genetic Susceptibility in Parkinson's Disease, How much phenotypic variation can be attributed to parkin genotype? *Ann. Neurol.*, 54, 176, 2003.

89. Bonifati, V. et al., Three parkin gene mutations in a sibship with autosomal recessive early onset parkinsonism, *J. Neurol. Neurosurg. Psychiatry*, 71, 531, 2001.

90. Munoz, E. et al., A new mutation in the parkin gene in a patient with atypical autosomal recessive juvenile parkinsonism, *Neurosci. Lett.*, 289, 66, 2000.

91. Yamamura, Y. et al., Autosomal recessive early-onset parkinsonism with diurnal fluctuation: clinicopathologic characteristics and molecular genetic identification, *Brain Dev.*, 22 suppl. 1, S87, 2000.

92. Khan, N. L. et al., Progression of nigrostriatal dysfunction in a parkin kindred: an [18F]dopa PET and clinical study, *Brain*, 125, 2248, 2002.

93. Thobois, S. et al., for the French Parkinson's Disease Genetics Study Group, Young-onset Parkinson disease with and without parkin gene mutations: a fluorodopa F 18 positron emission tomography study, *Arch. Neurol.*, 60, 713, 2003.

94. Broussolle, E. et al., [18 F]-dopa PET study in patients with juvenile-onset PD and parkin gene mutations, *Neurology*, 55, 877, 2000.

95. Hilker, R. et al., Positron emission tomographic analysis of the nigrostriatal dopaminergic system in familial parkinsonism associated with mutations in the parkin gene, *Ann. Neurol.*, 49, 367, 2001.

96. Pramstaller, P. P. et al., Parkin mutations in a patient with hemiparkinsonism-hemiatrophy: a clinical-genetic and PET study, *Neurology*, 58, 808, 2002.

97. Portman, A. T. et al., The nigrostriatal dopaminergic system in familial early onset parkinsonism with parkin mutations, *Neurology*, 56, 1759, 2001.

98. Dekker, M. et al., Clinical features and neuroimaging of PARK7-linked parkinsonism, *Mov. Disord.*, 18, 751, 2003.

99. Kuroda, Y. et al., Homozygous deletion mutation of the parkin gene in patients with atypical parkinsonism, *J. Neurol. Neurosurg. Psychiatry*, 71, 231, 2001.

100. Horstink, M. W. et al., Parkin gene related neuronal multisystem disorder, *J. Neurol. Neurosurg. Psychiatry*, 72, 419, 2002.

101. Mori, H. et al., Pathologic and biochemical studies of juvenile parkinsonism linked to chromosome 6q, *Neurology*, 51, 890, 1998.

102. Farrer, M. et al., Lewy bodies and parkinsonism in families with parkin mutations, *Ann. Neurol.*, 50, 293, 2001.

103. van Duijn, C. M. et al., PARK7, a novel locus for autosomal recessive early-onset parkinsonism, on chromosome 1p36, *Am. J. Hum. Genet.*, 69, 629, 2001.

104. Bonifati, V. et al., Localization of autosomal recessive early-onset parkinsonism to chromosome 1p36 (PARK7) in an independent dataset, *Ann. Neurol.*, 51, 253, 2002.

105. Bonifati, V. et al., Mutations in the DJ-1 gene associated with autosomal recessive early-onset parkinsonism, *Science*, 299, 256, 2003.

106. Mitsumoto, A. et al., Oxidized forms of peroxiredoxins and DJ-1 on two-dimensional gels increased in response to sublethal levels of paraquat, *Free Radic. Res.*, 35, 301, 2001.

107. Mitsumoto, A. and Nakagawa, Y., DJ-1 is an indicator for endogenous reactive oxygen species elicited by endotoxin, *Free Radic. Res.*, 35, 885, 2001.

108. Valente, E. M. et al., Identification of a novel primary torsion dystonia locus (DYT13) on chromosome 1p36 in an Italian family with cranial-cervical or upper limb onset, *Neurol. Sci.*, 22, 95, 2001.

109. Hague, S. et al., Early-onset Parkinson's disease caused by a compound heterozygous DJ-1 mutation, *Ann. Neurol.*, 54, 271, 2003.

110. Valente, E. M. et al., Localization of a novel locus for autosomal recessive early-onset parkinsonism, PARK6, on human chromosome 1p35-p36, *Am. J. Hum. Genet.*, 68, 895, 2001.

111. Valente, E. M. et al., and the European Consortium on Genetic Susceptibility in Parkinson's disease, PARK6-linked parkinsonism occurs in several European families, *Ann. Neurol.*, 51, 14, 2002.

112. Khan, N. L. et al., Clinical and subclinical dopaminergic dysfunction in PARK6-linked parkinsonism: an 18F-dopa PET study, *Ann. Neurol.*, 52, 849, 2002.

113. Hampshire, D. J. et al., Kufor-Rakeb syndrome, pallido-pyramidal degeneration with supranuclear upgaze paresis and dementia, maps to 1p36, *J. Med. Genet.*, 38, 680, 2001.

114. Najim al-Din, A. S. et al., Pallido-pyramidal degeneration, supranuclear upgaze paresis and dementia: Kufor-Rakeb syndrome, *Acta Neurol. Scand.*, 89, 347, 1994.

115. Hicks, A. A. et al., A susceptibility gene for late-onset idiopathic Parkinson's disease, *Ann. Neurol.*, 52, 549, 2002.

116. Sanders, J., Schenk, V. W. D. and van Veen, P., *A Family With Pick's Disease*, Noord-Hollandsche Uitg-Mij, Amsterdam, 1939.

117. The Lund and Manchester Groups, Clinical and neuropathological criteria for frontotemporal dementia, *J. Neurol. Neurosurg. Psychiatry*, 57, 416, 1994.

118. Wszolek, Z. K. et al., Rapidly progressive autosomal dominant parkinsonism and dementia with pallido-ponto-nigral degeneration, *Ann. Neurol.*, 32, 312, 1992.

119. Lynch, T. et al., Clinical characteristics of a family with chromosome 17-linked disinhibition-dementia-parkinsonism-amyotrophy complex, *Neurology*, 44, 1878, 1994.

120. Lanska, D. J. et al., Familial progressive subcortical gliosis, *Neurology*, 44, 1633, 1994.

121. Lendon, C. L. et al., Hereditary dysphasic disinhibition dementia: a frontotemporal dementia linked to 17q21-22, *Neurology*, 50, 1546, 1998.

122. Foster, N. L. et al., Frontotemporal dementia and parkinsonism linked to chromosome 17: a consensus conference, *Ann. Neurol.*, 41, 706, 1997.

123. Wijker, M. et al., Localization of the gene for rapidly progressive autosomal dominant parkinsonism and dementia with pallido-ponto-nigral degeneration to chromosome 17q21, *Hum. Mol. Genet.*, 5, 151, 1996.

124. Wilhelmsen, K. C. et al., Localization of disinhibition-dementia-parkinsonism-amyotrophy complex to 17q21-22, *Am. J. Hum. Genet.*, 55, 1159, 1994.

125. Heutink, P. et al., Hereditary frontotemporal dementia is linked to chromosome 17q21-22: a genetic and clinicopathological study of three Dutch families, *Ann. Neurol.*, 41, 150, 1997.

126. Wszolek, Z. K. et al., Hereditary tauopathies and parkinsonism, *Adv. Neurol.*, 91, 153, 2003.

127. Hutton, M. et al., Association of missense and 5'-splice-site mutations in tau with the inherited dementia FTDP-17, *Nature*, 393, 702, 1998.

128. Poorkaj, P. et al., Tau is a candidate gene for chromosome 17 frontotemporal dementia, *Ann. Neurol.*, 43, 815, 1998. Erratum in: *Ann. Neurol.*, 44, 428, 1998.

129. Spillantini, M. G. et al., Mutation in the tau gene in familial multiple system tauopathy with presenile dementia, *Proc. Natl. Acad. Sci. USA*, 95, 7737, 1998.

130. Werber, E. et al., Phenotypic presentation of frontotemporal dementia with Parkinsonism-chromosome 17 type P301S in a patient of Jewish-Algerian origin, *Mov. Disord.*, 18, 595, 2003.

131. Rosso, S. M. et al., A novel tau mutation, S320F, causes a tauopathy with inclusions similar to those in Pick's disease, *Ann. Neurol.*, 51, 373, 2002.

132. Spillantini, M. G., Bird, T. D. and Ghetti, B., Frontotemporal dementia and Parkinsonism linked to chromosome 17: a new group of tauopathies, *Brain Pathol.*, 8, 387, 1998.

133. Goedert, M., Neurofibrillary pathology of Alzheimer's disease and other tauopathies, *Prog. Brain Res.*, 117, 287, 1998.

134. Hong, M. et al., Mutation-specific functional impairments in distinct tau isoforms of hereditary FTDP-17, *Science*, 282, 1914, 1998.

135. Mailliot, C. et al., Phosphorylation of specific sets of tau isoforms reflects different neurofibrillary degeneration processes, *FEBS Lett.*, 433, 201, 1998.

136. Arvanitakis, Z. and Wszolek, Z. K., Recent advances in the understanding of tau protein and movement disorders, *Curr. Opin. Neurol.*, 14, 491, 2001.

137. Clark, L. N. et al., Pathogenetic implications of mutations in the tau gene in pallido-ponto-nigral degeneration and related neurodegenerative disorders linked to chromosome 17, *Proc. Natl. Acad. Sci., USA.*, 95, 13103, 1998.

138. Reed, L. A. et al., The neuropathology of a chromosome 17-linked autosomal dominant parkinsonism and dementia ("pallido-ponto-nigral degeneration"), *J. Neuropathol. Exp. Neurol.*, 57, 588, 1998.

139. Tsuboi, Y. et al., Clinical and genetic studies of families with the tau N279K mutation (FTDP-17), *Neurology*, 59, 1791, 2002.

140. Delisle, M. B. et al., A mutation at codon 279 (N279K) in exon 10 of the Tau gene causes a tauopathy with dementia and supranuclear palsy, *Acta Neuropathol. (Berl)*, 98, 62, 1999.

141. Yasuda, M. et al., A mutation in the microtubule-associated protein tau in pallido-nigro-luysian degeneration, *Neurology*, 53, 864, 1999.

142. Arima, K. et al., Two brothers with frontotemporal dementia and parkinsonism with an N279K mutation of the tau gene, *Neurology*, 54, 1787, 2000.

143. Pal, P.K. et al., Positron emission tomography in pallido-ponto-nigral degeneration (PPND) family (frontotemporal dementia with parkinsonism linked to chromosome 17 and point mutation in tau gene), *Parkinsonism Relat. Disord.*, 7, 81, 2001.

144. Reed, L. A., Wszolek, Z. K. and Hutton, M., Phenotypic correlations in FTDP-17, *Neurobiol. Aging*, 22, 89, 2001.

145. Bird, T. D. et al., A clinical pathological comparison of three families with frontotemporal dementia and identical mutations in the tau gene (P301L), *Brain*, 122, 741, 1999. Erratum in: *Brain*, 122, 1398, 1999.

146. Nasreddine, Z. S. et al., From genotype to phenotype: a clinical, pathological, and biochemical investigation of frontotemporal dementia and parkinsonism (FTDP-17) caused by the P301L tau mutation, *Ann. Neurol.*, 45, 704, 1999.

147. Spillantini, M. G., et al., Tau pathology in two Dutch families with mutations in the microtubule-binding region of tau, *Am. J. Pathol.*, 153, 1359, 1998.

148. Mirra, S. S. et al., Tau pathology in a family with dementia and a P301L mutation in tau, *J. Neuropathol. Exp. Neurol.*, 58, 335, 1999.

149. Dumanchin, C. et al., Segregation of a missense mutation in the microtubule-associated protein tau gene with familial frontotemporal dementia and parkinsonism, *Hum. Mol. Genet.*, 7, 1825, 1998.

150. Rizzu, P. et al., High prevalence of mutations in the microtubule-associated protein tau in a population study of frontotemporal dementia in the Netherlands, *Am. J. Hum. Genet.*, 64, 414, 1999.

151. Houlden, H. et al., Frequency of tau mutations in three series of non-Alzheimer's degenerative dementia, *Ann. Neurol.*, 46, 243, 1999.

152. Kodama, K. et al., Familial frontotemporal dementia with a P301L tau mutation in Japan, *J. Neurol. Sci.*, 176, 57, 2000.

153. Bugiani, O. et al., Frontotemporal dementia and corticobasal degeneration in a family with a P301S mutation in tau, *J. Neuropathol. Exp. Neurol.*, 58, 667, 1999.

154. Sperfeld, A. D. et al., FTDP-17: an early-onset phenotype with parkinsonism and epileptic seizures caused by a novel mutation, *Ann. Neurol.*, 46, 708, 1999.

155. Yasuda, M. et al., A Japanese patient with frontotemporal dementia and parkinsonism by a tau P301S mutation, *Neurology*, 55, 1224, 2000.

156. Goedert, M. et al., Tau gene mutation in familial progressive subcortical gliosis, *Nat. Med.*, 5, 454, 1999.

157. D'Souza, I. and Schellenberg, G. D., Tau exon 10 expression involves a bipartite intron 10 regulatory sequence and weak 5′ and 3′ splice sites, *J. Biol. Chem.*, 277, 26587, 2002.

158. Miyamoto, K. et al., Familial frontotemporal dementia and parkinsonism with a novel mutation at an intron 10+11-splice site in the tau gene, *Ann. Neurol.*, 50, 117, 2001.

159. Yasuda, M. et al., A novel mutation at position +12 in the intron following exon 10 of the tau gene in familial frontotemporal dementia (FTD-Kumamoto), *Ann. Neurol.*, 47, 422, 2000.

160. Baker, M., et al., Localization of frontotemporal dementia with parkinsonism in an Australian kindred to chromosome 17q21-22, *Ann. Neurol.*, 42, 794, 1997.

161. Dark, F., A family with autosomal dominant, non-Alzheimer's presenile dementia, *Aust. N. Z. J. Psychiatry*, 31, 139, 1997.

162. Hulette, C. M. et al., Neuropathological features of frontotemporal dementia and parkinsonism linked to chromosome 17q21-22 (FTDP-17): Duke Family 1684, *J. Neuropathol. Exp. Neurol.*, 58, 859, 1999.

163. Sima, A. A. et al., The neuropathology of chromosome 17-linked dementia, *Ann. Neurol.*, 39, 734, 1996.

164. Murrell, J. R. et al., Familial multiple-system tauopathy with presenile dementia is localized to chromosome 17, *Am. J. Hum. Genet.*, 61, 1131, 1997.

165. Ghetti, B., Hutton, M. and Wszolek, Z. K., Frontotemporal dementia and parkinsonism linked to chromosome 17 associated with tau gene mutations (FTDP-17), in *Neurodegeneration: the Molecular Pathology of Dementia and Movement Disorders*, Dickson, D., Ed., ISN Neuropath Press, Basel, Switzerland, 2003, 86.

166. Gwinn-Hardy, K. et al., Spinocerebellar ataxia type 2 with parkinsonism in ethnic Chinese, *Neurology*, 55, 800, 2000.

167. Shan, D. E. et al., Spinocerebellar ataxia type 2 presenting as familial levodopa-responsive parkinsonism, *Ann. Neurol.*, 50, 812, 2001.

168. Lu, C. S. et al., Dopa-responsive parkinsonism phenotype of spinocerebellar ataxia type 2, *Mov. Disord.*, 17, 1046, 2002.

169. Furtado, S. et al., SCA-2 presenting as parkinsonism in an Alberta family: clinical, genetic, and PET findings, *Neurology*, 59, 1625, 2002.

170. Kock, N. et al., Role of SCA2 mutations in early- and late-onset dopa-responsive parkinsonism, *Ann. Neurol.*, 52, 257, 2002.

171. Evidente, V. G. et al., Hereditary ataxias, *Mayo Clin. Proc.*, 75, 475, 2000.

172. Stevanin, G. et al., The gene for spinal cerebellar ataxia 3 (SCA3) is located in a region of approximately 3 cM on chromosome 14q24.3-q32.2, *Am. J. Hum. Genet.*, 56, 193, 1995.

173. Takiyama, Y. et al., The gene for Machado-Joseph disease maps to human chromosome 14q, *Nat. Genet.*, 4, 300, 1993.

174. Twist, E. C. et al., Machado Joseph disease maps to the same region of chromosome 14 as the spinocerebellar ataxia type 3 locus, *J. Med. Genet.*, 32, 25, 1995.

175. Tuite, P. J. et al., Dopa-responsive parkinsonism phenotype of Machado-Joseph disease: confirmation of 14q CAG expansion, *Ann. Neurol.*, 38, 684, 1995.

176. Subramony, S. H. and Currier, R. D., Intrafamilial variability in Machado-Joseph disease, *Mov. Disord.*, 11, 741, 1996.

177. Kramer, P. L. et al., Rapid-onset dystonia-parkinsonism: linkage to chromosome 19q13, *Ann. Neurol.*, 46, 176, 1999.

178. Lee, L. V. et al., Torsion dystonia in Panay, Philippines, *Adv. Neurol.*, 14, 137, 1976.

179. Lee, L. V. et al., The phenotype of the X-linked dystonia-parkinsonism syndrome: an assessment of 42 cases in the Philippines, *Medicine (Baltimore)*, 70, 179, 1991.

180. Waters, C. H., et al., Neuropathology of Lubag (X-linked dystonia parkinsonism), *Mov. Disord.*, 8, 387, 1993.

181. Evidente, V. G., et al., Phenomenology of "Lubag" or X-linked dystonia-parkinsonism, *Mov. Disord.*, 17, 1271, 2002.

182. Evidente, V. G., et al., X-linked dystonia ("Lubag") presenting predominantly with parkinsonism: a more benign phenotype? *Mov. Disord.*, 17, 200, 2002.

183. Wilhelmsen, K. C., et al., Genetic mapping of "Lubag" (X-linked dystonia-parkinsonism) in a Filipino kindred to the pericentromeric region of the X chromosome. *Ann. Neurol.*, 29, 124, 1991.

184. Haberhausen, G., et al., Assignment of the dystonia-parkinsonism syndrome locus, DYT3, to a small region within a 1.8-Mb YAC contig of Xq13.1. *Am. J. Hum. Genet.*, 57, 644, 1995.

185. Johns, D. R., Seminars in medicine of the Beth Israel Hospital, Boston: mitochondrial DNA and disease, *N. Engl. J. Med.*, 333, 638, 1995.

186. Langston, J. W., The etiology of Parkinson's disease with emphasis on the MPTP story. *Neurology*, 47 (suppl. 3), S153, 1996.

187. Parker, W. D., Jr., Boyson, S. J. and Parks, J. K., Abnormalities of the electron transport chain in idiopathic Parkinson's disease. *Ann. Neurol.*, 26, 719, 1989.

188. Schapira, A. H., et al., Mitochondrial complex I deficiency in Parkinson's disease, *Lancet*, 1, 1269, 1989.

189. Mizuno, Y. et al., Deficiencies in complex I subunits of the respiratory chain in Parkinson's disease, *Biochem. Biophys. Res. Commun.*, 163, 1450, 1989.

190. Schulz, J. B. and Beal, M. F., Mitochondrial dysfunction in movement disorders. *Curr. Opin. Neurol.* 7, 333, 1994.

191. Swerdlow, R. H., et al., Matrilineal inheritance of complex I dysfunction in a multigenerational Parkinson's disease family, *Ann. Neurol.*, 44, 873, 1998.

192. Simon, D. K., et al., Familial multisystem degeneration with parkinsonism associated with the 11778 mitochondrial DNA mutation, *Neurology*, 53, 1787, 1999.

193. Perry, T. L. et al., Hereditary mental depression and parkinsonism with taurine deficiency. *Arch. Neurol.*, 32, 108, 1975.

194. Perry, T. L, et al., Dominantly inherited apathy, central hypoventilation, and Parkinson's syndrome: clinical, biochemical, and neuropathologic studies of 2 new cases. *Neurology*, 40, 1882, 1990.

195. Elibol, B., Koboyashi, T. and Atac, F. B., Familial parkinsonism with apathy, depression, and central hypoventilation (Perry's syndrome), in *Mapping the Progress of Alzheimer's and Parkinson's Disease*, Mizuno, Y., Fisher, A., and Hanin, I., Eds., Kluwer Academic/Plenum Publishers, New York, 2002, 285.

196. Tsuboi, Y., et al., Japanese family with parkinsonism, depression, weight loss, and central hypoventilation, *Neurology*, 58, 1025, 2002.

197. Purdy, A. et al., Familial fatal Parkinsonism with alveolar hypoventilation and mental depression. *Ann. Neurol.*, 6, 523, 1979.

198. Roy, E. P., III, et al., Familial parkinsonism, apathy, weight loss, and central hypoventilation: successful long-term management. *Neurology*, 38, 637, 1988.

199. Lechevalier, B., et al., Familial parkinsonian syndrome with athymhormia and hypoventilation [French], *Rev. Neurol. (Paris)*, 148, 39, 1992.

200. Bhatia, K. P., Daniel, S. E. and Marsden, C. D., Familial parkinsonism with depression: a clinicopathological study. *Ann. Neurol.*, 34, 842, 1993.

201. Wszolek, Z. K., et al., German-Canadian family (Family A) with parkinsonism, amyotrophy, and dementia: longitudinal observations. *Parkinsonism Relat. Disord.*, 3, 125, 1997.

202. Wszolek, Z. K., Uitti, R. J. and Markopoulou, K., Familial Parkinson's disease and related conditions: clinical genetics. *Adv. Neurol.*, 86, 33, 2001.

15 Other Mutations: Their Role in Parkinson's Disease

Nathan Pankratz and Tatiana Foroud
Department of Medical and Molecular Genetics, Indiana University School of Medicine

CONTENTS

INTRODUCTION

Only a generation ago, the number of chromosomes in the human body was correctly determined to be 46. Then, in rapid succession, studies were begun to identify the genes underlying Mendelian, single-gene disorders. Initially, success was limited by the number of molecular markers that were available for analysis. However, during the past two decades, the field of gene mapping has been revolutionized by the advent of new molecular technologies that have made the rapid identification of genetic mutations possible. These developments have included the identification of thousands of molecular markers located throughout the human genome that can be easily genotyped and used to pinpoint the location of a disease gene to a small chromosomal segment. Then, through the careful examination of genes located within the narrowed critical interval, researchers have successfully identified the causative gene for nearly all of the common, Mendelian disorders such as cystic fibrosis and Duchenne muscular dystrophy.

The identification of disease genes was thought to provide the key to the development of improved therapies that would ameliorate clinical symptoms. The newly identified disease genes have proven amenable in many instances to molecular screening, which may be used presymptomatically, prenatally, or diagnostically. Unfortunately, for most genetic disorders, knowledge of the molecular mutation leading to disease has not resulted in substantially improved clinical outcome. Despite this initial disappointment, identification of disease genes is critical to the greater understanding of the causes of disease and has, in many cases, led to the recognition of novel disease pathways that would not have been identified without gene mapping studies.

The field of Parkinson's disease (PD) genetics research is only now beginning to reap the benefits of a decade of research designed to identify disease genes. As discussed in this chapter and elsewhere, there have been important discoveries that have revolutionized our conceptualization of disease pathogenesis. However, many more genes contributing toward the risk for PD must still be identified.

EVIDENCE FOR THE ROLE OF GENES IN PD

Evidence for a familial contribution to PD dates back over 100 yr, when Leroux[1-2] and Gowers[3] both noted that 15% of PD patients reported an affected family member. In the intervening century, the view of the scientific community has fluctuated with regard to the importance of genetics in the etiology of PD. Several studies provided additional evidence for a genetic role in disease causation.[4-7] In contrast, a few studies have strongly argued against a genetic role in PD. Most of the negative data have arisen from samples of monozygotic twins with low concordance rates for PD. The largest, a sample of World War II veteran twins, found greater concordance among twins with early-onset

PD but concluded that the genetic contribution to late-onset disease susceptibility was minimal.[8]

A major limitation of most twin studies is that they are usually cross-sectional in nature. In the case of PD, where the age at onset is quite variable, a cross-sectional study may fail to identify concordant twin pairs with widely differing ages of onset. In one instance, the age at onset of PD in a pair of monozygotic twins differed by 20 yr.[9] Functional imaging of the brain has suggested that some apparently normal co-twins actually have decreased function of the nigrostriatal dopaminergic system and may be presymptomatic, implying that the concordance rates for both monozygotic (MZ) and dizygotic (DZ) twins may be higher than previously estimated.[10,11]

Another established method used to better understand the role of genetics in a complex disease is the analysis of familial aggregation. When applied to PD, this approach compares the familial aggregation of PD in first and/or second-degree relatives of patients with the rate of disease observed in the general population. Most studies of this type will obtain information about the clinical status of the first and/or second degree relatives through family report rather than direct clinical evaluation. Despite the potential limitations of using family reported rates of PD to estimate familial aggregation, studies from around the world have provided evidence that genetic risk factors are involved in the pathogenesis of the idiopathic form of PD. Estimates of the increase in the relative risk to first-degree relatives of an affected individual range from 2.7 to 3.5 in the United States,[12–13] 2.9 in Finland,[14] 6.7 in Iceland,[15] 7.7 in France,[16] 3.2 in three centers within Europe,[17] 5.0 in Canada,[18] 13.4 in Italy,[19] and 7.1 in Germany.[20]

The growing body of evidence accumulating in the literature supports a role for genetics in the etiology of PD. It is also apparent that the genetics of PD is complex. Researchers have identified a subset of families in which PD appears to be inherited in a simple Mendelian fashion, in some cases autosomal dominant and in others autosomal recessive. Several other chapters in this book describe the current knowledge about the genes that have already been identified (i.e., alpha synuclein and parkin). Many of the causative genes have not yet been identified.

SINGLE GENE MUTATIONS RESULTING IN PD

One approach that has been employed to dissect the genetic etiology of late onset neurodegenerative disorders is the study of families segregating a mutation(s) in a single causative gene. Once families are identified with these strongly genetic inheritance patterns, it is critical that all family members are carefully evaluated neurologically so as to identify individuals with symptoms of disease and also to identify family members who lack any features of the disorder. In this way, the segregation of the mutant disease gene is characterized.

Once a sufficient number of families sharing a common pattern of inheritance have been identified, molecular studies are performed to determine the chromosomal region harboring the disease gene. In this approach, the entire genome is evaluated, typically by testing DNA sequences with a variable number of copies of a 2, 3, or 4 base pair sequence (termed microsatellite markers). If the marker being tested is in close physical proximity to a gene influencing the phenotype, then family members with the disease would be expected to have inherited the same marker allele, and the marker and the disease gene are "linked." Thus, the basic principle underlying linkage analysis is the detection, within a family, that a particular marker allele cosegregates with the disease phenotype. A strength of the genome-wide approach is that it allows susceptibility genes to be identified when we have only limited knowledge about the underlying pathophysiology of the disease process.

This genetic approach to the elucidation of disease genes has been successfully employed in the study of families with early onset, autosomal dominant Alzheimer disease. Studies performed by numerous laboratories using many different families led to the identification of mutations in three genes, amyloid precursor protein (APP), presenilin I (PS1), and presenilin II (PS2), which can cause AD. A similar approach has been applied to PD. At present, four genes or linkages have been implicated in autosomal dominant forms of parkinsonism, and three genes or linkages have been associated with autosomal recessive forms of parkinsonism (Table 15.1).

The first gene, alpha-synuclein, was identified by studying a large Italian kindred, in which PD was pathologically confirmed.[21] The same mutation in alpha-synuclein observed in the Italian kindred (Ala53Thr in exon 4) was later found in three Greek families, most of whom could trace their ancestry to a very small geographical area on the Peloponesos in Southern Greece.[21] Mutations in alpha-synuclein have been reported in eight additional individuals from six different families located in central and southwestern Greece.[22] Given the close historical ties to Southern Italy, these mutation results suggest the presence of a founder effect.[23] These individuals are very similar clinically and pathologically to idiopathic PD, with a response to levodopa and the presence of Lewy bodies; however, the age at onset is significantly earlier, with a mean of 46 yr. Two mutations in other parts of the gene (Ala30Pro in a German family[24] and Glu46Lys in a Spanish family,[92] both in exon 3) were later identified. Another autosomal dominant locus (PARK4 at chromosome 4p15)[25] was recently revealed to be a triplication of a large region that contained alpha-synuclein.[26] Still, since mutations in alpha-synuclein have not been identified in the large number of patients with sporadic or familial PD that have been screened,[27–30] alpha-synuclein is not

TABLE 15.1
Loci Linked to PD and Their Corresponding Genes, if Known

Locus	Chr. Region	Causative Gene	Mode of Inheritance[a]	Age of Onset[b]	Lewy Bodies
PARK1	4q21–23	*α-synuclein*	AD	Middle	Present
PARK2	6q25–27	*parkin*	AR	Juvenile	Absent
PARK3	2p13		AD	Typical	Present
PARK5	4p14	*UCH-L1*	AD	Middle	Unknown
PARK6	1p35–p36	*PINK1*	AR	Early	Unknown
PARK7	1p36	*DJ-1*	AR	Early	Unknown
PARK8	12p11–q13		AD	Middle	Absent

[a] AD = autosomal dominant; AR = autosomal recessive.

[b] Juvenile (mean age of onset <20 yr); early (mean age of onset 20–45 yr); middle (mean age of onset 46–60 yr); typical (mean age of onset 60 yr).

thought to be a major risk factor in familial PD. For more details, please see the chapter on alpha-synuclein.

A mutation in the ubiquitin carboxy-terminal hydrolase-L1 (UCH-L1) gene (Ile93Met in exon 4) was identified in a single sibling pair with PD in a German pedigree.[31] Symptoms for the two individuals were similar to cases with sporadic PD, with a response to levodopa and ages of onset at 49 and 51 yr; however, confirmation of Lewy body pathology is not yet available. As analysis of hundreds of other chromosomes has not identified any mutations in UCH-L1, this finding might be due to a coincidental polymorphism present in both siblings.

Despite the lack of identification of additional mutations in UCH-L1, this gene remains, for several reasons, a compelling candidate gene for PD. First, studies have shown that this mutation causes a 50% reduction in enzymatic activity when it is expressed in *Escherichia coli*.[32] Second, UCH-L1, found throughout the brain, belongs to a family of de-ubiquitinating enzymes and therefore plays an important role in the labeling of abnormal proteins in the ubiquitin-proteasome system.[33] It is possible that reduced labeling and impaired clearance of abnormal proteins could lead to the neurodegeneration seen in PD.

Synphilin-1 has also been implicated in the pathogenesis of PD for two reasons. First, similar to alpha-synuclein, it is a substrate of the gene product of *parkin*,[34] the ubiquitin E3 ligase that plays a role in the ubiquitin-proteasome system that clears abnormal proteins from the cell. Furthermore, it has been shown to interact directly with alpha-synuclein and is found, along with *parkin* and alpha-synuclein, in Lewy bodies. Modest evidence of linkage to the region of chromosome 5q that contains synphilin-1 has also been identified in the genome scans of three independent studies.[35-37] Recently, the same mutation (R621C) was reported in two apparently sporadic patients with late-onset idiopathic PD (ages of onset, 63 and 69 yr).[38] This mutation was the only polymorphism found in 328 German familial and sporadic patients that

was not seen in 351 control subjects. Both patients with this mutation reported no family history for PD. However, for five of the six microsatellite markers genotyped in the chromosomal region around the synphilin-1 gene, both patients share the same rare alleles, suggesting this variant was inherited from a common ancestor. Functional studies of the protein have shown that mutant synphilin-1 can form cytoplasmic inclusions in transfected cells and that cells transfected with the R621C polymorphism were more susceptible to apoptosis than cells expressing wild-type synphilin-1.

The PARK3 locus on chromosome 2p13 has been linked in a subset of families displaying an autosomal dominant mode of inheritance,[39] although the responsible gene has not yet been identified. Families linking to this region have symptoms similar to typical PD, a similar mean age at onset (59 yr) and display Lewy body pathology. All of the families in the study are of German ancestry, and two of these families share a common haplotype, indicating the possibility of a founder effect. The causative mutation in these families also appears to have a penetrance less than 40%, which raises the possibility that this locus might play a role in sporadic PD as well.

The PARK8 locus has been linked to chromosome 12p11.2–q13.1 in a large Japanese family with autosomal dominant parkinsonism.[40] The clinical features are typical of idiopathic PD, including a response to levodopa; however, the mean age at onset is slightly younger, at 51 yr. A disease haplotype was found among all 15 affected individuals as well as in 8 unaffected family members, indicating a mutation with lower penetrance, which is mediated by other genetic or environmental factors. Autopsy reports of four affected individuals showed "pure nigral degeneration," though Lewy bodies were not observed.

Autosomal recessive juvenile parkinsonism (ARJP) is a distinct clinical and genetic entity within familial PD. It is characterized by typical PD features and an early age at onset (<40 yr), slow progression of the disease, sustained response to levodopa, early and often severe levodopa-induced complications (such as fluctuations and dyskinesia), hyperreflexia, and mild dystonia, mainly in the feet.[41,42] Of the genes implicated in familial PD, the largest number of mutations have been found in the *parkin* gene,[43] and mutations in this gene might account for PD in as many as 50% of familial patients with ARJP.[44]

Mutations in the *parkin* gene appear to be the only causative gene with multiple reports of point mutations and exon rearrangements, including both deletions and duplications.[45–54] One study suggests that there is no discernible difference in the clinical phenotype between patients with missense mutations, truncating point mutations, or deletions.[52] However, there have been several reports of various *parkin* mutations having different modes of inheritance. A large pedigree from South Tyrol (a region of northern Italy) has been reported with adult-onset, clinically typical tremor-dominant parkinsonism of apparently autosomal dominant inheritance.[55] Other loci implicated in autosomal dominant PD had been excluded, suggesting that *parkin* may be important in the etiology of the more frequent late-onset typical Parkinson's disease. In a different study, Lewy body pathology typical of idiopathic Parkinson's disease was found at autopsy in a proband from a kindred with a single *parkin* mutation. This data suggests that, while compound heterozygous *parkin* mutations and loss of parkin protein may lead to early-onset parkinsonism, a single mutation may confer increased susceptibility to typical Parkinson's disease.[56] In both cases, it is possible that a second mutation was simply not detected or that the disease was caused by other factors. For more details, please see the chapter on parkin.

The PARK6 locus was originally mapped to chromosome 1p35–36 in a large Sicilian family,[57] with ages of onset ranging from 32 to 48. The gene responsible was recently determined to be PINK1.[93] Preliminary studies suggest that wild-type PINK1 protects against stress-induced mitochondrial dysfunction and apoptosis, whereas the mutant form negates this effect. Some of the typical features of autosomal recessive juvenile parkinsonism, like dystonia and improvement after sleep, were not seen. However, in the late-onset cases, the clinical presentation of symptoms was identical to that of idiopathic PD. While the locus was originally identified in a single, large family, linkage has since been observed in eight additional families from throughout Europe.[58] Although there was at least one individual with an age at onset below 45 in each of these families, the mean across the families was slightly higher 942.1 yr), with the latest onset not until 68 yr. Penetrance appears to be high in these families.

The PARK7 locus was mapped to chromosome 1p37 in a consanguineous pedigree from a genetically isolated population in the southwestern region of the Netherlands.[59] The gene underlying this locus was recently determined to be DJ-1, which had been previously associated with several different biological processes. It was first identified as an oncogene,[60] but it is also involved in the fertilization process of the rat and mouse.[61,62] DJ-1 was more recently identified as a hydroperoxide-responsive protein that shifts to a more acidic isoform under oxidative conditions. Thus, it appears DJ-1 functions as a sensor for oxidative stress.[63] Two types of mutations have been found in patients with PD.[64] One is a deletion of several of its exons, which prevents the synthesis of the protein. The other is a point mutation at a highly conserved residue (Leu166Pro) that makes the protein dramatically less stable and promotes degradation through the ubiquitin-proteasome system, thereby reducing the amount of DJ-1 to low or absent levels.[64]

IDENTIFICATION OF PD SUSCEPTIBILITY GENES

The study of families segregating Mendelian forms of PD has been critical to many of the recent scientific advances in the field of PD research. However, only a small proportion of individuals with PD have mutations in these genes. Rather, most individuals with PD have a more limited family history of the disease and are far more likely to have inherited polymorphisms in disease susceptibility genes that have increased the likelihood that they would develop PD.

To identify PD susceptibility genes, researchers can employ several different study designs. One commonly utilized strategy is family-based approaches. In this instance, families with multiple members affected with PD are recruited and carefully evaluated. In these individuals and families, it may be impossible to specify a particular inheritance pattern of the disease. Rather, researchers use another type of statistical approach called *nonparametric linkage analysis* to identify genes involved in the disorder. This statistical approach does not require a specific model of disease inheritance.

All nonparametric linkage analyses are based on a concept called *identity by descent (IBD)* marker allele sharing (see Figure 15.1). If related individuals inherit the same marker allele from the same ancestor, the allele is called IBD. This method is commonly used in the analysis of siblings, both of whom are affected with PD. In this instance, if the marker being tested is in close physical proximity to a gene influencing PD, then siblings who are both affected with the disease would be expected to share more marker alleles IBD. Conversely, siblings who are discordant for disease would be expected to inherit fewer marker alleles IBD near the gene influencing the trait. More recent extensions of this approach allow the inclusion of more extended families in the genetic analysis.

In linkage studies, the logarithm of the odds (LOD) for linkage is a commonly used statistical term employed

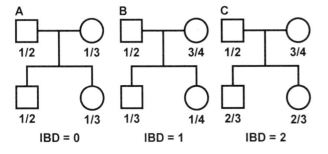

FIGURE 15.1 When relatives share alleles identical by descent (IBD), this means that they have inherited the same allele from a common ancestor. Panel 1B shows a pedigree in which both siblings have inherited one allele from a common ancestor (allele 1 from their father) and are said to have one allele IBD. The siblings in the pedigree in panel 1C have inherited the same two alleles from each of their parents (allele 2 from their father and allele 3 from their mother). Thus, the siblings have inherited two alleles IBD. The pedigree in Panel 1A illustrates two siblings who have inherited no alleles IBD. While both siblings have an allele 1, this allele is not IBD. This can be determined by examining from which parent each offspring has inherited allele 1. The individual with genotype 1/2 has received the 2 allele from his father; therefore, the 1 allele must have come from his mother. In contrast, the other sibling has genotype 1/3. She must have inherited the 3 allele from her mother and the 1 allele from her father. Thus, the son has inherited allele 1 from his mother and the daughter has inherited allele 1 from her father. Therefore, these two siblings do not have any alleles inherited IBD from a common ancestor for this marker and are said to have an IBD of 0. By estimating the number of alleles inherited IBD by affected relatives, it is possible to test for linkage to a chromosomal region. Under the null hypothesis, sibling pairs would be expected, on average, to share one allele IBD. By testing markers throughout the genome, a chromosomal region can be identified in which sibling pairs with PD from many different families consistently share more alleles IBD. This provides evidence that the marker and the putative disease gene do not segregate independently and a gene predisposing to PD may be somewhere in this chromosomal region.

to describe the evidence for linkage. While the exact calculation of a LOD score will vary depending on the assumptions made when linkage analyses are being performed, some basic properties of the LOD score are consistent across different types of analyses. First, a higher, positive LOD score is consistent with greater evidence in favor of linkage between a marker(s) being tested and a disease locus. Second, LOD scores greater than 3–3.6 are typically considered statistically significant.

Several research groups have used this experimental design to identify chromosomal regions that may harbor PD susceptibility genes. In the first report of the largest study collected, a genome screen was performed in a sample of 182 families, consisting of 203 affected sibling pairs.[35] Typical of many studies of complex disease, multiple disease definitions were employed. In Model I, only

those individuals meeting a more stringent diagnosis of Verified PD (96 sibling pairs from 90 families) were included in the analysis, while in Model II, all examined affected individuals, regardless of their final diagnostic classification (170 sibling pairs from 160 families), were employed in the genetic analyses. Under Model I, the highest LOD scores were observed on the X chromosome (LOD = 2.1) and on chromosome 2 (LOD = 1.9). Analyses performed utilizing all available sibling pairs (Model II) found even greater evidence of linkage to the X chromosome (LOD = 2.7) and to chromosome 2 (LOD = 2.5). Subsequent analyses were performed in an enlarged sample of 754 affected individuals, comprising 425 sibling pairs.[65] The same two disease models were employed in the analyses and the more stringent Model I resulted in a maximum LOD score of 3.4 on chromosome 2 and the broader Model II disease model yielded a LOD score of 3.1 on the X chromosome.

To improve the ability to identify the disease gene on chromosome 2, additional analyses were performed in only those families demonstrating the strongest genetic contribution to PD who still met criteria for Verified PD.[66] This subset was defined to include only those families having at least 4 members reported to have PD ($n = 40$) or a family with a sibling pair with PD who also had a parent diagnosed with disease ($n = 49$). The overlap between these two groups was substantial ($n = 24$). A total of 65 families with a strong history of PD were identified for subsequent analyses. Careful review of these families suggested that most were consistent with an autosomal dominant model of disease inheritance. Therefore, linkage analyses were performed employing an autosomal dominant model of disease with 80% penetrance, a disease allele frequency of 0.005, and a 3% phenocopy rate. Only those individuals meeting the stricter criteria of Verified PD were assumed affected with disease. This model yielded a maximum multipoint parametric LOD score of 5.1 near the markers D2S396, D2S206, and D2S338 (Figure 15.2).

Three other studies have reported linkage results for familial Parkinson's disease. The GenePD study,[67] analyzing data from 113 sibling pairs, has reported LOD scores between 0.9 and 1.3 on chromosomes 1, 9, 10, and 15. A second study, analyzing 174 extended pedigrees, including 378 affected individuals, has reported LOD scores between 1.5 and 2.5 on chromosomes 5, 8, 9, 14, 17, and X for late-onset families.[36] The study of Scott et al.[36] had its most significant linkage finding on chromosome 17, near the tau gene. The strength of the linkage finding was increased when analyses were limited to a subset of families in which at least one individual in the kindred was not responsive to levodopa treatment. Since a positive response to this dopamine precursor is very common among individuals with PD, they considered this potential phenotypic heterogeneity

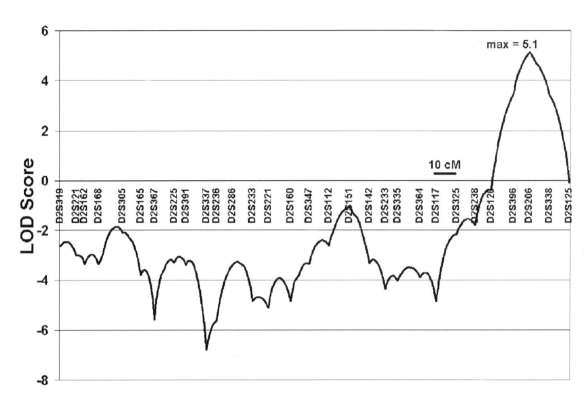

FIGURE 15.2 Results of multipoint linkage analysis performed in a sample of 65 multiplex PD families having a strong family history of PD. An autosomal dominant disease model was employed. (From Pankratz, N. et al., *Am. J. Hum. Genet.*, 72(4): 1053, 2003. With permission.)

to be indicative of genotypic heterogeneity. The most notable improvements to LOD scores were on chromosomes 3, 9, and 17.

The deCODE study is unique in that it has analyzed a relatively isolated and homogeneous population.[37] Using 117 affected individuals from Iceland, they report a LOD score of 4.9 at a region on chromosome 1p32 that they have named PARK10. No evidence of linkage near the centromere of chromosome 1 has been reported in the other three PD genome-wide studies, indicating that perhaps this putative gene is only at high frequency in the Icelandic population. This study has also limited evidence of linkage to chromosomes 5, 7, 13, 14, and X.

Three of these studies appear to report linkage to the same region of Xq21–25.[35–37,67] Interestingly, sex-linked dystonia parkinsonism (XDP), which has been reported at high incidence in Panay, Philippines, has been linked to Xq13.1.[68] It is an adult-onset, highly penetrant, X-linked disorder that primarily afflicts males (male-to-female ratio of 123:1). It is a severe, progressive disorder that typically onsets in the third or fourth decade with dystonic movements. Commonly, about a decade after disease onset, the dystonia coexists or is replaced by parkinsonism. Neuropathology reveals pronounced atrophy of the caudate and putamen, mostly in the cases with long standing illness.[68] The gene has been mapped to a <350 kb region in the DXS7117–DXS559 region.[69]

FROM LINKAGE TO THE IDENTIFICATION OF PD SUSCEPTIBILITY GENES

In the past few years, several family-based studies seeking to identify PD susceptibility genes have successfully recruited families with multiple members diagnosed with PD, which have then led to the detection of linkage to multiple chromosomal regions. This is a necessary first step in the identification of PD susceptibility genes. However, the broad linkage regions identified to date in most studies contain hundreds or thousands of genes. Therefore, it is essential that strategies be developed that can lead to the rapid identification of the susceptibility gene within this larger chromosomal interval.

A variety of different approaches have been proposed that may move researchers rapidly toward a PD susceptibility gene. Most involve the use of single nucleotide polymorphisms (SNPs), which are a single base pair change in the DNA sequence that can be found in either the coding or noncoding region of the DNA. SNPs are estimated to occur, on average, every 300 base pairs, suggesting that there are over 10 million SNPs in the human genome. Using SNPs, the evidence for linkage disequilibrium (LD; Figure 15.3) in various portions of the genome has been characterized.[70–73] Findings suggest that LD does not decay gradually, as was originally thought; rather, high levels of LD extend over blocks of variable physical

A. Linkage Equilibrium

	Allele 1 (frequency = 0.5)	Allele 2 (frequency = 0.5)	
Allele A (frequency = 0.5)	**Haplotype A1** (frequency = 0.25)	**Haplotype A2** (frequency = 0.25)	frequency = 0.5
Allele B (frequency = 0.5)	**Haplotype B1** (frequency = 0.25)	**Haplotype B2** (frequency = 0.25)	frequency = 0.5
	frequency = 0.5	frequency = 0.5	sum = 1.0

B. Linkage Disequilibrium

	Allele 1 (frequency = 0.5)	Allele 2 (frequency = 0.5)	
Allele A (frequency = 0.5)	**Haplotype A1** (frequency = 0.45)	**Haplotype A2** (frequency = 0.05)	frequency = 0.5
Allele B (frequency = 0.5)	**Haplotype B1** (frequency = 0.05)	**Haplotype B2** (frequency = 0.45)	frequency = 0.5
	frequency = 0.5	frequency = 0.5	sum = 1.0

FIGURE 15.3 Two loci are shown that are located physically near each other on the chromosome. At one locus, there are two alleles, termed A and B. At the other locus, there are also two alleles, termed 1 and 2. At each locus, the two alleles are equally frequent. *A. Linkage Equilibrium.* In this instance, the expected frequency of each of the four possible haplotypes is simply the product of the frequency of each of the two alleles composing the haplotype. For example, if the two loci are at equilibrium, a haplotype consisting of allele A at the first locus and allele 1 at the second locus (termed haplotype A1) would be expected to occur at a frequency of $1/2 \times 1/2 = 1/4$. Similarly, the frequency of the other three remaining haplotypes (A2, B1, and B2) are also 1/4. *B. Linkage Disequilibrium.* If two loci are not in equilibrium, then the frequency of the observed haplotypes is not equal to the product of the frequency of each of the alleles making up the haplotype. In the example shown, two haplotypes are more frequent then would be expected (A1 and B2), and two haplotypes are less frequent than would be expected (A2 and B1). These results indicate that these two loci are in linkage disequilibrium.

size, which are interrupted by short regions of extensive recombination ('hot spots').

Regions of high LD are typically shorter in African and African-American samples, with over half of the blocks estimated to be 22 kb or greater. In European samples, the blocks are larger, with over half the genome in blocks of 44 kb or greater.[74] Extensive study of selected chromosomal regions has consistently found that the boundaries of these haplotype blocks are highly correlated across populations. Within the blocks, a small number of common haplotypes, typically 3 to 5, can be identified that represent more than 90% of chromosomes.[74] Thus, it is essential within any candidate chromosomal region to rigorously delineate the haplotype block structure and identify the common SNP haplotypes. The characterization of SNP haplotype blocks throughout the genome is being accomplished rapidly and efficiently through the Haplotype Mapping project (HapMap).

Several approaches have been proposed to localize and isolate disease susceptibility genes through the use of LD and SNPs. Analyses can employ trios, consisting of an affected individual and their parents. In this type of analysis, a statistical test is performed to determine whether parents who are heterozygous at a marker (i.e., SNP) preferentially transmit a particular marker allele to their affected offspring.[75] This type of analysis is called the *transmission disequilibrium test (TDT)* and has been widely employed as a means to test the role of a SNP in a candidate gene. More recently, this method has been expanded to test for transmission distortion in multiple generations.[76] This test is termed the *pedigree disequilibrium test (PDT)*.

Since PD and other neurodegenerative disorders often onset later in life, after the parents of the affected individual are deceased, the TDT and PDT as initially proposed may have limited utility in genetic studies. Therefore, additional study designs have been developed that more efficiently exploit the family members most likely to be available for participation in these genetic studies of late onset disease. One such approach is the *discordant sib test*. While proposed independently by several different groups, the basic design premise is similar.[77–79] An affected and an unaffected sibling are required. A marker (or SNP) is genotyped in the discordant siblings, and a test is performed to determine if there has been preferential segregation of one marker allele to the affected siblings and a different marker allele to the unaffected siblings.

A common theme in all of these proposed association tests, commonly referred to as family-based tests of association, is the presence of linkage disequilibrium between the SNP or marker being tested and the disease susceptibility locus. It is now commonly appreciated that it may be difficult to identify a particular polymorphism as causing disease susceptibility. Rather, it may be that combinations of SNP alleles (i.e., haplotypes) may represent susceptibility alleles.

ASSOCIATION STUDIES WITH PD

Instead of prioritizing genes based on a linkage finding, a commonly employed experimental design to identify genes contributing to a disease is through the use of candidate gene analysis, which seeks to test the association between a particular allele of a candidate gene and disease risk. A candidate gene typically is chosen because it is suspected to play a role in the disorder, either because researchers have some information about the gene's function that might be related to the disease or because the gene lies in a DNA region that has already been linked to the disorder through linkage studies.

Many candidate gene studies use population-based association methods—that is, methods that compare the genes of groups of people. For example, in PD research, such analyses would involve two samples: a group of PD patients and a control group of individuals without PD. Ideally, the two groups would be matched with respect to numerous factors (e.g., age and ethnicity) so that they differ only in disease status. The investigators would then compare the frequencies of various alleles of a marker (e.g., an SNP) within or near the candidate gene. Evidence of differences in allele frequencies between the two groups would then typically be interpreted as evidence that the candidate gene contributes to PD susceptibility.

Several research groups have sought to identify PD susceptibility genes through the completion of population based case-control association studies of candidate genes, identified on the basis of their involvement in the dopamine pathway. Known genes investigated via association studies include the monoamine oxidase B (MAOB), dopamine D_2 receptor (DRD2), CYP2D6, CYP1A1, N-acetyltransferase 2 (NAT2), DAT1 and glutathione s-transferase M1 (GSTM1) genes.[80–82] Unfortunately, the few results from studies finding significant association between a candidate gene and PD have failed to replicate in other samples. This is most likely due to the susceptibility of association studies to yield spurious positive results due to population phenomena other than a causative relation (or close physical proximity) between the polymorphism being tested and a disease-causing mutation.[83] The most commonly cited phenomenon of this sort is population stratification, in which the population from which the sample is drawn is composed of two or more subgroups whose members tend to select mating partners within their own group. Frequency differences between subgroups for the alleles of the candidate polymorphism and the phenotype being investigated (e.g., Parkinson's disease) can easily produce false positive results for classical association tests, even when no trait-related gene exists at or near the candidate polymorphism tested.[83] Other possible explanations for differing results are poor sampling and the failure to use a correction for multiple tests.

There has been one association with PD that has eventually led to an identified mutation. Xu et al.[84] reported that an increase in homozygosity of a variant in the Nurr1 gene was associated with PD. More recently, an additional study in an independent sample suggested that heterozygosity for this polymorphism conferred an increased risk for PD.[85] Subsequently, Le et al.[86] described two distinct mutations in exon 1 in ten individuals with familial PD. The age at onset of disease and clinical features of these ten individuals did not differ from those of individuals with typical Parkinson's disease. Nurr1 mRNA levels were significantly decreased in transfected cell lines with the mutation, compared with those without the mutation, as well as in the lymphocytes of the affected individuals. Furthermore, expression levels of the dopamine biosynthesis enzyme tyrosine hydroxylase were also markedly decreased in the transfected cells with the mutations compared to the wild-type. These data suggest dopaminergic dysfunction can result from mutations in Nurr1.

Mitochondrial dysfunction, particularly with regard to complex I of the electron transport chain, has been implicated in the pathogenesis of Parkinson's disease. Since a majority of mitochondrial DNA is dedicated to the NADH complex I enzyme, van der Walt et al.[87] tested the hypothesis that mtDNA variation contributes to PD expression. Ten single-nucleotide polymorphisms (SNPs) that define the European mtDNA haplogroups were genotyped for white individuals with PD and for their matched control subjects. They found that individuals with either haplogroup J or K had a significantly lower risk of PD as compared with individuals carrying the most common haplogroup, H. These results suggest that variation in complex I proteins is an important risk factor in PD susceptibility among white individuals.

CONCLUSION

Studies have now consistently suggested that a family history of PD is an important risk factor for disease. This, in turn, led to a focus on genetic studies designed to identify the genes that increase the risk for PD. Great advances have now been made in the discovery of genes causing PD. In addition, numerous studies have now begun to detect linkage to PD susceptibility genes. These efforts will likely identify a number of genes that either increase or decrease the risk for PD.

It is important to recall that, while there are important genetic contributions to PD, other nongenetic factors also contribute to disease risk. Some factors, such as head trauma and rural living,[88–89] as well as lack of smoking or caffeine intake,[90–91] have already been shown to increase the risk for PD. It is likely that genes and environmental risk factors together determine which individuals will develop PD. Unfortunately, with only a few PD genes identified, studies have not yet been able to focus their efforts to determine which environmental factors most strongly modulate particular gene effects. This work will likely continue well into the current century.

The reward in understanding the genetic and environmental interplay at work in PD risk is enormous. Through a thorough characterization of these factors, it will be possible to determine an individual's genetic profile for PD and also evaluate environmental factors. From these data, high risk individuals might be identified and lifestyle alterations suggested which might decrease their risk for PD. Alternatively, individual specific therapies might be developed based on an individual's genetic profile which might delay PD symptom onset.

ACKNOWLEDGMENTS

This work was supported by R01 NS37167.

REFERENCES

1. Leroux, P. D., Contribution a l'etude des causes de laparalysie agitante, in These de Paris, 1880. Cited by Scarpalezos, 1948.
2. Scarpalezos, S., Sur la notion d'heredite similaire dans la maladie de Parkinson, *Rev Neurol*, 80, 184, 1948.
3. Gowers, W. R., A manual of the diseases of the nervous system, 2nd ed., Blakiston, Philadelphia, 636, 1902.
4. Martin, W. E., Young, W. I., and Anderson, V. E., Parkinson's disease: a genetic study, *Brain*, 96(3), 495, 1973.
5. Barbeau, A., Roy, M., and Boyer, L., Genetic studies in Parkinson's disease, *Adv. Neurol.*, 40, 333, 1984.
6. Alonso, M. E. et al., Parkinson's disease: a genetic study, *Can. J. Neurol. Sci.*, 13(3), 248, 1986.
7. Duvoisin, R. C. and Johnson, W. G., Hereditary Lewy-body parkinsonism and evidence for a genetic etiology of Parkinson's disease, *Brain. Pathol.*, 2(4), 309, 1992.
8. Tanner, C. M. et al., Parkinson's disease in twins: an etiologic study, *JAMA.*, 281, 341, 1999.
9. Dickson, D. et al., Pathology of PD in monozygotic twins with a 20-year discordance interval, *Neurology*, 56(7), 981, 2001.
10. Piccini, P. et al., Dopaminergic function in familial Parkinson's disease: a clinical and 18F-dopa positron emission tomography study, *Ann. Neurol.*, 41(2), 222, 1997.
11. Burn, D. J. et al., Parkinson's disease in twins studied with 18F-dopa and positron emission tomography, *Neurology*, 42(10), 1894, 1992.
12. Kuopio, A. et al., Familial occurrence of Parkinson's disease in a community-based case-control study, *Parkinsonism. Relat. Disord.*, 7(4), 297, 2001.
13. Payami, H. et al., Increased risk of Parkinson's disease in parents and siblings of patients, *Ann. Neurol.*, 36(4), 659, 1994.
14. Autere, J. M. et al., Familial aggregation of Parkinson's disease in a Finnish population, *J. Neurol. Neurosurg. Psychiatry*, 69(1), 107, 2000.
15. Sveinbjornsdottir, S. et al., Familial aggregation of Parkinson's disease in Iceland, *N. Engl. J. Med.*, 343(24), 1765, 2000.
16. Preux, P. M. et al., Parkinson's disease and environmental factors: matched case-control study in the Limousin region, France, *Neuroepidemiology*, 19(6), 333, 2000.
17. Elbaz, A. et al., Familial aggregation of Parkinson's disease: a population-based case-control study in Europe; EUROPARKINSON Study Group, *Neurology*, 52(9), 1876, 1999.
18. Uitti, R. J. et al., Familial Parkinson's disease: a case-control study of families, *Can. J. Neurol. Sci.*, 24(2), 127, 1997.
19. De Michele, G. et al., A genetic study of Parkinson's disease, *J. Neural. Transm. Suppl.*, 45, 21, 1995.
20. Vieregge, P. et al., (Multifactorial etiology of idiopathic Parkinson disease: a case-control study), *Nervenarzt*, 65(6), 390, 1994.
21. Polymeropoulos, M. H. et al., Mutation in the alpha-synuclein gene identified in families with Parkinson's disease, *Science*, 276(5321), 2045, 1997.
22. Bostantjopoulou, S. et al., Clinical features of parkinsonian patients with the alpha-synuclein (G209A) mutation, *Mov. Disord.*, 16(6), 1007, 2001.
23. Gasser, T., Genetics of Parkinson's disease, *J. Neurol.*, 248(10), 833, 2001.
24. Kruger, R. et al., Ala30Pro mutation in the gene encoding alpha-synuclein in Parkinson's disease, *Nat. Genet.*, 18(2), 106, 1998.
25. Farrer, M. et al., Low frequency of alpha-synuclein mutations in familial Parkinson's disease, *Ann. Neurol.*, 43(3), 394, 1998.
26. Singleton, A. et al., SNCA locus triplication causes Parkinson disease, *Science*, in press.
27. Pastor, P. et al., Analysis of the coding and the 5' flanking regions of the alpha-synuclein gene in patients with Parkinson's disease, *Mov. Disord.*, 16(6), 1115, 2001.

28. Chan, P. et al., Absence of mutations in the coding region of the alpha-synuclein gene in pathologically proven Parkinson's disease, *Neurology*, 50(4), 1136, 1998.

29. Farrer, M. et al., Low frequency of alpha-synuclein mutations in familial Parkinson's disease, *Ann. Neurol.*, 43(3), 394, 1998.

30. Vaughan, J. R. et al., Sequencing of the alpha-synuclein gene in a large series of cases of familial Parkinson's disease fails to reveal any further mutations; The European Consortium on Genetic Susceptibility in Parkinson's Disease (GSPD), *Hum. Mol. Genet.*, 7(4), 751, 1998.

31. Leroy, E. et al., The ubiquitin pathway in Parkinson's disease, *Nature*, 395(6701), 451, 1998.

32. Wintermeyer, P. et al., Mutation analysis and association studies of the UCHL1 gene in German Parkinson's disease patients, *Neuroreport.*, 11(10), 2079, 2000.

33. Pickart, C. M., Ubiquitin in chains, *Trends. Biochem. Sci.*, 25(11), 544, 2000.

34. Chung, K. K. et al., Parkin ubiquitinates the alpha-synuclein-interacting protein, synphilin-1: implications for Lewy-body formation in Parkinson disease. *Nat. Med.*, 7(10), 1144, 2001.

35. Pankratz, N. et al., Genome screen to identify susceptibility genes for Parkinson disease in a sample without parkin mutations, *Am. J. Hum. Genet.*, 71(1), 124, 2002.

36. Scott, W. K. et al., Complete genomic screen in Parkinson disease: evidence for multiple genes, *JAMA*, 286(18), 2239, 2001.

37. Hicks, A. A. et al., A susceptibility gene for late-onset idiopathic Parkinson's disease, *Ann. Neurol.*, 52(5), 549, 2002.

38. Marx, F. P. et al., Identification and functional characterization of a novel R621C mutation in the synphilin-1 gene in Parkinson's disease. *Hum. Mol. Genet.*, 12(11), 1223, 2003.

39. Gasser, T. et al., A susceptibility locus for Parkinson's disease maps to chromosome 2p13, *Nat. Genet.*, 18, 262, 1998.

40. Funayama, M. et al., A new locus for Parkinson's disease (PARK8) maps to chromosome 12p11.2-q13.1, *Ann. Neurol.*, 51(3), 296, 2002.

41. Yamamura, Y. et al., Paralysis agitans of early-onset with marked diurnal fluctuation of symptoms, *Neurology*, 23, 239, 1973.

42. Ishikawa, A. and Tsuji, S., Clinical analysis of 17 patients in 12 Japanese families with autosomal-recessive type juvenile Parkinsonism, *Neurology*, 47, 160, 1996.

43. Kitada, T. et al., Mutations in the parkin gene cause autosomal recessive juvenile parkinsonism, *Nature*, 392, 605, 1998.

44. Lucking, C. B. et al., Association between early-onset Parkinson's disease and mutations in the parkin gene; French Parkinson's Disease Genetics Study Group, *N. Engl. J. Med.*, 342(21), 1560, 2000.

45. Foroud, T. et al., Heterozygosity for a mutation in the parkin gene leads to later onset Parkinson disease, *Neurology*, 60(5), 796, 2003.

46. Oliveira, S. A. et al., Parkin mutations and susceptibility alleles in late-onset Parkinson's disease, *Ann. Neurol.*, 53(5), 624, 2003.

47. Tan, L. C. et al., Marked variation in clinical presentation and age of onset in a family with a heterozygous parkin mutation, Mov. Disord., 18(7), 758, 2003.

48. Nichols, W. C. et al., Linkage stratification and mutation analysis at the parkin locus identifies mutation positive, Parkinson disease families, *J. Med. Genet.*, 39, 489, 2002.

49. Kann, M. et al., Role of parkin mutations in 111 community-based patients with early-onset parkinsonism, *Ann. Neurol.*, 51(5), 621, 2002.

50. West, A. et al., Complex relationship between Parkin mutations and Parkinson disease, *Am. J. Med. Genet.*, 114(5), 584, 2002.

51. Hedrich, K. et al., Evaluation of 50 probands with early-onset Parkinson's disease for Parkin mutations, *Neurology*, 58(8), 1239, 2002.

52. Lucking, C. B. et al., Association between early-onset Parkinson's disease and mutations in the parkin gene; French Parkinson's Disease Genetics Study Group, *N. Engl. J. Med.*, 342(21), 1560, 2000.

53. Abbas, N. et al., A wide variety of mutations in the parkin gene are responsible for autosomal recessive parkinsonism in Europe; French Parkinson's Disease Genetics Study Group and the European Consortium on Genetic Susceptibility in Parkinson's Disease, *Hum. Mol. Genet.*, 8(4), 567, 1999.

54. Lucking, C. B. et al., Homozygous deletions in parkin gene in European and North African families with autosomal recessive juvenile parkinsonism; The European Consortium on Genetic Susceptibility in Parkinson's Disease and the French Parkinson's Disease Genetics Study Group, *Lancet*, 352(9137), 1355, 1998.

55. Klein, C. et al., Parkin deletions in a family with adult-onset, tremor-dominant parkinsonism: expanding the phenotype, *Ann. Neurol.*, 48(1), 65, 2000.

56. Farrer, M. et al., Lewy bodies and parkinsonism in families with parkin mutations, *Ann. Neurol.*, 50(3), 293, 2001.

57. Valente, E. M. et al., Localization of a novel locus for autosomal recessive early-onset parkinsonism, PARK6, on chromosome 1p35-p36, *Am. J. Hum. Genet.*, 68, 895, 2001.

58. Valente, E. M. et al., PARK6-linked parkinsonism occurs in several European families, *Ann. Neurol.*, 51(1), 14, 2002.

59. van Duijn, C. M. et al., PARK7, a novel locus for autosomal recessive early-onset parkinsonism, on chromosome 1p36, *Am. J. Hum. Genet.*, 69, 629, 2001.

60. Nagakubo, D. et al., DJ-1, a novel oncogene which transforms mouse NIH3T3 cells in cooperation with ras, Biochem. *Biophys. Res. Commun.*, 231(2), 509, 1997.

61. Wagenfeld, A., Gromoll, J., and Cooper, T.G., Molecular cloning and expression of rat contraception associated protein 1 (CAP1), a protein putatively involved in fertilization, Biochem. *Biophys. Res. Commun.*, 251(2), 545, 1998.

62. Okada, M. et al., DJ-1, a target protein for an endocrine disrupter, participates in the fertilization in mice, *Biol. Pharm. Bull.*, 25(7), 853, 2002.

63. Mitsumoto, A. and Nakagawa, Y., DJ-1 is an indicator for endogenous reactive oxygen species elicited by endotoxin, *Free. Radic. Res.*, 35(6), 885, 2001.

64. Bonifati, V. et al., Mutations in the DJ-1 gene associated with autosomal recessive early-onset parkinsonism, *Science*, 299(5604), 256, 2003.

65. Pankratz, N. et al., Genome-Wide Linkage Analysis and Evidence of Gene-by-Gene Interactions in a Sample of 362 Multiplex Parkinson Disease Families, *Hum. Mol. Genet.*, in press.

66. Pankratz, N. et al., Significant linkage of Parkinson disease to chromosome 2q36-37, *Am. J. Hum. Genet.*, 72(4), 1053, 2003.

67. DeStefano, A.L. et al., Genome-wide scan for Parkinson's disease: the GenePD Study, *Neurology*, 57(6), 1124, 2001.

68. Lee, L. V. et al., Sex linked recessive dystonia parkinsonism of Panay, Philippines (XDP), *Mol. Pathol.*, 54(6), 362, 2001.

69. Nemeth, A. H. et al., Refined linkage disequilibrium and physical mapping of the gene locus for X-linked dystonia-parkinsonism (DYT3), *Genomics.*, 60(3), 320, 1999.

70. Rioux, J. D. et al., Genetic variation in the 5q31 cytokine gene cluster confers susceptibility to Crohn disease, *Nat. Genet.*, 29, 223, 2001.

71. Daly, M. J. et al., High-resolution haplotype structure in the human genome, *Nat. Genet.*, 29, 229, 2001.

72. Johnson, G. C. et al., Haplotype tagging for the identification of common disease genes, *Nat. Genet.*, 29, 233, 2001.

73. Patil, N. et al., Blocks of limited haplotype diversity revealed by high-resolution scanning of human chromosome 21, *Science*, 294, 1719, 2001.

74. Gabriel, S. B. et al., The structure of haplotype blocks in the human genome, *Science*, 296, 2225, 2002.

75. Spielman, R. S., McGinnis, R. E., and Ewens, W. J., Transmission test for linkage disequilibrium: the insulin gene region and insulin-dependent diabetes mellitus (IDDM), *Am. J. Hum. Genet.*, 52(3), 506, 1993.

76. Martin, E. R. et al., A test for linkage and association in general pedigrees: the pedigree disequilibrium test, Am. *J. Hum. Genet.*, 67(1), 146, 2000.

77. Boehnke, M. and Langefeld, C. D., Genetic association mapping based on discordant sib pairs: the discordant-alleles test, *Am. J. Hum. Genet.*, 62(4), 950, 1998.

78. Horvath, S. and Laird, N. M., A discordant-sibship test for disequilibrium and linkage: no need for parental data, *Am. J. Hum. Genet.*, 63(6), 1886, 1998.

79. Curtis, D., Use of siblings as controls in case-control association studies, *Ann. Hum. Genet.*, 61 (Pt. 4), 319, 1997.

80. Mellick, G. D. et al., Variations in the monoamine oxidase B (MAOB) gene are associated with Parkinson's disease, *Mov. Disord.*, 14(2), 219, 1999.

81. Grevle, L. et al., Allelic association between the DRD2 TaqI A polymorphism and Parkinson's disease, *Mov. Disord.*, 15(6), 1070, 2000.

82. Nicholl, D. J. et al., A study of five candidate genes in Parkinson's disease and related neurodegenerative disorders; European Study Group on Atypical Parkinsonism, *Neurology*, 53(7), 1415, 1999.

83. Spielman, R. S. and Ewens, W. J., The TDT and other family-based tests for linkage disequilibrium and association, *Am. J. Hum. Genet.*, 59(5), 983, 1996.

84. Xu, P. Y. et al., Association of homozygous 7048G7049 variant in the intron six of Nurr1 gene with Parkinson's disease, *Neurology*, 58(6), 881, 2002.

85. Zheng, K., Heydari, B., and Simon, D. K., A common NURR1 polymorphism associated with Parkinson disease and diffuse Lewy body disease, *Arch. Neurol.*, 60(5), 722, 2003.

86. Le, W. D. et al., Mutations in NR4A2 associated with familial Parkinson disease, *Nat. Genet.*, 33(1), 85, 2003.

87. van der Walt, J. M. et al., Mitochondrial polymorphisms significantly reduce the risk of Parkinson disease, *Am. J. Hum. Genet.*, 72(4), 804, 2003.

88. Seidler, A. et al., Possible environmental, occupational, and other etiologic factors for Parkinson's disease: a case-control study in Germany, *Neurology*, 46(5), 1275, 1996.

89. Koller, W. et al., Environmental risk factors in Parkinson's disease, *Neurology.*, 40(8), 1218, 1990.

90. Louis, E. D. et al., Parkinsonian signs in older people: Prevalence and associations with smoking and coffee, *Neurology*, 61(1), 24, 2003.

91. Ross, G. W. et al., Association of coffee and caffeine intake with the risk of Parkinson disease, *JAMA*, 283(20), 2674, 2000.

92. Zarranz, J. J. et al., The new mutation, E46K, of alpha-synuclein causes Parkinson and Lewy body dementia, *Ann. Neurol.*, 55(2), 164, 2004.

93. Valente, E. M. et al., Hereditary early-onset Parkinson's disease caused by mutations in PINK1, *Science*, 304(5674), 1158, 2004.

16 Classical Motor Features of Parkinson's Disease

Kelvin L. Chou

Department of Clinical Neurosciences, Brown University School of Medicine, and Division of Neurology, Memorial Hospital of Rhode Island

Howard I. Hurtig

Parkinson's Disease and Movement Disorders Center, University of Pennsylvania Health System, Pennsylvania Hospital

CONTENTS

INTRODUCTION

James Parkinson's classic 1817 monograph, *An Essay on the Shaking Palsy,* marks the first detailed account of the disease that now bears his name.[1] This work was remarkable, because it was based solely on clinical interviews with three patients and street observations of three others, and his description of a clinical syndrome characterized by tremor, apparent weakness (due to a combination of bradykinesia and rigidity), and an abnormal gait remains unrivaled in its clarity and accuracy. Although our understanding of Parkinson's disease (PD) and its many variants has grown immensely since that time, no proven diagnostic procedures or laboratory tests for the diagnosis of PD have emerged, and the diagnosis of PD is still based on the clinical interpretation of signs and symptoms, obtained through a detailed history and thorough neurologic examination. If and when neuroprotection moves from optimis-

tic concept to therapeutic reality and is shown to prevent vulnerable neurons from degenerating, an early diagnosis of PD will be essential for initiating such life-saving treatment. Accuracy of diagnosis depends on a clear understanding of the varied clinical manifestations of PD, particularly the classical features that James Parkinson recorded in such lucid detail almost 200 years ago. Although it will be difficult for us to surpass his careful description, we have the advantage of time and a cumulative body of knowledge about the underlying pathophysiology of PD that Parkinson could only speculate about.

CLASSICAL MOTOR FEATURES AND DEFINITIONS

The classical or cardinal parkinsonian motor features are bradykinesia, rigidity, and rest tremor. A fourth feature, postural instability, is often mentioned, although it tends to occur in the later stages of the disease entity known in the modern era as idiopathic parkinsonism or PD. The combination of any of these classical features constitutes a cluster of signs and symptoms that has been termed "parkinsonism" or a "parkinsonian syndrome," which has many possible causes, including less common neurodegenerative disorders [progressive supranuclear palsy (PSP), multiple system atrophy (MSA), dementia with Lewy bodies (DLB), and cortical-basal ganglionic degeneration (CBGD)], adverse effects of therapeutic or recreational drugs (neuroleptics being the most common offender), environmental toxins (carbon monoxide, manganese), structural lesions in the basal ganglia (tumor, infarct or hemorrhage), and hereditary disorders [Huntington's disease (HD), Wilson's disease (WD), and neurodegeneration with brain iron accumulation (NBIA)]. A more comprehensive list of disorders that can present with parkinsonism is listed in Table 16.1. With the exception of psychogenic parkinsonism, all these disorders are believed to manifest parkinsonian features because of a primary disruption of the nigrostriatal dopaminergic pathway.

TABLE 16.1
Differential Diagnosis of Parkinsonism

Idiopathic
- Parkinson's disease

Symptomatic
- Drug-induced (neuroleptics, presynaptic dopamine blockers, certain anti-emetics)
- Hydrocephalus
- Metabolic (hepatocerebral degeneration, parathyroid disorders)
- Neoplasm
- Postencephalitic parkinsonism
- Post-traumatic parkinsonism
- Toxin (carbon monoxide, manganese, MPTP)
- Vascular parkinsonism

Neurodegenerative
- Alzheimer's disease
- Cortical-basal ganglionic degeneration
- Dementia with Lewy bodies
- Multiple System Atrophy (olivopontocerebellar atrophy, Shy-Drager syndrome, striatonigral degeneration)
- Parkinsonism-dementia-ALS complex of Guam
- Progressive supranuclear palsy

Hereditary Disorders
- Dentatorubral-pallidoluysian atrophy
- Huntington's disease
- Neurodegeneration with brain iron accumulation (Hallervorden-Spatz)
- Wilson's disease
- Spinocerebellar ataxias

Other
- Psychogenic

PD is by far the most common cause of parkinsonism, constituting approximately 75% of cases seen at large movement disorders centers.[2] It is a distinct clinical entity, characterized pathologically by loss of pigmented neurons in the substantia nigra (SN), reactive gliosis, and the presence of intracytoplasmic neuronal inclusions known as Lewy bodies.[3] The accepted criteria for a clinical diagnosis of PD include the presence of two out of the three classical motor features, unilateral onset (especially of rest tremor), a strong clinical response to levodopa, and the absence of features suggestive of any of the other disorders listed in Table 16.1.[4-8] Early literature suggested that the clinical diagnosis of PD was accurate only 75% of the time as compared with neuropathological diagnosis on autopsy, the accepted gold standard.[6,9] However, diagnostic accuracy increases to at least 90% when patients are followed long term by the same observers,[10,11] likely because some of the more distinctive clinical features of alternative diagnoses, such as eye movement abnormalities in PSP, autonomic insufficiency in MSA, or failure of levodopa therapy, do not manifest until later in the course of the disease. Hughes et al. looked at the diagnostic value of various parkinsonian features in predicting a final neuropathological diagnosis of PD.[12] The presence of two classical motor features resulted in a high sensitivity (99%) but extremely low specificity (8%). The specificity increased when other characteristics, such as asymmetrical signs or tremor-predominant disease, were included, but at the expense of sensitivity. Thus, a perfect set of clinical diagnostic criteria may be impossible to construct cause of the heterogeneity of PD and the disorders that imic it. An accurate biomarker or precise imaging techque could resolve this dilemma, but none currently ists.

LINICAL PHENOMENOLOGY

RADYKINESIA

generalized slowness of movement is arguably the fining feature of PD and other parkinsonian disorders, d this phenomenon has been termed *bradykinesia*. radykinesia is often used interchangeably with two other rms, *akinesia* (absence of movement) and *hypokinesia* overty of movement), and is a major cause of disability PD. It is eventually seen in all patients and is a require- ent for diagnosis of PD in many published diagnostic iteria.[5–7] Patients often have a difficult time describing mptoms of bradykinesia, instead using "weakness," ncoordination," and even "fatigue" or "tiredness" to scribe the difficulty and extreme effort of initiating ovement. In general, patients first notice a delay in the itiation of voluntary movement, with difficulty multi- sking or executing sequential actions. Family members ill first notice a decrease of spontaneous associated ovements, such as loss of gestures during conversation, creased eye blinking or facial masking, which may use family to think that the patient is unhappy, angry, not paying attention. The voice may become softer ypophonia), with the patient frequently needing to peat sentences. Swallowing may also be impaired but is ly rarely a source of major disability in early stage sease. Parkinsonian dysphagia is due to bradykinesia of e pharyngeal musculature, which can lead to pooling of liva in the mouth and drooling (sialorrhea),[13] even ough the amount of saliva produced in PD is normal.

Bradykinesia usually starts distally in the limbs, with creased dexterity of fingers and hand. One side is often ore affected than the other, and this pattern does not ange throughout the course of the disease.[14] Patients ten complain of difficulty buttoning buttons, tying shoe- ces, double clicking a computer mouse, or typing. andwriting is usually small (micrographia) and amped. Complaints of difficulty lifting the legs when alking, shuffling steps, and easy fatigability are com- on and plaintively expressed. Patients or family mem- ers observe that one or both arms become involuntarily exed when walking or that the arms hang by their side ithout swinging. A sensation of unsteadiness without lling is common in some cases as the disease progresses nd, in advanced stages, patients have difficulty standing p from a chair, getting out of a car, or maintaining pright posture without leaning or falling.

Freezing is a phenomenon that is poorly understood nd contributes to the gait difficulty seen in PD, but it can also be considered a manifestation of bradykinesia. It generally does not occur until the advanced stages of PD and may be seen as patients begin walking after standing up from a chair or bed (start hesitation), when open space is narrowed or constricted (e.g., approaching a doorway or in a crowd of people), or just before reaching a targeted destination (chair or bed).

In addition to careful clinical observation of the signs described above, bedside examination of bradykinesia includes evaluation of speed, amplitude, and rhythm of sequential movements on each side of the body through finger taps, opening and closing fists, pronation-supina- tion of the hands, and heel or toe tapping. In early disease, these tasks usually show mild slowing and decreased amplitude the longer the movements are performed. As the disease progresses, these movements become less coordinated, with frequent hesitation or arrests. Fortu- nately, bradykinesia responds well to dopaminergic medi- cations and deep brain stimulation (DBS) surgery. Stimulation of both the globus pallidus pars interna (GPi) and subthalamic nucleus (STN) seem to be equally effec- tive for bradykinesia.[15,16]

RIGIDITY

Tone in the context of the neurologic examination can be defined as the general resistance of a muscle to passive manipulation. *Rigidity* is a hypertonic state that is com- monly seen in disorders of the basal ganglia and is defined as unvarying increased resistance within the range of pas- sive movement about a joint.[17] Rigidity is therefore inde- pendent of the velocity used to manipulate the limb. This feature distinguishes it from *spasticity,* which is seen in disorders of the corticospinal tract, such as stroke or mul- tiple sclerosis, in which a limb can be moved passively at low velocity more easily than when moved rapidly. Rigid- ity can be difficult to distinguish from the "paratonia" or "gegenhalten" seen in encephalopathic disorders, in which the tone increases in proportion to the amount of force applied.

James Parkinson never mentioned "rigidity" as a main feature of the disease in his original essay,[1] though he did describe features that can be attributed to rigidity. Clinical series have reported that rigidity occurs in 89 to 99% of patients with PD.[18–21] The "cogwheel" phenomenon is a particular type of rigidity occurring in PD, and it refers to rhythmic brief increases in resistance that are palpated during passive movement.[17] Cogwheeling is thought to be tremor superimposed on rigidity.[22] It is a common mis- conception that a patient must have cogwheeling on examination to merit a diagnosis of PD. Many instead will have "lead pipe" rigidity, which is smooth throughout the entire range of motion. Rigidity, like tremor and bradykinesia, often presents unilaterally, progresses slowly to the other side of the body, and remains asymmetric throughout the disease.[23]

Clinical testing for rigidity involves passive manipulation of the arms and legs. Rigidity tends to be more prominent in biceps than triceps, quadriceps than hamstrings, and dorsiflexors of the ankle than plantar flexors, while flexors and extensors of the wrist are both equally affected.[22] When present, cogwheeling is most obvious at the wrist and can be brought out on exam by having the patient contract the muscles of the contralateral hand or perform mental arithmetic.

Rigidity affects both limb and axial musculature and contributes to the postural deformities seen in PD. The stooped or "simian" posture and lateral tilt of the trunk are common sequelae of axial rigidity. In some patients, severe forward flexion of the thoracolumbar spine occurs, a phenomenon known as "camptocormia."[24] It is controversial, however, whether camptocormia is due to rigidity of the iliopsoas and spinal muscles or a rare and extreme example of a typically parkinsonian postural deformity. A combination of rigidity and postural abnormality can result in involuntary flexion of the forearm, extension of the proximal and distal interphalangeal joints with flexion at the metacarpophalangeal joints (the "striatal hand"), and flexion at the knees. Common complaints of patients that can be attributed in part to rigidity include difficulty turning over in bed or standing up from a chair, and muscle cramps or pain

Rigidity, like bradykinesia, responds well to dopaminergic medications. Early trials with levodopa demonstrated that, of the three classical motor features of PD, rigidity seemed to respond best.[25,26] In addition, physical therapy has been shown to modestly improve rigidity,[27] and DBS of both the GPi and STN effectively suppress rigidity.[15,16]

TREMOR

Tremor is perhaps the most obvious manifestation of PD, causing James Parkinson to label the disease he was describing the "shaking palsy."[1] Tremor is defined as an involuntary rhythmic oscillation of a body region that is produced by alternating contractions of reciprocally innervated muscles.[28] Parkinson correctly observed the tremor as an "involuntary tremulous motion…in parts not in action and even when supported." The typical tremor in PD is a *rest* tremor, present mainly when the patient is relaxed and when the tremulous limb is not engaged in purposeful activities. This distinguishes it from tremors in other conditions, such as essential tremor (ET), in which the tremor is mainly postural and occurs with use of the hands, or multiple sclerosis, in which patients generally have an intention tremor due to cerebellar lesions. However, as PD progresses, it is not unusual to observe a postural or action component to the tremor and, in severe

cases, it may be difficult to tell whether the tremor primarily resting or postural.

Tremor is the most frequent presenting symptom PD, occurring in approximately 70% of patients.[18] Co monly described as "pill rolling" because of the unique to 6 Hz rhythmic oscillation of flexion and extensi movements of the thumb and fingers in the style of t early pharmacists who rolled or rounded hand-made pi into little balls by using the thumb and forefinger, parki sonian tremor at a similar frequency can also involve t legs, lips, jaw, and tongue, either alone or in combinati with tremors of other body parts.[29,30] Head tremor unusual in PD, and some experts in the field think th prominent head tremor in the setting of otherwise clas cal PD is evidence of coexistent ET.[31]

The tremor of early PD is usually intermittent, and may be so mild that it is not obvious to others. Appro mately half of all patients report episodes of invisible a uncomfortable *internal* tremulousness in the limbs body.[32] The typical tremor of PD tends to start unilatera in the hand, subsequently involving the ipsilateral leg the contralateral arm.[19] On average, the tremor sprea bilaterally six years after the onset of symptoms,[23] and t side initially affected continues to have more tremor th the contralateral side. However, exceptions to these ru are frequent, such as the patient whose tremor from t start is most prominent in one arm and the contralate leg or in both legs without involvement of either arm.

Rest tremor in PD disappears with REM sleep b may remanifest during light sleep and awaken t patient.[33] Anxiety, emotional excitement, and stressful s uations worsen the tremor. Since walking often accent ates tremor or uncovers latent tremor not otherwi apparent, the neurologic examiner should always inclu tremor inspection as part of the gait evaluation. Li rigidity, tremor in one limb can be brought out or rei forced by voluntary movement of the contralateral limb by having the patient concentrate on mental tasks.

The proportion of PD patients with rest tremor high. In several reports of patients with autopsy-prov PD, the percentage of patients with tremor of any ty (rest, action, postural) ranged from 79 to 100%.[18–21] Most of these had rest tremor alone. Action or postur tremor that often coexists with rest tremor in PD osc lates at a higher frequency of 7 to 12 Hz and, therefo may have a tremor generating mechanism different fro rest tremor.[30]

Most patients with prominent rest tremor want trea ment, not because it causes disability but because t tremor is annoying or embarrassing. However, whe action tremor is also present, disability becomes a rea ity. The response of tremor to treatment with drugs highly variable.[35] Amantadine hydrochloride, carb dopa/levodopa, the dopamine agonists, beta-blocke such as propranolol, clozapine, and anticholinergi such as trihexiphenidyl hydrochloride have all been shown to have some tremor-suppressing effect.[36–40] If the tremor is refractory to all medications, DBS of multiple targets in the basal ganglia and thalamus have been shown to be effective.[41–44]

POSTURAL INSTABILITY

Postural instability, or impairment of centrally mediated postural reflexes, is a fourth cardinal feature of PD, usually appearing for the first time several years after the onset of the first symptom. Patients with parkinsonian features who present with postural instability, falls, or gait difficulty should be evaluated for another parkinsonian disorder, such as PSP or MSA.[45] As the postural righting reflexes become impaired, falling and injury to head, joints, and peripheral nerves is a serious risk. Initially, postural instability may be subtle and found only on neurologic exam with the "pull" test, which evaluates the patient's ability to resist retropulsion. With further deterioration of postural reflexes, the gait may show signs of festination, defined by Parkinson as "an irresistible impulse to take much quicker and shorter steps, and thereby to adopt unwillingly a running pace."[1] Postural instability is the least treatable of the major motor features of PD,[46] especially late in the course when it ceases to respond to levodopa and may prevent walking altogether.

DISEASE ONSET AND CLINICAL EXPRESSION

AGE OF ONSET

PD is estimated to affect 2 to 3 percent of the population over the age of 65.[47,48] Age is the strongest and most consistent risk factor for the development of PD. In a recent study of PD using data from a large health maintenance organization, the estimated incidence increased rapidly over 60 yr of age, with a mean age at diagnosis of 70.5 yr.[49] PD is uncommon in people younger than 40, and when it occurs, Wilson's disease (WD), Huntington's disease (HD), or dopa-responsive dystonia (DRD) are the principal alternative diagnoses to consider. In the western hemisphere, only about 4 to 6 percent of patients with PD presenting to referral clinics are below age 40,[50–52] with approximately twice that number presenting in Japan.[53] By convention, patients with onset of symptoms between the ages of 21 and 40 have young-onset Parkinson's disease (YOPD), and patients with onset younger than age 21 have juvenile Parkinsonism (JP).[54] YOPD has the same underlying Lewy body (LB) pathology as the disease of later onset,[55] whereas JP may be a different disease, based on observations in a few autopsied cases showing nigrostriatal degeneration without LB pathology.[56,57]

EARLY NONSPECIFIC FEATURES

The dopamine deficiency of PD appears not to have much of a symptomatic impact until stores are depleted by at least 50%, according to studies using flourodopa PET and B-CIT SPECT imaging techniques.[58–60] As neuroprotective therapies are developed, early detection of the preclinical stage of neurodegeneration will be essential to prevent major symptoms from developing or to halt the manifest clinical disease before disability occurs. Symptoms that actually precede the classical motor signs of PD are often nonspecific and vague. Gonera et al., in a retrospective case-control study, found a prodromal phase prior to the onset of parkinsonism lasting 4 to 6 years, in which PD patients made more office visits to their general practitioners than control subjects.[60] Presenting symptoms in this prodromal period included fibromyalgia, pain and other sensory complaints in muscles and joints, hypertension, and mood disorders. These early nonspecific symptoms are often noted only in retrospect by patients whose first awareness of PD is with the appearance of one or more of the classical motor problems of PD. Other important early heralding features of PD include unilateral action dystonia[61] and rapid eye movement sleep behavior disorder (RBD), which may precede the appearance of classical signs by several years.[62] Most patients with PD lose their sense of smell—often without knowing it—at the earliest stage of the disease. Although this is a well documented finding, the relationship between olfactory system dysfunction and nigral degeneration is unknown.[63] Office evaluation with the handy University of Pennsylvania Smell Identification Test (UPSIT) can be useful in cases where the diagnosis of PD is in question, especially if the test results show normal olfactory function.[64]

The onset of motor symptoms can also be nonspecific. For example, patients may complain of leg dragging or limb weakness, leading to an extensive workup for neuropathy, radiculopathy, or stroke. Slowed thinking (bradyphrenia) or slowed response when talking may prompt a search for seizures or dementia. Oftentimes, the patient may not be aware that anything is wrong and may present to a physician only because a family member or colleague at work noticed cramped handwriting, stooped posture, subtle changes in speech, or loss of facial animation. The spectrum of presenting symptoms is similar for YOPD and JP patients compared with older onset patients, although dystonia is more common as an initial symptom in the two younger subgroups.[55,57,65]

SUBTYPES OF PD

The marked variability of clinical expression and rate of progression among patients with PD has led some investigators to conclude that PD is not one disease but a cluster of diseases subsumed under the umbrella of a single label.[66] This observation is somewhat contradicted by the limited spectrum of degenerative pathology found at autopsy, where the findings are more or less confined to the substantia nigra as the site most affected by neuronal loss and to a lesser extent other pigmented neurons of the brain stem (locus ceruleus and dorsal motor nucleus of the vagus). Clinical investigators have attempted to organize the heterogeneity of PD's natural history by creating somewhat artificial "subgroups," mainly to prognosticate outcome. Thus, age of onset (e.g., YOPD) and symptom predominance (tremor versus postural instability/gait disorder (PIGD) versus akinetic-rigid, and so forth) have become the principal headings in this taxonomy. Younger onset patients tend to progress more slowly and consequently have a longer course of illness.[18,55,67,68] They usually respond well to levodopa but are more likely to develop early levodopa-related motor complications (fluctuations and dyskinesias), usually within the first five years of starting levodopa therapy.[54,65,69] Dementia, which affects approximately one-third of all patients with PD, is reported to occur less commonly in young-onset patients, despite the longer disease duration.[54,65] Tremor predominance, like young age of onset, tends to be associated with a more benign course, with slower progression and less neuropsychologic impairment.[50,51,70,71] Although these subgroupings have no clear value in relation to disease mechanisms, they are broadly useful to the clinician as guides to management.

PATHOPHYSIOLOGY OF CLASSICAL MOTOR FEATURES

The underlying pathophysiology of motor symptoms in PD remains a mystery, despite all the advances brought about by research in the modern era. Degeneration of dopamine producing neurons in the substantia nigra (SN) is the constant, unvarying pathologic hallmark of PD, although the etiology of the insidious, progressive death of these cells is poorly understood. An in-depth discussion of basal ganglia physiology and processing is beyond the scope of this chapter, but, in brief, the currently popular model of basal ganglia function has normal voluntary movement regulated by a balance between two opposing physiologic circuits—the direct and indirect pathways—originating in the putamen and ultimately projecting to the internal segment of the globus pallidus (GPi).[72] Efferents from the GPi project in series to the motor nuclei of the thalamus and motor cortex, respectively. Activation of the direct pathway results in a decrease of the normal inhibitory outflow from the GPi on the thalamus and increases cortical motor activation, resulting in the execution of voluntary movement. Activation of the indirect pathway increases inhibitory outflow from the GPi onto the thalamus, resulting in decreased output to the motor

cortex and reduction in voluntary movement. In PD, degeneration of the dopaminergic nigrostriatal pathway and loss of dopaminergic input to the putamen results in overactivity of the indirect pathway, greater inhibitory output from GPi, and less voluntary movement or bradykinesia. In contrast, hyperkinetic disorders such as HD are the result of increased output from the direct pathway, less inhibition from GPi, and excessive movement that is involuntary. This model has been proven useful under experimental and clinical conditions but is oversimplified, because it does not account for all the features of PD, especially tremor. Moreover, new research is revealing more complex pathways and interactions of basal ganglia structures. For example, new data suggest that in the normal brain, information flows through independent and parallel circuits within the basal ganglia, while, in the parkinsonian brain, these circuits break down, and interconnections between these circuits become active and synchronized.[73]

BRADYKINESIA

Studies of reaction time in patients with PD have provided some insight into the pathophysiology of bradykinesia. Not surprisingly, reaction times are prolonged in PD,[74,75] which may be due to a problem either with motor programming or motor execution. To what extent motor programming affects reaction times is debatable,[76] but evidence suggests that execution of motor commands is impaired in PD and may play a stronger role in the overall mechanism of bradykinesia.[76,77] The motor cortex plays an essential role in movement execution, and cortical motor excitability has to build up to a certain level before movement can even be detected. Experiments using transcranial magnetic stimulation (TMS) have shown that this buildup is slower and starts earlier in patients with PD.[78,79] PET studies provide further evidence for this concept, showing impaired activation of the supplementary motor cortex (SMA) prior to the onset of voluntary movement in patients with PD, with an increase in activation of SMA and other premotor areas after treatment with the dopamine agonist apomorphine[80] or surgical ablation of GPi.[81]

Because the SMA is believed to have a role in the generation of complex movements, impairment of SMA activation might also explain why patients often have difficulty in sustaining sequential motor acts or performing simultaneous movements. When performing sequential tasks, the length of time needed to accomplish the two tasks together exceeds the amount of time necessary to perform each task alone.[82] Not only are the individual movements slower, there is also a significant pause in between tasks. Furthermore, movement time is increased when patients are asked to perform two different movements simultaneously.[83] It is hypothesized that, although the parkinsonian brain can process an individual motor program without significant difficulty, it lacks the processing capability to run two separate motor programs concurrently, perhaps because of poor SMA activation by the basal ganglia[82,84]

Another factor contributing to bradykinesia involves a problem in specifying the amplitude of movements. It has been shown that, when performing movements about a single joint, PD patients cannot produce sufficient EMG burst activity to accomplish a large, fast movement.[85] Instead, these patients add extra bursts of EMG to achieve enough force to reach the correct endpoint. Berardelli et al.[86] theorized that PD patients with bradykinesia underestimated the amount of burst activity required to complete a movement. Thus, the problem was one of matching or scaling an agonist burst appropriately to the movement size, rather than an inability to produce a large movement.

RIGIDITY

Some authors have suggested that changes in the mechanical properties of limbs play a part in the pathophysiology of rigidity. Dietz et al. showed that, although muscle action potentials as measured *qualitatively* by EMG were normal in the tibialis anterior muscle, PD patients had reduced dorsiflexion movement of the foot when walking, suggesting that increased *stiffness* of the muscles, as opposed to increased muscle *activity*, was responsible for rigidity.[87] Watts et al., using a *quantitative* measure of rigidity, demonstrated that muscle stiffness was increased when PD patients were in a relaxed state.[88] However, other investigators believe that neurally mediated stretch reflexes are more responsible for rigidity than any specific changes in the mechanical properties of the muscles themselves.[89]

The idea that long-latency stretch reflexes are abnormal in PD patients and contribute to rigidity in PD was first proposed by Tatton and Lee in 1975.[90] It is now generally well established that enhancement of the long latency reflex is a characteristic physiologic finding in PD.[89,91] The anatomic pathway of the long latency reflex is not entirely clear but is believed to be mediated through a transcortical loop.[77] How then does the current basal ganglia model account for an increase in long latency stretch reflexes? Increased inhibitory output from GPi on the thalamus likely decreases activation of cortical motor inhibitory circuits, which normally modulate these reflexes. Pallidotomy has been demonstrated to decrease the long latency stretch reflex in patients,[91] and transcranial magnetic stimulation studies have shown increased excitability of cortical motor inhibition pathways after GPi ablation.[92]

TREMOR

The typical 4 to 6 Hz rest tremor of PD is so uniform from patient to patient that investigators have believed for years that the nervous system must have a fixed "tremor generator." However, no definite site has been identified. Although there may be some mechanical stretch reflex component to tremor, it is likely that the main component comes from a central mechanism. If stretch reflex circuitry were solely responsible for the tremor in PD, we would expect higher tremor frequencies for proximal muscles as opposed to distal muscles because of added distances of the stretch reflex loop for distal muscles.[93] Instead, resting tremor frequency is uniform throughout the limbs and orofacial sites, such as the chin and lips, which have no stretch reflexes.[29]

How does loss of dopaminergic neurons in the SN cause tremor? One hypothesis holds that selective loss of dopaminergic neurons in the compact zone of the SN (SNc) is responsible.[94] Nigral neurons can be divided into ventral and dorsal tiers. In PD, the lateral ventral tier (VL) is thought to be affected first, followed by the ventromedial (VM) and then the dorsal tier.[58] Because PD can present with tremor, and because tremor often is reduced in advanced PD as rigidity and bradykinesia become more prominent, it may be that loss of dopaminergic neurons in the VL region is more responsible for tremor. As the disease progresses, a more extensive loss overall in the SNc occurs, accounting for the bradykinesia and rigidity.

On the other hand, research using fluorodopa PET as an index of nigrostriatal integrity has shown that the degree of nigrostriatal dopaminergic deficit correlates poorly with tremor.[95] Because there are also nigral dopaminergic projections to the STN and GPi, an alternative hypothesis is that tremor is not related to a particular pattern of neuronal degeneration in the SNc but to abnormal activation of STN and GPi by degeneration of the dopaminergic pathways to these structures.[96] The effectiveness of stereotactic lesions and suppression of output from GPi and STN by DBS in abolishing tremor lends support to this line of thinking.[15,16,97] Other supportive evidence for the role of STN and GPi in the generation of tremor includes the observation of 4 to 8 Hz rhythmic discharges from the GPi and STN, tightly linked to tremor, in parkinsonian monkeys treated with MPTP,[98] and 4 to 5 Hz bursting activity recorded from the STN in humans, time-locked to EMG bursts from the contralateral hand.[41,42]

Other circuits may also be involved in tremor production. Stimulation of the ventralis intermedius (Vim) nucleus of the thalamus, primarily a cerebellar relay nucleus without input from GPi or STN, is effective in reducing tremor, suggesting a role involving thalamo-cortical-cerebellar circuits.[44] It may be that lesions in the Vim of the thalamus are effective, because they interfere with pathways between the globus pallidus and other thalamic nuclei rather than the thalamus itself.[99] Complete suppression of tremor by thalamic stimulation is also associated with a decrease in cerebellar blood flow as measured by fluorodeoxyglucose PET, suggesting that tremor activation is connected to cerebellar hyperactivity.[100] Whatever the reasons, the exact basis for tremor remains unclear and likely arises from a complex network of neural connections in the brain involving the basal ganglia, thalamus, cerebellum and cerebral cortex.

POSTURAL INSTABILITY

Balance is a function that depends on several intact processes, including afferent sensory input from visual, vestibular, and somatosensory systems, and efferent control of muscle tone and peripheral movement. These separate sensory and motor systems are integrated through cortical and basal ganglia pathways and work synergistically to produce quick and coordinated compensatory responses to perturbations of posture. In the PD patient, postural stability is not only affected by a disruption of the pathways involved with postural control but also by bradykinesia and rigidity; however, the relative contributions of cardinal parkinsonian features and primary postural control mechanisms to postural imbalance in PD remain unknown. Current research suggests that postural instability in PD is not due to a primary afferent sensory defect but rather to profound deficits in the ability to scale the amplitude of postural leg responses appropriately to the size of perturbations, as well as an inability to quickly adapt to differing environmental conditions.[101–103]

CONCLUSIONS

Bradykinesia, rigidity, rest tremor, and postural instability are the classical motor features of PD. The astute clinician must be aware of the remarkable individual variability of PD as a complex illness affecting mobility and behavior in a mostly elderly population so as to make an early and accurate diagnosis. In particular, the earliest phase of the disease is often the most difficult to recognize, because the clinical picture may be dominated by subtle and non-specific complaints and findings. The pathophysiology of PD is an incomplete puzzle, but research is slowly filling in the missing pieces. Current working models suggest that nigral degeneration and the associated loss of dopaminergic input to the basal ganglia initiate a slowly progressive cascade of pathologic changes and interactions between basal ganglia, thalamus, cerebellum, and cortex.

REFERENCES

1. Parkinson, J., *An Essay on the Shaking Palsy,* Sherwood, Neely, and Jones, London, 1817.
2. Colcher, A. and Simuni, T., Clinical manifestations of Parkinson's disease, *Med. Clin. North Am.,* 83, 327, 1999.
3. Dickson, D. W., Neuropathology of parkinsonian disorders, in *Parkinson's Disease and Movement Disorders,* 4th ed., Jankovic, J. J. and Tolosa, E., Eds., Lippincott Williams & Wilkins, Philadelphia, 2002, 256.
4. Calne, D. B., Snow, B. J., and Lee, C., Criteria for diagnosing Parkinson's disease, *Ann. Neurol.,* 32 Suppl, S125, 1992.
5. Gelb, D. J., Oliver, E., and Gilman, S., Diagnostic criteria for Parkinson disease, *Arch Neurol,* 56, 33, 1999.
6. Hughes, A. J. et al., Accuracy of clinical diagnosis of idiopathic Parkinson's disease: a clinico-pathological study of 100 cases, *J Neurol. Neurosurg. Psychiatry,* 55, 181, 1992.
7. Langston, J. W. et al., Core assessment program for intracerebral transplantations (CAPIT), *Mov. Disord.,* 7, 2, 1992.
8. Ward, C. D. and Gibb, W. R., Research diagnostic criteria for Parkinson's disease, *Adv. Neurol.,* 53, 245, 1990.
9. Rajput, A. H., Rozdilsky, B., and Rajput, A., Accuracy of clinical diagnosis in parkinsonism—a prospective study, *Can. J. Neurol. Sci.,* 18, 275, 1991.
10. Hughes, A. J., Daniel, S. E., and Lees, A. J., Improved accuracy of clinical diagnosis of Lewy body Parkinson's disease, *Neurology,* 57, 1497, 2001.
11. Hurtig, H. I. et al., Alpha-synuclein cortical Lewy bodies correlate with dementia in Parkinson's disease, *Neurology,* 54, 1916, 2000.
12. Hughes, A. J. et al., What features improve the accuracy of clinical diagnosis in Parkinson's disease: a clinico-pathologic study, *Neurology,* 42, 1142, 1992.
13. Edwards, L. L., Quigley, E. M., and Pfeiffer, R. F., Gastrointestinal dysfunction in Parkinson's disease: frequency and pathophysiology, *Neurology,* 42, 726, 1992.
14. Lee, C. S. et al., Patterns of asymmetry do not change over the course of idiopathic parkinsonism: implications for pathogenesis, *Neurology,* 45, 435, 1995.
15. Deep-brain stimulation of the subthalamic nucleus or the pars interna of the globus pallidus in Parkinson's disease, *N. Engl. J. Med.,* 345, 956, 2001.
16. Burchiel, K. J. et al., Comparison of pallidal and subthalamic nucleus deep brain stimulation for advanced Parkinson's disease: results of a randomized, blinded pilot study, *Neurosurgery,* 45, 1375, 1999.
17. Deuschl, G., Bain, P., and Brin, M., Consensus statement of the Movement Disorder Society on Tremor. Ad Hoc Scientific Committee, *Mov. Disord.,* 13, suppl. 3, 2, 1998.
18. Hoehn, M. M. and Yahr, M. D., Parkinsonism: onset, progression and mortality, *Neurology,* 17, 427, 1967.
19. Hughes, A. J., Daniel, S. E., and Lees, A. J., The clinical features of Parkinson's disease in 100 histologically proven cases, *Adv. Neurol.,* 60, 595, 1993.
20. Martin, W. E. et al., Parkinson's disease. Clinical analysis of 100 patients, *Neurology,* 23, 783, 1973.
21. Louis, E. D. et al., Comparison of extrapyramidal features in 31 pathologically confirmed cases of diffuse Lewy body disease and 34 pathologically confirmed cases of Parkinson's disease, *Neurology,* 48, 376, 1997.

22. Lance, J. W., Schwarb, R. S., and Peterson, E. A., Action tremor and the cogwheel phenomenon in Parkinson's disease, *Brain*, 86, 95, 1963.

23. Scott, R. M. et al., Progression of unilateral tremor and rigidity in Parkinson's disease, *Neurology*, 20, 710, 1970.

24. Djaldetti, R. et al., Camptocormia (bent spine) in patients with Parkinson's disease—characterization and possible pathogenesis of an unusual phenomenon, *Mov Disord*, 14, 443, 1999.

25. Cotzias, G. C., Van Woert, M. H., and Schiffer, L. M., Aromatic amino acids and modification of parkinsonism, *N. Engl. J. Med.*, 276, 374, 1967.

26. Yahr, M. D. et al., Treatment of parkinsonism with levodopa, *Arch. Neurol.*, 21, 343, 1969.

27. Comella, C. L. et al., Physical therapy and Parkinson's disease: a controlled clinical trial, *Neurology*, 44, 376, 1994.

28. Zesiewicz, T. A. and Hauser, R. A., Phenomenology and treatment of tremor disorders, *Neurol. Clin.*, 19, 651, 2001.

29. Hunker, C. J. and Abbs, J. H., Uniform frequency of parkinsonian resting tremor in the lips, jaw, tongue, and index finger, *Mov. Disord.*, 5, 71, 1990.

30. Findley, L. J., Gresty, M. A., and Halmagyi, G. M., Tremor, the cogwheel phenomenon and clonus in Parkinson's disease, *J. Neurol Neurosurg. Psychiatry*, 44, 534, 1981.

31. Pal, P. K., Samii, A., and Calne, D. B., Cardinal features of early Parkinson's disease, in *Parkinson's Disease: Diagnosis and Clinical Management*, Factor, S. A. and Weiner, W. J., Eds., Demos Medical Publishing, Inc., New York, 41, 2002.

32. Shulman, L. M. et al., Internal tremor in patients with Parkinson's disease, *Mov. Disord.*, 11, 3, 1996.

33. Askenasy, J. J. and Yahr, M. D., Parkinsonian tremor loses its alternating aspect during non-REM sleep and is inhibited by REM sleep, *J. Neurol. Neurosurg. Psychiatry*, 53, 749, 1990.

34. Rajput, A. H., Rozdilsky, B., and Ang, L., Occurrence of resting tremor in Parkinson's disease, *Neurology*, 41, 1298, 1991.

35. Wasielewski, P. G., Burns, J. M., and Koller, W. C., Pharmacologic treatment of tremor, *Mov. Disord.*, 13, suppl. 3, 90, 1998.

36. Koller, W. C., Pharmacologic treatment of parkinsonian tremor, *Arch Neurol*, 43, 126, 1986.

37. Koller, W. C. and Herbster, G., Adjuvant therapy of parkinsonian tremor, *Arch. Neurol.*, 44, 921, 1987.

38. Friedman, J. H. et al. Benztropine versus clozapine for the treatment of tremor in Parkinson's disease, *Neurology*, 48, 1077, 1997.

39. Pogarell, O. et al., Pramipexole in patients with Parkinson's disease and marked drug resistant tremor: a randomised, double blind, placebo controlled multicentre study, *J. Neurol. Neurosurg. Psychiatry*, 72, 713, 2002.

40. Schrag, A., Keens, J., and Warner, J., Ropinirole for the treatment of tremor in early Parkinson's disease, *Eur. J. Neurol.*, 9, 253, 2002.

41. Krack, P. et al., Treatment of tremor in Parkinson's disease by subthalamic nucleus stimulation, *Movement Disorders*, 13, 907, 1998.

42. Rodriguez, M. C. et al., The subthalamic nucleus and tremor in Parkinson's disease, *Mov. Disord.*, 13, suppl. 3, 111, 1998.

43. Benabid, A. L. et al., Chronic electrical stimulation of the ventralis intermedius nucleus of the thalamus as a treatment of movement disorders, *J. Neurosurg.*, 84, 203, 1996.

44. Koller, W. et al., High-frequency unilateral thalamic stimulation in the treatment of essential and parkinsonian tremor, *Ann. Neurol.*, 42, 292, 1997.

45. Maher, E. R. and Lees, A. J., The clinical features and natural history of the Steele-Richardson-Olszewski syndrome (progressive supranuclear palsy), *Neurology*, 36, 1005, 1986.

46. Koller, W. C. et al., Falls and Parkinson's disease, *Clin. Neuropharmacol.*, 12, 98, 1989.

47. de Rijk, M. C. et al., Prevalence of parkinsonism and Parkinson's disease in Europe: the EUROPARKINSON Collaborative Study. European Community Concerted Action on the Epidemiology of Parkinson's disease, *J Neurol. Neurosurg. Psychiatry*, 62, 10, 1997.

48. Moghal, S. et al., Prevalence of movement disorders in elderly community residents, *Neuroepidemiology*, 13, 175, 1994.

49. Van Den Eeden, S. K. et al., Incidence of Parkinson's disease: variation by age, gender, and race/ethnicity, *Am J. Epidemiol.*, 157, 1015, 2003.

50. Zetusky, W. J., Jankovic, J., and Pirozzolo, F. J., The heterogeneity of Parkinson's disease: clinical and prognostic implications, *Neurology*, 35, 522, 1985.

51. Jankovic, J. et al., Variable expression of Parkinson's disease: a base-line analysis of the DATATOP cohort. The Parkinson Study Group, *Neurology*, 40, 1529, 1990.

52. Golbe, L. I., Young-onset Parkinson's disease: a clinical review, *Neurology*, 41, 168, 1991.

53. Yokochi, M. et al., Juvenile parkinsonism—some clinical, pharmacological, and neuropathological aspects, *Adv. Neurol.*, 40, 407, 1984.

54. Quinn, N., Critchley, P., and Marsden, C. D., Young onset Parkinson's disease, *Mov. Disord.*, 2, 73, 1987.

55. Gibb, W. R. and Lees, A. J., A comparison of clinical and pathological features of young- and old-onset Parkinson's disease, *Neurology*, 38, 1402, 1988.

56. Takahashi, H. et al., Familial juvenile parkinsonism: clinical and pathologic study in a family, *Neurology*, 44, 437, 1994.

57. Muthane, U. B. et al., Early onset Parkinson's disease: are juvenile- and young-onset different?, *Mov. Disord.*, 9, 539, 1994.

58. Fearnley, J. M. and Lees, A. J., Ageing and Parkinson's disease: substantia nigra regional selectivity, *Brain*, 114 (Pt. 5), 2283, 1991.

59. Marek, K. L. et al., [123I] beta-CIT/SPECT imaging demonstrates bilateral loss of dopamine transporters in hemi-Parkinson's disease, *Neurology*, 46, 231, 1996.

60. Gonera, e.g., et al., Symptoms and duration of the prodromal phase in Parkinson's disease, *Mov. Disord.,* 12, 871, 1997.

61. Poewe, W. H., Lees, A. J., and Stern, G. M., Dystonia in Parkinson's disease: clinical and pharmacological features, *Ann. Neurol.,* 23, 73, 1988.

62. Schenck, C. H., Bundlie, S. R., and Mahowald, M. W., Delayed emergence of a parkinsonian disorder in 38% of 29 older men initially diagnosed with idiopathic rapid eye movement sleep behaviour disorder, *Neurology,* 46, 388, 1996.

63. Doty, R. L., Deems, D. A., and Stellar, S., Olfactory dysfunction in parkinsonism: a general deficit unrelated to neurologic signs, disease stage, or disease duration, *Neurology,* 38, 1237, 1988.

64. Doty, R. L., Shaman, P., and Dann, M., Development of the University of Pennsylvania Smell Identification Test: a standardized microencapsulated test of olfactory function, *Physiol. Behav.,* 32, 489, 1984.

65. Schrag, A. et al., Young-onset Parkinson's disease revisited—clinical features, natural history, and mortality, *Mov. Disord.,* 13, 885, 1998.

66. Calne, D. B., Parkinson's disease is not one disease, *Parkinsonism Related Disord.,* 7, 3, 2000.

67. Birkmayer, W., Riederer, P., and Youdim, B. H., Distinction between benign and malignant type of Parkinson's disease, *Clin. Neurol. Neurosurg.,* 81, 158, 1979.

68. Goetz, C. G. et al., Risk factors for progression in Parkinson's disease, *Neurology,* 38, 1841, 1988.

69. Kostic, V. et al., Early development of levodopa-induced dyskinesias and response fluctuations in young-onset Parkinson's disease, *Neurology,* 41, 202, 1991.

70. Mortimer, J. A. et al., Relationship of motor symptoms to intellectual deficits in Parkinson disease, *Neurology,* 32, 133, 1982.

71. Hershey, L. A. et al., Tremor at onset. Predictor of cognitive and motor outcome in Parkinson's disease?, *Arch. Neurol.,* 48, 1049, 1991.

72. Wichmann, T. and DeLong, M. R., Functional neuroanatomy of the basal ganglia in Parkinson's disease, *Adv. Neurol.,* 91, 9, 2003.

73. Bergman, H. et al., Physiological aspects of information processing in the basal ganglia of normal and parkinsonian primates, *Trends Neurosci.,* 21, 32, 1998.

74. Kutukcu, Y. et al., Simple and choice reaction time in Parkinson's disease, *Brain Research,* 815, 367, 1999.

75. Evarts, E. V., Teravainen, H., and Calne, D. B., Reaction time in Parkinson's disease, *Brain,* 104, 167, 1981.

76. Berardelli, A. et al., Pathophysiology of bradykinesia in Parkinson's disease, *Brain,* 124, 2131, 2001.

77. Hallett, M., Parkinson revisited: pathophysiology of motor signs, *Advances in Neurology,* 91, 19, 2003.

78. Chen, R. et al., Impairment of motor cortex activation and deactivation in Parkinson's disease, *Clin Neurophysiol,* 112, 600, 2001.

79. Pascual-Leone, A. et al., Akinesia in Parkinson's disease. I. Shortening of simple reaction time with focal, single-pulse transcranial magnetic stimulation, *Neurology,* 44, 884, 1994.

80. Jenkins, I. H. et al., Impaired activation of the supplementary motor area in Parkinson's disease is reversed when akinesia is treated with apomorphine, *Ann. Neurol.,* 32, 749, 1992.

81. Ceballos-Baumann, A. O. et al., Restoration of thalamocortical activity after posteroventral pallidotomy in Parkinson's disease, *Lancet,* 344, 814, 1994.

82. Benecke, R. et al., Disturbance of sequential movements in patients with Parkinson's disease, *Brain,* 110 (Pt. 2), 361, 1987.

83. Benecke, R. et al., Performance of simultaneous movements in patients with Parkinson's disease, *Brain,* 109 (Pt. 4), 739, 1986.

84. Brown, R. G. and Marsden, C. D., Dual task performance and processing resources in normal subjects and patients with Parkinson's disease, *Brain,* 114, 215, 1991.

85. Hallett, M. and Khoshbin, S., A physiological mechanism of bradykinesia, *Brain,* 103, 301, 1980.

86. Berardelli, A. et al., Scaling of the size of the first agonist EMG burst during rapid wrist movements in patients with Parkinson's disease, *Journal of Neurology, Neurosurgery & Psychiatry,* 49, 1273, 1986.

87. Dietz, V., Quintern, J., and Berger, W., Electrophysiological studies of gait in spasticity and rigidity. Evidence that altered mechanical properties of muscle contribute to hypertonia, *Brain,* 104, 431, 1981.

88. Watts, R. L., Wiegner, A. W., and Young, R. R., Elastic properties of muscles measured at the elbow in man: II. Patients with parkinsonian rigidity, *J. Neurol. Neurosurg. Psychiatry,* 49, 1177, 1986.

89. Rothwell, J. C. et al., The behaviour of the long-latency stretch reflex in patients with Parkinson's disease, *Journal of Neurology, Neurosurgery & Psychiatry,* 46, 35, 1983.

90. Tatton, W. G. and Lee, R. G., Evidence for abnormal long-loop reflexes in rigid Parkinsonian patients, *Brain Research,* 100, 671, 1975.

91. Hayashi, R. et al., Relation between changes in long-latency stretch reflexes and muscle stiffness in Parkinson's disease—comparison before and after unilateral pallidotomy, *Clin. Neurophysiol.,* 112, 1814, 2001.

92. Young, M. S. et al., Stereotactic pallidotomy lengthens the transcranial magnetic cortical stimulation silent period in Parkinson's disease, *Neurology,* 49, 1278, 1997.

93. Rothwell, J. C., Physiology and anatomy of possible oscillators in the central nervous system, *Mov. Disord.,* 13, suppl. 3, 24, 1998.

94. Carr, J., Tremor in Parkinson's disease, *Parkinsonism Relat. Disord.,* 8, 223, 2002.

95. Otsuka, M. et al., Differences in the reduced 18F-Dopa uptakes of the caudate and the putamen in Parkinson's disease: correlations with the three main symptoms, *Journal of the Neurological Sciences,* 136, 169, 1996.

96. Hedreen, J. C., Tyrosine hydroxylase-immunoreactive elements in the human globus pallidus and subthalamic nucleus, *J. Comp. Neurol.,* 409, 400, 1999.

97. Lozano, A. M. et al., Effect of GPi pallidotomy on motor function in Parkinson's disease, *Lancet,* 346, 1383, 1995.

98. Bergman, H. et al., The primate subthalamic nucleus. II. Neuronal activity in the MPTP model of parkinsonism, *J. Neurophysiol.,* 72, 507, 1994.

99. Bergman, H. and Deuschl, G., Pathophysiology of Parkinson's disease: from clinical neurology to basic neuroscience and back, *Mov. Disord.,* 17, suppl. 3, S28, 2002.

100. Deiber, M. P. et al., Thalamic stimulation and suppression of parkinsonian tremor. Evidence of a cerebellar deactivation using positron emission tomography, *Brain,* 116, 267, 1993.

101. Horak, F. B., Frank, J., and Nutt, J., Effects of dopamine on postural control in parkinsonian subjects: scaling, set, and tone, *J. Neurophysiol.,* 75, 2380, 1996.

102. Horak, F. B., Nutt, J. G., and Nashner, L. M., Postural inflexibility in parkinsonian subjects, *J. Neurol. Sci.,* 111, 46, 1992.

103. Bronte-Stewart, H. M. et al., Postural instability in idiopathic Parkinson's disease: the role of medication and unilateral pallidotomy, *Brain,* 125, 2100, 2002.

17 Clinical Evaluation and Treatment of Gait Disorders in Parkinson's Disease

Yacov Balash
Movement Disorders Unit, Department of Neurology, Tel-Aviv Sourasky Medical Center

Jeffrey M. Hausdorff
Movement Disorders Unit, Department of Neurology, Tel-Aviv Sourasky Medical Center; Sackler School of Medicine, Tel-Aviv University; and Harvard Medical School

Nir Giladi
Movement Disorders Unit, Department of Neurology, Tel-Aviv Sourasky Medical Center and Sackler School of Medicine, Tel-Aviv University

CONTENTS

PATHOPHYSIOLOGY OF PARKINSONIAN GAIT

The capacity for upright standing and bipedal walking is a fundamental human function. The mechanism of posture and locomotion depends on sensory input (mainly visual, vestibular, and proprioceptive), cerebral central processing (identification and localization of target, planning of the action), and execution of the movement—robust involuntary and voluntary muscular responses with prompt adjustments of the standing and stepping according to environmental or internal conditions.[1]

Pathways from the cortical, basal ganglia, brain stem, and the cerebellum effect the final execution commands via the spinal motor neurons. These commands cause complex sequential muscles activation, rhythmic changes of joint position, linear and angular accelerations in the movements of the limbs, and, as a result, various modes of body equilibrium and locomotion.

Normal human gait requires the bilateral activation of the SMA, the primary sensorimotor cortex, striatum, visual cortex, and cerebellum.[2,3] In patients with Parkinson's disease (PD) and related posture and gait disorders, HMPAO SPECT demonstrated underactivation of the left medial frontal area, right precuneus, and left cerebellar hemisphere. In addition, overactivity in the left temporal cortex, right insula, left cingulated cortex, and cerebellar vermis was observed.[4] In other studies, selective right-side activation of the premotor area without synchronous activation of the primary motor cortex and cerebellum was seen in *kinesia paradoxa*—an improvement of hypokinetic gait after visual stimuli.[5,6] This overactivity within the cerebellar-parieto-lateral premotor loops in PD patients has been suggested to compensate for the mesial

0-8493-1590-5/05/$0.00+$1.50

premotor-prefrontal deficit. If, in normal subjects, gait execution activates dopaminergic neurons innervating the putamen, in PD patients increased uptake of dopa was found in the orbito-frontal cortex and the caudate nucleus.[7] Abnormal uptake of f[18]dopa in these areas was shown to significantly correlate with parkinsonian walking cadence. Thus, in PD, instead of the excitation of the classic (putaminal → GPi → thalamic) pathway during movement initiation, there is activation of less effective mesocortical-ventrostriatal dopaminergic system.[7]

Thus, the posture and gait disorders in PD are the result of dysbalance of cortical and subcortical activities with underactivation of the primary motor cortex, putamen, and cerebellum, which are replaced by over activation of nontypical brain areas. The probable role of basal ganglia is to act as "internal cue" or modulator and focusing of motor acts via the premotor cortex and the SMA.[8,9] The above-mentioned shifts in the functional condition of the thalamo-cortical system in PD might be compensated by a larger need for voluntary regulation of locomotion, use of attention, and, perhaps, increased susceptibility to distractions. This might be the basis for disorders of postural control and locomotion in PD.

CLINICAL STAGING OF GAIT DISORDERS IN PD

In accordance with progression of nigro-striato-pallidal dysfunction, gait disorders in PD can be staged as follows:

Stage 1.   Nonsignificant Gait Disturbances

In this first stage, functionally nonsignificant gait disturbances, such as decreased gait speed, shortened steps, decreased arm swing, and increased stride-to-stride variability as detected by gait dynamics analysis, may be observed.[10]

*Stage 2.   Mild to Moderate Functional Gait
             Disturbances*

In the second stage, shuffling with marked bradykinesia and hypokinetic gait becomes important functionally. Steps are short with decreased force of foot "push-off" or hip "pull-up" as well as flexed joints and stooped trunk. "Stop walking while talking"[11] as a presentation of advanced dual-tasking derangement can also be seen. Abnormally enlarged gait variability in response to distractions (dual tasking—subtracting of serial sevens) may also be observed.[12]

*Stage 3.   Significant Functional Gait
             Disturbances*

In this stage, significant episodic gait disorders (EGDs), including freezing of gait (FOG) and festinating gait

(FSG), take place on the background of continuous gait disturbances (e.g., slowed gait).[13] EGD typically appear unintentionally, break the background pattern of gait, and last for a period of seconds. FOG can be defined as intermittent episodes, which last for a few seconds (rarely exceeding 30 sec), of an inability to initiate or maintain locomotion or perform a turn. Typically, most FOG episodes can be overcome by motor, sensory, or mental tricks, but habituation has been described.[14,15] FOG episodes can be provoked most easily by asking the subject to turn around (turning hesitation).[16,17] However, in terms of ordinary motor behavior, 360° or even 180° turns are rare and, as a result, start hesitation is experienced more frequently in daily life.[17] Turning hesitation is important because of its possible contribution to falls during the act of turning. Other types of FOG occur while walking, passing through tight quarters, reaching a destination, and in stressful situations such as crossing the street ("open space") or entering an elevator ("tight quarters").[17] Festinating gait is an intermittent episode that lasts a matter of seconds and involves disturbed locomotion characterized by uncontrolled propulsion associated with rapid small steps. Patients frequently report a feeling that they were "pushed from behind."[18] The only clinical paper that characterized festination showed a significant association with FOG, in support of a possible pathophysiological linkage between the two.[19]

Previously EGD has been principally assessed by subjective measures and only recently was a validated questionnaire developed for evaluating freezing of gait (FOG).[20] A new objective method for quantifying FOG consists of an ambulatory gait analysis system with pressure sensitive insoles that continuously record walking, synchronized with a video recording.[21] Using this system, the episodic and unpredictable nature of FOG can be measured and assessed over several minutes. Given the transient nature of FOG, a longer walking evaluation period (i.e., minutes rather than just a few seconds) is preferred. In addition, the "tremor-like" shaking of the legs during FOG can be analyzed using time series analyses. This quantitative analysis demonstrated that the trembling during FOG is distinct from classical tremor—both in terms of frequency and complexity of the leg fluctuations—but also distinct from normal locomotion.[13] In FOG, the legs fluctuate in a complex pattern with much of the power centered around 2 to 4 Hz (Figure 17.1). Although the fluctuations may seem random, the leg movements fluctuate in a fairly organized pattern. One possibility is that the movements during FOG are generated by an independent generator or by misfiring oscillators that force the legs to move too fast for effective stepping. Using this method, it was shown that the center-of-pressure fluctuations during FOG are unique and very different from normal gait.[22]

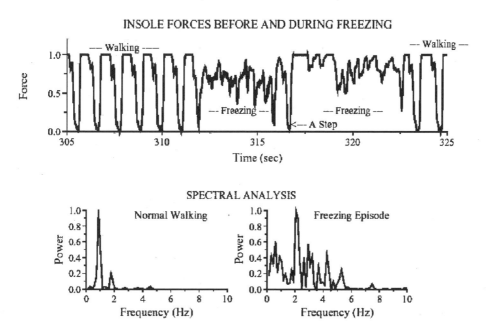

FIGURE 17.1   Example of freezing of gait in 77-year-old man with advanced PD (Hoehn and Yahr stage 3). The top panel shows the insole force before, during, and after a freezing episode. The bottom panel shows the results of spectral analysis for the walking period and for the freezing episode (the data before the breakthrough step were analyzed here). Note the large percentage of power in the 3 to 6 Hz band during freezing, but not during walking. (Adapted from Ref. 22.)

When stride-to-stride variation was compared during inter-episodic intervals (i.e., during regular walking, between two FOG episodes) among PD patients with FOG and those with no FOG in the "off" state, patients with FOG had significantly larger stride-to-stride variability.[23,24] This observation and the report by Nieuwboer et al.[25] about changes in stride length and cadence in the few steps prior to freezing episodes suggest primary disrhythmic locomotion in PD patients with FOG that may worsen until freezing appears. In other words, FOG might be the extreme form of a general continuous disrhythmicity.[13]

In the early stages of PD, a FOG episode usually lasts a second or two, occurring mainly in the form of turn or start hesitation and causing only minor disturbances in general function or quality of life, with tricks being needed only rarely to overcome the block. In a recent retrospect analysis of a group of patients who experienced FOG episodes early in the course of the disease, prior to any medical treatment, and who were followed in the Deprenyl and Tocopherol Antioxidative Therapy of Parkinsonism (DATATOP) cohort for a mean of 6.0 ± 1.4 years, early appearance of FOG was not associated with later development of atypical parkinsonism.[26]

As PD progresses, FOG episodes become one of the most disabling motor symptoms in the "off" state. In most cases, they respond to L-dopa treatment, which decreases their frequency, duration and akinetic nature.[17,23] At this stage, they are commonly associated with postural instability and, as a result, can lead to falls.[18,27–29]

Stage 4.   Appearance of Falls on the Background of Severe Continuous and Episodic Gait Disorders

Falls are the most serious complication of the gait disturbance in PD. Together with other accepted cardinal neurological signs of PD, falls may be recognized as a sign of disease progression and the result of decompensated postural instability and gait dysrhythmicity.[11] Falls become increasingly important and develop into one of the chief complaints among PD patients and their caregivers. They are the leading cause of physical trauma, fear of fallings, and restriction of day-to-day activity in PD patients.[10]

APPROACH TO TREATMENT

DRUG TREATMENT

L-dopa

L-dopa, the most effective and commonly used antiparkinsonian drug, has significant and long-lasting effects on parkinsonian gait. Shortly after levodopa was introduced for the treatment of parkinsonism in the late 1960s, its effect on gait velocity became clear.[30,31] L-dopa significantly improves stride length, velocity and synchronization of movements, double support time, and control of foot landing.[32–34] The effect of L-dopa on locomotion may occur through mechanisms involved in control of force and amplitude rather than rhythmicity or automaticity. [33] For example, improvement of muscle tone at the knee extensors can improve the angular excursion of the knee joint, which could provide better utilization of energy during propulsive phase and thus increase stride length.[35] L-dopa significantly decreased FOG frequency and the number of episodes with akinesia.[17] Stride time variability significantly improved in response to L-dopa, both in PD fallers and nonfallers, but remained higher in fallers (versus nonfallers).[23] The symptomatic benefit of L-dopa is better in younger onset (<40 yr) patients, suggesting again that gait disturbances in the older PD population are related in part to nondopaminergic mechanisms. L-dopa typically reduces rigidity, bradykinesia, and tremor and usually alleviates "off" state FOG,[23] but it is not very effective in preventing falls. In a prospective survey of fall circumstances during daily life,[36] most falls occurred when patients were in their best clinical condition ("on" state), possibly reflecting their increased mobility when symptoms are well controlled. In fact, one-third of the falls occurred when patients were experiencing dyskinesias, some of which may have been sufficiently severe to perturb the patients and cause a fall. Quantitative analyses of postural reactions to sudden movements of a supporting platform also indicate that dopaminergic medication provides insufficient relief of deficits, although some aspects partially improved.[37–39]

The motor complications of long-term treatment with levodopa, especially dyskinesia (dystonia and chorea), have serious implications with regard to gait. Severe painful foot or leg dystonia as well as violent generalized or focal chorea/ballism, frequently aggravated by walking, have become a major problem of advanced PD patients. Patients often have to choose between an akinetic "off" state with freezing or "on" with disabling dyskinesias.

Dopamine Agonists

Dopamine agonists are frequently used as monotherapy or as add-on therapy to improve L-dopa benefit or to spare L-dopa dosage. Agonist clinical benefit on gait, as measured by the Unified Parkinson's Disease Rating Scale (UPDRS)[40] and by other clinical assessment tools, has been shown by numerous studies. Dopamine agonists have also been shown to reduce FOG.[41] However, a case control study[42] and two long-term, prospective, double blind studies that compared dopamine agonists (ropinirole and pramipexole) to L-dopa in early PD patients[43,44] reported a trend toward increased frequency of FOG episodes in those patients who had been treated with dopamine agonists. Furthermore, a recently reported retrospective survey of 172 advanced PD patients showed that longer duration of treatment with dopamine agonist drugs tended to have an independent contribution to the risk of

developing FOG ($p = 0.07$).[29] In light of the current trend to substitute L-dopa with high-dose dopamine agonists, we suggest exercising caution when prescribing dopamine agonist drugs to those patients who experience FOG episodes in the early stages of PD until more long-term data is available.

MONOAMINE OXIDASE TYPE B (MAO-B) INHIBITOR

Selegiline as a MAO-B inhibitor is the only drug that has been shown to decrease FOG severity in a double blind, prospective study in the early stage of PD, prior to L-dopa treatment.[26] A long-term follow-up of the DATATOP study confirmed the symptomatic effect of selegiline; it decreased FOG severity even after 3 to 5 years of L-dopa treatment.[45]

Recently, a newly developed MAO-B inhibitor (Rasagiline) has also been shown to be effective for decreasing FOG in patients with advanced PD.[46] In a prospective double blind study, Rasagiline (1 mg/day) was better than Entacapone and placebo in improving FOG severity after 10 wk of treatment. The improvement was poorly correlated to decrease in "total daily OFF time," as measured by personal diaries and improvement in the UPDRS total or motor score, to suggest that Rasagiline improved FOG severity through other mechanisms than just increasing dopamine in the nigrostriatal dopaminergic synapse.

It is important to note that dopaminergic medication can cause balance and gait deficits in PD. For example, dopaminergic medication can lead to orthostatic hypotension and thereby cause syncopal falls.[47] Confusion, hallucinations, and psychosis are other adverse medication effects that may induce falls by causing recklessness or dangerous wandering behavior.

The resistance of many gait and balance problems to dopaminergic treatment suggests that nondopaminergic disturbances might be involved in its pathophysiology.[37,48] This could include the adrenergic locus coeruleus and the cholinergic/glutaminergic pedunculo-pontine nucleus. Quite logically, attempts have been made to ameliorate balance and gait deficits using drugs aimed at restoring these nondopaminergic deficits. L-threo-DOPS (a synthetic precursor of norepinephrine) has received most attention. Beneficial effects on freezing, gait impairment, retropulsion, and postural instability were noted in small, poorly controlled studies.[49,50] Pending further evidence, there is presently no place for L-threo-DOPS or other nondopaminergic compounds in daily clinical practice.

Deep Brain Stimulation

Stereotactic deep brain surgery aimed at the internal globus pallidus or subthalamic nucleus can partially alleviate gait impairment and postural instability in well selected patients with advanced PD.[51–64] Bilateral interventions seem to be more effective than unilateral interventions,[55,56] and the effects are generally greatest for "off" state axial symptoms. A recent meta-analysis showed that bilateral pallidal stimulation and bilateral subthalamic nucleus stimulation were significantly more effective than unilateral pallidotomy in reducing gait and balance impairment during the OFF state.[54] Both unilateral and bilateral STN stimulation can alleviate FOG,[55,58] but "on" state FOG may not improve much.[51] Thalamic surgery occasionally improves gait in PD, but the risk of postural deficits is considerable, particularly following bilateral approaches.[59] Pallidotomy also carries a risk of causing or aggravating gait and balance deficits, again mainly after bilateral procedures. Careful selection of appropriate candidates seems critical, because the effects on axial symptoms vary considerably among patients, and some do not improve at all. Younger patients with milder and dopa-responsive axial symptom seem to respond best.[60]

Physiotherapy

Posture, balance, gait, and transfers can be treated by physiotherapy.[14,61] Physical therapy may induce small but significant improvements in gait speed and stride length.[14] A sensory, cue-enhanced physical therapy program showed improvements lasting up to three months after the therapy had ended.[14,62] Examples of possibly useful interventions also include teaching of alternative motor strategies so as to make safer transfers,[63] gait training with external weight support,[64] and the use of exercises to improve stability, spinal flexibility, and general fitness.[65,66] Patients with FOG should be taught not to try to overcome their motor block during walking, as this may increase the risk of a fall. Physical therapy is best delivered in the domestic situation, as the effects of home treatment exceeded those of hospital-based interventions.[67] However, recent meta-analyses concluded that there is little evidence to support or refute the use of physical therapy, because of methodological flaws in published studies.[14,68,69] While there is much potential, further study is needed.

Other Measures

A recent study found that an intensive multidisciplinary rehabilitation program—focused on patients and care givers—significantly improved the patients' mobility and gait.[70] Interestingly, patients with more advanced disease at baseline benefited most from treatment. The care givers also responded positively. These findings should be expanded so as to design an optimal intervention program; to address such basic issues as cost-effectiveness, duration of effect, and compliance; and to distill the most effective components of the multidisciplinary rehabilitation program.

REFERENCES

1. Wolfson L. I., Whipple, R, Amerman, P., et al., Gait and balance in the elderly: two functional capacities that link sensory and motor ability to falls. *Clin. Geriatr. Med.,* 1: 649–659, 1985.
2. Fukuyama, H., Ouchi, Y., Matsuzaki, S., et al., Brain functional activity during gait in normal subjects: a SPECT study, *Neurosci. Lett.,* 228: 183–186, 1997.
3. Miyai, I., Tanabe, H. C., Sase, I. et al., Cortical mapping of gait in humans: a near-infrared spectroscopic topography study, *Neuroimage,* 14:1186–1192, 2001.
4. Hanakawa, T., Katsumi, Y., Fukuyam, H. et al., Enhanced lateral premotor activity during paradoxical gait in Parkinson's disease, *Brain,* 122:1271–1282, 1999.
5. Georgiou, N., Bradshow, J. L., Iansek, R. et al., Reduction in external cues and movement sequences in and Parkinson's disease, *J. Neurol. Neurosurg. Psychiatry,* 57: 368–370, 1994.
6. Azulay, J. P., Mesure, S., Amblard, B. et al., Visual control in Parkinson's disease, *Brain,* 122: 111–120, 1999.
7. Ouchi, Y., Kanno, T., Okada, H. et al., Changes in dopamine availability in the nigrostrial and mesocortical dopaminergic systems by gait in Parkinson's disease, *Brain,* 124: 784–792, 2001.
8. Marsden, C. D., What do the basal ganglia tell premotor cortical area? *Ciba Found. Symp.,* 132: 282–295, 1987.
9. Lewis, G. W., Byblow, W. D., Walt, S. E., Stride length regulation in Parkinson's disease: the use of extrinsic, visual cues, *Brain,* 123: 2077–2090, 2000.
10. Hausdorff, J. M., Balash, Y., Giladi, N., Effects of Cognitive Challenge on Gait Variability Patients with Parkinson's Disease, *J. Geriatr. Psychiatry Neurol.,* 2004. In press.
11. Bloem B. R., Grimbergen, Y. A., Cramer, M. et al., "Stops walking when talking" does not predict falls in Parkinson's disease. *Ann. Neurol.,* 48: 268, 2000.
12. Hausdorff, J. M., Schaafsma, J.D., Balash, Y. et al., Impaired regulation of stride variability in Parkinson's disease subjects with freezing of gait, *Exp Brain Res.,* 149: 187–94, 2003.
13. Giladi, M. D., Hausdorff, J. M., Balash, Y. et al., Episodic and Continuous Gait Disturbances in Parkinson's disease, in press.
14. Rubinstein, T., Giladi, N., Hausdorff, J. M., The power of cueing circumvent dopamine deficits: A brief review of physical therapy treatment of gait disturbances in Parkinson's disease, *Mov. Disord.,* 17:1148–1160, 2002.
15. Stern, G. M., Lander, C. M., Lees, A. J., Akinetic freezing and trick movements in Parkinson's disease, *J. Neural Transm.,* 137–141, 1980.
16. Giladi, N., Balash, Y., Hausdorff, J. M., Gait disturbances in Parkinson's disease, In Mizuno, Y., Fisher, A., Hanin, I., Eds., *Mapping the Progress of Alzheimer's and Parkinson's diseases,* Kluwer Academic/Plenum Publishers, New York, 329–335, 2002.
17. Schaafsma, J. D., Balash, Y., Gurevich, T. et al., Characterization of freezing of gait subtypes and the response

of each to levodopa in Parkinson's disease, *Eur. J. Neurol.*, 10: 391–398, 2003.

18. Giladi, N., McMahon, D., Przedborski, S. et al., Motor blocks in Parkinson's disease, *Neurology*, 42:333–339, 1992.

19. Bartels, A. L., Balash, Y., Gurevich, T. et al., Relationship between freezing of gait (FOG) and other features of Parkinson's: FOG is not correlated with bradykinesia, *J. Clin. Neurosci.*, 10: 584–588, 2003.

20. Giladi, N., Shabtai, H., Simon, E. S. et al., Construction of freezing of gait questionnaire for patients with parkinsonism, *Parkinsonism Relat. Disord.*, 6: 165–170, 2000.

21. Hausdorff, J. M., Ladin, Z., Wei, J. Y., Footswitch system for measurement of the temporal parameters of gait, *J. Biomech.*, 28:347–351, 1995.

22. Hausdorff, J. M., Balash, Y., Giladi, N., Time series analysis of leg movements during freezing of gait in Parkinson's disease: rhyme or reason? *Physica A.*, 321:565–570, 2003.

23. Schaafsma, J. D., Giladi, N., Balash, Y. et al., Gait dynamics in Parkinson's disease: relationship to Parkinsonian features, falls and response to levodopa, *J. Neurol. Sci.*, 212: 47–53, 2003.

24. Hausdorff, J. M., Balash, Y., Giladi, N., Effects of cognitive challenge on gait variability in patients with Parkinson's disease, *J. Geriatr. Psychiatry Neurol.*, 16:53–58, 2003.

25. Nieuwboer, A., Dom, R., De Weerdt, W. et al., Abnormalities of the spatiotemporal characteristics of gait at the onset of freezing in Parkinson's disease, *Mov. Disord.*, 16:1066–1075, 2001.

26. Giladi, N., McDermott, M. P., Fahn, S. et al., Freezing of gait in PD. Prospective assessment in the DATATOP cohort, *Neurology*, 56:1712–1721, 2001.

27. Giladi, N., Shabtai, H., Rozenberg, E. et al., Gait festination in Parkinson's disease, *Parkinsonism Relat. Disord.*, 7:135–138, 2001.

28. Gray, P., Hildebrand, K., Fall risk factors in Parkinson's disease, *J. Neurosci. Nurs.*, 32:222–228, 2000.

29. Giladi, N., Treves, T. A., Simon, E. S. et al., Freezing of gait in patients with advanced Parkinson's disease, *J. Neural Transm.*, 108:53–61, 2001.

30. Mones, R. J., An evaluation of L-dopa in Parkinson patients, *Trans. Am. Neurol. Assoc.*, 94:307–309, 1969.

31. Boshes, B., Blonsky, E. R., Arbit, J. et al., Effect of L-dopa on individual symptoms of parkinsonism, *Trans. Am. Neurol. Assoc.*, 94:229–231, 1969.

32. Bowes, S. G., Charlett, A., Dobbs, R. J. et al., Gait in relation to ageing and idiopathic parkinsonism, *Scand. J. Rehabil. Med.*, 24:181–186, 1992.

33. Ferrandez, A. M., Blin, O., A comparison between the effect of intentional modulations and the action of L-dopa on gait in Parkinson's disease, *Behav. Brain Res.*, 45:177–183, 1991.

34. Pedersen, S. W., Oberg, B., Larsson, L. E. et al., Gait analysis, isokinetic muscle strength measurement in patients with Parkinson's disease, *Scand. J. Rehabil. Med.*, 29: 67–74, 1997.

35. Knutsson, E., Martensson, A., Quantitative effects of L-dopa on different types of movements and muscle tone in Parkinsonian patients, *Scand. J. Rehabil. Med.*, 3:121–130, 1971.

36. Bloem, B. R., Grimbergen, Y. A., Cramer, M. et al., Prospective assessment of falls in Parkinson's disease, *J. Neurol.*, 248:950–958, 2001.

37. Bloem, B. R., Beckley, D. J., van Dijk, J. G. et al., Influence of dopaminergic medication on automatic postural responses and balance impairment in Parkinson's disease, *Mov. Disord.*, 11:509–521, 1996.

38. Beckley, D. J., Panzer, V. P., Remler, M. P. et al., Clinical correlates of motor performance during paced postural tasks in Parkinson's disease, *J. Neurol. Sci.*, 132:133–138, 1995.

39. Frank, J. S., Horak, F. B., Nutt, J., Centrally initiated postural adjustments in parkinsonian patients on and off levodopa, *J. Neurophysiol.*, 84:2440–2448, 2000.

40. Fahn, S., Elton, R. L., members of the UPDRS Development Committee, The unified Parkinson's disease rating scale, In: Fahn, S., Marsden, C. D., Calne, D. B., Goldstein, M., Eds, *Recent developments in Parkinson's disease,* Vol. 2., Macmillan Healthcare Information, Florham, Park, N. J., 153–163, 1987.

41. Linazasoro, G., The apomorphine test in gait disorders associated with parkinsonism, *Clin. Neuropharmacol.*, 19: 171–176, 1996.

42. Weiner, W. J., Factor, S. A., Sanches-Ramos, J. R. et al., Early combination therapy (bromocriptin and levodopa) does not prevent motor fluctuations in Parkinson's disease, *Neurology*, 43: 21–27, 1993.

43. Rascol, O., Brooks, D. J., Korczyn, A. D. et al., A five-year study of the incidence of dyskinesia in patients with early Parkinson's disease who were treated with ropinirole or levodopa, *N. Engl. J. Med.*, 342:1484–1491, 2000.

44. Holloway, R., Shoulson, I., Kieburtz, K. et al., Pramipexole vs. levodopa as initial treatment for Parkinson disease, A Randomized Controlled Trial, *JAMA*, 284:1931–1938, 2000.

45. Shoulson, I., Oakes, D., Fahn, S. et al., Impact of sustained Deprenyl (Selegiline) in Levodopa-treated Parkinson's disease: A randomized placebo-controlled extension of the Deprenyl and Tocopherol Antioxidative therapy of parkinsonism trial, *Ann. Neurol.*, 51:604–612, 2002.

46. Giladi, N., Rascol, O., Brooks, D. J. et al., and the European LARGO Study Group. Rasagiline treatment can improve freezing of gait in advanced Parkinson's disease; a prospective randomized, double blind, placebo and entacapone controlled study, *Neurology*, 62 (suppl.), p. 04.137, 2004.

47. Van Dijk, J G., Haan, J., Zwinderman, K. et al., Autonomic nervous system dysfunction in Parkinson's disease: relationships with age, medication, duration, and severity, *J. Neurol. Neurosurg. Psychiatry*, 56:1090–1095, 1993.

48. Bonnet, A. M., Loria, Y., Saint-Hilaire et al., Does long-term aggravation of Parkinson's disease result from non-dopaminergic lesions? *Neurology*, 37:1539–1542, 1987.

49. Narabayashi, N., Kondo, T., Results of a double-blind study of L-threo-DOPS in parkinsonism, in Fahn, S., Marsden, C. D., Goldstein, M., eds., *Recent Developments in Parkinson's Disease,* Macmillan, New York, 279–291, 1987.

50. Tohgi, H., Abe, T., Takahashi, S., The effects of L-threo-3, 4-dihydroxyphenylserine on the total norepinephrine and dopamine concentrations in the cerebrospinal fluid and freezing gait in parkinsonian patients, *J. Neural. Transm. Park. Dis. Dement. Sect.,* 5: 27–34, 1993.

51. Stolze, H., Klebe, S., Poepping, M. et al., Effects of bilateral subthalamic nucleus stimulation on parkinsonian gait, *Neurology,* 57:144–146, 2001.

52. Faist, M., Xie, J., Kurz, D. et al., Effect of bilateral subthalamic nucleus stimulation on gait in Parkinson's disease, *Brain,* 124:1590–1600, 2001.

53. Bronte-Stewart, H., M, Minn, A. Y., Rodrigues, K. et al., Postural instability in idiopathic Parkinson's disease: the role of medication and unilateral pallidotomy, *Brain,* 125:2100–2114, 2002.

54. Maurer, C., Mergner, T., Xie, J. et al., Effect of chronic bilateral subthalamic nucleus (STN) stimulation on postural control in Parkinson's disease, *Brain,* 126:1146–1163, 2003.

55. Yokoyama, T., Sugiyama, K., Nishizawa, S. et al., Subthalamic nucleus stimulation for gait disturbance in Parkinson's disease, *Neurosurgery,* 45:41–47, 1999.

56. Kumar, R., Lozano, A. M., Sime, E. et al., Comparative effects of unilateral and bilateral subthalamic nucleus deep brain stimulation, *Neurology,* 53:561–566, 1999.

57. Bakker, M., Esselink, R. A., Renooij, J. et al., Effects of stereotactic neurosurgery on postural instability and gait in Parkinson's disease, *Mov. Disord.,* 2004. In press.

58. Bejjani, B., Gervais, D., Arnulf, I. et al., Axial parkinsonian symptoms can be improved: the role of levodopa and bilateral subthalamic stimulation, *J. Neurol. Neurosurg. Psychiatry,* 68:595–600, 2000.

59. Speelman, J. D., *Parkinson's disease and stereotaxic neurosurgery,* Amsterdam, thesis, 1991.

60. Welter, M. L., Houeto, J. L., Tezenas du, M. S. et al., Clinical predictive factors of subthalamic stimulation in Parkinson's disease, *Brain,* 125:575–583, 2002.

61. Plant, R. D., Jones, D., Ashburn, A. et al., Physiotherapy for people with Parkinson's disease, UK Best Practice, Short Report, Newcastle upon Ty, Institute of Rehabilitation, 2001.

62. Nieuwboer, A., Feys, P., de Weerdt, W, et al., Is using a cue the clue to the treatment of freezing in Parkinson's disease? *Physiother. Res. Int.,* 2:125–132, 1997.

63. Kamsma, Y. P., Brouwer, W. H., Lakke, J. P., Training of compensation strategies for impaired gross motor skills in Parkinson's disease, *Physiother. Theory Pract.,* 11:209–229, 1995.

64. Schenkman, M., Cutson, T. M., Kuchibhatla, M. et al., Exercise to improve spinal flexibility and function for people with Parkinson's disease: a randomised, controlled trial, *J. Am. Geriatr. Soc.,* 46:1207–1216, 1998.

65. Morris, M. E., Movement disorders in people with Parkinson disease: a model for physical therapy, *Phys. Ther.,* 80:578–597, 2000.

66. Schenkman, M., Cutson, T. M., Kuchibhatla, M. et al., Exercise to improve spinal flexibility and function for people with Parkinson's disease: a randomised, controlled trial, *J. Am. Geriatr. Soc.,* 46:1207–1216, 1998.

67. Nieuwboer, A., de Weerdt, W., Dom, R. et al., The effect of a home physiotherapy program for persons with Parkinson's disease, *J. Rehabil. Med.,* 33: 266–272, 2001.

68. DeGoede, C. J. T., Keusm S. H. J., Kwakkelm G. et al., The effects of physical therapy in Parkinson's disease: a research synthesis, *Arch Phys. Med. Rehabil.,* 82:509–515, 2001.

69. Deane, K. H., Ellis-Hill, C., Jones, D et al., Systematic review of paramedical therapies for Parkinson's disease, *Mov. Disord.,* 17:984–991, 2002.

70. Trend, P., Kaye, J., Gage et al., Short-term effectiveness of intensive multidisciplinary rehabilitation for people with Parkinson's disease and their carers, *Clin. Rehabil.,* 16:717–725, 2002.

18 Sensory Symptoms and Sensorimotor Distortion in Parkinson's Disease

Padraig O'Suilleabhain
Department of Neurology, University of Texas Southwestern Medical School

CONTENTS

Parkinson's disease (PD) is, with good reason, conventionally considered a movement disorder. James Parkinson declared "the senses…uninjured" in the first paragraph of his monograph. However many patients with PD do experience unpleasant sensations, and for some these sensory symptoms are the biggest problem. The first half of this chapter focuses on these unpleasant somatic sensations occurring independent of or out of proportion to the cardinal motor symptoms. Vision and smell are not addressed, and pain as an extreme of dysesthesia is considered only to a limited extent. In the second half, attention is given to kinesthesia, which appears to be defective in PD. This is not a symptom as such, as patients are not conscious of it, but distortions of input and processing of interoceptive information may contribute to bradykinesia, dyskinesias, and postural instability, among other parkinsonian signs.

SENSORY SYMPTOMS

QUALITY

Numbness is the most common term used by patients with PD to describe unwanted somatic sensations. Other patients describe tingling, others burning, yet others itching or crawling. Some experience focal coldness as though the limb were going to sleep. The extent to which all these different descriptions reflect differing pathophysiology or semantics is uncertain. When asked what they think is the origin or depth of these types of sensations, most patients chose skin rather than muscle or bone. This class of symptoms will be referred to as *parasthesias* for the remainder of the chapter. These superficial parasthesias are common in the more parkinsonian arm or leg as compared with the contralateral side. They can last for hours at a time or can be continuous.

Another dysesthesia, fairly specific to PD, is the unpleasant restless sensation in muscle or skin that impels the patient to contract the muscle. The sensation mimics in character the experience of patients with restless legs syndrome, though the heredity, age of onset, and associated features are not identical.[1] When patients develop restless sensations in the years after PD onset and treatment with dopaminergic drugs, then the symptom usually occurs as an end of dose or off-state phenomenon.

Internal tremors, the subjective experience of rhythmic truncal vibrations in the absence of visible tremor, occur in PD patients only some of whom also have visible limb tremors.[2] Internal tremors appear to be distinct from palpitations, and while anxiety is common in these patients, the symptom is not likely a somatic manifestation of anxiety; anxious people who do not have PD do not describe it in terms of a tremor inside the trunk.

Occasionally, a patient will describe a brief shooting sensation, usually from the spine into an extremity. While these dysesthesias are not unique to PD, PD patients may be predisposed. In one case report, there was lancinating facial pain of trigeminal neuralgia, which was secondary to PD as indicated by its response to levodopa.[3]

Muscle aches, sometimes described as fatigue or heaviness, are not unusual. In some cases, the experience reaches a threshold for pain (although, in my clinic, when a patient with PD complains of an ache that I later call a pain, quite commonly I am pulled up: "It's not a pain, it's an ache."). Discomfort that sounds rheumatologic in nature was described in PD by Parkinson and by Charcot. Lumbar, cervical, and shoulder aches seem to be more common in people with PD than in others. Perhaps the chronic rigidity and paucity of spontaneous movements precipitate or worsen arthropathies.

Sensory symptoms generally follow, within a few years, the onset of motor impairment, although in about 20% of patients, dysesthesias come first.[4] The various sensory complaints described above are often fairly mild in severity. Motor and cognitive symptoms are usually of greater concern to the patient. However the shooting pains, the restlessness, and (less frequently) the heaviness and the parasthesias are very distressing to a significant minority of patients.[5]

Prevalence

Surveys of PD patients regarding dysesthesias find that roughly half of patients are affected. Snider et al. surveyed 101 consecutive patients in their clinic who had PD and who lacked an identifiable predisposition to neuropathy. Forty-three reported sensory complaints, compared with 8% of controls.[4] Koller surveyed 50 consecutive patients with PD and found that 9 had intermittent sensory complaints, and in 10 such symptoms were continuous. The most common dysesthesia in this cohort was parasthesia.[6] Shulman et al. asked 99 patients who had had PD for, on average, seven years and who were, for the most part, taking levodopa, "Do you notice any of the following sensations (pain, numbness, tingling or burning) related to your PD?" Sixty-three percent responded in the affirmative.[7] Gunal et al. surveyed 85 sequential PD patients, 72 of whom had motor fluctuations. Fifty-nine had sensory complaints, and, for many, these were parasthesias occurring specifically during the off period.[8]

Recognizing that dysesthesias can appear as one of a number of nonmotor fluctuations, a number of studies have addressed the prevalence of sensory symptoms specifically during the off state. Witjas et al. asked 50 patients with motor fluctuations a series of 54 yes/no questions about various sensory, autonomic, and neuropsychiatric symptoms occurring at specific stages of the dose cycle.[5] About 40% of the subjects described tightness or parasthesias, while others described sensations that the authors listed as autonomic phenomena (feeling warm all over, feeling cold distally, feeling hungry). Five

patients felt the off-state parasthesias or pains were more disabling than the motor component of the off state. However, less directed surveys provide lower estimates of prevalence and severity. Hillen et al. asked patients to volunteer those symptoms they associated with the "off" state. Of 130 consecutive patients with PD and motor fluctuations, only 3 described dysesthesias.[3]

Surveys about restlessness (or akathisia, used synonymously with restlessness by some authors) in PD produce a wide range of prevalence estimates, the variability probably due to the wording of the survey questions. The low estimates are 15 to 20%,[1,9] while other studies report prevalence of 45 to 54%.[5,10] Sixty-eight of 100 patients interviewed by Lang et al. described the frequent need to move, which was felt to be secondary to an uncomfortable sensation in most cases, though in a minority it was proactive (i.e., akathisia) rather than reactive.[11]

The surveys referenced in these past few paragraphs were performed at tertiary referral centers. None of the population studies of PD has systematically quantified the burden of sensory complaints.

Pathologic Basis

It is easy to explain radiculopathies and musculoskeletal pains as secondary to motor disruption without invoking direct parkinsonian involvement of the sensory system. The pathophysiology of parasthesias and restlessness in PD on the other hand is unknown. Clinical risk factors such as early age of PD onset and long duration of treatment and higher doses of levodopa[8] provide ambiguous clues at best. Despite the high prevalence of parasthesias, clinical exam signs and objective findings are uncommon. Deep tendon reflexes are usually intact. Exam of touch, vibration, temperature, and other sensory modalities typically do not uncover abnormalities. While challenging tests of proprioception and kinesthesia can reveal deficits as described later in the chapter, joint position sensation is generally intact as determined by bedside testing. There was a report of mildly impaired two-point discrimination in PD,[12] but this finding has not been replicated.

Nerve conduction studies are generally normal in uncomplicated PD even in the presence of parasthesias.[6] A single study reported electrophysiologic evidence of radiculopathies in 53% (cervical) and 73% (lumbar) of 26 patients with PD.[13] Electrophysiologic abnormalities can be found in approximately one-third of cases of juvenile and young-onset PD,[14] although it is suspected that this subpopulation differs from the general population of PD patients in the mix of etiologies and pathologies, including some which involve the peripheral in addition to the central nervous system.

Somatosensory evoked responses up to and including the parietal N20-P25 are generally normal in idiopathic PD[15–17]. Some but not all studies reported the frontal peaks N30-P45 are low in amplitude and can be boosted with dopaminergic medication.[18,19] Central aspects of the somatosensory evoked potentials can be delayed.[20]

While systematic histologic study of the sensory system in PD has not been undertaken as far as I am aware, the preceding observations imply integrity in the peripheral nerves and the primary somatosensory pathways extending from the spine to the thalamus and the primary sensory cortex. Logically, then, dysesthesias in PD are the result of either a central sensory distortion or the activation of sensory nerves secondary to motor or autonomic dysfunction. As outlined below, central mechanism seems plausible, although this has not been established, nor has peripheral activation of nerves been excluded.

We know that stimulation of the skin produces responses in the basal ganglia[21] and that the substantia nigra modulates sensory processing in thalamic nuclei.[22] It is reasonable to suspect that disruption of this extrapyramidal circuitry in PD will sometimes produce paresthesias and other sensory complaints.

Levodopa can relieve parasthesias in PD, and the fact that large neutral amino acids can interfere with the relief suggests that this is not a local action of dopamine at the extremity.[23] However, it is likely that more than simple dopamine deficiency is responsible; dopamine antagonists can produce uncomfortable sensations or a restless need to move but at a lower frequency than seen in PD, and chronic superficial parasthesias are not reported to occur widely in patients using neuroleptics.

Notwithstanding the tentative assumption that the parasthesias of PD are for the most part central in origin, some parkinsonian patients with dysesthesias have identifiable peripheral sources. Focal compression neuropathies and plexopathies can occur in patients frozen in abnormal positions for extended periods.[24] Iatrogenic neuropathy has been described as a result of amantadine toxicity.[25] There are uncommon neurodegenerations affecting the peripheral nervous system in combination with the extrapyramidal system. Some of these are genetic.[26,27] Parkinsonism developing at a young age probably has a different mix of genetic and biochemical etiologies as compared with PD arising later in life and, as noted in the previous section, extension of neuropathology into the peripheral nervous system seems more common. Polyglucosan body disease can affect both systems.[28] Some toxins[29–31] and Guamanian neurodegeneration,[32] which may be toxic, can cause neuropathy in combination with parkinsonism. Finally, sensory complaints are sometimes independent of the PD. Absent data to the contrary, it is assumed that PD patients have sensory and sensorimotor neuropathies to the same extent as age-matched non-PD controls. Thus, if symmetric sensory complaints and signs such as hyporeflexia and loss of vibration sensation are found in PD, they likely reflect a coincidental and independent neuropathy rather than a PD complication.

MANAGEMENT

There is no randomized controlled trial of treatment of sensory complaints in PD. That said, when patients with PD desire relief from dysesthesias, empiric treatment with dopaminergic drugs is generally practical and often rewarding. I usually adjust times and doses of dopaminergic drugs as a first step in addressing sensory complaints in PD. If the symptom tends to be better when drug levels tend to be higher, then no further investigation is needed for either cause or solution. Rarely, parasthesias, particularly burning sensations, are worsened by dopaminergic drugs.[4] There is little data on the relative efficacy of dopamine agonists as compared with levodopa for treating sensory symptoms, and each can be tried initially at low dose and then slowly increased as needed and tolerated. Post hoc analysis of the CALM-PD data revealed that, while those randomized to levodopa had more improvement in a number of motor and ADL items of the UPDRS, there was a trend toward better improvement of sensory item in the pramipexole arm.[33] As regards the time course of symptom relief, parasthesias and restlessness can improve as soon as a therapeutic dose of the dopaminergic drug is reached. By contrast, rheumatologic aches secondary to prolonged excessive rigidity and immobility often do not resolve for many weeks after hypertonicity is pharmacologically reduced. Musculoskeletal complaints can also be helped by physical therapies, including passive and active exercises, massage, and ultrasound.

Deep brain stimulation of the globus pallidus tends to improve sensory complaints; half of the 16 patients operated on by Loher et al. had dysesthesias, and in each case the dysesthesia essentially disappeared postoperatively.[34] The effect on sensory symptoms of subthalamic stimulation, now a more common intervention for PD, has not been systematically studied. High-frequency thalamic stimulation, sometimes used to alleviate parkinsonian tremor, in many patients will cause parasthesias at high settings. This iatrogenic sensory symptom is easily and instantly cured by reducing the stimulation. Alternatively, the stimulation can be delivered via adjacent lead contacts to reduce the field. In this way, side effects can be restricted without sacrificing tremor control. Parenthetically, thalamic (ventrocaudal) stimulation was found not to worsen somatosensory acuity.[35]

DEFECTIVE PROPRIOCEPTION AND KINESTHESIA IN PD

A MODEL EXPLAINING PD SIGNS AS CONSEQUENCES OF DISORDERED SENSORIMOTOR INTEGRATION

The precise function of the basal ganglia remains mysterious more than 20 years after Marsden's proposal that they "automatically execute learned motor plans."[36] Accurate execution requires sensory feedback from the muscles, joints, soft tissues, and skin. Three stages of a circuit might be considered: a sensory input, a black box sensorimotor integrator, and a motor output. Basal ganglia have often been considered in the "motor output" stage, but an argument can be made for placing it in the sensorimotor integration stage. The basal ganglia may function to scale movements depending, among other things, on its reading of muscle position and tension, titrating the output so that motion is sufficient but not excessive.[37] In the course of movement, it may participate in a comparator system in which kinesthetic feedback is analyzed with reference to a corollary discharge representing the size of the intended movement in terms of space, force, and effort.[38,39] The extrapyramidal system may, in addition, release successive stages of sequential movements based on the interoceptive feedback.[40] Sensory feedback about position and musculotendinous tension, and perhaps more importantly about their rates of change, would be important to movement and particularly the automatic movements for which the basal ganglia appear to have specialization. An incorrectly tuned system could program movements to be either bradykinetic and dampened or excessive and overflowing, depending on the nature of the disruption. The model might be extended to explain resting hypokinesis on the one hand and adventitious/dyskinetic/choreiform movement on the other, occurring while the patient is not engaged in purposeful activity; perhaps the brain desires a certain stream of information regarding the current deployment of body parts. If the extrapyramidal system overreads position and motion, then lower-than-normal spontaneous movement will provide the desired amount of information. If, on the other hand, the system lacks sensitivity toward interoceptive information, then amplified movements would be needed to supply the required level of input, and so adventitious movements would be unconsciously generated. In summary, then, extrapyramidal "movement" disorders might be considered sensorimotor disorders, because motor output cannot be divorced from sensory input and from the central system that interprets this input.

RESEARCH FINDINGS

Some empiric data support the role of basal ganglia in sensorimotor integration, and also reported are abnormalities in proprioception and kinesthesia in PD that would be predicted by the model to bias toward hypokinesis, bradykinesia, dyskinesias, or postural instability.[37,39,41] From the experiments described in the following paragraphs, a reasonably coherent picture emerges that proprioception has impaired discrimination value in PD and that, particularly in the off state, the system overestimates movements to be greater than they actually are.

While bedside joint position testing is intact ("is the finger pointing up or down?"), Demirci et al. found deficits using a more subtle and quantitative test in which the metacarpophalangeal joint was passively placed in a flexion angle of 0.5 to 19.5°, and the subject estimated the angle relative to a visual representation of a 10° flexion.[42] Both normal controls and "on" state PD patients underestimated the palmar angle (or equivalently overestimated the dorsal angle), particularly in the most flexed positions, but PD patients as a group were worse. Jobst et al. performed a test involving both position and motion sense, a test that incidentally required memory as subjects had to retain and reproduce test stimuli. Errors were worse in PD.[43] Zia et al. avoided the dependence on memory by performing small passive flexion/extension movements of both elbows on blindfolded subjects who were then asked to state which arm was the more flexed. Errors were more common in PD.[44] O'Suilleabhain et al. confirmed this finding and also found that dopamine acutely worsened the discrimination in PD.[45] Klockgether and his colleagues used a system that tested kinesthesia to a greater extent than proprioception.[46] Subjects (shortly before the second levodopa dose of the day) had their unseen arm passively and slowly extended. As the hand passed over a virtual target represented by a light in contralateral hemispace, the subject had to terminate the extension. Passive movements of this nature, as well as active movements by the subject, were universally short of the target, implying that the patient perceived the motion to be greater than it actually was. Movements were more accurate when visual feedback was used.[47] While the kinesthetic input is perceived to be higher than it actually is, the system appears to lack sensitivity to things that should change the kinesthetic input. Vibration of the muscle spindle normally activates afferents to produce an illusion of motion. Intact subjects receiving vibration to a contracting muscle perceive the motion to be farther and faster than it actually is. In PD patients in the "off" state, vibration has less than half the impact it has on control subjects. In the "on" state the misperception was closer to what occurs in normal subjects.[48]

Proprioception, then, as a conscious function, is modestly indiscriminating in unmedicated PD, while kinesthesia as a conscious function moderately overestimates the magnitude of a motion and almost seems to be stuck on "high," because external influences fail to reduce estimates. The unconscious influence of proprioception and kinesthesia on automatic movements (walking, shifting in a chair) is of even greater interest, though analysis is difficult as compared with conscious sensation; when motor performance is the outcome of interest (as it must be for such analyses), the sensorimotor integration process is difficult to separate from any coexisting parkinsonian impairments of motor output. One approach has been to use transcranial stimulation of the motor cortex while manipulating sensory feedback. In normal subjects, the motor-evoked potential becomes larger as a relaxed muscle is passively shortened. In "on" state patients with PD, the motor evoked potential is smaller, and it barely increases as the muscle is shortened.[49] The modulation of motor potential amplitude may be occurring at the cortex rather than the spinal level, as intracortical inhibition as determined using paired stimuli is less affected by passive movement in PD as compared with controls.[49] In addition, contingent negative variation, which is the preparatory slow depolarization of motor and premotor cortex in anticipation of initiating a cued movement, is low in amplitude in PD.[50–52]

Animal research has supported aspects of this model of disordered sensorimotor integration as a result of extrapyramidal disruption. The basal ganglia receive large sensory input primarily from deep rather than superficial somatic receptors.[53,54] In MPTP-treated primates, thalamic[55] and pallidal neurons have excessive and indiscriminate responses to joint movement.[56,57] When globus pallidus is cooled, monkeys have deficits in judging elbow movements, deficits that are corrected when visual input can compensate for the apparently disrupted kinesthetic input.[58,59]

INTERPRETATION OF RESEARCH FINDINGS IN RELATION TO PD SIGNS

All these findings could be interpreted as showing the central sensorimotor network misperceives proprioceptive and kinesthetic input as if it were biased toward a high setting, perhaps at the upper end of a sigmoid response curve. The system then perceives a small motion to be larger than it actually is. It also incorrectly estimates that a minor motor effort will move the limb farther than it actually will. The deficient contingent negative variation might be considered another consequence of sensorimotor integration that overreads sensory input; the premotor cortex gears up for a lower effort than is actually needed as a result and produces less preparatory depolarization. The movement when executed is smaller than is appropriate. The fact that the initial attempt at movement is too small and slow in PD corresponds with a reduced efference signal. The fact that iterative correction does not follow in PD corresponds with reduced impact of feedback. Both might be modeled as results of overread sensory data at the sensorimotor integrator. Interestingly, exteroceptive information, whose influence on motion seems less reliant on basal ganglia integrity, can compensate so as to release stalled movements in PD. Examples of this include the ability to trigger reflexive stepping with an inverted cane or by cueing with hand-clapping. A related phenomenon may occur with speech generation: patients with PD speak at low volumes but believe they are speaking loudly.[36] The Lee Silverman Voice Therapy method induces patients to project sound at objectively more normal volumes: "think loud, think shout." To be successful, the patient must accept the therapist's advice that the output is normal and suspend reliance on his own perception because, to his mind, the patient feels he is shouting.[60]

It has not been possible to determine if the overread of interceptive information proposed above is caused by abnormal fusimotor innervation[61] or by abnormal sensory gating at the level of the basal ganglia. The reported deficiency in frontal components of SEP despite normal parietal components might suggest a central gating abnormality, but that data have not been universally reproducible, and the interpretation is speculative.

In conclusion, there is evidence for a model of disrupted sensorimotor integration in the basal ganglia as a basis for some of the motor impairments of PD. However, research data supporting the model are not conclusive and do not indicate the degree to which bradykinesia, hypokinesia, or dyskinesia, among other phenomena, are functions of such disruption.

REFERENCES

1. Ondo, W. G., Vuong, K. D., Jankovic, J., Exploring the relationship between Parkinson disease and restless legs syndrome, *Arch Neurol.*, 59(3):421–424, 2002.
2. Shulman, L. M., Singer, C., Bean, J. A., Weiner, W. J., Internal tremor in patients with Parkinson's disease, *Mov. Disord.*, 11(1):3–7, 1996.
3. Hillen, M E., Sage, J. I., Nonmotor fluctuations in patients with Parkinson's disease, *Neurology*, 47(5):1180–1183, 1996.
4. Snider, S. R., Fahn, S., Isgreen, W. P., Cote, L. J., Primary sensory symptoms in parkinsonism, *Neurology*, 26(5):423–429, 1976.
5. Witjas, T., Kaphan, E., Azulay, J. P., Blin, O., Ceccaldi, M., Pouget, J. et al., Nonmotor fluctuations in Parkinson's disease: frequent and disabling, *Neurology*, 59(3):408–413, 2002.
6. Koller, W. C., Sensory symptoms in Parkinson's disease, *Neurology*, 34(7):957–959, 1984.
7. Shulman, L. M., Taback, R. L., Bean, J., Weiner, W. J., Comorbidity of the nonmotor symptoms of Parkinson's disease, *Mov. Disord.*, 16(3):507–510, 2001.
8. Gunal, D. I., Nurichalichi, K., Tuncer, N., Bekiroglu, N., Aktan, S., The clinical profile of nonmotor fluctuations in Parkinson's disease patients, *Can. J. Neurol. Sci.*, 29(1):61–64, 2002.
9. Tan, E. K., Lum, S. Y., Wong, M. C., Restless legs syndrome in Parkinson's disease, *J. Neurol. Sci.*, 196(1–2):33–36, 2002.
10. Comella, C. L., Goetz, C. G., Akathisia in Parkinson's disease, *Mov. Disord.*, 9(5):545–549, 1994.
11. Lang, A. E., Johnson, K., Akathisia in idiopathic Parkinson's disease, *Neurology*, 37(3):477–481, 1987.
12. Schneider, J. S., Diamond, S. G., Markham, C. H., Parkinson's disease: sensory and motor problems in arms and hands, *Neurology*, 37(6):951–956, 1987.
13. Lee, D. H., Seo, W., Koh, S. B., Kim, B. J., Park, M. K., Park, K. W., Sensory symptoms and radiculopathies

in Parkinson's disease, *Mov. Disord.,* 17 Suppl. 5:S228, 2002.

14. Taly, A. B., Muthanem U. B., Involvement of peripheral nervous system in juvenile Parkinson's disease, *Acta Neurol. Scand.,* 85(4):272–275, 1992.

15. Drory, V. E., Inzelberg, R., Groozman, G. B., Korczyn, A. D., N30 somatosensory evoked potentials in patients with unilateral Parkinson's disease, *Acta Neurol. Scand.,* 97(2):73–76, 1998.

16. Huttunen, J., Teravainen, H., Pre- and postcentral cortical somatosensory evoked potentials in hemiparkinsonism, *Mov. Disord.,* 8(4):430–436,1993.

17. Nakashima, K., Nitta, T., Takahashi, K., Recovery functions of somatosensory evoked potentials in parkinsonian patients, *J. Neurol. Sci.,* 108(1):24–31, 1992.

18. de Mari, M., Margari, L., Lamberti, P., Iliceto, G., Ferrari, E., Changes in the amplitude of the N30 frontal component of SEPs during apomorphine test in parkinsonian patients, *J. Neural. Transm.,* Suppl., 1995;45:171–176.

19. Rossini, P. M., Traversa, R., Boccasena, P., Martino, G., Passarelli, F., Pacifici, L. et al., Parkinson's disease and somatosensory evoked potentials: apomorphine-induced transient potentiation of frontal components, *Neurology,* 43(12):2495–2500, 1993.

20. Choi, S., Minn, Y. K., Lim, S. R., Lee, J. H., Somatosensory and motor evoked potentials in Parkinson's disease, *Mov. Disord.,* 17 suppl. 5:S128, 2002.

21. Albe-Fessard, D., Rocha-Miranda, C., Oswalde-Cruz, E., Activite evoquees dans le moyen caude du chat: en reponse a des types divers d'afferents, *Electroencephalogr. Clin. Neurophysiol.,* 12:649–661, 1960.

22. Bendrups, A., McKenzie, J., Suppression of extralemniscal thalamic unit responses by substantia nigra stimulation, *Brain Res.,* 80:131–134, 1974.

23. Nutt, J. G., Carter, J. H., Sensory symptoms in parkinsonism related to central dopaminergic function, *Lancet,* 2(8400):456–457, 1984.

24. Kurlan, R., Baker, P., Miller, C., Shoulson, I., Severe compression neuropathy following sudden onset of parkinsonian immobility, *Arch. Neurol.,* 42(7):720, 1985.

25. Shulman, L. M., Minagar, A., Sharma, K., Weiner, W. J., Amantadine-induced peripheral neuropathy, *Neurology,* 53(8):1862–1865, 1999.

26. Byrne, E., Thomas, P. K., Zilkha, K. J., Familial extrapyramidal disease with peripheral neuropathy, *J. Neurol. Neurosurg. Psychiatry,* 45(4):372–374, 1982.

27. Roy, M. K., Familial parkinsonism with peripheral neuropathy, *J. Assoc. Physicians India,* 49:944, 2001.

28. Robertson, N. P., Wharton, S., Anderson, J., Scolding N. J., Adult polyglucosan body disease associated with an extrapyramidal syndrome, *J. Neurol. Neurosurg. Psychiatry,* 65(5):788–790, 1998.

29. Bahiga, L. M., Kotb, N. A., El-Dessoukey, E. A., Neurological syndromes produced by some toxic metals encountered industrially or environmentally, *Z. Ernahrungswiss,* 17(2):84–88, 1978.

30. Choi, I. S., Carbon monoxide poisoning: systemic manifestations and complications, *J. Korean Med. Sci.,* 16(3):253–261, 2001.

31. Pezzoli, G., Strada, O., Silani, V., Zecchinelli, A., Perbellini, L., Javoy-Agid, F. et al., Clinical and pathological features in hydrocarbon-induced parkinsonism, *Ann. Neurol.,* 40(6):922–925, 1996.

32. Ahlskog, J. E., Litchy, W. J., Peterson, R. C., Waring, S. C., Esteban-Santillan, C., Chen, K. M. et al., Guamanian neurodegenerative disease: electrophysiologic findings, J. Neurol. Sci., 166(1):28–35, 1999.

33. Biglan, K. M., Holloway, R. G., Shoulson, I., Group tPS. An item analysis of the UPDRS in individual's with early PD treated initially with pramipexole or L-dopa: a subanalysis of the 4-year CALM-PD trial, *Mov. Disord.,* 17 Suppl. 5:S114, 2002.

34. Loher, T. J., Burgunder, J. M., Weber, S., Sommerhalder, R., Krauss, J. K., Effect of chronic pallidal deep brain stimulation on off period dystonia and sensory symptoms in advanced Parkinson's disease, *J. Neurol. Neurosurg. Psychiatry,* 73(4):395–399, 2002.

35. Abbassian, A. H., Shahzadi, S., Afraz, S. R., Fazl, A., Moradi, F., Tactile discrimination task not disturbed by thalamic stimulation, *Stereotact Funct. Neurosurg.,* 76(1):19–28, 2001.

36. Marsden, C. D., The mysterious motor function of the basal ganglia: the Robert Wartenberg Lecture, *Neurology,* 32(5):514–539, 1982.

37. Abbruzzese, G., Berardelli, A., Sensorimotor integration in Movement Disorders, *Mov. Disord.,* 18(3):231–240, 2002.

38. Gandevia, S. C., Burke, D., Does the nervous system depend on kinesthetic information to control natural limb movements? *Behav. Brain Sci.,* 15:614–632, 1992.

39. Moore, A. P., Impaired sensorimotor integration in parkinsonism and dyskinesia: a role for corollary discharges? *J. Neurol. Neurosurg. Psychiatry,* 50(5):544–552, 1987.

40. Georgiou, N., Iansek, R., Bradshaw, J. L., Phillips, J. G., Mattingley, J. B., Bradshaw, J. A., An evaluation of the role of internal cues in the pathogenesis of parkinsonian hypokinesia, *Brain,* 116 (Pt. 6):1575–1587, 1993.

41. Martin, P. M., The basal ganglia and posture, London: Pitman, 1967.

42. Demirci, M., Grill, S., McShane, L., Hallett, M., A mismatch between kinesthetic and visual perception in Parkinson's disease, *Ann. Neurol.,* 41(6):781–788, 1997.

43. Jobst, E. E., Melnick, M. E., Byl, N. N., Dowling, G.A., Aminoff, M. J., Sensory perception in Parkinson disease, *Arch. Neurol.,* 54(4):450–454, 1997.

44. Zia, S., Cody, F., O'Boyle, D., Joint position sense is impaired by Parkinson's disease, *Ann. Neurol.,* 47(2):218–228, 2000.

45. O'Suilleabhain, P., Bullard, J., Dewey, R. B., Proprioception in Parkinson's disease is acutely depressed by dopaminergic medications, *J. Neurol. Neurosurg. Psychiatry,* 71(5):607–610, 2001.

46. Klockgether, T., Borutta, M., Rapp, H., Spieker, S., Dichgans, J., A defect of kinesthesia in Parkinson's disease, *Mov. Disord.,* 10(4):460–465, 1995.

47. Klockgether, T., Dichgans, J., Visual control of arm movement in Parkinson's disease, *Mov. Disord.,* 9(1):48–56, 1994.

48. Rickards, C., Cody, F. W., Proprioceptive control of wrist movements in Parkinson's disease. Reduced muscle vibration-induced errors, *Brain*, 120 (Pt. 6):977–990, 1997.

49. Lewis, G. N., Byblow, W. D., Altered sensorimotor integration in Parkinson's disease, *Brain*, 125(Pt. 9):2089–2099, 2002.

50. Amabile, G., Fattapposta, F., Pozzessere, G., Albani, G., Sanarelli, L., Rizzo, P. A. et al., Parkinson disease: electrophysiological (CNV) analysis related to pharmacological treatment, *Electroencephalogr. Clin. Neurophysiol.*, 64(6):521–524, 1986.

51. Gerschlager, W., Alesch, F., Cunnington, R., Deecke, L., Dirnberger, G., Endl, W. et al., Bilateral subthalamic nucleus stimulation improves frontal cortex function in Parkinson's disease. An electrophysiological study of the contingent negative variation, *Brain*, 122 (Pt. 12):2365–2373, 1999.

52. Ikeda, A., Shibasaki, H., Kaji, R., Terada, K., Nagamine, T., Honda, M. et al., Dissociation between contingent negative variation (CNV) and Bereitschaftspotential (BP) in patients with parkinsonism, *Electroencephalogr. Clin. Neurophysiol.*, 102(2):142–151, 1997.

53. Crutcher, M. D., DeLong, M. R., Single cell studies of the primate putamen. I. Functional organization, Exp. Brain Res., 53(2):233–243, 1984.

54. Lidsky, T. I., Manetto, C., Schneider, J. S., A consideration of sensory factors involved in motor functions of the basal gangl, *Brain Res.*, 356(2):133–146, 1985.

55. Vitek, J. L., Ashe, J., DeLong, M. R., Alexander, G. E., Altered somatosensory response properties of neurons in the "motor" thalamus of MPTP treated parkinsonian monkeys, *Soc. Neurosci. Abstr.*, 16:425, 1990.

56. DeLong, M. R., Crutcher, M. D., Georgopoulos, A. P., Primate globus pallidus and subthalamic nucleus: functional organization, J. Neurophysiol., 53(2):530–543, 1985.

57. Filion, M., Tremblay, L., Bedard, P. J., Abnormal influences of passive limb movement on the activity of globus pallidus neurons in parkinsonian monkeys, *Brain Res.*, 444(1):165–176, 1988.

58. Hore, J., Meyer-Lohmann, J., Brooks, V. B., Basal ganglia cooling disables learned arm movements of monkeys in the absence of visual guidance, *Science*, 195(4278):584–586.

59. Hore, J., Vilis, T., Arm movement performance during reversible basal ganglia lesions in the monkey, *Exp. Brain Res.*, 39(2):217–228, 1980.

60. Ramig, L. O., Bonitati, C. M., Klemkkke, J. H., Hurii, Y., Voice treatment for patients with Parkinson's disease: development of an approach and preliminary efficacy data, *J. Med. Speech Lang. Pathol.*, 2:191–209., 1994.

61. Burke, D., Critical examination of the case for or against fusimotor involvement in disorders of muscle tone, in Desmedt, J. E., ed. Motor control mechanisms in health and disease, *Advances in Neurology,* 133–150, Raven Press, New York, 1983.

19 Pain in Parkinson's Disease

Blair Ford
Center for Parkinson's Disease, Columbia-Presbyterian Medical Center

CONTENTS

INTRODUCTION

Pain is an important but underrecognized symptom in Parkinson's disease (PD). The potential causes of chronic pain due to PD are protean, and an accurate diagnosis can be difficult. For most individuals with PD, the clinical features of bradykinesia, rigidity, tremor, postural instability, dementia, and other impairments dominate the clinical picture. Most patients, however, when questioned directly, will report painful symptoms or discomfort that they regard as connected to their Parkinson's disease.

Virtually every type of pain sensation has been described in PD. For many, painful symptoms fluctuate in parallel with the motor symptoms of the disorder, and are designated nonmotor sensory fluctuations.[1] While never held as a major feature of the disorder, pain is listed prominently in all of the earliest descriptions of PD.[2–7] James Parkinson wrote in his famous monograph that painful symptoms can be the first sign of the disorder.[2] In a minority of patients with PD, pain is so severe and intractable that it overshadows the motor symptoms of the disease.

Painful symptoms pose a considerable diagnostic and therapeutic challenge to the clinician. For diagnostic purposes, it may be helpful to classify into one or more of the following five categories: musculoskeletal pain, radicular or neuropathic pain, dystonia-related pain, akathitic discomfort, or primary, central parkinsonian pain. There is another category, namely pain that is *unrelated* to PD, which includes all of the other painful complaints that patients may experience, but this is outside of the scope of the present discussion.

PREVALENCE OF PAIN IN PD

Painful or unpleasant sensations in patients with PD are more common than one might expect, approximating 50% in most series.[1,8–11] The true prevalence of pain in PD remains unknown but is suspected to be higher than in the general population because of the rigidity, dystonia, physical restraints, and motor complications that the disease imposes. In five recent surveys of painful sensations in patients with PD, summarized in Table 19.1, the prevalence of pain has been estimated at between 38 and 54%.[1,8–11]

The challenge for the clinician is to recognize when a patient's complaint of pain requires further evaluation and to categorize the painful symptoms of PD into a framework for diagnosis and treatment. One recent study classified painful sensations by etiology.[11] In this survey, 43 of 95 patients with PD experienced pain. Muscle cramps occurred in 32 (74%), dystonia-associated pain occurred in 12 (28%), radicular or neuritic pain occurred in 6 (14%), and joint pains occurred in 6 (14%). In other series, there is a higher incidence of central or primary parkinsonian pain[10] or akathisia.[1]

TABLE 19.1
Surveys of Painful Sensations in Parkinson's Disease

Series	Prevalence	Pain Description	Comments
Sigwald, 1960	108 or 203 randomly selected PD patients described painful symptoms	All painful manifestations were divided into paresthesias and pain symptoms. Paresthesias were further divided into two categories: cramps (20% of patients) and restlessness (17%). Painful sensations included all types of discomforts, and were discussed by body region: legs (21 patients), arms (11 patients), neck (8), lumbar region (3), epigastrium (10) and abdomen (10).	This series is valuable because of the wide range of painful sensations it catalogs, but there was no systematic attempt to classify painful symptoms by etiology, as inferred from pain descriptors. Patients with burning pain, cold numbness, cramping, or "arthritic" pain were grouped together by affected body part.
Snider, 1976	43 of 101 PD patients had "sensory symptoms without apparent somatic etiology"	Sensory symptoms were classified as pain, tingling, burning, or numbness; many patients experienced several kinds of painful sensations: Pain (29 patients): intermittent, poorly localized, cramp-like or aching, no associated with increased muscle contraction, not affected by movement or pressure; sometimes correlated with "off" periods. Burning (11): burning sensations, sometimes related to dopaminergic therapy. Tingling/numbness (43): anesthetic feelings, formication, pins and needles; usually occurring in the extremities sometimes aggravated by immobility.	Patients with arthritis or diabetes were excluded from this survey. Painful muscle cramps or spasms were not counted as primary sensory symptom, and all other painful symptoms were regarded as primary parkinsonian sensations.
Koller, 1984	19 or 50 consecutive PD patients described sensory complaints	All abnormal sensations were considered primary sensory symptoms, and consisted of: numbness (12 patients), tingling (8), pain and achiness (6), coldness (6), and burning (1).	Patients with musculoskeletal disease, arthritis or diabetes were excluded.
Goetz, 1986	43 of 95 PD patients had pain directly related to PD	Sensory symptoms were classified by description into pain of muscular origin, dystonia, joint pain, radicular or neuritic pain, and akathitic discomfort: muscle cramps or tightness present in 32 (74%) patients with pain, painful dystonia in 12 (28%), joint pains in 6 (14%), radicular or neuritic pain in 6 (14%), and diffuse "akathitic" pain in 1 (2%).	Syndromes suggestive of thalamic pain were not described in this series.
Wijas, 2002	In a survey of 50 patients with motor fluctuations, 14 (28%) had sensory complaints	All patients experienced sensory symptoms that fluctuated with the motor symptoms, and were worse in the "off" state. Six patients stated that pain was the main cause of disability in the "off" state. Among 14 patients with sensory symptoms, 54% experienced akathisia, 42% had "tightening sensations," 38% had tingling sensations, 36% had diffuse pain, 18% had neuralgic pin, and 8% had burning sensations.	All of the sensory symptoms in this study occurred as nonmotor fluctuations.

SPECIFIC PAIN SYNDROMES IN PD

As a general approach, painful symptoms in PD should be considered in relation to the cardinal symptoms of tremor, rigidity, akinesia, dystonia, and akathisia that occur in PD. It is important to note whether antiparkinsonian medications induce, exacerbate, or relieve PD-associated pain. Most sensory symptoms are worse during "off" motor fluctuations and are considered as nonmotor fluctuations. However, not all pain in the parkinsonian "off" state represents a direct result of dopamine deficiency; for many patients, "off" pain can represent a secondary consequence of increased rigidity and immobility. Pain caused by dystonia can be diagnosed when there is visible twisting, cramping, and posturing of the painful extremity or body part. Dystonia may be painful in the "off" state, but medication-induced dystonia, occurring in the "on" state or during transitions between states, may be painful. Deep brain stimulation may induce painful, dystonic muscle spasms, sometimes attributed to spread of discharge to the corticospinal tract.

A careful appraisal of possible musculoskeletal or rheumatological pain mechanisms is important in patients with Parkinson's disease. Akathisia, while not painful, is intensely unpleasant and is a rare but distinctive symptom that occurs in PD. Primary parkinsonian pain, unrelated to a disturbance in motor function, is presumed to be of central origin; it may be inferred partly by the nature of its clinical features and partly through exclusion of other causes. Table 19.2 provides a classification of painful symptoms based on descriptions provided by the patients. In the sections that follow, categories of painful symptoms in PD are described in further detail.

TABLE 19.2
Types of Painful Sensations in Parkinson's Disease

Pain Category	Clinical Description
Musculoskeletal	Aching, cramping, arthralgic, myalgic sensations
	Associated findings may include muscle tenderness, arthritic changes, skeletal deformity, limited joint mobility may be exacerbated by parkinsonian rigidity, stiffness, immobility
	May improve with levodopa
	May fluctuate with medication dosing
	May improve with exercise
Radicular-neuritic	Pain in a root or nerve territory
	Associated with signs of nerve or root entrapment
Dystonia-related pain	Associated with dystonic movements and postures
	May fluctuate with medication dosing (see Table 19.3)
Central ("primary") pain	Burning, tingling, formication
	Bizarre quality
	Location not confined to root or nerve territory
	Not explained by rigidity, dystonia
	May fluctuate with medication effect
Akathisia and restlessness	Subjective sense of restlessness, often accompanied by an urge to move
	May improve with levodopa
	May fluctuate with medication effect

MUSCULOSKELETAL PAIN IN PD

Pain of musculoskeletal origin has long been described in PD, and, in some studies, appears related to the presence of parkinsonian rigidity and akinesia. In clinical practice, one observes deformities of posture, stiffness of limb movements and gait, and awkward mechanics for body motion and tasks. It is not hard to postulate that the abnormal postures and mechanics place unusual stresses on the musculoskeletal system. The aching, cramping, and joint pains in patients with PD are commonly held to result from a lack of mobility in affected limbs and joints. Supporting this notion is the observation in one study that parkinsonian discomfort tends not to occur during peak motor function but is maximal during periods of increased parkinsonism.[11]

Muscle cramps or tightness in PD typically affect the neck, arm, and paraspinal or calf muscles, while joint pains occur most frequently in the shoulder, hips, knees, and ankles.[11] One of the most common musculoskeletal afflictions in PD is shoulder stiffness, and a stiff shoulder may be the first sign of PD. The prevalence of frozen shoulder, also called periarthritis or adhesive capsulitis, in patients with PD is higher than in age-matched subjects without PD.[12] Among 150 consecutive patients with PD followed in a movement disorders clinic, 65 (43%) gave a history of shoulder disturbance of various causes, including shoulder trauma, that preceded the development of their parkinsonism. Six patients gave a history of spontaneous pain and restricted mobility about the shoulder joint. The peak incidence of frozen shoulder occurred in the two years *preceding* the first symptoms of PD, and, in almost all cases, the initial symptoms of PD developed in the upper limb ipsilateral to the frozen shoulder. Moreover, among patients with frozen shoulder, akinesia as the first symptoms was twice as frequent as tremor.[12]

Spinal deformities and arthritis are well described in PD. The typical posture of the parkinsonian is stooped forward with the neck held in flexion. Some patients have a fixed postural deformity, while others have an apparent truncal or neck dystonia that varies with posture and activity. In extreme cases, the spinal deformity of parkinsonism is termed camptocormia;[13] these individuals may have a curvature that forces the upper body to the horizontal and prevents upward gaze. A recent report described camptocormic posture in a patient due to focal myositis of the paraspinal muscles.[14]

There appears to be no specific or effective treatment for camptocormia. With advancing disease, the flexion deformity only worsens, despite treatment with antiparkinsonian agents. It is tempting to speculate that insertion of spinal rods might straighten the curvature, but anecdotal evidence suggests that this often fails due to hardware disruption or migration, or infection. Deep brain stimulation is not reported to have significant effects on severe truncal flexion.

Scoliosis occurs more frequently in PD than in the elderly general population,[15,16] and in one study was present in 62 of 103 (60%) patients with PD, the side of the convexity being unrelated to the side of maximal deficit.[17] Hand and foot deformities, consequences of the akinesia and dystonia in PD, have frequently been depicted.[18–20] A dystonic clenched fist is a complication of parkinsonism resulting from prolonged hand and finger immobility.[21]

Every type of rheumatological and orthopedic complaint is encountered in patients with PD, including temporomandibular joint disease, bursitis, arthritis, Baker's cyst, plantar fasciitis, stress fractures, cervical spondylosis, spinal stenosis, sciatica, ankylosing spondylitis, contractures, and others. The incidence of these conditions in PD has not been studied, and it is not possible to conclude a statistical or causal association with PD, because rheumatological conditions are common in the PD age range. The presence of osteoporosis, common in the female elderly, predisposes to pathological fractures, which are more likely to occur in a patient prone to falling.

Another important category of musculoskeletal pathology is painful contractures caused by immobility. Contractures, a pathologic shortening of muscle fibers, most often affect the Achilles tendon, hamstrings, hip adductors, finger flexors, biceps, brachioradialis, and neck—all of which result from the characteristic flexed attitude of the disease. The clenched fist that results from parkinsonian dystonia and immobility may progress to contracture.[21,22] It has been suggested that limb contractures may represent a side effect of bromocriptine,[23] an agent that may cause fibrosis, but there are too few cases on record to substantiate this. Contractures represent a complication of immobility and may form within a matter of weeks. The risk of developing contractures appears proportional to the amount of rigidity and bradykinesia.

Musculoskeletal pain requires a careful assessment of the muscles and tendons, bones, and joints. Painful symptoms must be considered in relation to parkinsonian signs, range of motion, posture, activity, and antiparkinsonian medication. It should be possible to arrive at an accurate diagnosis on the basis of the history and exam, but occasionally, ancillary testing, including serological tests, X-rays, bone scans, ultrasound, or rheumatological or orthopedic consultation, will be needed. The presence of joint deformities or a concurrent rheumatological condition should be obvious. Differentiating between parkinsonian

rigidity, painful cramping, contracture, dystonia, and a fixed skeletal deformity can all be done on clinical examination.

The treatment of musculoskeletal pain in PD depends on the cause. If the pain is due primary to parkinsonian rigidity, dopaminergic therapy, physical therapy, and an exercise program are indicated. The goal of treatment is to restore mobility, and, once achieved, an exercise program for most patients is an important way to prevent further musculoskeletal problems. Nonsteroidal anti-inflammatory drugs (NSAIDs) and analgesics are helpful for rheumatological and orthopedic conditions, in tandem with physical therapy. Passive range of motion exercises are important to prevent contractures in patients with limited mobility, but, once formed, a contracture will generally require surgical intervention.

RADICULAR AND NEURITIC PAIN

Pain and discomfort that is well localized to the territory of a nerve or nerve root is described as radicular or neuritic pain. Radicular or neuritic pain accounted for 14% of the pain syndromes experienced by patients with PD in one survey.[11] However, in most case reports and surveys of patients with PD, this category of pain is the least studied, because the descriptions do not provide adequate clinical information or neuroimaging data needed to confirm the pathological process. In some reports, the paresthetic sensations of coolness, numbness, or tingling may be mistakenly attributed to a central pain syndrome, when further evaluation could have revealed a compressive root or nerve lesion. As such, the incidence of these complaints in PD is unknown. It is possible to speculate that the postural deformities of PD may predispose to compressive radiculopathy, sciatica, or myelopathy. Immobility is a risk factor for a compressive focal neuropathy. A traumatic radial nerve palsy was described in one report.[24] It has been suggested that radicular and neuritic exacerbation may occur with peak effect dyskinesias.[11] Trigeminal neuralgia has also been described in PD.[25]

The evaluation and treatment of pains in this category begins with a careful clinical examination, supplemented if needed by electrodiagnostic studies and neuroimaging. Radicular pain can be treated with a judicious mobility program, NSAIDs, and pain medication. For a compressive peripheral neuropathy, avoidance of aggravating postures is needed, sometimes supplemented by splints or braces. In a recent report of two patients with Parkinson's disease and sciatica, drug-induced dyskinesias exacerbated the radicular pain.[26] Treatment with morphine not only reduced the sciatica pain but also suppressed the dyskinesias.[26] In the presence of refractory pain, or a severe or worsening neurological deficit that coincides with an abnormality on the radiological studies, decompressive surgery may be indicated.

Pain Associated with Dystonia

Dystonia describes a sustained, forceful twisting movement that leads to abnormal postures and deformities. Dystonia in PD may affect any limb, the trunk, neck, face, tongue, jaw, pharynx, and vocal cords, usually developing in sites most severely affected by parkinsonism. Dystonia may precede the development of parkinsonism, develop as a late feature, appear after the introduction of dopaminergic therapy, follow stereotactic neurosurgery, or be induced by deep brain stimulation. It is reasonable to classify the pattern of dystonia as focal, cranial, segmental, or generalized. It has been observed that off-period dystonia most commonly affects the feet, whereas drug-induced dystonia has a predilection for the neck, trunk, and cranial distribution.[27]

The best studied form of dystonia in PD is early-morning foot dystonia, which is reported in approximately 16% of patients with PD, usually causing foot or toe cramping and posturing.[28] Dystonic postures of the foot and hand, sometimes progressing to fixed deformity, have been well described.[18] Every variety of foot posture is possible: plantar flexion, dorsiflexion, foot inversion, curling of the toes, forced extension of the great toe ("striatal toe"), with associated stiffness of the calf muscles. In a study of 42 patients with PD and foot dystonia, 41 (97.5%) individuals experienced early-morning foot dystonia before the first dose of levodopa, with subsequent milder attacks during the late evening or in the night, suggesting that dystonia is intrinsic to the parkinsonism in most cases, representing a wearing-off phenomenon.[27]

It has been argued that early-morning dystonia represents a complication of long-term levodopa therapy, because it occurs more frequently in patients with a longer disease duration and in whom dyskinesias are present.[29–32] However, painful foot dystonia in PD was described long before the advent of levodopa and may often precede the other manifestations of the disorder.[33,34]

Dystonic spasms are among the most painful symptoms that a patient with PD may experience. The spasms may be paroxysmal, spontaneous, or triggered by movement or activity. They may be brief, lasting minutes, or prolonged, lasting hours, or even continuous, unrelieved by treatment attempts. The evaluation of dystonia requires especially careful consideration of its relationship to dopaminergic medication. As noted, dystonia may occur as an early-morning manifestation of dopaminergic deficiency or as a wearing-off phenomenon later in the day or in the middle of the night. In some patients, dystonia is a painful beginning-of-dose or end-of-dose phenomenon; in others, dystonia represents a peak dose effect of dopaminergic medication. The classic flowing, writhing, choreathetotic dyskinesias induced by dopaminergic medication are not sustained or painful and are generally to referred to as *dystonia*. Drug-induced dystonia, however,

are sustained, twisting postures that can be very painful. In uncertain cases, it may be helpful to observe the patient in the office for several hours so as to appreciate the relationship of the dystonia to the medication dose cycle. Classifying dystonia in relation to the levodopa dosing schedule provides a useful and rational framework for evaluating and treating painful dystonia in PD.[23,35]

Early-morning dystonia is typically relieved by activity, or it subsides shortly after the first dose of dopaminergic medication in the day. In some patients, early-morning dystonia is so severe that subcutaneous injections of apomorphine, with its onset of action in minutes, can be justified.[36] When dystonia occurs as a wearing-off effect during the day, the treatment is analogous to the treatment of wearing-off motor fluctuations and is aimed at reducing the duration of the off period. Dopaminergic agents, including long-acting levodopa, dopamine agonists, or apomorphine, can all be effective. The first line of therapy for off dystonia in PD is dopaminergic therapy, but dystonia in PD can be also treated by anticholinergics,[29,37] baclofen,[38] and lithium.[39] In patients with levodopa-induced dystonia, the treatment consists of reducing the dopaminergic stimulation, usually by decreasing the levodopa dosage or absorption, sometimes by substituting a less potent agonist.

Injections of botulinum toxin may also be helpful to treat focal dystonia in PD. In an open-label study of 30 patients with painful foot dystonia, in whom the dystonia was severely or completely disabling in 23 (77%), there was dramatic relief of pain and disability following injections with botulinum toxin A.[40] The injections were accomplished under EMG-guidance and were tailored to the specific appearance of the dystonic foot. The median dosage was 70 units (range 40 to 100 units).

The dystonic clenched fist in parkinsonism may begin as a dystonic posture but leads to contractures, usually within several months of sustained hand and finger flexion. The clenched fist is often extremely painful and leads to a loss of hand function, poor hand hygiene, and palmar infections.[21] The pain may result from sustained dystonic contractions of the fingers in flexion, from joint disease, or from the nails piercing the skin of the palm. Treatment with intramuscular botulinum toxin injections, given to the flexor digitorum superficialis or lumbricals, can relieve the dystonic contractures, sometimes for four months.[21] Active muscular contractions, as documented by EMG, are associated with a good result from botulinum toxin injections, whereas an absence of EMG activity, denoting contractures, may not improve.[21] The flexed neck posture that occurs in PD also responds poorly to botulinum toxin injections.

Advanced neurosurgical techniques may also reduce painful dystonia associated with PD. Pallidotomy has been reported to relieve painful dystonia in PD.[5,41] Painful "off" dystonia, present in 20 patients who underwent bilateral subthalamic nucleus stimulation, was completely relieved in 12 individuals and considerably improved in 4.[42] In a recent study that examined the effect of globus pallidus stimulation on parkinsonian pain,[43] all types of "off" period painful sensations were markedly reduced: dystonic pain, muscle cramping, dysesthesias, and "global pain," as measured using a rating scale. The benefit developed quickly after surgery and remained stable through the one-year follow-up interval. Patients with unilateral globus pallidus stimulation experienced pain reduction mainly on the contralateral body side, whereas bilateral stimulation produced bilateral reduction in pain.

Deep brain stimulation in the subthalamic nucleus or globus pallidus can induce acute painful, dystonic spasms, sometimes attributed to spread of current to the internal capsule. The necessary intervention is a change in the stimulator parameters, usually a reduction in voltage or pulse width, which promptly reverses the muscle spasms. Intrathecal baclofen, effective for spasticity of spinal or cerebral origin, has shown little effect on the dystonia associated with parkinsonism.[44]

CENTRAL PAIN SYNDROMES

Perhaps the most striking pain syndromes in patients with PD are those of central origin. Central pain in PD is presumed to be a direct consequence of the disease itself, and not the result of dystonia or other motor manifestations. Central pain is defined as pain produced directly by a lesion or abnormality of function within the central nervous system. Current concepts hold that central pain is generated by the brain itself and requires the presence of a lesion in the thalamus, its afferent, or efferent pathways.[45] Most of the clinical literature implicates the thalamus as the source of central pain, and specifically a lesion of the ventroposterior thalamus.[46]

The concept of primary parkinsonian pain was outlined in the seminal description of Souques in 1921,[5] in which he described 17 patients with Parkinson's disease or parkinsonism, some of whom were afflicted with pain syndromes that he believed were intrinsic to the PD. He listed many of the characteristics of his patients' pain: bizarre unexplained sensations of stabbing, burning, scalding, formication—all descriptions that have been associated with "neuropathic" pain originating in the central or peripheral nervous system. Souques noted that the presumed central pain syndromes in his patients typically afflicted the side of the body most affected by parkinsonism and could precede, even by years, the motor manifestations of the disorder. Using a conceptual framework outlined earlier by Dejerine,[47] Souques postulated a central origin of pain in PD due to a disturbance in signaling between the corpus striatum and thalamus.

The argument for a separate central pain syndrome in PD finds support in several unusual case descriptions.

While musculoskeletal conditions tend to occur in the limbs, muscles, and joints most afflicted with parkinsonism, there are several reports of unusual pain syndromes involving the face, head, epigastrium, abdomen, pelvis, rectum, and genitalia,[8,9,48,49] all areas in which painful dystonia or musculoskeletal conditions are unlikely or implausible. Sigwald described epigastric and abdominal pain as well as pharyngeal pain in his patients.[8] Schott described a patient with a cutaneous truncal burning sensation on the abdomen and back, improved ultimately by ECT. In a series of seven patients with parkinsonism and oral or genital pain,[44] oral pain involved the gums, teeth, tongue, inner cheek, face, and jaw. These oral pain syndromes, resembling the idiopathic "burning mouth syndrome,"[50] were described as burning, pulsating sensations that were often strikingly lateralized within the oral cavity. The pain tended to correlate with off periods but was not necessarily abolished by dopaminergic therapy. The genital pain in this series, occurring exclusively in women, was described as burning, numbness, or vibrating sensations. In all patients, the pain had a relentless, obsessional, distressing quality that overshadowed the patients' other parkinsonian symptoms.[44]

Further support for the notion that parkinsonian pain originates in the central nervous system is the observation that peripheral nerve blockade does not abolish the pain. One example of was a case of oral pain that was unaffected by a complete dental nerve block.[44] In a report of a patient with parkinsonism, dystonia, and severe leg pains, epidural anesthesia using chlorprocaine produced a complete sympathetic, sensory, and motor blockade, relieving the dystonia but not all elements of the patient's pain, suggesting that there was a central component to the pain, possibly due to deafferentation.[51]

While classic central pain is postulated to involve a lesion of the thalamus, primary parkinsonian pain is speculated to result from an abnormality of sensory pathways within the basal ganglia. The neuroanatomic and neurochemical substrate of sensory processing within the basal ganglia is summarized in an excellent review.[52] Somatosensory processing within the basal ganglia occurs in the substantia nigra, caudate, putamen, globus pallidus, thalamus, and their interconnections.[52] It has proposed that the basal ganglia perform an important gating role for nociceptive information before it reaches consciousness.[53] 6-hydroxydopamine lesions in the striatum or ventral tegmental area decrease the latencies of nociceptive reflexes in rats, suggesting that the dopaminergic system plays a role in modulating nociceptive information in the striatum and limbic system.[54] Noting that levodopa administration to 11 patients with PD increased heat pain threshold and tolerance, it has been speculated that dopamine, presumably acting through striatothalamic projections, modulates the peripheral inputs to the thalamus.[55]

The treatment of presumed central pain in PD is challenging, especially if dopaminergic agents, the first line of therapy for this disabling problem, are not effective. Conventional analgesics, opiates, tricyclics, and atypical neuroleptics, including clozapine, may be helpful.[49] In one report of a patient with intractable, recurrent, severe painful fluttering sensations in her left thoracic region, subcutaneous injections of apomorphine provided complete relief after all other classes of medication—dopaminergic, benzodiazepines, tricyclic antidepressants, opiates, baclofen, clozapine, intercostal nerve blocks—had failed.[56]

With the increased application of deep brain stimulation in advanced PD, it is possible that unusual painful or uncomfortable sensations of central origin may be reported. The stimulators can induce a variety of unpleasant sensations, including the "jolting" dysesthesias that transiently occur during stimulator programming sessions, and are reported in up to 70% of patients.

AKATHISIA AND RESTLESS LEGS SYNDROME

No discussion of uncomfortable symptoms in PD is complete without a mention of akathisia, or restlessness. Restlessness is a frequent and potentially disabling complaint in PD. Parkinsonian akathisia is defined as subjective inner restlessness, producing an intolerance of remaining still and manifesting as a constant need to move or change position. This complaint has long been observed in PD and is described in Gowers textbook of neurology.[4] In evaluating a complaint of restlessness, it is important to establish that the need to move that is not caused by symptoms of parkinsonism, somatic complaints or urges, dyskinesias, anxiety, depression, or claustrophobia.

Like many of the pain syndromes in PD in this discussion, restlessness is probably more common than expected if specifically inquired about. In one survey of 100 patients with PD, 68 (68%) complained of a periodic need to move. In 26 of these individuals (26%), this symptom represented genuine parkinsonian akathisia. In another study of 56 patients with PD, akathitic movements were present in 25 (45%), usually involving the legs and correlating with patients' own subjective descriptions of inner restlessness.

The definition of pure akathisia is meant to exclude additional neuropathic symptoms, but patients with akathisia often describe crawling sensations, burning, or tingling. Parkinsonian akathisia can be severe; patients may be unable to sit, drive a car, eat at a table, or attend social gatherings. Some patients remain in constant motion. In extreme cases, parkinsonian akathisia has driven individuals to suicide.[6] In about half of the reported cases of parkinsonian akathisia, the symptom fluctuated with levodopa dosing schedules and could often be relieved by additional dopaminergic treatment.

The link between akathisia and a dopaminergic deficit is well established, as the two other major causes of the syndrome are postencephalitic parkinsonism and neuroleptic-induced akathisia. It has been suggested that akathisia results from a dopaminergic deficiency involving the mesocortical pathway, which originates in the ventral tegmental area and is known to be affected in PD. Some indirect support for this notion is the observation that clozapine, which has a high affinity for D_4 receptors,[57] preferentially affecting the mesocortical and mesolimbic dopaminergic systems, can be remarkably effective in treating akathisia.

Restlessness is also a core element of restless legs syndrome (RLS), a disorder of unknown cause in which patients experience an intense irresistible urge to move the legs, accompanied by sensory complaints and motor restlessness. Characteristically, the symptoms are worse at rest, relieved with motion, and increase in severity in the evening or at night. True restless legs syndrome is rare in PD, although the fact that it may be dramatically relieved by levodopa or dopamine agonists suggests that RLS is a disorder of altered dopaminergic transmission.

HEADACHE IN PD

Headache is an important symptom that may occur in PD, but its relationship to the disease is uncertain. It does not fit into the pain categories described above but instead represents a painful symptom that often requires its own specific evaluation and treatment. In a survey of 71 patients with PD, headache was described in 25 individuals (35%).[58] Headaches were generally located in the nuchal region but did not correlate with a clinical assessment of nuchal rigidity. The character of the headaches ranged from dull, aching discomfort to sharp, squeezing or pulsatile pain. In a subsequent report, a specific early morning headache was described in three individuals, relieved within two hours of the first levodopa dose.[58] In another report, PD patients with headache scored significantly higher on measures of depression and anxiety than those without headache.[63] It is always important to note whether a headache in a patient with PD may represent an adverse effect of a medication, especially the dopaminergic ergot alkaloids pergolide and bromocriptine. If a headache develops or persists on a dopamine agonist, it is recommended to change to another agonist. A severe or unusual headache accompanied by neurological signs can never be attributed to PD and requires thorough neurological evaluation, usually with neuroimaging.

RELATIONSHIP BETWEEN PAIN, DEPRESSION, AND PD

Depression may influence the experience of pain in PD. Persistent pain of any cause can have a debilitating effect on an individual and can precipitate depression, which in turn may heighten the severity or intractability of the pain syndrome.[59] Depression may also alter the interpretation of painful symptoms in PD. The prevalence of major depression in PD is estimated to be 46%[60] and is unrelated to age, duration of disease, or disability. In a follow-up study to a survey on pain prevalence in PD, Goetz found that depression, as measured by the Beck Depression Inventory,[61] is more severe in patients with PD who experience pain[62] than among those who do not. Although there are no systematic data to guide the clinician, it is important that any assessment of pain in a PD patient take into account the potential contributing role of depression, which itself may require specific treatment.

SUMMARY

While not regarded as a cardinal feature of PD, pain can be an important complication of the disease that presents a diagnostic and therapeutic challenge. The first task is to decide whether the pain syndrome is likely related to PD or represents another, important medical condition requiring further evaluation. It is the nature of modern medical practice that a physician caring for a patient with PD may be the first to evaluate a new or unusual painful complaint, and defining the boundaries of neurological inquiry may not be simple. Chest pain or abdominal pain, for example,

may be due to PD but are best approached from a larger perspective, for which the appropriate referrals and testing may be necessary.

The diverse literature on pain in PD ranges from early clinical descriptions to systematic pain surveys, from speculations regarding the origin of central pain in PD to a growing understanding of the neurochemical substrate of pain in PD. In all series, approximately 40 to 50% of patients with PD experience pain, and, in a minority of these individuals, the problem is so distressing that it overshadows the motor symptoms and complications of parkinsonism. The various descriptions of painful sensations in the literature suggest that most parkinsonism-related pain can be assigned to one or more of five categories: musculoskeletal pain, neuritic or radicular pain, dystonia-associated pain, primary or central pain, and akathitic discomfort. The precise breakdown of painful symptoms into each category is not known, but it would appear that most pain in PD is related to a musculoskeletal cause, dystonia,[11] or restless sensations.[1] There are no diagnostic tests to guide the clinician, but the patients' own descriptions of pain and the associated clinical features should enable an accurate diagnosis of pain in most individuals. A summary of pain syndromes in PD, classified by etiology, is provided in Table 19.3. The treatment of the painful symptoms depends critically on the underlying cause and has been reviewed in the foregoing sections.

TABLE 19.3
Classification of Pain in Parkinson's Disease by Etiology

1. Musculoskeletal pain	Pain due to parkinsonian rigidity
	Pain due to rheumatological or musculoskeletal disorder or deformity
2. Neuritic-radicular pain	Pain due to roots lesion, focal, or peripheral neuropathy
3. Dystonic pain	Off-period painful dystonia (includes early-morning dystonia)
	Beginning of dose dystonia
	Peak-dose pain
	End-o-dose pain
4. Central pain	Off-period pain (includes early-morning pain)
	Beginning-of-dose pain
	Peak-dose pain
	End-of-dose pain
5. Akathisia	Off-period ("parkinsonian") akathisia
	Drug-induced akathisia

REFERENCES

1. Witjas, T., Kaphan, E., Azulay, J. P., Blin, O., Ceccaldi, M., Pouget, J., Poncet, M., Cherif, A.A., Nonmotor fluctuations inn Parkinson's disease. *Neurology* 2002; 59:408–413.

2. Parkinson, J., *An essay on the shaking palsy,* Sherwood, Neely, Jones, London, 1817.

3. Charcot, J. M., *Lectures on diseases of the nervous system,* Vol. I, The New Sydenham Society, London, 1877.

4. Gowers, W. R., *Diseases of the Nervous System,* Blakiston, Philadelphia, 1888.

5. Souques, M. A., Des douleurs dans la paralysie agitante, *Revue Neurologique,* Paris, 37:629–633, 1921.

6. Lewy, F. H., Die Lehre vom Tonus und der Bewegung Zugleich Systematische Untersuchungen zur Klinik,

Physiologie, Pathologie und Pathogenese der Paralysis Agitans, Springer, Berlin, pp. 28–29, 1923.

7. Wilson, S. A. K., *Neurology,* Hafner Publishing, New York, 1940.

8. Sigwald, J., Solignac, J., Manifestations douloureuses de la maladie de Parkinson et paresthesies provoquees par les neuroleptiques, *Sem. Hop. Paris,* 41:2222–5, 1960.

9. Snider, S. R., Fahn, S., Isgreen, W. P., Cote, L. J., Primary sensory symptoms in parkinsonism, *Neurology,* 34:957–9, 1984.

10. Koller, W. C., Sensory symptoms in Parkinson's disease, *Neurology,* 34:957–959, 1984.

11. Goetz, C. G., Tanner, C. M., Levy, M., Wilson, R. S., Garron, D. C., Pain in Parkinson's disease, *Mov. Disorders,* 1:45–49, 1986.

12. Riley, D., Lang, A. E., Blair, R. D. G., Birnbaum, A., Reid, B., Frozen shoulder and other shoulder disturbances in Parkinson's disease, *J. Neurol. Neurosurg. Psychiatry,* 52:63–66, 1989.

13. Djaldetti, R., Mosberg-Galili, R., Sroka, H., Merims, D., and Melamed, E., Camptocormia (bent spine) in patients with Parkinson's disease: characterization and possible pathogenesis of an unusual phenomenon, *Mov. Disorders,* 14:443–447, 1999.

14. Wunderlich, S., Csoti, I., Reiners, K., Gunther-Lengsfeld, T., Schneider, C., Becker, G., Naumann, M., Camptocormia in Parkinson's disease mimicked by focal myositis of the paraspinal muscles, *Mov. Disorders,* 17:598–600, 2002.

15. Duvoisin, R. C., Marsdenm C. D., Note on the scoliosis of parkinsonism, *J. Neurol. Neurosurg. Psychiatry,* 38:787–793, 1975.

16. Indo, T., Ando, K., Studies on the scoliosis of parkinsonism, *Clin. Neurol.,* 20:40–46, 1980.

17. Grimes, J. D., Hassan, M. N., Trent, G., Halle, D., Armstrong, G. W. D., Clinical and radiographic features of scoliosis in Parkinson's disease, *Adv. Neurol.,* 45:353–355, 1986.

18. Gortvai, P., Deformities of the hands and feet in parkinsonism and their reversibility by operation, *J. Neurol. Neurosurg. Psychiatry,* 26:33–36, 1963.

19. Reynolds, F. W., Petropoulos, G. C., Hand deformities in parkinsonism, *J. Chron. Dis.,* 18:593–595, 1965.

20. Bissonnette, B., Pseudorheumatoid deformity of the feet associated with parkinsonism, *J. Rheumatol.,* 13:825–826, 1986.

21. Cordivari, C., Misra, V. P., Catania, S., Less, A. J., Treatment of dystonic clenched fist with botulinum toxin, *Mov. Disorders,* 16:907–913, 2001.

22. Kyriakides, T., Langton Hewer, R., Hand contractures in Parkinson's disease, *J. Neurol. Neurosurg. Psychiatry,* 51:1221–1223, 1998.

23. Quinn, N. P., Ring, H., Honovar, M., Contractures of extremities in parkinsonian subjects: a report of three cases with a possible association with bromocriptine treatment, *Clin. Neuropharmacol.,* 11:268–277, 1988.

24. Pullman, S. L., Radial compression neuropathy in Parkinson's disease, *Neurology,* 44:1861–1864, 1994.

25. Hillen, M. E., Sage, J. I, Nonmotor fluctuations in patients with Parkinson's disease, *Neurology,* 47:1180–1183, 1996.

26. Berg, D., Becker, G., Reiners, K., Reduction of dyskinesia and induction of akinesia induced by morphine in two Parkinsonian patients with severe sciatica. *J. Neural. Trans.,* 106:725–728, 1999.

27. Poewe, W. H., Lees, A. J., Stern, G. M., Dystonia in Parkinson's disease: clinical and pharmacological features, *Ann. Neurol.,* 23:73–78, 1988.

28. Currie, J., Harrison, M. B., Trugman, J. M., Bennett, J. P., Jr., Wooten, G. F., Early morning dystonia in Parkinson's disease, *Neurology,* 51:283–285, 1988.

29. Duvoisin, R. C., Yahr, M. D., Lieberman, J., Antunes, J., Rhee, S., The striatal foot, *Trans. Am. Neurol. Assoc.,* 97:267, 1972.

30. Melamed, E., Early-morning dystonia: a late side effect of long term levodopa therapy in Parkinson's disease, *Arch. Neurol.,* 36:308–310, 1979.

31. Nausieda, P. A., Weiner, W. J., Klawans, H. L., Dystonic foot response of parkinsonism, *Arch. Neurol.,* 37:132–136, 1980.

32. Ilson, J., Fahn, S., Cote, L., Painful dystonic spasms in Parkinson's disease, *Adv. Neurol.,* 40:395–398, 1984.

33. Stewart, P., Paralysis agitans; with an account of a new symptom, *Lancet,* 2:1258–1260, 1898.

34. Lees, A. J., Hardie, R. J., Stern, G. M., Kinesigenic foot dystonia as a presenting feature of Parkinson's disease, *J. Neurol. Neurosurg. Psychiatry,* 47:885, 1984.

35. Quinn, N.P., Classification of fluctuations in patients with Parkinson's disease, Neurology, 51 (suppl. 2):S25–S29, 1998.

36. Pollak, P., Tranchant, C., Les autres symptomes de la phase evoluee de la maladie de Parkinson, *Rev. Neurol.,* 156 (suppl. 2):S2b165–173, 2000.

37. Poewe, W., Lees, A. J., Steiger, D., Stern, G. M, Foot dystonia in Parkinson's disease: clinical phenomenology and neuropharmacology, *Adv. Neurol.,* 45:357–360, 1986.

38. Lees, A. J., Shaw, K. M., Stern, G. M., Baclofen in Parkinson's disease, *J. Neurol. Neurosurg. Psychiatry,* 41:707–708, 1984.

39. Quinn, N. P., Marsden, C. D., Lithium for painful dystonia in Parkinson's disease, *Lancet,* 1:1366–1369, 1986.

40. Pachetti, C., Albani, G., Martignoni, E., Godi, L., Alfonsi, E., Nappi, G., "Off" painful dystonia in Parkinson's disease treated with botulinum toxin, *Mov. Disorders,* 10:333–336, 1995.

41. Laitinen, L. V., Bergenheim, A. T., Hariz, M. I., Leksell's posteroventral pallidotomy in the treatment of Parkinson's disease, *J. Neurosurg.,* 76:53–61, 1992.

42. Limousin, P., Krack, P., Pollak, P., Benazzouz, A., Ardouin, C., Hoffmann, D., Benabid, A. L., Electrical stimulation of the subthalamic nucleus in advanced Parkinson's disease, *New Engl. J. Med.,* 339:1105–1111, 1998.

43. Loher, T., Burgunder J. M., Weber, S., Sommerhalder, R., Krauss, J. K., Effect of chronic pallidal deep brain stimulation on off period dystonia and sensory symp-

toms in advanced Parkinson's disease, *J. Neurol. Neurosurg. Psychiatry,* 73:395–399, 2002.

44. Ford, B., Greene, P., Louis, E. D., Petzinger, G., Bressman, S. B., Goodman, R., Brin, M. F., Sadiq, S., Fahn, S., Use of intrathecal baclofen in the treatment of patients with dystonia, *Arch. Neurol.,* 53:1242–1246, 1996.

45. Casey, K. L., Pain and central nervous system disease: a summary and overview, in Casey, K. L., Ed., *Pain and Central Nervous System Disease,* Raven Press, New York, pp. 1–11, 1991.

46. Bogousslavsky, J., Regli, F., Uske, A., Thalamic infarcts: clinical syndromes, etiology and prognosis, *Neurology,* 38:837–848, 1988.

47. Dejerine, J., Roussy, G., Le syndrome thalamique, *Rev. Neurol.,* 14:521–528, 1906.

48. Quinn, N. P., Lang, A. E., Koller, W. C., Marsden, C. D., Painful Parkinson's disease, *Lancet,* 1:1366–1369, 1986.

49. Ford, B., Louis, E. D., Greene, P., Fahn, S, Oral and genital pain syndromes in Parkinson's disease, *Mov. Disorder,* 11:421–426, 1996a.

50. Grushka, M., Sessle, B. J., Burning mouth syndrome, *Dental Cl. N. Amer.,* 35:171–184, 1991.

51. Sage, J. I., Kortis, H. I., Sommer, W., Evidence for the role of spinal cord systems in Parkinson's disease-associated pain, *Clin. Neuropharm.,* 13:171–174, 1990.

52. Chudler, E. H., Dong, W. K., The role of the basal ganglia in nociception and pain, *Pain,* 60:3–38, 1995.

53. Lidsky, T. I., Manetto, C., Schneider, J. S., A consideration of sensory factors involved in motor functions of the basal ganglia, *Brain Research Rev.,* 9:133–146, 1985.

54. Saadé, N. E., Atweh, S. F., Bahuth, N. B., Jabbur, S. J., Augmentation of nociceptive reflexes and chronic deafferentation pain by chemical lesions of either dopaminergic terminals or midbrain dopaminergic neurons, *Brain Research,* 751:1–12, 1997.

55. Battista, A. F., Wolff, B., Levodopa and induced-pain response, *Arch Int. Med.,* 132:70–74, 1973.

56. Factor, S. A., Brown, D. L., Molho, E. S., Subcutaneous apomorphine injections as a treatment for intractable pain in Parkinson's disease, *Mov. Disorders,* 15:167–169, 2000.

57. Van Tol, H. H., Bunzow, J. R., Guen, H. C. et al., Cloning of the gene for a human D_4 receptor with a high affinity for the antipsychotic clozapine, *Nature,* 350:610–614, 1991.

58. Indo, T., Naito, A., Sobue, I., Clinical characteristics of headache in Parkinson's disease, *Headache,* 23:211–12, 1983.

59. Craig, K. D., Emotional aspects of pain, in Wall, P. D., and Melzack, R., Eds., *Textbook of Pain,* 3rd ed., Churchill Livingstone, Edinburgh, pp. 261–274, 1994.

60. Gotham, A. M., Brown, R. G., Marsden, C. D., Depression in Parkinson's disease: a quantitative and qualitative analysis, *J. Neurol. Neurosurg. Psychiatry,* 49:381–389, 1986.

61. Beck, A. T., *Depression: Causes and Treatment,* University of Pennsylvania Press, Philadelphia, PA, 1967.

62. Goetz, C. G., Wilson, R. S., Tanner, C. M., Garron, D. C., Relationships between pain, depression and sleep alterations in Parkinson's disease, *Adv. Neurol.,* 45:345–347, 1986a.

63. Meco, G., Frascarelli, M., Pratesi, L., Linfante, I., Rocchi, L., Formisano, R., Headache in Parkinson's disease, *Headache,* 128:26–29, 1988.

20 Fatigue: A Common Comorbidity in Parkinson's Disease

Carol Ewing Garber
Bouvé College of Health Sciences, Northeastern University; The Department of Medicine, Brown University School of Medicine; and The Department of Medicine, Memorial Hospital of Rhode Island

Joseph H. Friedman
The Department of Clinical Neurosciences, Brown University School of Medicine, and The Department of Neurology, Memorial Hospital of Rhode Island

CONTENTS

INTRODUCTION

Parkinson's disease (PD) is a neurobehavioral disorder defined clinically by its motor features.[1,2] Pathologically, it is defined by the loss of pigmented neurons in the brain stem, coupled with the presence of Lewy bodies in those degenerating centers.[1,3] However, advances in histology have led to the recognition of pathological changes in regions far more widespread than recognized even a decade ago, and there is every reason to believe that more surprises are in store in the near future.[4,5] The correlations between pathology and clinical phenomena have yet to be made for most brain regions, leaving our understanding of the mechanisms of the clinical features of the disease incomplete. The behavioral and nonmotor aspects of PD are particularly difficult to understand because of the major overlap among problems due to neuronal degeneration, psychological responses to progressive disability, iatrogenic complications, and the secondary effects of primary disorders, such as excessive daytime somnolence due to sleep disorder.[6–8]

One of the most common nonmotor symptoms associated with Parkinson's disease is fatigue.[6,7] "Fatigue is a complex and enigmatic entity;"[9] it is a symptom complex, rather than an isolated symptom or sign, and may be considered in a sense analogous to depression or even to the historical name of the disease itself, paralysis agitans. Patients with PD are never paralyzed, or even particularly weak, but they often feel weak and complain about it. Unlike weakness, which can be objectively measured, fatigue is, by its nature, an elusive concept.

In addition to being a problem in many if not most medical disorders,[10,11] fatigue also poses a problem in epistemology. What is fatigue? While we may know, as individuals, what it means to be fatigued, it is difficult to explain and more difficult to measure. Subjectively, it is often described as an "overwhelming sense of tiredness, lack of energy, or feeling of exhaustion."[12] Exemplifying the difficulty in defining what we mean by fatigue, many words have been used in its definition: "lassitude, over-tiredness, lacking in energy, weariness from bodily or mental exertion."[13] Synonyms of fatigue are similarly varied and imprecise and include "tired, debilitated, weary, enervated, languor, listlessness, heaviness, drowsiness, tedium, overtiredness."[14] In the medical literature, there are also many definitions of fatigue as shown in Table 20.1.

There is a physiological definition of fatigue, which refers to decreased function due to repeated use,[15] but this definition applies to isolated cells, organs, or physiological systems and not to the overall sensation of fatigue

TABLE 20.1
Fatigue Definitions

1. An overwhelming sense of tiredness, lack of energy and feeling of exhaustion. It is distinguished form symptoms of depression. Fatigue is also distinguished from limb weakness.[16]

2. A sense of physical tiredness and lack of energy, greater than expected for a usual task.[49]

3. A sense of tiredness, lack of energy or total body give out.[89]

4. A condition resulting from previous stress which leads to reversible impairment of performance and function. Affects the organic interplay of the functions and finally may lead to disturbance of the functional structure of the personality; it is generally accompanied by a reduction in readiness to work and heightened sensation of strain.[91]

5. A chronic form of tiredness, which is perceived by the patient as being unusual or abnormal, and absolutely disproportionate with respect to the amount of exercise or activity the subject has carried out and is not removed by resting or sleeping.[45]

6. Inability to maintain force. Sensation experienced when the effort to perform work, whether physical, mental or both, seems disproportionate to the task involved.[93]

7. A sense of physical tiredness and lack of energy, interfering with physical functioning and social life, distinct from mental exhaustion, sadness, sleepiness, and impaired motor function secondary to PD symptoms.[51]

8. A subjective lack of physical and/or mental energy that is perceived by the individual or caregiver to interfere with usual and desired activities.[94]

    • Physical fatigue [is the inability] to maintain the desired force during sustained or repeated exercise.[20]

    • A state with reduced capacity for work following a period of mental or physical activity.[21]

described by humans. For example, we may speak of fatigue as of the refractoriness to depolarization of a myofibril or neuron after repeated firing or due to the accumulation of metabolic by products.

Sleepiness, a distinct construct, complicates our interpretation of fatigue for several reasons. First of all, we use the word "tired" interchangeably with both sleepiness and fatigue, although one might be sleepy without feeling physically fatigued, or fatigued, such as from exercise, without feeling sleepy. In some circumstances, we feel both sleepy and fatigued; for example, after engaging in prolonged physical work. We may respond to fatigue by resting, which also has ambiguous meanings, encompassing sleeping and sitting or lying quietly without sleeping.

Human fatigue is often categorized into physical and mental components, with the mental component subcategorized into emotional and intellectual aspects.[16,17] Another classification distinguishes central and peripheral fatigue.[18] Peripheral fatigue refers to local muscular fatigue where an individual can no longer produce adequate force during repeated muscular contractions.[19,20] Even with peripheral fatigue, where there is an objective, measurable meaning, there is no agreement on the appropriate terminology. For example, Lou et al.[17] use the term "physical fatigue," and Schwid and colleagues[21] use the term "motor fatigue." Muscular fatigue has been identified in patients with Parkinson's disease,[17,22–24] but this is only one aspect of the persistent, disabling symptom complex that is experienced by so many PD patients.[6]

Further obscuring our understanding of fatigue is the possibility that the nature and etiology of fatigue may be different in different medical disorders. For example, the fatigue that is associated with multiple sclerosis is not necessarily the same as fatigue in Parkinson's disease.

Thus, knowledge of fatigue in one disorder may not apply to another.

In this chapter, we provide an overview of epidemiology and clinical features of fatigue in Parkinson's disease and discuss its measurement, potential causes, and treatments.

EPIDEMIOLOGY OF FATIGUE

Fatigue is a common problem that is pathological or normal, depending on circumstances. It is a common problem in primary care.[26–30] In a community study in Norway, Loge et al.[31] found that "substantial fatigue" lasting six months or more affected 11.4% of the population, aged 19 to 80. Other studies have found fatigue is the presenting complaint in 4 to 9% of primary care office visits.[27–29] In one study,[32] a sense of chronic fatigue was reported as a "major problem" by 25% of consecutive patients seen in a primary care clinic, with 75% of these patients suffering with it for at least one year. Fatigue is costly in terms of direct health care expenditures and indirect costs such as lost employment.[26] For instance, fatigue was estimated to account for 9.3% of formal health care expenditures in the UK.[26]

Fatigue is associated with worse physical and mental health[27,33,34] and often adds morbidity to common medical disorders such as diabetes,[35] chronic renal failure,[36] and cancer.[37] It is a diagnostic symptom of mood disorders such as depression and generalized anxiety disorders[38] and occurs in many psychiatric illnesses.[39] Fatigue is consistently associated with several neurological disorders[6,40–44] and affects 80 to 100% of patients with systemic lupus erythematosis.[45–47] In one study, fatigue was deemed the most bothersome symptom in more than one-fourth of MS patients.[48] Although fatigue is considered a hallmark symptom in multiple sclerosis patients, it was not recognized until 1984, when an influential paper by Freal et al.[49] reported that 78% of patients described fatigue as a symptom, making it both the single most common symptom as well as the one most likely to interfere with activities of daily life.

Fatigue is a prevalent and frequently disabling symptom in patients with Parkinson's disease.[6,7,16,50–54] Affecting up to two-thirds of all patients with PD,[7] 15 to 33% of all PD patients report fatigue to be their most disabling symptom, and more than 50% rank fatigue among their three worst symptoms.[6,7,52] In a recent study in southwest Norway,[54] every patient diagnosed with PD who might require medication was identified and, after excluding depressed and demented patients, 50% were found to suffer from fatigue. Of note here is the observation by Hoehn and Yahr in their classic 1967 paper[55] that fatigue was the presenting symptom in 2% of PD patients.

TYPES OF FATIGUE IN PD

Central and peripheral fatigue have been identified in PD patients.[16,18] Although some believe these are distinct types of fatigue, there is evidence that central mechanisms may underlie the accelerated muscle fatigue thought to be corroboration of peripheral fatigue.[17,18] Central fatigue is characterized by difficulty in initiating and sustaining mental and physical tasks in the absence of cognitive or motor impairment.[18] Mental fatigue has two subdivisions: mental lassitude induced either by hypo- or hypervigilance. The former occurs with repetitive and boring tasks. In PD patients, reduced stimulation due to physical dependence and social isolation consequent to the disease may result in a hypovigilant state. Sustained hypervigilance can also cause mental fatigue, for example, when keeping close track of breaking news stories and making complex decisions. Sustained emotional stressors, such as a critical illness in a close relative, may result in emotional fatigue.

IMPACT AND NATURE OF FATIGUE IN PD

Fatigue in PD was first mentioned in 1967 by Hoehn and Yahr,[55] which is notable because fatigue, rather than motor dysfunction, was the presenting complaint in a handful of patients. The study of fatigue was begun in earnest in 1993, with studies by Van Hilten et al.[51] and Friedman and Friedman.[6]

Van Hilten et al.[51] published the first report focused on fatigue in PD. They compared nondemented patients with PD to age-matched controls to test the hypothesis that fatigue in PD worsened over the course of the day.

Activity monitors were used to assess movement of the nondominant hand, and fatigue was assessed by frequency rather than severity and found to be "often," "very often," or "continuously present" in 31 of 65 patients. Of the fatigued patients, there was no discernible diurnal distribution of activity, nor any correlation between fatigue, motor activity, and bedtime.

Following shortly after the paper by Van Hilten et al., a survey of fatigue in PD patients conducted by Friedman and Friedman[6] was published. PD patients were compared to a same sex friend or relative without PD who was within five years of the subject's age. PD patients reported significantly greater levels of depression and fatigue than the control subjects, and one-third of patients reported fatigue as the single worst symptom of PD. More than half (58%) of the PD patients "agreed" or "strongly agreed" with question, "Fatigue is one of the three most disabling symptoms of PD." Most described fatigue as having a different quality than the fatigue experienced prior to the onset of PD. Although fatigue correlated with depression, not all patients with fatigue were depressed. Fatigue was not associated with motor dysfunction as hypothesized by other authors.[51,56]

Other studies have followed these seminal papers, but fatigue in PD remains an uncommon topic in the scientific literature. Most studies have described the nature and correlates of fatigue; none has evaluated treatments of fatigue in PD. What is clear from the literature is that there is strong, consistent evidence of a higher prevalence of fatigue in PD patients compared to healthy subjects. Several studies, using a variety of instruments and performed in different countries, have documented a higher frequency of fatigue in PD patients as compared to healthy age- and gender-matched control subjects.[6,17,51–54] These studies have shown consistently that about one-half of PD patients suffer from fatigue, but it is notable that fatigue is not a symptom common to all patients with PD. There is also some evidence[54] that the prevalence of fatigue in PD patients is higher that other patients suffering with some other chronic disease, although these findings need confirmation by others. A study by Herlofson and Larson[54] found that a higher proportion of patients with PD reported fatigue as compared to patients with diabetes and pre-hip surgery patients.

Friedman and Friedman,[57] following patients over nine years, found fatigue not only to be a consistent finding over time, but, in addition, the fatigue did not substantially change in severity in most patients, even with treatment and changes in disease severity. PD patients score higher on all dimensions of fatigue as compared with health controls.[17] Lou et al.[17] reported this, including physical fatigue, general fatigue, reduced motivation, reduced activity, mental fatigue, although mental fatigue was not significantly different between PD patients and control subjects. These results suggest that physical and mental fatigue are independent symptoms that should be evaluated separately.

PD patients often complain of weakness, which probably is a reflection of the fatigue of the skeletal muscles in performing repeated exercises.[17,22–24] Ziv et al.[23] observed that PD patients fatigued twice as quickly as healthy control subjects, and fatigue improved following a dose of carbidopa-levodopa, although the magnitude of this improvement was associated with disease severity. These results were confirmed by recent work by Lou et al.[25]

Recently, Hwang and Lin[22] used stimulated single fiber electromyography to evaluate the neuromuscular junction, hypothesizing that fatigue was due to cholinergic defect at the level of the muscle. Results showed that none of the patients had an abnormal individual mean consecutive difference in single fiber potentials, suggesting that the peripheral cholinergic system is intact in PD patients with fatigue. These results all suggest that peripheral muscle fatigue results from a central origin and perhaps is just one more manifestation of central fatigue. Interestingly, myasthenia, a paradigmatic example of peripheral fatigue, is also plagued by central fatigue. Supporting this possibility are several examples in other patient populations. Paul et al.,[58,59] studying patients with myasthenia gravis, found high levels of "cognitive fatigue," demonstrated by diminished cognitive performance correlated with self-perception of fatigue. This is somewhat surprising, because the weakness and neuropathology are limited to skeletal muscles, despite the known inflammatory process. Similarly, a study of fatigue in patients who had recovered from Guillain Barre syndrome found fatigue was not only a common sequellae but also inversely correlated with the degree of recovery.[60]

What is not in doubt is that fatigue in all of its manifestations occurs in PD, and it has a strong negative influence on quality of life and physical function in PD patients.[61–63] Fatigue has been reported to cause emotional distress and problems "in the areas of physical functioning, role limitation (physical), and social functioning and vitality" in nondepressed PD patients in multiple studies.[61–62] An inverse association between fatigue, habitual physical activity levels, physical function, and functional capacity in PD patients has been reported by Garber and Friedman.[63] Fatigued patients tend to be less active and have poorer functional capacity compared with those with lower levels of fatigue.[63] Van Hilten et al.[51] hypothesized that motor activity would have a diurnal pattern in PD patients, but their results documented only a lower activity level in the morning. After this "slow start," PD patients' activity increased and did not decline as expected as the day progressed. The pattern of activity also did not correlate with fatigue levels.

CAUSES OF FATIGUE IN PARKINSON'S DISEASE

There are multiple factors that probably contribute to the sense of fatigue in PD. One of the most obvious factors is disease severity. While studies show an association between more severe disease and fatigue,[51,52,54] none has found an independent association between disease severity and fatigue,[6,51–54,63] suggesting that disease severity in itself cannot explain fatigue. Furthermore, Friedman and Friedman[57] found that, even with progression of disease, fatigue declined but did not change substantially in PD patients followed clinically.

Other explanations for fatigue in PD include sleep dysfunction causing excessive daytime sleepiness, depression and other mood disorders, medication effects, and the motor dysfunction itself. Fatigue has, in fact, been associated with other comorbid conditions, including mood and sleep disorders in some, but not all, studies.

Drugs have been implicated both in exacerbating and reducing fatigue. For example, Pramiprexole has been associated with increased fatigue in several studies.[64,65] It should be considered, however, that most of the reports of increased fatigue related to drug therapy has come from randomized clinical trials of the efficacy of drugs in the treatment of motor symptoms of PD, and the data on fatigue has been collected as an adverse effect. Interpretation of the cause and effect relationships between a drug and adverse side effects is notoriously difficult,[66] so these results should be interpreted with caution, particularly because fatigue is a frequent complaint and a commonly reported adverse effect of many drugs.

Studies evaluating the effect of pharmacologic agents on fatigue in PD are rare. Carbidopa-Levodopa has been shown to reduce central and peripheral fatigue[23,25,67] but has also been found to be a predictor of fatigue in cross-sectional studies (e.g., Ref. 63). Recently, Abe et al.[68] compared fatigue in patients taking pergolide mesilate to bromocriptine and found that patients with pergolide had reduced levels of fatigue after taking the drug, but there was no change in patients taking bromocriptine. These results suggest the possibility that the D-1 receptor is involved in the sensation of fatigue in PD patients, but more work confirming these findings is needed.

There have been more studies evaluating drug therapy on fatigue in MS; however, none has been shown to have clear benefit.[69–71] Amantadine has been evaluated in several studies in the treatment of fatigue of MS (but not in PD); however, a Cochrane review[70] of its efficacy reports, "…[amantadine's] efficacy in reducing fatigue in people with MS is poorly documented and there is insufficient evidence to make recommendations to guide prescribing." More recently, trials of Prokarin,[71] Pemoline,[72] and 4-aminopyridine[73] have reported preliminary, although slightly promising, results in reducing fatigue in MS.

Whether these results will be verified by other studies, or the drugs might be used to treat fatigue in PD, remains to be seen.

Depression is often associated with the general feelings of tiredness and malaise that are often associated with fatigue,[74,75] but the link between depression and fatigue is complex. Lou et al.,[17] reporting on a sample of PD patients and healthy controls, found that PD patients on average had higher scores on the Profile of Mood States (POMS) and depression. Further depression correlated with all dimension of fatigue except physical fatigue. Karlsen et al.[52] also found an association between fatigue and depression as well as between fatigue and the use of sleeping pills. However, they also found fatigue to be equally prevalent in patients with and without depression and depression was not predictive of fatigue. Herlofson and Larsen,[54] using a multivariable analysis, also found that sleep disorders and pain were not independent predictors of fatigue. These results suggest that fatigue is an independent symptom of PD, overlapping with but not causally related to depression.

The same may be said about sleep disorders and fatigue. Sleep disorders are undeniably common in PD.[7,51,76–79] Although associated with fatigue, sleep disorders do not predict fatigue in PD patients.[54] Van Hilten and colleagues[51] reported that the prevalence of daytime sleepiness was similar in both PD patients and controls, with both having a diurnal pattern of sleepiness peaking in the early afternoon. On the other hand, in these subjects, fatigue was fairly constant throughout the day and more common in PD compared with controls.

Hogl et al.[80] evaluated daytime sleepiness in control subjects and in patients with PD. They found that, while daytime sleepiness was more common in PD patients compared with control subjects, in both groups, sleepiness was associated with heavy snoring, suggesting that daytime sleepiness reflects the presence of a sleep disorders. Other studies[81] have supported these findings, but not all. Fabbrini et al.[82] found no differences in daytime sleepiness in PD patients as compared with healthy control subjects, but they did find that sleepiness was associated with PD drug treatments, suggesting that sleepiness is a side effect of treatment rather than caused by the disease itself.

PD patients have higher resting energy utilization than age-matched controls.[83–85] The resting metabolic rate decreases with treatment in those patients who were stiff and, in general, there is a loose correlation between the improvement in rigidity and the decline in energy requirements.[83] L-dopa induced dyskinesias produce yet higher energy requirements.[83] Even early, untreated patients have an increased energy requirement.[87] Several authors have proposed that the increased energy expenditure may be a contributing factor to the weight loss that so commonly affects PD patients.[83,84] Data from Tzelepis et al.[88] have demonstrated that energy use for respiration is also increased in PD, which partly explains the increased resting energy requirements.

Given the increased energy requirements of PD patients at rest, it is a natural question to then ask, "Do PD patients who exercise less efficiently; that is, require more calories to perform a given exercise, suffer more from fatigue than those who are more efficient?" Preliminary data from our laboratory[87] suggest that PD patients use a higher proportion of their ventilation capacity as compared to health control subjects at a similar exercise workload. Thus, exercise may be more fatiguing in PD subjects due to a great ventilatory effort.

None of these factors discussed explain the phenomenon of fatigue in PD; they only assist researchers in identifying "clues" that may help to determine the underlying etiology of fatigue, which is not known. Several hypotheses have been proposed to explain fatigue in PD and other chronic diseases, but these remain largely untested.[11,18] Hypothesized mechanisms of fatigue range from altered activation of the hypothalamic-pituitary-adrenal axis due to prolonged stress, inflammatory processes, and alterations in neurotransmitters and neurotransmission within the CNS, including disruption of the nonmotor functions in the basal ganglia and dysfunction of the striato-thalamo-cortical loop.[11,18]

It remains a problem that fatigue, depression, sleep disorders, drug side effects, and other comorbid conditions are common in PD patients and, importantly, these symptoms and conditions are often not recognized by clinicians in their patients with PD patients.[8] To illustrate this problem, Shulman et al.,[8] studying patients with PD, found that 44% reported depression, 39% were anxious, 42% were fatigued, and 43% had sleep problems. On the other hand, these problems were often not diagnosed by the patient's neurologist. Of these same patients, 35% had been diagnosed with depression, 42% with anxiety disorder, 25% with fatigue, and 60% with sleep disturbance. The later is interesting, showing how patients often do not recognize sleep problems in themselves.

Although some authorities argue that better definitions and rating instruments are needed,[88] it is likely only that different rating instruments will be developed that are disease specific. It is unlikely, for example, that the fatigue of myasthenia gravis will be measured sensitively by an instrument aimed at multiple sclerosis or lupus, where effects of medications or polysystem diseases are important.

In summary, fatigue is one of most common disabling symptoms in patients with PD. The impact of fatigue on the quality of life of patients is substantial, although health care providers often underestimate it. Fatigue has two components that may be related but are likely independent: peripheral (local muscle) fatigue and mental fatigue. Fatigue is associated with sleepiness and depression, but patients with fatigue may not be depressed or have sleep disorders. Drugs may exacerbate or improve fatigue. Research of the causes of and treatments for fatigue are sorely needed.[89]

REFERENCES

1. Rowland, L. P., Ed., *Merritt's Neurology*, 10th ed., Lippincott, Williams and Wilkins, Philadelphia, 2000.

2. Fahn, S., Description of Parkinson's disease as a clinical syndrome, *Ann. N. Y. Acad. Sci.*, 991:1–14., 2003.

3. Hardy, J., The relationship between Lewy body disease, Parkinson's disease, and Alzheimer's disease, *Ann. N. Y. Acad. Sci.*, 991:167–170, 2003.

4. Maguire-Zeiss, K. A., Federoff, H. J., Convergent pathobiologic model of Parkinson's disease, *Ann. N. Y. Acad. Sci.*, 991:152–166, 2003.

5. Jellinger, K. A., Recent developments in the pathology of Parkinson's disease, *J. Neural. Transm.*, suppl. (62): 347–376, 2002.

6. Friedman, J., Friedman, H., Fatigue in Parkinson's disease, *Neurology*, 43(10): 2016–2018, 1993.

7. Shulman, L. M., Taback, R L., Bean, J, Weiner, W. J., Comorbidity of the nonmotor symptoms of Parkinson's disease, *Mov. Disord.*, 16(3):507–510, 2001.

8. Shulman, L. M., Taback, R. L., Rabinstein, A.A., Weiner, W. J., Non-recognition of depression and other non-motor symptoms in Parkinson's disease, *Parkinsonism Relat. Disord.*, 8(3):193–197, 2002.

9. Sullivan, P. F., Kovalenko, P., York, T. P. et al., Fatigue in a community sample of twins, *Psychol. Med.*, 33:263–281, 2003.

10. Ridsdale, L., Evans, A., Jerrett, W., Mandalia, S., Osler, K., Vora, H., Patients with fatigue in general practice: a prospective study, *B. M. J.*, 307(6896): 103–106, 1993.

11. Swain, M. G., Fatigue in chronic disease, *Clin. Sci., London*, 99(1): 1–8, 2000.

12. Abudi, S., Bar-Tal, Y., Ziv, L., Fish, M., Parkinson's disease symptoms—patient's perceptions, *J. Adv. Nurs.*, 25: 54–55, 1997.

13. *Websters Encyclopedic Unabridged Dictionary of the English Language*, 1996, Gramercy Books, New York.

14. Stein, J., Flexner, S. B., Eds., *The Random House Thesaurus College Edition*, Random House, New York, 1984.

15. Guyton, A. C., *Textbook of Medical Physiology*, 10th Ed., W. B. Saunders, Philadelphia, Pennsylvania, 2000.

16. Krupp, L. B., Pollina, D. A., Mechanisms and management of fatigue in progressive neurological disorders, *Curr. Opin. Neurol.*, 9: 456–460, 1996.

17. Lou, J. S., Kearns, G., Oken, B., Sexton, G., Nutt, J., Exacerbated physical fatigue and mental fatigue in Parkinson's disease, *Mov. Disord.*, 16(2):190–196, 22001.

18. Chaudhuri, A., Behan, P. O., Fatigue and basal ganglia, *J. Neurol. Sci.*, 179 (S1-2): 34–42, 2000.

19. Brooks, G. A., Fahey, T., White, T., *Exercise Physiology: Human Bioenergetics and Its Applications*, 3rd ed.,

Mayfield Publishing Company, Mountain View, CA, 2000.

20. Edwards, R. H. T., *Human muscle fatigue: physiological mechanisms*, Pitman Medical, London, 2000, pp. 1–18, 1981.

21. Schwid, S. R., Thornton, C. A., Pandya, S., Manzur, K. L., Sanjak, M., Petrie, M. D., McDermott, M. P., Goodman, A. D., Quantitative assessment of motor fatigue and strength in MS, *Neurology*, 53(4):743–750, 1999.

22. Hwang, W. J., Lin, T. S., Evaluation of fatigue in Parkinson's disease patients with stimulated single fiber electromyography, *Acta Neurol. Scand.*, 104(5): 271–274, 2001.

23. Ziv, I., Avraham, M., Michaelov, Y., Djaldetti, R., Dressler, R., Zoldan, J., Melamed, E., Enhanced fatigue during motor performance in patients with Parkinson's disease, *Neurology*, 51(6):1583–1586, 1998.

24. Abe, K., Takanashi, M., Yanagihara, T., Fatigue in patients with PD, *Behav. Neurol.*, 12: 103–106, 2000.

25. Lou, J. S., Benice, T., Kearns, G., Sexton, G., Nutt, J., Levodopa normalizes exercise related cortico-motoneuron excitability abnormalities in Parkinson's disease, *Clin. Neurophysiol.*, 114(5):930–937, 2003.

26. McCrone, P., Darbishire, L., Ridsdale, L., Seed, P., The economic cost of chronic fatigue and chronic fatigue syndrome in UK primary care, *Psychol. Med.*, 33(2):253–261, 2003.

27. Kirk, J., Douglass, R., Nelson, E., Jaffe, J. et al., Chief complaint of fatigue: a prospective study, *J. Fam. Practice*, 30:33–39, 1990.

28. David, A., Pelosi, A., McDonald, E., Stephens, D. et al., Tired, weak, or in need of rest: fatigue among general practice attenders, *Br. Med. J.*, 301:1199–1202, 1990.

29. Cathebras, P. J., Robbins, J. M., Kirmayer, L. J., Hayton, B. C., Fatigue in primary care: prevalence, psychiatric comorbidity, illness behavior, and outcome, *J. Gen. Int. Med.*, 7:276–286, 1992.

30. Fuhrer, R., Wessely, S., The epidemiology of fatigue and depression: a French primary-care study, *Psychol. Med.*, 25:895–905, 1995.

31. Loge, J. H., Ekeberg, O., Kaasa, S., Fatigue in the general Norwegian population: normative data and associations, *J. Psychosom. Res.*, 45(1 Spec No):53–65, 1998.

32. Kroenke, K., Wood, D. R., Mangelsdorff, E., Meier, N. J., Powell, J. B., Chronic fatigue in primary care: prevalence, patient characteristics and outcome, *JAMA*, 260:929–934, 1988.

33. Skapinakis, P., Lewis, G., Mavreas, V., One-year outcome of unexplained fatigue syndromes in primary care: results from an international study, *Psychol. Med.*, 33(5): 857–866, 2003.

34. Greden, J. F., Physical symptoms of depression: unmet needs, *J. Clin. Psychiatry*, 64 suppl. 7:5–11, 2003.

35. Konen, J. C., Curtis, L. G., Summerson, J. H., Symptoms and complications of adult diabetic patients in a family practice, *Arch. Fam. Med.*, 5:135–145, 1996.

36. Chang, W. K., Hung, K. Y., Huang, J. W., Wu, K. D., Tsai, T. J., Chronic fatigue in long-term peritoneal dialysis patients, *Am. J. Nephrol.*, 21(6):479–485, 2001.

37. Cella, D., Davis, K., Breitbart, W., Curt, G., Cancer-related fatigue: prevalence of proposed diagnostic criteria in a United States sample of cancer survivors, *J. Clin. Oncology,* 19:3385–3391, 2001.

38. First, M. B., Frances, A., Pincus, H. A., *DSM-IV-TR handbook of differential diagnosis,* American Psychiatric Press, Washington, D. C., 2002.

39. Gelder, M. G., Mayou, R., Gedes, J., *Psychiatry,* Oxford University Press, New York, 1999.

40. Jackson, C. E., Bryan, W. W., Amyotrophic lateral sclerosis, *Semin. Neurol.,* 18(1):27–39, 1998.

41. Miller, R. G., Role of fatigue in limiting physical activities in humans with neuromuscular diseases, *Am. J. Phys. Med. Rehabil.,* 81(11 suppl.): S99-107, 2002.

42. Paul, R. H., Cohen, R. A., Gilchrist, J. M., Ratings of subjective mental fatigue relate to cognitive performance in patients with myasthenia gravis, *J. Clin. Neurosci.,* 9(3):243–246, 2002.

43. Krupp, L. B., Christodoulou, C., Fatigue in multiple sclerosis, Curr. *Neurol. Neurosci. Rep.,* 1(3):294–298, 2001.

44. Sabin, T. D., An approach to chronic fatigue syndrome in adults, *Neurolog.,* 9(1):28–34, 2003.

45. Liang, M. H., Rogers, M., Larson, M., Eaton, H. M., Murawski, B. J., Taylor, J. E., Swafford, J., Schur, P. H., The psychosocial impact of systemic lupus erythematosus and rheumatoid arthritis, *Arthritis Rheum.,* 27(1): 13–19, 1984.

46. Wang, B., Gladman, D. D., Urowitz, M. B., Fatigue in Lupus Is Not Correlated with Disease Activity, *J. Rheumatol.,* 25:892–895, 1998.

47. Krupp, L. B., LaRocca, N. G., Muir-Nash, J., Steinberg, A. D., The fatigue severity scale. Application to patients with multiple sclerosis and systemic lupus erythematosus, *Arch. Neurol.,* 46(10): 1121–1123, 1989.

48. Ford, H., Trigwell, P., Johnson, M., The nature of fatigue in multiple sclerosis, *J. Psychosom. Res.,* 45: 33–38, 1998.

49. Freal, J. E., Kraft, G. H., Coryell, J. K., Symptomatic fatigue in multiple sclerosis, *Arch. Phys. Med. Rehabil.,* 65(3):135–138, 1984.

50. Friedman, J. H., Fernandez, H. H., The nonmotor problems of Parkinson's disease, *The Neurologist,* 6:8–27, 2000.

51. van Hilten, J. J., Weggeman, M., van der Velde, E. A., Kerkhof, G. A., van Dijk, J. G., Roos, R. A., Sleep, excessive daytime sleepiness and fatigue in Parkinson's disease, *J. Neural. Transm. Park Dis. Dement.,* Sect., 5(3):235–244, 1993.

52. Karlsen, K., Larsen, J. P., Trandberg, E., Kjell, Jorgensen, Fatigue in patients with Parkinson's disease, *Movement Disorders,* 14(2): 287–241, 1999.

53. Abudi, S., Bar-Tal, Y., Ziv, L., Fish, M., Parkinson's disease symptoms—patient's perceptions, *J. Adv. Nurs.,* 25: 54–59, 1997.

54. Herlofson, K., Larsen, J. P., The influence of fatigue on health-related quality of life in patients with Parkinson's disease, *Acta Neurol. Scand.,* 107(1):1–6, 2003.

55. Hoehn, M. M., Yahr, M. D., Parkinsonism: onset, progression and mortality, *Neurology,* 17(5): 427–442, 1967.

56. Schwab, R. S., England, A. C., Peterson, C., Akinesia in Parkinson's disease, *Neurology,* 9: 65–72, 1959.

57. Friedman, J. H., Friedman, H., Fatigue in Parkinson's disease: a nine-year follow-up, *Mov. Disord.,* 16(6): 1120–1122, 2001.

58. Paul, R. H., Cohen, R. A., Gilchrist, J. M., Ratings of subjective mental fatigue relate to cognitive performance in patients with myasthenia gravis, *J. Clin. Neurosci.,* 9(3): 243–246, 2002.

59. Paul, R. H., Cohen, R. A., Goldstein, J. M., Gilchrist, J. M., Fatigue and its impact on patients with myasthenia gravis, *Muscle Nerve,* 23(9): 1402–1406, 2000.

60. Merkies, I. S., Schmitz, P. I., Samijn, J. P., van der Meche, F. G., vanDoorn, P. A., Fatigue in immune-mediated polyneuropathies. European Inflammatory Neuropathy Cause and Treatment (INCAT) Group, *Neurology,* 53(8): 1648–1654, 1999.

61. Herlofson, K., Larsen, J. P., Measuring fatigue in patients with Parkinson's disease—the Fatigue Severity Scale, *Eur. J. Neurol.,* 9(6): 595–600, 2002.

62. Shrag, A., Jahanshahi, M., Quinn, N., What contributes to quality of life in patients with Parkinson's disease? *J. Neurol. Neurosurg. Psychiatry,* 69:308–312, 2000.

63. Garber, C. E., Friedman, J. H., Effects of fatigue on physical activity and function in patients with Parkinson's disease, *Neurology,* 60(7):1119–1124, 2003.

64. Pogarell, O., Gasser, T., van Hilten, J. J., Spieker, S., Pollentier, S., Meier, D., Oertel, W. H., Pramipexole in patients with Parkinson's disease and marked drug resistant tremor: a randomised, double blind, placebo controlled multicentre study, *J. Neurol. Neurosurg. Psychiatry,* 72(6):713–720, 2002.

65. Pinter, M. M., Pogarell, O., Oertel, W. H., Efficacy, safety, and tolerance of the non-ergoline dopamine agonist pramipexole in the treatment of advanced Parkinson's disease: a double blind, placebo controlled, randomised, multicentre study, *J. Neurol. Neurosurg. Psychiatry,* 66(4):436–441, 1999.

66. Riegelman, R. K., *Studying a Study and Testing a Test,* 4th ed., Lippincott Williams and Wilkins, Philadelphia, 2000.

67. Funkiewiez, A., Ardouin, C., Krack, P., Fraix, V., Van Blercom, N., Xie, J., Moro, E., Benabid, A. L., Pollak, P., Acute psychotropic effects of bilateral subthalamic nucleus stimulation and levodopa in Parkinson's disease, *Mov. Disord.,* 18(5):524–530.

68. Abe, K., Takanashi, M., Yanagihara, T., Sakoda, S., Pergolide mesilate may improve fatigue in patients with Parkinson's disease, *Behav. Neurol.,* 13(3-4):117–121, 2001–2002.

69. Rossini, P. M., Pasqualetti, P., Pozzilli, C., Grasso, M. G., Millefiorini, E., Graceffa, A., Carlesimo, G. A., Zibellini, G., Caltagirone, C., Fatigue in progressive multiple sclerosis: results of a randomized, double-blind, placebo-controlled, crossover trial of oral 4-aminopyridine, *Mult. Scler.,* 7(6):354–358, 2001.

70. Taus, C., Giuliani, G., Pucci, E., D'Amico, R., Solari, A., Amantadine for fatigue in multiple sclerosis, *Cochrane Database Syst. Rev.,* (2):CD002818, 2003.

71. Gillson, G., Richard, T. L., Smith, R. B., Wright, J. V., A double-blind pilot study of the effect of Prokarin on fatigue in multiple sclerosis, *Mult. Scler.,* 8(1):30–35, 2002.

72. Weinshenker, B. G., Penman, M., Bass, B., Eber, G. C., Rice, G. P., A double-blind, randomized, crossover trial of pemoline in fatigue associated with multiple sclerosis, *Neurology,* 42(8):1468–1471, 1992.

73. Rossini, P. M., Pasqualetti, P., Pozzilli, C., Grasso, M. G., Millefiorini, E., Graceffa, A., Carlesimo, G. A., Zibellini, G., Caltagirone, C., Fatigue in progressive multiple sclerosis: results of a randomized, double-blind, placebo-controlled, crossover trial of oral 4-aminopyridine, *Mult. Scler.,* 7(6):354–358, 2001.

74. Schrag, A., Jahanshahi, M., Quinn, N. P., What contributes to depression in Parkinson's disease? *Psychol. Med.,* 31: 65–73, 2001.

75. Dooneief, G., Mirabello, E., Bell, K., Marder, K., Stern, Y., Mayeux, R, An estimate of the incidence of depression in idiopathic Parkinson's disease, *Arch. Neurol.,* 49:305–307, 1992.

76. Garcia-Borreguero, D., Larrosa, O., Bravo, M., Parkinson's disease and sleep, *Sleep Med. Rev.,* 7(2):115–29, 2003.

77. Comella, C. L,. Sleep disturbances in Parkinson's disease, *Curr. Neurol. Neurosci. Rep.,* 3(2):173–80, 2003.

78. Stacy, M., Sleep disorders in Parkinson's disease: epidemiology and management, *Drugs Aging,* 19(10): 733–739, 2002.

79. Happe, S., Ludemann, P., Berger, K., FAQT study investigators. The association between disease severity and sleep-related problems in patients with Parkinson's disease, *Neuropsychobiology,* 46(2):90–96, 2002.

80. Hogl, B., Seppi, K., Brandauer, E., Glatzl, S., Frauscher, B., Niedermuller, U., Wenning, G, Poewe, W., Increased daytime sleepiness in Parkinson's disease: a questionnaire survey, *Mov. Disord.,* 18(3):319–323, 2003.

81. Chaudhuri, K. R., Pal, S., DiMarco, A., Whately-Smith, C., Bridgman, K., Mathew, R., Pezzela, F. R., Forbes, A., Hogl, B., Trenkwalder, C., The Parkinson's disease sleep scale: a new instrument for assessing sleep and nocturnal disability in Parkinson's disease, *J. Neurol. Neurosurg. Psychiatry,* 73(6):629–635, 2002.

82. Fabbrini, G., Barbanti, P., Aurilia, C., Vanacore, N., Pauletti, C., Meco, G., Excessive daytime sleepiness in de novo and treated Parkinson's disease, *Mov. Disord.,* 17(5):1026–1030, 2002.

83. Markus, H. S., Cox, M., Tomkins, A. M., Raised resting energy expenditure in PD and its relationship to muscle rigidity, *Clin. Sci.,* 83:199–204, 1992.

84. Lev, S., Coz, M., Lugon, M., Hodkinson, M., Tomkins, A., Increased energy expenditure in PD, *Br. Med. J.,* 301:1256–1257, 1990.

85. Brouselle, E., Borson, F., Gonzalez de Suso, J. M. et al., Augmentation de la depense energetique au cours de la maladie de Parkinson, *Rev. Neurolog.* (Paris), 147:46–51, 1991.

86. Tzelepis, G., McCool, F. D., Friedman, J. H., Hoppin, F. G., Jr., Respiratory muscle dysfunction in Parkinson's disease, *Am. Rev. Resp. Dis.,* 38:266–271, 1988.

87. Masoudi, O., Friedman, D., Garber, C., McCool, D., personal communication.

88. Schwid, S. R., Covington, M., Segal, B. M., Goodman, A. D., Fatigue in multiple sclerosis: Current understanding and future directions, *J. Rehab. Res. Develop.,* 39:211–224, 2002.

89. Djaldetti, R., Melamed, E., New drugs in the future treatment of Parkinson's disease, *J. Neurol.,* 249 suppl. 2:II30–II35, 2002.

90. Schwartz, J. E., Jandorf, L., Krupp, L. B., The measurement of fatigue: a new instrument, *J. Psychosom. Res.,* 37(7): 753–762, 1993.

91. Berrios, G. E., Feelings of Fatigue and psychopathology: A Conceptual History, *Comp. Psychiatry.,* 31:140–151, 1990.

92. Tavio, M., Milan, I., Tirelli, U., Cancer- related fatigue (a review), *Int. J. Oncol.,* 21:1093–1099, 2002.

93. Lane, R., Chronic fatigue syndrome: is it physical? *JNNP,* 69:289, 2000.

94. Shapiro, R. T., Schneider, D. M., Fatigue, in van den Noort, S. and Holland, N., Eds., *Multiple Sclerosis in Clinical Practice,* Chapter 3, Demos Medical Publishing, New York, 1993.

21 Sleep Disorders in Parkinson's Disease

Svenja Happe
Department of Clinical Neurophysiology, University of Göttingen

Claudia M. Trenkwalder
Paracelsus-Elena Klinik, Center of Parkinsonism and Movement Disorders Kassel, and University of Göttingen

CONTENTS

ABSTRACT

Almost all patients with Parkinson's disease (PD) develop sleep-wake disorders with a subsequently impaired quality of life. However, the frequent sleep complaints in PD and other parkinsonian syndromes (PS) have not been sufficiently acknowledged until recently. Sleep-wake disturbances in parkinsonism consist of many different factors, including psychological and physiological alterations with disease specific and treatment specific contributions. Motor symptoms [e.g., mild to violent movements in REM sleep behavior disorder (RBD) or periodic limb movements, tremor, or early morning dystonia] are characteristic features leading to sleep disturbances. Excessive daytime sleepiness (EDS) has recently been discussed in more detail and seems more a disease-related than a drug-related problem. Respiratory disturbances, including the sleep apnea syndrome, need to be considered as frequent causes for sleep disturbances and EDS as well.

A detailed medical and sleep history is the basis for differentiating the various sleep and daytime problems in PD. A recently developed and validated Parkinson's disease sleep scale (PDSS) may help to differentiate between various sleep disorders and to optimize treatment.

Treatment of sleep disorders in parkinsonism needs to be initiated according to their specific causes or focused on their diverse symptoms. A dopaminergic deficit can lead to nocturnal akinesia, which should be treated with increased dosages of dopaminergic medication at night, whereas nocturnal RBD should be treated with benzodiazepines. Hallucinations and psychosis require neuroleptic drugs such as clozapine or quetiapine. Nocturnal respiratory disorders need further investigations such as sleep apnea screening and polysomnography with initiation of specific treatment strategies such as nasal ventilation with continuous positive airway pressure (nCPAP).

Since sleep related problems have an important impact on quality of life in PD patients and their caregivers, it is important to develop new and improved management plans giving not only the patients but also their caregivers a good sleep and an improved quality of life.

A brief description of various sleep disorders in parkinsonism will be provided in this chapter, including a discussion on their clinical features, consequences, pathophysiology, and treatment options.

INTRODUCTION

James Parkinson, in 1817, noted that "sleep becomes much disturbed" and that the terminal stage of the disease is associated with "constant sleepiness, with slight delirium, and other marks of extreme exhaustion."[1] Later in the nineteenth century, Charcot described the impact of severe rigidity and bradykinesia on sleep.[2] However, sleep problems accompanying Parkinson's disease (PD) and other parkinsonian syndromes (PS) received relatively little attention until the middle of the twentieth century, when electroencephalography (EEG) was discovered and the first polysomnographic recordings (PSG), also in PD patients, were performed. In the second half of the twentieth century, more and more studies investigating different aspects of sleep and its disturbances in PD patients were published. Reports of excessive daytime sleepiness (EDS) with sudden sleep onset, so-called "sleep attacks," have drawn attention to sleep as an important issue for PD patients in recent years.[3] REM sleep behavior disorder (RBD) with complex movements during REM sleep may be a preclinical and premotor sign of PD or PS.[4] Therefore, the observation of early changes in sleep may be of particular importance and may gain relevance for the early diagnosis of PD, particularly if neuroprotective agents become available in the near future.

A variety of neurodegenerative, psychological, and pharmacological factors can lead to disturbances of the normal sleep in patients with parkinsonism. *First*, sleep disruption may be a consequence of the neurodegenerative process in the neurophysiological and neurochemical system that is responsible for sleep regulation itself. According to the six progressive stages described by Braak and co-workers, the neurodegenerative process starts in the lower brain-stem areas, where sleep-wake regulation gains important input.[5,6] The first signs of sleep-wake disruption may occur when Lewy bodies, the neuropathological equivalent of PD, intrude those brainstem areas relevant for sleep-wake regulation. *Second*, associated symptoms such as breathing-difficulties, akinesia, as well as depression can lead to nocturnal sleep problems. *Third*, the medication can induce new symptoms such as nightmares, psychosis, and nightly movements or increase wakefulness. *All* these effects on sleep and daytime alertness can have implications for choosing the right treatment strategy in patients with parkinsonism. Since sleep complaints and daytime sleepiness are common problems associated with parkinsonism leading to a subsequent impaired quality of life, they need sufficiently be acknowledged today.

PATHOPHYSIOLOGICAL CONSIDERATIONS

The most striking pathological features of PD are neuronal depletion and gliosis of pigmented areas of the basal ganglia leading to a marked reduction in dopamine content.[7] The loss of dopaminergic neurons of the substantia nigra is responsible for most of the motor symptoms of PD, whereas other brain abnormalities, including other neurotransmitter systems, may primarily account for some of the sleep-wake abnormalities. It is known that serotonergic neurons originating in the dorsal raphe nuclei, noradrenergic neurons originating in the region of the locus coeruleus, and cholinergic neurons originating in the pedunculopontine nucleus are also reduced in number in PD.[8,9] As the serotonergic, noradrenergic, and cholinergic systems are involved in the sleep-wake regulation, abnormalities in these systems may account for some of the sleep-wake disturbances in PD patients.

The mesocorticolimbic and the mesostriatal dopaminergic systems are additionally disturbed in PD and can contribute to sleep-wake disturbances.[10] The predominant dopamine receptor in the cerebral cortex, where dopaminergic neurons with cell bodies in the ventral tegmental area of the brain stem project to, is the D_1 type.[11] Dopamine D_1 receptor agonists produce EEG desynchronization and behavioral arousal,[12] whereas dopamine D_1 receptor antagonists lead to sedation.[13] This is consistent with an arousal function of the neurons in the ventral tegmental projection area. High dosages of dopamine D_1 and D_2 receptor agonists reduce total sleep time,[14] whereas very low dosages induce sleep and increase the amount of slow-wave sleep. Low dosages of dopamine D_1 and D_2 receptor agonists also induce sleep when injected directly into the ventral tegmental area. This effect is blocked by dopamine receptor autoantagonists, suggesting that dopamine D_2 autoreceptors play a role in the mediation of sleep due to an inhibition of the firing rate of ventral tegmental dopaminergic neurons.[15,16]

The regulation of rapid eye movement (REM) sleep may also be influenced by the dopaminergic system, mediated by D_1 and D_2 receptors. Dopamine D_1 receptor agonists suppress REM sleep, whereas dopamine D_1 receptor antagonists increase the amount and the duration of REM sleep.[15,17] High dosages of dopamine D_2 receptor agonists can also suppress REM sleep.[14] However, the exact effects of these agents on REM sleep and their sites of action are still not known. Levodopa, the biochemical precursor of dopamine and norepinephrine, and dopamine receptor agonists such as bromocriptine, pergolide, cabergoline, or nonergoline derivatives such as pramipexole and ropinirole are frequently used in the treatment of PD. All these mentioned drugs are known to enhance the activity of the dopaminergic system and to predominantly induce wake-

fulness in higher dosages. Therefore, they can also alter the sleep-wake regulation significantly. Levodopa is known to reduce the amount of REM sleep in healthy persons. This effect may be due to an increased activity of dopamine, norepinephrine, or both. Levodopa is also known to reduce the contents of serotonin, tryptophan, and tyrosine in the brain. Consequently, the effects of levodopa on sleep-wake regulation may be in part related to effects on serotonergic neurons also. Dopamine D_2 receptor agonists still alter sleep-wake patterns, although they have only little effect on noradrenergic or serotonergic neurons. This may be explained, presumably, through effects on mesocortical dopaminergic pathways. In Table 21.1, several other aspects of PD that can also contribute to sleep-wake disturbances are presented.

TABLE 21.1
Aspects of Parkinson's Disease that Contribute to Sleep-Wake Disturbances

Reasons	Predominant Impairments
Neurochemical alterations of cholinergic, serotonergic, and noradrenergic systems	Impaired sleep-wake control, reduced REM sleep
Bradykinesia and rigidity	Reduction of normal body shifts during sleep with problems of turn over in bed, leading to discomfort and awakenings; impaired ability to use the bathroom at night, respiratory muscle dysfunction
Periodic and non-periodic movements, tremor, dystonia, drug-induced myoclonus	Arousals, pain with discomfort laying still
Restless legs syndrome	Sleep onset problems, discrepancy between will of moving and immobility
REM sleep behavior disorder	Disrupted REM sleep, arousals, anxiety
Abnormal motor activity affecting respiratory and upper airway muscles	Breathing abnormalities during sleep such as sleep apnea syndrome
Drug-related effects	Reduced total sleep time with increased time awake, reduced REM sleep, reduced slow-wave sleep, nightmares, excessive daytime sleepiness
Depression and anxiety	Difficulty falling asleep, arousals, early morning awakening
Dementia and psychosis	Nocturnal confusion episodes, sun-downing with disturbed day-and-night rhythm
Nocturia	Increased time awake, problems of falling asleep again, nocturnal falls, discomfort

CLINICAL FEATURES

The most common sleep-related complaints in parkinsonism are difficulties with sleep initiation and sleep maintenance, nocturnal akinesia, and RBD with nocturnal vocalizations and complex movements. Sleep disturbances tend to increase with progression of the disease. Daytime drowsiness and EDS become increasingly common, and EDS is experienced in about 15% of PD patients as compared to 1% of healthy elderly.[18] Sudden sleep onset, so-called "sleep attacks," have been described recently and include sudden sleepiness, mostly related to drug intake. Severe sleep disruption is particularly common in older patients with on-off phenomena or halluci-

nations.[19,20] The sleep architecture can be almost normal in the early stages of PD. The abnormal sleep features associated with more advanced PD include a change of sleep stage patterns, abnormal motor activity, and disturbed breathing patterns.

ALTERATIONS IN THE ORGANIZATION OF SLEEP

Sleep maintenance seems to be the earliest and most frequent sign of sleep disruption in parkinsonism, leading to a fragmentation of the sleep profile.[21] PSG studies have demonstrated an increased sleep latency with frequent awakenings, leading to a reduced total sleep time with decreased sleep-efficacy. As much as 30 to 40% of the night are spent awake in the advanced stages.[22,23] Light sleep is relatively increased, slow-wave sleep is accordingly decreased, and REM-periods are shortened.[24–26] Although sleep may be normal in the milder stages of the disease, untreated PD patients already show a significant decrease of slow-wave sleep and an increased number of periodic limb movements.[27]

Not only the sleep macrostructure but also the microstructure of sleep may be changed. For example, the number of sleep spindles during slow-wave sleep may be reduced,[23,26] and REM-density may be increased with an increased alpha-power during REM-sleep.[25,28]

Sleep disturbances are not only a problem in PD but also in other PS. RBD is still more frequent in multiple system atrophy (MSA) than in PD[29] and can be seen as an early sign of neurodegeneration in PD patients as well as in MSA patients.[4] Patients with MSA may have an increased sleep latency and increased numbers of awakenings, along with reductions in REM sleep and slow-wave sleep as well as sleep efficacy.[30] Spindles are poorly formed or absent, slow wave activity is increased during REM sleep, REM sleep is decreased, and insomnia is often present in progressive supranuclear palsy (PSP). All these abnormalities tend to worsen with disease progression,[31] although RBD may improve during the progression of PD.

Nocturnal Motor Activity

Parkinsonian motor symptoms are most prominent during wakefulness; however, they are not completely abolished during sleep periods. Motor phenomena still occur to varying degrees in the different sleep stages and may include reoccurring tremor, nonperiodic and periodic movements, stiffness, and nocturnal akinesia with problems turning over in bed.[32–34] However, it is still controversial as to which degree these different motor symptoms contribute to sleep disturbances.

Usually, tremor disappears with the onset of stage 1 sleep, in some cases before alpha EEG activity is entirely gone.[35,36] However, tremor and dyskinesias may persist during light sleep stages and during short arousals, sleep stage changes, and body movements.[32,37] Tremor may also appear for a few seconds during bursts of rapid eye movements, and shortly before or after a REM period.[36] It rarely persists during slow-wave sleep, and there is no known association with K-complexes and sleep spindles. Its amplitude varies considerably during sleep and is usually less than 50% of the amplitude in the waking condition.

Abnormal simple and complex movements (as well as increased muscle tone) such as repeated blinking at the onset of sleep, rapid eye movements during nonREM sleep, blepharospasm at the onset of REM sleep, and pro-

longed tonic contraction of limb extensor or flexor muscles during nonREM sleep also frequently occur in PD patients.[25] Nocturnal akinesia with an inability to turn over in bed increases with the more advanced stages of PD.[38]

Periodic leg movements during sleep (PLMS) with a pathological index of more than five per hour of sleep (PLMS index >5) occur in up to one-third of patients with untreated PD and even more commonly with increasing age.[27] PLMS may be associated with arousals or awakenings and therefore can disturb sleep continuity. Symptoms of restless legs syndrome (RLS) are also common in patients with PD; however, except in patients with a family history of RLS, they seem to reflect a secondary phenomenon, and its prevalence is not higher than in the normal population.[39] To date, there is no evidence that RLS symptoms early in life predispose to a subsequent development of PD.[40] Fragmentary irregular nocturnal myoclonus is another motor phenomenon in PD patients that can be precipitated by levodopa. It occurs primarily during light nonREM sleep and is characterized by brief bursts of up to 150 ms in a random fashion without periodicity.[41] Repetitive muscle contractions followed by tremor may occur, particularly in the limb that is primarily affected by the disease. These muscle contractions can result in a painful extension of the great toe, finger, or foot as a sign for off-dystonia. Early-morning foot dystonia reflects the low concentration of dopamine after the last intake of dopaminergic medication at night and may occur just before waking or soon thereafter. Painful off-dystonia and early morning akinesia are frequent complaints of PD patients in an advanced stage of the disease that require adequate nocturnal treatment.[38]

Increased muscle tone, as well as abnormal simple and complex movements during sleep, are common in PD patients. It must be kept in mind that they can complicate the scoring of polysomnograms, in contrast to the quiescence of sleep in normal persons. Therefore, a special experience in analyzing sleep recordings of patients with parkinsonism is of importance.

Sleep Benefit

Patients with a so-called "sleep benefit" show a better morning motor function, although no alterations in the sleep pattern can be detected.[42] A sleep benefit is defined as "restoration of mobility on awakening from sleep prior to drug intake,"[43] about one-third of PD patients experience sleep benefit.[44]

Marked diurnal variations in rigidity and dystonia, with little rigidity soon after arising and worsening during the day, can be observed in some patients with familial early-onset Parkinson's disease. These symptoms also improve after naps.[45]

REM Sleep Behavior Disorder (RBD)

Schenck and co-workers described first criteria for RBD which are now defined[46] as complex, mild to harmful movements during REM sleep with a loss of skeletal muscle atonia (Figure 21.1, shows a typical PSG recording of RBD). Body movements are associated with dream mentation and nightmares; dreams appear to be "acted out," all leading to a disrupted sleep continuity. Although the underlying cause of RBD is still unknown, it most likely reflects a dysfunction in the brain-stem circuitry and the dorsolateral pontine tegmentum. In these areas, REM sleep without atonia can be induced in animal experiments. RBD may represent a preclinical marker of neurodegeneration in synucleinopathies such as PD and MSA and may precede motor symptoms for years.[47]

Fifteen[48] to 30%[49] of an unselected population of PD patients revealed RBD, more often in polysomnographically investigated patients. The percentage of RBD was up to 90% and more in patients with MSA, again increasing when investigated in the sleep laboratory as compared to "pure" clinical diagnosis. RBD preceded by more than one year the clinical onset of MSA in 44% of the cases, and PSG could differentiate between patients with pure autonomic failure and those developing MSA with autonomic failure.[50] In neuroimaging MRI studies, no specific region could be detected in relation to RBD.[51] Neuroimaging studies of dopamine receptors showed a significant reduction of striatal dopamine transporter binding in RBD patients as a sign of reduced dopamine release, indicating early or preclinical PD or MSA.[4,52] RBD seems to be particularly frequent in patients with PD and psychosis or hallucinations and is an important differential diagnosis of both, sometimes requiring PSG.[53]

Nocturnal Respiratory Disorders

Obstructive ventilatory deficits are common in moderate to severe PD patients during the day. They are apparently caused by a combination of upper airway obstruction, probably due to an abnormal tone in the upper airway muscles, and respiratory muscle weakness with decreased effective muscle strength.[54] Levodopa cannot improve these deficits, although they correlate to some extent with the severity of rigidity and tremor. Dyskinetic movements of glottic and supraglottic structures can also lead to intermittent airway closure in some patients.[55]

Disorganized patterns of respiration with central apneas, obstructive apneas, or episodes of hypoventilation can be observed in parkinsonism.[56,57] Patients with autonomic disturbances tend to have more severe respiratory abnormalities than those without. Daytime respiratory abnormalities include reduced ventilatory responses to hypercapnia and hypoxia.[58,59] PSG studies have demonstrated obstructive sleep apnea as well as other abnormal breathing patterns, including central sleep apnea, variable-amplitude respirations, and arrhythmic respirations during night.[60,61] Stridor, laryngeal stenosis, and obstruction may be caused by abnormal vocal cord function, which appears to be a major contributor to abnormal breathing during sleep in parkinsonism.[62]

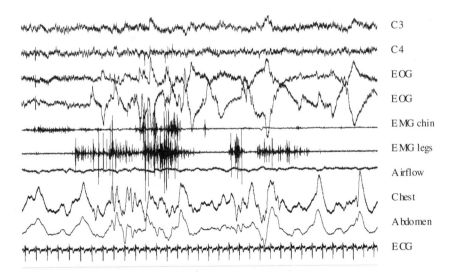

FIGURE 21.1 Thirty seconds polysomnographical recording of a 45-year-old male PD patient (Hoehn and Yahr, stage 2) with subclinical REM sleep behavior disorder. The chin EMG shows a highly elevated muscle tone, multiple motor activity occurs in the recording, and typical increased muscle tone also appears in the linked EMG channel of the legs. Channels from top down: 1–2, EEG; 3–4, EOG, eye movement showing typical REM sleep pattern; 5, chin-EMG with increased muscle tone; 6, linked EMG of both legs with increased muscle tone; 7–9, respiratory recording with airflow, thoracic, and abdominal belts, showing typical irregular breathing pattern during REM sleep; 10, ECG.

Excessive Daytime Sleepiness (EDS)

EDS is a well known phenomenon in PD patients. The neurodegenerative process itself, with disturbances of the reticular activating system, sleep disorders such as sleep apnea syndrome and narcolepsy, RBD, mood disorders as well as various drugs, can contribute to EDS.[21] Studies in recent years described an increased risk for causing motor vehicle accidents by sudden sleep onset, so-called "sleep attacks," in PD patients.[3] Those episodes were primarily attributed to the intake of non-ergot dopamine receptor agonists.[3] It was suggested that the sedating effect of non-ergot dopamine receptor agonists may be due to their stronger D_3 receptor activity as compared to other dopamine receptor agonists.[3,63] However, reports from more recent studies let suggest that sedation may be rather a class effect of dopaminergic medication in general, since EDS was generally more frequent in patients taking dopaminergic drugs.[64,65]

Sleepiness does not result only from pharmacotherapy or sleep abnormalities but is also related to the pathology of the disease itself.[66] Dopaminergic medication may exacerbate sleepiness in a subset of patients, but the primary pathology seems to be the greatest contributor to the development of EDS. Those patients may benefit from wake-promoting agents such as bupropion, modafinil, or traditional psychostimulants.[67]

The most serious question concerning EDS in parkinsonism is whether those patients are allowed to drive. It is necessary to advice PD patients that sudden sleepiness may occur in the course of the disease and may be attributed individually to specific dopaminergic drugs. However, the patients themselves are finally responsible for their ability to drive and may decide individually. Physicians who prescribe dopamine receptor agonists in PD patients must inform their patients about possible "sleep attacks" during the treatment.

It must be kept in mind that, in PD patients with complaints of EDS, diagnostic testings including PSG and daytime recordings of sleep preponderance are needed to exclude secondary causes of EDS as the basis for therapeutical interventions and to document and to quantify the severity of EDS.

DIAGNOSTIC EVALUATIONS

Importance of Medical History and Clinical Examination

A number of factors can contribute to the frequent sleep-wake disturbances in parkinsonism. Those factors include disease specific alterations, medication, nocturnal motor activity, sleep-wake rhythm abnormalities, and nocturnal respiratory disorders (see Table 21.1). However, sleep disturbances are rarely the presenting complaint of a patient with parkinsonism. To determine whether a sleep disorder exists and which factors are most important, the clinical history of the patient and desirable of partners or caregivers is extremely important. Questions concerning the sleep complaint should include all the features the physician would obtain from any patient with a sleep complaint. Additionally, it needs to include disease specific questions on nocturnal akinesia, daytime drowsiness, or EDS in relation to drug intake and psychiatric symptoms. The use of a disease specific questionnaire, the Parkinson's disease sleep scale (PDSS), can be helpful for diagnostic evaluation as for assessment of follow-up.[68] The bed partner needs to be asked for a careful description of the presence and frequency of movements during sleep as well as their timing, arousals and awakenings, snoring and episodes of apneas, and periods and durations of daytime sleepiness. The knowledge of the drug schedule is important: If dopaminergic drugs are not taken in the evening, nocturnal rigidity and akinesia may contribute to a relevant degree to the sleep disturbances; on the other hand, excessive evening doses of dopamine receptor agonists may be responsible for sleep-onset insomnia. The clinical examination may further hint to a specific sleep disturbance or motor pattern.

Technical Recordings of Sleep and Excessive Daytime Sleepiness (EDS)

During actigraphy, muscle activity is monitored by a small portable actimeter, usually worn at the wrist or the ankle. Actigraphic recordings can be performed for some days up to several weeks and may show an increased motor activity during night and a reduced motor activity during the day, reflecting, e.g., periods of EDS during the day. However, actigraphy cannot differ between sleep and wakefulness and is therefore only of limited value for analyzing sleep-wake disturbances in parkinsonism.[69]

Some patients with parkinsonism and sleep complaints require a further evaluation for the differential diagnosis of the underlying etiology. Polysomnography can be used as a last step for an exact evaluation and for analyzing nocturnal motor and breathing patterns.[69] In patients with the main complaint of EDS, a sleep apnea syndrome needs to be verified or excluded using a full cardio-respiratory PSG. Patients who present with symptoms that may underlie either an RBD or a psychosis need a PSG for the differential diagnosis as their treatment options are different. Simultaneous video monitoring and surface EMG recordings of all four limbs are often helpful for a better analysis of nocturnal motor patterns and their relevance for sleep disruptions.

If EDS is the most prominent complaint, daytime recordings are needed. The multiple sleep latency test (MSLT) and the maintenance of wakefulness test (MWT) are used to objectively determine its severity and circadian variation. It is best to do the recordings while the patient is under the usual medication schedule. However, if drugs appear to be a major factor for the sleep disturbances, definite diagnosis may require two or more nights of PSG with different medication schedules.

TREATMENT OPTIONS

The treatment of sleep disturbances in patients with parkinsonism is a great challenge for the physician, because treatment of the daytime motor symptoms in PD may affect sleep. Both, parkinsonism itself and its treatment with dopaminergic agents can lead to sleep fragmentation, nocturnal movements and vocalizations, abnormal muscle tone during sleep, as well as psychiatric symptoms such as depressive and psychotic behavior. The biphasic actions of dopaminergic drugs must be considered as low dosages may promote sleep, whereas high dosages may lead to increased waking effects, reduction of slow-wave sleep, and decreased sleep continuity.[70–72] Therefore, the variety of sleep disturbances in parkinsonism need different management strategies.

Nocturnal Akinesia

Nocturnal motor symptoms may disappear with adequate dopaminergic treatment during the day in the early and mild stages of the disease.[73] In more advanced cases, a controlled-release formulation containing 200 mg of levodopa plus 50 mg of a decarboxylase inhibitor given at bedtime or the introduction of a long-lasting dopamine receptor agonist therapy appear to improve sleep.[74–76] Early-morning akinesia is improved as well by these treatment schemes.

REM Sleep Behavior Disorder (RBD)

RBD with the occurrence of violent episodes during the night may disrupt the relationship between the patient and caregiver[48] and can be one of the main reasons for nursing home admittance. Therefore, and for the patient's and bed-partner's safety, violent or injurious behavior during sleep should be treated immediately. Loud sleep talking or screaming can be disturbing for the caregiver in a similar way. An early diagnosis is warranted, as RBD can easily be treated in most patients with small dosages of 0.5 to 2 mg clonazepam at bedtime.[77]

Periodic Limb Movements during Sleep (PLMS) and Restless Legs Syndrome (RLS)

Increased nocturnal dopaminergic stimulation, either by levodopa or dopamine receptor agonists, improves symptoms of PLMS and RLS.[76] A reduction of rigidity and bradykinesia during the day with consecutive improve-

ment of nocturnal mobility, reduced numbers of PLMS, disappearance of RLS symptoms, and normalization of sleep muscle activity are some factors that may contribute to an improved sleep due to an optimized dopaminergic treatment.[78] Nonperiodic movements and fragmentary myoclonus are best treated with benzodiazepines and may benefit from a reduction of dopaminergic medication.[47,79]

Insomnia

Assessment of psychosocial and behavioral factors that may contribute to sleep disturbances is the first step for treating insomnia. Concurrent psychiatric disorders need to be addressed: If nocturnal hallucinations occur, related drugs should be reduced as much as possible and therapy with clozapine or quetiapine should be started. Tricyclic antidepressants are drugs of first choice to improve depression in PD patients also suffering from problems of sleep initiation, sleep maintenance, and early awakenings.[80] In more advanced stages of the disease, the patient's spouse should be encouraged to sleep in a different bed or even a different room. Bad sleep for the patients also leads to bad sleep and inadequate rest for the spouses and can reduce quality of life in both groups.[81] Patients under high dosages of dopaminergic drugs may require further medication for insomnia. Next to tricyclic antidepressants with sedating properties such as amitryptiline (25 to 50 mg), the newer antidepressant mirtazapine (7.5 to 30 mg), which is well tolerated without influencing motor symptoms, is efficacious to improve sleep.[82] However, no controlled studies are available to date. The anticholinergic effects of tricyclic antidepressants may have therapeutic benefits for daytime parkinsonian symptoms and depression as well, but it must be kept in mind that they can also induce nocturnal delirium, particularly in patients with cognitive impairment. Dyskinetic nocturnal movements leading to insomnia may respond to a reduction of the evening dopaminergic medication. Nocturia is a common problem in patients with PD and is sometimes difficult to control. However, optimal urological treatment and easy provisions to go to the toilet during the night (e.g., with urinals or night-stools) should be provided. If insomnia is unresponsive to all these interventions, one may use benzodiazepines such as triazolam or clonazepam, or benzodiazepine receptor agonists such as zolpidem and zopiclone for a short period of time to normalize the sleep-wake schedule. Considerations of sleep hygiene and stimulus control mechanisms should always form the basis of getting a good night's sleep in all sleep-disordered persons.

Nocturnal Hallucinations, Psychosis, and Confusion

Unfortunately, dopaminergic drugs can induce entirely new sleep problems when used during the day or in the evening. Up to 30% of patients with parkinsonism taking levodopa, and an even higher number of those taking dopamine receptor agonists, experience vivid dreams, nightmares, and night terrors, particularly those with cognitive impairment.[83] These symptoms may necessitate a reduction of the afternoon or evening dosages of the dopaminergic drugs, which might worsen motor symptoms. Nocturnal confusion and hallucinations are often so disruptive for demented patients with PD that only low dosages of levodopa can be used with complete withdrawal of dopamine receptor agonists. The medication of first choice is still clozapine (starting with 6.25 to 25 mg), followed by quetiapine, as limited controlled data are available on quetiapine in PD patients with psychosis. Small dosages of clomethiazol at bedtime are sometimes also helpful in such cases. Low dosages of clozapine should be used and slowly increased until complete remission of nocturnal hallucinations is achieved.[84] Patients who do not tolerate clozapine may be switched to low dosages of quetiapine (12.5 to 50 mg).[85,86]

Nocturnal Respiratory Disturbances

In parkinsonism, the treatment of sleep-related respiratory disturbances is similar to their treatment in all other persons and patients. Nasal continuous positive airway pressure (nCPAP) ventilation offers the best chance of success in patients with obstructive sleep apnea syndrome. Upper airway surgery should be regarded with caution—it may help some patients with redundant palatal or pharyngeal tissue, but the abnormal motor activity of the upper airways is still present after surgery. Tracheostomy may be necessary in some patients with MSA and severe vocal cord dysfunction.

Appropriate nCPAP may improve the condition of a PD patient substantially. It can normalize nocturnal blood pressure, and neuropsychiatric symptoms, daytime drowsiness, and EDS may be improved immediately.

Excessive Daytime Sleepiness (EDS)

As a first step, respiratory disturbances, such as sleep apnea syndrome and snoring, and the relevance of pharmacologically induced EDS need to be excluded before EDS is treated with psychostimulants.[67,87] If possible, PSG should be performed for exclusion of treatable causes of EDS. If medication adjustments such as giving of amantadine or discontinuation of dopamine receptor agonists and benzodiazepines are not effective, agents specifically designed to promote daytime alertness may be beneficial. Previous reports have shown that amphetamines can increase daytime alertness in PD patients.[88] Therefore, psychostimulants may be considered as optional therapy for EDS in patients with PD.[67] Recently, it could be shown that EDS improved under treatment with modafinil, a psy-

chostimulant drug acting on postsynaptic alpha-1 adrenergic receptors, however, the exact mode of action of modafinil is still being debated. Modafinil (up to 200 mg) is the agent best investigated in PD patients, and a number of open and some controlled studies are available.[89–91] Some authors recommend modafinil even if EDS is drug-induced.[92] As long-term studies are still lacking, we do not know if the effects remain stable. PD patients suffering from cognitive impairment or psychotic episodes should only be very cautiously treated with modafinil, as the risk of side effects may be increased.

ABBREVIATIONS

EDS	excessive daytime sleepiness
EEG	electroencephalography
EMG	electromyography
EOG	electrooculogram
MRI	magnetic tomographic investigation
MSLT	multiple sleep latency test
MWT	maintenance of wakefulness test
nCPAP	nasal continuous positive airway pressure ventilation
PD	Parkinson's disease
PDSS	Parkinson's disease sleep scale
PS	Parkinsonian syndrome
PSG	polysomnography
RBD	REM sleep behavior disorder
REM	rapid eye movements
RLS	restless legs syndrome

REFERENCES

1. Parkinson, J., *Essay on the Shaking Palsy*, Sherwood, Neely, and Jones, London, England, 1817:17.
2. Charcot, J. M., Lectures on the diseases of the nervous system (trans. by G. Siegerson), *The New Sydenham Society*, Lecture V, London, England, 1877:147.
3. Frucht, S., Rogers, J. D., Greene, P. E. et al., Falling asleep at the wheel: motor vehicle mishaps in persons taking pramipexole and ropinirole, *Neurology*, 52:1908–1910, 1999.
4. Eisensehr, I., Linke, R., Noachtar, S. et al., Reduced striatal dopamine transporters in idiopathic rapid eye movement sleep behavior disorder. Comparison with Parkinson's disease and controls, *Brain*, 123:1155–1160, 2000.
5. Braak, H., R. B. U., Gai, W. P., Del Tredici, K., Idiopathic Parkinson's disease: possible routes by which vulnerable neuronal types may be subject to neuroinvasion by an unknown pathogen, *J. Neural. Transm.*, 110:517–536, 2003.
6. Del Tredici, K., R. B. U., De Vos, R. A. et al.,. Where does Parkinson's disease pathology begin in the brain? *Neuropathol. Exp. Neurol.*, 61:413–426, 2002.

7. Forno, L. S., Neuropathology of Parkinson's disease, *J. Neuropathol. Exp. Neurol.*, 55:259–272, 1996.

8. Jellinger, K., Pathology of parkinsonism, In Fahn, S., Marsden, C. D., Jenner, P. et al., (Eds.), *Recent Developments in Parkinson's disease*, New York, Raven Press. 33–66, 1986.

9. Zweig, R. M., Jankel, W. R., Hedreen, J. C. et al., The pedunculopontine nucleus in Parkinson's disease, *Ann. Neurol.*, 26:41–46, 1989.

10. Javoy-Agid, F., Agid, Y., Is the mesocortical dopaminergic system involved in Parkinson's disease? *Neurology*, New York, 30:1326–1330, 1980.

11. De Keyser, J., Ebinger, G., Vauquelin, G., Evidence for a widespread dopaminergic innervation of the human cerebral neocortex, *Neurosci. Lett.*, 104:281–285, 1989.

12. Ongini, E., Caporali, M. G., Massotti, M., Stimulation of dopamine-D-1 receptors by SKF 38393 induces EEG desynchronization and behavioural arousal, *Life Sci.*, 37:2327–2333, 1985.

13. Bo, P., Ongini, E., Giorgetti, A. et al., Synchronization of the EEG and sedation induced by neuroleptics depend upon blockade of both D_1 and D_2 dopamine receptors, *Neuropharmacology*, 27:799, 1988.

14. Chianchetti, C., Dopamine agonists and sleep in man, In Wauquier, A., Gaillard, J. M., Monti, J. M. et al., (Eds.), *Sleep: Neurotransmitters and Neuromodulators*, Raven Press, New York, 121–134, 1985.

15. Bagetta, G., De Sarro, G., Priolo, E. et al., Ventral tegmental area site through which dopamine D_2-receptor agonists evoke behavioural and electrocortical sleep in rats, *Br. J. Pharmacol.*, 95:860–866, 1988.

16. Svensson, K., Alfoldi, P., Hajos, M. et al., Dopamine autoreceptor antagonists: effects of sleep-wake activity in the rat, *Pharmacol. Biochem. Behav.*, 26:123–129, 1987.

17. Trampus, M., Ferri, N., Monopoli, A. et al., The dopamine D_1 receptor is involved in the regulation of REM sleep in the rat, *Eur. J. Pharmacol.*, 194:189–194, 1991.

18. Tandberg, E., Larsen, J. P., Karlsen, K., Excessive daytime sleepiness and sleep benefit in Parkinson's disease: a community-based study, *Mov. Disord.*, 922–927, 1999.

19. Menza, M. A., Rosen, R. C., Sleep in Parkinson's disease: the role of depression and anxiety, *Psychosomatics*, 36:262–266, 1995.

20. Comella, C. L., Tanner, C. M., Ristanovic, R. K., Polysomnographic sleep measures in Parkinson's disease patients with treatment-induced hallucinations, *Ann. Neurol.*, 34:710–714, 1993.

21. Chokroverty, S., Sleep and degenerative neurologic disorders, *Neurol. Clin.*, 4:807–826, 1996.

22. Kales, A., Ansel, R. D., Markham, C. H. et al., Sleep in patients with Parkinson's disease and normal subjects prior to and following levodopa administration, *Clin. Pharmacol. Ther.*, 12:397–406, 1971.

23. Bergonzi, P., Chiurulla, C., Gambi, D. et al., L-dopa plus dopadecarboxylase inhibitor: sleep organization in Parkinson's syndrome before and after treatment, *Acta Neurol. Belg.*, 75:5–10, 1975.

24. Puca, F., Bricolo, A., Rurella, G., Effect of L-dopa or amantadine therapy on sleep spindles in parkinsonism, *Clin. Neurophysiol.*, 35:327–330, 1973.

25. Mouret, J., Differences in sleep in patients with Parkinson's disease, *Electroencephalogr. Clin. Neurophysiol.*, 38:653–657, 1975.

26. Friedman, A., Sleep pattern in Parkinson's disease, *Acta Med. Pol.*, 21:193–199, 1980.

27. Wetter, T. C., Collado-Seidel, V., Pollm cher, T. et al., Sleep and periodic leg movement patterns in drug-free patients with Parkinson's disease and multiple system atrophy, *Sleep*, 23:361–367, 2000.

28. Brunner, H., Wetter, T. C., Högl, B., et al., Microstructure of the non-rapid eye movement sleep electroencephalogram in patients with newly diagnosed Parkinson's disease: effects of dopaminergic treatment, *Mov. Disord.*, 17:928–933, 2002.

29. Plazzi, G., Corsini, R., Provini, F. et al., REM sleep behaviour disorder in multiple system atrophy, *Neurology*, 48:1094–1097, 1997.

30. Martinelli, P., Coccagna, G., Rizzuto, N. et al., Changes in systemic arterial pressure during sleep in Shy-Drager syndrome, *Sleep*, 4:139–146, 1981.

31. Aldrich, M. S., Foster, N. L., White, R. F. et al., Sleep abnormalities in progressive supranuclear palsy, *Ann. Neurol.*, 25:577–581, 1989.

32. Fish, D. R., Sawyers, D., Allen, P. J. et al., The effect of sleep on the dyskinetic movements of Parkinson's disease, Gille de la Tourette syndrome, Huntington's disease, and torsion dystonia, *Arch. Neurol.*, 1991, 48:210–214.

33. Rye, D. B., Bliwise, D. L., Movement disorders specific to sleep and the nocturnal manifestations of waking movement disorders, in Watts, R. L., W. C., Koller, (Eds.), *Neurologic Principles and Practice*, McGraw-Hill, New York, 687–713, 1997.

34. van Hilten, J. J., Weggeman, M., van der Velde, E. A. et al., Sleep, excessive daytime sleepiness and fatigue in Parkinson's disease, *J. Neural. Transm.*, 5:235–244, 1993.

35. April, R. S., Observations on parkinsonian tremor in all-night sleep, *Neurology*, New York, 16:720–724, 1996.

36. Stern, M., Roffwarg, H., Duvoisin, R., The parkinsonian tremor in sleep, *J. Nerv. Ment. Dis.*, 147:202–210, 1968.

37. Askenasy, J. J. M., Yahr, M. D., Parkinsonian tremor loses its alternating aspect during non-REM sleep and is inhibited by REM sleep, *J. Neurol. Neurosurg. Psychiatry*, 53:749–753, 1990.

38. Lees, A. J., Blackburn, N. A., Campbell, V. L., The nighttime problems of Parkinson's Disease, *Clin. Neuropharmacol.*, 11:512–519, 1988.

39. Garcia-Borreguero, D., Odin, P., Serrano, C., Restless legs syndrome and PD: a review of the evidence for a possible association, *Neurology*, 61 (Suppl. 3): S49–55, 2003.

40. Ondo, W. G., Vuong, K. D., Jankovic, J., Exploring the relationship between Parkinson disease and restless legs syndrome, *Arch. Neurol.*, 59:421–424, 2002.

41. Broughton, R., Tolentino, M., Krelina, M., Excessive fragmentary myoclonus in NREM sleep. A report of 38 cases, *Electroencephalogr. Clin. Neurophysiol.*, 61:123–133. 1985.

42. Högl, B. E., Gomez-Arevalo, G., Garcia, S. et al., A clinical, pharmacologic, and polysomnographic study of

sleep benefit in Parkinson's disease, *Neurology*, 50:1332–1339, 1998.

43. Bateman, D. E., Levett, K., Marsden, C. D., Sleep benefit in Parkinson's disease, *J. Neurol. Neurosurg. Psychiatry*, 67:384–385, 1999.

44. Currie, L. J., Bennett, J. P., Jr., Harrison, M. B. et al., Clinical correlates of sleep benefit in Parkinson's disease, *Neurology*, 48:1115–1117, 1997.

45. Yamamura, Y., Sobue, I., Ando, K. et al., Paralysis agitans of early onset with marked diurnal fluctuation of symptoms, *Neurology*, New York, 23:239–244, 1973.

46. American Sleep Disorders Association, International classification of sleep disorders, revised: diagnostic and coding manual, *American Sleep Disorders Association*, Rochester, 1997.

47. Olson, E. J., Boeve, B. F., Silber, M. H., Rapid eye movement sleep behaviour disorder: demographic, clinical and laboratory findings in 93 cases, *Brain*, 123:331–339, 2000.

48. Comella, C. L., Nardine, T. M., Diederich, N. J. et al., Sleep-related violence, injury, and REM sleep behavior disorder in Parkinson's disease, *Neurology*, 51:526–529, 1998.

49. Gagnon, J. F., Bedard, M. A., Fantini, M. L. et al., REM sleep behavior disorder and REM sleep without atonia in Parkinson's disease, *Neurology*, 59:585–589, 2002.

50. Plazzi, G., Cortelli, P., Montagna, P. et al., REM sleep behaviour disorder differentiates pure autonomic failure from multiple system atrophy with autonomic failure, *J. Neurol. Neurosurg. Psychiatry*, 64:683–685, 1998.

51. Schenck, C., Mahowald, M., REM sleep behavior disorder: clinical, developmental, and neuroscience perspectives 16 years after its formal identification in sleep, *Sleep*, 25:120–138, 2002.

52. Eisensehr, I., Lindeiner, H., Jäger, M. et al., REM sleep behavior disorder in sleep-disordered patients with versus without Parkinson's disease: is there a need for polysomnography? *J. Neurol. Sci.*, 186:7–11, 2001.

53. Arnulf, I., Bonnet, A. M., Damier, P. et al., Hallucinations, REM sleep, and Parkinson's disease: a medical hypothesis, *Neurology*, 55:281–288, 2000.

54. Hovestadt, A., Bogaard, J. M., Meerwaldt, J. D. et al., Pulmonary function in Parkinson's disease, *J. Neurol. Neurosurg. Psychiatry*, 52:329–333, 1989.

55. Vincken, W. G., Gauthier, S. G., Dollfuss, R. E. et al., Involvement of upper-airway muscles in extrapyramidal disorders: a cause of airflow limitation, *N. Engl. J. Med.*, 311:438–442, 1984.

56. Hardie, R. J., Efthimiou, J., Stern, G. M., Respiration and sleep in Parkinson's disease, *J. Neurol. Neurosurg. Psychiatry*, 49:1326, 1986.

57. Apps, M. C. P., Sheaff, P. C., Ingram, D. A. et al., Respiration and sleep in Parkinson's disease, *J. Neurol. Neurosurg. Psychiatry*, 48:1240–1245, 1985.

58. Chokroverty, S., Sharp, J. T., Barron, K. D., Periodic respiration in erect posture in Shy-Drager syndrome, *J. Neurol. Neurosurg. Psychiatry*, 41:980–986, 1978.

59. McNicholas, W. T., Ruhterford, R., Grossman, R., et al., Abnormal respiratory pattern generation during sleep in patients with autonomic dysfunction, *Am. Rev. Respir. Dis.*, 128:429–433, 1983.

60. Guilleminault, C., Briskin J. G., Greenfield, M. S. et al., The impact of autonomic nervous system dysfunction on breathing during sleep, *Sleep*, 4:263–268, 1981.

61. Kenyon, G. S., Apps, M. C. P., Traub, M., Stridor and obstructive sleep apnea in Shy-Drager syndrome treated by laryngofissure and cord lateralization, *Laryngoscope*, 94:1106–1108, 1984.

62. Isozaki, E., Naito, A., Horiguchi, S. et al., Early diagnosis and stage classification of vocal cord abductor paralysis in patients with multiple system atrophy, *J. Neurol. Neurosurg. Psychiatry*, 60:399–402, 1996.

63. Ryan, M., Slevin, J. D., Wells, A., Non-ergot dopamine agonist-induced sleep attacks, *Pharmacotherapy*, 20:724–726, 2000.

64. Pal, S., Bhattacharya, K. F., Agapito, C. et al., A study of excessive daytime sleepiness and its clinical significance in three groups of Parkinson's disease patients taking pramipexole, cabergoline and levodopa mono and combination therapy, *J. Neural. Transm.*, 108:71–77, 2001.

65. Happe, S., Berger, K., The association of dopamine agonists with daytime sleepiness, sleep problems and quality of life in patients with Parkinson's disease. A prospective study, *J. Neurol.*, 248:1062–1067, 2001.

66. Arnulf, I., Konofal, E., Merino-Andreu, M. et al., Parkinson's disease and sleepiness: an integral part of PD, *Neurology*, 58:1019–1024, 2002.

67. Rye, D., Sleepiness and unintended sleep in Parkinson's disease, *Curr. Treat. Options, Neurol.*, 5:231–239, 2003.

68. Chaudhuri, K. R., Pal, S., DiMarco, A. et al., The Parkinson's disease sleep scale: a new instrument for assessing sleep and nocturnal disability in Parkinson's disease, *J. Neurol. Neurosurg. Psychiatry*, 73:629–633, 2002.

69. Happe, S., Trenkwalder, C., Movement disorders in sleep: Parkinson's disease and Restless Legs Syndrome, *Biomed. Tech.*, 48:62–67, 2003.

70. Leeman, A. L., O'Neill, C. J., Nicholson, P. W. et al., Parkinson's disease in the elderly: response to and an optimal spacing of night time dosing with levodopa, *Br. J. Clin. Pharmacol.*, 24:637–643. 1987.

71. Monti, J. M., Hawkins, M., Jantos, H. et al., Biphasic effects of dopamine D-2 receptor agonists on sleep and wakefulness in the rat, *Psychopharmacology*, 95:395–400, 1988.

72. Cantor, C. R., Stern, M. B., Dopamine agonists and sleep in Parkinson's disease, *Neurology*, 58:S71–78, 2002.

73. Askenasy, J. J., Yahr, M. D., Reversal of sleep disturbance in Parkinson's disease by antiparkinsonian therapy: a preliminary study, *Neurology*, 35:527–532, 1985.

74. Jansen, E. N., Meerwaldtt, J. D., Madopar, H. B. S., in nocturnal symptoms of Parkinson's disease, *Adv. Neurol.*, 53:527–531, 1990.

75. Koller, W. C., Hutton, J. T., Tolosa, E. et al., Immediate-release and controlled-release carbidopa/levodopa in PD: a 5-year randomized multicenter study, Carbidopa/Levodopa Study Group, *Neurology*, 53:1012–1019, 1999.

76. Högl, B., Rothdach, A., Wetter, T. C. et al., The effect of cabergoline on sleep, periodic leg movements in sleep, and early morning motor function in patients with Parkinson's disease, *Neuropsychopharmacology,* 28:1866–1870, 2003.

77. Schenck, C., Mahowald, M., A polysomnographic, neurologic, psychiatric and clinical outcome report on 70 consecutive cases with REM sleep behavior disorder (RBD): sustained clonazepam efficacy in 89,5 percent of 57 treated patients, *Clev. Clin. J. Med.,* 57:10–24, 1990.

78. Lang, A. E., Quinn, N., Brincat, S. et al., Pergolide in late-stage Parkinson disease, *Ann. Neurol.,* 12:243–247, 1982.

79. Lapierre, O., Montplaisir, J., Polysomnographic features of REM sleep behaviour disorder: Development of a scoring method, *Neurology,* 42:1371–1374, 1992.

80. Poewe, W., Seppi, K., Treatment options for depression and psychosis in Parkinson's disease, *J. Neurol.,* 248 (suppl. S3):12–21, 2001.

81. Happe, S., Berger, K., The association between caregiver burden and sleep disturbances in partners of patients with Parkinson's disease, *Age Ageing,* 31:349–354, 2002.

82. Gordon, P. H., Pullman, S. L., Louis, E. D. et al., Mirtazapine in Parkinsonian tremor, *Parkinsonism Relat. Disord.,* 9:125–126, 2002.

83. Scharf, B., Moskovitz, C., Lupton, M. D. et al., Dream phenomena induced by chronic levodopa therapy, *J. Neural. Transm.,* 43:143–151, 1978.

84. The Parkinson Study Group. Low-dose clozapine for the treatment of drug-induced psychosis in Parkinson's disease, *N. Engl. J. Med.,* 340:757–763, 1999.

85. Fernandez, H. H., Friedman, J. H., Jacques, C. et al., Quetiapine for the treatment of drug-induced psychosis in Parkinson's disease, *Mov. Disord.,* 14:484–487, 1999.

86. Reddy, S., Factor, S. A., Molho, E. S. et al., The effect of quetiapine on psychosis and motor function in parkinsonian patients with and without dementia, *Mov. Disord.,* 17:676-681, 2002.

87. Braga-Neto, P., Pereira da Silva-Junior, F., Sueli Monte, F., de Bruin, P. F., de Bruin, V. M., Snoring and excessive daytime sleepiness in Parkinson's disease, *J. Neurol. Sci.,* 217:41–45, 2004.

88. Parkes, J. D., Tarsy, D., Marsden, C. D. et al., Amphetamines in the treatment of Parkinson's disease, *J. Neurol. Neurosurg. Psychiatry,* 38:232–237, 1975.

89. Happe, S., Pirker, W., Sauter, C. et al., Successful treatment of excessive daytime sleepiness in Parkinson's disease with modafinil, *J. Neurol.,* 248:632–634, 2001.

90. Nieves, A. V., Lang, A. E., Treatment of excessive daytime sleepiness in patients with Parkinson's disease with modafinil, *Clin. Neuropharmacol.,* 25:111–114, 2002.

91. Högl, B., Saletu, M., Brandauer, E. et al., Modafinil for the treatment of daytime sleepiness in Parkinson's disease: a double-blind, randomized, crossover, placebo-controlled polygraphic trial, *Sleep,* 25:905–909, 2002.

92. Hauser, R. A., Wahba, M. N., Zesiewicz, T. A., McDowell, Anderson, W., Modafinil treatment of pramipexole-associated somnolence, *Mov. Disord.,* 15:1269–1271, 2000.

22 Visual Function in Parkinson's Disease

Robert L. Rodnitzky

Roy J. and Lucile A. Carver College of Medicine, University Hospitals, University of Iowa

CONTENTS

INTRODUCTION

Visual function is adversely affected in a great variety of ways in Parkinson's disease (PD).[1,2] Most of the visual dysfunction seen in this condition is relatively subtle from the patient's and physician's point of view. Although there are functional consequences under certain circumstances, the impairment resulting from visual symptoms in PD is seldom of sufficient severity to replace motoric dysfunction as the primary source of the patient's clinical disability. The visual abnormalities linked to PD are, for the most part, demonstrable in the very early clinical phase of the illness and possibly in the preclinical phase as well. Parkinson's disease is predominantly a disorder of the elderly, and patients in this age group commonly become aware of visual symptoms such as declining acuity, visual blurring, difficulty reading, impaired near vision, and abnormal light sensitivity. When these same symptoms occur in an elderly individual who also has PD, both the patient and the clinician may naturally wonder what contribution to these symptoms, if any, derives from the underlying neurological disorder. When visual complaints are formally solicited from PD patients, the most common are tired eyes or blurred vision when reading and diplopia.[3] Can the origin of such complaints be linked to the known pathophysiology of Parkinson's disease? To explore this possible relationship, this chapter will discuss the known aberrations of visual function that occur in PD as well as their pathogenesis. Since most forms of visual dysfunction in PD are clinically subtle, special attention will be paid to electrophysiologic and psychophysical techniques that are useful in demonstrating and quantifying them.

DOPAMINE AND VISION

An appreciation of the role of dopamine in the visual system and its abnormalities in PD is critical to understanding the aberrations of visual function seen in this condition. Dopamine is present in several anatomical structures that subserve vision. Most notable is its localization within the amacrine and interplexiform cells of the retina.[4] Several observations support the concept that dopamine subserves specific functions in the retina of primates. The chemical protoxin MPTP (1-methyl, 4-phenyl, 1-2-5-6-tetrahydropyridine) not only produces a clinical parkinsonian syndrome when injected into primates, it also significantly lowers retinal dopamine. This latter change is associated with abnormalities in the latency and amplitude of both the pattern visual evoked potential (VEP) and the electroretinogram, both of which can be reversed by the administration of the dopamine precursor levodopa.[5] Similarly, intravitreal injection of the neurotoxin 6-hydroxydopamine into aphakic monkeys results in abnormalities in both the phase and amplitude of the pattern electroretinogram (PERG) and the pattern VEP, especially for stimuli with higher spatial frequencies,[6] a finding that suggests a role for dopamine in retinal spatial tuning. In idiopathic PD as well, the visual evoked response[7] and pattern electroretinogram[8] are abnormal, and both can be improved by the administration of levodopa, especially the latter.[9] That these abnormalities in PD patients are related to retinal dopamine deficiency is supported by an autopsy study of PD patients in which retinal dopamine concentration was shown to be decreased;[10] however, in those patients who had received levodopa shortly before death, it was normal, suggesting that this therapy might be instrumental in reversing visual dysfunction related to retinal dopamine deficiency.

Dopaminergic innervation within the visual system has been demonstrated in structures other than the retina, including the lateral geniculate[11] and the visual cortex.[12] Single unit recordings in the lateral geniculate body of cats during simultaneous iontophoretic application of dopamine suggest that dopamine influences visual function in this structure by directly inhibiting relay cells through D_1 receptors and by both directly facilitating relay cell function and exciting inhibitory neurons through D_2 receptors.[13] Asymmetric primary visual cortex glucose hypometabolism has been demonstrated in PD, with the most severe abnormality appearing ipsilateral to the most severe motoric dysfunction.[14] The laterality of this abnormality suggests that it is more likely related to pathology in the nigrostriatal system than the retina, since the former structure is asymmetrically involved in PD, whereas the latter, even if asymmetrically affected, has bilateral input to the visual cortices and would be expected to result in symmetrical hypometabolism. Clinical evidence supporting possible cortical visual dysfunction in PD can be found in the observation that left hemi-Parkinson patients display a tendency to neglect the left upper visual field.[15]

Notwithstanding the potential widespread influence of dopamine within the visual system, its role in the retina seems to be most important. Dopamine content in the retina exhibits distinct circadian rhythms that can be driven by light/dark cycles or, in total chronic darkness, by the cyclic presence of melatonin.[16,17] Dopaminergic neurons are thought to subserve a modulatory role in the retina and may mediate center-surround functions that are important to receptive field organization.[18] An investigation in which the PERG spatial contrast response was recorded after administration of dopamine D_1 or D_2 antagonists or a D_1 agonist suggested that D_1 receptors may be most important for the surround organization of retinal ganglion cells, while D_2 receptors may play a role in center response amplification of other ganglion cells.[19] Within the D_2 receptor family, D_4 receptors predominate in the retina and appear to modulate the dopaminergic control of light sensitive cAMP.[20] Since dopamine receptors in the retina are not only found at synapses but at extrasynaptic sites as well, it appears that dopamine acts in this structure both as neurotransmitter and a neuromodulator.[21,22]

VISUAL ACUITY

It is generally accepted that there is not a severe or clinically impressive decline of visual acuity in PD, although careful group comparisons between PD patients and controls do reveal a difference. Repka et al.[3] tested high-contrast (Snellen chart) visual acuity in 39 PD patients and an equal number of age-matched controls. In this study, a small but statistically significant difference in visual acuity was found between PD patients (20/39) and controls (20/28). Visual acuity decline in PD patients correlated with increasing disease severity, in support of the notion that this abnormality is linked to the progressive pathology of the underlying Parkinson's disease. It is not certain whether loss of visual acuity in PD is related to retinal or cortical dysfunction, but the authors speculated that the reduction of retinal dopamine known to occur in PD might result in an increase in the receptive field size leading to the decrease in visual acuity. While the severity of visual acuity loss in PD may be related to advancing disease, it does not appear to be reversible with treatment, since high-contrast visual acuity has been found to be similarly impaired, whether patients are on or off dopaminergic drugs.[23] One other link between dopamine content and visual acuity is the clinical observation that administration of levodopa improves human amblyopia in both children and adults.[24] Whether this effect is exerted at the retinal or cortical level, or both, is still uncertain.[25]

Although not directly related to visual acuity, another common efferent visual problem in PD that can significantly reduce visual efficiency, is convergence insufficiency.[3] This condition, which is extremely common in PD, is associated with an abnormally distant near point of convergence, greater than 10 cm, and slow convergence amplitude. It is typically associated with the subjective complaint of asthenopia or eyestrain. Convergence insufficiency may also impair reading, especially in patients using bifocal eyeglasses for reading, since their proper use requires intact convergence. Impaired near vision in some affected patients may be amenable to correction with the use of prisms to compensate for impaired convergence or by instruction in the practice of monocular occlusion while reading. A recent report suggests that convergence insufficiency in PD can be improved by therapy with levodopa,[26] supporting the link between this form of dysfunction and dopamine deficiency.

COLOR VISION

Abnormal color discrimination has frequently been reported in patients with Parkinson's disease.[27,28] In many studies, this impairment has been found to be most prominent in the tritan (blue-yellow) axis.[29,30] Abnormalities of color perception have been demonstrated using both bedside clinical testing techniques such as the Farnsworth-Munsell (FM) 100-hue test[30] or more elaborate psychophysical means such as a computer-generated assessment of color contrast sensitivity.[29] Haug et al.[29] offered an explanation as to why the tritan contrast threshold is most affected in Parkinson's disease. In general, the blue cone system is preferentially affected in retinal disease, because its response range is limited, and it has the greatest vulnerability. The relatively selective involvement in PD can be explained by the fact that these short-wavelength-sensitive cones are relatively scarce in number in the retina and spaced widely apart, such that maintenance of their large receptive fields is dependent on interaction across considerable distances, a function mediated by the dopaminergic interplexiform and amacrine cells of the retina, the precise retinal elements that are most affected in PD. Involvement of these same retinal cells in PD may result in other forms of visual dysfunction but not necessarily related to the same pattern of impaired cellular connectivity. Pieri et at.[31] studied both color discrimination and contrast sensitivity in PD and found impairment of the two forms of visual dysfunction to be independent variables, suggesting that different retinal mechanisms underlie each.

The abnormality of color vision seen in PD can be demonstrated in very early patients who have not yet begun antiparkinson drug therapy. It can be reversed by treatment with levodopa.[32,33] Paradoxically, in one case, color vision was worsened after treatment with the dopamine agonist pramipexole.[34] Color discrimination testing in untreated, de novo PD patients has shown a significant correlation between the error score of the FM test and the severity of clinical parkinsonian signs as measured by the motor and activities of daily living subscales of the Unified Parkinson's Disease Rating Scale (UPDRS).[35] When PD patients are followed longitudinally over time, color discrimination scores decline progressively as the underlying disease worsens,[36–37] although, in one study, the decrementing scores only correlated with decline in the UPDRS activities of daily living (ADL) score,[37] and in another with both the UPDRS motor and ADL scores.[36] Despite the consistent correlation with disease severity by one measure or another, one investigation demonstrated that the magnitude of color vision abnormality in PD does not correlate with dopaminergic nigral degeneration as reflected by I^{123} β-CIT single photon emission tomography of the dopamine transporter. This observation s consistent with the prevailing notion that the visual abnormality in PD is largely extranigral in origin.[38] A plausible explanation of why color discrimination impairment does not correlate with nigral degeneration, yet parallels the clinical severity of PD, is that retinal dopamine depletion, although independent of nigral dopamine depletion, occurs contemporaneously at a relatively constant pace over time.

Regan et al.[39] questioned whether abnormalities uncovered during color discrimination testing in Parkinson's disease patients are just an epiphenomenon related to the motor disability of Parkinson's disease, since the FM test, used to demonstrate impaired color vision in many studies of PD, requires a motor response to correctly identify varying hues of color. They questioned whether it is the manual impairment of PD patients rather than a primary visual disorder that causes PD patients to fail this test and at the same time explains why levodopa, which corrects the motoric abnormality, improves the color discrimination score. These investigators utilized a computer-controlled test of color vision that did not require a motor response and found that PD patients performed as well as a control group. Their hypothesis, however, fails to explain why other investigators utilizing computer testing techniques[29] did uncover abnormalities of color vision in PD, or why most studies have revealed a preferential loss in the tritan color axis with little or no abnormality in the protan (red-green) axis, both of which should have been similarly affected were the abnormal test scores simply a reflection of parkinsonian motor impairment. There is additional evidence that supports the validity of a primary color vision abnormality in PD. Abnormalities of the visual evoked response produced by color pattern stimuli are more responsive to levodopa therapy than are those evoked by black-and-white stimuli.[40] Similarly, color contrast sensitivity in PD patients is most impaired along the tritan axis.[29] Lastly, other medical conditions characterized by impairment of dopaminergic transmission have been associated with abnormalities of color vision. In patients undergoing cocaine withdrawal, a relative hypodopaminergic state exists, and a similar tritan axis deficit in color discrimination has been noted. The same abnormality of color vision was not seen however during their hyperdopaminergic intoxication phase.[41] In schizophrenia, on the other hand, color discrimination abnormalities have been found to be general and not hue specific, leading to the hypothesis that axis-specific color discrimination abnormalities are a reflection of depletion of dopamine rather than its general dysregualtion.[42]

VISUAL CONTRAST SENSITIVITY

Visual contrast sensitivity (VCS) is a function that is not commonly tested by neurologists, yet it is an important sensory function that pervades many activities of daily living. It is probably a more meaningful reflection of functional vision than standard visual acuity tests as measured in most clinical settings. VCS has consistently been found to be abnormal in Parkinson's disease. VCS is measured by determining the minimal contrast required to distinguish objects from one another presented at a given spatial frequency. Visual targets spaced very closely together are said to have a high spatial frequency, and those spaced farther apart represent a low spatial frequency. Spatial frequency is expressed in cycles per degree of visual angle. The spectrum of contrast can be thought of as ranging from black on white (high contrast) to white on white (low contrast), with all shades of grey on black or grey on white in between. Another way to depict the concept of VCS is to ask how low in contrast adjacent images displayed at a given spatial frequency (distance apart) must be before they appear to be indistinguishable from a visually homogeneous field. The lower the contrast at which one can still detect a difference between adjacent objects, the higher the contrast sensitivity. Sinusoidal gratings of various spatial frequencies are among the most sensitive visual stimuli for testing VCS in humans. In this context, the term "sinusoidal" refers to the gradual diminution and then reconstitution of contrast between adjacent targets rather than a precipitous contrast change such as would be seen between adjacent black and white squares on a checkerboard. The peak of normal human contrast sensitivity is found at intermediate spatial frequencies. In Parkinson's disease, VCS is most reduced at these intermediate spatial frequencies.[43-45] This VCS abnormality is most exaggerated when the gratings are temporally modulated at medium frequencies of 4 to 8 Hz.[43] In addition, VCS is sometimes less attenuated at lower spatial frequencies in PD than it is in normal individuals.[46] These abnormalities are different from the declining VCS function associated with normal aging.[47] In some studies, VCS loss has been found to correlate with the overall severity of PD,[48] but in others it has not.[45] However, during the course of an individual day, there appears to be a more consistent correlation with the underlying severity of parkinsonian symptoms. VCS has been shown to exhibit a circadian variability that conforms to the common pattern of improved morning and worsened afternoon motoric disability seen in PD.[45] Recent evidence demonstrating a distinct circadian cycle of retinal dopamine content is consistent with this observation.[16,17] Similarly, VCS function can change in parallel to motor symptoms during transient "on" and "off" phases in fluctuating PD patients[49] and can be improved by the administration of levodopa.[50]

Whether the basic abnormality underlying abnormal VCS in PD resides in the retina, the visual cortex, or in both is still unclear. The fact that there are interocular differences in VCS[45,51] suggests the presence of retinal pathology. Moreover, the pattern electroretinogram, which largely reflects retinal ganglion cell activity, has been found to be abnormal in PD[52,53] with a characteristic amplitude loss at intermediate spatial frequencies similar to those associated with the greatest abnormality of VCS in PD.[52] As is the case with VCS, levodopa therapy improves the PERG abnormality in PD.[52,53] Langheinrich et al.[54] demonstrated that contrast discrimination threshold in PD patients correlated with frequency-specific PERG abnormalities (a retinal phenomenon) but not VEP impairment (a cortical phenomenon) and viewed these findings as further evidence that the VCS abnormality in PD is predominantly a result of retinal dysfunction. However, there is also evidence suggesting that cortical dysfunction may contribute to the VCS abnormality in PD. VCS impairment in PD patients has been found to be ori-

entation specific in that the VCS deficit is more severe for horizontally oriented patterns than those arrayed vertically.[43,44] Other dopamine deficiency syndromes, such as drug-induced parkinsonism, are also associated with VCS loss that is orientation dependent.[55] Although orientation specificity may be partially subserved by the lateral geniculate, [56] this perceptual function is felt to largely reside in the orientation-tuned receptive fields of the visual cortex.[57] While the presence of orientation specific VCS loss clearly raises the possibility of a central contribution to the VCS abnormality in PD, other investigators have noted that the cortically mediated function of contrast adaptation is preserved in PD and consider this finding evidence that cortical pathology is not significant in these patients and is not likely to play a major role with respect to reduced contrast sensitivity.[58]

Like the color vision abnormality in PD, VCS impairment progressively increases over time as the underlying neurologic condition worsens.[36] This worsening is accelerated at the intermediate spatial frequencies that are known to be most affected in PD, rather than at higher spatial frequencies, which would be expected to show the greatest rate of decline if the progressive worsening were solely due to aging.[59] As VCS worsens over time in PD, there is a contemporaneous progressive reduction in amplitude and lengthening of latency of the ERG, once again supporting the notion that abnormal VCS in this patient population is linked to retinal dysfunction.[60]

The use of low-contrast letter charts in patients with PD and other medical conditions has been found to detect visual dysfunction that was not appreciated through the use of standard visual acuity charts, which are confined to extremely high-contrast, high-spatial-frequency visual stimuli.[61,62] Parkinson's disease patients and their physician are usually unaware of this contrast sensitivity abnormality, but the patient may have noticed an inexplicable impairment in everyday visual tasks. This subtle abnormality, largely affecting VCS at the intermediate spatial frequencies, can impair such critical functions as facial recognition or proper and early identification of highway signs.[63] Additional functional correlates of this VCS deficit are possible.[64,65] Abnormal VCS might impair the ability to drive a motor vehicle in a low-contrast environment such as might exist at dusk or dawn. Intact spatiotemporal vision is functionally important on a day-to-day basis, since much of the visual world is periodic in array,[66] and is important for the normal perception of depth and depth discrimination.[67] It has been suggested that, in PD, abnormal contrast sensitivity might predispose to gait freezing. Mestre et al.[68] described a PD patient exhibiting increased contrast sensitivity to low and intermediate spatiotemporal frequencies who experienced gait freezing in the presence of environmental stripes arrayed at these low frequencies but not at higher spatial frequencies or with his eyes closed. They postulated that a hypersensitivity to low frequency visual stimuli resulted in an adaptive "braking" reflex leading to gait freezing. Of interest is the observation that levodopa therapy may preferentially increase VCS at these low spatial frequencies,[69] a fact that is consistent with the observation that dopaminergic therapy can paradoxically worsen gait freezing in some patients. Other investigators have demonstrated that the gait of PD patients improves in the presence of well illuminated periodic stimuli (lines) in the visual environment,[70] and that parameters of gait such as stride length are related to visual cues.[71]

VISUAL HALLUCINATIONS

Visual hallucinations occur commonly in advanced PD. Sanchez-Ramos et al.[72] recorded this complication in over 25% of 214 consecutive PD patients. In addition to known risk factors such as age, dementia, and drug therapy, visual loss can also contribute to the development of complex visual hallucinations in this patient population.[73–75] Visual hallucinations that occur in visually impaired but psychologically normal individuals are considered a form of the Charles Bonnet syndrome.[76–78] Patients afflicted with this syndrome are typically cognitively intact with retained insight such that this form of hallucinosis, which is usually devoid of personal meaning, tends to be somewhat less emotionally upsetting. Functional magnetic resonance imaging of patients with the Charles Bonnet syndrome has revealed increased activity in the ventral extrastriate region,[79] but whether this abnormal signal and the clinical syndrome with which it is associated reflect abnormal cortical excitation, a release phenomenon, or disrupted reentry signals is not yet known. While the Charles Bonnet syndrome is most typically associated with a significant loss of visual acuity,[74,77,80] in Parkinson's disease it has been associated with more covert visual abnormalities including abnormal color discrimination, reduced visual contrast sensitivity,[75] or impaired color contour perception.[73] In these studies, patients exhibiting the Charles Bonnet syndrome had otherwise normal visual acuity, confirming that any one of a wide range of visual abnormalities may be sufficient to predispose a PD patient to hallucinosis. The appearance of Charles Bonnet syndrome in PD patients and its predominance in elderly individuals has led some to postulate that some degree of underlying cerebral degeneration is critical to the development of the syndrome.[80] Treatment of the Charles Bonnet syndrome can be difficult. Therapy with neuroleptics that improve other forms of PD-related hallucinosis has been largely ineffective.[81] Improvement in the syndrome has been reported after institution of optical aids that result in improved functional vision,[82] raising the possibility that in some PD patients, whose hallucinations are predominantly related to abnormal VCS and/or color discrimination, treatment with dopaminergic drugs might reverse this symptom rather than exacerbate it.

CLINICAL UTILITY OF VISUAL TESTING IN PARKINSON'S DISEASE

Although, abnormalities such as impaired visual contrast sensitivity and abnormal color discrimination are unlikely to be apparent to the PD patient, it is still important that the clinician be aware that a wide variety of functional impairments as diverse as gait freezing, defective depth perception, impaired driving, and visual hallucinations might be related to these forms of visual impairment. Another clinically important question that arises is whether uncovering abnormalities of vision might be useful in identifying early or presymptomatic PD or in distinguishing PD from other parkinsonian syndromes. The possibility or differentiation between idiopathic Parkinson's disease and multiple system atrophy has been investigated in this regard, and distinct group differences between the two conditions have been identified in mean VEP latency and visual contrast thresholds.[58,51] However, it is not clear that these group differences would be useful in making a clinical distinction between the two conditions in individual patients. In progressive supranuclear palsy, mean VCS performance has been found to be more severely impaired than in PD but not so consistently abnormal in individual patients as to be useful in distinguishing this syndrome from other parkinsonian conditions.[54] In regard to the use of color testing as a diagnostic aid, Birch et al.[30] found that 23% of PD patients had tritan color vision deficits, while none of 40 age-matched controls were abnormal. These results suggest that the presence of impaired blue-yellow discrimination supports a diagnosis of Parkinson's disease, but normal function does not rule it out.

The prospect for using visual tests to identify PD in its earliest stage, or even prior to the onset of motoric symptoms, is slightly more promising but still not certain. In one study, color discrimination was found to be abnormal in mild de novo PD patients very early in the course of the illness, suggesting that the abnormality may have antedated the clinical diagnosis of Parkinson's disease.[28] However, a later investigation noted abnormal color discrimination in only a small percentage of such PD patients and no difference at all in mean performance compared to normal controls.[83] Perhaps the most useful application of VCS testing in the diagnosis of PD is its use in association with other assessments as part of a battery. Camicioli et al.[84] administered a battery consisting of tapping rate combined with either olfactory assessment or measurement of visual contrast sensitivity and found that it discriminated between mild PD patients and control subjects with greater than 90% accuracy.

SUMMARY

Involvement of the visual system in Parkinson's disease has been clearly demonstrated through electrophysiologic tests such as the electroretinogram or visual evoked potentials and by psychophysical tests of color discrimination and contrast sensitivity. There is abundant evidence that the visual system dysfunction seen, both in experimental parkinsonism and human Parkinson's disease is linked to retinal dopamine deficiency. The potential functional implications of the types of visual impairment found in PD are only beginning to be appreciated. The fact that convergence insufficiency, impaired VCS and reduced color discrimination all seem amenable to therapy with dopaminergic drugs provides hope that future advances in our understanding of the biology of dopamine in the visual system, and further development of neuroprotective or restorative therapies for Parkinson's disease can also effectively ameliorate the visual dysfunction associated with this condition.

REFERENCES

1. Rodnitzky, R. L., Visual dysfunction in Parkinson's disease, *Clinical Neuroscience*, 5(2):102–106, 1998.
2. Bodis-Wollner, I., Visualizing the next steps in Parkinson disease, *Archives of Neurology*, 59(8):1233–1234, 2002.
3. Repka, M. X., Claro, M. C., Loupe, D. N., Reich, S. G., Ocular motility in Parkinson's disease, *Journal of Pediatric Ophthalmology and Strabismus*, 33(3):144–147 1996.
4. Frederick, J. M., Rayborn, M. E., Laties, A. M., Lam, D. M., Hollyfield, J. G., Dopaminergic neurons in the human retina, *J. Comp. Neurol.*, 210(1):65–79, 1982.
5. Ghilardi, M. F., Chung, E., Bodis-Wollner, I., Dvorzniak, M., Glover, A., Onofrj, M., Systemic 1-methyl,4-phenyl,1-2-3-6-tetrahydropyridine (MPTP) administration decreases retinal dopamine content in primates, *Life Sciences*, 43(3):255–262, 1988.
6. Ghilardi, M. F., Marx, M. S., Bodis-Wollner, I., Camras, C. B., Glover, A. A., The effect of intraocular 6-hydroxydopamine on retinal processing of primates, *Annals of Neurology*, 25(4):357–364, 1989.
7. Bodis-Wollner, I., Yahr, M. D., Measurements of visual evoked potentials in Parkinson's disease, *Brain*, 101(4):661–671, 1978.
8. Peppe, A., Stanzione, P., Pierelli, F., Stefano, E., Rizzo, P.A., Tagliati, M. et al., Low contrast stimuli enhance PERG sensitivity to the visual dysfunction in Parkinson's disease, *Electroencephalography and Clinical Neurophysiology*, 82(6):453–457, 1992.
9. Peppe, A., Stanzione, P., Pierelli, F., De Angelis, D., Pierantozzi, M., Bernardi, G., Visual alterations in de novo Parkinson's disease: pattern electroretinogram latencies are more delayed and more reversible by levodopa than are visual evoked potentials, *Neurology*, 45(6):1144–1148, 1995.
10. Harnois, C., Di Paolo, T., Decreased dopamine in the retinas of patients with Parkinson's disease, *Investigative Ophthalmology and Visual Science*, 31(11):2473–2475, 1990.
11. Papadopoulos, G. C., Parnavelas, J. G., Distribution and synaptic organization of dopaminergic axons in the lateral geniculate nucleus of the rat, *Journal of Comparative Neurology*, 294:356–361, 1990.
12. Parkinson, D., Evidence for a dopaminergic innervation of cat primary visual cortex, *Neuroscience*, 30(1):171–179, 1989.
13. Zhao, Y., Kerscher, N., Eysel, U., Funke, K., D_1 and D_2 receptor-mediated dopaminergic modulation of visual responses in cat dorsal lateral geniculate nucleus, *Journal of Physiology*, 539(Pt. 1):223–238, 2002.
14. Bohnen, N. I., Minoshima, S., Giordani, B., Frey, K. A., Kuhl, D. E., Motor correlates of occipital glucose hypometabolism in Parkinson's disease without dementia, *Neurology*, 52(3):541–546, 1999.
15. Lee, A. C., Harris, J. P., Atkinson, E. A., Nithi, K., Fowler, M. S., Dopamine and the representation of the upper visual field: evidence from vertical bisection errors in unilateral Parkinson's disease, *Neuropsychologia*, 40(12):2023–2029, 2002.
16. Doyle, S. E., McIvor, W. E., Menaker, M., Circadian rhythmicity in dopamine content of mammalian retina: role of the photoreceptors, *J. Neurochem.*, 83(1):211–219, 2002.
17. Doyle, S. E., Grace, M. S., McIvor, W., Menaker, M., Circadian rhythms of dopamine in mouse retina: the role or melatonin, *Vis Neurosci.*, 19(5):593–601, 2002.
18. Bodis-Wollner, I., Tagliati, M., The visual system in Parkinson's disease, *Advances in Neurology*, 60:390–394, 1993.
19. Bodis-Wollner, I., Tzelepi, A., The push-pull action of dopamine on spatial tuning of the monkey retina: the effects of dopaminergic deficiency and selective D_1 and D_2 receptor ligands on the pattern electroretinogram, *Vision Research*, 38:1479–1487, 1998.
20. Patel, S., Chapman, K. L., Marston, D., Hutson, P. H., Ragan, C. I., Pharmacological and functional characterisation of dopamine D_4 receptors in the rat retina, *Neuropharmacology*, 44(8):1038–1046, 2003.
21. Ribelayga, C., Wang, Y., Mangel, S. C., Dopamine mediates circadian clock regulation of rod and cone input to fish retinal horizontal cells, *Journal of Physiology*, 544(3):801–816, 2002.
22. Puopolo, M., Hochstetler, S. E., Gustincich, S., Wightman, R. M., Raviola, E., Extrasynaptic release of dopamine in a retinal neuron: activity dependence and transmitter modulation, *Neuron*, 30(1):211–225, 2001.
23. Jones, R. D., Donaldson, I. M., Timmings, P. L., Impairment of high-contrast visual acuity in Parkinson's disease, *Movement Disorders*, 7(3):232–238, 1992.
24. Pandey, P. K., Chaudhuri, Z., Kumar, M., Satyabala, K., Sharma, P., Effect of levodopa and carbidopa in human amblyopia, *J. Pediatr. Ophthalmol. Strabismus*, 39(2):81–89, 2002.
25. Chao-I, Yang, Meng-Ling, Yang, Ju-Chuan, Huang, Yung-Liang, Wan, Ray, Jui-Fang, Tsai, Yau-Yau, Wai,

et al., Functional MRI of amblyopia before and after levodopa, *Neuroscience Letters*, 339:49–52, 2003.

26. Racette, B. A., Gokden, M. S., Tychsen, L. S., Perlmutter, J. S., Convergence insufficiency in idiopathic Parkinson's disease responsive to levodopa, *Strabismus*, 7(3):169–174, 1999.

27. Buttner, T., Kuhn, W., Klotz, P., Steinberg, R., Voss, L., Bulgaru, D. et al., Disturbance of colour perception in Parkinson's disease, *Journal of Neural Transmission, Parkinson's Disease and Dementia Section*, 6(1):11–15, 1993.

28. Buttner, T., Kuhn, W., Muller, T., Patzold, T., Heidbrink, K., Przuntek, H., Distorted color discrimination in "de novo" parkinsonian patients, *Neurology*, 45(2):386–387, 1995.

29. Haug, B. A., Kolle, R. U., Trenkwalder, C., Oertel, W. H., Paulus, W., Predominant affection of the blue cone pathway in Parkinson's disease, *Brain*, 118(Pt. 3):771–778, 1995.

30. Birch, J., Kolle, R. U., Kunkel, M., Paulus, W., Upadhyay, P., Acquired colour deficiency in patients with Parkinson's disease, *Vision Research*, 38(21):3421–3426, 1998.

31. Pieri, V., Diederich, N.J., Raman, R., Goetz, C. G., Decreased color discrimination and contrast sensitivity in Parkinson's disease, *Journal of the Neurological Sciences*, 172(1):7–11, 2000.

32. Buttner, T., Kuhn, W., Patzold, T., Przuntek, H., L-dopa improves colour vision in Parkinson's disease, *Journal of Neural Transmission, Parkinson's Disease and Dementia Section*, 7(1):13–19, 1994.

33. Buttner, T., Kuhn, W., Muller, T., Patzold, T., Przuntek, H., Color vision in Parkinson's disease: missing influence of amantadine sulphate, *Clinical Neuropharmacology*, 18(5):458–463, 1995.

34. Muller, T., Przuntek, H., Kuhlmann, A., Loss of color vision during long-term treatment with pramipexole, *Journal of Neurology*, 250(1):101–102, 2003.

35. Muller, T., Kuhn, W., Buttner, T., Przuntek, H., Distorted colour discrimination in Parkinson's disease is related to severity of the disease, *Acta Neurologica Scandinavica*, 96(5):293–296, 1997.

36. Diederich, N. J., Raman, R., Leurgans, S., Goetz, C. G., Progressive worsening of spatial and chromatic processing deficits in Parkinson disease, *Archives of Neurology*, 59(8):1249–1252, 2002.

37. Muller, T., Woitalla, D., Peters, S., Kohla, K., Przuntek, H., Progress of visual dysfunction in Parkinson's disease, *Acta Neurologica Scandinavica*, 105(4):256–260, 2002.

38. Muller, T., Kuhn, W., Buttner, T, Eising, E., Coenen, H., Haas, M. et al., Colour vision abnormalities do not correlate with dopaminergic nigrostriatal degeneration in Parkinson's disease, *Journal of Neurology*, 245(10):659–664, 1998.

39. Regan, B. C., Freudenthaler, N., Kolle, R., Mollon, J. D., Paulus, W., Colour discrimination thresholds in Parkinson's disease: results obtained with a rapid computer-controlled colour vision test, *Vision Research*, 38(21):3427–3431, 1998.

40. Barbato, L., Rinalduzzi, S., Laurenti, M., Ruggieri, S., Accornero, N., Color VEPs in Parkinson's disease, *Electroencephalography and Clinical Neurophysiology*, 92(2):169–172, 1994.

41. Desai, P., Roy, M., Brown, S., Smelson, D., Impaired color vision in cocaine-withdrawn patients, *Arch. Gen. Psychiatry*, 54(8):696–699, 1997.

42. Shuwairi, S. M., Cronin-Golomb, A., McCarley, R. W., O'Donnell, B. F., Color discrimination in schizophrenia, *Schizophr. Res.*, 55(1-2):197–204, 2002.

43. Regan, D., Maxner, C., Orientation-selective visual loss in patients with Parkinson's disease, *Brain*, 110(Pt. 2):415–432, 1987.

44. Bulens, C., Meerwaldt, J. D., van der Wildt, G. J., Effect of stimulus orientation on contrast sensitivity in Parkinson's disease, *Neurology*, 38(1):76–81, 1988.

45. Struck, L. K., Rodnitzky, R. L., Dobson, J. K., Circadian fluctuations of contrast sensitivity in Parkinson's disease, *Neurology*, 40(3 Pt. 1):467–470, 1990.

46. Bodis-Wollner, I., The visual system in Parkinson's disease, *Vision and the Brain*, Raven Press, 297–316, 1990.

47. Mestre, D., Blin, O., Serratrice, G., Pailhous, J., Spatiotemporal contrast sensitivity differs in normal aging and Parkinson's disease, *Neurology*, 40(11):1710–1714, 1990.

48. Hutton, J. T., Morris, J. L., Elias, J. W., Varma, R., Poston, J. N., Spatial contrast sensitivity is reduced in bilateral Parkinson's disease, *Neurology*, 41(8):1200–1202, 1991.

49. Bodis-Wollner, I., Marx, M. S., Mitra, S., Bobak, P., Mylin, L., Yahr, M., Visual dysfunction in Parkinson's disease. Loss in spatiotemporal contrast sensitivity, *Brain*, 110(Pt. 6):1675–1698, 1987.

50. Hutton, J. T., Morris, J. L., Elias, J. W., Levodopa improves spatial contrast sensitivity in Parkinson's disease, *Archives of Neurology*, 50(7):721–724, 1993.

51. Delalande, I., Hache, J. C., Forzy, G., Bughin, M., Benhadjali, J., Destee, A., Do visual-evoked potentials and spatiotemporal contrast sensitivity help to distinguish idiopathic Parkinson's disease and multiple system atrophy? *Movement Disorders*, 13(3):446–452, 1998.

52. Tagliati, M., Bodis-Wollner, I., Yahr, M. D., The pattern electroretinogram in Parkinson's disease reveals lack of retinal spatial tuning, *Electroencephalography and Clinical Neurophysiology*, 100(1):1–11, 1996.

53. Peppe, A., Stanzione, P., Pierantozzi, M., Semprini, R., Bassi, A., Santilli, A. M. et al., Does pattern electroretinogram spatial tuning alteration in Parkinson's disease depend on motor disturbances or retinal dopaminergic loss? *Electroencephalography and Clinical Neurophysiology*, 106(4):374–382, 1998.

54. Langheinrich, T., Tebartz, V. E., Lagreze, W. A., Bach, M., Lucking, C. H., Greenlee, M. W., Visual contrast response functions in Parkinson's disease: evidence from electroretinograms, visually evoked potentials and psychophysics, *Clinical Neurophysiology*, 111(1):66–74, 2000.

55. Bulens, C., Meerwaldt, J. D., van der Wildt, G. J., Keemink, C. J., Visual contrast sensitivity in drug-induced Parkinsonism, *Journal of Neurology, Neurosurgery and Psychiatry*, 52(3):341–345, 1989.

56. Xu, X., Ichida, J., Shostak, Y., Bonds, A. B., Casagrande, V. A., Are primate lateral geniculate nucleus (LGN) cells really sensitive to orientation or direction? *Vis. Neurosci.*, 19(1):97–108, 2002+.

57. Hubel, D. H., Wiesel, T. N., Stryker, M. P., Orientation columns in macaque monkey visual cortex demonstrated by the 2-deoxyglucose autoradiographic technique, *Nature*, 269:328–330, 1977.

58. Tebartz, V. E., Greenleem, M. W., Foleym, J. M., Lucking, C. H., Contrast detection, discrimination and adaptation in patients with Parkinson's disease and multiple system atrophy, *Brain*, 120(Pt. 12):2219–2228, 1997.

59. Kline, D. W., Ageing and the spatiotemporal discrimination performance of the visual system, *Eye,* 1, 323–329, 1987.

60. Ikeda, H., Head, G. M., Ellis, C. J., Electrophysiological signs of retinal dopamine deficiency in recently diagnosed Parkinson's disease and a follow up study, *Vision Research*, 34(19):2629–2638, 1994.

61. Regan, D., Neima, D., Low-contrast letter charts in early diabetic retinopathy, ocular hypertension, glaucoma, and Parkinson's disease, *British Journal of Ophthalmology*, 68(12):885–889, 1984.

62. Kupersmith, M. J., Shakin, E., Siegel, I. M., Lieberman, A., Visual system abnormalities in patients with Parkinson's disease, *Archives of Neurology*, 39(5):284–286, 1982.

63. Evans, D. W., Ginsburg, A. P., Contrast sensitivity predicts age-related differences in highway-sign discriminability, *Human Factors*, 27(6):637–642, 1985.

64. West, S. K., Rubin, G. S., Broman, A. T., Munoz, B., Bandeen-Roche, K., Turano, M., How does visual impairment affect performance on tasks of everyday life? The SEE project, Salisbury Eye Evaluation, *Archives of Ophthalmology*, 120(6):774–780, 2002.

65. Ginsburg, A. P., Contrast sensitivity and functional vision, *International Ophthalmology Clinics*, 43(2):5–15, 2003.

66. DeValois, R., DeValois, K., *Spatial Vision*, Oxford University Press, Inc., New York, 1988.

67. Rohaly, A. M., Wilson, H. R., The effects of contrast on perceived depth and depth discrimination, *Vision Research*, 39:9–18, 1999.

68. Mestre, D., Blin, O., Serratrice, G., Contrast sensitivity is increased in a case of nonparkinsonian freezing gait, *Neurology*, 42(1):189–194, 1992.

69. Giladi, N., Treves, T. A., Simon, E. S., Shabtai, H., Orlov, Y., Kandinov, B. et al., Freezing of gait in patients with advanced Parkinson's disease, *Journal of Neural Transmission—General Section*, 108(1):53–61, 2001.

70. Azulay, J. P., Mesure, S., Amblard, B., Blin, O., Sangla, I., Pouget, J., Visual control of locomotion in Parkinson's disease, *Brain*, 122(Pt. 1):111–120, 1999.

71. Lewis, G. N., Byblow, W. D., Walt, S. E., Stride length regulation in Parkinson's disease: the use of extrinsic, visual cues, *Brain*, 123(Pt. 10):2077–2090, 2000.

72. Sanchez-Ramos, J. R., Ortoll, R., Paulson, G. W., Visual hallucinations associated with Parkinson disease, *Archives of Neurology*, 53(12):1265–1268, 1996.

73. Buttner, T., Kuhn, W., Muller, T., Welter, F. L., Federlein, J., Heidbrink, K. et al., Visual hallucinosis: the major clinical determinant of distorted chromatic contour perception in Parkinson's disease, *Journal of Neural Transmission—General Section*, 103(10):1195–1204, 1996.

74. Lepore, F. E., Visual loss as a causative factor in visual hallucinations associated with Parkinson disease, *Archives of Neurology*, 54(7):799, 1997.

75. Diederich, N. J., Goetz, C. G., Raman, R., Pappert, E. J., Leurgans, S., Piery, V., Poor visual discrimination and visual hallucinations in Parkinson's disease, *Clinical Neuropharmacology*, 21(5):289–295, 1998.

76. Pfeiffer, R. F., Bodis-Wollner, I., Charles Bonnet syndrome, *J. Am. Geriatr. Soc.*, 44(9):1128–1129, 1996.

77. Antal, A., Pfeiffer, R., Bodis-Wollner, I., Simultaneously evoked primary and cognitive visual evoked potentials distinguish younger and older patients with Parkinson's disease, *Journal of Neural Transmission—General Section*, 103(89):1053–1067, 1996.

78. Teunisse, R. J., Cruysberg, J. R., Hoefnagels, W. H., Verbeek, A. L., Zitman, F. G., Visual hallucinations in psychologically normal people: Charles Bonnet's syndrome, *Lancet*, 347:794–797, 1996.

79. Ffytche, D. H., Howard, R. J., Brammer, M., David, A., Woodruff, P., Williams, S., The anatomy of conscious vision: an fMRI study of visual hallucinations, *Nat. Neurosci.*, 1(8), 738–742, 1998.

80. Manford, M., Andermann, F., Complex visual hallucinations. Clinical and neurobiological insights, *Brain*, 121(Pt. 10):1819–1840, 1998.

81. Batra, A., Bartels, M., Wormstall, H., Therapeutic options in Charles Bonnet syndrome, *Acta Psychiatr., Scand.*, 96, 129–133, 1997.

82. Pankow, L., Luchins, D. J., An optical intervention for visual hallucinations associated with visual impairment in an elderly patient, *Optometry and Vision Science*, 74(3):138–143, 1997.

83. Vesela, O., Ruzicka, E., Jech, R., Roth, J., Stepankova, K., Mecir, P. et al., Colour discrimination impairment is not a reliable early marker of Parkinson's disease, *Journal of Neurology*, 248(11):975–978, 2001.

84. Camicioli, R., Grossmann, S. J., Spencer, P. S., Hudnell, K., Anger, W. K., Discriminating mild parkinsonism: methods for epidemiological research, *Movement Disorders*, 16(1):33–40, 2001.

23 Visuocognitive Dysfunctions in Parkinson's Disease

Andrea Antal and Walter Paulus
Department of Clinical Neurophysiology, Georg-August University of Göttingen

Ivan Bodis-Wollner
Parkinson's Disease and Related Disorders Clinic, Center of Excellence,
State University of New York, Downstate Medical Center

CONTENTS

INTRODUCTION

Of all human senses, vision is crucial among sensory functions in shaping our perceptions and accurate actions. We need satisfactory visual processing for navigation; for recognizing faces, objects, buildings, and places; for writing and counting; and for a wide range of motor actions starting from eye movements ending with the execution of a motor response. Accurate visual information processing is also necessary for the satisfactory functioning of the visual memory. The processing and transfer of primary visual input to higher-order cortical areas is quick, almost automatic. Due to this efficient transfer process, most of our daily activities require not too much effort. However, when processing is impaired at a stage of the visual information flow, even simple actions may be significantly delayed and distorted. Indeed, in many neurological disorders, impaired vision is not the primary dominant symptom; however, unsatisfactory visual processing is likely to contribute to difficulties in daily living. Frequently studied relevant problems include consciously controlled visual information processing, sustained and selective attention, planning, problem solving, response selection, and decision making.

The clinical syndrome of parkinsonism is characterized by slowly progressive bradykinesia, rigidity, tremor, and postural changes that become disabling. Therefore, Parkinson's disease (PD) is generally known as a movement disorder due to the dopaminergic deficiency affecting the basal ganglia. However, in recent decades, several studies have demonstrated that beyond or parallel with the progressive motor impairments, nonmotor symptoms are also present, including visuospatial, visual perceptual, visuomotor, and visuocognitive dysfunctions. Visual abnormalities in PD are usually hidden and not likely to be uncovered during a routine neurological examination or even by ordinary high-contrast visual acuity (VA) testing. Most commonly, these nonspecific visual or visuocognitive symptoms are not considered as part of PD, known to be a "movement disorder." Furthermore, patients are not aware of sensory deficits, since they develop very slowly and, if present at all, may be attributed to the underlying aging process. To what extent a visual deficiency can go unnoticed may be evidenced in congenitally color blind subjects. For them, the outside world may look perfect, and they are confronted with their deficiency only when forced to make a direct comparison.

In this chapter, visual electrodiagnostic, psychophysical, and imaging data on visual and visuocognitive processing will be described in PD. Impairment of visual-spatial working memory and attentional set shifting are already implicated in the early stages of the disease. It will be shown that the dopaminergic dysregulation of pre-fronto-striatal circuits and the related posterior cortical areas in Parkinson patients lead to higher-level cognitive dysfunctions that are not passively caused by dopaminergic retinal or primary visual cortical impairments. Before the conclusion is made that these impairments represent higher-order cortical and subcortical dysfunctions, the relationship to lower-level visual impairments should be critically viewed.

PRIMARY VISUAL IMPAIRMENTS IN PD

THE ROLE OF DOPAMINE IN RETINAL PROCESSING

In the last decades pharmacological studies related to the electroretinogram (ERG) in normal human volunteers and in PD patients have suggested a specific role of dopamine (DA) in retinal processing of visual input.[21,37,93,99] Going beyond the limitations of human studies, extensive neuropharmacological and neurotoxicological studies affecting the dopaminergic system in the monkey and lower vertebrates have led to a more detailed understanding of visual impairment in PD. (For a review, see References 17 and 21.) Various types of DA receptors, broadly classified into D_1 and D_2 subtypes, are located on different neurons of the retina.[59] The dual physiologic action of DA on distal D_1 and D_2 receptors located on neuronal structures has been studied in detail only in the retina of lower vertebrates with larger neurones.[36] However, studying the effects of selective DA receptor ligands on massed electrophysiological retinal responses (ERG) in the monkey[22] has led to an understanding of the final retinal output in primates due to DA-s push-pull role in mediating center-surround interaction for establishing the receptive field structure of ganglion cells.[26] Visual electrophysiological and psychophysical abnormalities, originally observed in

PD patients,[15,16,62] have also been reported in neuroleptic treated normal volunteers,[12] in neuroleptics-induced parkinsonism in humans,[53,64] and also in parkinsonian animal models.[42–44] Taken together, the results of these studies suggest that dopaminergic deficiency, irrespective of the cause, results in characteristic visual impairment of spatial processing. The deficits are similar in experimental models and in idiopathic PD.

EVIDENCE FOR RETINAL AND CORTICAL DOPAMINERGIC DYSFUNCTION IN PD

It was originally reported by Bodis-Wollner and Yahr[15] that more than half of the examined PD patients had delayed visual evoked potentials (VEPs) (see Figure 23.1). This finding remained controversial until it became clear that the appropriate visual stimuli, preferentially Gabor filtered stimulus containing only one spatial frequency, should be used. The widely distributed checkerboard pattern, still ideal because of its robustness for a variety of patients, usually fails to reveal abnormalities in PD. (For reviews, see References 21, 25, and 26.) Now it is apparent that the VEP and pattern ERG (PERG) abnormality in PD is most evident for foveal stimuli of medium and high spatial frequencies (SFs) [above 2 cycles/degree (cpd)] where normal observers are most sensitive for the visual stimuli (see Figure 23.2).[18,95,100] Consistent with the results of the electrophysiological studies, in PD, contrast sensitivity (CS) is most reduced above 2 cpd.[19,21,29,33,34,70,88,95,102] However, reduced CS in PD goes undocumented in the majority of patients, as many vision care specialists are not aware of testing for a potentially profound CS deficit in a patient with near normal VA.

Contrast sensitivity loss in PD becomes more profound when the stimulus grating is temporally modulated at 4 to 8 Hz,[20,68,88] suggesting that a dopaminergic deficiency state also affects distal temporal processing.[67] It has been shown, however, that increasing stimulus strength can normalize some select temporal deficits seen in PD patients.[8] In summary, the spatial and temporal selectivity of visual losses detected with CS in PD is consistent with the results of electrophysiological tests (PERG and VEP). The interpretation of visual deficits in PD suggests that the disease process causes progressive, select pathology of dopaminergic neuronal processing in the human retina, leading to loss of spatio-temporal tuning and distorted retinal input to higher visual centers. An essential proof of visual system involvement in PD and the relationship of visual and motor changes was recently provided by a longitudinal study of visual dysfunction in PD patients: CS impairs in parallel with the worsening of motor score.[34] These results therefore suggest that the visual system shares with the motor system progressive degeneration of dopaminergic neurons and/or progressive failure of the effect of L-dopa therapy.

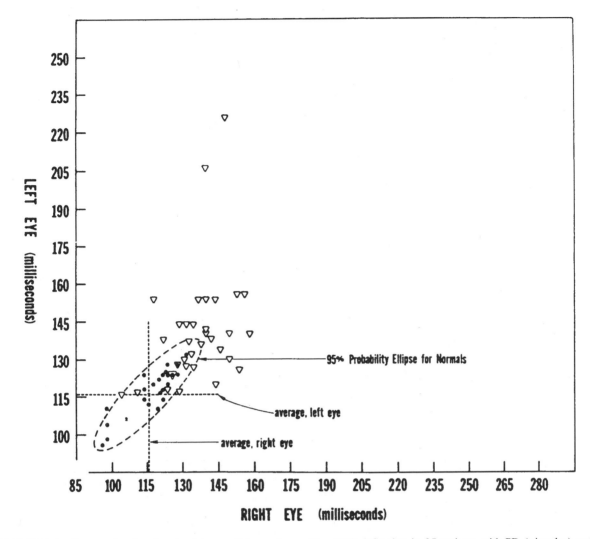

FIGURE 23.1 Scatter plot showing the latency of the major positive VEP deflection in 35 patients with PD (triangles) and 26 control subjects (dots). Numerals indicate the number of measurements falling on the same locus. Values for the left and right eye are shown on the ordinate and abscissa, respectively. An ellipse has been drawn within which 95% of the normal population would be expected to fall, based on the statistics of the control group (dark dots). Over two-thirds of the PD patients are outside the ellipse. (*Source:* Bodis-Wollner and Yahr,[15] with permission of *Brain.*)

The delay of the P100 component is observed in both de novo and also in treated PD patients using stimuli at middle (2 to 6 cpd) spatial frequencies.[49,75,80] It was reported that treated patients can exhibit longer delays.[75] This apparently paradoxical result is likely due to the more advanced disease in treated patients, which per se results in worse retinal visual responses.[34,100] While both ERG and VEP can improve with therapy, there is an apparent difference: levodopa therapy improves PERG abnormalities to a higher degree than it does VEP deficits.[80] One possible interpretation is that VEP changes in PD are secondary to retinal pathology and, at the cortical level, represent chronic and not exclusively dopaminergic alterations in visual processing. However, there is evidence of visual cortical dopaminergic innervation, even in the absence of retinal visual input.[86]

The question emerges: Is the visual dysfunction really an *integral* part of PD? It has been observed that the deficit fluctuates with motor symptoms in "on-off" patients[19] and worsens with the progression of motor symptoms.[34] While the role of DA deficiency is strongly implied by the above-mentioned studies, DA deficiency may not be *exclusively* responsible for visual changes in PD. For example, a higher onset/offset VEP amplitude ratio was found in PD patients compared to controls using sinusoidal grating as visual stimuli in on-off mode.[10] It is known that onset versus offset retinal responses may be separated using selective glutamate receptor blockers.[91] The relevance of dopaminergic deficiency or other neurotransmitter alteration, such as the involvement of selective glutamate receptor subtypes in the retina and beyond, in generating the "supernormal" offset VEP in PD is not yet

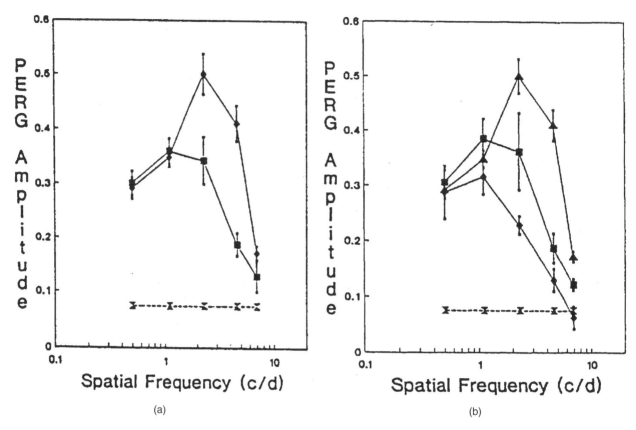

(a)   (b)

FIGURE 23.2 *(a)* The PERG tuning function in PD: PERG spatial transfer function obtained in patients (squares) and age-matched subjects (diamonds). The functions are parallel at lower SF and very close at the higher SF tested (6.9 cpd). Note lack of tuning of the PERG transfer function in PD. *(Source:* Tagliati et al.[100] with permission of *Clinical Neurophysiology.) (b)* Effects of L-dopa therapy on PERG amplitude. PERG amplitude obtained in age-matched subjects (triangles), PD patients receiving (squares) and not receiving (diamonds) L-dopa are plotted as a function of SF. PD patients receiving L-dopa show higher values and better tuning compared to untreated patients, although they rarely achieve normal values. The dashed line represents the mean noise level during recordings. Error bars indicate SE. *(Source:* Tagliati et al.[100] with permission of *Clinical Neurophysiology.)*

established. Although the findings appear robust and intriguing, no other studies have yet compared onset with offset responses on PD.

Are Visual Deficits in PD Solely Determined by Retinal Dopaminergic Dysfunction?

PERG changes in PD are definitely caused by retinal dopaminergic deficiency.[25,57] However, a retinal abnormality may passively cause visual deficits in subsequent processing, or other anatomical areas may also, independently of the retina may be affected in PD. The LGN[1] and the visual cortex also have dopaminergic innervation.[79,82,86,87] Asymmetrically lateralized primary visual cortex glucose hypometabolism has been demonstrated in PD with the most severe abnormalities contralateral to the most severe motoric dysfunction.[28] Although more confirmatory evidence is needed, it is possible that occipital hypometabolism indirectly reflects basal ganglia dysfunction or intrinsic cortical pathology.

Pattern orientation dependent CS losses have been reported in PD[88] more severe for horizontal than for vertical patterns[29] (see Figure 23.3). This finding cannot be due to retinal dopaminergic deficiency; rather, the deficit suggests the presence of intrinsic cortical pathology. However, contrast adaptation, which has a cortical origin, is spared in PD.[103]

VISUOCOGNITIVE PROCESSING IN PD

A correlation between cortical DA innervation and expression of cognitive capacities has been claimed (Nieoullon, 2002).[74] Impaired cognitive processing in PD is not surprising, due to the connections and loops[110] between the basal ganglia and various sensory cortical areas. However, DA is apparently involved in a more specific manner than just "gating" bottom up visual information flow. Several aspects of consciously controlled information processing, such as planning, problems solving, decision making, and response selection, are associated with the functions of fronto-striatal circuits.[41,46,50,65,76,77] A dopaminergic dysreg-

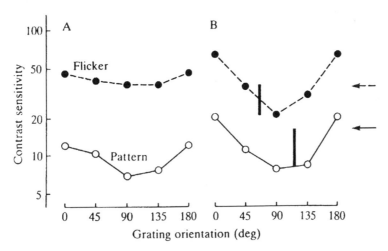

FIGURE 23.3 Effect of orientation on visual contrast sensitivity. Ordinates plot CS for flicker perception (filled dots) and for pattern perception (open dots) versus grating orientation (abscissa). Vertical is = deg. and 180 deg. on the abscissa, and horizontal is 90 deg. The vertical bars show the upper normal limits for orientational tuning, and the horizontal arrows show lower normal limits for absolute sensitivities (99% limits). The grating had a spatial frequency of 2 cpd and a temporal frequency of 8 Hz. A = left eye, B = right eye. (*Source:* Regan and Maxner,[88] with permission of *Brain.*)

ulation of this subcortico-cortical system in PD leads to higher-level cognitive dysfunctions.[31,41,69,77,78] Recent electrophysiological, neurophysiological, and functional imaging studies attempt to link cognitive symptoms and specific neuronal circuits of the basal ganglia and its connections.

ELECTROPHYSIOLOGY: THE RELATIONSHIP OF PRIMARY VEP-S AND THE CONCURRENTLY OBTAINED P300

Identifiable positive and negative deflections of event-related potentials (ERPs) provide indices for the timing in information processing including stimulus evaluation, response selection, and context updating.[63] ERPs are recorded in response to an external stimulus or event to which the subject is consciously paying attention. They are often elicited when the subject distinguishes one stimulus (target) from other stimuli (nontargets). The most extensively studied ERP component is the P300, appearing 300 to 400 ms after the onset of the target stimulus.[96] P300 amplitude is maximal at the midline electroencephalographic (EEG) electrodes (Cz and Pz) and is inversely related to the probability of the eliciting event.

Many visual ERP studies yielded a delayed P300 only in demented PD patients,[48,94,98,106,109] although other studies reported a delayed P300 in nondemented PD patients.[7,23,24,89,97] This suggests that the slowness of visual information processing may be independent or that it precedes global dementia.

LATENCY

Comparing the P100 and P300 of the concurrently obtained visual ERP resulted in a somewhat surprising finding in two independent and ethnically different groups of PD patients. A prolongation of the normalized P300 latency (P300–P100 latency difference, called central processing time) differentiated younger PD patients from controls.[7,89] These data suggest that younger PD patients could be differentiated from other types of PD using a concurrent VEP and visual P300 recording. Amantidine also shortened the latency of the visual P300 in PD with little or no effect on the primary VEP component.[11]

AMPLITUDE

Few studies have examined P300 amplitude in PD. In general, P300 amplitude increases when more attention is allocated, as in the case of unexpected or in complex tasks. However, it is conceivable that the interpretation of raw amplitude can be misleading, since a nonspecific, age-related, low-voltage EEG recording could cause low P300 amplitude.[7] Measuring the P300/P100 amplitude ratio therefore gives a more reliable measure on the nature of amplitude alterations.[7] This individually normalized P300 amplitude provided a significant distinction of younger nondemented PD patients from older patients and from age matched control subjects.[7,89]

N200 OF THE VISUAL ERP IN PD

Apparently, P100 and P300 are independently affected in PD. To localize the stage of visual processing at which this independence becomes established, earlier cognitive ERP components such as N200 were analyzed and showed that this component is also independently changing from P300.[7] The visual N200 probably represents a visual form of the auditory mismatch negativity.[101] This component is more negative for the infrequent stimuli and distributed over the extrastriate visual areas and the posterior-temporal cortex. N200 latency was delayed in nondemented PD patients, even when P300 was not prolonged using a simple visual paradigm.[7] In a semantic discrimination task, a similar result was found.[97] These data further suggest that visual deficits and processes indexed by the P300 may reflect processing that is either parallel to or well beyond the interface of bottom-up and top-down visual inputs.[32,60]

The Pharmacology of P300: Does the P300 Abnormality Represent only Dopaminergic Dysfunction?

A study in MPTP-treated monkeys suggests that levodopa therapy alone does not affect the visual P300,[44] however D_2 receptor blockade can influence the visual P300 in monkeys (see Figure 23.4).[6] Cellular electrophysiological evidence shows however that D_1 receptors are involved in visual working memory in the prefrontal area (for a review, Reference 47), which was also identified as one of the generators of P300.[51] It is therefore conceivable that the synergistic action of D_1 and D_2 receptors is necessary to improve the visual P300.

Levodopa treatment shortens the latency of P300 in PD.[92,94] However, some investigators have described a prolonged P300 latency in medicated patients.[52,85] One

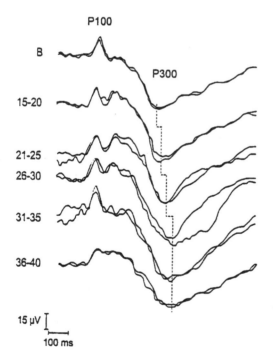

FIGURE 23.4 ERP traces illustrating the effects of Sulpiride (D_2 receptor antagonist) in different time intervals (B = baseline) at centro-parietal registration in a Cynomolgus monkey. Note the increasing latency and amplitude of P300. (*Source:* Antal et al.,[6] with permission of *Neurosci. Letters.*)

possibility is that medicated patients are more severely affected, and the P300 correlates with disease severity. Such correlation has not been studied in detail; this question therefore is still open.

The modulation of P300 by nondopaminergic agents has been frequently studied in healthy subjects.[35,40] Cognitive slowing in PD could also be caused by abnormalities of nondopaminergic systems,[83] although there is little direct evidence of correlation of the P300 in PD with cholinergic or other types of neurotransmitter alterations. Pretreatment delayed P300 improved in PD patients following treatment with amantidine, a low-affinity uncompetitive NMDA receptor antagonist.[11] Amantidine is closely related to memantine, advocated for the treatment of cognitive impairment in Alzheimer's disease. Amantidine's effect was noticeable not only as a monotherapy, but also in patients treated with levodopa. It is unknown how amantidine exercised this beneficial effect. It is often asserted that amantidine has DA-mimetic properties, and it cannot be therefore excluded that amantidine improves cognitive ERPs in PD as a DA-mimetic agent.

Impairment of "Cognitive Gamma Suppression" in PD

Cognitive processes require the interaction between distributed neuronal groups. The so called "binding hypothesis" (for a review, see Reference 39) essentially assumes that different brain areas have to be "bound" together within very short time intervals so as to solve perceptual-cognitive tasks, probably by synchronized or desynchronized activities of neuronal assemblies. The frequency range between 20 and 60 Hz is known as "gamma-band" activity. This rhythm exists spontaneously and/or can be evoked, induced, or emitted in different structures of the CNS in response to olfactory, auditory, somatosensory, and visual stimuli or in concomitance with attentional/perceptual-cognitive processes. In normal observers, gamma has been shown to accompany primary visual evoked responses and being suppressed during the P300.[25,26] Generally, this cortical suppression is thought to reflect competitive hippocampal gamma activation associated with P300 target processing,[13] and therefore hippocampal gamma activation may be due to short-term memory updating. However, this suppression does not exist in PD.[24] In PD patients the lack of "cognitive" gamma suppression may reflect visuocognitive processing deficits during the performance of the task.[24]

Cortico-cortical frequency coherence can be modified by L-dopa therapy in PD.[30] Using a simple visual tracking task, a coherence increase was found after levodopa therapy whereas, without levodopa, the coherence was much reduced when compared to age-matched normals. It appears that ascending dopaminergic projections from the mesencephalon may modulate the pattern and extent of cortico-cortical coupling in visuomotor tasks. Additionally, it seems that time-frequency analysis of visual ERPs might contribute to differentiate patients with and without hippocampal dysfunction or, more generally, it could help us to better understand of binding of different cortical areas in dysfunctional cognitive processing in PD.

Does P300 Abnormality Reflect Working Memory Impairment in PD?

Working memory (WM) refers to the short-term, attention-demanding maintenance and manipulation of information for purposeful actions.[9] WM is closely related to the notions of stimulus-representation matching and decision making. In the previously mentioned experiments, in which classic odd-ball paradigm has been used to elicit the P300 component, a target stimulus has to be stored in the active memory to compare that with subsequently presented stimuli for a same-different decision making. In addition, cortical areas identified as generators of P300 (dorsolateral prefrontal and parietal cortices) have also roles in WM processes.[51] One part of the WM system, the visuospatial sketchpad that related to the maintaining of visual information[9] shows a specific selective impairment in PD: while the visual subsystem responsible for the object-related visual analysis seems to be spared until the later stages of the disease, the visual processing of spatial location, motion, and three-dimensional properties is impaired.[66,72,77,84] In a delayed-response test, PD patients with mild symptoms were unable to briefly maintain the memory trace of spatial locations of irregular polygons, whereas they successfully kept on-line the shapes of the same stimuli.[84] PD patients also make significantly more errors in mental rotation of three-dimensional wire-frame figures.[72] Wang et al.[108] have combined the oddball paradigm with a delayed-response test (S1-S2 paradigm). In this procedure, first a simple geometric design is presented (S1), followed by another (S2) stimulus, which can be the same as S1. P300 is recorded only for S2 stimuli. It was shown that, when the time interval between S1 and S2 increases, nondemented PD patients show particular deficits, suggesting impaired working memory for visual shapes.

Visual Categorization Impairment in PD

Categorization, the evocative organization of our surrounding environment, plays a crucial role in our everyday life. Many neuroimaging and electrophysiological studies provided evidence for a discrete categorical organization of the human brain. In particular, there are specific representations of different categories in the occipito-temporal cortex and surrounding areas, such as faces in the occipito-temporal and the fusiform face area,[2] human body representation in the lateral occipito-temporal cortex,[38] animals

in the right fusiform cortex,[58] buildings in the right lingual sulcus,[3] man-made tools in the left posterior middle temporal cortex,[71] and plants in the right lateral occipital cortex.[58] Although previous studies have suggested that the visual subsystem responsible for the object-related visual analysis seems to be spared until the later stages of PD,[66,72,77,84] recent electrophysiological studies have found that it is not always the case.[4,5]

Categorization does not occur with the same latency as visual recognition. It was suggested that the basic visual feature encoding and initiating stages of perceptual categorization take place in the first 200 ms post-stimulus period, whereas conceptual and semantic properties are represented in later stages of information processing.[55,90] Thorpe and his associates found that non-animal scenes elicited more negative responses than images with animals, even at 150 ms following stimulus onset (N1)[105,107] In spite of relatively preserved P100, this difference was not observable in PD patients[4,5] (see Figure 23.5). The latter authors have hypothesized that the neostriatum may mediate feature weighting and extraction processes and the differential N1 may refer to this function. Consistent with this theory, multi-unit recordings from the basal ganglia of human volunteers revealed different neuronal responses when the subjects were asked to focus on distinct stimulus features. (For review, see Reference 61.) This suggests an attention-biased stimulus processing in the striatum, mediating the weighting and selecting of behaviorally relevant attributes. In PD, this weighting and selecting process is possibly dysfunctional, as reflected by the diminished differential N1.

Visuomotor Interaction in PD: The Possible Role of Impaired Saccades

Various, in particular saccadic, eye movement abnormalities have been described in PD.[56] Saccadic eye movements bring a new target into center of regard. Synchronized gamma range EEG rhythms of the occipital cortex have been recorded accompanying saccades in healthy subjects,[26] whereas, in nondemented PD patients, a desynchronization of the gamma rhythm was observed.[24] It is not clear whether this desynchronization occurs as a result of intrinsic visual cortical discrepancy or deficient subcortical input to the cortex.

CONCLUSIONS

In the last two decades, many specific and nonspecific visual abnormalities in patients with PD were found, such as abnormal PERG tuning,[81,94,100] longer latencies in visual evoked potentials,[15] and reduced CS mainly in the medium (2 to 5 cpd) spatial frequencies and 4 to 8 Hz temporal frequency range.[19,21,67,68] Improvement of these abnormalities by L-dopa therapy and the animal

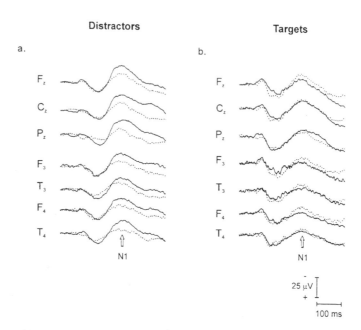

FIGURE 23.5 Grand averaged ERPs for non-animal/distractor and for animal/target stimuli in the control group (continuous line) and in the PD group (dotted line). Note that, while there is an amplitude difference of N1 component for distractors, there is no N1 difference for targets. (*Source:* Antal et al.,[4] with permission of *Cogn. Brain. Res.*)

models of this disorder have established a link between the visual symptoms observed in PD and dopaminergic dysfunction. Beyond these results, recent literature on the electrophysiology, neuropsychology, and functional neuroimaging of PD also suggest that dopaminergic dysregulation of the cortico-subcortical system in PD patients may lead to higher-level visuocognitive dysfunctions. However, the picture is far from complete. The experimental results are often controversial, probably due to the heterogeneity of patients, the different paradigms, and designs of the studies. Additionally, evidence suggests that L-dopa substitution ameliorates these visuocognitive symptoms only in the initial phase of the disease. As the disease advances, dopaminergic treatment seems to be less effective, possibly because of the progressive loss of dopaminergic neurons and because of nondopaminergic deficiencies (noradrenergic, serotonergic, and cholinergic deficits and cortical Lewy bodies) in PD.[77,26] Second, based on the available evidence, it is unlikely to find single anatomical loci to be responsible for complex visual deficits. In the last decades, it became obvious that, in visuocognitive processes, distributed parallel pathways, brain areas, and neuronal assemblies are involved according to well organized time plans. For instance, in attempting to explore the visual world, a sequence of visuospatial attentional shift and saccadic eye movements have been evidenced in psychophysical,[54] monkey,[104] and human fMRI[14] studies. In PD, in addition to (or as a consequence of) distal loss of stimulus strength,[8] the temporal time keeping may be defective in the visual system.[67] It is a challenging thought

that, in progressive disorders such as PD, once a critical number of neurons is lost, not only sensitivity to stimuli but also the synchrony of distributed neuronal groups lose the precision needed for task-related cooperation. In summary, it is plausible that, in PD, as a consequence of anatomical dopaminergic lesions, a significant part of visuocognitive impairments reflects temporal dysfunction of distributed neuronal assemblies among the visual, parietal, and frontal areas and the basal ganglia.

REFERENCES

1. Albrecht, D., Quaschling, U., Zippel, U., Davidowa, H., Effects of dopamine on neurons of the lateral geniculate nucleus: an iontophoretic study, *Synapse*, 23, 70–78, 1996.
2. Allison, T., Puce, A., Spencer, D. D., McCarthy, G., Electrophysiological studies of human face perception. I: Potentials generated in occipitotemporal cortex by face and non-face stimuli, *Cereb Cortex*, 9, 415–30, 1999.
3. Aguirre, G. K., Zarahn, E., D'Esposito, M., An area within human ventral cortex sensitive to "building" stimuli: evidence and implications, *Neuron*, 21, 373–83, 1998.
4. Antal, A., Keri, S., Dibo, G., Benedek, G., Janka, Z., Vecsei, L., Bodis-Wollner, I., Electrophysiological correlates of visual categorization: evidence for cognitive dysfunctions in early Parkinson's disease, *Cogn. Brain Res.*, 13, 153–158, 2002a.
5. Antal, A., Keri, S., Kincses, T., Kalman, J., Dibo, G., Benedek, G., Janka, Z., Vecsei, L., Corticostriatal cir-

cuitry mediates fast-track visual categorization, *Cogn. Brain Res.,* 13, 53–59, 2002b.

6. Antal, A., Keri, S., Bodis-Wollner, I., Dopamine D_2 receptor blockade alters the primary and cognitive components of visual evoked potentials in the monkey, Macaca fascicularis, *Neurosci. Lett.,* 232, 179–181, 1997.

7. Antal, A., Pfeiffer, R., Bodis-Wollner, I., Simultaneously evoked primary and cognitive visual evoked potentials distinguish younger and older patients with Parkinson's disease, *J. Neural. Transm.,* 103, 1053–1067, 1996.

8. Amick, A. M., Cronin-Golomb, A. and Gilmore, G. C., Visual processing of rapidly presented stimuli is normalized in Parkinson's disease when proximal stimulus strength is enhanced, *Vis. Res.,* 43, 2827–2835, 2003.

9. Baddeley, A., Recent developments in working memory, *Curr. Opin. Neurobiol.,* 8, 234–238, 1998.

10. Bandini, F., Pierantozzi, M., Bodis-Wollner, I., Parkinson's disease changes the balance of onset and offset visual responses: an evoked potential study, *Clin. Neurophysiol.,* 112, 976–983, 2001.

11. Bandini, F., Pierantozzi, M., Bodis-Wollner, I., The visuo-cognitive and motor effect of amantadine in non-Caucasian patients with Parkinson's disease. A clinical and electrophysiological study, *J. Neural. Transm.,* 109, 41–51, 2002.

12. Bartel, P., Bloom, M., Robinson, E., Van der Meyden, C., Soomers, D. K., Becker, P., Effects of chlorpromazine on pattern and flash ERG and VEPs compared to oxazepam and to placebo in normal subjects, *Electroencephalogr. Clin. Neurophysiol.,* 77, 330–339, 1990.

13. Basar-Eroglu, C., Basar, E., Endogenous components of event-related potentials in hippocampus: an analysis with freely moving cats, *Electroencephalogr. Clin. Neurophysiol.,* Suppl., 40, 440–444, 1987.

14. Berman, R. A., Colby, C. L., Genovese, C. R., Voyvodic, J. T., Luna, B., Thulborn, K. R., Sweeney, J. A., Cortical networks subserving pursuit and saccadic eye movements in humans: an FMRI study, *Hum. Brain Mapp.,* 8, 209–225, 1999.

15. Bodis-Wollner, I., Yahr, M. D., Measurements of visual evoked potentials in Parkinson's disease, *Brain,* 101, 661–671, 1978.

16. Bodis-Wollner, I., Yahr, M. D., Thornton, J., Visual evoked potentials and the severity of Parkinson's disease. In *Research Progress in Parkinson's Disease,* Rose, C. and Capildeo, R., Eds., Tunbridge Wells, Pitman Medical, pp. 126–137, 1981.

17. Bodis-Wollner, I., Yahr, M. D., Mylin, L. H., Non-motor functions of the basal ganglia. In *Advances in Neurology,* Vol. 40, Parkinson-Specific Motor and Mental Disorders, Hassler, R. G. and Christ, J. F., Eds., Raven Press, New York, p.p. 289–298, 1983.

18. Bodis-Wollner, I., Pattern evoked potential changes in Parkinson's disease are stimulus-dependent, *Neurology,* 35, 1675–1676, 1985.

19. Bodis-Wollner, I., Marx, M. S., Mitra, S., Bobak, P., Mylin, L., Yahr, M., Visual dysfunction in Parkinson's disease, Loss in spatiotemporal contrast sensitivity, *Brain,* 110, 1675–1698, 1987.

20. Bodis-Wollner, I., Piccolino, M., *Dopaminergic mechanisms in vision,* A. R. Liss, New York, 1988.

21. Bodis-Wollner, I., Visual deficits related to dopamine deficiency in experimental animals and Parkinson's disease patients, *Trends Neurosci.,* 13, 296–302, 1990.

22. Bodis-Wollner, I., Tagliati, M., The visual system in Parkinson's disease, *Adv. Neurol.,* 60, 390–394, 1993.

23. Bodis-Wollner, I., Borod, J. C., Cicero, B., Haywood, C. S., Raskin, S., Mylin, L., Sliwinski, M., Falk, A. and Yahr, M. D., Modality dependent changes in event related potentials correlate with specific cognitive function in non-demented patients with Parkinson's disease, *J. Neural. Transm.,* 9, 197–209, 1995.

24. Bodis-Wollner, I., Tzelepi, A., The push-pull action of dopamine on spatial tuning of the monkey retina: the effects of dopaminergic deficiency and selective D_1 and D_2 receptor ligands on the pattern electroretinogram, *Vis. Res.,* 38, 1479–1487, 1998.

25. Bodis-Wollner, I., Tzelepi, A., Sagliocco, L., Bandini, F., Mari, Z., Pierantozzi, A., Bezerianos, A., Ogliastro, E. C., Kim, J., Ko, Chr., and Gulzar, J., Visual Processing Deficit in Parkinson Disease, Y. Koga, K. Nagata and K Hirata, Eds., In *Brain Topography Today,* 606–611. Elsevier Science, B.V., 1998.

26. Bodis-Wollner, I., Tzelepi, A., Push-pull model of dopamine's action in the retina. In *Models of the Visual System,* Hung GK, Ciuffreda KC., Eds., Kluver Academic Publishers, p.p 191–214, 2002.

27. Bodis-Wollner, I., Von Gizycki, H., Avitable, M., Hussain, Z., Javeid, A., Habib, A., Raza, A., Sabet, M., Perisaccadic occipital EEG changes quantified with wavelet analysis, *Ann. N. Y. Acad. Sci.,* 956,464–467, 2002.

28. Bohnen, N. I., Minoshima, S., Giordani, B., Frey, K. A., Kuhl, D. E., Motor correlates of occipital glucose hypometabolism in Parkinson's disease without dementia, *Neurology,* 52, 541–546, 1999.

29. Bulens, C., Meerwaldt, J. D., van der Wildt, G. J., Effect of stimulus orientation on contrast sensitivity in Parkinson's disease, *Neurology,* 38, 76–81, 1988.

30. Cassidy, M., Brown, P., Task-related EEG-EEG coherence depends on dopaminergic activity in Parkinson's disease, *Neuroreport,* 12, 703–707, 2001.

31. Cools, R., Stefanova, E., Barker, R. A., Robbins, T. W., Owen, A. M., Dopaminergic modulation of high-level cognition in Parkinson's disease: the role of the prefrontal cortex revealed by PET, *Brain,* 125, 584–594, 2002.

32. Coull, J. T., Neural correlates of attention and arousal: insights from electrophysiology, functional neuroimaging and psychopharmacology, *Prog. Neurobiol.,* 55, 343–361, 1998.

33. Delalande, I., Hache, J. C., Forzy, G., Bughin, M., Benhadjali, J., Destee, A., Do visual-evoked potentials and spatiotemporal contrast sensitivity help to distinguish idiopathic Parkinson's disease and multiple system atrophy? *Mov. Disord.,* 13, 446–452, 1998.

34. Diederich, N. J., Raman, R., Leurgans, S., Goetz, C. G., Progressive worsening of spatial and chromatic processing deficits in Parkinson disease, *Arch. Neurol.,* 59, 1249–1252, 2002.

35. Dierks, T., Frolich, L., Ihi, R., Maurer, K., Event-related potentials and psychopharmacology, Cholinergic modulation of P300, *Pharmacopsychiatry*, 27, 72–74, 1994.

36. Djamgoz, M. B., Wagner, H. J., Localization and function of dopamine in the adult vertebrate retina, *Neurochem. Int.*, 20, 139–191, 1992.

37. Dowling, J. E., Functional and pharmacological organization of the retina: dopamine, interplexiform cells, and neuromodulation, *Res. Publ. Assoc. Res. Nerv. Ment. Dis.*, 67, 1–18, 1990.

38. Downing, P. E., Jiang, Y., Shuman, M., Kanwisher, N., A cortical area selective for visual processing of the human body, *Science*, 293, 2470–3, 2001.

39. Engel, A. K., Singer, W., Temporal binding and the neural correlates of sensory awareness, *Trends Cogn. Sci.*, 5, 16–25, 2001.

40. Frodl-Bauch, T., Bottlender, R., Hegerl, U., Neurochemical substrates and neuroanatomical generators of event-related P300, *Neuropsychobiol.*, 40, 86–94, 1999.

41. Gabrieli, J. D., Memory systems analyses of mnemonic disorders in aging and age-related diseases, *Proc. Natl. Acad. Sci., U.S.A.*, 93, 13534–13540, 1996.

42. Ghilardi, M. F., Chung, E., Bodis-Wollner, I., Dvorzniak, M., Glover, A., Onofrj, M., Systemic 1-methyl,4-phenyl,1-2-3-6-tetrahydropyridine (MPTP) administration decreases retinal dopamine content in primates, *Life Sci.*, 43, 255–262, 1988.

43. Ghilardi, M. F., Marx, M. S., Bodis-Wollner, I., Camras, C. B., Glover, A. A., The effect of intraocular 6-hydroxydopamine on retinal processing of primates, *Ann. Neurol.*, 25, 357–364, 1989.

44. Glover, A., Ghilardi, M. F., Bodis-Wollner, I., Onofrj, M., Alterations in event-related potentials (ERPs) of MPTP-treated monkeys, *Electroencephalogr. Clin. Neurophysiol.*, 71, 461–468, 1988.

45. Glover, A., Ghilardi, M. F., Bodis-Wollner, I., Onofrj, M., Mylin, L. H., Visual "cognitive" evoked potentials in the behaving monkey, *Electroencephalogr. Clin. Neurophysiol.*, 80, 65–72, 1991.

46. Goldman-Rakic, P. S., Lidow, M. S., Smiley, J. F., Williams, M. S., The anatomy of dopamine in monkey and human prefrontal cortex, *J. Neural. Transm.*, Suppl., 36, 163–177, 1992.

47. Goldman-Rakic, P., The cortical dopamine system: role in memory and cognition, *Adv. Pharmacol.*, 42, 707–711, 1998.

48. Goodin, D. S., Aminoff, L. M., Electrophysiological differences between demented and nondemented patients with Parkinson's disease, *Ann. Neurol.*, 21, 90–94, 1987.

49. Gottlob, I., Schneider, E., Heider, W., Skrandies, W., Alteration of visual evoked potentials and electroretinograms in Parkinson's disease, *Electroencephalogr. Clin. Neurophysiol.*, 66, 349–357, 1987.

50. Grossman, M., Zurif, E., Lee, C., Prather, P., Kalmanson, J., Stern, M. B., Hurtig, H. I., Information processing speed and sentence comprehension in Parkinson's disease, *Neuropsychology*, 16, 174–181, 2002.

51. Halgren, E., Marinkovic, K., Chauvel, P., Generators of the late cognitive potentials in auditory and visual odd-ball tasks, *Electroencephalogr, Clin, Neurophysiol.*, 106, 156–164, 1998.

52. Hansch, E. C., Syndulko, K., Cohen, S. N., Goldberg, Z. I., Potvin, A. R., Tourtellotte, W. W., Cognition in Parkinson disease: an event-related potential perspective, *Ann. Neurol.*, 11, 599–607, 1982.

53. Harris, J. P., Calvert, J. E., Leendertz, J. A., Phillipson, O. T., The influence of dopamine on spatial vision, *Eye*, 4, 806–812, 1990.

54. He, P., Kowler, E., The role of saccades in the perception of texture patterns, *Vision Res.*, 32, 2151–2163, 1992.

55. Hillyard, S. A., Teder-Salejarvi, W. A., Münte, T. F., Temporal dynamics of early perceptual processing, *Curr. Opin. Neurobiol.*, 8, 202–210, 1998.

56. Hodgson, T. L., Dittrich, W. H., Henderson, L., Kennard, C., Eye movements and spatial working memory in Parkinson's disease, *Neuropsychologia*, 37, 927–938, 1999.

57. Ikeda, H., Head, G. M., Ellis, C. J., Electrophysiological signs of retinal dopamine deficiency in recently diagnosed Parkinson's disease, *Vision Res.*, 34, 2629–2638, 1994.

58. Kawashima, R., Hatano, G., Oizumi, K., Sugiura, M., Fukuda, H., Itoh, K., Kato, T., Nakamura, A., Hatano, K., Kojima, S., Different neural systems for recognizing plants, animals, and artifacts, *Brain Res. Bull.*, 54, 313–317, 2001.

59. Kebabian, J. W., Calne, D. B., Multiple receptors for dopamine, *Nature*, 277, 93–96, 1979.

60. Kotchoubey, B., Lang, S., Parallel processing of physical and lexical auditory information in humans, *Neurosci. Res.*, 45, 369–374, 2003.

61. Kropotov, J. D., Etlinger, S. C., Selection and actions in the basal ganglia-thalamocortical circuits: review and model, *Int. J. Psychophysiol.*, 31, 197–217, 1999.

62. Kupersmith, M. J., Shakin, E., Siegel, I. M., Lieberman, A., Visual system abnormalities in patients with Parkinson's disease, *Arch. Neurol.*, 39, 284–286, 1982.

63. Kutas, M., McCarthy, G., Donchin, E., Augmenting mental chronometry: the P300 as a measure of stimulus evaluation time, *Science*, 197, 792–795, 1977.

64. Langston, J. W., Ballard, P., Tetrud, J. W., Irwin, I., Chronic Parkinsonism in humans due to a product of meperidine-analog synthesis, *Science*, 219, 979–980, 1983.

65. LeBras, C., Pillon, B., Damier, P., Dubois, B., At which steps of spatial working memory processing do stratiofrontal circuits intervene in humans? *Neuropsychologia*, 37, 83–90, 1999.

66. Lee, A. C., Harris, J. P., Calvert, J. E., Impairments of mental rotation in Parkinson's disease, *Neuropsychol.*, 36, 109–114, 1998.

67. Marx, M., Bodis-Wollner, I., Bobak, P., Harnois, C., Mylin, M., Temporal frequency-dependent VEP changes in Parkinson's disease, *Vision Res.*, 26, 185–193, 1986.

68. Masson, G., Mestre, D., Blin, O., Dopaminergic modulation of visual sensitivity in man, *Fundam. Clin. Pharmacol.*, 7, 449–463, 1993.

69. Mattay, V. S., Tessitore, A., Callicott, J. H., Bertolino, A., Goldberg, T. E., Chase, T. N., Hyde, T. M., Weinberger, D. R., Dopaminergic modulation of cortical

function in patients with Parkinson's disease, *Ann. Neurol.*, 51, 156–164, 2002.

70. Mestre, D., Blin, O., Serratrice, G., Pailhous, J., Spatiotemporal contrast sensitivity differs in normal aging and Parkinson's disease, *Neurology,* 40, 1710–1714, 1990.

71. Moore, H., West, A. R., Grace, A. A., The regulation of forebrain dopamine transmission: relevance to the pathophysiology and psychopathology of schizophrenia, *Biol. Psychiatry,* 46, 40–55, 1999.

72. Moreaud, O., Fournet, N., Roulin, J., Naegele, B., Pellat, J., The phonological loop in medicated patients with Parkinson's disease: a presence of phonological similarity and word length effects, *J. Neurol. Neurosurg. Psych.,* 62, 609–611, 1997.

73. Munk, M. H. J., Roelfsma, P. R., Koenig, P., Engel, A. K. and Singer, W., Role of reticular activation in the modulation of intracortical synchronization, *Science,* 272, 271–274, 1996.

74. Nieoullon, A., Dopamine and the regulation of cognition and attention, *Prog. Neurobiol.,* 67, 53–83, 2002.

75. Okuda, B., Tachibana, H., Kawabata, K., Takeda, M., Sugita, M., Visual evoked potentials (VEPs) in Parkinson's disease: correlation of pattern VEPs abnormality with dementia, *Alzheimer Dis. Assoc. Disord.,* 9, 68–72, 1995.

76. Owen, A. M., Downes, J. J., Sahakian, B. J., Polkey, C. E., Robbins, T. W., Planning and spatial working memory following frontal lobe lesions in man, *Neuropsychologia,* 28, 1021–1034, 1990.

77. Owen, A. M., Iddon, J. L., Hodges, J. R., Summers, B.A., Robbins, T. W., Spatial and non-spatial working memory at different stages of Parkinson's disease, *Neuropsychologia,* 35, 519–532, 1997.

78. Owen, A. M., James, M., Leigh, P. N., Summers, B. A., Marsden, C. D., Quinn, N. P., Lange, K. W., Robbins, T. W., Fronto-striatal cognitive deficits at different stages of Parkinson's disease, *Brain,* 115, 1727–1751, 1992.

79. Parkinson, D., Evidence for a dopaminergic innervation of cat primary visual cortex, *Neuroscience,* 30, 171–179, 1989.

80. Peppe, A., Stanzione, P., Pierelli, F., De Angelis, D., Pierantozzi, M., Bernardi, G., Visual alterations in de novo Parkinson's disease: pattern electroretinogram latencies are more delayed and more reversible by levodopa than are visual evoked potentials, *Neurology,* 45, 1144–1148, 1995.

81. Peppe, A., Stanzione, P., Pierelli, F., Stefano, E., Rizzo, P. A., Tagliati, M., Morocutti, C., Low contrast stimuli enhance PERG sensitivity to the visual dysfunction in Parkinson's disease, *Electroencephalogr. Clin. Neurophysiol.,* 82, 453–457, 1992.

82. Phillipson, O. T., Kilpatrick, I. C., and Jones, M. W., Dopaminergic innervation of the primary visual cortex in the rat, and some correlations with the human cortex, *Brain Res. Bull.,* 18, 621–633, 1987.

83. Pillon, B., Dubois, B., Cusimano, G., Bonnet, A. M., Lhermitte, F., Agid, Y., Does cognitive impairment in Parkinson's disease result from non-dopaminergic lesions? *J. Neurol. Neurosurg. Psychiatry,* 52, 201–206, 1998.

84. Postle, B. R., Jonides, J., Smith, E. E., Corkin, S., Growdon, J. H., Spatial, but not object, delayed response is impaired in early Parkinson's disease, *Neuropsychology,* 11, 171–179, 1997.

85. Prasher, D., Findley, L., Dopaminergic induced changes in cognitive and motor processing in Parkinson's disease: an electrophysiological investigation, *J. Neurol. Neurosurg. Psychiatry,* 54, 603–609, 1991.

86. Rakic, P. and Lidow, M., Distribution and density of monoamine receptors in the primate visual cortex devoid of retinal input from early embryonic stages, *J. Neurosci.,* 15, 2561–2574, 1995.

87. Reader, T. A., Quesney, L. F., Dopamine in the visual cortex of the cat, *Experientia,* 42, 1242–1244, 1986.

88. Regan, D., Maxner, C., Orientation-selective visual loss in patients with Parkinson's disease, *Brain,* 110, 415–432, 1987.

89. Sagliocco, L., Bandini, F., Pierantozzi, M., Mari, Z., Tzelepi, A., Ko, C., Gulzar, J., Bodis-Wollner, I., Electrophysiological evidence for visuocognitive dysfunction in younger non Caucasian patients with Parkinson's disease, *J. Neural. Transm.,* 104, 427–439, 1997.

90. Schendan, H. E., Ganis, G., Kutas, M., Neurophysiological evidence for visual perceptual categorizationg of words and faces within 150 ms, *Psychophysiol,* 35, 240–251, 1998.

91. Schiller, P. H., The ON and OFF channels of the visual system, *Trends Neurosci.,* 15, 86–92, 1992.

92. Sohn, Y. H., Kim, G. W., Huh, K., Kim, J. S., Dopaminergic influences on the P300 abnormality in Parkinson's disease, *J. Neurol. Sci.,* 158, 83–87, 1998.

93. Stanzione, P., Bodis-Wollner, I., Pierantozzi, M., Semprini, R., Tagliati, M., Peppe, A., Bernardi, G., A mixed D_1 and D_2 antagonist does not replay pattern electroretinogram alterations observed with a selective D_2 antagonist in normal humans: relationship with Parkinson's disease pattern electroretinogram alterations, *Clin. Neurophysiol.,* 110, 82–85, 1999.

94. Stanzione, P., Fattapposta, F., Giunti, P., D'Alessio, C., Tagliati, M., Affricano, C., Amabile, G., P300 variations in parkinsonian patients before and during dopaminergic monotherapy: a suggested dopamine component in P300, *Electroencephalogr. Clin. Neurophysiol.,* 80, 446–453, 1991.

95. Stanzione, P., Pierelli, F., Peppe, A., Stefano, E., Rizzo, P. A., Morocutti, C., and Bernardi, G., Pattern visual evoked potential abnormalities in Parkinson's disease: effects of L-dopa therapy, *Clin. Vis. Sci.,* 4, 115–127, 1989.

96. Sutton, S., Braren, M., Zubin, J., John, E. R., Evoked potentials correlate of stimulus uncertainty, *Science,* 150, 1187–1188, 1965.

97. Tachibana, H., Aragane, K., Miyata, Y., Sugita, M., Electrophysiological analysis of cognitive slowing in Parkinson's disease, *J. Neurol. Sci.,* 149, 47–56, 1997.

98. Tachibana, H., Toda, L., Sugita, M., Actively and passively evoked P3 latency of event-related potentials in Parkinson's disease, *J. Neurol. Sci.,* 111, 134–142, 1992.

99. Tagliati, M., Bodis-Wollner, I., Kovanecz, I., Stanzione, P., Spatial frequency tuning of the monkey pattern ERG depends on D_2 receptor-linked action of dopamine, *Vision Res.,* 34, 2051–2057, 1994.

100. Tagliati, M., Bodis-Wollner, I., Yahr, M., The pattern electroretinogram in Parkinson's disease reveals lack of retinal spatial tuning, *Electroencephalogr, Clin, Neurophysiol,* 100, 1–11, 1996.

101. Tales, A., Newton, P., Troscianko, T., Butler, S., Mismatch negativity in the visual modality, *Neuroreport,* 10, 3363–3367, 1999.

102. Tartaglione, A., Pizio, N., Bino, G., Spadavecchia, L., Favale, E., VEP changes in Parkinson's disease are stimulus dependent, *J. Neurol. Neurosurg. Psychiatry,* 47, 305–307, 1984.

103. Tebartz, van Elst, L., Greenlee, M. W., Foley, J. M. and Lucking, C. H., Contrast detection, discrimination and adaptation in patients with Parkinson's disease and multiple system atrophy, *Brain,* 120, 2219–2228, 1997.

104. Tehovnik, E. J., Lee, K., Schiller, P. H., Stimulation-evoked saccades from the dorsomedial frontal cortex of the rhesus monkey following lesions of the frontal eye fields and superior colliculus, *Exp. Brain Res.,* 98, 179–190, 1994.

105. Thorpe, S., Fize, D., Marlot, C., Speed of processing in the human visual system, *Nature,* 381, 520–522, 1996.

106. Toda, K., Tachibana, H., Sugita, M., Konishi, K., P300 and reaction time in Parkinson's disease, *J. Geriatr. Psychiatry Neurol.,* 6, 131–136, 1993.

107. VanRullen, R., Thorpe, S. J., The time course of visual processing: from early perception to decision-making, *J. Cogn. Neurosci.,* 13, 454–461, 2001.

108. Wang, L., Kuroiwa, Y., Kamitani, T., Takahashi, T., Suzuki, Y., Hasegawa, O., Effect of interstimulus interval on visual P300 in Parkinson's disease, *J. Neurol. Neurosurg. Psych.,* 67, 497–503, 1999.

109. Wang, L., Kuroiwa, Y., Li, M., Kamitani, T., Wang, J., Takahashi, T., Suzuki, Y., Ikegami, T., Matsubara, S., The correlation between P300 alterations and regional cerebral blood flow in non-demented Parkinson's disease, *Neurosci. Lett.,* 282, 133–136, 2000.

110. Wichmann, T., DeLong, M. R., Functional neuroanatomy of the basal ganglia in Parkinson's disease, *Adv. Neurol.,* 91, 9–18, 2003.

24 Olfactory Dysfunction in Parkinson's Disease and Parkinsonian Syndromes

Katerina Markopoulou
Department of Neurological Sciences, University of Nebraska Medical Center

CONTENTS

ABSTRACT

The olfactory system is a complex network whose organization and function depends on both peripheral and central input. It is commonly affected in Parkinson's disease (PD), parkinsonism-plus syndromes (PPS), other neurodegenerative disorders [e.g., Alzheimer's disease, (AD)], and in normal aging. Olfactory dysfunction usually appears early in the disease process. In PD, olfactory function is commonly impaired whereas, in PPS, olfactory function is only mildly impaired or preserved. Olfactory function is also impaired in familial forms of parkinsonism associated with a monogenic defect. In contrast to individuals with sporadic PPS, affected members of PPS kindreds do show olfactory impairment. Interestingly, olfactory dysfunction does not appear to be due to a dopamine deficiency. The neuropathological changes in the olfactory system appear to be disease specific. This suggests that olfactory dysfunction in neurodegenerative disorders may reflect a central rather than a peripheral process. The organization of the normal olfactory system is gradually being elucidated at the molecular, cellular, and system levels. The mechanisms underlying olfactory dysfunction in PD and other neurodegenerative diseases remain unknown.

OLFACTORY SYSTEM STRUCTURE AND ORGANIZATION

The olfactory system is composed of the olfactory epithelium, the olfactory nerves, the olfactory bulbs, the olfactory tracts, and the median and lateral olfactory striae that terminate in the contralateral hemisphere or the ipsilateral amygdaloid nucleus, septal nuclei, and hypothalamus. The olfactory epithelium is located on the superior-posterior aspect of the nasal septum and the lateral walls of the nasal cavity and contains the olfactory sensory neurons (OSNs). The OSNs are generated *in situ* from stem cells. Aging OSNs are replaced by cell division that persists into adulthood and throughout the adult life. The life span of an OSN is in the range of weeks to months. The OSNs are bipolar neurons, the axons of which form the olfactory nerves and pass through the cribiform plate and terminate in the olfactory bulb, where they synapse with second-order neurons and interneurons. In the olfactory bulb, the OSN axon terminates in a glomerulus, which is a globoid neural structure. The size of the glomerulus differs in different mammalian species. The axons of the second-order neurons form the olfactory tracts located in the orbital surfaces of the frontal lobes. As it courses centrally, the olfactory tract becomes divided into the median and lateral olfactory striae. In an organization analogous to that of the visual pathways, median stria fibers decussate through the anterior commisure, join fibers from the opposite olfactory tract, and terminate in the contralateral hemisphere, while lateral striae fibers reach the primary olfactory cortex (piriform cortex) and terminate in the ipsilateral amygdaloid nucleus, septal nuclei, and hypothalamus.[1]

In humans, odor detection of airborne odorants appears to be very efficient, but odor discrimination is considerably less efficient. Olfactory perception is initiated by the activation of odorant receptors by odorous ligands. Airborne odorants stimulate the olfactory sensory neurons (OSNs), contained in the olfactory epithelium. It is thought that the functional heterogeneity of the OSNs is derived from a very large number of odorant receptors (OR) that are expressed in the OSNs. In the last decade, approximately 1000 odorant receptor (OR) genes have been identified in humans. These represent approximately 1% of the human genome. Interestingly, a large subset (almost two-thirds) of these genes appears to be nonfunctional; i.e., they are pseudogenes. The OR genes are distributed in clusters on all chromosomes except chromosome 20 and the Y chromosome. This clustering has been observed in many different species, including mice, rats, zebrafish and humans. There does not appear to be any particular pattern to the clustering of the OR genes, and they can often be intermixed with other gene families such as the T-cell receptor and beta-globin genes. The OR genes are intronless and have open reading frames (ORF) of approximately 1 kb. Based on amino acid similarity they have been categorized into families and subfamilies. The predicted amino acid sequence indicates the presence of seven transmembrane domains, a characteristic of G-protein coupled receptors. Each OSN expresses a single allele of a single OR gene[2] and therefore the olfactory epithelium consists of distinct OSN populations (reviewed in References 82 through 84).

How is the sensitivity of the OSN translated into the specificity of individual smell? The principles underlying this specificity are still a matter of debate, but some interesting patterns are emerging. Both peripheral and central mechanisms seem to play an important role. In the periphery, specificity appears to be generated both by the OSN expressing a single allele of a single OR gene and by the pattern of connections that the OSN forms. All neurons expressing a single OR gene project axons that synapse in one medial and one lateral glomerulus of the olfactory bulb, which represents the first relay station of the olfactory pathway. It appears that the OR plays a role of organizing the connectivity of the olfactory map.[3] The glomerulus containing the second order neurons appears to serve as an "odorant feature detector" via mechanisms involving lateral feedback inhibition and excitation and temporal synchrony. Interestingly, in rodents, voltage-sensitive dye imaging has revealed that there are differences in the response latency and the response timecourse across different glomeruli in the olfactory bulb. The pattern of activity at the level of the glomerulus evolves over time and depends also on the identity of the different glomeruli. Both the temporal and the spatial context of the odor-evoked response is critical. The temporal patterning may be imposed both by the odor carrier medium, the sampling activity, or by the inherent neural dynamics of the cells comprising the olfactory bulb.[4-5] At the system level, it has been proposed that all odors are initially encoded as "objects" in the piriform cortex and that odor perception depends on higher cognitive functions such as memory and neural plasticity.[6]

Odors have long been thought to be linked to emotional responses, yet an association at the anatomical level has only recently been clearly demonstrated. Studies of patients with focal brain injuries suggest that the caudal orbitofrontal and medial temporal cortices are involved in odor perception. Using event-related fMRI,[7] researchers were able to show that responses could be identified in the piriform cortex in a rostrocaudal axis. The amygdala was activated bilaterally by all odors, regardless of valence. In the posterior orbitofrontal cortex, pleasant odors segregated in the medial aspects, whereas unpleasant odors segregated in the lateral aspects. fMRI studies have shown that odors can activate the cerebellum in a concentration-dependent manner.[8] These studies provide direct evidence in humans of the heterogeneity of brain regions involved in odor processing and that there is coupling between olfaction, emotion, and higher cognitive processes. At the same time, this heterogeneity of brain regions involved in normal nervous system function indicates its vulnerability in disease states and neurodegenerative processes.

In summary, the analysis of the olfactory system at the molecular, cellular, and system levels has identified a rather complicated organization that implicates both peripheral and central components in the function of the olfactory system. The characterization of the olfactory deficit in neurodegenerative disorders—specifically in Parkinson's disease and parkinsonian syndromes—has the potential to provide significant insights into the function of the olfactory system. Not only should it provide insight into its function in the normal and disease states, it should provide insight into the interplay of different aspects of central nervous system function in neurodegeneration.

ASSESSMENT OF OLFACTORY FUNCTION

In humans, different methods have been developed to assess distinct aspects of olfactory function, such as odor identification, threshold detection, and odor recognition memory. A number of these methods have achieved widespread use both in the research and clinical domain.

A commonly used test is the University of Pennsylvania Smell Identification Test (UPSIT), developed by Doty et al.[9–10] Its widespread use is based on the ease of administration, the relatively short completion time and its high test-retest reliability. This test uses 40 odorants that are released by using a pencil to scratch the surface of a strip containing a microencapsulated odorant. The subject is asked to identify each odorant by choosing among four items in a multiple-choice fashion. A simplified version of UPSIT is the CC-SIT developed by Cain and Rubin.[11] In that test, odor identification is combined with threshold testing. Threshold testing is performed using plastic squeeze-bottles containing successive dilutions of n-butanol in water, using 4% n-butanol as the highest concentration. For odor identification, the subjects sniff eight glass bottles containing different odorants and choose in a multiple-choice fashion from a uniform list of 16 items. More recently, Hummel et al. developed a new test using a pen-like odor-dispensing device.[12] This test assesses odor threshold (by using n-butanol in a stepwise presentation), odor discrimination (16 pairs of odorants and triple-forced choice), and odor identification (by using 16 common odorants and a multiple forced-choice from among four verbally stated options per odorant). Other tests include the San Diego Odor Identification Test,[13] the Scandinavian Odor Identification test,[14] the Viennese Olfactory Test Battery[15] and Smell Threshold Test.[16]

Olfactory event-related potentials (OERP) have been recorded in control and affected individuals in response to randomized stimulation with different odorants and the OERP latencies have been determined in control and affected individuals.[17] Statistical reliability of the OERP was established by Thesen et al.,[18] and it was shown that reliability of OERPs is comparable to that of visual and auditory evoked potentials. The generators for the OERP waveforms are not known. The early waveform (P1) is thought to originate in the olfactory bulb and the late waveform (P3) in the olfactory cortex.[19]

To determine whether anatomical changes are associated with olfactory dysfunction, endoscopic techniques have been developed to obtain olfactory epithelium from human subjects under either general or topical anesthesia.[21,22] The tissue is examined by light and/or electron microscopy and histochemistry.[23] The usefulness of this procedure is limited, since it is invasive and may require general anesthesia. The olfactory epithelium and the anterior olfactory nucleus can also be obtained postmortem and examined histologically and histochemically.[24]

In summary, a number of methods are currently available to assess olfactory function in humans. The choice of method depends on the ease of administration and on the aspect of olfactory function that is being assessed.

OLFACTORY DYSFUNCTION IN NORMAL AGING

A number of studies have shown that olfactory function is affected by aging.[25–27] Olfactory impairment associated with normal aging involves odor identification, threshold detection,[26] and odor recognition memory.[28] Olfactory function declines after age 65 and is severely affected after age 80. Interestingly, in women, olfactory impairment appears later than in men.[25]

It is useful to consider olfactory dysfunction contextually in light of findings that the olfactory epithelium undergoes continuous regeneration throughout development and adult life. A number of endogenous and exogenous mechanisms are implicated in maintaining a balance between regeneration and degeneration. Recently, Wu et al.[29] have shown that signals from neurons within the olfactory epithelium have the ability to inhibit the generation of new neurons by neural progenitors. In more general terms, it appears that neural repair of the mature CNS may be inhibited by the cellular and molecular microenvironment.[30] It is not clear what the role of these mechanisms is in aging or neurodegenerative disease. Cumulative exposure to environmental toxins, chemicals, upper respiratory viral infections, or head injury could contribute to gradual olfactory impairment by interfering with the endogenous mechanism of regeneration.

OLFACTORY DYSFUNCTION IN PARKINSON'S DISEASE (PD)

Olfactory dysfunction has been clearly demonstrated in sporadic PD. Olfactory dysfunction in this disorder includes impairment in odor identification, threshold detection, and odor recognition memory.[31] It has been shown that olfactory dysfunction is present early in the disease process and appears to remain stable as the disease progresses.[32] Studies have attempted to correlate olfactory dysfunction with disease parameters such as disease stage, duration, subtype, cognitive dysfunction, and therapy. Interestingly, olfactory dysfunction appears to be independent of disease stage and disease duration.[32] In contrast, olfactory dysfunction appears to be dependent on disease subtype, suggesting that disease subtype confers the specificity of the olfactory impairment. In a study by Stern et al., olfactory function was assessed in different PD subtypes.[33] Olfactory function was more impaired in advanced PD (Hoehn and Yahr stage III or greater) than early PD (Hoehn and Yahr stage II or less for four or more years). Both postural instability-gait disorder (PIGD) predominant PD (defined as UPDRS mean tremor score/mean PIGD score <1.0) and tremor-predominant PD (defined as UPDRS mean tremor score/mean PIGD score >1.5) subtypes exhibited olfactory impairment, but the impairment was more severe in the PIGD form than in the tremor-predominant form of PD. It is conceivable that differences in the degree of olfactory impairment between the disease subtypes may reflect different pathophysiological processes in the two disease subtypes. The olfactory deficits associated with PD appear to be independent of the cognitive dysfunction associated with the disease.[34] Olfactory dysfunction in PD is bilateral and does not respond to antiparkinsonian therapy.[35]

Olfactory impairment in PD has been attributed to the pathological changes, including neuronal loss and the presence of Lewy bodies identified in the olfactory cortex[24] and the amygdala.[36] Interestingly, sniffing impairment appears to contribute to the olfactory impairment in PD.[8]

OLFACTORY DYSFUNCTION IN PARKINSONISM-PLUS SYNDROMES (PPSs)

Olfactory function has also been assessed in multiple system atrophy (MSA), Shy-Drager syndrome,[37–38] progressive supranuclear palsy (PSP),[38–39] and the parkinsonism-dementia complex of Guam.[40] Wenning et al.[38] compared olfactory dysfunction in a large series of patients with either PD, MSA, corticobasal degeneration (CBD), or PSP. They showed that impairment of olfactory function was significantly more pronounced in PD than in PPS. In particular, olfactory impairment was mild in MSA, whereas olfactory function was preserved in CBD and PSP. The findings from a study by Muller et al. (2002) appear to confirm this difference in olfactory impairment between sporadic PD and PPS.[41] Additional studies have also demonstrated normal olfactory function in PSP.[39] This consistent difference in olfactory function can therefore be used as an aid in the differential diagnosis of PD and PPS.

Olfactory dysfunction has also been reported in the ALS-parkinsonism-dementia complex of Guam (PDC).[42] All four forms of the syndrome (ALS, pure parkinsonism, pure dementia, and parkinsonism-dementia complex) show impairment of olfactory function. This suggests a common mechanism of olfactory impairment in the different forms of the syndrome. There are no significant differences in the degree of the olfactory impairment in PD and PDC, making it impossible to distinguish these two entities on the basis of olfactory impairment.[40]

In contrast to what is seen in the sporadic forms of PPS, olfactory function is impaired in familial forms of PPS. Affected members of PPS kindreds show impairment similar to that seen in kindreds with idiopathic PD (IPD) phenotypes. Markopoulou et al.[43] assessed olfactory function in several multigenerational kindreds with an IPD phenotype as well as in kindreds with a PPS phenotype. Olfactory dysfunction appears to be a component

of the clinical phenotype in kindreds types of kindreds. No statistically significant differences in the degree of olfactory impairment were observed between these two phenotypes. Thus, it appears that, in regard to olfactory function, sporadic and familial forms of PD and PPS behave differently.

Three different genes are associated with the forms of parkinsonism assessed by Markopoulou et al. One is the *α-synuclein* gene (Family H),[43] a second is the gene for the microtubule-associated protein tau (pallido-ponto-nigral degeneration, PPND Family),[44] and the third is an as-yet unidentified gene on chromosome 2p13.[45] The expression of α-synuclein, along with that of its congeners β- and γ-synuclein, has been assessed in the olfactory mucosa of patients with PD, Lewy body disease, MSA, AD, and healthy controls.[46] While the synucleins are differentially expressed in the olfactory epithelium, and α-synuclein is the most abundantly expressed protein, there is no significant difference between affected individuals and healthy controls. However, it is conceivable that α-synuclein may play a role in the regeneration of the olfactory epithelium. This hypothesis is supported by other studies in which α-synuclein has been implicated in neuronal survival.[47–48]

To summarize, the presence of olfactory dysfunction in familial forms of parkinsonism associated with a monogenic defect suggests that genetic factors either directly or indirectly underlie olfactory dysfunction.

OLFACTORY DYSFUNCTION IN ATYPICAL PARKINSONIAN SYNDROMES (PPSs)

Olfactory function has been assessed in other atypical parkinsonian syndromes such as MPTP-induced parkinsonism. In this entity, olfactory function is preserved.[49] Olfactory function is also preserved in two syndromes that may be associated with PD, essential tremor,[50–51] and idiopathic restless leg syndrome.[52] While in sporadic forms of these syndromes, they appear to behave as independent disorders; in familial forms of PD, PD and essential tremor phenotypes appear to be associated at the genetic level and possibly reflect differential expressivity of the same monogenic defect.[53]

OLFACTORY DYSFUNCTION IN OTHER NEURODEGENERATIVE DISEASES

Perhaps not surprisingly, olfactory function is impaired in other neurodegenerative diseases such as Alzheimer's disease (AD),[15,19,31] motor neuron disease (MND),[54–56] and Huntington's disease (HD).[57–60]

In AD, the olfactory impairment appears to occur early in the disease process.[61] Interestingly, the ApoE epsilon-4 allele, a known risk factor for AD, appears to correlate with cognitive impairment and odor identification decline.[62,63] A meta-analysis of studies of olfactory function in AD and PD[31] suggests that olfactory impairment is relatively uniform in these diseases. This is consistent with the phenotypic overlap observed in the clinical manifestations of AD and PD. However, interesting differences exist in the olfactory impairment between AD and PD. In both PD and AD, odor identification is impaired,[64] but AD patients showed a higher olfactory threshold and poorer odor memory performance. In AD, olfactory impairment also appears to be a function of disease duration,[64] whereas this is not the case in PD.[32] In AD, the olfactory bulb, AON, piriform cortex, amygdala, and hippocampus show neurofibrillary tangles and amyloid plaques.[65–67] In PD, there is neuronal loss and Lewy bodies (LB) in the AON. The LB, however, resemble more the cortical than the nigral LB.[24] In addition, there are specific changes in the amygdala of PD patients.[68]

In motor neuron disease, the reports are somewhat conflicting. Some studies report olfactory impairment,[54–55] while others do not.[56] This could reflect selection bias and heterogeneity in the patient cohorts included in those studies.

In HD, the olfactory deficit is found only in affected individuals and not in gene-positive asymptomatic individuals.[58] The olfactory deficit involves primarily impairment of olfactory detection and odor identification but not odor recognition memory. As in AD and PD, the olfactory deficit in HD appears early in the disease process.[59–60]

A list of the neurodegenerative diseases associated with olfactory impairment associated discussed in this chapter is presented in Table 24.1.

TABLE 24.1
Olfactory Function in Neurodegenerative Diseases

Disease	Olfactory Function
Parkinson's disease	Impaired
Lewy body disease	Impaired
Familial Parkinson's disease (both IPD and PPS phenotypes)	Impaired
Progressive supranuclear palsy	Preserved
Multiple system atrophy	Mildly impaired
Corticobasal ganglionic degeneration	Preserved
Parkinsonism–dementia of Guam	Impaired
MPTP–induced Parkinsonism	Preserved
Essential tremor	Mildly–moderately impaired
Alzheimer's disease	Impaired
Motor neuron disease	Impaired/preserved
Huntington's disease	Impaired

OLFACTORY DYSFUNCTION IN THE CONTEXT OF CURRENT KNOWLEDGE OF NORMAL OLFACTION

The mechanism(s) underlying olfactory dysfunction in neurodegenerative diseases and normal aging have not yet been elucidated. However, a considerable body of information has accumulated over the last decade regarding the function of the olfactory system at the molecular, cellular, and system levels, both at the periphery and centrally. While several aspects of olfactory system function remain a mystery, a more complete understanding of the complex organization of the olfactory system is emerging from these analyses.

We now know that in the periphery, olfaction is initiated by binding of an odorous ligand to the ORs that are expressed in olfactory neurons (ORN), located within the olfactory epithelium. The ORs reflect the first organizational level at which specificity is established, as each neuron expresses only one receptor type. The spatial organization of the neurons that express one type of receptor in the olfactory epithelium reflects the second organizational level at which specificity is established. The third level of organization occurs at the olfactory bulb where the second-order neurons form connections in specific stereotypic sites in the olfactory bulb. The axons of first-order neurons form heterogeneous fascicles that defasciculate in the olfactory bulb and refasciculate with neurons expressing the same OR. Both permissive and inhibitory cues may contribute to this organizational process. This axon targeting may constitute another level of organization. Finally, behaviorally induced plasticity in the olfactory bulb may add yet another level of organizational complexity.[69] It will be important to understand whether the neurodegenerative process affects one or more levels of organization and the associated functions of the olfactory system.

Since the establishment of an association of olfactory dysfunction with PD and other neurodegenerative diseases, two broadly crafted, alternate hypotheses have been proposed to account for the nature of the olfactory deficits. According to the first hypothesis, the observed olfactory impairment is due to peripheral processes such as environmental insults to the olfactory system. According to the second hypothesis, the olfactory impairment is due to central processes. Support for the second hypothesis is provided by the fact that in both PD and AD the olfactory system appears to be affected in a disease-specific manner. In patients with autopsy-proven PD, the AON contains dystrophic neurites and Lewy bodies (LBs). These LB are morphologically more similar to cortical than to nigral LB.[70] In addition, there is considerable neuronal loss in the AON. The degree of neuronal loss correlates strongly with disease duration.[24] This is in apparent contradiction with the observation that olfactory dysfunction is independent of disease duration in PD. In AD, neurofibrillary tangles and amyloid plaques are seen in the AON. PD-specific pathology is also observed in the amygdala.[68] The amygdala is part of the limbic system and forms a large number of connections with the hippocampus and the entorhinal cortex as well as the neocortex. It is involved in memory, behavior, and regulation of endocrine and autonomic function and olfaction. In the amygdala, the neuropathological changes appear to accumulate slowly over time as the disease progresses. However, in PD patients, the severity of amygdala involvement appears to be independent of cognitive deficits.[68]

Furthermore, the olfactory bulb is rich in dopamine neurons, and a physiological role for dopamine in the olfactory bulb has been demonstrated in the rat olfactory system. Dopamine suppresses the electrical activity of mitral cells,[71] and the olfactory bulb is also rich in dopamine receptors (both D_1 and D_2). In the olfactory bulb, there is a differential distribution pattern of the dopamine receptors.[72] Recently, it has been demonstrated that, in the rat olfactory bulb, dopamine receptor subtypes can modulate the response of $GABA_A$ receptors and could be instrumental in odor detection, odor discrimination, and olfactory learning.[73] Interestingly, in clinical studies, the olfactory dysfunction observed in PD appears not to be a manifestation of dopamine deficiency.[74] Olfactory function was assessed in a small series of hyposmic PD patients before and after the administration of apomorphine, a potent, short-acting dopamine agonist, and no difference was observed. While the numbers of parkinsonian individuals tested in this study was small, the fact that olfactory dysfunction appears to be independent of disease stage or duration[32] provides indirect support for this hypothesis. However, this may also be explained by the fact that the early appearance of symptoms of olfactory dysfunction may reflect a threshold phenomenon that is achieved earlier in the olfactory system than in other areas of the CNS.

The complexity of the olfactory system's organization and its extensive connections to many cortical regions, the

basal ganglia, and cerebellum suggest that defects in any of a number of different molecular, cellular, or physiological processes may lead to olfactory dysfunction at the level of odor discrimination, recognition, and memory. In humans, many olfactory receptor genes (approximately 72%) are nonfunctional and are distributed on nearly all chromosomes. A large number of olfactory receptor genes are found in telomeric chromosomal regions.[75] Given the association of telomere length with senescence[76] as well as the known association of olfactory dysfunction with aging, it is tempting to speculate that the telomeric location of OR genes may make them more prone to deletion/inactivation that may in turn lead to age-dependent olfactory dysfunction. It is unclear whether such a process might play a role in the mechanisms underlying olfactory dysfunction associated with PD and neurodegenerative diseases.

System-level approaches have provided a valuable perspective on the role of central mechanisms in the development and function of the olfactory system. It is thought that the brain can determine which neurons are excited by analyzing a topographic map in the olfactory bulb.[78] Activity-dependent mechanisms and stimulus-specific synchronization of neuronal groups may be involved in olfactory processing.[79-80] Network dynamics can also be instrumental by creating odor representation and optimizing their distribution. Both slow, nonperiodic processes and fast, oscillatory processes may contribute to the coding that is inherent in the olfactory system.[81] It will be important to understand whether and how these central mechanisms are altered in neurodegenerative disease.

CONCLUDING REMARKS AND FUTURE DIRECTIONS

In conclusion, olfactory dysfunction is a consistent feature of PD and other neurodegenerative diseases. The molecular and cellular mechanisms underlying the disease-specific dysfunction remain unclear. The specificity of the neuropathological changes observed in the olfactory system in these diseases suggest that olfactory dysfunction is a result of the specific neurodegenerative process. It further suggests that disruption of the olfactory system at potentially different levels and by potentially different mechanisms can lead to the same end result.

In considering the complexity of the olfactory system and its connections, as well as the complexity of CNS involvement in PD and other neurodegenerative diseases, it seems a daunting task to systematically address the mechanism(s) underlying olfactory dysfunction and its relationship to neurodegeneration. For example, how does the different neuropathology of AD and PD result in a similar olfactory deficit? Given the specificity of neuropathological changes in the olfactory system in both PD and AD, a central process would appear more likely. A possible route of inquiry could make use of familial cases of PD, PPS, and AD where the genetic defect is known. A more systematic and extensive assessment of different aspects of olfactory function in such cases may provide significant insights into which particular olfactory deficit(s) are associated with which gene defect.

The study of the olfactory system in neurodegeneration offers a unique arena in which to employ a combination of analytical approaches at the molecular, cellular, physiological, and system levels. This uniqueness is not system-specific but, rather, the result of simultaneous advances in many scientific fronts. An important advance is the identification of the primary genetic defect in a number of neurodegenerative disorders. Another important advance is the development of genomic and proteomic analyses in which the simultaneous expression of thousands of genes in different tissues including brain tissue can be analyzed (e.g., using microarrays). Another advance is the analysis of the olfactory system by a dynamical systems approach that has led to significant insights into the organization and complexity of the olfactory system. Finally, the advent of functional imaging, including fMRI and PET, allows the *in vivo* functional characterization of olfaction and related higher cognitive processes in normal and diseased states. Understanding how the olfactory system is affected by neurodegeneration will require a synthesis of these conceptually different approaches. The field is in a particular moment in its development that a synthesis will open up new insights into both the functional understanding of the olfactory system as well as the neurodegenerative process.

REFERENCES

1. Brazis, P. W., Masdeu, J. C., Biller, J., *Localization in Clinical Neurology*, 3rd ed., Little, Brown and Company, Boston, MA, p.p. 109–114, 1996.
2. Buck, L. B., The molecular architecture of odor and pheromone sensing in mammals, *Cell,* 100:611–618, 2000.
3. Voshall, L. B., Putting smell on the map, *Trends in Neurosciences,* 26169–170, 2003.
4. Friedrich, R. W., Real time odor representations, *Trends in Neurosciences,* 25:487–489, 2002.
5. Spors, H., Grinvald, A., Spatio-temporal dynamics of odor representations in the mammalian olfactory bulb, *Neuron,* 34:301–315, 2002.
6. Wilson, D. A., Stevenson, R. J., The fundamental role of memory in olfactory perception, *Trends in Neurosciences,* 26:243–247, 2003.
7. Gottfried, J. A., Deichmann, R., Winston, J. S., Dolan, R. J., Functional heterogeneity in human olfactory cortex: an event-related functional magnetic resonance imaging study, *J. Neurosci.,* 22:10819–10828, 2002.
8. Sobel, N., Thomason, M. E., Stappen, I., Tanner, C. M., Tetrud, J. W., Bower, J. M., An impairment in sniffing contributes to the olfactory impairment in Parkinson's disease, *PNAS,* 98:4154–4159, 2001.
9. Doty, R. L., (1983), *The smell identification test administration manual,* Sensonics Inc., Hadden Height, NJ, 1983.
10. Doty, R. L., Shaman, P., Kimmelman, C. P., Dann, M. S., The University of Pennsylvania Smell Identification Test: a rapid quantitative olfactory function test for the clinic, *Laryngoscope,* 94:176–178, 1984b.
11. Cain, W. S., Rubin, M. D., Comparability of two tests of olfactory functioning, *Chem. Senses.,* 14:479–485, 1989.
12. Hummel, T., Sekinger, B., Wolf, S. R., Pauli, E., Kobal, G., "Sniffin' sticks": olfactory performance assessed by the combined testing of odor identification, odor discrimination and olfactory threshold, *Chem. Senses,* 22:39–52, 1997.
13. Anderson, J., Maxwell, L., Murphy, C., Odorant identification testing in the young child, *Chem. Senses,* 17:590, 1992.
14. Nordin, S., Bramerson, A., Liden, E., Bende, M., The Scandinavian odor-identification test: Development, reliability, validity and normative data, *Acta Otolaryngol.,* 118:226–234, 1999.
15. Lehrner, J., Deecke, L., Die Wiener olfaktorische Testbatterie (WOTB), *Akt. Neurol.,* 26:803–811, 1999.
16. Doty, R. L., *Odor Threshold Test™ Administration Manual,* Sensonics Inc., Hadden Heights, NJ, 2000.
17. Barz, S., Hummel, T., Pauli, E., Majer, M., Lang, C. J., Kobal, G., Chemosensory event-related potentials in response to trigeminal and olfactory stimulation in idiopathic Parkinson's disease, *Neurology,* 49:1424–1431, 1997.
18. Thesen, T., Murphy, C., Reliability analysis of event related brain potentials to olfactory stimuli, 39:733–738, 2002.
19. Sakuma, K., Nakashima, K., Takahashi, K., Olfactory evoked potentials in Parkinson's disease, Alzheimer's disease and anosmic patients, *Psychiatry Clin. Neurosci.,* 50:35–40, 1996.
20. Douek, E., Bannister, L. H., Dodson, H. C., Recent advances in the pathology of olfaction, *Proc. R. Soc. Med.,* 68:467–470, 1975.
21. Lanza, D. C., Moran, D. T., Doty, R. L., Trojanowski, J. Q., Lee, J. H., Rowley, J. C., 3rd, Crawford, D., Kennedy, D. W., Endoscopic human olfactory biopsy technique: a preliminary report," *Laryngoscope,* 103:815–819, 1993.
22. Lanza, D. C., Deems, D. A., Doty, R. L., Moran, D., Crawford, D., Rowley, J. C., 3rd, Sajjadian, A., Kennedy, D. W., The effect of human olfactory biopsy on olfaction: a preliminary report, *Laryngoscope,* 104:837–840, 1994.
23. Trojanowski, J. Q., Newman, P. D., Hill, W. D., Lee, V. M., Human olfactory epithelium in normal aging, Alzheimer's disease and other neurodegenerative disorders, *J. Comp. Neurol.,* 310:365–376, 1991.

24. Pearce, R. K., Hawkes, C. H., Daniel, S. E., The anterior olfactory nucleus in Parkinson's disease, *Mov. Disord.*, 10:283–287, 1995.

25. Doty, R. L., Shaman, P., Applebaum, S. L., Giberson, R., Sikorski, L., Rosenberg, L., Smell identification ability: changes with age, *Science,* 226:1441–1443, 1984a.

26. Doty, R. L., Influence of age and age-related diseases on olfactory function, *Ann. N. Y. Acad. Sci.,* 561:76–86, 1989.

27. Zucco, G. M., Zaglis, D., Wambsganss, C. S., Olfactory deficits in elderly subjects and Parkinson patients, *Percept. Mot. Skills,* 73:895–898, 1991.

28. Nordin, S., Murphy, C., Odor memory in normal aging and Alzheimer's disease, *Ann. N. Y. Acad. Sci.,* 855:686–693,1998.

29. Wu, H. H., Ivkovic, S., Murray, R. C., Jaramillo, S., Lyons, K. M., Johnson, J. E., Calof, A. L., Autoregulation of neurogenesis by GDF11, *Neuron.,* 37:197–207, 2003.

30. Hastings, N. B. and Gould, E., Neurons inhibit neurogenesis, *Nature Medicine,* 9:264–266, 2003.

31. Mesholam, R. I., Moberg, P. F., Mahr, R. N., Doty, R. L., Olfaction in neurodegenerative, 1998. Moberg, P. J., Pearlson, G. D., Speedie, L. J., Lipsey, J. R., Strauss, M. E., Folstein, S. E., Olfactory recognition: differential impairments in early and late Huntington's and Alzheimer's disease, *J. Clin. Exp. Neuropsychol.,* 9:650–664, 1987.

32. Doty, R. L., Deems, D. A., Stellar, S., Olfactory dysfunction in parkinsonism: a general deficit unrelated to neurological signs, disease stage, or disease duration, *Neurology,* 38: 1237–1244, 1988.

33. Stern, M. B., Doty, R. L., Dotti, M., Corcoran, P., Crawford, D., McKeown, D. A., Adler, C., Gollomp, S., Hurtig, H., Olfactory function in Parkinson's disease subtypes, *Neurology,* 44:266–268, 1994.

34. Doty, R. L., Riklan, M., Deems, D. A., Reynolds, C., Stellar, S., The olfactory and cognitive deficits of Parkinson's disease: evidence for independence, *Ann. Neurol.,* 25:166–171, 1989.

35. Doty, R. L., Stern, M. B., Pfeiffer, C., Gollomp, S. M., Hurtig, H. I., Bilateral olfactory dysfunction in early stage treated and untreated idiopathic Parkinson's disease, *J. Neurol. Neurosurg. Psychiatry,* 55:138–142, 1992a.

36. Harding, A. J., Stimson, E., Henderson, J. M., Halliday, G. M., Clinical correlates of selective pathology in the amygdala of patients with Parkinson's disease, *Brain,* 125:2431–2445, 2002.

37. Nee, L. E., Scott, J., Polinsky, R. J., Olfactory dysfunction in the Shy-Drager syndrome, *Clin. Auton. Res.,* 3:281–282, 1993.

38. Wenning, G. K., Shephard, B., Hawkes, C., Petruckevitch, A., Lees, A., Quinn, N., Olfactory function in atypical parkinsonian syndromes. *Acta Neurol. Scand.,* 91:247–250, 1995.

39. Doty, R. L., Golbe, L. I., McKeown, D. A., Stern, M. B., Lehrach, C. M., Crawford, D., Olfactory testing differentiates between progressive supranuclear palsy and idiopathic Parkinson's disease, *Neurology,* 43:962–965, 1993.

40. Doty, R. L., Perl, D. P., Steele, J. C., Chen, K. M., Pierce, J. D., Jr., Reyes, P., Kurland, L. T., Odor identification deficit of the parkinsonism-dementia complex of Guam: equivalence to that of Alzheimer's and idiopathic Parkinson's disease, *Neurology,* 41(Suppl. 2): 77–80, 1991.

41. Muller, A., Mungsdorf, M., Rechmann, H., Strehle, G., Hummel, T., Olfactory function in parkinsonian syndromes, *Journal of Clinical Neuroscience,* 9:521–524, 2002.

42. Ahlskog, J. E., Waring, S. C., Petersen, R. C., Esteban-Santillan, C., Craig, U. K., O'Brien, P. C., Plevak, M. F., Kurland, L. T., Olfactory dysfunction in Guamanian ALS, parkinsonism and dementia, *Neurology,* 51:1672–1677, 1998.

43. Markopoulou, K., Larsen, K. W., Wszolek, E. K., Denson, M. A., Lang, A. E., Pfeiffer, R. F., Wszolek Z. K., Olfactory dysfunction in familial parkinsonism, *Neurology,* 49:1262–1267, 1997.

44. Clark, L. N., Poorkaj, P., Wszolek, Z., Geschwind, D. H., Nasreddine, Z. S., Miller, B., Li, D., Payami, H., Awert, F., Markopoulou, K., Andreadis, A., D'Souza, I., Lee, V. M., Reed, L., Trojanowski, J. Q., Zhukareva, V., Bird, T., Schellenberg, G., Wilhelmsen, K. C., (1998): Pathogenic implications of mutations in the tau gene in pallido-ponto-nigral degeneration and related neurodegenerative disorders linked to chromosome 17, *Proc. Natl. Acad. Sci., U.S.A.,* 95:13103–13107, 1998.

45. Gasser, T., Muller-Myhsok, B., Wszolek, Z. K., Oehlmann, R., Calne, D. B., Bonifati, V., Bereznai, B., Fabrizio, E., Vieregge, P., Horstmann, R. D., A susceptibility locus for Parkinson's disease maps to chromosome 2p13, *Nat. Genet.,* 18:262–265, 1998.

46. Duda, J. E., Shah, U., Arnold, S. E., Lee, V. M., Trojanowski, J. Q., The expression of alpha- beta- and gamma synucleins in olfactory mucosa from patients with and without neurodegenerative disease, *Exp. Neurol.,* 160:515–522, 1999.

47. Kaplan, B., Ratner, V., Haas, E., Alpha-synuclein: its biological function and role in neurodegenerative diseases, *J. Mol. Neurosci.,* 20:83–92, 2003.

48. Lucking, C. B., Brice, A., Alpha-synuclein and Parkinson's disease, *Cell Mol. Life. Sci.,* 57:1894–1908, 2000.

49. Doty, R. L., Singh, A., Tetrud, J., Langston, J. W., Lack of major olfactory dysfunction in MPTP-induced parkinsonism, *Ann. Neurol.,* 32:97–100, 1992b.

50. Busenbark, K. L., Huber, S. J., Greer, G., Pahwa, R., Koller, W. C., Olfactory function in essential tremor, *Neurology,* 42:1631–1632, 1992.

51. Louis, E. D., Bromley, S. M., Jurewicz, E. C., Watner, D., Olfactory dysfunction in essential tremor, *Neurology,* 59:1631–1633, 2002.

52. Adler, C. H., Gwinn, K. A., Newman, S., Olfactory function in restless leg syndrome, *Mov. Disord.,* 13:563–565, 1998.

53. Farrer, M., Gwinn-Hardy, K., Muenter, M., Wavrant, DeVrieze F., Crook, R., Perez-Tur, J., Lincoln, S., Maraganore, D., Adler, C., Newman, S., MacElwee, K., McCarthy, P., Miller, C., Waters, C., Hardy, J., A chro-

mosome 4p haplotype segregating with Parkinson's disease and postural tremor, *Hum. Mol. Genet.,* 8:81–85, 1999.

54. Elian, M., Olfactory impairment in motor neuron disease: a pilot study, *J. Neurol. Neurosurg. Psychiatry,* 54:927–928, 1991.

55. Sajjadian, A., Doty, R. L., Gutnick, D., Chirurgi, R. J., Sivak, M., Perl, D. O.,(1994) Olfactory dysfunction in amyotrophic lateral sclerosis, *Neurodegeneration,* 3:153–157, 1994.

56. Hawkes, C. H., Shephard, B. C., Geddes, J. F., Body, G. D., Martin, J. E., Olfactory disorder in motor neuron disease, *Exp. Neurol.,* 50:248–253, 1998.

57. Nordin, S., Paulsen, J. S., Murphy, C., Sensory- and memory-mediated olfactory dysfunction in Huntington's disease, *J. Int. Neuropsychol. Soc.,* 1:281–290, 1995.

58. Bylsma, F. W., Moberg, P. F., Doty, R. L., Brandt, J., Odor identification in Huntington's disease patients and asymptomatic gene carriers, *J. Neuropsychiatry Clin. Neurosci.,* 9:598–600, 1997.

59. Moberg, P. J., Doty, R. L., Olfactory function in Huntington's disease patients and at-risk offspring, *Int. J. Neurosci.,* 89:133–139, 1997.

60. Hamilton, J. M., Murphy, C., Paulsen, G. S., Odor detection, learning and memory in Huntington's disease, *J. Int. Neuropsychol. Soc.,* 5: 609–615, 1999.

61. Christen-Zaech, S., Kraftsik, R., Pillevuit, O., Kiraly, M., Martins, R., Khalili, K., Miklossy, J., Early olfactory involvement in Alzheimer's disease, *Can. J. Neurol. Sci.,* 30:20–25, 2003.

62. Wang, Q. S., Tian, L., Huang, Y. L., Qin, S., He, L. Q., Zhou, J. N., Olfactory identification and apolipoprotein E epsilon 4 allele in mild cognitive impairment, *Brain Res.,* 951:77–81, 2002.

63. Graves, A. B., Bowen, J. D., Rajaram, L., McCormick, W. C., McCurry, S. M., Schellenberg, G. D., Larson, E. B., Impaired olfaction as a marker for cognitive decline: interaction with apolipoprotein E epsilon 4 status, *Neurology,* 53:1480–1487, 1999.

64. Lehrner, J. P., Brucke, T., Dal-Bianco, P., Gatterer, G., Kryspin-Exner, I., Olfactory functions in Parkinson's disease and Alzheimer's disease, *Chem. Senses,* 22:105–110, 1997.

65. Hyman, B. T., Arriagada, P. V., Van Hoesen, G. W., Pathologic changes in the olfactory system in aging and Alzheimer's disease, *Ann. N. Y. Acad. Sci.,* 640:14–19, 1991.

66. Reyes, P. F., Deems, D. A., Suarez, M. G., Olfactory-related changes in Alzheimer's disease: a quantitative neuropathologic study, *Brain Res. Bull.,* 32:1–5, 1993.

67. ter Laak, H. J., Renkawek, K., vanWorkum, F. P., The olfactory bulb in Alzheimer's disease: a morphologic study of neuron loss, tangles and senile plaques in relation to olfaction, *Alzheimer's Dis. Assoc. Disord.,* 8:38–48, 1994.

68. Braak, H., Braak, E., Yilmazer, D., deVos, R. A., Jansen, E. N., Bohl, J., Jellinger, K., Amygdala pathology in Parkinson's disease, *Acta Neuropath.,* (Berl.), 88: 493–500, 1994.

69. Brunjes, P. C., Green, C. A., Progress and directions in olfactory development, *Neuron,* 38:371–374, 2003.

70. Hawkes, C. H., Shephard, B. C., Daniel, S. E., Olfactory dysfunction in Parkinson's disease, *J. Neurol. Neurosurg. Psychiatry,* 62:436–446, 1997.

71. Duchamp-Viret, P., Coronas, V., Delaleu, J. C., Moyse, E., Duchamp, A., Dopaminergic modulation of mitral cell activity in the frog olfactory bulb: a combined radioligand binding-electrophysiological study, *Neuroscience,* 79:203–216, 1997.

72. Levey, A. I., Hersch, S. M., Rye, D. B., Sunahara, R. K., Niznik, H. B., Kitt, C. A., Price, D. L., Maggio, R., Brann, M. R., Ciliax, B. J., Localization of D_1 and D_2 dopamine receptors in brain with subtype-specific antibodies, *Proc. Natl. Acad. Sci., U.S.A.,* 90:8861–8865, 1993.

73. Brunig, I., Sommer, M., Hatt, H., Borman, J., Dopamine receptor subtypes modulate olfactory bulb g-aminobutyric acid type A receptors, *Proc. Natl. Acad., U.S.A.,* 96:2456–2460, 1999.

74. Roth, J., Radil, T., Ruzicka, E., Jech, R., Tichy, J., (1998): Apomorphine does not influence olfactory thresholds in Parkinson's disease, *Funct. Neurol.,* 13:99–103, 1998.

75. Rouquier, S., Taviaux, S., Trask, B. J., Brand-Arpon, V., van den Engh, G., Demaille, J., Giorgi, D., Distribution of olfactory receptor genes in the human genome, *Nat. Genet.,* 18:243–250, 1998.

76. Bodnar, A. G., Ouellette, M., Frolkis, M., Holt, S. E., Chiu, C. P., Morin, G. B., Harley, C. B., Shay, J. W., Lichtsteiner, S., Wright, W. E., Extension of life-span by introduction of telomerase into normal human cells, *Science,* 279:349–352, 1998.

77. Lin, D. M., Ngai, J., Development of the vertebrate main olfactory system, *Curr. Opin. Neurobiol.,* 9:74–78, 1999.

78. Ebrahimi, F. A., Chess, A., The specification of olfactory neurons, *Curr. Opin. Neurobiol.,* 8:453–457, 1998.

79. Laurent, G., Dynamical representation of odors by oscillating and evolving neural assemblies, *TINS,* 19:489–496, 1996.

80. Laurent, G., Olfactory processing: maps, time and codes, *Curr. Opin. Neurobiol.,* 7:547–553, 1997.

81. Laurent, G., Olfactory network dynamics and the coding of multidimensional signals, *Nature Reviews Neuroscience,* 3:884–895, 2002.

82. Mombaerts, P., The human repertoire of odorant receptor genes and pseudogenes, *Ann. Rev. Genomics Hum. Genet.,* 2-493–510), 2001.

83. Mombaerts, P., How smell develops, *Nature Neuroscience,* (Suppl.) 4:1192–1198, 2001.

84. Young, J. M., Trask, B. J., The sense of smell: genomics of vertebrate odorant receptors, *Human Molecular Genetics,* 11:1153–1160, 2002.

25 Gastrointestinal Dysfunction in Parkinson's Disease

Ronald F. Pfeiffer
Department of Neurology, University of Tennessee Health Science Center

CONTENTS

INTRODUCTION

Although gastrointestinal (GI) dysfunction is conventionally referred to as one of the "nonmotor" features of Parkinson's disease (PD), this is actually somewhat of a misnomer. Many (though certainly not all) aspects of GI function are clearly motor in character, and it is the more obscure sensory aspects of GI function that are often overlooked with regard to involvement in disease processes. What does distinguish GI dysfunction from the traditional motor features of PD is that the motor systems involved belong primarily, though not exclusively, to the autonomic and enteric, rather than somatic, nervous systems, and the muscles affected by the nervous system dysfunction are largely (though once again not exclusively) of the smooth, rather than striated, type.

Awareness of GI dysfunction in the setting of PD dates all the way back to James Parkinson and his remark-able 1817 treatise, in which he very clearly and concisely identified so many of the features of PD recognized today, including those involving the GI system. He described *drooling* (and even was aware that it is due to disordered swallowing), "...the saliva fails of being directed to the back part of the fauces, and hence it is continually draining from the mouth..."; *dysphagia,* "...food is with difficulty retained in the mouth until masticated; and then as difficultly swallowed..."; and *bowel dysfunction,* including both *constipation,* "...the bowels which had all along been torpid, now in most cases, demand stimulating medications of very considerable power..." and *defecatory dysfunction,* "...the expulsion of the feces from the rectum sometimes requiring mechanical aid."[1]

In the years that followed, however, other neurological masters paid perfunctory attention, at best, to the GI features of PD. Romberg briefly mentions difficulty chewing, drooling, dysphagia, and constipation in his

0-8493-1590-5/05/$0.00+$1.50

description of PD,[2] while Charcot provides only passing mention of dysphagia and drooling,[3] and neither Hammond nor Gowers, in their textbooks of neurology, mention the GI features of PD at all.[4, 5]

In fact, very little mention of GI dysfunction in PD surfaces in the post-Parkinson neurological literature until 1965, when Eadie and Tyrer [6] published their analysis of GI dysfunction in 107 patients with parkinsonism, 76 of whom had been diagnosed with idiopathic PD. A group of comparably aged persons with "acute orthopedic" disorders served as controls. Masticatory difficulty, drooling, dysphagia, frequent "heartburn," and constipation were found to be present more often in individuals with PD than in controls.

Study of the GI aspects of PD then devolved again into relative dormancy until 1991, when Edwards and colleagues resurrected interest in the topic with their survey of 98 patients with PD and 50 comparably aged spousal controls.[7] The GI features they identified in this and expanded upon in a subsequent series of reports[8–16] closely parallel those identified by both Eadie and Tyrer and by Parkinson himself, and include disordered salivation (drooling), dysphagia, nausea, constipation (decreased bowel movement frequency), and defecatory dysfunction (difficulty with the act of defecation). Other investigators have also subsequently contributed immensely to the expanding literature on the GI features of PD. This burgeoning literature is reviewed in this chapter.

GI ANATOMY AND PHYSIOLOGY

The GI system, like the nervous system, performs its vital tasks largely hidden from view and without conscious planning or effort (eating and evacuating being the obvious exceptions). It encompasses a rather astounding length and surface area between its oral and anal portals. Just as the study of the nervous system has been slowed by its complexity and inaccessibility, so has study of the GI system in many respects. While the function of the GI system seems quite straightforward—to process and absorb nutrients necessary for function, while eliminating waste—the control and coordination necessary between the nervous system and the GI system and within the GI system itself to perform this function is remarkably complex.

The oral cavity is the jumping off point for the journey food and drink take through the GI system. In the mouth, the teeth and masticatory muscles combine to grind and macerate food, mixing it with salivary gland secretions for lubrication and initiation of digestion. The tongue then forms the food into a bolus and propels it backward to the pharyngeal inlet where, in a piston-like action, it delivers the bolus into the pharynx.

The act of swallowing itself is a surprisingly intricate and complex action that is partly voluntary, partly reflex in character and is carried out by a combination of 26 pairs of pharyngeal and laryngeal muscles (not counting muscles used for chewing) under the control of 5 cranial nerves.[17] Swallowing is traditionally divided into three components: oral, pharyngeal, and esophageal. The oral component, described above, is largely under voluntary control, while the pharyngeal and esophageal phases are principally reflexive in character.

The reflex component of swallowing is coordinated at a brain-stem level by central pattern generators within the medial reticular formation of the rostral medulla and the reticulum adjacent to the nucleus solitarius, which contain the neural programs that conduct the symphony of swift, sequential movements of the oral, pharyngeal, and esophageal musculature that seal off the nasal passages and trachea while the upper esophageal sphincter (UES) relaxes and allows the food bolus to enter the esophagus.[18, 19] The volitional component of swallowing reflects involvement of the motor cortex and additional centers in the supplementary motor cortex, insula, and cerebellum.[20]

The esophagus is a tube, 18 to 26 cm in length, whose lumen is collapsed when not swallowing but can distend to several centimeters in diameter to accommodate food being swallowed.[21] Esophageal muscle composition, which consists of an inner circular muscle layer and an outer longitudinal layer, can be roughly divided into thirds: the proximal third (some report this as only 5%[22]) primarily striated muscle, the distal third smooth muscle, and the middle third of a mixture of the two. The UES is striated muscle, while the lower esophageal sphincter (LES) has a smooth muscle composition. Both are contracted at rest; contraction of the UES prevents air from entering the esophagus, while contraction of the LES prevents reflux of stomach contents into the esophagus. Esophageal skeletal muscle is under control of motor neurons originating in the nucleus ambiguus in the brain stem, while smooth muscle in the middle and distal esophagus is under autonomic direction with parasympathetic innervation from the dorsal motor nucleus of the vagus and sympathetic from the intermediolateral column in the spinal cord. The autonomic efferent fibers do not actually synapse on the muscle cells in the esophagus but, rather, on ENS neurons in Auerbach's plexus.

The stomach is a reservoir that serves several functions in the digestive process. It accommodates and stores food while grinding the solid particles down to appropriately small size, finally releasing the processed contents in a controlled fashion into the small intestine. In adults, the stomach typically has the capacity to hold 1.5 to 2.0 L of material. The grinding and propulsive movements of the stomach originate from "gastric pacemaker" cells at a site along the greater curvature of the stomach, which generate gastric slow waves at a rate of approximately three per minute.[23] The pacemaker cells have been identified as interstitial cells of Cajal (ICCs), which are a component of the ENS.

The next destination on the route through the GI tract is the small intestine, which is divided into three segments (duodenum, jejunum, and ileum) and in adults reaches the rather astounding length of approximately 6 m.[24] The small intestine is responsible for absorption of nutrients, salt, and water. Motility within the small intestine is produced by contractions of the circular and longitudinal muscle layers that compose the intestinal walls. As in the stomach, ICCs generate electrical slow waves that serve a pacemaker function and migrate in an aboral direction. When spike bursts are superimposed on the slow wave, actual muscle contraction occurs, which then travels for an undetermined, but probably short, distance in either direction along the small intestine. Two distinct patterns of small intestinal motor function have been identified.[25] The fed (postprandial) pattern, which appears within 10 to 20 min following a meal, is characterized by more segmental, and less propulsive, contractions that assist in the mixing of digestive enzymes with the chyme and maximize mucosal contact, thus promoting nutrient absorption. The second pattern, termed the fasting (interdigestive) pattern, appears 4 to 6 hr after a meal and is divided into three phases. The first is a period of relative motor quiescence, followed by increasingly prominent contractions in the subsequent two phases that presumably serve to "flush" solid residues from the small intestine, which prevents bezoar formation and minimizes bacterial accumulation. This complex pattern of small intestinal motility is under the direct control of the ENS but modulated by both autonomic and hormonal influences.

The colon, approximately 1.0 to 1.5 m in length in adults, is composed of the same two muscle layers—circular and longitudinal—found in the small intestine.[24,26] The ileocecal valve, which divides the colon from the small intestine, is not a true sphincter but still effectively regulates colonic filling and prevents colo-ileal reflux. The colon stores material marked for excretion and performs an important role in the regulation of fluid, electrolyte, and short-chain fatty acid absorption. It can increase fluid absorption up to fivefold in appropriate circumstances. As in the stomach and small intestine, ICCs perform a pacemaker function in the generation of the pressure waves that regulate colonic motility. For the most part, motor control of colonic motility is directly mediated through the ENS but modulated by the ANS. Parasympathetic innervation of the ascending and transverse colon is vagal in origin, while the descending and rectosigmoid regions receive their innervation via the pelvic nerves. Sympathetic supply to the colon originates in the thoracic spinal cord and reaches the colon via the inferior mesenteric and pelvic plexuses. Sympathetic activity produces vasoconstriction of mucosal and submucosal blood vessels, downregulates motility, and inhibits secretion (thus limiting water loss), while parasympathetic activity increases enteric motor activity and colonic motility.[26]

The rectum serves to store feces until a convenient opportunity to evacuate its contents is reached. Egress from the rectum is blocked by the internal and external anal sphincter muscles, which are tonically contracted. In the rectum, the longitudinal smooth muscle layer, which in the colon had been concentrated into the muscle bands called taenia, spreads out into an encircling sheath; the internal anal sphincter (IAS) consists of smooth muscle that is continuous with the circular muscle layer of the rectum.[26] In contrast, the external anal sphincter (EAS) is a band of striated muscle distal to the IAS. The IAS is under autonomic control via the pelvic plexus; the EAS under the control of motor neurons in the sacral spinal cord via the pudendal nerve. The puborectalis muscle also contributes to the maintenance of fecal continence by means of tonic contraction that pulls the rectum anteriorly, forming an anorectal angle that impedes rectal emptying.[27] The act of defecation is characterized by relaxation of the two anal sphincters and the puborectalis muscle, which results in a straightening or opening of the anorectal angle, and by contraction of the glottic, diaphragmatic, and abdominal wall muscles, which elevates intra-abdominal pressure and encourages evacuation of the rectal contents.

CLINICAL FEATURES

WEIGHT LOSS

Progressive weight loss is a frequent feature of PD. Although generally mild, in a minority of individuals, it can reach alarming proportions. In one study, weight loss was observed in 52% of PD patients, with loss of over 28 lb in 22%.[28] In another study, individuals with PD were four times as likely as controls to report weight loss of greater than 10 lb.[29] In some studies, the weight loss correlates with disease progression,[29] but in a recent study using data from two large prospective cohorts, Chen and colleagues discovered that weight loss in PD often begins even before the conventional motor features are identified.[30] In the 174 individuals with PD in their study, the average weight loss was 5.2 lb in the 10 years prior to diagnosis and 7.7 lb in the 8 years following diagnosis. The bulk of the prediagnosis weight loss occurred in the 4 years immediately prior to diagnosis.

Obscurity shrouds the explanation for weight loss in PD. For weight loss to occur in anyone, there must be either reduced energy intake or increased energy expenditure. In a dialog reminiscent of the well known beer commercial (more taste!—less filling!) theories and evidence favoring each mechanism have been advanced with regard to the PD patient.

Reduced energy intake can be the result of either reduced food intake or impaired food absorption. A bevy of factors, such as difficulty manipulating silverware, slowed chewing, and impaired swallowing, may incline PD patients to consume less food, while olfactory impairment may make food less enticing. However, when dietary intake has actually been assessed in PD patients, no significant differences from controls have been noted,[28,31] and increased energy intake has actually been noted in some studies.[30–32] Although intestinal absorption has not been extensively studied in PD, malabsorption has not been a generally recognized feature. However, using the differential sugar absorption test, Davies and colleagues found that PD patients displayed reduced mannitol, but not lactulose, absorption, suggesting impairment of nonmediated uptake across the enterocyte brush border membrane.[33]

Controversy has also swirled around the question of increased energy expenditure in PD. Energy expenditure can be divided into two components: energy expended at rest, which accounts for 60 to 80% of total expenditure, and energy expended during physical activity, which accounts for the remaining 20 to 40%.[31] Both increased energy expenditure due to parkinsonian rigidity[34] and increased expenditure as a consequence of dyskinesia[35] have been postulated. However, not all PD patients who lose weight suffer dyskinesia, and other investigators have demonstrated an overall 15% reduction in energy expenditure in PD patients, attributable to reduced physical activity.[31] Weight gain often occurs following stereotactic pallidal and subthalamic neurosurgical procedures for PD, and some investigators have attributed this to reduction in dyskinesia.[36] However, others have noted no correlation between post-pallidotomy weight gain and changes in dyskinesia.[37]

Neurochemical and hormonal factors could conceivably also be playing a role in parkinsonian weight loss. Dopamine has recently been shown to play a role in eating behaviors. In a series of studies utilizing[11] C-raclopride positron emission tomography (PET), Volkow, Wang, and their colleagues have demonstrated decreased D_2 dopamine receptor availability in obese individuals and have also documented significant increases of extracellular dopamine in dorsal but not ventral striatum in response to food stimulation, which correlated with self-reported hunger and desire for food.[38–40] Similar experiments have not been reported in PD patients, but both PD itself and antiparkinson medications could conceivably have an effect.

The possible role of hormonal influences on weight loss in PD is also largely uncharted. Leptin is a hormone produced in adipocytes and the hypothalamus that reduces food intake and increases energy expenditure. To investigate the hypothesis that leptin might be increased in patients with PD, leptin levels were measured in PD patients with and without weight loss, but no significant differences were found.[41] The possible role in parkinsonian weight loss of gastric-derived peptide hormones, such as ghrelin and peptide YY, or other substances such as α-melanocyte-stimulating hormone, neuropeptide Y, and agouti-related protein, that may play a role in long-term control of food intake and energy expenditure, has not been investigated.[42, 43]

ORAL DYSFUNCTION

DENTAL

It is generally perceived that individuals with PD are prone to develop dental dysfunction.[44–47] Difficulty with the repetitive motions necessary for teeth brushing, pooling of saliva in the mouth, or (alternatively) dry mouth, jaw muscle rigidity, difficulty retaining dentures, and involuntary jaw movements are problems that may be encountered by the person with PD.[48–50] The propensity of patients with PD to have a penchant for sweets has also led to concerns that this might promote dental decay, although such confectionary consumption does not seem to alter the oral microflora.[51]

It is, then, interesting to note that in at least two formal studies of dental problems, PD patients had significantly more teeth and fewer decayed, missing, or filled teeth than either a control group of comparable age[52] or compared to national statistics.[53] In the Japanese cohort described by Fukayo,[53] the superior dental status of the PD patients was attributed to the fact that an astounding 68% of the group (21/31) brushed their teeth three times a day. Whether this figure would be replicated in other populations is unknown, but the results of a study in Greek PD patients, in which 98% were found to be denture wearers and extensive oral problems were present in all participants,[54] might suggest that it would not be.

Other problems may also surface in PD patients. A burning sensation in the mouth was documented in 24% of PD patients surveyed in one questionnaire study.[55] Bruxism has also been reported, both as a presenting feature of PD[56] and as a complication of levodopa therapy.[57] Mandibular dislocation[58] and temporomandibular joint dysfunction[59] have also been described. Because patients with PD often have difficulty adjusting to complete dentures, the use of mandibular dental implants combined with overdentures has been advocated.[60] Concerns about potential mercury toxicity as a cause for PD[61] have led some individuals to have their amalgam fillings removed, although firm proof of such an association is lacking.

SALIVARY EXCESS

Excess saliva in the mouth, often with some degree of drooling, has been a recognized feature of PD since Parkinson's original description.[1] The frequency with which it occurs is reflected in the major survey studies, which record its presence in 70 to 78% of individuals with PD, compared with only 6% of control subjects.[6,7] Contrary to many patients' perceptions, saliva production is not increased in PD and is actually, in most instances, diminished.[62–64] Rather, the salivary excess is the consequence of inefficient and infrequent swallowing. Some treatment implications flow from this recognition.

Drooling is embarrassing and frustrating for PD patients and may produce a reluctance to go out in public. Although not dangerous in most instances, in individuals prone to dysphagia-related aspiration, the pooled saliva provides an ever-present source of aspirate. Improved swallowing efficiency can sometimes be attained by optimization of antiparkinson medication, but this is not invariably effective (see below). In social situations, the problem can be temporarily masked by chewing gum or sucking on hard candy, which projects a more voluntary component to the swallowing act but does not provide any permanent solution. Employment of a portable metronome brooch to cue swallowing has also been tested.[65] Most therapeutic attention, however, has focused on a number of more specific medical treatment techniques that have been advocated.

The traditional treatment approach has been to employ anticholinergic drugs, such as trihexyphenidyl or benztropine, to dry up the mouth. However, the adverse effect profile of these medications in elderly PD patients, including urinary retention, constipation, memory impairment, and even psychosis, renders them risky agents to use in this setting. An approach that may limit (though not eliminate) the potential for anticholinergic toxicity but still effectively reduce saliva production is the employment of one drop of 1% atropine ophthalmic solution sublingually twice daily.[66] Such prescriptions may, however, raise pharmacists' eyebrows.

Perhaps the most encouraging treatment for drooling in PD was introduced by Bhatia and colleagues in 1999 when they performed intraparotid injections of botulinum toxin type A (BTX) in four patients, one of whom had PD.[67] The possibility of using BTX in this fashion had first been suggested in 1997 by Bushara, who actually had patients with amyotrophic lateral sclerosis in mind when making the proposal.[68] Subsequent open label studies in PD patients confirmed subjective reduction of drooling in 67 to 88% of subjects and objective reduction in saliva production in 88 to 89% of patients tested.[69–71] Recent double-blind studies have further demonstrated the safety and efficacy of this technique.[72,73] BTX doses employed have varied considerably, from the 5 units per parotid employed by Friedman and Potulska[71] to the 225 units per side used by Mancini and colleagues, who injected both parotid and submandibular glands.[73] The latter investigators also performed their injections under ultrasonographic guidance, which they advocate as a means of improving accuracy and safety of the injections. The duration of effect of the injections has most often been

reported to be in the range of 6 to 8 weeks,[69,71,73] although improvement for as long as 4 to 7 months has also been described.[70] Although no significant complications have been reported in studies thus far, the potential for problems such as excessively dry mouth, dysphagia due to pharyngeal muscle weakness, and even facial nerve and artery damage from the injections has been emphasized by some investigators.[74]

Surgical approaches for the treatment of salivary excess in PD have also been employed. Tympanic neurectomy, in which both the chorda tympani and Jacobson's nerve are severed, is sometimes advocated for refractory drooling,[75] but formal studies of the procedure in PD patients have not been performed. Loss of taste in the anterior tongue accompanies tympanic neurectomy, and the reduction in salivation may not be permanent.

DYSPHAGIA

As noted earlier in this chapter, the act of swallowing requires multiple muscles in the mouth, throat, and esophagus to produce a precisely controlled and coordinated cascade of movement. This, perhaps not surprisingly, turns out to be difficult for the individual with PD. Survey studies reveal a rather broad range of positive responses when PD patients are asked whether they perceive difficulty swallowing. While the two large survey studies[6,7] each catalogued a subjective sense of dysphagia in approximately 50% of participants with PD, other studies have suggested that anywhere from 30 to 82% of PD patients may be aware of difficulty swallowing.[76–79] The reason for this broad range is uncertain but may simply reflect the degree of detail in the questionnaire.[79] While most attention regarding dysphagia in PD has centered on oropharyngeal abnormalities, it is clear that esophageal dysfunction may also play a role in the generation of dysphagia in some individuals.

OROPHARYNGEAL DYSPHAGIA

Objective studies of swallowing generally have demonstrated an even higher frequency of swallowing abnormalities than the subjective survey studies. The most frequently employed test has been the modified barium swallow (MBS) test, and various investigators have reported abnormalities on MBS in 75 to 97% of PD subjects tested.[80–83] An array of abnormalities in both the oral and pharyngeal phases of swallowing have been identified during MBS testing. Within the oral phase, alterations in lingual control and oral mobility, presumably due to rigidity and bradykinesia, result in abnormal bolus formation, delayed initiation of swallowing, repeated tongue pumping to accomplish the swallow, piecemeal deglutition, and the presence of residual material on the tongue or in the lateral or anterior sulci following swallowing.[84–89] An equally impressive roster of abnormalities have been identified in the pharyngeal phase, including pharyngeal dysmotility, misdirected swallows, pharyngeal and vallecular stasis, vallecular residue, and reflux of material from the vallecular and pyriform sinuses into the mouth.[81,84,85] With MBS testing, it has become abundantly clear that dysphagia can be present in individuals with PD, even if they have no symptoms to alert either patient or physician to its presence.

An important and potentially serious ramification of dysphagia in PD is the development of aspiration. Studies suggest that aspiration occurs in a significant proportion of PD patients, with the range of reported frequencies extending from 15 to 56%.[81,84,86,87,90,91] As is true with dysphagia itself, it has also become quite clear that symptoms alone are not sufficient predictors of the presence of aspiration in persons with PD. Entirely asymptomatic aspiration has been documented in 15 to 33% of PD patients.[81,86,92] Even this surprising figure may underestimate the potential problem. Bird and colleagues noted the presence of vallecular residue, an abnormality indicative of increased aspiration risk, in 88% of 16 PD patients they studied, all of whom were without any symptomatic dysphagia.[92] The high frequency of dysphagia and aspiration, both symptomatic and asymptomatic, in individuals with PD, and the recognition that the development of these abnormalities may be independent of disease severity, has led at least one investigator to suggest that screening videofluoroscopy be performed early in the course of PD to identify those patients with subclinical dysphagia and institute appropriate treatment measures.[91]

In the large survey studies, the presence of symptomatic dysphagia seemed to correlate with disease progression.[7] However, this correlation is not clearly evident when objective testing of swallowing is employed. Bushman and colleagues noted the presence of MBS abnormalities in approximately 50% of patients with early (Hoehn and Yahr stages 1 and 2) PD,[81] and other investigators have emphatically confirmed that the development of both dysphagia and aspiration is independent of both disease duration and severity.[90,91]

Cricopharyngeal muscle dysfunction is yet another source of swallowing impairment in PD. The cricopharyngeal muscle, serving as the UES, is tonically contracted, opening in response to the piston-like propulsive force of the tongue as it drives the food bolus into the pharynx. In individuals with PD those propulsive forces may be inadequate to trigger adequate cricopharyngeal relaxation, resulting in difficulty swallowing.[90] In some individuals, however, the cricopharyngeal muscle itself may be the source of the problem, with failure to relax resulting in cricopharyngeal bar formation.[90] In one study, 22% of PD patients referred for evaluation of dysphagia were found to have cricopharyngeal bars or Zenker's diverticula.[93] Zenker's diverticula, which form in Killian's

triangle and are felt to be the consequence of incomplete cricopharyngeal relaxation, are another structural source of dysphagia that, in addition to causing a sense of food hanging up in the throat, can also produce delayed regurgitation of undigested food that had been trapped in the diverticulum and halitosis. Perforation, especially during instrumentation such as nasogastric tube placement, with consequent mediastinitis, is a potentially life-threatening complication.

ESOPHAGEAL DYSPHAGIA

Esophageal dysfunction and its role in parkinsonian dysphagia has not been as extensively studied and thoroughly delineated as its oropharyngeal counterpart. However, videofluoroscopic abnormalities have been described in 5 to 86% of patients studied with PD [10,82,87,89,94,95] and disordered function during esophageal manometry in 61 to 73%.[96,97] An array of abnormalities have been observed, including slowed esophageal transit, segmental esophageal spasm, repetitive spontaneous contractions of the proximal esophagus, multiple simultaneous contractions producing diffuse esophageal spasm, ineffective or tertiary contractions with air trapping, aperistalsis, esophageal dilatation, and reduced pressure at the LES.[89,94–99] LES dysfunction has also been observed in PD, where advanced reflux disease with consequent esophagitis may produce dysphagia, in addition to the more typical gastroesophageal symptoms.[93]

TREATMENT OF DYSPHAGIA

Behavioral, pharmacological, and surgical treatment methods have all been employed in the treatment of dysphagia in PD. Behavioral management approaches may include compensatory techniques, such as postural strategies and swallowing maneuvers, and are useful for some individuals.

In contrast to the very predictable improvement in the conventional motor features of PD that occurs with levodopa treatment, there is conflicting evidence regarding the response of oropharyngeal dysphagia to standard antiparkinson medications. Improvement in dysphagia, sometimes striking, has been noted by some investigators, including Cotzias and colleagues in their pioneering studies.[100–103] However, more formal studies employing MBS have demonstrated objective improvement in only 33 to 50% of patients following levodopa or apomorphine administration.[81,83,104] Formal studies evaluating other dopamine agonists in this regard have not been published. The basis for this inconsistent response to dopaminergic medication is uncertain but may simply reflect the intricacy of movement necessary for normal swallowing and the difficulty reconstructing this with systemic drug administration, like trying to do calligraphy with a paint roller. Anticholinergic drugs have been reported to both improve[98] and exacerbate[105] dysphagia in PD patients.

Cricopharyngeal dysfunction may be amenable to more specific treatment maneuvers. Percutaneous injection with BTX has been successfully employed,[106] as has cricopharyngeal myotomy.[107] However, cricopharyngeal myotomy (and probably BTX injection also) should not be performed if the concomitant presence of esophageal dysmotility might leave the individual at greater risk for aspiration following myotomy.[107]

Improvement in esophageal dysphagia has been reported with apomorphine administration,[108,109] although extensive testing has not been undertaken. It has been suggested that sildenafil might be of benefit in the treatment of spastic esophageal motor disorders[110] because of its effect on nitric oxide, but there have been no published reports of its use for this purpose in PD. It is worth noting that transient esophageal obstruction has been attributed to levodopa, with resolution following drug discontinuation.[111]

It is very unusual for dysphagia in idiopathic PD to become sufficiently severe to require percutaneous endoscopic gastrostomy placement, but this procedure can be employed as a final step in patients in whom other treatment approaches have failed and both adequate nutritional support and medication administration have become impossible.[12,112]

GASTRIC DYSFUNCTION

Perhaps the first hint that patients with PD might have impaired gastric emptying (gastroparesis) can be found in the 1981 report of Evans and colleagues, who, although not specifically studying PD, identified delayed gastric emptying in a group of elderly individuals, 55% of whom did have PD, compared to young controls.[113] Subsequent investigators, using a variety of techniques, have confirmed that gastroparesis is, indeed, a common component of PD.[114–118] In one study, the average time to empty one-half of the gastric contents (GET1/2) was 59 min in a group of 28 untreated PD patients, compared with 43.4 min in a control group of slightly younger individuals.[115]

The clinical ramifications of delayed gastric emptying in PD have not been extensively explored. In patients with gastroparesis due to other sources of autonomic dysfunction, such as diabetes mellitus, early satiety, abdominal discomfort with a sense of bloating, nausea, vomiting, weight loss and even malnutrition may occur. Edwards and colleagues suggested that the nausea present in 24% of the PD patients they surveyed, including 16% of those not on PD treatment, might be due to impaired gastric emptying.[7] They also reported that almost 45% of the PD patients (including 43% of those untreated), experienced a sense of bloating, compared to 25% of the spousal controls, although this difference was not statistically significant.

Beyond the clinical symptoms described above, delayed gastric emptying also has potentially important pharmacokinetic implications for the PD patient taking levodopa. Absorption of levodopa takes place in the proximal small intestine.[119–121] Slowed gastric emptying might, therefore, be expected to result in delayed clinical response to levodopa doses. This supposition finds support in the report by Djaldetti and colleagues, who found the average GET1/2 to be 221 min in PD patients with motor fluctuations compared to 85 min in those without fluctuations,[116] but seems to be at odds with that of Hardoff and colleagues in which PD patients with motor response fluctuations had a shorter GET1/2 than those with a smooth response to levodopa.[115]

Several additional factors might further promote inconsistent responses to administered levodopa. Because aromatic amino acid decarboxylase is present in the gastric mucosa, delayed gastric emptying may also allow increased gastric conversion of levodopa to dopamine, rendering it unavailable for subsequent intestinal absorption. Moreover, the dopamine thus formed in the stomach could stimulate gastric dopamine receptors, which promote receptive relaxation of the stomach and inhibit gastric motility, and lead to further delays in gastric emptying.[122,123]

TREATMENT OF GASTROPARESIS

Drugs that block dopamine receptors accelerate gastric emptying, presumably by their action on the gastric dopamine receptors mentioned above. Both metoclopramide and domperidone are dopamine receptor antagonists and effective agents in the treatment of gastroparesis. However, because metoclopramide crosses the blood-brain barrier, it also can block striatal dopamine receptors and adversely impact motor function, rendering it contraindicated in persons with PD. Domperidone demonstrates little or no ability to cross the blood-brain barrier and, thus, can be safely and effectively used in PD patients,[124] although rare reports of domperidone producing extrapyramidal dysfunction can be found.[125] Domperidone is available throughout much of the world, including Canada, but is not approved for use in the U.S.A.

Other medications have also demonstrated efficacy as prokinetic agents. Cisapride, which stimulates acetylcholine release from myenteric cholinergic neurons,[126] was used successfully in PD patients but is no longer readily available, because of cardiac toxicity. While not specifically studied in PD, the macrolide antibiotic erythromycin, which also is a motilin agonist, accelerates gastric emptying in healthy volunteers[127] and has been reported to be superior to metoclopramide, domperidone, and cisapride as a prokinetic agent in patients with gastroparesis.[128] However, concerns about its long-term antibiotic effects have limited its use in chronic disease processes.

Drug delivery methods that bypass the stomach completely are another possible treatment for PD patients with severe gastroparesis. Direct jejunal infusion of drug via tube placement is effective[129] but impractical. Constant subcutaneous infusions of both apomorphine and lisuride have also been used very successfully,[130,131] but this technique is also technically challenging. Trials with the dopamine agonist rotigotine, administered transdermally, are ongoing.[132] Gastric pacemaker placement has been successfully utilized in individuals with severe refractory gastroparesis,[133] but no published reports of its use in PD exist.

SMALL INTESTINE DYSMOTILITY

The fate of small intestinal function in PD is largely unknown, since very few studies have directed their attention toward this most secluded and inaccessible component of the GI tract. Orocecal transit time, a measure of combined gastric and small intestine transit speed, was shown to be markedly prolonged in 15 patients with PD compared with 15 age- and sex-matched control individuals.[33] Abnormalities in small intestinal motor patterns in PD patients have also been demonstrated with small intestine manometry.[134] Small intestinal dilatation has also been observed.[135] In the laboratory disruption of the migrating myoelectric complex has been documented in rats administered MPTP, along with reduction in jejunal myenteric plexus dopamine levels.[136] Whether similar changes occur in PD is unknown.

The clinical consequences of small intestinal dysfunction in PD have not been documented. It is conceivable that the very uncomfortable bloating sensation experienced by some individuals with PD, primarily as an "off" phenomenon, might be related to small intestinal dysmotility, but no study has actually addressed this issue. If this is so, agents that accelerate small intestinal transit time, such as the serotonin-4 receptor agonist prucalopride,[137] might be beneficial for patients with these symptoms.

COLONIC DYSMOTILITY

When patients report the presence of bowel dysfunction, they typically use the term *constipation* as an all-inclusive descriptor that encompasses both decreased bowel movement frequency, often with hard stools, and difficulty with the act of defecation itself in the form of increased straining and sometimes incomplete evacuation.[138] From both a physiological and clinical standpoint, however, these two problems are quite different and a separate classification and discussion of each is warranted. Even within its more narrow medical definition as decreased bowel movement frequency, constipation (or colonic inertia) has undergone a redefinition in recent decades. In the past, it was standard

to label anything less frequent than a daily bowel movement as abnormal, but the current definition of constipation has been pegged as fewer than three bowel movements weekly.

Recognition of this definitional revision is important when reviewing reported frequencies of constipation in PD. Older studies indicate the presence of constipation in roughly 50 to 67% of PD patients,[6,139,140] but in more recent communications using the contemporary definition, frequencies of 20 to 29% have been reported.[7,141]

Multiple studies have documented that the physiological basis for decreased bowel movement frequency in PD is slowed colonic transit of fecal material. Colon transit studies, employing radiopaque markers, have indicated that as many as 80% of persons with PD may have abnormally prolonged transit times.[142] Reported colon transit times in PD patients have varied rather widely for reasons that are not clear. Jost and Schimrigk initially reported an average colon transit time (CTT) of 5 to 7 days (120 to 168 hr) in a group of 20 persons with PD,[142] and in a subsequent study of 22 subjects in whom CTT could be measured, the average time was 130 hr.[143] These times contrast with the report by Edwards and colleagues in a study of 13 PD participants, in which mean CTT was 44 hr, compared to 20 hr in spousal controls.[10] A more recent study further confirms that CTT is slowed in PD, although the times reported (82.4 min in PD patients and 39 min in controls) appear to be incorrectly labeled in minutes rather than hours.[144] Thus, despite the variance in average CTT in published reports, there seems to be ample agreement that CTT is prolonged in PD. Survey studies have suggested that constipation becomes more severe as PD progresses.[7,145] This is supported by another study by Jost and Schimrigk in recently diagnosed PD patients, where average CTT was 89 hr,[146] compared with the considerably longer CTT's reported in their earlier studies cited above, which included individuals with more advanced disease. Prolongation of the CTT in untreated individuals underscores that constipation is not simply medication-induced (although this certainly can occur), but part of the disease process itself. As with other aspects of GI dysfunction, objective abnormalities on testing do not necessarily translate into clinical symptoms, and prolongation of CTT has been noted in PD patients without symptomatic constipation.[147]

While constipation is most frequently recognized in persons who have already developed the motor features of PD, some individuals can retrospectively identify the presence of bowel dysfunction prior to the appearance of the more classical PD motor features.[11] A recent provocative epidemiological study has also identified an association between frequency of bowel movements and the risk of developing PD.[148] In the Honolulu Heart Program study, men who reported a bowel movement frequency of fewer than one per day were found to have a 2.7 times greater risk of developing PD than men who had daily bowel movements, and a fourfold higher risk than those with two or more bowel movements daily. While these findings may simply be a reflection that constipation may herald conventional PD motor features, other hypothetical explanations can also be advanced. Perhaps rapid transit of material through the GI tract, implied by frequent bowel movements, limits exposure to, and absorption of, some toxic substance capable of damaging dopaminergic neurons.

The pathophysiological basis of constipation in PD has not been completely clarified. Both central and peripheral factors may be operative. Animal studies employing intraventricular injection of dopaminergic agents have demonstrated that activation of central D_1 and D_2 receptors stimulates colonic motility by increasing colonic spike bursts.[149] It has also been suggested that this may be coordinated through Barrington's nucleus (also known as the pontine micturition center), which lies adjacent to, or possibly within,[150] the locus coeruleus in the pons.[144,151–153]

Support for a peripheral basis for slowed colonic transit in PD also arises from a number of sources. Kupsky and colleagues, in 1987, were the first to document the presence of Lewy bodies in the colonic myenteric and submucosal plexuses of individuals with PD.[154] This plentiful presence of colonic Lewy bodies was subsequently confirmed by several other groups,[155–157] who noted the Lewy bodies to be present both in dopaminergic neurons and in those containing vasoactive intestinal peptide. Using immunohistochemical methods, Singaram and colleagues were also able to demonstrate a very striking reduction in the number of dopaminergic neurons in the colonic myenteric plexus of PD patients, compared to both healthy controls and individuals with idiopathic constipation.[157]

Other abnormalities within colonic tissue have also been documented in individuals with constipation. Serotonin receptor immunoreactivity was recently found to be reduced in colonic tissue of individuals who underwent subtotal colectomy for treatment of colonic inertia.[158] Other studies of patients with chronic idiopathic intestinal pseudo-obstruction or slow transit constipation have documented a marked loss of interstitial cells of Cajal (ICC), which are believed to function as pacemaker cells in the gut.[159,160] Whether these abnormalities are also present in patients with PD suffering from constipation is unknown.

TREATMENT OF COLONIC DYSMOTILITY

The treatment of slow transit constipation in PD is largely empirical, since very few treatment modalities have undergone rigorous testing in this patient population.[161] Many studies have confirmed that increased dietary fiber reduces CTT in normal individuals,[162] most probably by increasing

bulk within the colonic lumen. A daily fiber intake of 15 g, along with at least 1.5 L of water, has been recommended.[163] Since daily fiber consumption is deficient in many PD patients,[7] increased dietary fiber or fiber supplements, such as psyllium, can be effective in increasing stool frequency.[147] Improved motor function, presumably reflecting increased levodopa bioavailability, has also been documented with increased fiber intake.[164] Adding a stool softener, such as docusate, can also be useful.

If these simple measures are insufficiently effective, an osmotic laxative, such as lactulose or sorbitol, can be a very useful next step. A lactulose dose of 30 ml once or twice daily can be used, with subsequent downward titration of dosage if necessary. Because sorbitol is less expensive than lactulose, it might be considered as a cost-effective alternative.[165] More recently, the effectiveness of polyethylene glycol electrolyte balanced solutions, well known as colon-cleansing agents prior to colonoscopy, administered on a regular or even daily basis in smaller amounts, has been demonstrated in PD patients.[166,167] Patients often turn to irritant laxatives, such as senna-containing compounds, that are available without prescription. These compounds can be effective, but daily use should probably be discouraged because of concerns of potential ENS damage from chronic use, even though such damage has not actually been definitively proven.

The role of prokinetic agents in the treatment of slow transit constipation is uncertain. Cisapride was reported to be effective in PD patients, at least in the short term,[168,169] but is no longer available. Prucalopride, a serotonin-4 agonist, has more recently been shown to be effective as a prokinetic agent in patients with severe chronic constipation,[137,170] but its effect in individuals with PD has not been specifically reported. Anecdotal reports have described the effectiveness of the cholinomimetic agents, pyridostigmine[171] and neostigmine,[172] in the treatment of constipation in PD, but no formal studies of these compounds have been reported. The efficacy of neurotrophin-3 in a small double-blind study of PD patients with constipation has been reported in abstract form.[173] Other potentially effective agents include misoprostol[174] and colchicine.[175]

Potentially life-threatening complications of slow transit constipation in PD include megacolon,[135,154,176,177] intestinal pseudoobstruction, volvulus, and even bowel perforation.[6,12,176,177] Surgical treatment in the form of colectomy may be necessary in such situations.

ANORECTAL DYSFUNCTION

Anorectal dysfunction, characterized by excessive straining and often accompanied by pain and a sense of incomplete evacuation, is actually the more prevalent form of bowel dysfunction in PD. In their survey study, Edwards and colleagues[7] differentiated between decreased bowel movement frequency and defecatory dysfunction and noted the latter in 67% of PD patients, compared with only 29% who reported decreased bowel movement frequency (see above). As with slow transit constipation, anorectal dysfunction can also appear early in the course of PD.[178]

Clinical neurophysiological and radiographic studies have shed considerable light on the pathophysiological basis for disordered defecation in PD. As described earlier, for effective defecation to occur the coordinated contraction and relaxation of a surprising array of muscles must take place. It is now clear from studies such as anorectal manometry, anorectal electromyography, and defecography that this does not always occur in individuals with PD and that dyscoordination may actually be the rule. In one study such abdominopelvic (or pelvic floor) dyssynergia was present in over 60% of PD patients.[178] Lower basal sphincter pressure and difficulty maintaining sphincter pressure have been noted on anorectal manometry in PD patients, as have some more distinctive abnormalities, including unusual phasic contractions of the sphincter muscles during voluntary contraction and a "paradoxical" hypercontractile response of the external anal sphincter and puborectalis muscles on rectosphincteric (rectoanal inhibitory) reflex testing, where sphincter relaxation, rather than contraction, is expected.[10,179–181] Failure of the external anal sphincter and puborectalis muscles to relax during attempted defecation, producing functional outlet obstruction, was originally observed by Mathers and colleagues[182,183] and subsequently confirmed by others.[10] It has been suggested that this is a focal dystonic phenomenon.[182,183] These abnormalities of anorectal muscle function appear to be distinctive for PD and not simply a reflection of constipation in general.[184] Moreover, fluctuation in the severity of the anorectal abnormalities in response to dopaminergic medications has been documented, with deterioration during "off" periods and improvement in function when patients are "on."[180] Evaluation of defecation with rectoanal videomanometry has provided objective confirmation of the subjective sense of incomplete emptying during defecation experienced by many PD patients by demonstrating that incomplete defecation with the presence of significant post-defecation residuals is common in PD.[144]

TREATMENT OF ANORECTAL DYSFUNCTION

The array of treatment options for anorectal dysfunction is somewhat limited. While softening the stool by various measures will make it easier to expel, such measures do not correct the fundamental defect in muscular coordination producing the problem. In fact, laxatives and other measures that hasten the arrival of stool to the rectum may sometimes accentuate the problem, producing a situation that might be likened to a crowd of frantic people trying to exit a burning building through a narrow, or even blocked, exit.

Some evidence suggests that dopaminergic medications may improve anorectal function in individuals with PD. As noted above, improvement in anorectal manometric and electromyographic measures of anorectal function during "on" periods, with deterioration during "off" episodes has been described,[183,185] and improvement following apomorphine injections has also been reported.[183,185] Occasional patients on levodopa will also report that it is easier for them to have a bowel movement when they are "on" than when they are "off."

Injection of BTX into the puborectalis muscle under transrectal ultrasonographic guidance has been successfully employed to treat defecatory dysfunction in PD patients.[186,187] The duration of benefit has not been thoroughly defined, but improvement lasting two to three months has been noted. Although these reports are encouraging, the risk for producing fecal incontinence is present with this procedure and perianal thrombosis has also been reported.[188]

Behavioral techniques, such as defecation training and biofeedback measures, have been successfully employed in the treatment of pelvic floor disorders, but they have not been specifically examined in PD patients. Sacral nerve stimulation is a technique that might conceivably have some application in PD patients, but has not yet been evaluated. Surgical treatment, such as colectomy, is rarely necessary in PD patients.

CONCLUSION

Recognition that nonmotor features, such as GI dysfunction, are an extremely important and frequent component of PD is rapidly growing. Such recognition, and the investigation that is prompted by it, will hopefully lead to a better understanding of the mechanisms responsible for such dysfunction and eventually to more effective treatment. For the individual with PD experiencing these problems, that time cannot come soon enough.

REFERENCES

1. Parkinson, J., *An Essay on the Shaking Palsy*, Whittingham and Rowland, London, 1817.
2. Romberg, M. H., *Nervous Diseases of Man*, Sydenham Society, London, 1853.
3. Charcot, J. M., *Lectures on the Diseases of the Nervous System*, Vol. 1, New Sydenham Society, London, 1877.
4. Hammond, W. A., *A Treatise on Diseases of the Nervous System*, Appleton and Company, New York, 1871.
5. Gowers, W. R., *Diseases of the Nervous System*, P. Blakiston and Company, Philadelphia, 1888.
6. Eadie, M. J. and Tyrer, J. H., Alimentary disorder in parkinsonism, *Aust. Ann. Med.*, 14, 13, 1965.

7. Edwards, L. L. et al., Gastrointestinal symptoms in Parkinson's disease, *Mov. Disord.*, 6, 151, 1991.

8. Edwards, L. L. et al., Gastrointestinal dysfunction in Parkinson's disease: frequency and pathophysiology, *Neurology*, 42, 726, 1992.

9. Edwards, L. L. et al., Gastrointestinal symptoms in Parkinson's disease: 18 month follow-up study, *Mov. Disord.* 8, 83, 1993.

10. Edwards, L. L. et al., Characterization of swallowing and defecation in Parkinson's disease, *Am. J. Gastroenterol.*, 89, 15, 1994.

11. Pfeiffer, R. F. and Quigley, E. M. M., Gastrointestinal motility problems in patients with Parkinson's disease: epidemiology, pathophysiology and guidelines for management, *CNS Drugs*, 11, 435, 1999.

12. Quigley, E. M. M., Gastrointestinal dysfunction in Parkinson's disease, *Semin. Neurol.*, 16, 245, 1996.

13. Quigley, E. M. M., Epidemiology and pathophysiology of gastrointestinal manifestations in Parkinson's disease, in *NeuroGastro-enterology*, Corazziari, E., Ed., deGruyter, Berlin, 1996, 167.

14. Pfeiffer, R. F., Gastrointestinal dysfunction in Parkinson's disease, *Clin. Neurosci.*, 5, 136, 1998.

15. Quigley, E. M. M., Gastrointestinal features, in *Parkinson's Disease. Diagnosis and Clinical Management*, Factor, S. A. and Weiner, W. J., Eds., Demos, New York, 87, 2002.

16. Pfeiffer, R. F., Gastrointestinal dysfunction in Parkinson's disease, *Lancet Neurol.*, 2, 107, 2003.

17. Wuttge-Hanning, A. and Hannig, C., Radiologic differential diagnosis of neurologically-induced deglutition disorders, *Radiologe*, 35, 733, 1995.

18. Hunter, P. C. et al., Response of parkinsonian swallowing dysfunction to dopaminergic stimulation, *J. Neurol. Neurosurg. Psychiatry*, 63, 579, 1997.

19. Miller, A. J., Neurophysiological basis of swallowing, *Dysphagia*, 1, 91, 1986.

20. Hamdy, S. et al., Identification of the cerebral loci processing human swallowing with $H_2^{15}O$ PET activation, *J. Neurophysiol.*, 81, 1917, 1999.

21. Long, J. D. and Orlando, R. C., Anatomy, histology, embryology, and developmental anomalies of the esophagus, in *Sleisenger and Fordtran's Gastrointestinal and Liver Disease*, 7th ed., Feldman, M., Friedman, L. S., and Sleisenger, M. H., Eds., Chap. 31, Saunders, Philadelphia, 2002.

22. Clouse, R. E. and Diamant, N. E., Esophageal motor and sensory function and motor disorders of the esophagus, in *Sleisenger and Fordtran's Gastrointestinal and Liver Disease*, 7th ed., Chap. 32, Feldman, M., Friedman, L.S., and Sleisenger, M.H., Eds., Saunders, Philadelphia, 2002.

23. Quigley, E. M. M., Gastric motor and sensory function, and motor disorders of the stomach, in *Sleisenger and Fordtran's Gastrointestinal and Liver Disease*, 7th ed., Chap. 37, Feldman, M., Friedman, L. S., and Sleisenger, M. H., Eds., Saunders, Philadelphia, 2002.

24. Keljo, D. J. and Gariepy C. E., Anatomy, histology, embryology, and developmental anomalies of the small and large intestine, in *Sleisenger and Fordtran's Gas-*

trointestinal and Liver Disease, 7th ed., Chap. 84, Feldman, M., Friedman, L.S., and Sleisenger, M.H., Eds., Saunders, Philadelphia, 2002.

25. Andrews, J. M. and Dent, J., Small intestinal motor physiology, in *Sleisenger and Fordtran's Gastrointestinal and Liver Disease*, 7th ed., Chap. 85, Feldman, M., Friedman, L.S., and Sleisenger, M.H., Eds., Saunders, Philadelphia, 2002.

26. Cook, I. J. and Brookes, S. J., Motility of the large intestine, in *Sleisenger and Fordtran's Gastrointestinal and Liver Disease*, 7th ed., Chap. 86, Feldman, M., Friedman, L.S., and Sleisenger, M.H., Eds., Saunders, Philadelphia, 2002.

27. Madoff, R. D., Williams, J. G., and Caushaj, P. F., Fecal incontinence, *N. Engl. J. Med.*, 326, 1002, 1992.

28. Abbott, R. A. et al., Diet, body size and micronutrient status in Parkinson's disease, *Eur. J. Clin. Nutr.*, 46, 879, 1992.

29. Beyer, P. L. et al., Weight change and body composition in patients with Parkinson's disease, *J. Am. Diet Assoc.*, 95, 979, 1995.

30. Chen, H. et al., Weight loss in Parkinson's disease, *Ann. Neurol.*, 53, 676, 2003.

31. Toth, M. J., Fishman, P. S., and Poehlman, E. T., Free-living daily energy expenditure in patients with Parkinson's disease, *Neurology*, 48, 88, 1997.

32. Davies, K. N., King, D., and Davies, H., A study of the nutritional status of elderly patients with Parkinson's disease, *Age Ageing*, 23, 142, 1994.

33. Davies, K. N. et al., Intestinal permeability and orocaecal transit time in elderly patients with Parkinson's disease, *Postgrad. Med. J.*, 72, 164, 1996.

34. Markus, H. S., Cox, M., and Tomkins, A. M., Raised energy expenditure in Parkinson's disease and its relationship to muscle rigidity, *Clin. Sci.*, 83, 199, 1992.

35. Kempster, P. A. and Wahlqvist, M. L., Dietary factors in the management of Parkinson's disease, *Nutr. Rev.*, 52, 51, 1994.

36. Gironell, A. et al., Weight gain after functional surgery for Parkinson's disease, *Neurologia*, 17, 310, 2002.

37. Ondo, W. G. et al., Weight gain following unilateral pallidotomy in Parkinson's disease, *Acta Neurol. Scand.*, 101, 79, 2000.

38. Wang, G. J. et al., Brain dopamine and obesity, *Lancet*, 357, 354, 2001.

39. Volkow, N. D. et al., "Nonhedonic" food motivation in humans involves dopamine in the dorsal striatum and methylphenidate amplifies this effect, *Synapse*, 44, 175, 2002.

40. Volkow, N. D. et al., Brain dopamine is associated with eating behaviors in humans, *Int. J. Eat. Disord.*, 33, 136, 2003.

41. Evidente, V. G. H. et al., Serum leptin concentrations and satiety in Parkinson's disease patients with and without weight loss, *Mov. Disord.*, 16, 924, 2001.

42. Batterham, R. L. et al., Inhibition of food intake in obese subjects by peptide YY[3-36,] *N. Engl. J. Med.*, 349, 941, 2003.

43. Korner, J. and Leibel, R. L., To eat or not to eat—how the gut talks to the brain, *N. Engl. J. Med.*, 349, 926, 2003.

44. Dirks, S. J. et al., The patient with Parkinson's disease, *Quintessence Int.*, 34, 379, 2003.

45. Clifford, T. and Finnerty, J., The dental awareness and needs of a Parkinson's disease population, *Gerodontology,* 12, 99, 1995.

46. Durham, T. M. et al., Management of orofacial manifestations of Parkinson's disease with splint therapy: a case report, *Spec. Care Dentist.*, 13, 155, 1993.

47. Jolly, D. E. et al., Parkinson's disease: a review and recommendations for dental management, *Spec. Care Dentist.*, 9, 74, 1989.

48. Chiappelli, F. et al., Dental needs of the elderly in the 21st century, *Gen. Dent.*, 50, 358, 2002.

49. Fiske, J. and Hyland, K., Parkinson's disease and oral care, *Dent. Update*, 27, 58, 2000.

50. Kieser, J. et al., Dental treatment of patients with neurodegenerative disease, *N. Z. Dent. J.*, 95, 130, 1999.

51. Kennedy, M. A. et al., Relationship of oral microflora with oral health status in Parkinson's disease, *Spec. Care Dentist.*, 14, 164, 1994.

52. Persson, M. et al., Influence of Parkinson's disease on oral health, *Acta Odontol. Scand.*, 50, 37, 1992.

53. Fukayo, S., Dental status in outpatients with Parkinson's disease, *Nippon. Eiseigaku Zasshi.*, 57, 585, 2002.

54. Anastassiadou, V. et al., Evaluating dental status and prosthetic need in relation to medical findings in Greek patients suffering from idiopathic Parkinson's disease, *Eur. J. Prosthodont. Restor. Dent.*, 10, 63, 2002.

55. Clifford, T. J. et al., Burning mouth in Parkinson's disease sufferers, *Gerodontology*, 15, 73, 1998.

56. Srivastava T. et al., Bruxism as presenting feature of Parkinson's disease, *J. Assoc. Physicians India,* 50, 457, 2002.

57. Magee, K. R., Bruxisma related to levodopa therapy, *JAMA*, 214, 147, 1970.

58. Sayama, S. et al., Habitual mandibular dislocation in two patients with Parkinson's disease, *Rinsho. Shinkeigaku*, 39, 849, 1999.

59. Minagi, S. et al., An appliance for management of TMJ pain as a complication of Parkinson's disease, *Cranio.*, 16, 57, 1998.

60. Heckmann, S. M., Heckmann, J. G., and Weber, H. P., Clinical outcomes of three Parkinson's disease patients treated with mandibular implant overdentures, *Clin. Oral Implants Res.*, 11, 566, 2000.

61. Ngim, C. H. and Devathasan, G., Epidemiologic study on the association between body burden mercury level and idiopathic Parkinson's disease, *Neuroepidemiology,* 8, 128, 1989.

62. Bagheri, H. et al., A study of salivary secretion in Parkinson's disease, *Clin. Neuropharmacol.*, 22, 213, 1999.

63. Bateson, M. C., Gibberd, F. B., and Wilson, R. S. E., Salivary symptoms in Parkinson's disease, *Arch. Neurol.*, 29, 274, 1973.

64. Eadie, M. J., Gastric secretion in parkinsonism, *Aust. Ann. Med.*, 12, 346, 1963.

65. Marks, L. et al., Drooling in Parkinson's disease: a novel speech and language therapy intervention, *Int. J. Lang. Commun. Disord.*, 36 (Suppl.), 282, 2001.

66. Hyson, H. C., Jog, M. S., and Johnson, A., Sublingual atropine for sialorrhea secondary to parkinsonism. *Parkinsonism Relat. Disord.*, 7 (Suppl.), 194, 2001 (abstract).

67. Bhatia, K. P., Munchau, A., and Brown, P., Botulinum toxin is a useful treatment in excessive drooling of saliva, *J. Neurol. Neurosurg. Psychiatry*, 67, 697, 1999.

68. Bushara, K. O., Sialorrhea in amyotrophic lateral sclerosis: a hypothesis of a new treatment—botulinum toxin A injections of the parotid glands, *Medical Hypothesis*, 48, 337, 1997.

69. Pal, P. K. et al., Botulinum toxin A as a treatment for drooling saliva in PD, *Neurology*, 54, 244, 2000.

70. Jost, W. H., Treatment of drooling in Parkinson's disease with botulinum toxin, *Mov. Disord.*, 14, 1057, 1999.

71. Friedman, A. and Potulska, A., Quantitative assessment of parkinsonian sialorrhea and results of treatment with botulinum toxin, *Parkinsonism Relat. Disord.*, 7, 329, 2001.

72. Lipp, A. et al., Treatment of drooling in Parkinson's disease and motorneuron disease with botulinum toxin A, *Naunyn Schmiedebergs Arch. Pharmacol.*, 365 (Suppl. 2), R38, 2002 (abstract).

73. Mancini, F. et al., Double-blind, placebo-controlled study to evaluate the efficacy and safety of botulinum toxin type A in the treatment of drooling in parkinsonism, *Mov. Disord.*, 18, 685, 2003.

74. O'Sullivan, J. D., Bhatia, K. P., and Lees, A. J., Botulinum toxin A as a treatment for drooling saliva in PD, *Neurology*, 55, 606, 2000.

75. Mullins, W. M., Gross, C. W., and Moore, J. M., Long-term follow-up of tympanic neurectomy for sialorrhea, *Laryngoscope*, 89, 1219, 1979.

76. Clarke, C. E. et al., Referral criteria for speech and language therapy assessment of dysphagia caused by idiopathic Parkinson's disease, *Acta Neurol. Scand.*, 97, 27, 1998.

77. Kurihara, K. et al., Dysphagia in Parkinson's disease, *Rinsho. Shinkeigaku*, 33, 150, 1993.

78. Hartelius, L. and Svensson, P., Speech and swallowing symptoms associated with Parkinson's disease and multiple sclerosis: a survey, *Folia. Phoniatr. Logop.*, 46, 9, 1994.

79. Leopold, N. A. and Kagel, M. C., Prepharyngeal dysphagia in Parkinson's disease, *Dysphagia*, 11, 14, 1996.

80. Logemann, J. A., Blonsky, E. R., and Boshes, B., Editorial: dysphagia in parkinsonism, *J.A.M.A.*, 231, 69, 1975.

81. Bushmann, M. et al., Swallowing abnormalities and their response to treatment in Parkinson's disease, *Neurology*, 39, 1309, 1989.

82. Leopold, N. A. and Kagel, M. C., Pharyngo-esophageal dysphagia in Parkinson's disease, Dysphagia, 12, 11, 1997.

83. Fuh, J. L. et al., Swallowing difficulty in Parkinson's disease, *Clin. Neurol. Neurosurg.*, 99, 106, 1997.

84. Nagaya, M. et al., Videofluorographic study of swallowing in Parkinson's disease, *Dysphagia*, 13, 95, 1998.

85. Silbiger, M. L., Pikielney, R., and Donner, M. W., Neuromuscular disorders affecting the pharynx. Cineradiographic analysis, *Invest. Radiol.*, 2, 442, 1967.

86. Robbins, J. A., Logemann, J. A., and Kirshner, H. S., Swallowing and speech production in Parkinson's disease, *Ann. Neurol.*, 19, 283, 1986.

87. Stroudley, J. and Walsh, M. Radiological assessment dysphagia in Parkinson's disease, *Br. J. Radiol.*, 64, 890, 1991.

88. Logemann, J., Blonsky, E. R., and Boshes, B., Lingual control in Parkinson's disease, *Trans. Am. Neurol. Assoc.*, 98, 276, 1973.

89. Blonsky, E. R. et al., Comparison of speech and swallowing function in patients with tremor disorders and in normal geriatric patients: a cinefluorographic, *J. Gerontol.*, 30, 299, 1975.

90. Ali, G.N. et al., Mechanisms of oral-pharyngeal dysphagia in patients with Parkinson's disease, *Gastroenterology*, 110, 383, 1996.

91. Mari, F. et al., Poor predictive value of clinical measures of dysphagia versus aspiration risk in parkinsonian patients, *Mov. Disord.*, 12 (Suppl. 1), 135, 1997.

92. Bird, M. R. et al., Asymptomatic swallowing disorders in elderly patients with Parkinson's disease: a description of findings on clinical examination and videofluoroscopy in sixteen patients, *Age Ageing*, 23, 251, 1994.

93. Byrne, K. G., Pfeiffer, R. F., and Quigley, E. M. M., Gastrointestinal dysfunction in Parkinson's disease: a report of clinical experience at a single center, *J. Clin. Gastroenterol.*, 19, 11, 1994.

94. Eadie, M. J. and Tyrer, J. H., Radiological abnormalities of the upper part of the alimentary tract in parkinsonism, *Aust. Ann. Med.*, 14, 23, 1965.

95. Gibberd, F. B. et al., Oesophageal dilation in Parkinson's disease, *J. Neurol. Neurosurg. Psychiatry*, 37, 938, 1974.

96. Bassotti, G. et al., Esophageal manometric abnormalities in Parkinson's disease, *Dysphagia*, 13, 28, 1998.

97. Castell, J. A. et al., Manometric abnormalities of the oesophagus in patients with Parkinson's disease, *Neurogastroenterol Motil*, 13, 361, 2001.

98. Penner, A. and Druckerman, L. J., Segmental spasms of the esophagus and their relation to parkinsonism, *Am. J. Dig. Dis.*, 9, 282, 1942.

99. Johnston, B. T. et al., Repetitive proximal esophageal contractions: a new manometric finding and a possible further link between Parkinson's disease and achalasia, *Dysphagia*, 16, 186, 2001.

100. Cotzias, G. C., Papavasiliou, P. S., and Gellene, R., Modification of parkinsonism—chronic treatment with L-dopa, *N. Engl. J. Med.* 280, 337, 1969.

101. Paulson, G. R. and Tafrate, R. H., Some minor aspects of parkinsonism, especially pulmonary function, *Neurology*, 20, 14, 1970.

102. Nowack, W. J., Hatelid, J. M., and Sohn, R. S., Dysphagia in parkinsonism, *Arch. Neurol.*, 34, 320, 1977.

103. Leiberman, A.N. et al., Dysphagia in Parkinson's disease, *Am. J. Gastroenterol.*, 74, 157, 1980.

104. Tison, F. et al., Effects of central dopaminergic stimulation by apomorphine on swallowing disorders in Parkinson's disease, *Mov. Disord.*, 11, 729, 1996.

105. Bramble, M. G., Cunliffe, J., and Dellipani, A. W., Evidence for a change in neurotransmitter affecting oesophageal motility in Parkinson's disease, *J. Neurol. Neurosurg. Psychiatry*, 41, 709, 1978.

106. Restivo, D. A., Palmeri, A., and Machese-Ragona, R., Botulinum toxin for cricopharyngeal dysfunction in Parkinson's disease, *N. Engl. J. Med.*, 346, 1174, 2002.

107. Born, L. J. et al., Cricopharyngeal dysfunction in Parkinson's disease: role in dysphagia and response to myotomy, *Mov. Disord.*, 11, 53, 1996.

108. Wang, S. J. et al., Dysphagia in Parkinson's disease. Assessment by solid phase radionuclide scintigraphy, *Clin. Nucl. Med.*, 19, 405, 1994.

109. Kempster, P. A. et al., Off-period belching due to a reversible disturbance of oesophageal motility in Parkinson's disease and its treatment with apomorphine, *Mov. Disord.*, 4, 47, 1989.

110. Bortolotti, M. et al., Effects of sildenafil on esophageal motility of patients with idiopathic achalasia, *Gastroenterology*, 118, 253, 2000.

111. Kellner, H. et al., Reversible esophageal dysfunction as a side effect of levodopa, *Bildgebung*, 63, 48, 1996.

112. Luman, W. et al., Percutaneous endoscopic gastrostomy—indications and outcome of our experience at the Singapore General Hospital, *Singapore Med. J.*, 42, 460, 2001.

113. Evans, M. A. et al., Gastric emptying rate in the elderly: implications for drug therapy, *J. Am. Geriatr. Soc.*, 29, 201, 1981.

114. Sulla, M. et al., Gastric emptying time and gastric motility in patients with untreated Parkinson's disease, *Mov. Disord.*, 11 (Suppl. 1), 167, 1996 (abstract).

115. Hardoff, R. et al., Gastric emptying time and gastric motility in patients with Parkinson's disease, *Mov. Disord.*, 16, 1041, 2001.

116. Djaldetti, R. et al., Gastric emptying in Parkinson's disease: patients with and without response fluctuations, *Neurology*, 46, 1051, 1996.

117. Krygowska-Wajs, A. et al., Gastric electromechanical dysfunction in Parkinson's disease, *Funct. Neurol.*, 15, 41, 2000.

118. Soykan, I. et al., Gastric myoelectrical activity in patients with Parkinson's disease: evidence of a primary gastric abnormality, *Dig. Dis. Sci.*, 44, 927, 1999.

119. Wade, D. N., Mearrick, P. T., and Morris, J., Active transport of L-dopa in the intestine, *Nature*, 242, 463, 1973.

120. Sasahara, K. et al., Dosage from design for improvement of bioavailability of levodopa: absorption and metabolism of levodopa in intestinal segment of dogs, *J. Pharm. Sci.*, 70, 1157, 1981.

121. Nutt, J. G. and Fellman, J. H., Pharmacokinetics of levodopa, *Clin. Neuropharmacol.*, 7, 35, 1984.

122. Valenzuela, J. E., Dopamine as a possible neurotransmitter in gastric relaxation, *Gastroenterology*, 71, 1019, 1976.

123. Berkowitz, D. M. and McCallum, R. W., Interaction of levodopa and metoclopramide on gastric emptying, *Clin. Pharmacol. Ther.*, 27, 414, 1980.

124. Soykan, I. et al., Effect of chronic oral domperidone therapy on gastrointestinal symptoms and gastric emptying in patients with Parkinson's disease, *Mov. Disord.*, 12, 952, 1997.

125. Barone, J. A., Domperidone: a peripherally acting dopamine2-receptor antagonist, *Ann. Pharmacother.*, 33, 429, 1999.

126. Wiseman, L. R. and Faulds, D., Cisapride, An updated review of its pharmacology and therapeutic efficacy as a prokinetic agent in gastrointestinal motility disorders, *Drugs*, 47, 116, 1994.

127. Boivin, M. A., Carey, M. C., and Levy, H., Erythromycin accelerates gastric emptying in a dose-response manner in healthy subjects, *Pharmacotherapy*, 23, 5, 2003.

128. Sturm, A. et al., Prokinetics in patients with gastroparesis: a systematic analysis, Digestion, 60, 422, 1999.

129. Syed, N. et al., Ten years' *experience* with enteral levodopa infusions for motor fluctuations in Parkinson's disease, *Mov. Disord.*, 13, 336, 1998.

130. Vaamonde, J., Luquin, M. R., and Obeso, J., Subcutaneous lisuride infusion in Parkinson's disease. Response to chronic administration in 34 patients, *Brain*, 114, 604, 1991.

131. Stibe, C. M. et al., Subcutaneous apomorphine in parkinsonian on-off oscillations, Lancet, 1, 403, 1988.

132. Behrens, S. and Sommerville, K., Non-oral drug delivery in Parkinson's disease: a summary from the symposium at the 7th International Congress of Parkinson's Disease and Movement Disorders. November 10–14, 2002, Miami, FL, U.S.A., *Expert Opin. Pharmacother.*, 4, 595, 2003.

133. McCallum, R. W. and George, S. J., Gastric dysmotility and gastroparesis, *Curr. Treat. Options Gastroenterol.*, 4, 179, 2001.

134. Bozeman, T. et al., Small intestinal manometry in Parkinson's disease, Gastroenterology, 99, 1202, 1990 (abstract).

135. Lewitan, A., Nathanson, L., and Slade, W. R., Megacolon and dilatation of the small bowel in parkinsonism, *Gastroenterology*, 17, 367, 1952.

136. Eaker, E.Y. et al., Chronic alterations in jejunal myoelectric activity in rats due to MPTP, *Am. J. Physiol.*, 253, G809, 1987.

137. Emmanuel, A. V. et al., Prucalopride, a systemic enterokinetic, for the treatment of constipation, *Aliment. Pharmacol. Ther.*, 16, 1347, 2002.

138. Stark, M. E., Challenging problems presenting as constipation, *Am. J. Gastroenterol.*, 94, 567, 1999.

139. Schwab, R. S. and England, A. C., Parkinson's disease, *J. Chron. Dis.*, 8, 488, 1958.

140. Pallis, C. A., Parkinsonism: natural history and clinical features, *B.M.J.*, 3, 683, 1971.

141. Siddiqui, M. F. et al., Autonomic dysfunction in Parkinson's disease: a comprehensive symptom survey, *Parkinsonism Relat. Disord.*, 8, 277, 2002.

142. Jost, W. H. and Schimrigk, K., Constipation in Parkinson's disease, *Klin. Wochenschr.*, 69, 906, 1991.

143. Jost, W. H. and Schimrigk, K., The effect of cisapride on delayed colon transit time in patients with idiopathic Parkinson's disease, *Wien Klin. Wochenschr.*, 106, 673, 1994.

144. Sakakibara, R. et al., Colonic transit time and rectoanal videomanometry in Parkinson's disease, *J. Neurol. Neurosurg. Psychiatry*, 74, 268, 2003.

145. Sakakibara, R. et al., Questionnaire-based assessment of pelvic organ dysfunction in Parkinson's disease, *Auton. Neurosci.*, 92, 76, 2001.

146. Jost, W. H. and Schrank, B., Defecatory disorders in de novo parkinsonians—colonic transit and electromyogram of the external anal sphincter, *Wien Klin. Wochenschr.*, 110, 535, 1998.

147. Ashraf, W. et al., Constipation in Parkinson's disease: objective assessment and response to psyllium, *Mov. Disord.*, 12, 946, 1997.

148. Abbott, R. D. et al., Frequency of bowel movements and the future risk of Parkinson's disease, *Neurology*, 57, 456, 2001.

149. Bueno, L. et al., Involvement of central dopamine and D_1 receptors in stress-induced colonic motor alterations in rats, *Brain Res. Bull.*, 29, 135, 1992.

150. Ding, Y. Q. et al., Localization of Barrington's nucleus in the pontine dorsolateral tegmentum of the rabbit, *J. Hirnforsch.*, 39, 375, 1999.

151. Pavcovich, L. A. et al., Novel role for the pontine micturition center, Barrington's nucleus: evidence for coordination of colonic and forebrain activity, *Brain Res.*, 784, 355, 1998.

152. Valentino, R. J., Miselis, R. R., and Pavcovich, L. A., Pontine regulation of pelvic viscera: pharmacological target for pelvic visceral dysfunctions, *Trends Pharmacol. Sci.*, 20, 253, 1999.

153. Vizzard, M. A., Brisson, M., and de Groat, W. C., Transneuronal labeling of neurons in the adult rat central nervous system following inoculation of pseudorabies virus into the colon, *Cell Tissue Res.*, 299, 9, 2000.

154. Kupsky, W. J. et al., Parkinson's disease and megacolon: concentric hyaline inclusions (Lewy bodies) in enteric ganglion cells, *Neurology*, 37, 1253, 1987.

155. Wakabayashi, K. et al., Parkinson's disease: an immunohistochemical study of Lewy-body containing neurons in the enteric nervous system, *Acta Neuropathol*, 79, 581, 1990.

156. Wakabayashi, K. et al., Lewy bodies in the visceral autonomic nervous system in Parkinson's disease, in *Parkinson's Disease. From Basic Research to Treatment (Advances in Neurology, Vol. 60)*, Narabayashi, H. et al., Eds., Raven Press, New York, 1993, 609.

157. Singaram, C. et al., Dopaminergic defect of enteric nervous system in Parkinson's disease patients with chronic constipation, *Lancet*, 346, 861, 1995.

158. Zhao, R. H. et al., Reduced expression of serotonin receptor(s) in the left colon of patients with colonic inertia, *Dis. Colon Rectum*, 46, 81, 2003.

159. Lyford, G. L. et al., Pan-colonic decrease in interstitial cells of Cajal in patients with slow transit constipation, *Gut*, 51, 496, 2002.

160. Jain, D. et al., Role of interstitial cells of Cajal in motility disorders of the bowel, *Am. J. Gastroenterol.*, 98, 618, 2003.

161. Wiesel, P. H., Norton, C., and Brazzelli, M., Management of faecal incontinence and constipation in adults with central neurological diseases, *Cochrane Database Syst. Rev.*, 4, CD002115, 2001.

162. Müller-Lissner, S. A., Effect of wheat bran on weight of stool and gastrointestinal transit time: a meta-analysis, *B.M.J.*, 296, 615, 1988.

163. Corazziari, E. and Badiali, D., Management of lower gastrointestinal tract dysfunction, *Semin. Neurol.*, 16, 289, 1996.

164. Astarloa, R. et al., Clinical and pharmacokinetic effects of a diet rich in insoluble fiber on Parkinson's disease, *Clin. Neuropharmacol.*, 15, 375, 1992.

165. Lederle, F. A. et al., Cost-effective treatment of constipation in the elderly: a randomized double-blind comparison of sorbitol and lactulose, *Am. J. Med.*, 89, 597, 1990.

166. Corazziari, E. et al., Small volume isosmotic polyethylene glycol electrolyte balanced solution (PMF-100) in treatment of chronic nonorganic constipation, *Dig. Dis. Sci.*, 41, 1636, 1996.

167. Eichhorn, T. E. and Oertel, W. H., Macrogol 3350/electrolyte improves constipation in Parkinson's disease and multiple system atrophy, *Mov. Disord.*, 16, 1176, 2001.

168. Jost, W. H. and Schimrigk, K., Cisapride treatment of constipation in Parkinson's disease, *Mov. Disord.*, 8, 339, 1993.

169. Jost, W. H. and Schimrigk, K., Long-term results with cisapride in Parkinson's disease, *Mov. Disord.*, 12, 423, 1997.

170. Coremans, G. et al., Prucalopride is effective in patients with severe chronic constipation in whom laxatives fail to provide adequate relief. Results of a double-blind, placebo-controlled clinical trial, *Digestion*, 67, 82, 2003.

171. Sadjadpour, K., Pyridostigmine bromide and constipation in Parkinson's disease, *J.A.M.A.*, 249, 1148, 1983.

172. Koornstra, J. J. et al., Neostigmine treatment of acute pseudo-obstruction of colon (Ogilvie syndrome), *Ned. Tijdschr. Geneeskd.*, 145, 586, 2001.

173. Pfeiffer, R. F. et al., Effect of NT-3 on bowel function in Parkinson's disease, *Mov. Disord.*, 17, S223, 2002 (abstract).

174. Roarty, T. P. et al., Misoprostol in the treatment of chronic refractory constipation: results of a long-term open label trial, *Aliment. Pharmacol. Ther.*, 11, 1059, 1007.

175. Sandyk, R. and Gillman, M. A., Colchicine ameliorates constipation in Parkinson's disease, *J. R. Soc. Med.*, 77, 1066, 1984.

176. Caplan, L. H. et al., Megacolon and volvulus in Parkinson's disease, *Radiology*, 85, 73, 1965.

177. Rosenthal, M. J. and Marshall, C. E., Sigmoid volvulus in association with parkinsonism. Report of four cases, *J. Am. Geriatr. Soc.*, 35, 683, 1987.

178. Bassotti, G. et al., Manometric investigation of anorectal function in early and late stage Parkinson's disease, *J. Neurol. Neurosurg. Psychiatry*, 68, 768, 2000.

179. Stocchi, F. et al., Anorectal function in multiple system atrophy and Parkinson's disease, *Mov. Disord.*, 15, 71, 2000.

180. Ashraf, W. et al., Anorectal function in fluctuating (on-off) Parkinson's disease: evaluation by combined anorectal manometry and electromyography, *Mov. Disord.*, 10, 650, 1995.

181. Normand, M. M. et al., Simultaneous electromyography and manometry of the anal sphincters in parkinsonian patients: technical considerations, *Muscle Nerve*, 19, 110, 1996.

182. Mathers, S. E. et al., Constipation and paradoxical puborectalis contraction in anismus and Parkinson's disease: a dystonic phenomenon? *J. Neurol. Neurosurg. Psychiatry*, 51, 1503, 1988.

183. Mathers, S. E. et al., Anal sphincter dysfunction in Parkinson's disease, *Arch. Neurol.*, 46, 1061, 1989.

184. Ashraf, W., Pfeiffer, R. F., and Quigley, E. M. M., Anorectal manometry in the assessment of anorectal function in Parkinson's disease: a comparison with chronic idiopathic constipation, *Mov. Disord.*, 9, 655, 1994.

185. Edwards, L. L. et al., Defecatory function in Parkinson's disease: response to apomorphine, *Ann. Neurol.*, 33, 490, 1993.

186. Albanese, A. et al., Severe constipation in Parkinson's disease relieved by botulinum toxin, *Mov. Disord.*, 12, 764, 1997.

187. Albanese, A. et al., Treatment of outlet obstruction constipation in Parkinson's disease with botulinum neurotoxin A, *Am. J. Gastroenterol.*, 98, 1439, 2003.

188. Jost, W. H. et al., Perianal thrombosis following injection therapy into the external anal sphincter using botulinum toxin, *Dis. Colon Rectum*, 38, 781, 1995.

26 Urinary Dysfunction in Parkinson's Disease

Carlos Singer
Department of Neurology, University of Miami School of Medicine

CONTENTS

INTRODUCTION

Parkinson's disease (PD) is defined by motor manifestations of tremor, bradykinesia, rigidity, gait disorder, postural instability and freezing. However, it is associated with multiple nonmotor problems including voiding difficulties. Urological symptoms in PD have a stereotypical presentation and character with its pathophysiology localized in the basal ganglia. A neurologist who obtains a thorough clinical history and who is acquainted with the neuro-urology of PD can play an important role in the management of this particular problem.

PREVALENCE OF URINARY SYMPTOMS IN PD

There is a surprising dearth of information covering the prevalence of urinary symptoms in the parkinsonian population at large.

Murnaghan[22] specifically investigated the presence of urological symptoms in 29 PD patients in the pre-levodopa era, who had been selected to undergo basal ganglia surgery. Eleven (38%) had urological symptoms. Porter and Bors[25] investigated a similar group of 62 patients being considered for basal ganglia surgery. Forty-four patients (71%) had urinary symptoms, but the group was primarily male. These figures have a selection bias. It is likely that older and feeble individuals were excluded.

Singer et al.[33] reported on a group of consecutive parkinsonian male patients attending a movement disorders clinic and compared the prevalence of autonomic symptoms with a group of healthy elderly controls. They found that the prevalence of urinary urgency (46%) and sensation of incomplete bladder emptying (42%) was significantly higher than in controls (3% and 16% respectively).

Sakakibara et al.[29] studied 115 PD patients (52 men and 63 women) and compared them to controls. All urinary symptoms were significantly higher in PD. Urgency (42% women, 54% men); daytime frequency (28% women, 16% men); nighttime frequency (53% women, 63% men); urge incontinence (25% women, 28% men); retardation in initiating urination (44% men); prolongation/poor stream (70% men); straining upon urination (28% women).

Lemack et al.[17] selected 80 men and 39 women with mild to moderate PD (Hoehn and Yahr lower than stage 3) and performed a questionnaire-based assessment using the American Urological Association Symptom Index (AUASI) in men and the Urogenital Distress Inventory-6 (UDI-6) in women. Men scored higher than age-matched controls with similar values to those of men with symptomatic benign prostatic hyperplasia. Results were less clear with PD women, who scored higher than non-age-matched volunteers but lower than an age-matched group of women (unaffected neurologically) presenting for urological evaluation.

DISTRIBUTION AND CHARACTERISTICS OF URINARY SYMPTOMS

Urinary symptoms are grouped either as *irritative*, encompassing frequency, urgency and urge incontinence, or as *obstructive*, represented by hesitancy and weak urinary stream.

Irritative symptoms invariably predominate. Murnaghan[22] found a proportion of 73% (8/11) irritative versus 27% (3/11) obstructive. Raz[27] examined 15 urologically symptomatic PD patients who were not on anticholinergics. Seventy three percent (11/15) had irritative symptoms and 36% (4/11) obstructive. Pavlakis et al.[24] reported a distribution of 57% irritative, 23% obstructive, and 20% mixed symptomatology in a group of 30 PD patients. Berger et al.[7] reported a distribution of 83% irritative and 17% obstructive in a group of 29 PD patients.

Eighty five per cent of the Chandiramani et al.[9] retrospective study of 41 PD patients (35/41) had urgency and frequency (but not incontinence), while only 15% (6/41) had troublesome incontinence as their main complaint, without mentioning if it was preceded by urgency.

Niimi et al.[23] studied seven patients with autonomic failure due to PD. Autonomic failure in PD was defined by the following criteria:

1. Progressive, systemic autonomic failure with predominant cardiovascular dysregulation, including orthostatic hypotension or postprandial hypotension
2. Parkinsonism as sole somatic neurologic manifestation
3. Responsiveness of parkinsonism to levodopa over a long period
4. Exclusion of drug-induced and other secondary forms of parkinsonism by neuroimaging and neurophysiological examinations
5. Absence of cerebellar and pontine atrophy on magnetic resonance imaging

All patients were assessed after being taken off medications for one week. Irritative symptoms were present in five, while obstructive symptoms were present in none.

Raudino[26] reported on the presence of nonmotor symptoms during off periods in 22 patients. Urinary symptoms of frequency and urgency were reported in two and three instances, respectively. Using depression as an example, the author suggested that it is necessary to determine whether any particular nonmotor symptom coincides with an off period or is independent of it.

Obstructive symptoms are less consistently present. They may sometimes be absent,[23,25] even in the presence of detrusor arreflexia,[25] a urodynamic finding well known for its association with hesitancy and weak urinary stream.

Certain urinary symptoms are mentioned in only a few reports. "Urinary retention" was reported in 10% of one series.[24] These symptoms may be related to the "sensation of incomplete bladder emptying" mentioned by others.[33] Post-void dribbling was seen in 7% of the same series.[24] The so-called "insensitive" incontinence is, in the opinion of one reviewer,[35] seen in patients with more advanced disease, but no additional data are provided.

Objective incomplete bladder emptying can occur in PD but is not a frequent finding—but perhaps it is not consistently searched for. Chandiramani et al.[7] reported that only 16% (5/32) of their PD patients had a post-void residual (PVR) of more than 100 ml. Specific correlation with obstructive symptoms was not offered.

There is limited information regarding the time of appearance of urinary symptoms in PD in relation to the motor symptoms. One of the few reports that specifically investigated the issue of time of appearance of urinary symptoms in relation to motor symptomatology was Chandiramani et al.[9] They found that urological symptoms in PD would usually follow the motor symptoms by an average of 5.75 years.

There is also data suggestive that disease severity correlates with presence of urological symptoms. In the Araki et al.[6] study of 70 PD urologically symptomatic patients, symptom index scores increased with disease severity, in particular the obstructive component of the score and its best urodynamic correlate was an elevated post-void residual.

Sakakibara et al.[30] studied ^{123}I-β-CIT SPECT scans of seven PD patients with urinary dysfunction and compared them to four PD patients free of urinary symptoms. The uptake was significantly reduced in the former group, suggesting a link between severity of the nigrostrial dopaminergic deficit and presence of urinary symptomatology. Although the seven urologically symptomatic PD patients had higher mean Unified Parkinson' Disease Rating scale (UPDRS) score and a higher mean Hoehn and Yahr stage, the difference did not achieve significance. No difference was noted in terms of duration of disease.

In conclusion, urological symptoms are more prevalent in the PD population at large, but especially in men. They tend to be irritative rather than obstructive. Urological symptoms may follow motor symptoms by a few years. Although they are commonly seen even in the early stages of the disease, they become more prevalent as the disease progresses.

DETRUSOR HYPERREFLEXIA

Detrusor hyperreflexia is a cystometric finding characterized by the presence of involuntary detrusor contractions in response to bladder filling that the patient is unable to inhibit, with pressure values exceeding 15 cm of water.[5,24,20] This "hyperactive" bladder is able to generate a subjective perception of fullness and a desire to void at an early stage in the course of filling the bladder.[25] A correlation has been described of irritative symptom scores with low maximum cystometric capacity and low volume at initial desire to void.[6] *Detrusor hyperreflexia therefore represents the immediate underlying cause of the irritative urinary symptoms.*

Cystometric studies have revealed a very high incidence of detrusor hyperreflexia in PD[6,7,22.25] and other etiologies of parkinsonism.[15] The reported prevalence of detrusor hyperreflexia among urologically symptomatic individuals ranges from 45% to 100%.[4,6,7,13,15,22,24,25,27] Araki et al.[6] assessed 70 PD consecutive PD patients referred for evaluation of urinary symptoms in whom obstructive etiologies had been excluded. Detrusor hyperreflexia was present in 67% (47/70) (free of additional abnormalities in the voiding phase).

Moreover, detrusor hyperreflexia may also be found in urologically asymptomatic PD patients. Murnaghan[22] reported that 25% (7/28) of PD patients in his study had unhibited contractions. Two of these patients had no urinary symptoms.[22] Stocchi et al.[36] reported a 37% (11/30) prevalence of detrusor hyperreflexia in 30 PD patients, of which 73%(8/11) had no urinary symptoms.

Raz reported a very close clinical correlation of irritative symptoms with detrusor hyperreflexia.[27] However, in a small numbers of cases detrusor hyperreflexia may coincide with obstructive symptoms (Murnaghan 1961; Galloway, 1983; Pavlakis et al., 1983; Fitzmaurice et al., 1985, Berger et al., 1987).[7,13,15,22,24] Araki et al. also found a similar correlation.[6]

There is limited information regarding conditions that predispose to the development of detrusor hyperreflexia. Stocchi et al.[36] reported that, of their 30 PD patients, those with a normal urodynamic pattern (36.6%) had significantly less *severity of disease* and a shorter *duration of disease* in years than those who had abnormal patterns. Araki et al.[6] studied 70 PD patients who had been referred for urological evaluation and who were free

of obstructive etiologies. Sixty seven percent (47/70) had pure hyperreflexia, with the majority (42/47) being in Hoehn and Yahr stage 3 or higher.

DETRUSOR ARREFLEXIA

Detrusor arreflexia, is a cystometrograhic finding where there is a decreased sensation during filling and an increased bladder capacity[25,27] on the order of 600 ml or higher, and a desire to void usually first experienced at a high filling volume.[3] The post-void residual volume is higher than 100 ml.[3] This results in hesitancy and weak urinary stream.[27]

DETRUSOR ARREFLEXIA IS UNCOMMON IN PD

Prevalence figures in series of urologically symptomatic patients[3,13,22,24] have ranged from 0% (0/9)[13] through 11%[6] to 27% (4/15).[27] The only report with unusually high prevalence figures for detrusor arreflexia comes from Porter and Bors, with a figure of 43% (19/44),[25] a discrepancy that remains unexplained.

Once faced with detrusor arreflexia in a PD patient, the clinician should consider anticholinergic effect, multiple system atrophy, and myogenic arreflexia.

According to some,[20] anticholinergics are the most common cause of detrusor arreflexia in PD patients. Although the concurrent use of anticholinergics is frequently mentioned in series reporting findings of detrusor arreflexia in PD, many studies lack careful detail in their clinical correlations.[7,22,24,25]

There are few reports of urologically symptomatic PD patients in which the confounding effect of anticholinergics has been factored in. Raz withdrew anticholinergics one week prior to the urodynamic investigations.[27] He still found a prevalence of 27% for detrusor arreflexia. On the other hand, Stocchi et al.[36] did not find detrusor arreflexia in any of their 30 PD patients—symptomatic or asymptomatic—who were studied with urodynamics with anticholinergics also having been withheld. Moreover, Araki et al.,[6] when reporting on 70 PD patients with urological symptoms, included 20 who were on anticholinergics and in whom they could not find a urodynamic correlation with atonia.

Once drug effect has been excluded, the next step is to consider benign prostatic hypertrophies and other forms of obstruction (see section below, "Coexistent Obstructive Uropathies") causing muscle fiber injury by overdistention, also known as "myogenic arreflexia."[35] However, a similar process may also occur in the absence of obstruction. Araki et al.[6] studied 70 PD patients referred for urological evaluation and who were free of obstructive etiologies. They found six patients (9%)—all stage 4—who had hyperreflexia with impaired contractile function.* To explain this finding, the authors raised a super-imposed myopathic process similar to what has been reported in the aged.[28] Perhaps this process explains older reports in which some cases of detrusor arreflexia were associated with irritative rather than obstructive symptomatology.[3,7,22,25]

Finally, detrusor arrefleixa, especially in the absence of anticholinergic effect or "myogenic arreflexia," should raise the possibility of multiple system atrophy. Please refer to the section, "Differentiation of PD from Multiple System Atrophy," in this chapter.

Although less clearly understood, detrusor arreflexia may also be asymptomatic. Murnaghan[22] studied 18 PD patients selected for basal ganglia surgery and who were asymptomatic from the urological standpoint. Three patients had detrusor arreflexia. However, this finding has not been reproduced in more recent series.

COEXISTENT OBSTRUCTIVE UROPATHIES

Obstructive uropathies (i.e., benign prostatic hypertrophy in the man, stenosis of the bladder neck in the woman) have been recognized as causes in their own right of both irritative and obstructive symptoms in the general population.[35] Such irritative symptoms associated with obstructive uropathies are equally the product of a detrusor hyperreflexia and indistinguishable from the purely neurogenic type.

Certain investigations have pointed to the presence of obstructive uropathies as contributing causes of urinary symptoms in some PD patients.[4,7,13,24] The prevalence figures vary from 17% to 33%.[7,13,24] However, correlation with specific obstructive symptoms is at times not outlined with sufficient clarity.[7,13,24]

DYSFUNCTION OF INFRAVESICAL MECHANISMS

The dysfunction of infravesical mechanisms (DIVM) encompasses the dysfunction of the striated urethral sphincter and that of the pelvic floor. Either dysfunction may occur alone or in combination. DIVM has been inconsistently reported in variable numbers,[3,22,24,36] including its complete absence.[13] Correlation with clinical symptomatology is frequently inadequate or lacking,[3,22,24] and therefore its clinical significance is unclear. Different kinds of dysfunctions have been described. In some cases, the descriptions are poorly characterized and may not be confirmed again in other reports.

The presence of DIVM was suggested by a number of early reports where an *elevated urethral pressure profile* was noted in a proportion of PD patients. Eighteen per cent of patients (2/11) in a series of PD patients with urinary symptoms had evidence of increased urethral resistance at the external sphincter level.[22] Adequate correlation with symptoms was not presented in this particular report.

Berger et al.[7,8] performed uroflow studies in 15 urologically symptomatic patients and demonstrated decreased flow in 10 (less than 12 ml per second). Only in 5 of these 10 patients was there evidence of an obstructive uropathy. A clear clinical correlation with obstructive symptoms or other manifestations was not available in this report. Although the possibility could be raised that the other five patients may have harbored a DIVM, no such activity was documented on sphincter EMG testing during voluntary detrusor contraction.

Raz was able to demonstrate changes in urethral pressure profile as a result of dopaminergic treatment in 66% (10/15) of a group of PD patients with urinary symptoms who were off anticholinergics and free of BPH.[27] Treatment with L-dopa reduced the closure pressure of the urethra as measured by urethral pressure. Interruption of treatment for a week resulted in an increase in the urethral pressure profile. Raz proposed that outlet dysfunction played an important role in the urinary symptomatology of PD by way of an increased tone of the external sphincter, the absence of a well coordinated pelvic floor relaxation during micturition and the lack of normal external sphincter function during interruption of micturition.

The most consistent DIVM consists of delayed relaxation of the striated urethral sphincter and pelvic floor, also known as *sphincter bradykinesia.* To understand this phenomenon, one must first understand that there is a normal guarding reflex where there is an increase in striated muscle activity during vesical filling before the onset of detrusor contraction. *Sphincter bradykinesia* would be an abnormality where *involuntary EMG activity persists through at least the initial part of the expulsive phase of the CMG.*[24]

In the Pavlakis et al. series,[24] 11% (3/28) had sphincter bradykinesia. Galloway[15] reported that 42% (5/12) of his urologically symptomatic patients were unable to relax the external urethral sphincter with voiding and were associated with low flow rates. Andersen et al.[3] studied 24 urologically symptomatic patients with parkinsonism (the words "Parkinson's disease" are not used). The same authors subsequently revised their data in a subsequent article.[5] They reported electromyographic findings in these 24 PD patients. The authors did not specify whether all 24 patients were symptomatic. Twenty-one per cent (5/24) had *impaired sphincter control defined as poor ability to contract or relax the sphincter on command.*

---

* Detrusor hyperreflexia with impaired contractile function was defined as an overactive bladder with uninhibited detrusor contractions associated with low maximum detrusor pressure during the voiding phase of less than 40 cm of water with a slow pressure increase and a large postvoid residual volume.[6]

Pavlakis et al.[24] also found *pseudodyssynergia* in two patients. This phenomenon was defined as "an attempt at continence by voluntary contraction of the pelvic musculature during an involuntary detrusor contraction."[35] These two patients were part of a group of ten in which the maximum flow rate was decreased. The clinical role of this phenomenon was not defined because of coexistent prostatic obstruction. Pseudodyssynergia has not been reported in any of the subsequent articles of this review. *Sphincter "tremor,"* described in 11/12 patients of Galloway's series,[15] has also not been confirmed in subsequent reports.

While Pavlakis et al.[24] called attention to the absence of vesicosphinter dyssynergia, Andersen et al.[3,5] also reported two patients with an abnormality they initially called "dyssynergia" in their first article[3] but later labeled "spasticity."[5] In the Araki et al.[6] series of 70 PD patients referred for urological evaluation and who were free of obstructive etiologies, they found 2 patients (3%) who had both hyperreflexia and *detrusor-sphincter dyssynergia* (2/70).

DIVM may also be asymptomatic. Berger et al.[7,8] studied 29 PD patients (24 men and 5 women) who were urologically symptomatic. They reported sporadic involuntary electromyography activity of the external sphincter during involuntary detrusor contractions in 61% (14/23 patients so tested) without any case resulting in obstruction. They termed this phenomenon *involuntary sphincteric activity*. Because the phenomenon was not associated with radiographic or manometric evidence of obstruction at the level of the membranous urethra, the authors concluded that it did not meet criteria for the definition of detrusor sphincter dyssynergia. This activity is reminiscent of pseudodyssynergia in that both occur in response to involuntary detrusor contractions, but pseudodyssynergia is seen as a voluntary or conscious act.

Stocchi et al.[36] studied 30 PD patients irrespective of presence of urological symptoms. They found that 27% (8/30) had an *inability to relax the perineal muscles* immediately and completely when asked to initiate micturition. This was their only abnormal finding (they had normal cystometrics). Not a single one is described with hesitancy or weak urinary stream. This subgroup of 8 patients had more severity and longer duration of disease than 11 patients with totally normal findings. An additional three patients (10%) had the same abnormality but associated with detrusor hyperreflexia. One of the three patients had urinary incontinence, diurnal and nocturnal, but the authors do not specify if preceded by urge incontinence (no mention of hesitancy or weak urinary stream).

Pavlakis et al.[24] reported "neuropathic potentials" in two of their patients. Andersen et al.[5] also reported two patients as having a "sphincter paralysis" in their 1976 report and as "flaccid sphincter" in their 1985 report. These cases most likely represent multiple system atrophy rather than PD.

EFFECT OF DOPAMINERGIC MEDICATION

One has to distinguish between effects during bladder emptying and effects during bladder filling. The effects on bladder emptying are frequently referred to as effects on voiding efficiency and include parameters such as bladder contractility, urethral pressure, and urethral flow. The effects on bladder filling are particularly focused on the effects on detrusor hyperreflexia, although effects on other parameters are also of interest such as urethral closure pressure.

Effect on Bladder Emptying Phase (a.k.a. Voiding Efficiency)

1. One of the earliest reports on the effect of dopaminergic treatment on urinary function of the parkinsonian patient suggested an "obstructive" effect of L-dopa.[21] The authors studied 24 PD patients (only one woman), questioning for urinary symptoms and performing air cystometrograms and excretory urography including voiding cystourethrograms (only 1 patient did not have a voiding cystourethrogram). They compared two subgroups: 18 patients receiving L-dopa (1.25 to 12 gr/day) and 6 patients not on L-dopa.

   These authors reported that 83% (15/18) of the 18 patients on L-dopa had radiographic evidence of bladder outlet obstruction, excluding a single patient with BPH. Specifically they reported "absence of the normal bladder neck funnel and a prominent lip of the posterior vesical neck, protruding into the urethral lumen." Ten of the 15 "radiologically obstructed" patients had obstructive symptoms (so did the one patient with BPH), 1 patient had irritative symptoms, and 5 patients were asymptomatic.

   Murdock et al.[21] contrasted this group with six patients who were not on L-dopa. One was demented and could not be adequately interviewed. Only one of the other five had obstructive symptoms but also had an enlarged prostate. The rest (two with irritative symptoms and two with no symptoms) had no radiographic evidence of obstruction.

   Murdock et al.[21] concluded that a pharmacological bladder neck obstruction could be caused by the alpha-adrenergic properties of the metabolites of levodopa. They postulated also that beta-adrenergic activity exerted by this drug could result in a decrease in bladder tone.

   We believe, however, that the postulated alpha-sympathomimetic activity of L-dopa metabolites as a cause of functional obstruction should be much less of a factor at present, given the concomitant use of dopa decarboxylase inhibitors in current practice. Moreover, Murdock et al.'s results are contradicted by all subsequent investigations on the subject.

2. Raz performed experiments in anesthetized dogs where urethral pressure profile (UPP) was measured before and after an infusion of L-dopa.[27] L-dopa produced a rapid and persistent drop in closure pressure of the urethra. Treatment with a curare-like striated muscle relaxant produced a similar effect without further enhancement with L-dopa. *L-dopa therefore appeared to exert its action on the distal part of the urethra (external sphincter) and probably has no effect on the smooth muscle component of the urethral closure mechanism.*[27]

   Raz also performed *in vitro* studies on excised urethra and bladder tissue of mongrel dogs.[27] The normal inhibition of basic rhythm and tone of bladder smooth muscle when exposed to a bath of noradrenaline was not modified by pre- or post-treatment with L-dopa. The normal increased rhythmic activity and basic tonus of the smooth muscle of the urethra when exposed to noradrenaline was similarly unaffected by the addition of L-dopa. This corroborated that there was no effect of L-dopa on the smooth muscle of these tissues.

   Raz then demonstrated a decrease in the UPP after treatment was instituted with L-dopa in 10 PD patients affected with urological symptoms.[27] In patients whose treatment with L-dopa was interrupted for one week (number of patients not specified), he reported an increase in UPP.

   In our opinion, Raz's important work has to be viewed as reflective of effect on PD patients when first exposed to L-dopa in contrast to subsequent work performed in more advanced cases such as fluctuators.[37]

3. In Stocchi et al.'s series of 30 PD patients,[36] studied irrespective of presence of urological symptoms), there were 11 with delayed or incomplete perineal floor relaxation. They all experienced greatly improved perineal muscle control after subcutaneous injection of apomorphine (4 mg). There was no effect on the detrusor hyperactivity.

4. In Christmas et al.'s[10] series of 10 PD patients with urinary symptoms, urodynamic studies were performed before and after subcutaneous administration of apomorphine, a dopamine receptor agonist. Voiding efficiency improved after apomorphine injection, with an overall decrease in bladder outflow obstruction. There was an increase in mean and maximum flow rate in nine patients and reduction in post-micturition residual volume in six patients. This was accompanied by fluoroscopic evidence of widening of the urethra at the level of the distal sphincter mechanism. Of additional interest was that three patients were unable to void during the off state despite considerable discomfort and a feeling of bladder fullness.

5. Uchiyama et al.[37] reported effects of a single dose of CD/LD 100 mg on urinary function of 18 PD patients who had severe wearing off. Patients were on L-dopa and dopamine agonists but not on anticholinergics. On one hand, there was an increase in detrusor contraction (force of contraction); on the other hand, there was an increase in urethral obstruction. The net effect favored the increase in bladder contraction with a decrease in residual volume. Consequently, one can say there was a consistent improvement in voiding efficiency.

In summary, the reports by Stocchi et al.,[36] Christmas et al.,[10] and Uchiyama et al.[37] have helped to buttress the idea of an active role of dopaminergic stimulation in improving voiding efficiency. The two first reports used apomorphine and demonstrated a decrease in bladder outflow resistance including promoting relaxation of the perineal floor during micturition. In contrast, the last report used levodopa and showed increase in bladder obstruction, but the net effect—via improved bladder contractility—was similarly an improvement in voiding efficiency.

Effect on Bladder Filling Phase (a.k.a. Bladder Storage)

1. Fitzmaurice et al.[13] reported on nine urologically symptomatic patients with detrusor hyperreflexia. The effects of levodopa were variable. Six patients had less severe detrusor hyperreflexia when off (including one patient whose hyperreflexia disappeared when off), while three were better when on levodopa. A description of impact of treatment on the actual symptoms was not provided.

2. In Christmas et al.'s report, detrusor function during filling and voiding was altered, albeit inconsistently, by apomorphine[10] with detrusor hyperreflexia improved in some cases and exacerbated in others. However, in three patients, poor detrusor contractility contributed to voiding dysfunction during the off state. After apomorphine injection, voiding detrusor pressure

in these three patients increased, while calculated bladder outflow resistance fell, resulting in considerable improvement in voiding. No information was provided as to whether these patients were on anticholinergics.

Christmas et al.[10] pointed out that, since their patients were all premedicated with domperidone, a peripheral dopamine antagonist, it follows that the effect both on smooth and striated musculature of the lower urinary tract are mediated by changes in the central dopaminergic transmission.

3. Uchiyama et al.,[37] in their study of 18 patients with PD who had severe wearing off, showed an unpredictable effect on bladder function during filling. Urinary urgency (with or without detrusor hyperreflexia or low compliance bladder) was aggravated in nine patients (50%), alleviated in three (17%) and unchanged in six (33%). Uchiyama's population represented a particular type of PD patient, namely advanced disease, with severe wearing off, and not receiving anticholinergic therapy.

In summary, the effect of dopaminergic stimulation on detrusor behavior during filling is not predictable, with both improvement and aggravation of detrusor hyperreflexia as possibilities. Additional factors must be playing a role.

DIFFERENTIATION OF PD FROM MULTIPLE SYSTEM ATROPHY

A proportion of patients with parkinsonism do not have PD but other forms of degenerative disease such as Lewy body disease, progressive supranuclear palsy, and multiple system atrophy (MSA).

In MSA, there is a progressive cell loss in the motor nuclei of the striated sphincters located in the S2–S4 segments of the spinal cord (Onuf's nucleus), a finding that has not been reported in PD.[14,38] MSA frequently courses with prominent urological symptoms. Since this disease carries a worse prognosis, early differentiation from PD may allow for more rational management.[38] Investigators have consequently searched for clinical, urodynamic and electrophysiological differences with PD.

Chandiramani et al.[9] performed a retrospective study of 52 patients with MSA and 41 patients with IPD. Sixty percent (31/52) of MSA patients had urinary symptoms preceding or coinciding with the diagnosis of the disease. Sixteen patients reported frequency, urgency, or incontinence before the onset of parkinsonism, and 15 patients developed urinary symptoms at the same time as parkinsonism. In contrast, in 94% of IPD patients, the urogenital

symptoms clearly followed the neurological diagnosis by a few years. The mean duration of IPD was 9 years of age (range 1 to 25) while the mean duration of urinary symptoms was 3.25 years (range 1 to 10). Only two patients had urinary symptoms at the time of diagnosis of IPD. Two other series identified in a review by Fowler also confirm a 60% prevalence of *early* urinary symptoms in MSA.[14]

The severity of the urinary symptoms, including incontinence in patients with MSA, is more marked than in PD.[14] In Chandiramani's series,[9] patients with MSA were more likely to suffer from troublesome incontinence (73%). They were also more likely to have elevated postvoid residuals than PD patients (66% versus 16%, respectively).

Among the males with MSA in Chandiramani et al's series, 93% had erectile dysfunction (ED), including 48% where this complaint preceded the diagnosis of MSA. However, ED can also be seen in PD, although the proportion of early ED is less.[32] One would also expect poor response to urological surgery targeting prostatism, even poorer than with PD. All 11 men with MSA in Chandiramani et al.'s series[9] who had a TURP were incontinent postoperatively. See the section below, *Effects of Urological Surgery.*

Fowler[14] has proposed the following five urogenital criteria as favouring the diagnosis of MSA:

1. Urinary symptoms preceding or presenting with parkinsonism
2. Male ED preceding or presenting with parkinsonism
3. Urinary incontinence
4. Significant post-micturition residue (>100 ml)
5. Worsening bladder control after urological surgery

However, none of these criteria is sufficiently specific and each requires analysis within the context of the individual case.

Berger et al.[8] reported that all patients with Shy-Drager-Syndrome (a variant of MSA with prominent orthostatic hypotension) evaluated with a voiding cystourethrogram had an *open bladder neck at rest.* In cases of PD, only those patients who had undergone a prior prostatectomy had this finding. Therefore, the presence of an open bladder neck during filling in someone who has not had prior surgery would point to the presence of sympathetic dysfunction and be suggestive of a diagnosis of MSA.

The cell loss in Onuf's nucleus reported in MSA has been associated with electromyographic changes of denervation (fibrillations and positive sharp waves) and reinnervation (abnormal polyphasic potentials of prolonged duration). Such urethral sphincter EMG abnormalities are also reflected in the anal sphincter,[11] a more easily accessible structure.

Stocchi et al.[36] found EMG to provide important differentiating data between MSA and PD. The main differentiating feature of the 32 MSA patients compared to 30 PD patients was abnormal sphincter EMG in 24/32 (75%) MSA patients as compared to none in the PD patients. Vodusek conducted a comprehensive review on the subject.[38] He concluded that anal sphincter EMG abnormalities could distinguish MSA from PD in the first five years after the onset of symptoms and signs, if other causes for sphincter denervation (such as surgeries) had been ruled out. With such criteria, however, as Vodusek readily admits, sphincter EMG offers a low sensitivity.

PATHOGENESIS OF VOIDING DYSFUNCTION IN PD

Voiding is a function of the autonomic nervous system with a core segmental representation in the spinal cord. As the bladder fills, afferent stimuli are conducted to the S2–S4 segments.

During filling, the external and internal urethral sphincters are tonically contracted, and there is an increased tone of striated musculature of the pelvic floor. At a certain level of bladder distention, a reflex efferent response is triggered by activated motor neurons, which stimulate the detrusor muscle via the pelvic nerve (parasympathetic) and relax the internal urethral sphincter via parasympathetic inhibition of sympathetic terminals that innervate the bladder neck. At the same time, inhibition of Onuf's nucleus and pudendal motor nuclei cause relaxation of the striated urethral sphincter and the perineal floor.

This segmentally organized function is subject to facilitatory and inhibitory impulses from higher neurologic centers that allow for voluntary control of the detrusor reflex. Specifically, impulses from the cortical micturition center in the mesial frontal lobes[36] would connect to the pontine-mesencephalic reticular formation. This pathway is further influenced by the basal ganglia, the thalamic nuclei, and the anterior vermis of the cerebellum (Pavlakis et al., 1983; Andersen et al., 1985).[5,24] Micturition is also influenced by the anterior cyngulate gyrus, the locus coeruleus, and the nucleus tegmento lateralis dorsalis.[36]

Based on a series of experiments and subsequent experience with basal ganglia surgery, *it is currently believed that the basal ganglia exert an inhibitory effect on the ponto-mesencephalic micturition center.* Lesions of basal ganglia, as in PD, would result in partial or total disconnection of the micturition reflex from voluntary control. The result would be unhibited detrusor contractions elicited at low volume threshold (detrusor hyperreflexia).[5] In PD, the presence of detrusor hyperreflexia with vesi-

cosphincter synergy is therefore suggestive of a suprapontine lesion. In contrast, in multiple sclerosis, the finding of detrusor-sphincter-dyssynergia denotes a lesion of the connections between the pontine micturition center and the spinal cord centers of micturition.

Lewin et al. performed pivotal experimental studies in cats that are still being cited as backbone for current theory on pathophysiology.[18,19] Lewin et al. stimulated the thalamus and different sites of the basal ganglia and found that the stimulation was inhibitory of detrusor contractions. The inhibition ranged from prolongation of intercontraction interval of the detrusor to occasional complete suppression of detrusor contractions with the activity only resuming after stimulation was stopped.

It is interesting to note that stimulation of the red nucleus, the subthalamic nucleus, and the substantia nigra was more inhibitory than that of the thalamus. This may suggest that current deep brain stimulation procedures may be more effective in improving voiding dysfunction if STN rather than the thalamus is the target. Stereotaxic thalamotomy in parkinsonian patients, on the other hand, demonstrated an increase in detrusor activity.[22,25]

The understanding of the pathophysiology of urethral sphincter dysfunction owes a lot to Raz' work.[27] Raz pointed out that, in the initiation of normal micturition, one of the important stages is relaxation occurring prior to maximal bladder contraction. In Parkinson's disease, there can be failure of the perineal muscle floor/shincter to relax rapidly before the detrusor contraction.[24] This delay in the normal relaxation of the pelvic floor would produce hesitancy and slow stream. This phenomenon of sphincter bradykinesia seems to be a condition peculiar to the parkinsonian patient, albeit not universally present. Pavlakis et al. believe that it represents a manifestation of skeletal muscle hypertonicity involving the perineal floor.

Studies in conscious rats suggest that D_1 receptors (linked to stimulation of adenylate cyclase[39]) tonically inhibit the micturition reflex.[31] Administration of mixed D_1/D_2 agonists in anesthetized MPTP-lesioned monkeys increased their pathologically reduced bladder volume threshold.[40] This effect could be antagonized by pretreatment with a D_1 antagonist. This inhibitory effect of the D_1 receptors would presumably be exerted via the forebrain system,[39] perhaps through a potentiation of the GABAergic system in the basal ganglia.[40] *The loss of D_1 activation in Parkinson's disease may therefore underlie the bladder overactivity in Parkinson's disease.*

On the other hand, similar studies in conscious rats suggest that D_2 receptors are involved in facilitation of the micturition reflex.[31] The pure D_2 agonist bromocriptine administered to MPTP-lesioned monkeys decreases their already pathologically reduced bladder volume threshold even further.[40] This excitatory effect of D_2 receptors on the micturition reflex would be exerted directly on the brain stem.[39]

This combination of effects would result in a D_1 effect during bladder filling and a D_2 effect during bladder emptying. We would expect a salutary effect of D_1 agonists on bladder control in Parkinson's disease. Studies comparing the effects of currently available "pure" D_2 versus mixed D_1/D_2 receptor agonists on the voiding dysfunction of PD patients would be of interest. We are only aware of one study (reported in abstract form) where patients with Parkinson's disease affected with urinary urgency and frequency while on bromocriptine experienced an improvement in their symptoms when switched to pergolide.[16]

TREATMENT

TREATMENT OF IRRITATIVE SYMPTOMS

The irritative symptoms in PD—themselves a manifestation of detrusor hyperreflexia—frequently respond to anticholinergics,[5,20] although there are no reports specifically evaluating the effectiveness and safety in PD.[34] Examples of commonly prescribed anticholinergics include oxybutinin (Ditropan®), propantheline bromide (Pro-banthine), hyoscyamine sulfate (Cystopaz® and other), flavoxate hydrochloride (Urispas®), and tolteridone tartrate (Detrol®).[1] Oxybutinin is possibly the most frequently used of these medications, with dosages ranging between 2.5 mg at hs and 5 mg t.i.d.

Side effects include the production of symptoms of obstructive type[20] such as hesitancy and weak urinary stream. Other well known side effects include dry mouth, difficulty with visual accommodation, constipation, and aggravation of glaucoma.

Tolterodine and its major active metabolite, DD 01, are muscarinic receptor antagonists that, in animals, are more active on the bladder than on the salivary glands. Based on an analysis of four 12-week double-blind studies in more than 1,000 patients, the dose of 2 mg PO BID has been proven effective. Tolterodine decreases the number of incontinence episodes per 24 hr, decreases by 20% the number of micturitions per day (same as oxybutinin), and increases the volume voided per micturition by 22% (oxybutinin increased it by 32%).[1]

Tolterodine may cause less side effects,[1] such as a lower incidence of severe dry mouth as compared to oxybutinin (4% versus 29%). Cardiac and cognitive effects are alleged to be less, but the original trials excluded those patients with history of "serious side effects" on oxybutinin. In addition, precautions related to narrow angle glaucoma or urinary retention remain relevant to tolterodine as they are to the other anticholinergics.

Some experts have suggested using the extended-release form of anticholinergics to prevent high serum levels during therapy with the idea that this may result in less likelihood of cognitive dysfunction.[34] Examples include tolterodine LA at doses of 2 to 4 mg QD and oxybutinin LA at doses of 5 to 30 QD.[1]

More recently, oxybutinin transdermal (Oxytrol®) has been released. This route avoids first-pass metabolism, resulting in a lower concentration of its active metabolite. Since this metabolite has a higher affinity *in vitro* for parotid cells than for bladder cells, it may explain the low incidence of dry mouth reported with transdermal oxybutinin.[2]

If therapy with a single anticholinergic agent proves to be suboptimal, the tricyclic antidepressant imipramine hydrochloride (Tofranil®) can be used in combination, since it has a different receptor site profile.[35]

TREATMENT OF OBSTRUCTIVE SYMPTOMS

The successful treatment of the "obstructive" symptoms of hesitancy and weak urinary stream begins with a careful drug history, searching for medications with an anticholinergic effect. Urodynamic studies should follow, investigating for the presence of detrusor arreflexia, DIVM, or an obstructive uropathy.

A frequent clinical setting for the development of detrusor arreflexia in PD occurs when symptomatic detrusor instability (hyperreflexia) is treated with anticholinergics. This may result in urodynamic findings of involuntary bladder contractions associated with incomplete emptying secondary to unsustained detrusor contractions.[35] In that case, management consists in combining anticholinergics with clean intermittent catheterization by self or others.[35] Successful management is will also help in preventing recurrent urinary tract infections.

The frequency with which catheterizations should be performed will depend on the degree of hesitancy or the volume of the post void residual.[35] The patient who attains continence at the cost of not being able to void at all will have to undergo catheterization every 5 to 8 hr, depending on the residual volume, which should be maintained below 500 cc.[35]

Finding detrusor arreflexia in the absence of anticholinergic treatment and in the absence of overdistention injury ("myogenic arreflexia") secondary to BPH (or other obstructive uropathies) raises the possibility of multiple system atrophy. Patients with MSA are also more likely to have poor bladder compliance and sphincter insufficiency.[8] This could result in episodes of incontinence, including overflow, and stress incontinence (in addition to hesitancy and weak stream). Intermittent catheterization with or without anticholinergics (i.e., oxybutinin) may be the initial treatment.[8,9] In some cases desmopressin spray may be of use.[9] Due to the motor dysfunction, treatment may evolve to permanent indwelling catheterization or suprapubic cystostomy.[8,34] Stress incontinence in females can be treated with urethral suspension or a sling procedure, but if there is concurrent detrusor hyperreflexia, the result may be suboptimal.[34]

Another possible cause of obstructive symptoms is represented by DIVM. In the cases of external urethral sphincter bradykinesia or pseudodyssynergia with high voiding pressures (above 90 cm H_2O), certain authors recommend anticholinergics and intermittent catheterization (similar to mixed detrusor hyperreflexia with incomplete bladder emptying).[35] The reason is that persistent high pressures are certain to result in damage to the bladder and, ultimately, to the upper urinary tract.[35] Sphincter bradykinesia has also been shown to be responsive to dopaminergic treatment,[10,27,36] while pseudodyssynergia may be correctable with biofeedback.[24]

EFFECT OF UROLOGICAL SURGERY

Obstructive uropathies coexistent with PD may also cause obstructive symptoms, and at the same time they may trigger detrusor hyperreflexia in their own right. The obstructive symptoms may be further enhanced by an overdistention injury to the bladder ("myogenic arreflexia"), which may gradually resolve after relief of the obstruction. (It should be noted that "myogenic arreflexia" may also be secondary to a temporary obstruction of the bladder outlet).[35]

The surgical relief of a well documented bladder outlet obstruction is well advised in the PD affected patient. However, the patient should understand clearly that such surgeries (i.e., prostatectomy) are primarily indicated for relief of the obstruction and to avoid the need of catheterization,[7] but they may not resolve the sometimes coexistent irritative symptoms.

Resolution of the detrusor instability can be expected in 60 to 70% of patients postoperatively if the instability is the result of prostatic obstruction.[35] Berger et al.[7] reported persistence of urge incontinence in eight PD men who had undergone prostatic surgery with evidence of detrusor hyperreflexia in seven. They could not find any urodynamic parameters to predict preoperatively which hyperreflexic bladder would stabilize after successful relief of the obstruction.[7] If urge incontinence persists after surgery, anticholinergic therapy may be added. If it still persists, a condom catheter drainage may be necessary. There are no urodynamic parameters capable of predicting preoperatively which hyperreflexic bladder will stabilize after successful relief of the obstruction.

Urologists should be aware of the need to rule out MSA prior to surgery. In Chandiramani et al.'s series,[9] postoperative results were very different for PD and MSA patients. These authors reported that three of the five IPD patients operated who underwent TURP reported a good result. One patient with an adequate flow rate had persistent urgency despite oral oxybutinin but improved considerably after intravesical oxybutinin. Another patient had a large PVR (post-void residual) after TURP and was said to have an atonic bladder of unknown etiology. Chandiramani et al. also reported that all 11 men with MSA who had a TURP were incontinent postoperatively. Nine (82%) had the problems immediately, and two (18%) eventually became incontinent within the year. Similarly, five anti-incontinence procedures in three women were unsuccessful.

EFFECTS OF BASAL GANGLIA SURGERY

There is very limited information on the effects of basal ganglia surgery on urological dysfunction. The few reports present contradictory results.

Murnaghan reported results of basal ganglia surgery on urological symptoms and urological findings in 29 PD patients. Eight complained of bladder disturbances, and 11 had abnormal cystometrograms. Eleven patients had cystometrograms performed pre- and postoperatively. and only five were unchanged postoperatively. Normal bladder function was converted into hyperreflexic bladder in two out of four patients examined before and after stereotaxic lesions on the thalamic nuclei, whereas stereotactic lesions of the posterior limb of the internal capsule normalized three out four uninhibited bladders. Murnaghan concluded that *thalamotomy may be associated with increased bladder tonus, pallidotomy with decreased bladder tonus and capsulotomy may decrease tonus but bladder sensation may be affected.*[22]

In 1971, Porter and Bors[25] also reported on the effects of thalamotomy on bladder function. They studied the effects of uni- and bilateral thalamotomy on 49 patients with PD (11 of whom had normal function). They concluded that neurogenic bladder dysfunction was more frequently seen in bilateral than in unilateral cases. It was only after bilateral stereotaxic surgery that improvement of bladder function could be consistently documented.

The same authors then followed up on the status of 40 patients over a "long term" (4 to 8 months after their last operation, uni- or bilateral). These patients had somatic manifestations that had been "significantly improved" after the surgery (no quantification provided). The results indicated to the authors that the neurogenic bladder of the parkinsonian patient was responsive to surgical therapy, although the response was not as prompt or as successful as the treatment of the somatic manifestations. Furthermore, the subjective response of the individual was often more pronounced than the objective evidence of improvement.

The authors also postulated that thalamotomy improved the post-void residual volume by relaxing the bladder floor and especially in the "hypoactive bladder," by increasing the activity of detrusor muscle.[25] This is consistent with the findings of Murnaghan. It would have been of interest to learn if the use of anticholinergics had decreased postoperatively as a possible alternative explanation to decrease in post-void residual.

Andersen et al.[4] examined 44 patients with parkinsonism, including 8 who had undergone thalamotomies. None of the eight patients had normal bladder function. The authors concluded that stereotactic operations on the thalamus could produce uninhibited bladder contractions with subsequent risks of urological disturbances.

To date, there is only one report of effect of basal ganglia surgery on parkinsonian voiding dysfunction stemming from the new era that started in the 1990s.[12] The authors studied five patients who had undergone bilateral implantation of subthalamic nucleus electrodes. These patients had not been assessed urologically preoperatively. Instead, they were studied urodynamically 4 to 9 months after surgery with comparisons made between the stimulator-on and stimulator-off states (no mention made as to being on or off levodopa during the procedures). The authors found consistent improvement in bladder capacity (bladder volume at which urinary leakage was observed or if leakage did not occur, bladder volume at unbearable desire to void) and reflex volume (bladder volume at first hyperflexic detrusor contraction).

REFERENCES

1. Abramovicz, M. et al., Tolterodine for Overactive Bladder, *Medical Letter,* 40: 101–103, 1998.
2. Abramovicz, M. et al., Oxybutinin Transdermal (Oxytrol) For Overative Bladder, *Medical Letter,* 45(1156): 38–39, 2003.
3. Andersen, J. T., Bradley, W. E., Cystometric, sphincter and electro-myelographic abnormalities in Parkinson's disease, *J. of Urol.,* 16: 75–78, 1976.
4. Andersen, J. T., Hebjorn, S., Frimodt-Moller, C., Walter, S., Worm-Petersen, J., Disturbances of Micturition in Parkinson's Disease, *Acta Neurol. Scandinav.,* 53: 161–170, 1976.
5. Andersen, J. T., Disturbances of Bladder and Urethral Function in Parkinson's Disease, *Int. Urol. Nephrol.,* 1: 35–41, 1985.
6. Araki, I., Kitahara, M., Tomoyuki, O., Kuno, S., Voiding Dysfunction and Parkinson's Disease: Urodynamic Abnormalities and Urinary Symptoms, *J. Urol.,* 164(5): 1640–1643, 2000.
7. Berger, Y., Blaivas, J. G., DeLaRocha, E.R., Salinas, J. M., Urodynamic findings in Parkinson's disease, *J. Urol.,* 138: 836–83, 1987.
8. Berger, Y., Salinas, J. N., Blaivas, J. G., Urodynamic differentiation of Parkinson Disease and the Shy Drager Syndrome, *Neurourology and Urodynamics,* 9:117–121, 1990.
9. Chandiramani, V. A., Palace, J., Fowler, C. J., How to recognize patients with parkinsonism who should not have neurological surgery, *Brit. J. Urol.,* 80: 100–104, 1979.
10. Christmas, T. J., Chapple, C. R., Lees, A. J., Kempster, P. A., Frankel, J. P., Stern, G. M., Milroy, E. J. G., Role of subcutaneous apomorphine in parkinsonian voiding dysfunction, *The Lancet,* pp. 1451–1454, December 24/31, 1998.
11. Eardley, I., Quinn, N. P., Fowler, C. J., Kirby, R. S., Parkhouse, H. F., Marsden, C. D., Bannister, R., The value of urethral sphincter electromyography in the differential diagnosis of parkinsonism, *Brit. J. Urol.,* 64:360–362, 1989.
12. Finazzi-Agrò, E., Peppe, A., D'Amico, A., Petta, F., Mazzone, P., Stanzione, P., Micali, F., Caltagirone, C., Effects of Subthalamic Nucleus Stimulation on Urodyanmic Findings in patients with Parkinson's Disease.
13. Fitzmaurice H., Fowler, C. J., Rickards, D., Quinn, N. P., Marsden, C. D., Milroy, E. J. G., Turner-Warwick, R. T., Micturition Disturbance in Parkinson's Disease, *Brit. J. Urol.,* 57: 652–656, 1985.
14. Fowler, C. J., Urinary Disorders in Parkinson's Disease and Multiple System Atrophy, *Funct. Neurol.,* 16: 277–282, 2001.
15. Galloway, N. T. M., Urethral Sphincter Abnormalities in Parkinsonism, *Brit. J. Urol.,* 55: 691–693, 1983.
16. Kuno, S., Mizuta, E., Yoshimura, N., Differential effects of D$_1$ and D$_2$ agonists on neurogenic bladder in parkinson's disease and MPTP-induced parkinsonian monkeys, *Mov. Disord.,* 12 (Suppl. 1): 63, 1997.
17. Lemack, G. E., Dewey, Jr., R. B., Roehrborn, C. G., O'Suilleabhain, P. E., Zimmern, P. E., Questionnaire-Based Assessment of Bladder Dysfunction in Patients with Mild to Moderate parkinson's Disease, *Urology,* 56: 250–4, 2000.
18. Lewin, R. J., Porter, R. W., Inhibition of spontaneous bladder activity by stimulation of the globus pallidus, *Neurology,* (Minneap) 15:1049–1052, 1965.
19. Lewin, R. J., Dillard, G. U., Porter, R. W., Extrapyramidal Inhibition of the Urinary Bladder. *Brain Research,* 4: 301–307, 1967.
20. Martignoni, E., Pacchetti, C., Godi, L., Micieli, G., Nappi, G., Autonomic Disorders in Parkinson's Disease, *J. Neural. Transm.,* (Suppl.) 45: 11–19, 1995.
21. Murdock, M. I., Olsson, C. A., Sax, D. S., Krane, R. J., Effects of levodopa on the bladder outlet, *J. Urol.,* 113:803–805, 1975.
22. Murnaghan, G. F., Neurogenic Disorders of the Bladder in parkinsonism, *Brit. J. Urol.,* 33:403–409, 1961.
23. Niimi, Y., Ieda, T., Hirayama, M., Koike, Y., Sobue, G., Hasegawa, Y., Takahashi, A., Clinical and physiological characteristics of autonomic failure with parkinson's disease, *Clin. Auton. Res.,* 9:139–144, 1999.
24. Pavlakis, A. J., Siroky, M. B., Goldstein, I., Krane, R. J., Neurourologic Findings in Parkinson's Disease, *J. Urol.,* 129:80–83, 1983.
25. Porter, R. W., Bors, E., Neurogenic bladder in parkinsonism: effect of thalamotomy, *J. Neurosurg.,* 34: 27–32, 1971.
26. Raudino, F., Non motor off in Parkinson's disease, *Acta Neurol. Scand.,* 104:312–313, 2001.

27. Raz, S., Parkinsonism and Neurogenic Bladder. Experimental and Clinical Observations, *Urol. Res.*, 4:133–138, 1976.

28. Resnick, N. M., Yalla, S. V., Detrusor hyperactivity with impaired contractile function. An unrecognized but common cause of incontinence in elderly patients, *JAMA*, 257(22): 3076–81, 1987.

29. Sakakibara, R., Shinotoh, H., Uchiyama, T., Sakuma, M., Kashiwado, M., Yoshiyama, M., Hattori, T., Questionnaire-based assessment of pelvic organ dysfunction in Parkinson's disease, *Autonomic Neuroscience: Basic and clinical*, 92: 76–85, 2001.

30. Sakakibara, R., Shinotoh, H., Uchiyama, T., Yoshiyama, M., Hattori, T., Yamanishi, T., SPECT imaging of the dopamine transporter with [^{123}I]-β-CIT reveals marked decline of nigrostriatal dopaminergic function in Parkinson's disease with urinary dysfunction, *J. Neurol. Sci.*, 187: 55–59, 2001.

31. Seki, S., Igawa, Y., Kaidoh, K., Ishizuka, O., Nishizawa, O., Andersson, K.E., Role of Dopamine D_1 and D_2 Receptors in the Micturition Reflex in Conscious Rats, *Neurology and Urodynamics*, 20:105–113, 2001.

32. Singer, C., Weiner, W. J., Sanchez-Ramos, J., Ackerman, M., Sexual dysfunction in men with Parkinson's disease, *J. Neurol. Rehab.*, 3(4):199–204, 1989.

33. Singer, C., Weiner, W. J., Sanchez-Ramos, J. R., Autonomic Dysfunction in Men with Parkinson's Disease, *Eur. Neurol.*, 32:134–140, 1992.

34. Siroky, M. B., Neurological disorders. Cerebrovascular disease and parkinsonism, *Urol. Clin. N. Am.*, 30: 27–47, 2003.

35. Sotolongo, J. R., Voiding Dysfunction in Parkinson's disease, *Seminars Neurol.*, 8:166–9, 1988.

36. Stocchi, F., Carbone, A., Inghilleri, M., Monge, A., Ruggieri, S., Berardelli, A., Manfredi, M., Urodynamic and neuro-physiological evaluation in Parkinson's disease and multiple system atrophy, *J. Neurol. Neurosurg. Psychiatry*, 62: 507–511, 1997.

37. Uchiyama, T., Sakakibara, R., Hattori, T., Yamanishi, T., Short-Term Effect of a Single Levodopa Dose on Micturition Distubance in parkinson's Disease Patients with Wearing-Off Phenomenon, *Mov. Disord.*, 18(5): 573–578, 2003.

38. Vodusek, D. B., Sphincter, E. M. G., Differential Diagnosis of Multiple System Atrophy, *Mov. Disord.*, 16: 600–7, 2001.

39. Yokoyama, O., Komatsu, K., Ishiura, Y., Akino, H., Kodama, K., Yotsuyanagi, S., Moriyama, N., Nagasaka, Y., Ito, Y., Namiki, M., Overactive bladder—Experimental Aspects, *Scand. J. Urol. Nephrol.*, Suppl., 210: 59–64, 2002.

40. Yoshimura, N., Mizuta, E., Yoshida, O., Kuno, S., Therapeutic Efficacy of Dopamine D_1/D_2 Receptor Agonists on Detrusor Hyperreflexia in 1-Methyl-4-Phenyl-1,2,3,6-Tetrahydropiridine-Lesioned Parkinsonian Cynomolus Monkeys, *J. Pharmacol. Exp. Therap.*, 286:228–233, 1998.

27 Sexual Dysfunction

Cheryl Waters and Janice Smolowitz
Division of Movement Disorders, Department of Neurology, Columbia University

CONTENTS

INTRODUCTION

Sexual interest and behavior may be altered in persons with Parkinson's disease (PD).[1-25] Impairment of sexual function may take the form of underactivity or impotence.[1-10] However, there exists a literature of anecdotal reports and uncontrolled studies describing resumption of sexual activity, increased interest in sex, and hypersexuality in PD patients as a result of antiparkinsonian therapy.[11-25] This chapter provides an overview of the literature to assist clinicians in identifying an aspect of disease that is not frequently discussed but greatly affects quality of life.

IMPAIRED SEXUAL FUNCTION

Impaired sexual function may result from emotional and physical illnesses as well as increasing age.[26-28] A variety of sexual functions and related variables have been studied in adults with PD using validated, self report questionnaires and interviews of men and women with PD, couples with one spouse affected by PD, men with PD, and women with PD. Comparison groups have included healthy adults matched for age and gender as well as age matched controls with chronic, nonneurological disease with motor impairment. There are no studies with quantitative measures that objectively evaluate sexual function.

The following three studies have described sexual function in men and women with PD. Thirty-six men and 14 women with idiopathic PD and no signs of mental deterioration completed a structured questionnaire that addressed sexual activity, function, and libido.[1] Partici-

pants mean age was 57.9 years, standard deviation 10.1 years. The duration of disease was 7.01 years, standard deviation 3.9 years. Sixty-eight percent of participants reported decreased sexual activity. Twenty-six percent described decreased libido. Erectile dysfunction (ED) was reported in 38.8% of men. ED was described more frequently in men over 61 years of age.

To determine whether adults with PD differed in sexuality from similarly aged healthy adults, 121 adults with PD and 126 controls matched for age and gender participated in a study comparing opinions about public sexual attitudes, emotion from personal sexual practice, personal sexual function, and general health perception.[2] Adults with PD were recruited from a PD self-support organization and physicians' patient lists. Controls were recruited for participation from a community registry. A physician investigator examined the adults with PD and reviewed their medical records. The physician completed the motor portion of the Unified Parkinson's Disease Rating Scale (UPDRS),[29] the Hoehn and Yahr[30] score, and interviewed participants about disease variables and sociodemographic data. In the presence of the investigator, participants completed a 33-item multiple-choice self-report questionnaire that addressed different aspects of sexuality,[31] a depression scale,[32] and the Wechsler Adult Intelligence scale[33] to measure the influence of education.

All subjects reported that they were currently living in heterosexual relationships. Frequency of intercourse did not differ between adults with PD and controls. The average age of adults with PD was 45 years. Adults with PD reported greater disagreement with present attitudes about sexuality than controls. Significantly, more adults with

PD were unemployed and depressed. Adults with PD indicated greater dissatisfaction with their personal sexual lives than controls. Those with depression expressed greater sexual dissatisfaction than those without depression. Men with PD reported greater dissatisfaction than women. Depressed, unemployed adults with PD were more often dissatisfied with their current sexual relationship, felt lonely more often, and were less able to enjoy small flirtations. Adults with PD were less satisfied with their lives, felt older than their stated age, and perceived their health to be poorer than controls.

Twenty-five patients with PD, 15 men and 10 women, younger than age 56, were interviewed on sexual function.[3] A female neurologist conducted the interviews and physical examinations. Participants' mean age was 50.3 years. The mean age of onset of disease was 44.7 years. Changes in libido were not statistically different between men and women, although women reported more marked changes in libido. Women reported more changes in sexual activity than men. Causes of sexual dysfunction in men included ED ($n = 3$), reduced libido after initiation of medication ($n = 2$), change in orgasm ($n = 2$), and lack of partner's acceptance ($n = 1$). Four women reported reduced libido after initiation of medication, three women reported change in orgasm, three women reported vaginal dryness, five women reported sexual dysfunction due to rigidity, and one woman reported lack of partner's acceptance. Four women and four men reported urinary incontinence. There was one female with major depression on the Beck Depression Inventory (BDI).[34] She was not sexually active. Fifty-five percent of the participants in this sample of optimally treated PD patients were found to have changes in sexual function.

Two studies have described patients' and spouses' perceptions of the affected partners' sexual function and aspects of the couples' relationship.[4,5] Thirty-six men and 14 women with PD, and their spouses, were recruited from a movement disorder clinic for participation in an investigation of the relationship of autonomic nervous system (ANS) dysfunction, depression, medication, motor disabilities, and sexual difficulties.[4] Patients and their spouses completed separate, self-report questionnaires. The patients were asked to answer the Geriatric Depression Scale (GDS),[35] a questionnaire of degree of sexual interest, arousal, and performance skills,[36] a medical history, a medication history, ANS function (increased sweating, constipation, or urinary difficulties). Spouses completed a questionnaire that addressed sexual interest, arousal, and performance of the affected spouse as well as their own sexual interests.

Patient mean age was 67.3 years, and the mean duration of disease was 6.96 years. Eighty percent of men stated that their sexual frequency had decreased since diagnosis of PD. Forty-four percent of men reported decreased sexual interest and drive. Fifty-four percent

were not able to achieve an erection. Fourteen percent reported they were able to maintain an erection. Depression was present in 19% of the male patients. Sexual dysfunction was present in 1.7% of these patients. Sixty-nine percent of male patients had ANS dysfunction. Of these, 70 percent reported problems with sexual function.

Seventy-nine percent of the female patients stated that their sexual frequency had decreased since diagnosis. Seventy-one percent of women patients reported a decrease in sexual interest. Thirty-eight percent of women were unable to achieve orgasm. Thirty-eight percent reported vaginal dryness during intercourse. Sixty-seven percent felt it was more difficult to be aroused. Seventy-five percent stated that frequency of orgasm was reduced since diagnosis. Depression was present in one woman. Seventy-eight percent of couples shared the same bed. A decrease in the affected partner's sexual interest was noted by 54% of the spouses.

Young onset PD patients and spouses, attending a weekend residential meeting in the United Kingdom, were surveyed for the purpose of estimating the prevalence of sexual dysfunction in patients with PD and their partners, describing the nature of sexual difficulties experienced and the relationship between sexual dysfunction, psychological morbidity, psychosocial stress, physical disability and autonomic dysfunction.[5] Forty-four couples attended the meeting. Thirty-four couples and four spouses of PD patients participated in the study. Twenty-three male and 11 female patients completed questionnaires. Data describing age of onset of PD, current medications, and physical disability were collected independently from patients and partners. Sexual function was assessed by the Golombok Rust Inventory of Sexual Satisfaction.[37] Marital function was assessed using the Golombok Rust Inventory of Marital Status.[38] Depression and anxiety in patients and spouses was assessed using the BDI[34] and the State Trait Scale Anxiety Inventory.[39] Patients completed an acceptance of illness scale.[40] Spouses completed a caregiver strain index.[41] Autonomic dysfunction was rated by questionnaire. Three neurologists rated the likelihood of autonomic dysfunction based on answers to the questionnaire.

Male patients (mean age 51.9 years, SD 8.9 years) were significantly older and had a later onset of disease than female patients (mean age 44.7 years, SD 7.2 years). A statistically significant difference was not found in the duration of illness or degree of disability for male and female patients. Sexual dissatisfaction and perception that sexual problems existed were greatest in couples where the patient was male. Marital dissatisfaction was highest in male patients and their partners. BDI scores were highest in the male and female patient groups. Thirty-six percent of the female patients and 29% of the male patients were depressed. Fifteen percent of female spouses were depressed. Female spouses demonstrated significantly greater trait anxiety than male spouses ($p < 0.01$). Significant differences were not demonstrated in caregiver strain and acceptance of illness. Thirty-nine percent of male and 54% of female patients were rated as having possible or probable autonomic dysfunction.

Singer reported on autonomic dysfunction, including sexual dysfunction in 48 men with PD.[6,7] The PD patients were compared with 32 healthy elderly men. ED affected 60.4% of men with PD versus 37.5% of controls. ED was not associated with other autonomic features, duration of levodopa therapy, or age.

The sexual function in men with PD and arthritis was compared in one study.[8] Sexual function and its relationship to age, severity of PD, and depression was described in 41 married men with PD. Twenty-nine married men with arthritis served as a comparison group. Men with history of dementia, illnesses, or use of medications known to cause impotence were excluded from participation. Men with PD were recruited from three neurology clinics. Men with arthritis were recruited from arthritis clinics at the same three hospitals. Providers of PD participants rated the patients' stage of disease using the Hoehn and Yahr scale[30] and the Columbia Parkinson scale.[42] Providers of patients with arthritis rated severity of disease using the Functional Capacity in Rheumatoid Arthritis Scale.[43] Participants completed the Zung Depression Scale[44] and the Sexual Functioning Questionnaire.[45]

The two groups were well matched for age but differed in duration of disease. The average duration of PD was 6 years, compared with 15 years for patients with arthritis. Similarities were found between the two groups. Total scores for sexual functioning, and subscores for desire, arousal, orgasm, satisfaction, and frequency of sex per month did not differ significantly between the two groups. Age was significantly related to total sexual function score (PD r = $-.40$, $p < 0.05$, arthritis r = $-.39$, $p < 0.05$). Sexual dysfunction increased with severity of illness. Sexual dysfunction, without depression, was found in both groups.

One report has exclusively addressed sexual function in women with PD.[9] Twenty-seven married women with PD and 27 age-matched, married women without a history of neurological disease participated in the study. Women with PD were assessed for presence of autonomic nervous system dysfunction by the presence of significant postural hypotension, and a history of urinary or fecal incontinence. To establish severity of disease, according to the Hoehn and Yahr scale,[30] neurological examinations of women with PD were conducted when they were in the "on" motor state. All participants completed the Brief Index of Sexual Functioning for Women (BISF-W)[46] and (BDI).[34]

Women with PD and women in the control group differed in employment and ethnicity. Twenty-two percent of PD patients were employed, 67% were retired, and 11% were unable to work, whereas 37% percent of control group participants were employed, and 63% were retired. Of the 27 PD patients, 23 were Caucasian, 2 were Asian, and 2 were Hispanic. In the control group, 19 women were Caucasian, 1 was Asian, 3 were Hispanic, 3 were African, and 1 woman described herself as "other." Approximately 50% of both samples were sexually active. PD patients reported less satisfaction with their sexual relationship than the control group. Women with PD reported greater anxiety or inhibition during sex ($p = 0.04$), more difficulty with vaginal tightness ($p = 0.03$), and more problems with involuntary urination ($p = 0.03$). PD patients were less satisfied with their partners than controls ($p = .005$). PD patients were significantly more depressed than community controls. In women with PD, the Hoehn and Yahr stage of disease was mildly correlated with change in satisfaction and change in sexual activity. In both groups, age was associated with change in sexual satisfaction and sexual activity.

THERAPEUTIC INTERVENTIONS FOR ADULTS WITH PD AND IMPAIRED SEXUAL FUNCTION

Impaired sexual function in PD patients is most likely multifactorial. Partners and couples should be individually assessed to guide therapeutic intervention. Depression, physical disability, and autonomic dysfunction may contribute to the increased incidence of ED in men with PD.[17] In men with PD, there is limited discussion of first line pharmacological therapies for ED. Treatments with testosterone gel[47] and sildenafil[48,49] have been described. Testosterone is thought to stimulate libido in the central nervous system.[50] Erections in response to erotic visual stimuli may be partially androgen dependent.[51] Animal and human studies have found that low normal range concentrations of testosterone are sufficient to maintain sexual activity.[52] Testosterone deficiency is found in 20 to 25% of men over age 60.[47] Testosterone deficiency can result in depression, fatigue, decreased libido, and decreased work performance. Okun et al.[47] retrospectively analyzed the effect of testosterone replacement therapy in five men with PD and evidence of plasma testosterone deficiency. The men had not clinically improved with antidepressants and antiparkinsonian medication. Four of the men were initially screened with the St. Louis Testosterone Deficiency Questionnaire (SLTDQ).[53] Men who met SLTDQ criteria were screened for total and free testosterone levels. Prostate specific antigen (PSA) and digital rectal exams were performed to exclude presence of prostate cancer. The UPDRS motor score was recorded. Patients with testosterone levels less than 70 pg/ml with no medical contraindications were treated with a topical application of testosterone gel. One month later, patients reported decrease in fatigue, depression, anxiety, and improved sexual function. To assess the prevalence of testosterone deficiency, total testosterone levels for 68 men enrolled in a PD registry were sent for evaluation. Thirty-five percent of the men had evidence of plasma testosterone deficiency. The risk of testosterone deficiency increased 2.8-fold per decade.

Sildenafil, a selective inhibitor of cGMP-specific phosphodiesterase type 5, enhances the effect of NO release into the corpora cavernosa from nonadrenergic-noncholinergic nerves of the parasympathetic system and vascular endothelium during sexual stimulation.[26] Sildenafil potentiates the hypotensive effect of nitrates and is absolutely contraindicated in men using nitrates.[54] Sildenafil may be hazardous to men with borderline low blood pressure, borderline low cardiac volume, or medications that can prolong its half-life.[55] Adverse effects include headache, flushing, nasal congestion, dyspepsia, abnormal vision, diarrhea, and dizziness.[26] To optimize the treatment outcome, sildenafil should be ingested on an empty stomach. Excessive alcohol consumption should be avoided.

To evaluate the efficacy and safety of sildenafil citrate in men with PD and ED, ten men participated in an eight-week, open-label pilot study.[48] The BDI,[34] UPDRS,[29] and a Sexual Health Inventory-M version questionnaire[56] were administered prior to the initiation of treatment and at the conclusion of the treatment period. Four 50-mg doses of sildenafil citrate were prescribed for use in four sexual encounters during the first month. At the conclusion of the first month, participants had telephone conversations with a urologist and a movement disorder neurologist. Participants were then permitted to increase the dose to 100 mg for each of four sexual encounters during the second month. All participants took eight doses of medication during the study period. Four men increased the dose to 100 mg during the second month. A statistically significant improvement in total SHI-M scores was demonstrated ($p = 0.01$). Significant improvement was demonstrated in overall sexual satisfaction, satisfaction with sexual desire, achievement of erection, maintenance of erection, and orgasm. One patient reported headache during three encounters. There were no reports of syncope or presyncope.

Twenty-four men with ED, 12 with PD, and 12 with multiple system atrophy (MSA) participated in a randomized, double-blind, placebo-controlled, crossover study of sildenafil.[49] Participants completed the International Index of Erectile Function questionnaire[57] and a quality of life questionnaire. Partners completed a brief questionnaire. The starting dose for active drug was 50 mg. The dose was titrated up to 100 mg or down to 25 mg at follow-up visits, depending on efficacy and tolerability. Ten of the 12 men with PD completed the study. Nine of the

ten men reported a good response to sildenafil. Eight men titrated up to 100 mg. One man titrated down to 25 mg. While one man reported lack of efficacy, most of the participants with PD reported significant improvement in ability to achieve and maintain erection and improvement in sex life with sildenafil. Partners' questionnaire responses confirmed the patients' reports. Men with PD demonstrated minimal change in blood pressure (BP). Six men with MSA were studied before recruitment was stopped. Four men received placebo first. Three men with MSA experienced significant postural fall in BP with symptoms of orthostatic hypotension one hour after receiving sildenafil. MSA patients reported improved sexual function and quality of sex life after receiving sildenafil. The authors recommended measurement of lying and standing BP as well as education about symptoms of hypotension before prescribing sildenafil for men with early parkinsonism, which may be difficult to distinguish from MSA.

A diagnostic framework and classification system for female sexual dysfunction has identified four categories of dysfunction[58] that are consistent with DSM IV: Diagnostic and Statistical Manual of Mental Disorders of the American Psychiatric Association and International Classification of Diseases-10 categories.[59]

A comprehensive approach to the evaluation of sexual function in women should include complete medical history, physical examination, pelvic examination, hormonal profile, and physiologic testing as indicated.[60] Therapeutic interventions for the treatment of impaired sexual function in women with PD were not identified in this literature review.

ANTIPARKINSONIAN THERAPY AND HYPERSEXUALITY

With the advent of antiparkinsonian therapy, reports of increased libido and sexual performance, hypersexual behavior (with or without concomitant hypomania), and rarely paraphilias have appeared in the literature.[11–25] Comparisons cannot be made among studies, as different criteria were used to collect data. Due to the small size of the studies, the incidence of increased sexual drive in patients receiving treatment with antiparkinsonian therapy cannot be determined. The data does not provide a basis for identifying patients who might develop adverse sexual behavior in response to antiparkinsonian medications. Since the initial reports were conducted, medication regimens and dosages have changed. This also affects applicability of reports to current patient care.

Not all early reviews found that levodopa increased sexual behavior.[61] A review of 152 patients treated for parkinsonism between 11/1/68 and 7/31//69 did not find significant changes in patients' sexual interests. Increased sexual awareness or activity was reported in 3 to 4 instances. The authors noted that the 5 men in the 40 to 60 year age group, with a relatively mild form of parkinsonism and 1 to 2 year history of impotence, demonstrated improved motor function with therapy but did not report a change in sexual function.

Other reviews described increase in sexual interest and or activity in some patients after L-dopa treatment. Barbeau[11] reported an increase in libido in four men after L-dopa therapy. However, erections were not sustained, and men had premature ejaculation. According to Yahr and Duvoisin,[12] 8% of 283 patients reported improvement in motor function with levodopa and increased sexual activity allowing a return to previous patterns. Hyyppa[13] reported that 10 of 41 patients, 7 males and 3 females, treated with 4 to 5 g of L-dopa per day for 2 to 9 months reported increased libido. Three men and two women had markedly increased sexual activity. Two patients reported sexual dreams. One person reported a decrease in libido when L-dopa was decreased from 5 to 3 g.

To assess the effect of L-dopa treatment on sexual behavior in 12 men and 7 women, treated for 3 to 15 months, semistructured interviews were conducted.[14] The interviewers assigned numerical values to interview responses. Six men and one woman (37%) reported activation of sexual behavior at some point during the therapy. There were strong negative trends between sexual activity and age of patient ($r = -.42$) and duration of parkinsonism ($r = -.44$). Three patterns of change in sexuality were described. In the first pattern, general improvement in overall function was accompanied by mild improvement in sexual function. This result depended on the patient's past sexual habits, age, and availability of a partner. In the second pattern, activation of sexual drive was independent of overall functional improvement. Three (16%) men demonstrated this effect, which was usually mild and did not persist despite continuation of levodopa. The third pattern was loss of sexual inhibition in patients who developed an acute brain syndrome.

Psychiatric interviews, sexual and affective rating scales, hormonal studies, and neurological assessment were used to evaluate L-dopa therapy in 7 men with PD, mean age 62, and mean duration of illness 4 years.[15] Four men reported increased sexual interest or activity related to treatment. One man also reported increased interest with placebo.

Levodopa's effect on mood was assessed in 20 patients followed during initiation and maintenance of therapy.[16] Six of nine men had spontaneous erections while taking 4 to 6.5 g of levodopa/day. Three of the men had been impotent for up to ten years prior to levodopa therapy and were generally puzzled and embarrassed. The erections were not related to sexual objects and were not accompanied by sexual fantasies. Three of the men

reported an increase in libido. One man who had been impotent was able to resume satisfactory sexual intercourse.

Erection has been reported in 5 of 15 men attending a movement disorder neurology clinic as a side effect of treatment with apomorphine injections.[17] Erections coincided with apomorphine administration in these patients with motor fluctuations who were offered the apomorphine to relieve severe symptoms. Four of the men had experienced ED before beginning apomorphine treatment. Two of the men reported improvement in their sexual relationship with their partner as a result of treatment. The one patient that had not experienced ED prior to beginning apomorphine treatment reported undesirable arousal associated with the erections.

Sandyk[18] reported on two men with PD and ED, ages 70 and 73, who experienced sexual arousal and nocturnal erections after receiving treatment for PD with transcranial administrations of AC-pulsed electromagnetic fields (EMFs) of 7.5 picotesla flux density. The first patient received EMF treatment for two consecutive days. He reported a decrease in parkinsonian symptoms after the first treatment and experienced sexual arousal and awakening during the night with several repetitive spontaneous erections lasting 15 to 20 min. During the second treatment, he experienced sexual arousal. The patient experienced nocturnal erections during the following three nights. The second patient had two successive EMF treatments for four days. This patient reported sexual arousal associated with nocturnal erection.

Hypersexuality is defined as a disturbance of sexuality in which there is a greatly or morbidly increased sexual activity.[62] Case reports describe increased masturbation, increased attempts at intercourse with the patients' partner, and initiation of extramarital affairs. Hypersexual behavior is one of the least common adverse psychiatric effects of antiparkinsonian therapy, occurring in less than 1% of patients.[25]

Hypersexual behavior in adults with PD has been reported as a complication of amantadine, levodopa, selegiline, bromocriptine, pergolide, thalamotomy, and high frequency-subthalamic stimulation. Hypersexual responses to therapeutic modalities occurred more often in men and persons with earlier than usual onset of PD. Hypomanic behavior has been associated with hypersexuality in some but not all patients.[11] In some men, excessive libido occurred despite ED. Hypersexual behavior was not always accompanied by improved motor function.

Transient mania with hypersexuality has been reported among 4 of 30 PD patients after high-frequency subthalamic nucleus implant surgery.[19] A 57-year-old woman with a history of hypomania during youth developed a marked increase in sexual drive which gradually appeared during the first month, lasted for 18 months, and then gradually resolved. A 54-year-old man with PD for ten years developed increased sexual interest and manic symptoms two months after implant that gradually subsided. Two men with young onset PD developed mania and hypersexuality several days after implant that gradually resolved after several months. When the stimulator was turned off, motor function deteriorated while symptoms of mania persisted. Two cases of transient hypersexuality following bilateral thalamotomy have been reported. Both patients were sexually overactive prior to thalamotomy.[20]

Among patients that exhibited hypersexual behavior as a complication of medication therapy, the behavior first appeared after initiation or increase in the dosage of a medication and resolved when the dosage was lowered or the medication was discontinued. Low-dose clozapine has been used to treat dopaminergic-induced psychiatric symptoms, including hypersexuality.[63,64] Clozapine improved control of sexual behavior without changing motor scores.[63]

CAUSES OF HYPERSEXUAL BEHAVIOR

Different mechanisms have been suggested to explain hypersexuality in response to antiparkinsonian therapy. The dopaminergic system, which is widely distributed in the central nervous system (CNS) and pelvic organs, is necessary for male sexual arousal and ejaculation, as documented in animal experiments and human studies.[65] The serotonergic system, which is also widely distributed in the CNS, has an inhibitory role in the sexual response cycle. Dopaminergic agents such as levodopa, bromocriptine, and pergolide may promote sexual behavior by activating the dopaminergic system and lowering serotonin concentrations at postsynaptic sites.[25]

Prolactin decreases libido. It has been hypothesized that dopaminergic therapy precipitates hypersexual behavior by decreasing serum prolactin levels.[11] Dopamine inhibits prolactin secretion from the anterior pituitary. Patients with hyperprolactinemia, treated with bromocriptine, demonstrate improved libido and decreased prolactin levels. Less than 1% of PD patients exhibit hypersexuality as a response to therapeutic interventions.[25] It has been postulated that hypersexuality is a symptom of dopaminergic overstimulation in susceptible individuals.[21]

PARAPHILIAS

Dopaminergic therapy may lead to paraphilias in predisposed individuals. Paraphilias are defined as disorders of specialized sexual fantasies, intense sexual urges, and practices that are repetitive and distressing to the individual.[23] The incidence of paraphilias in PD patients treated with dopaminergic therapy is not known.[66] Most PD

patients had not acted on their sexual interests prior to antiparkinsonian therapy. Usually, a delay was reported between the time antiparkinsonian treatment was initiated and the sexual behavior began.[11] A temporal, dose-dependent relationship has been reported between levodopa, pergolide, bromocriptine, selegiline, and paraphilic behavior.[22,23,66] Paraphilic behavior is treated by decreasing medication dosage or adding neuroleptic agents. Clozapine has been used to treat dopaminergic-induced paraphilias without reduction of antiparkinsonian medication so adequate motor function is maintained.[64,67]

Sexuality has been described as a holistic phenomenon that is more than and different from its physiological components.[68] Sexuality and sexual function are affected by the interaction of physiological, emotional, intellectual, spiritual, social, and cultural influences.[69] Life events, such as illness, affect sexuality.[26] PD and antiparkinsonian therapy can adversely affect sexual behavior. Patients and family members may be reluctant to discuss sexual concerns with health care providers.[64] Clinicians need to initiate and encourage discussion of sexual function. Patients should be questioned about sexual dysfunction at regular intervals and when medications are increased.

This chapter has overviewed sexual dysfunction in PD. Further research is needed to understand this complex phenomenon.

REFERENCES

1. Burguera, J. A., Garcia Reboll, L., and Martinez Agullo, E., Sexual dysfunction in Parkinson's disease, *Neurologia*, 9,176, 1994.
2. Jacobs, H., Vieregge, A., and Vieregge P., Sexuality in young patients with Parkinson's disease: a population based comparison with healthy controls, *J. Neurol. Neurosurg. Psychiatry,* 69, 550, 2000.
3. Wermuth, L., and Stenager, E., Sexual problems in young patients with Parkinson's disease, *Acta Neurol. Scand.*, 91, 453, 1995.
4. Koller, W. C. et al., Sexual dysfunction in Parkinson's disease, *Clinical Neuropharmacol.*, 13, 461, 1990.
5. Brown, R. G. et al., Sexual function in patients with Parkinson's disease and their partners, *J. Neurol. Neurosurg. Psychiatry*, 53, 480, 1990.
6. Singer, C. et al., Sexual function in patients with Parkinson's disease, *J. Neurol Neurosurg. Psychiatry*, 1991; 54:942.
7. Singer, C., Weiner, W. J., and Sanchez-Ramos J., Autonomic dysfunction in men with Parkinson's disease, *Eur. Neurol.*, 32,1992.
8. Lipe, H. et al., Sexual function in married men with Parkinson's disease compared to married men with arthritis, *Neurology*, 40, 1347, 1990.
9. Welsh, M., Hung, L., and Waters, C., Sexuality in women with Parkinson's disease, *Mov. Dis.* 12, 923, 1997.

10. Lambert, D., and Waters, C. Sexual dysfunction in Parkinson's disease, *Clin. Neuroscience*, 5, 73, 1998.

11. Barbeau, A., L-dopa therapy in Parkinson's disease: a critical review of nine years experience, *Canad. Med. Ass. J.*, 101, 59, 1969.

12. Duvoisin, R. C. and Yahr, M. D., Behavioral abnormalities occurring in Parkinsonism during treatment with L-dopa, in *L-dopa and Behavior*, Malitz, S., Ed, Raven Press, New York, 1972, 57.

13. Hyyppa, M., Rinne, U. K., and Sonninen, V., The activating effect of L-dopa treatment on sexual function and its experimental background, *Acta Neurologica Scandinavica*, 43, 223, 1970.

14. Bowers, M. B., Van Woert, M., and Davis, L., Sexual behavior during L-dopa treatment for Parkinsonism, *Am. J. Psychiat.*, 12, 127, 1971.

15. Brown, E. et al., Sexual function and affect in Parkinsonian men treated with L-dopa, *Am. J Psychiatry*, 135, 1552, 1978.

16. O'Brien, C. P., C. P. et al., Mental effects of high-dosage levodopa, *Arch. Gen. Psychiat.*, 24, 61, 1971.

17. O'Sullivan, J. D., and Hayes, A. J., Apomorphine induced penile erections in PD, *Mov. Dis.*, 13, 536, 1998.

18. Sandyk, R., AC pulsed electromagnetic fields-induced sexual arousal and penile erections in PD, *Intern. J. Neuroscience*, 99, 139, 1999.

19. Romito, L. M. et al., Transient mania with hypersexuality after surgery for high-frequency stimulation of the subthalamic nucleus in Parkinson's disease, *Mov. Dis.* 17, 1371, 2002.

20. Uitti, R. J. et al., Hypersexuality with antiparkinson therapy, *Clin. Neuropharm.*, 12, 375, *1989*.

21. Vogel, H. P. and Schiffter, R., Hypersexuality—a complication of dopaminergic therapy in Parkinson's disease, *Pharmacopsychiat.*, 16, 107, 1983.

22. Quinn, N. P. et al., Dopa-dependent sexual deviation, *Brit. J. Psychiat.*, 142, 296, 1983.

23. Riley, D. E., Transvestic fetishism in a man with Parkinson's disease treated with selegiline, *Clin. Neuropharm.*, 25, 234, 2002.

24. Bares, M., Pohanka M., and Rektor, I. Penile erections and hypersexuality induced by pergolide treatment in advanced fluctuating Parkinson's disease, *J. Neurol.*, 249, 112, 2002.

25. Shapiro, S. K., Hypersexual behavior complicating Levodopa therapy, *Minn. Med.*, 56, 58, 1973.

26. Shabsigh, R. and Anastasiadis, A. G., Erectile dysfunction, *Annu. Rev. Med.*, 54, 153, 2003.

27. Laumann, E. O., Paik, A., and Rosen, R. C., Sexual dysfunction in the United States: prevalence and predictors, *JAMA*, 281, 537, 1999.

28. Jonler, M. et al., The effect of age, ethnicity and geographical location on impotence and quality of life, *Br. J. Urol*, 75, 651, 1999.

29. Fahn, S. et al., Unified Parkinson's disease rating scale, in *Recent Developments in Parkinson's disease II*, Fahn, S., Marsden, C. D., and Goldstein, M., Eds., Macmillan, Florham Park, 1987,153.

30. Hoehn, M. and Yahr, M. D,. Parkinsonism: onset, progression and mortality, *Neurology*, 17, 427, 1967.

31. Schneider, H. D., Sexualverhalten in der zweiten lebenshalfte, in *Ergebnisse*.

32. *Sozialwissenschaftlicher Forschung*, Kohl-hammer Verlag, Stuttgart, 1980.

33. Zerssen, D. and Von Koeller, D. M., Paranoid-Depressivitats-Skala, Weinheim, Beltz Test, 1976.

34. Wechsler D., Handanweisung zum Hamburg-Weschler-Intelligenztest fur Erwachsene (HAWIE) Bern, Hans Huber Verlag, 1982.

35. Beck, A. T. and Beamesdorfer, A., Assessment of depression: the depression inventory, *Pharmacopsychiatr.*, 7, 151, 1974.

36. Yesavage, J. A. et al., Development and validation of a geriatric depression screening scale: a preliminary report, *J. Psychiatric. Res.*, 17, 37, 1983.

37. Othmer, E. and Othmer, S. C., Evaluation of sexual dysfunction, *J. of Clin. Psychiatry*, 48, 191,1987.

38. Rust, J. and Golombok, S., The Golombok Rust inventory of sexual satisfaction, Windsor: NFER-Nelson, 1986.

39. Rust, J., Bennun, I., et al., The Golombok-Rust inventory of marital state, *Sexual and Marital Therapy*, 1, 55, 1988.

40. Spielberger, C. D., Gorsuch, R. L., and Lushene, R. E., *Manual for the State-Trait Anxiety Inventor*, Consulting Psychologists Press, Palo Alto California 1970.

41. Felton, B. J. and Revenson, T A., Coping with chronic illness: a study of illness controllability and the influence of coping strategies on psychological adjustment, *J. Consult. Clin. Psychol.*, 2, 343, 1984.

42. Robinson, B. C., Validation of a caregiver strain index, *J. Gerontol.*, 38, 344, 1983.

43. Yahr, M. D. et al., Treatment of Parkinsonism with levodopa, *Arch. Neurol.*, 21, 343, 1969.

44. Steinbrocker, O., Traeyer, C. H., and Batterman, R. C., Therapeutic criteria in Rheumatoid Arthritis, *JAMA*, 140, 659, 1949.

45. Zung, W. K., A self-rating depression scale, *Arch. Gen. Psychiatry*, 12, 63, 1965.

46. Watts, R. J., Sexual functioning, health beliefs, and compliance with high blood pressure medication, *Nursing Res.*, 31, 278, 1982.

47. Taylor, J. F., Rosen, R. C., and Leiblum, S. R., Self-report assessment of female sexual function: psychometric evaluation of the Brief Index of Sexual Functioning for Women, *Arch. Sex. Behav.*, 23, 627, 1994.

48. Okun, M. S, McDonald, W. M., and DeLong, M. R. Refractory nonmotor symptoms in male patients with Parkinson disease due to testosterone deficiency, *Arch. Neurol.*, 59, 807, 2002.

49. Zesiewicz, T. A., Helal, M., and Hauser, R. A. Sildenafil citrate for treatment of erectile dysfunction in men with Parkinson's disease, *Mov. Dis.*, 25, 305, 2000.

50. Hussain, I. F. et al., Treatment of erectile dysfunction with sildenafil citrate (Viagra) in parkinsonism due to Parkinson's disease or multiple system atrophy with observations on orthostatic hypotension, *J. Neurol. Neurosurg. Psych.*, 71, 371, 2001.

51. Cohan, P. and Korenman, S. G., Erectile dysfunction, *J. Clin. Endocrinal. Metab.*, 86, 2392, 2001.

52. Carani, C. et al., The effects of testosterone administration and visual erotic stimuli on nocturnal penile tumescence in normal men, *Horm. Behav.,* 24, 435.

53. Bhasin, S., The dose-dependent effects of testosterone on sexual function and on muscle mass and function, *Mayo Clin. Proc.,* 75s, 70, 2000.

54. Morley, J. E. et al., Validation of a screening questionnaire for androgen deficiency in aging males, *Metabolism,* 49, 1239, 2000.

55. Padma-Nathan, F. and Giuliano, F., Oral drug therapy for erectile dysfunction, *Urol. Clin. No. Am.,* 28, 321, 2001.

56. Wespes, E. et al., Guidelines on erectile dysfunction, *Eur. Urol.,* 41, 1, 2002.

57. Rosen, R. C. et al., Development and evaluation of an abridged 5-item version of the International Index of Erectile Dysfunction (IIEF-5) as a diagnostic tool for erectile dysfunction, *Int. J. Impot. Res.,* 11, 319, 1999.

58. Rosen, R. C. et al., The international index of erectile function (IIEF): a multidimensional scale for assessment of ED, *Urology,* 49, 822, 1997.

59. Basson, R. et al., Report of the international consensus development conference on female sexual dysfunction: definitions and classifications, *J. Urol.,* 163, 888, 2000.

60. American Psychiatric Association, DSM-IV *Diagnostic and statistical manual of mental disorders,* 4th ed., American Psychiatric Press, Washington, D.C., 1994.

61. Berman, J. R., Berman, L., and Goldstein I, Female sexual dysfunction: incidence, pathophysiology, evaluation, and treatment options, *Urology,* 54, 385, 1999.

62. Mones, R. J., Elizan, T. S., and Siegek, G. J., Evaluation of L-dopa therapy in Parkinson's disease, *N.Y. State J. of Med.,* 448, 2309, 1970.

63. Campbell, R. J., *Psychiatric Dictionary,* 7th ed., Oxford University Press, New York, 1996, 335.

64. Ruggieri, S. Low dose clozapine in the treatment of dopaminergic psychosis in Parkinson's disease, *Clin. Neuropharm.,* 20, 204, 1997.

65. Fernandez, H. H. and Durso, R., Clozapine for dopaminergic-induced paraphilias in Parkinson's disease, *Mov. Dis.,* 13, 59, 1998.

66. Halaris, A. Neurochemical aspects of the sexual response cycle, CNS *Spectrums,* 8, 211, 2003.

67. Harvey, N. S., Serial cognitive profiles in levodopa induced hypersexuality, *Br. J. Psych.,* 153, 833, 1988.

68. Cummings, J. L., Behavioral complications of drug treatment of Parkinson's disease, *JAGS,* 39, 708, 1991.

69. Johnson, B. K., Older adults and sexuality: a multidimensional perspective. *J. of Gerontological Nursing,* 2, 6, 1996.

70. Woods, N. F., Toward a holistic perspective of human sexuality; alterations in sexual health and nursing diagnoses, *Holistic Nursing Practice,* 1, 1, 1987.

28 Cardiovascular Autonomic Dysfunction in Parkinson's Disease and Parkinsonian Syndromes

Christopher J. Mathias
Imperial College London and Institute of Neurology, University College London

CONTENTS

INTRODUCTION

The cardiovascular system is influenced by a variety of factors that include the autonomic nervous system. Through the arterial baroreceptor reflex, the sympathetic and parasympathetic components of the autonomic nervous system exert beat-by-beat control over the maintenance of blood pressure and both directly and indirectly influence the perfusion of various organs. In Parkinson's disease and various parkinsonian syndromes, cardiovascular dysfunction may occur for a variety of reasons. Autonomic failure is an integral component of parkinsonian syndromes such as multiple system atrophy (MSA), where orthostatic (postural) hypotension is an important clue to underlying cardiovascular autonomic failure and in the recognition of this disorder.[1] The majority of parkinsonian patients are over the age of 50, when the incidence of cardiovascular disorders increases regardless of associated disease. Many parkinsonian patients are on drugs (antiparkinsonian or for coincidental medical disorders) that may have cardiovascular side effects. Furthermore, with advancing years, there is impairment of autonomic function that may occur independently of the parkinsonian state.[2]

This chapter describes aspects of cardiovascular dysfunction resulting from autonomic impairment in PD and parkinsonian syndromes.

BLOOD PRESSURE CONTROL

Hypotension or hypertension may result from disruption of autonomic control. As organ function is dependent upon an adequate perfusion pressure, the symptoms arising from hypotension (such as syncope with head-up postural change) often are more prominent than those resulting from hypertension.

HYPOTENSION

Orthostatic Hypotension

In humans, standing upright results in considerable strains to the cardiovascular system as a result of gravitational forces. Maintaining blood pressure with head-up posture is essential to adequately perfuse organs, especially those above the heart such as the brain. Of the variety of factors

which help maintain blood pressure during gravitational stress, the sympathetic nervous system plays a key role.[3] A cardinal feature of failure of the sympathetic nervous system is orthostatic hypotension[4] (Figure 28.1). It is defined as a fall in systolic blood pressure of 20 mm Hg or more, or in diastolic blood pressure of 10 mm Hg or more, on either standing or head-up tilt to at least 60°.[5] Orthostatic hypotension reduces perfusion of organs. Hypoperfusion of the brain can result in dizziness, visual disturbances, and impaired cognition that often precede loss of consciousness (Table 28.1). These symptoms occur on assuming the upright posture, especially when getting out of bed in the morning, when patients often are at their worst. Many recognize the association between postural change and symptoms of cerebral hypoperfusion and either sit down or lie flat; some even assume curious postures, such as squatting or stooping, that now are rec-

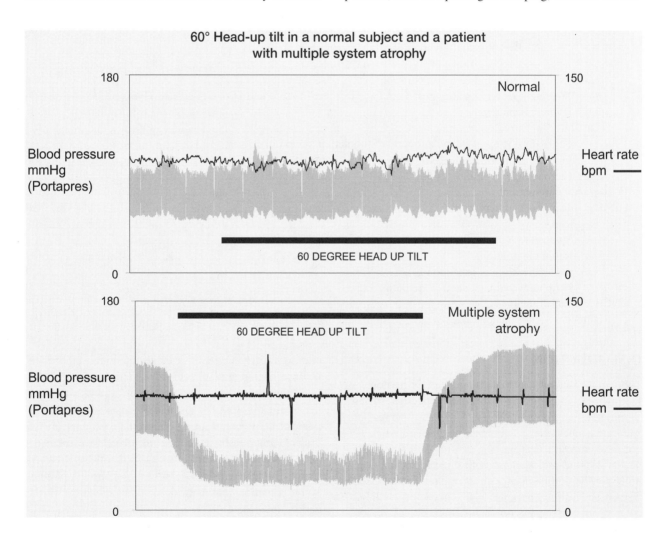

FIGURE 28.1 Blood pressure and heart rate before, during, and after head-up tilt in a normal subject (upper panel) and in a patient with autonomic failure due to multiple system atrophy (MSA, lower panel). In the normal subject, there is no fall in blood pressure during head-up tilt, unlike the patient with MSA, in whom blood pressure falls promptly and remains low, with a blood pressure overshoot on return to the horizontal. In the patient with MSA, there is only a minimal change in heart rate despite the marked blood pressure fall. In each subject, continuous blood pressure and heart rate was recorded with the Portapress II.

TABLE 28.1
Some Symptoms Resulting from Orthostatic (Postural) Hypotension

Cerebral hypoperfusion
  Dizziness
  Visual disturbances
    Blurred-tunnel
    Scotoma
    Graying out/blacking out
    Color defects
  Loss of consciousness
  Impaired cognition
Muscle hypoperfusion
  Paracervical and suboccipital ("coat hanger") ache
  Lower back/bullock ache
Cardiac hypoperfusion
  Angina pectoris
Spinal cord hypoperfusion
Renal hypoperfusion
  Oliguria
Nonspecific symptoms
  Weakness, lethargy, fatigue
  Falls

Source: Adapted from Mathias, 2003.[1]

TABLE 28.2
Factors Influencing Orthostatic (Postural) Hypotension

Speed of positional change
Time of day (worse in the morning)
Warm environment (hot weather, central heating, hot bath)
Raising intrathoracic pressure—micturition, defecation or coughing
Food and alcohol ingestion
Water ingestion[a]
Physical exertion
Maneuvers and positions (bending forward, abdominal compression, leg crossing, squatting, activating calf muscle pump)[b]
Drugs with vasoactive properties (including dopaminergic agents)

[a]Water raises seated blood pressure in autonomic failure and reduces orthostatic hypotension[8]

[b]These maneuvers usually reduce the postural fall in blood pressure, unlike the others.

Source: Adapted from Mathias, 2003.[1]

TABLE 28.3
Outline of Cardiovascular Autonomic Investigations in Autonomic Failure

Physiological
  Head-up tilt (45°);[a] standing;[a] Valsalva manoeuvre[a]
  Pressor stimuli—isometric exercise,[a] cutaneous cold,[a] mental arithmetic
  Heart rate responses—deep breathing,[a] hyperventilation,[a] standing,[a] head-up tilt,[a] 30:15 ratio
  Liquid meal challenge
  Exercise testing
  Carotid sinus massage
Biochemical
  Plasma noradrenaline—supine and head-up tilt or standing; urinary catecholamines; plasma renin activity and aldosterone
Pharmacological
  Noradrenaline-α-adrenoceptors—vascular
  Isoprenaline-ß-adrenoceptors—vascular and cardiac
  Tyramine-pressor and noradrenaline response
  Edrophonium-noradrenaline response
  Atropine-parasympathetic cardiac blockade

[a]Indicates screening tests used in our units.

Source: From Mathias and Bannister, 2002.[4]

ognized as reducing hypotension.[6] With time, symptoms may diminish for reasons that include improved cerebrovascular autoregulation. Occasionally, the blood pressure falls precipitously and syncope may occur rapidly, similar to a drop attack. In parkinsonian patients, the associated movement disorder may enhance the propensity to falls, from which injury may result. Seizures occasionally occur as a result of cerebral hypoxia. A variety of noncerebral symptoms that result from underperfusion of various organs are associated with orthostatic hypotension.[7] Many factors, which include the side effects of drugs and even previously-thought inert substances such as water,[8] can exacerbate or reduce orthostatic hypotension (Table 28.2). Orthostatic hypotension may be present without symptoms.

Measurement of blood pressure should be an integral component of the clinical evaluation of all parkinsonian patients. This should be performed both supine and standing, or sitting if it is difficult for the patient to stand. Laboratory testing using a tilt table is advisable, as this enables accurate evaluation of orthostatic hypotension if present and provides a baseline, ideally prior to initiation of drug therapy. In the laboratory, tilt table testing should be combined with autonomic screening tests (Table 28.3) to determine if orthostatic hypotension is present, and if so whether it is the consequence of failure of the autonomic nervous system or the result of nonneurogenic factors (Table 28.4). As previously emphasized (Table 28.5), the presence of orthostatic hypotension and such testing

itself does not separate MSA from other parkinsonian syndromes or disorders where autonomic dysfunction occurs,[9–11] which is the reason for the range of tests described in the different disorders.[4] Twenty-four-hour noninvasive blood pressure and heart rate monitoring, which can be obtained in the home environment, often provides valuable information (Figure 28.2), including on the effects of treatment.[12–14] In MSA, there often is a

TABLE 28.4

Non-neurogenic Causes of Orthostatic (Postural) Hypotension

Low intravascular volume
Blood/plasma loss—hemorrhage, burns, hemodialysis
Fluid/electrolyte
   Inadequate intake–anorexia nervosa
   Fluid loss—vomiting, diarrhea, losses from ileostomy
   Renal/endocrine–salt-losing nephropathy, adrenal insufficiency
   (Addison's disease), diabetes insipidus, diuretics
Vasodilatation
   Drugs—glyceryl trinitrate
   Alcohol
   Heat, pyrexia
   Hyperbradykinism
   Systemic mastocytosis
   Extensive varicose veins
Cardiac impairment
   Myocardial—Myocarditis
   Impaired ventricular—atrial myxoma, constrictive pericarditis
   filling
   Impaired output—aortic stenosis

Source: From Mathias and Bannister, 2002.[4]

TABLE 28.5

Possible Causes of Orthostatic Hypotension and Autonomic Dysfunction in a Patient with Parkinsonian Features

Side effects of antiparkinson therapy, including
L-DOPA, bromocriptine, pergolide
The combination of L-DOPA and COMT inhibitors (tolapone)
The MAO "b" inhibitor, selegiline
Coincidental disease causing autonomic dysfunction, e.g.,
   Diabetes mellitus
Coincidental administration of drugs for an allied condition
Antihypertensives
   α-adrenoceptor blockers (for benign prostatic hypertrophy)
   Vasodilators (for ischemic heart disease)
   Diuretics (for cardiac failure)
   Sildenafil (for erectile failure)
Multiple system atrophy (shy-Drager syndrome)
Parkinson's disease with autonomic failure
Diffuse Lewy body disease

Source: Adapted from Mathias, 1996,[9] and Mathias and Kimber, 1999.[10]

reversal of the circadian change in blood pressure; whether sleep abnormalities (sleep apnea, hypoxia, REM sleep disorders)[15–17] that are common in MSA account for this, in addition to autonomic impairment, is unclear.

There have been varying levels of resting blood pressure reported in PD. These include low basal levels,[18,19]

which was not confirmed in a large study in PD patients not on dopaminergic agents.[20] Various factors, ranging from age, stage of disease, drug treatment, and associated disorders, may contribute to these differences. In MSA where autonomic failure is a key component, orthostatic hypotension, especially due to neurogenic impairment, may occur early in the course of the disease. Its presence should lead to consideration of this disorder, especially when there are additional noncardiovascular features of autonomic failure. In MSA, unlike other primary autonomic failure syndromes such as pure autonomic failure (PAF), the autonomic lesions predominantly are central and preganglionic.[21,22] This explains, despite abnormalities on autonomic testing, especially in the early stages of MSA, why there is a near-normal basal level of plasma noradrenaline;[23] however, the levels do not rise appropriately with postural challenge, unlike normal subjects (Figure 28.3). In MSA, symptoms of orthostatic hypotension are common, although syncope occurs in less than 50% as compared to PAF patients with orthostatic hypotension.[7] The reasons for this difference are not entirely clear. The degree of autonomic failure often is more severe and possibly of more rapid onset in PAF than in MSA, and compensatory mechanisms, including those affecting the cerebral vasculature, are more likely to be operative in MSA than in PAF.

In PD, the reported prevalence and incidence of orthostatic hypotension varies considerably. Some consider it to be rare.[24] Others report a modest orthostatic fall (systolic mean fall of 11 mm Hg in 20 patients[25] and 10 mm Hg in 35 patients[26]) that do not fulfil current definition criteria of orthostatic hypotension. This differs from a high prevalence reported in other studies (43% of 80[27] and 58% of 91 patients.[28] Whether this reflects variations in the type of patient studied, the influence of the many factors that modify blood pressure control (such as increasing age, duration of the disorder and drug therapy), or differences in methods to evaluate orthostatic hypotension, is unclear.[29–34] Impaired mobility itself may contribute to autonomic dysfunction, as is known to occur, especially in elderly bed-bound patients.[35] Postmortem studies have emphasized the difficulties of *in vivo* diagnosis and in separating PD from non-PD disorders such as MSA that inadvertently and erroneously may have been included as PD patients.[36–38] However, this alone is unlikely to account for the considerably higher prevalence in some studies. In the PD study by Senard and co-workers,[28] of the 58% with orthostatic hypotension, 38.5% were symptomatic. However, there may be dissociation between symptoms suggestive of cerebral hypoperfusion and orthostatic hypotension; in a study by Turkka,[39] 12 out of 15 PD patients complained of dizziness, but orthostatic hypotension was present in only 4, suggesting that other factors were contributing.

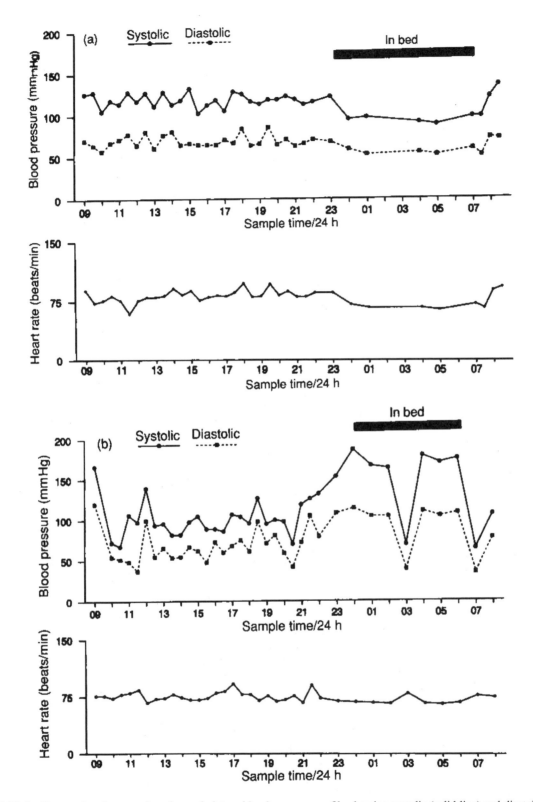

FIGURE 28.2 Twenty-four-hour noninvasive ambulatory blood pressure profile showing systolic (solid line) and diastolic (dotted line) blood pressure and heart rate at intervals through the day and night. (*a*) The changes in a normal subject with no postural fall in blood pressure; there was a fall in blood pressure at night while asleep, with a rise in blood pressure on wakening. (*b*) Marked fluctuations in blood pressure in a patient with autonomic failure. The marked falls in blood pressure are usually the result of postural changes, either sitting or standing. Supine blood pressure, particularly at night, is elevated. Getting up to micturate causes a marked fall in blood pressure (03.00 hr). There is a reversal of the diurnal changes in blood pressure. There are relatively small changes in heart rate, considering the marked changes in blood pressure. (*Source:* From Mathias & Bannister, 2002a.[4])

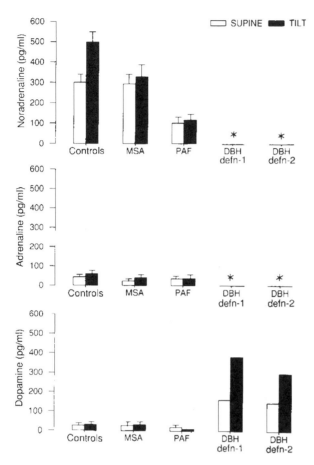

FIGURE 28.3 Plasma noradrenaline, adrenaline, and dopamine levels (measured by high-pressure liquid chromatography) in normal subjects (controls), patients with multiple system atrophy (MSA) and pure autonomic failure (PAF) while supine and after head-up tilt to 45° for 10 min. The asterisk indicates levels below the detection limits for the assay, which are less than 5 pg/ml for noradrenaline and adrenaline, and less than 20 pg/ml for dopamine. Bars indicate ± SEM. (Modified from Mathias et al., 1990.[23])

Postmortem studies of catecholaminergic neurons in the ventrolateral medulla show less atrophy in PD than in MSA.[40,41] In PD, the mechanisms responsible for orthostatic hypotension warrant discussion. Some studies exclude generalized autonomic failure and speculate on central lesions in the upper brain stem that affect postural control of blood pressure but spare other reflexes, such as the Valsalva manoeuvre.[42–46] Other studies suggest that autonomic failure in PD (PD + AF), is similar to that observed in pure autonomic failure (PAF). In PD + AF, there are subnormal levels of basal plasma noradrenaline, and an impaired plasma noradrenaline response to head-up postural challenge.[47–49] This may explain why such patients do not respond to certain drugs used to treat orthostatic hypotension, such as yohimbine, whose benefits are dependent on intact post-ganglionic sympathetic pathways.[50] In PD + AF,

drugs that act on alpha-adreno receptors, such as midodrine and its metabolite, are more likely to be effective. In MSA, the prodrug l-dihydroxyphenylserine (l-DOPS) reduces orthostatic hypotension.[51,52] However, in MSA there is no evidence of DBH deficiency or of variations in the DBH gene.[23,53,54]

In one of the early postmortem reports on PD with orthostatic hypotension, it was noted that the sympathetic ganglia were involved.[55] Lewy bodies have been observed in the autonomic ganglia of PD,[56,57] and also in PD with orthostatic hypotension,[58] and may indicate peripheral involvement of the autonomic nervous system. This differs from MSA, where sympathomimetic drugs that act through residual postganglionic sympathetic pathways, such as ephedrine often are effective.[10] The mechanisms and pathophysiological basis of orthostatic hypotension in the different parkinsonian disorders thus are of relevance for diagnosis and importantly for determining which drugs are more likely to provide effective benefit.

In progressive supranuclear palsy (PSP), previously it was unclear if autonomic function was affected. An asymptomatic fall in BP on orthostasis has been reported, with a systolic fall of 16 mmHg and diastolic of 7 mmHg,[59] and of 15 mmHg systolic.[26] Symptoms considered suggestive of orthostatic hypotension were reported in a large proportion of postmortem confirmed PSP (78%),[60] but a major limitation in this retrospective study was the lack of autonomic testing, raising the possibility that coincidental disorders may have contributed to such symptoms.[61] In a study of 35 PSP,[62] orthostatic hypotension was not recorded, and detailed autonomic investigations excluded autonomic failure; therefore, it was suggested that cardiovascular autonomic dysfunction should be an exclusion criterion in PSP.

Orthostatic hypotension may be a presenting feature in diffuse Lewy body disease (DLBD)[63] and was prominent enough in another case for the authors initially to consider a diagnosis of MSA.[64] In postmortem confirmed DLBD orthostatic symptoms have been reported;[65,66] in a series of 12, orthostatic dizziness was reported in 50% and syncopal attacks associated with standing in 33%.[67] Orthostatic hypotension has been described in a case with brain stem Lewy body disease presenting with various manifestations of autonomic failure.[68] Syncope, loss of consciousness, and repeated falls are supportive features to increase diagnostic sensitivity within the consensus guidelines for dementia with Lewy body disease,[69] a disorder known to be a synucleinopathy, as is Parkinson's disease. In a recent study of 29 patients with dementia with Lewy body disease,[70] 18 (62%) had severe autonomic failure, with 28 having autonomic dysfunction; 28% had episodic hypotension. In PAF, Lewy bodies have been described,[71–73] suggesting that this disorder also is an alpha-synucleinopathy. Whether PAF is a precursor state,[63] or a *forme fruste*, of Lewy body disease remains unclear.

In Huntington's disease, there is a greater fall in blood pressure on head-up tilt as compared to normal subjects, but investigations exclude generalized autonomic failure,[74] and it was concluded that the fall in blood pressure may have resulted from involvement of caudate nuclei and their putative influence on postural vasomotor mechanisms. Autonomic failure also may occur in Guam PD-dementia complex;[75] some patients also have diabetes mellitus, and the precise mechanisms responsible were unclear.

In Wilson's disease, there is recent evidence favoring autonomic dysfunction affecting both sympathetic and parasympathetic cardiovascular function.[76,77] This is thought to be central in origin. Autonomic dysfunction however, occurs in chronic hepatic disease[78,79] and needs to be considered also as a contributor to the dysautonomia described in Wilson's disease.

In male carriers of the Fragile X syndrome, there may be autonomic dysfunction (which currently includes urinary and bowel incontinence and impotence) as part of the Fragile X Permutation Tremor/Ataxia syndrome (FXTAS).[80,81] Whether there is significant cardiovascular autonomic involvement is not known.

Post-Prandial Hypotension

In normal subjects, food ingestion does not change systemic blood pressure, but there are a number of changes in gastrointestinal and pancreatic hormones accompanied by compensatory cardiac and regional hemodynamic responses.[82] In some parkinsonian patients, food ingestion may cause substantial hypotension even while supine (Figure 28.4), especially when they have autonomic failure as part of MSA.[83,84] The cardiovascular autonomic responses to food challenge can be tested using a liquid meal of mixed composition, with observations initially in the supine position, to avoid the additional effect of gravity.[85] The response to head-up postural challenge, using either head-up tilt or standing, before and after food ingestion, determines if food unmasks or exacerbates orthostatic hypotension (Figure 28.5). Recording the onset of, or increase in, symptoms of orthostatic hypotension on postural challenge after food ingestion is of importance.

When post-prandial hypotension is present, a variety of influencing factors need to be considered. These include the composition of food (as carbohydrate is more likely to lower blood pressure than fat or protein), the caloric load (a greater fall with a higher load), and the release of gastrointestinal and pancreatic peptides (espe-

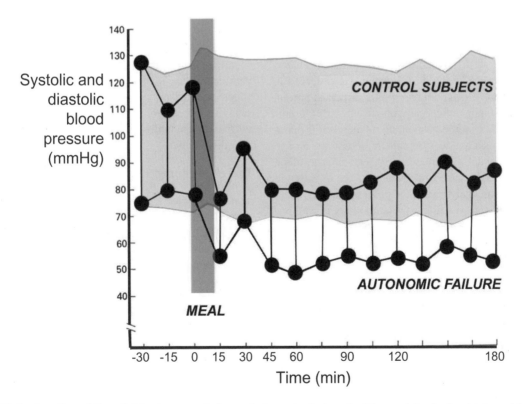

FIGURE 28.4 Systolic and diastolic blood pressure before and after a standard meal while remaining horizontal, to avoid the effects of gravity, in normal subjects (controls, stippled area) and in a patient with autonomic failure. Blood pressure does not change in normal subjects after a meal. In the patient, it rapidly falls to around 80/50 mmHg and remains low over three hours. (From Mathias, 2002.[3])

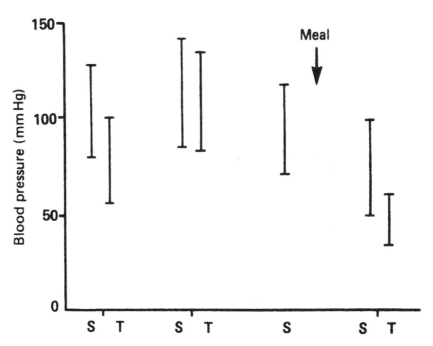

FIGURE 28.5 Systolic and diastolic blood pressure in a patient with multiple system atrophy while supine (S) and after 45° head-up tilt (T) on three occasions. On the first occasion, food intake was not controlled, and the patient had eaten earlier. On the second occasion, the patient had not eaten, and the postural blood pressure fall was negligible. The patient then had a liquid meal. Following food challenge, supine blood pressure fell and, on the third occasion of tilt, there was a considerably greater fall in blood pressure. These observations emphasize the importance of food intake in unmasking or exacerbating orthostatic hypotension (From Mathias et al., 1991.[85])

cially those with vasodilatatory effects).[82] In MSA, food ingestion causes a marked increase in superior mesenteric artery blood flow (as an indication of splanchnic blood flow) that is similar to the post-prandial rise observed in normal subjects.[86,87] Alcohol has similar effects in normal subjects and in autonomic failure due to MSA.[88,89] In normal subjects after food ingestion, blood pressure is maintained by an increase in sympathetic activity, a rise in cardiac output, and a decrease in skeletal muscle blood flow. Compensatory autonomic responses are impaired in MSA and are the likely reason for post-prandial hypotension,[84,90] which can be a major problem in some patients. Variations in the degree of post-prandial hypotension however, may occur even within the subgroups of MSA. In the cerebellar form (MSA-C), there is a greater degree of post-prandial hypotension while supine when compared to the parkinsonian (MSA-P) form. In MSA-C, a greater involvement of cerebellar and brain stem cardiovascular autonomic centers is a probable explanation.[91]

In PD, there are varying reports on the extent of post-prandial hypotension. In PD patients not on drug therapy, a small post-prandial fall in blood pressure while supine was reported without exacerbation of orthostatic hypotension post-meal;[92] these observations differ from those reported previously by Micielli et al.,[93] who noted a greater incidence and degree of hypotension. In elderly parkinsonian patients, post-prandial hypotension is more frequent (82% compared to 41% in controls) than orthostatic hypotension (13% versus 6% in controls); the combination of levodopa (125 mg) and benserazide did not significantly aggravate either.[94] In PD, the lowering of blood pressure by food ingestion may account for worsening motor control after a meal especially in patients on antiparkinsonian drugs that themselves may contribute to vasodilatation.[95]

Exercise-Induced Hypotension

In normal subjects, even a modest degree of exercise causes vasodilatation in working skeletal muscle. This is accompanied by activation of compensatory mechanisms that include an increase in sympathetic activity to the vasculature of nonexercising muscles and to other organs. In the presence of autonomic failure, as in MSA, exercise while supine (to avoid the effects of gravity) lowers blood pressure substantially[96] (Figure 28.6); there also usually is an increase in orthostatic hypotension post-exercise,[97] which explains why some patients become symptomatic during, or after, walking.[97,98] In some, there is a marked fall in blood pressure during exercise while upright (Figure 28.7); others can maintain their blood pressure while walking, but lower their blood pressure and have symptoms on cessation of walking.[99,98] This is presumably because the calf muscle pump that helps to maintain blood pressure while exercising is inoperative while standing still.

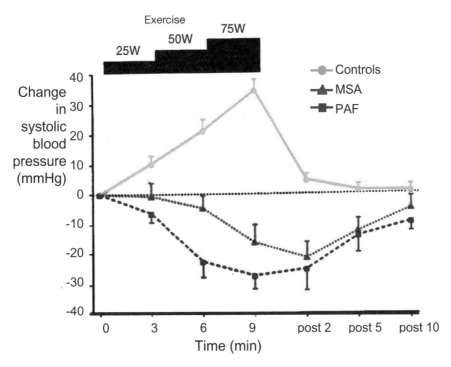

FIGURE 28.6 Changes in systolic blood pressure during horizontal bicycle exercise at three incremental levels in normal subjects (controls) and in patients with multiple system atrophy (MSA) and pure autonomic failure (PAF). In both MSA and PAF, unlike controls, there is a fall in blood pressure. (From Mathias, 2002,[3] with data from Smith et al., 1995.[96])

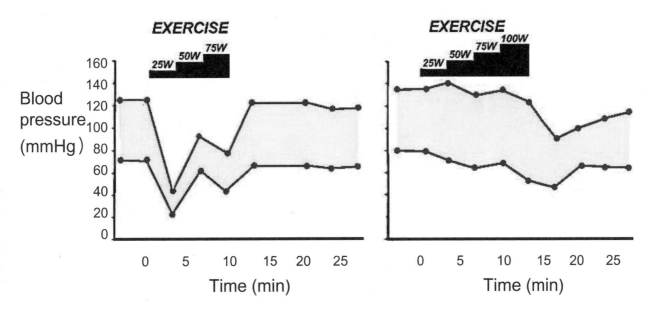

FIGURE 28.7 Systolic and diastolic blood pressure (top) and heart rate (bottom) in two patients with autonomic failure before, during, and after bicycle exercise performed with the patients in the supine position at different workloads, ranging from 25 to 100 W. In the patient on the left, there is a marked fall in blood pressure on initiating exercise; she had to crawl upstairs because of severe exercise induced hypotension. In the patient on the right, there are minor changes in blood pressure during exercise but a marked decrease soon after stopping exercise. This patient was usually asymptomatic while walking but developed postural symptoms when he stopped walking and stood still. It is likely that the decrease in blood pressure post-exercise was due to vasodilatation in exercising skeletal muscle, not opposed by the calf muscle pump. (From Mathias and Williams, 1994.[98])

Exercise-induced hypotension while supine also is greater in MSA-C than in MSA-P.[100] However, orthostatic hypotension post-exercise does not appear to be exacerbated in MSA-C, and this is similar to observations in these two groups after food challenge.[91] In PD, exercise causes minimal or no fall in blood pressure and does not appear to unmask orthostatic hypotension. The effect of exercise in other parkinsonian disorders is not known.

Drug-Induced Hypotension

Drugs may induce hypotension by causing autonomic dysfunction or as a result of the cardiovascular side effects of the drugs.[101] This applies to treatment with antiparkinsonian drugs (Table 28.5). In PD, there is variability in the ability of dopaminergic drugs to lower blood pressure. Some may cause orthostatic hypotension,[102–107] although other reports conclude that there has been no alteration of cardiovascular reflexes or enhancement of orthostatic hypotension.[34,108–110] The tendency to hypotension is likely to increase in patients who are older, have cardiovascular disease, and have either occult or evident autonomic failure, as part of MSA or PD with AF. Hypotension with dopaminergic drugs may result from vasodilatation induced by dopaminergic receptor stimulation. Thus, peripheral dopa-decarboxylase inhibitors may reduce the hypotensive effects of L-dopa, although the central effects of L-dopa or other dopaminergic agents may contribute.[111,112] Selegiline in PD may cause orthostatic hypotension by mechanisms that are unclear; they include the central effects of its metabolite methylamphetamine and interactions with L-dopa and ergoline derivatives.[113–116] The induction of marked hypotension following L-dopa challenge in a parkinsonian patient should necessitate consideration of autonomic failure and MSA[117] (Figure 28.8).

TREATMENT OF ORTHOSTATIC, POST-PRANDIAL, EXERCISE-INDUCED AND DRUG-ASSOCIATED HYPOTENSION

The prevention or reversal of hypotension is of importance, as it can result in considerable morbidity and may even contribute to death. There have been considerable advances in the nonpharmacological and pharmacological approaches used in the management of orthostatic hypotension[118,119] (Tables 28.6 and 28.7). These ideally should be coupled with appropriate autonomic investigations initially to evaluate the deficit,[4] both anatomically and functionally, so that management can be tailored to individual needs. Of importance is education of the patient and, where relevant, family and caregivers. There should be discussions of limitations of treatment and of realistic expectations, especially where parkinsonian features impair mobility and enhance susceptibility to the sequelae of fainting.

Investigations into the pathophysiological mechanisms causing post-prandial hypotension have resulted in therapeutic strategies that utilize both nonpharmacological and pharmacological methods to reduce hypotension[82] (Table 28.8). The management of exercise-induced hypotension currently is unsatisfactory. There is limited evidence that drugs such as octreotide[120] reduce it.

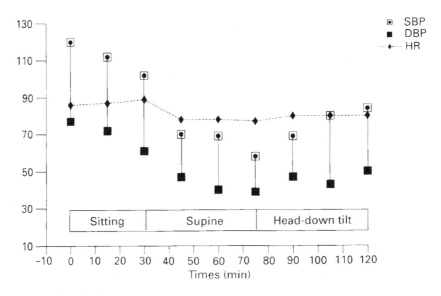

FIGURE 28.8 Systolic and diastolic blood pressure in a patient with parkinsonian features before and after a standard levodopa challenge (250 mg of levodopa along with 25 mg carbidopa). The patient was initially seated, but blood pressure fell and was so low that he needed to be laid horizontal and then head down. On further investigation, the patient had autonomic failure, and the final diagnosis was the parkinsonian form of multiple system atrophy. (From Mathias, 2000.[101])

TABLE 28.6
Outline of Nonpharmacological and Key Measures Used in the Management of Postural Hypotension Due to Neurogenic Failure

Nonpharmacological measures

To be avoided
   Sudden head-up postural change (especially on waking)
   Prolonged recumbency
   Straining during micturition and defecation
   High environmental temperature (including hot baths)
   "Severe" exertion
   Large meals (especially with refined carbohydrate)
   Alcohol
   Drugs with vasodepressor properties

To be introduced
   Head-up tilt during sleep
   Small, frequent meals
   High salt intake
   Judicious exercise (including swimming)
   Body positions and maneuvers

To be considered
   Elastic stockings
   Abdominal binders
   Water ingestion

Pharmacological measures

Starter drug: fludrocortisosne
Sympathomimetics: ephedrine, midodrine
Specific targeting: octreotide, desmopressin, erythropoietin

Source: From Mathias, 1003.[119]

It should be emphasized that non-neurogenic factors such as fluid loss due to vomiting or diarrhea substantially worsen neurogenic postural hypotension and will need to be rectified.

In drug-associated hypotension, it is necessary to achieve a balance between benefit and harm, which will need often to be considered individually, taking into account prognosis and quality of life among other factors.

HYPERTENSION

Hypertension often causes few, if any, symptoms, in contrast with the many symptoms that result from orthostatic hypotension. Hypertension may be sustained or intermittent. In parkinsonian patients, sustained hypertension may be a result of a variety of causes, the most common of which is essential (primary) hypertension. In normotensive PD patients treated with L-dopa, there may be a transient elevation of blood pressure without adverse effects, not requiring treatment.[121] Fluctuations of blood pressure occur during end-of-dose akinesia in PD, being higher in the "off" than the "on" phase.[122]

Parkinsonian patients with autonomic failure, such as with MSA, have pressor sensitivity to endogenous neurotransmitters (noradrenaline) and to a wide range of vasoactive agents, both pressor and depressor.[123,124] Knowledge of these supersensitive responses is of importance whenever vasoactive drugs are used for treatment and during anaesthetic procedures. Increased alpha-adrenergic pressor supersensitivity also occurs in PD[125] and in PD + AF (who have low plasma noradrenaline levels). This does not seem to occur in PD without orthostatic hypotension.[48,49] Presymptomatic, mainly transient hypertension with large fluctuation in blood pressure has been reported in 80% of PSP;[126] in contrast, hypertension occurred in 17% with PD, 19% with MSA, 33% with corticobasal degeneration, and 23% with drug-induced parkinsonism. Whether hypertension in PSP is due to involvement of brain stem autonomic nuclei or to other factors is unclear.

Supine hypertension may occur in patients with autonomic failure and orthostatic hypotension, as in MSA.[14,127] This may be a problem at night if such patients lie supine and horizontal, as demonstrated on the 24-hr ambulatory blood pressure profiles (Figure 28.2b). The probable mechanisms for supine hypertension include the movement of intra- and extravascular fluid from the peripheral to the central vascular compartment, impaired baroreflexes, and supersensitivity even to small amounts of circulating catecholamines or to pressor agents used for therapy.[4,14] Patients may report a fullness of the head or a throbbing headache while lying flat. It is unclear to what extent supine hypertension contributes to morbidity and mortality in such patients. In a few patients with autonomic failure (but not necessarily with MSA), intracerebral hemorrhages, myocardial failure, and aortic dissection have been reported, but the relationship to hypertension and its severity is unclear. Debate continues on the "safe" upper limits of supine hypertension and the use of antihypertensive drugs at night. However, even short-acting antihypertensives are likely to enhance orthostatic hypotension and increase vulnerability to falls of patients who need to get up at night to micturate, because of nocturia and urinary bladder dysfunction. The use of head-up tilt at night, together with a small nocturnal meal or alcoholic beverage (thus beneficially utilizing the hypotensive effects of food and alcohol), may transiently diminish supine hypertension.

HEART RATE CONTROL

Heart rate is controlled by the parasympathetic and sympathetic nervous systems, with the cardiac vagus playing a predominant role.

HEART RATE RESPONSES TO AUTONOMIC STIMULATION

A variety of stimuli, predominantly respiratory, are used to test parasympathetic cardiac activity (Table 28.3). These include the heart rate response to deep breathing

TABLE 28.7
Drugs Used in the Treatment of Postural Hypotension

Site of Action	Drugs	Predominant Action
Plasma volume expansion	Fludrocortisone	Mineralocorticoid effects—increased plasma volume
		Sensitization of α-adrenoceptors
Kidney–reducing diuresis receptors on renal	Desmopressin	Vasopressin$_2$-receptors on renal tubules
Vessels: vasoconstriction (adrenoceptor-mediated)	Ephedrine	Indirectly acting sympathomimetic
Resistance vessels	Midodrine,[a] phenylephrine methylphenidate	Directly acting sympathomimetics
	Tyramine	Release of noradrenaline
	Clonidine	Postsynaptic α$_2$-adrenoceptor agonist
	Yohimbine	Presynaptic α$_2$-adrenoceptor agonist
	DL–DOPS and L–DOPS	Pro-drug resulting in formation of noradrenaline
Capacitance vessels	Dihydroergotamine	Direct action on α-adrenoceptors
Vessels: vasoconstriction (non-adrenoceptor mediated)	Triglycyl-lysine-vasopressin (glypressin)	Vasopressin$_1$-receptors on blood (non-adrenoceptor mediated) vessels
Vessels: prevention of vasodilation	Propranolol	Blockade of ß-adrenoceptors
	Indoethacin	Prevents prostaglandin synthesis
	Metoclopromide	Blockade of dopamine receptors
Vessels: prevention of post-prandial hypotension	Caffeine	Blockade of adenosine receptors
	Octreotide	Inhibits release of vasodilator gut/pancreatic peptides
Heart: stimulation	Pindolol, xamoterol	Intrinsic sympathomimetic action
Red cell mass: increase	Erythropoietin	Stimulates red cell production, gut/pancreatic peptides

[a]Through its active metabolite.

Source: From Mathias, 1003.[119]

TABLE 28.8
Some of the Measures Used to Prevent or Reduce Postprandial Hypotension

Nonpharmacological
Small meals, more frequently
Reduce refined carbohydrate
Avoid alcohol
Do not stand or walk after meal

Pharmacological

Caffeine	Adenosine receptor blockade
Octreotide	Inhibition of gastro-intestinal/pancreatic peptide release
Indomethacin	Prostaglandin inhibition
Denopamine and midodrine	α-adrenoceptor agonist
L-Dihydroxyphenylserine	Prodrug converted to noradrenaline

(sinus arrhythmia), hyperventilation, the Valsalva maneuver, and head-up tilt or standing.[4] Heart rate changes are influenced by factors that include age and drug therapy. Impairment of parasympathetic cardiac control may not itself cause symptoms, and objective testing is the main means of determining abnormalities. In MSA, abnormalities are consistent with central involvement of the vagus.

In PD, normal and abnormal responses have been reported.[128–133] These differences may reflect duration of disease, age, and concomitant drug therapy[32,134,135] rather than a specific parasympathetic cardiac deficit. In PSP, small differences in heart rate responses to certain tests were attributed to central autonomic involvement;[59,136] this was not confirmed in a larger study.[62]

INVESTIGATION OF CARDIAC SYMPATHETIC INNERVATION

Control of heart rate exerted by the sympathetic nervous system is difficult to assess. An alternative, although this does not provide functional evaluation, is the use of imaging techniques that determine the integrity of cardiac sympathetic innervation. The physiological analogue of noradrenaline, meta-iodobenzylguanadine (MIBG) is actively transported into sympathetic nerve terminals by a noradrenaline transporter; its iodinated form ^{123}I-MIBG can be readily detected by myocardial scintigraphy. In MSA, consistent with a preganglionic lesion, noradrenaline transport is preserved, as in normal subjects (Figure 28.9).[137,138] In peripheral autonomic disorders, such as diabetes mellitus with a cardiac autonomic neuropathy, uptake is diminished.[139] In central disorders with spinocerebellar atrophy, such as Machado-Joseph disease, reduced MIBG uptake

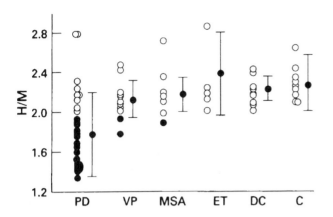

FIGURE 28.9 Comparison of the early phase heart/mediastinum ratio using [131]meta-iodo-benzylguanadine scintigraphy scanning in different neurological disorders and in controls. Open circles show normal and filled circles abnormal H/M ratios. (*Source:* From Orima et al., 1999.[137]) PD = Parkinson's disease, VP = vascular parkinsonism, MSA = multiple system atrophy, ET = essential tremor, DC = disease controls, and C = healthy controls.

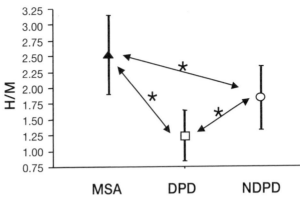

FIGURE 28.10 Values of heart-to-mediastinum activity ratios (H/M) calculated 240 min after [^{123}I]metaiodobenzylguanidine injection in patients with multiple system atrophy (MSA) or Parkinson's disease either with dysautonomia (DPD) or without nondysautonomia (NDPD) using clinical autonomic testing. Results are presented as mean ± SD. *P, 0.05. (From Courbon et al.[149])

also occurs,[140] and one explanation is peripheral denervation secondary to central and transynaptic degeneration. In PD there is impaired uptake, suggesting cardiac sympathetic denervation.[137,141–149] This occurs even in the absence of clinical features of autonomic dysfunction, suggesting autonomic failure at an early stage of PD.[149,150] Uptake in PD does not appear to be affected by dopaminergic drug therapy, and the results differ from vascular parkinsonism[137] and PSP.[151] It has been suggested that the defect is exclusive to PD but this needs further evaluation, as limited uptake also occurs in Machado-Joseph disease.[140] In the study by Orimo et al.,[137] heart rate variability was impaired only in a small proportion of PD (11%). The dissociation of functional autonomic testing with reduced cardiac MIBG uptake (Figure 28.10) is consistent with the suggestion that autonomic screening tests predominantly assess parasympathetic, and not sympathetic, cardiac activity. Sympathetic cardiac innervation also has been assessed with 18F-fluorodopamine and positron emission tomography scanning,[152,153] with similar results to those obtained from MIBG scanning; uptake in MSA is similar to normal subjects, with minimal uptake in PD with AF. Cardiac uptake, using this technique, as was described with MIBG scanning, also is impaired in PD without orthostatic hypotension and diminishes with time.[154,155]

Peripheral sympathetic denervation in the majority of PD, albeit only to the heart, appears surprising, considering that the manifestations of PD are central in origin. The combined MIBG and 18F FD cardiac scanning data in PAF and MSA, taken in conjunction with neuroendocrine and function neuroimaging mapping, is consistent with sparing of brain autonomic centres and intracerebral sym-

pathetic pathways in PAF and PD, and their involvement in MSA.[156–161] Confirmation that MIBG and 18F FD scanning provide a true indication of cardiac sympathetic denervation has been obtained from postmortem studies in a few patients with PD and MSA; in heart muscle of PD patients, in comparison with controls and MSA, there is diminished tyrosine hydroxylase activity.[162,163] Furthermore, in PD + AF, other organs, such as the adrenal medulla, appear spared, as levels of adrenaline and its metabolite, metanephrine, are within the normal range.[164] It remains unclear why selective cardiac sympathetic denervation should occur in PD even at a relatively early stage of the disease.

CARDIAC DYSRHYTHMIAS

Patients with PD usually are over 50 years of age, and some may be predisposed to cardiac dysrhythmias, especially when on antiparkinsonian drugs with cardiovascular effects. There also may be an intrinsic propensity to the development of arrhythmias in PD impaired who have impaired cardiac sympathetic innervation; the mechanisms may include denervation supersensitivity. The QT interval is the electrocardiographic description of ventricular depolarization and repolarization, and if abnormal may identify patients at risk of developing ventricular fibrillation and sudden cardiac death. In MSA the QT interval is prolonged; however, study of QT dispersion, which is thought to provide a better measure of the tendency to dysrhythmias, is unaffected.[165]

In PD, atrial and ventricular arrhythmias initially were reported with L-dopa,[166,167] but this was not confirmed in later studies.[168–170] Tremor artifacts on the ECG may result in an erroneous diagnosis of dysrhythmias as in PD.[171–173] Hypotension may induce a tachycardia and thus could increase the dysrhythmic potential. The introduction of dopa-decarboxylase inhibitors (such as carbidopa) was considered a means of avoiding potential dysrhythmogenic effects[174,175] by reducing hypotension secondary to increasing dopamine levels. There is no clear evidence that dopaminergic agents, such as pergolide increase the tendency to dysrhythmias.[176,177] Whether certain drugs predispose susceptible individuals (such as those with sympathetic cardiac denervation) to dysrhythmias is not known.

REGIONAL CIRCULATION

CEREBRAL CIRCULATION

The cerebral vasculature can be studied by noninvasive techniques that include the use of xenon inhalation, transcranial doppler, functional magnetic resonance imaging, and near infrared spectroscopy; they provide direct or indirect measurements of cerebral blood flow and oxygenation. In some patients with autonomic failure, despite severe orthostatic hypotension, there are few symptoms of cerebral hypoperfusion, and some do not develop syncope. This is thought to result from improved cerebrovascular autoregulation,[178–182] with cerebral blood flow maintained at a lower level of blood pressure than in normal subjects for reasons that are unclear; there are some studies that report the contrary.[183] Syncope occurs less frequently in MSA than in PAF despite substantial postural falls in blood pressure in both groups;[7] whether preservation of post-ganglionic pathways and a slower progression of the disease in MSA aids the ability to improve autoregulation of the cerebral vasculature is not known.

A wide variety of symptoms result from cerebral hypoperfusion and overlap with other disorders such as transient ischemic attacks, as in carotid artery stenosis.[184] Alternative causes to explain symptoms should be sought if relevant, especially in older patients[185] (Figure 28.11).

CORONARY CIRCULATION

Orthostatic hypotension may be associated with chest pain in MSA.[7] The possibility of coronary artery disease needs to be considered and, in some patients who have had coronary angiography, no abnormality may be found. In these patients, central chest pain may be the result of pericardiac and chest wall tissue hypoperfusion. In some, movement of arm muscles initiates symptoms, raising the possibility of a "steal phenomenon," as the blood supply to this region from the internal mammary artery arises from the subclavian artery. Chest pain in the majority of patients with parkinsonism, however, will warrant exclusion of coronary artery disease.

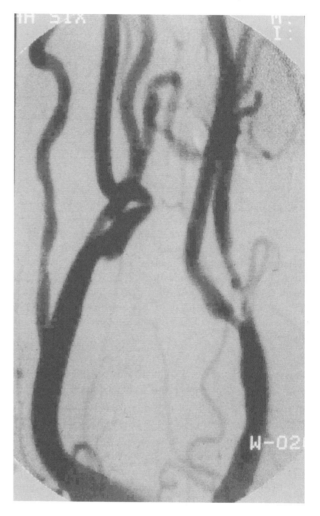

FIGURE 28.11 Intravenous digital subtraction angiogram of the cerebral vessels in a patient with hypertension and widespread atherosclerosis, indicating left carotid artery stenosis. She had symptoms that initially were considered to be transient ischemic attacks resulting from thrombo-embolism. The history, however, indicated symptoms closely associated with postural change. She had a postural fall in systolic blood pressure of only 10 mmHg, which presumably was sufficient to induce symptoms of cerebral ischemia because of cerebrovascular disease. A reduction in antihypertensive therapy abolished the small postural fall in blood pressure and also her symptoms. (From Mathias, 1998.[185])

SPLANCHNIC CIRCULATION

The splanchnic circulation takes a third of the cardiac output and may have 30% of the blood volume. Previously, measurement of splanchnic blood flow was dependent on invasive techniques. Noninvasive doppler techniques with imaging of the superior mesenteric artery (SMA) now provide information on the splanchnic vasculature. This has been studied in relation to a variety of stimuli that increase sympathetic neural activity; these include the responses to pressor tests, head-up postural change, food, and alcohol challenge. In MSA, neither pressor stimuli nor head-up tilt constrict the splanchnic vessels, unlike normal subjects.[186,187] With dynamic exercise, however, splanchnic vasoconstriction occurs later; this suggests that the initial response is neurally mediated, with the later response dependent on circulating hormones.[188]

Food ingestion, in patients with autonomic failure, causes SMA vasodilatation similar to that observed in normal subjects, suggesting that post-prandial hypotension results from lack of compensatory autonomic mechanisms in other regions rather than due to excessive splanchnic vasodilatation.[86,87] The role of vasodilator peptides released by food ingestion and alcohol has been emphasized by the ability of the peptide release inhibitor, octreotide, to prevent post-prandial splanchnic vasodilatation and effectively reduce post-prandial hypotension.[189]

RENAL CIRCULATION

Orthostatic hypotension may impair perfusion of the kidneys thus contributing to daytime oliguria. In MSA, the reverse, nocturia, occurs when the patient is supine and blood pressure is restored or elevated.[190,191] Vasopressin release is impaired in MSA and may contribute to abnormal circadian variation at night,[192] although changes in renal perfusion may be a dominant factor, as diuresis occurs even during the day when induced by recumbency.[193] The role of various hormones and peptides in influencing the renal circulation and its function remain to be further explored in parkinsonian disorders.[194] Renin release in response to tilt is impaired in MSA and diminished in PD compared to control subjects[(195)]. Vasopressin (antidiuretic hormone) levels in response to head-up tilt are impaired in MSA[196] but not in PD with AF.[197] The synthetic analogue of vasopressin, the V_2 agonist desmopressin, is effective when given nocturnally in reducing both nocturia and orthostatic hypotension.[191,198,199]

SKELETAL MUSCLE CIRCULATION

Basal levels of skeletal muscle blood flow, when measured in forearm or calf muscle, are higher in MSA than in normal subjects;[188] this may reflect impaired sympathetic vasoconstrictor activity. With exercise, there is a greater rise in blood flow in exercising muscle in MSA than in normal subjects, and the elevated levels remain elevated for longer after cessation of exercise, indicating impairment of corrective autonomic neural responses. The skeletal vasculature responses to exercise in other parkinsonian groups are not known. Understanding the mechanisms controlling the skeletal muscle vasculature may be of importance in the management of exercise-induced hypotension, for which there is no specific treatment.

PERIPHERAL AND CUTANEOUS CIRCULATION

In MSA a large proportion (76%) report cold sensitivity with initial pallor followed by cyanosis and, in some, the later onset of redness, affecting both hands and feet and thus similar to description of Raynaud's phenomenon[200]; similar mechanisms may account also for cold blue hands that are a frequent problem in MSA.[201] The mechanisms responsible for the abnormal vascular responses to cold are unclear. Cutaneous vasomotor function, in response to stimuli, is similar in controls and MSA.[202] Whether increased responses result from sympathetic denervation with pressor supersensitivity to small amounts of endogenous noradrenaline or to vasopressor drugs used for treatment are responsible, this remains a possibility. In PD, livedo reticularis or cutis marmorata has been reported with amantadine,[203] but the mechanisms are unknown.

CONCLUDING REMARKS

In Parkinson's disease and the parkinsonian syndromes, cardiovascular dysfunction may reflect autonomic involvement or cardiovascular disease as associated especially with an older population.[204,205] The control of blood pressure, heart rate, and various regional circulations may be affected. The presence of orthostatic hypotension should lead to consideration of MSA, especially in the presence of other features of autonomic failure. Orthostatic hypotension, however, also may occur in other parkinsonian disorders and in PD, especially with increasing age, severity of disease, and as a result of drug therapy, sometimes for associated disorders. In some, orthostatic hypotension may be unrelated to the parkinsonian disease process. Investigation of cardiovascular autonomic dysfunction in parkinsonism is important for a variety of reasons that include determining the precise diagnosis and in predicting prognosis (Figure 28.12). In parkinsonian disorders, understanding the pathophysiological basis of the abnormality aids targeting of therapy, improves management strategies, and should benefit such patients.

REFERENCES

1. Mathias, C. J., Autonomic diseases—clinical features and laboratory evaluation. *J. Neurol. Neurosurg. Psychiatry,* 74;31–41, 2003.
2. Lipsitz, L., Ageing and the autonomic nervous system, in *Autonomic Failure: A Textbook of Clinical Disorders of the Autonomic Nervous System.* Mathias, C. J., Bannister, R., Eds., 4th ed., Oxford University Press, Oxford, pp. 534–44, 2002.
3. Mathias, C. J., To stand on ones' own legs, *Clin. Med.,* 2:237–245, 2002.

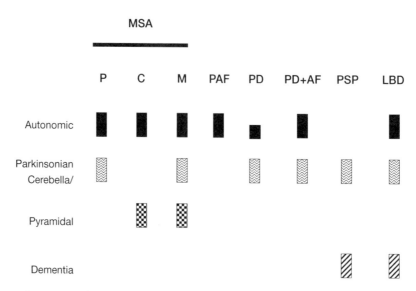

FIGURE 28.12 Schematic representation indicating the major clinical features in some of the parkinsonian disorders. These include, from the left, the three forms of multiple system atrophy (parkinsonian, MSA-P; cerebellar, MSA-C; and mixed, MSA-M), pure autonomic failure (PAF), Parkinson's disease (PD), Parkinson's disease with autonomic failure (PD+AF), progressive supranuclear palsy (PSP) and diffuse Lewy body disease (DLBD). The partially filled histogram in PD is to indicate that there is evidence for localized but not generalized AF, unlike PD+AF. (Adapted from Mathias, 1997[204] and 2002.[205])

4. Mathias, C. J., Bannister, R., Investigation of autonomic disorders, in *Autonomic Failure: A Textbook of Clinical Disorders of the Autonomic Nervous System,* Mathias, C. J., Bannister, R., Eds., 4th ed., Oxford University Press, Oxford, U.K., pp. 169–95, 2002.

5. Schatz, I. J., Bannister, R., Freeman, R. L., Jankovic, J., Koller, W. C., Low, P. A., Mathias, C. J., Polinsky, R. J., Quinn, N. P., Robertson, D., Streeten, D. H. P., Consensus Statement on the definition of orthostatic hypotension, pure autonomic failure and multiple system atrophy, *Clin. Auton. Res.,* 6:125–6, 1996.

6. Wieling, W., van Lieshout, J. J., van Leeuwen, A. M., Physical manoeuvres that reduce postural hypotension, *Clin. Auton. Res.,* 3:57–65, 1993.

7. Mathias, C. J., Mallipeddi, R., Bleasdale-Barr., K., Symptoms associated with orthostatic hypotension in pure autonomic failure and multiple system atrophy, *J. Neurol.,* 246;893–8, 1999.

8. Mathias, C. J., A 21st century water cure, *The Lancet,* 356;1046–8, 2000.

9. Mathias, C. J., Disorders affecting autonomic function in Parkinsonian patients, in *Parkinson's Disease,* Eds., Battistini, L., Scarlato, G., Caraceni, T., Ruggieri, S., Raven Press, New York, pp. 383–91, 1996.

10. Mathias, C. J., Kimber, J. R., Postural hypotension—causes, clinical features, investigation and management, *Ann. Rev. Med.,* 50:317–36, 1999.

11. Riley, D. E., Chelimsky, T. C., Autonomic nervous system testing may not distinguish multiple system atrophy from Parkinson's disease, *J. Neurol. Neurosurg. Psychiatry,* 74:56–60, 2003.

12. Senard, J. M., Chamentin, B., Rascol, A., Montastruc, J. L., (1992). Ambulatory blood pressure in patients with Parkinson's disease without and with orthostatic hypotension, *Clin. Auton. Res.,* 2:99–104, 1992.

13. Alam, M., Pavitt, D. V., Mathias, C. J., Cumulative sums of 24 hour blood pressure profiles in patients with sympathetic denervation, *J. Hypertens.,* 11, (suppl. 5):s286–287, 1993.

14. Alam, M., Smith, G. D. P., Bleasdale-Barr, K., Pavitt, D. V., Mathias, C. J., Effects of the peptide release inhibitor, Octreotide, on daytime hypotension and on nocturnal hypertension in primary autonomic failure, J. Hypertens., 13:1664–9, 1995.

15. Cortelli, P., Pierangeli, G., Provini, F., Plazzi, G., Lugaresi, E., (1996). Blood Pressure Rhythms in Sleep Disorders and Dysautonomia, *Annals New York Acad. Sci.,* 783:204–221, 1996.

16. Gilman, S., Chervin, R. D., Keoppe, R. A., Consens, F. B., Little, R., An, H., Junck, L., Heumann, M., Obstructive sleep apnoea is related to thalamic cholinergic deficit in MSA, *Neurology,* 61:35–9, 2003a.

17. Gilman, S., Keoppe, R. A., Chervin, R. D., Consens, F. B., Little, R., An, H., Junck, L., Heumann, M., REM sleep behaviour is related to striatal monoaminergic deficit in MSA, *Neurology,* 61:29–34, 2003b.

18. Yahr, M. D., General discussion on clinical effects of L-dopa upon blood pressure, in Barbeau, A., McDowell, F. H., eds., *L-dopa and parkinsonism,* Davis, F. A., Philadelphia, PA, pp. 266-8, 1970.

19. Barbeau, A., Mars, H., Gillo-Joffroy, Adverse clinical side effects of levodopa therapy, in Markham, F. H., C. H., Eds., *Recent Advances in Parkinson's Disease,* Blackwell, Oxford, pp. 203–37, 1971.

20. Aminoff, M. J., Gross, M., Laatz, B., Vakil, S. D., Petrie, A., Calne, D. B., Arterial blood pressure in patients with

Parkinson's disease. *J. Neurol. Neurosurg. Psychiatry,* 38:73–7, 1975.

21. Daniel, S. E., The neuropathology and neurochemistry of multiple system atrophy, in *Autonomic Failure. A textbook of clinical disorders of the autonomic nervous system.* Mathias, C. J. and Bannister, R., Eds., 4th ed. Oxford University Press, Oxford, pp. 321–8, 2002.

22. Matthews, M. R., Autonomic ganglia and preganglionic neurones in autonomic failure, in *Autonomic Failure,* A textbook of clinical disorders of the autonomic nervous system. Mathias, C. J. and Bannister, R., Eds., 4th ed., Oxford University Press, Oxford, pp. 329–39, 2002.

23. Mathias, C. J., Bannister, R., Cortelli, P., Heslop, K., Polak, J., Raimbach, S. J., Springall, D. B., Watson, L., Clinical autonomic and therapeutic observations in two siblings with postural hypotension and sympathetic failure due to an inability to synthesize noradrenaline from dopamine because of a deficiency of dopamine beta hydroxylase, *Q. J. Med.,* New Series, 75;278:617–33, 1990.

24. Colosimo, C., Albanese, A., Hughes, A. J., de Bruin, V. M. S., Lees, A. J., Some specific clinical features differentiate multiple system atrophy (striatonigral variety) from Parkinson's disease, *Arch. Neurol.,* 52:294–8, 1995.

25. Meco, G., Pratesi, L., Bonifati, V., Cardiovascular reflexes and autonomic dysfunction in Parkinson's disease, *J. Neurol.,* 238:195–99, 1991.

26. Sandroni, P., Ahiskog, E., Fealey, R. D, Low, P. A., Autonomic involvement in extrapyramidal and cerebellar disorders, *Clin. Auton. Res.,* 1:147–55, 1991.

27. Uono, Y., Parkinsonism and autonomic dysfunction, *Auton. Nerv. Syst.,* Tokyo, 10:163–70, 1973.

28. Senard, J. M., Rai, S., Lapeyre-Mestre, M., Brefel, C., Rascol, O., Rascol, A., Montastruc, J. L., (1997). Prevalence of orthostatic hypotension in Parkinson's disease, *J. Neurol. Neurosurg. Psychiatry,* 63:578–89, 1997.

29. Orskov, I., Jakobsen, J., Dupont, E., de Fine Olivarius, B., Christensen, N. J., Autonomic function in parkinsonian patients relates to duration of disease, *Neurology.* 37:1173–8, 1987.

30. Piha, S. J., Rinne, J. O., Rinne, J. O., Seppanen, A., Autonomic dysfunction in recent onset and advanced Parkinson's disease, *Clin. Neurol. Neurosurg.,* 90:221–6, 1988.

31. Mesec, A., Sega, S., Kiauta, T., The influence of the type, duration, severity and levodopa treatment of Parkinson's disease on cardiovascular autonomic responses, *Clin. Auton. Res.,* 3:339–44, 1993.

32. van Dijk, G. H., Haan, J., Zwinderman, K., Kremer, B., van Hilten, B. J., Roos, R. A., Autonomic nervous system dysfunction in Parkinson's disease: relationships with age, medication, duration and severity, *J. Neurol. Neurosurg. Psychiatry,* 56:1090–5, 1993.

33. Hillen, M. E., Wagner, M. L., Sage, J. I., "Subclinical" orthostatic hypotension is associated with dizziness in elderly patients with Parkinson disease, *Arch. Phys. Med. Rehabil.,* 77:710–712, 1996.

34. Kujawa, K., Leurgans, S., Raman, R., Blasucci, L., Goetz, C. G., Acute orthostatic hypotension when start-

ing dopamine agonists in Parkinson's disease, *Arch. Neurol.,* 57:1461–1463, 2000.

35. Kihara, M., Takahashi, M., Nishimoto, K., Okuda, K., Matsui, T., Yamakawai, T., Okumura, A., Autonomic dysfunction in elderly bedfast patients, *Age Ageing,* 27:551–5, 1998.

36. Rajput, A. H., Rozdilsky, B., Rajput, A., Ang, L., Levodopa therapy and pathological basis of Parkinson's syndrome, *Clin. Neuropharmacol.,* 13: 553–8, 1990.

37. Hughes, A. J., Daniel, S. E., Kilford, L., Less, A. J., Accuracy of clinical diagnosis of idiopathic Parkinson's disease: a clinicopathological study of 100 cases, *J. Neurol. Neurosurg. Psychiatry,* 55:181–184, 1992.

38. Litvan, I., Goetz, C. G., Jankovic, J., Wenning, G. K., Booth, V., Bartko, J. J., McKee, Jellinger K., Lai, E. C., Brandel, J. P., Verny, M., Ray, Chaudhuri K., Pearce, R. K. B., Agid, Y., What is the accuracy of the clinical diagnosis of multiple system atrophy? *Arch. Neurol.,* 54:937–44, 1997.

39. Turkka, J. T., Correlation of the severity of autonomic dysfunction to cardiovascular reflexes and to plasma noradrenaline levels in Parkinson's disease, *Eur. Neurol.,* 26:203–10, 1987.

40. Benarroch, E. E., Smithsoon, I. L., Low, P. A., Parisi, J. E., Depletion of catecholaminergic neurons of the rostral ventrolateral medulla in multiple system atrophy with autonomic failure, *Ann. Neurol.,* 43:156–63, 1998.

41. Benarroch, E. E., Schmeichel, A. M., Parisi, J. E., Involvement of the ventrolateral medulla in parkinsonism with autonomic failure, *Neurology,* 22: 963–8, 2000.

42. Appenzeller, O., Goss, J. E., Autonomic deficits in Parkinson's disease, *Arch. Neurol.,* 24:50–7, 1971.

43. Aminoff, M. J., Wilcox, C. S., Assessment of autonomic function in patients with a parkinsonian syndrome, *Brit. Med. J.,* 4:80–4, 1971.

44. Bannister, R., Oppenheimer, D. R., Degenerative diseases of the autonomic nervous system associated with autonomic failure, *Brain,* 95:457–74, 1972.

45. Gross, M., Bannister, R., Godwin-Austin, R., Orthostatic hypotension in Parkinson's disease, *Lancet,* 1:174–6, 1972.

46. Kuno, S., Parkinson's disease and its autonomic disorders, *Mod. Physician,* Tokyo, 9:1214–19, 1989.

47. Turkka, J. T., Juujarvi, K. K., Lapinlampi, T. O., Myllyla, V. V., Serum noradrenaline response to standing up in patients with Parkinson's disease, *Eur. Neurol.,* 25:355–61, 1986.

48. Senard, J. M., Valet, P., Durrieu, G., Berlan, M., Tran, M. A., Montastruc, J. L., Rascol, A., Montastruc, P., Adrenergic supersensitivity in parkinsonians with orthostatic hypotension, *Eur. J. Clin. Investig.,* 20:613–19, 1990.

49. Durrieu, G., Senard, J. M., Rascol, O., Tran, M. A., Lataste, X., Rascol, A., Montastruc, J. L., Blood pressure and plasma catecholamines in never-treated parkinsonian patients: effect of a D_1 agonist (CY 208-243), *Clin. Neuropharmacol.,* 16:70–6, 1990.

50. Senard, J. M., Rascol, O., Durrieu, G., Tran, M. A., Berlan, M., Rascol, A., Montastruc, J. L., Effects of

Yohimbine on plasma catecholamine levels in orthostatic hypotension related to Parkinson's disease or multiple system atrophy, *Clin. Neuropharmacol.*, 16:70–6, 1993.

51. Mathias, C. J., Senard, J., Braune, S., Watson, L., Aragishi, A., Keeling, J., Taylor, M., L-theo-dihydroxphenylserine (L-threo-DOPS; droxidopa) in the management of neurogenic orthostatic hypotension: a multi-national, multi-centre, dose-ranging study in multiple system atrophy and pure autonomic failure, *Clin. Auton. Res.*, 11;235-42, 2001.

52. Kaufmann, H., Saadia, D., Voustianiouk, A., Goldstein, D. S., Holmes, C., Yahr, M. D., Nardin, R., Freeman, R., Norepinephrine precursor therapy in neurogenic orthostatic hypotension, *Circulation,* 108:724–8, 2003.

53. Cho, S., Kim, C. H., Cubells, J. F., Zabetian, C. P., Hwang, D. Y., Kim, J. W., Cohen, H. M., Biaggioni, I., Robertson, D., Kim, K. S., Variations in the dopamine beta-hydroxylase gene are not associated with the autonomic disorders, pure autonomic failure or multiple system atrophy, *Am. J. Med. Genet.*, 120:234–6, 2003.

54. Mathias, C. J., Bannister, R., Dopamine β-hydroxylase deficiency—with a note on other genetically determined causes of autonomic failure, in *Autonomic Failure: A Textbook of Clinical Disorders of the Autonomic Nervous System*, Mathias, C. J., Bannister, R., Eds., 4th ed., Oxford University Press, Oxford, UK, pp. 387–401, 2002.

55. Fichefet, J. P., Sternon, J. E., Franken, L., Demanet, J. C., Vanderhaeghen, J. J., Etude anatomo-clinique d'un cas d'hypotension orthostatique "idiopathique." Considerations pathogeniques, *Act. Cardiol.*, Brux, 20:332–48, 1965.

56. Wakabayashi, K., Takahashi, H., Ohama, E., Takeda, S., Ikuta, F., Lewy bodies in the visceral autonomic nervous system in Parkinson's disease, *Adv. Neurol.*, 60:609–612, 1993.

57. Wakabayashi, K., Takahashi, H., Neuropathology of autonomic nervous system in Parkinson's disease, *Eur. Neurol.*, 38:2–7, 1997.

58. Rajput, A. H., Rozdilsky, B., Dysautonomia in parkinsonism: a clinicopathological study, *J. Neurol. Neurosurg. Psychiatry,* 39:1092–100, 1976.

59. van Dijk, J., Haan, J., Koenderink, M., Raymund, A., Roos, C., Autonomic nervous function in progressive supranuclear palsy, *Arch. Neurol.*, 48:1083–4, 1991.

60. Wenning, G. K., Scherfler, C., Granata, R., Bosch, S., Verny, M., Chaudhur, K. R., Jellinger, K., Poewe, W., Litvan, I., Time course of symptomatic orthostatic hypotension and urinary incontinence in patients with postmortem confirmed parkinsonian syndromes: a clinicopathological study, *J. Neurol. Neurosurg. Psychiatry,* 67:620–3, 1999.

61. Mathias, C. J., Can the early presence of autonomic dysfunction aid diagnosis in parkinsonism? *J. Neurol. Neurosurg. and Psychiatry,* 67; 566, 1999.

62. Kimber, J. R., Watson, L., Mathias, C. J., Physiological, pharmacological and neurohormonal assessment of autonomic function in progressive supranuclear palsy, *Brain*, 123:1422–1430, 2000.

63. Larner, A. J., Mathias, C. J., Rossor, M. N., Autonomic failure preceding dementia with Lewy bodies, *J. Neurol.*, 247: 229–31, 2000.

64. Pakiam, A. S. I., Bergeron, C., Lang, A. E., Diffuse Lewy body disease presenting as multiple system atrophy, *Can. J. Neurol. Sci.*, 26:127–31, 1999.

65. Dickson, D. W., Davies, P., Mayeuz, R. et al., Diffuse Lewy body disease. Neuropathological and biochemical studies of six patients, *Acta Neuropathol.*, 75:8–15, 1987.

66. Kosaka, K., Diffuse Lewy body disease in Japan, *J. Neurol.*, 237:197–204, 1990.

67. Kuzuhara, S., Yoshimura, M., Mizutani, T., Yamanouchi, H., Ihara, Y., Clinical features of diffuse Lewy body disease in the elderly: analysis of 12 cases, in Perry, R. H., McKeith, I. G., Perry, E. K., eds. Dementia with Lewy bodies, Cambridge University Press, New York, 153–60, Cambridge University Press, New York, 1996.

68. Hishikawa, N., Hashizume, Y., Hirayama, M. et al., Brainstem-type Lewy body disease presenting with progressive autonomic failure and lethargy, *Clin. Auton. Res.*, 10:139–43, 2000.

69. McKeith, I. G., Galasko, D., Kosaka, K., Consensus guidelines for the clinical and pathological diagnosis of dementia with Lewy bodies (DLB): Report of the consortium on DLB international workshop, *Neurology,* 47:1113–24, 1996.

70. Horimoto, Y., Matsumoto, M., Akatsu, H., Ikari, H., Kojima, H., Kojima, K., Yamamoto, T., Otsuka, Y., Ojika, K., Ueda, R., Kosaka, K., Autonomic dysfunction in dementia with Lewy bodies, *J. J. Neurol.*, 250:530–3, 2003.

71. van Ingelghem, E., van Zandijcke, M., Lammens, M., Pure autonomic failure. A new case with clinical, biochemical and necropsy data, *J. Neurol. Neurosurg. Psychiatry,* 57:745–7, 1994.

72. Hague, K., Lento, P., Morgello, S., Caro, S., Kaufmann, H., The distribution of Lewy bodies in pure autonomic failure: autopsy findings and review of the literature, *Acta Neuropathol.*, 94:192–6, 1997.

73. Arai, K., Kato, N., Kashiwado, K., Hattori, T., Pure autonomic failure in association with human alpha-synucleinopathy, *Neurosci. Lett.*, 296:171–3, 2000.

74. Aminoff, M. J., Gross, M., Vasoregulatory activity in patients with Huntington's chorea, *J. Neurol. Sci.*, 21:33–8, 1974.

75. Low, P. A., Ahlskog, J. E., Petersen, P. F., Waring, S. C., Esteban-Santiallas, C., Karsland, L. T., Autonomic failure in Guamanian neurodegenerative disease, *Neurology,* 49:1031–4, 1997.

76. Meenakshi-Sundaram, S., Taly, A. B., Kamath, V., Arunodaya, G. R., Rao, S., Swamy, H. S., Autonomic dysfunction in Wilson's disease—a clinical and electrophysiological study, *Clin. Auton. Res.*, 12:185–9, 2002.

77. Bhattacharya, K., Velickovic, M., Schilsky, M., Kaufmann, H., Autonomic cardiovascular reflexes in Wilson's disease, *Clin. Auton. Res.*, 12:190–2, 2002.

78. Oliver, M. I., Miralles, R., Rubies-Prat, J., Navarro, X., Espadaler, J. M., Sola, R., Andreu, M., Autonomic dys-

function in patients with non-alcoholic chronic liver disease, *Hepatol.,* 26:1242–8, 1997.

79. Chaudhry, V., Corse, A.M., O'Brian, R., Cornblath, D. R., Klein, A. S., Thuluvath, P. J., Autonomic and peripheral (sensorimotor) neuropathy in chronic liver disease: a clinical and electrophysiological study, *Hepatology,* 29:1698–1703, 1999.

80. Hagerman, R. J., Leehey, M., Heinrichs, W., Tassone, F., Wilson, R., Hills, J., Grigsbt, J., Gage, B., Hagerman, P. J., Intention tremor, parkinsonism, and generalized brain atrophy in male carriers of fragile X, *Neurology.* 57:127-30, 2001.

81. Jacquemont, S., Hagerman, R. J., Leehey, M., Grigsby, J., Zhang, L., Brunberg, J. A., Greco, C., DesPortes, V., Jardini, T., Levine, R., Berry-Kravis, E., Brown, W. T., Schaeffer, S., Kissel, J., Tassone, F., Hagerman, P. J., Fragile X permutation tremor/ataxia syndrome: Molecular, clinical and neuroimaging correlates, *Am. J. Hum. Genet.,* 72:869–878, 2003.

82. Mathias, C. J., Bannister, R., Postprandial hypotension in autonomic disorders, in *Autonomic Failure: A Textbook of Clinical Disorders of the Autonomic Nervous System,* Mathias, C. J., Bannister, R., Eds., 4th ed., Oxford University Press, Oxford, pp. 283–95, 2002b.

83. Seyer-Hansen, K., Post-prandial hypotension, *Br. Med. J.,* 2:1262, 1977.

84. Mathias, C. J., da Costa, D. F., Fosbraey, P., Bannister, R., Wood, S. M., Bloom, S. R., Christensen, N. J., Cardiovascular, biochemical and hormonal changes during food induced hypotension in chronic autonomic failure, *J. Neurol. Sci.,* 94: 255–69, 1989.

85. Mathias, C. J., Holly, E., Armstrong, E., Shareef, M., Bannister, R., The influence of food on postural hypotension in three groups with chronic autonomic failure: clinical and therapeutic implications, *J. Neurol. Neurosurg. Psychiatry,* 54:726–30, 1991.

86. Kooner, J. S., Peart, W. S., Mathias, C. J., The peptide release inhibitor, octreotide (SMS201-995) prevents the haemodynamic changes following food ingestion in normal human subjects, *Q. J. Experiment. Physiol.,* 74:569–72, 1989.

87. Kooner, J. S., Raimbach, S. J., Watson, L., Bannister, R., Peart, W. S., Mathias, C. J., Relationship between splanchnic vasodilatation and post-prandial hypotension in patients with primary autonomic failure, *J. Hypertens.,* 7: s40–1, 1989b.

88. Maule, S., Chaudhuri, K. R., Thomaides, T., Pavitt, D. V., McCleery, J., Mathias, C. J., Effects of oral alcohol on superior mesenteric artery blood flow in normal man, supine and tilted, *Clin. Sci.,* 84:419–25, 1993.

89. Chaudhuri, K. R., Maule, S., Thomaides, T., Pavitt, D., Mathias, C. J., Alcohol ingestion lowers supine blood pressure, causes splanchnic vasodilatation and worsens postural hypotension in primary autonomic failure, *J. Neurol.,* 241:145–52, 1994.

90. Hakusui, S., Sugiyama, Y., Iwase, S., Hasegawa, Y., Koike, Y., Mano, T., Takahashi, A., Post-prandial hypotension. Microneurographic analysis and treatment with vasopressin, *Neurology.* 41:712–15, 1991.

91. Smith, G. D. P., Von Der Thusen, J., Mathias, C. J., Comparison of the blood pressure response to food ingestion in two clinical subgroups of multiple system atrophy, *Parkinsonism Relat. Disord.,* 4:113–17, 1999.

92. Thomaides, T., Bleasdale-Barr, K., Chaudhuri, K. R., Pavitt, D. V., Marsden, C. D., Mathias, C. J., Cardiovascular and hormonal responses to liquid food challenge in idiopathic Parkinson's disease, multiple system atrophy and pure autonomic failure, *Neurology,* 43:900–4, 1993.

93. Micielli, G., Martignoni, E., Cavallini, A., Sandrini, G., Nappi, G.,Postprandial and orthostatic hypotension in Parkinson's disease, *Neurology.* 37:383–93, 1987.

94. Mehagnoul-Schipper, J., Boerman, R. H., Hoefnagels, W. H. L., Jansen, R. W. M. M., Effect of levodopa on orthostatic and post-prandial hypotension in elderly parkinsonian patients, *J. Gerontol. A. Biol. Sci. Med. Sci.,* 56A:M749–55, 2001.

95. Ray, Chaudhuri K., Love-Jones, S., Ellis, C. M., Watson, L., Mathias, C. J., Clift, S., Parkes, J. D., Postprandial hypotension and parkinsonian state in Parkinson's disease, *Mov. Disord.,* 12:877–84, 1997.

96. Smith, G. D. P., Watson, L. P., Pavitt, D. V., Mathias, C. J., Abnormal cardiovascular and catecholamine responses to supine exercise in human subjects with sympathetic dysfunction, *J. Physiol., London,* 485:255–65, 1995.

97. Smith, G. D. P., Bannister, R., Mathias, C. J., Post exercise dizziness as the sole presenting symptom in autonomic failure, *Brit. Heart. J.,* 69:359–361, 1993.

98. Smith, G. D. P., Mathias, C. J., Postural hypotension enhanced by exercise in patients with chronic autonomic failure, *Q. J. Med.,* 88: 251–6, 1995.

99. Mathias, C. J., Williams, A. C., The Shy Drager syndrome (and multiple system atrophy), in Neurodegenerative Diseases. Donald B. Calne. Ed., W. B. Saunders Company, Philadelphia, P. A., 43; 743–68, 1994.

100. Smith, G. D. P., Mathias, C. J., Differences in the cardiovascular responses to supine exercise and to standing post-exercise in two clinical subgroups of the Shy-Drager syndrome (multiple system atrophy), *J. Neurol. Neurosurg. Psychiatry,* 61: 297–303, 1996.

101. Tonkin, A. L., Frewin, D. B., Drugs, chemicals and toxins that alter autonomic function, in *Autonomic Failure: A Textbook of Clinical Disorders of the Autonomic Nervous System.* C. J. Mathias, R. Bannister, Eds., 4th ed., Oxford University Press, Oxford, pp. 527–33, 2002.

102. Calne, D. B., Brennan, A. S., Spiers, D., Stern, G. M., Hypotension caused by L-dopa, *Br. Med. J.,* 1:474–5, 1970.

103. Quinn, N., Illas, A., Lermitte, F., Agid, Y., Bromocriptine in Parkinson's disease: a study of cardiovascular effects, *J. Neurol. Neurosurg. Psychiatry,* 44:426–9, 1981.

104. Tanner, C. M., Goetz, C. G., Glantz, R. H., Glatt, S. L., Klawans, H. L., Pergolide mesylate and idiopathic Parkinson disease, *Neurology,* 32:1175–9, 1982.

105. Johns, D. W., Ayers, C. R., Carey, R. M., (1984). The dopamine agonist bromocriptine induces hypotension by venous and arteriolar dilatation, *J. Cardiovasc. Pharmacol.,* 6:582–7, 1984.

106. Piha, S. J., Rinne, J. O., Rinne, J. O., Seppanen, A., Autonomic dysfunction in recent onset and advanced Parkinson's disease, *Clin. Neurol. Neurosurg.*, 90:221–6, 1988.

107. Camerlingo, M., Ferrar, B., Gazzaniga, G. C., Casto, L., Cesana, B. M., Mamoli, A., Cardiovascular reflexes in Parkinson's disease: long-term effects of levodopa treatment on *de novo* patients, *Acta Neurol. Scand.*, 81:346–8, 1990.

108. Ballantyne, J. P., Early and late effects of levodopa on the cardiovascular system in Parkinson's disease, *J. Neurol. Sci.*, 19:97–103, 1973.

109. Sachs, C., Bergland, B., Kajser, L., Autonomic cardiovascular responses in parkinsonism: effect of levodopa with dopa-decarboxylase inhibition, *Acta Neurol. Scand.*, 71:37–42, 1985.

110. Goetz, C. G., Lutge, W., Tanner, C. M., (1986). Autonomic dysfunction in Parkinson's disease, *Neurology*, 36:73–5, 1986.

111. Calne, D. B., Reid, J. L., Vakil, S. D., George, C. F., Rao, S., Effects of carbidopa-levodopa on blood pressure in man, in *Advances in Neurology, Vol. 2., Treatment of Parkinsonism*, M. D. Yahr, Ed., Raven Press, New York, pp. 149–59, 1973.

112. Montastruc, J. L., Chamontin, B., Rascol, A., Parkinson's disease and hypertension: Chronic bromocriptine treatment, *Neurology*, 35:1644–7, 1985.

113. Karoum, F., Chuang, L. W., Eisler, T. et al., Metabolism of (-) deprenyl to amphetamine and metamphetamine may be responsible for deprenyl's therapeutic benefit: a biochemical assessment, *Ann. Neurol.*, 32:503–9, 1982.

114. Churchyard, A., Mathias, C. J., Lees, A. J., Autonomic effects of selegiline: possible cardiovascular toxicity in Parkinson's disease, *J. Neurol. Neurosurg. Psychiatry*, 63:228–34, 1997.

115. Turkka, J., Suominen, K., Tolonen, U., Sotaniemi, K., Myllyla, V. V., Selegiline diminishes cardiovascular autonomic responses in Parkinson's disease, *Neurology*, 48:662–7, 1997.

116. Churchyard, A., Mathias, C. J., Lees, A. J., Selegiline-induced postural hypotension in Parkinson's disease: a longitudinal study on the effects of drug withdrawal, *Mov. Disord.*, 2:246–51, 1999.

117. Mathias, C. J., Autonomic dysfunction, in *Oxford Textbook of Geriatric Medicine*. 2nd ed., Ed. J. Grimley-Evans, Oxford University Press, Oxford, pp. 833–52, 2000.

118. Mathias, C. J., Kimber, J. R., Treatment of postural hypotension, *J. Neurol. Neurosurg. Psychiatry*, 65:285–9, 1998.

119. Mathias, C. J., Autonomic diseases—Management, *J. Neurol. Neurosurg. Psychiatry*, 74:42–7, 2003b.

120. Smith, G. D. P., Alam, M., Watson, L. P., Mathias, C. J., Effects of the somatostatin analogue, octreotide, on exercise induced hypotension in human subjects with chronic sympathetic failure, *Clin. Sci.*, 89:367–73, 1995.

121. Cotzias, G. C., Papvasilou, P. S., Gellene, R., Modification of parkinsonism—chronic treatment with L-dopa, *N. Engl. J. Med.*, 280: 337–44, 1969.

122. Baratti, M., Calzetti, S., Fluctuation of arterial blood pressure during end-of-dose akinesia in Parkinson's disease, *J. Neurol. Neurosurg. Psychiatry*, 47:1241–3, 1984.

123. Mathias, C. J., Matthews, W. B., Spalding, J. M. K., Postural changes in plasma renin activity and response to vasoactive drugs in a case of Shy-Drager syndrome, *J. Neurol. Neurosurg. Psychiatry*, 2:147–56–124), 1977.

124. Polinsky, R., Neuropharmacological investigation of autonomic failure, in *Autonomic Failure: A Textbook of Clinical Disorders of the Autonomic Nervous System*. C. J, Mathias, R. Bannister, Eds., 4th ed., Oxford University Press, Oxford, pp. 232–44, 2002.

125. Wilcox, C. S., Aminof,f M. J., Blood pressure responses to noradrenaline and dopamine infusions in Parkinson's disease and the Shy-Drager syndrome, *Br. J. Clin. Pharmacol.*, 3:207–14, 1976.

126. Ghika, J., Bogousslavsky, J., (1997). Presymptomatic hypertension is a major feature in the diagnosis of progressive supranuclear palsy, *Arch. Neurol.*, 54:1104–8, 1997.

127. Goldstein, D. S., Pechnik, S., Holmes, C., Eldadah, B., Sharabi, Y., Association between supine hypertension and orthostatic hypotension in autonomic failure, *Hypertension*, 42:136–42, 2003.

128. Kuroiwa, Y., Shimada, Y., Toyokura, Y., Postural hypotension and low R-R interval variability in parkinsonism, spino-cerebellar degeneration and Shy-Drager syndrome, *Neurology*, 33:463–7, 1983.

129. Sachs, C., Bergland, B., Kajser, L., (1985). Autonomic cardiovascular responses in parkinsonism: effect of levodopa with dopa-decarboxylase inhibition, *Acta Neurol. Scand.*, 71:37–42, 1985.

130. Ludin, S. M., Steiger, U. H., Ludin, H. P., Autonomic disturbances and cardiovascular reflexes in idiopathic Parkinson's disease, *J. Neurol.*, 235:10–15, 1987.

131. Turkka, J. T., Tolonen, U., Myllya, V. V., Cardiovascular reflexes in Parkinson's disease, *Eur. Neurol.*, 26:104–12, 1987.

132. Piha, S. J., Rinne, J. O., Rinne, J. O., Seppanen, A., Autonomic dysfunction in recent onset and advanced Parkinson's disease, *Clin. Neurol. Neurosurg.*, 90:221–26, 1988.

133. Netten, P. M., de Vos, K., Horstink, M. I. M., Hoefnagels, W. H. L., Autonomic dysfunction in Parkinson's disease, tested with a computerized method using a Finapres device, *Clin. Auton. Res.*, 5:85–9, 1995.

134. Reid, J. L., Calne, D. B., George, C. F., Pallis, C., Vakil, S. D,. Cardiovascular reflexes in parkinsonism, *Clin. Sci.*, 41:63–7, 1971.

135. Merello, M., Pirtosek, Z., Bishop, S., Lees, A. J., Cardiovascular reflexes in Parkinson's disease: effect of domoperidone and apomorphine, *Clin. Auton. Res.*, 2:215–19, 1992.

136. Gutrecht, J. A., Autonomic cardiovascular reflexes in progressive supranuclear palsy, *J. Auton. Nerv. Syst.*, 39:29–35, 1992.

137. Orimo, S., Ozawa, E., Nakade, S., Sugimoto, T., Mizusawa, H., (1999). ^{123}I-metaiodobenzylguanidine

myocardial scintigraphy in Parkinson's disease, *J. Neurol. Neurosurg. Psychiatry*, 67:189–94, 1999.

138. Braune, S., Reinhardt, M., Schnitzer, R., Riedel, A., Lucking, C. H., Cardiac uptake of [123i]MIBG separates Parkinson's disease from multiple system atrophy, *Neurology*, 53:1020–6, 1999.

139. Claus, D., Meudt, O., Rozeik, C., Engelmann-Kempe, K., Huppert, P. E., Wietholtz, Prospective investigation of autonomic cardiac neuropathy in diabetes mellitus, *Clin. Auton. Res.*, 12:373–378, 2002.

140. Kazuta, T., Hayashi, M., Shimizu, T., Iwasaki, A., Nakamura, S., Hirai, S., Autonomic dysfunction in Machado-Joseph disease assessed by iodine[123]-labeled metaiodobenzylguanidine myocardial scintigraphy, *Clin. Auton. Res.*, 10:111–15, 2000.

141. Hakusui, S., Yasuda, T., Yanagi, T., Tohyama, J., Hasegawa, Y., Koike, Y., Hirayama, M., Takahashi, A., (1994). A radiological analysis of heart sympathetic functions with meta-[^{123}I] iodobenzylguanadine in neurological patients with autonomic failure, *J. Auton. Nerv. Syst.*, 49:81–4. 1994.

142. Kanzaki, N., Sato, K., Jayabara, T., Improved cardiac iodine-123 metaiodobenzyleguanidine accumulation after drug therapy in a patient with Parkinson's disease, *Nucl. Med. Commun.*, 22:697–9, 1998.

143. Braune, S., Reinhardt, M., Bathmann, J., Krause, T., Lehmann, M., Lucking, Ch., Impaired cardiac uptake of meta-[123]-iodobenzylguanidine in Parkinson's disease with autonomic failure, *Acta Neurol. Scand.*, 97:307–14, 1998.

144. Yoshita, M., Differentiation of idiopathic Parkinson's disease from striatonigral degeneration and progressive supranuclear palsy using iodine-123 and metaiodobenzylguanadine myocardial scintigraphy, *J. Neurol. Sci.*, 155:60–6, 1998.

145. Satoh, A., Serita, T., Seto, M. et al., Loss of 123I-MIBG uptake by the heart in Parkinson's disease: assessment of cardiac sympathetic denervation and diagnostic value, *J. Nucl. Med.*, 40:371–5, 1999.

146. Reinhardt, M. J., Jungling, F. D., Krause, T. M., Braune, S., Scintigraphic differentiation between two forms of primary dysautonomia early after onset of autonomic dysfunction; value of cardiac and pulmonary iodine-123 MIBG uptake, *Eur. J. Nucl. Med.*, 27:595–600, 2000.

147. Druschky, A., Hilz, M. J., Platsch, G. et al., Differentiation of Parkinson's disease and multiple system atrophy in early disease stages by means of I-123-MIBG-SPECT, *J. Neurol. Sci.*, 175:3–12, 2000.

148. Ohmura, M., Loss of 123I-MIBG uptake by the heart in Parkinson's disease: assessment of cardiac sympathetic denervation and diagnostic value, *J. Nucl. Med.*, 41:1594–5, 2000.

149. Courbon, F., Brefel-Courbon, C., Thalamas, C., Alibelli, M. J., Berry, I., Montastruc, J. L., Rascol, O., Senard, J. M., Cardiac MIBG scintigraphy is a sensitive tool for detecting cardiac sympathetic denervation in Parkinson's disease, *Mov. Disord.*, 18: 890–7, 2003.

150. Takatsu, H., Nishida, H., Matsuo, H. et al., Cardiac sympathetic denervation from the early state of Parkinson's disease: clinical and experimental studies with radiolabeled MIBG, *J. Nuc.l Med.*, 41:71–7, 2000.

151. Yoshita, M., Hayashi, M., Hirai, S., Decreased myocardial accumulation of 123I-metaiodobenzylguanidine in Parkinson's disease, *Nucl. Med. Commun.*, 19:137–42, 1998.

152. Goldstein, D. S., Holmes, C., Cannon, R. O., III, Eisenhofer, G., Kopin, I. J., Sympathetic cardioneuropathy in dysautonomias, *N. Engl. J. Med.*, 336:696–702, 1997.

153. Goldstein, D. S., Holmes, C., Li, S. T., Bruce, S., Metman, L. V., Cannon, R. O., Cardiac sympathetic denervation in Parkinson disease, *Ann. Intern. Med.*, 133:338–47, 2000.

154. Goldstein, D. S., Holmes, C., Dendi, R., Bruce, S., Li, S. T., Orthostatic hypotension from sympathetic denervation in Parkinson's disease, *Neurology*, 58:1247–55, 2002.

155. Li, S. T., Dendi, R., Holmes, C., Goldstein, D. S.,Progressive loss of cardiac sympathetic innervation in Parkinson's disease, *Ann. Neurol.*, 52:220–3, 2002.

156. Thomaides, T., Chaudhuri, K. R., Maule, S., Watson, L., Marsden, C. D., Mathias, C. J., The growth hormone response to clonidine in central and peripheral primary autonomic failure, *Lancet*, 340:263–6, 1992.

157. Kaufmann, H., Oribe, E., Miller, M., Knott, P., Wiltshire-Clement, M., Yahr, M. D., Hypotension-induced vasopressin release distinguishes between pure autonomic failure and multiple system atrophy with autonomic failure, *Neurology*, 42:590–3, 1992.

158. Kimber, J. R., Watson, L., Mathias, C. J., Distinction of idiopathic Parkinson's disease from multiple system atrophy by stimulation of growth hormone release with clonidine, *Lancet*, 349; 1877–1881, 1997.

159. Kimber, J. R., Mathias, C. J., Neuroendocrine responses to Levodopa in multiple system atrophy (MSA), *Mov. Disord.*, 14:981–7, 1999.

160. Critchley, H. D., Mathias, C. J., Dolan, R. J., Neuroanatomical basis for first-and second order representations of bodily states, *Nature Neurosci.*, 4:207–12, 2001.

161. Critchley, H. D., Mathias, C. J., Dolan, R. J., Fear conditioning in humans: the influence of awareness and autonomic arousal on functional neuroanatomy, *Neuron.* 33:653–63, 2002.

162. Orimo, S., Ozawa, E., Oka, T. et al., Different histopathology accounting for a decrease in myocardial MIBG uptake in PD and MSA, *Neurology*, 57:1140–41, 2001.

163. Orimo, S., Oka, T., Miura, H., Tsuchiya, K., Mori, F., Wakabayashi, K., Nagao, T., Yokochi, M., Sympathetic cardiac denervation in Parkinson's disease and pure autonomic failure but not in multiple system atrophy, *J. Neurol. Neurosurg. Psychiatry*, 73:776–78, 2002.

164. Goldstein, D. S., Holmes, C., Sharabi, Y., Brentzel, S., Eisenhofer, G., Plasma levels of catechols and metanephrines in neurogenic orthostatic hypotension, *Neurology*, 60:1327–32, 2003.

165. Lo, S. S., Mathias, C. J., St. John, Sutton, QT interval and dispersion in primary autonomic failure, *Heart*, 75:498–501, 1996.

166. Barbeau, A., L-dopa therapy in Parkinson's disease: nine years experience, *Can. Med. Assoc. J.*, 101:791–6, 1969.

167. McDowell, F. H., Lee, J. E., Swift, T., Sweet, R. D., Ogsburg, J. S., Kessler, J. T., Treatment of Parkinson's syndrome with L-dihydroxyphenalalinine (levo-dopa), *Ann. Intern. Med.,* 72:29–35, 1970.

168. Hunter, K. R., Hollman, A., Laurence, D. R., Stern, G. M., Levodopa in parkinsonian patients with heart disease, *Lancet,* 1:932–4, 1971.

169. Jenkins, R. B., Mendelson, S. H., Lamid, S., Klawans, H. L., Levodopa therapy of patients with parkinsonism and heart disease, *Br. Med. J.,* 3:512–14, 1972.

170. Tanner, C. M., Goetz, C. G., Klawans, H. L., Autonomic nervous system disorders, in *Handbook of Parkinson's Disease,* Koller, W. C., Ed., Marcel Dekker, pp. 145–70. 1987.

171. Pallis, C. A., Calne, D. B., (1970). A case of 'atrial flutter' due to tremor artifacts, *Lancet,* (Letter), 2:1313.

172. Freemon, F. R., Parkinsonism and cardiac arrhythmias, *Lancet,* 2:83–4, 1971.

173. Saint-Pierre, A., ECG artifacts simulating atrial flutter, *JAMA,* (Letter), 224:1534, 1973.

174. Mars and Krall, L-dopa and cardiac arrhythmias, *N. Engl. J. Med.,* (letter), 285:1437, 1971.

175. Desjacques, P., Moret, P., Gauthier, G., Effect cardio-vasculaires de las L-dopa et de l'inhibiteur de la decar-boxylase chez les malades atteints de la maladie de Parkinson, *Schweiz. Med. Wochenschr.,* 103:1783–5, 1973.

176. Lieberman, A. N., Goldstein, M., Gopinathan, G., Leibowitz, M., Neophytides, A., Walker, R., Hiesiger, E., Nelson, J., Further studies with pergolide in Parkinson disease, *Neurology,* 32:1181–4, 1982.

177. Tanner, C. M., Chhablani, R., Goetz, C. G., Klawans, H. L., Pergolide mesylate: lack of cardiac toxicity in patients with cardiac disease, *Neurology,* 35:918–21, 1985.

178. Depresseux, J. C., Rousseau, J. J., Franck, G., The autoregulation of cerebral blood flow, the cerebrovascular reactivity and their interaction in the Shy-Drager syndrome, *Eur. Neurol.,* 18:295–301, 1979.

179. Thomas, D. J., Bannister, R., Preservation of autoregulation of cerebral blood flow in autonomic failure, *J. Neurol. Sci.,* 44:205–12, 1980.

180. Briebach, T., Laubenberger, J., Fischer, P. A., Transcranial doppler sonographic studies of cerebral autoregulation in Shy-Drager syndrome, *J. Neurol.,* 236:349–50, 1989.

181. Titianova, E., Karakaneva, S., Velcheva, I., Orthostatic dysregulation in progressive autonomic failure – a transcranial doppler sonography monitoring, *J. Neuro. Sci.,* 146:87–91, 1997.

182. Brooks, D. J., Redmond, S., Mathias, C. J., Bannister, R., Symon, L., The effect of orthostatic hypotension on cerebral blood flow and middle cerebral artery velocity in autonomic failure, with observations on the action of ephedrine, *J. Neurol. Neurosurg. Psychiatry,* 52:962–6, 1989.

183. Meyer, J. S., Shimazu, K., Fukuuchi, Y., Ouchi, T., Okamoto, S., Koto, A., Ericsson, A. D., Cerebral dysautoregulation in central neurogenic orthostatic hypotension (Shy-Drager syndrome), *Neurology,* 23:262–73, 1973.

184. Akinola, A., Mathias, C. J., Mansfield, A., Thomas, D., Wolfe, J., Nicolaides, A. N., Tegos, T., Cardiovascular, autonomic and plasma catecholamine responses in unilateral and bilateral carotid artery stenosis, *J. Neurol. Neurosurg. Psychiatry,* 67:428–32, 1999.

185. Mathias, C. J., Autonomic disorders. In *Textbook of Neurology.* J. Bogousslavsky and M. Fisher, Eds., Butterworth Heinemann, MA, pp. 519–45, 1999.

186. Chaudhuri, K. R., Thomaides, T., Hernandez, P., Alam, M., Mathias, C. J., Non-invasive quantification of superior mesenteric artery blood flow during sympathoneural activation in normal subjects, *Clin. Aut. Res.,* 1:37–42, 1991.

187. Chaudhuri, K. R., Thomaides, T. N., Mathias, C. J., Abnormality of superior mesenteric artery blood flow responses in human sympathetic failure, *J. Physiol.,* London, 457;477–89, 1992.

188. Puvi-Rajasingham, S., Smith, G. D. P., Akinola, A., Mathias, C. J., Abnormal regional blood flow responses during and after exercise in human sympathetic denervation, *J. Physiol.,* 505:481–9, 1997.

189. Chaudhuri, K. R., Thomaides, T., Watson, L., Mathias, C. J., Octreotide reduces alcohol induced hypotension and orthostatic symptoms in primary autonomic failure, *Q. J. Med.,* 88:719–25, 1995.

190. Wilcox, C. S., Aminoff, M. J., Slater, J. D. H., Sodium homeostasis in patients with autonomic failure, *Clin. Sci.,* 53:321–8, 1977.

191. Mathias, C. J., Fosbraey, P., da Costa, D. F., Thornley, A., Bannister, R., The effect of desmopressin on nocturnal polyuria, overnight weight loss and morning postural hypotension in patients with autonomic failure, *Br. Med. J.,* 293: 353–4, 1986.

192. Ozawa, T., Tanaka, H., Nakano, R., Sato, M., Inuzuka, T., Soma, Y., Yoshimura, N., Fukuhara, N., Tsuji, S., Nocturnal decrease in vasopressin secretion into plasma in patients with multiple system atrophy, *J. Neurol. Neurosurg. Psychiatry,* 67:542–5, 1999.

193. Kooner, J. S., Frankel, H. L., Mirando, N., Peart, W. S., Mathias, C. J., Haemodynamic, hormonal and urinary responses to postural change in tetraplegic and paraplegic man, *Paraplegia,* 26:233–7, 1988.

194. DiBona, G. F., Wilcox, C. S., The kidney and the sympathetic nervous system., In: *Autonomic Failure. A Textbook of Clinical Disorders of the Autonomic Nervous System,* C. J. Mathias and R. Bannister, Eds., 4th ed., Oxford University Press, Oxford, pp. 143–50, 2002.

195. Plaschke, M., Schwarz, J., Dahlheim, H., Backmund, H., Trenkwalder, C., Cardiovascular and renin responses to head-up tilt tests in parkinsonism, *Acta Neurol. Scand.,* 96:206-10, 1997.

196. Puritz, R., Lightman, S. L., Wilcox, C. S., Forsling, M., Bannister, R., Blood pressure and vasopressin in progressive autonomic failure: response to postural stimulation, L-dopa and naloxone, *Brain,* 106:503–11, 1983.

197. Niimi, Y., da, T., Hirayama, M., Koike, Y., Sobue, G., Hasegawa, Y., Takahashi, A., Clinical and physiological characteristics of autonomic failure with Parkinson's disease, *Clin. Auton. Res.,* 9:139–44, 1999.

198. Sakakibara, R., Matsuda, S., Uchiyama, T., Yoshiyama, M., Yamanishi, T., Hattori, T., The effect of intranasal

desmopressin on nocturnal waking in urination in multiple system atrophy patients with nocturnal polyuria, *Clin. Auton. Res.,* 13:106–8, 2003.

199. Mathias, C. J., Young, T. M., Plugging the leak – benefits of the vasopressin-2 agonist, desmopressin in autonomic failure, *Clin. Auton. Res.,* 13:85–7, 2003.

200. Mallipedi, R., Mathias, C. J., Raynaud's phenomenon after sympathetic denervation in patients with primary autonomic failure: questionnaire survey, *Br. Med. J.,* 316;438–9, 1998.

201. Klein, C., Brown, R., Wenning, G. K., Quinn, N. P., (1997). The 'cold hands sign' in multiple system atrophy, *Mov. Disord.,* 12:514–8, 1997.

202. Asahina, M., Kikkawa, Y., Suzuki, A., Hattori, T., Cutaneous sympathetic function in patients with multiple system atrophy, *Clin. Auton. Res.,* 13:91–5, 2003.

203. Korczyn, A. D., Treatment of autonomic nervous system disturbances in Parkinson's disease, in Therapy of Parkinson's disease, W. C. Koller and G. Paulson, Eds., Marcel Dekker, pp. 405–11, 1990.

204. Mathias, C. J., Autonomic disorders and their recognition, *New. Eng. J. Med.,* 10;721–4, 1997.

205. Mathias, C. J., Neurodegeneration, parkinsonian syndromes and autonomic failure, *Aut. Neurosci: Basic and Clinical,* 96:50–8, 2002.

29 Disorders of Thermoregulation in Parkinson's Disease

Thomas L. Davis
Division of Movement Disorders, Vanderbilt University of School of Medicine

CONTENTS

INTRODUCTION

A healthy, nude human adult can be exposed to temperatures as low as 12°C or as high as 60°C in dry air and still maintain an almost constant core temperature of 37°C.[1] The body maintains this temperature through cooling (sweating) and heat conservation (peripheral vasoconstriction, shivering, and lack of sweating) regulated by the autonomic nervous system. This process is termed *thermoregulation*. Patients with Parkinson's disease often complain of temperature sensitivity that manifests as either intolerance to heat, intolerance to cold, or profuse periodic sweating (Table 29.1).

SWEATING

Stimulation of the anterior hypothalamus, especially the preoptic area, causes sweating.[2] The firing rate of these neurons may increase as much as tenfold from a 10°C increase in temperature. The signal from the hypothalamus travels through the autonomic pathways in the spinal cord and to the sympathetic outflow to the skin. Sweat glands are innervated by sympathetic cholinergic nerve fibers but can also be directly stimulated by circulating epinephrine or norepinephrine. Sweat glands of the hands and feet also have adrenergic innervation explaining why these areas preferentially sweat during times of stress or emotional states.

Eccrine sweat glands act to maintain body temperature in the presence of a hot environment or physical exertion. Eccrine glands number from 2 to 4 million per person and vary in density across the body surface. They are most dense on the palms and soles. Cooling is accomplished by the secretion of water onto the surface of the skin and subsequent evaporation. The rate of sweat secretion by the eccrine glands far surpasses that of other exocrine glands. The normal unacclimatized person can rarely produce more than 700 ml of sweat per hour. This rate can double in 1 to 6 weeks when the person is exposed to hot weather.

TABLE 29.1
Disorders of Thermoregulation in Parkinson's Disease

As part of the disease
Heat or cold intolerance
Excessive nighttime sweats[6]
Sweating abnormalities[8,10]
Predilection to accidental hypothermia[5]
Complication of medication
Sweating with reboxetine[14]
Anhidrosis secondary to therapeutic anticholinergics
Excessive sweating at threshold levels of L-dopa[21]
Excessive sweating with weak dose chorea
Parkinsonism hyperpyrexia syndrome
Drug holiday[15,18,22]
Change in medication[16]
Wearing-off response[17]
Hyponatremia[20]
Premenstrual[19]

DISORDERS OF THERMOREGULATION DUE TO PARKINSON'S DISEASE

It has long been recognized that abnormal tolerance to temperature is one of the nonmotor symptoms of Parkinson's disease. This temperature sensitivity was seen prior to the use of dopaminergics and was described by Gower in 1888.[3]

> Frequently also (in three-quarters of the cases) there is some abnormal sensation of temperature. The most common is a sense of heat, to which Charcot first directed attention. I have found this to be present in half the cases in which the point was investigated (twenty-two out of forty-seven), when slight, it may only cause the patient to dislike hot rooms or many bed clothes; but when considerable it is a source of considerable discomfort, and only the thinnest covering can be endured at night, even in the depth of winter.

> In other cases, instead of a feeling of heat, there is an abnormal sensation of cold: the patient always feels chilly. This is only half as common as the as a sensation of heat (eleven out of forty-seven cases). I have even known the sensation of cold and heat to alternate…

Episodic hyperthermia mimicking neuroleptic malignant syndrome may rarely occur in patients who are untreated with medications. VanHilton and Roos described a patient with mild parkinsonian features and dysautonomia, who developed two episodes of hyperthermia, increased rigidity and an elevated serum creatine kinase.[4] The patient was a 73-year-old man with a history of anhydrosis, xerostomia, and no prior exposure to neuroleptic or antiparkinsonian medications. No triggering cause could be found for his first episode of hyperthermia, and it resolved in three days with fluid replacement and a cooling. An allergic reaction to medication was presumed to precipitate the second episode. The authors felt that the anhydrosis likely contributed to the patient's susceptibility to these events. It has also been suggested that Parkinson's disease is a risk factor for the development of accidental hypothermia.[5]

Sleep disorders are a well recognized, common, nonmotor manifestation of Parkinson's disease that generally increase in prevalence with disease severity. Happe et al. recently reviewed the spectrum of sleep disorders seen in Parkinson's disease and prospectively studied quality of life in these patients.[6] Out of 108 patients answering the questionnaire, 30% complained of heavy sweating at night. No treatment options were mentioned.

In an attempt to quantify abnormalities in thermoregulation, Appenzeller and Gross formally studied autonomic function including thermoregulatory sweating in 18 patients with Parkinson's disease before and after the initiation of L-DOPA.[7] They found normal thermoregulatory sweating in only eight patients, three of whom were treated with anticholinergics. In the other patients, almost complete absence of sweating on the trunk and limbs was observed with apparent compensatory hyperhydrosis over the face. Normal reflex vasodilatation of the fingertips in response to heating was seen in seven patients. These observations were not effected by L-dopa therapy. In 12 of 17 patients, the absence or presence of normal reflex vasodilatation predicted the absence or presence of normal thermoregulatory sweating. In another study, Turkka and Myllyla measured sweating with an evaporimeter in 23 patients with idiopathic Parkinson's disease and in 11 age-matched control subjects before and after a heating stimulus.[8] Perspiration was increased significantly both before and after the heating provocation in the upper part of the body of Parkinson patients in comparison with the control subjects. The increase of perspiration correlated with the clinical severity of Parkinson's disease. The authors conclude that these results are consistent with autonomic nervous system dysfunction in some patients with Parkinson's disease.

Another measure of healthy temperature control is the presence of normal diurnal variation in core temperature. Pierangeli et al. compared 24-hr core temperature readings in 14 patients with probable multiple systems atrophy, to 7 subjects with Parkinson's disease and 8 healthy controls.[9] In this small sample, they found the circadian rhythm of body core temperature in those with Parkinson's disease to be identical to controls. Subjects with multiple systems atrophy showed a blunted nocturnal fall in core temperature. The authors did not feel that these results could be explained by differences in sleep structure and felt that they likely represented a pronounce impairment of central sympathetic activity in those with multiple systems atrophy.

Others have attempted to use the sympathetic skin response as a measure of sympathetic cholinergic activity to sweat glands. The sympathetic skin response measures the electrical potentials from electrodes placed on the hands and feet. Unfortunately, the amplitude and frequency of this response are highly variable. This generally necessitates the study of large numbers of subjects to determine population differences.

Mano et al. studied 38 patients with Parkinson's disease and 16 normal controls and reported that both the maximum and mean amplitude of the sympathetic skin response was significantly lower in Parkinson's disease compared to controls.[10] The authors also found a significant correlation between loss of amplitude of the sympathetic skin response and severity of the Parkinsonism. Hirashima and colleagues also found a similar relation between the severity of disease and extent of abnormality of the sympathetic skin response.[11] In this study, the sympathetic skin response was not affected by dopaminergic medications. It is important to note that, in both of these studies, some patients with Parkinson disease had normal sympathetic skin responses and that, unlike thermal sweating, the sympathetic skin response is considered to reflect emotional sweating and to depend on different afferent and efferent pathways.

The pathological explanation of the abnormalities in sweating and thermoregulation seen in some patients with Parkinson's disease is unknown. It is well documented that Lewy bodies may be found in the hypothalamus and sympathetic ganglia of patients with Parkinson's disease.[12] It is unclear, however, if these Lewy bodies are always associated with nerve cell loss. To further complicate the issue, in one patient, the coexistence of Lewy bodies in the hypothalamus and cytoplasmic inclusion in glia has recently been observed.[13] These inclusions were immunopositive for ubiquitin and α-synuclein identical to those classically associated with multiple systems atrophy.

DISORDERS DUE TO NONDOPAMINERGIC MEDICATIONS

Cholinergic antagonist have long been used as a treatment of Parkinson's and may be particularly effective in damping rest tremor. Their dose dependent adverse effects limit the therapeutic usefulness of these agents. These become more prevalent in older individuals and include anhidrosis leading to heat intolerance and heat stroke. This may also be seen with the more peripherally selective anticholinergics (oxybutynin and tolterodinel) sometime used to urinary frequency in patients with Parkinson's disease.

In a recent pilot study of the selective norepinephrine reuptake inhibitor reboxetine to treat depression in Parkinson's disease, increased sweating was noted in 3 of 15 subjects.[14] This observation was not appreciated in previous studies of reboxetine in patients without Parkinson's disease.

DISORDER IN PATIENTS ON DOPAMINERGIC THERAPY

A syndrome of fever, severe rigidity, and myoglobinuria (clinically identical to neuroleptic malignant syndrome) has been described as a rare complication of abrupt discontinuation of dopaminergics used to treat Parkinson's disease.[4,15] These are most often associated with "drug-holiday" but have also been reported with switching L-dopa to controlled release formulation[16] or as a dramatic wearing-off phenomenon.[17] The pathophysiology of this is poorly understood, but it is theorized that it is due to decreased dopaminergic stimulation of the hypothalamus. Since this syndrome is not always associated with neuroleptic use, some authors have suggested the term, *Parkinsonism hyperpyrexia syndrome.*[18] In patients with Parkinson's disease, this may represent the acute manifestation

of severe Parkinson's disease with rigidity and dysphagia. In these patients, a search for another cause of the fever such as aspiration should not be overlooked. Treatment includes hydration, supportive measures, and reinstitution of dopaminergics.

Rarely, the Parkinsonism hyperpyrexia syndrome has been reported as a complication of metabolic or hormonal abnormalities in the absence of changes in dopaminergic regimen.[19,20] In these cases, correction of the abnormality led to improvement in rigidity and fever.

Problems with thermoregulation may also be seen as in correlation with motor fluctuations. Sage and Mark closely followed four patients with Parkinson disease and severe intermittent drenching sweats.[21] In all patients, these tended to be end of dose phenomena. Serial plasma levels were collected in one patient. In this patient, the onset of sweating correlated with levels of L-dopa seen at the threshold of the motor response. The authors conclude that drenching sweats represent an early off sign in some patients.

Severe chorea or dystonia may be seen as a peak-dose response in some patients with Parkinson's disease. This activity is intensely physically and may appropriately be associated with tachycardia, shortness of breath, and sweating in proportion to the amount of energy expended in the movements.

CONCLUSION

Disorders of thermoregulation are common in Parkinson's disease and may occur as part of the disease itself or as a complication of treatment. The pathophysiology of these disorders remains poorly understood but is likely due in part to degeneration of the hypothalamus. Radical changes in antiparkinsonian medication may produce life threatening hyperpyrexia and rigidity and should be done under close supervision. Excessive sweating may represent side effects of medication or manifestation of drug induced motor fluctuations. Recognition of these can lead to effective treatment.

REFERENCES

1. Guyton, A. C. and J. E. Hall, *Metabolism and Temperature Regulation*, in *Textbook of Medical Physiology*, W. B. S. Company, Editor, pp. 916–922, 1996.
2. Wenzel, F. G. and T. D. Horn, *Nonneoplastic disorders of the eccrine glands, Journal of the American Academy of Dermatology*, 38(1), pp. 1–20, 1998.
3. Gowers, W. R., *Paralysis Agitans*, in *A Manual Diseases of the Nervous System: American Edition*, P. Blakiston, Son & Co., Philadelphia, PA, pp. 1005–1007, 1888.
4. van Hilten, B. J. and R. A. Roos, *The continuum of hyperthermic syndromes with impaired dopaminergic activity, Arch. Intern. Med.*, 152(8), pp. 1727, 1730, 1992.

5. Gubbay, S. S. and D. D. Barwick, Two cases of accidental hypothermia in parkinson's disease with unusual EEG findings, *J. Neurol. Neurosurg. Psychiat.*, 29: pp. 459–466, 1966.

6. Happe, S., Ludemann, P. and Berger, K., The association between disease severity and sleep-related problems in patients with Parkinson's disease, *Neuropsychobiology*, 46(2), 2002.

7. Appenzeller, O. and J. E. Goss, Autonomic deficits in Parkinson's syndrome, *Arch. Neurol.*, 24(1), pp. 50–57, 1971.

8. Turkka, J. T. and Myllyla, V. V., Sweating dysfunction in Parkinson's disease, *Eur. Neurol.*, 26(1), pp. 1–7, 1987.

9. Pierangeli, G., et al., Nocturnal body core temperature falls in parkinson's disease but not in multiple-system atrophy, *Mov. Dis.*, 16(2), pp. 226–232, 2001.

10. Mano, Y., et al., Sweat function in Parkinson's disease, J. Neurol., 241(10), pp. 573–576, 1994.

11. Hirashima, F., Yokota, T. and Hayashi, M., Sympathetic skin response in parkinson's disease, *Acta Neurol. Scand.*, 93, pp. 127–132, 1996.

12. Lowe, J. S. and Leigh, N., *Disorders of movement and system degeneration*, in *Greenfield's Neuropathology*, D. I. Graham and P. L. Lantos, Eds., Georgina Bentliff, pp. 334–336, 2002.

13. Mochizuki, A., Komatsuzaki, Y. and Shin'ichi, S, Association of Lewy bodies and glial cytoplasmic inclusions in the brains of Parkinson's disease, *Acta Neuropathol.*, 104, pp. 534–537, 2002.

14. Lemke, M. R., Effect of reboxetine on depression in Parkinson's disease patients. *J. Clin. Psychiatry*, 2002, 63(4), pp. 300–304, 2002.

15. Figa-Talamanca, L., et al., Hyperthermia after discontinuance of levodopa and bromocriptine therapy: impaired dopamine receptors a possible cause, *Neurology*, 35(2), pp. 258–261, 1985.

16. Cunningham, M. A., Darby, D. G. and Donnan, G. A., Controlled-release delivery of L-dopa associated with nonfatal hyperthermia, rigidity, and autonomic dysfunction, *Neurology*, 41(6), pp. 942–943, 1991.

17. Pfeiffer, R. F. and E. L. Sucha, "On-off"-induced lethal hyperthermia, *Mov. Disord.*, 4(4), pp. 338–341, 1989.

18. Granner, M. A. and Wooten, G. F., Neuroleptic malignant syndrome or parkinsonism hyperpyrexia syndrome. *Semin. Neurol.*, 11(3), pp. 228–235, 1991.

19. Mizuta, E., et al., Neuroleptic malignant syndrome in a parkinsonian woman during the premenstrual period, *Neurology*, 43(5), pp. 1048–1049, 1993.

20. Sechi, G. et al., Acute hyponatremia and neuroleptic malignant syndrome in Parkinson's disease, *Prog. Neuropsychopharmacol. Biol. Psychiatry*, 20(3), pp. 533–542, 1996.

21. Sage, J. I. and Mark, M. H., Drenching sweats as an off phenomenon in Parkinson's disease: treatment and relation to plasma levodopa profile, *Ann. Neurol.*, 1995. 37(1), pp. 120–122, 1995.

22. Reutens, D. C., Harrison, and Goldswain, P. R., Neuroleptic malignant syndrome complicating levodopa withdrawal, *Med. J. Aust.*, 155(1), pp. 53–54, 1991.

30 Respiratory Dysfunction

Holly Shill
Muhammad Ali Parkinson Research Center

CONTENTS

INTRODUCTION

James Parkinson, in *An Essay on the Shaking Palsy,* remarked on a man who "fetched his breath rather hard."[1] Parkinson's disease (PD) affects the respiratory tract at all levels from the upper airway down to the pulmonary tissue itself. Pneumonia and pulmonary embolism are the most frequent cause of death both in historical[2] and more contemporary series.[3–6] This chapter reviews the effects of PD on the pulmonary system at all levels and includes difficulties due to the disease itself, as well as complications related to medical and surgical therapy of PD.

HYPOPHONIA

Hypophonia is inventoried in the Unified Parkinson's Disease Rating Scale. It is graded by the patient as the difficulty in being understood by others. It is assessed by the clinician as the amount of monotonicty and intelligibility of speech during the examination. It can be considered a cardinal sign of PD, since it is a direct effect of bradykinesia and rigidity of the vocal cords and pharyngeal muscles.[7,8] It is a very common sign with more than 70% of patients experiencing problems with speech and voice.[9,10] Patients report that this can be one of the most disabling aspects of the disease. They are often unwilling to speak on the phone or go out in public places for fear of being unable to make themselves understood.

Hypophonia can respond to medication, but it may be incomplete. Standard speech therapy may be useful, particularly if combined with more specific voice therapy strategies.[11–13] The Lee Silverman Voice Therapy and variants of this may be particularly useful in augmenting speech. A small study of music therapy (singing and vocal exercises) failed to show improvement in voice,[14] but further study is being done.

Surgical procedures may be helpful in improving voice. Hypophonia may respond to percutaneous collagen injection into the vocal cords.[15,16] This procedure requires nasopharyngoscopy for direct visualization with injection through the tracheal structures. It is done under local anesthesia, requires minimal patient cooperation, and lasts about 12 weeks. The vocal cords may be injected with more permanent materials to produce longer-lasting benefit, but these are more complicated procedures. Laryngoplasty can also be performed; this requires general anesthesia.

Specific surgical treatments for PD, such as ablative procedures and deep brain stimulation of the thalamus, globus pallidus, and subthalamic nucleus, are generally thought to produce insubstantial benefit on the voice. Bilateral procedures and unilateral gamma knife may result in worsening of voice loudness and speech production. Patients should be counseled appropriately before undergoing these procedures.[17–20]

UPPER AIRWAY OBSTRUCTION

After hypophonia, upper airway obstruction is probably the most common respiratory abnormality in PD. In early studies, PD was thought effect the lower airway similar to chronic obstructive pulmonary disease,[21,22] although the higher frequency of smoking in this population confounds localization of this site as part of the pathology of PD.

Currently, the lower airway is not thought to be relevant in PD. This is primarily due to work by Vincken and co-workers who shifted focus to the upper-airway involvement.[23] These workers described two patterns on spirometry in treated patients with PD. Type A was characterized by regular acceleration and decelerations of the flow-volume loop that corresponded to the tremor frequency. This was termed *respiratory flutter* and was shown at endoscopy to be secondary to rhythmic contractions of the vocal cords and suproglottic structure rather than diaphragmatic oscillations. Type B was characterized by irregular and abrupt changes in flow, sometimes with complete obstructions. This was thought to be responsible for clinical manifestations of stridor and patient complaints of "running out of breath" while speaking. The study emphasized the high frequency of these findings in patients who were relatively asymptomatic. Twelve of 21 patients had a type A pattern, and 6 of 21 had a type B pattern. Seven patients met respiratory criteria for upper airway obstruction (UAO). Only four of these were overtly symptomatic with stridor.

In a subsequent study, completely asymptomatic patients with PD were studied by spirometry.[24] All patients had Hoehn and Yahr[2] staging greater than 3 and were excluded from study if they had any conditions that might be associated with pulmonary dysfunction. Patients were studied on levopdopa therapy. Only 8 of 31 patients had normal flow-volume loops. Four patients had a type A pattern and 16 had a type B pattern. Nine met criteria for UAO. The authors postulated that the higher frequency of a type B pattern was due to the more advanced stage of PD in their study. This study confirmed the earlier study by Vicken et al. and further demonstrated the very high occurrence of respiratory abnormalities in PD patients without pulmonary symptoms. Similar studies have gone on to support these initial findings with a relatively high frequency of obstructive abnormalities in aysmptomatic, treated PD patients.[25–27] Reasons for a lack of symptoms are unclear but may reflect the more sedentary lifestyle of advancing PD.

An additional piece of information that came from these early studies is that UAO obstruction responds to dopaminergic therapy. Patients studied on medication had frequencies of UAO ranging from 5 to 33%,[23,25,26] while 67% of those studied after medication was held had UAO.[24] Stridor may be a presenting symptom of PD and respond to levopdopa.[28] Acute respiratory failure as a result of UAO may result from abrupt changes or discontinuation of dopaminergic medications.[29–31] Systematic study of this concept has been done more recently. Subcutaneous apomorhine produced improvement in maximal flow rates in patients without their usual PD medication.[32] Oral levodopa produced similar findings in UAO.[33,34] These findings suggest that UAO is not simply a result of progressive degeneration of the motor control of the upper airway, but is dopamine responsive similar to the dysfunctions seen in other striatal muscles.

The clinical relevance of this is easy to see. Patients with PD have higher risk of postoperative pneumonia and longer hospital stays than controls after undergoing elective surgeries such as bowel resection or cholycystectomy.[35] It is probable that withdrawal of medications around the time of surgery may contribute to upper airway dysfunction and aspiration. Awareness of these complications of medication withdrawal will certainly help to reduce them. Further, change in management of those patients unable to take oral therapies, by use of subcutaneous apomorphine and rectal domperidone, may be helpful.[36]

Airway obstruction may also be a manifestation of more fixed, anatomical problems. Long-duration PD is associated with abnormal postures, which in turn contributes to kyphoscoliosis of the cervical spine. Studies have correlated the degree of pulmonary dysfunction with the amount of arthrosis on radiographic images and limitation of range of motion on passive testing.[27] The amount to which these changes are reversible and/or preventable is unknown. However, it is prudent to consider these factors in evaluating patients and addressing them when possible.

Finally, patients with autonomic failure, such as seen in the variant of idiopathic PD, multiple system atrophy (MSA), may be at higher risk for complications related to upper airway dysfunction.[37] These patients have non-dopaminergic sensitive UAO and may have fixed immobile vocal cords on direct laryngoscopy. Sleep studies done in this group of patients may show frequent obstructive and central sleep apneas.[38] Stridor occurring in these patients is considered a poor prognostic sign.[39] Patient with MSA and stridor are at risk for sudden death, presumably as a result of laryngeal obstruction. The pathology behind this complication is not completely understood but may reflect pathology in the brainstem respiratory structures as well as poor feedback control of respiration due to autonomic neuropathy.[40] If stridor is present, consideration should be given to further pulmonary evaluation and tracheostomy should be considered.

RESTRICTIVE ABNORMALITIES

In an early review of nine parkinsonian patients, with and without complaints of dyspnea, Nugent et al. found that vital capacity and airway pressures were reduced as compared to controls.[41] These authors compared the spirometric data to similar data collected from neuromuscular disease patients and postulated the restrictive pattern in parkinsonism to be secondary to "muscle weakness." Further studies emphasized the restrictive element of pulmonary changes in PD and, for many years, this was thought to be the primary changes seen in ventilatory function.[42,43] The restrictive changes were believed secondary to decreased chest wall compliance from muscle rigidity, and supported by electromyographic demonstration of respiratory musculature agonist and antagonist co-contraction.[44,45] It seems likely that restrictive changes also respond to dopaminergic therapy, since many of these patients were studied prior to levodopa therapy, and more contemporary studies do not support a primarily restrictive abnormality. The widespread use of medical therapy makes an advanced, untreated population difficult to find. It may be that the restrictive abnormalities come out only with very long-term withdrawal of dopaminergic therapy. An alternate explanation would be the potential for inclusion of larger numbers of post-encephalitic parkinsonism in the earlier studies. These patients have differing brain pathology and may have alternate respiratory system abnormalities. However, there is at least one study that supports the first hypothesis. Patients with PD were studied with repetitive, voluntary respiratory efforts as well as single respiratory efforts.[46] Patients did well with the single effort as long as they were treated. Untreated patients did not perform as well. All patients (treated or untreated) had problems with the repetitive tasks. It is likely, then, that the restrictive changes are a result of the fatigability of the respiratory pump that occurs with other coordinated muscles activities such as handwriting. Because these changes are more clinically subtle than obstruction, it may take longer to become manifest on isolated, single trial assessments and are seen only with advanced, untreated disease. With more advanced PD, the changes can mimic those seen in neuromuscular disorders.

Dopaminergic medications may also produce changes in the pulmonary tissue itself. All of the ergotamine-derived dopamine agonists have been reported to cause pleuropulmonary fibrosis with long-term therapy. Bromocriptine may do it in as many as 5% of patients,[47] although follow-up data suggest that this number may be much lower.[48] Pergolide, carbergoline, and lisuride seem to have a similar propensity to cause this complication.[49–53] Patients with this complication may present with shortness of breath, fatigue, and pleuritic chest pain. Chest radiograph may reveal pleural thickening, effusions, and mild infiltrates. Sedimentation rate may be elevated. Pleural fluid can show eosinophilia. If indicated, evaluation should be done for infectious and malignant causes of pleurisy. More recently, a link between this complication and prior asbestos exposure has been postulated.[54] Finally, reports of valvular heart disease with bromocriptine and pergolide should make the clinician more suspicious of shortness of breath in the congestive heart failure patient on these agents.[55,56] Pleuropulmonary fibrosis typically responds favorably to withdrawal of the offending agent. The newer dopamine agonists, pramipexole and ropinerole, are not thought to produce this complication. It is prudent to warn patients of these potential risks when using the ergot-derived dopamine agonists.

MOTOR FLUCTUATIONS

It is becoming increasingly well recognized that respiratory dysfunction may be part of motor fluctuations accompanying long-duration PD. Forty percent of patients with fluctuations experience dyspnea tied to their motor state, primarily as an "off" phenomena.[57] Stridor is relatively common during the off period.[58] Coughing may be seen. Advanced patients may report "panic attacks" that correlate with severity of motor fluctuations.[59] At least part of this syndrome may be related to the sudden appearance of chest wall tightness and upper airway restriction with the associated anxiety, choking sensations, and palpitations that may accompany these physical symptoms. Additionally, dyskinesia may effect the respiratory system leading to shallow, uncoordinated breathing, which results in shortness of breath.[60–62] Awareness of these entities is important, since an older patient complaining of chest tightness and shortness of breath invariably prompts and evaluation for cardiac and pulmonary causes for their symptoms. Management of the respiratory fluctuations is similar to the standard practices for typical motor fluctuations. Benzodiazepines are useful for the anxiety component. Behavioral strategies used for panic attacks may also be used.

EXERCISE AND RESPIRATION

Patients with PD do not coordinate breathing with locomotion as seen in normal individuals.[63] Combining this with fatigability of the respiratory pump during repetitive actions,[46] it is not surprising that patient with PD have prominent fatigue and poor exercise tolerance. However, patients are able to improve pulmonary function through a pulmonary rehabilitation program.[64] Furthermore, those who engage in regular aerobic exercise may maintain good pulmonary function.[65,66] These types of studies suggest that nonpharmacological measures may be as important in addressing pulmonary dysfunction as dopaminergic therapy. Because of limitations due to postural instability and freezing of gait, strategies to maintain aerobic capacity can be more challenging. Multidisciplinary care is often necessary. The PD population is increasingly using alternative medicine approaches. The effect of interventions such as Qigong, yoga, and tai chi on respiratory system are not entirely understood but certainly warrant further investigation.

CONCLUSION

Pneumonia remains the most common cause of death in the PD patient. The respiratory tract is effected at all levels from the voice and upper airway, down to the pulmonary tissue itself. Many of the ventilatory abnormalities in PD respond to dopaminergic medications similar to other motor dysfunctions. Awareness of these pulmonary complications is critical in attempts to prevent or delay aspiration pneumonia as a complication of long duration PD. Careful attention to medication timing is paramount, particularly during times of acute illness and elective surgery. Respiratory dysfunction may also respond to nonpharmacological interventions, and a further study of these treatments is warranted.

REFERENCES

1. Parkinson, J., *An essay on the shaking palsy*, Sherwood, Nealy and Jones, London, 1817.
2. Hoehn, M. M. and Yahr, M. D., Parkinsonism: onset, progression and mortality, *Neurology*, 17, 427, 1967.
3. Mosewich, R. K. et al., Pulmonary embolism: an under-recognized yet frequent cause of death in parkinsonism, *Mov. Disord.*, 9, 350, 1994.
4. Nakashima, K. et al., Prognosis of Parkinson's Disease in Japan, Tottori University Parkinson's Disease Epidemiology (TUPDE) Study Group, *Eur. Neurol.*, 38, 60, 1997.
5. Morgante, L. et al., Parkinson disease survival: a population-based study, *Arch. Neurol.*, 57, 507, 2000.
6. Beyer, M. K. et al., Causes of death in a community-based study of Parkinson's disease, *Acta Neurol. Scand.*, 103, 7, 2001.
7. Hanson, D. G., Gerratt, B. R. and Ward, P. H. Cinegraphic observations of laryngeal function in Parkinson's disease, *Laryngoscope*, 94, 348, 1984.
8. Baker, K. K. et al., Thyroarytenoid muscle activity associated with hypophonia in Parkinson disease and aging, *Neurology*, 51, 1592, 1998.
9. Logemann, J. A. et al., Frequency and co-occurrence of vocal tract dysfunction in the speech of a large sample of Parkinson patients, *J. Speech Hear. Disord.*, 43, 47, 1978.
10. Hartelius, L. and Svensson, P., Speech and swallowing symptoms associated with Parkinson's disease and multiple sclerosis: a survey, *Folia Phoniatrica Logopedica*, 46, 9, 1994.
11. Ramig, L. O. et al., Intensive voice treatment (LSVT) for patients with Parkinson's disease: a 2 year follow-up, *J. Neurol. Neurosurg. Psychiatry*, 71, 493, 2001.
12. Baumgartner, C. A., Sapir, S., and Ramig, T. O., Voice quality changes following phonatory-respiratory effort treatment (LSVT) versus respiratory effort treatment for individuals with Parkinson's disease, *J. Voice*, 15, 105, 2001.
13. Swart, B. J. M. et al., Improvement of voicing in patients with Parkinson's disease by speech therapy, *Neurology*, 60, 498, 2003.
14. Naneishi, E., Effects of a music therapy voice protocol on speech intelligibility, vocal acoustic measures and mood of individuals with Parkinson's disease, *J. Music Ther.*, 28, 273, 2001.
15. Berke, G. S.,et al., Treatment of Parkinson hypophonia with percutaneous collagen augmentation, *Laryngoscope*, 1098, 1295, 1999.
16. Kim, S. H., Kearney, J. J. and Atkins, J. P. Percutaneous collagen augmentation for treatment of parkinsonian hypophonia, *Otolaryngol Head Neck Surg.*, 126, 653, 2002.
17. Jankovic, J. et al., Outcome after stereotactic thalamotomy for parkinsonian, essential, and other types of tremor, *Neurosurgery*, 37, 680, 1995.
18. Lang, A. E. et al., Posteroventral medial pallidotomy in Parkinson's disease. *J. Neurol.*, Suppl., 2, 28, 1999.
19. Okun, M. S. et al., Complications of gamma knife surgery for Parkinson's disease, *Arch. Neurol.*, 58, 1995.
20. Romito, L. M. et al., Long-term follow up of subthlamic nucleus stimulation in Parkinson's disease, *Neurology*, 58, 15, 2002.
21. Neu, H. C. et al., Obstructive respiratory dysfunction in parkinsonian patients, *Am. Rev. Respir. Dis.*, 95, 33,1967.
22. Obenour, W. H. et al., The causes of abnormal pulmonary function in Parkinson's disease, *Am. Rev. Respir. Dis.*, 105, 382, 1972.
23. Vincken, W. G. et al., Involvement of upper-airway muscles in extrapyramidal disorders. A cause of airflow limitation, *N. Engl. J. Med.*, 311, 438, 1984.
24. Hovestadt, A. et al., Pulmonary function in Parkinson's disease, *J. Neurol. Neurosurg. Psychiatry*, 52, 329, 1989.
25. Izquierdo-Alonso, J. L. et al., Airway dysfunction in patients with Parkinson's disease, *Lung*, 172, 47, 1994.
26. Sabate, M. et al., Obstructive and restrictive pulmonary dysfunctions in Parkinson's disease, *J. Neurol. Sci.*, 138, 114, 1996.
27. Sabate, M. et al., Obstructive and restrictive pulmonary dysfunction increases disability in Parkinson disease, *Arch. Phys. Med. Rehabil.*, 77, 29, 1996.
28. Read, D. and Young, A. Stridor and parkinsonism, *Postgrad. Med. J.*, 59, 520, 1983.
29. Fink, M. E. et al., Acute respiratory failure during drug manipulation in patients with Parkinson's disease, *Neurology*, 39, 348, 1989.
30. Riley, D. E., Grossman, G., and Martin, L., Acute respiratory failure from dopamine agonist withdrawal, *Neurology*, 42, 1843, 1992.
31. Easdown, L. J., Tessler, M. J., and Minuk, J., Upper airway involvement in Parkinson's disease resulting in postoperative respiratory failure, *Can. J. Anaesth.*, 42, 344, 1995.
32. de Bruin, P. F. et al., Effects of treatment on airway dynamics and respiratory muscle strength in Parkinson's disease, *Am. Rev. Respir. Dis.*, 148, 1576, 1993.
33. Langer, H., and Woolf, C. R., Changes in pulmonary function Parkinson's syndrome after treatment with L-dopa, *Am. Rev. Respir. Dis.*, 104, 440, 1971.
34. Herer, B., Arnulf, I., and Housset, B., Effects of levodopa on pulmonary function in Parkinson's disease, *Chest.*, 119, 387, 2001.
35. Pepper, P. V. and Goldstein, M. K., Postoperative complications in Parkinson's disease, *J. Am. Geriatr. Soc.*, 47, 967, 1999.

36. Galvez-Jimenez, N. and Lang, A. E., Perioperative problems in Parkinson's disease and their management: apomorphine with rectal domperidone, *Can. J. Neurol. Sci.*, 23, 198, 1996.

37. Wenning, G. K., et al., Multiple system atrophy: a review of 203 pathologically proven cases, *Mov. Disord.*, 12, 133, 1997.

38. Apps, M. C. P. et al., Respiration and sleep in Parkinson's disease, *J. Neurol. Neurosurg. Psychiatry*, 48, 1240, 1985.

39. Silber, M. H. and Levine, S., Stridor and death in multiple system atrophy, *Mov. Disord.*, 15, 699, 2000.

40. Chester, C. S. et al., Pathophysiological findings in a patient with Shy-Drager and alveolar hypoventilation syndromes, *Chest*, 94, 212,1988.

41. Nugent, C. A. et al., Dyspnea as a symptom in Parkinson' syndrome, *Am. Rev. Tuberc.*, 78, 682, 1958.

42. Paulson, G. P. and Tafrate, R. H., Some "minor" aspects of Parkinsonism, especially pulmonary function, *Neurology*, 20, 14, 1970.

43. Bateman, D. N. et al., Levodopa dosage and ventilatory function in Parkinson's disease, *Br. Med. J. (Clin. Res. Ed.)*, 283, 190, 1981.

44. Hallett, M. and Khoshbin, S., A physiological mechanism for bradykinesia, *Brain*, 103, 301, 1980.

45. Estenne, M., Hubert, M. and De Troyer, A., Respiratory-muscle involvement in Parkinson's disease, *N. Engl. J. Med.*, 311, 1516, 1984.

46. Tzelepis, G. E. et al., Respiratory muscle dysfunction in Parkinson's disease, *Am. Rev. Respir. Dis.*, 138, 266, 1988.

47. Rinne, U. K., Pleuropulmonary changes during long-term bromocriptine treatment for Parkinson's disease, *Lancet*, 1, 44, 1981.

48. Krupp, P., Pleuropulmonary changes during long-term bromocriptine treatment for Parkinson's disease, *Lancet*, 1, 44, 1981.

49. McElvaney, N. G. et al., Pleuropulmonary disease during bromocriptine treatment of Parkinson's disease, *Arch. Intern. Med.*, 148, 2231, 1988.

50. Todman, D. H., Oliver, W. A., and Edwards, R. L., Pleuropulmonary fibrosis due to bromocriptine treatment for Parkinson's disease, *Clin. Exp. Neurol.*, 27, 79, 1990.

51. Bhatt, M. H. et al., Pleuropulmonary disease associated with dopamine agonist therapy, *Ann. Neurol.*, 30, 613, 1991.

52. Geminiani, G. et al., Cabergoline in Parkinson's disease complicated by motor fluctuations, *Mov. Disord.*, 11, 495, 1996.

53. Bleumink, G. S. et al., Pergolide-induced pleuropulmonary fibrosis, *Clin. Neuropharmacol.*, 25, 290, 2002.

54. Hillerdal, G. et al., Pleural disease during treatment with bromocriptine in patients previously exposed to asbestos, *Eur. Respir. J.*, 10, 2711, 1997.

55. Serratrice, J. et al., Fibrotic valvular heart disease subsequent to bromocriptine treatment, *Cardiol. Rev.*, 10, 334, 2002.

56. Prichett, A. M., Morrison, J. F., Edwards, W. D., et al., Valvular heart disease in patients taking pergolide, *Mayo Clin. Proc.*, 77, 1275, 2002.

57. Witjas, T. et al., Nonmotor fluctuations in Parkinson's disease, *Neurology*, 59, 408, 2002.

58. Corbin, D. O. and Williams, A. C., Stridor during dystonic phases of Parkinson's disease, *J. Neurol. Neurosurg. Psychiatry*, 50, 821, 1987.

59. Vazquez, A. et al., "Panic attacks" in Parkinson's disease. A long-term complication of levodopa therapy, *Acta Neurol. Scand.*, 87, 14, 1993.

60. Weiner, W. J. et al., Respiratory dyskinesias: extrapyramidal dysfunction and dyspnea, *Ann. Intern. Med.*, 88, 327, 1978.

61. Zupnick, H. M. et al., Respiratory dysfunction due to L-dopa therapy for parkinsonism: diagnosis using serial pulmonary function tests and respiratory inductive plethysmography, *Am. J. Med.*, 89, 109, 1990.

62. Rice, J. E., Antic, R. and Thompson, P. D., Disordered respiration as a levodopa-induced dyskinesia in Parkinson's disease, *Mov. Disord.*, 17, 524, 2002.

63. Scheiermeier, S. et al., Breathing and locomotion in patients with Parkinson's disease, *Plufers Arch.*, 443, 67, 2001.

64. Koseoglu, F. et al., The effects of a pulmonary rehabilitation program on pulmonary function tests and exercise tolerance in patients with Parkinson's disease, *Funct. Neurol.*, 12, 319, 1997.

65. Canning, C. G. et al., Parkinson's disease: an investigation of exercise capacity, respiratory function, and gait, *Arch. Phys. Med. Rehabil.*, 78, 199, 1997.

66. Bergen, J. L., Toole, T., Elliott, R. G., Wallace, B., Robinson, K., Maitland, C. G., Aerobic exercise intervention improves aerobic capacity and movement initiation in Parkinson's disease patients, *Neurorehabilitation*, 17: 161–168, 2002.

31 Depression in Parkinson's Disease

Steven P. Wengel, Daryl Bohac, and William J. Burke
Department of Psychiatry, University of Nebraska Medical Center

CONTENTS

IMPORTANCE OF DEPRESSION IN PD

Depression is the most common psychiatric complication affecting persons with Parkinson's disease (PD). Depressive symptoms may affect as many as half of all PD patients at some point in their illness, and, in many cases, depressive symptoms actually predate motor signs and symptoms. Aggressive surveillance for and treatment of depression is critical, since untreated depression produces a great deal of human suffering, and depression is usually treatable. In PD patients, depression may have even more impact on functional status than in other patients with depression. However, there is still uncertainty about many aspects of depression in PD despite an increasing amount of research devoted to this topic. In particular, issues revolving around differentiating depression from PD symptoms, the role of cognitive impairment from PD on mood symptoms, and unique treatment concerns in this population can be challenging for the clinician. This chapter provides an update on what is known about depression in PD, including its epidemiology, clinical features, neuropsychological features, and treatment, focusing on major themes of research in this field.

PREVALENCE OF DEPRESSION IN PD

The question of how frequently depression occurs in persons with PD is a striking example of how even ostensibly simple questions can be hard to answer when it comes to understanding depression in PD. Reported rates of depression have varied enormously from study to study. An overall depression rate of 43% is one of the most commonly cited figures.[1] However, this figure includes persons with major depressive disorder and minor depression or dysthymia. Another confounding factor with this estimate is the population being studied. Studies that have reported high rates generally used specialty populations, in contrast to the lower rates generally seen in community-based samples.

As part of a wider prevalence study of PD, Tandberg et al.[2] carefully assessed depressive symptoms in PD patients. They found that 7.7% of community dwelling patients met DSM-III criteria[3] for major depressive disorder (MDD). Additionally, they used the Montgomery-Åsberg depression rating scale (MADRS)[4] to obtain information on depressive symptoms. According to the MADRS, only 5.1% of patients were moderately to severely depressed, but another 45.5% had mild depressive symptoms. The rate of major depression was also strongly affected by cognitive impairment as defined by MMSE score.[5] Rates of MDD were 3.6% in patients with an MMSE ≥ 20 but increased to 25.6% in patients whose MMSE was below 20. Interestingly, rates were also higher in those with possible PD (18.8%) vs. those with probable PD (4.6%). The authors attribute these figures to a higher rate of dementia in those with possible PD and suggest that the higher rates in the cognitively impaired indicate more widespread cerebral involvement.[2]

Another study demonstrating relatively low rates of MDD used structured interviews to evaluate all cognitively intact PD patients in the community who reported depressive symptoms on the General Health Questionnaire (GHQ).[6,7] While depressive symptoms on the GHQ were reported in 34.2% of PD patients, only 2.7% met criteria for MDD.

REASONS FOR DISCREPANCIES IN RATES OF DEPRESSIVE SYMPTOMS

Probably the most salient reason for the wide range of figures quoted for rates of depression revolves around *how* depression is diagnosed. Conventional diagnostic tools, such as DSM-IV criteria,[8] can be difficult to apply to patients with PD, because it specifically excludes symptoms resulting from a general medical condition or a direct physiologic effect of a substance, such as a medication. In persons with PD, this presents obvious problems, particularly when trying to decide about the DSM-IV "somatic" symptoms such as changes in psychomotor activity, sleep, appetite, weight, and energy level. If one follows the "exclusive" directions of DSM-IV, many patients will end up without a primary mood disorder diagnosis in spite of appearing to meet criteria for major depression. In these cases, one is forced to use other diagnostic entities such as "mood disorder secondary to general medical condition."

With the inherent difficulty of separating primary from secondary causes of mood symptoms, a number of different approaches have been suggested. One such idea is to count symptoms as present or absent regardless of presumed causality. Another approach is to focus on the more "psychological" symptoms of depression.

To investigate this issue, Leentjens et al.[9] examined the individual items of the Hamilton[10] and Montgomery-Åsberg depression rating scales in a discriminant analysis. Nonsomatic symptoms were the most discriminating, but somatic symptoms also made meaningful contributions. Reduced appetite and early morning awakening were relatively low-prevalence symptoms that were useful in supporting a diagnosis of depression, whereas other somatic symptoms were not.

RISK FACTORS FOR DEPRESSION IN PD

Many attempts have been made to identify risk factors for developing depression in PD. However, these efforts have generally failed to consider factors known to predispose persons to depression in general. Leentjens et al. first considered general risk factors for depression (age, sex, prior history of depression, family history of depression, somatic comorbidity) in a PD population and found that these five risks factors predicted 75% of depression in their sample using a multivariate model.[11] When disease-specific markers were then included in the model, only right-sided onset of PD improved the model. Thus, established risk factors for depression in general may also be markers of depression in PD.

There is some evidence to suggest that depression in PD is more common in younger patients,[12] females,[13,14] and in persons with more bradykinesia and rigidity (as opposed to tremor-dominance).[15–17] Several authors have suggested that there is a bimodal distribution to the onset of depression in PD.[18–20] One peak seems to follow diagnosis and may be related to left hemisphere dysfunction while the other, later, peak occurs late in the course of depression and may relate to impaired activities of daily living.[18]

TABLE 31.1
Putative Risk Factors for Depression in Parkinson's Disease

- Younger age at diagnosis
- Female sex
- Early stage of illness
- Advanced stage of illness
- Prominent bradykinesia/rigidity
- Right-sided initial symptoms
- More rapid deterioration

PSYCHOSOCIAL RISK FACTORS

Attempts have been made to attribute depression to either psychological or biological sources. Given that biology of necessity underlies psychology, Brown and Jahanshahi provide a summary of the role of psychosocial factors that may contribute to depression in PD.[20] They suggest that certain patients are more vulnerable to depression including those who have an early age of onset, those in the earliest stages of the disease, those with more advanced disease and those with more rapidly progressive deterioration. While only a weak association has generally been reported between depression in PD and severity of illness, they suggest that a crucial related factor is the *rate* at which disability progresses. Those patients whose disability progresses slowly enough for them to adapt may show little depression or recover from a prior depression. Those with a more rapid progression may fail to adapt as easily and are then at higher risk of developing depression.

Social Support

These authors also raise the issue of why patients with apparently similar levels of physical illness and disability have distinctively different affective states. Factors that seem to explain some of this variability are the availability and quantity of social support as well as the strategies that individuals use to cope with stress. Patients with good social support, who are satisfied with that support, and who have good self-esteem and active coping mechanisms appear to be at lower risk of developing depression.[20,21]

Neurobiological Risk Factors

Neuropathological findings that may contribute to depression in PD are degeneration of the dopaminergic, serotonergic, noradrenergic, and cholinergic nuclei in the brainstem.[22] Neuronal loss in the locus coeruleus may be greater than in the substantia nigra.[23] A specific pathway that may play an important role is the mesocortical limbic pathway, which arises in the ventromedial tegmental area (VTA) and projects to areas critical for affect such as the cingulate, entorhinal, and orbitofrontal cortex as well as the subcortical portions of the limbic forebrain.[24,25] This pathway has been shown to be disrupted in patients with depression and PD. Additionally, PET studies have shown hypometabolism in the cingulate and frontal cortex in depressed PD patients compared to controls.[26]

IS PD AN INDEPENDENT RISK FACTOR FOR DEPRESSION?

Depression itself has been proposed as a preliminary symptom of Parkinson's disease. This association is supported by a retrospective study comparing the number of contacts with a general practitioner between persons who developed PD and a control group. There were substantially more visits to the GP in the two years preceding diagnosis in the developing PD group, and many of these visits were for mood symptoms.[27] Another study found that patients with a diagnosis of major depressive disorder had more than a twofold risk of receiving a diagnosis of PD compared to a control group of patients with osteoarthritis.[28]

With the preponderance of clinical studies showing elevated rates of depression in persons with PD, a critical issue becomes whether these rates are higher in patients with PD compared to those with other chronic illnesses producing a similar degree of disability. If depression is more common in persons with PD, a pathophysiologic link between the conditions is suggested.

Depression has been associated with a number of different general medical illnesses, yet the nature of the relationship is often unclear. In describing the difficulty of establishing causal relationships in comorbid illnesses, Krishnan et al.[29] suggested that this effort is particularly difficult in later life, because the lifetime prevalence of all conditions is steady or increases with age and, as such, there is a tendency to find a correlation between virtually all conditions. This "pseudocorrelation" is particularly observed in disorders where frequency increases with age, such as PD, Alzheimer's disease, and cardiovascular disease. As a result, many of these associations may be only a statistical artifact and not clinically relevant.[29]

In an attempt to answer the question about comorbidity of depression and PD, the Danish Psychiatric Central Register and Danish National Hospital Register were used.[30] The authors of this study determined rates of first admissions for depression in patients with PD, diabetes, and osteoarthritis. They found an increased incidence of depression in patients with PD as compared to those with the other conditions who had comparable degrees of disability. The risk of receiving a diagnosis of depression was highest in the six months following diagnosis of PD but remained elevated a year later, though to a lesser extent. The authors conclude that these findings support the idea that a common pathophysiology underlies these conditions.[30] A major limitation of this study is that it captures only a small slice of patients with PD who have depression severe enough to warrant hospitalization.

THE INTERPLAY OF MOOD AND COGNITION IN PD

Depression may mimic dementia, and vice versa. Determining if a patient is suffering from one or the other, or both, may be challenging, since both diagnoses are relatively common in PD. Furthermore, when the two entities coexist, depression may make existing cognitive deficits appear worse than they really are and lead to excess disability, that is, functional impairment greater than would be expected from PD or depression alone. Also, as cognitive impairment progresses, it can be increasingly difficult to recognize a depressive disorder, because of difficulty in accessing the individual's internal affective state.

To explore the complex relationship between mood and cognition in the context of PD, it is necessary to start with looking at how depression affects cognition in non-PD populations, then how PD itself affects cognition, and finally how the depression and cognitive impairment together can interact in PD.

Depression and Cognition

In patients with MDD but who do not have PD or dementia, depression often produces hesitancy in answering

TABLE 31.2
Effects of Depression on Cognition in PD Patients

- Overdiagnosis of dementia
- "Excess disability"
- Worsening of memory dysfunction
- Worsening of executive functioning deficits
- Worsening of other underlying cognitive deficits

questions and a tendency to "give up" easily while undergoing cognitive testing.[31,32] Depressed persons often answer "I don't know," when given challenging cognitive tasks, yet often respond correctly with a bit of encouragement. Depressed patients also tend to be inconsistent in their responses, which may produce lower than expected scores on a number of cognitive measures.

Rosenstein[33] offered these general guidelines about the cognitive functioning of the depressed patient. Memory and attention are not usually in the impaired range, although often slightly below expectation. Language functioning is almost always normal, as are intellectual functioning and visuospatial functioning. Psychomotor functions are often within normal limits but slightly below expectation. Finally, executive functioning efficiency appears to be reduced in depressed patients. For example, shifting cognitive sets may be slowed, but their overall performance may still be within normal limits.

PARKINSON'S DISEASE AND COGNITION

Cognition is quite variable in PD. Most patients will present with a pattern of cognitive impairment that does not rise to the level of dementia, while others may present with fairly clear-cut dementia, and others may have little or no cognitive impairment.[34] Moreover, other factors, including the age of the patient as well as the age of onset, motor symptom severity, side of onset, and medication effects,[35] also influence cognitive functioning. In general, a higher incidence of cognitive impairment and dementia can be expected with a later age of onset.

Cognitive impairments seen in PD are common and of varying severity. The most common specifically affected areas are free recall of previously learned information, visuospatial skills, and executive functions such as problem solving, planning, and flexibility.[26] There is considerable overlap of the effects of depression and PD on cognitive functioning.

Memory performance is characterized by inconsistent learning across trials and impaired free recall following delay, combined with normal recognition memory. In fact, the ability of the PD patient is often near normal for verbal learning, but free retrieval of the previously learned information is inefficient. In contrast,

when procedural learning, or implicit memory, is assessed, PD patients demonstrate poor ability to acquire and retain new skills.

Attention and concentration abilities are impaired in the PD patient, and it is especially when attentional complexity is increased that deficits may appear. This is likely related to impaired executive functioning in the PD patient. Complex attention is mediated by frontal systems and Bondi et al.[36] have shown that nondemented PD patients have frontal system impairments including problems with cognitive flexibility, planning, temporal ordering, and verbal fluency.

Language abilities are largely intact, but if deficits do appear, it is likely to be in the area of verbal fluency. As disease severity increases, complex language comprehension may be impaired, perhaps secondary to slowed speed of information processing. Generally, verbal intellectual functioning remains largely intact. In contrast, performance-based IQ, with its emphasis on speed of performance and visuospatial analysis, is often impaired relative to verbal IQ.

PARKINSON'S DISEASE AND DEPRESSION

Just as with non-PD patients, mild depression or dysthymia probably does not significantly impact cognition in PD. However, as the severity of depression increases, the likelihood that it will impact cognition increases. Recently, Norman et al.[37] demonstrated that depressed PD patients perform worse on measures of memory as compared to nondepressed PD patients. Depressed PD patients had memory impairment similar to depressed patients without PD, leading the authors to conclude that the memory dysfunction noted in their patients was due to depression alone.

The relationship between depression and PD and its influence on cognition is not well understood. Nevertheless, depression and PD do appear to have individual as well as overlapping influences on cognitive functioning. In the case of the nondemented PD patient who becomes depressed, the additive effect of depression on executive functioning and memory may lead one to suspect the patient has developed dementia. However, it is more likely that the cognitive dysfunction associated with PD alone is now exacerbated by the depression. A patient with mildly impaired cognitive flexibility prior to the onset of depression may now appear moderately impaired. In addition, memory functioning may be significantly worse due to encoding and consolidation problems and consequently the ability to even recognize previously learned information is reduced. Thus, timely treatment of depression in the PD patient may help to reduce the risk of developing excess disability and thereby help maintain a better quality of life.

TABLE 31.3
Cognitive Impairment from Depression vs. PD

Characteristic	Major Depressive Disorder	Parkinson's Disease
Speed of response	Hesitancy in answering	Psychomotor slowing
Demeanor	"I don't know" responses, but may respond to encouragement	Generally attempts to respond
Memory	Often within normal limits	Decreased recall of previously learned information, difficulty acquiring new skills
Consistency of responses	Inconsistent performance	Generally consistent
Attention	Normal to slightly impaired	Significant impairment, especially complex attention skills
Language	Normal	Decreased verbal fluency (early); decreased complex language comprehension (late)
Executive functioning	Decreased efficiency	Impairment in problem solving, planning
Visuospatial skills	Normal	Impaired

MOOD EFFECTS OF PARKINSON'S DISEASE TREATMENT

DEPRESSION

Conflicting reports exist for the effects of antiparkinsonian drugs on depressive symptoms. One explanation for this is that most studies of these agents have been designed to monitor effects of the drugs on motor symptoms rather than mood. Since the 1970s, levodopa has been reported to cause affective changes with both improvement and worsening of depressive symptoms noted. Some estimate the prevalence of levodopa-induced depressive symptoms at about 12%.[38] On the other hand, levodopa has also been reported to produce mild improvement in depressive symptoms, possibly due to improvement in motor symptoms rather than a true antidepressant effect *per se*.[38] Other agents, including bromocriptine and pergolide, have also been associated with depressive symptoms in PD patients. Overall, Factor and colleagues[38] conclude that "depression in PD patients more accurately reflects an issue of comorbidity rather than one of medication-induced side effects."

Subthalamic deep brain stimulation (DBS) has been reported to cause mood changes in PD patients. A fascinating case in point revolves around a 65-year-old woman with a 30-year history of PD who developed acute depression during high-frequency deep brain stimulation.[39] This woman went from a euthymic state to one of acute depression when her left basal ganglia was stimulated 2 mm below the site where stimulation relieved the signs of PD. The authors suggest that stimulation may have affected the activity of nigral GABAergic neurons innervating the ventral nuclei of the thalamus with projections to the prefrontal and orbitofrontal cortexes. This is an interesting region, since disruption of connections between the basal ganglia and frontal cortex have been reported to play a role in stroke-related depression, and disruption of these pathways by vascular disease has been proposed as an etiology for late life depression.[40] DBS was also reported to cause depressive symptoms in 6 of 24 consecutive PD patients.[41] Three of the six became transiently suicidal. In five of the six patients, depressive symptoms began within the first month postoperatively. All six required treatment with antidepressants.

MANIA

Dopaminergic agents have been reported to cause manic symptoms such as extreme optimism, spending sprees, and euphoria. Originally reported to occur in 1.5% of levodopa-treated patients,[38] manic symptoms have also been reported in patients treated with other dopaminergic agents. Bromocriptine has been reported to produce mania in PD patients, especially at higher doses, and also in non-PD patients such as postpartum women taking the drug to suppress lactation.[38] Manic symptoms were reported to begin within the first week of treatment and resolved on discontinuation of the drug. Selegiline has also been implicated in producing these symptoms.[38] Unlike depressive symptoms, manic symptoms are unlikely to occur in untreated PD patients but appear to be a consequence of treatment of PD symptoms.

Interestingly, although DBS has been implicated in causing depressive symptoms (see above), there have been cases of mania associated with this treatment also.[42] Three of 15 consecutive patients undergoing DBS developed manic symptoms after stimulation was initiated, and in the three cases, mania resolved after the stimulation was changed from the lower two electrodes to the higher pair of electrodes. The authors suggest that DBS may have triggered mania through stimulation of midbrain projections to the orbitofrontal or anterior cingulate striato-pallido-thalamo-cortical circuits.

APPROACHES FOR TREATING DEPRESSION IN PARKINSON'S DISEASE PATIENTS

The locus coeruleus and raphe nuclei are affected by Parkinson's disease, and levels of norepinephrine and serotonin may decrease as the illness progresses. This suggests that using agents to ameliorate deficiency states of these neurotransmitters could be therapeutic. However, a number of caveats apply when treating depressed PD patients. First, they may be susceptible to motor side effects of agents that do not ordinarily affect the motor systems of non-PD patients. Second, PD patients may be very sensitive to specific side effects such as anticholinergic side effects and orthostatic hypotension. Third, there are several important drug-drug interactions that need to be noted for PD patients.

Additionally, many PD patients report that their mood symptoms fluctuate in concert with their motor symptoms. When they are "off" motorically, they frequently experience abrupt dysphoric episodes. If such episodes occur, optimization of PD therapy should immediately follow rather than initially adding an antidepressant.[43,44] In fact, one author suggested that "optimized dopaminergic therapy is a prerequisite for a successful management of depression—particularly in patients with fluctuating PD."[45]

Treatment of depression in Parkinson's disease typically involves pharmacotherapy, electroconvulsive therapy, and/or psychosocial therapy. As in other depressive syndromes, the depression associated with PD appears to respond to conventional antidepressant (AD) drugs. To date, there has not been a randomized, placebo-controlled trial demonstrating their efficacy in PD.[46]

Several classes of antidepressants have been used to treat depression in PD. SSRIs are frequently prescribed for PD patients as first-line therapy. As a class, SSRIs have a number of attractive features:

1. Dosing is fairly straightforward, with once daily administration usually being adequate.
2. Titration often is not necessary, as the starting dose may also be the therapeutic dose.
3. SSRIs almost never produce orthostatic hypotension and rarely produce anticholinergic side effects (with the exception of paroxetine).

However, SSRIs may have an antagonistic effect on dopamine.[43] Case reports have been published in which PD patients experience a worsening of motor symptoms when an SSRI is added to their regimen, although this is a relatively uncommon. In an open trial of paroxetine, Ceravolo and colleagues[47] found no worsening of motor scores overall but did describe 1 of 33 subjects who experienced a worsening of tremor while taking paroxetine. This was completely reversed on discontinuation of the drug. In a review of 199 cases of PD patients taking at least one antidepressant, a group of French investigators found a rate of 4.5% of patients developing extrapyramidal side effects.[48] There was a trend of this effect being produced more frequently by SSRIs than other antidepressants, but this was not statistically significant. Another

uncommon but concerning event is the induction of "serotonin syndrome" when an SSRI is prescribed to a patient being treated with selegiline.[49] Symptoms of serotonin syndrome include myoclonus, delirium, tremors, fever, hyperreflexia, and diaphoresis.[45] Although most authors recommend avoiding this combination, others have reported using selegiline with an SSRI without causing the serotonin syndrome.[50] Caution, however, is still advised.

One other issue with SSRIs is their varying tendency to inhibit several cytochrome P450 enzymes. Probably of most interest is the inhibition of the 2D6 isoenzyme by fluoxetine and paroxetine.[51] Amitriptyline, a tricyclic antidepressant, is a substrate of this enzyme and often used by neurologists. Combining amitriptyline (as well as other tricyclics) with either of these two antidepressants can lead to clinically significant elevations in serum levels of the tricyclic, potentially producing toxicity.

Mirtazapine is a non-SSRI antidepressant that is an antagonist of alpha-2 noradrenergic autoreceptors and serotonin-2 and -3 receptors. It appears to be well tolerated in PD patients. At lower doses (less than 30 mg/day), it has prominent sedative and appetite stimulating effects. These side effects usually disappear at doses of 30 mg/day and higher, apparently due to predominance of alpha-2 antagonism at higher doses. Mirtazapine is usually given at bedtime due to sedation. However, a recent case series reported four PD patients who developed sleep-related behavioral problems, including nocturnal confusion, talking during sleep, and hallucinations, while taking mirtazapine.[52] In three of these patients, the dose was 30 mg at bedtime, while the fourth patient took 15 mg at bedtime. These symptoms resolved once mirtazapine was discontinued. Mirtazapine usually does not produce clinically relevant effects on the P450 system.

Bupropion is a novel antidepressant whose therapeutic mode of action is unknown. It is hypothesized to have pro-dopaminergic effects, although this has not been proven. It is usually well tolerated in patients with PD and does not produce orthostatic hypotension. In contrast, it may actually raise blood pressure in some patients. It tends to be more "activating" than many other antidepressants and thus may ameliorate fatigue. It is usually dosed twice daily with the time of the second administration being adjusted to avoid causing or aggravating insomnia. Bupropion does inhibit the P450 2D6 isoenzyme and has been reported to elevate TCA levels.[53]

Venlafaxine has a dual mechanism of action. At doses up to 150 mg/day, it functions much like an SSRI. At higher doses, it also acts as a norepinephrine reuptake inhibitor. Like most antidepressants, no controlled trials have been done with venlafaxine in PD patients. In clinical practice, it appears to be well tolerated in the PD patient. It may produce nausea initially, but this side effect usually attenuates after a week or two. Blood pressure should be monitored for those patients on 150 mg/day and higher, since venlafaxine may elevate blood pressure at higher doses. It usually has a minimal effect on the P450 system.

Tricyclic antidepressants (TCAs) were once the mainstay of treatment for depression. However, TCAs have seen a major decrease in use since the advent of the newer ADs. Nevertheless, they still have a role to play, particularly in patients who have not responded to several adequate trials of other ADs. Of the available TCAs, nortriptyline is preferred because of its lower risk of orthostatic hypotension and anticholinergic side effects. It can be started at 10 to 25 mg at bedtime and titrated up to 75 mg/day based on clinical response and serum levels. However, orthostatic hypotension is still a potential side effect in patients with PD who are more vulnerable due to the disease itself and the side effects of dopaminergic therapy. Patients should be counseled to consume adequate fluids and to exercise special care when rising from a chair or a bed. Because TCAs effect cardiac conduction, an electrocardiogram should be performed prior to initiation of a TCA. Additionally, this class of drugs should generally be avoided in the presence of atrioventricular or bundle-branch blocks. All TCAs may produce anticholinergic side effects, which may then cause constipation, a common problem in PD patients even when not taking an antidepressant. Tertiary amines, such as imipramine and most notably amitriptyline, are most likely to cause anticholinergic side effects and are best avoided in the context of PD.[54]

The use of testosterone has been a novel approach to treating PD. A case series described five men with PD who complained of fatigue, depression, anxiety, and sexual dysfunction. They were found to have low levels of free testosterone. Consequently, they all responded favorably to the application of a topical testosterone gel.[55]

Electroconvulsive therapy (ECT) has been reported to improve not only mood but motor symptoms in depressed PD patients.[56] It also decreases "off" time in nondepressed PD patients with severe on-off phenomenon.[57] These beneficial effects are transient, typically lasting several weeks. Thus, for the motor improvement to be maintained, ECT would need to be done at least monthly. While effective for depressive symptoms in this population, ECT may cause delirium in susceptible individuals. It also may contribute to an increase in dyskinesia or even psychotic symptoms in PD patients, possibility due to an increased permeability of the blood-brain barrier. Should either of these symptoms be seen during a course of ECT in a PD patient, the antiparkinsonian medication dosage should be reduced by approximately one-third.

Although ADs are effective and often necessary for treating depression in PD patients, psychosocial interventions should not be neglected. Psychotherapy can be very helpful in the cognitively intact PD patient and may be most helpful for those demonstrating depressive symptoms at the time of diagnosis.[45] Because of the chronic and progressive nature of PD, helping patients to develop effective coping strategies while providing support is therapeutic and beneficial. Involving the spouse in the treatment plan is preferable, as this person is often the key support person for the PD patient. Support groups should be strongly recommended, as they provide excellent support and encouragement for the PD patient and family.

TABLE 31.4
Overview of the Treatment of Depression in PD

- Maximize antiparkinsonian therapy first
- "Start low, go slow"
- Caution with combination of SSRIs and selegiline; may cause serotonin syndrome
- SSRIs may worsen motor function
- Avoid antidepressants with significant anticholinergic side effects
- Always consider psychosocial aspects
- Electroconvulsive therapy may benefit mood *and* motor function

---

REFERENCES

1. Cummings, J. L., Depression and Parkinson's disease: a review, *Am. J. Psychiatry,* 149(4):443–54, April, 1992.
2. Tandberg, E., Larsen, J. P., Aarsland, D., Cummings, J. L., The occurrence of depression in Parkinson's disease. A community-based study, *Arch. Neurol.,* 53(2):175–9, February, 1996.
3. American Psychiatric Association, *Diagnostic and statistical manual of mental disorder,* 3rd ed., Washington, DC, American Psychiatric Association, 1980.
4. Montgomery and Asberg, M., A new depression scale designed to be sensitive to change, *British Journal of Psychiatry,* 134:382–9, 1979.
5. Folstein, M. F., Folstein, S. E., McHugh, P. R., "Mini-Mental State: a practical method of grading the cognitive state of patients for the clinician," *J. of Psychiatric Research,* 12:189–98, 1975.
6. Goldberg, D. P., William, P., *A user's guide to the General Health Questionnaire,* Windsor, NFER-NELSON, 1988.
7. Hantz, P., Caradoc-Davies, G., Caradoc-Davies, T., Weatherall, M., Dixon, G., Depression in Parkinson's disease, *Am. J. Psychiatry,* 151(7):1010–4, July, 1994.
8. American Psychiatric Association, *Diagnostic and statistical manual of mental disorder,* 4th ed., Washington, DC,1994.
9. Leentjens, A. F. G., Marinus, J., Van Hilten, J. J., Lousberg, R., Verhey, F. R. J., The contribution of somatic symptoms to the diagnosis of depressive disorder in Parkinson's disease: a discriminant analytic approach, *J. Neuropsychiatry Clin. Neurosci.,* 15:(1):74–77, 2003.
10. Hamilton, M., A rating scale for depression, *Journal of Neurology, Neurosurgery and Psychiatry,* 23:56–62, 1960.

11. Leentjens, A. F. G., Lousber, R., Verhey, F. J. R., Markers for depression in Parkinson's disease, *Acta Psychiatr. Scand.*, 106:196–201, 2002.

12. Starkstein, S. E., Berthier, M. L., Bolduc, P. L., Preziosi, T. J., Robinson, R. G., Depression in patients with early versus late onset Parkinson's disease, *Neurology*, 39:1441–1445, 1989.

13. Gotham, A. M., Brown, R. G., Marsden, C. D., Depression in Parkinson's disease: a quantitative and qualitative analysis, *J. of Neurology, Neurosurgery Psychiatry*, 49:381–389, 1986.

14. Brown, R. C., MacCarthy, B., Psychiatric morbidity in patients with Parkinson's disease, *Psychol. Med.*, 20:77–87, 1990.

15. Huber, S. J., Paulson, G. W., Shuttleworth, E. C., Depression in Parkinson's disease, *Neuropsychiatry Neuropsychol. Behav. Neurol.*, 1:47–51, 1988.

16. Jankovic, J., McDermott, M., Carter, J. et al., Parkinson Study Group. Variable expression of Parkinson's disease: a baseline analysis of the DATATOP cohort, *Neurology*, 40:1529–1534, 1990.

17. Brown, G. L., Wilson, W. P., Parkinsonism and depression, *South Med. J.*, 65:540–545, 1972.

18. Starkstein, S. E., Preziosi, T. J., Bolduc, P. L., Robinson, R. G., Depression in Parkinson's disease, *J. Nerv. Ment. Dis.*, 178(1):27–31, January, 1990.

19. Celesia, G. G., Wanamaker, W. M., Psychiatric disturbances in Parkinson's disease.

20. Brown, R., Jahanshahi, M., Depression in Parkinson's disease: a psychosocial viewpoint, *Adv. Neurol.*, 65:61–84, 1995.

21. MacCarthy, B., Brown, R. G., Psychosocial factors in Parkinson's disease, *Br. J. Clin. Psychol.*, 28:41–52, 1989.

22. Oertel, W. H., Hoglinger, G. U., Caraceni, T., Girotti, F., Eichhorn, T., Spottke, A. E., Krieg, J. C. et al., Depression in Parkinson's disease, *Advances in Neurology*, 86:373–383, 2001.

23. Zarow, C., Lyness, S. A., Mortimer, J. A., Chui, H. C., Neuronal loss is greater in the locus coeruleus than nucleus basalis and substantia nigra in Alzheimer and Parkinson Diseases, *Arch. Neurol.*, 60:337–341, 2003.

24. Price, K. S., Farley, I. J., Hornykiewicz, O., Neurochemistry of Parkinson's disease: relation between striatal and limbic dopamine, *Adv. Biochem. Psychopharmacol.*, 19:293–300, 1978.

25. Javoy-Agid, F., Agid, Y., Is the mesocortical dopaminergic system involved in Parkinson's disease? *Neurology*, 30:1326–1330, 1980.

26. Baxter, L. R., Schwartz, J. M., Phelps, M. E. et al., Reduction of prefrontal cortex glucose metabolism common in three types of depression, *Arch. Gen. Psychiatry*, 46:243–250, 1989.

27. Gonera, vant't Hof, M., Berger, H. J., van Weel, C., Horstink, M. W., Symptoms and duration of the prodromal phase in Parkinson's disease, *Mov. Disorder*, 12:871–876, 1997.

28. Nilsson, F. M., Kessing, L. V., Bolwig, T. C., Increased risk of developing Parkinson's disease for patients with major affective disorder-a register study, *Acta Psychiatr. Scand.*, 104:380–386, 2001.

29. Krishnan, K. R., Delong, M., Kraemer, H., Carney, R., Spiegel, D. et al., Comorbidity of depression with other medical diseases in the elderly, *Biol. Psychiatry*, 52:559–588, 2002.

30. Nilsson, F. M., Kessing, L. V., Sorensen, T. M., Anderson, P. K., Bolwig, T. G., Major depressive disorder in Parkinson's disease: a register-based study, *Acta Psychiatr. Scand.*, 106:202–211, 2002.

31. desRosiers, G., Primary or depressive dementia: Psychometric assessment, *Clinical Psychology*, 12:307–343, 1992.

32. LaRue, A., *Aging and Neuropsychological Assessment*, Plenum, New York, 259–289, 1992.

33. Rosenstein, L. D., Differential diagnosis of the major progressive dementias and depression in middle and late adulthood: A summary of the literature of the early 1990s, *Neuropsychology Review*, 8:109–167, 1998.

34. Mahurin, R. K., Feher, E. P., Nance, M. L., Levy, J. K., Priozzolo, F. J., Cognition in Parkinson's disease and related disorders. In Parks, R. W., Zec, R. F., Wilson, R. S., Eds., *Neuropsychology of Alzheimer's Disease and Other Dementias*. Oxford University Press, New York, 308–349, 1993.

35. Levin, B. E., Katzen, H. L., Early cognitive changes and nondementing behavioral abnormalities in Parkinson's disease, in Weiner, W. J., Lang, A. E., Eds., *Advances in Neurology*, Vol. 65, Raven Press, New York, 85–95, 1995.

36. Bondi, M. W., Kaszniak, A. W., Bayles, K. A., Vance, K. T., Contributions of frontal system dysfunction to memory and perceptual abilities in Parkinson's disease, *Neuropsychology*, 7:89–102, 1993.

37. Norman, S., Troster, A. I., Fields, J. A., Brooks, R., Effects of depression and Parkinson's disease on cognitive functioning, *J. of Neuropsychiatry and Clinical Neurosciences*, 14:31–36, 2002.

38. Factor, S. A., Molho, E. S., Podskalny, G. D., Brown, D., Parkinson's disease: drug-induced psychiatric states, in *Behavioral Neurology of Movement Disorders*, Weiner, W. J., and Lang, A. E., eds., *Advances in Neurology*, Vol. 65, Raven Press, Ltd., New York, 1995.

39. Bejjani, B. P., Damier, P., Arnulf, I., Thivard, L., Bonnet, A. M. et al., Transient acute depression induced by high-frequency deep-brain stimulation, *NEJM*, 340(19): 1476–1480, 1999.

40. Alexopoulos, G., Kiosses, D., Klimstra, S., Kalayam, B., Bruce, M., Clinical presentation of the "depression-executive dysfunction syndrome" of late life, *Am. J. Geriatr. Psychiatry*, 10(1):98–106, 2002.

41. Berney, A., Vingerhoets, F., Perrin, A., Phil, L., Guex, P., Villemure, J. G., Burkhard, P. R., Benkelfat, C., Ghika, J., Effect on mood of subthalamic DBS for Parkinson's disease: a consecutive series of 24 patients, *Neurology*, 59:1427–1429, 2002.

42. Kulisevsky, J., Berthier, M. L., Gironell, A., Pascual-Sedano, B., Molet, J., Pares, P., Mania following deep brain stimulation for Parkinson's disease, *Neurology*, 59:1421–1424, 2002.

43. Mendis, T., Suchowersky, O., Lang, A., Gauthier, S., Management of Parkinson's disease: a review of current and new therapies, *Can. J. Neurol. Sci.,* 26:89–103, 1999.

44. Lieberman, A., Managing the neuropsychiatric symptoms of Parkinson's disease, *Neurology,* 50:S33–S38, 1998.

45. Poewe, W., Seppi, K., Treatment options for depression and psychosis in Parkinson's disease, *J. Neurol.,* 248:12–21, 2001.

46. Aarsland, D., Cummings, J. L., Depression in Parkinson's disease, *Acta Psychiatri. Scand.,* 106:161–162, 2002.

47. Ceravolo, R., Nuti, A., Piccinni, A., Dell'Agnello, G., Bellini, G., Gambaccini, G., Dell'Osso, L., Juri, L., Bonuccelli, U., Paroxetine in Parkinson's disease: effects on motor and depressive symptoms, *Neurology,* 55:1216–1218, 2000.

48. Gony, M., Lapeyre-Mestre, M., Montastruc, J. L., Risk of serious extrapyramidal symptoms in patients with Parkinson's disease receiving antidepressant drugs: a pharmacoepidemiologic study comparing serotonin reuptake inhibitors and other antidepressant drugs, *Clin. Neuropharmacol.,* 26(3):142–5, 2003.

49. Ritter, J. L., Alexander, B., Retrospective study of selegiline-antidepressant drug interactions and a review of the literature, *Ann. Clin. Psychiatry,* 9:7–13, 1997.

50. Waters, C. H., Fluoxetine and selegiline—lack of significant interaction, *Can. J. Neurol. Sci.,* 21(3):259–61, 1994.

51. Ereshefsky, L., Alfaro, C. L., Lam, Y. W., Treating depression: potential drug interactions, *Psychiatric Annals,* 27:244–258, 1991.

52. Onofrj, M., Luciano, A. L., Thomas, A., Iacono, D., D'Andreamatteo, G., Mirtazapine induces REM sleep behavior disorder (RBD) in parkinsonism, *Neurology,* 60: 113–115, 2003.

53. Weintraub, D., Nortriptyline toxicity secondary to interaction with bupropion sustained-release, *Depress Anxiety,* 13:50–52 2001.

54. Richard, I. H., Depression in Parkinson's disease, *Curr. Treat. Options. Neurol.,* 2(3):263–274, 2000.

55. Okun, M. S., McDonald, W. M., DeLong, M. R., Refractory nonmotor symptoms in male patients with Parkinson disease due to testosterone deficiency: a common unrecognized comorbidity, *Arch. Neurol.,* 59:807–811, 2002.

56. Burke, W. J., Peterson, J., Rubin, E. H., Electroconvulsive therapy in the treatment of combined depression and Parkinson's disease. *Psychosomatics,* 29:341–346, 1988.

57. Wengel, S. P., Burke, W. J., Pfeiffer, R. F., Roccaforte WH, Paige SR. Maintenance electroconvulsive therapy for intractable Parkinson's disease, *Am. J. Geriatr. Psychiatry,* 6:263–269, 1998.

32 Anxiety and Parkinson's Disease

Oksana Suchowersky
Department of Clinical Neurosciences, University of Calgary, Calgary Health Region

Michael Trew
Department of Psychiatry, University of Calgary, Calgary Health Region

CONTENTS

INTRODUCTION

Anxiety is a normal response felt in stressful or uncomfortable situations. However, when it exceeds what would be expected in intensity or duration, it becomes a disorder. In these situations, the individual experiences disturbances that can be categorized into three spheres:

1. Cognitive
2. Emotional
3. Physical

Physical symptoms include autonomic hyperactivity with sweating, increased heart rate, shortness of breath, and motor restlessness. Cognitive symptoms include trouble concentrating and negative, anxious, "what if" thoughts. These are accompanied by emotional feelings of distress and unwellness. All result in impaired ability to function. Anxiety disorders can be subdivided into the following categories: *panic disorder*—with or without agoraphobia, *specific phobias* (such as fear of heights or fear of spiders), *social phobias* (i.e., fear of public speaking or fear of eating in public), *post-traumatic stress disorder, generalized anxiety disorder, obsessive compulsive disorder,* and *anxiety disorder not otherwise specified* (used when symptoms do not fall into a specific subtype).[1]

Anxiety disorders are the most common of the mental disorders, with a lifetime risk of 15% in the general population.[2] In the majority, symptoms begin in young adulthood, and onset after the age of 50 is uncommon. Symptoms typically improve with age, with only 5% of well, older individuals having anxiety disorders. Prevalence is two to three times as frequent in women as compared to men.

When onset occurs later in life, the anxiety disorder is usually associated with chronic medical disease such as cardiac and pulmonary dysfunction, arthritis, and diabetes.[3,4] Chronic neurological disorders such as MS and vertigo have also been shown to have an increased incidence of anxiety disorders.[4,5] Reported rates of late onset anxiety disorders associated with chronic disease vary from 11 to 26% in the elderly populations studied. In both young and older onset anxiety disorders, depression is frequently present as a comorbidity, occurring in up to 40% of individuals studied.[2–5]

PREVALENCE OF ANXIETY DISORDERS IN PARKINSON'S DISEASE

In studies using formal psychiatric assessments and anxiety rating scales, 30 to 40% of all Parkinson's disease

(PD) patients have been shown to suffer from a definable anxiety disorder.[6,7] A large majority of these had no anxiety symptoms prior to development of PD. Thus, PD is the most common chronic medical disorder resulting in older-onset anxiety disorders.[6] The most prevalent categories in PD are panic disorder (with or without agoraphobia), generalized anxiety disorder (GAD), and social phobia, with obsessive-compulsive disorder being relatively uncommon (Table 32.1). In those patients with pre-existing anxiety disorder, the development of PD will exacerbate it significantly.[8]

TABLE 32.1
Anxiety Disorders in Parkinson's Disease

Generalized Anxiety Disorders
Panic disorder
• Panic disorder with agoraphobia
• Panic disorder without agoraphobia
Social phobias
Obsessive-compulsive disorder
Anxiety disorder, not otherwise specified

Over 90% of affected PD patients have coexistent depression.[7] Conversely, the majority of PD patients with depression (up to 67%) suffer from anxiety.[7,9] Several different types of anxiety disorders may also coexist in the same patient. An interesting observation is that PD individuals with predominantly left-sided symptoms have a five times higher prevalence of anxiety disorder as compared to those with right-handed symptoms.[10]

Anxiety disorders appear to be directly related to PD, as a study comparing parkinsonism due to multiple system atrophy with PD showed much lower rates of anxiety in multiple system atrophy.[11] Another study looking at corticobasiloganglionic degeneration also revealed a low prevalence of anxiety, although depression was common.[12] In hereditary Parkinsonism, a large number of affected individuals have been reported to be affected with anxiety disorders, although the prevalence of the different subtypes differed from that seen in idiopathic PD.[13]

CLINICAL FEATURES OF ANXIETY DISORDERS IN PD

In general, the clinical features of anxiety disorders in PD are phenomenologically indistinct from those seen in the general population. However, one distinguishing feature is that, in PD patients, the episodes of anxiety are frequently associated with fluctuations in motor function and can be a component of the "off" phenomenon. When anxiety symptoms are present, they will worsen the Parkinsonism and decrease effectiveness of medications.

PANIC DISORDER

Panic disorder consists of episodes with sudden onset of apprehension, malaise, and agitation. Individuals have a fear that symptoms will not improve and that death may be imminent. These are associated with autonomic overactivity including sweating, flushing, tachycardia, gastrointestinal distress, dyspnea, urinary urgency, and dizziness. The symptoms rapidly peak then wane within minutes, rarely lasting more than an hour. Dyspnea can be particularly prominent and distressing, sometimes leading to unnecessary pulmonary investigations.

In the general population, panic disorders are relatively uncommon, occurring in 1 to 2% of individuals. However, in PD, these appear to be the most common of the anxiety disorders, occurring in up to 24% of PD patients;[14] 90% of those with diagnosed anxiety disorders have panic disorder.[6]

Panic attacks tend to occur in patients with younger onset and more severe disease requiring higher doses of medications.[7,14] The panic disorder appears several years after initiation of therapy and appears to be associated with use of levodopa rather than dopamine agonists. In one study, panic attacks were directly related to motor fluctuations in 90% of patients and occurred in the "off" state.[14] Attacks can be alleviated by the next dose of levodopa. In some patients, this leads to overuse of levodopa, with patients taking increasingly higher doses to prevent these episodes.[15]

Occasionally the attack may be aborted by distraction, suggesting that nonpharmacological interventions may be helpful in some individuals.

GENERALIZED ANXIETY DISORDER (GAD)

A major feature of GAD is that anxiety is chronic but waxes and wanes over time. It is characterized by four groups of symptoms.

1. Unrealistic or excessive worry
2. Motor tension (i.e., trembling, muscle tension, easy fatigability)
3. Autonomic hyperactivity (i.e., shortness of breath, dry mouth, trouble swallowing)
4. Vigilance and scanning (i.e., exaggerated startle response, trouble concentrating)

Comorbidity with other anxiety and mood disorders is common.

GAD has received less attention than panic disorder in PD. In several studies, up to 12% of PD patients were diagnosed with GAD, representing 42% of all of those diagnosed with anxiety disorders.[6,7] No specific association has been shown with the "off" state.

PHOBIA

Phobias are characterized by irrational fear resulting in avoidance of the object or activity. These are the most common type of anxiety disorder in the general population (with a lifetime risk of 12.5%[2]), usually represented by specific phobia such as fear of heights or spiders.

In PD, specific phobias are rare, and patients typically suffer from social phobias with or without agoraphobia. These are characterized by anxiety in social settings, particularly if the individual is the focus of attention—for example, speaking in public or even writing a check. If the anxiety reaches a level at which the individual avoids going out in public, a diagnosis of agoraphobia is added.

The exact prevalence of social phobias is unknown, as studies have shown variable results. Stein et al.,[6] in a study of 24 patients, reported that 71% of the patients with anxiety disorders had social phobias and agoraphobia (with or without coexistent panic disorder). In another study, Menza et al.[7] reported a prevalence of only 8% among patients with anxiety disorders, with only 2% of all PD patients having this diagnosis. Recent prospective studies of PD patients has shown that with careful assessment, social phobia (and other anxiety disorders) may actually predate development of PD.[16,17]

In a number of patients, the development of agoraphobia or social phobias may be situational rather than a true psychiatric disorder. These are related to embarrassment or worry about disability related to tremor, gait, and freezing problems. Given the reality of physical disability, a clear distinction between "normal reactions" and "disorder" may be difficult in these situations.

OTHER ANXIETY DISORDERS

Although less common than the other types of anxiety disorders, a number of studies have reported the occurrence of obsessive-compulsive disorder[13,18] and anxiety disorder not otherwise specified[6]. Alegret et al.,[18] in a study of 72 PD patients and an equal number of controls, showed a significantly higher prevalence of obsessive—compulsive symptoms in the PD patients with advanced disease. As patients with mild disease did not differ from controls, the development of these symptoms appeared to be related to disease progression.

Up to 40% of PD patients experience anxiety that does not fulfill the psychiatric criteria of an anxiety disorder but nevertheless results in discomfort and interferes with daily functioning.[7]

COGNITIVE DEFICITS AND ANXIETY

To date, the majority of studies have not found a direct relationship between anxiety disorders and dementia or that patients with AD are more predisposed to developing dementia later in the course of their disease.[6,10,19] However,

as most studies of anxiety in PD excluded demented subjects, these conclusions should be considered preliminary.

Early dementia may be associated with agitation and symptoms of anxiety in stressful situations, and some patients with early dementia may appear to have anxiety disorder.[19] Neuropsychological assessment is helpful in determining if underlying cognitive changes are present.

DIAGNOSIS AND RELATIONSHIP TO FLUCTUATIONS

Anxiety disorders remain an unrecognized feature of PD. A recent study[20] has shown that neurologists frequently do not recognize symptoms of anxiety, or they misdiagnose them as related to motor or autonomic "off" symptoms. In this study, the correct clinical diagnosis was made less than 50% of the time. Thus, neurologists need to increase their awareness of the manifestations of these disorders; assessment by psychiatric colleagues is helpful in confirming the diagnosis.

It is now recognized that medication-related fluctuations in PD patients with advanced disease have prominent nonmotor features that result in significant morbidity.[21–23] These nonmotor fluctuations can be divided into autonomic, sensory, and cognitive/psychiatric categories. Under the cognitive/psychiatric heading, anxiety is reported frequently. It likely is the most common symptom of all nonmotor fluctuation symptomatology, occurring in up to 66% of patients studied.[14,21–23] Anxiety results in significant morbidity, which may be more distressing and disabling than the motor fluctuations.[23] Panic disorders are a particularly prominent feature of the "off" state.[14]

Several studies have shown that individuals feel less anxiety when they are in the "on" state as compared to "off."[24,25] A set of elegant studies correlated mood and anxiety changes with changes in motor improvement in PD patients receiving levodopa infusion. These showed that improvement in anxiety paralleled improvement in motor function with increasing levodopa levels and worsened with decreasing levels.[26] These effects were dose related.[27]

Anxiety disorders may occur in conjunction with symptoms of other nonmotor fluctuations. Also, symptoms of anxiety may mimic other nonmotor symptoms associated with sensory or autonomic phenomena.[28] For example, trembling and dyspnea may be features of both (see Table 32.2). Thus, care must be taken to differentiate among the different features and types of nonmotor fluctuations to arrive at the correct diagnosis and optimize treatment.

A number of anxiety scales are routinely used to aid in clinical diagnosis. These include the Hamilton Anxiety Scale,[29] Zung Anxiety Scale,[30] Beck Anxiety Scale,[31] Stait–Trait Anxiety Inventory,[32] and Sheehan Clinician-Rated Anxiety Scale,[33] all of which have been validated

TABLE 32.2
Differential Diagnosis of Anxiety Disorders in PD

"Off" phenomena
    Autonomic symptoms
    Akathesia
    Tremor
    Sensory symptoms
    Dyspnea
Other cognitive/psychiatric abnormalities
    Depression
    Situational anxiety
    Dementia

for use in general populations. The Hospital Anxiety and Depression Scale[34] was developed for use in the older population with medical problems. This scale excludes somatic symptoms of anxiety and depression that overlap with PD symptomatology and has been suggested to allow for more accurate diagnosis.[35] No studies have been done specifically looking at which scales are most effective in assessing the PD population.

SURGERY AND ANXIETY DISORDERS

Stereotactic surgery has become a standard option in the treatment of Parkinsonian symptoms. In many centers, patients undergo a detailed preoperative assessment including neuropsychological testing[36] to evaluate appropriateness for surgery. Evaluation for anxiety should be included in this testing, as patients with significant anxiety disorders make poor surgical candidates. Following surgery, patients with premorbid psychological and psychiatric problems may develop worsening of behavioral abnormalities.[37,38]

The effect of surgery on anxiety has been the subject of several studies. However, it should be noted that, in these studies, the diagnosis in most patients was adjustment reaction with anxiety rather that anxiety disorder, as patients with significant psychiatric problems were excluded from surgery.

Pallidal and thalamic surgeries have been reported to result in a reduction in anxiety following successful surgery.[39–42] This improvement cannot be explained by changes in medications, as these usually remain unchanged. No differences in anxiety reduction were found between left- and right-sided surgeries.[43] With subthalamic nucleus surgery, variable results have been reported, including a decrease,[44] no change,[45] or even an increase in anxiety.[38]

As follow-up in these studies has been limited (up to 18 months to date, with shorter follow-up in most studies), long-term effects are unknown. It has been suggested that anxiety may improve in the short term after surgery

but may recur with disease progression as the patient undergoes further deterioration.[45]

MECHANISMS OF DISEASE

Anxiety disorders appear to be an intrinsic part of the disease process of PD. Evidence supporting this includes the following:

1. Anxiety disorders are much more common in idiopathic and familial PD as compared to other parkinsonian disorders.[11–13]
2. Anxiety disorders may precede the development of PD by as much as 10 to 20 years.[16,17,19]

Although nigrostriatal degeneration has traditionally been felt to be the primary abnormality in PD, it is well recognized that the involvement in the brain is widespread, including noradrenergic neurons in the locus coeruleus, serotonergic neurons in the dorsal raphe, as well as mesolimbic and mesocortical dopaminergic systems.[46] In the general population, anxiety disorders have been postulated to arise from abnormalities in the central noradrenergic system.[47] Extrapolating to PD, it is felt that the most likely explanation for the high prevalence of anxiety disorders in PD is involvement of the noradrenergic system and the locus coeruleus that occurs as part of the disease process. In one study, the alpha–2–adrenergic receptor itself was also implicated.[48]

Serotonergic loss, also seen in PD and thought to result in the depression, may also play a role in anxiety.[49] Thus, the alteration of both neurotransmitter systems in PD would explain the high comorbidity of these two conditions.

In early PD, mesencephalic dopamine projections are decreased, resulting in decreased inhibition of the locus ceruleus. This would result in increased noradrenergic activity[50–52] and could lead to anxiety and anxiety disorders. This model would fit well with observations that fluctuations in dopamine levels, and particularly the "off" state, can result in increased anxiety.[53] Further support comes from the yohimbine challenge. Yohimbine, an alpha–2–antagonist, when given to patients with PD, results in the development of significant anxiety.[54]

According to this model, anxiety disorders would be expected to improve with advancing disease and with degeneration of the locus coreleus and loss of noradrenergic function. Although this has been shown to occur in one study on familial PD,[19] the majority of studies in idiopathic PD have shown anxiety to be more prevalent and severe in advanced disease.[7,14,18] Second, it has recently been suggested that degeneration in PD begins lower in the brain stem rather than in the substantia nigra.[46] Currently, the exact mechanism resulting in anxiety is still unclear, but it likely is a complex interplay among noradrenergic, serotonergic, and dopaminergic systems.

TREATMENT

Recommendations for treatment are currently based on anecdotal reports and experience drawn from the general population, as few studies of treatment of anxiety specific to the PD population have been conducted.[55,56]

Initial evaluation should involve assessment for type and severity of symptoms, identification of comorbidities, and exclusion of cognitive dysfunction. Psychiatric involvement is recommended to assist with accurate diagnosis and choice of medical therapy.

Careful evaluation of the role of dopaminergic medications should be performed. For patients with prominent symptoms related to fluctuations, these should be minimized as much as possible. At the same time, levodopa use should be controlled, due to the tendency for some patients to overmedicate themselves to prevent the "off" phenomenon.[15]

If anxiety severity warrants pharmacological treatment, individualization of therapy is recommended, with medications chosen that do not interfere with PD meds or worsen parkinsonian symptoms. As depression is frequently a comorbid condition, and serotonergic and noradrenergic mechanisms have both been implicated in pathogenesis of anxiety, antidepressants are the mainstay of treatment. (Table 32.3).

TABLE 32.3
Medical Therapy for Anxiety Disorders

Short-term

   Benzodiazepines

   Hypnotics

Long-term

   Selective serotonin reuptake inhibitors

   Cyclic antidepressants

   Atypical antipsychotics—quetiapine, clozapine

For short-term benefit, benzodiazepines can be helpful; long-term use should be avoided because of a potential for dependence. Also, high doses of benzodiazepines have been reported to worsen PD symptoms.[57] As anxiety may contribute to decreased sleep quality with increased nocturnal awakenings,[58] sleep quality should be assessed. Hypnotics may improve sleep abnormalities.

For chronic problems, selective serotonin reuptake inhibitors (SSRIs) are currently the first line of treatment.[59,60] All currently available SSRIs (fluoxetine, fluvoxamine, sertraline, paroxetine, and citalopram) have been shown to reduce symptoms of anxiety in the general population. Although it seems reasonable to assume that the sedating SSRIs (such as paroxetine) would be more helpful in treatment of anxiety disorders, this has not been borne out in trials. Thus, it remains a trial-and-error process to determine which is the most effective for each individual patient, and whether it improves or exacerbates the anxiety disorder. Citalopram, due to its low side-effect profile and low potential for interaction with other medications, is now frequently used in the PD population.[73]

The combination of selegiline and SSRIs has the potential of causing serotonin syndrome.[61] Although uncommon,[62] patients should be advised of the symptoms of this complication. SSRIs also have the theoretical risk of worsening motor symptoms,[63] although this is not usually a clinically significant problem in the majority of patients.

Tricyclic antidepressants can be helpful, as they deplete both serotonin and noradrenalin. It is recommended that ones with a low anticholinergic profile, such as nortriptaline or desipramine, be used to prevent cognitive side effects. Dosing should be kept in the low to moderate range and can be given once daily at bedtime.

Buspirone has been reported to have anxiolytic properties in the general population, although this has not been borne out in studies in PD patients.[64,65] If a trial of buspirone is being considered, dosing should be kept low, as worsening of Parkinsonism occurs at high doses.[64,65]

In cases of severe anxiety, a trial of clozapine or quetiapine can be considered. Although not studied specifically in anxiety disorders, both drugs have been shown to be helpful in the treatment of psychosis, agitation, and hallucinations and are generally well tolerated in the PD population.[66-68] Dosing is low as compared to usual psychiatric dosing; worsening of Parkinsonism can be seen with quetiapine at higher doses in demented PD patients.[68,69] Typical and other atypical neuroleptics, such as risperidone and olanzapine, should be avoided, as these have been shown to significantly worsen parkinsonism, even at low doses.[70,71]

Nonpharmacological approaches are equally important in management.[72] This includes explanation of symptomatology, its relationship to fluctuations, and counseling of the patient and family. Charting of the attacks can be helpful for the patient to understand timing and triggering factors for episodes. Use of relaxation and anxiety reduction techniques are frequently helpful. Regular exercise may decrease severity of the anxiety. Psychotherapy may provide a variety of compensatory and coping strategies to help the patient deal with the effects of the anxiety. While psychotherapy alone may be beneficial in mild disease, in more severe cases, it needs to be combined with medical management.

CONCLUSION

Anxiety disorders are a common feature of PD, occurring in 30 to 40% of patients. They are frequently associated with depression. Anxiety disorders result in significant morbidity in PD but, unfortunately, remain an unrecognized and underdiagnosed problem.

Treatment is hampered by lack of controlled trials. Current treatment recommendations include individualization of the treatment regimes using a multidisciplinary approach with pharmacological and nonpharmacological treatment strategies.

REFERENCES

1. American Psychiatric Association, *Diagnostic and Statistical Manual of Mental Disorders (DSM– IV)*, American Psychiatric Press, Washington, D. C., 1994.
2. Regier, D. A., Boyd, J. H., Burke, J. D., Jr., Rae, D. S., Myers, J. K., Kramer, M., Robins, L. N., George, L. K., Karno, M., and Locke, B. Z., One-month prevalence of mental disorders in the United States. Based on five Epidemiologic Catchment Area sites, *Arch. Gen. Psychiatry*, 45(11), 977–86, 1988.
3. Wells, K. B., Golding, J. M., and Burnam, M. A., Psychiatric disorder in a sample of the general population with and without chronic medical conditions, *Am. J. Psychiatry*, 145(8), 976–81, 1988.
4. Raj, B. A., Corvea, M. H., and Dagon, E. M., The clinical characteristics of panic disorder in the elderly: a retrospective study, *J. Clin. Psychiatry*, 54(4), 150–5, 1993.
5. Joffe, R. T., Lippert, G. P., Gray, T. A., Sawa, G., and Horvath, Z., Mood disorder and multiple sclerosis, *Arch. Neurol.*, 44(4), 376–8, 1987.
6. Stein, M. B., Heuser, I. J., Juncos, J. L., and Uhde, T. W., Anxiety disorders in patients with Parkinson's disease, *Am. J. Psychiatry*, 147(2), 217–20, 1990.
7. Menza, M. A., Robertson-Hoffman, D. E., and Bonapace, A. S., Parkinson's disease and anxiety: comorbidity with depression, *Biol. Psychiatry*, 34(7), 465–70, 1993.
8. Routh, L. C., Black, J. L., and Ahlskog, J. E., Parkinson's disease complicated by anxiety, *Mayo Clin. Proc.*, 62(8), 733–5, 1987.
9. Schiffer, R. B., Kurlan, R., Rubin, A., and Boer, S., Evidence for atypical depression in Parkinson's disease, *Am. J. Psychiatry*, 145(8), 1020–2, 1988.
10. Fleminger, S., Left-sided Parkinson's disease is associated with greater anxiety and depression, *Psychol. Med.*, 21(3), 629–38, 1991.
11. Fetoni, V., Soliveri, P., Monza, D., Testa, D., and Girotti, F., Affective symptoms in multiple system atrophy and Parkinson's disease: response to levodopa therapy, *J. Neurol. Neurosurg. Psychiatry*, 66(4), 541–4, 1999.
12. Litvan, I., Cummings, J. L., and Mega, M., Neuropsychiatric features of corticobasal degeneration, *J. Neurol. Neurosurg. Psychiatry*, 65(5), 717–21, 1998.
13. Lauterbach, E. C. and Duvoisin, R. C., Anxiety disorders in familial parkinsonism, *Am. J. Psychiatry*, 148(2), 274, 1991.
14. Vazquez, A., Jimenez-Jimenez, F. J., Garcia-Ruiz, P., and Garcia-Urra, D., "Panic attacks" in Parkinson's disease. A long-term complication of levodopa therapy, *Acta Neurol. Scand.*, 87(1), 14–8, 1993.
15. Nausieda, P. A., Sinemet "abusers," *Clin. Neuropharmacol.*, 8(4), 318–27, 1985.
16. Weisskopf, M. G., Chen, H., Schwarzschild, M. A., Kawachi, I., and Ascherio, A., Prospective study of phobic anxiety and risk of Parkinson's disease, *Mov. Disord.*, 18(6), 646–51, 2003.
17. Shiba, M., Bower, J. H., Maraganore, D. M., McDonnell, S. K., Peterson, B. J., Ahlskog, J. E., Schaid, D. J., and Rocca, W. A., Anxiety disorders and depressive disorders preceding Parkinson's disease: a case-control study, *Mov. Disord.*, 15(4), 669–77, 2000.
18. Alegret, M., Junque, C., Valldeoriola, F., Vendrell, P., Marti, M. J., and Tolosa, E., Obsessive-compulsive symptoms in Parkinson's disease, *J. Neurol. Neurosurg. Psychiatry*, 70(3), 394–6, 2001.
19. Lauterbach, E. C., The locus ceruleus and anxiety disorders in demented and nondemented familial parkinsonism, *Am. J. Psychiatry*, 150(6), 994, 1993.
20. Shulman, L. M., Taback, R. L., Rabinstein, A. A., and Weiner, W. J., Non-recognition of depression and other non-motor symptoms in Parkinson's disease, *Parkinsonism Relat. Disord.*, 8(3), 193–7, 2002.
21. Hillen, M. E. and Sage, J. I., Nonmotor fluctuations in patients with Parkinson's disease, *Neurology*, 47(5), 1180–3, 1996.
22. Raudino, F., Non motor off in Parkinson's disease, *Acta Neurol. Scand.*, 104(5), 312–5, 2001.
23. Witjas, T., Kaphan, E., Azulay, J. P., Blin, O., Ceccaldi, M., Pouget, J., Poncet, M., and Cherif, A. A., Nonmotor fluctuations in Parkinson's disease: frequent and disabling, *Neurology*, 59(3), 408–13, 2002.
24. Menza, M. A., Sage, J., Marshall, E., Cody, R., and Duvoisin, R., Mood changes and "on-off" phenomena in Parkinson's disease, *Mov. Disord.*, 5(2), 148–51, 1990.
25. Siemers, E. R., Shekhar, A., Quaid, K., and Dickson, H., Anxiety and motor performance in Parkinson's disease, *Mov. Disord.*, 8(4), 501–6, 1993.
26. Maricle, R. A., Nutt, J. G., and Carter, J. H., Mood and anxiety fluctuation in Parkinson's disease associated with levodopa infusion: preliminary findings, *Mov. Disord.*, 10(3), 329–32, 1995.
27. Maricle, R. A., Nutt, J. G., Valentine, R. J., and Carter, J. H., Dose-response relationship of levodopa with mood and anxiety in fluctuating Parkinson's disease: a double-blind, placebo-controlled study, *Neurology*, 45(9), 1757–60, 1995.
28. Berrios, G. E., Campbell, C., and Politynska, B. E., Autonomic failure, depression and anxiety in Parkinson's disease, *Br. J. Psychiatry*, 166(6), 789–92, 1995.
29. Bech, P., Kastrup, M., and Rafaelsen, O. J., Mini-compendium of rating scales for states of anxiety depression

mania schizophrenia with corresponding DSM-III syndromes, *Acta Psychiatr. Scand.*, Suppl., 326, 1–37, 1986.

30. Zung, W. W., A rating instrument for anxiety disorders, *Psychosomatics*, 12(6), 371–9, 1971.

31. Beck A, S. R., *Beck Anxiety Inventory Manual*, The Psychological Corporation, San Antonio, TX, 1993.

32. Speilberger, C., Gorsuch, R., and Lushene, R., *Manual for the State-Trait Anxiety Inventory*, Consulting Psychologists Press, Palo Alto, CA, 1970.

33. Sheehan, D. V., Coleman, J. H., Greenblatt, D. J., Jones, K. J., Levine, P. H., Orsulak, P. J., Peterson, M., Schildkraut, J. J., Uzogara, E., and Watkins, D., Some biochemical correlates of panic attacks with agoraphobia and their response to a new treatment, *J. Clin. Psychopharmacol.*, 4(2), 66–75, 1984.

34. Zigmond, A. S. and Snaith, R. P., The hospital anxiety and depression scale, *Acta Psychiatr. Scand.*, 67(6), 361–70, 1983.

35. Marinus, J., Leentjens, A. F., Visser, M., Stiggelbout, A. M., and Van Hilten, J. J., Evaluation of the hospital anxiety and depression scale in patients with Parkinson's disease, *Clin. Neuropharmacol*, 25(6), 318–24, 2002.

36. Saint-Cyr, J. A. and Trepanier, L. L., Neuropsychologic assessment of patients for movement disorder surgery, *Mov. Disord.*, 15(5), 771–83, 2000.

37. Saint-Cyr, J. A., Neuropsychology for movement disorders neurosurgery, *Can. J. Neurol. Sci.*, 30 Suppl., 1, S83-93, 2003.

38. Houeto, J. L., Mesnage, V., Mallet, L., Pillon, B., Gargiulo, M., du Moncel, S. T., Bonnet, A. M., Pidoux, B., Dormont, D., Cornu, P., and Agid, Y., Behavioural disorders, Parkinson's disease and subthalamic stimulation, *J. Neurol. Neurosurg. Psychiatry*, 72(6), 701–7, 2002.

39. Troster, A. I., Fields, J. A., Wilkinson, S. B., Pahwa, R., Miyawaki, E., Lyons, K. E., and Koller, W. C., Unilateral pallidal stimulation for Parkinson's disease: neurobehavioral functioning before and 3 months after electrode implantation, *Neurology*, 49(4), 1078–83, 1997.

40. Maeshima, S., Nakai, K., Nakai, E., Uematsu, Y., Ozaki, F., Terada, T., Nakakita, K., Itakura, T., and Komai, N., Effects on cognitive function and activities of daily living after stereotactic thalamotomy for Parkinson's disease, *No Shinkei Geka*, 23(5), 417–21, 1995.

41. Higginson, C. I., Fields, J. A., and Troster, A. I., Which symptoms of anxiety diminish after surgical interventions for Parkinson disease? *Neuropsychiatry Neuropsychol, Behav. Neurol.*, 14(2), 117–21, 2001.

42. Junque, C., Alegret, M., Nobbe, F. A., Valldeoriola, F., Pueyo, R., Vendrell, P., Tolosa, E., Rumia, J., and Mercader, J. M., Cognitive and behavioral changes after unilateral posteroventral pallidotomy: relationship with lesional data from MRI, *Mov. Disord.*, 14(5), 780–9, 1999.

43. Scott, R., Gregory, R., Hines, N., Carroll, C., Hyman, N., Papanasstasiou, V., Leather, C., Rowe, J., Silburn, P., and Aziz, T., Neuropsychological, neurological and functional outcome following pallidotomy for Parkinson's disease. A consecutive series of eight simultaneous

44. Daniele, A., Albanese, A., Contarino, M. F., Zinzi, P., Barbier, A., Gasparini, F., Romito, L. M., Bentivoglio, A. R., and Scerrati, M., Cognitive and behavioural effects of chronic stimulation of the subthalamic nucleus in patients with Parkinson's disease, *J. Neurol. Neurosurg. Psychiatry*, 74(2), 175–82, 2003.

45. Perozzo, P., Rizzone, M., Bergamasco, B., Castelli, L., Lanotte, M., Tavella, A., Torre, E., and Lopiano, L., Deep brain stimulation of subthalamic nucleus: behavioural modifications and familiar relations, *Neurol. Sci.*, 22(1), 81–2, 2001.

46. Braak, H., Del Tredici, K., Rub, U., de Vos, R. A., Jansen Steur, E. N., and Braak, E., Staging of brain pathology related to sporadic Parkinson's disease, *Neurobiol. Aging*, 24(2), 197–211, 2003.

47. Heninger, G. R. and Charney, D. S., Monoamine receptor systems and anxiety disorders, *Psychiatr. Clin. North Am.*, 11(2), 309–26, 1988.

48. Berlan, M., Rascol, O., Belin, J., Moatti, J. P., Rascol, A., and Montastruc, J. L., Alpha 2-adrenergic sensitivity in Parkinson's disease, *Clin. Neuropharmacol*, 12(2), 138–44, 1989.

49. Menza, M. A., Palermo, B., DiPaola, R., Sage, J. I., and Ricketts, M. H., Depression and anxiety in Parkinson's disease: possible effect of genetic variation in the serotonin transporter, *J. Geriatr. Psychiatry Neurol.*, 12(2), 49–52, 1999.

50. Cash, R., Dennis, T., L'Heureux, R., Raisman, R., Javoy-Agid, F., and Scatton, B., Parkinson's disease and dementia: norepinephrine and dopamine in locus ceruleus, *Neurology*, 37(1), 42–6, 1987.

51. Cedarbaum, J. M. and Aghajanian, G. K., Catecholamine receptors on locus coeruleus neurons: pharmacological characterization, *Eur. J. Pharmacol.*, 44(4), 375–85, 1977.

52. Iruela, L. M., Ibanez-Rojo, V., Palanca, I., and Caballero, L., Anxiety disorders and Parkinson's disease, *Am. J. Psychiatry*, 149(5), 719–20, 1992.

53. Erdal, K. J., Depressive symptom patterns in patients with Parkinson's disease and other older adults, *J. Clin. Psychol.*, 57(12), 1559–69, 2001.

54. Richard, I. H., Szegethy, E., Lichter, D., Schiffer, R. B., and Kurlan, R., Parkinson's disease: a preliminary study of yohimbine challenge in patients with anxiety, *Clin. Neuropharmacol.*, 22(3), 172–5, 1999.

55. Nutt, D., Management of patients with depression associated with anxiety symptoms, *J. Clin. Psychiatry*, 58 Suppl., 8, 11–6, 1997.

56. Lieberman, A., Managing the neuropsychiatric symptoms of Parkinson's disease, *Neurology*, 50(6 Suppl. 6), S33–8; discussion S44–8, 1998.

57. Suranyi-Cadotte, B. E., Nestoros, J. N., Nair, N. P., Lal, S., and Gauthier, S., Parkinsonism induced by high doses of diazepam, *Biol. Psychiatry*, 20(4), 455–7, 1985.

58. Menza, M. A. and Rosen, R. C., Sleep in Parkinson's disease. The role of depression and anxiety, *Psychosomatics*, 36(3), 262–6, 1995.

59. Kasper, S., Neurobiology and new psychopharmacological strategies for treatment of anxiety disorders, in *Current Therapeutical Approaches on Panic and Other Anxiety Disorders*, Racangi, B. Basel, Switzerland, 1995.

60. den Boer, J. A., Westenberg, H. G., Kamerbeek, W. D., Verhoeven, W. M., and Kahn, R. S., Effect of serotonin uptake inhibitors in anxiety disorders; a double-blind comparison of clomipramine and fluvoxamine, *Int. Clin. Psychopharmacol.*, 2(1), 21–32, 1987.

61. Suchowersky, O. and deVries, J., Possible interactions between deprenyl and prozac, *Can. J. Neurol. Sci.*, 17(3), 352–3, 1990.

62. Richard, I. H., Kurlan, R., Tanner, C., Factor, S., Hubble, J., Suchowersky, O., and Waters, C., Serotonin syndrome and the combined use of deprenyl and an antidepressant in Parkinson's disease, Parkinson Study Group, *Neurology*, 48(4), 1070–7, 1997.

63. Steur, E. N., Increase of Parkinson disability after fluoxetine medication, *Neurology*, 43(1), 211–3, 1993.

64. Ludwig, C. L., Weinberger, D. R., Bruno, G., Gillespie, M., Bakker, K., LeWitt, P. A., and Chase, T. N., Buspirone, Parkinson's disease, and the locus ceruleus, *Clin. Neuropharmacol.*, 9(4), 373–8, 1986.

65. Bonifati, V., Fabrizio, E., Cipriani, R., Vanacore, N., and Meco, G., Buspirone in levodopa-induced dyskinesias, *Clin. Neuropharmacol.*, 17(1), 73–82, 1994.

66. Dewey, R. B., Jr. and O'Suilleabhain, P. E., Treatment of drug-induced psychosis with quetiapine and clozapine in Parkinson's disease, *Neurology*, 55(11), 1753–4, 2000.

67. Klein, C., Gordon, J., Pollak, L., and Rabey, J. M., Clozapine in Parkinson's disease psychosis: 5-year follow-up review, *Clin. Neuropharmacol.*, 26(1), 8–11, 2003.

68. Reddy, S., Factor, S. A., Molho, E. S., and Feustel, P. J., The effect of quetiapine on psychosis and motor function in parkinsonian patients with and without dementia, *Mov. Disord.*, 17(4), 676–81, 2002.

69. Morgante, L., Epifanio, A., Spina, E., Di Rosa, A. E., Zappia, M., Basile, G., La Spina, P., and Quattrone, A., Quetiapine versus clozapine: a preliminary report of comparative effects on dopaminergic psychosis in patients with Parkinson's disease, *Neurol. Sci.*, 23 Suppl., 2, S89–90, 2002.

70. Goetz, C. G., Blasucci, L. M., Leurgans, S., and Pappert, E. J., Olanzapine and clozapine: comparative effects on motor function in hallucinating PD patients, *Neurology*, 55(6), 789–94, 2000.

71. Ellis, T., Cudkowicz, M. E., Sexton, P. M., and Growdon, J. H., Clozapine and risperidone treatment of psychosis in Parkinson's disease, *J. Neuropsychiatry Clin. Neurosci.*, 12(3), 364–9, 2000.

72. Marsh, L., Anxiety disorders in Parkinson's disease, *Int. Rev. Psychiatry*, 12, 307–318, 2000.

73. Trew, M., personal communication, 2003.

33 Dementia in Parkinson's Disease

Anne L. Barba, Eric S. Molho, Donald S. Higgins, Anthony J. Santiago, and Stewart A. Factor
Department of Neurology, Albany Medical Center

CONTENTS

INTRODUCTION

Although Parkinson's disease (PD) is traditionally thought of primarily as a motor disorder, cognitive dysfunction and frank dementia do occur in many patients. In fact, patients, families, and physicians alike have found that dementia occurring in PD can be the most frustrating and disabling consequence of the disease. Despite the dramatic advances that have been made in the treatment of motor symptoms in recent decades, PD patients who develop dementia still have a significantly worse prognosis than their nondemented counterparts and all too often lose the ability to live independently. Dementia is probably the greatest unmet need from a therapeutic standpoint in PD. In this chapter, we review the epidemiology, clinical features, neurochemical, neuroimaging, and pathological aspects of dementia in PD. We also review recent literature on the treatment of dementia and outline a practical approach to the management of this important complication of PD.

EPIDEMIOLOGY

The epidemiology of dementia in PD has been extensively studied, and several reviews have been published in recent years[1,2] The frequency of this extremely disabling complication has differed between studies, for several reasons. These include variation in the definition of dementia, the duration of PD, and method of ascertainment of subjects. Other methodological differences include the degree to which neuropsychological data was employed and the types of tests used, the choice of sample population, and the age of those studied. Although exact figures cannot be given, it is clear that dementia is a frequent problem that deserves increased attention.

James Parkinson[3] indicated in his initial essay that dementia was probably not a manifestation of the disease, although his report included only six patients, and three were observed on the street. This notion, however, probably survived for 150 years. In the 1970s and 1980s, though, the occurrence of dementia was increasingly recognized in prevalence studies generally carried out in clinic cohorts. It is possible that the frequency increased in the post-levodopa era, not because the drug caused dementia but because patients with PD were living longer. A meta-analysis of 17 such studies through 1984 indicated a prevalence of 15%.[4] Several studies in the mid to late 1980s also indicated low prevalence figures which ranged from 8 to 29%.[5–8] Studies examining community-based populations are more likely to provide realistic estimates of the prevalence of dementia in PD, and several of these studies have indicated higher rates than the older literature foretold, with figures of 18%,[9] 28%,[10] 37%,[11] and 41%[12] being reported. These are single point-prevalence estimates, however, which in all likelihood lead to an underestimation of actual prevalence rates.

It has been demonstrated that dementia prevalence rates vary depending on several factors, although the patients' age, duration of disease, and length of follow-up are likely most important. Several studies have examined the impact of age, establishing a significant difference in patients under age 70 as compared to those over 70.[12–14] The issues of disease duration and follow-up can be addressed only in longitudinal studies, of which there are few. Most recently, Aarsland et al.[15] performed a prospective study in a community-based population to examine the prevalence rates at baseline (mean duration of disease nine years) and after four and eight years of follow-up. Rates of dementia were 26% at baseline, 52% at four years, and 78% at eight years, and the mean duration of disease at the time of dementia development was approximately 14 years. These figures are higher than other point-prevalence studies, and they highlight the increasing frequency of dementia in relation to disease duration and advancing age. In addition, these figures may actually represent an underestimation, since, of the 224 PD patients evaluated at baseline, 65 died in the follow-up period, and it is not known how many of these had developed dementia. Another reason for underestimation is the elimination of patients with early dementia in this and other studies in an attempt to exclude those with dementia with Lewy bodies (DLB). However, since recent neuropathological studies indicate that PD with dementia and DLB may be the same disease,[2] their elimination from epidemiological studies probably leads to an underestimation of the frequency of dementia in PD. Longitudinal studies provide a more realistic measure of the frequency of dementia in PD and indicate that it is a much bigger problem than previously believed, as nearly all patients will develop dementia if they survive long enough.

Aarsland et al.[15] provide a cumulative risk of 78% after eight years of follow-up. This figure is similar to the cumulative risk of 65% by age 85 reported by Mayeux et al.,[16] but higher than the 38% of clinic patients seen in ten years by Hughes et al.[17] The estimated incidence of new dementia cases per year has also been reported and, in community-based population studies, has ranged from 4.2 to 9.5% per year.[11,18,19] Incidence rates also vary with age, from 3% for patients under age 60 to 15% for those over 80.[20] Several studies have also compared the risk of dementia in PD to that of healthy elderly populations. In general, results have revealed that PD patients have two to six times greater risk for developing dementia depending on the study.[9,11,15,18,19]

Dementia is a risk factor for increased mortality in PD patients. Epidemiological studies have demonstrated that having PD increases mortality by a factor of 1.5 to 2.5 times what is expected, and the development of dementia increases risk even further. In one clinic-based study and one population-based study where risk factors like age and disease duration were controlled, the mortality rate was increased twofold.[21,22] The occurrence of dementia is also associated with more severe motor features[1] and psychosis,[15,23] both of which are risk factors for increased mortality.

RISK FACTORS AND ASSOCIATED CONDITIONS

A discussion of disease-related risk factors generally includes an analysis of patient traits and environmental exposures, the presence of which affect the likelihood of developing a certain condition. In the case of PD, little is known about the existence of specific environmental exposures that might contribute to the risk of developing dementia. One study compared 43 demented and 51 nondemented PD patients and found that pesticide exposure was associated with a threefold greater risk of the presence of dementia, but only when combined with a specific genotype, namely, the presence of the CYP2D6 29B+ allele.[24] Neither factor, by itself, was predictive of the presence of dementia. In another community-based survey, Marder et al.[25] found that estrogen replacement therapy was "protective" against the development of dementia in women with PD. The rationale for this investigation was based on previous studies that found that estrogen replacement therapy was associated with a decreased risk of developing Alzheimer's disease (AD), and on the pathophysiological and clinical similarities between AD and PD dementia.[26] Clinical trials looking at the prospective use of estrogen replacement therapy in PD to prevent dementia are planned.

Recent efforts to identify genetic risk factors for the development of dementia in PD have attempted to build on the success of similar efforts to understand the genetic basis of AD. One of the genetic markers for AD relates to allelic variants the apolipoprotein E (ApoE) gene on chromosome 19. The ApoE gene has three common alleles, E2, E3, and E4, and in AD the E4 frequency is elevated, conferring an increased risk of developing the disease, while the E2 frequency is reduced. The increase in risk of developing AD is dose dependent. In addition, E4 carrier status relates to an earlier age of onset, while E2 leads to delayed onset, although the mechanism by which this effect occurs is unknown.[27] Due to the well known clinical and pathological overlaps between PD and AD, ApoE allele status has been examined in PD, and studies have demonstrated an allelic frequency similar to that of normal populations.[28,29] In addition, no association between dementia and ApoE 4 has been observed.[1] A few studies have implicated E2 in PD, but this finding has been inconsistent.[29,30] However, it does appear that the ApoE 4 allele is associated with an earlier age of onset of PD, a pattern resembling that seen in AD.[28,29] There was no interaction between family history and the carrier status of ApoE 4 in these studies to explain this association, but the effects were additive. Several small studies showed varied results on the relationship of ApoE 4 and earlier age of onset, but the largest confirmed this relationship.[29] Differences in results between studies may be the result of several factors, and Zareparsi et al.[29] delineated these in detail. The main issues related to sample size, ethnicity, and proportion of familial cases. The largest study was sufficiently powered to answer the question of a relationship between ApoE allele frequency and age of onset of PD. The authors examined 521 PD patients, and the patients with one or two E4 alleles (and no E2 alleles) had onset 3.5 years earlier than those with two E3 alleles. No clear relationship was found between the E2 allele and later age of onset.

PD has now been associated by linkage to ten loci, and five genes have been sequenced. The genes include alpha synuclein (Park 1), Parkin (Park 2), Ubiquitin carboxyl-terminal hydrolase L1 (Park 5), DJ1 (Park 7), and SCA2. Early attempts have been made to examine the effect that the genotype has on phenotype, and there is some information beginning to emerge regarding the occurrence of dementia in PD relating to particular genes. For example, in families with alpha synuclein parkinsonism, dementia has been described,[31] whereas Parkin does not seem to cause this problem.[32] It is likely that some types of hereditary PD will result in an increased risk for developing dementia, while others will have lower risk or no risk at all. Clearly, more work is needed before this information can be properly interpreted.

The two basic methodologies for looking at patient-associated phenotypic risk factors are case-control studies, which look retrospectively in a cross-sectional fashion at the association of putative risk factors and the incidence of a particular condition, and cohort studies, which look prospectively and longitudinally at baseline characteristics and the likelihood of the development of a condition over time. Each of these methodologies has been extensively employed to look at the patient and disease-related factors that might contribute to the risk of developing PD dementia. Older age and later age of onset have been found to be risks for dementia in PD in case-control studies.[9,10] This has been confirmed by several well designed, long-term prospective analyses.[17,18,33] However, in one ten-year study, Hughes et al.[17] found that only older age at evaluation, not age at onset, was a statistically significant risk, suggesting that early- and late-onset cases of PD may not be distinct as far as the risk for developing dementia. In another prospective analysis, Levy et al.[34] found that age was only an important risk for dementia when combined with worse severity of PD motor symptoms. The severity of motor symptomatology has been consistently implicated in prospective cohort studies as an important risk factor for developing PD-associated dementia.[11,17,18,33,34] Depression has also been shown to be a significant risk factor in numerous studies using both methodologies.[10,11,33,35–37] However, this relationship has been questioned by other investigators.[17,18,38] Ryder et al.[38] found that, when anxiety was separated from depression for the analysis, only anxiety was associated with the presence of cognitive dysfunction. The authors suggested that the inconsistency of other reports on the relationship between the presence of depression and the risk of developing dementia might be due to inadequate methods of separating out the overlapping signs and symptoms of anxiety and depression prior to performing statistical analysis.

The presence of several patient traits in early PD has also been suggested to represent a risk for developing dementia later in the course of illness. These include early cognitive dysfunction, such as an initial MMSE score less than 29 in one analysis[18] and, more specifically, the presence of deficits in verbal memory or executive function in another.[39] Interestingly, a higher educational level has not been found to have a protective influence against dementia in PD patients, unlike findings in AD patients.[17,40] Atypical motor presentations such as prominent facial masking,[33] symmetrical exam, early autonomic dysfunction, and poor response to levodopa[10] have also been implicated as risk factors for dementia.

The relationship between dementia and depression in PD is a complicated one. Depression is common in PD with estimates of the frequency of its occurrence ranging from 12 to 90%.[41] As a result, it often coexists with dementia. In addition, the signs and symptoms of each have significant overlap with the other, leading to frequent misdiagnosis. Pseudodementia can result from the effects of depression, and vice versa. Although it is generally accepted that depression is a risk factor for dementia in PD, the nature of this influence is unclear. Well designed, prospective studies have shown that patients with depression experience a more profound cognitive decline over time.[36,37] Does depression simply act as a marker for the impending onset of dementia, or does depression exert an independent negative influence on cognitive functioning? The beginnings of an answer may be provided by Tröster et al.,[42] who have shown that depression is associated with a pattern of memory impairment in PD that is distinguishable from that seen in AD. Furthermore, they have found that PD patients with dementia, relative to those without, have dysfunction in similar cognitive domains, but the magnitude of the difficulty is increased.[43] Since the pattern of cognitive dysfunction in PD is similar to what is seen in depression alone, they have suggested that aggressive treatment of depression in PD may actually help cognitive performance.[44]

The occurrence of drug-induced psychosis and dementia is also intimately related in PD patients. It is widely recognized that dementia is a major risk factor and may be a prerequisite for the development of drug-induced psychosis.[45] This is discussed more fully elsewhere in this text. It may also be true that the presence of psychotic symptoms in PD is a reliable marker for the development and progression of dementia. Factor et al.[23] evaluated the long-term outcome (26 months) of 59 patients initially treated with clozapine for drug-induced psychosis as part of a double-blind clinical trial. Of the 44 patients in the final analysis, 30 (68%) were demented. Other investigators have reported the high occurrence rate of dementia in PD patients experiencing psychosis as well.[33,46]

NEUROPSYCHOLOGICAL CHARACTERISTICS OF DEMENTIA IN PD

In the *Diagnostic and Statistical Manual of Mental Disorders,* 4th ed. (DSM-IV),[47] dementia is defined as the development of multiple cognitive deficits that affect memory and that result in aphasia, apraxia, agnosia, or executive dysfunction. Impairment must result in occupational or social dysfunction, must represent a decline from a previous level of ability, and must exist separate from delirium. Given that the bedside evaluation, the Mini-Mental State Examination,[48] and other common screening measures may not be sensitive to the presence of dementia in PD, an in-depth cognitive evaluation is often essential. In many cases, neuropsychological evaluation can assist in determining whether cognitive impairment or dementia

is present and can aid in quantifying the nature and degree of any cognitive dysfunction that is observed.

COGNITIVE DYSFUNCTION IN THE ABSENCE OF DEMENTIA

A proportion of individuals with PD will be spared clinically significant changes in cognition.[49] However, it is widely recognized that mild cognitive dysfunction occurs in patients, even early in the course of the disease[50,51] and in the absence of frank dementia.[52,53] When dementia is absent, cognitive dysfunction tends to be rather circumscribed, and executive and language functions, memory, and visuospatial skills are most often affected.[54] (Please refer to Table 33.1 for examples of tasks that assess these skill areas.)

TABLE 33.1
Tasks that Assess Skill Areas often Compromised in Those with PD

Skill Area	Assessment Tool
Executive function	Wisconsin Card Sorting Test (WCST); Booklet Category Test (BCT); Tower of London (TOL); Trailmaking Test–Part B (TMT–B); Stroop Color and Word Test
Language	Boston Naming Test (BNT); Controlled Oral Word Association Test (COWAT); Semantic Fluency
Learning and memory	California Verbal Learning Test–II (CVLT–II); Hopkins Verbal Learning Test–Revised (HVLT–R); Rey Auditory Verbal Learning Test (RAVLT); Rey Complex Figure Test (RCFT); Brief Visuospatial Memory Test-Revised (BVMT–R); Wechsler Memory Scale-III (WMS-III)
Visuospatial and visuoperceptual skill	Hooper Visual Organization Test (HVOT); Facial Recognition; Judgement of Line Orientation (JLO); Visual Form Discrimination (VFD); Wechsler Adult Intelligence Scale-III (WAIS-III): Block Design Subtest

EXECUTIVE FUNCTION

Executive function has been described as the capacity to plan and then initiate complex, goal-directed behavior. The ability to choose a correct response, sequence one's actions, benefit from feedback, monitor and modify behavior when appropriate, and inhibit one style of responding in favor of another are subsumed under these auspices[55,56] and are thought to rely heavily on the integrity of the prefrontal cortex.

Deficiencies on measures of executive function are thought to occur often in nondemented PD patients, including tasks that require temporal ordering,[57] cognitive sequencing,[50] adequate planning,[58] and the ability to inhibit a usual or non-novel response.[53] It seems, though, that the most commonly used paradigms assess a respondent's ability to form, maintain, and shift "set." This has been described as "a state of brain activity which predisposes a subject to respond in one way when several alternatives are available."[59] In fact, several authors[57,58,60–62] note that nondemented individuals with PD perform poorly on these paradigms, and difficulties in set formation,[50,52,63] maintenance,[52,60,64] and shifting[60,64–66] have often been revealed. In that vein, many[52,60,67] have suggested that nondemented individuals with PD tend to respond perseveratively on tasks of "set" (i.e., their performance is marred by an inability to disengage attention from a once-relevant dimension when shifting set is the objective), while others[68] note that "learned irrelevance" (i.e., an inability to reengage attention toward a once-irrelevant dimension) might contribute to often-observed decrements in set shifting capacity.

While difficulties in set shifting have often been revealed in nondemented PD patients, it seems that deficient performance on these tasks is not inevitable. Brown and Marsden[67,69] suggest that the nature of attentional control required by the task (whether internal or external) must be considered, as tasks that necessitate the use of an internal strategy (and are thus not guided by external cues) for the control of attention are deficient, while task performance tends not to suffer when external plans or cues are provided.

LANGUAGE FUNCTION

Although deficient performance on tasks of lexical and semantic fluency might be anticipated,[70] as both are thought to signal a disturbance in executive function,[71] support for this notion is at best inconsistent. While some investigators[72] have suggested that nondemented individuals with PD perform comparably to healthy volunteers on these measures, others[73,74] have demonstrated deficits in semantic, rather than lexical, fluency. However, subordinate category cueing may, in some instances, ameliorate these deficits.[75] On the other hand, poorer performance on quantitative (i.e., total word production) and qualitative (i.e., use of clustering and/or switching strategies) measures of lexical fluency have been reported,[70] although neither observation has received consistent support.[73,76]

Impaired action (or verb naming) fluency has been observed in PD,[77] although systematic support for this claim is also lacking, and Piatt et al.[72] did not find a significant difference in the task performance of individuals with PD as compared to healthy volunteers. Others[50,78] have examined performance on tasks of alternating fluency (which require that an individual alternate between given categories or letters of the alphabet) and suggest

that adequate performance (which requires that the patient shift set in the absence of an external cue) is deficient if compared to that of healthy volunteers.

While poor performance has also been observed on measures of confrontation naming in some studies,[52] others[50,51,79] suggest that nondemented individuals with PD perform comparably to healthy volunteers on these tasks. Others still contend that phonetic cueing may, in some instances, serve to normalize deficient performance on tasks of confrontation naming.[74]

LEARNING AND MEMORY

While a fair amount of evidence to date suggests that memory is relatively well preserved, especially early in the course of the disease, it remains that several investigators have observed memory deficiencies in nondemented PD patients.[80] While Taylor et al.[80] found that the performance of those with PD did not differ significantly from that of healthy volunteers on tasks of narrative memory function, their earlier work[63] revealed significant discrepancies on measures of immediate recall. In contrast, Cooper et al.[50] note that patients' performance was poorer on measures of both immediate and delayed narrative recollection, although they suggest that those with PD tended to recall as much of their immediate response as normal volunteers had after a delay. Performance on tasks of visual recollection has also been shown to be either normal[50,63] or deficient,[81] and the same is true of measures of paired associate learning.[50,63,82]

The most widely used measures of memory function have assessed supraspan word list learning, and many authors[63,83–85] have noted that, while recall is deficient in those with PD, performance on tasks of recognition is most often normal. Similarly, Taylor et al.[80] suggest that those with PD are less apt to cluster semantically when presented with word lists that contain an implicit embedded structure meant to guide recall. Thus, they reason that individuals with PD have difficulty when confronted with tasks that require "spontaneously organized subject-directed planning." Buytenhuijs et al.[84] observed the same tendency in their sample. Specifically, they noted that patients tended to perform poorly when forced to rely on internal strategies (as they must organize information in the absence of external cues) when recalling word list items but performed as well as volunteers on tasks of recognition (which are externally guided).

Procedural learning, which has been described as "knowing how" rather than "knowing that,"[86] is often deficient in those with PD. Poor performance has been observed on measures that require the acquisition of visuomotor (e.g., rotary pursuit[87] and serial reaction time tasks[88]) and cognitive (e.g., the Tower of Toronto task)[89] skills or routines. Deficiencies on measures of procedural learning are by no means universal. In fact, in their comparison of nondemented individuals with PD and healthy volunteers, Heindel et al.[90] did not observe a significant difference in performance on tasks of rotary pursuit and lexical priming (a measure of implicit memory). Furthermore, Harrington et al.[87] also did not find a significant difference on a task of visuoperceptual learning (i.e., mirror reading). In addition, on a task of rotary pursuit, only those with more advanced disease were found to be impaired.

VISUOSPATIAL SKILL

Visuospatial difficulties have often been cited and are considered a common consequence of PD, even when dementia is absent. Indeed, several have noted deficiencies on measures of visual-motor construction[7] and design copy.[50,81] Although adequate performance on these tasks relies to some extent on visual acuity, manual dexterity, and speed, when tasks' motor demands are diminished, and when speed requirements are erased, deficiencies on tasks of visual-motor construction remain.[91] In addition, poor performance has often been observed on measures that are neither timed nor dependent on motor skill, such as tasks of facial recognition,[51,91] perception of line orientation,[52,53,92] visual disembedding,[81] and mental object assembly.[91]

Several studies[91,93] suggest that adequate performance on tasks of visuospatial function might also rely heavily on other skills, such as sustained attention, memory function, or set shifting capacity. In fact, in their comparison of nondemented individuals with PD and elderly volunteers, Bondi et al.[57] initially observed a significant difference between groups on tasks of visual perception and construction. However, apparently deficient performance on these tasks was normalized when scores on measures of frontal system function was statistically covaried.

In summary, executive dysfunction is thought to be a hallmark of PD, as patients will often evidence impairment on tasks sensitive to the integrity of the prefrontal cortex.[63] Thus, poor performance has been noted on measures of set formation, maintenance, and shifting, cognitive sequencing, adequate planning, and the ability to inhibit an unusual or non-novel response. While several authors[57,71] suggest that frontal system dysfunction may also account for poorer than anticipated performance on other tasks (i.e., certain tasks of memory, language function, and visual perception and construction), this conclusion has not received universal support.[82]

NEUROPSYCHOLOGICAL PREDICTION OF INCIDENT DEMENTIA

Is there a pattern of neuropsychological impairment that might predict the onset of dementia in PD? Demographic risks (e.g., older age) have been identified, and risks related to the disease itself (e.g., motor symptom severity) have been acknowledged.[18,33] Research has also begun to highlight the presence of a pattern of neuropsychological risk, and several authors[71,94,95] have suggested that subtle executive dysfunction may precede the onset of dementia in PD.

For example Jacobs et al.[71] noted an association between poor baseline performance on measures of lexical and semantic fluency and the eventual occurrence of dementia. These authors suggest that executive dysfunction might underlie deficient task performance in this instance, as adequate performance on these measures likely requires the initiation of a systematic semantic memory search. Mahieux et al.[96] also noted an association between deficient lexical fluency and the subsequent development of dementia in PD. In addition, the authors noted that poor baseline performance on certain measures of sustained attention, visual perception, and executive function (i.e., the Picture Completion Subtest of the Wechsler Adult Intelligence Scale-Revised[97] and the interference component of the Stroop Test[98]) seemed to predict the presence of dementia at follow-up.

Levy et al.[39] observed an association between measures of both immediate and delayed verbal recall and subsequent development of dementia. Woods and Tröster[95] compared the performance patterns of nondemented patients who met diagnostic criteria for dementia after one year with those who did not. Their findings suggest that, relative to those who did not develop dementia, those who were demented after one year exhibited poor initial performance on certain tasks of executive function, complex attention, and word list-learning and recognition discriminability.

In summary, several investigators[71,95] have suggested that subtle executive dysfunction may precede the onset of dementia in PD. Interestingly, Piccirilli et al.[94] have maintained that the identification of a prodrome, or preclinical phase, of parkinsonian dementia is important in that it allows for the identification of at-risk individuals and may also permit the differentiation of various forms of dementing illness, as the pattern of neuropsychological performance observed in the preclinical stages of parkinsonian dementia differs from that observed in the preclinical stages of AD.

NEUROPSYCHOLOGICAL CHARACTERISTICS OF DEMENTIA

If present, dementia often emerges late in the course of the illness.[20] In addition, it is thought to be characterized by intellectual deterioration that is often less pronounced than that which accompanies AD and other "cortical" dementing illnesses.[99,100] This has led some to suggest that the pattern of deterioration, which is often described as "subcortical," can be distinguished by its clinical features.[99]

"Subcortical" dementing illnesses are typically characterized by slowed mentation (or bradyphrenia), alterations in personality and in mood, deficient manipulation of acquired knowledge, and forgetfulness or diminished retrieval of learned information. Frank amnesia, aphasia, apraxia, and agnosia, which are thought to be cardinal features of cortical illness, are often absent.[101] While some investigators[99] defend the use of this heuristic approach in characterizing the pattern of impairment that may accompany PD, others[102,103] note that there is little evidence to support its continued use. Specifically, they point to postmortem investigations that show that cortical pathology is present in those with PD. This would imply that its pathology extends beyond the subcortical nuclei. In fact, the dementia that may accompany PD has long been regarded as pathologically diverse, as postmortem evaluation has often revealed the presence of both senile plaques and neurofibrillary tangles which would be characteristic of AD[104] and cortical Lewy bodies, which are more characteristic of DLB.[105]

Considering that the dementia associated with PD is pathologically varied, it is not surprising that there has been debate about the nature of the neuropsychological impairment that typically characterizes it. Namely, some[7,106] suggest that the pattern of neuropsychological performance is best viewed as a continuum, and they note that those who are not demented may be distinguished from those who are on the basis of the depth or severity of their impairment alone. Others[107] maintain that PD is not a homogeneous entity and suggest that those who are truly demented can be distinguished on the basis of their unique clinical characteristics.

To this end, several investigators have attempted to examine and compare the neuropsychological performance patterns of those with varying degrees of intellectual impairment. In their examination of immediate, recent, and remote memory function, Huber et al.[108] noted that those with intellectual impairment performed poorly as compared to their nondemented counterparts on measures of remote memory function, and they concluded that this deficiency suggests that a qualitative distinction exists between those who are intellectually impaired and those who are not. Furthermore, McFadden et al.[109] compared the neuropsychological performance patterns of demented and nondemented individuals and suggested that the performance of those with dementia is unique and thus qualitatively different from the performance of nondemented controls. The authors concluded that dementia is the result of an "atypical development in Parkinson's disease" and is the consequence of a separate pathogenetic process.

To the contrary, Zakzanis and Freedman,[110] in their meta-analytic review, note that, while nondemented individuals tended to be most impaired on measures of delayed recall, those with dementia performed most poorly on tasks of manual dexterity, abstraction, and cognitive flexibility. In reviewing their findings, they concluded that the pattern of neuropsychological degeneration in PD is best viewed as a continuum, and they note that, as the disease progresses, performance initially deteriorates on tasks of delayed recall, and later worsens in measures of cognitive flexibility, manual dexterity, and abstraction. Furthermore, in their comparison of demented and nondemented PD patients, Girotti et al.[7] note that, while those with dementia were deficient on all tasks in a proportion greater than would be expected, they tended, in general, to be impaired above all on those measures that distinguished nondemented individuals from healthy volunteers. They therefore conclude that, while their impairment is more severe, the performance pattern of those with dementia typically mirrors that of their nondemented counterparts.

On first analysis, it appears that these investigators' conceptualizations of dementia are mutually exclusive. However, just as the dementia associated with PD is pathologically diverse, it seems that it may be neurobehaviorally heterogeneous as well, manifesting itself differently in different individuals.[20] While some may be categorized as suffering from a dementia that appears to be more subcortical in nature, others might exhibit a different[94] or more severe pattern of impairment—one that may, for instance, exhibit the pathologic and clinical features of AD.[111]

DIFFERENTIAL DIAGNOSIS AND PATHOLOGY

As is evident from Table 33.2, the differential diagnosis for an adult presenting with cognitive impairment is substantial. More specifically, dementia-associated parkinsonism includes several degenerative processes, the most important of which, for the purposes of this discussion, are PDD, DLB, and AD. There is an ongoing debate regarding whether the dementia associated with these three disorders is due to distinct pathologic processes, represents a continuum of a single disease process, or is some combination.[112] This section presents what is currently understood regarding the neuropathology of AD, PDD, and DLB and will attempt to formulate an understanding of the relationship between these closely related degenerative disorders. Discussion regarding movement disorders such as Huntington's disease, progressive supranuclear palsy (PSP), corticobasal ganglionic degeneration, and multiple system atrophy, which are not uncommonly associated with parkinsonism and dementia, is beyond the scope of this chapter.

Dementia is a common disorder, and it is estimated that at least 30% of the general population over age 65 are

TABLE 33.2
Differential Diagnosis of Dementia

Primary Degenerative Dementia

Alzheimer's disease
Down's syndrome
Dementia with Lewy bodies
Frontotemporal dementias
  Pick's disease
  Progressive aphasia
  Motor neuron disease with frontotemporal dementia
Thalamic dementia

Dementia Associated with Other Degenerative Processes

Parkinson's disease
Progressive supranuclear palsy
Corticobasal ganglionic degeneration
Prion disease
Huntington's disease
Amyotrophic lateral sclerosis

Vascular Dementia
Other Causes of Dementia

Infectious
  Human immunodeficiency syndrome
  Other chronic viral encephalitides
  Spirochetal disease
  Chronic bacterial meningitides
  Whipple's disease
Metabolic disorders
  Acquired
    Vitamin B12 deficiency
    Thiamine deficiency
    Hypothyroidism
    Chronic hypoglycemia
    Abnormalities of Calcium Homeostasis
    Corticosteroid disorders
    Renal impairment
    Hepatic disturbance
  Inherited
    Wilson's disease
    Metachromatic leukodystrophy
    Adult-onset neuronal ceroid-lipofuscinosis
    Lysosomal storage diseases
    Mitochondrial encephalomyelopathies
Neoplasms
Drugs and toxins
Trauma
Normal pressure hydrocephalus
Pseudodementia

afflicted.[113] As often as 50% of the time, the cause is AD.[114] AD often presents as impairment of memory, but depression, aphasia, spatial disorientation, or frontal lobe signs can be present as well.[115] Extrapyramidal features such as rigidity and akinesia may occur, but tremor is less common. It has been recognized that the presence of motor deficits, including gait difficulties, is associated with a more rapid decline in cognition and psychosis.[116]

Gross examination of the brain in AD typically shows widespread atrophy, most marked in association cortices of the frontal, temporal, and parietal lobes, with the medial portion of the temporal lobe most affected, including the hippocampus and adjacent entorrhinal cortex. Braak and Braak[117] have reported that changes first occur in the entorrhinal cortex, continue in to the hippocampus, and ultimately affect the neocortex. The basal ganglia and thalamus are usually unaffected. In AD, the substantia nigra is typically well pigmented, whereas the locus ceruleus shows loss of pigment.[114]

Consistent with the atrophy evident upon gross exam, microscopic examination reveals loss of neurons, gliosis, and loss of synapses.[114] The most conspicuous microscopic finding is that of abnormally phosphorylated tau protein, present in surviving neurons as paired helical filaments. These neurofibrillary tangles (NFTs) form within the cell body, while neuropil threads (NTs) are found in distal portions of dendritic processes. These are selectively found in cortical pyramidal projection cells.[118] Senile plaques are extracellular structures composed, in part, of amyloid and alpha-synuclein and are often associated with reactive glial changes. They are not specific to AD, as diffuse plaques have been found in nondemented elderly subjects.[119,120]

A number of quantification schemas have been developed, attempting to correlate pathologic changes with disease progression in AD. Braak and Braak[121] have described six stages with respect to the location of NFT-bearing neurons (entorhinal stages I-II: clinically silent; limbic stages III-IV: incipient dementia of the Alzheimer's type; neocortical stages V-VI: dementia of the Alzheimer's type). Other morphologic assessments include the Consortium to Establish a Registry for Alzheimer's Disease (CERAD), which designates cases as definite AD, probable AD, possible AD, and normal based on a combination of age-adjusted plaque scores and clinical history.[122] In another schema, the National Institute on Aging–Reagan Institute (NIA–Reagan) quantifies senile plaques as well as stressing topographic staging of neurofibrillary changes.[123] Newell et al.[124] applied the above measures to 84 brains, 33 with clinical and neuropathologically confirmed AD, 34 with non-AD dementing illnesses (including DLB and PSP), and 17 without neurodegenerative disease. Overall, the NIA–Reagan criteria correlated well with clinical history of AD as well as with the other measures but was more conservative in defining non-AD brains as AD than the other measures.

Historically, DLB has been defined as the onset of dementia within 12 months of parkinsonian symptoms. Clinical features include fluctuating cognitive impairment, variations in attentiveness and arousal, recurrent visual hallucinations, and spontaneous features of parkinsonism.[125] Once considered a rare entity, it is now believed to be the second most common degenerative dementia, estimated at 10 to 15% of cases.[125] A uniform set of pathological criteria has been harder to develop, due to extensive overlap with the pathological features of AD and PD. Consensus guidelines published in 1996 require only that brain stem and cortical Lewy bodies be present, although the authors also stressed that Lewy-related neuritis, AD pathology, and even spongiform changes are often present.[125] McKeith et al.[113] have pointed out that most patients thought to have DLB have AD pathology in the brain sufficient to meet CERAD criteria. Conversely, Lewy bodies occur in two-thirds of early-onset familial AD, sporadic AD, Down's syndrome, and Pick's disease. Thus, the relationship between AD and DLB is murky, and there does not appear to be a clear pathological delineation between the two.

PD is pathologically defined by neuronal loss, gliosis, and the presence of alpha-synuclein-immunopositive Lewy neurites (LNs) and Lewy body (LB) inclusions present in cellular processes and neuronal perikarya, respectively.[126] The most characteristic area of involvement is the substantia nigra, pars compacta, but extranigral pathology is also evident, including the dorsal motor nucleus of the glossopharyngeal and vagal nerves; subnuclei of the locus ceruleus, reticular formation, and raphe systems; and nuclei of the basal forebrain as well as subnuclei of the thalamus.[127] LBs are also seen in the cerebral cortex in layers V and VI of the limbic system, even in nondemented patients.[128] Brain stem and cortical LBs are morphologically distinct. LBs found in the brain stem are intraneuronal, round eosinophilic inclusions with a hyaline core and a pale peripheral halo, whereas cortical LBs are often less defined structures lacking a core and halo.[129]

Opinion regarding the pathological basis of dementia in PD has varied and basically falls into three camps: those who feel that brain stem pathology alone accounts for cognitive dysfunction, those who find that coexistent AD pathology is the salient finding, and those who propose cortical Lewy body pathology as the critical pathological change.[130] While the loss of nigrostriatal dopaminergic projection neurons are largely responsible for the motor signs seen in PD, they may also contribute to cognitive changes mainly through disruption of striatal/frontal connections.[131] Also, it has been seen that loss of cholinergic activity in the basal forebrain is greater (>70%) in patients with PDD as compared to PD patients without dementia.[132] Jellinger[133] has suggested that there may be a critical threshold of neuronal loss (65 to 80%) in the basal forebrain associated with cholinergic denervation to account for the presence of dementia in PD. It has also been proposed that degeneration of the thalamic nuclei involved in the limbic system may be responsible for not only cognitive dysfunction, but also autonomic and emotional disturbance in PD.[134]

Other investigators have contended that the presence of AD pathology is responsible for the occurrence of dementia in PD.[128,135–139] In one case series, 31 of 100 patients with pathologically confirmed PD had well documented dementia. Of these, 29% met pathologic criteria for AD, whereas the presence of Lewy bodies was less specific, being found to some degree in all cases, demented or not.[128] In a more recent study, regional NFT severity ratings were the best predictor of dementia in PD.[138] These findings have been corroborated by Jellinger et al.,[139] who reported a series of 200 consecutive autopsy cases of PD. Dementia was present in 33% and was highly correlated with AD pathology (97%).

Several recent investigations have suggested that LBs in the cortex and limbic system are the critical finding accounting for the occurrence of dementia in PD. Apayadin et al.[2] reviewed the postmortem examinations of 13 patients with a clinical diagnosis of PD in whom dementia developed at least 4 years after presentation of motor symptoms, and compared them with 9 PD patients not clinically defined as demented during their lifetime. Twelve of the 13 PDD patients showed a tenfold increase in neocortical and limbic median LB counts, compared with the nondemented group. One of the original 13 PDD patients was determined to have PSP, and one patient of the nine with PD met the criteria for intermediate probability of AD as defined by the NIA-Reagan criteria. Seven of the 12 patients with PDD were subsequently determined to have LB pathology primarily confined to the limbic areas with minimal neocortical change (so-called transitional Lewy body disease), with the remaining 5 showing widespread pathology (DLB). In another pathological analysis using recent advances in alpha-synuclein immunostaining to demonstrate the presence of LBs, Hurtig et al.[140] compared 22 demented and 20 nondemented PD patients. They found that the presence of cortical LBs were 91% sensitive and 90% specific for predicting dementia, while the finding of AD pathology by CERAD criteria was no better than 63% sensitive.

An interesting question is raised by the finding that senile plaques appear to be more frequently encountered in patients with a higher burden of neocortical LBs whether the clinical diagnosis is PDD or DLB.[141] Does the presence of tau pathology, associated with amyloid precursor protein and presenilin mutations, have an effect on the development of LBs?[142,143] Hardy[112] has made the assertion that perhaps it is the presence of amyloid pathology that may enhance cortical LB pathology. Although this theory may serve to integrate the seemingly dichotomous schools of thought regarding whether AD pathology or Lewy body pathology is responsible for dementia occurring in PD, to date, there are no reports demonstrating such.

In conclusion, as there are questions regarding the clinical features that differentiate AD, PDD, and DLB, so it is that the neuropathology also cannot fully define one entity from another. Further work is necessary to help elucidate the complex relationship between PDD and DLB, as well as AD with LB disorders.

NEUROCHEMISTRY OF DEMENTIA IN PD

Surprisingly, the neurochemical modifications that attend cognitive decline in PD have received limited direct exploration. Necropsy description of degeneration within select populations of neurons has provided the impetus to examine specific transmitter systems. While degeneration of dopaminergic neurons in the substantia nigra compacta (SNc) is the pathologic hallmark of PD, depletion within the lateral compartment, projecting to the putamen, accounts for characteristic motor dysfunction. Degeneration within the medial SNc, with projections primarily to the caudate nucleus, correlates with cognitive disturbance measured by a global scale.[144] In addition, diminished binding to the D_1 subtype of dopamine receptor in the caudate nucleus also correlates with the magnitude of cognitive dysfunction.[145] Cholinergic dysfunction, prominently involved in AD, is also observed in PDD. This is indicated by the finding that the affinity with which [^{3}H]quinuclidinylbenzilate (QNB) binds to muscarinic receptors is increased in the caudate nucleus and frontal cortex while receptor density is increased in the frontal cortex.[146] The activity of choline acetyltransferase (ChAT) is decreased in frontal cortex and hippocampus.[146] Diminished ChAT activity was also observed by Perry et al.[147] in frontal and temporal cortex with greater reduction in PDD than PD. Recent evaluation of ChAT also reported diminished activity in hippocampus, prefrontal cortex, and temporal cortex in PDD.[145] In this study, the density of LBs was negatively correlated with the degree of cognitive impairment.[145] Examination of ChAT activity in the basal nucleus of Meynert (nbM) revealed substantial diminution in PD relative to cortical regions.[148] In contrast, ChAT activity in AD was more substantially decreased in cerebral cortex than basal nuclei. These contrasting changes in cholinergic activity suggest distinct pathophysiologic modifications of cholinergic function in the cognitive disruption that attends these disorders.[148]

The role of other monoamine transmitters in PDD has received even less direct attention. Chan-Palay[149] has reported a reduction in norepinephrine (NE) synthesis and depletion of locus ceruleus (LC) neurons that correlates with "complex coexistent symptoms" that include depression and dementia. Concurrent depression and dementia was associated with the most significant alterations in NE content and neurons.[149] Recent examination of cell density in the LC, nbM, and SNc revealed the greatest depletion in the LC in PD and AD, suggesting that greater attention to noradrenergic function in these degenerative disorders is warranted.[150]

IMAGING OF DEMENTIA IN PD

In the evaluation and management of PD, neuroimaging tends to occupy an ancillary position. Widely available imaging modalities, as currently utilized, primarily examine anatomy and effectively evaluate for the presence of space-occupying lesions (i.e., malignancy, ischemia, and so on). Magnetic resonance (MR) imaging has largely supplanted computed tomography as the method of choice for this application. Functional imaging with positron emission tomography (PET) and single-photon emission computed tomography (SPECT) is not readily available and is less often used for diagnostic purposes. Refinements in MRI technology are expanding its application beyond simple assessment of structure. Imaging is being increasingly utilized to characterize the diverse manifestations of PD, their progression, and to determine the impact of treatment on disease natural history. These insights will hopefully lead to improved differentiation of the parkinsonian syndromes as well as contrast PDD with other causes of dementia (i.e., DLB and AD with parkinsonism).

Magnetic Resonance Imaging

Volumetrics

In the evaluation of dementia, the majority of the work has focused on AD. As a result, preliminary application to the study of PDD has embarked from this foundation. The extensive hippocampal pathology in AD prompted examination of this structure in PDD compared to PD, AD, vascular dementia, and matched controls.[151] While the volume of the hippocampus (measured by volumetric protocols) was diminished in all groups, this measure did not distinguish the etiology of cognitive decline. This observation was recently replicated by Camicioli et al.,[152] who also observed significant hippocampal atrophy in nondemented PD. No change was observed in parahippocampal gyrus or frontal, temporal, or parietooccipital cortex. Evaluation of the subcortical structures and the posterior fossa has revealed no difference between PD and matched controls.[153] Striatal and brain stem volume was found to be diminished in PSP and multiple system atrophy.[153] A comparison of AD to DLB and vascular dementia did not reveal significant change in caudate volume,[154] reinforcing the conclusion that volume alteration in this region does not discriminate the etiology of dementia. Reduction in the volume of the substantia innominata, encompassing the basal nucleus of Meynert, has been recently described. As was observed during the examination of the hippocampal formation, while PDD could be distinguished from matched control subjects, it was not possible to delineate distinct dementia etiologies.[155] A recent longitudinal examination of total brain volume and ventricular size in nondemented PD demonstrated more rapid decrease in both the absolute and relative rate of change in these measures.[156] Although PDD was not included, a significant correlation between volume parameters and performance and full-scale IQ was reported.

MAGNETIC RESONANCE SPECTROSCOPY

Magnetic resonance spectrosopy (MRS) is increasingly being utilized to examine metabolite content in discrete regions of brain through determination of n-acetyl aspartate (NAA), choline (CHO), lactate (LACT), and creatine (CR) in proton spectra. A significant increase in the LACT/NAA ratio has been reported in occipital cortex with the most prominent rise in PDD.[157] Enhanced glycolytic activity, reflected in the increased lactate content, is consistent with mitochondrial dysfunction as has been implicated in the pathogenesis of PD. Age, duration of disease, and medication exposure did not impact LACT/NAA.[157] Diminished NAA/CR has been reported in temporoparietal cortex in nondemented PD with the greatest change contralateral to maximum clinical impairment.[158] As CR content tends to change little, reduction in NAA/CR is believed to reflect decrease in the neuronal marker NAA. Whether this represents neuronal loss or dysfunction is a subject of speculation.[159] In this study, a significant correlation between the decrease in NAA/CR and measures of global cognitive dysfunction was also observed, independent of motor impairment.[158] Reduction in the absolute content of NAA has been reported in PDD occipital cortex with the magnitude of the decrease correlated to neuropsychological, but not motor, impairment.[160] Additional nuclei can be examined noninvasively with spectroscopy. Using phosphorus ([31P]) MRS bilateral increase in the inorganic phosphate (P_i)/β-ATP has been reported in temporoparietal cortex in nondemented PD.[161] Changes in the right hemisphere P_i/β-ATP correlate with reduction in performance IQ while left-sided increases correlated to full-scale and verbal IQ. Whether changes in oxidative metabolism reflect cortical pathology or loss of subcortical afferents awaits clarification.[161]

In addition to determination of metabolite content by MRS, change in blood flow/activation can be assessed with functional MRI (fMRI). Decrease in striatal and frontal signal intensity observed in PDD, not PD, during a working memory task suggested involvement of circuits distinct from those producing motor dysfunction.[162]

Magnetization transfer imaging is thought to provide information on the histological composition of tissues through determination of the exchange of magnetization between water and macromolecule bound protons. The magnetization transfer ratio (MTR), suggested to measure structural brain damage, is decreased in subcortical white matter in PDD. In contrast, the MTR is decreased in the globus pallidus and thalamus in PSP.[163]

SINGLE-PHOTON EMISSION COMPUTED TOMOGRAPHY

SPECT with technecium (^{99m}Tc)-HMPAO,[164,165] ^{123}Iodoampetamine (IMP),[166] and (^{99m}Tc)-ethyl cysteinate dimer (ECD)[167] has been commonly used to examine perfusion. While perfusion in PD was no different from controls, Spampinato et al.[164] described perfusion decreases in temporal, parietal, and occipital cortex in PDD. The reduction in blood flow reported by Firbank et al.[165] was more restricted (precuneus and infero-lateral parietal regions) and of a pattern similar to DLB. In addition to altered temporoparietal perfusion, Antonini et al.[167] identified a perfusion defect in frontal cortex in PDD. The pathogenesis of altered blood flow is not well understood but has been suggested to reflect cholinergic denervation. There has been limited use of SPECT to visualize the dopamine transporter in PDD. Comparison of PD and DLB revealed decreased striatal/cerebellar [^{123}I]- β-CIT relative to controls with the more profound reduction in DLB.[168] While diminished binding to the dopamine transporter in DLB suggests a role in cognitive decline, the implications of this observation await further study.

POSITRON EMISSION TOMOGRAPHY

As reported in other dementing disorders, cortical [^{18}F]deoxyglucose (FDG) uptake is diminished in PDD. Fairly widespread hypometabolism was reported in a single case, initially diagnosed with AD, corrected to PD at necropsy.[169] The larger series of Peppard et al.[170] also reported generalized hypometabolism with more prominent involvement of temporal cortex. Diminished [^{18}F]fluorodopa uptake in caudate and frontal cortex in PD is correlated with decreased performance on assessments of frontal lobe (executive) function (i.e., verbal fluency, working memory, and attention).[171] Compared to PD, more extensive basal ganglia involvement was observed in PDD, as well as extension into anterior cingulate cortex.[172] Correlation between dopaminergic denervation and frontal dysfunction has also been observed with [^{11}C]S-nomifensine labelling of the dopamine transporter.[173] Increased binding of [^{11}C]N-methyl-piperidyl benzilate (NMPB) to muscarinic cholinergic receptors in PD frontal cortex has also been reported.[174] While most participants (11 of 12) were cognitively intact, impaired performance on the Wisconsin Card Sorting Test suggested frontal lobe dysfunction. Up-regulation of muscarinic cholinergic receptors reflecting diminished activity of cholinergic projections from the basal forebrain may underlie these cognitive deficits. The activity of acetylcholinesterase, measured with [^{11}C]methylpiperidin-4-yl propionate (PMP) was reduced to a greater degree in PDD than PD or AD, consistent with cholinergic denervation in the pathogenesis of cognitive dysfunction.[175]

TREATMENT OF DEMENTIA IN PD

The treatment of dementia occurring in the setting of PD represents a major unmet need for patients and an ongoing source of frustration for clinicians. PD patients with dementia experience more severe side effects from their medications than those without dementia, including drug-induced psychosis.[23,176] In addition, patients with dementia often cannot tolerate adequate doses of medication to control their motor symptoms. Since the presence of dementia is considered a contraindication for surgical treatment of PD,[177] these patients do not even have the option of considering emerging therapies such as deep brain stimulation (DBS) for their disease. As a result, demented PD patients have a worse prognosis and a higher mortality than their nondemented counterparts.[178]

For many years, the treatment of dementia associated with PD consisted of merely shifting one's expectations regarding the risks and benefits of drug therapy and gradually reducing the doses of, and eventually eliminating, medications that could worsen cognitive function. Usually, adjunctive medications such as MAO-B inhibitors, dopamine agonists, and anticholinergic medications are eliminated first. Ultimately, doses of levodopa have to be reduced as well, at the expense of motor functioning. As patients worsen physically and cognitively over time, the main intervention has consisted of gradually increasing the amount of supervision and reducing the amount of freedom that patients have so as to promote their safety. All too often, this ultimately leads to nursing home placement.

One early controlled clinical trial attempted, unsuccessfully, to use phosphatidylserine to treat dementia in parkinsonian patients.[179] Sano et al.[180] used the nootropic (cognitive enhancer) piracetam to treat intellectual impairment in 20 PD patients in a double-blind trial and were equally unsuccessful. Marder et al.[25] have found that estrogen replacement therapy may be protective for the development of dementia in PD patients, but this has not been confirmed, and successful trials using estrogens to treat dementia in this setting have not been forthcoming. Although surgical treatments for PD such as pallidotomy, deep brain stimulation of the subthalamic nucleus, and dopaminergic cell transplant have not generally been associated with cognitive worsening in properly selected patients, there has also been no convincing evidence of cognitive improvements with these therapies.[177,181]

This is a grim picture and, until recently, there has been little cause for optimism. Lately, however, there has been growing interest in the use of cholinesterase inhibitors for dementia in PD, and positive results have been reported. These agents were originally developed for the treatment of AD, based on the finding that cholinergic transmission was reduced and cholinergic neurons depleted in the brains of AD patients.[182] Based on the results of well designed clinical trials, four agents are now approved for the treatment of AD—tacrine, donepezil, rivastigmine, and galantamine.[182] Both DLB and PD share pathological and neurochemical abnormalities with AD. In particular, deficits in cholinergic transmission and loss of cholinergic neurons in the medial forebrain have been demonstrated in all three disorders.[183–185] In fact, in DLB, the deficit may be more profound. These findings have provided the rationale for attempting to use cholinesterase inhibitors to treat dementia associated with parkinsonism. Initially, there was concern that these agents might worsen motor symptoms of parkinsonism. Because of the known benefit of anticholinergic agents in PD, it was thought that cholinesterase inhibitors might have the opposite effect neurochemically and clinically. Fortunately, this has not been the case in most early clinical trials.[186–189] In addition, no significant drug-drug interactions were found between donepezil and carbidopa/levodopa in a double-blind, pharmacokinetic and safety study.[190]

Unfortunately, not all reports have been satisfactory. One open-label trial using rivastigmine in 21 PD patients with dementia resulted in worsening of motor symptoms in six patients, even though the group, as a whole, did not experience a statistically significant worsening as measured by the UPDRS.[191] In another report of a single case, a patient with presumed DLBD was mistakenly given 25 mg of donepezil in 24 hr (2.5 times the recommended maximum dose) and experienced severe worsening of parkinsonian symptoms.[192] The patient's total UPDRS score increased from 27 to 96.

Building on the success of using cholinesterase inhibitors in AD, investigators also found that treatment with donepezil, rivastigmine, and galantamine was well tolerated and resulted in measurable cognitive improvements in patients with DLB.[186,193,194] However, most of these studies were small, uncontrolled trials. In one open trial, Kaufer et al.[195] found that the degree of improvement with donepezil in ten subjects with parkinsonism and dementia correlated with the degree of deficit in acetylcholinesterase activity as measured by PET. In the only large, double-blind trial, McKeith et al.[196] used rivastigmine over 20 weeks, in doses up to 12 mg per day, to treat 120 patients with DLBD. They reported significant improvements in several measures of neuropsychiatric and cognitive function compared to placebo. No motor worsening was seen.

Equally encouraging results have been reported in early trials looking at the utility of cholinesterase inhibitors in patients with idiopathic PD and dementia. In one open-label trial, Hutchinson et al.[188] reported dramatic improvements in hallucinations, mini-mental state scores, and even motor functioning in seven patients treated with tacrine up to 20 mg per day. These results have not been replicated in controlled clinical trials because of the risk of hepatic toxicity with this drug. Small open-label trials using donepezil and rivastigmine have also reported modest cognitive improvements.[187,191,197,198] In the only randomized, controlled trial, Aarsland et al.[189] treated 14 subjects over 20 weeks with donepezil up to 10 mg per day. Patients treated with donepezil experienced an average improvement in mini-mental state score of 2.1 points versus 0.3 points for placebo-treated patients. Again, no worsening of PD motor symptoms was seen, and donepezil was generally well tolerated.

With these results in mind, a rational approach to the treatment of dementia in PD would start with an investigation to rule out other coincident causes of cognitive dysfunction, such as metabolic or endocrine disturbance, and brain imaging to rule out structural causes such as chronic subdural hematoma, hydrocephalus, tumor, and multiple cerebral infarctions. True dementia must also be distinguished from pseudodementia, which might result from parkinsonian motor dysfunction, depression, or apathy. If the presence of true PD-associated dementia is established, then a review of the patient's medication regimen is in order. Anticholinergic medications such as trihexyphenidyl or amantadine are especially poorly tolerated in these patients and should be discontinued. However, caution should be used when stopping amantadine, since some patients may experience a paradoxical worsening of confusion.[199] Other adjunctive medication, such as dopamine agonists, MAO-B inhibitors, and COMT inhibitors, should also be reduced to the extent possible, depending on the degree of benefit individual patients obtain from these agents and the presence of other drug-induced side effects such as psychosis.

A trial of a cholinesterase inhibitor is reasonable and should be offered to any patient with cognitive dysfunction serious enough to impair quality of life. The best agent to use in this setting is still uncertain, as no head-to-head comparison trials have been reported in the literature. Although tacrine has been effective in this setting in open-label trials, its use has generally been abandoned secondary to the occurrence of serious liver dysfunction and the need for frequent blood testing. Thus, choosing between donepezil, rivastigmine, and galantamine will depend mostly on patient tolerance and ease of use. The clinically important side effects of these agents tend to be gastrointestinal, including nausea and vomiting, diarrhea, and anorexia. These problems tend to be most prominent with rivastigmine, and are least likely to occur with donepezil.[182] All three agents should be used with caution in patients taking high-dose nonsteriodal anti-inflammatory drugs or coumadin, because of the risk of gastrointestinal bleeding.[182] Donepezil would also seem easier to use, since it is dosed once per day, whereas the other two agents are taken twice per day. However, this advantage is likely to be inconsequential in most demented PD patients, since they are almost always taking other PD medications at least two to three times per day and will likely need supervision from a spouse or another caregiver to ensure that medications are taken reliably. If the first agent tried is not well tolerated, or if no clinical response is observed within three months, then switching to one of the other two medications is reasonable. We have certainly seen patients in our practice who seem to do much better on the second or third agent tried.

Unfortunately, there is no reliable data to guide the physician on how to use the cholinesterase inhibitors in the treatment of PD dementia over the long term. Since no data exist that support a neuroprotective or prognosis-modifying effect in PD, it would seem reasonable only to continue treatment in patients who experience a recognizable symptomatic benefit. In most cases, these drugs should be discontinued in patients who become severely demented, since these patients are more likely to experience side effects or drug interactions and less likely to retain a meaningful clinical benefit. There is open-label data suggesting that cholinesterase inhibitors may be useful in treating hallucinations and other symptoms of drug-induced psychosis in demented PD patients[197,198,200] Thus, it may be worthwhile continuing a cholinesterase inhibitor in some severely demented PD patients with psychotic symptoms to help control hallucinations and delusions, even if there is little evidence of specific cognitive benefits.

SUMMARY

Dementia is an all too common accompaniment to the physical features of PD that invariably leads to a significantly reduced quality of life. Although much has been learned about the nature of cognitive dysfunction in this setting, considerable controversy still exists regarding the causes and predictors of this often devastating complication. Fortunately, promising recent advances in the treatment of dementia may represent an important preliminary step in addressing this critical issue in the care of patients with PD.

ACKNOWLEDGMENT

This work was supported by the Albany Medical Center Parkinson's Research fund and the Riley Family Chair in Parkinson's Disease (SAF).

REFERENCES

1. Marder, K., Jacobs, D. M., Dementia, in Factor, S. A., Weiner, W. J., Eds., *Parkinson's Disease: Diagnosis and Clinical Management,* Demos Publishing, New York, 125–35, 2002.
2. Apaydin, H., Ahlskog, J. E., Parisi, J. E., Boeve, B. F., Dickson, D. W., Parkinson disease neuropathology: later-developing dementia and loss of the levodopa response, *Archives of Neurology,* 59:102–112, 2002.

3. Parkinson J., *An Essay on the Shaking Palsy,* Sherwood, Neely and Jones, London, 1817.

4. Brown, R. G., Marsden, C. D., How common is dementia in Parkinson's disease? *Lancet,* 1:1262–1265, 1984.

5. Taylor, A., Saint-Cyr, J. A., Lang, A. E., Dementia prevalence in Parkinson's disease, *Lancet,* 1:1037, 1985.

6. Elizan, T. S., Sroka, H., Maker, H., Smith, H., Yahr, M. D., Dementia in idiopathic Parkinson's disease: variables associated with its occurrence in 203 patients, *Journal of Neural Transmission,* 65:285–302, 1986.

7. Girotti, F., Soliveri, P., Carella, F., Piccolo, I., Caffarra, P., Musicco, M. et al., Dementia and cognitive impairment in Parkinson's disease, *Journal of Neurology, Neurosurgery, and Psychiatry,* 51:1498–1502, 1988.

8. Mayeux, R., Stern, Y., Rosenstein, R., Marder, K., Hauser, A., Cote, L. et al., An estimate of the prevalence of dementia in idiopathic Parkinson's disease, *Archives of Neurology,* 45:260–262, 1988.

9. Tison, F., Dartigues, J. F., Auriacombe, S., Letenneur, L., Boller, F., Alpérovitch, A., Dementia in Parkinson's disease: a population-based study in ambulatory and institutionalized individuals, *Neurology,* 45:705–708, 1995.

10. Aarsland, D., Tandberg, E., Larsen, J. P., Cummings, J. L., Frequency of dementia in Parkinson's disease, *Archives of Neurology,* 53:538–542, 1996.

11. Marder, K., Tang, M-X, Cote, L., Stern, Y., Mayeux, R., The frequency and associated risk factors for dementia in patients with Parkinson's disease, *Archives of Neurology.* 52:695–701, 1995.

12. Mayeux, R., Denaro, J., Hemenegildo, N., Marder, K., Tang, M-X., Cote, L. J. et al., A population-based investigation of Parkinson's disease with and without dementia: relationship to age and gender, *Archives of Neurology,* 49:492–497, 1992.

13. Martilla, R. J., Rinne, U. K., Dementia in Parkinson's disease, *Acta Neurologica Scandinavia,* 54:431–441, 1976.

14. Reid, W. G., The evolution of dementia in idiopathic Parkinson's disease: neuropsychological and clinical evidence in support of subtypes, *International Psychogeriatrics,* 4 Suppl., 2:147–160, 1992.

15. Aarsland, D., Anderson, K., Larsen, J. P., Lolk, A., Kragh-Sorensen, P., Prevalence and characteristics of dementia in Parkinson's disease: an 8-year prospective study, *Archives of Neurology,* 60:387–392, 2003.

16. Mayeux, R., Chen, J., Mirabello, E., Marder, K., Bell, K., Dooneief, G. et al., An estimate of the incidence of dementia in idiopathic Parkinson's disease, *Neurology,* 40:1513–1517, 1990.

17. Hughes, T. A., Ross, H. F., Musa, S., Bhattacherjee, S., Nathan, R. N., Mindham, R. H. et al., A 10-year study of the incidence of and factors predicting dementia in Parkinson's disease, *Neurology,* 54:1596–1602, 2000.

18. Aarsland, D., Andersen, K., Larsen, J. P., Lolk, A., Neilsen, H., Kragh-Sorensen, P., Risk of dementia in Parkinson's disease: a community based, prospective study, *Neurology,* 56:730–736, 2001.

19. Rajput, A. H., Offord, K. P., Beard, C. M., Kurland, L. T., A case-control study of smoking habits, dementia and other illnesses in idiopathic Parkinson's disease, *Neurology,* 37:226–232, 1987.

20. Tröster, A. I., Woods, S. P., Neuropsychological aspects of Parkinson's disease and parkinsonian syndromes, in Pahwa, R., Lyons, K. E., Koller, W. C., Eds., *Handbook of Parkinson's disease.* 3rd ed., Marcel Dekker, Inc., New York, 127–157, 2003.

21. Marder, K., Leung, D., Tang, M., Bell, K., Dooneief, G., Cote, L. et al., Are demented patients with Parkinson's disease accurately reflected in prevalence surveys? A survival analysis, *Neurology,* 41:1240–1243, 1991.

22. Louis, E. D., Marder, K., Cote, L., Tang, M., Mayeux, R., Mortality from Parkinson's disease, *Archives of Neurology,* 54:260–264, 1997.

23. Factor, S. A., Feustel, P. J., Friedman, J. H., Comella, C. L., Goetz, C. G., Kurlan, R. et al., Longitudinal outcome of Parkinson's disease patients with psychosis, *Neurology,* 60:1756–1761, 2003.

24. Hubble, J. P., Kurth, J. H., Glatt, S. L., Kurth, M. C., Schellenberg, G. D., Hassanein, R. E. et al., Gene-toxin interaction as a putative risk factor of Parkinson's disease with dementia, *Neuroepidemiology,* 17:96–104, 1998.

25. Marder, K., Tang, M-X., Alfaro, B., Mejia, H., Cote, L., Jacobs, D. et al., Postmenopausal estrogen use and Parkinson's disease with and without dementia, *Neurology,* 50:1141–1143, 1998.

26. Kawas, C., Resnick, S., Morrison, A., Brookmeyer, R., Corrada, M., Zonderman, A. et al., A prospective study of estrogen replacement therapy and the risk of developing Alzheimer's disease: the Baltimore longitudinal study of aging, *Neurology,* 48:1517–1521, 1997.

27. Tsuang, D. W., Bird, T. D., Genetics of dementia, *Medical Clinics of North America,* 86:591–614, 2002.

28. Zareparsi, S., Kaye, J., Camicioli, R., Grimslid, H., Oken, B., Litt, M. et al., Modulation of the age at onset of Parkinson's disease by apolipoprotein E genotypes, *Annals of Neurology,* 42:655–658, 1997.

29. Zareparsi, S., Camicioli, R., Sexton, G., Bird, T., Swanson, P., Kaye, J. et al., Age at onset of Parkinson disease and apolipoprotein E genotypes, *American Journal of Medical Genetics,* 107:156–161, 2002.

30. French Parkinson's Disease Study Group. Apolipoprotein E genotype in familial Parkinson's disease, *Journal of Neurology, Neurosurgery, and Psychiatry,* 63:394–395, 1997.

31. Golbe, L. I., Di Iorio, G., Sanges, G., Lazzarini, A. M., La Sala, S., Bonavita, V. et al., Clinical genetic analysis of Parkinson's disease in the Contursi kindred, *Annals of Neurology,* 40:767–775, 1996.

32. Khan, N. L., Graham, E., Critchley, P., Schrag, A. E., Wood, N. W., Lees, A. J. et al., Parkin disease: a phenotypic study of a large case series, *Brain,* 126:1279–1292, 2003.

33. Stern, Y., Marder, K., Tang, M-X., Mayeux, R., Antecedent clinical features associated with dementia in Parkinson's disease, *Neurology,* 43:1690–1692, 1993.

34. Levy, G., Schupf, N., Tang, M-X., Cote, L. J., Louis, E. D., Mejia, H. et al., Combined effect of age and severity

on the risk of dementia in Parkinson's disease, *Annals of Neurology,* 51:722–729, 2002.

35. Burn, D. J., Rowan, E. N., Minett, T., Sanders, J., Myint, P., Richardson, J. et al., Extrapyramidal features in Parkinson's disease with and without dementia and dementia with Lewy bodies: a cross-sectional comparative study, *Movement Disorders,* 18:884–889, 2003.

36. Starkstein, S. E., Bolduc, P. L., Mayberg, H. S., Preziosi, T. J., Robinson, R. G., Cognitive impairments and depression in Parkinson's disease: a follow up study, *Journal of Neurology, Neurosurgery, and Psychiatry,* 53:597–602, 1990.

37. Starkstein, S. E., Mayberg, H. S., Leiguarda, R., Preziosi, T. J., Robinson, R. G., A prospective longitudinal study of depression, cognitive decline, and physical impairments in patients with Parkinson's disease, *Journal of Neurology, Neurosurgery, and Psychiatry,* 55:377–382, 1992.

38. Ryder, K. A., Gontkovsky, S. T., McSwan, K. L., Scott, J. G., Bharucha, K. J., Beatty, W. W., Cognitive function in Parkinson's disease: association with anxiety but not depression, *Aging Neuropsychology and Cognition,* 9:77–84, 2002.

39. Levy, G., Jacobs, D. M., Tang, M-X., Cote, L. J., Louis, E. D., Alfaro, B. et al., Memory and executive function impairment predict dementia in Parkinson's disease, *Movement Disorders,* 17:1221–1226, 2002.

40. Pai, M-C., Chan, S-H., Education and cognitive decline in Parkinson's disease: a study of 102 patients, *Acta Neurologica Scandinavia,* 103:243–247, 2001.

41. Cummings, J. L., Depression and Parkinson's disease: a review, *American Journal of Psychiatry,* 149:443–454, 1992.

42. Tröster, A. I., Paolo, A. M., Lyons, K. E., Glatt, S. L., Hubble, J. P., Koller, W. C., The influence of depression on cognition in Parkinson's disease: a pattern of impairment distinguishable from Alzheimer's disease, *Neurology,* 45:672–676, 1995.

43. Tröster, A. I., Stalp, L. D., Paolo, A. M., Fields, J. A., Koller, W. C., Neuropsychological impairment in Parkinson's disease with and without depression, *Archives of Neurology,* 52:1164–1169, 1995.

44. Norman, S., Tröster, A. I., Fields, J. A., Brooks, R., Effects of depression and Parkinson's disease on cognitive functioning, *Journal of Neuropsychiatry and Clinical Neurosciences,* 14:31–36, 2002.

45. Molho, E. S., Psychosis and related problems, in Factor, S. A., Weiner, W. J., Eds., *Parkinson's Disease: Diagnosis and Clinical Management,* Demos Publishing, New York, 465–480, 2002.

46. Juncos, J. L., Jewart, R. D., Neparizde, N., Hanfelt, J., Long-term prognosis of hallucinating Parkinson's disease patients treated with quetiapine or clozapine [abstract], *Neurology,* 58:A435, 2002.

47. American Psychiatric Association. Diagnostic and statistical manual of mental disorders, 4th ed., Washington, D. C., American Psychiatric Association, 1994.

48. Folstein, M. F., Folstein, S. E., McHugh, P. R., Mini-Mental State: a practical method for grading the cognitive state of patients for the clinician, *Journal of Psychiatric Research,* 12:189–198, 1975.

49. Piccirilli, M., D'Alessandro, P., Finali, G., Piccinin, G. L., Neuropsychological follow-up of parkinsonian patients with and without cognitive impairment, *Dementia,* 5:17–22, 1994.

50. Cooper, J. A., Sagar, H. J., Jordan, N., Harvey, N. S., Sullivan, E. V., Cognitive impairment in early, untreated Parkinson's disease and its relationship to motor disability, *Brain,* 114:2095–2122, 1991.

51. Levin, B. E., Llabre, M. M., Weiner, W. J., Cognitive impairments associated with early Parkinson's disease, *Neurology,* 39:557–561, 1989.

52. Green, J., McDonald, W. M., Vitek, J. L., Evatt, M., Freeman, A., Haber, M. et al., Cognitive impairments in advanced PD without dementia, *Neurology,* 59:1320–1324, 2002.

53. Janvin, C., Aarsland, D., Larsen, J. P., Hugdahl, K., Neuropsychological profile of patients with Parkinson's disease without dementia, *Dementia and Geriatric Cognitive Disorders,* 15:126–131, 2003.

54. Taylor, A. E., Saint-Cyr, J. A., The neuropsychology of Parkinson's disease, *Brain and Cognition,* 28:281–296, 1995.

55. Elias, J. W., Treland, J. E., Executive function in Parkinson's disease and subcortical disorders, *Seminars in Clinical Neuropsychiatry,* 4:34–40, 1999.

56. Levin, B. E., Tomer, R., Rey, G. J., Cognitive impairments in Parkinson's disease, *Neurologic Clinics,* 10:471–485, 1992.

57. Bondi, M. W., Kaszniak, A. W., Bayles, K. A., Vance, K. T., Contributions of frontal system dysfunction to memory and perceptual abilities in Parkinson's disease, *Neuropsychology,* 7:89–102, 1993.

58. Owen, A. M., James, M., Leigh, P. N., Summers, B. A., Marsden, C. D., Quinn, N. P. et al., Fronto-striatal cognitive deficits at different stages of Parkinson's disease, *Brain,* 115:1727–1751, 1992.

59. Flowers, K. A., Robertson, C., The effect of Parkinson's disease on the ability to maintain mental set, *Journal of Neurology, Neurosurgery, and Psychiatry,* 48:517–529, 1985.

60. Alevriadou, A., Katsarou, Z., Bostantjopoulou, S., Kiosseoglou, G., Mentenopoulos, G., Wisconsin Card Sorting Test variables in relation to motor symptoms in Parkinson's disease, *Perceptual and Motor Skills,* 89:824–830, 1999.

61. Lees, A. J., Smith, E., Cognitive deficits in the early stages of Parkinson's disease, *Brain,* 106:257–270, 1983.

62. Tomer, R., Fishe, T., Giladi, N., Aharon-Peretz, J., Dissociation between spontaneous and reactive flexibility in early Parkinson's disease, *Neuropsychiatry, Neuropsychology, and Behavioral Neurology,* 15:106–112, 2002.

63. Taylor, A. E., Saint-Cyr, J. A., Lang, A. E., Frontal lobe dysfunction in Parkinson's disease: the cortical focus of neostriatal outflow, *Brain,* 109:845–883, 1986.

64. Farina, E., Gattellaro, G., Pomati, S., Magni, E., Perretti, A., Cannata, A. P. et al., Researching a differential

impairment of frontal functions and explicit memory in early Parkinson's disease, *European Journal of Neurology*, 7:259–267, 2000.

65. Cools, A. R., Van Den Bercken, J. H. L., Horstink, M. W. I., Van Spaendonck, K. P. M., Berger, H. J. C., Cognitive and motor shifting aptitude disorder in Parkinson's disease, *Journal of Neurology, Neurosurgery, and Psychiatry*, 47:443–453, 1984.

66. Richards, M., Cote, L. J., Stern, Y., Executive function in Parkinson's disease: set-shifting or set-maintenance? *Journal of Clinical and Experimental Neuropsychology*, 15:266–279, 1993.

67. Brown, R. G., Marsden, C. D., An investigation of the phenomenon of "set" in Parkinson's disease, *Movement Disorders*, 3:152–161, 1988.

68. Owen, A. M., Roberts, A. C., Hodges, J. R., Summers, B. A., Polkey, C. E., Robbins, T. W., Contrasting mechanisms of impaired attentional set-shifting in patients with frontal lobe damage or Parkinson's disease, *Brain*, 116:1159–1175, 1993.

69. Brown, R. G., Marsden, C. D., Internal versus external cues and the control of attention in Parkinson's disease, *Brain*, 111:323–345 1988.

70. Epker, M. O., Lacritz, L. H., Cullum, C. M., Comparative analysis of qualitative verbal fluency performance in normal elderly and demented populations, *Journal of Clinical and Experimental Neuropsychology*, 21:425–434, 1999.

71. Jacobs, D. M., Marder, K., Cote, L. J., Sano, M., Stern, Y., Mayeux, R., Neuropsychological characteristics of preclinical dementia in Parkinson's disease, *Neurology*, 45:1691–1696, 1995.

72. Piatt, A. L., Fields, J. A., Paolo, A. M., Koller, W. C., Tröster, A. I., Lexical, semantic, and action verbal fluency in Parkinson's disease with and without dementia, *Journal of Clinical and Experimental Neuropsychology*, 21:435–443, 1999.

73. Auriacombe, S., Grossman., M., Carvell, S., Gollomp, S., Stern, M. B., Hurtig, H. I., Verbal fluency deficits in Parkinson's disease, *Neuropsychology*, 7:182–192, 1993.

74. Matison, R., Mayeux, R., Rosen, J., Fahn, S., "Tip-of-the-tongue" phenomenon in Parkinson's Disease, *Neurology*, 32:567–70, 1982.

75. Randolph, C., Braun, A. R., Goldberg, T. E., Chase, T. N., Semantic fluency in Alzheimer's, Parkinson's, and Huntington's disease: dissociation of storage and retrieval failures, *Neuropsychology*, 7:82–88, 1993.

76. Tröster, A. I., Fields, J. A., Testa, J. A., Paul, R. H., Blanco, C. R., Hames, K. A. et al., Cortical and subcortical influences on clustering and switching in the performance of verbal fluency tasks, *Neuropsychologica*, 36:295–204, 1998.

77. Peran, P., Rascol, O., Demonet, J. F., Celsis, P., Nespoulous, J. F., Dubois, B. et al., Deficit of verb generation in nondemented patients with Parkinson's disease, *Movement Disorders*, 18:150–156, 2003.

78. Zec, R. F., Landreth, E. D., Fritz, S., Grames, E., Hasara, A., Fraizer, W. et al., A comparison of phonemic, semantic, and alternating word fluency in Parkinson's disease,

Archives of Clinical Neuropsychology, 14:255–264, 1999.

79. Berg, E., Bjornram, C., Hartelius, L., Laakso, K., Johnels, B., High-level language difficulties in Parkinson's disease, *Clinical Linguistics and Phonetics*, 17:63–80, 2003.

80. Taylor, A. E., Saint-Cyr, J. A., Lang, A. E., Memory and learning in early Parkinson's disease: evidence for a "frontal lobe syndrome," *Brain and Cognition*, 13:211–232, 1990.

81. Mohr, E., Juncos, J., Cox, C., Litvan, I., Fedio, P., Chase, T. N., Selective deficits in cognition and memory in high-functioning Parkinsonian patients, *Journal of Neurology, Neurosurgery, and Psychiatry*, 53:603–606, 1990.

82. Stefanova, E. D., Kostic, V. S., Ziropadja, L., Ocic, G. G., Markovic, M., Declarative memory in early Parkinson's disease: serial position learning effects, *Journal of Clinical and Experimental Neuropsychology*, 23:581–591, 2001.

83. Breen, E. K., Recall and recognition memory in Parkinson's disease, *Cortex*, 29:91–102, 1993.

84. Buytenhuijs, E. L., Berger, H. J. C., Van Spaendonck, K. P. M., Horstink, M. W. I. M., Borm, G. F., Cools, A. R., Memory and learning strategies in patients with Parkinson's disease, *Neuropsychologica*, 32:335–342, 1994.

85. Zakharov, V. V., Akhutina, T. V., Yakhno, N. N., Memory impairment in Parkinson's disease, *Neuroscience and Behavioral Physiology*, 31:157–163, 2001.

86. Cohen, N. J., Squire, L. R., Preserved learning and retention of pattern-analyzing skill in amnesia: dissociation of knowing how and knowing that, *Science*, 210:207–210, 1980.

87. Harrington, D. L., Haaland, K. Y., Yeo, R. A., Marder, E., Procedural memory in Parkinson's disease: impaired motor but not visuoperceptual learning, *Journal of Clinical and Experimental Neuropsychology*, 12:323–339, 1990.

88. Stefanova, E. D., Kostic, V. S., Ziropadja, L., Markovic, M., Ocic, G. G., Visuomotor skill learning on serial reaction time task in patients with early Parkinson's disease, *Movement Disorders*, 15:1095–1103, 2000.

89. Saint-Cyr, J. A., Taylor, A. E., Lang, A. E., Procedural learning and neostriatal dysfunction in man, *Brain*, 111:941–959, 1988.

90. Heindel, W. C., Salmon, D. P., Shults, C. W., Walicke, P. A., Butters, N., Neuropsychological evidence for multiple implicit memory systems: a comparison of Alzheimer's, Huntington's, and Parkinson's disease, *The Journal of Neuroscience*, 9:582–587, 1989.

91. Levin, B. E., Llabre, M. M., Reisman, S., Weiner, W. J., Sanchez-Ramos, J., Singer, C. et al., Visuospatial impairment in Parkinson's disease, *Neurology*, 41:365–369, 1991.

92. Montse, A., Pere, V., Carme, J., Francesc, V., Eduardo, T., Visuospatial deficits in Parkinson's disease assessed by Judgment of Line Orientation Test: error analyses and practice effects, *Journal of Clinical and Experimental Neuropsychology*, 23:592–598, 2001.

93. Brown, R. G., Marsden, C. D., Visuospatial function in Parkinson's disease, *Brain,* 109:987–1002, 1986.

94. Piccirilli, M., D'Alessandro, P., Finali, G., Piccinin, G., Early frontal impairment as a predictor of dementia in Parkinson's disease [letter], *Neurology,* 48:546–547, 1997.

95. Woods, S. P., Tröster, A. I., Prodromal frontal/executive dysfunction predicts incident dementia in Parkinson's disease, *Journal of the International Neuropsychological Society,* 9:17–24, 2003.

96. Mahieux, F., Fenelon, G., Flahault, A., Manifacier, M. J., Michelet, D., Boller, F., Neuropsychological prediction of dementia in Parkinson's disease, *Journal of Neurology, Neurosurgery, and Psychiatry,* 64:178–183, 1998.

97. Wechsler, D., Wechsler Adult Intelligence Scale-Revised, The Psychological Corporation, New York, 1981.

98. Stroop, J., Studies of interference in serial verbal reactions, *Journal of Experimental Psychology,* 18:643–662, 1935.

99. Cummings, J. L., Benson, D. F., Subcortical dementia: review of an emerging concept, *Archives of Neurology,* 41:874–879, 1984.

100. Mindham, R. H. S., Hughes, T. A., Cognitive impairment in Parkinson's disease, *International Review of Psychiatry,* 12:281–289, 2000.

101. Albert, M. L., Feldman, R. G., Willis, A. L., The "subcortical" dementia of progressive supranuclear palsy, *Journal of Neurology, Neurosurgery, and Psychiatry,* 37:121–130, 1974.

102. Mayeux, R., Stern, Y., Rosen, J., Benson, D. F., Is "subcortical dementia" a recognizable clinical entity? *Annals of Neurology,* 14:278–283, 1983.

103. Whitehouse, P. J., The concept of subcortical and cortical dementia: another look, *Annals of Neurology,* 19:1–6, 1986.

104. Boller, F., Mizutani, T., Roessmann, U., Gambetti, P., Parkinson's disease, dementia, and Alzheimer disease: clinicopathological correlations, *Annals of Neurology,* 7:329–335, 1980.

105. Yoshimura, M., Cortical changes in the parkinsonian brain: a contribution to the delineation of "diffuse Lewy body disease," *Journal of Neurology,* 229:17–32, 1983.

106. Pirozzolo, F. J., Hansch, E. C., Mortimer, J. A., Webster, D. D., Kuskowski, M. A., Dementia in Parkinson's disease: a neuropsychological analysis, *Brain and Cognition,* 1:71–83, 1982.

107. Piccirilli, M., Piccinin, G. L., Agostini, L., Characteristic clinical aspects of Parkinson patients with intellectual impairment, *European Neurology,* 23:44–50, 1984.

108. Huber, S. J., Shuttleworth, E. C., Paulson, G. W., Dementia in Parkinson's disease, *Archives of Neurology,* 43:987–990, 1986.

109. McFadden, L., Mohr, E., Sampson, M., Mendis, T., Grimes, J. D., A profile analysis of demented and non-demented Parkinson's disease patients, *Advances in Neurology,* 69:339–341, 1996.

110. Zakzanis, K. K., Freedman, M., A neuropsychological comparison of demented and nondemented patients with Parkinson's disease, *Applied Neuropsychology,* 6:129–146, 1999.

111. Lezak, M. D., Neuropsychological assessment. 3rd ed., Oxford University Press, New York, 1995.

112. Hardy, J., The relationship between Lewy body disease, Parkinson's disease and Alzheimer's disease, *Annals of the New York Academy of Science,* 991:167–170, 2003.

113. McKeith, I., Mintzer, J., Aarsland, D., Burn, D., Chiu, H., Cohen-Mansfield, J. et al., Dementia with Lewy bodies, *Lancet,* 3:19–28, 2004.

114. Dickson, D., Neuropathology of Alzheimer's disease and other dementias, *Clinics in Geriatric Medicine,* 17:209–228, 2001.

115. Reisberg, B., Clinical presentation, diagnosis, and symptomatology of age-associated cognitive decline and Alzheimer's disease, in Riesberg, B., Ed., *Alzheimer's Disease,* The Free Press, New York, 173–187. 1983.

116. Caligiuri, M. P., Peavey, G., Salmon, D. P., Galasco, D. R., Thal, L. J., Neuromotor abnormalities and risk for psychosis in Alzheimer's disease, *Neurology,* 61:954–958, 2003.

117. Braak, E., Braak, H., Alzheimer's disease: transiently developing dendritic changes in pyramidal cells of sector CA1 of the ammon's horn, *Acta Neuropathologica,* 93:323–325, 1997.

118. Braak, H., Braak, E., Demonstration of amyloid deposits and neurofibrillary changes in whole brain sections, *Brain Pathology,* 1:213–216, 1991.

119. Dickson, D., Pathogenesis of senile plaques, *Journal of Neuropathology and Experimental Neurology,* 56:321, 1997.

120. Giannalopoulus, P., Herrmann, F., Bussiere, T., Bouras, C., Kovari, E., Perl, D. et al., Tangle and neuron numbers, but not amyloid load, predict cognitive status in Alzheimer's disease, *Neurology,* 60:1495–1500, 2003.

121. Braak, H., Braak, E., Neuropathological staging of Alzheimer-related changes, *Acta Neuropathologica,* 82:239–259, 1991.

122. Mirra, S. S., Heyman, A., McKeel, D., Sumi, S. M., Crain, B. J., Brownlee, L. M. et al., The consortium to establish a registry for Alzheimer's disease (CERAD), Part II, Standardization of the neuropathologic assessment of Alzheimer's disease, Neurology, 41:479–486, 1991.

123. Jellinger, K., Bancher, C., Neuropathology of Alzheimer's Disease: a critical update, *Journal of Neural Transmission,* 54 Suppl:77–95, 1998.

124. Newell, K., Hyman, B., Growden, J., Hedley-Whyte, E., Application of the National Institute on Aging-Reagan Institute criteria for the neuropathologic diagnosis of Alzheimer's disease, *Journal of Neuropathology and Experimental Neurology,* 58:1147–1155, 1999.

125. McKeith, I. G., Galasko, D., Kosaka, K., Perry, E. K., Dickson, D. W., Hansen, L. A. et al., Consensus guidelines for the clinical and pathologic diagnosis of dementia with Lewy bodies (DLB): report of the consortium on DLB international workshop, *Neurology,* 47:1113–1124, 1996.

126. Forno, L. S., Neuropathology of Parkinson's disease, *Journal of Neuropathology and Experimental Neurology,* 55:259–272, 1996.

127. Damier, P., Hirsch, E. C., Agid, Y., Graybiel, A. M., The substantia nigra of the human brain. II. Patterns of loss of dopamine-containing neurons in Parkinson's disease, *Brain,* 122:1437–1448, 1999.

128. Hughes, A. J., Daniel, S. E., Blankson, S., Lee, A. J., A clinicopathological study of 100 cases of Parkinson's disease, *Archives of Neurology,* 50:140–148, 1993.

129. Galvin, J. E., Lee, V. M. Y., Baba, M., Mann, D. M., Dickson, D. W., Yamaguchi, H. et al., Monoclonal antibodies to purified cortical Lewy bodies recognize the mid-size neurofilament subunit, *Annals of Neurology,* 42:595–603, 1997.

130. Emre, M., What causes mental dysfunction in Parkinson's disease? *Movement Disorders,* 18 Suppl. 6:S63–S71, 2003.

131. DuBois, B., Pillon, B., Agid, Y., Deterioration of dopaminergic pathways and alterations in cognitive and motor functions, *Journal of Neurology,* 239 Suppl. 1:S9–S12, 1992.

132. McKeith, I., Burn, D., Spectrum of Parkinson's disease, Parkinson's disease with dementia and Lewy body dementia, *Neurology Clinics,* 18:865–902, 2000.

133. Jellinger, K. A., Structural basis of dementia in neurodegenerative disorders, *Journal of Neural Transmission,* 47:1–29, 1996.

134. Rub, U., Del Tredici, K., Schultz, C., Ghebremedhin, E., de Vos, R. A., Jansen, Steur E. et al., Parkinson's disease: the thalamic components of the limbic loop are severely impaired by alpha-synuclein immunopositive inclusion body pathology, *Neurobiology of Aging,* 23:245–254. 2002.

135. de Vos, R. A., Jansen, E. N., Stam, F. C., Ravid, R., Swaab, D. F., 'Lewy body disease': clinico-pathological correlations in 18 consecutive cases of Parkinson's disease with and without dementia, *Clinical Neurology and Neurosurgery,* 97:13–22, 1995.

136. Braak, H., Braak, E., Yilmazer, D., de Vos, R. A, Jansen, E. N., Bohl, J., New aspects of pathology in Parkinson's disease with concomitant incipient Alzheimer's disease, *Journal of Neural Transmission,* 48 Suppl:1–6, 1996.

137. Jellinger, K. A., Morphological substrates of dementia in parkinsonism. A critical update, *Journal of Neural Transmission,* 51 Suppl:57–82, 1997.

138. SantaCruz, K., Pahwa, R., Lyons, K., Troster, A., Handler, M., Koller, W. C. et al., Lewy body, neurofibrillary tangle and senile plaque pathology in Parkinson's disease patients with and without dementia, *Neurology,* 52 Suppl. 2:A476, 1999.

139. Jellinger, K. A., Seppi, K., Wenning, G. K., Poewe, W., Impact of coexistent Alzheimer pathology on the natural history of Parkinson's disease, *Journal of Neural Transmission,* 109:329–339, 2002.

140. Hurtig, H. I., Trojanowski, J. Q., Galvin, J., Ewbank, D., Schmidt, M. L., Lee, V. M. Y. et al., Alpha-synuclein cortical Lewy bodies correlate with dementia in Parkinson's disease, *Neurology,* 54:1916–1921, 2000.

141. Kosaka, K., Yoshimura, K., Ikeda, K., Budka, H., Diffuse type of Lewy body disease: progressive dementia with abundant cortical Lewy bodies and senile changes of varying degree—a new disease? *Clinical Neuropathology,* 3:185–192, 1994.

142. Hardy, J., Gwinn-Hardy, K., Neurodegenerative disease: a different view of diagnosis, *Molecular Medicine Today,* 5:514–517, 1996.

143. Duda, J., Concurrence of alpha-synuclein and tau brain pathology in the Contursi kindred, *Acta Neuropathologica,* 104:7–11, 2002.

144. Rinne, J. O., Rummukainen, J., Paljarvi, L., Rinne, U. K., Dementia in Parkinson's disease is related to neuronal loss in the medial substantia nigra, *Annals of Neurology,* 26:47–50, 1989.

145. Mattila, P. M., Roytta, M., Lonnberg, P., Marjamaki, P., Helenius, H., Rinne, J. O., Choline acetyltransferase activity and striatal dopamine receptors in Parkinson's disease in relation to cognitive impairment, *Acta Neuropathologica,* 102:160–166, 2001.

146. Ruberg, M., Ploska, A., Javoy-Agid, F., Agid, Y., Muscarinic binding and choline acetyltransferase activity in Parkinsonian subjects with reference to dementia, *Brain Research,* 232:129–139, 1982.

147. Perry, R. H., Tomlinson, B. E., Candy, J. M., Blessed, G., Foster, J. F., Bloxham, C. A. et al., Cortical cholinergic deficit in mentally impaired Parkinsonian patients, *Lancet,* 2:789–790, 1983.

148. Candy, J. M., Perry, R. H., Perry, E. K., Irving, D., Blessed, G., Fairbairn, A. F. et al., Pathological changes in the nucleus of Meynert in Alzheimer's and Parkinson's diseases, *Journal of the Neurological Sciences,* 59:277–289, 1983.

149. Chan-Palay, V., Depression and dementia in Parkinson's disease. Catecholamine changes in the locus ceruleus, a basis for therapy, *Advances in Neurology,* 60:438–446, 1993.

150. Zarow, C., Lyness, S. A., Mortimer, J. A., Chui, H. C., Neuronal loss is greater in the locus coeruleus than nucleus basalis and substantia nigra in Alzheimer and Parkinson diseases, *Archives of Neurology,* 60:337–341, 2003.

151. Laakso, M. P., Partanen, K., Riekkinen, P., Lehtovirta, M., Helkala, E-L., Hallikainen, M. et al., Hippocampal volume changes in Alzheimer's disease, Parkinson's disease with and without dementia, and in vascular dementia: an MRI study, *Neurology,* 46:678–681, 1996.

152. Camicioli, R., Moore, M. M., Kinney, A., Corbridge, E., Glassberg, K., Kaye, J. A., Parkinson's disease is associated with hippocampal atrophy, *Movement Disorders,* 18:784–790, 2003.

153. Schulz, J. B., Skalej, M., Wedekind, D., Luft, A. R., Abele, M., Voigt, K. et al., Magnetic resonance imaging-based volumetry differentiates idiopathic Parkinson's syndrome from multiple system atrophy and progressive supranuclear palsy, *Annals of Neurology,* 45:65–74, 1999.

154. Barber, R., McKeith, I., Ballard, C., O'Brien, J., Volumetric MRI study of the caudate nucleus in patients with dementia with Lewy bodies, Alzheimer's disease, and

vascular dementia, *Journal of Neurology, Neurosurgery, and Psychiatry,* 72:406–407, 2002.

155. Hanyu, H., Asano, T., Sakurai, H., Tanaka, Y., Takasaki, M., Abe, K., MR analysis of the substantia innominata in normal aging, Alzheimer disease, and other types of dementia, *American Journal of Neuroradiology,* 23:27–32, 2002.

156. Hu, M. T., White, S. J., Chaudhuri, K. R., Morris, R. G., Bydder, G. M., Brooks, D. J., Correlating rates of cerebral atrophy in Parkinson's disease with measures of cognitive decline, *Journal of Neural Transmission,* 108:571–580, 2001.

157. Bowen, B. C., Block, R. E., Sanchez-Ramos, J., Pattany, P. M., Lampman, D. A., Murdoch, J. B. et al., Proton MR spectroscopy of the brain in 14 patients with Parkinson disease, *American Journal of Neuroradiology,* 16:61–68, 1995.

158. Hu, M. T., Taylor-Robinson, S. D., Chaudhuri, K. R., Bell, J. D., Morris, R. G., Clough, C. et al., Evidence for cortical dysfunction in clinically non-demented patients with Parkinson's disease: a proton MR spectroscopy study, *Journal of Neurology, Neurosurgery, and Psychiatry,* 67:20–26, 1999.

159. Chard, D. T., Griffin, C. M., McLean, M. A., Kapeller, P., Kapoor, R., Thompson, A. J. et al., Brain metabolite changes in cortical grey and normal-appearing white matter in clinically early relapsing-remitting multiple sclerosis, *Brain,* 125:2342–2352, 2002.

160. Summerfield, C., Gomez-Anson, B., Tolosa, E., Mercader, J. M., Marti, M. J., Pastor, P. et al., Dementia in Parkinson disease: a proton magnetic resonance spectroscopy study, *Archives of Neurology,* 59:1415–1420, 2002.

161. Hu, M. T., Taylor-Robinson, S. D., Chaudhuri, K. R., Bell, J. D., Labbe, C., Cunningham, V. J. et al., Cortical dysfunction in non-demented Parkinson's disease patients: a combined (31)P-MRS and (18)FDG-PET study, *Brain,* 123:340–352, 2000.

162. Lewis, S. J., Dove, A., Robbins, T. W., Barker, R. A., Owen, A. M., Cognitive impairments in early Parkinson's disease are accompanied by reductions in activity in frontostriatal neural circuitry, *Journal of Neuroscience,* 23:6351–6356, 2003.

163. Hanyu, H., Asano, T., Sakurai, H., Takasaki, M., Shindo, H., Abe, K., Magnetisation transfer measurements of the subcortical grey and white matter in Parkinson's disease with and without dementia and in progressive supranuclear palsy, *Neuroradiology,* 43:542–546, 2001.

164. Spampinato, U., Habert, M. O., Mas, J. L., Bourdel, M. C., Ziegler, M., de Recondo, J. et al., (99mTc)-HM-PAO SPECT and cognitive impairment in Parkinson's disease: a comparison with dementia of the Alzheimer type, *Journal of Neurology, Neurosurgery, and Psychiatry,* 54:787–792, 1991.

165. Firbank, M. J., Colloby, S. J., Burn, D. J., McKeith, I. G., O'Brien, J. T., Regional cerebral blood flow in Parkinson's disease with and without dementia, *Neuroimage,* 20:1309–1319, 2003.

166. Sawada, H., Udaka, F., Kameyama, M., Seriu, N., Nishinaka, K., Shindou, K. et al., SPECT findings in Parkinson's disease associated with dementia, *Journal of Neurology, Neurosurgery, and Psychiatry,* 55:960–963, 1992.

167. Antonini, A., De Notaris, R., Benti, R., De Gaspari, D., Pezzoli, G., Perfusion ECD/SPECT in the characterization of cognitive deficits in Parkinson's disease, *Neurological Science,* 22:45–46, 2001.

168. Ransmayrl, G., Seppi, K., Donnemiller, E., Luginger, E., Marksteiner, J., Riccabona, G. et al., Striatal dopamine transporter function in dementia with Lewy bodies and Parkinson's disease, *European Journal of Nuclear Medicine,* 28:1523–1528, 2001.

169. Schapiro, M. B., Pietrini, P., Grady, C. L., Ball, M. J., DeCarli, C., Kumar, A. et al., Reductions in parietal and temporal cerebral metabolic rates for glucose are not specific for Alzheimer's disease, *Journal of Neurology, Neurosurgery, and Psychiatry,* 56:859–864, 1993.

170. Peppard, R. F., Martin, W. R., Carr, G. D., Grochowski, E., Schulzer, M., Guttman, M. et al., Cerebral glucose metabolism in Parkinson's disease with and without dementia, *Archives of Neurology,* 49:1262–1268, 1992.

171. Rinne, J. O., Portin, R., Ruottinen, H., Nurmi, E., Bergman, J., Haaparanta, M. et al., Cognitive impairment and the brain dopaminergic system in Parkinson disease: [18F]fluorodopa positron emission tomographic study, *Archives of Neurology,* 57:470–475, 2000.

172. Ito, K., Nagano-Saito, A., Kato, T., Arahata, Y., Nakamura, A., Kawasumi, Y. et al., Striatal and extrastriatal dysfunction in Parkinson's disease with dementia: a 6-[18F]fluoro-L-dopa PET study, *Brain,* 125:1358–1365, 2002.

173. Marie, R. M., Barre, L., Dupuy, B., Viader, F., Defer, G., Baron, J. C., Relationships between striatal dopamine denervation and frontal executive tests in Parkinson's disease, *Neuroscience Letters,* 260:77–80, 1999.

174. Asahina, M., Suhara, T., Shinotoh, H., Inoue, O., Suzuki, K., Hattori, T., Brain muscarinic receptors in progressive supranuclear palsy and Parkinson's disease: a positron emission tomographic study, *Journal of Neurology, Neurosurgery, and Psychiatry,* 65:155–163, 1998.

175. Bohnen, N. I., Kaufer, D. I., Ivanco, L. S., Lopresti, B., Koeppe, R. A., Davis, J. G. et al., Cortical cholinergic function is more severely affected in parkinsonian dementia than in Alzheimer disease: an *in vivo* positron emission tomographic study, *Archives of Neurology,* 60:1745–1748, 2003.

176. Holroyd, S., Currie, L., Wooten, G. F., Prospective study of hallucinations and delusions in Parkinson's disease, *Journal of Neurology, Neurosurgery, and Psychiatry,* 70:734–738, 2001.

177. Saint-Cyr, J. A., Trepanier, L. L., Kumar, R., Lozano, A. M., Lang, A. E., Neuropsychological consequences of chronic bilateral stimulation of the subthalamic nucleus in Parkinson's disease, *Brain,* 123:2091–2108, 2000.

178. Levy, G., Tang, M-X., Louis, E. D., Cote, L. J., Alfaro, B., Mejia, H. et al., The association of incident dementia with mortality in PD, *Neurology,* 59:1708–1713, 2002.

179. Funfgeld, E. W., Baggen, M., Nedwidek, P., Richstein, B., Mistlberger, G., Double-blind study with phosphati-

dylserine (PS) in parkinsonian patients with senile dementia of the Alzheimer's type (SDAT), *Progress in Clinical and Biological Research,* 317:1235–1246, 1989.

180. Sano, M., Stern, Y., Marder, K., Mayeux, R., A controlled trial of piracetam in intellectually impaired patients with Parkinson's disease, *Movement Disorders,* 5:230–234, 1990.

181. Trott, C. T., Fahn, S., Greene, P., Dillon, S., Winfield, H., Winfield, L. et al., Cognition following bilateral implants of embryonic dopamine neurons in PD: a double blind study, *Neurology,* 60:1938–1943, 2003.

182. Knopman, D., Pharmacotherapy for Alzheimer's disease: 2002, *Clinical Neuropharmacology,* 26:93–101, 2003.

183. Nakano, I., Hirano, A., Parkinson's disease: neuron loss in the nucleus basalis without concomitant Alzheimer's disease, *Annals of Neurology,* 15:415–418, 1984.

184. Dubois, B., Pillon, B., Lhermitte, F., Agid, Y., Cholinergic deficiency and frontal dysfunction in Parkinson's disease, *Annals of Neurology,* 28:117–121, 1990.

185. Perry, R. H., Irving, D., Blessed, G., Fairbairn, A., Perry, E. H., Senile dementia of Lewy body type. A clinically and neuropathologically distinct form of Lewy body dementia in the elderly, *Journal of the Neurolgical Sciences,* 95:119–139, 1990.

186. McKeith, I. G., Grace, J. B., Walker, Z., Byrne, E. J., Wilkinson, D., Stevens, T. et al., Rivastigmine in the treatment of dementia with Lewy bodies: preliminary findings from an open trial, *International Journal of Geriatric Psychiatry,* 15:387–392, 2000.

187. Werber, E. A., Rabey, J. M., The beneficial effect of cholinersterase inhibitors on patients suffering from Parkinson's disease and dementia, *Journal of Neural Transmission,* 108:1319–1325, 2001.

188. Hutchinson, M., Fazzini, E., Cholinesterase inhibition in Parkinson's disease [letter], *Journal of Neurology, Neurosurgery, and Psychiatry,* 61:324–325, 1996.

189. Aarsland, D., Laake, K., Larsen, J. P., Janvin, C., Donepezil for cognitive impairment in Parkinson's disease: a randomized controlled study, *Journal of Neurology, Neurosurgery, and Psychiatry,* 72:708–712, 2002.

190. Okereke, C. S., Kumar, D., Cullen, E. I., Pratt, R. D., Hahne, F., Pharmacokinetics and safety of concurrent donepezil HCL on levodopa/carbidopa administration in subjects with Parkinson's disease [abstract], *Neurology,* 56 Suppl. 3:A457, 2001.

191. Giladi, N., Shabtai, H., Benbunan, B., Gurevich, T., Anca, M., Sidis, S. et al., The effect of treatment with rivastigmine (Exelon) on cognitive functions of patients with dementia and Parkinson's disease [abstract], *Neurology,* 56 Suppl. 3:A128, 2001.

192. Onofrj, M., Thomas, A., Severe worsening of parkinsonism in Lewy body dementia due to donepezil, *Neurology,* 61:1452, 2003.

193. Shea, C., MacKnight, C., Rockwood, K., Donepezil for treatment of dementia with Lewy bodies: a case series of nine patients, *International Psychogeriatrics,* 10:229–238, 1998.

194. Edwards, K., Hershey, L., Lichter, D., Bednarczyk, E., Results of a 24-week, open-label, flexible-dose trial to assess the safety and efficacy of galantamine in patients with dementia with Lewy bodies [abstract], *Neurology,* 60:A193–194, 2003.

195. Kaufer, D. I., Bohnen, N. I., Lopresti, B., Mathis, C. A., Moore, R. Y., DeKosky, S. T. et al., PET acetylcholinesterase imaging correlates with clinical response to donepezil treatment in parkinsonian dementia [abstract], *Neurology,* 60 Suppl. 1:A414, 2003.

196. McKeith, I., Del Ser, T., Spano, P., Emre, M., Wesnes, K., Anand, R. et al., Efficacy of rivastigmine in dementia with Lewy bodies: a randomised, double-blind, placebo-controlled international study, *Lancet,* 356:2031–2036, 2000.

197. Reading, P. J., Luce, A. K., McKeith, I. G., Rivastigmine in the treatment of parkinsonian psychosis and cognitive impairment: preliminary findings from an open trial, *Movement Disorders,* 16:1171–1195, 2001.

198. Bullock, R., Cameron, A., Rivastigmine for the treatment of dementia and visual hallucinations associated with Parkinson's disease: a case series, *Current Medical Research and Opinion,* 18:258–264, 2002.

199. Factor, S. A., Molho, E. S., Brown, D. L., Acute delirium after withdrawals of amantadine in Parkinson's disease, *Neurology,* 50:1456–1458, 1998.

200. Bergman, J., Lerner, V., Successful use of donepezil for the treatment of psychotic symptoms in patients with Parkinson's disease, *Clinical Neuropharmacology,* 25:107–110, 2002.

34 Animal Models of Basal Ganglia Injury and Degeneration and Their Application to Parkinson's Disease Research

Giselle M. Petzinger and Michael W. Jakowec
Department of Neurology, Keck School of Medicine, University of Southern California

CONTENTS

INTRODUCTION

Starting in the late nineteenth century, neuroscientists began generating lesions using surgical, thermal, electrolytic, and toxicant means for scientific study of the effect of central nervous system injury on brain function. In addition to identifying anatomical regions of interest, scientists using early lesioning techniques began to recognize the behavioral similarities to known human neurological disorders and began to recognize the value of animal models. Since then, many new strategies have been identified to develop animal models with the hope of gaining insights into disease mechanism(s). It was hoped that through such knowledge of disease mechanisms, therapeutic modalities would emerge. Experience has now shown that an animal model need not replicate all neurological (i.e., behavioral, neurochemical, neuropathological) features of the human condition. In general, however, the overall utility of an animal model for a particular disease is often dependent on how closely the model replicates all or part of the human condition.

In Parkinson's disease (PD) and related parkinsonian disorders, there now exist a variety of animal models, including new invertebrate models, each of which makes a unique contribution to our understanding of the human condition. These models have been developed in a wide variety of species (i.e., mouse, rat, cat, and nonhuman primate and many other species including *Drosophila*) using a variety of techniques including:

1. Surgical lesioning
2. Administration of pharmacological agents
3. Administration of neurotoxicants
4. Development of transgenic animals

While many of these models are not identical to the human condition with respect to behavioral characteristics, brain anatomy, or disease progression, they have provided valuable insight into disease mechanism and have led to significant advancements in therapeutic treatment of movement disorders such as PD.

PD is characterized by bradykinesia, rigidity, postural instability, and resting tremor.[1] The primary pathological and biochemical features of PD are the loss of nigrostriatal dopaminergic neurons in the substantia nigra pars compacta (SNpc), the appearance of intracellular inclusions called Lewy bodies, and the depletion of striatal dopamine. Clinical features are apparent when striatal dopamine depletion reaches 80%, despite the fact that 40 to 60% of nigrostriatal dopaminergic neurons still remain.[2,3] Since the destruction of the nigrostriatal system and consequent depletion of striatal dopamine are key features in the human condition, attempts have been made in animal models to disrupt an analogous anatomical area through surgical, pharmacological, or neurotoxicant manipulation.

The purpose of this chapter is to provide a brief overview of the many different vertebrate and invertebrate models utilized in PD and basal ganglia research. Each model, when applicable, will be discussed with respect to its development, behavioral profile, biochemical and neuropathological alterations, and contribution to the field. Recently published reviews may provide additional details and complement references in this review.[4–13]

PHARMACOLOGICAL-INDUCED MODELS OF PARKINSON'S DISEASE

RESERPINE

The first animal model for PD was demonstrated by Carlsson in the 1950s using rabbits treated with reserpine.[14] Reserpine is a catecholamine-depleting agent that blocks vesicular storage of monoamines. The akinetic state, a consequence of reserpine-induced dopamine depletion in the caudate nucleus and putamen, led Carlsson to make the link that PD was the result of striatal dopamine depletion. This speculation was supported by the discovery of striatal dopamine depletion in postmortem brain tissue of PD patients and led to the subsequent use of L-dopa (L-dihydoxyphenylalanine) for symptomatic treatment of PD.[15–18] Thus, the conclusions derived from the behavioral and neurochemical features of this animal model led to the development of a clinical therapy for PD in the form of L-dopa replacement. In fact, L-dopa (in conjunction with carbidopa, a peripheral dopa-decarboxylase inhibitor) still remains the gold standard of treatment, 40 years later. For this contribution, Arvid Carlsson was awarded the Nobel Prize in Physiology or Medicine in 2000 along with Eric Kandel and Paul Greengard.

α-METHYL-para-TYROSINE (AMPT)

Although less commonly used, α-methyl-para-tyrosine (AMPT), like reserpine, serves as an effective catecholamine-depleting agent.[19] AMPT (sometimes used in combination with reserpine) directly inhibits tyrosine hydroxylase, the rate-limiting enzyme in dopamine biosynthesis. Thus, the nascent synthesis of dopamine in neurons of the substantia nigra pars compacta (SNpc) and ventral tegmental area (VTA) is prevented leading to the depletion of dopamine stores.

Both reserpine and AMPT have been used to discover new dopaminomimetics for the treatment of PD, but since their effects are transient (lasting only hours to days), these models are primarily useful for acute studies. In addition, neither agent can duplicate the extensive biochemical nor pathological changes seen in PD. Consequently, other models with long-lasting behavioral alterations have been sought using site-specific neurotoxicant injury directed toward the dopamine producing neurons of the SNpc.

NEUROTOXICANT-INDUCED MODELS OF PARKINSON'S DISEASE

6-Hydroxydopamine (6-OHDA)

6-Hydroxydopamine (6-OHDA or 2,4,5-trihydroxyphenyl-ethylamine), first described in 1959, is a specific catecholaminergic neurotoxin structurally analogous to both dopamine and noradrenaline.[20] Initial studies in cell culture and animals demonstrated the toxic effects of 6-OHDA on a variety of cell types including dopaminergic neurons.[21,22] Acting as a "false-substrate" 6-OHDA is rapidly accumulated in catecholaminergic neurons. The mechanism of 6-OHDA toxicity (both *in vivo* and *in vitro*) is complex and involves a number of biochemical features including (1) alkylation, (2) rapid auto-oxidation (leading to the generation of hydrogen peroxide, superoxide, and hydroxyl radicals), and (3) impairment of mitochondrial energy production.[12,23] Despite the toxicity of 6-OHDA, it is unable to cross the blood brain barrier. Therefore, to achieve toxicity in the brain, it is necessary to deliver 6-OHDA directly using stereotaxic surgery. The 6-OHDA induced rat model of PD was initially developed by Ungerstedt in 1968, using stereotaxic bilateral intracerebral injections into the substantia nigra or lateral hypothalamus (targeting the medial forebrain bundle).[24] The bilateral administration of 6-OHDA leads to catalepsy, generalized inactivity, aphagia and adipsia, and a high degree of animal morbidity and mortality. Consequently, the administration of 6-OHDA was modified to a unilateral intracerebral lesion (targeting the substantia nigra and/or medial forebrain bundle). With unilateral lesioning there was (1) minimal post-operative morbidity, (2) behavioral asymmetry, and (3) a nonlesioned side to serve as a control.[25,26]

An additional modification of 6-OHDA administration was the regimen of chronic low dose striatal injections. This led to progressive dopaminergic cell death that more closely resembled the human condition.[27] Similar to many other neurotoxicants (including MPTP, methamphetamine, and rotenone), the administration of 6-OHDA may lead to varied degrees of lesioning, molecular and morphological changes, and behavioral results depending on a number of parameters including the amount administered, target, rodent strain, gender, and age. For example, male rats are more susceptible to 6-OHDA than female rats, suggesting that estrogen may be neuroprotective.[28,29]

A distinctive behavioral feature of the unilateral lesion model is rotation that occurs either spontaneously or that which can be induced following pharmacological administration of drugs affecting dopamine neurotransmission.[30–32] This motor feature is due to asymmetry in dopaminergic neurotransmission between the lesioned and intact sides. Specifically, animals rotate away from the side of greater dopaminergic activity. Nomenclature describes the direction of rotation as either ipsilateral or contralateral with respect to the lesioned side. Initial reports of rotation examined both spontaneous and pharmacologically manipulated rotation. Spontaneous rotation consists of ipsilateral rotation (toward the lesioned side), while pharmacologically-induced rotation may be either contra- or ipsilateral. For example, apomorphine and other dopamine agonists induce contralateral rotation (away from the lesioned side). This is due to their direct action on "super-sensitized" dopaminergic receptors on the lesioned side. Conversely, d-amphetamine phenylisopropylamine (AMPH) induces ipsilateral rotation by blocking dopamine reuptake and increasing dopamine receptor activity on the nonlesioned side. In general, greater than 80% depletion of dopamine is necessary to manifest rotation in this model.[19,31] Circling behavior can be measured either by observation or by specially designed bowls called *rotometers*. The rate of rotation correlates with the severity of the lesion, and animals with more extensive striatal dopamine depletions are less likely to show behavioral recovery. This simple model of rotation away from the side with the most dopamine receptor occupancy has recently proven much more complex and less predictable than previously thought, especially in the context of various pharmacological treatments and neuronal transplantation. In addition to rotation, other behavioral assessments in the 6-OHDA model may include tests of (1) forelimb use, (2) bilateral tactile stimulation, (3) single-limb akinesia, (4) bracing, and (5) reaction time. Several reviews discussing behavior in this model have been published.[11,30,31,33–36]

The 6-OHDA-lesioned rat model has proven to be a valuable tool in evaluating the (1) pharmacological action of new drugs on the dopaminergic system, (2) mechanisms of motor complications, (3) neuroplasticity of the basal ganglia in response to nigrostriatal injury, and (4) the safety and efficacy of neuronal transplantation in PD. Extensive pharmacological studies have utilized the 6-OHDA-lesioned rat to investigate the roles of various dopamine receptor (D_1 through D_5) agonists and antagonists, and other neurotransmitter systems on the modulation of dopamine neurotransmission (including glutamate, adenosine, nicotine, or opioids).

These studies can elucidate the influence of these compounds on electrophysiological, behavioral, and molecular (signal transduction) properties of the basal ganglia. Discussions of the extensive pharmacological literature employing the 6-OHDA model can be found in several reviews.[11,12,30–32,37,38]

The 6-OHDA-lesioned rat model has also been an important tool in elucidating the mechanism(s) underlying motor complications, both "wearing-off" and dyskinesia. The chronic administration of L-dopa (over a period of weeks) to the 6-OHDA rat has been demonstrated to lead to a shortening response similar to the "wearing-off" complication in idiopathic PD.[39] L-dopa administration also leads to behavioral features in the 6-OHDA lesioned rats which resemble dyskinesia.[40] In the 6-OHDA, rat motor complications occur when greater than 95% of the nigrostriatal cells are lost. Studies using glutamate antagonists have demonstrated improvement in the "wearing-off" response and have implicated the role of glutamate receptor subtypes in the development of motor complications.[41,43] These findings have been supported by molecular studies that demonstrate alterations in the phosphorylation state of glutamate receptor subunits of the NMDA subtype after 6-OHDA-lesioning.[44]

In the context of neuroplasticity, the 6-OHDA-lesioned rat model demonstrates behavioral recovery and has been instrumental in characterizing the neurochemical, molecular, and morphological alterations within the basal ganglia in response to nigrostriatal dopamine depletion.[45] These mechanisms of neuroplasticity in surviving dopaminergic neurons and their striatal terminals include

1. Increased turnover of dopamine and its metabolites
2. Alterations in the expression of tyrosine hydroxylase, the rate-limiting step in dopamine biosynthesis
3. Decreased dopamine uptake through altered dopamine transporter expression
4. Alterations in the electrophysiological phenotype (both pattern and rate of neuronal firing) of substantia nigra neurons
5. Sprouting of new striatal dopaminergic terminals[46–52]

These molecular mechanisms may provide new targets for novel therapeutic interventions such as growth factors to enhance the function of surviving dopaminergic neurons. For example, the 6-OHDA rat has been useful for the evaluation of neurotrophic factor delivery (either stereotaxically or via vector carrier), including glial-derived neurotrophic factor (GDNF) and brain-derived neurotrophic factor (BDNF), to determine the capacity for neuroprotection and neurorestoration.[53,54]

The 6-OHDA-lesioned rat model has also been an important template for determining critical parameters for successful transplantation into the injured brain. These parameters include

1. Target site (striatum versus substantia nigra)
2. Volume of transplant tissue to the target site
3. Number of cells transplanted
4. Type and species of cells transplanted including fetal mesencephalon, engineered cell lines, and stem cells
5. Age of host and donor tissues
6. Pretreatment of transplant tissue or host with neurotrophic factors, antioxidants, immunosuppressive therapy, or neuroprotective pharmacological agents
7. Surgical techniques including needle design, cell suspension media, and transplant cell delivery methods[55,56]

The near absence of dopaminergic neurons and terminals within the striatum due to 6-OHDA lesioning provides a template for the assessment of sprouting axons and innervation from the transplant tissue. Measures of transplant success in this model include reduction in the rotational behavior and the survival, sprouting, and innervation (synapse formation) of dopaminergic fibers within the denervated striatum. The reduction of rotational behavior suggests increased striatal dopamine production originating from the transplanted tissue. Interestingly not all behavioral measures appear to respond to transplant. The advancements made in the 6-OHDA-lesioned rat provide a framework for the further testing of transplantation in nonhuman primates and future human clinical trials.

While the 6-OHDA-lesioned rat model has many advantages, it serves primarily as a model of dopamine dysfunction. Lesioning with 6-OHDA is highly specific for catecholaminergic neurons and does not replicate many of the behavioral, neurochemical, and pathological features of human PD. For example, the 6-OHDA-lesioned rat does not manifest alterations in the cholinergic and serotonergic neurotransmitter systems, which are commonly affected in PD. Stereotaxic injections of 6-OHDA to precise targets does not replicate the extensive pathology of PD where other anatomical regions of the brain (including the locus coeruleus, nucleus basalis of Meynert, and raphe nuclei) are affected. In addition, Lewy body formation, a pathological hallmark of PD, has not been reported in this model. Interestingly, a recent report using a regimen of chronic administration of 6-OHDA into the third ventricle did show a more extensive lesioning pattern reminiscent of human PD.[57] In addition to the rat, other species including the nonhuman primate have served as models for 6-OHDA lesioning.[58] Lesioning in nonhuman primates provides for the analysis of behaviors not observed in the rat, such as targeting and retrieval tasks of the arm and hand.

Overall, lesioning with 6-OHDA has provided a rich source of information regarding the consequences of precise dopamine depletion and its effects on rotational behavior, dopamine biosynthesis, and biochemical and morphological aspects of recovery, and it serves as an excellent template to study both pharmacological and transplantation treatment modalities for PD.

THE NEUROTOXIN MPTP (1-METHYL-4-PHENYL-1,2,3,6-TETRAHYDROPYRIDINE)

The inadvertent self-administration of MPTP by heroin addicts in the late1970s and early 1980s induced an acute form of parkinsonism whose clinical and biochemical features were indistinguishable from those of idiopathic PD.[59,60] The initial cohort of 7 MPTP-lesioned patients (from a population of over 400 heroin addicts exposed to MPTP) displayed severe bradykinesia, tremor, and impaired balance. Similar to patients with sporadic PD, this MPTP-lesioned cohort demonstrated an excellent response to L-dopa and dopamine agonist treatment. Positron emission tomography using ^{18}F-dopa showed severely reduced uptake similar to late-stage PD.[61–63] Also, similar to sporadic cases, the MPTP-induced parkinsonian cohort developed L-dopa-related motor complications (including dyskinesia). Interestingly, motor complications developed within a short period of time (weeks) of starting dopamine replacement therapy, compared to years in sporadic PD. The rapidity with which these motor complications appeared presumably reflected the severity of substantia nigra neuronal degeneration induced by MPTP exposure. Immediately following the identification of MPTP and its role in inducing an acute parkinsonian state in heroin addicts, MPTP was administered to both rodents and nonhuman primates, and some of the most valuable animal models of PD were developed (see below).

The acute effects of MPTP in humans have raised caution for its use in the laboratory setting. To date, there has been only one report of exposure to MPTP causing acute parkinsonism. This isolated report was described in a chemist who was working with unprotected full exposure to extremely large amounts of MPTP (in the kilogram range).[64] The degree to which humans are sensitive to MPTP and the minimum amount needed to induce a parkinsonian state in humans are unknown. The original cohort who had developed MPTP-induced parkinsonism had self-injected an estimated tens to hundreds of milligrams of MPTP.[65] Despite the uncertainty of the amount of exposure to MPTP required to develop a parkinsonian state in humans, established precautions are always taken when using MPTP in the laboratory.[66]

The subsequent administration of MPTP to a number of different species of animals has demonstrated that there exists a wide variety of sensitivity to the toxic effects of MPTP. These differences have been shown to be species, strain, and age dependent. For example, the nonhuman primate is the most sensitive to the toxic effects of MPTP with Old World monkeys much more sensitive than New World monkeys (see below). The mouse, cat, dog, and guinea pig are less sensitive, and the rat is the least sensitive. Even within species, there are strain differences. For example, the C57BL/6 mouse is the most sensitive of all mouse strains tested, while strains such as CD-1 and BALBc appear almost resistant.[67,68] Figure 34.1 shows the results of an experiment demonstrating MPTP-lesioning

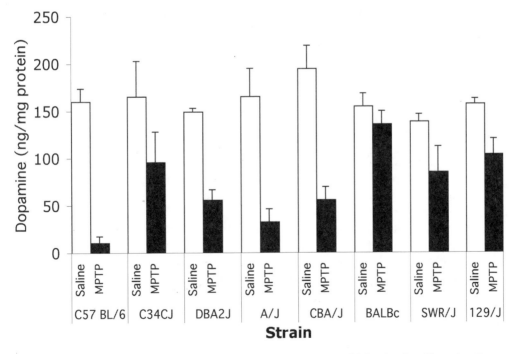

FIGURE 34.1 *Mice show strain-dependent susceptibility toward MPTP-lesioning.* Male mice 8 to 10 weeks of age were obtained from Jackson Laboratories (Bar Harbor, Maine) and administered MPTP in a series of four injections of 20 mg/kg free-base 2 hr apart. Saline injected mice acted as control. Striatal tissue was harvested seven days later, and the amount of dopamine determined using HPLC. Bars show comparison of the remaining striatal dopamine between the saline and MPTP-lesioned mice within each strain (*n* = 6 per saline or lesioned group). C57BL/6J were most sensitive while BALBc were the most resistant to striatal dopamine depletion.

susceptibility in different mouse strains. Eight different mouse strains (male 8 to 10 weeks of age), including C57BL/6J, C34CJ, DBA2J, A/J, CBA/J, BALBc, SWR/J, and 129J ($n = 6$ in each group), were obtained from Jackson Labs (Bar Harbor, Maine) and administered either (1) MPTP in a series of four injections 2 hr apart, at a concentration of 20 mg/kg free-base or (2) saline. Seven days later, the striatum was harvested and dopamine levels determined by HPLC analysis. These results show that mice strains can vary in their sensitivity to MPTP, ranging from marked striatal dopamine depletion (C57BL/6J), to intermediate depletion (C34CJ, 129J, or SWR/J), to modest or none (BALBc). Some strain differences have also been documented between different supply houses (Taconic, Charles River, and Jackson Laboratories). This may account for differences in MPTP sensitivity as reported in the literature with the Swiss Webster mouse.[69] Sensitivity to the neurotoxicant effects of MPTP is also influenced by the animal's age. Older mice, for example, are more sensitive to MPTP lesioning.[70,71] Studies suggest that age-dependent differences may be due in part to differences in MPTP metabolism, since older mice contain higher levels of MAO-B than younger mice.[72] Genetic analysis using quantitative trait loci (QTL) have identified a number of genetic loci that may also mediate MPTP-lesioning susceptibility.[73–75] These genes may be intrinsic to neurons, but studies have also shown a glia contribution of strain-dependent susceptibility.[76]

The mechanism of MPTP toxicity has been thoroughly investigated. Following systemic administration, MPTP is able to cross the blood-brain barrier. The meperidine analog 1-methyl-4-phenyl-1,2,3,6-tetrahydropyridine (MPTP) is converted to 1-methyl-4-pyridinium (MPP+) by monoamine oxidase B. MPP+ acts as a substrate of the dopamine transporter (DAT), leading to the inhibition of mitochondrial complex I, the depletion of ATP, and cell death of dopaminergic neurons. MPTP administration to mice and nonhuman primates destroys dopaminergic neurons of the SNpc, the same neurons affected in PD.[77] Similar to PD other catecholaminergic neurons, such as those in the ventral tegmental area (VTA) and locus coeruleus, may be affected to a lesser degree.[78–83] In addition, dopamine depletion occurs in both the putamen and caudate nucleus. The preferential lesioning of either the putamen or caudate nucleus may depend on animal species and regimen of MPTP administration.[84–86] Unlike PD, Lewy bodies have not been reported; however, eosinophilic inclusions (reminiscent of Lewy bodies) have been described in aged nonhuman primates.[80] The time course of MPTP-induced neurodegeneration is rapid and therefore represents a major difference with idiopathic PD, which is a chronic progressive disease. Interestingly, data from humans exposed to MPTP indicate that the toxic effects of MPTP may be more protracted than initially believed.[87]

Details of MPTP toxicity and utility are described in reviews.[66,88]

The MPTP-Lesioned Mouse Model

The administration of MPTP to mice results in the degeneration of SNpc neurons and striatal dopamine depletion. The severity of the lesioning depends on the amount of MPTP administered. For example, using a regimen of four intraperitoneal injections of 20 mg/kg (free base), 2 hr apart, for a total of 80 mg/kg in C57BL/6 mice, leads to a 90% depletion of striatal dopamine and 60 to 70% loss of nigrostriatal dopaminergic neurons.[77] The degree of injury can be titrated using lower levels of MPTP. A single injection of less than 10 mg/kg tends to show little or no injury. However, a more chronic administration regimen, such as a series of 5 or10 injections of MPTP (20 mg/kg free base), shows injury that may differ from the more acute regimen with respect to cell death with morphological features resembling apoptotic rather than necrotic pathways.[77,89] It should be noted that there is a wide range of administration regimens (with respect to the total amount administered, route, and time between injections) employed that vary in degree of dopamine depletion, SNpc cell death, behavioral deficits, and recovery.[77,90–92] Behavioral alterations in mice have been characterized and resemble some of the features of human parkinsonism. For example, hypokinesia, bradykinesia, and akinesia as well as altered balance can be observed through various behavioral analyses, including open-field activity monitoring, swim test, pole test, grip coordination, treadmill, and rotorod.[74,93] Whole body tremor and postural abnormalities have also been reported, but primarily in the acute phase.[74] In general, these behavioral alterations tend to be highly variable, with some mice showing severe deficits while others show little or no behavioral change (for review see Sedelis[74]). This behavioral variability may be due to a number of factors including the degree of SNpc lesioning, mouse strain, and time period after lesioning when recovery may occur. In addition, the reliability and validity of the behavioral analysis carried out by different investigators may also influence outcome measures.

The MPTP-lesioned mouse model has proven valuable to the investigation of potential mechanisms of neurotoxic induced dopaminergic cell death. For example, mechanisms under investigation have included mitochondrial dysfunction, energy (ATP) depletion, free-radical production, apoptosis, and glutamate excitotoxicity, and altered neurotransmission.[5,6,9,12,88,94–97] In addition to its utility in studying acute cell death, the MPTP-lesioned model also provides an opportunity to study injury-induced neuroplasticity. The MPTP-lesioned mouse displays the return of striatal dopamine several weeks to months after lesioning.[84,86,98,99] The molecular mechanism of intrinsic neuroplasticity of the injured basal ganglia is an area of investigation and appears to encompass alterations in both neurochemical and morphological components. In addition, work in our laboratory and those of others has shown that this intrinsic neuroplasticity may be enhanced through activity-dependent processes, including treadmill training.[100–102]

Administration of MPP+

The administration of MPP+ (1-methyl-4-pyridinium ion) is used for several reasons, including (1) to avoid the systemic toxicity of MPTP and (2) to achieve sufficient lesioning in animals who either lack CNS MAO-B (for conversion of MPTP to MPP+) or DAT (for uptake of dopamine). As outlined in the mechanisms of action of MPTP, MPTP itself is considered a pretoxin that crosses the blood brain barrier and is converted via MAO-B, in astrocytes, to the toxicant form MPP+. MPP+ does not cross the blood brain barrier. When systemically administered (via subcutaneous, intraperitoneal, or intravenous injection), MPTP may be converted by peripheral MAO-B to MPP+ causing systemic toxicity including alterations in blood pressure and liver function.[103–107] These alterations may compromise the health of the animal. To optimize the central toxicant effect of MPTP, MPP+ may be delivered stereotaxically to targets within and near the SNpc neurons. This strategy of MPP+ delivery is used primarily in the rat, since the rat, unlike the mouse, is resistant to the neurotoxic effects of MPTP. This resistance to MPTP may be due to biochemical differences in the rat, including the lack of CNS MAO-B, altered blood brain barrier permeability to MPTP, and differences in MPP+ uptake and storage.[108–113] MPP+ is also used in cell culture systems, since many cell types lack MAO-B for the conversion of MPTP to MPP+ or lack DAT necessary for uptake of MPTP.

Enhancing MPTP-Induced Toxicity in Mice

Several compounds have been shown to potentiate the neurotoxic effects of MPTP, including diethyldithiocarbamate (DDC), probenecid, acetalaldehyde, and ethanol.[114–117] DDC is a copper chelating agent that increases the bioavailability of MPP+ and enhances MPTP toxicity such that subthreshold levels of MPTP show enhanced lesioning with DDC coadministration.[118–122] Probenecid delays MPTP metabolism and reduces urinary and neuronal clearance of MPTP and its metabolites.[114,123] Mice treated for three weeks with MPTP and probenecid show small granular, nonfibrillary inclusions in the substantia nigra that ultrastructurally resemble the dense core of Lewy bodies, an intracytoplasmic inclusion considered a pathological hallmark of human PD. Acetaldehydes and ethanol as well as several other compounds may enhance MPTP susceptibility through a variety of mechanisms including alteration of dopamine and glutamate receptor expression, dopamine transporter, and mitochondrial func-

tion. Therefore, they can indirectly lower the threshold for MPTP-mediated toxicity as well as promote entry of MPP+ to its molecular targets.

MPTP-Lesioned Nonhuman Primate

The initial identification of MPTP-induced parkinsonism in humans who self-administered heroin derivatives[59,60] led to studies in the nonhuman primate including the squirrel monkey (*Saimiri sciureus*),[124] long-tailed macaque or cynomologus (*Macaca fascicularis*),[78] rhesus macaque (*Macaca mulatta*),[125,126] Japanese macaque (*Macaca fuscata*),[127,128] Bonnet monkey (*Macaca radiata*),[129] baboons (*Papio papio*),[130,131] African green monkey or vervet (*Chorocebus aethios,* formerly *Ceropithecus aethiops*),[132] and common marmoset (*Callithrix jacchus*).[128] The administration of MPTP to the nonhuman primate results in parkinsonian symptoms including bradykinesia, postural instability, freezing, stooped posture, and rigidity. Although postural and action tremor have been observed in many species after MPTP treatment, a resting tremor, characteristic of PD, is less commonly documented.[130,133,134] Similar to PD, the MPTP-lesioned nonhuman primate responds to traditional antiparkinsonian therapies such as L-dopa and dopamine receptor agonists. The degree of L-dopa response is dependent on the severity of the lesion and parkinsonian state. In general, the degree of motor deficits induced by MPTP-lesioning may vary at both the inter- and intraspecies level. Variability may be due to age and species phylogeny. For example, aged animals and Old World monkeys (such as rhesus, *Macaca mulatta or* African Green, *Cercopithecus aethiops*) tend to be more sensitive than young and New World monkeys (such as the squirrel monkey, *Saimiri sciureus* or marmoset, *Callithrix jacchus*).[13,135,136] Within a species of animals, behavioral variability is also observed. For example, in our studies, using a regimen of six subcutaneous injections of MPTP (2 mg/kg, free base) two weeks apart (for a total of 12 mg/kg), we have documented a range of symptoms from mild to severe (Petzinger, unpublished data). The molecular and morphological mechanism of this variability is complex and may be attributed to a combination of dopaminergic cell death and post-injury molecular adaptation (Petzinger, unpublished data).

Following the administration of MPTP, the nonhuman primate progresses through acute (hours), subacute (days), and chronic behavioral phases of toxicity that are due to the peripheral and central effects of MPTP. The acute phase occurs within minutes after MPTP administration and is characterized by sedation and a hyper-adrenergic state. This state may also include hyper-salivation, emesis, exaggerated startle, seizure-like activity, and dystonic posturing of trunk and limbs.[97,128,137–139] The subacute phase generally occurs within hours, persists for several days, and may be due to the peripheral actions of MPTP on the autonomic nervous system and peripheral organs such as the liver and heart.[97] Weight loss, altered blood pressure, and hypothermia may occur, requiring tube feeding and placement in an incubator to stabilize body temperature. In addition, elevated liver transaminases and creatinine phospokinase may develop, reflecting impaired liver function and muscle breakdown. Behaviorally, these animals may appear prostrate and cognitively impaired. Occasionally animals may demonstrate self-injurious behavior such as finger biting and hyperflexion of the neck and trunk with head banging. Assessment of parkinsonian features may be confounded by alterations in the general health of the animal. The chronic phase occurs within days to weeks after MPTP administration. It is characterized by the stabilization of body weight and temperature as well as the normalization of blood chemistries such as hepatic enzymes. Parkinsonian features clearly emerge and remain stable for weeks to months or longer. The degree of behavioral stability may be predicted in part by the initial degree of behavioral impairment as observed between the subacute to chronic phases. Those animals with greater behavioral impairments recover over a longer period of time. Behavioral recovery after MPTP administration has been reported in most species of nonhuman primates (see below).

The administration of MPTP through a number of different dosing regimens has led to the development of several distinct models of parkinsonism in the nonhuman primate. Each model is characterized by unique behavioral and neurochemical parameters. Reviews of these various models have recently been published.[6,10,94,96,140,141] Each of these models is different with respect to the modality and degree of lesioning as well as behavioral features. Accordingly, a variety of different models have been used to test different therapeutic interventions, including cell transplantation, gene delivery, and neuroprotective and neurorestorative therapies. For example, in some models, there is profound striatal dopamine depletion and innervation with little or no dopaminergic axons or terminals remaining. These models provide an optimal setting to test fetal tissue grafting, since the presence of any tyrosine hydroxylase positive axons or sprouting cells would be due to transplanted tissue survival. Other models have less extensive dopamine depletion and only partial denervation with a modest to moderate degree of dopaminergic axons and terminals remaining. This partially denervated model best resembles mildly to moderately affected PD patients. Therefore, sufficient dopaminergic neurons and axons as well as compensatory mechanisms are likely to be present. The effects of growth factors (inducing sprouting) or neuroprotective factors (promoting cell survival) are best evaluated in this situation. The following paragraphs review the most commonly used MPTP-lesioned nonhuman primate models, including the systemic, hemi-lesioned, bilateral intracarotid, over-lesioned, and chronic low-dose models.

The Systemic Model

In the *systemic lesioned* model, the administration of MPTP via intramuscular, intravenous, intraperitoneal, or subcutaneous injection leads to bilateral depletion of striatal dopamine and nigrostriatal cell death.[142–145] In this model, the degree of lesioning can be titrated, resulting in a range (mild to severe) of parkinsonian symptoms. The presence of clinical asymmetry is common, with one side more severely affected. L-dopa and dopamine administration leads to the reversal of all behavioral signs of parkinsonism in a dose-dependent fashion. After several days to weeks of L-dopa administration, animals develop reproducible motor complications, both "wearing-off" and dyskinesia. Animal behavior in this model and others may be assessed using (1) cage-side or video-based observation, (2) automated activity measurements in the cage through infrared-based motion detectors or accelerometers, and (3) examination of hand-reaching movements tasks. The principal advantage of this model is that the behavioral syndrome closely resembles the clinical features of idiopathic PD. The systemic model has partial dopaminergic denervation bilaterally and probably best represents the degree of loss seen in all stages of PD, including end-stage disease where some dopaminergic neurons are still present. This model is well suited for therapeutics that interact with remaining dopaminergic neurons, including growth factors, neuroprotective agents, and dopamine modulation. Dyskinesia is reproducible and permits extensive investigation regarding mechanism and treatment.[10,146] A disadvantage of this model is spontaneous recovery in mildly affected animals. Another is that bilateral severely affected animals may require extensive veterinary care and dopamine supplementation.

The Hemi-Lesioned Model

Administration of MPTP via unilateral intracarotid infusion has been used to induce a hemiparkinsonian state in the primate, called the *hemi-lesioned* model.[147] The rapid metabolism of MPTP to MPP+ in the brain may account for the localized toxicity to the hemisphere ipsilateral to the infusion. Motor impairments appear primarily on the contralateral side. Hemi-neglect, manifested by a delayed motor reaction time, also develops on the contralateral side. In addition, spontaneous ipsilateral rotation may develop. L-dopa administration reverses the parkinsonian symptoms and induces contralateral rotation. Substantia nigra neurodegeneration and striatal dopamine depletion (greater than 99%) on the ipsilateral side to the injection is more extensive than seen in the systemic model. The degree of unilateral lesioning in this model is dose dependent. Major advantages of the *hemi-lesioned* model include (1) the ability for animals to feed and maintain themselves without supportive care, (2) the availability of the unaffected limb on the ipsilateral side to serve as a control, and (3) the utility of the dopamine-induced rotation for pharmacological testing. In addition, due to the absence of dopaminergic innervation in the striatum, *the hemi-lesioned* model is well suited for examining neuronal sprouting of transplanted tissue. A disadvantage of this model is that only a subset of parkinsonian features are evident and are restricted to one side of the body, a situation never seen in idiopathic PD.

The Bilateral Intracarotid Model

The *bilateral intracarotid* model employs an intracarotid injection of MPTP followed several months later by another intracarotid injection on the opposite side. [148] This model combines the less debilitating features of the carotid model as well as creating bilateral clinical features, a situation more closely resembling idiopathic PD. The advantage of this model is its prolonged stability and limited interanimal variability. Similar to the hemi-lesioned model, where there is extensive striatal dopamine depletion and denervation, the bilateral intracarotid model is well suited for evaluation of transplanted tissue. However, L-dopa administration may result in only partial improvement of parkinsonian motor features and food retrieval tasks. This can be a disadvantage, since high doses of test drug may be needed to demonstrate efficacy, increasing the risk for medication related adverse effects.

The Over-Lesioned Model

A novel approach to MPTP-lesioning is the administration of MPTP via intracarotid infusion followed by a systemic injection.[140] This *over-lesioned* model is characterized by severe dopamine depletion ipsilateral to the MPTP-carotid infusion and a partial depletion on the contralateral side due to the systemic MPTP injection. Consequently, animals are self-sufficient and typically do not require extensive veterinary intervention due to a relatively intact side. The behavioral deficits consist of asymmetric parkinsonian feature. The more severely affected side is contralateral to the intracarotid injection.[149]

L-dopa produces a dose-dependent improvement in behavioral features; however, the complications of L-dopa therapy, such as dyskinesia, have not been as consistently observed. This model combines some of the advantages of both the systemic and intracarotid MPTP models, including stability. This model is suitable for both transplant studies, utilizing the more depleted side, and neuro-regeneration with growth factors, utilizing the partially depleted side where dopaminergic neurons still remain.[150–152]

The Chronic Low-Dose Model

Finally, the *chronic low-dose* model consists of intravenous injections of a low dose of MPTP administration over a 5- to 13-month period.[153] This model is characterized by cognitive deficits consistent with frontal lobe dysfunction reminiscent of PD or normal aged monkeys.

These animals have impaired attention and short-term memory processes and perform poorly in tasks of delayed response or delayed alternation. Since gross parkinsonian motor symptoms are essentially absent at least in early stages, this model is well adapted for studying cognitive deficits analogous to those that accompany idiopathic PD.

The MPTP-Lesioned Cat Model

Administration of MPTP to cats (*Felis catus*) leads to parkinsonian features (bradykinesia), striatal dopamine depletion, and dopaminergic cell death in the SNpc.[154–156] Histological features are not very distinct from those documented in mice and nonhuman primates. However, the MPTP-lesioned cat is worth noting, since it was one of the first species in which behavioral recovery following MPTP-lesioning was studied.[157–161] Investigations in the recovery of the MPTP-lesioned cat have proven important in investigating the mechanisms involved in intrinsic neuroplasticity of the injured basal ganglia. Later studies in the 6-OHDA-lesioned rat and MPTP-lesioned mouse and nonhuman primate have complemented these early analyses.[45–47,99,145,162–164]

MPTP-Lesion Models in Other Species

MPTP has been administered to a variety of species other than mice and nonhuman primates. These include the leech (*Hirudo medicinakis*),[165] the planarian flatworm (*Dugesia japonica*),[166,167] goldfish (*Carassius auratus*),[168–170] rainbow trout (*Oncorhynchus mykiss*),[171] frog (*Rana pipiens* and *R. clamitans*),[172,173] salamander (*Taricha torosa*),[174] snake (*Elaphe obsolete* and *Nerodia fasciata*),[175] lizard (*Anolis carolinensis*),[176] chicken (*Gallus gallus*),[177] rat (*Rattus rattus* and *Rattus norvegicus*),[109,178,179] guinea pig (*Cavia porcellus*),[180] rabbit (*Oryctolagus cuniculus*),[181] dog (*Canis familiaris*),[182] and pig (*Susscrofa domestica*).[183] There are no reports in the literature of MPTP administration to *C. elegans* or *Drosophila melanogaster*. While each of these different species has its own merits and drawbacks with respect to behavioral features, dopamine metabolism, MPTP toxicity, and recovery, their wide range application and utility have not been fully explored. The fruit fly and nematode have recently regained interest in PD research (see below), since transgene manipulations and large-scale genetic screens are possible, complementing studies in mice and avoiding many of the limitations present with nonhuman primates.

INSIGHTS FROM THE MPTP-LESIONED MOUSE AND NONHUMAN PRIMATE MODELS

The MPTP-lesioned mouse and nonhuman primate models have been useful in identifying mechanisms that may underlie degeneration of the substantia nigra neurons, and in investigating new pharmacological agents that may provide symptomatic, restorative, or neuroprotective treatment for PD. In addition, these models have provided insights into the mechanism of action of motor complications, such as dyskinesia, that are related to long-term use of L-dopa replacement therapy. The MPTP-lesioned nonhuman primate model has been particularly helpful in studying new surgical treatments such as deep brain stimulation and has provided the means to study novel surgical interventions, including cell replacement therapy and gene delivery.

Dyskinesia and Motor Complications

Dyskinesia is a motor complication resulting from long-term treatment with L-dopa, typically seen five to seven years after onset of L-dopa administration in sporadic PD.[184,185] There has been no documentation of dyskinesia-like behavior in the MPTP-lesioned mouse model. However, the 6-OHDA-lesioned rat does display movements in response to L-dopa administration that are considered analogous to dyskinesia but show limited anatomical distribution, such as the jaw.[186–188] The most dramatic display of L-dopa-induced dyskinesia is in the MPTP-lesioned nonhuman primate, where it has been best characterized in the systemic model of MPTP-lesioning in squirrel monkey and marmoset, as well in some macaques.[10,146,189–191] Dyskinesia manifests itself as chorea-like activity in both upper and lower limbs and trunk. There have been few reports of dyskinesia in other models of MPTP-lesioning possibly due to the extent of dopaminergic depletion. The squirrel monkey and marmoset retain a residual number of SNpc neurons and striatal dopamine integrity that may serve as a template to initiate dyskinesia. It is interesting to note that there have been reports of dyskinesia in non-lesioned monkeys receiving large amounts of L-dopa.[192–194]

The MPTP-lesioned nonhuman primate has provided a valuable tool in predicting which compounds are likely to elicit dyskinesia in humans. In addition, this model has provided a means to investigate potential mechanisms underlying motor complications. Studies thus far indicate that the capacity (storage and release) of the nigrostriatal neurons to handle L-dopa may play an important role in dictating the presence and severity of dyskinesia after L-dopa administration. The capacity of substantia nigra neurons may be compromised through neuronal and accompanying terminal loss, or through high-dose L-dopa administration, which overwhelms the cells' buffering system. Other factors that may also alter neuronal capacity may include age of animal as well as the degree of intrinsic adaptive molecular mechanisms (proteins involved in storage and uptake) of the cell after MPTP lesioning. Regarding pharmacological factors, work in the

MPTP nonhuman primate model have supported the idea that the tendency for dopaminergic drugs (L-dopa and dopamine agonists) to cause dyskinesia may be due in part to the half-life of the drug as well as the manner in which these agents are delivered. Animal studies have shown that high-dose intermittent L-dopa is more likely to induce dyskinesia than a dopamine agonist with a longer half-life. Alternatively, dopamine agonists with a short half-life, delivered in an intermittent fashion, have also been shown to induce dyskinesia. Although the etiology of dyskinesia is unknown, electrophysiological, neurochemical, molecular, and neuroimaging studies in the nonhuman primate models suggest that the pulsatile delivery of L-dopa may lead to

1. Changes in the neuronal firing rate and pattern of the globus pallidus and subthalamic nucleus
2. Enhancement of D_1- and/or D_2-receptor mediated signal transduction pathways
3. Super-sensitivity of the D_2 receptor
4. Alterations in the phosphorylation state and subcellular localization of glutamate (NMDA subtype) receptors
5. Modifications in the functional links between dopamine receptor subtypes (D_1 and D_2, and D_1 and D_3)
6. Changes in glutamate receptors (AMPA and NMDA receptor subtypes)
7. Enhancement of opioid-peptide mediated neurotransmission[195–199]

While the presence of a SNpc lesion has long been considered an important prerequisite for the development of dyskinesia in the MPTP model, recent studies demonstrate that even normal nonhuman primates, when given sufficiently large doses of L-dopa (with a peripheral decarboxylase inhibitor) over to 8 weeks, may develop peak-dose dyskinesia.[193] The high levels of plasma L-dopa in this dosing regimen may serve to exhaust the buffering capacity within the caudate and putamen of the normal animal and therefore lead to pulsatile delivery of L-dopa and priming of postsynaptic dopaminergic sites for dyskinesia.

Restorative Therapy in the MPTP-Lesioned Mouse and Nonhuman Primate

The MPTP-lesioned mouse and nonhuman primate have provided an important template to investigate a variety of restorative therapeutic modalities. These strategies are designed to restore the function of the basal ganglia by saving the surviving or injured SNpc neurons, introducing new dopaminergic neurons, or promoting differentiation of cells into a dopaminergic phenotype. These include

1. Cell transplantation
2. Surgical intervention
3. Neurotrophin delivery
4. Pharmacological targeting
5. Vector infusion

These strategies are not limited to the MPTP models but have been investigated to a great extent in the 6-OHDA-lesioned rat and marmoset. Discussion and references in the following paragraphs apply to both neurotoxicant models.

The administration of MPTP leads to SNpc cell death and striatal dopamine depletion in both the mouse and nonhuman primate, and both models have served as templates to investigate the transplantation of dopaminergic neurons. In addition, the 6-OHDA-lesioned rat model has also served as an excellent model to investigate transplantation as a means to restore basal ganglia function. A wide variety of cell sources have been investigated, including fetal mesencephalic tissue,[200–206] carotid bodies,[207] adrenal medullar tissue,[208–211] retinal cells,[212] as well as cultured cell lines including PC12 and others.[213,214] The recent characterization of embryonic, fetal, and adult stem cells has provided an additional source of material for transplantation into mice, rats, and nonhuman primates.[215–217] Transplantation of human stem cells into the MPTP-lesioned mouse has shown limited survival, migration, and phenotypic differentiation to TH neurons.[218] Studies in these animal models are critical for determining important parameters for successful transplantation, including age and stage of cells, number of cells and their volume, anatomical targets (caudate nucleus, putamen, or SNpc), pre- and post-transplantation treatment of cells with neurotrophins or antioxidants, and host treatment with immunosuppression therapy.[219–223] In these models, different outcome measures can be evaluated including the effect on behavior, dopamine biosynthesis, transplant phenotypes (glia versus neuron), neuron survival, and innervation. The goal of these studies is to restore or replace degenerative or dysfunctional cells in the substantia nigra with dopamine, producing cells either in the caudate putamen or SNpc. It is interesting to note that when human studies investigating cell transplantation (such as fetal mesencephalic tissue) are inconclusive, additional research studies in animal models may be useful to re-evaluate new treatment strategies.[224–226]

There has been a recent re-emergence of surgical treatment for PD and its motor complications.[227,228] The development of specific targets including the subthalamic nucleus, thalamus, and globus pallidus depend on studies in both the rodent and nonhuman primate. These studies will provide important information regarding the alterations in neuronal circuitry and electrophysiological properties of motor pathways within the basal ganglia. Not only do these models provide insights into the surgi-

cal techniques in humans (including deep brain stimulation), they also provide a means to map the circuitry of the basal ganglia and its regulatory features, including extrinsic inputs from the cerebral cortex, thalamus, and pedunculopontine nucleus.[229–233]

Both neurotrophic and pharmacological interventions to restore function in the MPTP-lesioned mouse and nonhuman primate primarily focus on stereotaxic infusion to either the SNpc or caudate-putamen. For example, infusion of caspase inhibitors, free-radical scavengers, dopamine and glutamate agonists and antagonists, and neurotrophic factors (including GDNF, BDNF, NGF, neurturin) have been evaluated for their restorative properties. Not only can compounds be delivered to specific anatomical targets, but extracellular constituents can be evaluated through microdialysis during different phases of these injury models.

An alternative means to intervene in basal ganglia function is the recent development of vectors based on infectious viruses. Earlier delivery strategies employed encapsulated cells or microspheres to target neurotrophic factors into the injured basal ganglia.[213,214,234] Viral vector delivery systems permit the targeted delivery of a number of different genes of interest including neurotrophic factors, inhibitory neurotransmitter proteins, and members of the dopamine biosynthetic pathway such as tyrosine hydroxylase (TH) and aromatic amino acid decarboxylase (AADC). A variety of different vectors have been constructed, including those from herpes simplex virus (HSV), adeno-associated virus (AAV), and lentiviruses (such as equine infectious anemia virus). Vectors (AAV and lentivirus) carrying GDNF have been shown to be both neuroprotective and neurorestorative in both rodent and nonhuman primate models.[235–238] The stereotaxic delivery of vectors carrying glutamic acid decarboxylase (GAD) to the subthalamic nucleus has provided an inhibitory phenotype to suppress a pathway hyperactive in PD and its models.[239] There are a number of advantages to using vector delivery as a restorative strategy, including (1) accurate stereotaxic delivery to the region of injury, (2) delivery to and integration into nondividing neurons, 3) avoidance of host immune response, and (4) the potential control of gene expression in vivo. In addition to providing potential benefit through vector delivery systems, the same technology can be used to deliver genes and proteins responsible for degenerative diseases.[240]

Whether this *in vivo* delivery system will become a therapeutic modality in the clinical setting will depend both on its safety and efficacy. Studies currently underway in our laboratory using a lentivirus-derived vector carrying green fluorescent protein (GFP) has shown that large numbers of putamen neurons (several hundred thousand) can be safely transdifferentiated to express GFP without adverse effects to the health and motor behavioral of the nonhuman primates. Studies are now underway to determine the efficacy of a lentivirus-based vector to alter behavioral motor deficits and dopamine biosynthesis in the MPTP-lesioned squirrel monkey. Complementary data from other labs have shown potential benefit from this strategy in rat and nonhuman primate models using three independent vectors carrying TH, aromatic L-amino acid decarboxylasae (AADC), and GTP-cyclohydrolase-1 (CH-1).[241,242] Delivery of these same genes in a tricistronic vector has also shown benefit in the 6-OHDA rat.[243] It should be kept in mind that the restoration of behavior in these models may involve more than the replacement of striatal dopamine. Other nondopaminergic pathways may prove important, including those that are glutamatergic or serotonergic and molecular changes downstream of the dopamine receptors may play important roles in functional restoration.

Recovery in the MPTP-Lesioned Animal Models

Behavioral recovery after MPTP-induced parkinsonism has been reported in both New and Old World nonhuman primates.[97,140,145,244–248] The degree and time course of behavioral recovery is dependent on age, species, and mode of MPTP administration.[97,136,249,250] The extent of the initial MPTP-induced dopamine depletion may be the most important factor determining the probability of spontaneous recovery.[250] The molecular and neurochemical mechanisms underlying behavioral recovery are unknown. However, two mechanisms have been proposed to play a major role. They are (1) biochemical compensation by surviving dopaminergic neurons and (2) the sprouting of new neuronal processes into dopamine depleted regions Results of our work and that of others have identified that the mechanisms underlying recovery may include

1. Alterations in dopamine biosynthesis (increased tyrosine hydroxylase protein and mRNA expression, and activity) and turnover
2. Down-regulation of dopamine transporter
3. Increased dopamine metabolism
4. Sprouting and branching of tyrosine hydroxylase fibers
5. Alterations of other neurotransmitter systems, including glutamate and serotonin
6. Alterations of signal transduction pathways in both the direct (D_1) and indirect (D_2) pathways[98,145,158,160,162,179,251–259]

Studies suggest that the recovery of TH activity (turnover rate) is more rapid than that of dopamine level in MPTP rodents and nonhuman primates. Figure 34.2 shows the partial recovery of dopamine storage as measured through 18F-dopa positron emission tomography (PET) imaging in the systemic MPTP-lesioned squirrel monkey. In addition to behavior, MPTP-lesioned mice show neurochemical recovery. It is interesting that behavioral recovery in the MPTP-lesioned mouse and nonhuman primate may be observed prior to significant return of dopaminergic parameters.[99,260,261] Increases in 5-HT levels and 5-HT hyperinnervation may also contribute to improve motor deficit in addition to the recovery of dopaminergic parameters.

METHAMPHETAMINE

Amphetamine and its derivatives, including methamphetamine (METH), provide an important model of neurotoxic injury to the basal ganglia, targeting dopaminergic and serotonergic neurotransmission.[262,263] Methamphetamine,

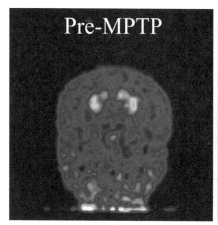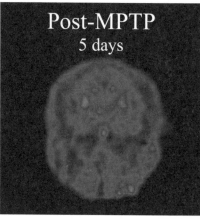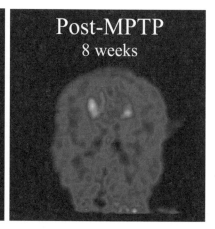

FIGURE 34.2 ^{18}F-dopa positron emission tomography (PET) imaging in the MPTP-lesioned squirrel monkey. The left panel shows an image in the coronal plane prior to MPTP administration. Note the high-intensity signal emitting from the caudate nucleus and putamen. The middle panel is an image from a monkey following a series of six injections of MPTP (2 mg/kg, free-base) with 2 weeks between each injection. Note the depletion of signal when imaged 5 days after the last injection of MPTP. The right panel shows the same monkey 8 weeks after the last injection of MPTP. Note the increased intensity of the ^{18}F-dopa PET image signal from the caudate nucleus and putamen. These studies were performed at the USC PET Imaging Center. Special thanks to Dr. Jim Bading and Michel Tohme. (*A color version of this figure follows page 518.*)

one of the most potent of these derivatives, leads to terminal degeneration of dopaminergic neurons in the caudate-putamen, nucleus accumbens, and neocortex. METH administration also leads to long lasting depletion of both dopamine and serotonin when administered to rodents and nonhuman primates.[264–270] In contrast to MPTP, the axonal trunks and soma of SNpc and VTA neurons are spared after METH administration.[271] However, there have been some reports of METH-induced cell death in the substantia nigra in rodents and possibly in nonhuman primates.[272–274] In general, the effects of severe METH lesioning are long lasting. Like many neurotoxicant models, there are strain and gender characteristics that influence the severity of lesioning. Young animals tend to be more resistant than older animals, different strains such as the BALBc mouse are less resistant than C57BL/6 mouse strains, and females are more resistant than males.[274–277] The mechanisms underlying these difference could be due to either expression of neuroprotective factors such as estrogen in females as well as differences in the metabolism of METH.[278] Here, there is also evidence of recovery of dopaminergic innervation depending on the METH regimen and species used. Recovery in the nonhuman primate has been reported and may involve mechanisms similar to those seen in recovery following MPTP-lesioning.[270,279] Despite the severe depletion of striatal dopamine, the motor behavioral alterations seen in rodents and nonhuman primates are typically subtle, with reports of akinesia and balance instability.[280]

The neurotoxic effects of METH are dependent on the efflux of dopamine, since agents that deplete dopamine or block its uptake are neuroprotective.[281,282] For example, α-methyl-para-tyrosine (AMPT), which blocks dopamine reuptake, attenuates METH toxicity, while reserpine, which promotes dopamine release, potentiates METH toxicity. In addition, altered dopamine storage can also influence METH toxicity, as seen in the vesicular monoamine transporter (VMAT-2) knockout mice, where METH toxicity is potentiated.[283] The metabolic mechanisms underlying METH-induced neurotoxicity involve the perturbation of antioxidant enzymes such as glutathione peroxidase or catalase, leading to the formation of reactive oxygen/nitrogen species, including H_2O_2, superoxide, and hydroxyl radicals.[262,284–287] The administration of antioxidant therapies or overexpression of superoxide dismutase (SOD) in transgenic mice models is neuroprotective against METH toxicity.[288,289] In addition, both glutamate receptors and nitric oxide synthase (NOS) are important to METH-induced neurotoxicity, since the administration of either NMDA receptor antagonists or NOS inhibitors are also neuroprotective.[290] Mice with nNOS knockout show protection from METH toxicity.[291,292] Other factors important to METH-induced neurotoxicity include the inhibition of both tyrosine hydroxylase, monoamine oxidase, and dopamine transporter activity and METH-induced hyperthermia.[287]

The administration of METH to adult animals has played an important role in testing the molecular and biochemical mechanisms underlying dopaminergic and serotonergic neuronal axonal degeneration, especially the roles of free radicals and glutamate neurotransmission. Understanding these mechanisms has led to the testing of different neuroprotective therapeutic modalities. An advantage of the METH model over MPTP is that the serotonergic and dopaminergic systems can be lesioned *in utero* during the early stages of the development of these neurotransmitter systems. Such studies have indicated that there is a tremendous degree of architectural rearrangement that occurs within the dopaminergic and serotonergic systems of injured animals as they develop. These changes may lead to altered behavior in the adult animal, including the development of self-injurious behaviors similar to *in utero* lesioning with 6-OHDA.[293,294] These results have implicated the dopaminergic systems role in Lesch-Nyhan syndrome, a disorder that includes self-injury behavior, and a possible relationship between this disorder and Parkinson's disease.[295–302] A recent study examined the effects of pretreatment with MPTP on the acute and long-term effects of METH in BALB/c mice and showed that the combination of METH and MPTP-lesioning in tandem has shown that initial lesioning with METH (resulting in the loss of nigrostriatal terminals) is protective against MPTP-induced nigrostriatal neuron cell death.[303] This result indicates that the mechanism of MPTP-induced SNpc cell death requires its uptake through the SNpc terminals and not cell bodies, supporting the role of terminal DAT in uptake of MPP+. The METH-lesioned rodent and nonhuman primate have also served as model systems to investigate recovery in response to injury, providing an avenue to study intrinsic basal ganglia neuroplasticity.[270,304]

In light of the toxic nature of these compounds in animals, studies in humans have suggested that abusers of METH and substituted amphetamines (including 3,4-methylenedioxymethamphetamine called MDMA or "ecstasy") may suffer from the long-lasting effects of these drugs.[305–307] Specifically, these individuals may be prone to develop parkinsonism.[308] Animals with neonatal exposure to METH and MDMA also show prolonged dopaminergic alterations into adulthood including TH fiber hyper-innervation.[309] A study by Ricaurte and colleagues showed that MDMA led to cell death within dopaminergic neurons of the SNpc.[274] Despite the retraction of their original study due to issues related to dosing compounds, the neurotoxic effects of MDMA and related compounds on dopaminergic neurons still remains an active area of research in both mice and nonhuman primates.

ROTENONE

Epidemiological studies have suggested that environmental factors such as pesticides may increase the risk for PD.[310] These include paraquat (1,1'-dimethyl-4,4'-bipyridinium) dieldrin, and maneb (manganese ethylenebisdithiocarbamate), all of which have structural or neurotoxicant similarities to MPTP and cause parkinsonian motor deficits in rodent and nonhuman primates.[311–319] The demonstration of specific neurochemical and pathological damage to dopaminergic neurons by the application of various pesticides such as rotenone (an inhibitor of mitochondrial complex I) have supported these epidemiological findings. For example, using a chronic rotenone infusion paradigm, Greenamyre and colleagues reported degeneration of a subset of nigrostriatal dopaminergic neurons, the formation of cytoplasmic inclusions, and the development of parkinsonian behavioral features (including hunched posture, rigidity, unsteady movement, and paw tremor) in the rat.[320] These animals also developed intracellular cytoplasmic inclusions that were immunoreactive for α-synuclein and ubiquitin. Some of these aggregates resembled Lewy bodies. Modification in rotenone administration from carotid infusion to the implantation of subcutaneous osmotic pumps has reduced the labor-intensive protocol.[321] Recent reviews outline some of the limitations of the rotenone model in the rat, including issues of cell death specificity and reproducibility.[5,322] Studies examining the effects of various pesticides and their applications in animal models may lead to insights into the mechanisms of neuronal death in Parkinson's disease and the factors that may trigger injury and disease onset.[318]

INTRANIGRAL INFUSION OF IRON

A large body of literature supports the concept that nigrostriatal neurodegeneration in PD as well as in models of neurotoxicant injury, including 6-OHDA and MPTP involve oxidative stress.[323–325] One of the numerous pathways leading to the formation of reactive oxygen species (ROS) involves the Fenton reaction, where iron catalyzes the conversion of hydrogen peroxide to hydroxyl radicals by the reaction $H_2O_2 + Fe^{2+} \Rightarrow OH^- + Fe^{3+}$. In addition, there have been reports of iron accumulation in neurons in PD, as well as neurotoxicant animal models.[326–328] In the rodent, the direct stereotaxic infusion of iron (in the form of ferrous citrate) to dopaminergic neurons of the SNpc leads to cell death and neurochemical changes, including dopamine depletion and increased turnover rate, reminiscent of PD.[251,329–331] This particular rodent model has been useful in elucidating the role of iron in mediating cell death. For example, iron-dependent aggregation of mutant forms of α-synuclein has been reported for both in vitro and in vivo systems.[332–334] An interesting transgenic mouse with altered expression of ferritin has supported the role of iron in models of SNpc neurodegeneration.[334]

1-BENZYL-1,2,3,4-TETRAHYDROISOQUINOLINE (1BNTIQ) MODEL

The advent of the discovery of MPTP-induced parkinsonism in humans and the potential link between causes of PD and the environment initiated the identification of endogenous MPTP-like neurotoxins in both humans and animals.[320,335,336] These compounds include β-carboline, salsolinol (SAL), tetrahydroisoquinolines (TIQ), and their derivatives.[337–339] One interesting compound is the dopamine derived MPTP-like compound 1-benzyl-1,2,3,4-tetrahydroisoquinoline (1BnTIQ). 1BnTIQ has been shown to be present in increased amounts in the cerebral spinal fluid (CSF) of PD patients[340] and has been reported to bind to dopamine receptors.[341] It utilizes the dopamine transporter and COMT for toxicity[342] and alters dopamine metabolism,[343] induces dopaminergic cell death in culture,[344] and induces PD-like symptoms (bradykinesia) in mice and nonhuman primates.[340] Despite the decrease in striatal dopamine, there appears to be no loss of nigrostriatal dopaminergic neurons.[345] This may indicate that the mode of action of compounds such as 1-BnTIQ and their induced motor behavioral deficits may differ from these in PD and MPTP-induced parkinsonism.

LIPOPOLYSACCHARIDE (LPS) MODEL

The activation of the immune response has been hypothesized as a potential contributor to PD and cell death in animal models of basal ganglia injury.[346,347] The substantia nigra is rich in microglia, as well as astrocytes, and the direct activation of these non-neuronal cells by compounds activating the immune response have been shown to lead to the production of agents such as superoxide, NO, and cytokines (TNF-α, IL-6, and IL-1b). These have been shown to play a role in mediating cell death of adjacent neurons such as nigrostriatal dopaminergic neurons. One means to induce the activation of an immune response is through injection of the bacterial-derived compound lipopolysaccharide (LPS). The administration of LPS to cells in culture or stereotaxic delivery to the SNpc in vivo leads to a selective and progressive destruction of dopaminergic neurons.[348–351] The LPS model can also be used to examine the synergistic effects of an activator of the inflammatory pathway with a mitochondrial inhibitor, such as rotenone, in potentiating neurotoxicity.[352]

QUINOLINIC ACID LESIONING MODEL

Excitatory amino acids have been employed to generate a range of basal ganglia lesions so as to replicate some of the pathological, morphological, and molecular changes seen in neurodegenerative diseases such as Huntington's disease, epilepsy, and multisystem atrophy.[353–355] A novel approach developed by Burke and colleagues, relevant to both the development of the basal ganglia and understanding of the mechanisms of SNpc dopaminergic neuron cell death, involves the injection of quinolinic acid (QA) into the ventricles or striatum of new born rats.[356–358] This destruction of striatal targets for SNpc dopaminergic neurons induces apoptotic cell death pathways similar to those seen in early postnatal life. These observations are similar to those seen in neonatal hypoxia-ischemia or striatal 6-OHDA lesioning and have been replicated in the mouse.[358–361] This model has been useful in investigating the developmental mechanisms involved in establishing basal ganglia circuitry and in defining characteristics of programmed cell death (apoptosis).[12,77,362,363]

GENETIC MODELS OF PARKINSON'S DISEASE

In addition to pharmacological and neurotoxicant models of PD, there are spontaneous rodent (such as the weaver mouse and AS/AGU rat) and transgenic mouse (including parkin, α-synuclein, β-synuclein, UCH-1, SOD, and GPx) models that provide important avenues to investigate basal ganglia injury, cell death, neuroprotection, and development.

SPONTANEOUS RODENT MODELS FOR PARKINSON'S DISEASE

There are several naturally occurring spontaneous mutations in rodents that are of particular interest in PD. Spontaneous rodent models include the weaver, lurcher, reeler, Tshrhyt, tottering, and coloboma mice and the AS/AGU and circling (ci) rat. These models possess unique characteristics that may provide insight into neurodegenerative processes of PD and related disorders. Several of these spontaneous rodent models display altered dopaminergic function or neurodegeneration and have deficits in motor behavior.[364] For example, the weaver mouse displays cell death of dopaminergic neurons, while the tottering mouse displays tyrosine hydroxylase hyper-innervation and altered MPTP toxicity.[365,366] The AS/AGU rat is a spontaneous model characterized by progressive rigidity, staggering gait, tremor, and difficulty in initiating movements.[367] Microdialysis in the AS/AGU rat model has revealed that, even prior to dopaminergic neuronal cell death, there is dysfunction in dopaminergic neurotransmission that correlates with behavioral deficit.[368,369] Another potentially interesting rodent model is the circling (ci) rat.[370] This animal model displays spontaneous rotational behavior as a result of an imbalance in dopaminergic neurotransmission despite the absence of asymmetric nigral cell death.

TRANSGENIC MOUSE MODELS

The development of transgenic animal models is dependent on the identification of genes of interest. A transgenic mouse is an animal in which a specific gene of interest has been altered through one of several techniques including (1) excision of the host gene (knock out), (2) introduction of a mutant gene (knock in), and (3) alteration of gene expression (knock down). In PD, one source of transgenic targeting is derived from genes identified through epidemiological and linkage analysis studies. α-synuclein and parkin are examples of genes that have been identified through linkage analysis in familial forms of parkinsonism. Other transgenic animals have been developed based on the identification of genes important for normal basal ganglia and dopaminergic function such as transgenics targeting the dopamine transporter, DAT. Once the transgene has been constructed, the degree of its expression and its impact on the phenotype of the animal depend on many factors, including the selection of sequence (mutant versus wild-type), site of integration, number of copies recombined, selection of transcription promoter, and upstream controlling elements (enhancers). Other important factors may include the background strain and age of the animal. These different features may account for some of the biochemical and pathological variations observed amongst transgenic mouse lines. Several examples of recent transgenic mouse lines are discussed.

α-Synuclein

Rare cases of autosomal dominant familial forms of PD (the Contursi and German kindreds) have been linked to point mutations in the gene encoding α-synuclein.[371] Originally, α-synuclein was identified in 1988 as a component of the electric organ in the eel (*Torpedo californica*).[372] The normal function of α-synuclein is unknown, but its localization and developmental expression suggest a role in neuroplasticity, with a role related to vesicle function as suggested by its lipid interactions and hydrophobicity.[373–375] In addition α-synuclein shares protein sequence identity with the chaperone 14-3-3 and may interact with other proteins, including protein kinases and their regulators.[376] The disruption of normal neuronal function may lead to the loss of synaptic maintenance and subsequent degeneration. It is interesting that mice with knockout of α-synuclein are viable suggesting that a "gain-in-function" phenotype, or other protein-protein interactions may contribute to neurodegeneration. Although no mutant forms of α-synuclein have been identified in idiopathic PD, its localization to Lewy bodies (including PD and related disorders) has suggested a pathophysiological link between α-synuclein aggregation and neurodegenerative disease.

To investigate the potential role of α-synuclein protein in PD, several groups have developed either transgenic mouse models or vector-mediated delivery systems of α-synuclein into the mouse or monkey. An interesting caveat is that the mutant allele of α-synuclein in the Contursi kindred is identical in sequence to the wild-type mouse, suggesting that protein expression and/or protein-protein interactions may be more important than simply loss of function due to missense mutation. This may have contributed to the fact that many different laboratories failed to report pathological or neurological deficits in their transgenic mouse constructs. Transgenic mouse models developed for α-synuclein have focused on altered protein expression through the use of different promoters (including promoters from genes such as tyrosine hydroxylase, platelet-derived growth factor or PDGF, prion gene, and Thy-1 genes) and gene cassette constructs (with respect to sequences that influence ribosomal entry, and transcription enhancers). However, studies are showing that protein processing, folding, and storage dysfunction may be as important as protein expression for α-synuclein-mediated toxicity.[377,378] Some transgenic mouse lines have shown pathological changes in dopaminergic neurons (including inclusions, decreased levels of striatal dopamine, and loss of striatal tyrosine hydroxylase immunoreactivity) and behavioral deficits (rotorod and attenuation of dopamine-dependent locomotor response to amphetamine), while other lines have shown no deficits.[379–381] A recent report documented behavioral deficits in mice expressing the A53T mutant form of α-synuclein as well as pathological changes including fibril formation.[382] No group has reported the progressive loss of substantia nigra dopaminergic neurons. This range of results, with different α-synuclein constructs from different laboratories, underscores the important link between protein expression (mutant versus wild-type alleles) and pathological and behavioral outcome. Despite the lack of consistent parkinsonian features in mouse transgenic models, there have been reports of SNpc specific inclusion-like bodies and cell death in rat and nonhuman primates stereotaxically administered α-synuclein to the midbrain.[383,384] Reviews of the development of α-synuclein models discuss many of these issues.[4,385–387]

Presently, the function of α-synuclein protein is unknown. Initial studies of α-synuclein identification and expression were focused on its role in developmental and post-injury neuroplasticity as well as its role in synaptic vesicle function.[388] The identification of α-synuclein in a familial form of parkinsonism and its localization to Lewy bodies prompted studies on the neurotoxic effects of α-synuclein.[377] However, these studies have shown that the mechanisms involving α-synuclein in neurodegeneration are very complex. The response of α-synuclein expression to neurotoxic injury including MPTP and 6-OHDA-lesioning is altered in SNpc neurons.[389–391] Studies both *in vivo* and *in vitro* have identified many candidate proteins involved in interactions with α-synuclein including its transport, localization, processing, folding, and protein-protein interactions.[377] For example, α-synuclein interacts with parkin protein and other proteins involved in ubiquitin-dependent proteasome degradation.[377,392] The selective toxicity to the SNpc may be explained by interactions between certain conformations of α-synuclein and dopamine metabolism including the formation of reactive oxygen species and catalysis of α-synuclein aggregation.[393]

Delivery of α-synuclein by viral vectors has the advantage of being site specific, it allows expression of this gene in animals other than mice, and its delivery can be performed in adult animals, thereby avoiding embryonic and early postnatal developmental complications. Both the adeno-associated virus (AAV) and a lentiviral vector have been used to introduce human wild-type or mutant A53T or A30P α-synuclein.[394] In these rodent models, varying degrees of dopaminergic neuronal loss, striatal dopamine depletion, and behavioral impairment have been observed. Recently, the administration of AAV expressing wild-type or mutant A53T α-synuclein was unilaterally administered to the SNpc of the marmoset.[395] Animals showed a 30 to 60% loss of dopamine neurons around the injection site, with an accompanying decrease in striatal dopamine innervation but no parkinsonian motor impairment.

Despite these advances in basic molecular mechanisms involving α-synuclein, there are still some very fundamental issues under debate. For example, what is the role of α-synuclein in sporadic PD, is the accumulation of α-synuclein protein in Lewy bodies a reflection of neurodegeneration or neuroprotection, what mediates the specificity of α-synuclein-dependent degeneration if at all, and what are the factors that may convert the normal intrinsic function of α-synuclein to ones implicated in neurodegenerative disease?

The development and characterization of α-synuclein-based animal models may be important to understanding the pathogenesis and mechanisms of cell death and provide a framework to test new therapeutic strategies. It is interesting to note that, as we begin to understand the role of α-synuclein, there appears to be a significant degree of commonality between PD and parkinsonism such as multisystem atrophy (MSA) and other neurodegenerative diseases including amyotrophic lateral sclerosis (ALS) and Alzheimer's disease.[396–399] Many of the disorders involving α-synuclein have been termed synucleinopathies.[400]

β-Synuclein

Investigations to identify mechanisms that protect against α-synuclein-induced neurodegeneration have implicated the related protein β-synuclein.[401] Coexpression of human α-synuclein and β-synuclein in transgenic mice showed no motor deficits, a reduced number of synuclein-positive inclusions, and increased striatal tyrosine hydroxylase immunoreactivity compared to mice overexpressing α-synuclein alone. *In vitro* studies have shown that β-synuclein reduced the aggregation of A53T α-synuclein in a dose-dependent manner.[402,403] Despite the fact that the function of β-synuclein is unknown, its interactions with related proteins such as α-synuclein may implicate one function in preventing aberrant protein accumulation.

Parkin

An autosomal recessive form of juvenile parkinsonism (AR-JP) led to the identification of a gene on chromosome 6q27 called parkin, now called PARK2.[404–406] Mutations in parkin may account for the majority of autosomal recessive familial cases of PD. Parkin protein has a large N-terminal ubiquitin-like domain and C-terminal cysteine ring structure, it is expressed in the brain, and it is present in Lewy Bodies.[407–409] Recent studies indicate that Parkin protein is an E3 ligase one of the many components of the complex protein degradation pathway in the ubiquitin/proteasome system.[410–413] Studies in both transgenic mice and *in vitro* are beginning to investigate the roles played by Parkin in the basal ganglia. Parkin plays a critical role in mediating interactions with a number of different proteins involved in the proteasome-mediated degradation pathway, including the E2-ubiquitin-conjugating protein, α-synuclein, and synphilin-1.[392,410,414,415] Mutations of the parkin gene have been introduced into transgenic mice. At present, there is very little known about pathological or behavioral alterations due to mutations in Parkin protein. However, parkin transgenic models enable investigation of the ubiquitin-mediated protein degradation pathway and its relationship to neurodegenerative disease.

UCH-1

Linkage analysis of another familial form of parkinsonism revealed a mutation in the gene encoding the protein, ubiquitin carboxy terminal hydrolase L1 (UCH-1).[416] UCH-1 mediates the release of ubiquitin from the polyubiquitin tail for reuse after digestion by the 26S proteasome. Altered UCH-1 function may lead to protein accumulation, including α-synuclein fibril formation. Therefore, transgenic mice targeting UCH-1 and other members of the ubiquitin/proteasome system will reveal potential mechanisms involving altered protein proteolysis and its resulting pathology. For example, a recent study employing a ubiquitin-green fluorescent protein fusion has been used to evaluate factors involved in proteasome function and may provide insight into pharmacological targeting

of this system.[417] These studies, as well as those dealing with α-synuclein, parkin, and related proteins, have shown that aberrations in protein function (including folding, protein-protein interactions, processing, and storage) may represent a common molecular mechanism between a number of neurodegenerative diseases and may reflect defective proteasomal function in many neurodegenerative diseases including PD.[400,418–420]

Transgenic Models Involving Free Radical Regulation

A variety of enzymes involved in scavenging free-radicals, including Cu/Zn-superoxide dismutase (SOD), glutathione peroxidase (GPx), glutathione reductase, and catalase, have been implicated in playing a key role in mechanisms involved in cell death of SNpc dopaminergic neurons.[5,12,323–325,328] For example, the identification of specific missense mutations encoding SOD protein in familial forms of amyotropic lateral sclerosis (ALS) has led to the development of SOD transgenic mice for studies not only in the mechanism of ALS but also PD.[421] For example, SOD transgenic mice have been shown to display degeneration of SNpc dopaminergic neurons.[422] Other transgenic mouse lines target several genes involved in free-radical scavenging and oxidative stress, including superoxide dismutase (SOD), glutathione reductase, glutathione peroxidase, and catalase. For example, transgenic mice expressing wild-type or mutant Cu/Zn-SOD have revealed important biochemical links between PD and ALS, as well as mechanisms (such as MPTP-lesioning) that may lead to midbrain dopaminergic cell death.[422–424] Extension of these studies using double transgenic mice (with mutations in both SOD and catalase) have led to important insights into the role of free radicals and injury due to reactive oxygen species in nigrostriatal dopaminergic neuronal cell death. Another interesting transgenic mouse model in this class of enzymes is one developed by Anderson and colleagues, having altered expression of GPx.[425,426] The GPx transgenic mouse displays altered levels of glutathione and, similar to the SOD and catalase transgenic models, has proven valuable in investigating the role of free radicals, mitochondrial dysfunction, energy depletion, and pharmacological and toxicant exposure.

Other Transgenic Mouse Models

A number of novel transgenic mouse models are emerging, with interesting results relevant to our understanding of basal ganglia development, function, and degeneration. Several of the genes involved encode proteins that show a high degree of specificity in the development of nigrostriatal dopaminergic neurons. For example, Nurr1, a transcription factor member of the zinc-finger nuclear receptor family, is expressed in the midbrain. Knockout of this gene leads to reduced striatal dopamine and loss of SNpc dopaminergic neurons. Nurr1 heterozygotes show increased susceptible to MPTP-lesioning compared to wild-type littermates.[427–429] Another interesting protein is NGFI-B (nerve growth factor inducible B) that is expressed in the striatum at the targets of the SNpc dopaminergic neurons. Neurotrophic factors that are important for basal ganglia development and function include GDNF (glial-derived neurotrophic factor) and BDNF (brain-derived neurotrophic factor). These transgenics have shown neuroprotection to different neurotoxicant paradigms and altered neuroplasticity in response to injury. The mouse model of the human ataxi-telangiectasia (ATm) also displays progressive loss of SNpc dopaminergic neurons.[430,431] Currently, the number of different transgenic models (either knockout, mutated, or altered expression) is growing at a tremendous rate. Recent transgenics include dopamine- and adenosine-3'5'-monophosphate regulated phosphoprotein of 32 kDa (DARPP-32), cAMP-responsive element binding protein (CREB), protein kinases, protein phosphatases, transcription factors, dopamine receptors, glutamate receptors, and members of other neurotransmitter systems. Mice lacking the dopamine transporter (DAT) have proven insightful into the function of DAT in behavior as well as revealing its role on TH expression, dopamine homeostasis, glutamate neurotransmission, vesicular storage, and susceptibility to MPTP.[432–436] The dopamine-deficient TH knockout transgenic mouse displays aberrant motor behavior due to altered dopamine receptor (D_1 and D_2) function.[437] Other important transgenic mouse models include those targeting monoamine oxidase (MAO), dopamine receptors (D_1, D_2), sonic hedgehog, caspases, neurotrophic factors (BDNF and GDNF), transcription factors like Pax6, and neurotransmitter receptors such as the glutamatergic receptor subtypes N-methyl-D-aspartate (NMDA) and α-amino-3-hydroxy-5-methyl-4-isoxaolepropionicacid (AMPA). These proteins are involved in various aspects of basal ganglia function including development, maintenance, susceptibility to neurotoxicant injury, and aging. The potential impact of these models on PD research will become evident in the not-too-distant future.[438]

INVERTEBRATE MODELS WITH α-SYNUCLEIN AND PARKIN

Recent developments of invertebrate transgenic models (such as in *Drosophila melanogaster*) for α-synuclein, parkin, and other genes of interest provide another avenue to investigate the function of proteins of interest in PD. In addition, the application of dopaminergic specific toxins such as 6-OHDA and MPTP/MPP+ to *Caenorhadbitis elegans,* may provide another tool for understanding mechanisms of cell death.[439] Unlike mammalian animal models, invertebrate models tend to be less expensive, and greater numbers can be generated in shorter periods of time. These advantages offer a means for high-volume screening of pharmacological agents for the treatment of PD.[440,441]

One of the first attempts to overexpress α-synuclein was made in *Drosophila melanogaster*, an organism that contains no endogenous α-synuclein or synuclein-like genes. The human wild-type and mutant (A30P and A53T) α-synuclein were overexpressed using the GAL4-responsive UAS pan-neuronal expression vector. In the transgenic *Drosophila*, the expression of high concentrations of mutant α-synuclein in the brain leads to the loss of dopamine neurons, the formation of α-synuclein-positive inclusions with morphology reminiscent of Lewy bodies and deficient motor output (which was reversible following treatment with antiparkinsonian medication).[440] This model has been helpful in examining the effects of different α-synuclein genotypes on dopamine survival, synuclein aggregation and behavior, as well as determining factors (such as the protein chaperone Hsp70) on α-synuclein mediated degeneration.[442] In mutant transgenic flies, there is selective depletion of dopaminergic neurons at 30 to 60 days of age. Transgenic flies also show loss of motor function, characterized by the impaired ability to climb. Impaired climbing behavior is manifested earlier in the more toxic A30P α-synuclein, compared to the wild-type or A53T protein. Studies investigating suppression of α-synuclein toxicity have identified proteins potentially involved in neuroprotection. For example, coexpression of the human chaperone protein heat-shock protein 70 (Hsp70) alleviated the toxicity of α-synuclein, including preventing the formation of protein aggregates.[442]

Drosophila transgenics expressing parkin have also been constructed.[443] Flies overexpressing human parkin showed protection of dopaminergic neurons from death compared to α-synuclein transgenics. They showed no evidence of inclusions or reduced ubiquitin in dopaminergic neurons. These results suggest that parkin protein may be neuroprotective by reducing the formation of protein accumulations in synuclein-related degeneration. Deletion of parkin in *Drosophila* did not show pathological features of parkinsonism including dopaminergic neuron cell loss. This may indicate that many of the necessary molecular factors needed for initiating dopaminergic specific neurodegeneration, including α-synuclein protein, protein processing pathways, and post-translational modifications (like glycosylation and phosphorylation), and protein degradation mechanisms, are either lacking or different in *Drosophila*.[440,444]

Despite the fact that invertebrates are not typically considered to develop human neurodegenerative diseases like Parkinson's disease or Alzheimer's disease, transgenics flies involving the synucleins, parkin, and tau provide an important link between *in vitro* molecular studies and models in phylogenetically higher animals such as rodents and nonhuman primates, and the human condition.[444–447]

CONCLUSION

Our understanding of Parkinson's disease and related disorders affecting the basal ganglia has been advanced through animal models using surgical, pharmacological, and neurotoxicant manipulation. The nonhuman primate, rodent, cat, fruit fly, and a wide variety of other species have provided models that have contributed to the development of symptomatic (dopamine modulation), neuroprotective (antioxidants, free-radical scavengers), and restorative (growth factors, cell transplantation, and gene delivery) therapies. In addition, these models have furthered our understanding of motor complications (wearing-off and dyskinesia), neuronal cell death, and neuroplasticity of the basal ganglia. Future direction in PD research is through the continued development of animal models with altered genes and proteins of interest. In conjunction with existing models, these genetic-based models may lead to the eventual cure of PD and related disorders.

ACKNOWLEDGMENTS

We would like to thank our colleagues at the University of Southern California for their support, including Kerry Nixon, Liz Hogg, Mark Liker, Beth Fisher, Mickie Welsh, Tom McNeill, Mark Lew, and Leslie Weiner. Studies in our laboratory were made possible through the generous support of the Parkinson's Disease Foundation, The Baxter Foundation, The Zumberge Foundation, The Ackerberg Foundation, The U.S. Army NETRP (to MWJ and to GMP), and NIH (NINDS RO144327-1 to MJ). Special thanks to the friends of the USC Parkinson's Disease Research Group for their generous support. Thank you to Nicolaus, Pascal, and Dominique for their patience and encouragement.

REFERENCES

1. Fahn, S. and Przedborski, S., Parkinsonism, in *Merritt's Neurology*, 10th ed., Rowland, L. P., Lippincott Williams and Wilkins, Philadelphia, PA, pp. 679, 2000.

2. Ma, S.Y. et al., Correlation between neuromorphometry in the substantia nigra and clinical features in Parkinson's disease using disector counts, *J. Neurol. Sci.*, *151* (1), 83, 1997.

3. Jellinger, K., Pathology of Parkinson's syndrome, in *Handbook of Experimental Pharmacology*, Calne, D. B., Spinger, Berlin, pp. 47, 1988.

4. Maguire-Zeiss, K. A. and Federoff, H. J., Convergent pathobiologic model of Parkinson's disease, *Ann. N. Y. Acad. Sci.*, 991, 152, 2003.

5. Dauer, W. and Przedborski, S., Parkinson's disease, mechanisms and models, *Neuron.*, 39 (6), 889, 2003.

6. Jenner, P., The contribution of the MPTP-treated primate model to the development of new treatment strategies for Parkinson's disease, *Parkinsonism Relat. Disord.*, 9 (3), 131, 2003.

7. Betarbet, R., Sherer, T. B., and Greenamyre, J. T., Animal models of Parkinson's disease, *Bioessays*,24 (4), 308, 2002.

8. Tolwani, R. J. et al., Experimental models of Parkinson's disease: insights from many models, *Lab. Anim. Sci.*, 49 (4), 363, 1999.

9. Schmidt, N. and Ferger, B., Neurochemical findings in the MPTP model of Parkinson's disease, *J. Neural. Transm.*, 108 (11), 1236, 2001.

10. Jenner, P., The MPTP-treated primate as a model of motor complications in PD: Primate model of motor complications, *Neurology*, 62 (6 Suppl. 3), S4, 2003.

11. Deumens, R., Blokland, A., and Prickaerts, J., Modeling Parkinson's disease in rats: an evaluation of 6-OHDA lesions of the nigrostriatal pathway, *Exp. Neurol.*, 175 (2), 303, 2002.

12. Blum, D. et al., Molecular pathways involved in the neurotoxicity of 6-OHDA, dopamine and MPTP: contribution to the apoptotic theory in Parkinson's disease, *Prog. Neurobiol.*, 6+5 (2), 135, 2001.

13. Gerlach, M. and Reiderer, P., Animal models of Parkinson's disease: an empirical comparison with the phenomenology of the disease in man, *J. Neural Transm.*, 103, 987, 1996.

14. Carlsson, A., The occurrence, distribution and physiological role of catecholamines in the nervous system, *Pharmacol. Rev.*, 11, 490, 1959.

15. Birkmayer, W. and Hornykiewicz, O., Der 1-3,4-dioxyphenylanin (l-DOPA)-effek bei der Parkinson-akinesia, *Klin. Wochenschr.*, 73, 787, 1961.

16. Ehringer, H. and Hornykiewicz, O., Verteilung von noradrenalin und dopamin (3-hydroxytyramin) in gehrindes menschen und ihr verhalten bei erkrankungen des extrapyramidalen systems, *Klin. Wochenschr.*, 38, 1238, 1960.

17. Cotzias, G. C., Papavasiliou, P. S., and Gellene, R., Modification of Parkinsonism-chronic treatment with L-dopa, *N. Engl. J. Med.*, 280 (7), 337, 1969.

18. Foley, P., The L-DOPA story revisited. Further surprises to be expected? *J. Neural Transm.*, Suppl., 60, 1, 2000.

19. Schultz, W., Depletion of dopamine in the striatum as an experimental model of Parkinsonism: direct effects and adaptive mechanisms, *Prog. Neurobiol.*, 18 (2–3), 121, 1982.

20. Senoh, S. et al., Chemical, enzymatic and metabolic studies on the mechanism of oxidation of dopamine, *J. Am. Chem. Soc.*, 81, 6236, 1959.

21. Stone, C.A. et al., Comparison of some pharmacologic effects of certain 6-substituted dopamine derivatives with reserpine, guanethidine, and metaraminol, *J. Pharmacol. Exp. Ther.*, 142, 147, 1963.

22. Porter, C. C., Totaro, J. A., and Stone, C. A., Effect of 6-hydroxydopamine and some other compounds on the

concentration of norepinephrine in the hearts of mice, *J. Pharmacol. Exp. Ther., 1*40, 308, 1963.

23. Glinka, Y. and Youdim, M. B. H., Mechanisms of 6-hydroxydopamine neurotoxicity, *J. Neural Transm., 50*, 55, 1997.

24. Ungerstedt, U., 6-Hydroxy-dopamine induced degeneration of central monoamine neurons, *Eur. J. Pharm., 5* (1), 107, 1968.

25. Ungerstedt, U. and Arbuthnott, G.W., Quantitative recording of rotational behavior in rats after 6-hydroxydopamine lesions of the nigrostriatal dopamine system, *Brain Res., 24*, 485, 1970.

26. Ungerstedt, U., Postsynaptic supersensitivity after 6-hydroxydopamine induced degeneration of the nigrostriatal dopamine system, *Acta Physiol. Scand.,* (Suppl.), 367, 69, 1971.

27. Sauer, H. and Oertel, W., Progressive degeneration of nigrostriatal dopamine neurons following intrastriatal terminal lesions with 6-hydroxydopamine: a combined retrograde tracing and immunocytochemical study in the rat, *Neurosci., 59* (2), 401, 1994.

28. Moroz, I. A. et al., Effects of sex and hormonal status on astrocytic basic fibroblast growth factor-2 and tyrosine hydroxylase immunoreactivity after medial forebrain bundle 6-hydroxydopamine lesions of the midbrain dopamine neurons, *Neurosci., 1*18 (2), 463, 2003.

29. Murray, H. E. et al., Dose- and sex-dependent effects of the neurotoxin 6-hydroxydopamine on the nigrostriatal dopaminergic pathway of adult rats: differential actions of estrogen in males and females, *Neurosci., 1*16 (1), 213, 2003.

30. Schwarting, R. K. et al., Relationships between indices of behavioral asymmetries and neurochemical changes following mesencephalic 6-hydroxydopamine injections, *Brain Res., 554* (1-2), 46, 1991.

31. Schwarting, R. K. and Huston, J. P., The unilateral 6-hydroxydopamine lesion model in behavioral brain research. Analysis of functional deficits, recovery and treatments, *Prog. Neurobiol., 20* (2-3), 275, 1996.

32. Schwarting, R. K. and Huston, J. P., Behavioral and neurochemical dynamics of neurotoxic meso-striatal dopamine lesions, *Neurotoxicol., 1*8 (3), 689, 1997.

33. Schallert, T. and Tillerson, J., Interventive strategies for degeneration of dopamine neurons in parkinsonism: Optimizing behavioral assessment of outcome, in *Central Nervous System Diseases*, Emerich, D., Dean, R., and Sanberg, P., Humana Press, Totawa, NJ, pp. 131, 2000.

34. Baunez, C., Nieoullon, A., and Amalric, M., In a rat model of parkinsonism, lesions of the subthalamic nucleus reverse increases of reaction time but induce a dramatic premature responding deficit, *J. Neurosci., 15* (10), 6531, 1995.

35. Tillerson, J. L. et al., Forced limb-use effects on the behavioral and neurochemical effects of 6-hydroxydopamine, *J. Neurosci., 21* (12), 4427, 2001.

36. Kirik, D., Rosenblad, C., and Bjorklund, A., Characterization of behavioral and neurodegenerative changes following partial lesions of the nigrostriatal dopamine

system induced by intrastriatal 6-hydroxydopamine in the rat, *Exp. Neurol., 152* (2), 259, 1998.

37. Reading, P. J. and Dunnett, S. B., 6-Hydroxydopamine lesoins of nigrostriatal neurons as an animal model of Parkinson's disease, in *Toxin-Induced Models of Neurological Disorders*, Woodruff, M. L. and Nonneman, A. J., Plenum Press, New York, pp. 89, 1994.

38. Schwarting, R. K. and Huston, J. P., Unilateral 6-hydroxydopamine lesions of meso-striatal dopamine neurons and their physiological sequelae, *Prog. in Neurobiol., 49* (3), 215, 1996.

39. Papa, S. M. et al., Motor fluctuations in levodopa treated parkinsonian rats: relation to lesion extent and treatment duration, *Brain Res., 662* (1-2), 69, 1994.

40. Henry, B., Crossman, A. R., and Brotchie, J. M., Characterization of enhanced behavioral responses to L-DOPA following repeated administration in the 6-hydroxydopamine-lesioned rat model of Parkinson's disease, *Exp. Neurol., 151* (2), 334, 1998.

41. Papa, S. M. et al., Reversal of levodopa-induced motor fluctuations in experimental parkinsonism by NMDA receptor blockade, *Brain Res., 701* (1-2), 13, 1995.

42. Chase, T. N., Engber, T. M., and Mouradian, M. M., Contribution of dopaminergic and glutamatergic mechanisms to the pathogenesis of motor response complications in Parkinson's disease, *Advan. Neurol., 69*, 497, 1996.

43. Chase, T. N., Konitsiotis, S., and Oh, J. D., Striatal molecular mechanisms and motor dysfunction in Parkinson's disease, *Adv. Neurol., 86*, 355, 2001.

44. Oh, J. D. et al., Enhanced tyrosine phosphorylation of striatal NMDA receptor subunits: effect of dopaminergic denervation and L-DOPA administration, *Brain Res., 813* (1), 150, 1998.

45. Zigmond, M. J. et al., Compensations after lesions of central dopaminergic neurons: some clinical and basic implications, *Trends Neurosci., 1*3 (7), 290, 1990.

46. Zigmond, M. J. et al., Neurochemical compensation after nigrostriatal bundle injury in an animal models of preclinical parkinsonism, *Arch. Neurol., 41*, 856, 1984.

47. Zigmond, M. J. et al., Compensatory responses to nigrostriatal bundle injury. Studies with 6-hydroxydopamine in an animal model of parkinsonism, in *Molec. Chem. Neuropath.*, Humana Press, Inc., pp. 185, 1989.

48. Zigmond, M. J., Abercrombie, E. D., and Stricker, E. M., Partial damage to nigrostriatal bundle: compensatory changes and the action of L-dopa, *J. Neural Transm.*, Suppl., 29, 217, 1990.

49. Zigmond, M. J., Do compensatory processes underlie the preclinical phase of neurodegenerative disease? Insights from an animal model of parkinsonism, *Neurobiol. Dis., 4* (3–4), 247, 1997.

50. Neve, K. A., Kozlowski, M. R., and Marshall, J. F., Plasticity of neostriatal dopamine receptors after nigrostriatal injury: relationship to recovery of sensorimotor functions and behavioral supersensitivity, *Brain Res., 244* (1), 33, 1982.

51. Meshul, C. K. et al., Time-dependent changes in striatal glutamate synapses following a 6-hydroxydopamine lesion, *Neurosci., 88* (1), 1, 1999.

52. Meshul, C. K. et al., Alterations in rat striatal glutamate synapses following a lesion of the cortico- and/or nigrostriatal pathway, *Exp. Neurol., 165* (1), 191, 2000.

53. Rosenblad, C. et al., In vivo protection of nigral dopamine neurons by lentiviral gene transfer of the novel GDNF-family member neublastin/artemin, *Mol. Cell. Neurosci., 15* (2), 199, 2000.

54. Rosenblad, C., Martinez-Serrano, A., and Bjorklund, A., Glial cell line-derived neurotrophic factor increases survival, growth and function of intrastriatal fetal nigral dopaminergic grafts, *Neurosci., 75* (4), 979, 1996.

55. Winkler, C. et al., Transplantation in the rat model of Parkinson's disease: ectopic versus homotopic graft placement, *Prog. Brain Res., 127*, 233, 2000.

56. Nikkhah, G. et al., A microtransplantation approach for cell suspension grafting in the rat Parkinson model: a detailed account of the methodology, *Neurosci., 63* (1), 57, 1994.

57. Rodriguez, M. et al., Dopamine cell degeneration induced by intraventricular administration of 6-hydroxydopamine in the rat: similarities with cell loss in Parkinson's disease, *Exp. Neurol., 169* (1), 163, 2001.

58. Annett, L. E. et al., Behavioral analysis of unilateral monoamine depletion in the marmoset, *Brain,115*, 825, 1992.

59. Davis, G. C. et al., Chronic parkinsonism secondary to intravenous injection of meperidine analogues, *Psychiatry Res., 1*, 249, 1979.

60. Langston, J. W. et al., Chronic parkinsonism in humans due to a product of meperidine-analog synthesis, *Science, 219*, 979, 1983.

61. Snow, B. J. et al., Pattern of dopaminergic loss in the striatum of humans with MPTP induced parkinsonism, *Neurol. Neurosurg. Psychiatry, 68* (3), 313, 2000.

62. Vingerhoets, F. J. et al., Positron emission tomographic evidence for progression of human MPTP-induced dopaminergic lesions, *Ann. Neurol., 36* (5), 765, 1994.

63. Calne, D. et al., Positron emission tomography after MPTP: observations relating to the cause of Parkinson's disease, *Nature, 317* (6034), 246, 1985.

64. Langston, J. W. and Ballard, P. A., Parkinson's disease in a chemist working with 1-methyl-4-phenyl-1,2,5,6-tertahydropyridine, *New Eng. J. Med., 309* (5), 310, 1983.

65. Langston, J. W. et al., MPTP-induced parkinsonism in humans: A review of the syndrome and observations relating to the phenomenon of tardive toxicity, in *MPTP: A Neurotoxin Producing a Parkinsonian Syndrome*, Markey, S. P., Castagnoli, N. J., Trevor, A. J., and Kopin, I. J., Academic Press, New York, pp. 9, 1986.

66. Przedborski, S. et al., The parkinsonian toxin 1-methyl-4-phenyl-1,2,3,6-tetrahydropyridine (MPTP): a technical review of its utility and safety, *J. Neurochem., 76* (5), 1265, 2001.

67. Muthane, U. et al., Differences in nigral neuron number and sensitivity to 1-methyl-4-phenyl-1,2,3,6-tetrahydropyridinium in C57/bl and CD-1 mice, *Exp. Neurol., 126*, 195, 1994.

68. Hamre, K. et al., Differential strain susceptibility following 1-methyl-4-phenyl-1,2,3,6-tetrahydropyridine

(MPTP) administration acts in an autosomal dominant fashion: quantitative analysis in seven strains of Mus musculus, *Brain Res., 828* (1–2), 91, 1999.

69. Heikkila, R. E., Differential neurotoxicity of 1-methyl-4-phenyl-1,2,3,6-tetrahydropyridine (MPTP) in Swiss-Webster mice from different sources, *Eur. J. Pharmacol., 117* (1), 131, 1985.

70. Jarvis, M. F. and Wagner, G. C., Age-dependent effects of 1-methyl-4-phenyl-1,2,3,6-tetrahydropyridine (MPTP), *Neuropharm., 24* (6), 581, 1985.

71. Ali, S. F. et al., MPTP-induced oxidative stress and neurotoxicity are age-dependent: evidence from measures of reactive oxygen species and striatal dopamine levels, *Synapse,18*, 27, 1994.

72. Saura, J., Richards, J., and Mahy, N., Age-related changes on MAO in Bl/C57 mouse tissues: a quantitative radioautographic study, *J. Neural Transm., 41* (Suppl.), 89, 1994.

73. Sedelis, M. et al., Chromosomal loci influencing the susceptibility to the parkinsonian neurotoxin 1-methyl-4-phenyl-1,2,3,6-tetrahydropyridine, *J. Neurosci., 23* (23), 8247, 2003.

74. Sedelis, M., Schwarting, R. K., and Huston, J. P., Behavioral phenotyping of the MPTP mouse model of Parkinson's disease, *Behav. Brain Res., 125* (1–2), 109, 2001.

75. Sedelis, M. et al., Evidence for resistance to MPTP in C57BL/6 x BALB/c F1 hybrids as compared with their progenitor strains, *Neuroreport,11* (5), 1093, 2000.

76. Smeyne, M., Goloubeva, O., and Smeyne, R. J., Strain-dependent susceptibility to MPTP and MPP(+)-induced parkinsonism is determined by glia, *Glia,34* (2), 73, 2001.

77. Jackson-Lewis, V. et al., Time course and morphology of dopaminergic neuronal death caused by the neurotoxin 1-methyl-4-phenyl-1,2,3,6-tetrahydropyridine, *Neurodegen., 4*, 257, 1995.

78. Mitchell, I. J. et al., Sites of the neurotoxic action of 1-methyl-4-phenyl-1,2,3,6-tetrahydropyridine in the monkey include the ventral tegmental area and the locus coeruleus, *Neurosci. Lett., 61*, 195, 1985.

79. Forno, L. S. et al., Neuropathology of MPTP-treated monkeys. Comparison with the neuropathology of human idiopathic Parkinson's Disease, in *MPTP: A Neurotoxin Producing a Parkinsonian Syndrome*, Markey, S.P., Castagnoli, N. J., Trevor, A. J., and Kopin, I. J., Academic Press, pp. 119, 1986.

80. Forno, L. S. et al., Locus ceruleus lesions and eosinophilic inclusions in MPTP-treated monkeys, *Ann. Neuol., 20*, 449, 1986.

81. Forno, L. S. et al., Neuropathology in MPTP-induced Parkinson in animals, *Proceedings of the Xth International Congress of Neuropathology*, 67, 1990.

82. Forno, L. S., Neuropathology of Parkinson's disease, *J. Neuropath. Exp. Neurol., 55* (3), 259, 1996.

83. Forno, L. S. et al., Similarities and differences between MPTP-induced parkinsonism and Parkinson's disease. Neuropathologic considerations, *Adv. Neurol., 60*, 600, 1993.

84. Ricaurte, G. A. et al., Fate of nigrostriatal neurons in young mature mice given 1-methyl-4-phenyl-1,2,3,6-

tetrahydropyridine: a neurochemical and morphological reassessment, *Brain Res.*, 376, 117, 1986.

85. Kalivas, P. W., Duffy, P., and Barrow, J., Regulation of the mesocortocolimbic dopamine system by glutamic acid receptor subtypes, *J. Pharmacol. Exp. Therap.*, 251 (1), 378, 1989.

86. Bezard, E. et al., Spontaneous long-term compensatory dopaminergic sprouting in MPTP-treated mice, *Synapse*, 38 (3), 363, 2000.

87. Langston, J. W. et al., Evidence of active nerve cell degeneration in the substantia nigra of humans years after 1-methyl-4-phenyl-1,2,3,6-tetrahydropyridine exposure, *Ann. Neurol.*, 46 (4), 598, 1999.

88. Royland, J. E. and Langston, J. W., MPTP: A dopamine neurotoxin, in *Highly Selective Neurotoxins*, Kostrzewa, R. M., Humana Press, Totawa, NJ, pp. 141, 1998.

89. Tatton, N. A. and Kish, S. J., *in situ* detection of apoptotic nuclei in the substantia nigra compacta of 1-methyl-4-phenyl-1,2,3,6-tetrahydropyridine-treated mice using terminal deoxynucleotidyl transferase labelling and acridine orange staining, *Neurosci.*, 77 (4), 1037, 1997.

90. Bezard, E. et al., Kinetics of nigral degeneration in a chronic model of MPTP-treated mice, *Neurosci. Lett.*, 234 (1), 47, 1997.

91. Bezard, E. et al., Effects of different schedules of MPTP administration on dopaminergic neurodegeneration in mice, *Exp. Neurol.*, 148 (1), 288, 1997.

92. Kuhn, K. et al., The mouse MPTP model: gene expression changes in dopaminergic neurons, *Eur. J. Neurosci.*, 17 (1), 1, 2003.

93. Tillerson, J. L. et al., Detection of behavioral impairments correlated to neurochemical deficits in mice treated with moderate doses of 1-methyl-4-phenyl-1,2,3,6-tetrahydropyridine, *Exp. Neurol.*, 178 (1), 80, 2002.

94. Collier, T. J., Steece-Collier, K., and Kordower, J. H., Primate models of Parkinson's disease, *Exp. Neurol.*, 183 (2), 258, 2003.

95. Fukuda, T., Neurotoxicity of MPTP, *Neuropath.*, 21 (4), 323, 2001.

96. Bingaman, K. D. and Bakay, R. A., The primate model of Parkinson's disease: its usefulness, limitations, and importance in directing future studies, *Prog. Brain Res.*, 127, 267, 2000.

97. Petzinger, G. M. and Langston, J. W., The MPTP-lesioned non-human primate: A model for Parkinson's disease, in *Advances in Neurodegenerative Disease.* Vol. I, *Parkinson's Disease*, Marwah, J. and Teitelbbaum, H. Prominent Press, Scottsdale, AZ, pp. 113, 1998.

98. Ho, A. and Blum, M., Induction of interleukin-1 associated with compensatory dopaminergic sprouting in the denervated striatum of young mice: model of aging and neurodegenerative disease, *J. Neurosci.*, 18 (15), 5614, 1998.

99. Jakowec, M. W. et al., Neuroplasticity in the MPTP-lesioned mouse and non-human primate, *Annals N. Y. Acad. Sci.*, 991, 298, 2003.

100. Fisher, B. et al., Activity-dependent plasticity in the MTPP-lesioned mouse, *Mov. Disord.*, 17 (Suppl. 5), S17, 2003.

101. Schallert, T. et al., CNS plasticity and assessment of forelimb sensorimotor outcome in unilateral rat models of stroke, cortical ablation, parkinsonism and spinal cord injury, *Neuropharm.*, 39 (5), 777, 2000.

102. Schallert, T. et al., Use-dependent structural events in recovery of function, *Adv. Neurol.*, 73, 229, 1997.

103. Arora P. K. et al., Chemical oxidation of 1-methyl-4-phenyl-1,2,3,6-tetrahydropyridine (MPTP) and its *in vivo* metabolism in rat brain and liver., *Biochem. Biophys. Res. Commun.*, 152 (3), 1339, 1988.

104. Gessner, W. et al., Studies on the mechanism of MPTP oxidation by human liver monoamine oxidase B, *FEBS Lett.*, 199 (1), 100, 1986.

105. Heikkila, R. E. et al., Studies on the oxidation of the dopaminergic neurotoxin 1-methyl-4-phenyl-1,2,5,6-tetrahydropyridine by monoamine oxidase B, *J. Neurochem.*, 45 (4), 1049, 1985.

106. Weissman, J. et al., Metabolism of the nigrostriatal toxin 1-methyl-4-phenyl-1,2,3,6-tetrahydropyridine by liver homogenate fractions, *J. Med. Chem.*, 28 (8), 997, 1985.

107. Fritz, R. R. et al., Metabolism of the neurotoxin in MPTP by human liver monoamine oxidase B, *FEBS Lett.*, 186 (2), 224, 1985.

108. Riachi, N. J. et al., On the mechanisms underlying 1-methyl-4-phenyl-1,2,3,6-tetrahydropyridine neurotoxicity. II. Susceptibility among mammalian species correlates with the toxin's metabolic patterns in brain microvessels and liver, *J. Pharmacol. Exp. Ther.*, 244 (2), 443, 1988.

109. Riachi, N. J., Behmand, R. A., and Harik, S. I., Correlation of MPTP neurotoxicity *in vivo* with oxidation of MPTP by the brain and blood-brain barrier *in vitro* in five rat strains, *Brain Res.*, 555 (1), 19, 1991.

110. Zuddas, A. et al., In brown Norway rats, MPP+ is accumulated in the nigrostriatal dopaminergic terminals but is not neurotoxic: a model of natural resistance to MPTP toxicity, *Exp. Neurol.*, 127, 54, 1994.

111. Giovanni, A. et al., Studies on species sensitivity to the dopaminergic neurotoxin 1-methyl-1,2,3,6-tetrahydropyridine. Part 1: systemic administration, *J. Pharmacol. Exp. Therap.*, 270, 1000, 1994.

112. Giovanni, A., Sonsalla, P. K., and Heikkila, R. E., Studies on species sensitivity to the dopaminergic neurotoxin 1-methyl-1,2,3,6-tetrahydropyridine. Part 2: central administration of 1-methyl-4-phenylpyridinium, *J. Pharmacol. Exp. Therap.*, 270, 1008, 1994.

113. Staal, R. G. et al., In vitro studies of striatal vesicles containing the vesicular monoamine transporter (VMAT2): rat versus mouse differences in sequestration of 1-methyl-4-phenylpyridinium, *J. Pharmacol. Exp. Ther.*, 293 (2), 329, 2000.

114. Petroske, E. et al., Mouse model of Parkinsonism: a comparison between subacute MPTP and chronic MPTP/probenecid treatment, *Neurosci.*, 106 (3), 589, 2001.

115. Zuddas, A. et al., MPTP-treatment combined with ethanol or acetaldhyde selectively destroys dopaminergic neurons in mouse substantia nigra, *Brain Res.*, 501, 1, 1989.

116. Zuddas, A. et al., Acetaldehyde directly enhances MPP+ neurotoxicity and delays its elimination from the striatum, *Brain Res.,* 501, 11, 1989.

117. Corsini, G. U. et al., 1-Methyl-4-phenyl-1,2,3,6-tetrahydropyridine (MPTP) neurotoxicity in mice is enhanced by ethanol or acetaldehyde, *Life Sci.,* 40 (9), 827, 1987.

118. Corsini, G. U. et al., 1-Methyl-4-phenyl-1,2,3,6-tetrahydropyridine (MPTP) neurotoxicity in mice is enhanced by pretreatment with diethyldithiocarbamate, *Eur. J. Pharmacol., 119* (1–2), 127, 1985.

119. Miller, D. B. et al., Diethyldithiocarbamate potentiates the neurotoxicity of *in vivo* 1-methyl-4-phenyl-1,2,3,6-tetrahydropyridine and of *in vitro* 1-methyl-4-phenylpyridinium, *J. Neurochem.,* 57 (2), 541, 1991.

120. Walters, T. L. et al., Diethyldithiocarbamate causes nigral cell loss and dopamine depletion with nontoxic doses of MPTP, *Exp. Neurol.,* 156 (1), 62, 1999.

121. McGrew, D. M., Irwin, I., and Langston, J. W., Ethylenebisdithiocarbamate enhances MPTP-induced striatal dopamine depletion in mice, *Neurotox.,* 21 (3), 309, 2000.

122. Irwin, I. et al., The effect of diethyldithiocarbamate on the biodisposition of MPTP: an explanation for enhanced neurotoxicity, *Eur. J. Pharmacol., 141* (2), 207, 1987.

123. Lau, Y. S. et al., Effects of probenecid on striatal dopamine depletion in acute and long-term 1-methyl-4-phenyl-1,2,3,6-tetrahydropyridine (MPTP)-treated mice, *Gen. Pharmacol.,* 21 (2), 181, 1990.

124. Langston, J. W. et al., Selective nigral toxicity after systemic administration of 1-methyl-4-phenyl-1,2,5,6,-tetrahydropyridine (MPTP) in the squirrel monkey, *Brain Res.,* 292, 390, 1984.

125. Burns, R. S. et al., A primate model of parkinsonism: selective destruction of dopaminergic neurons in the pars compacta of the substantia nigra by N-methyl-4-phenyl-1,2,3,6-tetrahydropyridine, *Proc. Natl. Acad. Sci. USA,* 80, 4546, 1983.

126. Chiueh, C. C. et al., Selective neurotoxic effects of N-methyl-4-phenyl-1,2,3,6-tetrahydropyridine (MPTP) in subhuman primates and man: a new animal model of Parkinson's disease, *Psychopharmac. Bull.,* 20, 548, 1984.

127. Crossman, A. R., Mitchell, I. J., and Sambrook, M. A., Regional brain uptake of 2-deoxyglucose in N-methyl-4-phenyl-1,2,3,6-tetrahydropyridine (MPTP)-induced parkinsonism in the macaque monkey, *Neuropharm.,* 24, 587, 1985.

128. Jenner, P. et al., MPTP-induced parkinsonism in the common marmoset: behavioral and biochemical effects, *Adv. Neurol.,* 45, 183, 1986.

129. Freed, C. R. et al., Fetal substantia nigra transplants lead to dopamine cell replacement and behavioral improvement in Bonnet monkeys with MPTP induced parkinsonism, In: *Pharmacology and Functional Regulation of Dopaminergic Neurons,* Beart, P. M., Woodruff, G. N., and Jackson, D. M. Macmillan Press, London, pp. 353, 1988.

130. Hantraye, P. et al., Stable parkinsonian syndrome and uneven loss of striatal dopamine fibres following chronic MPTP administration in baboons, *Neurosci.,* 53 (1), 169, 1993.

131. Moerlein, S. M. et al., Regional cerebral pharmacokinetics of the dopaminergic neurotoxin 1-methyl-4-phenyl-1,2,3,6-tetrahydropyridine as examined by positron emission tomography in a baboon is altered by tranylcypromine, *Neurosci. Lett.,* 66 (2), 205, 1986.

132. Taylor, J. R. et al., Behavioral effects of MPTP administration in the vervet monkey. A primate model of Parkinson's disease, in *Toxin-Induced Models of Neurological Disorders,* Woodruff, M. L. and Nonneman, A. J., Plenum Press, New York, pp. 139, 1994.

133. Raz, A., Vaadia, E., and Bergman, H., Firing patterns and correlations of spontaneous discharge of pallidal neurons in the normal and the tremulous 1-methyl-4-phenyl-1,2,3,6-tetrahydropyridine vervet model of parkinsonism, *J. Neurosci.,* 20 (22), 8559, 2000.

134. Tetrud, J. W. et al., MPTP-induced tremor in human and non-human primates, *Neurology,* 36 (Suppl.1), 308, 1986.

135. Ros, S. et al., Age-related effects of 1-methyl-4-phenyl-2,3,6-tetrahydropyridine treatment of common marmosets, *Eur. J. Pharm.,* 230, 177, 1993.

136. Ovadia, A., Zhang, Z., and Gash, D. M., Increased susceptibility to MPTP toxicity in middle-aged rhesus monkeys, *Neurobiol.Aging.,* 16 (6), 931, 1995.

137. German, D. C. et al., 1-Methyl-4-phenyl-1,2,3,6-tetrahydropyridine-induced parkinsonian syndrome in Macca fasicularis: which midbrain dopaminergic neurons are lost? *Neurosci.,* 24 (1), 161, 1988.

138. Jenner, P. and Marsden, C. D., MPTP-induced parkinsonism as an experimental model of Parkinson's disease, in *Parkinson's Disease and Movement Disorders,* Jankovic, J. and Tolosa, E., Urban and Schwarzenberg, Inc., Baltimore, pp. 37, 1988.

139. Irwin, I. et al., The evolution of nigrostriatal neurochemical changes in the MPTP-treated squirrel monkey, *Brain Res.,* 531, 242, 1990.

140. Oiwa, Y. et al., Overlesioned hemiparkinsonian non human primate model: correlation between clinical, neurochemical and histochemical changes, *Front. Biosci.,* 8, 155, 2003.

141. M., M., Experimental parkinsonism in primates., *Stereotact. Funct. Neurosurg.,* 77 (1–4), 91, 2001.

142. Tetrud, J. W. and Langston, J. W., MPTP-induced parkinsonism as a model for Parkinson's disease, *Acta Neurol. Scand., 126,* 35, 1989.

143. Elsworth, J. D. et al., MPTP-induced parkinsonism: relative changes in dopamine concentration in subregions of substantia nigra, ventral tegmental area and retrorubal field of symptomatic and asymptomatic vervet monkeys, *Brain Res.,* 513, 320, 1990.

144. Waters, C. M. et al., An immunohistochemical study of the acute and long-term effects of 1-methyl-4-phenyl-1,2,3,6-tetrahydropyridine in the marmoset, *Neurosci.,* 23 (3), 1025, 1987.

145. Eidelberg, E. et al., Variability and functional recovery in the N-methyl-4-phenyl-1,2,3,6-tetrahydropyridine model of parkinsonism in monkeys, *Neurosci., 18* (4), 817, 1986.

146. Langston, J. W. et al., Investigating levodopa-induced dyskinesias in the parkinsonian primate, *Annals Neurol.*, 47 (4 Suppl. 1), S79, 2000.

147. Bankiewicz, K. S. et al., Hemiparkinsonism in monkeys after unilateral internal carotid infusion of 1-methyl-4-phenyl-1,2,3,6-tetrahydropyridine, *Life Sci.*, 39, 7, 1986.

148. Smith, R. et al., Developing a stable bilateral model of parkinsonism in rhesus monkeys, *Neurosci.*, 52 (1), 7, 1993.

149. Eberling, J. L. et al., A novel MPTP primate model of Parkinson's disease: neurochemical and clinical changes, *Brain Res.*, 805 (1–2), 259, 1998.

150. Bankiewicz, K. S. et al., Transient behavioral recovery in hemiparkinsonian primates after adrenal medullary allografts, *Prog. Brain Res.*, 78, 543, 1988.

151. Bankiewicz, K S. et al., The effect of fetal mesencephalon implants on primate MPTP-induced parkinsonism. Histochemical and behavioral studies, *J. Neurosurg.*, 72 (2), 231, 1990.

152. Bankiewicz, K. S. et al., Fetal nondopaminergic neural implants in parkinsonian primates. Histochemical and behavioral studies, *J. Neurosurg.*, 74, 97, 1991.

153. Bezard, E. et al., A chronic MPTP model reproducing the slow evolution of Parkinson's disease: evolution of motor symptoms in the monkey, *Brain Res.*, 766 (1–2), 107, 1997.

154. Schneider, J. S. and Markham, C. H., Neurotoxic effects of N-methyl-4-1,2,3,6-tetrahydropyridine (MPTP) in the cat. Tyrosine hydroxylase immunohistochemistry, *Brain Res.*, 258, 1986.

155. Schneider, J. S., Yuwile, R. A., and Markham, C. H., Production of a parkinson-like syndrome in the cat with N-methyl-4 phenyl-1,2,3,6-tetrahydropyridine (MPTP): behavior, histology, and biochemistry, *Exp. Neurol.*, 91, 293, 1986.

156. Schneider, J. S., MPTP parkinsonism in the cat: pattern of neuronal loss may partially be explained by the distribution of MAO-A in the brain, in *The Basal Ganglia II: Concepts*, Carpenter, M.B. and Jayaraman, A. Plenum Press, New York, 1987.

157. Schneider, J. S. and Rothblat, D. S., Neurochemical evaluation of the striatum in symptomatic and recovered MPTP-treated cats, *Neurosci.*, 44 (2), 421, 1991.

158. Rothblat, D. S., Schroeder, J. A., and Schneider, J. S., Tyrosine hydroxylase and dopamine transporter expression in residual dopaminergic neurons: Potential contributors to spontaneous recovery from experimental Parkinsonism, *J. Neurosci. Res.*, 65 (3), 254, 2001.

159. Rothblat, D. S. and Schneider, J. S., Responses of caudate neurons to stimulation of intrinsic and peripheral afferents in normal, symptomatic, and recovered MPTP-treated cats, *J. Neurosci.*, 13 (10), 4372, 1993.

160. Frohna, P. A. et al., Alterations in dopamine uptake sites and D_1 and D_2 receptors in cats symptomatic for and recovered from experimental parkinsonism, *Synapse*, 19 (1), 46, 1995.

161. Rothblat, D. S. and Schneider, J. S., Regional differences in striatal dopamine uptake and release associated with recovery from MPTP-induced parkinsonism: an *in vivo* electrochemical study, *J. Neurochem.*, 72 (2), 724, 1999.

162. Cruz-Sanchez, F. F. et al., Plasticity of the nigrostriatal system in MPTP-treated mice. A biochemical and morphological correlation, *Mol. Chem. Neuropathol.*, 19 (1–2), 163, 1993.

163. Elsworth, J. D. et al., Striatal dopaminergic correlates of stable parkinsonism and degree of recovery in old-world primates one year after MPTP treatment, *Neurosci*, 95 (2), 399, 2000.

164. Sandyk, R., Mechanisms of recovery in MPTP-induced parkinsonism, *Neurosci.*, 27 (2), 727, 1988.

165. Altar, C. A. et al., 1-Methyl-4-phenylpyridine (MPP+): regional dopaminergic uptake, toxicity, and novel rotational behavior following dopamine receptor proliferation, *Eur. J. Pharmacol.*, 131, 199, 1986.

166. Kitamura, Y. et al., Inhibitory effects of antiparkinsonian drugs and caspase inhibitors in a parkinsonian flatworm model, *J. Pharmacol. Sci.*, 92 (2), 137, 2003.

167. Kitamura, Y., Kakimura, J., and Taniguchi, T., Protective effect of talipexole on MPTP-treated planarian, a unique parkinsonian worm model, *Jpn. J. Pharmacol.*, 78 (1), 23, 1998.

168. Youdim, M. B. et al., MPTP-induced "parkinsonism" in the goldfish, *Neurochem. Int.*, 20 (Suppl.), 275S, 1992.

169. Poli, A. et al., Effect of 1-methyl-4-phenyl-1,2,3,6-tetrahydropyridine (MPTP) in goldfish brain, *Brain Res.*, 534 (1–2), 45, 1990.

170. Poli, A. et al., Spontaneous recovery of MPTP-damaged catecholamine systems in goldfish brain areas, *Brain Res.*, 585 (1–2), 128, 1992.

171. Ryan, R. W. et al., Catecholaminergic neuronal degeneration in rainbow trout assessed by skin color change: a model system for identification of environmental risk factors, *Neurotox.*, 23 (4–5), 545, 2002.

172. Barbeau, A. et al., Comparative behavioral, biochemical and pigmentary effects of MPTP, MPP+ and paraquat in Rana pipiens, *Life Sci.*, 37 (16), 1529, 1985.

173. Barbeau, A. et al., Studies on MPTP, MPP+ and paraquat in frogs and *in vitro*, in *MPTP:A Neurotoxin Producing A Parkinsonian Syndrome*, Markey, S., Castagnoli, N. J., Trevor, A. J., and Kopin, I. J., Academic Press, New York, pp. 85, 1986.

174. Barbeau, A. et al., New amphibian models for the study of 1-methyl-4-phenyl-1,2,3,6-tetrahydropyridine (MPTP), *Life Sci.*, 36 (11), 1125, 1985.

175. Temple, J. G., Miller, D. B., and Barthalmus, G. T., Differential vulnerability of snake species to MPTP: a behavioral and biochemical comparison in ratsnakes (Elaphe) and watersnakes (Nerodia), *Neurotoxicol. Teratol.*, 24 (2), 227, 2002.

176. Lopez, K. H. et al., Catecholaminergic cells and fibers in the brain of the lizard Anolis carolinensis identified by traditional as well as whole-mount immunohistochemistry, *Cell Tissue Res.*, 270 (2), 319, 1992.

177. Sedlacek, J., Sensitivity of the generator of spontaneous motility in chick embryos to the acute and chronic administration of MPTP, *Physiol. Res.*, 41 (2), 109, 1992.

178. Weissman, E. M. et al., The effect of prenatal treatment with MPTP or MPP+ on the development of dopamine-

mediated behaviors in rats, *Pharmacol. Biochem. and Behav.,* 34, 545, 1988.

179. Chiueh, C. C. et al., Neurochemical and behavioral effects of systemic and intranigral administration of N-methyl-4-phenyl-1,2,3,6-tetrahydropyridine in the rat, *Eur. J. Pharmacol., 1*00 (2), 189, 1984.

180. Carvey, P. M., Kao, L. C., and Klawans, H. L., Permanent postural effects of MPTP in the guinea pig, in *MPTP: A Neurotoxin Producing a Parkinsonian Syndrome.*, Markey, S., Castagnoli, N. J., Trevor, A. J., and Kopin, I. J., Academic Press, New York, pp. 407, 1986.

181. Lermontova, N. et al., Relative resistance of rabbits to MPTP neurotoxicity, *Mol. Chem. Neuropathol.,* 25 (2–3), 135, 1995.

182. Parisi, J. E. and Burns, R. S., The neuropathology of MPTP-induced parkinsonism in man and experimental animals, in *MPTP:A Neurotoxin Producing a Parkinsonian Syndrome*, Markey, S., Castagnoli, N. J., Trevor, A. J., and Kopin, I. J. Academic Press, New York, pp. 141., 1986.

183. Wilms, H., Sievers, J., and Deuschl, G., Animal models of tremor, *Mov. Disord., 1*4 (4), 557, 1999.

184. Kostic, V. S. et al., The effect of stage of Parkinson's disease at the onset of levodopa therapy on development of motor complications, *Eur. J. Neurol.,* 9 (1), 9, 2002.

185. Colosimo, C. and De Michele, M., Motor fluctuations in Parkinson's disease: pathophysiology and treatment, *Eur. J. Neurol.,* 6 (1), 1, 1999.

186. Vitek, J. L. and Giroux, M., Physiology of hypokinetic and hyperkinetic movement disorders: model for dyskinesia, *Ann. Neurol.,* 47 (4 Suppl. 1), S131, 2000.

187. Lundblad, M. et al., Pharmacological validation of behavioural measures of akinesia and dyskinesia in a rat model of Parkinson's disease, *Eur. J. Neurosci., 1*5 (1), 120, 2002.

188. Cenci, M A., Whishaw, I. Q., and Schallert, T., Animal models of neurological deficits: how relevant is the rat?, *Nat. Rev. Neurosci.,* 3 (7), 574, 2002.

189. Clarke, C. E. et al., Drug-induced dyskinesia in primates rendered hemiparkinsonian by intracarotid administration of 1-methyl-4-phenyl-1,2,3,6-tetrahydropyridine (MPTP), *J. Neurol. Sci.,* 90, 307, 1989.

190. Clarke, C. E. et al., Levodopa-induced dyskinesia and response fluctuations in primates rendered parkinsonian with 1-methyl-4-phenyl-1,2,3,6-tetrahydropyridine (MPTP), *J. Neurol.l Sci.,* 78, 273, 1987.

191. Crossman, A. R., Primate models of dyskinesia: the experimental approach to the study of basal ganglia-related involuntary movement disorders, *Neurosci.,* 21 (1), 1, 1987.

192. Pearce, R. K. B. et al., Chronic L-DOPA administration induces dyskinesia in the 1-methyl-4-phenyl-1,2,3,6-tetrahydropyridine-treated common Marmoset (Callithrix jacchus), *Mov. Disord., 1*0 (6), 731, 1995.

193. Pearce, R. K. et al., L-dopa induces dyskinesia in normal monkeys: behavioural and pharmacokinetic observations, *Psychopharm. (Berl),* 156 (4), 402, 2001.

194. Togasaki, D. M. et al., Levodopa induces dyskinesias in normal squirrel monkeys, *Ann. Neurol.,* 50 (2), 254, 2001.

195. Bezard, E., Brotchie, J. M., and Gross, C. E., Pathophysiology of levodopa-induced dyskinesia: potential for new therapies, *Nat. Rev. Neurosci.,* 2 (8), 577, 2001.

196. Hurley, M. J., Mash, D. C., and Jenner, P., Dopamine D_1 receptor expression in human basal ganglia and changes in Parkinson's disease, *Brain Res. Mol. Brain Res.,* 87 (2), 271, 2001.

197. Papa, S. M. and Chase, T. N., Levodopa-induced dyskinesias improved by a glutamate antagonist in parkinsonian monkeys, *Ann. Neurol.,* 39, 574, 1996.

198. Bedard, P. J. et al., Levodopa-induced dyskinesia: facts and fancy. What does the MPTP monkey model tell us? *Can.J. Neurol. Sci., 1*9, 134, 1992.

199. Calon, F. et al., Alteration of glutamate receptors in the striatum of dyskinetic 1-methyl-4-phenyl-1,2,3,6-tetrahydropyridine-treated monkeys following dopamine agonist treatment, *Prog. Neuropsychopharmacol. Biol. Psychiatry,* 26 (1), 127, 2002.

200. Annett, L. E. et al., Behavioral assessment of the effects of embryonic nigral grafts in marmosets with unilateral 6-OHDA lesions of the nigrostriatal pathway, *Exp. Neurol., 1*25 (2), 228, 1994.

201. Armstrong, R. J. et al., Transplantation of expanded neural precursor cells from the developing pig ventral mesencephalon in a rat model of Parkinson's disease, *Exp. Brain Res., 1*51 (2), 204, 2003.

202. Elsworth, J. D. et al., Restoration of dopamine transporter density in the striatum of fetal ventral mesencephalon-grafted, but not sham-grafted, MPTP-treated parkinsonian monkeys, *Cell Transplant.,* 5 (2), 315, 1996.

203. Storch, A. et al., Long-term proliferation and dopaminergic differentiation of human mesencephalic neural precursor cells, *Exp. Neurol., 1*70 (2), 317, 2001.

204. Annett, L. E. et al., A comparison of the behavioural effects of embryonic nigral grafts in the caudate nucleus and in the putamen of marmosets with unilateral 6-OHDA lesions, *Exp. Brain Res., 1*03 (3), 355, 1995.

205. Di Porzio, U. and Zuddas, A., Embryonic dopaminergic neuron transplants in MPTP lesioned mouse striatum, *Neurochem. Int.,* 20 (Suppl.), 309S, 1992.

206. Sawamoto, K. et al., Visualization, direct isolation, and transplantation of midbrain dopaminergic neurons, *Proc. Natl. Acad. Sci. USA,* 98 (11), 6423, 2001.

207. Espejo, E. F. et al., Cellular and functional recovery in parkinsonian rats after intrastriatal transplantation of carotid body cell aggregates, *Neuron,* 20, 197, 1998.

208. Borlongan, C. V., Saporta, S., and Sanberg, P. R., Intrastriatal transplantation of rat adrenal chromaffin cells seeded on microcarrier beads promote long-term functional recovery in hemiparkinsonian rats, *Exp. Neurol.,* 151 (2), 203, 1998.

209. Fine, A. et al., Transplantation of adrenal tissue into the central nervous system, *Brain Res. Rev.,* 15, 121, 1990.

210. Watts, R. L., Mandir, A. S., and Bakay, R. A., Intrastriatal cografts of autologous adrenal medulla and sural nerve in MPTP-induced parkinsonian macaques: behavioral and anatomical assessment, *Cell Transplant.,* 4 (1), 27, 1995.

211. Fiandaca, M. S. et al., Adrenal medullary autografts into the basal ganglia of Cebus monkeys: injury-induced regeneration, *Exp. Neurol., 1*02 (1), 76, 1988.

212. Lund, R. D., Radel, J. D., and Coffey, P. J., The impact of intracerebral retinal transplants on types of behavior exhibited by host rats, *Trens. Neurosci., 1*4 (8), 358, 1991.

213. Aebischer, P. et al., Functional recovery in hemiparkinsonian primates transplanted with polymer-encapsulated PC12 cells, *Exp. Neurol., 1*26, 151, 1994.

214. Kordower, J. H. et al., Encapsulated PC12 cell transplants into hemiparkinsonian monkeys: a behavioral, neuroanatomical, and neurochemical analysis, *Cell Transplant.,* 4 (2), 155, 1995.

215. Isacson, O., Bjorklund, L. M., and Schumacher, J. M., Toward full restoration of synaptic and terminal function of the dopaminergic system in Parkinson's disease by stem cells, *Ann. Neurol.,* 53 (Suppl. 3), S135, 2003.

216. Bjorklund A et al., Neural transplantation for the treatment of Parkinson's disease., *Lancet Neurol.,* 2 (7), 437, 2003.

217. Redmond, D. E., Cellular replacement therapy for Parkinson's disease: where we are today?, *Neuroscientist,* 8 (5), 457, 2002.

218. Liker, M. A. et al., Human neural stem cell transplantation in the MPTP-lesioned mouse, *Brain Res.,* 971 (2), 168, 2003.

219. Svendsen, C. N. and Smith, A. G., New prospects for human stem-cell therapy in the nervous system, *Trends Neurosci.,* 22 (8), 357, 1999.

220. Reh, T., Neural stem cells: form and function, *Nat. Neurosci.,* 5 (5), 392, 2002.

221. Sommer, L. and Rao, M., Neural stem cells and regulation of cell number, *Prog. Neurobiol.,* 66 (1), 1, 2002.

222. Dunnett, S. B. and Bjorklund, A., Prospects for new restorative and neuroprotective treatments in Parkinson's disease, *Nature,* 399 (6838), A32, 1999.

223. Temple, S., Stem cell plasticity - building the brain of our dreams, *Nat. Rev. Neurosci.,* 2 (7), 513, 2001.

224. Freed, C. R. et al., Transplantation of embryonic dopamine neurons for severe Parkinson's disease, *N. Engl. J. Med.,* 344 (10), 710, 2001.

225. Freed, C. R., Will embryonic stem cells be a useful source of dopamine neurons for transplant into patients with Parkinson's disease? *Proc. Natl. Acad. Sci. USA,* 99 (4), 1755, 2002.

226. Lindvall, O. and Hagell, P., Cell therapy and transplantation in Parkinson's disease, *Clin. Chem. Lab. Med.,* 39 (4), 356, 2001.

227. Tasker, R. R., Lang, A. E., and Lozano, A. M., Pallidal and thalamic surgery for Parkinson's disease, *Exp. Neurol., 1*44, 35, 1997.

228. Limousin-Dowsey, P. et al., Thalamic, subthalamic nucleus and internal pallidum stimulation in Parkinson's disease, *J. Neurol.,* 246 (Suppl. 2), 1142, 1999.

229. Raz, A. et al., Activity of pallidal and striatal tonically active neurons is correlated in MPTP-treated monkeys but not in normal monkeys, *J. Neurosci.,* 21 (3), RC128, 2001.

230. Maurice, N. et al., Prefrontal cortex-basal ganglia circuits in the rat: involvement of ventral pallidum and subthalamic nucleus, *Synapse,* 29 (4), 363, 1998.

231. Parent, A. and Hazrati, L. N., Functional anatomy of the basal ganglia. II. The place of subthalamic nucleus and external pallidum in basal ganglia circuitry, *Brain Res. Brain Res. Rev.,* 20 (1), 128, 1995.

232. Parent, A. and Hazrati, L. N., Functional anatomy of the basal ganglia. I. The cortico-basal ganglia-thalamo-cortical loop, *Brain Res. Brain Res. Rev.,* 20 (1), 91, 1995.

233. Joel, D. and Weiner, I., The connections of the primate subthalamic nucleus: indirect pathways and the open-interconnected scheme of basal ganglia-thalamocortical circuitry, *Brain Res. Rev.,* 23, 62, 1997.

234. Gouhier, C. et al., Protection of dopaminergic nigrostriatal afferents by GDNF delivered by microspheres in a rodent model of Parkinson's disease, *Synapse,* 44 (3), 124, 2002.

235. McBride, J. L. and Kordower, J. H., Neuroprotection for Parkinson's disease using viral vector-mediated delivery of GDNF, *Prog. Brain Res., 1*38, 421, 2002.

236. Kordower, J. H., In vivo gene delivery of glial cell line—derived neurotrophic factor for Parkinson's disease, *Ann. Neurol.,* 53 (Suppl. 3), S120, 2003.

237. Kordower, J. H. et al., Neurodegeneration prevented by lentiviral vector delivery of GDNF in primate models of Parkinson's disease, *Science,* 290 (5492), 767, 2000.

238. Bilang-Bleuel, A. et al., Intrastriatal injection of an adenoviral vector expressing glial-cell-line-derived neurotrophic factor prevents dopaminergic neuron degeneration and behavioral impairment in a rat model of Parkinson disease, *Proc. Natl. Acad. Sc.i U.S.A.,* 94 (16), 8818, 1997.

239. Luo, J. et al., Subthalamic GAD gene therapy in a Parkinson's disease rat model, *Science,* 298 (5592), 425, 2002.

240. Kirik, D. and Bjorklund, A., Modeling CNS neurodegeneration by overexpression of disease-causing proteins using viral vectors, *Trends Neurosci.,* 26 (7), 386, 2003.

241. Shen, Y. et al., Triple transduction with adeno-associated virus vectors expressing tyrosine hydroxylase, aromatic-L-amino-acid decarboxylase, and GTP cyclohydrolase I for gene therapy of Parkinson's disease, *Hum. Gene Ther., 1*1 (11), 1509, 2004.

242. Muramatsu, S. et al., Behavioral recovery in a primate model of Parkinson's disease by triple transduction of striatal cells with adeno-associated viral vectors expressing dopamine-synthesizing enzymes, *Hum. Gene Ther., 1*3 (3), 345, 2002.

243. Azzouz, M. et al., Multicistronic lentiviral vector-mediated striatal gene transfer of aromatic L-amino acid decarboxylase, tyrosine hydroxylase, and GTP cyclohydrolase I induces sustained transgene expression, dopamine production, and functional improvement in a rat model of Parkinson's disease, *J. Neurosci.,* 22 (23), 10302, 2002.

244. Cruikshank, S. J. and Weinberger, N. M., Evidence for the Hebbian hypothesis in experience-dependent phys-

iological plasticity of neocortex: a critical review, *Brain Res. Rev.*, 22, 191, 1996.

245. Kurlan, R., Kim, M. H., and Gash, D. M., The time course and magnitude of spontaneous recovery of parkinsonism produced by intracarotid administration of 1-methyl-4-phenyl-1,2,3,6-tetrahydropyridine to monkeys, *Ann. Neurol.*, 29, 677, 1991.

246. Scotcher, K. P. et al., Mechanism of accumulation of the 1-methyl-4-phenyl-phenylpyridinium species into mouse brain synaptosomes, *J. Neurochem.*, 56, 1602, 1991.

247. Schneider, J. S. et al., Differential recovery of volitional motor function, lateralized cognitive function, dopamine-agonist-induced rotation and dopaminergic parameters in monkeys made hemi-parkinsonian by intracarotid MPTP infusion, *Brain Res.*, 672, 112, 1995.

248. Schneider, J. S., Rothblat, D. S., and DiStefano, L., Volume transmission of dopamine over large distances may contribute to recovery from experimental parkinsonism, *Brain Res.*, 643, 86, 1994.

249. Albanese, A. et al., Chronic administration of 1-methyl-4-phenyl-1,2,3,6-tetrahydropyridine to monkeys: behavioural, morphological and biochemical correlates, *Neurosci.*, 55 (3), 823, 1993.

250. Taylor, J.R. et al., Severe long-term 1-methyl-4-phenyl-1,2,3,6-tetrahydropyridine-induced parkinsonism in the vervet monkey (Cercopithecus aethiops sabaeus), *Neurosci.*, 81 (3), 745, 1997.

251. Rose, S. et al., Increased caudate dopamine turnover may contribute to the recovery of motor function in marmosets treated with the dopaminergic neurotoxin MPTP, *Neurosci. Lett.*, 101, 305, 1989.

252. Mori, S. et al., Immunohistochemical evaluation of the neurotoxic effects of 1-methyl-4-phenyl-1,2,3,6-tetrahydropyridine (MPTP) on dopaminergic nigrostriatal neurons of young adult mice using dopamine and tyrosine hydroxylase antibodies, *Neurosci. Lett.*, 90, 57, 1988.

253. Nishi, K., Kondo, T., and Narabayashi, H., Difference in recovery patterns of striatal dopamine content, tyrosine hydroxylase activity and total biopterin content after 1-methyl-4-phenyl-1,2,3,6-tetrahydropyridine (MPTP) administration: a comparison of young and older mice, *Brain Res.*, 489, 157, 1989.

254. Russ, H. et al., Neurochemical and behavioural features induced by chronic low dose treatment with 1-methyl-4-phenyl-1,2,3,6-tetrahydropyridine (MPTP) in the common marmoset: implications for Parkinson's disease?, *Neurosci. Lett.*, 123 (1), 115, 1991.

255. Rozas, G. et al., Sprouting of the serotonergic afferents into striatum after selective lesion of the dopaminergic system by MPTP in adult mice, *Neurosci. Lett.*, 245 (3), 151, 1998.

256. Morgan, S., Nomikos, G., and Huston, J. P., Changes in the nigrostriatal projections associated with recovery from lesion-induced behavioral asymmetry, *Behav. Brain Res.*, 46, 157, 1991.

257. Mitsumoto, Y. et al., Spontaneous regeneration of nigrostriatal dopaminergic neurons in MPTP-treated C57BL/6 mice, *Biochem. Biophys. Res. Comm.*, 248 (3), 660, 1998.

258. Wade, T.V., Rothblat, D. S., and Schneider, J. S., Changes in striatal dopamine D_3 receptor regulation during expression of and recovery from MPTP-induced parkinsonism, *Brain Res.*, 905 (1–2), 111, 2001.

259. Bezard, E. and Gross, C., Compensatory mechanisms in experimental and human parkinsonism: towards a dynamic approach, *Prog. Neurobiol.*, 55, 96, 1998.

260. Weihmuller, F. B., Hadjiconstantinou, M., and Bruno, J. P., Dissociation between biochemical and behavioral recovery in MPTP-treated mice, *Pharmacol. Biochem. Behav.*, 34, 113, 1989.

261. Rousselet, F. et al., Behavioral changes are not directly related to striatal monoamine levels, number of nigral neurons, or dose of parkinsonian toxin MPTP in mice, *Neurobiol. Dis.*, 14 (2), 218, 2003.

262. Davidson, C. et al., Methamphetamine neurotoxicity: necrotic and apoptotic mechanisms and relevance to human abuse and treatment, *Brain Res. Brain Res. Rev.*, 36 (1), 1, 2001.

263. Kita, T., Wagner, G. C., and Nakashima, T., Current research on methamphetamine-induced neurotoxicity: animal models of monoamine disruption, *J. Pharmacol. Sci.*, 92 (3), 178, 2003.

264. Ricaurte, G. A., Schuster, C. R., and Seiden, L. S., Long-term effects of repeated methylamphetamine administration on dopamine and serotonin neurons in the rat brain: a regional study, *Brain Res.*, 193, 153, 1980.

265. Ricaurte, G. A. et al., Dopamine nerve terminal degeneration produced by high doses of methylamphetamine in the rat brain, *Brain Res.*, 235 (1), 93, 1982.

266. Seiden, L. S., Fischman, M. W., and Schuster, C. R., Long-term methamphetamine induced changes in brain catecholamines in tolerant rhesus monkeys, *Drug Alcohol Depend.*, 1 (3), 215, 1976.

267. Fischman, M. W. and Schuster, C. R., Tolerance development to chronic methamphetamine intoxication in the rhesus monkey, *Pharmacol. Biochem. Behav.*, 2 (4), 503, 1974.

268. Wagner, G. C. et al., Long-lasting depletions of striatal dopamine and loss of dopamine uptake sites following repeated administration of methamphetamine, *Brain Res. Rev.*, 181 (1), 151, 1980.

269. Wagner, G. C. et al., Amphetamine induces depletion of dopamine and loss of dopamine uptake sites in caudate, *Neurology*, 30 (5), 547, 1980.

270. Harvey, D. C., Lacan, G., and Melega, W. P., Regional heterogeneity of dopaminergic deficits in vervet monkey striatum and substantia nigra after methamphetamine exposure, *Exp. Brain Res.*, 133, 349, 2000.

271. Kim, B. G. et al. Relative sparing of calretinin containing neurons in the substantia nigra of 6-OHDA treated rat Parkinsonian model, *Brain Res.*, 855 (1), 162, 2000.

272. Deng, X. et al., Methamphetamine causes widespread apoptosis in the mouse brain: evidence from using an improved TUNEL histochemical method, *Brain Res. Mol. Brain Res.*, 93 (1), 64, 2001.

273. Sonsalla, P. K. et al., Treatment of mice with methamphetamine produces cell loss in the substantia nigra, *Brain Res.*, 738 (1), 172, 1996.

274. Ricaurte, G. A. et al., Severe dopaminergic neurotoxicity in primates after a common recreational dose regimen of MDMA ("ecstasy"), *Science*, 297 (5590), 2260, 2002.

275. Yu, L. and Liao, P. C., Sexual differences and estrous cycle in methamphetamine-induced dopamine and serotonin depletions in the striatum of mice, *J. Neural Transm.*, 107 (4), 419, 2000.

276. Yu, Y. L. and Wagner, G. C., Influence of gonadal hormones on sexual differences in sensitivity to methamphetamine-induced neurotoxicity, *J. Neural Transm. Park. Dis. Dement. Sect.*, 8 (3), 215, 1994.

277. Wagner, G. C., Tekirian, T. L., and Cheo, C. T., Sexual differences in sensitivity to methamphetamine toxicity, *J. Neural Transm.*, 93, 67, 1993.

278. Caldwell, J., Dring, L. G., and Williams, R. T., Metabolism of (14 C) methamphetamine in man, the guinea pig and the rat, *Biochem. J.*, 129 (1), 11, 1972.

279. Melega, W. P. et al., Recovery of striatal dopamine function after acute amphetamine- and methamphetamine-induced neurotoxicity in the vervet monkey, *Brain Res.*, 766, 113, 1997.

280. Walsh, S. L. and Wagner, G. C., Motor impairments after methamphetamine-induced neurotoxicity in the rat, *J. Pharmacol. Exp. Ther.*, 263 (2), 617, 1992.

281. Westphale, R. I. and Stadlin, A., Dopamine uptake blockers nullify methamphetamine-induced decrease in dopamine uptake and plasma membrane potential in rat striatal synaptosomes, *Ann. N. Y. Acad. Sci.*, 914, 187, 2000.

282. Fumagalli, F. et al., Role of dopamine transporter in methamphetamine-induced neurotoxicity: evidence from mice lacking the transporter, *J. Neurosci.*, 18 (13), 4861, 1998.

283. Fumagalli, F. et al., Increased methamphetamine neurotoxicity in heterozygous vesicular monoamine transporter 2 knock-out mice, *J. Neurosci.*, 19 (7), 2424, 1999.

284. Cubells, J. F. et al., Methamphetamine neurotoxicity involves vacuolation of endocytic organelles and dopamine-dependent intracellular oxidative stress, *J. Neurosci.*, 14 (4), 2260, 1994.

285. Gluck, M. R. et al., Parallel increases in lipid and protein oxidative markers in several mouse brain regions after methamphetamine treatment, *J. Neurochem.*, 79 (1), 152, 2001.

286. Yamamoto, B. K. and Zhu, W., The effects of methamphetamine on the production of free radicals and oxidative stress, *J. Pharmacol. Exp. Ther.*, 287 (1), 107, 1998.

287. Imam, S. Z. et al., Methamphetamine-induced dopaminergic neurotoxicity: role of peroxynitrite and neuroprotective role of antioxidants and peroxynitrite decomposition catalysts, *Annals N. Y. Acad. Sci.*, 939, 366, 2001.

288. Cadet, J. L. et al., CuZn-superoxide dismutase (CuZn-SOD) transgenic mice show resistance to the lethal effects of methylenedioxyamphetamine (MDA) and of methylenedioxymethamphetamine (MDMA), *Brain Res.*, 655, 259, 1994.

289. Hirata, H. et al., Autoradiographic evidence for methamphetamine-induced striatal dopaminergic loss in mouse brain: attenuation in CuZn-superoxide dismutase transgenic mice, *Brain Res.*, 714 (1–2), 95, 1996.

290. Sonsalla, P. K., Riordan, D. E., and Heikkila, R. E., Competitive and noncompetitive antagonists at N-methyl-D-asparate receptors protect against methamphetamine-induced dopaminergic damage in mice, *J. Pharmacol. Exp. Therap.*, 256 (2), 506, 1991.

291. Imam, S. Z. et al., Peroxynitrite plays a role in methamphetamine-induced dopaminergic neurotoxicity: evidence from mice lacking neuronal nitric oxide synthase gene or overexpressing copper-zinc superoxide dismutase, *J. Neurochem.*, 76 (3), 745, 2001.

292. Itzhak, Y. and Ali, S. F., The neuronal nitric oxide synthase inhibitor, 7-nitroindazole, protects against methamphetamine-induced neurotoxicity *in vivo*, *J. Neurochem.*, 67 (4), 1770, 1996.

293. Kita, T. et al., Methamphetamine-induced neurotoxicity in BALB/c, DBA/2N and C57BL/6N mice, *Neuropharmacol.*, 37 (9), 1177, 1998.

294. Frost, D. O. and Cadet, J. L., Effects of methamphetamine-induced neurotoxicity on the development of neural circuitry: a hypothesis, *Brain Res. Brain Res. Rev.*, 34 (3), 103, 2000.

295. Breese, G. R., Criswell, H. E., and Mueller, R. A., Evidence that lack of brain dopamine during development can increase the susceptibility for aggression and self-injurious behavior by influencing D_1-dopamine receptor function, *Prog. Neuro-Psychopharmacol. Biol. Psychiat.*, 14, S65, 1991.

296. Breese, G. R. et al., A dopamine deficiency model of Lesch-Nyhan Disease: The neonatal-6-OHDA-lesioned rat, *Brain Res. Bull.*, 25, 477, 1990.

297. Breese, G. R. et al., Behavioral differences between neonatal and adult 6-hydroxydopamine-treated rats to dopamine agonists: relevance to neurological symptoms in clinical syndromes with reduced brain dopamine, *J. Pharmacol. Exp. Therap.*, 231 (2), 343, 1984.

298. Moy, S. S., Criswell, H. E., and Breese, G. R., Differential effects of bilateral dopamine depletion in neonatal and adult rats, *Neurosci. Biobehav. Rev.*, 21 (4), 425, 1997.

299. Lloyd, K. G. et al., Biochemical evidence of dysfunction of brain neurotransmitters in the Lesch-Nyhan syndrome, *N. Engl. J. Med.*, 305 (19), 1106, 1981.

300. Ye, H. J., Zheng, S., and Howard, B. D., Impaired differentiation of HPRT-deficient dopaminergic neurons: a possible mechanism underlying neuronal dysfunction in Lesch-Nyhan syndrome, *J. Neurosci. Res.*, 53 (1), 78, 1998.

301. Jinnah, H. A., Gage, F. H., and Theodore, F., Animal Models of Lesch-Nyhan Syndrome, *Brain Res. Bull.*, 25, 467, 1990.

302. Jinnah, H., Langlais, P., and Friedmann, T., Functional analysis of brain dopamine systems in a genetic mouse model of Lesch-Nyhan syndrome, *J. Pharmacol. Exp. Ther.*, 263 (2), 596, 1992.

303. Kita, T. et al., 1-Methyl-4-phenyl-1,2,3,6-tetrahydropyridine pretreatment attenuates methamphetamine-induced dopamine toxicity, *Pharmacol. Toxicol.*, 92 (2), 71, 2003.

304. Harvey, D. C. et al., Recovery from methamphetamine induced long-term nigrostriatal dopaminergic deficits without substantia nigra cell loss, *Brain Res.*, 871 (2), 259, 2000.

305. Wilson, J. M. et al., Striatal dopamine nerve terminal markers in human, chronic, methamphetamine users, *Nat. Med.*, 2 (6), 699, 1996.

306. McCann, U. D. et al., Reduced striatal dopamine transporter density in abstinent methamphetamine and methcathinone users: evidence from positron emission tomography studies with [11C]WIN-35,428, *J. Neurosci.*, 18 (20), 8417, 1998.

307. Paulus, M. P. et al., Behavioral and functional neuroimaging evidence for prefrontal dysfunction in methamphetamine-dependent subjects, *Neuropsychopharm.*, 26 (1), 53, 2002.

308. Guilarte, T. R., Is methamphetamine abuse a risk factor in parkinsonism?, *Neurotoxicol.*, 22 (6), 725, 2001.

309. Koprich, J. B. et al., Prenatal 3,4-methylenedioxymethamphetamine (ecstasy) alters exploratory behavior, reduces monoamine metabolism, and increases forebrain tyrosine hydroxylase fiber density of juvenile rats, *Neurotoxicol. Teratol.*, 25 (5), 509, 2003.

310. Tanner, C. M. et al., Parkinson disease in twins: an etiologic study, *JAMA*, 281 (4), 341, 1999.

311. Kitazawa, M., Anantharam, V., and Kanthasamy, A. G., Dieldrin-induced oxidative stress and neurochemical changes contribute to apoptopic cell death in dopaminergic cells, *Free Radic. Biol. Med.*, 31 (11), 1473, 2001.

312. Kitazawa, M., Anantharam, V., and Kanthasamy, A. G., Dieldrin induces apoptosis by promoting caspase-3-dependent proteolytic cleavage of protein kinase C delta in dopaminergic cells: relevance to oxidative stress and dopaminergic degeneration, *Neurosci,* 119 (4), 945, 2003.

313. Thiffault, C., Langston, J. W. and Di Monte, D. A., Increased striatal dopamine turnover following acute administration of rotenone to mice, *Brain Res.*, 885 (2), 283, 2000.

314. Sharma, R. P., Winn, D. S., and Low, B., Toxic, neurochemical and behavioral effects of dieldrin exposure in mallard ducks, *Arch. Environ. Contam. Toxicol.*, 5 (1), 43, 1976.

315. Sanchez-Ramos, J. et al., Toxicity of dieldrin for dopaminergic neurons in mesencephalic cultures, *Exp. Neurol.*, 150 (2), 263, 1998.

316. Heinz, G. H., Hill, E. F., and Contrera, J. F., Dopamine and norepinephrine depletion in ring doves fed DDE, dieldrin, and Aroclor 1254, *Toxicol. Appl. Pharmacol.*, 53 (1), 75, 1980.

317. Takahashi, R. N., Rogerio, R., and Zanin, M., Maneb enhances MPTP neurotoxicity in mice, *Res. Commun. Chem. Pathol. Pharmacol.*, 66 (1), 167, 1989.

318. Thiruchelvam, M. et al., The nigrostriatal dopaminergic system as a preferential target of repeated exposures to combined paraquat and maneb: implications for Parkinson's disease, *J. Neurosci.*, 20 (24), 9207, 2000.

319. Brooks, A. I. et al., Paraquat elicited neurobehavioral syndrome caused by dopaminergic neuron loss, *Brain Res.*, 823 (1-2), 1, 1999.

320. Betarbet, R. et al., Chronic systemic pesticide exposure reproduces features of Parkinson's disease, *Nat. Neurosci.*, 3 (12), 1301, 2000.

321. Sherer, T. B. et al., Subcutaneous rotenone exposure causes highly selective dopaminergic degeneration and alpha-synuclein aggregation, *Exp. Neurol.*, 179 (1), 6, 2003.

322. Trojanowski, J. Q., Rotenone neurotoxicity: a new window on environmental causes of Parkinson's disease and related brain amyloidoses, *Exp. Neurol.*, 179 (1), 6, 2003.

323. Fahn, S. and Cohen, G., The oxidant stress hypothesis in Parkinson's disease: evidence supporting it, *Ann. Neurol.*, 32 (6), 804, 1992.

324. Jenner, P. and Olanow, C. W., Oxidative stress and the pathogenesis of Parkinson's disease, *Neurology*, 47, Suppl., S161, 1996.

325. Jenner, P., Altered mitochondrial function, iron metabolism and glutathione levels in Parkinson's disease, *Acta Neurol. Scand*, 87 (Suppl., 146), 6, 1993.

326. Han, J. et al., Inhibitors of mitochondrial respiration, iron (II), and hydroxyl radical evoke release and extracellular hydrolysis of glutathione in rat striatum and substantia nigra: potential implications to Parkinson's disease, *J. Neurochem.*, 73 (4), 1683, 1999.

327. Morris, C. M. and Edwardson, J. A., Iron histochemistry of the substantia nigra in Parkinson's disease, *Neurodegen.*, 3, 277, 1994.

328. Friedman, A. and Galazka-Friedman, J., The current state of free radicals in Parkinson's disease. Nigral iron as a trigger of oxidative stress, *Adv. Neurol.*, 86, 134, 2001.

329. Arendash, G. W. et al., Intranigral iron infusion as a model for Parkinson's disease, in *Toxin-Induced Models of Neurological Disorders.*, Woodruff, M. L. and Nonneman, A.J., Plenum Press, New York, pp. 175, 1994.

330. Sossi, V. et al., Increase in dopamine turnover occurs early in Parkinson's disease: evidence from a new modeling approach to PET 18 F-fluorodopa data, *J. Cereb. Blood Flow Metab.*, 22 (2), 232, 2002.

331. Spina, M. B. and Cohen, G., Dopamine turnover and glutathione oxidation: Implications for Parkinson's disease, *Proc. Natl. Acad. Sci. USA*, 86, 1398, 1989.

332. Ostrerova-Golts, N. et al., The A53T alpha-synuclein mutation increases iron-dependent aggregation and toxicity, *J. Neurosci.*, 20 (16), 6048, 2000.

333. Galvin, J. E. et al., Neurodegeneration with brain iron accumulation, type 1 is characterized by alpha-, beta-, and gamma-synuclein neuropathology, *Am. J. Pathol.*, 157 (2), 361, 2000.

334. Kaur, D. et al., Genetic or pharmacological iron chelation prevents MPTP-induced neurotoxicity *in vivo*: a novel therapy for Parkinson's disease, *Neuron*, 37 (6), 899, 2003.

335. Giasson, B. I. and Lee, V. M., A new link between pesticides and Parkinson's disease, *Nat. Neurosci.*, 3 (12), 1227, 2002.

336. Rajput, A. H. et al., Geography, drinking water chemistry, pesticides and herbicides and the etiology of Par-

kinson's disease, *Can. J. Neurol. Sci., 14* (3 Suppl.), 414, 1987.

337. Musshoff, F. et al., A systematic regional study of dopamine and dopamine-derived salsolinol and norsalsolinol levels in human brain areas, *Forensic Sci. Int.,* 105 (1), 1, 1999.

338. Nagatsu, T. and Yoshida, M., An endogenous substance of the brain, tetrahydroisoquinoline, produces parkinsonism in primates with decreased dopamine, tyrosine hydroxylase and biopterin in the nigrostriatal regions, *Neurosci. Lett.,* 87 (1–2), 178, 1988.

339. Storch, A. et al., Selective dopaminergic neurotoxicity of isoquinoline derivatives related to Parkinson's disease: studies using heterologous expression systems of the dopamine transporter, *Biochem. Pharmacol.,* 63 (5), 909, 2002.

340. Kotake, Y. et al., Chronic administration of 1-benzyl-1,2,3,4-tetrahydroisoquinoline, an endogenous amine in the brain, induces parkinsonism in a primate, *Neurosci. Lett.,* 217 (1), 69, 1996.

341. Kawai, H., Kotake, Y., and Ohta, S., Inhibition of dopamine receptors by endogenous amines: binding to striatal receptors and pharmacological effects on locomotor activity, *Bioorg. Med. Chem. Lett., 10* (15), 1669, 2000.

342. Kawai, H., Kotake, Y., and Ohta, S., Dopamine transporter and catechol-O-methyltransferase activities are required for the toxicity of 1-(3',4'-dihydroxybenzyl)-1,2,3, 4-tetrahydroisoquinoline, *Chem. Res. Toxicol., 13* (12), 1294, 2000.

343. Antkiewicz-Michaluk, L. et al., Antidopaminergic effects of 1,2,3,4-tetrahydroisoquinoline and salsolinol, *J. Neural Transm., 107* (8–9), 1009, 2000.

344. Shavali, S. and Ebadi, M., 1-Benzyl-1,2,3,4-tetrahydroisoquinoline (1BnTIQ), an endogenous neurotoxin, induces dopaminergic cell death through apoptosis, *Neurotox.,* 24 (3), 417, 2003.

345. Abe, K. et al., Biochemical and pathological study of endogenous 1-benzyl-1,2,3,4-tetrahydroisoquinoline-induced parkinsonism in the mouse, *Brain Res.,* 907 (1–2), 134, 2001.

346. Hunot, S. and Hirsch, E. C., Neuroinflammatory processes in Parkinson's disease, *Ann. Neurol.,* 53, Suppl., 3, S49, 2003.

347. Orr, C. F., Rowe, D. B., and Halliday, G. M., An inflammatory review of Parkinson's disease, *Prog. Neurobiol.,* 68 (5), 325, 2002.

348. Le, W. et al., Microglial activation and dopaminergic cell injury: an *in vitro* model relevant to Parkinson's disease, *J. Neurosci.,* 21 (21), 8447, 2001.

349. Kim, W. G. et al., Regional difference in susceptibility to lipopolysaccharide-induced neurotoxicity in the rat brain: role of microglia, *J. Neurosci.,* 20 (16), 6309, 2000.

350. Herrera, A. J. et al., The single intranigral injection of LPS as a new model for studying the selective effects of inflammatory reactions on dopaminergic system, *Neurobiol. Dis.,* 7 (4), 429, 2000.

351. Castano, A. et al., Lipopolysaccharide intranigral injection induces inflammatory reaction and damage in nigrostriatal dopaminergic system, *J. Neurochem.,* 70 (4), 1584, 1998.

352. Gao, H. M. et al., Synergistic dopaminergic neurotoxicity of the pesticide rotenone and inflammmogen lipopolysaccharide: relevance to the etiology of Parkinson's disease, *J. Neurosci.,* 23 (4), 1228, 2003.

353. Chesselet, M. F. and Levine, M. S., Mouse models of Huntington's disease, in *Molecular Mechanisms of Neurodegenerative Diseases*, Chesselet, M. F., Humana Press, Totowa, N.J., pp. 327, 2000.

354. Ben-Ari, Y. and Cossart, R., Kainate, a double agent that generates seizures: two decades of progress, *Trends Neurosci.,* 23 (11), 580, 2000.

355. Scherfler, C. et al., Complex motor disturbances in a sequential double lesion rat model of striatonigral degeneration (multiple system atrophy), *Neurosci.,* 99 (1), 42, 2000.

356. Kelly, W. J. and Burke, R. E., Apoptotic neuron death in rat substantia nigra induced by striatal excitotoxic injury is developmentally dependent, *Neurosci. Lett.,* 220, 85, 1996.

357. Macaya, A. et al., Apoptosis in substantia nigra following developmental striatal excitotoxic injury, *Proc. Natl. Acad. Sci. USA,* 91, 8117, 1994.

358. Marti, M. et al., Early developmental destruction of terminals in the striatal target induces apoptosis in dopamine neurons of the substantia nigra, *J. Neurosci., 17* (6), 2030, 1997.

359. Jackson-Lewis, V. et al., Developmental cell death in dopaminergic neurons of the substantia nigra of mice, *J. Comp. Neurol.,* 424 (3), 476, 2000.

360. Marti, M. J. et al., Striatal 6-hydroxydopamine induces apoptosis of nigral neurons in the adult rat, *Brain Res.,* 958 (1), 185, 2002.

361. Oo, T. F., Henchcliffe, C., and Burke, R. E., Apoptosis in substantia nigra following developmental hypoxic-ischemic injury, *Neurosci.,* 69 (3), 893, 1995.

362. Tatton, W. G. et al., Apoptosis in Parkinson's disease: signals for neuronal degradation, *Ann. Neurol.,* 53, Suppl., 3, S61, 2003.

363. Lev, N., Melamed, E., and Offen, D., Apoptosis and Parkinson's disease, *Prog. Neuropsychopharmacol. Biol. Psychiatry,* 27 (2), 245, 2003.

364. Heintz, N. and Zoghbi, H. Y., Insights from mouse models into the molecular basis of neurodegeneration, *Ann. Rev. Physiol.,* 62, 779, 2000.

365. Kilbourn, M. R., Sherman, P., and Abbott, L. C., Reduced MPTP neurotoxicity in striatum of the mutant mouse tottering, *Synapse,* 30 (2), 205, 1998.

366. Blum, M., Weickert, C., and Carrasco, E., The weaver GIRK2 mutation leads to decreased levels of serum thyroid hormone: characterization of the effect on midbrain dopaminergic neuron survival, *Exp. Neurol., 160* (2), 413, 1999.

367. Payne, A. P. et al., The AS/AGU rat: a spontaneous model of disruption and degeneration in the nigrostriatal dopaminergic system, *J. Anat., 196* (Pt. 4), 629, 2000.

368. Campbell, J. M. et al., Extracellular levels of dopamine and its metabolite 3,4-dihydroxy-phenylacetic acid measured by microdialysis in the corpus striatum of con-

scious AS/AGU mutant rats, *Neurosci.*, 85 (2), 323, 1998.

369. Campbell, J. M. et al., Pharmacological analysis of extracellular dopamine and metabolites in the striatum of conscious as/agu rats, mutants with locomotor disorder, *Neurosci.*, *100* (1), 45, 2000.

370. Richter, A. et al., Immunohistochemical and neurochemical studies on nigral and striatal functions in the circling (ci) rat, a genetic animal model with spontaneous rotational behavior, *Neurosci.*, 89 (2), 461, 1999.

371. Polymeropoulos, M. et al., Mutation in the α-Synuclein Gene Identified in Families with Parkinson's Disease, *Science*, 276, 2045, 1997.

372. Maroteaux, L., Campanelli, J. T., and Scheller, R. H., Synuclein: a neuron-specific protein localized to the nucleus and presynaptic terminal, *J. Neurosci.*, 8 (8), 2804, 1988.

373. George, J. M. et al., Characterization of a novel protein regulated during the critical period for song learning in the zebra finch, *Neuron*, 15, 361, 1995.

374. Jakowec, M. W. et al., The postnatal expression of α-synuclein in the substantia nigra and striatum of the rodent, *Dev. Neurosci.*, 23 (2), 91, 2001.

375. Welch, K. and Yuan, J., Alpha-synuclein oligomerization: a role for lipids?, *Trends Neurosci.*, 26 (10), 517, 2003.

376. Ostrerova, N. et al., alpha-synuclein shares physical and functional homology with 14-3-3 proteins, *J. Neurosci.*, 19 (14), 5782, 1999.

377. Dev, K. K. et al., Part II: alpha-synuclein and its molecular pathophysiological role in neurodegenerative disease, *Neuropharmacol.*, 45 (1), 14, 2003.

378. Meredith, G. E. et al., Lysosomal malfunction accompanies alpha-synuclein aggregation in a progressive mouse model of Parkinson's disease, *Brain Res.*, 956 (1), 156, 2002.

379. Abeliovich, A. et al., Mice lacking alpha-synuclein display functional deficits in the nigrostriatal dopamine system, *Neuron*, 25 (1), 239, 2000.

380. Masliah, E. et al., Dopaminergic loss and inclusion body formation in alpha-synuclein mice: implications for neurodegenerative disorders, *Science*, 289 (5456), 1265, 2000.

381. Kahle, P. J. et al., Subcellular localization of wild-type and Parkinson's disease-associated mutant alpha-synuclein in human and transgenic mouse brain, *J. Neurosci.*, 20 (17), 6365, 2000.

382. Giasson, B. I. et al., Neuronal alpha-synucleinopathy with severe movement disorder in mice expressing A53T human alpha-synuclein, *Neuron*, 34 (4), 521, 2002.

383. Kowall, N. W. et al., MPTP induces alpha-synuclein aggregation in the substantia nigra of baboons, *Neuroreport*, 11 (1), 211, 2000.

384. Lo Bianco, C. et al., alpha-synucleinopathy and selective dopaminergic neuron loss in a rat lentiviral-based model of Parkinson's disease, *Proc. Natl. Acad. Sci. USA*, 99 (16), 10813, 2002.

385. Maries, E. et al., The role of alpha-synuclein in Parkinson's disease: insights from animal models, *Nat. Rev. Neurosci.*, 4 (9), 727, 2003.

386. Kahle, P. J. et al., Structure/function of alpha-synuclein in health and disease: rational development of animal models for Parkinson's and related diseases, *J. Neurochem.*, 82 (3), 449, 2002.

387. Kahle, P. J. et al., Physiology and pathophysiology of alpha-synuclein. Cell culture and transgenic animal models based on a Parkinson's disease-associated protein, *Ann. N. Y. Acad. Sci.*, 920, 33, 2000.

388. George, J. M., The synucleins, *Genome Biol.*, 3 (1), 3002.1, 2002.

389. Jackson-Lewis, V. et al., Increased expression of α-synuclein in the MPTP mouse model of Parkinson's disease, *Neurology*, 50 Suppl., 4, A96, 1998.

390. Neystat, M. et al., Alpha-synuclein expression in substantia nigra and cortex in Parkinson's disease., *Mov. Disord.*, 14 (3), 417, 1999.

391. Vila, M. et al., Alpha-synuclein up-regulation in substantia nigra dopaminergic neurons following administration of the parkinsonian toxin MPTP, *J. Neurochem.*, 74 (2), 721, 2000.

392. Shimura, H. et al., Ubiquitination of a new form of {alpha}-synuclein by parkin from human brain: Implications for Parkinson's disease, *Science*, 293, 263, 2001.

393. Junn, E. and Mouradian, M. M., Human alpha-Synuclein over-expression increases intracellular reactive oxygen species levels and susceptibility to dopamine, *Neurosci. Lett.*, 320 (3), 146, 2002.

394. Kirik, D. et al., Parkinson-like neurodegeneration induced by targeted overexpression of alpha-synuclein in the nigrostriatal system, *J. Neurosci.*, 22 (7), 2780, 2002.

395. Kirik, D. et al., Nigrostriatal alpha-synucleinopathy induced by viral vector-mediated overexpression of human alpha-synuclein: a new primate model of Parkinson's disease, *Proc. Natl. Acad. Sci. USA*, *100* (5), 2884, 2003.

396. Eisen, A. and Calne, D., Amyotropic lateral sclerosis, Parkinson's disease and Alzheimer's disease: phylogenetic disorders of the human neocortex sharing many characteristics, *Can. J. Neurol. Sci.*, 19, 117, 1992.

397. Heintz, N. and Zoghbi, H., α-Synuclein-a link between Parkinson's and Alzheimer's diseases? *Nat. Genet.*, *16*, 325, 1997.

398. Braak, H. et al., Neuropathological hallmarks of Alzheimer's and Parkinson's diseases, *Prog. Brain Res.*, 117, 267, 1998.

399. Mukaetova-Ladinska, E. B. et al., Alpha-synuclein inclusions in Alzheimer's and Lewy body diseases, *J. Neuropathol. Exp. Neurol.*, 59 (5), 408, 2000.

400. Trojanowski, J. Q. and Lee, V. M., Parkinson's disease and related synucleinopathies are a new class of nervous system amyloidoses, *Neurotox.*, 23 (4–5), 457, 2002.

401. Masliah, E. and Hashimoto, M., Development of new treatments for Parkinson's disease in transgenic animal models: a role for beta-synuclein, *Neurotox.*, 23 (4–5), 461, 2002.

402. Hashimoto, M. et al., beta-Synuclein inhibits alpha-synuclein aggregation: a possible role as an anti-parkinsonian factor, *Neuron*, 32 (2), 213, 2001.

403. Biere, A.L. et al., Parkinson's disease-associated alpha-synuclein is more fibrillogenic than beta- and gamma-synuclein and cannot cross-seed its homologs, *J. Biol. Chem.*, 275 (44), 34574, 2000.

404. Mizuno, Y. et al., Familial Parkinson's disease. Alpha-synuclein and parkin, *Adv. Neurol.*, 86, 13, 2001.

405. Kitada, T. et al., Mutations in the parkin gene cause autosomal recessive juvenile parkinsonism, *Nature*, 392, 605, 1998.

406. Hattori, N. et al., Molecular genetic analysis of a novel Parkin gene in Japanese families with autosomal recessive juvenile parkinsonism: evidence for variable homozygous deletions in the Parkin gene in affected individuals, *Ann. Neurol.*, 44 (6), 935, 1998.

407. Fallon, L. et al., Parkin and CASK/LIN-2 associate via a PDZ-mediated interaction and are co-localized in lipid rafts and postsynaptic densities in brain, *J. Biol. Chem.*, 277 (1), 486, 2002.

408. Huynh, D. P. et al., Differential expression and tissue distribution of parkin isoforms during mouse development, *Brain Res. Dev. Brain Res.*, 130 (2), 173, 2001.

409. Solano, S. M. et al., Expression of alpha-synuclein, parkin, and ubiquitin carboxy-terminal hydrolase L1 mRNA in human brain: genes associated with familial Parkinson's disease, *Ann. Neurol.*, 47 (2), 201, 2000.

410. Chung, K. K., Dawson, V. L., and Dawson, T. M., The role of the ubiquitin-proteasomal pathway in Parkinson's disease and other neurodegenerative disorders, *Trends Neurosci.*, 24, 11 Suppl, S7, 2001.

411. Bence, N. F., Sampat, R. M., and Kopito, R. R., Impairment of the ubiquitin-proteasome system by protein aggregation, *Science*, 292 (5521), 1552, 2001.

412. Moore, D. J., Dawson, V. L., and Dawson, T. M., Role for the ubiquitin-proteasome system in Parkinson's disease and other neurodegenerative brain amyloidoses, *Neuromolecular Med.*, 4 (1–2), 98, 2003.

413. Giasson, B. I. and Lee, V. M., Are ubiquitination pathways central to Parkinson's disease? *Cell*, 114 (1), 1, 2003.

414. Wakabayashi, K. et al., Immunocytochemical localization of synphilin-1, an alpha-synuclein-associated protein, in neurodegenerative disorders, *Acta Neuropathol. (Berl)*, 103 (3), 209, 2002.

415. Tanaka, K. et al., Parkin is linked to the ubiquitin pathway, *J. Mol. Med.*, 79 (9), 482, 2001.

416. Leroy, E. et al., The ubiquitin pathway in Parkinson's disease, *Nature*, 395 (6701), 451, 1998.

417. Lindsten, K. et al., A transgenic mouse model of the ubiquitin/proteasome system, *Nat. Biotechnol.*, 21 (8), 897, 2003.

418. Trojanowski, J. Q. and Lee, V. M., Transgenic models of tauopathies and synucleinopathies, *Brain Pathol.*, 9 (4), 733, 1999.

419. Lee, V. M. and Trojanowski, J. Q., Neurodegenerative tauopathies: human disease and transgenic mouse models, *Neuron*, 24 (3), 507, 1999.

420. McNaught, K. S. et al., Altered proteasomal function in sporadic Parkinson's disease, *Exp. Neurol.*, 179 (1), 38, 2003.

421. Gurney, M. E. et al., Motor neuron degeneration in mice that express a human Cu, Zn superoxide dismutase mutation, *Science*, 264, 1772, 1994.

422. Kostic, V. S. et al., Midbrain dopaminergic neuronal degeneration in a transgenic mouse model of familial amyotrophic lateral sclerosis, *Ann. Neurol.*, 41 (4), 497, 1997.

423. Przedborski, S. et al., Transgenic mice with increased Cu/Zn-superoxide dismutase activity are resistant to N-methyl-4-phenyl-1,2,3,6-tetrahydropyridine-induced neurotoxicity, *J. Neurosci.*, 12 (5), 1658, 1992.

424. Przedborski, S. et al., Superoxide dismutase, catalase, and glutathione peroxidase activities in copper/zinc-superoxide dismutase transgenic mice, *J. Neurochem.*, 58, 1760, 1992.

425. Klivenyi, P. et al., Mice deficient in cellular glutathione peroxidase show increased vulnerability to malonate, 3-nitropropionic acid, and 1-methyl-4-phenyl-1,2,5,6-tetrahydropyridine, *J. Neurosci.*, 20 (1), 1, 2000.

426. Bharath, S. et al., Glutathione, iron and Parkinson's disease, *Biochem. Pharmacol.*, 64 (5–6), 1037, 2002.

427. Tornqvist, N. et al., Generation of tyrosine hydroxylase-immunoreactive neurons in ventral mesencephalic tissue of Nurr1 deficient mice, *Brain Res. Dev. Brain Res.*, 133 (1), 37, 2002.

428. Zetterstrom, R. H. et al., Dopamine neuron agenesis in Nurr1-deficient mice, *Science*, 276, 248, 1997.

429. Law, S. W. et al., Identification of a new brain-specific transcription factor, NURR1, *Mol. Endocrinol.*, 6 (12), 2129, 1992.

430. Eilam, R. et al., Selective loss of dopaminergic nigrostriatal neurons in brains of Atm-deficient mice, *Proc. Natl. Acad. Sci. USA*, 95 (21), 12653, 1998.

431. Eilam, R. et al., Late degeneration of nigro-striatal neurons in ATM-/- mice, *Neurosci.*, 121 (1), 83, 2003.

432. Gainetdinov, R. R., Sotnikova, T. D., and Caron, M. G., Monoamine transporter pharmacology and mutant mice, *Trends Pharmacol. Sci.*, 23 (8), 367, 2002.

433. Gainetdinov, R. R. et al., Glutamatergic modulation of hyperactivity in mice lacking the dopamine transporter, *Proc. Natl. Acad. Sci. USA*, 98 (20), 11047, 2001.

434. Gainetdinov, R. R. et al., Dopamine transporter is required for *in vivo* MPTP neurotoxicity: evidence from mice lacking the transporter, *J. Neurochem.*, 69 (3), 1322, 1997.

435. Gerber, D. J. et al., Hyperactivity, elevated dopaminergic transmission, and response to amphetamine in M1 muscarinic acetylcholine receptor-deficient mice, *Proc Natl Acad Sci U S A*, 98 (26), 15312, 2001.

436. Jaber, M. et al., Differential regulation of tyrosine hydroxylase in the basal ganglia of mice lacking the dopamine transporter, *Eur. J. Neurosci.*, 11, (10), 3499, 1999.

437. Kim, D. S., Szczypka, M. S., and Palmiter, R. D., Dopamine-deficient mice are hypersensitive to dopamine receptor agonists, *J. Neurosci.*, 20 (12), 4405, 2000.

438. Ramsden, D. B. et al., The aetiology of idiopathic Parkinson's disease, *Mol. Pathol.*, 54 (6), 369, 2001.

439. Nass, R. et al., Neurotoxin-induced degeneration of dopamine neurons in Caenorhabditis elegans, *Proc. Natl. Acad. Sci. USA,* 99 (5), 3264, 2002.

440. Feany, M. B. and Bender, W. W., A Drosophila model of Parkinson's disease, *Nature,* 404 (6776), 394, 2000.

441. Pendleton, R. G. et al., Effects of pharmacological agents upon a transgenic model of Parkinson's disease in Drosophila melanogaster, *J. Pharmacol. Exp. Ther.,* 300 (1), 91, 2002.

442. Auluck, P. K. et al., Chaperone suppression of alpha-synuclein toxicity in a Drosophila model for Parkinson's disease, *Science,* 295 (5556), 865, 2002.

443. Feany, M. B. and Pallanck, L. J., Parkin: a multipurpose neuroprotective agent? *Neuron,* 38 (1), 13, 2003.

444. Shulman, J. M. et al., From fruit fly to bedside: translating lessons from Drosophila models of neurodegenerative disease, *Curr. Opin. Neurol.,* 16 (4), 443, 2003.

445. Scherzer, C. R. et al., Gene expression changes presage neurodegeneration in a Drosophila model of Parkinson's disease, *Hum. Mol. Genet.,* 12 (19), 2457, 2003.

446. Feany, M. B. and La Spada, A. R., Polyglutamines stop traffic: axonal transport as a common target in neurodegenerative diseases, *Neuron,* 40 (1), 1, 2003.

447. Muqit, M. M. and Feany, M. B., Modelling neurodegenerative diseases in Drosophila: a fruitful approach? *Na.t Rev. Neurosci.,* 3 (3), 237, 2002.

35 The Neuropathology of Parkinson's Disease and Other Parkinsonian Disorders

Christopher A. Robinson
Department of Pathology, Royal University Hospital, University of Saskatchewan

Ali H. Rajput
Department of Medicine, Royal University Hospital, University of Saskatchewan

CONTENTS

INTRODUCTION

Idiopathic Parkinson's disease (IPD) represents the most common cause of Parkinsonism,[1] the extrapyramidal movement disturbance that manifests as some combination of bradykinesia, rigidity, and resting tremor. However, many less-common causes of Parkinsonism are recognized, and clinical misclassification is not uncommon.[2] As a result, postmortem examination of the brain is required for confirmation of diagnoses. In this chapter, the neuro-pathology of IPD and a number of other disorders manifesting as Parkinsonism are discussed.

IDIOPATHIC PARKINSON'S DISEASE

It was James Parkinson, in his monograph *An Essay on the Shaking Palsy,* who first succinctly described the clinical features of Parkinsonism in 1817.[3] The cardinal clinical features of IPD are bradykinesia, rigidity, and resting

0-8493-1590-5/05/$0.00+$1.50

tremor, with the presence of at least two of the three required for establishing a clinical diagnosis. With time, impaired postural reflexes develop in most cases, and dementia, autonomic dysfunction, and dysphagia may also develop.[4,5] Although it is believed that Blocq and Marinesco first observed the anatomical site of the lesion responsible for Parkinsonian features in 1894, it was the extensive pathological studies of Tretiakoff that firmly established that pathological changes in the substantia nigra (SN) were invariably seen in patients with IPD.[6]

PATHOLOGICAL APPEARANCES

Gross examination of the external aspects of the brain in IPD reveals no distinguishing features. Sectioning of the brain reveals decreased pigmentation of the SN (Figure 35.1) and locus coeruleus (LC). Importantly, the globus pallidus, putamen, and caudate nucleus appear normal on gross examination.

On microscopic examination, the pathological hallmark of IPD is a loss of pigmented dopaminergic neurons of the pars compacta region of the SN, along with the presence of neuronal intracytoplasmic inclusions called Lewy bodies (LBs) (Figure 35.2a). Although Frederich H. Lewy first described these inclusions in the substantia innominata and dorsal motor nucleus of the vagus in 1912, it wasn't until 1919 that Tretiakoff established that intraneuronal LBs within the SN were a characteristic feature of IPD. In fact, LBs must be found to make a pathological diagnosis of IPD. If no LBs are found in the SN upon close examination of several sections of the midbrain, a diagnosis of IPD can usually be excluded, and other causes of Parkinsonism should be considered.[7] In addition to neuronal loss and LBs, the SN also shows the presence of variable astrocytic gliosis, extracellular neuromelanin and neuromelanin-laden macrophages (see Figure 35.2a), and pale bodies (Figure 35.2b), of which the latter have been suggested to represent precursors of LBs.[8]

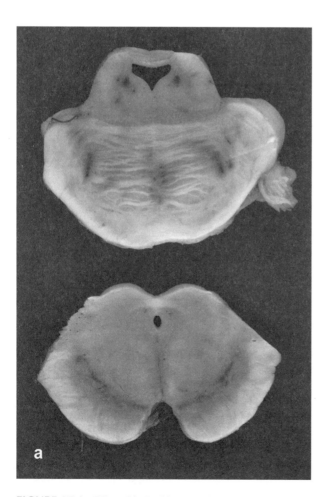

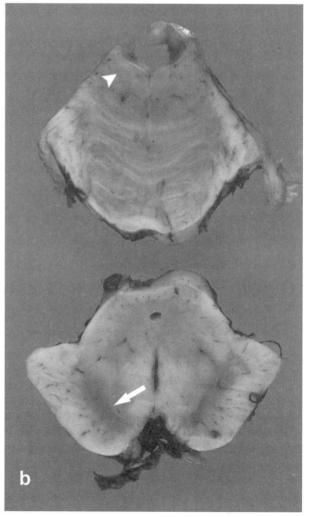

FIGURE 35.1 Idiopathic Parkinson's disease. Sections of midbrain and rostral pons from normal control (*a*) and IPD (*b*) brains. There is decreased pigmentation of the substantia nigra (arrow) and locus coeruleus (arrowhead) in the IPD brain.

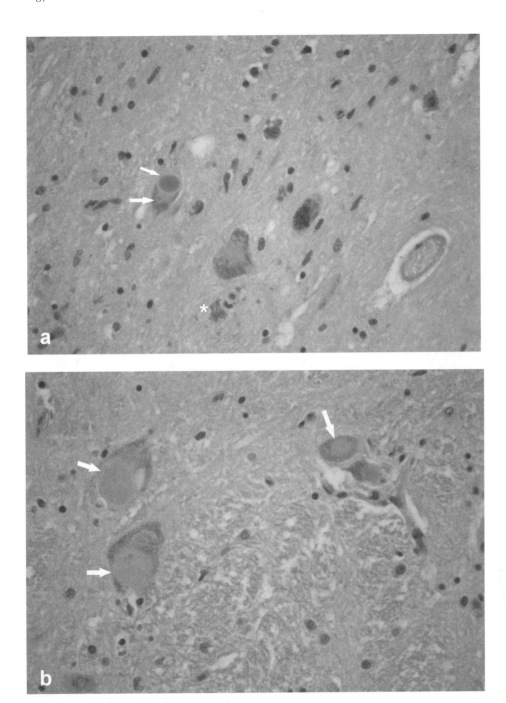

FIGURE 35.2 Idiopathic Parkinson's disease. (*a*) Two Lewy bodies within a pigmented neuron of the substantia nigra (arrows). Note the extracellular neuromelanin (asterisk) and neuromelanin-containing macrophage (arrowhead). (*b*) Pale bodies (arrows) in pigmented neurons of the substantia nigra. (*a* and *b*) Hematoxylin and eosin; original magnification ×400.

Additional brain areas that may show involvement by varying degrees of the aforementioned changes include the main source of noradrenergic innervation of the central nervous system, the LC, where the severity of changes often rivals that seen within the SN; the nucleus basalis of Meynert, which is the main source of cholinergic innervation of most forebrain areas; the dorsal motor nucleus of the vagus; the Edinger-Westphal nucleus; the paranigral nucleus, which is a component of the dopaminergic mesocortical limbic pathway; the intermediolateral cell column of the spinal cord; the seritonergic raphe nuclei; the pedunculopontine nucleus; the reticular formation of the pons and midbrain; the hypothalamus; and peripheral sympathetic ganglia.[9] As well, occasional LBs are typically seen within the neocortex of patients with IPD unassociated with dementia.

The postmortem neuropathological features of two patients from a large kindred with autosomal dominant Parkinson's disease associated with a mutation in the α-synuclein gene on chromosome 4q have been reported.[10] Changes typical of IPD were seen, with neuronal loss and LBs involving the SN and LC. The neuropathological findings from two patients of a different kindred with autosomal dominant Parkinson's disease and the same α-synuclein mutation have also recently been described.[11] In these patients, prominent neuronal loss, astrocytic gliosis, and LBs were found within the SN and LC. In addition, some astrocytic gliosis and neuronal loss, in the absence of LBs, were found within the neocortex, hippocampal CA2 and CA3 regions, putamen, and globus pallidus. Similar findings have been documented within additional kindreds with autosomal dominant Parkinson's disease in which the precise location of the chromosomal abnormality has yet to be identified.[12–14]

CLINICOPATHOLOGICAL CORRELATES

The pathological involvement of multiple neuronal systems in IPD accounts for the mixed clinical picture often seen in this patient population. Within a given patient, it is the severity of the pathological changes, coupled with the particular anatomical sites involved, which determine which clinical disturbances will become manifest.

Incidental Lewy Body Disease

A reduction of the pigmented neurons of the SN pars compacta by approximately 50% or more is necessary to produce contralateral Parkinsonian symptoms. Occasionally, Lewy bodies may be found incidentally in the brains of aged individuals without Parkinsonism, in the absence of a significant loss of pigmented neurons of the SN. It has been suggested that these individuals have presymptomatic Parkinson's disease, and for those patients who come to autopsy at this stage of disease progression, a diagnosis of "incidental Lewy body disease" has been recommended.[15]

Extrapyramidal Movement Disorder

Loss of pigmented neurons of the pars compacta of the SN occurs with a regional selectivity.[15] The pars compacta can be divided into ventral and dorsal tiers, which represent functionally distinct populations of neurons that project to different brain areas. Each tier can then be subdivided into three additional regions, from medial to lateral. In normal aging, neuronal loss is greatest in the dorsal tier, occurring at a rate of approximately 7% per decade, so that there is a 40 to 50% reduction in this neuronal population by the age of 65 years. In contrast, neuronal loss in IPD is greatest in the ventral tier, while loss from the dorsal tier does not differ significantly from that in normal aging. In IPD, neuronal loss is typically greatest in the lateral region of the ventral tier, where 70 to 90% of neurons may have been lost by the time a patient dies, while the medial region of the ventral tier is the next most affected. It has been suggested that, whereas neuronal loss in the akinetic-rigid subtype of IPD is most severe in the lateral region of the ventral tier, the tremor-dominant subtype of IPD shows less severe total neuronal loss in the pars compacta, as well as less severe neuronal loss in the lateral than in the medial region of the ventral tier,[16–18] but precise clinicopathological associations remain to be made.

Cognitive Dysfunction/Dementia with Lewy Bodies

Cognitive impairment in IPD is common, affecting up to 90% of patients,[19] often being characterized by cognitive slowing and impairment of reasoning and abstract thinking.[20] In most cases, however, these clinical features are quite subtle and do not interfere with activities of daily living. In a minority of patients, however, the cognitive impairment is more severe and may progress to dementia. In some cases, damage to subcortical structures may be extensive enough to cause dementia, as seen in IPD patients with dementia and little or no cortical damage present at autopsy.[21] In other patients in whom dementia is the prominent or presenting clinical feature, the pathological changes present in subcortical structures are milder, and Parkinsonism may be mild or even absent. When large numbers of neocortical Lewy bodies are found at postmortem examination in a patient with clinical dementia, for which no additional pathological cause is found, the disorder is known as "dementia with Lewy bodies" (DLB).[22] The cortical LBs found in DLB tend to be most numerous within the parahippocampal gyrus, cingulate gyrus, and temporal cortex; they can also be found in smaller numbers within the insular, frontal, and parietal cortices and cortical LBs tend to be found in neurons of the deeper cerebral cortical layers, particularly layers V and VI.

Autonomic Dysfunction

Autonomic disturbances are frequently seen in IPD and vary in type and severity, depending on the severity and anatomical sites of the associated pathological changes. For example, prominent LB pathology involving the intermediolateral cell columns of the spinal cord and peripheral sympathetic ganglia may manifest as progressive autonomic failure.[23] Dysphagia may be a prominent feature when pathological changes involving the dorsal vagal nuclei are prominent. Pathological changes involving the hypothalamus may also play a role in autonomic dysfunction.

LEWY BODIES

Lewy bodies are neuronal intracytoplasmic inclusion bodies of two types: the classical brain stem type and the cortical type.

Classical brain stem LBs are typically spherical inclusions that have a diameter of approximately 4 to 30 μm, with a hyaline eosinophilic core and a pale peripheral halo (see Figure 35.2a). Some LBs show a central dense core with pale concentric lamellae around the periphery. Although typically only one LB is present within a given neuron, several LBs may occasionally be seen within the same neuron. Ultrastructurally, classical LBs consist of radially arranged 7 to 20 nm intermediate filaments, associated with peripheral electron-dense granular material and vesicular profiles.[18,24] In IPD, classical LBs may be found within any of the aforementioned subcortical sites, and are typically easy to find in the SN and LC.

Cortical LBs are found in neurons of the cerebral cortex and are most numerous in the cortical sites mentioned previously. On H&E stained sections, cortical LBs appear as eosinophilic neuronal intracytoplasmic inclusions, which are often rounded but may be more irregular in shape (Figure 35.3a).[25] Very occasionally, a LB resembling the classical brain stem type may be encountered in the neocortex (Figure 35.3b). Cortical LBs may be difficult to identify on H&E and silver-stained sections, and in this situation immunocytochemistry is helpful (see below).[26,27] Ultrastructurally, cortical LBs are composed of intermediate filaments associated with granular electron-dense material.[18,24]

Both classical and cortical LBs demonstrate immunoreactivity with anti-ubiquitin (Figure 35.3c) and anti-α-synuclein antibodies (Figure 35.3d).[26–28] Ubiquitin, a heat shock protein that targets proteins for degradation, and α-synuclein, a protein normally involved in presynaptic vesicle homeostasis,[29,30] are major components of LBs, as are neurofilament proteins and αβ-crystallin. Additional components of LBs include chromogranin A, synaptophysin, amyloid precursor protein, microtubule-associated proteins, calbindin, tubulin, and tyrosine hydroxylase, among others.[31] Ubiquitin and α-synuclein immunocytochemistry also highlight pale bodies and Lewy neurites.[32,33] These latter structures are abnormal neurites that may be found in areas undergoing neuronal degeneration. Originally recognized in the hippocampal CA 2–3 region in cases of IPD and DLB (Figure 35.3e), they may also be found within the SN, dorsal motor nucleus of the vagus, and nucleus basalis of Meynert.

Although LBs are the pathological hallmark of both IPD and DLB, they are not restricted to these disorders. Lewy bodies have been described as a secondary pathology in a number of other disorders, such as progressive supranuclear palsy, corticobasal degeneration, multiple system atrophy, motor neuron disease, Hallervorden-

Spatz disease, sporadic and familial Alzheimer's disease, subacute sclerosing panencephalitis, and Down's syndrome.[18] The significance of these associations remains to be determined.

PROGRESSIVE SUPRANUCLEAR PALSY

Progressive supranuclear palsy (PSP) was first described in 1964 by Steele, Richardson, and Olszewski as a distinct clinicopathological entity, characterized by the presence of supranuclear ophthalmoplegia, rigidity, akinesia, pseudobulbar palsy, nuchal dystonia, and dementia.[34] Additional features include early disequilibrium with frequent falls, and corticobulbar and corticospinal pathology leading to dysarthria, dysphagia, brisk deep tendon reflexes, and extensor plantar responses.[35,36]

Gross examination of the brain may reveal atrophy involving areas within which the most severe pathological changes are present on microscopic examination. The external aspects of the cerebral hemispheres are usually unremarkable, but occasionally atrophy of the frontal and temporal regions, motor cortex, or hippocampus may be seen. The midbrain and pontine tegmentum often appear atrophic, with decreased pigmentation of the SN and LC, and variable atrophy of the globus pallidus and subthalamic nucleus may also be seen.

Microscopically, the characteristic pathological changes seen in affected areas are neuronal loss, astrocytic gliosis, and aggregates of the microtubule-associated protein tau within neurons, glial cells, and in neuronal and glial cell processes where they are known as neuropil threads. Demonstration of tau aggregates is facilitated through the use of a silver stain, such as the Gallyas technique, or immunocytochemistry with commercially available anti-tau antibodies.[37,38] In neurons, tau protein may be demonstrable as aggregates comprising neurofibrillary tangles (NFTs), or it may be dispersed more diffusely within the cytoplasm, where it is identifiable only with tau immunostaining. Neurofibrillary tangles within brain stem and basal ganglia structures often demonstrate a globose appearance (Figure 35.4a), while those in the neocortex are often flame-shaped or coiled. Ultrastructurally, the NFTs in PSP characteristically contain straight filaments 15 nm in diameter, but coexistence with paired helical filaments has also been described.[39–41] Accumulation of tau protein can also be demonstrated as inclusions within astrocytes, where they often highlight astrocytic processes (Figure 35.4b), and as coiled inclusions within oligodendroglial cells, where bundles of fibrils are coiled around the nucleus (Figure 35.4c).

Involved brain areas show varying degrees of these microscopic pathological changes, with the SN, LC, subthalamic nucleus, globus pallidus, periaquaductal grey matter, red nucleus, oculomotor nucleus, and cerebellar dentate nucleus often markedly affected. Additional brain

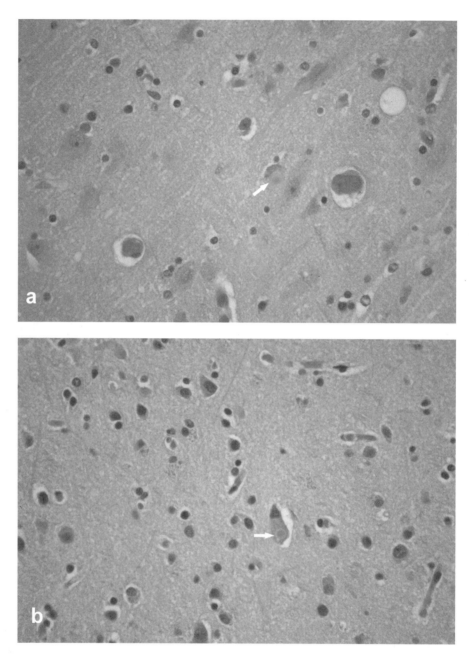

FIGURE 35.3 Dementia with Lewy bodies. (*a*) Cortical Lewy body in frontal lobe (arrow). (*b*) Occasionally, a cortical Lewy body may resemble the classical brain stem type (arrow). (*a*) and (*b*) Hematoxylin and eosin; original magnification ×400. *(Continues)*

areas which are commonly but variably involved include the putamen, caudate nucleus, pontine nuclei, inferior olivary nucleus, hippocampus, entorhinal cortex, cerebral cortex, and colliculi.[35,42]

CORTICOBASAL DEGENERATION

Corticobasal degeneration (CBD) was first described as a distinct clinicopathological entity by Rebeiz and coworkers in 1967,[43] who referred to this disorder as "corticodentatonigral degeneration with neuronal achromasia." Subse-

quent reports have referred to this disorder by a variety of titles, including cortical-basal ganglionic degeneration,[44] corticonigral degeneration,[45] and CBD.[46] The classical clinical picture is that of unilateral clumsiness or stiffness of a limb, followed after several years by dystonic rigidity, akinesia, apraxia, and stimulus-sensitive myoclonus of the involved limb. The alien limb phenomenon, in which the involved limb moves on its own, develops in many patients.[46,47] Dysarthria, dysphagia, supranuclear gaze palsy, mild cerebellar signs, and a cortical sensory disturbance may also develop. Cognitive dysfunction develops

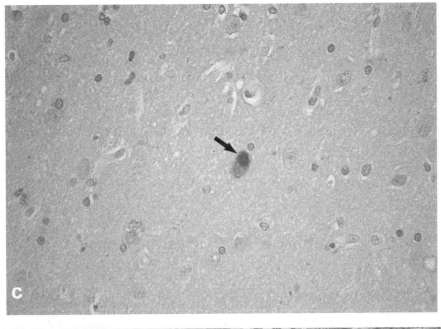

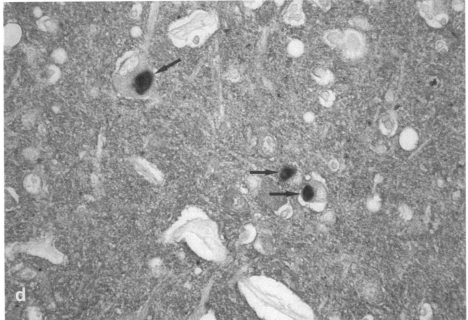

FIGURE 35.3 (Continued) (*c*) Anti-ubiquitin immunopositive Lewy body in frontal neocortex (arrow). Sigma ubiquitin immunoperoxidase, DAB/hematoxylin; original magnification ×400. (*d*) Anti-α-synuclein immunopositive Lewy bodies in cingulate gyrus (arrows). Sigma α-synuclein immunoperoxidase, DAB; original magnification ×400. *(Continues)*

in the later stages of the disease in many patients. It is now recognized, however, that pathologically proven CBD can occur with diverse clinical presentations, including an akinetic-rigid syndrome with involuntary movements, progressive aphasia, or frontal lobe dementia.[48–51] As well, several different pathological processes can manifest as an asymmetric akinetic-rigid syndrome leading to a mistaken clinical diagnosis of CBD.[52] The neuropathological criteria for a diagnosis of CBD have recently been proposed.[53]

On gross examination, the external aspects of the brain may appear relatively unremarkable, but more typically there is variable atrophy of cortical gyri of the parasagittal posterior frontal, peri-Rolandic, and parietal regions, which may be asymmetric. Cortical atrophy may be more generalized in cases presenting with dementia, with additional involvement of the inferior frontal and temporal lobes. Sectioning of the brain reveals decreased pigmentation of the SN and, often, the LC.[46,48,53]

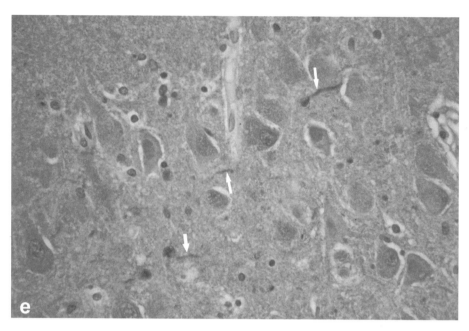

FIGURE 35.3 (Continued) (*e*) Anti-ubiquitin immunopositive Lewy neurites in the hippocampal CA2 region (arrows). Sigma ubiquitin immunoperoxidase, DAB/hematoxylin; original magnification ×400.

Microscopically, CBD is characterized by neuronal loss, astrocytic gliosis, swollen neurons, filamentous inclusions of tau protein within neurons, tau-containing glial cells, and tau-containing threads.[53–55] The swollen neurons contain neurofilament protein and can be demonstrated immunocytochemically with anti-neurofilament protein antibodies, while the tau protein aggregates in neurons, glial cells, and threads can be demonstrated with anti-tau antibodies, as well as the Gallyas silver stain.

Swollen neurons (also referred to as "achromatic" or "ballooned" neurons) are found scattered throughout affected cortical areas, where they are most frequent in layers III, IV, and VI. On H&E stained sections they appear as pale-stained eosinophilic to amphiphilic large neurons, which are often vacuolated, and may have an associated surrounding area of artifactual vacuolation. They often appear rounded, but may be more irregular or serpiginous in shape, particularly when there is also swelling of their proximal processes (Figure 35.5*a*). Ultrastructurally, the cytoplasm of swollen neurons contains accumulations of straight filaments 10 to 15 nm in diameter, as well as some larger straight filaments measuring 25 to 30 nm in diameter.[54,56,57] While swollen neurons are most frequently seen in involved cortical areas, they may also be found in the cingulate gyrus, insular cortex, claustrum, and amygdala.

The intraneuronal accumulations of tau protein range in appearance following tau immunostaining. In most involved neurons, cytoplasmic tau immunopositivity is diffuse or granular, representing so-called "pre-tangles" (Figure 35.5*b*). In other neurons, the tau immunopositiv-

ity is aggregated into small inclusions resembling NFTs (Figure 35.5*c*). In still other neurons, cytoplasmic tau immunopositivity is more filamentous and dispersed, or consists of skein-like NFTs. In brain stem nuclei, such as the SN and LC, intraneuronal tau immunopositive aggregates have an appearance similar to the globose NFTs found in Alzheimer's disease and PSP. Tau immunopositive astrocytes, in which aggregates of tau protein fill the astrocyte cell body and processes, are also seen (Figure 35.5*d*). A characteristic tau immunopositive astrocytic lesion which may be seen in grey matter in CBD is referred to as an "astrocytic plaque,"[58] in which a ring-like cluster of short tau immunopositive astrocytic processes surrounds a central nonstaining region. In addition, many tau immunopositive astrocytic processes are found scattered throughout the neuropil as threads (see Figure 35.5*b*), and tau immunopositive coiled inclusions may be seen in oligodendroglial cells.

Tau immunopositive pathological changes are quite widespread, involving both grey and white matter. Associated neuronal loss and astrocytic gliosis is variable in degree but is generally not marked, with the exception of the SN where there is often a prominent loss of pigmented dopaminergic neurons. Affected cortical areas often show variable microvacuolation, which may be limited to the superficial cortical layers, or may be transcortical in severely affected areas. Additional brain areas that may show variable degrees of involvement include the cerebellar dentate nucleus, LC, pontine nuclei, red nucleus, subthalamic nucleus, thalamus, globus pallidus, putamen, and caudate nucleus.[53]

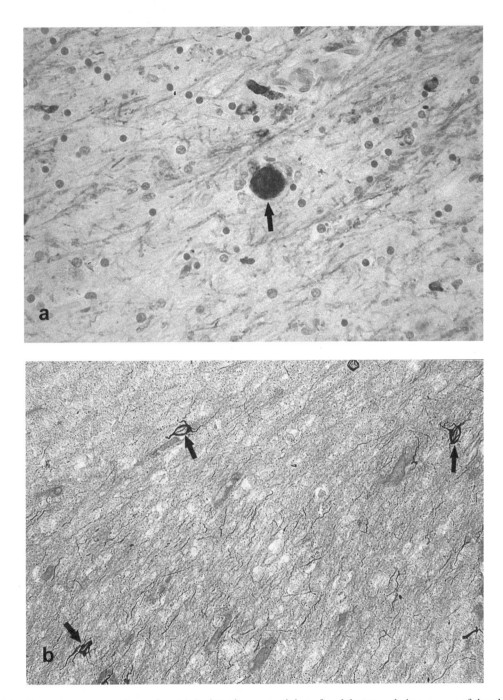

FIGURE 35.4   Progressive supranuclear palsy. (*a*) Anti-tau immunostaining of a globose tangle in a neuron of the globus pallidus interna (arrow). Sigma tau immunoperoxidase, DAB/hematoxylin; original magnification ×400. (*b*) Filamentous inclusions within astrocytes in white matter of the cerebellum (arrows). Gallyas; original magnification ×400. *(Continues)*

MULTIPLE SYSTEM ATROPHY

Graham and Oppenheimer introduced the term "multiple system atrophy" (MSA) in 1969 in a report of a patient who had developed autonomic failure followed by cerebellar and pyramidal signs.[59] At that time, the term MSA was used to refer to several sporadic and familial system degenerations that were previously considered to be separate neurological disorders. Today, however, the term MSA is used specifically to encompass cases of sporadic olivopontocerebellar degeneration, striatonigral degeneration, and the Shy-Drager syndrome. Although these three entities were formerly defined as separate disorders on the basis of clinical features and pathological changes limited

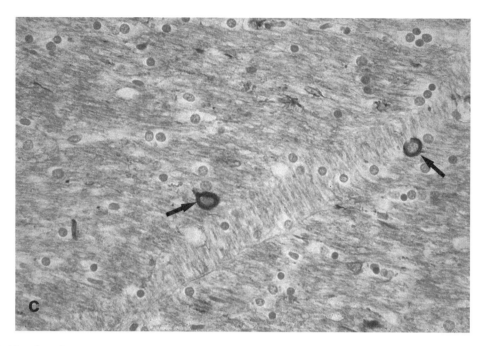

FIGURE 35.4 (Continued) (*c*) Coiled inclusions within oligodendroglial cells in the white matter of the anterior limb of the internal capsule (arrows). Sigma tau immunoperoxidase, DAB/hematoxylin; original magnification ×400.

to specific neuronal systems, in recent years glial cytoplasmic inclusions (GCIs) have been identified as a unifying pathological feature (see below).[60,61]

Patients may present with clinical symptoms and signs referable to any or all of the three disorders encompassed by the term MSA, dependent upon which neuronal systems are involved. It is uncommon for patients who present with symptoms and signs restricted to only one of the three disorders to maintain this clinical picture over time, as an overlap of symptoms typically develops as more neuronal systems become involved by the pathological process.

Glial cytoplasmic inclusions appear as flame or sickle-shaped structures within the cytoplasm of oligodendroglial cells and are well visualized with silver stains such as the Gallyas technique (Figure 35.6*a*), and anti-ubiquitin (Figure 35.6*b*) and anti-α-synuclein immunocytochemistry.[27,61] Ultrastructurally, GCIs are composed of a random arrangement of filaments having a diameter of 20 to 40 nm, along with associated granular material. Two classes of filaments have been identified: straight filaments, having a diameter of 10 nm, and twisted filaments, having a diameter alternating between 5 and 18 nm with a periodicity of 70 nm.[61–65] Other pathological changes accompanying the GCIs in involved areas include variable neuronal loss and astrocytic gliosis, as well as α-synuclein and ubiquitin immunopositive neuropil threads, neuronal cytoplasmic inclusions, and neuronal and oligodendroglial intranuclear inclusions.[60,61]

OLIVOPONTOCEREBELLAR ATROPHY

Dejerine and Thomas described the sporadic form of olivopontocerebellar atrophy in 1900.[66] The main clinical feature is cerebellar ataxia. Gross examination of the brain reveals atrophy of the cerebellum, middle cerebellar peduncles, and pons. Microscopic examination reveals GCIs and myelin pallor in these sites, with associated neuronal loss and astrocytic gliosis involving the cerebellar Purkinje cells, pontine nuclei, and inferior olivary nucleus.[60,67]

STRIATONIGRAL DEGENERATION

Adams and co-workers first described striatonigral degeneration in the early 1960s.[68] Clinically, patients manifest prominent Parkinsonian features, though they often lack a tremor. Typically, there is poor response to L-dopa. Gross examination of the brain reveals decreased pigmentation of the SN and LC, along with a variable degree of atrophy and grey-green discoloration of the putamen. Microscopic examination reveals GCIs in the putamen, globus pallidus, and caudate nucleus. Neuronal loss and astrocytic gliosis are most prominent within the putamen, SN, and LC and are milder within the globus pallidus and caudate nucleus.[60,61,67]

SHY-DRAGER SYNDROME

Shy and Drager described a syndrome characterized clinically by prominent autonomic dysfunction in 1960.[69]

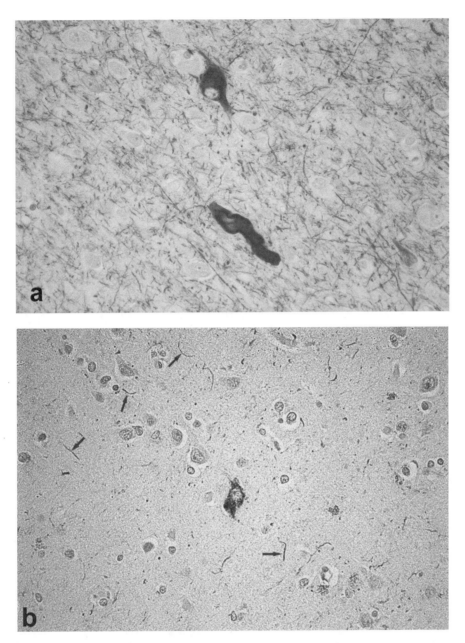

FIGURE 35.5 Corticobasal degeneration. (*a*) Anti-neurofilament protein immunopositive swollen neurons in frontal neocortex. Dako neurofilament protein immunoperoxidase, DAB/hematoxylin; original magnification ×400. (*b*) Diffuse granular cytoplasmic anti-tau immunopositivity in a neuron in temporal neocortex. Note the surrounding immunopositive thread-like structures (arrows). Sigma tau immunoperoxidase, DAB/hematoxylin; original magnification ×400. (*Continues*)

Clinical features of this syndrome may include orthostatic hypotension, urinary incontinence and atonic bladder, rectal incontinence and a loss of rectal sphincter tone, anhydrosis, and impotence. Gross pathological changes in brain and spinal cord areas associated with autonomic function are not identifiable. Microscopic examination, however, reveals the presence of GCIs within the intermediate zone of the spinal cord, and neuronal loss and astrocytic gliosis are seen within the intermediolateral cell columns of the spinal cord and the dorsal motor nucleus of the vagus.[60,67] Neuronal loss and astrocytic gliosis may also be seen in Onufrowicz's nucleus of the sacral spinal cord, which correlates with sphincter disturbances.[70,71]

Other brain areas may show varying degrees of these microscopic pathological changes in MSA, including frontal cortex and primary motor cortex and higher motor areas, which correlate with the frontal lobe dysfunction and pyramidal system involvement seen in some patients.[67,72–74]

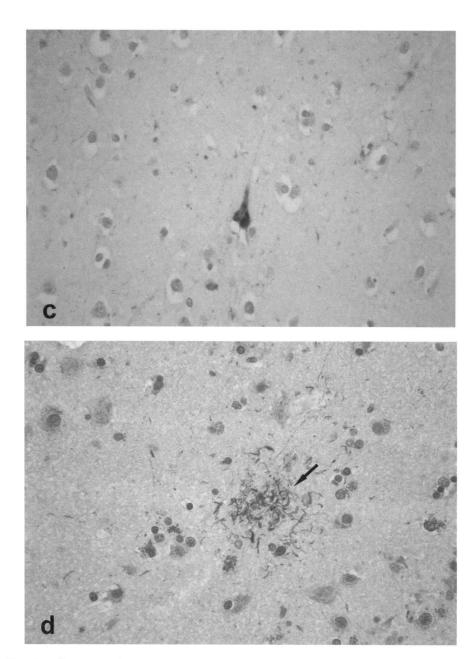

FIGURE 35.5 (Continued) (*c*) Neurofibrillary tangle-like collection of anti-tau immunopositive filaments in a neuron in temporal neocortex. (*d*) Collection of anti-tau immunopositive astrocytic processes in frontal neocortex (arrow). (*c*) and (*d*) Sigma tau immunoperoxidase, DAB/hematoxylin; original magnification ×400.

AMYOTROPHIC LATERAL SCLEROSIS-PARKINSONISM-DEMENTIA COMPLEX OF GUAM

Zimmerman, in 1945,[75] was the first to document aspects of the clinical and neuropathological features of the constellation of neurodegenerative disease now known as the amyotrophic lateral sclerosis-Parkinsonism-dementia complex (ALS/PDC) of Guam. The disease occurs with high incidence among the Chamorro people of the Mari-anas Islands of Guam and Rota in the Western Pacific Ocean. Although the ALS and PDC seen in this population were originally described as distinct clinical disease entities, it is now known that there is substantial clinical and neuropathological overlap between the two conditions. As a result, some have suggested that ALS and PDC represent the extremes of a spectrum of neurodegeneration, in which the relative degree of degeneration of the pyramidal system versus the extrapyramidal system is responsible for observed differences in clinical expression.[76]

The clinical and neuropathological features of this disorder have been recently reviewed.[77] Clinically, patients present with features of motor neuron disease and/or Parkinsonism associated with progressive cognitive deterioration.

Gross examination of the spinal cord from patients with clinical features of ALS shows atrophy of the ventral nerve roots as well as variable discoloration of the lateral columns. In patients with PDC, examination of the brain shows prominent cerebral atrophy with associated hydrocephalus ex vacuo and decreased brain weight, along with a prominent loss of pigmentation of the SN and LC. The neuropathological hallmark of both the ALS and PDC of Guam is the presence of prominent and widespread NFTs. In patients with dementia, and to a lesser extent in those with clinically intact cognitive function, numerous NFTs are present in the neocortices, basal forebrain, amygdala, hippocampus, and entorhinal cortex, associated with variable neuronal loss, astrocytic gliosis, and frequent extracellular ghost tangles. In patients with Parkinsonism, numerous NFTs can be found within the SN, LC, and periaquaductal grey matter, along with prominent neuronal loss and astrocytic gliosis. In patients with ALS, there is a loss of spinal cord anterior horn cells, with associated astrocytic gliosis. Many of the remaining viable anterior horn cells may appear shrunken, swollen, or contain Bunina bodies or, in cases with severe NFT formation, NFTs may be seen. Additional sites of NFT formation in cases with a prominent NFT burden include the cerebellar dentate nucleus and inferior olivary nucleus. The primary motor cortex often contains large numbers of NFTs,[78] and loss of upper motor neurons results in myelin pallor of the lateral and ventral corticospinal tracts. Numerous Hirano bodies and granulovacuolar degeneration can be seen in the hippocampus.

Ultrastructural, immunohistochemical, and biochemical studies have shown that the NFTs in ALS/PDC of Guam are essentially identical to those found in Alzheimer's disease. They are composed of paired helical filaments and straight filaments; react with antibodies to tau, beta-amyloid peptide, and ubiquitin; and contain paired helical filament-tau, which is abnormally phosphorylated.[77]

Although the etiology of ALS/PDC of Guam remains to be fully elucidated, exposure to a putative environmental etiological agent is suspected to be involved.[77] Its incidence has been reported to be in decline over recent years.[79]

POST-ENCEPHALITIC PARKINSONISM

Many cases of Parkinsonism appeared following the global pandemic of encephalitis lethargica (von Economo's disease), which began around the time of World War I. Hypersomnolence and ophthalmoplegia were characteristic clinical features of the acute phase of the encephalitic illness,

and a minority of patients showed Parkinsonian features.[80,81] Many patients surviving the acute encephalitic illness eventually developed Parkinsonism following an interval of several to many years. The Parkinsonian syndrome in postencephalitic Parkinsonism (PEP) is typically static or very slowly progressive, and additional findings may include dystonia, delirium, and pyramidal features, while dementia is unusual.[81,82] Cases of PEP are rare today, as most individuals affected by the pandemic have since died. Cases of sporadic encephalitis lethargica have been reported in recent years, however.[83] Although epidemiological evidence suggests an association between the acute encephalitic illness and an influenza virus, and the clinical and pathological features are suggestive of a viral infection, the definite etiology has not yet been ascertained.[84,85]

Gross examination of the PEP brain reveals variable but mild generalized cortical atrophy, as well as decreased pigmentation of the SN and LC.[86] Characteristic findings on microscopic examination include variable neuronal loss, which is usually severe in the SN, and astrocytic gliosis, as well as the accumulation of tau protein within neurons as NFTs, which can be demonstrated with appropriate silver stains or immunocytochemically with anti-tau antibodies. Other neurons may show diffuse cytoplasmic tau immunoreactivity, presumably representing a pre-tangle stage. Tau immunoreactive fibrillary inclusions within astrocytes are also seen in involved areas.[87] The molecular aspects of the tau protein that accumulates in PEP resemble those seen in Alzheimer's disease.[88] Similarly, the ultrastructural features of the NFTs found in PEP resemble those seen in Alzheimer's disease, consisting of typical paired helical filaments.[89]

The brain regions most commonly affected by the microscopic changes in PEP include the SN, LC, hippocampus, parahippocampal gyrus, and the nucleus basalis of Meynert. Other brain areas where changes may be found include the caudate nucleus, putamen, globus pallidus, thalamus, subthalamic nucleus, hypothalamus, cerebellar dentate nucleus, pontine nuclei, reticular formation, anterior horn cells, and insular, frontal, parietal, and temporal neocortices. Brain areas within which minor changes may be seen include the red nucleus, inferior olivary nucleus, superior and inferior colliculi, and the pedunculopontine nucleus.[86] There is considerable overlap between the pathological changes of PEP, PSP, and the ALS/PDC of Guam.[86] Clinical features, including a previous history of encephalitis, become important for establishing the diagnosis in this situation.

PARKINSONISM WITH NEUROFIBRILLARY TANGLES

A small number of patients have been reported in which Parkinsonism was associated with NFT pathology. Clinically, these patients were young at symptom onset, responded well to drug therapy, had a slowly progressive disease course, and had no history of previous encephalitis and no family history of Parkinsonism.[90] Postmortem examination revealed neuronal loss, astrocytic gliosis, and NFTs within the SN and LC; these changes were relatively inconspicuous elsewhere. The NFTs present showed tau immunopositivity, and ultrastructurally consisted predominantly of paired helical filaments. The relationship between these cases and PEP is uncertain.[90,91]

Similar pathological findings have been reported for an autopsied case of familial autosomal recessive juvenile-onset Parkinsonism linked to the parkin gene on chromosome 6q.[92] Neuronal loss and astrocytic gliosis were essentially restricted to the SN and LC, and small numbers of NFTs were found in these sites. Occasional NFTs were also present within the red nucleus, hypothalamus, and neocortex, while more frequent NFTs were seen within the hippocampus. Other reported cases of familial autosomal recessive Parkinsonism associated with parkin gene mutations have shown neuronal loss and astrocytic gliosis within the SN and LC without associated NFTs or LBs.[93–95]

MPTP-INDUCED PARKINSONISM

Langston et al.[96] reported the development of Parkinsonism, with bradykinesia, rigidity, and resting tremor, in a small number of young adults with a recent history of intravenous use of a "synthetic heroin" in 1983. The toxic agent was eventually identified as 1-methyl-4-phenyl-1,2,3,6–tetrahydropyridine (MPTP), a by-product of meperidine synthesis. MPTP is oxidized by the enzyme monoamine oxidase B in glial cells to form 1-methyl-4-phenyl pyridinium (MPP$^+$). The neurotoxic MPP$^+$ is concentrated by the neuronal synaptic system, with subsequent mitochondrial damage and neuronal death.[97] MPTP-induced Parkinsonism in primates and rodents is presently a popular experimental model of Parkinson's disease.

Postmortem examination of MPTP-treated experimental animals has revealed neuronal loss within the SN and LC as well as occasional eosinophilic inclusion bodies and α-synuclein immunopositive aggregates reminiscent of LBs.[98–100] Autopsy examination of three human cases of MPTP-induced Parkinsonism, with survival times ranging from three to 16 years, revealed moderate to severe loss of pigmented neurons of the SN with associated astrocytic gliosis, in the absence of LBs.[101]

METHANOL-INDUCED PARKINSONISM

Methanol intoxication most commonly occurs following accidental or intentional ingestion, often as an ethanol substitute, of a number of commercially available products such as gasoline antifreeze, windshield wiper fluid, wood alcohol, paint solvents, and perfume.[102] Methanol is oxidized to formaldehyde and formic acid by alcohol dehydrogenase in the liver, which leads to severe metabolic acidosis. As well, there is associated inhibition of cytochrome C oxidase with resultant mitochondrial dysfunction.[103] In patients surviving the acute intoxication, ophthalmological changes, including blindness, are common, and Parkinsonism associated with bradykinesia, rigidity, and resting tremor may occasionally develop.[104,105]

In the acute stages, gross examination of the brain reveals prominent cerebral edema and frequent petechial hemorrhages. With time, loss of retinal ganglion cells and optic nerve degeneration are seen. As well, symmetric necrosis of the putamen, which is often hemorrhagic, may be present, along with variable subcortical white matter necrosis. In patients with long-term survival, astrocytic gliosis and cystic degeneration are seen in previously necrotic areas.[104,105]

CYANIDE-INDUCED PARKINSONISM

Cyanide ingestion may occur accidentally or intentionally, as in attempted suicide. Cyanide inhibits cytochrome enzymes within mitochondria, thereby preventing oxidative phosphorylation. Clinically, acute cyanide intoxication can result in encephalopathy and coma, while in those who survive the acute stages severe Parkinsonism may develop. Neuroimaging findings have revealed extensive changes involving the basal ganglia, including hemorrhagic necrosis, as well as pseudolaminar necrosis of the cerebral cortex within six weeks of the acute event.[106–109]

Autopsy was performed on a 19-year-old man with a severe Parkinsonian syndrome who died 19 months following attempted suicide by cyanide ingestion.[110] The globus pallidus, putamen, and SN pars reticulata were shrunken, and microscopically revealed prominent neuronal loss and astrocytic gliosis. The subthalamic nucleus also showed some neuronal loss and astrocytic gliosis, resolved laminar necrosis was seen in the occipital arterial border zones, and there was a complete loss of Purkinje cells and a prominent reduction of internal granule cells in the cerebellum. The SN pars compacta and hippocampal regions were unremarkable.[110] Animal studies indicate that hypoxic/ischemic damage on the basis of associated hypotension, rather than direct toxicity, is the major cause of delayed damage after cyanide exposure.[111–113]

CARBON MONOXIDE-INDUCED PARKINSONISM

Carbon monoxide (CO) has an affinity for hemoglobin of approximately 200 times that of oxygen and binds irreversibly to hemoglobin for the life span of the red blood cell, thereby reducing the effective oxygen-carrying hemoglobin concentration.[114] Carbon monoxide may also bind directly to haem iron in brain regions rich in iron, such as the globus pallidus and SN pars reticulata. In combination with hypotension, brain areas selectively vulnerable to damage in CO poisoning include those rich in iron and those vulnerable to ischemic damage.[114,115]

The clinical features of CO intoxication are dependent upon the blood level of carboxyhemoglobin, which is dependent upon the concentration of inspired CO and duration of exposure, while the pathological features are dependent upon the severity of CO intoxication and length of survival. The clinical features seen with acute CO intoxication may range from psychomotor impairment through seizures and coma. Delayed neurological deterioration and encephalopathy may occur after a period of several days to weeks in those patients surviving the initial insult. Later on, many patients demonstrate Parkinsonian features.[116–118]

On gross examination, the brain of patients dying acutely reveals prominent pink discoloration due to the presence of carboxyhemoglobin, as well as variable edema. With time, ischemic damage involving selectively vulnerable areas, such as the hippocampus, cerebral cortex, and cerebellar Purkinje cells, may be seen. As well, well demarcated necrosis involving the globus pallidus, SN pars reticulata, and cerebral hemispheric white matter may be seen. Microscopically, damaged areas reveal features typical of ischemic injury. Cerebral hemispheric white matter may also show areas of demyelination with relative preservation of axons, a feature termed "Grinker's myelinopathy." The clinically evident delayed encephalopathy seen in some patients tends to correlate with the presence of cerebral hemispheric white matter changes.[114,115,119]

REFERENCES

1. Rajput, A. H. et al., Epidemiology of Parkinsonism: incidence, classification, and mortality, *Ann. Neurol.*, 16, 278, 1984.
2. Rajput, A. H., Rozdilsky, B., and Rajput, A., Accuracy of clinical diagnosis in parkinsonism—a prospective study, *Can. J. Neurol. Sci.*, 18, 275, 1991.
3. Parkinson, J., *An essay on the shaking palsy* (London, 1817), in *James Parkinson (1755–1824)*, Critchley, M., Ed., Macmillan, London, 1955, 147.
4. Olanow, C. W. and Tatton, W. G., Etiology and pathogenesis of Parkinson's disease, *Annu. Rev. Neurosci.*, 22, 123, 1999.
5. Hoehn, M. M. and Yahr, M. D., Parkinsonism: onset, progression, and mortality, *Neurology*, 17, 427, 1967.
6. Alvord, E. C. and Forno, L. S., Pathology, in *Handbook of Parkinson's Disease*, Koller, W. C., Ed., Marcel Dekker, New York, 209, 1987.
7. Gibb, W. R. G. and Lees, A., The significance of the Lewy body in the diagnosis of idiopathic Parkinson's disease, *Neuropathol. Appl. Neurobiol.*, 15, 27, 1989.
8. Dale, G. E. et al., Relationships between Lewy bodies and pale bodies in Parkinson's disease, *Acta Neuropathol.* (Berl.), 83, 525, 1992.
9. Jellinger, K. A., Pathology of Parkinson's Disease. Changes other than the nigrostriatal pathway, *Mol. Chem. Neuropathol.*, 14, 153, 1991.
10. Golbe, L. I. et al., A large kindred with autosomal dominant Parkinson's disease, *Ann. Neurol.*, 27, 276, 1990.
11. Spira, P. J. et al., Clinical and pathological features of a parkinsonian syndrome in a family with an Ala53Thr α-synuclein mutation, *Ann. Neurol.*, 49, 313, 2001.
12. Nicholl, D. J. et al., Two large British kindreds with familial Parkinson's disease: a clinico-pathological and genetic study, *Brain*, 125, 44, 2002.
13. Waters, C. H. and Miller, C. A., Autosomal dominant Lewy body parkinsonism in a four-generation family, *Ann. Neurol.*, 35, 59, 1994.
14. Grimes, D. A. et al., Large French-Canadian family with Lewy body parkinsonism: exclusion of known loci, *Mov. Disord.*, 17, 1205, 2002.
15. Fearnley, J. M. and Lees, A. J., Aging and Parkinson's disease: substantia nigra regional selectivity, *Brain*, 114, 2283, 1991.
16. Paulus, W. and Jellinger, K., The neuropathologic basis of different clinical subtypes of Parkinson's disease, *J. Neuropathol. Exp. Neurol.*, 50, 143, 1991.
17. Jellinger, K. A., Post mortem studies in Parkinson's disease—is it possible to detect brain areas for specific symptoms? *J. Neural. Transm.*, Suppl. 56, 1, 1999.
18. Jellinger, K. A., The pathology of Parkinson's disease, in *Parkinson's Disease: Advances in Neurology*, Vol. 86, Chapter 6, Calne, D. and Calne, S., Eds., Lippincott Williams & Wilkins, Philadelphia, 2001.
19. Pirozzolo, F. J. et al., Dementia in Parkinson disease: a neuropsychological analysis, *Brain Cogn.*, 1, 71, 1982.
20. Crystal, H. A. et al., Ante-mortem diagnosis of diffuse Lewy body disease, *Neurology*, 40, 1523, 1990.
21. Xuereb, J. H. et al., Cortical and subcortical pathology in Parkinson's disease: relationship to parkinsonian dementia, *Adv. Neurol.*, 53, 35, 1990.
22. McKeith, I. G., Perry, E. K., and Perry, R. H., Report of the second dementia with Lewy body international workshop, *Neurology*, 53, 902, 1999.
23. Rajput, A. H., and Rozdilsky, B., Dysautonomia in parkinsonism: a clinicopathological study, *J. Neurol. Neurosurg. Psychiat.*, 39, 1092, 1976.
24. Forno, L. S., Neuropathology of Parkinson's disease, *J. Neuropathol. Exp. Neurol.*, 55, 259, 1996.
25. Gibb, W., Esiri, M., and Lees, A., Clinical and pathological features of diffuse cortical Lewy body disease (Lewy body dementia), *Brain*, 110, 1131, 1985.
26. Love, S. and Nicoll, J. A., Comparison of modified Bielschowsky silver impregnation and anti-ubiquitin immunostaining of cortical and nigral Lewy bodies, *Neuropathol. Appl. Neurobiol.*, 18, 585, 1992.

27. Spillantini, M. G., Parkinson's disease, dementia with Lewy bodies and multiple system atrophy are α-synucleinopathies, *Parkinson. Rel. Disord.*, 5, 157, 1999.

28. Irrizary, M. C. et al., Nigral and cortical Lewy bodies and dystrophic nigral neurites in Parkinson's disease and cortical Lewy body disease contain α-synuclein immunoreactivity, *J. Neuropathol. Exp. Neurol.*, 57, 334, 1998.

29. Jo, E. et al., Alpha-synuclein membrane interactions and lipid specificity, *J. Biol. Chem.*, 275, 34328, 2000.

30. Perrin, R. J. et al., Interaction of human alpha-synuclein and Parkinson's disease variants with phospholipids. Structural analysis using site-directed mutagenesis, *J. Biol. Chem.*, 275, 34393, 2000.

31. Galvin, J. F. et al., Pathology of the Lewy body, in *Parkinson's Disease: Advances in Neurology*, Vol. 80, Stern, G., Ed., Lippincott Williams & Wilkins, Philadelphia, 1999, 313.

32. Braak, H. et al., Extensive axonal Lewy neurites in Parkinson's disease: a novel pathological feature revealed by alpha-synuclein immunocytochemistry, *Neurosci. Lett.*, 265, 67, 1999.

33. Gai, W. P., Blessing, W. W., and Blumbergs, P. C., Ubiquitin-positive degenerating neurites in the brain stem in Parkinson's disease, *Brain*, 118, 1447, 1995.

34. Steele, J., Richardson, J., and Olszewski, J., Progressive supranuclear palsy: a heterogeneous degeneration involving the brain stem, basal ganglia, and cerebellum with vertical gaze and pseudobulbar palsy, nuchal dystonia and dementia, *Arch. Neurol.*, 2, 473, 1964.

35. Rajput, A. and Rajput, A. H., Progressive supranuclear palsy: clinical features, pathophysiology and management, *Drugs Aging*, 18, 913, 2001.

36. Birdi, S. et al., Progressive supranuclear palsy diagnosis and confounding features: report on 16 autopsied cases, *Mov. Disord.*, 17, 1255, 2002.

37. Yamada, T., McGeer, P. L., and McGeer, E. G., Appearance of paired nucleated, tau-positive glia in patients with progressive supranuclear palsy brain tissue, *Neurosci. Lett.*, 135, 99, 1992.

38. Probst, A. et al., Progressive supranuclear palsy: extensive neuropil threads in addition to neurofibrillary tangles. Very similar antigenicity of subcortical neuronal pathology in progressive supranuclear palsy and Alzheimer's disease, *Acta Neuropathol.*, 77, 61, 1988.

39. Ghatak, N. R., Nochlin, D., and Hadfield, M. G., Neurofibrillary pathology in progressive supranuclear palsy, *Acta Neuropathol. (Berl.)*, 52, 73, 1980.

40. Takahashi, H. et al., Occurrence of 15-nm-wide straight tubules in neocortical neurons in progressive supranuclear palsy, *Acta Neuropathol. (Berl.)*, 79, 233, 1989.

41. Yagishita, S. et al., Ultrastructure of neurofibrillary tangles in progressive supranuclear palsy, *Acta Neuropathol. (Berl.)*, 8, 27, 1979.

42. Lowe, J. S. and Leigh, N., Disorders of movement and system degenerations, in *Greenfield's Neuropathology*, 7th ed., Vol. 2, Chapter 6, Graham, D.I. and Lantos, P.L., Eds., Arnold, London, 2002.

43. Rebeiz, J. J., Kolodny, E. H., and Richardson, E. P. Jr., Corticodentatonigral degeneration with neuronal achromasia: A progressive disorder of late adult life, *Trans. Am. Neurol. Assoc.*, 92, 23, 1967.

44. Riley, D. E. et al., Cortical-basal ganglionic degeneration. *Neurology*, 40, 1203, 1990.

45. Lippa, C. F., Smith, T. W., and Fontneau, N., Corticonigral degeneration with neuronal achromasia. A clinicopathologic study of two cases, *J. Neurol. Sci.*, 98, 301, 1990.

46. Gibb, W. R. G., Luthert, P. J., and Marsden, C. D., Corticobasal degeneration, *Brain*, 112, 1171, 1989.

47. Rinne, J. O. et al., Corticobasal degeneration. A clinical study of 36 cases, *Brain*, 117, 1183, 1994.

48. Bergeron, C. et al., Unusual clinical presentations of cortical-basal ganglionic degeneration, *Ann. Neurol.*, 40, 893, 1996.

49. Bergeron, C., Davis, A., and Lang, A. E., Corticobasal ganglionic degeneration and progressive supranuclear palsy presenting with cognitive decline, *Brain Pathol.*, 8, 355, 1998.

50. Schneider, J. A. et al., Corticobasal degeneration: neuropathologic and clinical heterogeneity, *Neurology*, 48, 959, 1997.

51. Kertesz, A. et al., Corticobasal degeneration syndrome overlaps progressive aphasia and frontotemporal dementia, *Neurology*, 55, 1368, 2000.

52. Boeve, B. F. et al., Pathologic heterogeneity in clinically diagnosed corticobasal degeneration, *Neurology*, 53, 795, 1999.

53. Dickson, D. W. et al., Office of rare diseases neuropathologic criteria for corticobasal degeneration, *J. Neuropathol. Exp. Neurol.*, 61, 935, 2002.

54. Mori, H. et al., Corticobasal degeneration: a disease with widespread appearance of abnormal tau and neurofibrillary tangles, and its relation to progressive supranuclear palsy, *Acta Neuropathol.*, 88, 113, 1994.

55. Horoupian, D. S. and Chu, P. L., Unusual case of corticobasal degeneration with tau/Gallyas-positive neuronal and glial tangles, *Acta Neuropathol.*, 88, 592, 1994.

56. Wakabayashi, K. et al., Corticobasal degeneration: etiopathological significance of the cytoskeletal alterations, *Acta Neuropathol. (Berl.)*, 87, 545, 1994.

57. Arima, K. et al., Corticonigral degeneration with neuronal achromasia presenting with primary progressive aphasia: ultrastructural and immunocytochemical studies, *J. Neurol. Sci.*, 127, 186, 1994.

58. Feany, M. B. and Dickson, D. W., Widespread cytoskeletal pathology characterizes corticobasal degeneration, *Am. J. Pathol.*, 146, 1388, 1995.

59. Graham, J. G. and Oppenheimer, D. R., Orthostatic hypotension and nicotine sensitivity in a case of multiple system atrophy, *J. Neurol. Neurosurg. Psychiat.*, 32, 28, 1969.

60. Papp, M. I., Khan, J. E., and Lantos, P. L., Glial cytoplasmic inclusions in the CNS of patients with multiple system atrophy (striatonigral degeneration, olivopontocerebellar atrophy and Shy- Drager syndrome), *J. Neurol. Sci.*, 94, 79, 1989.

61. Papp, M. I. and Lantos, P. L., Accumulation of tubular structures in oligodendroglial and neuronal cells as the

basic alteration in multiple system atrophy, *J. Neurol. Sci.*, 107, 172, 1992.

62. Spillantini, M. G. et al., Filamentous α-synuclein inclusions link multiple system atrophy with Parkinson's disease and dementia with Lewy bodies, *Neurosci. Lett.*, 251, 205, 1998.

63. Murayama, S. et al., Immunocytochemical and ultrastructural studies of neuronal and oligodendroglial cytoplasmic inclusions in multiple system atrophy. 2. Oligodendroglial cytoplasmic inclusions, *Acta Neuropathol. (Berl.)*, 84, 32, 1992.

64. Kato, S. et al., Argyrophilic ubiquitinated cytoplasmic inclusions of Leu-7-positive glial cells in olivopontocerebellar atrophy (multiple system atrophy), *Acta Neuropathol. (Berl.)*, 82, 488, 1991.

65. Abe, H. et al., Argyrophilic glial intracytoplasmic inclusions in multiple system atrophy: immunocytochemical and ultrastructural study, *Acta Neuropathol. (Berl.)*, 84, 273, 1992.

66. Dejerine, J. and Thomas, A. A., L'atrophie olivo-ponto-cerebelleuse, *Nouv. Icon. de le Salpet.*, 13, 330, 1900.

67. Papp, M. I. and Lantos, P. L., The distribution of oligodendroglial inclusions in multiple system atrophy and its relevance to clinical symptomatology, *Brain*, 117, 235, 1994.

68. Adams, R. D., van Bogaert, L., and van der Eecken, H., Striato-nigral degeneration, *J. Neuropathol. Exp. Neurol.*, 23, 584, 1964.

69. Shy, G. M. and Drager, G. A., A neurological syndrome associated with orthostatic hypotension: a clinical-pathologic study, *Arch. Neurol.*, 2, 511, 1960.

70. Mannen, T. et al., The Onuf's nucleus and the external anal sphincter muscles in amyotrophic lateral sclerosis and Shy-Drager syndrome, *Acta Neuropathol. (Berl.)*, 58, 255, 1982.

71. Konno, H. et al., Shy-Drager syndrome and amyotrophic lateral sclerosis. Cytoarchitectonic and morphometric studies of sacral autonomic neurons, *J. Neurol. Sci.*, 73, 193, 1986.

72. Konagaya, M. et al., Multiple system atrophy with remarkable frontal lobe atrophy, *Acta Neuropathol. (Berl.)*, 97, 423, 1999.

73. Wakabayshi, K. et al., Multiple system atrophy with severe involvement of the motor cortical areas and cerebral white matter, *J. Neurol. Sci.*, 156, 114, 1998.

74. Robbins, T. W. et al., Cognitive performance in multiple system atrophy, *Brain*, 1, 271, 1992.

75. Zimmerman, H. M., Monthly report to medical officer in command. *US Naval Medical Research Unit*, 2, 1945.

76. Perl, D. P., Amyotrophic lateral sclerosis and parkinsonism-dementia complex of Guam: two distinct entities or a single disease process with a spectrum of neuropathological and clinical expressions? In *Advances in Research on Neurodegeneration*, Vol. II, Etiopathogenesis, Mizuno, Y., Calne, D. B., and Horowski, R., Eds., Birkhauser, Boston, 209, 1994.

77. Perl, D. P., Amyotrophic lateral sclerosis-parkinsonism-dementia complex of Guam, in *Neuropathology of Dementing Disorders*, Chapter 14, Markesbery, W. R., Ed., Arnold, London, 1998.

78. Hof, P. R. and Perl, D. P., Neurofibrillary tangles in the primary motor cortex in Guamanian amyotrophic lateral sclerosis/parkinsonism-dementia complex, *Neurosci. Lett.*, 328, 294, 2002.

79. Stone, R., Guam: deadly disease dying out, *Science,* 261, 424, 1993.

80. von Economo, C. (translated by Newman, K. O.), *Encephalitis Lethargica: Its Sequelae and Treatment*, Oxford University Press, London, 1, 1931.

81. Calne, D. B. and Lees, A. J., Late progression of post-encephalitic Parkinson's syndrome, *Can. J. Neurol. Sci.*, 15, 135, 1988.

82. Gibb, W. R. G. and Lees, A. J., The progression of idiopathic Parkinson's disease is not explained by age-related changes. Clinical and pathological comparisons with post-encephalitic parkinsonian syndrome, *Acta Neuropathol. (Berl.)*, 73, 195, 1987.

83. Howard, R. S. and Lees, A. J., Encephalitis lethargica: a report of four recent cases, *Brain*, 110, 19, 1987.

84. Ravenholt, R. T. and Foege, W. H., 1918 Influenza, encephalitis lethargica, parkinsonism, *Lancet.*, ii, 860, 1982.

85. Elizan, T. S., Casals, J., and Swash, M., No viral antigens detected in brain tissue from a case of acute encephalitis lethargica and another case of post-encephalitic Parkinsonism [Lett.], *J. Neurol. Neurosurg. Psychiat.*, 52, 800, 1989.

86. Geddes, J. F. et al., Pathological overlap in cases of Parkinsonism associated with neurofibrillary tangles. A study of recent cases of postencephalitic Parkinsonism and comparison with progressive supranuclear palsy and Guamanian Parkinsonism-dementia complex, *Brain*, 116, 281, 1993.

87. Ikeda, K. et al., Anti-tau-positive glial fibrillary tangles in the brain of postencephalitic parkinsonism of Economo type, *Neurosci. Lett.* 162, 176, 1993.

88. Buee Scherrer, V. et al., Pathological tau proteins in postencephalitic parkinsonism: comparison with Alzheimer's disease and other neurodegenerative disorders, *Ann. Neurol.*, 42, 356, 1997.

89. Ishii, T. and Nakamura, Y., Distribution and ultrastructure of Alzheimer's neurofibrillary tangles in postencephalitic parkinsonism of Economo type, *Acta Neuropathol.*, 55, 59, 1981.

90. Rajput, A. H. et al., Parkinsonism and neurofibrillary tangle pathology in pigmented nuclei, *Ann. Neurol.*, 25, 602, 1989.

91. Forno, L. and Alvord, E., The pathology of parkinsonism. Part I. Some new observations and correlations, in *Recent Advances in Parkinson's Disease*, McDowell, F.H. and Markham, C.H., Eds., Blackwell, Oxford, 120, 1971.

92. Mori, H. et al., Pathologic and biochemical studies of juvenile parkinsonism linked to chromosome 6q, *Neurology*, 51, 890, 1998.

93. Gouider-Khouja, N. et al., Autosomal recessive parkinsonism linked to parkin gene in a Tunisian family. Clinical, genetic and pathological study, *Parkinson. Rel. Disord.*, 9, 247, 2003.

94. van de Warrenburg, B. P. et al., Clinical and pathologic abnormalities in a family with parkinsonism and parkin gene mutations, *Neurology*, 56, 555, 2001.

95. Yamamura, Y. et al., Autosomal recessive early-onset parkinsonism with diurnal fluctuation: clinicopathologic characteristics and molecular genetic identification, *Brain Dev.*, 22 Suppl. 1, S87, 2000.

96. Langston, J. W. et al., Chronic parkinsonism in humans due to a product of meperidine analog synthesis, *Science*, 219, 979, 1983.

97. Singer, T. P. and Ramsey, R. R., Mechanism of the neurotoxicity of MPTP. An update, *FEBS Lett.*, 274, 1, 1990.

98. Forno, L. S. et al., Locus ceruleus lesions and eosinophilic inclusions in MPTP-treated monkeys, *Ann. Neurol.*, 20, 449, 1986.

99. Forno, L. S. et al., Similarities and differences between MPTP-induced parkinsonism and Parkinson's Disease. Neuropathologic considerations. In *Advances in Neurology*, Vol. 60, 600, Narabayashi, H. et al., Eds., Raven Press, New York, 1993.

100. Kowall, N. W. et al., MPTP induces alpha-synuclein aggregation in the substantia nigra of baboons, *Neuroreport*, 11, 211, 2000.

101. Langston, J. W. et al., Evidence of active nerve cell degeneration in the substantia nigra of humans years after 1-methyl-4 phenyl-1,2,3,6-tetrahydropyridine exposure, *Ann. Neurol.*, 46, 598, 1999.

102. Winchester, J. F., in *Clinical Management of Poisoning and Drug Overdose*, 393, Haddad, L. M. and Winchester, J. F., Eds., W. B. Saunders, Philadelphia, 1983.

103. Liesivuori, J. and Savolainen, H., Methanol and formic acid toxicity: biochemical mechanisms, *Pharmacol. Toxicol.*, 69, 157, 1991.

104. McLean, D. R., Jacobs, H., and Mielke, B. W., Methanol poisoning: a clinical and pathological study, *Ann. Neurol.*, 8, 161, 1980.

105. Phang, P. T. et al., Brain hemorrhage associated with methanol poisoning, *Crit. Care Med.* 16, 137, 1988.

106. Rachinger, J. et al., MR changes after acute cyanide intoxication, *Am. J. Neuroradiol.*, 23, 1398, 2002.

107. Feldman, J. M. and Feldman, M. D., Sequelae of attempted suicide by cyanide ingestion: a case report, *Int. J. Psychiatry Med.*, 20, 173, 1990.

108. Grandas, F., Artieda, J., and Obeso, J. A., Clinical and CT scan findings in a case of cyanide intoxication, *Mov. Disord.*, 4, 188, 1989.

109. Rosenberg, N. L., Myers, J. A., and Martin, W. R., Cyanide-induced parkinsonism: clinical, MRI, and 6-fluorodopa PET studies, *Neurology*, 39, 142, 1989.

110. Uitti, R. J. et al., Cyanide-induced parkinsonism: a clinicopathologic report, *Neurology*, 35, 921, 1985.

111. MacMillan, V. H., Cerebral energy metabolism in cyanide encephalopathy, *J. Cereb. Blood Flow Metab.*, 9, 156, 1989.

112. Brierley, J. B. et al., Cyanide intoxication in *Macaca mulatta*, *J. Neurol. Sci.*, 31, 133, 1977.

113. Brierley, J. B., Brown, A. W., and Calverley, J., Cyanide intoxication in the rat—physiological and neuropathological aspects, *J. Neurol. Neurosurg. Psychiat.*, 39, 129, 1976.

114. Auer, R. N. and Sutherland, G. R., Hypoxia and related conditions, in *Greenfield's Neuropathology*, 7th ed., Vol. 1, Chapter 5, Graham, D. I. and Lantos, P. L., Eds., Arnold, London, 2002.

115. Okeda, R. et al., Comparative study on pathogenesis of selective cerebral lesions in carbon monoxide poisoning and nitrogen hypoxia in cats, *Acta Neuropathol. (Berl.)*, 56, 265, 1982.

116. Klawans, H. L. et al., A pure parkinsonian syndrome following acute carbon monoxide intoxication, *Arch. Neurol.*, 39, 302, 1982.

117. Chang, K. H. et al., Delayed encephalopathy after acute carbon monoxide intoxication: MR imaging features and distribution of cerebral white matter lesions, *Radiology*, 184, 117, 1992.

118. Hart, I. K. et al., Neurological manifestation of carbon monoxide poisoning, *Postgrad. Med. J.*, 64, 213, 1988.

119. Lapresle, J. and Fardeau, M., The central nervous system and carbon monoxide poisoning. II. Anatomical study of brain lesions following intoxication with carbon monoxide (22 cases), *Prog. Brain Res.*, 24, 31, 1967.

36 Pathophysiology of the Motor Disorder

Gaëtan Garraux and Mark Hallett
Human Motor Control Section, National Institute of Neurological Disorders and Stroke/National Institutes of Health

CONTENTS

INTRODUCTION

Parkinson's disease (PD) is classically characterized by bradykinesia, rigidity and tremor at rest. All features seem to be the result of the degeneration of the nigrostriatal pathway, but it has not been possible to define a single underlying pathophysiologic mechanism that explains everything. Nevertheless, there are considerable data that give separate understanding to each of the three classic features.

BRADYKINESIA

The most important functional disturbance in patients with PD is a disorder of voluntary movement prominently characterized by slowness. This phenomenon is generally called *bradykinesia*, although it has at least two components, which can be designated as *bradykinesia* and *akinesia*.[1] Bradykinesia refers to slowness of movement that is ongoing. Akinesia refers to failure of willed movement to occur. There are two possible reasons for the absence of expected movement. One is that the movement is so slow (and small) that it cannot be seen. A second is that the time needed to initiate the movement becomes excessively long.

While self-paced and tracking movements can give information about bradykinesia, the study of reaction time movements can give information about both akinesia and bradykinesia. In the reaction time situation, a stimulus is presented to a subject, and the subject must make a movement as rapidly as possible. The time between the stimulus and the start of movement is the *reaction time*, and the time from initiation to completion of movement is the *movement time*. Using this logic, prolongation of reaction time is akinesia, and prolongation of movement time is bradykinesia. Studies of patients with PD confirm that both reaction time and movement time are prolonged. However, the extent of abnormality of one does not necessarily correlate with the extent of abnormality of the other.[2] This suggests that they may be impaired by separable physiological mechanisms. In general, prolongation of movement time (bradykinesia) is better correlated with the clinical impression of slowness than is prolongation of reaction time (akinesia).

Some contributing features of bradykinesia are established. One is that there is a failure to energize muscles up to the level necessary to complete a movement in a standard amount of time. This has been demonstrated clearly with attempted rapid, monophasic movements at a single joint.[3] In this circumstance, normal movements of different angular distances are accomplished in approximately the same time by making longer movements faster. The EMG activity underlying the movement begins with a burst of activity in the agonist muscle of 50 to 100 ms, followed by a burst of activity in the antagonist muscle of 50 to 100 ms, followed variably by a third burst of activity in the agonist. This "triphasic" pattern has relatively fixed timing with movements of different distance, correlating with the similar total time for movements of different distance (Figure 36.1, left column). Different distances are accomplished by altering the magnitude of the EMG within the fixed duration burst. The pattern is correct in patients with PD, but there is insufficient EMG activity in the burst to accomplish the movement (Figure 36.1, middle and right columns). These patients often must go through two or more cycles of the triphasic pattern to

0-8493-1590-5/05/$0.00+$1.50

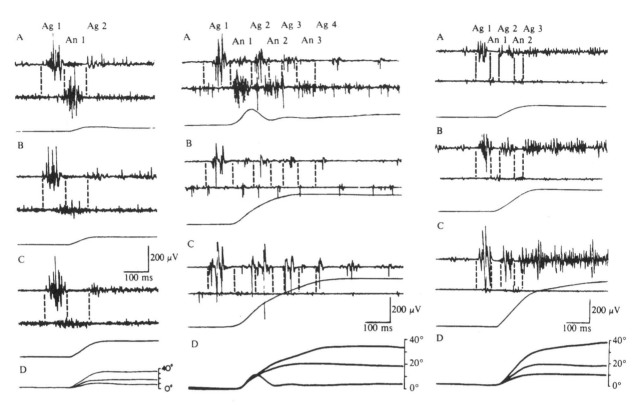

FIGURE 36.1 Ballistic elbow movements of 10 (*A*), 20 (*B*), and 40 (*C*) degrees by a normal subject (left column) and two patients with Parkinson's disease (middle and right columns). In *A*, *B*, and *C*, the traces correspond, from above downward, to biceps EMG, triceps EMG, and angular displacement at the elbow joint. (*D*) shows the three position traces superimposed. In each column, the figures are aligned so that the movements all began at the same time from the beginning of the traces. The left column illustrates a normal single cycle of a triphasic pattern characterized by alternating bursts of agonist (agonist 1, Ag1), antagonist (An), and agonist (agonist 2, Ag2) muscles. Note that, in this normal subject, all three movements were accomplished in approximately the same time, implying increasing velocities. Middle and right columns illustrate additional cycles of bursts in patients with Parkinson's disease. The right column illustrates that the amount of EMG activity in each burst could be increased by this patient but insufficiently to produce adequate velocity. (From Hallett, M. and Khoshbin, S.[3] Reproduced by permission of Oxford University Press.)

accomplish the movement. Interestingly, such activity looks virtually identical to the tremor at rest seen in these patients. The longer the desired movement, the more likely it is to require additional cycles. These findings were reproduced by Baroni et al., who also showed that L-DOPA normalized the pattern and reduced the number of bursts.[4]

Berardelli and colleagues showed that PD patients could vary the size and duration of the first agonist EMG burst with movement size and added load to some extent in the normal way.[5] However, there was a failure to match these parameters appropriately to the size of movement required (Figure 36.2). This suggests an additional problem in scaling of actual movement to the required movement. A problem in sensory scaling of kinesthesia was demonstrated by Demirci et al. PD patients used kinesthetic perception to estimate the amplitude of passive angular displacements of the index finger about the metacarpophalangeal joint and to scale them as a percentage of a reference stimulus.[6] The reference stimulus was

either a standard kinesthetic stimulus preceding each test stimulus (task K) or a visual representation of the standard kinesthetic stimulus (task V). The PD patients' underestimation of the amplitudes of finger perturbations was significantly greater in task V than in task K. Thus, when kinesthesia is used to match a visual target, distances are perceived to be shorter by the PD patients. Assuming that visual perception is normal, kinesthesia must be "reduced" in PD patients. This reduced kinesthesia, when combined with the well known reduced motor output and probably reduced corollary discharges, implies that the sensorimotor apparatus is "set" smaller in PD patients than in normal subjects.

In a slower, multijoint movement task, PD patients show a reduced rate of rise of muscle activity that also implies deficient activation.[7] Jordan, Sagar, and Cooper showed that release of force was just as slowed as increase of force and suggested that slowness to change, not deficient energization, was the main problem.[8] If termination of activity is an active process, then this find-

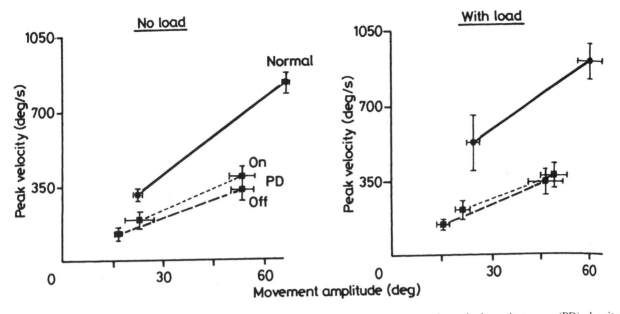

FIGURE 36.2 Abnormality of scaling of movements. Ballistic wrist flexion movements are slower in the patient group (PD), despite the fact that patients are able to increase movement velocity in the same amount as in normal subjects when 60° movements are compared with 15° movements. Similar results were found when flexion movements were performed against an opposing extensor torque (noted as load). (From Berardelli, A. et al.[5] Reproduction by permission of BMJ Publishing Group.)

ing, however, really does not argue against deficient energization.

A second physiologic mechanism of bradykinesia is that there is difficulty with simultaneous and sequential movements.[9] That PD patients have more difficulty with simultaneous movements than with isolated movements was first pointed out by Schwab, Chafetz, and Walker.[10] Quantitative studies show that slowness in accomplishing simultaneous or sequential movements is more than would be predicted from the slowness of each individual movement. With sequential movements, there is another parameter of interest—the time between the two movements, designated the interonset latency (IOL) by Benecke and colleagues.[9] The IOL is also prolonged in patients with PD. This problem, similar to the problem with simple movements, can also be interpreted as insufficient motor energy.

Akinesia would seem to be multifactorial, and a number of contributing factors are already known. As noted above, one type of akinesia is the limit of bradykinesia from the point of view of energizing muscles. If the muscle is selected but not energized, then there will be no movement. Such phenomena can be recognized on some occasions with EMG studies where EMG activity will be initiated but will be insufficient to move the body part. Another type of akinesia, again as noted above, is prolongation of reaction time; the patient is preparing to move, but the movement has not yet occurred. Considerable attention has been paid to mechanisms of prolongation of reaction time. One factor is easily demonstrable in

patients with rest tremor, who appear to have to wait to initiate the movement together with a beat of tremor in the agonist muscle of the willed movement.[11,12]

Another mechanism of prolongation of reaction time can be seen in circumstances when eye movement must be coordinated with limb movement.[13] In this situation, there is a visual target that moves into the periphery of the visual field. Normally, there is a coordinated movement of eyes and limb, the eyes beginning slightly earlier. In PD, some patients do not begin to move the limb until the eye movement is completed. This might be due to a problem with simultaneous movements, as noted above. Alternatively, it might be that PD patients need to foveate a target before they are able to move to it.

Many studies have evaluated reaction time quantitatively with neuropsychological methods.[14] The goal of these studies is to determine the abnormalities in the motor processes that must occur before a movement can be initiated. To understand reaction time studies, it is useful to consider, from a theoretical point of view, the tasks that the brain must accomplish. The starting point is the "set" for the movement. This includes the environmental conditions, initial positions of body parts, understanding the nature of the experiment and, in particular, some understanding of the expected movement. In some circumstances, the expected movement is described completely, without ambiguity. This is the "simple reaction time" condition. The movement can be fully planned. It then needs to be held in store until the stimulus comes to initiate the execution of the movement. In other circumstances, the set does not include a complete description of the required movement. It is intended that the description be completed at the time of the stimulus that calls for the movement initiation. This is the "choice reaction time" condition. In this circumstance, the programming of the movement occurs between the stimulus and the response. Choice reaction time is always longer than simple reaction, and the time difference is due to this movement programming.

In most studies, simple reaction time is significantly prolonged in patients as compared with normals.[14] On the other hand, patients appear to have normal choice reaction times, or the increase of choice reaction time over simple reaction time is the same in patients and normal subjects. Many studies in which cognitive activity was required for a decision on the correct motor response have shown that PD patients do not have apparent slowing of thinking, called *bradyphrenia*. We extended the study of choice reaction times by considering three different choice reaction time tasks that required the same simple movement but differed in the difficulty of the decision of which movement to make.[15] Comparing PD patients to normal subjects, the patients had a longer reaction time in all three conditions, but the difference was largest when the task was the easiest and smallest when the task was the most difficult. Thus, the greater the proportion of time there is in the reaction time devoted to motor program selection, the closer to normal are the PD results. Labutta et al. have shown that PD patients have no difficulty holding a motor program in store.[16] Hence, the difficulty must be executing the motor program. Execution of the movement, however, lies at the end of choice reaction time, just as it does for simple reaction time. How then can it be abnormal and choice reaction time be normal? The answer may be that in the choice reaction time situation both the motor programming and the motor execution can proceed in parallel.

Transcranial magnetic stimulation (TMS) can be used to study the initiation of execution. With low levels of TMS, it is possible to find a level that will not produce any motor evoked potentials (MEPs) at rest but will produce an MEP when there is voluntary activation. Using such a stimulus in a reaction time situation between the stimulus to move and the response, Starr et al. showed that stimulation close to movement onset would produce a response, even though there was still no voluntary EMG activity.[17] A small response first appeared about 80 ms before EMG onset and grew in magnitude closer to onset. This method divides the reaction time into two periods. In the first period, the motor cortex remains "unexcitable." In the second period, the cortex becomes increasingly "excitable" as it prepares to trigger the movement. We found that most of the prolongation of the reaction time was due to prolongation of the later period of rising excitability.[18] This result has been confirmed.[19] Our finding of

prolonged initiation time in PD patients is supported by studies of motor cortex neuronal activity in reaction time movements in monkeys rendered parkinsonian with MPTP.[20] In these investigations, there was a prolonged time between initial activation of motor cortex neurons and movement onset. Thus, an important component of akinesia is the difficulty in initiating a planned movement. This statement would not be a surprise to PD patients, who often say that they know what they want to do, but they just can't do it. Another factor that should be kept in mind is that patients appear to have much more difficulty initiating internally triggered movements than externally triggered movements. This is clear clinically in that external clues are often helpful in movement initiation. Examples include improving walking by providing an object to step over or playing march music. This can also be demonstrated in the laboratory with a variety of paradigms.[21–23]

The major problem in bradykinesia and akinesia seems to be a deficiency in activation of motor cortices. There are several additional lines of evidence for disturbed activation of cortical areas involved in motor control in PD.

Rossini et al. showed that the amplitude of the N30 of the median nerve somatosensory evoked potential (SEP) was diminished in PD. Other peaks of the SEP were normal, and the N30 had normal latency and topography. The origin of the N30 (as most of the waves of the SEP) is debated, but its decrease does suggest deficient cortical activation.[24]

The excitability of the motor cortex in PD patients has been assessed using TMS.[25] The threshold for a response was the same in normals and patients—there was a trend for the increase in MEP amplitude with stimulus intensity to be greater than normal, but the increase of the MEP amplitude with voluntary contraction was statistically less than normal. These results suggest that control of the excitability of the motor system is abnormal in PD patients, with enhanced excitability at rest and weak energization during voluntary muscle activation. There also appears to be slightly less intracortical inhibition in patients with PD. One study found reduced intracortical inhibition,[26] while another did not.[27] On the other hand, both studies found shortening of the TMS provoked silent period that lengthened with dopaminergic treatment. Studies of movement related cortical potentials (MRCPs) in patients with PD are controversial, but many studies show a decreased bereitschaftspotential (BP), a slowly rising negativity appearing during the 1 sec before self-paced voluntary movements.[28–30] In the study by Jahanshahi et al., the BP was deficient with self-paced movements but not externally triggered movements, suggesting a particular difficulty with internally triggered actions.[30]

Disturbed activation of cortical areas has also been demonstrated using functional imaging. One of the most consistent abnormalities in neuroimaging studies of PD patients during voluntary movements in the off state is a failure to activate putamen supplementary motor area (SMA) and caudate nucleus-dorsolateral prefrontal cortex (DLPFC) projections normally.[30,31] Changes in activation may, however, depend on the task performed by patients. For instance, in the study of Jahanshahi et al. where the neuroimaging was done together with EEG recording, it was found that there was a deficiency of activation of SMA in self-paced movements, but not in externally triggered movement.[30] Nevertheless, if the loss of dopaminergic activity is responsible for impaired activation of SMA and DLPFC, it should be possible to restore this functionally by systemic administration of a dopaminergic agent. Several functional imaging studies have tested that hypothesis and have shown a (partial) restoration of cortical activity after administration of apomorphine, a nonselective D_1/D_2 agonist.[31,32] Fetal mesencephalic transplant into caudate and putamen,[33] pallidotomy,[34] and pallidal and subthalamic stimulation[35] can also act to increase levels of SMA and DLPFC activation in PD along with resolution of bradykinesia.

Taken together, those data suggest that the dopaminergic system apparently provides energy to many different motor tasks, and the deficiency of this system in PD leads to both bradykinesia and akinesia.

Findings from animal studies and recent electrophysiological investigations in patients with Parkinson's disease, including intraoperative microelectrode single unit (neuron) studies and postoperative macroelectrode recordings (i.e., in the interval between implantation and subsequent connection to a stimulator), have considerably expanded our view on the mechanisms by which dopamine deficiency results in insufficient energization of the motor system. Single unit studies have shown that there is a tendency for discharge in the subthalamic nucleus (STN) and globus pallidus interna (GPi) to occur in three modes: irregular, bursting, and oscillatory.[36] Of these, only the oscillatory mode is accompanied by an established and strong tendency for synchronization between neurons. Three major frequency bands of oscillations have been identified in basal ganglia: <10 Hz, 11 to 30 Hz (beta band), and >60 Hz (gamma band). Analyses of coherence, which is a normalized measure of the linear and nonlinear correlation between two signals, have shown that each of these oscillations may be coupled to the activity in motor areas of the cerebral cortex.[37,38] Moreover, the level of dopaminergic stimulation appears to be critical for the expression of those oscillations.[38,39] Brown recently summarized evidence suggesting that different patterns of synchronization can affect movement in different ways, and proposed a model in which the low (<30 Hz) and high (>60 Hz) frequency modes of subthalamic-pallidal circuit oscillations could impair and promote motor function, respectively (Figure 36.3).[40] In untreated PD patients, there is considerable evidence suggesting an abnormal, subcortically driven, cortical rhythmicity: fast oscillations are replaced by low-frequency oscillations. Through the motor cortex, the synchronization at low frequency in PD would lead to suboptimal unfused motor unit contractions, thereby slowing the onset of voluntary action and decreasing force. It has also been proposed that trapping of cortical activity in synchronous oscillations of low frequency prevents corticocortical interactions in the gamma band, thereby contributing to akinesia and bradykinesia. This mechanism should be particularly evident in complex movements, which are especially difficult in Parkinson's disease, and

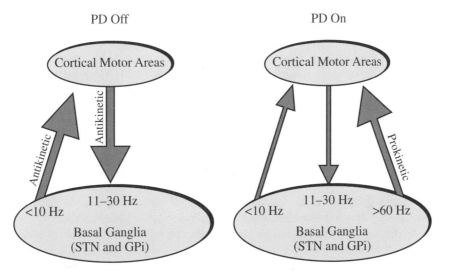

FIGURE 36.3 Schematic representation of a model of antikinetic and prokinetic activity of basal ganglia-cortical oscillatory interactions. The arrows show the dominating direction of connectivity in each frequency band. STN = subthalamic nuclei. GPi = globus pallidus interna. (From Brown, P.[40] Reproduction by permission of John Wiley & Sons, Inc.)

this has been confirmed in parkinsonian patients performing manual tracking or combined and sequential motor tasks on and off dopaminergic stimulation.[41,42] Thus, although the exact mechanisms are still under investigation, dopamine deficiency might lead to bradykinesia and akinesia by an absence of modulation of the pattern of synchronization of oscillatory activity in corticobasal ganglia-cortical networks at appropriate frequency bands during preparation and execution of voluntary movements.

RIGIDITY

Tone is defined as the resistance to passive stretch. Rigidity is one form of increased tone that is seen in disorders of the basal ganglia ("extrapyramidal disorders") and is particularly prominent in PD. Increased tone can result from changes in

1. Muscle properties or joint characteristics
2. Amount of background contraction of the muscle
3. Magnitude of stretch reflexes

There is evidence for all three of these aspects contributing to rigidity. For quantitative purposes, responses can be measured to controlled stretches delivered by devices that contain torque motors. The stretch can be produced by altering the torque of the motor or by altering the position of the shaft of the motor. The perturbation can be a single step or more complex, such as a sinusoid. The mechanical response of the limb can be measured: the positional change if the motor alters force, or the force change if the motor alters position. Such mechanical measurements can directly mimic and quantify the clinical impression.[43,44]

There are changes in the passive mechanical properties of muscle in patients with PD. The first suggestion that this might be true came from gait studies that showed reduced dorsiflexion movement of the ankle despite strong tibialis anterior activity and silent triceps surae.[45] Subsequently, using a quantitative measure, it was determined that the upper limb of patients was stiffer than normals in the totally relaxed state with no electromyographic activity present.[46] This phenomenon has been called into question by the findings of another group, which studied the lower leg and found normal contraction parameters (time-to-peak and half relaxation time), responses to short tetani, and resistance to stretch.[47] However, they found an increased resistance to passive stretch under static conditions, presumably elastic in origin. The results may be evidence against a contribution of altered muscle contractile properties to rigidity in PD, but they still reveal an increased totally passive component.

Patients with PD do not relax well and often have slight contraction at rest. This is a standard clinical as well as electrophysiological observation, and it is clear that this mechanism plays a significant part in rigidity.

There are increases in long-latency reflexes in PD patients. Generally, this is neurophysiologically distinct from the increases in the short-latency reflexes seen in spasticity, increase in tone of "pyramidal" type. The short-latency reflex is the monosynaptic reflex. Reflexes occurring at a longer latency than this are designated *long latency*. When a relaxed muscle is stretched, in general, only a short-latency reflex is produced. When a muscle is stretched while it is active, one or more distinct long-latency reflexes are produced following the short-latency reflex and prior to the time needed to produce a voluntary response to the stretch. These reflexes are recognized as separate because of brief time gaps between them, giving rise to the appearance of distinct "humps" on a rectified EMG trace. Each component reflex, either short or long in latency, has about the same duration—approximately 20 to 40 msec. They appear to be true reflexes in that their appearance and magnitude depend primarily on the amount of background force that the muscle was exerting at the time of the stretch and the mechanical parameters of the stretch; they do not vary much with whatever the subject might want to do after experiencing the muscle stretch. By contrast, the voluntary response that occurs after a reaction time from the stretch stimulus is strongly dependent on the will of the subject.

The short-latency stretch reflex can be easily measured with the tendon jerk or H reflex. To obtain a meaningful measure of the response, the amplitude of the maximal reflex must be compared with the amplitude of the EMG in maximum voluntary effort or the amplitude of the EMG produced by supramaximal stimulation of the nerve to that muscle (H/M ratio).[43,44] Unfortunately, there is a large interindividual variability that makes the measurement less useful than it might be. The H/M ratio is enhanced in spasticity but not in parkinsonian rigidity. Another clinically useful test is vibratory inhibition of the H reflex.[48] In normal subjects, the amplitude of the H reflex is markedly inhibited by vibration of the muscle. Vibratory inhibition is often dramatically reduced in spasticity but it is normal in parkinsonian rigidity.

Long-latency reflexes are best brought out with controlled stretches with a device such as a torque motor. While long-latency reflexes are normally absent at rest, they are prominent in PD patients (Figure 36.4).[43,44,49,50] Long-latency reflexes are also enhanced in PD with background contraction. Since some long-latency stretch reflexes appear to be mediated by a loop through the sensory and motor cortices, the enhancement of long-latency reflexes has been generally believed to indicate increased excitability of this central loop.

There is some evidence that at least one component of the increased long-latency stretch reflex in PD is a group II mediated reflex. This suggestion was first made

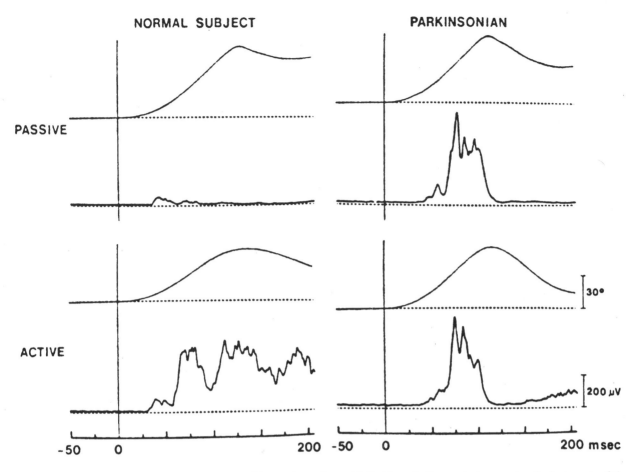

FIGURE 36.4 Long-latency reflex in a normal subject and a patient with Parkinson's disease during passive displacements of the wrist joint and active opposition to the imposed displacement. For each experimental condition (passive and active), the upper trace illustrates the angular displacement of the wrist, and the lower trace represents the EMG response in a wrist flexor. Passive: while in controls there is no long-latency EMG response, the parkinsonian patient shows markedly enlarged long-latency stretch reflex. Active: the long-latency EMG response is almost as large during stretch as when the patient is actively opposing the imposed displacement suggesting reduced ability to modulate long-latency reflex activity according to volitional intent. (From Lee, R. G. and Tatton, W. G.[49] Reproduced with permission of Elsevier.)

by Berardelli et al. on the basis of physiologic features including insensitivity to vibration.[51] It was subsequently supported by the observation that an enhanced late stretch reflex response could not be duplicated with a vibration stimulus.[52]

Some studies show a correlation between clinically measured increased tone and the magnitude of long-latency reflexes,[51] while others do not.[53,54] Long-latency reflexes contribute significantly to rigidity but are apparently not completely responsible for it.

The enhancement of long-latency reflexes can also be brought out by electrical stimulation of a mixed nerve. Such stimulation, while the limb is at rest, will produce only an M wave and F response in the muscles innervated by that nerve. If a mixed nerve is stimulated while the muscles are active, however, additional responses will be produced.[43,44] With mixed nerve stimulation, there is a short-latency response that seems analogous to the H

reflex (HR) and one or more long-latency responses. One of these long-latency responses, called LLRII by Deuschl and associates,[55] may have a transcortical pathway similar to some of the long-latency reflexes to stretch. A long-latency response, the LLRI, intermediate in latency between the HR and LLRII, is enhanced in about half of patients with PD.

Some spinal inhibitory reflexes such as reciprocal inhibition and Ib inhibition are deficient, and these mechanisms may also play a role. If inhibition is lacking, there will be excessive activity that could contribute to rigidity or failure to relax.

Reciprocal inhibition is a fundamental mechanism of motor control. There are multiple pathways for reciprocal inhibition, the simplest of which is the disynaptic pathway via the Ia inhibitory interneuron. In the arm, reciprocal inhibition has been studied looking at the effects of radial nerve stimulation upon the H reflex of the flexor carpi radialis (FCR).[56,57] Via various pathways, and therefore at various time intervals after the radial nerve stimulus, the radial afferent traffic can inhibit the motoneuron pools of the FCR. Normal subjects showed three periods of inhibition, reaching a peak at delays of 0, 10, and 75 ms. The first period of inhibition is caused by disynaptic Ia inhibition, the second period of inhibition is explained as a presynaptic inhibition, and, unfortunately, very little is known about the third period of inhibition, but the long latency (75 to 200 ms) appears to be compatible with a polysynaptic pathway. The first relative facilitation (at about 2 ms delay) is a function of Ib fiber actions, and indirect evidence indicates that the second facilitation (at about 50 ms delay) can be a function of cutaneous group II action.

Reciprocal inhibition is reduced in patients with dystonia, including those with generalized dystonia, writer's cramp, spasmodic torticollis, and blepharospasm.[57,58] Reciprocal inhibition is also abnormally reduced in PD patients.[59] On the other hand, short-latency reciprocal inhibition is increased in the lower extremities, the opposite of what is found in the upper extremities.[60]

That Ib inhibition can be demonstrated in the human was first demonstrated by the clever experiments of Pierrot-Deseilligny and colleagues.[61] They showed that stimulation of the nerve to the medial head of gastrocnemius (GM) provoked short-latency inhibition of the H reflex in soleus that was most consistent with Ib effects. Presumably, this is apparent because there are very few heteronymous Ia projections from the medial head of gastrocnemius onto soleus motoneurons. In patients with spasticity, Ib inhibition is absent and is replaced by facilitation.[62] The explanation for this inversion is not clear. Similarly, the Ib inhibition is diminished in PD and, when rigidity is more severe, the inhibition is replaced by facilitation. The authors explained this on the hypothesis of increased activity of the nucleus gigantocellularis of the brain stem.

Reduction of Ib inhibition was confirmed using a different method: electrical stimulation via skin electrodes placed over human tendons, resulting in a reflex inhibition of voluntary activity in the stimulated muscle.[63] The threshold of the inhibitory response was significantly increased in PD compared with controls. Also, the latency of the inhibitory wave was increased, and the duration of inhibition was increased in patients.

Inhibitory and excitatory reflex effects from stimulation of cutaneous nerves can be detected by recording changes in levels of tonic voluntary EMG activity of various hand muscles.[43,44] These reflexes consist of a series of bursts of EMG activity separated by periods of inhibition. The first excitatory component is generally agreed to be of spinal origin, while there is debate about a supraspinal or even a transcortical loop of the later reflex components. The first inhibitory component is produced by inhibition

above the level of the alpha motoneuron, but below the level of the cortex, and is diminished in PD.[64]

Recurrent inhibition can be studied using the complicated method developed by Pierrot-Deseilligny and colleagues.[65] While in spasticity, some patients show loss of inhibition, there is no loss of inhibition in PD.[59]

Delwaide et al. have suggested that the magnitude of audiospinal facilitation correlates with rigidity.[66] They compared audiospinal facilitation using the soleus H-reflex in control subjects and PD patients. In the patients, facilitation was significantly reduced during the 75 to 150 msec after the conditioning stimulation. This reduction was seen bilaterally even in patients with a hemisyndrome. It was corrected by L-dopa but not by anticholinergic agents. Facilitation at the 75-msec delay showed an inverse linear correlation with the bradykinesia intensity. The authors explain the results as a reduced excitability of the nucleus reticularis pontis caudalis from which a reticulospinal tract emanates as effector of audiospinal facilitation.

TREMOR AT REST

The so-called "tremor at rest" is the classic tremor of PD and other parkinsonian states, such as those produced by neuroleptics or other dopamine-blocking agents such as proclorperazine and metoclopramide.[44,67–69] It is present at rest, disappears with action, but may resume with static posture ("re-emergent tremor"). That the tremor may also be present during postural maintenance is a significant point of confusion in regard to naming this tremor "tremor at rest." It can involve all parts of the body and can be markedly asymmetrical, but it is most typical with a flexion-extension movement at the elbow, pronation and supination of the forearm, and movements of the thumb across the fingers ("pill-rolling"). Its frequency is 3 to 7 Hz but is most commonly 4 or 5 Hz, and EMG studies show alternating activity in antagonist muscles. PD is sometimes divided into two types—the akinetic-rigid form and the tremor-predominant form; the latter has a better prognosis.

Tremor at rest can also be seen in the parkinson plus disorders, but it is not as common as in PD itself. For example, rest tremor was seen in 29 of 100 patients thought to have multiple system atrophy, but only 9 had a "classic appearance."[70]

Some patients have rest tremor for a number of years without any other evidence of PD, and it has not been clear whether they really have PD. Eleven of these patients underwent 18F-dopa PET scan studies, and all showed reduced putaminal uptake, an abnormality characteristic of PD.[71] This result has been replicated in a double-blind fashion on five patients using MRI scanning. All five showed typical findings of PD with smudging or decreased distance between the substantia nigra and red nucleus.[72] However, several studies have indicated that the neuropathology of tremor dominant PD differs from the akinesia-rigidity dominant PD. The medial substantia nigra is more severely affected by dopaminergic cell degeneration in the tremor dominant form, in contrast to more severe damage of the lateral substantia nigra in the akinetic-rigid variant.[73]

In parkinsonian tremor at rest, there may be some mechanical-reflex component and some 8- to 12-Hz component, but the most significant component comes from a pathological central oscillator at 3 to 5 Hz. Evidence for the central oscillator includes the facts that the accelerometric record and the EMG are not affected by weighting, and small mechanical perturbations do not affect it. On the other hand, it can be reset by strong peripheral stimuli such as an electrical stimulus that produces a movement of the body part five times more than the amplitude of the tremor itself.[74] Where this strong stimulus acts is not clear, but does not have to be on the peripheral loop. Deafferentation changes the frequency of the parkinsonian tremor but does not suppress it.[75] This again argues against a major contribution of a peripheral component in the generation of the tremor.

There is strong evidence supporting a critical role of subcortico-cortical pathways in the generation of parkinsonian tremor.

First, there is evidence supporting a critical role of cerebello-thalamo-cortical pathways in the generation of parkinsonian tremor. An *in vivo* voxel-based morphometric study in patients with tremor dominant PD, compared with controls, showed increased grey matter in a thalamic region that may well correspond to the ventral intermediate (VIM) nucleus of the thalamus, a cerebellar relay nucleus.[77] However, the most striking point is that the tremor may be successfully treated with a stereotaxic lesion or deep brain stimulation of the VIM. Lenz and colleagues have been studying the physiological properties of cells in the VIM in relation to tremor production (Figure 36.5).[78] They have tried to see if the pattern of spike activity is consistent with specific hypotheses. They examined whether parkinsonian tremor might be produced by the activity of an intrinsic thalamic pacemaker or by the oscillation of an unstable long loop reflex arc. In one study of 42 cells, they found 11 with a sensory feedback pattern, 1 with a pacemaker pattern, 21 with a completely random pattern, and 9 that did not fit any pattern.[78] In another study of thalamic neuron activity, some cells with a pacemaker pattern were seen, but these did not participate in the rhythmic activity correlating with tremor.[79] These results confirm those of Lenz et al., suggesting that the thalamic cells are not the pacemaker.

Second, evidence also suggests participation of basal ganglia-thalamo-cortical pathways in the generation of tremor. A major point supporting this idea is that lesions and chronic stimulation of STN (and GPi to a lesser extent) are able to suppress parkinsonian tremor.[80-82] Tremor cells have been described in the STN of patients with PD, but some data suggest that those cells were most

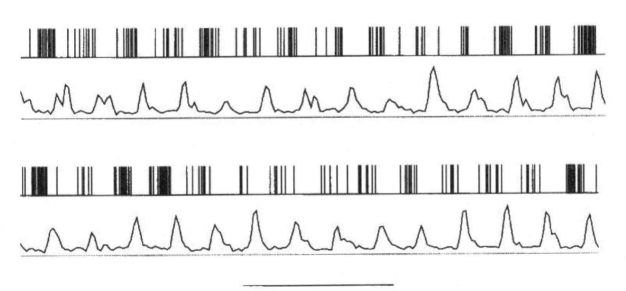

1 Sec.

FIGURE 36.5 Simultaneous human thalamic cell (spike trains, upper trace) and wrist flexor EMG (lower trace) recordings during spontaneous parkinsonian tremor. In the spike trace, vertical bars indicate the occurrence of action potentials. The lower trace illustrates filtered, rectified, and mean subtracted EMG activity from a wrist flexor. Simple visual inspection of the traces suggests a relationship between EMG and cell recording in the frequency domain. This was confirmed by a coherence analysis. However, results from more detailed investigations were not consistent with the presence of a tremor PMK in the thalamic cells that were studied, including those located in the VIM. (From Zirh, T. A. et al.[78] Reproduced with permission of Elsevier.)

sensitive to peripheral (kinesthesic) stimulation.[83] Single-unit recordings have shown synchronized discharges in STN and GPi predominantly at the frequency of rest and action tremor (3 to 10 Hz) in untreated PD patients undergoing functional neurosurgery,[84] but it remains debated whether synchronized oscillatory activity in STN and GPi at low frequency are confined to patients with tremor. Nevertheless, in reference to synchronized oscillatory activity in subcortico-cortical circuits (vide supra), available evidence is consistent with a net driving of motor cortical areas at tremor frequencies through STN-GPi-thalamo (ventralis anterior thalamus)-cortical pathway. The synchronization at low frequency in untreated PD patients would involve pyramidal neurons through its effect on the cortex, with consequent driving of muscles at low frequencies, manifest as parkinsonian rest and postural tremor.

Parkinsonian tremor can be thus successfully treated not only by intervention on the cerebello-thalamo-cortical but also the basal ganglia-thalamo-cortical pathway. This implies an interaction between the two loops. Although the exact nature of this interaction is unknown, one possibility is that the oscillator from one circuit is loosely coupled with the oscillator of the other in such a way that the basal ganglia loop oscillator would influence the cerebellar loop oscillator. The exact site of the interaction between the two loops is still undetermined. Since both pathways are believed to be separated up to the cortex, the interaction between the two should take place at the level of the cortex. Interestingly, cortical stimulation can also alleviate parkinsonian tremor.[85,86] The tremor can be reset by TMS,[74,76] presumably indicating a role of the motor cortex in the central processes that generate the tremor. In the studies of Pascual-Leone et al.,[76] using a relatively small stimulus, the tremor was reset with TMS, but not with transcranial electrical stimulation. Since TMS affects the intracortical circuitry more, this seems to be further evidence for a role of the motor cortex.

Wherever the pacemaker for the tremor, it is important to note while the tremor is synchronous within a limb, it is not synchronous between limbs.[87] Hence, a single pacemaker does not influence the whole body.

There are other types of tremor in PD, including an action tremor looking like essential tremor, but these have not been extensively studied.

ACKNOWLEDGMENTS

This review has been modified and updated from an original work of the U.S. government and has no copyright.[88]

REFERENCES

1. Berardelli A, Rothwell, J. C., Thompson, P. D. et al., Pathophysiology of bradykinesia in Parkinson's disease, *Brain*, 124:2131–2146, 2001.

2. Evarts, E. V., Teravainen, H., Calne, D. B., Reaction time in Parkinson's disease, *Brain*, 104:167–186, 1981.

3. Hallett, M., Khoshbin, S., A physiological mechanism of bradykinesia, *Brain*, 103:301–314, 1980.

4. Baroni, A., Benvenuti, F., Fantini, L. et al., Human ballistic arm abduction movements: effects of L-dopa treatment in Parkinson's disease, *Neurology*, 34:868–87, 1984.

5. Berardelli, A., Dick, J. P., Rothwell, J. C. et al., Scaling of the size of the first agonist EMG burst during rapid wrist movements in patients with Parkinson's disease, *J. Neurol. Neurosurg. Psychiatry*, 9:1273–1279, 1986.

6. Demirci, M., Grill, S., McShane, L. et al., A mismatch between kinesthetic and visual perception in Parkinson's disease, *Ann. Neurol.*, 41:781–788, 1997.

7. Godaux, E., Koulischer, D., Jacquy, J., Parkinsonian bradykinesia is due to depression in the rate of rise of muscle activity, *Ann. Neurol.*, 31:93–100, 1992.

8. Jordan, N., Sagar, H. J., Cooper, J. A., A component analysis of the generation and release of isometric force in Parkinson's disease, *J. Neurol. Neurosurg. Psychiatry*, 55:572–576, 1992.

9. Benecke, R., Rothwell, J. C., Dick, J. P. et al., Simple and complex movements off and on treatment in patients with Parkinson's disease, *J. Neurol. Neurosurg. Psychiatry*, 50:296–303, 1987.

10. Schwab, R. S., Chafetz, M. E., Walker, S., Control of two simultaneous voluntary motor acts in normals and in parkinsonism, *Arch. Neurol.*, 72:591–598, 1954.

11. Hallett, M., Shahani, B. T,. Young, R. R., Analysis of stereotyped voluntary movements at the elbow in patients with Parkinson's disease, *J. Neurol. Neurosurg. Psychiatry*, 40:1129–1135, 1977.

12. Staude, G., Wolf, W., Ott, M. et al., Tremor as a factor in prolonged reaction times of parkinsonian patients, *Mov. Disord.*, 10:153–162, 1995.

13. Warabi, T., Yanagisawa, N., Shindo, R., Changes in strategy of aiming tasks in Parkinson's disease, *Brain*, 111 (Pt. 3):497–505, 1988.

14. Hallett, M., Clinical neurophysiology of akinesia, *Rev. Neurol.* (Paris) 146:585–590, 1990.

15. Brown, V. J., Schwarz, U., Bowman, E. M. et al., Dopamine dependent reaction time deficits in patients with Parkinson's disease are task specific, *Neuropsychologia*, 31:459–469, 1993.

16. Labutta, R. J., Miles, R. B., Sanes, J. N. et al., Motor program memory storage in Parkinson's disease patients tested with a delayed response task, *Mov. Disord.*, 9:218–222, 1994.

17. Starr, A., Caramia, M., Zarola, F. et al., Enhancement of motor cortical excitability in humans by non-invasive electrical stimulation appears prior to voluntary movement, *Electroencephalogr. Clin. Neurophysiol.*, 70: 26–32, 1988.

18. Pascual-Leone, A., Valls-SolÈ, J., Brasil-Neto, J. et al., Akinesia in Parkinson's disease. I. Shortening of simple reaction time with focal, single pulse transcranial magnetic stimulation, *Neurology*, 44:884–891, 1994.

19. Chen, R., Kumar, S., Garg, R. R. et al., Impairment of motor cortex activation and deactivation in Parkinson's disease, *Clin. Neurophysiol.*, 112:600–607, 2001.

20. Watts, R. L., Mandir, A. S., The role of motor cortex in the pathophysiology of voluntary movement deficits associated with parkinsonism, *Neurol Clin.*, 10:451–469, 1992.

21. Majsak, M. J., Kaminski, T., Gentile, A. M. et al., The reaching movements of patients with Parkinson's disease under self-determined maximal speed and visually cued conditions, *Brain,* 121:755–766, 1998.

22. Curra, A., Berardelli, A., Agostino, R. et al., Performance of sequential arm movements with and without advance knowledge of motor pathways in Parkinson's disease, *Mov. Disord.,* 12:646–654, 1997.

23. Hanakawa, T., Fukuyama, H., Katsumi, Y. et al., Enhanced lateral premotor activity during paradoxical gait in Parkinson's disease, *Ann. Neurol.,* 45:329–336, 1999.

24. Rossini, P. M., Babiloni, F., Bernardi, G. et al., Abnormalities of short-latency somatosensory evoked potentials in parkinsonian patients, *Electroencephalogr. Clin. Neurophysiol.,* 74:277–289, 1989.

25. Valls-Solè, J., Pascual-Leone, A., Brasil-Neto, J. et al., Abnormal facilitation of the response to transcranial magnetic stimulation in patients with Parkinson's disease, *Neurology,* 44:735–741, 1994.

26. Ridding, M. C., Inzelberg, R., Rothwell, J. C., Changes in excitability of motor cortical circuitry in patients with Parkinson's disease, *Ann. Neurol.,* 37:181–188, 1995.

27. Berardelli, A., Rona, S., Inghilleri, M. et al., Cortical inhibition in Parkinson's disease. A study with paired magnetic stimulation, *Brain,* 119 (Pt. 1):71–77, 1996.

28. Tarkka, I. M., Reilly, J. A., Hallett, M., Topography of movement-related cortical potentials is abnormal in Parkinson's disease, *Brain Res.,* 522:172–175, 1990.

29. Dick, J. P., Rothwell, J. C., Day, B. L. et al., The Bereitschaftspotential is abnormal in Parkinson's disease, *Brain,* 112 (Pt. 1):233–244, 1989.

30. Jahanshahi, M., Jenkins, I. H., Brown, R. G. et al., Self-initiated versus externally triggered movements, I. An investigation using measurements of regional cerebral blood flow with PET and movement related potentials in normal and Parkinson's disease subjects, *Brain,* 118:913–933, 1995.

31. Jenkins, I. H., Fernandez, W., Playford, E. D. et al., Impaired activation of the supplementary motor area in Parkinson's disease is reversed when akinesia is treated with apomorphine, *Ann. Neurol.,* 32:749–757, 1992.

32. Rascol, O., Sabatini, U., Chollet, F. et al., Supplementary and primary sensory motor area activity in Parkinson's disease. Regional cerebral blood flow changes during finger movements and effects of apomorphine, *Arch Neurol.,* 49:144–148, 1992.

33. Piccini, P., Lindvall, O., Bjorklund, A. et al., Delayed recovery of movement-related cortical function in Parkinson's disease after striatal dopaminergic grafts, *Ann. Neurol.,* 48:689–695, 2000.

34. Grafton, S. T., Waters, C., Sutton, J. et al., Pallidotomy increases activity of motor association cortex in Parkinson's disease: a positron emission tomographic study, *Ann. Neurol.,* 37:776–783, 1995.

35. Limousin, P., Greene, J., Pollak, P. et al., Changes in cerebral activity pattern due to subthalamic nucleus or internal pallidum stimulation in Parkinson's disease, *Ann. Neurol.,* 42:283–291, 1997.

36. Levy, R., Dostrovsky, J. O., Lang, A. E. et al., Effects of apomorphine on subthalamic nucleus and globus pallidus internus neurons in patients with Parkinson's disease, *J. Neurophysiol.,* 86:249–260, 2001.

37. Williams, D., Tijssen, M., Van Bruggen, G. et al., Dopamine-dependent changes in the functional connectivity between basal ganglia and cerebral cortex in humans, *Brain,* 125:1558–1569, 2002.

38. Cassidy, M., Mazzone, P., Oliviero, A. et al., Movement-related changes in synchronization in the human basal ganglia, *Brain,* 125:1235–1246, 2002.

39. Salenius, S., Avikainen, S., Kaakkola, S. et al., Defective cortical drive to muscle in Parkinson's disease and its improvement with levodopa, *Brain,* 125:491–500, 2002.

40. Brown, P., Oscillatory nature of human basal ganglia activity: Relationship to the pathophysiology of Parkinson's disease, *Mov. Disord.,* 18:357–363, 2003.

41. Brown, P., Marsden, D., Bradykinesia and impairment of EEG desynchronization in Parkinson's disease, *Mov. Disord.,* 14:423–429, 1999.

42. Wang, H-C., Lees, A. J., Brown, P., Impairment of EEG desynchronisation before and during movement and its relation to bradykinesia in Parkinson's disease, *J. Neurol. Neurosurg. Psychiatry,* 66:442–446, 1999.

43. Hallett, M., Berardelli, A., Delwaide, P. et al., Central EMG and tests of motor control. Report of an IFCN committee, *Electroencephalogr. Clin. Neurophysiol.,* 90:404–432, 1994.

44. Hallett, M., Electrophysiologic evaluation of movement disorders, in Aminoff, M. J., Ed., *Electrodiagnosis in Electrophysiology,* Churchill Livingstone, New York, 365–380, 1999.

45. Dietz, V., Quintern, J., Berger, W., Electrophysiological studies of gait in spasticity and rigidity. Evidence that altered mechanical properties of muscle contribute to hypertonia, *Brain,* 104:431–449, 1981.

46. Watts, R. L., Wiegner, A. W., Young, R. R., Elastic properties of muscles measured at the elbow in man: II. Patients with parkinsonian rigidity, *J. Neurol. Neurosurg. Psychiatry,* 49:1177–1181, 1986.

47. Hufschmidt, A., Stark, K., Lucking, C. H., Contractile properties of lower leg muscles are normal in Parkinson's disease, *J. Neurol. Neurosurg. Psychiatry,* 54:457–460, 1991.

48. Bour, J. L., Ongerboer de Visser, B. W., Koelman, HTM et al., Soleus H-reflex tests in spasticity and dystonia: A computerized analysis, *J. Electromyogr. Kinesiol.,* 1:9–19, 1991.

49. Lee, R. G. and Tatton, W. G., Long loop reflexes in man: Clinical Applications, in Desmedt, J. E., Ed., Cerebral Motor Control in Man: Long Loop Mechanisms, *Prog. Clin. Neurophysiol.,* Vol. 4., pp. 320–333, 1978.

50. Rothwell, J. C., Obeso, J. A., Traub, M. M. et al., The behaviour of the long latency stretch reflex in patients

with Parkinson's disease, *J. Neurol. Neurosurg. Psychiatry,* 46:35–44, 1983.

51. Berardelli, A., Sabra, A. F., Hallett, M., Physiological mechanisms of rigidity in Parkinson's disease, *J. Neurol. Neurosurg. Psychiatry,* 46:45–53, 1983.

52. Cody, F. W., Macdermott, N., Matthews, P. et al., Observations on the genesis of the stretch reflex in Parkinson's disease, *Brain,* 109 (Pt. 2):229—249, 1986.

53. Meara, R. J., Cody, F. W., Stretch reflexes of individual parkinsonian patients studied during changes in clinical rigidity following medication, *Electroencephalogr. Clin. Neurophysiol.,* 89:261–268, 1993.

54. Bergui, M., Lopiano, L., Paglia, G. et al., Stretch reflex of quadriceps femoris and its relation to rigidity in Parkinson's disease, *Acta Neurol. Scand.,* 86:226–229, 1992.

55. Deuschl, G., Lucking, C. H., Physiology and clinical applications of hand muscle reflexes, *Electroencephalogr. Clin. Neurophysiol.* Suppl., 41:84–101, 1990.

56. Day, B. L., Marsden, C. D., Obeso, J. A. et al., Reciprocal inhibition between the muscles of the human forearm, *J. Physiol.,* 349:519–534, 1984.

57. Panizza, M. E., Hallett, M., Nilsson, J., Reciprocal inhibition in patients with hand cramps, *Neurology,* 39:85–89, 1989.

58. Panizza, M., Lelli, S., Nilsson, J. et al., H-reflex recovery curve and reciprocal inhibition of H-reflex in different kinds of dystonia, *Neurology,* 40:824–828, 1990.

59. Lelli, S., Panizza, M., Hallett, M., Spinal cord inhibitory mechanisms in Parkinson's disease, *Neurology,* 41:553–556, 1991.

60. Delwaide, P. J., Pepin, J. L., Maertens de Noordhout, A., Parkinsonian rigidity: clinical and physiopathologic aspects, *Rev. Neurol.,* (Paris), 146:548–554, 1990.

61. Pierrot-Deseilligny, E., Katz, R., Morin, C., Evidence of Ib inhibition in human subjects, *Brain Res,* 166:176–179, 1979.

62. Delwaide, P. J., Pepin, J. L., Maertens de Noordhout, A., Short-latency autogenic inhibition in patients with Parkinsonian rigidity, *Ann. Neurol.,* 30:83–89, 1991.

63. Burn, J. A., Lippold, O. C., Loss of tendon organ inhibition in Parkinson's disease, *Brain,* 119 (Pt. 4):1115–1121, 1996.

64. Fuhr, P., Zeffiro, T., Hallett, M., Cutaneous reflexes in Parkinson's disease, *Muscle Nerve,* 15:733–739, 1992.

65. Rossi, A., Mazzocchio, R., Presence of homonymous recurrent inhibition in motoneurones supplying different lower limb muscles in humans, *Exp. Brain Res.,* 84:367–373, 1991.

66. Delwaide, P. J., Pepin, J. L., Maertens de Noordhou, A., The audiospinal reaction in parkinsonian patients reflects functional changes in reticular nuclei, *Ann. Neurol.,* 33:63–69, 1993.

67. Elble, R. J., Koller, W. C., *Tremor,* Johns Hopkins University Press, Baltimore, MD, 1990.

68. Hallett, M., Classification and treatment of tremor, *JAMA,* 266:1115–1117, 1991.

69. Elble, R. J., The pathophysiology of tremor, in Watts, R. L., Koller, W. C., Eds., *Movement Disorders Neuro-*

logic Principles and Practice, McGraw-Hill, New York, 405–417, 1997.

70. Wenning, G. K., Ben-Shlomo, Y., Magalhaes, M. et al., Clinical features and natural history of multiple system atrophy, an analysis of 100 cases, *Brain,* 117:835–845, 1994.

71. Brooks, D. J., Playford, E. D., Ibanez, V. et al., Isolated tremor and disruption of the nigrostriatal dopaminergic system: an 18F-dopa PET study, *Neurology,* 42:1554–1560, 1992.

72. Chang, M. H., Chang, T. W., Lai, P. H. et al., Resting tremor only: a variant of Parkinson's disease or of essential tremor, *J. Neurol. Sci.,* 130:215–219, 1995.

73. Jellinger, K. A., Recent developments in the pathology of Parkinson's disease, *J. Neural. Transm. Suppl.,* 347–376, 2002.

74. Britton, T. C., Thompson, P. D., Day, B. L. et al., Modulation of postural wrist tremors by magnetic stimulation of the motor cortex in patients with Parkinson's disease or essential tremor and in normal subjects mimicking tremor, *Ann. Neurol.,* 33:473–479, 1993.

75. Pollock, L. J., Davis, L., Muscle tone in parkinsonian states, *Arch. Neurol. Psychiat., Chic.,* 23:303–319, 1930.

76. Pascual-Leone, A., Valls-Sole, J., Toro, C. et al., Resetting of essential tremor and postural tremor in Parkinson's disease with transcranial magnetic stimulation, *Muscle Nerve,* 17:800–807, 1994.

77. Kassubek, J., Juengling, F. D., Hellwig, B. et al., Thalamic gray matter changes in unilateral Parkinsonian resting tremor: a voxel-based morphometric analysis of 3-dimensional magnetic resonance imaging, *Neurosci. Lett.,* 323:29–32, 2002.

78. Zirh, T. A., Lenz, F. A., Reich, S. G. et al., Patterns of bursting occurring in thalamic cells during parkinsonian tremor, *Neuroscience,* 83:107–121, 1998.

79. Magnin, M., Morel, A., Jeanmonod, D., Single-unit analysis of the pallidum, thalamus and subthalamic nucleus in parkinsonian patients, *Neuroscience,* 96:549–564, 2000.

80. Krack, P., Pollak, P., Limousin, P. et al., Stimulation of subthalamic nucleus alleviates tremor in Parkinson's disease, *Lancet,* 350:1675, 1997.

81. Rodriguez, M. C., Guridi, O. J., Alvarez, L. et al., The subthalamic nucleus and tremor in Parkinson's disease, *Mov. Disord.,* 13 Suppl. 3:111–118, 1998.

82. Patel, N. K., Heywood, P., O'Sullivan, K. et al., Unilateral subthalamotomy in the treatment of Parkinson's disease, *Brain,* 126:1136–1145, 2003.

83. Hutchison, W. D., Allan, R. J., Opitz, H. et al., Neurophysiological identification of the subthalamic nucleus in surgery for Parkinson's disease, *Ann. Neurol.,* 44:622–628, 1998.

84. Levy, R., Hutchison, W. D., Lozano, A. M. et al., Synchronized neuronal discharge in the basal ganglia of parkinsonian patients is limited to oscillatory activity, *J. Neurosci.,* 22:2855–2861, 2002.

85. Woolsey, C. N., Erickson, T. C., Gilson, W. E., Localization in somatic sensory and motor areas of human cerebral cortex as determined by direct recording of

evoked potentials and electrical stimulation, *J. Neuro-surg.,* 51:476–506, 1979.

86. Canavero, S., Paolotti, R., Bonicalzi, V. et al., Extradural motor cortex stimulation for advanced Parkinson disease. Report of two cases, *J. Neurosurg.,* 97:1208–1211, 2002.

87. Hurtado, J. M., Lachaux, J. P., Beckley, D. J. et al., Inter-and intralimb oscillator coupling in parkinsonian tremor, *Mov. Disord.,* 15:683–691, 2000.

88. Hallett, M., Parkinson revisited: pathophysiology of motor signs, *Adv. Neurol.,* 91:19–28, 2003.

37 MPTP Disrupts Dopamine Receptors and Dopamine Transporters

Manuchair Ebadi and Jaturaporn Chagkutip
Departments of Pharmacology and of Neurosciences, University of North Dakota School of Medicine and Health Sciences

CONTENTS

ABSTRACT

The dopaminergic (DA) systems have been the focus of much research over the past three decades, mainly because several neurological diseases such as Parkinson's disease, Huntington disease, and Tourette's syndrome have been linked to malfunctioning of dopaminergic transmission. Furthermore, dopamine receptor agonists such as Ropinirole, pramipexole, and cabergoline have been used to treat Parkinson's disease, and dopamine receptor antagonists such as chlorpromazine, haloperidol, and risperidone have been used to treat schizophrenia.

The first evidence for the existence of separate dopamine receptors came from studies showing that dopamine was able to stimulate the activity of adenylyl cyclase. Initially, on the basis of pharmacological and biochemical evidence, it was proposed that two discrete populations of dopamine receptor existed, one positively liked to adenylyl cyclase and one independent of adenylyl cyclase. However, gene-cloning procedures revealed a higher degree of complexity within dopamine receptors, and the complementary DNAs of five distinct dopamine receptor subtypes (D_1–D_5) have been isolated and characterized.

Gene-targeting studies via homologous recombinate have been used to generate mice deficient in key molecules involved in dopaminergic transmission, including the dopamine transporters and the five subtypes of dopaminergic receptors. The dopamine transporter, a member of the family of Na^+, Cl^--dependent transporters, mediates uptake of dopamine into dopaminergic neurons by an electrogenic, Na^+ and Cl^- transport-coupled mechanism. Since its cloning, much information has been obtained regarding its structure and function. Binding domains for dopamine and various blocking drugs, including cocaine, are likely formed by interactions with multiple amino acid residues, some of which are separate in the primary structure but lie close together in the still unknown tertiary structure.

Dopamine and blockers of its uptake mechanism such as cocaine probably bind to both shared and separate domains on the transporter, which can be influenced dramatically by the presence of cations. Regulation of the dopamine transporter occurs both by chronic occupancy with the uptake blocker and by acute effects of D_2 dopamine receptors or second messengers such as diacylglycerol (protein kinase C) and arachidonic acid. The dopamine transporters are involved in the uptake of toxins, such as MPTP, generating Parkinson's disease. It is also an important target for psychostimulant drugs such as amphetamine.

INTRODUCTION

Dopamine is the youngest of three catecholamines that occur naturally in the mammalian organism (for a review and references, see Reference 1). The brain contains separate neuronal systems that utilize three different catecholamines—dopamine, norepinephrine, and epinephrine. Each system is anatomically distinct and serves separate but similar functional roles within their fields of innervation.[2]

Although dopamine originally was regarded only as a precursor of norepinephrine, assays of distinct regions of the CNS eventually revealed that the distributions of dopamine and norepinephrine are markedly different. In fact, more than half the CNS content of catecholamine is dopamine, and extremely large amounts are found in the basal ganglia (especially in the caudate nucleus), the nucleus accumbens, the olfactory tubercle, the central nucleus of the amygdala, the median eminence, and restricted fields of the frontal cortex.

Due to the availability of histochemical methods that can reveal all the catecholamines or immunohistochemical methods for enzymes that synthesize individual catecholamines, the anatomical connections of the dopamine-containing neurons are known with some precision, at least in the rodent brain. These studies indicate that dopaminergic neurons fall into three major morphological classes:

1. Ultrashort neurons within the amacrine cells of the retina and periglomerular cells of the olfactory bulb
2. Intermediate-length neurons within the tuberobasal ventral hypothalamus that innervate the median eminence and intermediate lobe of the pituitary, incertohypothalamic neurons that connect the dorsal and posterior hypothalamus with the lateral septal nuclei, and a small series of neurons within the perimeter of the dorsal motor nucleus of the vagus, the nucleus of the solitary tract, and the periaqueductal gray matter
3. Long projections between the major dopamine-containing nuclei in the substantia nigra and ventral tegmentum and their targets in the striatum, in the limbic zones of the cerebral cortex, and in other major regions of the limbic system except the hippocampus (for a review and references, see Reference 2)

At the cellular level, the actions of dopamine depend on receptor subtype expression and the contingent convergent actions of other transmitters to the same target neurons.

IDENTIFICATION OF DOPAMINE RECEPTORS

Two subtypes of dopamine receptors, D_1 and D_2, were identified on the basis of pharmacological and biochemical criteria.[3] The D_1 and D_2 dopamine receptors are pharmacologically distinct, displaying differing affinities not only for the endogenous ligand dopamine but also for several other compounds.[4] *Benzazepines,* including *SCH 23390,* bind to D_1 receptors with high affinity, and *benzamides,* such as *sulpride* or *raclopride,* and *butyrophenones,* such as *spiroperidol,* selectively label D_2 receptors.

At least five genes encoding subtypes of dopamine receptors have been isolated and categorized as D_1-like or D_2-like according to their nucleotide sequences and the pharmacological profile of the expressed proteins. The D_1-like receptors include the D_1 receptors (5–8), whereas the D_2-like receptors include the two isoforms of the D_2 receptor, differing in the length of their predicted third cytoplasmic loop, dubbed D_2 *short* (D_2S)[9] and D_2 long,[8,10–12] the D_3 receptors,[13] and the D_4 receptors.[14] The D_1 and D_5 receptors activate adenylyl cyclase. The D_2 receptors couple to multiple effector systems, including the inhibition of adenylyl cyclase activity, suppression of Ca^{2+} currents, and activation of K^+ currents (for a review, see Reference 15).

D_2 dopamine receptors have been implicated in the pathophysiology of schizophrenia and Parkinson's disease. A correlation exists between the average clinical dose of a neuroleptic and its affinity for brain dopamine receptors measured in studies of the binding of the D_2 antagonist ^{3}H-spiroperidol.[16] Although long-term administration of typical neuroleptics to human beings or experimental animals can lead to the development of extrapyramidal side effects, including Parkinsonian-like movement disorders and tardive dyskinesia, a group of antipsychotic drugs referred to as *atypical neuroleptics* has been reported to be effective in the treatment of psychiatric disorders while producing significantly fewer extrapyramidal side effects. Both typical and atypical neuroleptics bind to D_2, D_3, and D_4 receptors with affinity in the nanomolar range. This observation, together with the selective expression of D_3 receptor mRNA in the nucleus accumbens and olfactory tubercle and the high affinity of D_4 receptors for atypical neuroleptics like *clozapine,* which does not cause extrapyramidal side effects, has led to the hypothesis that the alleviation of psychoses by neuroleptics may be due, in part, to their ability to antagonize the stimulation of D_3 and/or D_4 receptors, while the motor dysfunction observed with long-term use of typical neuroleptics may be due to an increase in the density of D_2 receptors in the striatum.[17]

STRUCTURES OF DOPAMINE RECEPTORS

The distribution, functions, and structures of dopamine receptors have been reviewed recently.[18–21] The diverse physiological actions of dopamine are mediated by at least five distinct *G protein coupled receptor subtypes.* Two D_1-like receptor subtypes (D_1 and D_5) couple to the G protein G_S and activate adenylyl cyclase. The other receptor subtypes belong to the D_2-like subfamily (D_2, D_3, and D_4) and are prototypic of G protein-coupled receptors that inhibit adenylyl cyclase and activate K^+ channels. The genes for the D_1 and D_5 receptors are intronless, but pseudogenes of the D_5 exist. The D_2 and D_3 receptors vary in certain tissues and species as a result of alternative splicing, and the human D_4 receptor gene exhibits extensive polymorphic variation. In the central nervous system, dopamine receptors are widely expressed, because they are involved in the control of locomotion, cognition, and emotion and also affect neuroendocrine secretion.

In the periphery, dopamine receptors are present more prominently in kidney, vasculature, and pituitary, where they affect mainly sodium homeostasis, vascular tone, and hormone secretion. Numerous genetic linkage analysis studies have failed so far to reveal unequivocal evidence for the involvement of one of these receptors in the etiology of various central nervous system disorders. However, targeted deletion of several of these dopamine receptor genes in mice should provide valuable information about their physiological functions.[19]

The first evidence for the existence of DA receptors in the CNS came in 1972 from biochemical studies showing that DA was able to stimulate adenylyl cyclase (AC). In 1978, DA receptors were first proposed, on the basis of pharmacological and biochemical evidence, to exist as two discrete populations, one positively coupled to AC and the other independent of the adenosine 3´,5´-cyclic monophosphate (cAMP)-generating system. It was shown, in fact, that in the pituitary, DA inhibited *prolactin* secretion but did not stimulate cAMP formation and that, although the antipsychotic drug *sulpiride* was a DA antagonist when tested in the anterior pituitary, it was not able to block the striatal DA-sensitive AC. In 1979, Kebabian and Calne[3] summarized these observations and suggested calling D_1 the receptor that stimulated AC, and D_2 the one that was not coupled to this effector.

Subsequent studies confirmed this classification scheme, and D_1 and D_2 receptors were clearly differentiated pharmacologically, biochemically, physiologically, and by their anatomic distribution. Concurrently, in the late 1970s, by means of functional tests such as renal blood flow and cardiac acceleration measurements in the dog, the existence of specific peripheral receptors for DA was demonstrated. These receptors were named DA_1 and DA_2 on the basis of some pharmacological properties distinguishing them from their central counterparts. This led to a long-standing controversy concerning the identity or nonidentity of peripheral versus central receptors. However, subsequent biochemical and molecular biology studies in peripheral tissues pointed to extensive similarities between central and peripheral DA receptors so that the DA_1/DA_2 classification has been dropped.

For a decade, the dual receptor concept served as the foundation for the study of DA receptors. However, after the introduction of gene cloning procedures, three novel DA receptor subtypes have been characterized over the past 5 years. These have been called D_3, D_4, and D_5/D_{1b}.

Detailed structural, pharmacological, and biochemical studies pointed out that all DA receptor subtypes fall into one of the two initially recognized receptor categories. The D_1 and D_5/D_{1b} receptors share, in fact, a very high homology in their transmembrane domains. Similarly, the transmembrane sequences are highly conserved among D_2, D_3, and D_4 receptors. Pharmacologically, although the profiles of D_1 and D_2 receptors are substantially different, the D_5/D_{1b} receptor exhibits the classical ligand-binding characteristics of D_1 receptors, and the D_3 and D_4 receptors bind the hallmark D_2-selective ligands with relatively high affinity. In addition, the initial distinction between D_1 and D_2 receptors in terms of signaling events (that is, positive and negative coupling to AC) appears to apply, in broad terms, also to the novel members of the DA receptor family, the D_5/D_{1b} receptor being coupled to stimulation of AC and the D_3 and D_4 receptors to inhibition of cAMP formation.

The D_1/D_2 classification concept developed in the late 1970s thus is still valid, and D_1 and D_5/D_{1b} receptors are classified as D_1-like, and D_2, D_3, and D_4 receptor subtypes as D_2-like. The mammalian D_{1b} receptor, originally named on the basis of its high homology with the D_1 receptor, is now commonly referred to as the D_5 receptor (see Table 37.1).

The distribution of dopamine receptors in the brain is as follows:

- The cerebral cortex contains D_1, D_2, D_3, D_4, and D_5 receptors.
- The corpus striatum contains D_1 and D_2 receptors.
- The limbic system contains D_1, D_2, D_3, D_4, and D_5 receptors.
- The pituitary gland contains D_2 receptors.

MPTP AND DOPAMINE TRANSPORTER

The dopamine transporter or carrier, located on the plasma membrane of nerve terminals, transports dopamine across the membrane. By taking up synaptic dopamine into neurons, it plays a critical role in terminating dopamine neurotransmission and in maintaining dopamine homeostasis in the central nervous system (for review and references, see References 22 through 29).

Amphetamine, MPTP, and various sympathomimetic amines structurally resembling dopamine are transported (see Figure 37.1). The dopamine transporter is also a major molecular target for the addictive drug cocaine and, to a lesser extent, antidepressants. These drugs cannot be transported but can bind to the dopamine carrier to block dopamine transport. Therefore, interactions with the dopamine transporter protein can have profound neurobiological, pathophysiological, and pharmacological consequences.[28]

Methyl-4-phenyl pyridium (MPP^+), the active metabolite of the neurotoxin 1-methyl-4-phenyl-1,2,3,6-tetrahy-

TABLE 37.1

Characteristics of Dopamine Receptor Subtypes

| | D$_1$-like | | D$_2$-like | | |
	D$_1$	D$_5$	D$_{2\ (short)/(long)}$	D$_3$	D$_4$
Amino Acids	446(h,r)	477(h)	414/443(h)	400(h)	387(h,r)
		475(r)	415/444(r)	446(r)	
Pharmacological characteristics (Kd, nM)	*SCH 2390 (0.35)*	*SCH 23390*	*spiperone (0.05)*	*spiperone (0.61)*	*spiperone (0.05)*
	Dopamine (2340)	Dopamine (228)	raclopride (1.8)	raclopride (3.5)	raclopride (237)
			clozapine (56)	*clozapine (180)*	*clozapine (9)*
			dopamine (1705)	dopamine (27)	dopamine (450)
Homology					
with D$_1$ receptor	100	82	44	44	42
with D$_{2(short)}$	44	49	100	76	54
Receptor localization	caudate/putamen, nucleus accumbens, olfactory tubercle, hypothalamus, thalamus, frontal cortex	hippocampus, thalamus, lateral mamillary nucleus, stratum, cerebral cortex (all low)	caudate/putamen nucleus accumbens, olfactory tubercle, cerebral cortex (low)	nucleus accumbens, olfactory tubercle, islands of Calleja, cerebral cortex (low)	frontal cortex, midbrain, amygdala, hippocampus, hypothalamus, medulla (all low), retina
Response	adenylyl cyclase ↑	adenylyl cyclase ↑	adenylyl cyclase ↓	adenylyl cyclase ↓	adenylyl cyclase ↓
Organization of amino acid sequence					
putative third intracellular loop	short	short	long	long	long
carboxyl terminal tail	long	long	short	short	short

Several medical conditions are known to involve alterations in dopaminergic (DAergic) neurotransmission. They include neurological disorders such as Parkinson's disease, which is caused by a selective degeneration of midbrain nigrostriatal DAergic neurons, and Huntington's disease, which is due to a deterioration of dopaminoceptive projection neurons in the neostriatum. Moreover, drugs that alter DAergic neurotransmission are successfully employed in the management of certain psychiatric disorders. Neuroleptic drugs that predominantly block the D$_2$ class of DA receptors have powerful antipsychotic potencies, and drugs that regulate the tone of DAergic neurotransmission alleviate some of the symptoms that characterize attention deficit hyperactivity disorders. Finally, psychostimulants, such as cocaine, amphetamine, and opioid drugs, as well as nicotine and alcohol, have addictive properties that are thought to be (at least partially) due to their ability to alter DAergic neurotransmission.

The central actions of DA are mediated by five distinct receptors that are expressed in DA synthesizing neurons of the substantia nigra, ventral tegmental area, and the hypothalamus and/or in the targets of the DAergic pathways known as the nigrostriatal, mesolimbic, mesocortical, and tuberoinfundibular pathways that are implicated, respectively, in the modulation of locomotor behavior, motivated behavior, learning and memory, and the regulation of prolactin release. DA receptors are members of the large class of neurotransmitter receptors whose actions are mediated through the activation of heterotrimeric guanine nucleotide regulatory proteins (G-proteins). The five DA receptor subtypes identified to date by molecular cloning differ in their primary structure and show distinct affinities for DA receptor agonists and antagonists. Their current classification is based on the functional properties of these receptors. The D$_1$ class of receptors is composed of the receptor subtypes D$_1$ and D$_5$, which couple to the stimulatory G-proteins G$_s$ and G$_{olf}$ and activate adenylyl cyclase to increase cytosolic cyclic AMP (cAMP) levels. The D$_2$ class of receptors is composed of the subtypes D$_2$, D$_3$, and D$_4$, which couple to the inhibitory G-proteins G$_i$ and G$_o$ so as to modulate ion channel activity and/or depress adenylyl cyclase activity. The anatomical distribution and functional properties of the individual receptor subtypes have been reviewed extensively.

dropyridine (MPTP), mediates selective damage to dopaminergic neurons and has been widely used to generate a model of Parkinson's disease. However, the mechanisms of the neurotoxic action of MPP$^+$ are not fully understood. MPP$^+$ is transported into cells via the dopamine transporter (DAT) where it mediates cellular toxicity.[30–32] DAT is a presynaptic plasma membrane protein responsible for the regulation of extracellular dopamine levels and termination of its action by mediating the reuptake of dopamine.[33,34] Functional impairment of DAT alters many physiological and behavioral processes that

are mediated by dopamine. A dysfunction of dopamine transmission could consequentially interrupt motor neural circuits that control movement, as seen in Parkinson's disease (PD) (see Figure 37.2).

Recently, it has been shown that cellular mRNAs encoding DAT and vesicular monoamine transporters are decreased in PD.[35,36] Indeed, an alteration of dopaminergic neurotransmission by the modulation of DAT activity could have an important implication in the cellular events that lead to PD and could be a target for a potential therapeutic intervention.

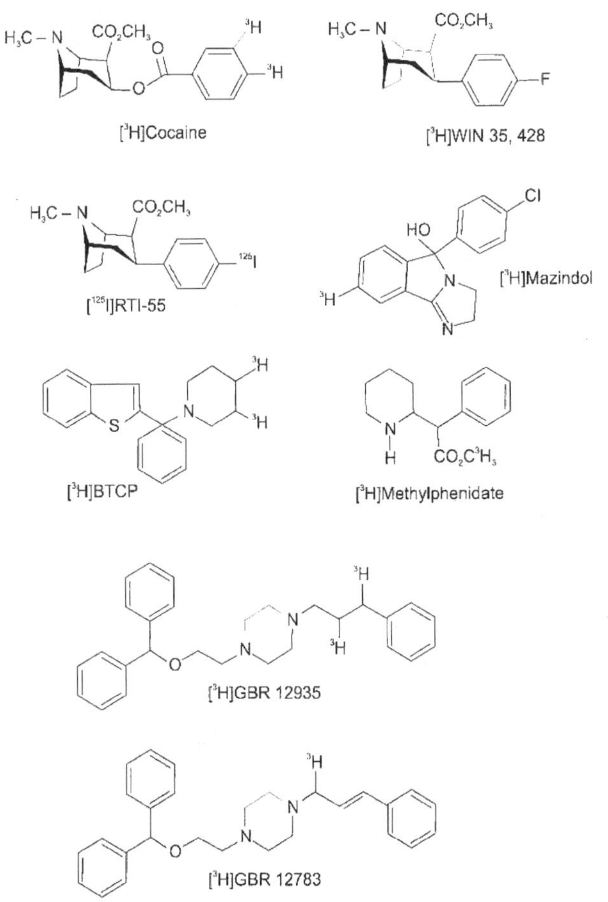

FIGURE 37.1 The reuptake of biogenic amine into the nerve terminal is a major mechanism for terminating the action of the neurotransmitter. It is believed that this carrier-mediated uptake mechanism requires sodium and energy and is blocked by a number of psychopharmacological agents, including cocaine. The dopamine uptake complex is located on presynaptic terminals and cell bodies of dopamine neurons. Because of the involvement of dopamine in the etiology of several neurological disorders such as Parkinson's disease and chorea, and psychiatric diseases such as schizophrenia, dopamine uptake complex is a major interest among neuroscientists. For example, the binding of [3H]cocaine is reduced to dopamine transporter in Parkinson's disease.

The dopamine uptake complex has been studied for many years using a variety of techniques including (a) dopamine uptake assays, (b) in vitro and in vivo homogenate binding assays, (c) in vitro and in vivo autoradiographic assays, (d) in vivo voltammetry, and (e) in vivo behavioral studies. Dopamine transporter has been characterized by in vivo administration of [3H]WIN35, 428, by utilization of [3H]GBR-12935 and membrane proteins, and by 3-Azido [3H]GBR-12935 to solubilized membrane proteins.

When MPP+ enters into the cells, it causes the release of dopamine from secretory vesicles and subsequently generation of free radicals.[37–40] Inside the cell, MPP+ disrupts cellular respiration by inhibiting mitochondrial complex I system,[41,42] reducing the level of ATP, and

hence contributing to degeneration of dopaminergic neurons (see Figure 37.3).

Several studies have shown that MPP+ causes a significant decrease in the activity of DAT.[43–46] However, it remained unclear whether the decrease in DAT activity resulted from the selective uptake of MPP+ through DAT, reduced dopaminergic neurons mediated by the toxin, or changes in the trafficking of the transporter molecules. To clarify this, we[47] have hypothesized that a reduction in cell surface expression of DAT may be an underlying mechanism for the down-regulation of DAT function by MPP+ and subsequent neurotoxicity. To test this hypothesis, we selected HEK-293 cells, which stably express human DAT so as to monitor specially the effects of MPP+ on DAT function.

We have shown that dopamine transporter (DAT) is required for MPP+ mediated cytotoxicity in HEK-293 cells stably transfected with human DAT. Furthermore, MPP+ produced a concentration- and time-dependent reduction in the uptake of [3H]dopamine. We observed a significant decrease in [3H]WIN 35428 binding in the intact cells with MPP+. The saturation analysis of the [3H]WIN 35428 binding (see Figure 37.1) obtained from total membrane fractions revealed a decrease in the transporter density (Bmax) with an increase in the dissociation equilibrium constant (Kd) after MPP+ treatment. Furthermore, biotinylation assays confirmed that MPP+ reduced both plasma membrane and intracellular DAT immunoreactivity. Taken together, these findings suggest that the reduction in cell surface DAT protein expression in response to MPP+ may be a contributing factor in the down-regulation of DAT function, while enhanced lysosomal degradation of DAT may signal events leading to cellular toxicity (see Reference 47 and Figure 37.3).

ACKNOWLEDGMENTS

The authors express their heartfelt appreciation to Mrs. Dani Stramer for her skill in typing this manuscript. Dr. Chagkutip is the recipient of the Thai Royal Golden Jubilee provided by the government of Thailand. The studies cited here have been supported by a contract from Counter Drug Technology Assessment Center, Office of National Drug Control Policy #DATM05-02-C1252 (M.E.).

REFERENCES

1. DiChiara, G., Dopamine in the CNS I and II, Springer-Verlag, Berlin, 2002.
2. Bjöklund, A. and Lindvall, O., Catecholaminergic brain stem regulatory systems, In, Handbook of Physiology, Vol. IV, Sect. 1. (Bloom, F. E., Ed.), American Physiological Society, Bethesda, MD, 155–236, 1986.
3. Kebabian, J. W. and Calne, D. B., Multiple receptors for dopamine, Nature, 277:93–96, 1979.

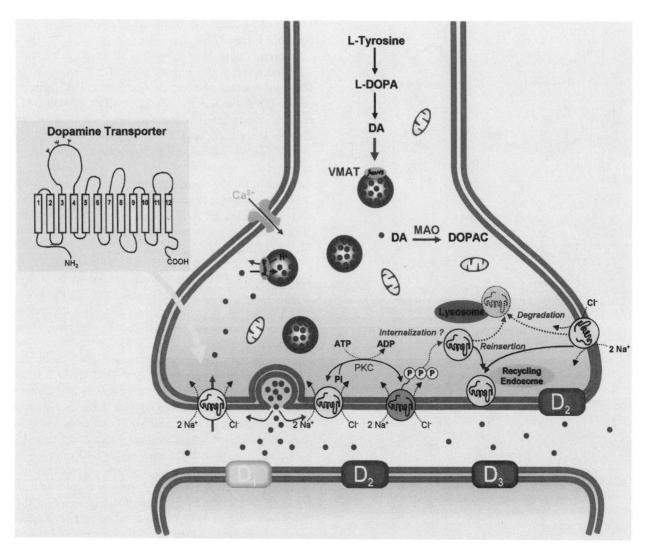

FIGURE 37.2 The sequence of dopamine transmission at a synaptic terminal. The biosynthetic pathway for dopamine starts with the amino acid tyrosine, which is converted into dihydroxyphenylalanine (L-DOPA) with tyrosine hydroxylase (TH), which is then catalyzed to dopamine by L-aromatic acid decarboxylase. Dopamine is stored in synaptic vesicles via vesicular monoamine transporter (VMAT), an ATP-dependent process linked to a proton pump, and released when the neuron is depolarized by Ca^{2+}-dependent exocytosis. Dopamine interacts with postsynaptic receptors as well as with presynaptic D_2 autoreceptors to produce its effects. These effects are terminated by reuptake into the presynaptic through a dopamine transporter (DAT) and metabolized to 3,4-dihydroxyphenylacetic acid (DOPAC) by monoamine oxidase B (MAO-B). The DAT is a plasma membrane transport protein, which is a member of a large family of Na^+/Cl^--dependent transporters, containing 12 transmembrane domains with cytoplasmic NH_2- and COOH-termini. The functional DAT is regulated by phosphorylation, which is coupled with the activity of PKC. Activation of PKC inhibits DAT function through the rapid sequestration/internalization of DAT protein.

4. Seeman, P., Dopamine receptors and the dopamine hypothesis of schizophrenia, *Synapse*, 1:133–152, 1987.

5. Dearry, A., Gingrich, J. A., Falardeau, P., Fremeau, R. T. Jr., Bates, M. D. and Caron, M. G., Molecular cloning and expression of the gene for a human D_1 dopamine receptor, *Nature*, 347:72–76, 1990.

6. Monsma, F. J., Jr., McVittie, L. D., Gerfen, C. R., Mahan, L. C. and Sibley, D. R., Multiple D_2 dopamine receptors produced by alternative splicing, *Nature*, 342:926–929, 1989.

7. Sunahara, R. K., Guan, H. C., O'Dowd, B. F., Seeman, P., Laurier, L. G., Ng, G., George, S. R., Torchia, J., Van

Tol, H. H. M. and Niznik, H. B., Cloning of the gene for a human dopamine D_5 receptor with higher affinity for dopamine than D_1, *Nature*. 350:614–619, 1991.

8. Zhou, Q. Y., Grandy, D. K., Thambi, L., Kushner, J. A., Van Tol, H. H. M., Cone, R., Pribnow, D., Salon, J., Bunzow, J. R. and Civelli, O., Cloning and expression of human and rat D_1 dopamine receptors, *Nature*, 347:76–80, 1990.

9. Bunzow, J. R., Van Tol, H. H. M., Grandy, D. K, Albert, P., Salon, J., Christie, M., Machida, C. A., Neve, K. A. and Civelli, O., Cloning and expression of a rat D_2 dopamine receptor cDNA, *Nature*, 336:783–787, 1988.

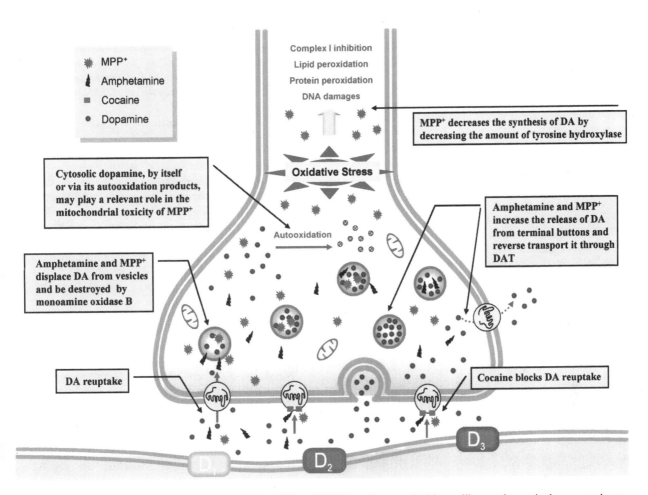

FIGURE 37.3 1-methyl-4-phenyl-1,2,3,6-tetrahydropyridine (MPTP) produces a Parkinson-like syndrome in human and non-human primates by causing a selective degeneration of dopaminergic neurons in the substantia nigra with eosinophilic inclusions. The neurotoxic effects of MPTP are thought to be mediated by 1-methyl-4-phylpyridinium ion (MPP+), which is biotransformed extraneuronally from MPTP by oxidation, catalyzed by monoamine oxidase type B. The selectivity of MPTP for dopaminergic neurons is considered to be a consequence of active uptake of MPP+ into dopaminergic neurons via the dopamine transporter. This hypothesis was confirmed by recent studies on mice lacking the dopamine transporter gene and *in vitro* investigations on neuronal cell lines transfected with the human dopamine transporter (hDAT). The molecular sequence of events that leads to MPP+-induced cell death is not fully understood. According to current hypotheses, intracellular MPP+ is accumulated within the inner mitochondrial membrane by an energy-dependent uptake where it inhibits NADH-Q reductase (mitochondrial complex 1), interrupts electron transport, and depletes intracellular ATP. This decrease of energy supply has been demonstrated in a wide variety of systems including brain mitochondria, human platelets, neuroblastoma cells, and isolated hepatocytes and astrocytes. Furthermore, reduction of mitochondrial complex 1 activity has also been observed in substantia nigra, non-nigral brain areas, platelets, muscle, and fibroblasts of patients with Parkinson's disease.

It has been established that the monoamine transporters [i.e., dopamine (DA), serotonin (5-HT), and norepinephrine (NE)] are the biological targets for stimulants such as *amphetamine*, a metabolite of *methamphetamine* (METH). These transporters are affected acutely by promoting the release of neurotransmitters via a carrier-mediated efflux, as well as through the reversal of the transporter activity, multiple high-dose administrations of METH cause a persistent loss of transporter protein and depress tyrosine hydroxylase activity. Furthermore, it has been shown that the acute effects caused by single and multiple administrations of METH are distinct from long-term transporter deficits presumably associated with nerve terminal degeneration. For instance, dopamine transporter (DAT) function rapidly and reversibly decreases after a single injection of METH, whereas multiple administrations of METH cause a rapid decrease in DAT activity that is partially reversed.

Clinical studies indicate that human *cocaine* users dying during exposure to cocaine have increased striatal [³H]WIN 35428 binding to the dopamine transporter, accompanied by up-regulation of dopamine uptake. Such functional alterations could be important, perhaps contributing to cocaine-induced binging, withdrawal symptoms, or craving. Beyond drug self-administration, dopamine neurons play a role in other rewarding phenomena, including sex and eating, suggesting that regulatory alterations in DAT function could have interesting implications for understanding the dynamics of a number of motivational and appetitive processes. Recent experiments in cell culture have determined that phosphorylative treatments or exposure to the stimulant d-amphetamine dynamically regulate DAT function by changing DAT cellular localization, perhaps invoking mechanisms that might be related to those activated by cocaine.[47,48]

10. Dal Toso, R., Sommer, B., Ewert, M., Herb, A., Pritchett, D. B., Bach, A., Shivers, B. D. and Seeburg, P. H., The dopamine D₂ receptor: two molecular forms generated by alternative spicing, *EMBO J.*, 8:4025–4034, 1989.

11. Giros, B., Sokoloff, P., Martres, M. P., Riou, J. F., Emorine, L. J. and Schwartz, J. C. Alternative splicing directs the expression of two D₂ dopamine receptor isoforms, *Nature*, 342:923–926, 1989.

12. Frothigham, L, Fischer, J. B., Burke-Howie, K. J., Bunzow, J. R., Server, A. C. and Civelli, O., Cloning of the cDNA and gene for a human D₂ dopamine receptor, *Proc. Natl. Acad. Sci. USA,* 86:9762–9766, 1989.

13. Sokoloff, P., Giros, B., Martres, M. P., Bouthenet, M. L. and Schwartz, J. C., Molecular cloning and characterization of a novel dopamine receptor (D₃) as a target for neuroleptics, *Nature*, 347:146–151, 1990.

14. Van Tol, H. H. M., Bunzow, J. R., Guan, H. C., Sunahara, R K., Seeman, P., Niznik, H. B. and Civelli, O., Cloning of the gene for a human dopamine D₄ receptor with high affinity for the antipsychotic clozapine, *Nature*, 350:610–614, 1991.

15. Vallar, L and Meldolesi, J. Mechanisms of signal transduction at the dopamine D₂ receptor, *Trends Pharmacol. Sci.*, 10:74–77, 1989.

16. Ebadi, M., Pfeiffer, R. F. and Murrin, L. C., Pathogenesis and treatment of neuroleptic malignant syndrome, *Gen. Pharmacol.*, 21:367–386, 1990.

17. Ebadi, M., Pfeiffer, R. F. and Murrin, L. C., Neuroleptic-induced movement disorders, in *Current Aspects of the Neurosciences,* Vol. 4., Ed., Osborne, N. N., Macmillan Press, pp. 159–203, 1992.

18. O'Dowd, B. F., Structures of dopamine receptors, *J. of Neurochem.*, 60:804–816, 1993.

19. Missale, C., Nash, S. R., Robinson, S. W., Jaber, M. and Caron M. G., Dopamine receptors: from structure to function, *Pharmacological Reviews*, 78:189–225, 1998.

20. Strange, P. G. Dopamine Receptors, *TOCR's Reviews*, 15:1–5, 2000.

21. Glickstein, S. B. and Schmauss, C., Dopamine receptor functions: lessons from knockout mice, *Pharmacology and Therapeutics,* 91:63–83, 2001.

22. Giros, B., Jaber, M., Jones, S., Wightman, R. M. and Caron, M. G., Hyperlocomotion and indifference to cocaine and amphetamine in mice lacking the dopamine transporter, *Nature,* 379, 606–616, 1996.

23. Jones, S. R., Gainetdinov, R. R., Jaber, M., Giros, B., Wightman, R. M. and Caron, M. G., Profound neuronal plasticity in response to inactivation of the dopamine transporter, *Proc. Natl. Acad. Sci. USA.*, 95:4029–4034, 1998.

24. Jones, S. R., Gainetdinov, R. R., Wightman, R. M. and Caron, M. G., Mechanisms of amphetamine action revealed in mice lacking the dopamine transporter, *J. Neurosci.,* 18:1979–1986, 1998.

25. Giros, B. and Caron, M. G., Molecular characterization of the dopamine transporter, *Trends in Pharmacol. Sci.*, 14:43–49, 1993.

26. Chen, N. and Reith, M. E. A., Structure and function of the dopamine transporter, *Eur. J. Pharmacol.*, 405: 329–339, 2000.

27. Reith, M. E. A., Xu, C. and Chen, N. H., Pharmacology and regulation of the neuronal dopamine transporter, *Eur. J. Pharmacol.*, 324:1–10, 1997.

28. Chen, N. H. and Reith, M. E. A. Structure and function of the dopamine transporter, *Eur. J. Pharamacol.*, 405: 329–339, 2000.

29. Reith, M. E. A., *Neurotransmitter transporters structure, function, and regulation,* 2nd ed., Human Press, New Jersey, 2002.

30. Tipton, K. F. and Singer, T. P., Advances in our understanding of the mechanisms of the neurotoxicity of MPTP and related compounds, *J. Neurochem.*, 61:1191–1206, 1993.

31. Gainetdinov, R. R., Fumagalli, F., Jones, S. R. and Caron, M. F., Dopamine transporter is required for *in vivo* MPTP neurotoxicity: evidence from mice lacking the transporter, *J. Neurochem,.* 69:1322–1325, 1997.

32. Javitch, J. A., D'Amato, R. J., Strittmatter, S. M. and Snyder, S., Parkinsonism-inducing neurotoxin, N-methyl-4-phenyl-1,2,3,6-tetrahydropyridine: uptake of the metabolite N-methyl-4-phenyl pyridium by dopamine neurons explains selective toxicity, *Proc. Natl. Acad. Sci. USA*, 82: 2173–2177, 1982.

33. Amara, S. G. and Kuhar, M. J., Neurotransmitter transporters: recent progress, *Ann. Rev. Neurosci.*, 16: 73–93, 1993.

34. Reith, M. E., Xu, C. and Chen, N. H., Pharmacology and regulation of the neuronal dopamine transporter, *Eur. J. Pharmacol.*, 324:1–10, 1997.

35. Uhl, G. R., Walther, D., Mash, D., Faucheux, B. and Javoy-Agid, F., Dopamine transporter messenger RNA in Parkinson's disease and control substantia nigra neurons, *Ann. Neurol.*, 35: 494–498, 1994.

36. Harrington, K. A., Augood, S. J., Kingsbury, A. E., Foster, O. J. and Emson, P. C., Dopamine transporter (DAT) and synaptic vesicle amine transporter (VMAT2) gene expression in the substantia nigra of control and Parkinson's disease, *Brain Res. Mol. Brain Res.*, 36: 157–162, 1996.

37. Speciale, S. G., Liang, C. L., Sonsalla, P. K., Edwards, R. H. and German, D. C., The neurotoxin 1-methyl-4-phenylpyridinium is sequestered within neurons that contain the vesicular monoamine transporter, *Neuroscience.* 84: 177–1185, 1998.

38. Chang, G. D. and Ramirez, V D., The mechanism of action of MPTP and MPP⁺ on endogenous dopamine release from the rat corpus striatum superfused *in vitro*, *Brain Res.*, 368: 134–140, 1986.

39. Obata, T. and Chiueh, C. C., *In vivo* trapping of hydroxyl free radicals in the striatum utilizing intracranial microdialysis perfusion of salicylate: effects of MPTP, MPDP⁺, and MPP⁺, *J. Neural Transm. Gen. Sect.*, 89:139–145, 1992.

40. Lotharius, J. and O'Malley, K. L., The parkinsonism-inducing drug 1-methyl-4 phenylpyridinium triggers intracellular dopamine oxidation. A novel mechanism of toxicity, *J. Biol. Chem.*, 275: 38581–38588, 2000.

41. Przedborski, S. and Jackson-Lewis, V., Mechanisms of MPTP toxicity, *Mov. Disord.*, 13: 35–38, 1998.

42. Ramsay, R. R., Salach, J. I. and Singer, T. P., Uptake of the neurotoxin 1-methyl-4-phenylpyridine (MPP$^+$) by mitochondria and its relation to the inhibition of the mitochondrial oxidation of NAD$^+$-linked substrates by MPP$^+$, *Biochem. Biophys. Res. Commun.*, 134: 743–748, 1986.

43. Barc, S., Page, G., Fauconneau, B., Barrier, L. and Huguet, F., A new *in vitro* approach for investigating the MPTP effect on DA uptake, *Neurochem. Int.*, 38:243–248, 2001.

44. Fonck, C. and Baudry, M., Toxic effects of MPP$^+$ and MPTP in PC12 cells independent of reactive oxygen species formation, *Brain Res.*, 905: 199–206, 2001.

45. Fonck, C. and Baudry, M., Rapid reduction of ATP synthesis and lack of free radical formation by MPP$^+$ in rat brain synaptosomes and mitochondria, *Brain Res.*, 975: 214–221, 2003.

46. Storch, A., Ludolph, A. C. and Schwarz, J., HEK-293 cells expressing the human dopamine transporter are susceptible to low concentrations of 1-methyl-4-phenylpyridine (MPP$^+$) via impairment of energy metabolism, *Neurochem. Int.*, 35: 393–403, 1999.

47. Chagkutip, J., Vaughan, R. A., Govitrapong, P. and Ebadi, M., 1-Methyl-4-phenylpyridinium-induced down-regulation of dopamine transporter function correlates with a reduction in dopamine transporter cell surface expression, *Biomedical and Biophysical Research Comm.*, Vol. 311, 49:49–54, 2003.

48. Richerson, G B. and Wu, Y. M., Dynamic equilibrium of neurotransmitter transporters: not just for reuptake anymore, *J. Neurophysiology*, Vol. 90, 1363–1374, 2003.

38 Pathophysiology of Parkinson's Disease, Neurochemical Pathology: Other Neurotransmitters

Yoshikuni Mizuno
Department of Neurology, Juntendo University School of Medicine

CONTENTS

0-8493-1590-5/05/$0.00+$1.50
© 2005 by CRC Press

ACETYLCHOLINE

Cholinergic Neurons in the Nervous System

Cholinergic neurons are frequently involved in Parkinson's disease (PD). The main cholinergic neurons in the brain include nucleus basalis of Meynert (NBM), pedunculopontine nucleus (PPN), dorsal motor nucleus of the vagal nerve, and the intrinsic neurons of the striatum. In the spinal cord, the preganglionic sympathetic neurons originating in the thoracic intermediolateral column and the preganglionic parasympathetic neurons of the sacral cord are cholinergic. In the peripheral autonomic nervous system, postganglionic parasympathetic neurons, including neurons in the Auerbach and Meissner plexuses in the gastrointestinal tract, are cholinergic. In addition, postganglionic sympathetic neurons innervating the sweat glands are cholinergic. Motor neurons of the brain stem and the anterior horn of the spinal cord are also cholinergic; these motor neurons are not involved in PD.

METABOLISM OF ACETYLCHOLINE

Acetylcholine (Ach) is synthesized from acetyl-CoA and choline by choline acetyltransferase (ChAT). Acetylcholine is stored in the synaptic vesicles of the cholinergic neurons. The released Ach is rapidly destroyed by acetylcholine esterase (AChE). ChAT is a marker enzyme for the study of cholinergic neurons. ChAT activity can be assayed in autopsy materials and ChAT immunostaining has been used to reveal cholinergic neurons.

CHOLINERGIC NEURONS IN PARKINSON'S DISEASE

Nucleus Basalis of Meynert (NBM)

NBM is located in the sublenticular region. NBM receives afferent input principally from the limbic system and projects to most of the cerebral cortical areas. This cortically projecting system is the largest diffusely projecting system to cortical areas from the basal area without going through the thalamus.[1] The nucleus is frequently involved in dementing neurodegenerative disorders including Alzheimer's disease (AD).[2,3] NBM has extensively been examined since 1983.

Mann and Yates[4] examined NBM from eight PD patients. They found loss of neurons in NBM in all the patients they examined, and the loss was more severe in three of eight patients with mental impairment. Arendt et al.[5] found neuronal loss in NBM in 77% of their 58 PD patients. Whitehouse et al.[6] found a selective loss of cells in NBM in demented patients out of 12 PD patients. Gaspar and Gray[7] examined 14 nondemented and 18 demented PD patients. They found neuronal loss and Lewy bodies in NBM in most cases (95%), and the loss was associated with significant reduction of ChAT activity both in NBM and the cortex. The lesion was more severe in subjects with dementia. Nakano and Hirano[8] found significant depletion of large neurons of NBM from 11 PD patients compared to the 13 controls. Interestingly, they did not find neuronal loss in NBM of majority of patients with AD.

Similar decreases in the number of NBM neurons in PD patients, particularly in demented PD patients, were reported by many investigators.[9–16] Kosaka[12] noted a good correlation between the presence of dementia and loss of NBM neurons in diffuse Lewy body disease (DLBD). Chan-Palay et al.[13] reported a decrease by 70% of the number of calbindin-D28-positive large neurons in NBM in two nondemented PD patients, and by 40% in one demented PD patient.

Thus, there is good agreement with regard to the loss of neurons in NBM in demented PD patients, unless there is no cause for dementia other than PD. Nondemented PD patients also show variable loss of NBM neurons to a lesser extent. Lewy bodies (LB) were frequently seen in NBM from PD patients. NBM neurodegeneration was particularly prominent in patients who fulfilled the pathologic criteria of DLBD.[13] Thus, neuropathologic changes of DLBD and PD with dementia (PDD) are essentially similar; the difference appears to be a matter of severity.

Neuronal loss in NBM has been reported in many other dementing disorders including AD,[2,3,5,9,13,16] parkinsonism-dementia complex of Guam,[17,18] PSP, Creutzfeldt-Jacob disease,[9] Korsakoff's syndrome,[5] dementia pugilistica, Pick's disease, and Down's syndrome.[2] Thus, NBM appears to be a very important basal structure that can cause significant impairment of intellectual functions and dementia. Particularly, the degeneration of NBM appears to be a very important pathologic substrate for the dementia of PD. The question is to what extent the NBM degeneration accounts for the dementia of PD and to what extent cortical changes such as Lewy neuritis and LB in the cortical areas in PD account for the dementia of PD. This question can be phrased in a different way, i.e., what symptoms of dementia are caused by NBM lesion, and what symptoms of dementia are caused by cortical lesions in PD? Further studies are necessary to answer these questions.

Pedunculopontine Nucleus

The pedunculopontine nucleus (PPN) is a small nucleus located in the dorso-lateral part of the ponto-mesencephalic tegmentum. The pars compact neurons in the PPN are cholinergic,[19] and those in the pars dissipatus are glutamatergic. The PPN is thought to be involved in the initiation and modulation of gait and other stereotyped movements. Glutamatergic neurons are thought to be important regulators of the basal ganglia and spinal cord. The cholinergic part constitutes a feedback loop from the spinal cord and limbic system back into the basal ganglia and thalamus.[20] The major afferents to PPN are the inputs from the internal segment of the globus pallidus (GPI) and the pars reticulata of the substantia nigra (SNR).

Patients with PD have significant loss of PPN neurons.[21–25] Hirsch et al.[21] first reported loss of cholinergic neurons in PPN from PD patients. Jellinger[22] found approximately 50% (range 6 to 39%) loss in cell number in pars compacta neurons of PPN in PD patients. Zweig et al.[24] counted the number of PPN pars compacta neurons from PD patients. They found significant reduction in cell numbers (average 40%) compared with the controls ($p <$

0.01). Gai et al.[25] found a 43% reduction in the number of substance P (SP)-positive neurons in PPN (SP co-localizes with Ach).

Thus, it is safe to say that the neurodegenerative change in the cholinergic neurons of PPN is a constant finding in PD, although the magnitude of the cell loss is less than that in substantia nigra pars compacta (SNC). The question is which symptoms of PD are related to the neurodegeneration in PPN. In this respect, it is interesting to note that PPN neurons are also reduced in progressive supranuclear palsy (PSP)[21,22] in which locomotion and balance are markedly impaired. Thus, it has been postulated that neurodegeneration in PPN might be related to postural instability and gait disturbance such as freezing of gait. But this hypothesis needs further investigation for the following reasons. First of all, at least 50% of dopaminergic neuronal loss is necessary before symptoms of PD appear. Second, postural instability is L-dopa responsive, and off-time freezing is L-dopa responsive. There is no evidence to indicate that cholinergic treatment is effective for postural instability and freezing of gain during an off period. Thus, the hypothesis that disturbances of postural instability and freezing of gait are related to loss of neurons in PPN should be tested by further studies. Some of L-dopa nonresponsive late manifestations of PD, such as on-time freezing and start hesitation, might be related to PPN degeneration. It should also be noted that PPN neurons are affected in AD, albeit to a lesser degree.[22,24,26]

Dorsal Motor Nucleus of the Vagal Nerve

Cholinergic neurons in the dorsal motor nucleus are almost constantly involved in PD. Gai et al.[27] studied the dorsal vagal nucleus of 8 PD patients in detail. They counted the number of the neurons in serial sections. They found marked neurodegenerative changes (Lewy bodies and neuronal loss) in this nucleus (13,637 ± 1,323 neurons in PD, 24,885 ± 1,157 in normal controls). Cells in the intermediate rostrocaudal part of the nucleus were most severely affected. They found a significant correlation between loss of vagal neurons and age at death in PD patients. Gai et al.[28] also studied ubiquitin-positive, tau-negative, degenerating neurites in PD brains. They were particularly striking in the dorsal motor nucleus of the vagus. In this nucleus, the density of degenerating neurites was inversely related to the duration of PD symptoms. They concluded that extensive ubiquitin-positive degenerating neurites might provide a clue to disease activity at the time of death.

Interestingly, Del Tredici et al.[29] reported that neuropathologic changes of PD commenced with the formation of the very first immunoreactive Lewy neurites and Lewy bodies in noncatecholaminergic neurons of the dorsal glossopharyngeus-vagus complex based on the observation on 30 incidental Lewy body disease patients.

Intermediolateral Column of the Thoracic Cord

The intermediolateral column neurons of the thoracic spinal cord give rise to preganglionic sympathetic fibers. They are cholinergic neurons. Wakabayashi and Takahashi[30] examined the number of neurons in the intermediolateral column (IML) from 25 PD patients. They found a decrease in the number of neurons by 69% at T2 and 57% at T9 compared with the controls. Lewy bodies were found in the remaining IML neurons in 24 of 25 cases of PD. These changes may in part be responsible for autonomic symptoms such as orthostatic hypotension and profuse sweating episodes in PD patients.

Meissner and Auerbach Plexuses

Neurons in the Meissner and Auerbach plexus are located within the muscular layers of the gastrointestinal tract representing postganglionic parasympathetic cholinergic neurons. Wakabayashi et al.[31] studied the enteric nervous system of the alimentary tract in seven PD patients. They found characteristic inclusions histologically and ultrastructurally identical to Lewy bodies in Auerbach's and Meissner's plexuses. They were most frequent in the Auerbach's plexus of the lower esophagus. Neurodegenerative changes in the Meissner and Auerbach plexuses may in part be responsible for some of the autonomic symptoms of PD patients such as constipation.

CHOLINERGIC MARKERS IN PARKINSON'S DISEASE

Choline Acetyltransferase (ChAT)

ChAT is most frequently used as a marker for cholinergic neurons. It is well known that ChAT is reduced in AD[32,33] and some other dementing disorders such as parkinsonism dementia complex of Guam,[34] and the decrease correlated well with loss of neurons in NBM.[3,7,35]

McGeer and McGeer[36,37] measured ChAT and AChE activities in various brain regions including the basal ganglia and the cortex of PD patients. They did not find any significant difference from the controls. Javoy-Agid et al.[38] measured ChAT activity in SNC and SNR in PD patients. The activity level was essentially unchanged compared with the controls.

On the other hand, Ruberg et al.[39] reported a significant decrease in the cortical ChAT activity in demented PD patients. Perry et al.[33] also reported a decrease in cortical ChAT activity in demented PD patients in comparison with AD patients. They found consistently lower cortical ChAT activities in demented PD patients than in the classical AD cases. Interestingly, two of the Lewy body cases with extremely low cholinergic activity were responders in therapeutic trials of the cholinesterase inhibitor, tacrine. Decrease in ChAT activity was also found in PD patients that included nondemented PD patients. Nishino et al.[40] measured ChAT activity in the caudate nucleus, putamen, pallidum, substantia nigra, and the cerebral cortex from L-dopa-treated Stage V PD patients. They found a significant decrease in the ChAT activity in all the regions they studied. Quirion[32] found a decrease in ChAT activity in various cortical areas of PD patients; this decrease was of the same magnitude as AD patients. The results are summarized in Table 38.1.

TABLE 38.1
Choline Acetyltransferase Activity in Parkinson's Disease

Authors	Results	References	Remarks
Cortex (frontal ad/or temporal)			
McGeer et al.	unchanged	36,37	
Ruberg et al.	decreased	39	Demented PD
Gaspar and Gray	decreased	7	
Nishino et al.	decreased	40	
Quirion	decreased	32	
Perry et al.	decreased	33	
Kuhl et al.	decreased	41	Vesicular Ach transporter
Putamen			
McGeer et al.	unchanged	36, 37	
Nishino et al.	decreased	40	
Globus pallidus			
Nishino et al.	decreased	40	
Substantia nigra pars compacta			
Javoy-Agid et al.	unchanged	38	
Nishino et al	decreased	40	

Kuhl et al.[41] studied cholinergic terminals using [123I]iodobenzovesamicol (IBVM) as an *in vivo* marker of the vesicular acetylcholine transporter for single photon emission computerized tomography (SPECT). They found extensive cortical binding decreases in demented PD subjects similar to early-onset AD.

Acetylcholinesterase in Parkinson's Disease

Acetylcholinesterase (AChE) is the main enzyme to destroy acetylcholine after its release from cholinergic nerve terminals. AChE is located in the extracellular space. Thus it is not a good marker for cholinergic neurons compared with ChAT. Not many studies have been done on AChE in PD. Rinne et al.,[42] measured AChE activity in various regions of the brain including the basal ganglia and the cortex. They found only modest decrease (20 to 30%) in the activity in PD patients compared with the controls.

Acetylcholine Receptors in Parkinson's Disease

Cortical Acetylcholine Receptors

Whitehouse et al.[43,44] first reported a profound reduction in nicotinic acetylcholine receptors (AChR) in various cortical areas from PD patients. They found similar reduction in AD brains also. Regarding muscarinic AChR, Nishisno et al.[40] first reported a normal amount in PD brains. Quirion[32] reported a decrease in the density of both muscarinic M2 and nicotinic AChR in various cortical areas in PD; the magnitude of the decrease was similar to that of AD. Shiozaki et al.[45] found a significant loss of the total amounts of muscarinic AChR in the temporal cortex of DLBD as well as AD patients. In both diseases, the proportion of the muscarinic 3 (m3) receptor in the frontal cortex was significantly increased, and that of the m4 receptor in the temporal cortex was significantly decreased as compared with the controls. Later, Shiozaki et al.[46] reported a decrease in m1 AChR in the temporal lobe of DLB patients as well.

Banerjee et al.[47] counted the number of nicotinic AChR using immunohistochemistry and stereological technique and found that the number of alpha7 subunit protein-expressing neurons in the AD temporal cortices amounted to approximately half of that of controls, while numbers in PD patients lay in between. They found no differences in the total number of neurons. Guan et al.[48] found significant decrease in the alpha3 mRNA and protein in the caudate nucleus and temporal cortex of PD patients. The level of the beta2 protein subunit in the temporal cortex and hippocampus and the beta2 mRNA in the temporal cortex was lowered. Both the levels of the alpha7 subunit protein and [125I]alpha-bungarotoxin binding were significantly increased in the temporal cortex of PD patients. Burghaus et al.[49] reported loss of cortical alpha4 and alpha7 subunits of nicotinic AChR in PD, similar to the findings for AD.

Asahina et al.[50] used PET to reveal muscarinic AChR in live patients. They used [11C]N-methyl-4-piperidyl benzilate ([11C]NMPB) as a tracer. They found significantly higher muscarinic AChR binding for the frontal cortex in PD patients (11 nondemented and 1 demented) than in the controls ($p < 0.01$). They ascribed the loss to neuronal loss in the NBM.

Results on cortical AChR in PD are summarized in Table 38.2. By reviewing the literature, there is apparent agreement in the decrease in nicotinic AChR in the cerebral cortex of PD patients. This result would indicate that the nicotinic AChR are located at least in part on the presynaptic terminals of cholinergic neurons from NBM. The neuronal nicotinic AChR are excitatory ligand-gated channels. They are widely expressed throughout the peripheral and central nervous system.[51] The nicotinic AChR have a pentameric structure composed of five membrane spanning subunits, of which nine different types have thus far been identified and cloned.[52]

Regarding muscarinic AChR, there is a controversy in the results. Muscarinic AChR may be localized both presynaptically and postsynaptically. Therefore, the state of denervation supersensitivity may influence the results. In

TABLE 38.2
Acetylocholine Receptors in Parkinson's Disease

Authors	Findings	References	Remarks
Cortex (frontal and/or temporal)			
(Nicotinic receptors)			
Whitehouse	decreased	43	
Quirion	decreased	32	
Banerjee et al.	decreased	47	
Guan, et al.	decreased	48	alpha3, beta2 subunits and mRNA
Guan et al.	increased	48	alpha7, [125] alpha–bungarotoxin
Burghaus et al.	decreased	49	alpha4 and alpha7 subunits
(Muscarinic receptors)			
Nishino et al.	unchanged	40	
Quirion	decreased	43	
Asahina et al.	increased	50	[11C]NMPB* PET
Shiozaki et al.	decreased	45,46	BLBD
Putamen			
(Nicotinic receptors)			
Court et al.	decreased	55	[3H]Nicotine binding
Martin-Ruiz et al.	decreased	56	[3H]Nicotine binding
Guan et al.	decreased	48	mRNA (caudate)
(Muscarinic receptors)			
Nishino et al.	unchanged	40	
Griffiths et al.	unchanged	53	QNB binding
Griffiths et al.	decreased	54	
Globus pallidus			
(Nicotinic receptors)			
Griffiths et al.	increased	53	GBI
(Muscarinic receptors)			
Nishino et al.	unchanged	40	
Substanita nigra pars compacta			
(Muscarinic receptors)			
Nishino et al.	unchanged	40	

*[11C]N-methyl-4-piperidyl benzilate

addition, the difference in the methods to measure the muscarinic AChR appears to be another source of the different results.

Striatal Cholinergic Receptors

Nishino et al.[40] measured muscarinic AChR binding in autopsied samples of the caudate nucleus, putamen, pallidum, substantia nigra, and the cerebral cortex from L-dopa-treated patients with Stage V PD patients using [3H]quinuclidinyl benzilate ([3H]QNB) as a ligand. They found no change in the muscarinic AChR bindings in PD. Griffiths et al.[53] measured muscarinic AChR in the basal ganglia of PD patients using QNB as a ligand for autoradiography. They found normal QNB binding in the caudate and putamen of PD patients. However, the binding in the medial pallidal segment was increased. They interpreted their finding as suggesting underactivity of the cholinergic pathway from PPN to GPI. Later, Griffiths et al.[54] reported decrease in muscarinic AChR receptor density in the rostral putamen.

Regarding the nicotinic AChR, Court et al.[55] found significant reduction in [3H]nicotine binding in both dorsal and ventral caudate and putamen in PD (43 to 67%, $n = 13$), AD (29 to 37%, $n = 13$), and DLB (50 to 61%, $n = 20$) compared to age-matched controls ($n = 42$). Martin-Ruiz et al.[56] studied alpha3, alpha4, alpha7, and beta2 subunit immunoreactivity of the nicotinic AChR in the putamen from PD patients; they found no loss in these subunits, despite a highly significant reduction in [(3)H]nicotine binding. They concluded that the loss of nicotine binding in the putamen in PD might involve nicotinic AChR subunits other than those investigated (e.g., alpha5 and/or alpha6).

SUMMARY OF CHOLINERGIC MARKERS IN PARKINSON'S DISEASE

Results of the ChAT activity in PD are summarized in Table 38.1. By reviewing the literature, there is no question about the significant decrease in the cortical ChAT activity in demented PD patients that correlated well with marked neuronal loss in NBM. Not every report clearly divided PD· patients into demented and nodemented patients; however, it appears to be safe to say that cortical ChAT activity is also reduced in nondemented PD patients. However, the magnitude of the decrease is less than that of demented PD patients. It is interesting to note that the decrease in the cortical ChAT activity in demented PD patients was even more prominent than that of AD patients, and those patients had responded well to a cholinesterase inhibitor, tacrine.[33]

Regarding the ChAT activity in the subcortical structures such as putamen and SNC, there is a controversy in the results as shown in Table 38.1. The number of reports is still small; further studies are needed before making a definite conclusion.

Regarding AchR, there is an agreement in the decrease in nicotinic AChR in the striatum of PD patients. As striatal neurons are not involved in PD, it appears to be likely that AChR are localized at least in part in the nerve terminals of the nigrostriatal dopaminergic neurons. Muscarinic AChR are generally unchanged in PD except for a part of the striatum.[54] It is likely that most of the muscarinic AChR are not located on the nigrostriatal dopaminergic terminals. In other subcortical areas, still the number of studies is small for making a definite conclusion.

SEROTONIN

SEROTONERGIC NEURONS IN THE NERVOUS SYSTEM

The major source of serotonergic innervation in the central nervous system is the raphe nucleus. Raphe nucleus is located in the central part of the midbrain to pontine tegmentum near the midline.[57] In addition scattered serotonergic neurons are found in the tegmentum through midbrain to medulla oblongata. Serotonergic neurons show widespread projections to the cortical areas, subcortical structures, including the basal ganglia, and the spinal cord. The striatum shows the highest concentration of 5HT in the brain. The raphe nucleus is composed of dorsal part and medial part; the former projects mainly to the striatum and the latter to the cortical areas.[58] The raphe nucleus is frequently involved in PD with neuronal loss and Lewy body formation,[59] and the serotonergic involvement may be important for nonmotor symptoms of PD such as depression.[60]

METABOLISM OF SEROTONIN

Serotonin (5HT) is synthesized from tryptophan (Figure 38.1). The initial step is 5-hydroxylation of tryptophan by tryptophan hydroxylase. This is the late-limiting step of 5HT synthesis. Then 5-hydroxytryptophan is decarboxylated by aromatic L-amino acid decarboxylase, the same enzyme that decarboxylates L-dopa, to produce 5-hydroxytryptamine (5HT = serotonin). 5HT is released from serotonergic nerve endings on excitation of the serotonergic neurons. 5HT binds to 5HT receptors and then reuptaken into the serotonergic nerve terminals through high-affinity 5HT uptake sites. Within neurons, 5HT is metabolized mainly by monoamine oxidase B (MAOB), which is located in the outer membrane of the mitochondria to form 5-hydroxyindoleacetic acid (5HIAA) that can be measured in the CSF.

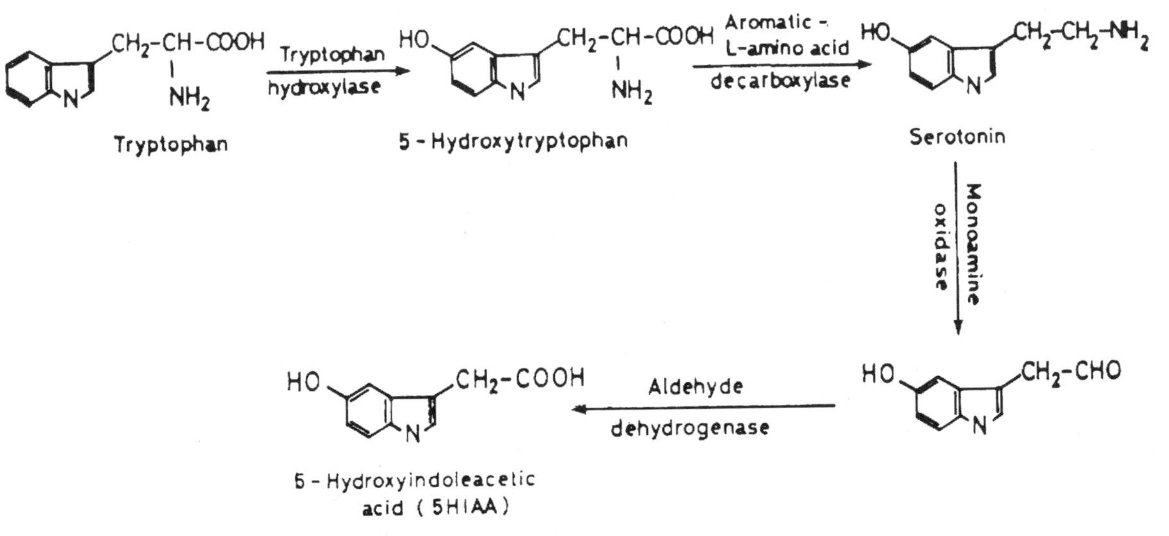

FIGURE 38.1 Biosynthesis and metabolism of serotonin. Serotonin (5HT) is synthesized from tryptophan. The initial step is 5-hydroxylation of 5HT by tryptophan hydroxylase followed by decarboxylation of 5-hydroxytriptophan by aromatic L-amino acid decarboxylase. 5HT is metabolized by monoamine oxidase B and aldehyde dehydrogease. The final metabolite is 5-hydroxyindoleacetic acid (5HIAA).

RAPHE NUCLEUS IN PARKINSON'S DISEASE

Halliday et al.[61] studied pontine cell groups from 4 PD patients. They found significant loss and degenerative changes of serotonergic neurons in PD patients. Paulus and Jellinger[62] studied 45 autopsy cases of PD and found more severe cell loss in the serotonergic dorsal raphe nucleus in PD patients with depression than in non-depressed ones.

Becker et al.[63] attempted to measure an alteration of the raphe nucleus by transcranial sonography. They studied 30 PD patients and 30 age and sex matched controls. They found significantly reduced echogenicity in depressed patients with PD compared with nondepressed patients with PD and control subjects.

SEROTONERGIC MARKERS IN PARKINSON'S DISEASE

Serotonin in the Brain

Bernheimer et al.[64] first reported 50% decrease in 5HT content in the striatum, thalamus, and hypothalamus from PD patients. Later, Scatton et al.[65] reported significant decrease in 5HT content in several cortical areas, hippocampus and caudate nucleus from PD patients. D'Amato et al.[66] also found similar reduction in 5HT in the cortex from PD patients. Ohara et al.[67] reported loss of 5HT in the striatum and the cerebral cortex of five DLBD patients. Shannak et al.[68] measured hypothalamic serotonin from PD patients. They found 26% reduction in 5HT content in the whole hypothalamic extract from PD patients compared with the controls.

Tryptophan Hydroxylase

Sawada et al.[69] analyzed tryptophan hydroxylase activity. They found significant reduction in the activity of this enzyme in the raphe nucleus of PD patients compared with controls ($p < 0.05$).

Serotonin Transporter

5HT transporters represent high affinity reuptake sites for 5HT after 5HT release and binding to 5HT receptors. Thus, 5HT transporters are expressed in 5HT neurons. Therefore, the 5HT transporter is a good marker to evaluate the state of 5HT neurons. Kienzl et al.[70] measured 5HT binding sites that include uptake sites in the frontal cortex from PD patients using 3H-5-HT as a ligand. They found significant reduction in the 5HT binding sites in PD patients. Cash et al.[71] measured 5HT reuptake sites in putamen and cortex from PD patients using [3H]imipramine as a ligand. They found a significant reduction in the density of the high affinity binding in the prefrontal cortex and putamen from PD patients. The decrease in [3H]imipramine binding was found predominantly in the

synaptosomal fractions. They concluded that the decrease in the 5HT reuptake site might be implicated in the depression of PD patients. D'Amato et al.[66] reported 40 to 50% decrease in the cortical uptake sites in PD patients using [3H] citalopram as a ligand. Chinaglia et al.[72] studied 5HT binding sites by autoradiography using [3H]citalopram as a ligand in PD patients. They found significant decrease in the binding sites both in the cortical areas and the basal ganglia of PD patients. The binding in the raphe nucleus was unchanged. There results would indicate that the terminal regions of the serotonergic system are more affected in PD.

5HT transporters can be visualized in live patients using SPECT. 2Beta-carboxymethoxy-3beta-(4-iodophenyl)tropane ([123I]beta-CIT) is a radio-ligand to label dopamine and serotonin uptake sites; its cortical and midbrain uptakes mainly represent serotonin reuptake sites. Brucke et al.[73] studied serotonin transporter in 2 PD patients using [123I]beta-CIT as a ligand for SPECT. They found only mild decrease in the activity in hypothalamus-midbrain areas. Uptake into cortical and cerebellar areas appeared to be unchanged in PD. Haapaniemi et al.[74] studied 27 PD patients using. ([123I]beta-CIT) SPECT. They found lower uptake of beta-CIT in the thalamic and frontal regions of PD patients compared with controls. The uptake ratio of the frontal area correlated with the Unified Parkinson's Disease Rating Scale. (UPDRS) subscore I. Murai et al.[60] also reported decrease of brain stem serotonin transporter sites in PD using [(123)I]beta-CIT SPECT and this decrease was correlated well with the mentation, behavior, and mood subscale of the UPDRS. But Kim et al.[75] found normal midbrain [(123)I]beta-CIT uptake in 7 depressed PD patients.

Serotonin and Its Metabolites in CSF

Davidson et al.[76] measured CSF 5HIAA in 54 untreated PD patients. They found no significant difference compared with the controls. Tabaddor et al.[77] measured ventricular 5HIAA in PD patients. They found no change compared with the controls. Tohgi et al.[78] measured total (free and conjugated) 5-HT concentration in the CSF of untreated PD patients. They found significant decrease in 5-HT compared with the controls. After treatment with L-dopa, CSF 5HT showed further decrease. This is an interesting observation suggesting the competition of aromatic L-amino acid decarboxylase between L-dopa and 5-hydroxytryptamine. Iacono et al.[79] measured ventricular CSF 5HT, 5-hydroxytryptophan (5HTP), and 5HIAA in PD patients. They found significant reduction in 5HT and significant increase in 5-HTP in PD patients with predominant postural instability and gait disturbance compared with tremor dominant PD patients. They concluded that gait and balance disorder of advanced PD patients might be in part due to impaired serotonergic transmission.

Serotonin Receptors

Not many studies have been published on 5HT receptors in PD. Perry et al.[80] reported 5HT-1 and 5HT-2 receptor bindings in cortical regions of demented PD patients. They found normal S2 sites but reduced S1 sites in cortical areas of demented PD patients. Maloteaux et al.[81] measured 5HT-2 receptors in the cortical areas of PD patients. They found decrease in the number of 5HT-S2 receptors in the temporal cortex but not in the frontal cortex and hippocampus of PD patients. There was no correlation between the decrease in number of 5HT-2 receptors and the degree of dementia. They concluded that those receptors were not directly involved in the deterioration of cognitive functions. Waeber and Palacios[82] studied 5HT-1 receptor in the basal ganglia from PD patients by autoradiography using 8-hydroxy-2-(di-n-propylamino)-tetralin. The majority of 5HT-1 sites belong to the 5HT1D class. They found no significant alteration of the density and distribution of these sites in PD brains. In contrast, a marked decrease in the density of the receptor binding was seen in the basal ganglia and SN from patients dying with Huntington's disease. They concluded that their results suggested that 5HT-1D receptors were expressed by cells intrinsic to the striatum projecting to the SNR.

SUMMARY OF SEROTONERGIC MARKERS IN PARKINSON'S DISEASE

The results on serotonergic markers in PD are summarized in Table 38.3. There is a good agreement in the loss of serotonergic neurons in the raphe nucleus and loss of 5HT uptake sites in the cerebral cortex and the basal ganglia. These two findings are closely linked. The results on the 5HT transporter on postmortem specimens are consistent with results of *in vivo* studies on 5HT transporter using beta-CIT SPECT. The loss of serotonergic neurons in the raphe nucleus is reflected in the decreases in 5HT in CSF and in brain. There is only one report on the 5HT synthesizing enzyme, tryptophan hydroxylase; the activity was significantly reduced in the raphe nucleus.[69] Regarding 5HT receptors, the number of the reports is still limited, and there is a controversy in the results. Further studies are needed to make any definite conclusion.

TABLE 38.3
Serotonergic Markers in Parkinson's Disease

Authors	Results	References	Remarks
Serotonin in brain			
Bernheimer et al.	decreased	64	striatum, hypothalamus
Scatton et al.	decreased	65	frontal cortex
D'Amato et al.	decreased	66	frontal cortex
Shannak et al.	decreased	68	hypothalamus
Ohara et al.	decreased	67	frontal and putamen, DLBD
Tryptophan hydroxylase			
Sawada et al.	decreased	69	raphe nucleus
Serotonin transporter			
Kienzl et al.	decreased	70	frontal cortex
Cash et al	decreased	71	frontal and putamen
D'Amato et al.	decreased	66	frontal cortex
Chinaglia et al.	decreased	72	frontal and putamen
Brucke et al.	decreased	73	SPECT, hypothalamus–midbrain
Brucke et al.	decreased	73	SPECT, cortex
Haapaniemi et al.	decreased	74	SPECT, cortex and hypothalamus
Murai et al.	decreased	60	brain stem
Kim et al.	normal	75	midbrain
Serotonin in CSF			
Tohgi et al.	decreased	78	
Lacono et al.	decreased	79	ventricular CSF
5HIAA in CSF			
Davidson et al.	unchanged	76	
Tabaddor et al.	unchanged	77	ventricular CSF
Serotonin receptors			
Perry et al.	decreased	80	cortical S1 receptor, demented PD
Perry et al.	unchanged	80	cortical S2 receptor, demented PD
Maloteaux et al.	decreased	81	temporal S2
Waeber and Palacious	unchanged	82	putamen S1

Loss of 5HT in PD is likely related to some of nonmotor symptoms of PD, particularly, anxiety and depression. Beneficial effects of selective serotonin reuptake inhibitors on depression of PD also support this hypothesis.

NOREPINEPHRINE

NORADRENERGIC NEURONS IN THE NERVOUS SYSTEM

The major noradrenergic neurons of the brain are located in the locus coeruleus (LC). LC is a small nucleus located in the tegmentum of the upper pons. LC neurons send axons to widespread areas of the cerebral cortex, subcortical areas including the thalamus, to the cerebellum, and to the spinal cord.[83] They utilize norepinephrine (NE) as a neurotransmitter. Scattered noradrenergic neurons are found in the tegmentum of the pons and the medullar oblongata. LC is constantly involved in PD showing, neuronal loss, Lewy body formation, and gliosis. In the peripheral autonomic nervous system, post-ganglionic fibers are noradrenergic except for the post-ganglionic fibers to the sweat gland.

METABOLISM OF NOREPINEPHRINE

NE is synthesized from tyrosine through dopamine. The final step of NE biosynthesis is mediated by dopamine beta-hydroxylase (DBH) (Fig. 38.2). NE is released upon

excitation of the noradrenergic neurons, and the released NE is reuptaken into the noradrenergic terminals through high affinity NE transporters. Within neurons, NE is metabolized by MAOB, which is located in the outer membrane of mitochondria, and catechole-O-methyltransferase (COMT). The major final product is 3-methoxy-4-hydroxy-phenylethylglycol (MHPG) that can be measured in CSF (Fig. 38.3).

CENTRAL NORADRENERGIC SYSTEM IN PARKINSON'S DISEASE

Mann and Yates[4] examined eight PD brains. They found LC neuronal loss in all the patients, and those changes were more severe in three of eight patients who had experienced mental impairment. Gasper and Gray[7] examined 22 brains of PD patients. LC was constantly involved and found that severe lesions of the LC were more frequent in demented PD patients.

Chan-Paly and Assan[84] examined LC NE neurons by immunocytochemistry against tyrosine hydroxylase (TH) in PD in comparison with AD. They carried out quantitations of neuronal parameters and cell numbers and three-dimensional reconstructions of the LC by a computer-assisted system. In PD patients, the rostrocaudal length (12.4 ± 1.5 mm) was shorter than in AD and controls. The neuronal morphology was more severely altered than in AD. Most cell bodies were swollen with Lewy bodies frequently, and the dendrites were short and thin, with absent

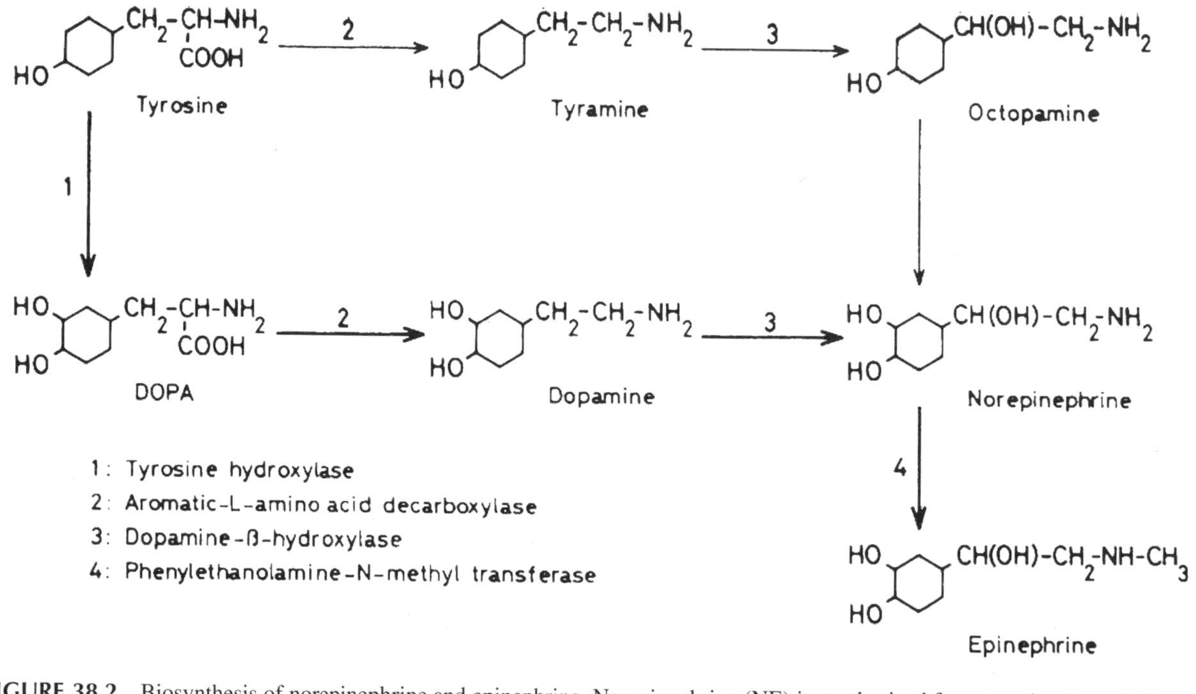

FIGURE 38.2 Biosynthesis of norepinephrine and epinephrine. Norepinephrine (NE) is synthesized from tyrosine through dopamine. The initial stem is 4-hydroxylation of tyrosine by tyrosine hydroxylase followed by decarboxylation of L-dopa by aromatic L-amino acid decarboxylase. Then dopamine is hydroxylated at the beta position by dopamine beta-hydroxylase. In epinephrine neurons, epinephrine is formed from NE by phenylethanolamine N-methyltransferase.

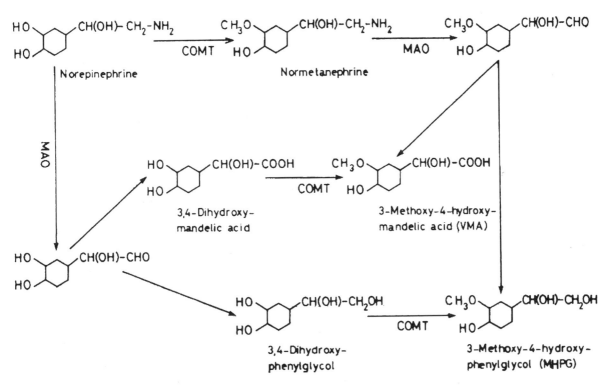

FIGURE 38.3   Metabolism of norepinephrine. Norepinephrine is metabolized by monoamine oxidase B and catechol-*O*-methyltransferase. The final metabolite is 3-methoxy-4-hydroxy-phenylethylglycol (MHPG). In the peripheral tissue, the main final product is vanylmandelic acid (VMA).

or reduced arborizations. Neuron numbers were more reduced in PD than in AD (between –26.4 and –94.4%).

Gasper et al.[85] examined cortical noradrenaergic terminals using DBH immunoreactivity in the primary motor, premotor, and prefrontal cortical regions in six PD and seven control subjects. Reductions of noradrenergic cortical innervations were observed, with similar magnitudes of reduction found in the motor and prefrontal regions of the cortex. Depletion of noradrenergic innervation was diffuse, involving all cortical laminae. Malessa et al.[86] studied adrenergic neurons in the medulla oblongata in three PD patients. TH-immunoreactive cells in the A1 and the A2 groups were not impaired in PD, in contrast to marked decrease in the immunoreactivity in MSA.

By reviewing the literature, there is a good agreement in the marked involvement of LC in PD patients. Changes in LC were more prominent in demented PD patients. Thus, LC may be responsible for some of the symptoms of dementia in PD patients. It is interesting to note that LC is also involved in AD to a lesser extent compared with PD. There is no evidence to indicate that noradrenergic neurons other than LC are involved in PD.

PERIPHERAL NORADRENERGIC SYSTEMS IN PARKINSON'S DISEASE

Post-ganglionic sympathetic fibers to the heart (noradrenergic) shows severe neurodegeneration, but post-gangli-

onic sympathetic fibers to the sweat glands (cholinergic) are spared.[87] Yoshita[88] studied the post-ganglionic cardiac sympathetic nerve terminals using [123I]metaiodobenzylguanidine (MIBG), a pharmacologically inactive analogue of noradrenalin, as a tracer for SPECT in 25 PD patients. A decrease in myocardial accumulation of MIBG was observed in the early stage of PD. They concluded that the measurement of MIBG was helpful in the diagnosis of early PD. Iwasa et al.,[89] Braune et al.,[90] and Orimo et al.[91] made essentially the same observation in PD patients. This is a useful test for the differential diagnosis of parkinsonism, as it is not decreased in PSP or MSA.[88,92] Orimo et al.[93] reported histopathological difference in the post-ganglionic cardiac sympathetic fibers between PD and MSA; in the latter, the cardiac sympathetic fibers are uninvolved. The reduction in cardiac MIBG uptake correlates well with Lewy body pathology in the brain, i.e., PD, DLB, and DLBD.[92]

Goldstein et al.[94] studied cardiac sympathetic innervation by visualizing cardiac sympathetic nerve terminals using 6-[(18)F]fluorodopamine as a tracer for PET. They studied 18 PD patients with and 23 PD patients without orthostatic hypotension. All the PD patients with orthostatic hypotension showed septal and lateral ventricular myocardial concentrations of 6-[(18)F]fluorodopamine-derived radioactivity >2 SD below the normal mean; 11 out of 23 PD patients without orthostatic hypotension also showed diffuse loss of 6-[(18)F]fluorodopamine-derived radioactivity.

In the literature, there is a good agreement in the involvement of post-ganglionic sympathetic fibers in PD. Degeneration of the cardiac sympathetic nerve fibers can be visualized by the absence of MIBG uptake using SPECT. Loss of MIBG uptake appears to be specific for Lewy body diseases (PD, DLBD, and pure autonomic failure with Lewy bodies). Usually, MIBG uptake is retained in MSA, PSP, and AD. Thus, this is a useful test for the differential diagnosis of parkinsonism and dementia. Examples of MIBG SPECT are shown in Fig. 38.4.

NORADRENERGIC MARKERS IN PARKINSON'S DISEASE

Norepinephrine in the Brain

NE is decreased in various parts of the brain from PD patients. Ehringer and Hornykiewicz[95] first reported loss of dopamine in the brain from PD patients. They also noted loss of NE in various parts of the brain from PD patients. Riederer et al.[96] also found decrease in NE in various brain areas from PD patients. Nagatsu et al.[97] measured the activity of DBH in various brain regions of PD patients. They found significant reduction in the DBH activity in hypothalamus of PD patients. Scatton et al.[65] reported decrease in NE in frontal cortex. Kish et al.[98] reported decrease in NE in the cerebellar cortex from PD patients.

Cash et al.[99] measured NE, MHPG and homovanillic acid (HVA) levels in LC in demented and nondemented PD patients. They were decreased in seven demented patients but within normal ranges in eight nondemented PD patients. Shannak et al.[68] reported decrease in hypothalamic NE (53% reduction compared with the controls) from PD patients. Ohara et al.[67] measured NE in the striatum and the cerebral cortex of five DLBD patients. NE was reduced in both regions compared with the controls.

Noepinephrine and Its Metabolite in CSF

There are many studies on the CSF NE, MHPG, and DBH. Davidson et al.[76] measured CSF MHPG in 54 untreated PD patients. They found no significant difference com-

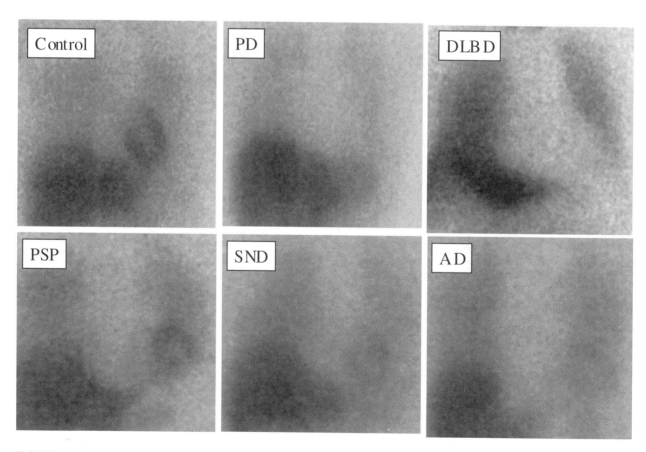

FIGURE 38.4 Cardiac MIBG SPECT. PD: Parkinson's disease, DLBD: diffuse Lewy body disease, PSP: progressive supranuclear palsy, SND: striatonigral degeneration, AD: Alzheimer's disease. In the control, clear cardiac MIBG uptake is seen indicating the presence of intact cardiac sympathetic terminals. In PD and DLBD, marked reduction in MIBG uptake is seen indicating the loss of cardiac sympathetic terminals. In PSP, SND, and in AD, cardiac images are clearly seen indicating the preservation of cardiac sympathetic fibers in these disorders. This test is very useful to differentiate PD and DLBD from PSP, MSA, and AD.

pared with the controls. Mitsui et al.[100] reported significantly decreased DBH activity in CSF of PD patients. Hurst et al.[101] reported 41% decrease in CSF dopamine beta-hydroxylase (DBH) activity in PD patients compared with the controls. Martignoni et al.[102] measured CSF NE and MEPH in 29 PD patients. They found significant decrease in NE in PD patients, however, MHPG levels were not unchanged compared with the controls. Tohgi et al.[103] examined CSF norepinephrine in PD and correlated with the severity of the disease. CSF norepinephrine was reduced in PD patients and the decrease correlated positively with the Hoehn and Yahr stage.

Adrenergic Receptors

Cash et al.[104] measured the amount of alpha 1, alpha 2, beta 1, and beta 2 adrenergic receptors in the prefrontal cortex of parkinsonian patients postmortem. Alpha 1 and beta 1

receptors were increased in number, particularly in demented parkinsonian patients, while alpha 2 receptors decreased. The affinity constants were unchanged. It is likely that increases in alpha 1 and beta 1 receptors represent denervation supersensitivity, and decease in alpha 2 receptor suggests that alpha 2 receptors are located mainly in the presynaptic terminals of noradrenergic fibers from LC.

Summary of Noradrenergic Systems in Parkinson's Disease

Results on noradrenergic markers are summarized in Table 38.4. In the literature, there is a good agreement in the involvement of LC noradrenergic neurons in PD. The involvement appears to be more prominent in demented PD patients. Loss of CSF NE and DBH appears to be good markers of the involvement of LC; however, MHPG level is not a good marker.

TABLE 38.4
Noradrenergic Markers in Parkinson's Disease

Authors	Results	References	Remarks
Norepinephrine in brain			
Ehringer and Hornykiewicz	decreased	95	striatum and cortex
Riederer et al.	decreased	96	striatum and cortex
Scatton et al.	decreased	65	frontal cortex
Kish et al.	decreased	98	cerebellar cortex
Cash et al.	decreased	99	locus coeruleus
Shannak et al.	decreased	68	hypothalamus
Ohara et al.	decreased	67	striatum and cortex, DLBD
Dopamine beta–hydroxylase (DBH)			
Nagatsu et al.	decreased	97	hypothalamus
Norepinephrine, MHP, and DBH in CSF			
Davidson et al.	unchanged	76	MHPG
Mitsui et al.	decreased	100	DBH
Hurst et al.	decreased	101	DBH
Martignoni et al.	decreased	102	NE
Martignoni et al.	unchanged	102	MHPG
Tohgi et al.	decreased	103	NE
Adrenergic receptors			
Cash et al.	increased	104	alpha–1, beta–1
Cash et al.	decreased	104	alpha–2

EPINEPHRINE

Epinephrine (EP) is synthesized from NE by phenylethanolamine N-methyltransferase (PNMT) (Figure 38.2); EP is metabolized by MAO and COMT to MHPG (Figure 38.5). EP is a neurotransmitter in limited numbers of neurons in the medulla oblongata (C1, C2, and C3).[105] Not many studies have been done on EP in PD. Nagatsu et al.[97] first measured the activity of PNMT in PD patients. They detected PNMT activity in hypothalamus, thalamus, and cerebellar nucleus of the control human brain, and the

PNMT activity was reduced in hypothalamus of PD patients.

Halliday et al.[61] studied brain stem nuclei in PD patients and found that the number of adrenaline-synthesizing and neuropeptide Y-containing neurons in the rostral ventrolateral medulla was reduced in PD patients., Gai et al.[106] studied C1, C2, and C3 adrenergic neurons in the medulla oblongata using antibody against PNMT and immunohistochemistry in 7 PD patients. The number of immuno-positive neurons was reduced by 47% in C1 and by 12% in C3 groups. They observed Lewy bodies in neu-

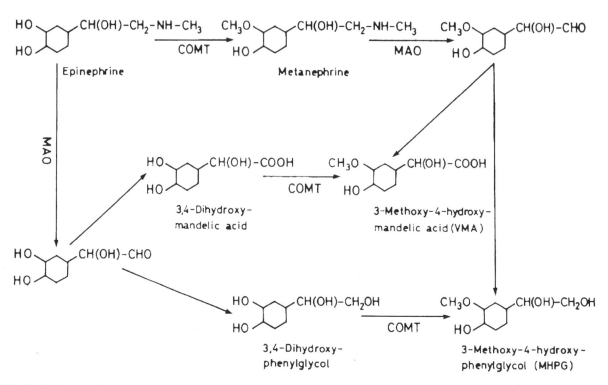

FIGURE 38.5 Metabolism of epinephrine. Epinephrine is metabolized by monoamine oxidase B and catechol-*O*-methyltransferase. The end products are essentially same as those of norepinephrine, i.e., MHPG and VMA.

rons positive for PNMT. C2 group neurons were not reduced. C1 is located in the ventrolateral part of the medulla oblongata, C2 is located in the region of the nucleus of the solitary tract, and C3 is located in the dorsal midline region near the hypoglossal nucleus.[105]

Tohgi et al.[103] examined CSF epinephrine and found that epinephrine was increased in demented PD patients compared with nondemented PD patients. Stoica and Enulescu[107] measured urinary EP and NE in PD patients. EP excretion was increased in PD patients compared with the controls. While NE excretion was unchanged. The major source of epinephrine in the urine is the adrenal medulla. Stoddard et al.[108] reported that the amount of total adrenal catecholamines were decreased in PD patients.

Thus, by reviewing the literature, we find that EP neurons of the brain are also involved in PD. The functions of these EP neurons have yet to be studied.

HISTAMINE

Histamine is synthesized from histidine by histidine decarboxylase. The cell bodies of histaminergic neurons are located in the hypothalamus and projecting to the widespread cortical, subcortical, and brain stem areas,[109] The central histaminergic system is one of the subcortical aminergic projection systems interacting extensively with the dopaminergic systems.[110]

Anichtchik et al.[110] examined the distribution of histaminergic fibers in SN from PD patients with a specific

immunohistochemical method. They found increase in the density of histaminergic fibers in the middle portion of SNC and SNR in PD brains. In PD the morphology of histaminergic fibers were thinner than in controls and had enlarged varicosities. Rinne et al.[111] measured histamine content in PD brains. They found significant increase in histamine in the putamen (to 159% of the control mean), SNC (to 201%), GPI (to 234%), and GPE (to 200%). They concluded that their finding might have implications in developing new drug therapies for PD.

Activities of histaminergic neurons appear to be increased. How such increase relates to symptoms of PD is yet to be studied. It is interesting to note that antihistamine drugs were used in the treatment of PD before the introduction of trihexyphenidyl and L-dopa. It has been claimed that the effects of antihistamines for PD may be mediated by their anticholinergic properties. But this proposal should be reinvestigated in the presence of new findings on histaminergic systems in PD.

GAMMA-AMINO BUTYRIC ACID

GABAᴇʀɢɪᴄ Nᴇᴜʀᴏɴs ɪɴ ᴛʜᴇ Nᴇʀᴠᴏᴜs Sʏsᴛᴇᴍ

Gamma-amino butyric acid (GABA) is a neutral amino acid that does not become a component of proteins. GABA is the major inhibitory neurotransmitter in the brain. GABAergic neurons are ubiquitously present in the brain, including the Purkinje cells in the cerebellum. The highest concentration of GABA is found in GPI and SNR. These areas receive GABAergic innervation from the striatum (putamen and caudate nucleus). Medium-sized spiny neurons are the major projecting neurons in the striatum. Substance P (SP) and dynorphin (DYN) co-localize with these striopallidal and strionigral pathways.[112,113] This strio-internal pallidal pathway has been called the direct pathway. Another set of medium sized spiny neurons in the striatum contains GABA and methionine-enkephalin (Met-Enk) as neurotransmitters and projects to the external segment of GP (GPE) making inhibitory synapses with GABAergic neurons. GABAergic neurons in GPE project to the subthalamic nucleus (STN) making inhibitory synapses. Glutamatergic neurons in STN project to GPI and making excitatory synapses with GABAergic neurons there. This striato-internal pallidal circuit through GPE and STN has been called the indirect pathway[112] (Figure 38.6). GABAergic neurons in GPI make inhibitory synapses with thalamic neurons in the ventrolateral nucleus and the anterior ventral nucleus.

Alexander and Crutcher[112] proposed that these parallel pathways between the striatum and GPI are very important regulating voluntary movements and the muscle tone. Also, pathologic states of these parallel pathways are considered to be responsible for the pathogenesis of bradykinesia, rigidity, and abnormal involuntary movements such as chore, dystonia, and ballisums.[112,114]

Nigrostriatal dopaminergic neurons make synapses with both of the GABAergic projecting neurons in the striatum (direct and indirect pathways). Dopamine is believed to make excitatory synapses with GABA-SP neurons and inhibitory synapses with GABA-Enk neurons.[112] Thus, in PD, the direct pathway is thought to be hypoactive and the direct pathway is thought to be hyperactive.

There is no evidence to indicate that striatal GABAergic neurons are morphologically abnormal in PD.

METABOLISM OF GABA

GABA is synthesized from glutamic acid by glutamic acid decarboxylase (GAD). Upon excitation of the GABAergic neurons, GABA is released into the synaptic space and binds to GABA receptors. The action of GABA as a transmitter is ceased by reuptake into the previous neurons or to glia cells through high affinity GABA transporters.[115] Intracellularly, GABA is metabolized by GABA transaminase that is an intracellular enzyme.

GABAERGIC MARKERS IN PARKINSON'S DISEASE

GABA Content in the Brain

Perry et al.[116] measured the content of GABA and the activities of GAD in whole putamen from 13 PD patients.

Mean GABA content was significantly elevated (by 28%) in the putamen of the PD patients. GAD activity was unchanged in PD. They interpreted their results as indicating increased activities of GABA-enkephalin neurons due to loss of dopamine. Kish et al.[117] also found increase in GABA level that was most prominent in the caudal subdivision of the putamen; this striatal subdivision also showed the most severe dopamine loss. They found, in the caudal putamen, a significant negative correlation between the (elevated) GABA and (reduced) dopamine levels. Rinne et al.[118] measured GABA content in the caudate nucleus and temporal cortex of demented and nondemented PD patients. They found no significant changes in GABA contents. Gerlach et al.[119] measured GABA content in the basal ganglia and the thalamocortical areas in PD patients and found decrease in GABA only in the centrum medianum nucleus of the thalamus compared with the controls.

SNR neurons are one of the major output neurons of the basal ganglia projecting to the thalamic neurons and neurons in the superior colliculus. They are GABAergic neurons. Hardman et al.[120] examined SNR neurons in PD using antibody against paralbumin that is expressed in SNR GABAergic neurons. There was a significant loss of paralbumin-immunoreactivity, though there was no evidence of actual cell loss.

Glutamic Acid Decarboxylase in Parkinson's Disease

Bernheimer and Hornykiewicz[121] first reported about 50% decrease in the GAD activity in putamen from PD patients. On the other hand, McGeer and McGeer[37] found normal or only slightly decreased GAD activity in striatum from PD patients. Lloyd and Hornykiewicz[122] found decrease in striatal GAD activity in PD patients who had not been treated with L-dopa and normal activity in patients who had been treated with L-dopa. Rinne et al.[123] also reported essentially the same results on GAD. Javoy-Agid et al.[38] measured GAD activity in SNC and SNR in PD patients. GAD activity was high and greater laterally and in the middle of the rostro-caudal extent in the controls. GAD activity was reduced to a uniformly low distribution in PD. Monfort et al.[124] found no difference in striatal and cortical GAD activity when 10 control and 9 parkinsonian brains were selected for an optimal premortem state. On the other hand, Nishino et al.[40] found significantly reduced GAD activities in the caudate nucleus and SN compared to normal controls, but these were normal when the values from protracted terminal illness cases were used as the controls.

Levy et al.[125] measured GAD67 mRNA by quantitative in situ hybridization in PD patients who had been treated with L-dopa. GAD67 mRNA expression was significantly decreased in all GABAergic neurons, in the

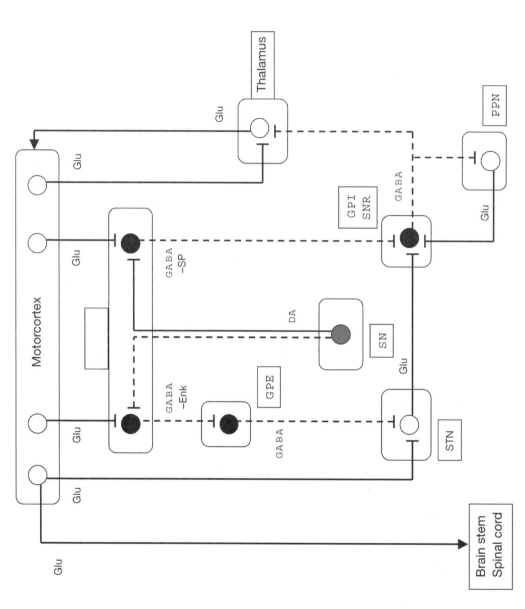

FIGURE 38.6 Parallel pathways between the striatum and GPI, adapted from Alexander and Crutcher, 1900.[103] Nigrostriatal dopaminergic neurons make excitatory synapses with GABA-SP neurons in the striatum and inhibitory synapses with GABA-Enk neurons in the striatum. The GABA-SP neurons make inhibitory synapses with GABAergic neurons in GPI. This pathway is called the direct pathway. The GABA-Enk neurons make inhibitory synapses with GABAergic neurons in GPE, and GPE neurons make inhibitory synapses with glutamatergic neurons in STN. The STN neurons make excitatory synapses with GABAergic neurons in GPI. This multisynaptic pathway is called the indirect pathway. Loss of striatal dopamine induces hypoactivity of GABA-SP neurons and hyperactivity of GABA-Enk neurons that results in hyperactivity of glutamatergic neurons of STN. The net results are marked hyperactivity of GABAergic neurons in GPI going to ventrolateral and anterior ventral nuclei of the thalamus. This marked inhibition of thalamic neurons going to motor areas is thought to be the pathophysiologic mechanism of bradykinesia in PD. (Adapted from Alexander and Crutcher, 1990.[112])

caudate nucleus (by 44%), putamen (by 43.5%), and ventral striatum (by 26%). In MPTP-treated monkeys, the expression of GAD67 mRNA was increased, and the increase was reversed by L-dopa treatment. Herrero et al.[126] measured GAD67 mRNA by quantitative *in situ* hybridization in GPI of PD patients who had been treated with L-dopa and in monkeys rendered parkinsonian by 1-methyl-4-phenyl-1,2,3,6-tetrahydropyridine (MPTP). In MPTP-treated monkeys, the expression of GAD67 mRNA was increased in cells from GPI, and this effect was abolished by L-dopa treatment. There were no differences in the levels of GAD67 mRNA between patients with PD, who were all treated with L-dopa, and control subjects. They interpreted their results as indicating that the level of GAD67 mRNA was increased in the cells of GPI after nigrostriatal dopaminergic denervation and that this increase could be reversed by L-dopa therapy. GABAergic neurons in GPI are believed to be overacting in untreated PD patients and the increase in GAD mRNA may be the result of this increased activity. Vila et al.[127] also found no difference in the expression of GAD67 mRNA in SNR of PD patients, including the basal ganglia. by *in situ* hybridization. They concluded that SNR GABAergic neurons were essentially unchanged in PD.

CSF GABA in Parkinson's Disease

There are many studies on CSF GABA content in PD. Manyam[128] measured CSF GABA in PD patients. The mean (±SD) CSF GABA levels were 200 ± 70 pmole/ml in controls and 121 ± 52 pmole/ml in PD patients. In the untreated PD patients, the CSF GABA level was 95 ± 31 pmole/ml ($n = 7$) and in those who were treated with L-dopa and carbidopa the level was 144 ± 53 pmole/ml ($n = 8$). Araki et al.[129] also found significant decrease in GABA levels in 14 PD patients. They found a positive correlation between the decreased GABA levels and severity of parkinsonism. Tohgi et al.[130] also found significant decrease in PD patients compared with controls. On the other hand, Jimenez-Jimenez et al.[131] reported higher than the control CSF GABA levels in 31 PD patients. In contrast, Bonnet et al.[132] reported normal CSF GABA level in PD. Perschak et al.[133] measured GABA level in the ventricular CSF from PD patients and the GABA levels were essentially unchanged in PD patients compared with the control subjects (cerebellar tremor or pain syndrome). Also, the ventricular GABA level was similar to that of lumber CSF levels reported in the literature.

GABA Receptors in Parkinson's Disease

Lloyd et al.[134] measured GABA binding using [3H]GABA and radioreceptor binding assay in various regions including the basal ganglia and cortical areas from PD patients. They found marked reduction in the [3H]GABA binding in SN but not in other brain areas examined. They concluded that GABA receptors were present in the nigral dopaminergic cell bodies. Nishino et al.[40] measured GABAA receptors using [3H]muscimol as a ligand for radioreceptor assay. GABAA receptor densities were significantly decreased in both the cortical and subcortical brain regions. They interpreted their results as indicating the loss of ascending monoaminergic neurons including nigral dopaminergic neurons. GABAA receptors are believed to be located at least in part in the presynaptic terminals of these monoaminergic neurons. On the other hand, Lloyd et al.[135] reported unchanged cortical GABA receptors in PD patients using [35S]TBPS (t-butylbicyclophosphorothionate) as a ligand for binding assay.

Calon and Di Paolo[136] correlated GABA receptor densities and motor fluctuations such as dyskinesias and wearing-off phenomenon. They found increased pre-proenkephalin expression in the putamen and increased GABA(A) receptors content in GPI in dyskinetic parkinsonian patients compared to nondyskinetic patients. They concluded that increased enkephalinergic activity in the putamen and increased sensitivity of GABA(A) receptors in the GPI were implicated in the pathogenesis of L-dopa-induced dyskinesias in PD. Calon et al.[137] studied striatal GABA(A) and GABA(B) receptor using (35)S-labeled t-butylbicyclophosphorothionate ([(35)S]TBPS) and [(3)H]flunitrazepam as ligands for GABA(A) receptors and [(125)I]CGP 64213 as a ligand for GABA(B) receptors in 14 PD patients, 10 of whom had developed motor fluctuations while receiving dopaminergic therapy. GABA(A) receptors were increased in the putamen of patients with wearing-off compared to those without. GABA(B) receptors were decreased in the putamen and GPE in PD patients.

SUMMARY OF GABAERGIC MARKERS IN PARKINSON'S DISEASE

Results on GABAergic markers in PD are summarized in Table 38.5. There is no morphological evidence to indicate neurodegenerative changes in the GABAergic neurons in the striatum and globus pallidus in PD, although the functional states of those neurons may be altered. GABA contents in putamen were reported as either unchanged or increased. GABA tends to increase postmortem.[138]

Regarding GAD activity, there appears to be an agreement indicating that striatal GAD is unchanged in PD patients. However, earlier studies reported decreased striatal GAD in L-dopa untreated PD patients. The decrease is likely secondary to dopaminergic degeneration. Therefore, there is no clear evidence to indicate involvement of GABAergic neurons in the striatum, GPE, and GPI in PD from the biochemical stand point of view.

Regarding CSF GABA, again there is controversy in the results. CSF GABA levels were reported as either decreased or unchanged.

TABLE 38.5
GABAergic Markers in Parkinson's Disease

Authors	Results	References	Remarks
GABA in brain			
Perry et al.	increased	116	putamen
Kish et al.	increased	117	putamen
Rinne et al.	unchanged	118	caudate and temporal cortex
Gerlach et al.	unchanged	119	putamen
Hardman et al.	unchanged	120	SNR, paralbumin (+) cell count
Glutamic acid decarboxylase (GAD)			
Bernheimer and Hornykiewicz	decreased	121	putamen
McGeer and McGeer	unchanged	37	putamen
Lloyd and Hornykiewicz	decreased	122	putamen, untreated PD
Lloyd and Hornykiewicz	unchanged	122	putamen, L–dopa treated PD
Rinne et al.	decreased	123	putamen, untreated PD
Rinne et al.	unchanged	123	putamen, L–dopa treated PD
Javoy–Agid et al.	decreased	38	SNC, SNR
Perry et al.	unchanged	116	putamen
Monfort et al.	unchanged	124	putamen and cortex
Nishino et al.	decreased	40	caudate and SN
Levy et al.	decreased	125	putamen and caudate, mRNA
Herrero et al.	unchanged	126	GPT, mRNA
GABA in CSF			
Manyam	decreased	128	GABA
Araki et al.	decreased	129	GABA
Tohgi et al.	decreased	130	GABA
Jimenez–Jimenez et al.	increased	131	GABA
Bonnet et al.	unchanged	132	GABA
Perschak et al.	unchanged	133	GABA
GABA receptors			
Iloyd et al.	unchanged	135	SN
Nishino et al.	decreased	40	putamen and cortex, GABAA
Lloyd et al.	unchanged	135	frontal cortex
Calon and DiPaolo	increased	136	GPI, GABAA, dyskinetic PD
Calon et al.	unchanged	137	putamen, GPE, GABAA
Calon et al.	decreased	137	putamen, GPE, GABAB

Overall evidence seems to indicating the absence of significant degenerative changes in GABAergic neurons in the striatum, GPE, and GPI. More information is needed to answer the question whether GABAergic neurons in SNR are involved in PD.

Regarding GABA receptors, once again there is controversy in the results. GABA(A) receptors are expressed in neurons in the striatum, GPE, and GPI as well as in cortical neurons. They are expressed in the putaminal GABA neurons as autoreceptors as well. They are also expressed in SNR neurons and in some of the SNC neurons.[139] Further studies are needed to make any definite conclusion about GABA receptors in PD.

GLUTAMIC ACID

Glutamic acid is the major excitatory neurotransmitter of the brain. In the basal ganglia, neurons in the subthalamic nucleus are glutamataergic, and hyperactive state of this glutamatergic input to GPI is considered to be a cause of bradykinesia in PD.[112] Also, excitatory input from the cerebral cortex to the basal ganglia are mostly glutamatergic.

GLUTAMATERGIC MARKERS IN PARKINSON'S DISEASE

Kish et al.[117] measured glutamate concentration in postmortem specimens of nine PD patients. They found elevated levels of striatal glutamate in three of the nine patients with PD. Rinne et al.[118] measured glutamate content in the caudate nucleus and temporal cortex of demented and nondemented PD patients. They found no significant changes in glutamate contents. On the other hand, aspartate content in the temporal cortex was increased in nondemented PD patients. Gerlach et al.[119] found no change in glutamate and aspartate in the basal ganglia and the thalamocortical areas in PD.

Glutamate + glutamine level can be estimated by magnetic resonance spectroscopy. Clarke et al.[140] found no difference in the striatum between PD patients and the controls by this method. Taylor-Robinson et al.[141] also found normal glutamate + glutamine content using magnetic resonance spectroscopy in dyskinetic and nondyskinetic PD patients. They found no evidence of increased striatal glutamate in either dyskinetic or nondyskinetic Parkinson's disease.

Tohgi et al.[130] reported reduction in CSF glutamate and aspartate in PD patients. Mally et al.[142] also reported a decrease in CSF glutamate in PD. On the other hand, Jimenez-Jimenez et al.[131] found normal CSF glutamate and aspartate levels in 31 PD patients. Perschak et al.[133] measured glutamate level in the ventricular CSF from PD patients, and the glutamate levels were essentially unchanged in PD patients compared with the control subjects (cerebellar tremor or pain syndrome). Also, the ventricular glutamate level was similar to that of lumber CSF levels reported in the literature.

GLUTAMATE RECEPTORS IN PARKINSON'S DISEASE

Difazio et al.[143] measured four subtypes of glutamate binding sites autoradiographically in PD midbrains. N-Methyl-D-aspartate (NMDA) binding sites were very low in control and were reduced in SNC from PD patients ($p < 0.02$). The alpha-amino-3-hydroxy-5-methylisoxazole-4-propi-onic acid (AMPA) binding sites were reduced in PD SNC. Nonkainate, nonquisqualate (NNKQ) sites were also reduced in total nigra. Metabotropic sites were unchanged. It is likely that AMPA and NNKQ binding sites are located on dopamine neurons. Ulsa et al.[144] measured glutamate receptors (NMDA, kainate, and AMPA receptors) using *in vitro* quantitative autoradiography in PD patients. PD patients showed substantially increased binding to NMDA receptors in the striatum and nucleus accumbens. But no statistically significant changes in binding to kinate and AMPA receptors were found. They concluded that the majority of striatal excitatory amino acid receptors were not located on dopaminergic nigrostriatal nerve terminals. Gerlach et al.[119] measured NMDA glutamate receptors using [3H]MK-801 as a ligand in the basal ganglia and the thalamocortical areas in PD patients and found that NMDA receptors were reduced in the head (–42%) and body (–38%) of the caudate nucleus.

SUMMARY OF GLUTAMATERGIC MARKERS IN PARKINSON'S DISEASE

Glutamatergic markers in PD are summarized in Table 38.6. There is a good agreement about absence of increase or decrease in glutamate contents in the striatum. There is a controversy in the CSF glutamate level. CSF glutamate may not be a good marker of the central glutamatergic neurons.

TABLE 38.6
Glutamatergic Markers in Parkinson's Disease

Authors	Results	References	Remarks
Glutamate content in brain			
Kish et al.	increased	117	striatum, 3 out of 8 patients
Rinne et al.	unchanged	118	caudate, temporal
Gerlach et al.	unchanged	119	putamen
Clarke et al.	unchanged	140	striatum, NMS
Taylor-Robinson et al.	unchanged	141	striatum, MNS
Glutamate in CSF			
Tohgi et al.	decreased	131	
Mally et al.	decreased	142	
Jimenez-Jimenez et al.	unchanged	131	
Perschak et al.	unchanged	133	ventricular CSF
Glutamate receptors			
Difazio et al.	decreased	143	SNC, NMDA, AMPA, NNKQ
Difazio et al.	unchanged	143	SNC, metabotrophic
Ulsa et al.	increased	144	striatum, NMDA
Ulsa et al.	unchanged	144	striatum, AMPA, Kinate
Gerlach et al.	decreased	119	caudate

Glutamate receptors are decreased in SNC. This finding is probably due to the loss of dopaminergic neurons indicating that glutamate receptors are located on the nigral dopaminergic neurons. Striatal glutamatergic receptors are unchanged in PD. There is no report on glutamatergic markers in the subthalamic nucleus (STN) or GPI in PD.

Glutamate has been implicated as a pathogenetic factor of nigral neurodegeneration;[145] however, there is no evidence to indicate that glutamatergic neurons are hyperactive in PD except for the glutamatergic neurons of STN. An interesting observation in PD is that drugs acting to inhibit glutamatergic transmission at the NMDA receptor can ameliorate L-dopa-induced dyskinesias.[146]

GLYCINE

Glycine binds to the glycine receptor located in the NMDA receptor,[147] one of the glutamate receptors. This site is strichinin insensitive. Glycine can modulate behavioral profile through modulation of NMDA receptors.

Not many reports are available on glycine in PD. Rinne et al.[118] measured glycine content in the caudate nucleus and temporal cortex of demented and nondemented PD patients. They found no significant changes in glycine contents.

Tohgi et al.[130] measured CSF glycine in PD patients and found significant decrease compared with controls. On the other hand, Jimenez-Jimenez et al.[131] found normal CSF glycine levels in 31 PD patients. Perschak et al.[133] measured glycine level in the ventricular CSF from PD patients, and the glycine levels were essentially unchanged in PD patients compared with the control subjects (cerebellar tremor or pain syndrome). Also, the ventricular glycine level was similar to that of lumber CSF levels reported in the literature.

BENZODIAZEPINE RECEPTOR

Benzodiazepine (BDZ) receptors are located on the GABAA receptors and BDZ enhance GABAergic neurotransmission. Reciprocally GABA enhances BDZ binding. BDZ receptors are expressed in cerebral cortical areas as well as subcortical structures such as basal ganglia where GABAergic neurons exist.[148]

Only limited numbers of studies are available. Uhl et al.[149] measured BDZ receptors in the substantia nigra in PD patients by autoradiography. They found reduction in BDZ receptor I but normal BDZ receptor II in the SN. Maloteaux et al.[148] reported an increase in BDZ receptor binding using [3H]flunitrazepam as a ligand in the caudate nucleus of PD patients compared with the controls. It was normal in the putamen. Griffiths et al.[53] measured BDZ receptor binding using flunitrazepam as a ligand for autoradiography and found reduced binding in GPE from PD patients. They interpreted their result as suggesting that the GABAergic pathway from the putamen to GPE was overactive in PD. Griffiths et al.[54] later reported a loss of BDZ receptors in the mid and caudal portions of the putamen.

Bonuccelli[150] measured peripheral BDZ receptors using platelets as a sample in 10 de novo and 18 L-dopa-treated PD patients. The binding assay was conducted using [3H]PK 11195, a specific ligand for peripheral BDZ receptors. A significant decrease in the density of [3H]PK

11195 binding sites was observed in PD patients with respect to controls ($p < 0.01$), but not between de novo and treated PD patients. Peripheral BDZ receptors are located in a variety of tissues, including platelets, in the nuclear and/or mitochondrial membranes.

SUBSTANCE P

Substance P (SP) is a peptide composed of 11 amino acids. SP has the highest concentration in SN and GPI. SP is also rich in hypothalamus, striatum, amygdala, posterior horns of the spinal cord, and the nucleus of the spinal tract of the trigeminal nerve.[151,152] SP co-localizes with GABA in the strio-pallidal (the direct pathway) and the strionigral pathways.

SUBSTANCE P IN PARKINSON'S DISEASE

Substance P Content by Radioimmunoassay

Mauborgne et al.[153] first measured SP immunoreactivity in PD brains. They found a 40 to 50% decrease in SP in GPE, SNR, and SNC from PD patients. Tenovuo et al.[154] measured immunoreactive-SP (SPI) content in various regions of the brain, including SN and basal ganglia from 42 PD patients. They found a significant decrease in SPI in SN from PD patients. There was also a significant decrease in the putamen of the PD patients who had not received L-dopa. The levels of SPI in the other brain regions studied did not show any difference. Clevens and Beal[155] also found significant reduction of SP by 44% in SNC from PD patients. There were no significant changes in GP. Sivam[156] found decrease in SP content in the striatum only when more than 80% of putaminal dopamine loss was present. Frontal SP was unchanged. They concluded that loss of SP in PD was likely due to secondary to dopamine loss. Fernandez et al.[157] measured SP content in the caudate and anterior putamen from PD patients. SP levels were reduced in caudate in PD but unchanged in putamen. Later, Fernandez et al.[158] reported reduction in SP level in the putamen in PD.

de Ceballos et al.[159] measured SP levels in GP and reported two subgroups of PD. In patients with >80% decrease in caudate nucleus DA content, they found a threefold increase in SP level in GPI. In contrast, in patients showing an approximately 50% reduction in DA content in caudate, SP was markedly reduced (approximately 80%) in GPI. SP level in GPE was unchanged in PD.

Two reports studied SP mRNA expression in PD. Levy et al.[160] studied the striatal expression of SP gene by in situ hybridization in PD patients. They found no significant difference in striatal expression of SP mRNAs compared with control subjects. Met-Enk mRNA also was not changed. This contrasts with animal models of parkinsonism, where expression of SP mRNA is decreased and Met-Enk mRNA increased, and these observations became the theoretical background of the parallel pathway theory on the motor loop of the basal ganglia.[112] Levy et al.[160] discuss that the difference between human and animals might be due to the effect of L-dopa treatment or to a chronic compensation mechanism. Nisbet et al.[161] studied preprotachykinin messenger RNA expression in the striatum. They found no change compared with the controls.

Substance P Immunostaining

Grafe et al.[162] studied SP immunostaining in the basal ganglia and SN from PD patients. They found intense immunoreactivity for SP in GP and SN. Met-Enk and SP fibers were distributed in essentially the same pattern. Waters et al.[163] studied SP immunocytochemistry in SN from PD patients. They found marked loss of SP immunoreactivity in SN. Gai et al.[164] examined brain stem SP containing neurons in PD. They found significant reductions in the total number of SP+ neurons in the pedunculopontine tegmental nucleus (loss 43%), in the laterodorsal tegmental nucleus (loss 28%), in the oral pontine reticular nucleus (loss 41%), and in the median raphe nucleus (loss 76%). de Ceballos and Lopez-Lozano[165] examined biopsied caudate specimens from 15 PD using immunohistochemistry. SP immunostaining was either normal or decreased in these patients.

Substance P in CSF

Cramer et al.[166] measured CSF SP in senile PD patients. They did not find significant difference compared with the controls. Cramer et al.[167] also measured ventricular CSF SP in PD patients. They found a 30% decrease in the SP immunoreactivity compared with patients without extrapyramidal disease.

Substance P Receptors

Tenovuo et al.[168] studied SP receptor in PD patients. They found significantly reduced SP receptor bindings in the NBM and parietal cortex as compared with controls, but not in the striatum or GP. Rioux and Joyce[169] studied SP receptors in striatum and pallidum. They found a significant increase in the SP receptors in the striatum of PD patients. No significant differences were observed for the GPI and GPE. Fernadez et al.[170] studied the specific binding of [3H]SP, membrane preparations of caudate nucleus, putamen, GP, and SN from PD patients and controls. The density of SP receptors was reduced in the putamen (anterior and posterior) and in GPE in PD. No difference in SP receptor binding was observed in SN from PD brains. The reductions in SP receptor density were less marked than the decrease in caudate and putamen content of DA and

its metabolites. They concluded that SP receptors were only partially localized to striatal dopamine terminals.

SUMMARY OF SUBSTANCE P IN PARKINSON'S DISEASE

Changes in SP are summarized in Table 38.7. By reviewing the literature, there is an agreement in the decrease of SP in SN and striatum, except for a few studies reporting contradictory results. There is controversy in SP contents in GPI and GPE. CSF SP is also decreased. Two papers reported unchanged SP mRNA expression in the striatum. Decrease in SP peptide level in the presence of normal mRNA expression suggest increased turnover of SP in PD brains. Regarding the SP receptors, there is controversy and more studies are needed. Many investigators believe that changes in SP in PD are secondary changes due to loss of DA. But the discrepancy of the behaviors between SP and GABA is a question. SP may be more vulnerable in the presence of nigrostriatal dopaminergic degeneration. Nevertheless, loss of SP in PD may produce a functional disturbance of the basal ganglia neuronal network. It is interesting to study whether or not loss of SP is responsible for some of the symptoms of PD.

MET-ENKEPHALIN

Methionine-enkephalin (Met-Enk) is a peptide composed of five amino acids; it belongs to the opioid peptide group. It has the highest concentration in GPI, but it is also found in striatum, SN, hypothalamus, and other places in the brain.[171] Met-Enk co-localizes with striatal GABAergic neurons projecting to GPE, the most important pathway in the pathogenesis of symptoms of PD and drug induced dyskinesias.

MET-ENKEPHALIN IN PARKINSON'S DISEASE

Radioimmunoassay of Met-Enkephalin

Taquet et al.[172] measured nigral Met-Enk in PD patients by radioimmunoassay. They found highly significant decreases in Met-Enk levels in SN and ventral tegmental area of PD patients. The decrease correlated well with the decrease of dopamine. Llorens-Cortes et al.[173] also found significant reduction in Met5-Enk levels (–72%) in SNC of PD patients. In addition, they found marked reduction in "enkephalinase" activity (–39%). In contrast, these markers were not significantly modified in caudate nucleus. Sivam[156] found decrease in caudatal Met-Enk in only those patients with more than 80% of caudatal dopamine loss. Frontal Met-Enk was unchanged. They concluded that loss of Met-Enk in PD was likely due to secondary to dopamine loss. Fernandez et al.[157] did not find any difference in Met-Enk content in the caudate and anterior putamen from PD patients. However, they found a significant correlation between DA and Met-Enk levels in caudate nucleus from PD.

TABLE 38.7
Substance P in Parkinson's Disease

Authors	Results	References	Remarks
Radioimmunoassay			
Mauborgne et al.	decreased	153	SNC, SNR, GPE
Tenovuo et al.	decreased	154	SN, putamen
Clevens and Beal	decreased	155	SNC
Clevens and Beal	unchanged	155	GP
Sivam	decreased	156	putamen, loss of DA > 80%
Fernandez et al.	decreased	157	caudate
de Ceballos et al.	increased	159	GPI, > 80% DA loss*
de Ceballos et al.	decreased	159	GPI, 50% DA loss
de Ceballos et al.	unchanged	159	GPE, 50% DA loss
Fernandez et al.	decreased	158	putamen
Levy et al.	unchanged	160	mRNA, striatum
Nisbet et al	unchanged	161	mRNA, striatum
Immunohistochemistry			
Grafe et al.	unchanged	162	SN, GP
Waters et al.	decreased	163	SN
Gai et al.	decreased	164	PPN, Raphe
de Ceballos and Lopez-Lozano	unchanged or decreased	165	caudate
Substance P in CSF			
Cramer et al.	decreased	166	lumbar
Cramer et al.	decreased	167	ventricular
Substance P receptors			
Tenovuo et al.	decreased	168	NBM, parietal cortex
Rioux and Joyce	increased	169	putamen
Rioux and Joyce	unchanged	169	GPE, GPI
Fernandez et al.	decreased	170	putamen, GPE
Fernandez et al.	unchanged	170	caudate, SN, GPI

*Decreased only in patients with more than 80% dopamine loss in the caudate nucleus.

de Ceballos et al.[159] measured Met-Enk levels in GP and reported two subgroups of PD. In patients with >80% decrease in caudate nucleus DA content, they found a threefold increase in Met-Enk level in GPI. In contrast, in patients showing an approximately 50% reduction in DA content in caudate, Met-Enk was markedly reduced (approximately 80%) in GPI. Met-Enk level in GPE was unchanged in PD. Fernandez et al.[158] found reduction in the levels of Met-Enk in the caudate nucleus, putamen, and substantia nigra in PD. Leu-enkephalin levels were decreased in the putamen and were undetectable in the substantia nigra in PD. They also analyzed samples from incidental Lewy body disease. The changes in basal ganglia peptide levels in incidental Lewy body disease generally followed a trend similar to those seen in PD but were less marked. They concluded that their data would suggest that they were an integral part of the pathology of the illness and not secondary to DA neuronal loss or a consequence of prolonged drug therapy.

There are four reports on Met-Enk mRNA expression in PD. Levy et al.[160] studied the striatal expression of Met-Enk together with SP genes by *in situ* hybridization in PD patients. They found no significant difference in striatal expression of these two neuropeptide mRNAs compared with control subjects. Nisbet et al.[161] found a statistically significant increase in preproenkephalin messenger RNA expression in the body of the caudate (109% increase, $P < 0.05$) and in the intermediolateral putamen (55% increase, $P < 0.05$) due to an increase in the level of gene expression per neuron rather than an increase in the number of neurons expressing preproenkephalin messenger RNA. No change in preprotachykinin messenger RNA expression was detected. They concluded that their findings demonstrated selective up-regulation of a striatal neuropeptide system in PD compatible with increased activity of the "indirect" striatopallidal pathway, which was thought to play a crucial role in the pathophysiology of akinesia and rigidity in this condition.

Calon and Di Paolo[136] found increased preproenkephalin expression in the putamen and increased GABAA receptors content in GPI in dyskinetic parkinsonian patients compared to nondyskinetic patients. They concluded that increased enkephalinergic activity in the putamen and increased sensitivity of GABAA receptors in the GPi are implicated in the pathogenesis of LD-induced dyskinesias in Parkinson's disease. Calon et al.[174] also found a significant increase of preproenkephalin mRNA levels in the lateral putamen of dyskinetic patients in comparison to controls (+210%; $p < 0.01$) and in comparison to nondyskinetic patients (+112%; $p < 0.05$).

Immunohistochemistry of Met-Enkephalin

Grafe et al.[162] studied Met-Enk immunostaining in the basal ganglia and SN from PD patients. They found intense immunoreactivity for Met-Enk in GP. Waters et al.[163] studied Enk immunocytochemistry in SN from PD patients. They found loss of Enk immunoreactivity in SN. Goto et al.[175] studied Met-Enk expression by immunohistochemistry in PD patients. They found strongly expressed Met-Enk-positive patches and Met-Enk-positive axon terminals in GPE of PD patients. de Ceballos and Lopez-Lozano[165] examined biopsied caudate specimens from 15 PD using immunohistochemistry. Met-Enk immunostaining was either unchanged or reduced. Low Met-Enk immunostaining tended to correlate with the severity of the disease as judged by higher Unified Parkinson's disease Rating Scale and gait scores.

Met-Enkephalin in CSF

Yaksh et al.[176] measured CSF Met-Enk in eight advanced PD patients. They found significant reduction in PD patients compared with the controls (166 ± 38 vs. 264 ± 44 pg/ml). Baronti et al.[177] measured CSF content of a proenkephalin, Met5 enkephalin-Arg6-Gly7-Leu8 (MERGL), in PD patients. They found a significantly lower level compared with the controls.

Met-Enkephalin Receptors

Rinne et al.[178] studied enkephalin receptors in postmortem brain samples of 27 patients with PD and of 26 controls by the radioligand-binding technique using 3H-Leu-Enk and 3H-Met-Enk. They found significant increases in the specific binding of both 3H-Leu- and 3H-Met-Enk in the caudate nucleus, putamen, nucleus accumbens, limbic cortex and hippocampus. Scatchard analysis showed an increase in the receptor number, but no significant changes in the mean dissociation constant. They interpreted their results as suggesting that there would be a supersensitivity of a population of enkephalin receptors in the striatum and limbic system,

SUMMARY OF MET-ENKEPHALIN IN PARKINSON'S DISEASE

Regarding the protein level of Met-Enk, there is a controversy about the striatal protein level. In the substantia nigra, two studies reported significant decrease. In GPI, decrease in Met-Enk was reported only in those patients who showed more than 80% dopamine loss in the caudate.[159] Many investigators felt that the decrease in Met-Enk was secondary to dopamine loss in the striatum in PD. Two studies reported increase in Met-Enk mRNA in the striatum only in dyskinetic PD patients.

Results of immunohistochemical studies are also controversial. There is an agreement in the decrease in CSF Met-Enk levels in PD. One study reported an increase in Met-Enk receptors in the striatum and the limbic cortex, which was interpreted as indicating denervation supersensitivity. But, generally, enkephalinergic neurons are morphologically uninvolved in PD.

In view of the parallel theory on the basal ganglia neuronal network according to Alexander and Cruthcer,[112] GABA-Met-Enk neurons in the striatum are believed to be hyperactive in PD because of a lack of inhibitory effect of dopamine. L-dopa treatment will improve this hyperactive state toward normal activity. Most of the PD patients studied in these reports had been taking L-dopa, except for the final agonal stage in some patients. The controversy in the results of Met-Enk may in part be due to the difference in the amount of L-dopa-derived dopamine remaining in the brain at the time of the studies.

DYNORPHIN

Dynorphin (DYN) is one of the co-transmitters of the striatonigral and striatopallidal pathways. It co-localizes with GABA and SP in the striatal projection neurons to GPI (the direct pathway) and SNR.[113]

Not many studies have been done on dynorphin in PD. Waters et al.[163] studied DYN immunocytochemistry in SN from PD patients. They found a loss of DYN immunoreactivity in SN. Sivam[156] measured dynorphin A (1–8) content in the striatum in the frontal cortex from PD patients by radioimmunoassay. They found no change compared with controls in striatal and frontal dynorphin. Baronti et al.[177] measured CSF dynorphin A(1–8) content in PD patients. They found no difference from the controls.

ENDORPHINS

Endorphins are a group of opioid peptides derived from the common precursor, proopiomelanocortin. There are four endorphins and they are named as alph-, beta-, gamma-, and delta-endorphin. Beta-endorphin is the major endorphin. The highest concentration of endorphins is found in the pituitary gland, but they are distributed widely in the brain.[179] Not many studies have been done on endorphins in PD. Pique et al.[180] measured beta-endorphin-immunoreactivities in whole hypothalamic extracts of normal subjects and PD patients. They found no significant difference between normal subjects and PD patients. Nappi et al.[181] measured CSF beta-endorphin content in untreated and treated PD patients. CSF beta-endorphin concentrations in untreated patients were lower than in 15 controls ($p < 0.005$), but values did not differ significantly in treated and untreated patients. Jolkkonen et al.[182] measured CSF beta-endorphin-like immunoreactivity in 36 PD patients. They found no significant difference in the level compared with the controls.

OPIOID RECEPTORS

Opioid receptors were found to be high-affinity binding sites for an exogenous opiate alkaloid, morphine.[183] Later on, endogenous ligands such as endorphins and enkephalins were found for opioid receptors.[184] Not many studies have been published on opioid receptors in PD. Llorens-Cortes et al.[173] studied opioid receptor binding in PD. They found marked reductions in opiate receptor binding (−42%) in SNC of PD patients. In contrast, opioid receptor binding was not significantly modified in caudate nucleus. Uhl et al.[149] measured densities of binding to opiate receptors in SN from PD patients by autoradiography. They found substantial reductions in mu-opiate and kappa-opiate receptors in SNC from PD patients. The reductions were less intense than the reduction in dopamine. Delay-Goyet et al.[185] found no difference from controls in the mu and delta opioid receptors in the cerebral cortex, amygdale, striatum, and in SNC from PD patients.

Fernadez et al.[170] reported reduction in the mu- and delta-opioid receptors in posterior putamen and reduction in mu receptors in caudate and putamen from PD patients. No differences in opioid receptor binding were observed in substantia nigra from PD brains. The reductions in neuropeptide receptor density were less marked than the decrease in caudate and putamen content of dopamine and its metabolites. They concluded that opioid receptors were only partially localized to striatal dopamine terminals. Yamada et al.[186] reported marked reduction in kappa opioid receptor mRNA in SNC. The kappa opioid receptor mRNA was expressed in melanized neurons in SNC

Opioid receptors can be visualized using PET. Burn et al.[187] studied striatal opioid receptor binding using [11C]diprenorphine with PET in eight PD patients and eight normal controls. They found no significant difference between mean ligand binding in the putamen and caudate of PD patients as compared with normals. Piccini et al.[188] studied *in vivo* opioid receptor binding in PD patients with and without L-dopa-induced dyskinesias using [11C] diprenorphine as a tracer. They found significantly reduced striatal and thalamic opioid binding in dyskinetic, but not in nondyskinetic PD patients.

Results on opioid receptors in PD are summarized in Table 38.8. By reviewing the literature, we see that controversy still exists in the results on opioid receptor changes in PD. Opioid receptors seem to be at least in part localized in the nigral dopaminergic neurons. A decrease in opioid receptors in the striatum only in dyskinetic PD patients is an interesting finding.[188] But how to put this finding into the parallel loop theory proposed by Alexander and Crutcher[112] is a question to be answered by further studies.

TABLE 38.8
Opioid Receptors in Parkinson's Disease

Authors	Results	References	Remarks
Llorens-Cortes et al	decreased	173	SNC
Llorens-Cortes et al.	unchanged	173	caudate
Uhl et al.	decreased	149	SNC, mu, kappa
Delay-Goyet et al.	unchanged	185	SNC, striatum, cortex, mu, delta
Fernandez et al.	decreased	170	putamen (delta, mu), caudate (mu)
Yamada et al.	decreased	186	SNC, mRNA, kappa
Burn et al.	unchanged	187	striatum, PET
Piccini et al.	decreased	188	striatum, PET, dyskinetic PD
Piccini et al.	unchanged	188	striatum, PET, nondyskinetic PD

NEUROTENSIN

Neurotensin (NT) is a peptide composed of 13 amino acids. It is rich in hypothalamus; in addition, NT is expressed in the nucleus accumbens, amygdala, and cortical areas. Low amount to NT is expressed in the striatum.[189] Bissette et al.[190] measured NT immunoreactivity

in various parts of the brain, including the basal ganglia from PD patients by radioimmunoassay. NT immunoreactivity was significantly reduced in the hippocampus but was not significantly altered in other areas examined. They also measured bombesin and found a significant decrease in the caudate nucleus and GP. Fernandez et al.[191] reported twofold increase in NT content in both SNC and SNR from PD patients. NT levels were unaltered in the caudate nucleus, putamen GPI, and GPE.

Regarding NT receptors, Uhl et al.[192] studied NT receptor binding in SN of PD patients by autoradiography. They found dense binding in normal human SN. Nigral receptor binding in PD patients was only about one-third of control values. Sadoul et al.[193] also found significant reduction in the NT receptors in SN from 17 PD patients. Saturation analysis using pooled substantia nigra demonstrated an almost complete loss of the high affinity component of the neurotensin receptor complex, yielding a 24% loss of the total binding capacity, with no alteration of the low-affinity component. Chinaglia et al.[194] found significant decreases in NT receptor bindings in SNC, SNR, and in the putamen from PD patients. Fernandez et al.[170] also reported decrease in NT binding in caudate nucleus, putamen, GPI, GPE, and SN from PD patients and controls.

Yamada et al.[195] studied NT receptor mRNA in SN. NT receptor mRNA was present in high levels in melanized neurons of SNC and the nucleus paranigralis. They noted a very low level of NT mRNA in SNC neurons in PD. This decrease appeared to be due to marked degenerative changes of the SNC neurons in PD.

Changes in NT in PD are summarized in Table 38.9. By reviewing the literature, we see that there is a fair agreement in the reduction of NT receptors in SNC. These results indicate that NT receptors are located at least in part on the nigral dopaminergic neurons; NT contents are unchanged in basal ganglia of PD patients.

TABLE 38.9
Neurotensin in Parkinson's Disease

Authors	Results	Reference	Remarks
Radioimmunoassay			
Bissette et al.	unchanged	190	striatum
Fernandez, et al.	increased	191	SNC, SNR
Fernandez, et al.	unchanged	191	striatum, GPE, GPI
Neurotensin receptors			
Uhl et al.	decreased	194	SN
Sadoul et al.	decreased	193	SN
Chinaglia et al.	decreased	194	SN, putamen
Fernandez et al.	unchanged	170	SN
Fernandez et al.	decreased	170	GPI, GPE
Yamada et al.	decreased	195	SNC

CHOLECYSTOKININ-8 (CCK-8)

Cholecystokinin (CCK) is a peptide composed of eight amino acids. CCK-8 co-localizes with dopamine (DA) in neurons of the mesolimbic-frontocortical and nigrostriatal DA system. CCK-8 is widely distributed in cortical as well as subcortical areas.[196] Not many studies are available on CCK-8 in PD. Studer et al.[197] measured immunoassayable

CCK-8 in 12 regions of control and PD brains. They found decrease in CCK-8 in SN but not in striatal and corticolimbic dopamine projecting areas. They concluded that the major proportion of dopaminergic neurons degenerated in PD might not contain the CCK-8 peptide. Fernandez et al.[157] measured cholecystokinin-8-S (CCK-8-S) content in the caudate and anterior putamen from PD patients. They found no difference from the controls.

SOMATOSTATIN

Somatostatin (SS) is a peptide composed of 14 amino acids. SS is distributed richly in the hypothalamus followed by cortical areas, amygdala, and the accumbens. SS is also expressed in the basal ganglia including the striatum.[198]

SOMATOSTATIN IN PARKINSON'S DISEASE

Radioimmunoassay

Epelbaum et al.[199] measured SS immunoreactivity in the cortex, hippocampus, and caudate nucleus of PD patients by radioimmunoassay. SS levels in the frontal cortex were significantly reduced in PD patients who were slightly or severely demented as compared to controls and to nondemented PD patients. Allen et al.[200] also found significant reduction in SS levels in cortical regions and hippocampus from demented PD patients. Beal et al.[201] also reported significant decrease in the cortical SS in demented PD patients. Later, Beal et al.[202] studied SS in more detail. They measured SSI in 22 cortical regions from 13 demented PD patients. They found 50 to 60% decrease in SSI in 17 out of 22 regions in demented PD patients. On the other hand, Whitford et al.[203] found no significant changes in SS immunoreactivity in the cortical regions compared with the controls, regardless of whether the PD patients were demented. Leake et al.[204] also found normal SS in temporal and occipital cortices from PD patients.

Eve et al.[205] measured SS mRNA expression in the striatum, medial medullary lamina (MML) of GP, and reticular thalamic nucleus from nine PD patients. They found a significant increase (82%) in SS mRNA expression in the MML in PD (56.5 microm2 of silver grain/cell, $n = 9$ cases) compared to controls (26.3 microm2/cell, $n = 13$ cases, $p < 0.01$, Student's t-test).

Somatostatin in CSF

There are many studies on CSF SS in PD. Dupon et al.[206] measured CSF SS immunoreactivity in 39 PD patients, CSF SS content was 88.0 ± 4.1 pg/ml, which was about 40% less than in controls (147.3 ± 5.1 pg/ml). Jolkkonen et al.[207] also found significantly lower SS in PD patients. The levels were lowest in demented PD patients. Cramer et al.[166] found significantly lower SS levels in PD patients.

Masson et al.[208] also found significantly lower SS in PD patients than in controls (P less than 0.02). Strittmatter et al.[209] also measured SS in the CSF from 35 PD patients. They found significant reduction in SS in PD in comparison to the control group ($p < 0.05$). The reduction was related to the progression of the disease.

On the other hand, Volicer et al.[210] reported normal CSF SS levels in PD patients, and Espino et al.[211] found significant increase in SS in 15 PD patients (107.9 ± 9.8 pg/ml) compared with the controls (73.5 ± 8.4 pg/ml).

There are two reports on ventricular CSF SS levels. Cramer et al.[212] measured ventricular CSF SSI in PD patients. They found significant decrease in SS in PD patients (mean ± SEM); 42.9 ± 2.9 fmol/ml) compared to patients with benign essential tremor (65.3 ± 9.7). Jost et al.[213] measured ventricular CSF SSI in PD patients. They found significantly lower concentration of SSI when compared with levels in control patients.

Somatostatin Receptors

Uhl et al.[149] measured densities of binding to SS in the substantia nigra in PD patients by autoradiography. They found substantial reduction in SS receptors, which was more modest compared to the loss of DA. Whitford et al.[203] measured SSI and SS binding in the frontal cortex of PD patients. They found no significant changes compared with the controls, regardless of whether the patients were demented. Epelbaum et al.[214] measured SSI and SS binding sites in demented and nondemented PD patients. Cortical SSI was significantly lower in demented PD patients, but the SS binding sites were not altered in both nondemented and demented PD patients, compared with the controls.

SUMMARY OF SOMATOSTATIN IN PARKINSON'S DISEASE

Results regarding SS in PD are summarized in Table 38.10. There is a good agreement in the loss of cortical SS in demented PD patients. Only one out of six studies on demented PD patients reported normal SS in the cortex. Also, there is a good agreement in the decreased levels of CSF SS in PD patients, demented or not. Only one out of nine studies reported a normal level, and one other study reported an increased level. On the other hand, cortical SS receptors are unchanged in PD. Nigral SS receptors are decreased suggesting that SS receptors are localized, at least in part, on the nigral dopaminergic neurons.

TABLE 38.10
Somatostatin in Parkinson's Disease

Authors	Results	References	Remarks
Radioimmunoassay			
Epelbaum et al.	decreased	199	cortex, caudate, demented PD
Allen et al.	decreased	200	cortex, hippocampus, demented PD
Beal et al.	decreased	201	cortex, demented PD
Beal et al.	decreased	201	17/22 cortical regions, demented PD
Whitford et al.	unchanged	203	cortex, demented, nondemented PD
Leake et al.	unchanged	204	cortex, PD
Eve et al.	increased	205	medial medullary lamina of GP
Somatostatin in CSF			
Dupon et al.	decreased	206	lumbar
Jolkkonen et al.	decreased	207	lumbar
Cramer et al.	decreased	166	lumbar
Masson et al.	decreased	208	lumbar
Strittmatter et al.	decreased	209	lumbar
Volicer et al.	unchanged	210	lumbar
Espino et al.	increased	211	lumbar
Cramer et al.	decreased	212	ventricular
Jost et al.	decreased	213	ventricular
Somatostatin receptors			
Uhl et al.	decreased	149	SNC
Whitford et al.	unchanged	203	cortex
Epelbaum et al	unchanged	214	cortex, demented, nondemented PD

NEUROPEPTIDE Y

Neuropeptide Y (NPY) is a peptide composed of 36 amino acids rich in tyrosine. It is one of the most abundant and widely distributed neuropeptides within the central nervous system, with particularly high concentrations in the hypothalamus and in several limbic regions. NPY co-localizes with somatostatin, galanin, GABA, noradrenalin, and adrenaline in discrete brain regions.[215]

Studies on NPY in PD are limited (Table 38.11). Allen et al.[200] measured NPY immunoreactivity in cortical regions from demented PD patients. They found no significant change compared with the controls. Beal et al.[202] measured NPY immunoreactivity in 22 cortical regions from 13 demented PD patients. They found a 20 to 30% decrease in NPY immunoreactivity in 9 out of 22 regions in demented PD patients.

TABLE 38.11
Neuropeptide Y in Parkinson's Disease

Allen et al.	unchanged	200	cortex
Beal et al.	decreased	202	cortex, demented
Cannizzaro et al.	increased	216	striatum
Yaksh et al.	unchanged	176	CSF
Martignoni et al.	decreased	102	CSF

Cannizzaro et al.[216] studied NPY mRNA expression in the basal ganglia from PD patients. The number of NPY mRNA-expressing cells was increased as was the density of the silver grains overlying each positive cell in PD patients compared with the controls. The increase was more pronounced in the nucleus accumbens and in the ventral part of the caudate nucleus. They interpreted their results as indicating the loss of dopaminergic tone on striatal NPY containing interneurons or the influence of chronic L-dopa therapy.

Yaksh et al.[176] measured CSF NPY in eight advanced PD patients. CSF NPY levels in PD patients were not changed as compared with the controls. On the other hand, Martignoni et al.[102] found a significant reduction in CSF NPY in PD patients compared with the controls.

VASOACTIVE INTESTINAL PEPTIDE

Vasoactive intestinal peptide (VIP) is a peptide composed of 28 amino acids. VIP is widely expressed in the brain with higher levels in cerebral cortex, amygdala, hypothalamus, and hippocampus and the lowest levels in basal ganglia including caudate nucleus, GPE, putamen, and SN.[217] Studies on VIP in PD are very limited. Jegou et al.[217] measured VIP in various brain regions of PD patients. They found no difference in the contents of VIP in PD patients compared with the controls.

CORTICOTROPIN RELEASING FACTOR

Corticotropin releasing factor (CRF) is a hypothalamic peptide that stimulates secretion of ACTH from the anterior pituitary gland. Studies on CRF in PD are very limited. Conte-Devolx et al.[218] measured hypothalamic CRF from PD patients. They found no difference from the controls. Whitehouse et al.[219] measured CRF-like immunoreactivity in the frontal, temporal, and occipital poles of the neocortex from demented PD patients. They found a significant decrease in the CRF-like immunoreactivity.

MELANOCYTE STIMULATING HORMONE

Melanocyte stimulating hormone (MSH) is a pituitary hormone released form the intermediate lobe of the pituitary gland stimulating melanocytes of the skin. Studies on MSH in PD are very limited. Rainero et al.[220] measured contents of immunoreactive MSH in CSF from nine PD patients. Mean CSF alpha-MSH-like immunoreactivity concentration was twofold greater in parkinsonian patients (44.1 ± 9.3 [SD] pg/ml) as compared with control subjects (21.8 ± 10.0 pg/ml).

SUMMARY

Many transmitters and transmitter-like substances were studied in PD in addition to dopamine and related substances. From a survey of the literature, it can be concluded that, in addition to loss of dopaminergic neurons and dopamine, locus coeruleus and noradrenalin are constantly involved in PD. Raphe nucleus and serotonin are also involved in PD, although the magnitude of degeneration is less than that of dopamine. Loss of serotonin in brain is likely responsible at least in part for depression in PD.

What is interesting is the change in cholinergic systems in PD. Degeneration of NBM and loss of cortical ACh are seen most of demented PD patients, and loss of Ach is responsible, at least in part, for symptoms of dementia in PD patients. Loss of neurons in NBM is also seen in nondemented PD patients; however, demented PD patients show more extensive neuronal loss. Other changes frequently seen in demented PD patients include loss of cortical SS and marked degeneration of LC and loss of cortical NE. Cholinergic neurons in the dorsal motor nucleus and the PPN are also degenerating in PD, but the loss of neurons in these nuclei is less than that of SN. Post-ganglionic cardiac sympathetic fibers are noradrenergic, and they also show extensive degeneration. Loss of cardiac sympathetic fibers can be visualized by MIBG SPECT. Loss of MIBG uptake correlates well with the presence of Lewy bodies in the brain. MIBG uptake is retained in other non-LB parkinsonism and dementia such as MSA, PSP, and AD. Thus, this is a very useful test for the clinical differential diagnosis.

GABAergic system and glutamatergic systems are generally preserved in PD. Regarding the neuropeptide systems, SP in striatum and SNC is reduced to a lesser extent compared with the loss of dopamine. Met-Enk seems to be reduced also in the striatum and SNC, but some controversy exists among the reports. These peptide changes are considered secondary to loss of dopaminergic neurons. SP and Met-Enk co-localize with respective sets of GABAergic neurons in the striatum. But only peptides are decreased. Therefore, decreases of these peptides may have some significance in the pathogenesis of symptoms of PD, and they can be considered as targets for new symptomatic drug treatment.

In conclusion, in addition to nigrostriatal dopaminergic neurons, many other transmitters show abnormalities in PD. It will be interesting to discover which symptoms of PD are due to changes of which neurotransmitters by further studies.

ABBREVIATIONS

Ach	acetylcholine
AchE	acetylcholine esterase
AchR	acetylcholine receptors
AD	Alzheimer's disease
AMPA	alpha-amino-3-hydroxy-5-methylisoxazole-4-propionic acid
BDZ	benzodiazepines
CCK-8	cholecystokinin-8
ChAT	choline acetyltransferase
COMT	catechol-*O*-methyltransferase
CRF	corticotropin releasing factor
CSF	cerebrospinal fluid
DA	dopamine
DBH	dopamine beta-hydroxylase
DLB	dementia with Lewy bodies
DLBD	diffuse Lewy body disease
DYN	dynorphin
EP	epinephrine
GABA	gamma-amino butyric acid
GAD	glutamic acid decarboxylase
GP	globus pallidus
GPE	external segment of globus pallidus
GPI	internal segment of globus pallidus
5HIAA	5-hydroxyindole acetic acid
5HT	5-hydroxytryptamine, serotonin
5HTP	5-hydroxytryptophan
HVA	homovanillic acid
LB	Lewy bodies
LC	locus coeruleus
MAO	monoamine oxidase
Met-Enk	methionine-enkephalin

MHPG	3-methoxy-4-hydroxy-phenylethylenglycol
MPTP	1-methyl-4-phenyl-1,2,3,6-tetrahydropyridine
MSH	melanocyte stimulating hormone
NBM	nucleus basalis of Meynert
NE	norepinephrine
NMDA	N-Methyl-D-aspartate
NPY	neuropeptide Y
NT	neurotensin
PD	Parkinson's disease
PDD	Parkinson's disease with dementia
PNMT	phenylethanolamine N-methyltransferase
PPN	pedunculopontine nucleus
PSP	progressive supranuclear palsy
SN	substantia nigra
SNC	substantia nigra pars compacta
SND	striatonigral degeneration
SNR	substantia nigra pars reticulata
SP	substance P
SPECT	single photon emission computerized tomography
STN	subthalamic nucleus
TH	tyrosine hydroxylase
UPDRS	Unified Parkinson's Disease Rating Scale
VMA	vanylmandelic acid

REFERENCES

1. Mesulam, M. M. and Van Hoesen, G. W., Acetylcholinesterase-rich projections from the basal forebrain of the rhesus monkey to neocortex, *Brain Res.*, 109, 152, 1976.
2. Ezrin-Waters, C. and Resch, L., The nucleus basalis of Meynert, *Can. J. Neurol. Sci.*, 13, 8, 1986.
3. Cummings, J. L. and Benson, D. F., The role of the nucleus basalis of Meynert in dementia: review and reconsideration, *Alzheimer Dis. Assoc. Disord.*, 1, 128, 1987.
4. Mann, D. M. and Yates, P. O., Pathological basis for neurotransmitter changes in Parkinson's disease, *Neuropathol. Appl. Neurobiol.*, 9, 3, 1983.
5. Arendt, T. et al., Loss of neurons in the nucleus basalis of Meynert in Alzheimer's disease, paralysis agitans and Korsakoff's Disease, *Acta Neuropathol. (Berl.)*, 61, 101, 1983.
6. Whitehouse, P. J. et al., Basal forebrain neurons in the dementia of Parkinson disease. *Ann. Neurol.*, 13, 243, 1983.
7. Gaspar, P. and Gray, F., Dementia in idiopathic Parkinson's disease. A neuropathological study of 32 cases, *Acta Neuropathol,. (Berl)*, 64, 43, 1984.
8. Nakano, I. and Hirano, A., Parkinson's disease: neuron loss in the nucleus basalis without concomitant Alzheimer's disease, *Ann. Neurol.*, 15, 415, 1984.

9. Rogers, J. D., Brogan, D., and Mirra, S. S., The nucleus basalis of Meynert in neurological disease: a quantitative morphological study, *Ann. Neurol.*, 17, 163, 1985.

10. Chui, H. C. et al., Bondareff, W. and Webster, D. D., Pathologic correlates of dementia in Parkinson's disease, *Arch. Neurol.*, 43, 991, 1986.

11. Yoshimura, M., Pathological basis for dementia in elderly patients with idiopathic Parkinson's disease, *Eur. Neurol.*, 28, Suppl. 1, 29, 1988.

12. Kosaka, K., Dementia and neuropathology in Lewy body disease, *Adv. Neurol.*, 60, 456, 1993.

13. Chan-Palay, V. et al., Calbindin D-28k and monoamine oxidase. An immunoreactive neurons in the nucleus basalis of Meynert in senile dementia of the Alzheimer type and Parkinson's disease, *Dementia*, 4, 1, 1993.

14. Jellinger, K. A., Morphological substrates of mental dysfunction in Lewy body disease: an update, *J. Neural. Transm.*, Suppl., 59, 185, 2000.

15. Tsuchiya, K. et al., Parkinson's disease mimicking senile dementia of the Alzheimer type: a clinicopathological study of four autopsy cases, *Neuropathology*, 22, 77, 2002.

16. Zarow, C. et al., Neuronal loss is greater in the locus coeruleus than nucleus basalis and substantia nigra in Alzheimer and Parkinson diseases, *Arch. Neurol.*, 60, 337, 2003.

17. Nakano, I. and Hirano, A., Neuron loss in the nucleus basalis of Meynert in parkinsonism-dementia complex of Guam, *Ann. Neurol.*, 13, 87, 1983.

18. Masullo, C. et al., The nucleus basalis of Meynert in parkinsonism-dementia of Guam: a morphometric study. Neuropathol, *Appl. Neurobiol.*, 15, 193, 1989.

19. Mesulam, M. M., Mufson, E. J., Wainer, B. H., and Levey, A. I., Central cholinergic pathways in the rat: an overview based on an alternative nomenclature, Chapter 1–Chapter 6), *Neuroscience*, 10, 1185, 1983.

20. Pahapill, P. A. and Lozano, A. M., The pedunculopontine nucleus and Parkinson's disease, *Brain*, 123, 1767, 2000.

21. Hirsch, E. C. et al., Neuronal loss in the pedunculopontine tegmental nucleus in Parkinson disease and in progressive supranuclear palsy, *Proc. Natl. Acad. Sci. USA*, 84, 5976, 1987.

22. Jellinger, K., The pedunculopontine nucleus in Parkinson's disease, progressive supranuclear palsy and Alzheimer's disease, *J. Neurol. Neurosurg. Psychiatry*, 51, 540, 1988.

23. Jellinger, K. A., Pathology of Parkinson's disease. Changes other than the nigrostriatal pathway, *Mol. Chem. Neuropathol.*, 14, 153, 1991.

24. Zweig, R. M. et al., The pedunculopontine nucleus in Parkinson's disease, *Ann. Neurol.*, 26, 41, 1989.

25. Gai, W. P. et al., Substance P-containing neurons in the mesopontine tegmentum are severely affected in Parkinson's disease, *Brain*, 114, 2253, 1991.

26. Mufson, E. J., Mash, D. C., and Hersh, L. B., Neurofibrillary tangles in cholinergic pedunculopontine neurons in Alzheimer's disease, *Ann. Neurol.*, 24, 623, 1988.

27. Gai, W. P. et al., Age-related loss of dorsal vagal neurons in Parkinson's disease, *Neurology*, 42, 2106, 1992.

28. Gai, W. P., Blessing, W. W., and Blumbergs, P. C., Ubiquitin-positive degenerating neurites in the brainstem in Parkinson's disease, *Brain*, 118, 1447, 1995.

29. Del Tredici, et al., Where does Parkinson disease pathology begin in the brain? *J. Neuropathol. Exp. Neurol.*, 61, 413, 2002.

30. Wakabayashi, K. and Takahashi, H., The intermediolateral nucleus and Clarke's column in Parkinson's disease, *Acta Neuropathol (Ber.l).*, 94, 287, 1997.

31. Wakabayashi, K. et al., Parkinson's disease: the presence of Lewy bodies in Auerbach's and Meissner's plexuses. *Acta Neuropathol (Berl.).*, 76, 217, 1988.

32. Quirion, R., Cholinergic markers in Alzheimer disease and the autoregulation of acetylcholine release, *J. Psychiatry Neurosci.*, 18, 226, 1993.

33. Perry, E. K. et al., Neocortical cholinergic activities differentiate Lewy body dementia from classical Alzheimer's disease, *Neuroreport*, 5, 747, 1994.

34. Masliah, E. et al., Cholinergic deficits in the brains of patients with parkinsonism-dementia complex of Guam, *Neuroreport*, 12, 3901, 2001.

35. Gibb, W. R. et al., A pathological study of the association between Lewy body disease and Alzheimer's disease. *J. Neurol. Neurosurg. Psychiatry*, 52, 701, 1989.

36. McGeer, P. L. and McGeer, E. G., Cholinergic enzyme system in Parkinson's disease, *Arch. Neurol.*, 25, 265, 1971.

37. McGeer, P. L. and McGeer, E. G., Enzymes associated with the metabolism of catecholamines, acetylcholine and GABA in human controls and patients with Parkinson's disease and Huntington's chorea, *J. Neurochem.*, 26, 65, 1976.

38. Javoy-Agid, F., Ploska, A., and Agid, Y., Microtopography of tyrosine hydroxylase, glutamic acid decarboxylase, and choline acetyltransferase in the substantia nigra and ventral tegmental area of control and Parkinsonian brains, *J. Neurochem.*, 37, 1218, 1981.

39. Ruberg, M. et al., Muscarinic binding and choline acetyltransferase activity in Parkinsonian subjects with reference to dementia, *Brain Res.*, 232, 129, 1982.

40. Nishino, N. et al., GABAA receptor but not muscarinic receptor density was decreased in the brain of patients with Parkinson's disease, *Jpn. J. Pharmacol.*, 48, 331, 1988.

41. Kuhl, D. E. et al., In vivo mapping of cholinergic terminals in normal aging, Alzheimer's disease, and Parkinson's disease, *Ann. Neurol.*, 40, 399, 1996.

42. Rinne, U. K. et al., Brain acetylcholinesterase in Parkinson's disease. *Acta Neurol. Scand.*, 49, 215, 1973.

43. Whitehouse, P. J. et al., Reductions in [3H]nicotinic acetylcholine binding in Alzheimer's disease and Parkinson's disease: an autoradiographic study, *Neurology*, 38, 720, 1988.

44. Whitehouse, P. J. et al., Reductions in acetylcholine and nicotine binding in several degenerative diseases, *Arch. Neurol.*, 45, 722, 1988.

45. Shiozaki, K. et al., Alterations of muscarinic acetylcholine receptor subtypes in diffuse Lewy body disease: relation to Alzheimer's disease, *J. Neurol. Neurosurg. Psychiatry*, 67, 209, 1999.

46. Shiozaki, K. et al., Distribution of m1 muscarinic acetylcholine receptors in the hippocampus of patients with Alzheimer's disease and dementia with Lewy bodies-an immunohistochemical study, *J. Neurol. Sci.*, 193, 23, 2001.

47. Banerjee, C. et al., Cellular expression of alpha7 nicotinic acetylcholine receptor protein in the temporal cortex in Alzheimer's and Parkinson's disease—a stereological approach, *Neurobiol., Dis.*, 7(6 Pt. B), 666, 2000.

48. Guan, Z. Z. et al., Selective changes in the levels of nicotinic acetylcholine receptor protein and of corresponding mRNA species in the brains of patients with Parkinson's disease, *Brain Res.*, 956, 358, 2002.

49. Burghaus, L. et al., Loss of nicotinic acetylcholine receptor subunits alpha4 and alpha7 in the cerebral cortex of Parkinson patients, *Parkinsonism Relat. Disord.*, 9, 243, 2003.

50. Asahina, M. et al., Brain muscarinic receptors in progressive supranuclear palsy and Parkinson's disease: a positron emission tomographic study, *J. Neurol. Neurosurg. Psychiatry*, 65, 155, 1998.

51. Weiland, S., Bertrand, D., and Leonard, S., Neuronal nicotinic acetylcholine receptors: from the gene to the disease. *Behav. Brain Res.*, 113, 43, 2000.

52. Paterson, D. and Nordberg, A., Neuronal nicotinic receptors in the human brain, *Prog. Neurobiol.*, 61, 75, 2000.

53. Griffiths, P. D. et al., Changes in benzodiazepine and acetylcholine receptors in the globus pallidus in Parkinson's disease, *J. Neurol. Sci.*, 100, 131, 1990.

54. Griffiths, P. D., Perry, R. H., and Crossman, A. R,. A detailed anatomical analysis of neurotransmitter receptors in the putamen and caudate in Parkinson's disease and Alzheimer's disease, *Neurosci. Lett.*, 169, 68, 1994.

55. Court, J. A. et al., Nicotine binding in human striatum: elevation in schizophrenia and reductions in dementia with Lewy bodies, Parkinson's disease and Alzheimer's disease and in relation to neuroleptic medication, *Neuroscience*, 98, 79, 2000.

56. Martin-Ruiz, C. M. et al., Alpha and beta nicotinic acetylcholine receptors subunits and synaptophysin in putamen from Parkinson's disease, *Neuropharmacology*, 39, 2830, 2000.

57. Dahlstöm, A, and Fuxe, K., Evidence for the existence of monoamine-containing neurons in the central nervous system. 1. Demonstration of monoamines in cell bodies of brain neurons, *Acta Physiol. Scand.*, 62, Suppl. 232, 1, 1964.

58. Ánden, N. E. et al., Ascending monoaminergic neurons to the telencephalon and diencephalon, *Acta Physiol. Scand.*, 67, 313, 1966.

59. Mochizuki, A., Komatsuzaki, Y., and Shoji, S., Association of Lewy bodies and glial cytoplasmic inclusions in the brain of Parkinson's disease, *Acta Neuropathol.*, 104, 534, 2002.

60. Murai, T. et al., In vivo evidence for differential association of striatal dopamine and midbrain serotonin systems with neuropsychiatric symptoms in Parkinson's disease. *J. Neuropsychiatry, Clin. Neurosci.*, 13, 222, 2001.

61. Halliday, G. M. et al. Neuropathology of immunohistochemically identified brainstem neurons in Parkinson's disease, *Ann. Neurol.*, 27, 373, 1990.

62. Paulus, W. and Jellinger, K., The neuropathologic basis of different clinical subgroups of Parkinson's disease, *J. Neuropathol. Exp. Neurol.*, 50, 743, 1991.

63. Becker, T. et al., Parkinson's disease and depression: evidence for an alteration of the basal limbic system detected by transcranial sonography, *J. Neurol. Neurosurg. Psychiatry*, 63, 590, 1997.

64. Bernheimer, H., Birkmayer H., and Hornykiewicz, O., Verteilung des 5-Hydroxytryptamins (Serotonin) im Gehirn des Menschen und sein Verhalten bei Patienten mit Parkinson-Syndrome, *Klin. Wochenschr.*, 39, 1056, 1961

65. Scatton, B. et al., Reduction of cortical dopamine, noradrenaline, serotonin and their metabolites in Parkinson's disease, *Brain Res.*, 275, 321, 1983.

66. D'Amato, R. J. et al., Aminergic systems in Alzheimer's disease and Parkinson's disease, *Ann. Neurol.*, 22, 229, 1987.

67. Ohara, K., Kondo, N., and Ohara, K., Changes of monoamines in post-mortem brains from patients with diffuse Lewy body disease, *Prog. Neuropsychopharmacol. Biol. Psychiatry*, 22, 311, 1998.

68. Shannak, K. et al., Noradrenaline, dopamine and serotonin levels and metabolism in the human hypothalamus: observations in Parkinson's disease and normal subjects, *Brain Res.*, 639, 33, 1994.

69. Sawada, M. et al., Tryptophan hydroxylase activity in the brains of controls and parkinsonian patients, *J. Neural. Transm.*, 62, 107, 1985.

70. Kienzl, E. et al., Transitional states of central serotonin receptors in Parkinson's disease, *J. Neural. Transm.*, 51, 113, 1981.

71. Cash, R. et al., High and low affinity [3H]imipramine binding sites in control and parkinsonian brains, *Eur. J. Pharmacol.*, 117, 71, 1985.

72. Chinaglia, G. et al., Serotoninergic terminal transporters are differentially affected in Parkinson's disease and progressive supranuclear palsy: an autoradiographic study with [3H]citalopram, *Neuroscience*, 54, 691, 1993.

73. Brucke, T. et al., SPECT imaging of dopamine and serotonin transporters with [123I]beta-CIT. Binding kinetics in the human brain, *J. Neural. Transm. Gen. Sec.*, 94,137, 1993.

74. Haapaniemi, T. H. et al., [123I]beta-CIT SPECT demonstrates decreased brain dopamine and serotonin transporter levels in untreated parkinsonian patients, *Mov. Disord.*, 16, 124, 2001.

75. Kim, S. E. et al., Serotonin transporters in the midbrain of Parkinson's disease patients: a study with 123I-beta-CIT SPECT, *J. Nucl. Med.*, 44, 870, 2003.

76. Davidson, D. L. et al., CSF studies on the relationship between dopamine and 5-hydroxytryptamine in Parkinsonism and other movement disorders. *J. Neurol. Neurosurg, Psychiatry*, 40, 1136, 1977.

77. Tabaddor, K., Wolfson, L. I., and Sharpless, N. S., Ventricular fluid homovanillic acid and 5-hydroxyindoleace-

tic acid concentrations in patients with movement disorders, *Neurology*, 28, 1249, 1978.

78. Tohgi, H. et al., Alterations in the concentration of serotonergic and dopaminergic substances in the cerebrospinal fluid of patients with Parkinson's disease, and their changes after L-dopa administration, *Neurosci. Lett.*, 159, 135, 1993.

79. Iacono, R. P. et al., Concentrations of indoleamine metabolic intermediates in the ventricular cerebrospinal fluid of advanced Parkinson's patients with severe postural instability and gait disorders, *J. Neural. Transm.*, 104, 451, 1997.

80. Perry, E. K. et al., Cortical serotonin-S2 receptor binding abnormalities in patients with Alzheimer's disease: comparisons with Parkinson's disease, *Neurosci. Lett.*, 51, 353, 1984.

81. Maloteaux, J. M. et al., Decrease of serotonin-S2 receptors in temporal cortex of patients with Parkinson's disease and progressive supranuclear palsy, *Mov. Disord.*, 3, 255, 1988.

82. Waeber, C. and Palacios, J. M., Serotonin-1 receptor binding sites in the human basal ganglia are decreased in Huntington's chorea but not in Parkinson's disease: a quantitative *in vitro* autoradiography study, *Neuroscience*, 32, 337, 1989.

83. Ungerstadt, U., Stereotaxic mapping of the monoamine pathways in the rat brain, *Acta Physiol. Scand.*, 82, [Suppl.] 367, 1, 1971

84. Chan-Palay, V. and Asan, E., Alterations in catecholamine neurons of the locus coeruleus in senile dementia of the Alzheimer type and in Parkinson's disease with and without dementia and depression, *J. Comp. Neurol.*, 287, 373, 1989.

85. Gaspar, P. et al., Alterations of dopaminergic and noradrenergic innervations in motor cortex in Parkinson's disease, *Ann. Neurol.*, 30, 365, 1991.

86. Malessa, S. et al., Catecholaminergic systems in the medulla oblongata in parkinsonian syndromes: a quantitative immunohistochemical study in Parkinson's disease, progressive supranuclear palsy, and striatonigral degeneration, *Neurology*, 40, 1739, 1990.

87. Sharabi, Y. et al., Neurotransmitter specificity of sympathetic denervation in Parkinson's disease, *Neurology*, 60, 1036, 2003.

88. Yoshita, M., Differentiation of idiopathic Parkinson's disease from striatonigral degeneration and progressive supranuclear palsy using iodine-123 meta-iodobenzylguanidine myocardial scintigraphy, *J. Neurol. Sci.*, 155, 60, 1998.

89. Iwasa, K. et al., Decreased myocardial 123I-MIBG uptake in Parkinson's disease, *Acta Neurol. Scand.*, 97, 303, 1998.

90. Braune, S. et al., Impaired cardiac uptake of meta-[123I]iodobenzylguanidine in Parkinson's disease with autonomic failure, *Acta Neurol. Scand.*, 97, 307, 1998.

91. Orimo, S. et al., (123)I-metaiodobenzylguanidine myocardial scintigraphy in Parkinson's disease, *J. Neurol. Neurosurg. Psychiatry*, 67, 189, 1999.

92. Orimo, S., et al., Sympathetic cardiac denervation in Parkinson's disease and pure autonomic failure but not

in multiple system atrophy, *J. Neurol. Neurosurg. Psychiatry*, 73, 776, 2002.

93. Orimo, S. et al., Different histopathology accounting for a decrease in myocardial MIBG uptake in PD and MSA, *Neurology*, 57, 1140, 2001.

94. Goldstein, D. S. et al., Orthostatic hypotension from sympathetic denervation in Parkinson's disease, *Neurology*, 58, 1247, 2002.

95. Ehringer, H. and Hornykiewicz, O., Verteilung von Noradrenalin und Dopamin (3-Hydroxytyramin) im Gehirn des Menschen und ihr Verhalten bei Erkrankungen des Extrapyramidalen systems, *Klin. Wochenschr.*, 38, 1236, 1960.

96. Riederer, P. et al., Brain-noradrenaline and 3-methoxy-4-hydroxyphenylglycol in Parkinson's syndrome, *J. Neural. Transm.*, 41, 241, 1977.

97. Nagatsu, T. et al., Phenylethanolamine N-methyltransferase and other enzymes of catecholamine metabolism in human brain, *Clin. Chim. Acta*, 75, 221, 1977.

98. Kish, S. J. et al., Cerebellar norepinephrine in patients with Parkinson's disease and control subjects, *Arch. Neurol.*, 41, 612, 1984.

99. Cash, R. et al., Parkinson's disease and dementia: norepinephrine and dopamine in locus coeruleus, *Neurology*, 37, 42, 1987.

100. Matsui, H. et al., Highly sensitive assay for dopamine-beta-hydroxylase activity in human cerebrospinal fluid by high performance liquid chromatography- electrochemical detection: properties of the enzyme, *J. Neurochem.*, 37, 289, 1981.

101. Hurst, J. H. et al., CSF dopamine-beta-hydroxylase activity in Parkinson's disease, *Neurology*, 35, 565, 1985.

102. Martignoni, E. et al., Cerebrospinal fluid norepinephrine, 3-methoxy- 4-hydroxyphenylglycol and neuropeptide Y levels in Parkinson's disease, multiple system atrophy and dementia of the Alzheimer type, *J. Neural. Transm. Park. Dis. Dement. Sect.*, 4, 191, 1992.

103. Tohgi, H., et al., Monoamine metabolism in the cerebrospinal fluid in Parkinson's disease: relationship to clinical symptoms and subsequent therapeutic outcomes, *J. Neural. Transm. Park. Dis. Dement. Sect.*, 5, 17, 1993.

104. Cash, R. et al., Adrenergic receptors in Parkinson's disease. *Brain Res.*, 322, 269, 1984.

105. Hökfelt, T. et al., Immunohistochemical evidence for the existence of adrenaline neurons in the brat brain. *Brain Res.*, 66, 235, 1974.

106. Gai, W. P. et al., Loss of C1 and C3 epinephrine-synthesizing neurons in the medulla oblongata in Parkinson's disease, *Ann. Neurol.*, 33, 357, 1993.

107. Stoica, E. and Enulescu, O., Abnormal epinephrine urinary excretion in Parkinsonians: correction of the disorder by levodopa administration, *J. Neurol. Sci.*, 38, 215–27, 1978.

108. Stoddard, S. L. et al., Decreased catecholamine content in parkinsonian adrenal medullae, *Exp. Neurol.*, 104, 22, 1989.

109. Watanabe, T. et al., Distribution of the histaminergic neuron system in the central nervous system of the rats:

a fluorescent immunohistochemical analysis with histidine decarboxylase as marker, *Brain Res.*, 295, 13, 1984.

110. Anichtchik, O. V. et al., An altered histaminergic innervation of the substantia nigra in Parkinson's disease, *Exp. Neurol.*, 163, 20, 2000.

111. Rinne, J. O., et al., Increased brain histamine levels in Parkinson's disease but not in multiple system atrophy, *J. Neurochem.*, 81, 954, 2002.

112. Alexander, C. E. and Crutcher, M. D., Functional architecture of basal ganglia circuits: neural substrates of parallel processing, *Trends Neurosci.*, 13, 266, 1990.

113. Steiner, H, and Gerfen, C. R., Role of dynorphin and enkephalin in the regulation of striatal output pathways and behavior, *Exp. Brain Res.*, 123, 60, 1998.

114. DeLong, M. R., Primate models of movement disorders of basal ganglia origin, *Trends Neurosci.*, 13, 281, 1990.

115. Itouji, A. et al., Neuronal and glial localization of two GABA transporters (GAT1 and GAT39 in the rat cerebellum, *Mol. Brain Res.*, 37, 309, 1996.

116. Perry, T. L. et al., Striatal GABAergic neuronal activity is not reduced in Parkinson's disease, *J. Neurochem.*, 40, 1120, 1983.

117. Kish, S. J. et al., Elevated gamma-aminobutyric acid level in striatal but not extrastriatal brain regions in Parkinson's disease: correlation with striatal dopamine loss, *Ann. Neurol.*, 20, 26, 1986.

118. Rinne, J. O. et al., Free amino acids in the brain of patients with Parkinson's disease, *Neurosci. Lett.*, 94, 182, 1988.

119. Gerlach, M. et al., A post mortem study on neurochemical markers of dopaminergic, GABA-ergic and glutamatergic neurons in basal ganglia-thalamocortical circuits in Parkinson syndrome, *Brain Res.*, 741, 142, 1996.

120. Hardman, C. D. et al., Substantia nigra pars reticulata neurons in Parkinson's disease, *Neurodegeneration*, 5, 49, 1996.

121. Bernheimer, H. and Hornykiewicz, O., Das Verhalten einigen Enzyme im Gehirn normaler und Parkinson-Kranken Menschen, *Naunyn-Schimiedeberg's Arch. Exp. Pathol. Pharmacol.*, 243, 295, 1962

122. Lloyd, E.G., and Hornykiewicz, O., L-glutamic acid decarboxylase in Parkinson's disease: effect of L-dopa therapy, *Nature*, 243, 521, 1970.

123. Rinne, U. et al., Brain glutamic acid decarboxylase activity in Parkinson's disease, *Eur. Neurol.*, 12, 1349, 1974.

124. Monfort, J. C. et al., Brain glutamate decarboxylase in Parkinson's disease with particular reference to a premortem severity index, *Brain*, 108, 301, 1985.

125. Levy, R. et al., Effects of nigrostriatal denervation and L-dopa therapy on the GABAergic neurons in the striatum in MPTP-treated monkeys and Parkinson's disease: an in situ hybridization study of GAD67 mRNA, *Eur., J. Neurosci.*, 7, 1199, 1995.

126. Herrero, M. T. et al., Consequence of nigrostriatal denervation and L-dopa therapy on the expression of glutamic acid decarboxylase messenger RNA in the pallidum, *Neurology*, 47, 219, 1996.

127. Vila, M. et al., Consequences of nigrostriatal denervation on the gamma-aminobutyric acidic neurons of sub-

stantia nigra pars reticulata and superior colliculus in parkinsonian syndromes, *Neurology*, 46, 802, 1996.

128. Manyam, B.V., Low CSF gamma-aminobutyric acid levels in Parkinson's Disease. Effect of levodopa and carbidopa, *Arch. Neurol.*, 39, 391, 1982.

129. Araki, K. et al., Alteration of amino acids in cerebrospinal fluid from patients with Parkinson's disease and spinocerebellar degeneration, *Acta Neurol. Scand.*, 73, 105, 1986.

130. Tohgi, H. et al., Significant reduction of putative transmitter amino acids in cerebrospinal fluid of patients with Parkinson's disease and spinocerebellar degeneration. *Neurosci. Lett.*, 126, 155, 1991.

131. Jimenez-Jimenez, F. J. et al., Neurotransmitter amino acids in cerebrospinal fluid of patients with Parkinson's disease, *J. Neurol. Sci.*, 141, 39, 1996.

132. Bonnet, A. M. et al., Cerebrospinal fluid GABA and homocarnosine concentrations in patients with Friedreich's ataxia, Parkinson's disease, and Huntington's chorea, *Mov. Disord.*, 2, 117, 1987.

133. Perschak, H. et al., Ventricular cerebrospinal fluid concentrations of putative amino acid transmitters in Parkinson's disease and other disorders, *Hum. Neurobiol.*, 6, 191, 1987.

134. Lloyd, G. K., Shemen, L., and Hornykiewicz, O., Distribution of high affinity sodium-independent [3H]gamma-aminobutyric acid [3H]GABA binding in the human brain: alterations in Parkinson's disease, *Brain Res.*, 127, 269, 1977.

135. Lloyd, G. K. et al., GABAA receptor complex function in frontal cortex membranes from control and neurological patients, *Eur. J. Pharmacol.*, 197, 33, 1991.

136. Calon, F. and Di Paolo, T., Levodopa response motor complications—GABA receptors and preproenkephalin expression in human brain, *Parkinsonism Relat. Disord.*, 8, 449, 2002.

137. Calon, F. et al., Changes of GABA receptors and dopamine turnover in the postmortem brains of parkinsonians with levodopa-induced motor complications, *Mov. Disord.*, 18, 241, 2003.

138. Perry, T. L., Hansen, S., and Gandham, S. S., Postmortem changes of amino compounds in human and rat brain, *J. Neurochem.*, 36, 406, 1981.

139. Petri, S. et al., Human GABA A receptors on dopaminergic neurons in the pars compacta of the substantia nigra, *J. Comp. Neurol.*, 452, 360, 2002.

140. Clarke, C. E., Lowry, M., and Horsman, A., Unchanged basal ganglia N-acetylaspartate and glutamate in idiopathic Parkinson's disease measured by proton magnetic resonance spectroscopy, *Mov. Disord.*, 12, 297, 1997.

141. Taylor-Robinson, S. D. et al., A proton magnetic resonance spectroscopy study of the striatum and cerebral cortex in Parkinson's disease, *Metab. Brain Dis.*, 14, 45, 1999.

142. Mally, J., Szalai, G., and Stone, T. W., Changes in the concentration of amino acids in serum and cerebrospinal fluid of patients with Parkinson's disease, *J. Neurol, Sci.*, 151, 159, 1997.

143. Difazio, M. C. et al., Glutamate receptors in the substantia nigra of Parkinson's disease brains, *Neurology*, 42, 402, 1992.

144. Ulas, J. et al., Selective increase of NMDA-sensitive glutamate binding in the striatum of Parkinson's disease, Alzheimer's disease, and mixed Parkinson's disease/Alzheimer's disease patients: an autoradiographic study, *J. Neurosci.*, 14, 6317, 1994.

145. Loopuijt, L. D. and Schmidt, W. J., The role of NMDA receptors in the slow neuronal degeneration of Parkinson's disease, *Amino Acids*, 14, 17, 1998.

146. Verhagen Metman, L. et al., Blockade of glutamatergic transmission as treatment for dyskinesias and motor fluctuations in Parkinson's disease, *Amino Acids*, 14, 75, 1998.

147. Kretschmer, B. D., Ligands of the NMDA receptor-associated glycine recognition site and motor behavior, *Amino Acids*, 14, 227, 1998.

148. Maloteaux, J. M. et al., Benzodiazepine receptors in normal human brain, in Parkinson's disease and in progressive supranuclear palsy, *Brain Res.*, 446, 321, 1988.

149. Uhl, G. R. et al., Parkinson's disease: nigral receptor changes support peptidergic role in nigrostriatal modulation, *Ann. Neurol.*, 20, 194, 1986.

150. Bonuccelli, U. et al., Platelet peripheral benzodiazepine receptors are decreased in Parkinson's disease, *Life Sci.*, 48, 1185, 1991.

151. Powell, L. T. et al., Radioimmunoassay for substance P, *Nature*, 241, 252, 1973.

152. Kanazawa, I. and Jessell, T., Postmortem changes and regional distribution of substance P in the rat and mouse nervous system, *Brain Res.*, 117, 462, 1976

153. Mauborgne, A. et al., Decrease of substance P-like immunoreactivity in the substantia nigra and pallidum of Parkinsonian brains, *Brain Res.*, 268, 167, 1983.

154. Tenovuo, O., Rinne, U. K., and Viljanen, M. K., Substance P immunoreactivity in the post-mortem parkinsonian brain, *Brain Res.*, 303, 113, 1984.

155. Clevens, R. A. and Beal, M. F., Substance P-like immunoreactivity in brains with pathological features of Parkinson's and Alzheimer's diseases, *Brain Res.*, 486, 387, 1989.

156. Sivam, S. P., Dopamine dependent decrease in enkephalin and substance P levels in basal ganglia regions of postmortem parkinsonian brains, *Neuropeptides*, 18, 201, 1991.

157. Fernandez, A. et al., Striatal neuropeptide levels in Parkinson's disease patients, *Neurosci. Lett.*, 145, 171, 1992.

158. Fernandez, A. et al., Alterations in peptide levels in Parkinson's disease and incidental Lewy body disease, *Brain*, 119, 823, 1996.

159. de Ceballos, M. L. et al., Parallel alterations in Met-enkephalin and substance P levels in medial globus pallidus in Parkinson's disease patients, *Neurosci. Lett.*, 160, 163, 1993.

160. Levy, R. et al., Striatal expression of substance P and methionine-enkephalin in genes in patients with Parkinson's disease, *Neurosci. Lett.*, 199, 220, 1995.

161. Nisbet, A. P. et al., Preproenkephalin and preprotachykinin messenger RNA expression in normal human basal ganglia and in Parkinson's disease, *Neuroscience,* 66, 361–76, 1995.

162. Grafe, M. R., Forno, L. S., and Eng, L. F., Immunocytochemical studies of substance P and Met-enkephalin in the basal ganglia and substantia nigra in Huntington's, Parkinson's and Alzheimer's diseases, *J. Neuropathol. Exp. Neurol.*, 44, 47, 1985.

163. Waters, C. M. et al., Immunocytochemical studies on the basal ganglia and substantia nigra in Parkinson's disease and Huntington's chorea, *Neuroscience*, 25, 419, 1988.

164. Gai, W. P. et al., Substance P-containing neurons in the mesopontine tegmentum are severely affected in Parkinson's disease, *Brain*, 114, 2253, 1991.

165. de Ceballos, M. L. and Lopez-Lozano, J. J., Subgroups of parkinsonian patients differentiated by peptidergic immunostaining of caudate nucleus biopsies, *Peptides*, 20, 249, 1999.

166. Cramer, H. et al., Immunoreactive substance P and somatostatin in the cerebrospinal fluid of senile parkinsonian patients, *Eur. Neurol.*, 29, 1, 1989.

167. Cramer, H. et al., Ventricular fluid neuropeptides in Parkinson's disease. II. Levels of substance P-like immunoreactivity, *Neuropeptides*, 18, 69, 1991.

168. Tenovuo, O. et al., Brain substance P receptors in Parkinson's disease, *Adv. Neurol.*, 53, 145, 1990.

169. Rioux, L. and Joyce, J. N., Substance P receptors are differentially affected in Parkinson's and Alzheimer's disease, *J. Neural. Transm. Park. Dis. Dement,* Sect., 6, 199, 1993.

170. Fernandez, A., et al., Neurotensin, substance P, delta and mu opioid receptors are decreased in basal ganglia of Parkinson's disease patients, *Neuroscience*, 61, 73, 1994.

171. Hong, J. S. et al., Determination of methionine enkephalin in discrete regions of rat brain, *Brain Res.*, 134, 383, 1977.

172. Taquet, H. et al., Microtopography of methionine-enkephalin, dopamine and noradrenaline in the ventral mesencephalon of human control and Parkinsonian brains, *Brain Res.*, 235, 303, 1982.

173. Llorens-Cortes, C. et al., Enkephalinergic markers in substantia nigra and caudate nucleus from Parkinsonian subjects, *J. Neurochem.*, 43, 874, 1984.

174. Calon, F. et al., Increase of preproenkephalin mRNA levels in the putamen of Parkinson disease patients with levodopa-induced dyskinesias, *J. Neuropathol. Exp. Neurol.*, 61, 186, 2002.

175. Goto, S., Hirano, A., and Matsumoto, S., Met-enkephalin immunoreactivity in the basal ganglia in Parkinson's disease and striatonigral degeneration, *Neurology*, 40, 1051, 1990.

176. Yaksh, T. L. et al., Measurement of lumbar CSF levels of met-enkephalin, encrypted met-enkephalin, and neuropeptide Y in normal patients and in patients with Parkinson's disease before and after autologous transplantation of adrenal medulla into the caudate nucleus, *J. Lab. Clin. Med.*, 115, 346, 1990.

177. Baronti, F. et al., Opioid peptides in Parkinson's disease: effects of dopamine repletion, *Brain Res.*, 560, 92, 1991.

178. Rinne, U. K. et al., Brain enkephalin receptors in Parkinson's disease, *J. Neural. Transm. Suppl.*, 19, 163, 1983.

179. Imura, H, et al., Biosynthesis and distribution of opioid peptides, *J. Endocrinol. Invest.*, 6, 139, 1983.

180. Pique, L. et al., Pro-opiomelanocortin peptides in the human hypothalamus: comparative study between normal subjects and Parkinson patients, *Neurosci. Lett.*, 54, 141, 1985.

181. Nappi, G. et al., beta-Endorphin cerebrospinal fluid decrease in untreated parkinsonian patients, *Neurology*, 35, 1371, 1985.

182. Jolkkonen, J. T., Soininen, H. S., and Riekkinen, P. J., beta-Endorphin-like immunoreactivity in cerebrospinal fluid of patients with Alzheimer's disease and Parkinson's disease, *J. Neurol. Sci.*, 77, 153, 1987.

183. Kuhar M. J., Pert, C. B., and Snyder, S. H., Regional distribution of opiate receptor binding in monkey and human brain, *Nature*, 245, 227, 1973

184. Huges, J. et al., Identification of two related pentapeptides from the brain with potent opiate agonist activity, *Nature*, 258, 577, 1975.

185. Delay-Goyet, P. et al., Regional distribution of mu, delta and kappa opioid receptors in human brains from controls and parkinsonian subjects, *Brain Res.*, 414, 8, 1987.

186. Yamada, M. et al., The expression of mRNA for a kappa opioid receptor in the substantia nigra of Parkinson's disease brain, *Brain Res. Mol. Brain Res.*, 44, 12, 1997.

187. Burn, D. J. et al., Striatal opioid receptor binding in Parkinson's disease, striatonigral degeneration and Steele-Richardson-Olszewski syndrome, A [11C]diprenorphine PET study, *Brain*, 118, 951, 1995.

188. Piccini, P., Weeks, R. A., and Brooks, D. J., Alterations in opioid receptor binding in Parkinson's disease patients with levodopa-induced dyskinesias, *Ann. Neurol.*, 42, 720, 1997.

189. Uhl, G. R., Kuhar, M. J., and Snyder, S. H., Neurotensin: immunohistochemical localization in rat central nervous system, *Proc. Natl. Acad. Sci. USA*, 74, 4059, 1977.

190. Bissette, G. et al., Alterations in regional brain concentrations of neurotensin and bombesin in Parkinson's disease, *Ann. Neurol.*, 17, 324, 1985.

191. Fernandez, A. et al., Characterization of neurotensin-like immunoreactivity in human basal ganglia: increased neurotensin levels in substantia nigra in Parkinson's disease, *Peptides*, 16, 339, 1995.

192. Uhl, G. R. et al., Parkinson's disease: depletion of substantia nigra neurotensin receptors, *Brain Res.*, 308, 186, 1984.

193. Sadoul, J. L. et al., Loss of high affinity neurotensin receptors in substantia nigra from parkinsonian subjects, *Biochem. Biophys. Res. Commun.*, 125, 395, 1984.

194. Chinaglia, G., Probst, A., and Palacios, J. M., Neurotensin receptors in Parkinson's disease and progressive supranuclear palsy: an autoradiographic study in basal ganglia, *Neuroscience*, 39, 351, 1990.

195. Yamada, M., Yamada, M., and Richelson, E., Heterogeneity of melanized neurons expressing neurotensin receptor messenger RNA in the substantia nigra and the nucleus paranigralis of control and Parkinson's disease brain, *Neuroscience*, 64, 405, 1995.

196. Emson, P. C., Lee, Rehfeld, J. F., and Rosser, M. N., Distribution of chorecystokinin-like peptide in the human brain, *J. Neurochem.*, 38,1177, 1982.

197. Studler, J. M. et al., CCK-8-Immunoreactivity distribution in human brain: selective decrease in the substantia nigra from parkinsonian patients, *Brain Res.*, 243, 176, 1982.

198. Epelbaum, J. et al., Effects of brain lesions and hypothalamic deafferentation on somatostatin distribution in the rat brain, *Endocrinology*, 101, 1495, 1977.

199. Epelbaum, J. et al., Somatostatin and dementia in Parkinson's disease, *Brain Res.*, 278, 376, 1983.

200. Allen, J. M. et al., Dissociation of neuropeptide Y and somatostatin in Parkinson's disease, *Brain Res.*, 337, 197, 1985.

201. Beal, M. F., Mazurek, M. F., and Martin, J. B., Somatostatin immunoreactivity is reduced in Parkinson's disease dementia with Alzheimer's changes, *Brain Res.*, 397, 386, 1986.

202. Beal, M. F., Clevens, R. A., and Mazurek, M. F., Somatostatin and neuropeptide Y immunoreactivity in Parkinson's disease dementia with Alzheimer's changes, *Synapse*, 2, 463, 1988.

203. Whitford, C. et al., Cortical somatostatinergic system not affected in Alzheimer's and Parkinson's diseases, *J. Neurol. Sci.*, 86, 13, 1988.

204. Leake, A. et al., Neocortical concentrations of neuropeptides in senile dementia of the Alzheimer and Lewy body type: comparison with Parkinson's disease and severity correlations, *Biol. Psychiatry*, 29, 357, 1991.

205. Eve, D. J. et al., Selective increase in somatostatin mRNA expression in human basal ganglia in Parkinson's disease, *Brain Res. Mol. Brain Res.*, 50, 59, 1997.

206. Dupont, E. et al., Low cerebrospinal fluid somatostatin in Parkinson disease: an irreversible abnormality, *Neurology*, 32, 312, 1982.

207. Jolkkonen, J. et al., Somatostatin-like immunoreactivity in the cerebrospinal fluid of patients with Parkinson's disease and its relation to dementia, *J. Neurol. Neurosurg. Psychiatry*, 49, 1374, 1986.

208. Masson, H. et al., Somatostatin-like immunoreactivity in the cerebrospinal fluid of aged patients with Parkinson's disease. The effect of dopa therapy, *J. Am. Geriatr Soc.*, 38, 19, 1990.

209. Strittmatter, M. et al., Somatostatin-like immunoreactivity, its molecular forms and monoaminergic metabolites in aged and demented patients with Parkinson's disease—effect of L-dopa, *J. Neural. Transm. Gen. Sect.*, 103, 591, 1996.

210. Volicer, L. et al., CSF cyclic nucleotides and somatostatin in Parkinson's disease, *Neurology*, 36, 89, 1986.

211. Espino, A. et al., CSF somatostatin increase in patients with early parkinsonian syndrome, *J. Neural. Transm. Park. Dis. Dement.*, Sect., 9, 189, 1995.

212. Cramer, H. et al., Ventricular somatostatin-like immunoreactivity in patients with basal ganglia disease, *J. Neurol.*, 232, 219, 1985.

213. Jost, S. et al., Ventricular fluid neuropeptides in Parkinson's disease. I. Levels and distribution of somatostatin-like immunoreactivity, *Neuropeptides*, 15, 219, 1990.

214. Epelbaum, J. et al., Somatostatin concentrations and binding sites in human frontal cortex are differentially affected in Parkinson's disease associated dementia and in progressive supranuclear palsy, *J. Neurol. Sci.*, 87, 167, 1988.

215. Wahlestedt, C., Ekman, R., and Widerlov, E., Neuropeptide Y (NPY) and the central nervous system: distribution effects and possible relationship to neurological and psychiatric disorders, *Prog. Neuropsychopharmacol. Biol. Psychiatry*, 13, 31, 1989.

216. Cannizzaro, C. et al., Increased neuropeptide Y mRNA expression in striatum in Parkinson's disease, *Brain Res. Mol. Brain Res.*, 110, 169, 2003.

217. Jegou, S. et al., Regional distribution of vasoactive intestinal peptide in brains from normal and parkinsonian subjects, *Peptides*, 9, 787, 1988.

218. Conte-Devolx, B. et al., Corticoliberin, somatocrinin and amine contents in normal and parkinsonian human hypothalamus, *Neurosci. Lett.*, 56, 217–22, 1985.

219. Whitehouse, P. J. et al., Reductions in corticotropin releasing factor-like immunoreactivity in cerebral cortex in Alzheimer's disease, Parkinson's disease, and progressive supranuclear palsy, *Neurology*, 37, 905, 1987.

220. Rainero, I. et al., Alpha-melanocyte-stimulating hormone like immunoreactivity is increased in cerebrospinal fluid of patients with Parkinson's disease, *Arch. Neurol.*, 45, 1224, 1988.

39 Metallothionein Isoforms Attenuate Peroxynitrite-Induced Oxidative Stress in Parkinson's Disease

Manuchair Ebadi, Sushil K. Sharma, Sawitri Wanpen, and Shaik Shavali
Departments of Pharmacology and of Neurosciences, University of North Dakota School of
Medicine and Health Sciences

CONTENTS

ABSTRACT

Parkinson's disease is the second most common neurodegenerative disorder after Alzheimer's disease, affecting approximately 1% of the population older than 50 years. There is a worldwide increase in disease prevalence due to the increasing age of human populations.

Parkinson's disease is characterized by a progressive loss of dopaminergic neurons in the substantia nigra zona compacta, and in other subcortical nuclei associated with a widespread occurrence of Lewy bodies. The causes of cell death in Parkinson's disease are still poorly understood, but a defect in mitochondrial oxidative phosphorylation and enhanced oxidative stress have been proposed.

Reactive oxygen species (ROS) denote superoxide, hydrogen peroxide, hydroxyl radical, and singlet oxygen, which are able to trigger both necrotic and apoptotic cell death. ROS are able to activate signaling pathways and transcription factors such as AP-1 and NF-κB. Antioxidants may modify gene expression in several ways. One possible mechanism for these observations is that antioxidants regulate the activation and/or binding of specific transcription factors to their cognate sites on DNA. Depletion of glutathione coupled with hydrogen peroxide treatment of cells increases the expression of metallothionein as well as metallothionein promotor expression vector-containing Sp1 sites.

The contribution of genetic factors to the pathogenesis of Parkinson's disease is increasingly recognized. A point mutation that is sufficient to cause a rare autosomal dominant form of the disorder has recently been identified in the α-synuclein gene on chromosome 4 in the much more common sporadic, or "idiopathic," form of

Parkinson's disease, and a defect of complex I of the mitochondrial respiratory chain was confirmed at the biochemical level. Disease specificity of this defect has been demonstrated for the parkinsonian substantia nigra. These findings and the observation that the neurotoxin MPTP (l-methyl-4-phenyl-l,2,3,6-tetrahydropyridine), which causes a Parkinson-like syndrome in humans, acts via inhibition of complex I have triggered research interest in the mitochondrial genetics of Parkinson's disease.

We have examined 3-morpholinosydnonimine (SIN-1)-induced apoptosis in control and metallothionein-overexpressing dopaminergic neurons, with a primary objective to determine the neuroprotective potential of metallothionein against peroxynitrite- induced neurodegeneration in Parkinson's disease. SIN-1 induced lipid peroxidation and triggered plasma membrane blebbing. In addition, it caused DNA fragmentation, α-synuclein induction, and intramitochondrial accumulation of metal ions (copper, iron, zinc, and calcium), and enhanced the synthesis of 8-hydroxy-2-deoxyguanosine. Furthermore, it down-regulated the expression of Bcl-2 and poly(ADP-ribose) polymerase but up-regulated the expression of caspase-3 and Bax in dopaminergic (SK-N-SH) neurons. SIN-1 induced apoptosis in aging mitochondrial genome knockout cells, α-synuclein-transfected cells, metallothionein double-knockout cells, and caspase-3-overexpressed dopaminergic neurons. SIN-1-induced changes were attenuated with selegiline or in metallothionein-transgenic striatal fetal stem cells. SIN-1-induced oxidation of dopamine to dihydroxyphenylacetaldehyde was attenuated in metallothionein-transgenic fetal stem cells and in cells transfected with a mitochondrial genome, and enhanced in aging mitochondrial genome knockout cells, in metallothionein double-knockout cells and caspase-3 gene-overexpressing dopaminergic neurons. Selegiline, melatonin, ubiquinone, and metallothionein suppressed SIN-1-induced down-regulation of a mitochondrial genome and up-regulation of caspase-3 as determined by reverse transcription-polymerase chain reaction.

The synthesis of mitochondrial 8-hydroxy-2-deoxyguanosine and apoptosis-inducing factors were increased following exposure to l-methyl-4-phenylpyridinium ion or rotenone. Pretreatment with selegiline or metallothionein suppressed l-methyl-4-phenylpyridinium ion-, 6-hydroxydopamine-, and rotenone-induced increases in mitochondrial 8-hydroxy-2-deoxyguanosine accumulation. Transfection of aging mitochondrial genome knockout neurons with mitochondrial genome encoding complex 1 or melanin attenuated the SIN-1-induced increase in lipid peroxidation. SIN-1 induced the expression of α-synuclein, caspase-3, and 8-hydroxy-2-deoxyguanosine, and augmented protein nitration. These effects were attenuated by metallothionein gene overexpression. These studies provide evidence that nitric oxide synthase activation and peroxynitrite ion overproduction may be involved in the etiopathogenesis of Parkinson's disease, and that metallothionein gene induction may provide neuroprotection.

INTRODUCTION

Although the exact etiopathogenesis of Parkinson's disease (PD) remains unknown, it has been hypothesized that the neuronal demise of nigrostriatal dopaminergic neurons (DA neurons) could occur due to the production of endogenous neurotoxins, such as *tetrahydroisoquinolines*, or by exposure to various environmental neurotoxins, such as rotenone. These neurotoxins produce a significant downregulation of mitochondrial complex 1 (ubiquinone NADH-oxidoreductase), as observed in the majority of PD patients. Furthermore, significantly reduced glutathione in the substantia nigra (SN) enhances the risk of free radical [mainly hydroxyl ($^{\cdot}$OH) and nitric oxide ($^{\cdot}$NO)] overproduction, leading to neuronal damage in PD.[2–7]

NO plays a critical role in mediating neurotoxicity associated with various neurological disorders, such as stroke, PD, HIV dementia, and multiple sclerosis. In the SN of PD patients, a significant increase in the density of glial cells expressing tumor necrosis factor-α, interleukin-1β, and interferon-γ has been observed. Although CD23 was not detectable in the SN of control subjects, it was found in both astroglial and microglial cells of parkinsonian patients, indicating the existence of cytokine/CD23-dependent activation pathways of inducible NO synthase (iNOS) and of proinflammatory mediators in glial cells and their involvement in the pathophysiology of PD.[8] In addition to NO, accumulation of iron in the SN has been implicated in the death of DA neurons in PD.[9–10]

Peroxynitrite (ONOO$^-$) ions, generated in the mitochondria by Ca^{2+}-dependent NO synthase (NOS) activation during oxidative and nitrative stress, readily react with lipids, aromatic amino acids, or metalloproteins, inhibiting mitochondrial respiratory complexes, and hence are thought to be involved in the etiopathogenesis of many diseases, including PD.[11]

Although NO has been shown to possess both apoptogenic and apoptostatic properties, its overproduction during oxidation and nitrative stresses induces deleterious consequences on mitochondrial complex 1 activity.[12–13] We have discovered recently that metallothionein (MT) gene overexpression in MT transgenic (MT$_{trans}$) mouse brain inhibited the l-methyl-4-phenyl-1,2,3,6-tetrahydropyridine (MPTP)-induced nitration of α-synuclein (α-Syn), and preserved mitochondrial coenzyme Q$_{10}$ levels, affording neuroprotection against nitrative and oxidative stress of aging brain.[14–15] In addition, MT isoforms are able to suppress 6-hydroxydopamine (6-OHDA)-induced $^{\cdot}$OH radical generation.[14] We have also reported that selegiline, a monoamine oxidase B inhibitor, provides neuro-protection via MT gene overexpression.[16]

As the involvement of oxidative and nitrative stresses is now advocated in the etiopathogenesis of PD, a detailed study was needed to explore the exact molecular mechanism of NO-mediated neurodegeneration and MT-induced neuroprotection in PD. 3-Morpholinosydnonimine (SIN-1), a vasorelaxant, a soluble guanylyl cyclase stimulator, and a potent $ONOO^-$ generator produced not only oxidative but also nitrative stresses in DA neurons, portending an important role in understanding the exact etiopathogenesis of PD. Therefore, we have investigated the extent of neuroprotection afforded by MT against SIN-1-induced lipid peroxidation. In addition, caspase-3 activation, and mitochondrial 8-hydroxy-2-deoxyguanosine (8-OH-2dG) synthesis in human DA (SK-N-SH) cell line and in MT-overexpressed striatal DA neurons (Figure 39.1), and DOPAL-induced oxidative damage (Figure 39.2) were studied. These studies have shown that selegiline affords neuroprotection by enhancing MT gene expression and by down-regulating α-Syn expression, a Lewy body molecular marker. MT pretreatment also attenuated SIN-1-induced intramitochondrial accumulation of metal ions (Fe^{3+}, and Ca^{2+}), known to be involved in the etiopathogenesis of PD, suggesting a possible neuroprotective potential for MT gene overexpression in attenuating neurotoxin-induced parkinsonism.[17–18]

Nitric oxide (NO), in excess, behaves as a cytotoxic substance mediating the pathological processes that cause neurodegeneration. The NO-induced dopaminergic cell loss causing Parkinson's disease has been postulated to include the following: an inhibition of cytochrome oxidase, ribonucleotide reductase, mitochondrial complexes I, II, and IV in the respiratory chain, superoxide dismutase, glyceraldehyde-3-phosphate dehydrogenase; activation or initiation of DNA strand breakage, poly(ADP-ribose) synthase, lipid peroxidation, and protein oxidation; release of iron; and increased generation of toxic radicals such as hydroxyl radicals and peroxynitrite. NO is formed by the conversion of L-arginine to L-citrulline by NO synthase (NOS). At least three NOS isoforms have been identified by molecular cloning and biochemical studies: a *neuronal NOS* or type 1 NOS (nNOS), an *immunologic NOS* or type 2 NOS (iNOS), and an *endothelial NOS* or type 3 NOS (eNOS). The enzymatic activities of eNOS or nNOS are induced by phosphorylation triggered by Ca^{2+} entering cells and binding to calmodulin. In contrast, the regulation of iNOS seems to depend on *de novo* synthesis of the enzyme in response to a variety of cytokines, such as interferon-γ and lipopolysaccharide. The evidence that NO is associated with neurotoxic processes underlying PD comes from studies using experimental models of this disease. NOS inhibitors can prevent 1-methyl-4-phenyl-1,2,3,6-tetrahydropyridine (MPTP)-induced dopaminergic

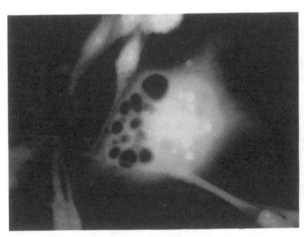

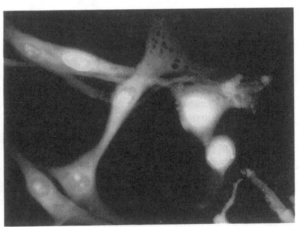

Controlwt   **MT**trans

FIGURE 39.1 A triple fluorochrome analysis of SIN-1-induced apoptosis illustrating genetic resistance of MT$_{trans}$ as compared with control$_{wt}$ striatal fetal stem cells. The apoptotic cells exhibited rounded appearance with reduced neuritogenesis as observed in control$_{wt}$ as compared with MT$_{trans}$ fetal stem cells. Triple fluorochrome analysis of apoptosis was performed using ethidium bromide, DAPI, and acridine orange. DAPI stained primarily intact nuclear DNA, ethidium bromide stained fragmented DNA, and acridine orange stained membrane proteins and RNA. Fluorescence images were captured after 24 h of exposure to SIN-1 (100 μM), using a SpotLite digital camera and ImagePro computer software. The images were merged to obtain detailed information about apoptosis as described in the text. (*Note:* Control$_{wt}$ cells exhibit typical membrane perforations, nuclear DNA fragmentation, and condensation in response to SIN-I.) Triple fluorochromes, acridine orange (green), ethidium bromide (red), and DAPI (blue), were merged to demonstrate SIN-1-induced membrane perforations, DNA condensation, and DNA fragmentation simultaneously. Magnification: 1,200×. (*A color version of this figure follows page 518.*)

DOPAL-INDUCED APOPTOSIS

[Striatal Fetal Stem Cells]   DOPAL: 10 uM

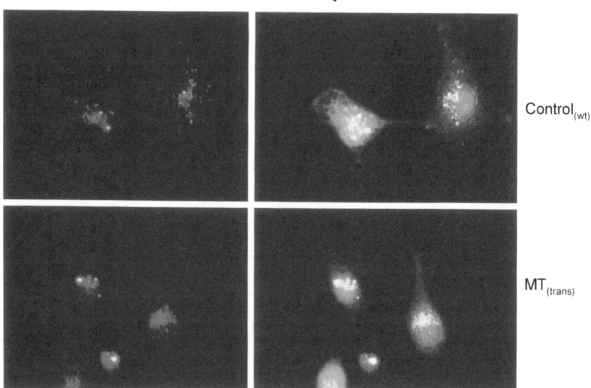

FIGURE 39.2 Digital fluorescence images demonstrating resistance to DOPAL-induced apoptosis in MT$_{trans}$ as compared with control$_{wt}$ striatal fetal stem cells within 24 h. DOPAL induced lipid peroxidation, membrane perforations, and DNA condensation and fragmentation, particularly in control$_{wt}$ cells. Fluorescence images were captured with a SpotLite digital camera and digitized using ImagePro computer software. Fluorochrome JC-l (red) was used as a mitochondrial marker, and FITC-conjugated MT I (green) was used to obtain the structural and functional relationship between mitochondrial membrane potential and MT I expression. Fluorescence images were merged (left panels) to obtain overall information about perinuclear accumulation of mitochondria, as well as membrane perforations in response to DOPAL (right panels). *(A color version of this figure follows page 518.)*

neurotoxicity. Furthermore, NO fosters dopamine depletion, and the said neurotoxicity is averted by nNOS inhibitors such as *7-nitroindazole* working on tyrosine hydroxylase-immunoreactive neurons in substantia nigra pars compacta. Moreover, mutant mice lacking the nNOS gene are more resistant to MPTP neurotoxicity as compared with wild-type littermates. Selegiline, an irreversible inhibitor of monoamine oxidase B, is used in PD as a dopaminergic function-enhancing substance. Selegiline and its metabolite, desmethylselegiline, reduce apoptosis by altering the expression of a number of genes, for instance, superoxide dismutase, Bcl-2, Bcl-xl, NOS, c-Jun, and nicotinamide adenine nucleotide dehydrogenase. The selegiline-induced antiapoptotic activity is associated with prevention of a progressive reduction of mitochondrial membrane potential in preapoptotic neurons. As apoptosis is critical to the progression of neurodegenerative disease, including PD, selegiline or selegiline-like compounds to

be discovered in the future may be efficacious in treating PD.[18]

MITOCHONDRIAL DYSFUNCTION AND NITRIC OXIDE (NO) IN NEURODEGENERATIVE DISEASES

Damage to the mitochondrial electron transport chain has been suggested to be an important factor in the pathogenesis of a range of neurological disorders, such as PD, Alzheimer's disease, multiple sclerosis, stroke, and amyotrophic lateral sclerosis. There is also a growing body of evidence to implicate excessive or inappropriate generation of NO in these disorders. It is now well documented that NO and its toxic metabolite, peroxynitrite (ONOO–), can inhibit components of the mitochondrial respiratory chain leading, if damage is severe enough, to a cellular energy deficiency state.[19]

Oxidative phosphorylation consists of five protein lipid enzyme complexes located in the mitochondrial inner membrane that contain flavins (FMN, FAD), quinoid compounds (coenzyme Q_{10}), and transition metal compounds (iron–sulfur clusters, hemes, protein-bound copper). These enzymes are designated *complex I* (NADH:ubiquinone oxidoreductase; EC 1.6.5.3), *complex II* (succinate:ubiquinone oxidoreductase; EC 1.3.5.1), *complex III* (ubiquinol:ferrocytochrome c oxidoreductase; EC 1.10.2.2), *complex IV* (ferrocytochrome c:oxygen oxidoreductase or cytochrome c oxidase; EC 1.9.3.1), and *complex V* (ATP synthase; EC 3.6.1.34). A defect in mitochondrial oxidative phosphorylation, in terms of a reduction in the activity of NADH CoQ reductase (complex I), has been reported in the striatum of patients with PD. The reduction in the activity of complex I is found in the SN, but not in other areas of the brain, such as globus pallidus or cerebral cortex. Therefore, the specificity of mitochondrial impairment may play a role in the degeneration of nigrostriatal dopaminergic neurons.

This view is supported by the fact that MPTP generating MPP[+] destroys dopaminergic neurons in the SN.[20–21] Lesions produced by the reversible inhibitor of succinate dehydrogenase (complex II), malonate, and the irreversible inhibitor 3-nitropropionic acid closely resemble the histologic, neurochemical, and clinical features of Huntington disease in both rats and nonhuman primates. The interruption of oxidative phosphorylation results in decreased levels of ATP. A consequence is partial neuronal depolarization and secondary activation of voltage-dependent NMDA receptors, which may result in excitotoxic neuronal cell death (secondary excitotoxicity, see also Figure 39.5). The increase in intracellular Ca^{2+} concentration leads to an activation of Ca^{2+}-dependent enzymes, including the constitutive neuronal nitric oxide synthase (cnNOS), which produces NO[•]. NO[•] may react with superoxide anion ($O_2^{•-}$) to form ONOO[−]. Schulz et al.[22] have shown that systemic administration of 7-nitroindazole (7-NI), a relatively specific inhibitor of cnNOS *in vivo*, attenuates lesions produced by striatal malonate injections or systemic treatment with 3-nitropropionic acid or MPTP. Furthermore, 7-NI attenuates increases in lactate production and hydroxyl radical ([•]OH) and 3-nitrotyrosine generation *in vivo*, which may be a consequence of ONOO[−] formation. These results suggest that neuronal nitric oxide synthase (nNOS) inhibitors may be useful in the treatment of neurologic diseases in which excitotoxic mechanisms play a role.[22]

In the CNS, NO may play important roles in neurotransmitter release, neurotransmitter reuptake, neurodevelopment, synaptic plasticity, and regulation of gene expression. However, excessive production of NO following a pathologic insult can lead to neurotoxicity. NO plays a role in mediating neurotoxicity associated with a variety of neurologic disorders, including stroke, PD, and HIV dementia.[4,23]

Due to its ability to modulate neurotransmitter release and reuptake, mitochondrial respiration, DNA synthesis, and energy metabolism, it is not surprising that NO is neurotoxic. Under conditions where NO is abnormally produced, such as when inducible nitric oxide synthase (iNOS) expression is induced in the CNS, dysregulation of normal physiologic activities of NO likely contributes to neuronal dysfunction and subsequently to neuronal death. However, acute toxicity mediated by NO appears to require production of *superoxide anion*.

NO in and of itself is a relatively nontoxic molecule that, in the absence of superoxide anion, will not kill cells, even at extremely high concentrations. In the presence of superoxide anion, however, NO is a potent neurotoxin. The reaction of NO with superoxide anion has the fastest biochemical rate constant currently known, resulting in the formation of the potent oxidant, ONOO[−], which is a lipid-permeable ion with a wider range of chemical targets than NO. It can oxidize proteins, lipids, RNA, and DNA. Neurotoxicity elicited by ONOO[−] formation may have a dual component. ONOO[−] can potently inhibit mitochondrial proteins. ONOO[−] inhibits the function of manganese superoxide dismutase, which could lead to increased superoxide anion formation and increased ONOO[−] formation. Additionally, ONOO[−] is an effective inhibitor of enzymes in the mitochondrial respiratory chain, resulting in decreased ATP synthesis. Second, ONOO[−] efficiently modifies and breaks DNA strands and inhibits DNA ligase, which increases DNA strand breaks. DNA strand breaks activate DNA repair mechanisms. One of the initial proteins activated by DNA damage is the nuclear enzyme poly (ADP-ribose) polymerase (PARP). PARP catalyzes the attachment of ADP-ribose units from NAD to nuclear proteins such as histone and PARP itself. PARP can add hundreds of ADP-ribose units within seconds to minutes of being activated. For every mole of ADP-ribose transferred from NAD, one mole of NAD is consumed and four free-energy equivalents of ATP are required to regenerate NAD to normal cellular levels. Activation of PARP can result in a rapid drop in energy stores. If this drop is severe and sustained, it can lead to impaired cellular metabolism and ultimately cell death.[4]

We[24] developed the multiple fluorescence Comet assay to examine mitochondrial, as well as nuclear, DNA damage simultaneously in a single neuron. The multiple fluorescence Comet assay is performed primarily on a single neuron using alkaline gel electrophoresis and digital fluorescence imaging microscopy. In these experiments, we have conducted the multiple fluorochrome Comet assay to determine MPP[+]-induced neurotoxicity in control$_{wt}$ and aging α-synuclein-overexpressed aging RhO$_{mgko}$ dopaminergic (SK-N-SH) neurons. Dominance of green fluorescence in control$_{wt}$ neurons suggests that, in control$_{wt}$ neurons, MPP$^+$-induced neuronal damage remains restricted primarily to the mitochondrial region [due to synthesis of mitochondrial DNA oxidation product, 8-hydroxy-2-deoxyguanosine (8-OH-2dG)], whereas in α-synuclein-overexpressed aging RhO$_{mgko}$ dopaminergic (SK-N-SH) neurons, the damage was observed in both mitochondrial and nuclear regions. These observations suggest that α-synuclein-overexpressed aging RhO$_{mgko}$ dopaminergic (SK-N-SH) neurons are highly susceptible to MPTP-induced neurotoxicity, which could involve both mitochondrial DNA and nuclear DNA. The damage in the nuclear DNA is illustrated with red fluorescence tails due to ethidium bromide staining, and the damage to mitochondrial DNA is represented by green fluorescence tails due to fluorescein isothiocyanate (FITC)-conjugated antibody to DNA oxidation product, 8-OH-2dG. In addition to nitrite, peroxide-dependent oxidation pathways of dopamine (DA) play a major role in NO neuronal degeneration.[25] Liu et al.[26] reported on the importance of opioids in NO-induced inflammation and neurodegeneration. Furthermore, they reported the beneficial effects of *naloxone*, the opioid receptor antagonist in the treatment of inflammation-related diseases. Inflammation in the brain primarily involves the participation of the two types of glial cells, microglia and astrocytes. Under physiologic conditions, microglia, the resident immune cells in the brain, serve a role of immune surveillance. Astrocytes, on the other hand, act to maintain ionic homeostasis, buffer the action of neurotransmitters, and secrete nerve growth factors. However, glia, especially microglia, readily become activated in response to immunologic challenge and injury. Activation of glia, a process termed *reactive gliosis,* has been observed during the pathogenesis of PD, Alzheimer's disease, multiple sclerosis, and AIDS dementia complex, as well as postneuronal death in cerebral stroke and traumatic brain injury. Activated astrocytes secrete trophic factors in an attempt to enhance neuronal survival. Activated microglia produce a variety of proinflammatory and neurotoxic factors, including the following: cytokines, such as tumor necrosis factor-α (TNF-α) and interleukin-lβ (IL-1β); fatty acid metabolites, such as *eicosanoids*; and free radicals, such as NO and superoxide. NO and IL-lβ are also produced by activated astrocytes. The production and release of these factors constitute a portion of the innate immunity that enables the host to destroy invading pathogens. However, excessive production and accumulation of these factors are deleterious to neurons, although the precise mechanism(s) and relative contribution of individual factors to neurodegeneration remain poorly understood.

Enzyme inhibitors, such as those of iNOS, are certainly valuable tools for studying the mechanism of action for inflammation-mediated neurotoxicity. However, side effects and toxicity often dampen the enthusiasm for their

eventual development into agents of therapeutic value. Analysis of the modulation of the inflammatory process by physiologically relevant agents, in contrast, may prove to be a critical step in the search for clinically useful remedies. One example is the study of opioids and related compounds in the regulation of brain inflammatory process and neuroprotection. Endogenous opioid peptides represent a family of peptides that include dynorphins, enkephalins, and β-endorphins. They are widely distributed throughout the CNS, as well as in peripheral tissues such as cardiac myocytes and lymphatic cells. Through interaction with G-protein distinct coupled transmembrane opioid receptors (δ, κ, and μ), these peptides regulate a wide range of biological activities, including respiration, immune responses, and ion-channel activity. Enkephalins are capable of reducing the lipopolysaccharide (LPS)-induced 1L-1β secretion from microglia. Subsequently, another group of endogenous opioid peptides, dynorphins, were found to inhibit LPS-induced NO production and to afford partial protection against LPS-induced neurotoxicity. A rather unexpected finding from these studies was the discovery that naloxone, a synthetic and nonselective opioid receptor antagonist, was capable of inhibiting LPS-induced microglial activation and production of NO and TNF-α. Most surprising of all was that (+)-naloxone, which is ineffective in blocking the stereospecific binding of opioids to their receptors, was as effective as the opioid receptor antagonist (−)-naloxone in inhibiting microglial activation and affording neuroprotection. This result suggests that the effect of naloxone isomers on microglial activation does not directly involve the opioid receptor mechanism. Analysis of the mechanism of action for naloxone isomers demonstrated that inhibition of microglial superoxide generation best correlated with their effectiveness in neuroprotection. Moreover, naloxone may have prevented the formation of ONOO$^-$ by inhibiting the production of both superoxide and NO. In a rodent model of inflammation-mediated parkinsonism, damage to dopaminergic neurons in the SN, induced by intracerebrally injected LPS, was reduced by systemically infused (−)-naloxone or (+)-naloxone. Naloxone, specifically the (+)-naloxone isomer, which does not bind conventional opioid receptors, may be a promising prototype for drugs with potential therapeutic value in the prevention and treatment of inflammation-mediated neurodegenerative disease such as PD.[26]

NO IN THE PATHOGENESIS OF PD

A potential role for excitotoxic processes in PD has been strengthened by the recent observations that there appears to be a mitochondrially encoded defect in complex I activity of the electron transport chain. An impairment of oxidative phosphorylation will enhance vulnerability to excitotoxicity. SN neurons possess NMDA receptors, and there are glutamatergic inputs into the SN from both the cerebral cortex and the subthalamic nucleus (STN). After activation of excitatory amino acid receptors, there is an influx of calcium followed by activation of nNOS, which can then lead to the generation of ONOO⁻. Studies with MPTP-induced neurotoxicity in both mice and primates have shown that inhibition of nNOS exerts neuroprotective effects (see Figure 39.5). Studies utilizing excitatory amino acid receptor antagonists have been inconsistent in mice but show significant neuroprotective effects in primates. These results raise the prospect that excitatory amino acid antagonists for nNOS inhibitors might be useful in the treatment of PD.[4]

Current concepts of the pathogenesis of PD center on the formation of reactive oxygen species and the onset of oxidative stress leading to oxidative damage to substantia nigra pars compacta (SN PC). Extensive postmortem studies have provided evidence to support the involvement of oxidative stress in the pathogenesis of PD. In particular, these include alterations in brain iron content, impaired mitochondrial function, alterations in the antioxidant protective systems [most notably superoxide dismutase (SOD) and GSR], and evidence of oxidative damage to lipids, proteins, and DNA.[20] Iron can induce oxidative stress, and intranigral injections of iron have been shown to induce a model of progressive parkinsonism. A loss of GSR is associated with incidental Lewy body disease and may represent the earliest biochemical marker of nigral cell loss. GSR depletion alone may not result in damage to nigral neurons but may increase susceptibility to subsequent toxic or free radical exposure. The nature of the free radical species responsible for cell death in PD remains unknown, but there is evidence of involvement of ·OH, ONOO⁻, and NO. Indeed, ·OH and ONOO⁻ formation may be critically dependent on NO formation. Central to many of the processes involved in oxidative stress and oxidative damage in PD are the actions of monoamine oxidase-B (MAO-B). MAO-B is essential for the activation of MPTP to MPP⁺, for a component of the enzymatic conversion of DA to H_2O_2 and for the activation of other potential toxins, such as isoquinolines and β-carbolines. Thus, the inhibition of MAO-B by drugs such as selegiline may protect against activation of some toxins and free radicals formed from the MAO-B oxidation of DA. In addition, selegiline may act through a mechanism unrelated to MAO-B to increase neurotrophic factor activity and up-regulate molecules such as GSH, SOD, catalase, and Bcl-2 protein, which protect against oxidant stress and apoptosis.[27] Consequently, selegiline may be advantageous in the long-term treatment of PD.[21] In addition to selegiline,[20] propargylamine[28] may rescue or protect dopaminergic neurons in PD.[28]

NO may play several roles in processes that lead to neurodegeneration. However, the mechanism by which NO kills cells, particularly neurons, is not fully understood. Toxicity may involve NO itself. NO inhibits a variety of enzymes, including the following: complexes I, II, and IV in the mitochondrial respiratory chain; aconitase, the citric acid cycle enzyme; ribonucleotide reductase, the rate-limiting enzyme in DNA replication; and glyceraldehyde-3-phosphate dehydrogenase in the glycolytic pathway. All these enzymes have a catalytically active nonheme iron–sulfur complex. Complex I activity is decreased in the SN PC of the PD brain and appears to be anatomically specific to the SN PC and disease-specific to PD. Because NO inhibits complex IV rather than complex I, it does not appear that NO accounts for the mitochondrial defects observed in PD. However, *in vitro* studies show that inhibition of complex IV can alter reversible mitochondrial dysfunction induced by agents that inhibit complex I, such as the l-methyl-4-phenyl-2,3-dihydropyridinium (MPDP⁺) ion. (See Figure 39.5.)

It is well known that mitochondrial complex I is down-regulated during the progression of neurodegeneration in PD. Sharma et al.[24] attempted to delineate the possible involvement of the mitochondrial genome in neuronal repair during aging. Aging RhO_{mgko} neurons were prepared by selective inactivation of the mitochondrial DNA with 5 μg/l DNA intercalating agent, ethidium bromide, for 6 to 8 weeks. The $control_{wt}$ neurons exhibited structurally intact neuronal morphology, with long axons and dendrites (Figure 39.3A). The aging Rho_{mgko} neurons exhibited enhanced granularity, mitochondrial aggregation, and elliptical appearance, caused by stunted neuritogenesis, down-regulation of oxidative phosphorylation, and reduced ATP generation (Figure 39.3B). Neuritogenesis and ATP production in aging RhO_{mgko} neurons were regained by transfecting the aging RhO_{mgko} neurons with the mitochondrial genome encoding complex I activity (Figure 39.3C). The transfection was done using pEGFP-N1 vector, Qiagen Effectine transfection reagent, and DNA enhancer as per the manufacturer's recommendations. The transfected neurons were selected using G-148, and enriched by limiting dilution technique. (See Figure 39.3.)

There is also evidence that NO can displace iron from its binding site on *ferritin*, an iron-storage protein, and consequently promote lipid peroxidation. NO also can influence iron metabolism at the post-transcriptional level by interacting with cytosolic aconitase. Cytosolic aconitase has dual functions that are regulated by NO. In the absence of NO, it functions as cytosolic aconitase, but in the presence of NO, it functions as the iron-responsive element binding protein to the iron-responsive element.[5]

The role of NO in 6-hydroxydopamine-induced parkinsonism has been established.[29] Riobo et al.[30] analyzed the potential reaction between 6-hydroxydopamine and nitric oxide. The results showed that NO reacts with the deprotonated form of 6-hydroxydopamine at pH 7 and 37°C with a second-order rate constant of 1.5×10^3 M⁻¹s⁻¹

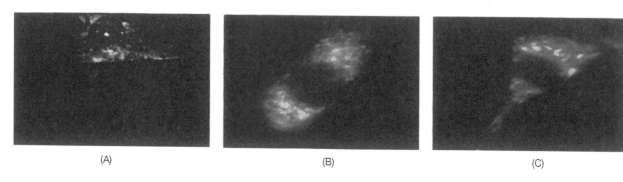

(A)   (B)   (C)

FIGURE 39.3 Multiple fluorochrome analysis of dopaminergic (SK-N-SH) neurons. Multiple fluorochrome analysis of control$_{wt}$ (*A*), aging RhO$_{mgko}$ (*B*), and aging RhO$_{mgko}$ dopaminergic (SK-N-SH) neurons, transfected with mitochondrial genome encoding complex I activity (*C*), is presented. Aging RhO$_{mgko}$ neurons were prepared as described earlier. Fluorescence images were captured using a SpotLite digital camera and analyzed with ImagePro Computer software. Selective knocking out of mitochondrial genome induced mitochondrial aggregation and changes in the neuronal morphology, represented by elliptical appearance due to reduced neuritogenesis (*B*). Neuritogenesis reappeared following transfection of aging RhO$_{mgko}$ neurons with mitochondrial genome (*C*). Red: JC-l (molecular marker of mitochondrial membrane potential); green: FITC-conjugated MT I; blue: nuclear stain DAPI.[24,31]

as calculated by the rate of NO decay measured with an amperometric sensor. Accordingly, the rates of formation of 6-hydroxydopamine quinone were dependent on NO concentration. The coincubation of NO and 6-hydroxydopamine with either bovine serum albumin or α-synuclein led to tyrosine nitration of the protein, in a concentration dependent-manner and sensitive to SOD. These findings suggest the formation of peroxynitrite during the redox reactions following the interaction of 6-hydroxydopamine with NO.

Hunot et al.,[31] using immunohistochemistry and histochemistry, analyzed the production systems of NO in the mesencephalon of four patients with idiopathic Parkinson's disease and three matched control subjects. Using specific antibodies directed against the inducible isoforms of NOS, they found that this isoform was present solely in glial cells displaying the morphological characteristics of activated macrophages. Immunohistochemical analysis performed with antibodies against the neuronal isoforms of nitric oxide synthase, however, revealed perikarya and processes of neurons but no glial cell staining. The number of NOS-containing cells was investigated by histoenzymology, using the NADPH-diaphorase activity of NOS. Histochemistry revealed (a) a significant increase in NADPH-diaphorase-positive glial cell density in the dopaminergic cell groups characterized by neuronal loss in PD and (b) a neuronal loss in PD that was twofold greater for pigmented NADPH-diaphorase-negative neurons than for pigmented NADPH-diaphorase-positive neurons.

These data suggest a potentially deleterious role of glial cells producing excessive levels of nitric oxide in PD, which may be neurotoxic for a subpopulation of dopaminergic neurons, especially those not expressing NADPH-diaphorase activity. However, it cannot be

excluded that the presence of glial cells expressing NOS in the SN of patients with PD represents a consequence of dopaminergic neuronal loss.[31]

Knott et al.[32] examined the cellular distribution of pro- and anti-inflammatory molecules in human parkinsonian and neurologically normal SN and caudate-putamen postmortem. An up-regulation of nitric oxide synthase- and cyclo-oxygenase-1-and-2-containing amoeboid microglia was found in parkinsonian, but not control, nigra. Astroglia contained low levels of these molecules in both groups. Lipocortin-1-immunoreactive amoeboid microglia were present within the astrocytic envelope of neurons adjacent to or within glial scars in parkinsonian nigra only. *Lipocortin-1* is known to have neuroprotective and anti-inflammatory properties. Up-regulation of nitric oxide synthase is generally associated with neurodestruction, whereas *prostaglandin* synthesis may be either neurodestructive or protective. The balance of these molecules is likely to be decisive in determining neuronal survival or demise. Barthwal et al.[33] showed an overexpression of nNOS in PD. Similarly, Kuiper et al.[34] demonstrated the levels of L-glutamate, L-arginine, and L-citrulline were altered in cerebrospinal fluids of patients with PD, multiple system atrophy, and Alzheimer's disease. Qureshi et al.[35] supported the possibility that cerebrospinal fluid concentration of nitrite was elevated in PD. However, the data by Molina et al.[36] suggested that the plasma levels of nitrate were unrelated to the risk of developing PD.

Lo et al.[37] hypothesized that inhibition of NOS can prevent the destruction of dopaminergic neurons in mammals. To determine if NOS gene polymorphism affects the 5′ flanking region that is immediately upstream of the transcription start site lying between the TATA element and CAATT boxes in PD, and differs significantly between patients with PD and normal controls, they studied genetic polymorphism in that region of the nNOS gene in Chinese patients with PD living in Taiwan. The results indicate that the allele size distribution in that region was statistically significant between patients with PD and normal subjects.

Oxidative stress is thought to be involved in the mechanism of nerve cell death in PD. Among several toxic oxidative species, NO has been proposed as a key element on the basis of the increased density of glial cells expressing inducible iNOS in the SN of patients with PD. However, the mechanism of iNOS induction in the CNS is poorly understood, especially under pathological conditions. Because cytokines and FcεRII/CD23 antigen have been implicated in the induction of iNOS in the immune system, Hunot et al.[31] investigated their role in glial cells *in vitro* and in the SN of patients with PD and matched control subjects. We show that, *in vitro*, interferon-γ (IFN-γ) together with interleukin-1β(Il-1β) and tumor necrosis factor-α (TNF-α) can induce the expression of CD23 in glial cells. Ligation of CD23 with specific antibodies resulted in the induction of iNOS and the subsequent release of NO. The activation of CD23 also led to an up-regulation of TNF-α production, which was dependent on NO release. In the SN of PD patients, a significant increase in the density of glial cells expressing TNF-α, Il-1β, and IFN-γ was observed. Furthermore, although CD23 was not detectable in the SN of control subjects, it was found in both astroglial and microglial cells in parkinsonian patients. These data demonstrate the existence of a cytokine/CD23-dependent activation pathway of iNOS and of proinflammatory mediators in glial cells and their involvement in the pathophysiology of PD.

Böckelmann et al.[38] examined postmortem putamen of PD patients and control subjects for distribution patterns of NOS-containing neurons, using the NADP-diaphorase technique). The ratio of positively stained neurons and the total number of cells (control: $1,120 \pm 69$ per mm^2, $n = 5$; PD: 575 ± 164 mm^2, $n = 5$) shows striking differences between controls and PD patients. Their findings give reason to conclude that NADPH-diaphorase positive structures may have pathogenetic importance in degenerative processes in PD putamen.

Eve et al.[39] studied the expression of nitric oxide synthase (NOS) mRNA in post mortem brain in putamen, globus pallidus, and STN of neurologically normal control subjects and patients with PD using *in situ* hybridization histochemistry. In PD, a significant increase in NOS mRNA expression was observed in the dorsal two-thirds of the STN with respect to the ventral one-third of the STN. A significant increase in NOS mRNA expression per cell in the medial medullary lamina of the globus pallidus was also observed in PD. NOS mRNA expression was significantly reduced in PD putamen. These findings provide evidence of increased activity of STN neurotrans-

mitter systems in PD and demonstrate in any species that basal ganglia nitric oxide systems can be selectively regulated in response to changes in dopaminergic input.

Although NO does not affect the survival of grafted dopaminergic neurons,[40] depletion of glutathione changes the neurotrophic effects of NO in midbrain cultures into neurotoxic. Under these conditions, NO triggers a programmed cell death with markers of both apoptosis and necrosis characterized by an early step of free radicals production followed by a late requirement for signaling on the soluble guanylate cyclase/cyclic GMP/PKG pathway.

ONOO⁻: A PUTATIVE CYTOTOXIN

The superoxide anion rapidly reacts with NO, yielding ONOO⁻, and this reaction occurs *in vivo* according to the following scheme:

$$NO^{\bullet} + O_2^{\bullet -} \rightarrow ONOO^-$$

The rate constant of this reaction is near the diffusion-controlled limit ($4-7 \times 10^9$ M^{-1} s^{-1}). The half-life of ONOO⁻ at 37°C and pH 7.4 is approximately 1 sec. ONOO⁻ is in equilibrium with peroxynitrous acid:

$$ONOO^- + H^+ \rightleftharpoons ONOOH$$

The ONOO⁻ anion itself is relatively stable, but peroxynitrous acid rapidly rearranges to form nitrate. Therefore, ONOO⁻ is practically stable in alkaline solutions. Although it has long been thought that peroxynitrous acid decomposes to form nitrate and hydroxy radicals, it is now believed that *peroxynitrous acid* (via an activated state: HOONO⁻) reacts with biological substrates in a hydroxyl radical-like way. Consequently, free radicals are probably not formed during the self-decomposition of peroxynitrous acid. Nonetheless, recent evidence suggests that the reaction of ONOO⁻ with carbon dioxide is the most important route for ONOO⁻ in biological environments, where carbon dioxide is relatively abundant. In short, ONOO⁻ reacts with carbon dioxide to form the nitrosoperoxycarbonate anion, which subsequently rearranges to form the nitrocarbonate anion,

$$ONOO^- + CO_2 \rightarrow ONOOCO_2^- \rightarrow O_2NOCO_2^-$$

The nitrocarbonate anion is postulated to be the proximal oxidant of ONOO⁻ mediated reactions in biological environments. Nitrocarbonate can undergo hydrolysis, can oxidize substrates via one- and two-electron transfers, and can nitrosylate substrates. Carbon dioxide concentration is therefore of crucial importance for ONOO⁻-mediated oxidation and nitrosylation. The exact biochemical fate of ONOO⁻ in biological systems, however, is very complex and is as yet not completely clear.[41]

ONOO⁻-MEDIATED LOSS OF DOPAMINERGIC NEURONS

Involvement of NO in the destruction of nigral dopaminergic neurons in PD has not been proven, but the finding of elevated iron levels in the SN PC, impaired complex I in the SN PC, increased lipid peroxidation, and DNA damage in the nigrostriatal system implicate NO as a mediator of neuronal oxidative damage in PD, although the hydroxyl free radical (˙OH) is also commonly associated with free radical mediated oxidative damage.

Cellular processes with which excess nitric oxide can interact to cause dopaminergic cell death are inhibition of cytochrome c oxidase, ribonucleotide reductase, mitochondrial complex I, II, IV, superoxide dismutase, glyceraldehyde-3-phosphate dehydrogenase; activation or initiation of DNA strand breakage, poly (ADP-ribose synthase, lipid peroxidation, and protein oxidation; release of iron (II), and increased generation of toxic radicals by fast reaction with the superoxide radical.[19,39]

Electron paramagnetic resonance spectroscopy has been used to study interactions of iron proteins in cells with NO. Nitrosyl complexes such as sodium nitroprusside, which are added as experimental NO generators, themselves produce paramagnetic nitrosyl species, which may be seen by EPR. Cammack et al.[42] have used this to observe the effects of nitroprusside on clostridial cells. After growth in the presence of sublethal concentrations of nitroprusside, the cells have been converted into other, presumably less toxic, nitrosyl complexes such as $(RS)_2Fe(NO)_2$. Nitric oxide is cytotoxic, partly due to its effects on mitochondria. This is exploited in the destruction of cancer cells by the immune system. The targets include iron–sulfur proteins. It appears that species derived from NO such as ONOO⁻ may be responsible. Addition of peroxynitrite to mitochondria led to depletion of the EPR-detectable iron–sulfur clusters. Paramagnetic complexes are formed *in vivo* from hemoglobin, in conditions such as experimental endotoxic shock. This has been used to follow the course of production of NO by macrophages. Cammack et al.[42] have examined the effects of suppression of NOS using biopterin antagonists. Another method is to use an injected NO-trapping agent, Fe-diethyldithiocarbamate, to detect accumulated NO by EPR. In this way, they have observed the effects of depletion of serum arginine by arginase. In brains from victims of PD, a nitrosyl species identified as nitrosyl hemoglobin has been observed in substantia nigra. This is an indica-

tion for the involvement of NO or a derived species in the damage to this organ.

Discoveries over the past 10 years indicate that crucial features of neuronal communication, blood vessel modulation, and immune response are mediated by a remarkably simple chemical, NO.[43–44] Endogenous NO is generated from arginine by a family of three distinct calmodulin-dependent NOS enzymes. NOS from endothelial cells (eNOS) and neurons (nNOS) are both constitutively expressed enzymes, whose activities are stimulated by increases in intracellular calcium. Immune functions for NO are mediated by a calcium-independent inducible NOS (iNOS). Expression of iNOS protein requires transcriptional activation, which is mediated by specific combinations of cytokines. All three NOS use NADPH as an electron donor and employ five enzyme cofactors to catalyze five-electron oxidation of arginine to NO with stoichiometric formation of citrulline.

The highest levels of NO throughout the body are found in neurons, where NO functions as a unique messenger molecule. In the autonomic nervous system, NO functions as a major nonadrenergic noncholinergic neurotransmitter. This nonadrenergic, noncholinergic pathway plays a particularly important role in producing relaxation of smooth muscle in the cerebral circulation and the gastrointestinal, urogenital, and respiratory tracts. Dysregulation of NOS activity in autonomic nerves plays a major role in diverse pathophysiological conditions including migraine headache. In the brain, NO functions as a neuromodulator and appears to mediate aspects of learning and memory.

Functions for NO in the brain remain less certain. Because NO is a uniquely diffusible mediator, it was proposed on theoretical grounds that NO may mediate neuronal plasticity, which underlies aspects of both development and information storage in brain. Evidence for NO involvement in synaptic plasticity has accumulated steadily. At the cellular level, NO signaling appears to be essential for two forms of neuronal plasticity: long-term potentiation in the hippocampus and long-term depression in the cerebellum.

NO appears to mediate synaptic plasticity by potentiating neurotransmitter release. In several model systems, NOS inhibitors such as nitroarginine blocked the release of neurotransmitters. In brain synaptosomes, the release of neurotransmitters evoked by stimulation of NMDA receptors is blocked by nitroarginine. Presumably, glutamate acts at NMDA receptors on NOS terminals to stimulate the formation of NO, which diffuses to adjacent terminals to enhance neurotransmitter release so that blockade of NO formation inhibits release. In addition to regulating glutamate release, NO can also regulate secretion of hormones and neuropeptides. Regulation of hormone secretion by NO has been most convincingly demonstrated in the hypothalamus.[45]

MESENCEPHALIC DOPAMINERGIC NEURONS ARE RESISTANT TO CYTOTOXICITY INDUCED BY NO

Sawada et al.[46] investigated the intracellular mechanism that protects dopaminergic neurons against NO toxicity in rat mesencephalic cultures. $ONOO^-$ anion, an active metabolite of NO, caused significant cytotoxic effects against dopaminergic and nondopaminergic neurons, but NO caused cytotoxic effects restricted to nondopaminergic neurons. In addition, they studied the effects of ascorbate, an antioxidant, on NO-induced neurotoxicity against dopaminergic neurons and found that coadministration of ascorbate failed to affect resistance against NO-induced neurotoxicity. These findings suggest that the protecting mechanism from NO neurotoxicity in dopaminergic neurons is based on inhibition of conversion of NO to $ONOO^-$ anion, is dependent on the NO redox state, and is possibly due to suppression of superoxide anion production. Furthermore, they investigated NO-induced neurotoxicity with or without pretreatment with sublethal doses of methylphenylpyridium ion (MPP^+). Following pretreatment with 1 µM MPP^+, which did not show significant cytotoxic effects against dopaminergic neurons, NO demonstrated significant cytotoxicity. Therefore, MPP^+ may inhibit the protecting systems from NO neurotoxicity in dopaminergic neurons.

GLIAL CELLS, INFLAMMATORY REACTIONS, AND THE PRODUCTION OF NO

NO is involved in LPS-induced dopamine cell loss. Gayle et al.[47] characterized the effects of the proinflammatory bacteriotoxin LPS on the number of tyrosine hydroxylase immunoreactive (THir) cells in primary mesencephalic cultures. LPS (10 to 80 µg/ml) selectively decreased THir cells and increased culture media levels of interleukin-1β (IL-1β) and tumor necrosis factor-α (TNF-α) as well as nitrite (an index of nitric oxide (NO) production). Cultures exposed to both LPS and neutralizing antibodies to IL-1β or TNF-α showed an attenuation of the LPS-induced THir cell loss by at least 50% in both cases. Inhibition of iNOS by N^ω-nitro-L-arginine methyl ester (L-NAME) did not affect LPS toxicity, but increased the LPS-induced levels of both TNF-α and IL-1β. These findings suggest that neuroinflammatory stimuli which lead to elevations in cytokines may induce DA neuron cell loss in a NO-independent manner and contribute to PD pathogenesis.

NO and other reactive nitrogen species appear to play several crucial roles in the brain. These include physiological processes such as neuromodulation, neurotransmission, and synaptic plasticity, and pathological processes such as neurodegeneration and neuroinflammation. There is increasing evidence that glial cells in the central nervous system can produce NO *in vivo* in response to stimulation by cytokines and that this production is mediated by iNOS. Although the etiology and pathogenesis of the major neurodegenerative and neuroinflammatory disorders (Alzheimer's disease, amyotrophic lateral sclerosis, PD, Huntington's disease, and multiple sclerosis) are unknown, numerous recent studies strongly suggest that reactive nitrogen species play an important role. Furthermore, these species are probably involved in brain damage following ischemia and reperfusion, Down's syndrome, and mitochondrial encephalopathies. Recent evidence also indicates the importance of cytoprotective proteins such as heat shock proteins, which appear to be critically involved in protection from nitrosative and oxidative stress.[48]

Perturbation of the cellular oxidant/antioxidant balance has been suggested to be involved in the neuropathogenesis of several disease states, including stroke, PD, and Alzheimer's disease, as well as "normal" physiological aging. Reactive oxygen species are constantly produced in the course of aerobic metabolism and, under normal conditions, there is a steady-state balance between prooxidants and antioxidants. Most of the reactive oxygen species produced by healthy cells results from "leakage" or short circuiting of electrons at several specific locations within the cell, which then become sources of free radical production. These include the mitochondrial respiratory chain, the enzyme *xanthine dehydrogenase* and, to a lesser extent, *arachidonic acid* metabolism and autooxidation of cathecholamines or hemoproteins. However, when the rate of free radical generation exceeds the capacity of antioxidant defenses, oxidative stress ensues, causing extensive damage to DNA, proteins, and lipids.

The brain has a large potential oxidative capacity due to the high level of tissue oxygen consumption. However, the ability of the brain to withstand oxidative stress is limited because of the following anatomical, physiological, and biochemical reasons:

1. High content of easily oxidizable substrates, such as polyunsaturated fatty acids and catecholamines
2. The relatively low levels of antioxidants, such as glutathione and vitamin E, and antioxidant enzymes, such as GSH peroxidase, catalase, and SOD
3. The endogenous generation of reactive oxygen free radicals via several specific reactions
4. The elevated content of iron in specific areas of the human brain, such as globus pallidus and SN

However, cerebrospinal fluid has very little iron-binding capacity because of its low content of transferrin. Furthermore, the CNS contains nonreplicating neuronal cells that, once damaged, may be permanently dysfunctional or committed to programmed cell death (apoptosis).[48]

ONOO⁻ is a powerful oxidant and can nitrate aromatic amino acid residues such as tyrosine to form nitrotyrosine. Nitration to form 3-nitrotyrosine can occur on either free or protein-bound tyrosine. Since the half-life of ONOO⁻ at physiological pH is short, the detection of 3-nitrotyrosine in tissues is often used as a biological marker of ONOO⁻ generation *in vivo*. Not only is 3-nitrotyrosine a marker for ONOO⁻ production, it also appears that the nitration of specific proteins by ONOO⁻ may be relevant to brain pathophysiology.

Inflammatory reaction is thought to be an important contributor to neuronal damage in neurodegenerative disorders such as Alzheimer's disease, PD, multiple sclerosis, amyotrophic lateral sclerosis, and the parkinsonism dementia complex of Guam. ONOO⁻ is a strong oxidizing and nitrating agent, which can react with all classes of biomolecules. In the CNS, it can be generated by microglial cells activated by proinflammatory cytokines or β-amyloid peptide, and by neurons in hyperactivity of glutamate neurotransmission, mitochondrial dysfunction, and depletion of L-arginine or tetrahydrobiopterin. The first two situations correspond to cellular responses to an initial neuronal injury and the ONOO⁻ formed only exacerbates the inflammatory process, whereas in the third situation the ONOO⁻ generated directly contributes to the initiation of the neurodegenerative process.[47]

ONOO⁻, generated in inflammatory processes, is capable of nitrating and oxidizing biomolecules, implying a considerable impact on the integrity of cellular structures. Cells respond to stressful conditions by the activation of signaling pathways, including receptor tyrosine kinase-dependent pathways such as mitogen-activated protein kinases and the phosphoinositide-3-kinase/Akt pathway. ONOO⁻ affects signaling pathways by nitration as well as by oxidation. Whereas nitration of tyrosine residues by ONOO⁻ modulates signaling processes relying on tyrosine phosphorylation and dephosphorylation, oxidation of phosphotyrosine phosphatases may lead to an alteration in the tyrosine phosphorylation/dephosphorylation balance. The flavanol (−)-epicatechin is a potent inhibitor of tyrosine nitration and may be employed as a tool to distinguish signaling effects due to tyrosine nitration from those that are due to oxidation reactions.[50]

Experimental evidence has implicated oxidative stress in the development of PD, amyotrophic lateral sclerosis, and other degenerative neuronal disorders. Recently, ONOO⁻, which is formed by the nearly diffusion-limited reaction of NO with superoxide, has been suggested to be a mediator of oxidant-induced cellular injury. The potential role of ONOO⁻ in the pathology of Parkinson's disease was evaluated by examining its effect on DOPA synthesis in PC12 pheochromocytoma cells. ONOO⁻ was generated from the compound 3-morpholinosydnonimine

(SIN-1), which releases superoxide and NO simultaneously. Exposure of PC12 cells to ONOO⁻ for 60 min greatly diminished their ability to synthesize DOPA without apparent cell death. The inhibition was due neither to the formation of free nitrotyrosine nor to the oxidation of DOPA by ONOO⁻. The inhibition in DOPA synthesis by SIN-1 was abolished when superoxide was scavenged by the addition of SOD. These data indicated that neither NO nor H_2O_2 generated by the dismutation of superoxide is responsible for the SIN-1-mediated inhibition of DOPA production. The inhibition of DOPA synthesis at high concentrations of SIN-1 persisted even after removal of SIN-1. The inactivation of the tyrosine hydroxylase may be responsible for the significant decline in DOPA formation by ONOO⁻. Inactivation of tyrosine hydroxylase may be part of the initial insult in oxidative damage that eventually leads to cell death.[51]

Various neurotoxins, such as 6-hyroxydopamine, *salsolinol*, MPP⁺, and *rotenone* induce mitochondrial and nuclear damage in the dopaminergic neurons. In view of the above, Ebadi and Hiramatsu[20] studied apoptosis in response to these neurotoxins in SK-N-SH neurons. All neurotoxins induced changes in neuronal morphology, represented by rounded appearance, and perinuclear aggregation of mitochondria, and caspase-3 activation. These data are interpreted to suggest that these agents induce neurotoxicity primarily through caspase-3 activation and mitochondrial degeneration (Figure 39.4).

Increased NO production has been implicated in many examples of neuronal injury such as those caused by methamphetamine and MPTP to dopaminergic cells, presumably through the generation of the potent oxidant ONOO⁻. DA is a reactive molecule that, when oxidized to DA quinone, can bind to and inactivate proteins through the sulfhydryl group of the amino acid cysteine. In a study, Lavoi and Hastings[52] sought to determine if ONOO⁻ could oxidize DA and participate in this process of protein modification. They measured the oxidation of the catecholamine by following the binding of [³H]DA to the sulfhydryl-rich protein alcohol dehydrogenase. The results showed that ONOO⁻ oxidized DA in a concentration- and pH-dependent manner. Furthermore, the resulting DA-protein conjugates were predominantly 5-cysteinyl-DA residues. In addition, it was observed that ONOO⁻ decomposition products such as nitrite were also effective at oxidizing DA. These data suggest that the generation of NO and subsequent formation of ONOO⁻ or nitrite may contribute to the selective vulnerability of dopaminergic neurons through the oxidation of DA and modification of protein.[52]

Immunophilin and NOS have been implicated in the pathogenesis of PD. Araki et al.,[53] using receptor autoradiographic technique, studied the sequential changes in FK-506 binding proteins, NOS and dopamine uptake sites in the brain 1 to 8 weeks after unilateral 6-hydroxy-

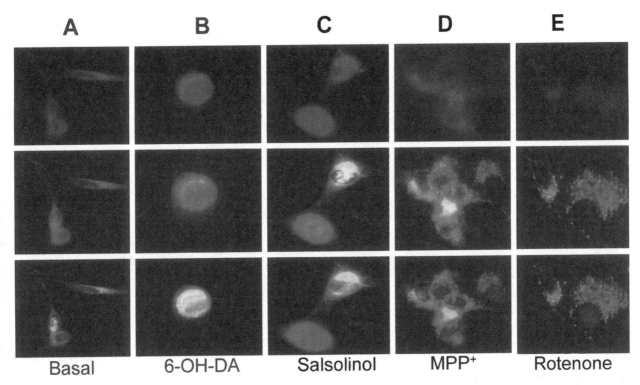

FIGURE 39.4 Multiple fluorochrome digital fluorescence imaging microscopic analysis of apoptosis in response to various parkinsonian neurotoxins is presented. Overnight exposure to parkinsonian neurotoxins induced apoptotic changes characterized by rounded appearance, enhanced caspase-3 activation, and mitochondrial aggregation in the perinuclear region, without any typical evidence of nuclear DNA fragmentation, suggesting that early changes occur most predominantly to down-regulate the mitochondrial genome. Fluorescence images were captured by a SpotLite digital camera equipped with ImagePro computer software. Parkinsonian neurotoxins: 6-hydroxydopamine (6-OH-DA), salsolinol, and MPP+, 100 μM each; rotenone, 100 nM. Green: FITC-conjugated anti-caspase-3; red: JC-1; blue: DAPI.[24]

dopamine injection of the medial forebrain bundle in rats. [³H]FK-506, [³H]L-N^G-nitro-arginine, and [³H]mazindol were used to label FK-506 binding proteins (immunophilin), NOS and dopamine uptake sites, respectively. [³H]FK-506 binding showed ~13 to 25% increase in the ipsilateral striatum from 2 to 8 weeks after degeneration of nigrostriatal pathway. However, no significant change in [³H]FK-506 binding was observed in the ipsilateral substantia nigra during the post lesion periods. In the contralateral side, [³H]FK-506 binding also showed ~13 to 25% increase in the striatum from 2 to 8 weeks post lesion. The SN showed a 21% increase in [³H] FK-506 binding only 2 weeks after the lesioning. On the other hand, [³H]L-N^G-nitro-arginine binding showed ~21 to 31% increase in the parietal cortex and striatum 1 to 2 weeks post lesion. In the contralateral side, a 21% increase in [³H]L-N^G-nitro-arginine binding was found in the dorsolateral striatum only 1 week post lesion. In contrast, degeneration of nigrostriatal pathway caused a conspicuous loss of [³H]mazindol binding in the ipsilateral striatum (87 to 96%), substantia nigra (36 to 73%), and ventral tegmental area (91 to 100%) during the post lesion periods. In the contralateral side, no sig-

nificant changes in [³H]mazindol binding were observed in these areas up to 8 weeks postlesion. The present study demonstrates that unilateral injection of 6-hydroxydopamine into the medial forebrain bundle of rats can cause a significant increase in [³H]FK-506 and [³H]L-N^G-nitro-arginine bindings in the brains. In contrast, a marked reduction in [³H] mazindol binding is observed in the brains after the lesioning, indicating severe damage to nigrostriatal dopaminergic pathway. These results suggest that immunophilin and NOS may play some role in the pathogenesis of neurodegenerative disorders such as PD.

SOD nNOS NEURONS FROM NO-MEDIATED NEUROTOXICITY

In the nervous system, nNOS is localized in discrete populations of neurons in the cerebellum, cortex, striatum, olfactory bulb, hippocampus, basal forebrain, and brain stem. Excess production of NO via nNOS has been implicated in various neurotoxic paradigms. Excess glutamate acting via NMDA receptors may mediate cell death in focal cerebral ischemia, trauma, and epilepsy and in neurodegenerative disease such as Huntington's disease, Alzheimer's disease, and PD.

Cu/Zn SOD is among the key cellular enzymes by which neurons and other cells detoxify free radicals and protect themselves from damage. Down-regulation of SOD causes apoptotic death of neurons. The postulated molecular mechanisms by which superoxide anions produce its toxicity are according to the following scheme:

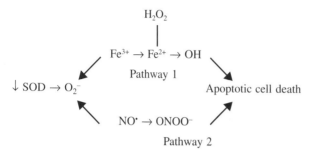

One pathway involves superoxide purely as a reducing agent for transition metal ions such as Fe^{3+}. In this scheme, the reduced metal ion catalyzes the conversion of H_2O_2 to the highly reactive and destructive hydroxyl radical ($^{\cdot}OH$). The other pathway invokes the interaction of superoxide with NO, leading to formation of $ONOO^-$. $ONOO^-$ then can be protonated and rapidly decomposed to a strong oxidant. The toxicity of $ONOO^-$ is hereby depicted.[54]

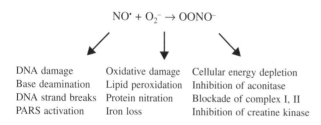

DNA damage	Oxidative damage	Cellular energy depletion
Base deamination	Lipid peroxidation	Inhibition of aconitase
DNA strand breaks	Protein nitration	Blockade of complex I, II
PARS activation	Iron loss	Inhibition of creatine kinase

PERGOLIDE, A DOPAMINE RECEPTOR AGONIST, PROVIDES PROTECTION AGAINST NITRIC OXIDE FREE RADICALS

Gómez-Vargas et al.[55] demonstrated that pergolide, a widely used DA agonist, has free radical scavenging and antioxidant activities. Using a direct detection system for NO radical ($NO^{\cdot}$) by electron spin resonance (ESR) spectrometry in an *in vitro* $^{\cdot}NO$-generating system, they examined the quenching effects of pergolide on the amount of $NO^{\cdot}$ generated. Pergolide dose-dependently scavenged $NO^{\cdot}$. In the competition assay, the IC_{50} value for pergolide was estimated to about 30 µM. Pergolide also dose-dependently attenuated the hydroxyl radical ($^{\cdot}OH$) signal in an *in vitro* $FeSO_4$-H_2O_2 ESR system with an approximate IC_{50} value of 300 µM. Furthermore, this agent significantly inhibited phospholipid peroxidation of rat brain homoge-

nates in *in vitro* experiments and after repeated administration (0.5 mg/kg/24 h, i.p. for 7 days). These findings suggest a neuroprotective role for pergolide on dopaminergic neurons due to its free radical scavenging and antioxidant properties. Similar results were found by Nishibayashi et al.[56]

THE ROLE OF nNOS IN MPTP-INDUCED DOPAMINERGIC NEUROTOXICITY

MPTP is used extensively in various mammalian species to produce an experimental model of PD, a common and disabling neurodegenerative disorder of unknown cause. In humans and nonhuman primates, MPTP induces irreversible and severe motor abnormalities almost identical to those observed in PD. In both monkeys and mice, MPTP replicates many of the biochemical and neuropathological changes in the nigrostriatal dopaminergic pathway found in PD. This includes a marked reduction in the levels of striatal DA and its metabolite dihydroxyphenylacetic acid (DOPAC) and homovanillic acid. In addition, as in patients with PD, animals that receive MPTP show significant reductions in the number of DA cell bodies in the SN PC. In monkeys, MPTP induces the formation of intraneuronal eosinophilic inclusions resembling Lewy bodies, a neuropathological hallmark of PD. These similarities provide appealing hints that the MPTP model may lead to important new insights into the pathogenesis of PD.[54] Smith et al.[58] showed that NO appears to be necessary for $^{\cdot}OH$ generation in MPP$^+$ toxicity and may play a role in neuronal generation in PD. Przedborski et al.[57] injected MPTP into mice in which NOS was inhibited by 7-NI in a time- and dose-dependent fashion. 7-NI dramatically protected MPTP-injected mice against indices of severe injury to the nigrostriatal dopaminergic pathway, including reduction in striatal DA contents, decreases in numbers of nigral tyrosine hydroxylase-positive neurons, and numerous silver-stained degenerating nigral neurons (see Figure 39.3). The resistance of 7-NI-injected mice to MPTP is not due to alterations in striatal pharmacokinetics or content of MPP$^+$, the active metabolite of MPTP. To study specifically the role of nNOS, MPTP was administered to mutant mice lacking the nNOS gene. Mutant mice are significantly more resistant to MPTP-induced neurotoxicity compared with wild-type littermates. These results indicate that neuronally derived NO mediates, in part, MPTP-induced neurotoxicity.

The similarity between the MPTP model and PD raises the possibility that NO may play a significant role in the etiology of PD. Moreover, Klivenyi et al.[59] showed that inhibition of nNOS protects against MPTP toxicity. Liberatore et al.[60] showed that, after administration of MPTP to mice, there was a robust gliosis in the SN PC associated with significant up-regulation of iNOS. These changes preceded or paralleled MPTP-induced dopamin-ergic neurodegeneration. They also show that mutant mice lacking the iNOS gene were significantly more resistant to MPTP than their wild-type littermates. This study demonstrates that iNOS is important in the MPTP neurotoxic process and indicates that inhibitors of iNOS may provide protective benefit in the treatment of Parkinson's disease. Obata and Yamanaka[61] examined the effect of L-NAME, a NOS inhibitor, on extracellular potassium ion concentration ($[K^+]_o$)-enhanced $\cdot$OH generation due to MPP^+ in the rat striatum. Rats were anesthetized, and sodium salicylate in Ringer's solution (0.5 nmol/μl per min) was infused through a microdialysis probe to detect the generation of $\cdot$OH as reflected by the nonenzymatic formation of 2,3-dihydroxybenzoic acid (DHBA) in the striatum. Induction of KCl (20, 70, and 140 mM) increased MPP^+-induced $\cdot$OH formation trapped as DHBA in a *concentration-dependent* manner. However, the application of L-NAME (5 mg/kg i.v.) abolished the $[K^+]_o$-depolarization-induced $\cdot$OH formation with MPP^+. DA (10 μM) also increased the levels of DHBA due to MPP^+. However, the effect of DA after application of L-NAME did not change the levels of DHBA. On the other hand, the application of allopurinol (20 mg/kg i.v. 30 min prior to study), a xanthine oxidase inhibitor, abolished both $[K^+]_o$- and DA-induced $\cdot$OH generation. Moreover, when iron (II) was administered to MPP^+ and then $[K^+]_o$ (70 mM)-pretreated animals, there was a marked increase in the level of DHBA. However, when corresponding experiments were performed with L-NAME-pretreated animals, the same results were obtained. Therefore, NOS activation may not be related to Fenton-type reaction via $[K^+]_o$ depolarization-induced $\cdot$OH generation. The present results suggest that $[K^+]_o$-induced depolarization augmented MPP^+-induced $\cdot$OH formation by enhancing NO synthesis. Furthermore, inhibition of nNOS prevented MPTP-induced Parkinsonism in baboons[62] and in mice.[63] Moreover, inhibition of nNOS by 7-NI protected against MPTP-induced neurotoxicity in mice.[64]

Cutillas et al.[65] found that a significant loss of dopamine was found in rat striatal slices incubated with MPP^+ at a concentration of 2 μM or higher. The addition of 7-NI, a specific inhibitor of nNOS, prevented this effect on dopamine when the concentration of MPP^+ was between 2 and 5 μM, but not at higher concentrations. This protection was reproduced with other less specific NOS-inhibitors, such as nitroarginine and nitroarginine methylester. 7-NI did not protect against the dopamine depletion caused by the nonspecific mitochondrial chain blocker rotenone. Neither MPP^+ nor rotenone significantly increased the nitrite concentration in striatal slices, measured as an index of NO production. The basal production of NO may be enough to trigger the DA depletion at very low concentrations of MPP^+, probably acting synergistically with cytosolic calcium increase.

Higher concentrations of MPP^+ are toxic by themselves without the mediation of NO. The inhibition of nNOS may protect against dopamine loss at early stages of a neurodegenerative process, and it could then be considered in the treatment of neurodegenerative human processes such as PD.

Muramatsu et al.[66] studied the effects of nNOS inhibitor 7-NI, nonselective NOS inhibitor L-NAME, and MAO inhibitor pargyline on MPTP-treated mice. The mice received four intraperitoneal injections of MPTP at one-hour intervals. A significant depletion in dopamine and DOPAC concentration was observed in the striatum from one day after MPTP treatment. The pretreatment of 7-NI and pargyline, but not L-NAME, dose-dependently protect MPTP-induced depletion in dopamine content 3 days after MPTP treatment. Their histochemical study also showed that 7-NI and pargyline can prevent a marked decrease in the nigral cells and a marked increase in astrocytes in striatum 7 days after MPTP treatment. The protective effect of 7-NI against MPTP-induced dopamine and DOPAC depletion in the striatum was not attenuated by intraperitoneal pretreatment with L-arginine. Furthermore, the post-treatment of 7-NI or pargyline protected against MPTP-induced depletion of dopamine content. These results demonstrate that the protective mechanism by which 7-NI counteracts MPTP neurotoxicity in mice may be due not only to inhibition of nNOS, but also to MAO-B inhibition. Furthermore, this study suggests that the post-treatment of 7-NI and pargyline can prevent a significant decrease in dopamine levels in the striatum of MPTP-treated mice. These findings have important implications for the therapeutic time window and choice of nNOS or MAO inhibitors in patients with PD (Figure 39.5).

Ferrante et al.[67] showed that increased nitrotyrosine immunoreactivity SN neurons in MPTP-treated baboons is blocked by inhibition of nNOS. However, Barc et al.[68] demonstrated that impairment of the neuronal dopamine transporter activity in MPP^+- treated rat was not prevented by treatments with nNOS or PARP inhibitors.

SELENIUM DEFICIENCY ENHANCES THE EXPRESSION OF NOS

Prabhu et al.[69] investigated the relationship between Se status, iNOS expression, and NO production in Se-deficient and Se-supplemented RAW 264.7 macrophage cell lines. The cellular GPx activity, a measure of Se status, was 17-fold lower in Se-deficient RAW 264.7 cells, and the total cellular oxidative tone, as assessed by flow cytometry with 2′, 7′-dichlorodihydrofluorescein diacetate, was higher in the Se-supplemented cells. Upon LPS stimulation of these cells in culture, we found significantly higher iNOS transcript and protein expression levels with

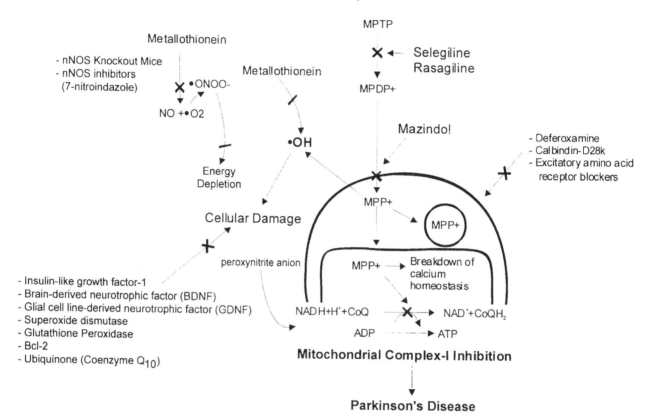

FIGURE 39.5   Free radicals and parkinsonism. MPTP became oxidized to l-methyl-4-phenyl-2,3-dihydropyridinium ion (MPDP⁺) and finally to MPP⁺, which generates free radicals and causes parkinsonism in human beings. A deficiency of NADH:ubiquinone oxidoreductase (EC 1.6.5.3, complex I) also causes striatal cell death. A deficiency of complex I may signify that an MPTP-like neurotoxin is generated endogenously, enhancing the vulnerability of striatum to oxidative stress reactions. The neurotoxic effects of MPP⁺ are blocked by metallothionein, a zinc-binding protein that scavenges ·OH; by deferoxamine, an iron-binding compound that inhibits Fenton reaction; and by *N-tert*-butyl-phenylnitrone, which traps free radicals. Neurons containing high concentrations of calbindin 28K are relatively resistant to MPP⁺. Excitatory amino acid receptor blockers such as dizocilpine attenuate MPP⁺-induced neurotoxicity. Amfonelic acid and mazindol, preventing the uptake of MPTP into dopaminergic neurons, and selegiline, preventing the formation of MPP⁺, prevent the neurotoxic effects of MPTP. After activation of excitatory amino acid receptors, there is an influx of calcium followed by activation of nNOS, which can then lead to the generation of ONOO⁻. Consistent with such a mechanism, studies of MPTP neurotoxicity in both mice and primates have shown that inhibition of nNOS exerts neuroprotection. The toxicity of MPTP is attenuated in nNOS knockout animals and following administration of 7-NI, which inhibits NOS.[61]

an increase in NO production in Se-deficient RAW 264.7 cells than in Se-supplemented cells. Electrophoretic mobility-shift assays, nuclear factor-κB (NF-κB)-luciferase reporter assays, and Western blot analyses indicate that the increased expression of iNOS in Se deficiency could be due to an increased activation and consequent nuclear localization of the redox sensitive transcription factor NF-κB. These results suggest an inverse relationship between cellular Se status and iNOS expression in LPS-stimulated RAW 264.7 cells and provide evidence for the beneficial effects of dietary Se supplementation in the prevention and/or treatment of oxidative-stress-mediated inflammatory diseases.

NO-MEDIATED ERYTHROPOIETIN PROTECTED AGAINST MPTP-INDUCED TOXICITY

EPO, produced by the kidney and fetal liver, is a cytokine-hormone that stimulates erythropoiesis under hypoxic conditions. It has been shown that EPO is produced in the central nervous system and its receptor is expressed on neurons. Since EPO has neuroprotective effects *in vitro* and *in vivo* against brain injury,[70–73] we investigated the effect of EPO treatment on locomotor activities of animals, survival of nigral dopaminergic neurons, and nitrate levels in substantia nigra and striatum in MPTP-induced animal model of parkinsonism in C57/BL mice. These findings suggest that EPO has protective and treating effects in MPTP-induced neurotoxicity in this mouse model of Parkinson's disease via increasing NO production.

SELEGILINE AS METALLOTHIONEIN MAY PROTECT AGAINST NO TOXICITY

MT isoforms are low-molecular-weight (6,000 to 7,000) zinc-binding proteins containing 60 to 61 amino acid residues, 25 to 30% cysteine, no aromatic amino acids or disulfide bonds, and binding 5 to 7 g zinc/mol of protein. The mammalian MT family consists of four similar but distinct isoforms, designated as MTI-IV. MTI and MT II isoforms were first identified in the rat brain. MT III containing 68 amino acids, also known as a growth inhibitory factor, and MT IV is expressed in stratified squamous epithelia. MT isoforms are found in glial cells as well as in neurons.

MT isoforms have been proposed to participate in the transport, accumulation, and compartmentation of zinc in various brain regions, including the areas that have extremely high concentrations of zinc, such as the hippocampus. Because, among its 61 amino acids, MT possesses 18 to 20 cysteine residues, it is the most abundant and important thiol source in the brain. In those cells that can express MT genes, they are transcriptionally regulated by metals, glucocorticoid hormones, and cytokines.[21]

By employing ESR spectroscopy,[74,75] we have demonstrated that rat hippocampal MT isoforms I and II were able to scavenge 1,1,-diphenyl-2-picrylhydrazyl radicals (DPPH), hydroxyl radicals ·OH generated in Fenton reaction, and superoxide anions generated by the hypoxanthine and xanthine oxidase system. In addition, MT I isoform protected the isolated hepatocytes from lipid peroxidation as determined by thiobarbituric acid bound malondialdehyde. MT antibodies scavenged DPPH radicals, hydroxyl radicals, and reactive oxygen species, but not superoxide anions. The results of these studies suggest that, although both isoforms of MT are able to scavenge free radicals, the MT-I appears to be a superior scavenger of superoxide anions and DPPH radicals. Moreover, antibodies formed against MT isoforms retain some, but not all, free radical scavenging actions exhibited by MT-I and MT-II (see Figure 39.3).

Thomas et al.[76] postulated that neuroprotectivity of selegiline L-deprenyl may be unrelated to inhibiting monoamine oxidase. As NO modulates activities including cerebral blood flow and memory, they examined the effect of L-deprenyl on NO. L-Deprenyl induced rapid increases in NO production in brain tissue and cerebral vessels. Vasodilation was produced by endothelial NO-dependent, as well as NO-independent, mechanisms in cerebral vessels. The drug also protected the vascular endothelium from the toxic effects of amyloid-peptide. These novel actions of selegiline may protect neurons from ischemic or oxidative damage and suggest new therapeutic applications for L-deprenyl in vascular and neurodegenerative disease. Similar results were reported by Sharma et al.[24]

CONCLUSIONS

Damage to the mitochondrial electron transport chain has been suggested to be an important factor in the pathogenesis of a range of neurological disorders, such as PD. It is now well documented that NO and its toxic metabolites, $ONOO^-$, can inhibit components of the mitochondrial respiratory chain leading, if damage is severe enough, to a cellular energy deficiency state. Within the brain, the susceptibility of different brain cell types to NO and $ONOO^-$ exposure may be dependent on factors such as the intracellular GSH concentration and an ability to increase glycolytic flux in the face of mitochondrial damage. Thus, neurons, in contrast to astrocytes, appear particularly vulnerable to the actions of these molecules. Following cytotoxins exposure, astrocytes can increase NO generation, due to de novo synthesis of the iNOS. Although the NO/$ONOO^-$ isoforms may not affect astrocyte survival, these molecules may diffuse out to cause mitochondrial damage, and possibly cell death, to other cells (such as neurons) in close proximity. Selegiline, selenium, coenzyme Q_{10}, naloxone, pergolide, and metallothionein provide neuroprotection against $ONOO^-$ induced damage to dopaminergic neurons.[77]

GSH deficiency, which causes accumulation of H_2O_2, leads to mitochondrial damage in brain. Moreover, coenzyme Q_{10} attenuates the MPTP-induced loss of striatal dopaminergic neurons. Deficiency of striatal GSH in PD fosters oxidative stress and causes apoptosis. γ–Glutamylcysteinylglycine assists in maintaining the intracellular reducing environment, protects protein thiol groups from oxidation, and participates as a coenzyme or cofactor in a wide variety of chemical reactions. GSH exerts its antioxidant activity synergistically with both vitamin C and vitamin E. Striatal GSH deficiency in PD enhances the susceptibility of SN to destruction by endogenous or exogenous neurotoxins. Moreover, treatment with lazaroid, which inhibits lipid peroxidation, prevents death of mesencephalic dopaminergic neurons following GSH depletion.[21]

The selective vulnerability and loss of certain neurons are remarkable characteristics of age-related degenerative disorders of the brain as seen in PD. Glutamate is the major excitatory neurotransmitter in the brain, and excitotoxicity plays a role in PD. Furthermore, growing evidence implicates oxidative stress as a mediator of excitotoxic cell death. Following activation of NMDA receptors, the generation of free radicals increases, oxidation damage to lipids occurs, and antioxidants prevent cell death. Dizocilpine, which blocks NMDA receptors, may provide neuroprotection in PD.[78–79]

ABBREVIATIONS

7-NI	7-nitroindazole
8-OH-2dG	8-hydroxy-2-deoxyguanosine
cnNOS	constitutive neuronal nitric oxide synthase
control$_{wt}$	wild-type control
DA	dopamine
DHBA	2,3-dihydroxybenzoic acid
DOPAC	dihydroxyphenylacetic acid
DPPH	1,1-diphenyl-2-picrylhydrazyl radicals
EPO	erythropoietin
EPR	electron paramagnetic resonance
ESR	electron spin resonance
FITC	fluorescein isothiocyanate
GSH	glutathione
H_2O_2	hydrogen peroxide
IL	interleukin
iNOS	inducible nitric oxide synthase
L-NAME	N^ω-nitro-L-arginine methyl ester
LPS	lipopolysaccharide
MAO	monoamine oxidase
MPDP+	1-methyl-4-phenylpyridinium
MPTP	1-methyl-4-phenyl-1,2,3,6-tetrahydropyridine
MT	metallothionein
NF-κB	nuclear factor-κB
NMDA	N-methyl-D-aspartate
NO	nitric oxide
NOS	endothelial nitric oxide synthase
NOS	neuronal nitric oxide synthase
NOS	nitric oxide synthase
ONOO⁻	peroxynitrite ion
PARP	poly(ADP-ribose) polymerase
PD	Parkinson's disease
RhO$_{mgko}$	mitochondrial genome knockout
SIN-1	3-morpholinosydnonimine
SN	substantia nigra
SN PC	substantia nigra pars compacta
SOD	superoxide dismutase
STN	subthalamic nucleus
THir	tyrosine hydroxylase-immunoreactive
TNF-α	tumor necrosis factor-α
˙OH	hydroxyl radical

ACKNOWLEDGMENTS

The studies cited in this review were supported in part by a grant from the USPHS, NINDS 2R01 NS34566-09, and NIA AG 17059-05. The authors express their heartfelt appreciation to Dani Stramer for typing this manuscript.

REFERENCES

1. Ebadi, M., Govitrapong, P., Sharma, S., Muralikrishnan, D., Shavali, S., Pellett, L., Schafer, R., Albano, C. and Eken, J., Ubiquinone (coenzyme Q_{10}) and mitochondria in oxidative stress of Parkinson's disease, *Biol. Signals Recept.*, 10:224–253, 2001.

2. Beckman, J. S., Peroxynitrite versus hydroxyl radical: the role of nitric oxide in superoxide-dependent cerebral injury, *Ann. N.Y. Acad. Sci.*, 738:69–75, 1994.

3. Bredt, D., Endogenous nitric oxide synthesis: biological functions and pathophysiology, *Free Rad. Res.*, 31: 577–596, 1999.

4. Dawson, V. and Dawson, T., Nitric oxide in neurodegeneration, *Prog. Brain Res.*, 118:215–229, 1998.

5. Gerlach, M., Blum-Dege, D., Lan, J. and Riederer, P., Nitric oxide in the pathogenesis of Parkinson's disease. In Parkinson's disease: Storm, G. M., (Ed)., *Adv. Neurol.*, 80:239–245, 1999.

6. Hirsch, E. C. and Hunot, S., Nitric oxide, glial cells and neuronal degeneration in parkinsonism, *Trends Pharmacol. Sci.*, 21:163–165, 2000.

7. Smith, K. J., Kapoor, R. and Felts, P. A., Demyelination: the role of reactive oxygen and nitrogen species, *Brain Pathol.*, 9:69–92, 1999.

8. Hunot, S., Dugas, N., Faucheux, B., Hartmann, A., Tardieu, M., Debr, P., Agid, Y., Dugas, B. and Hirsch, E., FcRII/CD23 is expressed in Parkinson's disease and induces, *in vitro*, production of nitric oxide and tumor necrosis factor in glial cells, *J. Neurosci.*, 19: 3440–3447, 1999.

9. Yodim, M. D., Ben-Schachar, D., Eshel, G., Finberg, J. P. and Riederer, P., The neurotoxicity of iron and nitric oxide. Relevance to the etiology of Parkinson's disease, *Adv. Neurol.*, 60:259–266, 1993.

10. Yodim, M. D., Lavie, L. and Riederer, P., Oxygen free radicals and neurodegeneration in Parkinson's disease: a role for nitric oxide, *Ann. N. Y. Acad. Sci.*, 738:64–68, 1994.

11. Torreilles, F., Salman-Tabcheh, S., Gurin, M. and Torreilles, J., Neurodegenerative disorders: the role of peroxynitrite, *Brain Res. Rev.*, 30:153–163, 1999.

12. Bringold, U., Ghafourifar, P. and Richter, C., Peroxynitrite formed by mitochondrial NO synthase promotes mitochondrial Ca^{2+} release, *Free Radic. Biol. Med.*, 29:343–348, 2000.

13. Ghafourifar, P., Bringold, U., Klein, S. and Richter, C., Mitochondrial nitric oxide synthase, oxidative stress and apoptosis, *Biol. Signals Recept.*, 10:57–65, 2001.

14. Sharma, S. K., Sangchot, P. and Ebadi, M., MT gene manipulation influences striatal mitochondrial ubiquinones and MPTP-induced neurotoxicity in dopaminergic neurons, *World Congress of Pharmacology*, 14:106–107, 2002.

15. Sharma, S. K., Shavali, S., El ReFaey, H. and Ebadi, M., Inhibition of α-Syn nitration and perinuclear aggregation by antioxidants in metallothionein transgenic and aging RhO (mgko) dopaminergic neurons, *FASEB J.*, 16:686–11, 2002.

16. Ebadi, M., Hiramatsu, M., Burke, W. J., Folks, D. G., and el-Sayed, M. A., MT isoforms provide neuroprotection against 6-hydroxy-dopamine-generated hydroxyl radicals and superoxide anions, *Proc. West Pharmacol. Soc.,* 41:155–158, 1998.

17. Sharma, S. K. and Ebadi, M., Metallothionein attenuates 3-morpholinosydnonimine (SIN-1)-induced oxidative and nitrative stress in dopaminergic neurons, *Antioxidants and Redox Signaling,* 5:231–264, 2003.

18. Ebadi, M. and Sharma, S. K., Peroxynitrite and mitochondrial dysfunction in the pathogenesis of Parkinson's Disease, *Antioxidants and Redox Signaling,* 5:319–335, 2003.

19. Heales, S., Bolaos, J., Stewart, V., Brookes, P., Land, J. and Clark, J., Nitric oxide, mitochondria and neurological disease, *Biochimica et Biophysica Acta,* 1410:215–228, 1999.

20. Ebadi, M. and Hiramatsu, M., Glutathione and metallothionein in oxidative stress of Parkinson's disease, in *Free Radicals in Brain Pathophysiology,* edited by Poli, G., Cadenas, E., and Packer, L., Marcel Dekker, Inc., California, pp. 427–465, 2000.

21. Ebadi, M., Srinivasan, S. K. and Baxi, M., Oxidative stress and antioxidant therapy in Parkinson's disease, *Prog. Neurobiol.,* 48:1–19, 1996.

22. Schulz, J., Matthews, R., Klockgether, T., Dichgans, J. and Beal, M., The role of mitochondrial dysfunction and neuronal nitric oxide in animal models of neurodegenerative disease, *Mol. Cell Biochem.,* 174:193–197, 1997.

23. Rose, S., MacKenzie, G. and Jenner, P., Nitric oxide and basal ganglia degeneration, in *Parkinson's Disease,* G. M. Storm, Ed., *Adv. Neurol.,* 80:247–257, 1999.

24. Sharma, S. K., Shavali, S., El ReFaey, H. and Ebadi, M., Inhibition of α-synuclein nitration and perinuclear aggregation by antioxidants in metallothionein transgenic and aging RhO (mgko) dopaminergic neurons, *FASEB J.* 16:696–711, 2002.

25. Palumbo, A., Napolitano, A., Barone, P. and d'Ischia, M., Nitrite- and peroxide-dependent oxidation pathways of dopamine: 6-nitrodopamine and 6-hydroxydopamine formation as potential contributory mechanisms of oxidative stress- and nitric oxide-induced neurotoxicity in neuronal degeneration, *Chem. Res. Toxicol.,* 12:1213–1222, 1999.

26. Liu, B., Gao, H., Wang, J., Jeohn, G., Cooper, C. and Hong, J., Role of nitric oxide in inflammation-mediated neurodegeneration, *Ann. N. Y. Acad. Sci.,* 962:318–331, 2002.

27. Sharma, S., Carlson, E. and Ebadi, M., Selegiline mediated neuroprotection in dopaminergic neurons are mostly unrelated to its inhibition of monoamine oxidase B, *13th International Symposium on the Autonomic Nervous System, Clinical Autonomic Research,* 12:307, 2002.

28. Naoi, M. and Maruyama, W., Future of neuroprotection in Parkinson's disease, *Parkinsonism Rel. Dis.,* 8:139–145, 2001.

29. Barthwal, M., Srivastava, N. and Dikshit, M., Role of nitric oxide in a progressive neurodegeneration model of Parkinson's disease in the rat, *Redox Report,* 6:297–302, 2001.

30. Riob, N., Schpfer, F., Boveris, A., Cadenas, E. and Poderoso, J., The reaction of nitric oxide with 6-hydroxydopamine: implications for Parkinson's disease, *Free Rad. Biol. Med.,* 32:115–121, 2002.

31. Hunot, S., Boissire, F., Faucheux, F., Brugg, B., Mouatt-Prigent, A., Agid, Y. and Hirsch, E. C., Nitric oxide synthase and neuronal vulnerability in Parkinson's disease, *Neuroscience,* 72:355–363, 1996.

32. Knott, C., Stern, G. and Wilkin, G. P., Inflammatory regulators in Parkinson's disease: iNOS, lipocortin-1, and cyclooxygenases-1 and –2, *Mol. Cell Neuro.,* 16:724–739, 2000.

33. Barthwal, M., Srivastava, N., Shukla, R., Nag, D., Seth, P., Srimal, R. and Dikshit, M., Polymorphonuclear leukocyte nitrite content and antioxidant enzymes in Parkinson's disease patients, *Acta Neurol. Scad.,* 100:300–304, 1999.

34. Kuiper, M. A., Teerlink, T., Visser, J. J., Bergmans, P., Scheltens, P. and Wolters, E., L-glutamate, L-arginine and L-citrulline levels in cerebrospinal fluid of Parkinson's disease, multiple system atrophy, and Alzheimer's disease patients, *J. Neural. Transm.,* 107:183–189, 2000.

35. Qureshi, G., Baig, S., Bednar, I., Sdersten, P., Forsberg, G. and Siden, A., Increased cerebrospinal fluid concentration of nitrite in Parkinson's disease, *NeuroReport,* 6:1642–1644, 1995.

36. Molina, J., Jimnez-Jimnez, F., Navarro, J., Ruiz, E., Arenas, J., Cabrera-Valdivia, F., Vzquez, A., Fernndez-Calle, P., Ayuso-Peralta, L., Rabasa, M. and Bermejo, F., Plasma levels of nitrates in patients with Parkinson's disease, *J. Neurol. Sci.,* 127:87–89, 1994.

37. Lo, H., Hogan, E. and Soong, B., 5'-flanking region polymorphism of the neuronal nitric oxide synthase gene with Parkinson's disease in Taiwan, *J. Neurol. Sci.,* 194:11–13, 2002.

38. Böckelmann, R., Wolf, G., Ransmayr, G. and Riederer, P., NADPH-diaphorase/nitric oxide synthase containing neurons in normal and Parkinson's disease putamen, *J. Neural. Transm.,* 7:115–121, 1994.

39. Eve, D., Nisbet, A., Kingsbury, A., Hewson, E., Daniel, S., Lees, A., Marsden, C. and Foster, O., Basal ganglia neuronal nitric oxide synthase mRNA expression in Parkinson's disease, *Mol. Brain Res.,* 63:62–71, 1998.

40. Van Muiswinkel, F., Steinbusch, H., Drukarch, B. and De Vente, J., Identification of NO-producing and receptive cells in mesencephalic transplants in a rat model of Parkinson's disease: a study using NADPH-*d* enzyme- and NOS*c*/cGMP immunocytochemistry, *Ann. N. Y. Acad. Sci.,* 738:289–305, 1994.

41. Muijsers, R., Folkerts, G., Henricks, P., Sadeghi-Hashjin, G. and Nijkamp, F., Peroxynitrite: a two-faced metabolite of nitric oxide, *Life Sci.,* 60:1833–1845, 1997.

42. Cammack, R., Shergill, J., Inalsingh, A. and Hughes, M., Applications of electron paramagnetic resonance spectroscopy to study interaction of iron proteins in cells

with nitric oxide, *Spectrochimica Acta*, Part A, 54:2393–2402, 1998.

43. Bredt, D. Endogenous nitric oxide synthesis: biological functions and pathophysiology, *Free Rad. Res.*, 31: 577–596, 1999.

44. Ghafourifar, P., Bringold, U., Klein, S. and Richter, C., Mitochondrial nitric oxide synthase, oxidative stress and apoptosis, *Biol. Signals Recept.*, 10:57–65, 2001.

45. Gatto, E., Riob, N., Carreras, M., Cheravsky, A., Rubio, A., Satz, M. and Poderoso, J., Over expression of neutrophil neuronal nitric oxide synthase in Parkinson's disease. *Nitric Oxide: Biology and Chemistry* 4:534–539, 2000.

46. Sawada, H., Shimohama, S., Kawamura, T., Akaike, A., Kitamura, Y., Taniguchi, T. and Kimura, J., Mechanism of resistance to NO-induced neurotoxicity in cultured rat dopaminergic neurons, *J. Neurosci. Res.*, 46:509–518, 1996.

47. Gayle, D., Ling, Z., Tong, C., Landers, T., Lipton, J. and Carvey, P., Lipopolysaccharide (LPS)-induced dopamine cell loss in culture: roles of tumor necrosis factor-, interleukin-1, and nitric oxide, *Devel. Brain Res.*, 133:27–35, 2002.

48. Calabrese, V., Bates, T. and Ginffrida-Stella, A., NO synthase and NO-dependent signal pathways in brain aging and neurodegenerative disorders: the role of oxidant/antioxidant balance, *Neurochem. Res.*, 9:1315–1341, 2000.

49. Torreilles, F., Salman-Tabcheh, S., Gurin, M. and Torreilles, J., Neurodegenerative disorders: the role of peroxynitrite, *Brain Res. Rev.*, 30:153–163, 1999.

50. Klotz, L. O., Schroeder, P. and Sies, H., Peroxynitrite signaling: receptor tyrosine kinases and activation of stress-responsive pathways, *Free Rad. BiolMed.*, 33:737–743, 2002.

51. Ischiropoulos, H., Duran, D. and Horwitz, J., Peroxynitrite-mediated inhibition of DOPA synthesis in PC12 cells, *J. Neurochem.*, 65:2366–2372, 1995.

52. LaVoie, M. and Hastings, T., Peroxynitrite- and nitrite-induced oxidation of dopamine: implications for nitric oxide in dopaminergic cell loss, *J. Neurochem.*, 73:2546–2554, 1999.

53. Araki, T., Tanji, H., Fujihara, K., Kato, H. and Itoyama, Y., Increases in [^{3}H] FK-506 and [^{3}H] L-N^G-nitro-arginine binding in the rat brain after nigrostriatal dopaminergic denervation, *Metabolic Brain Disease*, 14: 21–31, 1999.

54. Cammack, R., Shergill, J., Inalsingh, A. and Hughes, M., Applications of electron paramagnetic resonance spectroscopy to study interaction of iron proteins in cells with nitric oxide, *Spectrochimica Acta*, Part A, 54:2393–2402, 1998.

55. Gómez-Vargas, M., Nishibayashi-Asanuma, S., Asanuma, M., Kondo, Y., Iwata, E. and Ogawa, N., Pergolide scavenges both hydroxyl and nitric oxide free radicals *in vitro* and inhibits lipid peroxidation in different regions of the rat brain, *Brain Res.*, 790:202–208, 1998.

56. Nishibayashi, S., Asanuma, M., Kohno, M., Gmez-Vargas, M. and Ogawa, N., Scavenging effects of dopamine

agonists on nitric oxide radicals, *J. Neurochem.*, 67:2208–2211, 1996.

57. Przedborski, S., Jackson-Lewis, V., Yokoyama, R., Shibata, T., Dawson, V. and Dawson, T., Role of neuronal nitric oxide in 1-methyl-4-phenyl-1,2,3,6-tetrahydropyridine (MPTP)-induced dopaminergic neurotoxicity, *Proc. Natl. Acad. Sci.*, 93:4565–4571, 1996.

58. Smith, T., Swerdlow, R., Parker, W. and Bennett, J., Jr., Reduction of MPP$^+$ induced hydroxyl radical formation and nigrostriatal MPTP toxicity by inhibiting nitric oxide synthase, *NeuroReport*, 5:2598–2600, 1994.

59. Klivenyi, P., Andreassen, O., Ferrante, R., Lancelot, E., Reif, D. and Beal, M., Inhibition of neuronal nitric oxide synthase protect against MPTP toxicity, *NeuroReport*, 11:1265–1268, 2000.

60. Liberatore, G., Jackson-Lewis, V., Vukosavic, S., Mandir, A., Vila, M., McAuliffe, G., Dawson, V., Dawson, T. and Przedborski, S., Inducible nitric oxide synthase stimulates dopaminergic neurodegeneration in MPTP model of Parkinson disease, *Nat. Med.*, 12:1403–1409, 1999.

61. Obata, T. and Yamanaka, Y., Nitric oxide enhances MPP$^+$-induced hydroxyl radical generation via depolarization activated nitric oxide synthase in rat striatum, *Brain Res.*, 902:223–228, 2001.

62. Hantraye, P., Brouillet, E., Ferrante, R., Palfi, S., Dolan, R., Matthews, R. and Beal, M., Inhibition of neuronal nitric oxide synthase prevents MPTP-induced Parkinsonism in baboons, *Nat. Med.* 2:1017–1021, 1996.

63. Dehmer, T., Lindenau, J., Haid, S., Dichgans, J. and Schulz, J., Deficiency of inducible nitric oxide synthase protects against MPTP toxicity *in vivo*, *J. Neurochem.*, 74:2213–2216, 2002.

64. Schulz, J., Matthews, R., Muqit, M., Browne, S. and Beal, M., Inhibition of neuronal nitric oxide synthase by 7-nitroindazole protects against MPTP-induced neurotoxicity in mice, *J. Neurochem.*, 64:936–939, 1995.

65. Cutillas, B., Espejo, M. and Ambrosio, S., 7-nitroindazole prevents dopamine depletion caused by low concentration of MPP$^+$ in rat striatal slices, *Neurochem. Int.*, 33:35–40, 1998.

66. Muramatsu, Y., Kurosaki, R., Mikami, T., Michimata, M., Matsubara, M., Imai, Y., Kato, H., Itoyama, Y. and Araki, T., Therapeutic effect of neuronal nitric oxide synthase inhibitor (7-nitroindazole) against MPTP neurotoxicity in mice, *Meta Brain Dis.*, 17:169–182, 2002.

67. Ferrante, R., Hantraye, P., Brouillet, E. and Beal, M., Increased nitrotyrosine immunoreactivity in substantia nigra neurons in MPTP treated baboons is blocked by inhibition of neuronal nitric oxide synthase, *Brain Res.*, 823:177–182, 1999

68. Barc, S., Page, G., Barrier, L., Piriou, A. and Fauconneau, B., Impairment of the neuronal dopamine transporter activity in MPP+-treated rat was not prevented by treatments with nitric oxide synthase or poly(ADP-ribose) polymerase inhibitors, *Neuroscience Letters*, 314:82–86, 2001.

69. Prabhu, K., Zamamiri-Davis, F., Stewart, J., Thompson, J., Sordillo, L. and Reddy, C., Selenium deficiency

increases the expression of inducible nitric oxide synthase in RAW 264.7 macrophages: role of nuclear factor-B upregulation, *Biochem. J.,* 366:203–209, 2002.

70. Sadamoto, Y., Igase, K., Sakanaka, M., Sato, K., Otsuka, H. et al., Erythropoietin prevents place navigation disability and cortical infarction in rats with permanent occlusion of the middle cerebral artery, *Biochem. Biophys. Res. Commun.,* 253:26–32, 1998.

71. Sakanaka, M., Wen, T. C., Matsuda, S., Masuda, S., Morishita, E., Nagao, M. et al., *In vivo* evidence that erythropoietin protects neurons from ischemic damage, *Proc. Natl. Acad. Sci. USA.* 95: 4635–4640, 1998.

72. Dame, C., Juul, S. E., and Christensen R. D., The biology of erythropoietin in the central nervous system and its neurotrophic and neuroprotective potential, *Biol. Neonate,* 79: 228–235. 2001.

73. Morishita, E., Masuda, S., Nagao, M., Yasuda, Y. and Sasaki, R., Erythropoietin receptor is expressed in rat hippocampal and cerebral cortical neurons, and erythropoietin prevents *in vitro* glutamate-induced neuronal death, *Neuroscience,* 76:105–116. 1997.

74. Kumari, M., Hiramtsu, M. and Ebadi, M., Free radicals scavenging actions of metallothionein isoforms I and II, *Free Rad. Res.,* 29:93–101, 1998.

75. Kumari, M., Hiramtsu, M. and Ebadi, M., Free radical scavenging actions of hippocampal metallothionein isoforms and of antimetallothionein: an electron spin resonance spectroscopic study, *Cell Mol. Biolo.,* 46: 627–636, 2000.

76. Thomas, T., McLendon, C. and Thomas, G., L-deprenyl: nitric oxide production and dilation of cerebral blood vessels, *NeuroReport,* 9:2595–2600, 1998.

77. Jenner, P. and Olanow, W., Oxidative stress and the pathogenesis of Parkinson's disease, *Neurology,* 47:S161–S170, 1996.

78. Ebadi, M., Sharma, S. K., Shavali, S. and El Refaey, H., Neuroprotective actions of Selegiline, *J. Neurosci. Res.,* 67:285–289, 2002.

79. Ebadi, M., Sharma, S. K., Shavali, S., Sangchot, P., and Brekke, L., Neuroprotective actions of Selegiline in Parkinson's disease, *45th Annual Meeting of the Western Pharmacological Society and XXV Congreso Nacional de Farmacologia,* 45:77, 2002.

40 Excitotoxicity in Parkinson's Disease

James G. Greene and J. Timothy Greenamyre
Department of Neurology and Center for Neurodegenerative Disease, Emory University School of Medicine

CONTENTS

INTRODUCTION

The cardinal symptoms of Parkinson's disease (PD) are rest tremor, bradykinesia, rigidity, and postural instability. These features arise from the specific neurodegeneration of substantia nigra (SN) dopamine neurons in the midbrain. The resultant dopamine deficiency, predominantly in the striatum, causes a wave of circuitry changes in the basal ganglia, thus producing the characteristic movement disorder.

The dopamine deficit was initially described in PD decades ago, and the resulting elucidation of the pharmacology, neurochemistry, and functional anatomy of the basal ganglia has contributed to the discovery of several very effective symptomatic therapies for the disease.[1,2] Dopaminergic compounds are the most effective and widely used; these include levodopa and dopamine agonists. Anticholinergics, amantadine, and selegiline are also modestly effective symptomatic pharmacological agents.[3] In addition, recent advances in functional neurosurgery have made destructive lesions and deep brain stimulation of selected target nuclei a safe and successful therapeutic option.[4] All in all, the advances in PD treatment over the last 35 years have been amazing and have greatly improved quality of life for patients with the disease.

Nevertheless, there is still a significant amount of disability associated with PD, and the benefits of symptomatic therapy decrease over time as degeneration of the dopamine system continues.[5] In addition, there are many nonmotor symptoms associated with PD that are currently inadequately addressed, including cognitive impairment, psychiatric disturbances, autonomic dysfunction, and sleep disorders.[3] Even some of the motor problems, most notoriously postural instability, remain very difficult to treat.

The search for new and more efficient symptomatic treatments continues, but this strategy does not address the primary problem in PD, which is progressive neurodegeneration of dopamine neurons. The cause of dopamine neuron degeneration is unknown, but extensive investigation over the past two decades has provided new insights and intriguing hypotheses. One of the most widely studied rests on the duality of glutamate, an excitatory amino acid that functions in the CNS as not just a neurotransmitter, but also as a neurotoxin.

Although we will not discuss it in detail here, abnormal glutamate neurotransmission in the basal ganglia is believed to play a prominent role in the generation of PD symptoms (see the chapters on Physiological Neuropathology and Neurochemical Pathology). In fact, inhibition of glutamate activity is thought to be a promising therapeutic option for the symptoms of PD.[6]

In addition to the involvement of glutamate in the functional neuroanatomy of the basal ganglia under normal circumstances and in PD, glutamate may be a key mediator of neurotoxicity that results in loss of dopamine neurons. This hypothesis stems from the fact that glutamate neurotransmission is inherently demanding for neurons and the finding that excessive glutamate stimulation can be neurotoxic. This "excitotoxicity" may be part of a complicated scheme of neuronal vulnerability and neurodegeneration that ultimately causes specific degeneration of dopamine neurons in PD. Excitotoxicity is also intimately related to many other potential mechanisms of neurodegeneration recently proposed to be at work in PD. These include mitochondrial dysfunction, oxidative stress, and abnormal protein metabolism.

In this chapter, we briefly review the properties of glutamate as a neurotransmitter and neurotoxin. We review evidence for excitotoxicity in PD and examine the interrelationships between excitotoxicity and other proposed mechanisms of neurodegeneration in PD. We also discuss potential mechanisms of selective dopaminergic neurotoxicity related to glutamate. Finally, we briefly hypothesize about potential neuroprotective interventions in PD related to glutamate neurotransmission and neurotoxicity.

GLUTAMATE AS A NEUROTRANSMITTER

Glutamate is widely considered to be the predominant excitatory neurotransmitter in the central nervous system. It is the most abundant free amino acid in the CNS and is involved in a variety of reactions critical to intermediary metabolism. No one enzymatic pathway has been demonstrated unequivocally to be responsible for the production of the neurotransmitter pool of glutamate, and, as a consequence, there is no specific marker for the identification of glutamatergic neurons. However, almost every type of CNS neuron responds to glutamate stimulation. By employing a combination of experimental lesions, release and uptake studies, electrophysiology, and immunohistochemistry, a number of glutamatergic pathways have been identified. These include, but are not limited to, extensive connections in hippocampal, cortico-cortical, thalamo-cortical, and extrapyramidal circuitry.[6]

Glutamate plays a prominent role in the circuitry of the basal ganglia, and perturbation of this system has wide-ranging consequences for Parkinson's disease symptomatology and pathology (see the chapters on Physiological Neuropathology and Neurochemical Pathology). There are extensive glutamatergic projections from cortex to striatum and substantia nigra, subthalamic nucleus (STN) to substantia nigra and globus pallidus, midbrain extrapyramidal area to substantia nigra, and thalamus to cortex, among others. Of particular note concerning degeneration of dopamine neurons is the projection from STN to SN. This, in addition to cortical inputs, provides much of the excitatory input to the SN, and this input is noteworthy considering the potential toxic actions of glutamate. The STN is disinhibited in the Parkinson's disease brain, which produces even more excitatory drive at the level of the SN.[2,7] This persistent excitation may further damage dopaminergic neurons in the substantia nigra and may initiate a vicious cycle of excitation and neurodegeneration.[8]

As with most classical neurotransmitters, glutamate is specifically concentrated into synaptic vesicles in nerve terminals,[9,10] released during depolarization in a calcium-dependent manner,[11] and inactivated by reuptake.[12] However, unlike classical transmitters, glutamate uptake by glial cells is significant. Within glia, glutamate is transaminated by glutamine synthetase to form glutamine, which diffuses into nerve terminals and is converted back to glutamate by mitochondrial glutaminase.[13,14] In this way, glutamate is recycled from nerve terminal to glial cell and back to nerve terminal.

Glutamate exerts its actions via two main categories of receptors. "Ionotropic" receptors are linked to ion channels, and "metabotropic" receptors are coupled to second messenger systems by G-proteins. Ionotropic receptors are further divided into subtypes named initially for the agonists that most specifically stimulate them: N-methyl-D-aspartate (NMDA), α-amino-3-hydroxy-5-methyl-4-isoxzazolepropionic acid (AMPA), and kainate. Excitatory postsynaptic potentials resulting from glutamate stimulation generally involve both AMPA and NMDA receptor-mediated responses.[15]

AMPA and kainate receptors are thought to be the primary mediators of fast excitatory neurotransmission. The majority of these receptors are permeable only to sodium, but some endogenous AMPA receptors have been found to exhibit high calcium conductance.[16,17] Like most ion channel receptors, AMPA and kainate receptors are heteromeric complexes of subunits. At present, there are four cloned AMPA receptor subunits (GluR1-4) and five different kainate receptor subunits (GluR5-7, KA1-2), all of which have several splice variants.[15] Since the pharmacology, physiology, and anatomic distribution of individual subunits is distinct, it is hypothesized that there are a great many unique endogenous AMPA and kainate receptors. For instance, GluR1-4, GluR5, and KA1-2 are all expressed in midbrain dopaminergic neurons, leading to a multitude of potential combinations.

NMDA receptor pharmacology is exceedingly complex. The receptor has binding sites for glutamate and glycine, both of which must be occupied for receptor activation to occur.[18,19] It also has several other modulatory sites, which include binding sites for zinc, protons, polyamines, and a site modulated by oxidation-reduction.[15] All of these sites may have important consequences. For example, protons inhibit receptor activation, such that, when extracellular pH is low (e.g., during ischemia or intense neuronal activity), NMDA receptors are less likely to open.

In addition, the NMDA receptor ion channel is voltage gated as well as ligand gated.[20] At resting membrane potential, the ion channel is blocked by extracellular magnesium, which prevents ion transduction even when agonist is bound to the receptor. Since this magnesium blockade is voltage dependent, postsynaptic depolarization facilitates NMDA receptor activation. As a result, the NMDA receptor is activated only under conditions of coincident agonist binding and postsynaptic depolarization. This quality is vitally important for the physiological functions of the NMDA receptor, such as learning and memory, and also for the induction of NMDA receptor-mediated toxicity.[21,22]

Like AMPA and kainate receptors, endogenous NMDA receptors are thought to be composed of several different subunits (NMDAR1, NMDAR2A-D, NMDAR3A-B), the regional expression patterns of which result in a myriad of possible endogenous NMDA receptors with slightly different properties.[15,23] NMDAR1, NMDR2B, and NMDAR2C are all expressed in substantia nigra dopamine neurons.

Unlike most AMPA and kainate channels, the NMDA receptor ion channel has a very high calcium conductance,[24,25] and calcium is thought to be the primary mediator of both the physiological and toxic properties of the NMDA receptor.

In contrast to ionotropic glutamate receptors, metabotropic glutamate receptors are linked to G proteins. Depending on the cell and receptor subtype, metabotropic receptors may be linked to phosphoinositol turnover, arachidonic acid metabolism, or adenylate cyclase. Thus far, eight metabotropic receptor subunits have been cloned and designated mGluR1 through mGluR8; all have different anatomical distributions.[26] mGluR1 and mGluR3-4 are expressed in midbrain dopaminergic neurons. Finally, it is worth noting that mGluR activation can modulate ionotropic glutamate receptor activation.[27,28]

GLUTAMATE AS A NEUROTOXIN

Under certain circumstances, glutamate acts not only as a neurotransmitter but as a neurotoxin. The ability of glutamate analogs to cause neuronal excitation is correlated with their ability to cause neuronal death.[29] As such, the likelihood of neurotoxicity increases as the excitatory stimulation becomes more intense. This excitotoxicity is receptor mediated and blocked by glutamate receptor antagonists. Excessive stimulation of either ionotropic or certain metabotropic glutamate receptors can lead to excitotoxic neuronal death.[30,31] Extreme stimulation can result from neuronal activity (e.g., status epilepticus), nonsynaptic glutamate release (e.g., ischemia), or endogenous or exogenous excitotoxins (receptor agonists). Such extreme stimulation may also occur in the parkinsonian basal ganglia, where overactivation of the STN leads to excessive glutamatergic input to the substantia nigra, as mentioned above.

The post-receptor mechanisms of excitotoxicity have not yet been fully elucidated, but calcium influx appears to be crucial.[32–35] Due to its large calcium conductance, the NMDA receptor has been a focus of research into excitotoxicity, but non-NMDA receptors can also increase intracellular calcium, either directly[16,17] or indirectly via NMDA or voltage-dependent channels.[36] This calcium influx has a myriad of potential consequences that ultimately lead toward neuronal death, including mitochondrial dysfunction, oxidative stress, and induction of programmed cell death.

In certain models, mitochondrial calcium uptake is a requisite for excitotoxic neuronal death, and it has been suggested that calcium from NMDA receptors has privileged access to mitochondria.[37,38] The large quantities of calcium buffered by mitochondria during excitotoxic stimuli can cause mitochondrial dysfunction, thus impairing neuronal energy production and calcium buffering.[39] Excessive glutamate stimulation also causes mitochondrial production of oxygen radicals, such as superoxide ($O_2^{\cdot}$), hydrogen peroxide (H_2O_2), and hydroxyl radical ($OH^{\cdot}$), presumably through mitochondrial calcium overload.[40] These local free radicals may damage susceptible components of the electron transport chain, beginning a vicious cycle of further mitochondrial decompensation.[41,42]

Mitochondrial stress may also lead to activation of programmed cell death cascades during excitotoxicity. These pathways depend on release of mitochondrial intermembrane proteins that activate effectors of apoptotic cascades. While the details of these mechanisms are complicated and beyond the scope of this chapter, there are many potential initiating stimuli, such as free radicals, calcium accumulation, cytoskeletal disruption, and mitochondrial permeability transition (see chapter about Apoptosis; see also Reference 43).

In addition to mitochondrial toxicity, calcium influx can lead to generation of nitric oxide (NO) and cytotoxic radicals.[44,45] The elucidation of NO as an important downstream effector of NMDA receptor activation has led to description of several potential neurodegenerative mechanisms.[46,47] Calcium influx from NMDA receptors activates nitric oxide synthase (NOS) via a calmodulin interaction. Activation of NOS catalyzes the transformation of arginine to citrulline, with the concomitant liberation of NO. NO is involved in cellular signaling by activation of guanylate cyclase and protein nitrosylation. Additionally, NO may react with superoxide anion to produce peroxynitrite radical, a highly reactive species that can cause oxidative damage to proteins and DNA. DNA damage induces activation of poly-(ADP ribose) polymerase (PARP), which can ultimately cause depletion of mitochondrial cofactors (NAD$^+$) and mitochondrial dysfunction[48] In addition, dysregulated protein nitrosylation may also cause protein dysfunction and damage. Inhibition of NOS and PARP has been shown to protect against excitotoxicity *in vitro* and *in vivo*.[49–52]

Calcium also stimulates the production of arachidonic acid metabolites, which can lead to radical formation.[53,54] These free radicals can indiscriminately damage lipids, proteins, and nucleic acids, resulting in cellular damage. Extreme elevation of intracellular calcium may also result in the nonspecific activation of many other enzymes, such as proteases, nucleases, lipases, and those involved in cell signaling cascades. This may cause aberrant control of signaling pathways, decreased cytoskeletal integrity, and further cellular deterioration.[55]

There are numerous ways in which glutamate receptor activation can cause neurotoxicity. As the discussion above indicates, these mechanisms are not exclusive, and their interaction may be crucial to the ultimate expression of cellular damage. We will explore these interactions and their relationship to the pathogenesis of PD in later sections.

DIRECT AND INDIRECT EXCITOTOXICITY

Another important point about excitotoxicity must be made. Although *excessive* glutamate receptor stimulation is toxic, this is not the same as saying that *supranormal* glutamate receptor stimulation is toxic. There is a balance between the level of afferent glutamatergic stimulation and the ability of a neuron to cope with glutamatergic stimulation (Figure 40.1A). Neurons have a threshold for excitotoxic glutamate stimulation, but the threshold is not static. It varies based on many neuronal parameters, such as mitochondrial function, calcium buffering, and antioxidant defense mechanisms. Thus, under conditions of mitochondrial impairment, reduced calcium buffering, or increased oxidative stress, less glutamatergic stimulation is required to produce toxicity (Fig 40.1B). It follows then that the neuronal threshold for excitotoxic glutamate stimulation can be breached in two ways. Either the actual level of the glutamatergic stimulus can increase, or the ability of a neuron to cope with that stimulation can decrease. As such, *supranormal* glutamate receptor stimulation might not necessarily be *excessive*, and vice versa (Fig 40.1C).

Convention has defined "direct" excitotoxicity to be that which is caused by supranormal, excessive glutamate receptor stimulation, such as may occur following exposure to high concentrations of a glutamate agonist that subsequently overwhelms cellular neuroprotective mechanisms. The term "indirect" excitotoxicity is used for situations when impaired neuronal coping mechanisms result in excitotoxic degeneration in the face of normal levels of glutamate receptor stimulation. This separation is somewhat arbitrary, given the interrelationship between the two, but the key point is that some level of glutamate receptor activation is required.

For example, mitochondria and mitochondrial dysfunction are thought to play a central role in the toxicity resulting from excessive glutamate receptor stimulation. Conversely, mitochondrial dysfunction can sensitize neurons to excitotoxic death, even without supranormal glutamate. It has been convincingly shown that mitochondrial toxins can cause excitotoxic damage without elevation of extracellular glutamate concentration. The mechanisms for this type of indirect excitotoxicity are thought to involve impairment of all main aspects of mitochondrial function: energy metabolism, calcium buffering, free radical handling, and programmed cell death. Because of this, it has been hypothesized that mitochondrial dysfunction may produce excitotoxic neurodegeneration in a variety of acute and chronic neurological disorders, including PD.[56,57]

Alterations in other post-receptor mechanisms of excitotoxicity can also predispose neurons to an excitotoxic death. Over the remainder of this chapter, we will discuss the balance between glutamate receptor stimulation and excitotoxicity and explore how that balance may be implicated in the pathogenesis of PD.

EXCITOTOXICITY IN PD

Excitotoxicity has been implicated in a variety of human neurological diseases. These include acute syndromes, such as stroke, as well as epilepsy and neurodegenerative diseases (see Table 40.1).

TABLE 40.1
Excitotoxicity in Neurological Disease

Acute Disorders
  Stroke
  Hypoxic-ischemic injury
Chronic Disorders
  Alzheimer's disease
  Amyotrophic lateral sclerosis
  Epilepsy
  Huntington's disease
  Parkinson's disease

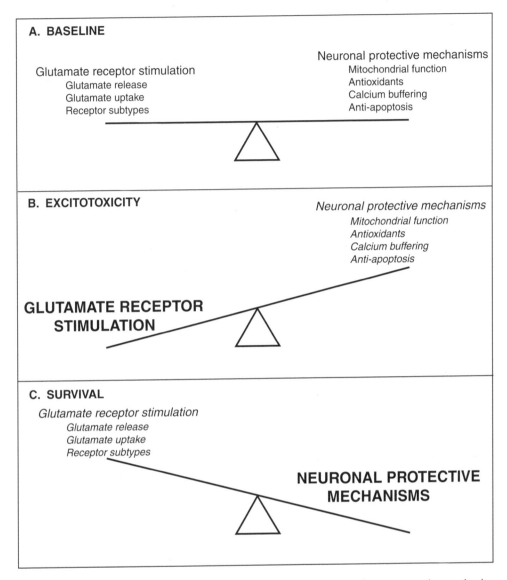

FIGURE 40.1 Schematic depiction of the balance between glutamate stimulation and neuroprotective mechanisms as relates to excitotoxicity. (*A*) Baseline balance between glutamate activation and protective mechanisms, (*B*) excitotoxic injury occurs when glutamate receptor stimulation increases dramatically or neuronal protective mechanisms are impaired, and (*C*) decreasing glutamate stimulation or enhancing protective mechanisms may result in neuroprotection.

PATHOLOGICAL EVIDENCE FOR EXCITOTOXICITY IN PD

Although the ultrastructural findings associated with active excitotoxic degeneration are quite specific, there are no specific pathological hallmarks of excitotoxicity. Nonetheless, there is a decrease of glutamate receptors in PD brains, suggesting that neurons expressing these receptors are vulnerable to neurodegeneration.[58,59] Another, less direct, indication of excitotoxicity in PD comes from the functional changes associated with induction of parkinsonism, specifically the increased glutamatergic tone in several basal ganglia pathways as discussed above. Although this evidence is not specific, it is intriguing, especially given results in laboratory models of PD.

EVIDENCE FOR TOXIC GLUTAMATE RECEPTOR ACTIVATION IN MODELS OF PD

One important finding implicating excitotoxicity in PD is that NMDA antagonists can prevent the neuropathological damage and behavioral consequences of MPTP and MPP+ administration.[60,61] MPTP (1-methyl-4-phenyl-1,2,3,6-tetrahydropyridine), is a well known neurotoxin originally produced as a by-product of illicit heroin synthesis that produced a clinical syndrome of parkinsonism in the unfortunate people who used it.[62] Administration of MPTP to rodents and primates causes behavioral parkinsonism and fairly specific degeneration of dopaminergic neurons.[63,64] Blockade of this toxicity by glutamate antagonists suggests that excitotoxicity is involved in the neurotoxicity in this model and possibly in idiopathic PD.

There is an increased incidence of PD welders, which may be related to manganese exposure.[65] Intracerebral injection of manganese produces a brain lesion that is pathologically similar to lesions produced by excitotoxins. In addition, this lesion is prevented by NMDA antagonists, providing more experimental evidence for excitotoxicity in PD.[66]

There is also the suggestion that several other dopaminergic neurotoxins used to model PD, such as 6-hydroxydopamine and methamphetamine, exert their toxic influence at least partially through excitotoxic stimulation of glutamate receptors.[67,68]

INDIRECT EXCITOTOXICITY AND PD PATHOGENESIS

The toxic action of glutamate may occur due to excessive agonist stimulation, reduced neuronal threshold for coping with normal stimulation, or a mixture of the two. Mitochondrial dysfunction, oxidative stress, nitric oxide, abnormal protein metabolism, and programmed cell death are thought to be several of the post-receptor mechanisms of excitotoxicity, and all have been implicated in PD pathogenesis. In this section, we explore in brief these mechanisms and examine how sensitization to glutamate excitotoxicity might occur.

EXCITOTOXICITY AND OXIDATIVE STRESS

There is mounting pathological and experimental evidence to suggest involvement of reactive oxygen species in the pathogenesis of PD. Excess iron deposition in the PD substantia nigra coupled with a decreased concentration of reduced glutathione may enhance free radical damage by increasing radical production and decreasing clearance, respectively.[69–71] This theory is bolstered by a recent report that iron chelation by genetic or pharmacological means decreases the dopaminergic toxicity associated with MPTP administration.[72] There is pathological evidence of oxidative damage in PD brains, including lipid peroxidation, protein oxidation, and oxidative damage to DNA.[73–77] Since excitotoxic glutamate receptor stimulation can cause generation of oxygen radicals, toxic synergy between enhanced glutamatergic stimulation and increased susceptibility to oxidative damage in the substantia nigra may contribute to their degeneration in PD. In fact, studies in cultured neurons demonstrate that glutamate receptor stimulation causes enhanced oxygen radical generation and that multiple mechanisms of free radical detoxification are neuroprotective *in vitro*.[40,78,79]

In vivo evidence for a synergistic interaction between excitotoxicity and oxidative stress is also quite robust. Toxicity of quinolinic acid, an NMDA receptor agonist, is exacerbated by free radical generation, indicating a role for oxidative stress in the neurotoxicity of this endogenous excitotoxin.[80] Furthermore, free radical spin traps attenuate the *in vivo* production of oxygen radicals induced by a variety of direct and indirect excitotoxins. These compounds also protect against the neurodegeneration associated with excitotoxin injection, as well as that caused by MPTP.[81,82]

EXCITOTOXICITY AND NITRIC OXIDE

There are several lines of experimentation implicating glutamate receptor activation and resultant NO production in PD pathogenesis.

Experiments in rodents and primates show increased levels of nitrosylated proteins in SN following MPTP administration, and one study has shown increased 3-nitrotyrosine in brains from PD patients.[83–85] These nitrosylated proteins are presumed to be a result of reaction with peroxynitrite, generated from NO and superoxide. If excessive, this form of protein modification can interfere with protein function. It may potentially impair subsequent protein degradation, possibly causing a feed-forward loop of protein malfunction.

Inhibitors of NOS, such as 7-nitroindazole, protect against MPTP toxicity in experimental animals, improving both the behavioral and neuropathological sequelae of toxin administration.[86,87] Inhibition of PARP has been reported to protect mice from MPTP toxicity, and PARP knockout mice are resistant to MPTP-induced damage.[88] These data provide experimental evidence to support the hypothesis that NO production induced by glutamate receptor activation is an important mechanism of degeneration in dopaminergic neurons.

EXCITOTOXICITY AND MITOCHONDRIAL DYSFUNCTION

Many of the effecter mechanisms of excitotoxicity are based on mitochondrial dysfunction. Oxidative stress, nitric oxide toxicity, calcium mishandling, and even protein processing problems can all be causes and effects of mitochondrial impairment. The mitochondrion appears to be centrally involved, not only as an effector of excitotoxic damage, but also as a susceptibility factor relating to excitotoxic damage.

Recently, mitochondrial dysfunction has been suggested to be intimately involved in PD pathogenesis. The early finding implicating a mitochondrial abnormality in PD was that MPP+ acts by inhibiting complex I of the mitochondrial electron transport chain.[89] There are also demonstrated deficits in complex I activity and immunostaining in PD substantia nigra.[90–92] Furthermore, a more generalized complex I abnormality is suggested by the finding that complex I activity is decreased in platelets and fibroblasts from patients with PD.[93] When PD mitochondria are combined with cells devoid of mitochondria, the resulting cybrid cell lines exhibit complex I deficiency, increased oxidative damage, and impaired calcium homeostasis.[94,95]

Complex I defects can have several deleterious consequences. First, decreased electron flux through the electron transport chain can result in decreased production of ATP, with resultant impairment of all cellular processes dependent on energy transfer. Second, complex I inhibition causes a backup of high-energy electrons that can nonspecifically reduce molecular oxygen and produce oxygen radicals. Third, decreased mitochondrial membrane potential causes dramatic perturbations in mitochondrial calcium handling. Finally, mitochondrial inhibition can cause release of factors central to the programmed cell death cascades, which may result in initiation of active neuronal death. All of these mechanisms can exacerbate, and in some cases initiate, excitotoxic death.

As mentioned above, the mitochondrial toxicity of MPTP and MPP+ can be ameliorated in experimental animals by glutamate receptor antagonists. This toxicity is associated with ATP depletion, oxidative damage, calcium dysregulation, and abnormal protein processing.[96,97] Additionally, the complex I inhibitor and organic pesticide rotenone also replicates several of the neuropathological and behavioral hallmarks of PD when administered systemically to rats.[98,99] Interestingly, unlike MPTP, which due to uptake via the dopamine transporter causes mitochondrial impairment specifically in dopamine neurons, rotenone produces a generalized inhibition of complex I activity yet still causes relatively specific neurodegeneration. This supports the idea that there is something specific to dopamine neurons that causes exquisite sensitivity to mitochondrial dysfunction and excitotoxicity. It also supports the hypothesis that a generalized complex I defect may cause specific neurodegeneration.

The link between mitochondrial function and excitotoxicity is further supported by multiple studies indicating that neurodegeneration caused by many different mitochondrial toxins is mediated at least partly by glutamate receptors and excitotoxicity.[100] Inhibition of energy metabolism with subsequent depletion of neuronal ATP can cause excitotoxic degeneration in several different ways. For example, ATP depletion results in decreased activity of the Na-K ATPase, which is responsible for maintenance of membrane polarization and sodium-potassium ion gradients. Normal activation of this pump utilizes approximately 40% of neuronal ATP production, and, during states of neuronal activation, this proportion is even higher.[101] Resultant membrane depolarization can cause increased calcium flux through NMDA receptors due to the voltage-dependent nature of NMDA receptor activation.[20] In addition, dissipation of the neuronal sodium gradient has profound consequences for glutamate and calcium homeostasis. Reuptake of glutamate is dependent on the sodium gradient, meaning that secondary active transport of glutamate is decreased without sufficient sodium as a driving source. This may result in increased synaptic glutamate concentrations and receptor overstimulation.[102] This would be particularly detrimental under conditions of increased glutamate transmission, such as in PD substantia nigra. Both extrusion of calcium at the plasma membrane and uptake of calcium into mitochondria are indirectly dependent on sodium-coupled calcium exchange, and impairment of these processes aggravates excitotoxic injury.[103]

Decreased complex I function may also affect mitochondrial calcium buffering capacity via depolarization of the mitochondrial membrane. This has several potential consequences, including that the capability of mitochondria to buffer excitotoxic calcium loads will be greatly diminished, causing a more rapid and severe loss of neuronal calcium homeostasis.[39,104,105] The apparent greater susceptibility of dopamine neurons with fewer calcium binding proteins to neurodegeneration is consistent with this hypothesis.[106,107]

A final area of mitochondria-excitotoxic interaction is programmed cell death. Mitochondria are central to neuronal programmed cell death cascades and excitotoxic mitochondrial dysfunction may initiate these cascades. For example, DNA fragmentation and caspase activation have been reported during glutamate-induced neurodegeneration in vitro.[108] Similar results have been noted after in vivo injection of glutamatergic compounds.[109,110] Additional in vivo results suggest that toxicity caused by intracerebral injection of NMDA agonists is mediated by a complex series of events that begins with NMDA receptor-induced NO production. This causes DNA damage and PARP activation, resulting in mitochondrial dysfunction, release of apoptosis inducing factor, and caspase-independent programmed cell death.[111,112] This pathway has also been implicated in the indirect excitotoxic action of MPTP, further strengthening the relationship of this mechanism to PD pathogenesis.[48]

EXCITOTOXICITY AND UPS FUNCTION

Abnormalities in protein handling and the ubiquitin-proteasome system (UPS) have become an attractive hypothesis to explain neurodegeneration in PD.[113] This theory has arisen from several lines of study, but there are two key underlying points. First, Lewy bodies are the pathological hallmarks of PD and are composed of an abnormal cytoplasmic accumulation of proteins. The UPS degrades damaged, misfolded, or disassembled proteins to maintain satisfactory function of cellular machinery and prevent accumulation of potentially toxic protein deposits.[114] The initiating signal for UPS recognition and degradation is thought to be polyubiquitination of the offending protein.

Since these ubiquitinated proteins are normally efficiently degraded by the UPS, the presence of ubiquitin in Lewy bodies has raised the question of whether there is some dysfunction of the UPS in PD that prevents recognition or degradation of abnormal ubiquitinated proteins. Further evidence for a pathological effect of abnormal UPS function comes from the recent elucidation that several familial forms of PD are the result of mutations involving UPS component genes, such as parkin and ubiquitin carboxy-terminal hydrolase 1 (UCHL1). Others, such as α-synuclein and DJ-1 appear to affect UPS function.[115]

Pharmacological inhibition of UPS function causes dopaminergic neuronal death and inclusion body formation in cultured midbrain neurons.[116] In addition, overexpression of α-synuclein in cultured neurons induces specific degeneration of dopaminergic cell types that appears to be in part related to endogenous dopamine and free radical production.[117] Supporting a potential excitotoxic interaction with UPS function is the finding that parkin protects against kainate excitotoxicity in cultured dopaminergic neurons, possibly by preventing accumulation of cyclin E. It should be noted, however, that parkin did not protect against MPP+ toxicity in this model.[118]

Interestingly, even though the function of the UPS is to degrade dysfunctional proteins, certain abnormal proteins can inhibit UPS function.[119,120] Additionally, accumulation of oxidative protein damage may put stress on UPS function, and oxidized or nitrosylated proteins might impair UPS function. Mutant α-synuclein may be toxic by a similar mechanism.[121]

Although excitotoxicity may cause UPS dysfunction, several studies have questioned the relationship between excitotoxicity and the UPS. One described enhanced toxicity from proteasome inhibition in the presence of NMDA receptor blockade, and another reported no proteasome activation following an excitotoxic insult.[122,123] Clearly, an interaction between excitotoxicity and UPS function requires further exploration to determine its exact contribution to neurodegeneration in PD.

EXCITOTOXICITY AND SELECTIVE VULNERABILITY OF DOPAMINE NEURONS

To this point, we have reviewed basic mechanisms of glutamate neurotransmission and excitotoxicity and explored potential mechanisms of excitotoxicity in PD. We have also noted that substantia nigra dopamine neurons receive glutamate afferents and possess a complement of receptors to respond, and pointed out that glutamate input to these dopamine neurons is aberrantly enhanced in parkinsonism. Thus, these neurons have the potential to undergo excitotoxic degeneration, a fact that has been repeatedly demonstrated experimentally, both *in*

vitro and *in vivo*.[124,125] However, a wide variety of neuronal subtypes in different brain regions are also susceptible to excitotoxic damage. The section that follows entails a brief discussion about how selective vulnerability of dopamine neurons in PD and excitotoxicity might be related. In general, the balance between receptor stimulation and neuronal resistance seems to be important for selective vulnerability, as it is for mechanisms of neurodegeneration (Figure 40.1).

In part, selective vulnerability of SN dopamine neurons might have an anatomical basis. Substantia nigra glutamate receptor activation is enhanced in PD due to an overactive STN, and, in rodents, the nigral toxicity associated with injection of the dopaminergic neurotoxin 6-hydroxydopamine can be attenuated by a concomitant toxic or pharmacologic lesion of the STN.[126,127] This lends credence to the theory that STN input to the nigra can be deleterious and cause excitotoxicity. On the other hand, basal ganglia circuitry models would predict that STN hyperactivity is a result of striatal dopamine depletion, as opposed to the primary cause. Furthermore, other regions receiving increased STN output (e.g., globus pallidus) are preserved in PD. Following this logic, while enhanced STN glutamate stimulation might exacerbate dopaminergic degeneration, it is less likely to be an initiating event.

It has been hypothesized that substantia nigra dopamine neurons are somehow more susceptible to excitotoxic damage than other neurons, and this is supported by several *in vitro* experiments. In mesencephalic culture systems, dopaminergic neurons appear to be somewhat more susceptible to glutamate toxicity than nondopaminergic neurons.[128] In addition, substantia nigra dopamine neurons appear to be particularly prone to degeneration from AMPA receptor stimulation.[129] Work in organotypic slice culture has shown similar results with both glutamate agonists and mitochondrial toxins.[130,131] There are several unique properties possessed by dopamine neurons that may predispose them to excitotoxic damage.

One obvious property is dopamine itself. It may be that dopaminergic neurotransmission puts distinctive stress on neurons such that they have less reserve capacity to respond to other insults. Dopamine has the potential to generate oxygen radicals, either enzymatically or spontaneously. These free radicals are typically readily detoxified by intracellular scavengers but in combination with excitatory stimulation may overwhelm these defenses. In certain *in vivo* models, dopamine does exacerbate excitotoxicity, and this is particularly obvious in the striatum, where a large amount of dopamine turnover occurs.[132–134] Additionally in a culture system, reduction of dopamine turnover and induction of scavenger pathways by D_2 autoreceptor stimulation reduced excitotoxic damage.[135]

Iron accumulation in PD substantia nigra may also contribute to oxidative stress in the region insofar that ferrous iron (Fe^{2+}) may undergo a Fenton reaction with hydrogen peroxide and result in hydroxyl radical generation. In addition, there appears to be a selective deficit of reduced glutathione (GSH) in the substantia nigra of PD patients.[69–71] This may further predispose this area to damage from oxidative stress, since the glutathione pathway is one of the main cellular mechanisms for detoxifying free radicals. In fact, there is extensive pathological evidence that there is increased oxidative damage to lipids, proteins, and nucleic acids in PD substantia nigra.[74–76] As such, the inherent oxidative dangers associated with dopaminergic neurotransmission, coupled with specific regional abnormalities in iron and glutathione metabolism, may result in an enhanced propensity for substantia nigra neurons to degenerate after glutamate stimulation.

As mentioned previously, calcium is intimately involved in glutamate neurotransmission and excitotoxicity. There is some evidence that some dopamine neurons have intrinsically less capacity to handle a calcium load and that this may make them more likely to undergo excitotoxic degeneration. This hypothesis comes predominantly from the finding that neuronal expression of certain calcium binding proteins is associated with relative resistance to toxic insults. More specifically, substantia nigra neurons that contain calbindin appear to be somewhat spared in experimental models and in human PD.[106,107] One might hypothesize that decreased calcium buffering capability, in the form of decreased calcium binding proteins, may lower the level of calcium influx required to cause excitotoxic damage.

One final interesting scheme for selective neurodegeneration involves indirect excitotoxicity from metabolic inhibition. As discussed above, systemic inhibition of mitochondrial metabolism at complex I can produce selective degeneration of the dopamine system in rats.[57,98] This raises the possibility that localized differences in mitochondrial function may influence selective toxicity to dopamine neurons. Specific clarification of such differences has not yet been described.

ANTI-EXCITOTOXIC STRATEGIES FOR NEUROPROTECTION IN PD

There are a myriad of potential ways that glutamate-mediated excitotoxicity may contribute to neurodegeneration in PD, and excitotoxicity may be associated in some way with many current theories of PD pathogenesis. Excitotoxicity may be involved in direct or indirect mechanisms of neurodegeneration and may be a central or a contributing factor to dopamine neuron death. In some respects, the complicated interactions between pathogenic mechanisms may be discouraging, as far as neuroprotection is concerned, because it is not clear which mechanism of neurotoxicity is the most important. However, this complex web of interactions provides the opportunity for mul-

tifocal intervention, and synergistic mechanisms of neurotoxicity may result in synergistic mechanisms of neuroprotection. As such, interference with these neuronal death cascades at many different levels may theoretically provide more effective neuroprotection than intervention at only one site.

Anti-excitotoxic strategies (Table 40.2) may be divided in a way similar to the division between direct and indirect excitotoxicity. As such, a direct anti-excitotoxic strategy might involve inhibition of glutamate neuron activation, inhibition of glutamate release, augmentation of glutamate uptake, or blockade of glutamate receptors. Conversely, an indirect anti-excitotoxic strategy could entail modulating intraneuronal calcium homeostasis, inhibiting NOS, free-radical scavenging, enhancing mitochondrial function, interfering with programmed cell death, or altering UPS function.

TABLE 40.2
Anti-Excitotoxic Strategies in PD

Inhibition of glutamate input
Inhibition of glutamate release
Augmentation of glutamate reuptake
Glutamate receptor blockade
Calcium buffering
NOS inhibition
PARP inhibition
Mitochondrial energization
Antioxidants
Caspase inhibition
Enhancement of UPS function

Both direct and indirect excitotoxic strategies have been successful in experimental animals. For example, inhibition of STN glutamatergic tone diminishes dopamine neuron degeneration in the substantia nigra.[126] This raises the possibility that STN lesioning or deep brain stimulation, which is already accepted as an effective symptomatic antiparkinsonian approach, may also be an effective neuroprotective paradigm.

To this point, the side effects associated with other direct anti-excitotoxic methods, such as NMDA receptor blockade, have limited attempts in human trials of neuroprotection even though these methods have been largely successful in animal studies. This may not be surprising, given the vital role of glutamate neurotransmission in normal CNS function. The few compounds that have been tried in human neuroprotectant trials, such as the NMDA antagonist remacemide and the glutamate release inhibitor lamotrigine, have not shown significant neuroprotective capabilities to this point.[136–138] Of course, such trials are in the nascent stages, and optimum dosing regimens remain unknown; these initial failures do not preclude eventual success. It is possible that a low level of glutamate antagonism will provide neuroprotection when combined with other strategies.

As already mentioned, NOS inhibition, free-radical scavenging, PARP inhibition, and other indirect anti-exci-totoxic strategies have been effective in minimizing damage in animal models of neurodegeneration in PD.[56] One recent suggestion that intervention of this type may be effective in human PD is the recent finding that a mitochondrial electron shuttle and antioxidant, coenzyme Q10, may slow the progression of PD.[139] This trial must be interpreted with caution due to its small size and recent suggestions that coenzyme Q10 may have some symptomatic benefit in PD, but the early results are encouraging.[140]

CONCLUSION

Evidence continues to mount implicating glutamate excitotoxicity in the neurodegeneration of dopamine neurons in PD. Even as multiple new factors related to PD pathogenesis are described, excitotoxic neurodegeneration persists as a plausible mechanism of neuronal death and compliments other pathogenic hypotheses. Further elucidation of the role of excitotoxicity in human PD, as well as exploration of its relationship with other proposed mechanisms of neurodegeneration, will likely lead to rational evaluation of anti-excitotoxic neuroprotective strategies in human PD.

REFERENCES

1. Ehringer, H. H. O., Verteilung von Noradrenalin und dopamin (3-Hydroxytyramin) im Gehirn des Menschen und ihr Verhalten bei Erkrankungen des extrapyramidalen Systems, *Klin Wochenschr*, 38, 1238, 1960.

2. DeLong, M. R., Primate models of movement disorders of basal ganglia origin, *Trends Neurosci.*, 13, 281, 1990.

3. Rascol, O. et al., Treatment interventions for Parkinson's disease: an evidence based assessment, *Lancet*, 359, 1589, 2002.

4. Olanow, C. W., Surgical therapy for Parkinson's disease, *Eur. J. Neurol.*, 9 Suppl., 3, 31, 2002.

5. Rascol, O. et al., A five-year study of the incidence of dyskinesia in patients with early Parkinson's disease who were treated with ropinirole or levodopa, 056 Study Group, *N. Engl. J. Med.*, 342, 1484, 2000.

6. Blandini, F., Porter, R. H., and Greenamyre, J. T., Glutamate and Parkinson's disease, *Mol. Neurobiol.*, 12, 73, 1996.

7. Bergman, H., Wichmann, T., and DeLong, M. R., Reversal of experimental parkinsonism by lesions of the subthalamic nucleus, *Science*, 249, 1436, 1990.

8. Rodriguez, M. C., Obeso, J. A., and Olanow, C. W., Subthalamic nucleus-mediated excitotoxicity in Parkinson's disease: a target for neuroprotection, *Ann. Neurol.*, 44, S175, 1998.

9. Naito, S. and Ueda, T., Characterization of glutamate uptake into synaptic vesicles, *J. Neurochem.*, 44, 99, 1985.

10. Naito, S. and Ueda, T., Adenosine triphosphate-dependent uptake of glutamate into protein I-associated synaptic vesicles, *J. Biol. Chem.*, 258, 696, 1983.

11. McMahon, H. T. and Nicholls, D. G., Transmitter glutamate release from isolated nerve terminals: evidence for biphasic release and triggering by localized Ca2+, *J. Neurochem.*, 56, 86, 1991.

12. O'Shea, R. D., Roles and regulation of glutamate transporters in the central nervous system, *Clin. Exp. Pharmacol. Physiol.*, 29, 1018, 2002.

13. Schousboe, A., Role of astrocytes in the maintenance and modulation of glutamatergic and GABAergic neurotransmission, *Neurochem. Res.*, 28, 347, 2003.

14. Nicklas, W. J., Glia-neuronal inter-relationships in the metabolism of excitatory amino acids., in *Excitatory Amino Acids*, J. S.-M. P. J. Roberts, and H. F. Bradford, Ed., Macmillan: London, p. 57, 1986.

15. Dingledine, R. et al., The glutamate receptor ion channels, *Pharmacol. Rev.*, 51, 7, 1999.

16. Brorson, J. R. et al., Calcium directly permeates kainate/alpha-amino-3-hydroxy-5-methyl-4- isoxazolepropionic acid receptors in cultured cerebellar Purkinje neurons, *Mol. Pharmacol.*, 41, 603, 1992.

17. Burnashev, N. et al., Calcium-permeable AMPA-kainate receptors in fusiform cerebellar glial cells, *Science*, 256, 1566, 1992.

18. Kleckner, N. W. and Dingledine, R., Requirement for glycine in activation of NMDA-receptors expressed in Xenopus oocytes, *Science*, 241, 835, 1988.

19. Johnson, J W. and Ascher, P., Glycine potentiates the NMDA response in cultured mouse brain neurons, *Nature*, 325, 529, 1987.

20. Nowak, L. et al., Magnesium gates glutamate-activated channels in mouse central neurones, *Nature*, 307, 462, 1984.

21. Riedel, G., Platt, B., and Micheau, J., Glutamate receptor function in learning and memory, *Behav. Brain Res.*, 140, 1, 2003.

22. Albin, R. L. and Greenamyre, J. T., Alternative excitotoxic hypotheses, *Neurology*, 42, 733, 1992.

23. Chatterton, J. E. et al., Excitatory glycine receptors containing the NR3 family of NMDA receptor subunits, *Nature*, 415, 793, 2002.

24. Dingledine, R., N-methyl aspartate activates voltage-dependent calcium conductance in rat hippocampal pyramidal cells, *J. Physiol.*, 343, 385, 1983.

25. MacDermott, A. B. et al., NMDA-receptor activation increases cytoplasmic calcium concentration in cultured spinal cord neurones, *Nature*, 321, 519, 1986.

26. Pin, J. P. and Acher, F., The metabotropic glutamate receptors: structure, activation mechanism and pharmacology, *Curr. Drug Target CNS Neurol. Disord.*, 1, 297, 2002.

27. Marino, M. J. and Conn, J. P., Modulation of the basal ganglia by metabotropic glutamate receptors: potential for novel therapeutics, *Curr. Drug Target CNS Neurol. Disord.*, 1, 239, 2002.

28. Marino, M. J. and Conn, P. J., Direct and indirect modulation of the N-methyl D-aspartate receptor, *Curr. Drug Target CNS Neurol. Disord.*, 1, 1, 2002.

29. Olney, J. W., Adamo, N. J., and Ratner, A., Monosodium glutamate effects, *Science*, 172, 294, 1971.

30. Wang, Y. et al., Glutamate metabotropic receptor agonist 1S,3R-ACPD induces internucleosomal DNA fragmentation and cell death in rat striatum, *Brain Res.*, 772, 45, 1997.

31. Aleppo, G. et al., Metabotropic glutamate receptors and neuronal toxicity, *Adv. Exp. Med. Biol.*, 318, 137, 1992.

32. Eimerl, S. and Schramm, M., Resuscitation of brain neurons in the presence of Ca2+ after toxic NMDA-receptor activity, *J. Neurochem.*, 65, 739, 1995.

33. Randall, R. D. and Thayer, S. A., Glutamate-induced calcium transient triggers delayed calcium overload and neurotoxicity in rat hippocampal neurons, *J. Neurosci.*, 12, 1882, 1992.

34. Choi, D. W., Calcium-mediated neurotoxicity: relationship to specific channel types and role in ischemic damage, *Trends Neurosci.*, 11, 465, 1988.

35. Choi, D. W., Glutamate neurotoxicity and diseases of the nervous system, *Neuron*, 1, 623, 1988.

36. Rothman, S. M., Excitotoxins: possible mechanisms of action, *Ann. N. Y. Acad. Sci.*, 648, 132, 1992.

37. Stout, A. K. et al., Glutamate-induced neuron death requires mitochondrial calcium uptake, *Nat. Neurosci.*, 1, 366, 1998.

38. Peng, T. I. and Greenamyre, J. T., Privileged access to mitochondria of calcium influx through N-methyl-D-aspartate receptors, *Mol. Pharmacol.*, 53, 974, 1998.

39. Gunter, T. E. and Gunter, K. K., Uptake of calcium by mitochondria: transport and possible function, *IUBMB Life*, 52, 197, 2001.

40. Dugan, L. L. et al., Mitochondrial production of reactive oxygen species in cortical neurons following exposure to N-methyl-D-aspartate, *J. Neurosci.*, 15, 6377, 1995.

41. Zhang, Y. et al., The oxidative inactivation of mitochondrial electron transport chain components and ATPase, *J. Biol. Chem.*, 265, 16330, 1990.

42. Flint, D. H., Tuminello, J. F., and Emptage, M. H., The inactivation of Fe-S cluster containing hydro-lyases by superoxide, *J. Biol. Chem.*, 268, 22369, 1993.

43. Ravagnan, L., Roumier, T., and Kroemer, G., Mitochondria, the killer organelles and their weapons, *J. Cell Physiol.*, 192, 131, 2002.

44. Bredt, D. S. and Snyder, S. H., Nitric oxide mediates glutamate-linked enhancement of cGMP levels in the cerebellum, *Proc. Natl. Acad. Sci. USA.*, 86, 9030, 1989.

45. Atlante, A. et al., Glutamate neurotoxicity, oxidative stress and mitochondria, *FEBS Lett.*, 497, 1, 2001.

46. Dawson, T. M. et al., Regulation of neuronal nitric oxide synthase and identification of novel nitric oxide signaling pathways, *Prog. Brain Res.*, 118, 3, 1998.

47. Dawson, V. L. and Dawson, T. M., Nitric oxide in neurodegeneration, *Prog. Brain Res.*, 118, 215, 1998.

48. Wang, H. et al., Apoptosis inducing factor and PARP-mediated injury in the MPTP mouse model of Parkinson's disease, *Ann. N. Y. Acad. Sci.*, 991, 132, 2003.

49. Dawson, V. L. et al., Nitric oxide mediates glutamate neurotoxicity in primary cortical cultures, *Proc. Natl. Acad. Sci. USA*, 88, 6368, 1991.

50. Eliasson, M. J. et al., Poly(ADP-ribose) polymerase gene disruption renders mice resistant to cerebral ischemia, *Nat. Med.*, 3, 1089, 1997.

51. Zhang, J. et al., Nitric oxide activation of poly(ADP-ribose) synthetase in neurotoxicity, *Science*, 263, 687, 1994.

52. Schulz, J. B. et al., Blockade of neuronal nitric oxide synthase protects against excitotoxicity *in vivo*, *J. Neurosci.*, 15, 8419, 1995.

53. Dumuis, A. et al., NMDA receptors activate the arachidonic acid cascade system in striatal neurons, *Nature*, 336, 68, 1988.

54. Gunasekar, P. G. et al., NMDA receptor activation produces concurrent generation of nitric oxide and reactive oxygen species: implication for cell death, *J. Neurochem.*, 65, 2016, 1995.

55. Nicotera, P. and Orrenius, S., Ca2+ and cell death, *Ann. N. Y. Acad. Sci.*, 648, 17, 1992.

56. Beal, M. F., Bioenergetic approaches for neuroprotection in Parkinson's disease, *Ann. Neurol.*, 53 Suppl. 3, S39, 2003.

57. Betarbet, R. et al., Mechanistic approaches to Parkinson's disease pathogenesis, *Brain Pathol*, 12, 499, 2002.

58. Difazio, M. C. et al., Glutamate receptors in the substantia nigra of Parkinson's disease brains, *Neurology*, 42, 402, 1992.

59. Penney, J. B. et al., Glutamate receptor genes in Parkinson's disease, *Adv. Neurol.*, 69, 79, 1996.

60. Turski, L. et al., Protection of substantia nigra from MPP+ neurotoxicity by N-methyl-D-aspartate antagonists, *Nature*, 349, 414, 1991.

61. Zuddas, A. et al., MK-801 prevents 1-methyl-4-phenyl-1,2,3,6-tetrahydropyridine-induced parkinsonism in primates, *J. Neurochem.*, 59, 733, 1992.

62. Langston, J. W. et al., Chronic Parkinsonism in humans due to a product of meperidine-analog synthesis, *Science*, 219, 979, 1983.

63. Burns, R. S. et al., A primate model of parkinsonism: selective destruction of dopaminergic neurons in the pars compacta of the substantia nigra by N-methyl-4-phenyl-1,2,3,6-tetrahydropyridine, *Proc. Natl. Acad. Sci. USA*, 80, 4546, 1983.

64. Heikkila, R. E., Hess, A., and Duvoisin, R. C., Dopaminergic neurotoxicity of 1-methyl-4-phenyl-1,2,5,6-tetrahydropyridine in mice, *Science*, 224, 1451, 1984.

65. Racette, B. A. et al., Welding-related parkinsonism: clinical features, treatment, and pathophysiology, *Neurology*, 56, 8, 2001.

66. Brouillet, E. P. et al, Manganese injection into the rat striatum produces excitotoxic lesions by impairing energy metabolism, *Exp. Neurol.*, 120, 89, 1993.

67. Olney, J. W. et al., Excitotoxicity of L-dopa and 6-OH-dopa: implications for Parkinson's and Huntington's diseases, *Exp. Neurol.*, 108, 269, 1990.

68. Sonsalla, P. K., Nicklas, W. J., and Heikkila, R. E., Role for excitatory amino acids in methamphetamine-induced

nigrostriatal dopaminergic toxicity, *Science*, 243, 398, 1989.

69. Dexter, D. T. et al., Increased nigral iron content in postmortem parkinsonian brain, *Lancet*, 2, 1219, 1987.

70. Sian, J. et al., Alterations in glutathione levels in Parkinson's disease and other neurodegenerative disorders affecting basal ganglia, *Ann. Neurol.*, 36, 348, 1994.

71. Perry, T. L. and Yong, V. W., Idiopathic Parkinson's disease, progressive supranuclear palsy and glutathione metabolism in the substantia nigra of patients, *Neurosci Lett.*, 67, 269, 1986.

72. Kaur, D. et al., Genetic or pharmacological iron chelation prevents MPTP-induced neurotoxicity *in vivo*: a novel therapy for Parkinson's disease, *Neuron*, 37, 899, 2003.

73. Dexter, D. T. et al., Basal lipid peroxidation in substantia nigra is increased in Parkinson's disease, *J. Neurochem.*, 52, 381, 1989.

74. Dexter, D. T. et al., Increased levels of lipid hydroperoxides in the parkinsonian substantia nigra: an HPLC and ESR study, *Mov. Disord.*, 9, 92, 1994.

75. Alam, Z. I. et al., Oxidative DNA damage in the parkinsonian brain: an apparent selective increase in 8-hydroxyguanine levels in substantia nigra, *J. Neurochem.*, 69, 1196, 1997.

76. Alam, Z. I. et al., A generalised increase in protein carbonyls in the brain in Parkinson's but not incidental Lewy body disease, *J. Neurochem.*, 69, 1326, 1997.

77. Yoritaka, A. et al., Immunohistochemical detection of 4-hydroxynonenal protein adducts in Parkinson disease, *Proc. Natl. Acad. Sci. USA*, 93, 2696, 1996.

78. Dykens, J. A., Stern, A., and Trenkner, E., Mechanism of kainate toxicity to cerebellar neurons *in vitro* is analogous to reperfusion tissue injury, *J. Neurochem.*, 49, 1222, 1987.

79. Lafon-Cazal, M. et al., NMDA-dependent superoxide production and neurotoxicity, *Nature*, 364, 535, 1993.

80. Behan, W. M. and Stone, T. W., Enhanced neuronal damage by co-administration of quinolinic acid and free radicals, and protection by adenosine A2A receptor antagonists, *Br. J. Pharmacol.*, 135, 1435, 2002.

81. Schulz, J. B. et al., Involvement of free radicals in excitotoxicity *in vivo*, *J. Neurochem.*, 64, 2239, 1995.

82. Lancelot, E. et al., alpha-Phenyl-N-tert-butylnitrone attenuates excitotoxicity in rat striatum by preventing hydroxyl radical accumulation, *Free Radic. Biol. Med.*, 23, 1031, 1997.

83. Good, P. F. et al., Protein nitration in Parkinson's disease, *J. Neuropathol. Exp. Neurol.*, 57, 338, 1998.

84. Ferrante, R. J. et al., Increased nitrotyrosine immunoreactivity in substantia nigra neurons in MPTP treated baboons is blocked by inhibition of neuronal nitric oxide synthase, *Brain Res.*, 823, 177, 1999.

85. Pennathur, S. et al., Mass spectrometric quantification of 3-nitrotyrosine, ortho-tyrosine, and o,o'-dityrosine in brain tissue of 1-methyl-4-phenyl-1,2,3, 6-tetrahydropyridine-treated mice, a model of oxidative stress in Parkinson's disease, *J. Biol. Chem.*, 274, 34621, 1999.

86. Hantraye, P. et al., Inhibition of neuronal nitric oxide synthase prevents MPTP-induced parkinsonism in baboons, *Nat. Med.*, 2, 1017, 1996.

87. Schulz, J. B. et al., Inhibition of neuronal nitric oxide synthase by 7-nitroindazole protects against MPTP-induced neurotoxicity in mice, *J. Neurochem.*, 64, 936, 1995.

88. Mandir, A. S. et al., Poly(ADP-ribose) polymerase activation mediates 1-methyl-4-phenyl-1, 2,3,6-tetrahydropyridine (MPTP)-induced parkinsonism, *Proc. Natl. Acad. Sci. USA*, 96, 5774, 1999.

89. Nicklas, W. J., Vyas, I., and Heikkila, R. E., Inhibition of NADH-linked oxidation in brain mitochondria by 1-methyl-4-phenyl-pyridine, a metabolite of the neurotoxin, 1-methyl-4-phenyl-1,2,5,6-tetrahydropyridine, *Life Sci.*, 36, 2503, 1985.

90. Schapira, A. H. et al., Mitochondrial complex I deficiency in Parkinson's disease, *Lancet*, 1, 1269, 1989.

91. Schapira, A. H. et al., Anatomic and disease specificity of NADH CoQ1 reductase (complex I) deficiency in Parkinson's disease, *J. Neurochem.*, 55, 2142, 1990.

92. Hattori, N. et al., Immunohistochemical studies on complexes I, II, III, and IV of mitochondria in Parkinson's disease, *Ann. Neurol.*, 30, 563, 1991.

93. Parker, W. D., Jr., Boyson, S. J., and Parks, J. K., Abnormalities of the electron transport chain in idiopathic Parkinson's disease, *Ann. Neurol.*, 26, 719, 1989.

94. Cassarino, D. S. et al., Elevated reactive oxygen species and antioxidant enzyme activities in animal and cellular models of Parkinson's disease, *Biochem. Biophys. Acta*, 1362, 77, 1997.

95. Sheehan, J. P. et al., Altered calcium homeostasis in cells transformed by mitochondria from individuals with Parkinson's disease, *J. Neurochem.*, 68, 1221, 1997.

96. Leist, M. et al., 1-Methyl-4-phenylpyridinium induces autocrine excitotoxicity, protease activation, and neuronal apoptosis, *Mol. Pharmacol.*, 54, 789, 1998.

97. Przedborski, S. and Vila, M., The 1-methyl-4-phenyl-1,2,3,6-tetrahydropyridine mouse model: a tool to explore the pathogenesis of Parkinson's disease, *Ann. N. Y. Acad. Sci.*, 991, 189, 2003.

98. Betarbet, R. et al., Chronic systemic pesticide exposure reproduces features of Parkinson's disease, *Nat. Neurosci.*, 3, 1301, 2000.

99. Sherer, T. B. et al., Subcutaneous rotenone exposure causes highly selective dopaminergic degeneration and alpha-synuclein aggregation, *Exp. Neurol.*, 179, 9, 2003.

100. Greene, J. G., Mitochondrial function and NMDA receptor activation: mechanisms of secondary excitotoxicity, *Funct. Neurol.*, 14, 171, 1999.

101. Astrup, J., Sorensen, P. M., and Sorensen, H. R., Oxygen and glucose consumption related to Na+-K+ transport in canine brain, *Stroke*, 12, 726, 1981.

102. Cousin, M. A., Nicholls, D. G., and Pocock, J. M., Modulation of ion gradients and glutamate release in cultured cerebellar granule cells by ouabain, *J. Neurochem.*, 64, 2097, 1995.

103. Andreeva, N. et al., Inhibition of Na+/Ca2+ exchange enhances delayed neuronal death elicited by glutamate in cerebellar granule cell cultures, *Brain Res*, 548, 322, 1991.

104. Kiedrowski, L. and Costa, E., Glutamate-induced destabilization of intracellular calcium concentration homeo-

stasis in cultured cerebellar granule cells: role of mitochondria in calcium buffering, *Mol. Pharmacol.*, 47, 140, 1995.

105. White, R. J. and Reynolds, I. J., Mitochondrial depolarization in glutamate-stimulated neurons: an early signal specific to excitotoxin exposure, *J. Neurosci.*, 16, 5688, 1996.

106. German, D. C. et al., Midbrain dopaminergic cell loss in Parkinson's disease and MPTP-induced parkinsonism: sparing of calbindin-D28k-containing cells, *Ann. N. Y. Acad. Sci.*, 648, 42, 1992.

107. McMahon, A. et al., Calbindin-D28k buffers intracellular calcium and promotes resistance to degeneration in PC12 cells, *Brain Res. Mol. Brain Res.*, 54, 56, 1998.

108. Brecht, S. et al., Caspase-3 activation and DNA fragmentation in primary hippocampal neurons following glutamate excitotoxicity, *Brain Res. Mol. Brain Res.*, 94, 25, 2001.

109. Qin, Z. H., Wang, Y., and Chase, T. N., Stimulation of N-methyl-D-aspartate receptors induces apoptosis in rat brain, *Brain Res.*, 725, 166, 1996.

110. Djebaili, M. et al., p53 and Bax implication in NMDA induced-apoptosis in mouse hippocampus, *Neuroreport*, 11, 2973, 2000.

111. Mandir, A. S. et al., NMDA but not non-NMDA excitotoxicity is mediated by Poly(ADP-ribose) polymerase, *J. Neurosci.*, 20, 8005, 2000.

112. Yu, S. W. et al., Mediation of poly(ADP-ribose) polymerase-1-dependent cell death by apoptosis-inducing factor, *Science*, 297, 259, 2002.

113. McNaught, K. S. and Olanow, C. W., Proteolytic stress: a unifying concept for the etiopathogenesis of Parkinson's disease, *Ann. Neurol.*, 53 Suppl. 3, S73, 2003.

114. Klimaschewski, L., Ubiquitin-dependent proteolysis in neurons, *News Physiol. Sci.*, 18, 29, 2003.

115. Cookson, M. R., Neurodegeneration: how does parkin prevent Parkinson's disease? *Curr. Biol.*, 13, R522, 2003.

116. McNaught, K. S. et al., Impairment of the ubiquitin-proteasome system causes dopaminergic cell death and inclusion body formation in ventral mesencephalic cultures, *J. Neurochem.*, 81, 301, 2002.

117. Xu, J. et al., Dopamine-dependent neurotoxicity of alpha-synuclein: a mechanism for selective neurodegeneration in Parkinson disease, *Nat. Med.*, 8, 600, 2002.

118. Staropoli, J. F. et al., Parkin is a component of an SCF-like ubiquitin ligase complex and protects postmitotic neurons from kainate excitotoxicity, *Neuron*, 37, 735, 2003.

119. Carrard, G. et al., Impairment of proteasome structure and function in aging, *Int. J. Biochem. Cell Biol.*, 34, 1461, 2002.

120. Bence, N. F., Sampat, R. M., and Kopito, R. R., Impairment of the ubiquitin-proteasome system by protein aggregation, *Science*, 292, 1552, 2001.

121. Petrucelli, L. et al., Parkin protects against the toxicity associated with mutant alpha-synuclein: proteasome dysfunction selectively affects catecholaminergic neurons, *Neuron*, 36, 1007, 2002.

122. Bobba, A. et al., Proteasome inhibitors prevent cytochrome c release during apoptosis but not in excitotoxic

death of cerebellar granule neurons, *FEBS Lett.*, 515, 8, 2002.

123. Snider, B. J. et al., NMDA antagonists exacerbate neuronal death caused by proteasome inhibition in cultured cortical and striatal neurons, *Eur. J. Neurosci.*, 15, 419, 2002.

124. Miranda, A. F. et al., Protection against quinolinic acid-mediated excitotoxicity in nigrostriatal dopaminergic neurons by endogenous kynurenic acid, *Neuroscience*, 78, 967, 1997.

125. Kikuchi, S. and Kim, S. U., Glutamate neurotoxicity in mesencephalic dopaminergic neurons in culture, *J. Neurosci. Res.*, 36, 558, 1993.

126. Blandini, F., Nappi, G., and Greenamyre, J. T., Subthalamic infusion of an NMDA antagonist prevents basal ganglia metabolic changes and nigral degeneration in a rodent model of Parkinson's disease, *Ann. Neurol.*, 49, 525, 2001.

127. Piallat, B., Benazzouz, A., and Benabid, A. L., Subthalamic nucleus lesion in rats prevents dopaminergic nigral neuron degeneration after striatal 6-OHDA injection: behavioural and immunohistochemical studies, *Eur. J. Neurosci.*, 8, 1408, 1996.

128. Sawada, H. et al., Different mechanisms of glutamate-induced neuronal death between dopaminergic and non-dopaminergic neurons in rat mesencephalic culture, *J. Neurosci. Res.*, 43, 503, 1996.

129. de Erausquin, G. A. et al., Nuclear translocation of nuclear transcription factor-kappa B by alpha-amino-3-hydroxy-5-methyl-4-isoxazolepropionic acid receptors leads to transcription of p53 and cell death in dopaminergic neurons, *Mol. Pharmacol.*, 63, 784, 2003.

130. Bywood, P. T. and Johnson, S. M., Mitochondrial complex inhibitors preferentially damage substantia nigra dopamine neurons in rat brain slices, *Exp. Neurol.*, 179, 47, 2003.

131. Bywood, P. T. and Johnson, S. M., Differential vulnerabilities of substantia nigra catecholamine neurons to excitatory amino acid-induced degeneration in rat midbrain slices, *Exp. Neurol.*, 162, 180, 2000.

132. Zhang, J. et al., Secondary excitotoxicity contributes to dopamine-induced apoptosis of dopaminergic neuronal cultures, *Biochem. Biophys. Re. Commun.*, 248, 812, 1998.

133. Garside, S., Furtado, J. C., and Mazurek, M. F., Dopamine-glutamate interactions in the striatum: behaviourally relevant modification of excitotoxicity by dopamine receptor-mediated mechanisms, *Neuroscience*, 75, 1065, 1996.

134. Chapman, A. G. et al., Excitotoxicity of NMDA and kainic acid is modulated by nigrostriatal dopaminergic fibres, *Neurosci. Lett.*, 107, 256, 1989.

135. Sawada, H. et al., Dopamine D2-type agonists protect mesencephalic neurons from glutamate neurotoxicity: mechanisms of neuroprotective treatment against oxidative stress, *Ann. Neurol.*, 44, 110, 1998.

136. Shinotoh, H. et al., Lamotrigine trial in idiopathic parkinsonism: a double-blind, placebo-controlled, crossover study, *Neurology*, 48, 1282, 1997.

137. Shoulson, I. et al., A randomized, controlled trial of remacemide for motor fluctuations in Parkinson's disease, *Neurology*, 56, 455, 2001.

138. Evaluation of dyskinesias in a pilot, randomized, placebo-controlled trial of remacemide in advanced Parkinson disease, *Arch. Neurol.*, 58, 1660, 2001.

139. Shults, C. W. et al., Effects of coenzyme Q10 in early Parkinson disease: evidence of slowing of the functional decline, *Arch. Neurol.*, 59, 1541, 2002.

140. Muller, T. et al., Coenzyme Q10 supplementation provides mild symptomatic benefit in patients with Parkinson's disease, *Neurosci. Lett.*, 341, 201, 2003.

41 Inflammation

Roger Kurlan
Department of Neurology, University of Rochester School of Medicine

CONTENTS

INTRODUCTION

Inflammation is a fundamental pathologic process consisting of a dynamic complex of cytologic and histological reactions that occur in affected blood vessels and adjacent tissues in response to an injury or other abnormal stimulation. Inflammation was historically presumed to play a beneficial role in tissue repair. Recently, however, it has become clear that the inflammatory response can be detrimental in disease, including neurodegenerative disorders. Thus, inflammation can be viewed as being a "double-edged sword" in the setting of neurodegeneration, with potentially beneficial neuroprotective effects and potentially harmful neurodestructive influences.[1,2] In this chapter, we review information regarding the occurrence of inflammation in Parkinson's disease (PD) and discuss current evidence regarding its effect on the disease process.

INFLAMMATION IN THE BRAIN

Glial cells represent the primary cytological components of inflammation in the brain. Glia are composed of macroglia (astrocytes and oligodendrocytes) and microglia. Oligodendrocytes, which are involved in the process of myelination, have not been implicated in PD and will not be discussed further. Microglia are the resident macrophage cell population within the brain, being the primary immunocompetent cells that deal with invasions by infectious agents and tumors and also remove cellular debris. Microglia are present in large numbers, consisting of 10 to 20% of the brain's glial cell population. In addition, perivascular microglia play a role in antigen recognition and processing at the blood-brain barrier. In the normal brain, astrocytes are critical for the homeostatic control of the neuronal extracellular environment. After an injury or damage to the brain, microglia and astrocytes undergo various phenotypic changes that allow them to respond to pathological processes. Recent information suggests that astrocytes play a key role in regulating the inflammatory response, perhaps via "death receptor" (e.g., tumor necrosis factor [TNF])-mediated apoptosis.[3] In normal brains, resting astrocytes and microglia are not evenly distributed.[4,5] The density of microglia is significantly higher in the substantia nigra (SN) compared to other midbrain regions and brain areas.[6] This observation, combined with the finding that SN neurons are particularly susceptible to activated microglial-mediated injury,[6] suggests that glial reactions may play an important role in PD.

INFLAMMATION IN PARKINSON'S DISEASE

Careful neuropathologic examination of postmortem PD brains consistently reveals evidence of a significant glial reaction, particularly robust in the SN, pars compacta (SNpc).[7–10] Interestingly, the magnitudes of the astrocytic and microglial reactions in parkinsonian brains are different.[5,10] At most, the SNpc exhibits a mild increase in the number of astrocytes, while a full-blown reactive astrocytosis is only rarely observed. Compared to the astrocytic response, the activation of microglial cells in PD is consistently robust.[8–10] This microglial response is most apparent in those subregions most affected by the neurodegenerative process.[8–10] Activated microglia are typically found in close proximity to free neuromelanin in the neuropil and to surviving neurons, onto which they aggregate to produce the appearance of neuronophagia.[8]

0-8493-1590-5/05/$0.00+$1.50

A recent study suggests that microglial reactions may be specific, with HLA microglial expression (which occurs in Alzheimer's disease) being absent in Lewy body diseases.[11]

Inflammatory glial reactions have been observed in a number of other parkinsonian disorders, including Guam-parkinsonism-dementia complex,[12] progressive supranuclear palsy (PSP),[13] striatonigral degeneration,[14] and familial forms of parkinsonism, including those linked to gene defects in alpha-synuclein[15] and parkin.[16] Similarly, inflammatory reactions are observed in neurotoxic animal models of PD, including 6-hydroxydopamine and MPTP.[17–20]

POTENTIAL PROTECTIVE EFFECTS OF INFLAMMATION IN PD

Recent information points to the possibility that the inflammatory response in PD could have a protective effect on the neurodegenerative process.[2] A variety of potential mechanisms have been suggested. One potential beneficial effect involves the production of trophic factors. It is well recognized that glial cells possess trophic properties that are essential for the survival of dopaminergic neurons. For example, the presence of oiligodendrocyte-type 2 astrocytes greatly improves the survival and phenotype expression of mesencephalic dopaminergic neurons in culture, at the same time reducing the apoptotic loss of these neurons.[21] Glial-derived neurotrophic factor (GDNF), which can be released by activated microglia, appears to be a potent factor in supporting SNpc dopaminergic neurons in culture.[22] Also, GDNF induces dopaminergic nerve fiber sprouting in the setting of striatal injury.[23] It has been shown that GDNF, delivered either by infusion or viral vectors, markedly reduces dopaminergic neuronal loss and enhances dopaminergic function in MPTP-treated animals.[24–26] Brain-derived neurotrophic factor (BDNF) is another trophic factor that is released by activated microglia and which can support dopaminergic neuronal terminals in the striatum.[23] Other potentially protective actions of inflammatory glial cells in PD include the following:

1. Scavenging toxic compounds released by dying neurons
2. Detoxifying reactive oxygen species and lowering oxidative stress[27]
3. Taking up extracellular glutamate to reduce the presumed harmful effects of the increased subthalamic excitotoxic input to the SN[28]

Although a variety of data suggests that inflammation could diminish the neurodegenerative process in PD, an actual neuroprotective role has yet to be established.

POTENTIALLY HARMFUL EFFECTS OF INFLAMMATION IN PD

Two general types of mechanisms have been proposed to explain how inflammatory actions may have direct neurotoxic sequelae and thereby contribute to the process of neurodegeneration in PD.[29,30] First is the *bystander lysis* mechanism. On cellular activation, which can occur by a diverse array of stimuli, microglia assume an amoeboid shape (for phagocytosis), and there is a dramatic genomic upregulation that leads to the production of a large number of potentially neurotoxic mediators, such as complement, cytokines (e.g., MAC, tumor necrosis factor-alpha [TNF-α]), reactive oxygen and nitrogen species, pro-inflammatory prostaglandins, and excitatory amino acids.[30] The mediators are critical for the normal "housekeeping" activities of microglia, and they are down-regulated once these activities have been completed. However, evidence suggests that in a number of clinical conditions (including neurodegenerative disorders) and in animal models for these diseases, microglia can remain in an active state for extended periods so that neurons in the local vicinity of a prolonged inflammatory response (bystanders) may be prone to lytic attack by the direct cytotoxic actions of microglial mediators.[30] Thus, neuronal injury or loss caused by some other mechanism may induce a normal inflammatory response to remove cellular debris, but it becomes prolonged and contributes to the death of these and other nearby neurons.

Growing evidence supports the possibility that inflammatory processes contribute to the disease process in PD. The presence of TNF-α receptors, and MAC has been shown on nigral neurons in PD.[20,31] A number of studies have identified increased levels of potentially toxic cytokines in the SNpc of postmortem brains or in cerebrospinal fluid of patients with PD, including interleukin (IL)-B, IL-2, IL-4, IL-6, IL-10, interferon-γ, transforming growth factor-a and TNF-α.[32–35] Current evidence indicates that elevations in these cytokines might result in a complex cascade of deleterious events that is toxic to dopaminergic neurons and also amplify and propagate the inflammatory reaction.[36] As mentioned, it has been demonstrated, in nigral lesions caused by administration of MPTP to animals, that there is a prominent inflammatory response.[37] The presence of a significant inflammatory response in the SN of postmortem brains from patients who had developed parkinsonism 3 to 16 years earlier from exposure to MPTP has also been observed, suggesting that a time-limited insult to the nigrostriatal system can set in motion a self-perpetuating process of neurodegeneration that includes an inflammatory component.[38]

The second mechanism proposed for how inflammation might contribute to a process of neurodegeneration in PD relates a disturbance of *buffering capacity*.[29,30] The large numbers of potentially neurotoxic mediators that are secreted by activated microglia are generally not neurotoxic under normal conditions, since they are presumably "buffered" or counteracted by an equally diverse array of inactivation and cytoprotective mechanisms. It is known, however, that a number of local microenvironmental compromises (e.g., stroke) and genetic factors will dramatically affect the degree of buffering capacity that is available to remove different potentially toxic mediators.[30] Thus, the concept of a disturbed buffering capacity, which might be determined by a number of complex inherited or environmental influences, is relevant to chronic, long-term degenerative processes that culminate in neuronal death, such as PD. In PD, the nigral buffering system for free radical species, for example, has been observed to be abnormal.[39] Damier et al. have shown that differential regional vulnerability of dopaminergic neurons to degeneration in PD is inversely correlated with the density of astroglial cells.[40,41] Thus, neurons may be less susceptible to the disease process when they are located in an astroglia-rich area, perhaps because of the ability of astroglia to buffer against free radicals and other neurotoxic substances.[36,42]

Even if not the initial, inciting cause of PD, a role for inflammation in the ongoing process of neurodegeneration may serve as a link, unifying most of the active theories of the diseases's pathogenesis (see Table 41.1).

TABLE 41.1
Inflammation and the Pathogenesis of PD

Pathogenetic Hypothesis	Roles of Inflammation
Oxidative stress	Microglia generate superoxide, other free radicals, NO
	Cytokines may make neurons vulnerable to oxidative stress
	COX activity generates superoxide, other free radicals, peroxynitrite, NO
Excitotoxicity	Microglia can secrete glutamate continuously, release quinolinic acid
	COX-2 co-localizes with glutamate in neurons
	COX-2 is induced by kainic acid
	COX-2 is induced by NMDA receptor activation
Apoptosis	COX-2 activation, prostaglandins, cytokines (especially TNF–α) can initiate programmed cell death
Iron	Nigral microglia in PD are activated and full of iron
	Microglia become activated when they acquire iron

In keeping with the *oxidative stress* hypothesis, microglia possess a large capacity to produce superoxide and other free radicals, which is sufficient, under conditions of prolonged microglial activation, to be neuro-

toxic.[43,44] Microglia can also generate nitric oxide (NO) at levels observed to be neurotoxic in cell culture[45,46] and might be responsible for induction of NO synthase (NOS), which has been reported in the SN of patients with PD[35] and MPTP-treated mice.[47,48] Upon induction, NOS can produce high amounts of NO[49] and superoxide radicals.[50] During inflammation, expression of cyclooxygenase type 2 (COX-2), an important prostaglandin synthesizing enzyme, can increase significantly in the brain and is involved in many of the cytotoxic effects of inflammation.[51,52] COX-2 promoter has many similarities to inducible NOS promoter, and so these two enzymes are often coexpressed in the setting of inflammation.[50] Both are expressed in SNpc glia in postmortem PD brains,[53] and the COX-2 product prostaglandin E_2 is also elevated.[54] The activity of COX-2 is linked to the generation of superoxide, other free radicals, peroxynitrite, and NO, which could contribute to oxidative stress of dopaminergic neurons in PD. It has been shown that interferon-gamma, which is released from microglia, induces the expression of CD-23, a glycoprotein on the cell surface of microphages that, when stimulated, provokes production of NOS.[55] This in turn results in the production of nitrates, which can activate the synthesis of cytokines. This cycle of pro-inflammatory processes elicited by microglia may be occurring in PD, since CD-23-expressing glia are seen in the SN of PD patients but not in the brain tissue of healthy controls.[36] As discussed above, many of the effects of microglia appear to be mediated by cytokines such as IL-B[56,57] or TNF-α,[30] which are reported to be increased in PD (see above). It is possible that cytokines released from microglia might contribute to the observation that neurons that degenerate in PD are particularly sensitive to oxidative stress.[36,39]

It has been hypothesized that a chronic state of "weak" *excitotoxicity* may be a pathogenetic mechanism in PD.[58] In this regard, it is of interest that microglia can secrete large quantities of glutamate continuously[59–61] and also release other excitotoxins such as quinolinic acid.[62,63] Thus, microglia may be a significant contributor to excitotoxins in the brain. Furthermore, COX-2 is induced by excitatory neurotransmission, including NMDA receptor activation and kainic acid.[52,53]

It has been shown in primary cultures of rat mesencephalon that activation of the TNF-α (a cytokine released by microglia) transduction pathway involves translocation of the transcription factor nuclear factor kappa B (NF-κB), resulting in almost complete neuronal degeneration by *apoptosis*.[64] Other studies have also suggested that TNF-α plays an important role in the cell death of neurons.[65] Bax, a potent pro-apoptotic substance, is up-regulated after MPTP administration, and its destruction prevents neuronal loss in this neurotoxin model.[66] Another important apoptotic protein, caspase-3, is activated in postmortem PD brain.[67] These findings suggest a possible link between inflammation and apoptosis, an active hypothesis in the pathogenesis of PD.

The *iron* hypothesis proposes that the observed selective increase in nigral iron and ferritin levels in PD is important in the disease pathogenesis, perhaps by the induction of oxidative stress and neuronal death due to their ability to promote formation of oxygen radicals.[68] It has been observed that the microglia identified in the substantia nigra of PD postmortem brains are activated and full of iron,[69] and it has become understood that microglia become activated and inflammatory because they acquire iron. Thus, the increased nigral iron content in PD may contribute to neurodegeneration by a variety of mechanisms, including an inflammatory process.

Some research results have suggested the possibility that inflammation and immune mechanisms may have a primary causative role in PD. Antibodies to dopamine neurons have been found in the cerebrospinal fluid of PD patients.[70–73] A recent study reported that sera from a subgroup of PD patients exhibited a specific IgG response to brain proteins modified by dopamine oxidation, suggesting that excessive oxidative nigral stress in PD could produce specific antigenic determinants that stimulate an autoimmune response.[74] Studies have demonstrated that immunization of guinea pigs with calf midbrain homogenates or with dopamine neuronal homogenates can result in damage to nigral dopaminergic neurons in the guinea pig.[75,76] One study suggests that immune mechanisms may initiate or amplify a cascade of selective neuronal injury. Chen et al. injected immunoglobulin (IgG) isolated from the blood of PD patients into the substantia nigra of rats and found a loss of dopamine neurons at and near the injection site.[77] Inflammation was observed to be present for at least four weeks after the injection. These results did not occur following injection of IgG from healthy controls or when PD IgG was injected into another site, such as the septal nucleus.

A ROLE FOR ANTI-INFLAMMATORY THERAPY IN PD?

The enzymes COX-1 and COX-2 convert arachidonic acid to prostaglandin (PG)G2 followed by conversion to PGH2. Tissue-specific isomerases further metabolize PGH2 into other PG isoforms or thromboxane A_2, compounds that are widely recognized as mediators of a variety of physiological processes in the nervous system, such as fever generation, modulation of the stress response, control of cerebral blood flow, and hyperalgesia. COX-1 is present in many cell types and, in general, is constitutively expressed. In contrast, COX-2 is induced as an immediate early gene in a wide variety of cell types and in response to a wide variety of stimuli. COX-2 has been shown to play a major role in many instances where PG

production is up-regulated, such as cellular reaction to injury and in response to pro-inflammatory cytokines. Factors known to induce COX-2 expression include growth factors, serum, pro-inflammatory cytokines (e.g., IL-B, TNF-α), lipopolysacaride, calcium entry, platelet activating factor, phorbol esters, and small peptide hormones.[78] Several anti-inflammatory factors (e.g., glucocorticoid hormones, IL-4, and IL-13) inhibit COX-2 expression.[78,79]

In most tissue where it is found, COX-2 is not expressed under resting conditions. Brain, however, is one of the few tissues that contains detectable levels of COX-2 mRNA under normal conditions.[80] It has been proposed that COX-2 is involved in the processing and integration of visceral and special sensory input in elaboration of autonomic, endocrine, and behavioral responses and generally in the process of synaptic signaling.[81] It has been confirmed that COX-2 is expressed by excitatory cortical neurons based on co-localization of the enzyme and glutamate.[82] Evidence suggests that NMDA receptor stimulation activates COX-2[82] and that there are COX-dependent steps in the NMDA activation of gene expression.[83]

Aside from its basal expression in neurons, COX-2 is established as an inducible gene in the brain. COX-2 is induced in animal models of seizures,[84–86] including electrically generated and kainic acid-induced (COX-2 induction is attenuated by glutamate antagonists and dexamethasone) and cerebral ischemia.[87,88] In these models, evidence suggests that neuronal COX-2 expression contributes to neuronal vulnerability and death, particularly in excitotoxin-dependent mechanisms.[89] Other mechanisms by which COX contributes to neuronal vulnerability have been discussed earlier and include generation of superoxide, other free radicals, peroxynitrite, NO, and stimulation of apoptosis. Overall, several lines of evidence point to a role for COX-2 activity as contributing to neuronal vulnerability in a variety of pathologic states, perhaps including neurodegeneration.

COX-2 activity has been directly implicated in the pathogenesis of PD. The enzyme has been shown to catalyze oxidation of dopamine to reactive dopamine quinones, potentially toxic reactive by-products.[54,90] In addition, COX activity was found to be increased in the SN from PD patients as compared to AD or control samples.[54] These results suggest the possibility that dopaminergic neurons may be selectively vulnerable due to COX activity.

In addition to the possibility that COX-2 activity may directly contribute to neurodegeneration in PD, growing evidence suggests that the enzyme may indirectly contribute to neuronal loss via brain inflammatory processes. As discussed above, a state of abnormal chronic inflammation may damage neighboring nerve cells, particularly if an underlying neuronal vulnerability exists. As in periph-

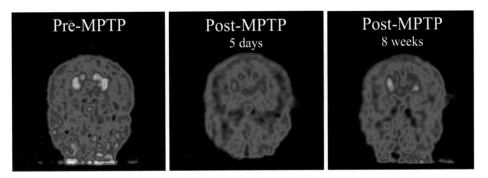

FIGURE 34.2 ¹⁸F-dopa PET in the MPTP-lesioned squirrel monkey. Left panel: coronal plane prior to MPTP administration showing high-intensity signal in the caudate nucleus and putamen. Middle panel: depletion of signal 5 days after the last injection in a series of six injections of MPTP (2 mg/kg, free-base, 2 weeks apart). Right panel: 8 weeks after the last injection of MPTP showing increased intensity of the ¹⁸F-dopa PET signal in the caudate nucleus and putamen. (A full caption accompanies this figure in Chapter 34.)

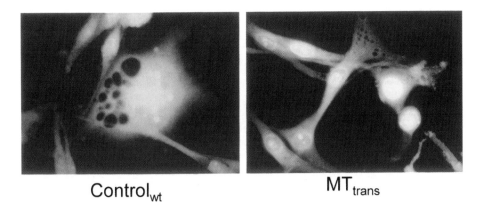

FIGURE 39.1 A triple fluorochrome analysis of SIN-1-induced apoptosis illustrating genetic resistance of MT-trans as compared with control-wt striatal fetal stem cells. (A full caption accompanies this figure in Chapter 39.)

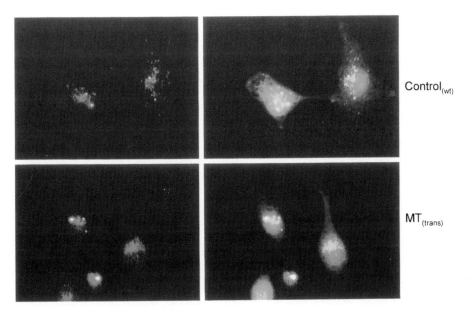

FIGURE 39.2 Digital fluorescence images demonstrating resistance to dopa-L-induced apoptosis in MT-trans as compared with control-wt striatal fetal stem cells within 24 hr. (A full caption accompanies this figure in Chapter 39.)

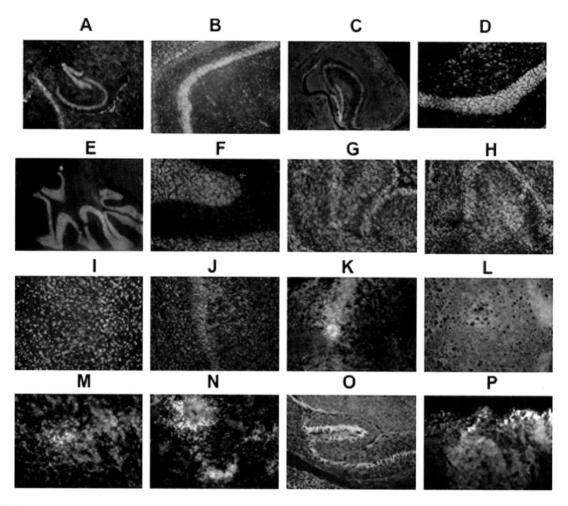

FIGURE 43.6 Progressive neurodegenerative changes in different brain regions of weaver mutant mice exhibiting movement disorders. (A full caption accompanies this figure in Chapter 43.

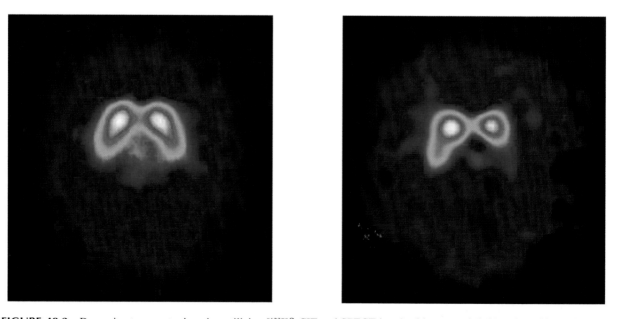

FIGURE 48.2 Dopamine transporter imaging utilizing [¹²³I]β-CIT and SPECT in a healthy control (left) and a subject with early Parkinson's disease (right).

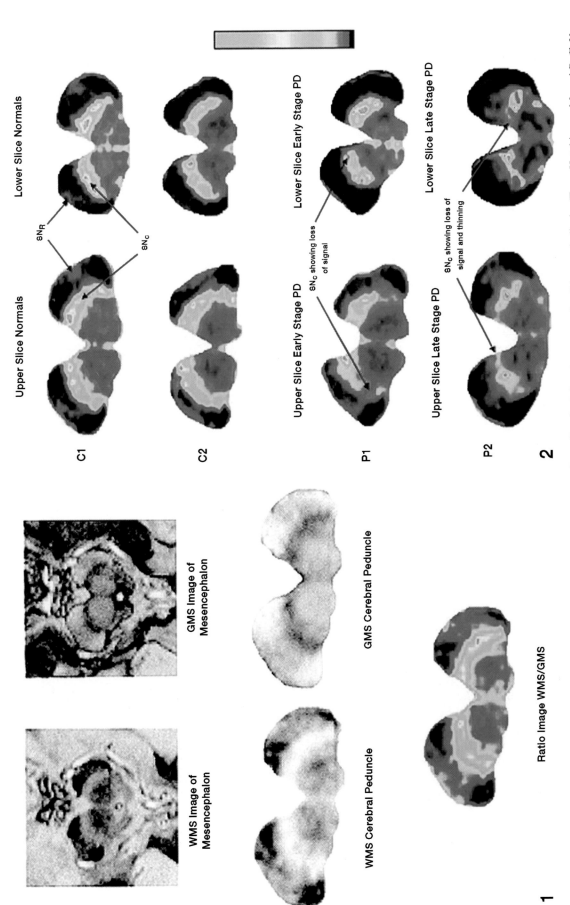

FIGURE 49.4 Computed ratio images from two different inversion recovery sequences that reflect both iron increase and cell loss and gliosis. (From Hutchinson. M. and Raff., U., *Am. J. Neuroradiol.*, 21, 697, 2000. With permission.) (A full caption accompanies this figure in Chapter 49.)

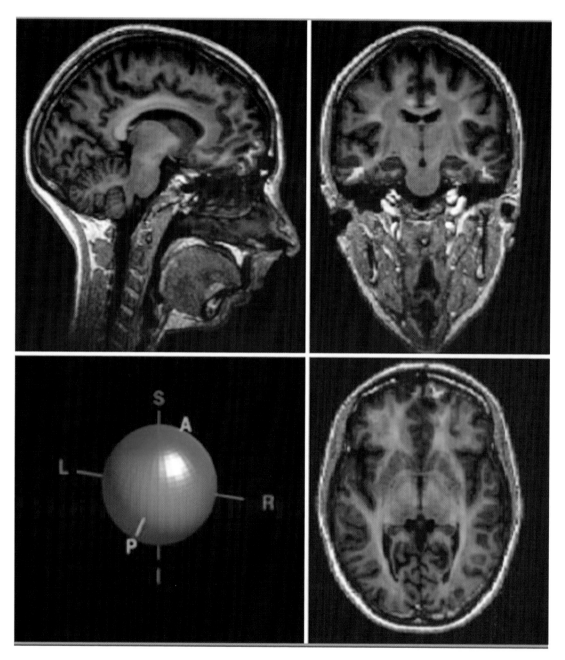

FIGURE 49.12 Diffusion tensor MR images that reflect directionality with color coding.

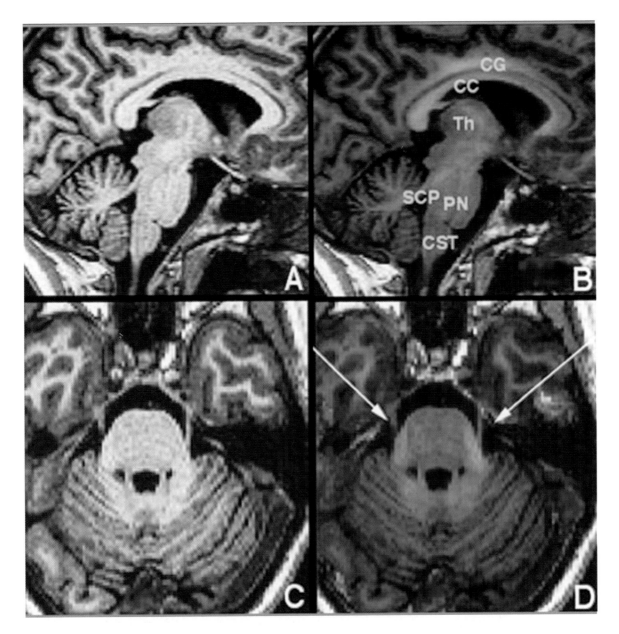

FIGURE 49.13 Diffusion tensor MR images that reflect directionality with color coding.

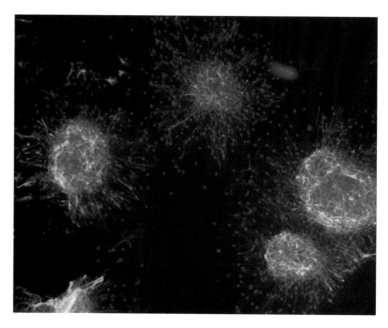

FIGURE 71.2 Neurosphere clones of cells derived from mouse neural stem/progenitor cells. (A full caption accompanies this figure in Chapter 71.)

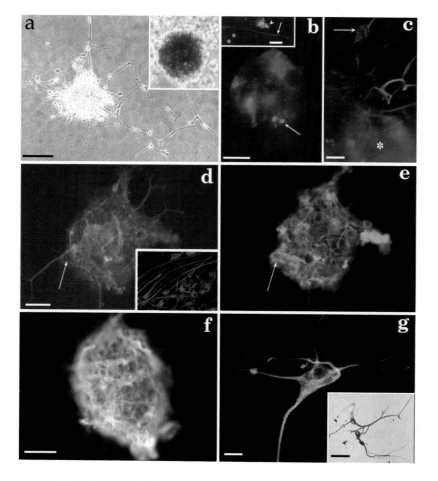

FIGURE 71.3 Human neurosphere clones, and cells derived from them, grown in tissue culture from stem/progenitor cells cultivated from adult human brain biopsy and autopsy specimens. (Adapted from References 5 and 22. With permission.) (A full caption accompanies this figure in Chapter 71.)

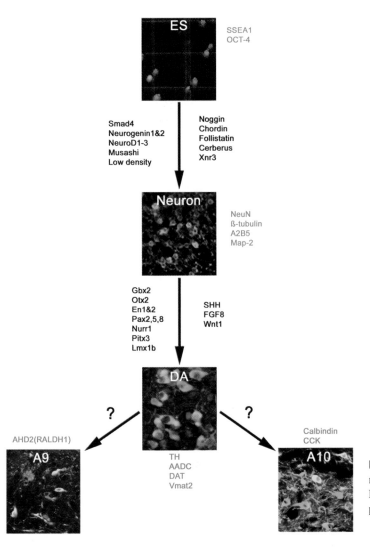

FIGURE 71.4 Midbrain dopamine neuronal development from embryonic stem (ES) cells. (Adapted from Reference 1. With permission.) (A full caption accompanies this figure in Chapter 71.)

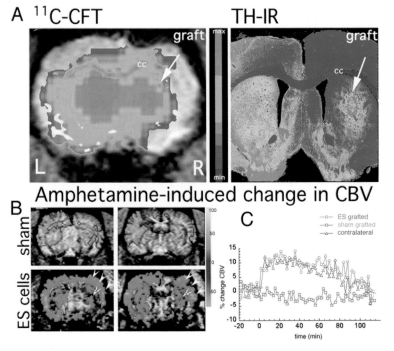

FIGURE 71.5 *In vivo* imaging of dopamine neurons after transplantation of mouse embryonic stem cells to the adult dopamine dennervated rat striatum. (Adapted from References 1 and 17; reprinted with permission of John Wiley & Sons. Copyright 2003 National Academy of Sciences.) (A full caption accompanies this figure in Chapter 71.)

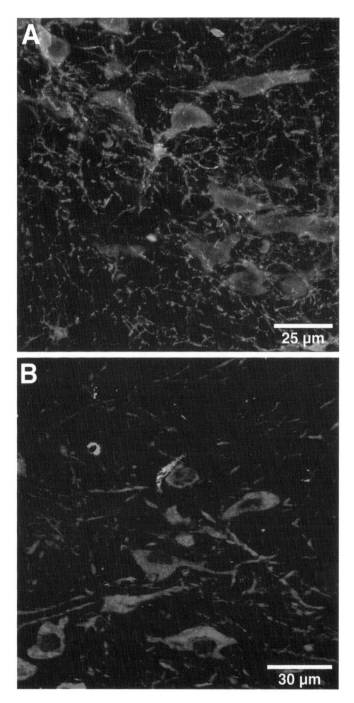

FIGURE 71.8 Phenotype of BrdU-positive cells (a thymidine analog that marks newly generated cells) in the substantia nigra after a 10 d BrdU pulse. (A) One-half of the BrdU-positive cells (green) express the glial progenitor marker NG2 (blue). (B) Some BrdU-positive cells (blue) that are not associated with blood vessels express the multipotent progenitor marker nestin (green). TH is shown in red. (Adapted from Reference 42 and reprinted with permission; copyright 2002 by the Society for Neuroscience.)

eral inflammation, evidence is accumulating for COX- and prostaglandin-dependent steps in the brain inflammatory cascade.[91–98]

Given the growing evidence that inflammatory processes might contribute to neuronal loss in PD, it is rational to consider an experimental therapeutic strategy to block inflammation, including COX-2 activity. Nearly all nonsteroidal anti-inflammatory drugs (NSAIDs) act by blocking access of the arachidonic acid substrate to the COX-2 active site. These drugs can generally be thought of as competing for binding to the active site. One major exception is aspirin, which, by a covalent modification, blocks access to the active site and permanently disables the enzyme. By inhibiting COX-2, NSAIDs prevent conversion of arachidonic acid to prostaglandin (PG) G_2 and eventually to other PG isoforms (PGE_2, PGD_2, PGF_2, PGI_2) and thromboxane A_2, thus interfering with important constituent of the inflammatory response. As indirect effects, it has been shown that COX-2 inhibition results in a major reduction in levels of pro-inflammatory cytokines, particularly IL-6.[99]

NSAID classification is generally based on selectivity for COX-1 or COX-2. Aspirin, for example, is 10 to 200 times more selective for COX-1 than COX-2. Most marketed NSAIDs are nonselective COX inhibitors, influencing both COX-1 and COX-2. Recently available are highly selective COX-2 inhibitors, such as celecoxib and rofecoxib, which have the theoretical advantages over nonselective inhibitors of not inhibiting platelet aggregation or causing renal or gastrointestinal side effects. A number of studies have shown that certain nonselective and selective COX inhibitors have good CNS penetration and act to reduce inflammatory responses in animal models.

Anti-inflammatory, immune-suppressing drugs have been shown to be protective in both the MPTP[17,19] and 6-hydroxy-dopamine[18] animal models of PD. A recent study demonstrated that deletion of the COX-2 gene in mice confers protection against MPTP-induced dopaminergic cell loss.[100] Another animal model in rats demonstrated that stereotaxic injection of bacterial endotoxin lipopolysaccharide (LPS) into the SNpc results in a strong activation of microglia followed by prominent degeneration of dopaminergic neurons.[101] This inflammation-induced degeneration can be blocked by pharmacologically inhibiting microglial activation. LPS-induced microglial activation has also been shown to injure dopaminergic neurons in cell culture.[102]

Based on available data and considering the recognized safety of chronic treatment with NSAIDs, it is reasonable to consider the study of this medication class as a potential neuroprotective therapy in PD. It is important to keep in mind, however, that blocking inflammation could negate potentially beneficial actions of the inflammatory process, as previously discussed.

REFERENCES

1. Hunot, S, Hirsch E., Neuroinflammatory processes in Parkinson's disease, *Ann. Neurol.,* 53:549–560, 2003.
2. Teismann, P., Thieu, K., Cohen, O. et al., Pathogenic role of glial cells in Parkinson's disease, *Mov. Disord.,* 18:121–129, 2003.
3. Dietrich, P. Y., Walker, P., Saas, P., Death receptors on reactive astrocytes. A key role in the fine tuning of brain inflammation? *Neurology,* 60:548–554, 2003.
4. Damier, P., Hirsch, E., Zhang, P., Agid, Y., Javoy-Agid, F., Glutathione peroxidase, glial cells and Parkinson's disease, *Neuroscience,* 52:1–6, 1993.
5. Lawson, L., Perry, V., Dri, P., Gordon, S., Heterogeneity in the distribution and morphology of microglia in the normal adult mouse brain, *Neuroscience,* 39:151–170, 1990.
6. Kim, W., Hohney, R., Wilson, B., Jeohn, G., Liu, B., Hong, J., Regional difference in susceptibility to lipopolysaccharide-induced neurotoxicity in the rat brain: role of microglia, *J. Neurosci.,* 20:6309–6316, 2000.
7. Forno, L., DeLaney, L., Irwin, I., DiMonte, D., Langston, J., Astrocytes and Parkinson's disease, *Prog. Brain Res.,* 94:429–436, 1992.
8. McGeer, P., Itagaki, S., Boyes, B., McGeer, E., Relative microglia are positive for HLA-DR in the substantia nigra of Parkinson's and Alzheimer's disease brains, *Neurology,* 38:1285–1291, 1988.
9. Banati, R., Daniel, S., Blunt, S., Glial pathology but absence of apoptotic nigral neurons in long-standing Parkinson's disease, *Mov. Disord.,* 13:221–227, 1998.
10. Mirza, B., Hadberg, H., Thomsen, P., Moos, T., The absence of reactive astrocytosis is indicative of a unique inflammatory process in Parkinson's disease, *Neuroscience,* 95:425–432, 2000.
11. Shepherd, C., Thiel, E., McCann, H. et al., Cortical inflammation in Alzheimer disease but not dementia with Lewy bodies, *Arch Neurol.,* 57:817–822, 2000.
12. Schwab, C., Steele, J., McGeer, P., Neurofibrillary tangles of Guam Parkinson-dementia are associated with reactive microglia and complement proteins, *Brain,* 707:196–205, 1996.
13. Steele, J., Richardson, J., Olszewski, J., Progressive supranuclear palsy, *Arch Neurol.,* 10:333–358, 1964.
14. Adams, R., Salam-Adams, M., Striatonigral degeneration, in Vinken, P., Bruyn, G., Klawans, H., Eds., Handbook of Clinical Neurology, *Extrapyramidal Disorders,* Elsevier, New York, 205–212, 1986.
15. Dwork, A., Balmaceda, C., Fazzini, E., MacCollin, M., Cote, L., Fahn, S., Dominantly inherited, early-onset parkinsonism: neuropathology of a new form, *Neurology,* 43:69–74, 1993.
16. Ishikawa, A., Takahashi, H., Clinical and neuropathological aspects of autosomal recessive juvenile parkinsonism, *J. Neurol. Neurosurg. Psychiatry,* 245:4–9, 1998.
17. Kitamura, Y., Itano, Y., Kubo, T., Nomura, Y., Suppressive effect of FK-506, a novel immunosuppressant, against MPTP-induced dopamine depletion in the stria-

tum of young C57BL/6 mice, *J. Neuroimmunol.*, 50:221–224, 1994.

18. Matsuura, K., Kabuto, H., Mikino, H., Ogawa, N., Cyclosporin A attenuates degeneration of dopaminergic neurons induced by 6-hydroxydopamine in the mouse brain, *Brain Res.*, 733:101–104, 1996.

19. Aubin, N., Curet, O., Deffois, A., Carter, C., Aspirin and salicylate protect against MPTP-induced dopamine depletion in mice, *J. Neurochem.*, 71:1635–1642, 1998.

20. Kurkowska-Jastrzebska, I., Wronska, A., Kohutnicka, M. et al., The inflammatory reaction following 1-methyl-4-phenyl-1,2,3,6-tetrahydropyridine intoxication in mouse, *Exp. Neurol.*, 156:50–61, 1999.

21. Sortwell, C., Daley, B., Pitzer, M., McGuire, S., Sladek, J., Collier, T., Oligodendrocyte-type 2 astrocyte-derived trophic factors increase survival of developing dopamine neurons through the inhibition of apoptotic cell death, *J. Comp. Neurol.*, 426:143–153, 2000.

22. Burke, R., Antonelli, M., Sulzer, D., Glial cell line-derived neurotrophic growth factor inhibits apoptotic death of postnatal substantia nigra dopamine neurons in primary culture, *J. Neurochem.*, 71:517–525, 1998.

23. Batchelor, P., Liberatore, G., Wong, J., Porritt, M., Frerichs, F., Donnan, G. et al., Activated macrophages and microglia induce dopaminergic sprouting in the injured striatum and express brain-derived neurotrophic factor and glial cell line-derived neurotrophic factor, *J. Neurosci.*, 19:1708–1716, 1999.

24. Batchelor, P., Liberatore, G., Porritt, M., Donnan, G., Howells, D., Inhibition of brain-derived neurotrophic factor and glial cell line-derived neurotrophic factor expression reduced dopaminergic sprouting in the injured striatum, *Eur. J. Neurosci.*, 12:3462–3468, 2000.

25. Gash, D., Zhang, Z., Ovadia, A., Cass, W., Yi., A., Simmerman, L. et al., Functional recovery in parkinsonian monkeys treated with GDNF, *Nature*, 380:252–255, 1996.

26. Kordower, J., Emborg, M., Bloch, J., Ma, S., Chu, Y., Leventhal, L. et al., Neurodegeneration prevented by lenti-viral vector delivery of GDNF in primate models of Parkinson's disease, *Science*, 290:767–773, 2000.

27. Hirsch, E., Hunot, S., Damier, P., Brugg, B., Faucheux, B., Michel, P. et al., Glial cell participation in the degeneration of dopaminergic neurons in Parkinson's disease, *Adv. Neurol.*, 80:9–18, 1999.

28. Benazzouz, A., Piallat, B., Ni, Z., Koudsi, A., Pollak, P., Benabid, A., Implication of the subthalamic nucleus in the pathophysiology and pathogenesis of Parkinson's disease, *Cell Transplant*, 9:215–221, 2000.

29. Kuhn, J., Mtiller, T., Nastos, I., Poehlau, D., The neuroimmune hypothesis in Parkinson's disease, *Rev. Neurosci.*, 8:29–34, 1997.

30. Wood, P., *Neuroinflammation: Mechanisms and Management*, Totowa, N. J., Humana Press, 1998.

31. McGeer, E., McGeer, P., The role of the immune system in neurodegenerative disorders, *Mov. Disord.*, 12:855–858, 1997.

32. Le, W. D., Rowe, D., Jankovic, J., Xie, W., Appel, S., Effects of cerebrospinal fluid from patients with Parkin-

son's disease on dopaminergic cells, *Arch Neurol.*, 56:194–200, 1999.

33. Blum-Degen, D., Muller, T., Kuhn, W., Gerlach, M., Przuntek, H., Riederer, P., Interleukin-1 beta and interleukin-6 an elevated in the cerebrospinal fluid of Alzheimer's and de novo Parkinson's disease patients, *Neurosci. Lett.*, 202:17–20, 1995.

34. Hunot, S., Betard, C., Faucheux, B., Agid, Y., Hirsch, E., Immunohistochemical analysis of interferon-γ and interleukin-10 in the substantia nigra of parkinsonian patients, *Mov. Disord.*, 12:20, 1997.

35. Nagatsu, T., Mogi, M., Ichinose, H., Togari, A., Changes in cytokines and neurotrophins in Parkinson's disease, *J. Neural. Transm.*, Suppl., 60:277–290, 2000.

36. Hirsch, E., Hunot, S., Damier, P., Faucheux, B., Glial cells and inflammation in Parkinson's disease: a role in neurodegeneration? *Ann. Neurol.*, 44:S115–S120, 1998.

37. Czlonkowska, A., Kohutnicka, M., Kurkowska-Jastrzebska, I., Czlonkowski, A., Microglial reaction in MPTP (1-methyl-4-phenyl-1,2,3,6-tetrahydropyride) induced Parkinson's disease mice model, *Neurodegeneration*, 5:137–143, 1996.

38. Langston, J., Forno, L., Tetrud, J. et al., Evidence of active nerve cell degeneration in the substantia nigra of humans years after 1-methyl-4-phenyl-1,2,3,6-tetrahydropyridine exposure, *Ann. Neurol.*, 46:598–605, 1999.

39. Jenner, P., Olanow, C., Oxidative stress and the pathogenesis of Parkinson's disease, *Neurology*, 47:S161–S170, 1996.

40. Damier, P., Agid, Y., Graybiel, A., Hirsch, E., Role of astroglial environment in selectivity of dopaminergic lesion in Parkinson's disease, *Soc. Neurosci. Abst.*, 22:219, 1996.

41. Damier, P., Hirsch, E., Zhang, P., Gluthathione peroxidase glial cells in Parkinson's disease, *Neurosci.*, 52:1–6, 1993.

42. Hirsch, E., Does oxidative stress participate in nerve cell death in Parkinson's disease? *Eur. Neurol.*, 33:522–559, 1993.

43. Colton, C., Gilbert, D., Microglia, an *in vivo* source of reactive oxygen species in the brain, *Adv. Neurol.*, 59:321–326, 1993.

44. Tanaka, M., Sotomatsu, A., Yoshida, T., Hirai S., Nishida, A., Detection of superoxide production by activated microglia using a sensitive and specific chemiluminescence assay and microglia-mediated PC12-cell death, *Neurochem.*, 63:266–270, 1994.

45. Boje, K., Arora, P., Microglial-produced nitric oxide and reactive nitrogen oxides mediate neuronal cell death, *Brain Res.*, 587:250–256, 1992.

46. Chao, C., Hu, S., Molitor, T., Shaskan, E., Peterson, P., Activated microglia mediate neuronal cell injury via a nitric oxide mechanism, *J. Immunol.*, 149:2736–2741, 1992.

47. Liberatore, G., Jackson-Lewis, V., Vukosavic, S., Mandir, A., Vila, M., McAuliffe, W. et al., Inducible nitric oxide synthase stimulates dopaminergic neurodegeneration in the MPTP model of Parkinson disease, *Nat. Med.*, 5:1403–1409, 1999.

48. Dehmer, T., Lindenau, J., Haid, S., Dichgans, J., Schulz, J., Deficiency of inducible nitric oxide synthase protects against MPTP toxicity *in vivo*, *J. Neurochem.*, 74:2213–2216, 2000.

49. Nathan, C., Xie, Q., Regulation of biosynthesis of nitric oxide, *J. Biol. Chem.*, 269:13725–13728, 1994.

50. Xia, Y., Zweier, J., Superoxide and peroxynitrite generation from inducible nitric oxide synthase in macrophages, *Proc. Natl. Acad. Sci., U.S.A.*, 94:6954–6958, 1997.

51. Seibert, K., Masferrer, J., Zhang, Y., Gregory, S., Olson, G., Hauser, S. et al., Mediation of inflammation by cyclooxygenase-2, *Agents Action Suppl.*, 46:41–50, 1995.

52. O'Banion, M., Cyclooxygenase-2: molecular biology, pharmacology, and neurobiology, *Crit. Rev. Neurobiol.*, 13:45–82, 1999.

53. Knott, C., Stern, G., Wilkin, G., Inflammatory regulators in Parkinson's disease: iNOS, lipocortin-1, and cyclooxygenases-1 and -2, *Mol. Cell Neurosci.*, 16:724–739, 2000.

54. Mattammal, M., Strong, R., Lakshmi, V., Chung, H., Stephenson, A., Prostaglandin H synthase-mediated metabolism of dopamine, implication for Parkinson's disease, *J. Neurochem.*, 64:1645–1654, 1995.

55. Hunot, S., Dugas, N., Faucheux, B., Hartmann, A., Tardieu, M., Debre, P. et al., FceRII/CD23 is expressed in Parkinson's disease and induces, *in vitro*, production of nitric oxide and tumor necrosis factor-alpha in glial cells, *J. Neurosci.*, 19:3440–3447, 1999.

56. Beasley, D., Schwartz, J., Brenner, B., Interleukin-1 induces prolonged L-arginine-dependent cyclic guanosine monophosphate and nitrate production in rat vascular smooth muscle cells, *J. Clin. Invest.*, 87:602–608, 1991.

57. Dinarello, C., Interleukin-I and interleukin-I antagonist, *Blood*, 77:1627–1652, 1991.

58. Beal, M., Does impairment of energy metabolism result in excitotoxic neuronal death in neurodegenerative illnesses? *Ann. Neurol.*, 31:119–130, 1992.

59. Piani, D., Frei, K., Quang, D. K., Cuenod, M., Fortana, A., Murine brain macrophages induce NMDA receptor mediated neurotoxicity *in vitro* by secreting glutamate, *Neurosci. Lett.*, 133:159–162, 1991.

60. Piani, D., Spranger, M., Frei, K., Schaffner, A., Fontana, A., Macrophage-induced cytotoxicity of N-methyl-D-aspartate receptor positive neurons involves excitatory amino acids rather than reactive oxygen intermediates and cytokines, *Eur. J. Immuno.*, 22:2429–2436, 1992

61. Patrizio, M., Levi, G., Glutamate production by cultured microglia: differences between rat and mouse, enhancement by lipopolysaccharide and lack of effect of HIV coat protein gp 120 and depolarizing agents, *Neurosci. Lett.*, 178:184–188, 1994.

62. Giulian, D., Corpuz, M., Chapman, S. et al., Reactive mononuclear phagocytes release neurotoxins after ischemic and traumatic injury to the central nervous system, *J. Neurosci. Res.*, 36:681–693, 1993.

63. Moffett, R., Espey, M., Gaudet, S., Namboodiri, M., Antibodies to quinolinic acid reveal localization in select

64. immune cells rather than neurons or astroglia, *Brain Res.*, 623:337–340, 1993.

64. Hunot, S., Brugg, B., Ricard, D., Nuclear translocation of NF-kB is increased in dopaminergic neurons of patients with Parkinson's disease, *Proc. Natl. Acad. Sci., U.S.A.*, 94:7531–7536, 1997.

65. Wallach, D., Cell death induction by TNF: A matter of self control, *Trends Biochem. Sci.*, 22:107–109, 1997.

66. Vila, M., Jackson-Lewis, V., Vukosavic, S., Djaldetti, R., Liberatore, G., Offen, D. et al., Bax ablation prevents dopaminergic neurodegeneration in the 1-methyl-4-phenyl-1,2,3,6-tetrahydropyridine mouse model of Parkinson's disease, *Proc. Natl. Acad. Sci., U.S.A.*, 98:2837–2842, 2001.

67. Hartmann, A., Hunot, S., Michel, P., Muriel, M., Vyas, S., Faucheux, B. et al., Caspase-3: a vulnerability factor and final effector in apoptotic death of dopaminergic neurons in Parkinson's disease, *Proc. Natl. Acad. Sci., U.S.A.*, 97:2875–2880, 2000.

68. Jellinger, K., Baulus, W., Grundke-Iqbal, I., Riederer, P., Youdim, M., Brain iron and ferritin in Parkinson's and Alzheimer's diseases, *J. Neural. Transm.*, [P–D Sect], 2:327–340, 1990.

69. McGeer, P., Itagaki, S., Boyes, B., McGeer, E., Reactive microglia are positive for HLA-DR in the substantia nigra of Parkinson's and Alzheimer's disease brains, *Neurology*, 38:1285–1291, 1998.

70. McRae-Degueurce, A., Rosengren, L., Haglid, K., Immunocytochemical investigations on the presence of neuron-specific antibodies in the CSF of Parkinson's disease cases, *Neurochem. Res.*, 13:679–684, 1998.

71. Loeffler, D., Brickman, C., Kapatos, G., Peter, J., LeWitt, P., Anti-neuronal antibodies and other markers of immune system activation in Parkinson's disease cerebrospinal fluid, *Neurodegen*, 1:145–153, 1992.

72. Klawans, H., Parkinson's disease ventricular CSF contains a growth-inhibitory factor directed at mesencephalic cultures, *Neurology*, 43:A388, 1993.

73. Dahlstrom, A., Wigander, A., Lundmark, K., Gottfries, C., Carvey, P., McRae, A., Investigations on auto-antibodies in Alzheimer's and Parkinson's disease using defined neuronal cultures, *J. Neural. Transm. Suppl.*, 29:195–206, 1990.

74. Rowe, D., Le, W., Smith, R., Appel, S., Antibodies from patients with Parkinson's disease react with protein modified by dopamine oxidation, *J. Neurosci. Res.*, 53:551–558, 1998.

75. Appel, S., Le, W., Tajti, J., Haverkamp, L., Engelhardt, J., Nigral damage and dopaminergic hypofunction in mesencephalon-immunized guinea pigs, *Ann. Neurol.*, 32:494–501, 1992.

76. Le, W. D., Engelhardt, J., Xie, W., Schneider, L., Smith, R., Appel, S., Experimental autoimmune nigral damage in guinea pigs, *J. Neuroimmunol.*, 57:45–53, 1995.

77. Chen, S., Le, W., Xie, W., Alexianu, M., Engelhardt, J., Siklos, L. et al., Experimental destruction of substantia nigra initiated by Parkinson's disease immunoglobulins, *Arch. Neurol.*, 55:1075–1080, 1998.

78. Crofford, L., COX-1 and COX-2 tissue expression: implications and predictions, *J. Rheumatol.*, 24:15–19, 1997.

79. O'Banion, M., Winn, V., Young, D., cDNA cloning and functional activity of a glucocorticoid-regulated inflammatory cyclooxygenase (griPGHS), *Proc. Natl. Acad. Sci., U.S.A.,* 89:4888–4892, 1992.

80. Seibert, K., Zhang, Y., Leahy, K., Hauser, S., Masferrer, J., Perkins, W. et al., Pharmacological and biochemical demonstration of the role of cyclooxygenase 2 in inflammation and pain, *Proc. Natl. Acad. Sci., U.S.A.,* 91:12013–12017, 1994.

81. Breder, C., DeWitt, D., Kraig, R., Characterization of inducible cyclooxygenase in rat brain, *J. Comp. Neurol.,* 355:296–315, 1995.

82. Kaufmann, W., Worley, P., Pegg, J., Bremer, M., Isakson, P., COX-2, a synaptically-induced enzyme, is expressed by excitatory neurons at postsynaptic sites in rat cerebral corte, *Proc. Natl. Acad. Sci., U.S.A.,* 93:2317–2321, 1996.

83. Lerea, L., Carslon, N., Simonato, M., Morrow, J., Roberts, J., McNamara, J., Prostaglandin F2alpha is required for NMDA receptor-mediated induction of *c-fos* mRNA in dentate gyrus neurons, *J. Neurosci.,* 17:117–124, 1997.

84. Chen, J., Marsh, T., Zhang, J., Graham, S., Expression of cyclooxygenase-2 in rat brain following kainate treatment, *Neuroreport,* 6:245–248, 1995.

85. Adams, J., Collaco-Moraes, Y., de Belleroche, J., Cyclooxygenase-2 induction in cerebral cortex: an intracellular response to synaptic excitation, *J. Neurochem.,* 13, 1996.

86. McCown, T., Knapp, D., Crews, F., Inferior collicular seizure generalization produces site-selective cortical induction of cycloxygenase-2 (COX-2), *Brain Res.,* 767:370–374, 1997.

87. Akins, P., Liu, P., Hsu, C., Immediate early gene expression in response to cerebral ischemia: friend or foe? *Stroke,* 27:1682–1687, 1996.

88. Sairanen, T., Raistimaki, A., Karjalainen-Lindsberg, L., Paetau, A., Kaste, M., Lindsberg, P., Cyclooxygenase-2 is induced globally in infarcted human brain, *Ann. Neurol.,* 43:738–747, 1998.

89. Hewett, S., Hewett, J., COX-2 contributes to NMDA-induced neuronal death in cortical cell cultures, *Soc. Neurosci. Abst.,* 23:1666, 1997.

90. Hastings, T., Enzymatic oxidation of dopamine, the role of protaglandin H synthase, *J. Neurochem.,* 65:919–924, 1995.

91. Blom, M., VanTwillet, M., DeVries, S., NSAIDs inhibit the IL-1 beta-induced IL-6 release from human post-mortem astrocytes: the involvement of prostaglandin E2, *Brain Res.,* 777:210–218, 1997.

92. Gottschall, P., Beta-amyloid induction of gelatinase B secretion in cultured microglia—inhibition by dexamethasone and indomethacin, *Neuroreport,* 7:3077–3080, 1996.

93. Janabi, N., Hau, I., Tardieu, M., Negative feedback between prostaglandin and alpha- and beta-chemokine synthesis in human microglial cells and astrocytes, *J. Immunol.,* 162:1701–1706, 1999.

94. Brambilla, R., Brunstock, G., Bonazzi, A., Ceruti, S., Cattabeni, F., Abbracchio, M., Cyclooxygenase-2 mediates P2Y receptor-induced reactive astrogliosis, *Brit. J. Pharmacol.,* 126:563–567, 1999.

95. Minghetti, L., Nicolini, A., Polazzi, E., Creminon, C., Levi, G., Up-regulation of cyclooxygenase-2 expression in cultured microglia by prostaglandin E21 Cyclic AMP, and non-steroidal anti-inflammatory drugs, *Eur. J. Neurosci.,* 9:934–940, 1997.

96. Minghetti, L., Nicolini, A., Polazzi, E., Crominon, C., Maclouf, J., Levi, G., Inducible nitric oxide synthase expression in activated rat microglial cultures in down regulated by exogenous prostaglandin E_2 and by cyclooxygenase inhibitors, *Glia,* 19:152–160, 1997.

97. Aloisi, F., Desimore, R., Columbia-Cabezas, S., Levi, G., Opposite effects on interferon-γ and prostaglandin E2 on tumor necrosis factor and interleukin-10 production in microglia: a regulatory loop controlling microglia pro- and anti-inflammatory cytokines, *J. Neurosci. Res.,* 56:571–580, 1999.

98. Fiebich, B., Hull, M., Lieb, K., Gyufko, K., Berger, M., Bauer, J., Prostaglandin E2 induces interleukin-6 synthesis in human astrocytoma cells, *J. Neurochem.,* 68:704–709, 1997.

99. Anderson, G., Hauser, S., McGeraity, K., Bremer, M., Isaakson, P., Gregory, S., Selective inhibition of cyclooxygenase (COX)-2 reverses inflammation and expression of COX-2 and interleukin-6 in rat adjuvant arthritis, *J. Clin. Invest.,* 97:2672–2679, 1996.

100. Teismann, P., Jackson-Lewis, V., Vila, M., Przedborski, S., Cycloxygenase-2 deficient mice are resistant to MPTP, *Soc. Neurosci. Abstr.,* 688:2, 2001.

101. Liu, B., Jiang, J., Wilson, B., Du, L., Yang, S., Wang, J. et al., Systemic infusion of naloxone reduces degeneration of rat substantia nigral dopaminergic neurons induced by intranigral injection of lipopolysaccharide, *J. Pharmacol. Exp. Ther.,* 295:125–132, 2000.

102. Le, W., Rowe, D., Xie, W., Ortiz, I., He, Y., Appel, S., Microglial activation and dopaminergic cell injury: an *in vitro* model relevant to Parkinson's disease, *J. Neurosci.,* 21:8447–8455, 2001.

42 Cell Death in Parkinson's Disease

Ciaran J. Faherty and Richard J. Smeyne
Department of Developmental Neurobiology, Saint Jude Children's Research Hospital

CONTENTS

INTRODUCTION

Parkinson's disease is a progressive neuropathological condition affecting two percent of people over 65 years and is visually characterized by muscle rigidity, resting tremor, and bradykinesia.[1] Following the initial description of these symptoms by James Parkinson, observations made by Brissaud, Tretiakoff, and Lewy led to the conclusion that the primary neuropathological lesion of PD is the degeneration of neuromelanin-containing dopamine neurons of the substantia nigra pars compacta (SNpc); characteristically associated with eosinophilic inclusions or Lewy bodies.[2] Although the neuropathology and symptomatology are well defined,[3] there is still no clear evidence to explain the etiology or the mechanism of dopamine cell death in PD. The etiology of PD is hypothesized to result from either a genetic predisposition, exposure to an environmental agent, or a combination of these factors.[4–7] There is evidence to suggest that the cell death of SNpc dopamine neurons in PD can occur through either an uncontrolled mechanism known commonly as *necrosis* or through a genetically defined series of programmed steps commonly known as *apoptosis*.[8,9]

WHAT ARE THE "TYPES" OF CELL DEATH?

NECROSIS

Necrosis, from the Greek meaning "dead body," is an uncontrolled form of cell death.[10] This form of cell death is characterized by excessive ionic flux across the plasma membrane resulting in the aberrant activation of proteases, critical depletion of cellular energy, and catastrophic swelling of organelles causing the cell to rupture. These cell effects are often seen during an inflammatory response. The critical depletion of cellular ATP during necrosis allows the cell to die independent of protein and RNA synthesis.[11] Although the nuclear material is largely preserved in necrosis, diffuse random degradation of DNA can occur. In general, when necrosis is seen in the brain, the effects occur among larger groups of cells.[6,10]

APOPTOSIS

Apoptosis, from the Greek meaning "falling off" (analogous to leaves from a tree), is a controlled form of cell death often termed *programmed cell death*. This form of cell death is characterized by chromatin condensation, preservation of plasma membrane integrity, cytoplasmic shrinkage, and cytoskeletal depolymerization.[12] Membrane phospholipid asymmetry, such as the translocation of phosphotidylserine (detected using Annexin V), may be an early signaling event during apoptosis functioning to elicit macrophage migration and phagocytosis.[13] Cellular ATP is required for protein synthesis and RNA transcription, and mitochondrial integrity is maintained because of the critical role this structure plays in the apoptotic cell death process.[14] Due to the controlled nature of the apoptotic process, the associated pathology can occur at the individual cell level.[11] Although a comprehensive overview of apoptosis is beyond the scope of the present discussion, excellent reviews can be found elsewhere.[15–17]

BIOCHEMICAL AND GENETIC PATHWAYS ACTIVATED IN APOPTOSIS

There is evidence to show that apoptosis can be instigated by diverse and unrelated mechanisms, from developmental to environmental cues, where these stimuli share downstream pathways that form a complex and unifying pattern of cell death.[15] Currently, it is thought that two main pathways exist (Figure 42.1) to expedite the apoptotic cell death signal, referred to as "extrinsic" and "intrinsic" pathways.[18] As a prelude to outlining these two pathways it is important to discuss a family of *cysteinyl aspartate*-specific protein*ases*, collectively termed *"Caspases,"* whose processing, activity, and function are central to the signal transduction pathways of apoptosis.[19] Genetic studies in the nervous system of *Caenorhabditis elegans* identified a critical component of the apoptotic apparatus, which was termed *ced*-3.[20] The product of the *ced*-3 gene, CED-3, was shown to be a cysteinyl aspartate-specific proteinase that has structural homology to the mammalian interleukin-1-β-converting enzyme (ICE), whose function involves maintenance and regulation of immunological cells and the inflammatory response.[20,21] The finding that CED-3, whose mammalian homolog is caspase-3, was related to the mammalian ICE (which has subsequently been termed *caspase-1*) led to an intense search for other related proteins influential in the apoptotic pathway.[19] The large family of caspases can be divided into three subdivisions. The first group is involved in cytokine activation (caspases 1/4/5/11/12/13/14). The second group is considered apoptotic effectors (caspases 3/6/7), and the third group is known as apoptotic initiators (caspases 2/8/9/10).[21] Many studies over the past few years have established the caspases as critical checkpoints within the cell death machinery and are normally activated by cleavage of a pro-domain (see Figure 42.1).[18,21]

EXTRINSIC PATHWAY OF CELL DEATH

Numerous physiological phenomena can induce apoptotic cell death. However, for the most part, only two pathways are activated when expediting this process. The "extrinsic" pathway of apoptosis involves a complex interaction of "death" receptors from the tumor necrosis factor (TNF) receptor family [e.g., TNF-R1, Fas and nerve growth factorR (NGFR-p75[NTR])], with their respective ligands.[15,22,23] Binding of a ligand to the death receptor induces structural changes (e.g., dimerization, trimerization) that function to bring together cytoplasmic death domains (DDs) that can sequester adapter molecules such as fas-associated protein with death domain (FADD/MORT1). Within the N-terminal region of FADD, the death effector domain (DED) is then responsible for the downstream signal transduction and binds to the initiator procaspase-8 via its two DED domains.[24] Therefore, the binding of ligand to the death receptor can bring about the formation of a plasma membrane linked; cytoplasmic pro-apoptotic complex termed the death-inducing signaling complex, or DISC.[22] Once pro-caspase-8 is bound to the DISC complex, the pro-domain is cleaved, and activated caspase-8 is released where it is free to proteolytically cleave its substrate, procaspase 3.[24]

INTRINSIC PATHWAY OF CELL DEATH

Within the cell, the "intrinsic" pathway involves the release of key contributors of the apoptotic pathway from the mitochondria. As a consequence of aberrant ionic flux, changes in pH, osmolarity, and transmembrane potential, possibly arising due to the production of reactive oxygen species (ROS), cytochrome c is released from the mitochondrial membrane.[14] The now cytosolic cytochrome c binds to the mammalian homolog of ced-4, the apoptosis protease activation factor (Apaf-1).[21] This cytosolic binding greatly increases the Apaf-1 affinity for deoxyadenosine triphosphate (dATP), which in turn triggers its oligomerization to form the apoptosome. The structural changes brought about by the formation of the apoptosome expose and thereby enable the caspase-recruitment domain (CARD) of Apaf-1 to bind to multiple procaspase-9 molecules and facilitate their autoactivation. The apoptosome-bound activated caspase-9 can now proteolytically activate procaspase-3 (Figure 42.1).

The release of cytochrome c is regulated by proteins encoded by the B-cell lymphoma (Bcl) family.[24] Members of this family include the anti-apoptotic proteins, Bcl-2, Bcl-x$_L$, and the pro-apoptotic proteins Bcl-x$_S$, Bax, Bad, Bid.[21] Members of the Bcl family either attach to or translocate to the outer membrane of the mitochondria during apoptosis. It is postulated that, within the mitochondria, the ratio of anti- and pro-apoptotic Bcl proteins regulate the function of the mitochondrial permeability transition pore (MPT). This ratio of Bcl proteins can act to either prevent the aberrant influx of ions (e.g., Bcl-2, Bcl-x$_L$) or facilitate the opening of the pore, leading to a large ionic influx, swelling of the mitochondrial membrane, and ulti-

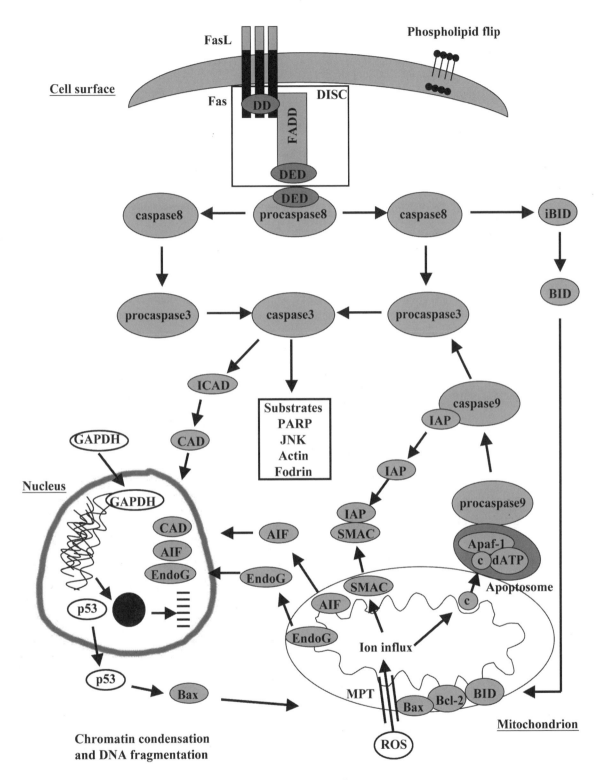

FIGURE 42.1 A brief overview of apoptosis illustrating the signaling through the receptor-mediated (extrinsic) and mitochondrial-based (intrinsic) cell death pathways. *Abbreviations:* AIF = apoptosis-inducing factor; Apaf-1 = apoptosis protease activation factor; Bax = Bcl-2-associated X protein; BID = BH3-interacting domain death; Bcl = B-cell lymphoma-2; c = cytochrome; CAD = caspase-associated Dnase; dATP = deoxyadenosine triphosphate; DD = death domains; DED = death effector domain; DISC= death inducing signaling complex; EndoG = EndonucleaseG; FADD = fas-associated protein with death domain; GAPDH = glyceraldehydes-3-phosphate dehydrogenase; IAP = inhibitor of apoptosis; ICAD = inhibitor of caspase-associated Dnase; JNK = c-jun N-terminal kinase; MPT = mitochondrial permeability transition pore; P53 = tumor suppressor gene/protein; PARP = poly(ADP-ribose) polymerase; ROS = reactive oxygen species; Smac = second mitochondrial derived activator of caspases.

mate release of cytochrome c (e.g., Bax, Bcl-x$_S$, Bid, Bad).[14,21,24]

Concurrent with the release of cytochrome c, other mitochondrial proteins are released that can contribute to the morphologic and biochemical consequences of apoptosis. One of the proteins that enter the cytosol is the second mitochondrial derived activator of caspases (Smac) or direct IAP (inhibitor of apoptosis) binding protein (Diablo).[14] Once in the cytosol, Smac binds to members of the IAP family of proteins. The IAPs function to inhibit active caspases (e.g. caspase-9) during normal conditions. Therefore, binding of Smac to the IAPs relieves the caspase-9 inhibition and allows further propagation of the apoptotic pathway. Two other mitochondrial proteins that can translocate to the nucleus during apoptosis, in a caspase-independent manner, are the apoptosis-inducing factor (AIF) and endonucleaseG.[14,25] Once in the nucleus these proteins contribute to chromatin condensation and DNA fragmentation (Figure 42.1).[14] The ensuing DNA damage instigated by caspase-mediated and independent mechanisms can ultimately lead to transcription of the p53 tumor suppressor gene. The p53 protein can then inhibit the progression of the cell cycle and increase the number of pro-apoptotic transcripts (e.g., Bax, Fas, Glyceraldehydes-3-phosphate dehydrogenase (GAPDH)), thereby inducing apoptosis.[16,26] Another signaling pathway that has been shown to participate in apoptosis and can affect the activities of a number of pro-apoptotic molecules is the c-jun N-terminal kinase (JNK) pathway.[27,28] The JNK pathway is induced by a range of stimuli including DNA damage, free radicals and cytokines.[29] This pathway involves a complex system of serine/threonine kinases that, through phosphorylation reactions, leads to nuclear translocation of JNK.[27,28] Nuclear JNK can induce or repress the expression of a number of target genes and thereby propagate an apoptotic signal.[30] Among the downstream targets of JNK are p53, Bcl-2, Bcl-xL, FasL, and caspases.[29]

Once the apoptotic signal transduction pathway proceeds to the activation of caspases, there are reportedly more than 60 potential substrates available for cleavage.[21] Molecules involved in DNA metabolism, such as the inhibitor of caspase-associated DNase (ICAD), poly-(ADP-ribose) polymerase (PARP), topoisomerase-1, and DNA-dependant protein kinase (DNA-PK), are all substrates for caspases. As the cell actively deconstructs, a number of cytoskeletal proteins are cleaved by caspase activity including lamins (involved in nuclear envelope structure), gelsolin, -fodrin, focal adhesion complex, actin, and actin regulatory enzymes. Other housekeeping proteins that are targeted by caspases include cell cycle regulators such as p21^{Cip1} and p27^{Kip1}, signaling molecules such as MEKK-1, anti-apoptotic proteins such as Bcl-2 and Bcl-x$_L$, and pro-inflammatory cytokines such as interleukin (IL)-1, IL-16, and IL-18.[21]

APOPTOSIS IN PARKINSON'S DISEASE

The numerous molecules involved in the apoptotic cell death pathway work together in a symphony of regulated containment, amplification, deconstruction, packaging, degradation, and elimination of cells. Apoptosis is proposed to be the primary mechanism of cell death in a number of neurologic diseases.[11] One of the neurodegenerative diseases postulated to involve apoptosis as a primary mechanism of cell death is Parkinson's disease.[26,29,31,32] Perhaps the most succinct approach to identifying potential mechanisms of dopaminergic cell death in Parkinson's disease can be obtained from analyzing postmortem tissue from human patients.

EVIDENCE OF APOPTOSIS AS A MECHANISM OF CELL DEATH IN PARKINSON'S DISEASE: POSTMORTEM TISSUE

As discussed earlier, one of the hallmarks of apoptosis is the distinct morphological changes that occur within the cell, including the condensation and degradation of nuclear material. Biochemical alterations within the apoptotic cell function to destabilize the nuclear material in an orderly fashion. This sequence of events is morphologically and biochemically amenable to detection using standard procedures.[33,34] Fragmentation of DNA during apoptosis occurs in two phases; first, the DNA is degraded into fragments of 50 to 300 kbp, followed by degradation into fragments of 180 to 200 bp.[34] The detection of low molecular weight fragments can be used for determining the presence of apoptosis, visualized by a "laddering" pattern when separated on conventional gel electrophoresis.[33] Another approach to analyzing structural changes within the nucleus of the apoptotic cell is to specifically label the fragmented DNA. As a result of these structural perturbations within the DNA molecule, the 3'OH groups are exposed and therefore become available for chemical reaction or labeling. The terminal deoxynucleotidyl transferase (TdT)-mediated dUTP nick-end labeling (TUNEL) procedure takes advantage of the exposed 3'OH groups within the fragmented DNA and enzymatically attaches a specific marker that can be used for visualization.[35,36]

At the ultrastructural level, using electron microscopy (EM), Anglade and colleagues used postmortem tissue from three PD patients and reported that, in 6% of the melanized dopaminergic neurons analyzed (169 total) throughout the SNpc, there was evidence of chromatin condensation along with convolution of the nuclear envelope.[37] In addition, the mitochondria and endoplasmic reticulum (ER) were compacted but maintained a normal morphology, while signs of phagocytosis by glia cells were observed. Another study, by

Hartmann and co-workers, used EM to analyze post-mortem tissue from one PD patient and showed that DA neurons in the SNpc exhibited condensation of the chromatin and ER.[38] In a study of Lewy body diseases, postmortem tissue from five PD patients all displayed characteristics of apoptotic cell death, namely chromatin condensation, formation of apoptotic-like bodies, and microglial phagosomes.[39]

Several studies also examined cell death using the immunohistochemical detection of "nicked" DNA (TUNEL), which is thought to be characteristic of apoptotic cells. Mochizuki et al.[40] analyzed the midbrain of postmortem brains from seven PD patients and found that only four (containing 0.6 to 4.2% of melanized neurons) were positive for TUNEL counted. Tompkins et al.[39] used TUNEL identification combined with EM analysis to verify apoptotic changes in five PD brains (0.38 to 10.13% TUNEL-positive melanized DA neurons). Kingsbury et al.[41] also analyzed the SN of postmortem brains from 16 PD patients using TUNEL and found that 10 (1 to 12.8% labeled melanized neurons) of the 16 brains had visible TUNEL labeling in neurons and glia. In this study, however, the control brains also displayed TUNEL labeling in 0.1 to 10.5% of melanized SNpc neurons, suggesting that the cell death may be due to antemortem hypoxia rather than a result of the PD.

Several other studies used TUNEL labeling in conjunction with DNA binding fluorescent markers, such as the cyanine dye YOYO-1, acridine orange (AO), Hoechst 33342, and DAPI in an attempt to biochemically detect and morphologically verify apoptotic nuclei in the same cell.[26,34,35,42–44] In one study, it was shown that TUNEL positive nuclei within the SN presented with morphology of apoptotic cell death where the condensation of chromatin was visualized using YOYO-1.[35]

While these studies support the induction of apoptosis as a mechanism of cell death in PD, other studies using the TUNEL labeling technique to investigate cell death in PD have failed to demonstrate evidence of apoptosis in postmortem PD brains. Kosel et al.[45] analyzed tissue from 22 PD patients using the TUNEL labeling technique. In approximately half of these brains, TUNEL positive labeling was observed in glia, but no consistent labeling was found within the neuronal population analyzed. Wullner et al.[46] also used TUNEL labeling to identify apoptotic cell death in three PD brains. They reported that no melanin-containing neurons were positive for TUNEL labeling, and there was no evidence of condensation or margination of chromatin. In an attempt to further sensitize the TUNEL labeling reaction, Wullner and colleagues prolonged the TdT incubation step but still did not observe any significant difference in positive labeling between the PD and control subjects.

As outlined in Figure 42.1, there are numerous pro-apoptotic molecules that form a complex signaling network, many of which have been identified as components in the cell death associated with PD. Postmortem studies have shown increases in the level of both TNF and the TNF receptor, TNFR1, in PD brains.[47,48] Polymorphisms within the TNF gene were reported in numerous cases of sporadic PD patients, purportedly influencing the onset of degeneration.[49,50] In addition to perturbations within the TNF family, postmortem studies have also reported increases in the nigrostriatal levels of both the Fas protein and soluble Fas in PD patients.[22,51–53] Further evidence supporting the role of Fas-mediated apoptosis in PD is the observation that FADD is significantly altered within the SNpc.[54] Hartmann and colleagues have suggested that FADD+ DA neurons within the nigrostriatal system may make these cells more vulnerable to degeneration in PD.

Other studies have shown increases in the activation of caspase-8 and -9 in postmortem PD brains,[9,55,56] as well as caspase-3.[38,57l,58] Hartmann et al.[38] analyzed the mesencephalons from five PD subjects and concluded that the activation of caspase-3 in these brains served to sensitize DA neurons to agents that cause degeneration.

Postmortem studies of PD brains have also shown compromised functions of mitochondrial complex I and NADH-cytochrome c reductase.[59,60] As a consequence of electron transport chain dysfunction, associated pro-apoptotic factors can interact with the mitochondrion to further expedite the apoptotic signal.[14] Analysis of sections through the SN from eight PD subjects revealed increases in the immunoreactivity for the pro-apoptotic protein, Bax.[58,61] This increase in Bax labeling was shown to be co-localized with activated caspase-3. Studies have also reported increases in the anti-apoptotic protein, Bcl-2, in midbrain tissue from PD subjects, possibly indicating a protective response to ongoing neuronal death.[51,62] In addition, the level of another anti-apoptotic member of the Bcl-2 family, Bcl-xL, was shown to be increased in melanized DA neurons in PD patients.[63] In this study, Hartmann and co-workers used EM and showed that localization of Bcl-xL to the intermembrane of the mitochondria occurred only in SNpc tissue obtained from PD subjects.

In addition to translocation and release from the mitochondria, molecules are shuttled to and from the nucleus during apoptosis (Figure 42.1).[16] One of the early events in neuronal apoptosis involves the translocation of the glycolytic enzyme GAPDH from the cytoplasm to the nucleus.[26] Once in the nucleus, GAPDH can suppress the expression of Bcl-2 and Bcl-xL, proteins that may be influential to the degenerative process of PD and other neurodegenerative diseases.[26,51,62–64] Nuclear accumulation of GAPDH has been reported to occur in PD brains, while some clinically effective drugs such as deprenyl, have been shown to modulate the availability of this cytoplasmic enzyme.[26,58] Further downstream, damage to the nuclear material can induce the expression of the p53 protein.[16] A p53-dependent apoptotic mechanism of cell death may be involved in PD as indicated by the observed increase in p53 protein in postmortem brains,[52] where p53 serves to increase the number of pro-apoptotic transcripts, thereby propagating the apoptotic cascade.

Although the observation that dopaminergic neurons display morphological hallmarks of apoptosis in postmortem tissue from PD patients, several caveats need to be kept in mind. There is clear evidence that apoptotic bodies can form as a result of normal aging.[65] In one study, tissue from four subjects without any known neurodegenerative disorders were analyzed, and 2% of melanized DA neurons displayed cellular shrinkage, chromatin condensation, shrunken mitochondria, and fragments of melanized neurons were found in glia. There is also disparity in the number of analyzed DA neurons between studies that show evidence for and against apoptotic cell death. Anglade et al.[37] reported evidence of apoptotic cell death by analyzing postmortem tissue from three PD patients. However, as few as 15 DA neurons were analyzed in one of these patients, and a total of 169 neurons for all three PD brains. In contrast, Wullner et al.[46] analyzed postmortem tissue from three PD patients and reported no evidence of apoptosis. However, in this study, approximately 10,000 DA neurons were analyzed per PD subject. As there are reportedly between 300,000 and 400,000 SNpc dopaminergic neurons before degeneration, which maybe reduced to approximately 70,000 to 140,000 in the later stages of PD (60 to 80% reduction), it is important to analyze a large cohort of affected cells.[32] Although there are many studies describing the biochemical manifestations of apoptotic cell death in the sequelae of PD degeneration, other groups have failed to find such evidence. These studies showed no differences in the level of Bcl-2, Bax[46,66–68] caspase-3, p53,[66] Fas or FasL[69] in PD brains compared to control subjects. Therefore, caution is needed when interpreting data from ultrastructural and biochemical analyses of DA neurons in postmortem PD brains. Attention must be given to confounding artifactual data possibly resulting from antemortem drug treatment, cause of death, postmortem interval, processing, fixation and variation in techniques, tissue pH, hypoxia, and the contribution of the normal ageing process.

Although postmortem tissue is a valuable basis from which to investigate and infer mechanisms of neuronal death in PD, these specimens are often difficult to obtain and limited in quantity. It is therefore invaluable to develop model systems that recapitulate many facets of the disease state and use this as a platform to investigate, infer, and speculate as to cause and potential treatment of PD. One such type of model system is available by using *in vivo* rodent models, the use of which has greatly accelerated and expanded our knowledge of PD.

EVIDENCE OF APOPTOSIS AS A MECHANISM OF CELL DEATH IN PARKINSON'S DISEASE: *IN VIVO* MODELS

One of the most widely used *in vivo* animal models for the study of PD is the systemic administration of the tertiary amine, 1-methyl-4-phenyl-1,2,3,6-tetrahydropyridine (MPTP).[70,71] Studies of young adult heroin addicts who unknowingly injected the synthetic heroin contaminant, MPTP, presented with symptoms indistinguishable from those of PD patients.[72] Subsequent investigations showed that administration of MPTP resulted in loss of dopamine neurons in the SNpc. *In vivo*, MPTP is metabolized to the free radical, 1-methyl-4-phenylpyridinium (MPP$^+$), a reaction catalyzed by monoamine oxidase B.[73] The active metabolite, MPP$^+$, is then transported into dopaminergic neurons via the dopamine transporter.[74] Although the precise mechanism that leads to SNpc cell death is still unclear, MPTP has been shown to inhibit mitochondrial function and other essential components of the cell's energy metabolism.[75] In particular, MPTP is reported to deplete cellular concentrations of ATP, inhibit glycolytic enzymes, and lead to the formation of free radical species.[71] The reported similarities in the proposed mechanisms leading to the changes in the nervous system following MPTP treatment and that of idiopathic Parkinson's disease make MPTP an attractive pharmacological model of PD.[6,71]

Tatton and Kish[76] used TUNEL labeling combined with AO staining to investigate the role of apoptotic cell death following a five-day systemic administration protocol of MPTP. Evidence of apoptosis was apparent in the SNpc 72 hours following the first injection of MPTP, demonstrated by confocal laser microscopy of AO-stained nuclei displaying chromatin condensation and nuclear clumping. This pattern of AO-stained DNA condensation was associated with intense nuclear TUNEL staining protracted to approximately three weeks following the last injection of MPTP. The observations by Tatton and Kish are supported by the findings of Spooren et al.[77] showing condensation and marked convolution of nuclear material in conjunction with TUNEL labeling in the SN following a single bolus injection of MPTP. In addition, Serra et al.[78] showed that MPTP induced apoptosis in the SNpc using TUNEL, AO staining, and transmission EM. The TUNEL labeling was associated with glial fibrillary acidic protein (GFAP) positive staining in the SNpc, suggesting the involvement of glia in the degenerative process following MPTP exposure.[79] Further staining of glial cells with Hoechst 33342 revealed pyknotic nuclei and chromatin condensation. However, some studies have shown evidence of mitochondrial swelling and deformation, and a lack of chromatin condensation and TUNEL labeling, findings that maybe contradictory to apoptotic cell death.[80,81]

In addition to structural aberrations in the DA neurons of the SNpc, there is also biochemical evidence of apoptotic cell death in the MPTP model of PD. Studies have shown that the pro-inflammatory cytokine, TNF-α, may be involved in regulation of DA homeostasis and the cell death process following MPTP exposure.[82] The expression of TNF-α increased as early as three hours following the systemic injection of MPTP, compared to vehicle-treated controls. Moreover, mice lacking the receptors for TNF-α were protected from loss of dopaminergic nerve terminals caused by MPTP.[82] Other studies support a role for TNF-α signaling in the cell death of DA neurons seen in the MPTP model of PD.[83,84]

In addition to the reported changes in TNF signaling, Viswanath et al.[56] showed that administration of MPTP to mice significantly increased the activation of caspase-8 within the SN. Others have shown that caspase-3 might also be involved in the SNpc DA neuronal death following the systemic administration of MPTP.[38] Here, they hypothesized that pro-caspase-3 within the SNpc may represent a "vulnerability factor" in the probabilistic loss of DA neurons following MPTP administration. However, further studies by Turmel and colleagues[85] have shown that caspase-3 is activated one to two days following MPTP injection, combined with neuronal pyknosis. Using cleavage of fluorogenic substrates to indirectly determine the activation and activity of caspases, Viswanath and co-workers showed that systemic injection of MPTP resulted in activation of caspases 3 and 9 within the SN.[56]

The activation of caspases 3 and 9 were associated with the release of cytochrome c from the mitochondria, cleavage and activation of the pro-apoptotic Bcl-2 family member, Bid (Figure 42.1).[21,56] Other members of the Bcl-2 family have also been implicated in MPTP-induced cell death (Hassouna et al., 1996; Vila et al., 2001; Yamada et al., 2001; Yang et al., 1998). Administration of MPTP increased the expression and immunoreactivity of Bax in the SN.[86,87] The Bax expression was highest at two days, while the peak increase in immunoreactivity occurred four days following MPTP treatment—a time corresponding to prominent apoptosis revealed by morphological criteria. The possible involvement of the pro-apoptotic Bax protein, in the neurotoxic response to MPTP, was further strengthened following the reported abrogation of cell death resulting from deletion of the Bax gene. The Bax-mediated cell death signal is likely to be compounded by the concurrent decrease in expression and immunoreactivity of the anti-apoptotic bcl-2 protein.[87] Conversely, MPTP-induced neurotoxicity is attenuated in mice overexpressing the human bcl-2 protein.[88,89] Additional evidence of caspases 3 activation, following MPTP exposure, is the changes observed in PARP activity.[90–93] Mandir et al. show that MPTP induces poly-(ADP-ribosyl)ation of nuclear proteins. The neurotoxic effects of MPTP can be prevented by deletion of the PARP gene and by the use of specific inhibitors such as benzamide.[91,92]

Further biochemical evidence to support the role of apoptotic cell death in the neurotoxicity of MPTP is the reported activation of the JNK and p53 pathways. MPTP induces the expression and phosphorylation of c-jun and activation of JNK and JNK kinase.[94–96] Furthermore, systemic injection or viral delivery of JNK-specific inhibitors protects mice against the neurotoxic effects of MPTP.[97,98] One of the consequences of JNK activation is the activation of the pro-apoptotic transcription factor, p53 (Milne et al., 1995). Ablation of the p53 gene as been shown to prevent MPTP-induced loss of DA neurons within the SNpc.[99] Furthermore, systemic administration of p53 inhibitors have been shown to protect midbrain dopaminergic neurons in a mouse model of PD, possibly by blocking the expression and subsequent translocation of Bax.[100]

Studies have also examined the modulation of pro- and anti-apoptotic proteins as they relate to degeneration of SN neurons following MPTP treatment. For example, the anti-apoptotic protein, AIF, is released from the mitochondria and is translocated to the nucleus within hours of MPTP exposure. This regulation of AIF is thought to be PARP dependent, since deletion of PARP abrogates the nuclear translocation of AIF.[90] Others groups have used viral delivery systems and transgenic approaches to study the potential involvement of pro and anti-apoptotic molecules in the MPTP model of PD.[101] Mochizuki et al.[102] showed, using adeno-associated viral delivery, that a dominant negative form of APAF prevented the nigrostriatal degeneration associated with MPTP. In addition, expression of the pro-apoptotic protein, IAP, can also attenuate the neurotoxicity of MPTP.[103]

A second neurotoxin that is widely used as a model of induced-PD is 6-hydroxydopamine (6-OHDA). This neurotoxin was isolated in 1959 by Senoh and is the hydroxylated form of the natural neurotransmitter, dopamine.[104] Initial studies showed that systemic 6-OHDA reduced noradrenaline concentration in the autonomic nervous system and destroyed nerve endings of sympathetic neurons.[32] However, due to inability to cross the blood-brain barrier, central administration of 6-OHDA into the striatum, SN, or ascending medial forebrain bundle (MFB) was necessary to recapitulate the neuronal loss and reduction of dopamine observed in PD patients.[105] Mechanistically, it has been postulated that the microenvironment of midbrain dopaminergic neurons, bringing into close proximity DA, hydrogen peroxide, and iron, may lead to the auto-oxidation and formation of the neurotoxin, 6-OHDA.[106] The in situ generation of this neurotoxin may explain the neurotoxicity of dopamine, possibly forming 6-OHDA via auto-oxidation and the reported 6-OHDA in urine of L-DOPA treated PD patients.[32,107]

He et al.[108] injected 6-OHDA into the MFB of rats and analyzed the associated cell death patterns in the midbrain up to 14 days post treatment. The dopaminergic neurons appeared shrunken and condensed or fragmented into small round bodies. These DA neurons were further identified as being apoptotic in nature by the presence of TUNEL positive labeling. Marti et al.[109] confirmed the presence of apoptosis using EM, and Zuch et al.[110] showed that MFB-injection of 6-OHDA caused degeneration within the SNpc of the rat brain by analyzing the pattern of FluoroJade staining, a marker used to identify degenerating neurons.[111] Although FluoroJade cannot distinguish the mechanism of cell death, the association with TUNEL labeling and chromatin condensation and fragmentation within the SNpc would infer an apoptotic form of death following 6-OHDA treatment.[110]

Further biochemical evidence that apoptotic cell death may be a component of 6-OHDA neurotoxicity comes from the observation that TNF-α protein[112] and activation of caspases 3[113] is observed following 6-OHDA lesions. In addition, there is evidence to suggest an involvement of the mitochondria in the neurotoxic profile of 6-OHDA.[114] Although the in vivo studies used to determine the mechanism of cell death of 6-OHDA are limited, there is clear evidence to indicate the involvement of an apoptotic-signaling pathway.[32,105]

While in vivo models are valuable for investigating the mechanism of cell death in PD, the cost to purchase and maintain animals can be expensive, they involve labor-intensive surgical procedures, and complications can arise due to strain differences and peripheral toxicity. Another alternate method in which to investigate and test hypotheses regarding mechanisms of cell death in PD is the use of cell culture systems.

EVIDENCE OF APOPTOSIS AS A MECHANISM OF CELL DEATH IN PARKINSON'S DISEASE: IN VITRO MODELS

Cell cultures represent an attractive model system for use in deciphering the components of the cell death mechanism of PD. These cell systems allow pharmacological manipulation, are often homogenous in cell type, and are amenable to transfection of test molecules. Among the many cell culture lines used as an in vitro model system of PD are primary cultures of mesencephalic dopaminergic neurons, the MN9D dopaminergic neuronal cell line, PC12 and SH-SY5Y cells.[115,116]

Dodel et al.[117] used primary ventral mesencephalic cultured neurons and showed that addition of MPP+ resulted in apoptosis. Morphologically, these neurons displayed nuclear condensation as measured by DAPI staining as well as TUNEL labeling. In addition, MPP+ was shown to cause chromatin condensation and DNA laddering when added (in large concentrations) to cerebellar

granule neurons.[118] Interestingly, Du and co-workers showed that higher concentration of MPP+ elicited a more rapid death independent of nuclear condensation, suggesting a switch to a nonapoptotic mode cell death. Lotharius et al. used murine mesencephalic dopaminergic cells to show evidence of early-stage apoptosis by measuring the membrane phospholipid asymmetry following administration of 6-OHDA. In this study, cells displayed translocation of phosphotidylserine, identified by an increase in Annexin-V staining, in combination with morphological attributes reflective of apoptosis using the Hoechst 33258 dye.[119] Using MN9D cells, 6-OHDA was shown to induce apoptotic cell death, as measured by the anatomical criterion of chromatin condensation and preservation of intracellular organelles without the apparent swelling of the mitochondria.[120] In addition, apoptosis cell death was also observed in primary cultured cortical neurons, SH-SY5Y and PC12 following 6-OHDA exposure.[121–123]

Biochemically, Han et al.[121] showed that both 6-OHDA and MPP+ caused the release of cytochrome c into the cytosol of MN9D cells. Further analyses revealed an increase in the immunoreactivity for activated caspase-3 and 9 following exposure to 6-OHDA.[122] The addition of a general caspase inhibitor attenuated the 6-OHDA-induced cell death, whereas no protection was afforded in the MPP+ model.[121] Other studies, however, have shown MPTP and MPP+ to induce cell death in association with caspase activation.[38,56,117,118,124] In vitro studies have shown, using SH-SY5Y cells, that both 6-OHDA and MPP+ induce cell death.[125] However, Storch and colleagues showed that, while both dopaminergic toxins led to the death of these cells, in contrast to MPP+, the 6-OHDA effects were independent of intracellular ATP depletion. Interestingly, a reconstitution assay showed that the addition of ATP to the cytosolic fraction of MPP+-treated cells was sufficient to activate caspase-3 and -9, suggesting an energy-dependant requirement for the propagation of apoptotic cell death within the cell.[121,126]

Although some studies have suggested that 6-OHDA-induced cell death is independent of mitochondrial function, others have suggested a role for pro- and anti-apoptotic mitochondrial-related proteins during cell death.[123,126–128] Studying the effects of 6-OHDA in PC12 cells, it was shown that within six hours of toxin administration, there was an increase in the pro-apoptotic Bax protein.[123] Conversely, primary cultures from transgenic mice[127] or in cells[129] overexpressing the anti-apoptotic bcl-2 protein were resistant to 6-OHDA-induced cell death. Overexpression of bcl-2 in vitro also afforded protection against MPTP.[127] It has also been reported that MPP+ increases the expression of cyclin dependant kinase, p21[waf1/Cip1], thereby possibly preventing the propagation of apoptotic signaling, a finding that was shown to independent of TUNEL labeling and DNA fragmentation.[130,131] Further analyses by Viswanath et al.[56] showed that MPP+ induced cleavage of the pro-apoptotic protein Bid in PC12, concurrent with biochemical and morphological evidence of apoptosis. In addition to influencing the expression of members of the bcl-2 family, 6-OHDA has also been reported to cause activation of the JNK pathway, translocate GAPDH to the nucleus and increase the level of the p53 protein. It has been suggested that post-translational modification, by PARP, may extend the half-life and therefore signaling capacity of the p53-induced apoptotic cell death.[132]

OTHER MECHANISMS OF CELL DEATH IN PARKINSON'S DISEASE

Although there is considerable evidence to support the role of apoptosis in PD, derived from postmortem PD tissue, *in vivo* animal models, and *in vitro* cell culture studies, there are also a number of conflicting reports regarding the mechanism of the associated DA cell death. One possible explanation for the disparity observed within these studies is the methodology employed by different investigators. One example of this is the use of the TUNEL assay. Although the TUNEL technique is a useful marker for DNA fragmentation by labeling the 3'OH groups, these can be generated during normal DNA repair mechanisms and random nuclear damage during necrosis. This labeling can lead to an erroneous interpretation of cell death mechanisms.[8] Therefore, morphological and other biochemical verification of apoptosis is required in conjunction with TUNEL labeling. One of the hallmarks of apoptosis is degradation of DNA into fragments of 180 to 200 bp.[34] However, analysis of the resultant DNA "laddering" is limited due to the estimated 10^5 cells required to be in the process of apoptosis to sufficiently visualize this by gel electrophoresis.[26,34] Therefore, this technique may be limited to cell culture systems where a larger cohort of cells is more readily available for analyses.

Many studies have based their biochemical validation of apoptotic cell death on the measurement of caspase activation such as caspase-3. However, caution should be taken, as studies have reported caspase-independent cell death involving endonuclease G, so that cell may still retain some classic hallmarks of apoptosis while being impervious to biochemical detection.[133]

The disparity between apoptosis and other forms of cell death in experimental Parkinsonism has led to the hypothesis that nigral neurons may die by a mechanism independent from apoptosis or necrosis but still display attributes of each—a process tentatively termed *aposklesis* from the Greek meaning to "wither."[134] Therefore, if necrosis and apoptosis represent opposing ends of a cell death scale, it is likely that these two mechanisms have some commonality at either the morphologic or biochemical levels and may overlap in DA cell death during PD.[135]

However, caution is warranted, as the absolute determination of the cell death mechanism has far-reaching consequences on the development of drugs to retard the degeneration and ameliorate the symptomatology in PD.[136,137]

REFERENCES

1. Parkinson, J., *An essay on shaking palsy*, Sherwood, Neeley and Jones, London, 1817.
2. Duvoisin, R. C., Overview of Parkinson's disease, *Ann. N. Y. Acad. Sci.*, 648, 187, 1992.
3. Hoehn, M. M. and Yahr, M. D., Parkinsonism: onset, progression and mortality, *Neurology*, 17, 427, 1967.
4. Warner, T. T. and Schapira, A. H., Genetic and environmental factors in the cause of Parkinson's disease, *Ann. Neurol.*, 53 Suppl. 3, S16, 2003.
5. Huang, Z., de la Fuente-Fernandez, R., and Stoessl, A. J., Etiology of Parkinson's disease, *Can. J. Neurol. Sci.*, 30 Suppl. 1, S10, 2003.
6. Olanow, C. W. and Tatton, W. G., Etiology and pathogenesis of Parkinson's disease, [Review] [133 refs], *Annual Review of Neuroscience*, 22, 123, 1999.
7. Di Monte, D. A., Lavasani, M., and Manning-Bog, A. B., Environmental factors in Parkinson's disease. *Neurotoxicology*, 23, 487, 2002.
8. Burke, R. E. and Kholodilov, N. G., Programmed cell death: does it play a role in Parkinson's disease? *Ann. Neurol.*, 44, S126, 1998.
9. Andersen, J. K., Does neuronal loss in Parkinson's disease involve programmed cell death? *Bioessays*, 23, 640, 2001.
10. Kanduc, D., et al., Cell death: apoptosis versus necrosis (review), *Int. J. Oncol.*, 21, 165, 2002.
11. Honig, L. S. and Rosenberg, R. N., Apoptosis and neurologic disease, *Am. J. Med.*, 108, 317, 2000.
12. Kerr, J. F., Wyllie, A. H., and Currie, A. R., Apoptosis: a basic biological phenomenon with wide-ranging implications in tissue kinetics, *Br. J. Cancer*, 26, 239, 1972.
13. Vanags, D. M., et al., Protease involvement in fodrin cleavage and phosphatidylserine exposure in apoptosis, *J. Biol. Chem.*, 271, 31075, 1996.
14. Wang, X., The expanding role of mitochondria in apoptosis, *Genes Dev.*, 15, 2922, 2001.
15. Konopleva, M. et al., Apoptosis. Molecules and mechanisms, *Adv. Exp. Med. Biol.*, 457, 217, 1999.
16. Geske, F. J. and Gerschenson, L. E., The biology of apoptosis, *Hum. Pathol.*, 32, 1029, 2001.
17. Kerr, J. F., History of the events leading to the formulation of the apoptosis concept, *Toxicology*, 181–182, 471, 2002.
18. Grutter, M. G., Caspases: key players in programmed cell death, *Curr. Opin. Struct. Biol.*, 10, 649, 2000.
19. Nicholson, D. W. and Thornberry, N. A., Caspases: killer proteases, *Trends Biochem. Sci.*, 22, 299, 1997.
20. Yuan, J. et al., The C. elegans cell death gene ced-3 encodes a protein similar to mammalian interleukin-1 beta-converting enzyme, *Cell*, 75, 641, 1993.

21. Chang, H. Y. and Yang, X., Proteases for cell suicide: functions and regulation of caspases, *Microbiol. Mol. Biol. Rev.*, 64, 821, 2000.

22. Nagata, S., Fas ligand-induced apoptosis, *Annu. Rev. Genet.*, 33, 29, 1999.

23. Dechant, G. and Barde, Y. A., Signalling through the neurotrophin receptor p75NTR, *Curr. Opin. Neurobiol.*, 7, 413, 1997.

24. Van Cruchten, S. and Van Den Broeck, W., Morphological and biochemical aspects of apoptosis, oncosis and necrosis, *Anat. Histol. Embryol.*, 31, 214, 2002.

25. Daugas, E. et al., Mitochondrio-nuclear translocation of AIF in apoptosis and necrosis, *Faseb. J.*, 14, 729, 2000.

26. Tatton, W. G. et al., Apoptosis in Parkinson's disease: signals for neuronal degradation, *Ann. Neurol.*, 53 Suppl. 3, S61, 2003.

27. Kuranaga, E. and Miura, M., Molecular genetic control of caspases and JNK-mediated neural cell death, *Arch. Histol. Cytol.*, 65, 291, 2002.

28. Weston, C. R. and Davis, R. J., The JNK signal transduction pathway, *Curr. Opin. Genet. Dev.*, 12, 14, 2002.

29. Eberhardt, O. and Schulz, J. B., Apoptotic mechanisms and antiapoptotic therapy in the MPTP model of Parkinson's disease, *Toxicol. Lett.*, 139, 135, 2003.

30. Leppa, S. and Bohmann, D., Diverse functions of JNK signaling and c-Jun in stress response and apoptosis, *Oncogene*, 18, 6158, 1999.

31. Lev, N., Melamed, E., and Offen, D., Apoptosis and Parkinson's disease, *Prog. Neuropsychopharmacol. Biol. Psychiatry*, 27, 245, 2003.

32. Blum, D. et al., Molecular pathways involved in the neurotoxicity of 6-OHDA, dopamine and MPTP: contribution to the apoptotic theory in Parkinson's disease, *Prog. Neurobiol.*, 65, 135, 2001.

33. Smyth, P. G. and Berman, S. A., Markers of apoptosis: methods for elucidating the mechanism of apoptotic cell death from the nervous system, *Biotechniques*, 32, 648, 2002.

34. Tatton, N. A. and Rideout, H. J., Confocal microscopy as a tool to examine DNA fragmentation, chromatin condensation and other apoptotic changes in Parkinson's disease, *Parkinsonism and Related Disorders*, 5, 179, 1999.

35. Tatton, N. A. et al., A fluorescent double-labeling method to detect and confirm apoptotic nuclei in Parkinson's disease, *Ann. Neurol.*, 44, S142, 1998.

36. Gavrieli, Y., Sherman, Y., and Ben-Sasson, S. A., Identification of programmed cell death in situ via specific labeling of nuclear DNA fragmentation, *J. Cell. Biol.*, 119, 493, 1992.

37. Anglade, P. et al., Apoptosis and autophagy in nigral neurons of patients with Parkinson's disease, *Histol. Histopathol.*, 12, 25, 1997.

38. Hartmann, A. et al., Caspase-3: A vulnerability factor and final effector in apoptotic death of dopaminergic neurons in Parkinson's disease, *Proc. Natl. Acad. Sci. USA*, 97, 2875, 2000.

39. Tompkins, M. M. et al., Apoptotic-like changes in Lewy-body-associated disorders and normal aging in substantia nigral neurons, *Am. J. Pathol.*, 150, 119, 1997.

40. Mochizuki, H. et al., Histochemical detection of apoptosis in Parkinson's disease, *J. Neurol. Sci.*, 137, 120, 1996.

41. Kingsbury, A. E., Mardsen, C. D., and Foster, O. J., DNA fragmentation in human substantia nigra: apoptosis or perimortem effect? *Mov. Disord.*, 13, 877, 1998.

42. Zelenin, A. V., Acridine Orange as a Probe for Cell and Molecular Biology, in *Fluorescent and Luminescent Probes for Biological Activity*, Mason, W. T., Editor, Academic Press, London. 1999.

43. Parish, C. R., Fluorescent dyes for lymphocyte migration and proliferation studies, *Immunol. Cell Biol.*, 77, 499, 1999.

44. Kapuscinski, J., DAPI: a DNA-specific fluorescent probe, *Biotech. Histochem.*, 70, 220, 1995.

45. Kosel, S. et al., On the question of apoptosis in the parkinsonian substantia nigra, *Acta Neuropathol. (Berl.)*, 93, 105, 1997.

46. Wullner, U. et al., Cell death and apoptosis regulating proteins in Parkinson's disease—a cautionary note, *Acta Neuropathol. (Berl.)*, 97, 408, 1999.

47. Boka, G. et al., Immunocytochemical analysis of tumor necrosis factor and its receptors in Parkinson's disease, *Neurosci. Lett.*, 172, 151, 1994.

48. Mogi, M. et al., Tumor necrosis factor-alpha (TNF-alpha) increases both in the brain and in the cerebrospinal fluid from parkinsonian patients, *Neurosci. Lett.*, 165, 208, 1994.

49. Nishimura, M. et al., Tumor necrosis factor gene polymorphisms in patients with sporadic Parkinson's disease, *Neurosci. Lett.*, 311, 1, 2001.

50. Kruger, R. et al., Genetic analysis of immunomodulating factors in sporadic Parkinson's disease, *J. Neural. Transm.*, 107, 553, 2000.

51. Mogi, M. et al., bcl-2 protein is increased in the brain from parkinsonian patients, *Neurosci. Lett.*, 215, 137, 1996.

52. de la Monte, S. M. et al., P53- and CD95-associated apoptosis in neurodegenerative diseases, *Lab Invest*, 78, 401, 1998.

53. Hartmann, A., Hunot, S., and Hirsch, E. C., CD95 (APO-1/Fas) and Parkinson's disease, *Ann. Neurol.*, 44, 425, 1998.

54. Hartmann, A. et al., FADD: A link between TNF family receptors and caspases in Parkinson's disease, *Neurology*, 58, 308, 2002.

55. Hartmann, A. et al., Caspase-8 is an effector in apoptotic death of dopaminergic neurons in Parkinson's disease, but pathway inhibition results in neuronal necrosis, *J. Neurosci.*, 21, 2247, 2001.

56. Viswanath, V. et al., Caspase-9 activation results in downstream caspase-8 activation and bid cleavage in 1-methyl-4-phenyl-1,2,3,6-tetrahydropyridine-induced Parkinson's disease, *J. Neurosci.*, 21, 9519, 2001.

57. Mogi, M. et al., Caspase activities and tumor necrosis factor receptor R1 (p55) level are elevated in the substantia nigra from parkinsonian brain, *J. Neural. Transm.*, 107, 335, 2000.

58. Tatton, N. A., Increased caspase 3 and Bax immunoreactivity accompany nuclear GAPDH translocation and

neuronal apoptosis in Parkinson's disease, *Exp. Neurol.*, 166, 29, 2000.

59. Mizuno, Y. et al., Deficiencies in complex I subunits of the respiratory chain in Parkinson's disease, *Biochem. Biophys. Res. Commun.*, 163, 1450, 1989.

60. Schapira, A. H. et al., Mitochondrial complex I deficiency in Parkinson's disease, *J. Neurochem.*, 54, 823, 1990.

61. Hartmann, A. et al., Is Bax a mitochondrial mediator in apoptotic death of dopaminergic neurons in Parkinson's disease? *J. Neurochem.*, 76, 1785, 2001.

62. Marshall, K. A. et al., Upregulation of the anti-apoptotic protein Bcl-2 may be an early event in neurodegeneration: studies on Parkinson's and incidental Lewy body disease, *Biochem. Biophys. Res. Commun.*, 240, 84, 1997.

63. Hartmann, A. et al., Increased expression and redistribution of the antiapoptotic molecule Bcl-xL in Parkinson's disease, *Neurobiol. Dis.*, 10, 28, 2002.

64. Mazzola, J. L. and Sirover, M. A., Alteration of intracellular structure and function of glyceraldehyde-3-phosphate dehydrogenase: a common phenotype of neurodegenerative disorders? *Neurotoxicology*, 23, 603, 2002.

65. Anglade, P. et al., Apoptosis in dopaminergic neurons of the human substantia nigra during normal aging, *Histol. Histopathol.*, 12, 603, 1997.

66. Jellinger, K. A., Cell death mechanisms in Parkinson's disease, *J. Neural. Transm.*, 107, 1, 2000.

67. Vyas, S. et al., Expression of Bcl-2 in adult human brain regions with special reference to neurodegenerative disorders, *J. Neurochem.*, 69, 223, 1997.

68. Tortosa, A., Lopez, E., and Ferrer, I., Bcl-2 and Bax proteins in Lewy bodies from patients with Parkinson's disease and Diffuse Lewy body disease, *Neurosci. Lett.*, 238, 78, 1997.

69. Ferrer, I. et al., Fas and Fas-L expression in Huntington's disease and Parkinson's disease, *Neuropathol. Appl. Neurobiol.*, 26, 424, 2000.

70. Langston, J. W. and Ballard, P., Parkinsonism induced by 1-methyl-4-phenyl-1,2,3,6-tetrahydropyridine (MPTP): implications for treatment and the pathogenesis of Parkinson's disease, *Can. J. Neurol. Sci.*, 11, 160, 1984.

71. Przedborski, S. et al., The parkinsonian toxin MPTP: action and mechanism, *Restorative Neurology and Neuroscience*, 16, 135, 2000.

72. Langston, J. W. et al., Chromic parkinsonism in humans due to a product of merperidine-analog synthesis, *Science*, 219, 979, 1983.

73. Chiba, K. et al., Studies on the molecular mechanism of bioactivation of the selective nigrostriatal toxin 1-methyl-4-phenyl-1,2,3,6-tetrahydropyridine, *Drug Metab. Disp.*, 13, 342, 1985.

74. Gainetdinov, R. R. et al., Dopamine transporter is required for *in vivo* MPTP neurotoxicity: evidence from mice lacking the transporter, *J. Neurochem.*, 69, 1322, 1997.

75. Tatton, W. G. and Olanow, C. W., Apoptosis in neurodegenerative diseases: the role of mitochondria, *Biochim. Biophys. Acta*, 1410, 195, 1999.

76. Tatton, N. A. and Kish, S. J., In situ detection of apoptotic nuclei in the substantia nigra compacta of 1-methyl-4-phenyl-1,2,3,6-tetrahydropyridine-treated mice using terminal deoxynucleotidyl transferase labelling and acridine orange staining, *Neuroscience*, 77, 1037, 1997.

77. Spooren, W. P., Gentsch, C., and Wiessner, C., TUNEL-positive cells in the substantia nigra of C57BL/6 mice after a single bolus of 1-methyl-4-phenyl-1,2,3,6-tetrahydropyridine, *Neuroscience*, 85, 649, 1998.

78. Serra, P. A. et al., The neurotoxin 1-methyl-4-phenyl-1,2,3,6-tetrahydropyridine induces apoptosis in mouse nigrostriatal glia. Relevance to nigral neuronal death and striatal neurochemical changes, *J. Biol. Chem.*, 277, 34451, 2002.

79. Smeyne, M., Goloubeva, O., and Smeyne, R. J., Strain-dependent susceptibility to MPTP and MPP+-induced Parkinsonism is determined by glia, *Glia*, 74, 73, 2001.

80. Jackson-Lewis, V. et al., Time course and morphology of dopaminergic neuronal death caused by the neurotoxin 1-methyl-4-phenyl-1,2,3,6-tetrahydropyridine, *Neurodegeneration*, 4, 257, 1995.

81. Cochiolo, J. A., Ehsanian, R., and Bruck, D. K., Acute ultrastructural effects of MPTP on the nigrostriatal pathway of the C57BL/6 adult mouse: evidence of compensatory plasticity in nigrostriatal neurons, *J. Neurosci. Res.*, 59, 126, 2000.

82. Sriram, K. et al., Mice deficient in TNF receptors are protected against dopaminergic neurotoxicity: implications for Parkinson's disease, *Faseb. J.*, 16, 1474, 2002.

83. Nagatsu, T. et al., Changes in cytokines and neurotrophins in Parkinson's disease, *J. Neural. Transm. Suppl.*, 277, 2000.

84. Mandel, S., Grunblatt, E., and Youdim, M., cDNA microarray to study gene expression of dopaminergic neurodegeneration and neuroprotection in MPTP and 6-hydroxydopamine models: implications for idiopathic Parkinson's disease, *J. Neural. Transm. Suppl.*, 117, 2000.

85. Turmel, H. et al., Caspase-3 activation in 1-methyl-4-phenyl-1,2,3,6-tetrahydropyridine (MPTP)-treated mice, *Mov. Disord.*, 16, 185, 2001.

86. Hassouna, I. et al., Increase in bax expression in substantia nigra following 1-methyl-4-phenyl-1,2,3,6-tetrahydropyridine (MPTP) treatment of mice, *Neurosci. Lett.*, 204, 85, 1996.

87. Vila, M. et al., Bax ablation prevents dopaminergic neurodegeneration in the 1-methyl- 4-phenyl-1,2,3,6-tetrahydropyridine mouse model of Parkinson's disease, *Proc. Natl. Acad. Sci., U.S.A.*, 98, 2837, 2001.

88. Offen, D. et al., Transgenic mice expressing human Bcl-2 in their neurons are resistant to 6-hydroxydopamine and 1-methyl-4-phenyl-1,2,3,6- tetrahydropyridine neurotoxicity, *Proc. Natl. Acad. Sci., U.S.A*, 95, 5789, 1998.

89. Yang, L.,et al., 1-Methyl-4-phenyl-1,2,3,6-tetrahydropyride neurotoxicity is attenuated in mice overexpressing Bcl-2, *J. Neurosci.*, 18, 8145, 1998.

90. Wang, H. et al., Apoptosis inducing factor and PARP-mediated injury in the MPTP mouse model of Parkinson's disease, *Ann. N. Y. Acad. Sci.*, 991, 132, 2003.

91. Mandir, A. S. et al., Poly(ADP-ribose) polymerase activation mediates 1-methyl-4-phenyl-1, 2,3,6-tetrahydropyridine (MPTP)-induced parkinsonism, *Proc. Natl. Acad. Sci., U.S.A.*, 96, 5774, 1999.

92. Cosi, C. et al., Poly(ADP-ribose) polymerase inhibitors protect against MPTP-induced depletions of striatal dopamine and cortical noradrenaline in C57B1/6 mice, *Brain Res.*, 729, 264, 1996.

93. Ha, H. C. and Snyder, S. H., Poly(ADP-ribose) polymerase-1 in the nervous system, *Neurobiol. Dis.*, 7, 225, 2000.

94. Nishi, K., Expression of c-Jun in dopaminergic neurons of the substantia nigra in 1-methyl-4-phenyl-1,2,3,6-tetrahydropyridine (MPTP)-treated mice, *Brain Res.*, 771, 133, 1997.

95. Saporito, M. S., Thomas, B. A., and Scott, R. W., MPTP activates c-Jun NH(2)-terminal kinase (JNK) and its upstream regulatory kinase MKK4 in nigrostriatal neurons *in vivo, J. Neurochem.*, 75, 1200, 2000.

96. Willesen, M. G., Gammeltoft, S., and Vaudano, E., Activation of the c-Jun N terminal kinase pathway in an animal model of Parkinson's disease, *Ann. N. Y. Acad. Sci.*, 973, 237, 2002.

97. Saporito, M. S. et al., CEP-1347/KT-7515, an inhibitor of c-jun N-terminal kinase activation, attenuates the 1-methyl-4-phenyl tetrahydropyridine-mediated loss of nigrostriatal dopaminergic neurons In vivo, *J. Pharmacol. Exp. Ther.*, 288, 421, 1999.

98. Xia, X. G. et al., Gene transfer of the JNK interacting protein-1 protects dopaminergic neurons in the MPTP model of Parkinson's disease, *Proc. Natl. Acad. Sci., U.S.A.*, 98, 10433, 2001.

99. Trimmer, P. A. et al., Dopamine neurons from transgenic mice with a knockout of the p53 gene resist MPTP neurotoxicity, *Neurodegeneration*, 5, 233, 1996.

100. Duan, W. et al., p53 inhibitors preserve dopamine neurons and motor function in experimental parkinsonism, *Ann. Neurol.*, 52, 597, 2002.

101. Eberhardt, O. et al., Protection by synergistic effects of adenovirus-mediated X-chromosome-linked inhibitor of apoptosis and glial cell line-derived neurotrophic factor gene transfer in the 1-methyl-4-phenyl-1,2,3,6-tetrahydropyridine model of Parkinson's disease, *J. Neurosci.*, 20, 9126, 2000.

102. Mochizuki, H. et al., An AAV-derived Apaf-1 dominant negative inhibitor prevents MPTP toxicity as antiapoptotic gene therapy for Parkinson's disease, *Proc. Natl. Acad. Sci., U.S.A.*, 98, 10918, 2001.

103. Crocker, S. J. et al., Attenuation of MPTP-induced neurotoxicity and behavioural impairment in NSE-XIAP transgenic mice, *Neurobiol. Dis.*, 12, 150, 2003.

104. Senoh, S. et al., Chemical, enzymatic and metabolic studies on the mechanism of oxidation of dopamine, *J. Am. Chem. Soc.*, 81, 6236 1959.

105. Grunblatt, E., Mandel, S., and Youdim, M. B., MPTP and 6-hydroxydopamine-induced neurodegeneration as

106. Jellinger, K. et al., Chemical evidence for 6-hydroxydopamine to be an endogenous toxic factor in the pathogenesis of Parkinson's disease, *J. Neural. Transm. Suppl.*, 46, 297, 1995.

107. Andrew, R. et al., The determination of hydroxydopamines and other trace amines in the urine of parkinsonian patients and normal controls, *Neurochem. Res.*, 18, 1175, 1993.

108. He, Y., Lee, T., and Leong, S. K., 6-Hydroxydopamine induced apoptosis of dopaminergic cells in the rat substantia nigra, *Brain Res.*, 858, 163, 2000.

109. Marti, M. J. et al., Striatal 6-hydroxydopamine induces apoptosis of nigral neurons in the adult rat, *Brain Res.*, 958, 185, 2002.

110. Zuch, C. L. et al., Time course of degenerative alterations in nigral dopaminergic neurons following a 6-hydroxydopamine lesion, *J. Comp. Neurol.*, 427, 440, 2000.

111. Schmued, L. C., Albertson, C., and Slikker, W., Jr., Fluoro-Jade: a novel fluorochrome for the sensitive and reliable histochemical localization of neuronal degeneration, *Brain Res.*, 751, 37, 1997.

112. Mogi, M. et al., Increase in level of tumor necrosis factor (TNF)-alpha in 6-hydroxydopamine-lesioned striatum in rats without influence of systemic L-DOPA on the TNF-alpha induction, *Neurosci. Lett.*, 268, 101, 1999.

113. Jeon, B. S. et al., Activation of caspase-3 in developmental models of programmed cell death in neurons of the substantia nigra, *J. Neurochem.*, 73, 322, 1999.

114. Yamada, M. et al., Herpes simplex virus vector-mediated expression of Bcl-2 prevents 6-hydroxydopamine-induced degeneration of neurons in the substantia nigra *in vivo, Proc. Natl. Acad. Sci., U.S.A.*, 96, 4078, 1999.

115. Shafer, T. J. and Atchison, W. D., Transmitter, ion channel and receptor properties of pheochromocytoma (PC12) cells: a model for neurotoxicological studies, *Neurotoxicology*, 12, 473, 1991.

116. Pahlman, S. et al., Human neuroblastoma cells in culture: a model for neuronal cell differentiation and function, *Acta Physiol. Scand. Suppl.*, 592, 25, 1990.

117. Dodel, R. C. et al., Peptide inhibitors of caspase-3-like proteases attenuate 1-methyl-4-phenylpyridinum-induced toxicity of cultured fetal rat mesencephalic dopamine neurons, *Neuroscience*, 86, 701, 1998.

118. Du, Y. et al., Involvement of a caspase-3-like cysteine protease in 1-methyl-4-phenylpyridinium-mediated apoptosis of cultured cerebellar granule neurons, *J. Neurochem.*, 69, 1382, 1997.

119. Lotharius, J., Dugan, L. L., and O'Malley, K. L., Distinct mechanisms underlie neurotoxin-mediated cell death in cultured dopaminergic neurons, *J. Neurosci.*, 19, 1284, 1999.

120. Choi, W. S. et al., Two distinct mechanisms are involved in 6-hydroxydopamine- and MPP+-induced dopaminergic neuronal cell death: role of caspases, ROS, and JNK, *J. Neurosci. Res.*, 57, 86, 1999.

121. Han, B. S. et al., Caspase-dependent and -independent cell death pathways in primary cultures of mesenceph-

alic dopaminergic neurons after neurotoxin treatment, *J. Neurosci.*, 23, 5069, 2003.

122. von Coelln, R. et al., Rescue from death but not from functional impairment: caspase inhibition protects dopaminergic cells against 6-hydroxydopamine-induced apoptosis but not against the loss of their terminals, *J. Neurochem.*, 77, 263, 2001.

123. Blum, D. et al., p53 and Bax activation in 6-hydroxy-dopamine-induced apoptosis in PC12 cells, *Brain Res.*, 751, 139, 1997.

124. Bilsland, J. et al., Caspase inhibitors attenuate 1-methyl-4-phenylpyridinium toxicity in primary cultures of mes-encephalic dopaminergic neurons, *J. Neurosci.*, 22, 2637, 2002.

125. Storch, A. et al., 6-Hydroxydopamine toxicity towards human SH-SY5Y dopaminergic neuroblastoma cells: independent of mitochondrial energy metabolism, *J. Neural. Transm.*, 107, 281, 2000.

126. Wu, Y. et al., Unlike MPP+, apoptosis induced by 6-OHDA in PC12 cells is independent of mitochondrial inhibition, *Neurosci. Lett.*, 221, 69, 1996.

127. Offen, D. et al., Transgenic mice expressing human Bcl-2 in their neurons are resistant to 6-hydroxydopamine and 1-methyl-4-phenyl-1,2,3,6- tetrahydropyridine neurotoxicity, *Proc. Natl. Acad. Sci., U.S.A.*, 95, 5789, 1998.

128. Choi, W. S. et al., Characterization of MPP(+)-induced cell death in a dopaminergic neuronal cell line: role of macromolecule synthesis, cytosolic calcium, caspase, and Bcl-2-related proteins, *Exp. Neurol.*, 159, 274, 1999.

129. Takai, N. et al., Involvement of caspase-like proteinases in apoptosis of neuronal PC12 cells and primary cultured microglia induced by 6-hydroxydopamine, *J. Neurosci. Res.*, 54, 214, 1998.

130. Soldner, F. et al., MPP+ inhibits proliferation of PC12 cells by a p21(WAF1/Cip1)-dependent pathway and induces cell death in cells lacking p21(WAF1/Cip1), *Exp. Cell Res.*, 250, 75, 1999.

131. Gartel, A. L. and Tyner, A. L., The role of the cyclin-dependent kinase inhibitor p21 in apoptosis, *Mol. Cancer Ther.*, 1, 639, 2002.

132. Mandir, A. S. et al., A novel *in vivo* post-translational modification of p53 by PARP-1 in MPTP-induced parkinsonism, *J. Neurochem.*, 83, 186, 2002.

133. van Loo, G. et al., Endonuclease G: a mitochondrial protein released in apoptosis and involved in caspase-independent DNA degradation, *Cell Death Differ.*, 8, 1136, 2001.

134. Graeber, M. B. et al., Nigral neurons are likely to die of a mechanism other than classical apoptosis in Parkinson's disease, *Parkinsonism and Related Disorders*, 5, 187, 1999.

135. Proskuryakov, S. Y., Konoplyannikov, A. G., and Gabai, V.L., Necrosis: a specific form of programmed cell death? *Exp. Cell. Res.*, 283, 1, 2003.

136. Kinloch, R. A. et al., The pharmacology of apoptosis. *Trends Pharmacol. Sci.*, 20, 35, 1999.

137. Dunnett, S. B. and Bjorklund, A., Prospects for new restorative and neuroprotective treatments in Parkinson's disease, *Nature*, 399, A32, 1999.

43 Weaver Mutant Mouse in Progression of Neurodegeneration in Parkinson's Disease

Manuchair Ebadi, Sushil Sharma, Amornpan Ajjimaporn, and Scott Maanum
Department of Pharmacology and of Neurosciences, University of North Dakota School of Medicine and Health Sciences

CONTENTS

KEY WORDS

wv/wv mice, GIRK channel, striatal DA, salsolinol, tyrosine nitration, serine (140)-phosphorylation, tyrosine hydroxylase, glutathione, caspase-3 activity, TNF-α synthesis, NF-κ-β activation, complex-1 nitration, α-syn index, Parkinson's disease, coenzyme Q_{10}

ABSTRACT

Developing weaver mutant (wv/wv) mice expressing point mutation (G156S) in G-protein-coupled inward rectifying potassium (GIRK) channel were used to explore the basic molecular mechanism of progressive neurodegeneration in Parkinson's disease. No overt clinical abnormality was observed within 7 days of postnatal life. Hypersensitivity to light, sound, touch, and electrical stimulation concomitant with jumping behavior was noticed within 14 days. Parkinsonism, characterized by stiff neck, drooping body posture, tremors, difficulty in walking, and ataxic body movements, was noticed within 28 days. Progressive neurodegenerative changes were observed primarily in the striatum, hippocampus, and cerebellar cortex. Striatal DA, ^{3}H-DA uptake, Zn^{+2}, and coenzyme Q_{10} were reduced, whereas homovalinic acid to DA ratios, Fe^{3+}, Ca^{2+}, lipid peroxidation, and α-syn indices were increased. MPTP or rotenone-induced DA release was reduced; whereas salsolinol and HVA release were increased. Various hallmarks of apoptosis including caspase-3, TNF-α, NF-κ-β, metallothionein (MT-1, 2), and complex-1 nitration were increased; whereas glutathione, complex-1, ATP, and Ser (40)-phosphorylation of tyrosine hydroxylase were reduced as compared to control wild-type (control$_{wt}$), α-synuclein knock out (α-syn$_{ko}$), metallothionein knock out (MT$_{dko}$), and metallothionein-transgenic (MT$_{trans}$) mice. Although MT$_{dko}$ mice were susceptible to MPTP-Parkinsonism, they did not exhibit progressive nigrostriatal neurodegeneration as observed in wv/wv mice. MT$_{trans}$ and α-syn$_{ko}$ mice were resistant to MPTP-Parkinsonism. Striatal neurons of wv/wv mice exhibited age-dependent increase in dense-cored intraneuronal inclusions, cellular aggregation, proto-oncogenes (c-fos, c-jun, caspase-3, and GAPDH) induction, internucleosomal DNA fragmentation, and progressive neuro-apoptosis. These data are interpreted to suggest that oxidative and nitrative stress may impair DA transporter, enhance α-syn indices, and proto-oncogene expression. Furthermore, mitochondrial dysfunction due to enhanced iron accumulation, lipid peroxidation, reduced complex-1 activity, and ATP synthesis may result in reduced Ser-40-phosphorylation of TH, and hence DA synthesis and cellular apoptosis. These early molecular events may trigger progressive neurodegeneration and hence Parkinsonism, which was ameliorated by coenzyme Q_{10} treatment in wv/wv mice.

INTRODUCTION

G-protein-activated inwardly rectifying K^+ channel protein (GIRK2) expression has been seen at multiple regions in the developing CNS, including the cerebellum and midbrain. During postnatal development, cerebellum, midbrain, and hippocampus exhibit significantly increased GIRK-2 expression.[1] Earlier studies have shown that GIRK-2 positive cells are most abundant in the substantia nigra pars compacta (SNpc). GIRK-2 expression may also have deleterious consequences outside of the SNpc and thus affect all GIRK-2-containing regions in the wv/wv mice.[2] Point mutation (glycine156serine) in the pore forming H5 region of GIRK2 induces hypothyroidism, inhibits insulin-like growth factor activity,[3] reduces striatal transforming growth factor-α (TGFα) and dopamine D_2 receptor-like immunoreactivity,[4] impairs synaptic integrity of the neurons in the substantia nigra, and increases caspase-3 activity.[4,5] These early molecular changes may affect the basal ganglia circuitry and are manifested as Parkinson-like symptoms in the wv/wv mice.[6,7,8]

Striatal DA transporter (DAT) is considered as a first target of weaver mutation as its activity was reduced within 3 days of postnatal life.[9] The electron microscopic analysis has revealed vacuolar and autophagic changes that were suggestive of a novel type of non-necrotic, non-apoptotic cell death in wv/wv mice.[10] To evaluate the possibility of a shared genetic defect in weaver mouse and PD, H5 pore region of GIRK-2 was analyzed in familial and sporadic cases of PD; however, the sequence was normal in all cases examined, suggesting a differing etiology of NS cell loss in PD and wv/wv mice.[11]

Recent studies have shown that GIRK-2 channels control cell proliferation, differentiation, neurogenesis, CNS hormone secretion, and neurotransmitter release, and may act as hypoxia sensors and regulate cerebrovascular tone.[12] In wv/wv mice, DA-ergic neurons of the NS pathway undergo spontaneous and progressive cell death. Dysfunction of GIRK channels in wv/wv mice may lead to severe phenotypes ranging from early postnatal death to an increased susceptibility to develop epileptic seizures, white matter disease, PD, loss of cerebellar granular cells[13,14] and midbrain DA-ergic neurons.[15–19] Homozygous (wv/wv) mice exhibit typical motor abnormalities including gait instability, uncoordinated muscular movements, and resting and intention tremors, consistent with the loss of midbrain DA-ergic neurons.[9,15,17,18] Forebrain DA and TH activities are significantly reduced in the adult striatum of wv/wv mice.[15,20,21] The DA-ergic deficit is more pronounced in the dorsal striatum as compared to the ventral striatum[21,22] and may result in both developmental defect and degeneration of mesotelencephalic neurons.[15–18,23] Previous studies[24] have reported that alterations in the NS-DA-ergic neurons occur as early as the day of birth. The striatal DA contents in wv/wv mice are significantly reduced during the second week of postnatal life[23,25] and the deficit remains persistent throughout life,[15,23] whereas deficit in DA uptake occurs as early as 7 days after birth.[25] Both striatal DA and the number of TH-positive cells in the substantia nigra remain normal at this age. A significant reduction in cerebellar weight occurs due to degeneration of granular neurons.[23,25]

The cells that are selectively affected by the GIRK-2 mutation in the wv/wv mice include the granule cells of the cerebellum and the DA-ergic neurons in the midbrain. Although it is known that progressive loss of NS DA-ergic neurons occurs in wv/wv mice, the basic molecular mechanism of selective neurodegeneration remains unknown. In the present study, we have estimated striatal mitochondrial metal ions (iron, zinc, and calcium), DA, DAT, Ser(40)-TH-phosphorylation, lipid-peroxidation, α-syn indices, coenzyme Q_{10}, MT-1, glutathione, ATP, and proto-oncogenes expression in wv/wv mice with a primary objective to explore the molecular mechanism of progressive neurodegeneration in PD. Various hallmarks of apoptosis, such as caspase-3, TNF-α, NF-κ-β, MT-1, salsolinol, and complex-1 nitration, were significantly increased, whereas coenzyme Q_{10}, glutathione, complex-1, ATP, and Ser(40) TH phosphorylation were significantly reduced, and proto-oncogenes (c-fos, c-jun, caspase-3, and GAPDH), TNF-α, and NF-κ-β expressions were significantly increased in the wv/wv mice striatum as compared to control$_{wt}$, MT$_{trans}$, MT$_{dko}$ and α-syn$_{ko}$ mice, suggesting that oxidative and nitrative stress downregulate DA transporter, reduce coenzyme Q_{10} synthesis and Ser-40-TH-phosphorylation, and enhance α-syn indices and tyrosine nitration of complex-1. These early molecular events may trigger neurodegeneration of NS-DA-ergic neurons and hence Parkinsonism in wv/wv mice that were attenuated by coenzyme Q_{10} treatment.

MATERIALS

Monoamines and coenzyme Q standards were purchased from Sigma Chemical Co. (St. Louis, MO). ^{35}S-methionine was purchased from NEN (Boston, MA). Nitrotyrosine, Ser-40-phosphorylated-anti-tyrosine hydroxylase, TNF-α, NF-κ-β, cytochrome-C, and α-syn antibodies were purchased from Chemicon International Inc. (Trimula, CA). Mouse anti-MT antibody was purchased from Zymed Laboratories (San Francisco, CA). Protein-A agarose was purchased from CalBiochem (La Jolla, CA). First strand cDNA kit including Taq-DNA polymerase, dNTPs, and murine leukemia virus (MMLV) reverse-transcriptase were purchased from Stratagene (La Jolla, CA). A caspase-3 assay kit was purchased from Pharmingen, Becton-Dickinson (Palo Alto, CA). Protein assay dye was purchased from Bio-Rad Laboratories (Hercules, CA). ECL-chemilumenescence kit and nitrocellulose were purchased from Amersham Bioscience Corp (Piscataway,

NJ). Nucleocytoplasmic fluorochromes, 4´, 6´diamidino-2-phenylindole dihydrochloride (DAPI), ethidium bromide, and fluorescein isothiocyanate (FITC)-conjugated anti-mouse IgG were purchased from Molecular Probes (Eugene, OR). Trizol, DNA-zol, and PCR primer sets were purchased from GIBCO-BRL Life Technologies (Rockville, MD). FITC-conjugated anti-mouse IgG was purchased from Sigma Chemical Co (St. Louis, MO). ATP-Lite bioluminescence kit was purchased from Packard (Meriden, CT). All other chemicals, including 1-methyl, 4-phenyl, 1,2,3,6-tetrahydropyridine (MPTP), rotenone, and diethyl, propyl carbonate (DEPC), were of reagent grade quality and were purchased from Sigma Chemical Co. (St. Louis, MO).

ANIMALS

Experimental animals were housed in temperature- and humidity-controlled rooms with 12-h day and 12-h night cycles and were provided with commercially prepared chow and water ad-libitum. The animals were acclimated to laboratory conditions for at least 4 days prior to experimentation. Care was taken to avoid any distress to animals during the period of experiment. Breeder pairs of control wild-type (control$_{wt}$) $C_{57}BL_{16}$, metallothionein double knock out (MT$_{dko}$), metallothionein-transgenic (MT$_{trans}$), α-synuclein knock out (α-syn$_{ko}$), and weaver mutant (wv/wv) mice were purchased from The Jackson Laboratory (Minneapolis, MN). Detailed information regarding these genotypes is available from The Jackson Laboratory web page (www.jax.org/jaxmice). They were bred in the central biomedical research facility. The breeder colony was maintained in an air-conditioned animal house facility in filtered cages with free access to water and chow. The zinc, copper, and iron contents in the chow were monitored by atomic absorption spectrometer to maintain their adequate supply. PCR analysis of the tail DNA was done for genotyping. Nonresponders were excluded from the present study.

METHODS

HIGH-PERFORMANCE LIQUID CHROMATOGRAPHY WITH ELECTROCHEMICAL DETECTION (HPLC-EC)

Striatal tissue (25 mg) was isolated and frozen at $-80°C$ before analysis. Monoamines were extracted in 0.4N perchloric acid by sonicating at low wattage for 30 sec. The homogenates were centrifuged at 14,000 rpm for 20 min at 4°C. The supernatant was Millipore-filtered with syringe tip filters (0.25 mμ). Twenty microliters of the filtered supernatant was injected in the HPLC loop in isocratic mode using mobile phase (composition: hepatane sulphonic acid 1.75 g, 2Na-EDTA 100 mg, triethyl amine 4.3 ml, and phosphoric acid 3.1 ml, acetonitrile 140 ml, and water to make 1 liter) and C_{18} reverse phase column. The data were analyzed employing ISCO 6000 computer software. Overlay chromatograms were prepared to confirm accuracy and reproducibility of the experimental data.

STRIATAL MICRODIALYSIS

Striatal microdialysis was performed as described in our recent report[26] using Bioanalytical Systems (BAS, West Lafayette, IN) microdialysis probe. Briefly, animals were anesthetized with 350 mg/kg i.p. tribromoethane and fixed steriotaxically. A small burr hole of 1-mm² diameter was made after determining the steriotaxic coordinates to approach the dorsal striatum (antero-posterior, +6 mm; dorso-ventral –4.2 mm, and lateral 2 mm, relative to bregma). Artificial CSF (composition: 155 mM Na+; 2.9 mM K+; 20.83 mM Mg+2; 132.76 mM Cl/pH 7.4) was prepared by making monovalent and divalent stock solutions. In 25 ml of monovalent stock solution, 0.0266 g of glucose was added, and pH was adjusted to 7.4 with 95% CO_2. After adding 50 µl of divalent stock solution, artificial CSF was loaded into the perfusion syringe. Monovalent stock solution (composition: NaCl, 3.693 g; $NaHCO_3$, 1.5551 g; KCl, 0.0895 g; Na_2SO_4, 0.0340 g; KH_2PO_4, 0.034 g) was prepared in 500 ml of deionized water and filtered through a 2-mµ Millipore filter. The divalent stock solution (composition: $CaCl_2$ $2H_2O$, 0.808 g $MgCl_2$ 6 H_2O, 0.8437 g) was prepared in 10 ml of de-ionized distilled water and filtered. Ringer Solution (0.5 ml; composition: 147.0 mM Na+; 2.4 mM Ca2+; 4.0 mM K+; 155.8 mM Cl pH 6.0) was injected i.v. after every 2 hr to avoid dehydration. Microdialysates were collected at a flow rate of 2 µl/min. Each fraction (40 µl) was collected in 300-µl sample collection vials in a semi-automated refrigerated fraction collector. The brain probes were calibrated in artificial brain (composition: 1 mM each of NE: 4.10 mg/10 ml; DA: 3.8 mg/10 ml; DOPAC: 3.36/10 ml; HVA: 3.64/10 ml; 5-HT: 3.64 mg/10 ml; 5-HIAA: 3.82 mg/10 ml). The temperature of the animals was maintained using the electronic blanket at 37°C. ECG and EEG were recorded to monitor the level of anesthesia. The microdialysates were analyzed by HPLC-EC detection to determine neurotransmitters release and metal ions with atomic absorption spectrometer under basal, and following 30 min of acute MPTP (30 mg/kg, i.p.) or rotenone (2.5 mg/kg, i.p.) intoxication.

COENZYME Q HOMOLOGS

Coenzyme Q homologs were estimated from 25 mg striatal tissue as described in our recent report.[27] Briefly, the striatal tissue was homogenized in 0.5 ml of 10% ethanol. One milliliter of ethanol:hexane (2:5) solution was added and the samples were shaken vigorously in amber centrifuge tubes. They were centrifuged at 14,000 rpm for 20 min at 4°C. The supernatant was dried in nitrogen and the residue was reconstituted in the mobile phase (25% hexane and 75% methanol). A 10-µl sample was injected in HPLC with UV detector set at 275 nm. Standard curves were prepared taking known concentrations of coenzyme Q homologs (Q_2, Q_4, Q_6, Q_9, and Q_{10}) to determine the exact concentrations of coenzyme Qs in the samples. Overlay chromatograms were prepared to authenticate accuracy and reproducibility of the experimental data employing JCL-6000 computer software as described above.

³H-DA UPTAKE

Striatal DA transporter (DAT) activity was assessed by estimating ³H-DA uptake in $control_{wt}$ and wv/wv mice. Twenty-five milligrams of the tissue homogenate, prepared in Dulbecco's-PBS, was treated with 1 µCi/ml of ³H-DA. The samples were incubated for 2 min at 30°C and filtered through GF-1B filters using a Millipore vacuum suction device. The filters were washed thrice to remove unspecific binding and placed in vials containing 5 ml of scintillation fluid. The samples were counted in a Perkin-Elmer Tricarb-4100 β-scintillation counter.

LIPID PEROXIDATION

Lipid peroxidation was estimated by taking 25 mg of striatal tissue and preparing homogenates in 1 ml of D-PBS. Five hundred microliters of thiobarbituric acid reagent was added (composition: 0.037% thiobarbituric acid, 0.15N Tris, 20 mg EDTA). The reaction mixture was placed in an 80°C water bath for 45 min, and the tubes were cooled in ice-cold water. The samples were centrifuged at 14,000 rpm for 20 min at 4°C. Two hundred microliters of the supernatant was read at 450 nm in a Hewlett-Packard microtiter plate reader. The data were analyzed using Soft-Max software. The pellet was dissolved in 1 M NaOH, and protein contents were estimated using Bradford's reagent.

RADIOIMMUNOPRECIPITATION

Animals were injected with 100 µCi/0.1 ml of ³⁵S-methinine i.p. After four hours, activity was stabilized by decapitation. Twenty-five milligrams of the striatal tissue was homogenized in 1 ml of D-PBS (pH 7.4) and lysed in 0.5 ml of lysis buffer containing sodium deoxycholate, Tris, and Triton-X100. To estimate protein tyrosine nitration, Ser(40)-TH phosphorylation, α-syn expression, and α-syn nitration, we used antibody dilutions per manufacturer's recommendations. To determine α-syn indices, first precipitation was done, by native α-syn antibody (1:5000) and 100 µl of protein-A agarose at 4°C for 48 hr. The samples were centrifuged at 14,000 rpm for 20 min at 4°C, and the immunoprecipitate was reconstituted in

100-μl of lysis buffer and counted above background by adding 5 ml of scintillation cocktail in a Perkin-Elmer TriCarb-4100 β-Scintillation counter. The second precipitation was done by nitrated tyrosine antibody (1:3000). α-syn indices were calculated by taking the ratios of counts derived from nitrated vs. native α-syn. For estimating complex-1 nitration, the first precipitation was performed using specific antibody to complex-1 and then with nitrotyrosine antibody as described above.

IMMUNOBLOTTING

The striatal lysates were prepared in Lammelli buffer with protease inhibitor cocktail as described in our recent report (Sharma and Ebadi, 2003). In brief, lysates containing 15 μg of protein were subjected to 12% SDS-poly-acrylamide gel electrophoresis. Proteins were transferred to nitrocellulose paper by electroblotting using 50 mA current-strength for 1.5 hr, and transfer efficiency was checked by Ponceau red stain. The blots were incubated overnight in 5% nonfat milk for nonspecific binding, washed thrice in PBS (pH 7.4), and subjected to recommended dilution of primary antibodies for 1 hr, washed thrice with PBS (pH 7.4), and exposed to 1:10,000 secondary antibody (HRP-labeled anti Mouse IgG) for 2 hr and washed thrice with PBS (pH 7.4) containing 0.1% Tween-20. An Amersham chemiluminescent kit was used for developing the autoradiograms. Immunoblots were scanned with a precalibrated Bio-Rad GS-810 high-resolution densitometric scanner and analyzed by Utility-1 software.

MITOCHONDRIAL IONS

Pooled from five animals, 250 mg of striatal tissue was taken to isolate mitochondria. The mitochondria were sonicated for 30 sec at low wattage using 0.4N perchloric acid. The homogenate was centrifuged at 14,000 rpm for 20 min at 4°C. The supernatant was passed through 2-μm syringe tip filters and subjected to Perkin-Elmer atomic absorption spectrometric analysis. A graphite furnace operating in argon environment was used to pyrolyze samples at 1300 to 1700°C and atomize at 2200 to 2800°C using optimal slit widths and spectral wavelengths. Perkin-Elmer (AA-analyst) equipped with WinLab software was used for the data analysis as described in our recent report.[26]

GLUTATHIONE ESTIMATION

The reduced glutathione in the mitochondrial extracts was estimated by the method of Jollow et al. (1974). Briefly, 0.3 ml of extract was mixed with 0.3 ml of sulphosylicylic acid (4%). The assay mixture contained 0.1 ml filtered aliquot, 1.7 ml of phosphate buffer (0.1 M, pH 7.4), and 0.2 ml 5,5'-dithio-bis-2-dinitrobenzoic acid (4 mg/ml) in a total volume of 2 ml. The samples were incubated at 4°C for 30 min and then subjected to centrifugation at 1200 g for 15 min. The supernatants were read at 412 nm employing a Hewlett-Packard UV spectrophotometer.

COMPLEX-1 ACTIVITY

Mitochondrial complex-1 activity was estimated by the colorimetric method as described earlier.[28]

MITOCHONDRIAL ATP

Mitochondrial ATP from the striatal lysates was estimated by ATP-*Lite* bioluminescence kit as described in our recent report.[29]

CASPASE-3 ACTIVITY

Caspase-3 activity was estimated in striatal lysates spectrofluorometrically by incubating striatal lysates in specific caspase-3 substrate AC-DEVD-AMC and inhibitor AC-DEVD-CHO as described earlier.[26]

TNF-α AND NFκ-β ESTIMATIONS

TNF-α and NFκ-β were estimated from striatal lysates by colometric microtiter ELISA as per manufacturer's recommendations.

DIGITAL FLUORESCENCE MICROSCOPY

Animals were anesthetized with 350 mg/kg body weight i.p. tribromoethane and perfused with 10% formaldehyde in PBS (pH 7.4, 0.15 M) at room temp. The brains were removed and embedded on the stubs using Cryo-M-Bed (Brite Instrument Co., Huntington, England) and kept frozen for 48 hr at −30°C. Eight-micrometer-thick frozen sections were cut in a Hacker-Brite microcryostat (Hacker Instruments Co., NJ). The sections were placed on poly-lysine-coated microscopic slides and exposed to 3% goat serum for 2 hr at room temperature to block unspecific binding, and washed three times in PBS. The sections were incubated for one hour in primary antibodies as per manufacturer's recommendations and were washed three times with PBS. They were then incubated in fluorescein-isothiocyanate (FITC)-conjugated IgG (1:10,000) secondary antibody for 2 hr at room temperature, washed three times, counterstained with 4', 6', diamidino-2-phenylindole dihydrochloride (DAPI) and ethidium bromide (5-nm) for 30 sec, and again washed three times in PBS. The sections were mounted in Fluor-Mounting Medium (Trevigen) and allowed to dry at room temperature in a dark chamber. The slides were examined under digital fluorescence microscope (Leeds Instruments, Minneapolis, MN) set at three wavelengths [blue for DAPI, green for fluorescein isothiocyanate (FITC), and red for ethidium bromide]. (DAPI stains preferentially structurally intact nuclear DNA, while ethidium bromide stains fragmented DNA.) The fluorescence images were captured using a *SpotLite* digital camera and analyzed by *ImagePro* software. Target accentuation and background inhibition software was employed to increase the image quality.

RT-PCR

RT-PCR was performed as described in our recent report.[26] Briefly, striatal tissue (25 mg) from different experimental genotypes was isolated to purify RNA using Triozol reagent. The RNA was quantitated by a Beckman UV spectrophotometer. Five micrograms of purified RNA was reverse transcribed by murine leukemia virus (MMLV) reverse transcriptase to prepare cDNA using a Stratagene first strand cDNA kit. *c-fos, c-jun*, MT-1, α-syn, and GAPDH genes were amplified using specific primer sets obtained from the Gene Bank. The cDNA was subjected to PCR with specific forward and reverse primer sets using 2.5U Taq DNA-polymerase and thermal cycling parameters including denaturation, annealing, and polymerization using a Programmable Brinkmann Eppendorf gradient thermal cycler. The PCR-amplified products were resolved simultaneously on 1% agarose gel at 70 V for 50 min (room temperature). The gels were visualized under a UV illuminator equipped with a UVP-Gel documentation system including video camera, *GelPro* software, and a Sony digital printer. The images were analyzed using a Bio-Rad calibrated GS-800 densitometer and Utility–1 software.

DNA Fragmentation (Apoptosis) Analysis

Striatal tissue (25 mg) was isolated from 10-, 20-, and 30-day-old wv/wv mice and kept on ice. The tissue was homogenized in DNA-zol and centrifuged at 14,000 rpm for 20 min at 4°C. DNA from the supernatant was precipitated with 1 ml isopropanol at –70°C for 1 hr. The DNA pellet was washed thrice with 70% ethanol, air-dried, and reconstituted in deionized DEPC-treated, autoclaved water. DNA was quantitated by Beckman UV spectrophotometer. DNA was loaded in 0.8% agarose gel in 1XTE buffer, electrophoresed at 80 V for 1 hr, stained with 5 μg/ml of ethidium bromide, and washed thrice in distilled water. The gels were visualized under UV illuminator and analyzed with UVP-Gel documentation system equipped with a Sony Digital Printer and *LabWorks* software. The intensity of 180-bp DNA fragment was quantitated using Bio-Rad calibrated GS-810 scanning densitometer and Utility-1 software.

STATISTICAL ANALYSIS

The data were analyzed with Sigma–Stat (version 3.02), employing repeated measures analysis of variance (ANOVA). p-values less than 0.05 were taken as statistically significant.

OBSERVATIONS AND RESULTS

Pregnancy

The first two pregnancies remained normal in wv/wv mice. During the third pregnancy, females either died or could not provide sufficient nutrients to the developing fetuses. Abortions and still births were also noticed during third pregnancy in wv/wv mice. Control$_{wt}$, MT$_{dko}$, MT$_{trans}$, and α-syn$_{ko}$ mice did not exhibit any gross abnormality during gestational and lactational periods.

Developing Pups

No apparent neurobehavioral or functional abnormalities were noticed in all the genotypes examined during birth. Food and water intake of lactating mothers was also similar. The body weight at birth did not vary significantly among different genotypes. Litter size, milk intake, and body weight also did not vary significantly among different genotypes during first 5 days of postnatal life. A picture of a lactating wv/wv mother along with pups at third postnatal day is presented in Fig. 43.1A. During 14 days of postnatal life, wv/wv mice exhibited neurobehavioral abnormalities represented by hypersensitivity to light (B), sound (C), touch (D), and electrical stimulation. The wv/wv mice could not run, exhibited hind limb ataxia, jumping behavior, and frequent urination and/or defecation upon handling. During weaning (28 days), wv/wv mice exhibited neurological symptoms of NS neurodegeneration. Adult (45 days) wv/wv mice exhibited reduced body and brain weight, immobility, morbidity, and eventually mortality. Parkinsonism (represented by stiff neck, drooping body posture and tremors, difficulty in walking, and ataxic body movements) was also noticed in adult wv/wv mice (E and F). Occasionally, wv/wv mice exhibiting Parkinsonism and also exhibited seizure activity, represented by generalized tonic-clonic convulsions (G, H). Body weight of adult wv/wv mice exhibiting typical Parkinsonism was also significantly reduced due to reduced food intake and reduced physical activity. Coenzyme Q$_{10}$ treatment (10 mg/kg, i.p.) for 7 days attenuated neurodegenertative and neurobehavioral abnormalities in adult wv/wv mice (I).

No overt symptoms of neurodegeneration were observed in adult MT$_{dko}$ and α-syn$_{ko}$ mice striatum. Adult MT$_{dko}$ mice exhibiting mild obesity did not exhibit progressive nigrostriatal neurodegeneration; however, they were genetically susceptible to MPTP-induced Parkinsonism (J). Chronic treatment of MPTP (30 mg/kg, i.p.) for 7 days induced severe Parkinsonism in MT$_{dko}$ mice as compared to MT$_{trans}$ and α-syn$_{ko}$ mice. MT$_{dko}$ mice were

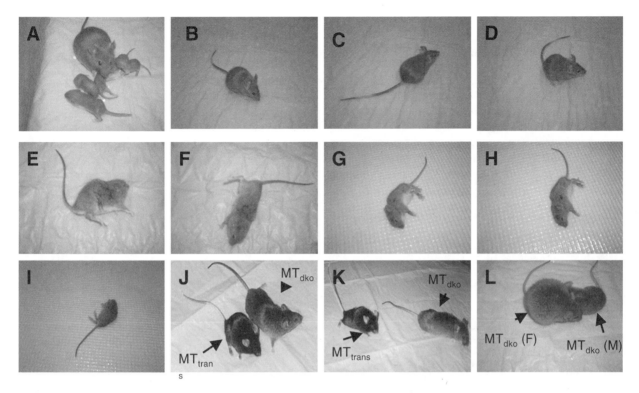

FIGURE 43.1 Pictures illustrating developmental neuropathogenesis of wv/wv mice. (*A*) Three-day-old developing pups exhibiting normal appearance and body weights without any overt symptoms of neurobehavioral abnormality. (*B*) Neurobehavioral abnormalities become apparent in wv/wv mice during the 14th day of postnatal life. These pups exhibit hyperactivity in response to external stimuli, such as (*C*) the sounds of footsteps and cages, (*D*) light flashes, and (*E*) electrical stimulation. (*F*) A mature (28 days) wv/wv mouse exhibiting typical parkinsonism, characterized by stiff neck and tail, (*G*) drooping posture and tremors, difficulty in walking, and (*H*) ataxic body movements. (*I*) Coenzyme Q_{10} (10 mg/kg, i.p.) for 7 days ameliorated progressive neurodegenerative symptoms in wv/wv mice. (*J*) Digital pictures of male MT_{trans} (black) and MT_{dko} (brown, mildly obese) mice. (*K*) Chronic treatment of MPTP (30 mg/kg, i. p.) for 7 days induces complete immobilization in MT_{dko} mouse (Brown), whereas MT_{trans} mouse can still walk with hind limbs spread out. (*L*) Aging (1.5 years) female MT_{dko} mouse exhibiting advanced obesity as compared to males. The pictures were made with a Hewlett-Packard digital camera and processed on the computer.

completely immobilized, whereas MT_{trans} and $\alpha\text{-syn}_{ko}$ mice could still walk with their stiff erect tail (*K*). Aging female MT_{dko} mice exhibited advanced obesity (*L*).

STRIATAL MONOAMINES AND THEIR METABOLITES

Striatal dopamine was significantly ($p < 0.05$) reduced, whereas HVA and salsolinol were significantly ($p < 0.1$) increased in wv/wv mice as compared to other experimental genotypes studied as presented in Table 43.1*a*. The DA-to-HVA ratio was also significantly ($p < 0.01$) reduced in wv/wv mice as compared to control$_{wt}$ mice. Striatal mitochondrial Zn^{2+} was significantly ($p < 0.05$) reduced, whereas Fe^{3+} and Ca^{2+} were significantly ($p < 0.01$) increased in wv/wv mice (Table 43.1*b*). Coenzyme Qs (coenzyme Q_9 and Q_{10}) was also significantly reduced in wv/wv mice as compared with control$_{wt}$, $\alpha\text{-syn}_{ko}$, MT_{trans}, and MT_{dko} mice (Table 43.1*c*).

STRIATAL MICRODIALYSIS

MPTP and rotenone enhanced striatal DA release within 30 min in both control$_{wt}$ and wv/wv mice. DA release was significantly (<0.01) reduced, while salsolinol, and HVA release was increased in wv/wv mice as compared to control$_{wt}$ mice as illustrated in the overlay chromatograms (Figure 43.2, left panel). Rotenone was more potent in inducing striatal DA release as compared to MPTP in both control$_{wt}$ and wv/wv mice.

DA METABOLISM

DA transporter activity, and Ser(40) phosphorylation of TH were significantly ($p < 0.5$) reduced, while lipid peroxidation was significantly ($p < 0.01$) enhanced. Histograms representing striatal DA, DAT, and lipid peroxidation in control$_{wt}$ and wv/wv mice are presented in Figure 43.3 (upper panels). Striatal mitochondrial ATP and glutathione levels were significantly ($p < 0.05$) reduced, whereas MT-1 immunoreactivity was signifi-

Table 43.1a
Monoamines and Their Metabolites

S. No.		Control$_{wt}$	wv/wv	alpha–Syn$_{ko}$	MT$_{trans}$	MT$_{dko}$
1.	NE	600 ± 5.0	446 ± 10	450 ± 8	650 ± 12	520 ± 9
2.	DA	1707 ± 15	$850 \pm 12^{\dagger}$	1250 ± 7	1750 ± 18	1250 ± 12
3.	DOPAC	500 ± 6	350 ± 6	450 ± 9	420 ± 10	420 ± 12
4.	HVA	300 ± 4	$1000 \pm 10^{\dagger}$	350 ± 7	320 ± 8	370 ± 9
5.	Salsolinol	50 ± 3	$150 \pm 4^{*}$	80 ± 5	55 ± 6	90 ± 7
6.	5–HT	400 ± 8	350 ± 9	219 ± 5	380 ± 7	430 ± 8
7.	5–HIAA	425 ± 7	300 ± 6	350 ± 6	450 ± 7	320 ± 5

Data are mean $\pm$ SD of eight determinations in each experimental group (ng/g wet tissue weight).

$^{*}p < 0.05$

$^{\dagger}p < 0.01$

The data were collected from 45-day-old animals from different experimental genotypes.

Table 43.1b
Mitochondrial Ions

S. No.	Genotype	Zn^{2+}	Fe^{+3}	Cu^{2+}	Ca^{2+}
1.	Control$_{wt}$	120 ± 3.5	150 ± 5.6	175 ± 4.5	250 ± 6.5
2.	wv/wv	$80 V 5.0^{*}$	$250 \pm 7.0^{*}$	120 ± 5.0	$380 \pm 8.6^{\dagger}$
3.	Alpha-syn$_{ko}$	100 ± 6.0	160 ± 4.5	180 ± 9.0	300 ± 7.0
4.	MT$_{trans}$	150 ± 8.0	175 ± 9.0	175 ± 10	275 ± 9.0
5.	MT$_{dko}$	90 ± 7.0	200 ± 12	185 ± 9.0	280 ± 10

Data are mean $\pm$ SD of six determinations in each experimental group (ng/mg protein).
$^{*}p < 0.05$

$^{\dagger}p < 0.01$

Table 43.1c
Coenzyme Qs

S. No.	Control$_{wt}$	wv/wv	alpha–Syn$_{ko}$	MT$_{trans}$	MT$_{dko}$
1. Coenzyme Q$_9$	5.7 ± 0.3	$2.8 \pm 0.2^{*}$	5.0 ± 0.5	6.5 ± 0.4	5.2 ± 0.5
2. Coenzyme Q$_{10}$	3.8 ± 0.4	$1.8 \pm 0.3^{\dagger}$	3.5 ± 0.6	5.8 ± 0.3	3.0 ± 11

Data are mean $\pm$ SD of seven determinations in each experimental group (ng/mg wet tissue weight).
$^{*}p < 0.05$
$^{\dagger}p < 0.01$

cantly ($p < 0.01$) increased in wv/wv mice as compared to control$_{wt}$ mice (Figure 43.3, lower panels).

Protein nitration, α-syn expression, α-syn nitration, and α-syn indices were also significantly ($p < 0.01$) increased in wv/wv mice as compared to control$_{wt}$ mice, as presented in Figure 43.4.

Caspase-3, TNF-α, NF-κ-β, and complex-1 nitration were significantly ($p < 0.05$) increased, whereas complex-1 activity and Ser(40) TH-phosphorylation were signifi-cantly ($p < 0.01$) reduced in wv/wv mice as compared to control$_{wt}$, α-syn$_{ko}$, MT$_{dko}$, and MT$_{trans}$ mice (Figure 43.5).

SERINE-(40) PHOSPHORYLATED TH-IMMUNOREACTIVITY

No significant difference in serine(40)-phosphorylated TH-immunoreactivity was observed in the hippocampal, striatal, and cerebellar regions of control$_{wt}$ and α-syn$_{ko}$

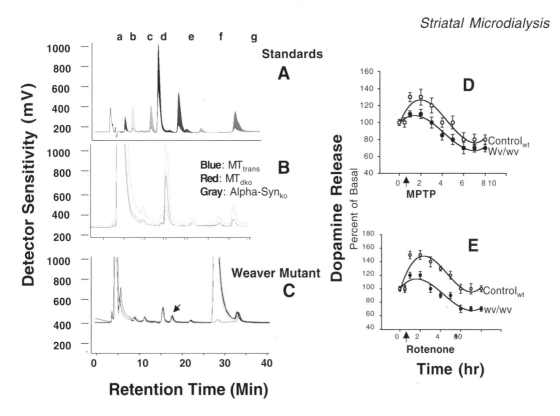

FIGURE 43.2 *Upper left panel:* overlay chromatograms of monoamine standards, a: MHPG, b: NE, c: DOPAC, d: DA, e: 5-HIAA, f: HVA, g: 5-HT (1.6 ng each). *Middle left panel:* striatal microdialysates of MT_{trans}, MT_{dko}, and α-Syn_{ko} mice; *Lower left panel:* striatal microdialysates from wv/wv mice employing HPLC-EC detection, demonstrating significantly reduced DA release, increased HVA release, and appearance of additional peak of salsolinol (confirmed by peak accentuation with known standard of salsolinol at 30.5 min retention time) in response to acute MPTP. The ratios of DA to HVA are significantly reduced in these animals. (Data are mean ± SD of five animals in each experimental group.) *A, upper left panel:* standards of monoamines and their metabolites: overlay chromatograms of striatal microdialysates obtained after 30 min of MPTP (30 mg/kg, i.p.). *D, upper right panel:* kinetic analysis of MPTP. *E, lower right panel:* rotenone-induced striatal DA release demonstrating significant reduction in DA release in wv/wv mice as compared to $control_{wt}$ mice. Data are mean ± SD of five determinations in each experimental group.

mice. $Control_{wt}$ and MT_{trans} mice striatal DA-ergic neurons had significantly high serine(40)-phosphorylated TH-immunoreactivity as compared to wv/wv, α-syn_{ko} and MT_{dko}. Progressive neurodegenerative changes during aging were associated with reductions in serine(40)-phosphorylated TH-immunoreactivity in wv/wv mice. Serine(40)-phosphorylated TH-immunoreactivity was significantly reduced in the hippocampal, striatal, and cerebellar regions. Progressive loss of serine(40)-phosphorylated TH-immunoreactivity in the hippocampal and striatal regions were associated with increase in α-syn immunoreactivity and apoptosis in wv/wv mice. Brain regional serine(40)-phosphorylated TH-immunoreactivity and apoptosis in the striatal, hippocampal, and cerebellar regions of $control_{wt}$ and wv/wv mice are presented in Figure 43.6. Ser-(140) phosphorylated TH immunoreactivity was particularly reduced in the degenerating apoptotic cells surrounding dense cores in the striatal regions of wv/wv mice.

PROGRESSIVE APOPTOSIS

Hippocampal CA-1, CA-3 and dentate gyrus, striatum, and cerebellar cortex regions exhibited progressive apoptosis, represented by increased ethidium bromide staining in wv/wv mice. Neuronal apoptosis in wv/wv mice was associated with age-dependent increase in intraneuronal inclusions. Intraneuronal inclusions were observed frequently in the hippocampal, striatal, and cerebellar regions. Cerebellar cortex exhibited trilamellar appearance with well-organized molecular layer, Purkinje cell layer, and granular layers in $control_{wt}$ mice. However, trilamellar structural organization of cerebellar cortex was obliterated due to apoptosis and the cells from the external granular layer were ectopically microdistributed in wv/wv mice.

PROGRESSIVE NEURODEGENERATION

$Control_{wt}$, α-syn_{ko}, MT_{dko}, and MT_{trans} mice did not exhibit neurodegenerative changes during brain development.

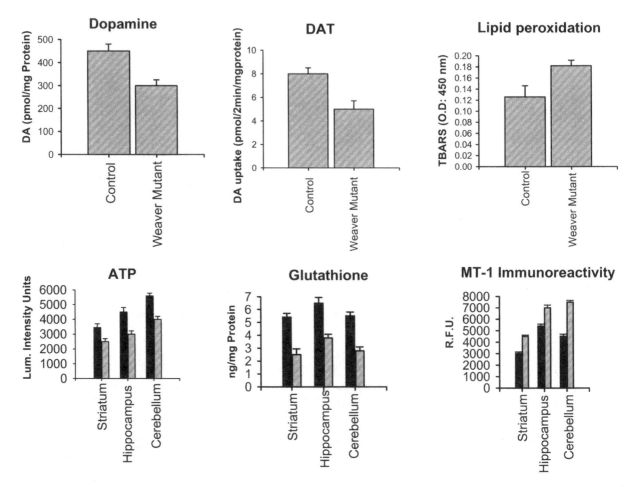

FIGURE 43.3 Upper panels: histograms illustrating DA content, DAT activity, and lipid peroxidation in control$_{wt}$ and wv/wv mice striatum. Data are mean ± SD of seven determinations in each experimental group. Lower panels: histograms illustrating mitochondrial ATP, Glutathione, and MT-1 Immunoreactivity in control$_{wt}$ and wv/wv mice. Data are mean ± SD of eight determinations in each group.

Progressive neurodegenerative changes, characterized by marked apoptosis, were observed primarily in the striatum, hippocampus, and cerebellar regions of wv/wv mice as illustrated in Figure 43.6A. Neurodegenerative changes in wv/wv mice ran concomitantly with severity of tremors, muscular rigidity, postural irregularities, ataxia, morbidity, and eventually mortality. Progressive neurodegeneration in wv/wv mice was also characterized by cellular aggregation in the dorsal striatum, hippocampal dentate gyrus, and cerebellar external granular layer as a function of age. A marked apoptosis was observed in these cellular aggregates. α-Syn immunoreactivity was also significantly high in the cellular aggregates. The incidence of α-syn-positive cellular aggregates and dense-cored structures was higher in the striatal region, as compared to hippocampal and cerebellar regions, and increased with age. Some of the cellular aggregates possessed Cyt-C positive dense cores surrounded by dead cells. In the cerebellar cortex, neurodegenerative changes were observed primarily in the external granular layer. Cell proliferation, migration, and differentiation were impaired in the cerebellar external

granular layer. Some of the cells from the external granular layer died at their origin, some of them on their way to the internal granular layer, and some at their target spot during migration. Purkinje cell layer was not very well organized in these animals. The number and size of Purkinje cells/unit area was also significantly ($p < 0.5$) reduced as compared to control$_{wt}$ mice. Developing control$_{wt}$, α-syn$_{ko}$, MT$_{dko}$, and MT$_{trans}$ mice did not exhibit such neurodegenerative changes. The incidence of α-syn-positive dense-cored structures was higher in the striatal region as compared to hippocampal and cerebellar regions of wv/wv mice. In the cerebellar cortex, progressive neurodegenerative changes were observed primarily in the external granular layer neurons. Cell proliferation, migration, and differentiation were all impaired in the cerebellar external granular layer of wv/wv mice.

SERINE-(40) PHOSPHORYLATION OF TH

An immunoblot representing Ser(40) TH-immunoreactivity along with densitometric analysis from control$_{wt}$,

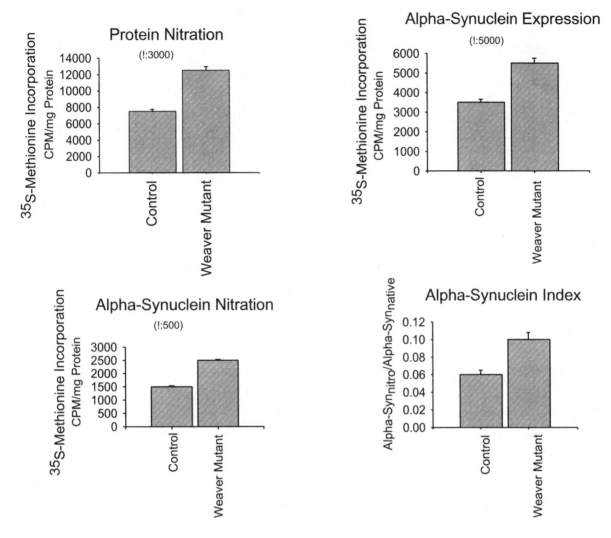

FIGURE 43.4   Histograms demonstrating significant increase in striatal protein nitration α-syn expression, α-syn nitration, and α-syn index in wv/wv mice as compared to control_{wt} mice.

wv/wv, MT_{trans}, MT_{dko}, and Syn_{ko} is presented in Figure 43.7. Serine (40) phosphorylation of TH was significantly ($p < 0.05$) reduced in wv/wv, MT_{dko}, and α-syn$_{ko}$ mice and increased in MT_{trans} mice striatum during adult age. Ser(40)-TH-immunoreactivity was significantly ($p < 0.05$) reduced in wv/wv mice striatum as compared to all other experimental genotypes. Ser(40)TH-immunoreactivity was high in aggregated cells and reduced in the degenerating apoptotic cells surrounding dense cores in the striatal and hippocampal regions of wv/wv mice.

PROTO-ONCOGENE EXPRESSION AND PROGRESSIVE NEURONAL APOPTOSIS

Various proto-oncogenes including *c-fos, c-jun,* GAPDH, and caspase-3 were significantly ($p < 0.05$) induced during degenerative phase in wv/wv mice striatum (Figure 43.8, upper panels). A semiquantitative analysis of proto-oncogene expression is presented as histograms (Figure 43.8*A* through *B,* lower panels). Although MT_{dko} mice were genetically susceptible to MPTP and rotenone, the striatal *c-fos, c-jun,* GAPDH, and caspase-3 expressions were not significantly altered in these genotypes. Striatal MT-1 expression was significantly ($p < 0.01$) enhanced in wv/wv mice as compared to control_{wt} and α-syn$_{ko}$ mice. Progressive neuronal apoptosis, characterized by internucleosomal DNA fragmentation was observed in wv/wv mice during brain development. DNA fragmentation was significantly ($p < 0.01$) increased in striatal DA-ergic neurons of wv/wv mice. Densitometric analyses of 180-bp fragment of DNA in 10, 20, and 30 days old wv/wv mice striatal DNA is presented in Figure 43.8*C* (lower panel). Other genotypes did not exhibit internucleosomal DNA fragmentation during brain development.

Striatum

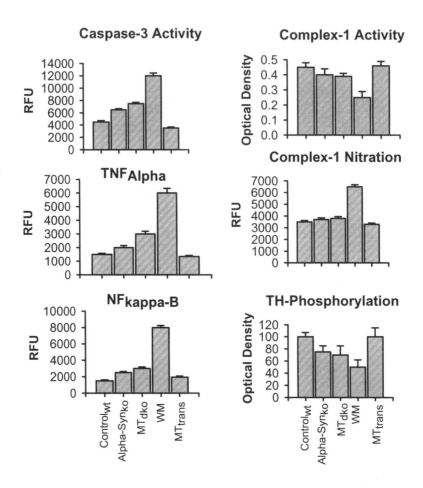

FIGURE 43.5 Histograms demonstrating significantly increased striatal caspase-3, TNF-α, NF-κ-β, complex-1, complex-1 nitration, and reduced Ser-(40)TH-phosphorylation in wv/wv mice as compared to other genotypes. Data are mean ± SD of five determinations in each group.

DISCUSSION

Since wv/wv mice exhibit age-dependent nigrostriatal dopaminergic neurodegeneration and Parkinsonism as observed in PD, even without any exogenous neurotoxic insult, they are highly suitable to explore the basic molecular mechanism of progressive neurodegeneration in PD. To explore the molecular mechanism of genetic susceptibility and progressive neurodegeneration in wv/wv mice, we have employed control$_{wt}$, MT$_{dko}$, MT$_{trans}$, and α-syn$_{ko}$ mice without any overt clinical symptoms of neurobehavioral abnormality as control genotypes. Although MTs gene ablation renders MT$_{dko}$ mice highly susceptible to PD, progressive neurodegeneration as observed in wv/wv mice is not seen in these animals. Significantly increased expression of α-syn and MT-1 indicates their involvement in neurodegeneration in wv/wv mice. We have used MT$_{dko}$

and α-syn$_{ko}$ animals, because PD occurs in old age when the antioxidant function is compromised due to oxidative and nitrative stress and impaired mitochondrial function. Since developmental neurodegenerative changes occur primarily in the striatal, hippocampal, and cerebellar cortex in wv/wv mice, we have estimated ATP, glutathione, and MT-1-immunoreactivity in these regions. We have also estimated mitochondrial ions, lipid peroxidation, DA transporter, protein nitration, α-syn expression and nitration, α-syn indices, caspase-3 activation, coenzyme Q$_{10}$, glutathione, MTs, and proto-oncogenes expression in the striatal neurons of wv/wv mice with a primary objective to explore the basic molecular mechanism of progressive neurodegeneration in PD.

Although weaver gene mutation is supposed to induce global impairment in DA-ergic neurotransmission, the exact molecular mechanism of specific neurodegeneration

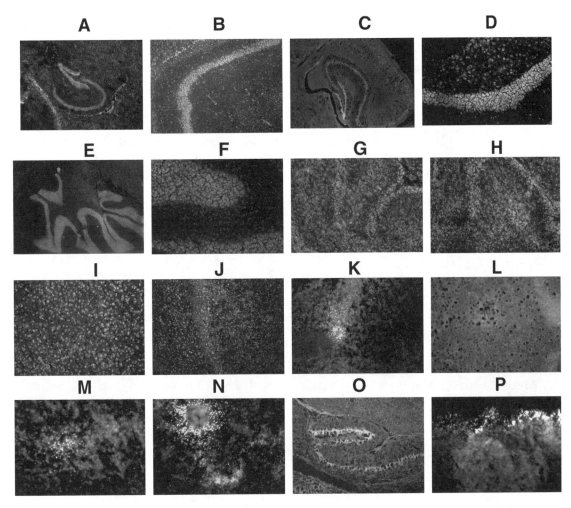

FIGURE 43.6 Multiple fluorochrome analysis demonstrating Ser (40) phosphorylation of TH and apoptosis in the hippocampal region at (*A*) low (100×) and at (*B*) high (400×) magnifications in control$_{wt}$ mice demonstrating normal appearance of CA1, CA3, and dentate gyrus neurons and in wv/wv mice at (*C*) low and (*D*) high magnifications, demonstrating thick dentate gyrus granular layer neurons. Cerebellar cortex at (*E*) low and (*F*) high magnifications demonstrating trilamellar structure and well-organized Purkinje cell layer in control$_{wt}$ mouse. Cerebellar cortex at (*G*) low and (*H*) high magnifications demonstrating obliterated trilamellar structure and disorganized Purkinje cell layer in wv/wv mouse. Striatal region of (*I*) control$_{wt}$, mouse demonstrating high Ser (40) phosphorylated TH immunoreactivity as compared to (*J*) wv/wv mouse. Progressive neurodegeneration, represented by loss of Ser (40) phosphorylation of TH and brain regional apoptosis in the (*L, M,* and *N*) striatal, (*O*) hippocampal dentate gyrus, and (*P*) cerebellar regions in adult (45 days) wv/wv mouse. Fluorescence images were captured by a Spot-Lite digital camera and analyzed by Image-Pro computer software. Target accentuation and background inhibition computer software was used to enhance the image quality. Digital fluorescence images were merged to determine Ser (40) phosphorylation of TH and apoptosis simultaneously. Fluorochromes: FITC-conjugated Ser (40) phosphorylated TH antibody; Blue: DAPI, Red: ethidium bromide. (*A color version of this figure follows page 518.*)

of NS-DA-ergic neurons, hippocampal dentate gyrus neurons, and cerebellar external granular neurons remains enigmatic. It seems that NS DA-ergic neurons and cerebellar external granular neurons are differentially susceptible to weaver gene mutation and are selectively damaged to induce Parkinsonism and ataxia in wv/wv mice. Maharajan et al.[19] have reported that, in addition to down-regulation of DAT and reductions in DA, TH activity is also reduced in wv/wv mice. We have now demonstrated that serine-40-phosphorylation of TH is also suppressed by weaver gene mutation, which could have serious patho-

physiological consequences. wv/wv mice may exhibit typical Parkinsonism due to significantly increased iron, lipid peroxidation, α-syn nitration, salsolinol, and reduced coenzyme Q_{10} synthesis, and TH-phosphorylation in the SN DA-ergic neurons. Although Lewy bodies are not observed in the striatal tissue, progressive neurodegeneration in the DA-ergic neurons was evident by significantly reduced DA synthesis, DA uptake, coenzyme $Q_{10,}$ glutathione, and ATP levels, and enhanced lipid peroxidation, tyrosine nitration, and cellular aggregation in wv/wv mice striatal and hippocampal regions. Furthermore, we have

TH (Ser40) Phosphorylation

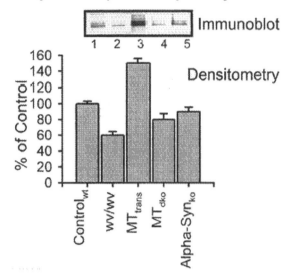

FIGURE 43.7 Upper panel: immunoblot representing significantly reduced serine (40) phosphorylation of TH in wv/wv, MT_{dko}, and α-syn$_{ko}$ mice as compared to MT_{trans} mice striatum during adult age (45 days). Lower panel: quantitative estimate of Ser(40) TH-immunoreactivity using densitometric analysis. Data are mean ± SD of five determinations in each experimental group.

estimated α-syn indices, which determine the extent of nitrative damage in the DA-ergic neurons and may be considered as sensitive hallmarks of neurodegeneration/neuroprotection.[30] α-Syn indices were enhanced in the striatal neurons of wv/wv mice as compared to control$_{wt}$ mice, indicating the involvement of oxidative and nitrative stress in PD.

Progressive DA-ergic cell death in wv/wv mice may be linked to an alteration in the DAT activity, because its cellular expression was significantly reduced in wv/wv mice. Striatal analysis of adult wv/wv mice compared to controls revealed a decrease in TH activity, DA, [3H]-DA uptake, and DA synthesis.[31] No loss of DA-ergic neurons was detected within the first 7 postnatal days, and reductions in the number of DA-ergic neurons occurred in the SN within 14 to 21 days.[32] By immunocytochemical labeling of TH and [3H]-thymidine autoradiography, it has been demonstrated that weaver mutation specifically targets late-generating DA-ergic neurons in the NS-pathways.[33] Elevations in striatal serotonin and increases in amphetamine-evoked fractional release of endogenous DA, indicated their increased susceptibility to amphetamine neurotoxicity.[9] The weaver mutation induced degeneration of pontine nuclei in addition to that of cerebellar granule cells and DA-ergic neurons. Degeneration of the pontine nuclei results in an elimination of pontocerebellar mossy fibers during cerebellar development.[34]

Down-regulation of D-2 receptor binding sites in wv/wv mice has been reported, but it does not reflect an important physiological mechanism through which they attempt to compensate loss of DA-ergic innervation. Almost complete loss of DA uptake sites in the neostriatum of wv/wv mice has been reported.[35]

Apoptosis appears to be a common pathway in the induced cell death and neurodegeneration in wv/wv mice. By digital fluorescence imaging microscopy and DNA fragmentation analysis, we have observed progressive apoptosis of NS-DA-ergic neurons, hippocampal dentate gyrus neurons, and cerebellar external granular neurons of wv/wv mice. Previous studies by Wullner et al.[36] have reported a six- to eightfold increase in the number of apoptotic cells in the external granular layer within postnatal day 9. Neuronal loss in the NS-DA-ergic neurons may induce progressive Parkinsonism, whereas impaired cell proliferation, migration, and differentiation of the external granular layer in the cerebellum may cause ataxia and postural irregularities in wv/wv mice as we have reported in this study.

By striatal microdialysis, we have demonstrated that MPTP and rotenone-induced striatal DA release is significantly reduced in wv/wv mice as compared to control$_{wt}$ mice, indicating reduced DA synthesis and impaired DA-ergic neurotransmission in these animals. Significantly increased striatal salsolinol and HVA release in wv/wv mice indicates the involvement of endogenous THIQs in the etiopathogenesis of PD. These observations have authenticated our previous hypothesis stating that endogenously synthesized THIQs, through DA oxidation, may induce apoptosis of DA-ergic neurons.[37] Recently, we have reported that insulin-like growth factor (IgF) inhibits salsolinol-induced apoptosis in SH-SY5Y cells;[38] therefore, cell death may occur due to early apoptotic signaling and insulin-like growth factor deprivation in wv/wv mice as reported recently by Zhong et al.[3]

Mitochondrial damage has been reported in wv/wv mice during CNS development.[39] Therefore, we estimated striatal coenzyme Q homologs to determine mitochondrial function and molecular mechanism of progressive neurodegeneration in developing wv/wv mice and to establish a functional correlation between coenzyme Q_{10} and cell survival. wv/wv mice exhibiting severe Parkinsonism had significantly reduced striatal coenzyme Q_{10}. Partial amelioration of progressive neurodegeneration and Parkinsonism by coenzyme Q_{10} treatment in wv/wv mice suggests its neuroprotective potential in PD. Coenzyme Q_{10} levels are also reduced in the striatal DA-ergic neurons, platelets, and in the CSF of PD patients.[40] Furthermore, exogenous administration of coenzyme Q_{10} increases CNS mitochondrial number, suggesting its role in neuroprotection.[41] Recently, we have reported that coenzyme Q homologs are heterogeneously microdistributed in the mouse brain and their concentration could

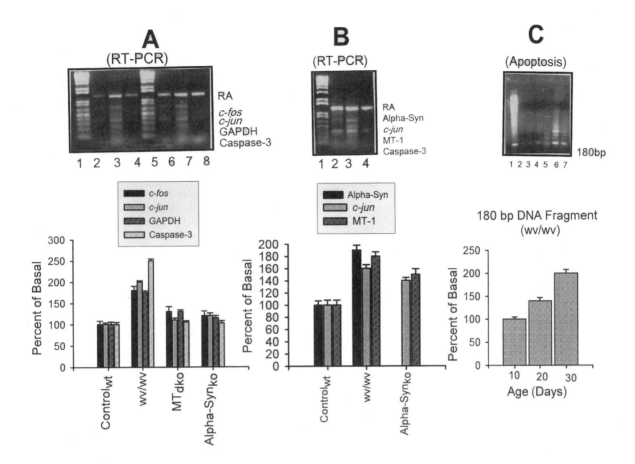

FIGURE 43.8 RT-PCR analysis of striatal *c-fos, c-jun,* GAPDH, and caspase-3 expression in control$_{wt}$, wv/wv, MT$_{dko}$, and α-syn$_{ko}$ mice. Lanes 1: molecular marker (1kDA DNA ladder), 2: Control$_{wt}$, 3: wv/wv, 4: MT$_{dko}$, 5: Molecular marker, 6: Control$_{wt}$, 7: wv/wv, 8: α-syn$_{ko}$. Histograms, illustrating densitometric analysis of retinoic acid receptor (RA) gene, c-fos, c-jun, GAPDH, and caspase-3. B. Lanes 1: molecular marker, 2: Control$_{wt}$, 3: wv/wv, 4: α-syn$_{ko}$. Histograms, illustrating densitometric analysis of α-syn, c-jun, MT-1, and caspase-3. Data are mean ± SD of seven determinations in each experimental group. RA gene expression was used to normalize the experimental data. (C) Ethidium bromide-stained 0.8% agarose gel electrophoresis of striatal DNA from 10-, 20-, and 30-day-old wv/wv mice exhibiting progressive internucleosomal DNA fragmentation and increase in 180 bp fragment as a function of brain development.

fluctuate by genetic manipulation and physiological demand. In particular, MT$_{dko}$ and α-syn$_{ko}$ mice had significantly reduced NS-coenzyme Q$_9$ and Q$_{10}$ levels as compared to control$_{wt}$ and MT$_{trans}$ mice. The NS-coenzyme Q$_{10}$ was significantly high in MT$_{trans}$ mice, and they were genetically resistant to MPTP and SIN-1-induced α-syn nitration and Parkinsonism,[27] suggesting the neuroprotective potential of coenzyme Q$_{10}$. Striatal coenzyme Q$_{10}$ levels also remained elevated in MT$_{trans}$ mice as compared to control$_{wt}$ and MT$_{dko}$ mice, confirming the neuroprotective potential of MTs gene overexpression. Thus, metallothioneins may provide coenzyme Q$_{10}$-mediated neuroprotection in PD.[42] Selegiline treatment also provided mitochondrial neuroprotection through enhanced MTs induction in SK-N-SH cells.[30]

We have now demonstrated that ser-40-TH-phosphorylation was reduced, whereas α-syn nitration was

enhanced, in wv/wv mice, indicating the involvement of oxidative and nitrative stress in the etiopathogenesis of PD. Recently[43] it has been reported that the density of striatal DA-ergic nerve terminals was increased following riluzole administration in wv/wv mice, suggesting its protective role in DA-ergic neurons. Transplantation of E-12 mesencephalic cells in the striatal region of DA-deficient wv/wv mice improved glutamatergic cortico-striatal and DA-ergic nigrostriatal pathways as estimated by ^{3}H-CNQX and ^{3}H-glutamate binding studies.[44] Dopamine D$_2$ receptor-like immunoreactivity was significantly reduced in the SNpc of wv/wv mice and the synaptic number was also reduced. Degenerative changes were observed in some of the D$_2$ receptor-like immunoreactive profiles, suggesting that the synaptic integrity of the DA-ergic system in the striatum is compromised in wv/wv mice, which in turn affects the functional efficacy of the basal ganglia circuitry. Thus altered NS-DA-ergic neurocircuitry is expressed in the Parkinson-like symptoms displayed in wv/wv mice.[7] Although wv/wv mice brains were calbindin-28K-deficient, they did not exhibit preferential degeneration upon MPTP treatment, indicating that endogenous calbindin-28kDA is not essential for the protection of DA-ergic neurons.[45]

Previous studies have shown that Lewy body synthesis is enhanced as a result of increased oxidative and nitrative stress in PD.[46] Oxidative and nitrative stress and excess iron could enhance α-syn nitration, which facilitates cross-bridging and hence its aggregation in Lewy bodies. We have now demonstrated that mitochondrial iron and calcium in the striatum is significantly increased, and zinc is significantly reduced in wv/wv mice. Recently, we have reported that α-syn is aggregated and translocated in the nuclear region in the presence of excess iron, whereas iron chelator deferoxamine significantly inhibited iron-induced α-syn aggregation and translocation in the nuclear region of DA-ergic (SK-N-SH) neurons.[29] Iron participates in the Fenton reaction to produce toxic OH radicals and enhances lipid peroxidation. Lipid peroxidation was significantly increased in wv/wv mice striatum as compared to control$_{wt}$ and other genotypes. Furthermore, MPTP-induced α-syn nitration could be suppressed by MT-gene over-expression.[47] MTs inhibited 3-morpholinosydnonimine (SIN-1)-induced caspase-3 activation, BCl-2 down-regulation, Bax activation, PARP down-regulation, and α-syn nitration, suggesting the neuroprotective role of MTs and the involvement of oxidative and nitrative stress in the etiopathogenesis of PD.[26] Previous studies[36] have shown that neurodegeneration may also occur due to down-regulation of anti-apoptotic, BCl-2 and up-regulation of pro-apoptotic, Bax in the cerebellar and NS-DA-ergic neurons of wv/wv mice.

Proto-oncogene (c-fos, c-jun, caspase-3, and GAPDH) expression was significantly increased in developing wv/wv mice. MT-1 expression was significantly increased during the proliferative phase and reduced during the degenerative phase in wv/wv mice. α-syn and Ser-40-TH-phosphorylation remained elevated in the cellular aggregates. The surrounding areas exhibited significantly reduced Ser-40-TH-phosphorylation. In general Ser-40-TH-phosphorylation remained reduced in the striatum, hippocampus, and cerebellum of wv/wv mice as compared to normal controls, suggesting that, during the neurodegenerative phase, α-syn degradation is reduced, and it starts accumulating in the cellular aggregates because α-syn over-expressing neurons were specifically degenerated through active apoptosis. Although proteolytic activity is enhanced due to induction of caspase-3, α-syn remains resistant to proteolytic degradation in wv/wv mice, which could participate in cellular aggregation and apoptosis during the degenerative phase. Significantly

reduced complex-1, ATP, and glutathione suggest the involvement of mitochondria in neurodegeneration of wv/wv mice.

The exact molecular basis of progressive apoptosis and its triggering mechanisms in wv/wv remains unknown. Apoptosis may be triggered due to enhanced proto-oncogene expression, which might be responsible for the progressive neurodegeneration in these animals. Recent studies have shown that caspase-3 activity is significantly enhanced in wv/wv mice, which could be prevented by caspase-3 inhibitors and provided neuroprotection in these animals.[3,5] wv/wv mice were zinc deficient; however, MT-1 expression was significantly increased, suggesting that these animals are under oxidative and nitrative stress. Mitochondrial iron levels were significantly high, which might participate in the Fenton reaction to produce toxic OH radicals and induce deleterious pathophysiological consequences. We have shown that lipid peroxidation is significantly increased in the striatum, hippocampus, and cerebellar cortex of wv/wv mice. Significantly enhanced lipid peroxidation may occur due to reduced glutathione, as we have reported in this study. MTs are induced during glutathione deficiency to cope with the neurodegenerative events occurring as a consequence of oxidative and nitrative stress. Significantly reduced zinc and glutathione and increased iron may enhance degenerative changes in wv/wv mice. Furthermore, zinc deficiency might be responsible for male infertility in these animals.

In conclusion, impaired metal ion homeostasis (particularly iron and zinc), reduced Ser(40)TH phosphorylation, glutathione, enhanced synthesis of salsolinol, and increased proto-oncogene (c-fos, c-jun, caspase-3, and GAPDH) expression may trigger progressive neurodegeneration in wv/wv mice. Furthermore, significantly increased α-syn indices and reduced mitochondrial complex-1 activity may contribute to the reduced brain regional DA-ergic neurotransmission and DAT, which remained depressed even upon MPTP and rotenone treatment, suggesting that synthesis, storage, and release of DA is impaired and is reflected as severe neurobehavioral deficits in wv/wv mice. These early molecular events might trigger apoptosis, enhance morbidity, mortality, and reduced life span in wv/wv mice. Early induction of proto-oncogenes may trigger neurodegeneration through progressive apoptosis in wv/wv mice. Furthermore, various inflammatory mechanisms that are routed through induction of TNF-α, NF-κ-β, and caspase-3 activation are yet to be explored in these animals. These molecular markers of neuro-inflammation may provide more information regarding the molecular mechanism(s) of neurodegeneration in developing wv/wv mice. Mass spectrometric analysis to determine post-translational modifications in α-syn and functional neuroimaging of complex-1 in vivo are in progress that may provide exact molecular mechanism of neurodegeneration, neuroprotective potential of coenzyme Q_{10}, and possible therapeutic strategies in PD.

ACKNOWLEDGMENTS

The authors express their heartiest appreciation to Dani Stramer for her excellent skills in typing the manuscript. These studies were supported in part by a Contract from Counter Drug Technology Assessment Center, Office of National Drug Control Policy, #DATMO5-02-C-1252 (ME).

ABBREVIATIONS

α-syn.	α-synuclein
α-syn$_{ko}$	α-synuclein knock out
ANOVA	analysis of variance
Cyt-C	cytochrome-C
DAPI	4′, 6′, diamidino-2-phenylindole dihydrochloride
DA	dopamine
DA-ergic neurons	dopaminergic neurons
FITC	fluorescein-isothiocyanate
GIRK	G-protein-activated inward rectifying K$^+$ channel protein
MMLV	murine leukemia virus
MPTP	1-methyl, 4-phenyl, 1,2,3,6-tetrahydropyridine
MT$_{dko}$	metallothionein double gene knock out
MT$_{trans}$	metallothionein transgenic
NF-κ-β	necrosis factor kappa β
PD	Parkinson's disease
RT-PC	Reverse-transcription polymerase chain reaction
SIN-1	3-morpholinosydnonimine
SNpc	substantia nigra pars compacta
TGFα	transforming growth factor-α
TH	tyrosine hydroxylase
TNFα	tumor necrosis factor alpha
wv/wv mice	homozygous weaver mutant mice

REFERENCES

1. Wei, J., Dlouhy, S. R., Bayer, S., Piva, R., Verina, T., Wang, Y., Feng, Y., Dupree, B., Hodes, M. E., Ghetti, B., In-situ hybridization analysis of GIRK2 expression in the developing central nervous system in normal and weaver mice, *J. Neuropathol. Exp. Neurol.*, 56: 762–771, 1997.
2. Schein, J. C., Hunter, D. D., Rofler-Tarlov, S., GIRK-2 expression in the ventral midbrain, cerebellum, and olfactory bulb and its relationship to the murine mutation weaver, *Dev. Biol.*, 204: 432–450, 1998.
3. Zhong, J., Deng, J., Ghetti, B., Lee, W. H., Inhibition of insulin-like growth factor I activity contributes to the premature apoptosis of cerebellar granular neurons in weaver mutant mice: *in vitro* analysis, *J. Neuroscience., Res.* 70: 36–45, 2002.
4. Blum, M., Weickert, C., Carrasco, E., The weaver GIRK2 mutation leads to decreased levels of serum thyroid hormone: characterization of the effect on midbrain DA-ergic neuron survival, *Exp. Neurol.*, 160: 413–424, 1999.
5. Peng, J., Wu, Z., Wu, Y., Hsu, M., Stevenson, F. F., Boonplueang, R., Roffler-Tarlov, S. K., Andersen, J. K., Inhibition of caspases protects cerebellar granular cells of the weaver mouse from apoptosis and improves behavioral phenotype, *J. Biol. Chem.*, 277: 44285–44291, 2002.
6. Surmeier, D. J., Mermelstein, P. G., Goldwitz, D., The weaver mutation of GIRK-2 results in a loss of inwardly rectifying K$^+$ current in cerebellar granule cells, *Proc. Natl. Acad. Sci., U.S.A.* 93:11191–11195, 1996.
7. Xu, S. G., Prasad, C., Smith, D. E., Neurons exhibiting DA D$_2$ receptor immunoreactivity in the substantia nigra of the mutant weaver mouse, *Neuroscience*, 89: 191–207, 1999.
8. Xu, S. G., Prasad, C,. Smith, D. E., Neurons exhibiting DA D$_2$ receptor immunoreactivity in the substantia nigra of the mutant weaver mouse, *Neuroscience*, 89: 191–207, 1999.
9. Simon, J. R., Richter, J. A., and Ghetti, B., Age-dependent alterations in DA content, tyrosine hydroxylase activity, and DA uptake in the striatum of the wv/wv mouse, *J. Neurochem.*, 62: 543–548, 1994.
10. Migheli, A., Piva, R., Wei, J., Attanasio, A., Casolino, S., Hodes, M. E., Dlouhy, S. R., Bayer, S. A. and Ghetti, B., Diverse cell death pathways result from a single missense mutation in weaver mouse, *Am. J. Pathol.*, 151:1629–1638, 1997.
11. Bandmann, O., Davis, M. B., Marsden, C. D. and Wood, N. W., The human homologue of the weaver mouse gene in familial and sporadic Parkinson's disease, *Neuroscience*, 72: 877–879, 1996.
12. Neusch, C., Weishaupt, J. H. and Bahr, M., Kir channels in the CNS: emerging new roles and implications for neurological diseases, *Cell Tissue Res.*, 311: 131–138, 2003.
13. Rakic, P. and Sidman, R. L. Sequence of developmental abnormalities leading to granule cell deficit in cerebellar cortex of wv/wv mice, *J. Comp. Neurol.*, 152: 103–132, 1973.
14. Harkins, A. B. and Fox, A. P., Cell death in weaver mouse cerebellum, *Cerebellum*, 1(3): 201–206, 2002.
15. Schmidt, M. J., Sawyer, B. D., Perry, K. W., Fuller, R. W., Foreman, M. M. and Ghetti, B., DA deficiency in wv/wv mouse, *J. Neurosci.*, 2: 376-380, 1982.
16. Gupta, M., Felten, D. L. and Ghetti, B., Selective loss of monoaminergic neurons in wv/wv mice: an immunocytochemical study, *Brain Res.*, 402: 379–382, 1987.

17. Triarhou, L. C., Norton, J. and Ghetti, B., Mesencephalic DA cell deficit involves areas A8, A9, and A10 in wv/wv mice, *Exp. Brain Res.*, 70: 256–265, 1988.

18. Ghetti, B. and Triarhou, L. C., Combined degeneration of cerebellar granular cells and the midbrain DA neurons in the wv/wv mouse, in *Progress in Parkinson's Disease Research-2* (Hefti, F., and Weiner, W. J., Eds.), pp. 375–388, Futura Publishing, Mount Kisco, NY.

19. Maharajan, P., Maharajan, V., Ravagnan, G. and Paino, G., The wv/wv mouse: a model to study the ontogeny of DA-ergic transmission systems and their role in drug addiction, *Prog. Neurobiol.*, 64: 269–276, 2001.

20. Lane, J. D., Nadi, N. S., McBride, W. J., Aprison, M. H. and Kusano, K., Contents of serotonin, norepinephrine, and DA in the cerebellum of the staggerer, weaver, and nervous, neurologically mutant mice, *J. Neurochem.*, 29: 349–350, 1977.

21. Roffler-Tarlov, S. and Grabiel, A. M., Weaver mutation has differential effects on the DA-containing innervation of the limbic and nonlimbic striatum, *Nature*, 307:62–66, 1984.

22. Simon, J. R. and Ghetti, B., Topographical distribution of DA uptake, choline uptake, cholin acetyl transferase and GABA uptake in the striata of wv/wv mice, *Neurochem Res.*, 17: 431–436, 1992.

23. Roffler-Tarlov, S. and Graybiel, A. M., The postnatal development of the DA containing innervation of dorsal and ventral striatum: effects of the weaver gene, *J. Neurosci.*, 7: 2364–2372, 1987.

24. Ghetti, B. and Triarhou, L. C., NS aberrations induced by weaver gene are present at birth, *Soc. Neurosci. Abstr.*, 18: 156, 1992b.

25. Roffler-Tarlov, S. and Graybiel, A. M., Genetic effects on DA uptake sites in the striatum, *Soc. Neurosci. Abstr.*, 17: 1289, 1991.

26. Sharma, S. K. and Ebadi, M., Metallothionein attenuates 3-Morpholinosydnonimine (SIN-1)-Induced Oxidative Stress in DA-rgic Neurons, *Antioxidants and Redox Signaling*, 5: 251–264, 2003.

27. Sharma, S., Kheradpezhou, Shavali S., Refaey, H. EI., Eken, J., Hagen, C. and Ebadi, M., Neuroprotective Actions of Coenzyme Q10 in Parkinson's disease, *Methods in Enzymology*, Vol. 382, 2003(in press).

28. Schapira, A. H., Cooper, J. M., Dexter, D., Clark, J. B., Jenner, P. and Marsden, C. D., Mitochondrial complex-1 deficiency in Parkinson's disease, *J. Neurochem.*, 54: 823–827, 1990.

29. Sangchot, P., Sharma, S., Chetsawang, B., Porter, J., Govitropong, P. and Ebadi, M., Deferoxamine attenuates iron-induced oxidative stress and prevents mitochondrial aggregation and α-syn translocation in SK-N-SH cells in culture, *Developmental Neurosci.*, 24: 143–153, 2002.

30. Sharma, S., Carlson, E. and Ebadi, M., Neuroprotective action of Selegiline in inhibiting 1-methyl, 4-phenyl, pyridinium ion (MPP+)-induced apoptosis in SK-N-SH neurons, *J. Neurocytol.*, 2003.

31. Adelgrecht, A., Agid, Y. and Raisman-Vozari, R., Effect of the weaver mutation on the expression of DA membrane transporter, tyrosine hydroxylase, and vesicular monoamine transporter in DA-ergic neurons of the substantia nigra and the ventral tegmental area, *Molecular Brain Res.*, 43: 291–300, 1996.

32. Verney, C., Febvert-Muzerelle, A. and Gasper, P., Early postnatal changes of the DA-ergic mesencephalic neurons in the wv/wv mouse, *Brain Res. Dev. Brain Res.*, 89: 115–119, 1995.

33. Bayer, S. A., Wills, K. V., Triarhou, L. C., Verina, T., Thomas, J. D. and Ghetti, B., Selective vulnerability of late-generated DA-ergic neurons of the substantia nigra in wv/wv mice, *Proc. Natl. Acad. Sci., U.S.A.*, 92: 9137–9140, 1995.

34. Ozaki, M., Hashikawa, T., Ikeda, K., Miyakawa, Y., Ichikawa, T., Ishihara, Y., Kumanishi, T. and Yano, R., Degeneration of pontine mossy fibers during cerebellar development in wv/wv mice, *Eur. J. Neurosci.*, 16:565–574, 2002.

35. Dewar, K. M., Paquet, M. and Sequeira, A., Apparent DA D1 and D2 receptors in the wv/wv mouse: receptor binding and coupling to adenyl cyclase, *J. Neural Transm.*, 106: 487–497, 1999.

36. Wullner, U., Weller, M., Schulz, J. B., Krajewski, S., Reed, J. C. and Klockgether, T., BCl-2, Bax, and BCl-X expression in neuronal apoptosis: a study of mutant weaver and lurcher mice, *Acta Neuropathol. (Berl.)*, 96: 233–238, 1998.

37. Hao, R., Norgren, R. B., Jr., Lau, Y. S. and Pfeiffer, R. F., Cerebrospinal fluid of Parkinson's disease patients inhibits the growth and function of dopaminergic neurons in culture, *Neurology*, 45: 138–142, 1995.

38. Shavail, S., Ren, J. and Ebadi, M., Insulin-like growth factor-1 protects dopaminergic SH-SY5Y cells from Salsolinol-Induced toxicity, *Neurosci. Lett.*, 340: 79–82, 2003.

39. Triarhou, Biology and Pathology of Weaver mutant mouse, *Adv. Exp. Med. Biol.*, 517: 15–42, 2002.

40. Ebadi, M., Govitropong, P., Sharma, S., Muralikrishnan, D., Shavali, S., Pellet, L., Schafer, R., Albano, C. and Ekan, J., Ubiquinone (Coenzyme Q10) and mitochondria in Oxidative Stress of Parkinson's Disease, *Biological Signals and Receptors*, 10:224–253, 2002.

41. Beal, M. F., Coenzyme Q10 in Neurodegenerative Diseases, *Ann. N. Y. Acad. Sci.*, 991: 120–131, 2003.

42. Ebadi, M., Muralikrishnan, D., Sharma, S., Shavali, S., Eken, J., Sangchot, P., Chetsawang, B. and Brekke, L., Metallothionein provides ubiquinone-mediated neuroprotection in Parkinson's disease, *Proceedings of the Western Pharmacol. Soc.*, 45: 36–38, 2002.

43. Douhou, A., Debier, T., Murer, M. G., Do, L., Dufour, N., Blanchard, V., Moussaoui, S., Bohme, G. A., Agid, Y. and Raisman-Vozari, R., Effect of chronic treatment with riluzole on the NS DA-ergic system in wv/wv mice, *Exp. Neurol.*, 176: 247–253, 2002.

44. Mitsacos, A., Tomiyama, M., Stasi, K., Giompres, P., Kouvelas, E. D., Cortes, R., Palacois, J. M., Mengod, G. and Triarhou, L. C., [3H] CNQX and NMDA-sensitive [3H]glutamate binding sites and AMPA receptor subunit RNA transcripts in the striatum of normal and wv/wv mice and affects of ventral mesencephalic grafts, *Cell Transplant*, 8: 11–23, 1999.

45. Airaksinen, M. S., Thoenen, H. and Meyer, M., Vulnerability of midbrain DA-ergic neurons in calbindin-D28k-deficient mice: lack of evidence for a neuroprotective role of endogenous calbindin in MPTP-treated and weaver mice, *Eur. J. Neurosci.,* 9: 120–127, 1997.

46. Giassen, B. I., Duda, J. E., Murray, I. V., Chen, Q., Souza, J. M., Hurtig, H. I., Ischiropoulos, H., Trojanowski, J. Q. and Lee, V. M., Oxidative damage linked to neurodegeneration by selective alpha-synuclein nitration in synucleinopathy lesions, *Science*, 290: 985–989, 2000.

47. Ebadi, M. and Sharma, S. K., Peroxynitrite and Mitochondrial Dysfunction in the Pathogenesis of Parkinson's disease, *Antioxidants and Redox Signaling,* 5: 319–335, 2003.

44 Differential Diagnosis of Parkinson's Disease

Serena W. Hung and Anthony E. Lang
Movement Disorders Centre, Toronto Western Hospital

CONTENTS

INTRODUCTION

There are a number of diseases that can mimic Parkinson's disease and make diagnosis difficult at times. The diseases that are most often discussed under the term "Parkinson-plus syndromes" include progressive supranuclear palsy, multiple system atrophy, corticobasal degeneration, and dementia with Lewy bodies. However, other diseases, such as Alzheimer's disease, drug-induced parkinsonism, and vascular parkinsonism, among many others, may present with clinical features similar to Parkinson's disease. In this chapter, we briefly describe some of the more common mimickers of Parkinson's disease. Then we outline clinical symptoms that are often helpful in approaching the differential diagnosis. Recently, the Scientific Issues Committee of the Movement Disorders Society commissioned an evaluation of the diagnostic criteria of the common neurodegenerative causes of Parkinsonism.[1] We refer extensively to these and provide the recommended criteria for several disorders in tabular form.

PROGRESSIVE SUPRANUCLEAR PALSY (PSP)

PSP is a progressive neurodegenerative disease first described in detail by Steele, Richardson, and Olszewski in 1964.[2] It is characterized by vertical supranuclear gaze palsy, prominent and early postural instability, cognitive changes and parkinsonism among other features (Table 44.1). Pathologically, characteristic features include neuronal loss associated with astrocytic gliosis, granulovacuolar degeneration, and neurofibrillary tangles in the globus pallidus, subthalamic nucleus, substantia nigra zona compacta, and some areas involved in ocular control (superior colliculus, periaqueductal gray, and pretectum).[3] This is now classified as a tauopathy with a variety of neuronal and glial tau positive changes primarily characterized by the presence of tau containing exon 10 (4 repeat tau).[4] Onset of PSP is usually between 55 and 70 years of age.[5] Prevalence is estimated to be 6.4 per 100,000 in the U.K.[6] At this time, there is no diagnostic test that is specific for PSP.

MULTIPLE SYSTEM ATROPHY (MSA)

MSA is another neurodegenerative disease of unclear etiology. It causes parkinsonism, cerebellar, autonomic and pyramidal dysfunction (see Table 44.2). MSA can be divided into MSA-C, the cerebellar variant (previously known as olivopontocerebellar atrophy), and MSA-P, the parkinsonian variant (previously known as striatonigral degeneration). Pathologically, MSA is characterized by oligodendroglial cytoplasmic inclusions (OCIs), neuronal loss, and astrogliosis that affect the striatonigral and olivopontocerebellar systems with involvement of spinal cord and brain stem autonomic nuclei accounting for prominent clinical features of dysautonomia.[7] α-synuclein is an important constituent of OCIs, and thus MSA is now classified as α-synucleinopathy. Prevalence is about 4.4 per 100,000 in the U.K.[6] There is no definitive diagnostic test available, but autonomic testing, sphincter EMG, and neuroimaging such as MRI, PET, and SPECT have been used to help determine the diagnosis.[8-11]

CORTICOBASAL DEGENERATION (CBD)

CBD was first described in 1968 by Rebeiz et al.[12] The major characteristics of the classical presentation that was first recognized include basal ganglia disturbances such as rigidity, akinesia, and dystonia and cortical signs including apraxia, cortical sensory loss, myoclonus, corticospinal tract signs, and alien limb phenomenon (see Table 44.3). Age of onset is in the seventh decade.[15,16] Histologically, swollen cortical neurons ("achromatic neu-

TABLE 44.1
NINDs-SPSP Clinical Criteria for the Diagnosis of PSP

Diagnostic Categories	Inclusion Criteria	Exclusion Criteria	Supportive Criteria
	For possible and probable: Gradually progressive disorder with age at onset 40 or later	*For possible and probable:* Recent history of encephalitis; alien limb syndrome; cortical sensory deficits; focal frontal or temporoparietal atrophy; hallucinations or delusions unrelated to dopaminergic therapy; cortical dementia or Alzheimer type; prominent, early cerebellar symptoms or unexplained dysautonomia; or evidence of other diseases that could explain the clinical features	Symmetric akinesia or rigidity, proximal more than distal; abnormal neck posture, especially retrocollis; poor or absent response of parkinsonism to levodopa; early dysphagia or dysarthria; early onset of congnitive impairment including > 2 of: apathy, impairment to abstract thought, decreased verbal fluency, utilization of imitation behavior, or frontal release signs
Possible	Either vertical supranuclear palsy or both slowing of vertical saccades and postural instability with falls < l yr of disease onset		
Probable	Vertical supranuclear palsy and prominent postural instability with falls within first year of disease onset*		
Definite	All criteria for possible or probably PSP are met and histopathologic confirmation at autopsy		

*Later defined as falls or the tendency to fall (patients are able to stabilize themselves).

NINDS-SPSP, National Institute of Neurological Disorders and Stroke and Society for Progressive Supranuclear Palsy, Inc.; PSP, progressive supranuclear palsy.

Source: Reprinted from Livan et al.[1] with permission from authors.

TABLE 44.2
Consensus Criteria for the Diagnosis of MSA

Clinical Domain	Features	Criteria
Autonomic and urinary dysfunction	Orthostatic (by 20 mm Hg systolic or 10 mm Hg diastolic)*; urinary incontinence or incomplete bladder emptying	Orthostatic fall in blood pressure (by 30 mm Hg systolic or 15 mm Hg diastolic)* and/or urinary incontinence (persistent, involuntary partial or total bladder emptying, accompanied by erectile dysfunction in men)
Parkinsonism	B, R, I, and T	1 or 3 (R, I, and T) and B
Cerebellar dysfunction	Gait ataxia; ataxic dysarthria; limb ataxia; sustained gaze-evoked nystagmus	Gait ataxia plus at least one other feature
Corticospinal tract dysfunction	Extensor plantar responses with hyperflexia	No corticospinal tract features are used in defining the diagnosis of MSA†

*Note the different figures for orthostatic hypotension depending on whether it is used as a feature or a criterion.

†In retrospect, this criterion is ambiguously worded. One possible interpretation is that, while corticospinal tract dysfunction can be used as a *feature* (characteristic of the disease), it cannot be used as a *criterion* (defining feature or composite of features required for diagnosis) in defining the diagnosis of MSA. The other interpretation is that corticospinal tract dysfunction cannot be used at all in consensus diagnostic criteria, in which case there is no point mentioning it.

MSA = multiple system atrophy, B = bradykinesia, R = rigidity, I = postural instability, T = tremor.

Source: Reprinted from Livan et al.[1] with permission from authors.

TABLE 44.3
Clinical Manifestations of CBD

Reference	Clinical Manifestations
Riley et al.[13]	*Basal ganglia signs:* Akinesia, rigidity; limb dystonia; athetosis; postural instability, falls; orolingual dyskinesias *Cerebral cortical signs:* Cortical sensory loss; alien limb phenomenon; dementia apraxia; frontal release reflexes; dysphasia *Other manifestations:* Postural-action tremor; hyperreflexia; impaired ocular motility; dysarthria; focal reflex myoclonus; impaired eyelid motion; dysphagia
Watts et al.[14]	*Major:* Akinesia, rigidity, postural/gait disturbance; action/postural tremor; alien limb phenomenon; cortical signs; dystonia; myoclonus *Minor:* Choreoathetosis; dementia; cerebellar signs; supranuclear gaze abnormalities; frontal release signs; blepharospasm

CBD = corticobasal degeneration.

Source: Reprinted from Livan et al.[1] with permission from authors.

rons" or "ballooned neurons") are found in the affected areas. Neuronal degeneration is evident in the cortex, basal ganglia, and brain stem. As with PSP, neuronal and glial pathological changes in CBD are characterized by involvement of 4 repeat tau, particularly the presence of the hallmark "astrocytic plaque" and extensive neuropil threads.[15,16] It is increasingly recognized that this pathological entity may result in other clinical presentations such as frontotemporal dementia or primary progressive aphasia. In addition, just as parkinsonism is not specific to Parkinson's disease, it is now recognized that the "classical" clinical presentation of CBD may be seen in other disorders such as PSP, frontotemporal dementia, and Alzheimer's disease (the so-called corticobasal syndrome).[19] Imaging studies can be helpful in establishing the diagnosis if there is presence of asymmetric cortical atrophy on MRI,[20] although this is clearly not specific for CBD. Other imaging studies that have been used include PET and SPECT.[21] However, none of the findings on imaging is pathognomonic.

DEMENTIA WITH LEWY BODIES (DLB)

DLB is the second most common cause of dementia, after Alzheimer's disease. It constitutes about 15 to 20% of cases in hospital autopsy series.[22] It is characterized by fluctuating cognition and level of consciousness, parkinsonism, progressive dementia, and visual hallucinations (see Table 44.4). Classical α-synuclein-positive Lewy bodies are found in the brain stem (where the pathological changes are indistinguishable from Parkinson's disease) as well as the limbic regions and the neocortex.[23] There has been discussion as to whether DLB and PD represent different ends of the spectrum of a single Lewy body disease. Certainly, aside from the relative times of onset of dementia and motor features, the clinical and pathological features of DLB are indistinguishable from Parkin-

TABLE 44.4
Consensus Criteria for the Clinical Diagnosis of Probable and Possible Dementia with Lewy Bodies[23]

Diagnostic Categories	Inclusion Criteria	Exclusion Criteria	Supportive Criteria

Possible	Progressive cognitive decline of sufficient magnitude to interfere with normal social or occupational function. Prominent or persistent memory impairment may not necessarily occur in the early stages but is usually evident with progression. Deficits on tests of attention and visuospatial ability may be especially prominent.
	
	One of these are features:
	  (a) Fluctuating cognition with pronounced variations in attention and alertness
	  (b) Recurrent visual hallucinations
	  (c) Parkinsonism
		For possible and probable:
		Stroke or evidence of any other brain disorder sufficient to account for the clinical picture
			Repeated falls, syncope, transient loss of consciousness, neuroleptic sensitivity, systematized delusions, hallucinations in other modalities*

Probable	Possible criteria plus one core feature		
Definite	Autopsy confirmation		

*Depression and rapid eye movement sleep behavior disorder have since been suggested as additional supportive features.[39]

Source: Reprinted from Livan et al.[1] with permission from authors.

son's disease with dementia (PD-D). Once again, there is no diagnostic test that can conclusively define DLB.

OTHER LESS COMMON DISEASES THAT CAN MIMIC PD

The diseases listed in Table 44.5 can also present with parkinsonism, sometimes clinically indistinguishable from Parkinson's disease, including response to L-dopa, with its accompanying motor complications.

DIAGNOSIS OF ATYPICAL PARKINSONISM—APPROACH TO PHYSICAL EXAMINATION

The application of the term *parkinsonism* is based on the presence of a combination of clinical features including the cardinal signs of Parkinson's disease: tremor, rigidity, akinesia, and postural instability. Subtle differences in these features may point to an alternative diagnosis, one of the "parkinson-plus" syndromes. There are also other features, for example, in facial expression, voice, distribution of clinical features, cognition, and other signs and symptoms that help make the diagnosis.

FACIAL EXPRESSION

The typical patient with Parkinson's disease is described as having "masked facies" with diminished facial expression, sometimes with the lips mildly parted. Sometimes patients present with a wide-eyed, anxious look that is especially common in PSP where brow furrowing and deepened nasolabial folds are also present.[25] Frequency of blinking is usually reduced. Other eyelid movement abnormalities may occur in PD but are more common in

the parkinson-plus syndromes. None are specific for any one disorder. Blepharospasm[26] and apraxia of eyelid opening[27,28] are more common in PSP. Occasionally, apraxia of eyelid closure and asymmetric slowing of eye opening can be seen in PSP.

EYE MOVEMENT ABNORMALITIES

Eye movement abnormalities are common in parkinsonian patients and are an important consideration in the differential diagnosis. The time taken to initiate a saccade, its speed, and the accuracy should be noted, as well as disturbances of pursuit eye movements. Slowing of vertical downward saccades is characteristic of early PSP, often accompanied by loss of downward fast components of optokinetic nystagmus (OKN).[29,30] This is followed by a loss of downward saccades and a clear vertical supranuclear gaze palsy, the downward gaze restriction being overcome by an oculocephalic maneuver. Patients with PSP also have decreased saccadic amplitude. Limitation of upgaze is rather nonspecific, occurring with normal aging, although in PSP this too may be overcome by the oculocephalic maneuver. Square-wave jerks that may disrupt steady fixation are common in PSP and may also be present in MSA[31] and other disorders with cerebellar dysfunction, such as the spinocerebellar atrophies (SCAs). Here, gaze-evoked nystagmus and hypermetric saccades (among other abnormalities) may also be evident. A supranuclear gaze palsy may be seen in other parkinsonian disorders, including Creutzfeld Jakob disease (CJD), Huntington's disease (HD), vascular parkinsonism, and even DLB. Probably the commonest, however, is CBD where, in the later stages, oculomotor disturbances can be identical to those of PSP. Early on, the saccadic latency rather than saccadic velocity is increased.[32] At this stage, major limitations in eye movement are often due to apraxia.

TABLE 44.5
Other Diseases that Mimic PD

Degenerative Diseases

Progressive supranuclear palsy

Multiple system atrophy

Corticobasal degeneration

Dementia with Lewy bodies

Huntington's disease (Westphal variant)

Frontotemporal dementia–Parkinsonism

Alzheimer's disease

X–linked dystonia–Parkinsonism (Lubag)

Neuroacanthocytosis

Spinocerebellar ataxia (SCA) e.g., SCA2, SCA3 (Machado–Joseph disease), SCA6

ALS–parkinsonism–dementia complex of guam

Progressive pallidal atrophy

Calcification of the Basal Ganglia (Fahr's disease)

Metabolic

Wilson's disease, Type 1 Neurodegeneration with brain iron accumulation [NBIA-1, formerly Hallervorden–Spatz disease, many cases have PANK 2 deficiency (PKAN)], aceruloplasminemia, neuroferritinopathy

Dopa-responsive dystonia

Vascular

Vascular parkinsonism ("lower-half parkinsonism")

Amyloid angiopathy

Post-hemorrhage due to other causes

Infectious

Post-encephalitic Parkinsonism

Creutzfeldt-Jakob disease

Subacute sclerosing panencephalitis

HIV and related opportunistic infections, e.g., toxoplasmosis, tuberculosis, PML

Other viruses

Toxic

MPTP, manganese, carbon monoxide, cyanide, mercury and others

Drugs

Neuroleptics, other dopamine receptor antagonists, e.g., metoclopramide, prochlorperazine, reserpine, tetrabenazine, lithium, alpha-methyldopa, cinnarizine, flunarizine, chemotherapeutic agents, e.g., cytosine arabinoside, cyclophosphamide, methotrexate

Neoplastic

Brain tumor and other mass lesions

Head Trauma

Post-Anoxia

Normal Pressure Hydrocephalus

Mitochondrial Cytopathies

Other—Essential Tremor, Multiple Sclerosis

ALS = amyotrophic lateral sclerosis

HIV = human immunodeficiency virus

MPTP = 1-methyl-4-phenyl-1, 2, 3, 6-tetrahydropyridine

PANK 2 = pantothenate kinase

PKAN = pantothenate kinase-associated neurodegeneration

PML = progressive multifocal leukoencephalopathy

VOICE AND BULBAR FUNCTION

The voice in parkinsonian disorders is monotonous, hypophonic, stuttering, and sometimes more rapid than normal ("tachyphemia"). In PSP, there is a pseudobulbar quality to the voice, with a more nasal, harsher quality to it,[30,33] and speed may become quite slow. Infrequently, low-pitched continuous moaning may be a feature seen in PSP. Palilalia and echolalia are most common in PSP[34] but can be present in other parkinsonian disorders. In MSA, the voice may be high-pitched, strained, and quivering. Cerebellar features (scanning speech) may be present here and in other diseases affecting the cerebellum in addition to the basal ganglia. Respiratory sighs and stridor are also common in MSA.[35,36] Occasionally, CBD manifests an unusual speech disturbance secondary to profound orolingual apraxia. Dysphagia is common in the late stages of all parkinsonian disorders. Early swallowing difficulties should raise concerns about a diagnosis other than PD (e.g., PSP, MSA, infarcts).

SYMMETRY (DISTRIBUTION)

In PD, the onset of symptoms and signs is frequently asymmetrical in the limbs. It is more common for features to be symmetrical in other parkinsonian disorders except CBD, where the classical phenotype typically presents in a very asymmetrical fashion. Another disorder that, by definition, has a unilateral presentation and always maintains profound asymmetrical features is hemiatrophy-hemiparkinsonism.[34] Evaluation of asymmetric limb size is critical to this diagnosis. Although symmetrical manifestation are more common in MSA and PSP than in PD, these disorders can also present with asymmetrical features.[35,36]

TREMOR

The typical tremor in PD is described as a 4 to 6 Hz, pill-rolling, pronation-supination rest tremor. It is frequently asymmetrical and starts in the hands more often than in the feet.[40] However, this type of tremor can also occur occasionally in MSA or even PSP, albeit uncommonly.[41,42] Rarely, if ever, is a classical resting tremor seen in CBD;[43] sometimes the myoclonus in CBD takes on a pattern that can mimic rest tremor.[44] In MSA-C, a slow (3-Hz) axial tremor is indicative of cerebellar disease. Postural and action tremor is a common, nonspecific feature of many parkinsonian disorders. In CBD and MSA, the "tremor" frequently looks "jerky," and electrophysiological studies may reveal that it is, in fact, a form of myoclonus.[44,45] Particularly in CBD, it is common to see the postural tremor "evolve" into better defined myoclonus.

RIGIDITY

Rigidity is an increase in muscle tone, usually equal in flexors and extensors and best appreciated using slow, passive movements. The distribution of rigidity can be helpful in the differential diagnosis of PD. The degree of limb rigidity in PD is usually asymmetrical,[40] but severe unilateral arm rigidity, often with pronounced paratonia, suggests CBD.[43] Nuchal and axial rigidity out of proportion to limb rigidity is common in PSP.[46] Sometimes, there is also spasticity, which can be difficult to detect in the presence of rigidity. A spastic catch with hyperreflexia is not uncommon in MSA, PSP, and CBD, although pyramidal weakness is usually not evident except occasionally in CBD and in the rare examples of parkinsonism accompanying motor neuron disease. Marked rigidity (with or without paratonia or spasticity) in the legs with little involvement in the arms might support the "lower-half parkinsonism" distribution most often found with vascular disease but also in hydrocephalus.

Akinesia/Bradykinesia

On examination, akinesia can be appreciated by observing the patient in conversation and watching for reduced use of hands, shifting of the body in the chair, or crossing of the legs. Bradykinesia can be tested by asking the patient to perform tasks such as finger tapping, opening and closing of the fist, pronation-supination of the forearm, and foot tapping. In addition to slowness and reduction of amplitude, there may be hesitation in starting the movements or arrests in the ongoing performance of the movements. Generally, the nature of bradykinesia is not very helpful in the differential diagnosis. In MSA, the extent of bradykinesia may be extreme.[41] In PSP, the axial slowness and generalized akinesia may be more impressive than the limb bradykinesia. In CBD, especially early on, the performance of simple motor tasks may be more impaired by limb-kinetic and then ideomotor apraxia than true bradykinesia. However, differentiating limb-kinetic apraxia from true bradykinesia may be extremely difficult.[47]

Posture

The typical posture of PD patients consists of head tilting forward and the body stooped, sometimes with kyphosis.[48] The arms are flexed at the elbows and wrists and the hands are sometimes ulnar-deviated, flexed at the metacarpophalangeal joints with partial extension at the interphalangeal joints, adduction of the fingers, and opposition of the thumb to the index finger.[49] Foot deformities that resemble rheumatoid arthritis, but without the swelling in the joints, can occur.[50] Another common foot deformity is the hammertoe-like appearance sometimes with the extension of the big toe ("striatal toe"). In the later stages, PD patients occasionally also demonstrate camptocormia, or severe flexion of the trunk at the waist.[51] In contrast, for the degree of general disability, PSP patients tend to be more

upright. Hyperextension of the neck is a characteristic but relatively uncommon feature of PSP.[52] In MSA, patients may have excessive flexion of the neck (anterocollis), but this has also been described in PD patients.[34] In MSA, profound anterocollis may be due either to dystonia or a focal myopathy involving posterior neck muscles.[53]

Postural Instability

In parkinsonism, patients usually demonstrate impaired postural reflexes and postural instability.[48] Patients may have trouble rising from a chair, requiring several attempts or the need to push off using the arms of a chair. They may be slow to get up, or fail to reposition their feet below the chair when arising. The combination of the above features may cause them to fall back into the chair. When checking for postural stability, the physician pulls the patients from behind, by the shoulders, into the upright position and evaluates the patients' ability to maintain or regain balance (the pull test).[54] The physician must stay close so as to catch the patients if necessary but also must provide enough room behind the patients to allow them to step back several steps before catching themselves. Patients may take extra steps backward to remain upright (retropulsion), or they may fall backward. In more severe cases, patients may not be able to move their feet at all to maintain balance and may make no other corrective movements to protect against a fall, resulting in falling "en bloc" like a plank of wood. In PSP, postural instability is an early manifestation (usually with falls within the first year of presentation).[29] Poor postural stability and frontal lobe dysfunction often combine in PSP to cause a failure to recognize or register the danger of attempting to arise and walk independently. This results in frequent falls as well as patient attempting to rapidly rise from a chair (rocketing or shooting out of the chair) followed immediately by a fall back into or beside the chair.

Gait

In PD, the gait is usually narrow-based, with small shuffling steps, and tandem gait is usually performed relatively well.[48] Freezing of gait is a nonspecific feature of many disorders, although its development very early in the course of the disease should suggest an alternative diagnosis such as PSP. If the gait is wide-based, MSA-C should be considered, especially if there are other signs of cerebellar dysfunction.[55] Some patients have a pattern of distribution of symptoms in which upper limb functions are relatively preserved while the lower limbs are predominantly impaired. This so-called "lower-half parkinsonism" results in early freezing or start hesitation and a magnetic gait or "marche a petit pas" as seen in normal pressure hydrocephalus, multiple infarctions, and Binswanger's disease.[56,57]

OTHER MOVEMENT DISORDERS UNRELATED TO LEVODOPA TREATMENT

In patients with PD, especially later in the course of the disease, dystonia associated with levodopa treatment is quite common. Dystonia most often occurs in the drug off state, e.g., early-morning dystonia typically involving the foot, although the striatal foot can be a presenting symptom of untreated PD.[58] Dystonia as an early manifestation, unrelated to dopaminergic treatment is more common in young-onset PD (<40 years),[59] including autosomal recessive parkinsonism due to mutations in the parkin gene. If an older patient presents with prominent dystonia prior to levodopa treatment, an alternative diagnosis should be considered. Patients with PSP[52] and CBD may have dystonia as one of their presenting symptoms. In CBD, the dystonia frequently involves one limb, with the arm more often affected than the leg.[15]

Myoclonus occurs in a variety of parkinsonian disorders, although PD patients may also have myoclonus as a toxic effect of medication and occasionally as an early manifestation unrelated to treatment,[60] and this is especially common when dementia is a feature. At least 18% of patients with DLB demonstrate myoclonus as well.[61] Stimulus-sensitive myoclonus in the limbs is especially common in CBD[43] and MSA.[45] In MSA, patients may have an axial stimulus-sensitive myoclonus as demonstrated by tapping the tip of the nose, which may result in nonfatiguing, pronounced backward jerks of the head, eyelid closure, facial contraction, and generalized body jerks akin to an excessive startle response. Myoclonus in response to visual stimulus has also been described in patients with MSA. Limb ataxia in a patient with parkinsonism should raise suspicion of MSA-C, an SCA or some other disorder affecting the cerebellum in addition to the basal ganglia.

DEMENTIA

Early cognitive decline should raise questions about the diagnosis of idiopathic Parkinson's disease. Typically, dementia occurs later in the course of PD[62] whereas, in PSP and CBD, it may be an earlier and more prominent feature, and it is a primary and mandatory manifestation of other disorders such as DLB, Alzheimer's disease (AD), and frontotemporal dementia with parkinsonism (FTD-P). In PSP[63] and FTD-P,[64] and to a lesser extent in classical CBD,[65] frontal lobe function is particularly impaired. This can be demonstrated on executive function tests or the frontal assessment battery (FAB), which is a short bedside tool that tests for conceptualization, mental flexibility, motor programming, sensitivity to interference, inhibitory control, and environmental autonomy.[66] PSP and FTD-P patients also demonstrate more prominent frontal release signs such as a grasp, snout, and rooting reflexes.[67] Of

note, CBD patients may demonstrate these prominent frontal release signs in the absence of dementia. Fluctuation in cognition and level of consciousness suggests DLB.[68] In patients with DLB, sensitivity to neuroleptic medications can be marked, causing profound parkinsonism, further impairment of consciousness level, and autonomic disturbances, all reminiscent of the neuroleptic malignant syndrome.[69]

OTHER CORTICAL SIGNS

CBD usually demonstrates signs that can be localized to the cortex.[16,43] These signs usually start asymmetrically but involve both sides as the disease progresses. Limb-kinetic and ideomotor apraxia, alien limb phenomenon, and cortical sensory loss such as simultagnosia, agraphesthesia, and astereognosis are features suggestive of a diagnosis of CBD. Again, it is important to remember that the corticobasal syndrome with varying degrees of apraxia may be seen in other pathological disorders. Involuntary limb levitation is a rather nonspecific feature that can be seen occasionally in several disorders (e.g., PSP[70] and HD[71]). Importantly, this simple movement should not be mistaken for the more complex behaviors of the alien limb phenomenon. Patients with very rigid or dystonic limbs in other parkinsonian conditions may also have difficulty with two-point discrimination and stereognosis testing. PD patients, despite common sensory complaints of numbness, tingling, or aching, do not demonstrate significant sensory loss on examination.[72] Language disorders such as a progressive dysphasia with paraphasic errors may be seen in CBD, frontotemporal dementias, and Alzheimer's disease, among others.

AUTONOMIC DYSFUNCTION

Prominent orthostatic hypotension, urinary incontinence, and impotence in men are often the first signs of MSA, commonly preceding parkinsonism.[35,70,71] Patients with MSA often have cold and dusky hands with poor capillary refill after blanching.[72] Other signs include gastrointestinal difficulties such as constipation and dysphagia,[73] and respiratory dysfunction, especially inspiratory stridor.[32,33] However, the presence of autonomic dysfunction does not preclude the diagnosis of idiopathic Parkinson's disease.[74] In fact, depending on the study performed, up to 80% of PD patients have autonomic incompetence during the course of their disease. It usually occurs later in the course and is not as prominent as in MSA. Occasionally, DLB patients can demonstrate prominent autonomic dysfunction prompting a consideration of a diagnosis of MSA;[75] however, MSA rarely if ever manifests prominent or early cognitive dysfunction. Urinary bladder dysfunction may be a prominent and early feature of PSP, sometimes causing diagnostic confusion with MSA.

SLEEP DISTURBANCES

REM behavior disorder is not uncommon in MSA[79] PD,[80] and DLB.[81] It may develop before the use of dopaminergic medication, and this may even be present in some patients for several years before signs of parkinsonism develop. It has also been described in SCA3[82] and PSP,[79] although much less frequently. Other parasomnias, including vivid dreams, nightmares, night terrors with nocturnal vocalization, and somnambulism, have also been described in PD. Restless legs syndrome and periodic limb movement in sleep are also more common in PD.[83,84] More recently described are excessive daytime sleepiness (EDS) and, less often, sudden onset of sleep (originally referred to as "sleep attacks") that can be due to PD itself or the use of dopaminergic medications.[85,86] The dopaminergic medications first implicated were pramipexole and ropinirole,[87] but EDS has now been described with all other dopamine agonists[88] and levodopa.[89]

RESPONSE TO LEVODOPA

Sustained response to levodopa with the development of motor fluctuations and peak-dose choreoathetotic dyskinesias involving the limbs and neck is characteristic of PD. Indeed, along with the presence of resting tremor and asymmetrical involvement, good and sustained response to levodopa is one of the best predictors of a pathological diagnosis of PD.[1] This pattern of response is typical of disorders with prominent degeneration of the substantia nigra pars compacta but with preservation of "downstream" areas from the striatal postsynaptic neurons and beyond. The poor or absent response to levodopa, which is characteristic of most other causes of parkinsonism, justifies the concept of "parkinson-minus" disorders (i.e., minus a response to levodopa) being equally important as the "parkinson-plus" designation. In MSA, patients may have good response to levodopa, but the response usually wanes, and prominent cranial dystonia (sometimes very asymmetrical) is the more common manifestation of dyskinesias.[90] Rare examples of MSA with nigral involvement but little or no striatal degeneration result in a pattern of levodopa response (as well as response to other dopaminergic drugs and surgical interventions such as STN DBS) more typical of that seen in PD. In MSA patients, psychiatric complications of dopaminergic drugs are much less common than in PD. PSP patients less often experience a good response to levodopa,[5,91] and good response is even rarer in CBD patients.[16] In DLB, response to levodopa is variable.[92] Dramatic response to levodopa without the development of dyskinesias (except occasionally upon initial exposure) may suggest an adult-onset presentation of dopa responsive dystonia.[93]

COURSE OF DISEASE/PROGRESSION

The onset of symptoms in PD is usually subtle, and progression is slow. Occasionally, patients can present more abruptly, for example, with tremor precipitated by physical or emotional upset or trauma. The course of the common parkinson-plus syndromes tends to be more rapid and aggressive than PD, and an alternative diagnosis should always be considered in cases of "malignant Parkinson's disease" (referring to disease course). Abrupt symptom onset should raise a consideration of other diagnoses such as vascular disease, encephalitis, CJD, rapid-onset dystonia parkinsonism, or Wilson's disease. The age of onset of PD is usually between the sixth and seventh decade, although some develop PD at a young age (as early as the thirties and forties). In PSP, the mean age of onset is in the mid-sixties with mean survival of 8.6 years (range 3 to 24 years).[5,34] The mean age of onset for MSA is earlier, in the early fifties, with mean survival of 7.5 years[34] (range 0.5 to 24 years[94]). The mean age of onset of CBD is early sixties, and mean survival is 7.9 years with a range between 2.5 to 12.5 years.[16] DLB has an age of onset in early to mid-sixties.[61,92] Juvenile parkinsonism (onset under age 20) raises other diagnostic considerations as outlined in Table 44.6.

Table 44.7 defines methods of investigation that may help in the differential diagnosis of PD.

CONCLUSION

There are numerous causes of a parkinsonian syndrome. Often, a detailed history and careful clinical examination will provide sufficient evidence for a definitive diagnostic alternative to Parkinson's disease, but at times only a suspicion can be raised, and further investigation or longer follow-up may provide the answer. A diagnosis of Parkinson's disease is most likely when atypical features (historical and on examination) are absent and when the clinical features are asymmetric, include resting tremor, and demonstrate a sustained good response to levodopa. An accurate diagnosis is obviously important in research evaluating the etiology and epidemiology of these disorders as well as in the evaluation of novel therapeutics, including putative neuroprotective agents. Independent of research, an accurate diagnosis is important in guiding day-to-day management, and particularly for providing patients and families an honest and realistic outlook and understanding of their illness so they can adequately plan for the future.

REFERENCES

1. Litvan, I. et al., SIC Task Force appraisal of clinical diagnostic criteria for parkinsonian disorders, *Mov. Disord.,* 18(5), 467–486, 2003.

TABLE 44.6
Differential Diagnosis of Juvenile Parkinsonism

Degenerative

Huntington's disease

Autosomal recessive parkinsonism [e.g., Parkin mutation (PARK2)]

Rapid-onset dystonia-parkinsonism

Pallidopyramidal degeneration

Structural

Tumor, e.g., basal ganglia, pineal gland

Infection

Postencephalitic parkinsonism

Other viruses, e.g., Japanese B encephalitis, Western Equine encephalitis, Coxsackie B type 2, measles, Murray Valley encephalitis, polio, influenza inoculation, herpes simplex, Epstein-Barr virus, St. Louis encephalitis

Intoxication (See Table 44.5)

Metabolic

Wilson's disease

Type 1 Neurodegeneration with Brain Iron Accumulation [NBIA-1, formerly Hallervorden-Spatz disease, many cases have PANK 2 deficiency (PKAN)]

Neuronal ceroid lipofuscinosis

Defects of the biosynthesis of dopamine (e.g., dopa-responsive dystonia, tyrosine hydroxylase deficiency, aromatic acid decarboxylase deficiency)

Niemann-Pick disease type C

Neuronal intranuclear hyaline inclusion disease

Mitochondrial cytopathies

Medications/Treatment

Neuroleptic medications

Amphoterocin B

Chemotherapeutic agents, e.g., cytosine arabinoside, cyclophosphamide methotrexate

Radiation therapy

PANK 2 = pantothenate kinase

PKAN = pantothenate kinase-associated neurodegeneration

2. Steele, J. C., Richardson, J. C., and Olszewski, J. Progressive supranuclear palsy, *Archives of Neurology,* 10 333–358, 1964.

3. Cervós-Navarro, J. and Schumacher, K., Neurofibrillary pathology in progressive supranuclear palsy (PSP), *J. Neural Transm.,* 98 Suppl., 42, 153–164, 1994.

4. Poorkaj, P. et al., An R5L tau mutation in a subject with a progressive supranuclear palsy phenotype, *Ann. Neurol.,* 52(4), 511–516, 2002.

5. Birdi, S. et al., Progressive supranuclear palsy diagnosis and confounding features: Report on 16 autopsied cases, *Movement Disorders,* 17(6), 1255–1264, 2002.

6. Schrag, A., Ben-Shlomo, Y., and Quinn, N. P., Prevalence of progressive supranuclear palsy and multiple system atrophy: a cross-sectional study, *Lancet,* 354(9192), 1771–1775, 1999.

7. Burn, D. J. and Jaros, E., Multiple system atrophy: cellular and molecular pathology, *Journal of Clinical Pathology: Molecular Pathology,* 54(6), 419–426, 2001.

8. Vodusek, D. B., Sphincter EMG and differential diagnosis of multiple system atrophy, *Mov. Disord.,* 16(4), 600–607, 2001.

9. Sakakibara, R. et al., Urinary dysfunction and orthostatic hypotension in multiple system atrophy: which is the more common and earlier manifestation? *Journal of Neurology, Neurosurgery and Psychiatry,* 68(1), 65–69, 2000.

10. Bhattacharya, K. et al., Brain magnetic resonance imaging in multiple-system atrophy and Parkinson disease—A diagnostic algorithm, *Archives of Neurology,* 59(5), 835–842, 2002.

11. Antonini, A. et al., Differential diagnosis of parkinsonism with [^{18}F]fluorodeoxyglucose and PET, *Movement Disorders,* 13(2), 268–274, 1998.

12. Rebeiz, J. J., Kolodny, E. H., and Richardson, E. P., Corticodentatonigral degeneration with neuronal achromasia, *Arch Neurol.,* 18, 20–33, 1968.

13. Riley, D. E. et al., Cortical-basal ganglionic degeneration, *Neurology,* 40:1203, 1990.

14. Watts, A. L. et al., Corticobasal degeneration, in Watts, R. L., Koller, W. C., Eds., *Movement Disorders: Neurologic Principles and Practice,* New York, McGraw-Hill, 611–621, 1967.

15. Vanek, Z. and Jankovic, J., Dystonia in corticobasal degeneration, *Movement Disorders,* 16(2), 252–257, 2001.

16. Wenning, G. K. et al., Natural history and survival of 14 patients with corticobasal degeneration confirmed at postmortem examination, *Journal of Neurology, Neurosurgery and Psychiatry,* 64(2), 184–189, 1998.

17. Dickson, D. W. et al., Office of Rare Diseases neuropathologic criteria for corticobasal degeneration, *J. Neuropathol. Exp. Neurol.,* 61(11), 935–946, 2002.

18. Houlden, H. et al., Corticobasal degeneration and progressive supranuclear palsy share a common tau haplotype, *Neurology,* 56(12), 1702–1706, 2001.

19. Kertesz, A. et al., The corticobasal degeneration syndrome overlaps progressive aphasia and frontotemporal dementia, *Neurology,* 55(9), 1368–1375, 2000.

20. Hauser, R. A. et al., Magnetic resonance imaging of corticobasal degeneration, *J. Neuroimaging,* 6(4), 222–226, 1996.

21. Lutte, I. et al., Contribution of PET studies in diagnosis of corticobasal degeneration, *European Neurology,* 44(1), 12–21, 2000.

22. Weiner, M. F., Dementia associated with Lewy bodies: dilemmas and directions, *Arch. Neurol.,* 56(12), 1441–1442, 1999.

23. Holmes, C. et al., Validity of current clinical criteria for Alzheimer's disease, vascular dementia and dementia with Lewy bodies, *Br. J. Psychiatry,* 174:45, 1999.

24. Armstrong, R. A., Cairns, N. J., and Lantos, P. L., The spatial patterns of Lewy bodies, senile plaques, and neurofibrillary tangles in dementia with Lewy bodies, *Experimental Neurology,* 150(1), 122–127, 1998.

25. Romano, S. and Colosimo, C., Procerus sign in progressive supranuclear palsy, *Neurology,* 57(10), 1928, 2001.

26. Krack, P. and Marion, M. H., "Apraxia of lid opening," a focal eyelid dystonia: clinical study of 32 patients, *Mov. Disord.,* 9(6), 610–615, 1994.

TABLE 44.7
Selected Investigations for Diseases that Mimic Parkinson's Disease

Type of Investigation	Disease	Findings
Laboratory work	Neuroacanthocytosis	Blood smear for acanthocytes
	SSPE	Increased titer of measles antibody in serum and CSF
	Wilson's disease	Decreased ceruloplasmin, increased 24h urine copper +/− increased serum unbound copper
Genetics/chromosomal studies	Autosomal recessive juvenile parkinsonism	Parkin mutation (chromosome 6)
	Dopa-responsive dystonia	GTP-cyclohydrolase gene (chromosome 14)
	FTDP-17	Chromosome 17 mutation
	Huntington's disease	Huntington/CAG repeats (chromosome 4)
	PKAN	PANK 2 mutation on chromosome 20
	SCA 2, 3, 6	CAG repeats
	X-linked Dystonia parkinsonism (Lubag)	DYT3 mutation (X chromosome)
Neuroimaging		
CT	Basal ganglia calcification	Increased signal intensity in basal ganglia
MRI	Amyloid angiopathy	Decreased signal on gradient echo in cortices
	Basal ganglia calcification	Increased T2 signal in basal ganglia
	Carbon monoxide poisoning	White matter hyperintensities (periventricular, cenrum semiovale), high signal in pallidum
	CBD	Asymmetrical atrophy in the posterior frontal and parietal regions, lateral ventricular dilatation, midbrain atrophy
	Creutzfeldt Jakob disease	"Cortical ribbon" on FLAIR and DWI, increased T2 signal in striatum
	MSA-C	Pontine and cerebellar atrophy ("hot cross bun sign")
	MSA-P	Putaminal atrophy and T2 hypointensity, putaminal "slit-like" hyperintensity
	Multiple sclerosis	White matter plaques
	PKAN	"Eye of the tiger" sign (bilateral area of hyperintensity within a region of hypointensity in the medial globus pallidus on T2-weighted images); prominent low signal in basal ganglia on T2 also seen in neuroferritinopathy and aceruloplasminemia
	Normal pressure hydrocephalus	Increased ventricular size
	PML	Diffuse white matter disease on T2
	PSP	Atrophy of midbrain, ballooned third ventricle, lateral ventricular dilatation
	Vascular parkinsonism	Evidence of old infarcts (white matter disease)
	Wilson's disease	Increase T2 signal in basal ganglia, less sos thalamus and brain stem, "face of the giant panda" midbrain sign, "bright claustrum sign," white matter changes
	X-linked Dystonia parkinsonism (Lubag)	Atrophy of the caudate and putamen or subtle putaminal abnormality
Others		
Slit lamp examination	Wilson's disease	Kayser-Fleischer ring
Biopsy	Wilson's disease	Liver biopsy with increased copper content
EEG	Creutzfeldt Jakob disease	Periodic lateralized discharges
	SSPE	Periodic discharges
Lumbar puncture–tap test	Normal pressure hydrocephalus	Improvement in symptoms, e.g., gait

CBD = corticobasal degeneration; DWI = diffusion weighted imaging; FLAIR = fluid-attenuated diversion recovery; FTDP-17 = frontotemporal dementia and parkinsonism linked to chromosome 17; MSA-C = multiple system atrophy, cerebellar variant; MSA-P = multiple system atrophy, parkinsonian variant; PANK2 = pantothenate kinase 2; PKAN = pantothenate kinase-associated neurodegeneration; PML = progressive multifocal leukoencephalopathy; PSP = progressive supranuclear palsy; SCA = spinocerebellar atrophy; SSPE = subacute sclerosing panencephalitis.

27. Nath, U. et al., Clinical features and natural history of progressive supranuclear palsy: A clinical cohort study, *Neurology,* 60(6), 910–916, 2003.

28. Lamberti, P. et al., Frequency of apraxia of eyelid opening in the general population and in patients with extrapyramidal disorders, *Neurol. Sci., 23,* Suppl. 2, S81–S82, 2002.

29. Litvan, I. et al., Which clinical features differentiate progressive supranuclear palsy (Steele-Richardson-Olszewski syndrome) from related disorders? A clinicopathological study, *Brain,* 120(1), 65–74, 1997.

30. Litvan, I. et al., Clinical features differentiating patients with postmortem confirmed progressive supranuclear palsy and corticobasal degeneration, *J. Neurol.,* 246, Suppl. 2, II1-II5, 1999.

31. Rascol, O. et al., Square wave jerks in parkinsonian syndromes, *J. Neurol. Neurosurg. Psychiatry,* 54(7), 599–602, 1991.

32. Rivaud-Pechoux, S. et al., Longitudinal ocular motor study in corticobasal degeneration and progressive supranuclear palsy, *Neurology,* 54(5), 1029–1032, 2000.

33. Kluin, K. J. et al., Perceptual analysis of speech disorders in progressive supranuclear palsy, *Neurology,* 43(3 Pt. 1), 563–566, 1993.

34. Testa, D. et al., Comparison of natural histories of progressive supranuclear palsy and multiple system atrophy, *Neurol. Sci.,* 22(3), 247–251, 2001.

35. Yamaguchi, M. et al., Laryngeal stridor in multiple system atrophy, *European Neurology,* 49(3), 154–159, 2003.

36. Merlo, I. M. et al., Not paralysis, but dystonia causes stridor in multiple system atrophy, *Neurology,* 58(4), 649–652, 2002.

37. Giladi, N. et al., Hemiparkinsonism-hemiatrophy syndrome: Clinical and neuroradiologic features, *Neurology,* 40, 1731–1734, 1990.

38. Colosimo, C. et al., Some specific clinical features differentiate multiple system atrophy (striatonigral variety) from Parkinson's disease, *Archives of Neurology,* 52 294–298, 1995.

39. Collins, S. J. et al., Progressive supranuclear palsy: Neuropathologically based diagnostic clinical criteria, *Journal of Neurology, Neurosurgery and Psychiatry,* 58, 167–173, 1995.

40. Fahn, S., Description of Parkinson's Disease as a Clinical Syndrome, *Ann. N. Y. Acad. Sci.,* 991, 1–14, 2003.

41. Tison, F. et al., Parkinsonism in multiple system atrophy: Natural history, severity (UPDRS-III), and disability assessment compared with Parkinson's disease, *Mov. Disord.,* 17(4), 701–709, 2002.

42. Masucci, E. F. and Kurtzke, J. F., Tremor in progressive supranuclear palsy, *Acta Neurol. Scand.,* 80(4), 296–300, 1989.

43. Kompoliti, K. et al., Clinical presentation and pharmacological therapy in corticobasal degeneration, *Arch Neurol.,* 55, 957–961, 1998.

44. Monza, D. et al., Neurophysiological features in relation to clinical signs in clinically diagnosed corticobasal degeneration, *Neurol. Sci.,* 24(1), 16–23, 2003.

45. Salazar, G. et al., Postural and action myoclonus in patients with parkinsonian type multiple system atrophy, *Movement Disorders,* 15(1), 77–83, 2000.

46. Tanigawa, A., Komiyama, A., and Hasegawa, O., Truncal muscle tonus in progressive supranuclear palsy, *J. Neurol. Neurosurg. Psychiatry,* 64(2), 190–196, 1998.

47. Lang, A. E., Parkinsonism in corticobasal degeneration, *Advances in Neurology,* 82, 83–89, 2000.

48. Giladi, N., Gait disturbances in advanced stages of Parkinson's disease, *Adv. Neurol.,* 86, 273–278, 2001.

49. Kyriakides, T. and Hewer, R. L., Hand contractures in Parkinson's disease, *J. Neurol. Neurosurg. Psychiatry,* 51(9), 1221–1223, 1988.

50. Uhrin, Z. and Stein, H., Rheumatoid-like deformities in Parkinson's disease, *Journal of Rheumatology,* 25(1), 177–179, 1998.

51. Schabitz, W. R. et al., Severe forward flexion of the trunk in Parkinson's disease: Focal myopathy of the paraspinal muscles mimicking camptocormia, *Mov. Disord.,* 18(4), 408–414. 2003.

52. Barclay, C. L. and Lang, A. E., Dystonia in progressive supranuclear palsy, *J. Neurol. Neurosurg. Psychiatry,* 62(4), 352–356, 1997.

53. Askmark, H. et al., Parkinsonism and neck extensor myopathy—A new syndrome or coincidental findings? *Archives of Neurology,* 58(2), 232–237, 2001.

54. Munhoz, R. P. et al., Evaluation of the pull test technique in assessing postural instability in Parkinson's disease, *Neurology,* 2003 (in press).

55. Wenning, G. K. et al., Cerebellar presentation of multiple system atrophy, *Movement Disorders,* 12(1), 115–117, 1997.

56. Kuba, H. et al., Gait disturbance in patients with low pressure hydrocephalus, *J. Clin. Neurosci.,* 9(1), 33–36, 2002.

57. Demirkiran, M., Bozdemir, H., and Sarica, Y., Vascular parkinsonism: a distinct, heterogeneous clinical entity, *Acta Neurol. Scand.,* 104(2), 63–67, 2001.

58. Winkler, A. S. et al., The frequency and significance of "striatal toe" in parkinsonism, *Parkinsonism Relat. Disord.,* 9(2), 97–101, 2002.

59. Schrag, A. et al., Young-onset Parkinson's disease revisited—clinical features, natural history, and mortality, *Mov Disord.,* 1998, 13(6), 885–894, 1998.

60. Caviness, J. N. et al., Cortical myoclonus in levodopa-responsive parkinsonism, *Movement Disorders,* 13(3), 540–544, 1998.

61. Louis, E. D. et al., Comparison of extrapyramidal features in 31 pathologically confirmed cases of diffuse Lewy body disease and 34 pathologically confirmed cases of Parkinson's disease, *Neurology,* 48(2), 376–380, 1997.

62. Aarsland, D. et al., Prevalence and characteristics of dementia in Parkinson disease—An 8-year prospective study, *Archives of Neurology,* 60(3), 387–392, 2003.

63. Dubois, B. et al., Slowing of cognitive processing in progressive supranuclear palsy. A comparison with Parkinson's disease, *Archives of Neurology,* 45(11), 1194–1199, 1988.

64. Lossos, A. et al., Frontotemporal dementia and parkinsonism with the P301S tau gene mutation in a Jewish family, *J. Neurol.*, 250(6), 733–740, 2003.

65. Pillon, B. et al., The neuropsychological pattern of corticobasal degeneration: Comparison with progressive supranuclear palsy and Alzheimer's disease, *Neurology*, 45, 1477–1483, 1995.

66. Dubois, B. et al., The FAB—A frontal assessment battery at bedside, *Neurology*, 55(11), 1621–1626, 2000.

67. Valls-Sole, J. et al., Distinctive abnormalities of facial reflexes in patients with progressive supranuclear palsy, *Brain*, 120 (Pt. 10), 1877–1883, 1997.

68. McKeith, I. G., Dementia with Lewy bodies, *Br. J. Psychiatry*, 180, 144–147, 2002.

69. McKeith, I. et al., Neuroleptic sensitivity in patients with senile dementia of Lewy body type, *British Medical Journal*, 305, 673–678, 1992.

70. Barclay, C. L., Bergeron, C. and Lang, A. E., Arm levitation in progressive supranuclear palsy, *Neurology*, 52(4), 879–882, 1999.

71. Louis, E. D. et al., Arm elevation in Huntington's disease: Dystonia or levitation? *Movement Disorders*, 14(6), 1035–1038, 1999.

72. Snider, S. R. et al., Primary sensory symptoms in parkinsonism, *Neurology*, 26(5), 423–429, 1976.

73. Wenning, G. K. et al., Clinicopathological study of 35 cases of multiple system atrophy, *J. Neurol. Neurosurg. Psychiatry*, 58(2), 160–166, 1995.

74. Wenning, G. K. et al., Multiple system atrophy: a review of 203 pathologically proven cases, *Mov. Disord.*, 12(2), 133–147, 1997.

75. Klein, C. et al., The "cold hands sign" in multiple system atrophy, *Mov. Disord.*, 12(4), 514–518, 1997.

76. Jost, W. H. and Schimrigk, K., Constipation in Parkinson's disease, *Klin. Wochenschr.*, 69(20), 906–909, 1991.

77. Siddiqui, M. F. et al., Autonomic dysfunction in Parkinson's disease: a comprehensive symptom survey, *Parkinsonism Relat. Disord.*, 8(4), 277–284, 2002.

78. Horimoto, Y. et al., Autonomic dysfunctions in dementia with Lewy bodies, *J. Neurol.*, 250(5), 530–533, 2003.

79. Olson, E. J., Boeve, B. F., and Silber, M. H., Rapid eye movement sleep behaviour disorder: demographic, clinical and laboratory findings in 93 cases, *Brain*, 123(2), 331–339, 2000.

80. Arnulf, I. et al., Hallucinations, REM sleep, and Parkinson's disease—A medical hypothesis, *Neurology*, 55(2), 281–288, 2000.

81. Boeve, B. F. et al., REM sleep behavior disorder and degenerative dementia—An association likely reflecting Lewy body disease, *Neurology*, 51(2), 363–370, 1998.

82. Friedman, J. H., Presumed rapid eye movement behavior disorder in Machado-Joseph disease (spinocerebellar ataxia type 3), *Movement Disorders*, 17(6), 1350–1353, 2002.

83. Krishnan, P. R., Bhatia, M., and Behari, M., Restless legs syndrome in Parkinson's disease: A case-controlled study, *Movement Disorders*, 18(2), 181–185, 2003.

84. Wetter, T. C. et al., Sleep and periodic leg movement patterns in drug-free patients with Parkinson's disease and multiple system atrophy, *Sleep*, 23(3), 361–367, 2000.

85. Factor, S. A. et al., Sleep disorders and sleep effect in Parkinson's disease, *Mov. Disord.*, 5(4), 280–285, 1990.

86. Gjerstad, M. D., Aarsland, D., and Larsen, J. P., Development of daytime somnolence over time in Parkinson's disease, *Neurology*, 58(10), 1544–1546, 2002.

87. Frucht, S. et al., Falling asleep at the wheel: Motor vehicle mishaps in persons taking pramipexole and ropinirole, *Neurology*, 52(9), 1908–1910, 1999.

88. Schlesinger, I. and Ravin, P. D., Dopamine agonists induce episodes of irresistible daytime sleepiness, *European Neurology*, 49(1), 30–33, 2003.

89. Pal, S. et al., A study of excessive daytime sleepiness and its clinical significance in three groups of Parkinson's disease patients taking pramipexole, cabergoline and levodopa mono and combination therapy, *Journal of Neural Transmission*, 108(1), 71–77, 2001.

90. Boesch, S. M. et al., Dystonia in multiple system atrophy, *J. Neurol. Neurosurg. Psychiatry*, 72(3), 300–303, 2002.

91. Litvan, I. and Hutton, M., Clinical and genetic aspects of progressive supranuclear palsy, *J. Geriatr. Psychiatry Neurol.*, 11(2), 107–114, 1998.

92. Hely, M. A. et al., Diffuse Lewy body disease: clinical features in nine cases without coexistent Alzheimer's disease, *J. Neurol. Neurosurg. Psychiatry*, 60(5), 531–538, 1996.

93. Nutt, J. G. and Nygaard, T. G., Response to levodopa treatment in dopa-responsive dystonia, *Archives of Neurology*, 58(6), 905–910, 2001.

94. Ben-Shlomo, Y. et al., Survival of patients with pathologically proven multiple system atrophy: A meta-analysis, *Neurology*, 48(2), 384–393, 1997.

45 Diagnostic Criteria for Parkinson's Disease

Daniel Tarsy
Harvard Medical School, Director, Movement Disorders Center, Beth Israel Deaconess Medical Center

CONTENTS

INTRODUCTION

Parkinson's disease (PD) is associated with a combination of characteristic clinical signs and symptoms which, in established cases, should allow for prompt and accurate diagnosis. Diagnosis is obviously essential for patient management as well as for epidemiologic studies and clinical trials of new medications, most of which are much more effective in PD than in other parkinson syndromes. In the absence of reliable biological markers for the diagnosis of PD diagnosis must rest on clinical criteria. The classical three cardinal signs of PD are tremor, rigidity, and akinesia. It is generally accepted that the presence of at least two of these signs is required for the diagnosis of PD. Postural instability is often considered a fourth cardinal sign but is less useful for early diagnosis, as it typically appears only later in the course of PD and is such a common feature of other forms of parkinsonism. Despite these simple criteria, underdiagnosis and misdiagnosis remain common. Regarding underdiagnosis, in one door-to-door survey, 24 percent of cases were detected for the first time at the time of the survey.[1] Underdiagnosis in the elderly occurs because bradykinesia and postural instability are common features of aging as well as chronic cerebrovascular disease. As a result, loss of facial expression, stooped posture, gait unsteadiness, and tremor are often attributed to aging by both laypersons and physicians. Misdiagnosis occurs because parkinsonism is a symptom complex as well as a disease and may be the result of a variety of underlying causes. The common use of qualifying terms such as parkinson syndrome, "atypical parkinsonism," or "Parkinson plus" indicates the general awareness that other forms of parkinsonism exist. Autopsy studies have shown that the clinical diagnosis of PD is incorrect in as many as 20 to 25% of cases.[2,3] This indicates that, although symptoms and signs of parkinsonism may be correctly identified in life, the diagnosis of idiopathic PD is often incorrect. Idiopathic PD can be definitively diagnosed only by neuropathological examination with findings of neuronal loss in the substantia nigra and the presence of Lewy bodies in substantia nigra and other brain nuclei such as locus ceruleus, nucleus basalis of Meynert, dorsal motor nucleus of the vagus, and hypothalamus. The most common alternative causes of parkinsonism that are discovered at autopsy, all of which lack characteristic Lewy bodies, are multiple system atrophy (MSA), progressive supranuclear palsy (PSP), Alzheimer's disease, and diffuse cerebrovascular disease.[3]

CLINICAL-PATHOLOGICAL CORRELATION IN PARKINSON'S DISEASE

Landmark studies at the United Kingdom Parkinson's Disease Society Brain Research Center using tissue from the United Kingdom Parkinson's Disease Society Brain Bank (UKPDSBB) have advanced awareness of pitfalls in the clinical diagnosis of PD.[3–8] An earlier clinical-pathological study of 59 patients followed by a single neurologist, basing his clinical diagnosis on the presence of 2 of the 3 cardinal signs of PD, had shown a 24% rate of incorrect diagnosis of PD.[2] In the UK studies, post-mortem tissue has been obtained from 710 of 1500 donors with PD in whom clinical data has been collected.[9] In the first 100 patients with a clinical diagnosis of PD, 70 were diagnosed using the UKPDSBB criteria. Only 76 proved to have pathological evidence of idiopathic PD with depletion of brain stem pigmented neurons and presence of Lewy bodies in surviving neurons.[3] Of the remaining 24 patients without Lewy bodies, 6 had PSP, 5 had MSA, 6 had Alzheimer's disease or Alzheimer-type pathology, 3 had vascular disease, 2 had isolated nigral atrophy without Lewy bodies, 1 had postencephalitic parkinsonism, and 1 patient with probable essential tremor had no pathological findings. After retrospective reanalysis of the clinical features, there were 11 patients who did not fulfill UKPDSBB criteria for idiopathic PD. Elimination of these 11 patients improved the accuracy rate to 82%. If very selected criteria such as asymmetric onset, no atypical features, and no other possible etiology for an alternative parkinson syndrome were used, the proportion of true PD cases increased to 92% but, at that point, 32% of pathological cases would have been rejected, thereby demonstrating increased specificity with reduced sensitivity.[9]

In a separate study, Hughes et al.[5] analyzed the clinical features of this same group of 100 patients to determine the specificity and sensitivity of clinical criteria thought to be useful in making the diagnosis of PD. As mentioned above, the retrospective use of the UKPDSBB diagnostic criteria eliminated 11 cases and improved diagnostic accuracy to 82%. Only asymmetric onset and absence of any atypical features were significantly different between pathologically proven PD and non-PD cases.[5] Response to L-dopa had little discriminatory power due to the fact that 16 of the 24 non-PD cases had shown a "marked" response to L-dopa. Tremor-dominant disease also failed to discriminate between the two groups.

In a more recent study by the same group, accuracy of diagnosis increased to 90% in a new population of 100 cases of clinically diagnosed PD, indicating significant improvement in diagnostic accuracy of PD over the last decade.[7] Retrospective application of current rigorous criteria for PD improved diagnostic accuracy only slightly to 91 to 92%. In the more recent series, MSA was the most frequent false positive diagnosis, as opposed to PSP in the previous series.

Hughes et al. also reversed the analysis by examining the clinical details of 100 pathologically confirmed cases of PD.[6] Although 90 of these cases were correctly diagnosed as PD by the time of death, 10 others carried clinical diagnoses of atypical parkinsonism, vascular parkinsonism, MSA, drug-induced parkinsonism, or post-traumatic parkinsonism. Disease onset was asymmetric in 72%, there was tremor at onset in 69%, and initial L-dopa response was good to excellent in 77% but none to poor in 6%. Dementia was present in 44%. Motor fluctuations occurred in 61% and dyskinesias in 60% of cases. Twelve patients had atypical features including severe early dementia, no response to L-dopa, early fluctuating confusion, myoclonus, apraxia, and early marked dysautonomia. These findings demonstrate that, in addition to a 10 to 20% rate of PD misdiagnosis, 10% of patients with pathologically proven PD have clinical features that lead to an alternative diagnosis.

Over the past ten years, greater awareness of alternative parkinson syndromes has led to improved diagnosis. Nonetheless, pathologic studies show that pitfalls in diagnosis remain, both in terms of incorrect, false positive diagnoses of PD[3,7] and incorrect, false positive diagnoses of parkinson syndrome.[6] As stated by Hughes et al.,[7] rigorous application of diagnostic criteria may reduce the number of false positive cases, but some cases of parkinson syndrome may still mimic PD enough to be diagnosed as PD. They have maintained that, based on current clinical diagnostic criteria, a diagnostic accuracy of greater than 90% may not be attainable.[7] However, their most recent study of 143 cases of parkinsonism, seen over 10 years by movement disorder neurologists at the National Hospital for Neurology and Neurosurgery in London,[8] showed a clinical accuracy for PD of 98.6%. Published criteria for the diagnosis of MSA[10] and PSP[11] have also served to improve diagnostic accuracy in patients with parkinsonism and will be discussed below. In addition, there has also been increased awareness of dementia with Lewy bodies (DLB) as a form of parkinsonism to be differentiated from PD. In one study, six neurologists were provided clinical vignettes from the first and last visits of autopsy studied patients with PD and similar disorders.[12] Inter-rater reliability for the diagnosis of PD was moderate for the first visit and substantial at the last visit, while agreement for the diagnosis of DLB was only fair for the first visit and slight for the last.[12,13]

It is generally agreed that studies based on brain bank material may not reflect more common community experience. Donor brain tissue is more likely to come from patients with unusual or atypical clinical manifestations without a clear clinical diagnosis and from patients dying in hospitals or institutions, thereby selecting for greater disability and dementia. However, similar results have come from a community-based clinical study in London.[14] In this study, 131 patients carrying a diagnosis of PD were carefully reviewed and followed for one year. The diagnosis of probable PD was confirmed in 109 of the 131 patients (83%), and the diagnosis was rejected in 20 patients (15%). Sensitivity was 88%, and specificity was 73%. Similar to pathologic series, alternative diagnoses were PSP, MSA, vascular parkinsonism, and tremor.

CLINICAL FEATURES OF PARKINSON'S DISEASE

Patients with PD typically present between ages 50 and 70 with complaints of tremor, which is usually accompanied by unilateral or asymmetric clumsiness of one hand and sometimes the leg on the same side. During initial evaluation, mild midline or contralateral signs often are also apparent. Most of these are manifestations of akinesia and may include reduced arm swing, reduced spontaneous gesturing, impaired repetitive movements such as toothbrushing, micrographia, mildly reduced vocal or facial expression (frequently mistaken for depression), reduced blink frequency, drooling, slowing in activities of daily living, hesitation arising from deep chairs, flexed posture, and shuffling gait while turning. The characteristic resting tremor of Parkinson's disease is present in about 75% of cases, but postural tremor may also be an early presenting sign. When tremor is absent, there is often a delay in appreciating the significance of the other early parkinsonian signs. In more advanced cases, signs of akinesia and postural disturbance predominate. Progressive gait disturbance, freezing, and falls due to loss of postural reflexes occur much later in idiopathic PD and make differentiation from other causes of parkinsonism more difficult than early in the disease.

The absence of resting tremor in a patient with akinesia, rigidity, and postural disturbance should serve as a clue that one may be dealing with one of the other akinetic-rigid syndromes. This is particularly true if symmetric and midline motor manifestations with postural instability appear early in the clinical course. However, as already mentioned, tremor may be absent in 25% of patients with early PD. Useful clues for other forms of parkinsonism are the presence of early gait and postural disturbance, pyramidal tract signs, cerebellar ataxia, oculomotor abnormalities, early or severe dementia, frontal release signs, sensory findings, or prominent autonomic disturbances early in the illness.

Parkinsonism due to MSA is typically the most difficult to distinguish on clinical grounds from idiopathic PD. MSA includes three disorders formerly known as Shy-Drager syndrome, striatonigral degeneration, and the sporadic form of olivopontocerebellar atrophy. Currently, Shy-Drager syndrome is designated as MSA-A, striatonigral degeneration as MSA-P, and olivopontocerebellar atrophy as MSA-C. These disorders share similar underlying neuropathology but exhibit a spectrum of variable and evolving neurologic findings that typically overlap to varying degrees, often depending on when the patient is seen in the course of the disease. The presence of orthostatic hypotension, neurogenic bladder, and impotence preceding the onset of significant motor disturbance is indicative of MSA-A. Predominant parkinsonism manifested by a rapidly progressive akinetic-rigid syndrome and relatively mild and variable tremor should suggest MSA-P. Cerebellar ataxia early in the course followed by parkinsonism and variable combinations of corticospinal tract deficits, oculomotor signs, severe dysarthria, and laryngeal stridor are the hallmarks of olivopontocerebellar atrophy. Clinical differentiation among these three forms of MSA is often difficult and somewhat arbitrary. Certain "red flags" have been cited as useful clues to the diagnosis of MSA.[15] In addition to the features mentioned above, these include stimulus sensitive focal myoclonus, severe anterocollis, peripheral neuropathy, and dusky cyanosis of the skin.

PSP is sometimes included in the category of Parkinson plus syndromes but is classified separately from MSA because of its distinctive clinical presentation and neuropathology. The condition is best known for paralysis of voluntary vertical and horizontal gaze, but it produces a number of other characteristic motor manifestations that often precede the abnormal eye movements. Early symptoms include gait instability, sudden unexplained falls, generalized slowing, visual complaints, speech and swallowing difficulty, sleep disturbance, and personality change. By contrast with PD, gait instability and falling are prominent early signs and are often the presenting manifestation. The supranuclear gaze palsy of PSP impairs downward more than upward gaze but also affects horizontal gaze. Initial eye findings may be subtle and limited to impaired saccadic movements. Early visual complaints include difficulty reading and descending stairs, trouble focusing, and diplopia. Oculocephalic eye movements are disinhibited consistent with a supranuclear gaze paralysis. Other eye findings may include blepharospasm, levator inhibition with impaired voluntary eye-lid opening, and square-wave jerks.

There is often a characteristic, sometimes astonished-appearing facial expression produced by the prominent stare, upper eyelid retraction, deep facial furrows, and impaired voluntary gaze, which is considerably different from the facial masking of PD. Gait is lurching or stumbling and associated with impaired balance and tendency to fall easily. Rigidity is more marked in axial than limb muscles, but limb spasticity and dystonia may be prominent late manifestations. Posture is erect and there may be hyperextension of the neck. Tremor is uncommon but may occur. There is usually a pseudobulbar palsy with spastic dysarthria, dysphagia, and emotional release. Bilateral corticospinal tract findings are present, including hyperreflexia, spasticity, and extensor plantar responses. Dementia is usually not prominent early in the disease, although the abnormal facial expression, severe bradykinesia, and marked dysarthria late in the disease often convey a false impression of cognitive deficit.

Acquired neurologic disorders that produce secondary parkinsonism can usually be differentiated from idiopathic PD by the history and examination. Drug-induced parkinsonism is relatively symmetric in distribution and more commonly associated with an action tremor than the typical resting tremor of idiopathic PD. However, asymmetry and rest tremor may also be observed. In such cases, often seen in older individuals, the possibility of underlying subclinical PD aggravated by dopamine-blocking medications should be considered. Parkinsonism due to toxin exposure is rare and is usually evident from the patient's medical history. Other common causes of secondary parkinsonism include cerebrovascular disease with multiple small infarcts, residual effects of anoxic encephalopathy, hydrocephalus, head trauma, brain tumor or arteriovenous malformation, and acquired hepatocerebral degeneration. A number of genetically mediated disorders may produce parkinsonism although usually in combination with other neurologic manifestations. The most important of these are the rigid form of Huntington's disease, Wilson's disease, familial olivopontocerebellar atrophy including Azorean disease, Hallervorden-Spatz disease, the pallidal degenerations, basal ganglia calcification, neuroacanthocytosis, dopa-responsive dystonia, juvenile parkinsonism, and rare metabolic and storage diseases usually seen in children or young adults such as GM1 gangliosidosis and Gaucher's disease.

CLINICAL DIAGNOSTIC CRITERIA FOR PARKINSON'S DISEASE

In the absence of reliable and readily available laboratory tests for the diagnosis of PD, the diagnosis currently relies on the clinical manifestations of the disease. Even for the most experienced neurologist, accurate diagnosis of PD can be particularly difficult to make early in course of the disease, although it typically improves with time. In the large prospective study of the effect of selegiline and tocopherol therapy in slowing progression of PD (DATATOP) in 800 patients,[16] the initial diagnosis of PD was changed in 8.1% over a follow-up period of 6.0 years.[17] Diagnostic criteria were actually not specified in this study, but diagnosis was made by an experienced expert in PD.[17] Mean duration of symptoms was 2.1 years, and mean Hoehn and Yahr stage was 1.6 years at randomization. Change in diagnosis was made on the basis of

lack of response to L-dopa, atypical neuroimaging, and atypical clinical features. Similar improvement in diagnostic accuracy during follow-up of patients with PD has been reported in other studies in which more stringent clinical criteria were used.[2,8,12]

There is general agreement that the cardinal signs of PD are akinesia, rigidity, and tremor, and that rest tremor, an excellent L-dopa response lasting at least five years, severe L-dopa induced dyskinesia, a progressive disorder lasting at least 10 years, and the absence of features suggestive of alternate forms of parkinsonism form the basis of the diagnosis of idiopathic PD.

Calne et al.[18] elaborated on these criteria by considering that resting tremor, rigidity, bradykinesia, and impaired postural reflexes make up the syndrome termed "idiopathic parkinsonism" (IP), while other supporting features, such as asymmetry, an excellent response to an adequate dose of L-dopa (1,500 mg/day with a peripheral decarboxyslase inhibitor), and progression, along with exclusion of other obvious causes of parkinsonism, allowed for the diagnosis of IP.

They proposed the following three categories of diagnosis:

- *Clinically possible IP* with presence of one of the following: resting or postural tremor, rigidity, or bradykinesia. Impaired postural reflexes was not included, because it was felt to be nonspecific.
- *Clinically probable IP* with any *two* of the cardinal features: resting tremor, rigidity, bradykinesia, or impaired postural reflexes. Alternatively, asymmetric resting tremor, rigidity, or bradykinesia were sufficient.
- *Clinically definite IP* with any combination of *three* of the four cardinal features. Alternatively sufficient were two of these features with one of the first three displaying asymmetry.

Calne's group has also provided a simpler and more practical definition of idiopathic PD:[19] a combination of the triad of akinesia, rigidity, and akinesia with no detectable cause; a response to dopaminergic drugs; absence of pyramidal, cerebellar, or lower motor neuron signs; eye signs limited to impaired upward gaze; and autonomic deficits limited to minor impairments.

Larsen et al.[20] proposed a similar but much more complex and cumbersome classification. *Definite clinical idiopathic PD* required the presence of resting tremor and at least two of the following: resting tremor (sic), rigidity, or postural abnormality along with unilateral onset of symptoms, asymmetric progression of disease, good to excellent response to dopaminergic medication, and absence of brain imaging and clinical signs that would suggest alternate causes of parkinsonism. *Proba-*

ble clinical idiopathic listed types A, B, and C as defined by various combinations of criteria including two of the four cardinal signs, moderate response to dopaminergic medication, absence of brain imaging and clinical signs of alternate causes of parkinsonism, and unilateral onset of symptoms. *Possible clinical idiopathic PD* was defined by at least two of the cardinal signs, a moderate response to dopaminergic medication, and absence of brain imaging or clinical signs of alternate causes of parkinsonism.

It should be emphasized that the diagnostic criteria utilizing categories of "definite, probable, and possible," which have been proposed[18,20] as suggested criteria, have no established validity and need to be subjected to prospective study. Both are of unknown validity and reliability and suffer from being cumbersome with a level of detail that doesn't appear to improve upon the more straightforward criteria of the UKPDSBB. Following a review of the literature, Gelb et al.[21] recently proposed a similar set of diagnostic criteria with categories of definite, probable, and possible PD. *Possible PD* required at least two of the four cardinal signs, which had to include either tremor or bradykinesia, no features suggestive of an alternative diagnosis, and a substantial and sustained response to dopaminergic medication. *Probable PD* required at least three of the cardinal signs, no exclusionary features, and a substantial and sustained response to dopaminergic medication. *Definite PD* required all of the criteria for possible PD and pathologic confirmation at autopsy. This appears to be a somewhat simpler set of criteria but is also unvalidated and, by requiring histologic confirmation, is not helpful for clinical diagnosis. Postural instability is commonly also included as a cardinal sign but is less useful for early diagnosis of PD, since it is a late clinical feature of PD and a much earlier feature of PSP and MSA. Clinical diagnostic criteria have been proposed by other authors,[4,5,22] but most have been based on retrospective data and not evaluated for validity or reliability.[13]

At present, the UKPDSBB criteria, which were published in the late 1980s,[4] remain the most widely accepted criteria for the diagnosis of PD.[13] Bradykinesia is required, plus the presence of either rigidity, rest tremor, or postural instability. Three or more supportive criteria and absence of a list of 16 exclusion criteria (see Table 45.1) are required.

TABLE 45.1
UK Parkinson's Disease Society Brain Bank Clinical Diagnostic Criteria

Inclusion Criteria	Exclusion Criteria	Supportive Criteria
Bradykinesia (slowness of voluntary movement with progressive reduction in speed and amplitude of repetitive actions)	History of repeated strokes with stepwise progression of parkinsonian features	(Three or more required for diagnosis of definite PD)
	History of repeated head injury	Unilateral onset
	History of definite encephalitis	Rest tremor present
	Oculgyric crises	Progressive disorder
And at least one of the following:	Neuroleptic treatment at onset of symptoms	Persistent asymmetry affecting side of onset most
	More than one affected relative	
• Muscular rigidity	Sustained remission	Excellent response (70 to 100%) to
• 4- to 6-Hz rest tremor	Strictly unilateral features after three years	levodopa
• Postural instability not caused by primary visual, vestibular, cerebellar, or proprioceptive dysfunction	Supranuclear gaze palsy	Severe levodopa–induced chorea
	Cerebellar signs	Levodopa response for 5 years or more
	Early severe autonomic involvement	Clinical course of 10 years or more
	Early severe dementia with disturbances of memory, language, and praxis	
	Babinski sign	
	Presence of cerebral tumor or communicating hydrocephalus on CT scan	
	Negative response to large doses of levodopa (if malabsorption excluded)	
	MPTP exposure	

Source: From Litvan, I., et al., *Mov. Disord.,* 18, 468, 2003. With permission.

CLINICAL DIAGNOSTIC CRITERIA FOR OTHER PARKINSON SYNDROMES

Criteria for the diagnosis of MSA, PSP, corticobasal degeneration (CBD), and DLB have also been published and have helped to differentiate idiopathic PD from other forms of parkinsonism. The Scientific Issues Committee (SIC) Task Force Appraisal of Clinical Diagnostic Criteria for Parkinsonian Disorders provides a useful critical review of these criteria.[13]

MULTIPLE SYSTEM ATROPHY

A consensus statement on the diagnosis of MSA[23] formalized previous criteria provided by Quinn[24] for the diagnosis of MSA-P and MSA-C. Several groups have attempted to identify factors that reliably differentiate MSA from PD.[25–27] Colosimo et al.[25] found that rapid progression, symmetric onset, absence of rest tremor, poor or no response to L-dopa, and orthostatic hypotension occurred more frequently in MSA than PD. However, MSA could not be differentiated from PSP in this study. Wenning et al.[27] identified four early features that differentiated MSA from PD: presence of autonomic signs, poor initial response to L-dopa, early fluctuations, and rigidity as a presenting sign. Not surprisingly, several studies analyzing the accuracy of criteria for diagnosis of MSA have found increased diagnostic accuracy in more advanced cases.[10,26] In the recent study by Hughes et al., at the National Hospital in London,[8] 30 of 34 cases of MSA were correctly diagnosed for a sensitivity of 88.2%. Diagnosis of MSA is made difficult by the fact that 10 to 20% of patients with MSA-P may have features of asymmetry, L-dopa responsiveness, dyskinesias, and motor fluctuations.[8,27] Similar to PD, none of the diagnostic criteria for MSA have been subjected to a prospective study with pathologic confirmation.

PROGRESSIVE SUPRANUCLEAR PALSY

In the case of PSP, there have been seven published sets of clinical criteria, which are described in the SIC Task force review,[13] although most of these were not derived systematically. A more rigorous approach led to the National Institute of Neurological Disorders and Stroke and Society for Progressive Supranuclear Palsy (NINDS-SPSP) diagnostic criteria[11] whose sensitivity, specificity, and positive predictive value were assessed retrospectively using a pathologic series and shown to be extremely high and superior to existing criteria.[13] The recent National Hospital of London study by Hughes et al. found a diagnostic accuracy rate of 80% for PSP.

CORTICOBASAL DEGENERATION

Diagnostic criteria have also been proposed for CBD but have not been formally validated.[13] In the National Hospital study,[8] there were very few cases of CBD, but only one of three cases diagnosed clinically was confirmed pathologically, and only one of four cases confirmed pathologically had been correctly diagnosed clinically. Larger series have shown variable results, with accuracy of clinical diagnosis ranging between 53 and 94% and only about 50% of pathologically proven cases being diagnosed in life.[8] When CBD presents with motor manifestations, it often resembles PSP more closely than PD. When CBD presents with dementia or aphasia, the issue becomes one of differentiation from other dementing illnesses.[13]

DEMENTIA WITH LEWY BODIES

DLB must be differentiated from PD as well as from other dementing illnesses such as Alzheimer's disease or vascular dementia. At the present time, it is unclear whether DLB is separable from PD or is a different clinical manifestation of the same disorder. There have been several attempts to define diagnostic criteria for DLB, with the most rigorous emerging from a DLB Consensus Conference.[28] The aim of these criteria was to differentiate DLB from other forms of dementia. Inclusion criteria were progressive cognitive decline with impaired frontal-subcortical functions and visuospatial ability along with one of three core features: fluctuating cognitive function and alertness, visual hallucinations, and parkinsonism. Supporting features were falls, syncope, neuroleptic sensitivity, delusions, and other hallucinations. Nine published studies examining the diagnostic accuracy of these criteria[13] have produced variable results. At present, it appears that the consensus criteria for probable DLB produce few false positives and therefore may be appropriate for diagnosis within a demented population but are limited in screening for DLB because of a high false negative rate.[13]

DOPAMINERGIC CHALLENGE TESTS

It is generally agreed that patients with idiopathic PD respond to treatment with L-dopa or dopamine agonists, while patients with other parkinson syndromes have a more variable response and, in many cases, show no clinical response. Response to L-dopa is therefore commonly used as a criterion for diagnosis of PD. With only rare exceptions,[29] the vast majority of patients with pathologically proven PD respond to L-dopa during life. However, about 30% of patients with MSA show an initial response to L-dopa that, in some cases, may be quite robust early in the disease,[30] and nearly one-third of patients with PSP show a partial albeit limited response to L-dopa.[13] L-dopa dyskinesias are more common in PD than PS, but craniocervical and limb dyskinesia may occur in MSA,[31] and dyskinesia very rarely does occur in PSP.[32]

In the interest of early and more accurate diagnosis of PD, acute pharmacologic challenge testing has been investigated in both clinical practice and research settings.[33] Acute dopaminergic challenge has been the subject of retrospective meta-analysis,[34,35] has been reviewed by a panel of experts in a consensus statement,[33] and has been included in guidelines to assess surgical treatment responses in patients with PD.[36,37]

In most cases, dopaminergic challenge testing has been carried out with single doses of L-dopa or apomorphine. For purposes of diagnosis, this is a more practical procedure than a prolonged course of treatment, but it assumes that a patient's response to a single dose of L-dopa or apomorphine will accurately predict the response to prolonged treatment. Only a few studies have compared responses to acute dopaminergic challenge with chronic L-dopa treatment. Several of these have been reviewed in meta-analyses.[34,35] Concerning patients with early parkinsonism, in three studies totaling 129 de novo patients, subjects were tested with apomorphine followed by 3 to 12 months of sustained L-dopa treatment.[38–40] These studies showed a sensitivity of 80%, a specificity of 90%, a positive predictive value of 96%, and an overall correct prediction rate of 83% for the apomorphine test.[34] However, there were significant numbers of false positive and false negative responses to apomorphine testing. Twenty percent of patients responding to sustained L-dopa had failed to respond to apomorphine, and only 43% with a negative response to apomorphine failed to respond to sustained L-dopa treatment. Meta-analysis of apomorphine testing in 222 patients with PD at all stages of disease duration showed very similar results with a sensitivity of 88%, a specificity of 88%, a positive predictive value of 97%, and an overall correct prediction rate of 88%.[34] In another meta-analysis[35] of two studies[41,42] in which acute L-dopa was compared to chronic L-dopa, 100 patients with PD responded to both, 12 responded to neither, 21 responded only to chronic L-dopa, and two responded only to acute L-dopa. The authors concluded that the results of acute testing were similar but not superior to chronic treatment.

A more recent study by Merello et al.[43] arrived at a somewhat different conclusion. Acute L-dopa was compared to one month of treatment with L-dopa and patients were followed up at 24 months. Sensitivity of the acute L-dopa challenge was 70.9%, specificity was 81.4%, and positive predictive value was 88.6%. When patients were stratified according to disease severity at time of initial testing, specificity was 100% for patients with mild disease. This is in contrast to the common impression that responses to acute challenges may be more difficult to document in patients with mild symptoms and signs.[33] The authors concluded that acute L-dopa challenge is a useful tool for separating PD from parkinson syndrome in less disabled patients.

As stated in the consensus statement on acute dopaminergic challenge,[33] challenge testing should be carried out early morning in the "practical off" state in which the patient has been free of dopaminergic medication overnight for a period of at least 12 hr. In drug-naïve patients, L-dopa should be given as the immediate release formulation in a dose of 250 mg with a peripheral dopa decarboxylase inhibitor. In patients already receiving L-dopa treatment, a suprathreshold dose of L-dopa (50% higher than the patient's usual dose) is sometimes used to assure maximal L-dopa response. Apomorphine is given subcutaneously in a single dose of 3 mg or incrementally beginning with 1.5 mg with additional 1.5 mg doses every 30 min, up to 9.0 mg. In drug-naïve patients, an antiemetic such as domperidone should be administered in a dose of 60 mg/day for 24 to 48 hours prior to testing.[33,34] Assessment should be carried out 10 to 30 minutes after apomorphine or 60 to 90 minutes after L-dopa using quantitative tests of motor function such as the Unified Parkinson's Disease Rating Scale (UPDRS) or other quantitative measurements such as timed motor testing recognizing that different motor symptoms may show different response thresholds. According to consensus guidelines, in drug-naïve patients a positive response to acute testing is defined as an improvement in motor scores of at least 20% compared to baseline.[33] There is no satisfactory definition of a positive response to chronic L-dopa treatment.[33] A minimum of 30% motor improvement has been considered clinically relevant.[44] It should be appreciated that patients with mild symptoms may show a clinically less impressive response than patients with more advanced disease.

In summary, acute dopaminergic challenge testing appears to be a reasonable optional tool for early and rapid prediction of therapeutic potential in patients with newly diagnosed parkinsonism. However, acute dopaminergic testing should not be used as a diagnostic test to reliably distinguish idiopathic PD from parkinson syndromes.[34] Even for prediction of therapeutic response, results of such testing should be viewed with considerable caution, given the substantial incidence of false positive and false negative responses. False negative rates in drug-naïve patients may be as high as 40%.[33] Also, some patients who fail to respond to apomorphine may respond to L-dopa.[39] In clinical practice, acute test results cannot supplant response to a more prolonged therapeutic trial.

By contrast, acute challenge testing may still play an important role in research settings such as in clinical drug trials, surgical treatment trials, and assessment and reassessment of patients in long-term follow-up studies.[33] It may also be useful in specific clinical situations such as estimating potential benefit of higher L-dopa doses, to reassess patients with apparently declining L-dopa responsiveness, and to assess specific symptoms with uncertain responses to L-dopa such as tremor or freezing.[34]

BRAIN IMAGING

The use of brain imaging in the differential diagnosis of parkinsonian syndromes is described in detail elsewhere in this volume. Since PD is associated with relatively limited nigral pathology, while other forms of parkinsonism are often associated with nigrostriatal pathology, brain imaging should theoretically be able to differentiate these disorders. However, in early stages of the disease, routine brain imaging is typically unhelpful in making this differentiation. In more advanced cases, magnetic resonance imaging (MRI) may show a mixture of high and low signal putamenal abnormalities in MSA-P, pontine or cerebellar atrophy in MSA-C, midbrain atrophy in PSP, asymmetric cortical atrophy in CBD, and a combination of subcortical ischemic changes and periventricular white matter changes in vascular parkinsonism.[45] Recent studies suggest that diffusion weighted imaging may be a more sensitive method to differentiate PD from MSA[46] or PSP[47] but require confirmation. MRI is more commonly used to identify less common causes of parkinsonism such as Wilson's disease, chronic liver disease, basal ganglia calcification, normal pressure hydrocephalus, and other structural brain abnormalities. In several studies, MR spectroscopy showed normal neuronal markers in striatum in PD and reduced levels in striatum in other forms of parkinsonism.[48,49] However, this has not been a universal finding,[50,51] and MRS is not a readily available diagnostic technique.

Functional brain imaging has been used to distinguish PD from other forms of parkinsonism but currently remains a relatively inaccessible tool with unproven value in early cases. Fluorodopa positron emission tomography (PET) studies show reduced uptake in putamen more than caudate nucleus in both PD and MSA-P. Single photon emission computed tomography (SPECT) using the dopamine transporter ligand β-CIT shows similar findings.[52,53] Both techniques show greater asymmetric abnormalities in PD than MSA, but this information adds little to the clinical examination, which is also capable of showing the same asymmetry. In addition, contrary to common belief, MSA is reported to be asymmetric in its initial presentation in 74% of patients.[30] PET studies of cerebral glucose metabolism show similar asymmetries and have also shown greater striatal metabolic abnormalities, including caudate nucleus involvement, in atypical parkinsonism than PD.[54,55] However these studies require special statistical analyses and are not generally available for routine clinical use. Since β-CIT SPECT labels presynaptic nigrostriatal dopamine terminals in striatum, it may be useful in differentiating PD from vascular parkinsonism in which the striatal neuronal pathology is largely postsynaptic.[56] PET and SPECT imaging in which postsynaptic dopamine receptors are labeled show increased uptake in early, untreated PD[57] and reduced uptake in primary striatal diseases such as MSA and PSP.[58]

In summary, a variety of brain imaging techniques may help to differentiate PD from other forms of parkinsonism. However, most studies in this area have compared carefully selected patients with clear-cut clinical features of these disorders. It remains unclear to what extent imaging can differentiate early cases with less well differentiated clinical manifestations.

FUTURE PROSPECTS

At present, the diagnosis of PD continues to rest on clinical criteria, while the criteria that have been proposed remain to be validated by appropriate clinicopathological studies. However, the pathological diagnosis of PD has recently become complicated by the identification of several genes associated with Parkinson's disease, including α-synuclein, Parkin, ubiquitin C-terminal hydrolase, and other gene loci. Although Lewy bodies contain all three of these proteins, young-onset PD due to Parkin is not associated with Lewy body formation. If neuropathological criteria are in a state of flux, the existing clinicopathological studies may be flawed.[13] As with any chronic neurological disorder that historically has been defined largely on clinical grounds, it is likely that accurate biological markers eventually will be identified that will allow for early and accurate detection of disease without the need for cumbersome and less reliable clinical criteria.

REFERENCES

1. deRijk, M. C., Tzourio, C., Breteler, M. M. B., et al., Prevalence of parkinsonism and Parkinson's disease in Europe: The Europarkinson Collaborative Study, *J. Neurol. Neurosurg. Psychiatry*, 62, 10, 1997.

2. Rajput, A. H., Rozdilsky, B., Rajput, A., Accuracy of clinical diagnosis in parkinsonism—A prospective study, *Can. J. Neurol. Sci.*, 18, 275, 1991.

3. Hughes, A. J., Daniel, S. E., Kilford, L., et al., Accuracy of clinical diagnosis of idiopathic Parkinson's disease: A clinico-pathological study of 100 cases, *J. Neurol. Neurosurg. Psychiatry*, 55, 181, 1992.

4. Gibb, W. R., Lees, A. J., The relevance of the Lewy body to the pathogenesis of idiopathic Parkinson's disease, *J. Neurol. Neurosurg. Psychiatry*, 51, 745, 1988.

5. Hughes, A. J., Ben-Shlomo, Y., Daniel, S. E., et al., What features improve the accuracy of clinical diagnosis in Parkinson's disease: A clinicopathologic study, *Neurology*, 42, 1142, 1992.

6. Hughes, A. J., Daniel, S. E., Blankson, S., et al., A clinicopathologic study of 100 cases of Parkinson's disease, *Arch Neurol.*, 50, 140, 1993.

7. Hughes, A. J., Daniel, S. E., Lees, A. J., Improved accuracy of clinical diagnosis of Lewy body Parkinson's disease, *Neurology*, 57, 1497, 2001.

8. Hughes, A. J., Daniel, S. E., Ben-Shlomo, Y., et al., The accuracy of diagnosis of parkinsonian syndromes in a specialist movement disorder service, *Brain*, 125, 861, 2002.

9. Lees, A. J., Problems in diagnosis, in *Parkinson's Disease. Diagnosis and Clinical Management*, Factor, S. A. and Weiner, W. J., Eds., Chapter 24, Demos, New York, 2002.

10. Osaki, Y., Ben-Shlomo, Y., Wenning, G., et al., Do published criteria improve clinical accuracy in multiple system atrophy? *Neurology*, 59, 1486, 2002.

11. Litvan, I., Agid, Y., Calne, D., et al., Clinical research criteria for the diagnosis of progressive supranuclear palsy (Steele-Richardson-Olszewski syndrome): report of the NINDS-SPSP international workshop, *Neurology*, 47, 1, 1996.

12. Litvan, I., MacIntyre, A., Goetz, C. G., et al., Accuracy of the clinical diagnoses of Lewy body disease, Parkinson disease, and dementia with Lewy bodies: A clinico-pathologic study, *Arch Neurol.*, 55, 969, 1998.

13. Litvan, I., Bhatia, K. P., Burn, D. J., et al., SIC Task Force appraisal of clinical diagnostic criteria of parkinsonian disorders, *Mov. Disord.*, 5, 467, 2003.

14. Schrag, A., Ben-Shlomo, Y., Quinn, N., How valid is the clinical diagnosis of Parkinson's disease in the community? *J. Neurol. Neurosurg. Psychiatry*, 73, 529, 2002.

15. Quinn, N., Multiple system atrophy—the nature of the beast. *J. Neurol. Neurosurg. Psychiatry*, Special supplement, 78, 1989.

16. Parkinson Study Group. Effect of deprenyl on the progression of disability in early Parkinson's disease, *N. Engl. J. Med.*, 321, 1364, 1989.

17. Jankovic, J., Rajput, A. H., McDermott, N. P., et al., The evolution of diagnosis in early Parkinson's disease, *Arch Neurol.*, 57, 369, 2000.

18. Calne, D. B., Snow, B. J., Lee, C., Criteria for diagnosing Parkinson's disease, *Ann. Neurol.*, 32, S125, 1992.

19. Pal, P. K., Samii, A., Calne, D. B., et al., Cardinal features of early Parkinson's disease, in *Parkinson's Disease. Diagnosis and Clinical Management,* Factor, S. A., Weiner, W. J., Eds., Chapter 6, Demos, New York, 2002.

20. Larsen, J. P., Dupont, E., Tandberg, E., Clinical diagnosis of Parkinson's disease. Proposal of diagnostic subgroups classified at different levels of confidence, *Acta Neurol. Scand.*, 89, 242, 1994.

21. Gelb, D. J., Oliver, E., Gilman, S., Diagnostic criteria for Parkinson disease, *Arch Neurol.*, 56, 33, 1999.

22. Ward, C., Gibb, W., Research diagnostic criteria for Parkinson's disease, in *Advances in Neurology: Parkinson's Disease: Anatomy, pathology, and therapy*, Streifler, M., Korczyn, A., Melamed, E., Youdim, M., Eds., Raven Press, New York, 1990.

23. Gilman, S., Low, P. A., Quinn, N., et al., Consensus statement on the diagnosis of multiple system atrophy, *J. Neurol. Sci.*, 163, 94, 1999.

24. Quinn, N., Multiple system atrophy, in *Movement Disorders 3*, Marsden, Chapter 13, C. D., Fahn, S., Eds., Butterworths, London, 1994.

25. Colosimo, C., Albanese, A., Hughes, A. J., et al., Some specific clinical features differentiate multiple system atrophy (striatonigral variety) from Parkinson's disease, *Arch Neurol.*, 52, 294, 1995.

26. Litvan, I., Booth, B., Wenning, G. K., et al., Retrospective application of a set of clinical diagnostic criteria for the diagnosis of multiple system atrophy, *J. Neural. Transm.*, 105, 217, 1998.

27. Wenning, G. K., Ben-Shlomo, Y., Hughes, A. J., et al., What clinical features are most useful to distinguish definite multiple system atrophy from Parkinson's disease? *J. Neurol. Neurosurg. Psychiatry*, 68, 434, 2000.

28. McKeith, I. G., Galasko, B., Kosaka, K., et al., Consensus guidelines for the clinical and pathologic diagnosis of dementia with Lewy bodies (DLB): report of the consortium on the DLB International Workshop, *Neurology*, 47, 1113, 1996.

29. Mark, M. H., Sage, J. I., Dickson, D. W., et al., Levodopa-non responsive Lewy body parkinsonism: clinicopathologic study of two cases, *Neurology*, 42, 1323, 1992.

30. Wenning, G. K., Ben-Shlomo, Y., Magalhaes, M., et al., Clinical features and natural history of multiple system atrophy, *Brain*, 117, 835, 1994.

31. Hughes, A. J., Colosimo, C., Kleedorfer, B., et al., The dopaminergic response in multiple system atrophy, *J. Neurol. Neurosurg. Psychiatry*, 55, 1009, 1992.

32. Kim, J. M., Lee, K. H., Choi, Y. L., et al., Levodopa-induced dyskinesia in an autopsy-proven case of progressive supranuclear palsy, *Mov. Disord.*, 17, 1089, 2002.

33. Albanese, A., Bonuccelli, U., Brefel, C., et al., Consensus statement on the role of acute dopaminergic challenge in Parkinson's disease, *Mov. Disord.*, 16, 197, 2001.

34. Hughes, A. J., Apomorphine test in the assessment of parkinsonian patients: A meta-analysis, in *Parkinson's Disease: Advances in Neurology*, Vol. 80, chapter 47, Stern, G. M., Ed., Lippincott Williams and Wilkins, Philadelphia, 1990.

35. Clarke, C. E. Davies, P., Systemic review of acute levodopa and apomorphine challenge tests in a diagnosis of idiopathic Parkinson's disease, *J. Neurol. Neurosurg. Psychiatry*, 69, 590, 2000.

36. Langston, J. W., Widener, H., Goetz, C. G., et al., Core assessment program for intracerebral transplantations (CAPIT), *Mov. Disord.*, 7, 2, 1992.

37. Defer, L., Widener, H., Marie, R. M., et al., Core assessment program for surgical interventional therapies in Parkinson's disease (CAPSIT-PD), *Mov. Disord.*, 14, 572, 1999.

38. Hughes, A. J., Lees, A. J., Stern, G. M., Challenge tests to predict the dopaminergic response in untreated Parkinson's disease, *Neurology*, 41, 1723, 1991.

39. Gasser, T., Schwarz, J., Arnold, G., et al., Apomorphine test for dopaminergic responsiveness in patients with previously untreated Parkinson's disease, *Arch Neurol.*, 49, 1131, 1992.

40. Bonuccelli, U., Piccini, P., Del Dotto, P., et al., Apomorphine test in de novo Parkinson's disease, *Funct. Neurol.*, 7, 295, 1992.

41. Hughes, A. J., Lees, A. J., Stern, G. M., Apomorphine test to predict dopaminergic responsiveness in parkinsonian syndromes, *Lancet*, 336, 32, 1990.

42. Zappia, M., Montesanti, R., Colao, R., et al., Short-term levodopa test assessed by movement time accurately predicts dopaminergic responsiveness in Parkinson's disease, *Mov. Disord.*, 12, 103, 1997.

43. Merello, N., Nouzeilles, M. I., Piran Arce, G., et al., Accuracy of acute levodopa challenge for clinical prediction sustained long-term levodopa response as a

major criterion for idiopathic Parkinson's disease diagnosis, *Mov. Disord.*, 17, 795, 2002.

44. Gilman, S., Low, P. A., Quinn, N., et al., Consensus statement on the diagnosis of multiple system atrophy, *J. Auton. Nerve Syst.*, 74, 189, 1998.

45. Lang, A. E., Lozano, A. M., Parkinson's disease, *N. Engl. J. Med.*, 339, 1044, 1998.

46. Schocke, M. F. H., Seppi, K., Esterhammer, R., et al., Diffusion-weighted MRI differentiates the Parkinson variant of multiple system atrophy from PD, *Neurology*, 58, 575, 2002.

47. Seppi, K., Schocke, M. F. H., Esterhammer, R., et al., Diffusion-weighted imaging discriminates progressive supranuclear palsy from PD, but not from the parkinson variant of multiple system atrophy, *Neurology*, 60, 922, 2003.

48. Davie, C. A., Wenning, G. K., Barker, G. J., et al., Differentiation of multiple system atrophy from idiopathic Parkinson's disease using proton magnetic resonance spectroscopy, *Ann. Neurol.*, 37, 204, 1995.

49. Tedeschi, G., Litvan, I., Bonavita, S., et al., Proton magnetic resonance spectroscopic imaging in progressive supranuclear palsy, Parkinson's disease and corticobasal degeneration, *Brain*, 120, 1541, 1997.

50. Holshouser, B. A., Komu, M., Moller, H. E., et al., Localized proton NMR spectroscopy in the striatum of patients with idiopathic Parkinson's disease: a multicenter pilot study, *Magn. Reson. Med.*, 33, 589, 1995.

51. Ellis, C. M., Lemmens, G., Williams, C. S. R., et al., Changes in putamen N-acetylaspartate and choline ratios in untreated and levodopa-treated Parkinson's disease: a proton magnetic resonance spectroscopy study, *Neurology*, 49, 438, 1997.

52. Schwarz, J., Linke, R., Kerner, M., et al., Striatal dopamine transporter binding assessed by [I–123] IPT and single photon emission computed tomography in patients with early Parkinson's disease, *Arch Neurol.*, 57, 205, 2000.

53. Parkinson Study Group. A multicenter assessment of dopamine transporter imaging with DOPASCAN/SPECT in parkinsonism, *Neurology*, 55, 1540, 2000.

54. Eidelberg, D., Takikawa, S., Moeller, J. R., et al., Striatal hypometabolism distinguishes striatonigral degeneration from Parkinson's disease, *Ann. Neurol.*, 33, 518, 1993.

55. Antonini, A., Kazumata, K., Feigin, A., et al., Differential diagnosis of parkinsonism with [18F] fluorodeoxyglucose and PET, *Mov. Disord.*, 13, 268, 1998.

56. Gerschlager, W., Bencsits, G., Pirker, W., et al., [123I] beta-CIT SPECT distinguishes vascular parkinsonism from Parkinson's disease, *Mov. Disord.*, 17, 518, 2002.

57. Antonini, A., Schwarz, J., Oertel, W. H., et al., Long term changes of striatal dopamine D-2 receptors in patients with Parkinson's disease: a study with positron emission tomography and [11C] raclopride, *Mov. Disord.*, 12, 33, 1997.

58. Schwarz, J., Tatsch, K., Arnold, G., et al., 123I—Iodobenzamide-SPECT in 83 patients with de novo parkinsonism, *Neurology*, 43, S17, 1993.

46 Neuroleptic-Induced Movement Disorders

Manuchair Ebadi

Departments of Pharmacology and of Neurosciences, University of North Dakota School of Medicine and Health Sciences

CONTENTS

0-8493-1590-5/05/$0.00+$1.50
© 2005 by CRC Press

ABSTRACT

Parkinsonism, tremor, chorea-ballismus, dystonia, tardive dyskinesia, myoclonus, tics, and akathisia can be induced by many drugs. The drugs that are most frequently implicated in movement disorders are antipsychotics, calcium channel antagonists, orthopramides and substituted benzamides (e.g., metoclopramide, sulpiride, clebopride, domperidone), CNS stimulants, antidepressants including the selective serotonin uptake inhibitors, anticonvulsants, antiparkinsonian drugs, and lithium. Moreover, extrapyramidal reactions (EPR) have also been reported to occur with the selective serotonin-reuptake inhibitors, and motor dysfunction is caused by *tacrine*.[1-17] It is possible for a single drug, such as one of the antipsychotics, to induce two or more types of movement disorders in the same patient. Movement disorders are not always reversible after drug withdrawal.

Strong positive correlations exist between hyperkinetic forms of four extrapyramidal syndromes (EPS), tardive dyskinesia, parkinsonism, akathisia, and tardive dystonia. More specifically, the probability of having akathisia, which is often neglected or misdiagnosed, is markedly increased in a patient suffering from tardive dyskinesia.

Furthermore, it is quite common for chronic psychiatric inpatients to suffer from combinations of EPS. Therefore, it is definitely advisable that neurologists and psychiatrists dealing with such patient groups should be familiar with treatment strategies for minimizing these EPS and should regularly check on the state of the EPS.

Another group of drugs that physicians must deal with, which are also commonly associated with neurologic complications, are the "street drugs." Most of these agents modulate central neurotransmitters, and some have direct cerebrovascular effects. These characteristics are the bases of their potential to produce neurologic symptoms. Of these agents, the most notorious for its neurologic effects and also one of the most commonly used is *cocaine*.

In addition to nonhemorrhagic infarctions, subarachnoid hemorrhage, intraparenchymal hemorrhage, and intraventricular hemorrhage;[18] single seizures, multiple seizures, and status epilepticus;[19,20] migraine-like headache caused by blockade of serotonin uptake mechanism;[21] optic neuropathy associated with osteolytic sinusitis;[22] and acute femoral neuropathy,[23] cocaine produces acute dystonia during administration[24] and after withdrawal.[24,25] In addition, cocaine-induced tics can occur in first-time and chronic users.[26] The tics may be multifocal and both vocal and motor in character. In some instances, the tics merely represent an uncovering of symptoms in a patient previously diagnosed as having *Tourette's syndrome*, but in others, tics occur for the first time during cocaine use. Abstinence from cocaine usually results in resolution of this syndrome.

Ecstasy is the name commonly used for 3,4-methylenedioxymethamphetamine (MDMA), a ring-substituted amphetamine derivative. Although patented in 1914 as an appetite suppressant, the drug did not become popular until the 1970s, when it was marketed as an adjunct to psychotherapy because of its effects in lowering the defensiveness of the patient, thus breaking the barriers between the patient and the therapist. Because of concerns for its abuse potential and reports of neurotoxicity in animal studies, it was declared illegal in the mid-1980s. Despite its well-established neurotoxicity in animals, the acute or chronic effects of this drug have not been well studied in humans. Its recreational use, especially on college campuses, and reports of cases associated with severe toxicity and death have increased awareness of the drug.

Overlapping symptoms of neuroleptic malignant syndrome and serotonin syndrome have occurred in patients taking MDMA.[27] Recognition of the potential neurologic complications of either prescription or illicit drugs is extremely important. Familiarity with the neurologic symptoms that can result from prescription drugs used to treat neuropsychiatric patients makes it easier to determine whether a given neurologic finding is a drug effect or part of an underlying syndrome for which the drug has been prescribed.

In this chapter, a brief description of the types of abnormal movements seen by physicians is provided, followed by a comprehensive discussion of drug-induced movement disorders.

TYPES OF ABNORMAL MOVEMENTS

Movement disorders (sometimes called *extrapyramidal disorders*) impair the regulation of voluntary motor activity without directly affecting strength, sensation, or cerebellar function. They include *hyperkinetic disorders* associated with abnormal, involuntary movement. Movement disorders result from dysfunction of deep subcortical gray matter structures termed the *basal ganglia*. While there is no universally accepted anatomic definition of the basal ganglia, for clinical purposes, they may be considered to comprise the caudate nucleus, putamen, globus pallidus, subthalamic nucleus, and substantia nigra. The putamen and globus pallidus are collectively termed the *lentiform nucleus*; the combination of lentiform nucleus and caudate nucleus is designated the *corpus striatum*.[28] Abnormal movements can be classified as *tremor, chorea, athetosis* or *dystonia, ballismus, myoclonus,* or *tics*.[29]

TREMOR

A tremor is a rhythmic oscillatory movement best characterized by its relationship to voluntary motor activity, i.e., according to whether it occurs at rest, during maintenance of a particular posture, or during movement. Tremor that occurs when the limb is at rest is generally referred to as *static tremor*, or *resting tremor*. If present during sustained posture, it is called a *postural tremor*. While this tremor may continue during movement, movement does not increase its severity. When present during movement but not at rest, a tremor is generally called an *intention tremor*. Both postural and intention tremors are also called *action tremors*. The causes of tremor are indicated in Table 46.1.

CHOREA

The word *chorea* denotes rapid, irregular muscle jerks that occur involuntarily and unpredictably in different parts of the body. In florid cases, the often forceful involuntary movements of the limbs and head and the accompanying facial grimacing and tongue movements are unmistakable. Voluntary movements may be distorted by the superimposed involuntary ones. In mild cases, however, patients may exhibit no more than a persistent restlessness and clumsiness. Power is generally full, but there may be difficulty in maintaining muscular contraction such that, for example, hand grip is relaxed intermittently (milkmaid grasp). The gait becomes irregular and unsteady, with the patient suddenly dipping or lurching to one side or the other (dancing gait). Speech often becomes irregular in volume and tempo and may be

---

TABLE 46.1
Major Causes of Tremor

Postural tremor

Physiologic tremor

Enhanced physiologic tremor

    Anxiety or fear

    Excessive physical activity or sleep deprivation

    Sedative drug or alcohol withdrawal

    Drug toxicity (e.g., lithium, bronchodilators, tricyclic antidepressants)

    Heavy metal poisoning (e.g., mercury, lead, arsenic)

    Carbon monoxide poisoning

    Thyrotoxicosis

Familial (autosomal dominant) or idiopathic (benign essential) tremor

    Cerebellar disorders

    Wilson's disease

Intention tremor

Brain stem or cerebellar disease

Drug toxicity (e.g., alcohol, anticonvulsants, sedatives)

Wilson's disease

Rest tremor

Parkinsonism

Wilson's disease

Heavy metal poisoning (e.g., mercury)

Four separate groups of symptoms are now described as part of the symptom complex of parkinsonism. These groups are *tremor, akinesia, rigidity,* and *loss of normal postural reflexes.*

The tremor consists of rhythmatically alternating contractions of a given muscle group and of its antagonists. It is insidious in onset. Most commonly, the tremor affects the distal parts of the extremity earlier and to a greater extent than the proximal parts. It is most prominent in the fingers, often less prominent in the wrists, and involves the forearm or upper arm only infrequently. The rate of the tremor averages about three to five oscillations per second. A number of descriptive terms such as pill-rolling or cigarette-rolling have been used to describe these movements, but the rotary component is often lacking, so the term *to-and-fro* is more applicable. The tremor can also involve the leg, where again it is usually more marked distally in the foot than it is proximally in the hip. The head, jaw, and pectoral structures can also become involved.

While the manifestations of the disease invariably involve the entire body, the symptoms can be markedly asymmetric. This is especially true of the tremor, which frequently begins in one arm or leg and can remain predominantly unilateral for several years. It can become disabling on one side while the other side is affected only slightly.

One of the classic features of this tremor is its presence during rest and its disappearance on purposeful movement. The tremor usually is aborted by the initiation of any willed act but tends to reappear a few moments later, despite the continuation of the action. In most cases, however, the tremor is less prominent on action than it is during rest. The tremor is characteristically absent during sleep. All extrapyramidal hyperkinesias stop during sleep with the exception of certain cases of hemiballismus. A number of psychologic factors increase the tremor. These include fatigue, cold, emotional stress of any sort, and almost anything that makes the patient nervous, including visits to his doctor. (For a review and reference, see Reference 29.)

explosive in character. In some patients, athetotic movements or dystonic posturing may also be prominent. Chorea disappears during sleep. Table 46.2 shows the major causes of chorea.

DYSTONIA AND ATHETOSIS

The term *athetosis* generally denotes abnormal movements that are slow, sinuous, and writhing in character. When the movements are so sustained that they are better regarded as abnormal postures, the term *dystonia* is used, and the terms are often used interchangeably. The abnormal movements and postures may be generalized or restricted in distribution. In the latter circumstance, one or more of the limbs may be affected (*segmental dystonia*), or the disturbance may be restricted to localized muscle groups (*focal dystonia*). The causes of dystonia and athetosis include static perinatal encephalopathy (cerebral palsy), Wilson's disease, Huntington's disease, Parkinson's disease, encephalitis lethargia drugs (levodopa, antipsychotic drugs), ischemic anoxia, focal intracranial disease, progressive supranuclear palsy, idiopathic torsion dystonia (hereditary, sporadic), and formes frustes or idiopathic torsion dystonia.

HEMIBALLISMUS

Hemiballismus is unilateral chorea that is especially violent because the proximal muscles of the limbs are involved. It is due most often to vascular disease in the contralateral subthalamic nucleus and commonly resolves spontaneously in the weeks following its onset. It is sometimes due to other types of structural disease, and it was an occasional complication of thalamotomy. Pharmacologic treatment is similar to that for chorea.

MYOCLONUS

Myoclonic jerks are sudden, rapid, and twitch-like muscle contractions. They can be classified according to their distribution, relationship to precipitating stimuli, or etiol-

---

TABLE 46.2
Causes of Chorea

Heredity
  Huntington's disease
  Benign hereditary chorea
  Wilson's disease
  Paroxysmal choreoathetosis
  Familial chorea with associated acanthocytosis
Static encephalopathy (cerebral palsy) acquired antenatally or perinatally (e.g., from anoxia, hemorrhage, trauma, kernicterus)
Sydenham's chorea
Chorea gravidarum
Drug toxicity
  Levodopa and other dopaminergic drugs
  Antipsychotic drugs
  Lithium
  Phenytoin
  Oral contraceptives
Miscellaneous medical disorders
  Thyrotoxicosis, hypoparathyroidism, or Addison's disease
  Hypocalcemia, hypomagnesemia, or hypernatremia
  Polycythemia vera
  Hepatic cirrhosis
  Systemic lupus erythematosus
  Encephalitis lethargia
Cerebrovascular disorders
  Vasculitis
  Ischemic or hemorrhagic stroke
  Subdural hematoma
Structural lesions of the subthalamic nucleus

ogy. *Generalized myoclonus* has a widespread distribution, while *focal* or *segmental myoclonus* is restricted to a particular part of the body. Myoclonus can be spontaneous, or it can be brought on by sensory stimulation, arousal, or the initiation of movement (*action myoclonus*). Myoclonus may occur as a normal phenomenon (*physiologic myoclonus*) in a healthy person, as an isolated abnormality (*essential myoclonus*), or as a manifestation of epilepsy (*epileptic myoclonus*). It can also occur as a feature of a variety of degenerative, infectious, and metabolic disorders (*symptomatic myoclonus*). The causes of general myoclonus are shown in Table 46.3.

TABLE 46.3
Causes of General Myoclonus

Physiologic myoclonus

  Nocturnal myoclonus

  Hiccup

Essential myoclonus

Epileptic myoclonus

Symptomatic myoclonus

  Degenerative disorders

    Dentatorubrothalamic atrophy (Ramsay Hunt syndrome)

    Storage diseases (e.g., Lafora body disease)

    Wilson's disease

    Huntington's disease

    Alzheimer's disease

  Infectious disorders

    Creutzfeldt-Jakob disease

    AIDS dementia complex

    Subacute sclerosing panencephalitis

  Metabolic disorders

Drug intoxications (e.g., penicillin, antidepressants, anticonvulsants)

    Drug withdrawal (ethanol, sedatives)

    Hypoglycemia

    Hyperosmolar nonketotic hyperglycemia

    Hyponatremia

    Hepatic encephalopathy

    Uremia

    Hypoxia

Myoclonic jerks are sudden, rapid, twitch-like muscle contractions. They can be classified according to their distribution, relationship to precipitating stimuli, or etiology. *Generalized myoclonus* has a widespread distribution, while *focal* or *segmental myoclonus* is restricted to a particular part of the body. Myoclonus can be spontaneous, or it can be brought on by sensory stimulation, arousal, or the initiation of movement (*action myoclonus*). Myoclonus may occur as a normal phenomenon (*physiologic myoclonus*) in healthy persons, as an isolated abnormality (*essential myoclonus*), or as a manifestation of epilepsy (*epileptic myoclonus*). It can also occur as a feature of a variety of degenerative, infectious, and metabolic disorders (*symptomatic myoclonus*). (For a review and reference, see Reference 29.)

TICS

Tics are sudden, recurrent, quick, coordinated abnormal movements that can usually be imitated without difficulty. The same movement occurs again and again and can be suppressed voluntarily for short periods, although doing so may cause anxiety. Tics tend to worsen with stress, diminish voluntary activity or mental concentration, and disappear during sleep. Tics can be classified into four groups, depending on whether they are simple or multiple and transient or chronic. *Transient simple tics* are very common in children, usually terminate spontaneously within 1 year (often within a few weeks), and generally require no treatment. *Chronic simple tics* can develop at any age but often begin in childhood, and treatment is unnecessary in most cases. The benign nature of the disorder must be explained to the patient. *Persistent simple* or *multiple tics* of childhood or adolescence generally begin before age 15 years. There may be single or multiple motor tics, and often vocal tics, but complete remission occurs by the end of adolescence. The syndrome of *chronic multiple motor and vocal tics* is generally referred to as *Gilles de la Tourette's syndrome*, after the French physician who was one of the first to describe its clinical features.[29]

BASAL GANGLIA AND MOVEMENT DISORDERS

Great strides have been made in the last five decades toward elucidating the neurochemistry of the pathways involved in motor function and movement disorders.[28,30–35] These pathways include those connecting motor cortex, brain stem, basal ganglia, and spinal cord. The basal ganglia play an important role in the control of movement and complex motor behavior. Figure 46.1 shows a simplified schematic diagram of the primary connections of the basal ganglia. Each area of cerebral cortex, from the most primitive olfactory structures to the most highly organized association cortex, has projections to one of a number of deep telencephalic gray matter nuclei, which include the putamen, caudate nucleus, nucleus accumbens, and the outer layers of the olfactory tubercle. These nuclei all have similar histochemical appearances and neurochemical properties. In addition, each of these nuclei also gets input from dopaminergic neurons in the midbrain and from the interlaminar nuclei of the thalamus. The caudate nucleus, putamen, nucleus accumbens, and outer tubercle can comprise the striatum.

The principal components of the basal ganglia are the striatum, the pallidum, the substantia nigra, and the subthalamic nucleus. The basal ganglia are neither a major sensory relay nor a coordinating neuronal system, such as the cerebellum, and they do not have direct access to the motor neurons of the spinal cord. Because

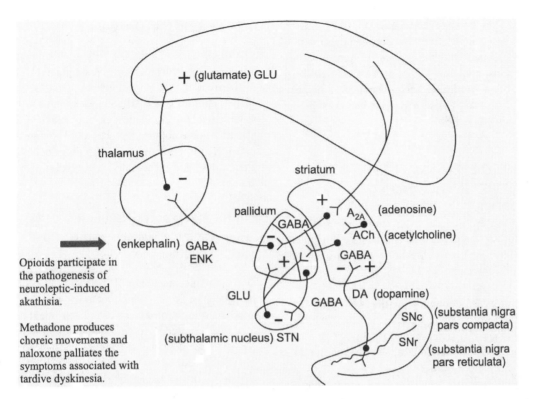

Opioids participate in the pathogenesis of neuroleptic-induced akathisia.

Methadone produces choreic movements and naloxone palliates the symptoms associated with tardive dyskinesia.

FIGURE 46.1   The basic circuit of basal ganglia. The major subcortical input to area 6 arises in a nucleus of the dorsal thalamus, called the *ventral lateral nucleus* (*VL*). The input to this part of VL, called VLo, arises from the basal ganglia buried deep within the telencephalon. The basal ganglia, in turn, are targets of the cerebral cortex, particularly the frontal, prefrontal, and parietal cortex. Thus, we have a loop where information cycles from the cortex through the basal ganglia and thalamus and then back to the cortex, particularly the supplementary motor area. One of the functions of this loop appears to be the selection and initiation of willed movements.

The basal ganglia consist of the *caudate nucleus*, the *putamen*, the *globus pallidus*, and the *subthalamus*. In addition, we can add the *substantia nigra*, a midbrain structure that is reciprocally connected with the basal ganglia of the forebrain. The caudate and putamen together are called the *striatum*, which is the target of the cortical input to the basal ganglia. The globus pallidus is the source of the output to the thalamus. The other structures participating in various side loops that modulate the direct path are:

Cortex → Striatum → Globus pallidus → VLo → Cortex (SMA)

The neurons of the striatum appear randomly scattered, with no apparent order such as that seen in the layers of the cortex. But this bland appearance hides a degree of complexity in the organization of the basal ganglia that we are only now beginning to appreciate. It appears that the basal ganglia participate in a large number of parallel circuits, only a few of which are strictly motor. Other circuits are involved in certain aspects of memory and cognitive function. (For reviews and references, see References 28 and 30 through 35.)

they lie just under the cerebral cortex and directly among the flow of corticifugal fibers, the basal ganglia are ideally located "to interact or act" in conjunction with the cerebral cortex.

Anatomically, this set of subcortical structures is involved in a closed corticobasal ganglia-thalamo-cortical loop, whose major axis is composed of sequentially arranged elements, namely the striatum, the globus pallidus or pallidum, the substantia nigra, and the ventral tier nuclei of the thalamus. Despite their privileged relationship with the cortex, it is not yet known how the basal ganglia complement the function of the cerebral cortex (for a review, see Reference 34). In addition to the anatomical substrate that allows information from the cerebral cortex to flow along the corticobasal ganglia-thalamo- cortical loops, there exist other structures that exert a profound modulatory influence upon the activity of the core structures of the basal ganglia. These structures are the subthalamic nucleus, the pars compacta of the substantia nigra, the centromedian/parafascicular thalamic complex, the dorsal raphe nucleus, and pedunculopontine tegmental nucleus (for a review, see Reference 36). These ancillary structures provide the basal ganglia with a wide variety of neurochemical inputs: (a) the subthalamic nucleus and the centromedian/parafascicular thalamic complex provide a glutamatergic entry;[37–39] (b) the dorsal raphe nucleus is the origin of a serotoninergic input;[40] (c) the pedunculopontine tegmental nucleus gives rise to a dual cholinergic and glutamatergic afferent;[41,42] and the substantia nigra is a major source of dopamine at basal ganglia levels.[43] Each element of the main axis of the basal ganglia is thus the focus of a highly complex interplay between these various chemospecific inputs and the striatal afferents that use GABA as a transmitter (see Figure 46.1). Diseases affecting one or more of the marked neurochemical imbalances are the hallmark of several basal ganglia disorders, including Parkinson's disease.[29,44]

DOPAMINERGIC TRANSMISSION INVOLVED IN MOVEMENT DISORDERS

Drugs causing movement disorders influence dopaminergic transmission. The main dopaminergic neurons in the brain are the following:

1. The ultrashort dopaminergic fibers, such as the interplexiform amacrine-like neurons, which link inner and outer plexiform layers of the retina, and the periglomerular dopamine cells of the olfactory bulb
2. The intermediate-length dopaminergic fibers, such as tuberohypophysial dopamine cells, incertohypothalamic neurons, and the medullary peri ventricular neurons
3. The long dopaminergic fibers linking the ventral tegmental and substantia nigra dopamine cells with three principal sets of targets: the neostriatum (principally the caudate and putamen); the limbic cortex (medial prefrontal, cingulate, and entorhinal areas); and other limbic structures (the regions of the septum, olfactory tubercle, nucleus accumbens septi, amygdaloid complex, and piriform cortex)

These latter two groups have been termed the mesocortical and mesolimbic dopamine projections, respectively.[45]

THE MESOLIMBIC- AND MESOLIMBIC-CORTICAL DOPAMINE PATHWAYS

These pathways originate primarily from the A10 dopamine neuron group (ventral tegmental area). The mesolimbic tract mainly innervates the nucleus accumbens and olfactory tubercle and is considered to be involved in arousal, locomotor activity, and motivational and affective states. The mesolimbic-cortical pathway innervates septum, hippocampus, amygdala, and many cortical regions (such as the prefrontal and cingulate cortices) and is important in higher cortical functions. It has been suggested that blockade of, in particular, limbic and prefrontal dopamine D_2 receptors might be the mode of action for the therapeutic effects of antipsychotic compounds.

THE NIGROSTRIATAL DOPAMINE PATHWAY

The nigrostriatal dopamine pathway originates from the A9 dopamine neuron group (substantia nigra) and projects primarily to the striatum (nucleus caudatus, putamen, and globus pallidus). The striatum is thought to be critically involved in the regulation of movement and may also subserve some cognitive processes. The nigrostriatal tract is believed to be associated with the production of extrapyramidal side-effects by antipsychotic drugs[46] (see Figure 46.2).

NEUROLEPTIC-INDUCED REGULATION OF DOPAMINE-RECEPTOR SUBTYPES AND ITS IMPLICATION IN SCHIZOPHRENIA

Dopamine receptors (DA-R) belong to the G protein-coupled receptor family, which includes many receptors such as adrenergic, serotoninergic, and neuropeptidergic receptors. The common structural features of the G protein-coupled receptors are (a) the seven hydrophobic transmembrane domains, (b) the extracellular N-terminus domain with glycosylation sites, (c) the cytoplasmic C-tenninus domain, and (d) the G protein coupling sites in the third cytoplasmic loop. All the known DA-R subtypes consist of a polypeptide chain containing about 400 amino acids (–50 kDa) and carbohydrate chains (for review, see References 47–52). The size of most receptor molecules detected with anti-D_2-R antibodies (anti-D_2-R) varies within the range of 90 to 120 kDa, depending on the tissues,[53,54] which is far larger than the molecular weight expected from the amino acid sequence. Therefore, D_2-R is likely to contain carbohydrate chains of various sizes. Although these carbohydrate chains are believed to have no effect on ligand affinity, it is important to clarify their roles in the receptor function.

Dopamine receptors were initially classified into various subtypes on the basis of pharmacological properties, but more recently they have been grouped into two types: the D_1-R group, which activates adenylate cyclase, and the D_2-R group, which inhibits (or has no effect on) the activity of adenylate cyclase.[55,56] Cloning of receptor genes in recent years has led to the identification of new subtypes not previously identified by conventional pharmacological and biochemical methods. At present, there are five DA-R subtypes, and they are classified into the D_1-R family (D_1-R and D_5-R) and D_2-R family (D_2-R, D_3-R and D_4-R) based on their structures and pharmacological features. The third cytoplasmic loop is "short" in the D_1-R family and "long" in the D_2-R family. It is generally believed that receptors with a short third cytoplasmic loop couple to stimulatory G proteins (G_s) and activate adenylate cyclase. On the other hand, receptors with a long third cytoplasmic loop react to G_1 and G_o, which inhibit adenylate cyclase, and G_q, which couples with phospholi-

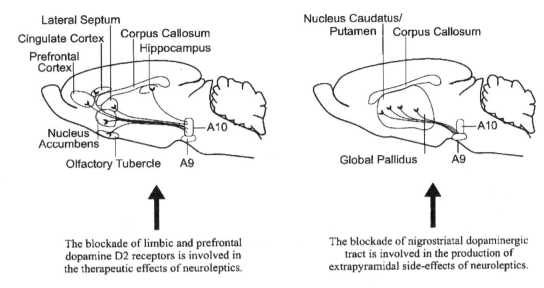

The blockade of limbic and prefrontal dopamine D2 receptors is involved in the therapeutic effects of neuroleptics.

The blockade of nigrostriatal dopaminergic tract is involved in the production of extrapyramidal side-effects of neuroleptics.

FIGURE 46.2 The main ascending dopaminergic pathways in rat brain. Dopaminergic neurons with intermediate-length axons include the tuberoinfundibular and hypophysial, incertohypothalamic cells, and the medullary peri ventricular group. The tuberoinfundibular neurons have a neurohumoral function; they secrete dopamine into a portal vascular system that supplies the anterior pituitary. This dopamine is responsible for inhibiting secretion of the anterior pituitary hormone, prolactin.

The final subdivision of dopaminergic neurons includes the midbrain groups from the substantia nigra and the ventral tegmental area. These systems have long axons that innervate the basal ganglia, parts of the limbic system, and the frontal cortex. The neostriatal system, which has cell bodies in the substantia nigra, innervates the caudate and putamen. This suggests that dopamine released from neostriatal areas has motor functions. The motor problems associated with Parkinson's disease are caused by a decrease in dopamine in these areas. Administration of the dopamine precursor L-Dopa bypasses tyrosine hydroxylase and alleviates some of the motor disturbances of Parkinson's disease. The specificity and complexity of the dopaminergic systems is further demonstrated by the mesolimbic system. These neurons originate in the ventral tegmental area of the midbrain, next to the substantia nigra. Long axons from these neurons project to many parts of the limbic system, including the nucleus accumbens, olfactory tubercle, septum, amygdala, and limbic cortex (e.g., frontal and cingulate cortex). These areas are associated with mood alterations and cognitive function, indicating another important role of central dopamine. The nucleus accumbens is involved with reward, and the release of dopamine in this area provides positive feelings of reinforcement. (For review and reference, see Reference 46.)

pase C.[57] D_2-R also activates K^+ channels.[57] While the structures of the extra- and intracellular loops of the DA-R vary with each receptor, the trans-membrane domains are highly homologous among most receptors. The subtypes belonging to the D_1-R and D_2-R families show overall sequence homology of about 50% within the families and 30% between the families. It is believed that an aspartate in the third transmembrane domain forms an ion pair with the protonated amine group of DA and that two serines in the fifth transmembrane domain form a hydrogen bonding interaction with two phenol groups of DA.[57,58] The latter interaction is specific for DA and its agonist. On the other hand, an aspartate in the second transmembrane domain of D_2-R has been shown to interact with antagonists.[59] In humans, the genes of the five DA-R subtypes are located on different chromosomes. In general, the genes for G protein-coupled receptors have no introns. The genes of the D_1-R family (D_1-R and D_5-R) also lack introns. On the other hand, a specific feature of the genes of the D_2-R family is the presence of introns in their coding regions; the D_2-R, D_3-R, and D_4-R genes have 6, 5, and 4 introns, respectively.[60,61] The presence of these introns strongly suggests that the gene products of each subtype of the D_2-R family undergo post-translational splicing, resulting in a greater number of receptor isoforms.

The dopaminergic hypothesis of schizophrenia postulates that an aberration of the brain's dopamine transmitter systems is key to the pathophysiology of schizophrenia. A cornerstone of this hypothesis, which has guided research in the field of neuropsychiatry for more than four decades, is the observation that therapeutic potency of antipsychotic drugs directly correlates with their affinity for dopamine D_2-R. This observation implies that the different antipsychotics achieve their therapeutic effects at doses that produce similarly high levels of D_2-R occupancy, an effect which, under chronic treatment conditions, can be expected to result in receptor up-regulation.

Indeed, antipsychotic drugs at clinically recommended doses occupy at least 70% of striatal D_2-R and significantly up-regulate these sites.[62] While the dopaminergic hypothesis of schizophrenia has not lost its currency, its original premise has been challenged by the discovery that therapeutically effective doses of the most beneficial atypical antipsychotic, *clozapine*, are significantly smaller than would be predicted on the basis of its relatively low affinity for D_2-R. The D_2- R occupancy of this drug in the striatum is only 50 to 66% of that produced by other antipsychotics, and clozapine does not up-regulate striatal D_2-R. Evidence from studies of receptor occupancy and regulation in postmortem brains of patients with neuropsychiatric disorders and in nonhuman primates is providing new leads in the ongoing quest to understand the pathophysiology and causes of schizophrenia and to develop more effective methods of treatment (Table 46.4).

These studies suggest that the cerebral cortex is the site of action of antipsychotic medications and indicate that chronic treatment with these drugs differentially reg-

TABLE 46.4
Effect of Chronic Treatment with Antipsychotics on the Levels of mRNAs Encoding Different Dopamine Receptor Subtypes in the Cortex and Neostriatum

Drugs	Chemical Class	Receptor Regulation				
		Striatum				
		D_2 long	D_2 short	D_4	D_1	D_5
Antipsychotics—typical						
Chlorpromazine	Phenothiazines	↑	↑	↑	←	←
Haloperidol	Butyrophernones	↑	↑	↑	←	←
Melindone	Indoles	↑	↑	←	←	←
Pimozide	Diphenylbutyl-piperidines	↑	↑	←	←	←
Antipsychotics—atypical						
Clozapine	Dibenzodiazepines	↑	↑	↑	←	←
Olanzapine	Thienobenzodiazepines	↑	↑	↑	←	←
Remoxipride	Substituted benzamides	←	←	←	←	←
Risperidone	Benzisoxazoles	←	←	↑	←	←
Nonantipsychotic D_2 receptor antagonist						
Tiapride	Substituted benzamides	↑	↑	↑	←	←
Antipsychotics—typical						
Chlorpromazine	Phenothiazines	↑	↑	↑	↓	↓
Haloperidol	Butyrophernones	↑	↑	↑	↓	↓
Melindone	Indoles	↑	↑	←	↓	↓
Pimozide	Diphenylbutyl-piperidines	↑	↑	←	↓	↓
Antipsychotics—atypical						
Clozapine	Dibenzodiazepines	↑	↑	↑	↓	↓
Olanzapine	Thienobenzodiazepines	↑	↑	↑	↓	↓
Remoxipride	Substituted benzamides	↑		←	↓	↓
Risperidone	Benzisoxazoles	↑	↑	↑	↓	↓
Nonantipsychotic D_2 receptor antagonist						
Tiapride	Substituted benzamides		←	↑	↓	↓

↑ = increase, ↓ = decrease, and ← = remains the same.

Data have been modified with permission from Lidow et al., 1997a, b.

For about three decades, the dopamine (DA) hypothesis of schizophrenia has been the reigning biological hypothesis of the neural mechanisms underlying this disorder. The DA hypothesis has undergone numerous revisions but has proven remarkably resistant to obliteration. In its original formulation, the hypothesis stated that schizophrenia is due to a central hyperdopaminergic state. This was based on two complementary lines of indirect pharmacological evidence: the DA releaser amphetamine as well as other DA-enhancing agents such as the DA precursor L-dopa or methylphenidate, produced and exacerbated schizophrenic symptoms, whereas drugs that were effective in the treatment of amphetamine-induced psychosis and schizophrenia [neuroleptics or antipsychotic drugs (APDs)] decreased DA activity, and their clinical potency was correlated with their potency in blocking D_2 receptors.

Recently, it has been suggested that schizophrenia may involve a hypodopaminergic state in the dorsal striatum coupled with a hyperdopaminergic state in the dorsal striatum coupled with a hyperdopaminergic state in the ventral striatum modes of DA activity within the prefrontal cortex, i.e., decreased phasic and increased tonic release. (For review and reference, see Reference 62.)

ulates both families of dopamine receptors in this structure. Up-regulation of the cortical dopamine D_2-R is accompanied by a down-regulation of the D_1 sites. Balancing the opposing actions of dopamine D_1 and D_2-R regulation may hold the key to optimal drug therapy and to understanding the pathophysiology of schizophrenia (see Reference 62 and Table 46.4).

THE MODULATORY ACTIONS OF ACETYLCHOLINE, ADENOSINE, GLUTAMATE, AND δ-OPIOID ON STRIATAL DOPAMINERGIC TRANSMISSION

The striatum is viewed as a structure performing fast neurotransmitter-mediated operations through somato-topically organized projections to medium-size spiny neurons. Modulatory influences act indirectly by setting the excitability of the neuron to incoming phasic input mediated by fast neurotransmitter actions. Modulatory influences have relatively long kinetics of action/desensitization, being related to modulation by voltage-operated ion channels as in the case of muscarinic and DA receptors or to operation of voltage-gated ion channels as in the case of N-methyl-D-aspartate (NMDA) receptors. Modulatory influences on neuronal excitability might not even be labeled as facilitatory (+) or inhibitory (−), their actual sign depending on the membrane potential. Modulatory actions can have long-lasting transcriptional effects that might be the basis for adaptive and plastic changes. The caudate-putamen is one of the areas of the brain rich in modulatory receptors such as acetylcholine, adenosine, GABA, glutamate, neurotensin, opioid, substance P, and somatostatin (for review, see References 35 and 63 through 65). A few examples will be cited to support this contention. Parkinson's disease is a disease of extrapyramidal motor function characterized by difficulties in initiating and smoothly sustaining motions. It is associated with severe loss of dopamine-containing neurons in the substantia nigra. Parkinson's disease can be treated with the dopamine precursor L-dopa, but this does not stop disease progression, its effectiveness ultimately decreases, and it may produce psychosis.[66] Parkinson's disease is also associated with a large (approximately 50%) loss of high-affinity nicotine binding sites from the brain.[67,68]

A central role in the modulatory operations taking place in the striatum is played by acetylcholine neurons. Acetylcholine neurons account for 1 to 2% of the striatal neuronal population. In all species examined, they are among the largest neurons of the striatum both for the size of the perikaryon (about 30 μm in its longest dimension) and the area of distribution of the dendritic tree (up to 0.5 mm²). Striatal acetylcholine neurons are interneurons, although a subpopulation of them also projects to neocortex. Striatal acetylcholine neurons receive three major synaptic inputs: (1) from intrinsic medium-size spiny neurons that use *substance P* and GABA as transmitters,

and project to the substantia nigra pars reticulata and entopeduncular nucleus (medial pallidal segment of primates), (2) from extrinsic DA neurons of the mesencephalic tegmentum (A_8, A_9, and A_{10} groups), and (3) from extrinsic excitatory (glutamate) neurons of the intralaminar thalamus (parafascicular complex) and, to a lesser extent, of the cortex. The output of acetylcholine neurons are the medium-size spiny neurons and the medium-size spiny interneurons containing somatostatin/neuropeptide Y (NPY) and neurotensin or GABA, which might be interposed between the acetylcholine neurons and medium-size spiny neurons (see Figure 46.3).

Stimulation of the striatum will evoke a short-latency inhibitory postsynaptic potential in zona compacta dopaminergic neurons and in zona reticulata neurons.[69] However, overstimulation of the striatum will actually cause an activation of DA neuron firing, and the said activation occurs with an inhibition of GABAergic neurons in the zona reticulata that normally inhibit the activity of dopaminergic neurons.[69,70] These GABAergic inhibitory neurons may represent collaterals of nigrothalamic neurons[71] or a short-axon interneuron located near the zona compacta.

Therefore, stimulation of striatum exerts two electrophysiological actions on DA neurons, which are a direct GABAergic inhibition and an indirect disinhibition.[72] The striatum is known to send a large number of GABAergic inhibitory projections to the globus pallidus, which in turn sends GABAergic fibers to the subthalamic nucleus.[73–75] Lesions of the nigrostriatal DA system activate the striato-pallidal pathway, thereby disinhibiting the subthalamus.[76,77] Single-pulse stimulation of the subthalamus has been shown to produce short-latency excitation of both dopaminergic and nondopaminergic neurons within the substantia nigra,[78–80] and glutamatergic excitatory postsynaptic potentials have been associated to DA neurons recorded *in vitro*[72,81,82] (see also Figure 46.3).

The operations performed by cortico-striatal projections and medium-size spiny neurons are regarded to be of a fast-neurotransmitter-like nature. Thus, cortically elicited fast EPSPs recorded from medium-size spiny neurons are mediated by glutamate receptors of the D, L-α-amino-3-hydroxy-5-methyl-4-isoxazole propionic acid (AMPA) subtype; in turn, these neurons use GABA as their fast inhibitory transmitter. Transmission through NMDA receptors can be regarded as modulatory in view of its voltage dependency and of its ability to act as a gain amplifier of excitatory phasic input, thus promoting burst firing. This transmission is largely inoperative in resting striatal medium-size spiny neurons, due to Mg^{2+} blockage of NMDA channels in hyperpolarized conditions. By contrast, acetylcholine neurons are already tonically active and depolarized to near threshold in basal conditions, making NMDA transmission fully operative in acetylcholine neurons and therefore capable of promotion burst fir-

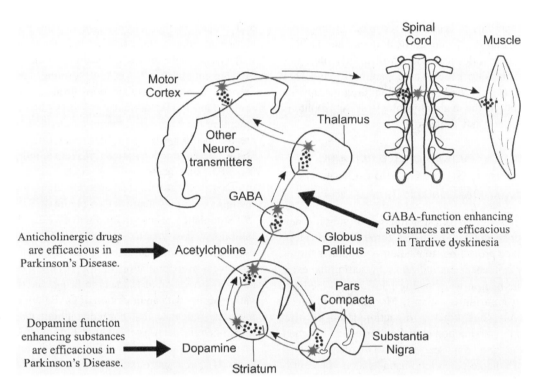

FIGURE 46.3 Striatal cholinergic transmission. Acetylcholine is the familiar transmitter at the neuromuscular junction, at synapses in autonomic ganglia, and at postganglionic parasympathetic synapses. Cholinergic interneurons also exist within the brain, in the striatum and the cortex, for example. In addition, there are two major diffuse modulatory cholinergic systems in the brain, one of which is called the basal forebrain complex. It is a "complex" because the cholinergic neurons lie scattered among several related nuclei at the core of the telencephalon, medial and ventral to the basal ganglia. The best known of these are the medial septal nuclei, which provide the cholinergic innervation of the hippocampus, and the basal nucleus of Meynert, which provides most of the cholinergic innervation of the neocortex. (For reviews and references, see References 69, 81, and 82.)

ing in response to low amplitude excitatory input. Fully active NMDA transmission might thus contribute to a major property of striatal acetylcholine neurons, that being, in contrast with medium-size spiny neurons, exquisitely sensitive to excitatory phasic input. This phenomenon supports the suggestion that, although medium-size spiny neurons express NMDA receptors and receive a massive glutamate projection, they are likely to be influenced by noncompetitive NMDA-receptor antagonists only indirectly as a result of a primary action of acetylcholine neurons.[64]

REGULATION OF CENTRAL DOPAMINERGIC NEURONS BY OPIOID RECEPTORS

Manzanares et al.[63] and Pan[65] have discussed the regulation of central dopaminergic neurons by opioid receptors. Three pharmacologically distinct subtypes of opioid receptors (μ, κ, and σ) have been identified in the CNS. Until recently, the study of the regulation of dopaminergic neurons by subtypes of opioid receptors has been limited by the lack of specific agonists and antagonists for these receptors. The development of selective μ–, κ–, and σ-opioid receptor agonists and antagonists has now permit-

ted exploration of the role played by each of these opioid receptor subtypes in the modulation of central dopaminergic neurons. To date, much of what is known regarding the effects of opioids on dopaminergic neurons has been based on studies using compounds that act at either μ- or σ-opioid receptors (see Figure 46.1).

Activation of μ-opioid receptors stimulates mesolimbic, nigrostriatal, and incertohypothalamic dopaminergic neurons; inhibits tuberoinfundibular dopaminergic neurons; and has no effect on periventricular-hypophysial dopaminergic neurons. On the other hand, activation of κ-opioid receptors inhibits the basal activity of peri ventricular-hypophysial dopaminergic neurons and the pharmacologically stimulated activities of mesolimbic, nigrostriatal, and tuberoinfundibular dopaminergic neurons. In contrast, comparatively less is known about the role of σ-opioid receptors in regulating dopaminergic neuronal systems in the brain. Administration of the σ-opioid selective agonist [D-Pen², D-Pen⁵] enkephalin (OPDPE) increases the release of DA in the nucleus accumbens but has no effect in the striatum. No information is available regarding the effects of activation or blockade of σ-opioid receptors on the activities of hypothalamic dopaminergic neurons.

The Striatal Blockade of the Adenosine A2A Receptor in Parkinson's Disease

The GABA-enkephalin neurons (see Figure 46.1) are excited by cortical inputs and inhibited by recurrent collaterals (via $GABA_A$ receptors). In Parkinson's disease, the feedback inhibition of striatal neurons by the recurrent collaterals may be insufficient to control the overactivity of these neurons. This overactivity probably arises from a reduction in the DA-R-mediated control of acetylcholine and glutamate release (from cortical and thalamic afferents) and in D_2-R-mediated inhibition of striatal neurons. Such overactivity may also be partly due to reduced D_1-R-mediated stimulation of striatal GABA release, and to the action of muscarinic acetylcholine receptors that stabilize an excitable state of the striatal neurons. It has been recently shown that blockage of the A_{2A} receptor increases the release of GABA from striatal synaptosomes and increases inhibitory input into medium spiny neurons. Therefore, A_{2A} receptor blockage serves to increase the GABA-mediated feedback control of the striatal output neurons.

This mode of action is intrinsically different from that of the DA-related modulators used in most parkinsonian therapies, as feedback inhibition is a function of the degree of excitation of individual striatal neurons. Thus, the effect of an A_{2A} receptor antagonist may be restricted to the control of the recurrent collaterals of those neurons that are highly active. The reduction in acetylcholine release caused by A_{2A} receptor blockage may also contribute to this inhibition of medium spiny neurons, by reducing cholinergic stimulation and thus the ability of acetylcholine to maintain the excitable state of the output neurons. In the absence of DA, the facilitatory drive of the D_1 receptors on the striatal GABAergic, substance P-containing neurons of the direct pathway is also lost, causing these cells to become less excitable. Although the A_{2A} receptor probably has no direct effect on this pathway, it opposes the behavioral effects of D_1-R stimulation, striatal neurons and the cholinergic interneurons.[35]

THE NEUROPATHOLOGY OF MOVEMENT DISORDERS

The term *extrapyramidal system* to be described in the following section was first coined by Samuel Alexander Kinnier Wilson in 1912 in describing the neurological disorder of hepatolenticular degeneration (*Wilson's disease*). Lesions in the pyramidal system, extrapyramidal system, or cerebellar system result in distinctive disturbances of motor activity.[29,83]

The Pyramidal System

The pyramidal tract derives its name from the fact that constituent fibers pass through the *medullary pyramid*, a prominent bulge on the ventral surface of the medulla oblongata. Its fibers arise in the cerebral cortex, principally from that area around the central sulcus that constitutes the motor cortex. The fibers then descend through the brain stem and, after a partial decussation in the medulla, continue through the spinal cord and finally terminate about the lower motor neurons. Injury to this tract produces paralysis of voluntary movement. As a result, the *pyramidal system is believed to be concerned with the initiation of voluntary movements.*

The Extrapyramidal System

The extrapyramidal system is made up of a number of paired nuclei and associated pathways. The major structures of this system include putamen, globus pallidus, substantia nigra, and subthalamic nuclei (corpus Luysii) (Figure 46.1). The caudate and putamen together are referred to as the *striatum* or *neostriatum*, whereas the globus pallidus is often referred to as the *pallidum*. The term *basal ganglia* is often used to refer to the extrapyramidal system. Strictly speaking, the basal ganglia are a number of large, paired masses of gray matter in the forebrain and include the caudate, putamen, and globus pallidus, as well as the amygdala. This latter structure is functionally a part of the limbic system so that the term *basal ganglia* usually refers only to the first three pairs of nuclei. Lesions of the extrapyramidal system often result in abnormal movements that usually are present at rest. Such lesions also result in abnormalities of station and postural reflexes. The extrapyramidal system then is thought to be concerned with maintenance of posture as opposed to initiation or coordination of voluntary movement.

The Cerebellar System

The cerebellar system is composed of the cerebellum and its afferent and efferent pathways, as well as associated structures such as the red nuclei and the inferior olives. Lesions of this system result in tremor with movement, incoordination, dyssynergia, and ataxia. *The cerebellar system is believed to be concerned with the coordination of movements* as opposed to the initiation of involuntary movement. Although clinical experience has demonstrated a high degree of interdependence among these three motor systems, the term *extrapyramidal disease,* however, still serves the useful purpose of tying together a number of clinically defined disease states of diverse etiology and obscure pathogenesis. Causing abnormal movements, these states share a number of related symptoms, and the major pathological changes noted in these diseases are all present within the extrapyramidal nuclei. The clinical signs and symptoms that help to tie these disease states together fall into the following groups: (a) *hyperkinesia, abnormal involuntary movements*, (b) *akinesia, slowness or poverty of spontaneous movement*, (c) *rigidity*, and (d) *loss of normal postural reactions*.[83] Movement disorders may be classified broadly as the syndrome of parkinsonism or an akinetic rigid syndrome and those disorders causing a variety of abnormal involuntary movements or "dyskinesias," including tremor, dystonia, athetosis, chorea, ballism, tics, and myoclonus. For a comprehensive classification of parkinsonism, diseases causing parkinsonism and dementia, differential diagnosis of tremor, etiological classification of chorea, features distinguishing tardive dyskinesia and Huntington's disease, causes of ballism, etiological classification of dystonia, classification of tics, etiological classification of myoclonus, differential diagnosis of paroxysmal dyskinesia, startle and related syndrome, and finally, abnormal involuntary movements in sleep, refer to a review by Lang and Weiner.[5]

DIVERSE CLASSIFICATION OF DRUGS CAUSING MOVEMENT DISORDERS

Drugs causing or aggravating movement disorders have diversified classification (Table 46.5) and mechanisms of actions.[14] For example: *Norpseudoephedrine* causes persistent dyskinetic syndromes such as spasmodic torticollis and cranial dystonia.[14] Tiagabine, an indirect GABA receptor agonist, has been shown to inhibit haloperidol-induced oral dyskinesias,[84]

A variety of neurological syndromes, involving particularly the extrapyramidal motor system, occur following the use of almost all antipsychotic drugs. These reactions are particularly prominent during treatment with the high-potency agents (tricyclic piperazines and butyrophenones). There is less likelihood of acute extrapyramidal side effects with *clozapine, thioridazine*, or low doses of *risperidone*. Six varieties of neurological syndromes are characteristic of antipsychotic drugs. Four of these (acute dystonia, akathisia, parkinsonism, and the rare neuroleptic malignant syndrome) usually appear soon after administration of the drug, and two (rare perioral tremor and tardive dyskinesias or dystonias) are late-appearing syndromes that occur following prolonged treatment.

Mianserin, a tetracyclic antidepressant, which increases the release of noradrenaline by blocking alpha-2 adrenoceptors,[85,86] has been shown to activate a latent involuntary movement disorder in predisposed persons.[87]

NEUROLEPTIC-INDUCED MOVEMENT DISORDERS

The treatment of schizophrenic patients with neuroleptics (antipsychotics) has had a decisive impact on psychiatry in that the number of patients hospitalized has decreased dramatically, and the number treated on an outpatient basis has increased steadily. However, a variety of neurological syndromes, involving particularly the extrapyramidal system, occur after either acute or chronic administration of neuroleptics, and the most serious syndromes include akathisia, dystonia, neuroleptic malignant syndrome, parkinsonism, and tardive dyskinesia.[88–90] Despite awareness that neuroleptics could produce extrapyramidal side effects, these drugs remain the most effective means of treating schizophrenic patients. Moreover, "atypical neuroleptics" such as clozapine (8-chloro-11 (4-methyl-1-piperazinyl)-5H-dibenzo (b,e)(1,4)diazepine) and *risperidone*, which cause substantially fewer numbers of potentially incapacitating side effects such as neuroleptic malignant syndrome, have been synthesized and marketed. In addition to accepting the involvement of dopaminergic receptors in the pathogenesis of neuroleptic-induced movement disorders, recent reports have implicated serotoninergic, GABAergic, glutamatergic, and peptidergic transmissions in the appearance, manifestation, and treatment of neuroleptic-induced movement disorders. By having a comprehensive understanding of pharmacokinetic and pharmacodynamic principles unique to antipsychotics, the neuroleptic-induced movement disorders may be minimized vastly and their potentially lethal neurotoxicity averted altogether.

NEUROLEPTIC-INDUCED AKATHISIA

Kathisia is a Greek word that may be translated as the act of sitting, and *akathisia* means literally an inability to remain seated.[91,92] Patients with neuroleptic-induced akathisia may describe vague feelings such as "inner tension," "emotional uneasiness," "all wound up like a spring," "unable to relax," "having a hurry-up feeling," or "uncomfortable in any position." In addition to an inability to sit still, akathisia is characterized by shifting of legs and tapping of feet while sitting, and by rocking and shifting weight while standing. Although the term akathisia was first used by Haskovec,[93] spontaneously occurring syndromes of restlessness were reported long before the introduction of neuroleptics in 1955. For example, in 1880, Beard described it as a fidgetiness and nervousness, inability to keep still—a sensation that amounts to pain-is sometimes unspeakably distressing. When the legs feel this way, the sufferer must get up and walk or run, even if he is debilitated. This was confirmed 75 years later by a warning that akathisia can be more difficult to endure than any of the symptoms for which the patient was originally treated.[94]

Akathisia may occur after administration of any neuroleptic but is especially found with more potent neuroleptics.[95–97] The prevalence has been reported as 12.5%, 20%,[96,98–99] or 75%[100] with more potent neuroleptics such as haloperidol. Indeed, by lowering the dose of a potent

TABLE 46.5
Examples of Drug-Induced Movement Disorders

1. *Drugs associated with induction of akathisia:*
   Metoclopramide
   Dopamine storage and transport inhibitors: a-methyltyrosine, reserpine, tetrabenazine
   Levodopa and dopamine agonists
   Antidepressants:
      selective serotonin reuptake
      inhibitors, tricyclic antidepressants
   Lithium
2. *Drugs associated with induction of chorea:*
   Dopamine antagonists (including antipsychotics)
   Dopamine agonists:
      levodopa, direct dopamine agonists
   CNS stimulants:
      amphetamines, pemoline, methylphenidate, cocaine, xanthines Anticholinergics
   HI antihistamines H_2 antihistamines
   Oral contraceptives
   Anticonvulsants:
      phenytoin
3. *Drugs that can induce myoclonus:*
   Antidepressants:
      cyclic antidepressants
      selective serotonin reuptake inhibitors, monoamine oxidase inhibitors, Levodopa
   Bismuth salts
   Anticonvulsants:
      valproic acid (sodium valproate), carbamazepine, phenytoin
   Lithium
   Morphine or its derivatives, antineoplastic drugs
   Bromocriptine
4. *Drugs associated with induction or aggravation of Parkinsonism*
   Antipsychotics
   Calcium channel antagonists:
      flunarizine, cinnarizine, diltiazem, verapamil, amlodipine, manidipine, orthopramides and substituted benzamides:
         metoclopramide, sulpiride, clebopride, cisapride, domperidone, veralipride, and Dopamine agonists
   Biogenice amine storage and transport inhibitors:
      reserpine, tetrabenazine
   Antiemetic/antivertiginous agents:
      thiethylperazine, prochlorperazine
   Methyldopa
5. *Drugs associated with the development of tardive dyskinesia*
   Antipsychotic drugs
   Orthopramides and substituted benzamides:
      metoclopramide, clebopride, sulpiride, veralipride
   Calcium channel antagonists:
      flunarizine, cinnarizine
   Antidepressants:
      cyclic antidepressants
6. *Drugs associated with induction of acute and/or tardive dystonia*
   Antipsychotic drugs
   Orthopramides and substituted benzamides:
      metoclopramide, sulpiride, tiapride, cisapride, domperidone, veralipride, and dopamine agonists: levodopa
   Direct dopamine agonists
   Antidepressants:
      selective serotonin reuptake inhibitors, tricyclic antidepressants, monoamine oxidase inhibitors
   Anticonvulsants:
      carbamazepine, phenytoin

Table 46.5 (continued)

---

7. Drugs associated with induction or aggravation of postural tremor
Anticonvulsant drugs
Tricyclic antidepressants, β-adrenergic agonists, levodopa
Amphetamines
Thyroxine
Antihyperglycemic drugs
Caffeine
Corticosteroids
Calcium channel antagonists:
    flunarizine, cinnarizine
Amiodarone

A variety of neurologic syndromes, involving particularly the extrapyramidal system, occur following shorter long-term use of neuroleptic (antipsychotic) drugs. These include akathisia, dystonia, neuroleptic malignant syndrome, parkinsonism, and tardive dyskinesia.

Akathisia is characterized by an inability to sit still, by shifting of the legs and tapping of feet while sitting, and by rocking and shifting of the weight while standing. Reducing the total dosage of neuroleptic medications and the addition of either an anticholinergic drug, one of the benzodiazepine derivatives, or propranolol have been shown to reduce the severity of akathisia.

Dystonia is characterized by an exaggerated posturing of the head, neck, or jaw; by spastic contraction of the muscles of the lips, tongue, face, or throat, which makes drinking, eating, swallowing, and speech difficult; by torticollis, retrocollis, opisthotonus, distress, and ultimately anoxia. Neuroleptic-induced dystonia, which may occur in children treated actively with phenothiazine derivatives for their antiemetic properties, disappears in sleep and is treated effectively with *diphenhydramine hydrochloride* (Benadryl), which possesses both anticholinergic and antihistaminic properties.

Parkinsonian symptoms may be characterized by postural instability, stooped posture, shuffling and festinating gate, or rigidity, due to enhanced muscle tone, with, at times, "cogwheel" or "ratchet" resistance to passive movements in any direction. There is also tremor at rest with regular rhythmic oscillations of the extremities especially in the hands and fingers as well as akinesia (poverty of movement) or bradykinesia (slowness in initiating volitional activities). These symptoms, which are due to blockade of dopaminergic receptor sites in the striatum, are lessened by reducing the dosage of neuroleptics and by the oral administration of anticholinergic compounds, such as *trihexyphenidyl hydrochloride* (Artane) or *benztropine mesylate* (Cogentin).

Tardive dyskinesia is characterized by abnormal involuntary movements frequently involving the facial, buccal, and masticatory muscles and often extending to the upper and lower extremities, including the neck, trunk, fingers, and toes. With continuous blockade, the dopaminergic receptors in the striatum up-regulate. Following the discontinued use of neuroleptics or a reduction in dosage, the dyskinesia becomes apparent. In the therapeutic management of neuroleptic-induced tardive dyskinesia, reserpine, lithium, diazepam, baclofen, and vigabatrin have all been used with unsatisfactory results. Therefore, in the absence of an effective treatment, the best prevention of tardive dyskinesia is to prescribe the neuroleptics at their lowest possible doses, have patients observe *drug-free holidays*, and avoid prescribing anticholinergic agents solely to prevent parkinsonism.

Neuroleptic Malignant Syndrome

Among the complications of neuroleptic chemotherapy, the most serious and potentially fatal complication is malignant syndrome, which is characterized by extreme hyperthermia; "lead pipe" skeletal muscle rigidity that causes dyspnea, dysphagia, and rhabdomyolysis; autonomic instability; fluctuating consciousness; leukocytosis; and elevated creatine phosphokinase levels.

The treatments of neuroleptic malignant syndrome consists of immediately discontinuing the neuroleptic agent and administering *dantrolene sodium* and dopamine function-enhancing substances such as *levodopacarbidopa*, *bromocriptine*, or *amantadine*. (For reviews and references, see References 3 and 4.)

---

neuroleptic,[99] or by switching the patient to a lower-potency neuroleptic,[96] it is possible to treat akathisia. The syndrome of akathisia is composed of both subjective feelings and the psychological experience of inner restlessness and objective motor signs such as jiggling or shaking of the legs when seated and rocking from foot to foot when standing.[92] In milder forms of akathisia, the patients may only have subjective complaints, whereas, in moderate and severe forms, both subjective feelings of restlessness and objective movements exist.[91,101] Akathisia can be a quite common and very troubling side effect of psychotropic treatment. Clinicians have become steadily more aware of this disorder, owing to descriptions of restless movement disorder originating in the first half of this century. Delineation of acute akathisia is crucial for providing patients with the best interventions.

The pathophysiology of akathisia is not completely understood but likely arises from complex interactions in subcortical and possibly spinal DA/norepinephrine systems. There are now valid and reliable methods to assess akathisia using standardized scales; doing so helps track the progress of treatment interventions. The secondary complications of akathisia are numerous. The most notable ones are noncompliance and assaultive or suicidal ideation or behavior. Causative agents of akathisia include all currently available neuroleptics, various other psychoactive medications, and occasionally other nonpsychotropics. Treatment first should include stopping the offending agent (if possible), lowering the dose, or changing to a lower-potency neuroleptic. If these strategies are not feasible, then there are a host of medications that are variably effective. The most common are β-adrenergic receptor blockers, anticholinergics, clonidine, or benzodiazepines. Less commonly prescribed agents such as opiates, amantadine, buspirone, piracetam, amitriptyline, and dopamine depleters can be tried in more treatment-refractory patients.

CONDITIONS RESEMBLING AKATHISIA

Ekbom (1944) described an idiopathic syndrome that he originally called "asthenia crurum paraesthetica" and later renamed "restless leg syndrome."[102] Similar to akathisia, the patients' descriptions are: *it feels like ants were running up and down in my bones, it feels like an internal itch, and it is a diabolical feeling that I would not wish on my worst enemy.* Akathisia-like syndromes have also been reported in encephalitis lethargica[103] and in patients with both postencephalitic and idiopathic parkinsonism.[104]

Nocturnal myoclonus, or periodic movements in sleep, causes intense and repetitive muscle jerking during sleep. Treatment with 100- to 200-mg L-dopa[105] or with L-dopa plus bromocriptine[106] has been reported to be beneficial in this movement disorder. Leg restlessness also occurs after withdrawal from narcotic analgesics, and this opioid-related restlessness responds to treatment with clonidine[107–109] or treatment with propranolol.[110] Propranolol is also effective in treating neuroleptic-induced akathisia.

CLASSIFICATION OF AKATHISIA

Akathisia has been divided into the three categories of chronic akathisia, acute akathisia, and pseudoakathisia:

1. *Chronic akathisia* (tardive akathisia). The term denotes that akathisia developed late in the course of neuroleptic therapy and both subjective restlessness and objective motor movements are present. Furthermore, it is frequently associated with tardive dyskinesia, does not respond to anticholinergic drugs[99] and, like tardive dyskinesia, is difficult to treat.[111]
2. *Acute akathisia.* The term *acute* denotes that akathisia developed recently (within 6 months)

and is related to an increase in the doses of neuroleptics. It is not associated with dyskinesia.
3. *Pseudoakathisia.* The term was coined by Munetz and Comes.[112] The condition resembles chronic akathisia without subjective feelings of inner restlessness.[113] This characteristic makes it difficult to differentiate among chronic akathisia, pseudoakathisia, and tardive dyskinesia.

In an effort to differentiate among these problems, it has been suggested that, if the patient is restless and therefore is moving, he suffers from akathisia. If the patient moves and therefore is restless, he suffers from tardive dyskinesia.[114]

DIFFERENTIAL DIAGNOSIS OF AKATHISIA

Anxiety and agitated depression are often associated with restlessness. In akathisia, unnatural or abnormal restlessness is confined mostly to the legs.[115,116] It is interesting that the anxiolytic agent *lorazepam* is less effective than *propranolol* in treating neuroleptic-induced akathisia.[92] Although five major rating scales have been developed to measure akathisia (see Reference 117 for review), the differentiation between subjective and objective akathisia is difficult. The Chouinard extrapyramidal rating scale[118] and the 23-item rating scale for akathisia developed by Braude et al.[99] do attempt to assess both subjective and objective akathisia.

TREATMENT OF NEUROLEPTIC-INDUCED AKATHISIA

Because the incidence of neuroleptic-induced akathisia is higher with more potent neuroleptics such as haloperidol, switching patients to a lower-potency neuroleptic such as chlorpromazine may improve akathisia.[96,97] Braude et al.[99] reported that the only consistently effective treatment of akathisia was a reduction in the dose of neuroleptic. In addition to this general guideline, the other agents used to treat akathisia are (a) antiparkinsonian agents, (b) anxiolytic agents, (c) alpha- and beta-adrenergic receptor blocking agents, and (d) to a limited extent, L-tryptophan.

ANTIPARKINSONIAN AGENTS

Unlike drug-induced parkinsonism (rigidity, akinesia, and tremor), neuroleptic-induced akathisia responds incompletely and unpredictably to anticholinergic medications. Clinical experience using benztropine and procyclidine,[119] benztropine,[120] benztropine or trihexyphenidyl,[100] and biperiden[121] indicates that the response is more satisfactory if akathisia coexists with parkinsonism, but less satisfactory if akathisia exists alone.

AMANTADINE

Amantadine, in a daily dose of 200 to 300 mg, has been shown to be as effective as, or more effective than, a daily dose of 2 to 4 mg of benztropine in treating neuroleptic-induced akathisia.[122–124] However, tolerance develops to the beneficial effects of amantadine.[125]

BENZODIAZAPINE DERIVATIVES

Moderate to marked improvement has been reported in treating neuroleptic-induced akathisia after daily administration of 5 to 15 mg diazepam,[126,127] 1.5 to 5.0 mg lorazepam,[128] and 0.5 mg clonazepam.[129]

BETA-ADRENERGIC RECEPTOR BLOCKING AGENTS

The most promising treatment of akathisia consists of beta-adrenergic receptor blocking agents. Lipinski et al.[130] were the first to treat 12 patients with propranolol (30 mg/day) and to report improvement in all patients. The beneficial effects of propranolol were confirmed by Kulik and Wilbur[131] and by Lipinski et al.[130] Adler et al.[92,132] compared the effectiveness of lorazepam (2 mg/day) with propranolol (20 to 30 mg/day) and concluded that propranolol decreased both subjective feelings of restlessness and objective motor signs, whereas lorazepam decreased subjective but not objective signs. Furthermore, they[133] showed that propranolol was far superior to benztropine in treating akathisia. Moreover, Reiter et al.[134] successfully treated a patient suffering from sinus bradycardia and akathisia with *pindolol*, a beta-adrenergic receptor blocking agent that possesses an intrinsic sympathomimetic activity. The unexplained selectivity of beta-adrenergic receptor blocking agents in treating akathisia may be cited by studies that showed *metoprolol* was less effective than propranolol,[135] whereas atenolol was ineffective.[136,137] Finally, propranolol has been used successfully for neuroleptic-induced akathisia resistant to treatment with anticholinergics and benzodiazepines.

Selective or nonselective, centrally acting beta-adrenergic antagonists reduce haloperidol-induced emotional defecation in rats, and this effect parallels the reported anti-akathisia effect of these drugs in humans.[139]

ALPHA-ADRENERGIC RECEPTOR BLOCKING AGENTS

Clonidine, a central alpha-2 adrenergic receptor agonist, was shown to be effective in treating akathisia in a daily dose of 0.2 to 0.8 mg[138] and in a daily dose of 0.15 to 0.40 mg.[117]

L-TRYPTOPHAN

After an initial report by Sandyk et al.,[140] Kramer et al.[141] reported that L-tryptophan "appeared to reduce" both the subjective and the objective components of akathisia in most of the patients in their study.

BUPROPION

Neuroleptic-induced akathisia responds dramatically to bupropion (300 mg/day/5 days) treatment.[142]

NEUROLEPTIC-INDUCED DYSTONIA

Dystonia is characterized by an exaggerated posturing of head, neck or jaw; by spastic contraction of muscles of the lips, tongue, face or throat, making drinking, eating, swallowing and speech difficult; by torticollis, retrocollis, opisthotonos, and oculogyric crisis; and by laryngeal and pharyngeal spasm potentially leading to respiratory distress and ultimately anoxia. The term *dystonia* was first coined by Oppenheim[143] and the full spectrum of the disease has been reviewed by Eldridge,[144] Marsden,[145] Lang,[146] McGeer and McGeer,[147] Dickey and Morrow,[148] and Lang and Weiner.[5] The acute dystonias are quite dramatic in their presentation. Usually precipitated by the neuroleptic phenothiazines and butyrophenone compounds, dystonic movements and postures take various forms.

Most commonly, these drug-induced dystonias occur in the acute phase of treatment with phenothiazines and butyrophenones and often affect patients under the age of 40, although this is not a hard-and-fast rule. The salient clinical features accompanying these neuroleptic-induced dystonias are oculogyric crisis and abnormal posturing of the head and neck. Patients with oculogyric crisis often complain of an inability to move their eyes in the vertical plane; they also complain of double vision and blurred vision, but rarely of pain on attempted gaze.

Most often, the eyes maintain a sustained upward gaze. The symptom complaints appear to result from the patient's attempt to maintain full visual field by manipulating head and neck musculature in an uncoordinated fashion. The abnormal posture of the head and neck including opisthotonos, in which the head and neck are in a retrocolic position, give the patient a most bizarre appearance. Other muscles may be involved in acute drug-induced dystonia, but these presentations are much less common than the opisthotonos and oculogyric crisis.[149]

Drug-induced dystonia is caused primarily by medications that affect dopaminergic mechanisms. For example, patients with Parkinson's disease treated with L-dopa might develop typical dystonic movements and postures.[150–153] In addition to dystonia caused by neuroleptics[154] or by metoclopramide,[155] dystonic-like reactions have also occurred with many other drugs including high doses of *carbamazepine*,[156] *phenytoin*,[157] or *propranolol*.[158] Neuroleptic-induced dystonia is divided into acute dystonia and tardive dystonia.[159] The two dystonic reactions may be differentiated by the fact that acute dystonia responds well to anticholinergic medications, whereas tardive dystonia does not, suggesting different underlying mechanisms.

IATROGENIC DYSTONIA

Dystonia in various forms, including torticollis, blepharospasm, and oromandibular movements, has occurred after administration of many drugs (Table 46.5[147]). These dystonic reactions are idiosyncratic, occurring in susceptible patients, with the susceptibility decreasing with age. Indeed, *diphenhydramine*, an antihistaminic substance with anticholinergic properties, and benztropine, an anticholinergic agent, have been used successfully in treating neuroleptic-induced acute dystonia. Yet, diphenhydramine[160,161] and benztropine[162] have caused dystonia in susceptible individuals. It is clearly seen that the complexity of these responses militates against provision of a unified concept regarding drug-induced dystonia. Therefore, only the most commonly occurring dystonias will be described.

INCIDENCE OF ACUTE DYSTONIA

The incidence of neuroleptic-induced dystonia has been reported to be as low as 2 to 3%[163] to as high as 50%.[164] However, it is generally agreed that incidence is considerably higher in children and young adults.[165–167]

ENHANCED SUSCEPTIBILITY TO DEVELOP DYSTONIA

Cocaine addiction,[168] hypoparathyroidism,[169] alcoholism,[170] excess stress,[171,172] childhood convulsion or birth trauma,[173] hypothyroidism,[174] and certain degenerative disorders of the central nervous system (e.g., neuronal ceroidlipofuscinosis)[175] are thought to enhance one's chances of developing neuroleptic-induced dystonia. Chronic administration of cocaine enhances the concentration of dopamine.[176]

TARDIVE DYSTONIA

Dystonias may be a group of disorders with multiple etiology in which both dopamine excess and deficiency are implicated. Furthermore, treatment of dystonias is generally empiric, defying generalization. The following comments in Table 46.6, are illustrative. A simplified presentation of the involvement of neurotransmitters in the genesis of dystonia is not possible at this time.[146,147,193–195]

IDIOPATHIC OROFACIAL DYSTONIA (MEIGE'S SYNDROME)

Ortiz[196] reported that a patient with Meige's syndrome responded best to a combination of haloperidol (a dopamine receptor blocking agent, hence reducing dopamine excess) and benztropine (acetylcholine receptor blocking agents, hence reducing acetylcholine excess).

This syndrome, usually occurring within 24 to 72 hr after initiation of neuroleptic therapy, must be differentiated from late-onset tardive dystonia.[197,198] Neuroleptic-induced tardive dystonia may or may not respond to anticholinergic agents,[199] whereas anticholinergic agents exacerbate neuroleptic-induced tardive dyskinesia.[200] Furthermore, in contrast to tardive dyskinesia, tardive dystonia decreases with age and shows no female preponderance.[201]

TABLE 46.6
Treatment of Dystonia

Potent neuroleptics such as haloperidol may cause dystonia, asphyxiation, and death.[177–181] Neuroleptic-induced dystonia is relieved by diphenhydramine or benztropine.[167,182,183]

Thioridazine-induced dystonia in a patient with neuronal ceroidlipofuscinosis was refractory to most drugs but improved with baclofen,[175] which acts by inhibiting the release of glutamate and other neurotransmitters.

In a double-blind, placebo-controlled crossover study, six patients with different forms of dystonia were treated with γ-vinyl GABA, an inhibitor of GABA-transaminase. γ-vinyl GABA therapy (2 g/day for 2 weeks) was compared with placebo by weekly assessments. There were no consistent changes in three evaluation scores. Based on this study, agents that augment central nervous system GABA are unlikely to benefit patients with generalized dystonia.[184]

Dystonia of the paroxysmal type, lasting only a few minutes (also called *basal ganglia epilepsy*), responds to anticonvulsants such as phenytoin, carbamazepine, or clonazepam.[147]

Juvenile dystonia parkinsonism[185] or progressive dystonia with marked diurnal fluctuations (also called *Segawa syndrome),* and sometimes aberrant juvenile parkinsonism,[186] responds well to L-dopa or anticholinergic agents.[187]

Botulinum toxin has been used for temporary relief of focal dystonia such as blepharospasm,[188–190] torticollis,[147] and writer's cramp.[191]

Pancuronium bromide, a skeletal muscle relaxant, has been used in patients with acute torticollis.[192]

Dystonias may be a group of disorders with multiple etiology in which both dopamine excess and deficiency are implicated. Furthermore, treatment of dystonias is generally empiric, defying generalization. A simplified presentation of the involvement of neurotransmitters in the genesis of dystonia is not possible at this time.

TREATMENT OF DYSTONIAS

High-dose anticholinergic therapy for childhood-onset dystonia, botulinum toxin injections for focal dystonia, and L-dopa for diurnal dystonia provide symptomatic relief in some patients.[147]

Baclofen in the Treatment of Dystonia

Greene,[202] by reviewing the literature, summarized the beneficial effects of *baclofen*, a presynaptically acting GABA receptor agonist, in the management of dystonia. Dramatic improvement in symptoms, especially in gait, was found in almost 30% of 31 children and adolescents with idiopathic dystonia in one retrospective study using doses ranging from 40 to 180 mg daily. The response to baclofen in adults with focal dystonia is less dramatic. One series of 60 adults with cranial dystonia found sustained benefit in only 18%, and smaller series have not consistently found significant benefit in adults. Baclofen has been used to treat several secondary dystonias; tardive dystonia has occasionally been reported to improve with administration of baclofen, and there are isolated reports of improvement in dystonia occurring in Parkinson's disease and in glutaric aciduria.

Botulinum Toxin in the Treatment of Dystonia

Botulinum toxin acts presynaptically at nerve terminals to prevent the calcium-mediated release of acetylcholine.[203] When botulinum toxin is injected locally, the effect is that of a chemical denervation.[204] Injected locally into extraocular muscles, botulinum toxin has been used to treat strabismus[205] and, injected subcutaneously over orbicularis oculi, has been used to treat *blepharospasm*.[189,206–213] Botulinum toxin has been used to treat hemifacial spasm.[214–216] In addition, botulinum toxin has been used to treat *spasmodic torticollis*,[217] *oromandibular-laryngeal cervical dystonia*,[218] and *stiff-person syndrome*.[219]

Phenylalanine in Dopa-Response Dystonia (DRD)

The syndrome of autosomal dominant dystonia that is exquisitely responsive to levodopa, termed *hereditary progressive dystonia* with diurnal fluctuation or dopa-responsive dystonia (DRD), is not well known outside the fields of pediatric neurology and of movement disorders and is likely to be underdiagnosed. The disease is classically present as a dystonic gait disorder in childhood, with an average age of symptom onset of 5 to 6 years, but the spectrum of clinical manifestations is broad. Some children have presented in the first 2 years of life with developmental motor delay. In older children, prominent upper motor neuron findings, including *spastic diplegia*, have led to the misdiagnosis of *cerebral palsy*. Because of this

overlap in clinical features, some have recommended that all children and young adults with dystonia and *athetoid-dystonic cerebral palsy* be given a trial of levodopa to exclude DRD. However, response to levodopa in patients with DRD is not always immediate, especially in early-onset and severe cases, and levodopa can provide significant symptomatic benefit in some patients with secondary dystonia who do not have DRD.

Hyland et al.[220] measured plasma levels of phenylalanine, tyrosine, biopterin, and neopterin at baseline, and 1, 2, 4, and 6 hr after an oral phenylalanine load (100 mg/kg). Seven adults with DRD, two severely affected children with DRD, and nine adult controls were studied. All patients had phenylalanine and tyrosine concentrations within the normal range at baseline. In the adult patients, phenylalanine levels were higher than in controls at 2, 4, and 6 hr post-load ($p < 0.0005$); tyrosine concentrations were lower than control levels at 1, 2, and 4 hr post-load ($p < 0.05$). Phenylalanine-to-tyrosine ratios were elevated in patients at all times post-load ($p < 0.0005$). *Biopterin* levels in the patients were decreased at baseline and 1, 2, and 4 hr post-load ($p < 0.005$). Pretreatment with *tetrahydrobiopterin* (7.5 mg/kg) normalized phenylalanine and tyrosine profiles in two adult patients. In the children with DRD, phenylalanine-to-tyrosine ratios were slightly elevated at baseline. Following phenylalanine loading, the phenylalanine profiles were similar to those seen in the adult patients, but there was no elevation in plasma tyrosine. Baseline biopterin levels were lower in the children with DRD than in the adult patients or the controls, and there was no increase in biopterin post-load. In both the children and adults with DRD, neopterin concentrations did not differ from control values at baseline or after phenylalanine load. These results are consistent with decreased liver phenylalanine hydroxylase activity due to defective synthesis of tetrahydrobiopterin in patients with DRD. The findings show that a *phenylalanine load* may be useful in the diagnosis of this disorder.

NEUROLEPTIC MALIGNANT SYNDROME

Among the complications of neuroleptic chemotherapy, the most serious and potentially fatal complication is malignant syndrome, which is characterized by extreme hyperthermia; "lead pipe" skeletal muscle rigidity causing dyspnea, dysphagia, and rhabdomyolysis; autonomic instability; fluctuating consciousness; leukocytosis; and elevated creatine phosphokinase (see References 3, 4, and 221 through 232).

Neuroleptic malignant syndrome, which is sometimes associated with abrupt withdrawal of anticholinergic agents,[233] may occur without muscular rigidity.[234] Furthermore, an acute imbalance of sodium in the central nervous system has been proposed to play a role in the pathophysiology of neuroleptic malignant syndrome,[235] and dehydration and hot weather are additional factors for the development of neuroleptic malignant syndrome.[236] Neuroleptic malignant syndrome associated with dysarthric disorders has been reported.[237] Pure akinesia is a syndrome characterized by akinesia without rigidity, tremor, and supranuclear gaze palsy. Yoshikawa et al.[238] reported a patient with pure akinesia-manifested in neuroleptic malignant syndrome. *Clozapine*, an atypical neuroleptic, was initially reported not to cause neuroleptic malignant syndrome. As a matter of fact, chronic schizophrenic patients should be considered for clozapine treatment.[239] However, it is evident that clozapine monotherapy may cause neuroleptic malignant syndrome.[240,241] Moreover, withdrawal from clozapine has caused catatonia.[242]

Neuroleptic malignant syndrome, as the most serious but the rarest and least known of the complications of neuroleptic chemotherapy,[243] is viewed as a triad of fever, movement disorder, and altered mentation.[226] Pulmonary abnormalities, including tachypnea, dyspnea, stridor, and pulmonary edema, are probably secondary complications resulting from movement disorder, alteration in the functions of the autonomic nervous system (tachycardia, diaphoresis, and labile blood pressure), and changes in mental status (stupor, lethargy, and coma).

In 115 cases of neuroleptic malignant syndrome studied by Addonizio et al.,[228] the primary psychiatric diagnoses, in descending order of occurrence, consisted of schizophrenia (44%), bipolar mania (26%), major depression (10%), schizoaffective disorder (6%), atypical psychosis (3%), alcohol abuse (3%), bipolar depression (2%), mental retardation (2%), organic mental syndrome (1%), Alzheimer's disease (1%), and sedative abuse (1%). Among these 115, there were 72 men (63%) and 43 women (37%). Furthermore, greater than 50% of the cases were patients 40 years or younger. Unlike the more familiar neuroleptic-induced movement disorders, which occur in 15 to 50% of patients, neuroleptic malignant syndrome is relatively rare, and the annual incidence of the syndrome has been reported to be 0.15%,[244] 0.4 to 1.4%,[223,245–247] or 2.4%.[227,247] Among the 115 cases reported by Addonizio et al.[228] and 52 cases analyzed by Kurlan et al.,[227] the incidence of neuroleptic malignant syndrome seems to be considerably higher with haloperidol than with any other neuroleptic.

DIFFERENTIAL DIAGNOSIS OF NEUROLEPTIC MALIGNANT SYNDROME

Many diseases and toxic reactions mimic the cardinal features of neuroleptic malignant syndrome, namely fever, muscular rigidity, changes in mental status, and autonomic dysfunction. Therefore, care must be taken to differentiate neuroleptic malignant syndrome from *malignant hyper-*

thermia, lethal catatonia, heat stroke, central anticholinergic toxicity, central nervous system infection, severe dystonic reaction, drug- and food-related allergic reactions, electrolyte imbalance, thyrotoxicosis, strychnine poisoning, rabies, tetanus, polymyositis, rhabdomyolysis, and stiff-person syndrome.[248]

EVENTS LEADING TO OR ENHANCING THE SEVERITY OF NEUROLEPTIC MALIGNANT SYNDROME

The contributing factors leading to and/or enhancing the incidence or severity of neuroleptic malignant syndrome are dehydration,[249] exhaustion,[250] pre-existing organic brain syndrome,[251] external heat load,[252] large dosage and rapid dose titration of neuroleptics,[253] excessive sympathetic discharge,[254,255] concurrent lithium therapy, and abrupt discontinuation of antiparkinsonian agents.

COMPLICATIONS OF NEUROLEPTIC MALIGNANT SYNDROME

In addition to respiratory failure, acute renal failure, and cardiovascular collapse, which occur frequently in inadequately treated or untreated patients with neuroleptic malignant syndrome, other serious complications have occurred, even in patients treated well with *dantrolene* and dopamine-enhancing substances. The reported complications are myocardial infarction,[256] periarticular ossification,[257] peripheral neuropathy,[258] respiratory distress syndrome and disseminated intravascular coagulation,[259] and necrotizing enterocolitis.[260] The cluster of aforementioned symptoms occurred in 52 patients who received neuroleptics for schizophrenia (24), for affective disorder (13), for other psychiatric disorders (13), for preinduction anesthesia (1), and for withdrawal from sedative-hypnotic drugs (1). Receiving more than one neuroleptic significantly increases the incidence of neuroleptic malignant syndrome. Furthermore, it is generally believed that the incidence of malignant syndrome is higher with high-potency neuroleptics than low-potency ones.

THE PATHOGENESIS OF NEUROLEPTIC MALIGNANT SYNDROME—THE ROLE OF DOPAMINE

Although the pathogenesis of neuroleptic malignant syndrome is not entirely clear, a blockade of dopaminergic receptors in the corpus striatum is thought to cause muscular contraction and rigidity-generating heat, and a blockade of dopaminergic receptors in the hypothalamus is thought to lead to impaired heat dissipation. Therefore, the excess heat production along with a lack of heat dissipation produces pronounced hyperthermia, which is the hallmark of the syndrome. Furthermore, a blockade of dopamine receptors in the spinal cord is thought to be responsible for dysautonomia. The involvement of dopamine in the genesis of malignant syndrome is further supported by the observation that, in addition to neuroleptics, which block dopamine receptors, dopamine-depleting agents such as *reserpine*, dopamine storage blocking agents such as *tetrabenazine*, and dopamine synthesis inhibitors such as α-methylparatyrosine might also produce neuroleptic malignant syndrome. The rapid (4-hr) reversal of hyperthermia of neuroleptic malignant syndrome by L-dopa/carbidopa also indicates that an alteration in the metabolism or function of dopamine and/or its receptors may be responsible for the hyperthermia. Because dopamine plays a role in central thermoregulation in mammals[261–263] and, because neuroleptics block dopamine receptor sites, the hyperthermia associated with neuroleptic malignant syndrome might result from a blockade of dopamine target sites within the preoptic-anterior hypothalamus.[264,265] A stereotaxic injection of dopamine into the preoptic-anterior hypothalamus causes a reduction in core temperature, and this effect is blocked by haloperidol.[262] *Histidylproline diketopiperazine* [cyclo(His-Pro)], an endogenously occurring cyclic dipeptide, shares certain properties with dopamine in that it causes hypothermia and is found in preoptic-anterior hypothalamus and in striatum.[266]

In addition to hyperthermia, the parkinsonism (98%) and alteration in mental status (8 to 27%) seen in patients with neuroleptic malignant syndrome may result from blockade of dopaminergic receptors in the nigrostriatal system and from disruption of dopaminergic function in the mesocortical system.[264] The gradual (1 to 5 days) disappearance of parkinsonism seen in patients with neuroleptic malignant syndrome and the resumption of normal physiological functions after treatment with L-dopa/carbidopa support this contention. In addition to *Sinemet*, other dopamine-function-enhancing drugs, such as *bromocriptine*[98,267–276] and *amantadine*[277–282] have shown efficacy in treating neuroleptic malignant syndrome.

TREATMENT OF NEUROLEPTIC MALIGNANT SYNDROME

GENERAL TREATMENTS

The most important factor in treatment of neuroleptic malignant syndrome is the early recognition of the incipient syndrome and prompt discontinuation of neuroleptic medication.[283] Allsop and Twigley[284] described a psychotic patient who developed neuroleptic malignant syndrome after administration of *fluphenthixol*. Because the treatment with dantrolene sodium was instituted too late, the patient died after massive intestinal hemorrhage, intra-abdominal sepsis, and disseminated intravascular coagulation. In addition to blocking dopamine receptors in the hypothalamus, basal ganglia, and spinal cord, neuroleptics

cause excessive release of calcium from the sarcoplasmic reticulum in peripheral muscle fibers, resulting in exaggerated muscular contraction and in enhanced thermogenesis.[223] Dantrolene or other muscle relaxants are useful[285] when used in conjunction with dopamine-enhancing substances. Furthermore, every attempt should be made to reduce morbidity and to avert mortality, which are related to the development of cardiac problems, pneumonia, pulmonary embolus, and renal failure, secondary to myoglobinuria.[286] Airway intubation and other supportive care might be required in some patients. Lack of fluid intake along with diaphoresis may result in dehydration requiring fluid supplementation. Furthermore, vigorous fluid therapy is needed to combat myoglobinuria. Although dialysis may improve renal failure, neuroleptic agents are protein bound and are not removed readily by dialysis. In treating neuroleptic malignant syndrome, it should be recalled that significant variations might occur in patients. For example, it is generally presumed that neuroleptic malignant syndrome lasts for 5 to 10 days after discontinuation of oral neuroleptic.[287] However, if the syndrome is caused by a long-acting neuroleptic, such as *fluphenazine enanthate*, a more prolonged and severe case may be anticipated.[281] Patients who are receiving neuroleptics and suffer from heat stroke may present with fever, rigidity, and elevated creatine kinase. However, their skin is dry, and their blood pressure is low or normal.[288,289] Neuroleptic malignant syndrome may occur in a milder form after the administration of weaker dopamine-depleting substances such as reserpine, and later develop fully in the same patient after the administration of a potent dopamine receptor blocking agent such as *haloperidol*.[290] Finally, neuroleptic malignant syndrome may be superimposed upon tardive dyskinesia,[291] making both diagnosis and treatment difficult.

SPECIFIC TREATMENTS

In neuroleptic malignant syndrome, central dopaminergic receptors are blocked, and altered levels of dopamine metabolites such as 3,4-dihydroxyphenylacetic acid and homovanillic acid have been found postmortem.[292] Therefore, the treatment of choice is to reverse the hypodopaminergic state by administration of L-dopa/carbidopa, bromocriptine, or amantadine.

L-DOPA/CARBIDOPA

L-dopa/carbidopa (Sinemet 25:100) is often effective in reversing hyperthermia and making the patient afebrile in hours. Treatment, however, may need to be continued for several days.[226] Harris et al.[293] reported a patient in whom malignant syndrome developed after taking haloperidol (15 mg/day), and therapy was initiated with dantrolene (10 mg/kg/24 hr). In this case, severe muscle rigidity resolved and temperature was reduced from 107 to 102 after the administration of dantrolene. Twenty-four hours after carbidopa-L-dopa (25:100) treatment was started, the temperature dropped to 100. When subsequent carbidopa-L-dopa was inadvertently not given, the temperature increased, despite continuous treatment with dantrolene.[293] Hirschorn and Greenberg[294] reported a case of L-dopa-induced myoclonus combined with L-dopa withdrawal-induced neuroleptic malignant syndrome, which was successfully treated with L-dopa/carbidopa along with 2 mg/day of *methysergide* (a serotonin receptor antagonist). It has been suggested that prolonged L-dopa therapy may result in deregulation of serotonergic transmission producing myoclonus.[295]

BROMOCRIPTINE

In 1983, several investigators[269,272,274,276] explored the beneficial effects of bromocriptine in reversing neuroleptic malignant syndrome. The recommended initial dose is 5.0 to 7.5 mg three times daily.[296] If the syndrome has progressed to the point at which the patient is unable to swallow the orally available bromocriptine, it is necessary to produce muscular relaxation by dantrolene (3 mg/kg four times daily). However, it should be stated that the hyperthermia in malignant syndrome does not respond to muscle relaxation alone.[297] Because dantrolene produces a rare but potentially fatal idiosyncratic hepatocellular injury, bromocriptine may be administered by a nasogastric tube in patients in whom a pre-existing liver disease may preclude the use of dantrolene.

Dhib-Jalbut et al.[269] reported that the amount of bromocriptine mesylate to be given to a patient depends on body temperature, autonomic and extrapyramidal signs, and symptoms. Therefore, in treating five patients with malignant syndrome, they used 7.5 to 45.0 mg/day in three divided doses for 10 days. In all five patients, significant improvement in vital signs and reduction in creatine kinase was noted within 24 to 72 hr after initiation of bromocriptine treatment. Resolution of confusion and mutism was noted within 24 to 48 hr, and resolution of extrapyramidal rigidity occurred within 1 week. In two patients, early discontinuation of bromocriptine resulted in a relapse of neuroleptic malignant syndrome, which responded to reinstitution of the drug.[269]

DANTROLENE SODIUM (DANTRIUM)

Dantrolene, a nitrophenylamino hydantoin derivative, is a unique skeletal muscle relaxant in that, unlike competitive neuromuscular blocking agents (e.g., *D-tubocurarine*), depolarizing blocking agents (e.g., *succinylcholine* and *decamethonium*), and agents enhancing or mimicking GABAergic transmission (e.g., diazepam and baclofen), dantrolene exerts its effects by direct action on excitation-contraction coupling and by reducing the amount of cal-

cium released from sarcoplasmic reticulum.[298] Dantrolene, which depresses the central nervous system, does not affect neuromuscular transmission, nor does it change the electric properties of skeletal muscle membranes.[299] Although the hepatotoxicity of dantrolene precludes its widespread and chronic usage, it has proven beneficial in treating patients with spasticity associated with stroke, cerebral palsy, spinal cord injury, and multiple sclerosis. Furthermore, it is effective in reducing the muscular rigidity associated with malignant hyperthermia[300,301] and in neuroleptic malignant syndrome.[285,302] In malignant hyperthermia, a dose of 2.4 mg/kg is given by intravenous infusion for prophylaxis or initial treatment of hyperthermia.[300] Britt[301] recommended that dantrolene should be administered at a rate of 1 mg/kg/min while monitoring electroencephalogram and until heart rate and temperature begin to fall and muscle stiffness starts to subside. If necessary, treatment may be repeated every minute, or a maintenance infusion of 1 to 2 mg/kg per 3 to 4 hours initiated until all evidence of hyperthermia has disappeared. Similar to L-dopa or bromocriptine, dantrolene sodium may produce rapid reversal of neuroleptic malignant syndrome.[246,248,272,296,303–306] The initial usual recommended dose of dantrolene, which may vary depending on the severity of the problem and the perceived need of the patient, to be decided by the attending physician (see Reference 307), is between 0.8 and 2.5 mg/kg given intravenously every 6 hr[308] or between 0.25 and 3.0 mg given intravenously every 6 hr.[272,304] When symptoms abate and the patient is able to swallow, oral doses in the range of 100 to 200 mg/day may be substituted.[305] Rapid resolution of symptoms (within 24 hr) is possible if treatment is begun early,[309] although the usual course of treatment is 5 to 10 days.[248]

AMANTADINE

Although universal agreement on its efficacy does not exist, amantadine has been tried in the management of neuroleptic malignant syndrome.[272,277–281]

ANTICHOLINERGIC AGENTS

Anticholinergic drugs such as benztropine (Cogentin) are usually ineffective in treating the rigidity of neuroleptic malignant syndrome and do not affect or may even aggravate the associated hyperthermia.[310] However, Schrehla and Herjanic[311] reported a schizoaffective patient who developed neuroleptic malignant syndrome and did not respond completely with bromocriptine (5 mg p.o. t.i.d.) but improved dramatically with 2 mg benztropine (i.m.) after initial treatment with bromocriptine.

BENZODIAZEPINE DERIVATIVES

Benzodiazepine derivatives, which enhance GABAergic function, have produced transient relief of symptoms.[222,214,264,306,312,313] Benzodiazepine derivatives have also been recommended to control "agitated" patients while being treated for neuroleptic malignant syndrome.[248,296]

NEUROLEPTIC-INDUCED PARKINSONISM

The cardinal features of Parkinson's disease are tremor at rest, rigidity, akinesia, and postural abnormalities. A small number of patients, however, might display only akinesia without rigidity or tremor.[314–317] Drug-induced parkinsonism is common but under-recognized.[318] A majority of patients initially diagnosed as having Parkinson's disease were subsequently shown to have drug-induced syndrome.[319,320] By far the most frequent causes of drug-induced parkinsonism are neuroleptics that block dopamine receptors in nigrostriatal dopaminergic pathways. Furthermore, there is a direct relationship between the potency of neuroleptics as antipsychotics and the incidence of parkinsonism. The more potent agents, such as haloperidol, produce more frequent pseudoparkinsonism than the less potent agents such as chlorpromazine.

Incidence of Parkinsonism

Although the severity of neuroleptic-induced parkinsonism varies with class of drug, potency of drug, dosage level, and the length of treatment, the factors enhancing one's susceptibility to develop parkinsonism are not known.[96,97,321] Some patients do not develop parkinsonism, even after taking potent neuroleptics for a long period of time, whereas others experience severe and disabling symptoms after a few doses of relatively weak neuroleptics. Neuroleptic-induced parkinsonism is reversible and disappears after discontinuation or lowering of drug dosage.

TREATMENT OF PARKINSONISM

Neuroleptic-induced parkinsonism is best treated by reducing doses of neuroleptic and by adding an anticholinergic agent such as benztropine or trihexyphenidyl. Anticholinergic drugs should not be used prophylactically, because the incidence of tardive dyskinesia is far higher in patients who receive neuroleptics and an anticholinergic drug given prophylactically to prevent parkinsonism. L-dopa is contraindicated, because it might aggravate the underlying psychiatric problem for which neuroleptic treatment was initiated. In animal models of Parkinson's disease, glutamate receptor antagonists diminish levodopa-associated motor fluctuations and dyskinesia.[322]

Antitremor Effects of Clozapine

Tremor at rest is a classic symptom of Parkinson's disease that causes significant disability and distress for the patient and is generally only weakly responsive to conventional treatment such as anticholinergic or amantadine. Jansen[323] reported that clozapine (18 mg/day) improved tremor.

Parkinsonism, Schizophrenia, and Dopamine

The coexistence of Parkinson's disease and schizophrenia is of theoretical interest and of therapeutic importance. Parkinson's disease is known to be a striatal dopamine deficiency syndrome. On the other hand, schizophrenic patients are thought to suffer from a dopaminergic hyperactivity state.[324–329] Clozapine along with L-dopa/carbidopa may be used in the management of patients with Parkinson's disease who also suffer from schizophrenia,[330–331] or in patients with L-dopa-induced psychosis.[332] Moreover, clozapine therapy may allow an increase in antiparkinson therapy leading to further amelioration of parkinsonism.[333] Clozapine has also been advocated to be of value in treating resistant schizophrenics[334] and in a paranoid subgroup of schizophrenics.[335]

NEUROLEPTIC-INDUCED TARDIVE DYSKINESIA

Persistent dyskinesia, which was initially described in 1956,[336,337] also became known as reversible and irreversible drug-related dyskinesia.[338] This neuroleptic-induced movement disorder is characterized by abnormal involuntary movements frequently involving the facial, buccal, and masticatory muscles and extending often to the upper and lower extremities, including the neck, trunk, fingers, and toes. The typical abnormal facial movements include opening and protrusion and retrieval of the tongue and closing of the mouth, chewing, licking, sucking, puckering, smacking, panting, and grimacing. Abnormal movements associated with the disorder, which might involve any part of the body, can be ataxic, myoclonic, dystonic, dyskinetic, or choreiform in nature. The neuroleptic-induced dyskinesias, which have been reported and studied extensively in adult patients,[339–341] occur also in children.[342]

Steen et al.[343] believe that autosomal inheritance of two polymorphic Ser 9 Gly alleles (2-2 genotype), but not homozygosity for the wild-type allele (1-1 genotype), is a susceptibility factor for the development of tardive dyskinesia.

Drugs and Conditions Causing Dyskinesia

In addition to neuroleptics[339,340–342] dyskinesias have been reported to occur after exposure to antidepressants,[344,345] anxiolytic agents,[346] anticonvulsants,[156] antihistaminics,[161,347] narcotics, L-dopa,[340] amphetamine,[348] lithium,[349] metoclopramide[155,350] used in the treatment of gastrointestinal motor dysfunctions, and allegedly nicotine,[351] inasmuch as the incidence of tardive dyskinesia is higher among smokers. However, because smoking induces the activity of the hepatic microsomal enzymes and enhances the metabolism of neuroleptics, it necessitates the use of higher-than-ordinary doses of neuroleptics for patients.[352] The reported higher incidence of tardive dyskinesia in smokers taking neuroleptics might be due to higher doses of neuroleptics themselves. In addition to drug-induced movement disorders, dyskinesias have been reported to take place occasionally in infantile autism, in Huntington's chorea, in some elderly individuals, and in some patients with mental retardation.[353] As a matter of fact, a positive correlation between cognitive impairment and tardive dyskinesia has been suggested.[354]

HETEROGENEITY OF TARDIVE DYSKINESIA

Tardive dyskinesia, a late-developing side effect syndrome arising as a consequence of chronic neuroleptic treatment, is pharmacologically heterogenous, and this heterogeneity is seen between individual patients (and groups of patients) as well as within body areas of individual patients (and groups of patients) as well as within body areas of individual patients.[355] For example, tardive dyskinesia most often involves the oral-lingual-facial regions but can affect all other body areas as well. The symptoms of tardive dyskinesia are most commonly choreoathetoid in nature but can have dystonic as well as other features (e.g., tics, akathisia, myoclonus). Symptoms of tardive dyskinesia vary widely in severity, but the majority of cases are mild. In addition, the syndrome varies in its duration with symptoms being transient in some patients, whereas, in others, they persist and in some cases may be irreversible. Although disturbances in dopamine and acetylcholine seem to be involved in these disorders, they do not in all cases exist in functionally opposite relationships. The observed pharmacologic heterogeneity in tardive dyskinesia response reflects the limitations of the dopamine/acetylcholine model of tardive dyskinesia, which oversimplifies the neuroanatomy of the basal ganglia (Figure 46.1) and the pathophysiology of tardive dyskinesia. The chemical and anatomical complexity of this region suggests that other neurotransmitter systems such as glutamate[356] and neuronal circuits within and extending from the basal ganglia might be disturbed in the pathogenesis (Figure 46.2). Consistent with these views is the observation that although trihexyphenidyl is effective in the treatment of neuroleptic-induced parkinsonism, a subpopulation of patients with tardive dyskinesia also shows improvement after anticholinergic treatment.[357]

TARDIVE DYSKINESIA AND DIABETES

Diabetic patients have been shown to exhibit a greater incidence of tardive dyskinesia.[358] Hyperglycemia suppresses the basal firing of dopamine-containing neurons,[359] and insulin reduces the severity of symptoms in tardive dyskinesia.[360]

L-DOPA-INDUCED DYSKINESIA

L-dopa-induced dyskinesias are a heterogeneous phenomenon, which might be difficult to explain on the basis of a single pathological mechanism. For example, Luquin et al.[361] classified L-dopa-induced dyskinesias into "on" dyskinesias, "diphasic dyskinesia,'" and "off" periods. Chorea, myoclonus, and dystonic movements occurred during the "on" period. Dystonic postures, particularly affecting the feet, were mainly present in the "off" period, but a few patients had a diphasic presentation. Repetitive stereotyped movements of the lower limbs always corresponded to diphasic dyskinesia. Moreover, Luquin et al.[361] have shown that dopamine agonists enhanced "on" dyskinesias and markedly reduced or abolished "off" period dystonia and diphasic dyskinesia. Dopamine receptor antagonists reduced all types of L-dopa-induced dyskinesia but also aggravated parkinsonism. These data indicate that L-dopa-induced dyskinesias in Parkinson's disease is a heterogeneous phenomenon difficult to explain on the basis of a single pathophysiological mechanism.[361]

TARDIVE OCULOGYRIC CRISIS

The syndrome of oculogyric crisis was originally described in association with epidemic encephalitis lethargica.[362] Oculogyric crisis is now most commonly seen as a side effect of neuroleptic medication.[363] It is recognized as a form of acute dystonia involving primarily the ocular muscles, although retrocollis, blepharospasm, contraction of the frontalis, jaw opening, and other movements might be associated with it. The importance of the syndrome lies in the fact that it is common, being reported in 10% to more than 60% of patients recently treated with neuroleptic medication, and very distressing to the patient and onlookers. It usually occurs within a few days of starting the drug or increasing its dose, with one report suggesting that 90% of patients experienced it in the first 4 days of drug treatment. Sachdev[364] reported on six patients with chronically recurring oculogyric crisis; three of the patients developed tardive side effects, and in one patient the episodes persisted for some months after cessation of neuroleptic.

TARDIVE DYSKINESIA AND TYPE II SCHIZOPHRENIA

Davis et al.[365] found a significant association between tardive dyskinesia, cognitive impairment, some negative symptoms, and formal thought disorders. These associations were independent of other illnesses and treatment variables. The severity of tardive dyskinesia correlated significantly with that of cognitive impairment.

MECHANISMS OF NEUROLEPTIC-INDUCED DYSKINESIA

Long-term administration of neuroleptics and other drugs causes tardive dyskinesia, which closely resembles L-dopa-induced dyskinesias and is brought about through complex mechanisms that are ill defined. It is generally believed that its pathogenesis involves blockade of dopamine receptor sites and that its pathophysiology results from a hypersensitivity of dopamine receptors (see Reference 366). This hypothesis, however, is not universally accepted, for the reasons presented in Table 46.7. The aforementioned studies, when examined collectively, indicate that the hypothesis proposing dopaminergic hyperactivity cannot explain completely the etiology of schizophrenia and/or the pathogenesis of tardive dyskinesia, hence making treatment difficult.

γ-Aminobutyric Acid in the Pathogenesis of Tardive Dyskinesia

GABA has been shown to interact with nigrostriatal dopaminergic neurons.[367–370] Defective GABAergic transmission has been advanced as a possible etiologic factor in the pathogenesis of tardive dyskinesia.[371,372] There is a reduction in the concentration of GABA in the cerebral spinal fluid of patients with tardive dyskinesia, and GABA-mimetic drugs improve dyskinetic symptoms. Moreover, it has been suggested that diminished neural activity in efferent GABAergic tracts from the substantia nigra pars reticulata to the thalamus underlies tardive dyskinesia in human beings (Figures 46.1 and 46.3). The repeated administration of neuroleptics reduces the activity of glutamic acid decarboxylase only in substantia nigra of animals developing dyskinesia.[373,374] Therefore, one of the recent therapeutic regimens advocates the administration of substances that functionally enhance GABAergic transmission, such as diazepam, or of agents that act as GABA receptor agonists, such as muscimol. GABA-related pharmacological treatment is based on the observation in experimental animals that the GABAergic efferent tract in the striatum constitutes a segment of the striatonigral feedback loop (Figure 46.3) with the ability to modulate the activity of dopaminergic cells in the substantia nigra.[375–378] Based on this concept, it has been shown that muscimol,[379] a GABA receptor agonist, and γ-acetylenic-GABA[380] or γ-vinyl-GABA,[195] substances that elevate the concentration of GABA by inhibiting GABA-transaminase, were effective in alleviating the severity of tardive dyskinesia. Singh et al.[381] have shown that diazepam significantly improved tardive dyskinesia and that

---

TABLE 46.7
The Pathogenesis of Tardive Dyskinesia and Chronic Blockade of Dopamine Receptor Sites

---

The pathophysiology of L-dopa-induced dyskinesia and of neuroleptic-induced dyskinesia is not identical because progabide, a GABA receptor agonist, is only beneficial in treating neuroleptic-induced (but not L-dopa-induced) dyskinesia.[401]

In addition to dopamine, noradrenergic overactivity might contribute to the pathogenesis of tardive dyskinesia.[402]

The hyperactivity of brain dopaminergic systems, especially in the cortical and limbic regions, has been postulated to play a definite role in the etiology of schizophrenia (reviewed in Reference 328). However, a study by Karoum et al.[403] suggests that the output of dopamine and its metabolites is lower in schizophrenia. Furthermore, at postmortem, no significant differences in D_1 and D_2 receptors have been found in schizophrenic patients with or without tardive dyskinesia.[404]

Because the blockade and hypersensitivity of dopamine receptors induced by neuroleptics are thought to play crucial roles in the etiology and manifestation of tardive dyskinesia, all pharmacotherapeutic interventions have concentrated on modifying the expression of dopaminergic transmission. Reserpine, which depletes dopamine in the brain and blocks its uptake into the intraneuronal storage vesicles, has been shown to benefit some patients with tardive dyskinesia.[405] Identical palliative effects have also been reported with tetrabenazine, which possesses reserpine-like properties.[406–409] A therapeutic regimen advocating step-wise and progressively smaller doses of neuroleptics to slowly desensitize the dopamine receptor sites has also been shown to be effective in many, but not all, schizophrenic patients with tardive dyskinesia.[410]

The neuroleptic-induced increase in the number of striatal dopamine receptors occurs rapidly within 1 week, whereas the appearance of tardive dyskinesia is observed months or years after initiation of neuroleptic treatment.

Dyskinesia is one of the main adverse events related to long-term dopa therapy in patients with Parkinson's disease. Generally, most drugs with reliable antidyskinetic properties, such as classical neuroleptics, also reduce the antiparkinsonian efficacy of dihydroxyphenyl-alanine (L-dopa), thus markedly limiting their clinical usefulness. L-dopa-induced dyskinesia is characterized by abnormal, involuntary movements such as chorea and dystonia. It can affect several body parts and can occur when the patient is on or off treatment. On-treatment dyskinesia can be apparent at the start of dose, at the peak dose, or at the end dose. Heterogeneity of the disorder has made it extremely difficult to determine the neural mechanisms underlying L-dopa-induced dyskinesia. In the past, it was suggested that L-dopa-induced dyskinesia might represent a symptom of the progression of Parkinson's disease, but data from 1-methy-4-phenyl-1,2,3,6-tetrahydropyridine (MPTP)-treated primates, which do not show progression of parkinsonian symptoms, strongly suggest that the appearance of L-dopa-induced dyskinesia results from the treatment. D_1 receptor blockade improves L-dopa-induced dyskinesia but worsens parkinsonism in MPTP monkeys.

some of the improvement persisted after discontinuation of the administration of diazepam. Studies by Sandyk[382] have shown that baclofen and clonazepam were effective in the treatment of neuroleptic-induced akathisia. These observations suggest, but do not confirm, a direct link between dopaminergic transmission, GABAergrc transmission, the pathogenesis of neuroleptic-induced dyskinesia, and the treatment of the dyskinesia with diazepam or GABA-mimetic agents.

An attempt has been made to demonstrate this association by studying the effects of treatment with diazepam alone, with haloperidol alone, and with haloperidol in combination with diazepam on the activity of glutamic acid decarboxylase and on the metabolism of dopamine and serotonin in discrete regions of the rat brain at the end of treatment with drugs and in brains of animals allowed to undergo a drug-free "holiday."[386] During a 3-day withdrawal period, after daily administration of 3 mg/kg of haloperidol (i.p.) for 3 weeks, the activity of glutamic acid decarboxylase in the striatum increased from 72.6 ± 7.8 to 92.6 ± 10.2 nmol $^{14}CO_2$/mg/protein/hr, and the concentration of dopamine in the striatum increased from 7.87 ± 0.23 to 8.86 ± 0.38 μg/g wet tissue. Diazepam (5 mg/kg, i.p.), given during the withdrawal period from haloperidol, was able to nullify the enhancement in the concentration of dopamine but not in the activity of glutamic acid decarboxylase in the striatum. The results of these studies were interpreted to indicate that the reported beneficial effects of diazepam and GABA-mimetic agents in ameliorating the symptoms of tardive dyskinesia might occur through a mechanism that does not necessarily link transmission involving both dopamine and GABA.[383] In another study, Mithani et al.[384] have shown that neuroleptic-induced chewing movements and decreases in nigral glutamic acid decarboxylase activity were not causally related.

DOPAMINE, PEPTIDES, SCHIZOPHRENIA, AND NEUROLEPTICS

The metabolism of neuropeptides including opioids, neurotensin, metenkephalin, substance P, and cholecystokinin in the spinal fluids of control and neuroleptic-treated schizophrenic patients has been reviewed.[385–394] The lack of detailed knowledge describing the exact nature of the interaction between dopamine and these neuropeptides militates against a comprehensive discussion of their involvement either in the pathogenesis of schizophrenia or in the pharmacodynamics of neuroleptics. Nevertheless, a few fragmentary yet interesting items will be outlined.

NEUROLEPTIC-CHOLECYSTOKININ INTERACTION

The possible involvement of cholecystokinin (CCK) in the pathogenesis of schizophrenia has been reviewed.[395] CCK,

a 33-amino acid peptide, originally characterized in the porcine gastrointestinal tract, was first detected in the vertebrate central nervous system by Vanderhaeghen et al.[396] In the brain, the majority of CCK-gastrin-like peptides exist as the sulfated form of CCK octapeptide (CCK8S).[397–400] CCK8S may serve as a neurotransmitter or a neuromodulator, influencing, among other functions, dopaminergic transmission.

The coexistence of dopamine and CCK8S in a subpopulation of mesolimbic dopaminergic neurons has been demonstrated,[411,412] suggesting that this peptide might modulate dopamine function. This observation is of interest in view of the suggested hyperactive dopaminergic transmission in schizophrenia; the beneficial antipsychotic effect of neuroleptics, which allegedly block the hyperactive dopamine receptors; and, as discussed previously, of neuroleptic-induced tardive dyskinesia, postulated to result from denervation supersensitivity of dopaminergic neurons in the striatum.[413,414] Other evidence pointing to a dopamine-CCK-linked transmission are the observations that CCK elevates the density of brain D_2 receptors[415] and that intrastriatally injected CCK is able to stimulate dopamine-mediated transmission.[416] In an attempt to study further the possibility of dopamine-CCK cotransmission, Hama and Ebadi[417] determined the CCK binding sites in the mouse brain. By using a synaptosomal fraction isolated from the mouse cerebral cortex and [propionyl-^{3}H]CCK8-sulfate ([^{3}H]CCK8S) as a ligand, a single binding site for [^{3}H]CCK8S with a K_D value of 1.04 nM and a B_{max} value of 42.9 fmol/mg/protein was identified. The competitive inhibition of [^{3}H]CCK8S binding by related peptides produced an order of potency of CCK8-sulfated (IC_{50} = 5.4 nM) > CCK8-unsulfated (IC_{50} = 40 nM) > CCK4 (IC_{50} = 125 nM). The regional distribution of [^{3}H]CCK8S binding in the mouse brain was highest in the olfactory bulb (34.3 ± 5.6 fmol/mg/protein) > cerebral cortex> cerebellum> olfactory tubercle> striatum> pons-medulla> midbrain> hippocampus> hypothalamus (12.4 ± 2.1 fmol/mg/protein).[417]

CCK peptides share certain properties with neuroleptics in that they induce catalepsy, antagonize conditioned avoidance behavior, antagonize stereotyped behavior, induce hypothermia, induce ptosis, and antagonize certain actions of amphetamine. In addition, ceruletide, CCK8, or CCK33 may produce rapid, effective, and persistent antipsychotic effects, especially in some neuroleptic-resistant patients.[414]

The aforementioned data led neuroscientists to study the effects of acute or chronic administration of neuroleptics, including haloperidol, on the concentrations of CCK and its receptor sites. The results of these studies provided interesting but inconclusive observations. The varied effects of haloperidol on the concentration of CCK[418–420] might depend on the varied mammalian species studied, the areas of brain examined, the nature of the experiments conducted, and especially the ligand used to determine either the content or the density of receptor sites for the octapeptide. Indeed, a study by Zetler[421] has shown that CCK-like peptides with neuroleptic activity were able to antagonize stereotyped behaviors caused by dopaminergic receptor agonists, but the mechanism of action of the peptides was not due to a simple clear-cut neuroleptic-like blockade of postsynaptic dopamine receptors. The coexistence of CCK peptides in the nigrostriatal and mesolimbic dopaminergic systems might modulate the synthesis, storage, and/or functions of dopamine and provide additional insight into the efficacy of neuroleptics and the psychopathology of schizophrenia. However, the nonuniform distribution of CCK8S receptors in the central nervous system signifies a broader function for the octapeptide than once anticipated, deserving further in-depth investigation.[416]

NEUROLEPTIC-OPIOID INTERACTION

Experimental evidence suggests close interaction between neuroleptic therapy and the endogenous opioid peptides[140] (see Table 46.8). The experimental evidence and clinical findings strongly support the contention that a modification in the metabolism and/or action of dopamine-opioid and dopamine-CCK transmission in part might have both beneficial and harmful effects with regard to the neuroleptic-induced movement disorders.

TREATMENT OF TARDIVE DYSKINESIA

BUSPIRONE IN L-DOPA-INDUCED DYSKINESIAS

Buspirone is an azaspirodecandeione drug with an anxiolytic efficacy comparable to that of the benzodiazepines, but without any sedative, muscle relaxant, or anticonvulsant effects.[422,423] Unlike the benzodiazepines, buspirone does not interact with GABA-benzodiazepine chloride channel complex, is thought to exert its neuropharmacological properties as an agonist for serotonin 5-HT$_{1A}$ receptor subtype, and blocks presynaptic dopamine D$_2$ receptors.[424] By stimulating 5-HT$_{1A}$ autoreceptors located on raphe neurons, buspirone inhibits the firing of serotonergic neurons, leading to a decrease of serotonin transmission in the brain. Moreover, it interacts directly with 5-HT$_{1A}$ postsynaptic receptors in the hippocampus, an action that has been invoked to explain, at least in part, its anxiolytic effects.[425] Bonifati et al.[425] reported that buspirone (10 mg orally twice a day) for 3 weeks significantly lessened the severity of the L-dopa-induced dyskinesia without worsening parkinsonism. Buspirone in relatively large doses of 180 mg/day (the recommended dosage of buspirone in anxiety is 20 to 60 mg/day) has been shown to be effective in the treatment of tardive dyskinesia.

TABLE 46.8

Experimental Evidence Suggesting Close Interaction between Neuroleptic Therapy and the Endogenous Opioid Peptides

Areas of the central nervous system, such as striatum and nucleus accumbens, contain high concentrations of both dopamine and opioid receptors.[430-434]

The interrelationship between dopaminergic and enkephalinergic neurons[435] is further extended by studies showing that the number of mesolimbic opioid binding sites is reduced after denervation of dopaminergic neurons.[436]

Chronic injection of haloperidol,[434] but not clozapine,[435] increased the concentration of enkephalins selectively in the striatum.

Neuroleptic-induced supersensitivity in the mesolimbic dopaminergic receptor is reduced by naloxone, an opioid receptor antagonist.[436]

Opioids might participate in the pathogenesis of neuroleptic-induced akathisia.[437,438]

Methadone, a narcotic used to detoxify individuals addicted to heroin, can produce choreic movements.[439] Conversely, naloxone, an opioid receptor antagonist, has been reported to palliate the symptoms associated with tardive dyskinesia.[440,441]

Cortical and basal ganglia levels of opioid receptor binding are altered in L-dopa-induced dyskinesia. Moreover, the fact that dyskinetic and nondyskinetic animals often show opposite changes in opioid radioligand binding suggests that the motor response to L-dopa is determined, at least in part, by compensatory adjustments of brain opioid receptors.

Improvement was also observed in neuroleptic-induced parkinsonism and akathisia. Although the dosages administered were considerably higher than those used in the treatment of anxiety, drug side effects were reported to be mild.[426] Although dyskinetic movements may improve with reduction in anxiety,[427] Moss et al.[426] believed that the observed antidyskinetic effect associated with buspirone treatment occurred independently of buspirone's effects on anxiety.

VITAMIN E AND DYSKINESIA

Vitamin E (1200 mg daily) for 1 month significantly ameliorated the severity of tardive dyskinesia.[428] Moreover, Dannon et al.[429] treated 16 patients with tardive dyskinesia with vitamin E in an open trial of on-off-on design. Abnormal involuntary movement scale (AIMS) ratings were performed in every phase of the study. The patients exhibited a significant reduction in their mean AIMS scores during vitamin E treatment. Thus, this finding may suggest a possible role for vitamin E in the treatment of tardive dyskinesia.

AMANTADINE IN TARDIVE DYSKINESIA

Angus et al.[442] reported that amantadine, initially 100 mg/day during the first week and then 300 mg/day during the third week, produced an improvement in dyskinesia without exacerbation of psychosis even with prolonged administration.

CLOZAPINE IN AXIAL TARDIVE DYSTONIA

Functionally disabling tardive dystonia is a well recognized subtype of tardive dyskinesia for which treatment is often ineffective.[159,443–445] Trugman et al.[446] reported a patient with severe axial tardive dystonia who showed improvement for 4 years after treatment with the atypical antipsychotic drug clozapine (625 mg/day). Clozapine differs from conventional neuroleptics in that it has higher affinity for dopamine D_1 and lower affinity for dopamine D_2 receptors than do conventional antipsychotics, which are relatively selective dopamine D_2 antagonists.

CHOLECYSTOKININ IN TARDIVE DYSKINESIA

CCK is known to modulate the nigrostriatal and mesolimbic dopamine neuronal system.[2,416] Kojima et al.[447] in a double-blind, placebo-controlled, and matched-pairs study, reported on the effectiveness of ceruletide (0.8 µg/kg/week), an analog of CCK, in suppressing the symptoms of neuroleptic-induced tardive dyskinesia. Global evaluation of the severity of tardive dyskinesia symptoms over the 8-week study period revealed a significant improvement with ceruletide as compared with placebo. Analysis of the therapeutic response to ceruletide over the course of treatment revealed a slow but long-lasting improvement of tardive dyskinesia symptoms. Side effects, which were mild and transient, consisted mainly of nausea and epigastric discomfort. The incidence of side effects did not differ between the ceruletide- and placebo-treated groups. Ceruletide appears to be a novel and practical treatment that can substantially alleviate the symptoms of dyskinesia.

RISPERIDONE AND TARDIVE DYSKINESIA

Risperidone is a novel benzisoxazole derivative that is characterized as a potent central serotonin receptor antagonist with less potent dopamine D_2 receptor antagonist properties.[448,449] The incidence of tardive dyskinesia with risperidone is low. In all studies to date, no cases of tardive dyskinesia have been conclusively attributed to risperidone. For example, in a Canadian multicenter, double-blind clinical trial of risperidone, 135 hospitalized chronic schizophrenic patients were randomly assigned to one of six parallel treatment groups for 8 weeks: risperidone, 2, 6, 10, or 16 mg/day, haloperidol, 20 mg/day; or placebo. Risperidone (6 to 16 mg)-treated patients showed significantly ($P < 0.05$) lower dyskinetic scores than those receiving placebo, whereas in haloperidol- and placebo-treated patients, no significant differences for dyskinetic symptoms were noted.[450]

The efficacy of risperidone versus haloperidol and amitriptyline in the treatment of patients with a combined psychotic and depressive syndrome has been studied.[451] In a multicenter, double-blind, parallel group trial, the efficacy of risperidone was compared with a combination of haloperidol and amitriptyline over 6 weeks in patients with coexisting psychotic and depressive symptoms with either a schizoaffective disorder, depressive type, a major depression with psychotic features, or a nonresidual schizophrenia with major depressive symptoms according to DSM-III-R criteria. A total of 123 patients (62 risperidone; 61 haloperidol and amitriptyline) were included; the mean daily dosage at endpoint was 6.9 mg risperidone versus 9 mg haloperidol combined with 180 mg amitriptyline. Efficacy results for those 98 patients (47 risperidone; 51 haloperidol/amitriptyline) who completed at least 3 weeks of double-blind treatment revealed, in both treatment groups, large reductions in the Positive and Negative Syndrome Scale-derived Brief Psychiatric Rating Scale (risperidone 37%; haloperidol/amitriptyline 51%) and the Bech-Rafaelsen Melancholia Scale total scores (risperidone 51%; haloperidol/amitriptyline 70%). The reductions in the Brief Psychiatric Rating Scale and the Bech-Rafaelsen Melancholia Scale scores in the total group were significantly larger in the haloperidol/amitriptyline group than in the risperidone group ($P < 0.01$), mostly because of significant differences in the subgroup of patients suffering from depression with psychotic features, whereas treatment differences in the other diagnostic subgroups were not significant.

The incidence of extrapyramidal side effects as assessed by the Extrapyramidal Symptom Rating Scale was slightly higher under risperidone (37%) than under haloperidol/amitriptyline (31%). Adverse events were reported by 66% of risperidone and 75% of haloperidol/amitriptyline patients. The results of this trial suggest that the therapeutic effect of haloperidol/amitriptyline is superior to risperidone in the total group of patients with combined psychotic and depressive symptoms. However, subgroup differences have to be considered.[451] Risperidone also causes neuroleptic malignant syndrome.[452–454]

CONCLUSIONS

A variety of neurological syndromes, involving particularly the extrapyramidal motor system, occur following the use of many drugs, but especially with almost all antipsychotic drugs. These drug-induced movement disorders are particularly prominent during treatment with the high-potency agents (tricyclic piperazines and butyrophenones). There is less likelihood of acute extrapyramidal side effects with *clozapine*, *thioridazine*, or low doses of *risperidone*.

Six varieties of neurological syndromes are characteristic of antipsychotic drugs. Four of these (acute dystonia, akathisia, parkinsonism, and the rare neuroleptic malignant syndrome) usually appear soon after administration of the drug, and two (rare perioral tremor and tardive dyskinesias or dystonias) are late-appearing syndromes that occur following prolonged treatment.

The pharmacodynamics of drug-induced movements are ill defined, and treatments are often unsatisfactory.

ACKNOWLEDGMENTS

The author gratefully acknowledges, appreciates, and admires the unique, dedicated, and excellent secretarial skills of Mrs. Dani Stramer. The studies cited in this paper have been supported by a grant from USPHS no. NS34566.

REFERENCES

1. Ebadi, M., Management of tremor by beta adrenergic receptor blocking agents, *Gen. Pharmacol.* 11, 257–260, 1980.

2. Ebadi, M. and Hama, Y., The possible involvement of dopamine, GABA and cholecystokinin octapeptide in neuroleptic-induced tardive dyskinesia, *J. Res. Commun. Psychol. Psychiatry Behav.,* 12, 225–226, 987.

3. Ebadi, M., Pfeiffer, R.F. and Murrin, L. C., Pathogenesis and treatment of neuroleptic malignant syndrome, *Gen. Pharmacol.* 21, 367–386, 1990.

4. Ebadi, M., Pfeiffer, R.F. and Murrin, L.C., Neuroleptic-induced movement disorders, *Current Aspects Neurosci.* 4, 159–203, 1992.

5. Lang, A. E. and Weiner, W. J. *Drug-induced Movement Disorders,* Futura Publishing Co., Mt. Kisco, NY, 1992.

6. Rodnitzky, R. L. and Keyser, D. L., Neurologic complications of drugs: Tardive dyskinesias, neuroleptic malignant syndrome, and cocaine-related syndromes, *Psychiat. Clinics North America* 15, 491–510, 1992.

7. Ebadi, M. and Srinivasan, S. K., Pathogenesis, prevention, and treatment of neuroleptic-induced movement disorders, *Pharmacal. Rev.* 47, 575–604, 1995.

8. Casey, D. E. Side effect profiles of new antipsychotic agents, *J. Clin. Psychiatry* 57, 40–45, 1996.

9. Casey, D. E., Extrapyramidal syndromes and new antipsychotic drugs: Findings in patients and non-human primate models, *Brit. J. Psychiatry* 168, 32–39, 1996.

10. Marti-Massó, J. F., Poza, J. J. and Lopez de Munain, A., Drugs inducing or aggravating parkinsonism: A review, *Therapie* 51, 568–577, 1996.

11. Kopala, L. C. Spontaneous and drug-induced movement disorders in schizophrenia, *Acta Psychiatr. Scan.* 94, 12–17, 1996.

12. Tu, J. B., Psychopharmacogenetic basis of medication-induced movement disorders, *Int. Clin. Psychopharmacol.* 12, 1–12, 1997.

13. Caley, C. F., Extrapyramidal reactions and the selective serotonin-reuptake inhibitors, *Ann. Pharmacother.* 31, 1481–1489, 1997.

14. Jiménez-Jiménez, F. J., Garcia-Ruiz, P. J. and Molina, J. A., *Drug Safety* 1, 180–204, 1997.

15. Van Harten, P. N., Hoek, H. W., Matroos, G. E., Koeter, M., and Kahn, R. S., The inter-relationships of tardive dyskinesia, parkinsonism, akathisia and tardive dystonia: The Curayao extrapyramidal syndromes study II, *Schizophrenia Res.* 26, 235–242, 1997.

16. Carriero, D. L., Outslay, G., Mayorga, A. J., Aberrnan, J., Gianutsos, G., and Salamone, J. D., Motor dysfunction produced by tacrine administration in rats, *Pharmacal. Biochem. Behav.* 5, 851–858, 1997.

17. Hallett, M., The neurophysiology of dystonia. *Archives of Neurology,* 55, 601–608, 1998.

18. Levine, S. R., Brust, J. C. M., and Futrell, N., Cerebrovascular complications of the use of the "crack" form of alkaloid cocaine, *N. Engl. J. Med.* 323, 699–704, 1990.

19. Alldredge, B. K., Lowenstein, D. H., and Simon, R.P., Seizures associated with recreational drug use, *Neurology* 39, 1037–1039, 1989.

20. Choy-Kwong, M. and Lipton, R. B. Dystonia related to cocaine withdrawal: A case report and pathogenic hypothesis, *Neurology* 39, 996–997, 1989a.

21. Satel, S. and Gawin, F. H., Migraine-like headache and cocaine use, *JAMA* 261, 2995–2996, 1989.

22. Newman, N. M., Diloretto, D. A. and Ho, J. T., Bilateral optic neuropathy and osteolytic sinusitis, Complications of cocaine abuse, *JAMA* 259, 72–74, 1988.

23. Kaku, D. and So, Y. T., Acute femoral neuropathy and iliopsoas infarction in intravenous drug abusers, *Neurology* 40, 1317–1318, 1990.

24. Kumor, K., Cocaine withdrawal dystonia (letter), *Neurology* 40, 863–864, 1990.

25. Choy-Kwong, M. and Lipton, R. B., Seizures in hospitalized cocaine users, *Neurology* 39, 425–427, 1989.

26. Pascual-Leone, A. and Dhuna, A., Cocaine-associated multifocal tics, *Neurology* 40, 999–1000, 1990.

27. Demirkiran, M., Jankovic, J., and Dean, J. M., Ecstasy intoxication: An overlap between serotonin syndrome and neuroleptic malignant syndrome, *Clin. Neuropharmacol.* 19, 157–164, 1996.

28. Chesselet, M. F. and Delfs, J. M., Basal ganglia and movement disorders: An update, *Trends Neurosci.* 19, 417–422, 1996.

29. Aminoff, M. J., Greenberg, D. A., and Simon, R. P., *Clinical Neurology,* 3rd ed., Appleton and Lange, Stanford, CT, pp. 212–233, 1996.

30. Young, A. B. and Penney, J. B. Neurochemical anatomy of movement disorders, *Neurol. Clin.* 2, 417–433, 1984.

31. Graybiel, A. M., Neurotransmitters and neuromodulators in the basal ganglia, *Trends Neurosci.* 13, 244–254, 1990.

32. Garrett, E. A. and Crutcher, M. D., Functional architecture of basal ganglia circuits: Neural substrates of parallel processing, *Trends Neurosci.* 13, 266–271, 1990.

33. Delong, M. R., Primate models of movement disorders of basal ganglia origin, *Trends Neurosci.* 13, 281–285, 1990.

34. Parent, A. and Hazrati, L. N., Functional anatomy of the basal ganglia, I. The corti co-basal ganglia-thalamo-cortical loop, *Brain Res. Rev.* 20, 91–127, 1995.

35. Richardson, P. J., Kase, H., and Jenner, P. G., Adenosine A_{2A} receptor antagonists as new agents for the treatment of Parkinson's disease, *Trends Pharmacal. Sci.* 18, 338–344, 1997.

36. Parent, A. and Hazrati, L. N., Functional anatomy of the basal ganglia, II. The place of subthalamic nucleus and external pallidum in basal ganglia circuitry, *Brain Res. Rev.* 20, 128–154, 1995.

37. Albin, R. L., Aldridge, J. W., Young, A. B., and Gilman, S., Feline subthalamic nucleus neurons contain glutamate-like but not GABA-like or glycine-like immunoreactivity, *Brain Res.* 491, 185–188, 1989.

38. Sadikot, A. F., Parent, A., and Franyois, C., Efferent connections of the centromedian and parafascicular thalamic nuclei in the squirrel monkey: A PHA-L study of subcortical projections, *J. Compo Neurol.* 315, 137–159, 1992.

39. Smith, Y. and Parent, A., Neurons of the subthalamic nucleus in primates display glutamate but not GABA immunoreactivity, *Brain Rev.* 45, 3353–356, 1988.

40. Lavoie, B. and Parent, A., Immunohistochemical study of the serotoninergic innervation of the basal ganglia in the squirrel monkey, *J. Compo Neurol.* 299, 1–16, 1990.

41. Lavoie, B. and Parent, A., The pedunculopontine nucleus in the squirrel monkey. Projections to the basal ganglia as revealed by anterograde tract-tracing methods, *J. Camp. Neurol.* 344, 210–231, 1994.

42. Lavoie, B. and Parent, A. The pedunculopontine nucleus in the squirrel monkey. Cholinergic and glutamatergic projections to the substantia nigra, *J. Compo. Neurol.* 344, 232–241, 1994.

43. Lavoie, B., Smith, Y., and Parent, A., Dopaminergic innervation of the basal ganglia in the squirrel monkey as revealed by tyrosine hydroxylase immunohistochemistry, *J. Compo. Neurol.* 289, 36–52, 1989.

44. Wichmann, T. and DeLong, M. R., Pathophysiology of parkinsonian motor abnormalities, *Adv. Neurol.* 60, 53–61, 1993.

45. Cooper, J. R., Bloom, F. E. and Roth, R. H., *The Biochemical Basis of Pharmacology,* 6th ed., Oxford University Press, New York, 1991.

46. Hietala, J., Lappalainen, J., Koulu, M., and Syvalahti, E., Dopamine D_1 receptor antagonism in schizophrenia: Is there reduced risk of extrapyramidal side-effects? *Trends Pharmacal. Sci.* 11, 406–410, 1990.

47. Jackson, D. M. and Westlind-Danielsson, A., Dopamine receptors: Molecular biology, biochemistry and behavioral aspects, *Pharmac. Ther.* 64, 291–369, 1994.

48. Ogawa, N., Molecular and chemical neuropharmacology of dopamine receptor subtypes, *Acta Med. Okayama* 49, 1–11, 1995.

49. Wise, R.A., D_1- and D_2-type contributions to psychomotor sensitization and reward: Implications for pharmacological treatment strategies, *Clin. Neuropharmacol.* 18, S74–S83, 1995.

50. Meador-Woodruff, J. H., Damask, S. P., Wang, J., Haroutunian, V., Davis, K.L., and Watson, S.J., Dopamine receptor mRNA expression in human striatum and neocortex, *Neuropsychopharmacology* 15, 17–29, 1996.

51. Gurevich, E. Y., Bordelon, Y., Shapiro, M., Arnold, S. E., Gur, R. E. and Joyce, J. N., Mesolimbic dopamine D_3 receptors and use of anti psychotics in patients with schizophrenia, A postmortem study, *Arch. Gen. Psychiatry* 54, 225–232, 1997.

52. Liegeois, J. F., Eyrolles, L., Bruhwyler, J., and Delarge, J., Dopamine D_4 receptors: A new opportunity for research on schizophrenia, *Current Med. Chem.* 577–100, 1998.

53. Farooqui, S. M., Brock, J. W., Hamdi, A., and Prasad, C., Antibodies against synthetic peptides predicted from the nucleotide sequence of D_2 receptor recognize native dopamine receptor protein in rat striatum, *J. Neurochem.* 57, 1363–1369, 1991.

54. Mc Vittie, L. D., Ariano, M. A., and Sibley, D. R., Characterization of antipeptide antibodies for the localization of 02 dopamine receptors in rat striatum, *Proc. Natl. Acad. Sci. USA* 88, 1441–1445, 1991.

55. Kebabian, L. W. and Caine, D. B. Multiple receptors for dopamine, *Nature* 277, 93–96, 1979.

56. Seeman, P., Dopamine receptors and the dopamine hypothesis of schizophrenia, *Synapse* 1, 133–152, 1987.

57. Schwartz, J. C., Giros, B., Martes, M. P., and Sokoloff, P., The dopamine receptor family: Molecular biology and pharmacology, *Semm. Neurosci.* 499–108, 1992.

58. Zhou, Q. Y., Grandy, D. K., Thambi, L., Kushner, J. A., Van, T. H., Cone, R., Pribnow, D., Salon, J., Bunzow, J. R., and Civelli, O., Cloning and expression of human and rat D_1 dopamine receptors, *Nature* 347, 76–80, 1990.

59. Neve, K. A., Cox, B. A., Henningsen, R. A., Spanoyannis, A., and Neve, R. L., Pivotal role for aspartate-80 in the regulation of dopamine D_2 receptor affinity for drugs and inhibition of adenylyl cyclase, *Mol. Pharmacal.* 39, 733–739, 1991.

60. Monsma, F. J., McVittie, L. D., Gerfen, C. R., Mahan, L. C., and Sibley, D. R., Multiple D_2 dopamine receptors produced by alternative RNA splicing, *Nature* 342, 926–929, 1989.

61. Giros, B., Sokoloff, P., Martres, M. P., Riou, J. F., Emorine, L. J., and Schwarz, J. C., Alternative splicing directs the expression of two D_2 dopamine receptor isoforms, *Nature* 342, 923–926, 1989.

62. Lidow, M. S., Williams, G. V. and Goldman-Rakic, P. S., The cerebral cortex: A case for a common site of action of antipsychotics, *Trends Pharmacal. Sci.* 19, 136–140, 1998.

63. Manzanares, J., Durham, R. A., Lookingland, K. J., and Moore, K. E., σ-opioid receptor-mediated regulation of central dopaminergic neurons in the rat, *Eur. J. Pharmacal.* 249, 107–112, 1993.

64. Di Chiara, G., Morelli, M., and Consolo, S., Modulatory functions of neurotransmitters in the striatum: ACh/dopamine/NMDA interactions, *Trends Neurosci.* 17, 228–233, 1994.

65. Pan, Z. Z., σ-opposing actions of the κ-opioid receptor, *Trend Pharmacal. Sci.* 19, 94–98, 1998.

66. Coleman, R., Current drug therapy for Parkinson's disease, A review, *Drugs Aging* 2, 112–124, 1992.

67. Whitehouse, P., Matino, A., Marcus, K., Zweig, R., Singer, H., Price, D., and Kellar, K., Reduction in acetylcholine and nicotine binding in several degenerative diseases, *Arch. Neurol.* 45, 722–724, 1988.

68. Lange, K., Wells, F., Jenner, P., and Marsden, P., Altered muscarinic and nicotinic receptor densities in cortical and subcortical regions in Parkinson's disease, *J. Neurochem.* 60, 197–203, 1993.

69. Grace, A. A. and Bunney, B. S., Opposing effects of striatonigral feedback pathways on midbrain dopamine cell activity, *Brain Res.* 333, 271–284, 1985.

70. Grace, A. A. and Bunney, B.S., Paradoxical GABA excitation of nigral dopaminergic cells: Indirect mediation through reticulata inhibitory interneurons, *Eur. J. Pharmacol.* 59, 211–218, 1979.

71. Deniau, J. M., Feger, J., and LeGuyader, C., Striatal evoked inhibition of identified nigrothalamic neurons, *Brain Res.* 104, 245–256, 1976.

72. Grace, A. A. and Bunney, B. S., Electrophysiological properties of midbrain dopamine neurons, In *Psychopharmacology,* Ed. by F. E. Bloom and D. J. Kupfer, Raven Press, NY, pp. 163–177, 1995.

73. Hollerman, J. R. and Grace, A. A., Subthalamic nucleus cell activity in the 6-OHDA-treated rat: Basal activity and response to haloperidol, *Brain Res.* 590, 291–299, 1992.

74. Kita, T., Chang, H. T. and Kitai, S. T., Pallidal inputs to subthalamus: Intracellular analysis, *Brain Res.* 264, 255–265, 1983.

75. Rouzaire-Dubois, B., Hammond, C., Hamon, B. and Feger, J., Pharmacological blockade of the globus pallidus-induced inhibitory response of subthalamic cells in the rat, *Brain Res.* 200, 321–329, 1980.

76. Miller, W. C. and DeLong, M. R., Altered tonic activity of neurons in the globus pallidus and subthalamic nucleus in the primate MPTP model of parkinsonism, in *The Basal Ganglia II-Structure and Function: Current Concepts,* Ed. by M. B. Carpenter and A. Jayaraman, Plenum Press, New York, pp. 415–427, 1987.

77. Mitchell, I. J., Clarke, C. E., Boyce, S., Robertson, R. F., Peggs, D., Sambrook, M. A., and Crossman, A. R., Neural mechanisms underlying parkinsonian symptoms based upon regional uptake of 2-deoxyglucose in monkeys exposed to 1-methyl-4-phenyl-1, 2, 3, 6-tetrahydropyridine, *Neuroscience* 32, 213–226, 1989.

78. Hammond, C., Deniau, J. M., Rizk, A. and Feger, J., Electrophysiological demonstration of an excitatory subthalamonigral pathway in the rat, *Brain Res.* 151, 235–244, 1978.

79. Nakanishi, H., Kita, H., and Kitai, S. T., Intracellular study of rat substantia nigra pars reticulata neurons in an *in vitro* slice preparation: Electrical membrane properties and response characteristics to subthalamic stimulation, *Brain Res.* 437, 45–55, 1987.

80. Smith, I. D. and Grace, A. A., The regulation of nigral dopamine neurons firing by afferents from the subthalamic nucleus, *Synapse* 12, 287–303, 1992.

81. Mereu, G., Costa, E., Armstrong, D. M. and Vicini, S., Glutamate receptor subtypes mediate excitatory synaptic currents of dopamine neurons in midbrain slices, *J. Neurosci.* 11, 1359–1366, 1991.

82. Johnson, S. W. and North, R. A., Two types of neuron in the rat ventral tegmental area and their synaptic inputs, *J. Physiol.* 450, 455–468, 1992.

83. Klawans, H. L. and Weiner, W. J., *Textbook of Clinical Neuropharmacology,* Raven Press Ltd., New York, 1981.

84. Gao, X. M., Kakigi, T., Friedman, M. B., and Tamminga, C. A., Tiagabine inhibits haloperidol-induced oral dyskinesias in rats, *J. Neural Transm.* 95, 63–69, 1994.

85. Maggi, A., U'Prichard, D. C., and Enna, S. J., Differential effects of antidepressant treatment on brain monoaminergic receptors, *Eur. J. Psychopharmacol.* 61, 91–98, 1980.

86. Tang, S. W. and Seeman, P., Effects of antidepressant drugs on serotonergic and adrenergic receptors, *Arch. Pharmacol.* 311, 255–261, 1980.

87. Bjorksten, K. S. and Walinder, J., Does mianserin induce involuntary movements in brain damaged patients? *Int. Clin. Psychopharmacol.* 8, 203–204, 1993.

88. Raja, M., Tardive dystonia: Prevalence, risk factors, and comparison with tardive dyskinesia in a population of 200 acute psychiatric inpatients, *Eur. Arch. Psychiatry Clin. Neurosci.* 245, 145–151, 1995.

89. Sachdev, P., The identification and management of drug-induced akathisia, *Contemp. Neurol. Ser. Drugs* 4, 28–45, 1995.

90. Jankovic, J., Tardive syndromes and other drug-induced movement disorders, *Clin. Neuropharmacol.* 18, 197–214, 1995.

91. Van Putten, T. and Marder, S. R., Behavioral toxicity of antipsychotic drugs, *J. Clin. Psychiatry* 48, 13–19, 1987.

92. Adler, L., Angrist, B., Peselow, E., Corwin, J., and Rotrosen, J., Efficacy of propranolol in neuroleptic-induced akathisia, *J. Clin. Psychopharmacol.* 5, 164–166, 1985.

93. Haskovec, L., Nouvelles remarques sur I'akathisie, *Nouv. Iconogr. Salpetriere* 16, 287–296, 1903.

94. Kalinowsky, L. B., Appraisal of the "tranquilizers" and their influences on other somatic treatment in psychiatry, *Am. J. Psychiatry* 115, 294–300, 1958.

95. Ayd, F. J., Drug-induced extrapyramidal reactions, Their clinical manifestations and treatment with akineton, *Psychosomatics* 1, 143–150, 1960.

96. Ayd, F. J., A survey of drug-induced extrapyramidal reactions, *J. Am. Med. Assoc.* 175, 1054–1060, 1961.

97. Ayd, F. J., Neuroleptics and extrapyramidal reactions in psychiatric patients, *Rev. Can. Bioi.* 20, 451–459, 1961.

98. Marsden, C. D. and Jenner, P., The pathophysiology of extrapyramidal side-effects of neuroleptic drugs, *Psychol. Med.* 10, 55–72, 1980.

99. Braude, W. M., Barnes, T. R. E., and Gore, S. M., Clinical characteristics of akathisia, *Br. J. Psychiatry* 143, 134–150, 1983.

100. Van Putten, T., May, P. R. A., and Marder, S. R., Akathisia with haloperidol and thiothixene, *Arch. Gen. Psychiatry* 41, 1036–1039, 1984.

101. Barnes, T. R. E. and Braude, W. M., Toward a more reliable diagnosis of akathisia (in reply), *Arch. Gen. Psychiatry* 43, 1016, 1986.

102. Ekbom, K. A., Restless legs syndrome, *Neurology* 10, 868–873, 1960.

103. Bing, R., Uber einige bemerkenswerte begleiterscheinungen der extrapyramidal en rigiditat (akathesie-mikrographie-kinesia paradoxica), *Schweiz. Med. Wochenschr.* 53, 167–171, 1923.

104. Hopkins, A., Movement disorders, in *Clinical Neurology,* Oxford University Press, pp. 208–239, 1993.

105. Montplaisir, R., Godbout, R., Poirier, G., and Bedard, M. A., Restless legs syndrome and periodic movements in sleep: Physiopathology and treatment with L-dopa, *Clin. Neuropharmacol.* 9, 456–463, 1986.

106. Akpinar, S., Restless legs syndrome treatment with dopaminergic drugs, *Clin. Neuropharmacol.* 10, 69–79, 1987.

107. Gold, M. S., Redmond, D. E., and Kleiber, H. D., Clonidine blocks the acute opiate withdrawal syndrome, *Lancet* 2, 403–405, 1978.

108. Washton, A. M. and Resnick, R. B., Clonidine in opiate withdrawal: Review and appraisal of clinical findings, *Pharmacotherapy* 1, 140–146, 1981.

109. Charney, D. S., Sternber, D. E., Kleber, H. D., Henninger, G. R., and Redmond, D.E., The clinical use of clonidine in abrupt withdrawal from methadone, *Arch. Gen. Psychiatry* 38, 1273–1277, 1981.

110. Roehrich, H. and Gold, M. S., Propranolol as adjunct to clonidine in opiate detoxification, *Am. J. Psychiatry* 144, 1099–1100, 1987.

111. Simpson, G. M., Neurotoxicity of major tranquilizers. In *Neurotoxicology,* Ed. by L. Roizin, H. Shiroki, and N. Grcevic, Raven Press, New York, p. 3, 1977.

112. Munetz, M. R. and Comes, C. L., Distinguishing akathisia and tardive dyskinesia: A review of the literature, *J. Clin. Psychopharmacol.* 3, 343–350, 1982.

113. Barnes, T. R. E. and Braude, W. M., Akathisia variants and tardive dyskinesia, *Arch. Gen. Psychiatry* 42, 874–878, 1985.

114. Munetz, M. R., Akathisia variants and tardive dyskinesia, *Arch. Gen. Psychiatry* 43, 1015, 1986.

115. Van Putten, T., The many faces of akathisia, *Compo Psychiatry* 16, 43–47, 1975.

116. Kendler, K. S., A medical student's experience with akathisia, *Am. J. Psychiatry* 133, 454–455, 1976.

117. Adler, L. A., Angrist, B., Reiter, S., and Rotrosen, J., Neuroleptic-induced akathisia: A review, *Psychopharmacology* 97, 1–11, 1989.

118. Chouinard, G., Annable, L., Ross-Chouinard, A., and Nestoros, J. N., Factors related to tardive dyskinesia, *Am. J. Psychiatry* 136, 79–83, 1979.

119. Kruse, W., Treatment of drug-induced extrapyramidal symptoms, *Dis. Nerv. Syst.* 21, 79–81, 1960.

120. Neu, C., DiMascio, A., and Demirgian, E., Antiparkinsonian medication in the treatment of extrapyramidal side-effects: Single or multiple daily doses? *Curro Ther. Res.* 14, 246–251, 1972.

121. Friis, T., Christensen, T. R., and Gerlach, J., Sodium valproate and biperiden in neuroleptic-induced akathisia, parkinsonism and hyperkinesia: A double-blind cross-over study with placebo, *Acta Psychiatr. Scand.* 67, 178–187, 1982.

122. Merrick, E. M. and Schmitt, P. P., A controlled study of the clinical effects of amantadine hydrochloride (Symmetrel), *Curro Ther. Res.* 15, 552–558, 1973.

123. Stenson, R. L., Donlon, P. T., and Meyer, J. E., Comparison of benztropine mesylate and amantadine Hcl in neuroleptic-induced extrapyramidal symptoms, *Compo Psychiatry* 17, 763–768, 1976.

124. DiMascio, A., Bernardo, D. L., Greenblatt, D., and Marder, J. E., A controlled trial of amantadine in drug-induced extrapyramidal disorders, *Arch. Gen. Psychiatry* 33, 559–602, 1976.

125. Zubenko, G. S., Barriera, P. and Lipinski, J. F. Development of tolerance to the therapeutic effect of amantadine on akathisia, *J. Clin. Psychopharmacol.* 4, 218–219, 1984.

126. Donlon, P., The therapeutic use of diazepam for akathisia, *Psychosomatics* 14, 222–225, 1973.

127. Gagrat, D., Hamilton, J., and Belmatier, R., Intravenous diazepam in the treatment of neuroleptic-induced dystonia or akathisia, *Am. J. Psychiatry* 135, 1232–1233, 1978.

128. Bartels, M., Heide, K., Mann, K., and Schied, H. W., Treatment of akathisia with lorazepam: An open clinical trial, *Pharmacopsychiatry* 20, 51–53, 1987.

129. Kutcher, S. P., Mackenzie, S., Galarraga, W., and Szalai, J., Clonazepam treatment of adolescents with neuroleptic-induced akathisia, *Am. J. Psychiatry* 144, 823–824, 1987.

130. Lipinski, J. F., Zubenko, G. S., Cohen, B. M., and Barriera, P. J., Propranolol in the treatment of neuroleptic-induced akathisia, *Am. J. Psychiatry* 141, 412–415, 1984.

131. Kulik, A. V. and Wilbur, R., Case report of propranolol (Inderal) pharmacotherapy for neuroleptic-induced akathisia and tremor, *Prog. Neuropsychopharmacol. Bioi. Psychiatry* 7, 223–225, 1983.

132. Adler, L., Angrist, B., Peselow, E., Corwin, J., Maslansky, R., and Rotrosen, J., A controlled assessment of propranolol in the treatment of neuroleptic-induced akathisia, *Br. J. Psychiatry* 149, 42–45, 1986.

133. Adler, L. A., Reiter, S., Corwin, J., Hemdal, P., Angrist, B., and Rotrosen, J., Differential effects of benztropine and propranolol in akathisia, *Psychopharmacol. Bull.* 23, 519–521, 1987.

134. Reiter, S., Adler, L., ErIe, S., and Duncan, E., Neuroleptic-induced akathisia treated with pindolol, *Am. J. Psychiatry* 144, 383–384, 1987.

135. Zubenko, G. S., Lipinski, J. F., Cohen, B. M. and Barriera, P. J., Comparison of metoprolol and propranolol in the treatment of akathisia, *Psychiatr. Res.* 11, 143–148, 1984.

136. Derome, E., Elinck, W., Buylaret, W., and Van der Straeten, M., Which beta-blocker for the restless leg? *Lancet* 857, 1984.

137. Reiter, S., Adler, L., Angrist, B., Corwin, J., and Rotrosen, J., Atenolol and propranolol in neuroleptic-induced akathisia, *J. Clin. Psychopharmacol.* 7, 279–280, 1987.

138. Zubenko, G. S., Cohen, B. M., Lipinski, J. F., and Jonas, J.M., Use of clonidine in the treatment of akathisia, *Psychiatr. Res.* 13, 253–259, 1984.

139. Sachdev, P. S. and Saharov, T., The effects of β-adreno-ceptor antagonists on a rat model of neuroleptic-induced akathisia, *Psychiatry Res.* 72, 133–140, 1997.

140. Sandyk, R., Consroe, P. F., and Iacono, R. P., L-tryptophan in drug-induced movement disorders with insomnia, *N. Eng. J. Med.* 314, 1257, 1986.

141. Kramer, M. S., DiJohnson, C., Davis, P., Dewey, D. A., and DiGiambattista, S., L-tryptophan in neuroleptic-induced akathisia, *Biol. Psychiatry* 27, 671–672, 1990.

142. Tanquary, J., Case report 1: Akathisia responsive to bupropion, *J. Drug Dev.* 6, 69–70, 1993.

143. Oppenheim, H., Uber eine eigenartige Kramptkrankheit des kindlischen und jugendichen Alters (Dysbasia lordotica progressiva, Dystonia musculorum deformans), *Neurolog. Centralblatt* 30, 1090–1107, 1911.

144. Eldridge, R., The torsion dystonias: Literature review and genetic and clinical studies, *Neurology* 20, 1–78, 1970.

145. Marsden, C. D., Dystonia: The spectrum of the disease, in *The Basal Ganglia,* Ed. By M. D. Yahr, Raven Press, New York, pp. 351–367, 1976.

146. Lang, A. E., Dopamine agonists in the treatment of dystonia, *Clin. Neuropharmacol.* 8, 38–57, 1985.

147. McGeer, E. G., and McGeer, P. L., The dystonias, *Can J. Neurol. Sci* 15, 447–483, 1988.

148. Dickey, W. and Morrow, J. I., Drug-induced neurological disorders, *Prog. Neurobiol.* 34, 331–342, 1990.

149. Rosenberg, R. N., *Neurology,* Grune and Stratton, New York, 1980.

150. Melamed, E., Early-morning dystonia, A late side effect of long-term levodopa therapy in Parkinson's disease, *Arch. Neurol.* 36, 308–310, 1979.

151. Muenter, M. D., Sharpless, N. S., Tyce, G. M., and Darley, F. L., Patterns of dystonia (>IDI= and >DID=) in response to L-dopa therapy by Parkinson's disease, *Mayo Clin. Proc.* 52, 163–174, 1977.

152. Lees, A. J. and Stern, G. M., Bromocriptine in treatment of levodopa-induced end-of-dose dystonia, *Lancet* 2, 215–216, 1980.

153. Ilson, J., Fahn, S., and Cote, L., Painful dystonic spasms in Parkinson's disease, *Adv. Neurol.* 40, 395–398, 1984.

154. Meldrum, B. S., Gill, M., Anlezark, G. M., and Marsden, C. D., Acute dystonia as an idiosyncratic response to neuroleptic drugs in baboons, *Brain* 100, 313–326, 1977.

155. Bateman, D. N., Rawlins, M. D. and Simpson, J. M., Extrapyramidal reactions with metoclopramide, *Br. Med. J.* 291, 930–932, 1985.

156. Chadwick, D., Reynolds, E. H. and Marsden, C. D., Anticonvulsant-induced dyskinesias: A comparison with dyskinesias induced by neuroleptics, *J. Neurol. Neurosurg. Psychiatry* 39, 1210–1218, 1979.

157. Critchley, E. M. R. and Phillips, M., Unusual idiosyncratic reaction to carbamazepine, *J. Neurol. Neurosurg. Psychiatry* 51, 1238, 1988.

158. Crawford, J. P., Dystonic reactions to high dose propranolol, *Br. Med. J.* 2, 1156–1157, 1977.

159. Burke, R. E., Fahn, S., and Jankovic, J., Tardive dystonia: Late-onset and persistent dystonia caused by antipsychotic drugs, *Neurology* 32, 1335–1346, 1982.

160. Lavenstein, B. L. and Cantor, F. K., Acute dystonia: An unusual reaction to diphenhydramine, *J. Am. Med. Assoc.* 236, 291, 1976.

161. Smith, R. E. and Domino, E. F., Dystonic and dyskinetic reactions induced by H_1 antihistaminic medication, in *Tardive Dyskinesia: Research Treatment,* Ed. by W. E. Fann, R. C. Smith, and J. M. Davis, Spectrum, New York, pp. 325–332, 1980.

162. Howrie, D. L., Rowley, A. H., and Krenzelok, E. P., Benztropine-induced acute dystonic reaction, *Ann. Emerg. Med.* 15, 141–143, 1986.

163. Rupniak, N. M. J., Jenner, P., and Marsden, C. D., Acute dystonia induced by neuroleptic drugs, *Psychopharmacology* 88, 403–419, 1986.

164. Sramek, J. J., Simpson, G. M., Morrison, R. L., and Heiser, J. F., Anticholinergic agents for prophylaxis of neuroleptic-induced dystonic reactions: A prospective study, *J. Clin. Psychiatry* 47, 305–309, 1986.

165. Swett, C., Drug-induced dystonia, *Am. J. Psychiatry* 132, 532–534, 1975.

166. Chiles, J. A., Extrapyramidal reactions in adolescents treated with high-potency antipsychotics, *Am. J. Psychiatry* 135, 239–240, 1978.

167. Keepers, G. A. and Casey, D. E., Clinical management of acute neuroleptic-induced extrapyramidal syndromes, *Curro. Psychiatr. Ther.* 23, 139–157, 1986.

168. Kumor, K., Haloperidol-induced dystonia in cocaine addicts. *Lancet* 2, 1341–1342, 1986.

169. Pratty, J. S., Ananth, J., and O'Brien, J.E., Relationship between dystonia and serum calcium levels, *J. Clin. Psychiatry* 47, 418–419, 1986.

170. Freed, E., Alcohol-triggered neuroleptic-induced tremor, rigidity and dystonia, *Med. J. Aust.* 445, 1981.

171. Sovner, R. and McGorrill, S., Stress as a precipitant of neuroleptic-induced dystonia, *Psychosomatics* 23, 707–709, 1982.

172. Malen, R. L., The role of psychological factors in reversible, drug-related dystonic reactions, *Mt. Sinai J. Med.* 43, 46–70, 1976.

173. Dick, D. J. and Saunders, M., Persistent involuntary movements after treatment with flupenthixol, *Br. Med. J.* 282, 1756, 1981.

174. Wood, G. M. and Waters, A. L., Prolonged dystonic reaction of chlorpromazine in myxedema coma, *Post grad. Med. J.* 56, 192–193, 1980.

175. Gospe, S. M., Jr. and Jankovic, J., Drug-induced dystonia in neuronal ceroidlipofuscinosis, *Pediatr. Neurol.* 2, 236–237, 1986.

176. Pettit, H. O., Pan, H. T., Parsons, L. H., and Justice, J. B., Jr., Extracellular concentrations of cocaine and dopamine are enhanced during chronic cocaine administration, *J. Neurochem.* 55, 798–804, 1990.

177. Flaherty, J. A. and Lahmeyer, H. W., Laryngeal-pharyngeal dystonia as a possible cause of asphyxia with haloperidol treatment, *Am. J. Psychiatry* 135, 1414–1415, 1978.

178. McDanal, C. E., Jr., Brief letter on case of laryngeal-pharyngeal dystonia induced by haloperidol, relieved by benztropine, *Am. J. Psychiatry* 138, 1262–1263, 1981.

179. Menuck, M., Laryngeal-pharyngeal dystonia and halo-peridol, *Am. J. Psychiatry* 138, 394–395, 1981.

180. Ravi, S. D., Borge, G. F., and Roach, F. L,. Neuroleptics laryngeal-pharyngeal dystonia, and acute renal failure, *J. Clin. Psychiatry* 43, 300, 1982.

181. Holmes, V. F., Adams, F., and Fernandez, F., Respiratory dyskinesia due to antiemetic therapy in a cancer patient, *Cancer Treat. Rep.* 71, 415, 1987.

182. Corre, K. A., Nieman, J. T., and Bessen, H. A., Extended therapy for acute dystonic reactions, *Ann. Emerg. Med.* 13, 194–197, 1984.

183. Gardos, G., Cole, J. O., Salomon, M. and Schniebolk, S., Clinical forms of severe tardive dyskinesia, *Am. J. Psychiatry* 144, 895–902, 1987.

184. Carella, F., Girotti, F., Scigliano, G., Caraceni, T., Joder-Ohlenbusch, A. M., and Schechter, P. J., Double-blind study of oral γ-vinyl GABA in the treatment of dystonia, *Neurology* 36, 98–100, 1986.

185. Nygaard, T., and Duvoisin, R., Hereditary dystonia-parkinsonism syndrome of juvenile onset, *Neurology* 36, 1424–1428, 1986.

186. Rondot, P. and Ziegler, M., Dystonia-L-dopa responsive or juvenile parkinsonism? *J. Neural Trans.* 19, 273–281, 1983.

187. Poewe, W. H. and Lees, A. J., The pharmacology of foot dystonia in parkinsonism, *Clin. Neuropharmacol.* 10, 47–56, 1987.

188. Faulstich, M. E., Carnrike, C. L. M., and Williamson, D. A., Blepharospasm and Meige syndrome: A review of diagnostic, aetiological and treatment approaches, *J. Psychosom. Res.* 29, 89–94, 1985.

189. Frueh, B. R., Felt, D. P., Wojno, T. H., and Musch, D. C., Treatment of blepharospasm with botulinum toxin, *Arch. Ophthalmol.* 102, 1464–1468, 1984.

190. Carruthers, J. and Stubbs, H. A., Botulinum toxin for benign essential belpharospasm, hemifacial spasm and age-related lower eyelid entropion, *Can. J. Neurol. Sci.* 14, 42–45, 1987.

191. Olney, J. W., Price, M. T., and Labruyere, J., Anti-parkinsonian agents are phencyclidine agonists and N-methyl-aspartate antagonists, *Eur. J. Pharmacol.* 142, 319–320, 1987.

192. Cremonesi, E. and Murata, K. N., Infiltration of a neuromuscular relaxant in diagnosis and treatment of torticollis, *Anesth. Anal.* 65, 1077–107, 1986.

193. Stahl, S. M. and Berger, P. A., Bromocriptine in dystonia, *Lancet* 745, 1981.

194. Stahl, S. M., Davis, K. L., and Berger, P. A., The neuropharmacology of tardive dyskinesia, spontaneous dyskinesia, and other dystonias, *J. Clin. Psychopharmacol.* 2, 321–328, 1982.

195. Stahl, S. M., Thornton, J. E., Simpson, M. L., Berger, P. A., and Napoliello, M. J., γ-Vinyl-GABA treatment of tardive dyskinesia and other movement disorders, *Biol. Psychiatry* 20, 888–893, 1985.

196. Ortiz, A., Neuropharmacological profile of Meige's disease: overview and a case report, *Clin. Neuropharmacol.* 6, 297–304, 1983.

197. Keegan, D. L. and Rajput, A. H., Drug induced dystonia tarda: Treatment with L-dopa, *Dis. Nerv. Syst.* 38, 167–169, 1973.

198. Bartels, M., Riffel, B., and Stohr, M., Tardive dystonie: Eine seltene nebenwirkung nach neuroleptika-langzeit-behandlung, *Nervenarzt* 53, 674–676, 1982.

199. Peatfield, R. C. and Spokes, E. G. S., Phenothiazine-induced dystonias, *Neurology* 34, 260, 1984.

200. Guy, N., Raps, A., and Assael, M., The Pisa syndrome during maintenance antipsychotic therapy, *Am. J. Psychiatry* 143, 1492, 1986.

201. Gimenez-Roldan, S., Mateo, D., and Bartolome, P., Tardive dystonia and severe tardive dyskinesia, *Acta Psychiatr. Scand.* 71, 488–494, 1985.

202. Greene, P., Baclofen in the treatment of dystonia, *Clin. Neuropharmacol.* 15, 276–288, 1992.

203. Kao, I., Drachrnan, D. B., and Price, D. L., Botulinum toxin: Mechanism of presynaptic blockade, *Science* (Washington, DC) 193, 1257–1258, 1976.

204. Lange, D. J., Brin, M. F., Fahn, S., and Lovelace, R. E., Distant effects of locally injected botulinum toxin: Incidence and course, *Adv. Neurology* 50, 609–613, 1988.

205. Scott, A. B., Botulinum toxin injection of eye muscles to correct strabismus, *Trans. Am. Ophthalmol. Soc.* 79, 734–770, 1981.

206. Tsoy, E. A., Buckley, E. G., and Dutton, J. J., Treatment of blepharospasm with botulinum toxin, *Am. J. Ophthalmol.* 99, 176–179, 1985.

207. Scott, A. B., Kennedy, R. A., and Stubbs, H. A., Botulinum, A toxin injection as a treatment for blepharospasm, *Arch. Ophthalmol.* 103, 347–350, 1985.

208. Shorr, N., Seiff, S., and Kopelman, J., The use of botulinum toxin in blepharospasm, *Am. J. Ophthalmol.* 99, 542–546, 1985.

209. Elston, J., and Russell, R., Effect of treatment with botulinum toxin on neurogenic blepharospasm, *Br. Med. J.* 290, 1857–1859, 1985.

210. Mauriello, J. A., Blepharospasm, Meige syndrome and hemifacial spasm: Treatment with botulinum toxin, *Neurology* 35, 1499–1500, 1985.

211. Fahn, S., List, T., Moskowitz, C., Brin, M. F., Bressman, S., Burke, R., and Scott, A., Double-blind controlled study of botulinum toxin for blepharospasm, *Neurology* 35, 271, 1985.

212. Perman, K., Baylis, H., Rosenblum, A., and Kirschen, D., The use of botulinum toxin in the medical management of benign essential blepharospasm, *Ophthalmology* 93, 1–3, 1986.

213. Elston, J. S. Botulinum toxin treatment of blepharospasm, *Adv. Neurol.* 50, 579–581, 1988.

214. Tsui, J. K., Eisen, A., Mak, E., Carruthers, J., Scott, A., and Calne, D. B., A pilot study on the use of botulinum toxin in spasmodic torticollis, *Can. J. Neurol. Sci.* 12, 314–316, 1985.

215. Savino, P., Sergott, R., Bosley, T., and Schatz, N., Hemifacial spasm treated with botulinum A toxin injection, *Arch. Ophthalmol.* 103, 1305–1306, 1985.

216. Brin, M. F., Fahn, S., Moskowitz, C., Friedman, A., Shale, H. M., Greene, P. E., Blitzer, A., List, T., Lange, D., Lovelance, R. E., and McMahon, D., Localized

injections of botulinum toxin for the treatment of focal dystonia and hemifacial spasm, *Adv. Neurol.* 50, 599–608, 1988.

217. Tsui, J. K. C., Eisen, A., and Calne, D. B., Botulinum toxin in spasmodic torticollis, *Adv. Neurol.* 50, 593–597.1988.

218. Jankovic, J., Blepharospasm and oromandibular-laryngeal-cervical dystonia: A controlled trial of botulinum A toxin therapy, *Adv. Neurol.* 50, 583–591, 1988.

219. Davis, D. and Jabbari, B., Significant improvement of stiff-person syndrome after paraspinal injection of botulinum toxin A, *Mov. Disord.* 8, 371–373, 1993.

220. Hyland, K., Fryburg, J. S., Wilson, W. G., Bebin, E. M., Arnold, L. A., Gunasekera, R. S., Jacobson, R. D., Rost-Ruffner, E., and Trugman, J. M., Oral phenylalanine loading in dopa-responsive dystonia: A possible diagnostic test, *Neurology* 48, 1290–1297, 1997.

221. Itoh, H., Ohtsuka, N., Ogita, K., Vagi, G., Miura, S., and Koga, Y., Malignant neuroleptic syndrome—Its present status in Japan and clinical problems, *Folia Psychiat. Neurol. Jap.* 31, 565–576, 1977.

222. Weinberger, D. R. and Kelly, M. J., Catatonia and malignant syndrome: a possible complication of neuroleptic administration, *J. Nerv. Ment. Dis.* 165, 263–268, 1977.

223. Geller, B. and Greydanus, D. E., Haloperidol-induced comatose state with hyperthermia and rigidity in adolescence: Two case reports with a literature review, *J. Clin. Psychiat.* 40, 102–103, 1979.

224. Caroff, S. N., The neuroleptic malignant syndrome, *J. Clin. Psychiatry* 41, 79–83, 1980.

225. Destee, A., Petit, H., and Warot, M., Le syndrome malin des neuroleptiques, *Nouv. Presse Med.* 19, 178, 1981.

226. Caroff, S. N., Rosenberg, H., and Gerber, J. C., Neuroleptic malignant syndrome and malignant hyperthermia, *Lancet* 244, 1983.

227. Kurlan, R., Hamill, R., and Shoulson, I., Neuroleptic malignant syndrome, *Clin. Neuropharmacol.* 7, 109–120, 1984.

228. Addonizio, G., Susman, V. L., and Roth, S. D. Symptoms of neuroleptic malignant syndrome in 82 consecutive inpatients, *Am. J. Psychiatry* 143, 1587–1590, 1986.

229. Addonizio, G., Susman, V. L., and Roth, S. D., Neuroleptic malignant syndrome: Review and analysis of 115 cases, *Biol. Psychiatry* 22, 1004–1020, 1987.

230. Kellam, A. M. P., The neuroleptic malignant syndrome, so-called a survey of the world literature, *Brit. J. Psychiat.* 150, 752–759, 1987.

231. Kaufmann, C. A. and Wyatt, R. J., Neuroleptic malignant syndrome, in *Psychopharmacology, The Third Generation of Progress,* Ed. by H. Y. Meltzer, Raven Press, New York, pp. 1421–1430, 1987.

232. Lee, T. H., and Tang, L. M., Neuroleptic malignant syndrome, *J. Neurol.* 235, 324–325, 1988.

233. Spivak, B., Gonen, N., Mester, R., Averbuch, E., Adlersberg, S., and Weizman, A., Neuroleptic malignant syndrome associated with abrupt withdrawal of anticholinergic agents, *Int. Clin. Psychopharmacol.* 11, 207–209, 1996.

234. Wong, M. M. C., Neuroleptic malignant syndrome: Two cases without muscle rigidity, *Australian New Zealand J. Psychiatry* 30, 415–418, 1996.

235. Sechi, G., Manca, S., Deiana, G., Corda, D. G., Pisu, A., and Rosati, G., Acute hyponatremia and neuroleptic malignant syndrome in Parkinson's disease, *Prog. Neuropsychopharmacol. Biol. Psychiatr.* 20, 533–542, 1996.

236. Kuno, S., Mizuta, E. and Yamasaki, S., Neuroleptic malignant syndrome in parkinsonian patients: Risk factors, *Eur. Neurol.* 38; 2, 56–59, 1997.

237. Kozian, R., Lesser, K., and Peter, K., Dysarthric disorders associated with the neuroleptic malignant syndrome, *Pharmacopsychiatry* 29, 220–222, 1996.

238. Yoshikawa, H., Oda, Y., Sakajiri, K., Takarnori, M., Nakanishi, I., Makifuchi, T., Ide, Y., Matsubara, S., and Mizushima, N., Pure akinesia manifested neuroleptic malignant syndrome: A clinical variant of progressive supranuclear palsy, *Acta Neuropathol.* 93, 306–309, 1997.

239. Spivak, B., Mester, R., Abesgaus, J., Wittenberg, N., Adlersberg, S., Gonen, N., and Weizman, A., Clozapine treatment for neuroleptic-induced tardive dyskinesia, parkinsonism, and chronic akathisia in schizophrenic patients, *J. Clin. Psychiatry* 58, 318–322, 1997.

240. Chatterton, R., Cardy, S., and Schramm, T. M., Neuroleptic malignant syndrome and clozapine monotherapy, *Australian New Zealand J. Psychiatry* 30, 692–693, 1996.

241. Amore, M., Zazzeri, N., and Berardi, D., Atypical neuroleptic malignant syndrome associated with clozapine treatment, *Neuropsychobiology* 35, 197–199, 1997.

242. Lee, J. W. Y. and Robertson, S., Clozapine withdrawal catatonia and neuroleptic malignant syndrome: A case report, *Ann. Clin. Psychiatry* 9; 3, 1997.

243. Delay, J. and Deniker, P., Drug-induced extrapyramidal syndromes, in *Handbook of Clinical Neurology: Diseases of the Basal Ganglia,* Ed. by P. J. Vinken and G. W. Bruyn, vol. 6, Elsevier North Holland, New York, pp. 248–266, 1968.

244. Sukanova, L., Maligni neurlepticky syndrom (Neuroleptic malignant syndrome), *Cesk. Psychiatrie* 81, 91–95, 1985.

245. Delay, J., Picot, P., Lemperiere, T., and Bailly, R., L'emploi des butyrophenones en psychiatrie: Etude statistique et psychmetrique, in *Proceedings of the Symposium Internazionale sull' Haloperidol e. Triperidol,* Milan, Inst. Luso Farmaco d'Italia, pp. 305–319, 1963.

246. Pope, H. G., Jr., Keck, P. E., Jr., and McElroy, S. L., Frequency and presentation of neuroleptic malignant syndrome in a large psychiatric hospital, *Am. J. Psychiatry* 143, 1227–1233, 1986.

247. Shalev, A. and Munitz, H., The neuroleptic malignant syndrome: agent and host interaction, *Acta Psychiatr. ScaM.* 73, 337–347, 1986.

248. Olmsted, T. R., Neuroleptic malignant syndrome: guidelines for treatment and reinstitution of neuroleptics, *South. Med. J.* 81, 888–891, 1988.

249. Wedzicha, J.A. and Hoftbrand, B. I. Malignant neuroleptic syndrome and hyponatraemia, *Lancet* 963, 1984.

250. Bernstein, R. A., Malignant neuroleptic syndrome: An atypical case, *Psychosomatics* 20, 845–846, 1979.

251. Diamond, J. M. and Hayes, D. D., A case of neuroleptic malignant syndrome in a mentally retarded adolescent, *J. Adolesc. Health Care* 7, 419–422, 1986.

252. Shalev, A., Hermesh, H., and Munitz, H., The role of external heat load in triggering the neuroleptic malignant syndrome, *Am. J. Psychiatry* 145, 110–111, 1988.

253. Gelenberg, A. J., Bellinghausen, B., Wojcik, J. D., Falk, W. E., and Sachs, G.S., A prospective survey of neuroleptic malignant syndrome in a short-term psychiatric hospital, *Am. J. Psychiatry* 145, 517–518, 1988.

254. Pearlman, C. A., Neuroleptic malignant syndrome, *J. Clin. Psychopharmacol.* 6, 257–273, 1986.

255. Fiebel, J. H. and Schiffer, R. B., Sympathoadrenomedullary hyperactivity in the neuroleptic malignant syndrome: A case report, *Am. J. Psychiatry* 138, 1115–1116, 1981.

256. Becker, D., Birger, M., Samuel, E., and Floru, S., Myocardial infarction: An unusual complication of neuroleptic malignant syndrome, *J. Nerv. Ment. Dis.* 176, 377–378, 1988.

257. Peylan, J., Goldberg, I., Retter, J., and Yosipovitch, Z., Articular ossification after malignant neurolepsis, A case of schizophrenia treated with phenothiazines, *Acta Orthop. Scand.* 58, 284–286, 1987.

258. Anderson, S. A. and Weinschenk, K., Peripheral neuropathy as a component of the neuroleptic malignant syndrome, *Am. J. Med.* 82, 169–170, 1987.

259. Johnson, M. D., Newman, J. H. and Baxter, J. W., Neuroleptic malignant syndrome presenting as adult respiratory distress syndrome and disseminated intravascular coagulation, *South. Med. J.* 81, 543–545, 1988.

260. Legras, A., Hurel, D., Dabrowski, G., Grenet, D., Graveleau, P., and Loirat, P., Protracted neuroleptic malignant syndrome complicating long-acting neuroleptic administration, *Am. J. Med.* 85, 875–878, 1988.

261. Kennedy, M. S. and Burks, T. F., Dopamine receptors in the central thermoregulatory mechanisms of the cat, *Neuropharmacology* 13, 119–128, 1974.

262. Cox, B. and Lee, T. F., Do central dopamine receptors have a physiological role in thermoregulation? *Br. J. Pharmacol.* 61, 83–86, 1977.

263. Cox, B., Kerwin, R., and Lee, T. F., Dopamine receptors in the central thermoregulatory pathways of the rat, *J. Physiol. (Lond.)* 282, 471–483, 1978.

264. Morris, H. H., McCormick, W. F. and Reinarz, J. A., Neuroleptic malignant syndrome, *Arch. Neurol.* 37, 462–463, 1980.

265. Henderson, V. W. and Wooten, G. F., Neuroleptic malignant syndrome: A pathogenetic role for dopamine receptor blockade? *Neurology* 31, 132–137, 1981.

266. Prasad, C., Neuropeptide-dopamine interactions, I. Dopaminergic mechanisms in cyclo (His-Pro)-mediated hypothermia in rats, *Brain Res.* 437, 345–348, 1987.

267. Ali, A. H. M., The neuroleptic malignant syndrome: Do we know enough? *Jefferson J. Psychiatry* 3, 45–49, 1985.

268. Bond, W. S., Detection and management of the neuroleptic malignant syndrome, *Clin. Pharmacol.* 3, 302–307, 1984.

269. Dhib-Jalbut, S., Hesselbrock, R., Brott, T., and Silbergeld, D., Treatment of the neuroleptic malignant syndrome with bromocriptine, *J. Am. Med. Assoc.* 250, 484–485, 1983.

270. Duke, M., Neuroleptic malignant syndrome, *Med. J. Aust.* 14, 198–199, 1984.

271. Figa-Talamanca, L., Gualandi, C., DiMeo, L., DiBattista, G., Neri, G., and LoRusso, F., Hyperthermia after discontinuance of levodopa and bromocriptine therapy: Impaired dopamine receptors a possible cause, *Neurology* 35, 258–261, 1985.

272. Granato, J. E., Stern, B. J., Ringel, A., Karim, A. H., Krumholz, A., Coyle, J., and Adler, S., Neuroleptic malignant syndrome: Successful treatment with dantrolene and bromocriptine, *Ann. Neurol.* 14, 89–90, 1983.

273. Levenson, J. L., Neuroleptic malignant syndrome, *Am. J. Psychiatry* 142, 1137–1145, 1985.

274. Mueller, P. S., Vester, J. W., and Fermaglich, J., Neuroleptic malignant syndrome, Successful treatment with bromocriptine, *J. Am. Med. Assoc.* 249, 386–388, 1983.

275. Rosse, R., and Ciolino, C., Dopamine agonists and neuroleptic malignant syndrome, *Am. J. Psychiatry* 142, 270–271, 1985.

276. Zubenko, G. and Pope, H. G., Management of a case of neuroleptic malignant syndrome with bromocriptine, *Am. J. Psychiatry* 140, 1619–1620, 1983.

277. Chayasirisobhan, S., Cullis, P., and Veeramasuneni, R. R., Occurrence of neuroleptic malignant syndrome in a narcoleptic patient, *Hosp. Comm. Psychiatry* 34, 548–550, 1983.

278. Gangadhar, B. N., Desai, N. G., and Channabasavanna, S. M., Amantadine in the neuroleptic malignant syndrome, *J. Clin. Psychiatry* 45, 526, 1984.

279. Lazarus, A., Neuroleptic malignant syndrome: Detection and management, *Psychiatr. Ann.* 15, 706–712, 1985.

280. Lazarus, A., Neuroleptic malignant syndrome and amantadine withdrawal, *Am. J. Psychiatry* 142, 1985.

281. McCarron, M. M., Boettger, M. L., and Peck, J. J., A case of neuroleptic malignant syndrome successfully treated with amantadine, *J. Clin. Psychiatry* 43, 381–382, 1982.

282. Woo, J., Tech, R., and Vallence-Owen, J., Neuroleptic malignant syndrome successfully treated with amantadine, *Postgrad. Med. J.* 62, 809–810, 1986.

283. Lazarus, A., Therapy of neuroleptic malignant syndrome, *Psychiatr. Dev.* 1, 19, 1986.

284. Allsop, P., and Twigley, A. J., The neuroleptic malignant syndrome, *Anaesthesia* 42, 49–53, 1987.

285. Goulon, M., de Rohan Chabot, P., Elkharrat, D., Gadjos, P., Bismuth, C., and Conso, F., Beneficial effects of dantrolene in the treatment of neuroleptic malignant syndrome: A report of two cases, *Neurology* 33, 516–518, 1983.

286. Smego, R. A. and Durack, D. T., The neuroleptic malignant syndrome, *Arch. Intern. Med.* 142, 1183–1185, 1982.

287. Sternberg, D. E., Neuroleptic malignant syndrome: The pendulum swings, *Am. J. Psychiatry* 143, 1273–1275, 1986.

288. Westlake, R. J. and Rastegar, A., Hyperpyrexia from drug combinations, *J. Am. Med. Assoc.* 225, 1250, 1973.

289. Mann, S. C. and Boger, W. P., Psychotropic drugs, summer heat and humidity and hyperpyrexia: A danger restated, *Am. J. Psychiatry* 135, 1097–1100, 1978.

290. McCarthy, A., Fatal recurrence of neuroleptic malignant syndrome, *Br. J. Psychiatry* 152, 558–559, 1988.

291. Haggerty, J. H., Jr. and Gillette, G. M., Neuroleptic malignant syndrome superimposed on tardive dyskinesia, *Br. J. Psychiatry* 150, 104–105, 1987.

292. Tollefson, G. D. and Garvey, M. J., The neuroleptic syndrome and central dopamine metabolites, *J. Clin. Psychopharmacol.* 4, 150–153, 1984.

293. Harris, M., Nora, L., and Tanner, C. M., Neuroleptic malignant syndrome responsive to carbidopa/levodopa: Support for a dopaminergic pathogenesis, *Clin. Neuropharmacol.* 10, 186–189, 1987.

294. Hirschorn, K. A. and Greenberg, H. S., Successful treatment of levodopa-induced myoclonus and levodopa withdrawal-induced neuroleptic malignant syndrome, A case report, *Clin. Neuropharmacol.* 11, 278–281, 1988.

295. Klawans, H. L., Goetz, C., and Bergen, D., Levodopa-induced myoclonus, *Arch. Neurol.* 32, 331–334, 1975.

296. Goldwasser, H. D. and Hooper, J. F., Neuroleptic malignant syndrome, *Am. Fam. Physician* 38, 211–216, 1988.

297. Birkhimer, L. J. and Devand, C. L., The neuroleptic malignant syndrome: Presentation and treatment, *Drug Intell. Clin. Pharmacol.* 18, 462–465, 1984.

298. Van Winkle, W. B., Calcium release from skeletal muscle sarcoplasmic reticulum: Site of action of dantrolene sodium, *Science* 193, 1130–1131, 1976.

299. Davidoff, A. R. Pharmacology of spasticity, *Neurology* 28, 46–51, 1978.

300. Flewellen, E. H., Nelson, T. E., Jones, W. P., Arens, J. F. and Wagner, D. L., Dantrolene dose response in awake man: Implications for management of malignant hyperthermia, *Anesthesiology* 59, 275–280, 1983.

301. Britt, B. A. Dantrolene, *Anesth. Soc. J.* 31, 61–75, 1984.

302. Bismuth, C., De Rohan-Chabot, P., Goulon, M., and Raphael, J. C., Dantrolene—a new therapeutic approach to the neuroleptic malignant syndrome, *Acta Neurol. Scand.* 100, 193–198, 1984.

303. Delacour, J. L., Daoudal, P., Chapoutot, J. L., and Rocq, B., Traitement du syndrome malin des neuroleptiques par i.e., dantrolene, *Nouv. Presse Med.* 10, 3572–3573, 1981.

304. Coons, D. J., Hillman, F. J. and Marshall, R. W., Treatment of neuroleptic malignant syndrome with dantrolene sodium: A case report, *Am. J. Psychiatry* 139, 944–945, 1982.

305. May, D. C., Morris, S. W., Stewart, R. M., Fenton, B. J., and Gaffney, F. A., Neuroleptic malignant syndrome: Response to dantrolene sodium, *Ann. Intern. Med.* 98, 183–184, 1983.

306. Greenberg, L. B., and Gujavarty, K., The neuroleptic malignant syndrome: Review and report of three cases, *Compo Psychiatry* 26, 63–70, 1985.

307. Sullivan, C. F., A possible variant of the neuroleptic malignant syndrome, *Br. J. Psychiatry* 151, 689–690, 1987.

308. Goekoop, J. G. and Cabaat, P. A., Treatment of NMS with dantrolene, *Lancet* 2, 49–50, 1982.

309. Khan, A., Jaffe, S. H., Nelson, W. H. and Morrison, B., Resolution of neuroleptic malignant syndrome with dantrolene sodium: Case report, *J. Clin. Psychiatry* 46, 244–246, 1985.

310. De Rohan Chabot, P., Elkharrat, D., Conso, F., Bismuth, C. H., and Goulon, M., Syndrome malin des neuroleptiques, Action benefique du dantrolene sur l'hyperthermie et la rigidite musculaire, *Nouv. Presse Med.* 11, 1067–1069, 1982.

311. Schrehla, T. J. and Herjanic, M., Neuroleptic malignant syndrome, bromocriptine and anticholinergic drugs, *J. Clin. Psychiatry* 49, 283–284, 1988.

312. Burke, R. E., Fahn, S., Mayeux, R., Weinberg, H., Louis, K., and Willner, J. H., Neuroleptic malignant syndrome caused by dopamine depleting drugs in a patient with Huntington's chorea, *Neurology* (NY) 31, 1022–1026, 1981.

313. Lew, T. and Tollefson, G., Chlorpromazine-induced neuroleptic malignant syndrome and its response to diazepam, *Biol. Psychiatry* 18, 141–146, 1983.

314. Narabayashi, H., Imai, H., Yokochi, M., Hirayama, K., and Nakamura, R., Cases of pure akinesia without rigidity and tremor and with no effect by L-dopa therapy. In *Advances in Parkinsonism,* Ed. by W. Birkmayer and O. Hornykiewicz, Editiones Roche, Basel, pp. 335–342, 1976.

315. Narabayashi, H., Clinical analysis of akinesia., *J. Neural Transm.* 16, 129–136, 1980.

316. Imai, H., Narabayashi, H., and Sakata, E., "Pure akinesia" and the later added supranuclear ophthalmoplegia, *Adv. Neurol.* 45, 207–212, 1986.

317. Quinn, N. P., Luthert, P., Honavar, M., and Marsden, C.D., Pure akinesia due to Lewy body Parkinson's disease: A case with pathology, *Mov. Disord.* 4, 85–89, 1989.

318. Pall, H. S. and Williams, A. C., Extrapyramidal disturbances caused by inappropriate prescribing, *Br. Med. J.* 295, 30–31, 1987.

319. Stephen, P. J. and Williamson, J., Drug-induced parkinsonism in the elderly, *Lancet* 2, 1082–1083, 1984.

320. Mutch, W. S., Dingwall-Fordyce, I., Downie, A. W., Paterson, J. G., and Roy, S. K., Parkinson's disease in a Scottish city, *Br. Med. J.* 292, 534–536, 1986.

321. Schiele, B. C., Symposium on side effects and drug toxicity, *Psychopharmacol. Bull.* 4, 56–61, 1967.

322. Verhagen Metman, L., Del Dotto, P., Blanchet, P. J., van den Munckhof, P. and Chase, T. N., Blockade of glutamatergic transmission as treatment for dyskinesias and motor fluctuations in Parkinson's disease. *Amino Acids* 14, 75–82, 1998.

323. Jansen, E. N. H., Clozapine in the treatment of tremor in Parkinson's disease, *Acta. Neurol. Scand.* 89, 262–265, 1994.

324. Davis, K. L., Davidson, M., and Mohs, R. L., Plasma homovanillic acid and the severity of schizophrenic illness, *Science* 227, 1601–1602, 1985.

325. Carlsson, A., Antipsychotic drugs, neurotransmitters, and schizophrenia, *Am. J. Psychiatry* 135, 164–173, 1978.

326. Carlsson, A., The current status of the dopamine hypothesis of schizophrenia, *Neuropsychopharmacology* 1, 179–186, 1988.

327. Wong, D. F., Wagner, N. H., and Tune, L. E., Positron emission tomography reveals elevated D_2 dopamine receptors in drug-naive schizophrenics, *Science* 234, 1558–1563, 1986.

328. Seeman, P., Brain dopamine receptors, *Pharmacol. Rev.* 32, 229–313, 1981.

329. Snyder, S. H., Psychotogenic drugs as models for schizophrenia, Comments on "the current status of the dopamine hypothesis of schizophrenia," *Neuropsychopharmacology* 1, 197–199, 1988.

330. Friedman, J. H., Max, J., and Swift, R., Idiopathic Parkinson's disease in a chronic schizophrenic patient: Long-term treatment with clozapine and L-dopa, *Clin. Neuropharmacol.* 10, 470–475, 1987.

331. Friedman, J. H. and Lannon, M. C., Clozapine in the treatment of psychosis in Parkinson's disease, *Neurology* 39, 1219–1221, 1989.

332. Pfeiffer, R. F., Kang, J., Graber, B., Hofman, R., and Wilson, J., Clozapine for psychosis in Parkinson's disease, *Mov. Disord.* 5, 239–242, 1990.

333. Wolters, E. C., Hurwitz, T. A., Peppard, R. F., and Calne, D.B., Clozapine: An antipsychotic agent in Parkinson's disease? *Clin. Neuropharmacol.* 12, 83–90, 1989.

334. Kane, J., Honigfeld, G., Singer, J., and Meltzer, H., Clozapine for the treatment-resistant schizophrenic. *Arch. Gen. Psychiatry* 45, 789–796, 1988.

335. Honigfeld, G. and Patin, J., Predictors of response to clozapine therapy, *Psychopharmacology* 99, S64–S67, 1989.

336. Hall, R. A., Jackson, R. B., and Swain, J., Neurotoxic reactions resulting from chlorpromazine administration, *J. Am. Med. Assoc.* 161, 214–218, 1956.

337. Kulenkampff, C. and Tarnow, G., Ein eigentumliches syndrom im oralen bereich bei megaphenapplikation, *Nervenarzt* 27, 178–180, 1956.

338. Uhrband, L. and Faurbye, A., Reversible and irreversible dyskinesia after treatment with perphenazine, chlorpromazine, reserpine, ECT therapy, *Psycho Pharmacologia* 1, 408–418, 1960.

339. Bannet, J. and Belmaker, R. H., *New Directions in Tardive Dyskinesia Research,* Karger, New York, 1983.

340. Campbell, M., Grega, D. M., Green, W. H., and Bennett, W. G., Neuroleptic-induced dyskinesia in children, *Clin. Neuropharmacol.* 6, 207–222, 1983.

341. DeVeaugh-Geiss, J., *Tardive Dyskinesia and Related Involuntary Movement Disorders, The Long-Term Effects of Antipsychotic Drugs,* John Wright, PSG, Boston, 1983.

342. Tarsy, D. and Baldessarini, R. J., The tardive dyskinesia syndrome, in *Clinical Neuropharmacology,* Ed. by H. L. Klawans, vol. 1, Raven Press, New York, pp. 29–61, 1976.

343. Steen, V. M., Lovlie, R., MacEwan, T., and McCreadie, R.G., Dopamine D_3-receptor gene variant and susceptibility to tardive dyskinesia in schizophrenic patients. *Mol. Psychiatry* 2, 139–145, 1997.

344. Fann, W. E., Sullivan, J. L., and Richman, B.W., Dyskinesia associated with antidepressants, *Br. J. Psychiatry* 128, 490–493, 1976.

345. Yassa, R., Camille, Y., and Belzile, L., Tardive dyskinesia in the course of antidepressant therapy: a prevalence study and review of the literature, *J. Clin. Psychopharmacol.* 7, 243–246, 1987.

346. Kaplan, S. R. and Murkofsky, C., Oral-buccal dyskinesia symptoms associated with low-dose benzodiazepine treatment, *Am. J. Psychiatry* 135, 1558–1559, 1978.

347. Thach, B. T., Chase, T. N., and Bosman, J. F., Oral facial dyskinesia associated with prolonged use of antihistaminic decongestants, *N. Eng. J. Med.* 293.486–487, 1975.

348. Rubovits, R. and Klawans, H. L., Implications of amphetamine-induced stereotyped behavior as a model for tardive dyskinesia, *Arch. Gen. Psychiatry* 27, 502–507, 1972.

349. Coffey, C. E., Ross, D. R., Massey, E. W., and Olanow, C.W., Dyskinesias associated with lithium therapy in parkinsonism, *Clin. Neuropharmacol.* 7, 223–229, 1984.

350. Beauclair, L. and Fontaine, R., Tardive dyskinesia associated with metoclopramide, *Can. Med. Assoc. J.* 134, 613, 1986.

351. Yassa, R., Lal, S., Korpassy, A., and Ally, J., Nicotine exposure and tardive dyskinesia, *Biol. Psychiatry* 22, 67–72, 1987.

352. Vinarova, E., Vinar, O., and Kalvach, Z., Smokers need higher doses of neuroleptic drugs, *Biol. Psychiatry* 19, 1265–1268, 1984.

353. Granacher, R. P., Differential diagnosis of tardive dyskinesia: An overview, *Am. J. Psychiatry* 138, 1288–1297, 1981.

354. Wade, J. B., Taylor, M. A., Kasprisin, A., Rosenberg, S., and Fiducia, D., Tardive dyskinesia and cognitive impairment, *Biol. Psychiatry* 22, 393–395, 1987.

355. Lieberman, J., Lesser, M., Johns, C., Pollack, S., Saltz, B., and Kane, J., Pharmacological studies of tardive dyskinesia, *J. Clin. Psychopharmacol.* 8, 57S–63S, 1988.

356. Meshul, C. K., Stallbaumer, R. K., Taylor, B., and Janowsky, A., Haloperidol-induced morphological changes in striatum are associated with glutamate synapses, *Brain Res.* 648, 181–195, 1994.

357. Wirshing, W. C., Freidenberg, D. L., Cummings, J. L., and Bartzokis, G., Effects of anticholinergic agents on patients with tardive dyskinesia and concomitant drug-induced Parkinsonism, *J. Clin. Psychopharmacol.* 9, 407–411, 1989.

358. Ganzini, L., Casey, D. E., Hoffman, W. F., and Heintz, R. T., Tardive dyskinesia and diabetes mellitus, *Psychopharmacol. Bull.* 28, 281–286, 1992.

359. Saller, C. F. and Chiodo, L. A., Glucose suppresses basal firing and haloperidol-induced increases in the firing rate of central dopaminergic neurons, *Science* 210, 1269–1271, 1980.

360. Mouret, J., Khomais, M., Lemoine, P., and Sebert, P.. Low doses of insulin as a treatment of tardive dyskinesia: Conjuncture or conjecture? *Eur. Neurol.* 312, 199–203, 1991.

361. Luquin, M. R., Scipioni, O., Vaamonde, J., Gershanik, O., and Obeso, J. A., Levodopa-induced dyskinesias in Parkinson's disease: Clinical and pharmacological classification, *Movement Disorders* 7, 117–124, 1992.

362. Jelliffe, S. E. Psychological components in postencephalitic oculogyric crises: Contributions to a genetic interpretation of compulsive phenomena, *Arch. Neurol. Psychiatry* 21, 491–532, 1929.

363. Dorevitch, A., Neuroleptics as causes of oculogyric crises, *Arch. Neurol.* 41, 15–16, 1984.

364. Sachdev, P., Tardive and chronically recurrent oculogyric crises, *Movement Disorders* 8, 93–97, 1993.

365. Davis, J. B., Borde, M., and Sharma, L. N., Tardive dyskinesia and type II schizophrenia, *Brit. J. Psychiatry* 160, 253–256, 1992.

366. Goetz, C. G., Weiner, W. J., Nausieda, P. A., and Klawans, H.L., Tardive dyskinesia: Pharmacology and clinical implications, *Clin. Neuropharmacol.* 5, 3–22, 1982.

367. Bartholini, G., Lloyd, K. G., Worms, P., Constantinidis, J., and Tissot, R., GAB A and GABA-ergic medication: Relation to striatal dopamine function and parkinsonism, *Adv. Neurol.* 24, 253–257, 1979.

368. Bartholini, G., Scatton, B., Zivkovic, B., and Lloyd, K. G., On the mode of action of SL 76002, a new GABA receptor agonist, in *GAB A Neurotransmitters: Pharmacochemical, Biochemical and Pharmacological Aspects, Proceedings of the 12th Alfred Benzon Symposium,* Ed. by P. Krogsgaard-Larsen, J. Scheel-Kruger, and H. Kofod, Munksgaard, Copenhagen, pp. 326–339, 1979.

369. Christensen, A. V. and Hyttel, J., Prolonged treatment with the GABA agonist THIP increases dopamine receptor binding more than it changes dopaminergic behavior in mice, *Drug Dev. Res.* 1, 255–263, 1981.

370. Christensen, A. V., Arnt, J., and Scheel-Kruger, J., Decreased antistereotypic effect of neuroleptics after additional treatment with a benzodiazepine, a GABA agonist or an anticholinergic compound, *Life Sci.* 24, 1395–1402, 1979.

371. Fibiger, H. C. and Lloyd, K. G., Neurobiological substrates of tardive dyskinesia: The GABA hypothesis, *Trends Neurosci.* 7, 462–464, 1984.

372. Cassady, S. L., Thaker, G. K., Moran, M., Birt, A., and Tamminga, C. A., GABA agonist-induced changes in motor, oculomotor, and attention measures correlate in schizophrenics with tardive dyskinesia, *Bioi. Psychiatry* 32, 302–311, 1992.

373. Gunne, L. M. and Haggstrom, J. E., Reduction of nigral glutamic acid decarboxylase in rats with neuroleptic induced oral dyskinesia. *Psychopharmacology* (Berlin) 81, 191–194, 1983.

374. Gunne, L. M., Haggstrom, J. E. and Sjoquist, B., Association with persistent neuroleptic induced dyskinesia

of regional changes in brain GABA synthesis, *Nature* 309, 347–349, 1984.

375. Bird, E. D., MacKay, A. V. P., Rayner, C. N., and Iversen, L.L., Reduced glutamic acid decarboxylase activity of post mortem brain in Huntington's chorea, *Lancet* 1, 1090–1092, 1973.

376. Carlsson, A. and Lindquist, M., Effect of chlorpromazine or haloperidol on formation of 3-methoxytyramine and normetanephrine in mouse brain, *Acta Pharmacal. Toxicol.* 20, 140–144, 1963.

377. McGeer, P. L., McGeer, E. G., and Fibiger, H. C., Choline acetylase and glutamic acid decarboxylase in Huntington's chorea, *Neurology* 23, 912–917, 1973.

378. Walters, J. R. and Chase, T. N., GABA systems and extrapyramidal function. in *Neurotransmitter Function: Basic and Clinical Aspects,* Ed. by W. S. Fields, Stratton Intercontinental, New York, 1977.

379. Tamminga, C. A., Crayton, J. W., and Chase, T.N., Improvement in tardive dyskinesia after muscimol therapy, *Arch. Gen. Psychiatry* 36, 595–598, 1979.

380. Tamminga, C. A., Thaker, G. K., and Goldberg, S. T., Tardive dyskinesia: GABA agonist treatment, in *Catecholamines: Neuropharmacology and Central Nervous System-Therapeutic Aspects,* Ed. by E. Usdin, A. Carlsson, A. Dahlstrom and J. Engel, Alan R. Liss, New York, pp. 69–72, 1984.

381. Singh, M. M., Becker, R. E., Pitman, R. K., Nasrallah, H. A., and Lal, H., Sustained improvement in tardive dyskinesia with diazepam: Indirect evidence for corticolimbic involvement, *Brain Res. Bull.* 11, 179–185, 1983.

382. Sandyk, R., Successful treatment of neuroleptic-induced akathisia with baclofen and clonazepam, A case report, *Eur. J. Neurol.* 24, 286–288, 1985.

383. Hama, Y. and Ebadi, M., The nullification by diazepam of haloperidol-induced increases in the level of striatal dopamine but not in the activity of glutamatic acid decarboxylase, *Neuropharmacology* 15, 1235–1242, 1986.

384. Mithani, S., Atmadja, S., Baimbridge, K. G. and Fibiger, H. C., Neuroleptic-induced oral dyskinesias: Effects of progabide and lack of correlation with regional changes in glutamic acid decarboxylase and choline acetyltransferase activities, *Psychopharmacology (Berlin)* 93, 94–100, 1987.

385. Berger, P. A., Endorphins in emotions, behavior and mental illness, in *Mind and Medicine: Emotions in Health and Illness,* Ed. by L. Temoshok, C. Van Dike, and L.S. Vegans, Grune and Stratton, New York, pp. 153–166, 1983.

386. Bissette, G., Nemeroff, C. B. and MacKay, A. V. P., Neuropeptides and schizophrenia, in *Progress in Brain Research,* Ed. by P. C. Emson, M. N. Rossor and M. Tohyama, Elsevier, Amsterdam, Vol. 66, pp. 161–174, 1986.

387. Buchsbaum, M. S., Davis, G. C., and van Kammen, D. P., Diagnostic classification and the endorphin hypothesis of schizophrenia. Individual differences and psychopharmacological strategies, in *Perspectives in Schizophrenia Research,* Ed. by C. Baxter and T.

Melnechuk, Raven Press, New York, pp. 177–191, 1980.

388. Davis, G. C., Buchsbaum, M. S., and Bunney, W. E., Research in endorphins and schizophrenia, *Schizophr. Bull.* 5, 244–250, 1979.

389. Koob, G., LeMoal, M., and Bloom, F. E., The role of endorphins in neurobiology, behavior and psychiatric disorders, in *Peptides, Hormones and Behavior,* Ed. by C. B. Nemeroff and A. J. Dunn, Spectrum, New York, pp. 349–384, 1984.

390. MacKay, A. V. P., Endorphins and the psychiatrist, *Trends Neurosci.* 4, R9–R11, 1981.

391. Terenius, L., The implications of endorphins in pathological states, in *Characteristics and Function of Opioids,* Ed. by J. M. van Ree and L. Terenius, North-Holland, Amsterdam, pp. 143–158, 1978.

392. Van Praag, H. M. and Verhoeven, W. M. A., Endorphins and schizophrenia, in *Hormones and the Brain,* Ed. by D. De Wied and P. A. Van Keep, University Park Press, Baltimore, pp. 141–153, 1980.

393. Van Ree, J. M. and De Wied, D., Endorphins in schizophrenia, *Neuropharmacology* 20, 1271–1277, 1981.

394. Vereby, K., Volavka, J., and Clouet, D., Endorphins in psychiatry, *Arch. Gen. Psychiatry* 35, 877–888, 1978.

395. Nair, N. P. V., Lal, S., and Bloom D. M., Cholecystokinin and schizophrenia, in *Progress in Brain Research,* Ed. by J. M. van Ree and S. Matthysse, vol. 65, Elsevier, Amsterdam, pp. 237–258, 1986.

396. Vanderhaeghen, J. J., Signeua, J. C., and Gepts, W., New peptide in the vertebrate CNS reacting with anti gastrin antibodies, *Nature* 257, 604–605, 1975.

397. Beinfeld, M. C., Meyer, D. K., Eskay, R. L., Jensen, R. T., and Brownstein, M. J., The distribution of cholecystokinin immunoreactivity in the central nervous system of the rat as determined by radioimmunoassay, *Brain Res.* 212, 51–57, 1981.

398. Dockray, G. J. and Gregory, R. A., Relations between neuropeptides and gut hormones, *Proc. R. Soc. Lond. Ser. B.* 210, 151–164, 1980.

399. Simon-Assmann, P. M., Yazigi, R., Greeley, G., Rayford, P. L., and Thompson, J. C., Biologic and radioimmunologic activity of CCK in regions of mammalian brains, *J. Neurosci. Res.* 10, 165–173, 1983.

400. Straus, E. and Yalow, R. S., Gastrointestinal peptides in the brain, *Fed. Proc.* 38, 2320–2324, 1979.

401. Ziegler, M., Fournier, V., Bathien, N., Morselli, P. L., and Rondot, P., Therapeutic response to progabide in neuroleptic- and L-dopa-induced dyskinesias. *Clin. Neuropharmacol.* 10, 238–246, 1987.

402. Kaufmann, C., Jeste, D. V., Shelton, R. C., Linnoila, M., Kafka, M., and Wyatt, R. J., Noradrenergic and neuro-radiological abnormalities in tardive dyskinesia, *Biol. Psychiatry* 21, 799–812, 1986.

403. Karoum, F., Karson, C. N., Bigelow, L. B., Lawson, W. B. and Wyatt, R. J., Preliminary evidence of reduced combined output of dopamine and its metabolites in chronic schizophrenia, *Arch. Gen. Psychiatry* 44, 604–607, 1987.

404. Casey, D. E., and Gerlach, J., Is tardive dyskinesia due to dopamine hypersensitivity? *Clin. Neuropharmacol.* 9, 134–136, 1986.

405. Duvoisin, R. C., Reserpine for tardive dyskinesia, *N Eng. J. Med.* 286, 611, 1972.

406. Brandrup, E., Tetrabenazine treatment in persisting dyskinesias caused by psychopharmaca, *Am. J. Psychiatry* 118, 551–552, 1961.

407. Kazamatsuri, H., Chien, C., and Cole, J. O., Treatment of tardive dyskinesias, I. Clinical efficacy of dopamine depleting agent, tetrabenazine, *Arch. Gen. Psychiatry* 27, 95–99, 1972.

408. Kazamatsuri, H., Chien, C., and Cole, J. O., Long-term treatment of tardive dyskinesia with haloperidol and tetrabenazine, *Am. J. Psychiatry* 130, 479–483, 1973.

409. MacCallum, W. A. G., Tetrabenazine for extrapyramidal motor disorders, *Br. Med. J. (Clin. Res.)* 1, 760, 1970.

410. Jus, A., Jus, K., and Fontaine, P., Long-term treatment of tardive dyskinesia, *J. Clin. Psychiatry* 30, 73–79, 1979.

411. Hokfelt, T., Skirboll, L., Rehfeld, J. F., Goldstein, M., Markey, K., and Dann, O., A subpopulation of mesencephalic dopamine neurons projecting to limbic areas contains a cholecystokinin-like peptide: Evidence from immunohistochemistry combined with retrograde tracing, *Neuroscience* 5, 2093–2124, 1980.

412. Skirboll, L. R., Crawley, J. N., and Hommer, D. W., Functional studies of cholecystokinin-dopamine coexistence: Electrophysiology and behavior, in *Progress in Brain Research,* Ed. by T. Hokfelt, K. Fuxe, and B. Pernow, Vol. 68, Elsevier, Amsterdam, pp. 357–370, 1986.

413. Nair, N. P. V., Lal, S., and Bloom, D. M., Cholecystokinin peptides, dopamine and schizophrenia, *Prog. Neuropsychopharmacol. Biol. Psychiatry* 9, 515–524, 1985.

414. Wang, R. Y., White, F. J., and Voigt, M. M., Cholecystokinin, dopamine and schizophrenia, *Trends Pharmacol. Sci.* 9, 436–438, 1984.

415. Dumbrille-Ross, A. and Seeman, P., Dopamine receptor elevation by cholecystokinin, *Peptides* 5, 1207–1212, 1984.

416. Worms, P., Martinez, J., Briet, C., Castro, B., and Biziere, K., Evidence for dopaminomimetic effect of intrastriatally injected cholecystokinin octapeptide in mice, *Eur. J. Pharmacal.* 121, 395–401, 1986.

417. Hama, Y. and Ebadi, M., Characterization of [^{3}H]cholecystokinin octapeptide binding to mouse brain synaptosomes, *Neurochem. Res.* 12, 729–737, 1987.

418. Govoni, S., Yang, H. Y. T., Bosio, A., Pasinetti, G., and Costa, E., Possible interaction between cholecystokinin and dopamine, *Adv. Biochem. Psychopharmacol.* 33, 437–444, 1982.

419. Chang, R. S. L., Lotti, V. J., Martin, G. E., and Chen, T. B., Increase in brain ^{125}I-cholecystokinin (CCK) receptor binding following chronic haloperidol treatment, intracisternal 6-hydroxydopamine or ventral tegmental lesions, *Life Sci.* 32, 871–878, 1983.

420. Frey, P., Cholecystokinin octapeptide levels in rat brain are changed after subchronic neuroleptic treatment, *Eur. J. Pharmacol.* 95, 87–92, 1983.

421. Zetler, G., Antistereotypic effects of cholecystokinin octapeptide (CCK-8), ceruletide and related peptides on apomorphine-induced gnawing in sensitized mice, *Neuropharmacology* 24, 251–259, 1985.

422. Eison, A. S. and Temple, D. L., Buspirone: Review of its pharmacology and current perspectives on its mechanism of action, *Am. J. Med.* 80, 1–9, 1986.

423. Goa, K. L. and Ward, A., Buspirone: A preliminary review of its pharmacological properties and therapeutic efficacy as an anxiolytic, *Drugs* 32, 114–129, 1986.

424. McMillen, B. A., Matthews, R. T., Sanghera, M. K., Shepard, P. D., and German, D. C., Dopamine receptor antagonism by the novel antianxiety drug, buspirone, *J. Neurosci.* 3, 733–738, 1983.

425. Bonifati, V., Fabrizio, E., Cipriani, R., Vanacore, N., and Meco, G., Buspirone in levodopa-induced dyskinesias, *Clin. Neuropharmacol.* 17, 73–82, 1994.

426. Moss, L. E., Neppe, V. M., and Drevets, W. C., Buspirone in the treatment of tardive dyskinesia, *J. Clin. Psychopharmacol.* 13, 204–209, 1993.

427. Sathananthan, G. L., Sanghvi, I., Phillips, N., and Gershon, S., MJ 9022: Correlation between neuroleptic potential and stereotype, *Curr. Ther. Res.* 18, 701–705, 1975.

428. Peet, M., Laughame, J., Rangarajan, N., and Reynolds, G. P., Tardive dyskinesia, lipid peroxidation, and sustained amelioration with vitamin E treatment, *Int. Clinc. Psychopharmacol.* 8, 151–153, 1993.

429. Dannon, P. N., Lepkitker, E., Iancu, I., Ziv, R., Horesh, N., and Kotler, M., Vitamin E treatment in tardive dyskinesia, *Human Psychopharmacol.* 12, 217–220, 1997.

430. Bloom, F., Battenberg, E., Rossier, J., Ling, N., and Guillemin, R., Neurons containing beta-endorphin in rat brain exist separately from those containing enkephalin: Immunocytochemical studies, *Proc. Natl. A cad. Sci. USA* 75, 1591–1595, 1978.

431. Chang, K. J., Cooper, N. R., Hazum, E., and Cuatrecasas, P., Multiple opiate receptors: Different regional distribution in the brain and differential binding of opiates and opioid peptides, *Mol. Pharmacol.* 1, 691–104, 1979.

432. Biggio, G., Casu, M., Corda, M. G., DiBello, C., and Gessa, G. L., Stimulation of dopamine synthesis in caudate nucleus by intrastriatal enkephalins and antagonism by naloxone, *Science* (Washington, DC) 200, 552–554, 1978.

433. Pollard, H., Llorens, C., Schwartz, J. C., Gros, C., and Dray, F., Localization of opiate receptors and enkephalins in the rat striatum in relationship with the nigrostriatal dopaminergic system: Lesion studies, *Brain Res.* 151, 392–398, 1978.

434. Hong, J. S., Yoshikawa, K., Kanamatsu, T., and Sabol, S. L., Modulation of striatal enkephalinergic neurons by antipsychotic drugs, *Fed. Proc.* 44, 2535–2593, 1985.

435. Sayers, A. C., Burki, H. R., Ruch, W., and Asper, H., Neuroleptic-induced hypersensitivity of striatal dopamine receptors in the rat as a model of tardive dyskinesias. Effects of clozapine, haloperidol, loxapine and chlorpromazine, *Psycho Pharmacologia* 4, 197–104, 1975.

436. Seeger, T. F., Nazzaro, J. M., and Gardner, E. L., Selective inhibition of mesolimbic behavioral supersensitivity by naloxone, *Eur. J. Pharmacol.* 65, 435–438, 1980.

437. Gillman, M. A., Sandyk, R., and Lichtigfeld, F. J., Evidence for under activity of the opioid system in neuroleptic-induced akathisia, *Psychiatry Res.* 13, 187, 1984.

438. Walters, A., Hening, W., and Chokroverty, S., Opioid responsiveness of neuroleptic-induced akathisia, *Ann. Neurol.* 18, 137, 1985.

439. Wasserman, S. and Yahr, M. D., Choreic movements induced by the use of methadone, *Arch. Neurol.* 37, 727–728, 1980.

440. Blum, I., Elizur, A., Segal, A., Ochshorn, N., and Stimantov, R., Effect of naloxone on the neuropsychiatric systems of a woman with partial adrenal 21-hydroxylase deficiency, *Am. J. Psychiatry* 140, 1058–1060, 1984.

441. Blum, I., Munitz, N., Shalev, A., and Roberts, E., Naloxone may be beneficial in the treatment of tardive dyskinesia, *Clin. Neuropharmacol.* 7, 265–267, 1984.

442. Angus, S., Sugar, J., Boltezar, R., Koskewich, S., and Schneider, N. M., A controlled trial of amantadine hydrochloride and neuroleptics in the treatment of tardive dyskinesia, *Clin. Psychopharmacol.* 17, 88–91, 1997.

443. Kang, U. J., Burke, R. E., and Fahn, S., Natural history and treatment of tardive dystonia, *Mov. Disord.* 1, 193–208, 1986.

444. Gardos, G., Dystonic reaction during maintenance antipsychotic therapy, *Am. J. Psychiatry* 138, 114–115, 1981.

445. Wojcik, J. D., Falk, W. E., Fink, J. S., Cole, J. O., and Gelenberg, A. J., A review of 32 cases of tardive dystonia, *Am. J. Psychiatry* 148, 1055–1059, 1991.

446. Trugman, J. M., Leadbetter, R., Zalis, M. E., Burgdorf, R. O., and Wooten, G. F., Treatment of severe axial tardive dystonia with clozapine: case report and hypothesis. *Movement Disorders* 9, 441–446, 1994.

447. Kojima, T., Yamauchi, T., Miyasaka, M., Koshino, Y., Nakane, Y., Takahashi, R., Shimazono, Y., and Yagi, G., Treatment of tardive dyskinesia with ceruletide: A double-blind, placebo-controlled study, *Psychiat. Res.* 43, 129–136, 1992.

448. Janssen, P. A. J., Niemegeers, C. J. E., Awouters, F. H. L., Schellekens, K. H. L., Megans, A. A. H. P., and Meert, T. F., Pharmacology of risperidone (R64766), a new antipsychotic with serotonin-S_2 and dopamine-D_2 antagonistic properties, *J. Pharmacol. Exp. Ther.* 244, 685–693, 1988.

449. Leysen, J. E., Gommeren, W., Eens, A., De Chaffoyde Courcells, D., Stoof, J. C., and Janssen, P. A. J., Biochemical profile of risperidone, a new antipsychotic, *J. Pharmacal. Exp. Ther.* 247, 661–670, 1988.

450. Chouinard, G., Effects of risperidone in tardive dyskinesia: An analysis of the Canadian multicenter risperidone study, *J. Clin. Psychopharmacol.* 15, 36S–44S, 1995.

451. Müller-Sicheneder, F., Müller, M. J., Hillert, A., Szegedi, A., Wetzel, H., and Benkert, O., Risperidone versus haloperidol and amitriptyline in the treatment of patients with a combined psychotic and depressive syndrome, *J. Clin. Psychopharmacol.* 18, 111–120, 1998.

452. Sharma, R., Trappler, B., Ng, Y. K., and Leeman, C.P., Risperidone-induced neuroleptic malignant syndrome, *Ann. Pharmacother.* 30, 775–778, 1996.

453. Bonwick, R. J., Hopwood, M. J. and Morris, P. L. P., Neuroleptic malignant syndrome and risperidone: A case report, *Australian New Zealand J. Psychiatry* 30, 419–421, 1996.

454. Gleason, P. P. and Conigliaro, R.L., Neuroleptic malignant syndrome with risperidone, *Pharmacotherapy* 17, 617–621, 1997.

47 The Placebo Effect in Parkinson's Disease

Sarah C. Lidstone, Raul de la Fuente-Fernandez, and A. Jon Stoessl
Pacific Parkinson's Research Centre

CONTENTS

INTRODUCTION

Since the first medical practices and healing rituals were performed in ancient civilizations, the ability of the mind to influence the healing of the body has been recognized across many cultures. Modern medical research has termed this the *placebo effect,* which is essentially the patient's ability to demonstrate improvement in condition in response to some type of "inert" treatment—whether it be a pill, an injection, or even sham surgery—but not from any properties that the treatment itself possesses. In 1811, *Hooper's Medical Dictionary* defined a placebo as "an epithet given to any medicine adapted more to please than to benefit the patient."[1] Ironically, scientific investigation, in the realization of this phenomenon, needed to account for the placebo effect in the interpretation of experimental results, and thus the placebo effect was largely considered to be a nuisance obscuring the true effects of the active treatment. However, with the growing amount of research available from clinical drug trials, the ability of placebos to produce therapeutic benefit in patients who suffer from various neurological disorders has proven to be real and effective. It is now accepted that a prominent placebo effect may be present in pain disorders, depression, and Parkinson's disease.[2-4] In the case of the latter, several randomized, placebo-controlled trials aimed at testing new pharmaceutical therapies have shown objective improvements in motor performance following placebo administration.[5] However, the precise neuropsychological and biochemical mechanisms underlying the placebo effect are only beginning to be unraveled. The original observation by Levine and colleagues in the late 1970s that placebo analgesia can be blocked by naloxone suggested that the placebo effect in pain disorders involves the release of endogenous opioids.[6] Following recent studies revealing direct biochemical evidence that a patient's expectation is central to the placebo response in Parkinson's disease,[7] research is currently directed at characterizing the psychological and biochemical links between the expectation of benefit and the improvement of motor function in patients. This "expectation theory" of the placebo effect is thought to depend on reward circuitry in the brain, and more specifically, as recent evidence suggests, to dopamine release in the ventral striatum.

PLACEBOS AND THE PLACEBO EFFECT

It is important to make the distinction between a placebo and the placebo effect. A large number of definitions for placebo and placebo effect have been put forth over the years,[8] each one slightly different from the next and emphasizing different aspects of this complex phenomenon. Certainly, a placebo is "inert," or devoid of any specific effect for the medical condition being treated. Wolf provides a straightforward and concise explanation, defining the placebo effect as "any effect attributable to a pill, potion, or procedure, but not to its pharmacodynamic properties."[9] Essentially, any sort of treatment can act as a placebo—pills, injections, or surgical procedures, for example—but it is the response of the patient to that treatment that determines whether there is an actual placebo effect. The placebo effect itself also depends on the type of placebo administered; it has been shown that the magnitude of the response to the placebo varies according to its supposed potency.[10] For example, placebo surgery seems to be more effective than a placebo pill,[1,11,12] and, as a recent study for arthroscopic knee surgery suggested, may produce the same outcome as the actual surgical procedure.[13] The term *nocebo effect* has also been used to describe the situation in which the patient exhibits a worsening condition in response to a placebo.[14] In this case, the placebo is referred to as a *nocebo,* since it produces a negative outcome. As an example, Benedetti et al. demonstrated that motor performance in Parkinsonian patients worsened with the induction of a negative verbal expectation, yet the induction of a positive verbal expectation blocked this "nocebo" effect.[15] In other words, the patient's expectation of improved motor performance reversed the motor worsening at the opposite (negative) suggestion. Another aspect that confounds the attempt to define the placebo effect is the lack of consistency that has been associated with it.[16] It has been shown that an individual may respond to a particular placebo at a given time yet fail to maintain a placebo effect on subsequent exposures to the same placebo or respond to a different placebo. In addition, the placebo effect can be very specific,[17] and this specificity depends on the information that is made available to the recipient (i.e., the recipient's expectation). For example, placebos have been shown to have opposite effects on heart rate or blood pressure, depending on whether they are given as tranquilizers or stimulants.[18]

Two alternative theories have developed with respect to the underlying psychological mechanisms of the placebo effect. The mentalistic or expectation theory previously mentioned proposes that the patient's expectation is the primary basis for the placebo effect, and the conditioning theory states that the placebo effect is essentially a conditioned response.[19] Original investigation into the placebo effect yielded models that supported one or the other of these theories.[17,20–22] In Parkinson's disease, and quite likely in pain and depression as well, recent research suggests that it is the expectation of clinical benefit that is directly related to the underlying biochemical mechanisms responsible for dopamine release.[7] A recent study demonstrated that hand movement velocity in Parkinsonian patients following subthalamic nucleus stimulation was affected by different expectations of motor performance.[23] This does not rule out the conditioning theory; naturally, it is possible that both mechanisms contribute to the placebo effect in any given patient, although their precise roles in different circumstances are, for the most part, largely unknown. The distinction may indeed be somewhat artificial, in that conditioning will enhance the level of expectation, and, particularly in sentient animals, it is the expectation itself rather than the final physiological effector response that may be conditioned. Benedetti and colleagues recently investigated the different roles of expectation and conditioning in different placebo responses.[15] In the study design, they compared a "conscious" or cognitive placebo response that would occur in ischemic arm pain in healthy subjects and motor performance in Parkinson patients versus the unconscious physiological process of hormone (growth hormone and cortisol) secretion. They found that verbally induced expectations of analgesia/hyperalgesia and motor improvement/worsening completely removed the effects of a conditioning procedure in the first two classes of patients, whereas verbally induced expectations had no effect on hormone secretion. These findings reveal that conscious expectation and unconscious conditioning play different roles in different circumstances in both the placebo and nocebo effect and, importantly, that when conscious perception is involved, expectation replaces conditioning.[15]

INVESTIGATION OF THE PLACEBO EFFECT

As previously emphasized, in accordance with the expectation model the patient's "belief" or "faith" that a treatment may be beneficial is the factor thought to determine the placebo effect.[24] The simple act of taking a pill or having a sham operation may only be regarded as the trigger of the placebo response.[25] Any meaningful investigation into the placebo effect must then be able to account for the participants' expectations and measure the resulting behavior of the participants. Thus, when selecting candidates for placebo research, certain requirements must be present within the participant population so as to detect the presence of a placebo effect with confidence. Patients must be conscious,[26] mental faculties must be preserved, and the disorder must result in symptoms of sufficient severity for the patient to be motivated to desire improvement if the researcher is to have the optimum chance of detecting a placebo effect.[25] There must be a reasonable expectation of obtaining benefit; thus, trials in which a placebo is tested against no treatment, or where there is at best a one-third chance of obtaining active treatment, are unlikely to demonstrate a significant placebo effect. In addition, it should be noted that, in contrast to pain and depression, Parkinson's disease is a disorder in which the response to treatment can be assessed directly by the examiner, and this direct measurability might allow a better evaluation of the placebo effect by clinicians.[25] This being said, it is equally important to emphasize that the clinical scales used for measuring motor function are subjective themselves. Also, patients may be less prone to report clinical changes than the clinicians are to observe them,[27] adding another dimension of subjectivity.

Aside from the selection of the experimental subjects, there is a host of problematic subtleties inherent in all placebo research. In studying depression, the placebo effect has proved to be a major issue in interpreting results, with some studies concluding that the entire observed response was due to the placebo effect.[28] Thus, study design is fundamental to ensuring that the researcher ends up investigating what is intended, whether it be an active drug, a surgical procedure, or the placebo effect itself. The literature on the placebo effect is largely based on standard two-group randomized controlled trials, where the changes in the placebo group with respect to baseline are attributed to the placebo effect. Many investigators have assumed that such a study design—a placebo group and an untreated group—should be ideal for detecting and quantifying the placebo effect.[26,29,30] Paradoxically however, with this study design and adequate informed consent, neither of the two groups will expect any benefit from the experiment and consequently, no full placebo intervention can be evaluated. The real placebo power in this scenario is lost, since a patient with no expectation of clinical benefit is not likely to manifest a placebo effect. Another approach is the three-group study[31] in which patients are randomly assigned to one of three groups: the active drug group, the placebo group, or the untreated group. However, even in this study design, the patients' expectation of benefit may be too low, because a fully informed patient may realize that there is only a one-in-three chance of getting some benefit. In addition, patients who volunteer for studies with low probability of benefit may have particularly low expectations and may therefore not be representative of the population as a whole. It is therefore unsurprising that many studies with these designs have failed to demonstrate a placebo effect.[29] In their meta-analysis of clinical trials involving two or three groups—including a placebo group and a no-treatment group—Hrobjartsson and Gotzsche concluded that, with the exception of pain disorders, placebos offered no beneficial clinical effects.[29] The subsequent letters published in response to that claim independently brought up the same point about the low expectations of benefit associated with three-group studies and the detection of the placebo effect. However, the results can be interpreted in another way; this observation shows that the simple act of being exposed to a placebo is not necessarily sufficient to provide clinical benefit to the patient.[25]

As noted above, patients with no expectation of benefit are not likely to manifest a placebo effect. Another psychological factor to consider is the patient's knowledge about the disease, the efficacy of available drugs, and the potential for placebos to affect the particular disease. The patient's knowledge about the possibility of receiving a placebo during the study might affect the placebo response. For example, the placebo effect may be greater in patients who have not been informed that they might receive a placebo during the study.[32] So, from a technical point of view, the best way to detect a placebo effect might be deliberately not to inform the patients that they may be receiving an inactive treatment, but this approach would clearly be unethical in most circumstances. How-

ever, it is interesting to see the patients' attitudes toward participating in placebo-controlled studies, for they volunteer for the study in full recognition that there is a chance they will not receive any treatment. Goetz and colleagues questioned Parkinson's disease patients after their completion of placebo-controlled trials and the revelation that they had received placebo treatment,[33] The patients' impressions were significantly more frequently positive than negative, and a large proportion (88%) of the respondents expressed that if another placebo-controlled trial were offered, they would be interested in participating.

Another confounding factor to consider when investigating the placebo effect is its interaction with the active drug. For example, the placebo effect associated with the simple act of ingesting a pill may potentiate the actual physiological effect of the drug (positive interaction). Conversely, the placebo effect could diminish the physiological response to the drug, especially in patients with a strong placebo effect (negative interaction). In both cases, if present, the placebo effect can obscure the actual results of the study and may jeopardize the randomized controlled trial if ignored in the statistical analysis.[34] This phenomenon strengthens the idea that the effects of placebos and active drugs may not summate in a simple fashion, and it is an important consideration in study design, as a negative interaction between the effect of an active drug and the placebo effect in Parkinson's disease may occur.[7,19] Research on antidepressants has shown a particularly strong placebo effect, which in some cases can demonstrate results indistinguishable from those of the active drug.[35] Kirsch and Sapirstein concluded, from their meta-analysis of 19 trials of antidepressants, that about 75% of the effectiveness of these drugs results from the placebo effect.[28] As previously mentioned, such a strong placebo effect may result in a negative interaction with the effect of the active drug, which would make detection of the effect related to active treatment very difficult.

In addition to depression, pain disorders have also shown significant susceptibility to the placebo effect. As mentioned earlier, the first direct demonstration of the physiology underlying the placebo effect was the observation that placebo analgesia could be blocked by naloxone, which indicated that the placebo could induce the release of endogenous opioids.[6] Indeed, subsequent research has revealed dopamine-opioid interactions in the mesolimbic and mesocortical pathways.[36] There is anatomical evidence that this relationship is bidirectional;[16] dopaminergic projections from the ventral tegmental area can control opioid release in the periaqueductal gray—a major center for pain regulation—and opioid release can in turn modulate dopamine release in the nucleus accumbens (described in detail in Reference 16). It has been shown that pain, opioids, and placebo analgesics activate cortical and subcortical areas that receive dopaminergic projections.[37,38] This implies that dopamine release can play a role in the transmission and perception of pain, which indicates a possible link between reward pathways and pain alleviation, and hence a potential mechanism for placebo-induced analgesia. Indeed, enhanced activity of dopamine in the nucleus accumbens seems to play a role in analgesia.[39]

THE RESULTS OF CLINICAL TRIALS

Several randomized controlled trials aimed at assessing the clinical efficacy of pharmaceutical therapies in Parkinson's disease have yielded evidence of a strong placebo effect. For example, in a double-blind trial of pergolide, a dopamine agonist used as a treatment for Parkinson's disease, Diamond and colleagues[40] found a significant improvement with respect to baseline in both the pergolide-treated group (17% improvement after 4 weeks and 30% after 24 weeks) and the placebo group (16% improvement after 4 weeks and 23% after 24 weeks). In fact, there was no significant difference between the drug and placebo groups. Shetty and colleagues[5] conducted a relevant review of 98 articles that were published between 1969 and 1996 and selected 36 that satisfied their inclusion criteria. Of these articles, 12 reported an improvement following placebo treatment in Parkinson's disease. The magnitude of improvement ranged from 9 to 59% of that seen in the active drug groups. Two other studies worth mentioning here were retrospective analyses of large, randomized, placebo-controlled clinical trials. The first study was an analysis of the placebo group from the large clinical trial of Deprenyl and Tocopherol Antioxidative Therapy of Parkinsonism (DATATOP), which found that 21% of patients demonstrated a blinded investigator-determined "objective" improvement in motor function during placebo therapy over a six-month period.[41] In this study, the predominant purpose was to determine whether selegiline had a neuroprotective effect; thus, the expectation of symptomatic benefit was not high. Furthermore, contrary to predictions, the placebo effect was not restricted to the early evaluations but was distributed approximately equally across the duration of the observations. Similarly, Goetz and colleagues reported that 14% of the patients enrolled in a six-month, randomized, multicenter, placebo-controlled clinical trial of ropinirole monotherapy achieved a 50% improvement in motor function while on placebo treatment.[42] In this particular study, all domains of parkinsonian disability were subject to the placebo effect—88% of the patients showing improvement in multiple domains—but, interestingly, bradykinesia and rigidity tended to be more susceptible than tremor, gait, or balance.

Pharmacological clinical trials have demonstrated a strong placebo effect in Parkinson's disease, but surgical intervention (e.g., transplantation of fetal mesencephalic tissue grafts) has yet to demonstrate a consistent placebo response in this disorder. The importance of including of a placebo group when investigating the efficacy of surgical procedures for treating Parkinson's disease has been emphasized by several authors[3] but also refuted by other authors.[43] The ethics of using sham surgery in the assessment of the efficacy of neural grafting in Parkinson patients continues to be a matter of debate. The inclusion of a placebo group to test surgical procedures arose to parallel the randomized, double-blind, placebo-controlled trials that have become the gold standard in biomedical research and evidence-based medicine,[44] yet the clinical benefits of such practices have been called into question. In a recent study for surgery to treat Parkinson's disease, the degree of motor performance improvement at 18 months was substantial, but the same after a sham operation as after stereotactic intrastriatal implantation of fetal porcine ventral mesencephalic tissue.[45] In one recently reported, multicenter, randomized, double-blind, sham surgery-controlled study of human fetal transplantation for Parkinson's, there was no significant clinical benefit of the transplant compared to sham surgery,[46] even though pilot studies performed using identical technique had demonstrated substantial benefit.[47] Although there was no improvement in the sham operated group, this does not necessarily mean that the disparate results could not arise from the placebo effect. Thus, patient expectations may conceivably have been much higher at the time of the pilot studies, resulting in an augmented placebo effect in the earlier, uncontrolled studies. Freed and colleagues also found the effect of an imitation operation to be modest.[27] Several factors could explain the differences in the magnitude of the placebo response between different trials. Variations in the information given to the patients, differences in group characteristics, and/or the surgical procedures could contribute to a range of placebo responses.[25] Naturally, ethical issues and consideration of the risks and benefits inherent in the conduct of the study will dictate whether a placebo treatment group is feasible. However, there is development in medical research involving surgical treatment for Parkinson's disease. In a recent small phase 1 clinical trial, Gill and colleagues[48] conducted direct intra-putamenal GDNF infusion in patients with Parkinson disease, resulting in significant increases in dopamine storage in three regions of the putamen. Placebo-controlled studies are now underway to confirm the benefit demonstrated in this pilot study.

DOPAMINE, EXPECTATION, AND REWARD

Recent investigation of the placebo effect in Parkinson's disease has linked it to the release of dopamine in areas of the brain related to reward mechanisms as well as motor control. The dopaminergic system has long been implicated in reward mechanisms.[49,50] Animal experiments have shown that the midbrain dopaminergic cell groups A8, A9, and A10 are activated by primary rewards, reward-predicting stimuli, and novel stimuli.[51,52] Although three major dopamine-containing pathways (the nigrostriatal, mesolimbic, and mesocortical pathways) participate in these responses, the projection to the nucleus accumbens has received the greatest attention.[50,51] In particular, it has been shown that most drugs of abuse (including cocaine, amphetamine, opioids, alcohol, and nicotine) increase dopamine levels in the nucleus accumbens, and, in fact, the basis of drug dependence seems to be related to dopamine release.[53] Of most relevance to the placebo effect is that dopamine release appears to be more related to the expectation of the reward than to the reward itself.[7] In an elegant experiment using an intracranial self-stimulation paradigm—in which an animal repeatedly presses a lever to stimulate its own dopaminergic projections—Garris and colleagues provided evidence that, although electrical intracranial stimulation in animals will occur only when the electrodes are positioned to stimulate dopamine release in response to experimenter-derived stimulation, self-stimulation itself does not result in the release of dopamine.[54] This result is consistent with pervious experiments showing that dopaminergic neurons are activated after stimuli that predict a reward.[52] Naturally, these observations led to the development of the following hypothesis: if the expectation of a reward triggers dopamine release, not only in the nucleus accumbens but also in the nigrostriatal pathway, the placebo effect in patients with Parkinson's disease could be related to the expectation of clinical benefit and could be mediated by the release of dopamine in the striatum. A link between the placebo effect, reward mechanisms, and dopamine release was demonstrated in a recent study using PET with [^{11}C] raclopride (RAC).[7] It was found that patients with Parkinson's disease release substantial amounts of dopamine in the dorsal striatum (i.e., caudate and putamen) in response to subcutaneous injections of saline (placebo). Changes in RAC binding between baseline and post-activation states—in this case, in response to placebo injection—reflected the release of endogenous dopamine, which displaces RAC binding to synaptic dopamine receptors. Placebo-induced changes in striatal RAC binding potential (17% in the caudate nucleus, 19% in the putamen) were of similar magnitude to those obtained after therapeutic doses of levodopa or amphetamine.[7,55] All patients showed a biochemical placebo effect (i.e., changes in RAC binding potential), but only half reported clinical benefit in motor function after placebo administration. Interestingly, the amount of dopamine release in the dorsal striatum was greater in those patients who perceived a placebo effect than in those who did not (22% and 12% decreases in RAC binding potential in the caudate nucleus, respectively; 24% and 14% decreases in the putamen, respectively). Given the known relationship

between dopamine levels in the putamen and motor function, it was concluded that the placebo effect in Parkinson's disease was triggered by the expectation of reward, in this case the reward being the clinical benefit. This idea was confirmed when the placebo-induced changes in the ventral striatum (nucleus accumbens) were analyzed; the region demonstrated a decline in RAC binding, similar to that seen in the caudate and putamen. However, in contrast to the results from the dorsal striatum, the magnitude of placebo-induced changes was not significantly different between patients who experienced clinical benefit and those who did not.[56] Because the perception of clinical benefit must be considered rewarding, this finding lends further support to the view that the release of dopamine is related to the expectation, and not the experience, of a reward.

Benedetti and colleagues[60] recently provided further evidence of a physiological underpinning for the placebo effect in Parkinson's disease. They recorded neuronal activity from single cells in the subthalamic nucleus (STN) at the time of electrode implantation for deep brain stimulation. In patients who had been preconditioned with apomorphine and then received an injection of placebo, the mean firing rate in the STN declined in response to placebo injection in those patients who showed a clinical response to placebo. Furthermore, the firing pattern of STN neurons in these patients changed from a bursting pattern to a more regular pattern. The authors concluded that their findings were compatible with striatal dopamine release in those patients who demonstrated a placebo response.

IMPLICATIONS FOR TREATMENT OF PARKINSON'S DISEASE

Can the power of placebos be harnessed to provide therapeutic benefit to Parkinson's disease patients? Clearly, the placebo effect can result in therapeutic benefit in some Parkinson's patients, so is it possible that, eventually, placebos could be used to augment the benefit derived from standard therapies? At this point, further research needs to be conducted into the precise mechanism by which placebos exert their positive effects. The relationship between the expectation that motor performance will improve and the actual improvement in physical function must be more clearly understood, as must be the means of maximizing the placebo effect. Thus, recent studies in monkeys suggest that dopaminergic activity is maximal when there is uncertainty regarding the likelihood of reward.[57] Placebos could potentially be involved in treatment in several different capacities. The most likely possibility is for placebos to supplement active medication; it is possible that, by reducing the requirement for active medication, some of the toxicity associated with use of the active drug might be diminished. This is not necessarily the case, however. In the case of Parkinson's disease, for example, a reduction in levodopa dose could potentially lead to reduced dyskinesias, but this might be offset by placebo-induced dopamine release, particularly if this were to occur in a pulsatile fashion. Furthermore, the interaction between the placebo effect and the effect of the active medication may vary among individuals and even within the same individual, and this may modify the response to therapy. It is as yet unresolved whether the response to placebo might be sustained over a prolonged period of time, but if so, it is almost certain that placebo substitution would have to be given according to a variable schedule. In addition, given that the ability to respond to a particular placebo is not associated with a specific psychological profile,[1,12] would that same response be maintained across a population of patients with similar profiles and levels of disease progression? It is likely that treatment regimens that incorporate a placebo would require individual tailoring.

Another possibility is for a placebo to act as a complete substitute for the active drug. This idea has already begun to be practiced in long-term substitution programmes for the treatment of drug addiction.[9] As it has been shown that placebos induce the release of endogenous dopamine in the nucleus accumbens, as do most drugs of abuse, it is then possible to use a placebo in lieu of the drug. There are reports of successful saline substitution for the active drug in morphine addicts,[58] and methadone substitution programs for heroin addicts can also benefit from placebo use.[59] However, care must be taken to avoid the addiction to "cross" from the drug to the placebo, for there is evidence to suggest that placebos can be addictive, causing withdrawal symptoms when treatment is discontinued.[1,9]

CONCLUSION

The placebo effect is a very real, widespread phenomenon with a significant role in medical history, and it occupies a prominent position in current clinical research, not only in Parkinson's disease but also in other neurological disorders. A great deal of progress has been made in identifying the areas of the brain that respond to placebos and give rise to a placebo response, and some of the biochemical bases of the placebo effect have already been elucidated.[16] In Parkinson's disease, the placebo effect has been associated with the release of endogenous dopamine in the striatum in response to the expectation of clinical benefit, and is likely secondary to activation of the reward circuitry of the brain. This raises the possibility that dopamine release may play a role in the placebo responses of other medical disorders, at least in part. Dopamine release has already been implicated in analgesia, and

dopamine-opioid interactions might mediate the placebo effect that has been observed in pain disorders.

The expectation theory of the placebo effect has strong implications for the design of future placebo studies. Certain elements must be present for the placebo effect to be detected to its full extent. Recognition of this is important not only in the design and interpretation of clinical trials but also for experiments designed to study the placebo effect itself. Thus, major advances in knowledge about the placebo effect will continue to stem from active drug trials, surgical procedure studies, and placebo-directed research. The results from this research will determine whether the placebo effect comes to represent a viable component of treatment of Parkinson's disease and other CNS disorders in clinical practice.

REFERENCES

1. Brody, H., *Placebos and the Philosophy of Medicine: Clinical, Conceptual, and Ethical Issues,* Chicago: University of Chicago Press, 1980.
2. Enserink, M., Can the placebo be the cure? *Science,* 284:238, 1999.
3. Freeman, T. B., Vawter, D. E., Leaverton, P. E., Godbold, J. H., Hauser, R. A., Goetz, C. G., et al. Use of placebo surgery in controlled trials of a cellular-based therapy for Parkinson's disease, *N. Engl. J. Med.,* 341(13): 988–992, 1999.
4. Turner, J. A., The importance of placebo effects in pain treatment and research, *J. Am. Med. Assoc.,* 271:1609, 1994.
5. Shetty, N., Friedman, J. H., Kieburtz, K., Marshall, F. J., Oakes, D., The placebo response in Parkinson's disease, Parkinson Study Group, *Clin. Neuropharmacol.,* 22(4):207–212, 1999.
6. Levine, J. D., Gordon, N. C., Fields, H. L., The mechanism of placebo analgesic, *Lancet,* ii:654, 1978.
7. de la Fuente-Fernandez, R., Ruth, T. J., Sossi, V., Schulzer, M., Calne, D. B., Stoessl, A. J., Expectation and dopamine release: mechanism of the placebo effect in Parkinson's disease, *Science,* 293(5532): 1164–1166, 2001.
8. Macedo, A., Farre, M., Banos, J. E., Placebo effect and placebos: what are we talking about? Some conceptual and historical considerations, *Eur. J. Clin. Pharmacol.,* 59:337, 2003.
9. Wolf, S., The pharmacology of placebos, *Pharmacol. Rev.,* 11:689, 1959.
10. de Craen, A. J., Tijssen, J. G., de Gans, J., Kleijnen, J., Placebo effect in the acute treatment of migraine: subcutaneous placebos are better than oral placebos, *J. Neurol.,* 247(3):183–188, 2000.
11. Kaptchuk, T.J., Goldman, P., Stone D, A., Stason, W. B., Do medical devices have enhanced placebo effects? *J. Clin. Epidemiol.,* 53(8):786–792, 2000.
12. Shapiro, A. K., Shapiro, E., *The placebo: is it much ado about nothing?* Cambridge, Massachusetts: Harvard University Press, 1997.

13. Moseley, J. B., O'Malley, K., Petersen, N. J., Menke, T. J., Brody, B. A., Kuykendall, D. H., et al., A controlled trial of arthroscopic surgery for osteoarthritis of the knee, *N. Engl. J. Med.*, 347(2):81–88, 2002.

14. Kennedy, W. P., The nocebo reaction, *Med. Exp. Int. J. Exp. Med.*, 95:203–205, 1961.

15. Benedetti, F., Pollo, A., Lopiano, L., Lanotte, M., Vighetti, S., Rainero, I., Conscious expectation and unconscious conditioning in analgesic, motor, and hormonal placebo/nocebo responses, *J. Neurosci.*, 23(10):4315–4323, 2003.

16. de la Fuente-Fernandez, R., Stoessl, A. J., The biochemical bases for reward, implications for the placebo effect, *Eval. Health. Prof.*, 25(4):387–398, 2002.

17. Kirsch, I., Specifying nonspecifics: psychological mechanisms of placebo effects, in Harrington, A., Ed., *The Placebo Effect: An Interdisciplinary Exploration*, Cambridge, Massachusetts: Harvard University Press, 166–186, 1997.

18. Flaten, M. A., Simonsen, T., Olsen, H., Drug-related information generates placebo and nocebo responses that modify the drug response, *Psychosom. Med.*, 61(2):250–255, 1999.

19. de la Fuente-Fernandez, R., Stoessl, A. J., The placebo effect in Parkinson's disease, *Trends Neurosci.*, 25(6):302–306, 2002.

20. Ader, R., The role of conditioning in pharmacotherapy, in Harrington, A., Ed., *The Placebo Effect: An Interdisciplinary Exploration*, Cambridge, Massachusetts: Harvard University Press, 138–165, 1997.

21. Evans, F. J., Expectancy, therapeutic instructions, and the placebo response, in White, L., Tursky, B., Schwartz, G. E., Eds., *Placebo: Theory, Research, and Mechanisms*, New York: Guilford Press, 215–228, 1985.

22. Wickramasekera, I. A., Conditioned response model of the placebo effect: predictions from the model, in White, L., Tursky, B., Schwartz, G. E., Eds., *Placebo: Theory, Research, and Mechanisms*, New York: Guilford Press, 255–287, 1985.

23. Pollo, A., Torre, E., Lopiano, L., Rizzone, M., Lanotte, M., Cavanna, A., et al., Expectation modulates the response to subthalamic nucleus stimulation in Parkinsonian patients, *NeuroReport*, 13(11):1383–1386, 2002.

24. Altman, D.G., *Practical Statistics for Medical Research*, London: Chapman & Hall, 1991.

25. Fuente-Fernandez, R., Schulzer, M., Stoessl, A. J., The placebo effect in neurological disorders, *Lancet Neurol.*, 1(2):85–91, 2002.

26. Ernst, E., Resch, K. L., Concept of true and perceived placebo effects., *BMJ*, 311(7004):551–553, 1995.

27. Freed, C.R., Greene, P. E., Breeze, R.E., Tsai, W. Y., DuMouchel, W., Kao, R., et al., Transplantation of embryonic dopamine neurons for severe Parkinson's disease, *N. Engl. J. Med.*, 344(10):710–719, 2001.

28. Kirsch, I., Sapierstein, G., Listening to Prozac but Hearing Placebo: A Meta-Analysis of Antidepressant Medications, *Prevention & Treatment*, 1(0002a), 1998.

29. Hrobjartsson, A., Gotzsche, P. C., Is the placebo powerless? An analysis of clinical trials comparing placebo with no treatment, *N. Engl. J. Med.*, 344(21):1594–1602, 2001.

30. Kaptchuk, T. J., Powerful placebo: the dark side of the randomised controlled trial, *Lancet.*, 351(9117): 1722–1725, 1998.

31. Rosenthal, R., Designing, analyzing, interpreting, and summarizing placebo studies, in White, L., Tursky, B., Schwartz, G. E., Eds., *Placebo: Theory, Research, and Mechanisms*, New York: Guilford Press, 110–136, 1985.

32. Kaptchuk, T. J., The double-blind, randomized, placebo-controlled trial: gold standard or golden calf? *J. Clin. Epidemiol.*, 54(6):541–549, 2001.

33. Goetz, C. G., Janko, K., Blasucci, L. M., Jaglin, J. A., Impact of placebo assignment in clinical trials of Parkinson's disease, *Mov. Disord.*, 18(10):1146–1149, 2003.

34. Kleijnen J., de Craen, A.J., van Everdingen, J., Krol, L., Placebo effect in double-blind clinical trials: a review of interactions with medications, *Lancet*, 344(8933): 1347–1349, 1994.

35. Mayberg, H. S., Silva, J. A., Brannan, S. K., Tekell, J. L., Mahurin, R. K., McGinnis, S., et al., The functional neuroanatomy of the placebo effect, *Am. J. Psychiatry*, 159(5):728–737, 2002.

36. Sesack, S. R., Pickel, V. M., Dual ultrastructural localization of enkephalin and tyrosine hydroxylase immunoreactivity in the rat ventral tegmental area: multiple substrates for opiate-dopamine interactions, *J. Neurosci.*, 12(4):1335–1350, 1992.

37. Petrovic, P., Kalso, E., Petersson, K. M., Ingvar, M., Placebo and opioid analgesia—imaging a shared neuronal network, *Science*, 295(5560):1737–1740, 2002.

38. Zubieta, J. K., Smith, Y. R., Bueller, J. A., Xu, Y., Kilbourn, M. R., Jewett, D. M., et al., Regional mu opioid receptor regulation of sensory and affective dimensions of pain, *Science*, 293(5528):311–315, 2001.

39. Altier, N., Stewart, J., The role of dopamine in the nucleus accumbens in analgesia., *Life Sci.*, 65(22): 2269–2287, 1999.

40. Diamond, S. G., Markham, C. H., Treciokas, L. J., Double-blind trial of pergolide for Parkinson's disease, *Neurology*, 35(3):291–295, 1985.

41. Goetz, C. G., Leurgans, S., Raman, R., Placebo-associated improvements in motor function: comparison of subjective and objective sections of the UPDRS in early Parkinson's disease, *Mov. Disord.*, 17(2):283–288, 2002.

42. Goetz, C. G., Leurgans, S., Raman, R., Stebbins, G. T., Objective changes in motor function during placebo treatment in PD, *Neurology*, 54(3):710–714, 2000.

43. Macklin, R., The ethical problems with sham surgery in clinical research, *N. Engl. J. Med.*, 341(13):992–996, 1999.

44. Dekkers, W., Boer, G., Sham neurosurgery in patients with Parkinson's disease: is it morally acceptable? *J. Med. Ethics.*, 27(3):151–156, 2001.

45. Watts, R. L., Freeman, T. B., Hauser, R. A., Bakay, R. A. E., Ellias, S. A., Stoessl, A. J., et al., A double-blind, randomised, controlled, multicenter clinical trial of the safety and efficacy of stereotaxic intrastriatal implantation of fetal porcine ventral mesencephalic tissue (Neurocell™-PD) vs. imitation surgery in patients with

Parkinson's disease (PD), *Parkinsonism & Related Disorders*, 7:S87, 2001.

46. Olanow, C. W., Goetz, C. G., Kordower, J. H., Stoessl, A. J., Sossi, V., Brin, M. F., et al., A double-blind controlled trial of bilateral fetal nigral transplantation in Parkinson's disease, *Ann. Neurol.*, 54(3):403–414, 2003.

47. Hauser, R. A., Freeman, T. B., Snow, B. J., Nauert, M., Gauger, L., Kordower, J. H., et al., Long-term evaluation of bilateral fetal nigral transplantation in Parkinson disease, *Arch. Neurol.*, 56(2):179–187, 1999.

48. Gill, S. S., Patel, N. K., Hotton, G. R., O'Sullivan, K., McCarter, R., Bunnage, M., et al., Direct brain infusion of glial cell line-derived neurotrophic factor in Parkinson disease, *Nat. Med.*, 9(5):589–595, 2003.

49. Phillips, A. G., Fibiger, H. C., Neuroanatomical bases of intracranial self-stimulation: untangling the Gordian knot, in Liebman, J. M., Cooper, S. J., Eds., *The Neuropharmacological Basis of Reward,* Clarendon Press, 66–105, 1989.

50. Wise, R. A., Rompre, P. P., Brain dopamine and reward, *Annu. Rev. Psychol.*, 40:191–225, 1989.

51. Rebec, G. V., Christensen, J. R., Guerra, C., Bardo, M. T., Regional and temporal differences in real-time dopamine efflux in the nucleus accumbens during free-choice novelty, *Brain Res.*, 776(1–2):61–67, 1997.

52. Schultz, W., Reward signaling by dopamine neurons, *Neuroscientist*, 7(4):293–302, 2001.

53. Robinson, T. E., Berridge, K. C., The neural basis of drug craving: an incentive-sensitization theory of addiction, *Brain Res. Brain Res. Rev.*, 18(3):247–291, 1993.

54. Garris, P. A., Kilpatrick, M., Bunin, M. A., Michael, D., Walker, Q. D., Wightman, R. M., Dissociation of dopamine release in the nucleus accumbens from intracranial self-stimulation, *Nature*, 398(6722):67–69, 1999.

55. de la Fuente-Fernandez, R., Lu, J. Q., Sossi, V., Jivan, S., Schulzer, M., Holden, J. E., et al., Biochemical variations in the synaptic level of dopamine precede motor fluctuations in Parkinson's disease: PET evidence of increased dopamine turnover, *Ann. Neurol.*, 49(3):298–303, 2001.

56. de la Fuente-Fernandez, R., Phillips, A. G., Zamburlini, M., Sossi, V., Calne, D. B., Ruth, T. J., et al., Dopamine release in human ventral striatum and expectation of reward, *Behav. Brain Res.*, 136(2):359–363, 2002.

57. Fiorillo, C. D., Tobler, P. N., Schultz, W., Discrete Coding of Reward Probability and Uncertainty by Dopamine Neurons, *Science*, 299(5614):1898–1902, 2003.

58. Leslie, A., Ethics and practice of placebo therapy, *Am. J. Med.,* 16(6):854–862, 1954.

59. Curran, H. V., Bolton, J., Wanigaratne, S., Smyth, C., Additional methadone increases craving for heroin: a double-blind, placebo-controlled study of chronic opiate users receiving methadone substitution treatment, *Addiction*, 94(5):665–674, 1999.

60. Benedetti, F., Colloca, L., Torre, E., Lanotte, M., Melcarne, A., Pesare, M., Bergamasco, B., and Lopiano, L., Placebo-responsive Parkinson patients show decreased activity in single neurons of subthalamic nucleus, *Nat. Neurosci.,* 7:587–588, 2004.

48 Dopamine Transporter Imaging Using SPECT in Parkinson's Disease

Danna Jennings, Ken Marek, and John Seibyl
The Institute for Neurodegenerative Disorders

CONTENTS

INTRODUCTION

Over the past decade, development of functional neuroimaging of the nigrostriatal dopaminergic system has improved our understanding of the natural history of pathophysiological changes in Parkinson's disease (PD). PD is characterized by degeneration of the nigral dopaminergic cells and their striatal terminals, resulting in decreased striatal dopamine and a loss in dopamine transporters. Functional neuroimaging uses radioactively labeled molecules called *ligands* as markers in conjunction with *single photon emission tomography (SPECT)* and *positron emission tomography (PET)* imaging to evaluate neurochemical systems in the brain. These methods offer a unique advantage in PD over structural imaging, such as computed tomography or magnetic resonance imaging. PET and SPECT provide a means to visualize the neurochemistry of the brain. Ligands targeting the dopamine transporter (DAT) have been developed as markers of dopaminergic neuronal cell loss. In this chapter, the role of DAT imaging as a marker for evaluating disease progression, severity, or stage of disease, and as a diagnostic tool in parkinsonian syndrome, is reviewed.

SINGLE PHOTON EMISSION COMPUTED TOMOGRAPHY OF THE DOPAMINERGIC SYSTEM

Dopaminergic system function in the brain can be visualized *in vivo* using either PET or SPECT imaging. Both PET and SPECT have been shown to be sensitive measures of the brain neurochemistry.[1,2] PET cameras have better resolution, but the availability of PET cameras and ligands are more limited due to the resources required to acquire and maintain a PET facility. In addition, the available PET ligands have a short half-life requiring an on-site cyclotron for successful synthesis and use. The widespread availability of SPECT cameras and the relatively long half-life of the radioligands used with SPECT makes it a more practical choice as a diagnostic tool in clinical practice and as a marker for disease progression in performing large clinical studies.

The strengths and limitations of *in vivo* neuroreceptor imaging studies depend on the imaging technology utilized to measure brain neurochemistry as well as the ligand or biochemical marker used to tag a specific brain neurochemical system. Currently available ligands have made it possible to visualize, *in vivo*, both the presynaptic nigrostriatal dopamine neurons and postsynaptic dopamine D$_2$ receptors using PET and SPECT ligands.[3,4] Specific ligands of the dopaminergic system have been developed to evaluate patients with PD, and the most extensively studied ligands include 18F-flurodopa,[3,5-9] 11C-vesicular monoamine transporter Type 2 (VMAT2),[10-12] and dopamine transporter (DAT)[13-17] ligands. DAT is a presynaptic protein located on the membrane of the dopaminergic neuron terminals. The function of DAT is to actively reuptake dopamine from the synaptic cleft after termination of its interaction with the postsynaptic dopamine receptors.[18-19] Imaging the DAT using specific ligands in conjunction with SPECT or PET offers the opportunity to measure the striatal uptake of the DAT ligand providing an *in vivo* assessment of the integrity of the presynaptic dopaminergic nerve terminals (Figure 48.1). Of the DAT and SPECT radioligands in development, [^{123}I]β-CIT has been the most widely evaluated ligand.[14,20] The unique binding kinetics of [^{123}I]β-CIT, characterized by a relatively long period of radiotracer uptake and slow elimination from the DAT sites in the striatum, allow reliable quantitative determination of the dopamine transporter density.

The Role of DAT Imaging in the Diagnosis of PD

The diagnosis of PD currently relies on clinical examination and is based on the identification of well recognized cardinal motor signs of rigidity, bradykinesia, and resting tremor. Long-term clinicopathologic studies of the diagnostic accuracy of PD demonstrate that the diagnoses most commonly mistaken for PD are PSP and MSA.[21,22] However, the diagnoses most commonly mistaken for PD early in its course include essential tremor, vascular parkinsonism, drug-induced parkinsonism, psychogenic parkin-

sonism, and Alzheimer's disease.[23,24] Factors making the diagnosis of PD more challenging include the subtlety of initial symptoms, variability of disease presentation, the slow rate of disease progression, and the lack of a convincing response to dopaminergic medications in some patients. In addition, the parkinsonian signs of bradykinesia and stiffness are relatively common in elderly subjects, making the diagnosis increasingly challenging in this population. In one series, 35% of individuals over the age of 65 years have been reported to have subtle extrapyramidal signs on neurological evaluation.[25,26] Prevalence estimates for clinically diagnosed parkinsonian syndromes in similarly aged subjects are much lower, at around 3%. Misdiagnosing other conditions as PD may lead to futile therapy with dopamine-replacing agents, often resulting in unnecessary side effects. In addition, significant resources are spent on medication trials and CT or MRI brain scans, which are performed in an attempt to clarify the diagnosis.

If the diagnosis of a parkinsonian syndrome is in question, the most common diagnostic approach is to perform serial examinations over several months to years until sufficient signs are present to determine a more definitive diagnosis. In many cases, a trial of dopaminergic replacement therapy is administered to clarify the diagnosis. Unfortunately, even a short trial of medications carries the risk of side effects related to dopaminergic therapy, and often the response to therapy is disappointingly unclear. This "wait and watch" approach has been the standard of practice; however, as disease modifying agents become available, identification of the disease state as early as possible will be essential. *In vivo* dopamine transporter imaging studies have demonstrated a reduction in dopamine transporter density in PD patients compared to healthy controls. The reduction in dopamine transporter density in PD is both region specific (putamen > caudate) and asymmetric, consistent with both pathologic assessment of the dopamine transporter loss and clinical presentation of PD. Similar to 18F Dopa and PET, dopamine transporter imaging using SPECT can discriminate patients with PD from control subjects with a sensitivity of greater than 95%.[9,20,27] The dopamine transporter density, quantitatively measured by [^{123}Iβ-CIT and SPECT imaging, has documented losses of 30 to 55% in early PD. The degree of loss of dopaminergic neurons is not as great as the loss of endogenous dopamine, reported in postmortem human tissue samples (>80%) to be in the range of in subjects with PD, but these are from subjects with more advanced PD.

Difficulty in accurately diagnosing individuals early in the course of PD clearly impacts the clinical care of individuals and may also have implications when recruiting subjects for early PD clinical trials. Two recent studies involving early, untreated parkinsonian subjects suggest that imaging may identify individuals without typical PD at the time of enrollment. In the REAL-PET

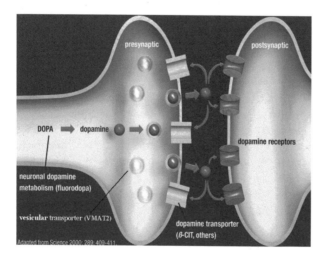

FIGURE 48.1 Dopamine pre- and postsynaptic neuronal receptors for radioligand targets. (Adapted from Marek, K. and Seibyl, J., *Science*, 289, 409–411, 2000. With permission.)

study, comparing ropinirole and levodopa as initial treatments in untreated patients, 11% (21/193) of enrolled subjects had scans without evidence of reduction in 18F-dopa uptake at baseline and after two years.[28] In the ELL-DOPA-CIT study,[29] comparing initial levodopa therapy to placebo in recently diagnosed patients, 14% (21/142) of enrolled subjects had scans without evidence of reduction in [123I]β-CIT uptake at baseline and again at 9 months (19/19, 2 terminated). The uncertainty of clinical diagnosis is an important factor in the design and critical analysis of clinical therapeutic trials. Inclusion of subjects who do not have PD increases estimates of disease frequency and confounds efficacy studies of agents that may alter the rate of progression of disease. Data from these studies underscores the difficulty in accurately diagnosing parkinsonian patients in the early stages, based solely on clinical evaluation. Dopamine transporter imaging offers an objective measure of the density of the presynaptic dopaminergic neurons. Several studies have shown that DAT ligand uptake is already reduced by about 50% when compared to age-corrected controls indicating a role for DAT and SPECT in confirming a diagnosis of PD in patients with early symptoms[30,31] (Figure 48.2).

In a recent blinded, prospective study, 35 patients with symptoms of suspected early parkinsonian syndrome (PS) were referred for DAT imaging, using [123I]β-CIT and SPECT, by community neurologists who were unsure of their diagnosis.[31] In this study, PS was defined as any condition expected to have a reduction in dopamine transporter density, including PD, PSP, MSA, DLBD, SND, and CBGD. To evaluate the accuracy of DAT imaging as a diagnostic tool in this population, patients were followed clinically over a six-month period. Two movement disorder experts assigned a clinical diagnosis at the time of referral. One movement disorder expert remained blind to the imaging data and evaluated and assigned a clinical diagnosis at the six-month interval. The six-month clinical diagnosis served as the "gold standard" diagnosis for the study. Figure 48.3 shows data from the subjects compared to healthy control database of 73 subjects. Based on this study, the sensitivity of the [123I]β-CIT and SPECT imaging diagnosis was 0.92, while the specificity of the imaging was 1.0 when compared to the clinical "gold standard" diagnosis at six-month follow-up. Two subjects referred with a questionable diagnosis of PS have a diagnosis of PS by the clinical "gold standard," while their imaging showed no deficit of DAT uptake. Longer follow-up of these subjects is necessary to clarify the diagnosis. In a similar study of subjects with an inconclusive diagnosis, [123I]FP-CIT and SPECT were performed, and subjects were followed over a two- to four-year period. In this study, the clinicians were aware of the imaging results and utilized this information in making a final diagnosis. In 9/33 subjects, dopaminergic neuronal

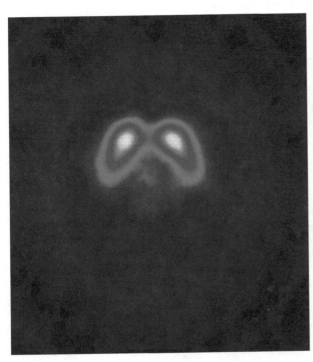 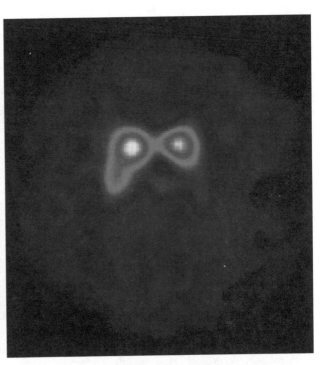

Healthy Subject — **Parkinson's disease Stage 1**

FIGURE 48.2 Dopamine transporter imaging utilizing [123I]β-CIT and SPECT in a healthy control (left) and a subject with early Parkinson's disease (right). (*A color version of this figure follows page 518.*)

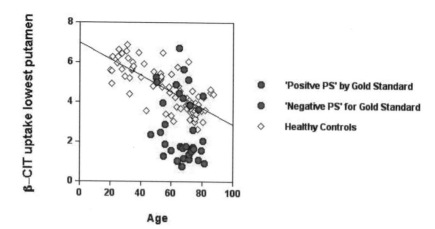

FIGURE 48.3 [¹²³I]β-CIT uptake in the putamen of 73 healthy controls and 35 subjects with suspected parkinsonian symptoms. The clinical diagnosis is congruent with the imaging diagnosis in 33/35 subjects. Two subjects with a clinical diagnosis of parkinsonian syndrome at six-month follow-up have putamenal uptakes in the range of the healthy controls.

degeneration was found, and in all cases a diagnosis of PS was confirmed clinically. In 24 subjects, there was no evidence of dopaminergic neuronal degeneration, and other non-PS diagnoses were assigned in 19 of these subjects at follow-up.[30] Both studies suggest that the positive predictive value of DAT imaging in the diagnosis of PS is high; however, the negative predictive value is lower. Combining data from DAT with the clinical evaluation improves the diagnostic accuracy of PS in difficult-to-diagnose cases.

DAT Imaging in the Differential Diagnosis of Parkinsonian Syndrome

Atypical Parkinsonian Syndromes

Distinguishing between PD and atypical parkinsonisms (such as progressive supranuclear palsy, striatonigral degeneration, diffuse Lewy body disease, or multiple systems atrophy) is important in offering information about prognosis and making appropriate treatment decisions. Differentiating PD from an atypical parkinsonian syndrome is difficult based on clinical exam alone and error rates can be as high as 25%.[21] Even in specialized movement disorders centers, the positive predictive value of a clinical diagnosis of PSP or MSA is only between 80 to 85%.[32] Most of the atypical parkinsonian syndromes are characterized pathologically by a loss of nigrostriatal dopaminergic neuronal loss in addition to other changes. Similar to PD, there is a reduction in striatal uptake of DAT tracers as a result of the pathology. The severity of DAT loss alone does not distinguish PD from the atypical parkinsonian syndromes. However, the pattern of loss in the atypical parkinsonisms is less region-specific than in idiopathic PD, with the caudate and putamen being more equally effected.[27,33,34] Although studies using DAT and SPECT imaging are unable to significantly differentiate

these atypical forms of parkinsonism from idiopathic PD, individuals with a relatively symmetric loss of DAT uptake have been shown to more likely have either PSP or MSA.[33,35] The more widespread pathology associated with the atypical parkinsonisms is more effectively evaluated with postsynaptic dopamine receptor imaging. Imaging of the postsynaptic dopamine receptors shows a decrease in ligand uptake in atypical parkinsonian syndromes, while patients with idiopathic PD have uptake in the postsynaptic receptors that appears to be similar to controls. The evaluation of presynaptic dopaminergic loss coupled with postsynaptic dopamine receptor imaging and clinical evaluation may improve our ability to distinguish PD from other atypical forms of parkinsonism.

Essential Tremor

Classic essential tremor (ET) with bilateral postural and action tremor of the limbs or head in the absence of any signs of rigidity or tremor can usually be differentiated from PD clinically. Diagnostic difficulties frequently arise when there is evidence of tremor that appears to be at rest, mild cogwheel rigidity, mild bradykinesia, or asymmetry of symptoms. Studies investigating the potential of DAT imaging to differentiate ET from PD have found a sensitivity and specificity of about 95% of DAT and SPECT for successfully discriminating between the two disorders.[27,36,37] DAT and SPECT imaging reliably and effectively distinguishes between individuals with PD or parkinsonian syndrome and ET.

Drug-Induced Parkinsonism

Parkinsonism secondary to drug exposure is common particularly in the elderly and in populations with psychiatric disorders. Dopamine receptor blockers, used primarily as antipsychotics and antiemetics, are most frequently the offending medications in drug-induced parkinsonism (DIP). Only a few patients exposed to dopamine receptor blockers develop parkinsonism, which suggests that there must be an individual susceptibility in those who develop parkinsonism. Differentiating DIP from a parkinsonian syndrome with nigrostriatal degeneration can be difficult clinically but has significant implications regarding treatment. Withdrawal from the dopamine receptor blocking medication, when possible, can require several months to reach full resolution of parkinsonian symptoms. Evaluating the integrity of the presynaptic dopamine neurons using DAT imaging can be useful in determining if there is a loss in nigrostriatal neurons and thus differentiating DIP from a parkinsonian syndrome. While there are few reports in the literature of patients with DIP who have undergone DAT imaging, 4 subjects from our center with DIP based on 6- to 12-month follow-up clinical examination all had [123I]β-CIT and SPECT imaging that was within the range of age-corrected healthy controls. DAT imaging appears to be a useful tool in evaluating whether parkinsonian symptoms are related to striatonigral dopaminergic neuronal loss in patients treated with dopamine receptor blocking medications.

Vascular Parkinsonism

Over the years, the term *vascular parkinsonism* has remained a poorly defined syndrome.[38,39] Diagnostic questions often arise when a patient presents with parkinsonism and diffuse white matter ischemic changes or lacunar lesions localized to the basal ganglia. Vascular parkinsonism typically presents with symptoms of rigidity and bradykinesia predominantly in the lower extremities, resulting in a frontal gait disorder and postural instability. When parkinsonian patients present with lower extremity predominant symptoms, it is difficult to distinguish PD from vascular parkinsonism, especially early in its course. Pathological studies have shown preservation of the presynaptic dopaminergic circuitry in vascular parkinsonism patients.[40,41] It has been hypothesized that deep periventricular white matter lesions disrupt connections between the primary motor cortex and the supplementary motor cortex with the cerebellum and the basal ganglia. A definitive diagnosis of vascular parkinsonism requires neuropathological evaluation postmortem. However, at least one study of 13 subjects who fulfilled the criteria for a clinical diagnosis of vascular parkinsonism demonstrated preservation of striatal binding and the putamen to caudate ratio with [123I]β-CIT and SPECT imaging.[38]

In studies from our group evaluating [123I]β-CIT and SPECT imaging in difficult-to-diagnose cases, it has become clear that patients presenting with lower body parkinsonism are diagnostically challenging. In our studies, a subset of 12 patients with lower body parkinsonism (LBP) have been referred for DAT to determine if there is

evidence of dopaminergic degeneration in these cases.[31] Three of the 12 LBP patients had a decrease in uptake of [123I]β-CIT and SPECT imaging consistent with PS, while 9 patients had scans without evidence of dopamine neuronal deficiency. A 12-month clinical follow-up demonstrated that two patients who had a DAT imaging with no evidence of dopaminergic neuronal loss at baseline did not have PS clinically. One patient was given a final diagnosis of NPH following a remarkable improvement with VP shunt. The second patient was diagnosed with vascular parkinsonism based on an MRI with marked white matter ischemic changes and a lack of response to levodopa. Given the long-term clinical follow-up resulting in the more definitive diagnoses of NPH and vascular parkinsonism in two cases correlating with a negative imaging diagnosis, we expect the imaging diagnosis at baseline will ultimately predict the 12-month follow-up clinical diagnosis. DAT imaging appears to be a particularly useful diagnostic tool in gait difficulties of the elderly.

Psychogenic Parkinsonism

Psychogenic parkinsonism (PsyP) is a rare form of secondary parkinsonism, which can be difficult to diagnose. Reported cases of PsyP reported in the literature have demonstrated a combination of parkinsonian symptoms that places them in the differential of parkinsonism; however, the parkinsonian symptoms may have atypical features. Recognizing these atypical features requires referral to a movement disorder specialist with considerable experience in the evaluation and treatment of PD. Experts often need to follow an individual patient during several months through treatment trials to definitively differentiate PsyP from PD or other parkinsonism with striatonigral degeneration. Adding to the complexity of this difficult to diagnose condition, previous studies have shown 10 to 25% of patients with psychogenic movement disorders had features of both organic and psychogenic disease.[42,43] In a study reported by Lang et al.,[44] one of their subjects had combined psychogenic and true parkinsonian features with decreased fluorodopa uptake on one side consistent with organic parkinsonism.

Dopaminergic neuronal degeneration has not been shown to occur in PsyP. Using DAT imaging in these cases provides additional objective information to help differentiate PsyP from parkinsonian syndrome related to striatal dopaminergic degeneration. There are a limited number of reports with small numbers of subjects with suspected PsyP who have undergone DAT imaging in patients with PsyP.[30,42] These reports suggest that those with a clinical diagnosis of PsyP have DAT imaging within the range of healthy controls.

In a pilot study, we have performed DAT imaging with [123I]β-CIT in ten patients with psychogenic features as described.[45] DAT imaging in these difficult-to-diagnose cases demonstrated a loss of [123I]β-CIT uptake consistent with PS in four of the ten subjects, while six of the ten subjects had scans without evidence of dopaminergic deficit. In the four subjects with a loss in [123I]β-CIT uptake, the loss of uptake was asymmetric correlating with the clinically most affected side and more pronounced in the putamen; patterns that have been described in PD.[37,46-48] Clinical follow-up of one patient with a reduction in [123I]β-CIT uptake has led to a more definitive diagnosis of PD, as there has been gradual progression of parkinsonian symptoms and a convincing response to dopaminergic therapy. In the patients with no evidence of a reduction in [123I]β-CIT uptake, three patients have been followed for a period of at least two years; two have had resolution of their symptoms, and one continues to have unusual clinical symptoms that appear to be psychogenic in nature. Further prospective studies involving long-term follow-up of patients with suspected PsyP are needed to confirm the diagnostic accuracy of DAT imaging in these patients.

Dopa-Responsive Dystonia

Dopa-responsive dystonia (DRD) is a dominantly inherited condition with a recognized mutation in the GTP-cyclohydrolase I gene.[49-51] This gene defect results in impaired synthesis of dopamine without degeneration of the striatonigral neurons.[52] The clinical presentations of DRD are broad, often making it difficult to differentiate diagnostically; however, initial symptoms usually include young-onset dystonia and parkinsonism. Discriminating DRD from adult-onset PD and juvenile parkinsonism (JP) can be especially challenging, given that all of these conditions respond to dopamine replacement therapy. Establishing the diagnosis has important implications for prognosis and long-term treatment. Ultimately, the diagnosis is clarified clinically through follow-up evaluations over months to years. Reports in the literature have shown no reduction in radiotracer uptake in patients with a clinical diagnosis of DRD,[53,54] consistent with the lack of striatonigral neuronal loss shown pathologically in this condition.[55] The use of DAT imaging in patients with dystonia and parkinsonism that is responsive to dopamine may be helpful in clarifying the diagnosis at an earlier stage of the illness.

PRECLINICAL DIAGNOSIS

An important goal in PD research is to develop biomarkers to identify individuals with neurochemical changes before the onset of symptoms. Preclinical identification of effected individuals is particularly important as we develop interventions that may slow or prevent disease progression. Several agents with disease modifying potential are being currently being tested in clinical trials. Both clinical and imaging data from longitudinal studies of patients with PD suggest that the preclinical phase of PD may be several years in duration. Specifically, DAT and SPECT imaging in patients with very early PD has improved our understanding of the duration between initiation of the pathophysiological process and the first appearance of clinical symptoms.

In most patients with PD, the initial presentation is characterized by a unilateral onset of motor symptoms, which progresses to affect the limbs bilaterally over time. Imaging studies of the DAT consistently demonstrate a decrease in uptake of the radiotracer in the striatum bilaterally, even in patients with unilateral symptoms.[13,46,56] In these patients, imaging of the DAT shows about a 50% reduction in the putamen contralateral to the symptomatic side and a 25% reduction in the putamen ipsilateral to the symptomatic side relative to healthy subjects. Based on these studies, DAT imaging appears to be a valuable test for the evaluation of presymptomatic PD, with the capability of identifying changes occurring in the brain before manifestation of clinical symptoms.

While it is not financially or logistically feasible to perform imaging studies on the population at large to identify individuals with evidence of early dopaminergic neuronal loss, identifying at-risk populations provides a more practical approach to identify preclinical PD. The recent identification of genes associated with the PD phenotype in familial PD provides an opportunity to evaluate an at-risk population, both clinically and with *in vivo* imaging studies.[57-60] Specific environmental risk factors are also being recognized, thus identifying individuals or populations of individuals that may be at risk for PD.[61-62] Performing sequential imaging studies to evaluate for dopaminergic neuronal cell loss and the rate of loss of time is essential in identifying and understanding characteristics of the preclinical phase. Another approach to more generally identify at-risk populations is the development of screening batteries, such as olfactory testing, neurocognitive evaluations, or mood and personality scales. Berendse et al.[63] studied relatives of subjects with PD, an abnormal reduction in striatal DAT binding was found in 4 out of 25 relatives who had a reduction in olfaction. Two of these individuals subsequently developed clinical parkinsonism. In the relatives with normal olfaction, none of the 24 individuals had abnormal DAT binding. This important study demonstrates that reductions in DAT binding can be detected in asymptomatic relatives of PD patients using [123I]β-CIT and SPECT. As additional at-risk populations are recognized, imaging will play an essential role in identifying individuals with early dopaminergic neuronal loss, a population that serves to gain the most benefit from neuroprotective agents as they become available.

DAT Imaging as a Measure of Disease Severity

The motor manifestations of PD can primarily be attributed to dopamine deficiency, which occurs as a result of progressive dopaminergic neuronal degeneration in the substantia nigra. Clinical examination and standardized rating scales are frequently used to measure the severity of PD; however, in the treated patient, the examination is often confounded by the masking affects of dopaminergic therapy. In addition, there is a variability in symptoms that occurs as a function of the time of day of the evaluation, anxiety, and other psychological factors that are not easily controlled. DAT imaging offers an objective marker of disease severity. Several cross-sectional studies have shown a significant correlation between severity of PD and DAT imaging.[14,20,64] In a study using [123I]β-CIT and SPECT, a correlation between both stage and severity of PD was demonstrated.[20] Correlation between striatal uptake and UPDRS scores was also found in a study using [123I] FP-CIT throughout Hoehn and Yahr stages I-IV.[36,65] Interestingly, when specific PD symptoms are compared, the loss of dopaminergic activity measured by imaging correlates best with severity of bradykinesia and a relatively poor correlation with tremor scores.[20,66] These correlations of DAT imaging with clinical ratings suggest that striatal uptake of DAT ligands is a useful marker of disease severity in PD, which enhances its utility as a measure of disease progression.

DAT Imaging as a Measure of Progression in Parkinson's Disease

PD is a progressive neurodegenerative disorder; however, the rate of progression for an individual is unpredictable and variable. One of the primary goals of PD research over the past several years has been to develop medications that slow disease progression. Several clinical trials are underway to evaluate whether neuroprotective candidate drugs may modify the rate of progression in PD, most of which rely on clinical outcome measures. The clinical ratings are useful tools to evaluate the subjects' status at the time of the examination; however, the examination is often confounded by the affects of medication. In addition, the timing of medications may influence clinical ratings during the study visits. Objective measures of disease progression are becoming imperative as neuroprotective agents for PD are developed and tested. *In vivo* imaging studies provide the opportunity to evaluate dopaminergic neuronal degeneration longitudinally through serial imaging scans.

Several studies utilizing neuroreceptor imaging have been performed to monitor progressive dopaminergic neuron loss in PD and to determine the rate of decline. These studies have been performed with DAT and 18F-flurodopa and have demonstrated loss from baseline of

approximately 4 to 10% per year of both DAT and 18F-fluorodopa in patients with early PD.[67–71] This rate of loss of dopaminergic neurons in PD is significantly greater than that of healthy controls, which has been shown to be 0 to 2.5% annually.[72]

Several longitudinal studies evaluating the disease-modifying potential of medications have already utilized neuroreceptor imaging as a secondary outcome measure. Recently, two studies that were similarly designed were conducted to compare the effect of initial treatment with dopamine agonist (pramipexole in the CALM-PD CIT trial[73]) and ropinerole in the REAL-PET trial[28]) or levodopa on the progression of PD as measured by [123I]β-CIT SPECT or 18F-DOPA PET imaging. These two clinical imaging studies targeting dopamine function with different imaging ligands and technology both demonstrate slowing in the rate of loss of [123I]β-CIT or 18F-DOPA uptake in early patients treated with dopamine agonists compared to levodopa. The relative reduction in the percent loss from baseline of [123I]β-CIT uptake in the pramipexole versus levodopa group was 47% at 22 months, 44% at 34 months, and 37% at 46 months after initiating treatment (Figure 48.4). The relative reduction of 18F-DOPA uptake in the ropinerole group versus the levodopa group was 35% at 24 months. These data suggest that treatment with a dopamine agonist may slow dopaminergic degeneration compared to treatment with levodopa. It is unclear whether this change represents a reduction in neuronal degeneration related to the dopamine agonist or accelerated neuronal loss as a result of levodopa treatment. However, questions have been raised regarding the potential for pharmacologic effects of the study drugs on the imaging outcome measures used in these studies that might provide an alternative explanation

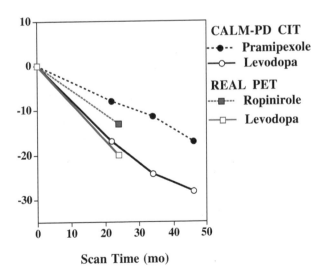

FIGURE 48.4 Percent change from baseline in [123I]β-CIT and 5F-dopa uptake by treatment assignment in the CALM-PD and REAL-PET studies.

for the imaging results. There is no clear evidence of a regulatory effect on the dopamine transporter as a result of exposure to levodopa or dopamine agonists in human studies;[69,73-76] however, larger studies are underway to more definitely address this question.

While the results from the CALM-PD CIT and REAL-PET studies have generated more questions than answers, we have begun to think more broadly about trial design and the potential for neuropharmacologic effects on both clinical and imaging outcome measures as a result of medication exposure. Ultimately, the upshot of these imaging results is initiation of additional studies to improve our understanding of dopaminergic neuronal function. Specifically, studies are underway to evaluate if current medications may have an effect on the uptake of radioligands by the dopamine transporter.

CONCLUSIONS

Dopamine transporter imaging with SPECT has become an important tool for clinical evaluation and research application in PD. The expanding utility of DAT imaging in improving the diagnostic accuracy of PD in early and difficult-to-diagnose cases, in the evaluation of at-risk populations, and in monitoring disease progression demonstrate the potential for the application of neuroreceptor imaging in answering key clinical questions. As screening tools become validated and more widely available for identification of early parkinsonian signs, DAT and SPECT imaging can be applied to establish disease onset at its earliest stages. As disease modifying treatments become available, it will become essential to identify individuals who would serve derive the most benefit from these potentially neuroprotective agents and monitor disease progression with time.

REFERENCES

1. Phelps, M., Positron emission tomography (PET), in *Clinical Brain Imaging: Principles and Applications*, J. Mazziota and S. Gilman, Eds., Philadelphia: F. A. Davis, pp. 71–107, 1992.
2. Lassen, N., Holm, S., Single photon emission computerized tomography (SPECT), in *Clinical Brain Imaging: Principles and Applications*, J. Mazziota and S. Gilman, Eds., Philadelphia: F. A. Davis, pp. 108–134, 1992.
3. Brooks, D. J., Advances in imaging in Parkinson's disease, *Curr. Opin. Neurol.*, 10:327–331, 1997.
4. Innis, R. B., Single photon emission computed tomography imaging of dopaminergic function: presynaptic transporter, postsynaptic receptor, and "intrasynaptic" transmitter, *Adv. Pharmacol.*, 42:215–219, 1998.
5. Leenders, K., Antonimi, A., PET 18F-Fluorodopa (FD) uptake and disease progression in Parkinson's disease, *Neurology*, 45:A220, 1995.
6. Eidelberg, D., Moeller, J. R., Ishikawa, T. et al., Early differential diagnosis of Parkinson's disease with 18F-fluorodeoxyglucose and positron emission tomography, *Neurology*, 45:1995–2005, 1995.
7. Snow, B. J., Tooyama, J., McGreer, E. G., et al., Human positron emission tomographic [18F]fluorodopa studies correlate with dopamine cell counts and levels, *Ann. Neurol.*, 34:324–330, 1993.
8. Piccini, P., Brooks, D. J., Etiology of Parkinson's disease: contributions from 18F-Dopa positron emission tomography, *Adv. Neurol.*, 80:227–231, 1999.
9. Sawle, G. V., Playford, E. D., Burn, D. J., et al., Separating Parkinson's disease from normality. Discriminate function analysis of fluordopa F-18 positron emission tomography data, *Arch. Neurol.*, 51:237–243, 1994.
10. Frey, K. A., Koeppe, R. A., Kilbourn, M. R., et al., Presynaptic monoaminergic and cholinergic vesicular transporters in the brain, *Adv. Pharmacol.*, 40:873–884, 1996.
11. Frey, K. A., Wieland, D. M., Kilbourn, M. R., Imaging of monoaminergic and cholinergic vesicular transporters in the brain, *Adv. Pharmacol.*, 42:269–272, 1998.
12. Frey, K. A., Koeppe, R. A., Kilbourn, M. R., Imaging the vesicular monoamine transporter, *Adv. Neurol.*, 86;237–247, 2001.
13. Booij, T., Tissingh, G., Boer, G., [123I]FP-SPECT shows a pronounced decline of striatal dopamine transporter labeling in early and advanced Parkinson's disease, *J. Neurol. Neurosurg. Psychiatry*, 62:133–140, 1997.
14. Brucke, T., Asenbaum, S., Pirker, W., et al., Measurement of the dopaminergic degeneration in Parkinson's disease with [123I]β-CIT and SPECT, *J. Neural. Transm. Suppl.*, 50:9–24, 1997.
15. Fischman, A. J., Bonab, A. A., Babich, J. W., et al., Rapid detection of Parkinson's disease by SPECT with altropane: a selective ligand for dopamine transporters, *Synapse*, 29:128–141, 1998.
16. Innis, R. B., Seibyl, J. B., Scanley, B. E., et al., Single photon emission computed tomographic imaging demonstrates loss of striatal dopamine transporters in Parkinson's disease, *Proc. Natl. Acad. Sci. USA*, 90:11965–11969 1993.
17. Tatsch, K., Schwarz, J., Mosley, P., Linker, R., Poglarell, O., Oertel, W., Fieber, R., Hahn, K., Kung, H., Relationship between clinical features of parkinson's disease and presynaptic dopamine transporter binding assessed with [123I]IPT and single-photon emission tomography, *Eur. J. Nucl. Med.*, 24:415–421, 1997.
18. Amara, G., Kuhar, M. J., Neurotransmitter transporters: recent progress, *Ann. Rev. Neurosci.*, 16:73–93, 1993.
19. Jaber, M., Jones, S., Giros, B., Caron, M. G., The dopamine transporter: a crucial component regulating dopamine transmission, *Mov. Disord.*, 12:629–633, 1997.
20. Seibyl, J. P., Marek, K. L., Quinlan, D., et al., Decreased single-photon emission computed tomographic [123I]β-CIT striatal uptake correlates with symptoms severity in Parkinson's disease, *Ann. Neurol.*, 38:589–598, 1995.
21. Hughes, A. J., Daniel, S. E., Kilford, L., Lees, A. J., The accuracy of clinical diagnosis in Parkinson's disease: a

clinicopathological study of 100 cases, *J. Neurol. Neurosurg. Psychiatry,* 55:181–4, 1992.

22. Rajput, A. H., Rozdilsky, B., Rajput, A., Accuracy of clinical diagnosis of parkinsonism—a prospective study, *Can. J. Neuol. Sci.,* 18:275–278, 1993.

23. Quinn, N., Parkinsonism—recognition and differential diagnosis, *Br. Med. J.,* 310:447–452, 1995.

24. Meara, J., Bhowmick, B., Hobson, P., Accuracy of diagnosis in patients with presumed Parkinson's disease, *Age and Ageing,* 28:99–102, 1999.

25. Bennett, D. A., Beckett, L. A., Murray, A. M., et al., Prevalence of parkinsonism signs and associated mortality in a community population of older people, *N. Engl. J. Med.,* 334:71–76, 1996.

26. Richards, M., Stern, Y., Mayeux, R., Subtle extrapyramidal signs can predict the development of dementia in elderly individuals, *Neurology,* 43:2184–2188, 1993.

27. Parkinson Study Group, A multicenter assessment of dopamine transporter imaging with DOPASCAN/SPECT in parkinsonism, *Neurology,* 55:1540–1547, 2000.

28. Whone, A., Remy, P., Davis, M. R., Sabolek, M., Nahmias, C., Stossel, A. J., Watts, R. L., Brooks, D. J., The REAL-PET study: slower progression in early Parkinson's disease treated with ropinerole compared with L-dopa, *Neurology,* 58 (Suppl. 3):A82–A83, 2002.

29. Fahn, S., Parkinson disease, the effects of levodopa and the ELLDOPA trial, *Arch. Neurol.,* 56:529–535, 1999.

30. Booij, J., Speelman, J. D., Horstink, M. W., Wolters, E. C., The clinical benefit of imaging striatal dopamine transporter with [123I]FP-CIT SPET in differentiating patients with presynaptic parkinsonism from those with other forms of parkinsonism, *Eur. J. Nucl. Med.,* 28(3):266–72, 2001.

31. Jennings, D. L., Seibyl, J. P., Oakes, D., Eberly, S., Murphy, J., Marek, K., [123I]β-CIT and SPECT Imaging versus clinical evaluation in parkinsonian syndrome: Unmasking an early diagnosis, *Arch. Neurol.,* 2004.

32. Hughes, A. J., Daniel, S. E., Ben-Shlomo, Y., Lees, A. J., The accuracy of diagnosis of parkinsonism syndromes in a specialist movements disorders service, *Brain,* 125(pt. 4):861–870, 2002.

33. Varrone, A., Marek, K. L., Jennings, D., Innis, R. B., Seibyl, J. P., [123I]beta-CIT SPECT imaging demonstrates reduced density of striatal dopamine transporter in Parkinson's disease and multiple systems atrophy, *Mov. Disord.,* 16(6):1023–32, 2001.

34. Pirker, W., Asenbaum, S., Bencsits, G., Prayer, D., Gershlager, W., Deecke, L., Brucke, T., [123I]beta-CIT SPECT in multiple systems atrophy, progressive supranuclear palsy, and corticobasal degeneration, *Mov. Disord.,* 15(6):1158–67, 2001.

35. Brucke, T., Asenbaum, S., Pirker, W., Djamshidian, S., Wenger, S., Wober, C., Muller, C., Podreka, I., *J. Neural Tram. Suppl.,* 50:9–24, 1997.

36. Benamer, H. T. S., Patterson, J., Wyper, D. J., Hadley, D. M., Macphee, G. J. A., Grosset, D. G., Correlation of Parkinson's disease severity and duration with 123I-FP-CIT SPECT striatal uptake, *Mov. Disord.,* 15(4):692–698, 2000.

37. Asenbaum, S., Pirker, W., Angelberger, P., Bencsits, G., Pruckmayer, M., Brucke, T., [123I]β-CIT and SPECT in essential tremor and Parkinson's disease, *J. Neural Transm.,* 105:1213–1228, 1998.

38. Gerschlager, W., Bencsits, G., Pirker, W., Bloem, B., Asenbaum, S., Prayer, D., Zijlmans, J., Hoffman, M., Brucke T., [123I]β-CIT SPECT distinguishes vascular parkinsonism from Parkinson's disease, *Mov. Disord.,* 17(3):518–523, 2002.

39. Winikates, J., Jankovic, J., Clinical correlates of vascular parkinsonism, *Arch. Neurol.,* 56;98–102, 1999.

40. Jellinger, K. A., Parkinsonism due to Binswanger's subcortical arteriosclerotic encephalopathy, *Mov. Disord.,* 11:461–462, 1996.

41. Yamanouchi, H., Nagura, H., Neurological signs and frontal white matter lesions in vascular parkinsonism, *Stroke,* 28: 965–969, 1997.

42. Factor, S. A., Podskalny, G. D., Molho, E. S., Psychogenic movement disorders: frequency, clinical profile, characteristics, *J. Neurol. Neurosurg. Psychiatry,* 59:406–412, 1995.

43. Ranawaya, R., Riley, D., Lang, A., Psychogenic dyskinesias in patients with organic movement disorders, *Mov. Disord.,* 5(2):127–133, 1990.

44. Lang, A. E., Koller, W. C., Fahn, S., Psychogenic parkinsonism, *Arch. Neurol.,* 52:802–810, 1995.

45. Williams, D. T., Ford, B., Fahn, S., Phenomenology and psychopathology related to psychogenic movement disorders, *Adv. Neurol.,* 65:231–257, 1995.

46. Marek, K., Seibyl, J., Scanley, B., Zea-Ponce, Y., Baldwin, R. M., Fussell, B., et al., [123I]β-CIT and SPECT imaging demonstrates bilateral loss of dopamine transporters in hemi-Parkinson's disease, *Neurology* 46:231–237, 1996.

47. Seibyl, J. P., Marek, K., Scheff, K., Zoghbi, S., Baldwin, R. M., Charney, D. S., vanDyck, C., Innis, R.B., Iodine-123-β-CIT and Iodine-123-FPCIT SPECT measurement of dopamine transporter in healthy subjects and Parkinson's patients, *J. Nucl. Med.,* 39:1500–1508, 1998.

48. Brooks, D. J., Ibanez, V., Sawle, G. V., et al., Differing patterns of striatal [18F]-Dopa uptake in Parkinson's disease, multiple systems atrophy and progressive supranuclear palsy, *Ann. Neurol.,* 28:547–555, 1990.

49. Ichinose, H., Ohye, T., Takahi, E., et al., Hereditary progressive dystonia with marked diurnal fluctuations caused by mutations in the GTP cyclohydrolase I gene, *Nat. Genet.,* 236–242, 1994.

50. Furukawa, Y., Shimadzu, M., Rajput, A. H., et al., GTP-cyclohydrolase I gene mutations in hereditary progressive and dopa-responsive dystonia, *Ann. Neurol.,* 39:609–17, 1996.

51. Hirano, M., Tamaru, Y., Ito, H., et al., Mutant GCP-cyclohydrolase I mRNA levels contribute to dopa-responsive dystonia onset, *Ann. Neurol.,* 40:796–798, 1996.

52. Rajput, A. H., Gibb, W. R. G., Zhong, X. H., et al., DOPA-responsive dystonia: pathological and biochemical observations in a case, *Ann. Neurol.,* 39:343–351, 1994.

53. Jeon, B. S., Jeong, J. M., Park, S. S., et al., Dopamine transporter density measured by [^{123}I]β-CIT single photon emission computed tomography is normal in dopa-responsive dystonia, *Ann. Neurol.*, 43:792–800, 1998.

54. Huang, C., Yen, T., Weng, Y., Lu, C., Normal dopamine transporter binding in dopa responsive dystonia, *J. Neurol.*, 249(8):1016–1020, 2002.

55. Gibb, W. R. G., Narabayashi, H., Yakochi, M., Iizuka, R., Lees, A. J., New observations in juvenile onset parkinsonism with dystonia, *Neurology*, 41:820–822, 1991.

56. Guttman, M., Burkholder, J., Kish, S. J., Hussey, D., Wilson, A., DaSilva, J., Houle, S., [11C]RTI-32 PET studies of the dopamine transporter in early dopa-naïve Parkinson's disease: implications for the symptomatic threshold, *Neurology*, 48(6):1578–1583, 1997.

57. Dekker, M. C., Bonifati, V., vanDuijn, C. M., Parkinson's disease: piecing together a genetic puzzle, *Brain*, 126(pt. 8):1722–33, 2003.

58. Baptista, M. J., Cookson, M. R., Miller, D. W., Parkin and alpha-synuclein: opponent actions in the pathogenesis of Parkinson's disease, *Neuroscientist*, 10(1):63–72, 2004.

59. Bertoli-Avella, A. M., Oostra, B. A., Heutink, P., Chasing genes in Alzheimer's and Parkinson's disease, *Hum. Genet.*, 114(5):413–438, 2004.

60. Pankratz, N., Nichols, W. C., Uniacke, S. K., Halter, C., Rudolph, A., Shults, C., Conneally, P. M., Foroud, T., Parkinson Study Group, Significant linkage of Parkinson's disease to chromosome 2q36–37, *Am. J. Hum. Genet.*, 72(4):1053–1057, 2003.

61. Di Monte, D. A., Lavasani, M., Manning-Bog, A. B., Environmental factors in Parkinson's disease, *Neurotoxicology*, 23(4–5)487–502, 2002.

62. Di Monte, D. A., The environment and Parkinson's disease: is the nigrostriatal system preferentially targeted by neurotoxins? *Lancet Neurol.*, 2(9):531–538, 2003.

63. Morrish, P. K., Sawle, G. V., Brooks, D. J., An [18F]Dopa-PET and clinical study of the rate of progression of Parkinson's disease, *Brain*, 119 (pt2):585–591, 1996.

64. Berendse, H. W., Booij, J., Francot, C. M., Bergnmans, P. L., Hijman, R., Stoof, J. C., Wolters, E. C., Subclinical dopaminergic dysfunction in asymptomatic Parkinson's disease patients' relatives with a decreased sense of smell, *Ann. Neurol.*, 50(1):34–41, 2001.

65. Ishikawa, T., Dhawan, V., Kazumata, K., et al., Comparative nigrostriatal dopaminergic imaging with iodine-123-β CIT-FP/SPECT and fluorine-18-FDOPA/PET, *J. Nucl. Med.*, 37:1760–1765, 1996.

66. Vingerhoets, F. J., Schulzer, M., Calne, D. B., Snow, B. J,. Which clinical sign of Parkinson's disease best reflects the nigrostriatal lesion? *Ann. Neurol.*, 41:58–64, 1997.

67. Marek, K., Innis, R., van Dyck, C., Fussell, B., Early, M., Eberly, S., Oakes, D., Seibyl, J., [^{123}I]β-CIT SPECT imaging assessment of the rate of Parkinson's disease progression, *Neurology*, 57:2089–2094, 2001.

68. Morrish, P., Rakshi, J., Bailey, D., Sawle, G., Brooks, D., Measuring the rate of progression and estimating the preclinical period of Parkinson's disease with [18F]dopa PET, *J. Neurol. Neurosurg. Psychiatry*, 64:314–319, 1998.

69. Nurmi, E., Bergman, J., Eskola, O., Solin, O., Hinnkka S, M., Sonninen, P., Rinne, J. O., Reproducibility and effect of levodopa on dopamine transporter function measurements: a [18F]CFT PET study, *J. Cereb. Blood Flow. Metab.*, 20:1604–1609, 2000.

70. Nurmi, E., Ruottinen, H., Kaasinen, V., Bergman, J., Haaparanta, M., Solin, O., Rinne, J., Progression in Parkinson's disease: a 6-[18F]fluoro-L-dopa PET study, *Mov. Disord.*, 16:608–615, 2001.

71. Pirker, W., Djamshidian, S., Asenbaum, S., Gerschlager, W., Tribl, G., Hoffman, M., Bruecke, T., Progression of dopaminergic degeneration in Parkinson's disease and atypical parkinsonism: a longitudinal β-CIT SPECT study, *Mov. Disord.*, 17:45–53, 2002.

72. vanDyck, C. H., Seibyl, J., Malison, R., Laurelle, M., Zoghbi, S., Baldwin, R. M., Innis. R, B., Age-related decline in dopamine transporter: analysis of striatal subregions, nonlinear effects, and hemispheric asymmetries, *Am. J. Geriatr. Psychiatry.*, 10(1):36–43, 2002.

73. Parkinson Study Group, Dopamine transporter brain imaging to assess the effects of Pramipexole vs. levodopa on Parkinson disease progression, *JAMA*, 287: 1653–1661, 2002.

74. Innis, R., Marek, K., Sheff, K., Zogbi, S., Castrnuovo, J., Feigin, A., Seibyl, J., Treatment with carbidopa/levodopa and selegiline on striatal transporter imaging with [^{123}I]β-CIT, *Mov. Disord.* 11999:4: 436–443.

75. Ahlskog, J. E., Uitti, R. J., O'Connor, M. K., et al., The effect of dopamine agonist therapy on dopamine transporter imaging in Parkinson's disease, *Mov. Disord.*, 4:940–946, 1999.

76. Guttman, M., Stewart, D., Hussey, D., Wilson, A., Houle, S., Kish, S., Influence of L-dopa and pramipexole on striatal dopamine transporter in early PD, *Neurology*, 56(11)1559–64, 2001.

49 MR Imaging of Parkinsonism

James B. Wood

Neuroradiology, Memphis VA Hospital and Radiology, University of Tennessee Health Science Center

CONTENTS

INTRODUCTION

Parkinson's disease is a slowly progressive neurodegenerative disorder of unknown origin. MRI is a powerful evolving imaging modality which is advancing in several directions to demonstrate this complex disease. With better understanding of the pathophysiology of the disease and the continuing advancement of MR technology, this modality will play a more active role in diagnosis and management of this disease. This chapter presents the progress to date.

The diagnosis of idiopathic Parkinson disease (IPD) is based on clinical criteria. Early diagnosis can be difficult. As demonstrated by Hughes, even adhering to strict clinical criteria, 24% of the time, the diagnosis was inaccurate when compared to pathologic specimens in 100 cases.[1] Hughes later showed, using current diagnostic criteria by movement disorder experts, that this number can be decreased to 10%.[2] The false positives are mainly the Parkinson plus diseases: multiple system atrophy (MSA), progressive supranuclear palsy (PSP), and cortical basal ganglionic syndrome (CBD).[1] However, toxins, medications, hydrocephalus, tumors, and vascular disease also can cause Parkinson-like features. MRI can help differentiate these diseases.

EVALUATION OF INHERENT VALUES OF TISSUES BY MRI

MRI allows evaluation of several inherent properties of tissue, including T1 and T2 relaxation times of the hydrogen nucleus, diffusion of water, and chemical composition.[3,4] The temporal change in these properties allows MR to evaluate blood flow. Blood flow not only demonstrates perfusion, it also correlates with neuronal function. Compare this with CT, which is only able to demonstrate the property of density. PET and SPECT with labeling metabolites demonstrate functioning tissue but with decreased spatial resolution. Fusion imaging with MRI helps locate the abnormalities. Also, these modalities have some limitations because of the ionizing effect on tissue, availability, and expense.

The process of imaging the brain has become very complicated. However, the objective of all modalities continues to be just one thing—contrast. This includes contrasting one normal structure from another, normal from abnormal tissue, and functioning from abnormally functioning tissue. Once contrast has been established, resolving the area of contrast into one point or two points is the next step. The contrast is derived from a signal from the object being scanned. The signal reflects an inherent property such as density in CT or T1 relaxation of the hydrogen nucleus in MR. In a digital world, the cube (volume element or voxel) of the tissue being analyzed has more signal if large. However, if the voxel is large, it cannot resolve small structures. If one makes the cube smaller, the resolution is better, but the signal goes down and may be lost in the background noise. Here is where technology has helped tremendously. Often, the signal can be increased without increasing the cube size, such as by increasing the magnet size from 1.5 to 3.0 Tesla. Also,

the speed of collecting the signal can be increased so that
several samples of the signal from the same small cube
can be collected and averaged, giving a more accurate
measure of the inherent property. Thus, better resolution
between very small objects, even with small differences
in contrast, can be obtained.

PATHOLOGY OF PARKINSONISM

Armed with these powerful tools, the challenge is to display the abnormalities of Parkinson's disease. Radiolo-
gists have attempted to demonstrate some of the main
pathological changes—the neuronal loss and gliosis and
also the changing iron deposits in the substantia nigra.
Because 70 to 80% of dopaminergic neurons projecting
from the substantia nigra to the striatum are lost before
symptoms develop, potentially asymptotic patients could
be demonstrated. To help differentiate from the false positives obtained with clinical exam, close evaluation of the
putamen, the cerebellum, brain stem, and frontal lobes is
also done. These findings help differentiate the Parkinson
plus diseases. Parkinson and the Parkinson plus diseases
involve many structures not mentioned above, and the lack
of function of all of these structures has a cascade affect
on other structures in the complex circuitry of the basal
ganglia. Therefore, pathologic change in these slowly progressive diseases with varied compensating affects is a
challenge to image. In the future, when a more robust
signal can be detected from the other pathologic changes,
increased sensitivity and specificity can be obtained. Other
less problematic diseases with Parkinson-like features can
be differentiated with well known radiographic features
of vascular disease, tumors, or hydrocephalus.[3,4]

T1 AND T2 FINDINGS

A routine screening MRI can demonstrate abnormalities
in some patients with Parkinson or Parkinson plus diseases.[5–22] IPD demonstrates a decrease in the width of the
pars compacta results from increased iron deposition in
this area adjacent to the pars reticulum, which normally
is already iron rich. The increased iron deposits are
thought to be associated with oxidative stress.[41,42] The
other border of the pars compacta is another normally
iron-rich-containing structure, the red nucleus. Iron
increases the T2 relaxation value of brain tissue. The best
way to demonstrate this has presented a challenge. Several
people demonstrated the resultant loss of signal on a T2
weighted image. Oikawa et al. concluded that iron changes
in the substantia nigra are better demonstrated on the
proton density weighted spin echo and fast STIR than on
the T2 weighted images (Figures 49.1, 49.2, and 49.3).[44]
Thus, the enlarging area of iron (normal and abnormal
deposits) causes the pars compacta to appear to shrink.
Some refer to this as *smudging* of the border the two pars

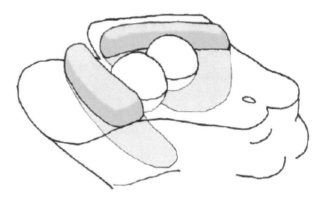

FIGURE 49.1 Schematic drawing of the three-dimensional
anatomy of the SN from the left superoposterolateral aspect. The
SN is located mainly beneath the red nucleus. (From Reference
44. With permission.)

of the substantia nigra. Unfortunately, overlap with controls is present in this small structure.

Another problem is normal aging also result in iron
deposits in the pars compacta. However, the progression
from medial to lateral is different from IPD, which is
from lateral to medial. Another problem described in
some people with IPD is loss of the normal low signal in
the pars reticulum, thought to be from depletion of normal iron by increased cellular metabolic activity or by
local cell death.[45] Using a more novel pulse sequence,
partial refocused interleaved multiple echo (PRIME),
Graham et al. accentuated increased iron concentration in
the substantia nigra.[23] However the amount of the
decreased width continues to overlap with controls and is
also seen in PSP and SND.[9,11]

The diminishing of pars compacta also reflects selective neural loss. Occasionally, patients with IPD have
hyperintense foci on T2 weighted images in SN, possibly
due to cell loss and gliosis.[10] Hutchinson and Raff[24] used
a combination of two different inversion recovery
sequences, one for white matter suppression (WMS) of
the crus cerebri and one for nigral gray matter suppression (GMS) to obtain a more robust signal reflecting both
iron increase and also cell loss and gliosis. A ratio image
(WMS/GMS) was then computed (Figure 49.4). Images
showed loss of signal in a lateral to medial gradient corresponding to the known neuropathologic pattern of degeneration in Parkinson's disease. Also, a radiologic index
was calculated to reflect this signal change (Figure 49.5).
The index was highly correlated with the Unified Parkinson's Disease Rating Scale score without overlap with
controls. Hu confirmed these results but concluded that F-
dopa PET was more reliable than inversion recovery MRI
in discriminating patients with moderately severe PD
from normal subjects. However, the structural changes
detected within the ratio image in his study did correlate
with measures of striatal dopaminergic function using[18]
F-dopa PET.[25] Hutchinson and Raff have recently refined

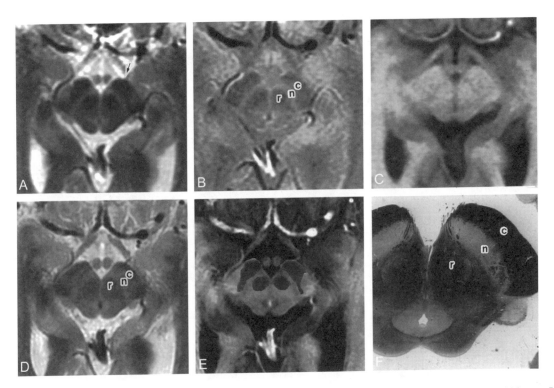

FIGURE 49.2 Axial MR images through the upper midbrain. (*A*) Axial T2-weighted image in a healthy control subject, a 55-year-old woman. A hypointense area, believed to be SNr, is located in the anteromedial part of the crus cerebri (*arrow*). No hyperintense gray matter area, representing the SN, is visible. (*B*) Proton density-weighted image in the same section as in (*A*). The SN (*n*) is clearly identified as an area of hyperintense gray matter surrounded by the hypointense red nucleus (*r*) and the crural fibers (*c*). (*C*) T1-weighted image in the same section as in *A*. The SN is not evident. (*D*) Fast STIR image in the same section as in (*A*). The SN (*n*) is readily identified as a structure with gray matter signal intensity. The red nucleus (*r*) with surrounding white matter and the crural fibers (*c*) are identified as areas with relatively low signal intensity. (*E*) Video-reversed fast STIR image onto which the hypointense areas on a T2-weighted image are superimposed (*shaded areas*). The hypointense area on the T2-weighted image includes the crural fibers and the anterior part of the SN. (*F*) Corresponding axial-section specimen obtained from a human cadaver. (From Reference 44. With permission.)

their technique enabling a more detailed assessment of the morphologic changes to increase specificity—segmented inversion recovery ratio imaging (SIRRIM).[26] To date, they have not yet completed a large-scale study.

Probably more useful on the screening MR (excluding the SIRRIM sequence) are specific findings in patient with Parkinson plus diseases. Multiple system atrophy with predominant parkinsonian features (MSA-P) has putaminal hypointensities and atrophy. Atrophy in the putamen is not associated with IPD. In addition, a hyperintense slit signal is seen in the lateral portion of the putamen with MSA-P. Certain sequences and field strengths affect the amount of this sign.[27] Macia et al., in a letter, describe their retrospective evaluation of 106 patients with PD and PD plus disease, finding similar hypointensities and slit signals in the putamen in 14 of 21 patients with MSA-C, 3 of 26 WITH PSP, 2 of 26 with CBD, but 0 of 33 with PD.[28] Thus, the sign may not be that specific for MSA-P. Another finding associated with MSA-P is the "hot cross bun" sign of the pons. Bhattacharyak[29] derived a useful diagnostic algorithm using the above findings on a screening MRI to dif-

ferentiate IPD from MSA (Figure 49.6). High-quality images in this article give an excellent demonstration of the findings of these diseases (Figures 49.7 through 49.11). Progressive supranuclear palsy demonstrates atrophy of the midbrain[30,31] and of the frontal lobe.[32] Asymmetrical atrophy in the posterior frontal and parietal regions contralateral to the side of the clinical manifestations is characteristic of corticobasal degeneration. Unfortunately, atrophy is either a subjective finding by an expert or an objective, time-consuming finding involving postprocessing exercises such as pixel counting (pixel are picture elements that represent cubes of tissue—voxels).

DIFFUSION IMAGING

Diffuse weight imaging demonstrates the property of random movement of water molecules in the brain. If the structure is homogeneous, the water can diffuse in all directions equally (isotropic). A drop of ink in a glass of water is one such example. Water in the extra cerebral space in a patient with vasogenic edema is another. How-

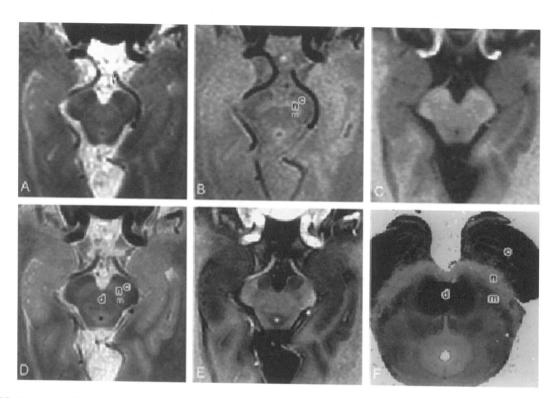

FIGURE 49.3   Axial MR images through the lower midbrain. (*A*) Axial T2-weighted image in the same control subject as in Figure 49.2. A hypointense area is visible on only the anteromedial end of the crus cerebri (*arrow*). An area with relatively high signal intensity suggestive of the SN is not depicted. (*B*) Proton density-weighted image in the same section as in (*A*.) The SN (*n*) is clearly depicted as a structure with hyperintense gray matter between the crural fibers (*c*) and the medial lemniscus (*m*). (*C*) T1-weighted image obtained at the same section as in (*A*). The SN is not visible. (*D*) Fast STIR image in the same section as in (*A*). The SN (*n*) is identified as an area of hyperintense gray matter posterior to the crural fibers (*c*). The medial lemniscus (*m*) and the decussation of superior cerebellar peduncle (*d*) show relatively low signal intensity. (*E*) Video-reversed fast STIR image onto which the hypointense areas on a T2-weighted image are superimposed (*shaded areas*). The hypointense areas on the T2-weighted image are located on the anteromedial part of the peduncular fibers, but they barely include the SN. (*F*) Axial-section specimen obtained in a human cadaver through the lower end of the midbrain. The SN (*n*) is not present on this section. It is located between the crural fibers cerebri (*c*) and medial lemniscus (*m*). (From Reference 44. With permission.)

ever, because the brain is not perfectly homogenous like a glass of water, the diffusion is referred to as an *apparent diffusion*. Thus, the diffusion coefficient is referred to as the apparent diffusion coefficient (ADC). In a brain with a two-hour-old infarct, the water collects in a swollen cell, because the sodium ATP pump has failed. The water cannot diffuse in any direction. The ADC value goes down, and an image can be made to demonstrate the contrast.[33]

Some normal structures like white matter fiber tracts allow diffuse in one direction but not another (anisotropic). Thus, a MR tensor image that reflects directionality with color coding can be created (Figures 49.12 and 49.13). Each cube of tissue (voxel) has a value, fractional anisotropy (FA) indicating the amount of directionality of water in different structures in the brain. Water in CSF is isotropic and has a FA of 0. Highly anisotropic (highly directional) structures have a FA approaching 1. An abnormal value indicates a focal lesion. However, different etiologies could give a similar appearance. A celery stalk hit

with a hammer would appear similar to a fungal infection, causing a focal loss of directionality (FA) on MR diffusion tensor imaging. A recent study by Yoshikawa et al.[34] was able to demonstrate early pathologic changes in the parkinsonian brain diffusion tensor MRI (Figure 49.14, Table 49.1). Seppi et al. demonstrated focal abnormal values of ADC in basal ganglia in PSP patients within a few years of onset, discriminating them from patients with PD with a sensitivity of 90% and a positive predictive value of 100% but not from those with MSA-P.[35]

SPECTROSCOPY

Spectroscopy allows the measurement of a specific biochemical in the brain.[3,4] Different molecules have specific frequencies in a given magnetic field. Changing the field strength (1.5 to 3.0 Tesla) also changes the frequency for a particular molecule. However, the change is proportional to the change in field strength. Therefore, a convention of parts per million (PPM) can be used in all field

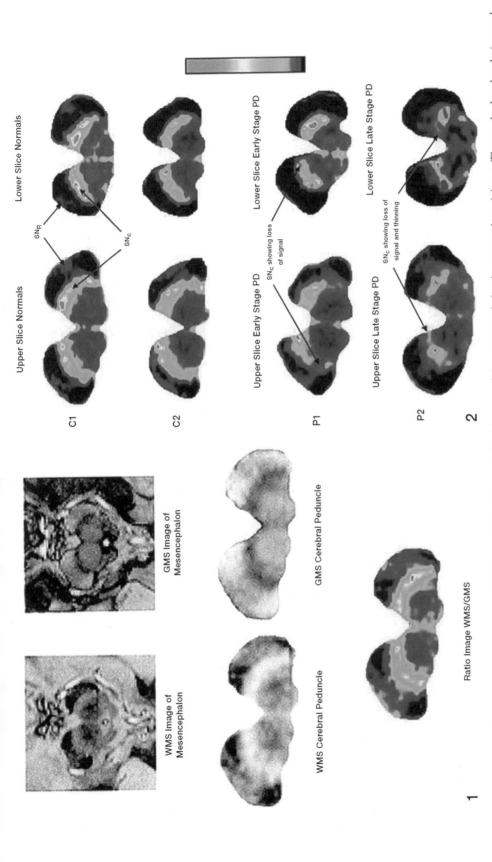

FIGURE 49.4 *Upper row* displays an example of axial WMS and GMS MR acquisition images of the mesencephalon in a control participant. The cerebral peduncle (*second row; left*) extracted from the WMS midbrain image serves as a template to extract the GMS image of the cerebral peduncle shown on the *right*. The SN_C is seen as a bright arch in the peduncular WMS image, whereas it appears as a dark band in the corresponding GMS image. Note also the substantia nigra pars reticulata (SN_R) reaching across the crus cerebri toward the SN_C. The ratio image (WMS/GMS) of the two images in the *second row* yields the color-coded ratio image displayed on the *bottom*. All black-and-white images are shown using a standard display of 256 gray levels. The color image uses a 256-pseudocolor lookup table. Ratio images of the cerebral peduncle displayed in pseudo colors show the morphologic characteristics of the SN_C in two control participants (C1 and C2) and the structural changes in two patients with Parkinson's disease (P1 and P2). The substantia nigra pars reticulata (SN_R) is indicated for participant C1. Notice that the SN_C in control participants reaches out toward the peduncular edge in the upper section, taking on the form of an arch. In the images of patient P1, who has Parkinson's disease, thinning and loss of signal can be seen in the lateral segment of the SN_C in the upper section. The lower section shows islands of cell loss on both sides of the SN_C. Note the considerable thinning and loss of signal in both upper and lower sections of the images of patient P2, who has late-stage Parkinson's disease. *Left* and *right sides* show two rims of preserved signal. (From Reference 24. With permission.) (*A color version of this figure follows page 518.*)

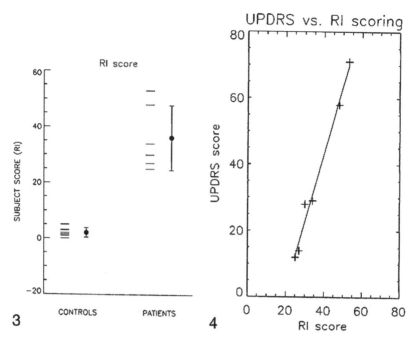

FIGURE 49.5 Radiologic indices are displayed for the six control participants and the six patients with Parkinson's disease. There is no overlap between the groups, which are distinct by Student's t test ($P < 0.00005$). The error bars represent one SD. Unified Parkinson's Disease Rating Scale scores for the six patients ranging from 12 to 71 are plotted versus radiologic indices. A linear regression analysis was conducted, yielding a linear correlation coefficient of $r = 0.99$. (From Reference 24. With permission.)

TABLE 49.1
Fractional Anisotropy (FA) Values in Five Circuits of the Extrapyramidal System

		Circuits (corresponding regions)			
	P(II+III+IV)	First (II+III)	Second (II)	Third (I+II+III)	Fourth (I+II+III+IV)
Control	0.490	0.483	0.583	0.542	0.542
PD12	0.451	0.445	0.548	0.493*	0.434*
PD345	0.456	0.456	0.537	0.495*	0.493*
PSP	0.403	0.410	0.449**	0.439**	0.432**

Parkinsonian patients showed a significant decrease in FA values in the third and fourth accessory circuits, regardless of clinical stage.

*$p < 0.05$

** $p < 0.01$

P, principal circuit; PD12, Parkinson's disease in the early stage group; PD345, Parkinson's disease in the advanced stage group; PSP, progressive supranuclear palsy.

Source: From Reference 34. With permission.

strengths to identify each molecule. Now, a contrast exists that not only demonstrates which chemicals are present but in what quantities. The significance of the different chemicals that can be measured continues to unfold. N-acetylaspartate (NAA) is a neuronal marker. If one loses neurons from any cause (infarct, near drowning, tumor, or degenerative disease), the amount of NAA goes down. Other significant hydrogen-containing metabolites includes creatine (CCR), a marker for general metabo-

lism; choline, a marker for the cell membranes; and myo-inositol (MI), contained by many glial cells. Ratios such as NAA/CR attempt to reflect the status of the neuron in a particular part of the brain. In other words, the lack of NAA in CSF is expected and reflected in the ratio of NAA/CR, whereas the lack of NAA in the occipital cortex is not expected and thus demonstrated in a low NAA/CR ratio compared to normal. The disadvantage of this strategy is if the CR, which is usually constant in most tissues,

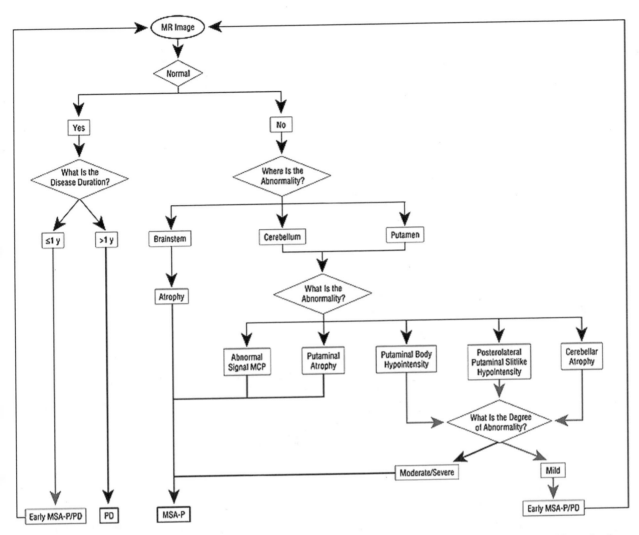

FIGURE 49.6 Algorithm for the magnetic resonance (MR) imaging diagnosis of multiple-system atrophy with predominant parkinsonism (MSA-P) and Parkinson's disease (PD). Red indicates MR imaging diagnosis of MSA-P; blue, MR imaging diagnosis of PD; green, MR imaging diagnosis of PD or early MSA-P; and MCP, middle cerebellar peduncles. (From Reference 29. With permission.)

is changed for whatever reason (trauma, hyperosmolar states, hypoxia, stroke, or tumors), the NAA/CR may falsely indicate neuronal damage.

Clarke showed that the only significant difference in any metabolite concentration was in the lentiform nucleus of patients with IPD compared with controls, with an increase in choline that led to a significant reduction in the NAA/choline ratio. The relevance of this finding is uncertain.[36] Helping with the differentiation of Parkinson's and MSA, Watanabe et al. used a 3.0-Tesla magnet, demonstrating elevated NAA/creatinine ratio in the pons and the putamen in MSA as compared to IPD and controls[37] (Figures 49.14 and 49.15). Decrease in NAA has been observed in the occipital lobes in advanced Parkinson's disease, with dementia helping differentiate from dementia associated with Alzheimer's disease.[38]

FUNCTIONAL IMAGING

At this time, evaluation of tissue function in parkinsonism is best demonstrated with nuclear medicine studies. Fusing the results with MRI aids in the location. This topic is covered in another chapter. Functional MRI with increased blood flow to activated areas of the brain is just now being evaluated[3,4] and shows some promise with early diagnosis of Parkinson's disease. J. Ross was able to define areas of brain activity by using functional MR imaging during imagery of a complex, coordinated motor task—a golf swing.[39] Because simulated movements, as well as executed movements, are slow in patients with Parkinson's disease,[40] abnormal results in such activities could indicate certain diseases such as movement disorders. An exciting new modality magne-

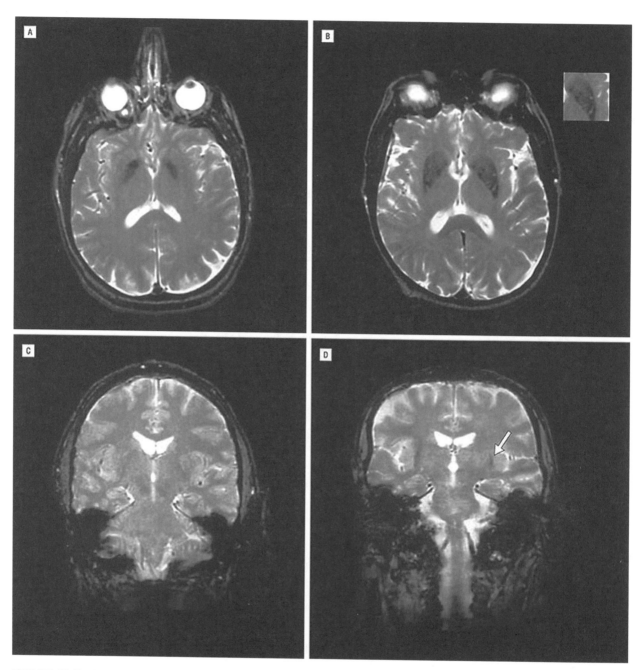

FIGURE 49.7   (*A*) Normal iron distribution, i.e., globus pallidum hypointensity relative to the putamen in a patient with a clinical diagnosis of Parkinson's disease (axial section, 1.5 T, T2 weighting). (*B*) Reversal of normal iron distribution with severe putaminal hypointensity relative to the globus pallidum (inset) in a patient with a clinical diagnosis of multiple-system atrophy with predominant parkinsonism (axial section, 1.5 T, T2 weighting). (*C*) Normal iron distribution in the globus pallidum relative to the putamen in a patient with a clinical diagnosis of Parkinson disease (coronal section, 1.5 T, T2 weighting). (*D*) Reversal of normal iron distribution with mild putaminal hypointensity relative to the globus pallidum (arrow) in a patient with a clinical diagnosis of multiple-system atrophy with predominant parkinsonism (coronal section, 1.5 T, T2 weighting). (From Reference 29. With permission.)

toencephalography (MEG) can also be fused with MRI to demonstrate function.[41] The breakthrough with this technique is much better resolution of the time involved in the brain activity in a particular location. Thus, slow or abnormal timing could indicate early Parkinson's disease.

SUMMARY

The challenge of imaging Parkinsonism and defining the exact pathology is proceeding in several different directions. Unfortunately, at this time, screening MRI does not provide accepted objective criteria to indicate Par-

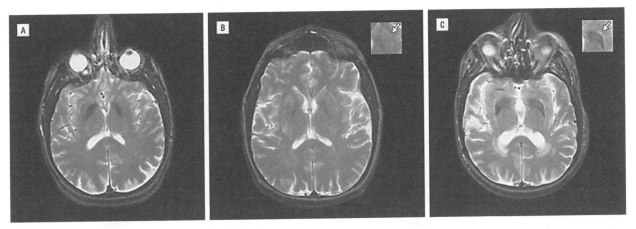

FIGURE 49.8 (*A*) Normal posterolateral putaminal margin in a patient with a clinical diagnosis of Parkinson disease (axial section, 1.5 T, T2 weighting). (*B*) Mild slit-like hyperintensity of the posterolateral putaminal rim (inset, arrow) in a patient with a clinical diagnosis of multiple-system atrophy with predominant parkinsonism (axial section, 1.5 T, T2 weighting). (*C*) Moderate slit-like hyperintensity of the posterolateral putaminal rim (inset, arrow) in a patient with a clinical diagnosis of multiple-system atrophy with predominant parkinsonism (axial section, 1.5 T, T2 weighting). (From Reference 29. With permission.)

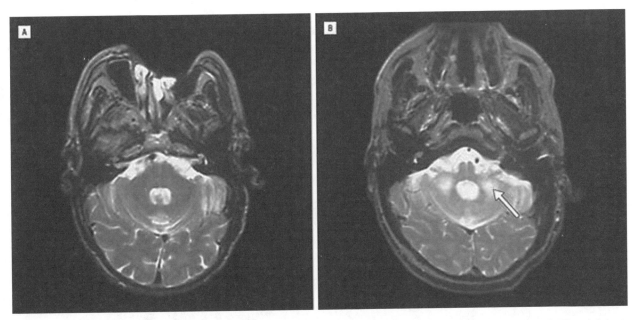

FIGURE 49.9 Normal cerebellum in a patient with a clinical diagnosis of Parkinson disease (axial section, 1.5 T, T2 weighting). B, Abnormal signal in the middle cerebellar peduncles (arrow) in a patient with a clinical diagnosis of multiple-system atrophy with predominant parkinsonism (axial section, 1.5 T, T2 weighting). (From Reference 29. With permission.)

kinson's disease. If MR diffusion tensor imaging is added, with computer-generated color maps for quick screening, IPD as well as many other diseases possibly could be detected. Then, the more time-consuming, more expensive, and less available techniques could be used to confirm the diagnosis if necessary. With the improving technology, a single imaging modality such as DTI, or a combination of modalities such as with fusion imaging with nuclear imaging, should emerge. Then, the most useful and most practical method for screening, diagnos-

ing, and monitoring of effective therapy of the disease can be performed.

REFERENCES

1. Hughes, A. J., Daniel, S. E., Kilford, L., Lee, A. J., Accuracy of clinical diagnosis of idiopathic Parkinson's disease: a clinicopathological study of 100 cases, *J. Neurol. Neurosurg. Psychiatry,* 55 181–184, 1992.

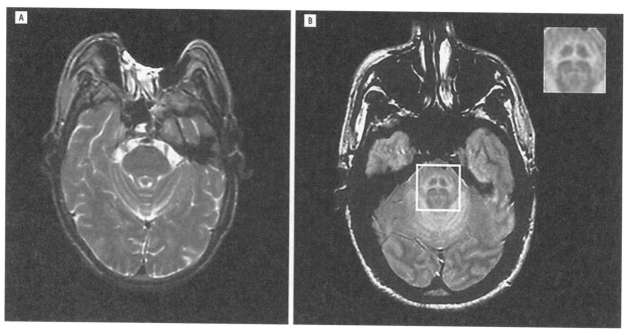

FIGURE 49.10 (*A*) Normal pons in a patient with a clinical diagnosis of Parkinson disease (axial section, 1.5 T, T2 weighting). (*B*) "Hot-cross bun" sign: cruciform degeneration of pontine fibers secondary to brain stem atrophy (inset) in a patient with a clinical diagnosis of multiple-system atrophy with predominant parkinsonism (axial section, 1.5 T, T2 weighting). (From Reference 29. With permission.)

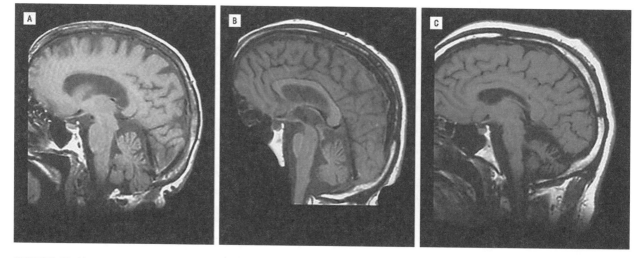

FIGURE 49.11 (*A*) Normal brain stem and cerebellum in a patient with a clinical diagnosis of Parkinson disease (sagittal section, 1.5 T, T1 weighting). (*B*) Mild brain stem and cerebellar atrophy in a patient with a clinical diagnosis of multiple-system atrophy with predominant parkinsonism (sagittal section, 1.5 T, T1 weighting). (*C*) Severe brain stem and cerebellar atrophy in a patient with a clinical diagnosis of multiple-system atrophy with predominant parkinsonism (sagittal section, 1.5 T, T1 weighting). (From Reference 29. With permission.)

2. Hughes, A. J., Daniel, S. E., Lees, A. J., Improved accuracy of clinical diagnosis of Lewy body Parkinson's disease, *Neurology,* Vol. 57, No. 8, October 23, 2001.

3. Stark, D., Bradley, W., *Magnetic Resonance Imaging,* Mosby, 3rd ed., 1999.

4. Atlas, S., *MRI of the Brain and Spine,* Lippincott, Williams and Wilkins, 3rd ed., 2002.

5. Drayer, B. P., Olanow, W., Burger, P., Johnson, G. A., Herfkens, R., Riederer, S., Parkinson plus syndrome: diagnosis using high field MR imaging of brain iron, *Radiology,* 159:493–498, 1986.

6. Drayer, B. P., Olanow, W., Burger, P., Johnson, G. A., Herfkens, R., Riederer, S., Parkinson plus syndrome: diagnosis using high field MR imaging of brain iron, *Radiology,* 159:493–498, 1986.

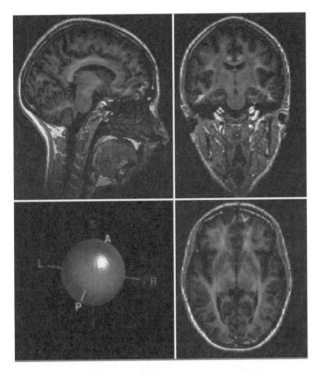

FIGURE 49.12 Diffusion tensor MR images that reflect directionality with color coding. *(A color version of this figure follows page 518.)*

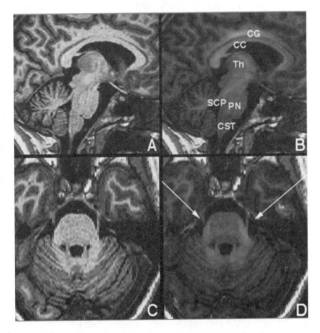

FIGURE 49.13 Diffusion tensor MR images that reflect directionality with color coding. *(A color version of this figure follows page 518.)*

7. Pastakia, B., Polinsky, R., Di Chiro, G., Simmons, J. T., Brown, R., Wener, L., Multiple system atrophy (Shy-Drager syndrome): MR imaging, *Radiology,* 159:499–502, 1986.

8. Duguid, J. R., De La Paz, R., DeGroot, J., Magnetic resonance imaging of the midbrain in Parkinson's disease, *Ann. Neurol.,* 20:744–747, 1986.

9. Stern, M. B., Braffman, B. H., Skolnick, B. E., Hurtig, H. I., Grossman, R. I., Magnetic resonance imaging in Parkinson's disease and parkinsonian syndromes, *Neurology,* 39:1524–1526, 1989.

10. Braffman, B. H., Grossman, R. I., Goldberg, H. I., et al., MR imaging of Parkinson disease with spin-echo and gradient-echo sequences, *Am. J. Roentgenol.,* 152:159–165, 1989.

11. Savoiardo, M., Strada, L., Girotti, F. et al., MR imaging in progressive supranuclear palsy and Shy-Drager syndrome, *J. Comput. Assis. Tomogr.,* 13:555–560, 1989.

12. Savoiardo, M., Strada, L., Girotti, F. et al., Olivopontocerebellar atrophy: MR diagnosis and relationship to multisystem atrophy, *Radiology,* 174(3, pt. 1):693–696, 1990.

13. O'Brien, C., Sung, J. H., McGeachie, R. E., Lee, M. C., Striatonigral degeneration: clinical, MRI, and pathologic correlation, *Neurology,* 40:710–711, 1990.

14. Konagaya, M., Konagaya, Y., Honda, H., Iida, M., A clinico-MRI study of extrapyramidal symptoms in multiple system atrophy—linear hyperintensity in the outer margin of the putamen (in Japanese), No To Shinkei, 45:509–513, 1993.

15. Testa, D., Savoiardo, M., Fetoni, V. et al., Multiple system atrophy: clinical and MR observations on 42 cases, *Ital. J. Neurol. Sci.,* 14:211–216, 1993.

16. Lang, A., Curran, T., Provias, J., Bergeron, C., Striatonigral degeneration: iron deposition correlates with the slit-like void signal of magnetic resonance imaging, *Can. J. Neurol. Sci.,* 21:311–318, 1994.

17. Wakai, M., Kume, A., Takahashi, A., Ando, T., Hashizume, Y., A study of parkinsonism in multiple system atrophy: clinical and MRI correlation, *Acta Neurol. Scand.,* 90:225–231, 1994.

18. Konagaya, M., Konagaya, Y., Iida, M., Clinical and magnetic resonance imaging study of extrapyramidal symptoms in multiple system atrophy, *J. Neurol. Neurosurg. Psychiatry,* 57:1528–1531, 1994.

19. Savoiardo, M., Girotti, F., Strada, L., Ciceri, E., Magnetic resonance imaging in progressive supranuclear palsy and other parkinsonian disorders, *J. Neural. Transm. Suppl.,* 42:93–110, 1994.

20. Schwarz, J., Weis, S., Kraft, E. et al., Signal changes on MRI and increases in reactive microgliosis, astrogliosis, and iron in the putamen of two patients with multiple system atrophy, *J. Neurol. Neurosurg. Psychiatry,* 60:98–101, 1996.

21. Schrag, A., Kingsley, D., Phatouros, C. et al., Clinical usefulness of magnetic resonance imaging in multiple system atrophy, *J. Neurol. Neurosurg. Psychiatry,* 65:65–7, 1998.

22. Schrag, A., Good, C. D., Miszkiel, K. et al., Differentiation of atypical parkinsonian syndromes with routine MRI, *Neurology,* 54:697–702, 2000.

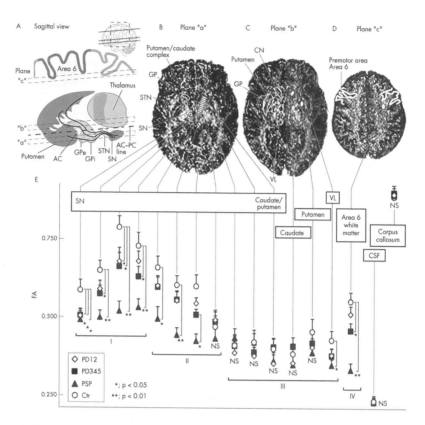

FIGURE 49.14 (From Reference 34. With permission.)

23. Graham, J.,Graham, M., Paley, N., Grunewald, M., Hoggard, N., Griffiths, P., Brain iron deposition in Parkinson's disease imaged using the PRIME magnetic resonance sequence, *Brain,* 123:12:2423–2431, 2000.

24. Hutchinson, M., Raff, U., Structural changes of the substantia nigra in Parkinson's disease as revealed by MR imaging, *Am. J. Neuroradiol.,* 21:697–701, 2000.

25. Hu, M. T. M., White, S. J., Herlihy, A. H., Chaudhuri, K. R., Hajnal, J. V., Brooks, D. J., A comparison of ^{18}F-dopa PET and inversion recovery MRI in the diagnosis of Parkinson's disease, *Neurology,* Vol. 56. No. 9, May 8, 2001.

26. Hutchinson, M., Raff, U., and Lebedev, S., MRI correlates of pathology in parkinsonism: segmented inversion recovery ratio imaging (SIRRIM), *NeuroImage,* 20, 1899–1902, 2003.

27. Watanabe, H., Fukatsu, H., Hishikawa, N., Hashizume, Y., Sobue, G., Field strengths and sequences influence putaminal MRI findings in multiple system atrophy, Neurology, 62(4): 671—671, Feb. 24, 2004.

28. Macia, F., Yekhlef, F., Ballan, G., Delmer, O., Tison, F. et al., T2-Hyperintense Lateral Rim and Hypointense Putamen are Typical but Not Exclusive of Multiple System Atrophy, *Arch Neurol.,* 58:1024–25, 2001.

29. Bhattacharya, K., Saadia, D., Eisenkraft, B., Yahr, M., Olanow, W., Drayer, B., Kaufmann, H., Brain Magnetic Resonance Imaging in Multiple-System Atrophy and Parkinson Disease: A Diagnostic Algorithm, *Arch Neurol.,* 59:835–842, 2002.

30. Aato, R., Akigughi, I., Masunaga, S., et al., Magnetic resonance imaging distinguishes progressive supranuclear palsy from multiple system atrophy, *J. Neural. Transm.,* 107:1427–36, 2000.

31. Warmuth-Metz, M., Naumann, M., Csati, I., et al., Measurement of the midbrain diameter on routine magnetic resonance imaging: a simple and accurate method of differentiating between Parkinson disease and progressive supranuclear palsy, *Arch Neuro.,* 58:1076–9, 2001.

32. Brennis, C., Seppi, K., Schocke, M., Benke, T., Wenning, G., Poewe, W., Voxel based morphometry reveals a distinct pattern of frontal atrophy in progressive supranuclear palsy, *J. Neuro. Neruosurg. Psychiatry,* 75:246–249, 2004.

33. Le Bihan, D., Turner, R., Douek, P. et al., Diffusion MR imaging: clinical applications, *Am. J. Roentgenol.,* 159:591–9, 1992.

34. Yoshikawa, K., Nakata, Y., Yamada, K., and Nakagawa, M., Early pathological changes in the parkinsonian brain demonstrated by diffusion tensor, *MRI J. Neuro. Neruosurg. Psychiatry,* 75:481–484, 2004.

35. Seppi, K., Schocke, M., Esterhammer, R., Kremser, C., Brenneis, C., Mueller, J., Boesch, S., Jaschke, W. Poewe, W., Wenning, G., Diffusion-weighted imagine discriminates progressive supranuclear palsy from PD, but not from the Parkinson variant of multiple system atrophy, *Neurol.,* 60:922–927, 2003.

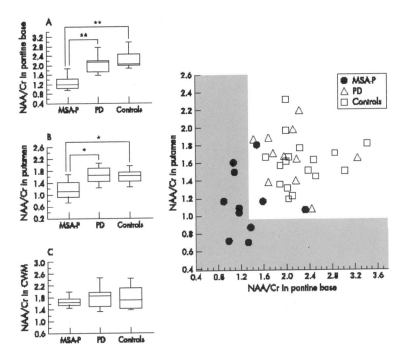

FIGURE 49.15 Box and whisker plot of the NAA/Cr ratio in the pontine base (*A*), putamen (*B*), and cerebral white matter (CWM, *C*) compared between MSA•P, PD, and controls. D is a scatter plot of the individual NAA/Cr data in the pontine base v putamen including MSA•P, PD, and control subjects. The scaled area corresponds to the mean − 2 SD of NAA/CR in the pontine base and putamen of control subjects. *$p = 0.02$, **$p < 0.0001$ by Scheffe's test, respectively. NAAA, N-acetylaspartate; CR, creatine; MSA•P, multiple system atrophy with parkinsonism predominant. (From Reference 37. With permission.)

36. Clarke, C. E., Lowry, M., Basal ganglia metabolite concentrations in idiopathic Parkinson's disease and multiple system atrophy measured by proton magnetic resonance spectroscopy, *Eur. J. Neuro.*, (6):661–5, Nov. 7, 2000.

37. Watnabe, H., Fukatsu, H., Katsuno, M., Sugiura, M., Hamada, K., Okada, Y., Hirayama, M., Ishigaki, T., Sobue, G., Multiple regional H-MR spectroscopy in multiple system atrophy: NAA/Cr reduction in pontine base as a valuable diagnostic marker, *J. Neuro. Neurosurg. Psychiatry,* 75:103–109, 2004.

38. Summerfield, C., Gomez-Anson, Tolosa, Mercader, J. M., Marti, M. J., Pastor, P., Junque, C., *Arch Neurol.,* 59:1415–1420, 2002.

39. Ross, Jeffrey S., Tkach, Jean, Ruggieri, Paul M., Lieber, Michael, and Lapresto, Eric, The Mind's Eye: Functional MR Imaging Evaluation of Golf Motor Imagery, *AJNR Am. J. Neuroradiol.,* 24: 1036–1044.

40. Dominey, P., Decety, J., Broussolle, E., Chazot, G., Jeannerod, M., Motor imagery of lateralized sequential task is asymmetrically slowed in hemi-Parkinson's patients, *Neuropsychologia,* 33:727–741, 14, 1995.

41. Roberts, Timothy P. L., Disbrow, Elizabeth A., Roberts, Heidi C., and Rowley, Howard A., Quantification and Reproducibility of Tracking Cortical Extent of Activation by Use of Functional MR Imaging and Magnetoencephalography, *Am. J. Neuroradiol.,* 21:1377–1387, August 2000.

42. McCord, J. M., Iron, Free Radicals, and Oxidative Injury, *Semin. Hematol.,* 35:5–12, 1998.

43. Chiueh, C. C., Iron Overload, Oxidative Stress, and Axonal Dystrophy in Brain Disorders, *Pediatr. Neurol.,* 25:138–47, 2001.

44. Hirobumi Oikawa, Makoto Sasaki, Yoshiharu Tamakawa, Shigeru Ehara and Koujiro Tohyama, The Substantia Nigra in Parkinson Disease: Proton Density-Weighted Spin-Echo and Fast Short Inversion Time Inversion-Recovery MR Findings, *Am. J. Neuroradiol.,* 23:1747–1756, November–December 2002.

45. Rutledge, J. N., Hilal, S. K., Silver, A. J. et al., Study of Movement Disorders and Brain Iron, by *MR, AJR.,* 149: 365–379, August 1987.

50 Clinical Batteries

Erwin B. Montgomery, Jr.
Department of Neurology, National Regional Primate Center, University of Wisconsin–Madison

CONTENTS

INTRODUCTION

Rapid advances in the neurobiology of Parkinson's disease generate great hope of one day preventing this disease. There are increasing numbers of models for environmental causes, including one agent in common use.[1] Furthermore, the genetic disorders now known to be associated with Parkinsonism may provide understanding of the biological events leading to the development of Parkinson's disease, which could lead to early detection and prevention. However, in some ways finding a prevention could be a social and political nightmare, as discussed below.

There is a sense of urgency. The "baby boom" generation is now entering the age for greatest risk for being diagnosed with Parkinson's disease. The prevalence could triple by the year 2015. What if a prevention was discovered that cost $10,000 per year and had to be taken every year? Who would get the prevention? As of yet, there are no predictive tests. Does this mean every person over some age, say 45 years, would be given the protective therapy? Could our country afford to pay for it? Would we allow only the rich to be protected? Maybe it would be better to hold off on developing protective therapy? This question is only partially facetious.

Every year, numerous manuscripts are submitted describing development of predictive or diagnostic tests for Parkinson's disease. Every year, numerous grants applications are submitted to funding agencies. Yet, the vast majority of them are go unpublished or unfounded, because they are flawed. Most are submitted by scientists who, for other reasons, develop tests that distinguish those with clinically definite disease from normal control subjects. These scientists think that it should be a relatively simple matter to extend these studies into a diagnostic or predictive test. However, the vast majority of such efforts fail to appreciate the complexities of diagnostic or predictive tests or follow well described methods and procedures for developing such tests.[2] Also, these manuscripts and grant applications fail to understand that the fundamental measure of any diagnostic or predictive test is not medical in nature. It is the social, ethical, moral, and economic consequences of making or failing to make a diagnosis or prediction.

The purpose of this chapter is to review efforts to develop diagnostic or predictive tests for the diagnosis of Parkinson's disease based on signs and symptoms usually in the domain of clinical assessment. Of the few papers published specifically discussing development of diagnostic or predictive tests based on clinical phenomena, based on a Medline search (March 2004),[3–10] virtually all failed the general guidelines for developing diagnostic or predictive tests.[2] Consequently, in hopes of spurring more successful efforts, this chapter reviews the issues related to developing such tests and the unique problems associated with diagnosing or predicting Parkinson's disease.

The implications go well beyond diagnostic or predictive tests for Parkinson's disease to virtually every decision made in clinical medicine. Indeed, a predictive or diagnostic for Parkinson's disease is a special case of more general approaches to clinical decision-making. When a clinician decides to use one form or treatment versus another, the physician is essentially making a diagnostic or predictive decision of whether that patient will have a satisfactory response. Therefore, analysis of methods used to develop diagnostic or predict tests of Parkinson's disease can illuminate the process by which clinicians make decisions.

It is a biological fact of life that humans, like nearly every other complex organism, display variability in response to disease or treatment. This variability translates into uncertainty when humans are assessed by diagnostic or predictive tests. Bertrand Russell said something to the effect that those who cannot stand uncertainty become religious, while those that can stand uncertainty become philosophical. This is not a value judgment regarding either religion or philosophy, but it clearly demonstrates the subjective impact that uncertainty has on individuals. Because the needs of patients compel clinicians to act on the patients' behalf, clinicians are compelled to deal with uncertainty.

Variability begets uncertainty, which in turn begets statistics. Fundamentally, statistics is about the management of uncertainty. This is not to be confused with the elimination of uncertainty. Rather, statistics allow us to define and measure uncertainty. It allows us to manipulate uncertainty by counterbalancing uncertainty in controlled studies. Statistics allows us to utilize uncertainty to dissect out relationships among multiple or complex variables using methods such as multivariate correlations or principal component analyses. Most importantly, statistics allows us to model the consequences of our decision-making processes, hopefully in a way that is predictive.

While statistics can be used to manage uncertainty, it is only through misuse of statistics that uncertainty can appear to be eliminated. For example, there is the tendency to utilize $p < 0.05$ as a de facto standard for statistical significance. Furthermore, a statistically significant difference is automatically assumed to be equal to clinical meaningfulness. These are both fallacious assumptions that can give rise to unacceptable consequences.[11]

Consider the following example. Assume for a moment that the prevalence of those at risk for Parkinson's disease in a community of 1000 older individuals is three percent. Furthermore, assume that we have a laboratory test that is 97 percent sensitive such that 97 percent of all of those individuals who truly are truly at risk can be identified by the test. Also, the test is 97 percent specific, meaning that 97 percent of normal individuals not at risk can be identified as normal by the test. The first impression is that this would be a very good diagnostic or predictive test. The community of older individuals to be tested will contain 30 individuals at risk for Parkinson's disease. Application of our tests will result in 29 of those 30 being identified. The other 970 individuals in the community will not be at risk for Parkinson's disease. However, application of our tests will result in 29 individuals falsely identified as at risk for Parkinson's disease. The net result this that, although the diagnostic test has a very high specificity and sensitivity, its application to this community of older individuals would result in as many individuals misdiagnosed with Parkinson's disease risk as those correctly identified. Maybe this test is not so good.

Assume further that we have a treatment that can prevent Parkinson's disease and that this cure costs pennies and has demonstrated side effects no more frequent or severe than a placebo. Now the social, economic, and health-care consequences of using this treatment are minimal, and thus the impact of this treatment on the 29 normal individuals who would be falsely diagnosed and treated is minimal. However, we would have prevented 29 individuals from getting Parkinson's disease. Now the impression is that maybe this test is a good idea.

Next assume that the performance of the test has a significant cost/or may be associated with nontrivial risks. The next impression would be at the test is not so good, because why apply a test when the social, economic, and health care cost of treating all 1000 individuals in the community is so much less than the test?

Now consider a scenario where we have a preventive treatment that is 100 percent effective with virtually no health side effects but costs $10,000 a year. The treatment has to be given continuously, so there is an annual recurring cost of $10,000. Who is going to get the treatment? Who is going to pay for it? It is unlikely that our society would adopt an egalitarian approach by which either everyone or no one receives the preventive treatment. What then would be the consequences to the fabric of our society if only those wealthy enough could be spared the disaster that is Parkinson's disease? The next impression would be that maybe the test is a good idea. Not because it is a very good diagnostic or predictive test but because it provides, at the least, some nonarbitrary way of deciding who would receive preventative treatment.

Finally, consider what happens after our test battery has been applied to our hypothetical community and we have identified all those at risk Parkinson's disease whose biomarker has reached a sufficient level for detection. There also may be those at risk for Parkinson's disease who will develop a biomarker of sufficient degree to be detected later. We must now consider follow-up testing to identify those individuals as they develop the disease. The working number now shifts from prevalence to incidence of risk, which is likely to be a much smaller percentage. Now, application of the tests at periodic intervals subsequent to the initial screening of the entire community is going to result in far more normal individuals being misdiagnosed than those who subsequently develop the disease and are diagnosed. The reader is left to consider the ramifications.

The purpose of the above discussion was to illustrate the complexities involved with clinical decision-making as it relates to the development of diagnostic or predictive tests. Demonstrating these complexities was not meant to engender feelings of nihilism—quite the contrary. Rather, it was meant to demonstrate the various factors that constitute the context within which any diagnostic or predictive test must be developed. Furthermore, the purpose of the remaining portion of this chapter is to clarify the complexities and discuss methodological approaches to increase the probability of success in developing predictive or diagnostic tests. As such, it is hoped that this chapter will be an antidote to both nihilism and naiveté.

Developing diagnostic or predictive tests based on measures that are clinical in nature gives rise to an important discussion that goes beyond semantics. Are we developing tests that are presymptomatic or preclinical? A presymptomatic diagnosis could imply the presumption that the patient is being identified prior to any meaningful irreversible injury. A predictive or diagnostic test that utilizes parameters typically in the domain of the clinical examination, such as reaction or movement times, depends by definition on the presence of symptoms, however minimal. Therefore, such predictive or diagnostic tests cannot be considered presymptomatic. The danger is the presumption that these patients already have "significant" injury, which could dampen the enthusiasm for developing such tasks. Furthermore, any therapies dependent on clinically based predictive or diagnostic tests would not be preventive—merely arresting. However, treatments that arrest the progression of the disease applied at a time at which symptoms, although present, are minimal would be tantamount to prevention and have the advantage of perhaps even being feasible.

There are three important elements in developing a diagnostic or predictive test. These are sensitivity, specificity, and prior probability. The latter is the incidence of the condition of interest; for example, being at risk for Parkinson's disease. As demonstrated above, the prior probabilities greatly influence the effects of sensitivity and specificity. However, it is exceptionally rare for investigators to include consideration of prior probability in their experimental design.[8]

The outcome of any diagnostic or predictive test typically is a dichotomous variable; that is, it is either positive or negative. Indeed, measures of specificity and sensitivity depend on a dichotomous description. However, most measures utilized by a diagnostic or predictive test are continuous, thus offering a range of values with a possible infinity of values in between. Thus, the question arises as to what value along this continuum to use as a cutoff. Some investigators will use two standard deviations from the mean value from a sample of results. Those individuals with test results greater than (or less than) two standard deviations are considered abnormal. However, this means that at least 4.6% of normal subjects, those without the condition of interest, will be falsely identified as having the condition of interest. If the consequences of a false positive are more severe than a false negative, using a result greater (or less) than two standard deviations may be inappropriate.

Usually, there is an overlap between the distributions of results for those with the condition of interest and those without. As can be seen, moving the threshold in one direction will result in more false positives but fewer false negatives, and moving the threshold in the other direction

will result in more false negatives but fewer false positives. Thus, the sensitivity and specificity depends on the cutoffs used, and there is a relationship between the sensitivity and specificity. This relationship is characterized by the *receiver-operator characteristics (ROC)* curve, and the area under the cure is a measure of diagnostic or prognostic utility (Figure 50.1). An area the ROC cure near 0.5 is a poor test, while an area under the cure of 1.0 would be a perfect test (Figure 50.2).

Ideally, the appropriate cut-off of a measure is determined by the full ramifications of a false positive versus a false negative result. However, examination of the ROC curve does provide some suggestion of an appropriate cut-off. This is usually the test result that is associated with a break or bend in the ROC curve. This often is associated with the best ratio of sensitivity to specificity.

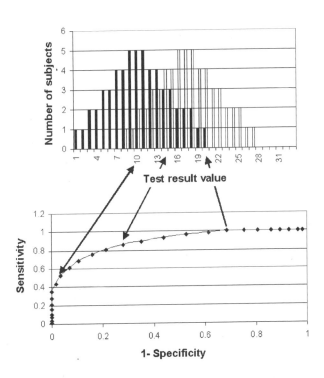

FIGURE 50.1 Demonstration of a receiver-operator characteristics curve. Typically, in developing diagnostic or predictive tests, there is an overlap in the test results between those with the condition of interest (filled bars) and those without the condition of interest (open bars). The question then becomes where to select a cutoff (top figure). The numbers of false positives and false negatives will vary depending on where the cutoff is selected. The receiver-operator characteristics curve (bottom figure) relates the specificity (true negatives) to the sensitivity (true positives) for different cutoffs (arrows running between the two figures). Thus, a decision of where to establish a cutoff can be drawn from examination of the receiver-operator characteristics curve by determining the consequences of false positives and false negatives. Furthermore, the area under the receiver-operator characteristics curve (bottom figure) also indicates the diagnostic value (see Figure 50.2).

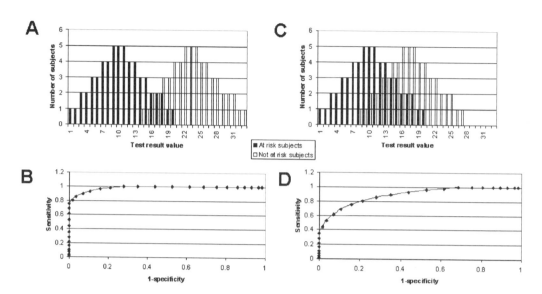

FIGURE 50.2 Demonstration of the effects of different overlaps on the receiver-operator characteristic curves. Graphs *A* and *C* show the interval histograms of hypothetical test results for a group at risk and a group not at risk. There is greater overlap in *A* compared to *B*, suggesting that the test used for *A* may be better than that used in *C*. This is reflected in their corresponding receiver-operator characteristic curves shown in *B* and *D*, respectively. As can be seen, the area under the receiver-operator characteristic curve is greater in *B* than *D*.

One of the fundamental approaches to the development of a diagnostic or prognostic test is correlation of some experimental variable with the presence or absence or degree of the condition of interest. In most cases, it is the presence or absence, and the correlational analysis typically used is logistic regression. This approach attempts to model the data consisting one variable that codes for diagnosis and the other variable(s) are the experimental measure(s), based on a sigmoid function. Thus, the data is parsed between the presence and absence of the condition of interest. The logistical regression analysis is then extended to the ROC curve and the area under the ROC curve.

The major limitation of correlational analysis is that regression is a mathematically optimizing procedure. In other words, the data are manipulated so as to produce a correlation if there is any correlation, even if spurious. It is possible that in the sampling process (that is, selection of test subjects), a correlation occurred solely by chance. Unfortunately, there is nothing in the correlational analysis that would tip one off that the correlation was spurious. Consequently, it is imperative that the regression equations that describe the correlations be applied prospectively to a second independent sample. Often, there is a marked reduction in the specificity and sensitivity when applied prospectively. Any scientist who does not repeat the analysis in a prospective manner is only doing half the job, and no one can have confidence in the half that has been done. Yet it is exceptionally rare to find studies that have prospectively tested the generalizability of their correlational analyses.[8] One measure to help ensure general-

izability during the development of the test is to be sure that there are a large number of subjects for each variable. One suggested ratio is 20 subjects for each variable.[12]

The issue of the number of subjects versus the number of tests becomes increasingly problematic when developing a battery of tests. Yet often, no individual test will have sufficient sensitivity and/or specificity to have a worthwhile positive or negative predictive value. Often, this requires a battery of tests. However, there are data reduction techniques that can reduce the number of variables, such a principle components analysis or multivariate regression.[7]

A major flaw in many studies and grant proposals is the tendency to use populations of convenience rather than the population of interest. For example, many diagnostic or predictive tests are developed on subjects known to have Parkinson's disease. Yet, this is not the population of interest, which is those that do not have or have not yet been diagnosed with Parkinson's disease. With respect to the test measure, these two populations may be very different, and merely extrapolating from one (those with known diagnoses) to the other will have spurious results. However, how does one find a group of subjects known to be at risk compared to those known not to be at risk if the intent is to develop predictive tests?

One approach has been to relate the experimental measure to the degree of a condition—for example, the severity or duration of Parkinson's disease. One then generates a regression line that relates the experimental measure to the degree of the condition (Figure 50.3). The regression lines are then extended or extrapolated back-

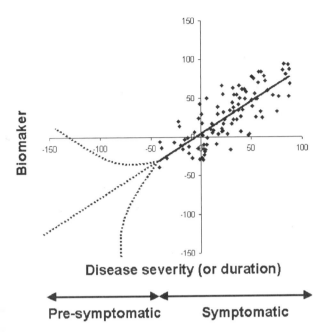

Biomaker

Disease severity (or duration)

Pre-symptomatic   **Symptomatic**

FIGURE 50.3 Demonstration of the difficulties associated with extrapolation compared to interpolation. Extrapolation requires extending statistical information, such as a trend line, from a region of data to a region without data. In this case, data from symptomatic subjects with varying results of a biomarker are extended to determine what the biomarker results would be in a presymptomatic group. As can be seen, extrapolation requires assumptions as the behavior of the trend line, such as linearity. Unfortunately, there are no data to determine whether the assumptions are correct.

ward in severity or time, and those values of the experimental measure are used to ascribe risk. However, this is extrapolation, not interpolation, and such extrapolation requires certain assumptions, specifically that the behavior of the correlation can be predicted. For example, the assumption is made for a linear extrapolation that the slope of the regression line derived with known data is the same as for the extrapolated or hypothesized data. While convenient, there is no *a priori* way to know what the behavior of the actual data in the range of the extrapolation would be if it were obtainable.

If the population of interest is those without, but at risk for, Parkinson's disease, there is no way to identify them so as to develop a predictive test. This is true; however, there are some methods that can be used to bring the slope of the regression line closer to the population of interest. For example, patients who are newly diagnosed or (even better) suspected of having Parkinson's disease can be studied. However, it may be difficult to have confidence in the development of the test when the illness is so mild and early. This can be compensated by adding subjects in whom there is confidence in the diagnosis. This combined sample has subjects with confident but later or more advanced disease and subjects with less confident

but earlier or milder disease. This combination helps finding a regression or relationship (those with definite diagnoses) while helping ensure generalizability to the population of interest (those with questionable diagnoses).[7]

Another major problem, particularly for diagnostic tests, is the question, "What is the gold standard?" Again, this relates to the population of interest. Is the issue the development of a diagnostic test that is more accurate than the clinical diagnosis by sophisticated physicians? If so, then what is to adjudicate which is better should there be a difference? It may not be sufficient to say that the diagnostic test is good because its results have a high correlation with clinical diagnoses by sophisticated physicians. Even if the correlation is perfect, this means that the diagnostic test is as good, or as bad, as the sophisticated physician. Then the question arises as to why we would develop such a diagnostic test. Why not just have the subject evaluated by the sophisticated physician? Is the issue the development of a diagnostic test that does not require sophisticated health care professionals? If so, then the clinical diagnosis of the sophisticated health care professional becomes the gold standard.

Currently, the best gold standard is pathological examination of brain tissue, but this is hardly helpful in developing diagnostic or predictive tests. Short of pathology, the best gold standard is the clinical examination by a sophisticated physician. However, it is possible to improve the standard over a single evaluation by adding the "test of time." Thus, serial evaluations by sophisticated physicians are the best gold standard by which to judge a diagnostic test. Indeed, the best population is those in whom the initial diagnosis was in doubt yet resolved in subsequent evaluations.[8,13]

Developing diagnostic or predictive tests where the prior probability of the condition of interest is low presents significant problems—particularly the high risk of false positive results. Indeed, the prior probability may be so low that it is highly unlikely that any test would have sufficient specificity to reduce the problem of false positives. However, there are methods that can be employed to help improve the specificity and sensitivities of diagnostic or predictive tests. These include use of a battery of tests. Taken sequentially, the result of one test changes the prior probabilities for the next test.

The tests in any battery must be carefully selected. If two or more tests are highly correlated, then the combined specificity and sensitivity actually will be less than the individual tests.[12] One method to help ensure that the different tests making up the battery will not be correlated is to test different domains in the condition of interest. For example, Parkinson's disease affects olfaction and mood as well as movement, and the presumed underlying mechanisms for each are different. Consequently, tests of olfaction, mood, and movement are not likely to be highly correlated, and combining these tests can improve specificity and/or sensitivity.[7]

Handwriting analysis often has been proposed for use as a diagnostic or prognostic test. However, there are many variables or measures in handwriting. The danger is that these variables are likely to be highly correlated. Combining these variables is likely to reduce specificity and/or sensitivity. However, different tests or aspects of motor function are not necessarily highly correlated. For example, different behavioral tasks have different susceptibility to the earliest effects of impending Parkinson's disease. For example, Parkinson's disease patients may have more difficulty moving to a target that is bounded by mechanical limitations compared to those with mechanical limitations compared to control subjects.[14] These differences can be utilized in developing a test battery.[7] Similarly, motor performance prior to administration of levodopa may not correlate with motor performance following levodopa, so a test battery including the response to levodopa may improve the specificity and sensitivity of the diagnostic or predictive test.

A wide range of clinical features have been considered for use as a diagnostic or predictive test, as described below.

LEVODOPA RESPONSE

Merello et al.[5] examined whether a response to a test dose of levodopa/carbidopa 250/50 would predict the long-term responsiveness to levodopa, which is taken as the gold standard for a diagnosis of idiopathic Parkinson's disease. An improvement in the Unified Parkinson Disease Rating Scale (UPDRS) of 30% was taken as the cutoff for a positive response to the levodopa/carbidopa test dose. A very high sensitivity (70.9%), specificity (81.4%), and positive predictive ratio (88.6%) were found. Even more interesting is the fact that those patients with the least motor involvement as detected by the UPDRS had even better specificity than those with greater disease. For those with a combined motor UPDRS less than or equal to 10, the sensitivity, specificity and positive predictive ratio was 71.4 and 100%, respectively. For those with a combined motor UPDRS of 11 to 20, the sensitivity and specificity were 75 and 75%, and for those with a combined motor UPDRS, the sensitivity and specificity were 36.4 and 87%, respectively. It would be interesting if the results would be less impressive in a large sample size, particularly for the subgroup analyses.

The question is, "What is this test detecting or predicting?" One can most confidently say that it is using a levodopa test to predict a levodopa response. The assumption is that a levodopa response for 24 months is a predictor of the pathological diagnosis. This is problematic. Given this assumption, and further assuming that the prior probability of nonidiopathic Parkinson disease in the

sample was 14%, then the test would identify 50 of the 71 subjects with idiopathic Parkinson disease but misidentified 21 as having nonidiopathic Parkinson disease. Furthermore, the test battery would have misidentified 2 of the 11 nonidiopathic Parkinson's disease subjects as having idiopathic Parkinson disease.

These results are clinically meaningless in themselves. The context in which the levodopa challenge test would be used is critically important. For example, assume that we want to develop a treatment for idiopathic Parkinson's disease, but we are concerned about inclusion of nonidiopathic Parkinson disease subjects that would skew the results. Applying the levodopa challenge test to the sample of 82 subjects would mean inclusion of 50 subjects with idiopathic Parkinson's disease and only 2 subjects with nonidiopathic Parkinson's disease, or 4%. This is a significant improvement over the assumed prior probability of 14%. However, consider using this test as a selection criterion for deep brain stimulation (DBS). While 2 of 11 nonidiopathic Parkinson disease subjects would have been subjected to DBS without benefit, 21 subjects with idiopathic Parkinson disease would have been denied DBS treatment. The answer to whether this is reasonable is not found in the numbers but in the social, ethical, moral, and economic consequences.

Finally, the question arises as to why a cutoff of 30% improvement with the levodopa challenge? What would have been the effect on specificity and sensitivity with a 20% cutoff; what about 40% or 50%? Answers to these questions would have allowed construction of a receiver-operator characteristic curves. As described above, the decision as to the appropriate cutoff could have been determined by the shape of the receiver-operator curve, although it would be better to base it on the social, ethical, moral, and economic consequences.

OLFACTION

A number of studies suggest that olfaction may be useful in developing diagnostic or predictive tests in asymptomatic patients. Studies have examined olfactory abnormalities in first-degree relatives.[3,4,6] The assumption is that more of these subjects are likely to be at risk for Parkinson's disease than those subjects with first-degree relatives with Parkinson's disease. Sobel et al.[15] suggest that the olfactory disturbances are actually due to motor impairments of sniffing. However, this would not explain the rate of olfactory disturbances in subjects tested when the motor symptoms were insufficient to warrant a diagnosis of Parkinson's disease, but who were later found to have Parkinson's disease.[8]

Olfaction tests also have been suggested for the differential diagnosis between idiopathic and nonidiopathic Parkinson's disease.[9] However, these studies amount to little more than a series of case reports without the application of critical epidemiological statistical evaluations.

CLINICAL NEUROPHYSIOLOGICAL STUDIES

Sabbahi et al.[10] examined motor neuron excitability in Parkinson's disease subjects. They studied at least 4 measures of the H-reflex in 30 subjects. Thus, there were nine subjects for each test raising questions about generalizability. There was no prospective testing.

REFERENCES

1. Betarbet, R., Sherer, T. B., MacKenzie, G., Garcia-Osuna, M., Panov, A. V., Greenamyre, J. T., Chronic systemic pesticide exposure reproduces features of Parkinson's disease, *Nature Neuroscience,* 3:1301–1306, 2000.

2. Wasson, J. R., Sox, H. C., Neff, R. K., Goldman, L., Clinical prediction rules: applications and methodological standards, *N. Engl. J. Med.,* 313:793–799, 1985.

3. Berendse, H. W., Booij, J., Francot, C. M. et al., Subclinical dopaminergic dysfunction in asymptomatic Parkinson's disease patients' relatives with a decreased sense of smell, *Ann. Neurol.,* 50:34–41, 2001.

4. Markopoulou, K., Larsen, K. W., Wszolek, E. K., et al., Olfactory dysfunction in familial parkinsonism, *Neurology,* 49:1262–67, 1997.

5. Merello, M., Nouzeilles, M. I., Arce, G. P., Leiguarda R., Accuracy of acute levodopa challenge for clinical prediction of sustained long-term levodopa response as a major criterion for idiopathic Parkinson's disease diagnosis, *Movement Disorders,* 17:795–798, 2002.

6. Montgomery, E. B., Jr., Baker, K. B., Lyons, K., Koller, W. C., Abnormal performance on the PD test battery by asymptomatic first-degree relative, *Neurology,* 52:757–762, 1999.

7. Montgomery, Jr., E. B., Koller, W. C., LaMantia, T. J. K., Newman, M. C., Swanson-Hyland, E., Kaszniak, A. W., Lyons, K., Early detection of probable idiopathic Parkinson's disease, I. Development of a diagnostic test battery, *Movement Disorders,* 15:467–473, 2000.

8. Montgomery, Jr., E. B., Lyons, K., Koller, W. C., Early detection of probable idiopathic Parkinson's disease, II. A Prospective Application of a Diagnostic Test Battery, *Movement Disorders,* 15:474–478, 2000.

9. Muller, A., Mungersdorf, M., Reichmann, H, Strehle, G., Hummel, T., Olfactory function in Parkinsonian syndromes, *Journal of Clinical Neuroscience,* 9:521, Sep.4, 2002.

10. Sabbahi, M., Etnyre, B., Al-Jawayed, I. A., Hasson, S., Jankovic, J., Methods of H-reflex evaluation in the early stages of Parkinson's disease, *Journal of Clinical Neurophysiology,* 19(1):67–72, 2002.

11. Montgomery, Jr., E. B., Turkstra, L. S., Evidenced based medicine: let's be reasonable, *Journal of Medical Speech Language Pathology,* 11:ix-xii, 2003.

12. Stevens, J., Applied Multivariate Statistics for the Social Sciences, 2nd ed., Lawerance Erlbaum Associates, New Jersey.

13. Adler, C. H., Hentz, J. G., Joyce, J. N., Beach, T., Caviness, J. N., Motor impairment in normal aging, clinically possible Parkinson's disease, and clinically probable Parkinson's disease: longitudinal evaluation of a cohort of prospective brain donors, *Parkinsonism and Related Disorders,* 9:103–110, 2002.

14. Montgomery, E. B., Jr., Nuessen, J., and Gorman, D. S., Reaction time and movement velocity abnormalities in Parkinson's disease under different conditions, *Neurology,* 41:1476–1481, 1991.

15. Sobel, N., Thomason, M. E., Stappen, I., et al., An impairment in sniffing contributes to the olfactory impairment in Parkinson's disease, *Proceedings of the National Academy of Sciences of the United States of America,* 98:4154–4159, 2001.

51 Rating Scales

Kathleen M. Shannon
Rush Medical College

CONTENTS

INTRODUCTION

Despite exhilarating advancements in our understanding of Parkinson's disease (PD), we remain dependent on clinical criteria for its diagnosis as well as for measuring changes in disease severity due to endogenous and exogenous influences. Clinical rating scales address the need to objectively describe and define the illness but evolved primarily to detect acute, subacute, and chronic symptomatic benefits of therapeutic interventions. Later, these rating scales were also applied to the detection of neuroprotective effects and to the validation of certain biomarkers, such as neuroimaging changes. However, rating scales have not proven useful for distinguishing between normal subjects and those with PD, or among patients with parkinsonism of different etiologies. In addition, rating scales do not well facilitate the characterization of mild, moderate, and severe disease, and they have other troublesome deficiencies. This chapter briefly reviews the history of rating scale development in PD, discusses rating scales most commonly used to measure the motor disorder in PD, and outlines goals for future scale development. The assessment of other aspects of PD, such as quality of life, affect, and cognition, is discussed elsewhere in the text.

HISTORY OF RATING SCALES IN PD

The history of rating scale development in PD largely follows the history of therapeutic intervention. Therapeutic trials in PD began with anticholinergic medications in the 1940s. In these early clinical trials, the therapeutic response was measured in terms of the clinical impressions of the investigator supplemented by the research subjects' own assessments. More formalized rating scales appeared in the context of clinical trials in the 1950s and 1960s. Some of these scales measured primarily objective signs of the illness (*impairment scales*), others addressed the impact of these signs on functional state (*disability scales*), and still others measured both impairment and disability (*multimodular scales*) (see Table 51.1). Because of the emphasis on detecting treatment effects, rating scales in PD focused almost exclusively on the motor

0-8493-1590-5/05/$0.00+$1.50
© 2005 by CRC Press

disorder, particularly the cardinal motor features *bradykinesia, rigidity, tremor, and postural instability*. Cognitive, affective, behavioral, and autonomic features of the illness were largely neglected. The recognition of delayed motor, behavioral, and psychiatric complications of pharmacotherapy necessitated the development of new assessment tools or modification of existing rating scales, and a newer emphasis on validity and reliability focused attention on clinimetric analyses of scales in common use and in development. Clinimetric assessments or validity and reliability are now considered essential in rating scale development. *Validity* is the extent to which a tool measures what is desired and fails to measure what is not desired. Validity encompasses the extent to which the relevant domains are measured, concordance with a standard, particularly a gold standard, and the degree to which a new instrument measures the same construct as is measured by another test. *Reliability* includes the extent to which all items in a given scale are measuring the same underlying construct and the reproducibility of the measure across investigators or over time (inter-rater and intra-rater reliability).

TABLE 51.1
Clinical Rating Scales for Parkinsonian Signs and Symptoms

Multimodular Rating Scales
Impairment Scales
New York University Rating Scale
Webster Scale
Columbia University Rating Scale
Parkinson's Disease Impairment Scale
Disability Scales
Schwab and England Disability Scale
Northwestern University Disability Scale
Intermediate Scale for the Assessment of Parkinson's Disease
Extensive Disability Scale
Multimodular Rating Scales
New York University Rating Scale
University of California Los Angeles Rating Scale
Unified Parkinson's Disease Rating Scale
Short Parkinson's Evaluation Scale
King's College Hospital Scale

IMPAIRMENT SCALES

Early impairment scales rated the cardinal and other signs of PD using numerical scores ranging from 0 to 3 or 0 to 100, with higher numbers indicating more severe impairment.[1,2] The first widely accepted PD rating scale was the *Webster scale*. The Webster scale scored nine impairment items—bradykinesia, rigidity, posture, arm swing, gait, tremor, facial expression, seborrhea, and speech—and one disability item (self-care)—each on a scale of 0 to 3, with 3 indicating more severe impairment.[3] The Webster scale

did not address all major aspects of PD, particularly postural instability. Studies of the Webster scale show poor to moderate inter-rater reliability.[4-6] *The Columbia University Rating Scale* was first introduced in 1969. This scale assessed facial expression, seborrhea, salivation, speech, tremor, rigidity, various limb-specific bradykinesia items, rising from a chair, standing, posture, postural stability, gait and global bradykinesia each on a 0 to 4 scale, with 0 representing normal and 4 representing severe difficulty performing the task. The scale showed moderate to good validity and reliability, enjoyed worldwide use, and later became the foundation for Part III, the motor impairment section of the *Unified Parkinson's Disease Rating Scale (UPDRS)* (see below).[7] In 1967, Hoehn and Yahr published the *Hoehn and Yahr Staging Scale (HY)*. This five-point staging system was developed in a large clinic-based PD population prior to the advent of dopaminergic therapy (Appendix A).[8] The HY describes five stages of PD beginning with unilateral disease and progressing through end-stage disease. The HY focuses on disease milestones such as the transition to generalized disease, the loss of postural reflexes and the loss of independent ambulation. Because each incremental increase in HY stage indicates a substantial qualitative deterioration, the HY is not generally used as a primary endpoint in clinical trials. However, it is the only impairment scale that specifically classifies the overall severity of the illness as mild, moderate or severe, and for this reason it is often used as a secondary endpoint and as the standard for validating other rating scales. The HY was later modified to allow half-point increments between Stages I and II and between Stages II and III (Appendix A).[9]

DISABILITY SCALES

The *Northwestern University Disability Scale (NUDS)* scored five types of activities of daily living—walking, dressing, hygiene, eating and feeding, and speech—on 0 to 10 point scales, with descriptors for each score. [10] Validity studies have shown good correlation with impairment scales such as the Webster and CURS, and the scale shows moderate to excellent inter-rater reliability.[7,11,12] The *Schwab and England Capacity for Daily Living Scale (SE)*, published in 1969, assigned a level of disability by percent decile ranging from complete independence (100%) to complete dependence (0%) with descriptive anchors for each level (Appendix B).[1] The SE can be completed by the subject or caregiver or by the investigator or clinician using historical data obtained from the patient. Separate scores can be given for function in the "on" and "off" state. The clinimetric properties of the SE have not been established, but based on its use as a comparator, it is believed to have moderate to very good validity and good reliability.[7]

MULTIMODULAR SCALES

The *New York University Scale (NYU)* assessed rigidity, tremor, bradykinesia, simultaneous movements, stance, postural stability and gait including timed walking as well as ten functional disability items, each on a 0 to 4 scale. A weighting system was used for the degree to which individual items were thought to influence overall severity. The NYU showed poor validity when compared to the HY.[13] The *University of California Los Angeles Scale (UCLA),* originally developed by McDowell and colleagues at Cornell University measured 14 signs and symptoms and seven activities of daily living each on a 0 to 2 (absent, present, marked) scale with weighting factors from 1 to 10 to adjust for the perceived contribution of items to global disability. The UCLA scale included some features of the illness, such as cognitive and mood disturbances that are not usually sensitive to dopaminergic therapy.[14] Although found to have moderate to good inter-rater reliability, the scale is not often used in clinical trials. [7] The *King's College Hospital Scale* evaluated 13 clinical signs and six disability items on a 0 to 3 scale. This scale had a unique emphasis on posture with separate ratings of posture for eight body areas.[15]

With the growth in PD research over time, it became apparent that the lack of a universally used rating scale was hampering the study of the disease, preventing the comparison of trial results across centers and across treatment interventions. There was scant data on inter- and intra-rater reliability and reliability was only fair to moderate for most assessed items.[5,16,17] Moreover, scales varied in their emphasis on the cardinal and other signs of the disease. Ramaker et al., reviewed seven impairment scales for the relative contribution of various items to the total impairment scale. She found, for example, that the contribution of bradykinesia items to the total impairment score ranged from 16% in the NYU to 40% in the Webster scale and that tremor contributed 10 to 33%, rigidity 0 to 20% and postural instability 0 to 10% of total impairment for various scales. Items other than those related to the cardinal signs of the illness contributed up to 57% of the total impairment score. In practice, the use of these different rating scales might lead to different conclusions about therapeutic response. For example, using one impairment and two disability scales, Markham and Diamond found quite different response profiles to dopamine agonist therapy in a group of patients followed over one year.[18] Most scales also fail to adequately address the manifestations of treated PD, particularly motor fluctuations, dyskinesias, and psychiatric complications of chronic dopaminergic therapy.

THE UNIFIED PARKINSON'S DISEASE RATING SCALE

In an attempt to create a more universally useful multimodular scale, a group of clinical investigators convened to develop the Unified Parkinson's Disease Rating Scale (UPDRS). First introduced in 1984, it underwent several revisions until its current iteration, version 3.0, was published in 1987.[17] Its authors intended the UPDRS to be a core assessment tool to be supplemented by other scales when more detailed information about some aspect of the illness was being sought.[19] Because the UPDRS is frequently supplemented by the HY and SE scales, there is a common misconception that the SE and HY are part of the UPDRS.

The UPDRS is a multimodular scale with four parts:

Part I   Mentation, behavior, and mood
Part II   Activities of daily living
Part III Motor
Part IV Complications of therapy (see Appendix C)

Part I includes four items: intellectual impairment, thought disorder due to dementia or drug intoxication, depression and motivation/initiative, each scored on a 0 to 4 scale. These items are rated by interview with the patient or research subject. Part II, the activities of daily living portion of the UPDRS, is also rated by interview; however, it can also be self-administered[20] or administered to caregivers.[21] Patients with a fluctuating response to dopaminergic medications may be asked to rate how they function in both the "on" and the "off" states. UPDRS Part III contains items assessing motor impairment. The scale was crafted from the Columbia University Rating Scale, with individual items that address speech, facial expression, rest and postural tremor, rigidity, bradykinesia, posture, gait, and postural stability. It has been used to detect the presence and evolution of parkinsonian signs in atypical parkinsonism, in nondemented elderly persons, and in Alzheimer's disease.[22–24] Nurses, research assistants, and field workers have been trained to use the scale with good inter-rater reliability,[25] and a published teaching tape can be used to standardize its application across investigators.[26] Part IV, Complications of therapy includes items that assess the predictability, duration, associated disability, and suddenness of "off" periods; the duration and severity of dyskinesias; the presence or absence of dystonia; and other complications of therapy, including anorexia/nausea/vomiting, sleep disturbance, and orthostatic hypotension. Part IV combines items rated on a 0 to 4 scale with items rated dichotomously, a troublesome feature that makes the total score less useful.

Use of the UPDRS has increased over time, and it is now the most popular scale for rating PD. In a survey of clinical trials in PD published between 1966 and 1998, use of the UPDRS increased from about 33% of publications between 1988 and 1993 to nearly 70% of studies published between 1994 and 1998 (see Figure 51.1).[27] It is used worldwide in patients and research subjects with

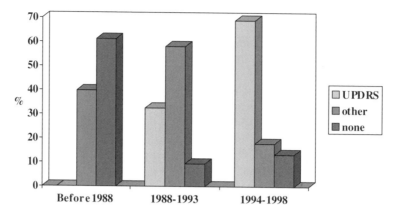

FIGURE 51.1   Evolution of rating scale usage in clinical trials in PD.

early to late disease and for the assessment of response to pharmacological, surgical, and other therapies. Longitudinal studies have shown that the scale is sensitive to increasing severity as well as to clinically relevant change, such as the need for pharmacological or other intervention, and that it detects symptomatic treatment effects. The scale is easy to administer, taking between 10 and 20 min. The UPDRS is also used as an index of disease severity, such as in studies of the development of surrogate neuroimaging markers.[28,29]

As is often the problem with composite scales, it may be difficult to detect changes in individual manifestations of a disease from the total score. There have thus been attempts to devise subscales by combining individual items from Part III or from Parts II and III. In some cases, items have been grouped empirically. For example, a tremor subscale has been derived using a sum of the two Part II tremor items, and five rest and two postural tremor items from Part III. Summing empirically derived items can also yield individual scores for rigidity and bradykinesia. Another score derived in this way from the UPDRS is the postural instability and gait disorder subscale (PIGD), which is the sum of the scores in Part I, for falling, freezing and walking (three items) as well as the scores for gait and postural instability (two items) from part III. From data collected using such an empirically derived subscale, for example, it has been suggested that higher PIGD relative to tremor subscales are associated with a more malignant form of PD.[30] Another way to define relevant subscales is to perform a factor analysis of the scale. Factor analysis is a statistical tool that can be used to identify groups of items within a scale that measure the same thing. Stebbins and Goetz performed a factor analysis of the motor UPDRS. They found that six factors (bradykinesia/gait, rest tremor, rigidity, right bradykinesia, left bradykinesia, and postural tremor) accounted for 78% of the total variance of the scale (see Table 51.2). This factor structure was stable in both the "on" and "off" states.[31,32]

TABLE 51.2
Clusters of Symptoms Identified by Factor Analysis of UPDRS Part III (Motor)

Factor	Part III Items
1. Bradykinesia/gait (7 items) 28 possible points	Speech
	Facial expression
	Arising from a chair
	Posture
	Gait
	Postural stability
	Body bradykinesia
2. Rest tremor (5 items) 20 possible points	Chin rest tremor
	Right hand rest tremor
	Lift hand rest tremor
	Right leg rest tremor
	Left leg rest tremor
3. Rigidity (5 items) 20 possible points	Neck rigidity
	Right arm rigidity
	Left arm rigidity
	Right leg rigidity
	Left leg rigidity
4. Right bradykinesia (4 items) 16 possible points	Right finger tapping
	Right hand movements
	Right pronation-supination movements
	Right leg agility
5. Left bradykinesia (4 items) 16 possible points	Left finger tapping
	Left hand movements
	Left pronation-supination movements
	Left leg agility
6. Postural tremor (2 items) 8 possible points	Right arm postural tremor
	Left arm postural tremor

LIMITATIONS OF THE UPDRS

Despite its widespread acceptance, a number of deficiencies in the UPDRS have become apparent. Part I items include a number of different cognitive and behavioral symptoms, relating to the primary degenerative illness, to comorbidities, and to adverse events of therapy. While useful for screening, the total score has little relevance, and changes over time may be difficult to interpret. Although billed as a disability scale, Part II includes a number of items eliciting subjective assessments of impairment (salivation, swallowing, falling, freezing, walking, tremor, and sensory complaints). These self-assessed impairment measures are less valid than self-assessed disability.[33]

Because Part III measures a number of different types of signs, the total score may not accurately reflect the severity of the illness. For example, a person with prominent tremor but little functional disability might have a UPDRS Part III score identical to that of a more disabled person whose score derives solely from bradykinesia. In addition, despite fairly good clinimetric performance for the scale as a whole, a number of Part III items have poor inter-rater reliability (facial expression, speech, action tremor, posture, and body bradykinesia). One particularly troublesome item in Part III is the assessment of postural stability, because it is difficult to standardize the administration of the pull test, and specific instructions for administering the test are not given.

Part IV mixes dichotomous yes–no items assessing the presence of various complications of therapy with items that address temporal aspects of drug-related motor fluctuations and complications and with items that address disability related to these complications, rendering the total Part IV score useless.

The UPDRS is not useful in assessing change in motor response fluctuations and dyskinesias over time or in response to therapeutic interventions, and supplemental information such as patient-completed diaries or special dyskinesia rating scales must be used in treated patients.

The UPDRS has a number of redundant items—particularly a large number of bradykinesia compared to other items. Indeed, some investigators have attempted to shorten the UPDRS by eliminating redundant items. For example, the *Short Parkinson's Evaluation Scale (SPES)* offers fewer items to rate and ratings on a 0 to 3 rather than 0 to 4 scale. A number of features are not addressed by the UPDRS, including anhedonia, bradyphrenia, anxiety, hypersexuality, sleep disorders, fatigue, dysautonomia, dysregulation, and quality of life.[19] One other problem is how to handle comorbidities, such as arthritis and neuromuscular disease, in this largely elderly population. As with most rating scales, floor and ceiling effects may limit the ability to accurately picture the early and late phases of the illness.[34] Finally, since published clinical trials include mainly a Caucasian population, possible biases related to race, ethnic group, gender, and age are likely.

The Movement Disorders Society Task Force on Rating Scales for Parkinson's Disease concluded that the UPDRS has substantive weaknesses and that it should be expanded and modified, focusing on improving clarity, resolving ambiguity, and providing ratings for milder disease. They recommend a new scale that retains the basic structure of the old scale but provides specific instructions for its administration, eliminates ambiguities, covers the full spectrum of the illness, removes cultural bias, has a strategy for dealing with comorbidities, includes more detailed information about dyskinesia, screens additional nonmotor disabilities, recommends additional scales as needed, and has an appropriate factor structure.[19]

SPECIAL CONSIDERATIONS

NEUROPROTECTION

Many intriguing putative neuroprotective therapies for PD are entering the testing pipeline. One challenge in designing clinical neuroprotective trials in PD is the selection of clinical endpoints. A change in UPDRS score is a potentially relevant clinical endpoint for such studies. However, the total UPDRS, and Parts II and III, are sensitive to symptomatic effects of dopaminergic therapy, limiting the population that can be studied. Although there are little available data on longitudinal changes in UPDRS over time, clinical trials in *de novo* subjects suggest an annual average increase in total UPDRS of 8 to 10 points. Patients are usually treated when the UPDRS score is around 35. Once potent dopaminergic therapy is initiated, there follows an average 40% decrease in UPDRS. The UPDRS then stabilizes over the next 1 to 2 years then gradually returns to pretreatment levels by about year 5 (see Figure 51.2).[35] Changes in UPDRS scores in subjects on dopaminergic therapy are thus a function of both the underlying disease progression and changes related to pharmacotherapy. In addition, treatment related side effects themselves might influence scores, particularly in disability ratings. Neuroprotection trials therefore typically use untreated subjects, which limits the study population as well as the ability to detect long-lasting effects. Newer clinical endpoints that might be more resistant to symptomatic effects, such as the development of postural instability, are being developed for use in neuroprotective studies.[35] Active exploration of biomarkers, such as neuroimaging changes, to supplement or supplant clinical measures, continues.

MOTOR FLUCTUATIONS AND DYSKINESIAS

Motor fluctuations and dyskinesias are a significant source of disability in more advanced PD patients. However, the clinic-based assessment of motor fluctuations and dyskinesias is difficult. Indeed, a recent study found that, even in clinical trials involving similar patient populations and treatment interventions, the reported incidence of dyskinesias ranged from 5 to 41%, depending on the method used to ascertain their presence.[36]

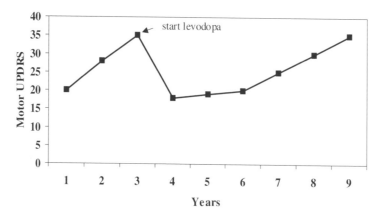

FIGURE 51.2 Representation of changes in UPDRS over time before and after levodopa initiation.

The UPDRS Part IV poorly assesses motor fluctuations. These data are especially important in assessing treatment interventions in more advanced patients on chronic dopaminergic therapy. Historical data on fluctuation-related disability can be obtained from collecting UPDRS Part II for both the "off" and "on" states. The SE can also be used to collect historical data on disability in the "off" and "on" states. The patient can complete hourly diaries—these are more reliable when the patient has been trained to accurately recognized different functional states.[37] Examining patients or research subjects in the "off" and "on" state captures the acute effect of dopaminergic therapy and is practiced in the detection of some pharmacological and surgical effects in PD (see below), but it is difficult to do in patients with advanced disability.

There are several available rating scales for assessing dyskinesias in treated PD. UPDRS Part IV assesses the duration and associated pain and disability of dyskinesia, using historical data. Larsen et al. used an assessment scale in which patients reported dyskinesia-related disability in the head/trunk and the limbs as well as the duration of dyskinesia expressed as a percentage of the waking day on a 0 to 4 scale. The total score was derived by multiplying severity and duration.[38] A scale developed by Obeso includes ratings of both the duration and severity of dyskinesia, with total score calculated as the arithmetic mean of these two items.[39] Goetz et al. revised the Obeso score to include observation of the patient performing three specific tasks—walking, drinking from a cup, and putting on a coat. Each task is rated on a 0 to 4 scale, with 0 indicated no dyskinesia present and 4 indicating severe dyskinesia rendering task completion impossible.[40] The Lang-Fahn Activities of Daily Living Dyskinesia Scale (LFADLDS) is modified from UPDRS Part II. The patient is asked to rate the effect of maximal dyskinesias on certain activities of daily living, including writing, eating, dressing, hygiene, and walking. Each task is rated on a 0 to 4 scale, with 4 indicating maximum interference with the task.[41] Hagell and Widner devised a scale rating differ-

ent body parts with separate ratings of dystonia and hyperkinesia. The scale showed good inter- and intra-rater reliability.[42] A comparison of the Modified Goetz Dyskinesia Scale (MGDS), the Lang-Fahn Activities of Daily Living Dyskinesia Scale (LFADLDS), and home and in-clinic diary ratings was performed within the context of a clinical trial in moderately disabled PD research subjects. This study suggested that the MDGDS agreed well with the home and in-clinic diary ratings and agreed poorly with the LFADLDS, suggesting that the impairment and disability dyskinesia scales may be complementary.

SURGICAL INTERVENTIONS

The advent of cellular restorative procedures for PD highlighted some of the inadequacies of the available rating scales for surgical trials in PD. In 1989, at a meeting of investigators studying adrenal medullary transplantation for PD, participants proposed that a standard system of patient diagnosis and evaluation be developed to establish a minimum standard evaluation for patients enrolled in neural transplantation trials. The product of the working committee, called the Core Assessment Program for Intrastriatal Transplantation (CAPIT), was published in 1992.[39] The CAPIT suggested clinical criteria for the diagnosis of idiopathic PD, as well as other patient inclusion criteria, outlined desired rating scales, included four-timed motor tests, and suggested the optimum follow-up schedule. Rating scales and timed motor testing were performed in the "practically defined off state" (in the morning after overnight withdrawal of levodopa) and the "best on" state (Appendix D).

Over time, the CAPIT was considered insufficient for a number of reasons:

1. New surgical procedures were developed.
2. The battery was overly laborious for the assessment of large numbers of patients.
3. Important features of the illness such as cognitive impairment were not adequately addressed.

The Core Assessment Program for Surgical Interventional Therapies in PD (CAPSIT-PD) was proposed to address these inadequacies. Additional inclusion criteria were proposed.[43] A "practically defined on" state was proposed, and the number of timed motor tests was reduced (Appendix E). However, despite these attempts to standardize data collection in surgical trials, a recent evidence-based review of more than 500 surgical trials combining more than 10,000 patients, failed to find reliable evidence of long-term safety and efficacy, suggesting that more work needs to be done in this area.[44]

CONCLUSIONS

Rating scales assessing impairment and disability in PD have provided the framework for studying the natural history of PD as well as for studying the efficacy of pharmacological, surgical and ancillary therapies. Monumental gains in our understanding of the disease and its potential treatments have unveiled the vulnerabilities in these familiar scales and have outlined the need to develop new ways to measure parkinsonian signs and symptoms.

APPENDIX A

HOEHN AND YAHR STAGING SCALES

Hoehn and Yahr Stage (Original Version)

Stage I. Unilateral involvement only, usually with minimal or no functional impairment.

Stage II. Bilateral or midline involvement, without impairment of balance.

Stage III. First sign of impaired righting reflexes. This is evident by unsteadiness as the patient turns or is demonstrated when he is pushed from standing equilibrium with the feet together and eyes closed. Functionally, the patient is somewhat restricted in his activities but may have some work potential, depending upon the type of employment. Patients are physically capable of leading independent lives, and their disability is mild to moderate.

Stage IV. Fully developed, severely disabling disease; the patient is still able to walk and stand unassisted but is markedly incapacitated.

Stage V. Confinement to bed or wheelchair unless aided.

Modified Hoehn and Yahr Stage

Stage 0.0. No signs of PD.
Stage 1.0. Unilateral involvement only.
Stage 1.5. Unilateral and axial involvement.
Stage 2.0. Bilateral involvement without impairment of balance.
Stage 2.5. Mild bilateral involvement with recovery on retropulsion (pull) test.
Stage 3.0. Mild to moderate bilateral involvement; some postural instability but physically independent.
Stage 4.0. Severe disability; still able to walk or stand unassisted.
Stage 5.0. Wheelchair bound or bedridden unless aided.

APPENDIX B

SCHWAB AND ENGLAND ACTIVITIES OF DAILY LIVING

100% = Completely independent. Able to do all chores without slowness, difficulty or impairment. Essentially normal. Unaware of any difficulty.

90% = Completely independent. Able to do all chores with some degree of slowness, difficulty and impairment. Might take twice as long. Beginning to be aware of difficulty.

80% = Completely independent in most chores. Takes twice as long. Conscious of difficulty and slowness.

70% = Not completely independent. more difficulty with some chores. Three to four times as long in some. Must spend a large part of the day with chores.

60% = Some dependency. Can do most chores, but exceedingly slowly and with much effort. Errors; some impossible.

50% = More dependent. Help with half the chores, slower, etc. Difficulty with everything.

40% = Very dependent. Can assist with all chores, but few alone.

30% = With effort, now and then does a few chores alone or begins alone. Much help is needed.

20% = Nothing alone. Can be a slight help with some chores. Severe invalid.

10% = Totally dependent, helpless. Complete invalid.

0% = Vegetative functions such as swallowing, bladder and bowel functions are not functioning. Bedridden.

APPENDIX C

UNIFIED PARKINSON DISEASE RATING SCALE

Part I—Mentation, behavior and mood

1. Intellectual impairment

0 = None.

1 = Mild. Consistent forgetfulness with partial recollection of events and no other difficulties

2 = Moderate memory loss, with disorientation and moderate difficulty handling complex problems. Mild but definite impairment of function at home with need of occasional prompting.

3 = Severe memory loss with disorientation for time and often to place. Severe impairment in handling problems.

4 = Severe memory loss with orientation preserved to person only. Unable to make judgments or solve problems. Requires much help with personal care. Cannot be left alone at all.

2. Thought disorder

0 = None.

1 = Vivid dreaming

2 = "Benign" hallucinations with insight retained.

3 = Occasional to frequent hallucinations or delusions; without insight; could interfere with daily activities.

4 = Persistent hallucinations, delusions or florid psychosis. Not able to care for self.

3. Depression

0 = Not present.

1 = Periods of sadness or guilt greater than normal, never sustained for days or weeks.

2 = Sustained depression (1 week or more).

3 = Sustained depression with vegetative symptoms (insomnia, anorexia, weight loss, loss of interest).

4 = Sustained depression with vegetative symptoms and suicidal thoughts or intent.

4. Motivation/initiative

0 = Normal.

1 = Less assertive than usual; more passive.

2 = Loss of initiative or disinterest in elective (non-routine) activities

3 = Loss of initiative or disinterest in day-to-day (routine) activities.

4 = Withdrawn, complete loss of motivation.

Part II—Activities of daily living

5. Speech

0 = Normal.

1 = Mildly affected. No difficulty being understood.

2 = Moderately affected. Sometimes asked to repeat statements.

3 = Severely affected. Frequently asked to repeat statements.

4 = Marked drooling, requires constant tissue or handkerchief.

6. Salivation

0 = Normal.

1 = Slight but definite excess of saliva in mouth; may have nighttime drooling.

2 = Moderately excessive saliva; may have minimal drooling.

3 = Marked excess of saliva with some drooling.

4 = Marked drooling, requires constant tissue or handkerchief.

7. Swallowing

0 = Normal.

1 = Rare choking.

2 = Occasional choking.

3 = Requires soft food.

4 = Requires NG tube or gastrostomy feeding.

8. Handwriting

0 = Normal

1 = Slightly slow or small.

2 = Moderately slow or mall; all words are legible.

3 = Severely affected; not all words are legible.

4 = The majority of words are not legible.

9. Cutting food/handling utensils

0 = Normal

1 = Somewhat slow and clumsy, but no help needed.

2 = Can cut most foods, although clumsy and slow; some help needed.

3 = Food must be cut by someone, but can still feed slowly.

4 = Needs to be fed.

10. Dressing

0 = Normal.

1 = Somewhat slow, but no help needed.

2 = Occasional assistance with buttoning, getting arms in sleeves.

3 = Considerable help required, but can do some things alone.

4 = Helpless.

11. Hygiene

0 = Normal.

1 = Somewhat slow, but no help needed.

2 = Needs help to shower or bathe; or very slow in hygienic care.

3 = Requires assistance for washing, brushing teeth, combing hair, going to the bathroom.

4 = Foley catheter or other mechanical aids.

12. Turning in bed/adjusting bed clothes

0 = Normal

1 = Somewhat slow and clumsy, but no help needed.

2 = Can turn alone or adjust sheets, but with great difficulty.

3 = Can initiate, but not turn or adjust sheets alone.

4 = Helpless.

13. Falling—unrelated to freezing

0 = None.

1 = Rare falling.

2 = Occasionally falls, less than once per day.

3 = Falls an average of once daily.

4 = Falls more than once daily.

14. *Freezing when walking*

0 = None
1 = Rare freezing when walking; may have start-hesitation.
2 = Occasional freezing when walking.
3 = Frequent freezing. Occasionally falls from freezing.
4 = Frequent falls from freezing.

15. *Walking*

0 = Nomal.
1 = Mild difficulty. May not swing arms or may tend to drag le.g.,
2 = Moderate difficulty, but requires little or no assistance.
3 = Severe disturbance of walking, requiring assistance.
4 = Cannot walk at all, even with assistance.

16. *Tremor*

0 = Absent.
1 = Slight and infrequently present.
2 = Moderate; bothersome to patient
3 = Severe; interferes with many activities.
4 = Marked; interferes with most activities.

17. *Sensory complaints related to parkinsonism*

0 = None.
1 = Occasionally has numbness, tingling, or aching; not distressing.
2 = Frequently has numbness, tingling, or aching; not distressing.
3 = Frequent painful sensations.
4 = Excruciating pain.

Part III—Motor examination

18. *Speech*

0 = Normal.
1 = Slight loss of expression, diction and/or volume.
2 = Monotone, slurred but understandable; moderately impaired.
3 = Marked impairment, difficult to understand.
4 = Unintelligible.

19. *Facial expression*

0 = Normal.
1 = Minimal hypomimia; could be normal "poker face."
2 = Slight but definitely abnormal diminution of facial expression.
3 = Moderate hypomimia; lips parted some of the time.
4 = Masked or fixed facies with severe or complete loss of facial expression; lips parted π inch or more.

20. *Tremor at rest*

a. Face
b. Left arm
c. Right arm
d. Left leg
e. Right leg

0 = Absent
1 = Slight and infrequently present.
2 = Mild in amplitude and persistent, or moderate in amplitude, but only intermittently present.
3 = Moderate in amplitude and present most of the time.
4 = Marked in amplitude and present most of the time.

21. *Action or postural tremor*

a. Left arm
b. Right arm

0 = Absent.
1 = Slight; present with action.
2 = Moderate in amplitude, present with action.
3 = Moderate in amplitude, with posture holding as well as action.
4 = Marked in amplitude, interferes with feeding.

22. *Rigidity*

a. Neck
b. Left arm
c. Right arm
d. Left leg
e. Right leg

0 = Absent.
1 = Slight or detectable only when activated by mirror or other movements.
2 = Mild to moderate.
3 = Marked, but full range of motion easily achieved.
4 = Severe, range of motion achieved with difficulty.

23. *Finger taps*

a. Left
b. Right

0 = Normal.
1 = Mild slowing and/or reduction in amplitude.
2 = Moderately impaired. Definite and early fatiguing. May have occasional arrests in movement.
3 = Severely impaired. Frequent hesitation in initiating movements or arrests in ongoing movement.
4 = Can barely perform the task.

24. *Hand movements (open and close hands in rapid succession)*

a. Left
b. Right

0 = Normal.
1 = Mild slowing and/or reduction in amplitude.

2 = Moderately impaired. Definite and early fatiguing. May have occasional arrests in movement.

3 = Severely impaired. Frequent hesitation in initiating movements or arrests in ongoing movement.

4 = Can barely perform the task.

25. *Rapid alternating movements (pronate and supinate hands)*

    a. Left

    b. Right

0 = Normal.

1 = Mild slowing and/or reduction in amplitude.

2 = Moderately impaired. Definite and early fatiguing. May have occasional arrests in movement.

3 = Severely impaired. Frequent hesitation in initiating movements or arrests in ongoing movement.

4 = Can barely perform the task.

26. *Leg agility (tap heel on ground, amplitude should be 3 inches)*

    a. Left

    b. Right

0 = Normal.

1 = Mild slowing and/or reduction in amplitude.

2 = Moderately impaired. Definite and early fatiguing. May have occasional arrests in movement.

3 = Severely impaired. Frequent hesitation in initiating movements or arrests in ongoing movement.

4 = Can barely perform the task.

27. *Arising from chair (arises with arms folded across chest)*

0 = Normal.

1 = Slow; or may need more than one attempt.

2 = Pushes self up from arms of seat.

3 = Tends to fall back and may have to try more than one time, but can get up without help.

4 = Unable to arise without help.

28. *Posture*

0 = Normal erect.

1 = Not quite erect, slightly stooped posture; could be normal for older person.

2 = Moderately stooped posture, definitely abnormal; can be slightly leaning to one side.

3 = Severely stooped posture with kyphosis; can be moderately leaning to one side.

4 = Marked flexion with extreme abnormality of posture.

29. *Gait*

0 = Normal.

1 = Walks slowly, may shuffle with short steps, but no festination or propulsion.

2 = Walks with difficulty, but requires little or no assistance; may have some festination, short steps, or propulsion.

3 = Severe disturbance of gait, requiring assistance.

4 = Cannot walk at all, even with assistance.

30. *Postural stability (retropulsion test)*

0 = Normal.

1 = Retropusions, but recovers unaided.

2 = Absence of postural response; would fall if not caught by examiner.

3 = Very unstable, tends to lose balance spontaneously.

4 = Unable to stand without assistance.

31. *Body bradykinesia/hypokinesia*

0 = None.

1 = Minimal slowness, giving movement a deliberate character; could be normal for some persons. Possibly reduced amplitude.

2 = Mild degree of slowness and poverty of movement which is definitely abnormal. Alternatively, some reduced amplitude.

3 = Moderate slowness, poverty or small amplitude of movement.

4 = Marked slowness, poverty or small amplitude of movement.

Part IV—Complications of therapy

A. Dyskinesias

32. *Duration: what proportion of the waking day are dyskinesias present?*

0 = None.

1 = 1-25% of day.

2 = 26-50% of day.

3 = 51-75% of day.

4 = 76-100% of day.

33. *Disability: how disabling are the dyskinesias (historical information—may be modified by office examination)*

0 = Not disabling.

1 = Mildly disabling.

2 = Moderately disabling.

3 = Severely disabling.

4 = Completely disabling.

34. *Painful dyskinesias: how painful are the dyskinesias*

0 = No painful dyskinesias.

1 = Slightly.

2 = Moderately.

3 = Severely.

4 = Markedly.

35. *Presence of early morning dystonia (historical information)*

0 = No.

1 = Yes.

B. Clinical fluctuations

36. *Are any "off" periods predictable as to timing after a dose of medication?*

    0 = No.
    1 = Yes.

37. *Are any "off" periods unpredictable as to timing after a dose of medication?*

    0 = No.
    1 = Yes.

38. *Do any "off" periods come on suddenly (e.g., within a few seconds?)*

    0 = No.
    1 = Yes.

What proportion of the waking day is the subject "off" on average?

    0 = None.
    1 = 1-25% of day.
    2 = 26-50% of day.
    3 = 51-75% of day.
    4 = 76-100% of day.

C. Other complications

40. *Does the subject have anorexia, nausea or vomiting?*

    0 = No.
    1 = Yes.

41. *Does the subject have any sleep disturbances (e.g., insomnia or hypersomnolence)?*

    0 = No.
    1 = Yes.

42. *Does the subject have symptomatic orthostasis?*

    0 = No.
    1 = Yes.

APPENDIX D

CAPIT—CORE ASSESSMENT PROGRAM FOR INTRACEREBRAL TRANSPLANTATIONS

Clinical Rating Scales

Unified Parkinson's Disease Rating Scale (UPDRS) (version 3.0)
Modified Hoehn & Yahr Staging
Dyskinesia Scale Score (from Obeso)
Self-reporting (diary)
Timed Testing (in both "practically defined off" and "best on" states)
Pronation-supination test (time to perform 20 successive cycles of alternating palm/dorsum tap on the knee)

Hand/arm movement between 2 points (time to perform 20 successive taps between 2 points placed 20 cm apart horizontally)
Finger dexterity (time to tap thumb with each finger in rapid succession 10 times)
Stand-Walk-Sit test (time to stand, walk 7 meters, turn, walk back to the chair and sit)
Pharmacological testing
Single-dose L-dopa test in "defined off" state

APPENDIX E

CAPSIT-PD—CORE ASSESSMENT PROGRAM FOR SURGICAL INTERVENTIONAL THERAPIES IN PARKINSON'S DISEASE

Clinical Rating Scales

Unified Parkinson's Disease Rating Scale
UPDRS III for evaluating dopamine responsiveness
Modified Hoehn and Yahr Stage
SF-36
PD-specific quality of life scale
Dyskinesia rating
dominant type of abnormal movement
response pattern
additional optional quantifications
Self-reporting (complete "off," partial "off," complete "on," and "on" with dyskinesias)

Timed tests

Hand-arm movement between 2 points 30 cm apart (number performed in 20 seconds)
Walking test (time to walk 7 meters back and forth including turning; note freezes)

Pharmacologic testing

Defined "off"
Defined "on"

Cognitive and behavioral assessment

General and behavioral
Mattis dementia rating scale
Montgomery and Asberg depression rating scale
Executive functions
Verbal fluency
Paced auditory serial addition test
Odd man out test
Modified Brown Peterson Paradigm
Explicit memory
Rey auditory verbal learning test
Visual mnesic battery of Signoret
Procedural memory
Tower of Hanoi, short version

REFERENCES

1. Schwab, R. S. and Chafetz, M. E., Kemadrin in the treatment of parkinsonism, *Neurol Clin,* 5, 273–277, 1955.

2. Strang, R. R., Experiences with cogentin in the treatment of parkinsonism, *Acta Neurol. Scand.,* 41, 413–418, 1965.

3. Webster, D. D., Critical analysis of the disability in Parkinson's disease, *Mod. Treat.,* 5, 257–258, 1968.

4. Geminiani, G., Cesana, B. M., Tamma, F., Contri, P., Pacchetti, C., Carella, F., Piolti, R., Martignoni, E., Giovannini, P., Girotti, F., and et al., Interobserver reliability between neurologists in training of Parkinson's disease rating scales. A multicenter study, *Mov. Disord.,* 6 (4), 330–5, 1991.

5. Ginanneschi, A., Degl'Innocenti, F., Magnolfi, S., Maurello, M. T., Catarzi, L., Marini, P., and Amaducci, L., Evaluation of Parkinson's disease: reliability of three rating scales, *Neuroepidemiology,* 7 (1), 38–41, 1988.

6. Martinez-Martin, P., Gil-Nagel, A., Gracia, L. M., Gomez, J. B., Martinez-Sarries, J., and Bermejo, F., Unified Parkinson's Disease Rating Scale characteristics and structure, The Cooperative Multicentric Group, *Mov. Disord.,* 9 (1), 76–83, 1994.

7. Ramaker, C., Marinus, J., Stiggelbout, A. M., and Van Hilten, B. J., Systematic evaluation of rating scales for impairment and disability in Parkinson's disease, *Mov. Disord.,* 17 (5), 867–76, 2002.

8. Hoehn, M. M. and Yahr, M. D., Parkinsonism: Onset, progression and mortality, *Neurology,* 17, 427–442, 1967.

9. Lang, A. E. and Fahn, S., Assessment of Parkinson's disease, in *Quantification of Neurologic Deficit,* Munsat, T. L. Butterworth-Heineman, Stoneham, M. A.,1989.

10. Canter, C. J., de la Torre, R., and Mier, M., A method of evaluating disability in patients with Parkinson's disease, *J. Nerv. Ment. Dis.,* 133, 143–147, 1961.

11. Henderson, B. T., Kenny, B. G., Hitchcock, E. R., Hughes, R. C., and Clough, C. G., A comparative evaluation of clinical rating scales and quantitative measurements in assessment pre and post striatal implantation of human foetal mesencephalon in Parkinson's disease, *Acta Neurochir. Suppl. (Wien),* 52, 48–50, 1991.

12. Martinez-Martin, P., Carrasco de la Pena, J. L., Ramo, C., Antiguedad, A. R., and Bermejo, F., [Inter-observer reproducibility of qualitative scales in Parkinson disease (I)], *Arch Neurobiol. (Madr)* 50 (5), 309–14, 1987.

13. Lieberman, A., Dziatolowki, M., Gopinathan, G., and et al., Evaluation of Parkinson's disease, in *Ergot Compounds and Brain Function: Neuroendocrine and Neuropsychiatric Aspects,* Goldstein, M., Raven, New York, 1980.

14. Treciokas, L. J., Ansel, R. D., and Markham, C. H., One to two year treatment of Parkinson's disease with levodopa, *Calif. Med.,* 114 (5), 7–14, 1971.

15. Parkes, J. D., Zilka, K. J., Calver, D. M., and Knill-Jones, R. P., Controlled trial of amantadine hydrochloride in Parkinson's disease, *Lancet,* 1, 259–262, 1970.

16. Montgomery, G. K., Reynolds, N. C., Jr., and Warren, R. M., Qualitative assessment of Parkinson's disease: study of reliability and data reduction with an abbreviated Columbia Scale, *Clin. Neuropharmacol.,* 8 (1), 83–92, 1985.

17. Fahn, S., Elton, R. I., and Members of the UPDRS Development Committee, The Unified Parkinson's Disease Rating Scale, in *Recent Developments in Parkinson's Disease,* Fahn, S., Marsden, C. D., Calne, D. B., and Goldstein, M. Macmillan Healthcare Information, Florham Park, pp. 153–163, 293–304, 1987.

18. Diamond, S. G. and Markham, C. H., A double-blind comparison of levodopa, madopar and sinemet in Parkinson's disease, *Ann. Neurol.,* 3, 267–272, 1978.

19. The Unified Parkinson's Disease Rating Scale (UPDRS): Status and recommendations, *Mov. Disord.,* 18 (7), 738–750, 2003.

20. Louis, E. D., Lynch, T., Marder, K., and Fahn, S., Reliability of patient completion of the historical section of the Unified Parkinson's Disease Rating Scale, *Mov. Disord.,* 11 (2), 185–92, 1996.

21. Martinez-Martin, P., Fontan, C., Frades Payo, B., and Petidier, R., Parkinson's disease: quantification of disability based on the Unified Parkinson's Disease Rating Scale, *Neurologia,* 15 (9), 382–7, 2000.

22. Wilson, R. S., Schneider, J. A., Bienias, J. L., Evans, D. A., and Bennett, D. A., Parkinsonianlike signs and risk of incident Alzheimer disease in older persons, *Arch Neurol.,* 60 (4), 539–44, 2003.

23. Wilson, R. S., Schneider, J. A., Beckett, L. A., Evans, D. A., and Bennett, D. A., Progression of gait disorder and rigidity and risk of death in older persons, *Neurology,* 58 (12), 1815–9, 2002.

24. Ballard, C., O'Brien, J., Swann, A., Neill, D., Lantos, P., Holmes, C., Burn, D., Ince, P., Perry, R., and McKeith, I., One year follow-up of parkinsonism in dementia with Lewy bodies, *Dement. Geriatr. Cogn. Disord.,* 11 (4), 219–22, 2000.

25. Bennett, D. A., Shannon, K. M., Beckett, L. A., Goetz, C. G., and Wilson, R. S., Metric properties of nurses' ratings of parkinsonian signs with a modified Unified Parkinson's Disease Rating Scale, *Neurology,* 49 (6), 1580–7, 1997.

26. Goetz, C. G., Stebbins, G. T., Chmura, T. A., Fahn, S., Klawans, H. L., and Marsden, C. D., Teaching tape for the motor section of the unified Parkinson's disease rating scale, *Mov. Disord.,* 10 (3), 263–6, 1995.

27. Mitchell, S. L., Harper, D. W., Lau, A., and Bhalla, R., Patterns of outcome measurement in Parkinson's disease clinical trials, *Neuroepidemiology,* 19 (2), 100–8, 2000.

28. Morrish, P. K., Sawle, G. V., and Brooks, D. J., An [18F]dopa-PET and clinical study of the rate of progression in Parkinson's disease, *Brain,* 119 (Pt. 2), 585–91, 1996.

29. Seibyl, J. P., Marek, K., Sheff, K., Zoghbi, S., Baldwin, R. M., Charney, D. S., van Dyck, C. H., and Innis, R. B., Iodine-123-beta-CIT and iodine-123-FPCIT SPECT measurement of dopamine transporters in healthy subjects and Parkinson's patients, *J. Nucl. Med.,* 39 (9), 1500–8, 1998.

30. Jankovic, J., McDermott, M., Carter, J., Gauthier, S., Goetz, C., Golbe, L., Huber, S., Koller, W., Olanow, C., Shoulson, I., and et al., Variable expression of Parkinson's disease: a base-line analysis of the DATATOP cohort. The Parkinson Study Group, *Neurology,* 40 (10), 1529–34, 1990.

31. Stebbins, G. T. and Goetz, C. G., Factor structure of the Unified Parkinson's Disease Rating Scale: Motor Examination section, *Mov. Disord.,* 13 (4), 633–6, 1998.

32. Stebbins, G. T., Goetz, C. G., Lang, A. E., and Cubo, E., Factor analysis of the motor section of the unified Parkinson's disease rating scale during the off-state, *Mov. Disord.,* 14 (4), 585–9, 1999.

33. Hariz, G. M., Lindberg, M., and Bergenheim, A. T., Impact of thalamic deep brain stimulation on disability and health-related quality of life in patients with essential tremor, *J. Neurol. Neurosurg. Psychiatry,* 72 (1), 47–52, 2002.

34. Vieregge, P., Stolze, H., Klein, C., and Heberlein, I., Gait quantitation in Parkinson's disease—locomotor disability and correlation to clinical rating scales, *J. Neural. Transm.,* 104 (2-3), 237–48, 1997.

35. Kieburtz, K., Designing neuroprotection trials in Parkinson's disease, *Ann. Neurol.,* 53 Suppl. 3, S100–7; discussion S107-9, 2003.

36. Marras, C. and Lang, A. E., Measuring motor complications in clinical trials for early Parkinson's disease, *J. Neurol. Neurosurg. Psychiatry,* 74 (2), 143–6, 2003.

37. Goetz, C. G., Stebbins, G. T., Blasucci, L. M., and Grobman, M. S., Efficacy of a patient-training videotape on motor fluctuations for on-off diaries in Parkinson's disease, *Mov. Disord.,* 12 (6), 1039–41, 1997.

38. Larsen, T. A., Calne, S., and Calne, D. B., Assessment of Parkinson's disease, *Clin. Neuropharmacol.,* 7 (2), 165–9, 1984.

39. Langston, J. W., Widner, H., Goetz, C. G., Brooks, D., Fahn, S., Freeman, T., and Watts, R., Core assessment program for intracerebral transplantations (CAPIT), *Mov. Disord.,* 7 (1), 2–13, 1992.

40. Goetz, C. G., Stebbins, G. T., Shale, H. M., Lang, A. E., Chernik, D. A., Chmura, T. A., Ahlskog, J. E., and Dorflinger, E. E., Utility of an objective dyskinesia rating scale for Parkinson's disease: inter- and intrarater reliability assessment, *Mov. Disord.,* 9 (4), 390–4, 1994.

41. Evaluation of dyskinesias in a pilot, randomized, placebo-controlled trial of remacemide in advanced Parkinson disease, *Arch Neurol.,* 58 (10), 1660–8, 2001.

42. Hagell, P. and Widner, H., Clinical rating of dyskinesias in Parkinson's disease: use and reliability of a new rating scale, *Mov. Disord.,* 14 (3), 448–55, 1999.

43. Defer, G. L., Widner, H., Marie, R. M., Remy, P., and Levivier, M., Core assessment program for surgical interventional therapies in Parkinson's disease (CAPSIT-PD), *Mov. Disord.,* 14 (4), 572–84, 1999.

44. Stowe, R. L., Wheatley, K., Clarke, C. E., Ives, N. J., Hills, R. K., Williams, A. C., Daniels, J. P., and Gray, R., Surgery for Parkinson's disease: lack of reliable clinical trial evidence, *J. Neurol. Neurosurg. Psychiatry,* 74 (4), 519–21, 2003.

52 Treatment of Parkinson's Disease with Anticholinergic Medications

Bahman Jabbari and Rene Pazdan
Department of Neurology, Uniformed Services University

CONTENTS

ABSTRACT

Synthetic anticholinergic drugs (ACDs) have been the leading agents for treatment of Parkinson's disease (PD) for decades before the introduction of L-dopa. These drugs are effective against Parkinsonian tremor and rigidity but fail to improve akinesia and postural changes. Their current indication in PD is for young patients with naïve resting tremor or tremor unresponsive to dopaminergic agents and for some selected cases of PD-associated dystonia. Their value as an adjunct medication to L-dopa therapy needs to be reassessed in view of emergence of newer anti-PD medications. Disturbing side effects, such as significant memory loss and blurred vision, limit the clinical utility of ACDs. Anticholinergic drugs should not be used in patients with close angle glaucoma, those prone to urinary retention, or the elderly with memory problems. The starting dose should be small and always increased in small increments (especially in the elderly). Slow withdrawal is also recommended so as to avoid exacerbation of Parkinsonian symptoms.

CHOLINERGIC SYSTEM OF BASAL GANGLIA AND ITS INTERACTIONS WITH DOPAMINERGIC SYSTEM

Acetylcholine (ACh) intuitively plays an important role within the striatum based on its abundant presence. In the cerebrum, the striatum has the highest expression of choline acetyltransferase (ChAT), used in ACh synthesis, and acetylcholinesterace (AchE), used in ACh breakdown.[1] ACh in the basal ganglia comes from two sources:

1. Long projection neurons from the pedunculopontine tegmental nucleus and the laterodorsal pontine tegmentum (brain stem ACh nuclei)
2. Giant interneurons of Kölliker (large interneurons)

The large interneurons account for only 1 to 3% of all striatal neurons but have extensive arborizations. Rat models have shown that these cholinergic interneurons provide a tonically active source of Ach[2] providing a constant release of Ach. It is postulated that just one or a few large cholinergic interneurons may synchronize dopamine (DA) release to a large amount of tissue.

ACh binds two very different classes of receptors: nicotinic and muscarinic, both of which have been identified within the striatum. To date, three subtypes of nicotinic acetylcholine receptors (nAChRs) have been identified:

1. Muscular nAChRs
2. Neuronal nAChRs composed of α–β subunit combinations ($\alpha2$–$\alpha6$, $\alpha10$ and $\beta2$–$\beta4$)
3. Homo-oligomeric neuronal nAChRs ($\alpha7$–$\alpha9$)[3]

A limited number of subunit combinations are favored in the formation of nAChRs, but others can also form. The most common and widely expressed nAChRs within the mammalian brain, including the striatum, contain $\alpha4$ and $\beta2$ subunits.[4] Based on recent studies of whole cell patch clamp recordings in striatal *in vitro* slices, nicotinic agonists increase GABAergic interneuron transmission, which subsequently inhibits medium spiny neurons and cholinergic interneurons.[5] The subtypes of nAChRs involved in this function remain unknown, and, in general, the function and even distribution of individual neuronal nAChR subtypes remain enigmatic.

Five distinct muscarinic acetylcholine receptors (mAChRs) have been identified, which are divided into two subtypes. The M1-like receptor subtype is composed of M1, M3, and M5 receptors, which are all coupled to phosphatidylinositol hydrolysis via Gq proteins. The M2-like receptor subgroup comprises M2 and M4 receptors that are coupled to Gi/Go proteins, which inhibit adenylyl cyclase and thereby reduce cyclic-AMP and inhibit Ca2+ channels. The inhibition of Ca++ channels serves to inhibit neurotransmitter release, and consequently M2 and M4 receptors have both been implicated as autoregulators in various parts of the brain.

Differential expression of the five distinct muscarinic receptors allows for variable effects. M1 and M4 receptor subtypes are abundant within the striatum, where only moderate amounts of M2 and hardly detectable levels of M3 and M5 are present.[6] M1 is expressed in 78% of neurons and is enriched in spiny dendrites and at postsynaptic densities.[7] M2 is expressed in the large cholinergic interneurons presynaptically and possibly acts as an ACh autoreceptor. Based on assessment of M2 and M4 knockout mice, M4 receptor appears to have predominantly an autoinhibitory effect within the striatum, while M2 acts (exerts its autoinhibitory influences) predominantly in the cortex and hippocampus.[8] M4 receptors are frequently colocalized with D_1 on medium spiny projection neurons, mainly in the direct pathway.[9] As M4 receptor activation decreases cAMP, and D_1 receptor activation increases cAMP, these opposing actions may be used in fine tuning modulatory control of the medium spiny neurons.

The cholinergic system of the basal ganglia is intimately related to the dopaminergic system. The previous model of ACh/DA interaction was one of balance, and in the setting of Parkinson's disease (PD), this balance is disrupted. Decreased DA production in the substantia nigra pars compacta (SNc) causes decreased dopamine levels within the striatum. ACh was presumed to remain either unchanged or increased, as early research showed inhibitory effects of DA on striatal ACh output. In this manner, the normal balance between ACh and DA was thrown off, and this imbalance was thought to account for the tremor, rigidity, and akinesia of PD. The current understanding of ACh/DA interaction reveals a much more complex picture, however, and the initial theory is constantly being modified and is in a state of evolution as new data become available. For example, the previous belief that ACh levels were unchanged in PD is now known to be inaccurate—studies have shown that, in PD, there is actually a loss of cholinergic neurons from the pedunculopontine nucleus.[10] Another effect of PD on the cholinergic system is an increase in the muscarinic binding sites in the globus pallidus internus (GPi), possibly representing a compensatory upregulation in response to a decrease in cholinergic activity.[11]

Opposing actions of various dopamine receptors also argue against a simple model of ACh versus DA balance. Inhibition of striatal ACh output via D_2 receptors and excitation of ACh via D_1 receptors has been demonstrated by systemic administration of selective DA compounds to normal animals. Extracellular ACh levels also alter the dopaminergic regulation of striatal ACh. For example, in normal physiologic conditions, D_2 inhibition is equal to or exceeds D_1 excitation of ACh release, but, in cases of increased extracellular ACh, the balance shifts toward D_1 predominance.[12] There is also evidence that nicotinic receptors have a role in PD. Several studies, both epidemiological and animal model-based, have shown that cigarette smoking and nicotine have a protective effect against development of PD.[13,14] However, other studies have shown no effect or a worsening of symptoms in PD with nicotine administration.[15,16] What has been substantiated, however, is that nicotinic stimulation is important in regulating dopamine release, although the specific mechanisms are not yet well understood. To date, nicotinic agonists have been of limited clinical value in the treatment of Parkinson's disease and other movement disorders. One difficulty in the production of therapeutic interventions has been in attempts to activate nAChRs without producing too much desensitization of these receptors. However, as various roles of neuronal nAChRs are being unraveled, particularly in enhancement of the presynaptic release of many neurotransmitters, the functional and anatomic interactions could be utilized in developing novel therapeutic approaches.

Research is continuously expanding our knowledge of the role of acetylcholine in the CNS and, specifically, the intimate interaction between ACh and dopamine in the striatum. As our understanding grows, clinical applications of cholinergic modifying agents will be better defined in patients with Parkinson's Disease.

ANTICHOLINERGIC DRUGS: CHEMICAL STRUCTURE AND PHARMACOKINETICS

Discovery of ACDs' effectiveness against the symptoms of PD is credited to one of the students of Charcot who published his observations with belladonna alkaloids (BAs).[17] For almost 100 years, BAs remained the leading agents for treatment of Parkinson's disease. In a recent review of the subject, Comella and Tanner[18] discussed the herbal nature of BAs and their limiting side effects.

In the 1940s, synthetic ACDs with fewer side effects were introduced for treatment of PD. Among these, trihexiphenidyl (Artane) was the first and most commonly used agent. Other commonly used ACDs include benztropine (Cogentin), biperidin (Akineton), procyclidine (Kemadrin), and ethopropazine (parsidol, Parsitan) (see Figure 52.1). Trihexyphenidyl and Biperiden bind selectively to M1 muscarinic receptors with trihexyphenidyl demonstrating a stronger receptor affinity.[19]

In addition to the aforementioned synthetic ACDs, a number of other drugs (e.g., diphyenhydramine, amitriptyline, and clozapine) used for treatment of emotional and psychiatric complications (depression, agitation, hallucinations, and so on) of PD also have anticholinergic properties.

Table 52.1 shows the dose of four commonly used anticholinergic agents in Parkinson's disease.

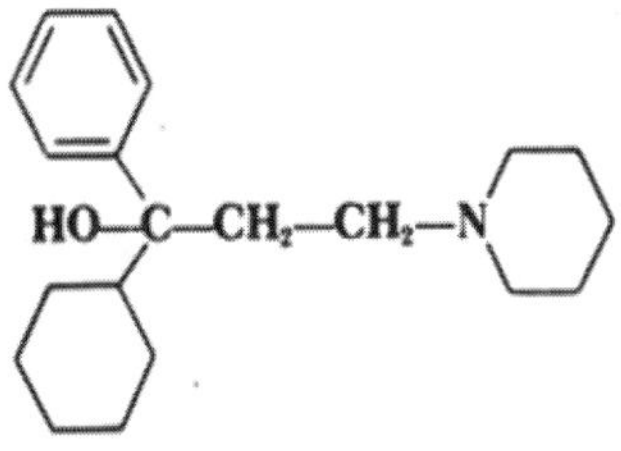

PHARMACOKINETICS

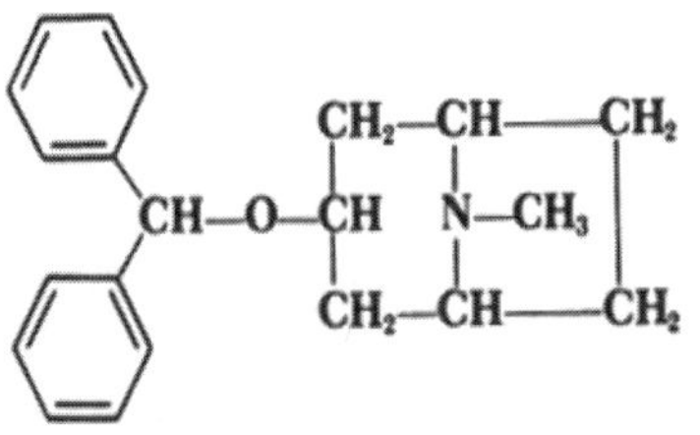

Brocks has recently reviewed the pharmacokinetic properties of anticholinergic agents.[20] As he has noted, surprisingly very little was done in this area during the almost 20 years when synthetic ACDs were the leading drugs for treatment of PD.

In general, anticholinergic drugs are absorbed rapidly through the gut, and their lag time to maximum plasma concentration (T-max) is 1.5 to 2.5 hr. The maximum plasma concentration (C-max) is dose dependent for all these agents with the exception of procyclidine, which, even in modest doses, produces high plasma levels (see Table 52.2). The drug half-life varies among the ACDs. Some earlier values were erroneously short, reflecting the inaccurate assessment technology of the time. For instance, in the case of trihexyphenidyl, the true half-life is approximately 33 hr rather than the earlier assessed value of 3.5 hr. According to Brock, this erroneous early value (most likely reflecting distribution phase rather than true half-life) was instrumental in a previous conclusion

FIGURE 52.1 Chemical structure of the two commonly used dugs for treatment of PD: trihexyphenidyl and benztropine.

TABLE 52.1
Four Anticholinergic Drugs Commonly Used in Parkinson's Disease

Name	Starting dose	Increments	Daily dose
Trihexyphenidyl (Artane) 2 and 5 mg tab/2 and 5 mL	1 mg q.d.	1 mg every 4th d, by 1 mg	2–3 mg, t.i.d.
Benztropine (Cogentin) 0.5, 1,2 mg tab–1 mg/mL injectable	0.5 mg q.d.	1 mg every 4th d, by 0.5–1 mg	2 mg, t.i.d.
Biperidin (Akineton) 2 mg tab–5/mL injectable	1 mg q.d.	1 mg every 4th d, by 0.5–1 mg	2–3 mg, t.i.d.
Ethopropazine (parsidol, Parsitan)	12.5 mg q.d.	12.5mg every 4th d, by 12.5 mg	50 mg t.i.d./q.i.d.

In elderly, slower increments are recommended.

TABLE 52.2
Pharmacokinetic Properties of Four Major Anticholinergic Agents

	C_{max} (ng/mL)	T_{max} (h)	AUC (ng.h/mL)	hl (h)
Trihexyphenidyl (Artane)	7.2	1.3	201	33
Benztropine (Cogentin)	2.5	7	–	–
Biperiden (Akineton)	3.9	1.5	27.2	18.4
Procyclidine (Kemadrin)	116	1.1	2007	12.6

on the unreliability of serum levels of trihexyphenidyl in the treatment of dystonia.[21]

The bioavailability of ACDs ranges from 30 to 70%. The tissue/plasma concentration of these drugs is large. For biperidine, a brain/plasma AUC ratio of 7:12 has been reported with maximal brain concentration reached within 3 to 10 min of IV infusion.[22] All ACDs are intensely lipophilic, a property that explains their high concentration in the brain. In case of trihexyphenidyl and biperidine, the high tissue uptake may be related to the intra-lysosomal uptake of the agents.[23] In animals, antipsychotic anticholinergics are highly bound to serum proteins. In humans, information on ACD protein binding is not available. Metabolism is through N-oxidation, N-dealkylation, and n-demethylation. Urine and bile are the major pathways for elimination of unchanged ACDs. After oral and IV administration of biperidin and procyclidine, serum plasma concentration correlates with degree of salivary flow and pupilary dilatation.[24,25]

In the elderly, due to larger composition of body fat and larger volume of distribution, brain concentration exceeds that of young individuals. This may explain the higher incidence of cerebral side effects in elderly taking anticholinergic medications.

CURRENT ROLE OF ANTICHOLINERGIC DRUGS IN TREATMENT OF PARKINSON'S DISEASE

Despite the widespread use of dopaminergic drugs and the steady development of newer antiparkinsonian agents, anticholinergic drugs still hold a place in treatment of Parkinsonian symptoms. They are useful in alleviation of the resting tremor and rigidity but have little effect on akinesia, bradykinesia, and postural changes. Their role in PD-associated dystonia and as an adjunct to L-dopa in treatment of difficult patients needs to be elucidated.

The exact mechanism(s) by which ACDs relieve some of the symptoms of PD is still unknown and may be multifactorial. The presence of muscaranic and nicotinic receptors on dopaminergic terminals suggests their modulatory effect. Reduced reuptake of dopamine has been shown after benztropine use in experimental studies.[26]

RESTING TREMOR

The original report on synthetic anticholinergic drug trihexyphenidyl claimed efficacy against tremor in 50% of the patients.[27] Postural and senile tremor were found less responsive to trihexyphenidyl. A subsequent open study conducted over a period of 5 years with benztropine in 302 patients yielded the similar success figure of 52% for tremor and rigidity.[28] Although some double blind studies reported no improvement of PD tremor with ACDs,[29] others confirmed the positive experience with the open observations.[30,31]

We have seen significant improvement of parkinsonian tremor with these agents and, like others,[32,33] we use these agents in younger patients in early stages of the disease and when L-dopa fails to improve the tremor. These drugs also have the advantage of being considerably cheaper than dopaminergic agents (monthly expense: Trihexyphenidyl, 6 mg/day, $15 to $17; Benztropine, 2 mg/day, $7 to $10; Sinemet, 25/100 to 600 mg/day, $113 to $149 and higher for newer direct agonists).[34] Like other areas of symptomatic treatment in movement disorders, the goal of treatment is not always total cessation of the abnormal movement but reducing it to a level that is acceptable to the patient.

The rapidly expanding practice of high-frequency, deep brain stimulation (thalamus and subthalamic nucleus) for parkinsonian tremor will undoubtedly influence the use of ACDs in this area. Anticholinergic drugs, however, remain an inexpensive and low-risk mode of treatment for young patients with Parkinsonian tremor.

PARKINSON-ASSOCIATED DYSTONIA

Anticholinergic drugs have been shown to improve many forms of human dystonia. In fact, their efficacy in both early onset and late onset human dystonias suggests that the cholinergic system plays a major role in neurochemistry of dystonia. Their use in late onset dystonias (blepharospasm, torticollis, and others) has been curtailed only recently by the introduction of botulinum toxin A to clinical medicine.

Patients with Parkinson's disease can have dystonic features early in the course of illness (drug naïve dystonia). More often, Parkinson-associated dystonia occurs either as wearing-off (end-dose) or peak-dose dystonia during L-dopa treatment. Wearing-off dystonias often take the form of early-morning foot dystonia, whereas peak-dose dystonias can involve the trunk and proximal limb muscles. Limb and trunk dystonia can also appear during severe motor fluctuations (biphasic dyskinesias).

Poewe et al.[35] investigated the role of anticholinergic drugs in patients with Parkinson-associated dystonia. Over a time span of 2 years, they observed 65 PD patients, 9 with drug naïve and 56 with dystonia secondary to L-dopa therapy. Administration of procyclidine 5 mg i.v. resulted in significant improvement of foot dystonia in eight of nine patients. In six patients, the drug totally abolished the foot dystonia. Oral anticholinergic therapy with trihexyphenidyl (10 to 12 mg/day) had a modest effect in only two of seven patients with drug naïve PD-associated dystonia.

We have seen improvement of PD associated dystonia (mostly wearing-off foot dystonia) with Trihexyphenidyl (6 to 10 mg/day). However, we have also seen patients with the same type of dystonia who did not respond to trihexyphenidyl, even at higher doses. Although foot dystonia may respond to dopaminergic manipulation, the condition can be stubborn and cause the patients significant discomfort. It is, therefore, reasonable to further explore the role of ACDs in PD-associated dystonia with careful prospective studies. Such studies may investigate also the combined effect of ACDs and botulinum toxin A, since the latter has been shown to produce partial relief of the foot dystonia.[36]

THE ROLE OF ACDs AS AN ADJUNCT TO DOPAMINERGIC TREATMENT

Some investigators have noted a synergistic effect between ACDs and L-dopa when the two were used in combination,[37] a view that is supported at least by one primate study.[38] A combination of low-dose ACDs with low-dose L-dopa has been advocated in early PD to reduce or delay the side effects of high-dose L-dopa therapy. One group of investigators reported a combination of trihexyphenidyl and amantidine being as effective in improving motor scores in early PD as L-dopa therapy alone.[39] On the other hand, a recent report claims chronic use of anticholinergic drugs may predispose the patients to dopa dyskinesias,[40] and another report suggests that combined therapy will cause a delay in gastric emptying of L-dopa, resulting in less predictable pattern of absorption.[41]

In the past few years, the introduction of new direct dopa agonists, COMT inhibitors, and subthalamic deep brain stimulation (effective against PD symptoms and dopa-dyskinesias) has reduced the role of anticholinergic drugs as an adjunct drug to L-dopa therapy.

SIDE EFFECTS OF ANTICHOLINERGIC DRUGS

The ACDs exert their function through binding to peripheral and central muscarinic receptors and therefore have both peripheral and central side effects.

Peripheral Side Effects

The peripheral side effects of ACDs result from their inhibitory effect on the parasympathetic system (parasympatholytic). The parasympathetic nervous system increases secretion of saliva and sweat glands and motility of the gut, constricts the pupils, slows the heart rate, and contracts the muscles of accommodation. The spectrum of peripheral effects of anticholinergic drugs is shown in Table 52.3.

TABLE 52.3
Peripheral Side Effects of Anticholinergic Drugs

Pupils	Dilation
Accommodation	Blurring of vision
Salivary glands	Decreased secretion, dryness of the mouth
Sweat glands	Decrease secretion, dry skin
Gut	Decreased motility, constipation
Bladder	Urinary retention
Heart	Increased heart rate
Cardiovascular	Orthostatic hypotension

Among these peripheral effects, dryness of the mouth is the common but rarely requires discontinuation of drug therapy. Drinking a lot of water or using artificial saliva helps. Blurring of vision is the next most common peripheral side effect and is related to the drug action on the muscles of accommodation. Urinary retention or severe constipation may occur but usually is seen in patients with a prior history of prostatic hypertrophy and gut motility disorders; ACDs should not be used in such patients. Patients with cardiac arrhythmias and increased intra-ocular pressure suspect for closed angle glaucoma should be excluded from ACD treatment. An electrocardiogram and a measurement of intra-ocular pressure should be performed in all patients taking anticholinergic drugs, preferably before initiation of the treatment.

In an study of 100 patients treated with high doses of trihexyphenidyl for different movement disorders (not PD),[42] the authors did not find any case of urinary retention, significant gut motility disorder, or increased intraocular pressure during two months of observation. (All patients were carefully screened for complications before initiation and at the completion of the study.) Trihexyphenydyl was started at 1 to 2 mg/day and increased to the tolerated dose or 60 mg/day over a period of a month (average dose 24 mg/day, with 25% of the group exceeding 40 mg/day). The most common side effects were dryness of the mouth, blurring of vision, and forget-fulness. The lack of the aforementioned side effects in this study, however, cannot be applied to the patients with PD, because PD patients often have subclinical autonomic nervous system dysfunction predisposing them to such complications.

Central Side Effects

— Memory loss and intermittent confusion are the most common central side effects of ACDs. The elderly are particularly prone to these side effects. Disorientation and hallucinations are less common but are not infrequent among elderly patients. Patients with a history of cognitive problems are at increased risk for developing these side effects. In one study, careful neuropsychiatric testing has shown a significant decline in recent memory, even with low doses of anticholinergics in nondemented PD patients.[43] In another study, conventional doses of trihexyphenidyl reduced rCBF and rCMR2 by 10 and 15% in the cortex and striatum of cognitively intact patients.[44] Frontal executive functions seem to be significantly affected by ACDs as in form of subcortico-frontal syndrome.[45] Receptor binding properties of ACDs seem to play a role in duration of memory problems with ACDs. In the case of trihexyphenidyl, reversible binding results in dose-dependent and short-duration memory deficit, whereas biperiden causes long-lasting cognitive deficits due to partial reversibility.[46] A recent neuropathological study of the brains of 120 elderly patients confirmed with PD suggested more amyloid plaques in the brains of those who have been chronic anticholinergic users.[47] The number of the plaques, however, has not been at the level seen with Alzheimer's disease. For the aforementioned reasons, the use of ACDs in elderly patients is discouraged.

— Anticholinergic drugs also can cause dyskinesias. These dyskinesias, unlike L-dopa-induced dyskinesias, are orobuccal in distribution, resembling tardive dyskinesias. We have seen a case of severe but reversible akathesia during anticholinergic therapy with trihexyphenidyl.

THE COCHRANE REVIEW (2003)

The authors reviewed previous clinical studies to determine the efficacy and tolerability of ACDs in Parkinson's disease.[48] The selection criteria consisted of randomized controlled trials versus placebo or no treatment in *de novo* or advanced PD. Both monotherapy and add-on trials were included in the review.

The initial search included 14 potentially eligible studies, but 5 had to be excluded (unmarketed drugs, poor methodology). The remaining 9 studies encompassed 221 patients. Different drugs were used in these studies. All of the studies but one reported significant improvement from the baseline among users of anticholinergic drug in at least one outcome measure. The difference between a placebo and the drug was statistically significant in four studies. Six of the nine studies reported neuropsychiatric or cognitive adverse effects. The reviewers concluded that monotherapy or adjunct therapy with anticholinergic drugs is superior to a placebo in improving motor function in Parkinson's disease. Results regarding a potentially better effect of anticholinergic drugs on tremor compared to other outcome measures were conflicting. The data were insufficient to allow a comparison in efficacy or tolerability among individual anticholinergic drugs.

REFERENCES

1. Zhou, F. M., Liang, Y., Dani, J. A., Endogenous nicotinic cholinergic activity regulates dopamine release in the striatum, *Nat. Neurosci.*, 4:1224–1229, 2001.

2. Wilson, C. J., Chang, H. T., Kitai, S. T., Firing patterns and synaptic potentials of identified giant aspiny interneurons in the rat neostriatum, *J. Neurosci.*, 10:508–519, 1990.

3. Zhou, F. M., Wilson, C., and Dani, J. A., Muscarinic and nicotinic cholinergic mechanisms in the mesostriatal dopamine systems, *Neuroscientist,* 9:23–36, 2003.

4. Wada, E., Wada, K., Boulter, J., Deneris, E., Heinemann, S., Patrick, J., and Swanson, L. W., Distribution of alpha 2, alpha 3, alpha 4, and beta 2 neuronal nicotinic receptor subunit mRNAs in the central nervous system: A hybridization histochemical study in the rat, *J. Comp. Neurol.*, 284:314–335, 1989.

5. Koos, T., Tepper, J. M., Dual cholinergic control of fast-spiking interneurons in the neostriatum, *J. Neurosci.*, 22:529–535, 2002.

6. Levey, A. I., Immunological localization of m1-m5 muscarinic acetylcholine receptors in peripheral tissues and brain, *Life Sci.*, 52:441–448, 1993.

7. Hersch, S. M., Gutekunst, C. A., Rees, H. D., Heilman, C. J., and Levey, A. I., Distribution of m1-m4 muscarinic receptor proteins in the rat striatum: Light and electron microscopic immunocytochemistry using subtype-specific antibodies, *J. Neurosci.*, 14:3351–3363, 1994.

8. Zhang, W., Basile, A. S., Gomeza, J., Volpicelli, L. A., Levey, A. I., and Wess, J., Characterization of central inhibitory muscarinic autoreceptors by the use of muscarinic acetylcholine receptor knock-out mice, *J. Neurosci.*, 22:1709–1717, 2002.

9. Ince, E., Ciliax, B. J., and Levey, A. I., Differential expression of D_1 and D_2 dopamine and m4 muscarinic acetylcholine receptor proteins in identified striatonigral neurons, *Synapse,* 27:357–366, 1997.

10. Hirsch, E. C., Graybiel, A. M., Duyckaerts, C., and Javoy-Agid, F., Neuronal loss in the pedunculopontine tegmental nucleus in Parkinson disease and in progressive supranuclear palsy, *Proc. Natl. Acad. Sci. USA,* 84:5976–5980, 1987.

11. Griffiths, P. D., Sambrook, M. A., Perry, R., and Crossman, A. R., Changes in benzodiazepine and acetylcholine receptors in the globus pallidus in Parkinson's Disease, *J. Neurol. Sci.*, 100:131–136, 1990.

12. DeBoer, P., and Aberrombie, E. D., Physiological release of striatal acetylcholine *in vivo:* Modulation by D_1 and D_2 dopamine receptor subtypes, *J. Pharmacol. Exp. Ther.*, 277:775–783, 1996.

13. Hernan, M. A., Takkouche, B., Caamano-Isorna, F., and Gestal-Otero, J. J., A meta-analysis of coffee drinking, cigarette smoking, and the risk of Parkinson's disease, *Ann. Neurol.*, 52:276–284, 2002.

14. Parain, K., Hapdey, C., Rousselet, E., Marchand, V., Dumery, B., and Hirsch, E. C., Cigarette smoke and nicotine protect dopaminergic neurons against the 1-methyl-4-phenyl-1,2,3,6-tetrahydropyridine Parkinsonian toxin, *Brain Res.*, 984:224–232, 2003.

15. Vieregge, A., Sieberer, M., Jacobs, H., Hagenah, J. M., and Vieregge, P., Transdermal nicotine in PD: a randomized, double-blind, placebo-controlled study, *Neurology*, 57:1032–1035, 2001.

16. Ebersbach, G., Stock, M., Muller, J., Wenning, G., Wissel, J., and Poewe, W., Worsening of motor performance in patients with Parkinson's disease following transdermal nicotine administration, *Mov. Disord.*, 14:1011–1013, 1999.

17. Ordenstein, L., Sur la Paralysie agitante et sclerose en plaque generalisee, Paris, Marinet, p. 32, 1667.

18. Comella, C. L. and Tanner, C. M., In: *Therapy of Parkinson's disease*, (2nd ed.), Koller and Paulson (Eds.), Marcel Dekker, 109–122, 1995.

19. Katayama, S., Ishizaki, F., Yamamura, Y., Khoriyama, T., Kito, S., Effects of anticholinergic antiparkinsonian drugs on binding of muscarinic receptor subtypes in rat brain, *Res. Commun. Chem. Pathol. Pharmacol.*, 69:261–70, 1990.

20. Brocks, D. R., Anticholinergic drugs used in Parkinson's disease: An overlooked class of drugs from a pharmacokinetic perspective, *J. Pharm. Pharmacol. Sci.*, 2:39–46, a review, 1999.

21. Burke, R. E., Fahn, S., Serum trihexyphenidyl levels in the treatment of torsion dystonia, *Neurology*, 35:1066–9, 1985.

22. Yokogawa, K., Nakashima, E., Ishizaki, J., Hasegawa, M., Kido, H., Ichimura, F., Brain regional pharmacokinetics of biperiden in rats, *Biopharm. Drug Dispos.*, 13:131–40, 1992.

23. Ishizaki, J., Yokogawa, K., Nakashima, E., Ohkuma, S., Ichimura, F., Influence of ammonium chloride on the tissue distribution of anticholinergic drugs in rats, *J. Pharm. Pharmacol.*, 50:761–6, 1998.

24. Whiteman, P. D., Fowle, A. S., Hamilton, M. J., Peck, A. W., Bye, A., Dean, K., Webster, A., Pharmacokinetics and Pharmacodynamics of procyclidine in man, *Eur. J. Clin. Pharmacol.*, 28:73–8, 1985.

25. Hollmann, M., Muller-Peltzer, H., Greger, G., Brode, E., Perucca, E., Grimaldi, R., Crema, A., Pharmacokinetic-dynamic study on different oral biperiden formulations in volunteers, *Pharmacopsychiatry*, 20:72–7, 1987.

26. Olanow, C., Koller, W. C., An algorithm for management of Parkinson's disease, *Neurology*, (Suppl.), 50:S14, 1988.

27. Corbin, K. B., Trihexyphenidyl: Evaluation of new agents in the treatment of Parkinsonism, *JAMA*, 141:377–382, 1949.

28. Doshay, L. J., Five-year study of benztropine (cogentin) methanesulfate, *JAMA*, 162: 1031–1034, 1956.

29. Norris, J. W., Vas, C. J., Mehixene hydrochloride and parkinsonian tremor, *Acta Neurol. Scand.*, 43:535–8, 1967.

30. Strang, R. R., Orphenadrine ("Disipal") in the treatment of Parkinsonism: a two-year study of 150 patients, *Med. J. Aust.*, 11:448–50, 1965.

31. Burns, D., DeJong, R. D., Solis-quiroga, O. H., Effects of trihexyphenidyl hydrochloride(artane) in Parkinson's disease, *Neurology*, 14: 134–23, 1964.

32. Koller, W. C., Pharmacologic treatment of parkinsonian tremor, *Arch Neurol.*, Feb., 43:126–7, 1986.

33. Obeso, J. A., Martinez-Lage, J. M., Anticholinergics and Amantiddine, In: Keller, W., Ed., *Handbook of Parkinson's Disease*, Marcel Dekker, New York, 309–316, 1987.

34. Ahlskog, E. J., Treatment options for mild and moderate Parkinson's disease, In: *Parkinson's Disease*, Course number 236, American Academy of Neurology, p. 35, 1996.

35. Poewe, W. H., Lees, A. J., Stern, G. M., Dystonia in Parkinson's disease: clinical and pharmacological features, *Ann. Neurol.*, 23:73–8, 1988.

36. Jankovic, J., Tintner, R., Dystonia and Parkinsonism. *Parkinsonism Relat. Disord.*, 8:109–21, 2001.

37. Tourtellotte, W. W., Potvin, A. R., Syndulko, K., Hirsch, S. B., Gilden, E. R., Potvin, J. H., Hansch, E. C., Parkinson's disease: Cogentin with Sinemet, a better response, *Prog. Neuropsychopharmacol. Biol. Psychiatry*, 6:51–5, 1982.

38. Domino, E. F., Ni, L., Trihexyphenidyl potentiation of L-DOPA: reduced effectiveness three years later in MPTP-induced chronic hemiparkinsonian monkeys, *Exp. Neurol.*, 152(2):238–42, August 1998.

39. Parkes, J. D., Baxter, R. C., Marsden, C. D., Rees, J. E., Comparative trial of benzhexol, amantadine, and levodopa in the treatment of Parkinson's disease, *J. Neurol. Neurosurg. Psychiatry*, 37:422–6, 1974.

40. Miller, R., Chouinard, G., Loss of striatal cholinergic neurons as a basis for tardive and L-dopa-induced dyskinesias, neuroleptic-induced supersensitivity psychosis and refractory schizophrenia, *Biol. Psychiatry*, review, 15, 34(10):713–38, November, 1993.

41. Roberts, J., Waller, D. G., von Renwick, A. G., O'Shea, N., Macklin, B. S., Bulling, M., The effects of co-administration of benzhexol on the peripheral pharmacokinetics of oral levodopa in young volunteers, *Br. J. Clin. Pharmacol.*, 41:331–337, 1996.

42. Jabbari, B., Scherokman, B., Gunderson, C. H., Rosenberg, M. L., Miller, J., Treatment of movement disorders with trihexyphenidyl, *Mov. Disord.*, 4(3):202–12, 1989.

43. Koller, W. C., Disturbance of recent memory function in parkinsonian patients on anticholinergic therapy, *Cortex,* 20(2):307–11, June, 1984.

44. Takahashi, S., Tohgi, H., Yonezawa, H., Obara, S., Yamazaki, E., The effect of trihexyphenidyl, an anticholinergic agent, on regional cerebral blood flow and oxygen metabolism in patients with Parkinson's disease, *J. Neurol. Sci.,* 167:56–61, 1999.

45. Bedard, M. A., Pillon, B., Dubois, B., Duchesne, N., Masson, H., Agid, Y., Acute and long-term administration of anticholinergics in Parkinson's disease: specific effects on the subcortico-frontal syndrome, *Brain Cogn.,* 40:289–313, 1999.

46. Kimura, Y., Ohue, M., Kitaura, T., Kihira, K., Amnesic effects of the anticholinergic drugs, trihexyphenidyl and biperiden: differences in binding properties to the brain muscarinic receptor, *Brain Res.,* 834:6–12, 1999.

47. Perry, E. K., Kilford, L., Lees, A. J., Burn, D. J., Perry, R. H., Increased Alzheimer pathology in Parkinson's disease related to antimuscarinic drugs, *Ann. Neurol.,* 54(2):235–238, August, 2003.

48. Katzenschlager, R., Sampaio, C., Costa, J., Lees, A., Anticholinergics for symptomatic management of Parkinson's disease, *Cochrane Review,* In: Cochrane Library, Issue 3, Oxford: update Software, 2003.

53 Amantadine

Kelly E. Lyons and Rajesh Pahwa
University of Kansas Medical Center

CONTENTS

INTRODUCTION

Amantadine (Symmetrel®) was first introduced as an antiviral agent for the treatment of influenza. It was recognized as a potential treatment for Parkinson's disease (PD) in 1968 when a woman treated for influenza noticed improvement in her parkinsonian symptoms and subsequent worsening of her PD symptoms when amantadine was discontinued.[1] In an open-label study, it was subsequently shown that two-thirds of 163 PD patients had some improvement in rigidity, bradykinesia, and tremor with amantadine.[1] Since this initial report, multiple studies have been conducted to examine the antiparkinsonian effects of amantadine as monotherapy in early PD, and as combination therapy in early and advanced PD. Currently, amantadine is the only antidyskinetic agent available for the treatment of levodopa induced dyskinsesia. The exact mechanism of action of amantadine is unknown, although it appears to have both dopaminergic and nondopaminergic properties. This chapter reviews the possible mechanisms of action, clinical studies, and safety of amantadine in PD.

PHARMACOKINETICS

Amantadine hydrochloride, 1-adamantanamine hydrochloride, is the salt of a primary amine. In the United States, it is available in 100-mg tablets or as a syrup containing 50 mg of amantadine per 5 mL. It is also available as an intravenous formulation (amantadine sulfate); however, this preparation is not available in the United States. Peak blood concentrations occur after 1 to 4 hr, and it is poorly metabolized, with 90% being excreted unchanged in the urine. The half-life of amantadine increases with age, with young healthy patients having a half-life of approximately 15 hr and elderly adults having a half-life of approximately 29 hr.[2,3] Amantadine is typically initiated at 100 mg/day and increased to a maximum recommended dose of 400 mg/day in divided doses in patients with normal renal function, with the most common therapeutic dose in PD being 100 mg twice a day.

MECHANISM OF ACTION

The mechanism of action of amantadine is currently unclear and is thought to include both dopaminergic and nondopaminergic systems. Presynaptically, amantadine is thought to increase dopamine release through an amphetamine-like action[4] and decrease the reuptake of dopamine.[5] Postsynaptically, amantadine acts as a direct dopamine agonist producing a high affinity state in dopamine D_2 receptors.[6] The combined presynaptic and postsynaptic dopamine receptor action of amantadine has been suggested to cause simultaneous interference with dopamine release and reuptake activity leading to antiparkinsonian effects.[6] Amantadine has also been shown to be a glutamatergic N-methyl-D-aspartate (NMDA) receptor antagonist.[7] The NMDA antagonism of amantadine allows the noncompetitive inhibition of acetylcholine release that is mediated by NMDA receptors[8] and may also be responsible for the indirect dopaminergic effects of amantadine by enhancing dopa decarboxylase activity and dopamine synthesis.[9] The glutamatergic inhibitory activity of aman-tadine suggests a potential neuroprotective effect of aman-tadine in PD.[7] Inhibition of NMDA receptors reduces glutamate excitotoxicity and consequent calcium influx, which in turn may prevent neuronal cell death.[10] It has also been suggested that oversensitized NMDA receptors play a critical role in the development of motor compli-cations in PD by altering basal ganglia responses to glutamate leading to irregular motor output and conse-quent motor fluctuations and/or dyskinesia.[11] Several ani-mal studies have demonstrated that glutamate antagonists can reduce motor complications in PD without interfering with the benefit of dopaminergic medications.[12,13]

CLINICAL STUDIES

After the initial serendipitous finding that amantadine may have beneficial effects in PD, Schwab and colleagues[1] conducted an open-label study of amantadine (200 mg/day) in 163 PD patients and observed improve-ments in rigidity and bradykinesia in two-thirds of the group, and some improvement in tremor as well. There was a decline in benefit after four to eight weeks; however, at eight months, patients were still improved as compared to baseline. In patients for whom amantadine was discon-tinued due to loss of effect, a worsening of symptoms occurred after discontinuation. Schwab et al.[14] subse-quently reported the results of 351 patients on amantadine for at least 2 months. They demonstrated that maximum benefit occurred within two to three weeks of treatment with amantadine, and 64% of the sample had improved by 2 months. Their results indicated that amantadine was beneficial both as monotherapy and as an adjunct to levodopa. Following the publication of these initial results, multiple studies of the antiparkinsonian effects of aman-tadine were conducted in early PD (Table 53.1), advanced PD as an adjunct to levodopa (Table 53.2), and advanced PD as a treatment for motor complications, primarily dys-kinesia (Table 53.2). This discussion includes only studies of idiopathic PD and is restricted to primarily double-blind, placebo-controlled studies.

AMANTADINE IN EARLY PD

In the 1970s, multiple studies were conducted to examine the efficacy of amantadine as monotherapy for PD[15–18] and as an adjunct treatment to anticholinergic medications[19–25] (Table 53.1). The majority of these studies reported a mild to moderate improvement in functional disability associ-ated with bradykinesia and rigidity and variable improve-ment in tremor with amantadine as both initial monother-apy and in combination with anticholinergics as compared to placebo. These studies were conducted in the 1970s and therefore had limited and variable study designs and methodology. They typically included a relatively small

number of patients, and often relied on timed tasks, patient self-report, and physician impressions to determine effi-cacy. In general, these studies were of short duration, lasting only a few weeks or months, making the long-term effects of amantadine difficult to evaluate. In a one-year study, Parkes and colleagues followed 26 PD patients on amantadine monotherapy in an open-label fashion.[15] At three months, total disability was reduced by 17%, and these effects were maintained for the duration of the study.

Although there are currently no prospective studies examining the potential of amantadine to slow disease progression, a retrospective study was reported. Uitti and colleagues[26] attempted to identify independent predictors of survival in 836 parkinsonian patients (92% PD) seen during a 22-year period. Amantadine was demonstrated to be an independent predictor of increased survival. There were 250 patients taking amantadine (200 mg/day) for an average of 37 months and 586 patients not taking amanta-dine. The groups were comparable in terms of age, gen-der, presence of dementia, type of parkinsonism, and Hoehn and Yahr staging. These results suggest that aman-tadine may have neuroprotective properties; however, a well controlled prospective study is necessary to confirm these results.

AMANTADINE IN ADVANCED PD

Several studies have examined the effects of amantadine in combination with levodopa therapy, with variable results[27–29] (Table 53.2). Millac et al.[27] reported no differ-ences in response when levodopa was added to amanta-dine therapy versus placebo. In addition, prior use of amantadine did not reduce the necessary levodopa dosage as compared to placebo. Other studies have shown that the addition of amantadine to stable patients on levodopa therapy provided a significant improvement in PD symp-toms of up to 80% as compared to placebo.[28,29]

Animal studies of amantadine have suggested that its NMDA antagonistic properties may indirectly lead to a reduction in motor fluctuations and dyskinesias.[11–13] Sev-eral studies have examined the efficacy of amantadine in advanced PD patients with levodopa-induced motor com-plications, primarily dyskinesia[30–34] (Table 53.2). Shannon and colleagues[30] reported a reduction in the severity of motor complications in 55% of patients receiving amanta-dine for at least two months. Improvement was sustained for an average of six months. Adler et al.[35] reported four PD patients with benefit from amantadine for up to two years. These patients had a reduction in predictable wear-ing off as well as a reduction in dyskinesia and dystonia. Their report also suggested that amantadine may be effica-cious in the treatment of the motor complications of advanced disease, even if it was not demonstrated to be effective in early disease. Several double-blind, placebo-controlled studies have examined the antidyskinetic

TABLE 53.1
Selected Studies of Amantadine as Monotherapy or in Combination with Nondopamineric Medications in Early Parkinson's Disease

Author, Yr.	Design	n	Duration	Drugs	Outcome	Adverse Events
Parkes et al., 1971[14]	Open label	26	1 year	Amantadine 200–600 mg (median 200mg/day)	Reduction of total disability of 17.3% at 3 months which was sustained at 1 year	Dry mouth, constipation, difficulty focusing
Mawdsley, et al., 1972[15]	Double-blind Crossover	42	2 weeks	Amantadine and Placebo	Significantly greater proportion improved with amantadine (76%) compared to placebo (43%)	Drowsiness, nausea, dry mouth, bad dreams
Parkes et al., 1974[16]	Double-blind Crossover	17	4 weeks	Amantadine (200 mg/day) Benzhexol (8 mg/day)	Amantadine monotherapy produced a 15% reduction in functional disability	Dry mouth, confusion, livedo reticularis
Fahn et al., 1975[17]	Randomized Crossover Placebo-control	23	2 weeks	Amantadine (200mg/day) and placebo	70% of patients were improved with amantadine compared to placebo	Insomnia, anorexia, dizziness, depression, light headedness, nervousness
Barbeau et al., 1971[18]	Double-blind Crossover Placebo-control	54	4 weeks	Current anticholinergics with add on amantadine and placebo	Improvement in functional ability was significantly greater with amantadine; 48% had moderate to good improvement	Mild, no serious adverse events
Jorgensen et al., 1971[19]	Multi-center Double-blind Crossover	149	3 weeks	Current anticholinergics with add on amantadine and placebo	Improvement in 56% of patients which was moderate to marked in 32%; improvement reported in tremor, rigidity and bradykinesia	Mild, no serious adverse events
Silver et al., 1971[20]	Double-blind	50	20 weeks	Current anticholinergics with add on amantadine and placebo	Significant improvement in PD symptoms peaking at 2–3 months and tapering thereafter	Dizziness, nausea, confusion, hallucinations, weakness, ataxia
Rinne et al., 1972[21]	Double-blind Placebo-control	38	4 weeks	Current anticholinergics with add on amantadine and placebo	Total disability scores significantly improved with amantadine compared to placebo; 60% had mild to moderate improvements	Dizziness, sweating, anxiety, insomnia
Walker et al., 1972[22]	Double-blind Placebo-control Crossover	42	3 weeks	Anticholinergics discontinued in 36/42 and maintained in 6; add on amantadine and placebo	64% had improvement with amantadine compared to 21% with placebo	Mild, no serious adverse events
Bauer et al., 1974[23]	Double-blind Placebo-control Crossover	48	3 weeks	Current anticholinergics with add on amantadine and placebo	10% reported improvement with amantadine compared to placebo; no changes in rigidity or tremor	Mild, no serious adverse events
Merry et al., 1974[24]	Double-blind Placebo-control	29	20 weeks	Current anticholinergics with add on amantadine or placebo	Seriously affected patients had a 47% improvement with amantadine compared to 8% with placebo	No serious adverse events

TABLE 53.2
Selected Studies of Amantadine as Adjunct Therapy to Levodopa in Advanced Parkinson's Disease

Author, Year	Design	n	Duration	Drugs	Outcome	Adverse Events
Millac et al., 1970[27]	Double-blind Placebo-control	32	4 months	Amantadine or placebo for 3 weeks followed by addition of levodopa	No differences between amantadine and placebo; no difference in optimal levodopa dose between amantadine and placebo	No serious adverse events
Fehling, 1973[28]	Double-blind Crossover	21	1 month	Current levodopa with add on amantadine and placebo	Amantadine significantly more efficacious than placebo	Dry mouth
Savery, 1977[29]	Double-blind Crossover Placebo-control	42	9 weeks	Current optimized levodopa with add on amantadine and placebo	Amantadine provided significant improvement in PD symptoms compared to baseline (90%) and placebo (80%)	Nervousness, nausea, confusion, livedo reticularis and mild blurred vision
Shannon et al., 1987[30]	Open label	20	3 months	Current optimized levodopa/other antiparkinsonian medication and add on amantadine (fluctuators)	Moderate improvement in motor fluctuations in 55% at 2 months and 65% at 3 months; 30% improvement in disability	Confusion, dizziness, worsening of chorea and foot dystonia
Verhagen et al., 1998[31]	Double-blind Crossover Placebo-control	18	3 weeks	Levodopa induced dyskinesias with add on of amantadine and placebo	Duration of off time significantly decreased with amantadine compared to placebo	No serious adverse events
Verhagen et al., 1998[31]	IV levodopa (end of each arm above)	14	7 hours	Levodopa infusion after 3 weeks of either amantadine or placebo	Amantadine reduced dyskinesia severity by 60% compared to placebo	Confusion, hallucinations, nausea, palpitations (pre-existing)
Metman et al., 1999[32]	IV levodopa Double-blind Placebo-control	13	1 year	Levodopa infusion after 1 year of amantadine; amantadine discontinued and amantadine or placebo given	Amantadine reduced dyskinesia severity by 56% after one year compared to baseline placebo group	No adverse events reported
Luginger et al., 2000[33]	Double-blind Crossover Placebo-control	10	2 weeks	Levodopa induced dyskinesia with add on amantadine or placebo; Oral levodopa challenge	No differences in on or off time per patient diaries with amantadine or placebo; Dyskinesia following L-dopa challenge reduced by 52% in amantadine group with no reduction in placebo group	Foot edema
Snow et al., 2000[34]	Double-blind Placebo-control Crossover	24	3 weeks	Levodopa induced dyskinesia with add on amantadine or placebo followed by levodopa challenge	Significant 24% reduction in dyskinesia after levodopa challenge in amantadine vs. placebo group	No adverse events reported

effects of amantadine by means of an acute levodopa challenge in which a large dose of levodopa was given to induce dyskinesia.[31–34] In a double-blind, placebo-controlled, crossover study, Verhagen and colleagues[31] examined the addition of a maximum of 400 mg/day of amantadine or placebo in 18 PD patients with levodopa-induced dyskinesia. At the end of three weeks, duration of off time was significantly reduced with amantadine compared to placebo. At the end of each treatment arm, 14 patients received a 7-hr intravenous infusion of levodopa. Patients had a 60% reduction in dyskinesia after amantadine therapy compared to placebo. Thirteen patients remained on amantadine therapy for one year. In a subsequent study of these 13 patients,[32] after one year of amantadine therapy, amantadine was discontinued, and patients received either amantadine or placebo for 10 days, followed by an intravenous infusion of levodopa. The results indicated that amantadine continued to reduce dyskinesias by 56% at one year compared to the initial baseline. These results were confirmed in two similar double-blind, placebo-controlled, crossover studies involving an oral levodopa challenge. In one study,[33] after 2 weeks of treatment with 300 mg/day of amantadine and again after treatment with placebo, 10 PD patients with levodopa-induced dyskinesias received an oral levodopa challenge in which antiparkinsonian medication was withheld overnight (12 hr) and either levodopa 25/100 or 50/200, depending on their usual dose, was administered at 8 A.M. in the clinic. After levodopa administration, dyskinesia was reduced by 52% with amantadine compared to placebo. In a second study, after 3 weeks of treatment with 200 mg/day of amantadine and again after treatment with placebo, a levodopa challenge (1.5 times the typical levodopa dose) was performed in 24 PD patients, and a 24% reduction in dyskinesia was observed with amantadine as compared to placebo.

The effect of intravenous amantadine on levodopa-induced dyskinesia has also been examined.[36] Nine PD patients with motor fluctuations and severe peak dose dyskinesia received their typical dose of levodopa followed by an intravenous infusion of either amantadine (200 mg) or placebo on two separate days. They were examined for three hours following the infusion while taking their typical antiparkinsonian medications. The results indicated that intravenous amantadine improved levodopa-induced dyskinesia by 50% compared to placebo and had no effect on the benefit received from typical levodopa therapy. These results, along with those from the levodopa challenge studies, provide support for the antidyskinetic properties of amantadine.

SIDE EFFECTS

Side effects of amantadine include nausea, dizziness, insomnia, dry mouth, constipation, drowsiness, irregular dreams, anorexia, depression, nervousness, confusion, hallucinations, and pedal edema.[15–18,21,22,29,30,31] These side effects are typically mild and resolve with continued use of the drug. Livedo reticularis, a reddish skin discoloration of the legs, has also been reported with amantadine.[37] Symptoms typically resolve in two to three weeks after drug discontinuation. Since amantadine is excreted unchanged in the urine, it should be used with caution in patients with abnormal renal function or reduced creatinine clearance.[2,3]

CONCLUSION

The antiparkinsonian effects of amantadine were first recognized in the late 1960s. Since that time, multiple studies have confirmed its mild to moderate benefit in early PD as well as its benefits as an adjunct to levodopa in advanced PD. More recently, the antidyskinetic properties of oral and intravenous amantadine have been demonstrated in several studies. The antiparkinsonian and antidyskinetic effects of amantadine have been reported to be sustained for up to two years. Although the mechanism of action of amantadine is not completely clear, it is thought to interact with the dopaminergic system, which may lead to its antiparkinsonian effects, as well as the glutaminergic system, which may account for its antidyskinetic effects and offers some suggestion of potential neuroprotective properties. Further research is necessary to determine if amantadine is neuroprotective in PD.

REFERENCES

1. Schwab, R. S., England, A. C., Poskanzer, D. C., et al., Amantadine in the treatment of Parkinson's disease, *JAMA*, 208:1168–1170, 1969.
2. Aoki, F. Y., Sitar, D. S., Clinical pharmacokinetics of amantadine hydrochloride, *Clin. Pharmacokinet.*, 14: 5135–5151, 1988.
3. Bleidner, W. E., Harmon, J. B., Hewes, W. E., et al., Absorption, distribution and excretion of amantadine hydrochloride, *J. Pharmacol. Exp. Ther.*, 150:484–490, 1965.
4. Stromberg, U., Svensson, T. H., Further studies on the mode of action of amantadine, *Acta Pharmacol. Toxicol.*, 30:161–171, 1971.
5. Von Voigtlander, P. F., Moore, K. E., Dopamine: release from the brain *in vivo* by amantadine, *Science*, 174:408–410, 1971.
6. Gianutsos, G., Chute, S., Dunn, J. P., Pharmacological changes in dopaminergic systems induced by long-term administration of amantadine, *Eur. J. Pharmacol.*, 110:357–361, 1985.
7. Kornhuber, J., Bormann, J., Hubers, M., et al., Effects of the 1-amino-adamantanes at the MK-801-binding site of the NMDA-receptor-gated ion channel: a human postmortem brain study, *Eur. J. Pharmacol. Mol. Pharmacol. Sect.*, 206:297–300, 1991.

8. Stoof, J. C., Booij, J., Drukarch, B., et al., The anti-parkinsonian drug amantadine inhibits the N-methyl-D-aspartic acid-evoked release of acetylcholine from rat neostriatum in a non-competitive way, *Eur. J. Pharmacol.,* 213:439–443, 1992.

9. Deep, P., Dagher, A., Sadikot, A., et al., Stimulation of dopa decarboxylase activity in striatum of healthy human brain secondary to NMDA receptor antagonism with a low dose of amantadine, *Synapse,* 34:313–318, 1999.

10. Blandini, F., Porter, R. H., Greenamyre, J. T., Glutamate and Parkinson's disease, *Mol. Neurobiol.,* 12:73–94, 1996.

11. Chase, T. N., Oh, J. D., Striatal dopamine- and glutamate-mediated dysregulation in experimental parkinsonism, *Trends Neurosci.,* Suppl., 23:S86–S91, 2000.

12. Papa, S. M., Boldry, R. C., Engber, T. M., et al., Reversal of levodopa-induced motor fluctuations in experimental parkinsonism by NMDA receptor blockade, *Brain Res.,* 701:13–18, 1995.

13. Blanchet, P. J., Konitsiotis, S., Chase, T. N., Amantadine reduces levodopa-induced dyskinesias in parkinsonian monkeys, *Mov. Disord.,* 13:798–802, 1998.

14. Parkes, J. D., Baxter, R. C., Curzon, G., et al., Treatment of Parkinson's disease with amantadine and levodopa. A one-year study, *Lancet,* 1:1083–1086, 1971.

15. Mawdsley, C., Williams, I. R., Pullar, I. A., et al., Treatment of parkinsonism by amantadine and levodopa, *Clin. Pharmacol. Ther.,* 13:575–583, 1972.

16. Parkes, J. D., Baxter, R. C., Marsden, C. D., Rees, J. E., Comparative trial of benzhexol, amantadine and levodopa in the treatment of Parkinson's disease, *J. Neurol. Neurosurg. Psychiatry,* 37;422–426, 1974.

17. Fahn, S., Isgreen, W. P., Long-term evaluation of amantadine and levodopa combination in parkinsonism by double-blind crossover analyses, *Neurology,* 25:695–700, 1975.

18. Barbeau, A., Mars, H., Botez, M. I., et al., Amantadine-HCl (Symmetrel) in the management of Parkinson's disease: a double-blind, crossover study, *Can. Med. Assoc. J.,* 105:42–47, 1971.

19. Jorgensen, P. B., Bergin, J. D., Haas, L., et al., Controlled trial of amantadine hydrochloride in Parkinson's disease, *N. Z. Med. J.,* 73:263–267, 1971.

20. Silver, D. E., Sahs, A. L., Double blind study using amantadine hydrochloride in the therapy of Parkinson's disease, *Trans. Am. Neurol. Assoc.,* 96:307–308, 1971.

21. Rinne, U. K., Sonninen, V., Siirtola, T., Treatment of Parkinson's disease with amantadine and L-dopa, *Eur. Neurol.,* 7:228–240, 1972.

22. Walker, J. E., Albers, J. W., Tourtellotte, W. W., et al., A qualitative and quantitative evaluation of amantadine in the treatment of Parkinson's disease, *J. Chronic. Dis.,* 25:149–182, 1972.

23. Bauer, R. B., McHenry, J. T., Comparison of amantadine, placebo and levodopa in Parkinson's disease, *Neurology,* 24:715–720, 1974.

24. Merry, R. T. G., Galbraith, A. W., A double-blind study of Symmetrel (amantadine hydrochloride) in Parkinson's disease, *J. Int. Med. Res.,* 2:137–141, 1974.

25. Schwab, R. S., Poskanzer, D. C., England, A. C., et al., Amantadine in Parkinson's disease: Review of more than two years experience, *JAMA,* 222:792–795, 1972.

26. Uitti, R. J., Rajput, A. H., Ahlskog, J. E., et al., Amantadine treatment is an independent predictor of improved survival in Parkinson's disease, *Neurology,* 50:1323–1326, 1998.

27. Millac, P., Hasan, I., Espir, M. L., Slyfield, D. G., Treatment of Parkinsonism with L-dopa and amantadine, *Lancet,* 2:720, 1970.

28. Fehling, C., The effect of adding amantadine to optimum L-dopa dosage in Parkinson's syndrome, *Acta Neurol. Scand.,* 49:245–251, 1973.

29. Savery, F., Amantadine and a fixed combination of levodopa and carbidopa in the treatment of Parkinson's disease, *Dis. Nerv. Syst.,* 38:605–608, 1977.

30. Shannon, K. M., Goetz, C. G., Carroll, V. S., et al., Amantadine and motor fluctuations in chronic Parkinson's disease, *Clin. Neuropharmacol.,* 10:522–526, 1987.

31. Verhagen, Metman L., Del Dotto, P., van den Munckhof, P., et al., Amantadine as treatment for dyskinesias and motor fluctuations in Parkinson's disease, *Neurology,* 50:1323–1326, 1998.

32. Metman, L. V., Del Dotto, P., LePoole, K., et al., Amantadine for levodopa-induced dyskinesias. A 1-year follow-up study, *Arch Neurol.,* 56:1383–1386, 1999.

33. Luginger, E., Wenning, G. K., Bosch, S., Poewe, W., Beneficial effects of amantadine on L-dopa-induced dyskinesias in Parkinson's disease, *Mov. Disord.,* 15:873–878, 2000.

34. Snow, B. J., Macdonald, L., McAuley, D., Wallis, W., The effect of amantadine on levodopa-induced dyskinesias in Parkinson's disease: a double-blind, placebo-controlled study, *Clin. Neuropharmacol.,* 23:82–85, 2000.

35. Adler, C. H., Stern, M. B., Vernon, G., et al., Amantadine in advanced Parkinson's disease: good use of an old drug, *J. Neurol.,* 244:336–337, 1997.

36. Del Dotto, P., Pavese, N., Gambaccini, G., et al., Intravenous amantadine improves levodopa-induced dyskinesias: an acute double-blind, placebo-controlled study, *Mov. Disord.,* 16:515–520, 2001.

37. Shealy, C. N., Weeth, J. B., Mercier, D., Livedo reticularis in patients with parkinsonism receiving amantadine, *JAMA,* 212:1522–1523, 1970.

54 The Role of MAO-B Inhibitors in the Treatment of Parkinson's Disease

John Bertoni
Department of Neurology, Creighton University

Lawrence Elmer
Department of Neurology, Medical College of Ohio

CONTENTS

SELEGILINE

INTRODUCTION

Selegiline was first discovered in the 1960s and has been lauded as a virtual fountain of youth due to its potential neuroprotective effects,[1] only to be later denigrated as a suspected cause of higher mortality in patients with Parkinson's disease (PD).[2] While neither of these extreme viewpoints seems reasonable now, selegiline nevertheless still invokes lively discussion regarding its role as a ther-

0-8493-1590-5/05/$0.00+$1.50
© 2005 by CRC Press

apeutic agent for PD and a variety of other neuropsychiatric illnesses. In this section, we review the historical development of selegiline, its mechanism of action, short- and long-term clinical trials, and conclude with current recommendations for its use in the management of PD.

HISTORICAL PERSPECTIVE

Monamine oxidase (MAO) inhibitors found use in the early 1950s during the development of isoniazid for treatment of tuberculosis.[1] Iproniazid, a derivative of isoniazid, was found to have psychic energizing effects thought to be related to its ability to inhibit monoamine oxidase.[1,4] Investigations of iproniazid led to widespread acceptance of MAO inhibition as therapy for depression. By the late 1950s, a variety of similar compounds were developed.

A catastrophic side effect of MAO inhibition was soon uncovered. Patients treated with MAO inhibitors were at risk of developing a hypertensive crisis with consumption of foods, typically cheeses, that contained tyramine.[5,6] Research then focused on agents that would inhibit monoamine oxidase activity without producing the so-called "cheese effect."

By the early 1960s, Knoll and colleagues set out to develop a methamphetamine derivative that would inhibit MAO. They developed a compound, racemic phenylisopropyl-methylpropargylamine HCl, or E-250, later named deprenyl.[7] The levorotatory form of deprenyl, L-deprenyl, demonstrated irreversible inhibition of MAO activity without increasing receptor sensitivity to tyramine,[3] suggesting that this compound would be the first MAO inhibitor free of the "cheese effect."

By 1968, Johnston synthesized clorgyline, which was chemically similar to deprenyl but with significantly different pharmacological actions.[8] Clorgyline was an irreversible inhibitor of a class of MAO activity that preferentially deaminated serotonin and norepinephrine rather than the synthetic substrate, benzylamine. In contrast, L-deprenyl inhibited a class of MAO activity that deaminated the synthetic substrates benzylamine and phenylethylamine rather than endogenous serotonin and norepinephrine. These differences in substrate specificity led to the concept of two separate classes of MAO activity. The MAO-A class of enzymatic activity preferentially deaminates serotonin and norepinephrine, while MAO-B activity preferentially deaminates benzylamine and phenylethylamine. Tyramine and dopamine are substrates for both.[3,9]

By then, pioneering studies of Ehringer and Hornykiewicz[10] implicated dopamine as the primary neurochemical deficiency in postmortem brains of patients afflicted with PD. As dopamine is an endogenous substrate for MAO, investigations shortly followed to determine whether MAO inhibitors could be used to increase dopamine levels and treat symptoms in patients with PD.

Unfortunately, the hypertensive crises seen in patients treated with nonspecific MAO inhibitors were also seen in the PD population. In the late 1960s, with the development of L-deprenyl, now referred to as L-selegiline (or simply selegiline; the levorotatory form is the only form in clinical use), investigators considered this compound as a far safer potential therapeutic agent for the treatment of PD[11] and depression[12] than the nonspecific MAO inhibitors.

MECHANISM OF ACTION/PHARMACOLOGY

Dopamine Metabolism

The basic pharmacological action of selegiline is inhibition of MAO-B activity. In the brain, dopamine is preferentially metabolized by the MAO-B system,[13] suggesting that selegiline or similar derivatives may be ideal inhibitors of dopamine degradation in PD. However, some of selegiline's metabolites are pharmacologically active including desmethylselegiline and methamphetamine[14] (see below). Inhibition of MAO-B is not the sole pharmacological action of selegiline. Selegiline and/or its metabolites also inhibit the reuptake of dopamine and increase the synthesis of dopamine by blocking the presynaptic dopamine autoreceptors that govern the synthesis rate of dopamine.[15]

Selegiline is rapidly absorbed from the gastrointestinal tract, with peak serum concentrations 1/2 to 2 hr after ingestion. It is lipophilic and distributes widely throughout the body. Studies with radiolabeled selegiline suggest an apparent volume of distribution of over 300 l with over 90% of the molecule bound to plasma proteins. In animal studies, selegiline has been shown to distribute more rapidly to the central nervous system than to the periphery. Positron emission tomography studies have demonstrated that selegiline binds in the thalamus, stratum, cortex, and brain stem, which are rich in MAO-B content.[16,17]

Since selegiline is a suicide inhibitor of MAO-B, the half-life of its effectiveness depends on synthesis of new enzyme, which usually requires weeks, especially after the majority of the endogenous MAO-B enzyme has been inactivated.[14] Inhibition of MAO-B activity in humans following oral administration of selegiline is rapid and complete. MAO-B inhibition in platelets exceeds 90% within hours after oral administration of either 5 or 10 mg of selegiline.[18,19] Postmortem examination of PD patients treated with 10 mg selegiline demonstrated almost complete MAO-B inhibition, while residual MAO-A activity was fairly unaffected.[20–22] These doses correspond to a brain concentration of 10^{-6} M. Doses significantly higher than this may also inhibit MAO-A activity and risk the "cheese effect."[14]

Five metabolites have been demonstrated in human plasma after the administration of oral selegiline.[23,24] The two main metabolites are desmethylselegiline and methamphetamine, which may then be metabolized to amphetamine. Methamphetamine and amphetamine can also be metabolized to pharmacologically inactive parahydroxy derivatives. The role of these metabolites and their potential contribution to the clinical efficacy of selegiline in dopamine-deficient states has been the object of extensive discussion.

At low doses, methamphetamine and amphetamine may release dopamine, which may contribute to therapeutic benefits by increasing striatal dopamine levels. Amphetamines—especially the dextro forms—also have MAO-A inhibitory properties.[25,26] However, the concentration of dextroamphetamines is in the micromolar range, which may be too low for significant clinical effects.[14]

NEUROPROTECTIVE EFFECTS

Selegiline is also thought to have neuroprotective effects. Selegiline blocks the neurotoxic effects of the synthetic narcotic derivative 1-methyl-4-phenyl-1,2,3,6-tetrahydropyridine (MPTP) on nigral neurons.[27–29] Intravenous drug users exposed to MPTP developed symptoms of PD nearly identical to sporadic forms of PD.[30,31] The toxicity of MPTP against nigral neurons requires conversion to a toxic derivative, 1-methyl-4-phenylpyridinium ion (MPP+), which is mediated by MAO-B activity. MPP+ is concentrated in dopaminergic neurons and blocks the mitochondrial respiratory chain by inhibiting complex I of the electron transfer chain, contributing to its high specificity for nigral neurons.[32] Selegiline blocks the conversion of MPTP to MPP+ and prevents nigral neurotoxicity.[33]

While outside the scope of this review, the list of possible neurodegenerative mechanisms that selegiline may influence favorably is extensive and includes the following:

1. Preventing conversion of environmental protoxins (by blocking MAO-B activity) into potential neurotoxic agents, analogous to the MPTP model of nigral neuronal cell death
2. Preventing excessive accumulation and toxicity of oxidative radicals through modification of dopamine metabolism and/or promoting endogenous antioxidant pathways and mechanisms
3. Mimicking endogenous neurotrophic growth factors, leading to repair and regrowth of damaged neurons
4. Blocking the presumed common pathway of cell death in most neurodegenerative disorders by acting as an antiapoptic agent and preventing

programmed cell death to occur (for reviews, see References 34 and 35)

CLINICAL TRIALS—DEPRESSION

In the management of depression, selegiline was found to be useful in patients with depression, with or without bipolar features.[12,36] Eleven out of 12 patients had at least 85% inhibition of platelet MAO-B activity after one week of therapy on 5 mg/day. After the first week of 5-mg therapy, patients were increased to 10 mg/day and then, a week later, given 15 mg/day. The authors found that a majority of the patients had greater than 50% reduction in their depression rating scales despite the absence of a clear correlation with MAO-B inhibition. One patient developed hypomania and had to be withdrawn from therapy. In this patient population, there was a small biphasic effect, with the first benefit seen within three days and the second benefit occurring three weeks later.[37]

CLINICAL TRIALS—PARKINSON'S DISEASE

Adjunctive Therapy

Also in the 1960s, levodopa (LD) was introduced as a dramatically effective treatment of PD.[38] However, LD had short-lived benefits and numerous central and peripheral side effects. Combination therapy with LD and nonselective inhibitors of MAO enhanced the symptomatic improvement seen with LD therapy but also led to the hypertensive crises described above.[39]

In 1975, Birkmayer and colleagues reported that selegiline given either parenterally or orally potentiated the antiparkinsonian effects of LD.[40,41] In 1977, they reported a clinical trial in 223 PD patients in which selegiline was used in combination with LD and a peripherally acting decarboxylase inhibitor (DCI).[42] The addition of selegiline to a regimen of LD/DCI resulted in a statistically significant benefit within 60 min after the oral dose of selegiline was given, and the benefit was prolonged for up to 1 to 3 days. Many parallel trials confirmed this clinical experience.[43,11]

Clinical trial data using selegiline accumulated in the years following selegiline's acceptance as a specific MAO-B inhibitor useful for the management of PD. In 1983, Birkmayer reviewed his experience using selegiline with LD and a peripheral DCI in almost 2000 patients over 9 years.[44] The primary clinical benefits included improvement of akinesia, decreased frequency and severity of on/off phases, and a reduction in daily fluctuations of disability and rigidity. Patients on selegiline were able to reduce their total daily doses of LD, and they had relatively limited side effects.

Also in 1983, Rinne reported on 45 PD patients with fluctuations following long-term LD therapy. In an open label trial of selegiline for 1 to 3 months, 5 to 10 mg of selegiline daily resulted in a significant improvement in up to 60% of the patients.[45]

Similarly, Gerstenbrand and colleagues reported that 48 patients taking LD and a DCI were improved with selegiline at a dose of 5 to 15 mg per day. Of 28 patients with an akinetic/rigid form of PD, 13 had a significant response to the administration of selegiline. In 14 patients with symptoms of tremor accompanying the akinetic/rigid syndrome, 8 responded in a significant fashion to the administration of selegiline. A third group of six patients with tremor-dominant PD showed no significant improvement in the tremor with the addition of selegiline.[46]

Many double-blind, placebo-controlled trials were published between 1977 and 1989 examining the LD enhancing effects of selegiline used adjunctively in patients with PD who were already receiving LD/DCI therapy.[47-61] Over 90% of these studies used 10 mg per day of selegiline in the active trial group. A total of 516 subjects are represented in these 16 trials, 380 (74%) of whom had motor fluctuations. The design of these studies examined either clinical improvement as measured by PD rating scales, reduction of motor fluctuations, reduction in total daily LD dosage, or any combination of these three outcome parameters.

When adjunctive selegiline therapy was statistically compared to placebo control, in 10 of 13 studies there was significant improvement in motor fluctuations, in 11 of 15 studies there was significant improvement in the clinical symptoms of PD, and in 9 there was a significant average reduction of 28% in daily LD dosage despite improved measures of clinical improvement. The most common side effect was the occurrence of dyskinesias, which are known to occur with LD (see References 15 and 62 for review).

Short-Term *de novo* Studies

Several large, double-blind, placebo-controlled clinical trials were performed to address the question of whether oral selegiline administration could slow the progression of PD in patients newly diagnosed with this disorder. In most of these studies, the time required until administration of LD/DCI rescue for control of PD symptoms was considered an indication of delay in the progression of the disease process. Delaying the time until LD/DCI therapy was required was also considered good clinical practice, as there was concern that LD/DCI administration by itself could accelerate the progression of PD due to generation of oxidative radicals during the metabolism of dopamine (see above).

The largest *de novo* study to date, the Deprenyl and Tocopherol Antioxidant Therapy of Parkinsonism (DATATOP) study performed by the Parkinson Study Group, reported interim results in 1989 demonstrating

that selegiline dramatically delayed the need for initiating LD. A total of 800 patients were randomized to receive selegiline (5 mg bid), tocopherol (2000 IU per day, selected for its antioxidant properties), a combination of selegiline and tocopherol, or placebo. The endpoint was when LD was required in the opinion of the investigator. After an average of 12 months of follow-up, only 97 patients receiving selegiline reached endpoint, whereas 176 patients not receiving selegiline reached endpoint. The blind was broken at that point, and the placebo arm was discontinued, even though the original design of the study was for 24 months of randomized follow-up.[63,64]

Similar results were obtained in a smaller double-blind, placebo-controlled study of 54 patients[65] measuring the time until LD therapy was required. Patients who were randomized to receive placebo averaged 312 days before they required LD, compared to an average of 549 days in the selegiline group. Just as in the DATATOP study, the difference between the two groups based on the Kaplan-Meier survival curves was highly significant in favor of selegiline, suggesting a reduction in the progression of the disease by 40 to 80% per year. Multiple similar studies corroborated these findings.[66-68]

But evidence accumulated that selegiline did have mild symptomatic benefits, either directly as selegiline, or perhaps through its metabolites, desmethylselegiline, methamphetamine, or amphetamine itself. Thoughtful reviews[34,69-73] argued there was a neuroprotective effect. Subsequent long-term analysis of the DATATOP cohort demonstrated differences in the rate of development of motor complications, supporting the use of selegiline and suggesting that long-term selegiline use in PD may actually alter the clinical course[74,75] (see below).

In another placebo-controlled trial, 157 *de novo* PD patients were randomized to receive either selegiline or placebo[76] until LD therapy became necessary. Then the selegiline or placebo was withdrawn for an eight-week washout period to evaluate possible symptomatic effects of selegiline administration. Selegiline was noted to have a wash-in effect, i.e., there was measurable symptomatic improvement compared to placebo. After the washout period, there was no significant difference in the degree of increased disability seen in the selegiline and placebo groups. So the clinical benefit in patients randomized to receive selegiline did not appear to be solely dependent on a symptomatic effect: the progression of symptoms was significantly slower in the selegiline group when the progression was adjusted by the time to reach the end point.

Semiannual progression of disability assessments (by different PD rating scales) showed statistically significant differences in favor of selegiline. The semiannual rate of progression of clinical disability decreased by 74 to 160% as measured by these different scales. A major concern of clinical trials designed to demonstrate a neuroprotective effect of selegiline is the appropriate length of the wash-out to eliminate residual symptomatic MAO-B inhibitory effects. In the original DATATOP study, homovanillic acid (HVA, an indirect marker of MAO activity) levels were still decreased at one month following selegiline treatment but returned to normal after two months. By two months, the authors in this paper argued that they were able to dissect out and eliminate the symptomatic effects of selegiline that may have confounded previous trials. Thus, there is strong evidence that selegiline either slows or favorably modifies PD's disease process.

A large, multicenter, open-label, prospective trial of selegiline in early PD by the Parkinson's Disease Research Group of the United Kingdom (PDRG-UK) in 1995[77] followed 520 patients randomized to receive LD/DCI selegiline plus LD/DCI, or bromocriptine, a dopamine agonist. Selegiline dramatically decreased the need for additional doses of LD/DCI during the four-year follow-up. The mean LD dose without selegiline was 635 mg versus 460 mg with selegiline. But the mortality rate in the selegiline group after five years of follow-up was significantly higher than the LD/DCI alone group. This resulted in a fear of the use of selegiline in early PD. Previous trial results were reexamined, mortality rates were calculated, and multiple publications followed, none of which demonstrated any similar data to support the controversial United Kingdom study.[78–84] No clear expla-nation for this observation of increased mortality with sel-egiline administration to patients with PD was ever discovered.[84]

Long-Term Analysis of *de novo* Studies

The Parkinson's Study Group reexamined the long-term effects of selegiline therapy in PD.[75] Data from 800 patients with early PD from the DATATOP clinical trial were reanalyzed. Seven percent of the patients had freez-ing at study entry, and 26% had it at the end. Risk factors for developing freezing of gait included onset of PD with a gait disorder, higher scores of rigidity, postural instabil-ity, bradykinesia and speech disturbance, and longer-term disease duration. Selegiline treatment was strongly pro-tective regarding risk for developing freezing of gait. Vita-min E had no such effect. Selegiline reduced the risk of developing freezing of gait by 53%. This effect only par-tially faded away after selegiline was stopped during a washout period, suggesting that selegiline may slow PD progression. However, this benefit was somewhat offset by an increased rate of developing dyskinesias, which occurred in 34% of the selegiline group compared to 19% of the placebo group.[75]

In another study, 44 patients who were initially ran-domized to receive either selegiline or placebo for *de novo* symptoms of PD were then examined for 5 years following the point at which all 44 patients required LD

rescue.[85] The assignment to placebo versus selegiline remained blinded during the five-year follow-up, and maximum therapeutic efficacy by adjusting LD dosages was maintained. At the end of five years of follow-up, on examination of subjects receiving either selegiline or pla-cebo as well as any additionally required LD therapy, the placebo group required an average of 725 mg of LD per day compared to 405 mg per day in the selegiline group. No significant differences in mortality were seen between the two groups.

Next, 116 *de novo* patients with PD were randomized in the SELEDO (selegiline plus LD) study to receive either selegiline/LD or placebo/LD and then followed for up to 5 years[86] The dose of LD (plus DCI in both groups) had been titrated to patient needs prior to randomization. The average time elapsed after randomization until the patients required an additional 50% increase in their dose of LD was statistically in favor of selegiline—4.9 years versus 2.6 years with the placebo group. Frequency of motor fluctuations was not statistically different between the two groups but showed a trend in favor of selegiline. There was no statistically significant difference in mortal-ity between the two groups.

Taken together, the data from these long-term, pro-spective, randomized, and blinded trials support the hypothesis that selegiline modifies and possibly slows the progression of PD before and/or after the clinical need for LD (plus DCI) arises. While the mechanism is still unknown, the benefit of using selegiline appears signifi-cant. The most common adverse side effect, dyskinesias, appears to be more common among those already receiv-ing LD therapy when selegiline is given as "add-on" ther-apy. This side effect may be ameliorated by lowering the total LD dosage. Finally, the possibility that selegiline contributes to increased mortality rates in PD now seems far less likely. Most PD researchers, including the lead author of the PDRG-UK paper[77] raising this possibility, now agree that the results from that study were probably an aberration.[84]

SELEGILINE USE IN OTHER DISEASES

Due to its potential neuroprotective, antidepressant, and amphetamine like effects, selegiline has been tried for many neurological and psychiatric conditions, particularly Alzheimer's disease.

In a recent meta-analysis of published trials using sel-egiline for Alzheimer's disease,[87] 14 trials that met the inclusion criteria for analysis demonstrated some evi-dence that selegiline improved cognition and activities of daily living, but the magnitude and the duration of the effect was not of great clinical importance. There was no evidence of long-term benefits supporting the use of sele-giline in Alzheimer's patients.

Some of the prospective studies of selegiline for other conditions are summarized in Table 54.1.

TABLE 54.1
Brief Summary of Other Neuropsychiatric Illnesses and Their Response to Selegiline Therapy

Disorder	Study Design	Dosage	Efficacy	Reference
Narcolepsy	Open label	20–30 mg daily	Equivalent to similar doses of dexamphetamine	88
Periodic limb movements of sleep (PLMS)	Open label, retrospective	10–30 mg daily	Significant improvement in PLMS	89
Cigarette addiction	Randomized, crossover trial	10 mg daily	Reduction in craving	90
Fronto-temporal dementia	Clinical observation	1.25 mg daily (sublingual)	Improvement in neuropsychometric parameters	91

CURRENT RECOMMENDATIONS FOR SELEGILINE

PD before Age 65

With accumulating evidence that selegiline delays the need for LD therapy and decreases the LD dose escalation over time, it is prudent to consider initiating selegiline early, possibly with dopamine agonists if dopaminergic therapy is needed. Selegiline is well tolerated in younger patients. Despite the controversy, selegiline appears to decrease the rate of progression of PD.[92]

Motor Fluctuations before Age 75

In patients who develop motor fluctuations, selegiline is a good alternative to increasing the frequency/amount of LD/DCI. Very elderly or debilitated patients may have cognitive and behavioral changes (confusion, agitation, hallucinations, sleep disturbance) with selegiline. These side effects are not as frequent as dyskinesias when selegiline is given with LD/DCI, but they may cause significant distress for patient and caregiver alike.[93]

Freezing

Freezing in PD is often difficult to treat. However, the DATATOP studies suggest that selegiline dramatically impacts freezing, either by delaying or preventing its development or by directly improving it.[74,75] In our patients, freezing has partially, and in rare cases completely, improved with use of selegiline (unpublished observations).

Published Treatment Guidelines for Selegiline

The recommendation for selegiline in early PD is outlined in the algorithm for the treatment of PD.[94] However, since the single UK report[77] of increased mortality in 1995, the use of selegiline in early PD has fallen. As reviewed above, this report has been refuted many times, and even the lead author agrees that it is probably erroneous. It is recommended, however, that selegiline be withdrawn if there is development of significant cognitive and/or behavioral changes that cannot be easily managed otherwise.

BUCCALLY OR ORALLY ABSORBED SELEGILINE

Another method of administering selegiline is in an orally disintegrating, orally absorbed form. This has several advantages. First, the avoidance of swallowing selegiline into the stomach and of corresponding "first pass" liver metabolism allows for a lower dose of selegiline and more economical delivery to the brain. Second, selegiline's metabolites, which include amphetamine and methamphetamine, are several-fold lower using the buccal route. Third, this approach is particularly helpful for PD patients who experience difficulty swallowing pills. Fourth, the blood levels of selegiline rise more rapidly through the buccal route. Fifth, once-daily dosing is convenient and promotes patient adherence to dose schedule.

In a randomized, double-blinded, multicenter, two-arm, placebo-controlled study of buccally absorbed selegiline (Zydis selegiline), patients with idiopathic PD with previous good response to levodopa/carbidopa, but with at least 3 hr of "off" time each day, were treated with buccally absorbed selegiline (1.25 or 2.5 mg) ($N = 94$) or placebo ($N = 46$) (95 submitted for publication). Patients in the active group initiated treatment at 1.25 mg once daily and were titrated to 2.5 mg once daily at week 6. Buccally absorbed selegiline proved to be significantly better than placebo at reducing the percentage of daily "off" time at 4 to 6 weeks ($p < 0.003$) and at 10 to 12 weeks ($p < 0.001$) (primary end point). Significant differences between active and placebo groups were found even at week 1 ($p < 0.05$). By the end of the 12-week study, the buccally absorbed selegiline group had 2.2 fewer hours of "off" time per day than at baseline versus 0.6 hr fewer than baseline in the placebo group. By the 12th week, the corresponding percentage of daily "on" time grew by 12% in the buccally absorbed selegiline group, versus only 3% in the placebo group ($p < 0.008$). The "on" time was increased by about 1.8 hr per day in the treatment group. There were no significant changes in the mean number of daily "asleep" hours, or the mean percentage of "asleep" time throughout the study. Adverse effects were apparently no different from those seen with placebo and were consistent with known adverse effects of LD. A second similar study of buccally absorbed selegiline showed comparable improvement in the treatment group, but, because of a large placebo effect, the results were not as statistically impressive.

A longer-term study[96] confirmed the long-term safety and efficacy of buccally absorbed selegiline (2.5 mg/day) in 254 PD patients already treated with LD. The most common side effects were dizziness, dyskinesias, insomnia, dry mouth, and nausea. Discontinuations were slightly higher in those not previously treated with the lower dose of buccally-absorbed selegiline, suggesting that it is better tolerated if more slowly titrated.

In conclusion, orally absorbed selegiline (Zydis selegiline) has been shown to be an effective and advantageous treatment in PD patients who take carbidopa/levodopa and have significant wearing off effect. It may be preferred in some or even most PD patients.

RASAGILINE

INTRODUCTION

Rasagiline, [N-propargyl-1R(+)aminoindan] is a unique, selective, and potent second-generation mitochondrial monoamine oxidase B inhibitor with distinctive neuroprotective as well as therapeutic properties for the treatment of PD. In this section, we review the historical development of rasagiline, its mechanism of action, its preclinical nonhuman animal and *in vitro* studies, its use in short- and long-term clinical trials, and conclude with current recommendations for its use in the management of PD.

HISTORICAL PERSPECTIVE: QUEST FOR THERAPIES WITH BOTH THERAPEUTIC AND NEUROPROTECTIVE EFFECTS

Through the pioneering work of Youdim and colleagues,[97] rasagiline was developed as an alternative to other agents with known or potential use in patients with PD and other central nervous system diseases. All therapies for PD have significant limitations, which can be summarized as incomplete effectiveness and/or unwanted side effects. Another challenge in chronic neurodegenerative diseases is that agents that once worked quite well lose their effectiveness over time. Much of this decrease in efficacy is attributed to loss or dysfunction of neurons, or supporting elements, or adaptive processes. The potential value of therapies that can halt or reverse these diseases can hardly be overestimated.

A great deal of information has been learned about the cascade of reactions and processes that lead to inflammation, oxidative stress, accumulation of intracellular toxins, or subcellular organelles, which may collectively result in apoptosis or necrosis. Just as all of these and other unknown processes cause damage to the neuron and its supporting elements in different ways, so also must neuroprotection require multiple approaches. It is no surprise, therefore, that the optimal management of neurodegenerative diseases like PD has become more complex and requires multiple drugs and multiple nondrug therapies. In the future, such therapy will likely become even more complex.

PHARMACOLOGY

MAO is present in the outer mitochondrial membrane and is found throughout the central nervous system and in platelets. Since both MAO-A and MAO-B oxidize dopamine, epinephrine, tyramine, tryptamine, but MAO-B also oxidizes phenylethylamine and benzylamine, MAO-B inhibitors like rasagiline prevent the destruction of dopamine.

Rasagiline, or TV3326, or [N-propargyl-1R(+)aminoindan], is a selective and highly potent second-generation mitochondrial monoamine oxidase B inhibitor.[97] Selegiline and rasagiline are compared in Figures 54.1 and 54.2 and Table 54.2, below.

As opposed to selegiline, rasagiline has quite a different metabolite profile. Selegiline's major metabolites are amphetamine and methamphetamine (Figure 54.1), while rasagiline's primary metabolite is aminoindan (Figure 54.2). While amphetamine and methamphetamine are potently addictive substances, they may promote alertness. Aminoindan has beneficial effects of its own and has no known adverse side effects.[98] In man, 1-mg daily dosages of rasagiline inhibit platelet MAO-B nearly completely.[99] Rasagiline has both therapeutic and protective properties.

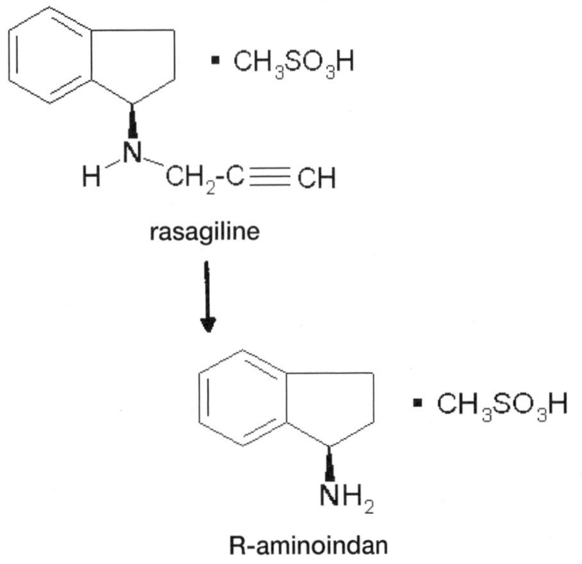

FIGURE 54.1 Selegiline and its metabolites.

TABLE 54.2
Comparison of Selegiline and Rasagiline

Characteristic	Selegiline	Rasagiline
Chemical structure		
Metabolites	Metabolized to N-desmethylselegiline, L-methamphetamine, and L-amphetamine	Major and only metabolite is aminoindan
Dosing	Dosed at 5 mg bid	Dosed at 0.5 to 2 mg q.d.
Efficacy (adjunctive)	Adjunct therapy in managing Parkinson's disease patients being treated with levodopacarbidopa	Greater *in vivo* potency (5 × greater) than selegiline
Efficacy (monotherapy)	Not indicated for monotherapy	Effective monotherapy
Neuroprotective properties	Demonstrated neuroprotective effects in cell and animal models, but is metabolized very rapidly N-desmethylselegiline, a minor metabolite, has demonstrated neuroprotective effects in cell models Amphetamine metabolites have potential to reduce, or counter, the neuroprotective effects	Demonstrated direct neuroprotective effect in cell and animal models Neuroprotective effects not due to its MAO-B inhibitory action

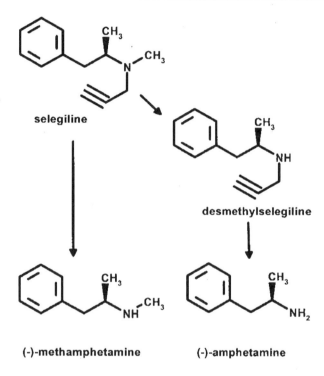

FIGURE 54.2 Rasagiline and its metabolite, R-aminoindan.

PRECLINICAL STUDIES—ANIMAL AND CELL CULTURE

Rasagiline has many neuroprotective properties in animal and cell culture models of PD as well as a variety of other conditions that are too numerous to delineate completely (see References 97 and 100 for review), so only some of these studies are mentioned here.

Some studies suggest that the neuroprotective effects of rasagiline are potentially separate and unrelated to its therapeutic effects.[97] For example, the *S*-enantiomer of rasagiline lacks the MAOB inhibitory and the antidepressant effects of rasagiline but shares with rasagiline its neuroprotective effects and its cholinesterase inhibitory effects.[101] This suggests that the neuroprotective effects of rasagiline are independent of its antidepressant and MAOB inhibitory effects. Rasagiline protects dopaminergic SH-SY5Y cells in culture by inhibiting mitochondrial permability transition, an apoptosis inducing effect, of an endogenous neurotoxin (*N*-methyl(*R*) salsolinol).[102] In the same cell model, rasagiline also *enhances* the expression of the Bcl-2 gene family which have anti-apoptotic effects.[103] Finally, rasagiline prevents the apoptosis and nuclear accumulation of glyceraldehyde-3-phosphate dehydrogenase induced by N-methyl(R)salsonilol in human dopaminergic SH-SY5Y cell cultures.[104] Since the SH-SY5Y cell system does not contain MAO, these protective effects must have other mechanisms. Rasagiline also is protective *in vivo* against MPTP (mice) and 6-OHDA (rats) (see Reference 100). In primate models of PD, rasagiline led to a return toward normal of cognitive and motor function, even in the absence of LD.[98]

In neuronal cell models (SH-SY5Y neuroblastoma cells and neuronally differentiated PC12 cells), rasagiline increases SOD activity and Bcl-2 expression, and it inhibits activation of capsase 3, prevents DNA laddering, inhibits toxin induced fall in mitochondrial membrane potential, prevents toxin induced apoptosis, and protects against cell death due to ischemia and glucose deprivation.[100] Rasagiline also has protective effects against stroke occurrence and promotes survival in spontaneously hypertensive rats.[105]

CLINICAL STUDIES

Rasagiline as Monotherapy for Early Untreated PD (Phase II Studies)

In a 10-week randomized, placebo-controlled trial of rasagiline in early untreated PD ($N = 56$), rasagiline at levels up to 4 mg/day were well tolerated.[106] The improvement in total Unified Parkinson's Disease Rating Scale (UPDRS) over the 10-week study was 0.5, 1.8, 3.6, and 3.6 units for 0 (placebo), 1, 2, and 4 mg/day rasagiline, respectively. Rasagiline was not associated with any EKG or hypertensive problems but actually tended to lower blood pressure a little. Adverse effects were minor and not significantly more in the treatment groups.

Rasagiline as Adjunctive Therapy in Fluctuating PD Patients

A multicenter, randomized, double-blind, placebo-controlled, parallel group (0, 0.5, 1, and 2 mg/day single dose) study of 12 weeks of rasagiline therapy was conducted on 70 patients with PD with motor fluctuations.[99] Although there was more improvement in UPDRS in the treatment group, the differences did not achieve statistical significance ($p = 0.078$). A large placebo effect confounded the results. However, the combined treatment group with fluctuations had more improvement in the UPDRS than the placebo group did (23% versus 8.5%). Rasagiline stopped "off" periods in 15% of the fluctuating patients, but the placebo didn't prevent fluctuations in any subject. Nearly complete platelet MAOB inhibition was achieved at each dose studied in a small subgroup ($N = 8$) of patients. Careful monitoring of blood pressure showed no hypertensive effect of rasagiline. Dose proportionality was observed for both rasagiline and aminoindan (the main metabolite). Kinetic studies showed that time to peak concentration of both rasagiline and aminoindan (T_{max}) was around 0.6 hr after an oral dose.

Rasagiline as Monotherapy for Early Untreated PD (Phase III, TEMPO Study)

In a 26-week, multicenter, parallel group randomized placebo-controlled trial of 404 nondemented patients over age 35 years with early untreated PD, rasagiline monotherapy significantly ($p < 0.001$) improved motor function as measured by the UPDRS at both the 1- and 2-mg daily dosages.[107] The 2-mg/day dosage was no more effective than the 1-mg/day dosage. The improvement in UPDRS related to rasagiline is relatively modest in comparison to dopamine agonists. The most plausible mode of action of

rasagiline, according to the authors, is slowed catabolism of endogenous dopamine. It was concluded that rasagiline is well tolerated, with reported adverse effects not significantly different from those of the placebo. Tyramine challenge tests to assess the safety of rasagiline in a subset of these patients showed no significant problems.[108] After six months, the patients receiving placebo were treated with rasagiline in an open-label extension, but they did not ever reach the same level of benefit as the group on rasagiline for the whole study. This suggests that rasagiline may have disease-altering benefits in *de novo* PD patients.

Rasagiline and Entacapone in More Advanced PD Patients (Phase III, LARGO Study)

In a study[109] of rasagiline versus placebo or entacapone with 687 advanced PD patients treated with LD who were already experiencing motor fluctuations, the primary efficacy outcome was the diary verified change in "off" time. Secondary endpoints were the number of responders with >1 hr reduction in daily "off" time, change in UPDRS, and change in global impression.

Both the rasagiline group and the entacapone groups experienced a reduction in mean daily "off" time of 1.2 hr, which was significantly ($p < 0.0001$) greater than the placebo group reduction of 0.4 hr. The mean daily total dose of LD was reduced by 24 mg/day in the rasagiline group, reduced by 17 mg/day in the entacapone group, and was *increased* by 5 mg/day in the placebo group. There were statistically more responders, better UPDRS scores, and better final global clinical assessments in those taking either rasagiline or entacapone than in the placebo group. In an ancillary study of 105 patients, rasagiline significantly improved motor UPDRS in the defined "off" periods, whereas entacapone did not.

Despite these significant clinical improvements, dyskinesias were not increased in the rasagiline or entacapone groups, and the frequency of adverse effects did not increase. This study established that a single daily 1-mg dosage of rasagiline was statistically effective in reducing "off" time in advanced LD-treated PD patients and was just as effective as multiple daily doses of entacapone.

Rasagiline in Advanced Fluctuating PD Patients (Phase III, PRESTO Study)

In a further study[110] of 472 PD patients with at least 2.5 hr of daily "off" time randomized to 0 (placebo), 0.5, or 1.0 mg rasagiline per day, patients had a mean daily "off" time of 6 hr. By the end of the study, the placebo group had 0.91 hr less daily "off" time than they had at baseline. Those taking 0.5 mg/day rasagiline had 1.41 hr less daily "off" time than they did at baseline or 0.49 hr better than with placebo (95% CI 0.08 to 0.91, $p = 0.02$). Those taking 1.0 mg/day rasagiline had 1.85 hr less daily "off" time than at baseline, or nearly 1 hr (0.94 hr) better than with placebo ($p < 0.0001$). Rasagiline also provided significant improvement in motor performance in the "on" state, in activities of daily living in the "off" state (as measured with the UPDRS), and in investigator assessment of clinical global impression.

Current Indications for Rasagiline

Rasagiline has a role in the treatment of PD by virtue of its proven ability to reduce the signs of PD in both the "on" and "off" states, and to improve global function. It appears to be of value in early stages of PD as well as after the appearance of clinical fluctuations in response to LD. As of this printing, it has not been given final approval by the United States Food and Drug Administration, and its official indications are not now available.

Rasagiline also has a promising but not fully explored potential to halt or slow down the progression of PD, as well as other clinical conditions. The accumulating evidence of rasagiline's neuroprotective effects in animal and cellular models is both intriguing and exciting. Further careful scientific basic and clinical studies and clinical experience are needed to establish the full therapeutic benefits of rasagiline for the treatment of PD.

CONCLUSIONS

The MAO-B inhibitors have a remarkable history as a treatment for PD and other neuropsychiatric illnesses. Initial widespread enthusiasm for selegiline gave way to fear of increased mortality, which has given way to scientifically supported indications as both first-line and adjunctive therapy for PD. Decisions to initiate selegiline therapy in early PD and at intermediate or advanced stages of the disease should be based on fact and the clinical situation after careful discussion with the patient and caregivers. The optimal indications for newer agents will also be developed over time and experience, but these principles remain.

When, why, and how to use MAO-B inhibitors for the best treatment of PD will remain important challenges as new delivery systems for selegiline and ever newer generations of MAO-B inhibitors continue to be developed in the future.

We wish to gratefully acknowledge the assistance of friend and colleague Dr. L. John Greenfield for his editorial comments.

REFERENCES

1. Knoll, J., The pharmacology of selegiline ((-)deprenyl), New aspects, *Acta Neurol. Scand.,* Suppl., 126:83–91, 1989.

2. Lees, A. J., Comparison of therapeutic effects and mortality data of levodopa and levodopa combined with selegiline in patients with early, mild Parkinson's disease. Parkinson's Disease Research Group of the United Kingdom, *B. M. J.,* 311(7020):1602–7, Dec. 16, 1995.

3. Knoll, J., Deprenyl (selegiline): the history of its development and pharmacological action, *Acta Neurol. Scand.,* Suppl., 95:57–80, 1983.

4. Zeller, E. A., Barsky, J., Fouts, J. E., Kirchheimer, W. F., Van Orden, I. S., Influence of isonicotinic acid hydrazide (INH) and 1-isonicotinic-2isopropyl hydrazide (IIH) on bacterial and mammalian enzymes, *Experientia.,* 8:349–350, 1952.

5. Blackwell, B., Hypertensive crisis due to monoamine oxidase inhibitors, *Lancet,* 11:849–851, 1963.

6. Horwitz, D., Lovenberg, W., Engelmann, K., Sjoerdsma, A., Monoamine oxidase inhibitors, tyramine and cheese, *JAMA,* 188:1108–1110, 1964.

7. Knoll, J., Ecseri, Z., Kelemen, K., Nievel, J., Knoll, B., Phenylisopropylmethylpropinylamine (E-250), a new spectrum psychic energizer, *Arch Int. Pharmacodyn. Ther.,* 155(1):154–64, May, 1965.

8. Johnston, J. P., Some observations upon a new inhibitor of monoamine oxidase in brain tissue, *Biochem. Pharmacol.,* 17(7):1285–97, July, 1968.

9. Riederer, P., Jellinger, K., Danielczyk, W., Seemann, D., Ulm, G., Reynolds, G. P., Birkmayer, W., Koppel, H., Combination treatment with selective monoamine oxidase inhibitors and dopaminergic agonists in Parkinson's disease: biochemical and clinical observations, *Adv., Neurol.,* 37:159–76, 1983.

10. Ehringer, H. Hornykiewicz, Verteilung von noradrenalin und dopamine (3-hydroxytyramin) im gehirn des menschen und ihr verhalten bei erkrankungen des extrapyramidalen systems, *Klin Wschr.,* 38:1236–1239, 1960.

11. Rinne, U. K., Recent advances in research on Parkinsonism, *Acta Neurol. Scand.,* Suppl., 67:77–113, 1978.

12. Mann, J., Gershon, S., L-deprenyl, a selective monoamine oxidase type-B inhibitor in endogenous depression, *Life Sci.,* 17;26(11):877–82, March, 1980.

13. Glover, V., Sandler, M., Owen, F., Riley, G. J., Dopamine is a monoamine oxidase B substrate in man, *Nature,* 6;265(5589):80–1, January 6, 1977.

14. Heinonen, E. H., Myllyla, V., Sotaniemi, K., Lamintausta, R., Salonen, J. S., Anttila, M., Savijarvi, M., Kotila, M., Rinne, U. K., Pharmacokinetics and metabolism of selegiline, *Acta Neurol. Scand.,* Suppl., 126:93–9, 1989.

15. Heinonen, E. H., Rinne, U. K., Selegiline in the treatment of Parkinson's disease, *Acta Neurol. Scand.,* Suppl., 126:103–11, 1989.

16. Fowler, J. S., MacGregor, R. R., Wolf, A. P., Arnett, C. D., Dewey, S. L., Schlyer, D., Christman, D., Logan, J., Smith, M., Sachs, H. et al., Mapping human brain monoamine oxidase A and B with 11C-labeled suicide inactivators and PET, *Science,* 3;235(4787):481–5, January 3, 1987.

17. Fowler, J. S., Wolf, A. P., MacGregor, R. R., Dewey, S. L., Logan, J., Schlyer, D. J., Langstrom, B., Mechanistic positron emission tomography studies: demonstra-

tion of a deuterium isotope effect in the monoamine oxidase-catalyzed binding of [11C]L-deprenyl in living baboon brain, *J. Neurochem.,* 51(5):1524–34, November, 1988.

18. Riederer, P., Youdim, M. B., Rausch, W. D., Birkmayer, W., Jellinger, K., Seemann, D., On the mode of action of L-deprenyl in the human central nervous system, *J. Neural. Transm.,* 43(3-4):217–26, 1978.

19. Riederer, P., Youdim, M. B., Birkmayer, W., Jellinger, K., Monoamine oxidase activity during (–)-deprenil therapy: human brain post-mortem studies, *Adv. Biochem. Psychopharmacol.,* 19:377–82, 1978.

20. Reiderer, P., Reynolds, G. P., Deprenyl is a selective inhibitor of brain MAO-B in the long-term treatment of Parkinsons's disease, *Br. J. Clin. Pharmacol.,* Jan., 9(1):98–9, 1980.

21. Riederer, P., Youdim, M. B., Monoamine oxidase activity and monoamine metabolism in brains of parkinsonian patients treated with l-deprenyl, *J. Neurochem.,* 46(5):1359–65, May, 1986.

22. Glover, V., Elsworth, J. D., Sandler, M., Dopamine oxidation and its inhibition by (-)-deprenyl in man, *J. Neural. Transm.,* Suppl., (16):163–72, 1980.

23. Reynolds, G. P., Elsworth, J. D., Blau, K., Sandler, M., Lees, A. J., Stern, G. M., Deprenyl is metabolized to methamphetamine and amphetamine in man, *Br. J. Clin. Pharmacol.,* 6(6):542–4, December, 1978.

24. Yoshida, T., Yamada, Y., Yamamoto, T., Kuroiwa, Y., Metabolism of deprenyl, a selective monoamine oxidase (MAO) B inhibitor in rat: relationship of metabolism to MAO-B inhibitory potency, *Xenobiotica,* 16(2):129–36, February, 1986.

25. Parkinson, D., Callingham, B. A., Substrate and inhibitor selectivity of human heart monoamine oxidase, *Biochem. Pharmacol.,* 15;28(10):1639–43, May, 1979.

26. Egashira, T., Yamamoto, T., Yamanaka, Y., Effects of d-methamphetamine on monkey brain monoamine oxidase, *in vivo* and *in vitro, Jpn. J. Pharmacol.,* 45(1):79–88, September, 1987.

27. Mytilineou, C., Cohen, G., Deprenyl protects dopamine neurons from the neurotoxic effect of 1-methyl-4-phenylpyridinium ion, *J. Neurochem.,* 45(6):1951–3, December 1985.

28. Lewin, R., Drug trial for Parkinson's, Science, 12;236 (4807):1420, June 12, 1987.

29. Fuller, R. W., Hemrick-Luecke, S. K., Perry, K. W., Deprenyl antagonizes acute lethality of 1-methyl-4-phenyl-1,2,3,6-tetrahydropyridine in mice, *J. Pharmacol. Exp. Ther.,* 247(2):531–5, November, 1988.

30. Davis, G. C., Williams, A. C., Markey, S. P., Ebert, M. H., Caine, E. D., Reichert, C. M., Kopin, I. J., Chronic parkinsonism secondary to intravenous injection of meperidine analogues, *Psychiatry Res.,* 1:249–254, 1979.

31. Langston, J. W., Ballard, P. A., Tetrud, J. W., Irwin, I., Chronic parkinsonism in humans due to a product of meperidine-analog synthesis, *Science,* 219:979–980, 1983.

32. Chiba, K., Trevor, A., Castagnoli, N. Jr., Active uptake of MPP+ a metabolite of MPTP by brain synapto-

somes, *Biochem. Biophys. Res. Commun.,* 120:574–578, 1984.

33. Heikkila, R. E., Manzino, L., Cabbat, F. S., Duvoisin, R. C., Protection against the dopaminergic neurotoxicity of 1-methyl-4-phenyl-1,2,3,6-tetrahydropyridine by monoamine oxidase inhibitors, *Nature,* 311:467–469, 1984.

34. Olanow, C. W., Deprenyl in the treatment of Parkinson's disease: clinical effects and speculations on mechanism of action, *J. Neural. Transm.,* Suppl., 48:75–84, 1996.

35. Ebadi, M., Sharma, S., Shavali, S., El Refaey, H., Neuroprotective actions of selegiline, *J. Neurosci. Res.,* 67(3):285–9, February 1, 2002.

36. Varga, E., Tringer, L., Clinical trial of a new type promptly acting psychoenergetic agent (phenyl-isopropyl-methylpropinyl-HCl, "E-250"), *Acta Med. Acad. Sci. Hung.,* 23(3):289–95, 1967.

37. Mann, J. J., Aarons, S. F., Wilner, P. J., Keilp, J. G., Sweeney, J. A., Pearlstein, T., Frances, A. J., Kocsis, J. H., Brown, R. P., A controlled study of the antidepressant efficacy and side effects of (-)-deprenyl. A selective monoamine oxidase inhibitor, *Arch Gen. Psychiatry,* 46(1):45–50, 1989.

38. Birkmayer, W., Hornykiewicz, O., Der L 3,4-dihydroxyphenylalanin (DOPA) effect bei der Parkinsonakinese, *Wien, Klin. Wochenschr.,* 73:787–788, 1961.

39. Gerstenbrand, F., Prosenz, P., On the treatment of Parkinson's syndrome with monoamine oxidase inhibitors alone and in combination with L-dopa, *Praxis,* 18, 54(46):1373–7, November 18, 1965.

40. Birkmayer, W., Danielczyk, W., Neumayer, E., Riederer, P., Dopaminergic supersensitivity in parkinsonism, *Adv. Neurol.,* 9:121–129, 1975.

41. Birkmayer, W., Riederer, P., Youdim, M. B., Linauer, W., The potentiation of the anti akinetic effect after L-dopa treatment by an inhibitor of MAO-B, Deprenil, *J. Neural. Transm.,* 36(3-4):303–326, 1975.

42. Birkmayer, W., Riederer, P., Ambrozi, L., Youdim, M. B., Implications of combined treatment with "Madopar" and L-deprenil in Parkinson's disease. A long-term study, *Lancet,* 1(8009):439–43, February 26, 1977.

43. Csanda, E., Anta,l J., Antony, M., Csanaky, A., Experiences with L-deprenyl in Parkinsonism, *J. Neural. Transm.,* 43(3–4):263–9, 1978.

44. Birkmayer, W., Deprenyl (selegiline) in the treatment of Parkinson's disease, *Acta Neurol. Scand.,* Suppl., 95:103–5, 1983.

45. Rinne, U. K., Deprenyl (selegiline) in the treatment of Parkinson's disease, *Acta Neurol. Scand.,* Suppl., 95:107–11, 1983.

46. Gerstenbrand, F., Ransmayr, G., Poewe, W., Deprenyl (selegiline) in combination treatment of Parkinson's disease, *Acta Neurol. Scand.,* Suppl., 95:123–6, 1983.

47. Lees, A. J., Shaw, K. M., Kohout, L. J., Stern, G. M., Elsworth, J. D., Sandler, M., Youdim, M. B., Deprenyl in Parkinson's disease, *Lancet,* 2(8042):791–5, October 15, 1977.

48. Stern, G. M., Lees, A. J., Sandler, M., Recent observations on the clinical pharmacology of (-) deprenyl, *J. Neural. Transm.,* 43(3–4):245–51, 1978.

49. Schachter, M., Marsden, C. D., Parkes, J. D., Jenner, P., Testa, B., Deprenyl in the management of response fluctuations in patients with Parkinson's disease on levodopa, *J. Neurol. Neurosurg. Psychiatry,* 43(11): 1016–21, November, 1980.

50. Goldstein, L., The "on-off" phenomena in Parkinson's disease—treatment and theoretical considerations, *Mt. Sinai J. Med.,* 47(1):80–4, January–February, 1980.

51. Eisler, T., Teravainen, H., Nelson, R., Krebs, H., Weise, V., Lake, C. R., Ebert, M. H., Whetzel, N., Murphy, D. L., Kopin, I. J., Calne, D. B., Deprenyl in Parkinson disease, *Neurology,* 31(1):19–23, January, 1981.

52. Presthus, J., Hajba, A., Deprenyl (selegiline) combined with levodopa and a decarboxylase inhibitor in the treatment of Parkinson's disease, *Acta Neurol. Scand.,* Suppl., 95:127–33, 1983.

53. Brodersen, P., Philbert, A., Gulliksen, G., Stigard, A., The effect of L-Deprenyl on on-off phenomena in Parkinson's disease, *Acta Neurol. Scand.,* 71(6):494–7, June, 1985.

54. Fischer, P. A., Baas, H., Therapeutic efficacy of *R*-(-)-deprenyl as adjuvant therapy in advanced parkinsonism, *J. Neural. Transm.,* Suppl., 25:137–47, 1987.

55. Golbe, L. I., Duvoisin, R. C., Double-blind trial of *R*-(-)-deprenyl for the "on-off" effect complicating Parkinson's disease, *J. Neural. Transm.,* Suppl., 25:123–9, 1987.

56. Lieberman, A. N., Gopinathan, G., Neophytides, A., Foo, S. H., Deprenyl versus placebo in Parkinson disease: a double-blind study, *N. Y. State J. Med.,* 87(12): 646–9, December, 1987.

57. Rascol, O., Montastruc, J. L., Senard, J. M., Demonet, J. F., Simonetta, M., Rascol, A., Two weeks of treatment with deprenyl (selegiline) does not prolong L-dopa effect in Parkinsonian patients: a double-blind crossover placebo-controlled trial, *Neurology,* 38(9): 1387–91, September, 1988.

58. Heinonen, E. H., Rinne, U. K., Tuominen, J., Selegiline in the treatment of daily fluctuations in disability of parkinsonian patients with long-term levodopa treatment, *Acta Neurol. Scand.,* Suppl., 126:113–8, 1989.

59. Teychenne, P. F., Parker, S., Double-blind, crossover placebo controlled trial of selegiline in Parkinson's disease—an interim analysis, *Acta Neurol. Scand.,* Suppl., 126:119–25, 1989.

60. Sivertsen, B., Dupont, E., Mikkelsen, B., Mogensen, P., Rasmussen, C., Boesen, F., Heinonen, E., Selegiline and levodopa in early or moderately advanced Parkinson's disease: a double-blind controlled short- and long-term study, *Acta Neurol. Scand.,* Suppl., 126:147–52, 1989.

61. C. D., Chouza, C., Aljanati, R., Scaramelli, A., De Medina, O., Caamano, J. L., Buzo, R., Fernandez, A., Romero, S., Combination of selegiline and controlled release levodopa in the treatment of fluctuations of clinical disability in parkinsonian patients, *Acta Neurol. Scand.,* Suppl., 126:127–37, 1989.

62. Golbe, L. I., Deprenyl as symptomatic therapy in Parkinson's disease, *Clin. Neuropharmacol.,* 11(5): 387–400, October, 1988.

63. The Parkinson Study Group, Effect of deprenyl on the progression of disability in early Parkinson's disease, *N. Engl. J. Med.,* 321(20):1364–71, November 16, 1989.

64. Parkinson Study Group, DATATOP: a multicenter controlled clinical trial in early Parkinson's disease, *Arch Neurol.,* 46(10):1052–60, October, 1989.

65. Tetrud, J. W., Langston, J. W., The effect of deprenyl (selegiline) on the natural history of Parkinson's disease, *Science,* 245(4917):519–22, August 4, 1989.

66. Myllyla, V. V., Sotaniemi, K. A., Vuorinen, J. A., Heinonen, E. H., Selegiline as a primary treatment of Parkinson's disease, *Acta Neurol. Scand.,* Suppl., 136:70–2, 1991.

67. Myllyla, V. V., Sotaniemi, K. A., Vuorinen, J. A., Heinonen, E. H., Selegiline as initial treatment in de novo parkinsonian patients, *Neurology,* 42(2):339–43, February, 1992.

68. Olanow, C. W., Hauser, R. A., Gauger, L., Malapira, T., Koller, W., Hubble, J., Bushenbark, K., Lilienfeld, D., Esterlitz, J., The effect of deprenyl and levodopa on the progression of Parkinson's disease, *Ann. Neurol.,* 38(5):771–7, November, 1995.

69. LeWitt, P. A., Deprenyl's effect at slowing progression of parkinsonian disability: the DATATOP study. The Parkinson Study Group, *Acta Neurol. Scand.,* Suppl., 136:79–86, 1991.

70. Shoulson, I., An interim report of the effect of selegiline (L-deprenyl) on the progression of disability in early Parkinson's disease, The Parkinson Study Group, *Eur. Neurol.,* 32 Suppl. 1:46–53, 1992.

71. Schulzer, M., Mak, E., Calne, D. B., The antiparkinson efficacy of deprenyl derives from transient improvement that is likely to be symptomatic, *Ann. Neurol.,* 32(6):795–8, December, 1992.

72. Brannan, T., Yahr, M. D., Comparative study of selegiline plus L-dopa-carbidopa versus L-dopa-carbidopa alone in the treatment of Parkinson's disease, *Ann. Neurol.,* 37(1):95–8, January, 1995.

73. Olanow, C. W., Mytilineou, C., Tatton, W., Current status of selegiline as a neuroprotective agent in Parkinson's disease, *Mov. Disord.,* 13 Suppl., 1:55–8, 1998.

74. Giladi, N., McDermott, M. P., Fahn, S., Przedborski, S., Jankovic, J., Stern, M., Tanner, C., Freezing of gait in PD: prospective assessment in the DATATOP cohort, *Neurology,* 56(12):1712–21, June 26, 2001.

75. Shoulson, I., Oakes, D., Fahn, S., Lang, A., Langston, J. W., LeWitt, P., Olanow, C. W., Penney, J. B., Tanner, C., Kieburtz, K., Rudolph, A., Impact of sustained deprenyl (selegiline) in levodopa-treated Parkinson's disease: a randomized placebo-controlled extension of the deprenyl and tocopherol antioxidative therapy of parkinsonism trial, *Ann. Neurol.,* 51(5):604–12, May, 2002.

76. Palhagen, S., Heinonen, E. H., Hagglund, J., Kaugesaar, T., Kontants, H., Maki-Ikola, O., Palm, R., Turunen, J., Selegiline delays the onset of disability in *de novo* parkinsonian patients, Swedish Parkinson Study Group, *Neurology,* 51(2):520–5, August, 1998.

77. Lees, A. J., Comparison of therapeutic effects and mortality data of levodopa and levodopa combined with

selegiline in patients with early, mild Parkinson's disease, Parkinson's Disease Research Group of the United Kingdom, *B. M. J.*, 16;311(7020):1602–7, December, 1995.

78. Olanow, C. W., Myllyla, V. V., Sotaniemi, K. A., Larsen, J. P., Palhagen, S., Przuntek, H., Heinonen, E. H., Kilkku, O., Lammintausta, R., Maki-Ikola, O., Rinne, U. K., Effect of selegiline on mortality in patients with Parkinson's disease: a meta-analysis, *Neurology*, 51(3):825–30, September, 1998.

79. Lees, A. J., Head, J., Shlomo, Y. B., Selegiline and mortality in Parkinson's disease: another view, *Ann. Neurol.*, 41(2):282–3, February, 1997.

80. Parkinson Study Group, Mortality in DATATOP: a multicenter trial in early Parkinson's disease, *Ann. Neurol.*, 43(3):318–25, March, 1998.

81. Oakes, D., Selegiline and excess mortality, *Clin. Neuropharmacol.*, 20(6):542, December, 1997.

82. Ben-Shlomo, Y., Churchyard, A., Head, J., Hurwitz, B., Overstall, P., Ockelford, J., Lees, A. J., Investigation by Parkinson's Disease Research Group of United Kingdom into excess mortality seen with combined levodopa and selegiline treatment in patients with early, mild Parkinson's disease: further results of randomized trial and confidential inquiry, *B. M. J.*, 18;316(7139):1191–6, April, 1998

83. Donnan, P. T., Steinke, D. T., Stubbings, C., Davey, P. G., MacDonald, T. M., Selegiline and mortality in subjects with Parkinson's disease: a longitudinal community study, *Neurology*, 55(12):1785–9, December 26, 2000.

84. Lees, A. J., Katzenschlager, R., Head, J., Ben-Shlomo, Y., Ten-year follow-up of three different initial treatments in de-novo PD: a randomized trial, *Neurology*, 13;57(9):1687–94, November, 2001.

85. Myllyl, V. V., Sotaniemi, K. A., Hakulinen, P., Maki-Ikol, O., Heinonen, E. H., Selegiline as the primary treatment of Parkinson's disease—a long-term double-blind study, *Acta Neurol. Scand.*, 95:211–218, 1997.

86. Przuntek, H., Conrad, B., Dichgans, J., Kraus, P. H., Krauseneck, P., Pergande, G., Rinne, U., Schimrigk, K., Schnitker, J., Vogel, H. P., SELEDO: a 5-year long-term trial on the effect of selegiline in early Parkinsonian patients treated with levodopa, *Eur. J. Neurol.*, 6(2):141–50, March, 1999.

87. Wilcock, G. K., Birks, J., Whitehead, A., Evans, S. J., The effect of selegiline in the treatment of people with Alzheimer's disease: a meta-analysis of published trials, *Int. J. Geriatr. Psychiatry*, 17(2):175–83, February, 2002.

88. Roselaar, S. E., Langdon, N., Lock, C. B., Jenne,r P., Parkes, J. D., Selegiline in narcolepsy, *Sleep*, 10(5): 491–5, October, 1987.

89. Grewal, M., Hawa, R., Shapiro, C., Treatment of periodic limb movements in sleep with selegiline HCl, *Mov. Disord.*, 17(2):398–401, March, 2002.

90. Houtsmuller, E. J., Thornton, J. A., Stitzer, M. L., Effects of selegiline (l-deprenyl) during smoking and short-term abstinence, *Psychopharmacology*, (Berl.), 163(2): 213–20, September, 2002.

91. Moretti, R., Torre, P., Antonello, R. M., Cazzato, G., Bava, A., Effects of selegiline on fronto-temporal dementia: a neuropsychological evaluation, *Int. J. Geriatr. Psychiatry*, 17(4):391–2, April, 2002.

92. Jankovic, J. J., Therapeutic strategies in Parkinson's disease, in Jankovic, J. J., Tolosa, E., Eds., *Parkinson's disease and Movement Disorders*, 4th ed., Lippincott, Philadelphia, P. A., 2002.

93. LeWitt, P., Therapies to extend duration of levodopa action, in Koller, W. C., Paulson, G., Eds., *Therapy of Parkinson's Disease*, 2nd ed., Marcel Dekker, New York, 77–90, 1995.

94. Olanow, C. W., Watts, R. L., Koller, W. C., An algorithm (decision tree) for the management of Parkinson's disease (2001): treatment guidelines, *Neurology*, 56(11 Suppl., 5):S1–S88, June, 2001.

95. Waters, C. H., Sethi, K. D., Hauser, R. A., Molho, E., Bertoni, J. M., Zydis selegiline reduces "off" time in Parkinson's disease patients with motor fluctuations: A 3-month, randomized, placebo-controlled study (submitted).

96. Longer-term Zydis selegiline study.

97. Youdim, M. B. H., Gross, A., Finberg, J. P. M., Rasagiline [*N*-propargyl-1*R*(+)-aminoindan], a selective and potent inhibitor of mitochondrial monoamine oxidase B, *British Journal of Pharmacology*, 132:500–506, 2001.

98. Speiser, Z., Levy, R., Cohen, S., Effects of *N*-propargyl-1-(*R*)aminoindan (rasagiline) in models of motor and cognition disorders, *J. Neural. Transm.*, Suppl., 52:287–300, 1998.

99. Rabey, J. M., Huberman, S. M., Melamed, E., Korczyn, A., Giladi, N., Inzelberg, R., Djaldetti, R., Klein, C., Berecz, G., Rasagiline mesylate, a new mao-B inhibitor for the treatment of Parkinson's disease: a double-blind study as adjunctive therapy to levodopa, *Clinical Neuropharmacology*, 23(6):324–330, 2000.

100. Youdim, M. B. H. and Weinstock, M., Novel neuroprotective anti-Alzheimer drugs with anti-depressant activity derived from the anti-parkinson drug, rasagiline, *Mech. Ageing Dev.*, 123(8):1081–6, April 30, 2002.

101. Youdim, M. B., Wadia, A., Tatton, W., Weinstock, M., The anti-parkinson drug rasagiline and its cholinesterase inhibitor derivatives exert neuroprotection unrelated to MAO inhibition in cell culture and *in vivo*, *Ann. NY Acad. Sci.*, 939:450–8, June, 2001.

102. Akao, Y., Maruyama, W., Shimizu, S., Yi, H., Nakagawa, Y., Shamoto-Nagai, M., Youdim, M. B., Tsujimoto, Y., Naoi, M., Mitochondrial permeability transition mediates apoptosis induced by N-methyl(R)salsolinol, an endogenous neurotoxin, and is inhibited by Bcl-2 and rasagiline, N-propargyl-1(R)-aminoindan, *J. Neurochem.*, 82(4):913–23, August, 2002.

103. Akao, Y., Maruyama, W., Yi, H., Shamoto-Nagai, M., Youdim, M. B., Naoi, M., An anti-parkinson's disease drug, N-propargyl-1(R)-aminoindan (rasagiline), enhances expression of anti-apoptotic bcl-2 in human dopaminergic SH-SY5Y cells, *Neurosci. Lett.*, 326(2):105–8, June 28, 2002.

104. Maruyama, W., Akao, Y., Youdim, M. B. H., Davis, B. A., Naoi, M., Transfection-enforced Bcl-2 overexpression and an anti-parkinson drug, rasagiline, prevent nuclear accumulation of glyceraldehydes-3-phosphate dehydrogenase induced by an endogenous dopaminergic neurotoxin, N-methyl®salsolinol, *Journal of Neurochemistry,* 78:727–735, 2001.

105. Eliash, S., Speiser, Z., Cohen, S., Rasagiline and its (*S*) enantiomer increase survival and prevent stroke in salt-loaded stroke-prone spontaneously hypertensive rats, *J. Neural. Transm.,* 108(8–9):909–23, 2001.

106. Marek, K., Friedman, J., Hauser, R., Juncos, J., LeWitt, P., Miyawaki, E., Tarsy, D., Olanow, C. W., Stern, M., Levy, R., Mulcahy, W., Phase II evaluation for rasagiline mesylate (TVP-1012), a novel anti-parkinsonian drug, in Parkinsonian patients not using levodopa/carbidopa, Presented at the PSG/MDS 11th Annual Symposium, *Movement Disorders,* 12(5):838, 1997.

107. Parkinson Study Group, A controlled trial of rasagiline in early Parkinson's disease, The Tempo Study, *Arch Neurol.,* 59:1937–1943, December, 2002.

108. Tyramine challenge subgroup of Tempo study, in preparation.

109. Rascol, O., Brooks, D., Melamed, E., Oertel, W., Poewe, W., Stocchi, F., Tolosa, E., A comparative randomized study of rasagiline versus placebo or entacapone as adjunct to levodopa in Parkinson's disease (PD) patients with motor fluctuations, LARGO Manuscript, LARGO Study Group, Abstract.

110. Parkinson Study Group, A controlled trial of rasagiline in Parkinson's disease patients with levodopa-related motor fluctuations (Presto Study), *American Neurological Association 128th Annual Meeting and the 6th Annual Neurology Outcomes Symposium,* San Francisco, CA, October, 19–22, 2003.

55 Catechol-O-Methyltransferase Inhibitors in the Treatment of Parkinson's Disease

Mervat Wahba, Theresa A. Zesiewicz, and Robert A. Hauser
Parkinson's Disease and Movement Disorders Center, Tampa General Healthcare, University of South Florida

CONTENTS

ABSTRACT

When levodopa is administered with a dopa decarboxylase inhibitor (DDCI), its predominant metabolism is via catechol-O-methyltransferase (COMT). The addition of a COMT inhibitor to levodopa/DDCI therapy extends the levodopa half-life ($t\frac{1}{2}$) and increases the concentration area under the curve (AUC). This makes more levodopa available for transport into the brain over a longer duration. COMT inhibitors are indicated for the treatment of Parkinson's disease patients who are experiencing motor fluctuations on levodopa/DDCI. In clinical trials, COMT inhibitors have been demonstrated to reduce "off" time, increase "on" time, and improve motor function and activities of daily living. Dopaminergic side effects such as an increase in dyskinesia are managed by lowering the levodopa dose. Tolcapone (Tasmar), a very powerful COMT inhibitor, is associated with rare fatal hepatotoxicity and is therefore reserved for patients with an inadequate response to other medications. Entacapone (Comtan) is not associated with hepatotoxicity and liver function test monitoring is not required. A new levodopa/carbidopa/entacapone product (Stalevo) is now available that offers patients the convenience of taking fewer pills per day. There is interest as to whether the use of COMT inhibition from the time levodopa is first introduced can reduce the development of motor fluctuations and dyskinesias.

KEY WORDS

Catechol-O-methyltransferase inhibitors, levodopa, motor fluctuations, dyskinesias, tolcapone, entacapone, Stalevo, clinical trials, Parkinson's disease.

INTRODUCTION

Parkinson's disease (PD) is a chronic, neurodegenerative disease associated with a progressive loss of nigrostriatal dopamine neurons and resultant reduced striatal dopamine concentration.[1] Levodopa, the chemical precursor of dopamine, remains the cornerstone of antiparkinsonian therapy, although its long-term use is hampered by the induction of motor complications including end-of-dose wearing off and dyskinesias.[2,3] Although levodopa has a short serum half-life, patients initially experience a sustained clinical response through the day, presumably due to the ability of remaining nigrostriatal neurons to take up levodopa, convert it to dopamine, and then store and release dopamine into the synaptic cleft over time.[4,5] As more and more dopamine neurons are lost, this storage and release capacity is diminished, and patients find that they improve for a few hours after taking levodopa and then the clinical benefit wears off. Ultimately, patients become dependent on a constant influx of levodopa into the brain, and clinical response fluctuates in concert with serum levodopa. In addition, many patients develop choreiform, twisting turning movements, called *dyskinesias*, that occur during dopamine peaks. The development of dyskinesias is related to both disease progression and exposure to medications, particularly levodopa.[6,7,8] It is thought that exposing postsynaptic dopamine receptors to fluctuating dopamine stimulation is a key factor in the development of dyskinesias.[9,10]

Levodopa has a serum half-life (t½) of approximately 60 min.[11,12] It is predominantly metabolized by the enzymes dopa-decarboxylase (DDC) and catechol-O-methyltransferase (COMT).[13] When orally administered by itself, only 1% of levodopa passes into the brain,[14] and there is a high incidence of nausea and vomiting due to the peripheral formation of dopamine.[15] When levodopa is combined with a DDC inhibitor such as carbidopa, the half-life of levodopa is extended to approximately 90 min, and the incidence of nausea and vomiting is greatly reduced.[16–18] Approximately 5 to 10% of orally administered levodopa reaches the brain when it is administered with a DDC inhibitor.[19]

In the presence of a DDC inhibitor, the main enzyme responsible for levodopa metabolism is COMT.[16,19] COMT metabolizes levodopa in the gut, liver, and systemic circulation by catalyzing the transfer of a methyl group from S-adenyosyl-L-methionine to a hydroxyl group of levodopa (O-methylation) to produce 3-O-methyldopa (3-OMD).[20–22] COMT inhibitors further inhibit the peripheral metabolism of levodopa, extending its central bioavailability and allowing a more continuous delivery of levodopa to the brain.

The development of first-generation COMT inhibitors was discontinued due to low efficacy and unacceptable toxicity.[23,24] The second-generation COMT inhibitors were introduced in the 1990s and are commercially available as tolcapone (Tasmar), and entacapone (Comtan, Comtess). The combination product levodopa/carbidopa/entacapone (Stalevo) was introduced in 2003.

This chapter reviews the COMT inhibitors tolcapone and entacapone, their effect on levodopa pharmacokinetics, and their clinical roles as antiparkinsonian agents.

TOLCAPONE

Tolcapone was the first COMT inhibitor to become available on the market. It is a very potent inhibitor of COMT but was found to be associated with liver toxicity in rare cases. For this reason, it is reserved for patients who do not experience an adequate response to other antiparkinsonian medications.

Tolcapone (4-methyl-3,4-dihydroxy-5-nitro-benzophenone) is a selective, reversible COMT inhibitor with peripheral and central effects. It is rapidly absorbed after oral administration and has a half-life of approximately two hours.[25] Approximately 65% of orally ingested tolcapone enters the general circulation, and less than 20% is lost during first-pass metabolism. Tolcapone is highly protein bound and is metabolized mainly in the liver by conjugation, including methylation, glucuronidation, sulphation, and acetylation.[27,28] Approximately 60% of tolcapone metabolites are found in urine, while 40% are excreted in the feces. Tolcapone has been shown to inhibit COMT activity in gut, liver, and brain in animal studies.[21] In human volunteers, tolcapone inhibits erythrocyte COMT by 80 to 90% at doses of 200 to 800 mg.[25,26]

Tolcapone increases the half-life and concentration area under the curve (AUC) of levodopa in healthy volunteers and in PD patients,[27] while reducing the AUC of 3-OMD by up to 80%.[26] In one study, tolcapone increased the AUC of levodopa in healthy volunteers in a dose-dependent fashion and, at a dose of 200 mg, the AUC of levodopa was doubled.[27,29,30] In another study, chronic administration of tolcapone 200 mg t.i.d. increased levodopa's half-life by 81% and AUC by 34%.[27]

TOLCAPONE SAFETY ISSUES

In preclinical toxicity testing, tolcapone did not cause hepatotoxicity. However, in clinical trials, liver chemistry tests were more than 3 times elevated in approximately 1% of patients taking tolcapone 100 mg t.i.d., and in almost 3% of patients taking tolcapone 200 mg t.i.d. Postmarketing surveillance studies revealed that tolcapone-induced fatal hepatitis in 3 out of 60,000 patients,[31] an incidence 10 to 100 times higher than that found in the general population.[32] The three stricken patients were elderly women who had not undergone liver function monitoring. These patients became ill 8 to 12 weeks following tolcapone initiation. For this reason, the drug was withdrawn from the market in Europe and Canada, and a black box warning was issued in the United States. Its use is currently reserved for PD patients with motor fluctuations who have not adequately responded to other therapy. Frequent monitoring of liver function tests is required, and written consent is recommended.

CLINICAL TRIALS EVALUATING TOLCAPONE

Tolcapone is indicated as an adjunct to levodopa for patients with motor fluctuations who have not been adequately controlled with other antiparkinsonian medications and in whom the benefits outweigh the potential risks. Several large, prospective, randomized clinical trials demonstrated that the addition of tolcapone to levodopa in patients with motor fluctuations reduces "off" time, increases "on" time, and improves motor function.

In a randomized, double-blind, placebo-controlled study, 202 PD patients with motor fluctuations were randomized to receive either tolcapone 100 or 200 mg t.i.d. or placebo.[33] Following 3 months of therapy, patients treated with tolcapone experienced a reduction in daily "off" time of 3.25 hr compared to baseline ($p < 0.01$). Motor function and overall efficacy were also significantly improved ($p < 0.01$). Adverse events included dyskinesias that worsened in 51% of patients taking tolcapone 100 mg t.i.d. and 64% of patients taking tolcapone 200 mg t.i.d., compared to 18% of patients taking placebo. Eighteen percent of patients withdrew due to side effects.

Myllyla et al.[34] studied the efficacy of different doses of tolcapone (50, 200, or 400 mg t.i.d.) over 6 weeks in 154 patients with wearing-off symptoms in a randomized, double-blind, placebo-controlled, parallel-group, multicenter trial. Tolcapone was effective in reducing "off" time at all three doses. The most effective dose was 200 mg t.i.d., which increased "on" time significantly from 37.9% of the waking day to 50.8% and significantly decreased "off" from 26.7 to 16.4%.

Baas et al.[35] studied the efficacy of tolcapone (100 or 200 mg t.i.d.) in 177 patients with "wearing-off" symptoms in a randomized, double-blind, placebo-controlled, parallel-group, multicenter trial. After 3 months, there was a significant reduction of "off" time by 31.5% (100 mg) and 26.2% (200 mg) in the tolcapone groups. "On" time increased by 21.3% (100 mg) and 20.6% (200 mg). Motor UPDRS scores improved with tolcapone 200 mg t.i.d.

Adler et al.[36] studied the efficacy of tolcapone (100 or 200 mg t.i.d. for 6 weeks) in 215 patients with predictable end of dose fluctuations in a randomized, double-blind, placebo-controlled, parallel-group, multicenter-trial. Both tolcapone doses significantly reduced "off" time, by 2hr/day and 2.5 hr/day, respectively, and increased "on" time by 2.1 and 2.3 hr/day, respectively.

Tolcapone was also demonstrated to provide mild symptomatic benefit in patients with a stable response to levodopa. The effects of tolcapone 100 mg and 200 mg t.i.d. were evaluated as adjunct therapy to levodopa in a double-blind, placebo-controlled study in 298 PD patients without motor fluctuations.[37] Both dosages of tolcapone produced significant improvements in UPDRS activities of daily living and the total scores at 6 and 12 months. Patients who received tolcapone had reductions in daily levodopa dosages at 6-month follow up (tolcapone 100 mg t.i.d., levodopa 20.8 mg; tolcapone 200 mg t.i.d., levodopa 32.3 mg) compared to patients who received placebo (mean increase of levodopa dose of 46.6 mg).

Tolcapone was well tolerated, and the principal adverse events were diarrhea and levodopa-related side effects. Nonetheless, tolcapone should not be used in nonfluctuating patients, as the risk of hepatotoxicity outweighs the potential benefit.

Side effects of tolcapone include increased dyskinesia, diarrhea, nausea, and hallucinations. Patients who have preexisting peak-dose dyskinesias often experience a rapid and substantial increase in dyskinesia with the introduction of tolcapone, necessitating a 25 to 50% reduction in levodopa dose. Alternatively, one can decrease the levodopa dose preemptively by about 25% at the time tolcapone is introduced in an effort to avoid an uncomfortable increase in dyskinesia. Once tolcapone is on board, the levodopa dose should be titrated to provide the best response and to minimize side effects. Other dopaminergic side effects such as nausea and hallucinations can also be managed by lowering the levodopa dose. Approximately 10% of patients on tolcapone experience diarrhea, and 3% discontinue tolcapone because of this side effect. Onset of diarrhea is usually delayed 4 to 12 weeks after initiation of therapy, but is uncommon after 6 months.

In clinical practice, tolcapone is administered at a dose of 100 or 200 mg t.i.d. The 200 mg t.i.d. dose should only be used if the additional benefit outweighs the potential risk. It is reasonable to introduce tolcapone at a dose of 100 mg with the first levodopa dose of the day and then titrate up by 100 mg per week to 100 mg t.i.d. This will allow reduction of the levodopa dose as necessary and may avoid a severe increase in dyskinesia. Liver function tests must be monitored.

ENTACAPONE

Entacapone ((E)-2-cyano-N,Ndiethyl-3-(3,4-dihydroxy-5-nitrophenyl-propenamide) is a selective, potent, reversible COMT inhibitor that acts peripherally.[38] The extent of COMT inhibition correlates with plasma concentration and was maximal at 82% after administration of the highest oral dose tested (800 mg).[39] It is a weak acid (pK of 4.5)[40] and possesses a nitrocatechol structure.[39,41] Entacapone is rapidly absorbed and has an elimination half-life of 1.6 to 3.4 hr.[39] Increases in C_{max} and AUC are directly proportional to dose. Food does not affect its absorption. Entacapone has a bioavailability of 29 to 46% following its oral administration.[42] Because of its short half-life, entacapone is administered with each levodopa dose (200 mg up to eight times per day in the U.S., and 10 times per day in the E.U.).

The major metabolite of entacapone is its Z-isomer, which is detectable in plasma and urine, and its main metabolic pathway is glucuronidation.[27] Entacapone and its metabolites are primarily eliminated via biliary excretion.[41] Eighty to 90% of entacapone is excreted in feces, while 10 to 20% is passed in the urine.

Effect of Entacapone on Levodopa Pharmacokinetics

~ Levodopa is an amino acid that is primarily metabolized by DDC. It is absorbed in the proximal small bowel by a facilitated transport system for large neutral amino acids.[42] ~ While almost 30% of orally administered levodopa enters the circulation, 98% is peripherally decarboxylated to dopamine by DDC and does not pass the blood-brain barrier.[43] The second most important route of levodopa metabolism is O-methylation by COMT to form 3-OMD. ~ Entacapone prolongs the elimination half-life and increases the bioavailability of levodopa by reducing its peripheral metabolism to 3-OMD. When entacapone 200 mg is administered with levodopa/carbidopa, the dual inhibition of DDC and COMT prolongs the half-life of levodopa from 1.3 to 2.4 hr, while the plasma bioavailability of levodopa (AUC) increases to 35 to 40%. This also results in a 30 to 50% reduction in the variability of levodopa plasma levels.[44] In single-dose models, entacapone does not affect levodopa's peak concentration (C_{max}) or the time it takes to reach maximum concentration (t_{max}).[45] In healthy volunteers, entacapone (doses 50 to 400 mg) administered with a single dose of levodopa/carbidopa (100/25) increased the AUC of levodopa dose-dependently, while the pharmacokinetics of carbidopa remained unchanged.[27,46]

The effects of entacapone on levodopa were tested in 25 PD patients with motor fluctuations over four 2-week treatment periods. Patients received 100, 200, or 400 mg of entacapone or a placebo with each levodopa intake in a randomized, double-blind, placebo-controlled trial with crossover design.[47,48] Entacapone decreased erythrocyte COMT activity by 25% (100 mg), 33% (200 mg), and 32% (400 mg). The elimination half life of levodopa was prolonged by 23% ($p < 0.05$), 26% ($p < 0.001$), and 48% ($p < 0.001$), and the levodopa AUC increased by 17% ($p < 0.05$), 27% ($p < 0.001$), and 37% ($p < 0.001$) with increasing doses. Daily "off" time was reduced by 11% (100 mg), 18% (200 mg), and 20% (400 mg), although these changes were not statistically significant in this small study. No severe adverse events were reported.

Clinical Trials of Entacapone

The Parkinson Study Group conducted a double-blind, placebo-controlled trial of entacapone in 205 PD patients with motor fluctuations, referred to as the SEESAW study (Safety and Efficacy of Entacapone Study Assessing Wearing Off).[20] Patients were randomized to receive either entacapone 200 mg with each levodopa dose or placebo

and were followed for 24 weeks. The primary efficacy measure was percent "on" time while awake, as recorded by home diaries. The use of entacapone increased "on" time by 5 percent (95% CI, 1.7 to 8.3; $p = 0.003$), or approximately 1 hr per day. "Off" time was reduced by 17.6% in patients taking entacapone, compared to 4.5% in placebo-treated patients ($p < 0.01$). 3-OMD levels in patients receiving entacapone declined from 6.5 μg/ml at baseline to 2.8 μg/ml at week 24, and the adjusted mean 3-OMD level was 57.8% lower in the entacapone group than in patients treated with placebo (95% CI, 53.7 to 61.5; $p < 0.0001$). Patients taking entacapone also reduced levodopa intake by about 100 mg/day, or 11.6% compared to 2.5% in the placebo group. There were significant improvements in UPDRS ADL, motor, and total scores in the entacapone group as well.

The Nomecomt Study Group evaluated the safety and efficacy of entacapone as an adjunct to levodopa in 171 PD patients with motor fluctuations in a 6-month double-blind, randomized trial.[49] Patients were randomized to receive placebo or entacapone 200 mg with each levodopa dose (4 to 10 per day). Entacapone increased mean (±SD) daily "on" time significantly (9.3 ± 2.2 to 10.7 ± 2.2 hr; $p < 0.01$), while decreasing "off" time (5.3 ± 2.2 to 4.2 ± 2.2 hr; $p < 0.001$). The daily levodopa dose was decreased in the entacapone group compared to the placebo group by 102 mg ($p < 0.01$). The mean duration of the beneficial effect after the first morning levodopa dose was 0.24 hr longer in patients taking entacapone ($p < 0.001$). Entacapone was well tolerated, and side effects were mild.

In a three-year open-label extension study of the Nomecomt study, the mean duration of benefit of a single dose of levodopa increased significantly from 2.1 to 2.8 hr at 3 months ($p < 0.01$) and remained increased throughout the remainder of the study.[50] The patients' global assessment indicated that 69% of patients improved when given entacapone, an effect that was maintained until the end of the study (64%). There was a significant worsening of disability upon withdrawal of entacapone.

The Celomen study evaluated the efficacy and safety of entacapone in PD patients with a suboptimal levodopa response in a double-blind, randomized trial. Three hundred and one PD patients were randomized to receive entacapone or placebo with each daily dose of standard or controlled-release (CR) levodopa/carbidopa and were followed for 24 weeks. "On" time increased by a mean of 1.7 hr while "off" time decreased by 1.5 hr, compared to placebo ($p < 0.05$). UPDRS motor and ADL scores were significantly improved in patients taking entacapone ($p < 0.05$). The daily levodopa dose was decreased by 54 mg in patients taking entacapone and increased by 27 mg in the placebo group ($p < 0.05$). Results were similar regardless of whether levodopa/carbidopa immediate-release (IR) or levodopa/carbidopa CR was used. Diarrhea occurred in 8% of patients taking entacapone, compared to 4% of patients taking placebo, and was rarely severe.[51]

There is no known hepatoxicity with entacapone use, and liver function tests are not required. Adverse events associated with entacapone are generally less severe than those associated with tolcapone. Side effects identified in clinical trials include increased dyskinesias, nausea, diarrhea, fatigue, urine discoloration, and abdominal pain.[52] An increase in dyskinesia, or other dopaminergic side effects, can potentially be resolved by a reduction of the levodopa dose. The likelihood of the need for a levodopa dose reduction following the addition of entacapone is higher in patients with preexisting dyskinesia and those on higher (>600 mg/day) daily levodopa doses.[32] Diarrhea is generally mild and may occur six weeks to six months after entacapone introduction. However, discontinuation of entacapone is rarely required due to diarrhea. It is possible that nausea is due in part to increased peripheral levodopa metabolism via DDC, and it may respond to increased DDC inhibition, but there are no data available on this. Urine discoloration is due to metabolites of entacapone and causes no medical problem. However, it may be annoying to patients if the discoloration occurs in sweat and stains clothes. There is one reported case of severe hypertension following ephedrine administration in a patient receiving entacapone.[53] Coadministration of entacapone with oral selegiline (an MAO-B inhibitor) at doses up to 10 mg per day has been shown to be safe, with no adverse events, drug interaction, or arrhythmias demonstrated in a double-blind study in 16 PD patients with end-of-dose motor fluctuations.[54] The effect of entacapone on the pharmacokinetics of warfarin was also studied in a randomized, double-blind, two-way crossover study in 12 healthy subjects who received either entacapone 200 mg 4 times daily or placebo for 1 week during individually optimized treatment with warfarin (INR 1.4–1.8).[25] Entacapone displayed a slight pharmacokinetic interaction with warfarin, but it appears that these medications can be safely used together.

In clinical practice, entacapone is administered at a dose of 200 mg with each levodopa intake, up to 8 times per day in the U.S. or 10 times per day in the E.U. In most patients, enatacapone is added "all at once," but one can consider upward titration, adding one entacapone tablet per week, especially in patients with moderate or severe dyskinesia. If dyskinesia or other dopaminergic side effects emerge, the levodopa dose should be reduced.

LEVODOPA/CARBIDOPA/ENTACAPONE (STALEVO)

Levodopa/carbidopa/entacapone (Stalevo) is now approved to treat patients who are experiencing motor fluctuations on levodopa/DDC inhibitor. Stalevo is available in 50-, 100-, and 150-mg strengths, indicating the dosage of levodopa contained in each tablet. Each tablet also contains carbidopa in a 1:4 mg ratio to levodopa and 200 mg of entacapone.

A series of pharmacokinetic studies demonstrated bioequivalence between Stalevo and corresponding dosages of levodopa/carbidopa plus entacapone.[55] The primary clinical advantage of Stalevo is convenience. Patients can take one combination pill rather than two (or more) separate tablets, and the Stalevo tablets are smaller and easier to swallow. These advantages may be particularly beneficial for patients taking many pills, those who have difficulty following complex medication regimens, and those with swallowing difficulty. A potential disadvantage of fixed combination pills is reduced flexibility in dosing.

Most patients taking levodopa/carbidopa IR plus entacapone can be directly switched to corresponding dosages of Stalevo. It is also possible to switch fluctuating PD patients receiving levodopa/carbidopa CR plus entacapone to Stalevo. The bioavailability of levodopa in the levodopa/carbidopa CR preparation is approximately 70 to 75% of levodopa/carbidopa IR.[56] This suggests that the levodopa AUC of levodopa/carbidopa CR 200/50 plus entacapone 200 mg should be comparable to Stalevo 150. However levodopa from Stalevo has a t_{max} similar to levodopa/carbidopa IR (0.5 to 1.5 hr), whereas the t_{max} is more delayed from levodopa/carbidopa CR (t_{max} = 1.5 to 3 hr).[54] These differences should be considered when switching patients from levodopa/carbidopa CR to Stalevo. In most patients, a shorter levodopa t_{max} is expected to be desirable to shorten the latency to onset of clinical benefit.

Not all levodopa/carbidopa plus entacapone regimens can be easily converted to Stalevo. There are no corresponding dose Stalevo tablets currently available for levodopa IR doses of 200 mg or more plus entacapone. In addition, the maximum recommended daily dose of levodopa from Stalevo alone is 1200 mg in the U.S. and 1500 mg in the E.U., due to limited experience with entacapone daily doses above 1600 mg/day in the U.S. and 2000 mg/day in the E.U. Patients on such regimens can be treated with levodopa/carbidopa plus entacapone or a combination of Stalevo and levodopa/carbidopa products.

For patients with motor fluctuations on levodopa/carbidopa without a COMT inhibitor, switching to the comparable levodopa-dose Stalevo tablet is analogous to adding entacapone. The FDA recommends that, for patients who are likely to require a reduction in levodopa dose, specifically those with preexisting dyskinesia or receiving more than 600 mg levodopa per day, entacapone should first be added, levodopa titrated as necessary, and then a switch can be made to the corresponding dose Stalevo tablets.

Use of COMT Inhibition when Levodopa Is First Introduced

Levodopa remains the most effective treatment for PD symptoms, but its use is associated with the development of long-term motor complications, specifically motor fluctuations and dyskinesias. Studies have demonstrated that using long-acting dopamine agonists to treat parkinsonian symptoms in 1-methyl-4-phenyl-1,2,3,6-tetrahydropyridine (MPTP)-treated primates results in a lower incidence of dyskinesias than levodopa.[57,58] Based on these studies, several clinical trials were conducted to compare the initial use of dopamine agonists to which levodopa can be added, to levodopa alone in patients with early PD. These studies demonstrated that initial dopamine agonist use was associated with a lower incidence of motor complications.[59,60] One hypothesis to explain these findings is the Continuous Dopaminergic Stimulation (CDS) hypothesis, which posits that symptomatic treatment strategies that provide more continuous dopaminergic stimulation reduce the emergence of motor fluctuations and dyskinesias.

Another strategy for providing CDS is to prolong the half-life of levodopa through the addition of a COMT inhibitor. Because tolcapone carries a small but finite risk of hepatotoxicity, it should not be considered for early use. However, because of its good safety profile, entacapone can be considered for early, off-label use.

Jenner and colleagues[61] compared the use of levodopa/carbidopa with levodopa/carbidopa plus entacapone using the MPTP primate model. Administration of levodopa/carbidopa on a three-hour dosing schedule (four administrations per day) resulted in a rapid induction of severe dyskinesias. However, when levodopa/carbidopa (at the same dose and on the same schedule) was administered with entacapone, significantly less dyskinesia occurred, with comparable improvement in parkinsonian signs over a greater duration. This study provides strong support for the CDS hypothesis, because the same agent (levodopa), when administered in a more continuous fashion (i.e., with entacapone) was demonstrated to cause less dyskinesia. The study suggests that the use of entacapone from the time levodopa is first introduced might reduce the emergence of motor complications in PD patients. Clinical trials to definitively evaluate this outcome are now in development. Important considerations for such a clinical trial include the optimal levodopa dose and interdose interval required to achieve CDS.

REFERENCES

1. Jellinger, K., Overview of morphological changes in Parkinson's disease, *Adv. Neurol.,* 45:1–18, 1987.
2. Fahn, S., Adverse effects of levodopa in Parkinson disease, in Calne, D. B., Ed., *Drugs for the treatment of Parkinson disease,* Berlin: Springer-Verlag, 385–409, 1989.
3. Marsden, C. D., Parkinson disease, *J. Neurol. Neurosurg, Psychiatry,* 57;672–-681, 1994.
4. Koller, W. C., Alternate day levodopa therapy in parkinsonism, *Neurology,* 32:324–326, 1982.
5. Sage, J. I., Mark, M. H., Basic mechanisms of motor fluctuations, *Neurology,* 44 (suppl. 6):S10–S14, 1994.
6. Ropinirole in the treatment of early Parkinson's disease: a 6-month interim report of a 5-year levodopa-controlled study, 056 Study Group, *Mov. Disord.,* 13:39–45, 1998.
7. Marek, K., Seibyl, J., Shoulson, I. et al., Dopamine transporter brain imaging to assess the effects of pramipexole vs. levodopa on Parkinson disease progression, *JAMA,* 287:1653–1661, 2002.
8. Whone, A. L., Rakshi, J. S., Watts, R. L., Brooks, D. J., Two trials demonstrating disease-slowing effects of ropinirole, compared with L-dopa, in early Parkinson's disease, *Mov. Disord.,* 17 Suppl., 5:85–86, 2002.
9. Engber, T. M., Susel, Z., Juncos, J., Chase, T., Continuous and intermittent levodopa differentially affect rotation induced by D1& D2 dopamine agonists, *Eur. J. Pharmacol.,* 168:291–298, 1989.
10. Juncos, I. L., Engber, T. M., Raisman, R., Chase, T. N., Continuous and intermittent levodopa differentially affect rotation induced by D1 and D2 dopamine agonists, *Ann. Neurol.,* 25:473–478, 1989.
11. Dunner, D. L., Brodie, H. K. H., Goodwin, Frederick K., Plasma DOPA response to levodopa administration in man: effects of peripheral decarboxylase inhibitor, *Clinical Pharmacology and Therapeutics,* 12:211–217, 1971.
12. Gancher, S. T., Nutt, G. J., Woodward, W. R., Peripheral pharmacokinetics of levodopa in untreated, stable and fluctuating parkinsonian patients, *Neurology,* 37: 940–944, 1987.
13. Reenila, Ilkka, Catechol-0-methyltransferase activity: Assay, distribution and pharmacologic modification, Institute of Biomedicine Department of Pharmacology and Toxicology, University of Helsinki, Finland, 1999.
14. Travainen, H., Rinne, U., Gordin, A., Catechol-O-methyl transferase inhibitors in Parkinson disease, *Advances in Neurology,* Vol. 86, Chapter 32, 311–325.
15. Nutt, J. G., Fellman, J. H., Pharmacokinetics of levodopa, *Clin. Neuropharmacol.,* 7:175–183, 1984.
16. Fahn, S., "On-off" phenomenon with levodopa therapy in parkinsonism, *Neurology,* 431–441, 1974.
17. Reid, J. L., Calne, D.B., Vakil, S. D., Allen, J. G., Davies, C. A., Plasma concentration of levodopa in parkinsonism before and after inhibition of peripheral decarboxylase, *J. Neurol. Sci.,* 17:45–51, 1972.
18. Mars, H., Modification of levodopa effect by systematic decarboxylase inhibition, *Arch Neurol.,* 28:91–95, 1973.
19. Mannisto, P. T., Kaakkola, S., Rationale for selective COMT inhibitors as adjuncts in the treatment of Parkinson disease, *Pharmacol. Toxicol.,* 66:317–323, 1990.
20. Parkinson Study Group, Entacapone improves motor fluctuations in levodopa-treated Parkinson's disease patients, *Ann. Neurol.,* 42:747–755, 1997.

21. Da Prada, M., Zurcher, G., Kettler, R., Colzi, A., New therapeutic strategies in Parkinson's disease: inhibition of MAO-B by Ro 19-6327 and of COMT by Ro 40-7592, *Adv. Behav. Biol.,* 39:723–732, 1991.

22. Mannisto, P. T., Tuomainen, P., Tuomainen, U. K., Different *in vivo* properties of three new inhibitors of catechol-o-methyltransferase in rat, *Br. J. Pharmacol.,* 105:569–574, 1992.

23. Guldberg, H. C., Marsden, C. A., Catechol-O-methyl transferase: pharmacological aspects and physiologic role, *Pharmacol. Rev.,* 27:135–206, 1975.

24. Tai, C. H. and Wu, R. M., Catchol-O-methyltransferase and Parkinson's disease, *Acta Medica Okayama,* 56:1-6, 2002.

25. Dingemanse, J., Jorga, K. M., Schmitt, M. et al., Integrated pharmacokinetics and pharmacodynamics of the novel catechol-o-methyltransferase inhibitor tolcapone during first administration to humans, *Clin. Pharmacol. Ther.,* 57:508–517, 1995.

26. Dupont, E., Burguner, J. M., Findley, L. J. et al., Tolcapone added to levodopa in stable parkinsonian patients: a double-blind placebo-controlled study, *Mov. Disord.,* 12:928–934, 1997.

27. Bonifati, V., Meco, G., New, selective catchol-O-methyltransferase inhibitors as therapeutic agents in Parkinson's disease, *Pharmacol. Ther.,* 81:1–36, 1999.

28. Lave, T. H., Dupin, S., Schmitt, M. et al., Interspecies scaling of tolcapone, a new inhibitor of catchol-O-methyltransferase (COMT). Use of *in vitro* data from hepatocytes to predict metabolic clearance in animals and humans, *Xenobiotica,* 26:839–851, 1996.

29. Dingemanse, J., Jorga, K., Zurcher, G. et al., Pharmacokinetics-pharmacodynamic interaction between the COMT inhibitor tolcapone and single-dose levodopa, *Br. J. Clin. Pharmacol.,* 40:253–262, 1995.

30. Sedek, G., Jorge, K., Schmitt, M. et al., Effect of tolcapone on plasma levodopa concentrations after coadministration with levodopa/carbidopa to healthy volunteers, *Clin. Neuropharmacol.,* 20:531–541, 1997.

31. Sedek, G., Jorge, K., Schmitt, M. et al., Parmacokinetics—pharmacodynamic interaction between the COMT inhibitor tolcapone and single-dose levodopa, *Br. J. Clin. Pharmacol.,* 40:253–262, 1995.

32. Assal, R., Spahr, L., Hadengue, A. et al., Tolcapone and fulminant hepatitis, *Lancet,* 352:958, 1998.

33. Hauser, R. A., Lyons, K. E., Pahwa, R., Zesiewicz, T. A., Golbe, L. I., *Parkinson's Disease: Questions and Answers,* 4th ed., Merit Publishing International, West Palm Beach, Florida, 2003.

34. Rajput, A. H., Marin, W., Saint-Hilaire, M. H. et al., Tolcapone improves motor function in parkinsonian patients with the "wearing-off" phenomenon: a double-blind, placebo-controlled, multicenter trial, *Neurology,* 49:1066–1071, 1997.

35. Myllyla, V. V., Jackson, M., Larsen, J. P. et al., Efficacy and safety of tolcapone in levodopa-treated Parkinson's disease patients with "wearing-off" phenomenon: a multicenter, double-blind, randomized, placebo-controlled trial, *Eur. J. Neurol.,* 4:333–341, 1997.

36. Baas, H., Beiske, A. G., Ghika, J. et al., Catachol-O-methytransferase inhibition with tolcapone reduces the "wearing off" phenomenon and levodopa requirements in fluctuations patients with PD, *J. Neurol. Neurosurg. Psychiatry,* 63:421–428, 1997.

37. Adler, C. H., Singer, C., OBrien, C. et al., Randomized, placebo-controlled study of tolcapone in patients with fluctuating Parkinson's disease treated with levodopa-carbidopa, *Arch Neurol.,* 55:1089–1095, 1998.

38. Waters, C. H., Kurth, M., Bailey, P. et al., Tolcapone in stable Parkinson's disease: efficacy and safety of long-term treatment, *Neurology,* 49:665–671, 1997.

39. Nissenen, E., Linden, I. B., Schultz, E. et al., Biochemical and pharmacological properties of a peripherally acting catechol-o-methyltransferase inhibitor entacapone, *Naunym-Schmiedebergs Arch Pharmacol.,* 346: 262–266, 1992.

40. Keranen, T., Gordin, A., Karlsson, M. et al., Inhibition of soluble catechol-O-methyltransferase and single-dose pharmacokinetics after oral and intravenous administration of entacapone, *Eur. J. Clin. Pharmacol.,* 446: 151–157, 1994.

41. Savolainen, J., Forsberg, M., Taipale, H. et al., Effects of aqueous solubility and dissolution characteristics on oral bioavailabilty of entacapone, *Drug Dev. Res.,* 49:238–244, 2000.

42. Lautala, P., Kivimaa, M., Salomies, H. et al., Glucuronidation of entacapone, nitecapone, tolcapone, and some other nitrocatechols by rat liver microsomes, *Pharm. Res.,* 14:1444–1448, 1997.

43. Wade, D. N., Mearrick, P. T., Morris, J. L., Active transport of L-dopa in the intestine, *Nature,* 242:463–465, 1973.

44. Calne, D. B., Karoum, F., Ruthven, C., Sandler, M., The metabolism of orally administered L-dopa in parkinsonism, *Br. J. Pharmacol.,* 37:57–68, 1969.

45. Nutt, J. G., Woodward, W. R., Beckner, R. M., Stone, C. K., Berggren, K., Carter, J. H., Gancher, S. T., Hammerstad, J. P., Gordin, A., Effect of peripheral catechol-O-methyltransferase inhibition on the pharmacokinetics and pharmacodynamics of levodopa in parkinsonian patients, *Neurology,* 44:913–919, 1994.

46. Kaakkola, S., Gordin, A., Mannisto, P. T., General properties and clinical possibilities of new selective inhibitors of catechol-o-methyltransferase, *General Pharmacol.,* 25:813–824, 1994.

47. Keranen, T., Gordin, A., Harjola, V. P. et al., The effects of catechol-O-methy transferase inhibition by entacapone on the pharmacokinetics and metabolism of levodopa in healthy volunteers, *Clin. Neuropharmacol.,* 16:145–156, 1993.

48. Lyytinen, J., Sovijarvi, A., Kaakkola, S. et al., The effect of catechol-O-methyltransferase inhibition with entacapone on cardiovascular autonomic responses in L-dopa-treated patients with Parkinson's disease, *Clin. Neuropharm.,* 24:50–57, 2001.

49. Heikkinen, H., Nutt, J. G., Le, Witt et al., The effects of different repeated doses of entacapone on the pharmacokinetics of L-Dopa and on the clinical response to

L-Dopa in Parkinson's disease, *Clin. Neuropharm.,* 24:150–157, 2001.

50. Rinne, U. K., Larsen, J. P., Siden, A., Worm-Petersen, J., Entacapone enhances the response to levodopa in parkinsonian patients with motor fluctuations. Nomecomt Study Group, *Neurology,* 51(5):1309–14, November, 1998.

51. Larsen, J. P., Worm-Petersen, J., Siden, A., Gordin, A., Reinikainen, K., Leinonen, M, NOMESAFE Study Group. The tolerability and efficacy of entacapone over 3 years in patients with Parkinson's disease, *Eur. J. Neurol.,* 10(2):137–46, March, 2003.

52. Poewe, W. H., Deuschl, G., Gordin, A. et al., Efficacy and safety of entacapone in Parkinson's disease patients with suboptimal levodopa response: a 6-month randomized placebo-controlled double-blind study in Germany and Austria (Celomen study), *Acta Neurol. Scand.,* 105:245–255, 2002.

53. Kaakola, S., Teravainen, H., Ahtila, S. et al., Effect of entacapone, a COMT inhibitor, on clinical disability and L-dopa metabolism in parkinsonian patients, *Neurology,* 44:77–80, 1994.

54. Renfrew, C., Dickson, R., Schwab, C. et al., Severe hypertension following ephedrine administration in a patient receiving entacapone, *Anesthesiology,* 93:1562, 2000.

55. Lyytinen, J., Kaakkola, S., Gordin, A. et al., Entacapone and selegiline with L-dopa in patients with Parkinson's disease: an interaction study, *Parkinsonism and Related Disorders,* 6:2125–222, 2000.

56. Hauser, R. A., Levodopa/carbidopa/entacapone (*Stalevo*), *Neurology,* Suppl., In press.

57. Gauthier, S., Amyot, D., Sustained release antiparkinson agents: controlled release levodopa, *Can. J. Neurol. Sci.,* Suppl., 153–5, 1992.

58. Pearce, R. K., Banerji, T., Jenner, P., Marsden, C. D., De novo administration of ropinirole and bromocriptine induces less dyskinesia than L-dopa in the MPTP-treated marmoset, *Movement Disorders,* 13:234–241, 1998.

59. Maratos, E. C., Jackson, M. J., Pearce, R. K., Jenner, P., Antiparkinsonian activity and dyskinesia risk of ropinirole and L-DOPA combination therapy in drug naive MPTP-lesioned common marmosets (Callithrix jacchus), *Movement Disorders,* 16:631–641, 2001.

60. Rascol, O., Brooks, D., Korczyn, A. D. et al., A five-year study of the incidence of dyskinesia in patients with early Parkinson's disease who were treated with ropinirole or levodopa, *N. Engl. J. Med.,* 342:1484–1491, 2000.

61. Parkinson Study Group, Pramipexole vs. levodopa as initial treatment for Parkinson's disease: A randomized controlled trial, *JAMA,* 284:1931–1938, 2000.

62. Jenner, P., Al-Barghouthy, G., Smith, L. et al., Initiation of entacapone with levodopa further improves antiparkinsonian activity and avoids dyskinesia in MPTP primate model of Parkinson's disease (abstract), *Neurology,* 58, Suppl., 58, A374–A375, 2002.

56 Symptomatic Treatment of Parkinson's Disease: Levodopa

John L. Goudreau
Michigan State University

J. Eric Ahlskog
Mayo Clinic

CONTENTS

INTRODUCTION

A revolution in the medical treatment of Parkinson's disease (PD) occurred with the recognition that levodopa (3,4 dihydroxy-L-phenylalanine, Figure 56.1), the direct biochemical precursor to dopamine (DA), replenishes neuronal dopamine stores and improves many of the motor features of PD. Since its introduction over three decades ago, levodopa remains the single most effective drug for alleviating the motor symptoms of PD and is the only drug that improves life span in this progressive disorder.[1–7]

Levodopa, unlike dopamine (DA), crosses the blood-brain barrier via large, neutral amino acid transporters (Figure 56.2). Once inside the neuron, levodopa undergoes rapid decarboxylation by aromatic amino acid decarboxylase to produce DA. Newly synthesized DA is then stored in synaptic vesicles for subsequent release. Since levodopa bypasses the rate-limiting enzyme, tyrosine hydroxylase, administration of this precursor accelerates DA synthesis through mass action, replenishing depleted central DA stores. Replacement of declining concentrations of synap-

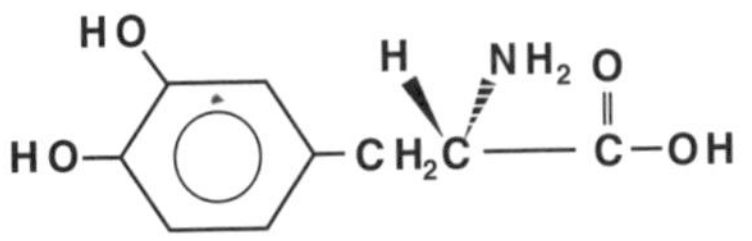

FIGURE 56.1 Levodopa.

0-8493-1590-5/05/$0.00+$1.50

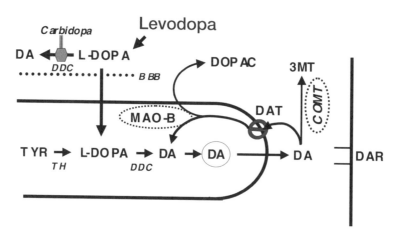

FIGURE 56.2 Schematic diagram of DA biosynthesis in a nigrostriatal nerve terminal. The rate-limiting enzyme, tyrosine hydroxylase (TH), converts the dietary amino acid tyrosine (TYR) into levodopa (L-dopa). L-dopa is then converted into dopamine (DA) by dopa-decarboxylase (DDC). DA is stored in vesicles for release into the synaptic cleft where it can activate postsynaptic dopamine receptors (DAR). DA may be brought back into the presynaptic nerve terminal via the dopamine transporter (DAT) or may be metabolized by mitochondria monoamine oxidase type B (MAO-B) or catechol-O-methyltransferase (COMT) into 3,4-dihydroxyphenylacetic acid (DOPAC) or 3-methoxytryamine (3-MT), respectively. Peripheral DDC-mediated catabolism of levodopa is blocked by coadministration of carbidopa, thus allowing appreciable amount of levodopa to pass across the blood-brain barrier (BBB) and enter the presynaptic nerve terminal, where it may be enzymatically converted into DA.

tic DA released from nigrostriatal neurons provides the rationale for the therapeutic benefit of levodopa.

Improvement in the motor features of PD with dopa was first reported in 1961 with intravenous administration and in 1962 with oral administration.[8,9] However, widespread use did not occur prior to 1967, when Cotzias and colleagues demonstrated profound improvement in the motor symptoms of PD following large oral doses of levodopa.[10] Early in the course of PD, the degree of improvement in tremor, rigidity, and bradykinesia can be profound when the dose is fully titrated and often returns patients to a nearly asymptomatic state. The dose-response curve for symptomatic benefit from levodopa can be steep and, as a result, there is often a threshold above which patients experience dramatic improvement in motor function.[11,12] Despite decades of research and development in the symptomatic treatment of PD, no other medication exceeds the therapeutic benefit in improving the motor features of PD.

PHARMACOKINETICS

Orally administered levodopa does not enter the bloodstream until passing into the proximal small intestine, where passage is via an active transport system for aromatic amino acids.[13] The rate and extent of absorption of levodopa is dependent on gastric emptying, gastric pH, and the length of time the compound is exposed to the degradative enzymes of the gastric and intestinal mucosa.[13] Levodopa, when administered alone (i.e., without a decarboxylase inhibitor), is extensively metabolized by peripheral decarboxylase, and less than 1% penetrates

the central nervous system.[14] Levodopa is carried across the blood-brain barrier by a large neutral aromatic amino acid transporter. Within the brain, the motor effects of levodopa are mediated by conversion to DA via intraneuronal dopa decarboxylase (L-aromatic amino acid decarboxylase); the pharmacologic activity of levodopa at DA receptors is nil. The DA receptors primarily responsible for the motor effects of levodopa are located on medium spiny neurons in the striatum. After release from presynaptic vesicles into the synaptic cleft, DA can be transported back into neurons via the DA transporter for recycling. Alternatively, DA may be metabolized by mitochondrial monoamine oxidase or extraneuronal catechol-O-methyl-transferase (COMT) into 3,4-dihydroxyphenyl acetic acid (DOPAC) or 3-methoxytryamine (3-MT) and eventually into 3-methoxy-4hydroxy-phenylacetic acid (HVA). These inactive metabolites are excreted primarily in the urine. The plasma half-life of levodopa is short (1 to 3 hr).[15–17] In late PD, the therapeutic benefit is time-locked to these levodopa plasma concentrations ("short-duration response"[16]), in contrast to early PD.

Prominent adverse side effects, especially nausea and vomiting, limited the utility of levodopa when it was first introduced. Vomiting and nausea occur via peripheral decarboxylation of levodopa to DA, which subsequently stimulates the chemoreceptive trigger zone in the brain stem area postrema. Although plasma dopamine cannot penetrate the blood-brain barrier, the area postrema is devoid of this barrier. Nausea can be abolished, however, by blocking peripheral dopa decarboxylase, which prevents premature conversion of levodopa to DA within the systemic circulation. The peripheral decarboxylase inhibitors, carbidopa and benserazide, were introduced in the early 1970s for this purpose.[18] They do not cross the blood-brain barrier and hence selectively block conversion of levodopa to DA outside the brain. Levodopa formulated with one of these two agents has been the standard of treatment of PD for over three decades. The pharmacokinetics of carbidopa and benserazide are almost identical;[18] carbidopa is approved for use within the U.S.A., whereas benserazide is available in other countries, including throughout Europe. It should be noted, however, that inhibition of peripheral decarboxylase is incomplete with carbidopa, and in some patients administration of supplemental carbidopa may be necessary to control side effects.[19,20]

Soon after the introduction of levodopa, pharmacokinetic interactions with dietary proteins were recognized.[16] Dietary protein digestion liberates large-neutral amino acids, which compete with levodopa for transport across the blood brain barrier and probably for entry into the neuron;[21,22] they may also compete with levodopa at the intestinal level, although this appears less important. Hence, meals containing proteins may impede the transport of levodopa to its site of action and contribute to loss of efficacy and quality of response. To counter the potential confounding effects of dietary proteins, levodopa should be taken on an empty stomach, if possible.

PHARMACODYNAMICS

The improvement in parkinsonism with levodopa treatment is relatively long-lasting in early PD, reflective of the "long-duration response."[16,23–25] Thus, once a stable response is achieved with chronic treatment, it persists for up to approximately two weeks if levodopa treatment is stopped.[25,26] Conversely, it takes a similar period of time for this long-duration response to fully develop after levodopa is initiated or the dose is changed. Patients with long duration responses will not experience improvement after the first few doses but will subsequently note a cumulative effect over approximately a week of continuous dosing and can often miss doses without any deleterious consequences. The long-duration response was initially thought to be due to vesicular storage of newly synthesized DA. However, it now appears the long-duration response is more likely a result of reduced turnover of synaptic vesicles or post-synaptic events in early PD.[25,27,28]

With passage of time, levodopa motor responses often become short lived (1 to 6 hr), resulting in motor fluctuations, i.e., short duration responses or "wearing off."[16,29] Among those with the short-duration response, the antiparkinsonian effect tends to follow fluxes in plasma and cerebrospinal fluid levodopa concentrations, although this correlation can be variable.[16,21,30] In these patients, the motor benefits of levodopa ebb and flow, time-locked to each dose of medication. Some patients may experience

little or no response to an individual dose of levodopa, i.e., a "dose failure," a phenomenon most likely due to interference from dietary proteins, a result of delayed gastric emptying time or a subthreshold dose of levodopa. Short-duration responses appear to primarily relate to the duration of PD rather than the duration of levodopa treatment.[31] Although motor fluctuations are rare in early PD, about 40% of PD patients experience motor fluctuations after four to six years of treatment with levodopa.[32]

OTHER FORMULATIONS AND ROUTES OF ADMINISTRATION

An extended-release formulation of levodopa (with a peripheral decarboxylase inhibitor) has been developed to prolong the duration of the plasma concentrations and hence improve treatment of short-duration levodopa responses. An erodable polymeric matrix slowly releases the pill contents, although this also delays absorption and time to onset of response.[33,34] The bioavailability of the extended-release drug is approximately 70% of the immediate-release preparation and, when used, the dose should be adjusted accordingly.[33] The extended release preparation may be of benefit to some, but not all, patients with fluctuating motor control.[34] An orally disintegrating tablet containing carbidodopa/levodopa has recently become available, but the dissolved contents must still be swallowed for intestinal absorption, and the pharmacokinetics are similar to the immediate-release formulation.

Other methods to deliver levodopa have also been employed. Levodopa may be readily dissolved in water (1 mg/ml) and used as an oral or intragastric liquid preparation.[35] Ascorbic acid may be added to prolong stability in solution for up to 24 hr but need not be added if the solution will be used within a few hours.[36] Dissolved in solution, levodopa is absorbed more rapidly than tablets, and this approach is sometimes helpful in patients with rapidly fluctuating motor responses and to "rescue" patients who develop sudden-onset off states. Intravenous administration of levodopa has limited practical utility and has largely been restricted to clinical research centers.[37] A rapid-release sublingual formulation is also under development.[38]

LEVODOPA SIDE EFFECTS

Most patients tolerate carbidopa/levodopa monotherapy. The most common levodopa-induced side effects are nausea, symptomatic orthostatic hypotension, hallucinations, somnolence, and dyskinesias (see Table 56.1). When nausea occurs, it is often mild and habituates with continued treatment. Supplemental carbidopa can ameliorate dose-limiting nausea in many cases. Symptomatic orthostatic hypotension may also occur as a dose-related side effect but can typically be managed by eliminating other offend-

TABLE 56.1
Common Levodopa Side Effects

Side Effect	Freq. (%)*	Treatment
Nausea	32%	Carbidopa; administration with non-protein-containing food
Somnolence	14%	Treat any underlying sleep disorder or nocturnal parkinsonism; discontinue DA agonists; consider amantadine or modafanil
Orthostatic hypotension	11%	Increase fluid and salt intake; elevate head of bed; fludrocortisone; midodrine
Hallucinations	4%	Minimize polypharmacy and other psychotropic medications; quetiapine, clozapine

*Data represent average frequencies reported in clinical studies comparing levodopa monotherapy with other treatments.[34,84,90,143]

ing medications (e.g., antihypertensives, diuretics), increasing blood volume (sodium chloride tablets, adequate hydration, elevating the head of bed, fludrocortisone), and increasing peripheral vasoconstriction (compressive stockings, midodrine). Hallucinations (typically visual) are rare in the treatment of early PD with levodopa monotherapy. They are much more likely when carbidopa/levodopa is administered with other antiparkinsonian drugs. They can be treated by simplifying the medication regimen, eliminating other psychoactive agents and, if necessary, adding an atypical antipsychotic medication (e.g., quetiapine). Levodopa reduction may be considered in the treatment of hallucinations, but this is often not tolerated because of decompensation of parkinsonism. Motor complications (dyskinesias and motor fluctuations) associated with levodopa treatment of patients with advancing PD and somnolence are discussed in detail below.

LEVODOPA: POTENTIAL TOXICITY

Concerns have been raised about the possibility that exogenously administered levodopa may contribute to the progression of PD.[39,40] This concern arises from the oxidative stress hypothesis of PD, which posits that degeneration of nigrostriatal DA neurons is a direct result of oxygen free-radical-mediated cell death. The metabolism of levodopa and DA occur via monoamine oxidase-mediated oxidation, and DA can also undergo spontaneous autoxidation.[41] Both processes could contribute to the generation of oxygen free-radical species and thus provide a mechanism for further neuronal damage. The microenvironment of the substantia nigra also favors the production of secondary, and more toxic-free, radical species due to the intrinsically high concentrations of iron and neuromelanin.[42]

Evidence for increased oxidative stress in the substantia nigra pars compacta has been demonstrated in postmortem studies. Products of lipid peroxidation, protein oxidation, and DNA damage, all sequelae of oxidative damage, are found in increased concentrations within the substantia nigra of patients with PD.[43–46] Levels of oxygen free-radical scavenger enzymes (peroxidase, catalase, glutathione peroxidase) are found in reduced concentrations in the brains of patients with PD.[42] It is not clear, however, if these changes reflect the primary pathogenesis or represent inevitable downstream consequences of the degenerative process.

Several *in vitro* studies have demonstrated that addition of levodopa or DA to cultured dopaminergic neurons increases cell death through free-radical-mediated mechanisms;[47–49] however, this data must be interpreted carefully. The concentrations of levodopa required to induce neuronal death in these experiments often exceed the concentrations that would be expected in the brains of patients treated with therapeutic doses. Concerns have also been raised that much of the *in vitro* evidence for DA-induced neurotoxicity could be attributed to nonphysiologic cell culture conditions and media.[50] Indeed, further studies have shown that levodopa can be neuroprotective, especially when neurons are co-cultured with supporting glial cells.[51–53] Moreover, *in vivo* studies in rodents suggest a neuroprotective effect of levodopa administration.[54,55] These latter findings would be in keeping with the observation that levodopa extends the life expectancy of patients with PD.[1–7]

Clearly, if oxidation of DA could contribute to the pathogenesis and progression of PD, this would have serious implications for symptomatic treatment. Concerns over levodopa toxicity have led to recommendations to delay levodopa administration until absolutely necessary and then using the lowest doses possible.[39,56] There is ample evidence, however, that levodopa is not toxic in PD, and a growing consensus that this highly effective treatment does not accelerate disease progression.[57]

There are multiple arguments against levodopa toxicity, including the following:

1. Levodopa use is associated with reduced mortality rates in PD.[1–7]
2. The postmortem histopathological changes seen in PD have not changed since the introduction of levodopa.[58]
3. Administration of huge doses of levodopa to mice for prolonged periods of time did not produce any evidence of neurotoxicity in the substantia nigra or other brain regions.[59–61]
4. Levodopa treatment of other disorders does not result in parkinsonism or neuropathological evidence of substantia nigra toxicity.[62,63]

5. Degeneration in PD is not restricted to DA or catecholamine neurons. In fact, dopaminergic neurodegeneration may represent an intermediate stage in the temporal course of PD, with the earliest (and latest) stages involving noncatecholamine neurons.[64–66]

CLINICAL TRIALS ASSESSING LEVODOPA INFLUENCE ON PD CLINICAL PROGRESSION

Levodopa has such a profound symptomatic effect that it is difficult to separate this from a direct influence on PD progression. Two clinical trials controlled for the symptomatic effect using a washout study design.[67,68] In each study, early PD patients were evaluated before treatment and again, one to two weeks after medications were discontinued. Compared to bromocriptine monotherapy, 14 months of carbidopa/levodopa treatment resulted in a similar rate of progression of parkinsonism.[67] In the more recent ELLDOPA trial, 40 weeks of carbidopa/levodopa treatment similarly did not accelerate the clinical progression of PD, compared to placebo.[68,69] In fact, progression may have been slowed by levodopa treatment in this trial, although a longer-lasting symptomatic effect could also have explained the results.

LEVODOPA TREATMENT AND FUNCTIONAL BRAIN IMAGING STUDIES

Two recent studies, using functional neuroimaging techniques, reported significantly faster PD progression with carbidopa/levodopa versus dopamine agonist treatment.[70–72] Perusal of the data suggests that if altered disease course is a true effect, then agonist treatment slowed progression, as opposed to levodopa accelerating progression. Regardless, these studies were confounded by the pharmacologic effects of these drugs on the regulation of proteins interacting with the radioligands used for functional neuroimaging in PD. Hence, PD progression may not truly have been measured.[73,74] Further adding to the uncertainty of these neuroimaging results are those of the ELLDOPA trial discussed above. In that study, the neuroimaging results were opposite to the clinical outcomes. Specifically, chronic carbidopa/levodopa treatment resulted in a slower rate of clinical progression after washout, compared to placebo; however, dopamine neuron imaging indicated a more rapid decline with levodopa treatment.

DELAYING LEVODOPA TREATMENT

Several retrospective studies have addressed the possible advantages of deferring levodopa treatment. Two early studies did suggest that motor fluctuations were less frequent among patients with PD in whom levodopa treatment was delayed.[75,76] However, selection bias in the levodopa-delayed group (i.e., patients with milder disease and

a slower inherent rate of progression) confounds the interpretation of the data in these retrospective studies.[77] Subsequent retrospective investigations failed to confirm any long-term benefit from delaying levodopa based on the risk of motor complications, dementia, or psychiatric symptoms.[77–81]

LEVODOPA DOSE RESTRICTION

Maintaining a relatively low dose of levodopa during the initial years of treatment has been suggested as a "levodopa sparing" strategy to stave off subsequent motor complications.[82] This approach was assessed in a single controlled trial in which low-dose levodopa therapy was compared with the "maximum tolerated dose" over six years of treatment.[83] The low-dose group experienced a lower frequency of dyskinesias and motor fluctuations, but the differences were not striking. More importantly, motor symptoms of PD were so poorly controlled in the low-dose group that the authors concluded that dose restriction was not advantageous.

INITIAL TREATMENT: LEVODOPA VS. DA AGONISTS

Initial treatment of disabling motor symptoms in patients with PD can be accomplished with either levodopa or DA agonists. Several studies have evaluated initial treatment of PD patients with levodopa versus DA agonists, comparing relative improvement in motor function along with the development adverse effects and longer-term motor complications.[84–90] Some who have reviewed these studies find support for choosing a DA agonist as initial treatment of PD, focusing on the reduced incidence of later-occurring dyskinesias and motor fluctuations.[91] However, others have expressed doubts about the potential long-term benefits of initial DA agonists treatment and the choice of initial therapy remains controversial.[74,92,93] This question is critical to the treating physician, and therefore we will touch on a few points that must be considered in this therapeutic decision.

There is no disputing that DA agonist monotherapy alone carries a low risk for subsequent motor fluctuations and dyskinesias.[94] However, DA agonist monotherapy is much less efficacious compared to levodopa treatment. Although agonist therapy alone may be sufficient for a year to a few years, few patients can be effectively managed on DA agonist monotherapy beyond this period of time. For example, in the five-year ropinirole trial, only 16% of patients initially started on ropinirole monotherapy remained on this drug alone by the end of the study.[84] Furthermore, these 16% were likely a very select group with much more benign PD. In fact, they may have been so select that, even with levodopa treatment, they may not have developed substantial motor complications. The lower efficacy of DA agonist monotherapy is the reason that all of the recent comparative clinical studies allowed those in the DA agonist group to add levodopa during the trial.[72,84,85,89,90] It is also clear from these studies that, once levodopa is added, the risk of dyskinesias and motor fluctuations increases markedly with this combination therapy, although it is not quite as high as with carbidopa/levodopa monotherapy.

One surprising finding in all of the recent comparative trials is that carbidopa/levodopa monotherapy was significantly more efficacious than combination DA agonist-carbidopa/levodopa.[72,84,85,89,90] In all four studies, patients treated with initial DA agonist therapy plus later optional levodopa had less satisfactory control of their parkinsonism than with levodopa alone. In fact, the lower incidence of dyskinesias and motor fluctuations in the DA agonist study arms of these trials has been attributed to less intense treatment.[74] Nonequivalent treatment may directly impact the development of motor complications, since the dose-response curves for desirable motor responses and induction of motor complications are steep and often overlap.[11,12] As such, small changes in treatment intensity may translate into significant differences in both motor performance and incidence of motor complications.

Why should the DA agonist groups have experienced less control of parkinsonism when they were allowed to add levodopa? This could have been due to unblinding of the investigators during the course of the study, as can occur when study arms involve treatments with noticeable differences in efficacy. The *a priori* prediction of fewer motor complications with DA agonist treatment may have led to bias; the agonist arm may have received less aggressive treatment if investigators could see through the blind. However, this seems unlikely to have occurred in all four studies. Another explanation is that chronic treatment with these selective D_2 agonists may have influenced the expression or turnover of proteins involved in dopamine neurotransmission.[73] Thus, chronic and selective stimulation of postsynaptic D_2 receptors may have led to selective receptor down-regulation and, hence, fewer motor fluctuations and lower efficacy. Regardless, it appears that seeking to achieve fewer motor complications with DA agonist treatment results in less satisfactory control of parkinsonism.

If we focus on levodopa motor fluctuations and dyskinesias, we need to know the longer-term risks and the implications for patient care. In a recent meta-analysis, the risk of dyskinesias after about five years of levodopa monotherapy was approximately 40%.[32] Similarly, the risk of motor fluctuations is about the same; i.e., 40% by five years of treatment.[32] These data represent incidence figures and included both problematic motor complications as well as mild dyskinesias and motor fluctuations. Presumably, many of these patients may have experienced mild motor complications that could be easily managed with levodopa adjustments. Thus, these five-year, 40% incidence figures overstate the true disability.

The design of the comparative trials of levodopa versus dopamine agonist focused on the initial drug treatment.[72,84,85,89,90] In these studies, starting with a DA agonist and subsequently adding carbidopa/levodopa ultimately resulted in a lower incidence of dyskinesias and motor fluctuations. What if treatment began with carbidopa/levodopa and the agonist was added only after the motor complications developed? After all, the conventional role for DA agonist drugs has been as later adjunctive therapy to counter levodopa motor complications.[95] With corresponding levodopa dose reduction, adjunctive DA agonist treatment has well recognized efficacy in controlling levodopa motor fluctuations and dyskinesias due to the longer half-life of agonists. None of the comparative trials has addressed the distinct possibility that the ultimate outcomes may have been the same if levodopa was started first and the DA agonist was added later.[72,84,85,89,90] It should also be noted that these studies counted only incident dyskinesias and motor fluctuations, without any crossover in the carbidopa/levodopa study arm. More important to clinicians is the prevalence and severity of motor complications after medication adjustments, which reflects the true disability.

Certainly later-developing motor fluctuations and dyskinesias deserve consideration. However, the potential for other common side effects must also be factored into the treatment decision. Hallucinations appear to be about three times as likely with ropinirole or pramipexole therapy compared to carbidopa/levodopa.[84,90] Somnolence was also significantly more likely with ropinirole or pramipexole in these studies, and the issue of "sleep attacks" has achieved some notoriety.[96] In addition, the ergot DA agonists (bromocriptine, pergolide and cabergoline) carry a risk of cardiac valvular, pericardial, pulmonary, and retroperitoneal inflammatory-fibrotic reactions.[95,97,98] Thus, complications of therapy are not limited to fluctuations and dyskinesias.

Initial use of levodopa for the symptomatic treatment of PD has certain advantages over DA agonists. Levodopa, when compared directly to DA agonists, provides significantly greater efficacy in alleviating the motor symptoms of PD and improving activities of daily living.[91,99,100] Levodopa is also the only medication that is associated with increased life span in patients with PD.[1–7,101] There is also a clear cost benefit to using levodopa; i.e., levodopa treatment can be implemented at 10 to 25% the cost of treatment with a DA agonist. Titration of levodopa to therapeutic levels can be accomplished in three to four weeks using a single-dose formulation. In contrast, DA agonist treatment is initiated at subtherapeutic doses and requires 6 to 8 weeks for titration to effective doses. Dose escalation is further complicated by the need to use DA agonist tablets with several different dose formulations. Finally, as mentioned, levodopa is much less likely to induce hallucinations, somnolence, or leg edema compared to DA agonists.[84,85]

The American Academy of Neurology practice parameter regarding the initial symptomatic treatment of PD (Table 56.2) concludes that it is appropriate to start treatment with either levodopa or DA agonist.[100] This systematic literature review comparing levodopa versus DA agonist treatment provided evidence (Level A, class I and II) that (a) levodopa is more effective at treating the motor symptoms and impaired activities of daily living features of PD, (b) agonist treatment results in fewer motor complications, and (c) agonist therapy is associated with more frequent adverse events that levodopa.

TABLE 56.2
American Academy of Neurology Practice Parameter Conclusions

- Levodopa is more effective in treating PD motor symptoms and improving ADL features of PD.
- Agonist treatment results in fewer motor complications (wearing off, on-off, dyskinesias).
- Agonist therapy is associated with more frequent adverse events than L-dopa (hallucinations, somnolence, leg edema).

As highlighted by the AAN practice parameter, the choice of initial treatment depends on the relative importance of improving motor disability and limiting adverse events (favoring levodopa) versus the possibility of lowering long-term motor complications (favoring DA agonists). An informed clinician and patient are in the best position to decide on initial treatment strategy on a case-by-case basis.

INITIAL DOSING

Replenishment of central nervous system DA with levodopa is an effective strategy for alleviating the motor symptoms of PD, plus a robust response to levodopa provides supportive clinical evidence in favor of a diagnosis of idiopathic PD. Levodopa, coadministered with carbidopa (or benserazide outside the United States), is available as an immediate-release standard tablet, orally disintegrating tablet, and a controlled-release preparation. The different formulations of levodopa contain varying amounts of carbidopa and levodopa (Table 56.3). The 10/100 and 25/250 formulations are appropriate for patients not experiencing nausea, whereas the formulations with a greater ratio of carbidopa to levodopa (e.g., 25/100) are advisable in patients with nausea and those undergoing initial titration. Of note, patients vary in the amount of carbidopa required to prevent nausea and, where necessary, additional carbidopa may be supplemented in 25-mg tablets (Lodosyn) for patients with persistent nausea.

The immediate-release formulation is a reasonable choice for initial treatment in most patients (Table 56.4).

TABLE 56.3
Levodopa Dose Formulations

Formulation	Carbidopa/Levodopa (mg)	Tablet Color
Immediate release*	25/100	Yellow
	10/100	Blue
	25/250	Blue
Controlled release	25/100	Pink
	50/200	Tan/peach

*Standard tablet and oral disintegrating tablet

The starting dose is usually one-half to one 25/100 tablet taken three times daily, taken one hour before meals to avoid the inhibiting effects of dietary proteins on levodopa absorption and distribution. Dose increments of one-half tablet for all doses (50 mg levodopa per dose) can be made on a weekly basis, since this period of time is often required to appreciate the full cumulative effect of a change in levodopa dose. Patients will usually achieve acceptable symptomatic improvement between 1 and 2.5 tablets three times daily (300 to 750 mg levodopa/day). The dose should be increased until optimal symptomatic control is attained or side effects develop. Ultimately, if several doses have similar efficacy, then the lowest dose may be used. Rare individuals may require higher doses, up to 300 to 350 mg levodopa three times daily (900 to 1050 mg/day), to appreciate optimal symptomatic improvement. If there is no response to maximal doses, taken on an empty stomach, then it is likely that the patient has a condition other than idiopathic PD.

Controlled-release levodopa preparations were initially developed for patients with advanced PD who had short-duration responses to levodopa, resulting in fluctuating clinical responses. Embedding levodopa in a polymeric matrix extends the therapeutic benefit of controlled release levodopa up to 90 min but can significantly delay the onset of therapeutic benefit (up to 2 hr).[34] The lower bioavailability and the slower release of the controlled-release formulation requires a 30 to 50% increase in the amount of levodopa administered each dose to achieve plasma concentrations similar to immediate-release levodopa.[34] While controlled-release levodopa is better absorbed when administered with food, competition with other dietary proteins for transport across the blood brain barrier makes it difficult to predict the clinical response with respect to meals.[33]

Some have proposed that initial treatment with controlled-release levodopa might lead to a lower subsequent frequency of levodopa-related motor complications (fluctuations and dyskinesias). This is based on the theory that the less pulsatile administration of levodopa with less fluctuation in synaptic DA concentrations should provide more constant and physiological stimulation of postsynaptic DA receptors (see below). However, two large multicenter trials found that the frequency of dyskinesias and

TABLE 56.4
Initial Dosing and Titration of Levodopa Therapy

- Increase the dose weekly, as tolerated, until substantial improvement in motor symptoms occurs or the maximum dose is reached.
- If several doses are equally effective, maintain the lowest dose.
- If side effects develop after an increment in dose, return to the previous dose and address side effects as discussed in Table 56.2 and the text before trying further increments.
- Administer the immediate-release formulation one hour before each of three meals (empty stomach). It is unclear when to administer controlled-release carbidopa/levodopa with respect to meals.[33]

Formulation	Starting Dose*	Weekly Increments	Maximum Dose†
Immediate-release carbidopa/levodopa	50–100 mg; three times daily	50 mg to all doses (150 mg/day total)	250–300 mg three times daily
Controlled-release carbidopa/levodopa	100 mg; two times daily	50 to 100 mg to all doses (100–200 mg total)	400 to 600 mg two times daily

*Doses are given in mg of levodopa

†For patients with no response to the maximum dose, further titration may be attempted using the immediate-release preparation by gradually increasing from 300 mg three times daily to 400 mg three times daily. If the patient is taking over 1000 mg/day of immediate-release levodopa on an empty stomach and has no response, then the patient probably does not have PD but, rather, another parkinsonian syndrome.

fluctuations was similar in patients receiving immediate-release versus controlled-release levodopa.[102,103]

The sustained-release preparation may be administered twice daily in patients with early PD, and this convenience may offset the substantial increased cost for some patients. In advanced PD, the sustained-release formulation may not provide optimal parkinsonism control in all patients due to the delayed response, plus variable bioavailability, absorption, and blood-brain barrier transport; these factors may result in unpredictable clinical responses.

MOTOR COMPLICATIONS OF LEVODOPA THERAPY

DYSKINESIAS

Dyskinesias are common features of advancing PD and present as involuntary chorea involving the limbs, trunk, or head. Often, the chorea is associated with a dystonic component, but the hallmark levodopa dyskinesias are predominantly choreiform. Although dyskinesias are not bothersome to many patients, they can be severe enough to interfere with goal directed voluntary movement and may be socially embarrassing. In this setting, dyskinesias can limit the dose of symptomatic therapy and complicate ongoing management of patients with advancing PD.

Dyskinesias occur in approximately 40% of patients receiving levodopa treatment for 4 to 6 years.[32] The frequency of dyskinesias in patients with young-onset PD, defined as PD symptom onset prior to age 40, is higher and approaches 100% by 5 years of levodopa treatment.[104,105] Although less commonly seen, dyskinesias can also occur with DA agonists. Dyskinesias are uncommon in early PD (approximately 7% after one year of levodopa treatment) and do not occur in untreated

patients.[32] In patients with idiopathic PD, dyskinesias tend to reflect the duration of PD rather than the duration of levodopa treatment.[31,32] Patients with long durations of D, before starting levodopa (e.g., in the pre-levodopa era) develop dyskinesias within the first few months; in contrast, dyskinesias surface much later with more recently developing PD (Figure 56.3). The correlation of dyskinesia frequency with PD duration probably reflects the severity of dopaminergic terminal loss, which is progressive. Dyskinesias have been observed early in the treatment of patients inadvertently exposed to 1-methyl-4-phenyl-1,2,3,6-tetrahydropyridine (MPTP) who have an acute severe insult to the dopaminergic system; this further supports the concept that the degree of nigrostriatal DA terminal damage is an important substrate for levodopa-induced motor complications.[106]

PROPOSED MECHANISMS OF LEVODOPA-INDUCED DYSKINESIAS

Both pre- and postsynaptic factors appear to play a role in the development of levodopa-induced dyskinesias. The severity of nigrostriatal DA neuronal loss and the mode of post-synaptic DA receptor stimulation appear to be critical elements. Preclinical experiments using putative animal models of PD have provided some insights into the mechanisms.[107–110] In these experiments, acute exposure to 6-hydroxyopamine or MPTP in rodents and nonhuman primates produces a substantial and rapid destruction of nigrostriatal DA neurons. Involuntary rotational movements occur after treatment with DA-mimetic agents (levodopa or DA agonists) in rodents with unilateral nigrostriatal DA destruction, and this has been proposed as rodent analog of dyskinesias.[111,112] Intermittent, chronic administration of levodopa produces an increase in rotational behavior, whereas chronic, continuous administration of levodopa does not alter the baseline level of rotational behavior. Marmosets with MPTP-induced destruction of nigrostriatal DA neurons rapidly develop involuntary movements after intermittent, initial treatment with levodopa (the "priming" effect). In contrast, initial treatment with a DA agonist does not induce involuntary movements in MPTP-lesioned marmosets.[113] In these studies, a severe depletion of dopaminergic presynaptic terminals is necessary for these phenomena to occur; limited dopaminergic terminal loss is insufficient.[114] Taken together, these studies suggest that the combination of extensive destruction of nigrostriatal DA neurons coupled with pulsatile levodopa administration can induce plastic changes in basal ganglia function and provide a substrate for the development of dyskinesias.

Pulsatile postsynaptic DA receptor occupancy has been proposed as a key factor in the induction of plastic changes in basal ganglia function.[107–110] Intermittent administration of levodopa may result in the bolus synthesis of DA resulting in fluctuations in synaptic DA concentrations. The resulting oscillatory stimulation of postsynaptic receptors is thought to precipitate a cascade of downstream plastic changes in the neurochemical architecture of the basal ganglia, thus setting the stage for dyskinesias to occur. In contrast, continuous levodopa treatment or administration of a DA agonist (with a longer half-life relative to levodopa) results in continuous, non-fluctuating postsynaptic DA receptor stimulation, thus avoiding the induction of plastic basal ganglia changes and avoiding the development of dyskinesias.

These animal studies have been used to promote the use of longer-acting preparations of levodopa (controlled release or adjunctive use of a COMT inhibitor) and DA agonists as initial treatment for patients with early, mild PD.[115] Although this research provides compelling mechanisms for the development of dyskinesias, extrapolation to patients with early, mild PD can be problematic. These animal models are based on acute destruction of nigrostriatal DA neurons, which is unlike the chronic progressive loss of DA neurons seen in patients with PD. The time course of DA neuronal loss could significantly alter the plastic changes in the basal ganglia. More importantly, these studies were conducted in animals with massive (>90%) destruction of nigrostriatal DA neurons. This degree of DA neuronal loss is not typical of patients with early, mild PD but can be seen in advanced PD of many years duration.[116] Indeed, animals with modest neurotoxin-induced destruction of nigrostriatal DA neurons do not exhibit plastic changes or involuntary movements following intermittent levodopa compared to animals with substantial nigrostriatal DA lesions.[114] Therefore, the importance of pulsatile stimulation of postsynaptic DA receptors may be more germane to patients with advanced PD with profound striatal DA denervation.

Also worth considering is the degree of pulsatile stimulation. Advocates of sustained-release levodopa therapy or adjunctive COMT inhibitor therapy have argued that these treatments are less pulsatile. However, they are pulsatile nonetheless,[33,117,118] and the frequency of motor complications is similar when comparing patients treated with immediate-release versus sustained-release levodopa.[102,103] Thus, it is unclear whether a pharmacologic strategy with only slightly less pulsatile stimulation of dopamine receptors translates into any meaningful reduction of levodopa motor complications.

DYSKINESIAS: CLINICAL FEATURES AND TREATMENT

The most common form of levodopa-induced dyskinesias is peak-dose chorea. This type of dyskinesia occurs at the time of peak levodopa symptomatic effect and plasma levodopa concentrations. Many patients actually feel their best when their medication dosage is adjusted to the point at which subtle peak-dose dyskinesias are present and often are unaware of the involuntary movements. When peak dose dyskinesias become severe or dose-limiting, they can complicate the management of patients with advanced PD.

If peak-dose dyskinesias are bothersome or severe, several treatment options are possible. A small (25 to 50 mg) reduction in the individual levodopa doses every three to four days, until dyskinesias abate, is often all that is necessary. If dose reduction produces intolerable worsening of the other motor symptoms of PD, adjunctive treatment with DA agonists may be helpful with subsequent reduction of levodopa, once the agonist dosage is therapeutic. Discontinuation of adjunctive medications that potentiate the pharmacological effects of levodopa (e.g., selegiline, tolcapone, entacapone) in favor of monotherapy with levodopa is sometimes helpful. Amantadine (200 to 400 mg per day in divided doses) has also been shown to reduce peak-dose dyskinesias.[119–122] Typical neuroleptic D_2 antagonists can suppress dyskinesias but result in unacceptable exacerbation of parkinsonism.[123,124] Clozapine, an atypical neuroleptic, may reduce dyskinesias without worsening other motor features of PD but requires close monitoring for agranulocytosis.[125] Several other adjunctive medications have been reported to alleviate levodopa-induced dyskinesias with varying degrees of success, e.g., dextromethorphan, buspirone, idazoxan, and propranolol.[126–130]

Chorea or dystonia that occurs at the beginning and end of each dose cycle (biphasic dyskinesias) is an uncommon pattern of dyskinesia that requires a different treatment approach from that of peak-dose dyskinesia.[131] In this dyskinesia-improvement-dyskinesia (D-I-D) response, an initial dyskinetic state occurs for about 10 to 30 min, just as the beneficial effects of levodopa begin and a late dyskinetic period develops as the symptomatic benefit of levodopa begins to wear off. There is typically an intervening satisfactory motor "on" response. Patients with the D-I-D response improve with shortening of the levodopa dose interval, thus producing overlapping dose effects. In some with biphasic dyskinesias, however, only 4 to 5 doses are tolerable; additional doses are associated with an adverse change in the pharmacodynamics as Muenter and colleagues described over 25 years ago.[131] Switching from controlled release to immediate release is usually necessary to facilitate careful dose and interval adjustments. Adjunctive use of a DA agonist may be helpful in some cases. Patients with the D-I-D response usually experience unavoidable dyskinesias at the beginning and end of each day.

MOTOR FLUCTUATIONS

In concert with advancing PD, patients may develop medication responses that are tightly time-locked to each levodopa dose, resulting in troublesome fluctuations in motor performance.[16,21] Motor fluctuations, reflective of short-duration levodopa responses, occur in approximately 40% of patients treated with levodopa for four to six years duration.[32] In the mildest form, motor fluctuations manifest as the "wearing-off" effect, in which the beneficial response to each dose of levodopa declines prior to next dose. In the most severe form, the response may be dramatic and sudden, with an abrupt transition from a mobile to immobile state, termed the "on-off" effect. Pronounced depletion of DA from presynaptic nigrostriatal terminals seems to be a prerequisite for the short-duration response,[114,132,133] but postsynaptic changes, such as downregulation of postsynaptic DA receptors, may also be involved.[25,134] Motor fluctuations are not usually seen in early PD, with a 3% incidence by the end of one year of treatment.[32] As with dyskinesias, motor fluctuations can complicate the symptomatic management of patients with PD.

Patients experiencing short-duration responses often report that their parkinsonism is "worse" and do not readily recognize the temporal relationship of motor fluctuations to their levodopa doses. Careful inquiry by the physician or dose-response daily diaries can be useful in identifying correlations between fluctuations and the timing of levodopa administration. When uncertainty remains, direct monitoring of patients throughout a levodopa response cycle may help resolve the issue. This can be done by having the patient present to the physician's office in the morning (before breakfast), just prior to taking the first medication dose of the day; a brief motor examination for parkinsonism is then done just before, and then serially after administration of medications. Careful documentation of the time to symptomatic onset, duration of beneficial effect, and temporal profile of adverse events can be invaluable in managing complicated patients.

Adjusting the dose interval to match levodopa response duration is the most direct method to treat motor fluctuations. For example, if a patient reports wearing off at three hours, then levodopa may be administered every three hours or slightly less so that the effects overlap. At times, however, the reductions in dose interval may not be acceptable to the patient due to the frequent need to take medication. Controlled-release levodopa was initially developed to address the short-duration response and can prolong the responses to levodopa by 60 to 90 min. Unfortunately, erratic absorption and variable clinical response can limit the utility of controlled-release levodopa in patients with advanced PD. Other strategies could include addition of a COMT inhibitor (tolcapone, entacapone), which can prolong the response duration by blocking the catabolism of levodopa and DA. Alternatively, adjunctive use of DA agonists, with a relatively long half-life, can be helpful in treating short-duration responses. When employing these latter strategies, a reduction in the dose of levodopa is often necessary to alleviate peak dose dyskinesias. The use of adjunctive medications in the treatment of PD is discussed in other chapters.

Freezing

Freezing refers to the inability to initiate movement, typically involving hesitancy in arising from a seated position, initiating gait or turning (gait freezing). Freezing can occur as an "off" phase phenomenon or reflect suboptimal levodopa dosage. Abrupt off states may be associated with profound freezing and is one of the most distressing events for patients with PD. In this setting, a "rescue dose" of levodopa may be administered by crushing and dissolving the patient's established dose of levodopa in several ounces of liquid (e.g., carbonated beverage or juice) and drinking the entire amount. This usually results in a rapid "on" response within about 20 min, although the symptomatic benefit of the "rescue dose" only persists for about 60 to 90 min. In rare cases, freezing gait can occur as a peak-dose effect (peak-dose freezing) and may require dose reduction of either levodopa or adjunctive medications (e.g., agonists). Unfortunately, for some patients, gait freezing is refractory to medication adjustments but may improve with physical therapy treatment modalities.

Functional Neurosurgery and Levodopa

Patients with advancing PD may develop disabling motor complications that become refractory to the medication treatment strategies outlined above. In select patients, particularly those in whom a previously satisfactory response to levodopa becomes limited by dyskinesias or fluctuations, functional neurosurgery may significantly improve motor function and quality of life.[135,136] Pallidotomy and deep brain stimulation of the pallidum and subthalamic nucleus are discussed in detail elsewhere.

NONMOTOR FEATURES OF PD AND LEVODOPA

A wide variety of nonmotor symptoms can occur as a manifestations of the levodopa "off" state.[137] These symptoms are reported to occur in up to 17% of patients who have motor fluctuations and often cause significant distress and disability for patients with PD.[138] The clinical spectrum includes autonomic, sensory/pain, cognitive and psychiatric symptoms, as listed in Table 56.5. Unexplained symptoms that occur in a time-locked fashion with the levodopa "off" state are suspicious for nonmotor manifestations of PD. Nonmotor symptoms often (75%) respond to appropriate adjustment of levodopa dose and dose interval (described above) to achieve optimum motor symptom control.[138] Recognition and treatment of these nonmotor manifestations of the levodopa "off" state can improve patient's quality of life while minimizing unnecessary tests and treatments.

TABLE 56.5
Nonmotor Features of Parkinson's Disease[137,138]

Sensory
  Pain: may be radicular, musculoskeletal, central or dystonia-related
  Paresthesias
  Akathisia
  Restless legs syndrome
Cognitive-Psychiatric
  Dementia
  Psychosis: hallucinations, delusions
  Anxiety/panic
  Depression
  Hypomania/mania
  Moaning/screaming
Hypersexuality
Sleep
  Insomnia
  Excessive daytime somnolence
Autonomic
  Cardiovascular: tachycardia, hypotension
  Gasrointestinal: constipation, bloating, belching, dysphagia, sialorrhea
  Genito-urinary: urinary frequency, urgency, incontinence, erectile dysfunction
  Dermatological: pallor, sweating, skin temperature changes
Respiratory
  Dyspnea
  Laryngeal stridor

INSOMNIA

Patients with PD may experience significant insomnia, and this symptom, along with the resultant increased daytime somnolence, can be a source of significant disability. Although there are numerous potential causes of insomnia in patients with PD, re-emergence of parkinsonism at bedtime or during the night due to inadequate nocturnal levo-

dopa coverage can be a significant factor. Nocturnal rigidity, immobility, tremor, and akathisia can contribute to insomnia, particularly in patients who have developed short-duration responses to levodopa.[139] Difficulty initiating sleep due to poorly controlled parkinsonism should respond to an evening dose of immediate-release levodopa administered one or two hours before bedtime. In addition, adding a bedtime dose of controlled-release levodopa may extend sleep; this is helpful for those who awaken several hours into the night because their levodopa effect has waned. In this setting, the delayed effect of controlled-release levodopa is advantageous. Finally, if patients awaken in the middle of the night in an uncomfortable state with prominent parkinsonian symptoms, then use of immediate-release levodopa is appropriate to obtain a rapid response.

Evening and nighttime doses of levodopa should be equivalent to the optimum individual daytime dose to achieve the best sleep-sustaining effect. Overnight doses of levodopa can also improve early morning functioning. Although nocturnal use of levodopa improves sleep in many patients, it does disrupt normal sleep architecture.[140] However, from a practical point of view, those with insomnia due to inadequate nocturnal levodopa coverage experience gratifying responses with optimal nocturnal levodopa dosage. Rarely, bedtime doses of levodopa lead to insomnia, primarily when it induces significant dyskinesias.

ANXIETY, AKATHISIA, AND PANIC

Anxiety and akathisia are common complaints of patients with PD and may represent a manifestation of the levodopa "off" state. These symptoms are frequently and incorrectly attributed to levodopa overdosage when, actually, the opposite is the case. A correlation to levodopa dosing should be suspected when symptoms of anxiety wax and wane over the course of the day, worsening at the end of each levodopa cycle. At times, the anxiety can be profound and present as a panic attack. Akathisia at night can mimic the symptoms of restless legs syndrome. Treatment of motor fluctuations, as described in the aforementioned sections, should prove effective if these symptoms are occurring as a nonmotor facet of PD. In this setting, optimization of levodopa coverage works better than anxiolytic therapy.

PAIN AND PARESTHESIAS

Pain and uncomfortable sensory disturbances are frequent complaints of patients in the age group commonly affected by PD.[138] Although a variety of common comorbid conditions can produce these symptoms (e.g., sciatica, peripheral neuropathy), they can also represent a nonmotor manifestation of under-treated PD. In one study, 46% of patients with PD reported a complaint of pain, and in about two-thirds of these cases, the pain correlated with motor fluctuations.[141] Descriptions of the pain range from superficial and burning to deep and boring pain. Even when the pain can be attributed to other causes, it can be exacerbated in the levodopa "off" state. A trial of levodopa dose or frequency adjustment may alleviate uncomfortable sensory symptoms and may avoid the need for other diagnostic tests or treatment modalities.

Painful dystonia, in the absence of more typical choreiform dyskinesias, can occur at the end of a levodopa cycle or in the early morning and typically represents a wearing-off effect. In this context, patients may complain about persistent pain, often cramp-like, in the affected extremity (often the distal lower leg or foot). Toe curling or extension are common manifestations of this phenomenon. This responds to strategies for motor fluctuations described above. Refractory dystonias sometimes diminish with anticholinergic therapy (e.g., trihexyphenidyl, 2 mg, three times per day) or may be treated with localized botulinum toxin injections.

Dyspnea

Dyspnea is an occasional complaint among patients with PD and obviously could be due to a comorbid primary pulmonary or cardiac condition. As such, a full cardiac and pulmonary evaluation is recommended. Among PD patients, however, two unique etiologies must be enertained. First, ergotamine-derived DA agonists (bromocriptine, pergolide, cabergoline) can induce inflammatory-fibrotic reactions that may lead to symptomatic cardiac valvular fibrosis, pleuropulmonary fibrosis, or constrictive pericarditis.[95,97,98,142] Although the fibrotic reactions to ergot-based agonists are rare, they can result in serious morbidity and always require discontinuation of the offending agent. Second, dyspnea may also be a nonmotor

manifestation of the levodopa "off" state. This usually becomes apparent when patients report resolution of dyspnea shortly after each dose of levodopa and recurrence at the end of each levodopa dose cycle. In the latter case, treatment is the same as any levodopa off-state symptom. In addition, dyspnea may be a persistent symptom in those who are generally undertreated with subtherapeutic levodopa doses.

SUMMARY

Levodopa is the most efficacious drug for symptomatic treatment of PD. Although theoretical concerns have been raised about possible toxicity, clinical evidence is lacking and is contradicted by evidence for reduced mortality with levodopa therapy. Later-developing levodopa motor complications appear to primarily reflect the inherent progression of PD with increasing loss of dopaminergic neurons. However, these are treatable in many cases with levodopa dosage adjustments and adjunctive therapy. Levodopa, when used appropriately, remains the mainstay for the symptomatic treatment of PD.

REFERENCES

1. Diamond, S. G. and Markham, C. H., Present mortality in Parkinson's disease: the ratio of observed to expected deaths with a method to calculate expected deaths, *J. Neural. Transm.*, 38, pp. 259–269, 1976.

2. Hoehn, M. M., Parkinsonism treated with levodopa: progression and mortality, *J. Neural. Transm.*, 19, Suppl. 1, pp. 253–264, 1983.

3. Joseph, C., Chassan, J. B., and Koch, M. L., Levodopa in Parkinson's disease: a long-term appraisal of mortality, *Ann. Neurol.*, 3, pp. 116–118, 1978.

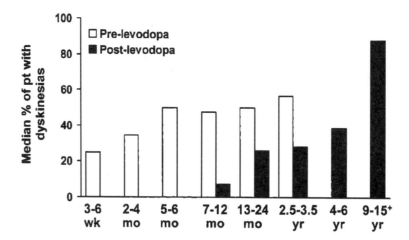

FIGURE 56.3 Comparison of the frequency of dyskinesias in the pre-levodopa versus post-levodopa eras. Median frequency (expressed as a nonweighted percentage) of dyskinesias from clinical series of patients who developed PD in the pre-levodopa era (open bars) compared to patients who developed PD in the post-levodopa era (solid bars). Patients in the pre-levodopa group had a longer duration of parkinsonism (6 to 9 years) than the pre-levodopa group (2 to 3 years) prior to the first exposure to levodopa. Data for dyskinesia frequency were derived from a systematic review of the literature from 1996 to September 2000 as previously described.[32]

4. Martilla, R. J. et al., Mortality of patients with Parkinson's disease treated with levodopa, *J. Neurol.*, 216, pp. 147–118, 1977.

5. Sweet, R. D. and McDowell, F. H., Five years' treatment of Parkinson's disease with levodopa: therapeutic results and survival of 100 patients, *Ann. Intern. Med.*, 83, pp. 456–463, 1975.

6. Zumstein, H. and Siegfried, J., Mortality among parkinson patients treated with L-dopa combined with a decarboxylase inhibitor, *Eur. Neurol.*, 14, pp. 321–327, 1976.

7. Uitti, R. J. et al., Levodopa therapy and survival in idiopathic Parkinson's disease: Olmsted County Project, *Neurology*, 43, pp. 1918–1926, 1993.

8. Birkmayer, W. and O. Hornykiewicz, The effect of L-3,4-dihydroxyphenylalanine on akinesia in Parkinsonism, *Wiener Klinische Wochenschrift*, 73, pp. 787–788, 1961.

9. Barbeau, A., Sourkes, T. L., and Murphy, G., *Les catecholamines dans la maladie de Parkinson*, in *Monoamines et systeme nerveux central*, Ajuriaguerra, J., Ed., Geneva and Masson, Paris, pp. 247–262, 1962.

10. Cotzias, G. C., Van Woert, M. H. and Schiffer, L. M., Aromatic amino acids and modification of parkinsonism, *N. Engl. J. Med.*, 276, pp. 374–379, 1967.

11. Nutt, J. G. and Holford, N. H., The response to levodopa in Parkinson's disease: imposing pharmacological law and order, *Ann. Neurol.*, 39(5), pp. 561–73, 1996.

12. Nutt, J. G., Clinical pharmacology of levodopa-induced dyskinesia, *Ann. Neurol.*, 47(4, Suppl. 1), p. S160–S164, discussion S164-6, 2000.

13. Nutt, J. G. and Fellman, J. H., *Pharmacokinetics of levodopa*, *Clin. Neuropharmacol.*, 7(1), pp. 35–49, 1984.

14. Goodman-Gillman, A., Ed., *Treatment of Parkinson's disease*, in *Goodman and Gilman's: The pharmacological basis of therapeutics*, McGraw Hill, New York, pp. 555–557, 2001.

15. Tyce, G. M., Sharpless, N. S. and Muenter, M. D., Free and conjugated dopamine in plasma during levodopa therapy, *Clinical Pharmacology and Therapeutics*, 16, pp. 782–788, 1974.

16. Muenter, M. D. and Tyce, G. M., L-dopa therapy of Parkinson's disease: plasma l-dopa concentration, therapeutic response, and side effects, *Mayo Clin. Proc.*, 46, pp. 231–239, 1971.

17. Rinne, U. K., Sonninen, V. and Siirtola, T., Plasma concentration of levodopa in patients with Parkinson's disease, *Eur. Neurol.*, 10(5), pp. 301–310, 1973.

18. Pinder, R. M. et al., Levodopa and decarboxylase inhibitors: a review of their clinical pharmacology and use in the treatment of parkinsonism, *Drugs*, 11, pp. 329–377, 1976.

19. Ward, C. D. et al., L-dopa decarboxylation in chronically treated patients, *Neurology*, 34, pp. 198–201, 1984.

20. Durso, R. et al., Variable absorption of carbidopa affects both peripheral and central levodopa metabolism. *Journal of Clinical Pharmacology*, 40(8), pp. 854–60, 2000.

21. Nutt, J. G. et al., The "on-off" phenomenon in Parkinson's disease: Relation to levodopa absorption and transport. *N. Engl. J. Med.*, 310, pp. 483–488, 1984.

22. Nutt, J. G., On-off phenomenon: relation to levodopa pharmacokinetics and pharmacodynamics. *Ann. Neurol.*, 22, pp. 535–540, 1987.

23. Zappia, M. et al., The long-duration response to l-dopa in the treatment of early PD, *Neurology*, 54, pp. 1910–1915, 2000.

24. Zappia, M. et al., Pharmacodynamics of the long-duration response to levodopa in PD, *Neurology*, 53: pp. 557–560, 1999.

25. Barbato, L. et al., The long-duration action of levodopa may be due to a postsynaptic effect, *Clin. Neuropharmacol.*, 20, pp. 394–401, 1997.

26. Hauser, R. A. et al., Time course of loss of clinical benefit following withdrawal of levodopa/carbidopa and bromocriptine in early Parkinson' s disease, *Mov. Disord.*, 15(3), pp. 485–9, 2000.

27. Melamed, E. and Hefti, F., Mechanism of action of short- and long-term l-dopa treatment in parkinsonism: role of the surviving nigrostriatal dopaminergic neurons, *Adv. Neurol.*, Christ, J. F., Ed., Raven Press, New York, N.Y., pp. 149–157, 1984.

28. de la Fuente-Fernandez, R. et al., Biochemical variations in the synaptic level of dopamine precede motor fluctuations in Parkinson's disease: PET evidence of increased dopamine turnover, *Ann. Neurol.*, 49, pp. 298–303, 2001.

29. Zappia, M. et al., Loss of long-duration response to levodopa over time in PD. Implications for wearing off, *Neurology*, 52, pp. 763–767, 1999.

30. Tolosa, E. S. et al., Patterns of clinical response and plasma dopa levels in Parkinson's disease, *Neurology*, 25(2), pp. 177–83, 1975.

31. Muenter, M. D. and Ahlskog, J. E., Dopa dyskinesias and fluctuations are not related to dopa treatment duration, *Ann. Neurol.*, 48, p. 464, 2000.

32. Ahlskog, J. E. and Muenter, M. D., Frequency of levodopa-related dyskinesias and motor fluctuations as estimated from the cumulative literature, *Movement Disorders*, 16, pp. 448–458, 2001.

33. Yeh, K. C. et al., Pharmacokinetics and bioavailability of Sinemet CR: a summary of human studies, *Neurology*, 39, Suppl., 2., pp. 25–38, 1989.

34. Ahlskog, J. E. et al., Controlled-release Sinemet (CR-4): a double-blind crossover study in patients with fluctuating Parkinson's disease, *Mayo Clin. Proc.*, 63, pp. 876–886, 1988.

35. Kurth, M. C. et al., Oral levodopa/carbidopa solution vs. tablets in Parkinson's disease with severe fluctuations: a pilot study, *Neurology*, 43, pp. 1036–1039, 1993.

36. Pappert, E. J. et al., The stability characteristics of levodopa solution, *Movement Disorders*, 9, p. 484, 1994.

37. Quinn, N., Parkes, J. D. and Marsden, C. D., Control of on/off phenomenon by continuous intravenous infusion of levodopa, *Neurology*, 34, pp. 1131–1136, 1984.

38. Djaldetti, R. and Melamed, E., Levodopa ethylester: a novel rescue therapy for response fluctuations in Parkinson's disease, *Ann. Neurol.*, 39, pp. 400–404, 1996.

39. Fahn, S., Is levodopa toxic? *Neurology*, 47, Suppl., 3, pp. S184–S195, 1996.

40. Olanow, C. W., Oxidation reactions in Parkinson's disease, *Neurology*, 40, 10 Suppl., 3, pp. 32–37, 1990.

41. Halliwell, B. and Gutteridge, J. M. C., Oxygen radicals and the nervous system, *TINS*, pp. 22–26, 1985.

42. Jenner, P. and Olanow, C. W., Oxidative stress and the pathogenesis of Parkinson's disease, *Neurology*, 47, 6 Suppl., 3, pp. S161–S170, 1996.

43. Dexter, D. T. et al., Basal lipid peroxidation in substantia nigra is increased in Parkinson's disease, *J. Neurochem.*, 52(2), pp. 381–389, 1989.

44. Dexter, D. T. et al., Increased levels of lipid hydroperoxides in the parkinsonian substantia nigra: an HPLC and ESR study, *Movement Disorders*, 9, pp. 92–97, 1994.

45. Alam, Z. I. et al., A generalised increase in protein carbonyls in the brain in Parkinson's but not incidental Lewy body disease, *J. Neurochem.*, 69(3), pp. 1326–1329, 1997.

46. Alam, Z. I. et al., Oxidative DNA damage in the parkinsonian brain: an apparent selective increase in 8-hydroxyguanine levels in substantia nigra, *J Neurochem*, 69(3), pp. 1196–1203, 1997.

47. Tanaka, M. et al., Dopa and dopamine cause cultured neuronal death in the presence of iron, *Journal of the Neurological Sciences*, 101(2), pp. 198–203, 1991.

48. Cheng, N. et al., Differential neurotoxicity induced by L-DOPA and dopamine in cultured striatal neurons, *Brain Res.*, 743(1–2): pp. 278–283, 1996.

49. Masserano, J. M. et al., Dopamine induces apoptotic cell death of a catecholaminergic cell line derived from the central nervous system, *Mol. Pharmacol.*, 50(5), pp. 1309–1315, 1996.

50. Clement, M. V. et al., The cytotoxicity of dopamine may be an artefact of cell culture, *J. Neurochem.*, 81(3), pp. 414–421, 2002.

51. Mena, M. A. et al., Glia protect fetal midbrain dopamine neurons in culture from L-DOPA toxicity through multiple mechanisms, *J. Neural Transm.*, 104, pp. 317–328, 1997.

52. Mytilineou, C., Han, S. K. and Cohen, G., Toxic and protective effects of L-dopa on mesencephalic cell cultures, *J. Neurochem.*, 61(4), pp. 1470–1478, 1993.

53. Mena, M. A., Davila, V. and Sulzer, D., Neurotrophic effects of L-DOPA in postnatal midbrain dopamine neuron/cortical astrocyte cocultures, *J. Neurochem.*, 69, pp. 1398–1408, 1997.

54. Murer, M. G. et al., Chronic levodopa is not toxic for remaining dopamine neurons, but instead promotes their recovery, in rats with moderate nigrostriatal lesions, *Ann. Neurol.*, 43(5), pp. 561–575, 1998.

55. Datla, K. P., Blunt, S. B. and Dexter, D. T., Chronic L-DOPA administration is not toxic to the remaining dopaminergic nigrostriatal neurons, but instead may promote their functional recovery, in rats with partial 6-OHDA or FeCl(3) nigrostriatal lesions, *Mov. Disord.*, 16(3), pp. 424–434, 2001.

56. Fahn, S., Levodopa-induced neurotoxicity: does it represent a problem for the treatment of Parkinson's disease? *CNS Drugs*, 8, pp. 376–393, 1997.

57. Agid, Y. et al., Levodopa in the treatment of Parkinson's disease: a consensus meeting, *Mov. Disord.*, 14(6), pp. 911–913, 1999.

58. Yahr, M. D. et al., Autopsy findings in parkinsonism following treatment with levodopa, *Neurology*, 22((Suppl)), pp. 56–71, 1972.

59. Cotzias, G. C. et al., Levodopa, fertility and longevity, *Science*, 196, pp. 549–551, 1977.

60. Hefti, F. et al., Long-term administration of l-dopa does not damage dopaminergic neurons in the mouse, *Neurology*, 31, pp. 1194–1195, 1981.

61. Sahakian, B. J. et al., Functional and structural consequences of long-term dietary l-dopa treatment in mice, *Commun. Psychopharmacol.*, 4, pp. 169–176, 1980.

62. Quinn, N. et al., Preservation of the substantia nigra and locus coeruleus in a patient receiving levodopa (2kg) plus decarboxylase inhibitor over a four-year period, *Movement Disorders*, 1, pp. 65–68, 1986.

63. Rajput, A. H. et al., Is levodopa toxic to human substantia nigra? *Movement Disorders*, 12, pp. 634–638, 1997.

64. Jellinger, K., Overview of morphological changes in Parkinson's disease, *Adv. Neurol.*, 45, pp. 1–18, 1987.

65. Braak, H. et al., Staging of brain pathology related to sporadic Parkinson's disease, *Neurobiol. Aging*, 24(2), pp. 197–211, 2003.

66. Del Tredici, K. et al., Where does parkinson disease pathology begin in the brain? *J. Neuropathol. Exp. Neurol.*, 61(5), pp. 413–26, 2002.

67. Olanow, C. W. et al., The effect of deprenyl and levodopa on the progression of Parkinson's disease, *Ann. Neurol.*, 38, pp. 771–777, 1995.

68. Parkinson-Study-Group, Does levodopa slow or hasten the rate of progression of Parkinson disease? The results of the Elldopa trial, *Neurology*, 60(Suppl. 1), p. A80, 2003.

69. Fahn, S. and Parkinson-Study-Group, Results of the ELLDOPA (Earlier vs. Later Levodopa) study, *Movement Disorders*, 17(Suppl. 5), pp. S13–S14, 2002.

70. Parkinson-Study-Group, Dopamine transporter brain imaging to assess the effects of pramipexole vs. levodopa on Parkinson disease progression, *JAMA*, 287, pp. 1653–1661, 2002.

71. Whone, A. L. et al., The REAL-PET study: slower progression in early Parkinson's disease treated with ropinirole compared with l-dopa, *Neurology*, 58(Suppl. 3), pp. A82–A83, 2002.

72. Whone, A. L. et al., Slower progression of Parkinson's disease with ropinirole versus levodopa: The REAL-PET study, *Ann. Neurol.*, 54(1), pp. 93–101, 2003.

73. Ahlskog, J. E., Slowing Parkinson's disease progression: recent dopamine agonist trials, *Neurology*, 60(3), pp. 381–389, 2003.

74. Albin, R. L. and K. A. Frey, Initial agonist treatment of Parkinson disease: a critique, *Neurology*, 60(3), pp. 390–394, 2003.

75. DeJong, G. J., Meerwaldt, J. D. and Schmitz, P. I. M., Factors that influence the occurrence of response variations in Parkinson's disease, *Ann. Neurol.*, 22, pp. 4–7, 1987.

76. Lesser, R. P. et al., Analysis of the clinical problems in parkinsonism and the complications of long-term levodopa therapy, *Neurology*, 29, pp. 1253–1260, 1979.

77. Cedarbaum, J. M., Gandy, S. E. and Mc Dowell, F. H., "Early" initiation of levodopa treatment does not promote the development of motor response fluctuations, dyskinesias or dementia in Parkinson's disease, *Neurology*, 41, pp. 622–629, 1991.

78. Blin, J., Bonnet, A. M. and Agid, Y., Does levodopa aggravate Parkinson's disease? *Neurology*, 38, pp. 1410–1416, 1988.

79. Caraceni, T., Scigliano, G. and Musicco, M., The occurrence of motor fluctuations in parkinsonian patients treated long-term with levodopa: role of early treatment and disease progression, *Neurology*, 41, pp. 380–384, 1991.

80. Roos, R. A. C., Vredevoogd, C. B. and van der Velde, E. A., Response fluctuations in Parkinson's disease, *Neurology*, 40, pp. 1344–1346, 1990.

81. Trabucchi, M. et al., Influence of treatment on the natural history of Parkinson's disease, in Parkinsonism and Aging, Calne, D. B., Ed., Raven Press, Ltd., New York, N. Y., pp. 239–254, 1989.

82. Rajput, A. H., Stern, W. and Laverty, W. H., Chronic low-dose levodopa therapy in Parkinson's disease: an argument for delaying levodopa therapy, *Neurology*, 34, pp. 991–996, 1984.

83. Poewe, W. H., Lees, A. J. and Stern, G. M., Low-dose l-dopa therapy in Parkinson's disease: a 6-year follow-up study, *Neurology*, 36, pp. 1528–1530, 1986.

84. Rascol, O. et al., A five-year study of the incidence of dyskinesia in patients with early Parkinson's disease who were treated with ropinirole or levodopa, *N. Engl. J. Med.*, 342, pp. 1484–1491, 2000.

85. Holloway, R. G. and Parkinson-Study-Group, Pramipexole versus levodopa as initial treatment for Parkinson disease: a four-year randomized controlled trial, *Neurology*, 58(Suppl., 3), pp. A81–A82, 2002.

86. Mizuno, Y., Kondo, T. and Narabayashi, H., Pergolide in the treatment of Parkinson's disease, *Neurology*, 45(Suppl., 3), pp. S13–S21, 1995.

87. Parkinson's Disease Research Group of the United Kingdom, Comparisons of therapeutic effects of levodopa, levodopa and selegiline, and bromocriptine in patients with early, mild Parkinson's disease: three year interim report, *B. M. J.*, 307, pp. 469–472, 1993.

88. Rinne, U. K., Early combination of bromocriptine and levodopa in the treatment of Parkinson's disease: a 5 year follow up, *Neurology*, 37, pp. 826–828, 1987.

89. Rinne, U. K. et al., Early treatment of Parkinson's disease with cabergoline delays the onset of motor complications. Results of a double-blind levodopa controlled trial, *Drugs*, 55(Suppl., 1), pp. 23–30, 1998.

90. Parkinson-Study-Group, A randomized controlled trial comparing the agonist pramipexole with levodopa as initial dopaminergic treatment for Parkinson's disease, *JAMA*, 284, pp. 1931–1938, 2000.

91. Rascol, O. et al., Treatment interventions for Parkinson's disease: an evidence based assessment, *Lancet*, 359(9317), pp. 1589–1598, 2002.

92. Katzenschlager, R. and Lees, A. J., Treatment of Parkinson's disease: levodopa as the first choice, *J. Neurol.*, 249, Suppl., 2, p. II,19–24, 2002.

93. Kondo, T., *Initial therapy for Parkinson's disease: levodopa vs. dopamine receptor agonists. J Neurol.*, 2002. 249, Suppl., 2, p. II, 25-9.

94. Rinne, U. K., Dopamine agonists as primary treatment in Parkinson's disease, in *Advances in Neurology*, Bergmann, K. J., Ed., pp. 519–523, Raven Press, New York, N. Y., 1986.

95. Uitti, R. Y. and Ahlskog, J. E., Comparative review of dopamine receptor agonists in Parkinson's disease, *CNS Drugs*, 5, pp. 369–388, 1996.

96. Frucht, S. et al., Falling asleep at the wheel: motor vehicle mishaps in persons taking pramipexole and ropinirole, *Neurology*, 52, pp. 1908–1910, 1999.

97. Ling, L. H. et al., Constrictive pericarditis and pleuropulmonary disease linked to ergot dopamine agonist therapy (cabergoline) for Parkinson's disease, *Mayo Clin. Proc.*, 1999. 74, pp. 371–375, 1999.

98. Pritchett, A. M. et al., Valvular heart disease in patients taking pergolide, *Mayo Clin. Proc.*, 77(12), pp. 1280–1286, 2002.

99. Agid, Y. et al., Levodopa in the treatment of Parkinson's disease: a consensus meeting, *Movement Disorders*, 14, pp. 911–913, 1999.

100. Miyasaki, J. M. et al., Practice parameter: initiation of treatment for Parkinson's disease: an evidence-based review. Report of the Quality Standards Subcommittee of the American Academy of Neurology, *Neurology*, 58, pp. 11–17, 2002.

101. Diamond, S. G. et al., Multi-center study of parkinson mortality with early versus later dopa treatment, *Ann. Neurol.*, 22, pp. 8–12, 1987.

102. Block, G. et al., Comparison of immediate-release and controlled release carbidopa/levodopa in Parkinson's disease, *Eur. Neurol.*, 1997. 37. pp. 23–27, 1997.

103. DuPont, E. et al., Sustained-release Madopar HBS compared with standard Madopar in the long-term treatment of de novo parkinsonian patients, *Acta Neurol. Scand.*, 1996. 93. pp. 14–20, 1996.

104. Quinn, N., Young onset Parkinson's disease, *Movement Disorders*, 2, pp. 73–91, 1987.

105. Schrag, A. et al., Young-onset Parkinson's disease revisited-clinical features, natural history, and mortality, *Movement Disorders*, 13, pp. 885–894, 1998.

106. Adler, C. H. and Ahlskog, J. E., Eds., *Parkinson's disease and movement disorders. Diagnosis and treatment guidelines for the practicing physician*, Humana Press, Totowa, N. J., 2000.

107. Chase, T. N., Levodopa therapy: consequences of the nonphysiologic replacement of dopamine, *Neurology*, 50(5 Suppl., 5), pp. S17–S25, 1998.

108. Chase, T. N. and Oh, J. D., Striatal mechanisms and pathogenesis of parkinsonian signs and motor complications, *Ann. Neurol.*, 47(4 Suppl., 1), pp. S122–S129; discussion S129-30, 2000.

109. Jenner, P., Pathophysiology and biochemistry of dyskinesia: clues for the development of non-dopaminergic treatments, *J. Neurol.*, 247, Suppl., 2, p. II, pp. 43–50, 2000.

110. Jenner, P., Factors influencing the onset and persistence of dyskinesia in MPTP-treated primates, *Ann. Neurol.*,

47(4 Suppl., 1), pp. S90–S99; discussion S99–S104, 2000.

111. Juncos, J. L. et al., Continuous and intermittent levodopa differentially affect basal ganglia function, *Ann. Neurol.*, 25, pp. 473–478, 1989.

112. Engber, T. M. et al., Continuous and intermittent levodopa differentially affect rotation induced by D-1 and D-2 dopamine agonists, *Eur. J. Pharmacol.*, 168, pp. 291–298, 1989.

113. Pearce, R. K. B. et al., De novo administration of ropinirole and bromocriptine induces less dyskinesias than l-dopa in the MPTP-treated marmoset, *Movement Disorders*, 13, pp. 234–241, 1998.

114. Papa, S. M. et al., Motor fluctuations in levodopa treated parkinsonian rats: relation to lesion extent and treatment duration, *Brain Res.*, 662, pp. 69–74, 1994.

115. Olanow, C. W., Watts, R. L. and Koller, W. C., An algorithm (decision tree) for the management of Parkinson's disease (2001): treatment guidelines, *Neurology*, 2001. 56(11 Suppl., 5), pp. S1–S88, 2001.

116. Guttman, M. et al., [11C]RTI-32 PET studies of the dopamine transporter in early dopa-naive Parkinson's disease: implications for the symptomatic threshold, *Neurology*, 48(6), pp. 1578–1583, 1997.

117. Bonifati, V. and G. Meco, G. New, selective catechol-O-methyltransferase inhibitors as therapeutic agents in Parkinson's disease, *Pharmacology and Therapeutics*, 81(1), pp. 1–36, 1999.

118. Mannisto, P. T. and S. Kaakkola, S., Catechol-O-methyltransferase (COMT): biochemistry, molecular biology, pharmacology, and clinical efficacy of the new selective COMT inhibitors, *Pharmacol. Rev.*, 51(4), pp. 593–628, 1999.

119. Snow, B. J. et al., The effect of amantadine on levodopa-induced dyskinesias in Parkinson's disease: a double-blind, placebo-controlled study, *Clin. Neuropharmacol.*, 23, pp. 82–85, 2000.

120. Verhagen Metman, L. et al., Blockade of glutamatergic transmission as treatment for dyskinesias and motor fluctuations in Parkinson's disease, *Amino Acids*, 14(1–3), pp. 75–82, 1998.

121. Verhagen Metman, L. et al., Amantadine as treatment for dyskinesias and motor fluctuations in Parkinson's disease [see comments], *Neurology*, 50(5): pp. 1323–1326, 1998.

122. Verhagen Metman, L. et al., *Amantadine for levodopa-induced dyskinesias*, Arch. Neurol., 56, pp. 1383–1386, 1999.

123. Klawans, H. L., Jr. and Weiner, W. J., Attempted use of haloperidol in the treatment of L-dopa induced dyskinesias, *J. Neurol. Neurosurg. Psychiatry*, 37(4), pp. 427–430, 1974.

124. Tarsy, D., Parkes, J. D. and Marsden, C. D., Metoclopramide and pimozide in Parkinson's disease and levodopa-induced dyskinesias, *J. Neurol. Neurosurg. Psychiatry*, 38(4), pp. 331–335, 1975.

125. Bennett, J. P., Jr. et al., Suppression of dyskinesias in advanced Parkinson's disease: moderate daily clozapine doses provide long-term dyskinesia reduction, *Mov. Disord.*, 9(4), pp. 409–414, 1994.

126. Verhagen Metman, L. et al., Dextromethorphan improves levodopa-induced dyskinesias in Parkinson's disease, *Neurology*, 51(1), pp. 203–206, 1998.

127. Bonifati, V. et al., Buspirone in levodopa-induced dyskinesias, *Clin. Neuropharmacol.*, 17(1), pp. 73–82, 1994.

128. Rascol, O. et al., Idazoxan, an alpha-2 antagonist, and L-DOPA-induced dyskinesias in patients with Parkinson's disease, *Mov. Disord.*, 16(4), pp. 708–713, 2001.

129. Manson, A. J., Iakovidou, E. and Lees, A. J., Idazoxan is ineffective for levodopa-induced dyskinesias in Parkinson's disease, *Mov. Disord.*, 15(2), pp. 336–337, 2000.

130. Carpentier, A. F. et al., Improvement of levodopa-induced dyskinesia by propranolol in Parkinson's disease, *Neurology*, 46(6), pp. 1548–1551, 1996.

131. Muenter, M. D. et al., Patterns of dystonia ("I-D-I" and "D-I-D") in response to l-dopa therapy for Parkinson's disease, *Mayo Clin. Proc.*, 52, pp. 163–174, 1977.

132. Ballard, P. A., J. W. Tetrud, and Langston, J. W., Permanent human parkinsonism due to 1-methyl-4-phenyl-1,2,3,6-tetrahydropyridine (MPTP): seven cases, *Neurology*, 35, pp. 949–956, 1985.

133. Leenders, K. L. et al., Brain dopamine metabolism in patients with Parkinson's disease measured with positron emission tomography, *J. Neurol. Neurosurg. Psychiatry*, 49, pp. 853–860, 1986.

134. Hwang, W. J. et al., Downregulation of striatal dopamine D2 receptors in advanced Parkinson's disease contributes to the development of motor fluctuation, *Eur. Neurol.*, 47(2), pp. 113–117, 2002.

135. Deuschl, G. et al., Deep-brain stimulation for Parkinson's disease, *J. Neurol.*, 249, Suppl., 3, p. III, 36–39, 2002.

136. Vitek, J. L., Deep brain stimulation for Parkinson's disease. A critical re-evaluation of STN versus GPi DBS, *Stereotact Funct. Neurosurg*, 78(3–4), pp. 119–131, 2002.

137. Riley, D. E. and A. E. Lang, The spectrum of levodopa-related fluctuations in Parkinson's disease, *Neurology*, 43, pp. 1459–1464, 1993.

138. Hillen, M. E. and Sage, J. I., Nonmotor fluctuations in patients with Parkinson's disease, *Neurology*, 47, pp. 1180–1183, 1996.

139. Factor, S. A. et al., Sleep disorders and sleep effect in Parkinson's disease, *Mov. Disord.*, 5(4), pp. 280–285, 1990.

140. Hogl, B. E. et al., A clinical, pharmacologic, and polysomnographic study of sleep benefit in Parkinson's disease, *Neurology*, 50(5), pp. 1332–1339, 1998.

141. Goetz, C. et al., Pain in Parkinson's disease, *Mov. Disord.*, 1: pp. 45–49, 1986.

142. Mear, J. Y. et al., Pergolide in the treatment of Parkinson's disease, *Neurology*, 34, pp. 983–986,1984.

57 Dopamine Agonists in Parkinson's Disease

Sandra Kuniyoshi and Joseph Jankovic
Parkinson's Disease Center and Movement Disorders Clinic, Baylor College of Medicine

CONTENTS

INTRODUCTION

Parkinson's disease (PD) is a neurodegenerative disorder characterized by the loss brain stem neurons and dysregulation of downstream basal ganglia motor pathways. Understanding the pathophysiology has led to the development of treatments that improve the symptoms of PD. Early pathological studies identified the loss of pigmented neurons in the substantia nigra of the parkinsonian brain.[1] The degenerated nigrostriatal tract produced the neurotransmitter dopamine. As dopamine does not cross the blood-brain barrier, L-dihydroxy-phenylalanine (LD) the metabolic precursor of dopamine was utilized to replenish the loss of dopamine. Though it was profoundly effective in alleviating symptoms of PD, LD was not considered an ideal therapy for PD.[2] LD had a short half-life. Conversion of LD was dependent on the presence of dopa decarboxylase, an enzyme that diminished in concentration in the degenerating nigrostriatal tract.[3] In addition, the supraphysiologic levels of dopamine nonspecifically stimulated dopamine receptors, resulting in severe limbic, sympathetic, and gastrointestinal adverse effects. Finally, *in vitro* studies suggested that LD promoted apoptosis and oxidative metabolites in nigrostriatal neurons.[4,5]

The natural history of Parkinson's disease changed after the advent of LD. Morbidity and mortality was positively impacted with widespread use of L-dopa, despite continued *in vitro* evidence of neuronal damage.[6] As degeneration of numerous nondopaminergic nuclei had also been reported, in retrospect, it is remarkable that replacement therapy works so well in compensating for impulse-driven release of endogenous dopamine. In no other syndrome or disease has simple exogenous replenishment of a neurotransmitter provided a therapeutic response. The pharmacologic treatment of PD remains dominated by strategies designed to compensate for the loss of dopaminergic neurons in the substantia nigra. Clinical studies continue to confirm the efficacy of LD as well as dopamine agonists in improving the symptoms of Parkinson's disease. Agonist therapy has not improved upon the potency of response obtained with L-dopa after the addition of the adjunctive medications designed to decrease the peripheral decarboxylation.

Symptomatic efficacy does not, however, indicate slowing of the progression or arrest of the disease process. Over time, the duration of action of an oral dose of LD diminishes, a phenomenon called *wearing off*.[7] This occurs partly because of the loss of striatal terminals and their diminished capacity to store and convert LD through the decarboxylase enzyme into dopamine presynaptically.[8] Efficacy then reflects the plasma LD level, which has a half-life of about 30 min. Downstream postsynaptic receptor loss contribute to the dysregulation of the basal ganglia and associated motor fluctuations as agonist therapy does not prevent fluctuations.[9,10]

Supraphysiologic fluctuations in the level of dopamine on the receptors caused by interval drug intake are associated with another effect termed *dyskinesia* or abnormal involuntary movements, phenomenologically termed *chorea* and *dystonia*.[11] These LD-induced dyskinesias (LID) can limit the dosage of LD tolerated and thereby the therapeutic effect. Recent investigations in continuous infusion therapies suggest that it is the severity and lengths of the troughs in serum levels rather than the peaks that trigger the development of on-off fluctuations.[9,11] Longer half-life, concentrations independent of the presence of AADC in dopamine terminals, and selective receptor specificity were mechanism whereby synthetic agonists could improve on the precursor.

The first dopamine agonist was discovered serendipitously. In determining the mechanism of ergot based toxins as chemotherapeutic agent in hypothalamic tumors, it was discovered that many of the ergots had dopamine agonist activity. Apomorphine was the first dopamine agonist successfully utilized for PD patients.[12] Significant adverse effects of nausea, vomiting, and hypotension, as well as poor oral bioavailability, limited its use in the clinical arena. Bromocriptine was the first oral dopamine agonist widely utilized clinically.[13] While it remains the oral agonist with the most mileage, it is seldom prescribed for Parkinson's disease today. Interest-

ingly, reports of the rapid therapeutic effect and potential neuroprotective effect suggest a good return reception of injectable apomorphine, which is now available in the U.S.[14]

The DA discussed in this chapter include (a) the ergot-based agents—apomorphine (APO), bromocriptine (BCP),[13] cabergoline,[15] dihydroergocryptine, lisuride, pergolide (PRG),[16] piribedil, and (b) the nonergots—pramipexole (PRAM)[17,18] and ropinirole (ROP). In the U.S., BCP, PRG, PRAM, and ROP[19] are licensed for the treatment of PD. While cabergoline (CAB) is available for the treatment of hyperprolactenemia, it is too expensive to use on an off-label basis for PD.

DOPAMINE RECEPTORS

Dopamine receptors were originally characterized by ligand specificity and coupling to Adenylate cyclase (D_1 and D_2). The D_1 class stimulated the adenylate cyclase pathway, while D_2 was inhibitory.[20] The synergistic effect of D_1 and D_2 activation is evident in human and animal models of Parkinson's disease.[21,22] The selective D_1 and D_2 agonists are clinically effective because of endogenous dopamine at remaining terminals.

Molecular cloning resulted in the development of subtypes (D_1 through D_5). The pharmacologically defined D_1 class contains the molecularly similar D_1 and D_5 subtypes, whereas the D_2 class is composed of subtypes D_2, D_3, and D_4.[23] D_2 short and long isoforms of D_2 exist, D_2S and D_2L. They differ in location and function in the basal ganglia.[24]

ANATOMY AND PHYSIOLOGY

Dopaminergic neurons originating in the substantia nigra primarily project axons to the striatum. The dominant "dual circuit" model suggests that the striatal output is composed of two projections, the direct and indirect pathways, to downstream basal ganglia nuclei. Medium spiny neurons containing D_1 receptors project prominently to the Gpi (globus pallidus interna), tonically inhibiting the primary output nuclei of the basal ganglia, whereas D_2 receptor stimulation disinhibits the Gpe (globus pallidus externa), allowing tonic inhibition of the subthalamic nucleus and the glutamatergic fibers, which project to the Gpi ("indirectly" providing inhibition).[25,26] Single-cell recordings and other physiological studies suggest that the effects of dopamine are considerably more complicated. There is evidence of direct tonic inhibition of the subthalamic nuclei from the substantia nigra and extensive striatofugal collateralization.[27,28] The primary actions of dopamine finally cannot be described in terms of simple inhibition and excitation, but modulation dependent on dose, as well as prior phasic and tonic activity.[29,30] Furthermore, dopaminergic denervation has been shown to result in a proliferation of D_2 receptors as well as a colocalization of D_1 and D_2 receptors.[30] Similarly, gap junctions, unaccounted for in the present model, allow rapid communications between striatal neurons and have been shown to increase dramatically in after dopaminergic denervation.[31,32]

A critical component in dopamine actions is its modulation of glutamate transmission. The modulatory actions of D_1 and D_2 receptors on prefrontal cortex pyramidal cell excitability are mediated by multiple intracellular mechanisms and by activation of GABA(A) receptors.[135] Single-cell recording suggests that the therapeutic effect of LD or dopamine agonists requires the reversal of glutamatergic overactivity and the normalization of hypersensitive D_2 dopamine receptors controlling glutamate release from corticostriatal terminals.[136] Finally, the number of proliferating cells in the subependymal zone and neural precursor cells in the subgranular zone and olfactory bulb are reduced in postmortem brains of individuals with Parkinson's disease. Recent investigations provide ultrastructural evidence that highly proliferative precursors in the adult subependymal zone express dopamine receptors and receive dopaminergic afferents. Depletion of dopamine decreases precursor cell proliferation in both the subependymal zone and the subgranular zone. Proliferation is restored by a selective agonist activity at the D_2 (D_2L) receptors.[137]

D_1 Receptors and Parkinsonism—D_1 Receptor Agonists

PRG and APO are commercially available products with active D_1 as well as D_2 activation. Investigations in selective D_1 agonists such as dihydrexidine [(6)-trans-10, 11-dihydroxy-5, 6, 6a,7,8,12bhexahydrobenzo[a]phenanthridine, A-77636, SKF 81297 [(R)-(1)-6-chloro-7,8-dihydroxy-1-phenyl-2,3,4,5-tetrahydro-1H-3-benzazepine], and A-86929 show a therapeutic response similar to other D_2 DA and LD.[33-36] Poor tolerance, attributed to D_1 internalization, short half-life, and decreased bioavailability have limited the clinical usefulness.[33-38]

D_3 Receptors

Studies on D_3 receptors have been problematic secondary to the lack of a suitable antagonist. D_3 receptors have a different distribution from that of D_1 and D_2, existing in high concentrations in the nucleus accumbens, a mesolimbic structure. Locomotion response appears to be biphasic and dose dependent.[39,40,138]

PRAM is a potent D_3 agonist and has been shown to have antidepressant effects, possibly mediated through the D_3 effect.[41,42] Preliminary data suggest that PRAM 0.375 to 1.0 mg/day adjunction to antidepressant treatment may be effective and well tolerated in patients with resistant major depression.

Receptor Affinities of DA Agonists

In a multivariate analysis of the binding profiles of antiparkinson agents at human D_2S, D_2L, D_3, and D_4 receptors and at α-(2A, 2B, 2C, and 1A-adrenoreceptors, as determined by guanosine $5'$-O-(3-[^{35}S]thio)triphosphate ((35)S]GTP-γS) binding, no ligand displayed "full" efficacy relative to dopamine at all "D(2)-like" sites.[43] However, at D_2S receptors, PRAM, ROP, PRG, and CAB were as efficacious as dopamine; APO was highly efficacious (79 to 92%); and piribedil, lisuride, and BCP, showed intermediate efficacy (40 to 55%). For all drugs, efficacies were lower than dopamine at $D_2(L)$ receptors. At D_3 receptors, efficacies ranged from 34% (piribedil) to 82% (APO), whereas, for D_4 receptors, highest efficacies (approximately 70%) were seen for ROP and quinpirole, whereas piribedil and terguride behaved as antagonists and BCP was inactive. Although efficacies at D_2S versus D_2L sites were highly correlated ($r = 0.79$), they correlated only modestly with D_3/D_4 sites ($r = 0.44$ to 0.59). In [^{35}S]-GTP-γS and [^{3}H] phosphatidylinositol depletion studies of α-adrenergic receptors, there was a diversity of results. The authors concluded that "antiparkinson agents display diverse agonist and antagonist properties at multiple subtypes of D_2-like receptor and α_1/α_2-adrenoreceptors, actions, which likely contribute to their contrasting functional profiles." (See Table 57.1.)

TABLE 57.1
Characteristics of Clinically Available Dopamine Agonists

	Apomorphine	Bromocriptine	Pergolide	Ropinirole	Pramipexole	Cabergoline
Metabolism	Hepatic	Hepatic	Hepatic	Hepatic	Renal	Hepatic
Half-life	35 min	6–8 hr	15–42 hr	6 hr	8–12 hr	65 hr
Therapeutic dose, mg/day	98	6.5–45	0.75–4.5	3–24	1.5–4.5	0.5–6
High dose	>100	50–100 mg	4.5–24 mg	18–34 mg	6 mg	>6 mg
Titration q week	—	1.25 mg	0.125 mg	0.25 mg t.i.d. until 3 mg then 0.5 mg t.i.d.	0.125 mg t.i.d. → 0.25 mg t.i.d. → 0.5 mg t.i.d.	0.5 mg

CLINICAL PHARMACOLOGY OF INDIVIDUAL DOPAMINE AGONISTS

BROMOCRIPTINE

BCP, a tetracyclic ergoline compound derived from plant alkaloids, was the first of the DA marketed for clinical treatment of PD. BCP is a D_2-like receptor agonist and a partial D_1-like receptor agonist (which means that it has some weak D_1 antagonistic effects on normosensitive receptors). Like most ergot derivatives, BCP has also 5-HT2 antagonist effects and mild adrenergic effects. The absolute oral bioavailability is less than 10%, since 90% of it undergoes first-pass hepatic metabolism. BCP plasma-elimination half-life is about 6 to 8 hr. Ninety percent is bound to plasma proteins.[13]

Titration regimens vary dramatically, ranging from 13.2 mg/d to over 50 mg a day. A randomized study investigated the effect of rapid, high dose (mean dose 22 mg, maximum dose 100 mg/d) versus slow titration low dose (mean dose 25 mg/d, maximum 55 mg/d) over 26 weeks. Approximately 50% of both groups experienced a therapeutic response and reached the improvement criteria (33% improvement in clinical rating score). Thirty-six percent of the patients in the high/fast group experienced adverse reactions of hallucinations, nausea/emesis, or orthostatic hypotension, which required discontinuation of the bromocriptine, whereas these were present in only 20% of the low/slow group. Discontinuation secondary to lack of therapeutic effect in the low/slow group equalized the poor tolerance.[44]

Typical doses required to demonstrate therapeutic efficacy as monotherapy are 20 to 30 mg per day. BCP has the typical dopaminergic side effects of the other ergot agents; i.e., nausea and vomiting, orthostatic hypotension, psychosis, drowsiness, and leg edema and painful, erythematous skin induration (erythromelalgia) and (rarely) alopecia.[13,45] In addition, it is associated with fibrosis of serosal membranes, including heart valves, seen with other ergots.[45–50] BCP is rarely used for PD anymore in the U.S. BCP continues to be prominent in treating non-PD disorders, i.e., gynecologic disorders (breast and brain tumors, infertility, menstrual disorders) and neuroleptic malignant syndrome, as well as for smoking and glycemic control. A transdermal formulation exists and awaits testing in humans.[51] The lack of quality clinical trials in PD and conflicting results regarding efficacy as compared to LD and other agonists limit the conclusions which can be drawn from the evidence.

PERGOLIDE

Pergolide is another ergoline compound, first marketed in 1989.[16] Although promoted as an agonist at the D_1 receptor, PRG is a very weak D_1 agonist. It predominantly activates D_2 receptors; it also is a mild agonist of D_3 and α-adrenergic receptors. PRG has a long half-life, between 15 and 42 hr, but its therapeutic half-life seems to be comparable to the other agents and so requires t.i.d. use. Plasma concentration peaks between 1 and 3 hr. The major route of excretion is the kidney.

PRG is one of the most frequently prescribed dopamine agonists worldwide. As other dopamine agonists, PRG is effective as monotherapy in *de novo* (patients never previously treated with dopaminergic drugs).

Three randomized trials have been accomplished in *de novo* Parkinson's disease comparing monotherapy PRG to LD or placebo. Two of the studies with a duration less than six months showed significant improvement relative to placebo.[52,53] In the study, which lasted for 12 months, greater UPDRS improvement was found in the LD group; however, the percentage of responders was similar in both groups.[54]

Randomized studies designed to compare the efficacy of PRG versus BCP found a similar efficacy in studies of *de novo* Parkinson's patients with a duration less than six months and greater improvement with a greater decrease in LD dose in those studies in patients with motor fluctuations after six months.[54,55]

The usual regimen of PRG involves a slow titration to 1 to 4 mg/d. Results from preliminary open-label studies suggest that off periods and dyskinesia are markedly reduced, and, on a high dose regimen of 8.25 ± 4.35 mg/d, a 50% reduction of LD dose could be managed.[56,57] (See Table 57.2.) There is evidence of sustained long-term improvement in patients on PRG treatment.[58]

TABLE 57.2
Dopamine Receptor Affinity[43]

	D_1	D_2S	D_21	D_3	D_4	D_5
BRC	+(ant)	+++	+++	+++	++	++
PRG	++	+++	+++	+++	+++	+++
CRB	++	++++	++++	++++	+++	+++
APO	++	+++	++	+++	+++	+++
PRAM		+++	++	+++	++	
ROP		+++	++	+++	++	

The side-effect profile of PRG is similar to that of other DA listed above and include nausea, vomiting, hypotension and psychosis. Complications such as sleep attacks and alopecia have been reported.[59,60] PRG as BCP is an ergot derived agonist and has 5 HT receptor affinities which may result in stimulation of immune factors and result in the development of rare but life-threatening symptoms of cardiac and pulmonary fibrosis and effusion.[139] Initial release of PRG was delayed because of concern about cardiac toxicity,[61] which was later largely dispelled.[62,63] More recently, three patients with cardiac valvular disease were reported. In an accompanying editorial, the premise that this was directly due to PRG was accepted, although a prospective controlled trial using rigorous cardiac morphometric criteria was proposed to determine the incidence rate. Annual chest X-ray and sedimentation rate are recommended for patients on long-term therapy, if more sensitive cardiac monitoring is not indicated. Very few cases have been reported for other ergot-derived agonists. Cases have been reported, though not published in the literature, in patients taking nonergot-derived agonist.[64]

CABERGOLINE

Cabergoline is a synthetic ergot derived DA with a high affinity for D(2) receptors and, as other ergots, has mild affinity for serotonergic and adrenergic receptors. In Europe and Asia, it is used to treat early and advanced PD and hyperprolactinaemic disorders but is approved only for hyperpolactinemia in the U.S. Peak plasma concentrations of CAB are reached within 2 to 3 hr. The elimination half-life of CAB is approximately 65 hr. CAB is moderately bound (around 40%) to human plasma proteins in a concentration-independent manner; administration of highly protein-bound drugs is unlikely to affect its efficacy. The absolute bioavailability of CAB is unknown. CAB is extensively metabolized by the liver, predominantly via hydrolysis of the acylurea bond of the urea moiety, and is excreted primarily through the bile and feces. Controversy surrounds the degree of cytochrome P450-mediated metabolism. A case report indicates 300% increase in plasma level after initiation of itraconazole, a potent inhibitor of CYP3A4.[65,66]

CAB is unique among the DA because of its long half-life. CAB is slowly titrated, starting from 1 mg once daily and increasing the dose by 0.5 to 1 mg at weekly or biweekly intervals until maximum therapeutic response or intolerable side effects occur. The recommended therapeutic dosage is 2 to 6 mg/d. Once-daily administration is recommended in patients with PD and twice-weekly administration in patients with hyperprolactinemia. In the U.S., since the indication is for hyperprolactinemia, which requires very low doses, the retail price is about $30 USD per 0.5-mg tablet; this would result in a cost of $120 to $360 per day for PD, effectively preventing off-label usage.

Few qualified clinical trials have been conducted to determine efficacy of CAB monotherapy. Placebo-controlled trials have provided evidence of a significant reduction "off" time and delay and decrease in motor complications [22% versus 34% (with LD)].[67] Adverse effects remain similar in quality and frequency to those reported for other ergot DA.

PRAMIPEXOLE

PRAM is a nonergot synthetic aminobenzathiazol derivative. It is a nearly pure DA with high selectivity for the D$_2$ dopamine receptor class. PRAM has a five- to seven-fold greater affinity for the D$_3$ receptor subtype with lower affinities for the D$_2$ and D$_4$ receptor subtypes. The drug has minimal α_2-adrenoreceptor activity and virtually no other receptor agonism or antagonism.

PRAM is rapidly absorbed after oral dosing, with a bioavailability >90% and a time to peak plasma concentration of 2 hr. Only 20% is protein bound, and it is excreted unmetabolized from the kidneys.[17,18] PRAM can be administered without regard to meals (protein load) or concern regarding interactions with drugs metabolized through the hepatic cytochrome P450 enzymes. The half-life ranges from 8 to 12 hr. The half-life is influenced by age, increased in the elderly (~12 hr), probably secondary to the decreased glomerular filtration rate. Dosing frequency should be reduced in patients with impaired renal function, i.e., b.i.d. in patients with a creatinine clearance of 35 to 59 ml/min and q.d. in patients with a creatinine clearance of 15 to 34 ml/min. PRAM can be titrated over three weeks to a conventional dose of 0.5 to 1.5 mg t.i.d.[68]

Randomized double-blind placebo-controlled trials with adequate power showed significant improvement in both part II and part III of the UPDRS. PRAM appears to be well tolerated and, in the majority of cases, upwards of two-thirds of the subjects completed the study.[69–72] In the Parkinson's study group, there was a 20% improvement in part II of UPDRS versus placebo. Initial PRAM treatment resulted in significantly less development of wearing off, dyskinesias, or on-off motor fluctuations (28%) compared with LD (51%) (hazard ratio, 0.45; 95% confidence interval [CI], 0.30 to 0.66; $P < 0.001$). Despite supplementation with open-label LD in both groups, the LD-treated group had a greater improvement in total UPDRS compared with the PRAM group (9.2 versus 4.5 points; $P < 0.001$). Somnolence was more common in PRAM-treated patients than in LD-treated patients (32.4% versus 17.3%; $P = 0.003$), and the difference was seen during the escalation phase of treatment.[72] In another open-label study, 47% patients (of the intention to treat group) had a LD dose reduction (adjusted) of more than 40% while maintaining or improving their level of efficacy, and 72.2% had a reduction of at least 20%. Motor fluctuations improved compared to baseline according to patient diaries and UPDRS Part IV.[73] Four-year follow-up studies comparing levodopa to PRAM suggested a significant reduction in the risk of freezing (25.3 versus 37.1%; hazard ratio, 1.7; 95% CI, 1.11–2.59; $P = 0.01$) and edema (42 versus 15%, $P < 0.001$). Mean changes in quality-of-life scores did not differ between the groups. Initial treatment with pramipexole continued to show a lower incidence of dyskinesias and wearing-off compared with initial treatment with levodopa. Both options resulted in similar quality of life.[140]

Despite the relatively low dropout rate, a significant proportion of subjects complained of adverse effects long term.[70–75] Gastrointestinal symptoms of nausea were prominent in almost every study and ranged from 20 to 39%. Hallucinations occurred more frequently in PRAM arm than in placebo or LD, ranging 7 to 14%. Significant somnolence in the PRAM versus placebo or LD arm was reported in many of the studies, with an incidence of 18 to 32%. Insomnia was reported in 8 to 39%. Peripheral edema occurred in 8 to 18% of the patients. Orthostatic hypotension was symptomatic in 25% of the PRAM arm and in 18% of the control.[74] Other studies reported an incidence of symptomatic orthostasis in the realm of 8%.

ROPINIROLE

ROP is a selective DA with a nonergoline structure. It has a high selectivity for D_2-like receptors, showing affinity for the $D_3 > D_2 > D_4$ receptor subtypes similar to dopamine with essentially no interaction with other neurotransmitter receptors (i.e., D_1, serotonergic, adrenergic).[19,76]

When taken as oral tablets, ROP is rapidly absorbed. The bioavailability is approximately 50%, and ROP plasma protein binding is only 30%. The drug is metabolized in the liver, and none of the major circulating metabolites has pharmacological activity. The principal metabolic enzyme is the cytochrome P450 (CYP) isoenzyme CYP1A2. ROP shows approximately linear pharmacokinetics when given as single or repeated doses, and is eliminated with a half-life of approximately six hours. The CYP1A2 inhibitor ciprofloxacin produced increases in the plasma concentrations of ROP when these two drugs were co-administered, but no interaction was seen with theophylline, which, like ROP, is also a substrate for CYP1A2. There is no obvious plasma concentration-effect relationship for ROP.[77] ROP may potentiate the anticoagulant effects of warfarin.[78]

Like the other DA, ROP should be slowly titrated initially to a usual minimum effective dose of about 3 mg t.i.d. Conventional doses of ROP are 8 to 24 mg, divided into two or three daily doses.[79] Six-month data from three trials of monotherapy in patients with PD showed that a therapeutic response may be expected at ≤7.5 mg/d, but there was further benefit with continued dose titration.[80] For some patients, the maximum recommended dose (24 mg/d) might be necessary. High-dose ROP has been examined in an open-label 12-month study in PD patients taking ROP and LD at conventional doses. The range of ROP was 28 to 50 mg, with the mean daily dose 34.7 mg. Twelve of the 36 patients recruited for the study dropped out within the first year, 7 for lack of efficacy, 3 for hallucinations, and 1 for somnolence.[81] There is a new sustained-release formulation of ROP (Requip CR), currently in Phase III trials, that allows once-per-day administration.

As in the case of PRAM, randomized double-blind placebo-controlled studies confirm the efficacy of ROP on symptoms of PD as assessed by the motor portion of the UPDRS. ROP-treated patients had a significantly greater percentage improvement in UPDRS motor score than patients who received placebo (+24% versus –3%; $p < 0.001$).[80]

In a five-year study, 85 of the 179 patients in the ROP group (47%) and 45 of the 89 patients in the LD group (51%) completed the study. In the ROP group, 29 of the 85 patients (34%) received no LD supplementation. The analysis of the time to dyskinesia showed a significant difference in favor of ROP (hazard ratio for remaining free of dyskinesia, 2.82; 95% confidence interval, 1.78 to 4.44; $P < 0.001$). At five years, the cumulative incidence of dyskinesia, regardless of LD supplementation, was 20% (36 of 177 patients) in the ROP group and 45% (40 of 88 patients) in the LD group. There was no significant difference between the two groups in the mean change in scores for activities of daily living among those who completed the study. Modest motor UPDRS improvement was sustained in the LD over the ROP arm. However, the small amplitude of the improvement and the power of the study diminished the clinical significance of the finding. Adverse events led to the early withdrawal from the study of 48 of 179 patients in the ROP group (27%) and 29 of 89 patients in the LD group (33%).[82]

In a three-year double-blind trial comparing BCP to ROP, approximately one-third withdrew prematurely. Occurrence of adverse experiences in both groups was similar. Emergence of dyskinesias was low. Both treatments provided comparable improvements in Unified Parkinson's Disease Rating Scale activities of daily living (ADL, Part II) and motor (Part III) scores. After three years, patients in the ROP group had a mean improvement in motor score of 31% compared with 22% in the BCP group ($p = 0.086$) and a significantly better ADL score (treatment difference 1.46 points, $p = 0.009$).[83]

Adverse effects for ROP remain similar to those reported for PRAM. In early therapy, peripheral adverse effects of nausea, vomiting, dyspepsia, somnolence, insomnia, and hypotension predominate. As the disease progresses, hallucinations and dyskinesia, and pedal edema, become significant; hypotension and nausea remain causes for discontinuation. Most adverse experiences were mild as most studies had similar withdrawal rate in the placebo group.

APOMORPHINE

Apomorphine is a dihydro-aporphine. APO is a mixed D_1 and D_2 agonist. It has D_1 and D_2 receptor affinity and has a high affinity for D_3 receptors.[64] Apomorphine is not a pure DA, and affinity for serotonergic and adrenergic receptors (adrenergic receptor affinity greater than D_1) are also described.[84]

APO is completely absorbed from the gastrointestinal tract but, due to pronounced first pass metabolism by the liver, bioavailability is poor. APO is 95% bound to proteins, and plasma half-life is 30 min. Hepatic glucuronidation, methylation, and demethylation may all inactivate the lipophilic compound. Severe adverse dopaminergic effects of hypersalivation, hypotension, nausea, vomiting, and sedation, as well as the technical complications of delivery, limited its usefulness in the clinic. Subcutaneous injection circumvents the rapid first-pass metabolism through the GI tract, and though there is a large variation among individuals, peak plasma concentration is commonly seen within 8 min.[84,85]

Pharmaceuticals that counteract the adverse effects and novel user-friendly delivery systems have been discovered. An injection "pen," similar to that used for insulin, is marketed in some countries. Rapid reversal of severe "off state" makes it an effective rescue medication in patients with advanced PD who struggle with motor fluctuations. It has also been utilized in clinic as a diagnostic tool in determining the response to dopaminergic agents and the diagnosis of Parkinson's disease.[86] However, in a recent study, 200 mg of LD was found to be more efficacious with improved sensitivity than APO.[87]

Placebo-controlled randomized studies of the injectable formulation support the usage of this DA.[88] Mean inpatient UPDRS motor scores were reduced by 23.9 and 0.1 points (62% and 1%) by apomorphine treatment and pH matched vehicle placebo, respectively ($P < 0.001$). The mean percentage of outpatient injections resulting in successful abortion of off-state events was 95% for apomorphine and 23% for placebo ($P < 0.001$). Inpatient response was significantly correlated with and predictive of outpatient efficacy ($P < 0.001$). The LD dose was not predictive of the apomorphine dose requirement. Frequent adverse events included dyskinesia, yawning, and injection site reactions.

Chronic infusion with an ambulatory mini-pump is available as well. Sublingual, intranasal, transdermal, and rectal preparations designed to avoid first-pass metabolism through the liver and the inconvenience of injections are being tested in clinical trials.[89–91] APO is currently available in the U.S.

LISURIDE

Lisuride is an α-amino-ergoline with D_2 receptor agonist properties with no apparent D_1 receptor effects. It is licensed in Europe as monotherapy and as an LD adjunct, and it is not available in the U.S. Similar to most ergotamine derivatives, lisuride also has 5-HT2 activity as well as affinity to 5HT1A.[92] Oral lisuride is absorbed completely from the gastrointestinal tract. Peak plasma levels are obtained within 60 to 80 min, though individual variation is high. Plasma elimination half-life for of lisuride is around 2 hr, which is shorter than most other DA. The long-duration effect, however, is longer than the longer half-life agents CAB and ROP.[93] Oral bioavailability of lisuride is low due to first-pass metabolism ranging from 10 to 20%; 60 to 70% of lisuride is bound to plasma proteins. Lisuride is extensively metabolized, with more than 15 metabolites identified.[94]

There are limited data on the clinical efficacy of oral lisuride. Its use as monotherapy or as an adjunct to LD in advanced PD patients is considered investigational. Lisuride is given t.i.d. at a dose ranging from 1.5 to 4.5 mg/d. Lisuride is associated with the usual side effects of ergot dopamine agonists, including serosal membranous fibrosis.[92]

Lisuride has solubility properties similar to APO and therefore can be administered subcutaneously and intravenously. Subcutaneous infusion of lisuride has been compared to conventional therapy with oral LD and dopamine agonists in a randomized, prospective, long-term (four-year) trial.[95] Patients receiving lisuride infusions experienced a significant reduction in both motor fluctuations and dyskinesia, which persisted for the four-year duration of the study. There was no significant change in the UPDRS scores in the "on" and "off" states between baseline and four years for patients in the lisuride group, but there was deterioration in patients in the LD group. The dopaminergic side effects were roughly comparable in the two groups, although there was no peripheral edema, which was seem in 4 of 20 patients in the conventional therapy group. Eleven of 20 patients receiving infusion therapy developed skin nodules. It is this mode of administration that makes lisuride a potentially useful agent.

PIRIBEDIL

Piribedil is a nonergot derivative, a D_2/D_3 agonist, and it has α_2-adrenergic antagonistic effects.[96] When administered orally, peak plasma level can be reached within 1 hr. There is a relatively long plasma elimination half-life (20 hr). Piribedil solubility allows intravenous and transdermal administration, though it has been used in this formulation only for experimental purposes to date.[97]

Placebo- or LD-controlled trials with piribedil are lacking. One recent double-blind placebo-controlled study on nonfluctuating patients insufficiently controlled with LD showed a modest therapeutic benefit with an oral dose of 150 mg/d relative to placebo.[98] The adverse event profile of Piribedil mirrors that of other nonergot DA. Sleep disorders, nausea, hallucinations, and other cognitive adverse effects predominate.

Experimental Dopamine Agonists

Sumanirole is a new, nonergot DA, currently investigated in Phase III clinical trials.[99] This DA is more selective at D_2 receptors than at the serotonergic 5HT-1A site, similar to BCP. It is essentially devoid of affinity for the other dopamine receptors. Sumanirole is not metabolized by the liver, and over half the administered drug is eliminated unchanged by renal excretion. It has been developed in an extended-release tablet formulation that increased its half-life from 1.7 to 5 hr. Doses of 8 to 24 mg/d on a BID regimen have been shown significantly superior to placebo in early clinical trials. At a dose of 48 mg/d, adverse cognitive effects were present. As with other DA, nausea, hypotension, and exacerbation of dyskinesias were demonstrated on combined LD and Sumanirole therapy. However, at the recommended dose, there has been no evidence of increased somnolence or psychosis to date. Thus, sumanirole has the potential of being particularly useful in patients who are at risk for developing dopaminergic psychiatric side effects. The putative neuroprotective effects are currently being explored in various *in vitro* studies. Since these effects are not blocked by D_2 receptor antagonists, this suggests that the drug may have some other, yet undefined, effects.

Rotigotine CDS, another new DA, represents an effective transdermal formulation. Previously known as N-0923, (-)-2-[N-propyl-N-(2-thienyl) ethyl-amino-5-hydroxy-tetralin]hydrochloride, it is a novel aminotetralin dopamine D_2 agonist whose i.v. formulation was shown to effectively reverse parkinsonian symptoms in nine patients with PD.[100] The short elimination half-life of N-0923 (90 min) made it unfavorable for oral maintenance therapy. However, because the drug is highly lipophilic, it was found to be particularly suited for transdermal delivery. An early clinical trial in patients on LD maintenance therapy demonstrated modest reductions in LD requirements (about 27%).[101] The patch was subsequently reformulated for higher delivery and renamed Rotigotine CDS.[102] This inpatient study consisted of a two-week dose escalation phase followed by a two-week dose maintenance phase at the highest dose (80 cm²). Data were evaluable in seven of ten subjects; two administrative dropouts and one individual were eliminated from the study because of recurrence of hallucinations. The median LD requirement (the primary efficacy variable) was reduced from 1,400 mg/d to 400 mg/d (i.e., a median reduction of 53%; range, 33 to 98%, $p = 0.018$). UPDRS motor scores were unchanged. Although diary variables improved in most individuals, only the reduction in "off" time attained statistical significance. Adverse effects were mild and consisted mainly of dopaminergic side effects and local skin reactions. This agent is currently in Phase 3 trials and is being studied in patients not on LD therapy. This is an excellent preparation to test the possible benefits of continuous dopaminergic stimulation.

USE IN EARLY AND LATE PD

Placebo-controlled clinical trials of BRC, PRG, ROP, and PRAM have demonstrated the efficacy of DA within the first six months of therapy in *de novo* PD patients.[13,53,69,80] In those investigative studies, which continued for one year or longer (PRG, BRC, ROP, CAB, and PRAM), adjunctive LD was required for adequate symptom control in the majority of patients.[67,71,82,83] Adequately dosed LD alone was found to be modestly more efficacious than monotherapy or combination (with supplemental LD) BRC, PRG, PRAM, ROP, though in the ROP study, it was felt that the difference was slight and the study insufficiently powered, which led the investigators to conclude that the change was negligible. In those studies with three- (PRAM, PRG) and five-year (ROP) follow-up, approximately one-third of the PD patients remaining in the DA arm were maintaining well on monotherapy. PD patients on combination and monotherapy CAB, PRG, PRAM, and ROP had a low incidence of motor fluctuations and dyskinesia (lower on monotherapy). These findings have led to a shift toward using agonists as monotherapy in early PD to delay the need for LD and thus prevent the development of LD-related motor complications. The most thorough and complete prospective studies have been performed with ROP and PRAM.[71,72,82]

Generally, advanced age and dementia are regarded as relative contraindications for DA, as the cognitive and hypotensive adverse effects are poorly tolerated in this population. As with all patients, the therapeutic strategy must be individualized and customized to the needs of the patient. Immediate LD monotherapy or combination therapy may be required in a young PD patient if a job is in jeopardy because of troublesome or disabling PD symptoms. Similarly, in young patients who cannot tolerate the adverse effects of the DA, LD is required. On the other hand, some elderly patients, particularly if cognitively intact, may be good candidates for DA therapy. Therapeutic dosages of DA are generally well tolerated by 46% of very elderly patients who received a trial of an agonist, indicating that dopamine receptor agonist therapeutic trials are warranted in selected elderly individuals.[103]

DO DOPAMINE AGONISTS HAVE DISEASE MODIFYING EFFECT?

One of the most controversial aspects of DA is whether these drugs favorably modify the natural history of the disease.[104,105] The controversy centers on the interpretation of imaging techniques designed to evaluate the integrity of dopamine terminals in patients treated with one of the

DA or LD. In the PRAM study, [^{123}I]β-CIT was used as a ligand for the dopamine transporter. In the ROP study, [^{18}F]-fluorodopa was used as a marker for nerve terminals that take up LD.

The ROP trial was a prospective, two-year, randomized, double-blind, multinational study that attempted to compare the rates of loss of dopamine-terminal function in *de novo* patients with clinical and (^{18})F-dopa PET evidence of early PD, randomized 1 to 1 to receive either ROP or LD. The primary outcome measure was reduction in putamen (^{18})F-dopa uptake (Ki) between baseline and two-year PET. Of 186, 162 randomized patients were eligible for analysis.[105] A blinded, central, region-of-interest analysis showed a significantly lower reduction in putamen Ki over two years with ROP (–13.4%; n = 68) compared with LD (–20.3%; n = 59; 95% CI, 0.65 to 13.06). Statistical parametric mapping localized lesser reductions in (^{18})F-dopa uptake in the putamen and substantia nigra with ROP. The greatest Ki decrease in each group was in the putamen (ROP, –14.1%; LD, –22.9%; 95% CI, 4.24 to 13.3), but the decrease was significantly lower with ROP compared with LD ($p < 0.001$). The conclusion was that ROP is associated with slower progression of PD than LD as assessed by [^{18}F]F-dopa PET.[109] An interim analysis involving only 45 patients followed for 2 years, the investigators found no significant overall difference in PD progression between ROP and LD groups. Although the magnitude of the apparent reduced decline in the imaging marker was about the same as the later larger group, this was not statistically significant, apparently because of the lower sample size and reduced the power.

In another other study, 82 patients with early PD were randomly assigned to receive PRAM, 0.5 mg 3 times per day with LD placebo (n = 42), or carbidopa/LD, 25/100 mg 3 times per day with PRAM placebo (n = 40).[104] For patients with residual disability, the dosage was escalated during the first ten weeks, and subsequently, open-label LD could be added. After 24 months of follow-up, the dosage of study drug could be further modified. The primary outcome variable was the percentage change from baseline in striatal [^{123}I]β-CIT uptake after 46 months. The percent changes and absolute changes in striatal, putamen, and caudate [^{123}I]β-CIT uptake after 22 and 34 months were also assessed. Sequential SPECT imaging showed a decline in mean (SD) [^{123}I]β-CIT striatal uptake from baseline of 10.3% ± 9.8% SD at 22 months, 15.3% ± 12.8% SD at 34 months, and 20.7% ± 14.4% SD at 46 months—approximately 5.2% per year. The mean percentage loss in striatal [^{123}I]β-CIT uptake from baseline was significantly reduced in the PRAM group compared with the LD group: 7.1% (9.0%) versus 13.5% (9.6%) at 22 months ($P = 0.004$); 10.9% (11.8%) versus 19.6% (12.4%) at 34 months ($P = 0.009$); and 16.0% (13.3%) versus 25.5% (14.1%) at 46 months ($P = 0.01$). The percentage loss

from baseline in striatal [^{123}I]β-CIT uptake was correlated with the change from baseline in UPDRS at the 46-month evaluation ($r = –0.40$; $P = 0.001$). Thus, patients initially treated with PRAM demonstrated a reduction in loss of striatal [^{123}I]β-CIT uptake, a marker of dopamine neuron degeneration, compared with those initially treated with LD during a 46-month period.

A study on 50 *de novo* PD patients investigated the mean annual change in the ratio of specific to nonspecific [(123)I]beta-CIT binding to the striatum, putamen, and caudate nucleus was used as the outcome measure. A decrease in [(123)I]beta-CIT binding ratios between the two images was found in all regions of interest. The average decrease in [(123)I]beta-CIT binding ratios was about 8% in the whole striatum, 8% in the putaminal region, and 4% in the caudate region. Power analysis indicated that to detect a significant ($p < 0.05$) effect of a neuroprotective agent with 0.80 power and 30% of predicted protection within 2 years, 216 patients are required in each group when the effects are measured in the whole putamen. Comparison of scans done in nine patients under two different conditions—in the off state and while on drug treatment—showed no significant alterations in the expression of striatal dopamine transporters as measured using [(123)I]beta-CIT SPECT.[106]

Limitations in design and interpretation of the results have been delineated.[107] As in the previous clinical studies,[72,82] the there were significant dropouts in the randomized patients, 20% in the CALM-PD study and 27% in the REAL-PET. Post hoc analysis in the REAL PET study revealed that 11% of the patients had F-dopa uptake within the normal range and were eliminated from the analysis. Furthermore, although only 14% of the REAL-PET ROP arm required L-dopa supplementation, by the end of the study, 75% of the PRAM arm required it. No placebo group was obtained in either study, although, as reported in the above study, subsequent reports indicated little short-term change in the DA and *de novo* group. In addition, the PRAM patients who required supplementation with L-dopa continued to have a reduced rate of signal loss relative to the monotherapy L-dopa group.

The difficulty in interpreting the findings is further confounded by *in vitro* and animal data suggesting that regulation of the cellular content of dopamine transporter, the protein to which the radioligand (b-CIT) binds, is highly susceptible to pharmacologic maneuvering and is not present in a fixed ratio on the neuron.[108] Regarding the REAL-PET study, a change in the concentration of dopa decarboxylase,[109] competition with the blood brain barrier transport,[110] and stimulation of dopamine autoreceptors[111] are all factors that could be uniquely affected by the agonist versus L-dopa long term and may not reflect accurately neuronal loss. Finally, despite the imaging findings, the therapeutic response was not improved in the patients on combination or monotherapy agonist. The patients on

L-dopa monotherapy consistently manifest an improved symptomatic response.

Despite the above caveats, the findings of two studies with unique imaging techniques and two different agonist argues in favor of a possible neuroprotective effect. Animal models of Parkinson's disease and *in vitro* studies suggested the neuroprotective effects of the agonist against toxic compounds long before they were marketed for human consumption. Further studies, appropriately powered, are needed to verify the significance of the improved imaging studies, and possibly more rigorous stratification of patient population should be obtained.

COMPARISONS BETWEEN DA AND SWITCHING

The Cochrane Reviews have been a continuing source of evaluations regarding the comparative effects of the DA, as few quality comparison trials of the best tolerated DA show significant differences in efficacy.[112–116]

COMPARISON OF ADVERSE EFFECTS OF DOPAMINE AGONISTS

In a meta-analysis comparing the adverse effects of ROP to PRAM, no significant difference was found in the risk of dizziness, nausea, or hypotension with either drug, individually or in combination as compared with LD. The risk of hypotension was approximately four times higher with ROP than PRAM when each drug was individually compared with placebo (6.46 [95% CI 1.47 to 28.28] for ROP, and 1.65 [0.88 to 3.08] for PRAM). The pooled RR (for PRAM and ROP combined) of hallucinations was 1.92 (95% CI 1.08 to 3.43) as compared with LD. Relative to placebo, PRAM had a significantly higher risk of hallucinations than ROP (PRAM 5.2 [95% CI 1.97 to 13.72] versus ROP 2.75 [95% CI 0.55 to 13.73]). There was no significant difference in the risk of somnolence between the two drugs when each was individually compared with LD. As compared with placebo, the pooled RR (PRAM and ROP combined) of somnolence was 3.16 (95% CI 1.62 to 6.13). Relative to placebo, the risk of somnolence was 2.01 (95% CI 2.17 to 3.16) with PRAM and 5.73 (95% CI 2.34 to 14.01) with ROP.[117]

The gastrointestinal symptoms can be often be ameliorated by using a peripheral dopamine receptor blocking agent that does not cross the BBB, such as domperidone.[118] Reports of favorable effects on hypotension have also been described.[119] Use of mineralocorticoids (fludrocortisone), sympathomimetics (ephedrine, and phenylpropanolamine), direct vasoconstrictors (midodrine hydrochloride), prostaglandin synthetase inhibitors (indomethacin) and prohemopoietic agents (erythropoietin) may be necessary if patients have no contraindications and do not respond to use of jobst stockings, increased salt intake, or adjustment of other medications.[120]

Confusion, hallucinations, or psychosis more commonly occurs in patients receiving DA therapy than in those receiving LD. Treatment of this is problematic, as dopamine receptor blockade may result in worsening of parkinsonism. There has been some success treating psychiatric complications of DA, such as hallucinations and psychosis, using atypical antipsychotics such as a olanzapine, quetiapine, and clozapine.[121,122]

Another side effect common to the recent, as well as the older, DA is peripheral edema.[123] The mechanism of edema associated with DA, particularly PRAM and ROP, has not been characterized. Treatment and diagnosis requires discontinuation of the DA as edema should subside after discontinuation. Symptoms reoccur with reinitiation typically.

A potentially serious complication of DA therapy is fibrosis of serosal membranes. This syndrome includes retroperitoneal, pleuropulmonary, pericardial, and valvular fibrosis. This adverse effect seems to occur exclusively with the ergot agents, such as BCP and PRG. Although fibrosis tends to occur in the first few years of use, we have reported a patient who developed it 11 years into treatment with PRG. His pleural fibrosis resolved when he was switched to PRAM, a nonergoline DA. In general, the fibrotic complications have been reversible, if recognized early enough, and patients have been successfully subsequently treated with the nonergot agents (i.e., ROP or PRAM).[124–126,139]

Somnulence as well as insomnia are commonly reported after initiating agonist therapy. Sleep disorders, particularly somnolence, attracted significant attention and debate following the publication of case reports of "sleep attacks" in PD patients.[127,128] Hypotension as sleepiness commonly occurs in PD, regardless of treatment.[120,129] In one study, conducted in Germany, LD monotherapy carried the lowest risk for sleep attacks with a relative risk of 2.9%, 1.7 to 4.0% (95% CI), followed by DA monotherapy (5.3%; 1.5 to 9.2%) and combination of L-dopa and a DA (7.3%; 6.1 to 8.5%).[130] Improvement can often be achieved by changing the dosing schedule or the amount of DA per dose, discontinuing the DA, or treating the sleepiness with CNS stimulants such as modafinil.[131] In a sleep study using standard polysomnographic techniques, the main risk factor associated with pathologic daytime sleep latency was high levodopa dosage equivalents (>867.5 mg; odds ratio, 4.2; 95% confidence interval, 1.3–13.7). Subjective accounts of daytime sleep and wakefulness, as indexed by scores on the Epworth Sleepiness Scale, were not related to impaired daytime sleepiness or wakefulness (chi(2)(1) [$n = 80$], 0.13; $P = 0.72$).[141]

Fatigue is a similar complaint frequently encountered among patients with PD. Interestingly, 41 patients with

PD and controls, after 5 weeks patients taking PRG, showed significant improvement in the fatigue scale (from 5.1 ± 0.7 SD to 4.4 ± 0.55 SD), but patients taking BCP did not (from 4.8 ± 0.9 SD to 4.7 ± 0.8 SD).[132]

TITRATING AND SWITCHING DOPAMINE AGONISTS

Slow titration of DA is recommended to prevent initial side effects such as nausea, hypotension, and drowsiness. As found in the BRC study, the rate of titration must be weighed against the patient's frustration during the subtherapeutic undermedicated state of the titration. Premedication with domperidone up to two days before and continued during the PRG titration period has been shown to effectively prevent gastrointestinal side effects. After rather quick titration of PRG with coadministration of domperidone, no symptomatic side effects were seen except for light-headedness in one patient, which disappeared after dose reduction.[119] In patients already on DA therapy, switch to an alternative DA can be made quickly, essentially overnight.[133,134]

Few systematic clinical trials have been performed with combination DA therapy. Complexity of study design and lack of pharmaceutical funding probably contribute. As DA and LD monotherapy appear to have less adverse effects of somnolence and simplify the medication regimen, it is probably reasonable to push DA to the point of asymptote of clinical benefit or side effects.

SUMMARY AND CONCLUSION

LD remains an imperfect solution to the progressive degeneration of the brain stem nuclei and downstream basal ganglia motor systems in Parkinson's disease. It does not delay the progression of the degeneration, and there is evidence that it may exacerbate the degenerative process. In contrast, there is evidence that DAs have a neuroprotective effect and may delay the progression of the disease. Although clinical data do not support the superiority of DA over LD, motor fluctuations and dyskinesia appear to a lesser degree in patients treated with DA. These findings suggest that DA monotherapy should be initiated in early PD, particularly in those with onset before age 50. Addition of LD should be initiated if and when required for symptom relief.

Advanced age and cognitive impairment are regarded as relative contraindications for DA, as the adverse effects of psychosis are poorly tolerated. Age alone should not prevent a trial of a dopamine agonist in a patient with intact cognitive function.

The nonergot-derived DA have a similar pharmacologic profile without the risk of serosal fibrosis. They are well tolerated and have similar adverse effect profiles.

Meta-analysis suggests that PRAM has a higher risk of hallucinations and ROP of hypotension.

The development of novel nonoral (i.e., transdermal, subcutaneous injection, intravenous, and intranasal) formulations of DA allow rapid distribution and a stable serum concentrations. Avoidance of the nonphysiologic pulsatile fluctuations associated with LD and DA with a shorter half-life should prevent the associating plastic changes in the striatum that result in wearing off and dyskinesia.

ABBREVIATIONS

APO	apomorphine
BCP	bromocriptine
CAB	cabergoline
CI	confidence interval
DA	dopamine agonists
LD	levodopa
LID	levodopa-induced dyskinesias
PD	Parkinson's disease
PRAM	pramipexole
PRG	pergolide
ROP	ropinirole
SD	standard deviation
UPDRS	Unified Parkinson Disease Rating Scale

REFERENCES

1. Ehringer, H., Hornykiewicz, O., Distribution of noradrenaline and dopamine (3-hydroxytyramine) in the human brain and their behavior in diseases of the extrapyramidal system, *Klin. Wochenschr.*, 38:1236–1239, 1960.

2. Goetz, C. G., Diederich, N. J., Dopaminergic agonist in the treatment of Parkinson's disease, *Neurologic Clinics*, 10: 527–540, 1992.

3. Ichinose, H., Ohye, T., Fujita, K., Pantucedk, F., Lanage, K., Riederer, P., Nagatsu, T., Quantification of mRNA of tyrosine hydroxylase and aromatic L-amino acid decarboxylase in the substantia nigra in Parkinson's disease and schizophrenia, *J. Neural. Transm. Park Dis. Dement. Sect.*, 8:149–58, 1994.

4. Blandini, F., Mangiagalli, A., Martignoni, E., Samuele, A., Fancellu, R., Tassorelli, C., Cosentino, M., Marino, F., Rasini, E., Calandrella, D., Riboldazzi, G., Colombo, C., Frigo, G. M., Nappi, G., Effects of dopaminergic stimulation on peripheral markers of apoptosis: relevance to Parkinson's disease, *Neurol. Sci.*, 24: 157–8, 2003.

5. Kristal, B. S., Conway, A. D., Brown, A. M., Jain, J. C., Ulluci, P. A., Li, S. W., Burke, W. J., Selective dopaminergic vulnerability: 3,4-dihydroxyphenylacetaldehyde targets mitochondria, *Free Radic. Biol. Med.*, 30:924–31, 2001.

6. Uitti, R. J., Ahlskog, J. E., Maraganore, D. M., Muenter, M. D., Atkinson, E. J., Cha, R. H., O'Brien, P. C.,

Levodopa therapy and survival in idiopathic Parkinson's disease. Olmsted County project, *Neurology,* 43:1918–1926, 1993.

7. Nutt, J. G., Carter, J. H., Lea, E. S., Sexton, G. J., Evolution of the response to levodopa during the first four years of therapy, *Ann. Neurol.,* 51:686–693. 2002.

8. Sage, J. I., Mark, M. H., Basic mechanisms of motor fluctuations, *Neurology,* 44:S10–4, 1994.

9. Barbato, L., Stocchi, F., Monge, A. et al., The long-duration action of levodopa may be due to a postsynaptic effect, *Clin. Neuropharmacol.,* 20:394–401, 1997.

10. Schrag, A., Quinn, N., Dyskinesias and motor fluctuations in Parkinson's disease. A community-based study, *Brain,* 123:2297–2305, 2000.

11. Stocchi, F., Vacca, L., Berardelli, A., De Pandis F., Ruggieri, S., Long-duration effect and the postsynaptic compartment: study using a dopamine agonist with a short half-life, *Mov. Disord.,* 16:301–305, 2001.

12. Cotzias, G. C., Papavasiliou, P. S., Fehling, C., Kaufman, B., Mena, I., Similarities between neurologic effects of L-dopa and of apomorphine, *N. Engl. J. Med.,* 282:31–3, 1970.

13. DA agonists—ergot derivatives: bromocriptine: management of Parkinson's disease, *Mov. Disord.,* 17:S53–67, 2002.

14. Gassen, M., Gross, A., Youdim, M. B., Apomorphine, a dopamine receptor agonist with remarkable antioxidant and cytoprotective properties, *Adv. Neurol.,* 80:297–302, 1999.

15. DA agonists—ergot derivatives: cabergoline: management of Parkinson's disease, *Mov. Disord.,* 17:S68–71, 2002.

16. DA agonists—ergot derivatives: pergolide: management of Parkinson's disease, *Mov. Disord.,* 17:S79–82, 2002.

17. DA agonists—non-ergot derivatives: pramipexole: management of Parkinson's disease, *Mov. Disord.,* 17:S93–97, 2002.

18. Biglan, K. M., Holloway, R. G., A review of pramipexole and its clinical utility in Parkinson's disease, *Expert Opin. Pharmacother.,* 3:197–210, 2002.

19. DA agonists—non-ergot derivatives: Ropinirole: management of Parkinson's disease, *Mov. Disord.,* 17:S98–102, 2002.

20. Missale, C., Nash, S. R., Robinson, S. W., Jaber, M., Caron, M. G., Dopamine receptors: from structure to function, *Physiol. Rev.,* 78:189–225, 1998.

21. Smith, L. A., Jackson, M. J., Al-Barghouthy, G., Jenner, P., The actions of a D-1 agonist in MPTP treated primates show dependence on both D-1 and D-2 receptor function and tolerance on repeated administration, *J. Neural. Transm.,* 109:123–140, 2002.

22. Blanchet, P., Bedard, P., Britton, D., Kebabian, J., Differential effect of selective D-1 and D-2 dopamine receptor agonists on levodopa-induced dyskinesia in 1-methyl-4-phenyl-1,2,3,6-tetrahydropyri, *J. Pharmacol. Exp. Ther.,* 267:275–279, 1993.

23. Sealfon, S. C., Dopamine receptors and locomotor responses: molecular aspects, *Ann. Neurol.,* 47:S12–9, 2000.

24. Seeman, P., Nam, D., Ulpian, C., Liu, I. S., Tallerico, T., New dopamine receptor, D2(Longer), with unique TG splice site, in human brain, *Brain Res. Mol. Brain Res.,* 76:132–41, 2000.

25. Yasumoto, S., Tanaka, E., Hattori, G., Maeda, H., Higashi, H., Direct and indirect actions of dopamine on the membrane potential in medium spiny neurons of the mouse neostriatum, *J. Neurophysiol.,* 87:1234–1243, 2002.

26. Wichmann, T., Delong, M. R., Pathophysiology of Parkinson's disease: the MPTP primate model of the human disorder, *Ann. N. Y. Acad. Sci.,* 991:199–213, 2003.

27. Svenningsson, P., Le Moine, C., Dopamine D1/5 receptor stimulation induces c-fos expression in the subthalamic nucleus: possible involvement of local D5 receptors, *Eur. J. Neurosci.,* 15:133–142, 2002.

28. Tofighy, A., Abbott, A., Centonze, D. et al., Excitation by dopamine of rat subthalamic nucleus neurones *in vitro*-a direct action with unconventional pharmacology, *Neuroscience,* 116:157–166, 2003.

29. Gerfen, C. R., Molecular effects of dopamine on striatal-projection pathways, *Trends Neurosci.,* 23(10 Suppl):S64–70, 2000.

30. Calabresi, P., Centonze, D., Bernardi, G., Electrophysiology of dopamine in normal and denervated striatal neurons, *Trends Neurosci.,* 23:S57–63, 2000.

31. Moore, H., Grace, A. A., A role for electrotonic coupling in the striatum in the expression of dopamine receptor-mediated stereotypes, *Neuropsychopharmacology,* 27:980–92, 2002.

32. Rufer, M., Wirth, S. B., Hofer, A., Dermietzel, R., Pastor, A., Kettenmann, H., Unsicker, K., Regulation of connexin-43, GFAP, and FGF-2 is not accompanied by changes in astroglial coupling in MPTP-lesioned, FGF-2-treated parkinsonian mice, *J. Neurosci. Res.,* 46:606–17, 1996.

33. Rascol, O., Nutt, J. G., Blin, O. et al., Induction by Dopamine D1 Receptor Agonist ABT-431 of Dyskinesia Similar to Levodopa in Patients With Parkinson Disease, *Arch Neurol.,* 58:249–254, 2001.

34. Grondin, R., Bedard, P., Britton, D., Shiosaki, K., Potential therapeutic use of the selective dopamine D1 receptor agonist, A-86929: An acute study in parkinsonian levodopa-primed monkeys, *Neurology,* 49:421–426, 1997.

35. Rascol, O., Blin, O., Thalamas, C., Descombes, S., Soubrouillard, C., Azulay, P., Fabre, N., Viallet, F., Lafnitzegger, K., Wright, S., Carter, J. H., Nutt, J. G., ABT-431, a D1 receptor agonist prodrug, has efficacy in Parkinson's disease, *Ann. Neurol.,* 45:736–741, 1999.

36. Blanchet, P., Fange, J., Gillespie, M., Sabounjian, L., Locke, K. W., Gammans, R., Mouradian, M. M., Chase, T. N., Effects of the full dopamine D1 receptor agonist dihydrexidine in Parkinson's disease, *Clin. Neuropharmacol.,* 21:339–343, 1998.

37. Blanchet, P., Grondin, R., Bedard, P., Shiosaki, K., Britton, D., Dopamine D1 receptor desensitization profile in MPTP-lesioned primates, *Eur. J. Pharmacol.,* 309:13–20, 1996.

38. Tomiyama, K., McNamara, F. N., Clifford, J. J., et al. Phenotypic resolution of spontaneous and D1-like agonist-induced orofacial movement topographies in congenic dopamine D1A receptor "knockout" mice, *Neuropharmacology*, 42:644–652, 2002.

39. Sibley, D. R., New insights into dopaminergic receptor function using antisense and genetically altered animals, *Annu. Rev. Pharmacol. Toxicol.*, 39:313–341, 1999.

40. Bezard, E., Ferry, S., Mach, U. et al., Attenuation of levodopa-induced dyskinesia by normalizing dopamine D3 receptor function, *Nat. Med.*, 9:762–767, 2003.

41. Corrigan, M. H., Denahan, A. Q., Wright, C. E., Ragual, R. J., Evans, D. L., Comparison of pramipexole, fluoxetine, and placebo in patients with major depression, *Depress Anxiety*, 11:58–65, 2000.

42. Lattanzi, L., Dell'Osso, L., Cassano, P. et al., Pramipexole in treatment-resistant depression: a 16-week naturalistic study, *Bipolar Disord.*, 4:307–314, 2002.

43. Newman-Tancredi, A., Cussac, D., Audinot, V. et al., Differential actions of antiparkinson agents at multiple classes of monoaminergic receptor. II. Agonist and antagonist properties at subtypes of dopamine D(2)-like receptor and alpha(1)/alpha(2)-adrenoceptor, *J. Pharmacol. Exp. Ther.*, 303:805–814, 2002.

44. Sampaio, C., Coelho, T., Castro-Caldas, A., Bastos-Lima, A., Levy, A., The response of "de novo" Parkinson's disease patients to bromocriptine in a "low and slow" regimen is predictive for prognosis, *J. Neural. Transm.*, Suppl., 45:197–202, 1995.

45. Blum, I., Leiba, S., Increased hair loss as a side effect of bromocriptine treatment, *N. Engl. J. Med.*, 303:1418, 1980.

46. Klaassen, R. J., Troost, R. J., Verhoeven, G. T., Krepel, H. P., Van Der Lely, A. J., Suggestive evidence for bromocriptine-induced pleurisy, *Neth. J. Med.*, 48:232–236, 1996.

47. Serratrice, J., Disdier, P., Habib, G., Viallet, F., Weiller, P. J., Fibrotic valvular heart disease subsequent to bromocriptine treatment, *Cardiol. Rev.*, 10:334–336, 2002.

48. Hely, M. A., Morris, J. G., Lawrence, S., Jeremy, R., Genge, S., Retroperitoneal fibrosis, skin and pleuropulmonary changes associated with bromocriptine therapy, *Aust. N. Z. J. Med.*, 21:82–84, 1991.

49. Barone, P., Bravi, D., Bermejo-Pareja, F. et al., Pergolide monotherapy in the treatment of early PD: a randomized, controlled study. Pergolide Monotherapy Study Group, *Neurology*, 53:573–579, 1999.

50. Varsano, S., Gershman, M., Hamaoui, E., Pergolide-induced dyspnea, bilateral pleural effusion and peripheral edema, *Respiration*, 67:580–582, 2000.

51. Degim, I. T., Acarturk, F., Erdogan, D., Demirez Lortlar, N., Transdermal administration of bromocriptine, *Biol. Pharm. Bull.*, 26:501–5, 2003.

52. Kulisevsky, J., Lopez-Villegas, D., Garcia-Sanchez, C. et al., A six month study of pergolide and levodopa in de novo Parkinson's disease patients, *Clin. Neuropharmacol.*, 21:358–62, 1998.

53. Barone, P., Bravi, D., Bermejo-Pareja, F. et al., Pergolide monotherapy in the treatment of early PD. A randomized, controlled study, *Neurology*, 53:573–9, 1999.

54. Lledo, A., Hundermer, H. P., Van Laar, T. et al., Long term efficacy of pergolide monotherapy in early stage Parkinson's disease. One year interim analysis of a 3 year double blind randomized study versus levodopa, *Movement Disord.*, 15:126, 2000.

55. Boas, J., Worm-Peterson, J., Dupont, E. et al., The levodopa dose sparing capacity of pergolide compared with that of bromocriptine in an open label crossover study, *Eur. J. Neurol.*, 3:44–9, 1996.

56. Facca, A., Sanchez-Ramos, J., High dose pergolide monotherapy in the treatment of severe levodopa induced dyskinesia, *Mov. Disord.*, 11:327–9, 1996.

57. Schwarz, J., Scheidtmann, K., Trenkwalder, C., Improvement of motor fluctuations in patients with Parkinson's disease following treatment with high doses of pergolide and cessation of levodopa, *Eur. Neurol.*, 37:236–8, 1997.

58. Jankovic, J., Long term study of pergolide in Parkinson's disease, *Neurology*, 35:296–9, 1985.

59. Tabamo, R. E., Di Rocco, A., Alopecia induced by dopamine agonists, *Neurology*, 12, 58:829–30, 2002.

60. Ulivelli, M., Rossi, S., Lombardi, C., Bartalini, S., Rocchi, R., Giannini, F., Passero, S., Battistini, N., Lugaresi, E., Polysomnographic characterization of pergolide-induced sleep attacks in idiopathic PD, *Neurology*, 58:462–5, 2002.

61. Pritchett, A. M., Morrison, J. F., Edwards, W. D. et al., Valvular heart disease in patients taking pergolide, *Mayo Clin. Proc.*, 77:1280–1286, 2002.

62. Tanner, C. M., Ghablani, R., Goetz, C. G., Klawans, H. L., Pergolide mesylate: lack of cardiac toxicity in patients with cardiac disease, *Neurology*, 35:918–921, 1985.

63. Rahimtoola, S. H., Drug-related valvular heart disease: here we go again: will we do better this time? *Mayo Clin. Proc.*, 77:1275–1277, 2002.

64. DA agonist-overview, *Mov. Disord.*, 17:S52, 2002.

65. Christensen, J., Dupont, E., Ostergaard, K., Cabergoline plasma concentration is increased during concomitant treatment with itraconazole, *Mov. Disord.*, 17:1360–2, 2002.

66. Del Dotto, P., Bonuccelli, U., Clinical pharmacokinetics of cabergoline, *Clin. Pharmacokinet.*, 42:633–645, 2003.

67. Rinne, U. K., Bracco, F., Chouza, C., Dupont, E., Gershanik, O., Marti Masso, J. F., Montastruc, J. L., Marsden, C. D., Early treatment of Parkinson's disease with cabergoline delays the onset of motor complications. Results of a double-blind levodopa controlled trial. The PKDS009 Study Group, *Drugs*, 55:23–30, 1998.

68. Parkinson Study Group. Safety and efficacy of pramipexole in early Parkinson's disease: a randomized dose ranging study, *JAMA*, 278:125–130, 1997.

69. Shannon, K. M., Bennett, J. P. J., Friedman, J. H., Efficacy of pramipexole, a novel dopamine agonist, as monotherapy in mild to moderate Parkinson's disease. The pramipexole study group, *Neurology*, 49:724–728, 1998.

70. Guttman, M., International pramipexole-bromocriptine study group. Double blind comparison of pramipexole

and bromocriptine with placebo in advanced Parkinson's disease, *Neurology,* 49:1060–1065, 1997.

71. Pinter, M. M., Pogarell, O., Oertel, W. H., Efficacy safety and tolerance of the non-ergoline dopamine agonist pramipexole in the treatment of advanced Parkinson's disease: a double blind, placebo-controlled randomized, multicentre study, *J. Neurol. Neurosurg. Psych.,* 66:436–441, 1999.

72. Parkinson Study Group, Pramipexole vs. levodopa as initial treatment for Parkinson's disease, *JAMA,* 284:1931–1938, 2000.

73. Pinter, M. M., Rutgers, A. W., Hebenstreit, E., An open-label, multicentre clinical trial to determine the levodopa dose-sparing capacity of pramipexole in patients with idiopathic Parkinson's disease, *J. Neural. Transm.,* 107:1307–23, 2000.

74. Hubble, J. P., Koller, W. C., Cutler, N. R. et al., Prami-pexole in patients with early Parkinson's disease, *Clin. Neuropharmacol.,* 18:338–347, 1995.

75. Weiner, W. J., Factor, S. A., Jankovic, J. et al., The long term safety and efficacy of Pramipexole in Parkinson's disease. Parkinsonism and related disorders, 7:115–120, 2001.

76. Tulloch, I. F., Pharmacologic profile of Ropinirole: a nonergoline dopamine agonist, *Neurology,* 49:S58–62, 1997.

77. Aye, C. M., Nicholls, B., Clinical pharmacokinetics of Ropinirole, *Clin. Pharmacokinet.,* 39:243–254, 2000.

78. Ir, J. D., Oppelt, T., Warfarin and Ropinirole interaction, *Ann. Pharmacother.,* 35:1202–1204, 2001.

79. Matheson, A. J., Spencer, C. M., Ropinirole: a review of its use in the management of Parkinson's disease, *Drugs,* 60:115–137, 2000.

80. Adler, C. H., Sethi, K. D., Hauser, R. A., Davis, T. L., Hammerstad, J. P., Bertoni, J., Taylor, R. L., Sanchez-Ramos, J., O'Brien, C. F., Ropinirole for the treatment of early Parkinson's disease. The Ropinirole Study Group, *Neurology,* 49:393–9, 1997.

81. Cristina, S., Zangaglia, R., Mancini, F., Martignoni, E., Nappi, G., Pacchetti, C., High dose Roponirole in advanced Parkinson's disease with severe dyskinesia, *Clin. Neuropharm.,* 26:146–150, 2003.

82. Rascol, O., Brooks, D. J., Korczyn, A. D., De Deyn, P. P., Clarke, C. E., Lang, A. E., for the 056 study group. A five year study of the incidence of dyskinesia in patients with early Parkinson's disease who were treated with Ropinirole or levodopa, *N. Engl. J. Med.,* 342:1448–1491, 2000.

83. Korczyn, A. D., Brunt, E. R., Larsen, J. P., Nagy, Z., Poewe, W. H., Ruggieri, S., A 3 year randomized trial of Ropinirole and bromocriptine in early Parkinson's disease, *Neurology,* 53:364–370, 1999.

84. Neef, C., Van Laar, T., Pharmacokinetic-pharmacody-namic relationships of apomorphine in patients with Par-kinson's disease, *Clin. Pharmacokinet,* 37:257–271, 1999.

85. Gancher, S., Pharmacokinetics of apomorphine in Par-kinson's disease, *J. Neural. Transm.,* Suppl., 45:137–141, 1995.

86. Bonuccelli, U., Piccini, P., Del Dotto, P., Rossi, G., Corsini, G. U., Muratorio A., Apomorphine test for dopaminergic responsiveness: a dose assessment study, *Mov. Disord.,* 8:158–64, 1993.

87. Muller, T., Benz, S., Bornke, C., Russ, H., Przuntek, H., Repeated rating improves value of diagnostic dopamin-ergic challenge tests in Parkinson's disease, *J. Neural. Transm.,* 110:603–9, 2003.

88. Dewey, R. B., Jr., Hutton, J. T., Lewitt, P. A., Factor, S. A., A randomized, double-blind, placebo-controlled trial of subcutaneously injected apomorphine for parkinso-nian off-state events, *Arch Neurol.,* 58:1385–1392, 2001.

89. Dewey, R. B., Jr., Maraganore, D. M., Ahlskog, J. E., Matsumoto, J. Y., A double-blind, placebo-controlled study of intranasal apomorphine spray as a rescue agent for off-states in Parkinson's disease, *Mov. Disord.,* 13:782–7, 1998.

90. Van Laar, T., Neef, C., Danhof, M., Roon, K. I., Roos, R. A., Pharmacokinetics and clinical efficacy of rectal apomorphine in patients with Parkinson's disease: a study of five different suppositories, *Mov. Disord.,* 10:433–439, 1995.

91. Priano, L., Albani, G., Calderoni, S. et al., Controlled-release transdermal apomorphine treatment for motor fluctuations in Parkinson's disease, *Neurol. Sci.,* 23, Suppl., 2, S99–100, 2002.

92. DA Agonists—Ergot derivatives: Lisuride, *Mov. Dis-ord.,* 17:S74–S78, 2002.

93. Stocchi, F., Vacca, L., Onofrj, M., Are there clinically significant differences between dopamine agonists, *Adv. Neurol.,* 91:259–266, 2003.

94. Marona-Lewicka, D., Kurrasch-Orbaugh, D. M., Selken, J. R. et al., Re-evaluation of lisuride pharmacology: 5-hydroxytryptamine(1A) receptor-mediated behavioral effects overlap its other properties in rats, *Psychophar-macology* (Berl.), 164:93–107, 2002.

95. Stocchi, F., Ruggiere, S., Vacca, L., Olanow, C. W., Prospective randomized trial of lisuride infusion versus oral levodopa in patients with Parkinson's disease, *Brain,* 125:2058–2066, 2002.

96. DA agonists—non-ergot derivatives: piribedil: manage-ment of Parkinson's disease, *Mov. Disord.,* 17:S90–2, 2002.

97. Smith L., Jackson, M., Bonhomme, C. et al., Transder-mal administration of piribedil reverses MPTP-induced motor deficits in the common marmoset, *Clin. Neurop-harmacol.,* 23:133–42, 2000.

98. Ziegler, M., Castrto-Caldas, A., Del Signore, S., Rasco, Lo., Efficacy of piribedil as early combination to levo-dopa in patients with stable Parkinson's disease: a 6-month, randomized, placebo-controlled study, *Mov. Dis-ord.,* 18:418–425, 2003.

99. De Paulis, T., Sumanirole Pharmacia, *Curr. Opin. Inves-tig. Drugs,* 4:77–82, 2003.

100. Calabrese, V. P. et al., N-0923, a novel soluble dopamine D2 agonist in the treatment of parkinsonism, *Mov. Dis-ord.,* 13:768–774, 1998.

101. Hutton, J. T., Metman, L. V., Chase, T. N. et al., Trans-dermal dopaminergic D(2) receptor agonist therapy in Parkinson's disease with N-0923 TDS: a double-blind,

placebo-controlled study, *Mov. Disord.,* 16:459–463, 2001.

102. The Parkinson Study Group, A controlled trial of rotigotine monotherapy in early Parkinson's disease, *Arch Neurol.,* 60:1721–8, 2003.

103. Shulman, L. M., Minagar, A., Rabinstein, A., Weiner, W. J., The use of dopamine agonists in very elderly patients with Parkinson's disease, *Mov. Disord.,* 15:664–668, 2000.

104. Parkinson Group Study, Dopamine transporter brain imaging to assess the effects of pramipexole vs. levodopa on Parkinson disease progression, *JAMA,* 287:1653–1661, 2002.

105. Whone, A. L., Watts, R. L., Stoessl, A. J. et al., Slower progression of Parkinson's disease with Ropinirole versus levodopa: The REAL-PET study, *Ann. Neurol.,* 54:93–101, 2003.

106. Winogrodzka, A., Bergmans, P., Booij, J., van Royen, E. A., Stoof, J. C., Wolters, E. C., [(123)I]beta-CIT SPECT is a useful method for monitoring dopaminergic degeneration in early stage Parkinson's disease, *J. Neurol. Neurosurg. Psychiatry,* 74:294–8, 2003.

107. Ahlskog, J. E., Slowing Parkinson's disease progression: recent dopamine agonist trials, *Neurology,* 60:381–389, 2003.

108. Meiergerd, S. M., Patterson, T. A., Schenk, J. O., D2 receptors may modulate the function of the striatal transporter for dopamine: kinetic evidence from studies *in vitro* and *in vivo, J. Neurochem.,* 61:764–767, 1993.

109. Berry, M. D., Juorio, A. V., Li, X. M., Boulton, A. A., Aromatic L-amino acid decarboxylase: a neglected and misunderstood enzyme, *Neurochem. Res.,* 21:1075–1087, 1996.

110. Kido, Y., Tamai, I., Uchino, H., Suzuki, F., Sai, Y., Tsuji, A., Molecular and functional identification of large neutral amino acid transporters LAT1 and LAT2 and their pharmacologic relevance at the blood brain barrier, *J. Pharm. Pharmacol.,* 53:497–503, 2001.

111. Ekesbo, A., Rydin, E., Torstenson, R., Sydow, O., Laengstrom, B., Westerberg, G., Tedroff, J., Dopamine autoreceptor effect is lost in advanced Parkinson's disease patients, *Neurology,* 52:120–125, 1999.

112. Clarke, C. E., Deane, K. H., Pergolide versus bromocriptine for levodopa-induced motor complications in Parkinson's disease, *Cochrane Database Syst. Rev.,* CD000236, 2000.

113. Clarke, C. E. and Deane, K. H., Ropinirole versus bromocriptine for levodopa-induced complications in Parkinson's disease, *Cochrane Database Syst. Rev.,* CD001517, 2001.

114. Clarke, C. E., Speller, J. M., Clarke, J. A., Pramipexole versus bromocriptine for levodopa-induced complications in Parkinson's disease, *Cochrane Database Syst. Rev.,* CD002259, 2000.

115. Clarke, C. E., Deane, K. D., Cabergoline versus bromocriptine for levodopa-induced complications in Parkinson's disease, *Cochrane Database Syst. Rev.,* CD001519, 2001.

116. Guttman, M., Double-blind comparison of pramipexole and bromocriptine treatment with placebo in advanced Parkinson's disease. International Pramipexole-Bromocriptine Study Group, *Neurology,* 49:1060–1065, 1997.

117. Etminan, M., Gill, S., Samii, A., Comparison of the risk of adverse events with pramipexole and ropinirole in patients with Parkinson's disease: a meta-analysis, *Drug Saf.,* 26:439–44, 2003.

118. Jansen, P. A., Herings, R. M., Samson, M. M. et al., Quick titration of pergolide in cotreatment with domperidone is safe and effective, *Clin. Neuropharmacol.,* 24:177–180, 2001.

119. Lang, A. E., Acute orthostatic hypotension when starting dopamine agonist therapy in parkinson disease: the role of domperidone therapy, *Arch Neurol.,* 58:835, 2001.

120. Kujawa, K., Leurgans, S., Raman, R., Blasucci, L., Goetz, C. G., Acute orthostatic hypotension when starting dopamine agonists in Parkinson's disease, *Arch Neurol.,* 57:1461–3, October, 2000.

121. Ondo, W. G., Levy, J. K., Vuong, K. D., Hunter, C., Jankovic, J., Olanzapine treatment for dopaminergic-induced hallucinations, *Mov. Disord.,* 17:1031–1035, 2002.

122. Morgante, L., Epifanio, A., Spina, E., Di Rosa, A. E., Zappia, M., Basile, G., La Spina, P., Quattrone, A., Quetiapine versus clozapine: a preliminary report of comparative effects on dopaminergic psychosis in patients with Parkinson's disease.

123. Tan, E. K., Ondo, W., Clinical characteristics of pramipexole-induced peripheral edema, *Arch Neurol.,* 57:729–732, 2000.

124. Oecsner, M. Groenke, L., Mueller, D., Pleural fibrosis associated with dihydroergocryptine treatment, *Acta Neurol Scand.,* 101:283–285, 2000.

125. Muller, T., Fritze, J., Fibrosis associated with dopamine agonist therapy in Parkinson's disease, *Clin. Neuropharmacol.,* 26:109–111, 2003.

126. Lund, B. C. Neiman, R. F., Perry, P. J., Treatment of Parkinson's disease with ropinirole after pergolide-induced retroperitoneal fibrosis, *Pharmacotherapy,* 19:1437–1438, 1999.

127. Erreira, J. J., Galitzky, M., Montastruc, J. L., Sleep attacks and Parkinson's disease treatment, *Lancet,* 355:1333–1334, 2000.

128. Frucht, S., Rogers, J. D., Greene, P. E., Gordon, M. F., Fahn, S., Falling asleep at the wheel: motor vehicle mishaps in persons taking pramipexole and Ropinirole, *Neurology,* 52:1908–1910, 1999.

129. Arnulf, I., Konofal, E., Merino-Andreu, M. et al., Parkinson's disease and sleepiness: an integral part of PD, *Neurology,* 58:1019–1024, 2002.

130. Paus, S., Brecht, H. M. Koster, J. et al., Sleep attacks, daytime sleepiness, and dopamine agonists in Parkinson's disease, *Mov. Disord.,* 18:659–667, 2003.

131. Nieves, A. V., Lang, A. E., Treatment of excessive daytime sleepiness in patients with Parkinson's disease with modafinil, *Clin. Neuropharmacol.,* 25:111–114, 2002.

132. Abe, K., Takanashi, M., Yanagihara, T., Sakoda, S., Pergolide mesilate may improve fatigue in patients with Parkinson's disease, *Behav. Neurol.,* 13:117–121, 2001.

133. Canesi, M., Antonini, A., Mariani, C. B. et al., An overnight switch to Ropinirole therapy in patients with Parkinson's disease. Short communication, *J. Neural. Transm.,* 106:925–929, 1999.

134. Goetz, C. G., Blasucci, L., Stebbins, G. T., Switching dopamine agonists in advanced Parkinson's disease: is rapid titration preferable to slow? *Neurology,* 52:1227–1229, 1999.

135. Tseng, K. Y., O'Donnell, P., Dopamine-glutamate interactions controlling prefrontal cortical pyramidal cell excitability involve multiple signaling mechanisms, *J. Neurosci.,* 2:5131–5139, 2004.

136. Picconi, B., Centonze, D., Rossi, S., Bernardi, G., and Calabresi, P., Therapeutic doses of L-dopa reverse hypersensitivity of corticostriatal D2-dopamine receptors and glutamatergic overactivity in experimental parkinsonism, *Brain,* 127:1661–1669, 2004.

137. Hoglinger, G. U., Rizk, P., Muriel, M. P., Duyckaerts, C., Oertel, W. H., Caille, I., and Hirsch, E. C., Dopamine depletion impairs precursor cell proliferation in Parkinson disease, *Nat. Neurosci.,* 7:726–735, 2004.

138. Golberg, J. F., Burdick, K. E., and Endick, C. J., Preliminary randomized, double-blind, placebo-controlled trial of pramipexole added to mood stabilizers for treatment-resistant bipolar depression, *Am. J. Psychiatry,* 16:564–566, 2004.

139. Titner, R., Manian, P., Gauthier, P., and jankovic, J., Pleuropulmonary fibrosis after chronic treatment with dopamine agonists for Parkinson's disease, *Arch. Neurol.,* 2004 (in press).

140. Parkinson Study Group, Pramipexole vs. levodopa as initial treatment for Parkinson disease: A 4-year randomized controlled trial, *Arch. Neurol.,* 61:1044–1053, 2004.

141. Razmy, A., Lang, A. E., and Shapiro, C. M., Predictors of impaired daytime sleep and wakefulness in patients with Parkinson disease treated with older (ergot) vs. newer (nonergot) dopamine agonists, *Arch. Neurol.,* 61:97–102, 2004.

58 Parkinson's Disease: Surgical Treatment—Stereotactic Procedures

Yasuhiko Baba, Robert E. Wharen, Jr., and Ryan J. Uitti
Mayo Clinic, Jacksonville

CONTENTS

INTRODUCTION

Surgical treatments for Parkinson's disease (PD) re-emerged in the 1990s after essentially two decades of relative obscurity. Prior to the late 1960s, treatment of PD consisted of the use of marginally beneficial pharmacological therapy and lesioning operations. Lesioning operations, including thalamotomy, were employed mainly in attempts to treat severe tremor. Major limitations encountered with surgical treatments in the first half of the 20th century included inconsistent targeting and side effects from lesioning. Inconsistent targeting occurred because of the lack of reliable radiological assistance for stereotactic procedures. Retrospective review of historic surgical records suggests that attempted thalamotomies, especially those deemed as particularly clinically successful for parkinsonism, were probably often subthalamotomies judging from notation of immediate, transient, postoperative chorea.* Lesioning side effects, even in instances with accurate targeting, such as dysphonia or dysarthria in bilateral lesioning operations, also limited enthusiasm for surgical treatment of PD.

With the introduction of levodopa in the 1960s, there was a precipitous decline in surgical treatment for PD. An expanding armamentarium of pharmacological agents, including dopamine (DA) agonists and inhibitors of DA catabolism, led to a two-decade period in which surgery was rarely employed.

Two factors led to re-emergence of surgery in the 1990s: (a) recognition of limitations in pharmacological therapy and (b) improvement in stereotactic neurosurgical technology.

Despite optimal pharmacological therapy, it is generally agreed that approximately 30% of PD patients experience motor complications after approximately five years of treatment.[1] While the severity of such complications varies tremendously, it is clear that a substantial proportion of PD patients eventually come to experience "intolerable" fluctuations in motor ability (by individual patients' definition). Unfortunately, some of the more disabling clinical problems related to PD, namely dementia and postural instability/gait difficulties, become increasingly prevalent with prolonged disease and are typically resistant or potentially exacerbated by pharmacological or surgical treatments aimed at improving motor symptoms. It is in the midst of these complicated clinical scenarios when surgical treatments are most commonly contemplated.

Improvements in radiology (computed tomography, CT, and magnetic resonance imaging, MRI) and stereotactic (software guidance systems) have dramatically reduced the frequency of adverse effects and increased the reliability of outcomes associated with all forms of surgery. However, as with all forms of new surgery, the apparent safety and efficacy of treatment may vary substantially on the basis of available equipment, expertise, and experience, not to mention patient selection.

Surgery for PD is presently thought of as predominantly symptomatic therapy, although it is conceivable that surgical treatments may have some influence on the natural progression of PD. The balance of the chapter discussion is divided into topics based on surgical type and target.

SURGICAL OPTIONS

It is convenient to subdivide surgery for PD into three groups:

1. Ablation/lesioning
2. High-frequency electrical stimulation/deep brain stimulation
3. Transplantation/neuroregenerative

Targets for these forms of treatment are generally shared between ablative and stimulation surgeries, although these modalities may well operate with different mechanism. The three main forms of surgery will be discussed sequentially, followed by conclusions regarding implementation and selection of specific procedures for individual PD patients.

LESIONING/ABLATION

Thalamotomy

Since Cooper[2] reported a patient who provided relief of tremor by ligation of anterior choroidal artery, surgical lesioning of the thalamus has been employed for treatment of tremor in patients with PD. In 1962, microelectrode recording techniques employed during the course of functional stereotactic surgery were introduced by Albe-Fessard et al.[3] These electrophysiologic studies helped to determine the ventral intermediate (VIM) nucleus of the thalamus as the optimal target for treatment of tremor. These techniques were used commonly with thousands of thalamotomies being carried out during the 1950s and 1960s.[4–6]

While elucidation of the anatomic basis of tremor in PD is still inadequate, Elble and others has suggested that central oscillation in the cortico-basal ganglia-thalamo-cortical loop plays an important role in the production of tremor.[7] Interrupting these centrally driven tremor activities, with destruction of the VIM nucleus of the thalamus, can effectively control tremor.

Additionally, VIM thalamotomy is useful for treatment of medically intractable tremor from not only parkinsonism, but also essential tremor, cerebellar, Holmes'

---

* Personal communication with Ross Miller, MD, one of the premier neurosurgeons of this era.

(midbrain), post-traumatic, and post-stroke tremor.[8–11] Thalamotomy has also been employed for treatment of other hyperkinetic disorders such as dystonia, hemiballism, and dyskinesias with various etiologies.[12,13]

Marked to complete relief of tremor is made by lesions as small as 2 mm located within the VIM nucleus. Tremor cells within VIM may be identified by neuronal firing at frequencies that coincide with electromyographic (EMG) tremor activity.[14] Thalamotomy can also improve levodopa-induced dyskinesias with optimal lesion sites reported as the ventral oral posterior (Vop) nucleus of the thalamus.[15] While other signs of parkinsonism (such as rigidity and bradykinesia) are not ameliorated by thalamotomy, this surgical treatment is often far more effective in reducing tremor than any pharmacological agent.

Pallidotomy

Pallidotomy has been performed with the intent to alleviate parkinsonism, especially tremor, since the 1950s.[16–20] Narabayashi and Okuma[17] and Cooper and Bravo[19,20] performed chemopallidotomy with injection of procaine and alcohol, or procaine oil. Initial pallidotomies targeted the anterodorsal region in the internal segment of the globus pallidus (GPi). However, lesioning in this location did not provide satisfactory results. Svennilson et al.[21] demonstrated that posteroventral pallidotomy could produce marked amelioration of parkinsonism. With the advent of the levodopa era and advances made with thalamotomy, pallidotomy became infrequently performed.[6] In 1992, Laitinen et al.[22,23] reports fostered new interest in pallidotomy as a procedure that resulted in long-term, marked or complete relief of tremor, rigidity, hypokinesia, and levodopa-induced dyskinesia. Based on these results, pallidotomy of the posteroventral portion of the GPi was the most frequently performed lesioning/ablation procedure in the 1990s.[6] Interestingly, Laitinen has argued that lesioning of the caudal portions of GPe may also produce good results (personal communication).

GPi, which receives projections from the striatum, external segment of globus pallidus (GPe), and subthalamic nucleus (STN), is a major output nucleus of the basal ganglia.[24] Additionally, GPi receives projections from the sensorimotor subloop arising from primary motor and sensory cortical areas in the basal ganglia-thalamocortical loop.[25] In models of PD, loss of DA cells in the pars compacta of the substantia nigra leads to alteration of neuronal activity in the GPi through the putamen. Subsequent excessive inhibitory outflow from GPi causes changes in the neuronal activity in the thalamus. Dysfunction of the thalamocortical circuit is considered responsible for the motor signs of parkinsonism.[24]

Unilateral posteroventral pallidotomy may lead to improvements in (a) contralateral parkinsonism (resting tremor, rigidity and akinesia/bradykinesia) and, to a lesser extent, (b) levodopa-induced side effects (dyskinesia and

"on/off" fluctuations), and (c) axial disabilities (postural instability and gait disturbance) in advanced and elderly patients with PD[22,23,26–36] Therefore, pallidotomy has more sweeping benefit than thalamotomy, which principally affects tremor.

Subthalamotomy

STN drives activity in the external segment of the globus pallidus (GPe), the pars compacta of the substantia nigra (SNc), and GPi/pars reticulata of the substantia nigra (SNr).[37] The STN is also a port of entry to the basal ganglia for cortical output. The dorsolateral STN is a sensorimotor territory by virtue of its afferents from motor cortex.[25]

Since the 1960s, subthalamotomy has been performed as a surgical procedure for PD. The location of the lesioning in the 1960s was probably in the vicinity of the fields of Forel and the zona incerta based on the technology available in that era.

In the 1980s, application of 1-methyl-4-phenyl-1,2,3,6-tetrahydropyridine hydrochloride (MPTP), which was discovered unexpectedly following illicit drug use, led to develop of an important PD animal model. Subsequently, Bergman et al.[38] demonstrated that abnormal neuronal activity in the STN as well as the GPi play a role in forming the parkinsonian animal model. In this model, lesioning of the STN provided significant improvement of parkinsonism and reversal of the increased neuronal activity in the GPi and SNr.[39–42] Furthermore, Bergman et al.[39] reported two MPTP-treated monkeys that showed improvements of contralateral tremor, rigidity, and akinesia after ibotenic acid injection in the ipsilateral STN. Guridi et al.[42] reported MPTP-treated monkeys that received kainic acid injections in the unilateral STN showed bilateral improvement of tremor, spontaneous activity, bradykinesia, and freezing. These monkeys also developed hemichorea after surgery. Additionally, PD patients were also reported to show amelioration of parkinsonism after STN hemorrhage.[43,44] Thus, a number of lines of evidence support the rationale for STN to be considered as a suitable target for functional surgery in PD.

The efficacy of subthalamotomy in PD patients has also been apparent.[45–47] Most patients show improvement of contralateral tremor, rigidity, and bradykinesia with reduction in the requirement of levodopa. Some of these patients experience hemichorea, hemiballism, and/or dyskinesia as a complication of the procedure, but these are less frequent than reported in the MPTP-treated animal model of parkinsonism.

DEEP BRAIN STIMULATION

Thalamic Stimulation

In the 1960s, shortly after introduction thalamotomy, it was recognized that high-frequency stimulation in the

VIM of the thalamus produced relief of tremor.[48,49] In the late 1980s, Benabid et al.[50] reintroduced high-frequency VIM stimulation as a surgical procedure for parkinsonian and essential tremor with similar efficacy and fewer complications. Thalamic stimulation is recommended as a treatment for medically refractory disabling tremor in PD, essential tremor, and multiple sclerosis (MS).[51] Thalamic stimulation is generally preferable to thalamotomy for tremor suppression.[51,52] Thalamic stimulation, as well as deep brain stimulation (DBS) in other targets, produces reversible changes and does not cause damage to adjacent brain parenchyma.[53] Because of lower risk for dysphasia and dysarthria, thalamic DBS is also practical to perform bilaterally.[54] Additionally, thalamic stimulation may be advantageous over static lesioning, as treatment can be modified over time. On the other hand, DBS procedures may carry a higher risk for infection, lead fracture, and hardware malfunction as well as greater equipment costs and need for periodic generator replacement.

It is assumed that high-frequency stimulation alters abnormal brain activity by one or more of the following mechanisms:

1. Depolarizing block
2. "Jamming" of neural activity
3. Channel blocking
4. Neuronal energy depletion
5. Synaptic failure
6. Antero- and/or retrograde effects
7. Activation of inhibitory and/or inactivation of excitatory neurotransmission
8. Effects on non-neuronal cells
9. Effects on local concentration of iron or neuro-active molecules[55]

However, the specific mechanism of action of DBS is not fully understood. Ceballos-Baumann et al.[56] investigated the functional effect of VIM DBS, and found that VIM DBS provides increased regional cerebral blood flow (rCBF) in the ipsilateral motor cortex. They suggested that the beneficial effect of VIM DBS is associated with increased synaptic activity in motor cortex, probably occurring as a result of activation of thalamocortical projection, or frequency-dependent neuroinhibition that overrides the abnormal periodic neuronal pattern underlying tremor. Perlmutter et al.[57] also found increased blood flow in the ipsilateral supplementary motor area, which is a terminal field of thalamocortical projections, in patients with VIM DBS.

Pallidal Stimulation

Achievement of thalamic stimulation brought a new technological achievement for the surgical treatment of movement disorders. Subsequently, targets of DBS for parkin-

sonism expanded to include VIM, GPi, and STN, on the basis of evidence that abnormal neural activity is present in GPi and STN in the MPTP-treated animal model. DBS in GPi and STN provided benefits for all forms of tremor and parkinsonism.

Pallidal stimulation shows similar clinical effects as pallidotomy and can be performed relatively safely, even in the context of bilateral surgery. Studies of efficacy and safety between unilateral pallidotomy and pallidal stimulation generally conclude these as being comparable.[58] Many studies have reported that pallidal stimulation improves all forms of parkinsonism and levodopa-induced dyskinesia.[59–63] The effects of pallidal stimulation are related to the location of the stimulating electrode within the GPi. Stimulation of dorsal GPi significantly improves parkinsonian features, including rigidity, akinesia, and gait disturbance, and can induce "off" state dyskinesia. On the other hand, stimulation of posteroventral GPi dramatically improves levodopa-induced dyskinesia and worsens gait and akinesia.[64] Stimulation of the most ventral part of the GPi leads to improvement of rigidity and complete relief of levodopa-induced dyskinesia but also produces severe akinesia. On the other hand, stimulation of the most dorsal part of the GPi leads to moderate improvement of akinesia and drug-induced dyskinesia.[65] Durif et al.[66] reported that stimulation of the anteromedial part of the GPi, which corresponds to the ventral pallidum and contains a greater number of fibers at the origin of the outflow pathways, is safe and improves cardinal parkinsonian signs and levodopa-induced dyskinesia. Motor performance improvements from pallidal stimulation are also reflected by increased of regional blood flow in the sensorimotor cortex, supplementary motor area, and anterior cingulate cortex on positron emission tomography (PET).[67]

Subthalamic Stimulation

High-frequency stimulation of the STN showed efficacy for amelioration of parkinsonism in MPTP-treated monkeys.[68] Subsequently, antiparkinsonian effects associated with subthalamic stimulation were proved also in PD patients;[69] the study of this procedure developed rapidly in the mid-1990s.

Subthalamic stimulation not only ameliorates cardinal parkinsonian signs and symptoms, it can also reduce the daily levodopa dose equivalent.[70,71] Bilateral procedures provide further benefits including improvement of axial instability.[72,73] In severely affected PD patients, subthalamic stimulation may be recommended unilaterally or bilaterally. Patients in whom unilateral or bilateral subthalamic stimulation has been performed may need to take antiparkinsonian medicine for ameliorating parkinsonism on the side ipsilateral to the implantation.

To date, although anecdotal reports exist that generally favor subthalamic stimulation,[74–77] there have been no large randomized studies comparing relative efficacy of subthalamic versus pallidal stimulation or lesioning procedures.

Electrostimulation procedures have also been performed following failed lesioning operations, with good results.[78–81] Moreover, it has been reported that patients in whom stimulation of the globus pallidus failed to give long-term relief may respond successfully to bilateral subthalamic stimulation.[77] Most experts suggest that patients with parkinsonism who do not benefit from levodopa therapy are poor candidates for subthalamic or pallidal stimulation. A patient with vascular parkinsonism, without response to levodopa, showed no beneficial response to bilateral subthalamic stimulation.[82] Pre- or intraoperative use of apomorphine, a short-acting DA agonist, has also been reported to be a good predictor of motor responsiveness to motor outcome from subthalamic stimulation.[83] Further patient selection criteria and technical specifics remain to be fully delineated.

Age, disease duration, and the severity of levodopa-related motor complications are not predictive factors for outcome of subthalamic stimulation. On the other hand, parkinsonian motor disability tends to be more improved in patients with younger age, shorter disease duration, and less axial symptoms. Therefore, older patients and/or patients with significant axial symptoms, such as gait disturbance and postural instability, who are poorly responsive to levodopa, may be not optimal surgical candidates for subthalamic stimulation.[84]

Cortical Stimulation

In 1979, Woolsey et al.[85] reported two PD patients who showed marked improvement of tremor and rigidity by primary motor cortex stimulation with subthreshold stimulation. Subsequently, repetitive transcranial magnetic stimulation (r-TMS) was introduced as a treatment for PD motor signs.[86] To date, extradural motor cortex stimulation (MCS) has been mainly used as a procedure for central and neuropathic pain with minimal morbidity/mortality risk,[87] but some PD patients have also been reported to benefit with MCS.[88,89]

Because motor symptoms of PD may reflect dysfunction of the thalamocortical circuit,[24] MCS may be logical treatment. In fact, modifications of motor cortex metabolism that probably accompany with improvement of motor symptom are demonstrated in PET studies of VIM, GPi, and STN DBS procedures.[56,90,91]

Extradural MCS has also been reported to reduce dyskinesia, and levodopa equivalent daily requirements.[88,89] However, reports concerning extradural MCS stimulation procedures in PD are few and describe anecdotal experiences. Therefore, much more systematic analysis of outcome with extradural MCS is required.

REGENERATIVE/GRAFTING

Fetal Mesencephalic Tissue Transplantation

PD is characterized by selective and progressive degeneration of neuronal cells, frequently leading to severe and uncontrollable disability. The goal of organ transplantation therapy is to regenerate specific neuronal cells or recreate the functional equivalent of those lost in the disorder. In 1979, Bjorklund and Steveni[92] and Perlow et al.[93] demonstrated that fetal mesencephalic tissue that was implanted in the striatum of 6-hydroxydopamine (6-OHDA) animal model reduced motor abnormalities in conjunction with neuronal survival. Implantation of fetal nigral cells to in the striatum of MPTP-treated monkeys provides manufacture of DA and amelioration of motor symptoms.[94,95] On the basis of these animal model studies and some anecdotal reports, clinical trials of transplants using adrenal autografts were performed in PD patients, with poor results.[96–98] Implantation of fetal nigral tissue in the MPTP-treated animals and PD patients provided modest benefits with effects varying between studies.[99–107] The reasons for inconstant results may reflect tissue collection, donor selection, grafting techniques, target site, tissue volume, and use of immunosuppression. Some researchers have limited fetal transplantation to young patients with parkinsonism responsive to DA or DA agonist, whose major symptoms are rigidity and hypokinesia.[108] Functional imaging (PET) demonstrates that dopaminergic grafts restore striatal dopaminergic function, with extracellular dynamics of DA that are different from those of intact striatum. These grafts can normalize ambient DA levels and permit transmission over an extended sphere.[109] Putaminal ^{18}F-fluorodopa uptake may also improve in selected patients, typically preceding improvements in clinical function.

GDNF Regeneration

Glial cell line-derived neurotrophic factor (GDNF) is a potent neurotrophic factor for dopaminergic neurons. GDNF may increase survival and growth and prevent apoptosis of dopaminergic neurons.[110] GDNF has also been studied in MPTP-treated monkeys[111] and PD patients. In animal models that were performed with intraventricular or intraputamenal infusion, GDNF treatment provides up to 60% improvements in rigidity, bradykinesia, and axial instability without significant adverse effect. Postmortem studies have demonstrated that nigral DA neuron cell size was increased by up to 30%, and DA metabolite levels in the striatum were increased by up to 70%. Moreover, DA levels in the periventricular striatum increased by 233%.

Human Amniotic Epithelial Cell Transplantation

Human amniotic epithelial (HAE) cells are generated from amnioblasts on the eighth day after fertilization. HAE cells release brain-derived neurotrophic factor (BDNF) and neurotrophin-3 (NT-3), which have trophic activities on cultured dopaminergic neurons.[112] In the 6-OHDA rat model, HAE cells were found to survive without evidence of overgrowth two weeks after midbrain infusion, and the number of nigral dopaminergic cells was significantly increased in the substantia nigra.[113] HAE cells may be part of further preventive or regenerative therapy for PD.

Porcine Mesencephalic Tissue Transplantation

Transplantation treatment using porcine tissue can overcome societal and ethical limitations and provide a large source of implantable tissue. It has been suggested that the pig is most suitable species for xenotransplantation.[114] Porcine fetal ventral mesencephalic transplantation provided functional recovery and reinnervation of striatum in PD animal models.[115] In recent studies in PD patients, however, functional recovery was minimal, and total Unified Parkinson's Disease Rating Scale (UPDRS) score in the "off" state was improved by 19% at 12 months after unilateral transplantation.[116,117] Moreover, a prospective, randomized, double-blind, surgical placebo-controlled study in PD reported that there was no significant statistical difference in total UPDRS score in the "off" state between control and treated groups at 18 months after transplantation.[118] A postmortem study in a single PD patient over seven months after fetal porcine transplantation was found to show minimal graft survival.[119] Additionally, ^{18}F-fluorodopa uptake after 12 months after surgery did not reveal significant changes on the side of transplant.[117]

There is a risk of transmission of porcine endogenous retroviruses to host cells. Transmission of porcine endogenous retroviruses, with graft survival, has been detected in animals with severe combined immunodeficiency (SCID) performed porcine islet transplants.[120] These potential problems with porcine tissue transplantation have led to concerns regarding immunosuppression used to prevent xenograft rejection with heteroplastic transplantation. Such concerns have slowed development and enthusiasm in this mode of regenerative therapy for PD.

METHODOLOGY

PATIENT SELECTION

Appropriate patient selection is the most important key to obtain a favorable outcome from any surgical procedure. Thorough neurological, neuropsychological, pharmaceutical, radiological, and systematic medical evaluations are best comprehensively performed by the surgical team: neurologist, neurosurgeon, neuroradiologist, and neuropsychologist.

The risks and potential benefits of brain surgery dictate that PD patients are best treated initially with pharmaco-

logical agents, with surgery reserved for circumstances in which medication fails to provide consistent benefit. Parkinsonian signs and symptoms that are not responsive to levodopa therapy rarely obtain significant benefit from any form of surgery. Consequently, patients who have little or no response from pharmacological therapy are generally considered poor surgical candidates. The one symptom/sign that does not follow this rule is tremor. Surgery may be effective in ameliorating tremor, even if minimally or unaffected by medication trials. Parkinsonian patients with neurodegenerative disorders other than PD (suggested by the presence of typical features such as supranuclear gaze palsy, predominant axial symptom, absence of tremor, pyramidal tract signs, and marked autonomic disturbance) respond poorly to surgical procedures. Relative contraindications for all surgery include major psychiatric disturbances (major depression, psychosis) and dementia. The surgical procedures themselves may lead to cognitive decline in elderly PD patients with early stage concomitant dementia.[121,122] Medication side effects, such as hallucinations or hypotension from dopaminergic agents, are not contraindications as long as they are modest and resolve with reduction in medication dosages. Pallidotomy and thalamotomy can result in postoperative dysarthria and dysphagia, especially when done bilaterally.[123–126] Therefore, significant abnormalities of speech and swallowing are relative contraindications, particularly for lesioning operations, although specific guidelines may vary between centers.[127] Chronic anticoagulation is not an absolute surgical contraindication, but careful perioperative management is required. The presence of severe hypertension, brain atrophy, and white matter signal changes that may increase the occurrence of intracerebral hemorrhage. Additionally, the presence of intracranial lesions on MRI or CT may compromise accurate radiological target determination.[127] Patients with other severe medical illness are generally not surgical candidates, because of higher risk for the surgical procedure and limited life span. Patients with cardiac pacemakers/defibrillators or spinal cord stimulators have generally not been treated with DBS. However, several patients treated with DBS have subsequently been successfully treated with cardiac pacemakers/defibrillators when required,[128] and therefore these devices are not absolute contraindications for DBS. There are no absolute contraindications to the surgical treatment of movement disorders in pregnancy, although the relative risks of the procedure contemplated must be weighed with the benefits. Because most movement disorders are chronic, it would be reasonable to defer surgery until the patient is postpartum.

SELECTIONS OF SURGICAL TARGET AND TYPE

Selection of target and surgical type must be selected by virtue of each individual patients need. Consequently, it is essential to delineate the patient's current and potential future needs that most influence daily life. While DBS can be performed with less neurologic complication than lesioning procedures, it does require potentially frequent adjustments in stimulation parameters and occasional pulse generator change, and it involves potential risk for infection and delayed complications with hardware. On the other hand, lesioning procedures are performed relatively easily, without high cost. While some patients have had DBS contralateral to lesioning procedures, it remains to be seen whether such a practice offers substantial benefits.

Tremor

A thalamic target provides beneficial results for parkinsonian tremor, but not for other cardinal parkinsonian features[51,52,129] (Figure 58.1). Therefore, PD patients with long-term tremor-dominant disability are good candidates for thalamic procedure. There is no significant difference in effect for tremor between thalamotomy and VIM DBS, but VIM DBS are performed with less adverse effects than thalamotomy.[51] Pallidal and subthalamic targets also provide good results for tremor. However, quantitative efficacy of thalamic target may be superior to those of pallidal or subthalamic target.

Rigidity, Bradykinesia/Akinesia, Axial Symptoms

Both pallidal and subthalamic targets provide benefit for rigidity, bradykinesia/akinesia, and axial symptoms such as gait disturbance and postural instability. Pallidotomy can improve moderately to markedly these signs. Long-term efficacy studies with unilateral pallidotomy demonstrate maintained contralateral benefit.[130–133] However, bilateral procedures carry risk for severe adverse effects.[124,134] Nevertheless, pallidotomy may be a reasonable option for a patient with severe, relatively unilateral parkinsonism and levodopa-induced dyskinesia. Subthalamotomy also produces improvement of cardinal symptoms of PD, and these effects endure.[45–47] Effects for axial symptoms are measurable but not long-lived with unilateral procedures.[47] GPi and STN DBS provide similar effects to lesioning procedure corresponding to each target but also with less complication, especially when per-

formed bilaterally. Effects for rigidity, bradykinesia/akinesia, and axial symptoms from both GPi and STN DBS are also similar, but efficacy in relieving bradykinesia/akinesia with STN DBS is likely mildly superior to GPi DBS.[65]

Dyskinesia

A pallidal target provides marked to complete and long-term relief of levodopa-induced dyskinesia.[63,135–141] Unilateral pallidotomy improves levodopa-induced dyskinesia, not only contralaterally but also ipsilateral to the side of the lesioning. Elimination of levodopa-induced dyskinesia is dramatic with bilateral GPi DBS. Subthalamic nucleus targets also result in amelioration of levodopa-induced dyskinesia. Both subthalamotomy and STN DBS allow marked to complete reduction of levodopa equivalent medication intake, which may account for dyskinesia relief.[32,45–47,70,71,142–145] The thalamic Vop target may also improve levodopa-induced dyskinesia.[15]

OPERATIVE METHODOLOGY

CT or MRI of the brain is obtained prior to and sometimes following the procedure to assess for hematoma and edema and to verify lesion placement. Because pallidotomy may result in visual field deficits, pre- and postoperative assessment of visual fields may be useful.[26] Prior to surgery, cerebral angiography may occasionally be performed to delineate surrounding blood vessels in the target area, although high-quality MRI usually would obviate the need for angiography.

Stereotaxic surgery was developed so as to better determine the relationship of the surgical target to nearby structures, which can be visualized radiographically, and then to direct an electrode or other probe to the target with minimal damage to surrounding structures.[146] The optimal means for targeting continues to be the subject of debate[147] but always includes neural imaging. Initial targeting is guided by integration of CT or MRI, with frames for stereotactic surgery, via specially designed computer software.[127,148] While MRI tends to be preferred, CT imaging may be acceptable, and the only randomized study evaluating this issue suggested no difference between MRI and CT for preoperative localization (in pallidotomy).[149]

Small statistical differences in MRI versus CT-derived targets have been identified and, although direct comparison with clinical outcomes have not been made, most institutions have concluded that MR-based target localization is superior to CT.[150] Comparisons of MRI-guided and ventriculography-based stereotactic surgery for PD have concluded that each results in similar clinical outcomes, concerning efficacy and complications.[151] Lesions in pallidotomy and thalamotomy are made using

FIGURE 58.1 Effects on tremor in VIM DBS, label. Left: presurgery. Drawing difficulty was seen for tremor. Right: postsurgery, "on."

stereotactic radio frequency ablation, often following electrophysiological recordings and stimulation procedures.[152] It has been determined that electrophysiologic recording typically leads to final placement of lesions usually within 2 to 3 mm of MRI targets, with the actual lesion overlapping the MRI theoretical target in 40 to 50% of patients.[153] Experienced centers using MRI and microelectrode recordings typically require only one or two trajectories for performance of pallidotomy.[154] Additionally, occasionally deep brain stimulation electrodes are employed in the process of directly making lesions.[155]

Surgical techniques for all types of surgical procedures vary between centers. This discussion provides descriptions of only some of the techniques commonly used. Pallidotomy, thalamotomy, and deep brain stimulation procedures are performed under local anesthesia so that the patient can be monitored with clinical criteria intraoperatively.[148,152] Usually, after the stereotactic frame is placed under local anesthesia, CT or MRI is performed to locate the coordinate of the anterior commissure (AC) and posterior commissure (PC). Computer programs allow for subsequent simulation with the patient's MRI of the precise trajectory and distance to target from the burr hole (Figure 58.2). Stereotactic devices are subsequently fixed to the frame, and microelectrode recording/stimulation ensues. Electrophysiological assessment of the activity of the target is used to ensure proper targeting and placement of the electrode. The tip of the electrode, which is used for recording, has a diameter of 0.01 to 0.02 mm and an impedance of 0.3 to 1 MΩ at 1,000 Hz. The sensorimotor region is delineated by recording the increases in neuronal discharge on passive manipulation of the limbs and during active movement. In the GPi and STN, the arm and face are in the most lateral region and leg slightly more medial. The reverse is the case for the VIM, where the leg is lateral to the arm.[156] After localization of the target by electrophysiological recording, in ablative procedures, lesioning is performed along the trajectory (Figure 58.3). In DBS procedures, electrode implantation is performed in the sensorimotor region of each target nucleus. The electrode is placed into the intended target with X-ray confirmation followed by intraoperative stimulation and characterization of stimulation effects (Figures 58.4 through 58.6).

In thalamotomy, a burr hole is made and, based on previously delineated coordinates, a monopolar microelectrode is advanced to identify somatotopy in the thalamic somatosensory nucleus, ventralis posterior. Moving anteriorly to sensory recordings made following tactile stimulation of the contralateral hand or face leads to stimulation of the ventralis intermedius nucleus, which has a characteristic spontaneous discharge coincident with tremor activity.[146] A separate, monopolar, stimulating electrode may subsequently be used, noting tremor suppression once the correct area is reached.[146] The operation is monitored clinically in terms of speech, manual and foot dexterity, sensation, tone, and tremor, and electri-

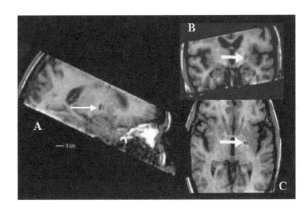

FIGURE 58.3 MRI findings in pallidotomy, label: T1 weighted parasagittal (A), coronal (B), and axial (C) imaging. Left pallidal lesion is confirmed (white arrow).

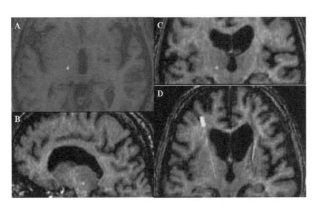

FIGURE 58.2 MRI stereotactic plan for VIM-thalamic target, label: T1 weighted axial (A), parasagittal (B), and coronal (C, D) imaging. The thalamic target is shown as light gray dots.

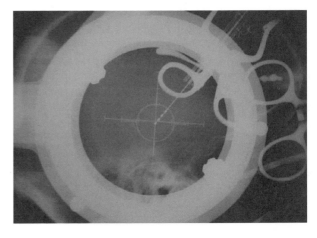

FIGURE 58.4 Implantation of deep brain stimulation electrode. Electrode positioning following MR targeting is confirmed with intraoperative lateral X-ray of the head.

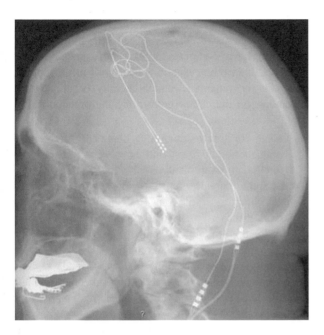

FIGURE 58.5 Lateral skull X-ray of bilateral subthalamic DBS electrodes (brain and extension leads and connections are visible).

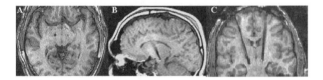

FIGURE 58.6 MRI findings of bilateral subthalamic deep brain stimulation electrode implant, label: T1 weighted axial (A), parasagittal (B), and coronal (C) imaging of bilateral subthalamic stimulation.

cally, in response to proprioceptive, kinesthetic, and electrical stimulation of the involved limb.[127,157,158] Once the appropriate coordinates are determined, a lesion is made with a lesioning electrode.[146,157] This electrode is hollow to allow the insertion of a thermocouple device, which uses radio frequency current to create the lesion.[146]

In pallidotomy, a microelectrode recording probe is introduced through a frontal burr hole and advanced to confirm somatotopic localization within the globus pallidus interna, which has been previously targeted with stereotactic information and CT or MRI.[26,148] Single-cell electrical recording is performed using tungsten-tipped, disposable microelectrodes and analyzed for responses during passive and active movement.[152] Electrical stimulation prior to lesioning is performed to prevent injury to the internal capsule and optic tracts adjacent to the globus pallidus interna.[152] Specific clinical signs have also been described that may occur and influence targeting decisions during surgery. For example, authors have reported a patient who had marked, sustained, contraversive eye

deviation by stimulation during pallidal surgery. This may occur as a result of excitation of internal capsular fibers by volume conducted current spread. Such conjugate eye deviation therefore may not necessarily indicate incorrect electrode placement.[159] If electrical stimulation does not result in weakness or visual field loss, a lesion is made.[148] The probe is then withdrawn several millimeters, with repeated lesioning, creating a three-dimensional lesion along the track (or tracks).[26,148]

Further lesions are performed based on clinical responses measured in the contralateral hemibody (i.e., reduction in parkinsonian signs). Some believe that bilateral pallidotomy can be safely done during the same procedure, and that the localization information from the first lesion is helpful in determining that of the second.[160] However, most centers favor performing only unilateral pallidotomy. Visual evoked potentials to photic stimulation of the eyes intraoperatively during pallidotomy are believed by some to facilitate the accuracy of the determination of the globus pallidus interna.[156,161] A consensus statement regarding pallidotomy has been reported suggesting, among other conclusions, that pallidotomy should be performed only at centers where a dedicated team of physicians has compiled substantial experience in the field.[162]

Lesions of the STN can be created in a similar fashion to those in pallidotomy and thalamotomy, but these lesions are technically more difficult, because neuroimaging techniques are less able to localize this target, and lesioning of the STN may cause hemiballism, chorea, or other adverse effects. However, stimulation of the STN is routinely performed in many experienced surgical centers and offers significant advantage in allowing the possibility of performing a bilateral procedure due to the relative lack of risk of irreversible dysphonia.

Issues still debated include the need for microelectrode recording,[136,152, 163,164] the number of lesions and lesion size, and the wisdom of making bilateral lesions. In a survey of 28 centers performing pallidotomy in North America, most centers were using MRI alone (50%) or with CT (n = 6) to localize the target. The median values of pallidal coordinates were stated as 2 mm anterior to the midcommissural point, 21 mm lateral to the midsagittal plane, and 5 mm below the intercommissural line, with a total of three permanent lesions placed 2 mm apart. According to the survey, lesions are typically made employing a median temperature of 75°C for 1 min. Microelectrode recording was performed by 50% of the centers surveyed, with the main target defining criteria being (1) the firing pattern of spontaneous neuronal discharge and (2) the response to passive manipulation of a limb. Proponents for microelectrode recordings indicate that such recordings altered the final target in almost every instance, with one of nine targets being more than 4 mm from the image-guided site.[154] Motor and visual evaluation was also done intraoperatively.[165]

Microvascular doppler evaluation, performed in order to identify intracerebral vessels in proximity to targets for thermocoagulation (in thalamotomy or pallidotomy), has been described as a means to minimize risk of vascular injury. A prominent vascular sound was identified in 3 of 13 cases in one series.[166] It is unclear whether use of this technique significantly impacts on safety in these lesioning operations.

In thalamic stimulation, placement of the stimulating quadripolar lead is performed under local anesthesia with subsequent implantation of an external, programmable stimulator being placed under general anesthesia.[167] Contrast ventriculography is used by some to allow advancement of the electrode through a burr hole toward the ventralis intermedius nucleus target.[167] Most centers employ MR- or CT-guided software to prepare a surgical targeting trajectory for placement of the stimulation lead. Electrophysiological confirmation of the target proceeds as with thalamotomy. When stimulation through the quadripolar lead suppresses tremor, the electrode is implanted and connected to a percutaneous lead that is tunneled to the implanted pulse generator.[167] A similar technique has been described for placement of the stimulator for pallidal and STN stimulation.[69] Microelectrode recordings in the STN suggest a somatotopic arrangement that may aid in electrode placement.[168]

The effects of intravenous anesthesia with propofol on intraoperative electrophysiologic monitoring were studied in patients during pallidotomy and thalamotomy. Infusion of this agent needed to be reduced to detect neural noise levels required for targeting in some patients but generally serves as a useful anesthetic agent for electrophysiological monitoring during functional neurosurgery.[169]

Gamma knife thalamotomy has been reported using stereotactic guidance.[170–172] Ohye and colleagues used 140 to 150 Gy and 4 mm collimators with microelectrode recording guidance similar to that described above in thalamotomy and pallidotomy.[170] Pan and colleagues have used slightly higher doses of radiation, 160 to 180 Gy maximum dose, also with 4 mm collimators.[171] Because of the delay in response and relative inaccuracy and unpredictability of radiation lesions, most specialists suggest other modes of surgical treatment be used in most instances.[173]

The surgical techniques used for fetal mesencephalic transplantation for PD vary between centers.[174] The ideal fetal age, amount of tissue, tissue handling methods, and preferred location of transplant remain in debate. Some researchers believe that storage of fetal tissue in liquid nitrogen is reasonable prior to transplantation, which would allow harvesting this scarce tissue source well in advance of transplantation.[175] Others perform fetal transplantation within 4 hours of the abortion, without freezing the tissue.[105] Some recommend the use of tissue from fetal cadavers of 6 to 11 weeks gestational age.[105,175] Others have used fetuses of 11 to 19 weeks' gestation.[176] Fetal transplants have been placed in the caudate nucleus, in the putamen, or in both (usually 1 fetal graft in the caudate and 3 in the putamen), and bilaterally or unilaterally, depending on the center performing the procedure.[99,100,105,175]

Tissue for transplantation is obtained from fetuses following suction abortion, and the mesencephalic region is dissected microscopically.[177] Testing of the maternal serum and recipient serum prior to transplantation includes studies to rule out HIV, hepatitis, and other infectious diseases.[108,178] ABO typing is done by some centers for donor/recipient compatibility.[177] Implantation is performed under general anesthesia.[108] Using stereotactic techniques, approximately 10 to 15 strands of fetal mesencephalic tissue are inserted via injection cannula into one or both lenticular nuclei of the recipient.[177] The importance of postoperative immunosuppression is not known, and its use varies between centers.[108,177–179] One study found that immunosuppression did not improve clinical outcome.[180] Use of fetal porcine tissue for transplantation is also now under study, as are issues relating to cell suspension versus solid graft transplant, use of neurotrophic factors, and routine immunosuppression.

Follow-up Care

Evaluation after surgery is of paramount important. Carefully follow-up observation is recommended to look for neurological deficit or procedure-related complications as well as to modify other pharmacological treatment. CT or MRI examination may be performed to confirm lesioning or implanted electrode site, and to find procedure-related complications. In DBS procedures, stimulators are often tuned "on" and programmed. Patients should be evaluated every one to three months for the first year following surgery and annually thereafter. At each visit, information regarding neurological examination, levodopa equivalent dosage, duration of "on" time, and severity of dyskinesia should be noted. Interviewing the patient's family is also important in discovering other psychological symptoms. Quantitative assessment following UPDRS, Dyskinesia Rating Scale (CAPIT) score, Parkinson's Disease Quality of Life Questionnaire (PDQL), and Mini-Mental State Examination (MMSE) may also be recorded.

DBS related complications or side effects may occur at various times following implantation and stimulation. These include disturbances in consciousness, seizures, confusion and bradyphrenia, and cerebral hemorrhage accompanying electrode implantation. Infection, skin erosion, and malfunction of brain and extension lead may occur as later complications. External magnetic devices and other electronic tools may cause inadvertent turning "off" of a pulse generator. Rarely, patients who suddenly lose stimulation may show sudden neurological deterioration, requiring emergency treatment.[181] One patient with STN DBS was left in a vegetative state from permanent brain stem lesioning after receiving pulsatile radio frequency diathermy for a dental condition.[182] As such, all surgical invasive procedures are best carried out only after consultation with the DBS device manufacturer.

OUTCOMES

LESIONING/ABLATION

Thalamotomy

Recent long-term follow up studies in PD patients who underwent unilateral thalamotomy report complete abolition of contralateral tremor in 86%[183,184] with reduced levodopa requirements. Immediate postoperative complications consist of mild contralateral weakness (34 to 42%), dysarthria (29 to 36%) and cognitive impairment (14 to 23%), with some side effects in approximately 58 to 67%. These complications usually resolve rapidly during the postoperative period. Potential persistent complications include contralateral hemiparesis, seizures, paresthesia, ataxia, apraxia, hypotonia, abulia, and gait disturbance. Hemorrhagic complications may accompany the procedure, causing serious or lethal morbidity in 1 to 2% of patients. However, the benefits in surgery usually outweigh potential complications at experienced surgical centers. Bilateral thalamotomy has a high incidence of complications including dysarthria, dysphagia, and hypophonia, typically on the order of 30%.

The mortality risk for thalamotomy in PD is less than 0.3%.[185] Causes of death include hemorrhage in deep gray matter, postoperative infection, and pulmonary embolism.

Thalamotomy vs. Thalamic Stimulation

Schuurman et al.[51] reported the effect of thalamotomy and thalamic stimulation on functional abilities of patients with severe tremor due to PD, essential tremor, and multiple sclerosis. After two years of follow-up, thalamotomy and thalamic stimulation showed equal effect for the suppression of tumor, but thalamic stimulation was associated with significantly fewer adverse effects than thalamotomy and resulted in greater functional improvement.

Schuurman et al.[186] demonstrated that thalamotomy and thalamic stimulation are associated with a small overall risk of cognitive deterioration. Worsening of verbal and reading tasks occurred after left-sided surgery in both thalamic procedures.

PALLIDOTOMY

The proportion of improvement in motor function varies in each study, with improvement of total UPDRS score ranging from 17.8 to 65% during short-term follow-up after surgery.[26–28,139,187] The UPDRS score in the "off" state reveals improvement ranging from 13.6 to 31%.[27,28,34,135,137,139,140,187,188]

Ameliorations of motor symptoms are striking contralateral to surgery. Some studies also report improvements of tremor, rigidity, bradykinesia, and tests of finger and foot tapping ipsilateral to surgery.[26,27,34,137,139,187] Such improvements are usually transient and undetectable six to nine months postoperatively. Axial disabilities such as postural instability and gait disturbance improve by 22 to 44% but, again, are not long-lived.[27,136,137,188] The most obvious and reliable beneficial effect of pallidotomy is improvement of levodopa-induced dyskinesia; the degree of reduction in dyskinesia ranges from 50 to 92% improvement in the contralateral side, and by 32 to 45% ipsilaterally.[135–137,139,140]

Reports of benefit of pallidotomy on activity of daily living (ADL) and quality of life (QOL) scores vary, ranging from 17 to 44%,[27,28,137,139,190] and from 35 to 77%,[139,190] respectively.

Laitinen et al.[22] reported that the dosage of levodopa could be reduced by 50 to 75% after pallidotomy. However, in most long-term follow-up studies, there is no change in dose of antiparkinsonian drug, including levodopa.[137] Pallidotomy provides around equivalent improvement of motor functioning in both younger and elderly PD patients.[28]

Improvement in cognitive function after pallidotomy emerges only in "off" state measures.[191] On the other hand, verbal fluency is reported to decrease after pallidotomy,[28,123,191,192] and left-sided lesioning produces more impairment in verbal fluency.[192] However, verbal fluency changes typically recover during long-term follow-up after surgery.[193] Additionally, the presence of cognitive impairment after pallidotomy does not correlate with volume of lesion.[192]

Long-term follow-up studies after unilateral pallidotomy over periods of 4 to 5.5 years report sustain improvements of parkinsonism and levodopa-induced dyskinesia in the contralateral side compared with baseline,[130–133,193] and improvement of tremor is strikingly preserved.[131,132,193] However, most other signs gradually return to preoperative levels, and signs deteriorate with disease progression.

Adverse effects with pallidotomy vary between studies. Because optic tract and internal capsule each reside adjacent to GPi, inaccurate targeting may cause visual and motor system problems. Adverse effects include facial paresis, dysarthria, acute confusion or somnolence, dysphagia, hemiparesis, changes in personality or behavior, worsening of handwriting, visual field defect, hypersalivation, and cerebrovascular accident. In these adverse effects, acute confusion or somnolence is likely to be a transient symptom after surgery. The occurrence rate of total adverse effects of unilateral pallidotomy is 30.2%, and of permanent adverse effects is about 13.8%.[194] Occurrence of adverse events may vary substantially between surgical centers. Some series report unilateral pallidotomy without serious permanent adverse effects.[28,190] Symptomatic brain infarction and hemorrhage are seen in 3.9%, and mortality in 1.2% of patients with unilateral pallidotomy.[194]

Bilateral pallidotomy also produces abolition of levodopa-induced dyskinesia and motor fluctuations. However, because of more frequent and severe complications, including neuropsychological and psychiatric changes, and corticobulbar syndromes in many patients,[123–126] such bilateral lesioning is usually avoided. Staged bilateral pallidotomy is also associated with increased risk of adverse effects, though most patients experience moderate benefit.[134]

SUBTHALAMOTOMY

Recent studies reported that unilateral subthalamotomy can improve all cardinal features of parkinsonism and reduce the required dose of levodopa.[45–47] Total scores of UPDRS II (motor score) and III (ADL score) in the "off" period show significant improvement after surgery, and these efficacies last for at least 6 to 24 months. Additionally, "on/off" fluctuations are reduced and tremor improved contralateral to the lesion.[45] In addition to improvements of rigidity and bradykinesia on the side contralateral to the lesion, transient benefit may occur ipsilateral to surgery.[45] Noteworthy ameliorations in axial disabilities, such as gait disturbance and postural instability, are obtained by unilateral subthalamotomy.[45,47] These improvements decrease gradually from about one year after surgery.[47] The levodopa equivalent daily intake reduces by 42 to 59%, and some patients may stop medical treatment after surgery,[45–47] subsequently resulting in significant reductions in dyskinesias.[46,47] Postoperative dyskinesia such as hemichorea and hemiballism are rarely seen,[46] ranging from 5 to 25%.[45–47] Other adverse effects include cerebrovascular accident with or without clinical symptoms. Patel et al.[46] report that combined lesioning of the dorsolateral and pallidofugal fibers (H2 field of Forel)/zona incerta is particularly effective for parkinsonism and dyskinesia.

DEEP BRAIN STIMULATION

Thalamic Stimulation

Thalamic stimulation produces significant reduction in both essential tremor and parkinsonian tremor contralateral to the side of stimulation in randomized controlled studies.[195,196] Koller et al.[195] reported that combined blinded tremor ratings (ranging from 0 to 4) of "on" stimulation scores in PD patients with unilateral thalamic stimulation were 0.6 compared to 3.2 for those of "off" stimulation; 58% of patients showed complete relief of tremor. Limousin et al.[196] reported resting tremor of the contralateral upper limb reduced in 85% of patients (Schwab and England scale, and UPDRS II items 6, 8). Double-blind long-term studies by Rehncrona et al.[197] demonstrated that thalamic stimulation provide significant improvement of the total motor score, not only by suppressing tremor but also by decreasing akinesia in the contralateral to the side of stimulation at the six- to seven-year follow-up evaluation.

Thalamic stimulation also can be beneficial in reducing midline tremor with bilateral thalamic stimulation being superior to unilateral.[128] All forms of tremor appear responsive to thalamic stimulation, with changes in postural tremor being associated with greatest functional improvement.[198] Thalamic stimulation has been recommended as treatment for disabling tremor in PD, essential tremor,[167] and multiple sclerosis. Based on experience with 118 patients, Benabid and colleagues found that complete arrest of tremor is frequently encountered.[167] Others authors have noted that repeated programming changes may be required in patients.[199]

Thalamic stimulation contributes benefit only for tremor, but it has been reported that relief of tremor by thalamic stimulation may also improve postural instability in PD patients.[83] However, in another study, gait comparisons in seven patients undergoing VIM stimulation for PD found no changes with stimulation.[200] Improvements in levodopa-induced dyskinesia have been shown to occur with thalamic stimulation inferior, medial, and more posterior to VIM, probably within the center median and parafascicularis complex.[201]

Significant cognitive decline is generally not encountered with thalamic stimulation.[202] It is reported that long-term cognitive functioning is maintained with improvement of QOL measures in patients with thalamic stimulation.[203] Some of patients with thalamic stimulation show improvement of depression score, semantic verbal fluency, or reaction time, with inconsistent worsening of immediate word recall, but not significantly.[202,204]

Sleep and sleep spindles do not appear to be affected by VIM stimulation, which theoretically might induce sleep because of the close proximity to thalamic reticular nuclei.[205] Another study reporting the influence of Vim stimulation on sleep found no modification of sleep quality or architecture between "on" (130 Hz, 2 to 3 V, unilateral and bilateral stimulation) and "off" states, suggesting that low-frequency stimulation of regions adjacent to the reticular nuclei do not induce sleep.[206] However, high-frequency stimulation of the STN in PD actually appeared to improve total sleep time in ten patients, likely on the basis of improved nighttime motor ability.[205]

Thalamic stimulation may lead to some adverse effects, which are commonly mild and acceptable to patients. Dysarthria is the most common, especially in bilateral procedures, and occurs in 20% of patients.[54,207] Other adverse effects include feelings of unsteadiness, limb numbness, muscle cramp, and dystonia.

Pallidal Stimulation

Pallidal stimulation, as well as pallidotomy, shows efficacy for amelioration of parkinsonian signs and symptoms, and dramatic effects on levodopa-induced dyskinesia. Bilateral pallidal stimulation appears to be more effective than unilateral pallidal stimulation for improvement of parkinsonism. It improves drug-induced dyskinesia by 80% and gait disturbance and postural disturbance by 40 to 50%.[208] Bilateral procedures improve "off" state UPDRS motor score by 40 to 50% and reduce the amount and severity of on/off fluctuations.[63,65,138] Unilateral pallidal stimulation appears to be relatively safe in either dominant or nondominant hemispheres from a cognitive perspective.[209] Bilateral pallidal stimulation does not bring a change of overall cognitive function in PD patients without dementia.[210] One case report exists showing a recurrent manic episode associated with pallidal stimulation.[211]

Bilateral pallidal stimulation has been shown to be beneficial two years following surgery, although the magnitude of benefit declined after one year.[63,138] Durif et al.[141] reported that significant improvement of dyskinesia severity and ADL was maintained at three years follow-up after surgery, but other efficacies induced by pallidal stimulation, including improvements of the "off" period overall UPDRS score (UPDRS I + II + III), UPDRS motor score, and mean daily "off" state duration were lost at three years follow-up after bilateral procedures. In addition to common hardware-related complication, adverse effects with pallidal stimulation include confusion, depression, increasing in akinesia, and induction of gait or speech disturbance.[212]

SUBTHALAMIC STIMULATION

While differences exist in study design, patient selection and evaluative method in each report, many investigators have reported the effects for cardinal parkinsonian signs and symptoms, and dyskinesia. In studies of 6 to 36 months follow-up after bilateral implantations, UPDRS motor scores in the "off" period are improved by 42 to 75%,[32,79,142–145,213,214] and UPDRS III (ADL) scores improved by 30 to 70%.[32,71,142–145,214] Beneficial effect for tremor is most striking improvement in the cardinal symptoms of PD, ranging between 55 and 90%.[32,70,142,190,214] Rigidity and bradykinesia are improved each by 52 to 72%,[32,70,142,190,214] with similar improvements being noted in axial symptoms.[142,190,214,215] Bejjani et al.[215] and Pinto et al.[216] also observed that bilateral subthalamic stimulation can improve not only limb motor function but also axial symptoms, including speech impairment.

The greatest benefit of DBS on motor function is that this form of therapy produces marked reduction (about 80%) in "off" time duration and leads to practical amelioration of motor fluctuations.[32,142,214] Reductions in levodopa-induced dyskinesia are clear; levodopa-induced dyskinesia is reduced by 65 to 90%,[142–144,190,214] and those effects last long-term after surgery. Bilateral subthalamic stimulation also provides complete or significant improvement of "off" state dystonia.[70,142]

The levodopa dose equivalent is reduced by 40 to 81% following subthalamic stimulation,[32,70,71,142–145] and some patients no longer require antiparkinsonian drugs after surgery. The decrease of levodopa is commonly recognized within a few months after surgery.

In an assessment of the impact of this procedure, using Parkinson's disease Quality of Life scale, by Lagrange et al.,[143] all subscales, including parkinsonian motor symptoms (+48%), systemic symptoms (+34%), emotional functioning (+29%), and social functioning (+63%) significantly improved with long-term follow-up after surgery.

Subthalamic stimulation also provides improvements in sleep.[71,218] These effects are probably due to increase nocturnal mobility and reduction of sleep fragmentation.

Potential adverse effects accompanying subthalamic stimulation include paresthesia, unilateral anisocoria, dysarthria/dysphonia, diplopia, eyelid opening apraxia, depressive mood, mania, delusion, and suicidal attempts.[32,70,71,142–144,213] Subthalamic stimulation may also produce ballism and chorea when the voltage is increased over a given threshold.

Pallidal Stimulation and Subthalamic Stimulation

Recent studies suggested that subthalamic stimulation might have advantage over pallidal stimulation in amelioration for overall disabilities with PD.[61,63,75,219] In prospective, nonrandomized, multicenter studies of patients with advanced PD treated with pallidal stimulation or subthalamic stimulation,[219,220] motor functions (compared with no stimulation) were greatly improved by 49% in subthalamic stimulation and 37% in pallidal stimulation. Between the baseline and six-month follow-up evaluation, the percentage of time during the day that patients had good mobility without involuntary movements increased from 27 to 74% with subthalamic stimulation and from 28 to 64% with pallidal stimulation. The daily levodopa dose equivalents were reduced by 37% in the subthalamic stimulation, with no decrease found with pallidal stimulation.

Other studies also report results that support the aforementioned study. Krack et al.[61] reported a six-month follow-up study in PD patients with young onset that UPDRS motor scores in the "off" state were improved by 71% in the subthalamic stimulation and 39% with pallidal stimulation. Rigidity and tremor showed equal improvement in both procedures, but efficacy for akinesia with subthalamic stimulation was superior to pallidal stimulation. While there was a marked improvement in levodopa-induced dyskinesia with pallidal stimulation, subthalamic stimulation led to an indirect reduction of levodopa-induced dyskinesia similar to that of pallidal stimulation because of the fact that the daily levodopa dose equivalents were reduced by 56% in the subthalamic stimulation. In a one-year follow-up study of PD patients treated with subthalamic stimulation or pallidal stimulation,[221] the motor symptoms in the "off" state were improved by 67% in the subthalamic stimulation and 54% in the pallidal stimulation, but speech and swallowing (UPDRS subscale) deteriorated significantly in patients with subthalamic stimulation. The investigators of this study suggested that significant reductions in levodopa dose equivalent with subthalamic stimulation also brought patient economic benefits.

A study of PD patients with unsatisfactory results ($n = 40$ out of 211) following pallidal or subthalamic stimulation concluded that the main causes for poor results were advanced age, abnormal MRI, and preoperative symptoms that were unresponsive to levodopa. Alternatively, misplacement of the electrode occurred in 5% of patients.

Bilateral subthalamic stimulation or pallidal stimulation affects neither memory nor executive functions three to six months after surgery.[210] In "on" stimulation, psychomotor speed and working memory are improved in patients treated with subthalamic stimulation. Except for a mild deterioration in lexical fluency, subthalamic or pallidal stimulation does not appear to affect cognitive performance.[222]

Cortical Stimulation

Only rare case reports have been reported to date, but extradural MCS may improve cardinal symptoms of PD. Canavero and Paolotti[88] reported an advanced PD patient with extradural MCS at cortical area corresponding to the left arm. The effects were seen in all limbs immediately after surgery. At three months after surgery, rigidity was improved markedly to completely. Choreiform dyskinesia, cogwheeling, and dysphagia were relieved completely. The patient became able to stand without assistance and walk. Execution of other movements also improved dramatically. Total UPDRS scores in the "off" state were improved by 46%, and levodopa equivalent dose was reduced by 80%. These benefits lasted about three years, with absence of rigidity and tremor in all limbs.[89] In another advanced PD patient extradural MCS at a cortical area corresponding to right arm (central sulcus),[89] at six months after surgery, tremor was relieved completely. Bradykinesia, gait disturbance, hypophonia, and dysar-

thria were also improved. Levodopa equivalent dose was reduced by 73%. Target-related adverse effects were not revealed in these case reports.

REGENERATIVE/GRAFTING

Fetal Mesencephalic Tissue Transplantation

Brundin et al.[223] reported that UPDRS motor scores and percent time "on" were improved each by a mean 40% and 25% at two years after transplantation. Levodopa dose equivalent was reduced by 54% compared with before surgery. PET showed a mean 61% increase of ^{18}F-fluorodopa uptake in the putamen. Hagell et al.[224] reported a retrospective, long-term follow-up (11 years after transplantation) study; maximum UPDRS motor score was improved by 39%, and levodopa dose equivalent was reduced by 36% during follow-up after surgery. These effects maintained several years after transplantation but gradually decreased thereafter. There was no significant amelioration of percent time "off" and percent time "on" dyskinesia.

Recently, the first double-blind, placebo-controlled study in advanced PD (at least seven years disease duration) was reported by Freed et al.[106] Forty patients were enrolled and stratified into younger than 60 years and older than 60 years. These patients were randomized to receive either four embryonic transplants or sham operation. At one year after surgery, there were no statistically significant differences between the mean global rating score of the transplanted and sham operated patients. The total UPDRS score in the "off" state also did not show statistically significant difference between the patient groups. However, in younger patients with transplant procedure, the UPDRS motor score in the "off" state (34% decrease) and the Schwab-England score (31% improvement) improved significantly compared with sham operated patients or older patients who received transplants. There are significant improvements in rigidity and bradykinesia in younger patients who received transplants, but no patient showed improvement of motor fluctuations and axial disability. In 19 transplanted patients who demonstrated that ^{18}F-fluorodopa uptake showed significant average improvement of 40% in the putamen, the result did not correlate with improvement of UPDRS scores except for in younger patients. At 12 months after surgery, severe dystonia and dyskinesia were seen in 9 patients. These late-onset dyskinesia were uncontrollable and did not response to levodopa.

A postmortem examination of another patient 18 months after transplantation demonstrated viability of large grafts obtained from 7 fetal mesencephalic donors.[225] Processes from these neurons had grown out of the grafts and provided extensive dopaminergic reinnervation to the striatum in a patch-matrix pattern. However, the neuronal processes extended only 5 to 7 mm away from the graft and reinnervated only 30% to 50% of the putamen (failing to reach any of the anterior putamen). Ungrafted regions of the putamen showed sparse dopaminergic innervation, and there was no evidence of any sprouting of host dopaminergic processes. This patient had also gained significant clinical improvement and enhanced ^{18}F-fluorodopa with uptake on PET scanning in association with survival of the grafts and dopaminergic reinnervation of the striatum.

Effects of fetal transplantation are quite variable in each study. There are patients with clear benefit and those with minimal or no benefit. These inconsistencies may reflect differences in patient selection, transplant location, preparation of transplant cells, and immunosuppressive treatment. The study of Freed et al.[106] also raised a new problem: off" state dyskinesia with fetal transplantation. Hagell et al.[224] suggested that "off" state dyskinesias probably did not result from excessive growth of grafted dopaminergic neurons, because the severity of dyskinesia was inversely correlated with uptake in the striatum. However, the cause of "off" state dyskinesia remains unknown, and these severe dyskinesias, which may lead to joint dislocation, have led to dampened enthusiasm for this mode of treatment.

RECOMMENDATIONS

To date, several stereotactic procedures have been established as surgical options for PD, while others await more systematic evaluation. Each procedure should be chosen on the basis of symptoms and individual background of each individual patient. Further studies will hopefully continue to clarify which surgical procedure is optimal for individual PD patients with particular clinical characteristics.

REFERENCES

1. Ahlskog, J. E. and Muenter, M. D., "Frequency of Levodopa-Related Dyskinesias and Motor Fluctuations as Estimated from the Cumulative Literature," *Mov. Disord.,* 16, 448, 2001.

2. Cooper, I. S., "Ligation of the Anterior Choroidal Artery for Involuntary Movements of Parkinsonism," *Arch Neurol.,* 75, 36, 1956.

3. Albe-Fessard, D. et al., "Derivations D'activites Spontanees Et Evoquees Dans Les Structures Cerebrales Profondes De L'homme," *Rev. Neurol.,* 109, 89, 1962.

4. Ohye, C., and Narabayashi, H., "Physiological Study of Presumed Ventralis Intermedius Neurons in the Human Thalamus," *J. Neurosurg.,* 50, 290, 1979.

5. Narabayashi, H., Maeda, T., and Yokochi, F. "Long-Term Follow-up Study of Nucleus Ventralis Intermedius

and Ventrolateralis Thalamotomy Using a Microelectrode Technique in Parkinsonism," *Appl. Neurophysiol.,* 50, 330, 1987.

6. Hariz, M. I., "From Functional Neurosurgery to "Interventional" Neurology: Survey of Publications on Thalamotomy, Pallidotomy, and Deep Brain Stimulation for Parkinson's Disease from 1966 to 2001," *Mov. Disord.,* 18, 845, 2003.

7. Elble, R. J., "Origins of Tremor," *Lancet,* 355, 1113, 2000.

8. Ohye, C. et al., "Vim Thalamotomy for the Treatment of Various Kinds of Tremor," *Appl. Neurophysiol.,* 45, 275, 1982.

9. Shahzadi, S., Tasker, R. R., and Lozano, A., "Thalamotomy for Essential and Cerebellar Tremor," *Stereotact. Funct. Neurosurg.,* 65, 11, 1995.

10. Kim, M. C. et al., "Vim Thalamotomy for Holmes' Tremor Secondary to Midbrain Tumour," *J. Neurol. Neurosurg. Psychiatry,* 73, 453, 2002.

11. Speelman, J. D. et al., "Stereotactic Neurosurgery for Tremor," *Mov. Disord.,* 17 Suppl., 3, S84, 2002.

12. Cardoso, F. et al., "Outcome after Stereotactic Thalamotomy for Dystonia and Hemiballismus," *Neurosurgery,* 36, 501, 1995.

13. Ohye, C. and Shibazaki, T., "Lesioning the Thalamus for Dyskinesia," *Stereotact Funct. Neurosurg.,* 77, 33, 2001.

14. Lenz, F. A. et al., "Statistical Prediction of the Optimal Site for Thalamotomy in Parkinsonian Tremor," *Mov. Disord.,* 10, 318, 1995.

15. Narabayashi, H., Yokochi, F., and Nakajima, Y., "Levodopa-Induced Dyskinesia and Thalamotomy," *J. Neurol. Neurosurg. Psychiatry,* 47, 831, 1984.

16. Guito, G. and Brion, S., "Traitement Des Mouvements Anormaux Par La Coagulation Pallidale. Technique Et Resulats," *Rev. Neurol.,* 89, 578, 1953.

17. Narabayashi, H. and Okuma, T., "Procaine-Oil Blocking of the Globus Pallidus for the Treatment of Rigidity and Tremor of Parkinsonism," *Psychiatr. Neurol. Jpn.,* 56, 471, 1954.

18. Talairach, J., "Actas Y Trabajos Del Vi Cong Latino-Am De Neurocir, Montevideo," *Impreuta Rosgal,* 1387, 866, 1955.

19. Cooper, I. S., and Bravo, G., "Chemopallidectomy and Chemothalamectomy," *J. Neurosurg.,* 244, 1958.

20. Cooper, I. S. and Bravo, G. J., "Implications of a 5-Years Study of 700 Basal Ganglia Operations," *Neurology,* 8, 701, 1958.

21. Svennilson, E. et al., "Treatment of Parkinsonism by Stereotactic Thermolesions in the Pallidal Region. A Clinical Evaluation of 81 Cases," *Acta Psychiatr. Neurol. Scand.,* 35, 358, 1960.

22. Laitinen, L. V., Bergenheim, A. T., and Hariz, M. I., "Leksell's Posteroventral Pallidotomy in the Treatment of Parkinson's Disease," *J. Neurosurg.,* 76, 53, 1992.

23. Laitinen, L. V., Bergenheim, A. T., and Hariz, M. I., "Ventroposterolateral Pallidotomy Can Abolish All Parkinsonian Symptoms," *Stereotact. Funct. Neurosurg.,* 58, 14, 1992.

24. Wichmann, T. and DeLong, M. R., "Functional and Pathophysiological Models of the Basal Ganglia," *Curr. Opin. Neurobiol.,* 6, 751, 1996.

25. Nakano, K. et al., "Neural Circuits and Functional Organization of the Striatum," *J. Neurol.,* 247 Suppl., 5, V1, 2000.

26. Dogali, M. et al., "Stereotactic Ventral Pallidotomy for Parkinson's Disease," *Neurology,* 45, 753, 1995.

27. Baron, M. S. et al., "Treatment of Advanced Parkinson's Disease by Posterior GPi Pallidotomy: 1-Year Results of a Pilot Study," *Ann. Neurol.,* 40, 355, 1996.

28. Uitti, R. J. et al., "Unilateral Pallidotomy for Parkinson's Disease: Comparison of Outcome in Younger Versus Elderly Patients," *Neurology,* 49, 1072, 1997.

29. Dalvi, A. et al., "Stereotactic Posteroventral Pallidotomy: Clinical Methods and Results at 1-Year Follow Up," *Mov. Disord.,* 14, 256, 1999.

30. Gross, R. E. et al., "Variability in Lesion Location after Microelectrode-Guided Pallidotomy for Parkinson's Disease: Anatomical, Physiological, and Technical Factors That Determine Lesion Distribution," *J. Neurosurg.,* 90, 468, 1999.

31. Jankovic, J. et al., "Movement and Reaction Times and Fine Coordination Tasks Following Pallidotomy," *Mov. Disord.,* 14, 57, 1999.

32. Kumar, R. et al., "Double-Blind Evaluation of Subthalamic Nucleus Deep Brain Stimulation in Advanced Parkinson's Disease," *Neurology,* 51, 850, 1998.

33. de Bie, R. M. et al., "Unilateral Pallidotomy in Parkinson's Disease: A Randomised, Single-Blind, Multicentre Trial," *Lancet,* 354, 1665, 1999.

34. de Bie, R. M. et al., "Outcome of Unilateral Pallidotomy in Advanced Parkinson's Disease: Cohort Study of 32 Patients," *J. Neurol. Neurosurg. Psychiatry,* 71, 375, 2001.

35. Bronte-Stewart, H. M. et al., "Postural Instability in Idiopathic Parkinson's Disease: The Role of Medication and Unilateral Pallidotomy," *Brain,* 125, 2100, 2002.

36. Vitek, J. L. et al., "Randomized Trial of Pallidotomy Versus Medical Therapy for Parkinson's Disease," *Ann. Neurol.,* 53, 558, 2003.

37. Wichmann, T. and DeLong, M. R., "Oscillations in the Basal Ganglia," *Nature,* 400, 621, 1999.

38. Bergman, H. et al., "The Primate Subthalamic Nucleus. Ii. Neuronal Activity in the MPTP Model of Parkinsonism," *J. Neurophysiol.,* 72, 507, 1994.

39. Bergman, H., Wichmann, T., and DeLong, M. R., "Reversal of Experimental Parkinsonism by Lesions of the Subthalamic Nucleus," *Science,* 249, 1436, 1990.

40. Aziz, T. Z. et al., "Lesion of the Subthalamic Nucleus for the Alleviation of 1-Methyl-4-Phenyl-1,2,3,6-Tetrahydropyridine (MPTP)-Induced Parkinsonism in the Primate," *Mov. Disord.,* 6, 288, 1991.

41. Wichmann, T., Bergman, H. and DeLong, M. R., "The Primate Subthalamic Nucleus, III. Changes in Motor Behavior and Neuronal Activity in the Internal Pallidum Induced by Subthalamic Inactivation in the MPTP Model of Parkinsonism," *J. Neurophysiol.,* 72, 521, 1994.

42. Guridi, J. et al., "Subthalamotomy in Parkinsonian Monkeys. Behavioural and Biochemical Analysis," *Brain,* 119 (Pt. 5), 1717, 1996.

43. Sellal, F. et al., "Contralateral Disappearance of Parkinsonian Signs after Subthalamic Hematoma," *Neurology,* 42, 255, 1992.

44. Vidakovic, A., Dragasevic, N. and Kostic, V. S., "Hemiballism: Report of 25 Cases," *J. Neurol. Neurosurg. Psychiatry,* 57, 945, 1994.

45. Alvarez, L. et al., "Dorsal Subthalamotomy for Parkinson's Disease," *Mov. Disord.,* 16, 72, 2001.

46. Patel, N. K. et al., "Unilateral Subthalamotomy in the Treatment of Parkinson's Disease," *Brain,* 126, 1136, 2003.

47. Su, P. C. et al., "Treatment of Advanced Parkinson's Disease by Subthalamotomy: One-Year Results," *Mov. Disord.,* 18, 531, 2003.

48. Hassler, R. et al., "Physiological Observation in Stereotactic Operations in Extrapyramidal Motor Disturbances," *Brain,* 83, 337, 1960.

49. Ohye, C. et al., "Ventrolateral and Subventralateral Thalamic Stimulation," *Arch Neurol.,* 11, 427, 1964.

50. Benabid, A. L. et al., "Combined (Thalamotomy and Stimulation) Stereotactic Surgery of the Vim Thalamic Nucleus for Bilateral Parkinson Disease," *Appl. Neurophysiol.,* 50, 344, 1987.

51. Schuurman, P. R. et al., "A Comparison of Continuous Thalamic Stimulation and Thalamotomy for Suppression of Severe Tremor," *N. Engl. J. Med.,* 342, 461, 2000.

52. Koller, W. C. et al., "Deep Brain Stimulation of the Vim Nucleus of the Thalamus for the Treatment of Tremor," *Neurology,* 55, S29, 2000.

53. Haberler, C. et al., "No Tissue Damage by Chronic Deep Brain Stimulation in Parkinson's Disease," *Ann. Neurol.,* 48, 372, 2000.

54. Benabid, A. L. et al., "Long-Term Suppression of Tremor by Chronic Stimulation of the Ventral Intermediate Thalamic Nucleus," *Lancet,* 337, 403, 1991.

55. Lozano, A. M. et al., "Deep Brain Stimulation for Parkinson's Disease: Disrupting the Disruption," *Lancet Neurol.,* 1, 225, 2002.

56. Ceballos-Baumann, A. O. et al., "Thalamic Stimulation for Essential Tremor Activates Motor and Deactivates Vestibular Cortex," *Neurology,* 56, 1347, 2001.

57. Perlmutter, J. S. et al., "Blood Flow Responses to Deep Brain Stimulation of Thalamus," *Neurology,* 58, 1388, 2002.

58. Merello, M. et al., "Unilateral Radiofrequency Lesion Versus Electrostimulation of Posteroventral Pallidum: A Prospective Randomized Comparison," *Mov. Disord.,* 14, 50, 1999.

59. Pahwa, R. et al., "High-Frequency Stimulation of the Globus Pallidus for the Treatment of Parkinson's Disease," *Neurology,* 49, 249, 1997.

60. Gross, C. et al., "High-Frequency Stimulation of the Globus Pallidus Internalis in Parkinson's Disease: A Study of Seven Cases," *J. Neurosurg.,* 87, 491, 1997.

61. Krack, P. et al., "Subthalamic Nucleus or Internal Pallidal Stimulation in Young Onset Parkinson's Disease," *Brain,* 121, 451, 1998.

62. Bejjani, B. P. et al., "Deep Brain Stimulation in Parkinson's Disease: Opposite Effects of Stimulation in the Pallidum," *Mov. Disord.,* 13, 969, 1998.

63. Volkmann, J. et al., "Bilateral High-Frequency Stimulation of the Internal Globus Pallidus in Advanced Parkinson's Disease," *Ann. Neurol.,* 44, 953, 1998.

64. Bejjani, B. et al., "Pallidal Stimulation for Parkinson's Disease. Two Targets?" *Neurology,* 49, 1564, 1997.

65. Krack, P. et al., "Opposite Motor Effects of Pallidal Stimulation in Parkinson's Disease," *Ann. Neurol.,* 43, 180, 1998.

66. Durif, F. et al., "Acute and Chronic Effects of Anteromedial Globus Pallidus Stimulation in Parkinson's Disease," *J. Neurol. Neurosurg. Psychiatry,* 67, 315, 1999.

67. Fukuda, M. et al., "Functional Correlates of Pallidal Stimulation for Parkinson's Disease," *Ann. Neurol.,* 49, 155, 2001.

68. Benazzouz, A. et al., "Reversal of Rigidity and Improvement in Motor Performance by Subthalamic High-Frequency Stimulation in MPTP-Treated Monkeys," *Eur. J. Neurosci.,* 5, 382, 1993.

69. Limousin, P. et al., "Effect of Parkinsonian Signs and Symptoms of Bilateral Subthalamic Nucleus Stimulation," *Lancet,* 345, 91, 1995.

70. Limousin, P. et al., "Electrical Stimulation of the Subthalamic Nucleus in Advanced Parkinson's Disease," *N. Engl. J. Med.,* 339, 1105, 1998.

71. Moro, E. et al., "Chronic Subthalamic Nucleus Stimulation Reduces Medication Requirements in Parkinson's Disease," *Neurology.* 53, 85, 1999.

72. Limousin, P. et al., "Bilateral Subthalamic Nucleus Stimulation for Severe Parkinson's Disease," *Mov. Disord.,* 10, 672, 1995.

73. Kumar, R. et al., "Comparative Effects of Unilateral and Bilateral Subthalamic Nucleus Deep Brain Stimulation," *Neurology,* 53, 561, 1999.

74. Hammerstad, J. et al., "Failure of Long-Term Pallidal Stimulation Corrected by Subthalamic Stimulation in PD," *Neurology,* 57, 566, 2001.

75. Krause, M. et al., "Deep Brain Stimulation for the Treatment of Parkinson's Disease: Subthalamic Nucleus Versus Globus Pallidus Internus," *J. Neurol. Neurosurg. Psychiatry,* 70, 464, 2001.

76. Scotto di Luzio, A. E. et al., "Which Target for DBS in Parkinson's Disease? Subthalamic Nucleus Versus Globus Pallidus Internus," *Neurol. Sci.,* 22, 87, 2001.

77. Houeto, J. L. et al., "Failure of Long-Term Pallidal Stimulation Corrected by Subthalamic Stimulation in PD," *Neurology,* 55, 728, 2000.

78. Galvez-Jimenez, N. et al., "Pallidal Stimulation in Parkinson's Disease Patients with a Prior Unilateral Pallidotomy," *Can. J. Neurol. Sci.,* 25, 300, 1998.

79. Moro, E. et al., "Bilateral Subthalamic Nucleus Stimulation in a Parkinsonian Patient with Previous Unilateral Pallidotomy and Thalamotomy," *Mov. Disord.,* 15, 753, 2000.

80. Mogilner, A. Y. et al., "Subthalamic Nucleus Stimulation in Patients with a Prior Pallidotomy," *J. Neurosurg.,* 96, 660, 2002.

81. Samii, A. et al., "Bilateral Subthalamic Nucleus Stimulation after Bilateral Pallidotomies in a Patient with Advanced Parkinson's Disease," *Parkinsonism Relat. Disord.,* 9, 159, 2003.

82. Krack, P. et al., "Ineffective Subthalamic Nucleus Stimulation in Levodopa-Resistant Postischemic Parkinsonism," *Neurology,* 54, 2182, 2000.

83. Pinter, M. M. et al., "Apomorphine Test: A Predictor for Motor Responsiveness to Deep Brain Stimulation of the Subthalamic Nucleus," *J. Neurol.,* 246, 907, 1999.

84. Welter, M. L. et al., "Clinical Predictive Factors of Subthalamic Stimulation in Parkinson's Disease," *Brain,* 125, 575, 2002.

85. Woolsey, C. N., Erickson, T. C. and Gilson, W. E., "Localization in Somatic Sensory and Motor Areas of Human Cerebral Cortex as Determined by Direct Recording of Evoked Potentials and Electrical Stimulation," *J. Neurosurg.,* 51, 476, 1979.

86. Cantello, R., Tarletti, R., and Civardi, C. "Transcranial Magnetic Stimulation and Parkinson's Disease," *Brain Res., Brain Res. Rev.,* 38, 309, 2002.

87. Canavero, S., and Bonicalzi, V. "Therapeutic Extradural Cortical Stimulation for Central and Neuropathic Pain: A Review," *Clin. J. Pain,* 18, 48, 2002.

88. Canavero, S., and Paolotti, R. "Extradural Motor Cortex Stimulation for Advanced Parkinson's Disease: Case Report," *Mov. Disord.,* 15, 169, 2000.

89. Canavero, S. et al., "Extradural Motor Cortex Stimulation for Advanced Parkinson Disease. Report of Two Cases," *J. Neurosurg.,* 97, 1208, 2002.

90. Ceballos-Baumann, A. O. et al., "A Positron Emission Tomographic Study of Subthalamic Nucleus Stimulation in Parkinson Disease: Enhanced Movement-Related Activity of Motor-Association Cortex and Decreased Motor Cortex Resting Activity," *Arch Neurol.,* 56, 997, 1999.

91. Fukuda, M. et al., "Networks Mediating the Clinical Effects of Pallidal Brain Stimulation for Parkinson's Disease: A Pet Study of Resting-State Glucose Metabolism," *Brain,* 124, 1601, 2001.

92. Bjorklund, A. and Stenevi, U., "Reconstruction of the Nigrostriatal Dopamine Pathway by Intracerebral Nigral Transplants," *Brain Res.,* 177, 555, 1979.

93. Perlow, M. J. et al., "Brain Grafts Reduce Motor Abnormalities Produced by Destruction of Nigrostriatal Dopamine System," *Science,* 204, 643, 1979.

94. Redmond, D. E. et al., "Fetal Neuronal Grafts in Monkeys Given Methylphenyltetrahydropyridine," *Lancet,* 1, 1125, 1986.

95. Bankiewicz, K. S. et al., "The Effect of Fetal Mesencephalon Implants on Primate MPTP-Induced Parkinsonism. Histochemical and Behavioral Studies," *J. Neurosurg.,* 72, 231, 1990.

96. Hirsch, E. C. et al., "Does Adrenal Graft Enhance Recovery of Dopaminergic Neurons in Parkinson's Disease?" *Ann. Neurol.,* 27, 676, 1990.

97. Forno, L. S. and Langston, J. W., "Unfavorable Outcome of Adrenal Medullary Transplant for Parkinson's Disease," *Acta Neuropathol (Berl.),* 81, 691, 1991.

98. Velasco, F. et al., "Autologous Adrenal Medullary Transplants in Advanced Parkinson's Disease with Particular Attention to the Selective Improvement in Symptoms," *Stereotact. Funct. Neurosurg.*, 57, 195, 1991.

99. Lindvall, O. et al., "Transplantation of Fetal Dopamine Neurons in Parkinson's Disease: One-Year Clinical and Neurophysiological Observations in Two Patients with Putaminal Implants," *Ann. Neurol.*, 31, 155, 1992.

100. Widner, H. et al., "Bilateral Fetal Mesencephalic Grafting in Two Patients with Parkinsonism Induced by 1-Methyl-4-Phenyl-1,2,3,6-Tetrahydropyridine (MPTP)," *N. Engl. J. Med.*, 327, 1556, 1992.

101. Hagell, P. et al., "Sequential Bilateral Transplantation in Parkinson's Disease: Effects of the Second Graft," *Brain*, 122 (Pt. 6), 1121, 1999.

102. Hagell, P. et al., "Health-Related Quality of Life Following Bilateral Intrastriatal Transplantation in Parkinson's Disease," *Mov. Disord.*, 15, 224, 2000.

103. Wenning, G. K. et al., "Short- and Long-Term Survival and Function of Unilateral Intrastriatal Dopaminergic Grafts in Parkinson's Disease," *Ann. Neurol.*, 42, 95, 1997.

104. Piccini, P. et al., "Dopamine Release from Nigral Transplants Visualized in vivo in a Parkinson's Patient," *Nat. Neurosci.*, 2, 1137, 1999.

105. Freed, C. R. et al., "Survival of Implanted Fetal Dopamine Cells and Neurologic Improvement 12 to 46 Months after Transplantation for Parkinson's Disease," *N. Engl. J. Med.*, 327, 1549, 1992.

106. Freed, C. R. et al., "Transplantation of Embryonic Dopamine Neurons for Severe Parkinson's Disease," *N. Engl. J. Med.*, 344, 710, 2001.

107. Hitchcock, E. R. et al., "A Series of Experimental Surgery for Advanced Parkinson's Disease by Foetal Mesencephalic Transplantation," *Acta Neurochir. Suppl. (Wien).* 52, 54, 1991.

108. Lindvall, O. et al., "Human Fetal Dopamine Neurons Grafted into the Striatum in Two Patients with Severe Parkinson's Disease. A Detailed Account of Methodology and a 6-Month Follow-Up," *Arch Neurol.*, 46, 615, 1989.

109. Cragg, S. J., Clarke, D. J., and Greenfield, S. A., "Real-Time Dynamics of Dopamine Released from Neuronal Transplants in Experimental Parkinson's Disease," *Exp. Neurol.*, 164, 145, 2000.

110. Clarkson, E. D., Zawada, W. M., and Freed, C. R., "GDNF Improves Survival and Reduces Apoptosis in Human Embryonic Dopaminergic Neurons in vitro," *Cell Tissue Resm.*, 289, 207, 1997.

111. Grondin, R. et al., "Chronic, Controlled GDNF Infusion Promotes Structural and Functional Recovery in Advanced Parkinsonian Monkeys," *Brain*, 125, 2191, 2002.

112. Hyman, C. et al., "Overlapping and Distinct Actions of the Neurotrophins BDNF, Nt-3, and Nt-4/5 on Cultured Dopaminergic and Gabaergic Neurons of the Ventral Mesencephalon," *J. Neurosci.*, 14, 335, 1994.

113. Kakishita, K. et al., "Implantation of Human Amniotic Epithelial Cells Prevents the Degeneration of Nigral Dopamine Neurons in Rats with 6-Hydroxydopamine Lesions," *Brain Res.*, 980, 48, 2003.

114. Dunning, J. J., White, D. J., and Wallwork, J., "The Rationale for Xenotransplantation as a Solution to the Donor Organ Shortage," *Pathol. Biol. (Paris)*, 42, 231, 1994.

115. Galpern, W. R. et al., "Xenotransplantation of Porcine Fetal Ventral Mesencephalon in a Rat Model of Parkinson's Disease: Functional Recovery and Graft Morphology," *Exp. Neurol.*, 140, 1, 1996.

116. Fink, J. S. et al., "Porcine Xenografts in Parkinson's Disease and Huntington's Disease Patients: Preliminary Results," *Cell Transplant*, 9, 273, 2000.

117. Schumacher, J. M. et al., "Transplantation of Embryonic Porcine Mesencephalic Tissue in Patients with PD," *Neurology*, 54, 1042, 2000.

118. Freeman, T. B. et al., "Placebo-Controlled Trial of Intrastriatal Transplantation of Fetal Porcine Ventral Mesencephalic Tissue (Neurocell-PD) in Subjects with Parkinson'S Disease [Abstract]," *Exp. Neurol.*, 175, 426, 2003.

119. Deacon, T. et al., "Histological Evidence of Fetal Pig Neural Cell Survival after Transplantation into a Patient with Parkinson's Disease," *Nat. Med.*, 3, 350, 1997.

120. van der Laan, L. J. et al., "Infection by Porcine Endogenous Retrovirus after Islet Xenotransplantation in Scid Mice," *Nature*, 407, 90, 2000.

121. Saint-Cyr, J. A. and Trepanier, L. L., "Neuropsychologic Assessment of Patients for Movement Disorder Surgery," *Mov. Disord.*, 15, 771, 2000.

122. Saint-Cyr, J. A. et al., "Neuropsychological Consequences of Chronic Bilateral Stimulation of the Subthalamic Nucleus in Parkinson's Disease," *Brain*, 123, 2091, 2000.

123. Scott, R. et al., "Neuropsychological, Neurological and Functional Outcome Following Pallidotomy for Parkinson's Disease. A Consecutive Series of Eight Simultaneous Bilateral and Twelve Unilateral Procedures," *Brain*, 121, 659, 1998.

124. Ghika, J. et al., "Bilateral Contemporaneous Posteroventral Pallidotomy for the Treatment of Parkinson's Disease: Neuropsychological and Neurological Side Effects. Report of Four Cases and Review of the Literature," *J. Neurosurg.*, 91, 313, 1999.

125. Favre, J. et al., "Outcome of Unilateral and Bilateral Pallidotomy for Parkinson's Disease: Patient Assessment," *Neurosurgery*, 46, 344, 2000.

126. Merello, M. et al., "Bilateral Pallidotomy for Treatment of Parkinson's Disease Induced Corticobulbar Syndrome and Psychic Akinesia Avoidable by Globus Pallidus Lesion Combined with Contralateral Stimulation," *J. Neurol, Neurosurg. Psychiatry*, 71, 611, 2001.

127. Tasker, R. R., "Thalamotomy," *Neurosurg. Clin. N. Am.*, 1, 841, 1990.

128. Obwegeser, A. A. et al., "Simultaneous Thalamic Deep Brain Stimulation and Implantable Cardioverter-Defibrillator," *Mayo Clin. Proc.*, 76, 87, 2001.

129. Jankovic, J. et al., "Late-Onset Hallervorden-Spatz Disease Presenting as Familial Parkinsonism," *Neurology*, 35, 227, 1985.

130. Fazzini, E. et al., "Stereotactic Pallidotomy for Parkinson's Disease: A Long-Term Follow-up of Unilateral Pallidotomy," *Neurology*, 48, 1273, 1997.

131. Baron, M. S. et al., "Treatment of Advanced Parkinson's Disease by Unilateral Posterior GPi Pallidotomy: 4-Year Results of a Pilot Study," *Mov. Disord.*, 15, 230, 2000.

132. Fine, J. et al., "Long-Term Follow-up of Unilateral Pallidotomy in Advanced Parkinson's Disease," *N. Engl. J. Med.*, 342, 1708, 2000.

133. Valldeoriola, F. et al., "Four Year Follow-up Study after Unilateral Pallidotomy in Advanced Parkinson's Disease," *J. Neurol.*, 249, 1671, 2002.

134. Intemann, P. M. et al., "Staged Bilateral Pallidotomy for Treatment of Parkinson Disease," *J. Neurosurg.*, 94, 437, 2001.

135. Lozano, A. M. et al., "Effect of GPi Pallidotomy on Motor Function in Parkinson's Disease," *Lancet*, 346, 1383, 1995.

136. Kishore, A. et al., "Efficacy, Stability and Predictors of Outcome of Pallidotomy for Parkinson's Disease. Six-Month Follow-up with Additional 1-Year Observations," *Brain*, 120, 729, 1997.

137. Lang, A. E. et al., "Posteroventral Medial Pallidotomy in Advanced Parkinson's Disease," *N. Engl. J. Med.*, 337, 1036, 1997.

138. Ghika, J. et al., "Efficiency and Safety of Bilateral Contemporaneous Pallidal Stimulation (Deep Brain Stimulation) in Levodopa-Responsive Patients with Parkinson's Disease with Severe Motor Fluctuations: A 2-Year Follow-up Review," *J. Neurosurg.*, 89, 713, 1998.

139. Samuel, M. et al., "A Study of Medial Pallidotomy for Parkinson's Disease: Clinical Outcome, MRI Location and Complications," *Brain*, 121, 59, 1998.

140. Schrag, A. et al., "Unilateral Pallidotomy for Parkinson's Disease: Results after More Than 1 Year," *J. Neurol. Neurosurg. Psychiatry*, 67, 511, 1999.

141. Durif, F. et al., "Long-Term Follow-up of Globus Pallidus Chronic Stimulation in Advanced Parkinson's Disease," *Mov. Disord.*, 17, 803, 2002.

142. Rodriguez-Oroz, M. C. et al., "Bilateral Deep Brain Stimulation of the Subthalamic Nucleus in Parkinson's Disease," *Neurology*. 55, S45, 2000.

143. Lagrange, E. et al., "Bilateral Subthalamic Nucleus Stimulation Improves Health-Related Quality of Life in PD," *Neurology*, 59, 1976, 2002.

144. Tavella, A. et al., "Deep Brain Stimulation of the Subthalamic Nucleus in Parkinson's Disease: Long-Term Follow-Up," *Neurol. Sci.*, 23 Suppl., 2, S111, 2002.

145. Thobois, S. et al., "Subthalamic Nucleus Stimulation in Parkinson's Disease: Clinical Evaluation of 18 Patients," *J. Neurol.*, 249, 529, 2002.

146. Andrew, J., "Surgical Treatment of Tremor." In: *Movement Disorders, Tremor*, Edited by Findley, Leslie and Capildeo, Rudy, 339, Oxford University Press, New York, N. Y., 1984.

147. Zonenshayn, M. et al., "Comparison of Anatomic and Neurophysiological Methods for Subthalamic Nucleus Targeting," *Neurosurgery*, 47, 282, 2000.

148. Laitinen, L. V., "Pallidotomy for Parkinson's Disease," *Neurosurg. Clin. N. Am.*, 6, 105, 1995.

149. Honey, C. R. and Nugent, R. A., "A Prospective Randomized Comparison of Ct and MRI Pre-Operative Localization for Pallidotomy," *Can J. Neurol. Sci.*, 27, 236, 2000.

150. Holtzheimer, P. E., 3rd, Roberts, D. W. and Darcey, T. M., "Magnetic Resonance Imaging Versus Computed Tomography for Target Localization in Functional Stereotactic Neurosurgery," *Neurosurgery*, 45, 290, 1999.

151. Meneses, M. S. et al., "Comparison of MRI-Guided and Ventriculography-Based Stereotactic Surgery for Parkinson's Disease," *Arq. Neuropsiquiatr.*, 55, 547, 1997.

152. Vitek, J. L. et al., "Microelectrode-Guided Pallidotomy: Technical Approach and Its Application in Medically Intractable Parkinson's Disease," *J. Neurosurg.*, 88, 1027, 1998.

153. Guridi, J. et al., "Stereotactic Targeting of the Globus Pallidus Internus in Parkinson's Disease: Imaging Versus Electrophysiological Mapping," *Neurosurgery*, 45, 278, 1999.

154. Alterman, R. L. et al., "Microelectrode Recording During Posteroventral Pallidotomy: Impact on Target Selection and Complications," *Neurosurgery*, 44, 315, 1999.

155. Oh, M. Y. et al., "Deep Brain Stimulator Electrodes Used for Lesioning: Proof of Principle," *Neurosurgery*, 49, 363, 2001.

156. Guridi, J. et al., "Targeting the Basal Ganglia for Deep Brain Stimulation in Parkinson's Disease," *Neurology*, 55, s21, 2000.

157. Kelly, P. J. et al., "Computer-Assisted Stereotactic Ventralis Lateralis Thalamotomy with Microelectrode Recording Control in Patients with Parkinson's Disease," *Mayo Clin. Proc.*, 62, 655, 1987.

158. Narabayashi, H., "Stereotaxic Vim Thalamotomy for Treatment of Tremor," *Eur. Neurol.*, 29, 29, 1989.

159. Anagnostou, E. et al., "Contraversive Eye Deviation During Deep Brain Stimulation of the Globus Pallidus Internus," *Neurology*, 56, 1396, 2001.

160. Iacono, R. P. et al., "The Results, Indications, and Physiology of Posteroventral Pallidotomy for Patients with Parkinson's Disease," *Neurosurgery*, 36, 1118, 1995.

161. Yokoyama, T. et al., "Visual Evoked Potential Guidance for Posteroventral Pallidotomy in Parkinson's Disease," *Neurol. Med. Chir. (Tokyo)*, 37, 257, 1997.

162. Bronstein, J. M., DeSalles, A., and DeLong, M. R., "Stereotactic Pallidotomy in the Treatment of Parkinson Disease: An Expert Opinion," *Arch Neurol.*, 56, 1064, 1999.

163. Eskandar, E. N. et al., "Stereotactic Pallidotomy Performed without Using Microelectrode Guidance in Patients with Parkinson's Disease: Surgical Technique and 2-Year Results," *J. Neurosurg.*, 92, 375, 2000.

164. Giller, C. A. et al., "Stereotactic Pallidotomy and Thalamotomy Using Individual Variations of Anatomic Landmarks for Localization," *Neurosurgery*, 42, 56, 1998.

165. Favre, J. et al., "Pallidotomy: A Survey of Current Practice in North America," *Neurosurgery*, 39, 883, 1996.

166. Kamiryo, T. and Laws, E. R., Jr., "Identification and Localization of Intracerebral Vessels by Microvascular Doppler in Stereotactic Pallidotomy and Thalamotomy: Technical Note," *Neurosurgery*, 40, 877, 1997.

167. Benabid, A. L. et al., "Chronic Electrical Stimulation of the Ventralis Intermedius Nucleus of the Thalamus as a Treatment of Movement Disorders," *J. Neurosurg.,* 84, 203, 1996.

168. Rodriguez-Oroz, M. C. et al., "The Subthalamic Nucleus in Parkinson's Disease: Somatotopic Organization and Physiological Characteristics," *Brain,* 124, 1777, 2001.

169. Fukuda, M. et al., "Intraoperative Monitoring for Functional Neurosurgery During Intravenous Anesthesia with Propofol," *No Shinkei Geka,* 25, 231, 1997.

170. Ohye, C. et al., "Gamma Thalamotomy for Parkinsonian and Other Kinds of Tremor," *Stereotact. Funct. Neurosurg.,* 66, 333, 1996.

171. Pan, L. et al., "Stereotactic Gamma Thalamotomy for the Treatment of Parkinsonism," *Stereotact. Funct. Neurosurg,* 66, 329, 1996.

172. Pollak, P. et al., "Treatment of Parkinson's Disease. New Surgical Treatment Strategies," *Eur. Neurol.,* 36, 400, 1996.

173. Niranjan, A. et al., "Functional Outcomes after Gamma Knife Thalamotomy for Essential Tremor and Ms-Related Tremor," *Neurology,* 55, 443, 2000.

174. Ahlskog, J. E., "Cerebral Transplantation for Parkinson's Disease: Current Progress and Future Prospects," *Mayo Clin. Proc.,* 68, 578, 1993.

175. Spencer, D. D. et al., "Unilateral Transplantation of Human Fetal Mesencephalic Tissue into the Caudate Nucleus of Patients with Parkinson's Disease," *N. Engl. J. Med.,* 327, 1541, 1992.

176. Henderson, B. T. et al., "Implantation of Human Fetal Ventral Mesencephalon to the Right Caudate Nucleus in Advanced Parkinson's Disease," *Arch Neurol.,* 48, 822, 1991.

177. Freed, C. R. et al., "Transplantation of Human Fetal Dopamine Cells for Parkinson's Disease. Results at 1 Year," *Arch Neurol.,* 47, 505, 1990.

178. Remy, P. et al., "Clinical Correlates of [18f]Fluorodopa Uptake in Five Grafted Parkinsonian Patients," *Ann. Neurol.,* 38, 580, 1995.

179. Freeman, T. B. et al., "Bilateral Fetal Nigral Transplantation into the Postcommissural Putamen in Parkinson's Disease," *Ann. Neurol.,* 38, 379, 1995.

180. Freed, C. R. et al., "Immunosuppressants May Not Improve Transplant Success after Human Fetal Dopamine Cell Implants for Parkinson's Disease," *Neurology,* 44, A323, 1994.

181. Hariz, M. I. and Bergenheim, A.T., "A 10-Year Follow-up Review of Patients Who Underwent Leksell's Posteroventral Pallidotomy for Parkinson Disease," *J. Neurosurg.,* 94, 552, 2001.

182. Nutt, J. G. et al., "DBS and Diathermy Interaction Induces Severe CNS Damage," *Neurology,* 56, 1384, 2001.

183. Fox, M. W., Ahlskog, J. E., and Kelly, P. J., "Stereotactic Ventrolateralis Thalamotomy for Medically Refractory Tremor in Post-Levodopa Era Parkinson's Disease Patients," *J. Neurosurg.,* 75, 723, 1991.

184. Jankovic, J. et al., "Outcome after Stereotactic Thalamotomy for Parkinsonian, Essential, and Other Types of Tremor," *Neurosurgery,* 37, 680, 1995.

185. Selby, G., "Stereotactic surgery," in *Handbook of Parkinson's Disease,* Koller, W. C., Ed., New York, Marcel Dekker, p. 421, 1987.

186. Schuurman, P. R. et al., "A Comparison of Neuropsychological Effects of Thalamotomy and Thalamic Stimulation," *Neurology,* 59, 1232, 2002.

187. Shannon, K. M. et al., "Stereotactic Pallidotomy for the Treatment of Parkinson's Disease. Efficacy and Adverse Effects at 6 Months in 26 Patients," *Neurology,* 50, 434, 1998.

188. Ondo, W. G. et al., "Assessment of Motor Function after Stereotactic Pallidotomy," *Neurology,* 50, 266, 1998.

189. Uitti, R. J. et al., "Neurodegenerative Overlap Syndrome: Clinical and Pathological Features of Parkinson's Disease, Motor Neuron Disease, and Alzheimer's Disease" *Parkinsonism and Related Disorders* (unpublished).

190. Martinez-Martin, P. et al., "Pallidotomy and Quality of Life in Patients with Parkinson's Disease: An Early Study," *Mov. Disord.,* 15, 65, 2000.

191. Alegret, M. et al., "Effects of Unilateral Posteroventral Pallidotomy on "on-Off" Cognitive Fluctuations in Parkinson's Disease," *Neuropsychologia,* 38, 628, 2000.

192. Junque, C. et al., "Cognitive and Behavioral Changes after Unilateral Posteroventral Pallidotomy: Relationship with Lesional Data from MRI," *Mov. Disord.,* 14, 780, 1999.

193. Alegret, M. et al., "Cognitive Effects of Unilateral Posteroventral Pallidotomy: A 4-Year Follow-up Study," *Mov. Disord.,* 18, 323, 2003.

194. de Bie, R. M. et al., "Morbidity and Mortality Following Pallidotomy in Parkinson's Disease: A Systematic Review," *Neurology,* 58, 1008, 2002.

195. Koller, W. et al., "High-Frequency Unilateral Thalamic Stimulation in the Treatment of Essential and Parkinsonian Tremor," *Ann. Neurol.,* 42, 292, 1997.

196. Limousin, P. et al., "Multicentre European Study of Thalamic Stimulation in Parkinsonian and Essential Tremor," *J. Neurol. Neurosurg. Psychiatry,* 66, 289, 1999.

197. Rehncrona, S. et al., "Long-Term Efficacy of Thalamic Deep Brain Stimulation for Tremor: Double-Blind Assessments," *Mov. Disord.,* 18, 163, 2003.

198. Hubble, J. P. et al., "Effects of Thalamic Deep Brain Stimulation Based on Tremor Type and Diagnosis," *Mov. Disord.,* 12, 337, 1997.

199. Montgomery, E. B. Jr., "Evaluation of Surgery for Parkinson's Disease: Report of the Therapeutics and Technology Assessment Subcommittee of the American Academy of Neurology," *Neurology,* 55, 154, 2000.

200. Defebvre, L. et al., "Effect of Thalamic Stimulation on Gait in Parkinson Disease," *Arch Neurol.,* 53, 898, 1996.

201. Caparros-Lefebvre, D. et al., "Improvement of Levodopa Induced Dyskinesias by Thalamic Deep Brain Stimulation Is Related to Slight Variation in Electrode Placement: Possible Involvement of the Centre Median and Parafascicularis Complex," *J. Neurol. Neurosurg. Psychiatry,* 67, 308, 1999.

202. Fields, J. A. and Troster, A. I., "Cognitive Outcomes after Deep Brain Stimulation for Parkinson's Disease:

A Review of Initial Studies and Recommendations for Future Research," *Brain Cogn.,* 42, 268, 2000.

203. Woods, S. P. et al., "Neuropsychological and Quality of Life Changes Following Unilateral Thalamic Deep Brain Stimulation in Parkinson's Disease: A One-Year Follow-Up," *Acta Neurochir (Wien),* 143, 1273, 2001.

204. Flament, D. et al., "Reaction Time Is Not Impaired by Stimulation of the Ventral-Intermediate Nucleus of the Thalamus (Vim) in Patients with Tremor," *Mov. Disord.,* 17, 488, 2002.

205. Arnulf, I. et al., "Effect of Low and High Frequency Thalamic Stimulation on Sleep in Patients with Parkinson's Disease and Essential Tremor," *J. Sleep Res.,* 9, 55, 2000.

206. Arnulf, I. et al., "Improvement of Sleep Architecture in PD with Subthalamic Nucleus Stimulation," *Neurology,* 55, 1732, 2000.

207. Pahwa, R. et al., "Bilateral Thalamic Stimulation for the Treatment of Essential Tremor," *Neurology,* 53, 1447, 1999.

208. Lozano, A. M., "Deep Brain Stimulation for Parkinson's Disease," *Parkinsonism Relat. Disord.,* 7, 199, 2001.

209. Vingerhoets, G. et al., "Cognitive Outcome after Unilateral Pallidal Stimulation in Parkinson's Disease," *J. Neurol. Neurosurg. Psychiatry,* 66, 297, 1999.

210. Ardouin, C. et al., "Bilateral Subthalamic or Pallidal Stimulation for Parkinson's Disease Affects Neither Memory nor Executive Functions: A Consecutive Series of 62 Patients," *Ann. Neurol.,* 46, 217, 1999.

211. Miyawaki, E. et al., "The Behavioral Complications of Pallidal Stimulation: A Case Report," *Brain Cogn.,* 42, 417, 2000.

212. Hariz, M. I., "Complications of Deep Brain Stimulation Surgery," *Mov. Disord.,* 17 Suppl., 3, S162, 2002.

213. Martinez-Martin, P. et al., "Bilateral Subthalamic Nucleus Stimulation and Quality of Life in Advanced Parkinson's Disease," *Mov. Disord.,* 17, 372, 2002.

214. Ostergaard, K., Sunde, N., and Dupont, E., "Effects of Bilateral Stimulation of the Subthalamic Nucleus in Patients with Severe Parkinson's Disease and Motor Fluctuations," *Mov. Disord.,* 17, 693, 2002.

215. Bejjani, B. P. et al., "Axial Parkinsonian Symptoms Can Be Improved: The Role of Levodopa and Bilateral Subthalamic Stimulation," *J. Neurol. Neurosurg. Psychiatry,* 68, 595, 2000.

216. Pinto, S. et al., "Bilateral Subthalamic Stimulation Effects on Oral Force Control in Parkinson's Disease," *J. Neurol.,* 250, 179, 2003.

217. Roberts-Warrior, D. et al., "Postural Control in Parkinson's Disease after Unilateral Posteroventral Pallidotomy," *Brain* (unpublished).

218. Iranzo, A. et al., "Sleep Symptoms and Polysomnographic Architecture in Advanced Parkinson's Disease after Chronic Bilateral Subthalamic Stimulation," *J. Neurol. Neurosurg. Psychiatry,* 72, 661, 2002.

219. "Deep-Brain Stimulation of the Subthalamic Nucleus or the Pars Interna of the Globus Pallidus in Parkinson's Disease," *N. Engl. J. Med.,* 345, 956, 2001.

220. Obeso, J. A., Olanow, C. W., and Lang, A., "Deep-brain stimulation in Parkinson's disease, *N. Engl. J. Med.,* 346, 452, 2002.

221. Volkmann, J. et al., "Safety and Efficacy of Pallidal or Subthalamic Nucleus Stimulation in Advanced PD," *Neurology,* 56, 548, 2001.

222. Pillon, B. et al., "Neuropsychological Changes between "Off" and "on" STN or GPi Stimulation in Parkinson's Disease," *Neurology,* 55, 411, 2000.

223. Brundin, P. et al., "Bilateral Caudate and Putamen Grafts of Embryonic Mesencephalic Tissue Treated with Lazaroids in Parkinson's Disease," *Brain,* 123 (Pt. 7), 1380, 2000.

224. Hagell, P. et al., "Dyskinesias Following Neural Transplantation in Parkinson's Disease," *Nat. Neurosci.,* 5, 627, 2002.

225. Kordower, J. H. et al., "Neuropathological Evidence of Graft Survival and Striatal Reinnervation after the Transplantation of Fetal Mesencephalic Tissue in a Patient with Parkinson's Disease," *N. Engl. J. Med.,* 332, 1118, 1995.

59 Neurotransplantation in Parkinson's Disease

Ronald F. Pfeiffer
Department of Neurology, University of Tennessee Health Science Center

CONTENTS

INTRODUCTION

The emergence of effective symptomatic treatment during the past four decades has made a tremendous impact on the lives of individuals with Parkinson's disease (PD), providing them with an almost normal life expectancy[1] and, for many, a reasonable quality of life. However, as described in other chapters in this text, the improvement afforded by the currently available pharmacological armamentarium is far from perfect. The combination of continued progression of PD and cumulative complications of medical treatment, particularly in the form of motor fluctuations and dyskinesia, eventually affects virtually all PD patients to some degree and poses significant difficulty for a distressing proportion of them.

Recognition of these therapeutic shortcomings has led a growing cadre of intrepid investigators to conceive and nurture the concept of neurotransplantation as a potential means of replenishing the supply of dopaminergic neurons lost as a fundamental part of the PD disease process. Developments in both the laboratory and the clinical arenas have generated excitement and anticipation within the scientific and medical communities. They have also attracted considerable media attention, which has sometimes led to misconceptions and unwarranted expectations by patients and family members, at times inadvertently advanced by organizations promoting fundraising for PD research.

One misconception that sometimes surfaces among patients (and even physicians) when the topic of neurotransplantation arises is that this technique will provide a "cure" for PD. In this context it is important to remember that PD involves more than simple dopamine deficiency due to nigrostriatal dopaminergic neuronal loss, although this is certainly the most prominent and striking feature of the pathological process. Other structures within the central nervous system (CNS), such as the locus ceruleus and the dorsal motor nucleus of the vagus, also sustain damage in the course of PD, and other neurotransmitter abnormalities are also present.[2–5] Furthermore, dopamine depletion itself in PD is not completely confined to the nigrostriatal system. Other dopaminergic neuronal populations are also affected by the pathological process, both within the CNS and beyond. Dopamine deficiency in PD has been identified in structures as varied as the retina[6] and the enteric nervous system in the gut.[7,8] Neurotransplantation of cells into either the substantia nigra or striatum (or both) cannot reasonably be expected to eliminate, or even affect, these components of PD.

Nevertheless, the possibility that neurotransplantation of cells into the nigrostriatal system might be able to both restore lost motor function and even provide some protection against further damage within that system is a tremendously exciting prospect that is reviewed in this chapter from the perspective of a practicing clinical neurologist.

BACKGROUND

The roots of cellular transplantation for the treatment of PD date back over 30 years to the reports of Olson and colleagues, who described the successful transplantation of chromaffin cells from the adrenal medulla into the anterior eye chamber of rats.[9] This procedure was then extended to the transplantation of fetal brain tissue to the same location.[10,11] These successes set the stage for subsequent undertakings involving transplantation of tissue to the brain proper.

In 1979, two groups of investigators reported the successful transplantation of fetal mesencephalic tissue containing dopaminergic neurons into the brains of rats that had been rendered hemiparkinsonian by unilateral 6-hydroxydopamine (6-OHDA) administration.[12,13] These initial procedures entailed the rather crude placement of solid transplanted tissue into either cortical cavities[12] or lateral ventricles,[13] and subsequent substantial refinements in experimental technique (cell preparation, graft location, tissue handling and storage) quickly followed.[14] These rodent studies and subsequent studies in nonhuman primates firmly documented the ability of grafts of fetal mesencephalic tissue to survive and reinnervate denervated striatum, with consequent improvement in motor function, and provided the proof of principle and justification for the next step to human transplantation studies.[14–16]

ADRENAL MEDULLARY TRANSPLANTATION

While the evaluation and refinement of fetal tissue transplantation was being methodically pursued, the first steps into the realm of neurotransplantation in humans with PD were taken using autologous adrenal medullary tissue. In 1985, Backlund and colleagues reported the results of the first implantation (performed in 1982) of adrenal medullary tissue into an individual with PD,[17] and in a second communication described short-term benefit in two additional patients.[18] A subsequent report of dramatic functional improvement in individuals with PD following adrenal medullary transplantation[19] created a flurry of excitement and led to a rush of additional clinical trials that were unable to duplicate or substantiate the earlier dramatic results, and also recorded significant morbidity

from the surgical procedure, which involved laparotomy for unilateral adrenalectomy in addition to the cerebral implantation.[20–24] These unfavorable reports, coupled with recognition that the grafted adrenal tissue generally did not survive, led to abandonment of adrenal transplantation as a treatment approach for PD.[25–27] Ideas regarding the potential use of adrenal medullary tissue obtained from liver transplantation donors also foundered on the recognition that the tissue was not viable.*

HUMAN FETAL MESENCEPHALIC TRANSPLANTATION

The first transplantation of fetal mesencephalic tissue into two patients with PD was undertaken by Lindvall and colleagues in 1987, in Sweden, and reported in 1989.[28] Cellular suspensions of tissue from three aborted fetuses (euphemistically termed "donors") were utilized and grafted into both caudate and putamen. Modest clinical improvement was documented, evident 4 to 6 months after grafting, but positron emission tomography (PET) with ^{18}F-fluorodopa (FD) did not demonstrate increased uptake, and the initial benefit had disappeared in one of the two patients by approximately one year.[29] Subsequent adjustments, improvements, and refinements in technical factors (amount of implanted tissue, tissue handling, tissue transport, surgical instrumentation) resulted in improved results in subsequent procedures performed on patients with PD and on individuals with MPTP-induced parkinsonism.[29–32]

The encouraging results from these Swedish trials were subsequently supplemented by reports from a number of additional centers.[33–43] Although it is difficult to arrive at an exact figure, it appears that, by 2001, approximately 300 individuals had received fetal tissue transplantation for treatment of PD,[44] and a more recent estimate asserts that more than 400 have undergone the procedure.[45] The vast majority of studies have been completed under nonblinded conditions with widely divergent trial designs, involving significant differences in variables such as patient selection, cell preparation, surgical implantation sites, surgical implantation techniques, immunosuppressive treatment, and assessment parameters[14–16,27,29,46–49] In light of this variability, it is, perhaps, not surprising that reported results from these clinical trials have also shown considerable variability.

Despite the difficulties imposed by the disparate data, assessment of available information from this collection of "open-label" clinical trials has led some investigators to draw a number of conclusions regarding the effectiveness of fetal mesencephalic transplantation for the treatment of PD. In a 1999 review, Lindvall provided some general assertions, stating that "grafted embryonic neu-

---

* *Source:* R. McComb and R. F. Pfeiffer, unpublished observations.

rons can exhibit short- and long-term survival in idiopathic PD" and that "grafts can give rise to long-term symptomatic relief of therapeutic value to patients with Parkinson's disease, but the symptomatic relief in most patients is incomplete with respect to both degree and pattern of functional recovery."[50] The evidence for the first assertion was based on the presence of increased [18]FD uptake on PET compared to preoperative values in several studies[42,51–53] and histopathologic evidence of surviving graft tissue upon autopsy studies of deceased individuals.[54–56]

Varying levels of investigator agreement have been achieved with regard to certain transplantation parameters. Some suggestions and evidence supporting the use of tissue from spontaneous abortions have been published,[57,58] but tissue from elective abortions has been utilized because of concerns regarding viability, potential defects, and availability of spontaneously aborted fetuses.[27] Optimal age of the fetus from whom tissue is recovered has been found to be between 5.5 and 8.0 weeks postconception for cell suspension grafts, and between 6.5 and 9.0 weeks for solid grafts.[48] The number of fetuses utilized in transplantation procedures has varied considerably. From animal studies and clinical trials it is known that only 5 to 20% of implanted dopamine neurons survive.[16,48,50] Extrapolating from this, investigators have suggested that, to achieve clinically meaningful improvement, it is necessary to implant tissue from three to five fetuses per side,[50,59–61] with some investigators recommending as many as six per side.[48] Differences of opinion regarding issues such as tissue storage and preparation and the necessity of immunosuppressive treatment posttransplantation have not been settled and have been the source of discussion and debate in the context of two recently completed double-blind, placebo surgery-controlled trials (discussed below).

The cumulative experience from nonblinded open clinical trials has demonstrated an overall improvement in motor scores on the Unified Parkinson's Disease Rating Scale in the range of 6 to 40% following transplantation.[14,50] Functional improvement typically does not begin until several months following the transplant procedure, and then may gradually grow over a period of up to 2 to 3 years, probably reflecting maturation and integration of the implanted neurons into the host striatum.[15,46] Placement of grafted tissue into the posterior putamen has produced results superior to caudate tissue placement; bilateral grafting is more effective than unilateral.[15,27,46,62] Graft survival and clinical improvement have been documented for periods over ten years in PD patients.[15,46,63,64]

The variable results reported in nonblinded clinical protocols have led to the performance of two double-blind, placebo surgery-controlled trials in the U.S.A.[65,66] In the first of the two to be completed, Freed and colleagues[65] studied 40 patients with advanced PD; 20

received bilateral putaminal transplantation of fetal mesencephalic tissue, and 20 underwent sham surgery in which burr holes were placed but the dura was not penetrated. Immunosuppressive treatment was not employed in this study. After one-year follow-up, there was no difference between the two groups in a subjective global rating of change in disease severity, which was the primary study outcome measure. However, some improvement in objective measures of motor function (UPDRS motor "off" score improvement of 34%, Schwab/England "off" score improvement of 31%) was evident in the portion of participants under age 60 in the transplantation group.[15,65] Improvement was not detectable in the over-age-60 group. Increased FD uptake was present in 85%[17,20] of transplanted patients at one year, consistent with graft survival. After one year, patients in the sham-surgery group were given the opportunity to undergo transplantation, and 14 chose to do so. Clinical improvement in this group was identical to that seen in the original transplant group.[15]

The second double-blind protocol was carried out by Olanow and colleagues.[66] In this study 34 patients were enrolled and randomized into one of three groups: 11 received sham surgery in which partial burr holes were placed but the inner table of the skull was not penetrated; 11 received putaminal placement of tissue from 1 fetus per side; 12 received putaminal placement of tissue from four fetuses per side. Participants were then followed for 24 months. In this study, the primary outcome measure utilized was the change in UPDRS motor "off" score. No significant differences were present between the three groups in this measure at 24 months. Unlike the earlier double-blind trial, no treatment effect was evident in patients younger than age 60, but patients with less severe disease (baseline UPDRS motor score of 49 or less) who received tissue from four fetuses per side demonstrated statistically significant improvement compared with those with less severe disease in the sham-surgery group

Both of the double-blind clinical trials were marked by the unexpected development of dyskinesia in some individuals that persisted despite reduction or even complete cessation of antiparkinson medication. Such dyskinesia, labeled by some as "runaway dyskinesia,"[67] developed in 15% of transplant recipients in the earlier study[65] and in 56% in the later trial.[66] Prompted by these developments, other investigators retrospectively reviewed their data and identified 14 individuals in whom off-phase dyskinesias also had developed.[68] Several hypotheses have been advanced to account for the development of these off-phase dyskinesias. It was originally suggested, based on PET data, that these dyskinesias might be the consequence of unbalanced regional increases in striatal dopaminergic function due to overgrowth of grafted dopaminergic neurons.[65,69] Other investigators, however, have proposed that culturing of the fetal tissue prior to transplantation might be in some way responsible[68] or that the off-phase dyskinesias represent a variant of diphasic dyskinesia and reflect partial graft survival.[66]

The two double-blind trials have also provoked some additional controversy. The decision by Freed and colleagues[65] not to employ immunosuppression has been criticized, as has their implantation technique, which utilized a strand, or "noodle," of tissue rather than cell suspension, and which entailed keeping the fetal tissue in culture for up to one month prior to implantation.[70,71] Olanow and colleagues employed immunosuppression with cyclosporine for six months following grafting and actually noted some functional deterioration that seemed to coincide with discontinuation of the immunosuppressive therapy.[66] Both criticism[72–76] and support[77] have also been voiced regarding the performance of sham surgery as a placebo procedure.

Significant complications from fetal tissue transplantation procedures are unusual but have been reported. A patient who received both intrastriatal implantation and intraventricular infusion of fetal tissue subsequently developed left lateral and fourth ventricular obstruction due to proliferation of intraventricular fetal tissue (containing ectodermal and mesenchymal, but not neural, elements) and died suddenly.[78] Another patient died as a result of a herniation syndrome secondary to formation of a large putaminal cyst, possibly containing choroid plexus tissue arising from the fetal graft.[79] One death has also been reported as a consequence of intraoperative surgical complications.[62] Other transplantation-related complications have included intracerebral hemorrhage, subdural hematoma, transient confusion, and enhanced psychiatric problems.[49] Brain abscess and partial motor seizures have also been described.[27]

The current status of transplantation of fetal mesencephalic tissue for treatment of PD is uncertain, clouded not only by the inability of the double-blinded studies to document clear-cut improvement, but also by the inconsistent benefit achieved by patients in the nonblinded studies, inconsistencies that are evident both between centers and among groups of patients within centers themselves.[15]

Two additional factors also stand in the way of routine clinical employment of fetal mesencephalic tissue transplantation. Ethical and religious objections to the use of fetal tissue are, and will continue to be, deeply and sincerely held by a significant segment of the population. Even if these moral reservations could be surmounted (which is unlikely), it has become clear that logistical limitations also are present in that the amount of fetal tissue available from elective abortions will never be sufficient to meet the demand, given the amount of fetal tissue currently required for successful grafting.[14–16,27,47,50,59,61,62,80–82]

In response to these barriers, a very active and innovative search for alternatives to direct fetal tissue transplantation has developed. Alternatives that have been proposed and studied include the use of other tissue sources, amplification of fetal tissue with trophic factors, use of embryonic stem cells or stem cells from other sources, gene therapy, and xenotransplantation. Stem cell strategies and gene therapy are addressed in other chapters of this text and therefore are not covered in detail here.

Adequate coverage of the many approaches that are being investigated with regard to manipulating and altering fetal tissue to increase viability and survival (such as cografting, trophic factor supplementation, antioxidant treatment, and many others) falls beyond the scope of this chapter. Instead, the remaining paragraphs focus on alternative tissue sources that are being investigated for potential treatment of PD.

OTHER TISSUE SOURCES FOR TRANSPLANTATION

CAROTID BODY GLOMUS CELLS

Carotid body glomus cells are physiologic arterial oxygen sensors that release dopamine in response to hypoxia.[83] This capability has prompted investigation of these cells as potential tissue for autologous transplantation into the striatum in persons with PD. Initial studies employing intrastriatal grafting of aggregates of these cells in rats produced improvement in motor function.[84] The cells retained the ability to secrete dopamine and, with time, dopaminergic fibers were found to reinnervate surrounding striatum. Subsequent studies on monkeys with MPTP-induced parkinsonism demonstrated similar findings.[85] Carotid body cells also express high levels of glial cell line-derived neurotrophic factor (GDNF), and it has recently been proposed that the motor improvement seen following intrastriatal grafting of carotid body cell aggregates into hemiparkinsonian rats is due to release of GDNF and its subsequent trophic action on surviving striatal neurons, rather than dopamine release by the implanted carotid body cells.[86] These animal studies paved the way for a pilot clinical trial in humans in which six patients underwent bilateral autotransplantation of carotid body cell aggregates into striatum.[87] Improvement was noted in five of six patients and was maximal at six months post-transplantation, with reduction in the off-motor score of the UPDRS ranging between 26 and 74%. At one year, improvement was maintained in three individuals, with the other two patients demonstrating a diminution, but not total loss, of benefit. The fact that both GDNF and dopamine are apparently secreted by the implanted carotid body cells, and the fact that the tissue is autologous, obviating the need for immunosuppression, are very appealing characteristics of this approach. However, the apparent reduction in benefit over time is troublesome and raises questions about graft survival. Much additional information regarding this procedure is needed, and further study certainly seems warranted.

CERVICAL SYMPATHETIC GANGLION CELLS

The knowledge that cells in the superior cervical sympathetic ganglion produce catecholamines, including dopamine, has prompted evaluation of these cells as potential tissue for CNS grafting in patients with PD. Successful autologous transplantation of the superior cervical ganglion (SCG) into the caudate nuclei of monkeys was reported in 1990; histofluorescence studies demonstrated survival of the graft.[88] That same year, apparently with little additional animal experimentation, SCG was grafted into the caudate nuclei of three patients with PD, and improved motor function was described.[89] By 1997, Itakura and colleagues had performed SCG autotransplantation on 35 patients with PD and noted that approximately half demonstrated some improvement in function, primarily in bradykinesia and gait dysfunction, but not in tremor and rigidity.[90] In a more recent report from the same group of investigators, four patients who had undergone unilateral intrastriatal autografting of SCV were studied one year post-surgery.[91] Motor scores were not improved, but "off" time had diminished. The authors speculated that this might be due to the ability of the grafted neurons to convert exogenous levodopa into dopamine and store it. The relatively sparse experimental data available regarding SCG transplantation does not appear to be as encouraging as other treatment approaches.

HUMAN RETINAL PIGMENT EPITHELIAL CELLS

Human retinal pigment epithelial (hRPE) cells are support cells that are found in the posterior layer of the retina, adjacent to the choroid and the neural elements of the retina.[80] They secrete dopamine or dopamine precursors and, thus, have caught the attention of investigators as a potential source of readily available tissue for transplantation in patients with PD. These cells also may have trophic functions and have been reported to produce platelet-derived growth factor, epidermal growth factor, and vascular endothelial growth factor.[80] Another very appealing feature, from the standpoint of potential utility in transplantation, is that they can be grown and expanded in tissue culture and survive prolonged storage.[80] Subramanian and colleagues studied the response of 6-OHDA-induced hemiparkinsonian rats to intrastriatal placement of hRPE cells attached to gelatin microcarriers and noted sustained reduction in apomorphine-induced rotations in animals receiving hRPE cells attached to the microcarriers, but not in animals who received hRPE cells alone.[92] Subsequent pathological examination demonstrated only a minimal host response to the implanted xenographic

tissue. Additional studies in three monkeys with MPTP-induced hemiparkinsonism also demonstrated improvement in motor function and minimal inflammatory response, and FD PET imaging in one of the monkeys revealed increased uptake at the transplantation site, suggesting that the implanted hRPE cells were actively taking up and metabolizing the FD.[80] In a follow-up study, the behavioral examiners were blinded as to whether the monkey subjects had undergone implantation of low dose hRPE, high dose hRPE, microcarrier implantation alone, or sham surgery.[80] Once again, hRPE-treated animals displayed improved motor function compared to those who had undergone sham surgery or had only received the microcarriers. With this background, human trials were initiated and carried out in an open-label fashion on six individuals with PD.[93,94] No immunosuppression was utilized. The UPDRS motor "off" scores were improved by 33% at six months ($n = 6$), 42% at nine months ($n = 6$), and 48% at 12 months ($n = 3$). No significant complications were encountered, and no "off" period dyskinesias were seen. The implantation of hRPE cells attached to gelatin microcarriers (Spheramine) thus displays promise as a potential treatment for PD. It is not a procedure that will reinnervate striatum, but rather will function as an implanted dopamine-producing "pump." Further studies are clearly needed before any firm assessment regarding the effectiveness of this technique can be made.

The potential use of encapsulated cell technology has also been promoted by other investigators,[95–97] and encapsulated PC12 cells, capable of secreting dopamine, have been transplanted successfully into monkeys.[98]

XENOTRANSPLANTATION

Humans have used animals for food and worn animal hides and other animal parts for clothing and adornment dating back to the mists and shadows of prehistory. The employment of animals or animal parts, ingested or externally applied, for medical treatment also has a long history. The idea of actually incorporating animal parts into humans as part of medical treatment, however, is relatively new and strange. To some, it is also a source of fear and consternation. The use of porcine skin grafts for cardiac valve replacement no longer raises eyebrows, but the idea of possibly using animal organs for transplantation purposes, such as hepatic and renal, conjures up more ambivalent feelings.[99] In recent years, the brain has become another frontier in the investigation of xenotransplantation, with human trials of porcine cell transplantation for the treatment of PD already undertaken.

PORCINE FETAL MESENCEPHALIC TISSUE

The pig has a number of characteristics that favor it as a potential source of tissue for xenografting procedures.[59,80]

The size of a pig brain is similar to that of a human, and there are similarities between their major histocompatibility complex (MHC) antigens. Pigs are also relatively easy to breed and have large litters, thus assuring a ready supply of tissue. Nevertheless, the potential for tissue rejection, the risk of zoonotic infection, and the issue of ultimate functional capacity are major questions confronting those investigating the prospect of using porcine tissue xenografting for the treatment of PD.[59] An abundance of animal studies have been carried out in evaluating the potential of porcine fetal mesencephalic tissue for transplantation procedures in the setting of PD.

When porcine fetal mesencephalic tissue is transplanted into rats, the tissue undergoes rejection over a period of approximately five weeks, due to both humoral and cellular mechanisms, the latter being more prominent.[100] Humans, however, possess high levels of naturally occurring antibodies to the glycoprotein α-galactosyl epitope, which is present on porcine neural cells and would likely incite a hyperacute rejection reaction if porcine neural tissue were to be transplanted into the human brain without immunosuppressive measures being taken.[15,59] It has been demonstrated in rats that cyclosporine alone does not fully protect against rejection;[101] presumably, this would also be so in humans. In an effort to deal with these rejection issues, genetically modified pigs that lack the α-galactosyl epitope have been developed, and more effective immunosuppressive therapies have been sought.[15] Nevertheless, rejection issues, particularly over the long term, remain a major barrier to the effective use of porcine xenografting in humans.

A second important issue with regard to xenografting of porcine tissue is the potential for introducing a zoonotic infection into humans. In particular, there is concern about the possibility that porcine endogenous retrovirus (PERV) might be transmitted to human cell lines following xenotransplantation, although this has not yet been documented.[102] This possibility might eventually be circumvented by cloning PERV-free pigs.

The ultimate functional capacity of porcine xenografts has been difficult to ascertain with confidence because of problems with tissue rejection, but improved function in lesioned animals has been documented, and microscopic studies have demonstrated some integration of porcine xenografts, with axonal extension and synapse formation.[15,59,103,104]

The jump to human testing of porcine xenografts has already been made. In an unblinded Phase I study, 12 individuals with PD underwent unilateral striatal implantation of porcine fetal mesencephalic tissue.[105] Six of the subjects received immunosuppressive treatment with cyclosporine, while the other six were treated with a monoclonal antibody directed against MHC class I. No serious operative complications were encountered. At one year post-transplantation, total UPDRS "off" scores had improved approximately 19%. The motor UPDRS "off" score had not improved, although a score assessing postural instability and gait did demonstrate improvement in the "off" state. Scores obtained in the "on" state did not reveal any improvement. Functional improvement did not correlate with changes on FD PET. One patient died from acute pulmonary emboli 7.5 months following transplantation. Neuropathological analysis with immunohistochemistry to tyrosine hydroxylase detected the presence of approximately 650 dopaminergic neurons in the grafts.

In a subsequent Phase II trial, 18 individuals with PD were enrolled and, in a double-blind fashion, assigned to undergo either porcine striatal xenografting or sham surgery.[107] At 18 months post-surgery, the 10 patients in the treatment group experienced a 24.6% improvement in the UPDRS, but the 8 subjects in the sham surgery group displayed a 21.6% improvement. There was no significant difference between the two groups due to the robust placebo response in the sham surgery group. Thus, there currently is no firm evidence of clinically meaningful benefit to transplantation of porcine fetal mesencephalic xenotransplantation in PD. The PET data and the neuropathologic examination of the one individual who died in the Phase I study would seem to indicate that insufficient survival of grafted tissue is responsible.

Work is progressing looking for methods to overcome the apparent immunologic barriers to porcine xenotransplantation.[102] Porcine expanded neural precursor cells may be less immunogenic and more advantageous than primary cells.[108,109] Combined treatment employing both caspase inhibition and complement inhibition might also enhance xenograft survival.[110] Other studies have suggested that testicular-derived Sertoli cells might provide protection to xenografts.[111] However, much needs to be accomplished in the laboratory before porcine xenografting returns to the clinical arena.

TISSUE FROM OTHER SOURCES

Even more exotic xenotransplantation studies have been described in isolated reports. Microencapsulated bovine chromaffin cells have been successfully transplanted into hemiparkinsonian rats, with a stable reduction in apomorphine-induced turning.[112] Finally, a mixture of human fetal brain and Notch Drosophila melanogaster neural embryonic tissues has been transplanted into the ventrolateral thalamic nucleus of patients with PD with subsequent sustained improvement in tremor.[113]

SUMMARY

The field of neurotransplantation is a rapidly advancing, constantly changing, enormously complex frontier of neuroscience that holds tremendous promise for improving our ability to effectively treat individuals with neurodegenerative disorders such as PD. While it is important to convey this excitement to patients with PD, it is equally important that patients recognize that, at the present time, neurotransplantation techniques and approaches are still within the realm of research rather than routine clinical care.

REFERENCES

1. Herlofson, K. et al., Mortality and Parkinson disease. A community based study, *Neurology,* 62, 937, 2004.

2. Zarow, C. et al., Neuronal loss is greater in the locus coeruleus than nucleus basalis and substantia nigra in Alzheimer and Parkinson diseases, *Arch. Neurol.,* 60, 337, 2003.

3. Halliday, G. M. et al., Neuropathology of immunohistochemically identified brainstem neurons in Parkinson's disease, *Ann. Neurol.,* 27, 373, 1990.

4. Kerenyi, L. et al., Positron emission tomography of striatal serotonin transporters in Parkinson disease, *Arch. Neurol.,* 60, 1223, 2003.

5. Srinivasan, J. and Schmidt, W. J., Potentiation of parkinsonian symptoms by depletion of locus coeruleus noradrenaline in 6-hydroxydopamine-induced partial degeneration of substantia nigra in rats, *Eur. J. Neurosci.,* 17, 2586, 2003.

6. Harnois, C. and Di Paolo, T., Decreased dopamine in the retinas of patients with Parkinson's disease, *Invest. Ophthalmol. Vis. Sci.,* 31, 2473, 1990.

7. Singaram, C. et al., Dopaminergic defect of enteric nervous system in Parkinson's disease patients with chronic constipation, *Lancet,* 346, 861, 1995.

8. Wakabayashi, K. et al., Parkinson's disease: an immunohistochemical study of Lewy-body containing neurons in the enteric nervous system, *Acta Neuropathol.,* 79, 581, 1990.

9. Olson, L. and Malmfors, T., Growth characteristics of adrenergic nerves in the adult rat. Fluorescence histochemical and ^{3}H-noradrenaline uptake studies using tissue transplantation into the anterior chamber of the eye, *Acta Physiol. Scand.,* 348, S1, 1970.

10. Olson, L. and Seiger, Å., Brain tissue transplanted to the anterior chamber of the eye. I. Fluorescence histochemistry of immature catecholamine and 5-hydroxytriptamine neurons innervating the iris, *Z. Zellforsch.,* 195, 175, 1972.

11. Olson, L., Seiger, Å. and Strömberg, I., Intraocular transplantation in rodents: a detailed account of the procedure and examples of its use in neurobiology with special reference to brain tissue grafting, *Adv. Cell. Neurobiol.,* 4, 407, 1983.

12. Björklund, A. and Stenevi, U., Reconstruction of the nigrostriatal dopamine pathway by intracerebral transplants, *Brain Res.,* 177, 555, 1979.

13. Perlow, M. J. et al., Brain grafts reduce motor abnormalities produced by destruction of nigrostriatal dopamine system, *Science,* 204, 643, 1979.

14. Dunnett, S. B., Björklund, A. and Lindvall, O., Cell therapy in Parkinson's disease—stop or go? *Nat. Rev. Neurosci.,* 2, 365, 2001.

15. Björklund, A. et al., Neural transplantation for the treatment of Parkinson's disease, *Lancet Neurol.,* 2, 437, 2003.

16. Clarkson, E. D. and Freed, C. R., Development of fetal neural transplantation as a treatment for Parkinson's disease, *Life Sci.,* 65, 2427, 1999.

17. Backlund, E. O. et al., Transplantation of adrenal medulla to the striatum in parkinsonism: first trials, *J. Neurosurg.,* 62, 169, 1985.

18. Lindvall, O. et al., Transplantation in Parkinson's disease: two cases of adrenal medulla grafting to the putamen, *Ann. Neurol.,* 22, 457, 1987.

19. Madrazo, I. et al., Open microsurgical autograft of adrenal medulla to the right caudate nucleus in two patients with intractable Parkinson's disease, *N. Engl. J. Med.,* 316, 831, 1987.

20. Goetz, C. G. et al., Multicenter study of autologous adrenal medullary transplantation to the corpus striatum in patients with advanced Parkinson's disease, *N. Engl. J. Med.,* 320, 337, 1989.

21. Olanow, C. W. et al., Autologous transplantation of adrenal medulla in Parkinson's disease: 18-month results, *Arch. Neurol.,* 47, 1286, 1990.

22. Allen, G. S. et al., Adrenal medullary transplantation to the caudate nucleus in Parkinson's disease. Initial clinical results in 18 patients, *Arch. Neurol.,* 46, 487, 1989.

23. Jankovic, J. et al., Clinical, biochemical, and neuropathologic findings following transplantation of adrenal medulla to the caudate nucleus for treatment of Parkinson's disease, *Neurology,* 39, 1227, 1989.

24. Goetz, C. G. et al., United Parkinson Foundation neurotransplantation registry on adrenal medullary transplants: presurgical and 1- and 2-year follow-up, *Neurology,* 41, 1719, 1991.

25. Quinn, N. P., The clinical application of cell grafting techniques in patients with Parkinson's disease, *Prog. Brain Res.,* 82, 619, 1990.

26. Ahlskog, J. E., Cerebral transplantation for Parkinson's disease: current progress and future prospects, *Mayo Clin. Proc.,* 68, 578, 1993.

27. Olanow, C. W., Freeman, T. B., and Kordower, J. H., Neural transplantation as a therapy for Parkinson's disease, *Adv. Neurol.,* 74, 249, 1997.

28. Lindvall, O. et al., Human fetal dopamine neurons grafted into the striatum in two patients with severe Parkinson's disease: a detailed account of methodology and a 6 months follow-up, *Arch. Neurol.,* 46, 615, 1989.

29. Widner, H., The case for neural tissue transplantation as a treatment for Parkinson's disease, *Adv. Neurol.,* 80, 641, 1999.

30. Lindvall, O. et al., Grafts of fetal dopamine neurons survive and improve motor function in Parkinson's disease, *Science,* 247, 574, 1990.

31. Lindvall, O. et al., Transplantation of fetal dopamine neurons in Parkinson's disease: one-year clinical and neurophysiological observations in two patients with putaminal implants, *Ann. Neurol.,* 31, 155, 1992.

32. Widner, H. et al., Bilateral fetal mesencephalic grafting in two patients with severe parkinsonism induced by 1-methyl-4-phenyl-1,2,3,6-tetrahydropyridine (MPTP), *N. Engl. J. Med.*, 327, 1556, 1992.

33. Huang, S. et al., Transplant operation of human fetal substantia nigra tissue to caudate nucleus in Parkinson's disease: first clinical trials, *Chin. J. Neurosurg.*, 5, 210, 1989.

34. Freed, C. R. et al., Transplantation of human fetal dopamine cells for Parkinson's disease: results at one year, *Arch. Neurol.*, 47, 505, 1990.

35. Henderson, B. T. H. et al., Implantation of human ventral mesencephalon to the right caudate nucleus in advanced Parkinson's disease, *Arch. Neurol.*, 48, 822, 1991.

36. Subrt, O. et al., Grafting of fetal dopamine neurons in Parkinson's disease: the Czech experience with severe akinetic patients, *Acta Neurochir. Suppl. (Wien)*, 52, 51, 1991.

37. Molina, H. et al., Transplantation of human fetal mesencephalic tissue in caudate nucleus as a treatment for Parkinson's disease: the Cuban experience, in *Intracerebral Transplantation in Movement Disorders*, Lindvall, O., Björklund, A., and Widner, H. Eds., Elsevier Science Publishing, Amsterdam, 99, 1991.

38. Spencer, D. D. et al., Unilateral transplantation of human fetal mesencephalic tissue into the caudate nucleus of patients with Parkinson's disease, *N. Engl. J. Med.*, 327, 1541, 1992.

39. Iacono, R. P. et al., Bilateral fetal grafts for Parkinson's disease: 22 months' results, *Stereotact. Funct. Neurosurg.*, 58, 84, 1992.

40. Peschanski, M. et al., Bilateral motor improvement and alteration of L-dopa effect in two patients with Parkinson's disease following intrastriatal transplantation of foetal ventral mesencephalon, *Brain*, 117, 487, 1994.

41. Zabek, M. et al., A long term follow-up of foetal dopaminergic neuronal transplantation into the brains of three parkinsonian patients, *Res. Neurol. Neurosci.*, 6, 97, 1994.

42. Freeman, T. B. et al., Bilateral fetal nigral transplantation as a treatment for Parkinson's disease, *Ann. Neurol.*, 38, 379, 1995.

43. Jacques, D. B. et al., Outcomes and complications of fetal tissue transplantation in Parkinson's disease, *Stereotact. Funct. Neurosurg.*, 72, 219, 1999.

44. Clarkson, E. D., Fetal tissue transplantation for patients with Parkinson's disease: a database of published results, *Drugs Aging*, 18, 773, 2001.

45. Linazasoro, G., Stem cells: solution to the problem of transplants in Parkinson's disease? *Neurologia*, 18, 74, 2003.

46. Isacson, O., The production and use of cells as therapeutic agents in neurodegenerative diseases, *Lancet Neurol.*, 2, 417, 2003.

47. Isacson, O., Bjorklund, L. M., and Schumacher, J. M., Toward full restoration of synaptic and terminal function of the dopaminergic system in Parkinson's disease by stem cells, *Ann. Neurol.*, 53(Suppl. 3), S135, 2003.

48. Olanow, C. W., Kordower, J. H., and Freeman, T. B., Fetal nigral transplantation as a therapy for Parkinson's disease, *Trends Neurosci.*, 19, 102, 1996.

49. No Author Listed. Surgical treatment for Parkinson's disease: neural transplantation, *Mov. Disord.*, 17(Suppl. 4), S148, 2002.

50. Lindvall, O., Neural transplantation: can we improve the symptomatic relief? *Adv. Neurol.*, 80, 635, 1999.

51. Sawle, G. et al., Transplantation of fetal dopamine neurons in Parkinson's disease: positron emission tomography [^{18}F]-6-L-fluorodopa studies in two patients with putaminal implants, *Ann. Neurol.*, 31, 166, 1992.

52. Remy, P. et al., Clinical correlates of [^{18}F] fluorodopa uptake in five grafted parkinsonian patients, *Ann. Neurol.*, 38, 580, 1995.

53. Lindvall, O. et al., Evidence for long-term survival and function of dopaminergic grafts in progressive Parkinson's disease, *Ann. Neurol.*, 35, 172, 1994.

54. Kordower, J. H. et al., Neuropathological evidence of graft survival and striatal reinnervation after the transplantation of fetal mesencephalic tissue in a patient with Parkinson's disease, *N. Engl. J. Med.*, 332, 1118, 1995.

55. Kordower, J. H. et al., Functional fetal nigral grafts in a patient with Parkinson's disease: chemoanatomic, ultrastructural, and metabolic studies, *J. Comp. Neurol.*, 370, 203, 1996.

56. Kordower, J. H. et al., Fetal grafting for Parkinson's disease: expression of immune markers in two patients with functional fetal nigral implants, *Cell Transplant.*, 6, 213, 1997.

57. Branch, D. W. et al., Suitability of fetal tissue from spontaneous abortions and from ectopic pregnancies for transplantation, *J.A.M.A.*, 273, 64, 1995.

58. Kondoh, T. et al., Functional effects of transplanted human fetal ventral mesencephalic brain tissue from spontaneous abortions into a rodent model of Parkinson's disease, *Transplant. Proc.*, 26, 335, 1994.

59. Barker, R A., Repairing the brain in Parkinson's disease: where next? *Mov. Disord.*, 17, 233, 2002.

60. Cochen, V. et al., Transplantation in Parkinson's disease: PET changes correlate with the amount of grafted tissue, *Mov. Disord.*, 18, 928, 2003.

61. Hagell, P. and Brundin, P., Cell survival and clinical outcome following intrastriatal transplantation in Parkinson disease, *J. Neuropathol. Exp. Neurol.*, 60, 741, 2001.

62. Freeman, T. B. et al., Neural transplantation in Parkinson's disease, *Adv. Neurol.*, 86, 435, 2001.

63. Piccini, P. et al., Dopamine release from nigral transplants visualized *in vivo* in a Parkinson's patient, *Nat. Neurosci.*, 2, 1137, 1999.

64. Barker, R. A. and Dunnett, S. B., Functional integration of neural grafts in Parkinson's disease, *Nat. Neurosci.*, 2, 1047, 1999.

65. Freed, C. R. et al., Transplantation of embryonic dopamine neurons for severe Parkinson's disease, *N. Engl. J. Med.*, 344, 710, 2001.

66. Olanow, C. W. et al., A double-blind controlled trial of bilateral fetal nigral transplantation in Parkinson's disease, *Ann. Neurol.*, 54, 403, 2003.

67. Piccini, P., Dyskinesias after transplantation in Parkinson's disease, *Lancet Neurol.*, 1, 472, 2002.

68. Hagell, P. et al., Dyskinesias following neural transplantation in Parkinson's disease, *Nat. Neurosci.*, 5, 627, 2002.

69. Ma, Y. et al., Dyskinesia after fetal cell transplantation for parkinsonism: a PET study, *Ann. Neurol.*, 52, 628, 2002.

70. Isacson, O., Bjorklund, L., and Pernaute, R. S., Parkinson's disease: interpretations of transplantation study are erroneous, *Nat. Neurosci.*, 4, 553, 2001.

71. Brundin, P. et al., Transplanted dopamine neurons: more or less? *Nat. Med.* 7, 512, 2001.

72. Macklin, R., The ethical problems with sham surgery in clinical research, *N. Engl. J. Med.*, 341, 992, 1999.

73. Dekkers, W. and Boer, G., Sham neurosurgery in patients with Parkinson's disease: is it morally acceptable? *J. Med. Ethics*, 27, 151, 2001.

74. Redmond, D. E. Jr., Sladek, J. R., and Spencer D. D., Transplantation of embryonic dopamine neurons for severe Parkinson's disease, *N. Engl. J. Med.*, 345, 146, 2001.

75. Boer, G. J. and Widner, H., Clinical neurotransplantation: core assessment protocol rather than sham surgery as control, *Brain Res. Bull.*, 58, 547, 2002.

76. London, A. J. and Kadane, J. B., Placebos that harm: sham surgery controls in clinical trials, *Stat. Methods Med. Res.*, 11, 413, 2002.

77. Albin, R. L., Sham surgery controls: intracerebral grafting of fetal tissue for Parkinson's disease and proposed criteria for use of sham surgery controls, *J. Med. Ethics*, 28, 322, 2002.

78. Folkerth, R. D. and Durso, R., Survival and proliferation of nonneural tissues, with obstruction of cerebral ventricles, in a parkinsonian patient treated with fetal allografts, *Neurology*, 46, 1219, 1996.

79. Mamelak, A. N. et al., Fatal cyst formation after fetal mesencephalic allograft transplant for Parkinson's disease, *J. Neurosurg.*, 89, 592, 1998.

80. Subramanian, T. Cell transplantation for the treatment of Parkinson's disease, *Semin. Neurol.*, 21, 103, 2001.

81. Freed, C. R., Will embryonic stem cells be a useful source of dopamine neurons for transplant into patients with Parkinson's disease? *Proc. Natl. Acad. Sci. U.S.A.*, 99, 1755, 2002.

82. Borlongan, C. V. and Sanberg, P. R., Neural transplantation for the treatment of Parkinson's disease, *Drug Discov. Today*, 7, 674, 2002.

83. Toledo-Aral, J. J. et al., Dopaminergic cells of the carotid body: physiological significance and possible therapeutic applications in Parkinson's disease, *Brain Res. Bull.*, 57, 847, 2002.

84. Espejo, E. F. et al., Cellular and functional recovery of Parkinsonian rats after intrastriatal transplantation of carotid body cell aggregates, *Neuron*, 20, 197, 1998.

85. Luquin, M. R. et al., Recovery of chronic parkinsonian monkeys by autotransplants of carotid body cell aggregates into putamen, *Neuron*, 22, 743, 1999.

86. Toledo-Aral, J. J. et al., Trophic restoration of the nigrostriatal dopaminergic pathway in long-term carotid body-grafted parkinsonian rats, *J. Neurol.*, 23, 141, 2003.

87. Arjona, V. et al., Autotransplantation of human carotid body cell aggregates for treatment of Parkinson's disease, *Neurosurgery*, 53, 321, 2003.

88. Nakai, M., Itakura, T., and Komai, N., Transplantation of autologous superior cervical ganglion into the brain of parkinsonian monkeys, *Stereotact. Funct. Neurosurg.*, 54–55, 337, 1990.

89. Horvath, M. et al., Autotransplantation of superior cervical ganglion to the caudate nucleus in three patients with Parkinson's disease (preliminary report), *Neurosurg. Rev.*, 13, 119, 1990.

90. Itakura, T. et al., Transplantation of autologous sympathetic ganglion into the brain with Parkinson's disease. Long-term follow-up of 35 cases, *Stereotact. Funct. Neurosurg.*, 69, 112, 1997.

91. Nakao, N. et al., Enhancement of the response to levodopa therapy after intrastriatal transplantation of autologous sympathetic neurons in patients with Parkinson's disease, *J. Neurosurg.*, 95, 275, 2001.

92. Subramanian, T. et al., Striatal xenotransplantation of human retinal pigment epithelial cells attached to microcarriers in hemiparkinsonian rats ameliorates behavioral deficits without provoking an immune response, *Cell. Transplant.*, 11, 207, 2002.

93. Watts, R. L. et al., Stereotaxic intrastriatal implantation of human retinal pigment epithelial (hRPE) cells attached to gelatin microcarriers: a potential new cell therapy for Parkinson's disease, *J. Neural Transm. Suppl.*, 65, 215, 2003.

94. Bakay, R. A. et al., Implantation of Spheramine in advanced Parkinson's disease (PD), *Front. Biosci.*, 9, 592, 2004.

95. Lanza, R. P., Hayes, J. L. and Chick, W. L., Encapsulated cell technology, *Nat. Biotechnol.*, 14, 1107, 1996.

96. Emerich, D. F. et al., A novel approach to neural transplantation in Parkinson's disease: use of polymer-encapsulated cell therapy, *Neurosci. Biobehav. Rev.*, 16, 437, 1992.

97. Chang, T. M. and Prakash, S., Therapeutic uses of microencapsulated genetically engineered cells, *Mol. Med. Today*, 4, 221, 1998.

98. Yoshida, H. et al., Stereotactic transplantation of a dopamine-producing capsule into the striatum for treatment of Parkinson disease: a preclinical primate study, *J. Neurosurg.*, 98, 874, 2003.

99. Lundin, S. and Widner, H., Attitudes to xenotransplantation: interviews with patients suffering from Parkinson's disease focusing on the conception of risk, *Transplant. Proc.*, 32, 1175, 2000.

100. Barker, R. A. et al., A role for complement in the rejection of porcine ventral mesencephalic xenografts in a rat model of Parkinson's disease, *J. Neurosci.*, 20, 3415, 2000.

101. Pakzaban, P. and Isacson, O., Neural xenotransplantation: reconstruction of neuronal circuitry across species barriers, *Neuroscience*, 62, 989, 1994.

102. Brevig, T., Holgersson, J., and Widner, H., Xenotransplantation for CNS repair: immunological barriers and strategies to overcome them, *Trends Neurosci.*, 23, 337, 2000.

103. Galpern, W. R. et al., Xenotransplantation of porcine fetal ventral mesencephalon in a rat model of Parkinson's disease: functional recovery and graft morphology, *Exp. Neurol.*, 140, 1, 1996.

104. LeBlanc, C. J. et al., Morris water maze analysis of 192-IgG-saporin-lesioned rats and porcine cholinergic transplants to the hippocampus, *Cell. Transplant.*, 8, 131, 1999.

105. Schumacher, J. M. et al., Transplantation of embryonic porcine mesencephalic tissue in patients with PD, *Neurology*, 54, 1042, 2000.

106. Deacon, T. et al., Histological evidence of fetal pig neural cell survival after transplantation into a patient with Parkinson's disease, *Nat. Med.*, 3, 350, 1997.

107. Freeman, T. B. et al., A prospective, randomized, double-blind, surgical placebo-controlled trial of intrastriatal transplantation of fetal porcine ventral mesencephalic tissue (Neurocell-PD) in subjects with Parkinson's disease, *Exp. Neurol.*, 175, 426, 2003

108. Armstrong, R. J. E. et al., Porcine neural xenografts in the immunocompetent rat: immune response following grafting of expanded neural precursor cells, *Neuroscience*, 106, 201, 2001.

109. Armstrong, R. J. E. et al., The potential for circuit reconstruction by expanded neural precursor cells explored through porcine xenografts in a rat model of Parkinson's disease, *Exp. Neurol.*, 175, 98, 2002.

110. Cicchetti, F. et al., Combined inhibition of apoptosis and complement improves neural graft survival of embryonic rat and porcine mesencephalon in the rat brain, *Exp. Neurol.*, 177, 376, 2002.

111. Emerich, D. F., Hemendinger, R., and Halberstadt, C. R., The testicular-derived Sertoli cell: cellular immunoscience to enable transplantation, *Cell. Transplant.*, 12, 335, 2003.

112. Xue, Y. et al., Microencapsulated bovine chromaffin cell xenografts into hemiparkinsonian rats: a drug-induced rotational behavior and histological change analysis, *Artif. Organs*, 25, 131, 2001.

113. Saveliev, S. V. et al., Chimeric brain: theoretical and clinical aspects, *Int. J. Dev. Biol.*, 41, 801, 1997.

60 The Role of Physical Therapy in Management of Parkinson's Disease

Rose Wichmann
Struthers Parkinson's Center

CONTENTS

INTRODUCTION

A 2001 research synthesis published in the *Archives of Physical Medicine and Rehabilitation* supported the hypothesis that Parkinson's disease (PD) patients benefit from physical therapy (PT) added to their standard regimen of medication.[1] An integral member of an interdisciplinary team approach to management of Parkinson's, the physical therapist plays an important role throughout the continuum of care from time of diagnosis to advanced stages of the disease. Parkinson's disease compromises patient mobility in a variety of ways, and physical therapy is helpful for patients experiencing difficulties with bed mobility, transfers, gait, or balance loss/falling. Referrals to physical therapy also are instrumental in the development of an individualized exercise program, posture awareness, and pain control. Patient/family education provided by physical therapists offers greater understanding of issues relating to safety, stress reduction, movement enhancement strategies, and compensation techniques.

PATIENT/CLIENT MANAGEMENT

The physical therapy patient management process begins with examination. A history is usually gathered through patient/client interview, providing a process for obtaining information as well as an opportunity for initial assessment of patient's communication and cognitive skills. It should be noted that facial masking and reduction of automatic movement seen in Parkinson's patients might affect the normal expressions, gestures, and body postures typically seen during an examination. Many health professionals frequently use these observations during the evaluation process and may misinterpret their absence as depression, confusion, or disinterest.

Asking patients to "describe an average day" provides great insight into workplace issues, leisure interests, activities of daily living (ADL) tasks, and activity levels. This description also allows the therapist to begin observing patterns in the patient's routine related to medication timing, motor fluctuations, and/or fatigue. It is important to clarify patient use of medical terminology and description of symptoms to prevent misunderstanding. Family member observations and input are valuable in providing further insight into a patient's daily function but should be "balanced" so patient input is not lost in the process. The patient will be asked to provide a listing of all medications, including herbs and other supplements. Physical therapy professionals need to familiarize themselves with potential side effects of common Parkinson's medications that impact patient mobility, including dyskinesia, dystonia, edema, orthostatic hypotension, confusion, and hallucinations.

A relevant systems review is included within the comprehensive examination. It should not be assumed that all reported symptoms are related to the diagnosis of Parkinson's disease. Many patients have other medical conditions that should be considered when developing appropriate interventions and treatment plans. The systems review also provides an opportunity for patient education, as those living with Parkinson's may mistakenly attribute all medical symptoms they experience to this chronic illness.

A variety of standardized scales are used by neurologists and researchers for assessment of Parkinson's disease symptoms, including the Hoehn and Yahr Rating Scale and the Unified Parkinson's Disease Rating Scale (UPDRS). Familiarity with and use of elements from these numerical scales provide a common language among health professionals when rating Parkinson's primary and secondary symptoms. A patient-scored questionnaire using the UPDRS ADL scale can provide a good overview of patient problems and concerns during the examination process. While it is helpful for physical therapists to be familiar with and use components of these scales for rating PD symptoms, they are usually not sensitive for use as a measure of functional impairments, or to determine progress based on treatment interventions. Specific tests and measures should be selected to establish a functional baseline, assess the level of impairment, and help accurately assess progress toward anticipated goals and expected outcomes during the course of treatment.

A variety of validated test measures can be used during physical therapy examination of individuals with Parkinson's disease. Functional tests, including the Five Times Sit to Stand,[2] Physical Performance Test,[3] Timed Up and Go,[4] Parkinson Activity Scale,[5] Berg Balance Test,[6] gait velocity and activity tolerance (i.e., two- or six-minute walk[7]) testing, may be used during the evaluation process. Goniometric measurement, sitting/standing blood pressure screening, sensory/proprioceptive testing, posture grids, digital photography, pain scales, vestibular screening, and other assessment tools also may be used, based on reported problems and concerns. A physical therapist will make clinical judgments based on data gathered during the examination to establish an appropriate physical therapy diagnosis, prognosis, and interventions.

Physical therapists must be aware of current research evidence when choosing specific tests and measures for the Parkinson's population. Some functional tests commonly used in physical therapy examination have been shown to be inappropriate for evaluating clients with PD. For example, the Tinetti gait assessment was studied and found not to be sensitive for detecting changes in persons with Parkinson's.[8] Functional reach testing also was found to be an insensitive instrument for determining fall risk within the Parkinson's population.[9]

There are a variety of mental and emotional factors to be considered when performing an examination of a patient with Parkinson's disease. Depression frequently is seen as a secondary symptom. When depressed, patients often describe sleep disturbances, appetite changes, or decreased activity levels, or complain of generalized weakness and fatigue. If untreated, depression may have a significant negative impact on physical therapy treatment. Patients experiencing depression should be referred to their physicians or other members of the interdisciplinary team as appropriate.

It also is important to consider mental/emotional factors that impact fall risk. A 2002 research study published in the *Journal of the American Geriatric Society* showed that individuals who fall develop a risk for fear of falling, which adds spiraling risk for additional falling, greater fear, and functional decline.[11] Physical therapists may wish to use balance "confidence scales" such as the Modified Falls Efficacy Scale[10] or the Activities Specific Balance Confidence Scale[11] to assess the fear of falling in their patients with Parkinson's disease.

Cognitive testing is needed to assess each individual's ability to learn and retain movement enhancement strategies prior to developing a treatment approach. Input from testing done by other members of the interdisciplinary team (i.e., occupational therapy, speech pathology, and/or neuropsychology) will be helpful in determining PT interventions. Noted impairments in executive function, short-term memory, orientation, or other cognitive skills necessitate more education and involvement of the family members/care partners during treatment. Visual perceptual changes alter the patient's ability to perform safe and efficient transfers, ambulation, or ADL tasks. Decreased contrast sensitivity causes distortion of the environment and adds to transfer difficulties, problems with stair climbing, and home safety concerns, especially in low-light conditions. Figure ground distortion creates a "clutter" of visual stimuli, causing increased difficulty and confusion during mobility tasks.

Psychosocial factors should also be considered during physical therapy evaluation of the PD patient. A physical therapist must be aware of the patient's living environment, current services, and care partner availability to develop the most appropriate plan. Factors including cultural diversity, financial resources, geography and access to services, personal beliefs, and education also may influence individual response to recommended interventions.

All data gathered throughout the physical therapy examination are evaluated and organized into clusters, syndromes, or categories to determine appropriate interventions and treatment. This process may include referrals to other members of the interdisciplinary team if identified problems are deemed outside the scope of physical therapy practice. A prognosis of anticipated physical therapy goals and expected outcomes is established. Interventions should be based on the impact Parkinson's symptoms create on patient function, and treatment goals should be written to accurately measure improvement of these daily functional tasks.

PHYSICAL THERAPY IN EARLY-STAGE PARKINSON'S DISEASE

The widely accepted Parkinson's treatment algorithm notes the importance of exercise in early stages of Parkinson's. Unfortunately, many PD patients do not receive a referral to physical therapy at this time. Patients may read PD literature that refers to the importance of movement and exercise yet remain unsure of what exercises would be most beneficial. Early referral to physical therapy offers an opportunity for the development of an individualized exercise program, as well as an introduction to effective use of an interdisciplinary team throughout the continuum of care. Patients may also have questions regarding posture changes, stress reduction, workplace issues, leisure interests, or pain at this stage of Parkinson's.

A multidimensional exercise routine is most effective at addressing deficits in balance, mobility, and fall risk.[13] Progressive changes in flexibility are often not observed by the patient until significant limitation is present. Exercises should include a foundation of stretching activities due to muscle rigidity and its accompanying potential for flexibility loss. Inclusion of exercises promoting spinal flexibility appears to be particularly needed in early stages of Parkinson's.[14] Even when made aware of the importance of regular stretching, patients frequently require instruction in proper stretching technique, including benefits of sustained stretches for 20 to 30 sec, maintaining deep breathing, and relaxing throughout the exercise activity. Additional flexibility activities, including tai chi or yoga, also may be recommended. Movement enhancement strategies with attention to making motions more mindful and complete can further enhance exercise performance.[15]

Many PD patients report feelings of muscle weakness, even when conventional muscle testing appears to show no significant deficits. It appears that bradykinetic movement combined with muscle rigidity may contribute to the PD patient's perceptions of lessened muscle strength. A research study published in the *American Journal of Physical Medicine and Rehabilitation* found no significant strength deficits between individuals with Parkinson's and normal subjects, with the exception of abdominal strength.[16] This seems to suggest a need for particular attention to strengthening core muscles of stability when designing the exercise program. Incorporating functional movements (i.e., practicing sit to stand from varying chair seat heights) into a strengthening exercise routine also proves beneficial and ensures regular follow-through.

Regular conditioning exercises are incorporated in a comprehensive routine to maintain activity tolerance and cardiovascular fitness. It has been found that PD patients benefit from aerobic exercise just as much as those without PD.[17] A variety of conditioning exercise activities or fitness equipment can be used, depending on availability and patient preferences. Special consideration should be given to the safety of certain types of exercise equipment (i.e., electric treadmills may move too quickly for safe operation in those with significant bradykinesia.) Mood and subjective reports of well-being also were shown to improve by participation in sports activities in early- to medium-stage PD patients.[18]

Family members may be the first to comment about observations of the patient's changing posture or gait pattern. A patient may first become aware of postural change when viewing recent photographs, or notice frequent tripping or "clumsiness" when walking on uneven terrain. Developing self-awareness of posture and gait is often advantageous to patients with early-stage Parkinson's disease. Performing frequent "posture checks" throughout the day promotes postural awareness and good alignment. Use of lumbar and/or cervical pillows help to improve sitting and sleeping postures. Exercises promoting axial extension, pelvic mobility, and back/abdominal strength also are helpful in posture training. Gait training is usually minimal in early-stage PD, though patients can benefit from instruction in self-monitoring skills and attentional strategies to increase velocity and heel strike in certain situations.

It is estimated that approximately 30% of individuals diagnosed with Parkinson's remain active in the work force.[19] Work site evaluation may prove helpful for determining areas of difficulty and recommendations for improved body mechanics, task performance, and safety. Physical or occupational therapists are appropriate referral sources for PD patients who have questions or concerns related to their work environment.

Instruction in appropriate relaxation activities should be included in the design of a comprehensive program. Reduced activity tolerance may necessitate dividing work, ADL, or leisure tasks into several components to allow adequate rest. The importance of stress reduction often must be emphasized, as most patients find that their PD symptoms exacerbate when they are under physical or emotional stress. Patients may benefit from tai chi, meditation, progressive relaxation, guided imagery, or other modes of relaxation based on individual preferences and interests. Referrals to other members of the interdisciplinary team or community resources may be included when designing appropriate relaxation activities.

Patient education is a primary component in all stages of Parkinson's disease. Physical therapy plays an important role in early-stage Parkinson's by answering patient and family member questions and providing information on treatment options beyond medications. The physical therapist also plays a supportive role in enhancing patient mobility by encouraging an active lifestyle that maximizes quality of life. Therapists should be knowledgeable about education and resource support services within their local communities, and offer this information to patients and families who are facing the sometimes overwhelming task of coping with a new diagnosis of Parkinson's disease.

PHYSICAL THERAPY IN MODERATE-STAGE PARKINSON'S DISEASE

As Parkinson's disease progresses, patients begin to experience greater difficulties with physical mobility skills. Clients may note greater difficulty with attempts to get out of bed, rise from a chair, or get into the car. Gait pattern changes become more pronounced, with increased shuffling, difficulty turning, or occurrence of festination and/or freezing. Many patients experience significant balance problems and frequently begin to report episodes of falling. Motor fluctuations may develop, further complicating mobility skills as patients experience variations in function throughout the day. Family care partners often need to become more involved in providing assistance with routine activities. Physical therapy can be helpful with all these mobility challenges and offer interventions and instruction to more effectively cope with changes in these daily tasks.

Many PD patients report problems turning in bed at night as one of the first difficulties noted with general mobility. Axial rigidity, with lack of dissociation between head and upper and lower trunk, combine to produce limited trunk rotation and difficulty rolling. Physical therapy can work with patients in breaking down the rolling sequence into a series of steps, focusing with conscious attention to detail during each individual movement. Mentally rehearsing each movement prior to performance often is beneficial. Compensatory strategies include wearing "slippery" fabric nightclothes or using a satin-based drawsheet through the middle of the bed. Installing a side rail also may be included within the physical therapy intervention plan. Clients may note difficulty with getting in or out of bed. Bed mobility instruction in proper body positioning and movement enhancement strategies improves transfer ease and allows the patient to retain maximized independence.

Chair, car, and toilet transfers become more effortful for a variety of reasons. Bradykinesia interferes with a patient's ability to generate upward momentum. Lack of flexibility or decreased muscle strength may result from a sedentary lifestyle and decreased activity levels. Most often, patients are observed attempting to stand using poor body mechanics, failing to place their center of gravity over their base of support. This results in reduced trunk flexion and improper foot placement when attempting to rise. Some clients also exhibit significant difficulty with body alignment as they attempt to sit down. Physical therapy offers transfer training, instruction in compensatory strategies, adaptive equipment, and appropriate exercise to improve these transfers.

Gait changes in Parkinson's disease include a narrowed base of support, decreases in step size, heel strike and arm swing, en bloc turns, and reduced gait velocity. Excessive dyskinesia or dystonic posturing also can negatively affect the gait pattern. It has been shown that patients experience increased gait difficulties when attempting to multitask, with added cognitive or motor tasks shown to be equally demanding.[20] Motor fluctuations cause some individuals with Parkinson's to experience only periodic deficits or to demonstrate different types of difficulties when on and off. Retropulsion, festination, and freezing are frequently seen at this stage of Parkinson's, requiring gait training through physical therapy to most effectively cope with these deficits.

There are a variety of techniques that can be used when working with clients who experience freezing. A physical therapist will provide assessment to help recognize individual freezing "triggers," determine locations where freezing is most likely to occur, and offer recommendations for appropriate environmental modifications. Other compensatory strategies, including visual, auditory, tactile, or kinesthetic cueing, are often helpful. The use of a laser pointer, step-over wand, or inverted walking stick is frequently used during gait training. Physical therapists may choose to work collaboratively with music therapists to provide gait training using rhythmic auditory stimulation, which has been shown to improve gait velocity, symmetry, stride length, and step cadence.[21]

Many patients with Parkinson's disease find it necessary to use a gait assistive device to maximize safety when ambulating. A referral to physical therapy is essential to receive recommendations regarding the most appropriate device and ensure proper sizing. Patients often are unaware of all assistive device options and make choices that do not offer maximized safety and support. As a general rule, four-post walkers and quad canes do not work well for those with Parkinson's. These devices interrupt the flow of movement and require divided attention/multitasking, which contributes to difficulties with balance stability. Single-end canes or walking sticks seem to work best for patients requiring unilateral support. Many patients benefit from the use of specialty walkers available with swivel casters, hand brakes, and a bench seat. These walkers offer more options for walker speed and control, improving turning stability. Some have an added "slow down" feature, especially helpful for those experiencing episodes of festination. The bench seat also is helpful for those with reduced activity tolerance or sudden "off" periods, although patients should receive instruction in safe transfers on and off these seats. Patients also require gait training to use these devices safely in a variety of situations, including varied floor surfaces and outdoor terrain. Recommendations for proper footwear and good foot health aid in improving gait stability. Physical therapists are also a resource for information about locations for equipment purchase and medical reimbursement.

Postural instability is the primary symptom of Parkinson's that is least responsive to available medications. Muscle rigidity and decreased flexibility combine with a narrowed base of support and decreased postural righting reflexes to produce balance changes. Many patients report frequent episodes of significant balance loss or falling. Retropulsion results in involuntary backward balance loss, worsened with attempts to reach overhead, open a door, or carry objects up against the body. Festination usually causes forward balance loss, as a patient experiences uncontrolled increases in gait velocity while step size declines. Physical therapy referrals for instruction in gait training, counterbalancing techniques, and movement enhancement strategies are helpful for patients experiencing these difficulties.

An emphasis on falls prevention is an integral part of the physical therapy intervention plan. A 2002 survey of 1061 patients with Parkinson's disease showed 55.4% reporting at least one fall within the past year, with 65.3% of fallers reporting injury, and 32.9% of fallers reporting a fracture.[21] Since a large number of patients with PD are diagnosed later in life, many have secondary complications such as osteoporosis, degenerative joint disease, or vision changes that increase fall injury risk. An assessment of the home environment may be needed to reduce barriers and remove potential hazards. Recommendations for furniture placement, visual cues, and installation of adaptive equipment also should be included. Some high-risk patients may benefit from use of protective kneepads or "hip saver" clothing to reduce potential for injury.

In the event of a fracture, illness, or other injury, a physical therapy referral should be initiated as soon as the patient is medically stable. It is typical to see Parkinson's symptoms exacerbate during times when a patient is under significant physical or emotional stress. Prolonged bedrest or inactivity significantly impairs mobility and complicates the rehabilitation process. Early physical therapy intervention allows timely mobilization and reduces risk of complications. It should be noted that the rehabilitation process might be slowed significantly for a patient with Parkinson's. Expected outcomes should reflect this slower progress, and interventions should be designed accordingly.

As PD symptoms progress, daily activities should be assessed to ensure safety and a balance of activity and rest in the daily routine. Modification of the exercise program usually is needed as balance declines. Patients may need to perform more of their exercise routine in seated or supine positions for maximized safety. Leisure interests also may need to be modified. Raised gardening beds, use of a stationary versus regular bike, and moving daily

walking from outdoor terrain to an indoor location are all examples of modifications that allow clients to remain safe while performing activities. Many patients enjoy involvement in community exercise groups for regular follow-through and support. Physicians and other members of the interdisciplinary team need to be alert to changes in patient mobility during each encounter, providing referrals as appropriate. An early introduction to physical therapy also allows patients to recognize changes or need for modification and seek referral for additional services as needed.

Care partner instruction should not be overlooked in this stage of Parkinson's disease. Many family care partners begin to offer assistance with transfers, exercises, or other daily cares without instruction in proper technique or body mechanics. The likelihood of care partner injury can be significant without proper instruction. If a patient is falling, instruction in safe techniques to get up from the ground is essential to minimize further injury risk. Instruction in providing clear, concise cues with a reduction of excessive verbal stimuli is particularly helpful and reduces frustration for both patient and care partner.

PHYSICAL THERAPY IN ADVANCED-STAGE PARKINSON'S DISEASE:

In advanced stages of Parkinson's disease, medications often become less effective for symptom control. Medication side effects also may become more prominent. Patients experience increasing immobility and require assistance with almost all activities of daily living. Although each individual experience with Parkinson's disease is unique, many clients develop cognitive changes that further complicate independence and safety. Care partners play a larger role in care of the patient and frequently need instruction, support, and respite care to cope with these complicated problems. Although some physicians and other members of the health care team may feel that an individual lacks "rehab potential," there is still a role for physical therapy in comprehensive management of individuals with advanced-stage Parkinson's disease. Physical therapy referrals are beneficial in areas of posture, positioning, pain control, and care partner assisted exercise. Continued instruction in care partner body mechanics is needed as patient care needs change and increase. The physical therapist also plays an important role in teaching other health care team members proper transfer techniques, compensatory strategies, and increased awareness of common challenges experienced by those with Parkinson's disease.

If gait and balance changes prevent safe ambulation, many clients begin to use a wheelchair for general mobility. A referral to physical therapy is helpful in determining proper wheelchair size, type, and features required. A lack of automatic movement and excessive muscle rigidity increase the risk of integumentary changes, and a proper cushion must be chosen to prevent skin breakdown. Seating systems offering reclining backrests, lateral trunk support, or elevating leg rests are often needed to achieve proper positioning. Consideration to chair size and width must be given to ensure that the wheelchair works within the home environment and that a care partner is able to lift the chair into the car trunk if needed. A physical therapist may work collaboratively with an occupational therapist in making these recommendations or decisions.

Instruction in proper positioning throughout the day and night is needed. Positioning for eating is especially important due to a significant risk of aspiration in many clients with advanced-stage Parkinson's. Some clients may use a recliner or other chair when not in a wheelchair. As posture declines, many patients begin to use excessive pillows under the head, shoulders, and knees while in bed, which may further promote flexed posture. Instruction in proper bed positioning offers maximized comfort and good body alignment. As patient immobility increases, family care partners or other assistants often need to return to physical therapy for additional instruction in proper body mechanics, use of a transfer belt, and safe methods to perform pivot transfers. Assessment for and instruction in the use of mechanical lifts may be needed for those with severe rigidity or balance impairments.

Clients with advanced-stage Parkinson's disease usually require assistance with performance of a daily exercise program. Emphasis on assisted range of motion and stretching helps to maximize flexibility and improve patient comfort. It is the author's experience that many family caregivers report receiving extensive, time-consuming home programs of assisted exercise for their family member with PD. These complicated routines often are impractical for families already overwhelmed with a variety of other caregiving tasks. Physical therapists should be mindful of these caregiver responsibilities and design simplified programs of exercise that can be more easily incorporated into the daily routine. Examples include adding a few extra arm and leg motions during assisted dressing and bathing, or performing assisted standing at a grab bar/counter to increase lower extremity weight bearing and ability to retain transfer skills. Involvement in adapted recreational tasks or household chores also may be successfully used as part of a movement and exercise program.

Pain may be reported more frequently in advanced-stage Parkinson's. This is often due to excessive rigidity, inability to change position independently, excessive dystonia/dyskinesia, or injuries sustained in falling. Crying, wandering, and agitation may be seen as pain-related behaviors in those with significant cognitive changes. Caregiver instruction in the use of superficial heat or cold,

repositioning, or massage is often helpful. Additional physical therapy, musculoskeletal evaluation, and interventions focusing on pain control also may be appropriate.

As care needs increase, many patients are faced with the transition to a new living environment. A move to assisted living, a skilled nursing facility, or other new environment can be extremely stressful for both patients and their care partners. Unfortunately, not all health care providers within these facilities are familiar with the symptoms or challenges of living with Parkinson's. Physical therapy evaluation of the new living environment is helpful to maximize patient safety and comfort. A physical therapist can help provide staff education for assisting patients experiencing fluctuating mobility, freezing, or other mobility challenges related to Parkinson's disease. Instruction in Parkinson's symptoms, as well as specific transfers and movement-enhancement strategies, aids staff understanding and improves patient care.

PHYSICAL THERAPY WITHIN A HOLISTIC MODEL OF CARE

Patients and family members often choose to incorporate complementary therapies into their comprehensive treatment plan. These specialty programs provide opportunities for a variety of creative physical therapy interventions. A physical therapist may successfully collaborate with a variety of professionals and/or programs, designing interventions to achieve desired outcomes while offering maximized quality of life.

For example, many clients enjoy gardening as a leisure interest. Therapeutic horticulture programs offer unique opportunities for physical therapy interventions. These include movement and exercise through gardening, transfer training opportunities to practice getting up and down from the ground, or instruction in balance safety strategies for walking on uneven terrain. Horticulture programs also offer methods for relaxation/stress management and exercise through performance of familiar tasks for clients with dementia.

Using animal-assisted therapy programs within physical therapy also can be successful for clients with Parkinson's disease. Working with therapy animals offers a wealth of opportunities for creative program design. Stroking, grooming, throwing, catching, tug of war, and other games with these pets are incorporated to achieve needed stretching motions and strengthening exercise. Therapy dogs also build confidence during gait and transfer training or may offer a "connection" to clients with dementia. It is recommended to work only with therapy animals certified through accrediting organizations such as Delta Society, and to work closely with the animal handlers when planning these programs. Animal assisted therapy is not appropriate for all clients, and careful screening of client interest, allergies, and past animal experiences should be conducted prior to initiating a program.

Access to music therapy offers opportunities for creative collaboration to achieve desired outcomes. It is helpful to work with board-certified music therapists trained in specific neurological music therapy techniques when working with clients with Parkinson's disease. Achieving exercise follow-through using therapeutic instrumental music performance (TIMP), and physical therapy gait training incorporating rhythmic auditory stimulation, are examples of potential collaborative efforts with music therapy. Music can also be successfully used to enhance movement during community exercise groups or incorporated into relaxation training sessions.

There are many other holistic programs and complementary therapies that offer collaborative opportunities for physical therapists. Working with these programs can help achieve anticipated goals and expected outcomes within the physical therapy plan of care. Physical therapy interventions also provide activity accommodation and adaptations allowing continued participation in a client's leisure interests and other activities designed to maximize quality of life.

THE INTERDISCIPLINARY TEAM

Comprehensive management of Parkinson's disease requires the skills of a full interdisciplinary team. Ongoing communication and understanding of each team member's role are vital to success of this approach. The patient, family members, health care professionals, and community resources must interact to effectively coordinate a plan designed to provide recommendations and treatment options focused on maximizing quality of life. Each team member must be able to recognize changes or areas of concern and be prepared to refer to other members of the team as needed.

The physical therapist plays an important role on the interdisciplinary team. Prompt communication of physical therapy evaluation results and planned interventions ensures coordination of needed patient services. Skilled observations obtained during a physical therapy session can be helpful to physicians in regard to optimizing patient medications or managing secondary symptoms. Posture training interventions and breath work designed in physical therapy work collaboratively with the efforts of a speech pathologist focused on improving communication and voice volume. Movement enhancement strategies learned in physical therapy benefit ADL training performed in occupational therapy. Information shared by a client or family member during a physical therapy session may indicate psychosocial concerns requiring referral to social services or other members of the interdisciplinary team.

The physical therapist serves an important role as educator for other team members who may be less familiar with the mobility challenges seen in individuals with Parkinson's disease. Instruction in helpful compensation techniques and movement enhancement strategies results in improved understanding and patient care from all disciplines. Physical therapy instructions apply to other allied health disciplines as well as to workers in home health care, adult day services, senior centers, assisted living, skilled nursing facilities, community exercise programs, or others who may be involved in the client's comprehensive care plan. Presentations for patient conferences, support groups, or other community events also prove useful. Education of all team members provides information needed to recognize patient changes, more effective transitions between skilled medical services and community programming, and prompt referrals back to physical therapy as new problems are identified.

EVIDENCE-BASED PRACTICE: RESEARCH AND PHYSICAL THERAPY

An increasing amount of research has been published that demonstrates the effectiveness of physical therapy for patients with Parkinson's disease. A continued focus on evidence-based practice is needed to establish benefits of treatment and best practice patterns for all physical therapy professionals. Research based practice also establishes credibility for the role of physical therapy in Parkinson's disease management and provides important information to referring physicians, consumers, policy makers, and insurers.

A comprehensive explanation of current physical therapy practice is available in the *Guide to Physical Therapist Practice,* a collaborative work published by the American Physical Therapy Association.[22] Physical therapists use the information developed for the *Guide* within their clinical practice, as well as for professional education purposes. The *Guide* defines physical therapy's scope of practice and provides preferred practice patterns grouped into several major categories.

REFERENCES

1. deGoede C., Keus, S., Kwakkel, G., Wagenaar, R., "The effects of physical therapy in Parkinson's disease: A research synthesis," *Arch Phys. Med. Rehabil.,* 82 (4), 2001.

2. Csuka, M., McCarty, D. J., "Simple Method for Measurement of Lower Extremity Muscle Strength," *Amer. J. Med.,* 78(1):77–81, 1985.

3. Reuben, D. B., Siu, Al, "An objective measure of physical dysfunction of elderly outpatients. The Physical Performance Test," *J. Amer. Ger. Soc.,* 38:1105–1112, 1990.

4. Shumway-Cook, A., Brauer, A., Woollacott, M., "Predicting the Probability for Falls in Community-Dwelling Older Adults Using the Time Up and Go Test," *Phys. Ther.,* September 2000.

5. Nieuwboer, A., DeWeerdt, W., Dom, R., Bogaerts, K., Nuyens, G., "Development of an Activity Scale for Individuals with Advanced Parkinson's Disease; Reliability and "On-Off" Variability," *Phys. Ther.,* November 2000.

6. Berg, K., Wood-Dauphinne, S., William, J. I. et al., "Measuring Balance in the Elderly: Preliminary Development of an Instrument," *Physiotherapy Canada,* 41:240B, 1989.

7. Light, K. E., Behrman, A. L., Thigpen, M., Triggs, W. J., "The 2-minute Walk Test: A Tool for Evaluating Walking Endurance in Clients with Parkinson's Disease," *Neurology Report,* Vol. 21.

8. Behrman, A. L., Light. K. F., Flynn, S. M., Thigpen, M. T., "Is the Functional Reach test useful for identifying falls risk among individuals with Parkinson's disease?" *Arch Phys. Med. Rehabil.,* 83(4):538–42, April 2002.

9. Behrman, A. L., Light. K. E., Miller, G. M., "Sensitivity of the Tinetti Gait Assessment for detecting change in individuals with Parkinson's disease," *Clin. Rehabil.,* 16(4):399–405, June 2002.

10. Friedman, S. M., Munoz, B., West, S., Rubin, G. S., Fried. L. P., "Falls and Fear of Falling: Which Comes First? A Longitudinal Model Suggests Strategies for Primary and Secondary Prevention," *J. Amer. Ger. Soc.,* 50(8):1329, August 2002.

11. Cheal, B., Clemson, L., "Older people enhancing self efficacy in fall risk situations," *Australian Occupational Therapy Journal,* 48: 80–91, 2001.

12. Powell, L. E., Myers, A. M., "The Activities Specific Balance Confidence (ABC) Scale," *Jour. Geren.,* 50A(1), M28–M34, 1995.

13. Shumway-Cook, A., Gruber, W., Baldwin, M., Liao, S., "The effect of multidimensional exercises on balance, mobility, and fall risk in community dwelling older adults," *Phys. Ther.,* 77;(1): 46–57, January 1977.

14. Schenkman, M., Morey, M., Kuchibhaita, M., "Spinal Flexibility and balance control among community dwelling adults with and without Parkinson's disease," *J. Gerontol. A. Biol. Sci.,* 55(8): M441–5, August 2000.

15. Morris, M., Movement Disorders in People with Parkinson's Disease: A Model for Physical Therapy," *Physical Therapy,* 80(6): 578–597, June 2000.

16. Scandalis, T. A., Bosak, A., Berliner, J. C., Helman, L. L., Wells, M. R., "Resistance training and gait function in patients with Parkinson's disease," *Am. J. Phys. Med. Rehabil.,* 80(1): 38–43, January 2001.

17. Bergen, J. L., Toole, T., Elliott, R. G., Wallace, B., Maitland, Robinson K., "Aerobic exercise intervention improves aerobic capacity and movement initiation in Parkinson's disease patients," *Neurorehabilitation,* 17(2):161–8, 2002.

18. Reuter, I., Engelhardt, M., Stecker, K., Baas, H., "Therapeutic value of exercise training in Parkinson's disease," *Med. Sci. Sports Exerc.,* 31 (11): 1544–9, November 1999.

19. O'Shea, S., Morris, M. E., Iansek, R., "Dual task interference during gait in people with Parkinson's disease. Effects of motor versus cognitive secondary tasks, *Phys. Ther.,* 82(9):888–97, September 2002.

20. McIntosh, G. C., Rice, R. R. and Thaut, M. H., "Rhythmic—auditory facilitation of gait patterns in Patients with Parkinson's disease," *Journal of Neurology, Neurosurgery and Psychiatry,* 62, 22–26, 1997.

21. Parashos, S., Wielinski, C., Erickson-Davis, C., Wichmann, R., Walde, Douglas M., "Injuries due to falls in Parkinsonian Patients," Abstract presented at *International Congress of Movement Disorders,* Miami, FL, 2002.

22. *The Guide to Physical Therapy Practice,* 2nd ed., American Physical Therapy Association.

61 Swallowing Function in Parkinson's Disease

Lisa A. Newman
Walter Reed Army Medical Center

CONTENTS

INTRODUCTION

Swallowing is the primary mode of maintaining one's nutritional status. One of the complications of Parkinson's disease (PD) is impaired swallowing, with an incidence reported to be as high as 95%.[1] Swallowing is an intricate and complex physiologic and neurologic process. Thus, when neurologic damage occurs, as in the case of PD, swallowing may be adversely affected.

Dysphagia is defined as impaired swallowing, which can occur anywhere from the mouth to the stomach, resulting from impaired function of the jaw, lips, tongue, velum, larynx, pharynx, upper esophageal sphincter, or esophagus[2,3] More specifically, oropharyngeal dysphagia refers to swallowing disorders involving the oral and pharyngeal cavities, which are distinguished from primary esophageal disorders. Before one can understand the effect of PD on swallowing, an understanding of normal swallowing is essential. This chapter incorporates the following aspects of swallowing: normal swallowing, diagnosis of swallowing disorders, swallowing disorders in PD, and treatment options for swallowing dysfunction in PD.

NORMAL SWALLOWING

Swallowing can be divided into four stages:[4]

1. The oral preparatory stage
2. The oral stage
3. The pharyngeal stage
4. The esophageal stage

The oral preparatory stage involves mastication of semisolid or solid food and formation of a bolus, which renders the food into an appropriate consistency for swallowing with the bolus being lubricated and chemically altered by mixing with saliva.[5] This stage involves lips closure, rotary and lateral motion of the jaw, buccal or facial tone, and rotary and lateral motion of the tongue.[4] Neurologically, this stage is under voluntary control of the patient; however, sensory information is processed from sensory receptors throughout the oral cavity.[4]

The oral stage of swallowing is also under the voluntary control of the patient and involves the transport of food to the posterior area of the oral cavity. Two distinct patterns of tongue movement during the oral stage have been described: the "tipper" and the "dipper."[6,7] The "tipper" is when the bolus is initially on the tongue dorsum with the tongue tip pressed against the posterior surface of the maxillary incisors. The "dipper" takes place when the bolus is initially in the anterior sublingual sulcus, which requires the tongue to elevate the bolus to a supralingual position, and the oral stage continues in a similar fashion to the "tipper" type swallow. The oral stage requires intact labial musculature that prevents drooling, intact buccal musculature preventing food from pocketing in the buccal cavity, normal palatal muscles, functional lingual musculature and movement, and the ability to breathe through the nose.[4] Physiologically, four events signal the onset of the oral phase of swallowing: tongue-tip movement, tongue-base movement, superior hyoid movement, and submental electromyographic (EMG) activity.[6] There is variability in tongue movement, depending on the type of bolus, suggesting that the sensory feedback of the oral phase is necessary to monitor the bolus and adjust tongue function.[8]

The pharyngeal stage of swallowing serves a twofold purpose: guiding the bolus through the pharynx into the esophagus and protection of the airway. It is an automatic phase of swallowing governed by the brain stem. Normal pharyngeal swallowing involves palatal closure, bolus transport through the pharynx, base of tongue propulsion to the posterior pharyngeal wall, glottal closure to prevent aspiration, and upper esophageal sphincter (UES) opening and transsphincteric fluid flow.[9] Transport of the bolus through the pharynx is mainly accomplished by the pressure applied by the tongue base directly onto the bolus in the oropharynx.[10] The anatomic and neurologic integrity of the tongue plays an important role in bolus transport during both the oral and pharynx stages of swallowing.

The pharyngeal stage is considered to be a swallowing response (as opposed to a swallowing reflex), as it varies with changes in bolus volume and/or viscosity. There are several phenomena that change during the pharyngeal phase. There is a progressive increase in the magnitude of superior hyoid movement with increases in the size of the bolus.[11] The upward motion of cricopharyngeus or upper esophageal sphincter (UES) increases with enlarging volumes of liquid. The diameter and duration of UES sphincter opening also increase with larger volumes.[12] Increasing viscosity to a thick paste results in greater magnitude of anterior hyoid displacement and greater diameter and duration of UES sphincter opening as compared to the same volumes of a low-density liquid.[13,14]

For food to pass into the esophagus, there must be an opening of the upper esophageal sphincter (UES). The mechanism of UES opening occurs during the pharyngeal phase of swallowing due to an interrelationship between hyoid/laryngeal movement and UES opening. Anatomically, the cricoid cartilage is linked to the hyoid by muscular and ligamentous connections, and it is an insertion point for the cricopharygeus muscle (the lower portion of the UES). As the hyoid moves superiorly and anteriorly during the swallow, the larynx also elevates, which in turn elevates and opens the cricopharygeus muscle. As a result, the passage of food across the upper esophageal sphincter occurs while the hyoid is at its highest and most anterior excursion.[11]

There must be coordination between respiration and swallowing, and this occurs at the level of the larynx. The larynx must be opened during respiration and closed during the pharyngeal stage of swallowing.[8] Laryngeal closure is a three-tier protective adduction. From superior to interior anatomic location, the first level of adduction involves the approximation of the aryepiglottic folds to cover the superior inlet of the larynx.[15] The downward, backward movement of the epiglottis completes the superior level of closure, preventing any material from entering the laryngeal vestibule. The second layer of adduction is the false vocal foods. The final layer is a forceful adduction at the level of the true vocal folds.[4,15,16]

The esophageal phase of swallowing is a reflexive stage that involves the descent of the bolus of food down the esophagus. This is accomplished by peristaltic action of the esophagus and relaxation of the lower esophageal sphincter. Above the bolus, the circular muscle contracts, and the longitudinal muscle relaxes. This combined action produces the bolus propulsion that travels the length of the esophagus.[17] The esophagus consists of both striated and smooth muscle. Neurologic innervation of the esophagus switches between motoneurons located in the CNS and autonomically located motoneurons in the periphery as the tissue changes from striated to smooth muscle.[8] Relaxation of the lower esophageal sphincter occurs during the esophageal phase of swallowing or with distention of the esophagus. Neurologically, it depends on descending motor fibers in the vagus nerve.[8]

EVALUATION OF OROPHARYNGEAL DYSPHAGIA

Swallowing disorders or dysphagia are common in patients with Parkinson's disease. Complications of swallowing disorders include malnutrition, dehydration, and respiratory complications of aspiration, all of which can be life threatening. Therefore, accurate diagnosis of the parameters that impair swallowing and management is absolutely vital. The swallowing process is complex and cannot be observed clinically. Swallowing is a rapidly moving process that is best assessed with a dynamic instrumental technique.

Two methods have been used most frequently for examining oropharyngeal swallowing: videofluoroscopy and fiber optic endoscopic evaluation of swallowing (FEES). This section describes both methods and focuses on videofluoroscopy, which is used most frequently.

Fiber optic endoscopic evaluation of swallowing (FEES) consists of passing an endoscope transnasally to view the larynx and pharynx while the patient swallows measured volumes of food and liquid dyed with food coloring.[18] The advantages of FEES are as follows:

1. The procedure can be done at bedside.
2. Real food is used.
3. There is no radiation exposure.
4. Fiber optic endoscopes are available in medical settings.

The disadvantages of FEES are the following:

1. It provides no ability to view the oral phase of swallowing.
2. Deflection of the epiglottis, which covers the laryngeal introitus, obstructs visualization of the pharyngeal response.
3. There is no visualization of the cervical esophageal stage of swallowing.

It is therefore difficult to diagnose the parameters that cause dysphagia, which may affect treatment options.[19]

Videofluoroscopy has been cited as the best method for evaluating the dynamic process of swallowing.[6,20,21] It is the only procedure designed to study the anatomy and physiology of the oral preparatory, oral, pharyngeal, and cervical esophageal stages of swallowing.[4]

Videofluoroscopy uses a fluoroscope, which permits the radiologic observation of the dynamic process of swallowing. A recording device, most recently digital recording instrumentation, is interfaced with the fluoroscope, allowing the radiographic procedure to be recorded. Swallowing is a very rapid process, and evaluation of the detailed parameters is impossible in real time. Therefore, recording the procedure to play back immediately or after the study is necessary. In the case of a generative disease process such as Parkinson's disease, previous studies can be archived to compare changes over time.

During videofluoroscopy, patients are viewed in the lateral and/or A-P projections on a tilt-table fluoroscope in the upright position.[22] Measured volumes of a liquid barium suspension are given to the patient. Smaller amounts of liquid are initially administered to minimize the risk of aspiration and allow visualization of the anatomic structures that can be obliterated by large quantities of radiopaque material.[23] Different viscosities of barium should be assessed; e.g., nectar thick liquid, honey thick liquids, and pudding. Finally, mastication of a solid material, e.g., a cookie covered with barium pudding, should be visualized. Changes in bolus volumes and viscosities also serve as factors for observing changes in swallowing function given different sensory stimuli and motor requirements for each bolus. Every study should also include strategies to improve swallowing function when appropriate, e.g., therapeutic maneuvers or position changes. If the patient demonstrates severe swallowing difficulties that are not amenable to treatment strategies, the study can be terminated, thus minimizing the danger of aspiration to the patient.

Following completion of the videofluoroscopy or FEES, two factors must be immediately considered in managing the patient with Parkinson's disease: can the patient swallow safely with or without therapeutic strategies, and can the patient consume adequate calories orally to maintain or improve nutritional status?[19] An instrumental assessment is the only method to evaluate swallowing reliably. Management of oropharyngeal dysphagia is possible only after the pathophysiology of the swallowing mechanism is diagnosed.

SWALLOWING DISORDERS IN PARKINSON'S DISEASE

Gastrointestinal dysfunction, which includes dysphagia, is a frequent and occasionally dominating symptom of Parkinson's disease, originally described by James Parkinson in 1817. Specifically, he described the swallowing disorder as "the food is with difficulty retained in the mouth until masticated; and then as difficulty swallows" and drooling as "the saliva fails of being directed to the back part of the fauces, and hence is continually draining from the mouth."[24] Complications of dysphagia are present in advanced stages of PD: aspiration, pneumonitis, weight loss, malnutrition, and dehydration. When unrecognized, dysphagia plays a role in weight loss and the recurrent airway infections in patients with Parkinson's disease.[25]

Excess saliva and drooling (as described by Parkinson himself) have been noted as a problem in patients with Parkinson's disease. Bagheri et al.[26] measured saliva production in PD patients across all stages of the disease and in some patients who were not receiving any dopaminergic drugs. This study revealed that saliva production was significantly lower in PD patients than age- and sex-matched controls across all stages of the disease.[26] One may question whether the administration of dopaminergic drugs may affect saliva production. In a small subset of patients not receiving any dopaminergic medication, saliva production was also lower than in age- and sex-matched controls.[26] The excess saliva and drooling experienced by Parkinson's disease patient is probably not the result of too much saliva, but rather a consequence of swallowing difficulties. When a patient swallows infrequently, there will be a buildup of saliva. The tendency for the mouth to drop open and posture to become more stooped as Parkinson's disease progresses further magnifies the problem.[24]

Patients with Parkinson's disease have a long period of time between disease onset and swallowing difficulties, with a mean latency of 130 months.[27] However, survival time between the onset of dysphagia and death was short, ranging from 15 to 24 months.[27] The late onset of dysphagia may be attributed to what is perceived and reported by the patient. Patients with Parkinson's disease may in fact have dysphagia; however, due to sensory deficits, they may not complain of difficulties until symptoms become severe.

Patients with all stages of Parkinson's disease, including those who are asymptomatic for dysphagia, have been shown to have swallowing deficits. Abnormal findings on

EMG and esophageal scintigraphy were noted during swallowing in PD patients who were both symptomatic and asymptomatic of dysphagia.[28] Clinical and videofluoroscopic examinations of patients with Parkinson's disease who were asymptomatic of dysphagia revealed abnormalities of swallowing[29] Early-stage Parkinson's patients with no symptoms of dysphagia had a high percentage of objective swallowing abnormalities demonstrated on videofluoroscopy,[30] and there was no relationship between complaints of swallowing difficulties and swallowing function.[31] Studies have also revealed that swallowing abnormalities do not correspond to the Hoehn and Yahr stages of Parkinson's disease or other clinical symptoms, e.g., tremor.[30,32–34]

There are clinical symptoms and signs of swallowing that can alert the clinician to the possibility of a swallowing disorder. A questionnaire sent to Parkinson's patients revealed that the risk of choking on food or drink is the most frequently noted problem, with 41% of the patients indicating that mastication and swallowing are more difficult than prior to their disease.[35] On clinical examination, the following abnormalities were observed in a group of 65 Parkinson's patients:[36]

1. Impaired mouth opening and palatal elevation (60%)
2. Poor lingual protrusion (70%)
3. Wet vocal quality after swallowing liquids (40%)
4. A cough after swallowing liquids (40%)

The cough is a mechanism to protect the airway from aspirated material. The absence of a cough upon aspiration is termed "silent aspiration." Silent aspiration occurs frequently in patients with Parkinson's disease[32] and more often than other neurologic disorders.[37] Silent aspiration makes it difficult to observe symptoms of aspiration in Parkinson's disease patients and demonstrates the necessity of an instrumental examination of swallowing.

Various studies have demonstrated swallowing deficits during all four stages of swallowing in patients with Parkinson's disease. Beginning with the oral preparatory and oral stage of swallowing, the task of taking liquid through a straw revealed reduced peak suction pressures and lower bolus volumes across all Hoehn and Yahr stages of patients with Parkinson's disease as compared to controls.[33] Other oral phase deficits included prolonged oral transit time, soft palate tremor, and abnormal lingual activity; specifically, premature loss of liquid, repetitive tongue pumping so that several tongue pumps are required to move the bolus posteriorly, tongue tremor, and piecemeal deglutition (several swallows of portion of the bolus rather than a single swallow or oral residue, which is cleared with several swallows), and increased occurrence of respiration during repetitive swallows.[32–34,38,39]

Deficits in lingual function have been attributed to rigidity and bradykinesia of the tongue.[32,33]

As stated earlier, bolus transport through the pharynx is mainly accomplished by the pressure applied by the tongue base directly onto the bolus in the oropharynx.[10] Therefore, rigidity and bradykinesia resulting in lingual deficits in the oral phase would also affect the pharyngeal phase of swallowing. Clearing of the space between the tongue base and epiglottis (valleculae) is largely the result of tongue base movement.[4] Residue in the valleculae on fluoroscopy would indicate problems with tongue base movement. Vallecular residue was noted to be one of the most significant findings in patients with Parkinson's disease in multiple studies.[30,34,38–40] Residue in the valleculae and pyriform sinuses would account for aspiration after the swallow. Residue in the pyriform sinuses was also a significant finding, as was aspiration after the swallow.[30,32,34,40] A delayed pharyngeal response and slow/reduced laryngeal closure will cause aspiration before and during the swallow respectively.[32] A delayed pharyngeal response with subsequent aspiration has been documented.[30,32,34,38,40] Bradykinesia may be responsible for a significantly prolonged reaction time to initiate the swallowing reflex as compared to controls.[41] Aspiration may also be silent due to decreased cough reflexes and lack of sensation, thus patients may not cough in response to material in the airway.[32] Other pharyngeal stage deficits include epiglottic dysmotility, pharyngeal constrictor dysfunction, and reduced mean sagittal UES diameter as compared to controls; slow laryngeal excursion, slow and/or incomplete vocal fold closure, and prolonged pharyngeal transit times.[34,39,40,42]

There has been limited examination of the esophageal stage of swallowing in patients with Parkinson's disease. However, there have been documented disorders during esophageal transit, including prolongation of transit time and retention in the lower part of the esophagus. These signs were more severe in PD patients with clinical symptoms of dysphasia than PD patients without clinical symptoms.[28] Prolonged esophageal transit may be the result of esophageal dysmotility. Specific documented esophageal dysmotility has included stasis, tertiary contractions, reverse peristalsis, reduced peristalsis, aperistalsis, and esophageal tortuosity or patulency. In addition, confirmed GE reflux was present in a group of 40 of 50 patients, and the LES failed to close in 21 of 50 patients with PD.[42]

The mean age of onset of Parkinson's disease is 54.8 yr in females and 55.6 yr in males, and patients live for a relatively long period of time with the disease.[43] Therefore, age-related changes might affect swallowing in addition to the disease process. In a group of 16 patients with Hoehn and Yahr Stage III-V, durational measures of oral and pharyngeal transit times were not significantly different from those of age-matched controls, although the PD patients had vallecular and pyriform sinus residue not seen in controls.[38] In a study of 19 patients with PD, both with and without dysphagia, the group with dysphagia tended to be older and had a shorter duration of the disease than the nondysphagia group, although this did not reach statistical significance.[34] However, one must consider that age-related changes may further exacerbate swallowing decompensation.

TREATMENT OPTIONS FOR SWALLOWING IN PARKINSON'S DISEASE

Treatment options for swallowing dysfunction in Parkinson's disease have been limited. One possible option to improve swallowing may be the pharmocologic treatment of Parkinson's. However, as the following studies illustrate, dopaminergic medications do not reliably improve dysphagia.[31,44] In three studies, antiparkinsonian medication was withheld for a minimum of eight hours, and videofluoroscopy was performed.[30,31,45] Bushman et al.[31] found that, after patients took their usual dose of levodopa, there was improvement in swallowing function in 5 out of 15 patients, with the greatest changes in decreasing vallecular residue and improving transit times for thick boluses.[31] Half of the 12 patients in study by Fuh et al.[30] showed swallowing function improved within 90 min after administration of 200 mg levadopa in combination with 50 mg bensarazide.[30] Specific oral and pharyngeal phase improvements included reduction in oral tremor, improved tongue elevation, and elimination of aspiration in two out of three patients. In each of these studies, there was one patient whose swallowing worsened after drug treatment.[30,31] Both of these studies included patients who were symptomatic and asymptomatic for swallowing difficulties. Hunter et al.[45] examined 15 patients who were symptomatic of dysphagia. Patients were first assessed after antiparkinsonian medication was withheld for at least eight hours, assessed again after a single oral dose of 250 mg levodopa and 25 mg carbidopa, and assessed a third time after a dose of apomorphine (mean dose 3.5 mg, range of 1.5 to 6 mg). The authors found few significant differences after levodopa or apomorphine. The only differences were reduced pharyngeal transit time with semisolids, less vallecular residue with solids after apomorphine, and fewer swallows to clear a solid bolus after levodopa. Surprisingly, there was an increase in oral transit time with solids after levodopa.[45]

Drooling remains a significant problem for the patient with Parkinson's disease. One study compared two therapeutic strategies to control drooling: botulinum toxin to both parotid glands, which reduced the amount of saliva produced, and a behavioral intervention program. The behavioral program, conducted by a speech-language pathologist, consisted of patient education; completion of a drooling awareness chart for dry swallows, three times a day for seven days; and a swallow reminder brooch that signaled the wearer to swallow at regular intervals by emitting a beep. The subject was instructed to wear the brooch for 30 min per day.[46] Both the botulinum toxin injections and the behavioral approaches were successful in reducing drooling as compared to baseline and aged-matched controls at one month. At three months, however, the degree of improvement was not fully maintained for the behavioral approach, while the botulinum toxin was still effective. The authors concluded that for the therapy program to be effective, reinforcement after the three-month period would need to be offered.[46]

Various swallowing parameters have been examined before and after surgical or behavioral treatment. One session of oral motor and swallowing exercises reduced the premotor time defined as the initiation time of the pharyngeal phase.[41] Cricopharyngeal myotomy showed mixed results in improving cricopharyngeal dysfunction in four case studies of Parkinson's disease patients with a Zenker's diverticulum, cricopharyngeal bar, or discoordination between pharyngeal contraction and cricopharyngeal relaxation.[44] A voluntary airway protection technique and variation of the supraglottic swallow technique[4] (consisting of holding breath, tilting the chin to the chest, swallow, cough, and then swallow again) eliminated silent aspiration in two out of three patients.[31]

Treatment programs used to improve voice and speech in patients with Parkinson's disease have also been shown to improve swallowing.[47,48] Patients reported an improvement in swallowing, although not documented instrumentally, after one month of treatment including increasing loudness, pushing exercises, and overarticulation.[47] The Lee Silverman Voice Treatment (LSVT) is a 1-month, 16-session program designed to improve the perceptual characteristics of voice through training high phonatory effort tasks that stimulate increased vocal fold adduction and respiratory support.[48] Patients underwent this therapy program to test its effects on swallowing function. Videofluoroscopy before and after treatment revealed an overall 51% reduction in the number of swallowing motility disorders, including the following significant improvements: reduction in oral transit time, reduction in oral residue after 3- and 5-ml liquid swallows, elimination of a delayed pharyngeal response during swallows of liquid, a 66% reduction in paste and cookie, reduced pharyngeal transit time, reduction in laryngeal penetration before the swallow, reduction in pharyngeal residue, and improvement in tongue base propulsion.[48] LSVT holds promise for significant improvements in swallowing function in patients with Parkinson's disease.

CONCLUSIONS

Swallowing is a complex physiologic and neurologic process, which has been categorized into four stages: oral preparatory, oral, pharyngeal, and esophageal. Dysphagia is impaired swallowing in any of these four stages and is common in patients with Parkinson's disease. Due to the physiologic complexity of swallowing, dysphagia cannot be observed or diagnosed on clinical examination. Therefore, an instrumental assessment is necessary to accurately diagnose the parameters that impair swallowing and is a mechanism for testing management techniques. Two instrumental techniques commonly used are videofluoroscopy and fiber optic endoscopic evaluation of swallowing (FEES).

Complications of dysphagia, including aspiration, pneumonitis, weight loss, malnutrition, and dehydration, play a role in weight loss and recurrent airway infections in patients with Parkinson's disease. Due to the swallowing disorders, patients with PD are less likely to swallow and therefore have a buildup of saliva, causing drooling. There is also a tendency for the mouth to drop open and posture to become more stooped as PD progresses, further magnifying the problem. Patients with PD are more likely to be asymptomatic of their swallowing difficulties until late in the disease process. However, asymptomatic patients have been shown to have swallowing deficits necessitating an instrumental evaluation of swallowing.

Parkinson's disease affects all four stages of swallowing. Oral preparatory and oral-stage swallowing deficits have included smaller bolus volumes, prolonged oral transit times, soft palate tremor, poor lingual control, repetitive tongue pumping, tongue tremor, and increased occurrence of respiration. Pharyngeal stage deficits have included a delayed swallowing response, prolonged pharyngeal transit times, reduced tongue base movement, slow laryngeal excursion, slow or incomplete vocal fold closure, slow or reduced laryngeal closure, residue in the valleculae and pyriform sinuses, and aspiration. Patients with PD may not have sensation to aspirated material or may have reduced cough reflexes and therefore will not have visible symptoms of aspiration. Esophageal disorders may be the result of esophageal dysmotility in addition to gastroesophageal reflux. As Parkinson's disease patients tend to be older and age during the course of the disease, age related changes might further exacerbate swallowing difficulties.

Treatment options for the swallowing disorders in the PD population have been limited. Dompaminergic medications have not reliably improved dysphagia. Botulinum toxin and behavioral therapy have been effective in improving drooling. Behavioral therapy techniques have shown some improvement in swallowing, specifically oral motor and swallowing exercises and airway protection techniques. Treatment programs used to improve voice and speech have also been successful in improving swallowing function in patients with Parkinson's disease and warrant further investigation.

Unfortunately, Parkinson's disease is a progressive degenerative disease. Swallowing function will usually mirror the progression of the disease despite all treatment attempts. However, treatment for swallowing may improve the quality of life and reduce dysphagia complications in the short term.

The opinions expressed herein are not to be construed as official or as reflecting the policies of either the Departments of the Army or Defense.

REFERENCES

1. Blonsky, E. R., Logemann, J. A., Boshes, B., Fisher, H. B., Comparison of speech and swallowing function in patients with tremor disorders and in normal geriatric patients: a cinefluorographic study, *J. Gerontol.,* 30(3):299–303, May, 1975.

2. Perlman, A., Schulze-Delrieu, K., *Deglutition and its disorders: anatomy, physiology, clinical diagnosis, and management,* Singular Publishing Group, Inc., San Diego, CA, 1997.

3. Horner, J., Massey, E. W., Managing dysphagia. Special problems in patients with neurologic disease, *Postgrad. Med.,* 89(5):203–206, 211–203, April 1991.

4. Logemann, J., *Evaluation and treatment of swallowing disorders,* 2nd ed., Pro-ed, Austin, TX, 1998.

5. Kennedy, J., Kent, R., Physiological substrates of normal deglutition, *Dysphagia,* 3:24–37, 1988.

6. Cook, I. J., Dodds, W. J., Dantas, R. O. et al., Timing of videofluoroscopic, manometric events, and bolus transit during the oral and pharyngeal phases of swallowing, *Dysphagia,* 4(1):8–15, 1989.

7. Dodds, W. J., Taylor, A. J., Stewart, E. T., Kern, M. K., Logemann, J. A., Cook, I. J., Tipper and dipper types of oral swallow, *Am. J. Roentgenol,.* 153(6):1197–1199, December, 1989.

8. Miller, A., *The Neuroscientific Principles of Swallowing and Dysphagia,* Singular Publishing Group, Inc., San Diego, CA, 1999.

9. Dodds, W. J., Stewart, E. T., Logemann, J. A., Physiology and radiology of the normal oral and pharyngeal phases of swallowing, *Am. J. Roentgenol.,* 154(5): 953–963, May 1990.

10. McConnel, F. M., Cerenko, D., Hersh, T., Weil, L. J., Evaluation of pharyngeal dysphagia with manofluorography, *Dysphagia,* 2(4):187–195, 1988.

11. Dodds, W. J., Man, K. M., Cook, I. J., Kahrilas, P. J., Stewart, E. T., Kern, M. K., Influence of bolus volume on swallow-induced hyoid movement in normal subjects, *Am J. Roentgenol.,* 150(6):1307–1309, June 1988.

12. Kahrilas, P. J., Dodds, W. J., Dent, J., Logemann, J. A., Shaker, R., Upper esophageal sphincter function during deglutition, *Gastroenterology,* 95(1):52–62, July 1988.

13. Dantas, R. O., Kern, M. K., Massey, B. T. et al., Effect of swallowed bolus variables on oral and pharyngeal phases of swallowing, *Am. J. Physiol.,* (258):G675–681, 1990.

14. Dantas, R., Dodds, W., Massey, B., Kern, M. K., The effect of high- vs. low-density barium preparations on the quantitative features of swallowing, *Am. J. Roentgenol,.* 153:1191–1195, 1989.

15. Sasaki, C., Isaacson, G., Functional anatomy of the larynx, *Otolaryngol. Clin. North Am.,* 21:595–611, 1988.

16. Ardran, G., Kemp, F., The mechanism of the larynx II, the epiglottis and closure of the larynx, *Br. J. Radiol.,* 40:372–389, 1967.

17. Hendrix, T. T., Coordination of peristalsis in the pharynx and esophagus, *Dysphagia,* 8:74–78, 1993.

18. Langmore, S., Schatz, K., Olson, N., Endoscopic and videofluoroscopic evaluations of swallowing and aspiration, *Ann. Otol. Rhinol. Laryngol.,* 100:678–681, 1991.

19. Newman, L., Cannito, M., Dysphagia and dysarthria: evaluation and management, in Gluckman, J., Ed., *Renewal of Certification Study Guide in Otolaryngology Head and Neck Surgery,* Kendall/Hunt Publishing Company, Dubuque, IA, 286–295, 1998.

20. Jones, B., Kramer, S., Donner, M., Dynamic imaging of the pharynx, *Gastroint. Radiol.,* 10:213–224, 1985.

21. Jing, B., The pharynx and larynx; roentgenographic technique, *Sem. Roentgenol.,* 9:259–265, 1974.

22. Beck, T., Gayler, B., Image quality and radiation levels in videofluoroscopy for swallowing studies: a review, *Dysphagia,* 5:118–128, 1990.

23. Logemann, J. A., *Manual for the Videofluorographic Study of Swallowing,* Pro-Ed, Austin, TX, 1993.

24. Pfeiffer, R., Gastrointestinal dysfunction in Parkinson's Disease, *Clinical Neuroscience,* 5(2):136–146, 2000.

25. Volonte, M., Porta, M., Comi, G., Clinical assessment of dysphagia in early phase of Parkinson's disease, *Neurol. Sci.,* 23:S121–S122, 2002.

26. Bagheri, H., Damase-Michel, C., Lapeyre-Mastre, M., et al., A study of salivary secretion in Parkinson's disease, *Clinical Neuropharmacol.,* 22(4):213–215, 1999.

27. Muller, J., G. K., W., Verny, M. et al., Progression of dysarthria and dysphagia in postmortem-confirmed Parkinsonian disorders, *Arch Neurol.,* 58(Feb.):259–264, 2001.

28. Potulska, A., Friedman, A., Krolicki, L., Spychala, A., Swallowing disorders in Parkinson's disease, *Parkinsonism and Related Disorders,* 9:349–353, 2003.

29. Bird, M., Woodward, M., Gibson, E., Phyland, D., Fonda, D., Asymptomatic swallowing disorders in elderly patients with Parkinson's disease: a description of findings on clinical examination and videofluoroscopy in sixteen patients, *Age and Ageing,* 23:251–254, 1994.

30. Fuh, J. L., Lee, R. C., Wang, S. J. et al., Swallowing difficulty in Parkinson's disease, *Clin. Neurol. Neurosurg,* 99(2):106–112, May 1997.

31. Bushmann, M., Dobmeyer, S. M., Leeker, L., Perlmutter, J. S., Swallowing abnormalities and their response to treatment in Parkinson's disease, *Neurology,* 39(10):1309–1314, October 1989.

32. Robbins, J. A., Logemann, J. A., Kirshner, H. S., Swallowing and speech production in Parkinson's disease, *Ann. Neurol.,* 19(3):283–287, March 1986.

33. Nilsson, H., Ekberg, O., Olsson, R., Hindfelt, B., Quantitative assessment of oral and pharyngeal function in Parkinson's disease, *Dysphagia,* 11:144–150, 1996.

34. Ali, G. N., Wallace, K. L., Schwartz, R., DeCarle, D. J., Zagami, A. S., Cook, I. J., Mechanisms of oral-pharyngeal dysphagia in patients with Parkinson's disease, *Gastroenterology,* 110(2):383–392, February 1996.

35. Hartelius, L., Svensson, P., Speech and swallowing symptoms associated with Parkinson's disease and multiple sclerosis: a survey, *Folia Phoniatr. Logop.,* 46(1):9–17, 1994.

36. Volonte, M. A., Porta, M., Comi, G., Clinical assessment of dysphagia in early phases of Parkinson's disease, *Neurol. Sci,.* 23 Suppl., 2:S121–122, September 2002.

37. Mari, F., Matei, M., Ceravolo, M. G., Pisani, A., Montesi, A., Provinciali, L., Predictive value of clinical indices in detecting aspiration in patients with neurological disorders, *J. N. N. P.,* 63(4):456–460, 1997.

38. Nagaya, M., Kachi, T., Yamada, T., Igata, A., Videofluorographic study of swallowing in Parkinson's disease, *Dysphagia,* 13(2):95–100, Spring 1998.

39. Bird, M. R., Woodward, M. C., Gibson, E. M., Phyland, D. J., Fonda, D., Asymptomatic swallowing disorders in elderly patients with Parkinson's disease: a description of findings on clinical examination and videofluoroscopy in sixteen patients, *Age Ageing,* 23(3):251–254, May 1994.

40. Leopold, N. A., Kagel, M. C., Pharyngo-esophageal dysphagia in Parkinson's disease, *Dysphagia,* 12(1):11–18; discussion 19–20, Winter 1997.

41. Nagaya, M., Kachi, T., Yamada, T., Effect of swallowing training on swallowing disorders in Parkinson's disease, *Scand. J. Rehabil. Med.,* 32(1):11–15, March 2000.

42. Leopold, N. A., Kagel, M. C., Laryngeal deglutition movement in Parkinson's disease, *Neurology,* 48(2):373–376, February, 1997.

43. Hoehn, M. M., Yahr, M. D., Parkinsonism: onset, progression, and mortality, *Neurology,* 17:427–442, 1967.

44. Born, L. J., Harned, R. H., Rikkers, L. F., Pfeiffer, R. F., Quigley, E. M., Cricopharyngeal dysfunction in Parkinson's disease: role in dysphagia and response to myotomy, *Mov. Disord.,* 11(1):53–58, January 1996.

45. Hunter, P. C., Crameri, J., Austin, S., Woodward, M. C., Hughes, A. J., Response of parkinsonian swallowing dysfunction to dopaminergic stimulation, *J. Neurol. Neurosurg. Psychiatry,* 63(5):579–583, November 1997.

46. Marks, L., Turner, K., O'Sullivan, J., Deighton, B., Lees, A., Drooling in Parkinson's disease: a novel speech and language therapy intervention, *Int. J. Lang. Commun. Disord.,* 36 Suppl., 282–287, 2001.

47. De Angelis, E. C., Mourao, L. F., Ferraz, H. B., Behlau, M. S., Pontes, P. A. L., Andrade, L. A. F., Effect of voice rehabilitation on oral communication of Parkinson's disease patients, *Acta Neurol. Scand.,* 96:199–205, 1997.

48. Sharkawi, A. E., Ramig, L., Logemann, J. A. et al., Swallowing and voice effects of Lee Silverman Voice Treatment (LSVT): A pilot study, *J. of Neurol. Neurosurg. Psychiatry,* 72(1):31–36, January 2002.

62 Restorative and Psychosocial Occupational Therapy in Parkinson's Disease

Surya Shah and Ann Nolen
University of Tennessee Health Science Center, College of Allied Health Services

CONTENTS

INTRODUCTION

This chapter focuses on neurological restitution and pathophysiological issues that contribute to neurological recovery and the associated psychosocial abilities that occupational therapists address to maximize capabilities of persons with Parkinson's. Management of these patients has changed in the last decade. Therefore, it is essential for occupational therapists to determine what and how the hallmark signs and symptoms and various therapeutic approaches discussed in earlier chapters affect function in Parkinson's. These would lead to utilizing occupation related assessments and intervention strategies for the stipulated deficits. Occupational therapists can then function as a member of the multidisciplinary team, and plan and implement novel solutions. Such occupational therapy approaches would maximize restitution (by considering the uniqueness of each individual), hasten their rehabilitation, and improve their quality of life.

HALLMARK ISSUES AFFECTING OCCUPATIONAL THERAPY

To determine occupational therapy intervention strategies, the manifestations of this chronic progressive condition are grouped as either negative or positive. Negative symptoms are the direct result of damage to the structures implicated in Parkinson's. The direct damage results in patients having difficulties in accessing motor plans and sequencing motor programs as a whole that organize the temporal progression of movements to eliminate robot-like sequencing effects, such as when movement control is acquired and learned by an infant from inherited mass patterns. Negative symptoms also include akinesia, bradykinesia, postural instability, gait disturbances, depression, and other psychosocial dysfunctions. Akinesia is difficulty in initiating movement and lack or loss of movement. Bradykinesia results from strial dopamine deficiency and results in slowness in executing movement from increased reaction time, delayed correction of an inaccurate attempt, prolonged time to recommence correct movement, and easy fatigue. These two negative manifestations are not due to a patient's perceptual difficulties but result from the paucity of self-generated voluntary movements.[1] Akinesia and bradykinesia are closely related but could be independent and present in absence of rigidity. Akinesia is also considered an extreme form of bradykinesia.[2] Due to deficits in movement preparation, patients are unable to transform general goals into specific motor actions. The prevailing predominance of movement execution difficulty is thought to result from the lack of control of muscle activation and regulation.[2] Once the reverberating circuits or their interlinking elements of corticonigral, nigropallidal, pallidothalamic, and thalamocortical are damaged, learned responses must be brought back to conscious effort. This damage makes the performance of a well rehearsed, learned, and automated task slow and deliberate. The lack of excitatory γ innervation from supraspinal segments not only results in a lack of dexterous movements, it requires increased and forceful efforts to drive the uninterrupted innervation. Such forceful efforts exhaust the remaining intact structures. There is also reduced ability to switch rapidly between sets of movements.[3] With akinesia, automatic execution of learned motor plans is altered, but with dyskinesia, the motor plan can be executed and completed by compensatory methods, indicating that the form of plan is preserved, but individual motor programs, which make up the motor plan, remain distorted.[4–5] The slowness of movement can be influenced by an external stimulus, followed by an optimal use of therapeutic activities that further influence these deficits to move to the required level of spontaneity in performance. Also, the cortical control is parallel and not hierarchical, thereby allowing intact structures to execute some of the functions.[6]

In contrast, positive symptoms (e.g., lead-pipe or cogwheel rigidity) are the result of overactivity of intact structures that have been released or escaped from the control of the implicated structures. Parkinson's rigidity is diffuse, pervasive, and not selective like spasticity. It affects both agonists and antagonists and is uniform throughout the passive range. In the limbs, this perceived resistance to passive motion is irrespective of joint position or direction of limb movement. The cogwheeling seen on passive range of movement is generally in the upper limbs and more pronounced in distal joints such the wrist. Lead pipe rigidity is felt when there is no tremor and predominates in the lower limb. The internal loop of the intrafusal muscle fibers with group II afferents contributes to rigidity by degenerating of inhibitory dopaminergic projections from the substantia to the putamen. This degeneration leads, in turn, to biased output from the globus pallidus to γ_2 efferents. Rigidity is also considered to be a release phenomenon normally suppressed by the globus pallidus, because a lesion of the medial internal zone of the globus pallidus abolishes rigidity. Rigidity is also the result of α excitation, so normalizing muscle tone is an important goal of occupational therapy.[7] Associated postural instability is attributed to structures connected with the basal ganglia but remote to subcortical areas such as the globus pallidus efferent to vestibular nuclei.

Resting tremor of 4 to 5.5 Hz is rhythmic, sinusoidal, and oscillatory from rhythmical contractions of agonists and antagonists and at times occurs when the same posture is maintained for a period of time. Resting tremor can be a side effect of drugs and can be exaggerated when the patient is emotionally excited, when conscious of being watched, and following exercise and fatigue. However, the tremor disappears during sleep. As a dorsal rhizotomy does not block the tremor, it appears that there is depressed γ activity and augmented α activity, as well as direct thalamic or cortical activation, contribute to the tremor.[8] The problem in reporting and quantifying clinical disorders resulting from damage to deeper structures is that conditions that affect those structures and cause movement disorders do not selectively damage one region but affect multiple areas. The neurological, pathological, and physiological implications of the symptoms and occupational therapy contributions affecting these are discussed in later sections.

OCCUPATIONS AND ACTIVITIES

Using purposeful, motivational, and preparatory activities in occupational therapy helps patients develop movements that lead to functional abilities that in turn enhance engagement in daily life tasks and occupations.[9–10] For example, occupations are tasks such as housekeeping tasks for a housewife, playing golf for a professional golfer, skilled manipulations for a machinist, manual labor for an unskilled laborer, or those tasks that have a unique meaning and purpose that influence how persons with Parkinson's spend time and how they make decisions.[11] The activities, on the other hand, are participation in tasks that are goal directed but not central to one's being. Occupational therapists believe that engaging in such therapeutic activities, or in occupations or elements that lead to occupations, helps maintain health status, prevents primary and secondary complications, encourages recovery from disease, and increases a person's ability to adapt despite Parkinson's.[12–13] Such therapeutic activities help minimize the effects of impairment, activities limitations, and restrictions imposed by Parkinson's.

Activities and occupations also facilitate the essential adaptive process. Adaptive abilities allow patients to change functional requirements as environmental demands alter and help individuals survive and self-actualize. Occupational therapy is participatory and requires patients to take responsibility for their personal well-being and to participate with the therapist in decision making.[14]

According to the American Occupational Therapy Association, there are three basic tiers of patient activities and occupations.[9,13] Top-level occupations allow patients to engage in age-appropriate activities that are unique to them and match their goals: a job for earning a living, dressing the lower body independently, purchasing their own groceries, doing their own laundry, managing money, or activities purely for pleasure, for example. At the next level are occupations that allow patients to engage in goal-directed activities during therapeutic interventions designed to maximize restitution, subsequently leading to the desired occupational functioning: washing vegetables for a meal, drawing nongeometrical designs that lead to arm and hand control, engaging in safe ways of getting in and out of car to go to work, or role playing to mimic real situations. These two levels of occupations and activities require ballistic (or open-loop) type and corrective (or closed-loop) type components of voluntary movements.[15] The third level refers to the preparatory activities that are meaningful interventions or are components of performance skills that prepare patients to isolate and develop mastery in performing essential ingredients of purposeful occupations and activities. Performing such tasks by patients is marred by increased reaction times, greater firing rate, and an increase in movement amplitude. These components cannot be varied according to task demands and are limited in spontaneity until the patient is systematically retrained during restorative occupational therapy. In acquiring control, patients would first understand the overall requirements for accomplishing the task before acquiring the necessary speed, force, and duration of each of the movement components. Finally, with repetitive practice, patients are freed from concentrating on the entire movement, which becomes automatic and spontaneous. Acquired control is then integrated with higher-level task requirements seen in levels 1 and 2 of the three tiers.[9,16–17] Learning of movements in the third preparatory level could be a fundamental, instinctive, primitive, natural, and unlearned mass patterns movement execution in an inherited diagonal pattern. Movements then progress to break away from such patterns of motion so as to allow isolated and precise execution to be developed. Use of exteroceptive or proprioceptive inputs and the use and application of various modalities facilitate the maximum response in such learned tasks. Functional and diagonal proprioceptive neuromuscular facilitation patterns need to commence in a shortened range and progress to the stretched positions, with progression beyond basic diagonal patterns to help maintain the required stretch, and to help build combinations of movements that help improve the repertoire of complex movement abilities. Possible lack of effectiveness of such attempts in rebuilding abilities to perform occupational tasks at levels 1 and 2 have been reported by Protas et al.[18] However, Protas was challenged by Murphy and Tickle-Degnen,[19] who concluded, based on their research, that interventions have positive effects. The research showed that 31% of patients improved without restorative occupational therapy, while 63% improved with restorative occupational therapy.

According to the model of occupational functioning,[20] and based on the systematic review by Murphy and Tickle-Degnen,[19] the capabilities and abilities level of control evident in preparatory tasks have certain outcomes such as motor control, co-ordination, dexterity,

and balance. The activities and tasks levels include outcomes of performance on a patient's ability to transfer, on performing activities of daily living, and on the ability to move. The roles level assists in classifying participants' perceived level of social resources and support (ICIDH-2 = levels of body functions and structure, activity, and participation). Finch et al.[21] and Christiansen and Townsend[11] recommend that the associated costs and utility as determined by self-report and quality of life become integral parts of considerations in determining occupational and activities choices and levels. Murphy and Tickle-Deanne[19] concluded positive occupation-centered treatment outcomes in 13 of 16 studies. In the same study, 10 of 16 studies showed positive effects on capabilities and abilities levels. Evidence also suggests that patients' participation in activities has a positive influence. Furthermore, patients engaging in these activities or their components need to have elements of a given skill mastered so as to perform the whole task meaningfully.[11] These sequential processes in occupational performance reinforce the globus pallidus facilitated functions of premovement programming, ability to initiate movement, the speed with which movement is executed, how body posture is maintained, and how movement elicitation occurs. These capabilities promote maximum adaptation.[21–22] Hinojosha and Youngsstrom[23] stated that a patient's ability to participate in self-chosen occupations for increased independence, with or without assistance, and achieving the desired goals contribute to increased independence achieved in a participatory, empowering, and satisfying manner. Control of complex goal-directed motor activities (such as writing letters of an alphabet, cutting paper with scissors, hammering a nail, shooting a basketball, shoveling dirt, throwing darts, punching a bag, and other ballistic movement tasks) become an integral part of occupational reeducation. Control of such goal-directed tasks facilitates participation by people with Parkinson's in actual occupations that are part of their own context and that match their performance expectations, and it further facilitates engagement in goal-directed activities.[10] Participatory, occupational therapy interventions enable patients to return home following rehabilitative care and remain actively functioning with improved quality of life. Goal-directed activities, such as combing hair and brushing teeth, would be further influenced when the disease-caused neural degeneration extends to the left frontal gyrus and intraparietal sulcus. These structures are responsible for control of goal-directed occupational tasks.[24] It is now conclusive that, when Lewy body dementia is diagnosed with Parkinson's, it would become necessary to look beyond occupational dysfunction and evaluate and plan to improve executive functions, such as memory, cognition, and psychological function to maximize restitution and improved quality of life.[25] While the environments are not the focus of this chapter, occupational therapists adapt and maximize the person–environment fit to promote wellness. Environments have a significant impact on performances; for example, the use of an aquarium with bright colored fish, gliding motion, and the sound of water trickling have been shown to be effective in pacifying patients, in improving nutritional status of elderly patients, and in cutting costs of rehabilitation.[26]

ASSESSMENTS

Assessment of patients with Parkinson's is complicated by many factors.[27] Because of the introduction of new therapeutic strategies, the development of more advanced surgical procedures, findings of oxygen extraction studies, and deep brain stimulation studies, occupational therapists should summate not only neurological restitution measures (vital signs, skin, mental status, cranial nerves, motor, sensory, superficial and deep reflexes, tone, tremor, cerebellar signs, gait, posture, and others), but also self-care ability and patients' ability to interact within their own environment in tasks such as meal preparation, shopping, managing money, cleaning, and going to a place of worship.[28] Lang[27] suggests that therapists undertaking to assess and quantify the findings should be aware of the following: progressive disease related consequences, medication-related fluctuations, psychosocial effects of the disease, the importance of having a "comments section" following self-care, and instrumental activities of daily living assessments to account for possible variations in items within the scales. With recent research reporting sudden and slow onset sleepiness in patients with Parkinson's, it is vital for the therapists to determine the times of the day that are most suitable for introducing assessments, treating executive functions, and commencing motor reeducation.[29–30]

When simple and chosen reaction times for movement preparation are examined to determine what would be required for a patient to select a movement or combination of movements among the repertoire of possibilities, the evaluation and intervention strategies become evident. Movement performance times are particularly important when pre-post-test changes need definitive evaluation. Reaction-time paradigm preparatory processes are also required when the therapist is evaluating the extent of premovement abnormalities, such as whether the patients are slower in planning a movement, slower in choosing between response alternatives from the repertoire of rehearsed options, or slower in the use of advance information from previous rehearsals. Electromyographic analysis is equally relevant when determining whether the deficiencies are due to force, displacement, or volitional control and whether the ballistic movements are conceptualized as agonist and/or antagonist or as an agonist sequence of muscle action.[31]

Kinematic analysis is evaluated if learned movements have one cycle of acceleration and deceleration or if there is irregularity, asymmetry, variation in frequency of cycles, or lack of precise control over termination or commencement trajectories.[32–34] A number of methods can be used. Chang et al.[35] uses video monitoring software to quantify posture and movement changes without the usual markers used in virtual reality. This video monitoring is 90% accurate, and further trials to improve accuracy and clinical utility are underway. Lang[27] developed a videotaping method and a videotape analysis system, while Graziano[36] developed a video to identify and address mobility problems. Spyers-Ashby et al.[37] utilize 3Space Fastrak® multidimensional movement analysis to differentiate limb tremor over six degrees of freedom. The system is capable of differentiating between postural tremor in unimpaired persons and persons with Parkinson's. It has a potential to be useful not only as an objective clinical tool to record progress but also as a diagnostic tool.

One of the measures used for measuring frailty and risk for falls is the Functional Reach Test. Multiple factors are involved in falls (e.g., progression, freezing, turning while attempting a task at hand, step length, cognitive and attention deficits,[38] and environmental factors including lighting, wet floors, loose carpet, flexibility, and weakness). The Functional Reach Test is recommended for identifying patients at risk of falling. A reach of 17.78 cm, according to this test, is regarded as a good marker of frailty. Also, people 70 years and over with a reach of 25.4 cm are at risk of recurrent falls.[38] Many preparatory and manipulatory tests are also in use (e.g., Purdue pegboard). When Purdue pegboard does not show finger dexterity improvement, physical and motor abilities improve, as does the disease staging ranking and thus quality of life following occupational therapy.

The Disability and Distress Index[39] has four domains of concern: self-care, social and personal relations, mobility, and usual activities. The responses obtained are combined and scores converted to a formula. The Euroqol System (EQ-5D)[40] has multidimensional items: self-care, mobility, pain and discomfort, usual activities, and anxiety and depression. Responses obtained on various items are combined to produce a summary score. The Health and Utilities Index II (HUI)[41] has items on sensation, emotion, cognition, self-care, mobility, pain, and fertility. Its scoring system converts responses to a numerical score. The Unified Parkinson's Disease Rating scale (UPDRS)[42] covers multidimensional items, and the scores are added to make a composite score. The scoring is negative, meaning the higher the score (199), the greater the disability. Items include cognition, activities of daily living, and motor function. Its use is widespread and is also recommended in occupational therapy studies,[43] yet the reliability, validity, and consequences of adding items of multidimensionality have not been adequately addressed.

Hohen and Yahr Scale[42] provide severity categories ranging from unilateral symptoms, unilateral and axial, bilateral, bilateral with impairment of postural reflexes, severe disability but ambulatory, to wheelchair-bound and bedridden. The PD Questionnaire-39 (PDQ-39)[44] is a disease-specific quality of life instrument with eight domains: self-care, mobility, emotional well-being, stigma, cognition, communication, bodily discomfort, and social support.[45] Beck Depression Inventory[46] is also frequently used. As evidenced by Siderowf et al.,[47] in a cost cutting climate, utilizing expensive neurosurgical and other interventions needs to be justified by outcome measures that are both reliable and valid, and it is vital that occupational therapists undertake cost-effectiveness analysis using preference-based measures for quality of life.

Other areas needing evaluation are energy conservation and work simplification; degree of contractures and prevention by positioning and alignment to counterbalance forces of gravity, imbalance, or faulty position; flexibility; control in antigravity muscles; degree of disuse and compensation; quality reciprocal movements; and the ability to utilize the trunk statically and dynamically in sitting and standing, rolling, turning, walking, and dynamic balancing.[43,48]

Other important and useful measures with biometric and psychometric qualities include the Barry Rehabilitation In-patient Screening of Cognition (BRISC), a brief neurological assessment battery with established psychometric properties.[49] It is a valid screening tool of cognition for those 39 years and older and requires less than 30 min. It has eight subtests that point to specific deficits that might need attention: reading, design copy, verbal concepts, orientation, mental imagery, mental control, initiation, and memory. The BRISC provides a maximum composite score of 135. A total score between 110 and 120 needs to be interpreted cautiously in older adults.

The computer-based Brain Train® software for carefully planned cognitive testing and training, developed by Sandford,[50] has many modules, e.g., attention skills, visual motor skills, conceptual skills, and higher faculty modules.

Along with self-care abilities, abilities to interact within the environment are vital to ensure maximum independent functioning in the community. The Moss Kitchen Assessment Revised[51] is a method of gauging community functioning of patients with Parkinson's. A succinct review of available meal preparation assessments for persons with neurological impairment and the importance of meal preparation ability as an instrumental activities of daily living, and as a basic task was published by Harridge and Shah.[28] The Moss Kitchen Assessment,[52] as revised by Harridge and Shah,[51] grades meal preparation performance on six hierarchical task levels in order of difficulty from serving and eating a cold meal to preparing a complex hot meal and cleaning. The performance on each task level can be evaluated on a five-point Likert-type scale from "unable" to "independent with or without assistive devices." The performance is then given ratings of relative contribution to each criterion and the ratings summed to obtain a score out of the 100 maximum points. These and the activities of daily living (ADL) scores help determine effectiveness of intervention, length of patient care, cost of rehabilitation, burden of care, housing requirements, likely outcomes, and other audit-based outcome comparisons.

The Barthel Index is considered to be comprehensive and the most researched ADL scale. It measures the individual's performance on ten ADL tasks. It is an empirically derived scale that measures performances in personal hygiene, bathing self, feeding, toilet, bowel control, bladder control, dressing, ambulation, stair climbing, and transfers. However, its sensitivity remained a concern. The Modified Barthel Index (MBI), by Shah et al.,[53] provides the required sensitivity in scoring those individuals who require assistance of some nature to perform the tasks. The increased sensitivity was achieved by expanding the number of categories used to record improvement in each ADL item. The modifications ensured that the minimum and maximum values assigned to each of the ten weighted tasks remained the same and that none of the underlying assumptions was violated or mutilated. Suggested changes and plausible explanations developed in 1989 were further modified, based on feedback from users, to include not only the amount of physical assistance required to perform the task but to accommodate cognitive aspects such as wandering, prompting, cuing, and standby supervision.[54] The MBI is valid and has with proven interobserver and test-retest reliability, internal consistency, portability, and appropriate psychometric and biometric qualities.[55] Corbett[56] used a number of measures to identify pre-post-test changes and found that the Functional Independence Measure did not identify any significant changes; therefore, it is not considered to be sensitive to detecting changes in functional ability.

A test consisting of basic nongraphic designs would help screen and evaluate perceptual deficits that might affect on Parkinson's patients' performance and daily function. Shah et al.[57] selected six reliable and valid domain items from the Burke Perceptual profile[58] and reported the normative data. The timed tests consist of six subtest sets and include picture sequences, body puzzle, 3-D space visualization, block design, figure-ground, and fine motor planning. The tests are unidimensional and provide specific domain strengths and weaknesses without possible nullifying from multidimensionality. Precise noting of functional difficulties during ADL evaluation is crucial, as patients report clumsiness, awkwardness, minimal unobservable lack of associated movements, and unsteadiness from tremor.

Careful observation of people reporting difficulties is of paramount importance for an early diagnosis of Parkinson's disease. Teive and Sa[59] reported a patient who was always late for appointments when his self-winding wristwatch was worn on the left wrist but was never late when he wore his watch on the right wrist or when he used a battery-operated watch. Such first signs of a lack of natural arm movement during walking, a slight foot drag leading to tripping, a slight intentional tremor when drinking water at a dining table, or difficulty during shaving become crucial in early diagnosis.[60]

POSTULATED FUNCTIONS, DEFICITS EXPERIENCED, AND RESOLUTIONS

Findings by Jones et al.[61] suggest that Parkinson's patients have difficulty in inhibiting a current movement and in preparing a new movement in the opposite hemisphere. These deficits are due to a lack of preprogramming of the new sequence. Stelmach and Phillips[62] identify arrest of motor activity, contralateral turning, circling around the base of support, limb flexion, chewing, licking, and swallowing as modifications of cortically induced movements. Firing of medial and lateral zones of globus pallidus neurons results in the onset of predictive, self-directed, and sequential movements. Released responses from the neostriatum, particularly anterior caudate, result from cues from the prefrontal cortex that prepare for tasks requiring initiation of movement or behavioral responses to environmental cues. Other parts of the caudate are involved in pattern-specific habituation from repeated visual stimuli, inattention, and in orientation to a changed visual stimulus pattern. These responses from the two parts of the caudate are the result of influences from prefrontal and inferior temporal cortical projections. For patients with globus pallidus disorders, once such purposeful activities are initiated, motor tasks requiring bilateral alternating movements such as pronation and supination, grasp and release, and reach and return followed by the ability to transfer occupations from one to the next hemisphere, should commence.[63]

The slowness of movements also affects participation in structured occupational tasks that require patients to initiate standing and walking, and it manifests difficulties in generating sequence of submovements, preparing for movement, and executing movement. Furthermore, damage to the pedunculopontine nucleus may be responsible for the deficits in initiating programmed movements.[64] The paucity of movements, which is not due to paralysis, becomes more pronounced when more than one act is requested, such as getting up from a chair and extending a hand to greet someone, talking on the phone and writing a message, preparing a meal and responding to a dialogue, or turning around a base of support to pick up an ingredi-

ent while stirring a hot meal. Paucity in such activities is attributed to difficulty in recruiting the peripheral nervous system following altered γ_2 innervation. Planned occupations excite intact reverberating circuits that activate γ_1 and γ_2 motor neurons, minimize inhibitory influences of the damaged neurons, and allow the intact cells to be efficiently utilized. In Parkinson's, the occupations and activities that are at the apex encourage goal direction, purpose, and a unique meaning to their being, and they activate a visual open-loop where amplitude, force, direction of movement, and accurate and precise termination of hand movements are determined. As these tasks are mastered, automatic corrections are performed using a closed-loop control where patients focus and visually ascertain each element of movement. As the akinesia improves in occupationally embedded tasks, movements become automatic and ballistic with decreased reaction time and increased spontaneity.[3,15]

The influence of exteroceptive stimuli, such as fast brushing and icing for preparing for a motor response via the reticular activating system, allows the diffuse and nonspecific central nervous system (CNS) to percolate and, via association areas, prepare the CNS to respond. A prepared, alert, and ready-to-receive CNS is further activated by a proprioceptive input such as a friction, rubbing, pressure, or a light static stretch while attempting to perform selected occupations while facilitating intrafusal muscle fibers to be activated to fire through 1a afferents.[62,65] These efforts maintain and increase range, maintain and increase strength, increase speed of muscle contraction, and prevent contractures, phlebitis, and other inactivity-related complications from developing.

In volition, movements and their components are superimposed, making volition a smooth and fluid skilled movement. However, the lack of ability to sequence movements and provide spontaneity by determining where the next muscle contraction should begin and where precisely to end the previous muscle contraction manifests robotic-type sequences, with one movement requiring completion before the emergence of the next. Hocherman and Aharon-Peretz[66] first identified difficulty in performing manual tracking tasks and the need for practice in the preparatory phase of task mastery. Carey et al.[67] further substantiated these findings and indicated that, in preparing patients to engage in productive occupations, occupational therapists need to provide practice in tracking tasks using compatible and incompatible hand and arm positions. There is also evidence to suggest that active participation in planned activities is paramount.[18] In holding objects still, no such deficits are evident, indicating intactness of static and tonic activity. Tonic and static contractions do not have such difficulties except for minimal slowness, the time taken to commence, and the delay in producing maximum force required for that task.

Fine finger facility, manipulative ability, individual, isolated, and combined finger movements; accuracy; and preciseness are improved with individualized, goal-directed activities such as spool knitting, link belts, cane work, bench and assembly work, toothpick-type projects, use of personal diaries, keyboard applications, computer keyboards and touch screen, and immersive virtual reality uses. Manual dexterity is further enhanced with purposeful activities such as using pegs in solitaire, placing washers, folding and putting a letter in an envelope, using scissors, hammering nails, dribbling and shooting a basketball, passing a football, shoveling dirt, and throwing darts. Patients need to participate in activities that require movement sequence as a whole, as it helps superimpose elements of movement components, maintain the computed force, initiate and increase variability of movements, and utilize visual guidance for ballistic movements. When difficulties are encountered, observing and copying another person, receiving verbal or tactile cues, imagining and/or visualizing an object may instill initiation required to perform the functional tasks.[68] Writing letters of the alphabet and tasks that require a quick and spontaneous response, progressively decreasing in reaction time, and increasing difficulty in providing alternating force within a movement sequence are improved by practicing nongeometrical designs. Such activities allow progression from unstructured control of the upper limb to precise manipulative writing skill. Introducing group or individual relaxation and activities using music, hand clapping to music, rhythm and movement, gentle rocking, and the rotation of trunk and extremities can improve the entire body range of motion. Hand clapping helps decrease cortical inhibition and abnormal EMG activity, and music facilitates movements. If the speed of movement is increased along with precision, dexterity, and coordinated tasks, overall performance is enhanced. Keeping patients as physically active as possible, practicing in learning to stop movements at a precise moment, and using ball games or musical chairs helps to maintain health status and agility. Games and leisure activities requiring spontaneity of movement (e.g., paying bills, balancing budget, writing checks, figuring taxes, using the internet and web sites) further advance patients' ability to master cognitive and daily skills. Heightening abilities thus facilitates coping with the progressive nature of the disease and improves the quality of life.

Experiments to produce peak forces reveal that Parkinson's does not impair intension, but it does impair the rate at which force develops and is produced, initiation, timing, and spacing; hence, there is a need to intentionally alter these. Patients perform a sequence of movements at a fast or a slow pace to improve the force of muscle contraction, to maximize recruitment of motor units, and to control amplitude. Selecting manipulatory tasks that require repetitive performance with occluded vision

makes provides good rehearsal for task performance. In micrographia, the ability to maintain the required amplitude rather than duration for which the force has to be maintained, the time between the generated pressure for a given word and the next, as well as the ability to maintain speed are vital. It might become necessary to increase pauses between movements for increased accuracy. Practicing nongeometrical shapes prior to controlled writing minimizes these difficulties. Based on inverted writing in nonmirror transformations, such tasks excite supplementary motor areas.[69]

Motor control of released movements by caudate enhances excitatory neurotransmitter glutamate and nigrostriatal inhibitory dopaminergic effects. Examining anatomical processing of stimuli relevant to motor actions and single neuron responses,[70] and the above motor release control, imply that striatum contributes to the resulting modulation on the anterior horn with corticospinal and other systems, and the firing of efferents from basal ganglia go to the supplementary motor area to primary cortex and back to the final pathway to pre-prepare for the task demands. The automaticity required in motor tasks is provided by the supplementary motor area's ability to prepare for the next movement that has been rehearsed.[71] Functions and deficits are somatotopically organized with leg and arm control in globus pallidus and neck and head control in substantia nigra.[72] Activation of the basal ganglia cells precedes limb movements, indicating their involvement in preparing for movement that requires alterations to axial postural muscles prior to activating limb movements.

Putamen, its cortical projections, and supplementary motor areas are functional system components. Putamen is involved in behavior of all movements that are associated with rewards. However, if there is no goal to be attained, or if a response is required without a motivation, then putamen activation is not evoked. Substantia nigra cells show no phasic activity to peripheral stimulation; however, the subthalamic nuclei show phasic activity somototopically. Globus pallidus firing is variable with some firing preceding phasic activity, some after movement has commenced[73] and the rest firing in relation to each discrete movement. This is termed a *set dependent response*. Globus pallidus and substantia nigra could be activated to stretch and other deep proprioceptors. However, cutaneous receptor stimulation does not evoke a response.

Thus, cortex, basal ganglia, cerebellum, and thalamus have highly specific afferent, efferent, internuncial, and commissural connections. This linking of data makes it possible to divide cortico-striato-pallido-thalamo-cortical loop into motor functions feedback circuit and the other loop for occulomotor and orbitofrontal functions.[74] A behavior generated and released by the cortex is focused by the basal ganglia on the desired motor actions. Once activities are released from striatum, the pallidum executes with spontaneity and swiftness. The pallidum executes learned motor plans and allows well rehearsed tasks to be implemented with ease. When damaged in Parkinson's, the individual reverts to a slower, deliberate, conscious, less accurate, and less automatic movement. In considering basal ganglia function and disorders, occupational therapists must consider motor manifestations and behavioral changes to hasten the process of remediation. Movements and occupational therapy interventions are most effective during the "on" periods.

While restorative occupational therapy is considered beneficial in Parkinson's, some[75] have suggested that motor learning itself could also be affected. However, it has been shown that manual pursuits and sequence learning can be learned. Platz et al.[75] investigated the question of such practices, leading to increased movement speed and its contribution in visually guided aiming movement. Platz et al.[75] and Sheridan et al.[76] found that training either to move fast or move accurately, but not both, quickly and accurately helped Parkinson's patients with difficulty in moving. The ability to use spatial, temporal, and force-monitored movements was learned or transferred via cerebellar and cortical areas and further facilitated by auditory cues during rehearsal.

Deficits in force requirements in acceleration and deceleration between agonist and antagonist forces and sensory-motor integration are facilitated by initial adjustment, early preplanning, and by fine tuning final corrective movement requirements. Rascol et al.[77] showed that Parkinson's patients could learn to decrease their movement time when visually guided occupations are used. The untrained opposite limb also shows significant improvement in anticipation of movement initiation. Auditory cues did not specifically help in this experiment, but humans are capable of activating α motor neurons and the final common pathways that are directly innervating the extrafusal skeletal muscles in gross movement and in strong muscle contractions. However, when fine, precise, and delicate, manipulatory movements are required,[7] the γ motor system has to be activated. The γ system has a central bag of nuclei with contractile elements at the two terminals. These centrally located nuclei are sensitized by a proprioceptive input such as a stretch put to the polar ends by active contraction or by a passive stretch. The Ia then fires and directly communicates with large α motor neurons. Thus, those muscles that are precise manipulators have a rich supply of intrafusal muscle fibers while muscles like lattisimus dorsi have a sparse supply.

Suprasegmental control and corticomotor connections influence the final common pathway in two ways. Direct influence is via α but, in most instances, via γ for precise augmentation, as γ innervation precedes α. In Parkinson's, the altered α and γ balance is altered: γ is depressed, and α is enhanced leading to freezing from tonic exaggeration. Reduced dopamine has the same effect on movement execution. While physical rehabilitation is vital in improving daily functioning,[78] individual and group interventions focus on neurological deficits (impairments) to maximize restitution such as reaching, grasping, relaxing, and breathing.

Tasks requiring simple to complex block designs, visual figure ground perceptions in extracting complex embedded tasks, and facial recognition are some of the first skills lost when premotor area is involved.[79] Lazaruk[80] reported difficulties in global visuospatial ability and changes in language, abstraction abilities, processing skills from complex stimuli, and visual memory from diminished ability in using the storage subsystem. These deficits emerge more frequently in later stages of Parkinson's. It is therefore vital to incorporate screening for such deficits, but not timed tests, because fatigue and medication can influence the outcome. Their effect on functional performance needs to be studied to measure functioning at home and work. A Canadian occupational and physical therapist designed an excellent self-referral plan for patients and their spouses to bring about health behavior changes by problem-solving the needed aspects of education and exercise. The program emphasizes self-management to affect quality of life by prioritizing activities so patients can adjust their life style in a manner that controls fatigue and encourages ease and endurance of daily tasks.[81–82] Instrumental activities facilitate developing interaction between volition, α and γ coactivation, cognition, and sensory-perceptual patterns. The goal-oriented retraining helps neurons to recruit and fire at a high level; develop synchronic patterns: and reinforce basic, fundamental, natural, instinctive central pattern generators. Functional tasks focus on daily living skills and mobility, hobbies and leisure, social interaction, and other activities to increase self-esteem and motivation to decrease dependence.[34]

Deane et al.[83] compared seven published trials and concluded that, because of methodology limitations and extreme variations in treatments, the data did not support or refute the effectiveness of therapeutic interventions. In another systematic review of 23 studies, researchers found that heterogeneous intervention methods and methodological flaws required further trials.[84] Manson and Caird[85] investigated sedentary and passive activities such as watching TV, listening to the radio, reading a newspaper or book, and knitting, and found that these activities were well preserved in most patients. Eighty-three percent did indoor gardening; 31% were active in the house or gardening and performed such manual dexterity activities such as sewing or knitting; and 29% required transport to leave the house, move and in and around the house, perform gardening indoors or outdoors, and engage in do-it-yourself activities. Thirty-three percent were active outside the house (e.g., going to the theater, cinema, restaurant, place of worship, and clubs). A considerable number did not have hobbies and interests apart from those of a sedentary and domestic kind, and they need further investigation.

APPLYING FUNCTIONAL ABILITIES TO AMBULATORY SKILLS

Festination (shuffling, hurried small steps, turning), freezing, lack of arm swing, decreased velocity, and stride length and width are some of the difficulties encountered by patients with Parkinson's.[34] When functional abilities are applied to ambulatory skills and novel postural adjustment situations, the difficulties are said to be due to rigidity, reduced preparatory postural adjustment, reweighing of sensory motor loops, and fluent adaptation to altered postural deviations.[86] Daily walks, dance routines such as ballroom and tap dance, confidence in shifting weight, just walking in a mall that has even terrain, and the ability to shop or go to a place of worship facilitate increased awareness of lower limbs. Swimming is considered good for toning and endurance and for transferring volitional effort to more automated tasks such as walking in open areas and using a stepping mechanism, which use the unaffected swing and stance times and enhance the ability to tilt forward and backward. Rocking from side to side, walking from side to side to free freeze, and inverting a cane and stepping over it prevent the arrest of walking. Line and tap dancing allow for fast and slow weight shift, whereas ballroom dancing facilitates increased awareness of lower extremities.

For Parkinson's patients, thinking of the steps of a complex task, such as walking, makes it a fall-free event as the task is brought under volition. Side walking, ascending and descending three steps, overstepping, and using heels rather than toes helps overcome freezing and falling. The judicious use of throw rugs and minimizing clutter eases walking and prevents falls. These types of ambulation facilitate confidence and prevent falls that result from a fear of falling and a lack of swiftness. A high incidence of hip fracture from falls in elderly women patients with Parkinson's have been reported by Johnell et al.[87] Learned, well rehearsed, and automated tasks facilitate increased awareness and minimize deliberation that lead to fragmenting performance. Leather shoes are considered superior to rubber or corrugated soles. The resulting socialization and participation can help minimize predisposition to fatigue.

Maintaining posture and learning to squat and kneel helps prevent tightness when the distribution of rigidity affects one group of muscles more than others. A wider base of support, learning to lean right or left, lifting rather than dragging the feet, and sliding heels to 90° and beyond when standing up counter freezing and facilitate

the initiation of movement. Walking, stair climbing, and other daily tasks can be rehearsed with or without external cues. Facilitatory stimuli, such as rhythm, result in improved speed of walking and stride length, and they increase muscle activity as evidenced by increased EMG activity. In EMG patterns, the number of bursts in a given time frame is thought to be due to peripheral control and is increased by therapy for Parkinson's patients who display slowness and irregularity to produce the required speed and range. While the EMG amplitude of burst is said to be due to central control, the observed jerkiness and periodic tremor are not.[6,31]

To improve accuracy, we employ neurorehabilitation approaches such as proprioceptive neuromuscular facilitation using visual cues and focusing on quality of movement rather than the end product. Auditory cues, such as counting, clapping, or verbalizing the steps in walking, have been reported to improve motor performance.[85] Brain activity using imagery in simple walking, repetitive movements, or programmed complex acts, or mentally rehearsing a complex movement sequences when integrated, have been effective therapies.[88] These authors showed increases in regional cerebral metabolism and regional cerebral blood flow provoked by visual imagery, in particular, with the strongest blood flow seen in the cortical areas involved in movement execution.

Miyai et al.[89] demonstrated that body-weight-supported treadmill training, using external cues such as attentional strategies, has a lasting effect in improving and maintaining stride length. Body weight support used during occupationally embedded tasks also ensures maintenance and an increase in stride length.[17,89] To make walking more meaningful, occupational therapists need to incorporate occupation-centered tasks that require change in direction, change in speed, shift in weight for reach and return, angle of foot placement, and modifiable body position to accommodate carpeted and linoleum type surfaces and terrains.[90] Well planned activities and games that involve patients walking in a corridor; entering and exiting doorways; walking on colored paper squares adhered to the floor; and turning, entering a confined space, and exiting help minimize freezing and improve initiation, velocity, acceleration, and the ability to constantly shift the center of gravity.[91] Practice facilitates the ability to anticipate component placing of leg and foot at a required angle while focusing on such occupational tasks as cooking, cleaning, sitting up, and rotating the body from a supportive phase to a new direction and position. Morris et al.[17] showed that the ability to generate a normal walking pattern is not lost in Parkinson's patients. The observed difficulties stem from an inability to activate the motor control system, which can be summoned to perform by appropriate cueing. Kinetic analyses during routine tasks help better understand the functional components that need to be emphasized.[92-93]

The required subconscious adaptation in all routine and automated tasks increases flexibility; modifies weight bearing; helps shift weight; improves arm swing and overall participation that requires walking as a component; and counters flexed posture with protracted shoulder joint complex, flexed elbows, flexed head and neck, dorsal, lumbar, and sacral spine, and flexed hip and knee.[94] The tendency to fall backward emanates from not lifting the feet, prolonged standing, and standing with the feet together. Freezing and falling are further addressed by stepping over an object, rocking medially and laterally to avoid sticking to the floor, and maintaining bilateral arm swing. Many activities contribute to improved functional walking and overall confidence, such as reaching overhead with one hand while the other helps to stabilize the person; installing a vertical grab bar on doors opening inward; forming a habit to carrying objects in one hand and close to the body; lifting a knee to move rather than backing away from stove, sink, dresser or turning to one side; feeling the back of a chair before reaching back for the armrest; moving slowly to change position; and breaking down the actions in steps by deliberately pausing for a few seconds.

There has been a growing trend to provide walkers to patients with Parkinson's to minimize freezing and to improve speed, balance, and visual guidance to facilitate walking. However, this practice has been refuted and criticized by Susman[95] and others. The authors state that the use of walkers and walkers equipped with wheels and other devices used to minimize freezing and to improve speed of walking are ineffective and actually impede ambulatory ability.

SURGERY, BRAIN STIMULATION, IMPLANTS, DRUGS, AND DIET

To participate in randomized, single-case, pre-post-test and other trials that estimate functional gains; that improve movement participation and quality of life in the short and long run for patients, family, and caregivers; and that determine which patients undergo which therapeutic procedures, occupational therapists need to understand current surgical procedures, current and planned therapeutic interventions, participatory inhibitory deep brain stimulation, and the implications of such procedures.[96–98]

Hubble and Berchou[104] suggested that patients treated with excessive levadopa, carbidopa, pargolide, bromocriptine, or dopamine agonist develop involuntary movements similar to chorea and athetosis and secondary changes of psychosis, vomiting, abdominal discomfort, loss of appetite, confusion, hallucination, low blood pressure, psychosis, motor fluctuations, and a marked inability to use volitional movements. Difficulty in judging force, amplitude, and duration is also evident. The dopamine D_2 receptor blockers, the butyrophenones, and the phenothiazines make rigidity, bradykinesia, and tremor worse. Dopamine deficiency blocks movements in one direction and increases the magnitude of choice reaction time. Levadopa with decarboxylase inhibitor helps with nausea, vomiting, abdominal discomfort, and entacapone reduces motor fluctuations.[105–106] Parkinson's patients benefit from anticholinergic drugs, while cholinergic agonists aggravate symptoms because of a decrease in dopamine and an increase in acetylcholine. The normal balance between the neurotransmitters acetylcholine (excitatory) and dopamine (inhibitory) is essential for normal muscle tone. Trickling of γ amino butyric acid from basal ganglia provides the stability of motor control.

Neurosurgical procedures to neutralize rigidity, akinesia, tremor, and drug-induced dyskinesia have included pyramidotomy and stereotactic surgery in the 1970s, stereotactic lesions of medial pallidum and ansa lenticularis renewed interest in pallidotomy in the 1980s, and for levadopa-induced movement dyskinesia and motor control fluctuations, transplanting of fetal dopamine cells, and other graft transplants. The consequences of these and other surgeries demand that occupational therapists move beyond traditional assessments and restorative care avenues to accommodate unforeseen changes.[99–101]

Kishore et al.[72] concluded that it is possible to affect "on phase" dyskinesias following pallidotomy, depending on the amount of necrosed volume excised. Thermal lesions, chemopallidotomy, and targeting the ventrointermediate nucleus of thalamus alleviate levadopa induced "on phase" dyskinesia. Surgery does not reverse the degenerative process and, hence, activating implicated cortical areas by implant counters the progression and helps modulate the output. Implantation of embryonic substantia nigra neurons from fetal dopamine and adrenal cells to alleviate rigidity has had some success. Kishore et al.[72] also concluded that there is a direct correlation between the amount of coagulated globus pallidus with the amount of improvement observed in levadopa-induced dyskinesia or "on signs." Montgomery and Rezai,[102] in discussing deep brain stimulation, caution therapists using diathermy, therapeutic ultrasound, shortwave, and other thermal modalities in patients with implanted neurostimulation systems, as these could cause severe injuries or death.

It is also important to recognize the lack of certainty of affecting features of Parkinson's following surgical lesions of the medial globus pallidus. Saint-Cyr and coworkers[100–101] found that the negative effects of such interventions could include memory impairment and diminished mental processing, encoding visuospatial information such as nongraphic designs, and motor speed and coordination. Another therapeutic side effect is decreased cognitive functioning following administration of coenzyme Q10 for improved mitochondrial function.[103]

Documenting adverse effects from therapeutic effects of drugs, surgery, and brain stimulation that complicate rehabilitation is also warranted. The importance of engaging persons in occupations to build strength and endurance following surgery and therapeutic interventions cannot be emphasized enough. Participation by patients at all levels in occupational therapy, activities of daily living, instrumental activities of daily living, and leisure and work activities is essential for a fuller and improved quality of life.[44] Research showing that pedunculopontine nucleus is significant as the newly found site for akinesia,[100–101,107] and the influences of glutamatergic pedunculopontine neurons on initiating programmed movements and the cholinergic pedunculopontine neurons on maintaining steady state,[64] are significant for occupational therapists to counter adverse influences.

A community-based study of 124 patients with Parkinson's demonstrated that dyskinesias are related to duration of levadopa treatment, and resulting motor fluctuations are strongly correlated to the length of Parkinson's from onset and quantity of levadopa administered.[106] Occupational therapists should also be aware of dietary implications for their patients, because diet can influence manifestations and their intensity. Contaminants in dairy products can adversely affect male patients' sexual performance.[67,108–109] Despite advances in therapeutic management, considerable disability and engaging in productive activities remain a challenge.[110] McNaught et al.[111] noted the importance of abnormal protein levels and resulting cytotoxicity, which can adversely influence the ability of the central cell bodies to respond to occupational therapy restitution attempts. Extra precautions in working with elderly women with Parkinson's are necessary because of both increased incidence of hip fracture from limited mobility and osteoporosis and reduced bone mineral deficiencies.[112] Botulinum toxin has been used successfully to therapeutically paralyze the rigid muscle in a graduated and reversible manner to facilitate self-care and other functions.[113–114] New drugs on the horizon include dopaminergic compounds and formulations for replenishing of depleted dopamine, nondopaminergic drugs from different neurochemical mechanisms, and neuroprotective and neurorestorative drugs that are said to reverse disease progression. Catechol-O-methyltransferase inhibitors and other new therapeutic agents such as oxidase inhibitors provide much-needed flexibility in minimizing secondary motor consequences with levadopa therapy.[115] The U.S. Food and Drug Administration has recently approved Stalevo[116] for Parkinson's. These new drug therapies should allow occupationally embedded tasks and therapeutic strategies to improve patient function and quality of life.[117]

FUNCTIONAL ISSUES IN DAILY LIVING

Ample information is available at the local, state, national, and international levels, from Parkinson's disease-related organizations and web sites, about functional issues such as activities of daily living, home adaptations, and prescription of devices and aids. Therefore, only a few such issues will be addressed here.

Driving is considered an important activity and, when compromised, it severely affects the Parkinson's patients. It has received *a priori*ty rating by the American Occupational Therapy Association.[118] Sleep attacks caused by dopamine drugs have been thought to contribute to car crashes. As a result, some provinces in Canada have restricted driving privileges.[119] The crashes occurred with nearly 23% of patients who experienced either slow or sudden onset of drowsiness or abrupt episodes of sleep during activities of daily living.[119] This problem was further complicated by a lack of awareness by passengers of the driver's reduced vigilance. However, McConnell[120] emphasizes that Parkinson's patients do not cause more road accidents than a matched control subjects and that limiting driving can be an improper decision, as it severely limits quality of life of persons with Parkinson's. Ondo et al.[119] analyzed 303 questionnaires to determine the factors contributing to daytime sleepiness and concluded that sleepiness correlated with a more advanced state of Parkinson's and the use of dopamine agonists. Male patients were more prone to such episodes. Clinical trials using Modafinil show that, on a subjective and a behavioral level, there is a significant improvement in daytime sleepiness. These findings also have implications for occupational therapy intervention.[121]

Whatever the level of participation, it is important that the focus of occupational therapy remain on engagement in activities and occupation. de Goede et al.[3] analyzed 19 studies and concluded that activities of daily living, ambulation, stride length, mobility, and transfers could be much improved by occupational therapy. Improving safety in kitchens and bathrooms, taking large steps, dropping a handkerchief to take a step, inverting a cane and stepping over it, and not succumbing to freezing are important daily skills. To reiterate, with limited functioning in bilateral cases, ambulatory devices requiring the use of upper limbs can help or hinder, depending on festination quality.[122] For spontaneity and easy rising from sitting, the use of lift chairs, satin sheets for rolling, and reclining bed with adjustable foot and head ends for transferring is essential. Palmar pockets help maintain the stability of foods when arm supination and pronation are used to reach to a plate for food and to reach the mouth. Long shoehorns for putting on shoes, a friction-suction mat in the bathtub, and a shower chair or bench can easily improve daily activities. Also, a shower is safer than a bath. Holding a toothpick or chewing gum and holding an an appointment card help minimize tremulousness. Some ways to facilitate independence include lip and tongue exercises, gently blowing candles, practice with speaking, learning not to speak toward the end of exhalation, breathing deeply from abdomen, swallowing semisolids then liquids to prevent food from entering the windpipe, and smelling strong pungent odors to encourage facial expression. Anticholinergics help reduce the volume of saliva and thus drooling. Posture should be as erect as possible to facilitate lung functioning. Guithier et al.[78] found that it was not the tremor at rest that affected activities of daily living, but the lack of postural stability that is resistant to levadopa that also led to difficulties at home and work. Impairment in functional activities increases with the progression of the disease. Trunk mobility and postural control are the first to interfere with functional independence, followed by increased difficulty in manipulative and skilled tasks. Based on the Barthel Index findings, the treatment group significantly improved in ADL following occupational therapy. Rigidity and bradykinesia can affect occupational performance such as cutting food, buttoning, writing, smiling, and interacting socially. α-synuclein, found in household pesticides, is a major contributor to the development of free radicals in the brain, and patients need to understand how to handle these with care and avoid setting off chemical reactions that intensify Parkinson's symptoms.[123]

Despite advances in pharmaceuticals that improve mobility and capabilities, Beattie[124] found that many patients continue to have functional limitations. On 6 to 18 months follow-up, the author found that 28% were independent, 26% were dependent and needed help, and 46% were dependent in self-care. In addition, 80% needed a bath rail and mat, and 10% needed a bath seat, board, rail, and mat. Thirteen percent benefited from a grab bar attached to the wall next to a toilet, and 5% needed a raised toilet seat. Feeding aids prevent scattering of food, dislodgement of plates, and difficulty using cups. Kitchen aids facilitate opening jars and cans. Miscellaneous other aids help with turning over, sitting in bed, and other tasks. These include a rope ladder, can opener, tap turner, trolley, telephone amplifier, doorbell, ramp, wheelchair, hand-held shower nozzle, walking frame, card holder, shoehorn, helping hand, raised chair, high-back chair, and incontinence pads. "On" and "off" phases lead to changes in motor control, akinesia, bradykinesia, and diffuse rigidity, all influencing movements and performance in self-care and environmental interactions and the quality of life for persons with Parkinson's. An independent means of transport is likely to prevent disengagement in activities and, although the amount and variety vary, participation could remain high despite self-consciousness. The lack of high participation or the lack of various activities can relate to dissatisfaction with life.[124] In nutritional matters, "eat well, stay well" with Parkinson's disease clearly highlights the food, fluid, and vitamin needs and requirements. AOTA[13,125] consumer reports advise patients that environment fitness is important, and occupational therapists contribute to modifications for increased accessibility in every room in the home. In the

bathroom, a hand-held shower head, large shower and bath controls, grab bars, and faucets with levers are such improvements. In the kitchen, they include adjusting sinks and countertops and designing the workspace to avoid twisting and turning while preparing a meal. In hallways and doorways, clearing clutter, providing unobstructed openings, securing carpet, and adding handrails will help. In the living room and bedroom, arranging furniture for clear passage, making the telephone accessible, and increased security are important. All these improvements will promote efficiency in completing daily tasks, promote energy conservation, and help prevent falls and other home injuries.[126–128]

It is important to recognize that advances, such as therapeutic interventions and surgical procedures, therapeutic use of drugs, and occupational therapy and rehabilitation strategies, are alone not enough. The importance of psychosocial management and perceived inner need and drive by Parkinson's patients contribute to their success.

PSYCHOSOCIAL OCCUPATIONAL THERAPY AND PARKINSON'S DISEASE

Among the domains composing the quality of life index, although no universal criteria exist, is the individual's psychosocial status.[47] To gain optimal understanding of patients' perception and self-evaluation of how PD has altered their social and emotional functioning, the occupational therapist uses a client-centered approach, empowering the patient to manage the PD.[129] One such tool that assists the patient in delineating life satisfaction is the Canadian Occupational Performance Measure (COPM), a semistructured interview that allows the patient to prioritize areas that are most important to maintain a quality of life and that often center around psychosocial needs, such as maintaining interpersonal interactions and continuing in family, religious, and leisure roles.[70,129]

Using a semistructured interview approach, Mainson and Caird[85] spoke with 74 PD patients who were living at home to determine the impact of the disease on the meaningful activities (hobbies) they had pursued prior to diagnosis, particularly relating to changes over the past 5 years. While 95% continued involvement in sedentary and passive activities around the house, such as watching TV or reading, only 31% maintained involvement in activities, such as knitting and sewing, that require fine motor dexterity. Eighty-three percent of indoor gardeners maintained their passion, but the interest in outdoor gardening saw a marked decline. Less than one-third of the PD patients continued to engage in community social activities—going to a place of worship, going out to lunch, or seeing a movie. While lack of mobility contributed to this decline, the embarrassment resulting from the tremor and social anxiety, experienced by two-thirds of the patients, were obstacles.[85] Not believing that one can maintain control over the illness impedes self-efficacy.[128] Individuals who took up a new hobby since the PD diagnosis were the exception; however, a few risked cycling and swimming.[85]

The overwhelming trend to go out less and engage in sedentary in-home activities was a concern of the researchers,[85] who conceded that the absence of motivation is inherent in the disease. They emphasized, however, the importance in continuing interests into one's mid-life. Concern about the participation and motivation of PD patients in social activities led Gauthier et al.[78] to implement a study of occupational therapy treatment groups. Using pretreatment evaluations, randomly assigning patients to either a treatment or control group, they followed up by evaluating approximately 30 participants in each group after therapy after 6 months and then at the end of 1 year. The experimental group received 20 hr of occupational therapy group treatment, 10 sessions, twice weekly over 5 weeks, consisting of mobility, functional, educational, and dexterity exercises, with interspersed with socialization. The Bradburn Index of psychological well-being showed that patients perceived greater psychological well-being after therapy, even noting a regression in their symptoms as compared with the control group that showed no change. Additionally, patients in the treatment group showed increased attention to grooming and demonstrated a clearer understanding of PD, which investigators linked to increased self-confidence. Over the ten, sessions the researchers also observed initial egocentric attitudes among the group participants changing to concern and interest in the well-being of others.[78] Davis[130] concluded that group activities designed to maintain a patient's level of functional activity and socialization, such as frisbee, shuffleboard, ball tossing and kicking, volleyball, hot potato, basketball, matching, sing-a-long with rhythm instruments, bean bag toss, and bowling, are optimal for patients to show encouragement and support for one another. While the effect of socialization on functional activity in the groups is speculative, Gauthier et al.[78] noted that one of the major motor systems (bradykinesia) improved along with psychological well-being and positive behavioral change. Through involvement ingroup activities, PD patients are able to remain active longer,[78,130] and the social support and improved motivation are by-products of the group participation. Maintenance of interpersonal relationships in friends and family may ultimately determine the PD patient's connection to occupation and the world at large.[85]

ACKNOWLEDGMENTS

The authors wish to thank David L. Armbruster, Ph.D., Scientific Publications, and Emeritus Professor Mary Ellis Gaston, University of Tennessee Health Science Center, for their editorial assistance with sections of this chapter.

REFERENCES

1. Hallett, M., Analysis of abnormal voluntary and involuntary movements with surface electromyography, *Adv. Neurol.,* 33, 387–414, 1983.
2. Paulson, H. L. and Stern, M. B., Clinical manifestations in Parkinson's disease, in *Movement Disorders: Neurologic Principles and Practice,* Watts, R. L. and Kohler, W. C., Eds., McGraw-Hill, New York, 184–1997.
3. De Goede, C. J. T. et al., The effects of physical therapy in Parkinson's disease, *Arch. Phys. Med. Rehabil.,* 82, 509–515, 2001.
4. Marsden, C. D., Parkinson's disease, *J. Neurol. Neurosurg., Psychiatry.,* 57, 672–681, 1994.
5. Penney, J. B., and Young, A. B., Speculations on the functional anatomy of BG disorders, *Annu. Rev. Neurosci.,* 6, 73–94, 1983.
6. Hallet, M. and Khoshbin, S., A physiological mechanism of Bradykinesia, *Brain,* 103, 301–322, 1980.
7. Jankovic, J. and Stacy, M., Movement disorders, in *Textbook of Clinical Neurology,* Goetz, C. G. and Pappert, E. J., Eds., Saunders, Philadelphia, PA, 655–679, 1999.
8. Alexander, G. E., DeLong, M. R., and Strick, P. L. Parallel organization of functionally segregated circuits linking basal ganglia and cortex, *Annu. Rev. Neurosci.,* 9, 357–381, 1986.
9. American Occupational Therapy Association. Occupational therapy practice framework: Domain and practice, *Am. J. Occup. Ther.,* 56, 609–639, 2002.
10. Law, M. Participation in the occupation of everyday life, *Am. J. Occup. Ther.,* 56, 640–649, 2002.
11. Christiansen, C. and Townsend, E. An introduction to occupation, in C. H. Christiansen, C. H. and Townsend, E. A., Eds., *Introduction to Occupation,* Prentice Hall, Upper Saddle River, New Jersey, 1–28, 2003.
12. Wilcock, A. A., The Dorris Sym Memorial Lecture: Developing a philosophy of occupation for health, *Br. J. Occup. Ther.,* 62. 192–198, 1999.
13. American Occupational Therapy Association. Modifying your home for independence, *AOTA Consumer Information—Tip Sheets,* AOTA.org, Maryland, 1–3, 2003.
14. Poglar, J. M. and Landry, J. E., Occupations as a means for individual and group participation in life, in Christiansen, C. H. and Townsend, E. A., Eds., *Introduction to Occupation,* Prentice Hall, Upper Saddle River, New Jersey, 197–220, 2003.
15. Flowers, K. A., Visual closed-loop and open-loop characteristics of voluntary movements in patients with Parkinsonism and intentional tremor, *Brain,* 99, 269–310, 1976.
16. Kannenberg, K. and Greene, S., Infusing occupation into practice, *OT Pract.,* 8, 27–34, 2003.
17. Morris, M. E. et al., Stride length regulation in Parkinson's disease, Normalization strategies and underlying mechanisms, *Brain,* 119, 551–568, 1996.
18. Protas, E. J., Stanley, R. K. and Jankovic, J., Exercise and Parkinson's disease, *Crit. Rev. Phys. Rehabil. Med.,* 8, 253–266, 1996.

19. Murphy, S. and Tickle-Degnen, L., The effectiveness of occupational therapy-related treatments for persons with Parkinson's disease: A meta-analytic review, *Am. J. Occup. Ther.,* 55,385–392, 2001.

20. Trombly, C. A., Conceptual foundations for practice, in *Occupational Therapy for Physical Dysfunction,* Trombly, C. A. and Radomski, M. A., Eds., 5th Ed., Lippincott, Williams and Wilkins, Philadelphia, PA, 1–16, 2002.

21. Finch, E. et al., *Physical Rehabilitation Outcome Measures: A Guide to Enhanced Clinical Decision Making,* 2nd Ed., Lippincott, Williams and Wilkins, Philadelphia, PA, 6–60, 2003.

22. World Health Organization. ICIDH-2 International classification of functioning, disability, and health, WHO, Geneva, 2001.

23. Hinojosha, J. and Youngsman, M. J. broadening the construct of independence, *Am. J. Occup. Ther.,* 56. 660, 2002.

24. Haaland, K. Y., Harrington, D. L. and Knight, R. Y., Neural representation of skilled movement, *Brain,* 123, 2306–2313, 2002.

25. Khotianov, N., Singh, R., and Singh, S., Lewy Body dementia: Case report and discussion, *J. Am. Board Fam. Pract.,* 15, 50–54, 2002.

26. Edwards, N., Aquariums may pacify Alzheimer's patients. Purdue University News, purduenews@uns.purdue.edu, 1–4, 1999.

27. Lang, A. E., Clinical rating scales and videotape analysis, in *Therapy for Parkinson's Disease,* Koller, W. C. and Paulson, G., Eds., Dekker, New York, 3–15, 1991.

28. Harridge, C. and Shah, S., The Moss Kitchen assessment revised, *N. Z. J. Occup. Ther.,* 46, 5–9, 1995b.

29. Ivanzo, A. et al., Sleep symptoms and polysomnographic architecture in advanced Parkinson's disease after chronic bilateral; subthalamic stimulation, *J. Neurol. Neurosurg. Psychiatry.,* 72, 661–664, 2002.

30. Short, R., Modafinil reduces sleepiness in Parkinson's disease, *Sleep,* 25, 905–909, 2002.

31. Hallett, M., Shahani, B. T. and Young, R. R., Analysis of stereotyped voluntary movement at the elbow in patients with Parkinson's disease, *J. Neurol. Neurosurg. Psychiatry.,* 40, 1129–1135, 1977.

32. Duncan, P. W. et al., Functional reach: predictive validity in a sample of elderly male veterans, *J. Gerontol.,* 47, M93–98, 1992.

33. Weiner, D. K. et al., Functional reach: A marker of physical frailty, *J. Am. Geriatr. Soc.,* 40. 203–207, 1992.

34. Melnick, M. E., Radtka, S., and Piper, M. P., Gait analysis and Parkinson's disease, *Rehab. Manag.,* 15, 46–48, 2002.

35. Chang, R., Guan, L., and Burne, J. A., An automated form of video image analysis applied to classification of movement disorders, *Disabil. Rehabil.,* 22, 97–108, 2002.

36. Graziano, M., Common mobility problems and how to address them, *Adv. Clin. Neurosci. Rehabil.,* 3, 38, 2003.

37. Spyers-Ashby, J. M. et al., Classification of normal and pathological tremors using a multidimensional electromagnetic system, *Med. Eng. Phys.,* 21. 713–723, 1999.

38. Behrman, A. L. et al., Is the functional reach useful for identifying falls risk among individuals with Parkinson's disease, *Arch. Phys. Med. Rehabil.,* 83, 538–542, 2002.

39. Rosser, R. M., and Kind, P., A scale of valuations of states if illness: is there a social conscious? *Int. J. Epidemiol.,* 7, 347–358, 1978.

40. The Euroqol Group. Euroqol: a new facility for the measurement of health related quality of life, *Health Policy,* 16, 199–208, 1990.

41. Glaser, A. W. et al., Applicability of the Health Utilities Index to a population of childhood survivors of CNS tumors in the U.K, *Eur. J. Cancer.,* 35, 256–261, 1999.

42. Stern, M. B., The clinical characteristics of Parkinson's disease and parkinsonian syndromes: diagnosis and assessment, in *The Comprehensive Management of Parkinson's Disease,* Stern, M. B. and Hurlig, H. I., Eds., PMA Publishing, New York, 3–50, 1987.

43. Gaudet, P., Measuring the impact of Parkinson's disease: An occupational therapy perspective, *Can. J. Occup. Ther.,* 69, 104–113, 2002.

44. Hoehn, M., and Yahr, M., Parkinsonism: onset, progression, and mortality, *Neurol.,* 17, 427–442, 1967.

45. Peto, V., Jenkinson, C., and Fitzpatrick, R., PDQ-39. A review of the development, validation, and application of a Parkinson's disease quality of life questionnaire and its associated measures, *J. Neurol.,* 245 (Suppl. 1), S10–S14, 1998.

46. Beck, A. T. et al., An inventory for measuring depression, *Arch. Gen. Psychiatry,* 4, 53–63, 1961.

47. Siderowf, A., Cianci, H. J. and Rorke, T. R., An evidence-based approach to management of early Parkinson's disease, *Hosp. Physician,* 37, 63–76, 2001.

48. Copperman, L. F., Forwell, S. J., and Hugos, L., Neurodegenerative diseases, in *Occupational Therapy for Physical Dysfunction,* Trombly, C. A. and Radomski, M. V., Eds., 5th ed., Lippincott, Williams and Wilkins, Philadelphia, PA, 895–908, 2002.

49. Barry, P. et al., Rehabilitation inpatient screening of early cognitive recovery, *Arch. Phys. Med. Rehabil.,* 70, 902–906, 1989.

50. Sandford, J. A., Browne, R. J. and Turner, A., *Software for Cognitive Training & Psychological Testing,* Brain Train, Richmond, VA, 2003.

51. Harridge, C. and Shah, S., A review of meal preparation ability as a measure of instrumental activities of daily living, *N. Z. J. Occup. Ther.,* 46, 5–12, 1995a.

52. Hays, C. A., Kassimir, J. and Parkin, J., Eds., *Sample Forms for Occupational Therapy,* American Occupational Therapy Association, Rockville, MD, 1996.

53. Shah, S., Vanclay, F. and Cooper, B., Improving the sensitivity of the Barthel Index for stroke rehabilitation, *J. Clin. Epidemiol.,* 42, 703–709, 1989.

54. Shah, S., Guidelines for the Barthel Index (Expanded) or Modified Barthel Index, in *Compendium of Quality of Life Instruments,* Salek, S., Ed., Wiley & Sons, New York, 1–9, 1998.

55. Shah, S., Biometric and psychometric qualities of the Barthel Index, P*hysiotherapy,* 80, 769–771, 1994.

56. Corbett, P. J., Focus on research: Functional assessment in Parkinson's disease: An investigation into the sensi-

tivity of commonly used assessments to determine change, pre-and postamorphine intervention, *Br. J.Occup. Ther.,* 61, 464, 1998.

57. Shah, S., Cooper, B., and Maas, F., Performance of perceptual tasks by neurologically unimpaired adults using the dominant right hand, *Aust. Occup. Ther. J.,* 40.165–178, 1993.

58. Figenson, J. S., Polkow, L. and Keegan, N., *Burke Perceptual Profile (BUPP),* American Neurological Foundation, New York, 1980.

59. Teive, H. A. G., and Sa, D. S., The Rolex sign first manifestation of Parkinson's disease, *Arq. Neuropsiquiatr.,* 58, 1–4, 2002.

60. Baylor Neurology Patient #33, http://www.hom.bcm.tmc.edu, Department of Neurology, Baylor College of Medicine, 2002.

61. Johnson, K. A. et al., Bimanual coordination in Parkinson's disease, *Brain,* 121, 743–753, 1998.

62. Stelmach, G. E. and Phillips, J. G., Movement disorders in Parkinson's disease, in *Physical Therapy Management of Parkinson's Disease,* Turnbull, G. I., Ed., Churchill Livingstone, New York, 37–48, 1992.

63. Laplane, D. et al., Clinical consequences of corticectomies involving the supplementary motor area in man, *J. Neurol. Sci.,* 34, 301–314, 1977.

64. Pahapill, P. A. and Lozano, A. M., The pedunculopontine nucleus in Parkinson's disease, *Brain,* 123, 1767–1783, 2000.

65. Stockmeyer, S. A., An interpretation of the approach by Rood to the treatment of neuromuscular dysfunction, *Am. J. Phys. Med.,* 46, 900–961, 1967.

66. Hocherman, A. and Aharon-Peretz, A., Two-dimensional tracing and tracking in patients with Parkinson's disease, *Neurol.,* 44, 111–116, 1994.

67. Carey, J. R. et al., Sex differences in tracking performance in patients with Parkinson's disease, *Arch. Phys. Med. Rehabil.,* 83, 972–977, 2002.

68. Quintyn, M. and Cross, E., Factors affecting the ability to initiate movement in Parkinson's disease, *Phys. Occup. Ther. Geriatr.,* 4, 51–60, 1986.

69. Chan, J. L. and Ross, E. D., Left-handed mirror writing following anterior cerebral artery infarction: evidence for non-mirror transformation of motor programs by right SMA, *Neurol.,* 38, 59–63, 1988

70. Connor, N. P. and Abbs, J. H., Sensorimotor contributions of the basal ganglia: Recent advances, *Phys. Ther.,* 70, 864–872, 1990.

71. Fahn, S., Hyperkinesia and hypokinesia, in *Textbook of Clinical Neurology,* Goetz, C. G. and Pappert, E. J., Eds., Saunders, Philadelphia, PA, 267–284, 1999.

72. Kishore, A. et al., Evidence of somatotopy in GPI from results of pallidotomy, *Brain,* 123, 2491–2500, 2000.

73. Weiner, W. J. and Singer, A. C., Parkinson's disease and non-pharmacological treatments, *J. Am. Geriatr. Soc.,* 37, 359–363, 1989.

74. DeLong, M. R., The neurophysiologic basis of abnormal movements in basal ganglia disorders, *Neurobehav. Toxicol. Teratol.,* 12, 366–375, 1983.

75. Platz, T., Brown, R. G. and Marsden, C. D., Training improves the speed aimed at movements in Parkinson's disease, *Brain,* 121, 505–514, 1998.

76. Sheridan, M. R., Flowers, K. A. and Hurrell, J., Programming and execution of movements in Parkinson's disease, *Brain,* 110, 1247–1271, 1987.

77. Rascol, O. et al., Cortical motor over activation in Parkinson's patients with L-dopa-induced peak-dose dysfunction, *Brain,* 121, 527–533, 1998.

78. Gauthier, L., Dalziel, S., and Gauthier, S., The benefits of group therapy for patients with Parkinson's disease, *Am. J. Occup. Ther.,* 41, 360–365, 1987.

79. Levin, B. et al., Visuospatial impairment in Parkinson's disease, *Neurol.,* 41, 365–369, 1991.

80. Lazaruk, L., Visuospatial impairment in persons with idiopathic Parkinson's disease: A literature review, *Phys. Occup. Ther. Geriatr.,* 12, 37–48, 1994.

81. Shah, S., Cooper, B. and Lyons, M., Investigation of the transient ischemia workload and its incidence: Implications for occupational therapy research, *Occup. Ther. J. Res.,* 12, 357–373, 1992.

82. Brownbridge, E., Unique Parkinson's program emphasizes self-management, *Ger. Aging,* 2, 3–27, 1999.

83. Deane, K. H. O. et al., Systematic review of paramedical therapies for Parkinson's disease, *Mov. Disord.,* 117, 984–991, 2002.

84. Deane, K. H. et al., Comparison of physiotherapy techniques for Parkinson's disease, *Cochran Database of Sys. Rev.,* 1, CD002815, 2001.

85. Mason, L., and Caird, F. I., Survey of the hobbies and transport of patients with Parkinson's disease, *Br. J. Occup. Ther.,* 48, 199–200, 1985.

86. Nieuwboer, A. et al., The effect of a home physiotherapy program for persons with Parkinson's disease, *J. Rehabil. Med.,* 33, 266–272, 2001.

87. Johnell, O. et al., Fracture risk in patients with Parkinson's: a population based study in Olmsted County, Minnesota, *Age Aging,* 21, 32–38, 1992.

88. Hummelsheim, H., Hauptmann, B. and Neumann, S., Influence of physiotherapeutic facilitation techniques on motor evoked potentials in centrally paretic hand extensor muscles, *Electroencephalogr. Clin. Neurophysiol.,* 97, 18–28, 1995.

89. Miyai, I. et al., Long-term effect of body weight-supported treadmill training in Parkinson's disease: a randomized control trial, *Arch. Phys. Med. Rehabil.,* 83, 1370–1373, 2002.

90. Grillner, S., Control of locomotion in bipeds, in *Handbook of Physiology,* Brooks, V. B., Ed., American Physiological Society, Bethesda, MD, 1179–1236, 1981.

91. Halliday, S. E. et al., The initiation of gait in young, elderly, and Parkinson's disease subjects, *Gait Posture,* 8, 8–14, 1998.

92. Bronstein, A. M. et al., Visual control of balance in cerebellar and parkinsonian syndromes, *Brain,* 113, 767–779, 1990.

93. Winter, D. A. and Eng, P., Kinetics: Our window into the goals and strategies of the central nervous system, *Behav. Brain Res.,* 67: 111–120, 1995.

94. http://www.geocities.com/parkinson.html/2003.

95. Susman, E., ANA: Walkers do not benefit freezing in Parkinson's disease patients, *Doctor's Guide,* http://www.docguide.com/news/content.nsf/new…/, P\S\L Consulting Group Inc., 2002.

96. Ahmad, S. O., Mu, K. L. and Scott, S. A., Meta-analysis of functional outcome in patients treated with unilateral pallidotomy, *Neurosci. Lett.,* 6, 2001.

97. de Bie, R. M. et al., Outcome of unilateral pallidotomy in advanced Parkinson's disease: Cohort study of 32 patients, *J. Neurol. Neurosur. Psychiatry.,* 71, 375–382, 2001.

98. Aziz, T. and Yianni, J., Surgical treatment of Parkinson's disease, *Adv. Clin. Neurosci. Rehabil.,* 2, 21–22, 2003.

99. The Deep-Brain Stimulation for Parkinson's Disease Study Group, Deep-brain stimulation of the subthalamic nucleus or the pars interna of the globus pallidus in Parkinson's disease, *N. Engl. J. Med.,* 345, 956–963, 2001.

100. Saint-Cyr, J. A. et al., Neuropsychological consequences of chronic bilateral stimulation of the subthalamic nucleus in Parkinson's disease, *Brain,* 123, 2091–2108, 2001.

101. Saint-Cyr, J. A. and Trepanier, L. L., Neuropsychologic assessment of patients for movement disorder surgery, *Mov. Disord.,* 15, 771–783, 2000b.

102. Montgomery, E. B., Jr. and Rezai, A. R., Deep brain stimulation for Parkinson's disease: Is it right for your patient? Cleveland: The Cleveland clinic center for functional and restorative neuroscience, www.clevelandclinicmeded.com

103. Shults, C. W. et al., Effects of coenzyme Q10 in ear of patients with Parkinson's disease, *Arch. Neurol.,* 59, 1541–1550, 2002.

104. Hubble, J. P. and Berchou, R. C., Parkinson's disease: Medications and side effects, http://www.parkinsonism.org, 2003.

105. Sylvester, B. Entacapone extends benefits of leva-dopa and improves condition of Parkinson's disease patients, *Doctor's Guide News,* 1–2, 2002.

106. Danisi, F. Parkinson's disease: Therapeutic strategies to improve patient function and quality of life, *Geriat.,* 57, 46–50, 2002.

107. Nandi, D. et al., Reversal of Akinesia in experimental Parkinsonism by GABA antagonist microinjections in the pedunculopontine nucleus, *Brain,* 125, 2418–2430, 2002.

108. Chen, H. et al., Dairy products and Parkinson's disease, *Ann. Neurol.,* 52, 793–801, 2002.

109. Siebert, C., Aging in place: implications for occupational therapy, *OT Pract.,* 8, CE-1-CE-7, 2003.

110. Lander, C., Parkinson's disease: Diagnosis and management, *Mod. Med. Aust.,* 5, 18–26, 1995.

111. McNaught, K. S. P. et al., Lewy bodies and aggresomes: altered protein handling and Parkinson's disease, *Eur. J. Neurosci.,* 16, 2136–2148, 2002.

112. Sato, Y. et al., Vitamin K deficiency and osteopenia in vitamin D- deficient elderly women with Parkinson's disease, *Arch. Phys. Med. Rehabil.,* 83, 86–91, 2002.

113. Pacchetti, C. et al., "off" painful dystonia in Parkinson's disease treated with botulinum toxin, *Mov. Disord.,* 10, 333–336, 1995.

114. Childers, M. K. et al., Treatment of painful muscle syndromes with botulinum toxin: A review, *J. Back Musculoskeletal Rehabil.,* 10, 89–96, 1998.

115. Lees, A. J., New advances in the management of late stage Parkinson's disease, *Adv. Clin. Neurosci. Rehabil.,* 1, 7–8, 2001.

116. Novatis., The US FDA approves Stalevo for treatment of Parkinson's, *Doctor's Guide,* 1–4, June 2003.

117. Schrag, A. and Quinn, N., Dyskinesias and motor fluctuations in Parkinson's disease, *Brain,* 123, 2279–2305, 2000.

118. Lee, H. C., Lee, A. H. and Cameron, D., Measuring visual attention skill of older drivers by using a driving simulator, *Am. J. Occp. Ther.,* 57, 324–328, 2003.

119. Ondo, W. G. et al., Daytime sleepiness and other sleep disorders in Parkinson's disease, *Neurol.,* 57, 1392–1396, 2001.

120. McConnell, H., No reason to stop driving by Parkinson's patients taking dopamine drugs, *Br. Med. J.,* 324, 1483–1487, 2002.

121. Sa, D. S. and Chen, R., Parkinson's disease: An update on therapeutic strategies, *Geriatr. Aging,* 5, 8–14, 2002.

122. Thaut, M. H. et al., Rhythmic auditory stimulation in gait training for Parkinson's disease patients, *Mov. Disord.,* 11, 193–200, 1996.

123. Spillantini, M. G. et al., Alpha-synuclein in Lewy bodies, *Nature,* 388, 839–840, 1997.

124. Beattie, A., Aids to daily living for the patient with Parkinson's disease, *Br. J. Occup. Ther.,* 44, 53–55, 1981.

125. Griffin, J. and McKenna, K., Influence on leisure and life satisfaction of elderly people, *Phys. Occup. Ther. Geriatr,* 15, 1–16, 1998.

126. Deane, K. H. O. et al., A survey of current occupational therapy practice for Parkinson's disease in the United Kingdom, *Br. J. Occp. Ther.,* 66, 193–197, 2003.

127. Majsak, M. J. et al., The reaching movements of patients with Parkinson's disease under self-determined maximal speed and visually cued conditions, *Brain,* 121, 755–766, 1998.

128. Montgomery, E. B., Jr. et al., Patient education and health promotion can be effective in Parkinson's disease: A randomized controlled trial, *Am. J. Med.,* 97: 429–435, 2001.

129. Cohn, E. S. et al., Introduction to evaluation and interviewing, in *Willard & Spackman's Occupational Therapy,* Crepeau, E. B. et al., Eds., 10th ed., Lippincott Williams and Wilkins, Philadelphia, PA., 249, 2003.

130. Davis, J. C., Team management of Parkinson's disease, *Am. J. Occup. Ther.,* 31: 305–308, 1977.

63 Music Therapy for People with Parkinson's

Sandra L. Holten
Struthers Parkinson's Center

CONTENTS

INTRODUCTION

Music therapy is an established, allied health profession in which music is used within a therapeutic relationship to address the physical, psychological, cognitive, spiritual, and social needs of individuals with an illness or disability. Music therapists complete a bachelors or masters music therapy degree and a minimum of 1200 hr combined clinical practice and internship. To have board certified credentials, the music therapist is required to pass a certification examination offered by the Certification Board for Music Therapists. Board certified (BC) music therapists must complete 100 continuing music therapy education credits (CMTE) within a five-year cycle to maintain their board-certified status. Continuing education may include institutes that provide specialized training in a specific area of music therapy.

In the United States, music therapy was established as a profession in 1950 as a result of work in which music was used with patients in veteran's hospitals following World War II. Today, more than 5,000 music therapists are employed throughout the U.S. in settings such as hospitals, clinics, community centers, day care centers, substance abuse facilities, schools, nursing homes, rehabilitation centers, and private practices.[1]

Recent research has provided evidence that validates music therapy in a way that it has not previously. As a result of research done by the Center for Biomedical Research in Music in Fort Collins, Colorado, music therapy techniques have been more clearly defined and standardized into an approach called *neurologic music therapy (NMT)*. NMT is defined as the therapeutic application of music to cognitive, sensory, and motor dysfunction due to neurologic disease of the human nervous system. It is based on a neuroscience model of music perception and production and the influence of music on functional changes on nonmusical brain and behavior functions. NMT treatment techniques are based on scientific research and are directed toward functional goals. They are standardized and applied to therapy as *therapeutic music interventions* that are adaptable to the patient's needs. In addition to music therapy training, a neurologic music therapist is educated in the areas of neuroanatomy/physiology, brain pathologies, medical terminology, and rehabilitation of cognitive and/or motor functions.[2] NMT applies directly in addressing the symptoms of Parkinson's.

Within the scope of providing music therapy services, necessary steps are followed in planning and facilitating interventions. Music therapists assess patient needs, identify goals, and then translate and apply the qualities of music in a focused, intentional manner to plan and facilitate interventions, which improve the quality of life and optimize the functioning level of the people with whom they work. In the monograph, *A Scientific Model of Music in Therapy and Medicine,* Michael Thaut discusses an approach called *transformational design model (TDM)* that "provides a system for the therapist to immediately translate the scientific model into functional clinical practice."[3] He delineates five basic steps of the TDM as follows:

1. Diagnostic and functional assessment of the patient
2. Development of therapeutic goals/objectives
3. Design of functional, nonmusical therapeutic exercises and stimuli
4. Translation of Step 3 into functional, therapeutic music experiences
5. Transfer of therapeutic learning to real-world applications

All of the steps, excluding the fourth, are shared by all therapy disciplines. Thaut notes that "the crucial clinical process for music therapists occurs in Step 4. Here, a role for music therapists emerges that is unique to the profession: translating functional and therapeutic exercises and stimuli into functional therapeutic music exercises and stimuli...." The correspondence and translation between the musical and nonmusical elements or exercises is direct.

Music therapists providing services for people with Parkinson's face a unique and rewarding challenge. Parkinson's is a disease that affects every fiber of a person's being. The qualities of music have the potential to affect humans by eliciting responses that facilitate change and positively affect rehabilitation and healing process. Music therapy is an exciting, effective modality that holds great promise in improving the quality of life for people with Parkinson's. This chapter explores the applications and related research for music therapy within the areas of sensorimotor function, speech and cognition, and psychosocial challenges for people with Parkinson's.

NMT TECHNIQUES FOR SENSORIMOTOR TRAINING

NMT techniques for sensorimotor training address gait, posture, and arm and trunk training. Three techniques have been developed and standardized based on a body of research. These techniques are *rhythmic auditory stimulation (RAS), patterned sensory enhancement (PSE),* and *therapeutic instrumental music performance (TIMP).*

Rhythmic auditory stimulation is a specific technique of rhythmic motor cueing to facilitate training of movements that are intrinsically and biologically rhythmical. Because the most important type of these movements in humans is gait, RAS is almost exclusively used for gait rehabilitation to aid in the recovery of functional, stable, and adaptive walking patterns in patients with significant gait deficits. The underlying mechanism in RAS is rhythmic entrainment, which enhances gait. In this process, rhythm is an external timekeeper. Through the anticipation of the functional movement patterns, rhythmic cues are provided to synchronize and change existing to desired movement frequencies. This entrainment retrains the motor programs through anticipatory cueing of functional movement patterns. Rhythmic cues are provided to synchronize and change existing to desired movement frequencies. "Through frequency entrainment of motor patterns, rhythm stabilizes the timing, kinematic control, and force applications in movement."[2]

There are two ways in which RAS can be used. First, it can be used as an immediate entrainment stimulus providing rhythmic cues during the movement.[2] Given the difficulties for people with Parkinson's with "freezing" episodes, this is an effective coping strategy to assist patients in the initiation of movement. Utilization of rhythm through verbalization of a cadence or counting, singing or playing a music that encourages movement, such as a march, provides an excellent compensation strategy. Although people with Parkinson's or their care partners may already use this method, the music therapist who is trained in using these techniques will be able to go beyond the obvious in the execution of RAS. Therapists will capitalize on and enhance the rhythm with dynamics, force, and pitch as needed specifically by the patient.

The second way in which RAS can be employed is as a facilitating stimulus for training where patients train with RAS for a certain period of time so as to achieve more functional gait patterns, which they then transfer to walking without rhythmic facilitation.[2] When utilizing RAS in this way, the music therapist will follow a specific treatment protocol. The length of treatment time is determined by the patient needs and may be done in an inpatient setting or with a home program on an outpatient basis. The music therapist may co-treat with a physical therapist. Because their approaches are different, collaboration can help to solidify the patient's learning and rehabilitation in gait training. Neurologic music therapists specialized training provides them the ability to work with patients in gait assessment and training.

From the early 1990s to the present, the body of research that demonstrates the effectiveness of both applications with different gait disorders continues to grow. Thaut et al.[4] studied the effect of RAS gait training with people with Parkinson's. Study participants involved in the RAS program significantly ($p < 0.05$) improved their gait velocity by 25%, stride length by 12%, and step cadence by 10% more than subjects who self-paced, improving their velocity by 7%, and those who received no training, whose velocity decreased by 7%. EMG patterns in the anterior tibialis and vastus lateralis muscles were significantly influenced by RAS ($p < 0.05$). This study also found that RAS improved velocity ($p < 0.01$), cadence ($p < 0.02$), and stride length ($p < 0.03$) for patients with Parkinson's on medication, for those off medication, and for normal (non-Parkinson's) controls, with RAS 10% faster than baseline cadence.[5] In addition, a short-term continuance effect of RAS on the gait patterns was maintained during an uncued walking condition in all three groups. It is vital to stress the importance of maintaining medication management as an integral piece in the treatment of Parkinson's. However, an interesting finding in this study was that rhythmic entrainment mechanisms were still able to improve gait patterns in the absence of dopaminergic medications. This may be an interesting point for future research.

McIntosh et al.[6] demonstrated long-term carryover effects of RAS. The study showed, "After an improvement of about 18% in gait velocity after three weeks of training, the improved gait parameters were retained for three weeks. At week four, follow-up testing showed a first decline of about 10%, and at week five, gait performance had declined almost to pretest values."

The far-reaching implications for people with Parkinson's, when these techniques are properly applied by a trained neurologic music therapist, are important to consider when managing this disease. When velocity and stride length are improved, a person with Parkinson's experiences more gait stability. With this stability comes a decreased fall risk, creating a safer ambulation for the patient. In addition, there is solid evidence that music can be accessed even when a patient is not receiving dopaminergic medication and for a potential three-week carryover even when the music is not present. There is great potential in improving functional mobility and then further improving a person's independence and quality of life.

PATTERNED SENSORY ENHANCEMENT

Patterned sensory enhancement (PSE) is a technique that has a broader application than RAS in that it is used for movements that, unlike gait, may not be intrinsically or biologically rhythmic. However, "the underlying neurologic mechanisms for RAS extend to the technique of PSE. Music provides a sensory cue that temporally structures movement tasks and enhances movement patterns, thus accelerating the rehabilitation process."[2] Music, like movement, is multidimensional, and thus it is uniquely positioned to provide the support and motivation to cue and to truly facilitate the spatial, temporal, and force elements of movement. This is achieved through practice and repetition. It is important to note, "The clinical goal addressed through PSE is the practice of functional movements of daily life activity or the motor patterns underlying these movements. Functional movement organized in repetitive patterns facilitates active training and learning, a common and effective strategy of motor rehabilitation."[2]

PSE can be used to directly address functional improvement of deficits caused by symptoms of Parkinson's. Posture may improve with exercises such as stretching arms high above a person's head. When PSE is employed in within community and day program exercise groups, patients are able to extend their arms above the head by spatially cueing them with an ascending pitch line. The stretch can be maximized by cueing the patient to increase the stretch by moving further away from their body with higher pitch and with force cueing of volume and possibly dissonance. The music provides the direction in which patients will move their arms with either an ascending or descending musical line as well as providing a point of reference for the size of the movement. If a patient's goal is to work on weight shift to improve functional gait, this might be cued temporally by the music therapist choosing an appropriate meter and by emphasizing the beats within the meter that will facilitate the weight shift.

Anecdotal reports from patients indicate that, when PSE is used, they experience the feeling of "freedom" in their movements. They state that the movements are "easier" and they feel that, because music supports their movements, they are able to increase the range of the movements and the duration of their exercises. This in turn increases the repetitions necessary to motor rehabilitation and improves the patient's endurance. PSE supports passive range of motion exercises for patients confined to bed with severe muscle rigidity. Therapists report a dramatic increase in patient's range of motion, and the patients report decreased pain with the exercises whenever PSE was provided in support of the treatment.

The important concept is that functional goals are met through music by supporting and driving the movement. Along with maximizing movements, PSE can increase the repetitions of the movements by minimizing the perception of fatigue and discomfort. Increasing the number of repetitions results in increased benefits in the area of improved endurance and functional movement. Learning and rehabilitation are solidified through this process.

THERAPEUTIC INSTRUMENTAL MUSIC PERFORMANCE

Therapeutic instrumental music performance (TIMP) uses the playing of musical instruments to further accomplish rehabilitative goals. Playing musical instruments involves a spectrum of movement. For example, appropriate instrumental performance optimizes gross and fine motor skills and often requires repetitive movements. Consequently, therapeutic instrument playing in motor rehabilitation assists the patient in improving movements and increasing strength, endurance, and range of motion. It is further possible to enhance functional hand movements, finger dexterity, and limb coordination.

As in the case of PSE, TIMP interventions can directly work on functional physical goals that address symptoms of Parkinson's. Although this might be accomplished through playing an instrument in a traditional manner, the music therapist may adapt the instrument or the way in which the patient plays the instrument to achieve the desired outcome. In circumstances where patients who enjoy activities that require fine motor control, like sketching, but have a decreased sensation and range of motion in their fingers, the traditional playing of an omnichord (or electronic autoharp) might be used. This instrument requires fine motor functions in the fingers to change chords and to strum along a textured fingerboard. Positioning the omnichord on a flat surface in front of and sufficiently distanced away from the patient's body also improves range of motion by stretching toward the instrument. Posture is also addressed in this exercise, since erect posture needs to be maintained to reach the instrument. Patients have reported that consistent use of this or other instruments that demand fine motor exercises have provided them increased sensation and range of motion in their fingers and improved ability to perform other fine motor tasks.

Posture changes for many patients diagnosed with Parkinson's include forward head position, stooped shoulders with rounding or the upper back and a forward trunk position with increased bending of hip and knee joints. This impairs balance, walking, breathing, and the ability to display eye contact for communication. It poses challenges for swallowing food and saliva, as gravity fights these efforts. In a TIMP session, the music therapist might use a drum in a nontraditional way by holding it above patient's head or fastening it to an IV pole to position an instrument at eye level or higher in front of a patient. Playing the instrument in this position repetitively creates a healthier pattern for the position of the muscles. Anecdotally, observations following TIMP sessions note posture improving by as much as a 45° angle. Another nontraditional use of an instrument might be to instruct the patient to strike a heel onto a tambourine placed on the floor, simulating the raising of the leg and encouraging heel strikes to practice healthy gait patterns instead of the shuffling gait that is often seen in people with Parkinson's. Again, when these techniques are skillfully employed, positive carryover from the interventions has been observed in a day program setting.

In addition to creating different ways to play the instruments, adaptations may need to be made to the instrument itself if a patient is unable to play the instrument because of restricted motor ability due to Parkinson's or another condition. For individuals with involuntary neurological impulses, missing limbs, and restricted motor abilities, adaptive equipment or assistance is often necessary.[7] Examples might include using a foam grip on a mallet so it is more easily held or Velcro® attached to an instrument so it can be attached to a brace. It is interesting to note that involuntary movements like resting tremor or dyskinesia do not usually precipitate challenges in playing instruments. Rather, the action of playing and instrument actually appears to decrease the amplitude of these involuntary movements.

ADDRESSING VOICE CONCERNS WITH MUSIC THERAPY

It is estimated that "at least 70% of patients with PD manifest speech and voice disorders."[8] The use of music, specifically rhythm and singing, is an effective rehabilitative tool in this area. Posture can be improved with NMT techniques previously mentioned. Good posture is essential to the breathing needed for clear phonation and for volume. NMT has also provided standardization and a common language in this area. These techniques also directly address the difficulties in speech encountered by people with Parkinson's.

One of the primary issues is soft, fading volume. Breath, the primary source of energy behind phonation, is severely affected by poor posture. Improved posture will maximize deep abdominal breathing. It also allows for improved ability to ration and wisely use the respiratory energy during the exhalation of the breath. After helping the patient to achieve these goals, the therapist can use *therapeutic singing*[2] techniques. It is important to distinguish that this is not "just a sing-along." Important steps are followed to maximize the potential for achieving positive outcomes and meeting patient goals during this type of session. Deeper breaths taken while singing allow the patient to practice obtaining the energy/air that is needed to counteract the soft fading volume. Therapeutic singing may also be used to practice the initiation of speech. This might also be practiced through *musical speech stimulation*.[2] In this technique, the therapist uses "musical and sound patterns to stimulate nonpropositional speech. For example, when completion of familiar sentences is stimulated through familiar tunes, poems, or obvious completion of melodic phrases." Examples of this might be "Let Me call you (sweetheart)" or "How are you (today)?"

Voice quality changes are also associated with Parkinson's. The voice can become raspy due to the bowing of the vocal folds. With singing, a person can practice not only bringing the vocal folds together but also sustaining this action. The tendency for patients to slip into a glottal fry voice also affects voice quality. Utilization of resonance, respiratory control, forward focus of sound, and "head" voice is effective in counteracting this issue. In addition to therapeutic singing, *vocal intonation therapy*[2] can again accomplish the goal of improving voice quality. This technique employs the use of "intoned phrases simulation the prosody, inflection, and pacing of normal speech and vocal exercises to train all aspects of voice control, i.e., inflection, pitch breath control, timbre, loudness, etc." This has been observed to be especially effective when the music therapist gradually moves the phonation from the singing realm into the speech realm.

At a day program for people with Parkinson's, the staff has observed a carryover through much of the remainder of the program day. Utilization of this technique, with the focus for the patient on the sensation of the desired outcome, has been successful for these same patients. Verbal cues of remembering "how it felt" can be enough for most patients to replicate the improved phonation. This aspect is important, considering how our own perception of our voice is often inaccurate, and because the increased effort that is required for persons with Parkinson's to be heard feels uncomfortable to them. Monotone voice is also improved with therapeutic singing and vocal intonation therapy. Singing can address this because of the pitch range that is used.

Also important to verbal communication is facial expression, which is often masked in people with Parkinson's due to muscle rigidity. When singing, facial animation increases and becomes more expressive. *Oral motor exercises*[2] also help to counteract the muscle rigidity in facial muscles. Using different musical elements (melody, rhythm, and dynamics) to practice muscular control of speech apparatus can facilitate the production of specific sounds. This may include vocal exercises such as overarticulating phrases like "Red leather, yellow leather" or "Mama made me mash my M&Ms." It also may include playing wind instruments such as recorders or kazoos, which can be helpful in working on rationing air for speech. Fast, irregular rate of speech and imprecise and blurred articulation are also addressed using this technique.

Another especially effective technique is the use of *rhythmic speech cueing*.[2] This technique is the "use of rhythmic cueing by tapping hand, drum etc. to control the initiation and the rate of speech cueing and pacing, e.g., with apraxia, dysarthria, fluency disorders." Thaut et al. found that auditory rhythm increases the intelligibility rates in patients with hypokinetic dysarthric speech due to Parkinson's by 19.7% ($p = 0.004$).[9] The level of improvement appeared to be directly related to the level of severity in the deficit. Severely impaired speakers' intelligibility increased 92.5% ($p = 0.001$), moderately impaired speakers' intelligibility increased nonsignificantly by 7.7%, and the mildly affected showed no improvement. The researchers stated "that rigid, temporally constraining techniques such as rhythmic cueing appear to benefit severe intelligibility impairments more that less severe impairments of speech functioning.... Higher levels of speech functioning may benefit more from therapeutic techniques that are less externally constraining.[9, p.170] They conclude, "The differential benefits of rhythmic cueing depend on rate selection as well as on the patient's level of speech functioning. Rate reduction frequencies close to intrinsic speech rhythms should be avoided due to possible entrainment interference."[9, p.171]

In a separate research article published by the *Journal of Music Therapy* in 2001,[8] Haneishi found a significant increase in speech intelligibility according to caregiver rating ($p = 0.0410$) and in pre- to post-test vocal intensity ($p = 0.032$). Haneishi developed a *music therapy voice protocol* that includes opening conversation, warm-up, vocal exercises, singing exercises, sustained vowel phonation, review and speech exercises, and closing conversation. This protocol works well with patients. However, in the interest of maintaining common language, it is advantageous to utilize the NMT terminology that was standardized after Haneishi research was completed. The addition of rhythmic speech cueing is an important component to a music therapy treatment when working with severe speech/voice deficits.

To achieve the best possible outcome for a patient, it is helpful for the music therapist to collaborate with the speech language pathologist. The SLP and music therapist facilitate work toward achieving overlapping goals using different techniques. Providing the patient with this consistency in care with different approaches solidifies and enriches the learning and rehabilitation process.

MUSIC-ASSISTED RELAXATION

Just as rhythm has the potential to impact the intrinsically, biological rhythmical function of gait, other intrinsically rhythmical functions can be affected. The theory of entrainment also works in changing heart rate, breathing rate, and blood pressure. Spingte and Droh[10] discussed their findings in using anxiolytic music with nearly 50,000 patients undergoing medical procedures over a period of eight years (1977 to 1985). They cite the following "preconditions" be fulfilled to categorize the music as anxiolytic:

1. Musical works should be selected according to duration, instrumentation, dynamics, and interpretation. It is important that there be no extremes in rhythm, melody, or dynamics; more lower than higher frequencies; a frequency range from 100 to 10,000 Hz; and that instru-

mental (preferably strings) rather than vocal music be chosen.

2. Patients should make their own musical selections.
3. The effects of individual pieces and combinations of pieces should be tested in ongoing clinical settings.
4. Recordings should be of high quality yet technically simple and reliable.

Their studies "indicate that patients listening to music of their choice during certain medical procedures experience better psychophysical effects that those who do not have musical stimuli…These results are replicated in all our studies and show a statistically significant reduction ($p < 0.01$) of stress response in the cardiovascular and endocrinological systems."

Stress can exacerbate the symptoms of Parkinson's. Involuntary movements such as tremor, as well as the side effect of dyskinesia, increase in amplitude when a person with Parkinson's perceives stress. Freezing episodes increase, as a patient may be "rushed" by a family member or friend. Music-assisted relaxation can effectively address these issues. In assisting patients in their choices for relaxation music, selection is of sedative music that is at 50 to 60 beats per minute (heart rate at rest); is instrumental with minimal extremes in dynamics, melody, and rhythm; and has lower frequencies. Either live or recorded music can be used, and sung human voice without the use of lyrics can be effective. Anecdotally, patients and care partner's report that they experience/observe feelings of increased relaxation during and after the sessions. In most cases, patient's tremor and dyskinesias subside, and in some cases stop altogether. There is usually a reduction in involuntary movements; however, carryover after the session is short.

Active involvement in playing music can also have relaxation effects. Playing instruments creates action in the parts of the body exhibiting tremor. Since a resting tremor is seen in Parkinson's, the action of playing an instrument can provide relief from the symptom. When a person experiencing dyskinesia plays a floor drum, such as a tubano, the amplitude of the dyskinesia significantly decreases.

An involuntary movement like tremor and dyskinesias entrain to the rhythm of music that is played. Decreased amplitude of involuntary movements allows patients to participate in activities they may not be able to otherwise. Activities may range from activities of daily living to creative and artistic endeavors. When balance is compromised, music can decrease the amplitude of dyskinesias, enabling a patient to walk with minimal assistance and unimpaired balance. When the music is stopped, dyskinesia returns, and patient fall risk is increased.

Symptoms can be exacerbated by the stress caused by pleasant and unpleasant emotions. For example, tremor may increase in a person who is hearing a favorite song. So, even though symptoms may increase, the perceived relaxation benefit of listening to favorite, familiar music outweighs the short-term effects of increased symptoms.

ADDRESSING PSYCHOSOCIAL CONCERNS THROUGH MUSIC

There are many psychosocial concerns for those who are dealing with a progressive illness like Parkinson's. Estimates of the number of people with Parkinson's who suffer from depression vary. A survey conducted by Levin, Llabre, and Weiner[11] found a range of 37 to 90%. An explanation in the variability could be the differences in measures used to assess depression in people with Parkinson's. In addition to this, there is variability in the level of depression reported, ranging from some depression to depression that is significant enough to require treatment.

Sometimes contained within this range are two concepts that differ in their definitions of depression but are frequently not differentiated. Pauline Boss identifies a state or condition called *ambiguous loss*.[12] Ambiguous loss can be described as a loss that is unclear, confusing, and continuous. This stems from the changes that are experienced by a person with a progressive disease like Parkinson's. Loss is experienced even during times that may appear to be a plateau, as the fear of the unknown future is always present in a disease that affects each individual in a unique way.

The second helpful concept, *chronic sorrow,* was first identified by Olshansky[13] to characterize the recurring grief that is felt by parents of mentally retarded children as they continue to grieve the loss of normal development in their child. In 1992, Lindgren, Burke, Hainsworth, and Eakes began further clarifying this concept and how it applies to those challenged by progressive diseases like Parkinson's. "Chronic sorrow is a continuous grief that occurs in a cyclical pattern of resurging feelings of sorrow interspersed with periods of calmer emotions. The resurging feelings of sorrow can increase in intensity with the accumulation of new losses and remembrance of old losses. This pattern of peaks and valleys of sorrowful feelings does not abate."[13] Lindgren studied the occurrence of chronic sorrow in those diagnosed with Parkinson's and their spouse caregivers. She found that with the reminder of each "major change," there was resurgence in the feelings of sadness. She noted, "Changes encompassed more than a diagnoses. It resulted in a wholly different life, with continual distress from the unrelenting demands of the disability that the diseases produced and a continual coping to keep from being consumed mentally and physically by the illness."

People with Parkinson's frequently report experiencing feelings of isolation. Relationships with family and friends change as cognitive or communication challenges increase. A frequent ongoing theme in patient's verbalizations is that of anger and frustration in how they are perceived by others. The physical, emotional, cognitive, and spiritual changes that are experienced also create a change in how people may perceive themselves. Self-concept may be severely affected.

Many people with Parkinson's experience impaired thinking and memory. In addition to this, they may experience medication-induced difficulties such as hallucinations. These changes, along with impaired speech, add to the challenges in communication.

Music therapy literature specific to addressing the psychosocial concerns of those diagnosed with Parkinson's is limited. Pachetti et al.[14] researched the effects of "active music therapy" on people with Parkinson's. "Active music therapy" included listening and relaxation; use of the voice and body, and singing; free body movement with rhythmic cadence; sound-gesture synchronization; rhythmic exercises using the body and instruments; ensemble music; improvisation; and listening and free body expression (creative movement). They found that "active music therapy" with 16 patients with Parkinson's improved hypokinesia ($p < 0.0001$) and emotional well-being (as measured by the Parkinson's Disease Quality of Life Questionnaire) ($p < 0.0001$). They also documented significant changes in overall effect on daily performance ($p < 0.0001$). Specific areas in which they found improvement were in the ability to cut food, to dress, and in decreased falls ($p < 0.0001$) and freezing ($p < 0.0005$).

The same facility conducted a study comparing active music therapy to physical therapy interventions.[15] They reported that music therapy significantly changed bradykinesia (as measured by the Unified Parkinson's Disease Rating Scale, $p < 0.034$) and created improvement in emotional functions (as measured by the Happiness Measure ($p < 0.0001$) and improvement in activities of daily living and quality of life (as measured by the Parkinson's Disease Quality of Life Questionnaire ($p < 0.0001$), especially in emotional ($p < 0.0001$) and social ($p < 0.0001$) functioning scores. Their physical therapy group improved rigidity ($p < 0.0001$).

In the Haneishi study,[8] mood changes that occurred as her small sample size of patients participated in *music therapy voice protocol* were discussed. Along with finding the aforementioned increased intelligibility and increased pre- to post-test vocal intensity, she found higher mean mood scores. It was hypothesized that these improvements in speech may decrease frustration in patients and their family members, and may increase functional use of voice and increase social interaction, all important aspects when dealing with psychosocial needs.

In the book, *Music in Therapy,* E. Thayer Gaston outlined fundamental considerations of man in relation to music.[16] One of these is that "music is derived from the tender emotions." Music provides many ways to express tender emotions. Music can provide needed validation of feelings, a vehicle for realization and expression of feelings. Altschuler developed a specific technique in which music is matched with the patient's mood and gradually altered to affect the desired mood state.[17]

As the field of music therapy moves forward, attempts to standardize terminology is increasingly important. Within NMT, a standardized term for most of the interventions discussed addressing psychosocial concern is *Musical Incentive Training for Behavior Modification.* This is defined in the *Training Manual for Neurologic Music Therapy* as "techniques employing guided music listening, musical role playing, expressive improvisation or composition exercises using musical performance to address issues of mood control, affective expression, cognitive coherence, reality orientation, appropriate social interaction, etc. to facilitate psychosocial functions. Techniques are based on models derived from, e.g., affect modification, associative network theory of mood and memory, social learning theory, classical and operant conditioning, iso principle, etc."[2-3]

Within a clinic, adult day services, and community classes, music therapy techniques successfully address a spectrum of psychosocial issues. Music therapy sessions have afforded the staff at the Struthers Parkinson's Center in Golden Valley, Minnesota, a unique glimpse from the perspective of those challenged by Parkinson's. This can be accomplished through a variety of interventions. In discussing these, the reader will hopefully see and feel the humanity and soul that is bigger than this disease.

SONG DISCUSSION

Song discussion can facilitate work on a number of goals. Although communication challenges and bradyphrenia can affect the speed at which persons with Parkinson's may be able to express themselves, this is greatly helped by providing the "gift of time." The song itself may provide the words for which those with word-find difficulties may be searching. After song discussions, group participants have chosen to bring home song lyrics to share with their family, sometimes giving the lyrics to someone in their support system as the message that they are unable to articulate. The outcome for the patient and family can be unifying. One wife wrote back that receiving this communication from her husband was meaningful and encouraging, since she rarely heard utterances from her husband. Songs centering on changes and losses are especially powerful, providing the needed "marker" to define, express, and possibly ritualize a loss or a group of losses. This is especially important when dealing with a loss that is ambiguous.

GROUP THERAPEUTIC SINGING

Group therapeutic singing is an important strategy that also accomplishes speech goals as mentioned earlier in this chapter. Singing with a defined purpose improves mood and decreases isolation. For some people who are singers, this intervention can also surface feelings of another loss, as their voices are not the same as pre-Parkinson's. However, by reframing singing as now being therapeutic rather than performance oriented is empowering. Although sadness regarding the loss is acknowledged, music via singing is now a vehicle for coping with the change.

EXPRESSIVE IMPROVISATION

Expressive improvisation is most often used to facilitate affective expression. Facial masking, coupled with communication difficulties, often makes expression of feelings a difficult task. The importance of this is paramount as the issues of depression are considered. In expressive improvisation, patients are able to extract and express the emotions that they are experiencing. After these sessions, outcomes expressed by patients include connecting with anger at having PD and then being able to experience the sadness and loss that they were avoiding. Another observation is that patients play and then express grief at the loss of their friends who have Parkinson's. In some expressive improvisations, patients have orchestrated what they think Parkinson's sounds like or what they would like to say to Parkinson's. Along with expression of feelings, the patient is allowed to feel empowered by the expression.

Cognitive issues are also addressed by the therapist, offering elements like instrument choice, mood of improvisation, and musical form, which may present decisions. Within the milieu of the adult day services, expressive improvisation may be used for mood control and appropriate social interaction. When a person has the added challenge of dementia, expression of emotional and physical discomfort often comes out in the form of anxious, loud, and/or aggressive behavior. This can be disruptive to the milieu and pose a great discomfort for the patient involved. Utilizing Altschuler's iso-principle in mirroring the observed behavior musically and then gradually altering the music to a calmer mood is an effective way to work with this challenge. Patients then engage in the musical improvisation, changing their expression from behavioral to musical. In some cases, the expression has become a milieu that is appropriate verbalization of the discomfort they are experiencing.

Another of the fundamental considerations proposed by Gaston is that "music is structured reality."[16] In its most basic definition, music is organized sound and silence. When working with those challenged by the hallucinations that occur for some diagnosed with Parkin-

son's, the writer has observed that any and all of the interventions mentioned throughout this chapter have the potential to help the patient reconnect with reality. While observing a patient hallucinate, the writer has effectively utilized singing as well as playing music to help establish a shared reality with a patient. As hallucinations and paranoia are often accompanied by anxiety, both passive listening or playing along with the therapist is effective in calming a patient. The writer has observed that some patient's can then openly discuss the hallucination. In any milieu in which the patient is found, experiencing issues with hallucinations, paranoia, and anxiety can be disruptive. Music therapy interventions offer a respectful alternative that honors a patient's dignity and state of being while moving them to a place that is more physically and emotionally more comfortable.

SONGWRITING

Songwriting is a powerful form of expression. It is imperative that the music therapist provide adequate response time. In songwriting, the perspective of those who deal with Parkinson's is evident. The outcome is vitally important. Patients report that they feel like they have received more than what they have contributed and that, through the process of writing the lyrics and choosing the sound, tempo, dynamics, and meter of the piece, they experience a renewal of their own sense of identity outside of Parkinson's. With the permission of Struthers Parkinson's Center, sharing the following song will allow the reader to experience only the lyrical power of songwriting, however it will give the glimpse of the humanity that truly is bigger than the disease. In these lyrics the writers refer to Parkinson's as a "thief" who has taken many valued pieces of life from them. The writers also show you their courage and determination to draw upon the strengths and essence of themselves that remains despite the disease.

> Why'd You Have To Come?
> Why'd you have to come and take away some fun?
> Now I cannot run. Why'd you have to come?
> Now I had to move away from home.
> It's a changing life, the loss is big and tough.
> Tears fill my eyes but I can still arise
> To the challenge of the day, though the sky is gray.
> Thief, you know, he couldn't take away the day.
> So we look into his eyes and say:
> Still have the faith, still have the hope, still have the love.
> I know the One that made me is smiling down upon me from above.
> I know that love will save me. I won't let that darkness take me
> Oh my passion is strong, you know I'll laugh the day along.

With the help from God and family and my friends
They all save me and give me strength to live.
You know they give me strength to live.

CONCLUSION

First you adapt, then you adept.

Robert W. Terry

Adaptation is a process of self-discovery that can take many forms: it applies to our spiritual, cognitive, and physical self. The losses associated with Parkinson's begin with the diagnosis and the impact on a person's personal and professional life. As the disease progresses, new challenges are continuously presented. In negotiating the journey through this process, a variety of strategies are needed to assist the patient to adapt and become adept. Music therapy is poised to facilitate this by addressing the needs and issues of the whole person challenged by Parkinson's disease in unique, well grounded ways.

Music therapy can be applied in individual or group sessions, depending on the needs of the patient and what best addresses their goal. Music therapy interventions can be especially powerful in a group setting where patients are motivated by each other as stated by E. Thayer Gaston, "The potency of music is greatest in a group."[16, p.27]

Also critical to all music therapy techniques is the importance of the quality of the music. Thaut describes this well in the *Training Manual for Neurologic Music Therapy* when he states,

> While much research had investigated the impact of isolated musical components on human behavior, the therapeutic value of music also depends on its overall effect or artistic quality. The auditory stimulus ought to be 'musical' to elicit the desired therapeutic response. That is, patients must perceive the music as organized, meaningful, and pleasurable, whether they are hearing it or producing it. This consideration calls upon clinicians to carefully design and select musical materials, while keeping in mind the patient's needs and functioning level. Therapists must take into account not only the physical/motor needs of the patient, but the cognitive deficits as well. A client with lower cognitive functioning, for instance, may find simple auditory stimuli far more appealing than complex musical structure.[2]

This points to the need for the specialized training of the music therapy clinician.

Practicing music therapy with people diagnosed with one disease may be considered by some to be limiting; however, in this case, the contrary is true. The nature of Parkinson's stretches the music therapist music to be knowledgeable and use music in rehabilitation of physi-cal, speech, and cognitive needs and to address the psychosocial challenges posed by the disease. Others may consider this work to be a negative experience for the therapist, also. Although it is difficult to witness persons undergoing the changes and losses caused by Parkinson's, their hopeful attitude and drive to maintain abilities makes one feel privileged to walk on their journey with them.

ACKNOWLEDGMENTS

I would like to thank all of my colleagues at the Struthers Parkinson's Center, especially Rose Wichmann, Catherine Wielinski, and Marjorie Johnson, for their incredible support of me in completing this project. I also acknowledge the assistance from former music therapy intern, Melissa Hirokowa, for her assistance with the literature review. Above all, I thank the patients who have allowed me to walk with them on their Parkinson's journey; you are the best teachers and my heroes.

REFERENCES

1. *American Music Therapy Association Member Sourcebook,* American Music Therapy Association, Silver Spring, MD, p. vi, 2002.
2. Thaut, Michael, *Training Manual for Neurologic Music Therapy,* Center for Biomedical Research in Music, Colorado State University, Fort Collins, 1999.
3. Thaut, Michael H., *A Scientific Model of Music in Therapy and Medicine,* Institute for Music Research Press, Chapter 3, The University of Texas at San Antonio, San Antonio, TX, 2000.
4. Thaut, M. H., McIntosh, G. C., Rice, R. R., Miller, R. A., Rathbun, J., Rhythmic Auditory stimulation in gait training for Parkinson's disease patients, *Mov. Disord.,* 11,193, 1996.
5. McIntosh, G. C., Brown, S H., Rice, R. R., and Thaut, M. H., Rhythmic auditory-motor facilitation of gait patterns in patients with Parkinson's disease, *J. Neurol., Neurosurg., and Psych.,* 62, 22, 1997.
6. McIntosh, G. C., Rice, R. R., Hurt, C. P., Thaut, M. H., Long-term training effects of rhythmic auditory stimulation on gait in patients with Parkinson's disease, *Mov. Disord.,* 13, (Suppl. 2), 212, 1998.
7. Clark, C., Chadwick, D., *Clinically Adapted Instruments for the Multiply Handicapped,* Magnamusic-Baton, St. Louis, MO, 1980.
8. Haneishi, E., Effects of a music therapy voice protocol on speech intelligibility, vocal acoustic measures, and mood of individuals with Parkinson's disease, *J. Music Ther.,* 38(4), 273, 2001.
9. Thaut, M. H., McIntosh, G. C., Hoemberg, B., Auditory rhythmicity enhances movement and speech motor control in patients with Parkinson's disease, *Functional Neurol.,* (16) 2, 163, 2001.

10. Spingte, R., Droh, R., Effects of anxiolytic music on plasma levels of stress hormones in different medical specialties, in *The Fourth International Symposium on Music Rehabilitation and Human Well-Being,* Pratt, R. B., University Press of America, Chapter III, p. 88, 1985.

11. Levin, B. E., Llabre, M. M., Weiner, W. J., Parkinson's disease and depression: psychometric properties of the Beck Depression Inventory, *J. Neurol., Neurosurg., and Psych.,* 51, 1401, 1988.

12. Boss, P., *Ambiguous Loss Learning to Live With Unresolved Grief,* Harvard University Press, Cambridge, MA, 2000.

13. Lindgren, C. L., Chronic sorrow in persons with Parkinson's and their spouses, *Scholarly Inquiry for Nursing Practice: An International Journal,* 10(4), 351, 1996.

14. Pachetti, C., Alglieri, R., Mancini, F. Martignoni, E. Nappi, G., Active music therapy and Parkinson's disease: methods, *Func. Neurol.,* 13(1), 57, 1998.

15. Pachetti, C., Mancini, F., Algieri, R. Fundaro, C., Martignoni, E., Nappi, G., Active music therapy in Parkinson's disease: an integrative method for motor and emotional rehabilitation, *Psychosom. Med.,* 62, 386, 2000.

16. Gaston, E. T., *Music In Therapy,* Chapter 1, Macmillan Publishing Company, Inc., New York, 1958.

17. Davis, W., Gfeller, K., and Thaut, M., *An Introduction to Music Therapy: Theory and Practice,* 2nd ed., McGraw-Hill College, 1999.

18. For further information on Neurologic Music Therapy, contact the Center for Biomedical Research in Music at www.colostate/edu/depts/cbrm.

64 Pathogenesis of Motor Response Complications in Parkinson's Disease: "Mere Conjectural Suggestions" and Beyond

Leo Verhagen Metman
Rush University Medical Center

CONTENTS

THE REMAINING THERAPEUTIC CHALLENGE

In his 1817 monograph, James Parkinson expressed the wish that his essay would excite others to "extend their researches" to this disease so that they might *"point out the most appropriate means of relieving a tedious and most distressing malady."*[1] It was not until 1967 that his wish was granted, but the advent of levodopa was well worth the wait.[2] Currently, levodopa still stands out among other therapies in neurology as a rational neurotransmitter replacement strategy that confers great benefit to its target population. Unfortunately, Parkinson did not have the foresight to wish that *"the relief of this distressing malady"* should last indefinitely. Soon after the introduction of levodopa, it became clear that its chronic use is associated with motor fluctuations and dyskinesias that even-

tually occur in the majority of patients with advanced PD and can be major causes of disability in their own right.[2,3] Thus, paraphrasing Parkinson's words, we are now confronted with the remaining challenge *"to point out appropriate means to prevent or relieve the long-term complications of levodopa treatment."*

MOTOR RESPONSE COMPLICATIONS

While patients with early Parkinson's disease characteristically enjoy a good response to levodopa, the combination of disease progression and chronic levodopa therapy eventually compromises this benefit in several ways.[4]

Over the years, the antiparkinson effect of each levodopa dose lasts progressively shorter, necessitating increasingly more frequent levodopa administration. At first, when the shortening of motor benefit occurs in a pre-

0-8493-1590-5/05/$0.00+$1.50
© 2005 by CRC Press

dictable and gradual fashion, patients are said to have wearing-off fluctuations. Subsequently, when the effect of an individual dose ceases in an unpredictable and abrupt manner, the term on-off phenomenon is used.

In addition to these fluctuations in motor function, patients also develop involuntary movements called *dyskinesias*. When these first arise, they are usually associated with high levodopa levels and may be prevented or minimized by lowering the levodopa dose. Later on, however, the therapeutic window of levodopa narrows progressively, and dyskinesias occur at a plasma levodopa level equal to that needed to induce an antiparkinson effect (obligatory dyskinesias). Dyskinesias most commonly are choreiform and occur when plasma levodopa levels are high (peak-dose dyskinesias) or, in more advanced disease, throughout the levodopa-induced motor benefit (square-wave dyskinesias); they may also be more dystonic or ballistic in appearance and occur when levodopa levels are rising or falling (diphasic dyskinesias). Combined, motor fluctuations and dyskinesias are referred to as motor response complications (MRCs) and occur in 40 to 50% of patients after five years of levodopa treatment.[4,5]

Even though our current understanding of the pathogenesis of motor response complications, to some degree is based, in James Parkinson's self-depreciating words, on *"mere conjectural suggestions,"*[1] recent investigations have begun to identify several pathogenetic mechanisms associated with the complications of chronic levodopa therapy.

PHARMACOKINETICS OF LEVODOPA

Once ingested, levodopa passes through the stomach to reach its primary absorption site, the duodenum.[6] The stomach itself does not contribute significantly to levodopa absorption, but prolonged exposure of levodopa to the gastric mucosa, which contains aromatic amino acid decarboxylase (AAAD), decreases the amount of drug that eventually reaches the duodenum.[7] Therefore, factors interfering with the transport through the stomach have an impact on the timing and amount of levodopa absorption and consequently on the latency and magnitude of the motor response to levodopa. Gastric emptying may be delayed by large meals, low pH, and anticholinergic drugs.[7–9] Bypassing the stomach completely, following gastrectomy[12] or with duodenal levodopa infusions,[11–13] results in faster and higher peaks in levodopa plasma levels. Accelerating gastric transport using oral liquid levodopa formulations[14,15] or drugs that increase gastric motility[9] also results in faster and increased absorption. In the duodenum, absorption of levodopa occurs via a saturable active transport system shared with other large neutral amino acids (LNAAs), and therefore dietary amino acids may compete with levodopa for its absorption into the bloodstream.[6,16]

Levodopa clearance from the plasma compartment follows a biphasic pattern.[17] The initial phase lasts five to ten minutes and represents redistribution of levodopa from plasma to various tissue reservoirs. Quantitatively, skeletal muscles form the most important reservoir. The second or elimination phase involves levodopa metabolism and has a half-life of about 90 min.[17–19]

In contrast to dopamine, levodopa readily passes the blood-brain barrier (BBB) mediated by a saturable carrier system for LNAAs similar to that in the gut.[20,21] Thus, the BBB is the second site where dietary LNAAs compete with levodopa on its way to the brain.[20,22–25] Consequently, not only plasma levodopa levels, but also the concentrations of other LNAAs, determine the extent of levodopa access to the brain.[21,26] Increased plasma levels of LNAA can reduce levodopa's clinical efficacy,[23] whereas lowering dietary protein and circulating amino acids reduces levodopa requirements.[22–24,27] These observations have formed the basis for the so-called protein redistribution diets, which limit protein intake at breakfast and lunch but provide the daily protein requirement at dinner,[27] allowing patients to be more predictably in the mobile "on" state when most needed.

Levodopa is decarboxylated by l-AAAD to form dopamine.[28–30] This process starts and is quantitatively most prominent in the gut, but it continues in other tissues that contain l-AAAD throughout the body, including liver, kidney, and even cerebral endothelium. Peripheral conversion of levodopa to dopamine, however, poses two problems. First, circulating dopamine causes side effects such as nausea, vomiting, and cardiac disturbances. Second, only a fraction of the levodopa taken crosses the BBB unscathed. The use of peripheral decarboxylase inhibitors (carbidopa or benzeraside) decreases levodopa metabolism in the gut and thus increases its bioavailability, allowing significant reduction in levodopa dose. Carbidopa prolongs levodopa elimination half-life only modestly and has no significant impact on clinical efficacy duration.[17,31,32]

The second most important pathway of levodopa metabolism is its catalytic conversion to 3-O-methyl-dopa by the enzyme catechol-O-methyl-transferase. Under experimental conditions, 3-O-methyl-dopa can compete with levodopa for carrier sites at the BBB and adversely affect its efficacy.[33–38] Because of its long plasma half-life of over 15 hr, 3-O-methyl-dopa tends to accumulate in the body.[31,39] However, in patients chronically treated with levodopa, 3-O-methyl-dopa makes up only a small proportion of LNAAs, and its plasma concentration is not high enough to significantly impede levodopa transport into the brain.[38,40] Inhibitors of catechol-O-methyl-transferase have been developed in recent years. They can act peripherally (entacapone) as well as centrally (tolcapone) to enhance levodopa bioavailability and prolong its anti-parkinson efficacy.

Peripheral pharmacokinetic parameters thus determine the rate and extent of levodopa's arrival at its target site, the striatum. It is therefore conceivable that motor response complications occurring with long-term levodopa therapy are due to changes in one of these "supply" variables. Several studies, however, have failed to demonstrate any differences in pharmacokinetic parameters among patients who were never treated with levodopa, those with a stable response to levodopa, and those with a fluctuating response.[19,41,42] Thus, while peripheral pharmacokinetic factors can and obviously do contribute to motor fluctuations in patients with advanced PD, they are not thought to play an important role in the *pathogenesis* of the altered response to levodopa. Instead, investigative attention has increasingly focused on central pharmacodynamic processes that are induced first by nigrostriatal denervation and subsequently by chronic intermittent levodopa therapy.

INTERMITTENT, NONPHYSIOLOGIC, DOPAMINERGIC STIMULATION

Under normal conditions, the nigrostriatal pathway is tonically active, i.e., nigral neurons most of the time fire at a fairly constant rate of about 5 Hz.[43,44] Thus, under physiological conditions, dopamine release associated with neuronal depolarization is an essentially continuous process, leading to relatively constant stimulation of striatal dopamine receptors. In Parkinson's disease, neurotransmitter replacement therapy ideally should mimic this process, but the therapeutic reality is much different. Oral levodopa has a short half-life, and standard dosing schedules lead to wide variations in plasma concentrations. In the striatum of patients with *mild* PD, these swings in levodopa levels may be buffered by the compensatory effort of surviving nigrostriatal neurons.[45–48] Thus, uptake of exogenous levodopa, conversion to dopamine, and vesicular storage and release under neuronal control presumably will initially continue to provide more or less constant dopamine receptor activation. This is evidenced by a long-duration antiparkinsonian effect that outlasts the half-life of the drug, enabling a stable motor state even when patients occasionally skip their levodopa dose.[49] In more *advanced* disease, associated with a further demise of nigrostriatal cells, dopamine synthesis from exogenous levodopa largely takes place in other dopa-decarboxylase-containing cells that are not equipped to store and release dopamine in a controlled fashion.[50] As a result, intrasynaptic dopamine levels no longer remain constant but now mirror the broad swings in plasma levodopa levels associated with the intermittent administration of this rapidly metabolized prodrug. Dopaminergic transmission in the

nigrostriatal system is converted from a tonic to a phasic process.[51] This process was recently visualized through a [¹¹C]-raclopride PET study, where patients with wearing-off fluctuations one hour after levodopa had threefold higher (estimated) synaptic dopamine levels than stable responders.[52]

The periodic high-intensity stimulation of postsynaptic dopamine receptors is now believed to contribute to the development of the motor response complication syndrome.[53] Although this concept has become widely accepted,[54] direct clinical evidence is still lacking. However, several lines of evidence from animal models as well as clinical studies strongly support the continuous stimulation hypothesis.

Indirect evidence comes from the 6-OHDA-lesioned rat model of Parkinson's disease. When these animals are given levodopa twice daily, a significant shortening of the motor response to a test dose of levodopa occurs within three weeks. Moreover, the frequency of off-responses to an ordinarily effective dose of levodopa substantially increases.[55–57] These changes in response have been likened to wearing-off and on-off fluctuations, respectively, in PD patients.[53] They occur only when more than 95% of nigrostriatal cells are lost.[56] When, on the other hand, the same daily dose of levodopa is administered in a continuous fashion instead of intermittently, none of the above changes occurs.[55,58]

In MPTP-exposed nonhuman primates, intermittent dopaminergic stimulation readily induces dyskinesia. Analogous to the above findings in the 6-OHDA-lesioned rat model, nigrostriatal damage also appears to be required for the induction of dyskinesias in squirrel monkeys,[59] and dyskinesias are most readily invoked in MPTP-exposed monkeys with the most severe motor impairment[60] associated with >95% dopamine depletion.[61] It should be noted, however, that recent data indicate that very high-dose levodopa can induce dyskinesias, even in normal macaque monkeys, suggesting that nigrostriatal denervation may be a risk factor rather than an absolute requirement for the development of dyskinesia.[62]

In the MPTP-exposed monkeys, a differential behavioral effect of continuous versus pulsatile administration of a D_2 agonist has been demonstrated as well.[63] Thus, intermittent subcutaneous injection induced significant dyskinesias, whereas animals treated with continuous delivery through an osmotic pump at identical doses manifested no or only transient involuntary movements. Furthermore, in the MPTP-treated marmoset, dopamine agonists such as bromocriptine and ropinerole were shown to induce less dyskinesia than levodopa, presumably because their long half-lives provided more continuous dopaminergic stimulation than levodopa, with a shorter half-life of about 90 min.[64] However, when, in a recent study, the dyskinesiogenic potential of levodopa was compared with that of the very short-acting agonist apomorphine and the long-acting agonist pergolide in drug-naïve MPTP-lesioned marmosets, it was found that both agonists caused significant less dyskinesia than levodopa at equally effective antiparkinsonian doses. This suggests that factors other than half-life may also determine the dyskinesiogenic potential of a dopaminergic agent.[65]

Marked striatal denervation also appears to be a risk factor for the occurrence of dyskinesias in humans, as they occur predominantly in Parkinson's disease and not in conditions where the nigrostriatal pathway is intact.[66–69] Additionally, in asymmetrically affected parkinsonian patients, dyskinesias usually appear first on the most affected side.[70] And in patients with MPTP-induced parkinsonism, dyskinesias develop on average six months after initiation of levodopa therapy, which is much sooner than in patients with idiopathic Parkinson's disease, presumably due to the more severe denervation of the former group.[71] In humans, prophylactic studies have been limited by the lack of practical means to administer effective dopaminergic agents in a continuous, noninvasive fashion for an extended period of time to drug-naïve patients. Controlled-release levodopa/carbidopa, while designed for this purpose, does not provide steady plasma levels with commonly used dosing schedules and indeed was not successful in preventing or slowing down the development of MRCs.[72,73] More convincing are findings from recent studies showing a lower incidence and severity of MRCs when antiparkinsonian therapy is initiated with dopamine agonists such as cabergoline, pergolide, pramipexole, and ropinerole, first as monotherapy and later (as needed) in combination with levodopa, compared with levodopa therapy alone.[74–78] The longer half-life of these dopamine agonists compared to that of levodopa has been credited for these results, as it provides more continuous dopaminergic stimulation than shorter acting agents.

In contrast to the prophylactic benefits, the palliative effects of continuous dopaminergic stimulation on MRCs, once they have developed, have been well documented. Continuous intravenous infusions of levodopa, or subcutaneous administration of the dopamine agonists apomorphine and lisuride, have been shown to diminish response fluctuations as well as dyskinesias, indirectly making a case for the potential benefit of early institution of continuous therapy.[79–83]

Based on the foregoing arguments, it appears reasonable to accept a causative relationship between nonphysiological, pulsatile dopaminergic stimulation of a denervated striatum and the development of MRCs.[54] However, the pathophysiological substrate of MRCs and the intricacies of the cellular mechanisms by which they come about are only beginning to be unraveled.

Early investigations into pathogenetic mechanisms of MRCs have frequently employed intravenous levodopa. Differences in motor responses to levodopa between groups with mild versus those with advanced disease led to the belief that "presynaptic" processes were directly responsible for the occurrence of MRCs.[84] In other words, the progressive loss of nigral neurons with the associated dwindling capacity to take up levodopa, decarboxylate it, store dopamine in vesicles, and gradually release dopamine, would, according to this "storage hypothesis," be clinically expressed as motor fluctuations. Subsequently, the above studies were repeated with the dopamine agonist apomorphine instead of levodopa.[85,86] Patients were again classified into a dopa-naïve, stable response, wearing-off, and on-off group, in identical fashion as in the earlier levodopa studies. Differences in motor responses to parenterally administered apomorphine between these groups were strikingly similar as those found with levodopa. Apomorphine, in contrast to levodopa, does not require presynaptic processing but instead directly activates postsynaptic dopamine receptors. This implies that the anatomical substrate for the development of MRCs is situated postsynaptic to the nigral neuron, at the striatal level or possibly farther downstream.[85–87] Since elements downstream from the nigrostriatal neuron are presumably intact when MRCs first appear, this offers the possibility that the alterations leading to MRCs are functional changes amenable to correction through pharmaceutical intervention.

The occurrence of motor response complications has been associated with a variety of alterations throughout the cortical-basal ganglia-thalamo-cortical loop. Thus, decreased neuronal activity in globus pallidus internus (GPi) and subthalamic nucleus (STN) was demonstrated in dyskinetic monkeys[88,89] and humans[90,91] through electrophysiological recordings; decreased metabolic activity in the pallidal-receiving area of the thalamus (VA/VL) was found in dyskinetic primates using 2-deoxyglucose;[92] and increased activity in the motor cortex was demonstrated in parkinsonian patients with $[H_2{}^{15}O]$-PET and $[{}^{133}Xe]$-SPECT studies.[93,94] Because the motor loop is, by definition, a closed circuit, it remains speculation as to where the primary/causative abnormality is in the production of MRCs and which factors play a secondary role. However, as the striatum is the first station downstream from the degenerating nigrostriatal pathway, it seems plausible to propose a primary role for the striatum in the generation of MRCs.

STRIATAL MEDIUM SPINY NEURONS

Within the striatum, medium-sized spiny neurons account for over 90% of striatal nerve cells. They can be subdivided into two main subtypes based on their projection target. One subtype provides a projection axon to the internal segment of the globus pallidus and SNR, whereas the other subtype provides a projection axon that arborizes extensively in the external part of the globus pallidus and does not project beyond that nucleus. The former type of medium spiny neuron gives rise to the "direct" pathway, as it provides input directly into the main output nuclei of the basal ganglia (GPi and SNR), whereas the second subtype gives rise to the "indirect" pathway, named because of its indirect connection with the output nuclei of the basal ganglia.[95] Both subtypes use GABA as their neurotransmitter. However, neurons of the direct pathway express dopamine D_1 receptors (as well as the neuropeptides dynorphin and substance P), while projection neurons of the indirect pathway express dopamine D_2 receptors (and the neuropeptide enkephalin, see below). Dopamine D_1 and D_2 receptors are coupled to different G-proteins to respectively stimulate and inhibit adenylate cyclase signal transduction cascades. A balance between these opposing actions of the direct and indirect pathways is essential for normal basal ganglia function, as both pathways contribute to the ultimate output of the basal ganglia, the GABA-ergic projection to motor thalamus.[95,96] Medium spiny neurons receive synaptic input not only from the dopaminergic nigrostriatal pathway but also from glutamatergic corticostriatal pathways, as well as a number of other systems both intrinsic and extrinsic to the striatum, including adenosine, adrenergic, and serotonergic pathways. Their projections provide the major output nuclei of the basal ganglia, the internal segment of the globus pallidus, and the pars reticulata of the substantia nigra, with GABA-ergic input.[97–99] Medium spiny neurons may thus serve as integrative stations that process the flow of cortical and nigral information into the basal ganglia and have been implicated in striatal synaptic plasticity (below). Not surprisingly, the development of MRCs has been associated with numerous alterations at the level of the medium spiny neuron, as described below.

DOPAMINE SYSTEM

The focus of investigative efforts into the nature of the above striatal changes has, logically, been on dopaminoceptive systems. Dopamine receptor supersensitivity secondary to striatal denervation is frequently offered as an explanation for the MRC phenomenon.[100] There is, however, little evidence to support this presumption. It is true that supersensitivity of dopamine receptors after nigrostriatal denervation has been demonstrated in the Ungerstedt rat model of PD.[101,102] Similarly, in *untreated* parkinsonian patients, most studies evaluating dopamine receptor binding in postmortem tissue or with PET-imaging suggest an increase in striatal D_2 binding.[103–106] The same, however, does not appear to be the case in the striatum of *levodopa-treated* patients, where either normal or decreased dopamine receptor binding/density has been reported.[105–109] Furthermore, in a recent PET study of levodopa-treated PD patients with and without dyskinesias, no differences in D_1 and D_2 receptor binding between the two groups were observed.[106]

Other recent studies, however, suggest alternative mechanisms whereby dopamine receptor function may be altered as a result of levodopa therapy. Studies in normal rats,[110] 6-OHDA rats,[111] and humans[112] with PD all showed that activation of D_1 dopamine receptors provoked a modification of the localization of D_1 receptors by their internalization into cytoplasmic endosomes. This may interfere with their function, as it is assumed that the receptors must be localized at the plasma membrane and be coupled to second messenger transduction pathways to be fully functional.

In another interesting report, investigators found that levodopa treatment induced expression of the D_3 dopamine receptor in the dorsal striatum of the 6-OHDA-lesioned rat.[113] These findings were initially not reproduced in studies with MPTP-treated monkeys[114] and parkinsonian patients,[115] but subsequent studies did find overexpression of the D_3 receptor in levodopa-treated, *dyskinetic* monkeys, and, in addition, correlation of putaminal D_3 receptor binding with the occurrence and severity of levodopa induced dyskinesia.[116] Taken together, these data indicate that levodopa treatment and levodopa-induced dyskinesias are associated with changes in dopamine receptor expression and localization. The exact contribution of these changes to the pathogenesis of MRCs remains to be clarified.

GLUTAMATE SYSTEM

The close proximity of dopaminergic and glutamatergic terminals on the dendrites of medium spiny neurons[99,117] provides the anatomical ambiance for functional interactions between the two transmitter systems and raises the possibility that enhanced glutamatergic transmission may have relevance for the development of the MRC syndrome. Since protein phosphorylation serves as a major regulatory mechanism for glutamate receptors, the phosphorylation state of different subunits of NMDA and AMPA receptors has recently been investigated in the 6-OHDA-lesioned rat. These studies indicate that nonphysiological stimulation of DA receptors on striatal spiny neurons leads to changes in the subunit phosphorylation pattern of coexpressed ionotropic glutamatergic receptors, i.e., the N-methyl-D-aspartate (NMDA) subtype and amino-3-hydroxy-5-methyl-4-isoxazole proprionic acid (AMPA) subtype receptors. It now appears that these alterations in the phosphorylation state of striatal NMDA and AMPA receptors reflect the aberrant activation of signaling cascades linking DA and glutamate receptors expressed along the dendrites of medium spiny neurons.[118,119] More specifically, changes in the balance between specific spiny neuron kinase and phosphatase activity appear to affect the degree and pattern of phosphorylation.[119–122]

NMDA receptors are heteroligomers assembled to form ligand-gated ion channels from one or two NR1 subunits, expressed in eight currently recognized splice variants (a through h), and two or three NR2 subunits composed of four homologous isoforms (A through D).[123,124] In rat striatum, medium spiny neurons express NR1 variants along with NR2B and, to a lesser extent, NR2A subunits.[125] Protein phosphorylation serves as a major regulatory mechanism for these receptors.[126,127] Current evidence suggests that the chronic nonphysiological stimulation of rat DA receptors activates various kinases responsible for direct subunit phosphorylation as well as for synaptic clustering. These include serine kinases, such as cyclic AMP-protein kinase A (PKA), calcium/calmodulin-dependent protein kinase II (CaMKII), and calcium-activated protein kinase (PKC), as well as src or fyn tyrosine kinases. The phosphorylation of tyrosine residues has been reported to modulate channel opening probability[128,129] and receptor trafficking to the postsynaptic membrane,[130] while serine/threonine phosphorylation by calcium/phospholipid-stimulated or cAMP-stimulated protein kinases appears to affect their subcellular distribution, plasma membranes anchoring,[131,132] and synaptic clustering.[133] Recently, PKC has also been shown to influence NMDA currents by direct serine phosphorylation of the NR2B tail at residues S1303 and S1323[134] or by direct tyrosine phosphorylation of the NR2A and NR2B subunits.[135] The intrastriatal administration of inhibitors of certain of these serine and tyrosine kinases alleviates the motor response alterations in this rat model.

The AMPA receptors for glutamate are multimeric protein channels that mediate fast excitatory responses.[136] In rat striatum, somatodendritic AMPA glutamate receptors on medium spiny neurons are of GluR1, GluR2, or GluR3 type, and activation of these receptors results in a significant calcium influx. In addition, all neuronal classes in the striatum appear to express GluR5 or GluR6 and/or GluR7 AMPA subunits in addition to kainate (KA) subunits.[137] Phosporylation of AMPA receptors is an important mechanism for the modification of their channel function and is thought to play an important role in synaptic plasticity in this brain region.[136]

With respect to striatal AMPA receptor subunits, changes in the phosphorylation state of serine residues by a PKC signaling cascade may also affect motor function. Gene transfer of constitutively active protein kinase C into striatal neurons was recently found to accelerate the onset of levodopa-induced motor response alterations in parkinsonian rats in part, by the phosphorylation of AMPA receptor subunits and consequent modification of the strength of corticostriatal glutamatergic input.[138]

Taken together, differential activation of signal transduction pathways within spiny neurons leads to characteristic changes in the phosphorylation state of NMDA and AMPA glutamate receptors and thus in their sensitivity to corticostriatal synaptic input. As a consequence of these molecular and cellular events, striatal output changes in ways that contribute to the motor complications associated with levodopa therapy.

Attenuation of the enhanced glutamatergic function, and thus of the motor response complications, can be provided by drugs that target signaling proteins within striatal spiny neurons[139] or by pharmaceutical agents that interact extracellularly with glutamate (NMDA or AMPA) receptors. At this point in time, the former class of pharmaceuticals is not available for clinical use, and therefore most investigative efforts have focused on drugs that are active at the cell surface receptors, in particular the NMDA receptor.

GLUTAMATE ANTAGONISTS

If enhanced NMDA receptor sensitivity were to play a significant role in the expression of motor response complications, then NMDA receptor blockade can be predicted to confer benefit. Current evidence from studies in the 6-OHDA-lesioned rat, MPTP intoxicated monkey, and parkinsonian man suggests this is indeed the case.

In the 6-OHDA rat model of motor fluctuations, systemically administered drugs that selectively block NMDA receptors, such as MK-801, act palliatively as well as prophylactically to suppress or even prevent the shortening of the motor response to levodopa that otherwise occurs within a few weeks of intermittent levodopa treatment.[140–142] Striatal stereotactic injection studies suggest that NMDA antagonists exert their action primarily in the striatum.[142] Reversal of the shortened duration of the motor response to levodopa has subsequently also been demonstrated with the nonselective noncompetitive NMDA antagonist amantadine,[143] the NR2B selective antagonist ACEA 10-1244,[144] the competitive NMDA antagonist CPP, the inhibitor of glutamatergic transmission riluzole, and the AMPA antagonist NBQX.[145] Attenuation of levodopa-induced abnormal involuntary movements in 6-OHDA-lesioned rats,[146] comparable with MPTP-primate and human dyskinesias, has also been shown with amantadine.[147]

Chronic intermittent levodopa therapy in the MPTP-lesioned monkey leads to choreiform dyskinesias within several weeks.[148] Systemic administration of some, but not all, NMDA antagonists reduces chorea at doses that have no effect on the antiparkinson benefit from levodopa. In cynomolgus and rhesus monkeys, the competitive NMDA antagonist LY 235959, co-administered subcutaneously with a dyskinetic dose of levodopa, reduced choreic dyskinesia by 68% without diminishing the antiparkinsonian benefit of levodopa.[149] At the highest dose given, however, chorea suppression continued, but dystonic posturing tended to increase. Earlier studies with the noncompetitive NMDA antagonist MK-801 had led to quite different results. In cynomolgus monkeys, MK-801 blocked both the antiparkinson and dyskinetic effects of levodopa.[150] In squirrel monkeys, MK-801 again blocked the antiparkinson effect of levodopa and also converted chorea into dystonia.[151] Yet other results were obtained with the competitive NMDA receptor antagonist MDL 100,453. This agent had an antiparkinson effect of its own and increased dyskinesia severity when co-administered with levodopa.[152] In contrast, the noncompetitive allosteric site NMDA antagonist Co 101244/PD 174494 did not change motor activity by itself and reduced levodopa-induced dyskinesias by 70%.[152] Also in the cynomolgus monkey, the weak channel blocker amantadine improved dyskinesia by 30%.[153] At the highest doses, the antiparkinson effect was mildly antagonized. Memantine, another aminoadamantane with higher affinity for the NMDA receptor than amantadine, had no effect on dyskinesias in the same model (unpublished observation). NMDA receptor subtype selectivity may be one of the determinants of the behavioral effects of NMDA antagonists, with some evidence pointing to the relative importance of the NR2B subunit for antidyskinetic efficacy. However, other factors, such as the mechanism of antagonistic action (competitive antagonism, channel blocking, allosteric site inhibition), affinity, binding kinetics, and efficacy of antagonism, may all play a role as well.[152] In addition, differences in species, drug dose ranges, routes of administration, and experimental paradigms limit direct comparison of results across the aforementioned studies. As far as AMPA receptors is concerned, LY300164, a drug acting at an allosteric modulation site to block AMPA receptors, not only potentiated the motor response to low- and medium-dose levodopa by 50%, but also decreased the levodopa-induced chorea by 40%.[154] In the same study, administration of CX516, an ampakine that enhances glutamatergic transmission, did not enhance motor activity but increased dyskinesia.

In patients with Parkinson's disease, only a limited number of glutamate modulating agents have been studied so far. The noncompetitive NMDA antagonist dextrorphan hydrochloride was given to two patients with levodopa-induced dyskinesias under controlled conditions.[155] The antiparkinson response to levodopa remained intact, and dyskinesias were substantially reduced. The parent compound of dextrorphan, dextromethorphan, is also a noncompetitive NMDA antagonist. Its effects on motor response complications was studied first in 18 subjects.[156] In 12 participants, side effects occurred, including decreased levodopa efficacy. A double-blind, placebo-controlled, crossover study, consisting of two arms of two weeks each, separated by a one week washout, was conducted in the remaining six individuals. With the addition of DM, dyskinesia scores during an eight-hour in-hospital observation period were 25% lower than with placebo, while motor fluctuations improved as well. The antidyski-

netic effect was confirmed in another double-blind, placebo-controlled, crossover study with DM.[157] In this study, subjects received DM or placebo for two to three weeks each. At the end of each arm, they were admitted to the hospital for intravenous levodopa dose response studies. At the highest LD dose, maximum dyskinesia scores were 54% lower with DM than with placebo. It thus appears that DM can decrease dyskinesia in some PD patients. A majority of the subjects, however, did not tolerate DM, and a subset showed decreased levodopa efficacy when DM was co-administered, reminiscent of some of the nonhuman primate studies with NMDA antagonists. It should be kept in mind that these were acute studies, and the long-term effects were not formally studied.

Amantadine is a noncompetitive NMDA antagonist with a long history in the treatment of mild Parkinson's disease. The NMDA antagonistic properties were only recently recognized,[158] providing a safe opportunity to test the hypothesis that NMDA antagonists diminish dyskinesias under clinical conditions. Eighteen patients with choreiform dyskinesias received 300 to 400 mg amantadine or placebo for three weeks each, in a double-blind crossover study.[159] At the end of each arm of the study, patients received a continuous intravenous levodopa infusion at the same, individually determined, infusion rate. Five hours into the infusion, after steady-state conditions had presumably been achieved, parkinsonian symptoms and dyskinesia severity were scored every ten minutes for two hours. Dyskinesia scores were more than 50% lower, whereas parkinson scores were unaffected. Two other double-blind, placebo-controlled studies have since confirmed the impressive acute antidyskinetic effect of amantadine.[160,161] Another study has provided long-term follow-up data indicating that amantadine's antidyskinetic benefit is maintained for at least one year.[162] The related compound memantine was found, as in the MPTP monkey, to have no effect on dyskinesia in a double-blind study; it did improve parkinsonian symptoms.[163] There may be a subset of individuals, as was the case with dextromethorphan, in whom memantine does afford antidyskinetic benefit, as indicated by a recent case report.[164] It remains unclear which pharmacodynamic or pharmacokinetic factors explain the interindividual variability of the antidyskinetic effects. The inhibitor of glutamatergic transmission riluzole, in accordance with the results obtained in rat, was found to reduce the presence and severity of dyskinesias in an open study of six PD patients.[165]

ADENOSINE SYSTEM

Striatal medium spiny neurons express adenosine A1 and A2a receptors.[166] A2a receptors are predominantly expressed on GABA-ergic neurons that co-express D_2 dopamine receptors, giving rise to the indirect pathway projecting to GPe. Adenosine receptors may play a role in the control of motor behaviors through modulation of striatal output.[166,167] Recent studies have shown that selective adenosine A2a receptor antagonists confer antiparkinson activity in the 6-OHDA-lesioned rat[168] and MPTP monkey[169,170] without inducing significant dyskinesia. Similarly, A2a receptor blockade in patients with Parkinson's disease not only potentiated and prolonged[171] the response to levodopa, it also reduced the severity of dyskinesias.[172] Interestingly, a recent report demonstrated increased adenosine A2a mRNA levels in the brain of severely dyskinetic monkeys but not in the brains of animals exposed to the same high levodopa dose but who had developed no dyskinesias.[173] Several mechanisms of action have been proposed. One theory centers on the finding that A2a receptor activation modulates GABA release in the striatum,[174–176] in the opposite direction of D_2 receptors.[174] A second hypothesis is that A2a receptor activation modulates the activity of striatal glutamate neurotransmission,[177] possibly through activation of kinases that regulate the phosphorylation state of ionotropic glutamate receptors.[172,178] In support of this hypothesis, treatment of parkinsonian rats with a selective A2a antagonist reduced the hyperphosphorylation of AMPA receptors.[179] The latter explanation would be compatible with the aforementioned hypothesis that hyperphosphorylation of gluatamatergic receptors contributes to the development of motor response complications and that NMDA and AMPA antagonists reduce dyskinesia severity in animals and PD patients.[120]

SEROTONIN SYSTEM

In Parkinson's disease, the serotonergic system is not spared,[180] and serotonergic dysfunction may be involved in some of the clinical features of Parkinson's disease and complications of levodopa therapy.[181] Drugs that alter serotonergic transmission, such as the antidepressant fluoxetine[182] and the anxiolytic buspirone,[183] have been reported to reduce levodopa-induced dyskinesias in open label studies. On the other hand, enhanced serotonergic transmission has been associated with increased parkinsonism and tardive dyskinesias.[184] More recently, the 5-HT1A agonist sarizotan was studied in 6-OHDA-lesioned rats and MPTP monkeys.[185] In both species, sarizotan reduced levodopa-induced response complications. In the rat, the shortening in motor response duration, induced by repeated administration of levodopa, was reversed. In the parkinsonian monkey, levodopa-induced dyskinesias were markedly diminished. These effects were blocked by co-administration of a selective 5-HT1A antagonist,[185] suggesting that the antidyskinetic effect was indeed specific to 5-HT1A stimulation. This effect could be mediated through stimulation of 5-HT1A autoreceptors located on the soma and dendrites of 5-HT neurons, leading to reduced 5-HT neuronal firing activity and reduced sero-

tonin release. Since serotonergic striatal terminals may serve as an important source of decarboxylation of levodopa and therefore as a source of dopamine in the absence of dopaminergic terminals, the inhibition of 5-HT neurons could lead to a more sustained release of dopamine and, as a consequence, to a reduction in fluctuations and dyskinesias.[185] Clinical studies with sarizotan are currently underway.

Other investigations have explored the effects of serotonin receptor blockade. Ritanserin, a 5-HT2 antagonist, was reported to improve levodopa induced dyskinesias in a small study.[186] The same investigators recently reported that mirtzapine, a 5-HT2 antagonist, 5-HT1A agonist, and alpha-2 antagonist, was effective in reducing levodopa-induced dyskinesias.[187] In the same vein, clozapine (active at 5-HT2A and other serotonin as well as dopamine receptors) was found to decrease dyskinesias.[188,189] Quetiapine, another atypical neuroleptic with 5-HT2A/C and D2/3 antagonistic activity, was recently examined in the 6-OHDA rat and MPTP monkey model of Parkinson's disease.[190] In the rat, quetiapine reversed the shortening of response to levodopa produced by twice daily injection of levodopa for three weeks. In the monkey model, co-administration of quetiapine with levodopa substantially reduced levodopa-induced dyskinesias without affecting the response to levodopa. Given the potential of quetiapine to bind to multiple receptors including D_2/D_3 and adrenergic receptors, the antidyskinetic effects cannot be attributed to 5-HT2A/C antagonism with certainty.[190] As an interesting aside, it should be noted that serotonin receptor stimulation activates certain signaling kinases that are responsible for phosphorylation of striatal ionotropic glutamate receptors, leading to their increased sensitivity to corticostriatal glutamatergic synaptic input. The latter, as discussed above, may lead to motor response complications.

ADRENERGIC SYSTEM

Several reports suggest that alpha-2 adrenoceptor antagonists can exert antiparkinsonian as well as antidyskinetic activity. Co-administration of yohimbine reduces levodopa-induced abnormal involuntary movements in MPTP treated monkeys,[191] and idazoxan reduces dyskinesia while enhancing the antiparkinsonian action of levodopa in the latter animal model.[192–194] In patients with Parkinson's disease idazoxan may also have a modest effect,[195] although another small clinical trial found no antidyskinetic benefit.[196] The underlying mechanism for the antidyskinetic effect of alpha-2 adrenoceptor antagonists remains unclear. However, rodent studies using double-labeled immunohistochemistry demonstrate that alpha2C adrenoceptors are abundantly localized on striatal GABA-ergic medium-sized spiny projection neurons,[197] where they may modulate GABA release.

OPIOID SYSTEM

In recent years, the relation between changes in the opioid system and the development of LID has been a focus of investigation. There are three major classes of opioid receptors (the mu, kappa, and delta receptors), all of which are G-protein coupled receptors, present in the striatum and other nuclei of the basal ganglia.[198] Their ligands, the endogenous opioid peptides, include the enkephalins, beta-endorphins, and dynorphins.[199,200] These neuropeptides are formed by cleavage of their precursor proteins preproenkephalin-A (PPE-A) (enkephalin) and preproenkephalin-B (PPE-B) (dynorphin, beta-endorphin) and show different affinities for the different classes of opioid receptors, with enkephalins binding mainly to delta receptors and dynorphins to kappa receptors.[201] Dynorphin and enkephalin are co-transmitters in the GABA-ergic striatal output pathways. Dynorphin is co-expressed on medium spiny neurons of the direct pathway that mainly express dopamine D_1 receptors and project to the GPi/SNR, whereas enkephalin is expressed on medium spiny neurons of the indirect pathway that bear D_2 receptors and project to GPe.[202] These opioid peptides are known to act as neuromodulators and may modulate neurotransmitter release in the basal ganglia. Enkephalin has been shown to reduce GABA release in GPe.[203] Thus, enhanced opioid neurotransmission may cause diminished inhibition of GPE leading to increased inhibition of STN. Underactivity of STN and, as a consequence, diminished activity in the output nuclei of the basal ganglia, the GPi/SNR, is thought to be associated with the appearance of involuntary movements.[204,205] Alternatively, enhanced opioid transmission along the direct pathway could lead to increased stimulation of mu, delta, or kappa subtype opioid receptors in GPi and SNR and thereby reduce the activity in these output regions of the basal ganglia.[206]

Many studies have indeed demonstrated changes in the expression of opioid peptide precursors in animal models after lesioning of the nigrostriatal pathway with 6-OHDA (rat[207–211]) or MPTP (mice[212], monkey[213–215]). In general, these studies agree that dopaminergic denervation increases striatal expression of PPE-A, while expression of PPE-B is reduced or unchanged.

Subsequent intermittent levodopa treatment has no effect (rats[209,210], monkeys[213,215]) or further increases PPE-A mRNA[207] but substantially increases striatal PPE-B mRNA levels.[206-211,213,215] Repeated bromocriptine administration, on the other hand, does not increase PPE-B expression and normalizes the lesion induced increase in PPE-A expression.[207,215]

Several recent studies are now suggesting that alterations in the opioid system are not only associated with levodopa administration but also correlate with dyskinesia severity. Using opioid receptor-binding autoradiography in a rat model of levodopa-induced dyskinesia,[216] it was found that dyskinetic rats had considerably lower striatal and nigral levels of kappa binding densities compared to nondyskinetic rats. These levels were negatively correlated with dyskinesia scores and with the striatal expression of opioid precursor mRNAs.[208] In another study, involving unlesioned monkeys that nonetheless had developed dyskinesias after high-dose levodopa treatment,[62] animals with marked LID had significantly higher PPE-A mRNA levels than animals with little or no dyskinesia.[173] Similarly, in MPTP-lesioned macaques striatal PPE-B mRNA expression was significantly higher in dyskinetic monkeys than in nondyskinetic parkinsonian animals or controls.[206] In situ hybridization studies were also performed on postmortem striatal tissue from PD patients, some of whom had experienced dyskinesias while others had not. Significantly higher preproenkephalin mRNA levels were observed in the lateral putamen of the dyskinetic patients in comparison to nondyskinetic patients[217] or age-matched controls.[206] A recent neuroimaging study supports the above findings. When dyskinetic and nondyskinetic patients with PD were studied with 11C-diprenorphine PET, the dyskinetic group showed a reduction in striatal and thalamic opioid site availability, suggesting that dyskinesias are associated with raised levels of endogenous opioid ligands in these nuclei.[218,219] Finally, injection of endogenous mu opioid receptor ligand endomorphin-1 in the globus pallidus of rats induced orofacial dyskinesia, an effect that could be blocked using a selective mu antagonist.[220]

Regardless whether enhanced opioid transmission is a cause or consequence of the basal ganglionic changes leading to motor response complications, it offers a novel strategy to interfere with dyskinesias. Several studies have explored the antidyskinetic potential of opioid receptor antagonists. The non-subtype-selective opioid antagonist naloxone reduced levodopa-induced dyskinesias in MPTP-treated marmosets[221] and blockade of the related nonselective opioid antagonist naltrexone, the mu opioid receptor selective antagonist cyprodime, or the delta-opioid receptor selective antagonist naltrindole reduced LID in the MPTP-treated marmoset.[222] Small studies in dyskinetic MPTP-lesioned monkeys[150] and PD patients[223,224] however, showed no significant reduction in LID. Higher potency, longer duration of action, and subtype selectivity may be important to achieve an antidyskinetic effect.[222]

Taken together, these findings suggest that increased synthesis of preproenkephalin in medium spiny neurons of the direct and indirect pathways contributes to the development of LID.[206,217]

CANNABINOID SYSTEM

Neurotransmission in the basal ganglia is also modulated through signaling at cannabinoid receptors, and it has been proposed that changes in cannabinoid transmission are associated with parkinsonian symptomatology and motor response complications.[225] However, the actions of endogenous cannabinoids are only beginning to be understood, and alterations in the cannabinoid system occurring in PD and with chronic levodopa treatment are complex and even farther from being elucidated.

Of the three currently known cannabis receptors, the CB_1 receptor is found in high levels in the basal ganglia, in particular the striatum, globus pallidus and SNR.[226] Their endogenous ligands, the endocannabinoids, are also predominantly expressed in the basal ganglia.[227] In general, cannabinoid receptors and their ligands are thought to act as inhibitors of neurotransmitter release.[228] However, multiple and often opposite actions have recently been ascribed to endocannabinoids.

Within the striatum, stimulation of presynaptic CB_1 receptors may reduce glutamate and GABA release,[229,230] and it has been proposed that CB_1 activation is essential for long-term depression (LTD) in corticostriatal synapses and thus for glutamatergic striatal plasticity.[231] Postsynaptic interactions may also occur between striatal CB_1 receptors and other G-protein-coupled receptors, perhaps due to their ability to activate adenylyl cyclase and protein kinases.[232,233] CB1 receptors are localized on GABA-ergic neurons[234] where they are co-expressed with, among others, D_1 and D_2 receptors,[235] and signal transduction convergence between CB_1 and DA receptors has been demonstrated in rat and monkey striatum, suggestive of interactions between these receptors at the level of G-protein/adenylyl cyclase signal transduction.[236] In addition, dopamine regulates striatal CB_1 receptor mRNA,[237] and activation of CB_1 receptors increases firing rates of dopaminergic neurons in conscious rats,[238] suggesting that CB_1 receptor activation could enhance DA release in striatum.[239]

In the globus pallidus and SNR, CB_1 receptors are located presynaptically on terminals of striatal GABA-ergic projection neurons,[226] well positioned to modulate GABA transmission. In the GPe, cannabinoids enhance GABA transmission by reducing GABA re-uptake.[240,241] In GPi and SNR, CB_1 activation blocks GABA-ergic input from the striatum as well as glutamatergic input from the subthalamic nucleus.[242]

Experimental parkinsonism has been associated with increased striatal endocannabinoid levels[243] as well as increased striatal CB_1 receptor binding and CB_1 receptor-G-protein coupling.[244] Subsequent chronic administration of levodopa has been reported to reverse these changes in the MPTP-lesioned marmoset.[244] In contrast, levodopa was found to increase striatal CB1 mRNA in 6-OHDA-lesioned rats.[245] In the GP(e) of the rat, reserpine-induced parkinsonism is associated with a sevenfold increase of cannabinoid levels, which is reversed by subsequent dopamine agonist treatment.[246]

It will be evident from the foregoing that the cannabinoid system could play a prominent role in the control of movement under parkinsonian and dyskinetic conditions, but the exact mechanisms of action are in need of further clarification. However, this has not precluded exploration of the effects of exogenous cannabinoid agonists or antagonists in animal models of parkinsonism and levodopa-induced dyskinesias.[247,248]

Cannabinoid receptor modulation has been investigated in the reserpine-treated rat model of Parkinson's disease. Both a cannabinoid CB_1 agonist and antagonist improved levodopa induced hyperkinetic behavior.[248] Furthermore, in the MPTP-lesioned marmoset, the CB_1 receptor agonist nabilone alleviated levodopa-induced dyskinesias,[247] as did the selective cannabinoid antagonist SR 141716A. Obtaining antidyskinetic benefit with both agonists and antagonists has been proposed to be due to actions at different target sites in the brain.[225,247] Finally, a small but controlled crossover study suggested that nabilone can also reduce dyskinesia severity in patients with PD.[249]

Possible explanations for the antidyskinetic effect of CB_1 modulating compounds have been sought at the level of GPe and the striatum. In GPe, CB_1 receptor stimulation reduces GABA-ergic uptake. This results in enhanced GABA-ergic transmission, leading in turn to increased inhibition of GPe and, as a consequence, reduced inhibition of STN. The latter is thought to result in reduced dyskinesia production.[250] Alternatively, at the level of the striatum, stimulation of cannabinoid receptors could reduce synaptic glutamate release.[229] As discussed before, overactivity of striatal glutamatergic transmission is thought to be involved in the generation of motor response complications,[118] as evidenced by changes in phosphorylation state of striatal NMDA and AMPA receptors.[119,144] Interestingly, cannabinoids are known to activate protein kinase C,[233] which is one of the kinases responsible for NMDA and AMPA subunit phosphorylation.[251] The finding that cannabinoid CB_1 receptor activation is a key element for LTD in corticostriatal synapses[231] suggests the intriguing possibility that CB_1 modulating agents could act by modifying striatal plasticity. As discussed below, levodopa induced dyskinesia has been proposed to represent a form of pathological learning.[252]

STRIATAL PLASTICITY

The striatum plays a role in motor learning.[253] Long-term activity-dependent changes in the efficacy of excitatory synaptic transmission are considered fundamental to the process of information storage in different areas of the brain. In the striatum, corticostriatal excitatory, glutamatergic transmission can express two opposite forms of neuronal plasticity: long-term potentiation (LTP) and long-term depression (LTD).[254–256] Both require activation of glutamate receptors (NMDA and AMPA receptors, respectively) as well as dopamine receptors. Dopaminergic denervation induces hyperactivity of corticostriatal glutamate mediated transmission and interferes with the generation of corticostriatal plasticity. Subsequent nonphysiological dopamine replacement may modify the balance between LTD and LTP in ways that lead to impaired motor performance and altered responses to dopaminergic therapy.[256]

The hypothesis that LID may result from abnormal plastic processes finds support from studies of immediate-early gene (IEG) expression.[257,258] i.e.,Gs encode proteins that show rapid changes in level of transcription after certain types of neuronal stimulation and may initiate cellular mechanisms of neuronal plasticity. Studies in rats with a unilateral 6-OHDA lesion showed that the i.e.,G response to D_1 dopamine agonists is different in the denervated compared to the intact striatum. Thus, where doses of a selective D_1 agonist showed little or no i.e.,G induction in the normal striatum, a robust i.e.,G response was present in the dopamine depleted striatum. Furthermore, whereas repeated dopaminergic stimulation results in a diminished i.e.,G response in the normal striatum, such an adaptive response was absent in the dopamine-depleted striatum. The persistence of the increased i.e.,G response suggests that adaptive mechanisms operating normally in the striatum to reverse neuronal plasticity resulting from excessive dopamine receptor activation are ineffective.[257]

Of particular interest is a recent report[259] showing that changes in corticostriatal plasticity associated with dopamine denervation and levodopa treatment differ, depending on whether dyskinesias are present in the experimental animal. In this study, 6-OHDA-lesioned rats were injected with low-dose levodopa twice daily to induce a gradual development of abnormal involuntary movements. Fifty percent of the animals developed dyskinesias. The animals with dyskinesias and those without dyskinesias had comparable levels of dopamine denervation. Using a corticostriatal slice preparation to measure LTP from striatal spiny neurons, the authors found that 25 min of high-frequency stimulation (HFS) of corticostriatal afferents induced LTP in sham operated but not 6-OHDA-lesioned rats, confirming earlier reports that dopaminergic denervation blocks LTP.[260] Next, they showed that chronic levodopa treatment restored LTP in all dopamine denervated rats, dyskinetic or nondyskinetic. Normally, subsequent exposure of corticostriatal afferents to low-frequency stimulation (LFS) reverses the increased synaptic strength in a process called *synaptic depotentiation*.[261] This process has also been described in the hippocampus[262] and may be a mechanism through which efficiency of information storage can be increased by "erasing" irrelevant information.[261] In the 6-OHDA-lesioned rats, however, an intriguing difference between

the two groups existed.[259] LFS fully reversed LTP in the nondyskinetic rats, whereas dyskinetic rats showed no capacity for depotentiation. This loss of depotentiation could lead to storage of abnormal motor information, impeding the selection and automatic execution of learned motor programs and, instead, leading to the development of abnormal motor patterns and behaviors, such as dyskinesias.

CONCLUSION

Nigrostriatal degeneration and subsequent intermittent dopamine replacement therapy are associated with widespread changes throughout the cortical-basal ganglia-thalamo-cortical loop, and modifications anywhere along this loop potentially contribute to the development of an altered response to levodopa over time. The striatum appears to play a prominent role in this process, as it is positioned directly downstream from the degenerating nigra and serves as an integrative station processing the flow of cortical and nigral information into the basal ganglia. The close proximity of multiple extrinsic and intrinsic synaptic inputs in the striatal medium spiny neuron provides an ambiance for crosstalk between different receptors, their neurotransmitters, and neuromodulators, which may explain why manipulation of so many different messenger systems can modify motor behaviors. Increased synaptic efficacy of corticostriatal glutamatergic synaptic transmission, associated with changes in medium spiny neuron signaling pathways, and, perhaps, with impaired down-regulation of striatal synaptic plasticity, may contribute to the development of motor response complications.

While some of the recent findings described here may still be in a stage of, in James Parkinson's words, *"mere conjectural suggestions,"* they are slowly but surely starting *"to point out the most appropriate means of relieving this tedious and most distressing malady,"* indefinitely.

REFERENCES

1. Parkinson, J., *An essay on the shaking palsy*, Sherwood, Neely and Jones, London, 1817.
2. Cotzias, G. C., Van, W. M., and Schiffer, L. M., Aromatic amino acids and modification of parkinsonism., *N. Engl. J. Med.*, 276, 374, 1967.
3. Barbeau, A., The clinical physiology of side effects in long-term L-DOPA therapy, *Adv. Neurol.*, 5, 347, 1974.
4. Marsden, C., Problems with long term levodopa therapy for Parkinson's disease, *Clin. Neuropharmacol.*, 17, S32, 1994.
5. Ahlskog, J. E. and Muenter, M. D., Frequency of levodopa-related dyskinesias and motor fluctuations as estimated from the cumulative literature, *Mov. Disord.*, 16, 448, 2001.

6. Wade, D. N., Mearrick, P. T., and Morris, J. L., Active transport of L-dopa in the intestine, *Nature*, 242, 463, 1973.

7. Rivera, C. L. et al., L-dopa treatment failure: explanation and correction, *Br. Med. J.*, 4, 93, 1970.

8. Nutt, J. G. et al., The on-off phenomenon in Parkinson's disease. Relation to levodopa absorption and transport. *N. Engl. J. Med.*, 310, 483, 1984.

9. Fermaglich, J. and O'Doherty, D. S., Effect of gastric motility on levodopa, *Dis. Nerv. Syst.*, 33, 624, 1972.

10. Rivera, C. L. et al., Absorption and metabolism of L-dopa by the human stomach, *Eur. J. Clin. Invest.*, 1, 313, 1971.

11. Kurlan, R. et al., Duodenal and gastric delivery of levodopa in parkinsonism. *Ann. Neurol.*, 23, 589, 1988.

12. Sage, J. I. et al., Long-term duodenal infusion of levodopa for motor fluctuations in parkinsonism, *Ann. Neurol.*, 24, 87, 1988.

13. Bredberg, E. et al., Intraduodenal infusion of a water-based levodopa dispersion for optimisation of the therapeutic effect in severe Parkinson's disease, *Eur. J. Clin. Pharmacol.*, 45, 117, 1993.

14. Kurth, M. C. et al., Oral levodopa/carbidopa solution versus tablets in Parkinson's patients with severe fluctuations: a pilot study, *Neurology*, 43, 1036, 1993.

15. Verhagen Metman, L. V. et al., Fluctuations in plasma levodopa and motor responses with liquid and tablet levodopa/carbidopa, *Mov. Disord.*, 9, 463, 1994.

16. Wooten, G. F., Pharmacokinetics of levodopa, in *Movement Disorders 2,* Marsden, C. D. and Fahn, S., Eds., Butterworths, London, 1987, 231.

17. Nutt, J. G., Woodward, W. R., and Anderson, J. L., The effect of carbidopa on the pharmacokinetics of intravenously administered levodopa: the mechanism of action in the treatment of parkinsonism, *Ann. Neurol.*,18, 537, 1985.

18. Nutt, J. G., and Fellman, J. H., Pharmacokinetics of levodopa, *Clin. Neuropharmacol.*, 7, 35, 1984.

19. Fabbrini, G. et al., Levodopa pharmacokinetic mechanisms and motor fluctuations in Parkinson's disease, *Ann. Neurol.*, 21, 370, 1987.

20. Wade, L. A., and Katzman, R., Synthetic amino acids and the nature of L-DOPA transport at the blood-brain barrier, *J. Neurochem.*, 25, 837,1975.

21. Pardridge, W. M., and Oldendorf, W. H., Transport of metabolic substrates through the blood-brain barrier, *J. Neurochem.*, 28, 5, 1977.

22. Cotzias,G. C., Van, W. M., and Schiffer L. M., Aromatic amino acids and modification of parkinsonism, *N. Engl. J. Med.*, 276, 374, 1967.

23. Nutt, J. G. et al., The on-off phenomenon in Parkinson's disease. Relation to levodopa absorption and transport, *N. Engl. J. Med.*, 310, 483, 1984.

24. Mena, I., and Cotzias, G. C., Protein intake and treatment of Parkinson's disease with levodopa, *N. Engl. J. Med.*, 292, 181, 1975.

25. Leenders, K. L. et al., Inhibition of L-[18F]fluorodopa uptake into human brain by amino acids demonstrated by positron emission tomography, *Ann. Neurol.*, 20, 258, 1986.

26. Pardridge, W. M., Kinetics of competitive inhibition of neutral amino acid transport across the blood-brain barrier, *J. Neurochem.*, 28, 103, 1977.

27. Pincus, J. H., and Barry, K., Protein redistribution diet restores motor function in patients with dopa-resistant "off" periods, *Neurology*, 38, 481, 1988.

28. Rivera, C. L. et al., L-dopa treatment failure: explanation and correction, *Br. Med. J.*, 4, 93, 1970.

29. Rivera, C. L. et al., Absorption and metabolism of L-dopa by the human stomach, *Eur. J. Clin. Invest.*, 1, 3, 1971.

30. Rivera, C. L. et al., L-3,4-dihydroxyphenylalanine metabolism by the gut *in vitro*, *Biochem. Pharmacol.*, 20, 3051, 1971.

31. Fahn, S., "On-off" phenomenon with levodopa therapy in Parkinsonism. Clinical and pharmacologic correlations and the effect of intramuscular pyridoxine, *Neurology*, 24, 431, 1974.

32. Nutt, J. G., On-off phenomenon: relation to levodopa pharmacokinetics and pharmacodynamics, *Ann. Neurol.*, 22, 535, 1987.

33. Muenter, M. D. et al., 3-O-methyldopa, L-dopa, and trihexyphenidyl in the treatment of Parkinson's disease, *Mayo. Clin. Proc.*, 48, 173, 1973.

34. Wade, L. A., and Katzman, R., 3-O-methyldopa uptake and inhibition of L-dopa at the blood-brain barrier, *Life Sci.*, 17,131, 1975.

35. Reches, A., Mielke, L. R., and Fahn S., 3-o-methyldopa inhibits rotations induced by levodopa in rats after unilateral destruction of the nigrostriatal pathway, *Neurology*, 32, 887, 1982.

36. Reches, A., and Fahn S., 3-O-methyldopa blocks dopa metabolism in rat corpus striatum, *Ann. Neurol.*, 12, 267, 1982.

37. Gervas, J. J. et al., Effects of 3-OM-dopa on monoamine metabolism in rat brain, *Neurology*, 33, 278, 1983.

38. Nutt, J. G. et al., 3-O-methyldopa and the response to levodopa in Parkinson's disease, *Ann. Neurol.*, 21, 584, 1987.

39. Muenter, M. D., Sharpless, N. S., and Tyce, G. M., Plasma 3-0-methyldopa in L-dopa therapy of Parkinson's disease, *Mayo. Clin. Proc.*, 47, 389, 1972.

40. Fabbrini, G. et al., 3-O-methyldopa and motor fluctuations in Parkinson's disease, *Neurology*, 37, 856, 1987.

41. Gancher, S. T., Nutt, J. G., and Woodward, W. R., Peripheral pharmacokinetics of levodopa in untreated, stable, and fluctuating parkinsonian patients, *Neurology*, 37, 940, 1987.

42. Contin, M. et al., Response to a standard oral levodopa test in parkinsonian patients with and without motor fluctuations, *Clin. Neuropharmacol.*, 13, 19, 1990.

43. Bunney, B. S. et al., Dopaminergic neurons: effect of antipsychotic drugs and amphetamine on single cell activity, *J. Pharmacol. Exp. Ther.*,185, 560, 1973.

44. Schultz, W., Behavior-related activity of primate dopamine neurons, *Rev. Neurol.*, 150, 634, 1994.

45. Agid, Y., Javoy, F., and Glowinski, J., Hyperactivity of remaining dopaminergic neurones after partial destruction of the nigro-striatal dopaminergic system in the rat, *Nat. New. Etiol.*, 245, 150, 1973.

46. Zigmond, M. J. et al., Compensations after lesions of central dopaminergic neurons: some clinical and basic implications, *TINS,* 13, 290, 1990.

47. Hirsch, E. C., Nigrostriatal system plasticity in Parkinson's disease: effect of dopaminergic denervation and treatment, *Ann. Neurol.,* 47, S115, 2000.

48. Kumar, A., Huang, Z., and de la Fuente-Fernandez R., Mechanisms of motor complications in treatment of Parkinson's disease, *Adv. Neurol.,* 91,193, 2003.

49. Nutt, J. G. and Holford, N. H. G., The response to levodopa in Parkinson's disease: imposing pharmacological law and order, *Ann. Neurol.,* 39, 561, 1996.

50. Melamed, E., Hefti, F., and Wurtman, R. J., Nonaminergic striatal neurons convert exogenous L-dopa to dopamine in parkinsonism, *Ann. Neurol.,* 8, 558, 1980.

51. Chase, T. N., Engber, T. M., and Mouradian, M. M., Contribution of dopaminergic and glutamatergic mechanisms to the pathogenesis of motor response complications in Parkinson's disease, *Adv. Neurol.,* 69, 497, 1996.

52. de la Fuente-Fernandez, R. et al., Biochemical variations in the synaptic level of dopamine precede motor fluctuations in Parkinson's disease: PET evidence of increased dopamine turnover, *Ann. Neurol.,* 49, 298, 2001.

53. Chase, T. N., Levodopa therapy: Consequences of the non-physiologic replacement of dopamine, *Neurology,* 50, S17, 1998.

54. Fahn, S., Parkinson disease, the effect of levodopa, and the ELLDOPA trial, *Arch. Neurol.,* 56, 529, 1999.

55. Engber, T. M. et al., Continuous and intermittent levodopa differentially affect rotation induced by D-1 and D-2 dopamine agonists, *Eur. J. Pharmacol.,* 168, 291, 1989.

56. Papa, S. M. et al., Motor fluctuations in levodopa treated parkinsonian rats: relation to lesion extent and treatment duration, *Brain Res.,* 662, 69, 1994.

57. Papa, S. M., Reversal of levodopa-induced motor fluctuations in experimental parkinsonism by NMDA receptor blockade, *Brain Res.,* 701, 13, 1995.

58. Gnanalingham, K. K. and Robertson, R. G., The effects of chronic continuous versus intermittent levodopa treatments on striatal and extrastriatal D1 and D2 dopamine receptors and dopamine uptake sites in the 6-hydroxydopamine lesioned rat—an autoradiographic study, *Brain Res.,* 640, 185, 1994.

59. Boyce, S. et al., Nigrostriatal damage is required for induction of dyskinesias by L-DOPA in squirrel monkeys, *Clin. Neuropharmacol.,* 13, 448, 1990.

60. Crossman, A. R., A hypothesis on the pathophysiological mechanisms that underlie levodopa- or dopamine agonist-induced dyskinesia in Parkinson's disease: implications for future strategies in treatment, *Mov. Disord.,* 5, 100, 1990.

61. Schneider, J. S., Levodopa-induced dyskinesias in parkinsonian monkeys: relationship to extent of nigrostriatal damage, *Pharmacol. Biochem. Behav.,* 34, 193, 1989.

62. Pearce, R. K. et al., L-dopa induces dyskinesia in normal monkeys: behavioural and pharmacokinetic observations, *Psychopharmacol,* 156, 402, 2001.

63. Blanchet, P. J. et al., Continuous administration decreases and pulsatile administration increases behavioral sensitivity to a novel dopamine D2 agonist (U-91356A) in MPTP-exposed monkeys, *J. Pharmacol. Exp. Ther.,* 272, 854, 1995.

64. Pearce, R. K. et al., De novo administration of ropinirole and bromocriptine induces less dyskinesia than L-dopa in the MPTP-treated marmoset, *Mov. Disord.,* 13, 234, 1998.

65. Maratos, E. C. et al., Both short- and long-acting D-1/D-2 dopamine agonists induce less dyskinesia than L-DOPA in the MPTP-lesioned common marmoset (Callithrix jacchus), *Exper. Neurol.,* 179, 90, 2003.

66. Nygaard, T. G., Marsden, C. D., and Fahn, S., Doparesponsive dystonia: long-term treatment response and prognosis, *Neurology,* 41, 174, 1991.

67. Mones, R. J., Elizan, T. S., and Siegel, G. J., Analysis of L-dopa induced dyskinesias in 51 patients with Parkinsonism, *J. Neurol. Neurosurg. Psychiatry,* 34, 668, 1971.

68. Markham, C. H., The choreoathetoid movement disorder induced by levodopa, *Clin. Pharmacol. Ther.,* 12, 340, 1971.

69. Chase, T. N., Holden, E. M., and Brody, J. A., Levodopa-induced dyskinesias. Comparison in Parkinsonism-dementia and amyotrophic lateral sclerosis, *Arch. Neurol.,* 29, 328, 1973.

70. Horstink, M. W. et al., Severity of Parkinson's disease is a risk factor for peak-dose dyskinesia, *J. Neurol. Neurosurg. Psychiatry,* 53, 224, 1990.

71. Ballard, P. A., Tetrud, J. W., and Langston, J. W., Permanent human parkinsonism due to 1-methyl-4-phenyl-1,2,3,6-tetrahydropyridine (MPTP): seven cases, *Neurology,* 35, 949, 1985.

72. Block, G. et al., Comparison of immediate-release and controlled release carbidopa/levodopa in Parkinson's disease, *Eur. Neurology,* 37, 23, 1997.

73. Dupont, E. et al., Sustained-release Madopar HBS compared with standard Madopar in the long-term treatment of de novo parkinsonian patients, *Acta Neurol. Scand.,* 93, 14, 1996.

74. Rinne, U. K. et al., Cabergoline in the early treatment of Parkinson's disease: results of the first year of treatment in a double-blind comparison of cabergoline and levodopa. The PKDS009 Collaborative Study Group, *Neurology,* 48, 363, 1997.

75. Rinne, U. K. et al., Early treatment of Parkinson's disease with cabergoline delays the onset of motor complications. Results of a double-blind levodopa controlled trial. The PKDS009 Study Group, *Drugs,* 55, 23, 1998.

76. Rascol, O. et al., A five-year study of the incidence of dyskinesia in patients with early Parkinson's disease who were treated with ropinirole or levodopa, *N. Engl. J. Med.,* 342, 1484, 2000.

77. Lledo, A. et al., Long-term efficacy of pergolide monotherapy in early-stage Parkinson's disease. One-year interim analysis of a 3-year double-blind randomized study versus levodopa, *Mov. Disord.,* 15, 126, 2000.

78. Parkinson Study Group. Pramipexole versus levodopa as initial treatment for Parkinson disease: A randomized

controlled trial. Parkinson Study Group, *J.A.M.A.,* 284, 1931, 2000.

79. Quinn, N., Parkes, J. D., and Marsden, C. D., Control of on/off phenomenon by continuous intravenous infusion of levodopa, *Neurology,* 34, 1131, 1984.

80. Mouradian, M. M. et al., Modification of central dopaminergic mechanisms by continuous levodopa therapy for advanced Parkinson's disease, *Ann. Neurol.,* 27, 18, 1990.

81. Baronti. F. et al., Continuous lisuride effects on central dopaminergic mechanisms in Parkinson's disease, *Ann. Neurol.,* 32, 776, 1992.

82. Colzi, A., Turner, K., and Lees, A. J., Continuous subcutaneous waking day apomorphine in the long term treatment of levodopa induced interdose dyskinesias in Parkinson's disease, *J. Neurol. Neurosurg. Psychiatry,* 64, 573, 1998.

83. Manson, A. J., Turner, K., and Lees, A. J., Apomorphine monotherapy in the treatment of refractory motor complications of Parkinson's disease: long-term follow-up study of 64 patients, *Mov. Disord.,* 17,1235, 2002.

84. Mouradian, M. M. et al., Motor fluctuations in Parkinson's disease: central pathophysiological mechanisms, Part II, *Ann. Neurol.,* 24, 372, 1988.

85. Bravi, D. et al., Wearing-off fluctuations in Parkinson's disease: contribution of postsynaptic mechanisms, *Ann. Neurol.,* 36, 27, 1994.

86. Verhagen-Metman, L. et al., Apomorphine responses in Parkinson's disease and the pathogenesis of motor complications, *Neurology,* 48, 369. 1997.

87. Colosimo, C. et al., Motor response to acute dopaminergic challenge with apomorphine and levodopa in Parkinson's disease: implications for the pathogenesis of the on-off phenomenon, *J. Neurol. Neurosurg. Psychiatry,* 60, 634, 1996.

88. Papa, S. M. et al., Internal globus pallidus discharge is nearly suppressed during levodopa-induced dyskinesias, *Ann. Neurol.,* 46, 732, 1999.

89. Boraud, T. et al., Dopamine agonist-induced dyskinesias are correlated to both firing pattern and frequency alterations of pallidal neurones in the MPTP-treated monkey, *Brain,* 124, 546, 2001.

90. Levy, R. et al., Effects of apomorphine on subthalamic nucleus and globus pallidus internus neurons in patients with Parkinson's disease, *J. Neurophysiol.,* 86, 249, 2001.

91. Stefani, A. et al., Effects of increasing doses of apomorphine during stereotaxic neurosurgery in Parkinson's disease: clinical score and internal globus pallidus activity, *J. Neural Transm.,* 104, 895, 1997.

92. Mitchell, I. J. et al., Neural mechanisms underlying parkinsonian symptoms based upon regional uptake of 2-deoxyglucose in monkeys exposed to 1-methyl-4-phenyl-1,2,3,6-tetrahydropyridine, *Neurosci.,* 32, 213, 1989.

93. Brooks, D. J. et al., Neuroimaging of dyskinesia. *Ann. Neurol.,* 47(Suppl. 1), S154, 2000.

94. Rascol, O. et al., Cortical motor overactivation in parkinsonian patients with L-dopa-induced peak-dose dyskinesia, *Brain,* 121, 527, 1998.

95. Gerfen, C. R. and Wilson, C. J., The basal ganglia, *Hndbk. Chem. Neuroanatomy,* 13, 365, 1996.

96. Gerfen, C. R., Dopamine-mediated gene regulation in models of Parkinson's disease, *Ann. Neurol.,* 47, S42, 2000.

97. Gerfen, C. R., The neostriatal mosaic: multiple levels of compartmental organization in the basal ganglia, *Ann. Rev. Neurosci.,* 15, 285, 1992.

98. Graybiel, A. M. et al., The basal ganglia and adaptive motor control. *Science,* 265, 1826, 1994.

99. Kotter, R., Postsynaptic integration of glutamatergic and dopaminergic signals in the striatum, *Prog. Neurobiol.,* 44, 163, 1994.

100. Klawans, H. L. et al., Levodopa-induced dopamine receptor hypersensitivity, *Trans. Am. Neurol. Assoc.,* 102, 80, 1977.

101. Ungerstedt, U., Postsynaptic supersensitivity after 6-hydroxydopamine induced degeneration of the nigrostriatal dopamine system, *Acta Physiol. Scand.,* S367, 69, 1971.

102. Schultz, W. and Ungerstedt, U., Striatal cell supersensitivity to apomorphine in dopamine-lesioned rats correlated to behaviour, *Neuropharmacol.,* 17, 349, 1978.

103. Lee, T. et al., Receptor basis for dopaminergic supersensitivity in Parkinson's disease, *Nature,* 273, 59, 1978.

104. Seeman, P. et al., Human brain D1 and D2 dopamine receptors in schizophrenia, Alzheimer's, Parkinson's, and Huntington's diseases, *Neuropsychopharmacol.,* 1, 5, 1987.

105. Brooks, D. J.et al., Striatal D2 receptor status in patients with Parkinson's disease, striatonigral degeneration, and progressive supranuclear palsy, measured with 11C-raclopride and positron emission tomography, *Ann. Neurol.* 31, 184, 1992.

106. Turjanski, N., Lees, A. J., and Brooks, D. J., In vivo studies on striatal dopamine D1 and D2 site binding in L-dopa-treated Parkinson's disease patients with and without dyskinesias, *Neurology,* 49, 717, 1997.

107. Ahlskog, J. E. et al., Reduced D2 dopamine and muscarinic cholinergic receptor densities in caudate specimens from fluctuating parkinsonian patients, *Ann. Neurol.,* 30, 185, 1991.

108. Rinne, J. O. et al., A post-mortem study on striatal dopamine receptors in Parkinson's disease, *Brain Res.,* 556, 117, 1991.

109. Pierot, L. et al., D1 and D2-type dopamine receptors in patients with Parkinson's disease and progressive supranuclear palsy, *J. Neurol. Sci.,* 86, 291, 1988.

110. Dumartin, B. et al., Internalization of D1 dopamine receptor in striatal neurons *in vivo* as evidence of activation by dopamine agonists, *J. Neurosci.,* 18, 1650, 1998.

111. Muriel, M. P. et al., Levodopa induces a cytoplasmic localization of D1 dopamine receptors in striatal neurons in Parkinson's disease, *Ann. Neurol.,* 46, 103, 1999.

112. Muriel, M. P., Orieux, G., and Hirsch, E. C., Levodopa but not ropinirole induces an internalization of D1 dopamine receptors in parkinsonian rats, *Mov. Disord.,* 17, 1174, 2002.

113. Bordet, R. et al., Induction of dopamine D3 receptor expression as a mechanism of behavioral sensitization to levodopa, *Proc. Natl. Acad. Sci., U.S.A.,* 94, 3363, 1997.

114. Hurley, M. J. et al., Dopamine D3 receptors in the basal ganglia of the common marmoset and following MPTP and L-DOPA treatment, *Brain Res.*, 709, 259, 1996.

115. Hurley, M. J. et al., D3 receptor expression within the basal ganglia is not affected by Parkinson's disease, *Neurosci. Lett.*, 214, 75, 1996.

116. Bezard E. et al., Attenuation of levodopa-induced dyskinesia by normalizing dopamine D3 receptor function, *Nat. Med.*, 9, 762, 2003.

117. Cepeda, C. and Levine, M. S., Dopamine and N-methyl-D-aspartate receptor interactions in the neostriatum, *Dev. Neurosci.*, 20, 1, 1998.

118. Chase, T. N. and Oh, J. D., Striatal mechanisms and pathogenesis of parkinsonian signs and motor complications, *Ann. Neurol.*, 47, S122, 2000.

119. Chase, T. N. and Oh, J. D., Striatal dopamine- and glutamate-mediated dysregulation in experimental parkinsonism, *Trends Neurosci.*, 23, S86, 2000.

120. Chase, T. N., Oh, J. D. and Blanchet, P. J., Neostriatal mechanisms in Parkinson's disease, *Neurology*, 51, S30, 1998.

121. Oh, J. D., Dotto, P. D. and Chase, T. N., Protein kinase A inhibitor attenuates levodopa-induced motor response alterations in the hemi-parkinsonian rat, *Neurosci. Lett.*, 228, 5, 1997.

122. Oh, J. D. et al., Enhanced tyrosine phosphorylation of striatal NMDA receptor subunits: Effects of dopaminergic denervation and levodopa administration, *Brain Res.*, 813, 150, 1998.

123. Ozawa, S., Kamiya, H., and Tsuzuki, K., Glutamate receptors in the mammalian central nervous system, *Prog. Neurobiol.*, 54, 581, 1998.

124. Wollmuth, L. P. et al., Differential contribution of the NR1- and NR2A-subunits to the selectivity filter of recombinant NMDA receptor channels, *J. Physiol.*, 491, 779, 1996.

125. Chen, Q. and Reiner, A., Cellular distribution of the NMDA receptor NR2A/2B subunits in the rat striatum, *Brain Res.*, 743, 346, 1996.

126. Gurd, J. W., Protein tyrosine phosphorylation: Implications for synaptic function, *Neurochem. Intl.*, 31, 635, 1997.

127. Suen, P. C. et al., NMDA receptor subunits in the postsynaptic density of rat brain: expression and phosphorylation by endogenous protein kinases, *Mol. Brain Res.*, 59, 215, 1998.

128. Wang, J. H., Ko, G. Y. and Kelly, P. T., Cellular and molecular bases of memory: synaptic and neuronal plasticity, *J. Clin. Neurophysiol.*, 14, 264, 1997

129. Yu, X. M. et al., NMDA channel regulation by channel-associated protein tyrosine kinase, *Src. Science*, 275, 674, 1997

130. Dunah, A. W. and Standaert, D. G. Dopamine D1 receptor-dependent trafficking of striatal NMDA glutamate receptors to the postsynaptic membrane, *J. Neurosci.*, 21, 5546, 2001.

131. Hisatsune, C. et al., Phosphorylation-dependent regulation of N-methyl-D-aspartate receptors by calmodulin, *J. Biol. Chem.*, 272, 20805, 1997.

132. Tingley, W. G. et al., Characterization of protein kinase A and protein kinaseC phosphorylation of the N-methyl-D-aspartate receptor NR1 subunit using phosphorylation site specific antibodies, *J. Biol. Chem.*, 72, 5157, 1997.

133. Crump, F. T., Dillman, K. S. and Craig, A. M., cAMP-dependent protein kinase mediates activity-regulated synaptic targeting of NMDA receptors, *J. Neurosci.*, 21, 5079, 2001.

134. Liao, G. Y. et al., Evidence for direct protein kinase-C mediated modulation of N-methyl-D-aspartate receptor current, *Mol. Pharmacol.*, 59, 960, 2001.

135. Grosshans, D. R. and Browning, M. D., Protein kinase C activation induces tyrosine phosphorylation of the NR2A and NR2B subunits of the NMDA receptor, *J. Neurochem.*, 76, 737, 2001.

136. Carvalho, A. L., Duarte, C. B. and Carvalho, A. P., Regulation of AMPA receptors by phosphorylation. *Neurochem. Res.*, 25, 1245, 2000.

137. Stefani, A. et al., Physiological and molecular properties of AMPA/Kainate receptors expressed by striatal medium spiny neurons, *Dev. Neurosci.*, 20, 242, 1998.

138. Oh, J. D. et al., Gene transfer of constitutively active protein kinase C into striatal neurons accelerates onset of levodopa-induced motor response alterations in parkinsonian rats, *Brain Res.*, 971, 18, 2003.

139. Chase, T. N., Oh J. D., and Konitsiotis, S., Antiparkinsonian and antidyskinetic activity of drugs targeting central glutamatergic mechanisms, *J. Neurol.*, 247, S36, 2000.

140. Engber, T. M. et al., NMDA receptor blockade reverses motor response alterations induced by levodopa, *Neuroreport*, 5, 2586, 1994.

141. Marin, C. et al., MK801 prevents levodopa-induced motor response alterations in parkinsonian rats, *Brain Res.*, 736, 202, 1996.

142. Papa, S. M. et al., Reversal of levodopa-induced motor fluctuations in experimental parkinsonism by NMDA receptor blockade, *Brain Res.*, 701, 13, 1995.

143. Karcz-Kubicha, M., Quack, G. and Danysz, W., Amantadine attenuates response alterations after repetitive L-DOPA treatment in rats, *J. Neural Transm.*, 105, 1229, 1998.

144. Oh, J. D. et al., Enhanced tyrosine phosphorylation of striatal NMDA receptor subunits: Effects of dopaminergic denervation and levodopa administration, *Brain Res.*, 813, 150, 1998.

145. Marin, C. et al., Non-NMDA receptor-mediated mechanisms in levodopa-induced motor response alterations in parkinsonian rats, *Synapse*, 36, 267, 2000.

146. Andersson, M., Hilbertson, A., and Cenci, M. A., Striatal fosB expression is causally linked with l-DOPA-induced abnormal involuntary movements and the associated upregulation of striatal prodynorphin mRNA in a rat model of Parkinson's disease, *Neurobiol. Dis.*, 6, 461, 1999.

147. Lundblad, M. et al., Dyskinetic versus anti-dyskinetic effects of L-DOPA and bromocriptine in a rat model of Parkinson's disease: modulation by non-dopaminergic drugs, *Mov. Disord.*, 15, 33, 2000.

148. Blanche, P. J. et al., Continuous administration decreases and pulsatile administration increases behavioral sensitivity to a novel dopamine D2 agonist (U-91356A) in MPTP-exposed monkeys, *J. Pharmacol. Exp. Ther.,* 272, 854, 1995.

149. Papa, S. M. and Chase, T. N., Levodopa-induced dyskinesias improved by a glutamate antagonist in Parkinsonian monkeys, *Ann. Neurol.,* 39, 574, 1996.

150. Gomez-Mancilla, B. and Bedard, P. J., Effect of non-dopaminergic drugs on L dopa-induced dyskinesias in MPTP-treated monkeys, *Clin. Neuropharmacol.,* 16, 418, 1993.

151. Rupniak, N. M. J.et al., Dystonia induced by combined treatment with L-Dopa and MK-801 in parkinsonian monkeys, *Ann. Neurol.,* 32, 103, 1992.

152. Blanchet, P. J. et al., Differing effects of N-methyl-D-aspartate receptor subtype selective antagonists on dyskinesias in levodopa-treated 1- methyl-4-phenyl-tetrahydropyridine monkeys, *J. Pharmacol. Exp. Ther.,* 290, 1034, 1999.

153. Blanchet, P. J., Konitsiotis, S., and Chase, T. N., Amantadine reduces levodopa-induced dyskinesias in parkinsonian monkeys, *Mov. Disord.,* 13, 798, 1998.

154. Konitsiotis, S. et al., AMPA receptor blockade improves levodopa-induced dyskinesia in MPTP monkeys, *Neurology,* 54, 1589, 2000.

155. Blanchet, P. J. et al., Acute pharmacologic blockade of dyskinesias in Parkinson's disease, *Mov. Disord.,* 11, 580, 1996.

156. Verhagen Metman, L. et al., A trial of dextromethorphan in parkinsonian patients with motor response complications, *Mov. Disord.,* 13, 414, 1998.

157. Verhagen Metman, L. et al., Dextromethorphan improves levodopa-induced dyskinesias in Parkinson's disease, *Neurology,* 51, 203, 1998.

158. Kornhuber, J., Effects of the 1-amino-adamantanes at the MK-801-binding site of the NMDA receptor-gated ion channel: a human postmortem brain study, *Eur. J. Pharmacol.,* 206, 297, 1991.

159. Verhagen Metman, L. et al., Amantadine as treatment for dyskinesias and motor fluctuations in Parkinson's disease, *Neurology,* 50, 1323, 1998.

160. Del Dotto, P. et al., Intravenous amantadine improves levadopa-induced dyskinesias: An acute double-blind placebo-controlled study, *Mov. Disord.,* 16, 515, 2001.

161. Luginger, E., Beneficial effects of amantadine on L-Dopa-induced dyskinesias in Parkinson's disease, *Mov. Disord.,* 15, 873, 2000.

162. Verhagen Metman, L., Amantadine for levodopa-induced dyskinesias: a one year follow-up study, *Arch. Neurol.,* 56, 1383, 1999.

163. Merello, M. et al., Effect of memantine (NMDA antagonist) on Parkinson's disease: a double-blind crossover randomized study, *Clin. Neuropharmacol.,* 22, 273, 1999.

164. Hanagasi, H. A. et al., The use of NMDA antagonist memantine in drug-resistant dyskinesias resulting from L-Dopa, *Mov. Disord.,* 15, 1016, 2000.

165. Merims, D. et al., Riluzole for levodopa-induced dyskinesias in advanced Parkinson's disease, *Lancet,* 353, 1764, 1999.

166. Schiffmann, S. N. et al., Distribution of adenosine A2 receptor mRNA in the human brain, *Neurosci. Lett.,* 130, 177, 1991.

167. Hettinger, B. D. et al., Ultrastructural localization of adenosine A2A receptors suggests multiple cellular sites for modulation of GABAergic neurons in rat striatum, *J. Comp. Neurol.,* 431, 331, 2001.

168. Koga, K. et al., Adenosine A(2A) receptor antagonists KF17837 and KW-6002 potentiate rotation induced by dopaminergic drugs in hemi-Parkinsonian rats, *Eur. J. Pharmacol.,* 408, 249, 2000.

169. Kanda, T. et al., Adenosine A2A antagonist: a novel antiparkinsonian agent that does not provoke dyskinesia in parkinsonian monkeys, *Ann. Neurol.,* 43, 507, 1998.

170. Grondin, R. et al., Antiparkinsonian effect of a new selective adenosine A2A receptor antagonist in MPTP-treated monkeys, *Neurology,* 52, 1673, 1999.

171. Hauser, R. A. et al., Randomized trial of the adenosine A2A receptor antagonist istradefylline in advanced PD, *Neurology,* 61, 297, 2003.

172. Bara-Jimenez, W. et al., Adenosine A2A receptor antagonist treatment of Parkinson's disease, *Neurology,* 61, 293, 2003.

173. Zeng, B. Y. et al., Alterations in preproenkephalin and adenosine-2a receptor mRNA, but not preprotachykinin mRNA correlate with occurrence of dyskinesia in normal monkeys chronically treated with L-DOPA, *Eur. J. Neurosci.,* 12,1096, 2000.

174. Dayne Mayfield, R. et al., Opposing actions of adenosine A2a and dopamine D2 receptor activation on GABA release in the basal ganglia: evidence for an A2a/D2 receptor interaction in globus pallidus, *Synapse,* 22, 132, 1996.

175. Mori, A. et al., The role of adenosine A2a receptors in regulating GABAergic synaptic transmission in striatal medium spiny neurons, *J. Neurosci.,* 16, 605, 1996.

176. Kurokawa, M. et al., Inhibition by KF17837 of adenosine A2A receptor-mediated modulation of striatal GABA and ACh release, *Br. J. Pharmacol.,* 113, 43, 1994.

177. Nash, J. E. and Brotchie, J. M., A common signaling pathway for striatal NMDA and adenosine A2a receptors: implications for the treatment of Parkinson's disease, *J. Neurosci.,* 20, 7782, 2000.

178. Cunha, R. A. and Ribeiro, J. A., Adenosine A2A receptor facilitation of synaptic transmission in the CA1 area of the rat hippocampus requires protein kinase C but not protein kinase A activation, *Neurosci. Lett.,* 289, 127, 2000.

179. Bibbiani, F. et al., A2A antagonist prevents dopamine agonist-induced motor complications in animal models of Parkinson's disease, *Exp. Neurol.,* 184, 285, 2003.

180. Hornykiewicz, O., Brain monoamines and parkinsonism. *National Institute on Drug Abuse: Research Monograph Series,* 3, 13, 1975.

181. Melamed, E. et al., Involvement of serotonin in clinical features of Parkinson's disease and complications of L-DOPA therapy, *Adv. Neurol.,* 69, 545, 1996.

182. Durif, F. et al., Levodopa-induced dyskinesias are improved by fluoxetine. *Neurology,* 45, 1855, 1995.

183. Bonifati, V. et al., Buspirone in levodopa-induced dyskinesias. *Clin. Neuropharmacol.*, 17, 73, 1994.

184. Rascol, O., L-dopa-induced peak-dose dyskinesias in patients with Parkinson's disease: a clinical pharmacologic approach, *Mov. Disord.*, 14(Suppl), 1, 19, 1999.

185. Bibbiani, F., Oh, J. D., and Chase, T. N., Serotonin 5-HT1A agonist improves motor complications in rodent and primate parkinsonian models, *Neurology*, 57, 1829, 2001.

186. Meco, G. et al., Controlled single-blind cross-over study of ritanserin and placebo in L-dopa-induced dyskinesias in Parkinson's disease, *Curr. Ther. Res.*, 43, 262, 1988.

187. Meco, G. et al., Mirtazapine in L-dopa-induced dyskinesias, *Clin. Neuropharmacol.*, 26, 179, 2003.

188. Durif, F. et al., Low-dose clozapine improves dyskinesias in Parkinson's disease, *Neurology*, 48, 658, 1997.

189. Bennett, J. P., Jr. et al., Suppression of dyskinesias in advanced Parkinson's disease: moderate daily clozapine doses provide long-term dyskinesia reduction, *Mov. Disord.*, 9, 409, 1994.

190. Oh, J. D., Bibbiani, F., and Chase, T. N., Quetiapine attenuates levodopa-induced motor complications in rodent and primate parkinsonian models, *Exper. Neurol.*, 177, 557, 2002.

191. Gomez-Mancilla, B. and Bedard, P. J., Effect of non-dopaminergic drugs on L-dopa-induced dyskinesias in MPTP-treated monkeys, *Clin. Neuropharmacol.*, 16, 418, 1993.

192. Henry, B. et al., The alpha2-adrenergic receptor antagonist idazoxan reduces dyskinesia and enhances antiparkinsonian actions of L-dopa in the MPTP-lesioned primate model of Parkinson's disease, *Mov. Disord.*, 14, 744, 1999.

193. Grondin, R. et al., Noradrenoceptor antagonism with idazoxan improves L-dopa-induced dyskinesias in MPTP monkeys, *Naunyn-Schmiedebergs Arch. Pharmacol.*, 361, 181, 2000.

194. Fox, S. H. et al., Neural mechanisms underlying peak-dose dyskinesia induced by levodopa and apomorphine are distinct: evidence from the effects of the alpha(2) adrenoceptor antagonist idazoxan, *Mov. Disord.*, 16, 642, 2001.

195. Rascol, O. et al., Idazoxan, an alpha-2 antagonist, and L-DOPA-induced dyskinesias in patients with Parkinson's disease, *Mov. Disord.*, 16, 708, 2001.

196. Manson, A. J., Iakovidou, E., and Lees, A. J., Idazoxan is ineffective for levodopa-induced dyskinesias in Parkinson's disease, *Mov. Disord.*, 15, 336, 2000.

197. Holmberg, M. et al., Adrenergic alpha2C-receptors reside in rat striatal GABAergic projection neurons: comparison of radioligand binding and immunohistochemistry, *Neurosci.*, 93,1323, 1999.

198. Mansour, A. et al., Anatomy of CNS opioid receptors, *Trends Neurosci.*, 11, 308, 1988.

199. Lord, J. A. et al., Endogenous opioid peptides: multiple agonists and receptors, *Nature*, 267, 495, 1977.

200. Goldstein, A. et al., Dynorphin-(1-13), an extraordinarily potent opioid peptide, *Proc. Natl. Acad. Sci., U.S.A.*, 76, 6666, 1979.

201. Zhang, S. et al., Dynorphin A as a potential endogenous ligand for four members of the opioid receptor gene family, *J. Pharmacol. Exper. Therapeut.*, 286, 136, 1998.

202. Gerfen, C. R. et al., D1 and D2 dopamine receptor-regulated gene expression of striatonigral and striatopallidal neurons, *Science*, 250, 1429, 1990.

203. Maneuf, Y. P. et al., On the role of enkephalin cotransmission in the GABAergic striatal efferents to the globus pallidus, *Exper. Neurol.*, 125, 65, 1994.

204. Crossman, A. R., Primate models of dyskinesia: the experimental approach to the study of basal ganglia-related involuntary movement disorders, *Neurosci.*, 21, 1, 1987.

205. Bezard, E., Brotchie, J. M., and Gross, C. E., Pathophysiology of levodopa-induced dyskinesia: potential for new therapies, *Nat. Rev. Neurosci.*, 2, 577, 2001.

206. Henry, B. et al., Increased striatal pre-proenkephalin B expression is associated with dyskinesia in Parkinson's disease, *Exper. Neurol.*, 183, 458, 2003.

207. Henry, B., Crossman, A. R., and Brotchie, J. M., Effect of repeated L-DOPA, bromocriptine, or lisuride administration on preproenkephalin-A and preproenkephalin-B mRNA levels in the striatum of the 6-hydroxydopamine-lesioned rat, *Exper. Neurol.*, 155, 204, 1999.

208. Johansson, P. A. et al., Alterations in cortical and basal ganglia levels of opioid receptor binding in a rat model of l-DOPA-induced dyskinesia, *Neurobiol. Dis.*, 8, 220, 2001.

209. Westin, J. E. et al., Persistent changes in striatal gene expression induced by long-term L-DOPA treatment in a rat model of Parkinson's disease, *Eur. J. Neurosci.*, 14, 1171, 2001.

210. Engber, T. M. et al., Levodopa replacement therapy alters enzyme activities in striatum and neuropeptide content in striatal output regions of 6-hydroxydopamine lesioned rats, *Brain Res.*, 552, 113, 1991.

211. Cenci, M. A., Lee, C. S., and Bjorklund, A., L-DOPA-induced dyskinesia in the rat is associated with striatal overexpression of prodynorphin—and glutamic acid decarboxylase mRNA, *Eur. J. Neurosci.*, 10, 2694, 1998.

212. Gross, C. E. et al., Pattern of levodopa-induced striatal changes is different in normal and MPTP lesioned mice, *J. Neurochem.*, 84,1246, 2003.

213. Herrero, M. T. et al., Effects of L-DOPA on preproenkephalin and preprotachykinin gene expression in the MPTP-treated monkey striatum, *Neurosci.*, 68, 1189, 1995.

214. Morissette, M. et al., Differential regulation of striatal preproenkephalin and preprotachykinin mRNA levels in MPTP-lesioned monkeys chronically treated with dopamine D1 or D2 receptor agonists, *J. Neurochem.*, 72, 682, 1999.

215. Tel, B. C. et al., Alterations in striatal neuropeptide mRNA produced by repeated administration of L-DOPA, ropinirole or bromocriptine correlate with dyskinesia induction in MPTP-treated common marmosets, *Neurosci.*, 115, 1047, 2002.

216. Lundblad, M. et al., Pharmacological validation of behavioural measures of akinesia and dyskinesia in a rat

model of Parkinson's disease, *Eur. J. Neurosci.*, 15, 120, 2002.

217. Calon, F. and Di Paolo, T., Levodopa response motor complications-GABA receptors and preproenkephalin expression in human brain, *Parkins. Rel. Disord.*, 8, 449, 2002.

218. Piccini, P., Weeks, R. A., and Brooks, D. J., Alterations in opioid receptor binding in Parkinson's disease patients with levodopa-induced dyskinesias, *Ann. Neurol.*, 42, 720, 1997.

219. Brooks, D. J. et al., Neuroimaging of dyskinesia, *Ann. Neurol.*, 47, S154, 2000.

220. Mehta, A. et al., Endomorphin-1: induction of motor behavior and lack of receptor desensitization, *J. Neurosci.*, 21, 4436, 2001.

221. Klintenberg, R. et al., Naloxone reduces levodopa-induced dyskinesias and apomorphine-induced rotations in primate models of parkinsonism, *J. Neural Trans.*, 109, 1295, 2002.

222. Henry B. et al., Mu- and delta-opioid receptor antagonists reduce levodopa-induced dyskinesia in the MPTP-lesioned primate model of Parkinson's disease, *Exper. Neurol.*, 171, 139, 2001.

223. Rascol, O. et al., Montastruc, J. L., Rascol, A., Naltrexone, an opiate antagonist, fails to modify motor symptoms in patients with Parkinson's disease, *Mov. Disord.*, 9, 437, 1994.

224. Manson, A. J. et al., High dose naltrexone for dyskinesias induced by levodopa, *J. Neurol. Neurosurg. Psychiatry*, 70, 554, 2001.

225. Brotchie, J. M., CB1 cannabinoid receptor signalling in Parkinson's disease, *Curr. Opin. Pharmacol.*, 3, 54, 2003.

226. Mailleux, P., Parmentier, M., and Vanderhaeghen, J. J., Distribution of cannabinoid receptor messenger RNA in the human brain: an in situ hybridization histochemistry with oligonucleotides, *Neurosci. Lett.*, 143, 200, 1992.

227. Di Marzo, V. et al., Enhanced levels of endogenous cannabinoids in the globus pallidus are associated with a reduction in movement in an animal model of Parkinson's disease, *FASEB*, 14, 1432, 2000.

228. Wilson, R. I. and Nicoll, R. A., Endocannabinoid signaling in the brain, *Science*, 296, 678, 2002.

229. Gerdeman, G. and Lovinger, D. M., CB1 cannabinoid receptor inhibits synaptic release of glutamate in rat dorsolateral striatum, *J. Neurophys.*, 85, 468, 2001.

230. Szabo, B. et al., Inhibition of GABAergic inhibitory postsynaptic currents by cannabinoids in rat corpus striatum, *Neurosci.*, 85, 395, 1998.

231. Gerdeman, G. L., Ronesi, J., and Lovinger, D. M., Postsynaptic endocannabinoid release is critical to long-term depression in the striatum, *Nat. Neurosci.*, 5, 446, 2002.

232. Maneuf, Y. P. and Brotchie, J. M., Paradoxical action of the cannabinoid WIN 55,212-2 in stimulated and basal cyclic AMP accumulation in rat globus pallidus slices, *Br. J. Pharmacol.*, 120, 1397, 1997.

233. Hillard, C. J. and Auchampach, J. A., In vitro activation of brain protein kinase C by the cannabinoids, *Biochimica et Biophysica Acta*, 1220, 163, 1994.

234. Hohmann, A. G. and Herkenham, M., Localization of cannabinoid CB(1) receptor mRNA in neuronal subpopulations of rat striatum: a double-label in situ hybridization study, *Synapse*, 37, 71, 2000.

235. Hermann, H., Marsicano, G., and Lutz, B., Coexpression of the cannabinoid receptor type 1 with dopamine and serotonin receptors in distinct neuronal subpopulations of the adult mouse forebrain, *Neurosci.*, 109, 451, 2002.

236. Meschler, J. P. and Howlett, A. C., Signal transduction interactions between CB1 cannabinoid and dopamine receptors in the rat and monkey striatum, *Neuropharmacol.*, 40, 918, 2001.

237. Giuffrida, A. et al., Dopamine activation of endogenous cannabinoid signaling in dorsal striatum, *Nat. Neurosci.*, 2, 358, 1999.

238. Melis, M., Gessa, G. L., and Diana, M., Different mechanisms for dopaminergic excitation induced by opiates and cannabinoids in the rat midbrain, *Progress in Neuro-Psychopharmacology Biological Psychiatry*, 24, 993, 2000.

239. Voruganti, L. N. et al., Cannabis induced dopamine release: an in-vivo SPECT study, *Psych. Res.*, 107, 173, 2001.

240. Maneuf, Y. P. et al., Activation of the cannabinoid receptor by delta 9-tetrahydrocannabinol reduces gamma-aminobutyric acid uptake in the globus pallidus, *Eur. J. Pharmacol.*, 308, 161, 1996.

241. Romero, J. et al., The activation of cannabinoid receptors in striatonigral GABAergic neurons inhibited GABA uptake, *Life Sciences*, 62, 351, 1998.

242. Sanudo-Pena, M. C., Tsou, K., and Walker, J. M., Motor actions of cannabinoids in the basal ganglia output nuclei, *Life Sciences*, 65, 703, 1999.

243. Gubellini, P. et al., Experimental parkinsonism alters endocannabinoid degradation implications for striatal glutamatergic transmission, *J. Neurosci.*, 22, 6900, 2002.

244. Lastres-Becker, I. et al., Increased cannabinoid CB1 receptor binding and activation of GTP-binding proteins in the basal ganglia of patients with Parkinson's syndrome and of MPTP-treated marmosets, *Eur. J. Neurosci.* 14, 1827, 2001.

245. Zeng, B. Y. et al., Chronic L-DOPA treatment increases striatal cannabinoid CB1 receptor mRNA expression in 6-hydroxydopamine-lesioned rats, *Neurosci. Lett.*, 276, 71, 1999.

246. Di Marzo, V. et al., Enhanced levels of endogenous cannabinoids in the globus pallidus are associated with a reduction in movement in an animal model of Parkinson's disease, *FASEB*, 214, 1432, 2000.

247. Fox, S. H. et al., Stimulation of cannabinoid receptors reduces levodopa-induced dyskinesia in the MPTP-lesioned nonhuman primate model of Parkinson's disease, *Mov. Disord.*, 17, 1180, 2002.

248. Segovia, G. et al., Effects of CB1 cannabinoid receptor modulating compounds on the hyperkinesia induced by high-dose levodopa in the reserpine-treated rat model of Parkinson's disease, *Mov. Disord.*, 18, 138, 2003.

249. Sieradzan, K. A. et al., Cannabinoids reduce levodopa-induced dyskinesia in Parkinson's disease: a pilot study, *Neurology*, 57, 2108, 2001.

250. Mitchell, I. J.et al., A 2-deoxyglucose study of the effects of dopamine agonists on the parkinsonian primate brain. Implications for the neural mechanisms that mediate dopamine agonist-induced dyskinesia, *Brain*, 115, 809, 1992.

251. Verhagen Metman, L. and Oh, J. D., Dyskinesias in Parkinson's disease, in *Ionotropic glutamate receptors as therapeutic targets*, Lodge, D., Danysz, W., Parsons, C. G., Eds., F. P. Graham Publishing Co, Johnson City, TN, 2002.

252. Calabresi, P. et al., Levodopa-induced dyskinesia: a pathological form of striatal synaptic plasticity? *Ann. Neurol.*, 47, S60, 2000.

253. Graybiel, A. M. et al., The basal ganglia and adaptive motor control. *Science*, 265, 1826, 1994.

254. Calabresi, P. et al., Abnormal synaptic plasticity in the striatum of mice lacking dopamine D2 receptors, *J. Neurosci.*, 7, 4536, 1997.

255. Calabresi, P. et al., The corticostriatal projection: from synaptic plasticity to dysfunctions of the basal ganglia, *Trends Neurosci.*, 19, 19, 1996.

256. Calabresi, P., Centonze, D., and G Bernardi., Electrophysiology of dopamine in normal and denervated striatal neurons, *Trends Neurosci.*, 23, S57, 2000.

257. Gerfen, C. R., Dopamine-mediated gene regulation in models of Parkinson's disease, *Ann. Neurol.*, 47, S42, 2000.

258. Canales, J. J. and Graybiel, A. M., Patterns of gene expression and behavior induced by chronic dopamine treatments, *Ann. Neurol.*, 47, S53, 2000.

259. Picconi, B. et al., Loss of bidirectional striatal synaptic plasticity in L-DOPA-induced dyskinesia, *Nat. Neurosci.*, 6, 501, 2003.

260. Centonze, D. et al., Unilateral dopamine denervation blocks corticostriatal LTP, *J. Neurophysiol.*, 82, 3575, 1999.

261. Huang, C. C. and Hsu, K. S., Progress in understanding the factors regulating reversibility of long-term potentiation, *Rev. Neurosci.*, 12, 51, 2001.

262. O'Dell, T. J. and Kandel, E. R., Low-frequency stimulation erases LTP through an NMDA receptor-mediated activation of protein phosphatases, *Learning and Memory*, 1, 129, 1994.

65 Treatment of Early Parkinson's Disease

John C. Morgan and Kapil D. Sethi
Medical College of Georgia and Veterans Affairs Medical Center

CONTENTS

INTRODUCTION

From the time of the original description of Parkinson's disease (PD)[1] until the 1960s, the pharmacotherapy of PD was not too complex a matter: anticholinergics were the cornerstone of treatment. The introduction of levodopa in the 1960s, and the proof of its effectiveness in the treatment of PD,[2] have resulted in vastly improved quality of life and reduction of disability in PD patients. Today, carbidopa/levodopa (or benserazide/levodopa) and dopamine agonists (DA) have become the major drugs used in the treatment of PD. The optimum initial pharmacological treatment of early PD is debatable,[3–8] and the goal of this chapter is to provide a sound rationale for initial pharmacotherapy in early PD based on current scientific evidence and personal experience.

Ideally, therapy in early PD should focus on slowing progression of the disease as well as relieving symptoms. While carbidopa/levodopa (CD/LD) is the most efficacious agent in our armamentarium for the symptomatic control of PD, its effect on disease progression is unknown. In spite of recent controlled trials, it remains unknown if CD/LD slows, does not alter, or accelerates the progression of PD.[9,10] DA are thought by some to slow PD progression, based on functional neuroimaging in recent studies; however, the interpretation of these results remains controversial.[11–15]

ANTICHOLINERGICS

The anticholinergic effects of belladonna alkaloids were initially employed by Charcot in the later half of the nineteenth century to treat drooling in PD patients.[16] It was soon discovered that other features of PD were ameliorated by anticholinergics, and, even today, these drugs remain a viable option in the treatment of early PD. In the same journal issue as Hoehn and Yahr's seminal paper[17] on parkinsonism, there was actually an advertisement for procyclidine, an anticholinergic agent frequently used for treatment of PD at that time. Trihexyphenidyl and benztropine are the two most commonly employed anticholinergics used in the treatment of PD today. These medications are lipophilic and readily enter into the CNS, binding to muscarinic acetylcholine receptors.[18] Their precise mechanism of action is unknown, but they are thought to help correct the dysequilibrium between striatal dopamine and acetylcholine activity.[18] Benztropine also has the ability to block dopamine re-uptake in central dopaminergic neurons.[18]

In the 1993 American Academy of Neurology Practice Parameter on the treatment of PD, the authors reported that anticholinergics appeared more efficacious for tremor relative to CD/LD.[19] More recently, the Movement Disorder Society conducted an evidence-based systematic review of anticholinergics, and they concluded that anticholinergics were "likely efficacious" in the symptomatic control of PD.[18] Most recently, a Cochrane database systematic review conducted by Katzenschlager et al.[20] examined nine randomized double-blind cross-over studies of anticholinergics versus placebo in monotherapy or adjunctive treatment of early to advanced PD. The analysis revealed that anticholinergics were more effective than placebo in improving motor function in PD. While anticholinergics such as trihexyphenidyl and benztropine are typically used as a CD/LD-sparing treatment of tremor in younger PD patients,[4] Katzenschlager et al.[20] found that there was no strongly supported effect of anticholinergics on individual features of PD, including tremor. In clinical practice, it appears that these agents do not benefit akinesia, unlike other features of PD.

Neuropsychiatric and cognitive side effects often limit therapy with these drugs in elderly patients; however, younger patients can tolerate high doses of anticholinergics. Cognitive side effects were a more common cause for anticholinergic withdrawal rather than lack of efficacy.[20] In early PD, the cognitive side effects from anticholinergics are probably minimal in patients without baseline cognitive problems.[21] Other common side effects from anticholinergics include dry mouth (beneficial for PD patients with sialorrhea), constipation (a problem for the majority of PD patients), blurry vision, and urinary hesitancy. Trihexyphenidyl and benztropine are very inexpensive in generic form.

In clinical practice, a typical scenario would be treating a cognitively intact PD patient with anticholinergics early in disease. As the patient ages, and if the patient develops baseline confusion or cognitive problems, anticholinergic drugs would be one of the first drugs removed from the therapeutic regimen. In our practice, these drugs are sometimes continued as long-term therapy in patients with tremor; however, there is little data to support their long-term efficacy.[18] There is no significant data to support the efficacy of one anticholinergics drug over another in early PD patients. Table 65.1 outlines typical starting and maintenance doses for anticholinergics and other drugs used in the treatment of early PD.

TABLE 65.1
Medications Used in the Initial Therapy of Early Parkinson's Disease in the U.S.

Generic Name	Mechanism(s)	Initial Dose	Maintenance Dose	Major Side Effects
Amantadine	Multiple	100 mg b.i.d.	100 mg b.i.d.-t.i.d.	Nausea, confusion, hallucinations, leg edema, livedo reticularis
Benztropine	Anticholinergic	0.5 mg q.h.s.	1 mg b.i.d.-t.i.d.	Dry mouth, confusion, urinary retention, hallucinations, blurry vision, constipation
Bromocriptine	Dopamine agonist	1.25 mg q.d.-b.i.d.	5–10 mg b.i.d.-t.i.d.	Nausea, hallucinations, orthostatic hypotension, fibrosis, sleep attacks
Carbidopa/ levodopa	Converted to dopamine	25 mg/100 mg b.i.d.	Variable	Nausea, orthostatic hypotension, confusion, hallucinations, dizziness
Pergolide	Dopamine agonist	0.05 mg q.d.-b.i.d.	0.75–1 mg t.i.d.	Nausea, hallucinations, orthostatic hypotension, fibrosis, sleep attacks, leg edema, valvular heart disease
Pramipexole	Dopamine agonist	0.125 mg t.i.d.	1.5 mg t.i.d.	Nausea, hallucinations, orthostatic hypotension, sleep attacks, leg edema
Ropinirole	Dopamine agonist	0.25 mg t.i.d.	2–8 mg t.i.d.	Nausea, hallucinations, orthostatic hypotension, sleep attacks, leg edema
Selegiline	MAO-B Inhibitor	5 mg q.d.-b.i.d.	5 mg b.i.d.	Confusion, hallucinations, orthostatic hypotension, insomnia, nausea, serotonin syndrome, benign cardiac arrythmias
Trihexyphenidyl	Anticholinergic	1–2 mg q.d.	2 mg b.i.d.-t.i.d.	Dry mouth, confusion, urinary retention, hallucinations, blurry vision, constipation

MONOAMINE OXIDASE-TYPE B (MAO-B) INHIBITORS

Monoamine oxidase (MAO) is the major metabolizing enzyme of dopamine. Preventing dopamine breakdown in PD should theoretically reduce metabolism of both endogenous dopamine and dopamine produced from ingestion of levodopa. Early attempts at MAO inhibition resulted in nonselective inhibition of both Type A and Type B forms of MAO, leading to the "cheese effect"—a hypertensive crisis that can be life threatening when a patient ingests food rich in tyramine.[22]

The introduction of selegiline as a relatively selective MAO-B inhibitor resulted in re-emergence of the idea of MAO-B inhibition to treat PD.[23] Selegiline is antiapoptotic, potentiates the effect of dopamine, prevents dopaminergic neurotoxicity due to 6-OH-dopamine and MPTP, acts as an antioxidant, and has a multitude of other effects that could contribute to its putative neuroprotective mechanism of action.[24]

The first large-scale attempt of achieving neuroprotection and slowing disease progression in PD was a comparison of selegiline and α-tocopherol (vitamin E) versus placebo in the DATATOP trial.[25] Eight hundred patients with early, untreated PD were randomized in a 2 × 2 factorial design to receive selegiline (a relatively selective MAO-B inhibitor at a dose of 10 mg/day), α-tocopherol (2000 IU/day), both drugs, or placebo in a two-year trial.[25] Patients who were treated with placebo required CD/LD treatment of disability due to PD 11 months earlier than patients treated with selegiline (with or without α-tocopherol).[26] This was initially thought to be due to neuroprotection; however, selegiline was found to have an unexpected symptomatic benefit in a post-hoc analysis of the trial.[27] In the early 1990s, prescribing practices changed after the publication of these preliminary results, and selegiline became a standard initial therapy for many patients with early PD, based on these results.

A subsequent trial by the Parkinson's Disease Research Group of the U.K. reported a 60% higher relative mortality rate in early PD patients who were treated with selegiline and decarboxylase inhibitor/levodopa (DCI/LD) versus DCI/LD alone.[28] In contrast, mortality data in the DATATOP cohort revealed that patients treated with selegiline, α-tocopherol, or both drugs combined had similar mortality after eight years of observation as compared to the age- and gender-matched U.S. population without PD.[29] Also in contrast to the U.K. group, a meta-analysis of five randomized trials of selegiline in untreated PD patients revealed there was no evidence of excess mortality in patients treated with selegiline, with or without concomitant DCI/LD therapy.[30] We believe that there is not excess mortality in PD patients treated with selegiline and DCI/LD, similar to the findings of another recent clinical trial comparing DCI/LD versus DCI/LD plus selegiline.[31]

Today, most clinicians use selegiline as a levodopa-sparing option and to provide mild symptomatic benefit for patients with early PD.[4] The Movement Disorder Society's evidence-based review of selegiline revealed that it was efficacious as monotherapy in PD, but there was insufficient evidence to make a conclusion regarding selegiline's putative neuroprotective effect.[24]

Rasagiline, a selective MAO-B inhibitor, was recently compared to placebo as monotherapy in PD (TEMPO).[32] In this trial, 404 patients with early, untreated PD were randomized to receive 1 mg/day of rasagiline, 2 mg/day of rasagiline, or placebo. Patients were followed on rasagiline monotherapy versus placebo for 26 weeks. The average total Unified Parkinson's Disease Rating Scale (UPDRS) score was 4.2 units less for patients treated with 1mg/day of rasagiline compared to placebo ($p < 0.001$).

Similar results were obtained in patients treated with 2 mg/day of rasagiline. These results may be related to symptomatic benefit versus neuroprotection—this trial was not designed to address this issue.[32] A staggered wash-in design clinical trial was subsequently performed, and patients who were treated with placebo never caught-up with the group of patients treated with rasagiline suggesting disease-slowing activity. Rasagiline is currently awaiting approval by the United States Food and Drug Administration.

In early PD patients without significant disability, MAO-B inhibitors such as selegiline or rasagiline are acceptable initial therapeutic choices for symptomatic benefit. Side effects include nausea, dizziness, confusion, hallucination, insomnia, and cardiovascular changes. While there is some concern that this medication may precipitate serotonin syndrome in patients taking concomitant serotonin-reuptake inhibitors, this is very rare.[33] MAO-B inhibitors are frequently continued, even after the patient is treated with DA or CD/LD.

AMANTADINE

Amantadine was originally used as an antiviral agent for Asian influenza[34] until the serendipitous discovery by Schwab et al.[35] that amantadine was beneficial in PD. Amantadine's mechanism of action in PD is uncertain; however, it appears that it may act presynaptically by enhancing the release of dopamine from intact dopaminergic terminals and by inhibiting catecholamine reuptake at presynaptic terminals.[36] Postsynaptically, amantadine alters dopamine receptor affinity.[37] A recent study also demonstrated that 10 to 14 days of amantadine treatment increased striatal ^{11}C-raclopride binding, possibly indicating an increase in D_2 dopamine receptor expression.[38] Amantadine may also have anticholinergic action[39] and NMDA glutamate receptor blockade.[40,41] This combination of actions leads to a net effect of increased dopaminergic stimulation.

A double-blind, placebo-controlled, crossover study of amantadine versus placebo revealed modest sustained benefit in tremor and rigidity in PD patients over one year.[42] The 1993 AAN practice parameter concluded that amantadine had a modest effect on all features of PD and had a low adverse effect profile.[19] One study found that amantadine was actually an independent predictor of improved survival using a Cox regression model.[43] The Movement Disorder Society evidence-based review concluded that amantadine was likely efficacious for symptomatic control in PD; however, the duration of benefit is unknown.[44] Most recently, Crosby et al.[45] reviewed six randomized controlled trials of amantadine as monotherapy or as adjunctive therapy in PD, and the authors concluded there was insufficient evidence that amantadine was effective or safe in therapy of PD.

Our experience with amantadine is that it serves a role in early disease prior to therapy with CD/LD or DA. It will also benefit patients later in their disease course when they suffer levodopa-induced dyskinesias.[44,46]

Amantadine is usually prescribed at a dose of 100 mg b.i.d.-t.i.d., and major side effects include nausea, light-headedness, insomnia, confusion, hallucinations, leg edema, and levido reticularis. Amantadine should never be discontinued suddenly, as this may precipitate neuro-leptic malignant syndrome in rare cases.[47,48] In addition, the dose of amantadine should be lowered appropriately (see package insert) in patients with renal failure, given that the drug is mainly excreted in the urine. Amantadine may worsen congestive heart failure, so prudent use and frequent monitoring in PD patients with this condition are important. Amantadine is very inexpensive in generic form.

LEVODOPA

Levodopa is converted to dopamine in the brain by aro-matic amino acid decarboxylase (AADC). Only 1% of levodopa alone and 5% of levodopa with a DCI actually makes it into the brain to undergo conversion to dopamine by AADC. Levodopa combined with a peripheral DCI (carbidopa or benserazide) became the gold standard of PD treatment in the early 1970s. Recently, peripheral inhi-bition of another levodopa metabolizing enzyme (cate-chol-O-methyl transferase—COMT) has proven success-ful in increasing the levodopa available to the brain. Thirty years after its introduction into clinical practice, CD/LD is still the most efficacious drug in the treatment of PD and remains the first line of therapy for many patients. While CD/LD therapy clearly ameliorates the cardinal symptoms of PD (bradykinesia, tremor, rigidity, and pos-tural instability), it remains unknown if CD/LD slows or adversely affects progression of PD. This led to the pro-posal for and conduction of the ELLDOPA trial.[9,10]

The recent publication of the ELLDOPA trial results in abstract form[10] has created further controversy as to the role of CD/LD therapy in the treatment of early PD. The design and performance of this clinical trial by the Par-kinson Study Group was undertaken to determine if CD/LD therapy in early PD patients not requiring treat-ment had any effect on the progression of PD.[9,10] Patients were randomized into three groups (placebo, 300 mg of CD/LD/day, and 600 mg CD/LD/day) and treated for 40 weeks. After two weeks of washout, the UPDRS score was calculated for each subject, and the treatment arms were compared. Patients enrolled in the study also under-went striatal SPECT imaging after administration of the dopamine transporter ligand, β-CIT.

β-CIT was used as a possible surrogate marker of PD progression, given that the amount of β-CIT binding in the striatum declines as PD progresses.[14]

Despite a half-life of 90 min to 2 hr, the CD/LD-treated group of patients did better than the placebo-treated group after two, or even after four, weeks of wash-out.[10] This trial was also quite interesting in that patients who were treated with CD/LD, while having better UPDRS scores, actually had a greater reduction of β-CIT binding in the striatum on SPECT imaging,[10] indicating reduced dopamine transporter binding in patients treated with CD/LD. This makes one wonder if functional neu-roimaging of the nigrostriatal system with β-CIT can be used as a current and future surrogate marker of the clini-cal status of PD patients. Future clinical trials are planned, and ELLDOPA-2 is currently being designed to answer these questions.

Even though CD/LD markedly improves the motor symptoms of PD, it was recognized early on that CD/LD can also lead to significant motor fluctuations.[49] Within 5 years of starting CD/LD therapy, approximately 50% of PD patients experience motor fluctuations and dyskine-sias, and 70% of patients are affected after 15 years of treatment.[50] This is especially true in patients with younger ages of PD onset—these patients are more likely to develop motor fluctuations and dyskinesias.[51–53] Recent evidence suggests that quality of life is reduced in younger PD patients with dyskinesias and motor fluctua-tions greater than in older patients.[54] Some studies have found that dyskinesias and motor fluctuations have a sig-nificant negative impact on quality of life in PD,[55,56] while others have demonstrated that patients with PD do not have significant differences in quality of life due to motor complications after four years of disease,[54] six years of disease,[57] or even after nine years of disease in one study.[58] The prevention of motor complications is an important argument between those who favor DA versus those who use CD/LD in the treatment of early PD.

The half-life of levodopa is on the order of 90 min to 2 hr, so controlled-release CD/LD was developed to reduce end-of-dose bradykinesia or the "wearing off" effect.[59,60] The controlled-release formulation of CD/LD has a plasma half-life approximately two hours longer than standard CD/LD;[61] however, it has approximately 30% less bioavailability relative to equivalent doses of immediate-release CD/LD.[61,62] The pharmacokinetics also appear to be more erratic with the controlled-release prep-aration relative to immediate-release CD/LD after the first two hours of metabolism.

A five-year randomized, controlled trial compared immediate-release CD/LD and controlled-release CD/LD and demonstrated that the drugs were similar in efficacy and rates of motor complications.[63] Given the higher cost of controlled-release CD/LD, its comparable efficacy to immediate release CD/LD, and its reduced bioavailability, this medication has a limited role in our practice.

CD/LD administration on a typical t.i.d. schedule the-oretically results in pulsatile dopaminergic stimulation of striatal dopamine receptors, and this may play a role in the development of dyskinesias and motor fluctuations in PD. The idea of continuous dopaminergic stimulation (CDS) as a way to avoid levodopa-induced motor fluctuations and dyskinesias has subsequently emerged.[64,65] Achieving CDS in clinical practice is difficult. It remains to be proven that initiating therapy with CD/LD coupled with a catechol-O-methyltransferase (COMT) inhibitor such as entacapone[64,65] will provide CDS. To date, no trials exist to support this concept. COMT irreversibly converts levodopa into 3-O-methyldopa, a compound that cannot be utilized further in dopamine synthesis. This therapeutic approach resulted in enhanced efficacy and a reduction of dyskinesias in MPTP-lesioned monkeys compared to levodopa alone when these drugs were given q.i.d.[66] This hypothesis will be tested in clinical trials in the future and perhaps the combination of CD/LD with entacapone can provide the best of both worlds in early PD—the efficacy of CD/LD coupled with a reduction in subsequent motor fluctuations by providing CDS.

In the 1993 AAN practice parameter, CD/LD was considered the most effective of all drugs for the symptoms of PD, especially for bradykinesia and rigidity.[19] CD/LD was also considered as likely more efficacious than monotherapy with anticholinergics, amantadine, or bromocriptine according to a systematic review by the Movement Disorder Society.[61] Both CD/LD and DA were considered equally sound choices for the initial therapy of PD, based on the most recent AAN guidelines.[4]

The major side effects of levodopa include nausea, orthostatic hypotension, dyskinesias, confusion, hallucinations, and dizziness. If levodopa is abruptly stopped, patients can develop a neuroleptic malignant-like syndrome. Levodopa (coupled with benserazide or carbidopa) is inexpensive and readily available worldwide. Typically CD/LD is titrated to one tablet t.i.d., and PD patients frequently experience a "honeymoon period" lasting two to three years where the dose response is steady and smooth, and there are no motor fluctuations.[67]

The most effective ratio of carbidopa to levodopa is 25 mg/100 mg for prevention of peripheral metabolism and side effects (nausea) and for the optimum entry of remaining levodopa into the brain. If nausea is a significant problem when initiating therapy, there are multiple options to improve tolerability. First, taking the CD/LD with food or on a full stomach may help. However, patients need to be aware that amino acids from high-protein meals can interfere with CD/LD absorption due to competition for the amino acid transporter in the gut. A second option is to pretreat the patient for one to two weeks with carbidopa or to give the patient supplemental doses of carbidopa. A third option is to treat the patient with domperidone (not available in the U.S.), which has the rare side effect of galactorrhea and is usually taken at a dose of 10 to 20 mg 30 min prior to each CD/LD dose.

Trimethobenzamide is also a viable but somewhat less effective option. Trimethobenzamide is dosed at 250 mg three to four times daily and can be sedating for some patients.

DOPAMINE AGONISTS

DA have found significant favor in recent years as alternatives to CD/LD, since they were shown to cause fewer dyskinesias and motor fluctuations over time relative to CD/LD. These compounds bind to intact dopamine receptors in the striatum, bypassing the needed step of levodopa conversion to dopamine. Bromocriptine was the first commercially available agonist, becoming available in 1974. Pergolide followed and became available in 1988. Both pergolide and bromocriptine are ergot-derived DA with predominant side effects of nausea, hallucinations, dizziness, somnolence, abnormal involuntary movements, and leg edema. Both bromocriptine and pergolide are associated with both retroperitoneal and pleuropulmonary fibrosis in some patients,[68,69] and pergolide was recently associated with valvular heart disease as well.[70] Two nonergot-derived, D_2/D_3 dopamine receptor agonists (ropinirole and pramipexole) were introduced in 1997, and these drugs have become the most commonly used DA in the treatment of PD.

Two trials that have certainly influenced treatment of early PD are the CALM-PD trial (pramipexole versus CD/LD)[71] and the Rascol et al.[72] ropinirole versus benserazide/levodopa trial. CALM-PD was a randomized, double-blind trial of pramipexole monotherapy versus CD/LD monotherapy in PD patients ($n = 301$) requiring dopaminergic therapy.[71] Time to the first occurrence of any of three dopaminergic complications (wearing off, dyskinesias, on-off motor fluctuations), change in UPDRS score, and change in striatal dopamine transporter (DAT) binding by β-CIT on SPECT imaging ($n = 82$) were the primary endpoints.[71] Patients treated with pramipexole had less development of the three motor complications listed above versus CD/LD (28% versus 51%, $p < 0.001$). Both drugs improved total UPDRS scores relative to baseline (untreated) scores; however, patients in the CD/LD treatment arm had greater improvement in total UPDRS scores relative to pramipexole-treated patients (9.2 versus 4.5 points, $p < 0.001$).[71] This occurred despite this trial allowing open-label CD/LD in patients with emerging disability in either treatment arm. It is possible that patients in this trial were "undertreated" in the pramipexole arm, and this may have contributed to a higher incidence of motor complications in CD/LD-treated patients. The allowance of supplemental open-label CD/LD in either treatment arm should have theoretically prevented this difference, however. It is doubtful that this marginal difference in UPDRS scores between the groups was exclusively responsible for decreased motor complications.

There was no difference in striatal β-CIT binding when comparing the pramipexole and CD/LD-treated groups at two years.[71] This trial was continued out to 46 months, and the decline in striatal β-CIT binding in the CD/LD treatment arm was greater relative to the prami-pexole arm (25.5% decline from baseline with CD/LD compared to 16.0% decline from baseline with pramipex-ole, $p = 0.01$).[11] Interestingly, at 46 months, the mean change in UPDRS scores for the two treatment arms was not significantly different in the subset of patients ($n = 65$) who remained in the imaging portion of the study.[11] Similarly, quality of life indices were better for the CD/LD arm at 2 years but better in the pramipexole arm at 46 months.

A five-year randomized (2 ropinirole:1 levodopa/DCI), double-blind trial of ropinirole versus benser-azide/levadopa as monotherapy in early PD demonstrated that ropinirole treatment resulted in fewer dyskinesias.[72] The study enrolled 268 patients, and approximately half of the patients in each treatment arm completed the study. Thirty-four percent of the patients remaining in the ropin-irole treatment arm at five years remained on monother-apy. Dyskinesias occurred in only 20% in the ropinirole-treated group, regardless of benserazide/levodopa supple-mentation and in 45% of patients in the benserazide/levo-dopa monotherapy group.[72] Activities of daily living scores were similar between the two treatment groups; however, the UPDRS motor scores were better in the benserazide/levodopa arm, as was seen in the CALM-PD trial.[72]

Both pramipexole and ropinirole appear to have neu-roprotective effects based on functional neuroimaging of the nigrostriatal pathway as a surrogate marker of PD pro-gression. Assessment of the nigrostriatal pathway with ^{18}F-dopa PET also demonstrated greater striatal uptake in patients treated with ropinirole relative to CD/LD.[13] This reflects greater amounts of AADC in nigral neurons, or more of these neurons are remaining. These results have to be interpreted cautiously, however, since DA may alter expression of the dopamine transporter or the AADC enzyme.[15] The ELLDOPA trial also suggests that CD/LD may have regulatory effects on striatal dopamine trans-porter expression as well.[10]

Slow initiation of DA therapy is important for maxi-mal tolerance in PD patients. The makers of both ropin-irole and pramipexole have excellent starter packs for patients to achieve this goal. Patients who are treated with DA should be warned about possible "sleep attacks," which can occur with DA therapy and appear to be best correlated with the total dose of dopaminergic therapy.[73,74] Patients on DA can also develop compulsive behaviors, so this should be monitored periodically.[75] While DA are effective in the treatment of early PD symptoms, patients

who have developed postural instability should be ini-tially treated with CD/LD and not DA.

COENZYME Q$_{10}$

Coenzyme Q$_{10}$ (CoQ$_{10}$ or ubiquinone) is the electron acceptor for mitochondrial complexes I and II in the res-piratory chain and is an antioxidant that also enhances complex I activity.[76] Given that complex I activity is reduced in PD patients,[77] and CoQ$_{10}$ is protective in the MPTP model of PD,[76] a clinical trial was designed to determine if treatment with high doses of CoQ$_{10}$ slowed progression in PD.[78] In this trial, 80 early, untreated PD patients were randomized to placebo ($n = 16$), 300 mg ($n = 21$), 600 mg ($n = 20$), or 1200 mg ($n = 23$) of CoQ$_{10}$ therapy for 10 months. The change from the baseline UPDRS score was the primary outcome measure. This study demonstrated a dose-dependent trend for slowing of PD progression ($p = 0.09$).[78] There was a statistically significant difference in UPDRS scores between those patients receiving 1200 mg/day of CoQ$_{10}$ and those receiv-ing placebo ($p = 0.04$). The adjusted mean total UPDRS score was actually 5.3 points lower for the 1200 mg/day group relative to those treated with placebo.[78] One month after initiation of CoQ$_{10}$ therapy, however, there was a reduction in the UPDRS ADL score, suggesting some symptomatic benefit of CoQ$_{10}$.[78]

While this Phase II clinical trial clearly demonstrated the safety of high dose CoQ$_{10}$, the efficacy of the drug in slowing PD progression needs further proof. A larger Phase III trial is planned, and doses of CoQ$_{10}$ higher than 1200 mg/day may be used. A more recent four-week, pla-cebo-controlled, double-blinded trial of 360 mg/day of CoQ$_{10}$ ($n = 14$) versus placebo ($n = 14$) in stable, treated PD patients was conducted.[79] There was actually a reduc-tion in the total UPDRS score by 2.3 points in CoQ$_{10}$-treated patients at four weeks ($p = 0.012$), indicating that CoQ$_{10}$ may provide a mild symptomatic benefit in PD patients.[79]

While we do not advise against CoQ$_{10}$ therapy in our patients (if they choose to do so), we are not convinced that the current available data support widespread treat-ment of early PD patients with this drug. Taking CoQ$_{10}$ at a dose of 1200 mg/day using the same formulation (with 300 IU of vitamin E per 300 mg of CoQ$_{10}$) as in the PSG trial[78] is very expensive. If CoQ$_{10}$ does prove to be a drug that slows progression of PD, then, hopefully, less expen-sive formulations will become available, and they will be equally effective as the form used in the PSG trial.

NONMOTOR PROBLEMS IN EARLY PD

While the motor problems of early PD are bothersome, nonmotor problems (mood disturbances, sleep problems, constipation, autonomic dysfunction) can also significantly affect the quality of life in early PD patients. Depression can precede the diagnosis of PD by years and occurs in approximately half of PD patients at some point in their illness.[80] It is usually effectively treated with selective serotonin reuptake inhibitors or tricyclic antidepressants in our personal experience, but a recent systematic review revealed that there is little data on the efficacy and safety of antidepressants in PD.[81] More controlled trials are needed in this setting

Sleep problems abound throughout the course of PD, and REM sleep behavior disorder can even present years before the onset of PD.[82] Sleep is impaired in PD as a result of the disease process itself (vivid dreams), due to motor symptoms of PD (akinesia, early morning dystonia, and so on), and due to insomnia.[83] Medications used to treat PD also have significant effects on sleep architecture: (1) DA may cause excessive daytime sleepiness, and (2) selegiline and amantadine are converted to stimulant-like metabolites.[83] Restless legs syndrome may occur more often in PD than the general population, as well.[84]

Good sleep hygiene is paramount in PD due to the multitude of other problems that the patient may not be able to control (dopaminergic therapy, early morning dystonia, and so forth). Insomnia may respond to benzodiazepines such as clonazepam, zolpidem, or tricyclic antidepressants. REM sleep behavior disorder is usually treated with clonazepam at bedtime, and this is frequently effective. DA are the treatment of choice for restless legs syndrome.

Autonomic problems in PD are also very common. Constipation and reduced bowel movement frequency are common in PD, and infrequent bowel movements (<1 per day) were actually associated with an elevated risk of future PD in one study.[85] Constipation is treated with good diet, adequate fluid intake, stool softeners, and laxatives. Patients with impotence related to PD appear to benefit from and safely tolerate sildenafil with subsequent improvement in depression scores as well.[86] Orthostatic hypotension can occur secondary to medications or due to PD, and patients should be screened for orthostatic symptoms at each visit.

NONPHARMACOLOGIC THERAPIES

Sensitive education about PD and its implications is important in patients presenting with early PD. Both the National Parkinson Foundation [NPF, (800) 327-4545] and the American Parkinson Disease Association [APDA, (800) 223-2732] serve as excellent resources for a multitude of patient education resources. Local support groups can also help patients with a new diagnosis of PD cope with this condition that affects approximately one million people in the United States. Nutrition, exercise, and paramedical therapies (physical therapy, speech therapy, and occupational therapy) are also useful adjuncts to improve function as PD progresses. Referral to PD-experienced therapists is important for maximal patient benefit.

RECOMMENDATIONS FOR TREATMENT OF EARLY PD

In reality, treatment of early PD is more controversial today than ever. Based on current evidence, multiple agents are acceptable for initial pharmacotherapy in PD, depending on the age of the patient, stage of disease, urgency of obtaining therapeutic benefit, comorbidities, and economic factors. Table 65.1 summarizes various treatment options for symptomatic therapy of early PD.

We hope that patients with early PD, without significant disability and comorbidities, will participate in clinical trials aimed at slowing the progression of PD. Current examples include the ongoing NINDS futility trial of multiple putative neuroprotective agents (NET-PD) and the planned QE3 trial of high dose CoQ$_{10}$. These and other large-scale, multicenter clinical trials will hopefully identify treatments that clearly impede PD progression on both clinical rating scales (UPDRS) and on surrogate markers such as functional neuroimaging (^{18}F-dopa PET and β-CIT SPECT).

If patients do not qualify for, or refuse entry into, ongoing clinical trials, then evaluation of the patient's composite presentation is essential for deciding initial therapy. If the patient has no disability (physical or social) due to PD symptoms and does not want to initiate pharmacotherapy, then we follow the patient until PD causes problems with the patient's occupation or activities of daily living. When this occurs, or if the patient wants to be treated with medications on initial presentation, then multiple factors come into play in our decision-making process: age of the patient, cognitive status, comorbidities, level of disability, and cost of therapy for the patient.

In our practice, we use the ages of 50 and 70 as relative cut-offs to guide our therapeutic choices. In an otherwise healthy, 46-year-old patient with PD, we would frequently use CD/LD-sparing therapy, given that a younger patient is demographically the most likely to develop dyskinesias and motor fluctuations on CD/LD therapy. CD/LD-sparing options include anticholinergics (we typically use trihexyphenidyl), selegiline, or amantadine. Alternatively, we would often initiate this type of patient on a DA for symptomatic control, titrating-up gradually to achieve maximal antiparkinsonian effects.

If the patient is already taking anticholinergics, selegiline, or amantadine, then these medications are continued as well. We do not typically initiate patients on bromocriptine, due to higher cost, and we typically use pergolide as a second line agonist (predominantly due to cost) if patients do not tolerate ropinirole or pramipexole.

After a few years of titrated DA therapy, the majority of patients require the addition of CD/LD due to disease progression and further disability. It is important to include nonpharmacological therapies (physical therapy, occupational therapy, and speech therapy) in the care plan for these patients as their PD progresses. We continue DA and add CD/LD to the therapeutic regimen at lower doses, usually one 25/100 tablet of CD/LD titrated from once daily to t.i.d. Circumstances that would modify our choice of therapy include comorbidities—if this patient had cancer, significant congestive heart failure, or other conditions with significantly reduced life expectancy, then we would use CD/LD as the first choice for the patient. Also, if patients are unable to afford DA therapy and do not qualify for patient assistance programs, then we use CD/LD after CD/LD-sparing therapies (anticholinergics, selegiline) are tried.

In the other patient group, those greater than 70 years of age, we are more likely to initiate CD/LD, given that life expectancy is shorter, and PD may be more aggressive in older patients. In addition, these patients frequently do not tolerate DA well, and they experience confusion and hallucinations when treated with these drugs. CD/LD-sparing therapies are tried first; however, patients frequently need CD/LD or have difficulty tolerating anticholinergics. Alternatively, in the case of patients older than 70 years of age, cognitively intact, and otherwise healthy, then we may attempt to treat them with a DA on the assumption that they may live into their 80s or longer with PD. Older patients with cognitive impairment are initiated on CD/LD. We typically avoid anticholinergics, selegiline, and amantadine in older patients given the risk of hallucinations, confusion, and cognitive disturbances.

In our practice, if a relatively healthy man or woman between the ages of 50 and 70 presents with early PD, the patient is typically treated with a DA (pramipexole or ropinirole). We feel that the reduction of motor complications and the possible delay in PD progression afforded by DA warrants their use in the majority of early PD patients who are experiencing disease-related disability. While cost is an issue, the patient assistance programs available for both drugs should make these medications available for the majority of patients who need them.

Surgeries including deep brain stimulation are viable, effective options for patients with moderate to advanced PD today. We do not feel, however, that the risk of surgical intervention is warranted in early PD, given that early PD patients typically have dramatic improvement with medical therapy early in the course of disease.

SUMMARY AND FUTURE DIRECTIONS

In summary, the foundation of our therapeutic practice in PD is consistent with the most recent Movement Disorder Society and AAN practice guidelines—either CD/LD or DA, are acceptable initial therapies in the treatment of early PD.[4,18,24,44,61] The appropriate initial pharmacotherapy, however, is dependent on multiple circumstances as outlined above. If further evidence of disease-modifying therapy with DA or with other drugs arises in current and future clinical trials, then certainly these would become the drug of choice for the early treatment of PD. Until then, there will be continued debate and controversy surrounding the initial pharmacotherapy of early PD. The development of reliable surrogate markers in addition to following the UPDRS in PD patients will allow researchers to answer these questions over time.

With the pending approval of rasagiline (a newer MAO-B inhibitor) in the U.S. and the recent release of CD/LD/entacapone for treating patients with motor fluctuations, options for early PD pharmacotherapy will change. The delineation of rasagiline's effects beyond six months of therapy in early PD will be important for clinicians. While not typical in our practice, some neurologists use CD/LD/entacapone as first line therapy for early PD patients given that this compound may help provide CDS. Further clinical trials are necessary to determine the safety and efficacy as well as long-term complications of combined CD/LD/entacapone therapy.

Rotigotine (a transdermally administered dopamine agonist) was recently shown to be effective in ameliorating the signs and symptoms of PD over 11 weeks.[87] This DA may allow for CDS (in theory) and may reduce future motor complications while simultaneously providing symptomatic benefit. The DA apomorphine and cabergoline may also one day have use in early PD in the United States, though not currently approved for this indication.

Multiple forms of neuroprotective therapies have shown promise in animal models of PD: iron chelators, free radical scavengers, antioxidants, neuroimmunophilin ligands, MAO-B inhibitors, calcium channel antagonists, glutamate antagonists, and trophic factors. It remains to be seen which of these therapies will slow the progress of PD in humans.

REFERENCES

1. Parkinson J., *An Essay on the Shaking Palsy*, Sherwood, Neely, and Jones, London, England, 1817.
2. Cotzias, G. C., Van Woert, M. H., Schiffer, L. M., Aromatic amino acids and modification of parkinsonism, *N. Engl. J. Med.*, 276:374–379, 1967.
3. Wooten, G. F., Agonists vs. levodopa in PD: the thrilla of whitha, *Neurology*, 60:360–362, 2003.
4. Miyasaki, J. M., Martin, W., Suchowersky, O., Weiner, W. J., Lang, A. E., Practice parameter: initiation of treatment for Parkinson's disease: An evidence-based review, *Neurology*, 58:11–17, 2002.

5. Katzenschlager, R., Lees, A. J., Treatment of Parkinson's disease: levodopa as the first choice, *J. Neurol.*, 249 (Suppl. 2):II19–24, 2002.

6. Kondo T., Initial therapy for Parkinson's disease: levodopa vs. dopamine receptor agonists, *J. Neurol.*, 249 (Suppl. 2): II25–29, 2002.

7. Ahlskog, J. E., Parkinson's disease: is the initial treatment established? *Curr. Neurol. Neurosci. Rep.*, 3:289–295, 2003.

8. Albin, R. L., Frey, K. A., Initial agonist treatment of Parkinson disease: A critique, *Neurology*, 60:390–394, 2003.

9. Fahn, S., Parkinson disease, the effect of levodopa, and the ELLDOPA trial, *Arch Neurol.*, 56:529–535, 1999.

10. Parkinson Study Group, Does levodopa slow or hasten the rate of progression of Parkinson disease? The results of the ELLDOPA trial, *Neurology*, 60 (Suppl. 1):A80–A81, 003.

11. Parkinson Study Group, Dopamine transporter brain imaging to assess the effects of pramipexole vs. levodopa on Parkinson disease progression, *JAMA*, 287:1653–61, 2002.

12. Marek, K., Jennings, D., Seibyl, J., Do dopamine agonists or levodopa modify Parkinson's disease progression? *Eur. J. Neurol.*, 9 (Suppl. 3):15–22, 2002.

13. Whone, A. L., Watts, R. L., Stoessl, A. J., Davis, M., Reske, S., Nahmias, C., Lang, A. E., Rascol, O., Ribeiro, M. J., Remy, P., Poewe, W. H., Hauser, R. A., Brooks, D. J., REAL-PET Study Group. Slower progression of Parkinson's disease with ropinirole versus levodopa: The REAL-PET study, *Ann. Neurol.*, 54:93–101, 2003.

14. Morrish, P. K., How valid is dopamine transporter imaging as a surrogate marker in research trials in Parkinson's disease? *Mov. Disord.*, 18 (Suppl. 7):S63–S70, 2003.

15. Ahlskog, J. E., Slowing Parkinson's disease progression: Recent dopamine agonist trials, *Neurology*, 60:381–389, 2003.

16. Lang, A. E., Blair, R. D. G., Anticholinergic drugs and amantadine in the treatment of Parkinson's disease, In: Calne, D. B., ed., *Handbook of Experimental Pharmacology*, Vol. 88: Drugs for treatment of Parkinson's disease, Berlin Heidelberg, Springer-Verlag, 1989.

17. Hoehn, M. M., Yahr, M. D., Parkinsonism: onset, progression, and mortality, *Neurology*, 17:427–442, 1967.

18. Anticholinergic therapies in the treatment of Parkinson's disease, *Mov. Disord.*, 17 (Suppl. 4):S7–S12, 2002.

19. Quality Standards Subcommittee of the American Academy of Neurology, Practice parameters: initial therapy of Parkinson's disease (summary statement), *Neurology*, 43:1296–1297, 1993.

20. Katzenschlager, R., Sampaio, C., Costa, J., Lees, A., Anticholinergics for symptomatic management of Parkinson's disease, *Cochrane Database Syst. Rev.*, (2):CD003735, 2003.

21. Schelosky, L., Benke, T., Poewe, W. H., Effects of treatment with trihexyphenidyl on cognitive function in early Parkinson's disease, *J. Neura.l Transm.*, Suppl., 33:125–132, 1991.

22. Anderson, M. C., Hasan, F., McCrodden, J. M., Tipton, K. F., Monoamine oxidase inhibitors and the cheese effect, *Neurochem. Res.*, 18:1145–9, 1993.

23. Knoll, J., Deprenyl [selegiline]: the history of its development and pharmacological action, *Acta Neurol. Scand.*, 95:57–80, 1983.

24. MAO-B Inhibitors for the treatment of Parkinson's disease, *Mov. Disord.*, 17 (Suppl. 4), S38–S44, 2002.

25. Parkinson Study Group, DATATOP: a multicenter controlled clinical trial in early Parkinson's disease, Parkinson Study Group, *Arch Neurol.*, 46:1052–1060, 1989.

26. Parkinson Study Group, Effect of deprenyl on the progression of disability in early Parkinson's disease, *N. Engl. J. Med.*, 321:1364–1371, 1989.

27. Olanow, C. W., Calne, D., Does selegiline monotherapy in Parkinson's disease act by symptomatic or protective mechanisms? *Neurology*, 42:13–26, 1991.

28. Lees, A. J., On behalf of the Parkinson's Disease Research Group of the United Kingdom. Comparison of therapeutic effects and mortality data of levodopa and levodopa combined with selegiline in patients with early, mild Parkinson's disease, *B.M.J.*, 311:1602–1607, 1995.

29. Parkinson Study Group, Mortality in DATATOP: a multicenter trial in early Parkinson's disease, *Ann. Neurol.*, 43, 318–325, 1998.

30. Olanow, C. W., Myllyla, V. V., Sotaniemi, K. A., Larsen, J. P., Palhagen, S., Przuntek, H., Heinonen, E. H., Kilkku, O., Lammintausta, R., Maki-Ikola, O., Rinne, U. K., Effect of selegiline on mortality in patients with Parkinson's disease: a meta-analysis, *Neurology*, 51: 825–830, 1998.

31. Donnan, P. T., Steinke, D. T., Stubbings, C., Davey, P. G., MacDonald, T. M., Selegiline and mortality in subjects with Parkinson's disease: a longitudinal community study, *Neurology*, 55, 1785–1789, 2000.

32. Parkinson Study Group, A controlled trial of rasagiline in early Parkinson disease: the TEMPO study., *Arch Neurol.*, 59:1937–1943, 2002.

33. Heinonen, E. H., Myllyla, V., Safety of selegiline (deprenyl) in the treatment of Parkinson's disease, *Drug Saf.*, 19:11–22, 1998.

34. Davies, W. L., Grunert, R. R., Haff, R. F., McGahen, J. W., Neumayer, E. M., Paulshock, M., Watts, J. C., Wood, T. R., Hermann, E. C., Hoffmann, C. E., Antiviral activity of 1-adamantanamine (amantadine), *Science*, 144:862–863, 1964.

35. Schwab, R. S., England, A. C., Jr., Poskancer, D. C., Young, R. R., Amantadine in the treatment of Parkinson's disease, *JAMA*, 208:1168–1170, 1969.

36. Von Voigtlander, P. F., Moore, K. E., Dopamine: release from the brain *in vivo* by amantadine, *Science*, 174:408–410, 1971.

37. Gianutsos, G., Chute, S., Dunn, J. P., Pharmacological changes in dopaminergic systems induced by long-term administration of amantadine, *Eur. J. Pharmacol.*, 110:357–361, 1985.

38. Moresco, R. M., Volonte, M. A., Messa, C., Gobbo, C., Galli, L., Carpinelli, A., Rizzo, G., Panzacchi, A., Franceschi, M., Fazio, F., New perspectives on neurochemical effects of amantadine in the brain of parkin-

sonian patients: a PET - [(11)C]raclopride study, *J. Neural. Transm.,* 109:1265–74, 2002.

39. Nastuck, W. C., Su, P. C., Doubilet, P., Anticholinergic and membrane activities of amantadine in neuromuscular transmission, *Nature,* 264:76–79, 1976.

40. Greenamyre, J. T., O'Brien, C. F., N-methyl-D-aspartate antagonists in the treatment of Parkinson's disease, *Arch Neurol.,* 48:977–981, 1991.

41. Stoof, J. C., Booij, J., Drukarch, B., Amantadine and N-methyl-D-aspartic acid receptor antagonist. New possibilities for therapeutic application? *Clin. Neurol. Neurosurg.,* 94 (Suppl):S4–S6, 1992.

42. Butzer, J. F., Silver, D. E., Sahs, A. L., Amantadine in Parkinson's disease. A double-blind, placebo-controlled, crossover study with long-term follow-up, *Neurology,* 25:603–606, 1975.

43. Uitti, R. J., Rajput, A. H., Ahlskog, J. E., Offord, K. P., Schroeder, D. R., Ho, M. M., Prasad, M., Rajput, A., Basran, P., Amantadine treatment is an independent predictor of improved survival in Parkinson's disease, *Neurology,* 46:1551–1556, 1996.

44. Amantadine and other antiglutamate agents, *Mov. Disord.,* 17 (Suppl. 4):S13–S22, 2002.

45. Crosby, N., Deane, K. H., Clarke, C. E., Amantadine in Parkinson's disease, *Cochrane Database Syst. Rev.,* (1):CD003468, 2003.

46. Luginger, E., Wenning, G. K., Bosch, S., Poewe, W., Beneficial effects of amantadine on L-dopa-induced dyskinesias in Parkinson's disease, *Mov. Disord.,* 15:873–878, 2000.

47. Weller, M., Kornhuber, J., Amantadine withdrawal and neuroleptic malignant syndrome, *Neurology,* 43:2155, 1993.

48. Ito, T., Shibata, K., Watanabe, A., Akabane, J., Neuroleptic malignant syndrome following withdrawal of amantadine in a patient with influenza A encephalopathy, *Eur. J. Pediatr.,* 160:401, 2001.

49. Marsden, C. D., Parkes, J. D., On-off effects in patients with Parkinson's disease on chronic levodopa therapy, *Lancet,* 1:292–296, 1976.

50. Miyawaki, E., Lyons, K., Pahwa, R. et al., Motor complications of chronic levodopa therapy in Parkinson's disease, *Clin. Neuropharmacol.,* 20:523–530, 1997.

51. Quinn, N., Critchley, P., Marsden, C. D., Young onset Parkinson's disease, *Mov. Disord.,* 2:73–91, 1987.

52. Golbe, L. I., Young-onset Parkinson's disease: a clinical review, *Neurology,* 41:168–173, 1991.

53. Schrag, A., Ben-Shlomo, Y., Brown, R., Marsden, C. D., Quinn, N., Young-onset Parkinson's disease revisited—clinical features, natural history, and mortality, *Mov. Disord.,* 13:885–94, 1998.

54. Marras, C., Lang, A., Krahn, M., Tomlinson, G., Naglie, G. and The Parkinson Study Group, Quality of life in early Parkinson's disease: impact of dyskinesias and motor fluctuations, *Mov. Disord.,* 19:22–28, 2004.

55. Damiano, A. M., McGrath, M. M., Willian, M. K., Snyder, C. F., Le Witt, P. A., Richter, R. R., Evaluation of a measurement strategy for Parkinson's disease: patient health-related quality of life, *Qual. Life Res.,* 9:87–100, 2000.

56. Pechevis, M., Clarke, C. E., Vieregge, P. et al., Direct and costs of Parkinson's disease and L-dopa induced dyskinesias: a prospective European study, *Parkinsonism Relat. Disord.,* 7 (Suppl):106, 2001.

57. Schrag, A., Quinn, N., Dyskinesias and motor fluctuations in Parkinson's disease: a community based study, *Brain,* 123:2297–2305, 2000.

58. Siderowf, A., Ravina, B., Glick, H. A., Preference-based quality of life in patients with Parkinson's disease, *Neurology,* 59:103–108, 2002.

59. Nutt, J. G., On-off phenomenon: relation to levodopa pharmacokinetics and pharmacodynamics, *Ann. Neurol.,* 22:535–540, 1987.

60. Chase, T. N., Engber, T. M., Mouradian, M. M., Palliative and prophylactic benefits of continuously administered dopaminomimetics in Parkinson's disease, *Neurology,* 44 (Suppl. 6), S15–S18, 1994.

61. Levodopa. *Mov. Disord.,* 17 (Suppl. 4):S23–S37, 2002.

62. Pahwa, R., Lyons, K., McGuire, D., Silverstein, P., Zwiebel, F., Robischon, M., Koller, W. C., Comparison of standard carbidopa-levodopa and sustained-release carbidopa-levodopa in Parkinson's disease: pharmacokinetic and quality-of-life measures, *Mov. Disord.,* 12:677–681, 1997.

63. Koller, W. C., Hutton, J. T., Tolosa, E., Capilldeo, R., Immediate-release and controlled-release carbidopa/levodopa in PD: a 5-year randomized multicenter study. Carbidopa/Levodopa Study Group, *Neurology,* 53:1012–1019, 1999.

64. Olanow, C. W., Stocchi, F., COMT inhibitors in Parkinson's disease: can they prevent and/or reverse levodopa-induced motor complications? *Neurology,* 62(1 Suppl. 1), S72–S81, 2004.

65. Jenner, P., Avoidance of dyskinesia: preclinical evidence for continuous dopaminergic stimulation, *Neurology,* 62(1 Suppl. 1), S47-S55, 2004.

66. Jenner, P., Al-Barghouthy, G., Smith, L., Kuoppamaki, M., Jackson, M., Rose, S., Entacapone combined with L-dopa enhances antiparkinsonian activity and avoids dyskinesia in the MPTP-treated primate model of Parkinson's disease (PD), *Mov. Disord.,* 17 (Suppl. 5), P146, Abstract, 2002.

67. Rascol O., Payoux, P., Ory, F., Ferreira, J. J., Brefel-Courbon, C., Montastruc, J. L., Limitations of current Parkinson's disease therapy, *Ann. Neurol.,* 53 (Suppl. 3), S3–S12; discussion S12–S15, 2003.

68. Ward, C. D., Thompson, J., Humby, M. D., Pleuropulmonary and retroperitoneal fibrosis associated with bromocriptine treatment, *J. Neurol. Neurosurg. Psychiatry,* 50:1706–1707, 1987.

69. Jimenez-Jimenez, F. J., Lopez-Alvarez, J., Sanchez-Chapado, M., Montero, E., Miquel, J., Sierra, A., Gutierrez, F., Retroperitoneal fibrosis in a patient with Parkinson's disease treated with pergolide, *Clin. Neuropharmacol.,* 18:277–9, 1995.

70. Pritchett, A. M., Morrison, J. F., Edwards, W. D., Schaff, H. V., Connolly, H. M., Espinosa, R. E., Valvular heart disease in patients taking pergolide, *Mayo Clin. Proc.,* 77:1280–1286, 2002.

71. Parkinson Study Group, Pramipexole vs. levodopa as initial treatment for Parkinson disease: A randomized controlled trial. Parkinson Study Group, *JAMA,* 284:1931–1938, 2002.

72. Rascol, O., Brooks, D. J., Korczyn, A. D., De Deyn, P. P., Clarke, C. E., Lang, A. E., A five-year study of the incidence of dyskinesia in patients with early Parkinson's disease who were treated with ropinirole or levodopa, 056 Study Group, *N. Engl. J. Med.,* 342, 1484–1491, 2000.

73. Homann, C. N., Wenzel, K., Suppan, K., Ivanic, G., Kriechbaum, N., Crevenna, R., Ott, E., Sleep attacks in patients taking dopamine agonists: review, *B.M.J.,* 324:1483–1487, 2002.

74. Razmy, A., Lang, A. E., Shapiro, C. M., Predictors of impaired daytime sleep and wakefulness in patients with Parkinson disease treated with older (ergot) vs. newer (nonergot) dopamine agonists, *Arch Neurol.,* 61:97–102, 2004.

75. Driver-Dunckley, E., Samanta, J., Stacy, M., Pathological gambling associated with dopamine agonist therapy in Parkinson's disease, *Neurology,* 61:422–423, 2003.

76. Beal, M. F., Coenzyme Q_{10} as a possible treatment for neurodegenerative diseases, *Free Radic. Res.,* 36:455–60, 2002.

77. Parker, W. D., Jr., Boyson, S. J., Parks, J. K., Abnormalities of the electron transport chain in idiopathic Parkinson's disease, *Ann. Neurol.,* 26:719–23, 1989.

78. Shults, C. W., Oakes, D., Kieburtz, K., Beal, M. F., Haas, R., Plumb, S., Juncos, J. L., Nutt, J., Shoulson, I., Carter, J., Kompoliti, K., Perlmutter, J. S., Reich, S., Stern, M., Watts, R. L., Kurlan, R., Molho, E., Harrison, M., Lew, M., Parkinson Study Group, Effects of coenzyme Q_{10} in early Parkinson disease: evidence of slowing of the functional decline, *Arch Neurol.,* 59:1541–1550, 2002.

79. Müller, T., Buttner, T., Gholipour, A. F., Kuhn, W., Coenzyme Q_{10} supplementation provides mild symptomatic benefit in patients with Parkinson's disease, *Neurosci. Lett.,* 341:201–204, 2003.

80. McDonald, W. M., Richard, I. H., DeLong, M. R., Prevalence, etiology, and treatment of depression in Parkinson's disease, *Biol. Psychiatry,* 54:363–375, 2003.

81. Shabnam, G. N., Th, C., Kho, D. H. R., Ce, C., Therapies for depression in Parkinson's disease, *Cochrane Database Syst. Rev.,* (3):CD003465, 2003.

82. Tan, A., Salgado, M., Fahn, S., Rapid eye movement sleep behavior disorder preceding Parkinson's disease with therapeutic response to levodopa, *Mov. Disord.,* 11:214–216, 1996.

83. Chaudhuri, K. R., Nocturnal symptom complex in PD and its management, *Neurology,* 61(6 Suppl. 3):S17–23, 2003.

84. Ondo, W. G., Vuong, K. D., Khan, H., Atassi, F., Kwak, C., Jankovic, J., Daytime sleepiness and other sleep disorders in Parkinson's disease, *Neurology,* 57:1392–139, 2001.

85. Abbott, R. D., Petrovitch, H., White, L. R., Masaki, K. H., Tanner, C. M., Curb, J. D., Grandinetti, A., Blanchette, P. L., Popper, J. S., Ross, G. W., Frequency of bowel movements and the future risk of Parkinson's disease, *Neurology,* 57:456–62, 2001.

86. Raffaele, R., Vecchio, I., Giammusso, B., Morgia, G., Brunetto, M. B., Rampello, L., Efficacy and safety of fixed-dose oral sildenafil in the treatment of sexual dysfunction in depressed patients with idiopathic Parkinson's disease, *Eur. Urol.,* 41:382–386, 2002.

87. Parkinson Study Group, A controlled trial of rotigotine monotherapy in early Parkinson's disease, *Arch Neurol.,* 60:1721–1728, 2003.

66 Moderate Parkinson's Disease

John E. Duda and Matthew B. Stern
University of Pennsylvania School of Medicine and Parkinson's Disease Research, Education and Clinical Center, Philadelphia Veterans Affairs Medical Center

CONTENTS

INTRODUCTION

Moderate Parkinson's disease (PD) can be defined as the phase of the illness that occurs when PD symptoms are no longer completely controlled, and complications of drug therapy and disease progression emerge. This phase includes the appearance of motor fluctuations and dyskinesias, the initiation of adjunctive therapy, and the emergence of many nonmotor symptoms. The moderate phase of PD may encompass years of illness and usually represents the longest phase of an individual's battle with PD. During this period, appropriate care often involves frequent follow-up with regular adjustments of the medical regimen, a reliance on rational polypharmacy, referral to movement disorder specialists, and the use of the myriad of ancillary health services.

NATURAL HISTORY OF PARKINSON'S DISEASE

In discussing PD, it is useful to recognize four distinct phases of the illness. The first is a prodromal phase that is poorly characterized but probably lasts several years and ends when the typical motor symptoms of PD emerge. While it is often difficult to determine with any degree of certainty, neuropathological, neuroimaging, and clinical studies have attempted to determine the average duration of this phase and have extrapolated numbers ranging from 3 to 30 years. Much of the variability in the preclinical phase may reflect the uncertainty in rapidity of PD progression, i.e., whether PD progresses more rapidly in its early or later stages. Using nonlinear models that account for this variability, most studies have estimated a duration of 3 to 10 years for the prodromal phase.[3] Similarly difficult to determine with certainty is the range of symptoms that represent prodromal PD. Candidate signs and symptoms are outlined in Table 66.1. Interestingly, in a recent attempt to characterize the progression of PD pathology in the brain, Braak and colleagues demonstrated that pathological alterations do not begin in the substantia nigra or basal ganglia but rather in the medulla oblongata

---

TABLE 66.1
Possible Signs and Symptoms of the Prodromal Phase of PD

Psychiatric disturbances:
  PD personality
  Mood disorders, esp. depression
Sensory disturbances:
  Pain
    Fibromyalgia
    Shoulder pain
  Non-specific sensory disturbances
Autonomic disturbances:
  Decreased olfaction
  Constipation/obstipation
  Diarrhea
  QT interval prolongation
  Hypertension
Sleep disturbances:
  REM sleep behavior disorder

0-8493-1590-5/05/$0.00+$1.50

and olfactory system.[4,5] Presumably, these early sites of involvement are responsible for many of the clinical symptoms observed during the prodromal phase, including olfactory dysfunction, mood disturbances, and sleep dysfunction. As this phase of PD becomes better characterized, future research efforts will focus neuroprotective therapies that can slow or halt disease progression on this stage of the illness, before symptoms become more overt.

The second phase, or early clinical phase, begins with the appearance of the cardinal signs and symptoms of PD including tremor, bradykinesia, and rigidity. Symptoms are often unilateral and are usually easily controlled with medical therapies. The modern goal of early PD therapy is adequate control of symptoms while minimizing the risk of motor complications. The early clinical phase can be thought of as ending when these early therapies begin to fail.

The third, or moderate, phase of PD, the subject of this chapter, includes the appearance of motor complications, the need for adjunctive therapy, and the emergence of many nonmotor symptoms that can be as disabling as motor symptoms. The advanced phase of PD includes the period when advanced medical and surgical therapies are considered for the management of symptoms and/or secondary complications and continues until the time of death. This phase is discussed in detail in the following chapter.

RATE OF PROGRESSION OF PARKINSON'S DISEASE

Physicians who care for many PD patients recognize that no two patients are identical concerning symptom progression. There is a wide range of variation in progression, and it is often very difficult to predict, for any given patient, what that rate will be. Perhaps the best predictor of future progression is the rate of progression in the past. In a study of 442 PD patients, Rinne and Martilla[6] showed that, while there was great variation in the population for time of progression between Hoehn and Yahr stages for a given individual, the rate of progression from Hoehn and Yahr stage 1 to stage 2 and from stage 3 to stage 4 was relatively fixed. In addition, several clinical factors have been associated with either a slower or faster rate of progression.[7–9] These factors are useful to consider when addressing the prognosis for a given patient and are outlined in Table 66.2.

EARLY TREATMENT FAILURE

While initial therapeutic interventions in PD were discussed at length in the previous chapter, the first therapeutic decision in the management of moderate PD entails the point at which initial therapy begins to fail and trou-

TABLE 66.2
Clinical Factors Related to Rate of Progression in PD

Factors associated with a slower rate of progression:
- Younger age at onset
- Tremor predominant phenotype

Factors associated with a faster rate of progression:
- Older age at onset
- Akinetic-rigid predominant phenotype
- Cognitive decline
- Concomitant major depression

blesome symptoms return. It is important to consider that this does not include patients who fail to achieve adequate symptom relief from the initial intervention, but only patients who have been free of troublesome symptoms for a period of time. It is also important to recognize the difference between troublesome or disabling symptoms and nontroublesome or nondisabling symptoms. It is essential that the physician discuss this concept early in the course of therapy with the patient—that the goal is not necessarily to alleviate every trace of PD symptoms that arise. This approach often produces medication side effects that are more problematic than the symptoms being treated. Instead, the therapeutic goal is to alleviate symptoms that are troublesome or disabling to the patient, i.e., prohibiting normal function in daily life. Therefore, minor symptoms of decreased arm swing, tremor during ambulation, or mild dyskinesias that are hardly noticeable to the patient are not necessarily indications for augmenting medical therapy. However, it is important that the patient remain an active participant in this decision-making process, because symptoms that appear innocuous (e.g., a mild tremor) may be the cause of sufficient social embarrassment to the patient that they would limit social interactions, thus impairing normal function.

Alternatively, it is also important to realize that all disabling or troublesome symptoms should be treated adequately, with levodopa, if necessary. Concerns of levodopa toxicity remain unproven and should not dictate therapeutic interventions. In practice, levodopa is virtually always administered in combination with a peripheral dopa decarboxylase inhibitor (carbidopa or benserazide) so that when levodopa is mentioned in this article as a therapeutic option, it should be as a combination of the two agents. In the future, this may be further refined because of the availability of a triple combination pill of levodopa, carbidopa, and entacapone (Stalevo, Novartis).[10] With the advent or return of any symptom, the first objective is to rule out any cause that is not necessarily related to disease progression. These include changes in the way that medications are taken, concomitant illnesses, or medications that can exacerbate PD and are outlined in Table 66.3.

TABLE 66.3
Clinical Factors Related to Symptom Worsening in PD

Factors related to medication effectiveness:

- Change in medication formulation (e.g., change from brand name to generic formulation or from immediate release to controlled-release formulation)
- Change in timing of medication in relation to meals that may affect intestinal absorption
- Change in dietary preferences, especially a change to a high-protein, low-carbohydrate diet that may interfere with intestinal absorption more than a well balanced diet
- Problems with compliance (a frequent cause early in the disease course)
- Addition of ancillary medications that may affect medication effectiveness by competing for absorption or metabolism

Factors that can worsen PD symptoms:

- Addition of a medication known to block dopamine receptors and worsen PD symptoms (e.g., typical neuroleptics, anti-nausea medications, including metoclopramide and prochlorperazine)
- *Concomitant illnesses that can worsen PD symptoms:*
  Infectious processes including urinary tract infections, pneumonia
  Sleep disorders
  Depression

CONSIDERATIONS FOR ADJUNCTIVE THERAPY

Once it becomes clear that PD has progressed and current treatment is no longer effective, there are a number of options. If a patient is on monotherapy, the dose of the single agent can be increased, or adjunctive medications can be added. Sometimes the knee-jerk reaction early in the course of treatment for patients with occasional disabling symptoms is to prescribe an as-needed dose of medication to control these symptoms. However, one current theory concerning the cause of secondary complications of levodopa therapy may justify alternative therapeutic options. The theory, involving the concept of continuous dopaminergic stimulation,[11–14] suggests that under normal conditions the postsynaptic dopamine receptor is exposed to tonic stimulation from the continuous slow release of dopamine into the synaptic cleft. However, as dopaminergic neuron loss proceeds in the course of the illness, the remaining neurons have less capacity to store dopamine and therefore lose the ability to maintain a tonic stimulation of the postsynaptic receptors. Therefore, postsynaptic stimulation mirrors the fluctuating blood levels that follow oral dopamine replacement therapy, and it is this pulsatile pattern of stimulation that eventually leads to secondary complications. While this concept remains theoretical, it is preferable to avoid this pattern and to devise a medication regimen that prevents the symptom altogether. For any given patient, there is often an equally valid argument for either increasing the dose of current medications or adding adjunctive medications. The decision-making process should be guided by an assessment of the timing of symptoms, the symptom type, the initial therapeutic intervention, and patient characteristics.

Perhaps the most important component of the decision-making process in determining how to treat moderate PD is the assessment of the timing of the bothersome symptom and whether the patient is describing a gradual wearing-off of drug effect. The wearing-off phenomenon entails development of a shortened response to dopaminergic therapy so that symptom control becomes inadequate prior to the next dose of medication.[15,16] It is important to examine the timing of symptoms in relation to medication administration so as to determine if the symptom is a manifestation of the wearing off phenomenon. If the symptom in question occurs most prominently in the period before the next dose or even in the first 30 to 45 minutes after the next dose (due to delayed onset), it is likely a wearing-off symptom. Therapeutic strategies designed specifically for wearing-off symptoms are outlined in Table 66.4. These include shortening the frequency between levodopa doses or adding adjunctive agents that either have longer half-lives or delay the elimination of levodopa. In particular, adjunctive therapy with a catechol-O-methyltransferase (COMT) inhibitor—either tolcapone (Tasmar, Hoffman-La Roche) or entacapone (Comtan, Novartis)—that prolongs the half-life of levodopa by eliminating one catabolic pathway can be useful. In fact, due to concerns of hepatic toxicity with tolcapone, entacapone has become the most commonly used agent in this class and has an FDA-approved indication for end-of-dose wearing off.[17–19] It should be recognized that many patients starting COMT inhibitor therapy will require reductions in their levodopa dosage to prevent the onset or worsening of dyskinesias.

TABLE 66.4
Therapeutic Options for Wearing-Off Symptoms

- Increase the frequency of medications (e.g., changing from t.i.d. dosing to q.i.d. dosing with the last dose still during the day rather than at bedtime)
- Change from immediate release levodopa to controlled-release levodopa
- Adjunctive therapy with a COMT inhibitor
- Adjunctive therapy with a dopamine agonist
- Adjunctive therapy with an MAO-B inhibitor

In addition, some symptoms may present as a wearing-off symptom or as a peak-dose symptom, and management is different for the two. For example, dystonia is a common end-of-dose phenomenon but can also occur as a peak-dose symptom.[20] Treating end-of-dose dystonia by augmenting dopaminergic therapy can exacerbate peak-dose dystonia. When the relationship to medications is not clear, it is often helpful to have patients complete symptom and dosing diaries.[21,22] Symptom diaries are completed by the patient and include a record of exactly when each dose of medication is taken, what the patient's condition is for every half hour of the day, and when troublesome symptoms occur. Symptom diaries are particularly useful in determining patterns of symptoms in patients who are taking multiple medications frequently throughout the day, but are also useful early in the disease course to identify wearing off symptoms.

Symptom type is also important to consider when making therapeutic interventions, because each symptom has particular patterns of medication responsiveness. The first aspect to consider is whether the symptom is a re-emergence of a previously well treated symptom or a novel symptom. In the case of a re-emergent symptom, an argument can be made for simply increasing the amount or frequency of the current medication as a symptom that has responded to a particular agent in the past should respond to higher doses just as adequately. If the symptom is novel, then particular patterns of symptom responsiveness should be considered. For example, tremor may respond well to anticholinergic medications [trihexyphenidyl (Artane), benztropine (Cogentin)], or amantadine (Symmetrel, DuPont Pharma), whereas bradykinesia and rigidity respond better to dopaminergics.

Adjusting treatment for moderate PD will depend on initial PD therapy. Dopamine agonists are the frequent choice for initial therapy in PD; however, DA require generally slow titration over weeks to months. Patients and physicians often stop short of safe and effective doses. Increasing the DA dose might therefore be an appropriate first step in alleviating new symptoms. Furthermore, if a patient is currently on levodopa, even in low doses, adding a DA would be prudent, considering the long-term benefits of DA and lower levodopa doses compared with higher doses of levodopa alone. Titrating DA to maximal recommended doses or, in some patients, to higher-than-recommended doses[23,24] improves efficacy, although side effects can be limiting. In the event of treatment failure at adequate doses, the likelihood that switching to a second dopamine agonist will achieve adequate symptom control is low and is probably not warranted. However, if adequate doses are not achieved because of side effects, then a trial with a second dopamine agonist may be justified. If symptoms do not respond, or troublesome side effects occur, then adjunctive therapy with levodopa should be considered.

If the patient was started on levodopa, then dosage escalation is appropriate for patients on low doses or infrequent doses. Adjunctive therapy with long-acting levodopa, dopamine agonists, or a COMT inhibitor is appropriate to consider after a patient has been treated with at least 300 mg levodopa equivalents (100 mg of controlled-release levodopa is equivalent to approximately 70 mg immediate-release levodopa) on a t.i.d. schedule.

The final consideration for modifying therapy for treatment failures is an assessment of patient characteristics that would favor use of one therapy over others. The first characteristic to consider is the functional age of the patient, which is an attempt to assess the patient's longevity. It is considered to the same extent that it is considered in determining initial therapy in relation to the probability of surviving until the onset of secondary complications. Other patient characteristics to consider include cognitive status, autonomic symptoms, and payment method for medications. In general, dopamine agonists are more likely to cause cognitive and psychiatric complications (e.g., hallucinations, delusions, delirium) and autonomic symptoms (e.g., orthostatic hypotension, peripheral edema) than levodopa, and therefore patients who have underlying cognitive or autonomic deterioration may tolerate levodopa better. In addition, the financial burden to the patient should be considered in the decision-making process. Many PD medications are expensive, and the financial burden will vary depending on financial resources and medication cost to the patient. As we have seen, there are few absolutes in PD therapy, and often a less expensive regimen is just a valid as a more expensive regimen.

MANAGEMENT OF EARLY SECONDARY COMPLICATIONS

While the wearing off phenomenon is often easily managed with medication adjustments, managing other secondary complications, including on-off fluctuations and dyskinesias, can be more difficult with a vast number of treatment alternatives. An understanding of the normal physiological response to antiparkinsonian therapy will empower the clinician to make rational decisions with a greater likelihood of success. Early in the management of PD, it is common for patients taking levodopa to deny any definite differences in symptoms before or after taking PD medications. They do not feel the medications "kick in," and they can even miss a dose or two and have no appreciable change in symptoms, because of the long-duration response to medication in early PD.[25-26] As PD evolves, a short-duration response predominates, and patients become clearly aware of the onset and end of response for each dose on their symptoms. Eventually, the patient will be able to recognize periods when the medications are controlling his symptoms well, referred to as "on time," and other periods when his symptoms are not controlled, referred to as "off time."

Frequently, the first "off time" is experienced after the longest delay between two doses of medication, which is usually between the last dose in the evening and the first dose in the morning. In addition to a recurrence of the first symptoms that the patient developed, many patients will develop dystonia of their lower extremities as a manifestation of early morning "off-time."[27] Adding a DA or a single dose of long-acting levodopa, with or without a COMT inhibitor, at bedtime will often alleviate early-morning "off-time." There are many other options for the management of motor fluctuations as outlined in Table 66.5. Recently, the results of a large multi-center, placebo-controlled study of the use of rasagiline (Teva), which is an irreversible MAO-B inhibitor, in the treatment of motor fluctuations were presented at the 128th annual meeting of the American Neurological Association, in October 2003. The PRESTO study revealed significant decreases in "off time" in patients with optimized dosages of levodopa and two doses of rasagiline versus placebo. In addition to decreasing the amount of time spent in the "off state," rasagiline also significantly improved the patients' ability to perform activities of daily living in the "off" state as well as motor function in the "on" state. Therefore, rasagiline is a promising new agent in the management of patients with motor fluctuations. Similarly, a new orally absorbed preparation of selegiline (Zelapar, Amarin) has shown promise in clinical trials[28] and should be available in the near future.

TABLE 66.5
Therapeutic Options for Motor Fluctuations

- Increase the frequency of medications (e.g., changing from t.i.d. dosing to q.i.d. dosing with the last dose still during the day rather than at bedtime)
- Change from immediate-release levodopa to controlled-release levodopa
- Adjunctive therapy with a COMT inhibitor
- Adjunctive therapy with a dopamine agonist
- Adjunctive therapy with an MAO-B inhibitor (esp., rasagiline)
- Use liquid levodopa for "rescue therapy"
- Surgery

Patients whose fluctuations become sudden transitions between the on and off states, the "on-off phenomenon," are more difficult to manage and often require surgical approaches such as deep brain stimulation (see Chapter 67, "Management of Advanced Parkinson's Disease"). If surgical therapy is not an option, with the patient switching rapidly from an "on" state to an "off" state of being virtually "frozen," it can be useful to recommend a liquid levodopa preparation.[29–31] To prepare this, the patient's caregiver must crush and suspend a 25/100 tablet of either regular or controlled-release levodopa (crushing controlled-release levodopa negates the controlled-release aspect) into a glass of water or juice. If the

patient can drink the whole solution, he can expect liquid levodopa to "kick-in" in 15 to 20 min and last for 60 to 90 min.

Another challenging symptom that frequently occurs in moderate PD is dyskinesia that can present as chorea, dystonia or, more rarely, as myoclonus (reviewed in References 32 and 33). As with other secondary complications, a determination must be made as to whether a particular patient's dyskinesias are troublesome or disabling. Often, mild dyskinesias are not even recognized by the patient until pointed out by someone else and, in many cases, do not require therapeutic intervention. In addition, if given a choice, most patients would prefer to be mildly dyskinetic than "off." Similarly, as with other secondary symptoms, the timing of dyskinesias relative to drug dosing must be established so as to differentiate between different patterns of involuntary movements, including peak-dose dyskinesias, diphasic dyskinesias, and wearing-off dyskinesias. By far the most common pattern is the peak-dose dyskinesias that, if troublesome or disabling, can be managed with a variety of different interventions as outlined in Table 66.6. These interventions include shifts to agents with longer half-lives, discontinuation of agents that slow levodopa elimination, and adjunctive therapy with two agents that block glutamate receptors, amantadine[34–38] and riluzole (Rilutek, Aventis).[39] Diphasic dyskinesias consist of a dyskinetic period immediately following a medication dose followed by a period of improvement and then a final period of dyskinesias. In the case of diphasic dyskinesias, therapeutic intervention in designed to completely avoid wearing off and therefore mainly consists of increasing the frequency of medication doses or continuous apomorphine infusion.[40]

TABLE 66.6
Therapeutic Options for Peak-Dose Dyskinesias

- Decrease the dose of medications (e.g., decrease a levodopa dose by 50 mg or decrease the dose of a dopamine agonist) and increasing the frequency of doses, if necessary
- Reduce levodopa dose and increase dopamine agonist dose
- Change to multiple small doses of immediate release levodopa throughout the day
- Discontinue a COMT inhibitor
- Adjunctive therapy with a dopamine agonist
- Discontinue an MAO-B inhibitor
- Adjunctive therapy with amantadine
- Adjunctive therapy with riluzole
- Surgery

In many aspects, the management of peak-dose dyskinesias involves interventions that are in opposition to treatment options for the wearing-off phenomenon, and the resulting interplay of therapeutic strategies often characterizes the transition from moderate to advanced PD. A period of guided experimentation is often beneficial for patients and may involve gradually introducing or increasing DA while trying to taper the total levodopa dose. In patients with complicated motor fluctuations, we often recommend discontinuing monoamine oxidase inhibitors, anticholinergics, and other "nonessential" antiparkinson drugs unless proven effective for a given patient. Continuous apomorphine administration[41–44] is an option where available, but, for most of these patients, surgical intervention becomes a viable alternative (see Chapter 67, "Management of Advanced Parkinson's Disease").

RATIONAL POLYTHERAPY

Often, in today's medical culture, with limited time dedicated to each office visit, there is a pressure to treat each new symptom with a new medication. Many patients who are referred to movement disorder specialists when this approach fails to achieve adequate symptom control are best served by eliminating medications. However, in moderate PD, rational polypharmacy has become the mainstay of treatment. Most of the current PD medications were developed and approved as adjuncts to levodopa and therefore are appropriate in combination with levodopa. As long as the patient and physician understand the effects of a particular medication on an individual patient, it may be totally appropriate for a PD patient with moderate disease to be on levodopa, a COMT inhibitor, a MAO-B inhibitor, amantadine, and a dopamine agonist. In addition, immediate- and controlled-release formulations of levodopa are often used in combination. Therapeutic options that have little or no rational support include a COMT inhibitor as monotherapy, adding a COMT inhibitor to dopamine agonist monotherapy, and using multiple dopamine agonists in the same patient.

Of current interest is the combination of levodopa, carbidopa, and entacapone (Stalevo, Novartis) that was recently approved by the FDA for the treatment of PD.[10] Stalevo is available in three strengths, referred to as Stalevo 50, 100, and 150, with a combination of 200 mg of entacapone and 12.5, 25, or 37.5 mg of carbidopa and 50, 100, and 150 mg of levodopa, respectively. It is reasonable to substitute equivalent levodopa doses of Stalevo for patients already taking a combination of levodopa and entacapone or changing to Stalevo when wearing-off symptoms dictate a need for adjunctive COMT inhibition. Whether using this new combination tablet from the outset of levodopa therapy lessens the risk of developing motor complications because of more continuous dopaminergic stimulation will depend on the results of well designed clinical trials.

REFERENCES

1. Gonera, E. G. et al., Symptoms and duration of the prodromal phase in Parkinson's disease, *Mov. Disord.*, 12, 871, 1997.

2. Koller, W. C., When does Parkinson's disease begin? *Neurology*, 42 Suppl. 4, 27, 1992.

3. Morrish, P. K., Parkinson's disease is not a long-latency illness, *Mov. Disord.*, 12, 849, 1997.

4. Braak, H. et al., Staging of the intracerebral inclusion body pathology associated with idiopathic Parkinson's disease (preclinical and clinical stages), *J. Neurol.*, 249 Suppl. 3,1, 2002.

5. Braak, H. et al., Staging of brain pathology related to sporadic Parkinson's disease, *Neurobiol. Aging*, 24, 197, 2003.

6. Martilla, R. J. and Rirme, U. K., Disability and progression of Parkinson's disease, *Acta Neurol. Scand.*, 56, 159, 1997.

7. Goetz, C. G. et al., Risk factors for progression in Parkinson's disease, *Neurology*, 38,1841, 1998.

8. Roos, R. A., Jongen, J. C., and van der Velde, E. A., Clinical course of patients with idiopathic Parkinson's disease, *Mov. Disord.*, 11, 236, 1996.

9. Zetusky, W. J., Jankovic, J., and Pirozzolo, F. J., The heterogeneity of Parkinson's disease: Clinical and prognostic implications, *Neurology*, 35, 522, 2004.

10. Hauser, R. A., Levodopa/carbidopa/entacapone (Stalevo), *Neurology*, 62 Suppl. 1, 64, 2004.

11. Chase, T. N. et al., Rationale for continuous dopaminomimetic therapy of Parkinson's disease, *Neurology*, 39, 7, 1989.

12. Olanow, C. W. and Obeso, J. A., Preventing levodopa-induced dyskinesias, *Ann. Neurol.*, 47 Suppl. 1, 167, 2000.

13. Olanow, C. W. and Obeso, J. A., Pulsatile stimulation of dopamine receptors and levodopa-induced motor complications in Parkinson's disease: implications for the early use of COMT inhibitors, *Neurology*, 55 Suppl. 4, 72, 2000.

14. Sage, J. I. and Mark, M. H., The rationale for continuous dopaminergic stimulation in patients with Parkinson's disease, *Neurology*, 42 Suppl. 1, 23, 1992.

15. Nutt, J. G. et al., Short- and long-duration responses to levodopa during the first year of levodopa therapy, *Ann. Neurol.*, 42, 349,1997.

16. Zappia, M. et al., Loss of long-duration response to levodopa over time in PD: implications for wearing-off. *Neurology*, 52, 763,1999.

17. Kurth, M. C. and Adier, C. H., COMT inhibition: a new treatment strategy for Parkinson's disease, *Neurology*, 50 Suppl. 5, 3, 1998.

18. Lang, A. E. and Lees, A., COMT inhibitors: management of Parkinson's disease, *Mov. Disord.*, 17 Suppl. 4, 45, 2002.

19. Schapira, A. H., Obeso, J. A., and Olanow, C. W., The place of COMT inhibitors in the armamentarium of drugs for the treatment of Parkinson's disease, *Neurology*, 55 Suppl. 4, 65, 2000.

20. Jankovic, J. and Tintner, R., Dystonia and parkinsonism, *Parkinsonism Relat. Disord.*, 8, 109,2001.

21. Defer, G. L. et al., Core Assessment Program for Surgical Interventional Therapies in Parkinson's Disease (CAPSIT-PD), *Mov. Disord.*, 14, 572, 1999.

22. Montgomery, G. K. and Reynolds, N. C., Jr., Compliance, reliability, and validity of self-monitoring for physical disturbances of Parkinson's disease, The Parkinson's Symptom Diary, *J. Nerv. Ment. Dis.*, 178, 636, 1990.

23. Cristina, S. et al., High-dose ropinirole in advanced Parkinson's disease with severe dyskinesias, *Clin. Neurophamacol.*, 26,146, 2003.

24. Facca, A. and Sanchez-Ramos, J., High-dose pergolide monotherapy m the treatment of severe levodopa-induced dyskinesias, *Mov. Disord.*, 11, 327, 1996.

25. Nutt, J. G., Carter, J. H., and Woodward, W. R., Long-duration response to levodopa, *Neurology*, 45, 1613,1995.

26. Nutt, J. G. and Holford, N. H., The response to levodopa in Parkinson's disease: imposing pharmacological law and order, *Ann. Neural.*, 39, 561, 1996.

27. Currie, L. J. et al., Early morning dystonia in Parkinson's disease, *Neurology*, 51, 283, 1998.

28. Clarke, A. et al., A new low-dose formulation of selegiline: clinical efficacy, patient preference and selectivity for MAO-B inhibition, *J. Neural Transm.*, 110,1257, 2003.

29. Kurth, M. C., Using liquid levodopa in the treatment of Parkinson's disease. A practical guide, *Drugs Aging*, 10, 332, 1997.

30. Metman, L. V. et al., Fluctuations in plasma levodopa and motor responses with liquid and tablet levodopa/carbidopa, *Mov. Disord.*, 9, 463, 1994.

31. Pappert, E. J. et al., Liquid levodopa/carbidopa produces significant improvement in motor function without dyskinesia exacerbation, *Neurology*, 47, 1493, 1996.

32. Fahn, S., The spectrum of levodopa-induced dyskinesias, *Ann. Neural.*, 47 Suppl. 1, 2, 2000.

33. Luquin, M. R. et al., Levodopa-induced dyskinesias in Parkinson's disease: clinical and pharmacological classification, *Mov. Disord.*, 7, 117, 1992.

34. Metman, L. V. et al., Amantadine for levodopa-induced dyskinesias: a 1-year follow-up study. *Arch. Neurol*, 56, 1383, 1999.

35. Paci, C., Thomas, A., and Onofrj, M., Amantadine for dyskinesia in patients affected by severe Parkinson's disease, *Neural. Sci.*, 22, 75, 2001.

36. Ruzicka, E. et al., Amantadine infusion in treatment of motor fluctuations and dyskinesias in Parkinson's disease, *J. Neural Transm.*, 107, 1297, 2000.

37. Snow, B. J. et al., The effect of amantadine on levodopa-induced dyskinesias in Parkinson's disease: a double-blind, placebo-controlled study, *Clin. Neuropharmacol.*, 23, 82, 2000.

38. Verhagen, M. L. et al., Amantadine as treatment for dyskinesias and motor fluctuations in Parkinson's disease. *Neurology*, 50, 1323, 1998.

39. Merims, D. et al., Riluzole for levodopa-induced dyskinesias in advanced Parkinson's disease, *Lancet*, 353, 1764, 1999.

40. Durif, F. et al., Apomorphine and diphasic dyskinesia, *Clin. Neuropharmacol.*, 17, 99, 1994.

41. Durif, F., Treating and preventing levodopa-induced dyskinesias, *Drugs Aging*, 14, 337, 1999.

42. Kanovsky, P. et al., Levodopa-induced dyskinesias and continuous subcutaneous infusions of apomorphine: results of a two-year, prospective follow-up, *Mov. Disord.*, 17, 188, 2002.

43. Manson, A. J. et al., Intravenous apomorphine therapy in Parkinson's disease: clinical and pharmacokinetic observations, *Brain*, 124, 331, 2001.

44. Manson, A. J., Turner, K., and Lees, A. J., Apomorphine monotherapy in the treatment of refractory motor complications of Parkinson's disease: long-term follow-up study of 64 patients, *Mov. Disord.*, 17, 1235, 2002.

67 Management of Advanced Parkinson's Disease

James W. Tetrud
Movement Disorders Treatment Center, The Parkinson's Institute

CONTENTS

INTRODUCTION

In the early stages of Parkinson's disease (PD), motor symptoms generally predominate and, in most cases, patients experience a robust response to dopaminergic therapy. Indeed, the lack of such a response often portends one of the more aggressive "atypical" parkinsonian syndromes such as multiple system atrophy (MSA), progressive supranuclear palsy (PSP), cortical basilar ganglionic degeneration (CBG), or vascular Parkinson's (VP).[1] Following the initiation of dopaminergic therapy, patients may maintain a relatively stable course for several years with little change in their independence. However, inevitably, as the disease progresses, therapeutic strategies that originally focused on motor symptoms alone will begin to include a complex compilation of nonmotor symptoms as well (Table 67.1). Indeed, the medical management of patients with advanced PD can be one of the most challenging faced by the primary care physician, general neurologist, and/or movement disorder specialist. This chapter discusses treatment issues in patients with advanced-stage Parkinson's disease. Many of the topics discussed here are covered throughout this book;

TABLE 67.1
Symptoms of Advanced PD

Motor	Nonmotor
Bradykinesia	Cognitive decline
Postural instability	Psychoses
Tremor	Depression and anxiety
End-of-dose wearing-off	Autonomic dysfunction
Freezing	Skeletal changes
Dyskinesia	Pain
Dystonia	Sleep disorders
Myoclonus	Visual dysfunction

thus, the goal of this chapter is to provide a more condensed, but comprehensive, overview of management issues in the patient with advanced PD.

PROGRESSION OF PARKINSON'S DISEASE

Parkinson's disease is inevitably progressive, yet the clinical course can vary. Following the initiation of dopamin-

0-8493-1590-5/05/$0.00+$1.50

ergic therapy, some patients (particularly younger individuals) develop complex motor complications, others evolve a steady loss of medication response and, yet another subgroup becomes intolerant to these drugs, primarily due to cognitive decline.[2] Some patients with presumed PD may experience little or no response to antiparkinsonian drugs and eventually evolve one of a variety of atypical parkinsonian syndromes (PS) such as MSA, PSP, VP, DLB (dementia with Lewy bodies), or PFG (primary freezing of gait).[3] Even an experienced movement disorder specialist may be unable to distinguish some of these patients from those with typical PD. Indeed, patients with late-stage PD and those with atypical PS can face similar disease-related symptoms that include autonomic dysfunction, impaired cognition, dysarthria, dysphagia, sleep disorders, postural instability, freezing of gait, and disorders of the musculoskeletal system.[4]

The variability of patients' responsiveness to dopaminergic therapy and overall clinical course appears to be associated with a number of factors. These include the clinical subtype,[5] age at disease onset,[6–8] evolution of dementia,[9,10] depression,[10,11] and perhaps a host of as yet unknown genetic and environmental risk factors. For example, it is well recognized that, in general, patients with tremor-onset PD seem to have a slower progression than those with the so-called "akinetic-rigid" form of the disease.[5] This clinical observation appears to correlate with pathological findings. In a clinicopathological correlation study of autopsy confirmed PD, Jellinger[5] compared pathological changes in the substantia nigra (SN) from these two clinical subtypes of PD. He noted that, in patients with akinetic-rigid PD, cell loss was more prominent in the ventrolateral region of the SN whereas, in the tremor-dominant form, cell loss was more prominent in the medial SN, suggesting differing pathophysiological mechanisms for these two clinical subtypes. In a study comparing age of onset with evolution of PD, older-age onset patients appeared to evolve motor impairment more rapidly than younger-age onset PD patients.[12] Dementia, which may reach a prevalence as high as 75% in the late stages of PD, appears to be associated with the akinetic-rigid form of the disease[13] and has been reported to be one of the strongest predictors of death in a large study of nursing home residents.[14] Also, in a study of 97 PD patients assessed with the Beck Depression Inventory, severe depression was relatively common in patients with PD, occurring in nearly 20%.[11] In this study, the degree of depression was associated with disease severity and clinical deterioration as well as the occurrence of falls. Finally, comorbid diseases (e.g., diabetes, cerebrovascular disease, and injuries) may also modify the progression of this disease.[15]

One issue that remains controversial is whether pharmacological agents can modify the progression of PD. To date, studies aimed at assessing putative neuroprotective drugs have been inconclusive due to confounding symptomatic effects of the drug (e.g., selegiline),[16,17] insufficient prospective clinical data (e.g., amantadine),[18,19] or questions regarding reliability of imaging data (e.g., ropinirole and pramipexole).[20,21] On the other hand, there has been a persistent concern that levodopa might actually contribute to the progression of PD by means of enhanced oxidative stress and other mechanisms.[22,23] This issue remains controversial, although, to date, there is no convincing clinical evidence to support the notion that levodopa is neurotoxic.[24,25] Furthermore, in a clinical trial assessing the impact of levodopa on disease progression and motor complications, the motor and activity of daily living (ADL) scores from the Unified Parkinson's Disease Rating Scale (UPDRS) demonstrated less disability in the levodopa-treated groups compared to the placebo group, even after a four-week washout.[26] Nonetheless, it is widely recognized that dopaminergic drugs, particularly levodopa, contribute to the development of motor complications, one of the more complex treatment challenges of advanced PD.

MOTOR COMPLICATIONS IN ADVANCED PD

Several clinical trials in patients with PD have assessed the time to onset of motor complications following the introduction of levodopa and the dopamine agonists. These studies have shown that over 30% of patients will begin to experience end-of-dose wearing-off (EODWO) and/or dyskinesia within three years following initiation of levodopa therapy.[27,28] However, a large proportion of patients with advanced PD may not experience fluctuations. For example, in a community-based study of 245 PD patients, nearly 80% did not experience motor fluctuations. Of the nonfluctuators, age of onset was older, and there was a higher prevalence of dementia.[29] Nonetheless, in many patients, motor complications tend to become increasingly complex and difficult to manage as the disease advances.

Several studies have shown, rather conclusively, that initiating dopamine agonists (DA) in the early stages of the disease as monotherapy can reduce the risk of motor complications.[28,30] However, in advanced PD, nearly all patients will require levodopa with or without DA and/or other adjunctive drugs. Levodopa exerts a more potent symptomatic benefit and is less likely to induce psychiatric and other side effects compared to DA.[31] Fluctuations generally begin with EODWO but will often evolve into a more complex pattern that includes: rapid "on-off" fluctuations, delayed drug response, lack of levodopa response related to food intake, diurnal fluctuations with end-of-day worsening, and freezing episodes.[26] These complex motor fluctuations in the patient with advanced PD present an ever-increasing challenge for the clinician.[32]

Dyskinesia often emerges about the same time as motor fluctuations following the initiation of dopaminergic drugs.[27,28] Involuntary twisting of the foot on the affected side is often the first manifestation of impending dyskinesia, with subsequent spread to involve the upper extremities, trunk, head, and face, but there is variation from patient to patient.[33] The most common form of dyskinesia is associated with the time of maximum dopaminergic effect (peak-dose dyskinesia); however, dyskinesia can occur during the beginning and/or end of the dose cycle (diphasic dyskinesia) or even be associated with the "off" period, concomitant with the return of parkinsonian features (usually "off" dystonia).[34] In one study of early-morning off dyskinesia in 68 PD patients, 11 exhibited focal dystonia, and 1 exhibited chorea.[35] Interestingly, patients frequently complain of an increase in dyskinesia toward the end of the day (e.g., at dinner time).[34] During EODWO, patients can experience chorea or even ballism, but "off" dystonia occurs more frequently.[34] Myoclonus is another form of dyskinesia that can occur in advanced PD and in certain cases appears to be associated with impending cognitive decline.[36]

In addition to motor fluctuations and dyskinesia, nonmotor fluctuations can also occur and may be even more disabling for the patient (Table 67.2). In a study assessing nonmotor symptoms in 50 PD patients experiencing EODWO, Witjas and colleagues[37] reported that 66% experienced anxiety, 64% drenching sweats, 58% slowness of thinking, 56% fatigue, and 52% irritability, and nearly 30% indicated that nonmotor symptoms were more disabling than the motor fluctuations. Other symptoms reported to occur during EODWO include subjective shortness of breath, blurred vision, abdominal bloating, depression, lowered pain threshold, and akathisia.[38,39]

TABLE 67.2
Nonmotor Symptoms during "End-of-Dose Wearing-Off"

Anxiety	Abdominal bloating
Depression	Akathisia
Panic	Slowed cognition
Shortness of breath	Bladder urgency
Decreased pain threshold	Drenching sweats
Blurred vision	

The goal in managing patents with complex motor complications is to maximize "on" time while minimizing dyskinesia and other drug-related side effects (Table 67.3). However, prior to changing a patient's medication schedule, it is important for the clinician to fully understand the pattern of motor fluctuations and dyskinesia. The patient and/or caregiver will need to understand the basic reasons for the fluctuating response to medica-

TABLE 67.3
Treatment Strategies for Motor Complications

End-of-dose wearing-off	• Add dopamine agonist
	• Add COMT-inhibitor
	• Shorten dose interval
	• Add extended release CD/LD
	• Add MAO-B inhibitor
	• Take CD/LD 30–45 min. prior to meals
	• "Liquid" solution of CD/LD
	• Patient home diary
Peak-dose dyskinesia	• Add amantadine (with cognitive impairment)
	• Lower levodopa dose and add dopamine agonist, COMT inhibitor or shorten dose interval
	• Switch from extended to immediate release CD/LD
	• Discontinue selegiline
Diphasic dyskinesia	• Add amantadine
	• Add dopamine agonist
	• Consider DBS
Random and sudden on/off	• Booster dose of immediate release levodopa (dissolved in water): "liquid CD/LD"
Impaired gait and balance	• Physical therapy
• Start/stop/turn hesitation	• Ambulatory aid
• Freezing of gait ("on" and "off")	• Sensory cues
	• Home health assessment
• Postural instability	
Myoclonus	• Lower CD/LD
	• Clonazepam
	• Levetiracetam

tion and the origin of dyskinesia. For example, some patients may believe that dyskinesia is a symptom of the underlying disease process or may confuse tremor and dyskinesia. Thus, educating the patient and caregiver is often the first step in the management of these motor complications. Also, having the patient record motor fluctuations on a diary[40] can improve communication between patient and clinician. Once the pattern of motor fluctuations is understood, the clinician will need to devise a strategy individualized for the patient.

The concomitant use of dopamine agonists, levodopa, amantadine, and a COMT inhibitor may be required to maximize "on" time and minimize dyskinesia. However, because of the risk of inducing psychiatric side effects, this complex combination of drugs should be used with caution in older PD patients and those with cognitive impairment. Younger individuals are more likely to develop troublesome dyskinesia and are better able to tolerate this combination of drugs.[41] Extended-release carbidopa/levodopa may be helpful in patients experiencing wearing-off symptoms during the night; however, in patients with a "narrow therapeutic window," it may be preferable to use immediate-release carbidopa/levodopa rather than extended-release carbidopa/levodopa, since this formulation seems to provide a less predictable response (personal observation). Immediate-release carbidopa/levodopa will provide a more predictable "on" response, especially if the patient delays protein intake for 30 to 45 min following dosing or waits 60 to 90 min after a protein containing meal.[42,43] Another strategy that can be helpful for patients with complex motor fluctuations is the use of "liquid carbidopa/levodopa" whereby carbidopa/levodopa is dissolved in water or juice with a ratio of 100 mg levodopa to 100 cc of water. If this is used in larger volumes throughout the day, a simple formula is 1000 mg levodopa dissolved in 1000 cc of water along with 1000 mg ascorbic acid (to stabilize levodopa). This solution can then be used to aliquot the desired doses of levodopa using the formula 1 mg levodopa/cc. This appears to provide a more rapid absorption, presumably due to more rapid passage through the pyloric valve into the duodenum.[44] In many patients with advanced PD, it may be necessary to reduce DA or adjunctive drugs such as amantadine and/or anticholinergics and compensate by increasing levodopa and/or adding an inhibitor of catechol-O-methyl transferase (COMT).

Several studies have reported a reduction of dyskinesia in patients treated with amantadine.[45–47] In one study, patients with peak-dose and/or diphasic dyskinesia showed a mean dyskinesia reduction of 38%.[48,49] The use of 100 to 300 mg of amantadine should be considered in patients with troublesome dyskinesia while monitoring for side effects such as confusion, psychosis, and leg edema.[50] Treatment of motor complications is most difficult for those patients experiencing diphasic dyskinesia. Some have suggested using higher doses of levodopa;[51] however, this strategy can enhance peak-dose dyskinesia and has not been a widely accepted strategy.[34] The use of higher doses of DA may provide some benefit, but in many cases has been unsatisfactory.[52] Apomorphine has been reported to be of benefit in these cases due to the drug's rapid onset.[53] In selected cases, these difficult-to-treat motor complications may respond to deep brain stimulation (DBS).[54] For the patients with excessive myoclonus, clonazepam, sodium valproate, and levetiracetam have been reported to be beneficial.[55]

FREEZING OF GAIT

Freezing in PD is a phenomenon manifested by the abrupt cessation of movement. In patients with advanced PD, freezing of gait (FOG) is one of the more troublesome and perplexing symptoms, often resulting in falls and injury.[56] In a study of 172 consecutive patients 5 or more years following the diagnosis of PD, 53% reported FOG.[57] This phenomenon appears to be associated with gait initiation, change of gait direction, and turning.[58] Although FOG is often associated with EODWO, as the disease progresses, patients often experience this phenomenon even during the "on" state. The underlying pathophysiology of FOG remains elusive and relatively resistant to therapeutic intervention. In a clinicopathological correlation study of patients with atypical parkinsonism (PSP, MSA, CBD, and DLB), 47% exhibited FOG on their last examination but, at autopsy, no consistent neuropathological substrate was detected.[59]

There is no specific pharmacological treatment proven effective for this phenomenon. However, some novel strategies have been attempted. In a pilot study, injection of botulinum toxin into the calf[60] was reported to be of some benefit. Also, treatment with pharmacological agents such as L-threo-dops[61,62] and selegiline[57] has been reported to reduce the incidence of FOG. In a prospective assessment of the DATATOP cohort, the use of deprenyl (selegiline), in the absence of levodopa, was associated with a lower incidence of freezing.[57] A subsequent review of patients in the same study cohort following initiation of levodopa also revealed a reduction of FOG in the selegiline treated group.[63]

Nonpharmacological strategies such as the use of sensory cues have been reported to benefit some patients. The use of auditory signals such as a metronome or even taped march music can help the patient focus on cadence.[64,65] Tactile stimuli may also be of some benefit. For example, touching a patient lightly on the shoulder or arm may facilitate the initiation of a step (personal observation). Dogs trained to place a paw on the dorsum of a patient's foot to unlock a FOG episode was observed to reduce falls from FOG.[66] Dietz and colleagues[67] reported the effect of walking time and freezing episodes in PD patients using an inverted walking stick. They concluded that a subpopulation of patients showed continued reduction of freezing episodes using this technique. However, in a similar study using a "modified inverted stick" and a "visual laser beam stick," Kompoliti and colleagues[68] studied episodes of FOG and walking time in PD patients during the "on" state. They concluded that these visual cues did not provide significant reduction of "on" freezing. Finally, some reports suggest a reduction in FOG episodes after deep brain stimulation (DBS) of the subthalamic nucleus; however, whether this beneficial effect includes "on" as well as "off" FOG remains controversial.[69,70]

COGNITIVE DECLINE

There are conflicting reports on the incidence and prevalence of dementia in PD; however, one of the grim realities of the disease is that a substantial proportion of patients will experience cognitive decline to some degree as the disease progresses. Based on a 10-year longitudinal study,

Hughes and colleagues reported dementia occurring in 17 of 83 patients (about 20%) who were nondemented at baseline.[71] In a community-based, prospective study of 171 Norwegian PD patients who were nondemented on entry into the study, 43 (25%) were classified as demented (DSM-III-R criteria) 4.2 years later, with a nearly sixfold increased risk compared to controls subjects.[72] This same group published results from an 8-year longitudinal study of 224 patients with PD without cognitive impairment at disease onset and detected a prevalence of 26% dementia at baseline, but 78.2% dementia after 8 years.[13] These data suggest that cognitive decline may affect the majority of PD patients in the later stages of the disease. Even in patients with advanced PD without dementia or depression, comprehensive neuropsychological measures detect a range of cognitive defects, particularly impaired frontal lobe function.[73] The pathological substrate of dementia in PD appears to be complex with Alzheimer's pathology admixed with cortical Lewy body pathology.[74,75]

As with any patient exhibiting signs of cognitive decline, factors other than PD should be ruled out (Table 67.4). Importantly, the patient's mood should be addressed, since depression is common in PD, and severe depression may further complicate and even mimic dementia.[76,77] In this regard, comprehensive neuropsychological testing can be helpful in assessing the type and degree of cognitive impairment as well as the presence of depression. Brain imaging, basic metabolic screening, B12, and heavy metal assay should be considered. The patient's medication regimen should be evaluated to determine if any drugs might be affecting cognitive func-tion. Certain antiparkinsonian drugs such as selegiline, amantadine, and anticholinergics may need to be tapered in favor of a simpler regimen of levodopa alone.[78,79] Other medications commonly used in PD patients such as tri-cylic antidepressants, oxybutynin or tolterodine, and ben-zodiazepams may need to be avoided. Insomnia is another common symptom of PD that can have an impact on cog-nitive function.[80] In such cases, a trial of mild sedatives may be of some benefit, but, in more severe cases, a for-mal sleep study should be considered. Hearing impair-ment is common in the elderly, often unrecognized by the patients and a condition that can add to disability in patients with cognitive impairment.[81] The clinician should encourage such patients to undergo a hearing evaluation and use of a hearing aid, if needed.[81]

Despite the high prevalence of dementia in PD, there is a relative paucity of data on therapeutic interventions for this devastating feature of late-stage PD.[82,83] The use of cholinesterase inhibitors (donezepil, rivastigmine, galan-tamine) would seem reasonable in view of the fact that this class of drugs seems to be effective in DLB[84] and Alzhe-imer's disease.[85] Memantine, an NMDA receptor antago-nist, has been reported to be of benefit in moderate to advanced Alzheimer's disease and might be a candidate for clinical trials in PD patients with cognitive impairment.[86]

PSYCHIATRIC SYMPTOMS

PSYCHOSES

Psychotic symptoms are generally induced by antiparkin-sonian medication but are of much higher prevalence in patients who are cognitively impaired.[87] Amantadine, anti-cholinergic, and DA are more likely associated with psy-chotic symptoms than levodopa alone.[83] In nearly all cases, visual hallucinations and/or illusions will predom-inate, often with the patient expressing insight.[88,89] These "benign" hallucinations and illusions are often well toler-ated but should alert the clinician that a change in anti-parkinsonian medication or treatment with an antipsy-chotic medication may be required. These psychotic symptoms are often associated with cognitive decline and should prompt close observation by the clinician and car-egiver to look for more flagrant psychotic symptoms. In patients with DLB, psychotic symptoms may occur with or without antiparkinsonian medication and are often asso-ciated with fluctuations in cognition, disturbance in con-sciousness, and parkinsonism.[90]

Treatment of psychotic symptoms should start with adjustment of the antiparkinsonian drugs. For example, amantadine and anticholinergics may need to be discon-tinued and DA reduced. The use of atypical antipsychot-ics such as clozapine or quentiapine can be quite effective in small doses ranging from 6.25 to 100 mg[91] and, because these drugs are sedating, they are generally better

---

TABLE 67.4
Treatment of Cognitive and Psychiatric Symptoms of Advanced PD

Cognitive decline	• Assess and treat depression
	• Simplify dopaminergic therapy (use CD/LD only)
	• Taper anticholinergic drugs and amantadine
	• Consider discontinuing oxybutynin or tolterodine
	• Add cholinesterase inhibitor
	• Psychological counseling for patient and caregiver
Depression	• SSRIs
	• Bupropion
	• Tricyclic antidepressants
	• Psychiatric/psychological counseling
Anxiety	• Benzodiazepines
	• Treat EODWD
Psychoses	• Taper dopamine agonists
	• Taper amantadine
	• Taper anticholinergics
	• Discontinue MAO-B inhibitors
	• Switch to CD/LD
	• Clozapine (6.25–50 mg HS)
	• Quentiapine (12.5–100 mg HS)

tolerated if administered at bedtime. If psychosis becomes more severe, the use of small amounts of these drugs during the day may be required; however, sedation, particularly in patients with cognitive impairment, will be a limitation. Patients with PD who undergo hospitalization for a variety of reasons appear to be at risk for developing psychiatric symptoms.[92] Thus, preemptive reduction of dopaminergic medication should be considered, particularly in patients with cognitive impairment. Other atypical antipsychotics such as olanzapine, resperidone, and aripiprazole may be required in certain patients, but these drugs are more likely to worsen parkinsonism.[93,94]

DEPRESSION, ANXIETY, AND AGITATION

Although the reported prevalence of depression varies widely, the generally accepted figure is about 40%.[95] Generally, depression in patients with PD is considered mild to moderate; however, a community-based assessment of depression in 97 PD patients using the Beck Depression Inventory revealed a nearly 20% prevalence of severe depression that correlated with advancing disease.[11] Thus, recognition of this symptom and discussion with the patient and family regarding appropriate therapy is an important aspect of managing patients with advanced PD. There is a wide range of antidepressant pharmacotherapy available, including the tricyclic antidepressants, selective serotonin reuptake inhibitors (SSRIs), and monoamine oxidase type A and B inhibitors. With the exception of nortriptyline, there is a paucity of reliable efficacy data on most of these drugs.[96] Nonetheless, the use of SSRIs and tricycliclic antidepressants is routine in clinical practice.[97] Psychiatric or psychological consoling can be helpful for the patient and caregiver and, since the caregiver is an important member of the treatment team, the clinician should always be sensitive to and supportive of caregiver burnout.[98] In severe cases, ECT has been reported to be of some benefit,[99] and repetitive transcranial magnetic stimulation (rTMS) may be as effective, but with fewer side effects.[100,101]

AUTONOMIC DYSFUNCTION

Autonomic dysfunction is also a variable symptom in PD patients (Table 67.5). Pathologically, Lewy bodies appear to be widespread throughout the autonomic nervous system and have been found in the dorsal vagal nucleus, sympathetic ganglion, enteric nervous system of the alimentary tract, the parasympathetic submandibular ganglion, cardiac plexus, pelvic plexus, and adrenal medulla.[102–104] One of the most common manifestations of autonomic dysfunction is bladder urgency, often leading to incontinence. Medications such as tolteridine and oxybutinin can be of benefit, particularly for patients experiencing nighttime urgency; however, these drugs have anti-

TABLE 67.5
Autonomic Symptoms in PD

Bladder urgency	• Oxybutynin
	• Tolterodine
Constipation	• Stool softener (docusate sodium), etc.
	• Adequate fluid intake
	• Ample fruits and vegetables
	• Exercise
	• Psyllium fiber
	• Lactulose
	• Polyethylene glycol
Orthostatic hypotension	• Flurocortisone 0.1 mg with sup K+
	• Midodrine 2.5–10 mg/day
Diaphoresis, heat intolerance	• Small amounts of propranolol
Sexual dysfunction (males)	• Sildenafil
Sialorrhea	• Glucopyrrolate
	• Botulinum toxin
Weight loss	• Protein supplementation
	• Reassurance

cholinergic properties and should be used with caution in patients with cognitive impairment.[105]

Another common symptom associated with autonomic dysfunction is constipation, often occurring long before the diagnosis of PD,[106] a symptom that is often exacerbated by most antiparkinsonian drugs.[107] Treatment for constipation may require lifestyle changes for some patients. For example, changing to a high-fiber diet, the use of a stool softener, attention to adequate fluid intake, and exercise may be necessary to minimize this problem.[108,109] However, some patients may require more rigorous measures such as lactulose, polyethylene glycol 3350, or enemas.[110]

Orthostatic hypotension, although generally not as severe a symptom as in MSA, may be problematic for some patients with PD. Dopaminergic drugs may exacerbate underlying autonomic dysfunction and may need to be adjusted. Drugs such as fludrocortisone or midodrine may be required in some cases.[111] More aggressive measure such as the use of alpha-adrenergic agonist clonadine, indomethacin, and ephedrine are usually reserved for patients with MSA and rarely required in typical PD.[111] Patients with sexual dysfunction may experience benefit from sildenafil, although this drug is contraindicated in patients treated with nitrates or symptomatic orthostatic hypotension.[112] Other autonomic symptoms may include cold limbs, diaphoresis, seborrhea, and sialorrhea.[112] Weight loss is common in PD and a frequent concern among family members, patients, and primary caregivers. The underlying pathophysiology of this phenomenon is poorly understood but does not appear to be related to reduced food intake.[114] Perhaps the widespread gastrointestinal dysfunction that occurs in PD may be a factor.[108] Diaphoresis can occur in some patients, but it is usually associated with the "off" state.[115,116]

SLEEP DISORDERS

Disorders of sleep are quite common in patients with PD, with a reported prevalence between 70 and 90%.[117,118] Sleep disorders in PD range from difficulty with sleep initiation and maintenance to parasomnias and excessive daytime sleepiness (EDS). It is estimated that about 20% of PD patients experience restless leg syndrome (RLS), a syndrome that may precede the diagnosis of PD and is generally responsive to dopaminergic mediations.[119] Other medications used to treat RLS include opiates and benzodiazepines.[120] Insomnia is complicated by bladder urgency, but other factors such as medication effect should be considered.[117] REM sleep behavioral syndrome (RBD) seems to be extremely common in PD patients and generally is more annoying for the spouse than the patient. Symptoms can range from talking during sleep to complex movements and even striking out. RBD appears to be the result of disinhibition of the usual motor paralysis that occurs during REM sleep and likely related to dream content.[121,122] Although it is believed that RBD is associated with PD pathophysiology, the addition of antiparkinsonian medication may contribute to or exacerbate this syndrome.[123] Since the patient is generally unaware of RBD, treatment is often aimed at helping the patient's partner. Benzodiazepines such as clonazepam have been reported to reduce RBD activity.[120]

Another sleep-related disorder in patients with PD is EDS. This symptom is likely multifactorial, related to the lack of adequate nighttime sleep and to PD medications.[124] All dopaminergic drugs can contribute to EDS,[125] although DA appear to be more likely than levodopa.[126] For example, in patients enrolled in two clinical trials of DA compared to levodopa, the percentage of daytime somnolence was 32% for pramipexole 32.4% versus 17.3% for LD,[28] and 27.5% for Ropinirole 27.4% versus 19.1% for LD.[30] A major concern of EDS is falling asleep while driving. Thus, the patient and family should be urged to carefully monitor driving habits. In some cases, driving may need to be curtailed and medication adjustments pursued. Reduction of DA, switching from one DA to another, and/or reducing levodopa may be required. In certain cases, DA may need to be discontinued.[127] The use of caffeine may provide some benefit, although insomnia may result. A trial of modafinil, a drug designed to treat narcolepsy, may also be helpful.[127]

DYSPHAGIA AND DYSARTHRIA

Dysphagia is common in advanced PD and appears to be associated with impaired triggering of the swallowing reflex as well as prolonged laryngeal movement.[128] In patients experiencing choking, frequent coughing, or difficulty tolerating larger pieces of food, a comprehensive dysphagia evaluation should be undertaken to determine the origin and severity of dysphagia as well as the proper treatment strategy.[129,130] With the high incidence of gastroesophageal reflux disease (GERD), esophageal stricture should be considered along with other esophageal disorders.[131] Even in asymptomatic PD patients, a regular assessment of swallowing ability may be warranted because of the serious consequences of aspiration.[129,130] In more severe cases, video fluoroscopy using varying consistencies of food may be required to determine optimal intervention.[132] A speech therapist can provide the patient with certain maneuvers such as keeping the head flexed when swallowing and/or learning to exert a preemptive cough after swallowing, to minimize aspiration. Another treatment modality that may be effective is "sour taste stimulation" of the pharynx and larynx, but clinical trials will be needed to determine the efficacy of this technique. In anesthetized rats, acetic acid facilitated reflex swallowing via the superior laryngeal nerve and glossophayngeal nerve.[133] Drooling is another common symptom of late-stage PD associated with dysfunctional swallowing.[129] Sublingual atropine and injections of botulinum toxin have been reported to reduce drooling in some studies.[134,135] The use of oral anticholinergics should be used with caution in patients with impaired cognitive function.

Dysarthria affects the great majority of patients with advanced PD and can be the source of social isolation in some patients.[136] Although voice changes have commonly been described as monotone, patients more often will experience a hoarse, raspy quality with low volume and impaired articulation.[137,138] Even so, patients may not realize that others have difficulty understanding them.[139,140] Speech therapy is frequently advocated for PD patients;[140] however, there is no clear objective evidence that a specific technique is either superior or provides long-term benefit.[141] One of the most widely advocated techniques is the Lee Silverman voice therapy program (LSVT), designed to increase vocal cord adduction and loudness. In a study of 35 PD patients over a 13-month period, improvement in speech volume persisted at follow-up evaluations of 6 and 12 months.[142] Although the long-term benefit of speech therapy is unknown, it seems reasonable to encourage voice exercises for PD patients in an attempt to maintain tone in the muscles of voice production. A simple program that appears to have benefited some patients includes reciting from a text, singing, or even creating a speech training station at home with a tape recorder, mirror, several paragraphs of text, and a metronome (personal observation). Finally, percutaneous laryngeal collagen augmentation has been reported to benefit some patients.[143]

MUSCULOSKELETAL CHANGES

A commonly ignored condition in advanced PD is the development of musculoskeletal symptoms. Progressive bradykinesia can reduce the daily repertoire of stretching and joint movement, resulting in shortening of ligaments and muscle tissue, placing the patient at risk for injury to muscles and joints.[144] Thus, it is not surprising that shoulder, hip, and knee pain are common in PD patients. In addition, the evolution of scoliosis appears to be quite common in advanced PD, although there is a paucity of data on the prevalence or incidence of this phenomenon. In an outpatient evaluation of 21 patients with parkinsonism, Duvoisin and Marsden[145] and Marsden and Duvoisin[146] noted scoliosis in 19 patients and commented that the direction of postural deviation correlated with the major signs of parkinsonism. Camptocormia, a rare condition manifested by an exaggerated stooped posture, occasionally occurs in PD. Djaldetti and colleagues[147] reported eight cases of severe forward flexion in PD patients and noted that the symptoms were most prominent while walking, but resolved when recumbent, suggesting that this phenomenon may be a form of dystonia.[148] The use of a staff tends to bring the posture more upright in some patients (personal observation). These musculoskeletal changes clearly affect the quality of life in PD patients and need to be better characterized.

VISUAL DYSFUNCTION

Visual symptoms are common in patients with advanced PD and can be the source of great concern for patient and caregiver, although routine ophthalmological evaluations are unlikely to detect the cause. A number of studies have documented an underlying defect in retinal function. For example, levodopa administration has been reported to normalize abnormal visual evoked responses in PD patients.[149] Also, in a study of 39 patients with PD, a loss of contrast sensitivity was detected in 64%, suggesting "widespread neurotransmitter deficiency."[150] Other studies documented "orientation-selective visual loss" suggesting visual cortex involvement in PD.[151–153] Diplopia with visual blurring is commonly mentioned by patients and seems to be associated with the "off" state (personal observation). This symptom might be related to an underlying dopamine-sensitive ocular motility disorder.[154] Thus, visual dysfunction is another issue faced by the clinician in managing advanced PD. Unfortunately, there are, as yet, no specific therapies available other than optimizing the patient's "on" state. Nonetheless, it is reasonable to have these symptoms reviewed by a neuro-ophthalmologist to rule out other causes.

PSYCHOSOCIAL ISSUES

Late-stage PD not only impacts on the patient, but the caregiver and family members as well.[155] Ideally, the services of a social worker, psychiatrist, and/or psychologist should be sought; however, the patient's primary care physician and neurologist are often the only source of support and advice regarding many psychosocial issues. One issue that inevitably arises is driving competency. If there is a question about the patient's driving skills, it is often helpful to have the family arrange a driving test from a local driving school. This is generally nonthreatening for the patient and can provide valuable insight for the family regarding the patient's driving skill.

Another issue that emerges at this stage in the disease is the need for either home care assistance or an assisted care facility for the patient and respite for the overwhelmed caregiver. At this point, it is often helpful to have a discussion with members of the patient's family, if possible, to determine the degree of support available to the patient and caregiver. A family conference can be particularly helpful when both the patient and caregiver are in poor health and need assisted living. Fortunately, there is a well organized network of Parkinson's disease support groups throughout North America and other countries that can provide support and advise for patient and caregiver. These psychosocial issues can be complicated and difficult to resolve, but the primary care physician and neurologist can play a major role in easing the burden.

SUMMARY

The treatment of PD in the advanced stage is likely one of the most daunting challenges faced by the neurologist. Because of the ever-changing sequence of medical and social problems faced by the patient and caregiver as the disease progresses, the medical management becomes complex and demanding. Fortunately, there is a growing appreciation and understanding of the multifaceted nature of this disease, and a number of treatment strategies that have become available to treat both the motor and nonmotor symptoms (see Table 67.6). Over the past decade, drugs that reduce motor fluctuations, dyskinesia, depression, psychosis, sleep disorders, and that enhance cognition, have been developed. Nonetheless, certain aspects of the disease, such as impaired balance, freezing, dementia, and autonomic dysfunction, are poorly treated. Furthermore, the motor and nonmotor features are admixed with social and psychological issues for the patient and caregiver. Therefore, the treatment of advanced PD is not for the faint of heart and can be challenging for even the most experience clinician. Hopefully, more effective therapeutic strategies will emerge out of the ongoing surgical and pharmacological clinical trials currently underway to improve the quality of life for our patients with advanced Parkinson's disease.

TABLE 67.6
Miscellaneous Symptoms of Advanced PD

REM sleep behavioral syndrome	• Clonazepam 0.25–1 mg
Excessive daytime sleepiness	• Caffeine • Modanifil • Reduce dopamine agonist
Restless leg syndrome	• Dopamine agonists • Levodopa • Opiates • Benzodiazepines
Insomnia	• Temazepam • Zolpidem • Zaleplon
Dysphagia	• Video esophageal fluoroscopy • Head flexed while swallowing • Preemptive cough following swallow • Soar stimulation of pharynx?
Dysarthria	• Speech therapy • Voice exercises • Aerobic exercises
Camptocormia	• Use of upright walker or staff
Drooling	• Sublingual atropine • Botulinum toxin injections

REFERENCES

1. Poewe, W., Wenning, G., The differential diagnosis of Parkinson's disease, *Eur. J. Neurol.*, 9 (Suppl. 3), 23, 2002.

2. Fahn, S., *Consensus? How to proceed in treatment in treatment today. Conclusions,* in Rinne, U. K., Nagatsu, T., Horowski, R., Eds., International workshop Berlin Parkinson's disease, Bussum, The Netherlands, Medicom Europe, 368, 1991.

3. Factor, S. A., Jennings, D. L., Molho, E. S., Marek, K. L., The natural history of the syndrome of primary progressive freezing gait, *Arch Neurol.*, 59, 1778, 2002.

4. Brooks, D. J., Diagnosis and management of atypical parkinsonian syndromes, *J. Neurol. Neurosurg. Psychiatry.*, 72 (Suppl. 1), I10, 2002.

5. Jellinger, K. A., Post mortem studies in Parkinson's disease—is it possible to detect brain areas for specific symptoms? *J. Neural. Transm.,* Suppl., 56:1, 1999.

6. Gibb, W. R., Lees, A. J., A comparison of clinical and pathological features of young- and old-onset Parkinson's disease, *Neurology,* 38, 1402, 1988.

7. Gasparoli, E., Delibori, D., Polesello, G., Santelli, L., Ermani, M., Battistin, L., Bracco, F., Clinical predictors in Parkinson's disease, *Neurol. Sci.*, 23, (Suppl. 2), S77, 2002.

8. Wang, X., You, G., Chen, H., Cai, X., Clinical course and cause of death in elderly patients with idiopathic Parkinson's disease, *Chin. Med. J. (Engl).*, 115, 1409, 2002.

9. Jellinger, K. A., Recent developments in the pathology of Parkinson's disease. *J. Neural. Transm.,* Suppl., 62, 347, 2002.

10. Poewe, W. H., Wenning, G. K., The natural history of Parkinson's disease, *Ann. Neurol.*, 44 (Suppl. 1), S1, 1998.

11. Schrag, A., Jahanshahi, M., Quinn, N. P., What contributes to depression in Parkinson's disease? *Psychol. Med.*, 31, 65, 2001.

12. Diederich, N. J., Moore, C. G., Leurgans, S. E., Chmura, T. A., Goetz, C. G., Parkinson disease with old-age onset: a comparative study with subjects with middle-age onset, *Arch Neurol.*, 60, 529, 2003.

13. Aarsland, D., Andersen, K., Larsen, J. P., Lolk, A., Kragh-Sorensen P., Prevalence and characteristics of dementia in Parkinson disease: an 8-year prospective study, *Arch Neurol.*, 60, 387, 2003.

14. Fernandez, H. H., Lapane, K. L., Predictors of mortality among nursing home residents with a diagnosis of Parkinson's disease, *Med. Sci. Monit.*, 8, CR241, 2002.

15. Pressley, J. C., Louis, E.D., Tang, M. X., Cote L., Cohen, P. D., Glied, S., Mayeux, R., The impact of co-morbid disease and injuries on resource use and expenditures in parkinsonism, *Neurology,* 14, 60, 87, 2003.

16. The Parkinson Study group, Effect of deprenyl on the progression of disability in early Parkinson's disease, *N. Engl. J. Med.*, 321, 1364, 1989.

17. Parkinson Study Group, Effects of tocopherol and deprenyl on the progression of disability in early Parkinson's disease, *N. Eng.l J. Med.,* 328, 176, 1993.

18. Uitti, R. J., Rajput, A. H., Ahlskog, J. E., Offord, K. P., Schroeder, D. R., Ho, M. M., Prasad, M., Rajput, A., Basran, P., Amantadine treatment is an independent predictor of improved survival in Parkinson's disease, *Neurology,* 46, 1551, 1996.

19. Kieburtz, K., Designing neuroprotection trials in Parkinson's disease, *Ann. Neurol.*, 53 (Suppl. 3), S100, 2003.

20. Ahlskog, J. E., Slowing Parkinson's disease progression: recent dopamine agonist trials, *Neurology,* 60, 381, 2003.

21. Albin, R. L., Frey, K. A., Initial agonist treatment of Parkinson disease: a critique, *Neurology,* 60, 390, 2003.

22. Agid, Y., Olanow, C. W., Mizuno, Y., Levodopa: why the controversy? *Lancet,* 360, 575, 2002.

23. Olanow, C. W., Stocchi, F., Why delaying levodopa is a good treatment strategy in early Parkinson's disease, *Eur. J. Neurol.*, 7 (Suppl. 1), 3, 2000.

24. Rajput, A. H., Fenton, M., Birdi, S., Macaulay, R., Is levodopa toxic to human substantia nigra? *Mov. Disord.*, 12, 634, 1997.

25. Schapira, A. H., Olanow, C. W., Rationale for the use of dopamine agonists as neuroprotective agents in Parkinson's disease, *Ann. Neurol.*, 53 (Suppl. 3), S149, 2003.

26. Fahn, S., Description of Parkinson's disease as a clinical syndrome, *Ann. N. Y. Acad. Sci.*, 991, 1, 2003.

27. The Parkinson Study Group, Impact of deprenyl and tocopherol treatment on Parkinson's disease in DATATOP subjects not requiring levodopa, *Ann. Neurol.*, 39, 29, 1996.

28. Parkinson Study Group, Pramipexole vs. levodopa as initial treatment for Parkinson's disease: a randomized controlled trial, *JAMA*, 284, 1931, 2000.

29. Larsen, J. P., Karlsen, K., Tandberg, E., Clinical problems in non-fluctuating patients with Parkinson's disease: a community-based study, *Mov. Disord.*, 15, 826, 2000.

30. Roscole, O., Brooks, D., Korczyn, A. D. et al., A five-year study of the incidence of dyskinesia in patients with early Parkinson's disease treated with ropinirole or levodopa, *N. Eng. J. Med.*, 342, 1484, 2000.

31. Kuzuhara, S., Drug-induced psychotic symptoms in Parkinson's disease, Problems, management and dilemma, *J. Neurol.*, 248 (Suppl. 3), III28, 2001.

32. Nutt, J. G., Motor fluctuations and dyskinesia in Parkinson's disease, *Parkinsonism Relat. Disord.*, 8, 101, 2001.

33. Vidailhet, M., Bonnet, A. M., Marconi, R., Durif, F., Agid, Y., The phenomenology of L-dopa-induced dyskinesias in Parkinson's disease, *Mov. Disord.*, 14 (Suppl. 1), 13, 1999.

34. Fahn, S., The spectrum of levodopa-induced dyskinesias, *Ann. Neurol.*, 47 (Suppl. 1), S2, 2000.

35. Cubo, E., Gracies, J. M., Benabou, R., Olanow, C. W., Raman, R., Leurgans, S., Goetz, C. G., Early morning off-medication dyskinesias, dystonia, and choreic subtypes, *Arch Neurol.*, 58, 1379, 2001.

36. Caviness, J. N., Adler, C. H., Newman, S., Caselli, R. J., Muenter, M. D., Cortical myoclonus in levodopa-responsive parkinsonism, *Mov. Disord.*, 13, 540, 1998.

37. Witjas, T., Kaphan, E., Azulay, J. P., Blin, O., Ceccaldi, M., Pouget, J., Poncet, M., Cherif, A. A., Nonmotor fluctuations in Parkinson's disease: frequent and disabling, *Neurology*, 59, 408, 2002.

38. Burkhard, P. R. and Tetrud, J. W., The spectrum of nonmotor fluctuations in Parkinson's disease, *Mov. Disord*, 13 (Suppl. 2) (abstract), 1998.

39. Raudino, F., Non motor off in Parkinson's disease, *Acta Neurol. Scand.*, 104, 312, 2001.

40. Hauser, R. A., Friedlander, J., Zesiewicz, T. A., Adler, C. H., Seeberger, L. C., O'Brien, C. F., Molho, E. S., Factor, S. A., A home diary to assess functional status in patients with Parkinson's disease with motor fluctuations and dyskinesia, *Clin. Neuropharmacol.*, 23, 75, 2000.

41. Jankovic, J., Tintner, R., Dystonia and parkinsonism, *Parkinsonism Relat. Disord.*, 8, 109, 2001.

42. Carter, J. H., Nutt, J. G., Woodward, W. R., Hatcher, L. F., Trotman, T. L., Amount and distribution of dietary protein affects clinical response to levodopa in Parkinson's disease, *Neurology*, 39, 552, 1989.

43. Contin, M., Riva, R., Martinelli, P., Albani, F., Baruzzi, A., Effect of meal timing on the kinetic-dynamic profile of levodopa/carbidopa controlled release [corrected] in parkinsonian patients, *Eur. J. Clin. Pharmacol.*, 54, 303, 1998

44. Kurth, M. C., Tetrud, J. W., Tanner, C. M., Irwin, I., Stebbins, G. T., Goetz, C. G., Langston, J. W., Double-blind, placebo-controlled, crossover study of duodenal infusion of levodopa/carbidopa in Parkinson's disease patients with "on-off" fluctuations, *Neurology*, 43 1698, 1993.

45. Rajput, A. H., Rajput, A., Lang, A. E., Kumar, R., Uitti, R. J., Galvez-Jimenez, N., New use for an old drug: amantadine benefits levodopa-induced dyskinesia, *Mov. Disord.*, 13, 851, 1998.

46. Chase, T. N., Oh, J. D., Konitsiotis, S., Antiparkinsonian and antidyskinetic activity of drugs targeting central glutamatergic mechanisms, *J. Neurol.*, 247 (Suppl. 2), II36, 2000.

47. Snow, B. J., Macdonald, L., Mcauley, D., Wallis, W., The effect of amantadine on levodopa-induced dyskinesias in Parkinson's disease: a double-blind, placebo-controlled study, *Clin. Neuropharmacol.*, 23, 82, 2000.

48. Paci, C., Thomas, A., Onofrj, M., Amantadine for dyskinesia in patients affected by severe Parkinson's disease, *Neurol. Sci.*, 22, 75, 2001.

49. Blanchet, P. J., Metman, L. V., Chase, T. N., Renaissance of amantadine in the treatment of Parkinson's disease, *Adv. Neurol.*, 91, 251, 2003.

50. Verhagen, Metman L., Del Dotto, P., van den Munckhof, P., Fang, J., Mouradian, M. M., Chase, T. N., Amantadine as treatment for dyskinesias and motor fluctuations in Parkinson's disease, *Neurology*, 50, 1323, 1998.

51. Lhermitte, F., Agid, Y., Signoret, J. L., Onset and end-of-dose levodopa-induced dyskinesias. Possible treatment by increasing the daily doses of levodopa, *Arch Neurol.*, 35, 261, 1978.

52. Quinn, N., Drug treatment of Parkinson's disease, *B. M. J.*, 310, 575, 1995.

53. Durif, F., Deffond, D., Dordain, G., Tournilhac, M., Apomorphine and diphasic dyskinesia, Clin. *Neuropharmacol.*, 17, 99, 1994.

54. The Movement Disorder Society Task Force, (Goetz, C. G., Koller, W. C., Poewe, W., Roscol, O., Sampaio, et al.,) Surgical treatment for Parkinson's disease: Deep brain surgery in Management of Parkinson's Disease: An evidence-based review, *Mov. Disord.* 17 (Suppl. 4), S128, 2002.

55. Van Zandijcke, M., Treatment of myoclonus, *Acta Neurol. Belg.*, 103, 66, 2003.

56. Giladi, N., Treves, T. A., Simon, E. S., Shabtai, H., Orlov, Y., Kandinov, B., Paleacu, D., Korczyn, A. D., Freezing of gait in patients with advanced Parkinson's disease, *J. Neural. Transm.*, 108, 53, 2001a.

57. Giladi, N., McDermott, M. P., Fahn, S., Przedborski, S., Jankovic, J., Stern, M., Tanner, C., Parkinson Study Group. Freezing of gait in PD: prospective assessment in the DATATOP cohort, *Neurology*, 56, 1712, 2001b.

58. Nieuwboer, A., Dom, R., De Weerdt, W., Desloovere, K., Fieuws, S., Broens-Kaucsik, E., Abnormalities of the spatiotemporal characteristics of gait at the onset of freezing in Parkinson's disease, *Mov. Disord.*, 16, 1066, 2001.

59. Muller, J., Seppi, K., Stefanova, N., Poewe, W., Litvan, I., Wenning, G. K., Freezing of gait in postmortem-confirmed atypical parkinsonism, *Mov. Disord.*, 17, 1041, 2002.

60. Giladi, N., Gurevich, T., Shabtai, H., Paleacu, D., Simon, E. S., The effect of botulinum toxin injections to the calf muscles on freezing of gait in parkinsonism: a pilot study, *J. Neurol.*, 248, 572, 2001c.

61. Narabayashi, H., [L-threo-DOPS therapy and parkinsonism], *No To Shinkei*, 38(1):60–2, January, 1986.

62. Kondo, T., L-threo-DOPS in advanced parkinsonism, *Adv. Neurol.*, 60, 660, 1993.

63. Shoulson, I., Oakes, D., Fahn, S., Lang, A., Langston, J. W., LeWitt, P., Olanow, C. W., Penney, J. B., Tanner, C., Kieburtz, K., Rudolph, A., Parkinson Study Group. Impact of sustained deprenyl (selegiline) in levodopa-treated Parkinson's disease: a randomized placebo-controlled extension of the deprenyl and tocopherol antioxidative therapy of parkinsonism trial, *Ann. Neurol.*, 51, 604, 2002.

64. McIntosh, G. C., Brown, S. H., Rice, R. R., Thaut, M. H., Rhythmic auditory-motor facilitation of gait patterns in patients with Parkinson's disease, *J. Neurol. Neurosurg. Psychiatry*, 62, 22, 1997.

65. Enzensberger, W., Oberlander, U., Stecker, K., Metronome therapy in patients with Parkinson disease, *Nervenarzt*, 68, 972, 1997.

66. Stern, M., personal communication.

67. Dietz, M. A., Goetz, C. G., Stebbins, G. T., Evaluation of a modified inverted walking stick as a treatment for parkinsonian freezing episodes, *Mov. Disord.*, 5, 243, 1990.

68. Kompoliti, K., Goetz, C. G., Leurgans, S., Morrissey, M., Siegel, I. M., "On" freezing in Parkinson's disease: resistance to visual cue walking devices, *Mov. Disord.*, 15, 309, 2000.

69. Stolze, H., Klebe, S., Poepping, M., Lorenz, D., Herzog, J., Hamel, W., Schrader, B., Raethjen, J., Wenzelburger, R., Mehdorn, H. M., Deuschl, G., Krack, P., Effects of bilateral subthalamic nucleus stimulation on parkinsonian gait, *Neurology*, 57, 144, 2001.

70. Su, P. C., Tseng, H. M., Gait freezing and falling related to subthalamic stimulation in patients with a previous pallidotomy, *Mov. Disord.*, 16, 376, 2001.

71. Hughes, T. A., Ross, H. F., Musa, S., Bhattacherjee, S., Nathan, R. N., Mindham, R. H., Spokes, E. G., A 10-year study of the incidence of and factors predicting dementia in Parkinson's disease, *Neurology*, 54, 1596, 2002.

72. Aarsland, D., Andersen, K., Larsen, J. P., Lolk, A., Nielsen, H., Kragh-Sorensen, P., Risk of dementia in Parkinson's disease: a community-based, prospective study, *Neurology*, 56, 730, 2001.

73. Green, J., McDonald, W. M., Vitek, J. L., Evatt, M., Freeman, A., Haber, M., Bakay, R. A., Triche, S., Sirockman, B., DeLong, M. R., Cognitive impairments in advanced PD without dementia, *Neurology*, 59, 1320, 2002.

74. Apaydin, H., Ahlskog, J. E., Parisi, J. E., Boeve, B. F., Dickson, D. W., Parkinson disease neuropathology: later-developing dementia and loss of the levodopa response, *Arch Neurol.*, 59, 102, 2002.

75. Perl, D. P., Olanow, C. W., Calne, D., Alzheimer's disease and Parkinson's disease: distinct entities or extremes of a spectrum of neurodegeneration? *Ann. Neurol.*, 44, (Suppl. 1), S19, 1998.

76. Gainotti, G., Marra, C., Villa, G., Parlato, V., Chiarotti, F., Sensitivity and specificity of some neuropsycholog-ical markers of Alzheimer dementia, *Alzheimer Dis. Assoc. Disord.*, 12, 152, 1998.

77. Gimenez-Roldan, S., Mateo, D., Dobato, J. L., [Depressive pseudodementia in early Parkinson's disease: lessons from a case with long-term follow-up], *Neurologia*, 12, 130, 1997.

78. Playfer, J. R., The therapeutic challenges in the older Parkinson's disease patient, *Eur. J. Neurol.*, 9 (Suppl. 3), 55, 2002.

79. Rabinstein, A. A., Shulman, L. M., Management of behavioral and psychiatric problems in Parkinson's disease, *Parkinsonism Relat. Disord.*, 7, 41, 2000.

80. Cricco, M., Simonsick, E. M., Foley, D. J., The impact of insomnia on cognitive functioning in older adults, *J. Am. Geriatr. Soc.*, 49,1185, 2001.

81. Allen, N. H., Burns, A., Newton, V., Hickson, F., Ramsden, R., Rogers, J., Butler, S., Thistlewaite, G., Morris, J., The effects of improving hearing in dementia, *Age Ageing*, 32, 189, 2003.

82. Emre, M., Dementia associated with Parkinson's disease, *Lancet Neurol.*, 2, 229, 2003.

83. The Movement Disorder Society Task Force, (Goetz, C. G., Koller, W. C., Poewe, W., Roscol, O., Sampaio, et al.), Management of Parkinson's disease: an evidence-based review. Drugs to treat dementia and psychosis: management of Parkinson's disease, *Mov. Disord.*, 17 (Suppl. 4), S120, 2002.

84. McKeith, I., Del Ser, T., Spano, P., Emre, M., Wesnes, K., Anand, R., Cicin-Sain, A., Ferrara, R., Spiegel, R., Efficacy of rivastigmine in dementia with Lewy bodies: a randomised, double-blind, placebo-controlled international study, *Lancet*, 356, 2031, 2000.

85. Thal, L. J., Therapeutics and mild cognitive impairment: current status and future directions, *Alzheimer Dis Assoc. Disord.*, 17 (Suppl. 2), S69, 2003.

86. Reisberg, B., Doody, R., Stoffler, A., Schmitt, F., Ferris, S., Mobius, H. J., Memantine Study Group., Memantine in moderate-to-severe Alzheimer's disease, *N. Engl. J. Med.*, 348, 1333, 2003.

87. D'Souza, C., Gupta, A., Alldrick, M. D., Sastry, B. S., Management of psychosis in Parkinson's disease, *Int. J. Clin. Pract.*, 57, 295, 2003.

88. D'Souza, S., Barnett, S. K., Rangu, S., Rowland, F. N., Treatment of Psychosis in Parkinson's Disease. *J. Am. Med. Dir. Assoc.*, 1, 211, 2000.

89. Rascol, O., Payoux, P., Ory, F., Ferreira, J. J., Brefel-Courbon, C., Montastruc, J. L., Limitations of current Parkinson's disease therapy, *Ann. Neurol.*, 53 (Suppl. 3), S3, 2003.

90. Ballard, C. G., O'Brien, J., Lowery, K., Ayre, G. A., Harrison, R., Perry, R., Ince, P., Neill, D., McKeith, I. G., A prospective study of dementia with Lewy bodies, *Age Ageing*, 27, 631, 1998.

91. Friedman, J. H., Fernandez, H. H., Atypical antipsychotics in Parkinson-sensitive populations, *J. Geriatr. Psychiatry Neurol.*, 15, 156, 2002.

92. Danielczyk, W., Mental disorders in Parkinson's disease, *J. Neural. Transm.*, Suppl. 38, 115, 1992.

93. Ondo, W. G., Levy, J. K., Vuong, K. D., Hunter, C., Jankovic, J., Olanzapine treatment for dopaminergic-induced hallucinations, *Mov. Disord.*, 17, 1031, 2002.

94. Friedman, J. H., Factor, S. A., Atypical antipsychotics in the treatment of drug-induced psychosis in Parkinson's disease, *Mov. Disord.*, 15, 201, 2000.

95. Cummings, J. L., Masterman, D. L., Depression in patients with Parkinson's disease, *Int. J. Geriatr. Psychiatry*, 14, 711, 1999.

96. The Movement Disorder Society Task Force, (Goetz, C. G., Koller, W. C., Poewe, W., Roscol, O., Sampaio, et al.), Management of Parkinson's disease: an evidence-based review. Treatment of Depression in idiopathic Parkinson's disease, *Mov. Disord.*, 17 (Suppl. 4), S112, 2002

97. Burn, D. J., Depression in Parkinson's disease, *Eur. J. Neurol.*, 9 (Suppl. 3), 44, 2002.

98. Kasuya, R. T., Polgar-Bailey, P., Takeuchi, R., Caregiver burden and burnout. A guide for primary care physicians, *Postgrad. Med.*, 108, 119, 2000.

99. Moellentine, C., Rummans, T., Ahlskog, J. E., Harmsen, W. S., Suman, V. J., O'Connor, M. K., Black, J. L., Pileggi, T., Effectiveness of ECT in patients with parkinsonism, *J. Neuropsychology. Clin. Neurosci.*, 10, 187, 1998.

100. Wassermann, E. M., Lisanby, S. H., Therapeutic application of repetitive transcranial magnetic stimulation: a review, *Clin. Neurophysiol.*, 112, 1367, 2001.

101. Kobayashi, M., Pascual-Leone, A., Transcranial magnetic stimulation in neurology, *Lancet Neurol.*, 2, 145, 2003.

102. Gibb, W. R., Scott, T., Lees, A. J., Neuronal inclusions of Parkinson's disease, *Mov. Disord.*, 6, 2, 1991.

103. Takeda, S., Yamazaki, K., Miyakawa, T., Arai, H., Parkinson's disease with involvement of the parasympathetic ganglia, *Acta Neuropathol. (Berl.)*, 86, 397, 1993.

104. Wakabayashi, K., Takahashi, H., Neuropathology of autonomic nervous system in Parkinson's disease, *Eur. Neurol.*, 38 (Suppl. 2), 2, 1997.

105. Singer, C., Urinary dysfunction in Parkinson's disease. *Clin Neurosci.*, 5, 78, 1998.

106. Abbott, R. D., Petrovitch, H., White, L. R., Masaki, K. H., Tanner, C. M., Curb, J. D., Grandinetti, A., Blanchette, P. L., Popper, J. S., Ross, G. W., Frequency of bowel movements and the future risk of Parkinson's disease, *Neurology*, 57, 456, 2001.

107. Jost, W. H., Gastrointestinal motility problems in patients with Parkinson's disease. Effects of antiparkinsonian treatment and guidelines for management, *Drugs Aging*, 10, 249,1997.

108. Pfeiffer, R. F., Gastrointestinal dysfunction in Parkinson's disease, *Clin. Neurosci.*, 5,136,1998.

109. Stark, M. E., Challenging problems presenting as constipation, *Am. J. Gastroenterol.*, 94, 567,1999.

110. Eichhorn, T. E., Oertel, W. H., Macrogol 3350/electrolyte improves constipation in Parkinson's disease and multiple system atrophy, *Mov. Disord.*, 16, 1176, 2001.

111. Senard, J. M., Brefel-Courbon, C., Rascol, O., Montastruc, J. L., Orthostatic hypotension in patients with Parkinson's disease: pathophysiology and management, *Drugs Aging*, 18, 495, 2001.

112. Hussain, I. F., Brady, C. M., Swinn, M. J., Mathias, C. J., Fowler, C. J., Treatment of erectile dysfunction with sildenafil citrate (Viagra) in parkinsonism due to Parkinson's disease or multiple system atrophy with observations on orthostatic hypotension, *J. Neurol. Neurosurg. Psychiatry*, 71, 371, 2001.

113. Micieli, G., Tosi, P., Marcheselli, S., Cavallini, A., Autonomic dysfunction in Parkinson's disease, *Neurol. Sci.*, 24 (Suppl. 1), S32, 2003.

114. Chen, H., Zhang, S. M., Hernan, M. A., Willett, W. C., Ascherio, A., Weight loss in Parkinson's disease, *Ann. Neurol.*, 53, 676, 2003.

115. Sage, J. I., Mark, M. H., Drenching sweats as an off phenomenon in Parkinson's disease: treatment and relation to plasma levodopa profile, *Ann. Neurol.*, 37, 120, 1995.

116. Goetz, C. G., Lutge, W., Tanner, C. M., Autonomic dysfunction in Parkinson's disease, *Neurology*, 36, 73, 1986.

117. Kumar, S., Bhatia, M., Behari, M., Sleep disorders in Parkinson's disease, *Mov. Disord.*, 17, 775, 2002.

118. Larsen, J. P., Sleep disorders in Parkinson's disease, *Adv. Neurol.*, 91, 329, 2003.

119. Hobson, D. E., Lang, A. E., Martin, W. R., Razmy, A., Rivest, J., Fleming, J., Excessive daytime sleepiness and sudden-onset sleep in Parkinson disease: a survey by the Canadian Movement Disorders Group, *JAMA*, 287, 455, 2002.

120. Ondo, W. G., Vuong, K. D., Jankovic, J., Exploring the relationship between Parkinson disease and restless legs syndrome, *Arch. Neurol.*, 59, 421, 2002.

121. Silber, M. H., Sleep disorders, *Neurol. Clin.*, 19, 173, 2001.

122. Comella, C. L., Sleep disturbances in Parkinson's disease, *Curr. Neurol. Neurosci. Rep.*, 3, 173, 2003.

123. Wetter, T. C., Collado-Seidel, V., Pollmacher, T., Yassouridis, A., Trenkwalder, C., Sleep and periodic leg movement patterns in drug-free patients with Parkinson's disease and multiple system atrophy, *Sleep*, 23, 361, 2000.

124. Gagnon, J. F., Montplaisir, J., Bedard, M. A., Rapid-eye-movement sleep disorders in Parkinson's disease, *Rev. Neurol. (Paris)*, 158, 135, 2002.

125. O'Suilleabhain, P. E., Dewey, R. B., Jr., Contributions of dopaminergic drugs and disease severity to daytime sleepiness in Parkinson disease, *Arch Neurol.*, 59, 986, 2002.

126. Brodsky, M. A., Godbold, J., Roth, T., Olanow, C. W., Sleepiness in Parkinson's disease: A controlled study, *Mov. Disord.*, 18, 668, 2003.

127. Paus, S., Brecht, H. M., Koster, J., Seeger, G., Klockgether, T., Wullner, U., Sleep attacks, daytime sleepiness, and dopamine agonists in Parkinson's disease, *Mov. Disord.*, 18, 659, 2003.

128. Adler, C. H., Caviness, J. N., Hentz, J. G., Lind, M., Tiede, J., Randomized trial of modafinil for treating subjective daytime sleepiness in patients with Parkinson's disease, *Mov. Disord.*, 18, 287, 2003.

129. Potulska, A., Friedman, A., Krolicki, L., Spychala, A., Swallowing disorders in Parkinson's disease, *Parkinsonism Relat. Disord.*, 9, 349, 2003.

130. Volonte, M. A., Porta, M., Comi, G., Clinical assessment of dysphagia in early phases of Parkinson's disease, *Neurol. Sci.*, 23 (Suppl. 2), S121, 2002.

131. Fuh, J. L., Lee, R. C., Wang, S. J., Lin, C. H., Wang, P. N., Chiang, J. H., Liu, H. C., Swallowing difficulty in Parkinson's disease, *Clin. Neurol. Neurosurg.*, 99, 106, 1997.

132. Mantynen, T., Farkkila, M., Kunnamo, I., Mecklin, J. P., Juhola, M., Voutilainen, M., The impact of upper GI endoscopy referral volume on the diagnosis of gastroesophageal reflux disease and its complications: a 1-year cross-sectional study in a referral area with 260,000 inhabitants, *Am. J. Gastroenterol.*, 97, 2524, 2002.

133. Miyazaki, Y., Arakawa, M., Kizu, J., Introduction of simple swallowing ability test for prevention of aspiration pneumonia in the elderly and investigation of factors of swallowing disorders, *Yakugaku Zasshi*, 122, 97, 2002.

134. Kajii, Y., Shingai, T., Kitagawa, J., Takahashi, Y., Taguchi, Y., Noda, T., Yamada, Y., Sour taste stimulation facilitates reflex swallowing from the pharynx and larynx in the rat, *Physiol. Behav.*, 77, 321, 2002.

135. O'Sullivan, J. D., Bhatia, K. P., Lees, A. J., Botulinum toxin A as treatment for drooling saliva in PD, *Neurology*, 55, 606, 2000.

136. Marks, L., Turner, K., O'Sullivan, J., Deighton, B., Lees, A., Drooling in Parkinson's disease: a novel speech and language therapy intervention, *Int. J. Lang. Commun Disord.*, 36 (Suppl.), 282, 2001.

137. Stewart, C., Winfield, L., Hunt, A., Bressman, S. B., Fahn, S., Blitzer, A., Brin, M. F., Speech dysfunction in early Parkinson's disease, *Mov. Disord.*, 10, 562, 1995.

138. Sapir, S., Ramig, L. O., Hoyt, P., Countryman, S., O'Brien, C., Hoehn, M., Speech loudness and quality 12 months after intensive voice treatment (LSVT) for Parkinson's disease: a comparison with an alternative speech treatment, *Folia Phoniatr Logop*, 54, 296, 2002.

139. Gamboa, J., Jimenez-Jimenez, F. J., Nieto, A., Montojo, J., Orti-Pareja, M., Molina, J. A., Garcia-Albea, E., Cobeta, I., Acoustic voice analysis in patients with Parkinson's disease treated with dopaminergic drugs, *J. Voice*, 11, 314, 1997.

140. Ramig, L. O., Dromey, C., Aerodynamic mechanisms underlying treatment-related changes in vocal intensity in patients with Parkinson disease., *J. Speech Hear Res.*, 39, 798, 1996.

141. Schulz, G. M., The effects of speech therapy and pharmacological treatments on voice and speech in Parkin-

son s disease: a review of the literature, *Curr. Med. Chem.*, 9, 1359, 2002.

142. The Movement Disorder Society Task Force, (Goetz, C. G., Koller, W. C., Poewe, W., Roscol, O., Sampaio, et al.), Management of Parkinson's disease: an evidence-based review., Speech therapy in Parkinson's disease, *Mov. Disord.*, 17 (Suppl. 4), S163, 2002.

143. Ramig, L. O., Countryman, S., O'Brien, C., Intensive speech treatment for patients with Parkinson' disease, *Neurology*, 47, 1496, 1996.

144. Kim, S. H., Kearney, J. J., Atkins, J. P., Percutaneous laryngeal collagen augmentation for treatment of parkinsonian hypophonia, *Otolaryngol Head Neck Surg.*, 126, 653, 2002.

145. Ford, B., Pain in Parkinson's disease, *Clin. Neurosci.*, 5, 63, 1998.

146. Duvoisin, R. C., Marsden, C. D., Note on the scoliosis of Parkinsonism, *J. Neurol. Neurosurg. Psychiatry*, 38, 787, 1975.

147. Marsden, C. D., Duvoisin, R., Scoliosis and Parkinson's disease, *Arch Neurol.*, 37, 253, 1980.

148. Djaldetti, R., Mosberg-Galili, R., Sroka, H., Merims, D., Melamed, E., Camptocormia (bent spine) in patients with Parkinson's disease—characterization and possible pathogenesis of an unusual phenomenon, *Mov. Disord.*, 14, 443, 1999.

149. Slawek, J., Derejko, M., Lass, P., Camptocormia as a form of dystonia in Parkinson's disease, *Eur. J. Neurol.*, 10, 107, 2003.

150. Bodis-Wollner, I., Yahr, M. D., Mylin, L., Thornton, J., Dopaminergic deficiency and delayed visual evoked potentials in humans, *Ann. Neurol,*. 11, 478, 1982.

151. Bulens, C., Meerwaldt, J. D., van der Wildt, G. J., Keemink, C. J., Contrast sensitivity in Parkinson's disease, *Neurology*, 36, 1121, 1986.

152. Regan, D., Maxner, C., Orientation-selective visual loss in patients with Parkinson's disease, *Brain*, 110, (Pt. 2), 415, 1987.

153. Bulens, C., Meerwaldt, J. D., Van der Wildt, G. J., Effect of stimulus orientation on contrast sensitivity in Parkinson's disease, *Neurology*, 38, 76, 1988.

154. Rodnitzky, R. L., Visual dysfunction in Parkinson's disease, *Clin. Neurosci.*, 5, 102, 1998.

155. Repka, M. X., Claro, M. C., Loupe, D. N., Reich, S. G., Ocular motility in Parkinson's disease, *J. Pediatr. Ophthalmol. Strabismus*, 33, 144, 1996.

156. Bhatia, K., Brooks, D. J., Burn, D. J., Clarke, C. E., Grosset, D. G., MacMahon, D. G., Playfer, J., Schapira, A. H., Stewart, D., Williams, A. C., Parkinson's Disease Consensus Working Group. Updated guidelines for the management of Parkinson's disease, *Hosp. Med.*, 62, 456, 2001.

68 Future Symptomatic Therapy in Parkinson's Disease

Ruth Djaldetti and Eldad Melamed
Rabin Medical Center and Sackler Faculty of Medicine

CONTENTS

INTRODUCTION

The treatment of Parkinson's disease (PD) is continuously evolving. The last three decades have witnessed an explosion of new drugs with different mechanisms of action for improving not only the motor symptoms of PD but also the autonomic, psychiatric, and cognitive symptoms that negatively affect daily living. Treatment trends have shifted from surgery (before the revolutionary discovery of levodopa) to medical treatment and back to surgery (with the advances in stereotactic procedures and transplantation and disappointment with the poorly manageable side effects of levodopa). The current medical therapy for the motor symptoms of the disease includes levodopa preparations, dopamine agonists, dopamine extenders such as monoamine oxidase B (MAO-B) inhibitors, and catechol-O-methyl transferase (COMT) inhibitors. Anticholinergic, serotoninergic, noradrenergic, and gabaergic medications are used to treat the secondary long-term dopaminergic symptoms and the nonmotor symptoms (Table 68.1). The future holds much promise for PD patients. This chapter envisions what the next decades may look like in terms of the symptomatic medical treatment of PD.

TREATMENT OF MOTOR SYMPTOMS

EARLY STAGES

Oral Dopaminergic Treatment

Dopamine Agonists

The introduction of levodopa for the treatment of PD was initially greeted with much enthusiasm, owing to the rapid and successful improvement noted in motor symptoms. With time, however, clinicians noted that the peripheral pharmacokinetic and central pharmacodynamic mechanisms led to an inadequate response to the drug and complications, namely on-off fluctuations and dyskinesias.[1-3] These adverse reactions appeared even earlier (within 1 to 2 years) in patients with disease onset before age 45 years.[4-6] To overcome this problem, adjunctive therapy with MAO-B inhibitors, dopamine agonists, or COMT inhibitors was added to levodopa when the motor compli-

TABLE 68.1
Current and Future Symptomatic Treatment of Motor Symptoms in Parkinson's Disease

- Levodopa preparations
- Sinemet (levodopa + carbidopa)
- Sinemet CR (controlled release)
- Levopar (levodopa + benserazide)
- Madopar (levodopa + benserazide)
    - Levodopa ethylester (soluble levodopa + carbidopa)
    - Dual release levodopa (sinemet CR + sinemet)
- MAO-B inhibitors
- Selegiline (deprenyl)
- Rasagiline
- Dopamine agonists
- Pergolide
- Ropinirole
- Cabergoline
- Pramipexole
    - Apomorphine (pumps, subcutaneous injections)
- Rotigotine (transdermal, currently in clinical trials)
    - Piribedil
    - Sumanirole
- COMT inhibitors
- Entacapone
- Tolcapone (not approved by FDA)
- NMDA receptor antagonists
    - Amatadine
    - Memantine
    - Budipine
    - Riluzole

cations appeared. Later, when underlying mechanisms became clear, this strategy was abandoned in favor of starting with dopamine agonist monotherapy before levodopa. In the normal brain, dopamine is released from dopaminergic neurons into the synapse and stimulates the dopaminergic receptors in a constant, tonic manner. Excess dopamine is recycled by dopamine transporter into the neurons, stored in vesicles, and then released into the synapse for reuse. The dopaminergic receptors are activated by the smooth and constant stimulation. In the parkinsonian brain, pre- and postsynaptic changes cause the surviving neurons to lose their buffer capacity to maintain the physiologic tonic release of dopamine in the synapse. Moreover, the excessive amounts of exogenous dopamine, which can no longer be stored in the vesicles, stimulate the dopaminergic receptors in a burst activity, followed by an extreme drop and, later, an acute rise. This pulsatile stimulation of the striatal receptors induced by the combination of disease progression and chronic levodopa treatment may lead to further secondary "downstream" postsynaptic changes, rendering the brain susceptible to motor complications. Therefore, researchers hypothesized that stable and continuous stimulation of these receptors

with dopamine agonists would prevent these biochemical and molecular changes.

Indeed, the first double-blind, multicenter placebo-controlled trials with cabergoline,[7,8] and then with ropinirole[9,10] and pramipexole,[11,12] showed that patients with early-stage PD could be maintained on monotherapy for several years (although with less effectiveness than with levodopa preparations) and that the appearance of fluctuations and dyskinesias was postponed, as was the progression of the disease relative to levodopa treatment. These findings were supplemented with nuclear imaging studies.[13,14] Yet, oral dopamine agonists did not lead to the same dramatic improvement in motor symptoms as subcutaneous injection of apomorphine, a highly potent selective D_2 receptor dopamine agonist, and patients treated with these agents alone were less well managed than those treated with levodopa alone. The dopamine agonists were also associated with a wide range of adverse effects already in the early phase of dose escalation, especially nausea, orthostatic hypotension, constipation, and hallucinations. Furthermore, the single photon emission computed tomography (SPECT) and positron emission tomography (PET) studies themselves have been influenced by the drugs, although this issue requires additional research. Nevertheless, the ability of apomorphine to relieve motor symptoms is a promising indication for the future ability of researchers to develop an ideal dopamine agonist agent with the potency of apomorphine, the long half-life of cabergoline, and minimal adverse effects.

These agents will probably have D_2 receptor selectivity, perhaps with partial agonism for other dopaminergic receptors. This might aid in the prevention of late motor complications.

COMT Inhibitors

Another promising approach is to initiate treatment with a combination of levodopa, dopa decarboxylase, and COMT inhibitor in the same tablet to achieve less pulsatile stimulation of the dopaminergic receptors and thereby prevent or delay complications. The coadministration of a COMT inhibitor provides more stable levodopa levels than levodopa alone; it extends the plasma half-life of levodopa without increasing its maximum concentration (C_{max}) or time of maximum concentration (T_{max}),[15,16] so that levodopa is delivered to the brain in a more continuous fashion. Studies in monkeys treated with 1-methyl-4-phenyl-1,2,3, 6-tetrahydropyridine (MPTP) showed that pretreatment with the COMT inhibitor entacapone enhanced the duration and intensity of the locomotor response and reversed the disability induced by levodopa.[17] In another experimental study, the administration of levodopa b.i.d. resulted in an antiparkinsonian effect with severe dyskinesia, whereas four smaller doses of levodopa combined with entacapone reduced the occurrence and severity of the dyskinesia, without decreasing the favorable antiparkinsonian effect.[18] If this strategy can be replicated in patients, they would gain not only the unrefuted beneficial motor effect of levodopa but also a delay in motor complications. So far, a preliminary trial by the Celomen study group[19] in 172 patients with motor complications and 25 patients without, yielded a drop in activities of daily living (ADL) from 11.3 to 10.3 and a drop in motor scores from 28.0 to 25.7 in the group without motor complications. Levodopa dosage was reduced in this subgroup by 5%, but, owing to its small size, no statistically significant differences were noted as compared to patients receiving placebo. A major drawback of entacapone is its short half-life, necessitating frequent administration with each dose of levodopa. Tolcapone, a peripheral and central COMT inhibitor, is more effective than entacapone owing to its longer half-life and its ability to block the enzyme in the central nervous system. However, its use is restricted by hepatotoxicity. Researchers are still seeking other centrally acting COMT inhibitors with a long half-life, long-term action, and no adverse effects.

Transdermal Treatment

Although dopamine agonists are considered the first-line treatment for PD by most specialists, there is not enough evidence indicating that they completely mimic the physiologic stimulation of endogenous dopamine. The half-life of most of the drugs is less than 24 hr, requiring at least 3 doses during the day (only cabergoline has a longer half-life of 65 to 110 hr). Also, dopamine agonists alleviate motor symptoms, but they do not necessarily prevent them. As their symptomatic effect is insufficient, most patients require the addition of levodopa preparations, and eventually, 75% of patients reach the phase of motor complications. Therefore, other routes for smoother and longer-lasting delivery of plasma dopamine concentrations are needed. So far, intravenous and intraduodenal levodopa infusion;[20–22] rectal, sublingual, intranasal, and subcutaneous apomorphine administration;[23–26] and pumps providing slow and constant release of dopamine agonists[27] have all been found to be inconvenient, impractical, and applicable only for short-term use in patients with acute-phase deterioration. The transdermal route is the most appealing. Dermal patches are comfortable and user friendly; the applied drug dissociates through the epidermis into the blood capillaries and is released in a slow and continuous manner. Their major advantage is that they bypass the problems of gastrointestinal tract absorption and first-pass liver metabolism. Gastric atony is one of the symptoms caused by autonomic dysfunction in PD. It might present as a primary phenomenon, i.e., degeneration of the Meissner and Auerbach plexuses in the gastrointestinal tract, or secondary to drugs that block dopaminergic and/or cholinergic receptors within the

stomach. The reduced gastric motility delays absorption of levodopa from the stomach, leading to motor fluctuations. Once absorbed, the first-pass metabolism in the liver renders most medications less available to the central nervous system. The dermal route of delivery overcomes these barriers and allows the medications to reach their target.

Dopamine Agonists

Transdermal delivery has already been tested for various dopamine agonists with different degrees of success. Naxagolide, a naphthoxazine derivative, was the first to be introduced as a dermal patch, but the minor beneficial effect gained was outweighed by the drug's possible toxicity by this route.[28] Piribedil failed to have a pharmacological or clinical effect with the tested doses,[29,30] and the patch size necessary to supply sufficient amounts of the drug was unfeasible. Rotigotine, a lipid-soluble, nonergot aminotetraline derivative, is currently in the final stages of dermal patch testing in a multicenter, double-blind, placebo-controlled trial. Several small, preliminary studies found the drug efficacious in patients with PD who were receiving levodopa at a reduced dosage;[31–33] a few were able to discontinue levodopa completely. One monotherapy trial reported improved motor function.[34] It seems that intensive research is required for the development of more potent agonists that will be suitable for dermal delivery.

Monoamine Oxidase B (MAO-B) Inhibitors

Selegiline has been tested for transdermal administration to avoid its first-path metabolism. So far, studies in in vitro models have been encouraging.[35] The potential of selegiline to prevent freezing (Fahn, personal communication) should prompt further investigations. Rasagiline, another novel MAO-B inhibitor, also holds promise as a monotherapy agent in the early stage of disease.[36]

Iontophoretic Delivery

There are already a few reports of the iontophoretic transport of the dopamine agonist apomorphine.[37–39] The use of electric current to produce electrorepulsive and electroosmotic forces was found to facilitate the drug's absorption through the skin and into the plasma; passive absorption was not noted. Further supportive studies are expected to promote the development of a patch delivery system. Iontophoretic delivery will also enable individual titration of the drug, not only according to dosage but also by current density.

Nondopaminergic Treatments

Nicotine

Thus far, knowledge on nondopaminergic agents for the treatment of mild and moderate PD is limited. Studies have reported that the brains of patients with PD show a substantial loss of nicotine receptors that may be involved in the characteristic motor, cognitive, and behavioral changes.[40,41] However, clinical studies with nicotine yielded conflicting results. Administration of intravenous nicotine in 15 patients followed by chronic administration of up to 14 mg/d by transdermal patch improved some motor and cognitive performance for up to 1 month.[42] By contrast, 16 patients exposed to 35 mg of transdermal nicotine showed worsening of symptoms compared to placebo.[43] Future studies are needed to resolve this question.

Trophic Factors

Intracerebral administration of glial cell-line-derived neurotrophic factor (GDNF) was found to have neuroprotective and neurorestorative effects in rodents and primates, with improvement of parkinsonian symptoms.[44–46] In clinical studies, however, GDNF not only failed to improve parkinsonian symptoms, it caused major sensory (paresthesias and Lhermitte sign), gastrointestinal (anorexia and vomiting), and behavioral and autonomic side effects.[47,48] There may be several explanations for the treatment failure: inadequate doses, exceeding those given in the animal studies; lack of penetration of GDNF into brain parenchyma, so that it does not reach its target; and mismatch of brain and cerebrospinal fluid parameters between humans and primates. All these factors should be considered in further studies with trophic factors. One recent breakthrough that may bring this potential therapy one step forward is the use of Lentiviral vector to carry GDNF to its target.[49] Another option is direct nigral injections. In a recent study, GDNF was delivered through a catheter directly into the putamen of five patients with PD. After a year of treatment, there was an improvement in motor scores and activities of daily living, with no serious clinical side effects. In one patient, the catheter had to be repositioned. These procedures are invasive, however, bearing the unavoidable complications of foreign objects in brain parenchyma. If GDNF or other trophic factors prove successful in treating parkinsonian symptoms, other means of delivery that avoid the problem of direct intraparenchymal administration will need to be developed. GDNF analogs that can be delivered orally or at least subcutaneously would represent a major advance in the treatment of PD.

Genetic Factors

Like other neurodegenerative diseases with genetic impact, PD may also be influenced by multiple genes.[51,52] Five genes have already been mapped (PARK 1, 2, 5, 7, and 10), and there are five more known linkage loci (PARK 3, 4, 6, 8, and 9). Some of these genes are involved in the protein-degradation pathway within the cell, and at least two are biochemically related.[53] Proteins within cells are in danger of assuming an incorrect conformation, which could threaten cell function and viability.

The mechanism underlying the selective degeneration of dopaminergic neurons in the substantia nigra involves dopamine-induced accelerated accumulation of α-synuclein protofibrils and their aggregation within Lewy bodies.[54] Therefore, compounds that inhibit the formation of α-synuclein protofibrils or prevent their aggregation may be of therapeutic value in PD.[55] Another possibility is the reduction of α-synuclein toxicity by overexpression of heat-shock proteins.[56,57] The degeneration of the dopaminergic cells may also impair their metabolism of ubiquitin, another protein that aggregates in the Lewy bodies. By determining the mechanism of the toxicity causing mutations or misfolding of aggregation-prone proteins, clinicians may disclose alternative methods of treating PD. Agents that block apoptosis, inhibit the induction of caspases, inhibit the formation of aggregates, enhance proteasome activity, and degrade toxic proteins may all be of enormous importance. An exciting tactic attempting in the treatment of Alzheimer's disease is vaccination with the pathologic protein so as to induce autoantibodies that attack and destroy the pathologic accumulation of the offending pathogen. Vaccination with amyloid β was successful in mice models of Alzheimer's disease[58] (liquefying amyloid pluques), but clinical trials in humans were stopped because of serious adverse effects.[59] An analogous possibility involves vaccinating with a concentrate of Lewy body to dismantle the pathologic aggregation of proteins within the substantia nigra.

ADVANCED-STAGE PD

Dopaminergic Treatments

Eventually, all patients with PD reach the point at which the motor disability compels them to initiate levodopa use. Therefore, there is a need for novel strategies to treat fluctuations and dyskinesias, not only to prevent them.

Motor fluctuations are caused mainly by pharmacodynamic ("wearing off," "on-off," off periods, and early-morning dystonia) or pharmacokinetic ("delayed on" and "no on") mechanisms. When present, the "wearing off" and "on-off" fluctuations are partly ameliorated by dopamine agonists,[60–65] COMT inhibitors,[66,67] and MAO-B inhibitors.[68,69] Continuous parenteral or transdermal administration has been found effective, and rotigotine patches have had modest success.[28]

Rasagiline, a novel long-acting MAO-B inhibitor that is not metabolized to amphetamines, has also shown promising results, with a decrease in total scores on the Unified Parkinson Disease Rating Scale (UPDRS).[70] A recent multicenter, double-blind, placebo-controlled study confirmed the ability of rasagiline to reduce daily "off" hours (unpublished data). *In vitro* studies demonstrated a potential neuroprotective effect of the drug in culture and in animal models.[71,72]

In patients with PD, the autonomic nervous system in the gastrointestinal tract is denervated, leading to decreased gastric motility. The increased latencies to "on" are believed to be caused by stagnation of the insoluble levodopa tablets in the atonic parkinsonian stomach, combined with delayed absorption of levodopa from the duodenum. Therefore, one way to control fluctuations is to bypass the oral route; intravenous or intraduodenal delivery of levodopa, as well as subcutaneous delivery of apomorphine, can ameliorate the "delayed on" and "no on" types of response fluctuations. However, while patients can be maintained for at least five years on subcutaneous delivery of apomorphine,[73,74] most find this route difficult to adhere to, and researchers continue to seek means to improve levodopa's oral absorption. Levodopa ethylester (LDEE), a highly soluble prodrug of levodopa, was designed to solve this problem. It was found to be efficacious in reducing latencies to "on" and in decreasing the number of dose failures in patients with response fluctuations.[75] A larger multicenter, double-blind study with replacement of all daily doses of levodopa with LDEE is still needed to reach definitive conclusions.

Nondopaminergic Therapy

Along with the reduction in "off" hours, all the above treatments are associated with an increase in dyskinesias due to the excess levodopa concentrations in plasma and brain. There are currently two main pathophysiological theories of the emergence of dyskinesias: pulsatile stimulation of DA receptors and imbalance between the major striatal glutamatergic output pathways. The first theory was supported by the recent finding that D_3 receptor expression increased in MPTP-treated monkeys with levodopa-induced dyskinesias and decreased in monkeys without dyskinesias. This led to the assumption that limiting the fluctuations in D_3 receptor function would decrease the dyskinesias without loss of the beneficial effect of levodopa. Initial studies with BP 897, a partial D_3 receptor agonist, normalized D_3 receptor function in primates.[76] A small amount of the drug promoted the action of levodopa, but it acted as an antagonist when there was too much levodopa. Some authors suggest that dopamine itself regulates the excitation of striatal synapses, producing a decrease in glutamate release;[77] therefore, therapy should combine the existing strategies with nondopaminergic agents that block glutamate receptors. However, remacemide, a weak N-methyl-D-asparate (NMDA) receptor antagonist, tested in 39 patients with PD and disabling dyskinesias for 2 weeks in a double-blind, placebo-controlled trial, failed to induce any change in the dyskinesia scale.[78] Other trials yielded similar results.[79,80] Yet, acute or chronic administration of amantadine, a noncompetitive NMDA receptor antagonist, or rimantadine, a derivative of amantadine, attenuated levodopa-induced dyskinesias

by 50 to 60% without deterioration in motor scores.[81–84] The use of other NMDA receptor antagonist compounds is restricted by their potential neurotoxicity. Riluzole, an agent that blocks glutamatergic receptors without any toxic side effects, has been tried successfully in a few patients with PD[85] and should be tested in a larger controlled trial.

Using another pathway, studies in rodents have shown that the activation of cannabinoid receptors, which are scattered throughout the neostriatum, may inhibit glutamatergic transmission.[86] These preliminary data may point to a potential role for cannabinoids in ameliorating fluctuations and dyskinesias. More rigorous studies will probably be conducted in the future.

A range of opioid antagonists, selective muscarinic and nicotinic cholinergic antagonists, 5-hydropytryptamine (5-HT)-2c antagonists, and α-2 antagonists have been found to be efficacious in increasing locomotor activities in PD-primed rodents and primates, without provoking a dyskinetic response.[87–89] Other downstream receptors serve as a potential target for treating symptoms of PD and reducing the magnitude of dyskinesias. In one Phase II clinical trial, idazoxan, an α-2 antagonist, improved levodopa-induced dyskinesias without exacerbating motor symptoms.[90] These results were not replicated in another controlled study (unpublished data). PET studies showing a decrease in [11C] diprenorphine, an opioid receptor-binding ligand, in patients with dyskinetic PD[91] led to the assumption that opioid antagonists may not only prevent but also improve dyskinesias. However, naloxone had only a mild beneficial effect,[92] and high-dose naltrexone only a mild subjective effect.[93] Finally, some authors noted that the nonphysiologic stimulation of dopamine receptors activates signal transduction pathways associated with upregulation of striatal kinases,[94] leading to dyskinesias. Subsequent trials inhibiting serine kinase and tyrosine kinase were successful in reducing dyskinesias in rodents.[95,96] Apparently, dyskinesias cannot be adequately controlled with drugs treating only one mechanism of action. Future therapy should include medications that activate or inhibit different types of receptors in combination with drugs that inhibit downstream factors involved in the mechanism of dyskinesias.

NONMOTOR SYMPTOMS

PSYCHIATRIC SYMPTOMS

Atypical Neuroleptics

Another grave prognostic milestone in PD is the development of psychosis, which is the most common reason for transferring patients to a nursing home. Family members who, despite great difficulties, continue to cope with the many incapacitating aspects of this chronic disease may become helpless and discouraged when psychosis emerges. More importantly, parkinsonian psychosis is the predominant limiting factor for optimal antiparkinsonian drug therapy. Its presence prevents an increase in levodopa dosage or the addition of other drugs to improve motor function, particularly in advanced stages of the illness, which are almost invariably associated with intolerable worsening of the psychotic phenomena. The introduction of atypical neuroleptic agents revolutionized the treatment of psychotic symptoms in PD. Clozapine, quetiapine, and (with somewhat less efficacy and more parkinsonian side effects) risperidone and olanzapine can control hallucinations and behavioral changes without aggravating other parkinsonian symptoms.[97,98] Improved novel atypical neuroleptic agents that refine the balance between the beneficial antipsychotic effect of these drugs and their dopaminergic receptor blockade are being sought. They need to possess unique features to prevent a deterioration in motor parkinsonian symptoms yet be efficacious in controlling hallucinations and aggressive behavior. This might be achieved by drugs with more selective affinity or partial agonism to different subtypes of dopaminergic and serotoninergic receptors. The new medications should also be devoid of bothersome adverse effects such as leukopenia, diarrhea, and sleepiness.

Cholinesterase Inhibitors

Cholinergic neurotransmission also seems to be involved in the appearance of cognitive, behavioral, and psychic symptoms in PD. Cholinesterase inhibitors (donepezil, rivastigmine) bypass the obstacle of dopaminergic blockade, making them promising therapeutic agents for PD psychosis. This assumption is supported by their effectiveness in patients with diffuse Lewy body disease.[99] Preliminary studies in patients with PD reported a reduced severity of the psychotic symptoms and improved social behavior.[100,101] Other cholinesterase inhibitors currently under study for Alzheimer's disease (galantamine) will probably also be tested for use in PD.

Hormonal Replacement Therapy

It has been shown that there is a correlation between low levels of testosterone and nonmotor symptoms of PD.[102] Thus, hormonal replacement therapy in men and women with PD might offer a new approach. Treatment of testosterone-deficient men with a single daily dose of transdermal testosterone gel was found to improve energy level, enjoyment of life, and libido; however, there was no improvement in scores on the motor part of the UPDRS or in cognitive function.[103] In women, there are as yet no prospective studies of adjunct hormonal therapy, but indirect epidemiologic evidence links hormonal replacement therapy with disease prevention or amelioration of motor symptoms.[104,105] Future studies will shed more light on this intriguing issue.

MISCELLANEOUS

As PD progresses, neurotransmitter systems other than dopamine degenerate and aggravate the motor, vegetative, and cognitive symptoms, leading to depression, panic attacks, cognitive deterioration, and gait problems (freezing, postural instability, falls). Some symptoms, such as sialorrhea, constipation, urinary frequency, and impotence, are not reported by the patients, as they seem negligible compared to the devastating motor symptoms. However, they should be considered by the caring physician, as they cause a significant loss of quality of life, and more important, they can be helped. For example, excess salivary secretion can be alleviated with sublingual atropine and intraparotid botulinum toxin injection[106,107] (anticholinergic agents should not be used, owing to their side effects in elderly patients). Fatigue might be treated with provigil,[108] and panic attacks with cholecystokinin receptor antagonists.[109] Unfortunately, there is still no progress in the treatment of gait problems (freezing, postural instability, and falls). Table 68.2 presents the currently inadequately treated symptoms of PD and the possible future strategies to alleviate them.

TABLE 68.2
Presently Unmet Needs in the Treatment of Parkinson's Disease

Tremor	Potent dopamine agonists
	Potent serotonin reuptake blockers
Falls	Adrenoreceptor antagonists
Freezing gait	Adrenoreceptor antagonists
	Partial adrenoreceptor agonists
	Regulation of cholinergic transmission
Autonomic failure	Enhancement of gastric motility
	Maintaining adequate blood pressure by atrial pressors
	Long acting phosphodiesterase inhibitors for erectile dysfunction
	Psychostimulants
Fatigue	Improve cardiac output
	Nutritional care
Weight loss	Regulation of weight-reducing hormones (leptin, ghrelin)

ACKNOWLEDGMENT

The authors thank Gloria Ginzach and Charlotte Sachs of the Editorial Board, Rabin Medical Center, Beilinson Campus, for their assistance.

REFERENCES

1. Verhagen-Metman, I., Konitsiotis, S., and Chase, T. N., Pathophysiology of motor response fluctuations in Parkinson's disease: Hypotheses on the why, where and what, *Mov. Disord.*, 15,3, 2000.
2. Obeso, J. A. et al., Pathophysiology of levodopa-induced dyskinesias in Parkinson's disease: problems with the current model, *Ann. Neurol.*, 47, S22, 2000.
3. Obeso, J. A. et al., The evolution and origin of motor complications in Parkinson's disease, *Neurology*, 55, S13, 2000.
4. Kostic, V. et al., Early development of levodopa-induced dyskinesia and response fluctuations in young-onset Parkinson's disease, *Neurology*, 41, 202, 1991.
5. Quinn, N., Cirtchley, P., and Parsden, C. D., Young onset Parkinson's disease, *Mov. Disord.*, 2,73, 1987.
6. Schrag, A. and Quinn, N., Dyskinesias and motor fluctuations in Parkinson's disease: a community-based study, *Brain*, 123, 2297, 2000.
7. Rinne, U. K. et al., Cabergoline in the treatment of early Parkinson's disease: results of the first year of treatment in a double-blind comparison of cabergoline and levodopa. The PKDS009 Collaborative Study Group, *Neurology*, 48, 363, 1997.
8. Rinne, U. K. et al., and the PKDS009 study group. Early treatment of Parkinson's disease with cabergoline delays the onset of motor complications, *Drugs*, 55,23, 1998.
9. Sethi, K. D. et al., for the Ropinirole Study Group. Ropinirole for the treatment of early Parkinson's disease: a 12-month experience, *Arch. Neurol.*, 55, 1211, 1998.
10. Rascol, O. et al., A five-year study of the incidence of dyskinesia in patients with early Parkinson's disease who were treated with ropinirole and levodopa. 056 Study Group, *N. Engl. J. Med.*, 342, 1484, 2000.
11. Shannon, K. M., Bennett, J. P. Jr., and Friedman, J. H., Efficacy of pramipexole, a novel dopamine agonist, as monotherapy in mild to moderate Parkinson's disease. Pramipexole Study Group, *Neurology*, 49, 724, 1997.
12. Parkinson Study Group, Pramipexole vs. levodopa as initial treatment for Parkinson disease—a randomized controlled trial, *J. Am. Med. Assoc.*, 284, 1931, 2000.
13. Marek, K. et al., Dopamine transporter brain imaging to assess the effects of pramipexole vs. levodopa on Parkinson disease progression, *J. Am. Med. Assoc.*, 287, 1653, 2002.
14. Whone, L. et al., Slower progression of Parkinson's disease with ropinirole versus levodopa: the REAL-PET study, *Ann. Neurol.*, 54, 93, 2003.
15. Nutt, J. G., Effect of COMT inhibition on the pharmacokinetics and pharmacodynamics of levodopa in parkinsonian patients, *Neurology*, 55, S33, 2000.
16. Olanow, C. W. and Obeso, J. A., Pulsatile stimulation of dopaminergic receptors and levodopa-induced motor complications in Parkinson's disease: implications for the early use of COMT inhibitors, *Neurology*, 55, S72, 2000.
17. Smith, L. A. et al., Entacapone enhances levodopa-induced reversal of motor disability in MPTP-treated common marmosets, *Mov. Disord.*, 12, 935, 1997.
18. Smith, L. A. et al., Effect of pulsatile administration of L-dopa on dyskinesia induction in drug naive MPTP-treated common marmosets: effect of dose, frequency of administration and brain exposure, *Mov. Disord.*, in press.
19. Poewe, W. H. et al., and the Celomen study group, Efficacy and safety of entacapone in Parkinson's disease patients with suboptimal levodopa response: a 6-month randomized placebo-controlled double-blind study in German and Austria, *Acta Neurol. Scand.*, 105, 245, 2002.
20. Quinn, M., Parkes, J. D., and Marsden, C. D., Control of on/off phenomenon by continuous intravenous infusion of levodopa, *Neurology*, 34, 1131, 1984.
21. Mouradian, M. M. et al., Modification of central dopaminergic mechanisms by continuous levodopa therapy for advanced Parkinson's disease, *Ann. Neurol.*, 27, 18, 1990.
22. Nillson, D., Nyholm, D., and Aquilonius, S. M., Duodenal levodopa infusion in Parkinson's disease - long-term experience, *Acta Neurol. Scand.*, 104, 343, 2001.
23. Van Laar, T. et al., Pharmacokinetics and clinical efficacy of rectal apomorphine in patients with Parkinson's disease: a study of five different suppositories, *Mov. Disord.*, 10, 443, 1995.
24. Ondo, W. et al., A novel sublingual apomorphine treatment for patients with fluctuating Parkinson's disease, *Mov. Disord.*, 14, 664, 1999.
25. Dewey, R. B. et al., A double-blind, placebo-controlled study of intranasal apomorphine spray as a rescue agent for off-states in Parkinson's disease, *Mov. Disord.*, 13, 782, 1998.
26. Colzi, A., Turner, K., and Lees, A. J., Continuous subcutaneous waking day apomorphine in the long term treatment of levodopa induced interdose dyskinesias in Parkinson's disease, *J. Neurol. Neurosurg. Psychiatry*, 64, 573, 1998.
27. Vaamonde, J., Luquin, M. R., and Obeso, J. A., Subcutaneous lisuride infusion in Parkinson's disease. Response to chronic administration in 34 patients, *Brain*, 114, 601, 1991.
28. Ahlskog, J. E. et al., Parkinson's disease monotherapy with controlled release MK-458(PHNO): double-blind study and comparison to carbidopa/levodopa, *Clin. Neuropharmacol.*, 14, 214, 1991.
29. Montastruc, J. L. et al., A randomized, double-blind study of a skin patch of a dopaminergic agonist, piribedil, in Parkinson's disease, *Mov. Disord.*, 14, 336, 1999.
30. Rondot, P. and Ziegler, M., Activity and acceptability of piribedil in Parkinson's disease: a multicentre study, *J. Neurol.*, 239, 28, 1992.
31. Hutton, J. T. et al., Transdermal dopaminergic D(2) receptor agonist therapy in Parkinson's disease with N-0923 TDS: a double-blind, placebo-controlled study, *Mov. Disord.*, 16, 459, 2001.
32. Verhagen-Metman, L. et al., Continuous transdermal dopaminergic stimulation in advanced Parkinson's disease, *Clin. Neuropharmacol.*, 24, 163, 2001.

33. Quinn, N., for the SP 511 Investigators. Rotigotine transdermal delivery system (TDS) (SPM-962): a multicenter, double-blind, randomized, placebo-controlled trial to assess the safety and efficacy of rotigotine TDS in patients with advanced Parkinson, *Parkinsonism Relat. Disord.*, 7, S66, 2001.

34. Fahn, S., Parkinson Study Group, Rotigotine transdermal system (SPM-962) is safe and effective as monotherapy in early Parkinson, *Parkinsonism Relat. Disord.*, 7, S55, 2001.

35. Rohatagi, S. et al., Integrated pharmacokinetic and metabolic modeling of selegiline and metabolites after transdermal administration, *Bipharm. Drug Dispos.*, 18, 567, 1997.

36. Parkinson Study Group, A controlled trial of rasagiline in early Parkinson disease: the TEMPO Study, *Arch. Neurol.*, 59, 1937, 2002.

37. Van der Geest, R. et al., Iontophoretic delivery of apomorphine. I. In vitro optimization and validation, *Pharm. Res.*, 14, 1798, 1997.

38. Van der Geest, R. et al., Iontophoretic delivery of apomorphine. II. An *in vitro* study in patients with Parkinson's disease, *Pharm. Res.*, 14, 1804, 1997.

39. Dayan, N. and Touitou, E., Carriers for skin delivery of trihexyphenidyl HCl: enthosomes vs. liposomes, *Biomaterials*, 21, 1879, 2000.

40. Smith, C. J. and Giacobini, E., Nicotine, Parkinson's and Alzheimer's disease, *Rev. Neurosci.*, 3, 25, 1992.

41. Kelton, M. C. et al., The effects of nicotine on Parkinson's disease, *Brain Cogn.*, 43, 274, 2000.

42. Vieregge, A. et al., Transdermal nicotine in PD: a randomized, double-blind, placebo-controlled study, *Neurology*, 57, 1032, 2000.

43. Ebersbach, G. et al., Worsening of motor performance in patients with Parkinson's disease following transdermal nicotine administration, *Mov. Disord.*, 14, 1001, 1999.

44. Bjurklund, A., Rosenblad, C., and Winkler, C., Studies on neuroprotective and regenerative effects of GDNF in a partial lesion model of Parkinson's disease, *Neurobiol. Dis.*, 4, 186, 1997.

45. Gash, D. M. et al., Functional recovery in parkinsonian monkeys treated with GDNF, *Nature*, 380, 252, 1996.

46. Zhang, Z. et al., Dose response to intraventricular glial cell-line derived neurotrophic factor administration in parkinsonian monkeys, *J. Pharmacol. Exp. Ther.*, 282, 1396, 1997.

47. Kordower, J. H. et al., Clinicopathological findings following intraventricular glial-derived neurotrophic factor treatment in a patient with Parkinson's disease, *Ann. Neurol.*, 46, 419, 1999.

48. Nutt, J. G. et al., Randomized, double-blind trial of glial cell line-derived neurotrophic factor (GDNF) in PD, *Neurology*, 60, 69, 2003.

49. Kordower, J. H. et al., Neurodegeneration prevented by Lentiviral vector delivery of GDNF in primate models of Parkinson's disease, *Science*, 290, 767, 2000.

50. Gill, S. S. et al., Direct brain infusion of glial cell line derived neurotrophic factor in Parkinson's disease, *Nat. Med.*, 9, 589, 2003.

51. Gwinn-Hardy, K., Genetics in parkinsonism, *Mov. Disorder.*, 17, 645, 2002.

52. Foltynie, T. et al., The genetic basis of Parkinson's disease, *J. Neurol. Neurosurg. Psychiatry*, 73, 363, 2002.

53. Shimura, H. et al., Ubiquitination of a new form of alpha-synuclein by parkin from human brain: implications for Parkinson's disease, *Science*, 293, 263, 2001.

54. Taylor, J. P., Hardy, J., and Fischbeck, K. H., Toxic proteins in neurodegenerative disease, *Science*, 296, 1991, 2002.

55. Conway, K. A. et al., Kinetic stabilization of the alpha-synuclein protofibril by a dopamine-alpha-synuclein adduct, *Science*, 294, 1346, 2001.

56. Warrick, J. M. et al., Suppression of polyglutamine-mediated neurodegeneration in Drosophila by the molecular chaperone HSP70, *Nat. Genet.*, 23, 425, 1999.

57. Auluck, P. K. et al., Chaperone suppression of alpha-synuclein toxicity in a Dorosphila model for Parkinson's disease, *Science*, 295, 865, 2002.

58. Schenk, D. et al., Immunization with amyloid-beta attenuates Alzheimer-disease-like pathology in the PDAPP mouse, *Nature*, 400, 173, 1999.

59. Janus, C., Vaccines for Alzheimer's disease: how close are we? *CNS Drugs*, 17, 457, 2003.

60. Inzelberg, R. et al., Double-blind comparison of cabergoline and bromocriptine in Parkinson's disease patients with motor fluctuations, *Neurology*, 47, 785, 1996.

61. Geminiani, G. et al., Cabergoline in Parkinson's disease complicated by motor fluctuations, *Mov. Disord.*, 11, 495, 1996.

62. Rascol, O. et al., Ropinirole in the treatment of levodopa-induced motor fluctuations in patients with Parkinson's disease, *Clin. Neuropharmacol.*, 19, 234, 1996.

63. Lieberman, A. et al., A multicenter trial of ropinirole as adjunct treatment for Parkinson's disease, *Neurology*, 51, 1057, 1998.

64. Lieberman, A., Ranhosky, A., and Korts, D., Clinical evaluation of pramipexole in advanced Parkinson's disease: results of a double-blind, placebo-controlled, parallel group study, *Neurology*, 49, 162, 1997.

65. Pinter, M. M., Pogarell, O., and Oertel, W. H., Efficacy, safety, and tolerance of the non-ergoline dopamine agonist pramipexole in the treatment of advanced Parkinson's disease: a double-blind, placebo-controlled, randomized, multicentre study, *J. Neurol. Neurosurg. Psychiatry*, 66, 436, 1999.

66. Parkinson Study Group, Entacapone improves motor fluctuations in levodopa-treated Parkinson's disease patients, *Ann. Neurol.*, 42, 747, 1997.

67. Rinne, U. K., Larsen, J. P., Siden, A., and the NomeComt Study Group, Entacapone enhances the response to levodopa in parkinsonian patients with motor fluctuations. *Neurology*, 51, 1309, 1998.

68. Lieberman, A. N. et al., Deprenyl versus placebo in Parkinson's disease. A double-blind study, *N.Y. State J. Med.*, 87, 646, 1987.

69. Golbe, L. I. et al., Deprenyl in the treatment of symptom fluctuations in advanced Parkinson's disease, *Clin. Neuropharmacol.*, 11, 45, 1989.

70. Rabey, R. M. et al., Rasagiline mesylate, a new MAO-B inhibitor for the treatment of Parkinson's disease: a double-blind study as adjunctive therapy to levodopa, *Clin. Neuropharmacol.*, 23, 324, 2000.

71. Youdim, M. B. et al., The anti-Parkinson drug rasagiline and its cholinesterase inhibitor derivatives exert neuroprotection unrelated to MAO inhibition in cell culture and *in vivo*, *Ann. N. Y. Acad. Sci.*, 939, 450, 2001.

72. Maruyama, W. et al., Neuroprotection by propargylamines in Parkinson's disease: suppression of apoptosis and induction of postviral genes, *Neurotoxicol. Teratol.*, 24, 675, 2002.

73. Stocchi, R. et al., Subcutaneous continuous apomorphine infusion in fluctuating patients with Parkinson's disease: long-term results, *Neurol. Sci.*, 22, 93, 2001.

74. Kanovsky, P. et al., Levodopa-induced dyskinesias and continuous subcutaneous infusions of apomorphine: results of a two-year, prospective follow-up, *Mov. Disord.*, 17, 188, 2002.

75. Djaldetti, R. et al., Oral solution of levodopa ethylester for treatment of response fluctuations in patients with advanced Parkinson's disease, *Mov. Disord.*, 17, 297, 2002.

76. Bezard, E., et al., Attenuation of levodopa-induced dyskinesias by normalizing dopamine D3 receptor function, *Nat. Med.*, 9, 762, 2003.

77. Tang, K. et al., Dopamine-dependent synaptic plasticity in striatum during *in vivo* development, *Proc. Natl. Acad. Sci. USA.*, 98, 1255, 2001.

78. Parkinson Study Group, Evaluation of dyskinesias in a pilot, randomized, placebo-controlled trial of remacemide in advanced Parkinson disease, *Arch. Neurol.*, 58, 1660, 2001.

79. Clarke, C. E., Copper, J. A., and Holdich, T. A., TREMOR study group. A randomized, double-blind, placebo-controlled, ascending-dose tolerability and safety study of remacemide as adjuvant therapy in Parkinson's disease with response fluctuations, *Clin. Neuropharmacol.*, 24, 133, 2001.

80. Shoulson, I. et al., A randomized, controlled trial of remacemide for motor fluctuations in Parkinson's disease, *Neurology*, 56, 455, 2001.

81. Luginger, E. et al., Beneficial effect of amantadine on L-dopa induced dyskinesia in Parkinson's disease, *Mov. Disord.*, 15, 873, 2000.

82. Snow, B. J. et al., The effect of amantadine on levodopa-induced dyskinesia in Parkinson's disease: a double-blind, placebo-controlled study, *Clin. Neuropharmacol.*, 23, 82, 2000.

83. Del Dotto, P. et al., Intravenous amantadine improves levodopa-induced dyskinesias: an acute double-blind placebo-controlled study, *Mov. Disord.*, 16, 515, 2001.

84. Evidente, V. G. et al., A pilot study on the motor effects of rimantadine in Parkinson's disease, *Clin. Neuropharmacol.*, 22, 30, 1999.

85. Merims, D. et al., Riluzole for levodopa-induced dyskinesias in advanced Parkinson's disease, *Lancet*, 353, 1764, 1999.

86. Gerdeman, G. and Lovinger, D. M., CB1 cannabinoid receptor inhibits synaptic release of glutamate in rat dorsolateral striatum, *J. Neurophysiol.*, 85, 468, 2001.

87. Pearce, R. K. et al., The monoamine reuptake blocker reverses akinesia without dyskinesia in MTPT treated and levodopa primed common marmosets, *Mov. Disord.*, 17, 877, 2002.

88. Klintenberg, R. et al., Naloxone reduces levodopa-induced dyskinesias and apomorphine-induced rotations in primate models of parkinsonism, *J. Neurol. Transm.*, 109, 1295, 2002.

89. Jenner, P., Pathophysiology and biochemistry of dyskinesia: clues for the development of non-dopaminergic treatments, *J. Neurol.*, 247, 43, 2000.

90. Rascol, O. et al., Idazoxan, an alpha-2 antagonist, and L-dopa-induced dyskinesias in patients with Parkinson's disease, *Mov. Disord.*, 16, 708, 2001.

91. Piccini, P., Weeks, R. A., and Brooks, D. J., Alterations in opioid receptor binding in Parkinson's disease patients with levodopa-induced dyskinesias, *Ann. Neurol.*, 42, 720, 1997.

92. Trabucchi, M., Bassi, S., and Frattola, L., Effects of naloxone on the "on-off" syndrome in patients receiving long-term levodopa therapy, *Arch. Neurol.*, 39, 120, 1982.

93. Rascol, O. et al., Naltrexone, an opiate antagonist, fails to modify motor symptoms in patients with Parkinson's disease, *Mov. Disord.*, 9, 437, 1994.

94. Oh, J. D., Vaughan, C. L., and Chase, T. N., Effect of dopamine denervation and dopamine agonist administration on serine phosphorylation of striatal NMDA receptor subunits, *Brain Res.*, 821, 433, 1999.

95. Oh, J. D. et al., Enhanced tyrosine phosphorylation of striatal NMDA receptor subunits: effect of dopaminergic denervation and L-DOPA administration, *Brain Res.*, 813, 150, 1998.

96. Oh, J. D., Del Dotto, P., and Chase, T. N., Protein kinase A inhibitor attenuates levodopa-induced motor response alterations in the hemi-parkinson rat, *Neurosci. Lett.*, 228, 5, 1997.

97. Wolters, E. C. and Berendse, H. W., Management of psychosis in Parkinson's disease, *Curr. Opin. Neurol.*, 14, 499, 2001.

98. Poewe, W. and Seppi, K., Treatment options for depression and psychosis in Parkinson's disease, *J. Neurol.*, 248, 12, 2001.

99. McKeith, I., et al., Efficacy of rivastigmine in dementia with Lewy bodies: a randomised, double-blind, placebo-controlled international study, *Lancet*, 356, 2031, 2000.

100. Bergman, J. and Lerner, V., Successful use of donepezil for the treatment of psychotic symptoms in patients with Parkinson's disease, *Clin. Neuropharmacol.*, 25, 107, 2002.

101. Fabbrini. G. et al., Donepezil in the treatment of hallucinations and delusions in Parkinson's disease, *Neurol. Sci.*, 23, 41, 2002.

102. Okun, M. S., McDonald, W. M., DeLong, M. R., Refractory nonmotor symptoms in male patients with Parkinson's disease due to testosterone deficiency: a common unrecognized comorbidity, *Arch. Neurol.*, 59, 807, 2002.

103. Okun, M. S. et al., Beneficial effect of testosterone replacement for the nonmotor symptoms of Parkinson's disease, *Arch. Neurol.*, 59, 1750, 2002.

104. Tsang, K. L., Ho, S. L., and Lo, S. K., Estrogen improves motor disability in parkinsonian postmenopausal women with motor fluctuations, *Neurology*, 54, 2292, 2000.

105. Blanchet, P. J. et al., Short-term effect of high-dose 17beta-estradiol in postmenopausal PD patients: a cross-over study, *Neurology*, 53, 91, 1999.

106. Pak, P. K. et al., Botulinum toxin A as treatment for drooling saliva in PD, *Neurology*, 54, 244, 2000.

107. Hyson, H. C., Johnson, A. M., and Jog, M. S., Sublingual atropine for sialorrhea secondary to parkinsonism – a pilot study, *Mov. Disord.*, 16, 1318, 2002.

108. Happe, S. et al., Successful treatment of excessive daytime sleepiness in Parkinson's disease with modafinil, *J. Neurol.*, 248, 632, 2001.

109. Bradwejn, J. and Koszycki, D., Cholecystokinin and panic disorder: past and future clinical research strategies, *Scand. J. Clin. Lab. Invest.*, 234, 19, 2001.

69 Restorative Therapy in Parkinson's Disease

Joel M. Trugman
University of Virginia

CONTENTS

INTRODUCTION

Restoration means "to bring back to a former or original state, to repair or renew."[1] In Parkinson's disease (PD), a restorative therapy should return anatomical and functional integrity to the nigrostriatal dopamine system and reverse clinical symptoms and signs. There are currently no restorative therapies in PD, but two new classes of drugs, the neuroimmunophilin ligands and glial cell line-derived neurotrophic factor, hold great promise. In *in vitro* experiments and in animal models of PD, these drugs have the ability to promote axonal regeneration and repair the damaged nigrostriatal system. This chapter reviews the mechanism of action, preclinical, and clinical studies of these two agents.

NEUROIMMUNOPHILIN LIGANDS

BIOLOGY AND MECHANISM OF ACTION

Neuroimmunophilin ligands are a class of drugs that has been developed over the past ten years, based on a series of remarkable insights from cell biology, chemistry, and neuroscience. Immunophilins are cytosolic binding proteins that are receptors for the new immunosuppressant drugs cyclosporin A and FK506 (tacrolimus). The term "immunophilin ligand" refers to drugs that bind to immunophilins and includes both the clinically used immunosuppressants (cyclosporin A, tacrolimus, and rapamycin) and the novel nonimmunosuppressant drugs, which are being developed for neurological therapeutic indications (neuroimmunophilin ligands). Immunophilins are present in the central and peripheral nervous system in concentrations 10 to 30 times greater than in the immune system, and immunophilin ligands promote nerve regeneration in models of peripheral nerve and CNS injury. The chemical synthesis of nonimmunosuppressant immunophilin ligands that retained the ability to promote nerve regeneration was a key discovery leading to the current interest in these drugs as potential restorative therapy in PD.[2–4]

Immunophilins were first identified in T lymphocytes as cytoplasmic binding proteins, or receptors, for immunosuppressive drugs. In 1984, Handschumacher and colleagues showed that cyclosporin A binds to a 18 kDa cytoplasmic protein termed cyclophilin A.[5] This was the first indication that the immunosuppressive activity of cyclosporin A is mediated by an intracellular rather than a membrane-associated mechanism. Other members of the cyclophilin family have since been identified.[6] In 1989, it was found that FK506 binds to a different cytosolic protein, termed FK506 binding protein (FKBP).[7] Like the cyclophilins, the FKBPs constitute a family of related proteins including FHBP12, FKBP13, FKBP25, and

FKBP52. The cyclophilins and the FKBPs are the two main classes of immunophilins.

The next key observation was that immunophilins are present in significant concentration in neural tissue of the central and peripheral nervous systems. Steiner et al.[8] showed that the levels of FKBP and FKBP mRNA are high in rat brain, and their regional localizations are nearly identical to those of calcineurin. In rat brain, [^{3}H]-FK506 binding sites are present in highest concentration in the striatum, substantia nigra pars reticulata, and the hippocampus. Concentrations of binding sites in these regions are approximately 10 to 20 times greater than in the thymus or spleen. Quinolinic acid lesions of the striatum markedly reduced [^{3}H]-FK506 binding in the striatum and substantia nigra pars reticulata, indicating that the FKBP is present mainly in striatonigral neurons (striatal output neurons).[8,9] In the cerebellum, FKBP is present in granule cells and their processes.

Immunophilin ligands can have both immunosuppressant and nerve regenerating actions. Immunosuppression is mediated by calcineurin inhibition. Complexes of cyclosporin A-cyclophilin A and FK506-FKBP12 act by binding to calcineurin, a calcium/calmodulin-activated protein phosphatase, and inhibiting its activity. In T lymphocytes, inhibition of calcineurin phosphatase activity, via a cascade of events, prevents the expression of cytokines such as interleukin-2 and leads to the failure of T-cell activation.[10,11]

The intracellular mechanisms mediating the neuroregenerative actions of immunophilin ligands have been difficult to pinpoint. FK506 has two binding domains, one mediating inhibition of FKBP isomerase activity and the other mediating calcineurin inhibition.[11] Using principles of structure-based drug design, FK506 derivatives were synthesized that retained the ability to bind to the enzymatic site but not inhibit calcineurin (therefore lacking immunosuppressant activity). The FK506 derivatives GPI-1046 and V-10,367 are the prototypic nonimmunosuppressant neuroimmunophilin ligands.[12,13] The separation of the nerve-regenerating properties from the immunosuppressant properties of these ligands was a crucial breakthrough from a mechanistic/drug development point of view. Binding to FKBP12 and calcineurin inhibition have been excluded as mechanisms mediating nerve regeneration; recent work has implicated FKBP-52, a component of mature steroid receptor complexes.[14] A model has been proposed whereby ligands that bind to FKBP-52 disrupt steroid receptor complexes leading to activation of the mitogen activated protein kinase/extracellular signal-related kinase pathway.[15]

PRECLINICAL STUDIES

The potential of immunophilin ligands to promote nerve regeneration was first detected in cell culture systems.

Growth-associated protein (GAP) 43 is a neural specific protein involved in axon growth during nerve development and regeneration and is also a substrate for calcineurin phosphatase activity.[16–18] In models of facial and sciatic nerve regeneration following injury, levels of FKBP increase in parallel with increases in GAP-43, suggesting a role for FKBP in neurite outgrowth.[19] Lyons et al.[20] examined the effects of FK506 in rat pheochromocytoma (PC12) cells and in rat sensory ganglia and showed that it enhances nerve growth factor-induced outgrowth in both cell systems. The drug was highly potent, effective at subnanomolar concentrations.

Immunophilin ligands were next shown to promote nerve regeneration in *in vivo* models of peripheral nerve injury. Gold et al.[12,21] treated rats with daily subcutaneous injections of FK506 (1.0 to 10.0 mg/kg) for 18 days following a sciatic nerve crush injury. There was a maximal effect of a 34% increase in the rate of axonal regeneration in sensory fibers. This was accompanied by a significant increase in the caliber (mean axonal area) of the regenerating axons. This was the first demonstration that systemic administration of an immunophilin ligand can promote axonal regeneration. In contrast, cyclosporin A failed to increase axonal caliber or functional recovery, showing that the nerve regenerating property of FK506 is independent of calcineurin inhibition.[22] Injections of the nonimmunosuppressant FKBP-12 ligand V-10,367 also accelerate nerve regeneration in the rat sciatic nerve crush model, and this drug is orally active.[12,23]

There is solid experimental evidence from several laboratories to indicate that neuroimmunophilin ligands can reverse some of the abnormalities in the nigrostriatal system in rodent models of PD. Treatment with GPI-1046 seems to promote striatal reinnervation by sprouting of residual dopamine neurons. Ross et al.[24] studied rats with unilateral 6-hydroxydopamine nigral lesions and treated the rats with GPI-1046 beginning two months after the lesion. Treatment with GPI-1046 (10 mg/kg/day for two weeks) resulted in a two- to threefold increase in the density of tyrosine hydroxylase (TH) positive fibers in the ipsilateral striatum. Interestingly, there were extensive regions in the dorsolateral striatum that remained completely denervated with no indication of reinnervation. Treatment with GPI-1046 did not alter the TH positive fiber innervation in the contralateral, normally innervated striatum. These results are consistent with a regenerative process mediated by GPI-1046-induced sprouting of residual TH positive fibers, which were spared by the 6-hydroxydopamine lesion. In a different model, in which 6-hydroxydopamine was injected directly into the striatum, the orally active immunophilin ligand V-10,367 prevented the retrograde death of dopamine neurons in the substantia nigra, enhanced TH positive fiber innervation of the lesioned striatum, and improved behavioral measures of dopamine denervation.[25]

Treatment with GPI-1046 reverses behavioral and electrophysiological indexes of dopamine denervation. Zhang et al.[26] treated rats with stable unilateral 6-hydroxydopamine nigral lesions with GPI-1046 (10 mg/kg/day for one week). GPI-1046 did not increase the number of nigral neurons. GPI-1046 reduced amphetamine-induced rotations by 75%, a behavioral marker for asymmetric dopamine innervation in this model. Dopamine denervation prevents the normally occurring long-term potentiation induced by high-frequency stimulation of the corticostriatal pathway. Treatment with GPI-1046 for one week restored long-term potentiation in this model. This is a striking result, because long-term potentiation in the corticostriatal pathway may underlie dopamine-mediated synaptic plasticity and play a key role in dopamine effects on movement.[27]

Not all studies in rats have found a beneficial effect of GPI-1046. Harper et al.[28] reported that two weeks of treatment with GPI-1046 did not improve circling behavior or the density of TH positive fiber innervation in rats with stable 6-hydroxydopamine nigral lesions. It is possible that these rats were too extensively denervated, and the beneficial effects of GPI-1046 may depend on having a sufficient number of residual dopamine neurons available to reinnervate the striatum.

Regenerative/restorative effects of the neuroimmunophilin ligands have not yet been demonstrated in primate models of PD. Monkeys with unilateral MPTP lesions, achieved through intracarotid infusions of the toxin, were treated with GPI-1046 (10 mg/kg/day) for eight weeks.[29] GPI-1046 did not improve clinical parkinsonism scores, striatal dopamine transporter density, or the density of TH immunoreactive fibers in the striatum. At this time, it is not clear why the immunophilin ligands appear to be effective in rodent but not primate models of PD. The dose given to primates may not have been sufficient, or there may be species differences in the biological effects of this compound, but clearly more work is needed in primates to establish efficacy.

CLINICAL TRIALS

The effect of the neuroimmunophilin ligand GPI-1485 was tested in a phase 2, 6-month, clinical trial of 300 patients with PD.[30,31] Subjects with PD of less than five years duration and no significant motor fluctuations were treated with either placebo, low-dose GPI 1485 (200 mg p.o.q. i.d.), or high-dose GPI 1485 (1000 mg p.o. q.i.d.). One hundred and five subjects had [123I]β-CIT/SPECT scans performed at baseline and at six months to determine if the neuroimmunophilin ligand had an effect on dopamine transporter density. GPI-1485 was well tolerated except for nausea and mild dyspepsia. The clinical results were unremarkable: treatment with GPI-1485 at low or high dose did not improve the UPDRS motor score in the

"defined off" condition. There was improvement in the Hoehn and Yahr score, however, when the high-dose treatment group was compared to placebo. There were several limits to this trial, the main one being that six months of treatment is probably too short a time to see a clinical improvement. In addition, subjects with a broad distribution of PD severities (H and Y stages 1 through 3) were enrolled, and the study was not controlled for the modification of antiparkinsonian medications during the treatment period.

[123I]β-CIT/SPECT imaging of dopamine transporter density showed interesting results: in the placebo group, striatal β-CIT uptake declined 0.6% at six months compared to baseline, whereas there was an increase of 1.1% and 2.2% in striatal uptake in the subjects treated with low-dose and high-dose GPI-1485. These differences were not statistically significant but suggest that there may be a dose-dependent effect of GPI-1485 on striatal dopamine nerve terminal density.[31]

To pursue these preliminary results, Guilford Pharmaceuticals Inc. has initiated a new phase 2 clinical trial to evaluate the safety, pharmacokinetics, and efficacy of GPI-1485 in patients with mild to moderate PD.[32] Two hundred subjects, treated with dopamine agonists but not levodopa at baseline, will receive either placebo or GPI-1485 (1000 mg p.o. q.i.d.). Interestingly, the primary endpoint of this new study will be a neuroimaging measure, a 50% or greater reduction in the loss of dopamine transporters as assessed by [123I]β-CIT/SPECT. Also of interest is that the dose being tested in this trial is approximately 50 mg/kg/day, which is five times greater than the dose tested in monkeys with MPTP lesions.[29]

GLIAL CELL LINE-DERIVED NEUROTROPHIC FACTOR

BIOLOGY AND MECHANISM OF ACTION

Glial cell line-derived neurotrophic factor (GDNF) was discovered and cloned in 1993 and reported to be a potent trophic factor for midbrain dopamine neurons.[33,34] The existence of dopaminergic neurotrophic activity in media derived from primary glial cells had been known; Lin and colleagues purified GDNF to homogeneity from the supernatant of the rat B49 glial cell line. In embryonic midbrain cultures, GDNF increased the number of dopamine neurons about threefold and increased the average cell body size more than 50%. GDNF dramatically increased TH positive neurite outgrowth. In the original publication, the authors suggested that this newly identified neurotrophic factor might be useful in the treatment of PD.

Although it was originally thought to have specificity only for dopamine neurons, GDNF was later found to have effects on other neuronal populations, including spinal motoneurons, a subset of dorsal root ganglia cells, as well as sympathetic, parasympathetic, and enteric neurons.[35,36] GDNF also is important for the development of the kidney and regulates the differentiation of spermatogonia. Molecules related to GDNF have subsequently been discovered including neurturin, artemin, and persephin, and, collectively, these are referred to as GDNF family ligands (GFLs). GFLs belong to the transforming growth factor-β superfamily and function as homodimers.[35,37]

Neurotrophic factors are molecules that are secreted by target tissues and required for the development and maintenance of innervating neurons.[38] After intracellular processing, GDNF is secreted as a glycosylated protein of 134 amino acid residues and acts as a trophic factor for substantia nigra dopamine neurons. The mRNA for GDNF is expressed in striatal astrocytes,[39] and retrograde transport of GDNF from the striatum to nigral dopamine neurons has been demonstrated.[40] Expression of GDNF receptor components (GFRα1 and RET) has been studied as well and all neurons responsive to GDNF *in vivo* express both receptor components.[41]

The receptor and signalling mechanisms that mediate the actions of GDNF are complex.[35,37,42–44] GDNF acts via a multicomponent receptor complex composed of the GDNF family receptor α1 subtype (GFRα1), which is linked to the plasma membrane by a glycosyl phosphotidylinositol anchor, and the transmembrane RET receptor that incorporates an intracellular tyrosine kinase domain. The binding of GDNF, via the activation of the intracellular tyrosine kinase, ultimately activates intracellular signalling pathways, leading to neuronal differentiation and survival.[42]

PRECLINICAL STUDIES

A large number of studies indicate that GDNF can partially restore anatomical and functional integrity to the nigrostriatal system in animal models of PD. Initial studies in rats showed that direct nigral injection of GDNF, as well as intraventricular injection, increase dopamine levels and promote sprouting of TH positive neurons in the striatum and functional recovery.[45–47] GDNF was also found to protect dopaminergic neurons from retrograde cell death after axotomy.[48]

Bjorklund and colleagues have studied extensively the partial lesion model of unilateral striatal injection of 6-hydroxydopamine in the rat.[49–52] In this model, the injury to dopamine axons in the striatum leads to the gradual death of cell bodies in the substantia nigra. The protective properties of GDNF can be examined in the acute phase of injury and restorative/regenerative properties can be examined in the later phase when some dopamine neurons persist in a dysfunctional state. Using this model, it has been shown that GDNF delivered to the substantia nigra is not sufficient to restore striatal innerva-

tion. Functional recovery in this model is best achieved when GDNF is delivered continuously to the striatum over a period of months.

Grondin and colleagues[53] have published the definitive study to date on the effects of GDNF in parkinsonian monkeys, following up on the initial observations of Gash et al.[54] Rhesus monkeys with unilateral MPTP lesions induced by intracarotid injection were treated with continuous infusion of GDNF by the intracerebroventricular (ICV) route (5 μg per day) or by direct infusion into the dopamine-denervated putamen (15 μg per day). Treatment was not begun until three months after MPTP lesion, and the animals were treated for three months. The animals were studied behaviorally using a nonhuman primate parkinsonism scale, and they were extensively examined neurochemically and histochemically for dopaminergic indexes. The aim of the study was to assess the restorative potential of GDNF in a primate model of PD.

GDNF treatment improved parkinsonism by both ICV and intraputamenal routes of infusion, and the routes were not compared. Overall, there was a 36% improvement in parkinsonism score compared to vehicle-treated controls after 12 weeks of treatment. At postmortem, there was substantial evidence for restoration of the dopamine system. The number of dopamine neurons expressing tyrosine hydroxylase increased by 20% on the unlesioned side and nearly doubled on the lesioned side. Dopaminergic neuron cell body size increased about 30% bilaterally. Interestingly, these findings are similar to the first reported effects of GDNF on embryonic dopamine neurons in culture.[33] In the GDNF-treated monkeys there was a threefold increase in dopamine levels in the medial, periventricular striatum and a fivefold increase in TH positive fiber density in this same striatal region on the lesioned side. In this model, the dopamine fibers in the periventricular striatal regions (the medial caudate and nucleus accumbens) are relatively spared by the MPTP lesion and the findings suggest that some residual dopaminergic innervation is necessary for GDNF-induced recovery.

The mechanisms that underlie the restorative effects of GDNF are not yet clear and may be multiple. There appear to be bilateral effects on nigral cell number and size. This study also emphasized the importance of dose and timing of GDNF delivery in optimizing the benefit from this drug.

CLINICAL TRIALS

There have been two published studies of GDNF treatment for PD. In a placebo-controlled trial of GDNF given by the ICV route, no benefit was observed.[55] In a small, open-label pilot trial of GDNF infused directly into the putamen, clear antiparkinsonian effects were seen.[56] The drug used in these clinical trials is a recombinant protein with an amino acid sequence identical to native human GDNF with the addition of an amino terminal methionine (r-metHuGDNF). These studies will be reviewed in detail.

Fifty subjects with advanced PD were enrolled in a phase 1 to 2 placebo-controlled dose-escalation trial of GDNF administered by monthly ICV injection.[55] Successive cohorts of subjects received 25, 75, 150, 300, and 500 to 4000 µg of GDNF, or placebo, for 6 to 8 months. Total and motor UPDRS scores in both "on" and "off" states were not improved by GDNF at any dose. Nausea, vomiting, and anorexia were common for several days after the injection of GDNF. Weight loss, on occasion severe, occurred in most subjects receiving ≥75 µg of GDNF per injection. Sensory symptoms, including paresthesias and Lhermitte's sign, were common in GDNF-treated subjects, perhaps reflecting the effect of GDNF on dorsal root ganglion neurons.[36] Hyponatremia occurred in more than half of the subjects who received doses of ≥75 µg of GDNF; in two subjects, serum sodium levels fell to the 115 to 120 mmol/L range. The investigators concluded that GDNF is biologically active, as evidenced by the spectrum of side effects but that GDNF did not improve parkinsonism when given by the ICV route. It was speculated that GDNF may not have reached the target tissues of the putamen or the substantia nigra when given by this route. This view was supported by a single case report in which one of the subjects in this study died, and there was no postmortem evidence of significant regeneration of nigrostriatal neurons.[57]

Five subjects with advanced PD were studied in a pilot, phase 1 safety trial of GDNF delivered directly into the putamen by chronic infusion.[56] One subject had unilateral, and four subjects had bilateral, infusions into the posterior-dorsal putamen. Areas of low [18F]dopa uptake in the putamen visualized by PET were targeted for injection of GDNF, which was delivered by continuous infusion using pumps (14 µg per putamen per day at 6 µl per hour). The subjects were followed for 12 months.

In this study, PD symptoms and signs began to improve within three months of treatment. After 12 months of GDNF, total UPDRS in the clinically defined "off" phase was 48% lower than at baseline. In addition, there was a 45% reduction in total UPDRS in the defined "on" phase, a noteworthy finding since improvement in the "on" phase is not typically seen in surgical treatments of PD. Periods of severe immobility were markedly attenuated. Improvements were observed in the activities of daily living and motor subsections of the UPDRS. Dyskinesias were reduced by 64%, and off-medication dyskinesias were not seen. Timed motor tasks such as pronation/supination and finger dexterity also improved in the off and on medication states. Smell, taste, and sexual function partially improved.

[18F]dopa uptake increased by 16 to 25% in the putamen bilaterally and right substantia nigra. Increased [18F]dopa uptake in the substantia nigra suggested that nigral cell bodies were responding to GDNF delivered to the nerve terminals in the putamen.

The results of this preliminary trial are impressive and suggest that intraputamenal delivery of GDNF may well be an effective treatment for PD. A placebo-controlled trial of intraputamenal GDNF is currently underway in North America to investigate this further.

CONCLUSION

As can be seen from this brief review, the potential for restorative therapy in PD is real, here and now. For both the neuroimmunophilin ligands and GDNF, the time from laboratory discovery to the initiation of clinical trials was less than five years, an incredible speed. However, as noted by Brundin,[58] the path from discovery of a new molecule to testing in cell systems, animal models of PD, and clinical trials is neither simple nor linear. Research at each of these levels collectively should bring progress. The fact that both of the compounds described in this chapter are currently in clinical trial in PD is most encouraging.

REFERENCES

1. *The Merriam-Webster Dictionary*, G & C Merriam Co., Springfield, MA, 1974.
2. Hamilton, G. S. and Steiner, J. P., Immunophilins: beyond immunosuppression, *J. Med. Chem.*, 41(26), 5119, 1998.
3. Snyder, S. H., Lai, M. M., and Burnett, P. E., Immunophilins in the nervous system, *Neuron*, 21(2), 294, 1998.
4. Gold, B. G. and Nutt, J. G., Neuroimmunophilin ligands in the treatment of Parkinson's disease, *Curr. Opin. Pharmacol.*, 2(1), 82, 2002.
5. Handschumacher, R. E. et al., Cyclophilin: a specific cytosolic binding protein for cyclosporin A, *Science*, 226(4674), 544, 1984.
6. Stamnes, M. A. and Zuker, C. S., Peptidyl-prolyl cis-trans isomerases, cyclophilin, FK506-binding protein, and ninaA: four of a kind, *Curr. Opin. Cell Biol.*, 2(6), 1107, 1990.
7. Harding, M. W. et al., A receptor for the immunosuppressant FK506 is a cis-trans peptidyl-prolyl isomerase, *Nature*, 341(6244), 758, 1989.
8. Steiner, J. P. et al., High brain densities of the immunophilin FKBP colocalized with calcineurin, *Nature*, 358 (6387), 584, 1992.
9. Dawson, T. M. et al., The immunophilins, FK506 binding protein and cyclophilin, are discretely localized in the brain: relationship to calcineurin, *Neuroscience*, 62(2), 580, 1994.
10. Liu J. et al., Calcineurin is a common target of cyclophilin-cyclosporin A and FKBP-FK506 complexes, *Cell*, 66(4), 807, 1991.

11. Schreiber, S. L., Chemistry and biology of the immunophilins and their immunosuppressive ligands, *Science,* 251(4991), 283, 1991.

12. Gold, B. G. et al., A nonimmunosuppressant FKBP-12 ligand increases nerve regeneration, *Exp. Neurol.,* 147(2), 269, 1997.

13. Steiner, J. P. et al., Neurotrophic immunophilin ligands stimulate structural and functional recovery in neurodegenerative animal models, *Proc. Natl. Acad. Sci. U.S.A.,* 94(5), 2019, 1997.

14. Gold, B. G. et al., Immunophilin FK506-binding protein 52 (not FK506-binding protein 12) mediates the neurotrophic action of FK506, *J. Pharmacol. Exp. Ther.,* 289(3), 1202, 1999.

15. Gold, B. G. and Villafranca, J. E., Neuroimmunophilin ligands: the development of novel neuroregenerative/neuroprotective compounds, *Curr. Top. Med. Chem.,* 3(12), 1368, 2003.

16. Liu, Y. C. and Storm. D. R., Dephosphorylation of neuromodulin by calcineurin, *J. Biol. Chem.,* 264(22), 12800, 1989.

17. Skene, J. H., Axonal growth-associated proteins, *Annu. Rev. Neurosci.,* 12, 127, 1989.

18. Tetzlaff, W. et al., Axonal transport and localization of B-50/GAP-43-like immunoreactivity in regenerating sciatic and facial nerves of the rat, *J. Neurosci.,* 9(4), 1303, 1989.

19. Lyons, W. E. et al., Neuronal regeneration enhances the expression of the immunophilin FKBP-12, *J. Neurosci.,* 15(4), 2985, 1995.

20. Lyons, W. E. et al., Immunosuppressant FK506 promotes neurite outgrowth in cultures of PC12 cells and sensory ganglia, *Proc. Natl. Acad. Sci. U.S.A.,* 91(8), 3191, 1994.

21. Gold, B. G., Katoh, K., and Storm-Dickerson, T., The immunosuppressant FK506 increases the rate of axonal regeneration in rat sciatic nerve, *J. Neurosci.,* 15(11), 7509, 1995.

22. Wang, M. S., Zeleny-Pooley, M., and Gold, B. G., Comparative dose-dependence study of FK506 and cyclosporin A on the rate of axonal regeneration in the rat sciatic nerve, *J. Pharmacol. Exp. Ther.,* 282(2), 1084, 1997.

23. Gold, B. G. et al., Oral administration of a nonimmunosuppressant FKBP-12 ligand speeds nerve regeneration, *Neuroreport,* 9(3), 553, 1998.

24. Ross, D. T. et al., The small molecule FKBP ligand GPI 1046 induces partial striatal re-innervation after intranigral 6-hydroxydopamine lesion in rats, *Neurosci. Lett.,* 297(2), 113, 2001.

25. Costantini, L. C. et al., Immunophilin ligands can prevent progressive dopaminergic degeneration in animal models of Parkinson's disease, *Eur. J. Neurosci.,* 13(6), 1085, 2001.

26. Zhang, C. et al., Regeneration of dopaminergic function in 6-hydroxydopamine-lesioned rats by neuroimmunophilin ligand treatment, *J. Neurosci.,* 21(15), RC156, 2001.

27. Arbuthnott, G. W., Ingham, C. A., and Wickens, J. R., Dopamine and synaptic plasticity in the neostriatum, *J. Anat.,* 196(Pt. 4), 587, 2000.

28. Harper, S. et al., Analysis of the neurotrophic effects of GPI-1046 on neuron survival and regeneration in culture and *in vivo, Neuroscience,* 88(1), 267, 1999.

29. Eberling, J. L. et al., The immunophilin ligand GPI-1046 does not have neuroregenerative effects in MPTP-treated monkeys, *Exp. Neurol.,* 178(2), 236, 2002.

30. Guilford Pharmaceuticals Inc., Guilford Pharmaceuticals announces completion of NIL-A phase II clinical trial for Parkinson's disease, press release, July 26, 2001, at http://www.corporate-ir.net/ireye/ir_site.zhtml?ticker= GLFD&script=410&layout=-6&item_id=195083.

31. Seibyl, J. P. et al., 123I beta-CIT SPECT imaging assessment of Parkinson disease patients treated for six months with a neuroimmunophillin ligand Nil-A, *Neurology,* 58(7)(Suppl. 3), A203, 2002.

32. Guilford Pharmaceuticals Inc., Guilford Pharmaceuticals initiates Phase II trial of GPI 1485 for the treatment of Parkinson's disease, press release, Nov. 6, 2002 at http://www.corporate-ir.net/ireye/ir_site.zhtml?ticker= GLFD&script=410&layout=-6&item_id=353663.

33. Lin, L. F. et al., GDNF: a glial cell line-derived neurotrophic factor for midbrain dopaminergic neurons, *Science,* 260(5111), 1130, 1993.

34. Lin, L. F. et al., Purification and initial characterization of rat B49 glial cell line-derived neurotrophic factor, *J. Neurochem.,* 63(2), 758, 1994.

35. Airaksinen, M. S. and Saarma, M., The GDNF family: signalling, biological functions and therapeutic value, *Nat. Rev. Neurosci.,* 3(5), 383, 2002.

36. Bennett, D. L. et al., A distinct subgroup of small DRG cells express GDNF receptor components and GDNF is protective for these neurons after nerve injury, *J. Neurosci.,* 18(8), 3059, 1998.

37. Baloh, R. H. et al., The GDNF family ligands and receptors—implications for neural development, *Curr. Opin. Neurobiol.,* 10(1), 103, 2000.

38. Oppenheim, R. W., Cell death during development of the nervous system, *Annu. Rev. Neurosci.,* 14, 453, 1991.

39. Schaar, D. G. et al., Regional and cell-specific expression of GDNF in rat brain, *Exp. Neurol.,* 124(2), 368, 1993.

40. Tomac, A. et al., Retrograde axonal transport of glial cell line-derived neurotrophic factor in the adult nigrostriatal system suggests a trophic role in the adult, *Proc. Natl. Acad. Sci. U.S.A.,* 92(18), 8274, 1995.

41. Trupp, M. et al., Complementary and overlapping expression of glial cell line-derived neurotrophic factor (GDNF), c-ret proto-oncogene, and GDNF receptoralpha indicates multiple mechanisms of trophic actions in the adult rat CNS, *J. Neurosci.,* 17(10), 3554, 1997.

42. Tansey, M. G. et al., GFRalpha-mediated localization of RET to lipid rafts is required for effective downstream signaling, differentiation, and neuronal survival, *Neuron,* 25(3), 611, 2000.

43. Trupp, M. et al., Functional receptor for GDNF encoded by the c-ret proto-oncogene, *Nature,* 381(6585), 785, 1996.

44. Sariola, H. and Saarma, M., Novel functions and signalling pathways for GDNF, *J. Cell Sci.,* 116(Pt. 19), 3855, 2003.

45. Hudson, J. et al., Glial cell line-derived neurotrophic factor augments midbrain dopaminergic circuits *in vivo*, *Brain Res. Bull.*, 36(5), 425, 1995.

46. Hebert, M. A. et al., Functional effects of GDNF in normal rat striatum: presynaptic studies using *in vivo* electrochemistry and microdialysis, *J. Pharmacol. Exp. Ther.*, 279(3), 1181, 1996.

47. Lapchak, P. A. et al., Glial cell line-derived neurotrophic factor attenuates behavioural deficits and regulates nigrostriatal dopaminergic and peptidergic markers in 6-hydroxydopamine-lesioned adult rats: comparison of intraventricular and intranigral delivery, *Neuroscience*, 78(1), 61, 1997.

48. Beck, K. D. et al., Mesencephalic dopaminergic neurons protected by GDNF from axotomy-induced degeneration in the adult brain, *Nature*, 373(6512), 339, 1995.

49. Bjorklund, A. et al., Studies on neuroprotective and regenerative effects of GDNF in a partial lesion model of Parkinson's disease, *Neurobiol. Dis.*, 4(3–4), 186, 1997.

50. Rosenblad, C., Martinez-Serrano, A., and Bjorklund, A., Intrastriatal glial cell line-derived neurotrophic factor promotes sprouting of spared nigrostriatal dopaminergic afferents and induces recovery of function in a rat model of Parkinson's disease, *Neuroscience*, 82(1), 129, 1998.

51. Kirik, D., Rosenblad, C., and Bjorklund, A., Preservation of a functional nigrostriatal dopamine pathway by GDNF in the intrastriatal 6-OHDA lesion model depends on the site of administration of the trophic factor, *Eur. J. Neurosci.*, 12(11), 3871, 2000.

52. Kirik, D. et al., Delayed infusion of GDNF promotes recovery of motor function in the partial lesion model of Parkinson's disease, *Eur. J. Neurosci.*, 13(8), 1589, 2001.

53. Grondin, R. et al., Chronic, controlled GDNF infusion promotes structural and functional recovery in advanced parkinsonian monkeys, *Brain*, 125(Pt. 10), 2191, 2002.

54. Gash, D. M. et al., Functional recovery in parkinsonian monkeys treated with GDNF, *Nature*, 380(6571), 252, 1996.

55. Nutt, J. G. et al., Randomized, double-blind trial of glial cell line-derived neurotrophic factor (GDNF) in PD, *Neurology*, 60(1), 69, 2003.

56. Gill, S. S. et al., Direct brain infusion of glial cell line-derived neurotrophic factor in Parkinson disease, *Nat. Med.*, 9(5), 589, 2003.

57. Kordower, J. H. et al., Clinicopathological findings following intraventricular glial-derived neurotrophic factor treatment in a patient with Parkinson's disease, *Ann. Neurol.*, 46(3), 419, 1999.

58. Brundin, P., GDNF treatment in Parkinson's disease: time for controlled clinical trials? *Brain*, 125(Pt. 10), 2149, 2002.

70 Gene Therapy for Parkinson's Disease

Ronald J. Mandel, Edgardo Rodriguez, Fredric P. Manfredsson, and Carmen S. Peden
Department of Neuroscience, Powell Gene Therapy Center, McKnight Brain Institute, University of Florida College of Medicine

M. Angela Cenci
Wallenburg Neuroscience Centre, University of Lund

Stuart E. Leff
Emory University

CONTENTS

INTRODUCTION

Gene therapy involves the introduction of a novel genetic sequence to a specific population of cells, either *ex vivo* or *in vivo*, for the purpose of *de novo* synthesis of proteins or various peptides. An engineered vector containing the desired genetic sequence facilitates the transduction of the target cells. Gene therapy for Parkinson's disease (PD) can be used to continuously deliver large molecules like dopamine (DA), its precursors, enzymes, and various trophic factors inside the blood-brain barrier (BBB) without the use of indwelling mechanical devices such as pumps or reservoirs. A neurosurgical procedure is still used to deliver the gene therapy vector to a localized delivery area. Therefore, the treatment of PD continues to be a major focus for neurological gene therapists, because it will be sufficient to deliver a transgene or reduce expression of a gene in a particular anatomical region to the exclusion of global delivery. The present chapter will review the current gene therapy strategies under investigation for the treatment of PD and will introduce gene therapies that are potentially on the clinical horizon.

The progressive nature of PD and the current limitations in its treatment have opened the door for a variety of gene therapy applications. PD advances from early to late stages in both its idiopathic and inherited forms. Early and late stages can be differentiated by time from disease onset, the severity of symptoms, and response to medications. Gene therapy strategies for PD also separate along these lines, and many therapies are being investigated to either prevent progression in early PD or to ameliorate symptoms and side effects of medicinal therapy seen in late PD.

In the earliest stage, PD patients are newly diagnosed and often have unilateral symptom presentation. In these early stages of PD, medical management of patients is optimal. L-dopa can replace lost function and restore speed and flexibility at doses that produce minimal adverse effects. This is often referred to as the "honeymoon period." At this point, approximately 50% of the dopaminergic neurons in the substantia nigra are lost, and those remaining will continue to die in a linear fashion. SPECT scanning in these early, treatment naïve PD patients reveals an average reduction of 8% of striatal DA receptor activity from baseline per year.[238] For this reason, and the relative success in treating early PD patients, preventing further cell loss and subsequent decline in function is an appealing strategy. Of all the neuroprotection strategies being investigated, gene therapy strategies exhibit a great deal of promise in reducing, preventing, and in some cases reversing the pathology. Existing results from neuroprotection drug trials are subject to confounding interference, and gene therapy neuroprotection studies are still in the preclinical stages. Therefore, at this time, early PD inevitably progresses to late-stage PD.

Late-stage PD is usually reached when patients have been taking L-dopa for more than five years. Symptoms are manifested on both sides of the body and are usually much more debilitating. Likewise, previously effective doses of L-dopa are ineffective, and increasing doses and additional palliative drugs are required. L-dopa resistance includes both motor and nonmotor symptoms as well as side effects resulting from the increased dose. Reduction in the levels of L-dopa will reduce side effects but can result in loss of therapeutic effect. Fluctuations in DA levels may be responsible, while constant delivery via gene therapy may be able to ameliorate these unwanted effects.

GENE THERAPY BACKGROUND

Ex Vivo Gene Therapy

Foreign genes can be delivered into the CNS via two distinct methods: *ex vivo* or *in vivo* gene transfer. *Ex vivo* gene transfer involves the transduction, or genetic engineering, of cells that are grown in culture to express a foreign gene. Following transduction, cells are characterized and implanted into the specific target region where gene expression is required. This approach conveys advantages, because the precise quantity of transduced cells implanted can be controlled, concomitant with the dose of foreign gene delivered. Another advantage attributed to *ex vivo* gene transfer is the potential to genetically modify neural stem cells that are capable of both replacing neural cell types that are lost during disease, as well as promoting the survival and function of remaining endogenous cells via secretion of neurotrophic factors.[1,245]

Most of the research has been focused on the use of neural stem cells or modified fibroblasts as the cells of choice for *ex vivo* gene therapy.[1,72] However, it has proven difficult to achieve a high level of transduction in cultured neural stem cells. Additionally, while cell types, such as modified fibroblasts, can support unlimited gene expression while in culture, it has been difficult to demonstrate long-term gene expression *in vivo* after transplantation.[9,62,131,142] However, the use of different cell types as well as new regulatory sequences in the gene expression cassette could overcome this lack of long-term *in vivo* gene expression.

Ex vivo gene therapy has many conceptual as well as practical safety advantages that make this strategy attractive for clinical use. Many of the technical difficulties associated with *ex vivo* gene transfer was thought to be simplified by encapsulating cells. The main advantage of this strategy is the use of a biocompatible membrane that allows for the secretion of therapeutic factors while isolating the implanted cells from the surrounding tissue. Encapsulation results in the protection of the implant from immune surveillance as well as offering the possibility of implant removal if necessary.[2] However, technical challenges such as long-term transgene expression and survival of the cells for long periods has limited the usefulness of this method.

In Vivo Gene Therapy

Viral as well as non-viral vectors can be used to deliver therapeutic genes directly into the CNS in a process termed *in vivo* gene transfer. Viral vectors have been designed to exploit the natural ability of viruses to infect, transfer and express their genetic material in host cells. Several viral vectors have been designed, each with advantages and disadvantages. When deciding on the potential clinical use of engineered viral vectors for the CNS, several aspects of their biology must be considered. First, binding and entry into host cells can elicit immunological responses that can lead to neutralization of the virus or toxicity. Also, the engineered viral vector must have the ability to integrate its genome in order to achieve long-term expression of the transferred gene. Finally, it is important to utilize viruses that have a sufficient genetic payload capacity to include regulatory elements to control the expression of the therapeutic gene.

Nonviral vectors include the use of naked DNA as well as DNA that has been placed inside cationic lipids (liposomes) to facilitate fusion and crossing of the cell membrane.[167] Although this strategy circumvents the potential toxic problems associated with the binding and entry of viral vectors, it has been difficult to achieve and maintain a high level of gene expression. Nonviral vectors hold promise in the field of gene therapy and will eventually replace the use of viral vectors in the clinic. However, our current knowledge of this technology is limited and will require breakthroughs that can address the problems with expression before it can enter initial clinical trials.

Gene Transfer Vectors

The advent of viral vector technology is due in great part to the intrinsic evolutionarily developed ability of wild-type viruses to efficiently transfer foreign genes to host cells.[177] The successful development of a wild-type virus into an efficient viral vector depends on the removal of as much viral genetic material as possible and the inhibition of viral replication without perturbing the natural processes of viral assembly, infection, and transduction. Because neurons are postmitotic cells, it is imperative that viral vectors intended for use in the CNS be able to transfer and express therapeutic genes in these nondividing cells. Adenoviruses (Ad), adeno-associated viruses (AAV), herpes simplex viruses (HSV), and lentiviruses (Lv) are the most common vectors being currently used in animal models of CNS disorders (see Table 70.1).

Recombinant Adenoviruses

Adenovirus is a relatively large, double-stranded DNA virus that has been mainly associated with respiratory infections.[208] The recombinant version of Ad (rAd) can transduce postmitotic neurons and glia and leads to high levels of transgene expression in the CNS.[3] However, this first-generation rAd leads to a significant host inflammatory response due to expression of viral genes that remain in the vector, preventing the use of this particular vector in the clinic (see Table 70.1). In fact, the use of this first-generation rAd led to an ill-fated event in a gene therapy clinical trial to correct ornithine transcarbamylase deficiency.[154]

RAd-mediated inflammatory response, which is also observed in the brain,[139,140,219,240] has been the subject of extensive research and has led to the development of newer versions of rAd vectors that are completely depleted of viral genes.[86,175,191,220] These "gutless" rAd vectors show significant decrease in toxicity and normal tropism. However, their transduction efficiency is somewhat reduced compare to first-generation rAd.[220] The main advantage of rAd vectors is their large payload capacity as well as their ability to transduce both neurons and glia. However, problems with toxicity and long-term gene expression will have to be addressed in order to bring this vector into the clinic.

Recombinant Adeno-Associated Virus (rAAV)

Wild type AAV is a relatively small, single-stranded DNA virus. It is a nonpathological member of the *parvoviridae* family.[179] Following infection, the viral genome is integrated into the host's DNA. This lack of pathogenicity and the tendency to integrate into the host genome make recombinant versions of AAV (rAAV) an excellent vector for gene transfer to the CNS.[157,179] However, this vector has a small payload capacity that could result in a limited use of rAAV in the treatment of CNS disorders.

Recombinant versions of AAV have been shown to transduce primarily neurons in the CNS.[8,148,157] However, recent studies have focused on the diverse tropism shown by the six different AAV serotypes that have been clinically isolated.[52,73] This variety in capsid structure and receptor binding has enabled the potential to engineer rAAV for efficient transduction of varied target tissue. Another advantage of having these different serotypes is the theoretical ability to avoid immune surveillance given that 80% of the human population is seropositive to antibodies against AAV2 (the most common serotype currently in use for gene transfer).[20,21,63]

Recombinant Herpes Simplex Virus

Herpes simplex virus (HSV) is a large, linear, double-stranded DNA virus associated with cold sores, corneal blindness, and encephalitis.[196] Wild-type HSV can infect a wide variety of cells where it can establish latency. HSV-1 can infect both neurons and glia. HSV's inherent neural tropism, its ability to be retrogradely transported, as well as its large transgene capacity make it a good candidate for CNS gene delivery.[68,80] However, its inherent cytotoxicity as well as long-term gene expression problems are hurdles that remain to be cleared prior to clinical use.[68,80]

The genetically engineered versions of this virus consist of recombinant HSV or amplicon vectors. In the recombinant version, most of the viral genes contained within the 150kb genome have been kept, and only a portion of the genes have been deleted to prevent viral reproduction.[71] Amplicon vectors have been produced that carry a very limited amount of viral genes. Current efforts are being undertaken to remove additional viral genes that

TABLE 70.1
Catalog of Properties of the Different Recombinant Viral Vectors*

	Viral Gene Deletions	Vector Type	CNS Cells Transduced	Fate of Provirus	Transgene Packaging Size	In Vivo Transgene Expression	Potential Safety Issues
rAd[128]							
rAd	E1a, E1b, E3	Double-stranded DNA	Neurons, glia	Episomal	~8kb	Produces large protein quantities, variable duration of expression, depending on promoter and immune response of the species used	Inflammatory response to capsid and expressed viral proteins
Gutless	All structural	Double-stranded DNA	Neurons, glia	Episomal	<37kb	Low-moderate protein production long-term	Capsid/infection related inflammation
rAAV[106,128,179,214]							
rAAV2	Rep, cap	Single-stranded DNA	Neurons	Episomal/integrated	~4.7kb	Very high protein expression depending on transcriptional cassette, very long-term	Insertional mutagenesis
r HSV[23,27]							
Replication-defective	ICP4, ICP22, ICP27, ICP47	Double-stranded DNA	Neurons, glia	Episomal/latent	~20kb	High protein production short-term	Moderate toxicity
Amplicon	All except *oris* and *pac*	Double-stranded DNA	Neurons, glia	Episomal/latent	~20kb	Low protein production short-term	Low toxicity
rLV[37]							
1st-generation rLV	*gag, pol, env*	RNA	Neurons	Integrated	~9kb	High protein production long-term	Potential toxicity from VSV-g, pseudotyped envelope, possible HIV sero-conversion, possible recombination, insertional mutagenesis
2nd-generation rLV	*gag, pol, env,* vif, vpr, vpu, nef	RNA	Neurons	Integrated	~9kb	High protein production long-term	Potential toxicity from VSV-g, pseudotyped envelope, possible HIV sero-conversion, possible recombination, insert ional mutagenesis
3rd-generation rLV	*gag, pol, env;* vif, vpr, vpu, nef tat, rev	RNA	Neurons	Integrated	~9kb	High protein production long-term	VSV-g, seroconversion, recombination, insertional mutagenesis
Self-inactivating	All above and 5' LTR inactivation	RNA	Neurons	Integrated	~9kb	High protein production long-term	VSV-g, seroconversion, insertional mutagenesis

*Each vector has advantages and disadvantages, and as these vectors come into clinical testing, they are likely to find their own niche. For example, rAD, while currently suffering from some toxicity issues, might be very useful in situations where high protein production is necessary for a short period of time. HSV vectors may be most useful to transfer very large genes. The brief descriptions given for each vector are only generalizations. Various laboratories are working to improve each of these vector systems, and it is entirely possible that any and all of these vectors will be perfected as clinically relevant gene therapy vectors.

are thought to be involved in cytotoxicity. Insights into HSV long-term gene expression as well as improvements on its toxicity should lead to the development of vectors that can be used to treat the targets they naturally infect such as the ganglion cells of the retina.

Lentiviral Vectors

Lentivirus is a RNA retrovirus capable of infecting both mitotic as well as nonmitotic cells.[184] The most widely studied lentivirus is the human immunodeficiency virus type I (HIV-1).[212] HIV-1 naturally transduces neurons and it integrates its reverse-transcribed DNA into the host genome. This naturally led to the engineering and production of recombinant versions of HIV-1 to be used as a gene delivery vector.[182–184] The HIV-1 genome in its wild type form is composed of three coding regions: *gag*, *pol*, and *env*. These coding regions encode core proteins (*gag*), virion-associated enzymes (*pol*), and envelope glycoproteins (*env*). The viral genome is flanked by long terminal repeats (LTRs) that are crucial for viral integration, transcription, and polyadenylation of mRNA.

The recombinant version of HIV-1 is produced by pseudotyping the envelope glycoproteins with the vesicular somatitis virus glycoprotein (VSV-G).[183] This allows for more diverse tropism as well as increased stability of the virion particle. The biosafety concerns associated with the use of HIV-1 as a gene therapy vector have been extensively addressed. One approach is to remove all of the genes present in the viral genome that are not involved in vector propagation (*vif, vpr, vpu and nef*).[57,249,250] In fact, removal of these genes may reduce levels of toxicity associated with HIV-1. Another approach is to remove the promoter and enhancer sequences present in the 5´LTR of the viral genome. The possibility of recombination events that could potentially generate wild-type replication competent HIV-1 is greatly reduced in these self-inactivating vectors.[249]

Recent advances in vector production and purification are making it possible to generate high titer vector preparations of recombinant pseudotyped HIV-1. Additionally, lentiviruses have been repeatedly shown to have great tropism for neurons and are capable of sustaining transgene expression for long periods of time. Finally, these vectors have a moderate payload capacity. Although there are safety concerns associated with this virus, recombinant HIV-1 vectors hold great promise in the field of CNS gene therapy. In addition, evolutionarily related immunodeficiency viruses have been engineered into viral vector which may also have a better safety profile.[76,186,202]

Intracerebral Delivery of Vectors

To maximize the area of transduced CNS tissue, many have described intracarotid, intraventricular, or intrathecal injections of the viral vectors in the presence or absence of solutions that induce hyperosmolar disruptions of the BBB.[13,54-56,151] However, these injections lead to minimal transduction of the parenchyma. At present, the most reliable way to introduce viral vectors into the brain is via MRI-guided stereotaxic injections. With regard to PD, the neurosurgical experience with fetal transplantation indicates that the neurosurgical aspect of the gene therapy procedure will not be a major obstacle.

GENE THERAPY STRATEGIES FOR EARLY-STAGE PD

Of primary concern in the treatment of PD is to halt the progression of neuronal death rather than treating the symptoms themselves, and although the underlying etiology of PD still remains unknown, there are a few proteins that may serve as potential agents to inhibit cell death. Proteins that inhibit the apoptotic cascade are potentially important for this strategy.[84] However, the most promising gene therapeutic approach to slow the advance of PD is the overexpression of neurotrophic factors.

GENE TRANSFER OF GLIAL CELL LINE-DERIVED NEUROTROPHIC FACTOR

RATIONALE FOR INTRASTRIATAL GDNF

Neurotrophic factors are an attractive candidate for the treatment of neurodegenerative disease as they can potentially block both apoptotic and necrotic modes of cell death, as well as facilitate neuronal recovery following various forms of cellular damage.

Striving to retard the progressive neurodegeneration in PD, glial cell line-derived neutrophic factor (GDNF) has been the strongest candidate in terms of neurotrophic factors facilitating DA neuron survival.[16] GDNF is a member of the transforming growth factor-β super family with a narrow range of target cells, one of them being the nigral neurons which express the GDNF receptor, GFR-α.[25,79,156,201,224,230] Injections of GDNF protein have been shown to improve DA neuron survival both *in vitro*[138] and *in vivo*. Thus, GDNF injections can ameliorate 6-hydroxy-dopamine (6-OHDA)- or MPTP-induced damage as well as improving the functional integrity of the nigrostriatal system (see Reference 19 for review). Recently, promising results of continuous intrastriatal GDNF protein administration have been reported in a small clinical trial in PD patients,[77] although previous trials using intraventricular GDNF injections have been less successful.[122,188]

To deliver GDNF to the striatum of PD patients, a small cannula was implanted and attached to a pump.[77] The complications and risks associated with this type of delivery method may not be desirable for some PD patients who might benefit from GDNF delivery. Therefore, gene transfer of GDNF may be an attractive strategy for GDNF delivery, given that several vector systems appear to support extremely long-term transgene expression, and a single surgical session should be sufficient for long-term treatment.[18] Longevity of transgene expression for treatment of PD is important, since PD results in the progressive loss of neurons over decades.[67,158]

GDNF Gene Therapy for PD: Progress

Progress toward a GDNF-based gene therapy for PD has recently been reviewed in detail.[16,120] Numerous groups have reported positive results in the rodent and primates by using viral vectors expressing GDNF in the striatum and the substantia nigra.[16,43,44,46,114,121,123,149,150,231] Figure 70.1 shows the effect of rAAV-mediated striatal GDNF in the 6-OHDA rat model of PD.[114] In this rAAV study, several potentially advantageous properties of GDNF were demonstrated. First, at the dose of GDNF delivered in that study, as had been shown for GDNF protein injections,[96] GDNF increased TH activity and DA-mediated behaviors in intact (nonlesioned) rats.[114] Second, rAAV mediated GDNF expression spared DA neurons,[114] as has been shown using other vector systems.[10,15,43] Finally, functionally significant sprouting of spared DA axons was demonstrated[114] (Fig. 70.1). Similar positive functional results were obtained using lentiviral vectors to deliver GDNF to the rhesus monkey in the MPTP-lesion model of PD and in aged monkeys.[121] All these GDNF gene transfer studies establish that it is the striatal DA innervation that leads to functional improvements in motor symptoms, not merely saving DA neurons.[114,121] Thus, most studies agree that vector mediated GDNF expression must be directed at the striatum to support and protect striatal dopaminergic innervation but it is unclear whether both the substantia nigra and the striatum should be transduced as well.[16]

Hurdles Remaining for GDNF Gene Therapy

While the gene therapy data gathered thus far using GDNF are very encouraging, there does appear to be expression levels of GDNF that can adversely affect DA neuron function. Hagg and colleagues first reported that GDNF protein injections, while they spare DA neurons from cell death, reduce TH expression in substantia nigra.[141] Subsequently, it was reported that GDNF regulates TH expression at the mRNA level,[166] and overexpression of GDNF in transgenic mice leads to an apparent reduction in nigral DA neurons, which may result from reduced TH expression levels.[45] Using lentiviral vectors to deliver intrastriatal GDNF, Georgievska et al.[74] found that GDNF induced a down-regulation of nigrostriatal TH and induced aberrant sprouting of dopaminergic striatal fibers in partially DA depleted rats. Later, Rosenblad et al.[197] demonstrated that maintained GDNF expression over a period of 13 months in normal rats also resulted in a drop in TH levels in treated striatum, although abnormal sprouting was not observed. These data are not necessarily at odds with the reports of

beneficial GDNF effects. It is highly likely that there is a dose-response relationship for GDNF on TH activity where higher doses, such as those reported in Georgievska et al.[74] and Rosenblad et al.,[197] induce the opposite effect on TH activity. These data merely point out the absolute requirement for the ability to regulate the GDNF transgene expression prior to the onset of clinical trials. Using a tetracycline transcriptional regulation system[82,83,117] may enable regulated GDNF production using rAAV vectors. GDNF is a very stable protein in the brain,[198] and even very low baseline levels (leakiness) of GDNF expression in the absence of tetracycline treatment leads to significant accumulation in the brain (Mandel, Burger, and Muzyczka, unpublished data). Thus, prior to the initiation of clinical trials for this extremely promising gene therapy strategy, more technical hurdles must be resolved.

Brain-Derived Neurotrophic Factor

A second neurotrophic factor of interest is brain-derived neurotrophic factor (BDNF), a trkB ligand, which is normally expressed in the striatum and substantia nigra.[99] BDNF has been demonstrated to be reduced in PD as observed in postmortem studies.[98] Similarly to GDNF, BDNF has also been shown to provide a protective effect on dopaminergic neurons both *in vivo* and *in vitro*, in the 6-OHDA rat model of PD,[210,85] and in the MPTP monkey model.[225] Several *ex vivo* gene therapy studies report that BDNF-producing cells transplanted into the neostriatum can block 6-OHDA induced nigrostriatal dopaminergic degeneration.[135,246,247] In addition, one rAAV-BDNF study also reports some positive effect of nigral BDNF expression on a small intrastriatal 6-OHDA lesion, but this effect was extremely limited.[118] Although it is generally agreed that BDNF has some trophic actions on dopaminergic cells, these effects are not as robust as that seen with GDNF. Therefore, to date no efforts have been made in clinical trials evaluating its efficiency to reduce degeneration.

Antiapoptotic Strategies

Increasing amounts of data strongly support apoptosis as a mode of nigral cell death in PD.[5,26,146,170,180,199,221] Both the 6-OHDA and the MPTP models of PD also are likely to ultimately kill DA neurons via apoptosis.[26,155] Given that rAAV vectors have been shown to have natural targeting for the nigrostriatal DA neurons,[112,115,119,150] this vector might be very useful for an antiapoptotic strategy. However, it has been noted that, identical to data from GDNF studies in PD models, sparing of nigral neurons via apoptotic inhibition without saving the striatal DA terminals is not functionally effective.[60,228]

Several studies have reported that inhibition of apoptosis by overexpressing of antiapoptotic proteins is efficacious in animal models of PD. For example, HSV-1-mediated overexpression of bcl-2 in substantia nigra has

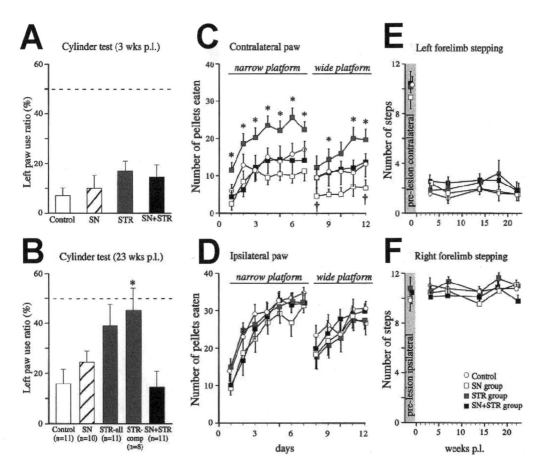

FIGURE 70.1 There were four experimental groups in this experiment, named according to where rAAV-GDNF was injected. Thus, "controls" were either untreated or received rAAV-GFP injections, "SN" animals received rAAV-GDNF injections four weeks prior to 6-OHDA in the substantia nigra, "STR" animals received identically timed vector injections in striatum only, and the "SN+STR" received identical vector injections in both the substantia nigra and striatum as described in Kirik et al.[114] (reprinted with the permission of the Society for Neuroscience). *Analysis Spontaneous Behaviors.* (A) Spontaneous limb use was evaluated using the cylinder test[204] three weeks after the 6-OHDA lesion. There was a highly significant ipsilateral-side bias present in all groups at this time point, as indicated by only 10 to 15% of contralateral limb use in this test ($p = 0.52$) (p.l. refers to "post-lesion"). (B) The animals were tested again in the cylinder test 23 weeks after the 6-OHDA lesion. At this time point, there was a general improvement across all groups (effect of time $p = 0.001$). In the STR group, the improvement was more pronounced, but this trend did not reach significance (group effect $p = 0.07$). However, considering only the animals from the STR group that displayed full compensation in the amphetamine-induced rotation test (8/11 animals, *STR-comp*), these animals used their contralateral paw to contact the sides of the cylinder at near normal levels (~45%), and they performed significantly better than the uncompensated animals and the controls ($p < 0.0001$). (C) Staircase (skilled limb use) test,[172] data from the contralateral paw. The first period of testing (days 1 through 7) was performed using the standard narrow platform. The groups performed significantly differently from one another ($p = 0.003$). The groups improved over the course of the testing (effect of training; $p < 0.0001$), but the rate of learning did not differ between the groups (time 3 group interaction, $p = 0.058$). However, the STR group was able to successfully retrieve significantly more pellets than the other three groups on all days (*asterisks*; simple main effects, $p < 0.02$), whereas the control, SN, and SN+STR groups did not perform differently from one another (simple main effects, $p > 0.1$ on each individual day). Beginning on day 8, the test was made more difficult by using a wider platform. In this part of the test, the groups differed from each other in their ability to successfully retrieve pellets ($p = 0.035$). The STR group again performed significantly better than all the other groups (*asterisks*, simple main effects, $p < 0.05$, except on day 8 and day 10, where $p > 0.05$). The SN group performed significantly worse than the controls and the SN1STR on day 8 and day 12 (†, simple main effects, $p < 0.05$), whereas the control group and the SN1STR groups did not perform differently at any time point ($p > 0.4$ on all days). (D) Staircase (skilled limb use) test, data from the ipsilateral paw. Performance with the ipsilateral paw was significantly better than that of the contralateral paw for all four groups ($p < 0.0001$). (E, F) Performance of the contralateral (E) and the ipsilateral forelimb (F) in the stepping test. In the prelesion testing (*gray shaded column*) there was no difference between the groups on either side ($p = 0.82$). The lesion severely affected the number of steps on the contralateral side in all groups (effect of side; $p < 0.0001$), whereas the performance on the ipsilateral side was unaffected. No improvement was observed over time in any of the groups in this test. The legend in the *bottom right corner* refers to the symbols representing each group and applies to C through F. All values are means ± SEM. *(continues)*

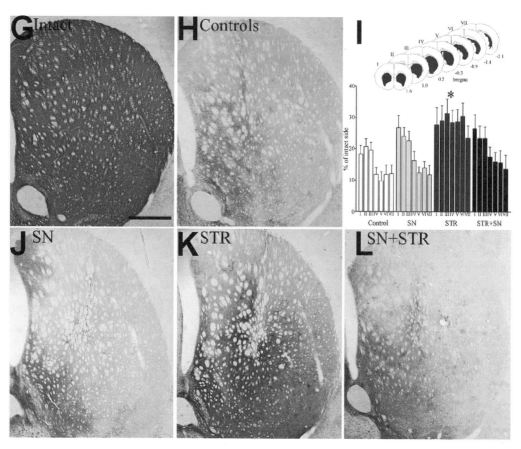

FIGURE 70.1 *(continued)* (*G* through *L*) TH immunocytochemical staining of the central striatum. The 6-OHDA lesion resulted in degeneration of the TH-positive fibers in the controls (*H*). Note the higher intensity of TH-positive fiber innervation in the STR (*K*) group. The innervation in the SN (*J*) and SN+STR (*L*) groups was not different from the controls. The scale bar in *G* represents 500 μm and applies also to photomicrographs in *G* through *L*. Panel *I* shows the density of TH-positive fibers in the striatum measured at seven rostrocaudal levels, as shown in the sketch. The 6-OHDA lesion induced extensive degeneration of the striatal TH-positive innervation in the control group. There was a highly significant difference in striatal TH-positive fiber density between the groups (*p* < 0.0001). Anterior regions (levels I through III) were relatively more spared compared with posterior regions (levels IV through VII) across all groups (*p* < 0.0001). However, there was not a significant level 3 group interaction (*p* = 0.92); therefore, *post hoc* tests between groups were performed regardless of level. The STR group displayed significantly higher TH fiber density across all rostrocaudal levels compared with all other groups (**p* < 0.05; Tukey HSD). All values are means ± SEM.

been reported to save DA neurons from 6-OHDA-induced cell death.[244] rAAV-mediated overexpression of an apoptotic protease-activating factor-1 dominant negative mutant has likewise been reported to have similar effects in the mouse MPTP model.[171] Finally, the neuronal apoptosis inhibitor protein has been used in the substantia nigra of rats via adenoviral vector delivery to protect against 6-OHDA lesion.[50]

In addition to the use of apoptosis inhibition by itself two studies, have combined GDNF overexpression with some form of antiapoptosis. One report combined striatal Ad-GDNF and nigral Ad- over-expressing X-linked inhibitor of apoptosis (XIAP) to show good functional effects in the 6-OHDA rat model of PD.[60] Similarly, Natsume et al., who used HSV vectors to deliver both bcl-2 and GDNF, reported an additive effect of both genes in the rat 6-OHDA.[185] Although these strategies are very

promising, this body of work is still in its early stages, and more detailed experiments in primates are required prior to consideration of clinical trials.

FUTURE DIRECTIONS: PROTEOSOMAL PROCESSING

Recently, rare familial forms of Parkinson's disease have been uncovered that point to proteosomal processing of proteins as a potential etiological factor in idiopathic PD.[5,223] As in other neurological disorders, accumulation of protein aggregates may be a toxic event in nigral DA neurons. α-synuclein, the first familial PD gene, has been found in Lewy bodies.[211] Thus, strategies aimed at reducing α-synuclein, may be beneficial[51] (see also references 103 and 152). Viral vectors are capable of delivering various agents that reduce protein expression such as dominant negative proteins,[171] anti-sense oligonucleotides,[242] ribozymes,[129] and siRNA.[241] Parkin, an E3 ligase that functions to polyubiquinate proteins and is mutated in juvenile forms of familial PD,[209] may also be protective in PD.[192] However, to date, no studies have been published using gene transfer to modulate proteosomal processing in PD models.

GENE THERAPY STRATEGIES FOR LATE-STAGE PD TREATMENTS

TRANSMITTER REPLACEMENT STRATEGY

Intrastriatal Continuous L-Dopa Delivery

The first strategy proposed as a gene therapeutic treatment for PD was striatal L-dopa delivery by *ex vivo* expression of tyrosine hydroxylase (TH).[72] The initial focus on the *ex vivo* delivery of TH was based on the known efficacy of peripheral L-dopa treatment combined with the success of fetal ventral mesencephalic DA grafting in rodents and humans.[17] As the ability to reliably and durably express transgenes progressed, it became clear that this strategy was actually more complicated than originally thought.

Rationale for Striatal Specific L-Dopa Delivery to Treat PD

When considering a gene therapy strategy to treat any human disease, two major unknowns must be surmounted. These are as follows:

1. Lack of experience with gene transfer itself
2. Efficacy of the transgene product

In the specific striatal delivery of L-dopa strategy, the clinical efficacy of L-dopa and its side-effect profile are well known. This fact removes one major uncertainty for this particular gene therapy strategy. Site-specific L-dopa delivery would also be expected to remove any side effects that result from global decarboxylation of L-dopa in nontherapeutic compartments.

In addition, any new treatment, especially a highly experimental strategy such as gene therapy, must be shown to be superior to current standard therapy. The potential for site-specific intrastriatal L-dopa delivery to be superior to current peripheral L-dopa regimens is supported by ample data indicating that many of the side effects associated with L-dopa in late-stage PD can be relieved by continuous L-dopa delivery.[39,40,42,189,205] In addition, continuous, nonpulsatile stimulation of striatal DA receptors does not produce the abnormal changes in striatal gene expression that are associated with L-dopa-induced dyskinesia.[53,61,234] Continuous L-dopa administration instantly relieves wearing-off simply due to the static levels of L-dopa. In addition, "on-off" motor fluctuations and peak-dose dyskinesias are also inhibited by continuous optimal-dose intravenous infusion of L-dopa after about nine days of treatment, with a gradual decline of

benefit after cessation of the continuous treatment.[205,41] Thus, based on these data, continuous site-specific L-dopa delivery via gene therapy should extend the period of L-dopa efficacy in PD patients experiencing motor fluctuations.

Ideal Features Required for L-Dopa Gene Therapy

Because PD is a long-term degenerative disorder and the gene delivery itself will probably require neurosurgery, a single surgical procedure that results in long-term therapy will likely be required. This means that the gene transfer should lead to very long-term or permanent gene expression. Since known side effects are associated with L-dopa treatment, and permanent transgene expression is necessary, the ability to regulate the transgene expression will also be required. Thus, the gene therapy reagent to be tested in humans should be proven to produce therapeutic and regulatable levels of L-dopa for a near-permanent length of time. To date, *ex vivo* strategies have not shown the ability to provide long-term expression. Table 70.2 shows the advantages and disadvantages of various viral vector systems.

Another complication associated with the L-dopa strategy is a controversy in the field concerning which genes are actually necessary to express L-dopa in the DA-depleted PD striatum.[110] This issue has been reviewed in detail previously.[147] Briefly, the controversy revolves around the availability of the essential cofactor for TH activity, tetrahydrobiopterin (BH4)[187] in the DA-depleted striatum.

Is Co-expression of BH4 Necessary?

It is universally accepted that BH4 is absolutely required for TH enzymatic activity to occur (Figure 70.2). A simple demonstration of this fact is that all the references listed in Table 70.2 that report *in vitro* L-dopa levels added BH4 to the media (with the exception of astrocytes, which may produce low levels of BH4). Since BH4 can act as a TH cofactor transcellularly (*in vitro* data from Table 70.3), and[131] it is controversial as to whether BH4 levels in the PD striatum are sufficient to allow L-dopa expression by transferring the gene for TH alone to the denervated striatum. As can be seen in Tables 70.2 and 70.3, most gene therapy studies have transferred only the TH gene. However, at present, the best data indicate that, in the complete 6-OHDA lesion model in the rat, there is not sufficient BH4 available to support TH activity (e.g., Figure 70.3). Despite one publication that indicates that about 30% of normal BH4 levels remain in the PD striatum,[134] the remaining levels of BH4 in the PD striatum is still an open question. Admittedly, almost every study utilizing only striatal TH gene transfer reports significant behavioral improvement manifested by reduced apomorphine-induced rotational behavior (see Table 70.1). This issue greatly complicates the evaluation of this field, because

TABLE 70.2
Ex Vivo Gene Transfer for L-Dopa Production

Source	Cell Type	Vector	Transgenes Used	Demonstration of Transgene Expression	Long-Term Expression?	In vivo Striatal L-Dopa?	Striatal Transcriptional Correction?	Reduction of Apomorphine Rotational Behavior?	Nondrug-Induced Behaviors?
Wolff et al., 1989[239]	Fibroblasts	Retrovirus	TH	ND	ND	ND	ND	Yes	ND
Horellou et al., 1990[93]	Neurblastoma line and Att20 neroendocrine lines (tumors)	Plasmid	TH	TH staining	15 days (limited by tumor growth)	ND	ND	Yes	ND
Fisher et al., 1991[69]	Fibroblasts	Retrovirus	TH	TH mRNA	7 weeks	ND	ND	Yes	ND
Horellou et al., 1991[92]	3T3 cells RIN cells (tumors)	Retrovirus	TH	ND	10 days	Tissue levels and microdialysis	ND	Yes	ND
Uchida et al., 1992[226]	Fibroblasts	Retrovirus	TH	TH staining	ND	Microdialysis with biopterin infusion	ND	Yes	ND
Jiao and Wolff, 1992[105†]	Myoblasts		TH	—	—		—	—	—
Lundberg et al., 1996[143]	Astrocytes	Retrovirus	TH	Yes	2 weeks?	Yes, check	ND	No check	ND
Bencsics et al., 1996[9]	Fibroblasts	Retrovirus	TH ± GTPCHI	TH staining check	ND	Yes, only with GTPCH1 co-expression	ND	No	ND
Tomatore et al., 1996[222]	SVG, glial cell line	Plasmid transfection	TH	No	ND	ND	ND	Yes	ND
Leff et al., 1998[131]	Fibroblasts and 9L glioblastoma cells	Retrovirus	TH ± GTPCHI	TH staining check	10 day	Yes, only with GTPCH1 co-expression	ND	No	ND
Segovia et al., 1998[206]	Astrocytes	Plasmid transfection	TH	TH staining rt-PCR	3 weeks	ND	ND	Yes	ND
Fitoussi et al., 1998[70]	Human astrocytes	Plasmid transfection	TH	TH staining	11 days	ND	ND	Yes	ND
Hida et al., 1999[91]	Astrocytes	Ad	TH	TH staining rt-PCR	4–6 weeks	ND	ND	Yes	ND
Cortez et al, 2000[47]	Astrocytes	Retrovirus	TH	TH staining	4 weeks	ND	ND	Yes	ND
Cao et al.., 2000[29]	Myoblasts	Plasmid transfection	TH	TH staining rt-PCR	3 months	ND (have DA data)	ND	Yes	ND

*This table is intended to compile almost all of the publications using *ex vivo* gene therapy to deliver L-dopa in rodent models of PD. The main point of this table is to show that many desirable measurements have never been attempted in most studies (ND = not done). The other point is that everyone finds reductions of apomorphine-induced rotational behavior but, as discussed in the text, this is not a good measure of the functional effects of L-dopa delivery. If a measurement was taken, and a positive result was found, a "yes" is placed in a cell. If the measurement was taken, but a negative result was found, "no" is placed in a cell.

†Paper retracted.[105]

TABLE 70.3
In Vivo **Gene Transfer for L-Dopa Production***

Source	Vector	Transgenes Used	Demonstration of Transgene Expression?	Long-Term Expression?	*In Vivo* Striatal L-Dopa?	Striatal Transcriptional Correction?	Reduction of Apomorphine Rotational Behavior?	Non-drug Induced Behaviors?
Kaplitt, et al., 1994[111]	rAAV	TH	TH staining?	4 months	ND	ND	Yes	ND
During et al., 1994[59]	HSV	TH	TH staining	1 year	Yes[†]	ND	Yes	ND
Isacson, 1995[101]	HSV[T]	TH	No	Check	ND	ND	Yes, but related to damage, not transgene expression	ND
Horellou et al., 1994[95]	Ad	TH	TH staining	1 month check	ND	ND	Yes	ND
Imaoka et al., 1998[100]	Plasmid	TH[†]	TH staining	7 days	ND	ND	Yes	ND
Mandel et al., 1998[148]	rAAV	TH ± GTPCHI	TH staining GTPCH1 staining	1 year	Yes, only with GTPCH1 co-expression	ND	No	ND
Corti et al., 1999[49]	Ad	TH	TH staining	17 weeks	Yes, only with biopterin infusion	ND	Yes, but related to inflammation, not transgene expression	ND
Leone et al., 2000[133]	Ad	TH	TH	6 weeks	ND	ND	Yes	ND
Kirik et al., 2002[113]	rAAV	TH ± GTPCHI	TH staining	15 weeks	Yes, only with GTPCH1 co-expression	ND	Yes	Yes, but only with L-dopa above threshold
This chapter	Lentivirus	TH ± GTPCHI	TH staining GTPCH1 staining	14 weeks	Yes, only with GTPCH1 co-expression	No	No	No
Carlsson et al., in prep.[32]	rAAV	TH ± GTPCHI	TH staining	12 weeks	Yes, only with GTPCH1 co-expression	ND		
PD-related model (DA-/-mice)								
Szczypka et al., 1999[217]	rAAV	TH ± GTPCHI	TH staining DA content	> 1 year	Yes, only with GTPCH1 co-expression	ND	NA	Yes, feeding and drinking
Szczypka et al., 2001[216]	rAAV	TH ± GTPCHI	TH staining DA content	ND	ND	ND	NA	Yes, feeding and drinking

*This table is intended to compile almost all of the publications using in vivo gene therapy to deliver L-dopa in rodent models of PD. The main point of this table is the same as Table 70.1 (ND = not done). If a measurement was taken and a positive result was found, a 'yes' is placed in a cell. If the measurement was taken but a negative result was found, 'no' is placed in a cell. The other point is that more of the function of L-dopa in striatum is being elucidated as evidenced by the more sophisticated measurements used by a number of the papers shown toward the bottom of the table (more recent.)

†A partial striatal lesion was used here and very few cells were transduced, therefore, the source of L-dopa might be questionable.

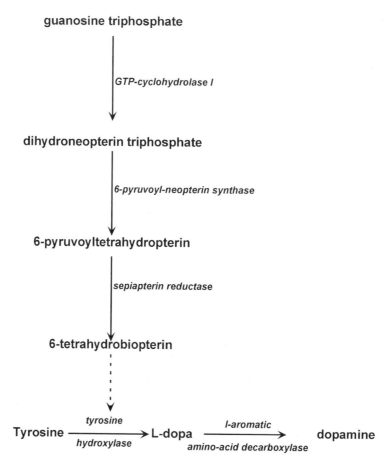

FIGURE 70.2 The dopamine biosynthetic pathway and the biosynthetic pathway for the essential cofactor for TH, BH4. Most cells constitutively express 6-pyruvoyl-neopterin synthase and sepiapterin reductase.[233] Therefore, gene expression of the rate-limiting enzyme, GTP-cyclohydrolase 1 is all that is necessary for BH4 production. The dashed arrow indicates that BH4 is needed for TH enzyme activity. BH4 also appears to stabilize the TH protein.[131,213]

apomorphine-induced rotational behavior is a highly confounded behavioral assay.[109,147] Any treatment that reduces striatal DA receptors will also inhibit apomorphine-induced rotational behavior (see Mandel et al.[147] for review). This effect is clearly demonstrated using rAd-TH by Corti et al.[48]

Production of BH4 in the striatum requires the expression of a second gene, GTP-cyclohydrolase I, the primary synthetic enzyme for BH4[233](GTPCH1, Figure 70.2). When the further requirement for the ability to regulate the levels of striatal L-dopa is added to the probable requirement for the expression of two transgenes (TH and GTPCH1), the complexity of the technical aspects of this strategy can clearly be appreciated.

In summary, a major advantage of choosing striatal L-dopa as a gene therapy strategy is that the safety profile of L-dopa is well known. However, durable regulated gene expression that produces levels of striatal L-dopa that are therapeutic must be achieved. In addition, the combination of genes to be used is still controversial. Furthermore, especially because a PD patient may live several decades

after the gene therapy procedure, the safety of the chosen gene therapy vector should be established as well as possible in animal models. Clear attainment of some of these goals is hampered by the absence of a truly valid animal model that displays long-term progressive nigrostriatal degeneration that is also L-dopa responsive. This issue is particularly important, because the response to L-dopa therapy wanes over time as the disease progresses in humans.

Therapeutic Efficacy of Intrastriatal rAAV-Mediated L-Dopa Delivery

Until recently, the apomorphine-induced rotational behavior paradigm was the most accepted test for functional recovery in the rat 6-OHDA PD model. However, several recent nondrug-induced functional tests have been shown to be sensitive to striatal DA depletions and to respond to therapeutic treatments including L-dopa. Thus, stepping behavior, a measure of forelimb akinesia,[203] and the cylinder test, a measure of spontaneous forelimb usage,[204] are deficient in response to 6-OHDA lesions and respond

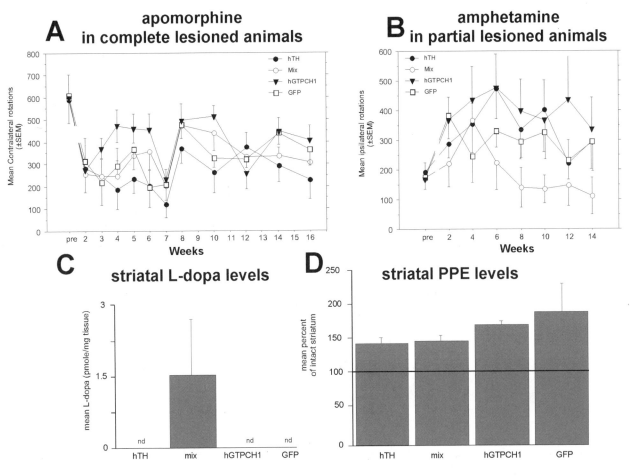

FIGURE 70.3 Intrastriatal Lentiviral vectors expressing L-dopa synthesizing enzymes in the 6-OHDA PD rat model. (*A*) Apomorphine-induced rotational behavior (0.1 mg/kg i.p.) in complete lesioned (MFB) animals from the experimental groups indicated in the figure legend. There was no significant difference between any of the groups in this test ($p > 0.05$). Each point represents seven to eight animals. Error bars = ±1 SEM. (*B*) Amphetamine-induced rotational behavior (2.5 mg/kg, i.p.) in partial lesioned animals that received identical intrastriatal lentiviral vector injections to that shown in *A*. Again, there was no significant effect of expressing striatal L-dopa in this group; however, the mixed (1:1 mixture of lenti-hGTPCH1 and lenti-hTH) group's reduction in rotational behavior approached significance ($p = 0.07$). Each point represents five to six animals. (*C*) Striatal L-dopa levels from striatal punches taken from the animals in *A*. Detectable L-dopa levels were observable only in the mix vector group. Note that the mean L-dopa was below 1.5 pmole per mg striatal tissue (nd = not detectable). (*D*) Relative striatal preproenkephalin (PPE) mRNA levels in partial lesioned animals shown in *B*. Striatal PPE mRNA levels were quantified by densitometry from striatal sections that underwent *in situ* hybridization. Striatal PPE mRNA levels normally rise in response to DA denervation and have been shown to be reduced after L-dopa treatment.[90,173,174,248] Intrastriatal lentiviral mediated L-dopa production had no effect on the 6-OHDA induced increase in PPE mRNA levels. (*continues*)

positively to intrastriatal fetal transplants[190,116] and L-dopa treatment.[116,144]

More than doubling the dose of the same vectors used in Mandel et al.[148] allowed significantly greater coverage of striatum yet produced no behavioral recovery in either stepping or cylinder tests (Mandel, Björklund, and Kirik, unpublished observations). One obvious hypothesis, given that peripheral L-dopa injections can significantly ameliorate these behaviors, is that insufficient striatal L-dopa levels were being produced.

To investigate this issue, rats were administered a peripheral dose of L-dopa known to improve function and

their striatal L-dopa tissue levels were determined (Figure 70.4).[113] The data indicate that L-dopa levels must be greater than 1.5 pmoles above the DA depleted background levels. New, high-titer rAAV vectors expressing TH and GTPCH1 under the control of a very strong promoter[243] were injected in five sites in striatum in order to reach this level of striatal L-dopa (Figure 70.4). Using this injection procedure in both completely DA denervated rats (MFB lesions) or partially DA depleted rats (n. accumbens DA innervated) revealed that, when striatal L-dopa levels exceeded the defined threshold, both partially and complete 6-OHDA lesioned rats significantly

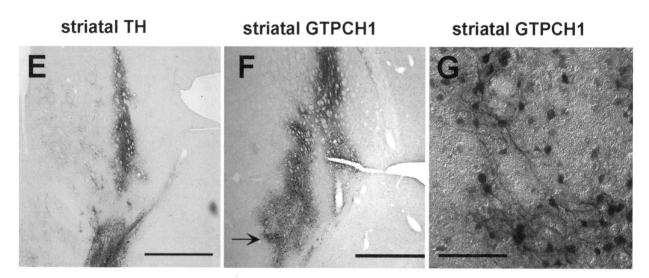

FIGURE 70.3 *(continued)* (*E*) Striatal TH immunostaining from an animal receiving Lenti-TH alone, scale bar = 1 mm. (*F*) Striatal GTPCH1 immunostaining in an animal that received a mixed vector injection, scale bar = 1 mm. Arrow shows area of magnification shown in *G*. (*G*) Higher magnification of GTPCH1+ cells shown in *E*. Scale bar = 100 μm. These data are in agreement with the idea that subthreshold striatal L-dopa levels do not affect behavior and also suggest that 6-OHDA-induced striatal molecular alterations also require high L-dopa expression levels for functional reversal.

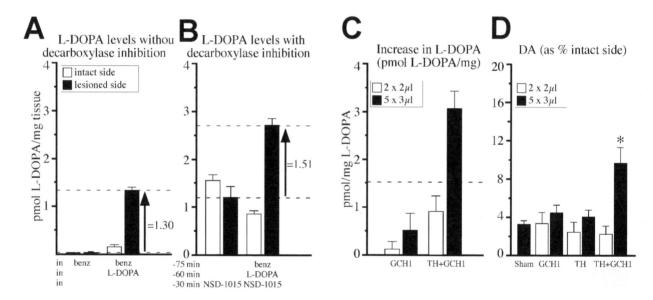

FIGURE 70.4 *(A, B)* Striatal L-dopa levels after i.p. injection. *(A)* Two groups of partially lesioned animals received injections of benzeraside (benz, 15 mg/kg i.p.) or L-dopa (6.25 mg/kg i.p). In the benz group, low levels of L-dopa were recovered on the lesion or the intact sides. Injection of L-dopa caused accumulation of 1.30 pmol/mg of L-dopa on the lesion side above that of benz alone (arrow). *(B)* When central decarboxylation is blocked for 30 min before killing, 1.2 pmol/mg of L-dopa was recovered on the lesion side (NSD-1015 alone group). L-dopa injection increased striatal L-dopa to 2.75 pmol/mg. Thus, we estimated that accumulation of 1.51 pmol/mg of L-dopa was induced by the injection (arrow). *(C, D)* Striatal levels of L-dopa and dopamine in completely lesioned animals injected with rAAV vectors. *(C)* Striatal TH enzyme activity was estimated 30 min after NSD-1015 by measuring the accumulation of L-dopa. Increased rAAV-mediated striatal L-dopa in the five-site 1:1 mix group was above the threshold value obtained with peripheral injection (dashed line). In contrast, animals receiving the TH vector alone or the 1:1 mixed vectors in only two sites did not produce this threshold L-dopa level. *(D)* Dopamine (DA) levels were reduced greatly on the lesioned side in all groups. Striatal dopamine was increased only in the 5 × 3 × 3-μl mixed-vector group. *(continues)*

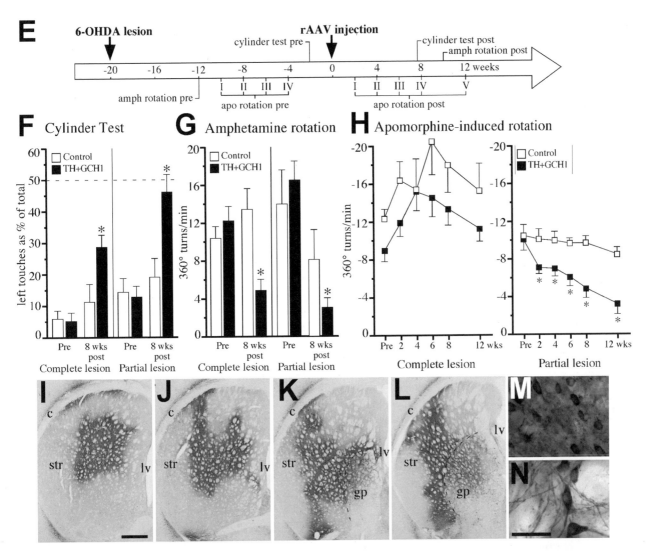

FIGURE 70.4 *(continued)* (*E–H*) Behavioral effects of L-dopa delivery. (*E*) Experimental time course. The 1:1 vector mix group showed clear improvements in the cylinder (*F*) and amphetamine-rotation (*G*) tests. Reduction in the apomorphine rotation (compared with the fourth pretest; prevalue in *H*) was observed only in the partially lesioned animals (* = significantly different from the control group and their baseline values before vector injection). (*I–N*) TH immunohistochemistry in the striatum at three weeks after injection of the 1:1 rAAV-THyrAAV-GCH1 vector mix. In large parts of the striatum (*I–L*) and globus pallidus (gp, *K* and *L*), cell bodies and fibers were TH+. Nearly all infected cells had neuron-like morphology both in striatum (*M*) and globus pallidus (*N*). There were about 300,000 striatal neurons expressing human TH as estimated by unbiased stereological methods and this number of transgene expressing striatal neurons remained stable over the 15-week period of the experiment (ac = anterior commissure; cc = corpus callosum; lv = lateral ventricle; str = striatum). (The scale bar shown in *I* = 1 mm for *I–L*; the scale bars in *M* and *N* = 40 μm). *Reprinted from Kirik et al.[113] with the permission of the National Academy of Sciences.*

recovered. Functional recovery, as defined by amphetamine-induced rotational behavior (index of restored striatal DA release[227] forelimb akinesia and skilled limb use, was improved in both PD models (Figure 70.4) but, in all cases, partially lesioned rats improved to a greater extent (Figure 70.4). This difference in vector delivered L-dopa efficacy in less severely DA depleted rats can be interpreted to model the situation in PD where L-dopa has a greater therapeutic window earlier in the natural history of the disease in humans.[64,65,176]

These data are very encouraging, but they do not alone demonstrate any therapeutic advantage over standard L-dopa therapy. Fortunately, a highly reproducible and validated model of L-dopa-induced dyskinesias has been recently developed for the rat.[35,36,144,237] rAAV delivered intrastriatal L-dopa in this model produced highly significant reduction in the number of rats that developed dyskinesias by intrastriatal L-dopa delivery.[32] Very significantly, molecular changes that occur in striatum in dyskinetic rats such as increased ΔFosB and prodynorphin were also reversed.[32] Moreover, extremely long-term expression of L-dopa synthetic enzymes supported by rAAV vectors.[148,217] Thus, at least in rodents, several key demonstrations of efficacy have been completed, and marked advantages over standard L-dopa therapy have been observed.

Remaining Questions for the L-Dopa Strategy

The results reviewed above are extremely encouraging for suggesting that site-specific rAAV-mediated striatal L-dopa delivery might be a useful strategy for treating PD. However, externally regulated L-dopa has not been demonstrated in this model, and this is an absolute requirement considering the different dose levels that are needed for peripheral L-dopa in PD patients and the potential for side effects. Moreover, it would be very prudent to obtain similar positive data in primate models of PD. Primate data will be especially useful in determining if there is a different threshold of striatal L-dopa levels needed to treat the MPTP lesioned monkey as compared to the 6-OHDA lesioned rat.

Finally, while wt-AAV is nonpathogenic, and rAAV does not induce significant inflammation in rodents, no significant data regarding rAAV-induced immune response are available from primates or from animals pre-exposed to wt-AAV. As stated above, much of the human population has been exposed to wt-AAV at some point during their life span. Therefore, the effect of circulating anti-AAV antibodies on the immune response to intracerebral administration of rAAV must be investigated prior to the initiation of human trials.

Intrastriatal Gene Transfer of AADC as a PD Treatment

Rationale for the AADC Strategy

As the nigrostriatal pathway continues to degenerate in PD, the therapeutic window for L-dopa narrows. The idea that supplementing striatal levels of AADC, that converts L-dopa to form DA (Figure 70.2), might be therapeutic in PD was first proposed by Kang and colleagues in 1993.[108] Increased striatal AADC levels may allow use of lower L-dopa doses that might widen the therapeutic window in late-stage PD patients.[108,132] This gene therapy strategy is particularly attractive, because modulation of the peripheral L-dopa dose will control the amount of striatal DA and therefore regulated transgene expression is not necessary.

While the rationale for the use of intrastriatal AADC gene transfer is clear and simple, the actual situation in PD is not quite so straightforward. While AADC activity is reduced in both the unilateral 6-OHDA rat model and in PD, there is an unidentified pool of residual AADC activity in the DA depleted striatum.[87,88,160–165,181] Moreover, because AADC is a highly active enzyme, even residual striatal AADC levels are capable of efficiently decarboxy-

lating exogenously applied L-dopa. Indeed, it is residual striatal AADC activity that allows L-dopa therapy to be efficacious.[31,97] If AADC is rate limiting in the PD striatum, then L-dopa's conversion to DA would not be therapeutic. While the therapeutic response to L-dopa does wane over the course of PD, a major limitation to L-dopa therapy is largely due to overwhelming L-dopa-induced dyskinetic side effects, and possibly not due to a loss of decarboxylation of L-dopa.

Finally, the AADC gene therapy strategy relies on the production of AADC in nondopaminergic striatal neurons. In contrast to the normal situation in the PD striatum, where peripheral L-dopa is decarboxylated in the remaining nigrostriatal DA neurons and DA is released in a physiological manner, AADC-expressing neurons release DA in response to peripheral L-dopa doses in a completely unregulated fashion.[200] The therapeutic efficacy of unregulated striatal DA release should be essentially similar to the clinical efficacy of direct DA agonists.[81] While DA agonists do abate PD symptoms, they are not as effective as the best response to L-dopa. These caveats raise the possibility that the AADC-based gene therapy strategy may not be better than currently available standard therapy.

Gene Transfer Can Restore Striatal AADC Activity

Contrary to the situation with L-dopa-based gene therapy strategies, there is little disagreement in the AADC gene therapy literature with regard to the effect of overexpressing striatal AADC on L-dopa decarboxylation. Both intrastriatal transplanted retrovirally transduced AADC-expressing fibroblasts and rAAV mediated striatal transduction have been conclusively shown to functionally increase decarboxylation of peripheral L-dopa.[107,132,200,229] In one rAAV study, increased AADC activity was shown to last for at least six months.[132] Similar results have been achieved using rAAV in rhesus monkeys.[7] In an elegant study, Kang and colleagues have shown that AADC expressed along with the vesicular monoamine transporter (VMAT2) overexpression in fibroblasts can prolong the time-course of striatal conversion of L-dopa to DA.[130] These experiments show that addition of VMAT2 to AADC in nondopaminergic cells may confer more regulated release of DA.

Functional Effects of Intrastriatal AADC Gene Transfer

If striatal overexpression of AADC increases the efficiency of decarboxylation of peripheral L-dopa, then lower doses of L-dopa should induce rotational behavior in animal models of PD, i.e., the L-dopa dose response curve should be shifted to the left. At a low dose of peripheral L-dopa, in one study, intrastriatally rAAV-AADC transduced 6-OHDA-lesioned rats responded with rotational behavior, whereas control rats did not.[200] This is in direct contrast to the behavioral results obtained in an earlier study.[132] The behavioral data were not shown in the Leff et al. (132) study, however, they are shown in Figure 70.5. No L-dopa dose, including one that is quite close to the dose used by Sanchez-Pernaute et al.[200] (2.5 mg/kg versus 3.0 mg/kg), produced a significant increase in L-dopa-induced rotations in animals that received rAAV-AADC in striatum.

AADC Gene Therapy: Conclusion

Regardless of the caveats involved with this gene therapy strategy, intrastriatal AADC gene therapy remains a very interesting idea due to its safety profile and the clear ability to overexpress AADC in striatum. In addition, the development of fluoro-meta-tyrosine as a positron emission tomography ligand[78,127] enables the detection of the AADC transgene product *in vivo*.[200] The ability to noninvasively monitor the transgene adds to the scientific interest in this strategy. A small carefully designed AADC-based clinical trial might be warranted.

Gene Transfer-Mediated DA Synthesis

As stated above with regard to the AADC gene therapy strategy, it is unclear whether it will be beneficial to produce unregulated striatal DA as a treatment for PD. However, PD symptoms are probably caused by the specific loss of DA neurons; therefore, it is a simple concept that replacing lost DA might be therapeutic in PD.

To synthesize DA, at least three genes are necessary (Figure 70.2)—GTPCH1 to produce BH4, TH to produce L-dopa, and AADC to produce DA. Positive results in animal models of PD have been reported using various vectors to produce striatal DA including rAAV[178,207] and lentivirus.[6] There are also studies utilizing only TH and AADC that report striatal DA production and behavioral correction in animal models of PD.[66,215] However, as reviewed above, it is unclear if there are sufficient striatal BH4 levels to support TH enzyme activity in these animal models. Given the relatively few gene therapy reports available that aim to deliver striatal DA, no firm conclusions can be drawn regarding the potential to treat PD in this manner.

Gene Transfer of Glutamic Acid Decarboxylase (GAD) to the Subthalamic Nucleus (STN) as a Treatment for PD

Recent physiological and anatomical analysis of the motor function of basal ganglia circuitry suggested that the STN could be a critical structure for controlling the motor symptoms of PD.[11,12,235,236] These data have led to the development of deep brain stimulation (DBS) of the STN as an effective treatment for late-stage PD patients.[102,124–126,137,194] The precise physiological basis for the effectiveness of STN DBS is currently under debate.

However, STN lesions improve symptoms and nigral DA neuron survival in animal models of PD.[11,22,33,89,159,193] These data have lead to the idea that reduction of activity in STN would be therapeutic in PD.

Along these lines, rAAV mediated overexpression of GAD, the synthetic enzyme for gamma-amino-butyric acid (GABA), in the STN has been reported to reduce symptoms in the rat model of PD.[145] In further support of this idea, muscimol (a GABA agonist) has been injected into the STN of human PD patients in an area reported to be related to tremor activity.[136] Acute muscimol injections were reported to reduce tremor in these patients ($n = 2$).[136] These data, have, in turn, lead to the advent of a Phase I clinical trial to test rAAV mediated GAD expression in STN of PD patients.[58]

While DBS is clearly effective in controlling symptoms of PD, the rAAV vector proposed in the rAAV-GAD clinical trial is not externally regulated. Due to the safety concerns associated with unregulated transgene expression, it is questionable whether this gene transfer strategy will be superior to current version of STN DBS, which has an excellent safety profile. Moreover, chronic stimulation of GABA receptors has been shown to induce tolerance;[168,169] however, this phenomenon has not been studied using rAAV-GAD gene delivery. Finally, if symptomatic improvements are observed in this trial, since STN lesions can theoretically improve PD symptoms, there will be no way to determine whether GAD gene expression or toxicity is responsible for the positive results. Indeed, in the human study where STN muscimol injections improved tremor, lidocaine dramatically improved the PD symptoms when injected locally in the STN, indicating that any treatment that reduced STN firing rates should be clinically effective.[136] Given the potential hazards and the outstanding conceptual issues associated with this strategy, it would be desirable to expand the current basic research base before proceeding.

Future Directions: L-Dopa-Induced Dyskinesia

The clinical response to L-dopa in PD, even after the "honeymoon" period, is still quite good. One of the main limitations of L-dopa therapy after side effects begin is the onset of peak-dose dyskinesias. Therefore, blockade of the onset of L-dopa-induced dyskinesia would represent a significant advance in PD therapy, and there has been virtually no attention paid to this avenue in the gene therapy field.

The mechanisms underlying the development of dyskinesia and motor fluctuations in PD are not fully understood but are believed to depend on the intermittent elevation of brain DA levels caused by pulsatile L-dopa administration in severely DA-denervated subjects.[14,38] The consequent intermittent stimulation of brain DA receptors is thought to cause maladaptive plastic changes

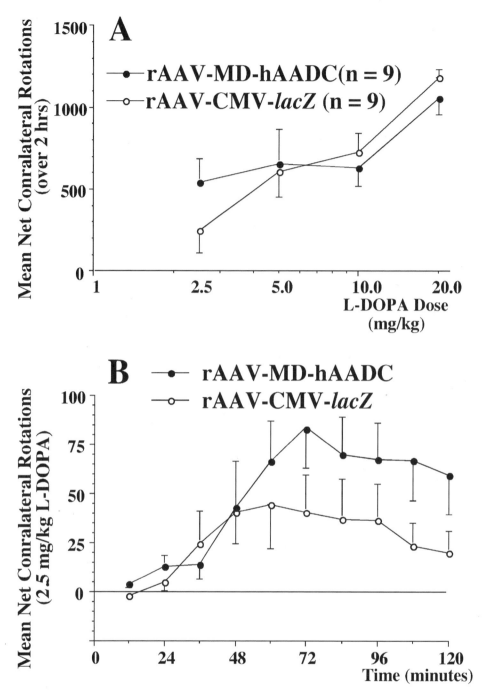

FIGURE 70.5 L-dopa-induced rotational behavior from the animals reported on in Leff et al.[132] The animals were tested on L-dopa rotational behavior once per week following a balance Latin square design to control for sensitization and/or carryover effect, none of which was observed. (*A*) Dose-response curve for L-dopa (+25 mg/kg carbidopa in all cases) induced rotational behavior from 6-OHDA lesioned animals expressing 80% of normal levels of striatal AADC after rAAV injections versus control animals expressing about 18% of normal striatal AADC levels. There was no significant differences between controls and rAAV-AADC injected rats. However, the 2.5 mg/kg group approached significance ($p = 0.06$). (*B*) To rule out the possibility that there was a difference in L-dopa-induced rotational behavior between the vector injected groups at the 2.5 mg/kg dose, the time course of rotational behavior was examined for this dose. There was no time point at which the rotational behavior was statistically different. In a separate experiment, 10 more rAAV-AADC transduced animals were tested at the 2.5 mg/kg dose in a crossover experiment with 1.25 mg/kg L-dopa (data not shown). Again, at 2.5 mg/kg L-dopa, there was no significant effect ($p > 0.05$); when both experiments were combined ($n = 19$), the statistical significance remained unchanged. These data indicate that a fourfold increase in striatal AADC activity in the rat 6-OHDA model was insufficient to affect this particular behavior in contrast to the effect reported for 3.0 mg/kg L-dopa.[200]

in neurons of the basal ganglia. Current knowledge of such changes has mainly been obtained from studies performed in DA-denervated rodents. Rats that develop dyskinesia in response to chronic L-dopa treatment show exceedingly high levels of ΔFosB-like transcription factors, prodynorphin and glutamic acid decarboxylase (GAD67) mRNA in the striatum, while L-dopa-treated nondyskinetic animals show normal levels of these gene products.[4,35] Similar findings have been obtained in nonhuman primate models of L-dopa-induced dyskinesia[24,28,53,218] and postmortem tissue from dyskinetic parkinsonian patients.[34] Knowledge of these transcriptional changes in the denervated striatum that only occur in dyskinetic animals suggests a potential therapeutic target for gene therapy. As indicated above, there are methods available to inhibit very specific mRNAs. Thus, blockade of the molecular changes known to be coincident with the expression of peak dose L-dopa-induced dyskinesias may be worth pursuing.

CONCLUSION

The continued development of viral vectors for gene transfer in the brain has provided the research community with vectors that now transduce a relatively large number of neurons and are capable of supporting very long-term gene expression in the brain. This fact allows the study of various ameliorative gene therapy strategies in animal models of PD. Moreover, especially given the exciting and high profile successes reported in this field,[121,145] clinical trials have begun[58] and more are being considered. With regard to these "mature" gene therapy strategies, the proteins that are being expressed are secreted, and therefore the viral vectors do not have to transduce every cell in the target population. Indeed, in transmitter replacement gene therapy and GDNF gene therapy, gene transfer is simply providing a technically advanced drug delivery method.

In contrast, other gene therapy strategies discussed in this chapter, such as antiapoptotic treatments or the expression of transgenes designed to have molecular effects on target neurons of the striatum, require that very large populations of neurons be transduced by viral vectors. At present, transduction of nearly every neuron in the human neostriatum is probably beyond any vector's capabilities. However, rapid improvements in viral vector design and production, for example new rAAV serotypes that are far more efficient in the brain[52] (Burger, Mandel, and Muzyczka, unpublished data), will eventually enable the consideration of clinical trials using molecular therapeutic strategies.

While the technical advances in the development of vectors for use in gene therapy protocols for neurological disorders are encouraging, safety considerations are paramount when considering the use of viral vectors in clinical protocols. The development of regulated vector systems is a prerequisite prior to human use (except in noted cases). Currently, all transcriptional regulation systems require pharmacological agents that may pose additional safety concerns. While promising, these systems have not been validated for clinical use.[153,195,232] Moreover, in some cases, most notably with regard to rAAV, the study of the potential immune response in the CNS to viral vectors has not been adequately studied. Certainly, all the technical challenges that remain for the eventual widespread use of gene therapy to treat PD can be resolved; nevertheless, it remains important for clinical research only to proceed when all conceivable safety avenues have been proven in animal models.

REFERENCES

1. Aboody, K. S., Brown, A., Rainov, N. G., Bower, K. A., Liu, S., Yang, W., Small, J. E., Herrlinger, U., Ourednik, V., Black, P. M., Breakefield, X. O., Snyder, E. Y., From the cover: neural stem cells display extensive tropism for pathology in adult brain: evidence from intracranial gliomas, *Proc. Natl. Acad. Sci., U.S.A.,* 97, 12846–12851, 2000.

2. Aebischer, P., Winn, S. R., Galletti, P. M., Transplantation of neural tissue in polymer capsules, *Brain Res.,* 448, 364, 1988.

3. Akli, S., Caillaud, C., Vigne, E., Stratford-Perricaudet, L. D., Poenaru, L., Perricaudet, M., Kahn, A., Peschanski, M. R., Transfer of a foreign gene into the brain using adenovirus vectors, *Nat. Genet.,* 3, 224, 1993.

4. Andersson, M., Hilbertson, A., Cenci, M. A., Striatal fosB expression is causally linked with l-DOPA-induced abnormal involuntary movements and the associated upregulation of striatal prodynorphin mRNA in a rat model of Parkinson's disease, *Neurobiol. Dis.,* 6, 461, 1999.

5. Anglade, P., Vyas, S., Javoy-Agid, F., Herrero, M. T., Michel, P., P., Marquez, J., Mouatt-Prigent, A., Ruberg, M., Hirsch, E. C., Agid, Y., Apoptosis and autophagy in nigral neurons of patients with Parkinson's disease, *Histol. Histopathol.,* 12, 25–31, 1997.

6. Azzouz, M., Martin-Rendon, E., Barber, R. D., Mitrophanous, K. A., Carter, E. E., Rohll, J. B., Kingsman, S. M., Kingsman, A. J., Mazarakis, N. D. Multicistronic lentiviral vector-mediated striatal gene transfer of aromatic L-amino acid decarboxylase, tyrosine hydroxylase, and GTP cyclohydrolase I induces sustained transgene expression, dopamine production, and functional improvement in a rat model of Parkinson's disease, *J. Neurosci.,* 22, 10302, 2002.

7. Bankiewicz, K. S., Eberling, J. L., Kohutnicka, M., Jagust, W., Pivirotto, P., Bringas, J., Cunningham, J., Budinger, T. F., Harvey-White, J., Convection-enhanced delivery of AAV vector in parkinsonian monkeys, *in vivo* detection of gene expression and restoration of dopaminergic function using pro-drug approach, *Exp. Neurol.,* 164, 2–14, 2000.

8. Bartlett, J. S., Samulski, R. J., McCown, T. J., Selective and rapid uptake of adeno-associated virus type 2 in brain, *Hum. Gene Ther.,* 9, 1181, 1998.

9. Bencsics, C., Wachtel, S. R., Milstien, S., Hatakeyama, K., Becker, J. B., Kang, U. J., Double transduction with GTP cyclohydrolase I and tyrosine hydroxylase is necessary for spontaneous synthesis of L-dopa by primary fibroblasts, *J. Neurosci.,* 16, 4449–4456, 1996.

10. Bensadoun, J. C., Deglon, N., Tseng, J. L., Ridet, J. L., Zurn, A. D., Aebischer, P., Lentiviral vectors as a gene delivery system in the mouse midbrain: cellular and behavioral improvements in a 6-OHDA model of Parkinson's disease using GDNF, *Exp. Neurol.,* 164, 15, 2000.

11. Bergman, H., Wichmann, T., DeLong, M. R., Reversal of experimental Parkinsonism by lesions of the subthalamic nucleus, *Science,* 249, 1436–1438, 1990.

12. Bergman, H., Wichmann, T., Karmon, B., DeLong, M. R., The primate subthalamic nucleus.2. neuronal activity in the MPTP model of parkinsonism, *J. Neurophysiol.,* 72, 507–520, 1994.

13. Betz, A. L., Shakui, P., Davidson, B. L., Gene transfer to rodent brain with recombinant adenoviral vectors: effects of infusion parameters, infectious titer, and virus concentration on transduction volume, *Exp. Neurol.,* 150, 136, 1998.

14. Bezard, E., Brotchie, J. M., Gross, C. E., Pathophysiology of levodopa-induced dyskinesia: potential for new therapies, *Nat. Rev. Neurosci.,* 2, 577, 2001.

15. Bilang-Bleuel, A., Revah, F., Colin, P., Locquet, I., Robert, J. J., Mallet, J., Horellou, P., Intrastriatal injection of an adenoviral vector expression glial-cell-line-derived neurotrophic factor prevents dopaminergic neuron degeneration and behavioral impairment in a rat model of Parkinson disease, *Proc. Natl. Acad. Sci. U.S.A.,* 94, 8818, 1997.

16. Björklund, A., Kirik, D., Rosenblad, C., Georgievska, B., Lundberg, C., Mandel, R. J., Towards a neuroprotective gene therapy for Parkinson's disease: use of adenovirus, AAV, and lentivirus vectors for gene transfer of GDNF to the nigrostriatal system in the rat Parkinson model, *Brain Res,.* 886, 82, 2000.

17. Björklund, A., Lindvall, O., Cell replacement therapies for central nervous system disorders. *Nat. Neurosci.,* 3, 537, 2000.

18. Björklund, A., Lindvall, O., Parkinson disease gene therapy moves toward the clinic, *Nat. Med.,* 6, 1207, 2000.

19. Björklund, A., Rosenblad, C., Winkler, C., Kirik, D., Studies on neuroprotective and regenerative effects of GDNF in a partial lesion model of Parkinson's disease. *Neurobiol. Dis.,* 4, 186, 1997.

20. Blacklow, N. R., Hoggan, M. D., Rowe, W. P., Serologic evidence for human infection with adenovirus-associated viruses, *J. Natl. Cancer Inst.,* 40, 319, 1968.

21. Blacklow, N. R., Hoggan, M. D., Sereno, M. S., Brandt, C. D., Kim, H. W., Parrott, R. H., Chanock, R. M., A seroepidemiologic study of adenovirus-associated virus infection in infants and children, *Am. J. Epidemiol.,* 94, 359, 1971.

22. Blandini, F., Nappi, G., Greenamyre, J. T., Subthalamic infusion of an NMDA antagonist prevents basal ganglia metabolic changes and nigral degeneration in a rodent model of Parkinson's disease, *Ann. Neurol.,* 49, 525, 2001.

23. Bowers, W. J., Olschowka, J. A., Federoff, H. J., Immune responses to replication-defective HSV-1 type vectors within the CNS: implications for gene therapy, *Gene Ther.,* 10, 941, 2003.

24. Brotchie, J. M., Henry, B., Hille, C. J., Crossman, A. R., Opioid peptide precursor expression in animal models of dystonia secondary to dopamine-replacement therapy in Parkinson's disease, *Adv. Neurol.,* 78, 41, 1998.

25. Burazin, T. C., Gundlach, A. L. Localization of GDNF/neurturin receptor (c-ret, GFRalpha-1 and alpha-2) mRNAs in postnatal rat brain: differential regional and temporal expression in hippocampus, cortex and cerebellum, *Brain Res. Mol. Brain Res.,* 73, 151, 1999.

26. Burke, R. E., Kholodilov, N. G., Programmed cell death: does it play a role in Parkinson's disease? *Ann. Neurol.,* 44, S126, 1998.

27. Burton, E. A., Fink, D. J., Glorioso, J. C., Gene delivery using herpes simplex virus vectors, *DNA Cell Biol.,* 21, 915, 2002.

28. Calon, F., Grondin, R., Morissette, M., Goulet, M., Blanchet, P. J., Di Paolo, T., Bedard, P. J., Molecular basis of levodopa-induced dyskinesias, *Ann. Neurol.,* 47, S70, 2000.

29. Cao, L., Zhao, Y. C., Jiang, Z. H., Xu, D. H., Liu, Z. G., Chen, S. D., Liu, X. Y., Zheng, Z. C., Long-term phenotypic correction of rodent hemiparkinsonism by gene therapy using genetically modified myoblasts, *Gene Ther.,* 7, 445, 2000.

30. Cao, L., Zheng, Z. C., Zhao, Y. C., Jiang, Z. H., Liu, Z. G., Chen, S. D., Zhou, C. F., Liu, X. Y., Gene therapy of parkinson disease model rat by direct injection of plasmid DNA-lipofectin complex, *Hum. Gene Ther.,* 6, 1497–1501, 1995.

31. Carlsson, A., Biochemical and pharmacological aspects of Parkinsonism., *Acta Neurol. Scand.,* Suppl. 51, 11–42, 1972.

32. Carlsson, T., Winkler, C., Burger, C., Muzyczka, N., Mandel, R. J., Cenci-Nilsson, M. A., Bjorklund, A., Kirik, D., Reversal of dyskinesias in a rat model of PD by continuous levodopa delivery using AAV vector-mediated gene transfer of TH and GTPCH1, *Brain* (in press).

33. Carvalho, G. A., Nikkhah, G., Subthalamic nucleus lesions are neuroprotective against terminal 6-OHDA-induced striatal lesions and restore postural balancing reactions, *Exp. Neurol.,* 171, 405, 2001.

34. Cenci, M. A., Transcription factors involved in the pathogenesis of L-DOPA-induced dyskinesia in a rat model of Parkinson's disease, *Amino Acids,* 23, 105, 2002.

35. Cenci, M. A., Lee, C. S., Björklund, A., L-DOPA-induced dyskinesia in the rat associated with striatal overexpression of prodynorphine-and glutamic acid decarboxylase mRNA, *Eur. J. Neurosci.,* 10, 2694–2706, 1998.

36. Cenci, M. A., Tranberg, A., Andersson, M., Hilbertson, A., Changes in the regional and compartmental distribution of FosB- and JunB-like immunoreactivity induced in the dopamine-denervated rat striatum by acute or chronic L-dopa treatment, *Neuroscience,* 94, 515, 1999.

37. Chang, L. J., Gay, E. E., The molecular genetics of lentiviral vectors—current and future perspectives, *Curr. Gene Ther.,* 1, 237, 2001.

38. Chase, T. N., Levodopa therapy: consequences of the nonphysiologic replacement of dopamine, *Neurology* 50, S17, 1998.

39. Chase, T. N., Baronti, F., Fabbrini, G., Heuser, I. J., Juncos, J. L., Mouradian, M. M., Rationale for continuous dopaminomimetic therapy of Parkinson's disease, *Neurol.,* 39, 7–10, 1989.

40. Chase, T. N., Baronti, F., Fabbrini, G., Heuser, I. J., Juncos, J. L., Mouradian, M. M., Rationale for continuous dopaminomimetic therapy of Parkinson's disease, *Neurology,* 39, 7–10, 1989.

41. Chase, T. N., Engber, T. M., Mouradian, M. M., Palliative and prophylactic benefits of continuously administered dopaminomimetics in Parkinson's disease, *Neurology,* 44, S15–S18, 1994.

42. Chase, T. N., Mouradian, M. M., Engber, T. M., Motor response complications and the function of striatal efferent systems, *Neurol.,* 43, 23–27, 1993.

43. Choi-Lundberg, D. L., Lin, Q., Chang, L. N., Chiang, L., Hay, C. M., Mohajeri, H., Davidson, B. L., Bohn, M. C., Dopaminergic neurons protected from degeneration by GDNF gene therapy, *Science,* 275, 838–841, 1997.

44. Choi-Lundberg, D. L., Lin, Q., Schallert, T., Crippens, D., Davidson, B. L., Chang, Y. N., Chiang, Y. L., Qian, J., Bardwaj, L., Bohn, M. C., Behavioral and cellular protection of rat dopaminergic neurons by an adenoviral vector encoding glial cell line-derived neurotrophic factor, *Exp. Neurol.,* 154, 261–275, 1998.

45. Chun, H. S., Yoo, M. S., DeGiorgio, L. A., Volpe, B. T., Peng, D., Baker, H., Peng, C., Son, J. H., Marked dopaminergic cell loss subsequent to developmental, intranigral expression of glial cell line-derived neurotrophic factor, *Exp. Neurol.,* 173, 235, 2002.

46. Connor, B., Kozlowski, D. A., Schallert, T., Tillerson, J. L., Davidson, B. L., Bohn, M. C., Differential effects of glial cell line-derived neurotrophic factor (GDNF) in the striatum and substantia nigra of the aged Parkinsonian rat, *Gene Ther.,* 6, 1936, 1999.

47. Cortez, N., Trejo, F., Vergara, P., Segovia, J. Primary astrocytes retrovirally transduced with a tyrosine hydroxylase transgene driven by a glial-specific promoter elicit behavioral recovery in experimental Parkinsonism, *J. Neurosci. Res.,* 59, 39, 2000.

48. Corti, O., Sanchez-Capelo, A., Colin, P., Hanoun, N., Hamon, M., Mallet, J., Long-term doxycycline-controlled expression of human tyrosine hydroxylase after direct adenovirus-mediated gene transfer to a rat model of Parkinson's disease, *Proc. Natl. Acad. Sci., U.S.A.,* 96, 12120, 1999.

49. Corti, O., Sanchez-Capelo, A., Colin, P., Hanoun, N., Hamon, M., Mallet, J., Long-term doxycycline-controlled expression of human tyrosine hydroxylase after direct adenovirus-mediated gene transfer to a rat model of Parkinson's disease, *Proc. Natl. Acad. Sci., U.S.A.,* 96, 12120, 1999.

50. Crocker, S. J., Wigle, N., Liston, P., Thompson, C. S., Lee, C. J., Xu, D., Roy, S., Nicholson, D. W., Park, D. S., MacKenzie, A., Korneluk, R. G., Robertson, G. S., NAIP protects the nigrostriatal dopamine pathway in an intrastriatal 6-OHDA rat model of Parkinson's disease, *Eur. J. Neurosci.,* 14, 391, 2001.

51. Dauer, W., Kholodilov, N., Vila, M., Trillat, A. C., Goodchild, R., Larsen, K. E., Staal, R., Tieu, K., Schmitz, Y., Yuan, C. A., Rocha, M., Jackson-Lewis, V., Hersch, S., Sulzer, D., Przedborski, S., Burke, R., Hen, R., Resistance of alpha -synuclein null mice to the parkinsonian neurotoxin MPTP, *Proc. Natl. Acad. Sci., U.S.A.,* 99, 14524, 2002.

52. Davidson, B. L., Stein, C. S., Heth, J. A., Martins, I., Kotin, R. M., Derksen, T. A., Zabner, J., Ghodsi, A., Chiorini, J. A., Recombinant adeno-associated virus type 2, 4, and 5 vectors: transduction of variant cell types and regions in the mammalian central nervous system, *Proc. Natl. Acad. Sci., U.S.A.,* 97, 3428, 2000.

53. Doucet, J. P., Nakabeppu, Y., Bedard, P. J., Hope, B. T., Nestler, E. J., Jasmin, B. J., Chen, J. S., Iadarola, M. J., Stjean, M., Wigle, N., Blanchet, P., Grondin, R., Robertson, G. S., Chronic alterations in dopaminergic neurotransmission produce a persistent elevation of delta FosB-like protein(s) in both the rodent and primate striatum, *Eur. J. Neurosci.,* 8, 365–381, 1996.

54. Driesse, M. J., Esandi, M. C., Kros, J. M., Avezaat, C. J., Vecht, C., Zurcher, C., vanderVelde, I., Valerio, D., Bout, A., Sillevis-Smitt, P. A., Intra-CSF administered recombinant adenovirus causes an immune response-mediated toxicity, *Gene Ther.,* 7, 1401, 2000.

55. Driesse, M. J., Kros, J. M., Avezaat, C. J., Valerio, D., Vecht, C. J., Bout, A., Smitt, P. A., Distribution of recombinant adenovirus in the cerebrospinal fluid of nonhuman primates, *Hum. Gene Ther.,* 10, 2347, 1999.

56. Driesse, M. J., Vincent, A. J., Sillevis, S., Kros, J. M., Hoogerbrugge, P. M., Avezaat, C. J., Valerio, D., Bout, A., Intracerebral injection of adenovirus harboring the HSVtk gene combined with ganciclovir administration: toxicity study in nonhuman primates, *Gene Ther.,* 5, 1122, 1998.

57. Dull, T., Zufferey, R., Kelly, M., Mandel, R.J., Nguyen, M., Trono, D., Naldini, L., A third-generation lentivirus vector with a conditional packaging system, *J. Virol.,* 72, 8463, 1998.

58. During, M. J., Kaplitt, M. G., Stern, M. B., Eidelberg, D., Subthalamic GAD gene transfer in Parkinson disease patients who are candidates for deep brain stimulation, *Hum. Gene Ther.,* 12, 1589, 2001.

59. During, M. J., Naegele, J. R., O'Malley, K. L., Geller, A. I., Long-term behavioral recovery in parkinsonian rats by an HSV vector expressing tyrosine hydroxylase, *Science,* 266, 1399–1403, 1994.

60. Eberhardt, O., Coelln, R. V., Kugler, S., Lindenau, J., Rathke-Hartlieb, S., Gerhardt, E., Haid, S., Isenmann, S., Gravel, C., Srinivasan, A., Bahr, M., Weller, M., Dichgans, J., Schulz, J. B., Protection by synergistic effects of adenovirus-mediated X-chromosome- linked inhibitor of apoptosis and glial cell line-derived neurotrophic factor gene transfer in the 1-methyl-4-phenyl-1,2,3,6-tetrahydropyridine model of Parkinson's disease, *J. Neurosci.*, 20, 9126, 2000.

61. Engber, T. M., Susel, Z., Kuo, S., Gerfen, C. R., Chase, T. N., Levodopa replacement therapy alters enzyme activities in striatum and neuropeptide content in striatal output regions of 6-hydroxydopamine lesioned rats, *Brain Res.*, 552, 113–118, 1991.

62. Englund, U., Ericson, C., Rosenblad, C., Mandel, R. J., Trono, D., Wictorin, K., Lundberg, C., The use of a recombinant lentiviral vector for *ex vivo* gene transfer into the rat CNS, *Neuroreport*, 11, 3973, 2000.

63. Erles, K., Sebokova, P., Schlehofer, J. R., Update on the prevalence of serum antibodies (IgG and IgM) to adeno-associated virus (AAV), *J. Med. Virol.*, 59, 406, 1999.

64. Fabbrini, G., Juncos, J., Mouradian, M. M., Cerrati, C., Chase, T. N., Levodopa pharmacokinetic mechanisms and motor fluctuations in Parkinson's disease, *Ann. Neurol.*, 21, 370–376, 1987.

65. Fabbrini, G., Mouradian, M. M., Juncos, J. L., Schlegel, J., Mohr, E., Chase, T. N., Motor fluctuations in Parkinson's disease: central pathophysiological mechanisms, Part I., *Ann. Neurol.*, 24, 366–371, 1988.

66. Fan, D. S., Ogawa, M., Fujimoto, K. I., Ikeguchi, K., Ogasawara, Y., Urabe, M., Nishizawa, M., Nakano, I., Yoshida, M., Nagatsu, I., Ichinose, H., Nagatsu, T., Kurtzman, G. J., Ozawa, K., Behavioral recovery in 6-hydroxydopamine-lesioned rats by cotransduction of striatum with tyrosine hydroxylase and aromatic L-amino acid decarboxylase genes using two separate adeno-associated virus vectors, *Hum. Gene Ther.*, 9, 2527–2535, 1998.

67. Fearnley, J. M., Lees, A. J., Ageing and Parkinson's disease: substantia nigra regional selectivity, *Brain*, 114, 2283–2301, 1991.

68. Fink, D. J., Glorioso, J. C., Engineering herpes simplex virus vectors for gene transfer to neurons, *Nat. Med.*, 3, 357, 1997.

69. Fisher, L. J., Jinnah, H. A., Kale, L. C., Higgins, G. A., Gage, F. H., Survival and function of intrastriatally grafted primary fibroblasts genetically modified to produce L-DOPA, *Neuron*, 6, 371–380, 1991.

70. Fitoussi, N., Sotnik-Barkai, I., Tornatore, C., Herzberg, U., Yadid, G., Dopamine turnover and metabolism in the striatum of parkinsonian rats grafted with genetically-modified human astrocytes, *Neuroscience*, 85, 405, 1998.

71. Fraefel, C., Jacoby, D. R., Breakefield, X. O., Herpes simplex virus type 1-based amplicon vector systems, *Adv. Virus Res.*, 55, 425, 2000.

72. Gage, F. H., Wolff, J. A., Rosenberg, M. B., Xu, L., Yee, J. K., Shults, C., Friedmann, T., Grafting genetically modified cells to the brain: possibilities for the future, *Neurosci.*, 23, 795–807, 1987.

73. Gao, G. P., Alvira, M.R., Wang, L., Calcedo, R., Johnston, J., Wilson, J. M., Novel adeno-associated viruses from rhesus monkeys as vectors for human gene therapy, *Proc. Natl. Acad. Sci., U.S.A.*, 99, 11854, 2002.

74. Georgievska, B., Kirik, D., Bjorklund, A., Aberrant sprouting and downregulation of tyrosine hydroxylase in lesioned nigrostriatal dopamine neurons induced by long-lasting overexpression of glial cell line derived neurotrophic factor in the striatum by lentiviral gene transfer, *Exp. Neurol.*, 177, 461, 2002.

75. Giasson, B. I., Lee, V. M., Are ubiquitination pathways central to Parkinson's disease? *Cell*, 114, 1, 2003.

76. Gilbert, J. R., Wong-Staal, F., HIV-2 and SIV vector systems, *Somat. Cell Mol. Genet.*, 26, 83, 2001.

77. Gill, S. S., Patel, N. K., Hotton, G. R., O'Sullivan, K., McCarter, R., Bunnage, M., Brooks, D. J., Svendsen, C. N., Heywood, P., Direct brain infusion of glial cell line-derived neurotrophic factor in Parkinson disease, *Nat. Med.*, 9, 589, 2003.

78. Gjedde, A., Leger, G. C., Cumming, P., Yasuhara, Y., Evans, A. C., Guttman, M., Kuwabara, H., Striatal L-DOPA decarboxylase activity in Parkinson's disease *in vivo*– implications for the regulation of dopamine synthesis, *J. Neurochem.*, 61, 1538–1541, 1993.

79. Glazner, G. W., Mu, X., Springer, J. E., Localization of glial cell line-derived neurotrophic factor receptor alpha and c-ret mRNA in rat central nervous system, *J. Comp. Neurol.*, 391, 42, 1998.

80. Glorioso, J., Bender, M. A., Fink, D., DeLuca, N., Herpes simplex virus vectors, *Mol. Cell Biol. Hum. Dis. Ser.*, 5, 33–63, 1995.

81. Goetz, C. G., Diedrich, N. J., Dopaminergic agonists in the treatment of Parkinson's disease, *Neurologic Clinics*, 10, 527–540, 1992.

82. Gossen, M., Bujard, H., Tight control of gene expression in mammalian cells by tetracycline-responsive promoters, *Proc. Natl. Acad. Sci.*, 89, 5547, 1992.

83. Gossen, M., Freundlieb, S., Bender, G., Muller, G., Hillen, W., Bujard, H., Transcriptional activation by tetracyclines in mammalian cells, *Science*, 268, 1766–1769, 1995.

84. Goyal, L., Cell death inhibition: keeping caspases in check, *Cell*, 104, 805, 2001.

85. Hagg, T., Neurotrophins prevent death and differentially affect tyrosine hydroxylase of adult rat nigrostriatal neurons *in vivo*, *Exp. Neurol.*, 149, 183, 1998.

86. Hardy, S., Kitamura, M., Harris-Stansil, T., Dai, Y., Phipps, M. L., Construction of adenovirus vectors through Cre-lox recombination, *J. Virol.*, 71, 1842, 1997.

87. Hefti, F., Melamed, E., Wurtman, R. J., The decarboxylation of DOPA in the parkinsonian brain: *in vivo* studies on an animal model, *J. Neural. Transm.*, Suppl. 16, 95–101, 1980.

88. Hefti, F., Melamed, E., Wurtman, R. J., The site of dopamine formation in rat striatum after L-Dopa administration, *J. Pharmacol. Exp. Ther.*, 217, 189–197, 1981.

89. Henderson, J. M., Annett, L. E., Ryan, L. J., Chiang, W., Hidaka, S., Torres, E. M., Dunnett, S. B., Subthalamic nucleus lesions induce deficits as well as benefits

in the hemiparkinsonian rat, *Eur. J. Neurosci.,* 11, 2749, 1999.

90. Henry, B., Crossman, A. R., Brotchie, J. M., Effect of repeated L-DOPA, bromocriptine, or lisuride administration on preproenkephalin-A and preproenkephalin-B mRNA levels in the striatum of the 6-hydroxydopamine-lesioned rat, *Exp. Neurol.,* 155, 204, 1999.

91. Hida, H., Hashimoto, M., Fujimoto, I., Nakajima, K., Shimano, Y., Nagatsu, T., Mikoshiba, K., Nishino, H., Dopa-producing astrocytes generated by adenoviral transduction of human tyrosine hydroxylase gene: *in vitro* study and transplantation to hemiparkinsonian model rats, *Neurosci. Res.,* 35, 101, 1999.

92. Horellou, P., Lundberg, C., LeBourdelles, B., Wictorin, K., Brundin, P., Kalen, P., Björklund, A., Mallet, J., Behavioural effects of genetically engineered cells releasing dopa and dopamine after intracerebral grafting in a rat model of parkinson's disease, *J. Physiol. (Paris),* 85, 158–170, 1991.

93. Horellou, P., Marlier, L., Privat, A., Mallet, J., Behavioural Effect of Engineered Cells that Synthesize l-dopa or Dopamine after Grafting into the Rat Neostriatum, *Eur. J. Neurosci.,* 2, 116, 1990.

94. Horellou, P., Vigne, E., Castel, M. N., Barneoud, P., Colin, P., Perricaudet, M., Delaere, P., Mallet, J., Direct intracerebral gene transfer of an adenoviral vector expressing tyrosine hydroxylase in a rat model of Parkinson's disease, *Neuro. Rep.,* 6, 49–53, 1994.

95. Horellou, P., Vigne, E., Castel, M. N., Barneoud, P., Perricaudet, M., Delaire, P., Mallet, J., Transplantation of cells genetically modified and direct intracerebral gene transfer with an adenoviral vector expressing tyrosine hydroxylase in a rat model of parkinson's disease, *Restor. Neurol. Neurosci.,* 8, 63, 1995.

96. Horger, B., Nishimura, M., Armanini, M., Wang, L. C., Poulsen, K., Rosenblad, C., Kirik, D., Moffat, B., Simmons, L., Johnson, E., Milbrandt, J., Rosenthal, A., Björklund, A., Vandlen, R., Hynes, M., Phillips, H., Neurturin exerts potent actions on survival and function of midbrain dopaminergic neurons, *J. Neurosci.,* 18, 4929–4937, 1998.

97. Hornykiewicz, O., The mechanisms of action of L-DOPA in Parkinson's disease, *Life Sci.,* 15, 1249–1259, 1993.

98. Howells, D. W., Porritt, M. J., Wong, J. Y., Batchelor, P. E., Kalnins, R., Hughes, A. J., Donnan, G. A., Reduced BDNF mRNA expression in the Parkinson's disease substantia nigra, *Exp. Neurol.m* 166, 127, 2000.

99. Hyman, C., Hofer, M., Barde, Y. A., Juhasz, M., Yancopoulos, G. D., Squinto, S. P., Lindsay, R. M., BDNF is a neurotrophic factor for dopaminergic neurons of the substantia nigra, *Nature,* 350, 230–232, 1991.

100. Imaoka, T., Date, I., Ohmoto, T., Nagatsu, T., Significant behavioral recovery in Parkinson's disease model by direct intracerebral gene transfer using continuous injection of a plasmid DNA-liposome complex, *Hum. Gene Ther.,* 9, 1093, 1998.

101. Isacson, O., Behavioral effects and gene delivery in a rat model of Parkinson's disease, *Science,* 269, 856–857, 1995.

102. Jahanshahi, M., Ardouin, C. M., Brown, R. G., Rothwell, J. C., Obeso, J., Albanese, A., Rodriguez-Oroz, M. C., Moro, E., Benabid, A. L., Pollak, P., Limousin-Dowsey, P., The impact of deep brain stimulation on executive function in Parkinson's disease, *Brain,* 123, 1142, 2000.

103. Jensen, P. J., Alter, B. J., O'Malley, K. L., Alpha-synuclein protects naive but not dbcAMP-treated dopaminergic cell types from 1-methyl-4-phenylpyridinium toxicity, *J. Neurochem.,* 86, 196, 2003.

104. Jiao, S., Gurevich, V., Wolff, J. A., Long-term correction of rat model of Parkinson's disease by gene therapy, *Nature,* 380, 734,1996.

105. Jiao, S., Wolff, J. A., Long-term survival of autologous muscle grafts in rat brain, *Neurosci. Lett.,* 137, 207–210, 1992.

106. Jooss, K., Chirmule, N., Immunity to adenovirus and adeno-associated viral vectors: implications for gene therapy, *Gene Ther.,* 10, 955, 2003.

107. Kaddis, F. G., Clarkson, E. D., Weber, M. J., Vandenbergh, D. J., Donovan, D. M., Mallet, J., Horellou, P., Uhl, G. R., Freed, C. R., Intrastriatal grafting of Cos cells stably expressing human aromatic L-amino acid decarboxylase: neurochemical effects, *J. Neurochem.,* 68, 1520–1526, 1997.

108. Kang, U. J., Fisher, L. J., Joh, T. H., O'Malley, K. L., Gage, F. H., Regulation of dopamine production by genetically modified primary fibroblasts, *J. Neurosci.,* 13, 5203, 1993.

109. Kang, U. J., Lee, W. Y., Chang, J. W., Gene therapy for Parkinson's disease: determining the genes necessary for optimal dopamine replacement in rat models, *Hum. Cell,* 14, 39, 2001.

110. Kang, U. J., Lee, W. Y., Chang, J. W., Nakamura, K., Ahmed, M., Barr, E., Leiden, J. M., Kang, U. J., Gene therapy for Parkinson's disease: determining the genes necessary for optimal dopamine replacement in rat models: The localization and functional contribution of striatal aromatic L-amino acid decarboxylase to L-3,4-dihydroxyphenylalanine decarboxylation in rodent parkinsonian models, *Hum. Cell: Cell Transplant,* 14. 9, 39, 567, 2001.

111. Kaplitt, M. G., Leone, P., Samulski, R. J., Xiao, X., Pfaff, D. W., O'Malley, K. L., During, M. J., Long term gene expression and phenotypic correction using adeno associated virus vectors in the mammalian brain, *Nat. Genet.,* 8, 148, 1994.

112. Kirik, D., Annett, L. E., Burger, C., Muzyczka, N., Mandel, R. J., Bjorklund, A., Nigrostriatal alpha-synucleinopathy induced by viral vector-mediated overexpression of human alpha-synuclein: a new primate model of Parkinson's disease, *Proc. Natl. Acad. Sci., U.S.A.,* 100, 2884, 2003.

113. Kirik, D., Georgievska, B., Burger, C., Winkler, C., Muzyczka, N., Mandel, R. J., Bjorklund, A., Reversal of motor impairments in parkinsonian rats by continuous intrastriatal delivery of l-dopa using raav-mediated gene transfer, *Proc. Natl. Acad. Sci., U.S.A.,* 99, 4708, 2002.

114. Kirik, D., Rosenblad, C., Björklund, A., Mandel, R. J., Long-term rAAV mediated gene transfer of GDNF in

the rat Parkinson's model: intrastriatal but not intranigral transduction promotes functional regeneration in the lesioned nigrostriatal system, *J. Neurosci.,* 20, 4686, 2000.

115. Kirik, D., Rosenblad, C., Burger, C., Lundberg, C., Johansen, T. E., Muzyczka, N., Mandel, R. J., Bjorklund, A., Parkinson-like neurodegeneration induced by targeted overexpression of alpha-synuclein in the nigrostriatal system, *J. Neurosci.,* 22, 2780, 2002.

116. Kirik, D., Winkler, C., Bjorklund, A., Growth and functional efficacy of intrastriatal nigral transplants depend on the extent of nigrostriatal degeneration, *J. Neurosci.,* 21, 2889, 2001.

117. Kistner, A., Gossen, M., Zimmermann, F., Jerecic, J., Ullmer, C., Lubbert, H., Bujard, H., Doxycycline-mediated quantitative and tissue-specific control of gene expression in transgenic mice, *Proc. Natl. Acad. Sci., U.S.A.,* 93, 10933–10938, 1996.

118. Klein, R. L., Lewis, M. H., Muzyczka, N., Meyer, E. M., Prevention of 6-hydroxydopamine-induced rotational behavior by BDNF somatic gene transfer, *Brain Res.,* 847, 314, 1999.

119. Klein, R. L., Meyer, E. M., Peel, A. L., Zolotukhin, S., Meyers, C., Muzyczka, N., King, M. A., Neuron-specific transduction in the rat septohippocampal or nigrostriatal pathway by recombinant adeno-associated virus vectors, *Exp. Neurol.,* 150, 183, 1998.

120. Kordower, J. H., In vivo gene delivery of glial cell line—derived neurotrophic factor for Parkinson's disease, *Ann. Neurol.,* 53 Suppl. 3, S120, discussion S132-4, 2003.

121. Kordower, J. H., Emborg, M. E., Bloch, J., Ma, S. Y., Chu, Y., Leventhal, L., McBride, J., Chen, E. Y., Palfi, S., Roitberg, B. Z., Brown, W. D., Holden, J. E., Pyzalski, R., Taylor, M. D., Carvey, P., Ling, Z., Trono, D., Hantraye, P., Deglon, N., Aebischer, P., Neurodegeneration prevented by lentiviral vector delivery of GDNF in primate models of Parkinson's disease, *Science,* 290, 767, 2000.

122. Kordower, J. H., Palfi, S., Chen, E. Y., Ma, S. Y., Sendera, T., Cochran, E. J., Cochran, E. J., Mufson, E. J., Penn, R., Goetz, C. G., Comella, C. D., Clinicopathological findings following intraventricular glial-derived neurotrophic factor treatment in a patient with Parkinson's disease, *Ann. Neurol.,* 46, 419, 1999.

123. Kozlowski, D. A., Connor, B., Tillerson, J. L., Schallert, T., Bohn, M. C., Delivery of a GDNF gene into the substantia nigra after a progressive 6-OHDA lesion maintains functional nigrostriatal connections, *Exp. Neurol.,* 166, 1–15, 2000.

124. Krack, P., Benazzouz, A., Pollak, P., Limousin, P., Piallat, B., Hoffmann, D., Xie, J., Benabid, A. L., Treatment of tremor in Parkinson's disease by subthalamic nucleus stimulation, *Mov. Disord.,* 13, 907–914, 1998.

125. Krack, P., Limousin, P., Benabid, A. L., Pollak, P., Chronic stimulation of subthalamic nucleus improves levodopa-induced dyskinesias in Parkinson's disease (letter), *Lancet,* 350, 1676, 1997.

126. Krack, P., Pollak, P., Limousin, P., Benazzouz, A., Benabid, A. L., Stimulation of subthalamic nucleus alleviates tremor in Parkinson's disease (letter), *Lancet,* 350, 1675, 1997.

127. Kuwabara, H., Cumming, P., Yasuhara, Y., Leger, G.C., Guttman, M., Diksic, M., Evans, A. C., Gjedde, A., Regional striatal DOPA transport and decarboxylase activity in parkinson's disease, *J. Nucl. Med.,* 36, 1226–1231, 1995.

128. Lai, C. M., Lai, Y. K., Rakoczy, P. E., Adenovirus and adeno-associated virus vectors, *DNA Cell Biol.,* 21, 895, 2002.

129. LaVail, M. M., Yasumura, D., Matthes, M. T., Drenser, K. A., Flannery, J. G., Lewin, A. S., and Hauswirth, W. W., Ribozyme rescue of photoreceptor cells in P23H transgenic rats: long-term survival and late-stage therapy [In Process Citation], *Proc. Natl. Acad. Sci., U.S.A.,* 97(21), 11488–11493. 2000.

130. Lee, W. Y., Chang, J.W., Nemeth, N. L., Kang, U. J., Vesicular monoamine transporter-2 and aromatic L-amino acid decarboxylase enhance dopamine delivery after L-3, 4- dihydroxyphenylalanine administration in Parkinsonian rats, *J Neurosci.,* 19, 3266, 1999.

131. Leff, S. E., Rendahl, K. G., Spratt, S. K., Kang, U. J., Mandel, R. J., In vivo L-dopa production by genetically modified primary rat fibroblast or 9L gliosarcoma cell grafts requires co-expression of GTP-cyclohydrolase I with tyrosine hydroxylase, *Exp. Neurol.,* 151, 249, 1998.

132. Leff, S. E., Spratt, S. K. S., Mandel, R. J., Long-term restoration of striatal L-aromatic amino acid decarboxylase activity using recombinant adeno-associated virus in an animal model of Parkinson's disease, *Neurosci.,* 92, 187, 1999.

133. Leone, P., McPhee, S. W., Janson, C. G., Davidson, B. L., Freese, A., During, M. J., Multi-site partitioned delivery of human tyrosine hydroxylase gene with phenotypic recovery in Parkinsonian rats, *Neuroreport,* 11, 1145, 2000.

134. Levine, R. A., Miller, L. P., Lovenberg, W., Tetrahydrobiopterin in striatum: localization in dopamine nerve terminals and role in catecholamine synthesis, *Science,* 214, 919–921, 1981.

135. Levivier, M., Przedborski, S., Bencsics, C., Kang, U. J., Intrastriatal implantation of fibroblasts genetically engineered to produce brain-derived neurotrophic factor prevents degeneration of dopaminergic neurons in a rat model of parkinson's disease, *J. Neurosci.,* 15, 7810–7820, 1995.

136. Levy, R., Lang, A. E., Dostrovsky, J. O., Pahapill, P., Romas, J., Saint-Cyr, J., Hutchison, W. D., Lozano, A. M., Lidocaine and muscimol microinjections in subthalamic nucleus reverse Parkinsonian symptoms, *Brain,* 124, 2105, 2001.

137. Limousin, P., Pollak, P., Benazzouz, A., Hoffmann, D., Lebas, J. F., Broussolle, E., Perret, J. E., Benabid, A. L., Effect on parkinsonian signs and symptoms of bilateral subthalamic nucleus stimulation, *Lancet,* 345, 91–95, 1995.

138. Lin, L. F., Doherty, D. H., Lile, J. D., Bektesh, S., Collins, F., GDNF: a glial cell line-derived neurotrophic factor for midbrain dopaminergic neurons, *Science,* 260, 1130–1132, 1993.

139. Lowenstein, P. R., Immunology of viral-vector-mediated gene transfer into the brain: an evolutionary and developmental perspective, *Trends Immunol.,* 23, 23, 2002.

140. Lowenstein, P. R., Castro, M. G., Inflammation and adaptive immune responses to adenoviral vectors injected into the brain: peculiarities, mechanisms, and consequences, *Gene Ther.,* 10, 946, 2003.

141. Lu, X., Hagg, T., Glial cell line-derived neurotrophic factor prevents death, but not reductions in tyrosine hydroxylase, of injured nigrostriatal neurons in adult rats, *J. Comp. Neurol.,* 388, 484–494, 1997.

142. Lundberg, C., Engineered cells and *ex vivo* gene transfer, in *Neuromethods: Neural Transplantation Methods,* Dunnett, S. B., Boulton, A. A., Baker, G. B., Eds., Humana Press, Inc., Totowa, NJ, 89, 2000.

143. Lundberg, C., Horellou, P., Mallet, J., Björklund, A., Generation of DOPA-producing astrocytes by retroviral transduction of the human tyrosine hydroxylase gene: *in vitro* characterization and *in vivo* effects in the rat parkinson model, *Exp. Neurol.,* 139, 39–53, 1996.

144. Lundblad, M., Andersson, M., Winkler, C., Kirik, D., Wierup, N., Cenci, M. A., Pharmacological validation of behavioural measures of akinesia and dyskinesia in a rat model of Parkinson's disease, *Eur. J. Neurosci.,* 15, 120, 2002.

145. Luo, J., Kaplitt, M. G., Fitzsimons, H. L., Zuzga, D. S., Liu, Y., Oshinsky, M. L., During, M. J., Subthalamic GAD gene therapy in a Parkinson's disease rat model, *Science,* 298, 425, 2002.

146. Macaya, A., Munell, F., Gubits, R. M., Burke, R. E., Apoptosis in substantia nigra following developmental striatal excitotoxic injury, *Proc. Natl. Acad. Sci.,* 91, 8117–8121, 1994.

147. Mandel, R. J., Rendahl, K. G., Snyder, R. O., Leff, S. E., Progress in direct striatal delivery of L-dopa via gene therapy for treatment of Parkinson's disease using recombinant adeno-associated viral vectors, *Exp. Neurol.,* 159, 47, 1999.

148. Mandel, R. J., Rendahl, K. G., Spratt, S. K., Snyder, R. O., Cohen, L. K., Leff, S. E., Characterization of intrastriatal recombinant adeno-associated virus mediated gene transfer of human tyrosine hydroxylase and human GTP-cyclohydroxylase I in a rat model of Parkinson's disease, *J. Neurosci.,* 18, 4271, 1998.

149. Mandel, R. J., Snyder, R. O., Leff, S. E., Recombinant adeno-associated viral vector-mediated glial cell line-derived neurotrophic factor gene transfer protects nigral dopamine neurons after onset of progressive degeneration in a rat model of Parkinson's disease, *Exp. Neurol.,* 160, 205, 1999.

150. Mandel, R. J., Spratt, S. K., Snyder, R. O., Leff, S. E., Midbrain injection of recombinant adeno-associated virus encoding rat glial cell line-derived neurotrophic factor protects nigral neurons in a progressive 6-hydroxydopamine-induced degeneration model of Parkinson's disease in rats, *Proc. Natl. Acad. Sci., U.S.A.,* 94, 14083, 1997.

151. Mannes, A. J., Caudle, R. M., O'Connell, B. C., Iadarola, M. J., Adenoviral gene transfer to spinal-cord neurons: intrathecal vs. intraparenchymal administration, *Brain Res.,* 793, 1, 1998.

152. Manning-Bog, A. B., McCormack, A. L., Purisai, M. G., Bolin, L. M., Di Monte, D. A., Alpha-synuclein overexpression protects against paraquat-induced neurodegeneration, *J. Neurosci.,* 23, 3095, 2003.

153. Mansuy, I. M., Bujard, H., Tetracycline-regulated gene expression in the brain, *Curr. Opin. Neurobiol.,* 10, 593, 2000.

154. Marshall, E., FDA halts all gene therapy trials at Penn, *Science,* 287, 565, 567, 2000.

155. Marti, M. J., Saura, J., Burke, R. E., Jackson-Lewis, V., Jimenez, A., Bonastre, M., Tolosa, E., Striatal 6-hydroxydopamine induces apoptosis of nigral neurons in the adult rat, *Brain Res.,* 958, 185, 2002.

156. Matsuo, A., Nakamura, S., Akiguchi, I., Immunohistochemical localization of glial cell line-derived neurotrophic factor family receptor alpha-1 in the rat brain: confirmation of expression in various neuronal systems, *Brain Res.,* 859, 57, 2000.

157. McCown, T. J., Xiao, X., Li, J., Breese, G. R., Samulski, R. J., Differential and persistent expression patterns of CNS gene transfer by an adeno-associated virus (AAV) vector, *Brain Res.,* 713, 99–107, 1996.

158. McGeer, P. L., Itagaki, S., Akiyama, H., McGeer, E. G., Rate of cell death in parkinsonism indicates active neuropathological process, *Ann. Neurol.,* 24, 574, 1988.

159. Meissner, W., Harnack, D., Paul, G., Reum, T., Sohr, R., Morgenstern, R., Kupsch, A., Deep brain stimulation of subthalamic neurons increases striatal dopamine metabolism and induces contralateral circling in freely moving 6-hydroxydopamine-lesioned rats, *Neurosci. Lett.,* 328, 105, 2002.

160. Melamed, E., Hefti, F., Liebman, J., Schlosberg, A. J., Wurtman, R. J., Serotonergic neurones are not involved in action of L-dopa in Parkinson's disease, *Nature,* 283, 772, 1980.

161. Melamed, E., Hefti, F., Pettibone, D. J., Liebman, J., Wurtman, R. J., Aromatic L-amino acid decarboxylase in rat corpus striatum: Implications for action of L-dopa in parkinsonism, *Neurol.,* 31, 651–655, 1981.

162. Melamed, E., Hefti, F., Wurtman, R. J., Nonaminergic striatal neurons convert exogenous L-DOPA to dopamine in Parkinsonism, *Ann. Neurol.,* 8, 558–563, 1979.

163. Melamed, E., Hefti, F., Wurtman, R. J., Decarboxylation of exogenous L-DOPA in rat striatum after lesions of the dopaminergic nigrostriatal neurons: the role of striatal capillaries, *Brain Res.,* 198, 244–248, 1980.

164. Melamed, E., Hefti, F., Wurtman, R. J., Diminished decarboxylation of L-DOPA in rat striatum after intrastriatal injections of kainic acid, *Neuropharmacol.,* 19, 409–411, 1980.

165. Melamed, E., Hefti, F., Wurtman, R. J., Nonaminergic striatal neurons convert exogenous L-dopa to dopamine in parkinsonism, *Ann. Neurol.,* 8, 558–563, 1980.

166. Messer, C. J., Son, J. H., Joh, T. H., Beck, K. D., Nestler, E. J., Regulation of tyrosine hyroxylase gene transcription in ventral midbrain by glial cell line-derived neurotrophic factor, *Synapse,* 34, 241, 1999.

167. Miller, A. D., The problem with cationic liposome/micelle-based non-viral vector systems for gene therapy, *Curr. Med. Chem.*, 10, 1195, 2003.

168. Miller, L. G., Greenblatt, D. J., Barnhill, J. G., Shader, R. I., Tietz, E. I., Rosenberg, H. C., Chronic benzodiazepine administration. I. Tolerance is associated with benzodiazepine receptor downregulation and decreased gamma-aminobutyric acid A receptor function: Behavioral measurement of benzodiazepine tolerance and GABAergic subsensitivity in the substantia nigra pars reticulata, *J. Pharmacol. Exp. Ther. and Brain Res.*, 438, 170, 41, 1988.

169. Miller, L. G., Woolverton, S., Greenblatt, D. J., Lopez, F., Roy, R. B., Shader, R. I., Miller, L. G., Greenblatt, D. J., Barnhill, J. G., Shader, R. I., Tietz, E. I., Rosenberg, H. C., Chronic benzodiazepine administration. IV. Rapid development of tolerance and receptor downregulation associated with alprazolam administration, *Biochem. Pharmacol.*, 38, 3773, 1989.

170. Mochizuki, H., Goto, K., Mori, H., Mizuno, Y., Histochemical detection of apoptosis in Parkinson's disease, *J. Neurol. Sci.*, 137, 120, 1996.

171. Mochizuki, H., Hayakawa, H., Migita, M., Shibata, M., Tanaka, R., Suzuki, A., Shimo-Nakanishi, Y., Urabe, T., Yamada, M., Tamayose, K., Shimada, T., Miura, M., Mizuno, Y., An AAV-derived Apaf-1 dominant negative inhibitor prevents MPTP toxicity as antiapoptotic gene therapy for Parkinson's disease, *Proc. Natl. Acad. Sci., U.S.A.*, 98, 10918, 2001.

172. Montoya, C. P., Campbell-Hope, L. J., Pemberton, K. D., Dunnett, S. B., The "staircase test": a measure of independent forelimb reaching and grasping abilities in rats, *J. Neurosci. Meth.*, 36, 219–228, 1991.

173. Morissette, M., Goulet, M., Soghomonian, J. J., Blanchet, P. J., Calon, F., Bedard, P. J., Di Paolo, T., Preproenkephalin mRNA expression in the caudate-putamen of MPTP monkeys after chronic treatment with the D2 agonist U91356A in continuous or intermittent mode of administration: comparison with L-DOPA therapy, *Brain Res. Mol. Brain Res.*, 49, 55, 1997.

174. Morissette, M., Grondin, R., Goulet, M., Bedard, P. J., Di Paolo, T., Differential regulation of striatal preproenkephalin and preprotachykinin mRNA levels in MPTP-lesioned monkeys chronically treated with dopamine D1 or D2 receptor agonists, *J. Neurochem.*, 72, 682, 1999.

175. Morsy, M. A., Gu, M., Motzel, S., Zhao, J., Lin, J., Su, Q., Allen, H., Franlin, L., Parks, R. J., Graham, F. L., Kochanek, S., Bett, A. J., Caskey, C. T., An adenoviral vector deleted for all viral coding sequences results in enhanced safety and extended expression of a leptin transgene, *Proc. Natl. Acad. Sci., U.S.A.*, 95, 7866–7871, 1998.

176. Mouradian, M. M., Juncos, J. L., Fabbrini, G., Schlegel, J., Bartko, J. J., Chase, T. N., Motor fluctuations in Parkinson's disease: central pathophysiological mechanisms, Part II, *Ann. Neurol.*, 24, 372–378, 1988.

177. Mulligan, R. C., The basic science of gene therapy, *Science*, 260, 926, 1993.

178. Muramatsu, S., Fujimoto, K., Ikeguchi, K., Shizuma, N., Kawasaki, K., Ono, F., Shen, Y., Wang, L., Mizukami, H., Kume, A., Matsumura, M., Nagatsu, I., Urano, F., Ichinose, H., Nagatsu, T., Terao, K., Nakano, I., Ozawa, K., Behavioral recovery in a primate model of Parkinson's disease by triple transduction of striatal cells with adeno-associated viral vectors expressing dopamine-synthesizing enzymes, *Hum. Gene Ther.*, 13, 345, 2002.

179. Muzyczka, N., Berns, K. I., Parvoiridae, The viruses and their replication. In: *Fields Virology*, Knipe, D. M., Howley, P. M., Eds., Lippincott, Williams and Wilkins, New York, 2327, 2001.

180. Nagatsu, T., Parkinson's disease: changes in apoptosis-related factors suggesting possible gene therapy, *J. Neural. Transm.*, 109, 731, 2002.

181. Nakamura, K., Ahmed, M., Barr, E., Leiden, J. M., Kang, U. J., The localization and functional contribution of striatal aromatic L-amino acid decarboxylase to L-3,4-dihydroxyphenylalanine decarboxylation in rodent parkinsonian models.,*Cell Transplant*, 9, 567, 2000.

182. Naldini, L., Lentiviruses as gene transfer agents for delivery to non-dividing cells, *Curr. Opin. Biotechnol.*, 9, 457, 1998.

183. Naldini, L., Blömer, U., Gallay, P., Ory, D., Mulligan, R., Gage, F. H., Verma, I. M., Trono, D., In vivo gene delivery and stable transduction of nondividing cells by a lentiviral vector, *Science*, 272, 263, 1996.

184. Naldini, L., Verma, I. M., Lentiviral vectors, *Adv. Virus Res.*, 55, 599, 2000.

185. Natsume, A., Mata, M., Goss, J., Huang, S., Wolfe, D., Oligino, T., Glorioso, J., Fink, D. J., Bcl-2 and GDNF delivered by HSV-mediated gene transfer act additively to protect dopaminergic neurons from 6-OHDA-induced degeneration, *Exp. Neurol.*, 169, 231, 2001.

186. Negre, D., Cosset, F. L., Vectors derived from simian immunodeficiency virus (SIV), *Biochimie*, 84, 1161, 2002.

187. Numata, Y., Kato, T., Nagatsu, T., Sugimoto, T., Matsuura, S., Effects of stereochemical structures of tetrahydrobiopterin on tyrosine hydroxylase, *Biochim. Biophys. Acta*, 480, 104–112, 1977.

188. Nutt, J. G., Burchiel, K. J., Comella, C. L., Jankovic, J., Lang, A. E., Laws, E. R., Jr., Lozano, A. M., Penn, R. D., Simpson, R. K., Jr., Stacy, M., Wooten, G. F., Randomized, double-blind trial of glial cell line-derived neurotrophic factor (GDNF) in PD, *Neurology*, 60, 69, 2003.

189. Obeso, J. A., Grandas, F., Herrero, M. T., Horowski, R., The role of pulsatile versus continuous dopamine receptor stimulation for functional recovery in parkinson's disease, *Eur. J. Neurosci.*, 6, 889–897, 1994.

190. Olsson, M., Nikkhah, G., Bentlage, C., Björklund, A., Forelimb akinesia in the rat parkinson model: differential effects of dopamine agonists and nigral transplants as assessed by a new stepping test, *J. Neurosci.*, 15, 3863–3875, 1995.

191. Parks, R. J., Chen, L., Anton, M., Sankar, U., Rudnicki, M. A., Graham, F. L., A helper-dependent adenovirus vector system: removal of helper virus by Cre-mediated excision of the viral packaging signal, *Proc. Natl. Acad. Sci., U.S.A.*, 93, 13565, 1996.

192. Petrucelli, L., O'Farrell, C., Lockhart, P. J., Baptista, M., Kehoe, K., Vink, L., Choi, P., Wolozin, B., Farrer, M., Hardy, J., Cookson, M. R., Parkin protects against the toxicity associated with mutant alpha-synuclein: protea-some dysfunction selectively affects catecholaminergic neurons, *Neuron*, 36, 1007, 2002.

193. Piallat, B., Benazzouz, A., Benabid, A. L., Neuroprotec-tive effect of chronic inactivation of the subthalamic nucleus in a rat model of Parkinson's disease, *J. Neural. Transm.*, Suppl. 55, 71, 1999.

194. Pollak, P., Benabid, A. L., Gross, C., Gao, D. M., Lau-rent, A., Benazzouz, A., Hoffmann, D., Gentil, M., Per-ret, J., Effects of subthalamic nucleus stimulation in parkinsons disease, *Rev. Neurol.*, 149, 175–176, 1993.

195. Rivera, V. M., Clackson, T., Natesan, S., Pollock, R., Amara, J. F., Keenan, T., Magari, S. R., Phillips, T., Courage, N. L., Cerasoli, F., Holt, D. A., Gilman, M., A humanized system for pharmacologic control of gene expression, *Nat. Med.*, 2, 1028, 1996.

196. Roizman, B., Sears, A. E., An inquiry into the mecha-nisms of herpes simplex virus latency, *Annu. Rev. Micro-biol.*, 41, 543, 1987.

197. Rosenblad, C., Georgievska, B., Kirik, D., Long-term striatal overexpression of GDNF selectively downregu-lates tyrosine hydroxylase in the intact nigrostriatal dopamine system, *Eur. J. Neurosci.*, 17, 260, 2003.

198. Rosenblad, C., Kirik, D., Bjorklund, A., Sequential administration of GDNF into the substantia nigra and striatum promotes dopamine neuron survival and axonal sprouting but not striatal reinnervation or functional recovery in the partial 6-OHDA lesion model, *Exp. Neu-rol.*, 161, 503, 2000.

199. Ruberg, M., Brugg, B., Prigent, A., Hirsch, E., Brice, A., Agid, Y., Is differential regulation of mitochondrial transcripts in Parkinson's disease related to apoptosis? *J. Neurochem.*, 68, 2098–2110, 1997.

200. Sanchez-Pernaute, R., Harvey-White, J., Cunningham, J., Bankiewicz, K. S., Functional effect of adeno-asso-ciated virus mediated gene transfer of aromatic L-amino acid decarboxylase into the striatum of 6-OHDA-lesioned rats, *Mol. Ther.*, 4, 324, 2001.

201. Sarabi, A., Hoffer, B. J., Olson, L., Morales, M., GFRal-pha-1 mRNA in dopaminergic and nondopaminergic neurons in the substantia nigra and ventral tegmental area, *J. Comp. Neurol.*, 441, 106, 2001.

202. Sauter, S. L., Gasmi, M., FIV vector systems, *Somat. Cell Mol. Genet*, 26, 99, 2001.

203. Schallert, T., Norton, D., Jones, T. A., A clinically rel-evant unilateral rat model of Parkinsonian akinesia, *J. Neural. Transplant Plast.*, 3, 332–333, 1992.

204. Schallert, T., Tillerson, J. L., Intervention strategies for degeneration of dopamine neurons in Parkisonism: opti-mizing behavioral assessment of outcome, In: *Innova-tive Models of CNS Disease: From Molecule to Therapy*, Anonymous Humana Press, Clifton, NJ, 131, 1999.

205. Schuh, L. A., Bennett, J. P., Suppression of dyskinesias in advanced parkinson's disease.1. continuous intrave-nous levodopa shifts dose response for production of dyskinesias but not for relief of parkinsonism in patients

with advanced parkinson's disease, *Neurology*, 43, 1545–1550, 1993.

206. Segovia, J., Vergara, P., Brenner, M., Astrocyte-specific expression of tyrosine hydroxylase after intracerebral gene transfer induces behavioral recovery in experimen-tal parkinsonism, *Gene Ther.*, 5, 1650–1655, 1998.

207. Shen, Y., Muramatsu, S. I., Ikeguchi, K., Fujimoto, K. I., Fan, D.S., Ogawa, M., Mizukami, H., Urabe, M., Kume, A., Nagatsu, I., Urano, F., Suzuki, T., Ichinose, H., Nagatsu, T., Monahan, J., Nakano, I., Ozawa, K., Triple transduction with adeno-associated virus vectors expressing tyrosine hydroxylase, aromatic-L-amino-acid decarboxylase, and GTP cyclohydrolase I for gene therapy of Parkinson's disease, *Hum. Gene Ther.*, 11, 1509, 2000.

208. Shenk, T., Adenoviridae: the viruses and their replica-tion. In: *Fundamental Virology*, Fields, B. N., Knipe, D. M., Howley, P. M., Eds., Lippincott-Raven Co., New York, 979, 1996.

209. Shimura, H., Hattori, N., Kubo, S. I., Mizuno, Y., Asakawa, S., Minoshima, S., Shimizu, N., Iwai, K., Chiba, T., Tanaka, K., Suzuki, T., Familial Parkinson disease gene product, parkin, is a ubiquitin-protein ligase, *Nat. Genet.*, 25, 302–305, 2000.

210. Shults, C. W., Kimber, T., Altar, C. A., BDNF attenuates the effects of intrastriatal injection of 6-hydroxydopam-ine, *NeuroRep.*, 6, 1109–1112, 1995.

211. Spillantini, M. G., Schmidt, M. L., Lee, V. M., Trojan-owski, J. Q., Jakes, R., Goedert, M., Alpha-synuclein in Lewy bodies [letter], *Nature*, 388, 839, 1997.

212. Stevenson, M., HIV-1 pathogenesis, *Nat. Med.*, 9, 853, 2003.

213. Sumi-Ichinose, C., Urano, F., Kuroda, R., Ohye, T., Kojima, M., Tazawa, M., Shiraishi, H., Hagino, Y., Nagatsu, T., Nomura, T., Ichinose, H., Catecholamines and serotonin are differently regulated by tetrahydro-biopterin. A study from 6-pyruvoyltetrahydropterin syn-thase knockout mice, *J. Biol. Chem.*, 276, 41150, 2001.

214. Sun, J. Y., Anand-Jawa, V., Chatterjee, S., Wong, K. K., Immune responses to adeno-associated virus and its recombinant vectors, *Gene Ther.*, 10, 964, 2003.

215. Sun, M., Zhang, G. R., Kong, L., Holmes, C., Wang, X., Zhang, W., Goldstein, D. S., Geller, A. I., Correction of a rat model of Parkinson's disease by coexpression of tyrosine hydroxylase and aromatic amino Acid decar-boxylase from a helper virus-free herpes simplex virus type 1 vector, *Hum. Gene Ther.*, 14, 415, 2003.

216. Szczypka, M. S., Kwok, K., Brot, M. D., Marck, B. T., Matsumoto, A. M., Donahue, B. A., Palmiter, R. D., Dopamine production in the caudate putamen restores feeding in dopamine-deficient mice, *Neuron*, 30, 819, 2001.

217. Szczypka, M. S., Mandel, R. J., Donahue, B. A., Snyder, R. O., Leff, S. E., Palmiter, R. D., Viral gene therapy selectively restores feeding and prevents lethality of dopamine-deficient mice, *Neuron*, 22, 167–178, 1999.

218. Tel, B. C., Zeng, B. Y., Cannizzaro, C., Pearce, R. K., Rose, S., Jenner, P., Alterations in striatal neuropeptide mRNA produced by repeated administration of L-DOPA, ropinirole or bromocriptine correlate with dys-

kinesia induction in MPTP-treated common marmosets, *Neuroscience,* 115, 1047, 2002.

219. Thomas, C. E., Schiedner, G., Kochanek, S., Castro, M. G., Lowenstein, P. R., Peripheral infection with adenovirus causes unexpected long-term brain inflammation in animals injected intracranially with first-generation, but not with high-capacity, adenovirus vectors: toward realistic long-term neurological gene therapy for chronic diseases, *Proc. Natl. Acad. Sci., U.S.A.,* 97, 7482, 2000.

220. Thomas, C. E., Schiedner, G., Kochanek, S., Castro, M. G., Lowenstein, P. R., Peripheral infection with adenovirus causes unexpected long-term brain inflammation in animals injected intracranially with first-generation, but not with high-capacity, adenovirus vectors: toward realistic long- term neurological gene therapy for chronic diseases, *Proc. Natl. Acad. Sci., U.S.A.,* 97, 7482, 2000.

221. Tompkins, M. M., Basgall, E. J., Zamrini, E., Hill, W. D., Apoptotic-like changes in Lewy-body-associated disorders and normal aging in substantia nigra neurons, *Am. J. Pathol.,* 150, 119–131, 1997.

222. Tornatore, C., Bakercairns, B., Yadid, G., Hamilton, R., Meyers, K., Atwood, W., Cummins, A., Tanner, V., Major, E., Expression of tyrosine hydroxylase in an immortalized human fetal astrocyte cell line, *in vitro* characterization and engraftment into the rodent striatum, *Cell Transplant.,* 5, 145–163, 1996.

223. Trojanowski, J. Q., Lee, V. M., Parkinson's disease and related alpha-synucleinopathies are brain amyloidoses, *Ann. N. Y. Acad. Sci.,* 991, 107, 2003.

224. Trupp, M., Belluardo, N., Funakoshi, H., Ibanez, C. F., Complementary and overlapping expression of glial cell line-derived neurotrophic factor (GDNF), c-ret proto-oncogene, and GDNF receptor-alpha indicates multiple mechanisms of trophic actions in the adult rat CNS, *J. Neurosci.,* 17, 3554, 1997.

225. Tsukahara, T., Takeda, M., Shimohama, S., Ohara, O., Hashimoto, N., Effects of brain-derived neurotrophic factor on 1-methyl-4- phenyl-1,2,3,6-tetrahydropyridine-induced parkinsonism in monkeys, *Neurosurgery,* 37, 733–739, 1995.

226. Uchida, K., Tsuzaki, N., Nagatsu, T., Kohsaka, S., Tetrahydrobiopterin-dependent functional recovery in 6-hydroxydopamine-treated rats by intracerebral grafting of fibroblasts transfected with tyrosine hydroxylase cDNA, *Dev. Neurosci.,* 14, 173–180, 1992.

227. Ungerstedt, U., Striatal dopamine release after amphetamine or nerve degeneration revealed by rotational behaviour, *Acta Physiol. Scand.,* Suppl. 367, 51–68, 1971.

228. von Coelln, R., Kugler, S., Bahr, M., Weller, M., Dichgans, J., Schulz, J. B., Rescue from death but not from functional impairment: caspase inhibition protects dopaminergic cells against 6-hydroxydopamine-induced apoptosis but not against the loss of their terminals, *J. Neurochem.,* 77, 263, 2001.

229. Wachtel, S. R., Bencsics, C., Kang, U. J., Role of aromatic L-amino acid decarboxylase for dopamine replacement by genetically modified fibroblasts in a rat model of Parkinson's disease, *J. Neurochem.,* 69, 2055–2063, 1997.

230. Walker, D. G., Beach, T. G., Xu, R., Lile, J., Beck, K. D., McGeer, E.G., McGeer, P. L., Expression of the proto-oncogene Ret, a component of the GDNF receptor complex, persists in human substantia nigra neurons in Parkinson's disease, *Brain Res.,* 792, 207, 1998.

231. Wang, L., Muramatsu, S., Lu, Y., Ikeguchi, K., Fujimoto, K., Okada, T., Mizukami, H., Hanazono, Y., Kume, A., Urano, F., Ichinose, H., Nagatsu, T., Nakano, I., Ozawa, K., Delayed delivery of AAV-GDNF prevents nigral neurodegeneration and promotes functional recovery in a rat model of Parkinson's disease, *Gene Ther.,* 9, 381, 2002.

232. Wang, Y., O'Malley, B. W., Jr., Tsai, S. Y., O'Malley, B. W., A regulatory system for use in gene transfer, *Proc. Natl. Acad. Sci., U.S.A.,* 91, 8180, 1994.

233. Werner, E. R., Werner-Felmayer, G., Fuchs, D., Hausen, A., Reibnegger, G., Yim, J. J., Pfeiderer, W., Wachter, H., Tetrahydrobiopterin biosynthetic activities in human macrophages, fibroblasts, THP-1, and T 24 cells, *J. Biol. Chem.,* 265, 3189–3192, 1990.

234. Westin, J. E., Andersson, M., Lundblad, M., Cenci, M. A., Persistent changes in striatal gene expression induced by long-term L-DOPA treatment in a rat model of Parkinson's disease, *Eur. J. Neurosci.,* 14, 1171, 2001.

235. Wichmann, T., Bergman, H., DeLong, M. R., The primate subthalamic nucleus.3. changes in motor behavior and neuronal activity in the internal pallidum induced by subthalamic inactivation in the MPTP model of parkinsonism, *J. Neurophysiol.,* 72, 521–530, 1994.

236. Wichmann, T., Bergman, H., DeLong, M. R., The primate subthalamic nucleus. I. functional properties in intact animals, *J. Neurophysiol.,* 72, 494–506, 1994.

237. Winkler, C., Kirik, D., Bjorklund, A., Cenci, M. A., L-DOPA-induced dyskinesia in the intrastriatal 6-hydroxydopamine model of parkinson's disease: relation to motor and cellular parameters of nigrostriatal function, *Neurobiol. Dis.,* 10, 165, 2002.

238. Winogrodzka, A., Bergmans, P., Booij, J., van Royen, E. A., Stoof, J. C., Wolters, E. C., [(123)I]beta-CIT SPECT is a useful method for monitoring dopaminergic degeneration in early stage Parkinson's disease, *J. Neurol. Neurosurg. Psychiatry,* 74, 294, 2003.

239. Wolff, J. A., Fisher, L. J., Xu, L., Jinnah, H. A., Langlais, P. J., Iuvone, P. M., O'Malley, K. L., Rosenberg, M. B., Shimohama, S., Friedmann, T., Gage, F. H., Grafting fibroblasts genetically modified to produce L-dopa in a rat model of Parkinson's disease, *Proc. Natl. Acad. Sci.,U.S.A.,* 86, 9011, 1989.

240. Wood, M. J., Charlton, H. M., Wood, K. J., Kajiwara, K., Byrnes, A. P., Immune responses to adenovirus vectors in the nervous system, *Trends. Neurosci.,* 19, 497, 1996.

241. Xia, H., Mao, Q., Paulson, H. L., Davidson, B. L., siRNA-mediated gene silencing *in vitro* and *in vivo, Nat. Biotechnol.,* 20, 1006, 2002.

242. Xiao, X., McCown, T. J., Li, J., Breese, G. R., Morrow, A. L., Samulski, R. J., Adeno-associated virus (AAV) vector antisense gene transfer *in vivo* decreases GABA(A) alpha1 containing receptors and increases

inferior collicular seizure sensitivity, *Brain Res.*, 756, 76–83, 1997.

243. Xu, L., Daly, T., Gao, C., Flotte, T. R., Song, S., Byrne, B. J., Sands, M. S., Ponder, K. P., CMV-beta-actin promoter directs higher expression from an adeno-associated viral vector in the liver than the cytomegalovirus or elongation factor 1alpha promoter and results in therapeutic levels of human factor X in mice, *Hum. Gene Ther.*, 12, 563, 2001.

244. Yamada, M., Oligino, T., Mata, M., Goss, J. R., Glorioso, J. C., Fink, D. J., Herpes simplex virus vector-mediated expression of Bcl-2 prevents 6- hydroxy-dopamine-induced degeneration of neurons in the substantia nigra *in vivo*, *Proc. Natl. Acad. Sci., U.S.A.*, 96, 4078, 1999.

245. Yandava, B. D., Billinghurst, L. L., Snyder, E. Y., "Global" cell replacement is feasible via neural stem cell transplantation: evidence from the dysmyelinated shiverer mouse brain, *Proc. Natl. Acad. Sci., U.S.A.*, 96, 7029, 1999.

246. Yoshimoto, Y., Lin, O., Collier, T., Frim, D., Breakefield, X. O., Bohn, M. C., Astrocytes genetically altered to express brain derived neurotrophic factor BDNF ame-

liorate amphetamine induced rotation following grafting into the striatum of the partially lesioned hemiparkinsonian rat, *American Society for Neural Transplantation*, 1, 21,1994.

247. Yoshimoto, Y., Lin, Q., Collier, T. J., Frim, D. M., Breakefield, X. O., Bohn, M. C., Astrocytes retrovirally transduced with BDNF elicit behavioral improvement in a rat model of parkinson's disease, *Brain Res.*, 691, 25, 1995.

248. Zeng, B. Y., Jenner, P., Marsden, C. D., Partial reversal of increased preproenkephalin messenger ribonucleic acid (mRNA) and decreased preprotachykinin mRNA by foetal dopamine cells in unilateral 6- hydroxydopamine-lesioned rat striatum parallels functional recovery, *Movement. Disord.*, 11, 43–52, 1996.

249. Zufferey, R., Dull, T., Mandel, R. J., Bukovsky, A., Quiroz, D., Naldini, L., Trono, D., Self-inactivating lentivirus vector for safe and efficient In vivo gene delivery, *J. Virol.*, 72, 9873–9880, 1998.

250. Zufferey, R., Nagy, D., Mandel, R. J., Naldini, L., Trono, D., Multiply attenuated lentiviral vector achieves efficient gene delivery *in vivo*, *Nat. Biotechnol.*, 15, 871–875, 1997.

71 Translating Stem Cell Biology to Regenerative Medicine for Parkinson's Disease

Dennis A. Steindler

Departments of Neuroscience and Neurosurgery, McKnight Brain Institute, University of Florida College of Medicine

CONTENTS

INTRODUCTION

This chapter is supposed to focus on potential stem cell therapies for Parkinson's disease (PD). Since there have been only a few such encouraging stem cell studies in animal models of the disease, the preclinical studies are likewise short in number, and, furthermore, there is perhaps only one human PD patient transplanted to date with human stem or progenitor cells, this chapter will instead concentrate on the new field of regenerative medicine as a target for stem cell biology to help translate new approaches for both protecting and replacing the compromised dopaminergic nigrostriatal pathway.

There are a plethora of recent reviews on cell replacement therapies for PD, including position papers on the first human transplant trials.[1,2] Despite certain drawbacks in the fetal human mesencephalon transplantation trials for PD, there is still hope for transplant and stem cell therapies to help ameliorate symptoms and possibly even cure the disease. That said, learning from past trials and tribulations is also relevant, and the paradigm shifts accompanying the now burgeoning fields of stem cell biology and regenerative medicine have also helped move this tech-

nology into the forefront of translation and even include potentially ongoing and best-is-yet-to-come clinical trials. For example, there have been reports of autologous transplants of "brain marrow"-derived cells in at least one human PD patient,[3] some growth factor/gene therapies (e.g., GDNF[4]), and possibly by the time this book appears, other cell/molecular therapies may have charted their initial courses toward human clinical protocols. But, without question, the best is yet to come. The enthusiasm for applying the fruits of stem cell biology, including use of the biogenic factors derived from the field (including new in addition to well characterized neurotrophic/tropic growth factors), though justified, still warrants some tempering in light of the need for more understanding of the biology of these cellular and molecular therapeutic reagents. This has been discussed in recent reviews and position papers[5] and therefore will not be extensively reiterated here, but includes an ominous, newly discovered connection between stem cell biology and oncology.[6–8]

So what are the most promising candidates for regenerative medicine-generated cell/molecular therapies for PD right now? There are many. Some of them are controversial. Governments, ethicists, clergy, and the lay public have begun to voice their opinions on the use of controversial reagents tied to anything to do with human embryos and fetuses. This is perhaps justified to certain degrees, but the emotional, personal, moral, and religious debates must be separate from the science. The science is so promising as to mandate creative applications of the technologies without compromising any of the opinions, not solely relying on animal reagents because they may bypass most of the controversy, but also including use of any postmortem human materials for research that allows the development of new therapeutics for human diseases.[9] This is extremely difficult to accomplish if the debates on human embryonic stem (ES) cells, derived from blastocysts, continue on their current course, but the world is beginning to come to grips with this, and it is hopefully generally assumed that science and scientists are sensitive to the issues and the ethical/moral/religious ramifications of all of this work to the point of not conceding any standards of humanity. Unfortunately, the world has become a bit distracted with the debates surrounding stem cell biology, and even the scientific community must get involved, taking stances to try and ensure that the science will not suffer from any guilt by association generated by either the lay press or stated positions of journals and editors. This is exemplified in the following quotes from recent publications.

> The appropriately honest and accurate response to the false statement that, if adult stem cells could do everything embryonic stem cells can do, then embryonic stem cell research is unnecessary would be that the issue is not a simple "either/or" but instead is more complicated: maximizing the therapeutic potential of either cell population inevitably would require studying both cell populations. Information for making use of one cell population would inevitably be provided by studying the other, and furthermore it was entirely premature to predict which cell population would ultimately produce the most cost-effective therapeutic options. However, the response of the pro-embryonic stem cell lobby, including scientists, journalists, and politicians, was swift, unequivocal, and unfortunate. Instead of attempting to convey a more complex message, they largely accepted these overly simplistic terms of the debate, responding in kind: embryonic stem cell research must go forward because adult stem cell research is not convincing. Thus, we have evolved to the current state in which support of embryonic stem cell research now requires that adult stem cell plasticity be repeatedly cast in a negative light....[10]

Drazen, editor of the *New England Journal of Medicine,* recently wrote,

> There are two distinct uses of embryonic stem cells. The first, for which there is no support among members of the scientific and medical communities, is the use of stem cells to create a genetically identical person. There is a de facto worldwide ban on such activities, and this ban is appropriate. The second use is to develop genetically compatible materials for the replacement of diseases tissues in patients with devastating medical conditions, such as diabetes or Parkinson's disease. This is important work that must and will move forward....[11]

At least the author of this chapter has not seen a breach of this in his experiences and interactions with the field. We can therefore assume that all of the debates will lead to guidelines that help to govern an emerging field like regenerative medicine to not enrage patients or the general public, but rather to bring them into the excitement and possibilities the discoveries offer—discoveries that will so change our lives, and those of our children and their children, with the awesome potential of self-repair.[5] Again, the cliché that regenerative medicine offers the potential of immortality can be used in awesome scenarios, but the vision should focus more on longer quality-of-life goals with reducing the human suffering from disease and injury, and fewer examples of outrageous applications (e.g., cloning armies of servants or warriors). Again, this author is tremendously optimistic about applying the many recent examples of stem cell therapies to many debilitating neurological diseases, including PD, multiple sclerosis, and traumatic injuries. This will contribute to true restorative neurology, with attainable rebuilding of any compromised complex central nervous system (CNS) circuit to restore lost function. Such an advance would represent the translation of stem cell biology to regenerative medicine for Parkinson's disease—the goal being to completely regenerate or reconstitute lost cells and connections, as well as protect cells that are vulnerable in the disease, utilizing stem cell and other technologies in the most powerful of regeneration protocols. Whether it be in toto or partial, topographically appropriate (i.e., replacing cells in the midbrain) or inappropriate (e.g., grafting cells to the striatum), the protection and reconstitution of dopaminergic innervation of the striatum is an attainable goal of both stem cell biology and regenerative medicine for PD.

FETAL MIDBRAIN GRAFTS IN PD DID CONTAIN STEM CELLS

Among the list of possible explanations for disappointing outcomes from the fetal mesencephalic transplant trials in Parkinson's disease are poor survival and integration of the grafts; exuberant production of dopamine from the fetal cells; the generation of a supernumerary and possibly unbefitting basal ganglia structure following the intrastriatal rather than intranigral grafting of midbrain cells; and the unintended but difficult to surmount inclusion of non-dopamine neurons from other midbrain structures in the grafts that could contribute to anomalous behaviors, including dyskinesias. Many investigators and studies have suggested that the best approach for treating PD would be to rescue at-risk dopamine neurons and completely reconstitute the dopaminergic nigrostriatal axis. Complexities and problems associated with the human fetal mesencephalic trials, including variance in the different protocols applied at different institutions (e.g., preparation of the grafts, additions of certain growth/survival factors) and differing viewpoints for graft placement, are nicely summarized in a recent review article by Gene Redmond.[2] Included within this and many recent review and position papers is the discussion of how stem cell therapy is now a focus of cell replacement therapies because of the potential to generate only dopamine neurons using differentiation protocols that nonetheless still represent a great deal of work in progress. Issues in stem cell therapy include which is the best cell, neural versus nonneural stem cell and embryonic versus fetal versus adult sources; and what are the best priming approaches for preparing such cells to generate requisite numbers of dopamine neurons that innervate target striatal cells following grafting?

The clinical trials in PD that used fetal tissue did achieve some degree of success, and the tissue that was transplanted did contain neural stem and/or progenitor cells. In animal model studies using grafts of fetal mesencephalon, it was certainly appreciated early on that immature neurons, precursor cells, and even perhaps stem cells were included in the grafts. This is exemplified in one of our early studies[12] where very immature cells were meant to be the grafted reagent so they could insinuate themselves in a very complex circuitry because of their youth (e.g., lacking receptors to the hostile environment of even the normal adult CNS, not to mention a compromised circuit as in the case of the PD nigrostriatal pathway). We showed the presence of very immature cells in these fetal mouse nigral grafts, placed either in the midbrain or neostriatum, that expressed markers of primitive radial glia including the RC-2 protein, that we now know can also be expressed by multipotent astrocytic stem cells[13] that are present during developmental neurogenesis and secrete developmentally-regulated proteins that support their neuropoiesis.[14] It is presumed that stem cells in the tissue grafts would either eventually differentiate into neurons and glia or die as a result of altered molecular environments in the mature brain that are not conducive for stemness. A supportive environment is artificially created *in vitro* with the addition of certain growth factors including epidermal growth factor (EGF), fibroblast growth factors (FGF), brain derived neurotrophic factor (BDNF), glial cell line derived growth factor (GDNF), serum, and other factors that keep stem/progenitor cells in an immature and proliferative mode.* Regardless of whether stem or progenitor cells had any positive or negative contribution to the results of the fetal midbrain transplant trials, these cells should quickly differentiate or die if they do not get the right combinations of growth factors that are normally

present in the developing brain, or that are supplemented in the *in vitro* experiments that promote expansion of these cells under particular growth conditions. Finally, in other non-PD human fetal tissue transplants, e.g., the trials for syringomyelia performed by a University of Florida team,[15] some of the positive outcomes observed in the eight or so patients grafted with fetal human spinal cord might in part relate to the presence of not only young neurons and glia in these grafts, but also the presence of undifferentiated regeneration-friendly cells including stem/progenitor cells that normally reside in the developing CNS.

A question arises as to whether the environment of a solid tissue graft or dissociated fetal human ventral mesencephalon was any better or worse for the survival and functional integration of stem and progenitor cells present in those grafts, versus homogeneous primary cells or lines of dopaminergic precursor cells for ameliorating the symptoms of PD. The field is certainly expounding that enriched populations of stem cells and dopaminergic neuron precursors are better for reintroducing dopamine to the depleted basal ganglia than the fetal tissue grafts because of homogeneity (e.g., lack of nondopaminergic nigral neurons, including glia and cells from surrounding midbrain structures that were certainly included in both the animal and human transplant studies[1,2]), presumed better efficiency, consistency, and maybe even survival and dopaminergic innervation compared to the fetal tissue and cell grafts. But at the present time, it is still uncertain as to which stem, stem-like, or progenitor cell will best accomplish cell replacement or circuitry protection in brain injury or neurodegenerative diseases including PD.

GRAFTS OF ES, STEM CELL LINES (E.G., THE C17.2 RODENT AND NOW HUMAN IMMORTALIZED NEURAL PRECURSOR LINES), AND OTHER STEM/PROGENITOR CELLS IN ANIMAL MODELS OF PD (FOR RESCUE AND REPLACEMENT)

EMBRYONIC STEM (ES) CELLS

There are several choices of stem, stem-like, progenitor, and dopaminergic neuron precursor cells for grafting into the striatum or midbrain of PD patients. These include

---

* It should be defined here that a stem cell is a self-renewing cell that can give rise to different progeny, is able to survive and keep its self-renewing ability following serial transplantation, and can respond to injury or disease with repopulating prowess. A progenitor cell is a more committed cell that can give rise to more differentiated cells, e.g., neurons or glia, and the term "stem/progenitor cell" is used where the degree of stemness of the cell in question has not been resolved. See Reference 14 for details.

adult "brain marrow" (see below, but defined as those periventricular regions of the adult brain that exhibit persistent expression of developmentally regulated proteins including extracellular matrix, and the presence of cycling, neurogenic stem, and progenitor cells; this includes the periventricular subependymal zone, SEZ, of the forebrain lateral ventricle, its migratory extension into the olfactory bulb termed the rostral migratory stream, RMS, and the hippocampus, see Figure 71.1), fetal sources, bone marrow, cord blood or mesenchymal-derived cells, cell lines including those derived from human cancers, indigenous cycling precursor cells that do inhabit the nigra (see below), and embryonic stem cells. ES cells are controversial for a variety of reasons, as already mentioned, but they offer potential for combining cell and gene therapies for a variety of neurological and other disorders. The advantages of ES cells as donors for replacement or rescue therapies include their pluripotency, defined by their ability to generate cells exhibiting the phenotype of most or all tissues in the body; the potential for unrestricted proliferation; their amenability to genetic modification; and, finally, the possibility for controlled fate choice and differentiation such that it might be possible to obtain highly purified neural cell populations.[16–18] ES cell-derived neural precursor cells have been efficiently derived from both rodent and human ES cells. Upon transplantation, these cells incorporate widely throughout the CNS and differentiate into neurons, astrocytes, and oligodendrocytes. Transplantation of neural precursor cells represents an alternative route to replace lost or damaged neurons in the adult CNS. Although this approach can be developed to a clinical scale,[19] it is currently complicated by its dependency on donor tissue, e.g., from *in vitro* fertilization clinics. Transplanted neural precursors derived from primary CNS tissue or cultured ES cells can integrate into the developing brain and differentiate into mature neurons and glia. The successful integration of these cells can be established using morphological and immunophenotypic labeling methods.

There is a paucity of electrophysiological data on the functional integration of embryonic stem cell derived neural precursors following transplantation, although very recent studies[16,18] have shown functional integration of neuronal precursor cells that were derived from ES cells. One might suggest that complete functional integration may be mandatory for complete functional restoration of the dopaminergic nigrostriatal pathway. There is some possibility that dopamine-replacement alone from these cells might be therapeutic, much like L-dopa is. One would assume that reconstitution of the cell replacement topography and precise afferent and efferent innervation patterns is likewise important for complete functional reconstitution. These ultimate goals fall on one end of a spectrum of emerging palliative as well as cure—attempting protocols that use dopamine and nondopamine cells

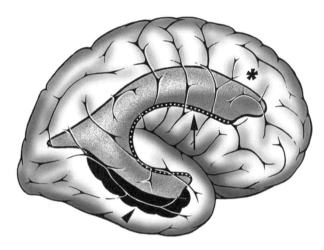

FIGURE 71.1 Sites of persistent neurogenesis, or neuropoiesis, within the adult human brain. Neural stem cells (NSCs), and a potentially heterogeneous population thereof, reside within adult "brain marrow"—the periventricular subependymal zone (SEZ, "long arrow") and the hippocampus (arrowhead). In addition, there are small numbers of cycling cells found throughout the neuraxis (e.g., asterisk in the cerebral cortex) that could exhibit stem cell-like behaviors under certain conditions. Adult NSCs are clonogenic and multipotent, able to give rise to all three types of CNS cells (also see Figures 71.2 and 71.3). It is still not clear that these cells can give rise to all types of neurons and glia of the CNS (they may have a propensity to generate local circuit cells). We do not understand why cells persist in the senescent and even cadaveric brain, but their position may indicate something about their nature. Most proliferative or clonogenic cells appear to reside within the periventricular SEZ and hippocampus, but recent evidence indicates the possibility for dedifferentiating glial cells to exhibit stem cell-like behaviors, suggesting the possibility that other areas besides the SEZ and hippocampus may be neurogenic. The SEZ and hippocampal cells may be involved in steady-state neurogenesis for the replacement of cells in the constantly remodeling olfactory bulb, and in the hippocampus. In the latter, the functional implications of new granule cells in the dentate gyrus have not been clearly elucidated, although recent reports indicate a loss of these cells in experimental animal models impacts short term memory. One possibility is that a small stem cell population within the SEZ (a vestigial remnant of the proliferative germinal matrix of the embryonic forebrain) represents "leftover bricks at a construction site"; this should not imply that the cells that inhabit the SEZ are the same cells that built the brain during development, rather, they should be considered nonprogrammed neural cells that can be induced to exhibit stem cell-like behaviors. (Adapted from Steindler and Pincus, Reference 5. With permission.)

derived from ES, fetal, neural, and nonneural stem/progenitor cell graft therapies. These cells might supply dopamine or other sustaining neurotrophic factors that help to promote survival of at-risk nigral dopamine neurons as well as attempt cell replacement. New designer therapies might combine a variety of cellular and genetic therapies in combinations to generate cells that will

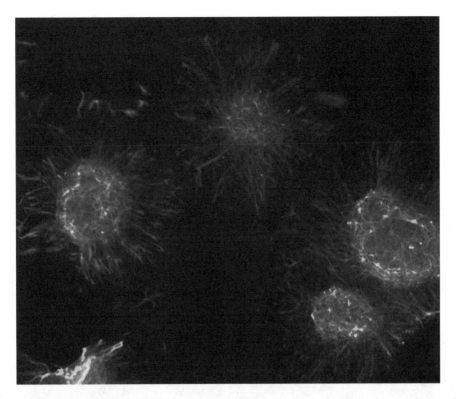

FIGURE 71.2 Neurosphere clones of cells derived from mouse neural stem/progenitor cells. Neurospheres reflect the ability to expand small numbers of neural stem/progenitor cells into large numbers of these densely packed cellular aggregates that contain neuronal, in addition to the clonogenic stem cell, glial progenitor cells, neurons, and glia in different states of differentiation. The neurons in these clones are stained with a red fluorescent neuronal marker, the glia are green, and the neurospheres were counterstained with a blue fluorescent nuclear marker. The diameter of the neurospheres is approximately 150 μm. *(A color version of this figure follows page 518.)*

accomplish both of the aforementioned goals, but there will be chaperone and replacement cell therapies for PD that owe their development to evolving understandings of the molecular cell biology of dopaminergic precursor cells.

Two studies in particular show the promise as well as the uncertainty of using ES cells as PD therapeutics at this point in the time. Lars Bjorklund with the Isacson group, and Ron McKay and collaborators[16,17] transplanted low doses of undifferentiated ES cells into the normal and 6-hydroxydopamine (OHDA)-lesioned adult rat striatum. Following an injection of 1 to 2000 mouse ES cells, with cyclosporine A immunosuppression, after 14 to 16 weeks, 2059 ± 626 tyrosine hydroxylase positive cells were identified in the graft site, and these cells expressed other markers of midbrain dopamine cells including aromatic amino acid decarboxylase and calretinin. The Isacson lab has been a very strong proponent of ES cell therapies for PD, and at the same time has cautioned the field about the need for precise molecular differentiation protocols (see Figure 71.4 and Reference 1), and proper assessment of differentiated cell phenotype that matches the desired neuron fate choice (e.g., A9 versus A10 dopamine groups, Vental Tegmental Area phenotypy that can be resolved

using the appropriate molecular markers and fingerprinting; the list of both neuralization and dopaminization factors, often requiring presentation in combinations and at particular stages of the neuron and dopamine neuron generating protocols, reads like a list of morphogenetic genes used to build a fruit fly, e.g., sonic hedgehog, Pax's, Lmx, Smad4, neurogenin, noggin, chordin, FGFs, Wnts, and on and on, see Reference 1). In their study of undifferentiated ES cell grafts in the normal and lesioned striatum, in addition to the very positive outcomes of apparent ventral mesencephalon dopamine neuron differentiation and improved "behavioral restoration of dopamine-mediated motor asymmetry," some astrocyte differentiation, and some apparent generation of mesodermal cells (via the observed expression of desmin/myosin and keratin), 5 of the 25 ES cell-injected animals also became severely ill and had developed "teratoma-like structures" at the graft site. This should not be surprising in light of their choice of undifferentiated ES cells, but their positive outcomes were quite impressive and supported the notion that ES cells can generate dopamine neurons that release dopamine and improve motor behaviors in the 6-OHDA-lesioned rat (see Figure 71.5).

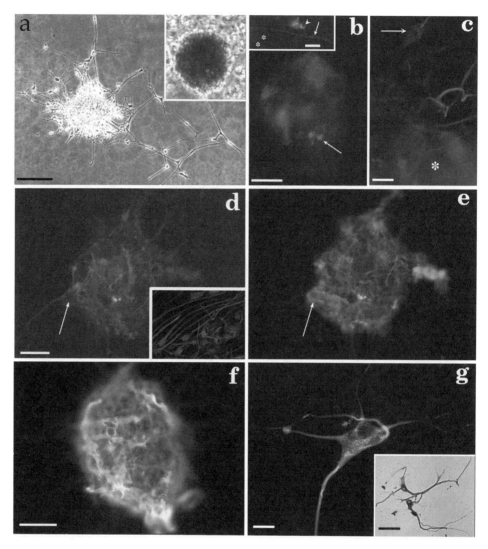

FIGURE 71.3 Human neurosphere clones, and cells derived from them, grown in tissue culture from stem/progenitor cells cultivated from adult human brain biopsy and autopsy specimens. A neurosphere is a tissue culture-generated clone of cells that are in different states of differentiation (e.g., maturing glia and neurons), all presumed to arise from a single multipotent stem/progenitor cell. Phase (a) and immunocytochemical (b–g) images of differentiated human neurospheres after plating on laminin. (a) Phase microscopy of a neurosphere plated on laminin shows how the sphere begins to differentiate following plating, and cells migrate out from the sphere. Scale bar = 100 μm. (b) Immunofluorescence for the intermediate filament protein nestin in a single sphere, showing many large nestin-labeled cells, in addition to small cells (arrow). Inset shows a cluster of nestin-positive cells (arrowhead) that have migrated from a neurosphere (asterisks), as well as a labeled long process projecting from the sphere (arrow). Scale bar = 50 μm in (b) and 10 μm in the inset. (c) Immunolabeling of a sphere (asterisk) shows dense vimentin expression by neurosphere cells, including those and their processes that have emigrated from the sphere (e.g., arrow). Scale = 50 μm. (d) GFAP immunolabeling of astrocytes within a neurosphere, and processes emanating from the neurosphere. Arrows in (d) and (e) point to the same landmark in this neurosphere stained with GFAP vs. beta III tubulin markers. Inset in (d) shows a cluster of vimentin positive radial-like glia and immature astrocytes that have migrated away from a neurosphere. Scale = 50 μm for (d) and (e). (e) Neuronal beta III tubulin immunolabeling of the neurosphere shows many labeled neurons. (f) Double labeling for GFAP (blue) and beta III tubulin (green) shows both astrocytes and neurons residing within a plated sphere. Scale bar = 50 μm. (g) De novo generated neurons, with distinct morphologies, that have grown out from a plated neurosphere stained for beta III tubulin (brown, peroxidase in the inset) and using immunofluorescence (g). Scale bar = 25 μm in (g), 100 μm in the inset. (Adapted from References 5 and 22. With permission.) *(A color version of this figure follows page 518.)*

Another ES cell-Parkinson's rat model grafting study published in 2002 from Ron McKay's lab[16] utilized the five-stage method of ES cell differentiation into neurons, and also exploited a well appreciated dopaminization fac-

tor, Nurr1, and manipulating exposure of the cells to the fibroblast growth factors (FGFs), Leukemia Inhibitory Factor, and sonic hedgehog (SHH) to attempt enrichment of the tyrosine hydroxylase (TH)-expressing population.

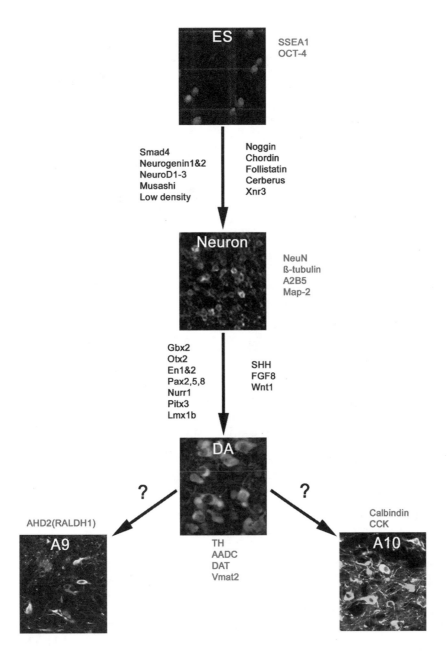

FIGURE 71.4 Midbrain dopamine neuronal development from embryonic stem (ES) cells. Schematic illustration of known developmental factors involved in the identification and generation of midbrain dopamine (DA) neurons from mouse ES cells. ES cells can be identified by expression of embryonic markers such as stage specific embryonic antigen 1 (SSEA-1)or OCT-4. ES cells can adopt neuroectodermal fate through a "default" mechanism involving inhibition of transforming growth factor (TGF)-related molecules such as activin and bone morphogenetic proteins (BMPs) as well as TGF-beta downstream targets like smad4. Factors known to be of importance for "default" neuralization are the so-called "BMP inhibitors," Noggin, Chordin, Follistatin, Cerberus, and Xnr3 as well as culturing or transplanting ES cell in low density to avoid autocrine and paracrine TGF- signaling. Cells of neuroectodermal lineage are believed to adopt a neuronal fate under the influence of basic helix-loop-helix (bHLH) factors such as NeuroDs and Neurogenins or other factors such as Musashi 1 and 2. Neuronal phenotype can be identified by expression of Beta III tubulin, neuronal nuclear antigen (NeuN), A2B5 antigen, or microtubule-associated protein 2 (MAP-2). Midbrain DA neurons are generated at the intersection between midbrain and hindbrain in response to a ventral-dorsal gradient of floor plate-derived sonic hedgehog (SHH) and a anterior-posterior gradient of fibroblast growth factor 8 (FGF-8). Factors known to be of importance for proper development of midbrain DA neurons are Gbx-2, Otx-2, Pax 2, 5, 8, Wnt-1, Nurr1, Pitx-3, and Lmx1b. Midbrain DA neurons that express tyrosine hydroxylase (TH), L-aromatic amino acid decarboxylase (AADC), the dopamine transporter (DAT), and the vesicular monoamine transporter 2 (Vmat-2) will, through yet unknown mechanisms, develop into functionally and anatomically distinct subpopulations such as the A9 (aldehyde dehydrogenase 2, also known as retinaldehyde 1 [Raldh1] expressing) and A10 (calbindin and cholecystokinin [CCK]) expressing cells. (Adapted from Reference 1. With permission.) *(A color version of this figure follows page 518.)*

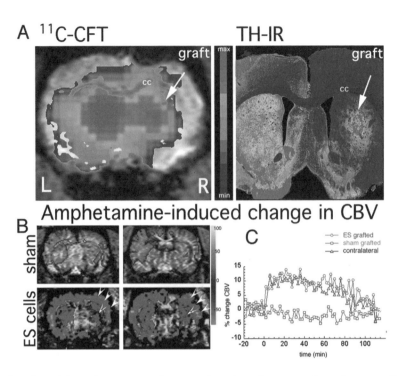

FIGURE 71.5 *In vivo* imaging of dopamine neurons after transplantation of mouse embryonic stem cells to the adult dopamine denervated rat striatum. (*A*) positron emission tomography (PET) imaging using the specific dopamine transporter (DAT) ligand carbon-11-labeled 2–carbomethoxy-3–(4-fluorophenyl)tropane (11 C-CFT) showing specific binding in the right grafted striatum, as shown in this brain slice (*A*, left panel) acquired 26 min after tail vein injection of the ligand. Color-coded (activity) PET images were overlaid with magnetic resonance imaging images for anatomical localization. The increased 11 C-CFT binding in the right striatum correlated with postmortem presence of TH-immunoreactive (IR) neurons in the graft (*A*, right panel). (*B*) Animals receiving embryonic stem (ES) grafts showed restored DA release mediated neuronal activation in response to amphetamine (2 mg/kg). Color-coded maps (percentage change) in relative cerebral blood volume (rCBV)in an animal with an ES cell-derived DA graft are shown in three slices spanning the striatum. Only ES cell-grafted animals showed recovery of signal change in motor and somatosensorial cortex (arrows), and this was also seen, although to a minor extent, in the striatum. (*C*) Graphic representation of signal changes over time in the same animal as shown in (*B*.) The response in grafted (red line) and normal (blue line) striata was similar in magnitude and time course, whereas no changes were observed in sham-grafted dopamine-lesioned animals (green line). Baseline was collected for 10 min before and 10 min after MION injection, and amphetamine was injected at time 0. cc = corpus callosum. (Adapted from References 1 and 17. With permission.) *(A color version of this figure follows page 518.)*

It was found that SHH and FGF8 further enhanced the production of TH-expressing cells from the Nurr1 overexpressing ES cells to almost 80% of the neurons generated. Furthermore, when slice electrophysiological studies were performed, it was found that the "…ES-derived neurons develop functional synapses and show electrophysiological properties expected of mesencephalic neurons…," including those of TH+ and TH- mesencephalic neurons (see Figure 71.6). These animals, in behavioral studies, also showed significant improvement in rotational, adjusting step, cylinder and paw-reaching tests. This prompted McKay and collaborators to say, "Our results encourage the use of ES cells in cell-replacement therapy for Parkinson's Disease…."[16]

It is at this point that, in light of the Isacson and McKay group ES cell transplant findings, that stem cell tumorigenesis and incomplete neuronal differentiation[18] issues should be addressed, since Isacson's group

reported the generation of tumors following their undifferentiated ES cell grafts, and the McKay group reported apparently normal synaptic activities of their Nurr1 overexpressing ES-derived midbrain-dopamine-like neurons. Of course, the world has appreciated the roles of hematopoietic stem cells in the leukemias (see Reference 5 for review), but it has only been recently that three studies have shed a great deal of light on the potential roles of stem/progenitor cells in solid tumorigenesis in the brain[6] and breast.[8] In both of these studies, the behaviors of abnormal stem-like cells associated with neoplasia and the generation of diverse progeny have been noted, with the suggestion that a primitive cell that normally gives rise to limited numbers of progeny with characteristic lineage diversity now gives rise, due to genetic and/or epigenetic anomalies, to a diverse set of abnormal progeny that constitute gliomas and other solid tumor types. Thus, the existence of a "cancer stem cell"[7] suggests that we need to

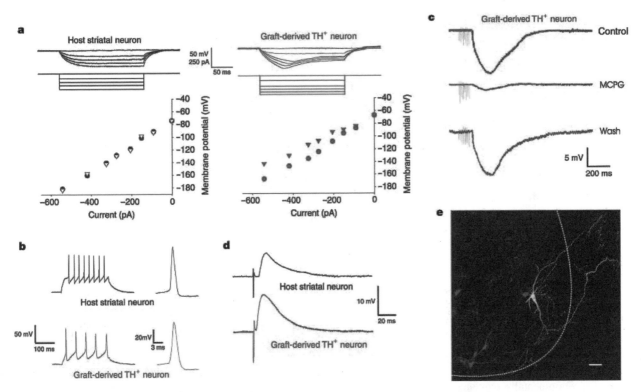

FIGURE 71.6 Electrophysiological properties of TH+ neurons. Simultaneous recordings were performed from neurons in the graft located on the graft-host border and neurons in the host striatum. (*a*) Representative current-voltage relationship for a host striatal TH– neuron and a TH+ neuron in the graft. The TH+ neurons display the time-dependent anomalous rectifier characteristic of dopamine-synthesizing cells after a hyperpolarizing pulse. Circles, the full extent of the immediate reduction in the membrane potential; triangles, the sustained membrane potential. (*b*) Spike train profiles of a host striatal neuron and a graft-derived TH+ neuron. TH+ neurons fired broader action potentials at a lower frequency than TH– neurons. (*c*) ES-derived TH+ neurons in the graft displayed a unique evoked IPSP mediated by activation of the metabotropic glutamate receptors. (*d*) Extracellular stimulation in the center of the graft resulted in an EPSP in both a graft-derived TH+ neuron and a host striatal neuron, indicating the presence of graft-to-host and graft-to-graft synapses. (*e*) Confocal micrograph illustrating a biocytin-filled (green) TH+ neuron in the graft in close proximity to other nonfilled, graft-derived TH+ neurons (red). The neuron was filled during recording. The filled neuron extends processes well into the host striatum. The dotted line shows the host-graft interface. Scale bar, 50 mm. (Adapted from Reference 16. With permission.)

be cautious before transplanting any potent cell into the nervous system, and this is further reinforced by the recent study from our group that showed transplantation of an indigenous stem cell population from the postnatal and adult mouse brain, the SEZ multipotent astrocytic stem cell, back into the vicinity of brain marrow can give rise to hyperplasias that have attributes of tumors.[20] Whether the *ex vivo* manipulations of the clonogenic cells, supporting dedifferentiation programs and possibly making these cells susceptible to genetic malformations during the excessive growth factor exposures, or other factors or conditions contributed to the potential for hyperplasia, the potential for neoplasia must be studied further before pronouncing a stem or progenitor cell population ready for human use. Likewise, pronouncing any stem cell, including ES, as giving rise to a completely differentiated and "normal" populations of desired neurons (e.g., dopamine neurons for PD), based on electrophysiological and molecular phenotyping, also requires a fair

amount of diligence, since it is possible that we might miss incomplete differentiation as well as the retention of subtle attributes of the starting cell population (e.g., in the case of ES cells, undifferentiated or non-neural differentiated states), since our protocols for characterizing phenotype and differentiation are not yet perfect. Most groups admit that certainly less than 100% of their populations, even following the best neuralization protocols, are neural; we probably do not want to graft these non-neural and undifferentiated non-neural cells into the brain along with the neuronal precursors of choice. Even the best of differentiation protocols of ES-derived neuronal progenitors still give rise to populations of cells that do not express the full repertoires of neurotransmitter-related receptors and channels (e.g., see Benninger et al.,[18] who elegantly showed many normal neurophysiological behaviors of their ES-derived neurons, but they appear to lack NMDA currents that are so rudimentary to the functions of the hippocampal neurons they were studying). It is certainly possible that a cell that does not grow up in the nervous system does not have all of the appropriate programs for generating molecular and behavioral hallmarks of fully-differentiated neurons. We must therefore devise protocols and assays to test for this and surmount this. Perhaps exposing such primitive cells in search of identity to morphogenetic factors and conditions that might favor full differentiation sequalae can compensate for the lack of exposure to critical and sequential morphogenetic programs and factors that are normally present *in situ*.

Adult Neural Stem Cells

Adult human brain neuropoiesis[5] supports suggestions that adult brain-derived stem cells[21,22] might also see a place in PD transplant therapies, especially if we might be able to modify their maturation with factors including telomerase.[23] At the end of a neurodevelopmental critical period (roughly the second postnatal week in rodents), extracellular matrix (ECM) and other developmentally regulated "recognition" molecules are down-regulated to negligible levels, except for in a few brain areas including the periventricular subependymal zone (SEZ).[14,22] The presence of developmentally regulated ECM and other molecules in the SEZ, along with cycling cells as seen using tritiated thymidine- or bromodeoxyuridine-labeling, suggested the possibility of persistent neurogenesis, or "neuropoiesis" throughout life in the mammalian forebrain, and also prompted comparisons of this neurogenic region to hematopoietic bone marrow (thus, "neuropoietic brain marrow"[5,14]). The long-standing axiom within neuroscience that, with few exceptions, the postnatal mammalian CNS is capable of little or no *de novo* neurogenesis, includes now the well accepted findings of newly generated neurons in the periventricular SEZ as well as other CNS regions in rodents and primates. In addition to *in vivo* demonstrations of neurogenesis, cells have also been isolated from postnatal and adult CNS that can grow as proliferative clones, termed "neurospheres" (see Figure 71.2), that are multipotent, since they can give rise to the three major classes of neural cells: neurons, astrocytes, and oligodendrocytes from a variety of fetal, postnatal, and adult mammalian (including human, see Figure 71.3) brain marrow sources. Neurosphere-generating cells, or the tissue-specific "neural stem cell," have been isolated from the SEZ[24] and other regions, including spinal cord, of rodents as well as from adult human SEZ and hippocampus.[22] When fetal human neural stem cells were transplanted into the neostriatum of rats with unilateral dopaminergic lesions, even after 20 weeks, there were surviving cells in the grafts, and some of these cells expressed tyrosine hydroxylase and seemed to attempt innervation of the host striatum.[25] The fetal and adult human neural stem cell studies obviously drove the first autologous stem cell transplants for human PD by Levesque and colleagues

who have described in lay press publications[3] the grafting of a PD patient's own "neural stem cell population," obtained from a brain biopsy, into their neostriatum. Despite the apparent initial expression of dopamine in these grafts, they reported that, after one year, these levels returned to presurgery levels, but "...an 83% reduction in symptoms, such as tremor, has inexplicably persisted.... may be due to other cells in the transplanted mixture...." It would seem reasonable to further explore the proof of principle of autologous stem/progenitor cell grafts in animal models before taking these cells to additional human trials.

The ability to purify NSC populations from postnatal and adult CNS is particularly important for studies attempting to characterize, expand, and use these cells in cell-replacement therapies. Not surprisingly, the search for the NSC has concentrated on the periventricular SEZ of the rostral forebrain. The adult SEZ seems to be a special region, since it represents the vestigial embryonic germinal zone, displays a high level of constitutive proliferation, and is likely to contain the greatest density of putative NSCs. Work from our laboratory[13] has shown that a cell exhibiting characteristics of an astrocyte, from the developing brain until the end of the second postnatal week, and from the SEZ throughout life, is a multipotent stem cell that can give rise to neurospheres containing both glia and neurons. Detailed ultrastructural analyses of this cells has also been performed by Alvarez-Buylla's group.[26] Studies recently conducted in our laboratory have shown that neurospheres derived from multipotent SEZ astrocytes (the type "B" cell of Alvarez-Buylla's group), in the presence of epidermal growth factor (EGF) and FGF -2, can differentiate into cells with antigenic profiles of neurons, astrocytes, and oligodendrocytes. These studies support a notion of dedifferentiation ability of mature, differentiated populations of neural cells (e.g., a subclass of the pervasive astrocyte population), and also support recent observations of a potential lack of replicative senescence (the possible overturning of the "Hayflick limit," whereby it was suggested that cells can undergo only so many population doublings, e.g., 50 divisions, before "aging"[27]) of these clonogenic, multipotent cells (see Reference 5 for review). Despite holes in our understanding of the nature of adult neural stem cells, they do offer a potential source of cells for transplantation in PD. The finding that both adult mouse and human neural stem cells can survive with rather protracted postmortem intervals[28] even suggests that the cadaveric human brain is a source of cells that could be manipulated for therapeutic applications in diseases including PD.

Hematopoietic Stem Cells

In addition to the neuropoietic sources in brain, the hematopoietic system is another potential source of multipotent cells that might be able to be coaxed into neuronal and glial differentiation. Hematopoietic stem cells (HSCs), including the highly touted, presumably extremely potent umbilical cord blood stem cells, retain their pleuripotent differentiative capacity throughout the life span of an organism, and homeostasis is maintained by a constant, ordered, and tightly regulated developmental cascade. HSCs differentiate through a hierarchical array of multipotent and monopotent progenitor cells to form all blood cell types, including lymphocytes, granulocytes, monocytes, erythrocytes, and megakaryocytes. The regulatory pathways that control hematopoiesis consist of several interconnected mechanisms. One level of regulation is the transcriptional control of hematopoietic-specific gene expression. A second mechanism involves the interaction of hematopoietic growth factors with their cognate cell-surface receptors. Cell-cell signaling mediated by adhesion molecules also plays an important role in regulating hematopoiesis. The hematopoietic system also offers a well characterized, adult model of reconstitution following ablation.

Despite a recent study by our collaborators here at the University of Florida,[29] raising the question of "cell fusion" involved in the so-called recent stem cell plasticity studies, there is a considerable body of literature regarding the possible ability of hematopoietic cells to participate in the formation of neural tissue. *In vitro* studies indicate that conditions can be found to coax hematopoietic cells to express neural characteristics. For example, Sanchez-Ramos and collaborators, and Black and collaborators[30,31] showed that bone marrow stromal cells in culture can express neural markers of astrocytes and neurons when exposed to particular growth conditions and factors. It also has been shown that bone marrow stromal cells, injected into neonatal mouse ventricles, could adopt neural fates as evidenced by their expression of glial fibrillary acidic protein (GFAP, an intermediate filament protein specific to astrocytes) and a neuron-specific neurofilament protein.[32] When Mezey and colleagues[33] transplanted bone marrow from male wild-type mice into hematopoietically compromised PU.1 null female neonatal mice, they found that numerous cells containing the Y-chromosome were present in the brain, and some of these expressed antigens specific for mature neurons. Finally, Brazelton and colleagues[34] transplanted adult-derived bone marrow from green fluorescent protein (GFP) transgenic mice into lethally irradiated adult wild-type mice, and found that GFP-expressing cells entered numerous brain regions, and some of the grafted cells co-expressed neuronal, but not glial, antigens. However, both of these latter studies failed to show complete differentiation and integration of donor cells in the host brain, as would be typified by elaboration of extensive neuritic arbors and the generation of synapses, and the yield of donor cells that exhibited transdifferentiation was quite

low. The goal of rebuilding functional brain tissue following injury or disease could be achieved through the directed homing of bone marrow-derived stem cells to the injured brain. This would certainly overcome the many obstacles of trying to access, isolate, and expand one's neural stem cell population for cell replacement therapies in neurological disease. There is very little known about potential homing and trafficking factors (see Figure 71.9) involved in CNS stem cell recruitment, unlike our increasing knowledge of "Homing and Trafficking of Hemopoietic Progenitor Cells."[35] It is interesting that we have recently found, looking at postmortem brain specimens from patients receiving bone marrow transplants, that chimerism can occur in the hippocampus, with bone marrow-derived cells apparently homing to the possibly radiation/chemotherapy-affected forebrain whereby they transdifferentiated into neurons and glia without cell fusion involved.[36]

IMMORTALIZED NEURAL PRECURSOR CELLS

As recently reviewed,[37] the year 1992 was an important year for restorative neuroscience with the discovery of clonogenic stem cells in the adult brain by Reynolds and Weiss, Bartlett and colleagues, and the report of an immortalized neural stem cell line able to participate in mouse cerebellar development by Evan Snyder and the Cepko lab. The most utilized cell line, the C17.2 cell line, has demonstrated extreme potential and plasticity, and these "stem-like" cells exhibit a remarkable ability to insinuate themselves within established and compromised rodent central nervous system CNS structures and circuits, home to injured areas, migrate, exhibit diverse lineage potential, release neurohumoral factors, and overall offer the possibility of reconstituting neural circuitry in a variety of models of neurological disease (see Figure 71.7). These cells are "stem-like" cells because the starting cell population (neonatal mouse cerebellar external granular layer cells) was immortalized by retrovirus-mediated transduction of avian myc (V-myc) that produced a clonal multipotent progenitor cell line. These cells are therefore somewhat "transformed," which probably helps contribute to their impressive plasticity. Nonetheless, the cells offer insights into the control of fate and integration of cells as therapeutics for a variety of neurological disorders.

This cell line has been used to generate an interesting chaperone effect for rescuing TH expression by injured cells in the 1-methyl-4-phenyl-1,2,3,6-tetrahydropyridine (MPTP)-injured midbrain substantia nigra of aged mice (see Figure 71.7). Transplantation of the C17.2 cells in the vicinity of the MPTP-injured dopamine neurons in the midbrain resulted in some new dopamine neurons being generated, but in addition, it was found that most of the TH (the rate-limiting enzyme in dopamine synthesis) expression was associated with rescued host cells. In this

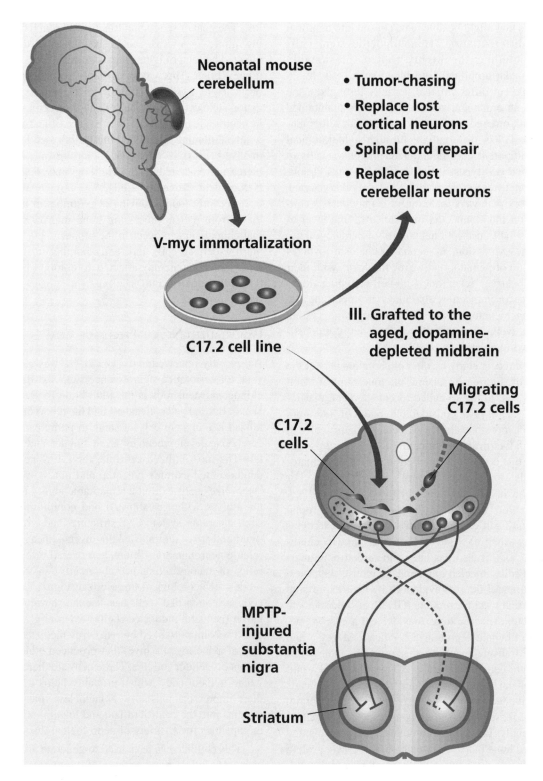

FIGURE 71.7 Immortalized cerebellar precursor cells for dopamine neuron rescue. The C17.2 cell line was generated by V-myc immortalization of neonatal mouse cerebellar precursor cells. This cell line has been used in previous studies to "chase" tumor cells, replace lost cells in the injured cerebral cortex, help ameliorate some of the behavioral deficits following spinal cord injury, and replace lost cerebellar neurons in a mutant mouse. Grafts of C17.2 cells within the dopamine neuron-injured midbrain of adult mice that have a "chaperone" effect on the rescuing of dopamine neurons in a rodent MPTP model of Parkinson's disease. In addition to an apparent neurotrophic rescue of the injured dopaminergic neurons and their nigrostriatal axonal projections, C17.2 cells also gave rise to small numbers of newly generated dopamine neurons, and they also were found to migrate to disparate CNS sites. These grafts also appeared to improve motor behaviors of the MPTP-injured animals. (Adapted from Reference 37. With permission.)

study,[38] in addition to the C17.2 cells rescuing midbrain dopamine neurons in the MPTP-lesioned animals (as seen using immunocytochemical and biochemical assays), in a functional assay D-amphetamine-evoked rotational behavior also regressed.

In addition to this immortalized mouse neural precursor cell line seeming to offer another potential source of cells for protection and replacement of dopamine neurons in rodent PD models, a human neural precursor cell line has been generated with the help of Evan Snyder and Angelo Vescovi,[39] again using a v-myc immortalization protocol and human fetal neural stem cells. Even though all examples of the immortalized precursor and stem cell-like lines indicate a strong propensity to protect and rebuild damaged brains, their zealous ability to migrate, and their potential for harboring an abnormal genome that could conceivably generate neoplastic transformation, must make us take pause when considering the use of such cells for human transplantation therapeutics in diseases including PD without exhaustive analyses in animal models.

OTHER CELLS AND MORPHOGENETIC FACTORS AS POTENTIAL THERAPEUTICS FOR PD

A paper was published in 2001 using Evan Snyder's C17.2 immortalized neural precursor cell line to provide the potent dopaminergic neuron growth/protective factor glial-cell line-derived neurotrophic factor (GDNF) to at risk neurons in the 6-hyrdroxydopamine mouse model of PD.[40] This built on numerous previous observations of the potential of GDNF in gene therapy protocols from Ron Mandel, Jeff Kordower, Marty Bohn, and many others, showing that this growth factor could be used to protect and rescue at risk dopamine neurons in these lesion models. In addition to the GDNF-expressing C17.2 cells insinuating themselves into the compromised striatum and generating neurons and glia, including neurons with dopaminergic traits, they "...gave rise to therapeutic levels of GDNF in vivo, suggesting a use for NSCs engineered to release neuroprotective molecules in the treatment of neurodegenerative disorders, including Parkinson's disease...." Thus, as depicted in Figure 71.7, in addition to stem cells being used to replace lost dopaminergic neurons in models of PD, they hold the potential to act as sources of neurotrophic, neuroprotective molecules to be released to possibly rescue at-risk cells during the course of disease. Clive Svendsen and collaborators[41] also showed that genetically modified neurospheres, engineered to release GDNF using lentiviral constructs, when cotransplanted with primary dopamine neurons in 6-OHDA-lesioned animals, seemed to support the survival of these cografted cells. Obviously, some of this work, along with the previous GDNF successes in rodent and primate models by other investigators, led to the recent human trials by Svendsen and collaborators.[4]

Perhaps one of the most exciting therapeutic approaches for PD will utilize the findings of a pervasive network of multipotent glial cells that exist into adulthood in the CNS, even in the substantia nigra. Such cycling glial cells, which we originally described as being able to generate both glia and neurons throughout the neuraxis during a critical period in postnatal development,[13] have now been described as cells that persist in the adult rodent and have many attributes of "satellite" cells whose potency and repopulation prowess have been studied in other tissues including muscle and vasculature (e.g., the elusive pericyte). The cycling glial stem cell data from the Gage group (see Figure 71.8 and Reference 42), and the neurogenesis findings of nigral glial cells by the Gallo lab,[43] suggest that certain substantia nigra glia may be capable of sustaining and/or making dopamine neurons. With the idea being set forth that glial cells can exhibit stem cell-like behaviors (e.g., Reference 44), a goal now is to establish how to get cells like the one recently discovered in the adult rodent substantia nigra to be neurogenic in vivo, as it has been demonstrated in vitro,[42] since it has been shown to only give rise to glial cells in vivo even when manipulated using a variety of approaches. If what appears to be a truly mutlipotent, cycling glial cell does inhabit the human substantia nigra, and if the rodent data is also extended to the human, and this cell can only generate neurons when manipulated ex vivo (or following grafting into forebrain brain marrow or hippocampal[42] structures), then we must figure out ways to encourage true stem cell abilities of such cells so they can attempt neuronal, e.g., dopaminergic neuronal, repopulation in PD.

The potential for dedifferentiation of cycling glial cells in the mature and compromised brain offers numerous repair possibilities and is now being considered by the field.[44] It is also possible that these cells could be exploited in novel protocols that combine stem cell therapy with gene therapy to attempt protection and replacement of at-risk populations of dopamine neurons in both early and late stages of PD. Most genetic therapy protocols have concentrated on growth factor, e.g., GDNF, rescue approaches,[45] but these technologies could be combined with stem cell grafting or recruitment approaches in PD. Recent advances in identifying new factors for controlled expansion of dopaminergic neurons, as well as fate control of non-neural cells (e.g., pigmented epithelia, see Reference 46), support optimism for the discovery of new factors that could be targets of novel combined gene therapy-stem cell approaches.

Finally, although somewhat controversial, it might also be possible to use nonhuman sources of stem/progenitor cells as starting populations for xenografting in circuit reconstruction paradigms for PD.[47]

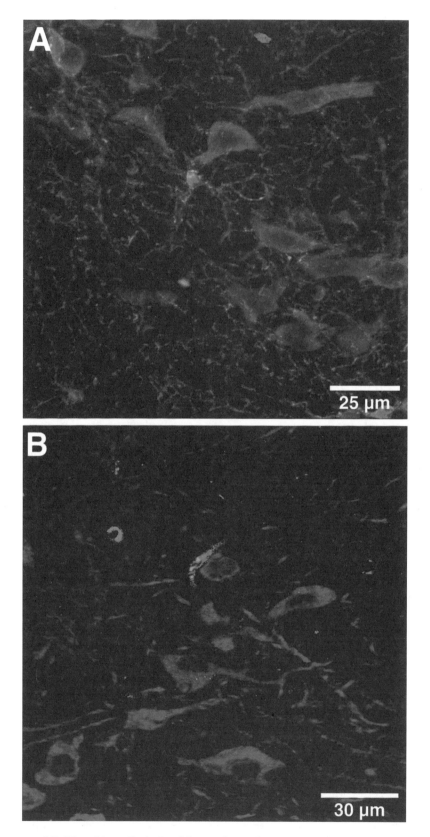

FIGURE 71.8   Phenotype of BrdU-positive cells (a thymidine analogue that marks newly-generated cells) in the substantia nigra after a 10 d BrdU pulse. (*A*) One-half of the BrdU-positive cells (green) express the glial progenitor marker NG2 (blue). (*B*) Some BrdU-positive cells (blue) that are not associated with blood vessels express the multipotent progenitor marker nestin (green). TH is shown in red. (Adapted from Reference 42. With permission.) *(A color version of this figure follows page 518.)*

HOPES FOR NEW REGENERATIVE MEDICINE DESIGNER THERAPIES FOR PD

Based on accumulating evidence in support of stem cell therapies for PD, one can envision a variety of dream scenarios for new therapeutic approaches that might evolve with increased therapeutic options for both early and late stage PD neuron rescue and replacement. Basic mechanisms of cellular de- and transdifferentiation must continue to be studied, with the idea that old brain cells might be coaxed into young, repair-interested cells,[13,44] and nonneural (e.g., bone marrow stem cells) ones could be readily harvested from the patient and gene therapy applied, *ex vivo* if necessary, and reintroduced into the patient via either intracerebral or systemic injections to seek out the damaged basal ganglia circuitry (Figure 71.9). In addition to the cells mentioned above, there is evidence that skin and fat cells might be able to be neuralized, and some of these other epithelial and mesenchymal cell sources may also provide significant growth and survival

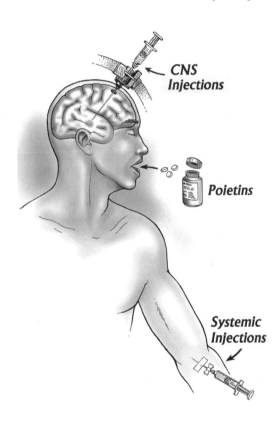

CNS Injections

Poietins

Systemic Injections

FIGURE 71.9 Once the choice of source (e.g., bone vs. brain marrow) has been made, and particular stem/progenitor cells have been characterized (e.g., phenotypy issues have been resolved), one can imagine several scenarios for the eventual use of the cells deemed best suited for neurorepair. With the scenario of allografting of donor stem/progenitor cells, large-scale expansion of the select population, enrichment, purification, and resolution of pathogen-free cells (good medical practice issues), the best root of introduction must be determined (stereotaxic placement vs. systemic administration to reach the compromised regions). Precise image-guided approaches are required for discrete and specific circuitry repair. The use of microcarriers and degradable biomatrices will also be used to support the initial survival and integration of intracerebral grafts. Whether stem/progenitor cell homing occurs, or the possibility of undesired migration might take place once in the adult CNS, might determine which "root of entry" strategy is the best. Stem/progenitor cells must then make short and long distance axonal connections. There is right now only limited evidence of long distance connections of newly generated neurons in the injured adult rodent CNS. There may be the need to simultaneously encourage axon growth with growth promoting factors and inhibit the growth-inhibiting molecules known to be upregulated in injured brain areas so as to facilitate axon growth and synaptogenesis. These events may also require adjunctive behavioral modification and therapies to help recapitulate particular ontogenetic events involved in circuitry formation. Self-repair also could be induced or augmented with "poietins," small, selective growth factors that trigger repair processes by one's own indigenous populations of stem/progenitor cells. Such a CNS regeneration dream scenario might utilize systemic administration of such factors that specifically home to or selectively affect areas of the damaged CNS where normally quiescent stem/progenitor cells reside. These factors might also induce dedifferentiation of mature cells that could then replace cells lost to disease or injury. Until this is possible, functional microrepair approaches are needed to provide new cells and support circuit repair. It is reasonable to predict that all of these scenarios are possible, in light of rapid breakthroughs in functional genomics and proteomics that should lead to the discovery of new factors that induce neuro- and synaptogenesis. (Adapted from Reference 5. With permission.)

factors as well. Of course, as previously described,[5] we need much better and efficient methods for the isolation and purification of different stem cell populations on the way to producing higher yields of dopaminergic neural precursors (e.g., immunopanning, gradient centrifugation, fluorescence activated cell sorting), but other methods must be developed to generate highly enriched populations of specific stem/progenitor cell populations for massive expansion and specific neuronal fate determination and differentiation. Furthermore, we need to determine why there seems to be a poor ability of SEZ, hippocampal, and nigral stem/progenitor cells to proliferate, migrate, and home to areas of injury and depletion in the adult CNS. Even if ES cell-derived, fetal, or adult brain stem/progenitor cell populations exhibit the propensity to integrate within neurogenic migratory pathways as well as established circuitries in the adult brain, there is still a great deal of information needed on the potential establishment of long-distance axonal projections from newly generated neurons in the mature brain. We must be able to place dopaminergic neuron precursors in a hostile environment of the injured nigrostriatal pathway, either in the midbrain (optimal, for appropriate afferent and efferent driving of their biochemical, physiological, and behavioral phenotypes) or in the dopamine-depleted neostriatum, have them survive and functionally integrate, and achieve the ability to innervate their distant target structures in the same topographic patterns as that achieved originally during development. This may require conjunctive neurite-growth support presented in distinct spatial and temporal sequences and patterns as that seen during normal ontogeny. This is an extremely daunting possibility and suggests that achieving dopamine neuron repopulation in the PD brain is but the first "obstacle" in the process of rebuilding a functional nigrostriatal dopaminergic pathway. There is still limited evidence for long distance axonal growth and reestablishment of circuits even after possibly successful integration of the progeny of a grafted or indigenous stem/progenitor cell population.

It is worth reiterating (see Reference 5 for review) that stem cells have already been used as experimental therapeutics for various rodent models of human disease, including metabolic disorders, muscular dystrophy, global CNS cell replacement, spinal cord repair, and brain tumors, but, in light of their potential for overproduction, hyperplasia, we must be cautious to not underestimate the abilities of a prolific population of cells to generate brain tumors.

> Ideally, one would like to understand the biology of fetal and adult stem cells a bit more before translating them into therapeutics; yet, morbidity and mortality associated with devastating human diseases and injuries argue for exploiting stem cell therapeutics without clearly understanding their nature...."[5]

Furthermore, if autologous repair is not possible, and heterologous stem/progenitor cells must be supplied from donor sources, immunorejection must be considered[2] with the need for haplotype-matching or immunosuppressants, although a very recent report suggests the possibility that, unlike stem cells from other sources (e.g., blood), neural stem cells do not usually express MHC class I and II, therefore lacking immunogenecity and possessing resistance to destruction as allografts.[48]

Figure 71.9 provides a summary of some scenarios for PD repair that may be on the horizon. Indigenous and poised-for-repair stem/progenitor cells attempt repair following injury or disease. If, for example, indigenous cycling stem cell populations in and around the substantia nigra are not able to meet the demands for rescue or replacement, cells from other "poietic regions" (e.g., brain marrow or bone marrow cells) might be able to home to the midbrain via cues provided by sick and dying cells. The release of trophic and guidance factors following injury might contribute to the reestablishment of both short and long distance axonal connections that lead to functional and behavioral recovery, but other growth and guidance factors may have to be introduced, along with cellular and/or biomatrix support systems to help rebuild connections within an already complex and growth-inhibitory neuraxis that is well respected for its inability to support growth compared to the more compact and regeneration-friendly developing brain. Thus, self-repair might also be augmented through noncellular, e.g., drug therapies, since there are studies that show peripheral growth factor injections can expand indigenous neural stem cell populations. "Neuropoietins" and guidance molecules that help support proliferation, fate choice, and neurite growth are clearly important elements to be further studied in nigrostriatal circuitry rebuilding paradigms.

To quote one of our recent position papers,

> Despite potential rapid progress from the bench to the bedside for restorative neurology and neurosurgery, we must still employ a tremendous amount of caution in applying exciting new stem cell findings to human therapeutics. Continued progress is dependent upon the prudent use of new technologies in justifiably highly scrutinized new therapeutic approaches. There are still examples of an impressive ability of different stem cells and immortalized stem cell lines to survive, release neurotransmitters, and other factors to overcome various molecular deficiencies, and perhaps even integrate in different and compromised mature brain circuits. But as of right now, we are not certain which potent cell population, either indigenous or donor derived, will safely and efficiently protect and replace vulnerable and lost neurons; we are also not certain that molecular therapeutics developed with insights gained from stem cell biology and regenerative neuroscience, might augment or even act on their own to achieve a cure in PD. Nonetheless, we might

take one lesson from the hematopoietic stem cell field where their concern over the use of immortalized, transformed stem/progenitor cells stems from their comprehensive knowledge of the leukemias, and another lesson from the Parkinson's disease cell-replacement field which early on called for "patience rather than patients."[5]

ACKNOWLEDGMENTS

The author is supported by NIH grants NS37556 and HL70143.

REFERENCES

1. Isacson, O., Bjorklund, L.M., and Schumacher, J.M., Toward full restoration of synaptic and terminal function of the dopaminergic system in Parkinson's disease by stem cells, *Ann. Neurol.,* 2003, 53, S135, 2003.
2. Redmond, D.E. Jr., Cellular therapy for Parkinson's disease: Where are we today? *The Neuroscientist,* 8, 457, 2002.
3. Weiss, R., Stem cell transplant works in a California Case; Parkinson's Traits Largely Disappear," *Washington Post,* p. A8, April 9, 2002 (in referring to work by Dr. M. Levesque).
4. Gill, S. S., et al., Direct brain infusion of glial cell line-derived neurotrophic factor in Parkinson disease, *Nat. Med.* 9, 589, 2003.
5. Steindler, D.A. and Pincus, D., Stem cells and neuropoiesis in the adult human brain, *Lancet,* 359, 1047, 2002.
6. Ignatova, T., et al., Human cortical glial tumors contain stem-like cells expressing astroglial and neuronal markers *in vitro, GLIA,* 39, 193, 2002.
7. Reya, T., et al., Stem cells, cancer, and cancer stem cells, *Nature,* 414:105, 2001.
8. Al-Hajj, M. et al., Prospective identification of tumorigenic breast cancer cells, *Proc. Natl. Acad. Sci. U.S.A.,* 100:3983, 2003.
9. Capron, A. M., Stem cells: Ethics, law, and politics, *Biotech Law Report,* 20, 678, 2001.
10. Theise, N. D., Stem cell research: Elephants in the room, *Mayo Clin. Proc.* 78, 1004, 2003.
11. Drazen, J. M., Legislative myopia on stem cells, *N. Engl. J. Med.,* 349:300, 2003.
12. Gates, M. A., Laywell, E. D., Fillmore, H., and Steindler, D. A., Astrocytes and extracellular matrix in adult mice following intracerebral transplantation of embryonic ventral mesencephalon or lateral ganglionic eminence, *Neuroscience,* 74, 579, 1996.
13. Laywell, E. D. et al., Identification of a multipotent astrocytic stem cell in the immature and adult mouse brain, *Proc. Natl. Acad. Sci. U.S.A.,* 97, 13883, 2000.
14. Scheffler, B., et al, Marrow-mindedness: a perspective on neuropoiesis, *Trends in Neurosciences,* 22, 348, 1999.
15. Wirth, E. D. 3rd et al., Feasibility and safety of neural tissue transplantation in patients with syringomyelia, *J. Neurotrauma,* 18, 911, 2001.
16. Kim, J. H. et al., Dopamine neurons derived from embryonic stem cells function in an animal model of Parkinson's disease, *Nature,* 418, 50, 2002.
17. Bjorklund, L. M. et al., Embryonic stem cells develop into functional dopaminergic neurons after transplantation in a Parkinson rat model, *Proc. Natl. Acad. Sci. U.S.A.,* 99, 2344, 2002.
18. Benninger, F. et al., Functional integration of embryonic stem cell-derived neurons in hippocampal slice cultures, *J. Neurosci.,* 23, 7075, 2003.
19. Bjorklund A, Lindvall, O., Cell replacement therapies for central nervous system disorders, *Nat. Neurosci.,* 3, 537, 2000.
20. Zheng, T., Steindler, D. A., and Laywell, E. D., Transplantation of an indigenous neural stem cell population leading to hyperplasia and atypical integration, *Cloning and Stem Cells,* 4, 3, 2002.
21. Kirschenbaum, B. et al., In vitro neuronal production and differentiation by precursor cells derived from the adult human forebrain, *Cereb. Cortex.* 4, 576, 1994.
22. Kukekov, V. G. et al., Multipotent stem/progenitor cells with similar properties arise from two neurogenic regions of adult human brain, *Experimental Neurology* 156, 333, 1999.
23. Ostenfeld, T. et al., Human neural precursor cells express low levels of telomerase *in vitro* and show diminishing cell proliferation with extensive axonal outgrowth following transplantation, *Exp. Neurol.,* 164, 215, 2000.
24. Reynolds, B. A., and Weiss, S., Generation of neurons and astrocytes from isolated cells of the adult mammalian central nervous system, *Science,* 255,1707, 1992.
25. Svendsen, C.N., et al., Long-term survival of human central nervous system progenitor cells transplanted into a rat model of Parkinson's disease, *Exp. Neurol.,* 135, 1997.
26. Doetsch, F., et al., Subventricular zone astrocytes are neural stem cells in the adult mammalian brain, *Cell* 97, 703, 1999.
27. Brautbar, C., Payne, R., and Hayflick, L., Fate of HL-A antigens in aging cultured human diploid cell strains, *Exp. Cell Res.,* 75:31, 1972.
28. Laywell, E. D., Kukekov, V. G., and Steindler, D. A., Multipotent neurospheres can be derived from forebrain subependymal zone and spinal cord of adult mice after protracted post-mortem intervals, *Experimental Neurology* 156, 430, 1999.
29. Terada, N. et al., Bone marrow cells adopt the phenotype of other cells by spontaneous cell fusion, *Nature,* 416, 542, 2002.
30. Sanchez-Ramos, J. et al., Adult bone marrow stromal cells differentiate into neural cells *in vitro, Exp.Neurol.,* 164, 247, 2000.
31. Woodbury, D. et al., Adult rat and human bone marrow stromal cells differentiate into neurons, *J. Neurosci. Res.,* 61, 364, 2000.
32. Kopen, G. C., Prockop, D. J., and Phinney, D. G., Marrow stromal cells migrate throughout forebrain and cerebellum, and they differentiate into astrocytes after

injection into neonatal mouse brains, *Proc. Natl. Acad. Sci. U.S.A.,* 96, 10711, 1999.

33. Mezey, E. et al., Turning blood into brain: cells bearing neuronal antigens generated *in vivo* from bone marrow, *Science*, 290,1779, 2000.

34. Brazelton T. R. et al., From marrow to brain: expression of neuronal phenotype in adult mice, *Science*, 290, 1775, 2000.

35. Papayannopoulou, T., and Craddock, C., Homing and trafficking of hemopoietic progenitor cells, *Acta Haematol.,* 97, 97, 1997.

36. Cogle, C. et al., Bone marrow transdifferentiation in the brain following transplantation, *The Lancet,* 363: 1432–1437, 2004.

37. Steindler, D. A., Neural stem cells, scaffolds, and chaperones, *Nature Biotechnol.,* 20,1091, 2002.

38. Ourednik, J. et al., Neural stem cells display an inherent mechanism for rescuing dysfunctional neurons, *Nat. Biotechnol.,* 20, 1103, 2002.

39. Villa, A., et al., Establishment and properties of a growth factor-dependent, perpetual neural stem cell line from the human CNS, *Exp. Neurol.,*161, 67, 2000.

40. Akerud, P. et al., Neuroprotection through delivery of glial cell line-derived neurotrophic factor by neural stem cells in a mouse model of Parkinson's disease, *J. Neurosci.,* 21, 8108, 2001.

41. Ostenfeld, T. et al., Neurospheres modified to produce glial cell line-derived neurotrophic factor increase the survival of transplanted dopamine neurons, *J. Neurosci. Res.,* 69, 955, 2002.

42. Lie, D. C., et al., The adult substantia nigra contains progenitor cells with neurogenic potential, *J. Neurosci.,* 22, 6639, 2002.

43. Belachew, S. et al., Postnatal NG2 proteoglycan-expressing progenitor cells are intrinsically multipotent and generate functional neurons. *J. Cell Biol.,*161, 169, 2003.

44. Steindler, D. A., and Laywell E. D., Astrocytes as stem cells: nomenclature, phenotype, and translation, *Glia,* 43, 62, 2003.

45. Kirik, D. et al., Long-term rAAV-mediated gene transfer of GDNF in the rat Parkinson's model: intrastriatal but not intranigral transduction promotes functional regeneration in the lesioned nigrostriatal system, *J. Neurosci.,* 20, 4686, 2000.

46. Kawasaki, H. et al., Generation of dopaminergic neurons and pigmented epithelia from primate ES cells by stromal cell-derived inducing activity, *Proc. Natl. Acad. Sci. U.S.A.,* 99,1580, 2002.

47. Armstrong, R. J., The potential for circuit reconstruction by expanded neural precursor cells explored through porcine xenografts in a rat model of Parkinson's disease, *Exp. Neurol.,* 175, 98, 2002.

48. Hori, J. et al., Neural progenitor cells lack immunogenicity and resist destruction as allografts, *Stem Cells,* 21, 405, 2003.

72 Alternative Drug Delivery in the Treatment of Parkinson's Disease

Pierre J. Blanchet
Department of Stomatology, University of Montreal and University of Montreal Hospital Center

CONTENTS

INTRODUCTION

Even though oral levodopa (combined with a peripheral dopa decarboxylase inhibitor) remains the most efficacious symptomatic drug treatment for Parkinson's disease (PD), it is not devoid of problems. The emergence of motor-response complications in levodopa-treated PD patients has puzzled and fascinated neurologists for over 30 years. Sooner or later, most patients eventually experience a predictable (so-called "wearing-off" effect) and less often unpredictable return in parkinsonian disability during the day, associated with nonmotor symptoms (neuropsychiatric, cognitive, or dysautonomic) and various dyskinesias, for reasons that remain elusive. The very short plasma half-life, poor bioavailability, and erratic absorption of oral levodopa are consistently present throughout the illness, and unequivocal changes in the peripheral pharmacokinetic handling of levodopa have never been demonstrated. Thus, attention shifted toward acquired changes in central pharmacokinetic properties (so-called "buffering capacity") and additional pharmacodynamic factors to explain and eventually correct the fluctuations of the levodopa motor response. Experimental attempts have tried to match the fairly good correlation between stable plasma levels of levodopa and sustained motor response with ways to continuously supply the dopamine precursor by bypassing the oral route.

The proof of concept of the remarkable efficacy of constant intravenous infusions of levodopa to sustain stable and adequate plasma levodopa levels and rapidly maintain fluctuating patients in the "on" state was made by Woods et al.[1] and Shoulson et al.,[2] and later confirmed and extended by Quinn et al.[3] Initially, the duration of levodopa infusions lasted up to 8 hr, and hourly infusion rates were from 70 to 143 mg (mean, 109 mg/hr) in most subjects.[4] Patients with biphasic dyskinesias also improved but experienced "breakthrough" dyskinesias during intravenous levodopa, transiently responding to a further increase in the rate of infusion. The clinical results of more sustained infusions were initially thought to be less satisfactory, with reduced mobility and severe dyskinesias reported.[5-6] However, repeated but interrupted daytime infusion for 12 to 14 hr for 5 consecutive days in three patients with the "on-off" phenomenon was shown to be an effective strategy.[7] In two cases, motor benefit was sustained during the day without severe dyskinesias or a need to modify the rate of infusion. The third patient also had a favorable, albeit less stable, response, with daily "on" time increasing from 42% on oral treatment to 82% during the infusion. A consistent threshold of plasma levodopa level to maintain the "on" state could not be determined. The importance of the infusion paradigm and necessity to keep drug-free nights to maintain the clinical response and avoid severe dyskinesias were highlighted. Nonetheless, continuous, around-the-clock intravenous infusions of levodopa gradually lessened the fluctuations in motor status during the day over 10 days and raised the antiparkinsonian and dyskinetic threshold plasma levodopa levels, and the severity of the complications typical of the oral treatment only progressively returned to baseline after the end of the infusion.[8] This response profile highlighted the importance of the continuous mode of dopamine receptor occupancy as a more physiological therapeutic strategy in PD, and the potential reversibility of the "priming" process underlying motor response complications provided dopamine receptors are stimulated correctly. The distinct behavioral effects resulting from the pulsatile and continuous delivery of levodopa and dopamine agonists have been studied in animal models of PD in the last decade, using subcutaneous osmotic minipumps. In one experiment, the rotational response to the nonselective dopamine agonist apomorphine was clearly enhanced following intermittent but not continuous levodopa administration in hemiparkinsonian rats.[9]

Non-oral drug administration has several potential advantages (Table 72.1). Constant, rate-controlled drug delivery through the subcutaneous or transdermal route is expected to reduce drug toxicity and fluctuations in clinical status during the day attended by the excessive maximal plasma concentrations (C_{max}) and variations around the mean plasma concentration observed with oral dosing. Sustained relief of parkinsonian symptoms during sleep may also desirable, particularly in more advanced stages of the illness. Moreover, the delay of onset of response to standard oral dopaminergic drugs represents a serious limitation when emergency rescue therapy is required to quickly reverse an akinetic crisis. A peripheral route of delivery is also impractical for drugs unable to cross the blood-brain barrier. The experience with the alternative routes of drug administration explored over the years (Table 72.2) to obviate such problems is reviewed.

TABLE 72.1
Advantages of Non-Oral Drug Delivery

- Constancy of drug delivery
- Avoidance of first-pass effect
- Rapid drug delivery for emergency use (subcutaneous)
- Lesser toxicity (lower peak levels)
- Rate-controlled delivery (patch, pump infusion)
- Targeting of central nervous system (intracranial)
- Avoidance of blood-brain barrier (intracranial)
- Good patient acceptability (patch)

TABLE 72.2
Alternative Modes of Drug Delivery Studies in Parkinson's Disease

Intravenous
- Bolus injections
- Continuous pump perfusion

Subcutaneous
- Bolus injections
- Continuous pump perfusion

Enteric
- Nasoduodenal
- Nasojejunal
- Duodenal (gastrostomy)

Intranasal

Sublingual

Rectal

Transdermal
- Passive
- Enhanced (chemical, iontophoretic)

Intracranial
- Ventricular
- Site-specific

ENTERIC INFUSION

Since the mid 1980s, a number of clinical investigators have attempted to lessen fluctuations in levodopa plasma levels and motor status by infusing levodopa continuously beyond the pylorus. Open-label observations on the effects of continuous daytime nasoduodenal delivery of levodopa at an hourly rate between 10 and 96 mg greatly improved mobility in three patients with the "on-off" phenomenon, and, in one of these to a greater extent than intermittent infusion.[10] On occasion, a transient loss of motor benefit was recorded when the distal end of the tube in the proximal duodenum slipped back into the stomach. Further work revealed that continuous duodenal delivery was comparatively more beneficial than intermittent duodenal and continuous gastric infusions in improving motor ratings and fluctuations in a different set of patients.[11] The long-term experience of two patients under daytime continuous duodenal levodopa infusions via a gastrostomy for 145 and 240 days was also favorable.[12] The rate of infusion adjusted to keep patients mobile gradually diminished over the first two months and remained constant thereafter. The daily levodopa intake rapidly declined initially, then more gradually during the first two months before it stabilized. Motor fluctuations virtually disappeared, and dyskinesias decreased in severity. Plasma levodopa levels just sufficient to maintain the "on" state dropped from 5.6 ± 1.2 µg/ml (on oral tablets) to 3.5 ± 0.6 µg/ml on infusion day 63. Thus, erratic gastric emptying may contribute to fluctuating plasma levels and motor response complications. Single nasoduodenal dosing has also been shown to shorten the time to normalize motor function and time to reach maximal plasma levels (0.5 hr in both instances) compared to oral dosing (1.71 and 2.0 hr, respectively), and to prolong the mean duration of clinical benefit from 1.2 hr (oral dosing) to 3.2 hr (duodenal dosing).[13]

The acute benefit of continuous duodenal infusion compared to intermittent oral levodopa treatment was subsequently evaluated in a double-blind, placebo-controlled, crossover trial.[14] Ten patients were randomly assigned to one of six four-day protocols alternating between continuous duodenal delivery for eight hours daily and standard intermittent oral therapy. The daily amount of levodopa used during the seven-hour observation period was comparatively higher on the infusion than on oral therapy (501 ± 253 mg versus 420 ± 205 mg). The variability or deviation from the daily mean plasma levodopa levels decreased by 47% on infusion days, from $38 \pm 11\%$ on oral therapy to $17 \pm 9\%$ on duodenal infusion. Seven patients showed improved "on" time, increasing on average from 2.88 ± 1.43 hr to 4.51 ± 2.06 hr. The number of "off" episodes declined in five subjects, from 3.8 ± 1.4 to 2.7 ± 1.6. The patients primarily experiencing dyskinesia reduced their total intake of levodopa while on duodenal infusion. In contrast, those with the "wearing-off" phenomenon increased their total dose during the infusion without worsening dyskinesia. Five patients elected to continue daily duodenal infusion therapy for at least one year after completing the study, and four of these eventually consented to a jejunostomy.

Some investigators have suggested that the problem with duodenal infusion levodopa therapy lies in its low water solubility. In one experiment,[15] oral levodopa/carbidopa tablets were milled and dispersed in a 1.8% aqueous methylcellulose solution (20 mg/ml). The solution is stable for one week, requires no added antioxidants, and can be kept refrigerated overnight and redispersed in the morning. It was delivered continuously during daytime through a nasoduodenal tube. The effects of this duodenal dispersion were tested on two nonconsecutive days in five patients with advanced PD and compared six weeks apart to their usual oral treatment. The intra-individual variation in levodopa plasma concentrations declined from three- to tenfold on oral therapy to twofold at the most on the duodenal dispersion, with parallel reductions in motor response fluctuations.

The highly soluble methylester of levodopa has also been used to reduce the volume of solution necessary for daytime infusions. At a dilution of 250 mg/ml, one patient showed dramatic motor improvement following continuous daytime nasoduodenal and jejunal administration of a solution of methylester at a rate of 180 mg/hr through a micropump with a 10-ml capacity.[16] An oral peripheral dopa decarboxylase inhibitor was co-administered. This approach kept him "on" throughout the day and improved

dyskinesias, enabling him to lower the infusion rate to 140 mg/hr.

Thus, continuous enteric infusion, albeit impractical for long-term use, produced gradual changes in clinical response and in levodopa dose requirements, suggestive of an inducible shift in sensitivity by the mode of receptor occupancy. Support for this hypothesis can be found in the experience of one parkinsonian patient receiving continuous nighttime jejunal infusions, who noticed some carryover benefit on daytime fluctuations and dyskinesias in spite of the maintenance of a standard oral levodopa regimen during the day.[17] In another single case report,[18] continuous around-the-clock enteric infusions failed to improve motor status in spite of increases in the infusion rate. Nighttime interruptions of the infusion to "reset" the receptors led to a gradual and considerable improvement of the situation. This apparent dopamine receptor downregulation with raised motor response threshold induced by the constant supply of the dopamine precursor should be borne in mind in the design of experimental therapies aiming to provide continuous dopamine receptor stimulation. Receptor internalization or uncoupling may be at play.

SUBCUTANEOUS BOLUS INJECTIONS

By virtue of the rapid onset of action it provides and bioavailability of certain dopamine agonists, the subcutaneous route of drug administration has been the main alternative route tested. In addition to the usual pharmacokinetic, pharmacodynamic, and safety issues that are part of any experimental drug development, practical issues related to handling and local skin tolerability must also be carefully addressed. For over half a century now, the nonselective, direct-acting dopamine agonist apomorphine has been the gold standard for this route of delivery. It is almost completely bioavailable and has shown efficacy in the treatment of PD following subcutaneous dosing.[19] In fact, it is equipotent to levodopa and produces virtually identical motor and dyskinetic responses.[20] The therapeutic benefit is of rapid onset (5 to 15 min) with T_{max} averaging 16 ± 11 min.[21] The mean duration of motor response to a single suprathreshold injection is 60 min (range, 20 to 120).[22] In view of this short duration of action, frequent adverse effects (including nausea, vomiting, hypotension and sedation), poor oral bioavailability, and dose-dependent, reversible uremia when effective oral doses up to 1500 mg per day were used,[23-24] it fell in disrespect for years. The use of domperidone to improve the tolerability of apomorphine[25] led to renewed interest for apomorphine as a rescue drug in levodopa-responsive parkinsonian patients with resistant response fluctuations.[6,26-27]

Since the individual standard levodopa dosage is not predictive of apomorphine dose requirements,[28] the optimal effective dose must be titrated and adjusted for each patient. The threshold dose is predetermined with incremental test doses under domperidone and a standard dose (e.g., double threshold dose) given subcutaneously in the abdomen or thigh using an insulin syringe mounted in a "penject" system.[22] The goal of this add-on approach is to help reducing the severity and duration of "off" periods during daytime, not to eradicate them. Patients are instructed to inject apomorphine as soon as the "off" period has set in. Single doses average 40 to 50 µg/kg (between 1 and 3 mg), and the protocol may be repeated 3 to 8 times during the day. Benefit usually lasts between 40 and 90 min.[27,29] A reduction in daily "off" time of approximately 50% is often reported with intermittent apomorphine add-on therapy, usually with maintenance of the same daily levodopa dose.[22,27,30–32] Some patients have used apomorphine at night to rapidly improve nocturnal akinesia or other unpleasant, sleep-disrupting "off" period disabilities.[22] Single apomorphine injections can also quickly relieve unusual and intractable "off" period symptoms such as pain,[22,33] panic symptoms,[33] freezing,[34] functional bladder outlet obstruction,[35] anismus,[36] and impotence.[37] This strategy has also been used successfully in the perioperative management of PD patients undergoing abdominal surgery.[38]

In spite of its clinical efficacy, subcutaneous apomorphine is felt to be underused.[39–40] The reasons for this include the following:

1. Perception that the technique is too cumbersome and that the patient will be unable to learn how to self-inject and overcome the technical demands
2. Fear that apomorphine frequently causes severe abdominal wall panniculitis
3. Misconception that neuropsychiatric and dyskinetic complications worsen and constitute contraindications to this strategy
4. Ambiguity regarding long-term efficacy

These sources of concern are only partly justified. Stibe and colleagues[27] commented on the easy use and acceptability of penjects by patients who are able to predict impending "off" periods. In 14 out of 22 PD patients evaluated in a double-blind, placebo-controlled study,[31] 13 were able to self-inject apomorphine by the end of the maintenance phase of 8 weeks, and 11 found this easy to handle. Rare patients develop an aversion for self-injection.[30] Supervision and support by a dedicated nurse practitioner and neurologist help solve most technical issues in motivated patients. In some countries, easy-to-use, prefilled single-use pen injectors are also commercially available.[31] Advanced patients with profound "off" episodes who cannot self-inject apomorphine for rescue may also benefit from daytime continuous apomorphine infusion (see below).

The treatment is well tolerated by most patients, and the adverse event profile is considered favorable with co-administration of oral domperidone.[27] Intermittent injections cause minor local skin reactions with small itchy nodules at some of the injection sites, which usually disappear within 48 hr.[30] Patients with advanced PD are at risk to experience neuropsychiatric complications with any dopaminergic drug, and apomorphine is no exception. However, under chronic intermittent apomorphine injections, psychotic features have been conspicuously absent[27] or have occurred in 12% (3 out of 24)[32] and up to 22% (11 out of 49) of individuals,[30] but their intensity was usually insufficient to require discontinuation of treatment. Since intermittent apomorphine is used as add-on therapy in patients with motor response complications, it is not surprising that almost 50% of patients maintained on long-term bolus apomorphine as part of a polypharmacy displayed a gradual worsening in peak dose dyskinesias,[30] representing a 67% increase in the mean daily duration of involuntary movements.[31] In another study, dyskinesias worsened in only 12% (3 out of 24) of subjects.[32] Adjustment of the antiparkinsonian co-medications may help solve the problem. The use of apomorphine injections to counteract biphasic dyskinesias is more controversial. Some case reports suggest that intermittent subcutaneous apomorphine is antidyskinetic in such cases[22,33,41–42] but eventually becomes a much less effective modality[22,43] in spite of upward dose titration. However, tachyphylaxis to the usual peripheral adverse effects of apomorphine (nausea, vomiting, postural hypotension, sedation) is normally seen over time, and only a few patients require long-term domperidone administration.

CONTINUOUS SUBCUTANEOUS INFUSION

These results have encouraged several groups, mainly in Europe, to infuse apomorphine continuously during the daytime through a needle inserted subcutaneously into the abdominal wall and connected to a portable minipump automated system. This strategy produces striking improvement in motor function in many PD patients with severe motor complications. The first study by Stibe and colleagues[27] showed great improvement in 11 severely disabled patients using an apomorphine pump for several months at a mean hourly infusion rate of 40 µg/kg (range, 20 to 70) during waking hours (mean daily dose of 77 mg). The quality of the motor benefit was comparable to that seen with oral levodopa, and two patients reported a greater sense of well-being on apomorphine. Mean reductions in daily "off" time (by 62%) and levodopa dose (by 19.5% or 209 mg) were more significant than those resulting from apomorphine injections. Residual "off" episodes were also less severe. A confirmatory account of the experience of 7 patients also infused for several months was soon published,[44] in which a lower mean daily apomorphine dose of 29.7 mg (range, 15 to 55) produced even more dramatic reductions in mean daily "off" time (by 85%) and levodopa dose (by 39% or 348 mg). The mean infusion rate is 3.3 mg/hr or 0.05 mg/kg/hr (range, 1.25 to 5.5 mg/hr or 0.02-0.08 mg/kg/hr), the majority of patients requiring between 2–4 mg/hr.[22] In a cohort of 17 patients with severe motor fluctuations switched to continuous apomorphine infusion, the mean hourly infusion rate was 97 ± 44 μg/kg during daytime, and 39 μg/kg overnight in six patients.[45] Marked motor improvement was observed with reductions in mean "off" time (by 61%) and mean daily levodopa dose (by 53% or 420 mg). Residual "off" episodes still occurred but were felt to be less severe in some cases. Additional 3-mg boluses (4 on average) of apomorphine were used during the day and co-treatment with oral levodopa was necessary in all but one patient. Overall, the infusion program was maintained beyond one year in 41% (7/17) of patients.

Eleven open-label long-term infusion studies with follow-up extending between 1 and 8 years published since 1990 were reviewed.[22,30,32,45–52] Collectively, the experience of these 263 patients indicates highly favorable motor benefit (Figure 72.1). The mean daily apomorphine dose was 114 ± 33 mg, with a range extending from 31.4 mg up to 162 mg due to different treatment approaches regarding attempts to lessen cutaneous complications or wean patients off oral levodopa, or maintenance of the infusion during waking hours only or around the clock. While low infusional rates of apomorphine (<40 mg/day) maximize "on" time as part of a polypharmacy, only higher rates (>70 mg/day) allow the discontinuation of oral levodopa which was achieved in 30% of cases. The global reduction of apomorphine infusion on mean daily "off" time and oral levodopa dose was 60 ± 11% and 47 ± 24%, respectively (Figure 72.1). These figures are in many instances superior to those resulting from the combination of levodopa with oral dopamine agonists[40] that only exceptionally allow patients to discontinue levodopa treatment completely. The short-term replacement of continuous subcutaneous apomorphine for levodopa has shown efficacy in advanced patients following major surgery.[53] Nocturnal apomorphine infusion also constitutes a sensible alternative for patients with refractory severe sleep fragmentation due to motor symptoms or restless legs manifestations.[54]

The outcome of levodopa-induced dyskinesias under continuous apomorphine infusion is variable. In one study, the severity of dyskinesias was either spared, decreased in 4 out of 11 subjects, or initially increased in 2 others.[27] However, the cumulative experience of 234 chronically infused patients reveals a favorable antidyskinetic effect contrasting with the impact of add-on intermittent injections. Over half of the patients with peak

dose dyskinesias have noticeably improved, with striking benefit in some cases in spite of the improvement in "on" time. Some authors expressed less consistent improvement in those with severe dyskinesias.[55] The initial reduction in peak-dose dyskinesia intensity generally paralleled the decrease in daily levodopa dose.[45,51–52] However, the observation of a more gradual and sustained antidyskinetic benefit taking place over a mean interval of six months[47] or longer[52] following the onset of apomorphine infusion argued in favor of the development of a genuine "depriming" effect. Few studies have quantified this antidyskinetic effect. Some patients able to achieve apomorphine monotherapy have obtained better results, with 65% reduction in peak-dose dyskinesia severity reported in two studies.[47,52] Those treated with apomorphine infusion as part of a polypharmacy showed a mean reduction in peak-dose dyskinesia severity of 30%[52] and reduction in the duration of dyskinetic periods averaging 61%.[51] In these patients, dyskinesias may gradually return over time, the mean antidyskinetic benefit decreasing from 45% at one year to a respectable 27% at five years in a long-term study.[50] The presence of biphasic dyskinesias is not an absolute contra-indication to apomorphine infusion since they worsened in six cases[22,45] but disappeared in four others.[46]

The main limitation with long-term continuous apomorphine infusions is the consistent development of local cutaneous reactions with bruises, rash, or nodules occurring in virtually all cases. However, the nodules are often small and not problematic in up to 62% of cases.[52] Nodular formation may be minimized by a combination of standard approaches such as aseptic, frequent rotation of the infusion sites, dilution of the apomorphine solution with saline from 10 to 5 mg/ml, restricting infusion times, and use of silicone gel patches or of ultrasound on indurated areas. Nonetheless, moderate to severe skin reactions have occurred in one-third of the 263 patients maintained on long-term infusions, and are more likely to manifest during the second and third year of treatment.[30] The nodules can bleed or get infected, forming abscesses that may require surgical debridement and antibiotics, and the abdominal wall skin can become necrotic and fibrotic to produce large indurated areas conceivably affecting drug absorption. A biopsy of a nodule revealed subepidermal edema, mild dermal perivascular inflammation, and patchy inflammation with an eosinophil response extending into subcutaneous fat, compatible with panniculitis.[27]

Various neuropsychiatric disorders can arise *de novo* or worsen under apomorphine infusion. Such complications have collectively affected 21% of the 263 infused patients, but these have generally been mild and rarely constituted a cause for concern and discontinuation of the therapy. These have included visual illusions and hallucinations, nightmares, mild confusion, diurnal agitation, and frank paranoid psychosis. In one study,[52] apomor-

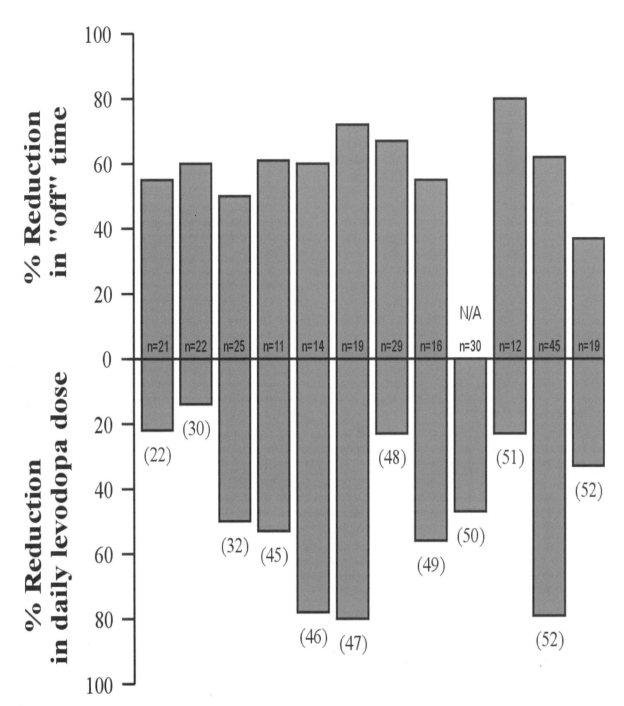

FIGURE 72.1 Benefit of long-term continuous subcutaneous apomorphine infusions (≥1 year) in patients with Parkinson's disease. Data concerning the percentage reduction in mean daily "off" time (top histogram) and daily levodopa dose (bottom histogram) compared to baseline are provided. One study compared subjects who achieved apomorphine monotherapy ($n = 45$) or polytherapy ($n = 19$).[52] The number of subjects (within bars) and the reference numbers (in parentheses under bars) are indicated.

phine demonstrated the potential to alleviate neuropsychiatric problems, patients under apomorphine infusion monotherapy improving by 40% and those in the polypharmacy group by 16%. One can only speculate about the reasons underlying this effect, which may be similar to those underlying the reported antidyskinetic effect, including reduction in daily levodopa dose, sustained dopamine receptor occupancy responsible for a new functional state, or direct specific pharmacodynamic properties of the drug through its piperidine moiety.[56] Sedation and hyperlibidinous behavior have also occurred. Weight gain was documented in 60% of patients in one series,[52] perhaps secondary to improvement in dyskinesias or some other pharmacological effects.

Peripheral blood eosinophilia, often transient, has been documented in a total of 27 patients with abdominal nodules.[22,46,48] A positive Coombs test,[52] associated with an autoimmune hemolytic anemia in eight subjects,[22,46,49,52] has also been reported. An early transient elevation in pancreatic enzymes has been documented in 5 out of 29 subjects in a single study.[48] Tachyphylaxis to the peripheral adverse effects of apomorphine (nausea, vomiting, orthostatic hypotension) often develops to eventually allow the discontinuation of domperidone in most cases.[57] Some patients also withdraw treatment prematurely due to practical technical issues. The penject needle may break and remain in subcutaneous tissue.[27]

Lisuride [N-(D-6-methyl-8-isoergolenyl)-N',N'-diethylcarbamide hydrogen maleate] is another water-soluble ergoline derivative that has been administered to patients with PD. Single intravenous doses between 0.1 and 0.2 mg produced motor benefit within 5 min and improved motor scores by 40%, comparable to oral levodopa.[58] Tremor appeared particularly sensitive. The effect was maximal for 10 to 30 min and disability gradually returned to baseline after 120 to 180 min. All subjects were given domperidone 50 mg orally one hour prior to lisuride dosing. Adverse effects included yawning, hiccups, nausea and vomiting, hypotension, and dyskinesias. In another experiment,[59] intravenous lisuride was continuously infused at a mean rate of 76 ± 31 Hg/hr (range, 33 to 125) for a maximum of 12 hr as co-treatment with the usual oral antiparkinsonian regimen in 10 patients with advanced disease and severe motor fluctuations. Domperidone was administered intravenously 30 min prior to the infusion. This drug combination greatly improved daily "off" time and strikingly reduced dyskinesias in five patients. Only one subject showed worsening in dyskinesias during lisuride infusion. Oral levodopa could not be discontinued with the dose of lisuride selected. Hypotension, nausea, sweating, and malaise were reported in one-half of the subjects. Similar results were obtained in three patients maintained on subcutaneous lisuride infusion for 4 to 7 months administered as co-treatment at a rate of 83 to 129 Hg/hr through a thin needle rotated every 5 days and connected to a portable minipump.[60] Cutaneous and systemic tolerability was excellent. Biphasic dyskinesias were abolished, but peak-dose dyskinesias became apparent by the second or third month, requiring a reduction in daily levodopa dose. One patient developed transient hallucinosis.

Few studies have addressed the long-term efficacy and adverse event profile of lisuride infusions. Long-term experience with 12- or 24-hr subcutaneous lisuride infusion in advanced PD confirmed its favorable impact on motor response fluctuations and severe nocturnal akinesia.[61-62] Daytime infusions (rate, 55 to 80 µg/hr) for 3 to 14 months allowed 6 patients to stop levodopa completely and to remain "on" most of the time with less dyskinesias;

6 others required supplemental levodopa and 24-hour lisuride infusion to achieve similar control of fluctuations, but dyskinesias persisted at a lesser intensity and hallucinosis became apparent after five months in one patient.[61] The daytime infusion protocol was felt to be better tolerated. No tolerance occurred. All patients showed local red nodules that spontaneously recovered within days and never justified withdraw from the protocol. In another study of 38 patients infused at a mean hourly rate of 111.3 ± 29.5 µg together with supplemental oral levodopa (729.6 ± 452.1 mg/day), three were withdrawn due to psychiatric complications, and one more patient could not pursue the protocol for technical reasons.[62] The others were infused for a mean period of 21 months (range, 6 to 45) at a mean daily dose of 2.6 mg or 109 ± 26.5 µg/hr. Two portable, battery-driven, infusion pumps were used. The 22-gauge needle was replaced every two days by the majority of patients. Eleven of 34 patients showed an excellent response, with minimal "off" time and no worsening in dyskinesias. The mean daily dose of levodopa was reduced by 47% in that group. In another subgroup of 18 subjects, "off" periods were reduced and the mean daily dose of oral levodopa decreased by 32%, but dyskinesia intensity increased by more than 50%. A last subgroup of 5 patients were considered therapeutic failures due to loss of benefit within 15 months and side effects. Twenty-six patients dropped out of the protocol during the first 2 years for various reasons, most frequently because of psychiatric reactions (16 cases). Under lisuride, daily "off" time fell from 4 to 12 hr to 1 hr or less, and the severity of "off" periods was greatly ameliorated to maintain 50% of these patients fully independent throughout the day. An unstable response was still observed in eight patients, with a partial "off" state often emerging in the early afternoon perhaps due to pharmacokinetic problems. A mean reduction in daily levodopa dose by 37% was possible, and while 2 patients almost achieved levodopa withdrawal, none of the 24 subjects was able to stop levodopa completely. Biphasic dyskinesia and "off" period dystonia largely resolved. However, problematic beginning-of-dose choreic dyskinesia increased or emerged in five patients. Neuropsychiatric complications developed in over one-third of the subjects, including visual and/or auditory hallucinosis (nine), confusional state (six), paranoia (five), and constituted the main reason for permanent withdrawal in five patients. Subcutaneous nodules developed in all patients, enough to interfere with lisuride absorption in five patients. Stocchi and colleagues[63] reported their observations in 13 patients, 10 of whom were kept under a 12-hr lisuride infusion regimen for 3 to 15 months. Four patients were able to stop levodopa (mean infusion rate of 0.064 mg/hr, range, 0.06 to 0.08) and the mean daily dose of oral levodopa was reduced by 45% in the others (mean infusion rate of 0.08 mg/hr, range, 0.05 to 0.12). The amount of

"off" time fell under one hour in most cases. Three patients required nocturnal infusion at a lower rate to reduce adverse effects. A single patient under a 24-hr infusion experienced hallucinosis and progressed to a frank paranoid state even after the infusion regimen was converted back to 12 hr. Red nodules were observed to regress spontaneously. In another report, 10 out of 12 patients, known for mental side effects with prior antiparkinsonian medications, displayed psychotic features and stopped lisuride infusion.[64] The serotonin effects of the ergot drug were thought to be involved. Bittkau and Przuntek[65] documented the possibility to infuse lisuride in parkinsonian patients with previous psychiatric disturbances with a low-dose regimen (mean dose of 0.94 ± 0.65 mg/day) in five cases, yet producing improvement in motor scores, a mean reduction in daily levodopa dosing by 74%, and reduction in daily "off" time by 54%. Interestingly, this treatment combination completely alleviated dyskinesias. Nodule formation greatly improved in two patients after replacing metal needles by plastic catheters. Thus, treatment with constant subcutaneous lisuride pump infusion provides good motor benefit, but dose-limiting and troublesome psychotic features may develop.

In another report,[66] 29 patients with severe and often unpredictable fluctuations were studied prospectively in an open-label trial for periods extending up to 36 months. Lisuride was infused at a mean daily dose of 1 mg (range, 0.3 to 2). The hourly rate was either kept constant for 24 hr (8 patients) or reduced by 30% overnight (21 patients). The mean daily levodopa dose (512 ± 249 mg) administered at baseline did not change appreciably. Thirteen patients pursued the protocol, while 16 others abandoned treatment after 0.5 to 30 months. Motor scores improved, and 11 of 17 disabled patients regained the ability to walk during "off" periods. The amount of daily "off" time improved by 35% at 3 months. Only 7 subjects were available for assessment at 12 months of treatment. The assessment of peak-dose dyskinesias at 3 months revealed improvement in 11, no change in 2, worsening in 4, and de novo dyskinesias in 4 patients. Biphasic dyskinesias were abolished in one patient and persisted in two others. Psychiatric disturbances, often benign, occurred in 11 out of 13 patients maintained on protocol, leading to a reduction in lisuride dose (in 10) and addition of promethazine (in 3). Only one patient stopped lisuride because of psychosis. Painless nodules were observed in all cases, but two patients required surgical drainage for abscess formation.

A recent prospective, randomized open-label trial in 40 fluctuating PD patients compared the clinical efficacy of subcutaneous lisuride infusion (with supplemental oral levodopa) to standard oral antiparkinsonian regimens over a 4-year period.[67] Only two lisuride patients withdrew early during the study due to compliance difficulties. The mean hourly infusion rate of lisuride was 0.91 ±

0.17 µg/hr maintained for 12 hr during the daytime. The patients were instructed to rotate needle sites every 2 to 5 days, but many preferred to remove the needle every night. After four years, mean daily "off" time showed a persistent drop by 59% from baseline to 1.2 ± 0.7 hr in the lisuride group, compared to a 21% increase to 5.1 ± 0.7 hr in the levodopa group. The outcome on the Abnormal Involuntary Movements Scale scores strikingly differed between the groups, with a sustained improvement by 49% in the lisuride group and worsening by 59% in the levodopa group. The mean daily levodopa dose was decreased by half (from 688.2 ± 133.4 mg to 333.3 ± 89.9 mg) in the lisuride group, and increased by half (from 675 ± 180.6 mg to 1032.5 ± 144.4 mg) in the levodopa group. The alleviation of dyskinesias was not felt to be the result of this reduction in daily levodopa dose, since this reduction took place prior to lisuride infusion, while dyskinesias gradually lessened over weeks to months. The results support the view that continuous dopamine receptor stimulation is more physiological and produces less functional anomalies in the basal ganglia than the repeated, short-lived, "pulsatile" stimulation attended by standard oral levodopa treatment. No patient withdrew because of adverse effects. Eleven out of 18 patients under lisuride showed generally mild skin nodules, 3 had psychiatric complications, and 3 experienced hypersexuality. According to these authors, the patients under lisuride infusions showed almost normal daily functioning, with minimal motor complications. The absence of problematic psychiatric reactions and skin nodules was attributed to the 12-hr infusion schedule. Skin nodules are also reduced by removal of the needle at night and infusion of a small volume of fluid made possible with lisuride. No fibrotic complications occurred.

In three patients comparing the effects of apomorphine (0.04 to 0.06 mg/kg/hr) and lisuride (1.79 to 2.5 µg/kg/hr) infusions at least 72 hr apart, the mean daily "off" time was 2.8 ± 0.3 hr with apomorphine and 7.6 ± 2.6 hr with lisuride.[68] In addition, the side-effect profile was less favorable with lisuride, including nightmares, depression, and vomiting. The patients continued to receive apomorphine for up to five months with sustained benefit and no evidence of tolerance. The comparative efficacy of continuous subcutaneous 24-hr infusions of apomorphine and lisuride was also evaluated at 6, 12, and 24 months in 20 patients with PD randomly assigned to one treatment arm.[69] Patients under lisuride infusion at a mean rate of 84 µg/hr required additional oral levodopa at a mean dose of 297 ± 193 mg/day; those under apomorphine required a mean rate of 3.2 mg/hr with supplemental oral levodopa at a mean dose of 327 ± 324 mg/day. The reduction in mean daily oral levodopa requirements compared to baseline was equivalent (46 and 47.5% with lisuride and apomorphine co-treatment, respectively). No upward titration in infusional rate was necessary over two

TABLE 72.3
Experimental Nonsubcutaneous Routes of Administration of Apomorphine in Parkinson's Disease

Route	N	Dose (mg)	Delay of Onset (min)	Duration (min)	Motor Response	Adverse Effects	Ref.
Nasal	8	6 mg (0.6 ml)	8.9 (6–15)	44 (36–55)	Off score ↓ 50%	None	73
	5	4.3 mg (2–6)	5–15	30–60	Off time ↓ 24%	• Transient nasal congestion or burning sensation • Vestibulitis requiring discontinuation (N = 2) • Hypotension (N = 4)	74
	7	1–10 (1–2 mg in 4)	8.8 (5–15)	49 (26–90)	Magnitude equal to s.c. dosing	• Mild vestibulitis (congestion, crusting) (N = 3) • Severe vestibulitis with 10 mg (N = 1)	71
	7	5.3	18.1	61.0	"On" response achieved	n/a	75
	11	2–5	7.5	60 (–90)	Similar to oral levodopa	• Vomiting (N = 2) • Nausea (N = 3) • Hypotension • Yawning • Bitter taste • Dyskinesia	145
	9	4.1	11	50	Off time ↓ 19%	• Dyskinesia (N = 4) • Nausea (N = 2) • Yawning (N = 2) • Hypotension (N = 3) • Nasal irritation, severe (N = 3); mild (N = 2)	146
	9 (chronic dosing)	3	11.3	50.5	Off time ↓, 45% No tolerance	• Nasal crusting (N = 5) within 4–6 weeks • Severe nasal crusting with pain and bleeding (N = 3) • Infection (N = 1)	72
Sublingual	9	10 × 3 mg tab. single dose	43 (30–55)	73 (30–110)	Comparable to s.c. APO and oral levodopa	• Mild sedation (N = 6) • Nausea (N = 2) • Yawning (N = 2) • Hypotension (N = 1)	147
	8 (chronic dosing)	3 × 3 mg tab t.i.d. to 10 × 3 mg tab t.i.d.	20–40	60–90	Off time ↓ 56%	• Stomatitis with ulcers and loss of taste (N = 4)	79
	7	0.5–3 × 6 mg tab	20–40	15–100 (no response in 2)	Off time ↓45% (one patient)	• Yawning (N = 2) • Flushing, diaphoresis, nausea (N = 1)	77
	10 (acute dosing)	57 mg	25 (12–53)	118 (60–200)	Full	• Nausea (N = 2) • Sedation (N = 2) • Unpleasant taste (N = 8)	81

TABLE 72.3
Experimental Nonsubcutaneous Routes of Administration of Apomorphine in Parkinson's Disease (continued)

Route	N	Dose (mg)	Delay of Onset (min)	Duration (min)	Motor Response	Adverse Effects	Ref.
Sublingual	3 (chronic dosing)	57, 114, or 228 mg up to t.i.d.	—	—	Similar to s.c. APO No tolerance	• No renal impairment	81
	10	40 mg t.i.d. (N = 8) or 20 mg 6 times a day (N = 2)	<20 min		↓ daily levodopa dose (32%), ↑"on" time, 8% (0–20) UPDRS-II, III improved	• Unpleasant taste (N = 8) • Nausea (N = 2) • No buccal irritation at 3 months • Lightheadedness (N = 5) • Sedation (N = 4)	148
	5	10–30 (0.1–0.3 ml)	19	65	Comparable to oral levodopa	• Yawning (N = 5) • Disorientation with marked postural tremor (N = 1)	149
	7	40 mg tab	28 (17–45)	N/A	Off time ↓ 68% No tolerance	• Bitter taste (N = 7) • Stomatitis with burning tongue after 2 months (N = 2)	80
	13	10	25	88	Effective in 56% of patients	• No renal impairment • Yawning (N = 4) • Bitter taste and hypersalivation (N = 2) • Nausea (N = 1)	78
	8	6 × 3 mg tab	30 (20–35)	120 (85–200)	Comparable to s.c. APO, off score ↓ 50%	• Bitter taste	150
	7	23 ± 1.3	21.7 ± 4.7	24.2 ± 14.6	Off scores ↓ 53% (4 responders)	• Dyskinesia (N = 4) • Yawning (N = 3) • Dizziness with sedation (N = 1)	151
	6	45.5 ± 2.6	18.3 ± 3.7	86.7 ± 14.9	Off scores ↓ 86% (all responded)	• Excitation (N = 1)	
Rectal	11	200 mg suppos.	32 (20–55) in 5/11	195 (60–350)	• Comparable to s.c. APO in 5/11 • Partial in 3 • None in 3	• Mild sedation (N = 2) • Nausea (N = 2) • Hypotension (N = 1)	82
	3	Enema 10–25 mg (10 mg/ml)	Short (15 min)	35–60	As effective as s.c.	N/A	83
	5 (acute dosing)	10–100 (3 forms: solution, gelatin suppos.. Witepsol-H15 suppos.)	14–27.5	50 ± 13 (enema), 156 ± 43 (Witepsol-H15, 100 mg)	Efficacy comparable to s.c.	• Rectal irritation (N = 1) • Hypotension (N = 1) • Sedation, yawning (N = 4) (Witepsol-100)	84

years. The reduction in daily "off" time and severity of dyskinesias was similar between treatment arms. Transient nausea and dizziness occurred with both drugs. Almost all patients experienced inflammatory nodules at infusion sites. Psychiatric adverse events were observed in only one patient in each arm. Thus, lisuride and apomorphine were similarly effective in alleviating motor response complications. The authors felt that apomorphine is more reliable when given in monotherapy and more flexible given the possibility for additional acute bolus injections to overcome severe "off" periods. On the other hand, apomorphine requires bulkier infusion devices.

INTRANASAL ADMINISTRATION

The intranasal administration of solutions using a metered dose nebulizer has been considered very promising with respect to convenience, rapidity, and quality of response for drugs penetrating mucous membranes, such as apomorphine.[70] Seven open-label studies in small groups (N = 5 to 11 subjects) conducted in a total of 56 patients have been published (Table 72.3). Effective doses have been variable (1 to 10 mg), with intranasal dosing considered equivalent to the subcutaneous route in some reports,[71-72] but doses between 1.5 and 2 times[73-74] and up to 5 times[75] the subcutaneous dose were necessary in others. The relative bioavailability has varied from 45%[75] to 90 to 100%.[71] The onset of response has generally occurred between 5 and 15 min, and duration of benefit lasted about 50 to 60 min. Motor improvement on the order of 50% and reduction in daily "off" time between 19 and 45% have been reported. Marked interindividual variations in susceptibility to local adverse effects have been observed. In chronic studies involving 30 subjects, nasal adverse events have occurred from mild irritation with congestion, rhinorrhea, burning, or itching to vestibulitis with crusting within 4 to 6 weeks, considered severe in 10 cases with pain, nosebleed, secondary infection, and eventually discontinuation of the drug. Others complained of the bitter taste of the solution. Although this attractive route of delivery for emergency drug intake has not been fruitful thus far, other medications may eventually prove to be well tolerated and effective in the future.

Besides intranasal solutions, mucoadhesive delivery systems have been of interest for certain drugs. The rapid nasal mucociliary clearance is an important limiting factor in that regard, restricting the time for complete and sustained drug absorption following nasal delivery. Mucoadhesive polymers such as cellulose derivatives and degradable starch microspheres (DSM) have been tested to enhance the intranasal bioavailability of apomorphine.[76] Two carboxymethylcellulose (CMC)-based powders [15% (w/w) or 30% (w/w) apomorphine loading] were compared to DSM and a lactose mixture of a drug loading of about 30% (w/w) each following intranasal insufflation in rabbits. Apomorphine absorption occurred within 10 min with all formulations. The lactose and DSM formulations were shown to be immediate release preparations with a mean absorption half-life for the lactose mixture of 1.8 min, compared to 4 to 5 min for DSM and CMC 30% (w/w), and nearly 10 min for the CMC 15% (w/w). The T_{max} values were approximately 10 min and equivalent between the lactose mixture, DSM, and subcutaneous route, and 1.5 to 2 times longer in the case of the CMC powders. Plasma apomorphine concentrations were better sustained with CMC, particularly with CMC (15% w/w), which maintained values within 50% of the C_{max} for 70 min, compared to 20 min for lactose and 35 min for subcutaneous injection. Apomorphine was still detectable in the plasma at four hours only following CMC dosing. All formulations yielded excellent relative bioavailability values between 82 to 100%. Thus, lactose and mucoadhesive DSM displayed a pharmacokinetic profile that could be advantageous for emergency drug release in PD. On the other hand, CMC-based powders may offer a reasonable compromise between rapid drug absorption and more sustained plasma drug concentrations to reduce the number of intranasal insufflations needed daily and thereby reduce potential toxicity. Further investigations seem worthwhile.

SUBLINGUAL ADMINISTRATION

Little experience with a single drug has been generated with this route in patients with PD. Ten open-label studies in small groups of patients (N = 5 to 13) using different formulations of apomorphine have been published (Table 72.3). Individual tablets of 3, 6, 10, 40, and 57 mg have been used, and apomorphine (100 mg/ml) was dissolved in glycerol and antioxidants in another study. The time for tablet dissolution and swallowing of part of the dose have delayed the response and produced large interindividual variations in response and inadequate relative bioavailability between 10 and 22%,[77-78] not improved with vitamin C.[78] The dose administered has been 10 times the subcutaneous dose.[79-80] Tablet dissolution has taken place within 10 min[77] and 21 min (range, 6 to 50),[81] and the onset of response occurred shortly thereafter (15 to 45 min). The clinical response was felt to be equivalent to oral levodopa or subcutaneous apomorphine in many cases, for a duration of benefit generally extending between 60 and 90 min, up to a mean of 118 min (range, 60 to 200) with the 57-mg tablets.[81] In two long-term studies involving 15 patients, sublingual apomorphine reduced daily "off" hours by 56%[79] and 68%.[80] Only one of these patients displayed apparent tolerance to the drug over five months. Stomatitis with ulcers and loss of taste (four cases) and burning tongue (two cases) occurred after two to six months, requiring discontinuation of the drug.

Adverse events abated within two weeks. Almost all patients complained of the unpleasant bitter taste of the medication. No renal impairment was documented following a mean daily dose of 150 mg,[80] even following individual doses of 228 mg up to thrice daily in a long-term study over 3 to 8 months.[81] Clearly, the hope for a rapid absorption in emergency situations was not fulfilled with sublingual tablets of apomorphine. The mean onset of response was more favorable with the glycerol preparation. Perhaps a Zydis fast-dissolving dosage form would enhance the pharmacokinetic profile of the medication and reduce the risk of stomatitis at the same time.

RECTAL ADMINISTRATION

Only two groups have explored the clinical potential of this route of delivery with enema or suppository formulations of apomorphine in small patient samples (Table 72.3). In one study, a single 200-mg suppository has produced a good clinical response after 32 min in half the patients tested.[82] Much smaller doses of 10 to 25 mg of apomorphine given in an enema (10 mg/ml) produced a response after 15 min for a motor benefit lasting 18 to 60 min.[83] Three other pharmaceutical formulations (enema of 10 or 15 mg, and two different suppositories of 25, 50, or 100 mg) were found to be quickly absorbed (onset between 14 and 28 min) and produced a standard response duration.[84] However, the Witepsol-H15 suppository of 100 mg generated a mean response of 156 min. The relative bioavailability of these formulations varied between 18 and 40%. Some rectal irritation was seen in one out of five patients. Long-term results are missing. Evidently, this route is not convenient for severely akinetic patients with sudden "off" episodes but provides yet another way to partly avoid the first-pass effect and improve drug bioavailability relative to the oral route.

TRANSDERMAL "PATCH" DELIVERY

This approach has long been recognized as a satisfactory means to achieve constant and controlled drug delivery in different areas of medicine. Passive diffusion over a large enough skin surface through the lipid phase of the stratum corneum can reach effective plasma levels for a few drugs only, but technologic advances in skin delivery systems are widening the classes of candidate drugs. This route by-passes the stomach and avoids variable and excessive peak plasma concentrations associated with adverse effects. The patch can be rapidly removed if such effects occur. It improves compliance and avoids the necessity to wake up at night to prevent nocturnal or early morning akinesia and dystonia. The application is noninvasive and painless, but issues related to the number and size of patches, the difficulty in maintaining skin contact (particularly during hygiene), and the technical demands in

manipulating and opening the sealed pocket need consideration. In view of their solubility and stability, dopamine agonists have mainly been examined for their suitability for transdermal application.

The first dopamine agonist considered was the potent and highly selective D_2 agonist (+)-4-propyl-9-hydroxynaphthoxazine (PHNO or naxagolide). It has mixed aqueous and lipid solubility and easy skin absorption, as suggested by rat experiments[85] and accidental exposure in humans.[86] The transdermal delivery system for (+)-PHNO free base was made of an outer covering membrane, a drug reservoir, a rate-controlling membrane, and an adhesive contact surface covered by a disposable release liner. The patch was made of silicone elastomeric components allowing an estimated drug release rate of 2.6 µg/cm²/hr in vitro. In 4 to 7 1-methyl-4-phenyl-1,2,3,6-tetrahydropyridine (MPTP)-exposed parkinsonian squirrel monkeys, the application of (+)-PHNO over 48 hr normalized motor activity after a delay of 9 hr, representing the time taken to permeate the epidermis and reach the dermal capillaries, correlating with T_{max} values on the order of 12 to 16 hr.[87] Steady-state plasma levels were achieved after 20 to 24 hr and remained stable for the following 24 hr. Exposure of a skin surface area of 4.78 cm², providing plasma levels between 100 and 500 pg/ml, was necessary to restore locomotor activity, a larger skin surface providing no further benefit. Compared with the in vitro drug release rate, the systemic availability was estimated to be no greater than 20%. The profile of response in 4 PD patients, co-administered oral levodopa and transdermal (+)-PHNO applied for 12 to 24 hr over chest skin surface areas rising from 2 to 20 cm², was less favorable.[88] In two, patch application over the maximum skin area improved "on" time after a delay of 4 to 6 hr. Plasma levels started to rise after 3 to 4 hr, but it took over 8 hr to achieve steady state. In two others, bedtime application of a 15-cm² or 20-cm² patch failed to turn patients "on" the next morning, and plasma steady-state levels between 200 and 300 pg/ml were approached by 20 to 24 hr after application. In one patient, plasma levels continued to rise and doubled even after the patch was removed, suggesting drug storage in dermal depots. Local skin irritation was not observed. This pharmacokinetic profile precluded the use of larger doses for monotherapy and increased the chance of overdosing, and testing was abandoned.

Developed in the 1970s, piribedil (1-[3,4-methylenedioxybenzyl)]-4-[(2-pyrimidinyl)]piperazine; ET 495) is a dopamine D_3/D_2 agonist also exerting D_1 agonist activity through its catechol derivative S 584.[89–90] Oral dosing between 200 and 300 mg/day reduced parkinsonian symptoms by 25 to 30%, but its antiparkinsonian efficacy was limited by extensive hepatic first-pass effect, short duration of action (–2 hr) and side-effect profile,[91–93] all potentially correctable using a transdermal approach (Smith et al., 2000).[94] Different amounts of piribedil

(2.5, 5.0, and 10.0 mg) were dissolved in an ethanol/water/ethylene glycol mixture to form a paste molded in 16-cm² squares and taped over an area of 25 cm² of abdominal skin, to be left in place for 24 hr. In addition, two types of self-adhesive 20-cm² patches impregnated with micronized piribedil were used: (1) monolayer patches had a single coating comprising the drug (67.5 mg), matrix, enhancer, and adhesive, and (2) bilayer patches contained 68.3 mg of active drug and two coatings of the same ingredients with twice as much adhesive as monolayer patches. These were left in place for 48 hr. Piribedil paste application in four MPTP-exposed parkinsonian marmosets produced rapid, complete, and long-lasting (~10 hr after application of 5 or 10 mg) improvement in motor status in a dose-dependent fashion. The motor response produced by the patch application was even better, with a slight advantage for the bilayer patches. No nausea or drowsiness occurred. After patch removal, the animals turned "off" within 2 to 3 hr, suggesting limited dermal storage. Measured plasma drug levels averaged 64 ng/ml in four animals 24 hr after patch application and remained high at 2 and 4 hr after patch removal, to gradually decline over the following 10 to 12 hr.

Montastruc and colleagues[95] tested a 30-cm² matrix patch containing 50 mg micronized piribedil and an adhesive strip in 27 levodopa-treated PD patients with disease duration ranging from 2 to 10 years. They were randomized into three parallel arms to either apply two active patches, two matching placebo patches, or one of each. Patches were applied on the arm and replaced daily over three weeks. No clinical efficacy was demonstrated, but mean plasma levels were <10 ng/ml in those who applied two active patches. Patch adhesion and local skin tolerance were reportedly good with mild, transient erythema observed in less than 20% of cases. Six patients experienced adverse effects, including nausea, vomiting, and malaise. Development was stopped in view of the skin area necessary to optimize drug delivery.

The latest dopamine agonist to undergo extensive testing is the nonergolinic aminotetralin derivative N–0923 ((-)-5,6,7,8-tetrahydro-6-[propyl-[2-(2-thienyl) ethyl]amino]-1-naphthalenol hydrochloride), the (–)-enantiomer of the racemic agonist N–0437.[96-97] It is 15-fold selective at D_2 versus D_1 receptors and displays antiparkinsonian efficacy in animal models and PD patients when given parenterally.[98–99] Following intravenous infusion, its calculated elimination half-life is approximately 71 min in humans.[99] Although N–0923 is not orally bioavailable in view of extensive hepatic first-pass effect,[100] its lipid solubility and ability to penetrate the skin[101] have prompted the development of a transdermal patch delivering the drug across human skin *in vitro* at a rate of 10 to 15 µg/cm/hr.[102] A Phase II, multicenter, clinical trial was then conducted in 85 levodopa-treated PD patients

to evaluate the effectiveness of various doses of transdermal N–0923 at replacing levodopa.[103] Patients were randomized to receive either placebo or one of four doses of medication: 8.4 mg (5 cm²), 16.8 mg (10 cm²), 33.5 mg (20 cm²), or 67 mg (40 cm²). Patches were applied every morning for 21 days and removed at bedtime for approximately 14 hr of skin application daily. The intent-to-treat population and the efficacy evaluable group included 82 and 67 patients, respectively. Patients under the two highest dosages achieved only a mild reduction in daily levodopa intake by 25 to 30%, but six patients under N–0923 and one under placebo required no levodopa throughout the study period. Importantly, no deterioration in motor status was documented. Two patients withdrew from the study because of postural hypotension (N–0923 group) and a moderate skin reaction (placebo group). Adverse effects were typically dopaminergic and mild in intensity, not clearly dose dependent except for somnolence.

The transdermal delivery system was later modified to allow a threefold increase in plasma concentrations of N–0923 compared with the earlier patch preparation of similar size (unpublished data, 1999, internal report, Schwarz Pharma AG, Germany). The new system, called Rotigotine CDS, was applied daily on the abdomen over four weeks by ten PD patients and doses titrated individually over the first two weeks starting with 9 mg (20 cm²), escalated by 4.5 mg (10 cm²) every two to three days up to the maximum tolerated dose or 36 mg (80 cm²) maintained for two more weeks.[104] In seven patients deemed evaluable, a significant decrease in "off" time was noted in six patients in spite of a mean reduction in daily levodopa intake by 53% (range, 33 to 98%), often improving dyskinesia. Mild, transient dopaminergic side effects were common, along with local itching and erythema. Plasma rotigotine levels from two patients 23 hr after application of the highest dose were between 2 and 3.5 ng/ml. The authors commented about the lack of a plateau effect on levodopa dose reduction within the Rotigotine CDS dose used.

Subsequently, two Phase II industry-sponsored, international, randomized, placebo-controlled trials of Rotigotine CDS were completed, but only preliminary results are available. The first (SP 506) 5-arm, parallel-group study enrolled 400 recently diagnosed PD patients not requiring levodopa or dopamine agonist treatment.[105] A total of 329 patients were randomized to target doses of 4.5, 9.0, 13.5, or 18.0 mg/day rotigotine or placebo, escalating the dose over four weeks by weekly increments of 4.5 mg/day up to the target dose maintained for seven additional weeks, using combinations of placebo and active 4.5 mg (10 cm²) patches to always apply four patches daily over the abdomen. A linear dose-dependent improvement in UPDRS parts II+III scores was observed, significant for the three higher dosages, reaching an apparent ceiling effect between 13.5 mg/day and 18.0 mg/day. A reduction in UPDRS scores of ~30% was documented in approximately 40 to 45% of subjects receiving 13.5 mg/day and 18.0 mg/day. Adverse events were mainly dopaminergic, but only local skin reactions occurred in a dose-dependent fashion, affecting 19% of subjects in the placebo group up to 46% at the highest effective dosage level. Eight subjects (3.6% of all actively treated patients) under Rotigotine CDS and none under placebo withdrew from the study because of skin reactions. The second (SP 511) 4-arm, parallel-group study enrolled 383 patients with advanced PD and motor fluctuations; 324 were randomized to treatment and 274 completed.[106] This time, combinations of placebo and active 10- and 20-cm^2 patches were used to escalate the weekly dose of Rotigotine over five weeks up to a selected target dose of (0, 9, 18, 27 mg/day). Although daily "off" time and UPDRS-III scores improved under Rotigotine treatment, the significance of the results was blurred by a large placebo effect. The number of "off" episodes was reduced particularly with Rotigotine at 27 mg. Thus, the current patch appears safe and superior to the initial formulation. However, higher doses may be needed in some patients.

Nondopaminergic drugs may also be administered transdermally to alleviate specific problems in PD. Several recent clinical trials have assessed the potential of different transdermal delivery systems of nicotine to correct motor and cognitive deficits in parkinsonian patients. These systems vary in terms of their *in vivo* transport rates with widely different C_{max} and T_{max},[107] affecting treatment outcome and side effects. A single open-label study testing 7-mg and 14-mg patches in 15 patients reported good tolerability and improvement in attention, processing speed, and motor speed.[108] Another study with a better design evaluated nicotine (17.5 and 35 mg) and placebo patches as add-on therapy in 32 PD patients. Adverse events were mild, but a lack of motor benefit was documented, possibly due to a "ceiling effect."[109] Two other studies showed untoward effects with motor deterioration[110] and poor tolerability with an unacceptable dropout rate,[111] independent of age, body mass index, disease duration, and severity. Specific nicotinic agonists may eventually provide a suitable alternative.

The drug absorption through human skin can be optimized in many ways.[112] For instance, transdermal delivery can be enhanced by iontophoresis, whereby an external electrical current is applied to produce electrorepulsive or electro-osmotic forces promoting drug flux into tissue. This concept was tested in PD with open-chamber TransQ-E® patches (Iomed Inc., Utah, U.S.A.), containing Ag/AgCl electrodes, with a skin exposed surface area of 20 cm^2.[113] The anodal compartment was filled with an apomorphine solution and the cathodal compartment filled with saline. The two patches (anode and cathode) were applied on the forearm, and patients carried a (pulsed) constant current power supply testing two different current densities. No drug absorption took place with passive drug delivery (no current applied). When apomorphine was applied by iontophoresis, clinical improvement was observed in two patients at low current density (250 μA.cm^{-2}) and in three patients at high current density (375 μA.cm^{-2}). Plasma concentrations of apomorphine increased at both current densities in a current-dependent fashion and doubled with the higher current (2.5 ng/ml at 1 hr). Local tingling sensation and transient erythema were reported, possibly in relation to transdermal iontophoresis rather than apomorphine. A reduction in current was necessary in a single patient. Thus, this strategy may afford individual titration by patch size and current density. Long-term local skin toxicity and damage to hair follicles in particular, which constitute a low resistance route, will need due consideration. Chemical enhancers and electrically assisted methods can also be combined to promote drug flux. In one study, a solution combining ethanol 40% and L-menthol 2% greatly influenced the profile of levodopa skin permeation *in vitro* compared to an ethanol 20% and L-menthol 2% solution.[114] A pharmacokinetic study was then conducted *in vivo* in rats with a cutaneous hydrogel preparation containing 5 g of levodopa in 40% ethanol and 2% L-menthol. The levodopa plasma levels gradually rose until 180 min to reach 45 ng/ml, whereas dopamine reached a plateau level of 1.2 ng/ml at 30 min maintained until 180 min. In comparison, plasma concentrations (ng/ml) of levodopa and dopamine following a single intravenous injection of levodopa (2.5 mg/kg) reached 664 and 21 at 5 min, and 200 and 5 at 15 min, respectively. The potential of vesicular carriers has also been investigated.[112] Although liposomes (colloidal particles made of phospholipids and cholesterol) do not generally penetrate intact skin, ethosomes (liposomes high in ethanol content) more effectively enhance drug delivery.[115] A novel ethosomal system carrying trihexyphenidyl has been developed to reduce C_{max} and adverse events.[116] The effectiveness and applicability of these approaches remain undetermined in PD.

INTRACRANIAL APPLICATION

Intracerebroventricular (ICV) or direct site-specific infusion, albeit invasive and generally impractical, affords targeting of the central nervous system, access to drugs not penetrating the blood-brain barrier, and rate-controlled delivery. Over 15 years ago, one patient with advanced PD and neuropsychiatric complications benefited from chronic ICV pump infusion of dopamine hydrochloride at a rate of 10 to 15 mg/day.[117] Motor improvement was sustained between pump refill every three weeks, but the patient showed fluctuating hallucinosis and delusional state. The ICV route was also used to administer glial cell line-derived neurotrophic factor (GDNF), a 134 amino acid peptide displaying neurotrophic activity for dopaminergic neurons. In MPTP-lesioned parkinsonian monkeys, bolus ICV injections[118] and chronic ICV pump infusion (SynchroMed™, Medtronic, Minneapolis, U.S.A.) of GDNF (5 or 15 μg/day)[119] promoted neurorestorative effects together with functional motor improvement. However, the ICV GDNF Study Group reported no apparent motor benefit in 50 subjects with PD treated for 8 months with monthly ICV administration of GDNF at single doses extending from 25 to 4000 μg.[120] The cerebrospinal fluid volume and distribution may have contributed to a dilutional effect that limited drug penetration into the brain parenchyma to reach the striatum and substantia nigra.

Site-specific tissue delivery is technically more demanding, but it offers the advantages of "surgical" manipulations of brain circuits, reminiscent of high-frequency deep brain stimulation protocols, with low toxicity. In MPTP monkeys, chronic intraputamenal pump infusion and ICV administration of GDNF yielded comparable results.[119] However, in a pilot trial, five PD patients implanted with SynchroMed™ pumps, refilled monthly to continuously deliver GDNF in the putamen bilaterally for one year, showed much better results than those in the ICV trial.[121] Using a daily delivery rate of GDNF of 10.8 to 14.4 μg, all patients were improved after three months, and "off" episodes completely vanished after six months. The mean total UPDRS score in the practically defined "off" state was reduced by 48% at one year compared to baseline, together with a reduction in the intensity of drug-induced dyskinesias. The sense of smell was improved in three patients, with intermittent dysosmic sensations. [18F]fluorodopa uptake in the putamen assessed by positron emission tomography increased by 28% around the tip of the infusion catheter after 18 months. The adverse event profile was favorable, with the Lhermitte's phenomenon as the only consistent finding. A high signal of unknown meaning developed around the catheter tip on magnetic resonance imaging when the daily delivery rate of GDNF was increased to 43.2 μg. Its disappearance with dose reduction suggests that it might reflect protein deposition or local vasogenic edema. One subject needed catheter repositioning in the dorsal putamen, and another developed a wound infection related to the pumps and connection tubing, requiring temporary explantation of the extracranial hardware and antibiotics. These unblinded results certainly warrant further investigation.

For obvious reasons, little effort has been made previously to deliver symptomatic antiparkinsonian medications *in situ* in the basal ganglia, with few research teams attempting to infuse dopamine directly in the striatum in animal models.[122] Advances in stereotaxic brain surgery and interventions such as pallidotomy and electrode

placement procedures for deep brain stimulation now allow to conduct neuronal recordings and acute pharmacological experiments in the operating room setting in the awake patient. The GABA$_A$ agonist muscimol (2.5 μg), injected into the ventral posterior globus pallidus pars interna of a single PD patient prior to a standard pallidotomy, increased movement speed and amplitude and abolished rigidity starting 10 min after the microinjection.[123] Interestingly, resting tremor persisted until the pallidotomy lesion abolished it. A spatial effect due to the volume of substance covered or the destruction of multiple circuits with pallidotomy, including not only pallidal tissue but also fibers passing through from the putamen, may explain the difference. The size of the subthalamic nucleus (STN) would certainly be advantageous for such protocols. Indeed, local microinjections of muscimol ($N = 2$) and lidocaine ($N = 4$) to focally inactivate the STN produced a marked antiparkinsonian effect within five to ten minutes, including tremor.[124] Lidocaine also induced choreic and dystonic dyskinesias. Simultaneous neuronal recordings revealed a suppression of neuronal activity at distances of <0.9 mm from the lidocaine injection site. In MPTP-lesioned parkinsonian monkeys, local microinjections of the GABA$_A$ antagonist bicuculline in the pedunculopontine nucleus produced an antiakinetic effect comparable to oral levodopa in two animals, suggesting that overinhibition of that structure in the parkinsonian brain contributes to akinesia.[125] The need for frequent pump refills and associated risk of infection make it unlikely that intraparenchymal drug delivery will ever replace deep brain stimulation as a viable therapeutic strategy.

CONTINUOUS DOPAMINERGIC DRUG DELIVERY: EVIDENCE FOR TOLERANCE?

As new drugs and technologies will soon make continuous dopamine receptor stimulation a reality, the question as to whether this strategy will induce plastic changes to favorably modify the functional state and response of the basal ganglia in drug-naïve and advanced PD remains largely unanswered. The clinical studies addressing the impact of continuous stimulation have been open-label trials involving a small number of subjects from single institutions, without controls, using different drugs (levodopa or agonists such as apomorphine, lisuride), dosing schedules (around the clock or waking hours only), and time courses [day(s) to 5 years].[126] Apparent acute changes in dopamine receptor sensitivity have been suggested by the reduced responsiveness to apomorphine when administered following overnight exposure to levodopa[127] and the decreasing response to a pulsatile treatment involving repetitive, closely spaced, low effective doses of apomorphine during one day in PD patients.[128] The latter results were not confirmed by others using a different pulsatile protocol.[129,130] The clinical response to a bolus of apomorphine was spared following continuous subcutaneous lisuride 24-hr infusion for 18 months.[131] However, evidence for tolerance was suggested in another experiment examining the response to test boluses of apomorphine before and after short (6 hr) or long (24 hr) continuous fixed-rate infusions of apomorphine, a greater reduction in acute response occurring following the long infusion.[132] Interestingly, this phenomenon was reversed when apomorphine was stopped overnight. The briefer duration of response observed after discontinuing a 21-hr levodopa infusion compared to a 2-hr infusion in 19 PD patients is also coherent with this interpretation.[133] In these experiments, apparent tolerance following continuous apomorphine or levodopa but not lisuride infusions may stem from the constancy of dopamine D_1 receptor occupancy provided by the former drugs, both mixed D_1/D_2 agents (see below).

Only some pharmacodynamic indices of central dopaminergic function have suggested the development of partial tolerance in clinical studies of continuous dopamine receptor stimulation. For instance, around-the-clock intravenous levodopa infusions for 7 to 12 days have required an 8% increase in the optimal levodopa dose so as to keep the patients "on," raised the threshold antiparkinsonian levodopa dose by 50%, and produced a shift to the right in the dose-response curve to levodopa test doses, all suggestive of tolerance.[134] However, the therapeutic efficacy of levodopa was maintained following the infusion and the duration of antiparkinsonian action even slightly prolonged by 30%. In another experiment, the dose-response curve to levodopa did not change appreciably following an around-the-clock levodopa infusion protocol for 3 to 5 days, although the infusion rates slowly increased in most patients to keep an optimal motor response.[135] One patient, under a continuous enteric infusion of levodopa, required higher infusion rates for a suboptimal motor response during daytime when maintained on an around-the-clock schedule, while overnight interruption of the infusion progressively reduced levodopa intake for a better motor response.[18] The profile of response with continuous synthetic dopamine agonists may well differ, the lack of endogenous metabolic machinery providing more sustained occupancy of central dopamine receptors compared to the precursor levodopa. While infusion studies using dopamine agonists did not produce clinically significant tolerance, save for a reduction in dyskinesia scores in some cases (see sections on apomorphine and lisuride), a sustained release oral formulation of the highly selective D_2 agonist naxagolide, administered as adjunctive therapy to 12 levodopa-treated PD patients up to a maximal dose of 30 mg twice daily, produced partial tolerance within 12 weeks, only 5 subjects completing the 24-week study.[136] That dosage form was estimated to provide efficacious plasma levels for 8 to 12 hr, so the profile of occupancy of central D_2 receptors was unlikely reset-permitting. No clinical studies with continuous D_1 agonists are available for comparison.

Pharmacological studies in animal models of parkinsonism have clearly shown evidence in favor of the development of tolerance or desensitization in the setting of continuous dopamine receptor stimulation, following a dopamine receptor subtype-dependent profile. In 6-hydroxydopamine rats, continuous levodopa therapy, unlike intermittent "pulsatile" dosing, does not diminish the rotational response[137] and electrophysiological response of substantia nigra pars reticulata neurons[138] to a selective D_1 agonist. The picture is dramatically different in the case of continuously administered dopamine agonists. Winkler and Weiss provided evidence that continuously infusing D_1 agonists with subcutaneous minipumps in 6-hydroxydopamine mice produced rapid loss of efficacy (within a few days) and nearly abolished the response to a test dose of D_1 agonist, while sparing the response to a D_2 agonist.[139] The extreme vulnerability of the D_1 receptor to desensitization was supported in MPTP-exposed parkinsonian monkeys.[140] Receptor internalization in the cytoplasm has been proposed as a mechanism.[141] Less profound and rapid development of tolerance occurs during continuous dopamine D_2 receptor agonist therapy with partial desensitization of striatal D_2 receptors.[142–144]

Although the fundamental point concerning the physiological nature of continuous dopamine receptor stimulation has not been extensively studied, particularly in relation to the distinct impact of levodopa and synthetic dopamine agonists, the observations available indicate that it is less detrimental than pulsatile dopaminergic therapy.[126] Tolerance is a risk but highly depends on the profile of receptor occupancy and targeted dopamine receptor subtype. The available evidence suggests that a reset at the receptor level is necessary to maintain an optimal response.

CONCLUSION

Advances in our understanding of the induction process conducive to levodopa-related motor response complications, the needs to provide access to drugs not penetrating the blood-brain barrier, and to reduce drug toxicity through rate-controlled delivery drive the research for alternative modes of drug delivery in PD. Several novel approaches are now made possible thanks to recent advances in biomaterials and drug delivery technology, supporting the development of new symptomatic and neuroprotective/neurorestorative strategies. Among them, transdermal delivery systems may soon meet success.

REFERENCES

1. Woods, A. C., Glaubiger, G. A., and Chase, T. N., Sustained-release levodopa, *Lancet*, 1, 1391, 1973.

2. Shoulson, I., Glaubiger, G. A., and Chase, T. N., On-off response, *Neurology*, 25, 1144, 1975.

3. Quinn, N., Parkes, J. D., and Marsden, C. D., Complicated response fluctuations in Parkinson's disease: response to intravenous infusion of levodopa, *Lancet*, 2, 412, 1982.

4. Quinn, N., Parkes, D., and Marsden, C. D., Control of on/off phenomenon by continuous intravenous infusion of levodopa, *Neurology*, 34, 1131, 1984.

5. Nutt, J. G. et al., The "on-off" phenomenon in Parkinson's disease: relation to levodopa absorption and transport, *N. Engl. J. Med.*, 310, 483, 1984.

6. Hardie, R. J., Lees, A. J., and Stern, G. M., On-off fluctuations in Parkinson's disease. A clinical and neuropharmacological study, *Brain*, 107, 487, 1984.

7. Marion, M. H. et al., Repeated levodopa infusions in fluctuating Parkinson's disease: clinical and pharmacokinetic data, *Clin. Neuropharmacol.*, 9, 165, 1986.

8. Mouradian, M. M. et al., Motor fluctuations in Parkinson's disease: central pathophysiological mechanisms, Part II, *Ann. Neurol.*, 24, 372, 1988.

9. Juncos, J. L. et al., Continuous and intermittent levodopa differentially affect basal ganglia function, *Ann. Neurol.*, 25, 473, 1989.

10. Kurlan, R. et al., Duodenal delivery of levodopa for on-off fluctuations in parkinsonism: preliminary observations, *Ann. Neurol.*, 20, 262, 1986.

11. Kurlan, R. et al., Duodenal and gastric delivery of levodopa in parkinsonism, *Ann. Neurol.*, 23, 589, 1988.

12. Sage, J. I. et al., Long-term duodenal infusion of levodopa for motor fluctuations in parkinsonism, *Ann. Neurol.*, 24, 87, 1988.

13. Deleu, D., Ebinger, G., and Michotte, Y., Clinical and pharmacokinetic comparison of oral and duodenal delivery of levodopa/carbidopa in patients with Parkinson's disease with a fluctuating response to levodopa, *Eur. J. Clin. Pharmacol.*, 41, 453, 1991.

14. Kurth, M. C. et al., Double-blind, placebo-controlled, crossover study of duodenal infusion of levodopa/carbidopa in Parkinson's disease patients with "on-off" fluctuations, *Neurology*, 43, 1698, 1993.

15. Bredberg, E. et al., Intraduodenal infusion of a water-based levodopa dispersion for optimisation of the therapeutic effect in severe Parkinson's disease, *Eur. J. Clin. Pharmacol.*, 45, 117, 1993.

16. Ruggieri, S. et al., Jejunal delivery of levodopa methyl ester, *Lancet*, 2, 45, 1989.

17. Sage, J. I. and Mark, M. H., Nighttime levodopa infusions to treat motor fluctuations in advanced Parkinson's disease: preliminary observations, *Ann. Neurol.*, 30, 616, 1991.

18. Cedarbaum, J. M., Silvestri, M., and Kutt, H., Sustained enteral administration of levodopa increases and interrupted infusion decreases levodopa dose requirements, *Neurology*, 40, 995, 1990.

19. Schwab, R. S., Amador, L. V., and Lettwin, J. Y., Apomorphine in Parkinson's disease, *Trans. Am. Neurol. Assoc.*, 76, 251, 1951.

20. Kempster, P. A. et al., Comparison of motor response to apomorphine and levodopa in Parkinson's disease, *J. Neurol. Neurosurg. Psychiatry*, 53, 1004, 1990.

21. Nicolle, E. et al., Pharmacokinetics of apomorphine in parkinsonian patients, *Fundam. Clin. Pharmacol.*, 7, 245, 1993.

22. Frankel, J. P. et al., Subcutaneous apomorphine in the treatment of Parkinson's disease, *J. Neurol. Neurosurg. Psychiatry*, 53, 96, 1990.

23. Cotzias, G. C. et al., Similarities between neurologic effects of L-dopa and of apomorphine, *N. Engl. J. Med.*, 282, 31, 1970.

24. Düby, S. E. et al., Injected apomorphine and orally administered levodopa in parkinsonism, *Arch. Neurol.*, 27, 474, 1972.

25. Corsini, G. U. et al., Therapeutic efficacy of apomorphine combined with an extracerebral inhibitor of dopamine receptors in Parkinson's disease, *Lancet*, 1, 954, 1979.

26. Yahr, M. D., Clough, C. G., and Bergmann, K. J., Cholinergic and dopaminergic mechanisms in Parkinson's disease after long term levodopa administration, *Lancet*, 2, 709, 1982.

27. Stibe, C. M. et al., Subcutaneous apomorphine in parkinsonian on-off oscillations, *Lancet*, 1, 403, 1988.

28. Dewey, R. B., Jr. et al., A randomized, double-blind, placebo-controlled trial of subcutaneously injected apomorphine for parkinsonian off-state events, *Arch. Neurol.*, 58, 1385, 2001.

29. van Laar, T. et al., A double-blind study of the efficacy of apomorphine and its assessment in "off"-periods in Parkinson's disease, *Clin. Neurol. Neurosurg.*, 95, 231, 1993.

30. Hughes, A. J. et al., Subcutaneous apomorphine in Parkinson's disease: response to chronic administration for up to five years, *Mov. Disord.*, 8, 165, 1993.

31. Ostergaard, L. et al., Pen injected apomorphine against off phenomena in late Parkinson's disease: a double blind, placebo controlled study, *J. Neurol. Neurosurg. Psychiatry*, 58, 681, 1995.

32. Pietz, K., Hagell, P., and Odin, P., Subcutaneous apomorphine in late stage Parkinson's disease: a long term follow up, *J. Neurol. Neurosurg. Psychiatry*, 65, 709, 1998.

33. Factor, S. A., Brown, D. L., and Molho, E. S., Subcutaneous apomorphine injections as a treatment for intractable pain in Parkinson's disease, *Mov. Disord.*, 15, 167, 2000.

34. Corboy, D. L., Wagner, M. L., and Sage, J. I., Apomorphine for motor fluctuations and freezing in Parkinson's disease, *Ann. Pharmacother.*, 29, 282, 1995.

35. Christmas, T. J. et al., Role of subcutaneous apomorphine in parkinsonian voiding dysfunction, *Lancet*, 2, 1451, 1988.

36. Mathers, S. E. et al., Constipation and paradoxical puborectalis contraction in anismus and Parkinson's disease: a dystonic phenomenon?, *J. Neurol. Neurosurg. Psychiatry*, 51, 1503, 1988.

37. O'Sullivan, J. D. and Hughes, A. J., Apomorphine-induced penile erections in Parkinson's disease, *Mov. Disord.*, 13, 536, 1998.

38. Gálvez-Jiménez, N. and Lang, A. E., Perioperative problems in Parkinson's disease and their management: apomorphine with rectal domperidone, *Can. J. Neurol. Sci.*, 23, 198, 1996.

39. Chaudhuri, K. R. and Clough, C., Subcutaneous apomorphine in Parkinson's disease, *B. M. J.*, 316, 641, 1998.

40. Poewe, W. and Wenning, G. K., Apomorphine: an underutilized therapy for Parkinson's disease, *Mov. Disord.*, 15, 789, 2000.

41. Luquin, M. R. et al., Levodopa-induced dyskinesias in Parkinson's disease: clinical and pharmacological classification, *Mov. Disord.*, 7, 117, 1992.

42. de Saint-Victor, J. F. et al., Levodopa-induced diphasic dyskinesias improved by subcutaneous apomorphine, *Mov. Disord.*, 7, 283, 1992.

43. Durif, F. et al., Apomorphine and diphasic dyskinesia, *Clin. Neuropharmacol.*, 17, 99, 1994.

44. Chaudhuri, K. R. et al., Subcutaneous apomorphine for on-off oscillations in Parkinson's disease, *Lancet*, 2, 1260, 1988.

45. Pollak, P. et al., External and implanted pumps for apomorphine infusion in parkinsonism, *Acta Neurochir. Suppl. (Wien)*, 58, 48, 1993.

46. Poewe, W. et al., Continuous subcutaneous apomorphine infusions for fluctuating Parkinson's disease. Long-term follow-up in 18 patients, *Adv. Neurol.*, 60, 656, 1993.

47. Colzi, A., Turner, K., and Lees, A. J., Continuous subcutaneous waking day apomorphine in the long term treatment of levodopa induced interdose dyskinesias in Parkinson's disease, *J. Neurol. Neurosurg. Psychiatry*, 64, 573, 1998.

48. Pinter, M. M. et al., Transient increase of pancreatic enzymes evoked by apomorphine in Parkinson's disease, *J. Neural Transm.*, 105, 1237, 1998.

49. Wenning, G. K. et al., Effects of long-term, continuous subcutaneous apomorphine infusions on motor complications in advanced Parkinson's disease, *Adv. Neurol.*, 80, 545, 1999.

50. Stocchi, F. et al., Subcutaneous continuous apomorphine infusion in fluctuating patients with Parkinson's disease: long-term results, *Neurol. Sci.*, 22, 93, 2001.

51. Kaňovsky, P. et al., Levodopa-induced dyskinesias and continuous subcutaneous infusions of apomorphine: results of a two-year, prospective follow-up, *Mov. Disord.*, 17, 188, 2002.

52. Manson, A. J., Turner, K., and Lees, A. J., Apomorphine monotherapy in the treatment of refractory motor complications of Parkinson's disease: Long-term follow-up study of 64 patients, *Mov. Disord.*, 17, 1235, 2002.

53. Broussolle, E., Marion, M. H., and Pollak, P., Continuous subcutaneous apomorphine as replacement for levodopa in severe parkinsonian patients after surgery, *Lancet*, 340, 859, 1992.

54. Reuter, I., Ellis, C. M., and Ray, C. K., Nocturnal subcutaneous apomorphine infusion in Parkinson's disease

and restless legs syndrome, *Acta Neurol. Scand.*, 100, 163, 1999.

55. Gancher, S. T., Nutt, J. G., and Woodward, W. R., Apomorphine infusional therapy in Parkinson's disease: clinical utility and lack of tolerance, *Mov. Disord.*, 10, 37, 1995.

56. Ellis, C. et al., Use of apomorphine therapy in parkinsonian patients with neuropsychiatric complications to oral treatment, *Parkinsonism Relat. Disord.*, 3, 103, 1997.

57. Lees, A. J., Dopamine agonists in Parkinson's disease: a look at apomorphine, *Fundam. Clin. Pharmacol.*, 7, 121, 1993.

58. Parkes, J. D. et al., Lisuride in parkinsonism, *Ann. Neurol.*, 9, 48, 1981.

59. Obeso, J. A., Luquin, M. R., and Martínez Lage, J. M., Intravenous lisuride corrects oscillations of motor performance in Parkinson's disease, *Ann. Neurol.*, 19, 31, 1986.

60. Obeso, J. A., Luquin, M. R., and Martínez-Lage, J. M., Lisuride infusion pump: a device for the treatment of motor fluctuations in Parkinson's disease, *Lancet*, 1, 467, 1986.

61. Ruggieri, S., Stocchi, F., and Agnoli, A., Lisuride infusion pump for Parkinson's disease, *Lancet*, 2, 348, 1986.

62. Vaamonde, J., Luquin, M. R., and Obeso, J. A., Subcutaneous lisuride infusion in Parkinson's disease. Response to chronic administration in 34 patients, *Brain*, 114 (Pt. 1B), 601, 1991.

63. Stocchi, F. et al., Subcutaneous lisuride infusion in Parkinson's disease: clinical results using different modes of administration, *J. Neural Transm. Suppl.*, 27, 27, 1988.

64. Critchley, P. et al., Psychosis and the lisuride pump, *Lancet*, 2, 349, 1986.

65. Bittkau, S. and Przuntek, H., Chronic s.c. lisuride in Parkinson's disease—motor performance and avoidance of psychiatric side effects, *J. Neural Transm. Suppl.*, 27, 35, 1988.

66. Heinz, A. et al., Long-term observation of chronic subcutaneous administration of lisuride in the treatment of motor fluctuations in Parkinson's disease, *J. Neural Transm. Park. Dis. Dement. Sect.*, 4, 291, 1992.

67. Stocchi, F. et al., Prospective randomized trial of lisuride infusion versus oral levodopa in patients with Parkinson's disease, *Brain*, 125, 2058, 2002.

68. Stibe, C., Lees, A., and Stern, G., Subcutaneous infusion of apomorphine and lisuride in the treatment of parkinsonian on-off fluctuations, *Lancet*, 1, 871, 1987.

69. Stocchi, F. et al., Apomorphine and lisuride infusion. A comparative chronic study, *Adv. Neurol.*, 60, 653, 1993.

70. Walton, R. P. and Lacey, C., Absorption of drugs through the oral mucosa, *J. Pharmacol. Exp. Ther.*, 54, 61, 1935.

71. van Laar, T. et al., Intranasal apomorphine in parkinsonian on-off fluctuations, *Arch. Neurol.*, 49, 482, 1992.

72. Muñoz, J. E. et al., Long-term treatment with intermittent intranasal or subcutaneous apomorphine in patients with levodopa-related motor fluctuations, *Clin. Neuropharmacol.*, 20, 245, 1997.

73. Kapoor, R. et al., Intranasal apomorphine: a new treatment in Parkinson's disease, *J. Neurol. Neurosurg. Psychiatry*, 53, 1015, 1990.

74. Kleedorfer, B. et al., Intranasal apomorphine in Parkinson's disease, *Neurology*, 41, 761, 1991.

75. Sam, E. et al., Apomorphine pharmacokinetics in parkinsonism after intranasal and subcutaneous application, *Eur. J. Drug Metab. Pharmacokinet.*, 20, 27, 1995.

76. Ugwoke, M. I. et al., Intranasal bioavailability of apomorphine from carboxymethylcellulose-based drug delivery systems, *Int. J. Pharm.*, 202, 125, 2000.

77. Gancher, S. T., Nutt, J. G., and Woodward, W. R., Absorption of apomorphine by various routes in parkinsonism, *Mov. Disord.*, 6, 212, 1991.

78. van Laar, T. et al., A new sublingual formulation of apomorphine in the treatment of patients with Parkinson's disease, *Mov. Disord.*, 11, 633, 1996.

79. Montastruc, J. L. et al., Sublingual apomorphine in Parkinson's disease: a clinical and pharmacokinetic study, *Clin. Neuropharmacol.*, 14, 432, 1991.

80. Deffond, D., Durif, F., and Tournilhac, M., Apomorphine in treatment of Parkinson's disease: comparison between subcutaneous and sublingual routes, *J. Neurol. Neurosurg. Psychiatry*, 56, 101, 1993.

81. Hughes, A. J. et al., Sublingual apomorphine in the treatment of Parkinson's disease complicated by motor fluctuations, *Clin. Neuropharmacol.*, 14, 556, 1991.

82. Hughes, A. J. et al., Rectal apomorphine in Parkinson's disease, *Lancet*, 337, 118, 1991.

83. van Laar, T. et al., Rectal apomorphine: a new treatment modality in Parkinson's disease, *J. Neurol. Neurosurg. Psychiatry*, 55, 737, 1992.

84. van Laar, T. et al., Pharmacokinetics and clinical efficacy of rectal apomorphine in patients with Parkinson's disease: a study of five different suppositories, *Mov. Disord.*, 10, 433, 1995.

85. Koller, W., Herbster, G., and Gordon, J., PHNO, a novel dopamine agonist, in animal models of parkinsonism, *Mov. Disord.*, 2, 193, 1987.

86. Stahl, S. M. and Wets, K. M., Recent advances in drug delivery technology for neurology, *Clin. Neuropharmacol.*, 11, 1, 1988.

87. Rupniak, N. M. J. et al., Antiparkinsonian efficacy of a novel transdermal delivery system for (+)-PHNO in MPTP-treated squirrel monkeys, *Neurology*, 39, 329, 1989.

88. Coleman, R. J. et al., The antiparkinsonian actions and pharmacokinetics of transdermal (+)-4- propyl-9-hydroxynaphthoxazine (+PHNO): preliminary results, *Mov. Disord.*, 4, 129, 1989.

89. Creese, I., Behavioural evidence of dopamine receptor stimulation by piribedil (ET495) and its metabolite S584, *Eur. J. Pharmacol.*, 28, 55, 1974.

90. Cagnotto, A., Parotti, L., and Mennini, T., In vitro affinity of piribedil for dopamine D3 receptor subtypes, an autoradiographic study, *Eur. J. Pharmacol.*, 313, 63, 1996.

91. Sweet, R. D., Wasterlain, C. G., and McDowell, F. H., Piribedil, a dopamine agonist, in Parkinson's disease, *Clin. Pharmacol. Ther.*, 16, 1077, 1974.

92. Chase, T. N., Woods, A. C., and Glaubiger, G. A., Parkinson disease treated with a suspected dopamine receptor agonist, *Arch. Neurol.*, 30, 383, 1974.

93. Rondot, P. and Ziegler, M., Activity and acceptability of piribedil in Parkinson's disease: a multicentre study, *J. Neurol.*, 239 (Suppl. 1), S28, 1992.

94. Smith, L. A. et al., Transdermal administration of piribedil reverses MPTP-induced motor deficits in the common marmoset, *Clin. Neuropharmacol.*, 23, 133, 2000.

95. Montastruc, J. L. et al., A randomized, double-blind study of a skin patch of a dopaminergic agonist, piribedil, in Parkinson's disease, *Mov. Disord.*, 14, 336, 1999.

96. Beaulieu, M. et al., N,N-disubstituted 2-aminotetralins are potent D-2 dopamine receptor agonists, *Eur. J. Pharmacol.*, 105, 15, 1984.

97. Van der Weide, J. et al., Pharmacological profiles of three new, potent and selective dopamine receptor agonists: N-0434, N-0437 and N-0734, *Eur. J. Pharmacol.*, 125, 273, 1986.

98. Belluzzi, J. D. et al., N-0923, a selective dopamine D2 receptor agonist, is efficacious in rat and monkey models of Parkinson's disease, *Mov. Disord.*, 9, 147, 1994.

99. Calabrese, V. P. et al., N-0923, a novel soluble dopamine D_2 agonist in the treatment of parkinsonism, *Mov. Disord.*, 13, 768, 1998.

100. Swart, P. J. and de Zeeuw, R. A., Extensive gastrointestinal metabolic conversion limits the oral bioavailability of the dopamine D2 agonist N-0923 in freely moving rats, *Pharmazie*, 47, 613, 1992.

101. Loschmann, P. A. et al., Stereoselective reversal of MPTP-induced parkinsonism in the marmoset after dermal application of N-0437, *Eur. J. Pharmacol.*, 166, 373, 1989.

102. Chiang, C. M. et al., A two-phase matrix for the delivery of N-0923, a dopamine agonist, *Proc. Intl. Symp. Contr. Rel. Bioact. Mat.*, 22, 710, 1995.

103. Hutton, J. T. et al., Transdermal dopaminergic D_2 receptor agonist therapy in Parkinson's disease with N-0923 TDS: a double-blind, placebo-controlled study, *Mov. Disord.*, 16, 459, 2001.

104. Metman, L. V. et al., Continuous transdermal dopaminergic stimulation in advanced Parkinson's disease, *Clin. Neuropharmacol.*, 24, 163, 2001.

105. Fahn, S., Rotigotine transdermal system (SPM-962) is safe and effective as monotherapy in early Parkinson's disease, *Parkinsonism Relat. Disord.*, 7, S55, 2001.

106. Quinn, N., Rotigotine transdermal delivery system (TDS) (SPM 962)-A multicenter, double-blind, randomized, placebo-controlled trial to assess the safety and efficacy of Rotigotine TDS in patients with advanced Parkinson's disease, *Parkinsonism Relat. Disord.*, 7, S66, 2001.

107. Fant, R. V. et al., A pharmacokinetic crossover study to compare the absorption characteristics of three transdermal nicotine patches, *Pharmacol. Biochem. Behav.*, 67, 479, 2000.

108. Kelton, M. C. et al., The effects of nicotine on Parkinson's disease, *Brain Cogn.*, 43, 274, 2000.

109. Vieregge, A. et al., Transdermal nicotine in PD. A randomized, double-blind, placebo-controlled study, *Neurology*, 57, 1032, 2001.

110. Ebersbach, G. et al., Worsening of motor performance in patients with Parkinson's disease following transdermal nicotine administration, *Mov. Disord.*, 14, 1011, 1999.

111. Lemay, S. et al., Poor tolerability of a transdermal nicotine treatment in Parkinson disease, *Clin. Neuropharmacol.*, 2003 (in press).

112. Barry, B. W., Novel mechanisms and devices to enable successful transdermal drug delivery, *Eur. J. Pharm. Sci.*, 14, 101, 2001.

113. Van der Geest, R. et al., Iontophoretic delivery of apomorphine. II: An *in vivo* study in patients with Parkinson's disease, *Pharm. Res.*, 14, 1804, 1997.

114. Sudo, J. et al., Transdermal absorption of L-dopa from hydrogel in rats, *Eur. J. Pharm. Sci.*, 7, 67, 1998.

115. Touitou, E. et al., Ethosomes - novel vesicular carriers for enhanced delivery: characterization and skin penetration properties, *J. Control. Release*, 65, 403, 2000.

116. Dayan, N. and Touitou, E., Carriers for skin delivery of trihexyphenidyl HCl: ethosomes vs. Liposomes, *Biomaterials*, 21, 1879, 2000.

117. Horne, M. K. et al., Intraventricular infusion of dopamine in Parkinson's disease, *Ann. Neurol.*, 26, 792, 1989.

118. Gash, D. M. et al., Functional recovery in parkinsonian monkeys treated with GDNF, *Nature*, 380, 252, 1996.

119. Grondin, R. et al., Chronic, controlled GDNF infusion promotes structural and functional recovery in advanced parkinsonian monkeys, *Brain*, 125, 2191, 2002.

120. Nutt, J. G. et al., Randomized, double-blind trial of glial cell line-derived neurotrophic factor (GDNF) in PD, *Neurology*, 60, 69, 2003.

121. Gill, S. S. et al., Direct brain infusion of glial cell line-derived neurotrophic factor in Parkinson disease, *Nature Medicine*, 9, 589, 2003.

122. Hood, T. W., Domino, E. F., and Greenberg, H. S., Possible treatment of Parkinson's disease with intrathecal medication in the MPTP model, *Ann. N.Y. Acad. Sci.*, 531, 200, 1988.

123. Penn, R. D. et al., Injection of GABA-agonist into globus pallidus in patient with Parkinson's disease, *Lancet*, 351, 340, 1998.

124. Levy, R. et al., Lidocaine and muscimol microinjections in subthalamic nucleus reverse parkinsonian symptoms, *Brain*, 124, 2105, 2001.

125. Nandi, D. et al., Reversal of akinesia in experimental parkinsonism by GABA antagonist microinjections in the pedunculopontine nucleus, *Brain*, 125, 2418, 2002.

126. Nutt, J. G., Obeso, J. A., and Stocchi, F., Continuous dopamine-receptor stimulation in advanced Parkinson's disease, *Trends Neurosci.*, 23, S109, 2000.

127. Vaamonde, J., Luquin, M. R., and Obeso, J. A., Levodopa consumption reduces dopaminergic receptor responsiveness in Parkinson's disease, *Clin. Neuropharmacol.*, 12, 271, 1989.

128. Grandas, F. and Obeso, J. A., Motor response following repeated apomorphine administration is reduced in Parkinson's disease, *Clin. Neuropharmacol.*, 12, 14, 1989.

129. Hughes, A. J., Lees, A. J., and Stern, G. M., The motor response to sequential apomorphine in parkinsonian fluctuations, *J. Neurol. Neurosurg. Psychiatry*, 54, 358, 1991.

130. Gervason, C. L. et al., Reproducibility of motor effects induced by successive subcutaneous apomorphine injections in Parkinson's disease, *Clin. Neuropharmacol.*, 16, 113, 1993.

131. Vaamonde, J., Luquin, M. R., and Obeso, J. A., Dopaminergic responsiveness to apomorphine after chronic treatment with subcutaneous lisuride infusion in Parkinson's disease, *Mov. Disord.*, 5, 260, 1990.

132. Gancher, S. T., Nutt, J. G., and Woodward, W. R., Time course of tolerance to apomorphine in parkinsonism, *Clin. Pharmacol. Ther.*, 52, 504, 1992.

133. Nutt, J. G. et al., Does tolerance develop to levodopa? Comparison of 2- and 21-H levodopa infusions, *Mov. Disord.*, 8, 139, 1993.

134. Mouradian, M. M. et al., Modification of central dopaminergic mechanisms by continuous levodopa therapy for advanced Parkinson's disease, *Ann. Neurol.*, 27, 18, 1990.

135. Schuh, L. A. and Bennett, J. P., Jr., Suppression of dyskinesias in advanced Parkinson's disease, *Neurology*, 43, 1545, 1993.

136. Cedarbaum, J. M. et al., Sustained-release (+)-PHNO [MK-458 (HPMC)] in the treatment of Parkinson's disease: evidence for tolerance to a selective D_2-receptor agonist administered as a long-acting formulation, *Mov. Disord.*, 5, 298, 1990.

137. Engber, T. M. et al., Continuous and intermittent levodopa differentially affect rotation induced by D-1 and D-2 dopamine agonists, *Eur. J. Pharmacol.*, 168, 291, 1989.

138. Weick, B. G. et al., Responses of substantia nigra pars reticulata neurons to GABA and SKF 38393 in 6-hydroxydopamine-lesioned rats are differentially affected by continuous and intermittent levodopa administration, *Brain Res.*, 523, 16, 1990.

139. Winkler, J. D. and Weiss, B., Effect of continuous exposure to selective D_1 and D_2 dopaminergic agonists on rotational behavior in supersensitive mice, *J. Pharmacol. Exp. Ther.*, 249, 507, 1989.

140. Blanchet, P. J. et al., Dopamine D1 receptor desensitization profile in MPTP-lesioned primates, *Eur. J. Pharmacol.*, 309, 13, 1996.

141. Muriel, M. P., Orieux, G., and Hirsch, E. C., Levodopa but not ropinirole induces an internalization of D1 dopamine receptors in parkinsonian rats, *Mov. Disord.*, 17, 1174, 2002.

142. Alexander, G. M. et al., Dopamine receptor changes in untreated and (+)-PHNO-treated MPTP parkinsonian primates, *Brain Res.*, 547, 181, 1991.

143. Blanchet, P. J. et al., Continuous administration decreases and pulsatile administration increases behavioral sensitivity to a novel dopamine D_2 agonist (U-91356A) in MPTP-exposed monkeys, *J. Pharmacol. Exp. Ther.*, 272, 854, 1995.

144. Blanchet, P. J. et al., Regulation of dopamine receptors and motor behavior following pulsatile and continuous

dopaminergic replacement strategies in the MPTP primate model, *Adv. Neurol.*, 86, 337, 2001.

145. Dewey, R. B., Jr. et al., Intranasal apomorphine rescue therapy for parkinsonian "off" periods, *Clin. Neuropharmacol.*, 19, 193, 1996.

146. Dewey, R. B., Jr. et al., A double-blind, placebo-controlled study of intranasal apomorphine spray as a rescue agent for off-states in Parkinson's disease, *Mov. Disord.*, 13, 782, 1998.

147. Lees, A. J. et al., Sublingual apomorphine and Parkinson's disease, *J. Neurol. Neurosurg. Psychiatry*, 52, 1440, 1989.

148. Ondo, W. et al., A novel sublingual apomorphine treatment for patients with fluctuating Parkinson's disease, *Mov. Disord.*, 14, 664, 1999.

149. Panegyres, P. K. et al., Sublingual apomorphine solution in Parkinson's disease, *Med. J. Aust.*, 155, 371, 1991.

150. Durif, F., Deffond, D., and Tournilhac, M., Efficacy of sublingual apomorphine in Parkinson's disease, *J. Neurol. Neurosurg. Psychiatry*, 53, 1105, 1990.

151. Durif, F. et al., Relation between clinical efficacy and pharmacokinetic parameters after sublingual apomorphine in Parkinson's disease, *Clin. Neuropharmacol.*, 16, 157, 1993.

73 The Role and Designs of Clinical Trials for Parkinson's Disease

Andrew Siderowf
University of Pennsylvania

Stanley Fahn
Columbia University

CONTENTS

INTRODUCTION

A randomized clinical trial is the "gold standard" through which potential new therapies are tested. The purpose of clinical trials is to demonstrate the effects, both good and bad, of a given intervention in an objective manner and ultimately to improve clinical care. In the case of Parkinson's disease (PD), results from trials may be used for several purposes, including

1. To compare treatment choices for existing therapies
2. To demonstrate the efficacy of new symptomatic treatments
3. To evaluate potentially neuroprotective therapies

In the first part of this chapter, we review the role of clinical trials. We discuss the conditions that should exist to justify conducting a trial; the process by which, after a trial is completed, the knowledge gained is adopted by physicians and applied to individual patients; and the eth-ically sensitive relationship between clinical trials and clinical practice. In the second part of the chapter, we describe design paradigms for clinical trials in PD and review some frequently used outcome measures. The chapter may serve as a starting point for interpreting existing trials and possibly for designing new trials for emerging therapies for PD.

ROLE OF CLINICAL TRIALS

APPROPRIATE SETTING FOR CLINICAL TRIALS

Since clinical trials are scientific experiments on human subjects, they should only be performed when the knowledge to be gained from the trial is important, and when the information cannot be obtained through other research designs. Although observational studies may approximate the results of subsequent clinical trials,[1,2] only randomized trials reliably control for unidentifiable differences between subjects in separate treatment arms (which often include a placebo arm) and provide unbiased estimates of the effects of treatment.[3] Even when the treatment effects seen in open studies appear unmistakable, it is quite possible for those results to be a result of a placebo effect or other source of bias, and uncontrolled studies may be followed by randomized, controlled studies showing no effect, or even harmful effects, from treatment. The case of fetal cell transplantation for PD is an example of this problem. The promising results from observational studies[4,5] were not borne out by subsequent randomized controlled trials.[6,7] The perspective that the randomized, controlled trial is the most reliable type of medical "evidence" has been reinforced by evidence rating systems[8] that strongly favor clinical trials relative to observational methods. Thus, it is reasonable to conduct a randomized trial when the information to be gained from the trial is important, and the potential for bias may invalidate observational study designs.

A precondition for the conduct of a clinical trial is that a state of *equipoise* exists between the treatments in the various arms of the trial. In its most simple form, equipoise can be defined as a state of "genuine uncertainty" about the relative merits of the treatment alternatives offered in the trial. If equipoise is disturbed, and one treatment is perceived as superior, then an ethical clinician-investigator must recommend the superior treatment. In the extreme, equipoise can be considered to exist only when the evidence for the alternatives is perfectly balanced. Freedman[9] refers to this situation as *theoretical equipoise*. He points out that this formulation of equipoise is extremely fragile. Any preclinical or preliminary human study showing the promise of a new treatment would disturb equipoise and preclude further, definitive studies. As a solution to this problem, Freedman proposes the concept of *clinical equipoise*. He defines clinical equipoise as a state of "genuine disagreement" within the expert community about the comparative merits of the alternatives to be tested. Clinical equipoise is much more easily achieved and more difficult to disturb than theoretical equipoise. It is only disturbed when the evidence in favor of one of the alternatives is strong enough to create consensus within the expert community. As a result, clinical equipoise provides greater ethical latitude for the conduct of clinical trials and has become the ethical framework in which modern clinical trials are conducted.[10]

IMPACT OF TRIAL INFORMATION ON CLINICAL PRACTICE

Once a trial is conducted, what is the process by which the information generated by the trial is incorporated into clinical practice? The result from a clinical trial must cross three hurdles for it to affect treatment of patients. The first hurdle is that the results from a trial must be accepted and approved by the government's regulatory agency. The second hurdle is that the results must be adopted by physicians as part of their therapeutic armamentarium. Trial results may reach physicians through a number of different routes, including presentation at meetings, publication in professional journals, being discussed by colleagues, and dissemination at continuing medical education activities. This process of adopting new information has been extensively studied by social scientists interested in the diffusion of technology. The third hurdle comes once a clinician has adopted a technology. At that point, the clinician needs to decide how to apply the technology to individual patients. These decisions depend on the applicability of the trial result.

IMPACT OF TRIAL INFORMATION ON PHYSICIANS—DIFFUSION OF TECHNOLOGY

Historically, health care providers have been remarkably slow to adopt innovations. Several authors[11,12] have referred to the delay between the generation of medical knowledge and its implementation as the "practice gap." To understand why this practice gap exists, one may turn to the work of social scientists who study the diffusion of innovation through populations. (Rogers[13] provides an excellent discussion of the issues related to adopting new technology.) The rate at which an innovation, such as the information from a clinical trial, is adopted by practitioners depends on three major factors:

1. Attributes of the innovation as perceived by potential adopters
2. The characteristics of the people who adopt the innovation
3. Contextual factors including communications, incentives, leadership, and management

The way that health care providers view a new technology is the first factor that determines whether it will be adopted. Berwick[11] cites five perceptual factors that determine the likelihood that a new technology will be adopted. These factors include the perceived benefit, how well it fits into existing conceptual frameworks, its simplicity, its trialability (whether the intervention can be tested on a small scale), and its observability (whether potential adopters can observe others try the innovation).

Perhaps the most important of these factors is the perceived benefit of a new innovation. In the case of a clinical trial, the perceived benefit is strongly related to the magnitude of effect observed in the trial. It is not uncommon to hear the concepts of the *magnitude* of an effect observed in a trial and its *statistical significance* used interchangeably. Although the magnitude of an effect observed in a trial contributes to its statistical significance, these concepts are quite distinct. The magnitude of an effect is simply how much change was observed in the trial. Important changes that are likely to matter to patients and doctors are sometimes referred to as "clinically meaningful." By contrast, statistical significance describes the probability that the effect observed in a trial was due to chance. Significance is reported as a "p-value," with lower values representing lower likelihood that a result is due to chance. However, a very small p-value does not necessarily mean that a drug has a strong effect, but rather that the effect observed in a trial was measured precisely and is not due to chance. For a trial to have any impact on practice, both conditions should be satisfied: the effect should be clinically meaningful, and one should be relatively certain that the observed benefit was not the result of chance.

The factors described by Berwick can be observed at work in the adoption of deep brain stimulation (DBS) surgery for PD. The results for DBS of the subthalamic nucleus have been reported in many open trials,[14-16] and a randomized, controlled trial is underway. Based on results from the available open trials, the perceived benefit of DBS is great, and so is its observability, since descriptions of patients, complete with video showing remarkable improvement, are available in the professional literature and on television. By contrast, while it is trialable at academic centers, DBS cannot be considered simple or trialable by individual practitioners. Similarly, DBS fits only moderately into a therapeutic culture outside of academic centers that is dominated by medical therapy, but may be more consistent with the culture at academic centers familiar with surgical approaches to PD. As a result, DBS is practiced at academic centers, and by a relatively small number of specialized surgeons, but is not commonly available in the community setting.

Calculating the perceived benefit of a new intervention depends on the interpretability of the trial result, and this has been an issue for PD trials. In the framework described above, interpretability is related both to the perceived benefit of an intervention and the ability to fit the intervention into available conceptual frameworks. Several large-scale, carefully designed trials have produced results that have been difficult to interpret, and this problem has limited the impact of these trials. The deprenyl and tocopherol antioxidative therapy of parkinsonism (DATATOP) trial[17] provides the seminal example of the challenges in interpreting a trial result. The trial showed a striking and statistically significant treatment effect in favor of deprenyl compared to placebo in the time to reaching disability sufficient to require dopaminergic therapy. Although the authors were careful to conclude only that deprenyl delayed the onset of disability, some readers interpreted the treatment effect as possibly slowing down the progression of PD. The authors had also cautioned that the mild symptomatic benefit that was observed when deprenyl was introduced renders the interpretation with uncertainty. The perceived benefit of a pos-

sible neuroprotective benefit by readers was great, and the results fit very well into the existing conceptual framework for treating PD in which slowing disease progression was, and remains, the most important unmet therapeutic need. Deprenyl was also simple to use and could easily be tried by practicing doctors. Hence, there was a great increase in the use of deprenyl to treat PD at all stages of disease.

Because of the mild symptomatic benefit observed with the introduction of deprenyl, and because the benefit seemed to wane with continuing treatment with deprenyl, critics of the DATATOP study suggested that the results of the trial were not due to a "neuroprotective" effect.[18] This conflicting opinion on how to interpret the DATATOP results led to decreased confidence in the benefit of deprenyl and confusion about where to place the drug in the existing conceptual framework. As a result, this led to a reduction in its use by practicing physicians. However, interpretation of deprenyl's effect is coming round in full circle, as a subsequent controlled trial of deprenyl in the presence of levodopa therapy showed that deprenyl slowed the clinical worsening of PD compared to placebo in the presence of levodopa to provide its symptomatic benefit.[19]

More recently, several trials comparing levodopa to placebo[20] and to a dopamine agonist[21,22] using both clinical and neuroimaging biomarker endpoints have resulted in problematic interpretation. In both cases, the clinical endpoints favor levodopa, whereas the biomarker endpoints favor the alternative (either placebo or dopamine agonist). Because of limited interpretability of these trials, the benefits of the alternatives tested are difficult to assess, and the impact of these trials on practice may be blunted.

The uptake of a new technology also depends on the characteristics of the people who will be adopting it. Adoption of a health care innovation is a social process and depends on networks of interconnected individuals. In the simplest case, diffusion theorists have promulgated a two-step model in which information is passed from mass media or other sources of knowledge generation to opinion leaders, and then the information is passed on to potential adopters by opinion leaders.[13] The characteristics of opinion leaders are crucial to this process. In general, opinion leaders tend to be have more contact with information sources, have higher social or educational status, and to be more innovative than followers. Opinion leaders need to be similar enough to followers that they can be seen as credible role models, but not so similar that their contact with sources of innovation is limited. Once adoption of an innovation has been passed to potential adopters, it is gradually taken up until a critical mass of individuals have adopted the innovation, and soon the remainder of the potential adopters follow suit.

Social scientists considered the adoption of new medical interventions as a quintessential example of this process.[23] New information becomes available that is either published in a medical journal or presented at a meeting. Then the information diffuses through local communities of physicians beginning with the most influential members who may be more likely to attend meetings or read journals. Clearly, the opinion leader is a key link in this process, since he may play a dual role as generator and disseminator of the new information and as an example of early adoption. Because of the great influence the opinion leader has in modifying practice, a great burden falls on them to evaluate emerging therapies in a responsible and disinterested fashion.

IMPACT OF TRIAL INFORMATION ON PATIENTS—APPLICABILITY OF RESULTS

The concept of *generalizability* (or external validity) has been used to describe the likelihood that a given patient will benefit from an intervention that has been shown to be effective in a clinical trial. Formally, generalizability can be defined as is the extent to which the results of a trial provide a correct basis for generalizations to other circumstances.[24] Patients who participate in clinical trials may be selected based on factors that make them good study subjects and may not necessarily be representative of the overall population of patients with a given disease. For example, patients who have been included in reported case series describing the effectiveness of deep brain stimulation are younger than most patients with advanced, medically refractory PD who would be considered candidates for DBS. In this case, the results of the study must be extrapolated to patient groups, such as older patients, that may benefit from treatment but were not explicitly included in the trial. Likewise, the treatment milieu of a trial may not be easily reproduced outside of specialized centers. Therefore, the results of the trial may not be applicable in some treatment settings, even though the patients there have similar signs and symptoms. Phase IV, or post-marketing studies that study the "real life" application of new treatments, can be useful in determining how such treatments work outside of the clinical trial setting.

Expanding on the idea of generalizability, Dans and colleagues[25] have put forward the broader concept of *applicability* in deciding whether the results of a clinical trial are likely to be relevant to a specific patient. The concept of "applicability" is closely related to generalizability but places greater emphasis on nonbiological factors. Dans and colleagues suggest six points to consider, divided into categories of biologic, social and economic, and epidemiologic, when considering whether a trial results applies to a specific patient (Table 73.1). The bio-

logical factors relate to differences in the disease characteristics between the study population and an individual patient. In the case of PD, these may be highly relevant, since several pathogenic mechanisms may lead to the same clinical presentation and may either enhance or diminish the expected response to treatment. The social and economic factors have to do with the likelihood of patient compliance and the ability of the provider to deliver the intervention in the same manner as in the trial. The last set of factors consider whether a patient has comorbid conditions, or an increased risk of adverse outcomes due to factors other than the biological characteristics of the disease.

THE RELATIONSHIP OF TRIALS AND CLINICAL PRACTICE

In academic centers, and increasingly in private physician's offices, patients are offered the opportunity to participate in clinical trials. In this context, the role of clinical trials may appear to be an extension of clinical practice. In fact, the relationship between clinical practice and trial participation is complex and has been the subject of ongoing debate among ethicists and clinicians involved in trials. Traditionally, there had been a free mixing of standard practice and clinical research.[26] Even now, informal clinical trials continue to take place when physicians attempt to treat conditions that do not have approved therapies by using medications in "off-label" indications, and the trial-and-error process that forms clinical experience could be considered a type of clinical research. Moreover, formal clinical investigations take place in the same setting as clinical practice and involve a very similar relationship between doctors and patients. Because of these factors, it is not surprising that the division between clinical research and practice can become indistinct.

Nonetheless, there are crucial differences between clinical trials and clinical practice. Proponents of the distinction between clinical trials and practice stress that trials differ from practice in their purpose, methods, and justification of risk.[27] Unlike standard care, the primary purpose of treatment in a clinical trial is to test the efficacy of a new treatment rather than to provide treatment for individual patients participating in the trial. Many research participants do not recognize this distinction. This confusion is referred to as *therapeutic misconception*. Second, clinical trials are often structured with standardized dosing regimens, assessments for the effects of treatment, and safety monitoring by physical examination and laboratory tests. This lack of individualization of treatment is required to preserve the scientific validity of the trial but does not contribute to patient care and may even pose risks or discomforts that would not be encountered in standard practice. Finally, the justification of risk and the doctor-patient relationship in trials are distinct

TABLE 73.1

Factors to Consider in Deciding Whether a Trial Result Is Applicable to a Specific Patient*

Category	Questions to Ask before Applying Trial Result	Hypothetical Example Relevant to Parkinson's Disease
Biologic	Are there differences in pathophysiologic aspects of the illness under study that may alter the treatment response for this patient?	A patient with atypical or secondary parkinsonism may not respond to a dopaminergic therapy tested in patients with idiopathic PD.
	Are there differences in the genetic or physiologic make-up of a patient that may alter the response to treatment?	A patient with impairment of hepatic or renal function may respond differently, or have increased risk of toxicity, from a treatment than was tested in patients without these problems.
Social and economic	Are there differences in patient compliance that may alter treatment response?	A patient with cognitive symptoms may not be able to comply with a complex titration schedule needed to initiate a medication that has side-effects when started abruptly.
	Are there differences in provider compliance, resources or expertise that may alter treatment response?	A surgeon who performs deep brain stimulation surgery infrequently may not have the same results as a highly experienced surgeon.
Epidemiologic	Are there comorbid conditions that alter the risk-benefit ratio of treatment?	A patient with a severe comorbid illness (e.g., cancer) that is likely to limit life-expectancy may not need a treatment strategy that avoids long-term problems like motor complications.
	Will differences in untreated patient's risk of adverse outcomes effect the efficacy of treatment?	A patient with atypical parkinsonism and a low risk of developing motor complications may not need a therapeutic strategy designed to avoid these problems.

*This table shows questions to ask divided into three categories: biologic, social and economic, and epidemiologic. Each question is illustrated with an example that might be encountered in deciding whether a trial result applies to a patient with Parkinson's disease. *Adapted from Dans et. al.*[25]

from clinical practice. In clinical care, the principles of *beneficence* and *nonmalficence* require that the physician provide the greatest benefit to an individual patient while minimizing risk. In the context of a clinical trial, formulation of risk and benefit extends to the potential good that can come to future patients as a result of the trial. Because of this additional factor, it may be reasonable to allow subjects participating in a trial to be exposed to a less favorable risk-benefit ratio, as long as they understand the risk and benefits of trial participation and have given informed consent.

While there are clearly differences between the structure and purposes of a clinical trial and practice, it is disingenuous to suggest that there is nothing therapeutic about participating in a clinical trial.[28] The arguments for a separation between trials and practice are based on the condition that receiving experimental treatment is no better than receiving placebo—the condition of theoretical equipoise. In reality, most trials are conducted under the condition of clinical equipoise, and there is often reasonable possibility that research participants may benefit from experimental treatment. Empirical studies have confirmed that patients who participate in trials tend to outperform those receiving standard medical care.[29] The benefits may be the result of the study intervention or simply the "therapeutic milieu" of the trial. A second problem with the nontherapeutic view of trial participa-

tion is that it places too great an emphasis on altruism as a motivation for trial participation. If there were no expectation of personal benefit, it would be extremely difficult to recruit adequate numbers of subjects to conduct trials, and trials could not realistically be conducted. Thus, there needs to be at least some "therapeutic orientation" to clinical trials. Finding an approach to trials that addresses the scientific requirements of the trial while providing at least standard care to research participants remains a substantial challenge and is likely to be an ongoing topic of spirited debate.

DESIGN OF CLINICAL TRIALS

For the purpose of drug registration, the Food and Drug Administration (FDA) has divided clinical trials into four phases (Table 73.2).[30] The FDA defines Phase I clinical trials as investigations that are typically closely monitored and are often carried out utilizing normal volunteers, but in some circumstances may also be conducted in patients. The purposes of Phase I studies are to determine the metabolism and pharmacologic actions of drugs in humans, the side-effects associated with increasing doses, and, if possible, gain early evidence on effectiveness. The total number of subjects included in Phase I studies varies with the drug, but it is generally in the range of 20 to 80.

TABLE 73.2
Phases of Clinical Investigation as Defined by the Code of Federal Regulations (CFR)[30]

Phase	Description
I	Small studies, usually involving 20 to 80 patients, for the purpose of determining safety and clinical pharmacology.
II	Larger studies, involving up to several hundred subjects, to further explore safety and to determine effective dosage for a specific indication.
III	Still larger studies, involving up to several thousand subjects, for the purpose of gathering additional information about safety, efficacy, and dosage that is needed to determine the overall benefit-risk relationship of the drug and to characterize the drug for is intended use.

Phase II clinical studies are intended to evaluate the effectiveness of a drug for a particular indication or indications in patients with the disease or condition under study. Phase II trials are also done to determine common short-term side effects and risks associated with the drug. Phase II studies are typically controlled, closely monitored, and conducted in a relatively small number of patients, usually involving no more than several hundred subjects. Identification of the most well tolerated and biologically active dose continues to take place in Phase II trials.

Phase III studies are performed after preliminary evidence suggesting effectiveness of the drug has been obtained and are attempts to gather additional information about effectiveness and safety, which is needed to evaluate the overall benefit-risk relationship of the drug and provide a basis for physician labeling. Phase III studies usually include from several hundred to several thousand subjects and are intended to be of a sufficient scope to identify adverse effects of treatment that occur with a relatively low frequency. Phase IV studies are conducted after the treatment has been approved by the FDA. These studies are intended to further characterize its effectiveness in patient subgroups and in different clinical settings. Phase IV trials enable sponsors to evaluate the drug in populations that may not have been well represented in Phase III trials.

Clinical trials may also be classified based on the quality of evidence that they provide, depending on the research methods they employ. The spectrum of research quality moves from observational designs including case-series, case-control studies, and cohort studies to experimental designs. Although evidence of the effectiveness of an intervention can come from cohort or case-control studies, three designs are more commonly used to evaluate new therapies in prospective interventional studies—case series, crossover studies, and parallel group trials. In this "hierarchy" of research designs, studies that employ the key features of *randomization* of treatment assignment and *blinding* of both research participants and investigators to treatment assignment produce the most reliable evidence for the effectiveness of an intervention.[31]

Although the case-series is the least methodologically rigorous type of study, it may be the simplest to perform. In this type of trial, research subjects are evaluated before an intervention and again while on treatment. The crucial limitation of case-series (and other observational designs) is that this type of trial does not offer protection against confounding such as concomitant use of beneficial therapies, bias such as unrecognized selection of the fittest patients (who may have good outcomes regardless of treatment), or placebo effects. Because the case-series design is susceptible to bias, it is not appropriate for testing the efficacy of a new intervention if other designs are available. The most appropriate use of case-series trials is to test the feasibility of delivering a new intervention and to identify frequent side effects. Case-series trials may also give some indication of the variability of response to a therapy, information that can be used in planning the sample size requirements for future controlled trials.

In contrast to case-series trials, crossover trials avoid most problems with confounding and bias. (Pocock[32] provides a detailed description of this type of trial.) In the first phase of a crossover trial, half of the subjects are randomly assigned to receive treatment, and the other half receive placebo. During the second phase, the two groups are crossed over so that the group initially receiving placebo now receives treatment, and the group initially receiving treatment receives placebo. The crossover trial design has several advantages. Because each subject serves as his own control in a crossover study, the sample-size requirements are one quarter of a parallel group study. Second, because it is a randomized design, the potential sources of bias are equally distributed between the treatment groups, and thus most types of bias are minimized. However, crossover trials are also subject to two specific sorts of bias: *period effects* and *carryover effects*. Period effects occur when a subject's underlying condition changes or the ability to respond to therapy is different in one phase of the trial from the other. Carryover effects occur when the treatment given in the first period continues to exert an effect during the second period. Carryover effects can sometimes be overcome by inserting a washout period between the two treatment periods. Because of these limitations, the FDA does not view crossover trials as acceptable for drug registration.

The most compelling type of trial design is the randomized, controlled, parallel group trial. In this type of trial, patients are randomly assigned either to treatment or a control state and are followed over time. In one typical analysis, the change in clinical status of patients receiving treatment is compared to that of patients who received the control treatment. Alternatively, the frequency of a clinically relevant event (such as mortality) may be compared between groups. Compared to crossover studies, parallel group trials are not susceptible to period or carryover effects and are usually of longer duration. In some cases, parallel group trials may be able to provide information about the long-term effects of therapy.

One major threat to the validity of a randomized trial is the problem of differential dropout of subjects over the course of observation. If subjects randomly assigned to one treatment group drop out of the trial at a higher rate than those in the other group, the benefits of randomization in controlling bias may be compromised. To minimize this problem, formal procedures for handling subjects who do not complete the entire trial should be part of a well designed trial. The most common procedure is to analyze the data according to the intent to treat (ITT) principle. ITT means that every subject who is randomized is included in the analysis in some fashion, and the missing values for subjects who do not complete the trial are imputed. Traditionally, in PD trials, noncompleters had their last available observation carried forward to the end of the trial. More recently, strategies in which missing values are imputed using statistical modeling have gained wider acceptance.[33,34]

A second potential weakness of this type of trial is its generalized ability or external validity. As discussed earlier, subjects enrolled in randomized, parallel group trials can be carefully chosen for their ability to comply with study procedures and may not be represent patients in general with a given medical condition. In spite of these limitations, randomized, parallel group trials are generally accepted as providing the strongest clinical evidence of efficacy of an intervention.

SPECIAL ISSUES RELATED TO PARKINSON'S DISEASE FOR CLINICAL TRIAL DESIGN

The model of the double-blind, randomized, controlled trial is applicable for any medical intervention. However, understanding the details of a specific disease and the possible effects, both intended and unintended, of a specific intervention is as important to the success of a clinical trial as following the principles of randomization and blinding. This is clearly the case in PD. First, the validity of a PD trial depends on having sufficient knowledge of the disease to choose appropriate outcome measures (Table 73.3). Second, in the particular case of trials intended to identify a "neuroprotective" effect, design elements intended to separate symptomatic effects from true neuroprotective effects are crucial to the interpretability of the trial.

TABLE 73.3
Possible Measurement Tools for Clinical Trials in PD*

Patient Group	Target Symptom	Specific Measurement Instrument
Early PD	Symptom severity	Change in UPDRS[35]
	Development of motor complications	Time to dyskinesias Time to motor fluctuations
	Disease progression	Change in UPDRS Time to dopaminergic therapy Time to postural instability Change in functional imaging biomarkers (beta-CIT SPECT or fluorodopa PET)[57,58]
Advanced PD	Symptom severity	Change in UPDRS
	Wearing-off	On-off diary
	Dyskinesias	Lang-Fahn ADL dyskinesia scale[48]
	Cognitive impairment	ADAS-cog[51] Mattis Dementia Rating Scale[50]
	Visual hallucinations	Scale for assessment of positive symptoms[53] Brief psychiatric rating scale[52]
All	Quality of life	PDQ-39[72] PDQUALIF[73]

*This table lists outcome measures commonly used in clinical trials for anti-parkinsonian therapies. Severity of the core features of parkinsonism are usually measured using the UPDRS. Other symptoms of PD, including motor complications, cognitive or psychiatric symptoms, and quality of life, may be measured using a range of instruments. Functional imaging, such as PET or SPECT, may be used as a biomarker.

SYMPTOMATIC TRIALS

Clinical trials to investigate new symptomatic therapies are conducted in patients at all stages of disease. For patients with early disease, subjects who are currently receiving no therapy are randomly assigned to receive either the new medication or placebo. In trials involving patients with more advanced disease, the new therapy or placebo is added to the medications the subject is already receiving. Because the effect of symptomatic treatment is maximal after a relatively short period of time, these trials are usually six months or less in duration.

The primary endpoint in symptomatic trials depends on the target symptom that the intervention is intended to treat. In studies in patients with early PD, the primary goal is to reduce the severity of the cardinal features of PD—tremor, rigidity, bradykinesia, as well as postural instability—and the standard measure of disease severity is the Unified Parkinson's Disease Rating Scale (UPDRS). The UPDRS was developed by editing and adopting several existing PD rating scales with new additions.[35–38] It consists of 36 items rated on a 0 to 4 scale covering the cardinal features of PD, and it is divided into three sections assessing mental/behavioral, historical symptoms, and descriptions of the ability to carry out daily activities and physical exam features of the disease. In some cases, a fourth section, assessing complications of therapy, is added. The psychometric properties of the UPDRS have been demonstrated,[39–41] and this scale has been responsive to the effects of various pharmacological interventions.[42–43]

Several studies have assessed the long-term progression of UPDRS scores. In early, untreated disease, the total UPDRS score deteriorates by about 8 to 10 points per year.[44,45] In clinical trials, the rate of progression of UPDRS in untreated subjects appears to accelerate immediately before the initiation of symptomatic therapy.[46] However, this phenomenon may be confined to measurement within a clinical trial. The UPDRS is very sensitive to dopaminergic treatment. After the initiation of treatment, UPDRS scores in the "on" state may remain at or above pretreatment levels for up to five years.[22,47] There are limited data on the long-term performance of the UPDRS in patients with more advanced disease.

In symptomatic trials involving patients with advanced PD, there is more variability in the choice of primary endpoint. Change in overall disease severity is still evaluated in advanced PD using the UPDRS. Motor complications (on-off fluctuations and dyskinesias) are also important target symptoms. Reduction in "off" time can be captured by direct observation over a prolonged period of time but is more often measured by patient diaries in which research subjects record whether they were experiencing an adequate response to medication for each 30-min or 1-hr interval for several days prior to study visits. The other important motor complication, dyskinesia, is measured both in terms of frequency and severity. Frequency is measured by means of diaries in a manner analogous to measuring motor fluctuations. Severity of dyskinesia and interference with normal function are measured by rating scales including the Lang-Fahn Activities of Daily Living Dyskinesia scale[48] and the Modified Goetz Dyskinesia Rating scale.[49] Neuropsychiatric features of PD can be measured with cognitive scales such as the Mattis Dementia Rating Scale[50] or the Alzheimer's Disease Assessment Scale (ADAS-cog)[51] or psychiatric scales such as a modified version of the Brief Psychiatric Rating Scale [52] or the Scale for Assessment of Positive Symptoms.[53]

NEUROPROTECTION TRIALS

Finding approaches that separate short-term, symptomatic effects from more durable neuroprotective effects has been the central challenge for clinical investigators working to develop neuroprotection trials. The methodological efforts have focused on two main fronts: selection of appropriate endpoints, including biomarkers, and development of novel study designs.

There has been considerable debate over the most suitable measure to use as a primary endpoint in neuroprotection trials. Although the UPDRS is commonly used in symptomatic trials to measure changes in disease severity, sensitivity of the UPDRS to the effects of medications that make it less appropriate for measuring changes in the rate of disease progression in trials in which subjects are likely to be receiving symptomatic treatments. Studies looking at disease progression have tended to measure effectiveness in terms of clinical milestones. Clearly, mortality would be the most clinically meaningful milestone. However, progression of PD is gradual enough, with life expectancy at the time of diagnosis of 10 years or more,[54] that measuring mortality is impractical even in a very long trial. Because mortality is not a practical endpoint, a number of other milestones have been used or suggested. In the DATATOP trial, the milestone of interest was development of the need for symptomatic treatment with levodopa. The onset of postural instability, as measured by inability to recover from a retropulsive stress ("pull test"), has been suggested as another measure of disease progression for use in neuroprotection trials.[55] The advantages of this measure are that it is generally considered to be a true clinical measure of progression rather than a surrogate, and that impairment of postural reflexes is relatively refractory to dopaminergic therapy.

Two recent studies[56,57] used the endpoint of time to development of motor complications in comparisons of levodopa to a dopamine agonist as initial monotherapy for patients with early PD. In these studies, the time to development of motor complications was longer in patients randomly assigned to receive the dopamine agonist; however, short-term motor performance, measured by the UPDRS, was better for patients assigned to levodopa. Although time to the milestone of motor complications was slower in patients treated with an agonist, these studies do not necessarily show a "neuroprotective" effect for the agonist relative to levodopa. The reasons for this interpretation are as follows: (1) Motor complications are not necessarily a function only of disease progression, but are also a result of treatment. (2) In these trials, the effect of treatment on the cardinal features of disease (UPDRS) was in the opposite direction of the effect on the incidence of motor complications. (3) Without a placebo control, one cannot be certain whether the effects observed in these trials represent deviations from the natural history of the underlying disease.

Because of these issues, compounds that affect the onset of motor complications but do not necessarily reduce progression of the cardinal signs of PD are described as "disease-modifying" but not neuroprotective.

Imaging biomarkers are increasingly viewed as potential endpoints that may be incorporated in neuroprotection trials. The most commonly used biomarkers are fluorodopa PET[58] and beta-CIT SPECT.[59] Fluorodopa PET measures metabolic (decarboxylase), activity and beta-CIT measures dopamine transporter binding. Both are indirect measures of the integrity of the dopaminergic system. These imaging modalities represent a major technological advance for clinical trials methods, and have the potential to support a neuroprotective effect in the right clinical setting. However, it is important to bear in mind that biomarkers cannot necessarily substitute for clinically meaningful outcomes. To be a valid, they must be subject to the same rigorous psychometric process as other measures (e.g., the UPDRS). Specifically, they must be shown to predict (and not merely be correlated with) important clinical outcomes.

A number of design strategies have been employed to separate potential neuroprotective effects from symptomatic effects. These design techniques include periods of washout of drug at the end of trials and periods of wash-in or delayed start of treatment at the beginning of trials. Washout periods have been used in a number of trials.[20,45,46,60] However, the results of these trials have been difficult to interpret, because the pharmacodynamic time to washout of some PD medications is either very long, as is the case with irreversible enzyme inhibitors, or very difficult to estimate, as is the case with levodopa. In addition, many subjects cannot tolerate prolonged washout from symptomatic medication. The second design option is to delay the initiation of a medication for some subjects and then evaluate all subjects on medication at the end of the trial. This "wash-in" design is often referred to as a *randomized start design* (see Figure 73.1). The idea behind a randomized-start design is that exposure to a neuroprotective compound over a longer period of time will produce greater benefit. Symptomatic effects, if present, will be balanced between groups at the end of the trial. This design has been used in one recent trial[61] that showed a small but statistically detectable difference for the group receiving active treatment for a longer period of time. Current small clinical trials being conducted by NIH for potential neuroprotective drugs are designed as "futility" trials. If the results indicate that there is no statistical chance that the drug will be successful, no further expenditure of time or money will be invested in the agent.

The goal of these designs is to identify a neuroprotective effect in a relatively short period of time. Another approach that will be implemented as part of a long-term effort by the National Institutes of Health (NIH) to identify neuroprotective compounds for PD is the "large simple trial" design.[62] In a large simple trial, relatively brief evaluations are made over a long period of time, possibly up to ten years. The idea behind a large, simple trial is that over time, symptomatic effects will be small relative to the effect of disease progression.

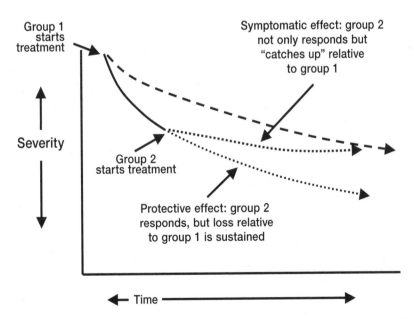

FIGURE 73.1 Schematic illustrating the "randomized-start" trial design. In this design, one group receives active treatment from the time of randomization, and the other group receives placebo initially, resulting in a delay in the initiation of active treatment. At the time of the final evaluation, all subjects are receiving active treatment, and thus the symptomatic effects of treatment are presumably balanced between groups. If differences in function remain at the end of the trial, these differences may be due to something other than a short-term symptomatic effect of treatment. (From Siderowf, A. D. and Stern, M. D., in *Parkinson's Disease: Diagnosis and Current Management,* Demos Medical Publishing, New York, 2002. With permission.)

Appropriate Use of Placebo in Clinical Trials

The appropriate use of placebo in controlled trials has been a topic of intense debate in many areas of medicine, but it is particularly relevant to PD, because it is one of the rare therapeutic areas in which a placebo-controlled surgery trial has been conducted.[6] There are two main settings in which a placebo may not be appropriate: when ethical considerations preclude the use of placebo, and when the research question focuses on helping clinicians choose between competing therapies rather than showing that one or the other is efficacious by itself. This latter type of trial has recently been referred to as a "practical clinical trial."[63]

The primary ethical conflict in conducting placebo-controlled trials is between the imperatives to provide research subjects with at least standard therapy and the imperative to produce the most valid results that will benefit other patients in the future. The imperative to protect research subjects from harm is clearly spelled out in the Declaration of Helsinki, which states that, "In any medical study, every patient, including those of a control group, if any, should be assured of the best proven diagnostic and therapeutic method."[64] It is clear that research participants should not be deprived of standard therapies for extended periods of time. However, some ethicists including Rothman[65] and Macklin[66] have interpreted this statement to mean that no patient should suffer unnecessary risk or be exposed to anything less than standard therapy, even for very brief periods of time. Others[67] point out that, if the declaration of Helsinki is interpreted in this manner, no clinical trial could be conducted. Virtually all trials involve blood tests or even inconvenient additional visits to the study center that present "harms" that would not occur in standard practice.

Placing the emphasis entirely on eliminating excess risk for trial subjects confuses the ethics of clinical research with that of clinical practice. As was described earlier in this chapter, the ethical responsibility of the clinician in clinical practice is only to the patient. In research, the responsibility of the clinician-investigator is both to the individual research subject and to future patients who may benefit from the knowledge gained from a properly conducted trial. This ethical balance is incorporated into the common law,[68] which stresses that institutional review boards should weigh three factors: the benefit to research subjects, the risk to research subjects, and the value of the knowledge to be gained from the research. Of course, it is impossible to know the exact value of the knowledge to be gained until the trial is completed; however, in the case of the trial of fetal cell transplantation, many patients with advanced PD were probably saved from having unnecessary and possibly harmful surgery based on the trial results.

By contrast, there are situations in which the use of an active control instead of placebo is dictated by the research question being studied rather than ethical imperatives. Practical clinical trials, which are a class of investigations that directly address choices between two treatment alternatives faced by decision makers (either physicians or policy makers), illustrate this point. Practical clinical trials are specifically designed to answer questions faced by decision makers (either physicians or policy makers). The characteristic features of practical clinical trials is that they (1) select clinically relevant alternative interventions to compare, (2) include a diverse population of study participants, (3) make use of a variety of practice settings, and (4) collect data across a broad range of health outcomes.

Most practical trials would be considered Phase IV trials. Such trials have been rare in PD; however, recent comparisons between levodopa and dopamine agonists as initial dopaminergic therapy[21,22] are examples of practical clinical trials. Historically, the major funding sources for clinical trials have not focused on practical trials. However, the data from these trials are necessary to guide treatment decisions for individual patients and for quality improvement activities such as formulating practice guidelines.

Quality of Life and Pharmacoeconomic Trials

There has been increasing interest in studying both the global effects of therapy on quality of life in Parkinson's disease and also the economic consequences of treatment given increasingly scarce health care resources.[69–71] Because most therapy that is available for Parkinson's disease presently is symptomatic rather than neuroprotective, quality of life is a logical construct to measure in clinical trials. In fact, several instruments have been validated to measure quality of life in patients with PD, including the PDQ-39[72] and the PDQUALIF.[73] In trials, there has been less focus on quality of life, perhaps because current tools for measuring it appears to be less sensitive to interventions than physiologic measures. Quality of life also may be affected by factors including comorbid medical conditions and a patient's health beliefs and attitudes. To correct these deficiencies, the National Institutes of Health is currently undertaking a large-scale project to develop a quality-of-life instrument to be used in neurological clinical trials.

Likewise, pharmacoeconomic evaluations remain somewhat at the periphery of clinical trials. However, the methodology for conducting such pharmacoeconomic evaluations in conjunction with clinical trials continues to improve.[74] This fact, coupled with the relentless rise in health care expenditures, may lead to greater emphasis on pharmacoeconomic evaluations in future trials.

SUMMARY

The primary role of clinical trials in PD is to provide evidence for physicians to select and effective therapies in their efforts to make the best treatment decisions for their patients. Before recommending a new therapy based on a trial result, however, physicians must first adopt the treatment as part of their overall therapeutic armamentarium and then weigh whether the trial result is applicable to the specific patient. To fulfill these roles, trials must be carefully designed. Several different trial designs can be used to test new treatments for PD. Of these, the most powerful is the parallel-group, randomized, controlled trial. Design of trials to evaluate neuroprotective compounds presents a particular challenge. This challenge centers around the problem of separating symptomatic effects from effects on underlying disease progression. Finding solutions to this problem is a necessary precondition to bringing neuroprotective treatments to patients with PD.

ACKNOWLEDGMENTS

Dr. Siderowf is supported by grant number K-08 HS00004 from the Agency for Healthcare Research and Quality (AHRQ). Dr. Fahn is the recipient of a grant from the Parkinson's Disease Foundation.

REFERENCES

1. Concato, J., Shah, N., Horwitz, R. I., Randomized, controlled trials, observational studies, and the hierarchy of research designs, *N. Engl. J. Med.,* 342:1887–1892, 2000.
2. Benson, K., Hartz, A. J., A comparison of observational studies and randomized, controlled trials, *N. Engl. J. Med.,* 342:1878–1886, 2000.
3. Pocock, S. J., Elbourne, D. R., Randomized trials or observational tribulations? *N. Engl. J. Med.,* 342:1907–1909, 2000.
4. Freeman, T. B., Olanow, C. W., Hauser, R. A. et al., Bilateral fetal nigral transplantation into the postcommissural putamen in Parkinson's disease, *Ann. Neurol.,* 38:379–388, 1995.
5. Lindvall, O., Sawle, G. V., Widner, H., Evidence for long-term survival and function of dopaminergic grafts in progressive Parkinson's disease, *Ann. Neurol.,* 35:172–180, 1994.
6. Freed, C. R., Greene, P. E., Breeze, R. E. et al., Transplantation of embrionic dopamine neurons for severe Parkinson's disease, *N. Engl. J. Med.,* 344:710–719, 2000.
7. Olanow, C. W., Goetz, C. G., Kordower, J. H. et al., A double-blind controlled trial of bilateral fetal nigral transplantation in Parkinson's disease, *Ann. Neurol.,* 54:403–414, 2003.
8. Guyatt, G. H., Cook, D. J., Sackett, D. L., Eckman, M., Pauker, S. G., Grades of recommendation for antithrombotic agents, *Chest,* 114:441S–444S, 1998.
9. Freedman, B., Equipoise and the ethics of clinical research, *N. Engl. J. Med.,* 317:141–145, 1987.
10. Edwards, S. J., Lilford, R. J., Braunholtz, D. A., Jackson, J. C., Hewison, J., Thornton, J., Ethical issues in the design and conduct of randomised controlled trials, *Health Technology Assessment,* 2:1–6, 1998.
11. Berwick, D. M., Disseminating innovations in health care, *JAMA,* 289:1969–1975, 2003.
12. Lenfant, C., Shattuck lecture—clinical research to clinical practice—lost in translation? *N. Engl. J. Med.,* 349:868–874, 2003.
13. Rogers, E. M., *Diffusion of Innovations,* 4th ed., The Free Press, New York, New York, 1995.
14. Simuni, T., Jaggi, J. L., Mulholland, H. et al., Bilateral stimulation of the subthalamic nucleus in patients with Parkinson disease: a study of efficacy and safety, *J. Neurosurg.,* 96:666–672, 2002.
15. Kumar, R., Lozano, A. M., Sime, E., Halket, E., Lang, A. E., Comparative effects of unilateral and bilateral subthalamic nucleus deep brain stimulation, *Neurology,* 53:561–566, 1999.
16. Limousin, P., Krack, P., Pollak, P. et al., Electrical stimulation of the subthalamic nucleus in advanced Parkinson's disease, *N. Engl. J. Med.,* 339:1105–1111, 1998.
17. Parkinson Study Group, Effect of deprenyl on the progression of disability in early Parkinson's disease, *N. Engl. J. Med.,* 321:1364–1371, 1989.
18. Schulzer, M., Mak, E., Calne, D. B., The antiparkinson efficacy of deprenyl derives from transient improvement that is likely to be symptomatic, *Ann. Neurol.,* 32:795–798, 1992.
19. Shoulson, I., Oakes, D., Fahn, S., Lang, A., Langston, J. W., LeWitt, P., Olanow, C. W., Penney, J. B., Tanner, C., Kieburtz, K., Rudolph, A., Parkinson Study Group. Impact of sustained deprenyl (selegiline) in levodopa-treated Parkinson's disease: a randomized placebo-controlled extension of the deprenyl and tocopherol antioxidative therapy of parkinsonism trial, *Ann. Neurol.,* 51(5):604–612, 2002.
20. Fahn, S., Parkinson Study Group, Does levodopa slow or hasten the rate of progression of Parkinson disease? *Neurology,* 60:A80, 2003.
21. Rascol, O., Brooks, D., Korczyn, A. et al., A five-year study of the incidence of dyskinesia in patients with early Parkinson's disease who were treated with ropinirole or levodopa, *N. Engl. J. Med.,* 342:1484–1491, 2000.
22. Parkinson Study Group, Pramipexole vs. Levodopa as initial treatment for Parkinson's disease, *JAMA,* 284:1931–1938, 2000.
23. Coleman, J. S., Katz, E., Menzel, H., The diffusion of an innovation among physicians, *Sociometry,* 20:253–270, 1957.
24. Bucher, H. C., Guyatt, G. H., Cook, D. J., Holbrook, A., McAlister, F. A., Users' guides to the medical literature: XIX. Applying clinical trial results. A. How to use an article measuring the effect of an intervention on surro-

gate end points, Evidence-Based Medicine Working Group, *JAMA,* 282:771–778, 1999.

25. Dans, A. L., Dans, L. F., Guyatt, G. H., Richardson, S., Users' guides to the medical literature: XIV. How to decide on the applicability of clinical trial results to your patient, Evidence-Based Medicine Working Group, *JAMA,* 279:545–549, 1998.

26. Reiser, S. J., Human experimentation and the convergence of medical research and patient care, *Ann. Am. Acad. Polit. Sci.,* 437:18, 1978.

27. Miller, F. G., Rosenstein, D. L., The therapeutic orientation to clinical trials, *N. Engl. J. Med.,* 348:1383–1386, 2003.

28. Grunberg, S. M., Cefalu, W. T., The integral role of clinical research in clinical care, *N. Engl. J. Med.,* 348: 1386–1388, 2003.

29. Braunholtz, D. A., Edwards, S. J., Lilford, R. J., Are randomized clinical trials good for us (in the short term)? Evidence for a "trial effect," *J. Clin. Epidemiol.,* 54:217–224, 2001.

30. Investigational New Drug Application, Code of Federal Regulations 21 Part 312.21. Revised as of 4-1-2004. Washington, D.C., U.S. Government Printing Office, 2004.

31. Guyatt, G. H., Sackett, D. L., Cook, D. J., Users' guides to the medical literature. II. How to use an article about therapy or prevention. B. What were the results and will they help me in caring for my patients? Evidence-Based Medicine Working Group, *JAMA,* 271:59–63, 1994.

32. Pocock, S. J., *Crossover Trials. Clinical Trials: A Practical Approach,* Chichester: John Wiley and Sons, 110–122, 1983.

33. Rubin, D. B., *Multiple Imputation for Nonresponse in Surveys,* John Wiley and Sons, New York, 1987.

34. Little, R. J. A., Rubin, D. B., *Statistical Analysis with Missing Data.* John Wiley and Sons, Inc, New York, New York, 1987.

35. Fahn, S., Elton, R. L., Unified Parkinson's disease rating scale. In: Fahn, S., Marsden, C. D., Calne, D., Goldstein, M., Eds., *Recent Developments in Parkinson's disease.* Macmillan Health Care Information, Florham Park, NJ, 153–164.1987.

36. Canter, C. J., de la Torre, R., Mier, M., A method of evaluating disability in patients with Parkinson's disease, *J. Nervous Ment. Dis.,* 133:143–147, 1961.

37. Duvoisin, R. C., The evaluation of extrapyramidal disease. In: de Ajuriaguerra, J., Gauthier, G., Eds., *Monoamines Noyaux Gris Centraux et Syndrome de Parkinson,* Georg and Cie, Geneva, 313–325, 1971.

38. Webster, D. D., Clinical analysis of the disability in Parkinson's disease, *Modern Treatment,* 5:257–282, 1968.

39. Martinez-Martin, P., Gil-Nagel, A., Gracia, L. M., Gomez, J. B., Martinez-Sarries, J., Bermejo, F., Unified Parkinson's Disease Rating Scale characteristics and structure. The Cooperative Multicentric Group, *Mov. Disord.,* 9:76–83, 1994.

40. van Hilten, J. J., van der Zwan, A. D., Zwinderman, A. H., Roos, R. A., Rating impairment and disability in Parkinson's disease: evaluation of the Unified Parkin-

son's Disease Rating Scale, *Mov. Disord.,* 9:84–88, 1994.

41. McDermott, M. P., Jankovic, J., Carter, J., Fahn, S., Gauthier, S., Goetz, C. G. et al., Factors predictive of the need for levodopa therapy in early, untreated Parkinson's disease, *Arch Neurol.,* 52:565–570, 1995.

42. Parkinson Study Group, Safety and efficacy of pramipexole in early Parkinson's disease, *JAMA,* 278: 125–130, 1997.

43. Parkinson Study Group, A controlled trial of Rasagiline in early Parkinson's disease, *Arch Neurol.,* 59: 1937–1943, 2003.

44. Parkinson Study Group, Effects of tocopherol and deprenyl on the progression of disability in early Parkinson's disease, *N. Engl. J. Med.,* 328:176–183, 1993.

45. Parkinson Study Group, Effect of lazabemide on the progression of disability in early Parkinson's disease, *Ann. Neurol.,* 40:99–107, 1996.

46. LeWitt, P., Oakes, D., Cui, L., and the Parkinson Study Group, The need for levodopa as an endpoint of Parkinson's disease progression in a clinical trial of selegiline and alpha-tocopherol, *Mov. Disord.,* 12:183–89, 1997.

47. Larsen, J. P., Boas, J., Erdal, J. E. and the Norwegian-Danish Study Group, Does selegiline modify the progression of early Parkinson's disease? Results from a five-year study, *Euro. J. Neurol.,* 6:539–547, 1999.

48. Parkinson Study Group, Evaluation of dyskinesias in a pilot, randomized, placebo-controlled trial of remacemide in advanced Parkinson disease, *Arch Neurol.,* 58:1660–1668, 2001.

49. Goetz, C. G., Stebbins, G. T., Shale, H. M. et al., Utility of an objective dyskinesia rating scale for Parkinson's disease: inter- and intrarater reliability assessment, *Mov. Disord.,* 9:390–394, 1994.

50. Dementia Rating Scale, Psychological Assessment Resources, Odessa, FL, 1973.

51. Rosen, W. G., Mohs, R. C., Davis, K. L., A new rating scale for Alzheimer's disease, *Am. J. Psychiatry,* 141:1356–1364, 1984.

52. Overall, J. E., Gorham, D. R., Introduction —The Brief Psychiatric Rating Scale (BPRS): recent developments, *Psychopharmacol. Bull.,* 21:97–99, 1988.

53. Parkinson Study Group, Low-dose clozapine for the treatment of drug-induced psychosis in Parkinson's disease, *N. Engl. J. Med.,* 340: 757–763, 1999.

54. Louis, E. D., Marder, K., Cote, L., Tang, M., Mayeux, R., Mortality from Parkinson disease, *Arch Neurol.,* 54:260–264, 1997.

55. Kieburtz, K., Design of neuroprotection trials in Parkinson's disease, *Ann. Neurol.,* 53: S100–S109, 2003.

56. Whone, A. L., Watts, R. L., Stoessl, A. J., Davis, M., Reske, S., Nahmias, C. et al., Slower progression of Parkinson's disease with ropinirole versus levodopa: The REAL-PET study, *Ann. Neurol.,* 54:93–101, 2003.

57. Parkinson Study Group, Dopamine transporter brain imaging to assess the effects of pramipexole vs. levodopa on Parkinson disease progression, *JAMA,* 287:1653–1661, 2002.

58. Morrish, P. K., Sawle, G. V., Brooks, D. J., Regional changes in [18F]dopa metabolism in the striatum in

Parkinson's disease, *Brain,* 119(Pt 6):2097–2103, 1996.

59. Seibyl, J. P., Marek, K., Sheff, K. et al., Iodine-123-beta-CIT and iodine-123-FPCIT SPECT measurement of dopamine transporters in healthy subjects and Parkinson's patients, *J. Nucl. Med.,* 39:1500–1508, 1998.

60. Olanow, C. W., Hauser, R. A., Gauger, L. et al., The effect of deprenyl and levodopa on the progression of Parkinson's disease, *Ann. Neurol.,* 38:833–834, 1995.

61. Parkinson Study Group, A controlled, randomized, delayed-start study of Rasagiline in early Parkinson's disease, *Arch Neurol.,* 2004 (in press).

62. Ravina, B., Fagan, S., Hart, R. G. et al., Neuroprotective agents for clinical trials in Parkinson's disease: a systematic assessment, *Neurology,* 60:1234–1240, 2003.

63. Tunis, S. R., Stryer, D. B., Clancy, C. M., Practical clinical trials: Increasing the value of clinical research for decision making in clinical and health policy, *JAMA,* 290:1624–1632, 2003.

64. Levine, R. J., The need to revise the Declaration of Helsinki, *N. Engl. J. Med.,* 341:531–534, 1999.

65. Rothman, K. J., Michels, K. B., The continuing unethical use of placebo controls, *New Engl. J. Med.,* 331:394–398, 1994.

66. Macklin, R., The ethical problems with sham surgery in clinical research, *N. Engl. J. Med.,* 341:992–996, 1999.

67. Temple, R. T., Ellenberg, S. S., Placebo-controlled trials and active-control trials in the evaluation of new treatments. Part 1: Ethical and scientific issues, *Ann. Int. Med.,* 133:464–470, 2000.

68. Protection of Human Subjects. Code of Federal Regulations, 45 Part 46.111-119. Revised as of 11-13-2001. Washington, DC, U.S. Government Printing Office, 2001.

69. Palmer, C. S., Nuijten, M. J., Schmier, J. K., Subedi, P., Snyder, E. H., Cost effectiveness of treatment of Parkinson's disease with entacapone in the United States, *Pharmacoeconomics,* 20:617–628, 2002.

70. Davey, P., Rajan, N., Lees, M., Aristides, M., Cost-effectiveness of pergolide compared to bromocriptine in the treatment of Parkinson's disease: a decision-analytic model, *Value in Health,* 4:308–315, 2001.

71. Martinez-Martin, P., An introduction to the concept of "quality of life in Parkinson's disease," *J. Neurol.,* 245:1–6, 1998.

72. Peto, V., Jenkinson, C., Fitzpatrick, R., PDQ-39: a review of the development, validation and application of a Parkinson's disease quality of life questionnaire and its associated measures, *J. Neurol.,* 245:S10–S14, 1998.

73. Welsh, M., McDermott, M. P., Holloway, R. G. et al., Development and testing of the Parkinson's disease quality of life scale, *Mov. Disord.,* 18:637–645, 2003.

74. Drummond, M. F., McGuire, A., *Economic Evaluation in Health Care,* Oxford University Press, London, 2001.

74 Positron Emission Tomography in Parkinson's Disease

John R. Adams and A. Jon Stoessl
University of British Columbia

CONTENTS

HISTORICAL PERSPECTIVE

Positron-emission tomography (PET) has developed into a leading technology for the *in vivo* assessment and investigation of physiological function and pathophysiology. The existence of positrons was initially postulated and confirmed in the early 1930s.[1,2] Subsequent demonstration of artificially producible radioactive atoms and the invention of the cyclotron paved the way for the development of positron-emission tomography.[3,4] Invention of the scintillation radiation detector was quickly followed by the development of the first PET scanners in the late 1950s. The first medical cyclotron was introduced at the Hammersmith Hospital (London, U.K.) (see Ter-Pogossian, M., and Wagner, H., for reviews).[5,6]

The basis of PET derives from the release of a positron (positive electron) from a decaying radiotracer. The subsequent annihilation reaction resulting from collision with an electron results in the production of two photons. These photons, travelling in opposite directions, are subsequently detected and used for source localization via reconstruction algorithms. The result is the PET image. The applicability of PET, in studying physiological and pathophysiological processes, stems from the ability to incorporate radiotracers in to various biological substances. Thus, the fate of such substances can be tracked and information derived regarding the pathways involved. Therefore, the advantage of PET is its ability to study function in contrast to more conventional imaging technologies that focus on structure. Recent and ongoing technological advances, along with novel ideas, have broadened the scope of usefulness and feasibility of PET, allowing a predominantly research-oriented tool to be increasingly applicable clinically.

GENERAL PET USAGE

CARDIOLOGY

The role of PET as a diagnostic tool has been expanding over recent years, with the technique having particular prominence in the fields of cardiology, oncology, and neurology. Cardiac patients have benefited from the application of PET technology in a variety of areas. PET has been used in the study of coronary artery disease via determinations of myocardial blood flow using oxygen-15 (^{15}O)

labeled water or nitrogen-13 (^{13}N) labeled ammonia. By studying fluorodeoxyglucose (FDG) myocardial uptake, PET can determine myocardial metabolic activity by quantifying glucose utilization. Using this technique, ratios of myocardial metabolism to perfusion, or the relation to akinetic segments on echocardiography, have been employed to identify hibernating myocardium amenable to revascularization. Neurocardiological autonomic function can also be examined with PET as demonstrated in the study of congestive heart failure and following cardiac transplantation. For a more detailed discussion on the role of PET in cardiology, readers are referred to a recent review by Camici.[7]

ONCOLOGY

The indications for PET in oncology have recently been reviewed by Bomanji et al.[8] Suggested primary indications to date have included preoperative staging of non-small-cell carcinoma, staging of recurrent lymphoma and colorectal cancer, and evaluation of stage II melanoma and solitary pulmonary nodules. Secondary indications have included staging of head and neck and recurrent breast cancer, grading of brain tumors, and determining tumor recurrence versus scar or tissue necrosis. Additional roles for PET in oncology currently being investigated include the evaluation of response to therapy, gene expression, and the pharmacodyanmics and pharmacokinetics of various drugs.[9–11]

NEUROLOGY

Neuro-oncology, cerebrovascular disorders, dementia, and epilepsy represent several of the neurological subdisciplines in which PET has been applied.[12] PET has been used for differentiating intracranial tumor recurrence from postradiation changes; differentiating CNS lymphoma, toxoplasmosis, and other infections in immunocompromised patients; and detecting cerebrovascular perfusion abnormalities in stroke and migraine. Other uses have included the evaluation of dementia based on patterns of reduced cerebral blood flow (CBF) or regional glucose metabolism, epileptic foci identification, and the investigation of cortical visual loss. In the subspecialty of movement disorders, PET has been used in the evaluation of blepharospasm, dystonia, and prominently in the akinetic-rigid syndromes, most notably Parkinson's disease (PD).

PET AND PARKINSON'S DISEASE

PRESYNAPTIC NIGROSTRIATAL IMAGING: ^{18}F-DOPA

By labeling various biological substrates with radiotracers, the integrity and function of various *in vivo* biochemical systems or pathways can be assessed. 6-^{18}F-fluoro-L-dopa (^{18}F-dopa) has until relatively recently been the gold standard technique for the *in vivo* assessment of the nigrostriatal dopaminergic pathway.[13] ^{18}F-dopa activity reflects uptake of the radiolabeled dopamine (DA) precursor, its subsequent decarboxylation to fluorodopamine (FDA) by L-aromatic amino acid decarboxylase (AAAD), and storage within presynaptic vesicles. In established Parkinson's disease, striatal ^{18}F-dopa activity is reduced compared to age-matched normal controls indicating a presynaptic dopaminergic nerve terminal deficit.[14,15] Consistently, the pattern of reduced striatal uptake follows a rostro-caudal gradient, with the putamen predominantly affected contralateral to the patient's more severely afflicted side.[16] This distribution is consistent with postmortem neurochemical findings.[17] Comparative analysis of ^{18}F-dopa uptake in clinically affected and unaffected striatum estimates a range of 47 to 62% of normal uptake as the threshold for the development of clinical symptoms.[18]

^{18}F-dopa uptake has been shown to correlate well with postmortem nigral dopamine cell counts in humans and nonhuman primates, suggesting a role for this tracer as a marker for the severity of nigral injury.[19,20] Using N-methyl-4-phenyl-1,2,3,6-tetrahydropyridine (MPTP)-treated monkeys, Yee and colleagues recently used PET to evaluate the relationship of ^{18}F-dopa uptake and decarboxylation with nigral cell counts and striatal concentrations of dopamine and its metabolites.[21] This study demonstrated excellent correlation between ^{18}F-dopa uptake and decarboxylation with striatal levels of DA and its metabolites. However, nigral cell counts did not correlate with ^{18}F-dopa uptake/decarboxylation or striatal DA levels, although a weak correlation between ^{18}F-dopa uptake/decarboxylation and nigral levels of DA and its metabolites was seen. The discrepancy between this study and the postmortem findings noted above has been attributed to the assessment of monkeys with only a minor degree of cell loss. A stronger correlation with striatal DA levels was predicted, if a broader range of dopaminergic cell loss had been examined.

Caution is required in interpreting the results of long scans using ^{18}F-dopa. Studies longer than two hours have demonstrated declining tissue activity as a function of FDA turnover, reflecting egress from synaptic vesicles and metabolic conversion to nontrapped metabolites.[22] However, changes reflecting the metabolism of ^{18}F-dopa can be used to advantage in the estimation of DA turnover, as demonstrated in MPTP-treated monkeys and PD patients. In early MPTP-treated monkeys, DA turnover was increased.[23] Using a four-hour ^{18}F-dopa study protocol, Sossi et al. demonstrated a similar increase in DA turnover in early PD patients.[24] The authors postulated that the increase in DA turnover was a compensatory mechanism delaying the development of clinically apparent disease.

Presynaptic Nigrostriatal Imaging: DAT and VMAT2

Nigrostriatal dopaminergic nerve terminal integrity can also be evaluated using the presynaptic dopamine membrane (DAT) and vesicular monoamine (VMAT2) transporters. Numerous radiotracers, most commonly using ^{123}I, have been developed for use in single-photon emission-computed tomography (SPECT) studies of the DAT.[25] Recently, the more readily available tracer technitium-99 (^{99m}Tc) has been employed as ^{99m}Tc-TRODAT-1 for DAT imaging.[26] The usefulness of ^{99m}Tc-TRODAT-1 has been demonstrated in hemiparkinsonian PD patients who showed reduced uptake contralateral to the affected side with most profound reduction in the posterior striatum.[27] Furthermore, less markedly reduced uptake was also demonstrated in the putamen contralateral to the clinically asymptomatic side. The observed reductions in ^{99m}Tc-TRODAT-1 DAT binding have demonstrated good correlation with symptom severity in PD patients when assessed by the Hoehn and Yahr and Unified Parkinson's Disease Rating Scales (UPDRS).[28,29] ^{99m}Tc-TRODAT-1, when studied in idiopathic PD patients and subjects with vascular parkinsonism, was capable of differentiating the two disorders; striatal binding was significantly lower and more asymmetric in the idiopathic PD patient population.[30] Comparative studies of ^{99m}Tc-TRODAT-1 SPECT and ^{18}F-dopa PET in clinically diagnosed PD patients and in the 6-hydroxydopamine (6-OHDA)-lesioned monkey PD model have revealed similar patterns of reduced striatal uptake.[31,32]

Several tropanes (cocaine analogs) have been introduced for PET study of the DAT, including the carbon-labeled tracers ^{11}C-WIN 35,428 (CFT) and ^{11}C-RTI-32,[33–36] and the fluorinated tracers ^{18}F-CFT, ^{18}F-FPCIT, and ^{18}F-FECNT.[37–39] ^{18}F-FECNT exhibits much higher binding affinity for DAT than for the serotonin transporter, excellent kinetics, and the highest peak uptake striatum-to-cerebellum ratios when compared to other fluorine-labeled DAT ligands.[37] The nontropane analog [11]C-*d-threo*-methylphenidate (MP) also has excellent kinetics and a high degree of selectivity for the DAT as compared to other monoamine transporters.[40]

These DAT tracers have shown similar patterns of reduced striatal uptake as seen with ^{18}F-dopa in PD patients.[18,33,34–36,39] Using MP, a threshold of 29 to 44% of normal uptake has been estimated for symptom manifestation in PD.[18] A similar estimate was provided by Ma et al. using ^{18}F-FPCIT.[38] Despite numerous potential ligands for the DAT, there are limitations to its study, and caution must be used when interpreting the results. This largely reflects the susceptibility of the DAT to regulatory and compensatory mechanisms, potentially making its use less appealing for studying DA nerve terminal integrity.

In contrast to the DAT, the VMAT2 does not appear to be as susceptible to pharmacological modulation, particularly by dopaminergic compounds.[41,42] However, ovarian steroids appear to reduce VMAT2 mRNA expression and receptor density in the rat striatum.[43] This apparent steroid effect notwithstanding, ^{11}C-dihydrotetrabenazine (DBTZ) binding to VMAT2 may more specifically identify monoaminergic nerve-terminal loss. Unfortunately, the VMAT2 is not specific for dopaminergic neurons; it provides a similar role in vesicular packaging of monoamines in serotonergic and noradrenergic nerve terminals. As a consequence, one cannot distinguish between the different monoaminergic terminals using VMAT2 binding. However, as the majority of VMAT2-related activity in the striatum derives from dopaminergic terminals, it remains a useful marker of nigrostriatal nerve terminal integrity, relatively free of the uncertainty presented by the effects of modulation. An asymmetric reduction, with a similar rostro-caudal gradient as seen in ^{18}F-dopa studies, of striatal ^{11}C-DTBZ binding has been demonstrated in Parkinson's disease patients. Reduction of ^{11}C-DBTZ uptake to 38 to 48% of normal levels has been estimated to correspond to symptom threshold.[44,18]

Postsynaptic Nigrostriatal Imaging: Dopamine Receptors

[11]C-raclopride acts as a competitive inhibitor of dopamine binding at the postsynaptic type-2 dopamine (D_2) receptor. Increased ^{11}C-raclopride binding has been demonstrated in Parkinson's disease patients with more prominent uptake contralateral to the clinically more severely affected side.[45,46] In a study of untreated PD patients, Sawle and colleagues correlated the highest levels of putaminal ^{11}C-raclopride uptake with the most severely reduced ^{18}F-dopa binding.[47] They attributed their findings to greater post-synaptic D_2 receptor up-regulation in response to more severely damaged substantia nigra. By demonstrating similarly increased uptake of the reversible ligand ^{11}C-raclopride and the irreversible D_2 ligand ^{11}C-N-methylspiperone in PD patients, Kaasinen et al. corroborated the view that increased ^{11}C-raclopride uptake was due to post-synaptic receptor up-regulation, rather than simply a manifestation of reduced receptor occupancy by endogenous dopamine (see below).[48] These findings are also in keeping with earlier postmortem studies demonstrating an initial up-regulation of D_2 receptors.[49] Even as symptoms worsen, the up-regulation of D_2 receptors appears to persist, with increased ^{11}C-raclopride binding still present on repeat investigation six months from baseline in untreated PD patients.[50]

In a longitudinal ^{11}C-raclopride PET longitudinal study, the effect of L-dopa and the dopamine agonist, lisuride, on D_2 receptors was studied in previously untreated PD patients. Three to four months of oral L-dopa therapy resulted in no change in ^{11}C-raclopride binding. Similarly, no change was elicited with an acute intravenous L-dopa infusion. Oral lisuride therapy resulted in a nonsignificant trend to reduced ^{11}C-raclopride uptake that reversed within days of drug withdrawal. Intravenous lisuride, in a healthy rhesus monkey, resulted in reduced striatal ^{11}C-raclopride uptake. The authors proposed a blocking effect on ^{11}C-raclopride binding at D_2 receptors by lisuride and suggested that D_2 receptor density was not affected by oral therapy with L-dopa or lisuride.[51] However, in the same patients, a follow-up study three to five years later demonstrated a significant reduction in the previously increased (at baseline) putaminal ^{11}C-raclopride binding levels to within the normal, age-matched control range. Furthermore, whereas baseline caudate uptake had been normal, it was reduced at repeat testing. The authors postulated a down-regulation of postsynaptic D_2 receptors due to chronic dopaminergic therapy or to postsynaptic structural adaptation of the dopaminergic system.[52] In an interim cross-sectional study of PD patients ranging from early to advanced disease, Antonini demonstrated a reduction of putaminal ^{11}C-raclopride binding, toward normal levels, in more advanced patients.[53] These findings mirrored previous postmortem studies that demonstrated a return of up-regulated D_2 receptors to normal levels following dopaminergic therapy.[49] Thus, PET imaging in PD patients using ^{11}C-raclopride has been able to demonstrate an early up-regulation of D_2 receptors within the striatum that reflects the severity of nigral injury. Furthermore, the postsynaptic D_2 receptor appears to be down-regulated by dopaminergic therapy.

[11]C-raclopride, although highly specific for D_2/D_3 receptors, has a comparatively low affinity, and consequently its binding is subject to competition from endogenous dopamine.[54,55] By exploiting this feature, ^{11}C-raclopride binding studies can be used to estimate changes in dopamine release. In a study of a single, intravenous L-dopa infusion in PD patients, there was a significant reduction of ^{11}C-raclopride binding, as compared to baseline, that was more pronounced in the posterior putamen and correlated with drug-free disability. The authors suggested an accelerated dopamine turnover in dopamine-depleted striatal tissue based on a positive correlation between striatal dopaminergic nerve-terminal deficiency and the capacity for levodopa to increase synaptic dopamine and consequently block ^{11}C-raclopride binding.[56] De la Fuente-Fernandez et al. used ^{11}C-raclopride binding in a similar capacity to demonstrate rapid DA turnover in PD patients who later go on to develop motor fluctuations compared to those who maintain a stable response.[57] By stimulating dopamine release with amphetamine and observing reduced ^{11}C-raclopride binding, in combination with demonstrating reduced ^{18}F-dopa uptake, Piccini et al. correlated dopamine release with reduced dopamine storage capacity in PD patients. Furthermore, significant dopamine release could still be stimulated in advanced PD patients.[58] Restoration of neurochemical function following fetal mesencephalic transplantation for Parkinson's has been associated with increased ^{18}F-dopa uptake and by a concurrent restoration of amphetamine-induced reductions in ^{11}C-raclopride binding.[59]

Exogenously administered drugs acting on the dopaminergic system will also compete for ^{11}C-raclopride binding to the D_2 receptor. Application of this feature has been used to estimate dopamine receptor occupancy by neuroleptic medications[60] and has been used to demonstrate reductions in ^{11}C-raclopride binding in L-dopa-treated PD patients after treatment with pergolide.[61] However, ^{11}C-raclopride studies must be interpreted with caution, given that ^{11}C-raclopride binding reflects not only direct receptor occupancy but also changes in occupancy by endogenous dopamine. In stable PD patients, de la Fuente-Fernandez et al. used a novel method for measuring synaptic dopamine levels to demonstrate reduced dopamine release following administration of subcutaneous apomorphine. They proposed this response was mediated by activation of presynaptic D_2/D_3 autoreceptors.[62] Dopamine release, as demonstrated by reduced ^{11}C-raclopride binding, has also been reported in healthy controls and PD patients during prelearned sequential motor tasks, in response to placebo administration and with the expectation of reward.[63–65] It is clear that ^{11}C-raclopride binding studies provide an important component to the study of the nigro-striatal pathway, but study planning and interpretation must be undertaken with care to avoid misleading results.

Cerebral Blood Flow and Metabolism in Parkinson's Disease

PET activation studies have provided insight into the broader effects of nigrostriatal dysfunction throughout the cerebral hemispheres, thus shedding light on clinicopathological correlation. Playford et al. used PET with C1502 to study regional cerebral blood flow following activation by simple, freely chosen motor tasks in PD patients and age-matched controls.[66] For the controls, they demonstrated activation of the left primary sensorimotor cortex, left premotor cortex, left putamen, right dorsolateral prefrontal cortex and supplementary motor area, anterior cingulate area, and parietal association areas bilaterally. In the PD group, compared with the rest condition, there was impaired activation of the contralateral putamen, the anterior cingulate, supplementary motor area, and dorsolateral prefrontal cortex. Normal activation in the left sensorimotor and premotor cortices was observed. They suggested that impaired activation of medial frontal areas accounted for the difficulties PD patients have in movement initiation. A companion study examining the effects of subcutaneous apomorphine infusion revealed no effect on focal or global cerebral blood flow under resting conditions. However, activation of the supplementary motor area significantly improved in association with reversal of akinesia by apomorphine.[67] Similar improvement has been demonstrated following pallidotomy and with subthalamic nucleus (STN) stimulation.[68–70] Recent studies have suggested a regional pattern of differential activation within the SMA of PD patients. Thobois et al. demonstrated reduced activation within the rostral SMA (pre-SMA) with normal levels of activation seen within the caudal SMA (SMA proper).[70] Using a task designed to robustly activate the SMA, Cunnington et al. used motor imagery in a PET study of PD patients to reveal a similar pattern.[71] The impaired activation of SMA and DLPFC has been postulated as the clinicopathological correlate of akinesia and as a consequence of abnormal pallidal output. In a study of sustained, unilateral high-frequency vibratory stimulation, Boecker et al. have demonstrated abnormalities within the somatosensory system of PD patients.[72] Reduced rCBF activation was seen in the contralateral primary sensorimotor, secondary sensory, posterior cingulate, lateral premotor cortex, and the contralateral basal ganglia and bilateral prefrontal cortex in response to the sensory stimulus. Furthermore, ipsilateral activation was apparent in the primary sensory, secondary sensory, and insular cortex. Thus, they provided evidence of impaired somatosensory processing in PD patients, including a possible ipsilateral compensatory component not seen in normal controls.

Compensatory overactivation of lateral premotor and inferolateral parietal regions has been demonstrated in a PET study of simple sequenced finger movements in PD patients.[73] The longer the motor sequences the greater the compensatory changes according to a study by Catalan et al.[74] This overactivity in parietal and premotor regions may underlie the facilitation of movement initiation by external cues seen in PD patients.

Using ^{18}F-fluorodeoxglucose (FDG) and a form of principal-components analysis, Eidelberg and colleagues have developed an approach to identify a Parkinson's disease-related pattern (PDRP) of covariance.[75] In PD patients, the pattern has been characterized by relative hypometabolism of lateral premotor cortex, supplementary motor cortex, the dorsolateral prefrontal cortex, and the parietooccipital association areas. Furthermore, hypermetabolic covariance was demonstrated in pallidal, thalamic, and pontine regions. This pattern implicates cerebral regions responsible for the planning and performance of voluntary movement, which have demonstrated functional abnormalities in Parkinson's disease (see previous section). The PDRP has been demonstrated in early PD patients.[76] The pattern has consistently shown positive correlation with motor disability and disease severity.[77] Application of this technique permitted differentiation of early PD patients from normal controls, PD patients from

those with striatonigral degeneration, and atypical parkinsonism in general.[77–79]

FAMILIAL PARKINSON'S DISEASE

Reduced ^{18}F-dopa uptake, reduced binding of the DAT and VMAT2, and normal to increased ^{11}C-raclopride binding within the striatum are consistent with the nigrostriatal injury of established sporadic Parkinson's disease. Similar findings have been demonstrated in familial Parkinson's disease patients. Thobois et al. studied young-onset Parkinson's disease patients with and without *parkin* mutations in comparison with normal age-matched controls using ^{18}F-dopa. Although ^{18}F-dopa uptake was reduced in the PD patients with the characteristic rostro-caudal gradient and asymmetry, there was no significant difference between those with and those without the *parkin* mutation.[80] These findings have been documented in other studies.[81,82] However, some studies have suggested findings different from sporadic PD. Reduced striatal ^{11}C-raclopride uptake has been demonstrated in PD patients with *parkin* mutations in comparison to asymptomatic family members, sporadic idiopathic PD patients, and controls, suggesting postsynaptic dysfunction in this group.[83] In a study of two brothers with *parkin* mutations, Portman and colleagues demonstrated significantly reduced striatal ^{18}F-dopa uptake with the caudate more severely affected than expected in sporadic PD.[84] The apparent excess in caudate involvement was attributed to particularly long disease duration rather than as a novel finding in the familial group. PET has also been used to suggest slower disease progression in *parkin*-related PD than in idiopathic PD.[85] In studies of familial Parkinson's disease related to mutations within the α-*synuclein* gene, the striatal PET findings have been similar to idiopathic disease.[86,87] Finally, in a pallido-ponto-nigral degeneration family, ^{18}F-dopa uptake was much more severely affected than in idiopathic PD patients.[88] In contrast, clinically affected patients with PARK 6 mutation have been shown to have reductions in striatal ^{18}F-dopa similar to those of idiopathic PD patients.[89] The similarities and differences detected on PET studies of familial and sporadic PD patients may reflect the phenotypic similarities and differences seen clinically and pathologically. The further study of these patient groups using PET will provide a better understanding of the underlying pathophysiological mechanisms that contribute to the various syndromes.

EARLY DETECTION AND PROGRESSION

Of particular interest to patients, investigators, and pharmaceutical companies is the possibility of halting the process of degeneration and clinical deterioration. Effective neuroprotective intervention has been elusive to date, despite promising animal and *in vitro* experimental evidence.[90–93] Selegiline, a monoamine oxidase B inhibitor, was initially employed as a neuroprotective agent after the DATATOP study suggested delayed disease progression.[94] Further analysis revealed that selegiline has mild anti-parkinsonian properties that may have accounted for the apparent neuroprotective effect.[95] Furthermore, increased mortality rates have been reported with this agent, although subsequent meta-analysis failed to confirm this.[96,97] The early detection of nigrostriatal degeneration, preferably prior to symptom onset, is desirable if the pathway of degeneration is to be thwarted. It is likely that the false positive result arising from the initial analysis of the DATATOP study could have been prevented had neuroimaging been used as a biomarker of disease progression. However, more recent studies have suffered from the opposite problem. Claims have been made with respect to differential effects of L-dopa and dopamine agonists on disease progression based on imaging, but there was no clinical correlate. In the REAL-PET study, early PD patients not previously treated with L-dopa or dopamine agonist were randomized to therapy with carbidopa/levodopa or ropinirole in a two-year study of neuroprotection. The primary outcome measure was putaminal ^{18}F-dopa uptake as an estimate of loss of nigrostriatal dopaminergic nerve-terminal function. Secondary outcomes included the motor component of the UPDRS and the Clinical Global Impression score (CGI). In the ropinirole group, there was a 13.4% reduction in ^{18}F-dopa uptake compared with 20.3% in the L-dopa group at two years. Despite this apparent reduction in loss of dopaminergic function, the UPDRS motor score increased by 0.7 for the ropinirole group compared to a reduction of 5.6 in the L-dopa group.[98] In the CALM-PD study, pramipexole was compared to L-dopa in a randomized, blinded study evaluating changes in DAT binding using ^{123}I-β-CIT SPECT. Again, as with ropinirole, there was a reduced rate of decline of the imaging marker over the course of the study with pramipexole versus L-dopa. However, clinical severity scores were worse in the dopamine agonist group than in the L-dopa group at two years, and similar at four years.[99] Thus, in both of these studies, there is a discrepancy between imaging and clinical findings, emphasizing the need for caution when interpreting studies with incongruous results.

Reduced ^{18}F-dopa activity has been demonstrated in asymptomatic individuals exposed to MPTP and in asymptomatic concordant twins of Parkinson's disease patients.[100–102] A portion of these patients went on to develop symptoms, suggesting that reduced ^{18}F-dopa activity can be seen prior to symptom onset. At-risk, asymptomatic relatives of familial PD patients with proven *parkin* mutations have demonstrated reduced ^{18}F-dopa activity when compared to controls.[85] Some of these patients also went on to develop symptoms. Finally, studies of hemiparkinsonian patients have revealed reduced ^{18}F-dopa binding both contralateral and ipsilateral to the

affected side.[103] DAT ligand binding is also reduced in early PD patients. ^{11}C-WIN[35,428] (CFT) uptake was reduced in Hoehn and Yahr stage II PD patients and ^{11}C-RTI-32 reduced with a rostro-caudal gradient in dopa-naive patients.[104,33] Similar findings have been seen in more recent studies using ^{18}F-FPCIT and ^{18}F-CFT.[38,105,106] A recent study using ^{18}F-FECNT showed reduced DAT binding in striatum contralateral to the unaffected side in early PD patients suggesting a role for this ligand in identifying presymptomatic disease.[37] DBTZ has similarly shown early binding reduction in patients with PD.[44,18] Comparing [76]Br-FE-CBT and ^{18}F-dopa, Ribeiro et al. demonstrated greater reduction in [76]Br-FE-CBT binding than ^{18}F-dopa in putamen of early PD patients with no difference in binding between the two radioligands in advanced PD patients.[107] These findings suggest up-regulation of ^{18}F-dopa uptake or, conversely, down-regulation of DAT. Motor dysfunction correlated with ^{18}F-dopa uptake but not [76]Br-FE-CBT binding. Lee et al. have studied the compensatory presynaptic changes in nigrostriatal neurons in Parkinson's disease using ^{18}F-dopa, ^{11}C-DTBZ, and ^{11}C-MP. Putaminal ^{18}F-dopa uptake was reduced in all subgroups of PD patients. Significantly greater reduction was seen in Hoehn and Yahr stages 2 and 3 versus stage 1 patients and in treated versus drug-naïve patients. Analogous findings were seen with ^{11}C-DTBZ and ^{11}C-MP. The ratio of normalized ^{18}F-dopa to ^{11}C-DTBZ was increased in the putamina of early (Hoehn and Yahr stage 1) and drug-treated patients when compared with normal controls. ^{11}C-MP/^{11}C-DTBZ ratios were significantly lower in the intermediate and caudal putamen in PD patients than in normal controls. The findings suggest compensatory changes in the presynaptic dopaminergic striatal nerve terminals in response to cell loss. These changes include up-regulation of aromatic acid decarboxylase and down-regulation of the dopamine transporter in an attempt to maintain adequate striatal dopamine levels.[18] Sossi et al., using prolonged (four-hour) ^{18}F-dopa PET scans and employing a graphical method to determine effective dopamine turnover, demonstrated increased dopamine turnover as an earlier change in PD patients than reduced ^{18}F-dopa uptake and storage.[24] Compensatory mechanisms may also involve nigro-pallidal projections. Whone and colleagues used ^{18}F-dopa PET in early PD patients to demonstrate increased ^{18}F-dopa uptake in the internal segment of the globus pallidus (GPi). They proposed a modulating effect on the pattern of pallidal output to ventral striatum and motor cortex. The increased binding was not seen in PD patients with advanced disease.[108]

Postsynaptic changes in early, untreated PD patients and normal controls were studied by Rinne et al. using ^{11}C-raclopride. They demonstrated an increase in the number of D$_2$ receptors in PD patients when compared to normal age-matched controls and a 33% increase in the number of putaminal D_2 receptors contralateral to the predominantly affected side within the PD patient group. No side-to-side difference was seen in the caudate.[109] In a preliminary study, van Netten et al. studied compensatory changes for nigro-striatal cell loss in members of a family with autosomal dominantly inherited Parkinson's disease. Using ^{18}F-dopa, DTBZ, and MP, they demonstrated findings similar to established idiopathic PD in an affected member. A recently diagnosed member showed ^{18}F-dopa uptake, DTBZ binding, and MP values of 73, 51, and 40% of normal, respectively. In an asymptomatic member, they recorded DTBZ and MP values of 54 and 35% of normal, respectively. However, ^{18}F-dopa values were considered within the normal range at 90% of normal.[110] Thus, increased dopamine turnover, down-regulated DAT activity, increased D_2 receptors, and augmented aromatic decarboxylase activity, as detected by PET, appear to be the earliest markers of nigrostriatal dysfunction.

The pathophysiological changes associated with progression of Parkinson's disease have also been investigated using SPECT and PET. Using PET, ^{18}F-dopa uptake has been shown to decline as disease progresses. Morrish et al. estimated an annual rate of decline of 4.7% for ^{18}F-dopa uptake in the putamen of PD patients using a graphical analysis method when repeated PET scans were performed over an 18-month period. Ratio analysis revealed no significant changes compared to controls in this study. They suggested multiple time graphical analysis of ^{18}F-dopa uptake as a more sensitive method than ratio analysis for detecting progression.[111] In a five-year study, Nurmi et al. estimated reduced ^{18}F-dopa uptake of 10.5, 8.3, and 5.9% per year for the posterior putamen, anterior putamen, and caudate, respectively. They also estimated a 6.5-year preclinical period of detectable ^{18}F-dopa uptake reduction for the posterior putamen. Thus, they suggested PD first affects the posterior putamen, then the anterior putamen, and subsequently the caudate nucleus, with stable absolute rates of decline thereafter.[112] In a cross-sectional study of ^{18}F-dopa uptake in normal controls and PD patients looking at subregions of the striatum, the dorsal putamen appeared to be affected most severely followed by ventro-caudal and then ventro-rostral putaminal regions.[113]

Studies using tracers for the DAT have demonstrated similar findings to ^{18}F-dopa, as discussed earlier. Subregional analysis of the striatum, using PD patients and normal controls with scans an average of 2.2 years apart, showed annual rates of decline for ^{18}F-CFT binding of 5.6, 3.3, and 5.6% for the anterior putamen, posterior putamen, and caudate, respectively. At baseline, binding was most markedly reduced in the posterior putamen at 27% of normal, with the anterior putamen and caudate at 45 and 71% of normal, respectively. The absolute rate of decline of DAT binding was greater ipsilateral to the most severely affected side. The findings were suggestive of a

slower progression in more severely affected regions.[39] However, despite a study suggesting that ^{18}F-CFT binding was unchanged after three months of L-dopa therapy, other studies have suggested caution in using DAT tracers in studies of disease progression, due to potential modulation of the DAT by pharmaceutical intervention.[114,115]

In a recent cross-sectional ^{11}C-DTBZ PET study of progression, Lee et al. evaluated the regional selectivity of dopamine nerve terminal loss. they reported a greater extent of terminal loss in posterior versus anterior putamen between normal controls and early PD (effectively the preclinical period); whereas the extent of loss of ^{11}C-DTBZ BP was similar between anterior and posterior putamen once disease had become clinically manifest. A biphasic kinetic model of progression was suggested with regional selectivity explained by an initial causative factor and subsequent secondary mechanisms, common to multiple degenerative conditions, responsible for ongoing cell death.[157]

As discussed in a previous section, elevated ^{11}C-raclopride binding persisted after six months in untreated PD patients.[50] In a longer-duration study of treated PD patients, a return to normal levels for ^{11}C-raclopride binding was demonstrated after three to five years. The effects may have been due to chronic dopaminergic therapy or postsynaptic structural adaptation of the dopaminergic system.[52]

In summary, functional imaging studies of Parkinson's disease have been able to demonstrate a focal, asymmetric pathophysiological process affecting only the dorsal putamen in its earliest (and preclinical) phase. Thereafter, ventral putamen and caudate dysfunction can be detected with the less severely affected regions progressing at a greater rate. The disease process extends beyond the basal ganglia and is not restricted to the motor pathways, reflecting the diverse spectrum of clinical abnormalities characteristic of PD. Furthermore, various compensatory patterns of cerebral activation suggesting alternate pathway recruitment for task completion are apparent. Differences between sporadic and familial cases have provided insight into the pathogenesis of these disorders; however, the familial disorders as a rule behave in a fashion very similar to sporadic disease. A greater understanding of the compensatory mechanisms at a cellular level has resulted from study of early and asymptomatic cases. As early markers of nigrostriatal dysfunction become established and consistent patterns of progression are elucidated, studies of neuroprotective interventions will become increasing feasible. As with many investigational tools, interpretation of results must be undertaken with caution. Since PD patients are unlikely to remain without treatment for substantial periods, studies using tracers for the DAT, raclopride, and ^{18}F-dopa, which may be susceptible to modulation by pharmaceutical intervention, will be limited in their interpretability.

PET AND DIFFERENTIAL DIAGNOSIS

Clinical diagnostic accuracy for Parkinson's disease has been estimated to have a predictive value of approximately 90% using a variety of diagnostic criteria. Sensitivity of these various criteria, applied retrospectively, ranged from 67 to 90%.[116] An adjunctive diagnostic tool would be welcome to guide discussions regarding prognosis and expected therapeutic efficacy, and to permit research opportunities. Ultrasonography, MRI, SPECT, and PET have all been investigated as diagnostic imaging tools.

Ultrasonography, detecting hyperechogenic substantia nigra (SN), has been proposed as such a test.[158] Hyperechogenicity of the SN has been documented in PD patients and postulated to result from increased iron deposition.[117–119] However, up to 9% of healthy adults and 23% of patients with non-PD cerebral disorders demonstrate hyperechogenicity of the SN, thereby reducing its sensitivity and specificity for PD.[120,121] Furthermore, ultrasound can be limited by technical factors such as inadequate bone windows.[122]

MRI has also been employed in the differential diagnosis of atypical parkinsonian syndromes and has recently been reviewed.[123] Previously, MRI in idiopathic Parkinson's disease was generally accepted as normal, although more recent studies have demonstrated abnormalities in up to 67% of cases.[124] Various manipulations of MRI technique, including modification of slice thickness, image weighting, and volumetry, have been used to identify PD and differentiate other conditions such as multiple system atrophy (MSA) and progressive supranuclear palsy (PSP).[124–129,159] Differentiation of PSP from MSA remains a challenge with significant overlap of MRI findings. However, recently, Asato et al. have suggested that antero-posterior measurements of rostral and caudal midbrain tegmentum may distinguish the two conditions.[130] Magnetization transfer[160] and diffusion tensor imaging techniques[161,162] have recently demonstrated potential for the differential diagnosis of parkinsonism. Despite promising results, MRI studies have not yet been able to consistently demonstrate the ability to discriminate atypical parkinsonian syndromes.

SPECT and PET have also been investigated as tools for the differential diagnosis of parkinsonism. The advantage of these techniques lies in the functional information generated from pathophysiological alterations in different neurochemical pathways. Readers are referred to the chapter on SPECT imaging elsewhere in this text for further discussion of this modality.

[18]F-dopa PET does not reliably differentiate idiopathic PD from MSA. Reduced striatal ^{18}F-dopa has been demonstrated in both conditions.[126,131,132] ^{18}F-dopa and S-^{11}C-nomifensine (NMF), a DA reuptake blocker, were employed to detect presynaptic dopaminergic differences between 8 patients with IPD, 10 patients with MSA-P, 7

with pure autonomic failure (PAF), and 13 normal controls. Whereas both MSA-P and PD subjects had decreased striatal ^{18}F-dopa and NMF uptake, only PD patients showed a rostro-caudal gradient, while MSA-P displayed a more homogeneous pattern of caudate-putamen abnormality.[133] Similar findings were reported using ^{18}F-dopa by Brooks et al.[134] However, the rostro-caudal gradient classically associated with PD has also been seen in MSA, with overlapping caudate to putamen uptake ratios seen in some cases.[131,132] Raclopride binding to the postsynaptic D_2 receptor has been shown to be reduced in MSA and PSP patients and may help differentiate these patients from those with PD but not from each other.[45,126,131] Glucose metabolism, as studied by FDG PET, may differentiate MSA from PD with reduced striatal values demonstrated in the former.[131] Antonini and collaborators used FDG PET in 56 individuals with idiopathic Parkinson's disease (IPD) and 48 patients with nonspecific atypical parkinsonism (APD), as defined by autonomic failure (orthostatic hypotension) or poor response to levodopa. Significant abnormalities of glucose metabolism were demonstrated in the striatum and thalamus of 75% of the APD patients.[78] However, in another study using FDG, despite different overall patterns, the values for glucose metabolism, particularly within the putamen, overlapped between MSA and PD patients, thus reducing the specificity of this technique.[132] The study of networks of covariant glucose metabolism may increase the specificity and diagnostic utility of FDG PET.[75,76] Autonomic dysfunction in MSA has been related to brain stem hypoperfusion findings in ^{18}F-FDG PET.[135] Cardiac PET, using 6-[18]Fluorodopamine or N-methyl-^{11}C-meta-hydroxyephedrine, to study cardiac innervation has shown reduced myocardial radioactivity in PD with orthostatic hypotension but not in MSA, making it a potentially useful diagnostic test.[136,137]

PET studies of PSP have demonstrated reduced striatal ^{18}F-dopa uptake and a global reduction in cortical cerebral blood flow more pronounced in the superior frontal lobes.[45,138–143] This pattern of hypometabolism was considered a reliable marker for PSP, but such findings have not been consistently replicated.[144] Furthermore, whereas no FDG abnormalities are generally expected in idiopathic PD patients, at least one SPECT study using Tc99m-HMPAO suggested a similar degree of left frontal hypoperfusion in PSP and PD patients.[145] A recent study by Juh et al. suggested that application of statistical parametric mapping analysis to ^{18}F-FDG PET may help differentiate parkinsonian disorders.[163]

In corticobasal degeneration (CBD), the asymmetry of parkinsonian—mainly rigidity and akinesia—and parietal cortical findings is rarely reflected as asymmetric contralateral atrophy on structural imaging. On the other hand, PET studies on cerebral activation show striking asymmetry of blood flow reduction, particularly in inferior parietal, posterior frontal, and superior temporal areas, as well as thalamus and striatum.[146] Early changes studied in six individuals with CBD included hypoperfusion of the frontal (premotor, primary motor, supplementary motor, and prefrontal cortices) and parietal lobes (primary sensory and associative cortices), with sparing of the occipital lobes as usually expected in the disease. Striatal reduction in ^{18}F-dopa uptake has also been demonstrated.[147]

Changes in cerebral glucose metabolism profile have been studied in several additional dementia subtypes, many of which may manifest parkinsonism and present a diagnostic dilemma. An ability to differentiate PSP from PD with dementia (PDD) using cerebral blood flow and PET was demonstrated by Karbe and colleagues. In this study, PD patients with no cognitive impairment had normal cerebral blood flow, those with PDD were similar to Alzheimer's disease (AD) patients, and patients with PSP showed significantly reduced metabolic activity in frontal and subcortical regions.[141] Turjanski et al. demonstrated a pattern of fronto-parieto-temporal hypometabolism for patients with dementia with Lewy bodies (DLB) similar to that seen for patients with AD and PDD. They suggested this pattern could differentiate these clinical entities from PSP and CBD, but not each other.[148] In another study with specific attention paid to posterior cortical metabolic activity, PDD patients demonstrated a greater degree of visual cortex hypometabolism when compared to AD patients matched for dementia severity.[149] Occipital hypometabolism may also differentiate DLB from AD patients with the former showing reduced activity.[150] These findings were pathologically confirmed and may well be pathognomonic. Thus, specific regional analysis of cerebral metabolic activity with attention to posterior cortical areas may provide a means of differentiating AD from clinically similar Lewy body disorders.

Several additional disease entities that may present diagnostic difficulty have been studied using PET. Dopa-responsive dystonia (DRD) patients have demonstrated normal striatal ^{18}F-dopa uptake in contrast to the reduced uptake seen in early PD patients presenting with dystonia.[20] An uncommon, late-onset parkinsonism clinical phenotype of DRD may also be differentiated from idiopathic PD using ^{18}F-dopa PET.[151] Przedborski et al. used FDG and [18]fluoroethylspiperone (FESP) to study glucose metabolism and striatal postsynaptic D_2 receptor integrity, respectively, in a single case of hemiparkinsonism-hemiatrophy syndrome (HP-HA). They demonstrated reduced metabolism within the basal ganglia and, to lesser extent, in the fronto-parietal cortex contralateral to the clinically involved side. This was in contrast to idiopathic PD in which no abnormalities were detected. FESP binding in the HP-HA case and PD patient controls remained normal. Thus, they proposed FDG PET as a potential diagnostic tool to differentiate HP-HA from idiopathic PD.[152]

In an ^{18}F-dopa study of a 76-year-old woman with long-standing postencephalitic parkinsonism, a rostro-caudal gradient of reduced uptake similar to idiopathic PD was seen.[153] Reduced ^{18}F-dopa uptake with a rostro-caudal gradient has also been reported for patients with acute postencephalitic parkinsonism.[154] In contrast, Ghaemi et al. studied a 74-year-old woman with parkinsonism consequent to acute viral encephalitis and reported ^{18}F-dopa and blood flow studies different from those seen in idiopathic PD patients.[155] Finally, post-traumatic parkinsonism, when evaluated with ^{18}F-dopa and compared to PD, showed decreased striatal uptake with putamen and caudate equally affected, disclosing a probable distinct pathophysiological process.[156]

CONCLUSION

Functional imaging with PET is an invaluable investigational tool for understanding the pathophysiology, detecting at-risk individuals, and monitoring disease progression in Parkinson's disease. Although PET imaging is beginning to find a home in clinical medicine in a variety of disciplines, its application in Parkinson's disease and related disorders remains predominantly that of a research tool. As a diagnostic tool for Parkinson's disease, PET, using a variety of tracers, can generally separate dopamine deficiency states from an intact nigrostriatal system. However, while the use of a combination of tracers to assess both pre- and postsynaptic nigrostriatal integrity and cerebral glucose utilization may narrow the diagnostic possibilities, PET has not generally been able to reliably differentiate between the different parkinsonian syndromes. Furthermore, cost and the limited availability of scanners and radiotracers severely curtail accessibility. Both PET and SPECT have been reported normal in slightly more than 10% of patients who carry a clinical diagnosis of early PD, and while this could at least partially reflect misdiagnosis, the full meaning of this phenomenon is not yet clear.[98] Findings on PET may be susceptible to pharmacological intervention, making interpretation difficult. With our growing understanding, potential interventions will become available, necessitating the need to make timely and accurate diagnoses. In the meantime, clinical judgement remains the standard for premortem diagnosis. PET will continue to play a valuable role in understanding the pathophysiology of disease and complications of therapy.

REFERENCES

1. Dirac, P. A. M., A theory of electrons and protons, *Proc. Roy. Soc,* A126, 360, 1930.
2. Anderson, C. D., Energies of cosmic-ray particles, *Phys. Rev,* 40, 405, 1932.

3. Curie, I. and Joliot, F., Artificial production of a new kind of radioactive element, *Nature,* 133, 201, 1934.

4. Lawrence, E. O. and Livingston, M. S., Production of high speed light ions without the use of high voltage, *Phys. Rev,* 40, 19, 1932.

5. Ter-Pogossian, M. M., The origins of positron emission tomography, *Sem. Nucl. Med.,* 22, 140, 1992.

6. Wagner, H. N., A brief history of positron emission tomography (PET), *Sem. Nucl. Med.,* 28, 213, 1998.

7. Camici, P. G., Positron emission tomography and myocardial imaging, *Heart,* 83, 475, 2000.

8. Bomanji, J. B., Costa, D. C., and Ell, P. J., Clinical role of positron emission tomography in oncology, *Lancet Oncol.,* 2, 157, 2001.

9. Blasberg, R., PET imaging of gene expression, *Eur. J. Cancer,* 38, 2137, 2002.

10. Cohade, C. and Wahl, R. L., PET scanning and measuring the impact of treatment, *Cancer J.,* 8, 119, 2002.

11. Gupta, N., Price, P. M., and Aboagye, E. O., PET for *in vivo* pharmacokinetic and pharmacodynamic measurements, *Eur. J. Cancer,* 38, 2094, 2002.

12. Blake, P., Johnson, B., and VanMeter, J. W., Positron emission tomography (PET) and single photon emission computed tomography (SPECT): clinical applications, *J. Neuroophthal.,* 23, 34, 2003.

13. Martin, W. R. et al., Nigrostriatal function in humans studied with positron emission tomography, *Ann. Neurol.,* 26, 535, 1989.

14. Nahmias, C. et al., Striatal dopamine distribution in parkinsonian patients during life, *J. Neurol. Sci.,* 69, 223, 1985.

15. Leenders, K. L. et al., Brain dopamine metabolism in patients with Parkinson's disease measured with positron emission tomography, *J. Neurol. Neurosurg. Psychiatry,* 49, 853, 1986.

16. Garnett, E. S. et al., A rostrocaudal gradient for aromatic acid decarboxylase in the human striatum, *Can. J. Neurol. Sci.,* 14, 444, 1987.

17. Kish, S. J., Shannak, K., and Hornykiewicz, O., Uneven pattern of dopamine loss in the striatum of patients with idiopathic Parkinson's disease. Pathophysiologic and clinical implications, *N. Engl. J. Med.,* 318, 876, 1988.

18. Lee, C. S. et al., In vivo positron emission tomographic evidence for compensatory changes in presynaptic dopaminergic nerve terminals in Parkinson's disease, *Ann. Neurol.,* 47, 493, 2000.

19. Pate, B. D. et al., Correlation of striatal fluorodopa uptake in the MPTP monkey with dopaminergic indices, *Ann. Neurol.,* 34, 331, 1993.

20. Snow, B. J. et al., Positron emission tomographic studies of dopa-responsive dystonia and early-onset idiopathic parkinsonism, *Ann. Neurol.,* 34, 733, 1993.

21. Yee, R. E. et al., Novel observations with FDOPA-PET imaging after early nigrostriatal damage, *Mov. Disord,* 16, 838, 2001.

22. Holden, J. E. et al., Graphical analysis of 6-fluoro-L-dopa trapping: effect of inhibition of catechol-O-methyltransferase, *J. Nucl. Med.,* 38, 1568, 1997.

23. Doudet, D. J. et al., 6-[18F]Fluoro-L-DOPA PET studies of the turnover of dopamine in MPTP-induced parkinsonism in monkeys, *Synapse,* 29, 225, 1998.

24. Sossi, V. et al., Increase in dopamine turnover occurs early in Parkinson's disease: evidence from a new modeling approach to PET 18 F-fluorodopa data, *J. Cereb. Blood Flow Metab.,* 22, 232, 2002.

25. Marek, K., Jennings, D., and Seibyl, J., Single-photon emission tomography and dopamine transporter imaging in Parkinson's disease, *Adv. Neurol.,* 91, 183, 2003.

26. Kung, H. F. et al., Imaging of dopamine transporters in humans with technetium-99m TRODAT-1, *Eur. J. Nucl. Med.,* 23, 1527, 1996.

27. Mozley, P. D. et al., Binding of [99mTc]TRODAT-1 to dopamine transporters in patients with Parkinson's disease and in healthy volunteers, *J. Nucl. Med.,* 41, 584, 2000.

28. Bao, S. Y. et al., Imaging of dopamine transporters with technetium-99m TRODAT-1 and single photon emission computed tomography, *J. Neuroimaging,* 10, 200, 2000.

29. Huang, W. S. et al., Evaluation of early-stage Parkinson's disease with 99mTc-TRODAT-1 imaging, *J. Nucl. Med.,* 42, 1303, 2001.

30. Tzen, K. Y. et al., Differential diagnosis of Parkinson's disease and vascular parkinsonism by (99m)Tc-TRODAT-1, *J. Nucl. Med.,* 42, 408, 2001.

31. Huang, W. S. et al., Crossover study of (99m)Tc-TRODAT-1 SPECT and (18)F-FDOPA PET in Parkinson's disease patients, *J. Nucl. Med.,* 44, 999, 2003.

32. Huang, W. S. et al., 99mTc-TRODAT-1 SPECT in healthy and 6-OHDA lesioned parkinsonian monkeys: comparison with 18F-FDOPA PET, *Nucl. Med. Commun.,* 24, 77, 2003.

33. Guttman, M. et al., [11C]RTI-32 PET studies of the dopamine transporter in early dopa-naive Parkinson's disease: implications for the symptomatic threshold, *Neurology,* 48, 1578, 1997.

34. Ouchi, Y. et al., Alterations in binding site density of dopamine transporter in the striatum, orbitofrontal cortex, and amygdala in early Parkinson's disease: compartment analysis for beta-CFT binding with positron emission tomography, *Ann. Neurol.,* 45, 601, 1999.

35. Ouchi, Y. et al., Presynaptic and postsynaptic dopaminergic binding densities in the nigrostriatal and mesocortical systems in early Parkinson's disease: a double-tracer positron emission tomography study, *Ann. Neurol.,* 46, 723, 1999.

36. Ouchi, Y. et al., Changes in dopamine availability in the nigrostriatal and mesocortical dopaminergic systems by gait in Parkinson's disease, *Brain,* 124, 784, 2001.

37. Davis, M. R. et al., Initial human PET imaging studies with the dopamine transporter ligand 18F-FECNT, *J. Nucl. Med.,* 44, 855, 2003.

38. Ma, Y. et al., Parametric mapping of [18F]FPCIT binding in early stage Parkinson's disease: a PET study, *Synapse,* 45, 125, 2002.

39. Nurmi, E. et al., Progression of dopaminergic hypofunction in striatal subregions in Parkinson's disease using [18F]CFT PET, *Synapse,* 48, 109, 2003.

40. Ding, Y. S. et al., Pharmacokinetics and *in vivo* specificity of [11C]dl-threo-methylphenidate for the presynaptic dopaminergic neuron, *Synapse,* 18, 152, 1994.

41. Vander Borght, T. M. et al., The vesicular monoamine transporter is not regulated by dopaminergic drug treatments, *Eur. J Pharmacol.,* 294, 577, 1995.

42. Kilbourn, M. R. et al., Effects of dopaminergic drug treatments on *in vivo* radioligand binding to brain vesicular monoamine transporters, *Nucl. Med. Biol.,* 23, 467, 1996.

43. Rehavi, M. et al., Regulation of rat brain vesicular monoamine transporter by chronic treatment with ovarian hormones, *Brain Res. Mol. Brain Res.,* 57, 31, 1998.

44. Frey, K. A. et al., Presynaptic monoaminergic vesicles in Parkinson's disease and normal aging, *Ann. Neurol.,* 40, 873, 1996.

45. Brooks, D. J. et al., Striatal D_2 receptor status in patients with Parkinson's disease, striatonigral degeneration, and progressive supranuclear palsy, measured with 11C-raclopride and positron emission tomography, *Ann. Neurol.,* 31, 184, 1992.

46. Rinne, U. K. et al., Positron emission tomography demonstrates dopamine D_2 receptor supersensitivity in the striatum of patients with early Parkinson's disease, *Mov. Disord.,* 5, 55, 1990.

47. Sawle, G. V. et al., Asymmetrical pre-synaptic and post-synaptic changes in the striatal dopamine projection in dopa naive parkinsonism. Diagnostic implications of the D_2 receptor status, *Brain,* 116 (Pt. 4), 853, 1993.

48. Kaasinen, V. et al., Upregulation of putaminal dopamine D_2 receptors in early Parkinson's disease: a comparative PET study with [11C] raclopride and [11C]N-methyl-spiperone, *J. Nucl. Med.,* 41, 65, 2000.

49. Guttman, M. et al., Dopamine D_2 receptor density remains constant in treated Parkinson's disease, *Ann. Neurol.,* 19, 487, 1986.

50. Rinne, J. O. et al., PET study on striatal dopamine D_2 receptor changes during the progression of early Parkinson's disease, *Mov. Disord.,* 8, 134, 1993.

51. Antonini, A. et al., [11C]raclopride and positron emission tomography in previously untreated patients with Parkinson's disease: Influence of L-dopa and lisuride therapy on striatal dopamine D_2-receptors, *Neurology,* 44, 1325, 1994.

52. Antonini, A. et al., Long-term changes of striatal dopamine D_2 receptors in patients with Parkinson's disease: a study with positron emission tomography and [11C]raclopride, *Mov. Disord.,* 12, 33, 1997.

53. Antonini, A. et al., Complementary positron emission tomographic studies of the striatal dopaminergic system in Parkinson's disease, *Arch Neurol.,* 52, 1183, 1995.

54. Dewey, S. L. et al., Striatal binding of the PET ligand 11C-raclopride is altered by drugs that modify synaptic dopamine levels, *Synapse,* 13, 350, 1993.

55. Seeman, P., Guan, H. C., and Niznik, H. B., Endogenous dopamine lowers the dopamine D_2 receptor density as measured by [3H]raclopride: implications for positron emission tomography of the human brain, *Synapse,* 3, 96, 1989.

56. Tedroff, J. et al., Levodopa-induced changes in synaptic dopamine in patients with Parkinson's disease as measured by [11C]raclopride displacement and PET, *Neurology,* 46, 1430, 1996.

57. Fuente-Fernandez, R. et al., Biochemical variations in the synaptic level of dopamine precede motor fluctuations in Parkinson's disease: PET evidence of increased dopamine turnover, *Ann. Neurol.,* 49, 298, 2001.

58. Piccini, P., Pavese, N., and Brooks, D. J., Endogenous dopamine release after pharmacological challenges in Parkinson's disease, *Ann. Neurol.,* 53, 647, 2003.

59. Piccini, P. et al., Dopamine release from nigral transplants visualized *in vivo* in a Parkinson's patient, *Nat. Neurosci.,* 2, 1137, 1999.

60. Kapur, S., Zipursky, R. B., and Remington, G., Clinical and theoretical implications of 5-HT2 and D_2 receptor occupancy of clozapine, risperidone, and olanzapine in schizophrenia, *Am. J. Psychiatry,* 156, 286, 1999.

61. Linazasoro, G. et al., Modification of dopamine D_2 receptor activity by pergolide in Parkinson's disease: an *in vivo* study by PET, *Clin. Neuropharmacol.,* 22, 277, 1999.

62. Fuente-Fernandez, R. et al., Apomorphine-induced changes in synaptic dopamine levels: positron emission tomography evidence for presynaptic inhibition, *J. Cereb. Blood Flow Metab.,* 21, 1151, 2001.

63. Goerendt, I. K. et al., Dopamine release during sequential finger movements in health and Parkinson's disease: a PET study, *Brain,* 126, 312, 2003.

64. Fuente-Fernandez, R. et al., Expectation and dopamine release: mechanism of the placebo effect in Parkinson's disease, *Science,* 293, 1164, 2001.

65. Fuente-Fernandez, R. et al., Dopamine release in human ventral striatum and expectation of reward, *Behav. Brain Res.,* 136, 359, 002.

66. Playford, E. D. et al., Impaired mesial frontal and putamen activation in Parkinson's disease: a positron emission tomography study, *Ann. Neurol.,* 32, 151, 1992.

67. Jenkins, I. H. et al., Impaired activation of the supplementary motor area in Parkinson's disease is reversed when akinesia is treated with apomorphine, *Ann. Neurol.,* 32, 749, 1992.

68. Ceballos-Baumann, A. O. et al., A positron emission tomographic study of subthalamic nucleus stimulation in Parkinson disease: enhanced movement-related activity of motor-association cortex and decreased motor cortex resting activity, *Arch Neurol.,* 56, 997, 1999.

69. Samuel, M. et al., Pallidotomy in Parkinson's disease increases supplementary motor area and prefrontal activation during performance of volitional movements an H2(15)O PET study, *Brain,* 120 (Pt. 8), 1301, 1997.

70. Thobois, S. et al., Effects of subthalamic nucleus stimulation on actual and imagined movement in Parkinson's disease: a PET study, *J. Neurol.,* 249, 1689, 2002.

71. Cunnington, R. et al., Motor imagery in Parkinson's disease: a PET study, *Mov. Disord.,* 16, 849, 2001.

72. Boecker, H. et al., Sensory processing in Parkinson's and Huntington's disease: investigations with 3D H(2)(15)O-PET, *Brain,* 122 (Pt. 9), 1651, 1999.

73. Samuel, M. et al., Evidence for lateral premotor and parietal overactivity in Parkinson's disease during

sequential and bimanual movements. A PET study, *Brain,* 120 (Pt. 6), 963, 1997.

74. Catalan, M. J. et al., A PET study of sequential finger movements of varying length in patients with Parkinson's disease, *Brain,* 122 (Pt. 3), 483, 1999.

75. Eidelberg, D. et al., The metabolic topography of parkinsonism, *J. Cereb. Blood Flow Metab.,* 14, 783, 1994.

76. Eidelberg, D. et al., Early differential diagnosis of Parkinson's disease with 18F-fluorodeoxyglucose and positron emission tomography, *Neurology,* 45, 1995.

77. Eidelberg, D. et al., Assessment of disease severity in parkinsonism with fluorine-18-fluorodeoxyglucose and PET, *J. Nucl. Med.,* 36, 378, 1995.

78. Antonini, A. et al., Differential diagnosis of parkinsonism with [18F]fluorodeoxyglucose and PET, *Mov. Disord,* 13, 268, 1998.

79. Eidelberg, D. et al., Striatal hypometabolism distinguishes striatonigral degeneration from Parkinson's disease, *Ann. Neurol.,* 33, 518, 1993.

80. Thobois, S. et al., Young-onset Parkinson disease with and without parkin gene mutations: a fluorodopa F 18 positron emission tomography study, *Arch. Neurol.,* 60, 713, 2003.

81. Broussolle, E. et al., [18 F]-dopa PET study in patients with juvenile-onset PD and parkin gene mutations, *Neurology,* 55, 877, 2000.

82. Hilker, R. et al., The striatal dopaminergic deficit is dependent on the number of mutant alleles in a family with mutations in the parkin gene: evidence for enzymatic parkin function in humans, *Neurosci. Lett.,* 323, 50, 2002.

83. Hilker, R. et al., Positron emission tomographic analysis of the nigrostriatal dopaminergic system in familial parkinsonism associated with mutations in the parkin gene, *Ann. Neurol.,* 49, 367, 2001.

84. Portman, A. T. et al., The nigrostriatal dopaminergic system in familial early onset parkinsonism with parkin mutations, *Neurology,* 56, 1759, 2001.

85. Khan, N. L. et al., Progression of nigrostriatal dysfunction in a parkin kindred: an [18F]dopa PET and clinical study, *Brain,* 125, 2248, 2002.

86. Kruger, R. et al., Familial parkinsonism with synuclein pathology: clinical and PET studies of A30P mutation carriers, *Neurology,* 56, 1355, 2001.

87. Samii, A. et al., PET studies of parkinsonism associated with mutation in the alpha-synuclein gene, *Neurology,* 53, 2097, 1999.

88. Pal, P. K. et al., Positron emission tomography in pallido-ponto-nigral degeneration (PPND) family (frontotemporal dementia with parkinsonism linked to chromosome 17 and point mutation in tau gene), *Parkinsonism Relat. Disord.,* 7, 81, 2001.

89. Khan, N. L. et al., Clinical and subclinical dopaminergic dysfunction in PARK6-linked parkinsonism: an 18F-dopa PET study, *Ann. Neurol.,* 52, 849, 2002.

90. Cohen, G. et al., Pargyline and deprenyl prevent the neurotoxicity of 1-methyl-4-phenyl-1,2,3,6-tetrahydropyridine (MPTP) in monkeys, *Eur. J. Pharmacol.,* 106, 209, 1984.

91. Collier, T. J. et al., Cellular models to study dopaminergic injury responses, *Ann. N.Y. Acad. Sci.,* 991, 140, 2003.

92. Heikkila, R. E. et al., Protection against the dopaminergic neurotoxicity of 1-methyl-4-phenyl-1,2,5,6-tetrahydropyridine by monoamine oxidase inhibitors, *Nature,* 311, 467, 1984.

93. Olanow, C. W., Selegiline: current perspectives on issues related to neuroprotection and mortality, *Neurology,* 47, S210, 1996.

94. Parkinson Study Group, Effects of tocopherol and deprenyl on the progression of disability in early Parkinson's disease. The Parkinson Study Group, *N. Engl. J Med.,* 328, 176, 1993.

95. Olanow, C. W. and Calne, D., Does selegiline monotherapy in Parkinson's disease act by symptomatic or protective mechanisms? *Neurology,* 42, 13, 1992.

96. Lees, A. J., Comparison of therapeutic effects and mortality data of levodopa and levodopa combined with selegiline in patients with early, mild Parkinson's disease. Parkinson's Disease Research Group of the United Kingdom, *B. M. J.,* 311, 1602, 1995.

97. Olanow, C. W. et al., Effect of selegiline on mortality in patients with Parkinson's disease: a meta-analysis, *Neurology,* 51, 825, 1998.

98. Whone, A. L. et al., Slower progression of Parkinson's disease with ropinirole versus levodopa: The REAL-PET study, *Ann. Neurol.,* 54, 93, 2003.

99. Parkinson Study Group, Dopamine transporter brain imaging to assess the effects of pramipexole vs. levodopa on Parkinson disease progression, *JAMA,* 287, 1653, 2002.

100. Burn, D. J. et al., Parkinson's disease in twins studied with 18F-dopa and positron emission tomography, *Neurology,* 42, 1894, 1992.

101. Calne, D. B. et al., Positron emission tomography after MPTP: observations relating to the cause of Parkinson's disease, *Nature,* 317, 246, 1985.

102. Laihinen, A. et al., Risk for Parkinson's disease: twin studies for the detection of asymptomatic subjects using [18F]6-fluorodopa PET, *J. Neurol.,* 247 Suppl. 2, II110, 2000.

103. Nagasawa, H. et al., 6-[18F]fluorodopa metabolism in patients with hemiparkinsonism studied by positron emission tomography, *J. Neurol. Sci.,* 115, 136, 1993.

104. Frost, J. J. et al., Positron emission tomographic imaging of the dopamine transporter with 11C-WIN 35,428 reveals marked declines in mild Parkinson's disease, *Ann. Neurol.,* 34, 423, 1993.

105. Nurmi, E. et al., Progression in Parkinson's disease: a positron emission tomography study with a dopamine transporter ligand [18F]CFT, *Ann. Neurol.,* 47, 804, 2000.

106. Rinne, J. O. et al., Striatal uptake of a novel PET ligand, [18F]beta-CFT, is reduced in early Parkinson's disease, *Synapse,* 31, 119, 1999.

107. Ribeiro, M. J. et al., Dopaminergic function and dopamine transporter binding assessed with positron emission tomography in Parkinson disease, *Arch Neurol.,* 59, 580, 2002.

108. Whone, A. L. et al., Plasticity of the nigropallidal pathway in Parkinson's disease, *Ann. Neurol.,* 53, 206, 2003.

109. Rinne, J. O. et al., Increased density of dopamine D$_2$ receptors in the putamen, but not in the caudate nucleus in early Parkinson's disease: a PET study with [11C]raclopride, *J. Neurol. Sci.,* 132, 156, 1995.

110. van Netten, H. G. et al., Compensatory mechanisms in the dopamine system of patients with dominantly inherited parkinsonism, *Ann. Neurol.,* 54, S19,2 003.*abstract*

111. Morrish, P. K. et al., Measuring the rate of progression and estimating the preclinical period of Parkinson's disease with [18F]dopa PET, *J. Neurol. Neurosurg. Psychiatry,* 64, 314, 1998.

112. Nurmi, E. et al., Rate of progression in Parkinson's disease: a 6-[18F]fluoro-L-dopa PET study, *Mov. Disord.,* 16, 608, 2001.

113. Morrish, P. K., Sawle, G. V., and Brooks, D. J., Regional changes in [18F]dopa metabolism in the striatum in Parkinson's disease, *Brain,* 119 (Pt. 6), 2097, 1996.

114. Guttman, M. et al., Influence of L-dopa and pramipexole on striatal dopamine transporter in early PD, *Neurology,* 56, 1559, 2001.

115. Nurmi, E. et al., Reproducibility and effect of levodopa on dopamine transporter function measurements: a [18F]CFT PET study, *J. Cereb. Blood Flow Metab.,* 20, 1604, 2000.

116. Hughes, A. J., Daniel, S. E., and Lees, A. J., Improved accuracy of clinical diagnosis of Lewy body Parkinson's disease, *Neurology,* 57, 1497, 2001.

117. Becker, G. et al., Degeneration of substantia nigra in chronic Parkinson's disease visualized by transcranial color-coded real-time sonography, *Neurology,* 45, 182, 1995.

118. Berg, D., Siefker, C., and Becker, G., Echogenicity of the substantia nigra in Parkinson's disease and its relation to clinical findings, *J. Neurol.,* 248, 684, 2001.

119. Berg, D. et al., Echogenicity of the substantia nigra: association with increased iron content and marker for susceptibility to nigrostriatal injury, *Arch Neurol.,* 59, 999, 2002.

120. Berg, D. et al., Vulnerability of the nigrostriatal system as detected by transcranial ultrasound, *Neurology,* 53, 1026, 1999.

121. Walter, U. et al., Substantia nigra echogenicity is normal in non-extrapyramidal cerebral disorders but increased in Parkinson's disease, *J. Neural. Transm.,* 109, 191, 2002.

122. Berg, D. et al., Relationship of substantia nigra echogenicity and motor function in elderly subjects, *Neurology,* 56, 13, 2001.

123. Savoiardo, M., Differential diagnosis of Parkinson's disease and atypical parkinsonian disorders by magnetic resonance imaging, *Neurol. Sci.,* 24 Suppl. 1, S35, 2003.

124. Bhattacharya, K. et al., Brain magnetic resonance imaging in multiple-system atrophy and Parkinson disease: a diagnostic algorithm, *Arch Neurol.,* 59, 835, 2002.

125. Righini, A. et al., Thin section MR study of the basal ganglia in the differential diagnosis between striatonigral degeneration and Parkinson disease, *J. Comput. Assist. Tomogr.,* 26, 266, 2002.

126. Ghaemi, M. et al., Differentiating multiple system atrophy from Parkinson's disease: contribution of striatal and midbrain MRI volumetry and multi-tracer PET imaging, *J. Neurol. Neurosurg. Psychiatry,* 73, 517, 2002.

127. Schulz, J. B. et al., Magnetic resonance imaging-based volumetry differentiates idiopathic Parkinson's syndrome from multiple system atrophy and progressive supranuclear palsy, *Ann. Neurol.,* 45, 65, 1999.

128. Seppi, K. et al., Diffusion-weighted imaging discriminates progressive supranuclear palsy from PD, but not from the parkinson variant of multiple system atrophy, *Neurology,* 60, 922, 2003.

129. Warmuth-Metz, M. et al., Measurement of the midbrain diameter on routine magnetic resonance imaging: a simple and accurate method of differentiating between Parkinson disease and progressive supranuclear palsy, *Arch Neurol.,* 58, 1076, 2001.

130. Asato, R. et al., Magnetic resonance imaging distinguishes progressive supranuclear palsy from multiple system atrophy, *J. Neural. Transm.,* 107, 1427, 2000.

131. Antonini, A. et al., Complementary PET studies of striatal neuronal function in the differential diagnosis between multiple system atrophy and Parkinson's disease, *Brain,* 120 (Pt. 12), 2187, 1997.

132. Otsuka, M. et al., Differentiating between multiple system atrophy and Parkinson's disease by positron emission tomography with 18F-dopa and 18F-FDG, *Ann. Nucl. Med.,* 11, 251, 1997.

133. Brooks, D. J. et al., The relationship between locomotor disability, autonomic dysfunction, and the integrity of the striatal dopaminergic system in patients with multiple system atrophy, pure autonomic failure, and Parkinson's disease, studied with PET, *Brain,* 113 (Pt. 5), 1539, 1990.

134. Brooks, D. J. et al., Differing patterns of striatal 18F-dopa uptake in Parkinson's disease, multiple system atrophy, and progressive supranuclear palsy, *Ann. Neurol.,* 28, 547, 1990.

135. Taniwaki, T. et al., Cerebral metabolic changes in early multiple system atrophy: a PET study, *J. Neurol. Sci.,* 200, 79, 2002.

136. Berding, G. et al., [N-methyl 11C]meta-Hydroxyephedrine positron emission tomography in Parkinson's disease and multiple system atrophy, *Eur. J. Nucl. Med. Mol. Imaging,* 30, 127, 2003.

137. Goldstein, D. S. et al., Sympathetic cardioneuropathy in dysautonomias, *N. Engl. J Med.,* 336, 696, 1997.

138. Bhatt, M. H. et al., Positron emission tomography in progressive supranuclear palsy, *Arch Neurol.,* 48, 389, 1991.

139. Foster, N. L. et al., Cerebral hypometabolism in progressive supranuclear palsy studied with positron emission tomography, *Ann. Neurol.,* 24, 399, 1988.

140. Goffinet, A. M. et al., Positron tomography demonstrates frontal lobe hypometabolism in progressive supranuclear palsy, *Ann. Neurol.,* 25, 131, 1989.

141. Karbe, H. et al., Positron emission tomography in degenerative disorders of the dopaminergic system, *J. Neural. Transm. Park Dis. Dement. Sect.,* 4, 121, 1992.

142. Piccini, P. et al., Familial progressive supranuclear palsy: detection of subclinical cases using 18F-dopa and 18fluorodeoxyglucose positron emission tomography, *Arch Neurol.*, 58, 1846, 2001.

143. Taniwaki, T. et al. Positron emission tomography (PET) in "pure akinesia," J. Neurol. Sci., 107, 34, 1992.

144. Santens, P. et al., Cerebral oxygen metabolism in patients with progressive supranuclear palsy: a positron emission tomography study, *Eur. Neurol.*, 37, 18, 1997.

145. Defebvre, L. et al., Tomographic measurements of regional cerebral blood flow in progressive supranuclear palsy and Parkinson's disease, *Acta Neurol. Scand.*, 92, 235, 1995.

146. Brooks, D. J., Functional imaging studies in corticobasal degeneration, *Adv. Neurol.*, 82, 209, 2000.

147. Laureys, S. et al., Fluorodopa uptake and glucose metabolism in early stages of corticobasal degeneration, *J. Neurol.*, 246, 1151, 1999.

148. Turjanski, N. and Brooks, D. J., PET and the investigation of dementia in the parkinsonian patient, *J. Neural. Transm.*, Suppl., 51, 37, 1997.

149. Vander Borght T. M. et al., Cerebral metabolic differences in Parkinson's and Alzheimer's diseases matched for dementia severity, *J. Nucl. Med.*, 38, 797, 1997.

150. Minoshima, S. et al., Alzheimer's disease versus dementia with Lewy bodies: cerebral metabolic distinction with autopsy confirmation, *Ann. Neurol.*, 50, 358, 2001.

151. Takahashi, H. et al., Clinical heterogeneity of dopa-responsive dystonia: PET observations, *Adv. Neurol.*, 60, 586, 1993.

152. Przedborski, S. et al., Brain glucose metabolism and dopamine D_2 receptor analysis in a patient with hemiparkinsonism-hemiatrophy syndrome, *Mov. Disord.*, 8, 391, 1993.

153. Caparros-Lefebvre, D. et al., PET study and neuropsychological assessment of a long-lasting post-encephalitic parkinsonism, *J. Neural. Transm.*, 105, 489, 1998.

154. Lin, S. K. et al., Isolated involvement of substantia nigra in acute transient parkinsonism: MRI and PET observations, *Parkinsonism Relat. Disord.*, 1, 67, 1995.

155. Ghaemi, M. et al., FDG- and Dopa-PET in postencephalitic parkinsonism, *J. Neural. Transm.*, 107, 1289, 2000.

156. Turjanski, N., Lees, A. J., and Brooks, D. J., Dopaminergic function in patients with posttraumatic parkinsonism: an 18F-dopa PET study, *Neurology*, 49, 183, 1997.

157. Lee, C. S., Schulzer, M., de la Fuente-Fernandez, R., Mak, E., Kuramoto, L., Sossi, V., Ruth, T. J., Calne, D. B., and Stoessl, A. J., Lack of regional selectivity during the progression of Parkinson's disease: implications for pathogenesis, *Arch. Neurol.*, 2004 (in press).

158. Walter, U., Niehaus, L., Probst, T., Benecke, R., Meyer, B. U., and Dressler, D., Brain parenchyma sonography discriminates Parkinson's disease and atypical parkinsonian syndromes, *Neurology*, 60, 74, 2003.

159. Schocke, M. F., Seppi, K., Esterhammer, R., Kremser, C., Jaschke, W., Poewe, W., and Wenning, G. K., Diffusion-weighted MRI differentiates the Parkinson variant of multiple system atrophy from PD, *Neurology*, 58, 575, 2002.

160. Eckert, T., Sailer, M., Kaufmann, J., Schrader, C., Peschel, T., Bodammer, N., Heinze, H. J., and Schoenfeld, M.A., Differentiation of idiopathic Parkinson's disease, multiple system atrophy, progressive supranuclear palsy, and healthy controls using magnetization transfer imaging, *Neuroimage*, 21, 229, 2004.

161. Schocke, M. F., Seppi, K., Esterhammer, R., Kremser, C., Mair, K. J., Czermak, B. V., Jaschke, W., Poewe, W., and Wenning, G. K., Trace of diffusion tensor differentiates the Parkinson variant of multiple system atrophy and Parkinson's disease, *Neuroimage*, 21, 1443, 2004.

162. Yoshikawa, K., Nakata, Y., Yamada, K., and Nakagawa, M., Early pathological changes in the parkinsonian brain demonstrated by diffusion sensor MRI, *J. Neurol. Neurosurg. Psychiatry*, 75, 481, 2004.

163. Juh, R., Kim, J., Moon, D., Choe, B., and Suh, T., Different metabolic patterns analysis of parkinsonism on the ^{18}F-FDG PET, *Eur. J. Radiol.*, 51, 223, 2004.

75 Economics of Parkinson's Disease

Katia Noyes and Robert G. Holloway
University of Rochester School of Medicine

CONTENTS

INTRODUCTION

There are many reasons why we should care about the cost of Parkinson's disease (PD). Therapeutic advances have occurred in PD care, but these have come at an increased cost—to payers, providers, and patients—and some have even questioned the value of some of these advances.[1] In 2001, the U.S. spent over $1.4 trillion on providing health care to its citizens, and the fastest growing components are prescription and hospital costs. The U.S. prescription drug market is a $140 billion dollar industry and has been growing at a rate of 15 to 18% each year.[2] Given the aging of the population and the continued technological advances likely to occur over the next decade, managing patients with PD will likely become much more costly than it is today. At the same time, patient out-of-pocket costs will likely increase, and there will be growing pressures to contain costs and more efficiently manage care and societal resources. Economic studies in PD will allow us to know how the money is being spent and assist in determining more effective ways to spend it.

USES OF ECONOMIC INFORMATION IN PD

Economic information about PD serves many purposes and has audiences beyond providers and patients. Other audiences include decision-makers at various policy levels who help establish clinical, health, and public policies. Specific examples include practice guideline developers, pharmacy formulary committees, managed care companies, and federal and state legislators.

Economic information can inform decisions of health care coverage, access, and cost, which ultimately can lead to improvements in quality of care. In addition, economic information can provide cost predictions so that providers can be paid appropriate rates for caring for groups of people including patients with PD. Patient advocacy groups, lobbyists, and researchers often use economics information (i.e., estimates of economic burden of disease) to support or further their cause of obtaining more funding for social programs and research.

One of the most common purposes of economic information about PD is to inform research on cost-effectiveness. The main objectives of cost-effectiveness research are to determine the value of new or expensive interventions compared to the standard of care and to improve the efficiency in the delivery of health care services. For example, deep brain stimulation for PD is a costly upfront investment in terms of time and money, and many would rightly ask if this investment is worth the long-term improvements in health gained. The methods of cost-effectiveness research provide the analytic framework to make such assessments.[3]

KEY (IF NOT UNIQUE) CONCEPTS WHEN PERFORMING AND INTERPRETING ECONOMIC STUDIES

There are several key concepts when performing and interpreting data on the economics of PD that are not part of usual patient-oriented research practice. These include

- Perspective
- Collecting data on health care utilization
- Costing health care resources
- Opportunity costs
- Direct costs and productivity costs
- Analyzing data on utilization and cost
- Inflation and discounting
- Health preference measurement
- Cross-national studies

Each of these is discussed briefly below.

PERSPECTIVE

Central to any study on cost is the study's perspective.[4] The perspective of the cost study refers to the perspective in which the costs are being gained or lost. Commonly used perspectives include the "patient," "third-party payer," "provider," and "societal." From the patient perspective, cost is measured in terms of out-of-pocket losses. From the third-party payer perspective, the cost of disease and treatments is measured in terms of claims paid. A study from the societal perspective would include all costs gained or lost.

COLLECTING DATA ON HEALTH CARE UTILIZATION

Estimating the cost of illness or the impact that a treatment has on the cost of an illness can be a painstakingly tedious process. Most cost studies employ a two-step process, first by collecting data on utilization (i.e., resource use) and second by costing each resource use (i.e., unit pricing) to arrive at a total estimated cost. Collecting utilization data can occur alongside many study designs, including cohort studies, case series, and randomized trials. In addition, the process of collecting utilization data can be done naturalistically as health insurance members submit claims.

A variety of approaches to collect data on utilization exist, and these include subject interviews, subject surveys, provider surveys, medical record reviews, health care utilization diaries, and insurance claims data.[5] Collecting data on utilization can be viewed as a detailed accounting exercise. Comprehensive lists of resource categories exist to serve as reminders of possible resources to include in the data collection exercise.[6] The objectives, the perspective, the costing detail required, and budgetary constraints often dictate the data collection method chosen for a study. However, much more research is needed to validate the accuracy of these different collection methods.

COSTING HEALTH CARE RESOURCE USE

Costing resource units should be viewed as a research exercise in itself, and it usually occurs after the collection of medical resources. Cost estimates for resource units are not organized for research purposes and, therefore, are contained in many different sources used for commercial purposes. This decentralization of cost information and a lack of a research-based "cost-coding dictionary" can make the costing exercise challenging, tedious, and, if we are not careful, inaccurate. Unit pricing decisions and sources used should be prespecified within the protocol. Important considerations when deciding on sources of unit cost will be data availability, perspective of the analysis, country of origin, and data purchasing costs. In the U.S., there are a variety of methods to obtain estimates of hospital costs.[7] The Medicare Reimbursement Schedule is becoming an increasingly used method to obtain cost estimates for physician visits and procedures by Current Procedural Terminology (CPT) codes.[8] In addition, the *Red Book* can provide average wholesale prices for prescription drugs.[9] Gender, age, and occupation-specific productivity costs can be obtained from information from the Bureau of Labor statistics.[10]

OPPORTUNITY COSTS

The opportunity cost of a good or service is its true market value and is defined as the value of that good or service in its next best use.[11] Opportunity costs are reflected as the price in a perfectly competitive marketplace. No marketplace is "perfect," however, and the health care marketplace has many distinguishing features (e.g., information asymmetries, market distortions, and cross subsidies) that make it less "perfect" than other markets. Therefore, routinely used prices of health care goods and services (e.g., charges and reimbursements) are not true opportunity costs. At their best, health care market prices can be viewed as "proxy" costs, which can be either higher or lower than opportunity costs.

Therefore, cost estimates used in economic studies may be far removed from opportunity costs, and there are methods to convert certain available prices to better reflect costs (e.g., hospital cost-to-charge ratios).[7,12] Research is needed to better understand how these market deviations affect cost estimates and to develop better methods to adjust for these deviations, if necessary.

DIRECT COSTS AND PRODUCTIVITY COSTS

The long list of cost categories can be rolled up into two discrete resource categories: direct costs and productivity costs.[13] Total cost estimates often combine both cost categories, but it is useful to also report them separately. Direct costs reflect the dollar burden of the medical care and nonmedical care expenditures made in response to disease. The cost of pharmaceuticals is one type of direct medical cost. Other types of direct medical costs include hospitalizations, physician visits, tests and procedures, and durable medical equipment. Direct nonmedical costs include cost to caregivers or the valued time in dollar terms in caring for a loved one. Productivity costs reflect the dollar value of the lost work induced by disease or its treatment. Therefore, productivity cost is especially important for studies conducted from the societal perspective.

INFLATION AND DISCOUNTING

Cost data need to be adjusted for inflation and time preferences. For example, if the source of cost information is obtained from an earlier year, and one would like to express the value of these costs in current dollar terms, then one needs to adjust for inflation. Inflation is related to the change in the value of money due to economic development over time. Methods to adjust for inflation should be based on the Consumer Price Index, its health care components, or one of its subcomponents.[14] If, on the other hand, a study includes costs gained or lost in future years, then these costs should be discounted to present value. Discounting reflects how people value goods and services obtained today versus the same goods and services they could buy in the future. Discount rates can vary, but the most commonly recommended discount rate is 3%.[15]

ANALYZING DATA ON UTILIZATION AND COST

Utilization and cost data have several characteristics that make them challenging to analyze.[16] Utilization data and cost data in any given population are often skewed, since a small minority of subjects often contribute to a disproportionately large percentage of the total costs. Therefore, assumptions of normality do not hold, simple data transformation may not suffice, and more sophisticated statistical methods are often required. In addition, many of the available utilization and cost estimates are from nonsampled data (i.e., deterministic rather than stochastic), and therefore, traditional methods of statistical analysis are not appropriate. Therefore, alternative methods to investigate the implications of uncertainty are needed (e.g., sensitivity analysis). Finally, despite the best intentions and best implementation of a research design, some cost information is unavailable, and missing data need to be imputed or possibly modeled.[17]

HEALTH PREFERENCE MEASUREMENT

Health preferences are numeric ratings of the desirability of health states and should be distinguished from measures of health status.[18] Health status measures classify patients into specific health states; for example, Hoehn and Yahr stage 2 or "advanced PD." On the other hand, health preference measures have individuals value the desirability of health states. The value scale is usually from death, anchored at "0," to the best imaginable health states, anchored at "1." In addition, some rating systems explicitly include health state valuations worse than death.

Health preference measures are often part of cost studies, particularly when the research is to inform decisions about cost-effectiveness or resource allocation, as the end result is often an estimate of cost per quality-adjusted life year (QALY) gained. In the calculation of a QALY, estimates of health preferences are needed. There are several methods of transforming health status values into quality of life weights to be used in the estimation of a QALY. These include scaling methods such as visual analog systems, and choice methods such as the time-trade-off and the standard gamble. In addition, quality-of-life weights can be obtained from prescored health state classifications systems, including the Health Utilities Index, EuroQol-5Q/D, Quality Well Being, and the SF-6D.[19–22] These classification systems include a series of health status questions that generate a numeric estimate of the desirability of a health state based on quality-of-life weighting obtained from other populations. Caution should be exercised, however, when using different methods of generating preference estimates, since research has shown that the different elicitation techniques yield different estimates of QALYs gained or lost.[23]

CROSS-NATIONAL STUDIES

Economic studies in PD have been performed in many different countries, and methods are available to make adjustments across countries using the purchasing power parity. In general, most cost estimates should be viewed as country-specific, given the differences in national health care systems and financing (e.g., different ratio of specialists to generalists, payment mechanisms, and availability of treatments across countries). Exceptions to this rule can be made, however, for multinational clinical trials and if methods are used to account for this variation.[24]

SOURCES OF ECONOMIC STUDIES IN PD

Economic studies can be obtained from several sources. In addition to Pubmed, the National Health Service Economic Evaluation Database is an excellent source to locate published studies and contains keyword search options that allow searches specific to disease and treatment options.[25] A registry of cost-effectiveness analyses also exists, as well as a cataloging of published preference estimates for a growing list of health states.[26,27]

COST OF ILLNESS STUDIES

CROSS-SECTIONAL STUDIES

Cross-sectional studies assess a defined and relatively heterogeneous population of patients over time. The population scope may include a country, a defined geographical area, an insured population, or a clinic population. Cross-sectional cost-of-illness studies take a snapshot of the defined population and ask the question "What is the cost of providing PD care across the population of patients over a defined period of time?" These types of analyses can provide "burden of disease estimates" and provide insights into utilization and cost relationships with demographic characteristics, access to care, and quality.

LONGITUDINAL STUDIES

Longitudinal studies assess a defined and relatively homogenous population of patients over time. The population scope may be by stage of disease (e.g., *de novo* or early-stage patients) or other patient characteristics (e.g., age) pertinent to the analysis. Longitudinal cost-of-illness studies address the question, "What is the cost to care for an individual with PD or a homogenous cohort of individuals over the course of study (e.g., duration of a clinical trial or patient's lifetime)?" These epidemiologically based models require a precise knowledge of the natural history of disease and the probability of risk factors and complications over time. Therefore, to track the economic stream of events over time, one needs a thorough and detailed knowledge of the clinical events over time.

SUMMARY OF PUBLISHED STUDIES

Table 75.1 summarizes the available cost of illness studies. There have been 16 published studies from 9 coun-

tries.[28–43] The studies involve different populations of patients and use a variety of different data collection methods to estimate resource use. Most of the studies include both direct medical costs and productivity costs, and no lifetime cost estimates have yet been published. The different countries of origin, patient populations, and data collection methods make comparisons across studies difficult if not impossible. However, self-evident themes emerge. Examples include cost increases with more severe disease, in patients with motor fluctuations, and in patients with worse quality of life. In addition, prescription costs are the most costly aspect of care, particularly in the early stages of illness, and PD patients are more costly than age-matched controls.

COST EFFECTIVENESS RESEARCH

OVERVIEW OF METHODOLOGY

Cost-effectiveness research is the comparative analysis of two or more alternative interventions in terms of both their health effects and cost. Many research designs can be used to obtain data on health effects and costs, but the two most common designs are clinical trials and decision analytic modeling. Clinical-economic trials collect cost outcomes in addition to the health outcomes, and general reviews are available regarding important considerations when designing, implementing, and analyzing clinical trials that include an economic component.[44] Decision-analytic models structure evidence on clinical and economic outcomes in a form that can help to inform decisions about clinical practice and health care resource allocation. In contrast with clinical-economic trials, models synthesize evidence on health consequences and costs from many different sources, including data from clinical trials, observational studies, insurance claim databases, registries, and preference surveys, and they link these data to outcomes that are of interest to health care decision makers.[45] The two designs, modeling and clinical trials, can also be used in combination where one models the long-term health and costs consequences beyond the time horizon of the clinical trial.[46]

Figure 75.1 is a graphical illustration of how to analyze the cost-effectiveness of a new intervention compared to an old intervention; for example, pramipexole compared to levodopa in the management of early PD. The cost-effectiveness plane, with a cost dimension (x-axis) and an effects dimension (y-axis), shows the incremental change in effects and in costs in comparing the initial pramipexole compared with initial levodopa. The plane can be divided into four quadrants: northwest, southeast, southwest, and northeast. The northwest quadrant is the cost-effectiveness space where pramipexole results in greater QALYs and more costs compared to the levodopa strategy; in this instance, pramipexole is said to

TABLE 75.1
Cost of Illness in Parkinson's Disease

Country	Patient Population	Method of Collecting Resource Use	Type of Medical Costs	Reference
U.S.A.	149 PD subjects from outpatient clinics	Subject interview	Productivity costs	28
	43 PD subjects from the 1987 National Medical Expenditure Survey	Mailed subject survey	Direct medical costs Productivity costs	29
	109 PD subjects from a variety of sources	Subject interviews	Direct medical costs Productivity costs	30
	193 patients from outpatient clinic	Subject interviews and questionnaires, medical record review	Direct medical costs Productivity costs	31
	790 PD subjects from the National Long-Term Care Survey	Insurance claims data	Comorbidity odds ratios and cost ratios	32
	70 PD subjects from Central North Carolina	Subject interviews	Direct medical costs Productivity costs	33
Canada	Population of Ontario, Canada, 1996–98	Insurance claims data (Ontario Ministry of Health)	Direct medical costs	34
	Population of Ontario, Canada, 1992–98	Insurance claims data (Ontario Ministry of Health)	Direct medical costs	35
Germany	409 PD subjects from outpatient clinics	Medical record review	Prescription drugs	36
Finland	260 PD subjects from outpatient clinics	Subject questionnaire	Direct medical costs Productivity costs	38
Sweden	127 PD subjects from outpatient clinic	Mailed subject questionnaire	Direct medical costs Productivity costs	39
UK	72 PD subjects from outpatient clinic	Subject and caregiver interviews	Direct medical costs Productivity costs	40
Japan	104 PD subjects from inpatient and outpatients clinic	Subject interviews	Direct medical costs Productivity costs	41
Netherlands	235 PD subjects patients from Parkinson's Disease Society	Mailed subject survey	Direct medical costs	42
France	294 PD subjects from outpatient clinics	Interview-administered questionnaires	Direct medical costs	43

"dominate" levodopa. The southeast quadrant is the cost-effectiveness space where pramipexole results in fewer QALYs and more costs compared to the levodopa strategy; in this instance, pramipexole is said to be "dominated by" levodopa. The southwest and northeast quadrant provide results that are not as straightforward. The southwest quadrant is the cost-effectiveness space where pramipexole treatment generates fewer QALYs and fewer costs compared to levodopa, and the question becomes, "Is the loss in health worth the savings in dollars?" The northeast quadrant is the cost-effectiveness space pramipexole has greater QALYs and greater costs compared to levodopa, and the question becomes, "Is the gain in health worth the additional cost in dollars?"

The two-dimensional cost-effectiveness plane can also be dissected into regions of cost-effectiveness based on the socially acceptable cost of one QALY. For example, the upper shaded region is that space where the estimate for a new technology compared to an old technology would cost less than $50,000 per QALY. The lighter region, in the center, is the space that is between $50,000 and $100,000 per QALY. Finally, the lower shaded region is the region where the cost effectiveness estimate is greater than $100,000 per QALY.

EVIDENCE FROM COST-EFFECTIVENESS RESEARCH

Evidence from cost-effectiveness research depends on the methods used to generate the point-estimate for the incremental cost-effectiveness ratios. When the cost and effects data are generated from the same population as in a clinical trial, one can obtain 95% confidence estimates around the point estimate.[47] For example, in Figure 75.1, the point estimate is indicated by the star and generates a cost-effectiveness ratio of $106,000 per QALY gained. Using the variance and covariance between the cost and effects estimates, one can generate confidence ellipses, and the figure shows the 50, 80, 90, and 95% confidence ellipses around the point estimate.

Although graphically appealing, however, the confidence ellipses do not fully address the question of cost-effectiveness, which requires an estimation of the probability of falling within a region of cost-effectiveness. There are methods to determine the probability within which the true cost-effectiveness ratio lies.[48] For example, using bootstrap techniques one can regenerate cost-effectiveness estimates shown in the figure as the superimposed scatterplot. One can then determine the probability of being below some prespecified cost-effectiveness threshold (e.g., <$50,000 per QALY, <$100,000 per

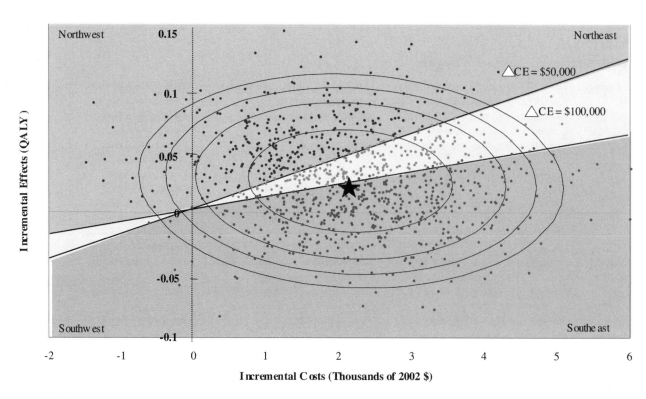

FIGURE 75.1 Incremental cost-effectiveness and confidence ellipses. The point estimate for the incremental cost-effectiveness ratio is highlighted as a magnified star. Fifty, 80, 90, and 95% confidence ellipses are superimposed on the distribution of 1,000 bootstrapped data points. Bootstrapped distributions of the estimates can be used to determine the probability that the true value of the cost-effectiveness ratio lies within each of the regions of cost-effectiveness.

QALY, or <$200,000 per QALY) by counting the number of scatterplot points within the region of cost-effectiveness. Another method is to use the net benefit approach and generate acceptability curves as shown in Figure 75.2.[49] The acceptability curve plots the probability of the new technology being cost-effective based on different valuations of a QALY, ranging from $0 to $600,000. If decision-makers could prespecify thresholds of cost-effectiveness, these methods will become increasingly useful, and recent proposals have recommended that the societal threshold for considering a new technology to be cost-effective should be $200,000 per QALY.[50]

The evidence obtained from most models differs from that generated from clinical trials, since models are a logical mathematical framework that permits the integration of facts and values, and many of the model inputs are from nonsampled data; therefore, the cost and effects outputs do not have known variances and covariances that allow confidence intervals to be estimated. Nonetheless, uncertainty can be evaluated by using sensitivity analysis and other techniques. It is important to realize that the purpose of models is to help produce informed decisions, not necessarily to predict future events, and the standards of inferential statistics used in clinical research decision making (i.e., $p < 0.05$) are not necessary. Models are commonly used in other policy arenas including envi-

ronmental policy and defense policy, and they can be valuable for health-care decision makers.[51] However, it is the responsibility of the research conducing modeling studies to adhere to best practice standards of quality, to communicate the results with sufficient disclosure of assumptions, and to ensure that the model results are completely conditional upon those assumptions. This is particularly important, given that multiple biases can exist in the model construction, data inputs, data analysis, and interpretation.

SUMMARY OF PUBLISHED STUDIES

Table 75.2 presents a summary of the published cost-effectiveness studies.[52–63] There have been 12 published studies: 9 that addressed new medications, 2 that addressed deep brain stimulation surgery, and 1 that addressed a community nurse specialist intervention. Although the actual cost-effectiveness point estimates are not presented in the table, all but one of the studies concluded that the intervention was cost-effective compared with the alternative.[52] Ten of the 12 studies were funded all or in part by pharmaceutical companies, and caution should be exercised when interpreting the studies' conclusions, given the potential bias in research funded by major stakeholders.[64]

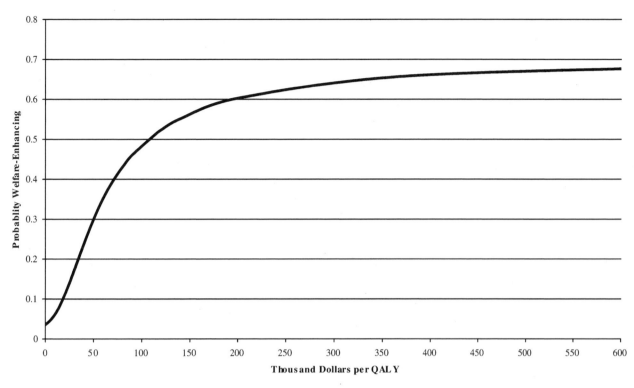

FIGURE 75.2 Probability that the "new" intervention is welfare-enhancing, given different economic values of a QALY, ranging from 0 to $600,00. Data obtained from a clinical-economic trial comparing pramipexole to levodopa in the treatment of early PD.[50]

TABLE 75.2
Cost-Effectiveness Studies in Parkinson's Disease

Patient Population	Country	Study Design	Outcomes	Intervention	Alternative	Time Horizon	Ref.
HY* I–III	U.S.	Clinical trial	Costs and effects	Pramipexole	Levodopa	2 years	52
Patients with motor fluctuations	U.S.	Before/after study	Costs	Sinemet CR	Sinemet	1 year	53
Early patients/late patients	U.S.	Model	Costs and effects	Pramipexole	No levodopa Levodopa	Life expectancy	54
HY stages I	Australian	Model	Costs and effects	Pergolide	Bromocriptine	10 years	55
HY stages II to V	Japan	Model	Costs and effects	Bromocriptine Pergolide	Levodopa	10 years	56
Patients with motor fluctuations	Netherlands	Model	Costs and effects	Entacapone	No entacapone	5 years	57
Late stage patients	U.S.	Model	Costs and effects	Deep brain stimulation	No deep brain stimulation	Life expectancy	58
Late stage patients	Germany	Before/after study	Costs and effects	Deep brain stimulation	No deep brain stimulation	1 year	59
All PD patients on anti-PD medications	England	Clinical trial	Costs and effects	Community-based nurse specialist	Usual care	2 years	60
HY stages 1.5 and 2.5	U.S.	Model	Costs and effects	Entacapone	No entacapone	5 years	61
Early patients	Canada	Model	Costs only	Ropinirole	Levodopa	5 years	62
Early patients	Germany	Model	Costs and effects	Cabergoline	Levodopa	10 years	63

*HY = Hoehn and Yahr.

HOW TO IMPROVE THE USE OF ECONOMIC INFORMATION IN PD

Table 75.3 provides suggestions to improve the use of economic information in PD. First, one should become familiar with study designs and methods to perform and appraise cost-of-illness and cost-effectiveness research.[65,66] Along with this should be the recognition that, although the evidence from cost-effectiveness research is different from that of clinical trials, it can be useful to informed decision-making. Second, there needs to be more research on the most reliable and accurate ways to collect and measure resource use and on natural history studies to serve as the clinical event backbone on which lifetime costs can be estimated. Third, more Phase III and IV clinical trials should include an economic component, and information about cost-effectiveness should be incorporated into clinical practice guidelines and practice parameters. Fourth, collaborative partnerships should be strengthened with health economists, quality of life researchers, and health policy researchers. Finally, support for general advances within the field, such as to periodically update and revise cost per QALY benchmarks to account for budget increases and inflation, will improve the fields' credibility and acceptance by many health care decision makers. Only then will economic information regarding PD stand beside clinical information to help guide decisions to improve the quality and efficiency of PD care.

TABLE 75.3
How to Improve the Use of Economic Information in PD

- Become familiar with study designs and methods to perform cost of-illness and cost-effectiveness research
- Become familiar with how to appraise cost-effectiveness research
- Recognize that the evidence from cost-effectiveness research is different from that of clinical trials but can be useful to inform decision-making
- Research on most reliable and accurate ways to collect resource use and measure costs
- More natural history studies to serve as the clinical backbone of lifetime cost estimates
- More Phase III and IV clinical trials should include an economic component
- Incorporate information about cost-effectiveness into clinical practice guidelines and practice parameters
- Work collaboratively with health economists and quality of life researchers
- Update and periodically revise cost per QALY benchmarks to account for budget increases and inflation

---

REFERENCES

1. Relman, A. S. and Angell, M., How the drug industry distorts medicine and politics. America's other drug problem, *The New Republic,* 27, 2002.
2. Frank, R. G., Government commitment and regulation of prescription drugs, *Health Aff.,* 22, 46, 2003.
3. Panel on Cost-Effectiveness in Health and Medicine. Cost-Effectiveness in Health and Medicine, 1st ed., Oxford University Press, New York, 1996.
4. Davidoff, A. J. and Powe, N. R., The role of perspective in defining economic measures for the evaluation of medical technology, *Int. J. Technol. Assess Health Care,* 12, 9, 1996.
5. Goossens, M. E., Rutten-van Molken, M. P. Vlaeyen, J. W., and van der Linden, S. M., The cost diary: a method to measure direct and indirect costs in cost-effectiveness research, *J. Clin. Epidemiol.,* 53, 688, 2000.
6. Luce, B. R. and Elixhauser, A., Estimating costs in the economic evaluation of medical technologies, *Int. J. Technol. Assess Health Care,* 6, 57, 1990.
7. Mushlin, A. I. Hall, W. J. Zwanziger, J.,Gajary, E., Andrews, M., Marron, R., Zou, K. H., and Moss, A. J., The cost-effectiveness of automatic implantable cardiac defibrillators, *Circulation,* 97, 2129, 1998.
8. http://www.cms.gov/medicare/, 2003.
9. *The Red Book Drug Topics,* Medical Economics Company Inc., Montvale, NJ, 2001.
10. http://www.census.gov/hhes/www/income.html, 2003.
11. Russell, L. B., Opportunity costs in modern medicine, *Health Aff.,* 11, 162, 1992.
12. Finkler, S. A., The distinction between cost and charges, *Ann. Intern Med.,* 96, 102, 1982.
13. Luce, B. R., Manning, W. G., Siegel, J. E., and Lipscomb, J., Estimating costs in cost-effectiveness analysis. In: Gold, M. R., Siegel, J. E., Russell, L. B., and Weinstein, M. C., *Cost-Effectiveness in Health and Medicine,* Oxford University Press, New York, 1996: 176–213.
14. http://www.bls.gov/cpi, 2003.
15. Lipscomb, J., Weinstein M. C., and Torrance, G. W., Time Preference, In: Gold, M. R., Siegel, J. E., Russell, L. B., and Weinstein, M. C., *Cost-Effectiveness in Health and Medicine,* Oxford University Press, New York, 1996, 214–246.
16. Diehr, P., Yanez, D., Ash, A.; Hornbrook, M., and Lin, D. Y., Methods for analyzing health care utilization and costs, *Annu. Rev. Public Health,* 20, 125, 1999.
17. Little, R. J. A. and Rubin, D. B., *Statistical Analysis with Missing Data,* Toronto, Canada, John Wiley, 1987.
18. Neumann, P., Goldie, S. J., and Weinstein, M. C., Preference-based measures in economic evaluation in health care, *Ann. Rev. Public Health,* 21, 587, 2000.
19. The Euro Qol Group, EuroQol*—a new facility for the measurement of health-related quality of life, *Health Policy,* 16, 199, 1990.
20. Kaplan, R., Bush, J. W., and Berry, C. C., Health status: types of validity and the index of well-being, *Health Serv. Res.,* 11, 478–507, 1976.

21. Brazier, J., Usherwood, T., Harper, R., and et al., Deriving a preference-based single index from the UK SF-36 health survey, *J. Clin. Epidemiol.*, 51, 1115, 1998.

22. Feeny, D., Furlong, W., and Barr, R.,D., A comprehensive multiattribute system for classifying the health status of survivors of childhood cancer, *J. Clin. Oncol.*, 10, 923, 1992.

23. Lenert, L. and Kaplan, R., Validity and interpretation of preference-based measures of health-related quality of life, *Med. Care*, 38, II–138, 2000.

24. Schulman, K., Burke, J., Drummond, M., Davies, L., Carlsson, P, Gruger, J., Harris, A., Lucioni, C., Gisbert, R., Llana, T, Tom, E, Bloom, B., Willke, R., Glick, H., and Soikos, S. L., Resource costing for multinational neurologic clinical trials: methods and results, *Health Econ.*, 7, 629, 2003.

25. http://nhscrd.york.ac.uk/nhsdhp.htm.

26. www.hsph.harvard.edu/cearegistry/.

27. http://www.hcra.harvard.edu/pdf/preferencescores.pdf

28. Singer, E., Social costs of Parkinson's disease, *J. Chronic. Dis.*, 26, 243, 1973.

29. Rubenstein, L. M.; Chrischilles, E. A., and Voelker, M. D., The impact of Parkinson's disease on health status, health expenditures, and productivity, *Pharmacoeconomics*, 12, 486, 1997.

30. Whetten-Goldstein, K., Sloan, F.A., Kulas, E., Cutson, T. and Schenkman, M., The burden of Parkinson's disease on society, family, and the individual, *JAGS*, 45, 844, 1997.

31. Chrischilles, E. A., Rubenstein, L. M., Voelker, M. D., Wallace, R. B., and Rodnitzky, R. L., The Health burden's of parkinson's disease, *Mov. Disord.*, 13, 406, 1998.

32. Pressley, J. C., Louis, E.D., Tang, M.X., Cote, L., Cohen, P. D., Glied, S., and Mayeux, R., The impact of comorbid disease and injuries on resource use and expenditures in parkinsonism, *Neurology*, 60, 87, 2003.

33. Schenkman, M.; Wei, Z. C.; Cutson, T. M., and Whetten-Goldstein, K., Longitudinal evaluation of economic and physical impact of Parkinson's disease, *Parkinsonism and Related Disorders*, 8, 41, 2001.

34. Muir, T. and Zegarac, M., Societal costs of exposure to toxic substances: economic and health costs of four case studies that are candidates for environmental causation, *Neurol*, 60, 885, 2001.

35. Guttman, M., Slaughter, P.M., Theriault, M.E., DeBoer, D. P., and Naylor, C. D., Burden of parkinsonism: a population-based study, *Mov. Disord.*, 18, 313, 2003.

36. Dodel, R. C., Eggert, K. M., Singer, M. S., Eichhorn, T.E., Pogarell, O., and Oertel, W. H., Costs of drug treatment in Parkinson's disease, *Mov. Disord.*, 13, 249, 1998.

37. Dodel, R. C., Singer, M. S., and Kohne-Volland, R., The economic impact of Parkinson's disease, *Pharmacoeconomics*, 14, 299, 1998.

38. Keranen, T., Kaakkola, S., Sotaniemi, K., Laulumaa, V., Haapaniemi, T., Jolma, T., Kola, H., Ylikoski, A., Satomaa, O., Kovanen, J., Taimela, E.; Haapaniemi, H., Turunen, H., and Takala, A., Economic burden and qual-

ity of life impairment increase with severity of PD, *Parkinsonism and Related Disorders*, 9, 163, 2003.

39. Hagell, P., Nordling, S., Reimer, J., Grabowski, M., and Persson, U., Resource use and costs in a Swedish cohort of patients with Parkinson's disease, *Mov. Disord.*, 17, 1213, 2002.

40. Clarke, C. E., Zobkiw, R. M., and Gullaksen, E., Quality of life and care in Parkinson's disease, *Br. J. Clin. Pract.*, 49, 288, 1995.

41. Fukunaga, A., Kasai, T., and Yoshidome, H., Clinical findings, status of care, comprehensive quality of life, daily life therapy and treatment at home in patients with Parkinson's disease, *Eur. Neurol.*, 38, 64, 1997.

42. De Boer, A. G., Sprangres, M. A., Speelman, H. D., and de Haes, H. C., Predictors of heath care use in patients with Parkinson's disease: a longitudinal study, *Mov. Disord.*, 14, 772, 1999.

43. LePen, C., Wait, S., Moutard-Martin, F., Dujardin, M., and Ziegler, M., Cost of illness and disease severity in a cohort of French patients with Parkinson's disease, *Pharmacoeconomics*, 16, 59, 1999.

44. Ramsey, S. D., McIntosh, M., and Sullivan, S. D., Design issues for conducting cost-effectiveness analyses alongside clinical trials, *Ann. Rev. Public Health*, 22, 129, 2001.

45. Weinstein, M. C., O'Brien, B., Hornberger, J., Jackson, J., Johannesson, M., McCabe, C., Luce, B.R., and ISPOR Task Force on Good Research Practices, Principles of good practice for decision analytic modeling in health-care evaluation: report of the ISPOR Task Force on Good Research Practices—Modeling Studies, *Value In Health*, 6, 9, 2003.

46. National Emphysema Treatment Trial Research Group, Cost effectiveness of lung-volume-reduction surgery for patients with severe emphysema, *N. Engl. J. Med.*, 348, 2092, 2003.

47. Briggs, A. H., A Bayesian approach to stochastic cost-effectiveness analysis. An illustration and application to blood pressure control in type 2 diabetes, *Int. J. Technol. Assess Health Care*, 17, 69, 2001.

48. Hoch, J. S., Briggs, A. H., and Willan, A. R., Something old, something new, something borrowed, something blue: a framework for the marriage of health econometrics and cost-effectiveness analysis, *Health Econ.*, 11, 415, 2002.

49. Fenwick, E., Claxton, K., and Sculpher, M., Representing uncertainty: The role of cost-effectiveness acceptability curves, *Health Econ.*, 779, 2003.

50. Ubel, P. A., Hirth, R. A., Chernew, M. E., and Fendrick, A., What is the price of life and why doesn't it increase at the rate of inflation? *Arch. Intern. Med.*, 163, 1637, 2003.

51. Weinstein, M. C., Toy, E. L., Sandberg, E., Neumann, P. J., Evans, J. S., Kuntz, K. M., Graham, J. D., and Hammitt, J. K., Modeling for health care and other policy decisions: Uses, roles, and validity, *Value in Health*, 4, 348, 2001.

52. Noyes, K., Dick, A. W., and Holloway, R. G., Pramipexole vs. levodopa as initial treatment for Parkinson's

disease: A randomized clinical-economic trial (in press), *Med. Decis. Making,* 2003.

53. Hempel, A. G.; Wagner, M. L.; Maaty, M. A., and Sage, J. I., Pharmacoeconomic analysis of using Sinemet CR over standard Sinemet in parkinsonian patients with motor fluctuations, *Ann. Pharmacother.,* 32, 878, 1998.

54. Hoerger, T. J., Bala, M. V., Rowland, C., Greer, M., Chrischilles, E. A., and Holloway, R. G., Cost effectiveness of pramipexole in Parkinson's disease in the US, *Pharmacoeconomics,* 14, 541, 1998.

55. Davey, P., Rajan, N., Lees, M., and Aristides, M., Cost-effectiveness of pergolide compared to bromocriptine in the treatment of Parkinson's disease: a decision-analytic model, *Value in Health,* 4, 308, 2001.

56. Shimbo, T., Hira, K., Takemura, M., and Fukui, T., Cost-effectiveness analysis of dopamine agonists in the treatment of parkinson's disease in Japan, *Pharmacoeconomics,* 19, 875, 2001.

57. Nuijten, M. J., van Iperen, P., Palmer, C., van Hilten, B. J., and Snyder, E., Cost-effectiveness analysis of entacapone in Parkinson's disease: a Markov process analysis, *Value in Health,* 4, 316, 2001.

58. Tomaszweski, K. and Holloway, R. G., Deep brain stimulation in the treatment of PD: A cost-effectiveness analysis, *Neurology,* 57, 663, 2001.

59. Spottke, E. A., Volkmann, J., Lorenz, D., Krack, P., Smala, A. M., Sturm, V., Gerstner, A., Berger, K., Hellwig, D., Deuschl, G., Freund, H.J., Oertel, W. H., and Dodel, R. C., Evaluation of healthcare utilization and health status of patients with Parkinson's disease treated with deep brain stimulation of the subthalamic nucleus, *J. Neurol.,* 249, 759, 2002.

60. Jarman, B., Hurwitz, B., Cook, A., Bajekal, M., and Lee, A., Effects of community based nurses specialising in Parkinson's disease on health outcome and costs: randomised controlled trial, *B. M. J.,* 324, 1072, 2002.

61. Palmer, C. S., Nuijten, M. J., Schmier, J. K., Subedi, P., and Snyder, E. H., Cost effectiveness of treatment of Parkinson's disease with entacapone in the United States, *Pharmacoeconomics,* 20, 617, 2002.

62. Iskedjian, M. and Einarson, T. R., Cost analysis of ropinirole versus levodopa in the treatment of Parkinson's disease, *Pharmacoeconomics,* 21, 115, 2003.

63. Smala, A. M., Spottke, E. A., Machat, O., Siebert, U., Meyer, D., Kohne-Volland, R., Reuther, M., DuChane, J., Oertel, W. H., Berger, K. B., and Dodel, R. C., Cabergoline versus levodopa monotherapy: A decision analysis, *Mov. Disord.,* 18, 898, 2003.

64. Kassirer, J. P. and Angell, M., The *Journal's* policy on cost-effectiveness analyses, *N. Engl. J. Med.,* 331, 669, 1994.

65. Drummond, M. F., Richardson, W. S., O'Brien, B. J., Levine, M., and Heyland, D., Users' guides to the medical literature. XIII. How to use an article on economic analysis of clinical practice A. Are the results of the study valid? Evidence-Based Medicine Working Group, *JAMA,* 277, 1552, 1997.

66. O'Brien, B. J., Heyland, D., Richardson, W. S., Levine, M., and Drummond, M. F., Users' guides to the medical literature. XIII. How to use an article on economic analysis of clinical practice B. What are the results and will they help me in caring for my patients? Evidence-Based Medicine Working Group, *JAMA,* 277, 1802, 1997.

76 Informal Caregivers: A Valuable Part of the Health Care Team

Kirsten Maier
Parkinson Society British Columbia

Susan Calne
Pacific Parkinson's Research Centre, University Hospital, University of British Columbia

CONTENTS

INTRODUCTION

Over the past several decades, there has been an ever increasing interest in the plight of people caring at home for a family member with a chronic or serious illness. Much has been written about the burden and family changes that accompany declining health in one family member. The impact is felt at both the individual and the family level. An illness is an unwelcome, indefinite guest in the home that has the potential to cause anxiety, sadness, and chaos. Health professionals have a responsibility to be aware of the possible impact of these changes and to become a facilitator for the family social network. In this way the caregiver and the family can be empowered to make decisions and changes that are right for them. In this chapter, we outline some of the changes caregivers can experience and propose practical strategies for health professionals as a formal support for informal caregivers.

The word *caregiver* originated in the U.S.A. and appeared in a 1966 paper on mental illness by Mackey.[1] The first time *caregiver* appears as a key word specifically applied to Parkinson's disease (PD) in Pub Med is 1986.[2] The *American Heritage Dictionary* 2000 edition contains somewhat paternalistic definitions that include the following:

1. "An individual, such as a physician, nurse, or social worker, who assists in the identification, prevention, or treatment of an illness or disability."
2. "An individual, such as a parent, foster parent, or head of a household, who attends to the needs of a child or dependent adult."

The term *caregiver* is now often loosely applied to anyone living with someone who has a chronic illness, almost regardless of the amount of daily care this person provides. In the public consciousness, caregivers are almost invariably deemed to be "burdened" and therefore finding it difficult to cope. Recent research is finding that, as with many life situations, the personal experience of caregivers is more ambiguous.[3,4] Research may have focused on the burden, due to the somewhat selective group of caregivers that health care professionals had contact with—namely, those who were having the most difficulty providing care or whose partners needed the most medical help. However, Greenberger and Litwin found that burdened caregivers could provide good care if adequate resources, self-esteem, and support were present.[5]

Pearlin et al. described three transitions in the caregiving career.[3] The initial transition takes place when the family is providing care at home. It is during this period that families access community health services such as social workers, home care, respite, day care programs, and in-home rehabilitation assessments. The second transition occurs when the disease has progressed to the point at which the person being cared for requires more formal and continuous care in a long- term care facility. Such a separation is often traumatic for the caregiver (a form of divorce or death). When a patient moves into a facility, this does not mean an end to caregiving, but a shift in focus as the caregiver works with the care facility to provide ongoing support. Pearlin's third and final transition is the bereavement caused by the death of the person with the chronic illness. If the person dies at home, clearly the caregiver will experience only transitions 1 and 3. Maier proposes another transition, prior to the period of at-home care, which follows the diagnosis when the family is coming to terms with the fact that a chronic illness has moved into the family setting.[6] Each of these transitions requires a different response from both the caregiver and those who provide support.

A more descriptive term, in the beginning and middle stages of an illness, may be *care partner*. This term has been adopted by some self-help groups, as it implies a less one-sided relationship than the label *caregiver*. This term suggests a relationship that is still reciprocal, where the care partner is still receiving something from the family member being helped. This reciprocity continues as the disease process progresses. However, Carter et al., in their PD caregiver study, found that reciprocity had begun to decline by the time patients reached Hoehn and Yahr stage 2.[7,8] Therefore, the term may become inappropriate even before the patient reaches stages 4 or 5. In PD, the line between being a care partner and a caregiver blurs before it is finally crossed. The "blurred" stage occurs when a PD patient's drugs fluctuate in their effectiveness ("wearing off" and "on off") or with episodes of confusion and/or concomitant illness. This leads to a fluctuating need for help, which in turn can create tensions when the care partner is unsure of how much help to offer and when it is appropriate to do this.

What brings about the increasing attention that is being paid to the role of the caregiver? For example, in the first three months of 2003 alone, a Pub Med search using *caregiver* and *2003* as key words produced an astounding 119 results. There are multiple reasons—some social, some economic, and some political. As public funding for health care continues to decrease and the population ages, more individuals are finding themselves in the position of providing physical and emotional care for a family member with a chronic illness. This reality alone creates an awareness of the patient's need for care, the needs of the informal caregiver, and preparation for "best practices" in this area a worthwhile study for those providing formal care to patients and families living with a chronic illness such as PD.

CHANGES IN THE CAREGIVING LANDSCAPE

Social shifts in work patterns and advances in treatment have led to changes and choices in caregiving. Effective treatment for PD was introduced only at the end of the 1960s.[9,10] Until that time, one can perhaps assume that most people with PD would have had a diagnostic consultation and then disappeared into their communities, as there was little that could be done for them. A few PD patients would have been treated with anticholinergic drugs, and even fewer would have had surgery.[11,12] Their numbers and plight largely went unrecorded. Levodopa not only improved quality of the patient's life, but the duration as well.[13] At the same time, the allied health professions of nursing, social work, and rehabilitation were beginning their evolution into more autonomous professions with their own goals. With this change, health care delivery became less paternalistic, and patients, including those with PD, began to be proactive consumers with regard to their health. At about the same time, large numbers of married women returned to school and entered the work force, not just for income but also for professional satisfaction.

Improvements in public health and medicine have also been a factor in the changing landscape of caregiving. Since World War II, there has been increasing access to improved nutrition, sanitation, immunization, and medical care; this has led to a greater expectation that health care will be provided. As a result, larger numbers of people now grow very old and, in spite of medical advances, are often in poor health. Elderly people put enormous strain on health care systems, and health care systems in many countries have begun to buckle under the strain. Medical practice has moved from a public service model toward more of a traditional business model, leading to more bureaucrats and fewer hospital beds.[14] There is a chronic shortage of nurses, which some have attributed to the phasing out of more easily accessible diploma nursing programs.[15] In addition, rehabilitation therapy and social work services are being reduced or eliminated. The number of beds in long-term care facilities is decreasing when it should be increasing. In a paper comparing the ratio of beds and nurses to need in Britain, U.S.A., and Germany, the U.S.A. lags behind Britain. Both are followed by Germany, which relies far more on the voluntary sector for service.[16]

Historically, politicians have viewed the elderly as expendable, as they have had few influential advocates and were no longer valued as contributing members of society. This is changing and, increasingly, the elderly are better educated, more financially secure, and are beginning to advocate for themselves.[16,17]

It is interesting to examine who is becoming the informal caregiver. While the *American Heritage Dictionary* asserts that caregiving is usually provided by the "head of the household" (traditionally the male), Attias Dornfut surveyed 2000 families and demonstrated that, regardless of whether men or women were being looked after, women provided the bulk of the care.[18] Only 14% of men provided care to elderly fathers, with the remaining care provided by women. Women also provided the most care to elderly mothers (Figure 76.1). This finding is echoed in numerous studies on caregiving in which the majority of participants is found to be women.[5,19–21]

There is some evidence that there are further gender differences in the caregiving experience. Faison et al. found that caregiving sons reported significantly less burden than did daughters or other relatives.[22] For most men, providing physical care is a new role. Male caregivers may find it easier than female counterparts to problem solve, delegate, and seek help. However, Jette et al. found that the odds of being institutionalized if the caregiver was male were twice those for female caregivers.[23] They postulated that this is because men provide assistance with tasks that are not as relevant to the ability to remain at home as those dispensed by the female caregiver, who provides more housekeeping and personal care. Many women, on the other hand, have provided care to children and elderly parents for much of their lives and may view a new caring role with dismay.[24]

For the most part, the huge social and medical gains made since World War II have conspired to make modern informal caregiving a challenge. Social changes have left us with many more women of all ages in the workforce (but not nurses), creating smaller more scattered families. These same gains are resulting in large numbers of elderly people needing care but having fewer family members at home to provide it. Financial cutbacks have meant fewer hospital beds and social services to support them. When combined, these factors leave increasing numbers of partners and adult children with dwindling resources to care for a PD relative.

THINKING AHEAD: CAREGIVING AND PUBLIC POLICY

In 1989, Zarit wrote an editorial asking, "Do we need another caregiver and stress paper study?"[25] He presented his first paper on family burden at a meeting in 1979. His was the only paper on this topic. In 1989, he reported that family caregiving papers were now dominating geriatric meetings and journals. At that time, he argued that "the point that caregiving is stressful is now well established" (p. 147), with the drawback that the point was largely supported by descriptive papers. He proposed a more rigorous approach to analyzing the topic, using multivariate models, controlled studies, and large sample sizes, as just three examples. Since then, there has been a steady stream of more controlled studies on PD caregiver "plight," "stress," and "burden."[7,19–21,26–42]

Despite the fact that the subject of PD caregivers has been addressed since 1989, there have not been enough controlled studies of sufficient size to influence public policy on caregiver needs. The largest well controlled

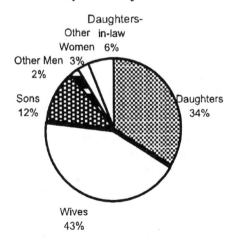

Who helps elderly men ?

Daughters-in-law
Other Women 6%

Other Men 3%
2%

Sons 12%

Daughters 34%

Wives 43%

Women help elderly men in 86% of the cases

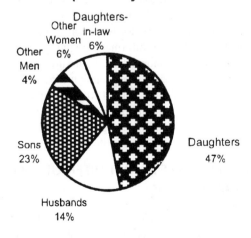

Who helps elderly women ?

Daughters-in-law
Other Women 6%

Other Men 4%

Sons 23%

Daughters 47%

Husbands 14%

Women help elderly women in 59% of the cases

FIGURE 76.1   Taken from Attias Donfut, C., The dynamics of elder support, *Zeitschrift für Gerontologie und Geriatrie* 34:9–15, 2001, with permission of Steinkoppf.

study is by Aarsland et al., who studied the effect of mental symptoms in PD patients had on 94 caregivers and compared them with the same number of caregivers of patients with diabetes mellitus, and with healthy aged controls.[26]

The logistics of recruiting large numbers of elderly people who provide care to different disease groups, some of whom are also in poor health, make these studies difficult and expensive.[28] While it is easier to conduct controlled studies with smaller numbers of subjects, it is argued that results from small sample sizes lack the scientific "power" to provide the impetus for political change. Almost invariably, the observations in the discussion section of these small sample papers point to the need for further controlled studies with larger numbers. Those determining government health policy and expenditure on PD are therefore unwilling to be influenced by either small or uncontrolled studies.

Although larger, controlled research studies in this area are advisable before public policy can be changed, they may not always be necessary if a study is well designed. For example Dura et al. compared two small groups of 23 caregivers of patients with PD and Alzheimer's disease (AD) and found no statistical difference between the degree of depression in the two groups.[28] If one adjusts their data for age and sex, it is apparent that increasing the number of subjects would not increase the predictive power of the findings. Dura's results could therefore be generalized to these caregiver groups as a whole and used by agencies in developing policy.

This relationship between research and policy leads to a challenge in providing service to caregivers of PD patients. In the absence of the studies, and the resulting changes in public policy, those who provide support to PD caregivers are left to react on an *ad hoc* basis to individual caregiver needs. If they are fortunate, they can rely on the local voluntary sector to provide more structured support for them.[14,16,43] Informal caregiving is a modern reality. It has been shown that, with sufficient support, caregivers can maintain family relationships while saving health care dollars.[5] Therefore, it would seem prudent for government agencies to implement policies that provide adequate support to address the growing need.

WHY DO PEOPLE FIND IT DIFFICULT TO BE CAREGIVERS?

Caring for a family member with chronic illness can be challenging. Carter et al. studied 380 spouses of PD patients at all stages of the illness.[7] They found that caregiver strain was present at all levels of disease and had tripled by the time the patient was at stage 4/5 Hoehn and Yahr.[8] It is when the role of caring begins to have a negative impact on the caregiver's quality of life and

health that the experience of burden becomes significant. Burden can take many forms, including loss of personal time and autonomy, loss of financial stability, and loss of personal relationships. Robinson, in a study of female primary caregivers of severely demented patients, found that poor financial status, difficulty asking for help, and past unhappy marriages were good predictors of caregiver burden.[44] Neufeld and Harrison found that criticism and disparaging comments by family and friends about the caregiver's performance, together with conflict over the needs of patient within the family, lowered self esteem, caused resentment, and increased the likelihood of depression.[45]

Berry suggests that the health of elderly caregivers must be a factor in decision making and care plans; the time and energy involved in providing care puts them at risk for a decline in health and well-being.[19] The majority of PD caregivers studied are female spouses close in age to the PD patient. They may have their own medical problems. Fatigue, depression, and anxiety are common.[26,28,31] Faison et al. found a positive correlation between the amount of care provided and caregiver burden, and also suggested that caregivers be included in patient care plans.[22]

Caregivers may have difficulty accessing support. Attias-Donfut found that, when women cared for men, they received less help from local health units and support services.[18] There can also be personal barriers to accessing the available services. Severely affected patients may be anxious and resent being left alone and, on the other hand, may also dislike the idea of programs outside the home.[46] Caregivers may be ambivalent, facing tensions between the need for rest and a lack of confidence in available programs or facilities.[47] Personal beliefs about dignity, pride, and independence can negatively influence decisions about accessing services, even when they could be helpful. Often, a natural affinity develops between the caregiver and the family member, based on spending extended periods of time together. This affinity can lead caregivers to feel that no one else will be able to care for a family member as well as they can. This belief may be a badge as well as a burden for the caregiver. Financial concerns or a lack of awareness about what is offered can also be limitations. Alternatively, caregivers may have expectations about how much help they can expect to receive—expectations that are not actually met. Such expectations may be unrealistic. Family or friends may promise help that never materializes or offer help with conditions that do not fit with the caregiver's needs.[45]

It has been well documented that caregiving and chronic illness have a strong impact on the caregiver. In a study to validate the Parkinson's Impact Scale (PIMS)[48] in PD caregivers, Calne et al. examined the impact that living with someone with PD had on 134 spousal caregivers caring full time for someone with the illness.[21] The caregivers' PIMS scores showed a significant sensitivity to disease severity and a high degree of responsiveness to changes in the caregiver situation. PIMS allows PD patients' fluctuations in mobility to be taken into account, and subjects can assign scores to a "stable" column or "best" and "worst" columns. PIMS captured changes in caregiver "best" and "worst" scores reflecting a deterioration or improvement in their partner's symptoms. Caregivers (110 female partners versus 24 male) rated sexuality, travel, and leisure as the areas of their lives most adversely affected by their partner's PD (see Table 76.1). The results regarding loss of sexuality or satisfying sexual relationships confirmed the results of Basson.[49] She reported that women found it difficult be sexual beings when they were also caregivers and had difficulty with the fact that their PD partners looked and felt different. Their partners, in turn, felt that "sick people" could not be sexual beings. Some of the PD patients were able to have sex only when their antiparkinson drugs were working and they were mobile. Consequently, the female caregivers felt that it was the drugs that were driving the desire.

TABLE 76.1
Factor Loadings on Each of the Four Factors Derived from the Ten-Item PIMS

	Factor 1	Factor 2	Factor 3	Factor 4
Sex	0.87			
Travel	0.67			
Leisure	0.64			
Financial		0.92		
Family		0.63		
Work	0.56			
Self Positive			0.88	
Self Negative		0.87		
Safety				0.90
Friendship				0.56

Factor analysis (varimax rotation) revealed four factors, which together accounted for 78.3% of the total variation in the data. Factor 1 represented "sex," "travel," and "leisure." Factor 2 accounted for "financial security," "family relationships," and "work." Factor 3 encompassed "self positive" and "self negative." Factor 4 included "safety" and "community relationships."[21]

In the later stages of PD, patients not only become increasingly immobile, they may also be cognitively impaired. (For a review of late-stage PD, see Calne and Kumar, 2003.) In a preliminary study, a comparison of caregivers (matched for duration of care) caring for AD patients and demented PD patients showed these two

groups to be equally and more depressed than age-matched controls.[28] In agreement with this, in a cross-sectional, population-based, controlled study of 94 subjects, Aarsland et al. examined the emotional and social distress of caring for a patient with PD, looking particularly at the impact of mental symptoms in PD on their caregiver's situation.[26] Spousal caregivers had more severe depression, with a higher proportion reporting tiredness, sadness, and less satisfaction with life as compared with healthy elderly subjects. Calne et al. found a significant correlation between the caregivers' total PIMS score and the length of time since diagnosis.[21] The longer the duration of disease, the greater impact it had on the caregiver. This confirmed the findings of Carter et al., who also found that the severity of the disease correlated with burden.[7]

Von Kanel et al., in a study of spousal Alzheimer caregivers, have shown them to be at twice the risk for cardiovascular disease than their controls, due to the added stress.[50] The caregivers of PD patients with dementia might share the same risk. Poor health and mood and psychological stress may leave them ill equipped to provide appropriate care. Psychiatric disturbances in patients with PD (particularly delusions, agitation, depression, and inappropriate behavior) contribute significantly to caregiver stress.[26,51]

Grief and loss are also part of the caregiving experience, both in the form of anticipatory grief and in the form of bereavement following the death of the person with PD. Anticipatory grief is due to the ongoing process of loss that occurs during a progressive chronic condition. It is the grief experienced on a daily basis as the relationship between the patient and caregiver changes and as the patient's ability to function decreases. The course of grief and bereavement is individual to each person and, therefore, the time a caregiver takes to grieve will vary.[52] One caregiver summarized her experience with the following, "First we grieve the husband we lost, then we grieve the person we cared for."[53, p. 108] A caregiver who has experienced a long period of anticipatory grief may be ready to move on to a new stage in life before other family members and friends have finished their grieving process. When social expectations of grief and the caregiver's experience conflict, it can compound the difficulty of the caregiver's grief experience.

In summary, some of the key issues in caregiving that can cause it to be a difficult experience include the length and severity of the partner's illness, difficulty accessing services, changes in sexual relations, loss of travel and leisure time, cognitive impairment in the partner, continual decline in the partner's daily functioning, and the experience of anticipatory grief. It is important for health professionals to keep these issues in mind when providing support and assistance to caregivers, as these issues can have a negative impact on the their health.

WHAT CONTRIBUTES TO A POSITIVE CAREGIVING EXPERIENCE?

Hooker et al. set out to determine why, in similar circumstances, there is such variability among caregivers' ability to cope.[20] They found that two variables were significant. The physical health of the patient does not always correlated positively with the caregiver's ability to cope. However, the degree of poor mental health in the partner is an indicator of how well the caregiver was able to cope. A caregiver with an optimistic personality and self-confidence, who feels competent as a caregiver, is also more likely to cope well.[5] Skaff et al. found that greater burden was not a predictor of lower levels of self-esteem; rather, the reverse was true—burdened caregivers who develop a sense of mastery can be more self-confident.[54]

Berry et al. reported that a pre-existing happy relationship (be it spousal partnership or parent/child relationship) often had a positive influence on how caregivers provide care.[19] Interestingly, this is not dependent on economic status. The duration of the relationship is also a factor in how successfully a caregiver assumes the role.

Naturally, the person with PD also plays a role. Caring for a person with an optimistic personality, whatever the level of disability, may also be a factor in easing the role of the caregiver. Andersen promotes the notion of a patient "breaking through" rather than "breaking down."[55] He suggests that, after an appropriate period of grieving, the patient accepts the challenge and moves forward. He defines health as equaling "the good life," confidence in life, good relations, self-confidence, having meaning, the experience of controlling one's own life, and succeeding by doing things. To quote the title of the paper by Hooker et al., it is evident that "personality counts for a lot."[20]

Greenberger and Litwin examined caregiver resources, burden, and competence as predictors of ability to facilitate health care on behalf of older patients.[5] They found a positive correlation between (a) the presence of personal and social resources and (b) a caregiver's ability to facilitate health care. Good resources correlated positively with reduced levels of burden. The greater the ability to cope and the higher the self-esteem, the more likely that caregivers were able to seek and acquire help. They, like Berry and Murphy, found that this was not related to financial status.[19] They suggest that quality and therefore successful caregiving can coexist with burden, provided that ample caregiver resources are present. The most important resources were a caregiver sense of competence and support from the professional health care provider.

The meaning that the caregiver attributes to the situation and the expectations around family relationships are also significant in shaping the caregiving experience. Health care professionals can obtain valuable information by determining these meanings and expectations.[4,56,57] These beliefs shape choices around caring, such as how long the person with Parkinson's should live at home and what types of social or medical interventions may be necessary. For example, when should a caregiver insist on taking responsibility for administration of medications? How far should a caregiver go to prevent an unsafe PD patient from driving? A reshaping of expectations allows people to continue to love their family member in new and flexible ways as the disease process progresses.

Positive meanings attributed to caregiving have included a reported increase in caring capacity and altruism[58] as well as increased emotional closeness and personal competency.[59] Other benefits reported by caregivers in a small qualitative study were a feeling of being needed, the moral attribution of being a good human being, enjoyment of the ongoing relationship, and practicing useful self-care strategies.[53]

Two challenges caregivers face are dealing with the large range of negative emotions they may experience and dealing with role changes as family relationships change. Emotions such as anger, frustration, and guilt can be seen as messengers. They provide information that the caregiver can use as a resource when making choices about changing roles and the family situation. Framing emotions in this light gives the person the freedom to experience natural feelings without guilt and to use these feelings in a positive way.[60]

Encouraging self-care strategies for maintaining good physical and mental health is also a necessary part of the caregiving process. The needs of the caregiver are as important as those of the person with PD. It takes continuing compromise and creativity on the part of all family members to balance the demands on time and activities. Similarly, encouraging self-care in the person with PD allows a sense of autonomy and independence for as long as possible. As mentioned earlier, determining where this line should be drawn can be difficult, as the nature of PD means that the patient's mobility can fluctuate rapidly. Self-care strategies for both caregiver and partner are important in maintaining the best possible quality of life. This focus on self-care is supported by the findings of Skaff and Pearlin, who documented role engulfment and the experience of loss of self as risks of informal caregiving.[61]

Promoting the use of community resources is also important. Many factors influence how caregivers access available services. Health professionals, most often nurses, who take the time to listen to frustrated, exhausted caregivers allow them to vent their feelings to an independent party without fear of repercussion. They can promote good physical and mental health and reinforce the caregiver's need for appropriate respite. This could be in the form of short-term admission for residential care, day care programs, part-time paid respite, a volunteer visitor, or a family visit.[62] When caregivers or patients have concerns about access support, it can be helpful to encourage the idea that the family and health care professionals will work together to strive for *excellence*, rather than *perfection*, in care. This emphasis on making the teamwork for excellence provides space for natural human limitations without compromising the quality of life of those involved. Gilmour suggests that caregivers must be included in developing respite care plans and that facilities have an obligation to maintain the home routine as much as possible as the patient is a temporary visitor.[47]

Health professionals cannot assume that the health of the PD patient will deteriorate more quickly than that of the caregiver. Spouses can and do die first. Contingency plans for spousal sickness, such as respite, long-term care health directives, and wills, can be introduced into discussions about future care and needs.

Maintaining a strong social network is seen as important in combating the isolation that can occur due to chronic illness in the family as well as in supporting both physical and mental well-being. However, research findings on social isolation are mixed. Maier[6] found that caregivers experienced a loss of long-time friends, a shift in friendships to single friends, and a decrease in friendships in general. The participants in this study saw this as a loss in their lives. Calne et al.,[21] on the other hand, found female caregivers largely unconcerned by the loss of community support. These differences may be the result of sample size or of personality, as described by Hooker et al.[20]

Education, assistance, listening, and counseling as provided by all health professionals form the foundation of caring for caregivers and their families. These activities take time but are a worthwhile investment. Validating the individual's experience allows the care giver to normalize what is happening and make the best choices under the circumstances. Practicing active listening, wherein both the content and the feeling of what is being said are acknowledged by the health professional, is essential. Encouraging appropriate boundaries to helping can support the ongoing health of all family members while still fulfilling the families' caring goals. This is made easier by strengthening family communication skills and encouraging space for personal growth. For more details on these and other interventions, see Egan.[63] Caring for the caregiver is achieved by building a professional relationship whereby the caregiver is acknowledged to be the expert provider of informal care, and the health professional has the medical training and experience needed to provide assistance and support.

SUMMARY

Caregiving can be viewed as a personal life transition wherein those involved must be seen first and foremost as human beings. At this level, the health professional is in an ideal position to help families who are taking on the role of providing informal care. This person brings both professional expertise and personal human experience to bear in working with people dealing with illness in the family.

While it is not easy to alleviate caregiver burden, members of the health profession can do several things to help. Appropriate local services can be identified and referrals made. Professionals can listen to caregivers without criticism and can also provide positive reinforcement and bolster self-esteem. Greenberger and Litwin found that the most important resources were caregiver sense of competence and support from the professional health care provider.[5,p. 339] They argue that "nurses, who are orientated to holistic family centered care, are especially well-suited for this important intervention."

Caregivers are both family members and health care providers. They are increasingly becoming a financial and social necessity in the present climate of health care. It is essential and prudent to include them in the creation of care plans for the person with PD. Building systems that provide them with the support they need, and using them as resources in making health care plans, will encourage the health and well-being of the families we work with and the health care system as a whole.

REFERENCES

1. Mackey, R. A., The meaning of mental illness to caregivers and mental health agents, *Catholic University of America. Studies in Social Work*, 43, 1966.

2. Hurwitz, A., Home visiting by nursing students to patients with Parkinson's disease, *J. Neurosci. Nurs.*, 18, 344–348, 1986.

3. Pearlin, L. I., Mullan, J. T., Semple, S. J., et al., Caregiving and the stress process: an overview of concepts and their measures, *Gerontologist*, 30, 583–594, 1990.

4. Noonan, A. E. and Tennstedt, S. L., Meaning in caregiving and its contribution to caregiver well-being, *Gerontologist*, 37, 785–794, 1997.

5. Greenberger, H. and Litwin, H., Can burdened caregivers be effective facilitators of elder care-recipient health care? *J. Advanced Nursing*, 41, 332–341, 2003.

6. Maier, K. Creating a collective narrative with caregivers of family members with dementia, unpublished master's thesis, University of British Columbia, 2001.

7. Carter, J. H., Stewart, B. J., Archbold, P. G., et al., Living with a person who has Parkinson's disease: the spouse's perspective by stage of disease. *Mov. Disord.*, 13, 20–28, 1998.

8. Hoehn, M. M. and Yahr, M. D., Parkinsonism: onset, progression, and mortality, *Neurology*, 17, 427–442, 1967.

9. Cotzias, G. C., L-dopa in Parkinson's disease, *Hospital Practice*, 4 Sept., 35, 1969.

10. Yahr, M. D., Duvoisin, R. C., Schear, M. J. et al., Treatment of Parkinsonism with levodopa, *Arch. Neurol.,* 21, 343–354, 1969.

11. Charcot, J., *Leçons sur les maladies du systeme nerveux,* V.a. Delahaye, Paris, 1867.

12. Guiot, G., Le Trâitment des syndromes parkinsoniens par la destruction du pallidum interne, *Neurochirurgica,* 1, 94–98, 1958.

13. Rajput, A. H., Levodopa prolongs life expectancy and is non-toxic to substantia nigra, *Parkinsonism Relat. Disord.,* 8, 95–100, 2001.

14. Barodawala, S., Community care: the independent sector, *BMJ,* 740–743, 313, 1996.

15. Berliner, H. S. and Ginzberg, E., Why this hospital nursing shortage is different, *JAMA,* 288, 2742–2744, 2002.

16. Alber, J., Residential care for the elderly, *J. Health Polit. Policy Law,* 17, 929–957, 1992.

17. Paul, P., Make Room for Granddaddy, *American Demographics,* 24, 40–45, 2002.

18. Attias-Donfut, C., The dynamics of elderly support: The transmission of solidarity patterns between generations, *Zeitschrift für Gerontologie und Geriatrie,* 34, pp. 9–15, 2001.

19. Berry, R. A. and Murphy, J. F., Well-being of caregivers of spouses with Parkinson's disease, *Clin. Nurs. Res.,* 4, 373–386, 1995.

20. Hooker, K., Monahan, D. J., Bowman, S. R.,et al., Personality counts for a lot: Predictors of mental and physical health of spouse caregivers in two disease groups, *J. Gerontol. B. Psychol. Sci. Soc. Sci.,* 53, 73–85, 1998.

21. Calne, S. M., Mak, E., Hall, J., Fortin, M. J. et al.,Validating a quality-of-life scale in caregivers of patients with Parkinson's disease: Parkinson's Impact Scale (PIMS), *Adv. Neurol.,* 91, 115–122, 2003.

22. Faison, K. J., Faria, S. H., and Frank, D., Caregivers of chronically ill elderly: perceived burden, *J. Community Health Nurs.,* 16, 243–253, 1999.

23. Jette, A. M., Tennstedt, S., and Crawford, S., How does formal and informal community care affect nursing home use? *J. Gerontol. B. Psychol. Sci. Soc. Sci.,* 50, S4–S12, 1995.

24. Calne, S., Nursing care of patients with idiopathic parkinsonism, *Nursing Times,* 90, 38–39, 1994.

25. Zarit, S. H., Do we need another "stress and caregiving" study? *Gerontologist,* 29, 147–148, 1989.

26. Aarsland, D., Larsen, J. P., Karlsen, K., et al., Mental symptoms in Parkinson's disease are important contributors to caregiver distress, *Int. J. Geriatr. Psychiatry,* 14, 866–874, 1999.

27. Bannister, P., Parkinson's disease in the elderly, *Home Health Care Consultant,* 7, 20–24, 2000.

28. Dura, J. R., Haywood-Niler, E., Kiecolt-Glaser, J. K., Spousal caregivers of persons with Alzheimer's and Parkinson's disease dementia: a preliminary comparison, *Gerontologist,* 30, 332–336, 1990.

29. Edwards, N. E. and Ruettiger, K. M., The influence of caregiver burden on patients' management of Parkinson's disease: implications for rehabilitation nursing, *Rehabil. Nurs.,* 27, 182–186, 2002.

30. Edwards, N. E. and Scheetz, P. S., Predictors of burden for caregivers of patients with Parkinson's disease, *J. Neurosci. Nurs.,* 34, 184–190, 2002.

31. Fernandez, H. H., Tabamo, R. E., David, R. R.et al., Predictors of depressive symptoms among spouse caregivers in Parkinson's disease, *Mov. Disord.,* 16, 1123–1125, 2001.

32. Habermann, B., Spousal perspective of Parkinson's disease in middle life, *J. Adv. Nurs.,* 31, 1409–1415, 2000.

33. Happe, S. and Berger, K., The association between caregiver burden and sleep disturbances in partners of patients with Parkinson's disease, *Age Ageing,* 31, 349–354, 2002.

34. Lindgren, C. L., Chronic sorrow in persons with Parkinson's and their spouses, *Sch. Inq. Nurs. Pract.,* 10, 351–366, 1996.

35. Lyons, K. S., Zarit, S. H., Sayer, A. G., et al., Caregiving as a dyadic process: perspectives from caregiver and receiver, *J. Gerontol. B. Psychol. Sci. Soc. Sci.,* 57, 195–204, 2002.

36. McRae, C., Sherry, P., and Roper, K., Stress and family functioning among caregivers of persons with Parkinson's disease, *Parkinsonism Relat Disord,* 5, 69–75, 2001.

37. Monahan, D. J. and Hooker, K., Caregiving and social support in two illness groups, *Soc. Work,* 42, 278–287, 1997.

38. O'Reilly, F., Finnan, F., Allwright, S., et al., The effects of caring for a spouse with Parkinson's disease on social, psychological and physical well-being, *British Journal of General Practice,* 46, 507–512, 1996.

39. Pal, P. K., Calne, S. M., Flemming, J. et al., Sleep disturbances in Patients with Parkinson's Disease and their Caregivers. *Parkinsonism Relat Disord.,* 7[S1], 353, 2001.

40. Sirapo-Ngam, Y., Stress, caregiving demands, and coping of spousal caregivers of Parkinson's patients, *University of Alabama at Birmingham ** D,* S. N. 1994.

41. Smith, M. C., Ellgring, H., and Oertel, W. H., Sleep disturbances in Parkinson's disease patients and spouses, *J. Am. Geriatr. Soc.,* 45, 194–199, 1997.

42. Speer, D. C., Parkinson's disease patient and caregiver adjustment: preliminary findings, *Behaviour, Health and Aging,* 3, 139–146, 1993.

43. Baker, M., The role of the voluntary sector: pump primer or pit prop? *J. Neurol. Neurosurg. Psychiatry,* 55 Suppl., 45–46, 1992.

44. Robinson, K. M., Predictors of burden among wife caregivers, *Sch. Inq. Nurs. Pract.,* 4, 189–203, 1990.

45. Neufeld, A. and Harrison, M. J., Unfulfilled expectations and negative interactions: nonsupport in the relationships of women caregivers, *J. Adv. Nurs.,* 41, 323–331, 2003.

46. Henderson, R., Kurlan, R., Kersum, J. M. et al., Preliminary examination of the co-morbidity of anxiety and depression in Parkinson's disease, *J. Neuropsychiatry Clinical Neuroscience,* 4, 257–264, 1992.

47. Gilmour, J. A., Dis/integrated care: family caregivers and in-hospital respite care, *J. Adv. Nurs.,* 39, 546–553, 2002.

48. Calne, S., Schulzer, M., Mak, E. et al., Validating a Quality of Life Rating Scale For Idiopathic Parkinsonism: Parkinson's Impact Statement (PIMS), *Parkinsonism and Related Disorders,* 2, 55–61, 1996.

49. Basson, R., Sexuality and Parkinson's disease, *Parkinsonism Relat Disord.,* 177–185, 1996.

50. von Kanel, R., Dimsdale, J. E., Patterson, T. L. et al., Association of negative life event stress with coagulation activity in elderly Alzheimer caregivers, *Psychosom. Med.,* 65, 145–150, 2003.

51. Thommessen, B., Aarsland, D., Braekhus, A. et al., The psychosocial burden on spouses of the elderly with stroke, dementia and Parkinson's disease, *Int. J. Geriatr. Psychiatry,* 17, 78–84, 2002.

52. Reeves, N. C., A Path Through Loss: A guide to writing your healing and growth. Victoria, B. C. Victoria British Columbia, Grieving Issues Publications, 2000.

53. Maier, K. Grief and Loss. Master's Thesis, University of British Columbia, Canada, 2001.

54. Skaff, M. M., Pearlin, L. I., and Mullan, J. T., Transitions in the caregiving career: effects on sense of mastery, *Psychol. Aging,* 11, 247–257, 1996.

55. Andersen, S., Patient perspective and self-help, *Neurology,* 52, S26–S28, 1999.

56. Perry, J. A. and Olshansky, E. F., A family's coming to terms with Alzheimer's disease, *West J. Nurs. Res.,* 18, 12–28, 1996.

57. Sheehan, N. W. and Dornofio, L. M., Efforts to create meaning in the relationship between aging mothers and their caregiving daughters, *Journal of Aging Studies,* 13, 161–176, 1999.

58. Bar-David, G., Three phase development of caring capacity in primary caregivers for relatives with Alzheimer's disease, Journal of Aging Studies, 13, 177–197, 1999.

59. Walker, A. J., Pratt, C. C., and Eddy, L., Informal caregiving to aging family members, *Family Relations,* 44, 402–411, 1995.

60. Greenberg, L. S., Emotion-focused Therapy: Coaching clients to work through their feelings, American Psychological Association., Washington, D. C., 2002.

61. Skaff, M. M. and Pearlin, L. I., Caregiving: role engulfment and the loss of self, *Gerontologist,* 32, 656–664, 1992.

62. Calne, S. M. and Kumar, A., Nursing care of late stage Parkinson's disease, *Journal of Neuroscience Nursing,* 35, 242–251, 2003.

63. Egan, G., The Skilled Helper: A Problem Management Approach to Helping, Thomson Learning, New York, 2003.

77 Quality of Life in Parkinson's Disease: A Conceptual Model

Mickie D. Welsh
Keck School of Medicine, University of Southern California

CONTENTS

INTRODUCTION

In the years to come, the U.S. population will include larger numbers of older individuals, likely including the number of people with Parkinson's disease (PD) and other chronic and degenerative diseases. Demographic studies of PD suggest that the associated changes in functional status and dependency rise dramatically with each year of life.[3] Deterioration in functional status in individuals with PD increases the likelihood of their dependence on families, communities, and health care resources. Epidemiological studies suggest that, in diseases such as PD, improving *mortality* alone leads to a higher prevalence of dependent individuals in the life table population; improving *morbidity* alone leads to a lower percentage of individuals with problems in functioning, in both the years of life and the proportion of years of dependent life.[4]

In the last decade, dramatic improvements in the medical and surgical management of Parkinson's disease motor symptoms and complications of therapy have occurred. Control of motor dysfunction remains the major focus of PD treatment plans. PD patients also experience significant changes in other aspects of their lives. Patients suffer from depression, decreased social interaction, anxiety, insomnia, fatigue, role alteration, and cognitive changes, to name a few. However, we still understand little about which of these nonmotor factors (individually or combined) have the greatest impact on morbidity or quality of life in PD.[5] *Health-related quality of life* (HRQOL) refers specifically to an individual's sense of well-being, purpose in life, autonomy, and the ability to maintain worthwhile roles and participate in significant relationships.[1] It is noteworthy that recently published comprehensive articles addressing current concepts in the management of PD give little attention to the role of nonmotor symptoms or their influence on overall patient well-being, or HRQOL.[6–9]

Functional ability, independence, and the quality of life of patients and their families are constantly threatened by the degenerative nature of PD. The visible motor symptoms are frequently exacerbated by the less obvious yet profound demands of a changing body.[2] PD symptoms and their unpredictable presentation may endanger one's self-esteem, identity, role fulfillment, social relationships, and financial security. To reduce morbidity and improve quality of life in patients with PD, we need a better understanding of the influence and interrelatedness of the non-motor symptoms such as anxiety, depression, sleep disturbances, and fatigue and their complex relationships with motor function.

Over the past 25 years, there has been a growing interest in measuring the impact of chronic conditions and therapies on a person's health-relate quality of life. In neurology, the use of quality of life measures has markedly increased, and HRQOL indicators are now regularly used as secondary outcomes in clinical trials. In PD clinical trials, change in quality of life is usually reported as a summary score. A need exists to better understand how the component parts of the total score contribute to overall quality of life. Design of interventions to improve quality of life in PD patients is dependent on understanding which target areas are most important.

Despite the increased interest in this area, to date, little research is focused on characterizing the specific components of HRQOL, which are most influential in PD. Identifying the components that influence quality of life seems especially important. As therapeutic efforts focus more on improving functional performance and well-being, as well as motor performance, the need to understand these complex relationships increases.[10]

The goal of medical treatment is to make the manifest effects of an illness go away. In a sense, the quality of life measure represents the final common pathway of all the physiological, psychological, and social inputs into the therapeutic process.[11] If we are to direct specific interventions at improving quality of life, a clear understanding of the most influential factors that contribute to HRQOL in PD, and the relationship among them, must be described.

Understanding the complex relationships requires a more cohesive approach than now exists. We propose development of a conceptual model of quality of life knowledge development in PD as a framework for empirical testing and knowledge development. Testing the model would clarify how different mediating and outcome variables are interrelated and the weight associated with each variable. This approach would integrate two different paradigms of health outcomes: the medical model and the outcomes model.[12]

MEDICAL MODEL

Traditionally, the medical model emphasizes diagnosis and depends on measures of disease process. Clinical medicine has typically been focused on outcomes such as mortality, presence and absence of disease symptoms, duration of survival, and duration of disease-free interval.[11] Utilizing the medical model approach, the majority of outcomes are objective: biological, physiological, and pathological findings, and disease symptoms. The acute disease model dominates how we have developed health care delivery. The focus of this model has not yet been amended to include the major burden on our health care system today: chronic disease.[12] Changing social conditions, improvements in disease management, scientific discoveries, and therapeutic options have resulted in a reduction in episodic and contagious illnesses. Reducing the numbers of these "acute" diseases drastically reduced mortality, increased life span and eventually led to the current prevalence of chronic illnesses. The needs and outcomes of patients with chronic diseases are very different from those with acute illnesses.

THE OUTCOMES MODEL

The outcomes model emphasizes dimensions of individual patient functioning and overall well-being, or HRQOL. Outcomes models rely on patient self-reports of functioning, well-being, behaviors, and feelings, and they focus more on a continuum of disease state rather than on disease state at isolated points in time. The outcomes model also places a greater emphasis on epidemiological data.[12]

CONTRASTING THE MODELS

The differences emphasized by the two models reflect the changing landscape of health care, which is a landscape of people living longer with chronic and comorbid conditions—all of which highlight the need for understanding the pertinent but abstract concepts such as disability, handicap, and quality of life.

An important distinction between the traditional medical model and the outcomes model is the value placed on patients self-reports. The traditional model emphasizes disease pathology and treatment, and the outcomes model emphasizes the impact of detection and treatment. Individuals with chronic diseases typically have other chronic diseases as well. In the Medical Outcomes Study, 90% of the participants had other chronic conditions.[13] Because chronic conditions are not cured, patients must adapt to life with their disease. In adaptation, psychological and social factors are of key importance. Patient interpretation of the condition and adaptation to it cannot be ignored.

A CONCEPTUAL MODEL FOR PD QUALITY OF LIFE

Development and testing of a conceptual model is one way to bring clarity to a complex phenomenon. Conceptual models are useful for the organization they provide, as well as for clarification, analysis, and interpretation. Conceptual models also give direction to the search for relevant questions about phenomena, pointing out solutions to practical problems, and giving direction for future research and model development. Furthermore, they provide general criteria for knowing when a problem has been solved.[14] In this chapter, we propose developing a model of HRQOL in Parkinson's disease. Based on existing literature, this conceptualization is but a starting point. The basic assumption, however, is crucial—quality of life for a patient with Parkinson's disease is multifactorial. Motor symptoms alone do not account for the morbidity of this disease. Complex interrelationships and numerous factors may contribute to quality of life. Describing and manipulating any of these variables could potentially result in improvements in quality of life.

To begin, we must focus on the continuum of Parkinson's disease patients' quality of life (PDQOL) is illus-

trated in Figure 77.1. The quietly changing biological landscape results in the emergence of the symptoms of PD. In most cases, these initial changes are subtle; over time, increasing symptoms bring the patient to a doctor to determine their meaning. Gradually, symptoms that are more obvious replace the subtle symptoms, and the process of change in quality of life begins. Potential threats to quality of life may begin anywhere along this continuum and are, of course, specific to each individual.

Figure 77.2 illustrates a proposed framework of the model. The patient is the focus of the PDQOL model. A diagnosis of Parkinson's disease potentially threatens one's quality of life. Influencing the patient's quality of life are five domains: the patient's general health, the patient's Parkinson's disease state and medication ther-

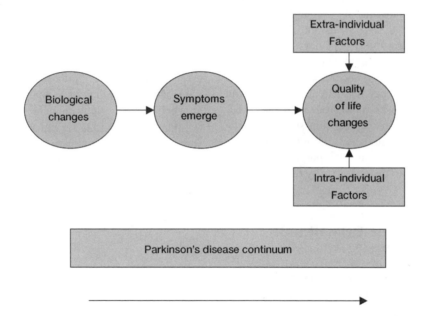

FIGURE 77.1   Parkinson's disease quality of life (PDQOL).

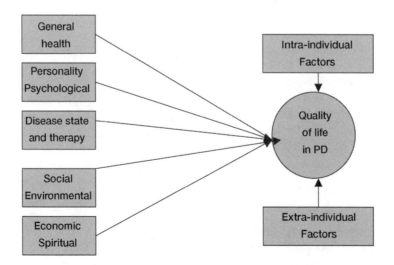

FIGURE 77.2   Quality of life in Parkinson's disease.

apy, the patient's social and environmental influences, individual personality and psychological characteristics, and spiritual and economic influences. Each of the domains is hypothesized to contain multifactorial components. The relationships of the domain concepts and subconcepts to each other are unknown at this time. Empirical testing of the model will be necessary to provide support for the dimensions and subdimensional concepts and give insight into the weight of individual or groups of factors within the domains.

Empirical testing of a model such as this provides a framework for practice, research, and knowledge development. Stuifbergen studied 786 persons with multiple sclerosis (MS) to test an explanatory model of variables influencing health promotion and quality of life.[15] The final model supported Stuifbergen's hypothesis that quality of life is the outcome of a complex interplay among contextual factors: severity of illness, antecedent variables, and health-promoting behaviors. The strength of direct and indirect paths suggested that interventions to enhance social support, decrease barriers, and increase specific self-efficacy for health behaviors would result in improved health-promoting behaviors and quality of life. Disease-specific information such as Stuiffbergen's study provides guidance for future studies in Parkinson's disease as well as MS, possibly revealing areas for interventions to enhance quality of life in many chronic illnesses.

INFLUENCE OF MODEL DOMAINS ON QUALITY OF LIFE

Within the five suggested domains of PDQOL, there are also subdomains. Many factors influence quality of life for the patient with Parkinson disease. In our schema of the continuum of life with PD, quality of life within the five domains is further influenced by two other sources, those within the individual and those outside of the individual. Figure 77.2 proposes the relationships of model variables as unidirectional. The actual direction, reciprocal relationship, and individual weight of model variables would be confirmed after empirical testing. Control over all these influences is an important issue for patients with chronic disease. Helping them with this control can empower patients; empowerment is a necessary step in assisting them toward adaptation to the chronic disease.

GENERAL HEALTH

For the individual with Parkinson's disease, one's underlying health status plays a key role in influencing overall quality of life. As an example, the patient with severe degenerative disease of the back and chronic pain finds himself with decreased motor function and increased off-time exacerbated by the presence of this stressful, comorbid condition. Comorbid health conditions place additional stressors on the patient and can be responsible for increasing PD symptoms and interfering with the efficacy of therapeutic treatments.

Many conditions such as lack of exercise, fatigue, lack of energy, and poor nutrition contribute to health status. Sleep disturbances and bowel and bladder difficulties commonly accompany PD and may contribute to changes in overall health.

Shulman's study of 105 PD patients found that many patient general health issues such as depression, anxiety, insomnia, and fatigue had not been previous identified (or treated) by the physician[16]. Findings from this study emphasize the importance of attention to nonmotor symptoms, which can be responsible for enhancing greater PD severity. Comorbidity of nonmotor symptoms in PD raises the question of whether the presence of these conditions might act as a feedback loop that, when acting synergistically, increases suffering and disability and lowers the quality of life. Furthermore, if these symptoms go unrecognized and untreated, a potentially mediating variable becomes a complicating one.

Intra-individual factors can act overtly to act as stressors, and enhance symptom activity and impact health and well-being. Lifestyle and behavior changes (at any time in the disease course) can maximally alter quality of life. Healthy lifestyle patterns, such as diet and exercise, are established contributors to overall health. Repeated exercise in any form is beneficial for maintaining function, regaining function and limiting the long-term effects of Parkinson's disease.

DISEASE STATE AND THERAPY

PD motor symptoms are well described in clinical practice and research. The decision to treat and the choice of treatment depends on individual patient symptom presentation. In the earliest stages of the disease, many patients require minimal pharmacological treatment and may experience little change in their quality of life due specifically to motor symptoms or to effects of treatment.[6–9] However, the patient's concerns for their future and the threats posed from having a chronic degenerative disease are psychologically stressful, and the subtle but continuous changes in function and symptoms haunt the patient, both of which can negatively affect HRQOL.

Over time, as the disease progresses and the patient becomes symptomatic, increased polypharmacy is required.[6–9] Response to treatment, symptom presentation, and some complications of therapy are captured using the UPDRS Sections I through IV. Until only a few years ago, clinical research data addressed only disease-specific symptoms expressed by scores on the UPDRS

Sections I through IV. Even now, few reports of quality of life assessment in clinical practice document the relationship of disease state and changes in quality of life.

Complications of medication therapy have a significant impact on the life of a patient with Parkinson's disease. Development of dyskinesias affect the ability to perform activities of daily living, impair gait and balance, and may increase dependence on others. Dystonic posturing can be very painful and limiting to functional ability. An array of medication side effects can also ensue, including constipation, nausea, vomiting, changes in mental status and memory, hallucinations, and psychosis. The presence of side effects such as hallucination and psychosis can limit therapeutic options to control motor symptoms.

SOCIAL AND ENVIRONMENTAL INFLUENCES

Research studies provide evidence that social networks strongly affect health outcomes and even survival. The direct costs of the disease often reflect a small portion of the burden. In PD, the hidden costs in the form of lost wages, informal care, and changing roles are substantial.[17] An eight-year study of community-based elderly lent support for the importance of social relationships in maintaining health. Being "embedded" in a social network of relatives and friends reduced risk for activity of daily living (ADL) disability as well as enhancing recovery from ADL disability.[18] Motor changes from PD can threaten independence and greatly reduce one's ability to maintain social and environmental contacts.

Social engagement, the maintenance of many social connections, and a high level of participation in social activities have been thought to prevent cognitive decline in elderly persons. Bassuk and colleagues followed 2812 noninstitutionalized elderly patients over a 12-year period. Compared with individuals with four to five social ties, those with no ties were at increased risk for cognitive decline after controlling for numerous other variables. Enhancing engagement in social activities is a critical to maintaining health in both the physical and psychological realms.[19]

Identifying PD patients at risk for threats in this domain are critical. A diagnosis of PD in someone with family and financial responsibilities can pose a threat to continued employment, insurance benefits, and long-term financial security. PD studies suggest that family relationships suffer changes early in the disease course, highlighting the importance of recognition of these threats.[20] Providing early referrals to services such as social workers and counseling are essential to helping families cope. Individuals within a family setting may or may not have an extant environment of support.

PERSONALITY ATTRIBUTES AND PSYCHOLOGICAL INFLUENCES

Extremely difficult to deal with for families, changes in cognition and behavior are regrettably common in PD. Neuropsychiatric symptoms are a frequent feature of advancing Parkinson's disease. There is a high prevalence of affective comorbidity in advanced PD.[21] Cognitive impairment, depression, hallucinations, and psychosis threaten individual and family quality of life. Psychosis in PD is often associated with older age, coexistence of dementia, protracted sleep disturbances, and use of a dopamine agonist.

Medication side effects of compulsive behaviors such as gambling and sexual hyperactivity can have severe consequences for the patient as well as the family. However, compulsive behaviors frequently go undiagnosed and are reported only when a family is in crisis.

Depression prevalence varies by study but may be as high as 40 to 50% and may even antedate motor symptoms.[22] Frequently identified as highly predictive of lower QOL, depression can go unrecognized. Numerous studies address depression. Karlsen studied 233 subjects with PD and compared them with 100 healthy elderly. PD patients had higher distress scores; more depression, insomnia, and lack of energy; and a lower degree of independence than the healthy elderly.[23] All of these variables were more strongly predictive of a higher Nottingham Health profile score (worse) than severity of Parkinson's symptoms. Depression may also present as changes in cognitive function masking the underlying problem.

PERSONALITY ATTRIBUTES

Individual personality characteristics help promote successful adaptation to chronic illness. Studies of coping in different chronic illnesses have found more similarities than differences between different disease conditions. Individual characteristics such as mastery, health beliefs, values, and health perceptions contribute to overall well-being and quality of life. Health perceptions are personal beliefs and evaluation of general health status.[24] People's perceptions of their health were found to be the best predictors of utilization of general medical services in the Health Insurance study performed by RAND.[13] Helping patients use personality attributes, or dispelling altered perceptions, is an important component in facilitating adaptation to chronic illness.

Despite the difficulties involved in living with a chronic illness, some individuals appear to adapt to their illnesses and maintain an enhanced qualify of life. Increasing attention is being paid to personality characteristics that play a role in enhancing health among individuals with chronic disease. Shifren's study of PD patients' optimism suggested that individuals with higher levels of optimism reported less need for assistance with basic functional abilities than less optimistic individuals.[25] Assisting chronically ill individuals to develop coping tasks so they can preserve their integrity, restore or maintain a positive concept of self, and function effectively in relationships and life roles is key to living well with PD.

Multidisciplinary approaches to rehabilitation applied to PD and MS patients over a 10-day period found that both groups experienced significant improvements. While the MS patients' strongest improvement was in physical mobility, the PD group had higher emotional reactions.

ECONOMIC AND SPIRITUAL INFLUENCES

The financial impact of neurological disease has received significant attention in public health studies but much less in clinical neurological assessments. Determining the economic costs of chronic illness is a specialized area of assessment and is addressed in a separate chapter. The importance of recognizing and assessing the presence of these potential burdens should be included in clinical practice and not limited to clinical trials.

Spirituality is more than the sum of the client's religious preference, religious beliefs, and religious practices. It relates to the totality of man's inner resources, the ultimate concerns around which all other values are focused, that guides conduct and influences all individual social behavior.[26]

CONCLUSIONS

Helping individuals with Parkinson's disease cope and adapt is an enormous challenge. The goal of all therapeutics for PD is to limit disability to whatever degree possible. The multifactorial nature of HRQOL in PD requires that we fully understand the impact of these factors if we intend to make real improvements in patient's quality of life. Gage and colleagues studied 14,530 Parkinson's veterans and found that the negative impact of PD (in both physical and psychological domains) on HRQOL was greater than in 8 other chronic conditions, including spinal cord injury, congestive heart failure, stroke, and diabetes.[27] The findings of this large study emphasize the high degree of burden of Parkinson's disease and reinforces the case for identifying all the factors that need to be targeted to improve patients' quality of life.

The model proposed in this chapter is intended to provide but a starting point with respect to better understanding the concepts and constructs that impact quality of life. Empirical testing of the model will no doubt surface where interventions can be targeted. It is hoped that this

work will ultimately help us to improve quality of life in patients with PD.

REFERENCES

1. Spilker, B., *Quality of life Studies: Definitions and Conceptual Issues,* in Schipper, H., Lippincott-Raven, Philadelphia, 11, 1996.
2. Koplas, P. A. et al., Quality of Life and Parkinson's Disease, *J. Gerontology,* 54A, 197, 1999.
3. Tison, F., Barberger-Gateau, P., and Dubroca,B., Dependency in Parkinson's disease: A population-based survey in don-demented elderly subjects, *Mov. Dis.,* 12, 910, 1997.
4. Crimmins, E. M., Hayward, M. D., and Saito, Y., Changing mortality and morbidity rates and the health status of older adults, *Demorgraphy,* 31, 159, 1994.
5. Janca, A., A report of the WHO working group on parkinson's disease, *Neuroepidemiology,* 18, 240, 1999.
6. Lang,.A. E., Lozano, A. M., Parkinson's disease: second of two parts, *NEJM.,* 339, 1130, 1998.
7. Rascol, O. et al., Limitations of current parkinson's disease therapy, *Ann. Neurology,* 53, Suppl. 3, 3, 2003.
8. Clarke, C. E., Medical management of parkinson's disease. *J. Neurol. Neuosurg. Psych.,* 72, Suppl. 1, 22, 2003.
9. Guttman, M., Kish, S., and Furukawa, Y., Current concepts in the diagnosis and management of Parkinson's disease. *CMA, JAM,* 3168, 293, 2003.
10. Wilson, I. B. and Cleary, P. D., Linking clinical variables with health-related quality of life: a conceptual model of Patient outcomes, *JAMA,* 273, 59, 1995.
11. Kaplan, R. M., The significance of quality of life in health care, *Quality of Life Research*, 12, 3, 2003.
12. Hobart, J. C., Lamping, D. L., and Thompson A. J., Evaluating neurological outcome measures: the bare essentials, *J. Neurol. Neurosurg. and Psych.,* 127, 1999 and *N. E. J. M.,* 339, 1130, 1998.
13. Stewart, A. L. and Ware, J. E., *Measuring Functioning and Well-Being: The Medical Outcomes Study Approach,* Duke University Press, London 141, 1992.
14. Fawcett, J., *Analysis and Evaluation of Conceptual Models of Nursing* 1st ed., F. A. Davis, Philadelphia, Chapter 2, 1984.
15. Stuifbergen, A. K., Seraphine, A., and Roberts, G., An Explanatory model of health promotion and quality of life in chronic disabling conditions, *Nursing Research,* 49, 122, 2000.
16. Shulman, L. M. et al., Comorbidity of the Non-motor symptoms of Parkinson's disease, *Mov. Dis.,* 16, 507, 2001.
17. Whettenn-Goldstein, K, The burden of Parkinson's disease on society, family and the individual, *J. Am. Geriatrics Soc.,* 45, 844, 1997.
18. Mendes de Leon, C. F., Social networks and disability transitions across eight intervals of yearly data in the New Haven EPESE study, *J. Gerontology,* 54, 162, 1999.

19. Bassuk, S. S., Glass, T. A., and Berkman L. F., Social disengagement and incident cognitive decline in community-dwelling elderly persons, *Ann. Intern. Med.,* 131, 165, 1999.

20. Glass, T. A. et al., Population based study of social and productive activities as predictors of survival among elderly Americans, *B. M. J.,* 319, 478, 1999.

21. Juncos, J. L., Management of Psychotic aspects of Parkinson's disease, *J. Clin. Psych.,* 60, 42, 1999.

22. Poewe, W. and Seppi, K., Treatment options for depression and psychosis in Parkinson's disease, *J. Neurology,* 248, 12, 2001.

23. Karlsen, K., Larsen, J. P., Tandberg, E. Maeland, J. G., Influence of clinical and demographic variables on quality of life in patients with Parkinson's disease, *J. Neurol., Neurosurg. and Psych.,* 66, 431, 1999.

24. Miller, J. F., Analysis of coping with illness, In *Coping with Chronic Illness: Overcoming Powerlessness,* Miller J. E., Ed., F. A. Davis, Philadelphia, 3, 1992.

25. Shifren, K., Individual differences in the perception of optimism and disease severity: A study among individuals with Parkinson's disease, *J. Behav. Med.,* 19, 241, 1995.

26. Moberg, D. O., Development of social indicators of spiritual well-being for quality of life research, in *Spiritual Well-Being Sociological Perspectives,* Moberg, D. O., Ed., Washington University Press of America, 13, 1979.

27. Gage, G. et al., The relative health related quality of life of veterans with Parkinson's disease, *J. Neurol. Neurosurg. Psych.,*74, 163, 2003.

78 The Evolution and Potential of the National Parkinson Organizations: A Brief Overview

Ruth A. Hagestuen
The National Parkinson Foundation

CONTENTS

INTRODUCTION

There has been and continues to be a clear, viable role for the private, Parkinson focused, nonprofit organizations. Since their inception beginning as early as the late 1950s, contributions to research for Parkinson's disease, the development of a broad range of support services, provision of educational resources and conferences, and an ever-growing voice of advocacy on behalf of persons with Parkinson's disease have been significant.

The current movement has been toward organizations developing with a more specific mission. A collaborative process has begun wherein organizations are beginning to mobilize around specific goals so as to reduce duplication and maximize potential. There is continued need for collaboration around the development of products and services to better meet the variable, diverse, and complex needs of people with Parkinson's and their families.

This chapter describes contributions of some of the major Parkinson's organizations in the U.S. by presenting a very brief history and description of their mission and some of their work. The purpose is also to highlight some of the similarities and differences as well as to show the progress in collaboration among the organizations. This is not meant to be an exhaustive listing of all organizations or their activities.

Four of the five organizations referenced in this chapter were founded because of a personal connection to Parkinson's disease. Perseverance and dedication are hallmarks of each organization, and, as will be discussed, each organization remains true to its original mission.

THE NATIONAL PARKINSON FOUNDATION

The National Parkinson Foundation (NPF) was founded in 1957 by Mrs. Jeanne Levey, whose husband was diagnosed with Parkinson's disease. At that time, little was known about the disease. She established a facility in the same location where NPF is today, in Miami, Florida, across from the University of Miami's School of Medicine. NPF still operates under the same charter Mrs. Levey created in 1957.

Nathan Slewett, Jeanne Levey's successor, has led the National Parkinson Foundation in a continual process of expansion to become a leading international Parkinson organization. The National Parkinson Foundation describes its mission as dedicated to finding the cause and cure of Parkinson's disease as well as supporting a broad array of educational, support, and care initiatives designed to improve the quality of life for people with Parkinson's and their caregivers.

A core component of the NPF research and care delivery system is its Centers of Excellence (COE). Conceptualized as the local area or regional "hub" for Parkinson's disease research, services, education, and outreach, COEs are expected to assume a leadership position in the provision of innovative models of service and in the development of community relations to support health promotion efforts in Parkinson's disease. There are currently 65 affiliated COEs in the international network.

The NPF awards individual research grants that are reviewed and recommended by their scientific advisory board as particularly promising. They also designate research funds for qualifying Centers of Excellence, which are engaged in bench or translational research.

NPF chapters are community organizations, consisting primarily of volunteers who are dedicated to helping fulfill the larger, mutual mission. Chapters function within their geographic areas as representatives, delivering products and services, and they are also the representatives of their region to the organization as a whole. There are currently 35 affiliated NPF chapters.

NPF provides educational and medical information through its series of publications, professional education, support group development, caregiver and young onset programming, as well as other services designed to impact quality of life in Parkinson's.

NPF's newest initiative, *Community Partners for Parkinson Care* (CPP), is an outreach program that is funded, in part, through a grant from the Medtronic Foundation. This is in response to the identified need to address health disparities that exist in the delivery of care for persons with PD and their caregivers. The program is designed to raise community awareness of Parkinson's disease and help some of the most diverse communities around the country to find information, support, and access to informed services to meet their needs. This program is being conducted initially in six regions of the U.S. with the hope to develop this model more widely.

THE PARKINSON'S DISEASE FOUNDATION

The Parkinson's Disease Foundation (PDF) was founded in the same year as NPF, 1957, by William Black. Black was a New York City businessman and philanthropist who learned that the controller of his company, Chock Full o' Nuts, had been diagnosed with Parkinson's disease. Black quickly realized that there was little to help control Parkinson's disease symptoms and even less to educate and assist patients. As a result, he made two major gifts to Columbia University: one to help build a research facility that now houses one entire floor dedicated to PD research, and one to endow the support of research. Black then established the Parkinson's Disease Foundation.

PDF still maintains a primary relationship with Columbia University although, over time, it has expanded the scope of its activities. In 1999, PDF merged with a Chicago-based Parkinson's organization, the United Parkinson Foundation. The head office remains in New York City, with a regional office in Chicago.

PDF describes itself as a national nonprofit organization devoted to education, advocacy and the funding of research. The PDF program goals are as follows:

1. Advancing Parkinson's science through research awards and training fellowships
2. Sponsorship of professional and lay conferences
3. Meeting the needs of the Parkinson's community for information and education
4. Engaging in advocacy for heightened public interest in Parkinson's and for increased government allocation of funds for research into the causes of Parkinson's disease and the cure

Other services include:

1. Support groups
2. Clinical neurologists specializing in diagnosis and treatment of PD symptoms
3. A variety of not-for-profit and for-profit resources available to persons with Parkinson's (PWP)

THE AMERICAN PARKINSON'S DISEASE ASSOCIATION

The American Parkinson's Disease Association (APDA) was founded in 1961 to "ease the burden and find a cure." It is headquartered in New York. APDA focuses its energies on research, patient services, education, and raising public awareness of the disease. At the time of its founding, APDA Executive Director Joel Gerstel pointed out that no one ever thought about raising money, and there weren't very many treatments for Parkinson's disease. Physical therapy, according to Gerstel, was the only way those with Parkinson's disease were being treated. Gerstel, in a recent phone interview, reflected on the progress made in clinical treatment, therapies, and surgical techniques. "Obviously," he said, "we have come a long way."

Annually, the APDA scientific advisory board reviews grant requests and submits recommendations for the funding of researchers whose work shows promise of yielding new breakthroughs or improved treatments for PD.

Its national office also coordinates the efforts of chapters and Information and Referral Centers (I & R) across the nation. Chapters and I & R Centers are of equal stature in the APDA structure.

Chapters are made up of grassroots volunteers who are given the assignment of getting the word out about Parkinson's disease, making people aware of the disease and how it's being handled. Gerstel believes that it is the grassroots approach through its 67 chapters across the U.S. that makes APDA unique. These chapters are completely staffed by volunteers. According to Gerstel, the best way to accomplish that feat is to get a support group interested in becoming more than a support group. As they grow, they want to do more to fight the disease.

The APDA's goal of promoting a better quality of life for the Parkinson's community is actualized through their network of I & R centers. The unique contribution of these centers is their ability to respond to the particular needs of persons affected by Parkinson's disease and their caregivers through education, referral, support, and public awareness programs. The centers serve as regional "hubs' that process requests received by mail, phone, or on-site visits. They are funded through grants given to the institutions which house them.

THE PARKINSON'S ACTION NETWORK

Founded in 1991 by the Joan Samuelson four years after she was diagnosed with Parkinson's disease, the Parkinson's Action Network (PAN) is the unifying advocacy arm of the Parkinson's community, fighting for a cure. Through education and interaction with the Parkinson's community, scientists, lawmakers, opinion leaders, and the public at large, PAN works to increase awareness about Parkinson's disease and advocates for increased federal support for Parkinson's research.

The Parkinson's Action Network goals are to

1. Increase public awareness about Parkinson's disease
2. Educate members of Congress and other federal decision makers
3. Advocate for increased federal funding
4. Enhance the effectiveness of the Parkinson's community through coordination with the leadership of other national Parkinson's organizations

5. Encourage federal support for research on causes, treatments, and a cure
6. Empower the Parkinson's community to become effective advocates for Parkinson's disease

The Parkinson Action Network recently opened an office in downtown Washington, DC, and developed a Communications Program with increased emphasis on becoming the coordinating point and unified voice for the Parkinson's community in Washington, DC.

Jeff Martin, executive director of Parkinson Action Network, is very clear about his vision for the future of Parkinson Action Network. PAN, he said, is basically run by a group of people who either have the disease or care very much about somebody who has it. It is not a large organization. There are no thoughts of trying to build for the future in terms of organizational structure. "What we want," declared Martin, "is to go out of business ASAP. Our legacy is not going to be a better organization. We hope our legacy is going to be that we were a part of the team that made efforts such as ours unnecessary."

THE PARKINSON ALLIANCE

Shortly after Congress approved appropriations or funding for the Udall bill in 1999, Martin and Margaret Tuchman founded the Parkinson Alliance and the Tuchman Foundation. Prior to establishing the Parkinson Alliance, the Tuchman's had been actively involved for four years in advocating for the passage of the Morris K. Udall Parkinson's Research and Education Act. It was a very personal campaign for the Tuchman's, since Margaret had been diagnosed with Parkinson's disease 20 years earlier.

Headquartered in Princeton, NJ, the Parkinson Alliance's mission is to fund PD research. The Tuchman Foundation's main objective is supporting the Parkinson Alliance, whose mission is raising money for pilot study programs that allow researchers to qualify for major funding from NIH. The Alliance has also been collaborating with the other private Parkinson organizations and the NIH in joint funding of the R21 Fast Track Grant Program as well as other research initiatives that they see as most promising. The Tuchman Foundation matches all donations to the Parkinson Alliance (www.parkinsonalliance.net).

The Tuchman's and the Parkinson Alliance have played an important role in developing the connection with the NIH. In an interview featuring Margaret Tuchman in a "New Hope Profile" on Medtronic's web site, Parkinson's advocate Jim Mauer said, "The connection with the NIH is probably the single most important contribution of the Alliance."

MICHAEL J. FOX FOUNDATION FOR PARKINSON'S RESEARCH

The newest funding organization is the Michael J. Fox Foundation, devoted to funding Parkinson's Research (michaeljfox.org). It was established in May 2000 by actor Michael J. Fox, two years after he disclosed that he had been diagnosed with young-onset Parkinson's disease seven years earlier.

The Michael J. Fox Foundation for Parkinson's Research (MJFF), headquartered in New York City, is dedicated to ensuring the development of a cure within this decade through an aggressively funded research agenda. "The Foundation seeks to hasten progress further through the awarding grants that help guarantee that new and innovative research avenues are thoroughly funded and explored."

The Foundation is made up of a variety of programs. In August 2002, the Foundation established a $1.2 million Research Fellowship program designed to encourage promising young scientists and clinicians to enter the field of Parkinson's research.

The response was very strong, with more than two-hundred applicants for the first round. The goal was to fund projects that might be considered higher risk/higher reward than those typically funded through the National Institutes of Health.

In May, 2003, the MJFF announced a new research program, Linked Efforts to Accelerate Parkinson's Solutions (LEAPS). This program, composed of fewer but larger grants, has been described as a new paradigm to jump-start progress through collaborative, multidisciplinary, and holistic research efforts that translate into new treatments or otherwise have a tangible impact on Parkinson's disease. Deborah W. Brooks, MJFF executive director, called the LEAPS program "the next step in the Foundation's research funding strategy."*

A concern repeatedly expressed by scientists is that one barrier to fast progress is the time it takes to apply for and receive funding. Responding to those concerns, the Michael J. Fox Foundation made its application shorter and pledged to make the turnaround time much shorter. The Fast Track Program is funded within 6 months of announcement, as opposed to the more traditional time of up to 18 months for some other programs.

DEVELOPING A COLLABORATIVE PROCESS FOR ADVOCACY AND RESEARCH

It was only with the emergence of people who took the case to Washington and subsequently the passage of the Udall Act in 1997, as well as other steps such as the

---
* Via telephone interview.

research agenda at the NIH, that federal investment began in a significant way. Parkinson's disease research has enjoyed substantially increased support from the federal government, with the NIH doubling it's funding over the last five years. Awareness of the need to increase federal investment so as to benefit from the scientific opportunities to find a cure for Parkinson's has been greatly heightened for the whole community in the past few years. In a parallel process, the Parkinson organizations are moving toward a more intentional collaborative process to accomplish their shared mission.

In May 2003, the Michael J. Fox Foundation proposed the formation of a representative scientific body to share broad research-related information among the Parkinson's disease research funding foundations and their scientific advisory boards. Calling it the Parkinson's Community Research Advisory Council, this small work group of five to seven scientists will assess a detailed yearly inventory of the community's research program and will offer suggestions to improve coordination of the community-wide effort. The belief is that better understanding of the community-wide investment will allow individual organizations to better target their resources, help reduce redundancy, and increase coordination of separate programs that share common goals. The Council will not be a decision-making body but will facilitate information sharing and consensus building.

One of the questions that continue to surface in the Parkinson's community is the appearance of duplication of efforts. All the foundations except PAN raise and designate funds for research. The NPF, APDA, PDF, and Fox Foundation all have a scientific advisory board. Past concerns in this regard are beginning to be addressed.

A second research collaborative in the Community Fast Tract 2003. This is an MJFF $4 million grant program supported by the Foundation and other Parkinson's organizations, including the Parkinson's Disease Foundation, The National Parkinson Foundation, and Parkinson's Unity Walk/the Parkinson Alliance. The idea for the program came from the Fox Foundation's Scientific Advisory Board.

The Community Fast Track is an investigator-initiated, peer-reviewed program that considers a broad range of research applications relevant to the cure, cause, prevention, or improved treatment of Parkinson's disease. The intent of this program is to stimulate novel, innovative, and/or high-impact approaches to the field of Parkinson's disease as well as to fill funding gaps missed by more conventional funding sources.

In a statement announcing the establishment of the Community Fast Track 2003, the MJFF declared, "Over time, we hope that Community Fast Track serves as a springboard to great collaboration within the Parkinson's community, as we all strive toward our common goal of curing Parkinson's disease."

The collaborative research funding process now being initiated through the coordination of the Fox Foundation, combined with the individual funding mechanisms of each organization, simply broadens the field of opportunity and expands the potential. It is felt that the current level of communication and sharing of information will minimize redundancies, encourage greater collaborations, and speed the progress to better quality of life and the cure for Parkinson's disease.

ADVOCACY

The combined efforts of all the PD organizations have increasingly coalesced behind the Parkinson Action Network (PAN). PAN has become the primary representative voice including, as appropriate, the most influential and eloquent individuals representing the Parkinson's community in congressional and other public appearances. As a more united front, good progress is being made in the fight for increased investment in biomedical research by the federal government, including research funded by the National Institutes of Health (NIH) and the Department of Defense (DOD). The developing public-private partnerships contribute to the larger process of education, communication, and collaboration. In 2003, the Department of Defense was a sponsor of PAN's Research and Education Forum, where renowned scientists and researchers who are working on Parkinson's disease are brought together. The Defense Department has a Parkinson's disease research program of its own, which is run by the Army. A great deal of research, which they are conducting on ways to protect soldiers from chemical agents, radiological agents, and other types of environmental toxins, turns out to be applicable to Parkinson's research, which also looks at the effects of toxins and environmental agents on the neurological system.

In development is an effort to establish a grassroots network of advocates across the country. The PAN Director of Outreach is now working on selecting coordinators in every state. Through a developing partnership with leaders of NPF and APDA chapters, expanded potential and sustainability of a shared infrastructure for advocacy will be established to continue to identify issues and educate and mobilize the whole community. The job of state coordinators is developing networks within their own states, speaking to support groups, and encouraging people at the grassroots level to become advocates.

SUPPORT, EDUCATION, AND CARE INITIATIVES

For organizations such as the APDA and the NPF, whose mission includes providing services that affect quality of life, the establishment and maintenance of support groups for persons with Parkinson's and caregivers has been and continues to be a basic component in the array of services offered. The coordination for development and facilitation of these groups is most commonly offered through their affiliated centers and chapters.

The offering of support services has been a priority not only for these Parkinson organizations but also for other regional organizations that have developed. Parkinson's support groups vary in frequency of meeting as well as membership. Some are closed groups that meet weekly for a specified period, others are ongoing and meet monthly. Some combine patients and caregivers and others split the groups. Some are peer led, and others are professionally led. The idea is to provide a safe environment to discuss and learn more about the disease and ways to cope, in the midst of others going through similar experiences. There are also many Parkinson groups that have an educational focus and invite speakers for their monthly meetings.

There are clear benefits to participating in a support group. The challenge in the development and maintenance of these groups is continuing to provide appropriate guidance and training for the facilitators.

EDUCATIONAL SERVICES FOR THE PARKINSON COMMUNITY

The development and distribution of educational materials designed to inform and assist persons whose lives are affected by Parkinson's disease, sponsorship of educational conferences, and the establishment of web resources are also basic services provided through the work of these organizations. The web sites as well as regional conferences combine to provide the information needed by persons who are integrating the reality of Parkinson's disease into their lives and seeking to maintain the best possible quality of life.

ADDRESSING THE NEEDS OF THE FAMILY CAREGIVER

The needs of family caregivers of persons with Parkinson's disease are not being adequately addressed through the health care system. The voices of caregivers in search of support, respite, and assistance in navigating the system of services are often directed toward the private foundations in search of an answer. As mentioned previously, support groups have been traditionally established, and national networks of caregivers can be helpful.

Financial support for respite care services, education of professionals in facilities providing respite care services for persons with Parkinson's, and educational series for caregivers are being developed and offered in many locations with a variety of results.

The national organizations have the opportunity and have begun to assemble providers, help discern best practices, and provide education and resources to expand the scope of services and level of support to alleviate the burden now shouldered by caregivers. In particular, the National Parkinson Foundation is facilitating the launch of a program called *Caregiver Connections* in which existing services are being evaluated and resources assembled to bring caregiver services to the next level.

YOUNG-ONSET PARKINSON'S

Persons with young onset Parkinson's disease have become simultaneously much more visible and vocal. The process of being diagnosed and coming to terms with Parkinson's disease has historically been an isolating experience for younger persons, and many have found it difficult to make connections with the larger community of young persons with common young-onset and lifestyle concerns.

Many of the strongest voices advocating for research to find the cure as well as care and support systems to better meet their needs are those of the young onset. The American Parkinson Disease Association and the National Parkinson Foundation both support young-onset programming, and there are a number of young-onset chat rooms and support groups in existence. The development of age specific, stage specific materials and programs is a continual need and role for these organizations.

PROFESSIONAL EDUCATION FOR ALLIED HEALTH PROFESSIONALS

There is concern throughout the Parkinson community that the majority of professionals who provide care do not have sufficient understanding Parkinson's or the clinical expertise specific to Parkinson's to able to provide optimum care.

A recent innovation introduced by NPF is a model training program, the Allied Team Training for Parkinson (ATTP). The program, designed for allied health professionals, brings together professionals and students nearing graduation in the fields of music therapy, occupational therapy, physical therapy, social work, and speech-language pathology for 4.5 days of training in the specialized care of Parkinson's disease.

Goals of this new program are, first and foremost, to raise the level of knowledge and quality of care provided by allied health practitioners to persons with PD and their care partners. A second goal is to convey the message that interdisciplinary care is optimal care. And, finally, the program has a third goal of addressing issues of cultural diversity in health care delivery.

POTENTIAL CONTRIBUTIONS OF THESE ORGANIZATIONS

The greatest potential of the private, nonprofit Parkinson organizations will be attained through a continued process of communication and collaboration. This needs to move beyond the organizations discussed in this chapter. There are numerous independent regional organizations, some of whom have developed excellent resources and programs. The organizations have an opportunity to collectively

- Identify and capitalize on opportunities for collaboration
- Establish goals that minimize duplication so as to maximize the potential of available resources
- Develop an expanded nationwide network of communications
- Identify and document unmet needs and then work to effect change
- Mobilize as a community at large to advocate for research funding and also to facilitate change in the health care system to better serve people whose lives are affected by Parkinson's disease

The opportunities for increased collaboration do not stop with the voluntary organizations, as has been illustrated in the preceding discussion. The current collaboration with NIH, as well as with the Department of Defense and the Veterans Administration with the establishment of the Parkinson's Disease Research Education and Clinical Centers (PADRECCs) is an example. All of the organizations devoted to the combined mission of finding a cure for Parkinson's disease and providing optimum care and services, whether public or private, will make the greatest contribution if it can be done in concert with each other.

CONTACT INFORMATION

The American Parkinson's Disease Association
1250 Hylan Boulevard, Suite 4B
Staten Island, NY 10305-1946
(800) 223 2732 or (718) 981 8001
http://www.apdaparkinson.com
info@apdaparkinson.org

Michael J. Fox Foundation for Parkinson's Research
Grand Central Station
P.O. Box 4777
New York, NY 10163
(800) 708-7644
http://www.michaeljfox.org

National Parkinson Foundation, Inc.
1501 NW Ninth Avenue/Bob Hope Road
Miami, FL 33136-1494
(800) 327-4545 or (305) 547-6666
http://www.parkinson.org
mailbox@parkinson.org

The Parkinson's Action Network
1000 Vermont Avenue, NW, Suite 900
Washington, D.C. 20005
(800) 850 4726
(202) 842-4101
http://www.parkinsonsaction.org
info@parkinsonsaction.org

Parkinson's Disease Foundation
William Black Medical Building
Columbia-Presbyterian Medical Center
710 West 168th Street
New York, NY 10032-9982
(800) 457-6676 or (212) 923-4700
http://www.pdf.org/index.cfm
info@pdf.org

The Parkinson Alliance
211 College Road East, 3rd Floor
Princeton, NJ 08520
(609) 688-0870 or (800) 579-8440
http://www.parkinsonalliance.net/home.html
admin@parkinsonalliance.net

Index